Imaging Practice and Radiation Protection in Pediatric Radiology

Michael Seidenbusch
Veronika Rösenberger • Karl Schneider

Imaging Practice and Radiation Protection in Pediatric Radiology

Conventional Radiography

Springer

Michael Seidenbusch
Department of Paediatric Radiology
Dr. von Hauner Children's Hospital
University of Munich
Munich
Germany

Veronika Rösenberger
Department of Paediatric Radiology
Dr. von Hauner Children's Hospital
University of Munich
Munich
Germany

Karl Schneider
Department of Paediatric Radiology
Dr. von Hauner Children's Hospital
University of Munich
Munich
Germany

ISBN 978-3-030-18502-2 ISBN 978-3-030-18504-6 (eBook)
https://doi.org/10.1007/978-3-030-18504-6

This Springer imprint is published by the registered company Springer Nature Switzerland AG.
The registered company address is: Gewerbestrasse 11, 6330 Cham, Switzerland

This book is dedicated to
Karla
with my sincerest thanks for all her love and patience
and to
my parents
for rendering possible my path of life
and my professional history.

Munich, Summer 2019

Michael Seidenbusch

Foreword

In neonates, organs and tissues of infants and young children are localized closer to each other than in adults and—according to radio-biological investigations—children are more sensitive to ionizing radiation. Quality criteria for diagnostic radiographic imaging needed to be specifically adapted to paediatric radiology. After the important work in 1979 of *Rosenstein* on the calculation of organ doses in paediatric age groups, this book closes a wide gap on this topic.

The authors, internationally renowned in the field, provide new normalized organ doses for conventional paediatric X-ray examinations in all age groups. Although the mathematical phantoms had some inaccuracies concerning the location, size, and shape of the organs in the different age groups, the anatomy of the developing child is detailed enough to get reliable dose calculations. Clinical indications, different field settings, exposure parameters, image quality criteria, and normalized radiation doses have been investigated. The combination of these criteria, especially with various field settings, offers new aspects of dose reduction and risk calculation.

The book is well organized with clear figures and similar structure and formation of the tables for all investigated regions.

Imaging practice and radiation protection should be the reference tool for everybody who works in paediatric radiology.

Don't forget, children are our future.

Cologne, Germany Gabriele Benz-Bohm

Preface

"Es hat nie einen Mann gegeben, der für die Behandlung von Einzelheiten so begabt gewesen wäre. Wenn er sich mit den kleinsten Dingen abgab, so tat er das in der Überzeugung, daß ihre Vielheit die großen zuwege bringt."—"There has never been a man who was more capable to address details. If he had to do with the minor details he did it believing that a lot of minor things would produce the great ones."

Friedrich II. about Friedrich Wilhelm I., both kings of Prussia

Children are amongst the most radiation-sensitive living creatures on earth. Thus, radiation protection of young patients might be one of the most important issues in paediatric diagnostic radiology. On ethical grounds, children also need greater protection against diseases compared to adults. Thus, imaging of the child in paediatric radiology is an essential part of medicine. But trying to optimize radiation exposure and image quality simultaneously may make the paediatric radiologist feel himself like Ulysses between Scylla and Charybdis.

Indeed, especially in paediatric radiology, there are numerous interacting exposure parameters which influence radiation exposure and image quality, a fact that makes the optimization process in paediatric radiology so complicated. The authors of this book would like to contribute some facts to help clarify this situation. This book addresses mainly to all radiologists and technologists who are involved in paediatric imaging and who wish to improve the radiation protection of children by choosing adequate field settings and exposure parameters. The book addresses also to medical physicists and epidemiologists who want to reconstruct organ doses achieved in patients during conventional paediatric X-ray examinations.

The authors are aware that they perform a balancing act when trying to bring together two quite distinct areas of paediatric radiology, the exposure technique and the radiation dosimetry, each of them comprehensive enough to fill an entire book. However, the authors believe that it is this combination that may help the reader to assess the dosimetric consequences of the applied exposure practice.

The book is divided into four parts. In the first part, a very short introduction will be given to the topics of radiation exposure and radiation risk. In the second part, a short introduction to radiation physics and radiation dosimetry will be given. However, the first two parts are no prerequisites at all for understanding the third part as the main part of this book. In the third part, a short introduction will be given on how to calculate organ doses from clinically measured dose indicators. Each chapter in this part deals with a specific radiographic technique and provides a lot of tables subsuming clinical indications, image quality requirements, field settings, exposure parameters, and normalized organ doses which result from various field settings and exposure conditions. Finally, the fourth part of this book provides additional information on the state of the art of dose reconstruction in paediatric radiology.

In preparation for this book, the authors were greatly assisted by several people. First of all, the authors are grateful to Mrs. Corinna Hauser, to Mrs. Anna Lena Buchholz, and to Mrs. Wilma McHugh, Springer Company, for all their advice and patience. The authors want also to express their gratitude to Mrs. Saanthi Shankhararaman and Mr. Kumar Srinivasan Deepak, Springer Company, and their team for their patience and accuracy when preparing the layout of the book. Special thanks go to Prof. Dr. Stefan Milz, anatomist at the University of Munich and editor-in-chief of the Journal of Anatomy, as well as to M. A. Franz Jürgen Götz, librarian of the national library of Bavaria, for supporting the literature research. Thanks go also to our colleague Dr. med. Birgit Kammer, head of the section for paediatric radiology at Dr. von Hauner Children's Hospital, and to Mrs. Sieglinde "Sigi" Eberle, chief radiographer of the section for paediatric radiology, who have provided photographs of X-ray devices. And last but not least, the authors want to thank Mrs. Tanja Smith and Mrs. B. A. Isabella Körbl for performing the linguistic review of this book.

The authors want herewith to express their hope that this book might support a short step on the long way of radiation protection in paediatric radiology.

Munich, Germany
Summer 2019

Michael Seidenbusch
Veronika Rösenberger
Karl Schneider

Contents

Part I
Introduction

"Wir werden ja sehen, was wir sehen."—"We shall see what we will see."
***Wilhelm Conrad Röntgen,* 28.12.1895**

1 Radiation Risk and Radiation Protection in Paediatric Radiology

1.1 Paediatric Radiology

Immediately after *Wilhelm Conrad Röntgen*'s publication about X-rays[1] in December 1895 [58], the evolution of diagnostic radiology as a new medical discipline started [30], and so did the evolution of paediatric radiology. By 1896, within a very short period of time, 21 reports had been published on X-ray examinations in children [67]. Now, in rapid succession, departments for paediatric radiology were founded all over the world: In 1897, possibly for the first time ever, a department of paediatric radiology was established by the German physician and bacteriologist *Theodor Escherich* in the Anna Children's Hospital in Graz, Austria, [2, 16, 26], and in 1899, in the United States of America, the first department of paediatric radiology was founded in the Children's Hospital of Boston by the civil war surgeon *Francis Henry Brown* [8, 26]. These foundations can be traced back to the fact that, within a little while, diagnostic radiology had become an essential diagnostic tool not only in adult medicine [30] but also in paediatrics [1, 30, 57]. Nevertheless, child-specific radiological methods had still to be developed in the following decades [55, 59].

[1] In Germany, Austria, and Switzerland, the new kind of radiation discovered by *Roentgen* has been called "Röntgenstrahlung" (Roentgen rays), in other countries "x-rays" with "x" standing for "unknown".

1.2 Radiation Risk in Paediatric Radiology

At that time, typical radiographies required radiation exposures of several minutes [7, 14, 37]. Thus, radiation doses achieved in patients during X-ray examinations were about three orders of magnitude higher than today [38, 41]. As a consequence, adverse biological effects could be observed in clinical practice already within the first year of X-ray diagnostics. As an example, a case of hair-loss after a paediatric X-ray examination was described by *Daniel* [11] as an "interesting observation [of] a physiological effect of the x-rays" in April of 1896, but, of course, due to the lack of knowledge about the true nature of X-rays in former times, no correct pathomechanism could be identified.

Today, after more than 100 years of world-wide radiation research, the risks associated with ionizing radiation can be divided into deterministic and stochastic biological effects [39]. While deterministic effects do occur only after exceeding a tissue-specific threshold dose below which the effect is not observed, there is, according to the present state of research, no threshold for stochastic effects [39], of which radiation-induced cancerogenesis is the most terrifying one. Thus, since it is impossible to avoid stochastic radiation effects completely, these effects do play a crucial role in radiation protection especially in diagnostic radiology where ionizing radiations are intended to be

M. Seidenbusch et al., *Imaging Practice and Radiation Protection in Pediatric Radiology*,
https://doi.org/10.1007/978-3-030-18504-6_1

clinically applied in order to avert further deterministic pathologic damage of patients.

Unfortunately, children have to be considered as being much more radiation-sensitive than adults. Radiation exposure in children under 10 years has been assumed to increase the stochastic radiation risk by a factor of 3–4 [33]. And in fact, radio-biological investigations have shown in recent years that organs and tissues in children seem to be much more sensitive to ionizing radiation than those of adults [64, 66]. This fact is of utmost importance in paediatric radiology as most of the paediatric X-ray examinations are performed in neonates and young infants [62, 63] who may be even more radiosensitive than children of other age groups.

To add insult to injury, the collective radiation exposure of the European and US American population is mainly determined by medical imaging [13, 48, 56] and is still going up due to the increasing use of computed tomography in medical diagnostics [5]. Therefore, it can be assumed that this will also apply to children who represent a percentage of about 15–25% of the European and US American populations [20, 26]. Indeed, up to 10% of all conventional X-ray examinations are performed in children in Western Europe [17], and up to 15% in the United States [48], a lot of them in neonates and young infants [54]. Accordingly, *Brenner* et al. have warned at the beginning of the new millennium that the stochastic radiation risk of children might be considerably increased through paediatric computed tomographic (CT) examinations [3, 4]. Moreover, direct epidemiologic proof has been given recently of a significantly heightened risk for neoplastic diseases after CT examinations in the age range under 22 years [40, 52]. Nevertheless, diagnostic imaging of children by means of ionizing radiation is one of the most important diagnostic techniques in paediatrics and there is a clear common agreement that children's health is a primary consideration [65].

1.3 Radiation Protection in Paediatric Radiology

Against this background, the radiation exposure during X-ray examinations ought to be minimized especially in children and young adolescents. This requires a rational use of ionizing radiation [21–23, 28]. Today, there exist three fundamental principles of rational use of ionizing radiation in medicine [44] which are based on each other: justification, optimization, and limitation [26, 31, 35, 43, 63].

1.3.1 Justification

The fundamental and most effective principle of radiation protection in paediatric radiology is the principle of justification which represents one of the main tasks of the paediatric radiologist. Justification involves the question whether the particular procedure will actually have any benefit for the individual patient [26]. Thus, by observing the principles of efficacy [23, 26], many paediatric X-ray examinations might be avoided and replaced by alternative diagnostic procedures of similar diagnostic yield [50].

1.3.2 Optimization

If a paediatric X-ray examination is indicated according to the principle of justification, the radiation exposure ought to be kept as low as reasonably achievable. This demand has been summarized in the so-called ALARA concept, where ALARA is an acronym for the phrase "as low as reasonably achievable" [68]. This concept can be traced back to the phrase that radiation exposure should be kept "at the lowest practical level" in a report of the American National Committee on Radiation Protection (NCRP) which was published in 1954 [42, 47]. The ALARA concept may be also derived from the ALARP principle that has arisen in 1974 from legislation in the United Kingdom [51], where ALARP is the acronym for "as low as reasonably practicable".

Probably for the first time in paediatric radiology the principle of minimization was postulated by the German paediatrician *Reyher* in his paediatric radiology textbook entitled "Das Röntgenverfahren in der Kinderheilkunde" ("Roentgenology in paediatrics") [49, 57].

In Europe, systematic efforts were made in order to improve image quality and radiation exposure of patients in diagnostic radiology [46] since the 1960s. Based on these efforts and on the concept of efficacy which was defined by the World Health Organization (WHO) in 1977, *Helmut Fendel*, a paediatric radiologist at the Dr. von Hauner Children's Hospital, University of Munich, introduced a new era of radiation protection in paediatric radiology [21, 22, 24] considering the radiation doses as well as the image quality achieved during paediatric X-ray examinations [23, 25]. His publications anticipated the principles of justification and optimization introduced by the International Commission on Radiological Protection (ICRP) [31, 32]. In 1989, *Fendel* founded the so-called Lake Starnberg Group (Fig. 1.1, Table 1.1), a group of researchers of the European Society of Paediatric Radiology (ESPR) that held their meetings at the Lake Starnberg, Bavaria. In these years, efforts have been made by international working groups in order to optimize image quality and radiation exposure during conventional X-ray examinations [15, 45, 46]. After *Fendel*'s death in 1991, his successor *Karl Schneider* performed fundamental investigations on the image quality as well as the radiation dose in paediatric radiology all over Europe [61]. In 1996, the European Network of Pediatric Radiology (ENPR) that emerged from the former Lake Starnberg Group published the European Guidelines on quality criteria in paediatric radiology [9, 53]. These studies were considered in a recommendation of the German Commission for Radiation Protection [12] even many years later. In the United Kingdom, *Valmai Cook* published guidelines on the best imaging practice in paediatric imaging in 1998 [10]. The German Medical Association published in 2007 guidelines for all radiographic and fluoroscopic examinations with specific comments on the imaging procedures of paediatric patients [29], and so did the Austrian Ministry of Health [27].

Fig. 1.1 Helmut Fendel, founder of the Lake Starnberg Group, at the opening ceremony of the annual meeting of the European Society of Paediatric Radiology (ESPR), Munich 1990

Table 1.1 Members of the Lake Starnberg Group and the European Network of Paediatric Radiologists

Rosemary Arthur	Leeds, UK
Giampiero Beluffi	Pavia, Italy
Valmai Cook	London, UK
Clément Fauré ┼	Paris, France
Helmut Fendel ┼	Munich, Germany
Eldad Horwitz	Würzburg, Germany
Peter Kramer ┼	Utrecht, The Netherlands
Jean-Philippe Montagne	Paris, France
Noemi Perlmutter	Brussels Belgium
Karl Schneider	Munich, Germany
Betty Sweet	Glasgow, UK
Paul Thomas	Belfast, UK

Annotations:
┼ = deceased

1.3.3 Limitation

The concept of diagnostic reference levels (DRL) was introduced in radiation protection in 1997 [34]. This concept was developed and used for a long time in the United Kingdom. The DRL were defined as the third quartile dose values—either the entrance surface dose or the dose-area product values—received from large field studies in country. The very meaning is that radiology departments or private practices which exceed the third quartile values make at least one significant mistake in the daily routine imaging [36]. In Germany, DRL were introduced in 2003 [6] using the data from several European field studies published in European Guidelines [9] and have been revised three times, last amended in 2016 [60]. In the meantime, dose reference levels of many European countries have been published by the national authorities [9, 18, 19]. However, the DRL were not implemented in all European member states. Furthermore, they differ by a factor between 5 and 15 for chest X-rays which is the most radiographic examination in paediatrics. The reasons are diverse, but can be mainly attributed to ignore the recommendations on good radiographic practice outlined in the European guidelines on quality criteria for diagnostic radiographic images in paediatrics [9].

References

1. Benz-Bohm G, Richter E. Chronik der Kinderradiologie. Berlin, Heidelberg: Springer; 2012.
2. Benz-Bohm G, Stöver B, Riccabona M. 50 years Gesellschaft für Pädiatrische Radiologie—a challenge. Fortschr Röntgenstr. 2013;185:807–9.
3. Brenner DJ. Estimating cancer risks from pediatric CT: going from the qualitative to the quantitative. Pediatr Radiol. 2002;32:228–31.
4. Brenner DJ, Elliston CD, Hall EJ, Berdon WE. Estimated risks of radiation-induced fatal cancer from pediatric CT. AJR. 2001;176:289–96.
5. Brenner DJ, Hall EJ. Computed tomography: an increasing source of radiation exposure. N Engl J Med. 2007;357:2277–84.
6. Brix G. Bekanntmachung der diagnostischen Referenzwerte für radiologische und nuklearmedizinische Untersuchungen: Bundesamt für Strahlenschutz, Salzgitter; 2003.
7. Bürgerspital Basel. Jahresbericht pro 1899. Basel; 1899.
8. Caffey J. The first sixty years of pediatric roentgenology in the United States—1896 to 1956. AJR. 1956;76:437–54.
9. Commission of the European Communities. European guidelines on quality criteria for diagnostic radiographic images in paediatrics. EUR 16261, SBN 92-827-7843-6; 1996.
10. Cook JV, Shah K, Pablot S, Kyriou J, Pettett A, Fitzgerald M. Guidelines on best practice in the x-ray imaging of children. A manual for all x-ray departments. London: Queen Mary's Hospital for Children, Surrey, and The Radiological Protection Centre; 1998.
11. Daniel J. The x-rays. Science. 1896;3:562–3.
12. Deutsche Strahlenschutzkommission. Bildgebende Diagnostik beim Kind—Strahlenschutz, Rechtfertigung und Effektivität. Empfehlung der Strahlenschutzkommission, Deutschland; 2006.
13. Deutscher Bundestag, 18. Wahlperiode. Unterrichtung durch die Bundesregierung. Umweltradioaktivität und Strahlenbelastung im Jahr 2014. Drucksache 18/9600; 2016.
14. Dommann M. Durchsicht, Einsicht, Vorsicht. Eine Geschichte der Röntgenstrahlen, 1896-1963. Zürich. Chronos-Verlag; 2003.
15. Drexler G, Eriskat H, Schibilla H, Haybittle JL, Secretan LF. Criteria and methods for quality assurance in medical x-ray diagnosis. Br J Radiol. 1985;Suppl 18.
16. Escherich T. Die diagnostische Verwertung des Röntgenverfahrens bei Untersuchung der Kinder. Fortschr Röntgenstr. 1899;3:119.
17. European Commission. Radiation Protection 154. European guidance on estimating population doses from medical x-ray procedures. Directorate-General for energy and transport, Directorate H—Nuclear energy, Unit H.4—Radiation protection; 2008.
18. European Commission. Radiation Protection 180. Diagnostic reference levels in thirty-six European countries. Directorate-General for energy, Directorate D—Nuclear safety & fuel cycle, Unit D.3—Radiation protection; 2014.
19. European Commission. Radiation Protection 185. European guidelines on diagnostic reference levels for paediatric imaging. Directorate-General for Energy, Directorate D—Nuclear Energy, Safety and ITER, Unit D3—Radiation Protection and Nucl Saf; 2018.
20. Eurostat. Population structure and ageing. ISSN 2443-8219; 2016.
21. Fendel H. Die zehn Gebote des Strahlenschutzes bei der Röntgenuntersuchung im Kindesalter. Pädiat Prax. 1976;17:339–46.
22. Fendel H, Schneider K. Vernünftige diagnostische Bildgebung in der Kinderurologie. Krankenhaus-Arzt. 1988;61:513–22.
23. Fendel H, Schneider K, Bakowski C, Glas J, Drews K, Kohn MM. The efficacy of diagnostic radiation in paediatrics. Report of the Federal Ministery for Environment, Environment Protection and Reactor Safety of the Federal Republic of Germany, BMU-1987-161; 1987.
24. Fendel H, Stieve FE, editors. Radiation protection in pediatric radiology. NCRP report 68; 1983.
25. Fendel H, Stieve FE, editors. Vernünftige diagnostische Bildgebung bei Kindern. Bericht einer Studiengruppe der Weltgesundheitsorganisation. Berlin: Hoffmann Verlag; 1990.
26. Fendel H, Sweet EM, Thomas PS, editors. The status of paediatric radiology in Europe. Berlin: Schering AG; 1991. ISBN 3-921817-43-9.

27. Fotter R, Sinzig M, Jost G, Unterweger O, Weissensteiner S. Qualitätskriterien für Röntgenaufnahmen bei Kindern. Ein Leitfaden der AG Kinderradiologie: Österreichisches Bundesministerium für Gesundheit.
28. Fuchs W. Simple recommendations on how to avoid radiation harm. Western Electrician. 1896;12.
29. German Medical Association. Guidelines of the German Medical Association for quality assurance in diagnostic radiology. 2007.
30. Gocht H. Die Röngten-Literatur. II. Teil: Sachregister. Stuttgart: Enke Verlag; 1912.
31. International Commission on Radiological Protection. Protection against ionizing radiation from external sources. ICRP report 15; 1969.
32. International Commission on Radiological Protection. Recommendations of the International Commission on Radiological Protection. ICRP publication 26; 1977.
33. International Commission on Radiological Protection. 1990 recommendations of the International Commission on Radiological Protection. ICRP publication 60; 1990.
34. International Commission on Radiological Protection. Radiological protection and safety in medicine. ICRP publication 73; 1997.
35. International Commission on Radiological Protection. Radiological protection in paediatric diagnostic and interventional radiology. ICRP publication 121; 2013.
36. International Commission on Radiological Protection. Diagnostic reference levels in medical imaging. ICRP publication 135; 2017.
37. Kaufmann HJ, Ringertz H, Sweet E, editors. The first 30 years of the ESPR. The history of pediatric radiology in Europe. Berlin, Heidelberg: Springer; 1993. ISBN 978-3-540-56541-3.
38. Kemerink M, Dierichs TJ, Dierichs J, Huynen HJM, Wildberger JE, van Engelshoven JMA, Kemerink GJ. Characteristics of a first-generation x-ray system. Radiology. 2011;259:534–9.
39. Little MP. Risks associated with ionizing radiation. Br Med Bull. 2003;68:259–75.
40. Mathews JD, Forsythe AV, Brady Z, Butler MW, Goergen SC, Byrnes GB, Giles GG, Wallace AB, Anderson PR, Guiver TA, McGale P, Cain TM, Dowty JG, Bickerstaffe AC, Darby SC. Cancer risk in 680 000 people exposed to computed tomography scans in childhood or adolescence: data linkage study of 11 million Australians. BMJ. 2013;346:f2360.
41. Melo DR, Miller DL, Chang L, Moroz B, Linet MS, Simon SL. Organ doses from diagnostic medical radiography—trends over eight decades (1930 to 2010). Health Phys. 2016;111:235–55.
42. Miller DL, Schauer D. The ALARA principle in medical imaging. AAPM Newslett. 2015;40:38–40.
43. Moores BM. A review of the fundamental principles of radiation protection when applied to the patient in diagnostic radiology. Radiat Prot Dosim. 2017;175:1–9.
44. Moores BM. The psychology of decision making and its relevance to radiation protection of the patient in medicine. Radiat Prot Dosim. 2018;178:245–53.
45. Moores BM, Stieve FE, Eriskat H, Schibilla H, editors. Technical and physical parameters for quality assurance in medical diagnostic radiology. British Institute of Radiology, BIR report 18; 1989.
46. Moores BM, Wall BF, Eriskat H, Schibilla H, editors. Optimization of image quality and patient exposure in diagnostic radiology. British Institute of Radiology, BIR report 20; 1989.
47. National Council on Radiation Protection and Measurements. Permissible dose from external sources of ionizing radiation. NRCP report no. 17; 1954.
48. National Council on Radiation Protection and Measurements. Ionizing radiation exposure of the population of the United States. NRCP report no. 160; 2009.
49. Oestreich AE. ALARA 1912: "as low a dose as possible" a century ago. Radiographics. 2014;34: 1457–60.
50. Oikarinen H, Meriläinen S, Pääkkö E, Karttunen A, Nieminen MT, Tervonen O. Unjustified CT examinations in young patients. Eur Radiol. 2009;19:1161–5.
51. Parliament of the United Kingdom. Health and safety at work etc act. UK; 1974.
52. Pearce MS, Salotti JA, Little MP, McHugh K, Lee C, Kim KP, Howe NL, Ronckers CM, Rajaraman P, Craft AW, Parker L, Berrington de González A. Radiation exposure from CT scans in childhood and subsequent risk of leukaemia and brain tumours: a retrospective cohort study. Lancet. 2012;380:499–505.
53. Perlmutter N, Arthur RJ, Beluffi G, Cook V, Horwitz EA, Kramer P, Montagne JP, Thomas PS, Schneider K. The quality criteria for diagnostic radiographic images in paediatrics. Radiat Prot Dosim. 1998;80:45–8.
54. Portelli JL, McNulty JP, Bezzina P, Rainford L. Frequency of paediatric medical imaging examinations performed at a European teaching hospital over a 7-year period. Eur Radiol. 2016;26:4221–30.
55. Pouders E, Van Tiggelen R. Milestones in paediatric radiology. JBR-BRT. 2007;90:15–7.
56. Regulla D, Eder H. Patient exposure in medical x-ray imaging in Europe. Radiat Prot Dosim. 2005;114:11–25.
57. Reyher P. Das Röntgenverfahren in der Kinderheilkunde. Berlin: Hermann Meusser; 1912.
58. Röntgen WC. Ueber eine neue Art von Strahlen. (Vorläufige Mittheilung.) Sitzungsberichte der Würzburger Physik. medic. Gesellschaft. 1895:137–47.
59. Rotch TM. Living anatomy and pathology; the diagnosis of diseases in early life by the Roentgen method. Philadelphia, London: Lippincott Company; 1910.
60. Schegerer A. Diagnostische Referenzwerte für diagnostische und interventionelle Röntgenanwendungen. Salzgitter: Bundesamt für Strahlenschutz; 2016.

61. Schneider K. Evolution of quality assurance in paediatric radiology. Radiat Prot Dosim. 1995;57:119–23.
62. Seidenbusch MC, Schneider K. Radiation exposure of children in pediatric radiology. Part 1: referral criteria and x-ray examination frequencies at a university children's hospital between 1976 and 2003. Fortschr Röntgenstr. 2008;180:410–22.
63. Seidenbusch MC, Schneider K. Strahlenhygienische Aspekte bei der Röntgenuntersuchung des Thorax. Radiologe. 2015;55:580–7.
64. Stephan G, Schneider K, Panzer W, Walsh L, Oestreicher U. Enhanced yield of chromosome aberrations after CT examinations in paediatric patients. Int J Radiat Biol. 2007;83:281–7.
65. Unites Nations. Convention of the rights of the child; 1989.
66. United Nations Scientific Committee on the Effects of Atomic Radiation (UNSCEAR). UNSCEAR 2013 Report, Volume II: scientific findings on effects of radiation exposure of children, Scientific Annex B: effects of radiation exposure of children. ISBN 978-92-1-142293-1, New York: United Nations; 2013.
67. Willich E, Ebel KD. The history of paediatric radiology in Germany. In: Kaufmann HJ, Ringertz H, Sweet E, editors. The first 30 years of the ESPR. Berlin, Heidelberg: Springer; 1995.
68. Willis CE, Slovis TL. The ALARA concept in pediatric CR and DR: dose reduction in pediatric radiographic exams—a white paper conference executive summary. Radiology. 2005;234:343–4.

2 The Concept of Normalized Organ Doses for Dose Reconstruction in Paediatric Radiology

2.1 Normalized Organ Doses

The estimation of the radiation exposure of the patient during X-ray examinations is an essential requirement in order to assess the stochastic radiation risk of the patient. In Chap. 3, an outline of the fundamentals of paediatric dosimetry will be given.

In recent years, the estimation of the organ doses received during conventional radiographic X-ray examinations has been usually performed by applying normalized organ doses (synonymously: conversion factors, conversion coefficients) to basic dose indicators like the incident air kerma or the dose area product which can be measured. This concept of normalized organ doses is highly applicable in clinical radiology as normalized organ doses for conventional radiographs are given in numerous individual publications and in a few large systematic reference tables which have been published in recent years (see Chap. 29). Thus, organ doses can be estimated easily and quickly from basic dose indicators like the incidence air kerma or the dose area product.

2.2 Normalized Organ Doses Considering Varying Field Settings

In the reference tables published up to now, the selection of exposure parameters and standardized radiation field settings (size and position of the radiation field on the patient) has been performed through the institutions publishing the reference tables of normalized organ doses. However, exposure parameters and radiation field settings applied in clinical practice may differ from the standard values given in the reference tables. Beyond that, the cited reference tables are lacking quantitative information regarding the impact of the variation of exposure parameters and field settings on the radiation exposure especially in such organs and tissues which are localized close to the radiation field. Thus, the normalized organ doses listed in these tables are probably not generally applicable to a particular clinical examination situation, or are only applicable with limitations.

For this reason in particular and for considering the varying radiation field settings in clinical routine, new sets of normalized organ doses for calculating organ doses from measured incidence doses have been provided by the authors of this book in recent years, basing on optimal and suboptimal radiation field settings and considering different age groups of 0,[1] 1, 5, 10, 15, and 30 years for the most frequent X-ray examinations, i.e. the skull, the thorax, the abdomen, the segmental spine, the

[1]The age group "0" refers to children having an age less than 1 year. Although this term can be criticized from the paediatric point of view, it is very appropriate for mathematically describing anthropometric properties.

M. Seidenbusch et al., *Imaging Practice and Radiation Protection in Pediatric Radiology*,
https://doi.org/10.1007/978-3-030-18504-6_2

pelvis, and the thoraco-abdominal radiograph in neonates and young infants [1–6].

However, due to the limited space available in scientific publications, only a small set of exposure parameters and field settings could be considered in these publications. As normalized organ doses are significantly affected by the radiation quality, the normalized organ doses listed in these publications may be not complete enough for any kind of clinical application. In addition, no standard operating procedures regarding the patient preparation, the field setting, and the exposure parameters depending on the patient's age could be provided. As the field setting and the exposure parameters do have strong impact on the organ doses achieved in a patient, an integrated presentation of clinical and dosimetric aspects seems to be essential in order to achieve a maximum of clinical applicability.

Therefore, the intention of this book has been to provide new normalized organ doses for all common conventional paediatric X-ray examinations in paediatric patients of all age groups considering various aspects of field setting, exposure parameters, image quality requirements, and radiation doses.

References

1. Seidenbusch MC, Regulla D, Schneider K. Radiation exposure of children in pediatric radiology. Part 2: the PAEDOS algorithm for computer-assisted dose reconstruction in pediatric radiology and results for x-ray examinations of the skull. Fortschr Röntgenstr. 2008;180:522–39.
2. Seidenbusch MC, Regulla D, Schneider K. Radiation exposure of children in pediatric radiology. Part 3: conversion coefficients for reconstruction of organ doses achieved during chest x-ray examinations. Fortschr Röntgenstr. 2008;180:1061–81.
3. Seidenbusch MC, Regulla D, Schneider K. Radiation exposure of children in pediatric radiology. Part 6: conversion coefficients for reconstruction of organ dose in abdominal radiography. Fortschr Röntgenstr. 2009;181:945–61.
4. Seidenbusch MC, Regulla D, Schneider K. Radiation exposure of children in pediatric radiology. Part 7: conversion factors for reconstruction of organ dose during thoracoabdominal babygrams. Fortschr Röntgenstr. 2010;182:415–21.
5. Seidenbusch MC, Schneider K. Conversion coefficients for determining organ doses in paediatric spine radiography. Pediatr Radiol. 2014;44:434–56.
6. Seidenbusch MC, Schneider K. Conversion coefficients for determining organ doses in paediatric pelvis and hip joint radiography. Pediatr Radiol. 2014;44:1110–23.

Part II

Fundamentals of Radiation Dosimetry in Conventional Paediatric Radiology

"Ἀεὶ ὁ ***θεὸς γεωμετρεῖ.***"—"*God is ever doing geometry.*"

Platon

3 Specific Problems of Paediatric Radiology

Children are not to be considered as "little adults" [10, 20, 93]. Thus, medical examinations in children will have to be different from those in adults [19]. This is particularly true for paediatric X-ray examinations [10]. Children differ from adults regarding

- anthropometry
- anatomy and physiology
- psychology
- radiation biology
- radiation risk

3.1 Anthropometry

"Anthropometry is the measurement of human size, shape, and physical capabilities" [37]. Anthropometry is as old as humanity.[1] In ancient Egypt and Greece, it was a common matter in knowledge of artists and philosophers that there inevitably had to exist a divine relationship between anthropometry and geometry. Probably one of the first of the ancient artists who described the divine proportions in humans was *Polyklet* [24, 56]. Once again in the Renaissance, artists like *Leonardo da Vinci* tried to find a relationship between human body measures and geometrical figures [13]. Across the millennia, there have been a lot of motivations for the geometric approach to human anthropometry: theology [13], natural philosophy [47], visual arts [13, 24, 39, 44, 56, 79, 86], architecture and ergonomics [15, 74], and last but not least medicine [1, 3, 6, 7, 22, 23, 29, 33, 36, 40, 41, 42, 46, 51, 54, 82, 85, 89, 92]. Indeed, anthropometry is an important measure in paediatrics e.g. for diagnosing disorders in the somatic evolution of children [3, 22, 36, 85]. But also in paediatric radiology, the anthropometry of children does play a crucial role as radiation exposure during conventional radiographs is an exponential function of patient diameter (see Chap. 4.6) [1, 29, 38, 40, 41, 42, 46] and thus has to be considered when planning paediatric CT protocols [55]. Beyond that, the effects of scatter radiation, collimation, and shielding are strongly affected by the children's body measures as well. As children's anthropometry does vary over a wide range not only during ontogenesis but also within a defined age group [e.g. [54, 85]], determining exposure parameters for performing paediatric X-ray examinations or reconstructing radiation doses achieved during X-ray examinations in children is rather extensive.

There are various publications about paediatric anthropometry. Modern relations between human's age and anthropometry have been established e.g. by various authors in the seventeenth century [39], in the eighteenth century [47], in the nineteenth century [63, 78], and in the twentieth century [3, 22, 54, 86, 92]. These findings

[1] During the national-socialistic period in Germany (1933–1945), anthropometry was misused for racist purposes. The German authors of this book distance themselves strongly from this approach.

M. Seidenbusch et al., *Imaging Practice and Radiation Protection in Pediatric Radiology*,
https://doi.org/10.1007/978-3-030-18504-6_3

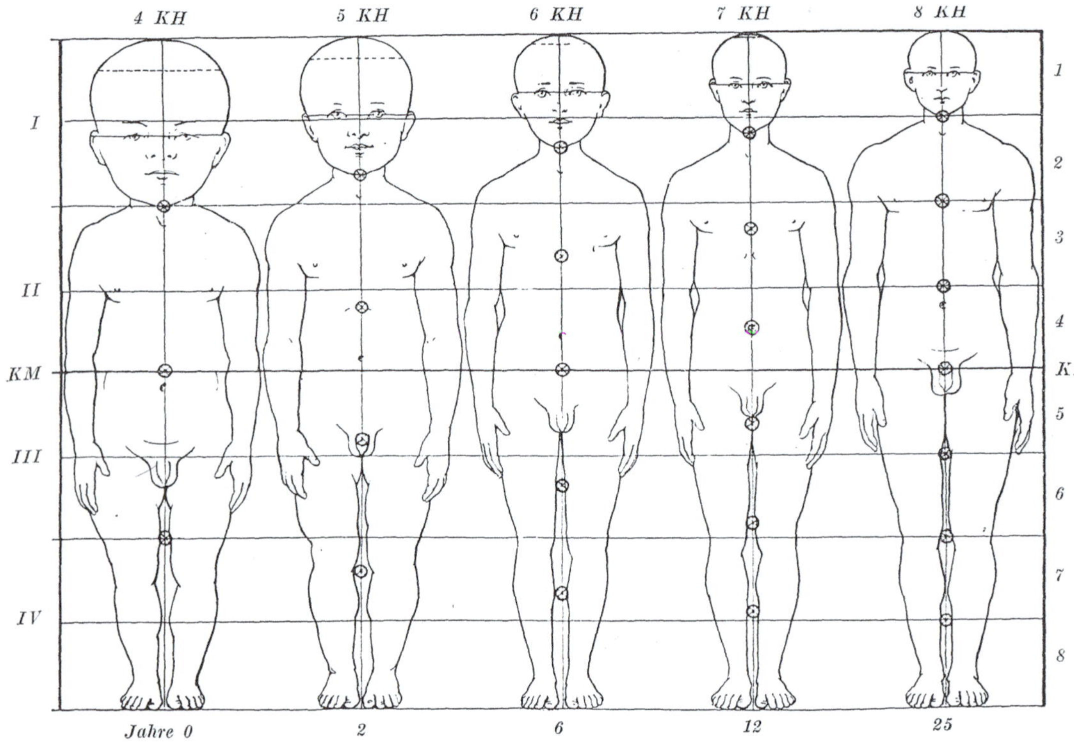

Fig. 3.1 Variation of human anthropometry over the years of life. Body measures are given in units of skull heights (KH = Kopfhöhe (German) = skull height, KM = Körpermittelpunkt (German) = midpoint of the body in the longitudinal axis) (From [54])

have also been transferred in textbooks of graphic art[2] [44, 86].

3.1.1 Body Height in Units of Head Heights

Figure 3.1 [54] shows the variation of human body measures as a function of age with the head height as measuring unit as it is common in visual arts [79] and probably was used first by *Polyklet* and his predecessors in ancient Greece and Egypt. A simple equation[3] can be used to describe the patient's total body height h_{Body} as a linear function of patient's age a in years (0–18 years) in terms of patient's head height h_{Head}:

$$h_{\text{Body}} = \left(\frac{2}{9} \cdot a + 4 \right) \cdot h_{\text{Head}} \qquad (3.1)$$

3.1.2 Anthropometric Parameters of Anatomical Regions

Detailed anthropometric parameters of children are given in various publications, e.g. in the tables of *Bohmann* [6, 7] and Hart et al. [29]. More data have been provided e g. by *Snyder et al.* [74] and by the National Health and Nutrition Examination Surveys (NHANES) database [34]. In recent years, especially diameters of anatomical regions have been determined through several clinical studies. Numerical values of diameters of various anatomical regions being examined in paediatric patients are given also in the publications mentioned above [6, 7, 29, 36]. In many X-ray examinations, however, the trunk diameter of children does play a crucial role. As the human trunk region has an elliptical cylindrical shape, the

[2] It seems to be somewhat strange that data of human body proportions can be taken from textbooks of graphic art rather than from books of human anatomy, but indeed, anthropometry seems to be more a topic of visual arts and architecture than a topic of human medicine [49].

[3] This equation has been derived by the British painter *Martin Dace.*

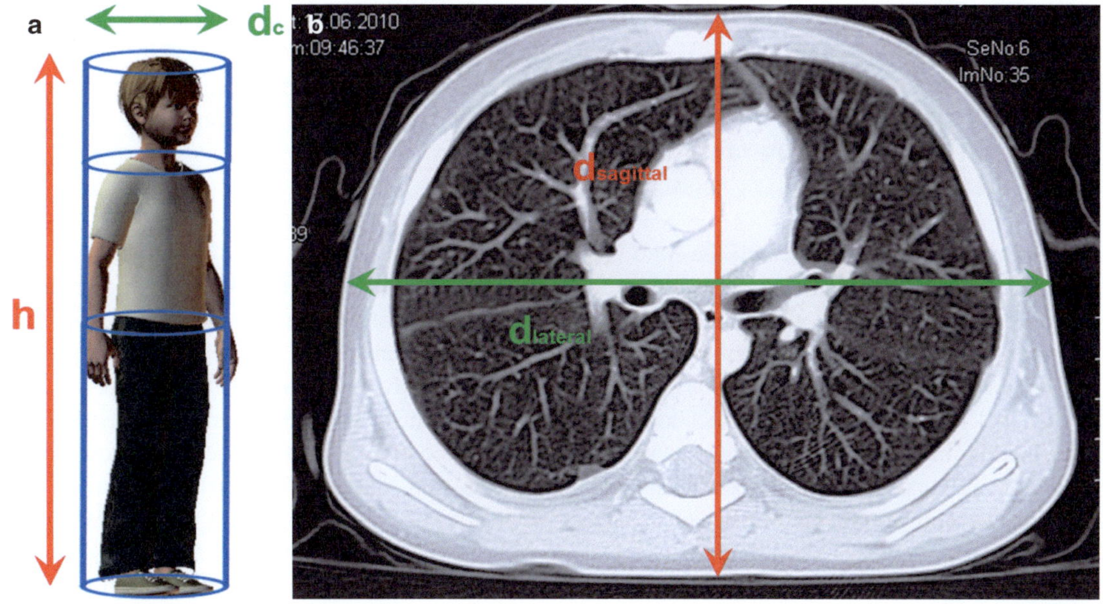

Fig. 3.2 Diameters of the trunk. (**a**) Concept of Lindskoug's equivalent cylindrical diameter. The equivalent diameter of the chest according to Lindskoug is derived from the patient's height h and the patient's mass m assuming the patient's body as a cylinder. The body density is assumed to be constant in all body regions. This approximation of the body of a paediatric patient by a cylinder is valid for children in their first 2 years of life. (**b**) Concept of the effective elliptical diameter. The effective elliptical diameter is the geometric mean of the sagittal and the lateral trunk diameter of the patient. Using computed tomography the diameters can be exactly measured [52, 64, 67]. (**b**) shows a chest CT image of a 6-month-old infant (© M. Seidenbusch 2018)

thickness of the human trunk can be described by its anatomical sagittal and lateral diameters d_{sagittal} and d_{lateral}.[4] For dosimetric purposes, it can be advantageous to calculate a virtual mathematical diameter representing these both real anatomical diameters.

Beyond that, also the chemical composition of organs and tissues determines the dosimetric properties of patients. Detailed chemical compositions are listed e.g. in [31].

3.1.3 Anthropometric Parameters of the Trunk Region

3.1.3.1 Equivalent Cylindrical Diameter of the Trunk Region

An approach to calculate a mathematical diameter from clinical anthropometric parameters is *Lindskoug's* concept of the equivalent cylindric diameter (in short: equivalent diameter) d_c (Fig. 3.2) [40, 41, 42, 46] which can be calculated from the patient's height h and mass m approximating the patient's shape by a cyclic cylinder with height h, diameter d_c, and medium density ρ:

$$d_c = \frac{2}{\sqrt{\pi \cdot \rho}} \cdot \sqrt{\frac{m}{h}} \qquad (3.2)$$

3.1.3.2 Effective Elliptical Diameter of the Trunk Region

Another approach is the calculation of an effective elliptical diameter (in short: effective diameter) d_e from the real sagittal diameter d_{sagittal} and lateral diameter d_{lateral} considering the fact that the transversal section of human body has an elliptical shape [1, 9]:

$$d_e = \sqrt{d_{\text{sagittal}} \cdot d_{\text{lateral}}} \qquad (3.3)$$

As can be seen e.g. in Chap. 4 from Table 4.3, the equivalent cylindrical diameter is closer to the

---

[4] By the way, during the adolescence, the ratio of the lateral to the sagittal diameter converges to the golden section [72].

sagittal diameter and the effective elliptical diameter is closer to the lateral diameter.

3.1.4 Anthropometric Parameters of the Skull

The anthropometric parameters of the skull (sagittal diameter, lateral diameter, head circumference, skull surface, brain weight, brain volume and many more) have been described since several centuries in numerous publications [e.g. [30, 54, 65, 89]]. In order to perform correct dosimetric calculations in paediatric patients, a realistic model of skull and brain are necessary especially in neonates and young infants because of the short distance between skull and trunk. Only then, scattered radiation originating from skull and chest is appropriately taken into account. In Chap. 4 will be shown that the skull volume has been considerably underestimated in the mathematical phantoms used for calculating the normalized organ doses in this book.

3.1.5 Somatograms

In paediatrics, for describing the age-related development of children, so-called somatograms are used which depict the dependence of body height ad body weight as functions of age [23, 51, 85]. As can be seen from somatograms derived from patients' anthropometric data acquired in the Dr. von Hauner Children's Hospital of the University of Munich during 1976–1998 [68], anthropometric properties such as body height and body weight show a wide range even within a defined age group (Figs. 3.3 and 3.4).

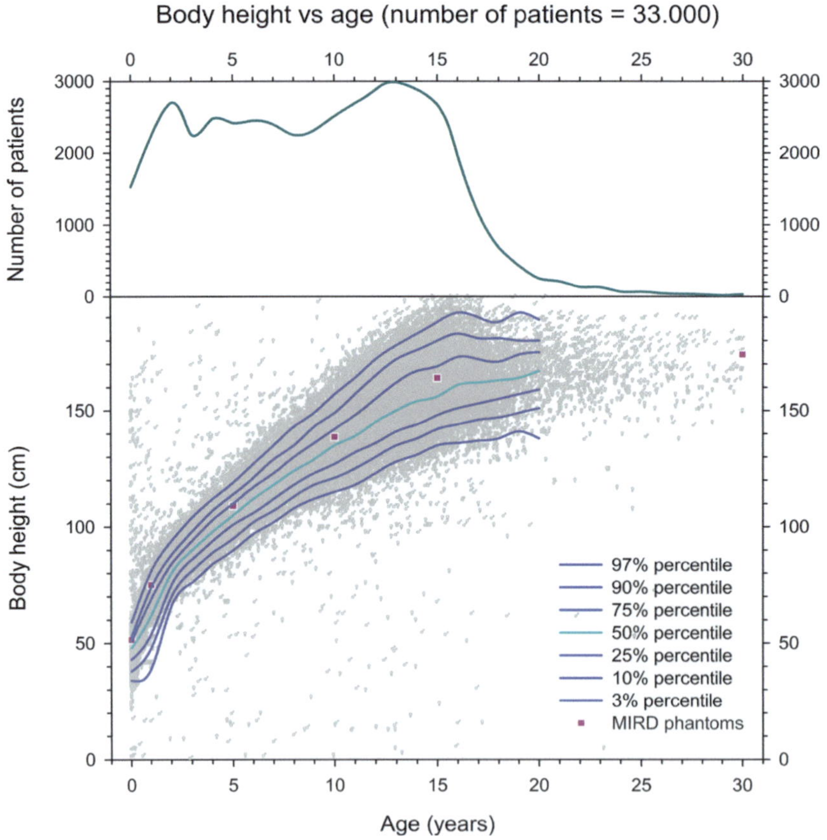

Fig. 3.3 Somatogram for body height of children in the Dr. von Hauner Children's Hospital, University of Munich, 1976–1998 [68]. Blue lines = 3%, 10%, 25%, 75%, 90%, and 97% percentiles; green line = 50% percentile = median; red squares = body heights of mathematical STUK MIRD phantoms. Body heights of STUK MIRD phantoms do match very well with the median body heights of the patient collective. The symbols represent original, uncorrected values as taken from the database of the paediatric radiology department (© M. Seidenbusch 2003)

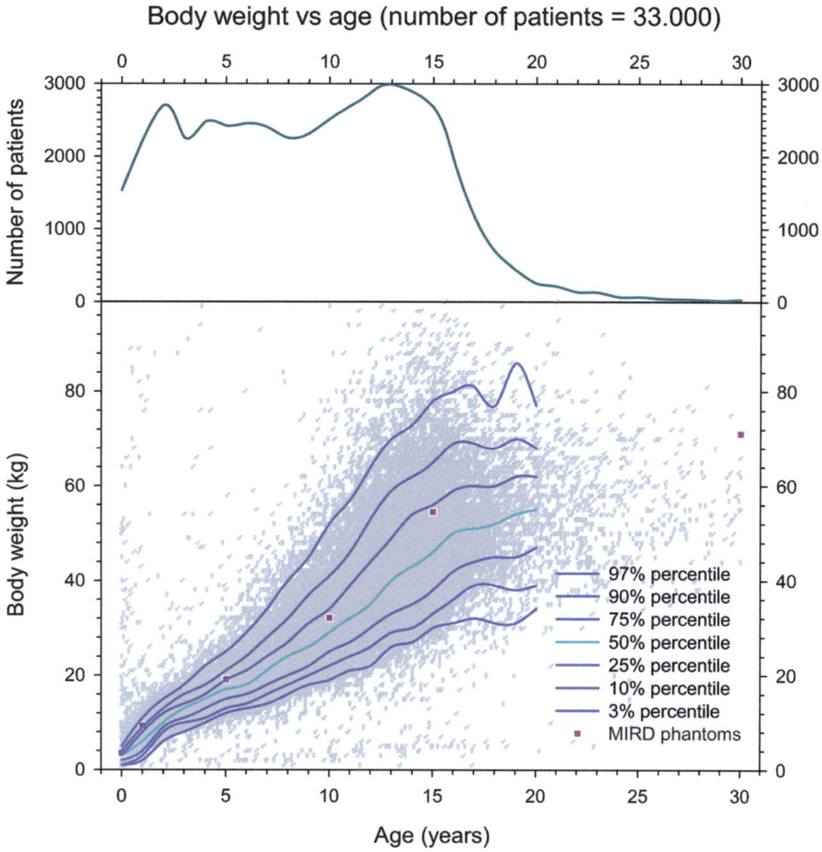

Fig. 3.4 Somatogram for body weight of children in the Dr. von Hauner Children's Hospital, University of Munich, 1976–1998 [68]. Blue lines = 3%, 10%, 25%, 75%, 90%, and 97% percentiles; green line = 50% percentile = median; red squares = body weights of mathematical STUK MIRD phantoms. Body weights of STUK MIRD phantoms do match very well with the median body weights of the patient collective. The symbols represent original, uncorrected values as taken from the database of the of the paediatric radiology department (© M. Seidenbusch 2003)

3.1.6 Age Groups[5]

The simplest way to assign normalized organ doses to a paediatric patient is characterizing the patient's anthropometry by his age. However, despite of the fact that patients' anthropometries may vary over a wide range even in the same age group, the definition of age groups may be reasonable not only for practical reasons.

There are various possibilities to define age groups for children [82]. From the medical point of view, the so-called traditional age groups can be defined (Table 3.1, [83]). Beyond that, according to numerous anthropometric considerations, several systems of age group intervals have been proposed (Table 3.2, [82]).

[5]The age group "0" refers to children having an age less than 1 year. Although this term can be criticized from the paediatric point of view, it is very appropriate for mathematically describing anthropometric properties.

Table 3.1 Traditional age groups [83]

Age group	Age range
Newborn	0–3 months
Infant	0–12 months
Toddler	–3 years
Preschool	–6 years
School age	–12 years
Adolescent	–18 years

Table 3.2 Examples for different age group systems [82]

Age group interval (years)				
Various countries [14, 26, 29, 47]	UK [32, 71, 72]	Japan [28]	Portugal [52]	Australia [7]
–	–	–	0–0.25	–
0–1	<1	<1	0.25–3	0–4
1–5	5	1–7	3–8	
5–10	10	8–12	8–15	5–14
10–15	–	13–19	15–19	–
–	–	>19	–	–

These age group intervals can be assigned to body weight intervals (Table 3.3, [84]).

Another approach is calculating age groups by defining equidistant trunk or body height intervals which can be derived from the anthropometric data published in somatograms (Tables 3.4 and 3.5 [83]).

In another approach, age group intervals are chosen in a way that within an age group organ sizes do not vary more than 15% (Table 3.6, [83]).

Overall, it can be stated that the age groups and age group intervals defined in Table 3.6 correspond to minimal variations of trunk or body height and of organ dimensions within an age group.

Table 3.3 Age groups corresponding to body weight intervals [84]

Age group interval (years)	Body weight interval (kg)
0–1	<10
1–5	10–20
5–10	20–35
10–15	>35

Table 3.4 Age groups corresponding to nearly equal trunk height intervals [83]

Age group (years)	Age range (years)	Trunk height range (cm)
0	0–0.5	10–24
1	0.5–3	24–30
5	3–7.5	30–36
10	7.5–12.5	36–42
15	12.5–17	42–50

Considering the medium percentile of a somatogram of trunk height vs. age, age ranges were derived from somatogram values corresponding to trunk height intervals between the age groups of about 6 cm

Table 3.5 Age groups corresponding to body height intervals [83]

Age group (years)	Age range (years)	Height range (cm)
0	0–0.5	0–66
1	0.5–3	66–95
5	3–7	95–124
10	7–13	124–156
15	13–17	156–168

Considering the medium percentile of a somatogram of body height vs. age, age ranges were derived from somatogram values corresponding to body height intervals of about 30 cm

Table 3.6 Age groups corresponding to variation of organ dimensions [83]

Age group (years)	Age range
0	0–0.5
1	0.5–3
5	3–7
10	7–13
15	13–17

Age groups are chosen in a way that within an age group organ sizes do not vary more than 15%

3.2 Anatomy and Physiology

Children differ from adults not only in anthropometry [33] but also in anatomy and physiology [54]. There are several implications arising from these differences: The age-depending anatomical situs of organs, the size of organs and tissues have an impact on X-ray findings. As in infants and young children organs are localized closer together than in older children, radiation protection through collimation or shielding is much more sophisticated. Beyond that, the age-related specific peculiarities of the physiology, e.g. respiratory rate, have to be considered. By this means, the exposure time should be as low as possible in uncooperative patients. It should be kept below 4 ms for ap chest X-rays in infants to avoid motion and minimize pulsation artefacts of small intrapulmonary interstitial pulmonary structures [10]. In this paragraph, reference will be made to some age-specific properties of some organs and tissues insofar they are important for radiation protection. The following paragraphs refer also to the STUK MIRD phantom family described in Chap. 4.8.

3.2.1 Skull

The ocular lenses are not inserted in the STUK-phantoms. Therefore, no calculations of the lens dose are possible.

In first year of life, a considerable proportion of the active bone marrow (about 30%) is located in the cranial vault and the facial bones. This proportion of bone marrow decreases continuously over the years. Post puberty it is in the range of 10% of the total.

3.2.2 Thymus

The thymus varies greatly in size and shape up to the age of 2 years. The cervical lobe of the thymus—a frequent anomaly in infancy—is not incorporated in the STUK MIRD phantoms. This variant underestimates the thymus dose even if the skull ap radiograph is collimated tightly (optimal field setting).

3.2.3 Chest

Size and shape of the human chest depend on patient's age [6, 7, 61, 72]. Thus, the sagittal and lateral chest diameters are a function of age. In elder children, the increase of the lateral chest diameter has been overestimated when the phantom family of PCXMC program was created [72].

3.2.4 Lungs

The lung density and thus the value of the effective linear mass-attenuation coefficients (see Chap. 4) of the lungs decrease with patient's age [35, 43, 72, 77]. This can be explained by the enlargement of the air spaces with simultaneous decrease of the interstitial tissue of the developing lungs. This process, the continuing increase in number and size of the alveoli occurs within the first 7 years of life. At the same time, the patient's diameters grow preferably in the horizontal and longitudinal direction with age. These two characteristics implicate that the radiation doses required for chest radiographs can be kept nearly constant over the age range from infants till the age of 10 years, a physical effect which at first glance may seem somewhat strange but which is well known in clinical routine [71].

3.2.5 Mammae

Post-partum maternal hormones can sensitize the breast tissue of full-term newborns leading to a transient increase of the mammary tissue. This increase of glandular tissue mass in the first 3–6 months of life has not been respected when the phantom family of PCXMC program was created [69]. The glandular breast tissue in this age period is probably more vulnerable to ionizing radiation as the mitotic activity is increased [90].

3.2.6 Gonads

The location of ovaries varies considerably in all age groups of females. In neonates they are often located higher in the pelvic region than in older girls or adolescents. One reason for the high position is a long mesovar which allows the ascension of ovaries even into the upper abdomen [2, 14, 18, 21, 58]. The size of the urinary bladder can also change the position of the ovaries considerably.

The male gonads descend to the scrotum. This process is not finished before the school age. In premature male babies born before the 28th gestational week, they are located within the abdomen.

3.2.7 Active Bone Marrow

The active bone marrow with its stem cells is amongst the most radiation-sensitive tissues in human body. Therefore, the radiation exposure of active bone marrow will contribute significantly to the total stochastic radiation risk. The estimation of the active bone marrow dose requires detailed quantitative information about its regional distribution in the different compartments of the human body. This regional distribution of red bone marrow cells, however, is depending on age [11, 16, 62, 66, 87, 88, 91]. For example, in premature babies with a gestational age of 28 weeks, erythropoietic stem cells are mainly found in the liver with a percentage of 50% and in the spleen with a percentage of 15% [50]. Even in mature neonates, during the first weeks of life, not all stem cells will have migrated to the active bone marrow. It must be stated here that it remains questionable whether these aspects have been considered sufficiently when constructing the mathematical MIRD phantoms forming the basis of the normalized organ doses listed in this book.

As active bone marrow cells are enclosed by bone cells and as the linear mass absorption coefficients (see Chap. 4) of bone are higher than these of red bone marrow, interface effects may play an important role during radiation exposure of active bone marrow cells [75] resulting in a local dose enhancement which likely cannot be calculated through the conventional phantom models used in radiation dosimetry.

3.3 Psychology

It may be somewhat surprising at first glance that even children's psychology should play an important role in paediatric radiology [5, 80]. However, already in 1911, the German paediatrician *Reyher* reported that X-ray examinations create additional distress for the child [60]. Indeed, for a child, "admission to hospital is a kind of trap from which there is no escape" [20], and "there is always the fear of unknown" [20]. Moreover, "large [X-ray] machines are intimidating, and most children do not like the feeling that they are trapped under something which may fall on them" [20]. Therefore, children often will not cooperate during medical examinations or even try to escape, especially during X-ray examinations. Of course, a restless child will enhance the number of motion artefacts during X-ray examinations [59]. Thus, from the dosimetric point of view, this behaviour of children may result in repeated radiographs or in extended fluoroscopy durations. As a consequence, the radiation burden in uncooperative children may be increased compared to the radiation doses achieved in cooperating children.[6,7]

[6] Moreover, also the parents' psychology may be taken into consideration as the perception of radiation risk is rather different in different population groups [17, 45] and thus affects radiation risk communication in clinical practice.

[7] More literature about this subject has been provided in [80].

3.4 Radiation Biology

As mentioned in the Introduction section, children have to be considered as being much more radiation-sensitive than adults. The radiation exposure in children under 10 years has been assumed to increase the stochastic radiation risk by a factor of 3–4 [32]. Radio-biological investigations have shown in recent years that at least some organs and tissues in children seem to be much more sensitive to ionizing radiation than those of adults [76, 81]. In particular, children younger than 10 years seem to be more radiation-sensitive than adolescents. Finally, this factor may be of wider importance as a lot of paediatric X-ray examinations are performed in neonates and young infants [70, 73]. The reason for the enhanced radiosensitivity of children may be the fact "that radiosensitivity of tissues depends upon the number of undifferentiated cells which the tissue contains, the degree of mitotic activity in the tissue, and the length of time that the cells of the tissue stay in active proliferation" (law of *Bergonié* and *Tribondeau*) [4, 57] and all of these factors are enhanced in children.

3.5 Radiation Risk

Against this background, the question comes up whether children have an increased stochastic radiation risk through paediatric X-ray examinations. For the dose range of conventional X-ray examinations, no significant enhancement of stochastic radiation risk could be found in epidemiologic studies [e.g. [25–28]]. However, *Brenner et al.* assumed that the stochastic radiation risk of children might be considerably increased by paediatric CT examinations [8]. According to this, direct epidemiologic proof of a significantly heightened risk for neoplastic diseases after CT examinations in the age range under 22 years has been given [48, 53].

References

1. American Association of Physicists in Medicine. Size-specific dose estimates (SSDE) in pediatric and adult body CT examinations. AAPM report 204; 2011.
2. Bardo DME, Black M, Schenk K, Zaritzky MF. Location of the ovaries in girls from newborn to 18 years of age: reconsidering ovarian shielding. Pediatr Radiol. 2009;39:253–9.
3. Bakwin H, Bakwin RM. Body build in infants: V. Anthropometry in the new-born. Human Biol. 1934;6:612–26.
4. Bergonié J, Tribondeau L. Interprétation de quelques résultats de la radiothérapie et essai de fixation d'une technique rationnelle. C R Acad Sci. 1906;143:983–5.
5. Björkman B. Children in the radiology department—a study of anxiety, pain, distress and verbal interaction. Thesis, Jönköping University, Dissertation series no. 51; 2014.
6. Bohmann I. Ermittlung des Durchstrahlungsdurchmessers bei Säuglingen, Kindern und Jugendlichen zur Aufstellung von Belichtungswerten in der Röntgendiagnostik und Abschätzung der Organdosiswerte bei typischen Röntgenuntersuchungen. Dissertation, Universität München; 1990.
7. Bohmann I. Ermittlung des Durchstrahlungsdurchmessers bei Säuglingen, Kindern und Jugendlichen zur Aufstellung von Belichtungswerten in der Röntgendiagnostik und Abschätzung der Organdosiswerte bei typischen Röntgenuntersuchungen. GSF-Bericht 16/90; 1990.
8. Brenner DJ, Elliston CD, Hall EJ, Berdon WE. Estimated risks of radiation-induced fatal cancer from pediatric CT. AJR. 2001;176:289–96.
9. Burton CS, Szczykutowicz TP. Evaluation of AAPM reports 204 and 220: estimation of effective diameter, water-equivalent diameter, and ellipticity ratios for chest, abdomen, pelvis, and head CT scans. J Appl Clin Med Phys. 2018;19:228–38.
10. Commission of the European Communities. European guidelines on quality criteria for diagnostic radiographic images in paediatrics. EUR 16261, SBN 92-827-7843-6; 1996.
11. Cristy M. Active bone marrow distribution as a function of age in humans. Phys Med Biol. 1981;26:389–400.
12. Daldrup-Link HE, Gooding CA, editors. Essentials of pediatric radiology. New York: Cambridge University Press; 2010.
13. Da Vinci L. In: Pacioli L, editor. De divina proportione. Venice: Paganini; 1509.
14. Dawood RM, Hall CM. Too much radiation for too many children? Br Med J. 1988;296:1277–8.
15. Dreyfuss H. The measure of man. Human factors in design. New York: Whitney Library of Design; 1967.
16. Ellis RE. The distribution of active bone marrow in the adult. Phys Med Biol. 1961;5:255–8.
17. European Commission. Radiation protection No 184—EU Scientific Seminar 2015 "Risk communication". Directorate-General for Energy, Directorate D—Nuclear Energy, Safety and ITER, Unit D3—Radiation Protection and Nuclear Safety; 2016.
18. Featherstone C, Harnett AN, Brunt AM. Ultrasound localization of the ovaries for radiation-induced ovarian ablation. Clin Oncol. 1999;11:393–7.
19. Fendel H, Schneider K, Bakowski C, Glas J, Drews K, Kohn MM. The efficacy of diagnostic radiation in paediatrics. Report of the Federal Ministry for Environment, Environment Protection and Reactor Safety of the Federal Republic of Germany, BMU-1987-161; 1987.
20. Fendel H, Sweet EM, Thomas PS, editors. The status of paediatric radiology in Europe. Berlin: Schering AG; 1991. isbn:3-921817-43-9.
21. Fochem K, Pape R. Problematik des Ovarialschutzes bei Röntgenaufnahmen des Beckens. Fortschr Röntgenstr. 1962;97:785–79.
22. Friedenthal HWC. Allgemeine und spezielle Physiologie des Menschenwachstums. Berlin: Springer; 1914.
23. Fryar CD, Gu Q, Ogden CL. Anthropometric reference data for children and adults: United States, 2007–2010. National Center for Health Statistics. Vital Health Stat. 2012;(252):1–48.
24. Galenos of Pergamon A. De placitis Hippocratis et Platonis 5, 449.
25. Hammer GP, Seidenbusch MC, Regulla DF, Spix C, Zeeb H, Schneider K, Blettner M. Childhood cancer risk from conventional radiographic examinations for selected referral criteria: results from a large cohort study. AJR. 2011;197:217–23.
26. Hammer GP, Blettner M, Schneider K, Seidenbusch MC, Regulla DF. Kohortenstudie zur Abschätzung des Krebsrisikos durch diagnostische Strahlenexposition im Kindesalter (RICC-Studie). Ergebnisbericht, Forschungsvorhaben StSch 4432, Bundesamt für Strahlenschutz, urn:nbn:de:0221-2009011214, Salzgitter; 2009.
27. Hammer GP, Seidenbusch MC, Schneider K, Regulla DF, Zeeb H, Spix C, Blettner M. A cohort study of childhood cancer incidence after postnatal diagnostic x-ray exposure. Radiat Res. 2009;171:504–12.
28. Hammer GP, Seidenbusch MC, Schneider K, Regulla DF, Zeeb H, Spix C, Blettner M. Cancer incidence rate after diagnostic x-ray exposure in 1976–2003 among patients of a university children's hospital. Fortschr Röntgenstr. 2010;182:404–14.
29. Hart D, Wall BF, Shrimpton PC, Bungay DR, Dance DR. Reference doses and patient size in paediatric radiology. Didcot: National Radiological Protection Board NRPB-R318; 2000. isbn:0-85951-448-X.
30. Huschke E. Schaedel, Hirn und Seele des Menschen und der Thiere nach Alter, Geschlecht und Race, dargestellt nach neuen Methoden und Untersuchungen. Jena: Friedrich Mauke Verlag; 1854.
31. International Commission on Radiological Protection. Report of the task group on reference man: anatomi-

cal, physiological and metabolic characteristics. ICRP publication 23. Oxford: Pergamon Press; 1975.
32. International Commission on Radiological Protection. 1990 recommendations of the International Commission on Radiological Protection. ICRP publication 60. Oxford: Pergamon Press; 1990.
33. International Commission on Radiological Protection. Basic anatomical and physiological data for use in radiological protection: reference values. ICRP publication 89. Oxford: Pergamon Press; 2003.
34. Johnson PB, Whalen SR, Wayson M, Juneja B, Lee C, Bolch WE. Hybrid patient-dependent phantoms covering statistical distributions of body morphometry in the U.S. adult and pediatric population. Proc IEEE. 2009;97:2060–75.
35. Kalender WA, Rienmüller R, Seissler W, Behr J, Welke M, Fichte H. Measurement of pulmonary parenchymal attenuation: use of spirometric gating with quantitative CT. Radiology. 1990;175:265–8.
36. Kleinman PL, Strauss KJ, Zurakowski D, Buckley KS, Taylor GA. Patient size measured on CT images as a function of age at a tertiary care children's hospital. AJR. 2010;194:1611–9.
37. Klinich KD, Reed MP. Pediatric anthropometry. In: Crandall JR, Myers BS, Meaney DF, Zellers-Schmidtke S, editors. Pediatric injury biomechanics. New York: Springer company; 2013. p. 1–31.
38. Ladia AP, Skiadopoulos SG, Karahaliou AN, Messaris GAT, Delis HB, Panayiotakis GS. The effect of increased body mass index on patient dose in paediatric radiography. Eur J Radiol. 2016;85:1689–94.
39. Lautensack H. Dess Cirkelss und Richtscheyts, auch der Perspectiva, und Proportion der Menschen und Rosse. Simon Schamberger; 1618.
40. Lindskoug BA. The reference man in diagnostic radiology dosimetry. Br J Radiol. 1992;65:431–7.
41. Lindskoug BA. Reference man in diagnostic radiology dosimetry. Radiat Prot Dosim. 1992;43:111–4.
42. Lindskoug BA. Exposure parameters in x ray diagnostics of children, infants and the newborn. Radiat Prot Dosim. 1992;43:289–92.
43. Long FR, Williams RS, Castile RG. Inspiratory and expiratory CT lung density in infants and young children. Pediatr Radiol. 2005;35:677–83.
44. Loomis A. Figure drawing for all it's worth. London: Chapman & Halls; 1943.
45. Luetzelschwab J. The 3 R's: radiation, risk, and reason. Carlisle: Dickinson College, Department of Physics; 1996.
46. Martin CJ, Farquhar B, Stockdale E, Macdonald S. A study of the relationship between patient dose and size in paediatric radiology. Br J Radiol. 1994;67:864–71.
47. Martinet JF. Katechismus der natuur. Amsterdam: Joanne van Selm; 1777.
48. Mathews JD, Forsythe AV, Brady Z, Butler MW, Goergen SC, Byrnes GB, Giles GG, Wallace AB, Anderson PR, Guiver TA, McGale P, Cain TM, Dowty JG, Bickerstaffe AC, Darby SC. Cancer risk in 680 000 people exposed to computed tomography scans in childhood or adolescence: data linkage study of 11 million Australians. BMJ. 2013;346:f2360.
49. Milz S. Private communication. Chair of anatomy. Munich: University of Munich; 2018.
50. Nathan DG, Oski FA. Hematology of infancy and childhood, vol. I. Philadelphia: Holt-Saunders; 1981.
51. Neuhauser H, Schienkiewitz A, Schaffrath Rosario A, Dortschy R, Kurth BM. Referenzperzentile für anthropometrische Maßzahlen und Blutdruck aus der Studie zur Gesundheit von Kindern und Jugendlichen in Deutschland (KiGGS). Berlin: Robert Koch-Institut; 2013. isbn:978-3-89606-218-5.
52. Parikh RA, Wien MA, Novak RD, Jordan DW, Klahr P, Soriano S, Ciancibello L, Berlin SC. A comparison study of size-specific dose estimate calculation methods. Pediatr Radiol. 2018;48:56–65.
53. Pearce MS, Salotti JA, Little MP, McHugh K, Lee C, Kim KP, Howe NL, Ronckers CM, Rajaraman P, Craft AW, Parker L, Berrington de González A. Radiation exposure from CT scans in childhood and subsequent risk of leukaemia and brain tumours: a retrospective cohort study. Lancet. 2012;380:499–505.
54. Peter K, Wetzel G, Heiderich F, editors. Handbuch der Anatomie des Kindes. München: Verlag von J. F. Bergmann; 1938.
55. Phillips GS, Stanescu AL, Alessio AM. Relationships of pediatric anthropometrics for CT protocol selection. AJR. 2014;203:W85–91.
56. Philon of Byzanz. Mechanike syntaxis 4, 1, 49.
57. Pizzarello DJ, Witcofski RL. Basic radiation biology. Philadelphia: Lea & Febiger; 1967.
58. Poretti GG. Radiation exposure of a population due to diagnostic x-ray examinations: some critical remarks. Phys Med Biol. 1985;30:1017–27.
59. Poznanski EO. Some psychologic factors affecting the child in the radiology department. In: Poznanski AK, editor. Practical approaches to pediatric radiology. Chicago: Year Book Medical Publishers; 1976. p. 55–9.
60. Reyher P. Das Röntgenverfahren in der Kinderheilkunde. Berlin: Hermann Meusser; 1912.
61. Riccabona M, Beer M, Mentzel H-J, editors. Bildgebung des Thorax bei Neugeborenen und Kleinkindern. Berlin: Springer; 2019.
62. Ricci C, Cova M, Kang YIS, Yang A, Rahmouni A, Scott WW, Zerhouni EA. Normal age-related patterns of cellular and fatty bone marrow distribution in the axial skeleton: MR imaging study. Radiology. 1990;177:83–8.
63. Richer P. Description des formes extérieures du corps humain au repos et dans les principaux mouvements. Paris: Imprimeurs-éditeurs Plon, Nourrit et Cie; 1890.
64. Richter E, Lierse W. Radiologische Anatomie des Neugeborenen. Gesicht, Hals und Thorax. München: Urban & Schwarzenberg; 1990. p. 68–146.
65. Schienkiewitz A, Schaffrath Rosario A, Dortschy R, Ellert U, Neuhauser H. German head circumference references for infants, children and adolescents in comparison with currently used national and international references. Acta Paediatr. 2011;100:e28–33.
66. Schneider C, Montz R. Die quantitative Verteilung des erythropoetischen Knochenmarks beim Menschen gemessen mit Radioeisen. J Mol Med. 2005;16:969–73.

67. Schneider K. The normal child: growth and development of the infant and the child; frequent and important variants. In: Daldrup-Link HE, Gooding CA, editors. Essentials of pediatric radiology. New York: Cambridge University Press; 2010. p. 1–19.
68. Seidenbusch M. Rekonstruktion von Organ- und Effektivdosen bei konventionellen Röntgenuntersuchungen am Dr. von Haunerschen Kinderspital der Universität München mit einer Berechnung neuer Konversionsfaktoren für die pädiatrische Radiologie. Dissertation, LMU München: Medizinische Fakultät; 2006.
69. Seidenbusch M, Kammer B, Specht T, Stahl R, Schneider K. Zur Rekonstruktion der Brustdrüsendosis bei Kindern in der pädiatrischen Radiologie: Gewicht, Dichte und Volumen der Mammagewebe bei Neugeborenen, Säuglingen, Kleinkindern und Adoleszenten und die dosimetrischen Konsequenzen. Fortschr Röntgenstr. 2018;190:869–70.
70. Seidenbusch MC, Schneider K. Radiation exposure of children in pediatric radiology. Part 1: referral criteria and x-ray examination frequencies at a university children's hospital between 1976 and 2003. Fortschr Röntgenstr. 2008;180:410–22.
71. Seidenbusch MC, Schneider K. Radiation exposure of children in pediatric radiology. Part 4: entrance doses achieved during the x-ray examination of the chest. Fortschr Röntgenstr. 2008;180:1082–103.
72. Seidenbusch MC, Schneider K. Zur Strahlenexposition von Kindern in der pädiatrischen Radiologie. Anthropometrische Merkmale von Kindern und den zur Dosisrekonstruktion verwendeten mathematischen Phantomen (ISIMEP-Projekt des Bundesministeriums für Bildung und Forschung, Förderkennzeichen 02NUK016A). Fortschr Röntgenstr. 2012;184:A13.
73. Seidenbusch MC, Schneider K. Strahlenhygienische Aspekte bei der Röntgenuntersuchung des Thorax. Radiologe. 2015;55:580–7.
74. Snyder RG, Schneider LW, Owings CL, Reynolds HM, Golomb DH, Schork MA. Anthropometry of infants, children, and youths to age 18 for product safety design. Report UM-HSRI-77-17. Michigan: Highway Safety Research Institute, University of Michigan; 1977.
75. Spiers FW. Transition-zone dosimetry. Radiat Dosimetry. 1969;3:809–67.
76. Stephan G, Schneider K, Panzer W, Walsh L, Oestreicher U. Enhanced yield of chromosome aberrations after CT examinations in paediatric patients. Int J Radiat Biol. 2007;83:281–7.
77. Stein JM, Walkup LL, Brody AS, Fleck RJ, Woods JC. Quantitative CT characterization of pediatric lung development using routine clinical imaging. Pediatr Radiol. 2016;46:1804–12.
78. Stratz CH. Der Körper des Kindes und seine Pflege. Stuttgart: Enke Verlag; 1903.
79. Stratz CH. Die Darstellung des menschlichen Körpers in der Kunst. Berlin: Springer; 1914.
80. Thacker PG, Collins HR, Hill J. Distress experienced during pediatric VCUGs—a granular, prospective assessment using the brief behavioral distress scale. Pediatr Radiol. 2016;46:660–5.
81. United Nations Scientific Committee on the Effects of Atomic Radiation (UNSCEAR). UNSCEAR 2013 Report, Volume II: Scientific findings on effects of radiation exposure of children, Scientific Annex B: Effects of radiation exposure of children. ISBN 978-92-1-142293-1, United Nations, New York; 2013.
82. Vassileva J, Rehani M. Patient grouping for dose surveys and establishing diagnostic reference levels in paediatric computed tomography. Radiat Prot Dosimetry. 2015;65:81–5.
83. Varchena V. Pediatric phantoms. Pediatr Radiol. 2002;32:280–4.
84. Verdun F, Gutierrez D, Vader JP, Aroua A, Alamo-Maestre LT, Bochud F, Gudinchet F. CT radiation dose in children: a survey to establish age-based diagnostic reference levels in Switzerland. Eur Radiol. 2008;18:1980–6.
85. Voigt M, Fusch C, Olbertz O, Hartmann K, Rochow N, Renken C, Schneider KTM. Analysis of the neonatal collective in the Federal Republic of Germany. 12th report: presentation of detailed percentiles for the body measurement of newborns. Geburtsh Frauenheilk. 2006;66:956–70.
86. Von Froriep A. Anatomie für Künstler. Kurzgefasstes Lehrbuch der Anatomie, Mechanik, Mimik und Proportionslehre des menschlichen Körpers. Leipzig: Johann Ambrosius Barth Verlag; 1937.
87. Waitches G, Zawin JK, Poznanski AK. Sequence and rate of bone marrow conversion in the femora of children as seen on MR imaging: are accepted standards accurate? AJR. 1994;162:1399–406.
88. Wall BF, Shrimpton PC. Bone marrow dose in chest radiography: the posteroanterior vs. anteroposterior projection. Letter to the editor. Radiology. 1980;137:258–9.
89. Welcker H. Wachsthum und Bau des menschlichen Schädels. Leipzig: Wilhelm Engelmann Verlag; 1862.
90. Weninger WJ, Meng S, Geyer SH, Weninger SUG. Morphologie und funktionelle Anatomie des wachsenden Thorax. Radiologe. 2003;43:1036–44.
91. Wilpshaar J, Jokes EC, Lim FTH, van Leuwwen GHM, van den Boogaard PJ, Kanhai HHH, Willemze R, Falkenburg JHF. Magnetic resonance imaging of fetal bone marrow for quantitative definition of the human fetal stem cell compartment. Blood. 2002;100:451–7.
92. World Health Organization. WHO child growth standards. Length/height-for-age, weight-for-age, weight-for-length, weight-for-height and body mass index-for-age. Geneva: World Health Organization; 2006. isbn: 92 4 154693 X.
93. World Health Organization. WHO training package for the health sector, children's health and the environment: children are no little adults. Geneva: World Health Organization; 2008.

4 Radiation Dose and Dose Reconstruction in Conventional Paediatric Radiology

4.1 Introduction

Radiation dosimetry is the science of determining the fraction of the energy which is absorbed in matter during interaction with ionizing radiation [5]. Due to the fact that children are much more sensitive to ionizing radiation than adults, radiation dosimetry may be considered as one of the most important topics in paediatric radiology. In this chapter, the fundamentals of radiation dosimetry against the background of paediatric radiology will be presented. Therefore, the description of radiation physics will be focus on clinical paediatric aspects. For a more physical description, see e.g. [1, 5, 13, 17, 33, 40, 46].

Organ doses achieved in patients during conventional paediatric X-ray examinations depend on many influence parameters. Thus, radiation dosimetry and dose reconstruction are the most sophisticated subjects in paediatric radiology. In the following section, a short overview over the fundamentals of radiation dosimetry in paediatric radiology shall be given in order to enable the reader to perform dose reconstructions in children using this book. For more details in clinical radiation physics, see e.g. the books of *Bushberg* et al. [13] and *Dance* et al. [14].

4.2 Technical Principle of a Conventional X-Ray Examination

In order to achieve optimal image quality under optimized radiation exposure to the patient during an X-ray examination, the technical components of an X-ray unit and the physical exposure parameters have to be simultaneously adapted to each other with a lot of care. Figure 4.1 exemplifies the technical principle of a standard paediatric X-ray examination by a chest radiography in a 5-year-old patient and the exposure parameters associated to the technical components of the X-ray unit. From the technical point of view, a conventional X-ray examination works as follows.

4.2.1 X-Ray Tube

An electric field between the cathode and the anode is generated in the X-ray tube (depicted in Fig. 4.2) by connecting the cathode material with the negative pole and the anode material with the positive pole of a high voltage source with voltage V. Electrons with elementary electric charge e are released from a glow filament connected with the cathode and are accelerated in the electric field up to a kinetic energy E_{kin} when hitting the anode:

M. Seidenbusch et al., *Imaging Practice and Radiation Protection in Pediatric Radiology*,
https://doi.org/10.1007/978-3-030-18504-6_4

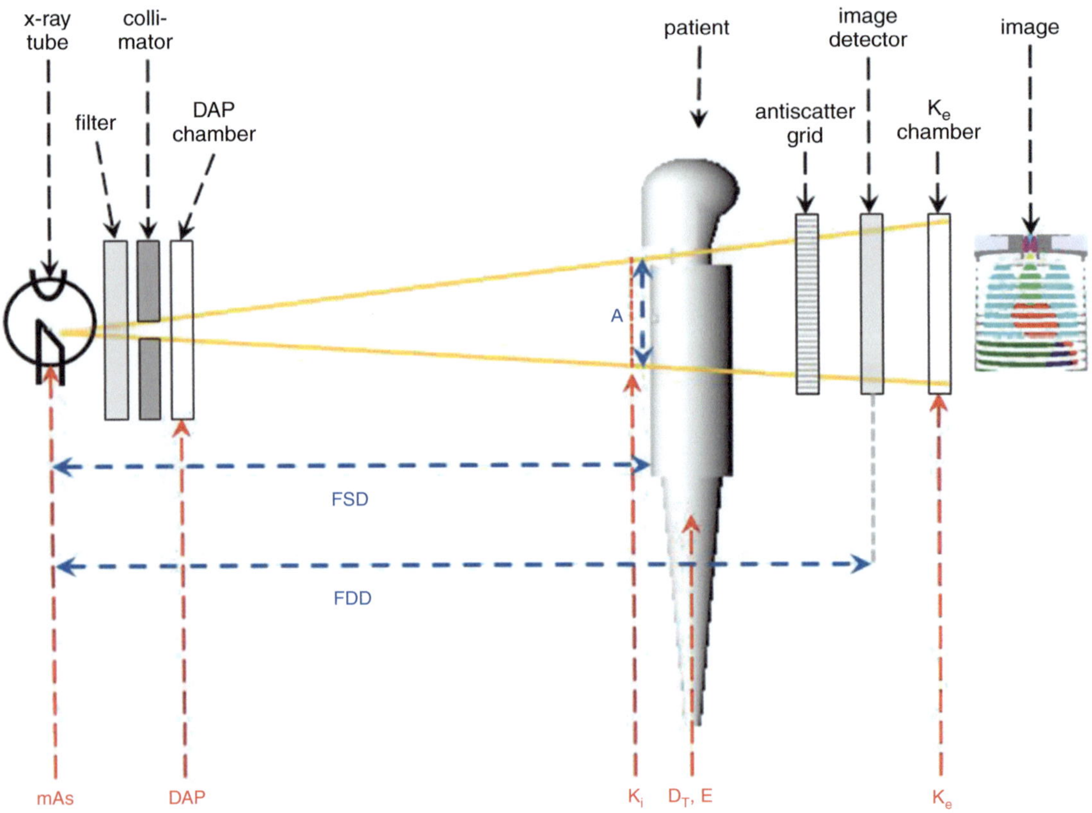

Fig. 4.1 Technical principle of an ap chest X-ray and dose indicators corresponding to the technical components of the X-ray unit. The pictures of phantom and X-ray image were taken from the user interface of the PCXMC software used for the calculations in this book [58, 59] (© M. Seidenbusch 2018). Dose indicators (red): mAs = mAs product; DAP = Dose area product; K_i = Incidence air kerma; D_T = Organ dose in tissue T; E = effective dose; K_e = Exit air kerma. Irradiation geometry (blue): FSD = Focus-to-skin distance; FDD = Focus-to-detector distance; A = Field size

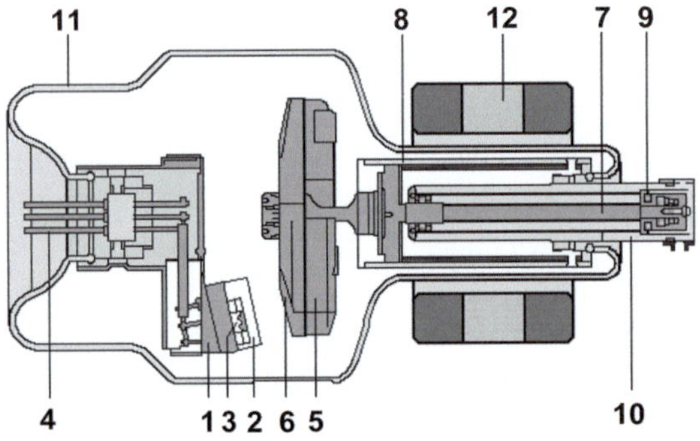

Fig. 4.2 Technical principle of an X-ray tube. (The picture of the X-ray tube is a modified picture from the book of *Hoxter* and *Schenz* [30].) (1) Cathode, (2) *Wehnelt* electrode, (3) Glow filaments, (4) High voltage connectors, (5) Rotating anode, (6) Focal track, (7) Molybdenum spindle, (8) Rotor, (9) Ball-bearing of the rotating anode, (10) Outer part of the rotating anode, (11) Bulb, (12) Stator

$$E_{\text{kin}} = e \cdot V \tag{4.1}$$

This kinetic energy is transformed into energy of bremsstrahlung that is released from the electrons during their slowing-down within the electric fields of the atomic nucleus of the anode material. As the slowing-down of electrons is a probabilistic process, bremsstrahlung will not have a defined energy but show an energy distribution which is called *bremsstrahlung spectrum* or *X-ray spectrum* with a maximum energy $E_{\max}$ which equals to the kinetic energy of the electrons:

$$E_{\max} = E_{\text{kin}} \tag{4.2}$$

Typical X-ray spectra generated by diagnostic X-ray tubes when selecting a tube voltage of

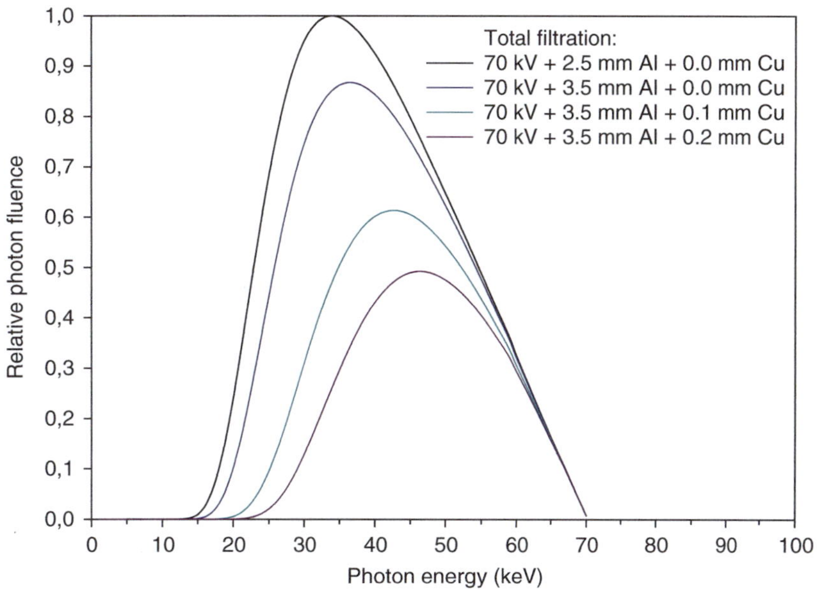

Fig. 4.3 X-ray spectra resulting from various total filtrations. The maximal photon energy is defined by the tube voltage and equals in this example to 70 keV. (From [53], © Springer Verlag 2015, courtesy Springer Verlag)

70 kV are depicted in Fig. 4.3. It should be stated here that the shape of X-ray spectra is also affected a little bit by the anode angle, but the contribution of the anode angle is to be neglected for practical clinical purposes.

The percentage of electric energy being transformed into radiation energy is called the *yield Y* of the X-ray tube. As the dose rate of an X-ray unit is proportional to the yield of the tube and the yield is proportional to the tube current, any kind of radiation dose quantity D is a linear function of the current-time product or mAs product:

$$D \sim Y \sim \text{mAs product} \quad (4.3)$$

However, as the yield of an X-ray tube must be known in order to estimate the dose rate of an X-ray tube from the tube current, the mAs product cannot be considered as an appropriate dose indicator for the reconstruction of organ and effective doses.

4.2.2 Filters and Total Filtration

X-ray spectra originating from X-ray tubes are composed of bremsstrahlung photons with an energy range from zero to the maximal photon energy defined by the tube voltage. As low energy X-rays are absorbed in the patient's skin surface and thus increase the patient's skin dose without contributing to the diagnostic imaging process, X-ray spectra originating from the X-ray tube should not be applied for X-ray diagnostics in their native form. However, the filtering of X-ray spectra by placing metallic filters between the X-ray tube and the patient will reduce the percentage of low energy photons and thus will reduce the exposure of the patient. The *total filtration* of an X-ray unit is defined as the sum of the *inherent filtration* of the X-ray tube and the *additional filtration* that can be added depending on the clinical situation. Figure 4.3 depicts various diagnostic X-ray spectra resulting from different total filtrations when selecting a tube voltage of 70 kV. As can be seen, an increasing total filtration goes along with a shift of the maximum of the X-ray spectrum towards higher photon energies but also with a decrease of the relative photon fluence and thus with a decrease of the dose rate. As a consequence, the tube current has to be increased when increasing the total filtration in order to achieve the same image detector dose.

4.2.3 Collimator

Collimation of the primary X-ray field originating from the X-ray tube to the anatomical region of interest is necessary when performing diagnostic X-ray examinations. Collimation has two effects: First, it is one of the most effective measures in order to reduce the radiation exposure of the patient. Second, reducing the irradiated patient volume will reduce the amount of scattered photons and thus reduce the image noise and therefore improve the image quality. Please note that by now collimating is the unique method in diagnostic radiology for optimizing radiation exposure and image quality simultaneously.

Depending on the patient's age and the clinical situation, optimal field sizes which can be achieved under optimal clinical conditions can be defined. In clinical application, field sizes are visually defined by harmonizing the boundaries of the light field with specific anatomical landmarks of the patient. However, small deviations of an optimal field collimation may considerably increase the field size (Fig. 4.4). Thus, not only optimal but also suboptimal field settings should be considered in dose calculations - one of the main concerns of this book.

Of course, uncertainties in the visual estimation of the radiation field through the light field do also play a role when defining the field size. A very important prerequisite for determining an optimal field size is, of course, that there is congruency of light field and radiation field [29]. Nevertheless, the field sizes applied during paediatric X-ray examinations have been found to differ considerably between different institutions and even within radiological departments [7, 18, 19, 23, 27, 29, 41, 56, 57, 60, 66].

4.2.4 KAP Meter

The KAP meter (synonymously: DAP chamber) is mounted at the X-ray unit after the collimation system and serves for measurement of the kerma area product (KAP) or dose area product

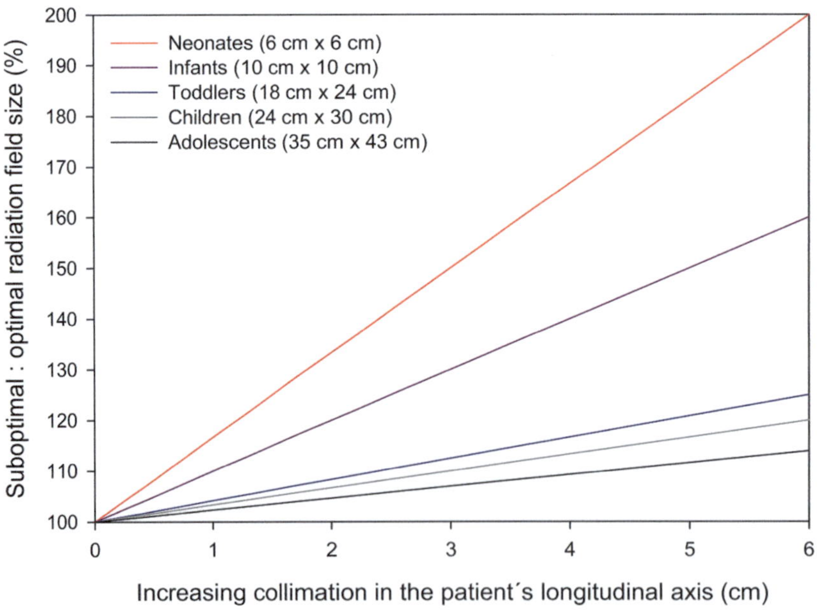

Fig. 4.4 Ratio of optimal and suboptimal field size of an ap/pa chest X-Ray due to additional increase of field size in longitudinal axis of the patient. Naturally, small optimal field sizes are applied in neonates and young infants and large optimal field sizes are applied in adolescents accordingly. Thus, an additional increase of field height due to suboptimal collimation in longitudinal direction will not considerably increase the field size in adolescents but will increase the field size in neonates and young infants by more than 100%. (From [53], © Springer Verlag 2015, courtesy Springer Verlag)

(DAP) (see Sect. 4.3.2) as a dose indicator for the radiation exposure of the patient. The dose area product is the most important dose quantity for the reconstruction of the organ and the effective doses achieved and thus for assessing radiation risks during conventional X-ray examinations [32].

In daily routine, no entrance surface dose will be measured but the dose area product. In Germany, recording of the dose measured during paediatric X-ray examinations is mandatory [63]. Especially in neonatology, dose meters must be sensitive enough in order to be able to measure entrance surface doses of 10 μGy or less. Highly sensitive dose meters are offered by several manufacturers of medical dosimetry systems which are, however, unable to capture all relevant dose quantities. It is particularly important in paediatric radiology that clinical KAP meters have been calibrated accurately [43].

4.2.5 Patient

The patient is the most variable component of the X-ray equipment in paediatric radiology as his diameter varies over a wide range depending on age group, beam projection, and body part to be examined. As X-rays are attenuated exponentially when passing through the patient, the radiation exposure of the patient is strongly affected by the patient's size and shape.

4.2.6 Antiscatter Grid

The purpose of antiscatter grids is to reduce scattered radiation originating from the patient and thus reducing image noise. Antiscatter grids consist of a system of lead lamellae that are aligned according to the irradiation geometry. Therefore, antiscatter grids have to be adjusted carefully considering the focus-to-detector distance. However, not only scattered radiation is absorbed by antiscatter grids but also a fraction of the useful beam. Thus, the use of antiscatter grids goes along dependent on patient age and type of X-ray examination with an increase of radiation exposure of the patient up to a factor of three to five [13]. Overall, the use of antiscatter grids in paediatric radiology is discussed controversial [3, 15, 24, 34, 48, 49]. Nevertheless, the authors of this book recommend using of antiscatter grids in paediatric X-ray examinations of children beyond the age of 8 years for chest X-rays (Fig. 4.5). For skull radiographs, an antiscatter grid is already used in infants. Details on the use of antiscatter grid were published in many guidelines (e.g. [15]).

4.3 Dose Quantities

In this section, dose quantities as being used in this book shall be presented short from the clinical point of view. For more precise physical definitions, see e.g. the books of *Attix* [5] or *Jones* and *Cunningham* [40].

The interaction between ionizing radiation and material can be divided into four phases [46]: The physical phase of first order, the physical phase of second order and the physico-chemical and biological phase. In the physical phase of first order, radiation energy from the incident X-ray beam is transferred into kinetic energy of secondary electrons through the photo effect, the *Compton* effect, and the pair formation effect. In the physical phase of second order, the kinetic energy of the secondary electrons is absorbed in matter, generating ions and radicals in the chemical phase. Ions and radicals cause chemical reactions in material, as well as radiation damages like DNA strand brakes in biological environment during the biological phase of radiation interaction. To each of these phases, a dose quantity can be attributed: the Kerma (Kerma = kinetic energy released in material) to the physical phase of first order, the absorbed dose (synonymously: energy dose) to the physical phase of second order, the equivalent dose to the chemical phase, and the effective dose to the biological phase. Figure 4.6 depicts the four phases of radiation interaction with matter and the dose quantities associated.

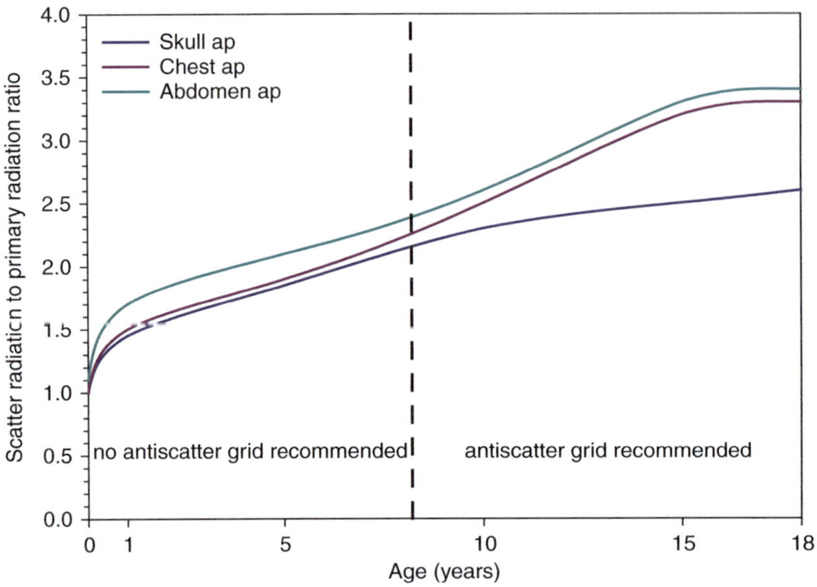

Fig. 4.5 Ratio of scatter radiation to primary radiation as a function of patient age and type of X-ray examination (using the scatter ratios of *Bushberg* et al. [13]). The ratio of scatter is highest for abdominal X-rays even in early childhood. As could be expected, the fraction of scattered radiation is increasing with age for chest and abdominal X-rays. For the skull, the slope of scatter curve decreases during adolescence. According to these data, an antiscatter grid is recommended for all X-rays in children older than 8 years of age. (© M. Seidenbusch 2018)

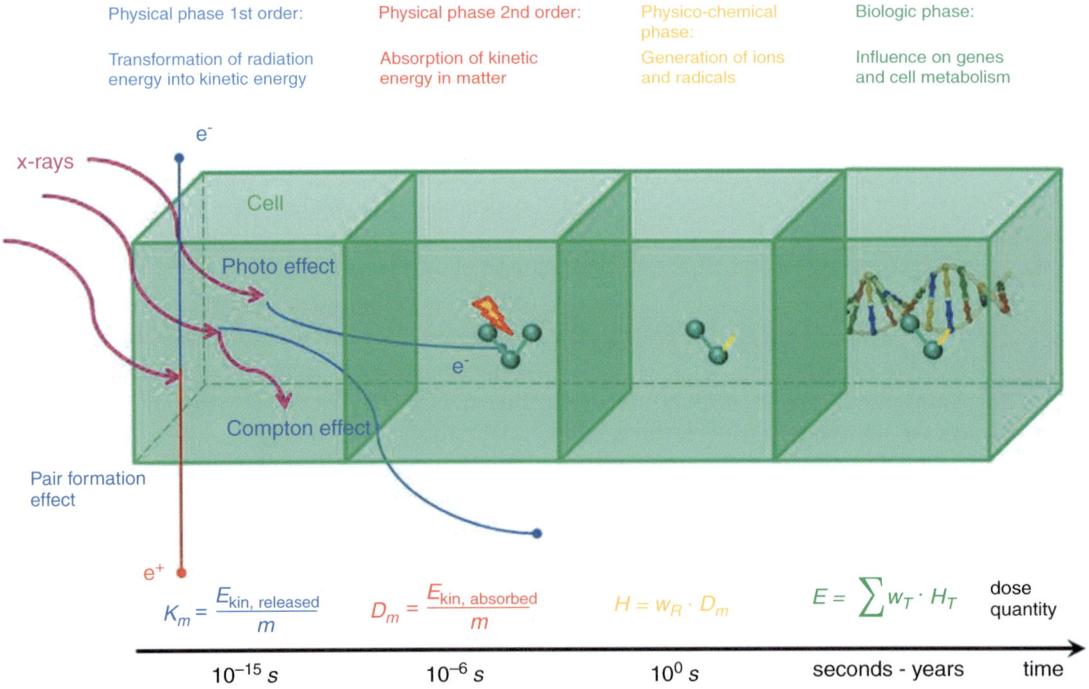

Fig. 4.6 The four phases of radiation interaction with living matter [46] and corresponding dose quantities. Green cubes symbolize living cells. Purple arrows symbolize X-ray photons, blue and red lines secondary electrons (e^-) and secondary positrons (e^+), respectively. The four phases of interaction are depicted in chronological order in four cells. For details, see the text. (© M. Seidenbusch 2018)

Table 4.1 Backscatter factors B for various radiation qualities, field sizes, and focus-to-skin distances [45]

	Total filtration F				Backscatter factor B for ICRU tissue		
Voltage U (kV)	Al (mm)	Cu (mm)	HVL (mm)	Mean energy (keV)	10 × 10 cm²	20 × 20 cm²	25 × 25 cm²
50	**2.5**	–	1.74	32.0	1.25	1.27	1.28
60	**2.5**	–	2.08	35.8	1.28	1.32	1.32
70	**2.5**	–	2.41	39.3	1.31	1.36	1.36
	3.0	–	2.64	40.0	1.32	1.37	1.38
	3.0	**0.1**	3.96	44.0	1.39	1.47	1.47
80	**2.5**	–	2.78	42.9	1.33	1.39	1.39
	3.0	–	3.04	43.7	1.34	1.40	1.41
	3.0	**0.1**	4.55	48.2	1.40	1.50	1.51
90	**2.5**	–	3.17	46.3	1.34	1.41	1.42
	3.0	–	3.45	47.0	1.36	1.43	1.44
	3.0	**0.1**	5.12	51.7	1.41	1.51	1.53
100	**2.5**	–	3.24	48.1	1.34	1.41	1.42
	3.0	–	3.88	50.0	1.37	1.45	1.46
	3.0	**0.1**	5.65	54.8	1.42	1.53	1.55
120	**2.5**	–	–	–	–	–	–
	3.0	–	4.73	55.4	1.38	1.45	1.49
	3.0	**0.1**	6.62	60.1	1.42	1.54	1.56
150	**2.5**	–	4.79	59.1	1.36	1.46	1.48
	3.0	–	6.80	64.9	1.39	1.51	1.53
	3.0	**0.1**	8.50	69.2	1.41	1.54	1.57

Annotations: HVL = half value length of the X-rays of a certain radiation quality

4.3.1 Kerma

If material M is hit by an X-ray photon field, secondary electrons are released within the material by the photoelectric effect, by the *Compton* effect, and by the pair formation effect. The amount of the radiation energy of the incident photon field that is transferred to kinetic energy E_{kin} of secondary electrons within a given mass element m is named the *kerma* (KERMA = kinetic energy released in material) K_M:

$$K_M = \frac{dE_{\text{kin}}}{dm} \tag{4.4}$$

The physical SI[1] unit of the kerma is 1 Gray = 1 Gy = 1 J kg^{-1}.

The *incidence air kerma* K_i is defined as the kerma in air in the entrance plane of the incident radiation field at a patient's position but without presence of the patient:

$$K_i = K_{\text{air}} \tag{4.5}$$

When a patient is present, the corresponding dose quantity is the *entrance surface dose* K_s that is defined as the sum of incidence air kerma and the backscatter radiation from the patient. The entrance surface dose can be calculated by multiplying the incidence air kerma with a *backscatter factor B*:

$$K_s = B \cdot K_i \tag{4.6}$$

Backscatter factors depend on the parameters of the beam quality (X-ray tube voltage V and total filtration F [12]) and the parameters of the beam geometry (beam projection p, field size A, and focus-to-skin distance FSD):

$$B = B(V, F, p, A, \text{FSD}) \tag{4.7}$$

Backscatter factors for diagnostic photon energies and various field sizes and focus-to-film distances are given in [45] and are provided in Table 4.1 and Fig. 4.7.

4.3.2 Dose Area Product (DAP), Kerma Area Product (KAP)

The *Dose Area Product (DAP)* or strictly speaking: *air Kerma Area Product* (KAP) [32, 42] is the product of incident air kerma K_i at any given

[1] SI = Système international d'unités.

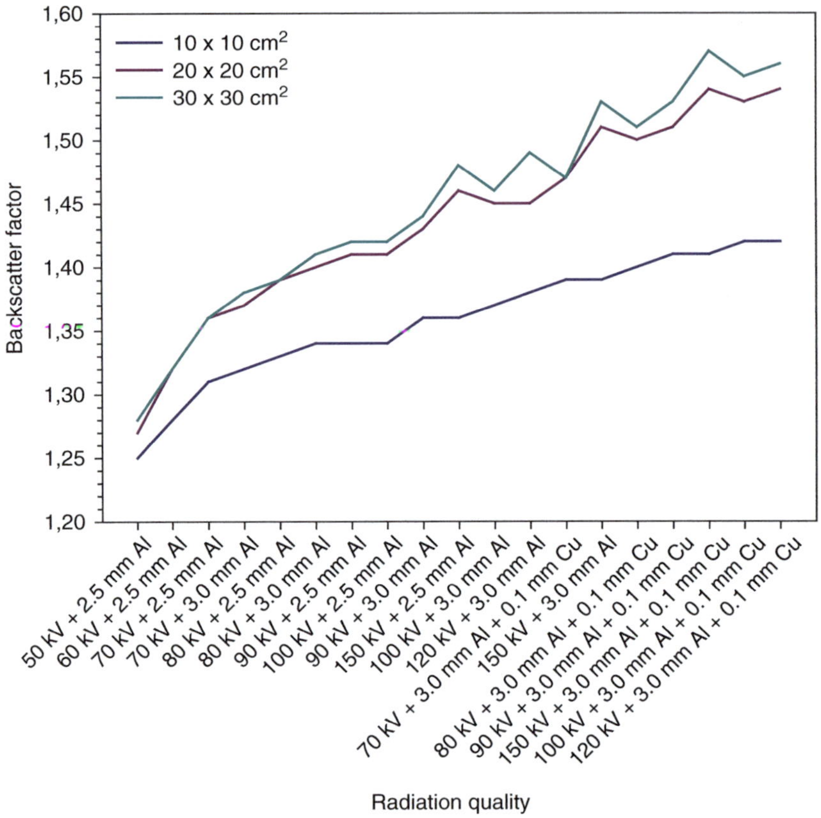

Fig. 4.7 Backscatter factors for various radiation qualities and field sizes. Visualization of the data in Table 4.1. Generally, backscatter factors go up with increasing effective energy of the radiation field. (© M. Seidenbusch 2018)

Table 4.2 Conversion coefficients *C* for various DAP measuring units

	mR cm²	cR cm²	dR cm²	R cm²	mGy cm²	cGy cm²	μGy m²	*B* = *A*/*C*
mR cm²	1	0.1	0.01	0.001	0.0084	0.00084	0.00084	
cR cm²	10	1	0.1	0.01	0.084	0.0084	0.0084	
dR cm²	100	10	1	0.1	0.84	0.084	0.084	
R cm²	1000	100	10	1	8.4	0.84	0.84	
mGy cm²	119	11.9	1.19	0.119	1	0.1	0.1	
cGy cm²	1190	119	11.9	1.19	10	1	1	
μGy m²	1190	119	11.9	1.19	10	1	1	
A = *C B*	C							

Note that DAP values in column A can be calculated from DAP values in row B by the equation $A = C\,B$ and vice versa

point of the X-ray beam and the field size *A* at this point:

$$\mathrm{KAP} = \mathrm{DAP} = K_i \cdot A \tag{4.8}$$

The physical SI unit of the dose area product is 1 Gy m².

As an X-ray source is usually a point source, the inverse square law is valid for the irradiation geometry, and the dose area product remains nearly constant over the whole focus-to-skin distance. Furthermore, the dose area product is a dose indicator which represents a measure of the radiation exposure of the patient. The dose area product is usually measured by an ionization chamber that is fixed at the beam collimator.

In diagnostic radiology, there are various dose measuring units in practical use for the DAP as shown in Table 4.2. Table 4.2 provides conversion coefficients *C* for transforming the DAP measuring units into each other.

4.3.3 Absorbed Dose

The part E of kinetic energy of secondary electrons E_{kin} that is transferred to a mass element m of a material M is named the *absorbed dose* (in Europe synonymously: *energy dose*) D_M:

$$D_M = \frac{dE}{dm} \quad (4.9)$$

The physical SI unit of the absorbed or energy dose is 1 Gray = 1 Gy = 1 J kg^{-1}.

In case of secondary electron equilibrium which can be assumed to be always achieved in the photon energy range of X-ray diagnostics, the absorbed dose equals the kerma.

If an absorbed dose or energy dose is estimated for an organ or tissue T, the absorbed dose or energy dose is also called *organ absorbed dose* (in Europe synonymously: *organ energy dose*) D_T:

$$D_T = D_M \text{ achieved in tissue T} \quad (4.10)$$

4.3.4 Equivalent Dose

In order to allow for the fact that different radiation qualities differ in energy transfer mechanisms, the *equivalent dose* H_M can be calculated from an absorbed or energy dose by multiplying the absorbed or energy dose D_M with a radiation-specific *radiation weighting factor* w_R (almost synonymously: *quality factor* Q) as follows:

$$H_M = w_R \cdot D_M = Q \cdot D_M \quad (4.11)$$

The physical SI unit of the equivalent dose is 1 Sievert = 1 Sv = 1 J kg^{-1}.

Within the photon energy range applied in X-ray diagnostics, the radiation weighting factor has been defined to be $w_R = 1$.

If an equivalent dose is estimated for an organ or tissue T, the equivalent dose is also called *organ equivalent dose* or, in short, *organ dose* H_T:

$$H_T = H_M \text{ achieved in tissue T} \quad (4.12)$$

Note that the organ dose is defined as the average dose achieved within an organ or tissue.

4.3.5 Effective Dose

In order to allow for the fact that different organs and tissues T differ in radiation sensitivity, the *effective dose E* has been defined as the weighted sum of all organ doses, each of them weighted with a tissue-specific *tissue weighting factor* w_T representing the specific radiation sensitivity of organs and tissues T as follows:

$$H_T = \sum_T w_T \cdot H_T \quad (4.13)$$

The physical SI unit of the effective dose is 1 Sievert = 1 Sv = 1 J kg^{-1}.

As the effective dose is intended to represent the virtual whole body dose that would have to be achieved in order to cause the same stochastic risk as a really applied dose distribution in some organs or tissues, the condition

$$\sum_T w_T = 1 \quad (4.14)$$

has to be fulfilled when defining the tissue weighting factors w_T. For this reason, the effective dose has sometimes also been called the "weighted sum of stochastic risks". However, the effective dose has been intended to serve as a risk indicator in occupational health and has not been defined for describing the radiation exposure of patients originally [39], particularly not for children. Therefore, its application for describing radiation exposure of patients in diagnostic radiology has to be estimated critically [14, 21, 22]. Although attempts have been made to derive tissue weighting factors for children [2, 8, 44] in recent years, the concept of effective dose ought to be considered as to be inapplicable for paediatric patients at least in the opinion of the authors of this book.

4.4 Dose Reconstruction Algorithms

Reconstruction of the organ and the effective doses achieved during conventional X-ray examinations can be performed through various mathematical methods [1, 20]. In recent years, the conversion coefficient concept, the X-ray source concept, and the image detector concept have been developed. The decision which concept is to be used best will depend on the known set of exposure data.

4.4.1 Conversion Coefficient Concept

The conversion coefficient concept represents the methodic basis of this book and will be described in the following section. In short, organ and effective doses can be calculated from direct dose indicators such as entrance air kerma or dose area product by applying specific so-called conversion coefficients (synonymously: conversion factors, normalized organ doses). Entrance air kerma or dose area product will usually be measured during an X-ray examination with ionization chambers placed in the X-ray unit after the beam collimator. If neither an entrance air kerma nor a dose area product has been measured, these dose indicators can be estimated from indirect exposure parameters by applying the X-ray source concept or the image detector concept.

4.4.2 X-Ray Source Concept

The X-ray source concept consists of an estimation of entrance air kerma K_{E} from characteristics of the X-ray source as follows:

$$K_{\mathrm{E}} = Y_R \cdot I \cdot t \cdot \left(\frac{R}{\mathrm{FSD}} \right)^2 \quad (4.15)$$

with Y_{R} being the tube yield (synonymously: tube output) in reference distance R from the X-ray tube and FSD being the focus-to-skin distance. The term $(I\,t)$ is the mAs product that can be taken from the exposure parameter settings at the console of the X-ray unit. Numerical values of the yield Y_{R} should be taken from the technical documentation of the X-ray unit. Typical values for the yield have been also provided e.g. in [1, 20]. However, as the tube output is strongly dependent on the technical properties of the high voltage generator and the X-ray tube and is affected by the total filtration as well, the uncertainty of this dose estimation concept may be considerable. Thus, it is strongly recommended here to apply the X-ray source concept only alternatively.

4.4.3 Image Detector Concept

The image detector concept consists of an estimation of entrance air kerma K_{E} from the kerma K_{ID} required by the image detector system ID as follows:

$$K_{\mathrm{E}} \approx K_{\mathrm{ID}} \cdot \mathrm{e}^{\mu_{\mathrm{P}} \cdot d_{\mathrm{P}}} \cdot m_{\mathrm{g}} \cdot m_{\mathrm{e}} \cdot (1-\alpha) \cdot \left(1 + \frac{d_{\mathrm{P}}}{\mathrm{FSD}} \right)^2 \quad (4.16\mathrm{a})$$

with μ_{P} being the average linear attenuation coefficient of the anatomical region of the patient P, d_{P} the patient's diameter, m_{g} the attenuation factor of an antiscatter grid g, m_{e} the attenuation factor of the X-ray equipment e (Bucky table or chest stand), α the percentage of scattered radiation, and FSD the focus-to-skin distance. Numerical values of the image detector dose K_{ID} should be taken from the technical documentation of the X-ray unit. Typical numerical values of m_{g}, m_{d}, and α can be taken e.g. from [20]. Linear attenuation coefficients have been provided e.g. in the tables of *Hubbell* [31]. However, linear attenuation coefficients are defined for narrow beam geometry. As in clinical practice broad beams are applied, these linear attenuation coefficients may be replaced by so-called *effective* linear attenuation coefficients μ_{eff} as listed by *Hart* et al. [28] which are a function of the tube voltage and the field size A:

$$\mu_{\mathrm{eff}} = \mu_{\mathrm{eff}}(V, A) \quad (4.16\mathrm{b})$$

By the way, linear attenuation coefficients can be considered as the sum of partial linear attenuation coefficients for the photo effect μ_{Photo}, the *Compton* effect μ_{Compton}, and the pair formation effect μ_{Pair}[2]:

[2]The exponential attenuation law is due to the fact that the photo, Compton, and pair formation effects are stochastic effects occurring with a probability being proportional to the irradiated volume.

$$\mu = \mu_{\text{Photo}} + \mu_{\text{Compton}} + \mu_{\text{Pair}} \tag{4.16c}$$

As the entrance air kerma estimated through the image detector concept is an exponential function of the patient's diameter, the uncertainty of the dose estimation is mainly an exponential function of the error in estimation of the patient diameter. As patient diameters vary over a wide range even within a defined age group, the uncertainty of this dose estimation concept may be considerable and is recommended to be applied alternatively only.

4.5 The Concept of Normalized Organ Doses

The concept of normalized organ doses has become the standard method for reconstructing organ doses in patients in recent years.

Generally, so-called conversion coefficients (almost synonymously: conversion factors) establish a multiplicative relationship between two physical quantities. In the case of completely different physical quantities with different physical units one would talk about conversion coefficients in the stricter sense of the word and in the case of similar physical quantities with identical physical units one would usually talk about conversion factors. In the special case of conversion coefficients being provided for calculating organ doses from measurable dose indicators, the conversion coefficients which are given as ratios of organ dose and measurable dose indicator are also named *normalized organ doses*. As the term *normalized organ dose* seems most appropriate to clinical applications of conversion coefficients, we will use this term in this book.

The concept of normalized organ doses consists of the reconstruction of organ doses D_T achieved in organs or tissues T by means of the application of conversion coefficients $C_{K\to DT}$ or $C_{DAP\to DT}$ to measurable dose indicators like the incidence air kerma K_i or the dose area product DAP:

$$D_T = C_{K\to DT} \cdot K_i = C_{DAP\to DT} \cdot \text{DAP} \tag{4.17a}$$

Thus, the conversion coefficients $C_{K\to DT}$ and $C_{DAP\to DT}$ are defined as organ doses D_T which were normalized to the measurable dose indicators K_i and DAP and thus can also be called *normalized organ doses*:

$$C_{K\to DT} = \frac{D_T}{K_i} \tag{4.17b}$$

$$C_{DAP\to DT} = \frac{D_T}{\text{DAP}} \tag{4.17c}$$

Normalized organ doses C depend on various variables such as the type of X-ray examination X, the reference organ or reference tissue T for which the organ dose D_T is to be calculated, the parameters of the beam quality (X-ray tube voltage V and total filtration F [12]), the parameters of the beam geometry (beam projection p, field size A, and focus-to-skin distance FSD) and, most important, the patient's anthropometry (body height h_P and body mass m_P of the patient P):

$$C = C\left(X, T, V, F, p, A, \text{FSD}, h_P, m_P\right) \tag{4.18}$$

Body size and body mass are a function of the patient's age a_P, respectively:

$$h_P = h_P\left(a_P\right) \tag{4.19a}$$

$$m_P = m_P\left(a_P\right) \tag{4.19b}$$

In paediatrics, these functions are usually described by somatograms. In the general population, there is also a functional dependence between patient's age and patient's anthropometric parameters. Thus, normalized organ doses can be considered as to approximately depend on the patient's age provided that the patient is a member of the general population:

$$C \approx C\left(X, T, V, F, p, A, \text{FSD}, a_P\right) \tag{4.20}$$

As normalized organ doses are a function of patient's anthropometry, the consideration of anthropometric characteristics of paediatric patients are of utmost importance in paediatric radiation dosimetry.

4.6 Human Anthropometry

4.6.1 Entrance Surface Dose, Exit Dose, and Patient Diameter

The anthropometric measures of the patient are the most important determinants of the radiation exposure. Assuming a decrease of dose with distance from the X-ray tube focus which follows the inverse square law and an exponential attenuation of an X-ray field through the patient, the relationship between the entrance air kerma K_E and the exit dose K_{exit} is given by

$$K_{exit} = K_E \cdot \left(\frac{FSD}{FSD + d_P} \right)^2 \cdot e^{-\mu_P \cdot d_P} \quad (4.21)$$

in which FSD is the focus-to-skin distance, μ_P is the linear attenuation coefficient of the patient's body region being examined, and d_P is the diameter of the patient's body region [25]. This function is also the essential term in (4.16a).

The required exit dose is determined by the sensitivity of the imaging system, the required image quality, and the facultative use of an antiscatter grid. As the relationship between entrance surface dose and exit dose is described by an exponential function, the radiation exposure of the patient that corresponds to the entrance surface dose will exponentially increase with the diameter of the anatomical region being examined. Thus, the radiation exposure of the patient is strongly determined by the human anthropometry and therefore by the age of the patient. As in paediatric radiology all age groups from the neonate till the young adult are to be examined and as children's anthropometry does vary over a wide range depending on the age group (Chap. 3), the patient has to be considered as the most variable component of the imaging system in paediatric radiology.

4.7 Mathematical and Anthropomorphic Phantoms

Normalized organ doses for a given X-ray examination can be determined by measurements in physical phantoms or by simulations in mathematical phantoms [65]. Today, the most powerful method is the conversion coefficient concept that has become more important in recent years.

Conversion coefficients for a given X-ray examination can be measured in physical phantoms or be calculated in mathematical simulations. However, the absorption of radiation energy in the human body is substantially affected by the patient's anatomy and anthropometry [61, 62]. As the human body consists of a very large number of organs and tissues with heterogeneous chemical composition, the possibilities for reproducing human anatomy for physical radiation measurements in order to perform dose estimations are extremely limited and rather extensive. Therefore, there has been an increase in the application of the mathematical method of the Monte Carlo simulation to mathematical phantoms in recent years. An overview over the mathematic phantoms constructed for paediatric dosimetry is given in Chap. 28 of this book. Broadly speaking, two groups of mathematical phantoms have been developed in recent years, mathematical stylized phantoms and anthropomorphic voxel phantoms. In mathematical stylized phantoms, inner organs and tissues of the human body are described by mathematical expressions representing planes or cylindrical, conical, elliptical, or spherical surfaces. In anthropomorphic voxel phantoms, the phantoms have been constructed from voxel datasets originating from post-mortem computed tomographic (CT) or magnetic resonance tomographic (MRT) slices of real patients.

Thus, anthropomorphic voxel phantoms refer to the patients' individual anatomy and mathematical stylized phantoms do refer to an average anatomy of the age group and gender represented by the phantom. As children's anatomy and anthropometry show a high interindividual variability (Figs. 3.3 and 3.4 in Sect. 3.1), dose calculations using mathematical phantoms may be more appropriate than using those using anthropomorphic voxel phantoms. As also shown in Figs. 3.3 and 3.4 in Sect. 3.1 paediatric mathematical phantoms of the MIRD type represent the median anthropometric properties (body height and body weight) of paediatric patients very well and may therefore be more adequate for determining normalized organ doses than anthropometric phantoms. However, as these phantoms represent the anthropometric properties of the American and European population, the normalized organ doses presented in this book may at least a priori not be transferable to patients of other nationalities [64].

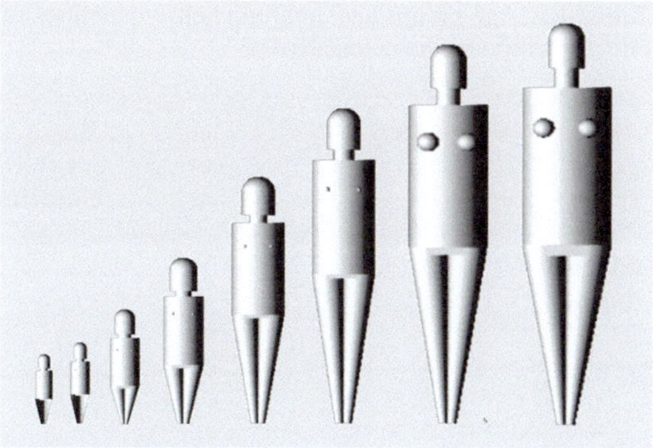

Fig. 4.8 MIRD family of hermaphrodite mathematical phantoms implemented into the PCXMC algorithm, version 2.0. The extended phantom family consists of two premature neonates (24 and 28 gestational weeks old), a neonate, a 1-year-old, a 5-year-old, a 10-year-old and a 15-year-old child and a young adult. (Taken from the user interface of the PCXMC algorithm)

4.8 Monte Carlo Method

4.8.1 Description

As mentioned above, there has been an increase in the application of the mathematical method of Monte Carlo simulation in order to determine normalized organ doses in diagnostic radiology in recent years. The Monte Carlo method is based on the simulation of stochastically distributed interactions of radiation particles with matter of a known chemical composition. In X-ray diagnostics it follows the individual photon histories, starting from the impact of the photon on the phantom surface, throughout the scattering and absorption processes and up to the deposition of energy in the phantom or detector material [4, 47]. The normalized organ doses listed in this book were determined through Monte Carlo simulations, using the commercially available PCXMC algorithm, which will be described in the following section.

4.8.2 PCXMC Algorithm

All normalized organ doses listed in this book were determined through Monte Carlo simulations using the commercially available PC software PCXMC, Version 2.0, developed by the Finnish Centre for Radiation and Nuclear Safety Säteilyturvakeskus (STUK) [55, 58, 59]. The algorithm consists of three parts and permits the determination of the absorbed doses through external photon fields of any spectral composition and any examination configuration of conventional paediatric X-ray examinations in 40 human reference organs and tissues. For this purpose, a complete phantom family of mathematical hermaphroditic MIRD phantoms at age levels 0, 1, 5, 10, 15, and 18 years [54] has been implemented into the algorithm in accordance with the models given by *Cristy* [16] and the ICRP standard [35, 37]. The phantom family is depicted in Fig. 4.8, and their anthropometric parameters are listed in Table 4.3.

In the first part of the algorithm, the parameters of the beam geometry (beam projection, X-ray tube tilt angle, focus-to-skin distance or source to image receptor distance, and position and size of the radiation field in the plane of incidence) are to be defined at the surface of the six mathematical phantoms for the X-ray examinations to be simulated. The adjusted X-ray exposure parameters and the radiation field settings are checked by means of a graphic display in the program's user interface showing the radiation field settings at the phantom surface and the virtual radiograph resulting from these field settings [58]. Furthermore, the number of individual photon histories to be tracked is to be entered here. In addition to that, size and shape of the mathematical phantoms can be adjusted here. For this purpose, scaling factors (x_P/x_0), (y_P/y_0) and (z_P/z_0) for the length of the lateral (x), transversal (y), and longitudinal (z) axis of the mathematical phantoms are calculated by

Table 4.3 Age groups and anthropometric quantities of STUK MIRD phantoms [59] and equivalent cylindrical and effective elliptical diameters derived

Skull region							
Age group (years)	**Weight (kg)**		**Height (cm)**	**Maximal sagittal diameter (cm)**	**Lateral diameter (cm)**	**Equivalent cylindrical diameter (cm)**	**Effective elliptical diameter (cm)**
24. gw[a]	–		6.1	6.2	4.6	–	5.3
28. gw[a]			7.1	7.4	5.7		6.5
0			11.1	11.3	8.8		10.0
1			15.3	15.3	11.8		13.4
5			17.6	17.6	14.2		16.0
10			18.0	18.0	14.4		16.1
15			19.2	18.6	15.2		16.8
18			19.8	19.2	15.4		16.9
Neck region							
Age group (years)	**Weight (kg)**		**Height (cm)**	**Sagittal diameter (cm)**	**Lateral diameter (cm)**	**Equivalent cylindrical diameter (cm)**	**Effective elliptical diameter (cm)**
24. gw[a]	–		1.0	2.6	2.6	–	2.6
28. gw[a]			1.2	3.4	3.4		3.4
0			1.6	5.3	5.3		5.3
1			1.8	7.0	7.0		7.0
5			3.0	7.2	7.2		7.2
10			5.6	8.8	8.8		8.8
15			6.8	9.6	9.6		9.6
18			6.8	10.5	10.5		10.5

Whole body region								
Age group (years)	**Whole body weight (kg)**	**Whole body height (cm)**	**Trunk height (cm)**	**Sagittal diameter of trunk region (cm)**	**Lateral diameter of trunk region (cm)**		**Equivalent cylindrical diameter (cm)**	**Effective elliptical diameter (cm)**
					With arms	**Without arms**		
24. gw[a]	0.7	30.0	12.8	5.2	6.9	6.0	5	6
28. gw[a]	1.0	35.0	14.8	6.0	8.2	7.1	6	7
0	3.40	50.9	21.6	9.8	12.6	10.9	9	10
1	9.20	74.4	30.7	13.0	17.1	15.1	13	14
5	19.0	109.1	40.8	15.0	22.6	19.6	15	17
10	32.4	139.8	50.8	16.8	27.3	23.8	17	20
15	56.3	168.1	63.1	19.6	34.4	29.6	20	24
18	73.2	178.6	70.0	20.0	38.9	34.4	23	26

[a]*gw* gestational weeks (in neonates). Premature neonates were defined by the authors of this book using the scaling function of the PCXMC software [55, 58, 59]

the algorithm from pre-defined values of the patient's weight w_P, the patient's height h_P, and the corresponding standard values w_0, h_0 through *Lindskoug's* equivalent cylindrical diameters d (3.2) as follows:

$$\frac{x_P}{x_0} = \frac{d_P}{d_0} = \sqrt{\frac{m_P}{m_0} \cdot \frac{h_0}{h_P}} \tag{4.22a}$$

$$\frac{y_P}{y_0} = \frac{d_P}{d_0} = \sqrt{\frac{m_P}{m_0} \cdot \frac{h_0}{h_P}} \tag{4.22b}$$

$$\frac{z_P}{z_0} = \frac{h_P}{h_0} \tag{4.22c}$$

Thus, organ masses m_P are adjusted by the algorithm according to

Table 4.4 Age groups and anthropometric quantities of children in a German survey [5, 6] and equivalent cylindrical and effective elliptical diameters derived

Skull region								
Age group (years)	**Weight (kg)**		**Height (cm)**	**Sagittal diameter (cm)**	**Lateral diameter (cm)**		**Equivalent cylindrical diameter (cm)**	**Effective elliptical diameter (cm)**
24. gw[a]	–		–	–	–		–	–
28. gw[a]				–	–			–
0				11.9	9.3			10.5
1				16.2	12.4			14.2
5				17.9	13.8			15.7
10				18.4	14.1			16.1
15				18.9	14.6			16.6
18				–	–			–
Whole body region								
Age group (years)	**Whole body weight (kg)**	**Whole body height (cm)**	**Trunk height (cm)**	**Mean sagittal diameter of trunk region (cm)**	**Mean lateral diameter of trunk region (cm)**		**Equivalent cylindrical diameter (cm)**	**Effective elliptical diameter (cm)**
					With arms	**Without arms**		
24. gw[a]	–	–	–	–	–	–	–	–
28. gw[a]				–		–		–
0				9.1		10.3		9.7
1				12.3		15.2		13.7
5				13.5		18.1		15.6
10				15.6		21.6		18.4
15				17.7		25.2		21.1
18				–		–		–

[a] *gw* gestational weeks (in neonates). Neonates were defined by the authors of this book using the scaling function of the PCXMC software [55, 58, 59]

$$\frac{m_\mathrm{P}}{m_0} = \frac{\rho_\mathrm{P}}{\rho_0} \cdot \frac{x_\mathrm{P}}{x_0} \cdot \frac{y_\mathrm{P}}{y_0} \cdot \frac{z_\mathrm{P}}{z_0} = \left(\frac{d_\mathrm{P}}{d_0}\right)^2 \cdot \frac{h_\mathrm{P}}{h_0} \tag{4.23}$$

in which $\rho_\mathrm{P} = \rho_0$ is the mean density of the organ or tissue. In this book, mathematical PCXMC phantoms of neonates in 24th and 28th gestational week were adjusted using these scaling factors.

In the second part of the algorithm, ten individual Monte Carlo simulations are carried out for each energy level of the photon radiation field with the stipulated number of individual photon histories. The organ doses achieved in the individual reference organs and tissues of the mathematical phantoms are determined by the algorithm from the mean values of the corresponding spectral energy depositions.

Finally, in the third part of the algorithm, a photon spectrum is calculated from the parameters of the beam quality (X-ray tube voltage, total filtration, and tube anode angle [12]) in accordance with the theory of *Birch* and *Marshall* [9]. The photon spectrum is defined by the distribution of the number N_j of photons over the photon energy j. This photon spectrum of the incident X-ray field with field size A_i is used to determine the incidence air kerma from the spectral kerma factors k_j and the spectral energy deposition ε_j in an organ or tissue T with mass m_T. The organ doses D_T calculated in the second part are normalized to the total incidence air kerma K_i and put out in a table as normalized organ doses:

$$C_{\mathrm{K}\rightarrow\mathrm{DT}} = \frac{D_\mathrm{T}}{K_i} = \left[A_i \cdot \sum_{\mathrm{j=1\,keV}}^{150\,\mathrm{keV}} \varepsilon_i \cdot N_\mathrm{j}\right] \cdot \left[m_\mathrm{T} \cdot \sum_{\mathrm{j=1\,keV}}^{150\,\mathrm{keV}} k_\mathrm{i} \cdot N_\mathrm{j}\right]^{-1} \tag{4.24}$$

The calculations of the normalized organ doses for the effective dose are based on the tissue weighting factors as defined by ICRP 60 [36] and ICRP 103 [38].

4.9 Comparison of the Mathematical MIRD Phantoms with Real Patients

It is obvious that the anatomy of stylized mathematical phantoms does represent an approximation to the real anatomy of an average patient. However, it has been shown that this approximation is quite sufficient for clinical purpose [26]. Nevertheless, in order to enable the reader of this book to assess the validity of normalized organ doses provided in this book, inconsistencies between the stylized mathematical MIRD phantoms and the real human anatomy shall be discussed in this section.

4.9.1 Body Height and Body Weight

Body height and body weight of the mathematical MIRD phantoms do agree quite well with the median body height and the median body weight of a real patient collective (patient collective of the Dr. von Hauner Children's Hospital, University of Munich, Figs. 3.3 and 3.4 in Sect. 3.1). Small deviations can be observed for adolescents whose somatic development may be retarded in the real patient collective compared to the average European population. Thus, the normalized organ doses provided in this book refer more to the average population than to paediatric patients at least for adolescent patients.

4.9.2 Head Diameter and Head Circumference

The head circumference is a function of age [50] and serves as a clinical screening parameter in paediatrics. As the mathematical phantoms of various ages as used for the calculation of normalized organ doses in this book were constructed through isometric transformation of the anthropomorphic trunk data (see Sect. 4.8), the question comes up whether the phantoms' head sizes do represent the human head size adequate over all age groups. However, a comparison of the measured head circumference of children of various ages [50] with the calculated head circumference of the mathematical phantoms which have been used for calculating the normalized organ doses listed in this book shows clearly that the head size of the mathematical phantoms has been formed too small especially in the first years of life (Fig. 4.9). This is also valid for the effective elliptical diameters. Thus, the radiation field sizes defined in this book for skull radiographs in neonates and young infants will underestimate the field sizes required for skull radiographs in real patients. However, the impact of this error on the validity of the normalized organ doses calculated with these field sizes is assumed by the authors to be negligible.

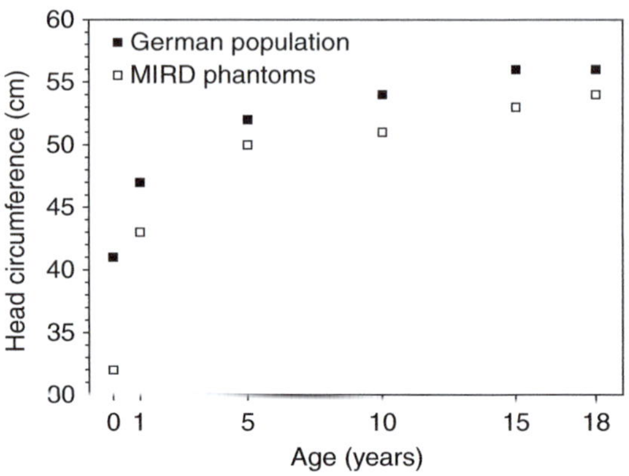

Fig. 4.9 Comparison of the head circumferences of the mathematical phantoms used for the calculations in this book with real children from the German population. Head circumferences of the mathematical phantoms have been calculated from the sagittal and lateral skull diameter and compared to the measured head circumferences of an anthropometric study in Germany [50]. The deviation of the circumferences is in the range of 20% within the first 2 years of life. (© M. Seidenbusch 2018)

4.9.3 Trunk Diameter

The effective trunk diameters of the mathematical MIRD phantoms do agree quite well with

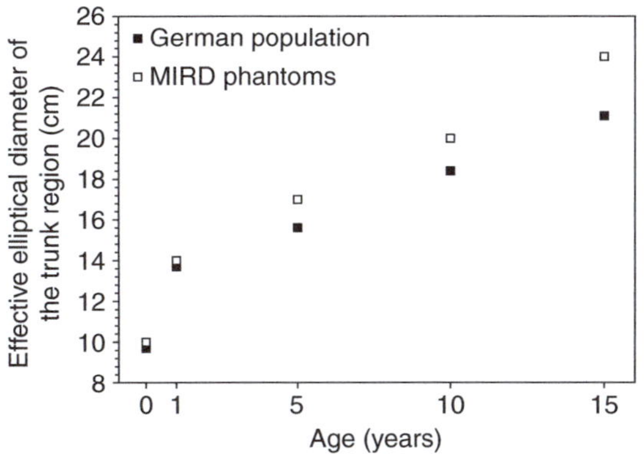

Fig. 4.10 Comparison of the equivalent elliptical diameter of the trunk region of the mathematical phantoms used for the calculations in this book with that of real children from the German population. Equivalent elliptical diameters have been calculated from the sagittal and lateral trunk diameter given for the mathematical phantoms and an anthropometric survey in Germany [10, 11]. The deviation of the diameters is increasing from about 3% in the first 2 years of life till about 12% in adolescents. (© M. Seidenbusch 2018)

the average trunk diameter of the German population (Fig. 4.10, Tables 4.3 and 4.4). However, in elder children, the increase of the lateral chest diameter has been overestimated when the phantom family of PCXMC program was created [52].

4.9.4 Organs and Tissues

As mentioned in Sect. 3.2, children differ considerably in anatomy from adults depending of age. This fact complicates the construction of mathematical phantoms for paediatric radiology. Conversely, the construction of the mathematical MIRD phantoms basing on geometrical structures means greatly simplifying the paediatric anatomy. As a consequence, there are a lot of inconsistencies between the geometrical anatomy of the mathematical phantoms and the real human anatomy (Table 4.5). This also applies to organs and tissues of the mathematical phantoms.

Table 4.5 Inconsistencies between the mathematical MIRD phantoms and real human anatomy

Anatomical region	Inconsistencies in the mathematical MIRD phantoms
Head	No eyes, nose, mouth, and ears have been inserted
	No eye lens has been inserted
Chest	No variation of lung density with age
	Mass of glandular tissue is underestimated in newborn infants
Abdomen	No navel has been inserted
Gonads	No variation of gonad situs with age
Active bone marrow	Active bone marrow within the liver and spleen is not inserted in newborn infants
Skeleton	The shape of the spine is unrealistic (no lordosis, no kyphosis considered), the vertebral column is not clearly depicted. The transverse processes are not outlined
	Ribs are drawn as complete circles and thicker than in real humans, the course of the ribs is horizontal
	No xiphoid has been inserted
	No acetabulum, no symphysis pubis have been inserted
Extremities	Position of the extremities cannot be changed
	No hands and feet have been inserted
	No shoulder, elbow, and wrist joints have been inserted
	Arms are constructed too long in the ages 0, 1, and 5 years
	No hip joint, no knee and no ankle joint have been inserted

For details, see the text

4.9.4.1 Eye Lens

Although the eye lens is considered as rather sensitive to radiation, no eye lens has been inserted into the mathematical MIRD phantoms till today. Therefore, no normalized dose for the eye lens can be provided in this book.

4.9.4.2 Thymus

The size of the thymus within the first 2 years varies considerably. This fact may have not been considered adequately when constructing the mathematical MIRD phantoms.

4.9.4.3 Lungs

The decrease of lung density with patient's age has not been considered when constructing the mathematical MIRD phantoms [52].

4.9.4.4 Mamma

As mentioned in Sect. 3.2, post-partum maternal hormones can sensitize the breast tissue of young infants leading to a transient increase of the mammarian tissue. This increase of breast tissue in the first 3–6 months of life has not been respected when the phantom family of PCXMC program was created, and the proportion of glandular tissue in the various age groups has been assessed too low [51].

4.9.4.5 Gonads

The location of ovaries varies considerably in all age groups. Naturally, ovaries and testes have been inserted into the mathematical MIRD phantoms considering their average situs in the adult patient.

4.9.4.6 Active Bone Marrow

As mentioned in Sect. 3.2, it must be stated here that it remains questionable whether the age-dependent regional distribution of active bone marrow in the children of various age groups has been considered sufficiently when constructing the mathematical MIRD phantoms forming the basis of the normalized organ doses listed in this book.

References

1. Aichinger H, Dierker J, Joite-Barfuß S, Säbel M. Radiation exposure and image quality in x-ray diagnostic radiology. Heidelberg, Dordrecht, London, New York: Springer; 2012.. ISBN 978-3-642-11241-6
2. Almén A, Mattsson S. On the calculation of effective dose in children and adolescents. J Radiol Prot. 1996;16:81–9.
3. Alzen G, Benz-Bohm G. Radiation protection in pediatric radiology. Dtsch Arztebl. 2011;108:407–14.
4. Andreo P. Monte Carlo techniques in medical radiation physics. Phys Med Biol. 1992;36:861–920.
5. Attix FH. Introduction to radiological physics and radiation dosimetry. New York, Chichester, Brisbane, Toronto, Singapore: Wiley; 1986.
6. Australian Radiation Protection and Nuclear Safety Agency. Australian national paediatric diagnostic reference levels for MDCT. Australian Radiation Protection and Nuclear Safety Agency; 2012.
7. Bader D, Datz H, Bartal G, Juster AA, Marks K, Smolkin T, Zangen S, Kugelman A, Hoffmann C, Shani G, Ben-Shlomo A, Margaliot M, Sadetzki S. Unintentional exposure of neonates to conventional radiography in the neonatal intensive care units. J Perinatol. 2007;27:579–85.
8. Beninson D, Sowby D. Age and sex dependent weighting factors for medical irradiation. Radiat Prot Dosim. 1985;11:57–60.
9. Birch R, Marshall M. Computation of bremsstrahlung x-ray spectra and comparison with spectra measured with a Ge(Li) detector. Phys Med Biol. 1979;24:505–17.
10. Bohmann I. Ermittlung des Durchstrahlungsdurchmessers bei Säuglingen, Kindern und Jugendlichen zur Aufstellung von Belichtungswerten in der Röntgendiagnostik und Abschätzung der Organdosiswerte bei typischen Röntgenuntersuchungen. Dissertation, Universität München; 1990.
11. Bohmann I. Ermittlung des Durchstrahlungsdurchmessers bei Säuglingen, Kindern und Jugendlichen zur Aufstellung von Belichtungswerten in der Röntgendiagnostik und Abschätzung der Organdosiswerte bei typischen Röntgenuntersuchungen. GSF-Bericht 16/90; 1990.
12. Boone JM. The three parameters of equivalent spectra as an index of beam quality. Med Phys. 1988;15:304–10.
13. Bushberg JT, Seibert JA, Leidholdt EM, Boone JM. The essential physics of medical imaging: Lippincott Williams and Wilkins; 2012.
14. Carlsson GA, Carlsson CA. Effective energy in diagnostic radiology. A critical review. Phys Med Biol. 1984;29:953–8.
15. Commission of the European Communities. European guidelines on quality criteria for diagnostic radiographic images in paediatrics. EUR 16261, ISBN 92-827-7843-6; 1996.
16. Cristy M, Eckerman KF. Specific absorbed fractions of energy at various ages from internal photon sources. Oak Ridge National Laboratory, Health and Safety Research Division, ORNL/TM-8381/V1; 1987.
17. Dance DR, Christofides S, Maidment ADA, McLean ID, Ng KH, editors. Diagnostic radiology physics. A handbook for teachers and students. Vienna: International Atomic Energy Agency (IAEA); 2014. ISBN 978–92–131010–1.
18. Datz H, Ben-Shlomo A, Bader D, Sadetzki S, Juster-Reicher A, Marks K, Smolkin T, Zangen S, Margaliot M. The additional dose to radiosensitive organs caused by using under-collimated x-ray beams in neonatal intensive care radiography. Radiat Prot Dosim. 2008;130:518–24.
19. Dell'Agnolo E, Seidenbusch M, Genzel-Boroviszeny O, Münch HG, Schneider K. Der Neugeborenen-Thorax I—Analyse von Feldgröße, Bildgüte und anderer Qualitätsparameter. Fortschr Röntgenstr. 2013;185:V30.
20. Deutsche Gesellschaft für Medizinische Physik. Pränatale Strahlenexposition aus medizinischer Indikation. Dosisermittlung, Folgerungen für Arzt und Schwangere. DGMP-Bericht Nr.7. Deutsche Gesellschaft für Medizinische Physik; 2002.

21. Deutsche Strahlenschutzkommission. Use of the effective dose in medical examinations. In: Veröffentlichungen der Strahlenschutzkommission, Band 41. Gustav Fischer Verlag; 1998. ISBN 3-437-21438-1.
22. Drexler G, Panzer W, Petoussi N, Zankl M. Effective dose—how effective for patients? Radiat Environ Biophys. 1993;32:209–19.
23. Fichtner C, Schneider K, Freidhof C, Endemann B, Horwitz AE, Kohn MM, Fendel H. Critical analysis of field size in chest x-rays of infants—a EC-wide survey in children's clinics. Eur Radiol. 1993;Suppl 3:389.
24. Fritz S, Jones AK. Guidelines for anti-scatter grid use in pediatric digital radiography. Pediatr Radiol. 2014;44:313–21.
25. Galanski M, Nagel HD, Stamm G. Paediatric CT exposure practice in the federal republic of Germany: results of a nationwide survey in 2005–2006. Hannover: Medizinische Hochschule Hannover; 2007.
26. Golikov V, Barkovsky A, Wallström E, Cederblad A. A comparative study of organ doses assessment for patients undergoing conventional x-ray examinations: phantom experiments vs. calculations. Radiat Prot Dosim. 2018;178:223–34.
27. Griebl G. Variation von Feldgröße und Dosis bei Röntgen-Thoraxaufnahmen beim 5 Jahre alten Kind—kritische Analyse einer Feldstudie in europäischen Kinderkliniken. Dissertation, Universität München; 2001.
28. Hart D, Wall BF, Shrimpton PC, Bungay DR, Dance DR. Reference doses and patient size in paediatric radiology. National Radiological Protection Board NRPB-R318; 2000. ISBN 0-85951-448-X.
29. Horwitz AE, Schweighofer-Berberich K, Schneider K, Kohn MM, Bakowski C, Stein E, Freidhof C, Fendel H. Selected image quality parameters in a survey using a test phantom in radiological departments and offices in the Federal Republic of Germany. Radiat Prot Dosim. 1993;49:79–82.
30. Hoxter EA, Schenz A. Röntgenaufnahmetechnik. Siemens AG; 1991. ISBN 3-8009-1566-9.
31. Hubbell JH. Photon mass attenuation and energy-absorption coefficients from 1 keV to 20 MeV. Int J Appl Radiat Isot. 1982;33:1269–90.
32. Huda W. Kerma-area product in diagnostic radiology. AJR. 2014;203:W565–9.
33. International Atomic Energy Agency. Dosimetry in diagnostic radiology for paediatric patients. IAEA human health series no. 24; 2013.
34. International Atomic Energy Agency. Radiation protection in paediatric radiology. IAEA safety reports series no. 71; 2012.
35. International Commission on Radiological Protection. Report of the task group on reference man: anatomical, physiological and metabolic characteristics. ICRP publication 23. Oxford: Pergamon Press; 1975.
36. International Commission on Radiological Protection. 1990 recommendations of the International Commission on Radiological Protection, ICRP publication 60; 1991.
37. International Commission on Radiological Protection. Basic anatomical and physiological data for use in radiological protection: reference values. ICRP publication 89; 2003.
38. International Commission on Radiological Protection. The 2007 recommendations of the International Commission on Radiological Protection, ICRP publication 103; 2007.
39. Jacobi W. The concept of the effective dose—a proposal for the combination of organ doses. Radiat Environ Biophys. 1975;12:101–9.
40. Jones HE, Cunningham JR. The physics of radiology. Springfield: Charles C Thomas; 1981.
41. Klinzmann M. Qualitätsbeurteilung von Röntgenthoraxaufnahmen auf den pädiatrischen Intensivstationen in Abhängigkeit von den ausführenden Berufsgruppen. Dissertation, Justus-Liebig-Universität Gießen; 2015.
42. Löster W, Drexler G, Stieve FE. Die Messung des Dosisflächenproduktes in der diagnostischen Radiologie als Methode zur Ermittlung der Strahlenexposition. Berlin: Hoffmann; 1995.
43. Malusek A, Sandborg M, Carlsson GA. Accurate KAP meter calibration as a prerequisite for optimisation in projection radiography. Radiat Prot Dosim. 2016;169:353–9.
44. Mettler FA, Davis M, Moseley RD, Kelsay CA. The effect of utilising age and sex dependent factors for calculating detriment from medical irradiation. Radiat Prot Dosim. 1986;15:269–71.
45. Petoussi-Henß N, Zankl M, Drexler G, Panzer W, Regulla D. Calculation of backscatter factors for diagnostic radiology using Monte Carlo methods. Phys Med Biol. 1998;43:2237–50.
46. Reich H. Dosimetrie ionisierender Strahlung. Stuttgart: Teubner; 1990.
47. Rogers DWO. Fifty years of Monte Carlo simulations for medical physics. Phys Med Biol. 2006;51:R287–301.
48. Šabič I, Ključevšek D, Thaler M, Žontar D. The effect of anti-scatter grid on radiation dose on chest radiography in children. Paediatr Today. 2016;12:75–80.
49. Sandborg M, Dance DR, Carlsson GA, Persliden J. The choice of anti-scatter grids in diagnostic radiology: the optimization of image quality and absorbed dose. University of Linköping; 1993.
50. Schienkiewitz A, Schaffrath Rosario A, Dortschy R, Ellert U, Neuhauser H. German head circumference references for infants, children and adolescents in comparison with currently used national and international references. Acta Paediatr. 2011;100:e28–33.
51. Seidenbusch M, Kammer B, Specht T, Stahl R, Schneider K. Zur Rekonstruktion der Brustdrüsendosis bei Kindern in der pädiatrischen Radiologie: Gewicht, Dichte und Volumen der Mammagewebe bei Neugeborenen, säuglingen, Kleinkindern und Adoleszenten und die dosimetrischen Konsequenzen. Fortsch Röntgenstr. 2018;190:869–70.

52. Seidenbusch MC, Schneider K. Zur Strahlenexposition von Kindern in der pädiatrischen Radiologie. Anthropometrische Merkmale von Kindern und den zur Dosisrekonstruktion verwendeten mathematischen Phantomen (ISIMEP-Projekt des Bundesministeriums für Bildung und Forschung, Förderkennzeichen 02NUK016A). Fortschr Röntgenstr. 2012;184:A13.
53. Seidenbusch MC, Schneider K. Strahlenhygienische Aspekte bei der Röntgenuntersuchung des Thorax. Radiologe. 2015;55:580–7.
54. Servomaa A, Rannikko S, Nikitin V, Golikov V, Ermakov I, Madarskyi L, Saltukova L. A topographically and anatomically unified phantom model for organ dose determination in radiation hygiene. STUK-A87. Helsinki: Finnish Centre for Radiation and Nuclear Safety; 1989.
55. Servomaa A, Tapiovaara M. Organ dose calculation in medical x-ray examinations by the program PCXMC. Radiat Prot Dosim. 1998;80:213–9.
56. Soboleski D, Theriault C, Acker A. Unnecessary irradiation to non-thoracic structures during pediatric chest radiography. Pediatr Radiol. 2006;36:22–5.
57. Stollfuss J, Schneider K, Krüger-Stollfuss I. A comparative study of collimation in bedside chest radiography for preterm infants in two teaching hospitals. Eur J Radiol Open. 2015;2:118–22.
58. Tapiovaara M, Lakkisto M, Servomaa A. PCXMC. A PC-based Monte Carlo program for calculating patient doses in medical x-ray examinations. Finnish Centre for Radiation and Nuclear Safety, Säteilyturvakeskus (STUK), Report STUK A-139; 1997.
59. Tapiovaara M, Siiskonen T. PCXMC. A Monte Carlo program for calculating patient doses in medical x-ray examinations. Säteilyturvakeskus (STUK), Report STUK-A231; 2008.
60. Tschauner S, Marterer R, Gübitz M, Kalmar PI, Talakic E, Weissensteiner S, Sorantin E. European guidelines for ap/pa chest x-rays: routinely satisfiable in a paediatric radiology division? Eur Radiol. 2016;26:495–505.
61. Veit R, Zankl M. Influence of patient size on organ doses in diagnostic radiology. Radiat Prot Dosim. 1992;43:241–3.
62. Veit R, Zankl M. Variation of organ doses in paediatric radiology due to patient diameter, calculated with phantoms of varying voxel size. Radiat Prot Dosim. 1993;49:353–6.
63. Verordnung über den Schutz vor Schäden durch Röntgenstrahlen (Röntgenverordnung—RöV). BGBl. 1987;3:114–33.
64. White DR. The design and manufacture of anthropomorphic phantoms. Radiat Prot Dosim. 1993;49:359–69.
65. Zankl M. Methods for assessing organ doses using computational models. Radiat Prot Dosim. 1998;80:207–12.
66. Zeiss A. Feldgrößen- und Dosisvariation von Röntgen-Thorax-Aufnahmen beim zehnjährigen Kind—kritische Analyse einer Feldstudie in europäischen Kinderkliniken. Dissertation, Universität München; 2002.

5 Image Quality in Paediatric Radiography

5.1 General Remarks

“Image quality” in diagnostic radiology has been defined differently by physicists and radiologists. Physicists usually evaluate the objective image quality using standardized phantom measurements under various exposure conditions, i.e. the optical density, the spatial and contrast resolution of images. Thereby, they are able to calculate and quantify the resultant objective image quality, which can be defined through the signal to noise ratio or the modulation transfer function [21]. The imaging capacity of digital X-ray imaging depends on several components: pixel size, dynamic range to cover a wide range of intensities, detector sensitivity in the low dose range, and the read-out [24]. The most common characteristic of digital X-ray imaging used is the term digital quantum efficiency (DQE) [13, 40], which describes the overall system performance. This term is mainly dependent on the detector material, the kilovoltage, additional filtration, and the dose at the detector, and is mostly correlated with spatial resolution. However, the final image quality seen on a high resolution monitor depends on the X-ray converter (storage phosphor, photoconductor, scintillator) but also on the coupling state (optical glue, direct deposition, lens/fibre optic) and finally on the pixel read-out [24]. Then a long chain of imaging processing follows which may greatly differ amongst the manufacturers of current digital detector systems [15, 18]. Today, two digital X-ray imaging systems are available: computed radiography (CR) using storage phosphor image plates and direct digital radiography (DR) using flat-panel systems with two variants, semiconductor or scintillator technology, as conversion layers [15, 18, 24].

The assessment of the subjective image quality of a radiographic image is in the personal charge of the radiographer and the radiologist who evaluate distinct anatomical structures of the X-ray image, which are important for the radiological diagnosis [4, 9]. For this reason, radiologists are primarily trained to recognize certain anatomical image details, which are of relevance to achieve the correct interpretation of the radiographs [38]. In this context, the type of display monitors—medical grade, consumer grade, and tablet devices—as well as the ambient light play a significant role in recognizing subtle anatomic details in chest radiographs and bone X-ray images. The newest generation of tablets, even calibrated with DICOM-GSDF (Digital Imaging and Communications in Medicine-Grayscale Standard Display Function), is not suitable for evaluating digital images [19]. However, radiologists often do not understand the relation between objective image degrading effects, the dose to the patient, and the image quality. This is especially true for direct digital radiography and computed radiography with their wider dynamic range, but limited resolution power of CR systems [18, 47]. Occasionally, the radiologists can even misinterpret image details, if the whole system is not appropriately optimized for all paediatric X-ray examinations.

M. Seidenbusch et al., *Imaging Practice and Radiation Protection in Pediatric Radiology*,
https://doi.org/10.1007/978-3-030-18504-6_5

5.2 Image Quality Criteria

Image quality criteria of adult radiology are not at all or at least not reliably transferable to paediatrics. Therefore, the European Commission published the European guidelines on quality criteria for diagnostic radiographic images in paediatrics in 1996 [4]. These recommendations include “positioning criteria” as well as criteria of “important image details” for the most frequent conventional radiographs in paediatrics [4, 25].

Three levels of image criteria have been defined in this document, which refer to specific anatomical structures:

- Visualization: Characteristic features are detectable but details are not fully reproduced; features are just visible
- Reproduction: Details of anatomical structures are visible but not necessarily clearly defined; details are emerging
- Visually sharp reproduction: Anatomic details are clearly defined; details are clear

Bilateral opacities of lung parenchyma or mediastinal disease can sometimes prevent the use the aforementioned criteria, which evaluate the “important image details” like the vascular pattern, the heart border, the diaphragm or the trachea and large bronchi. However, pathological changes are mostly confined only to one lung or affect the mediastinal structures asymmetrically.

Figure 5.1a, b show chest X-rays of young infants with different noise levels with normal and pathological findings. Standardized radiographic technique for both X-rays followed the national and EU-Guidelines. DAP values were far below the German dose reference levels [31]. The exposure technique was similar with the exception that different image detectors were used. In Fig. 5.1b positioning criteria of the EU-Guidelines are fulfilled, but not in Fig. 5.1a. More important, not all relevant image details for the newborn chest ap of these guidelines were fulfilled in Fig. 5.1a. For example, the contours of the thymus and trachea are not reproduced as “visually sharp” and the vascular pattern was not at all visible. Nevertheless, in Fig. 5.1a with a

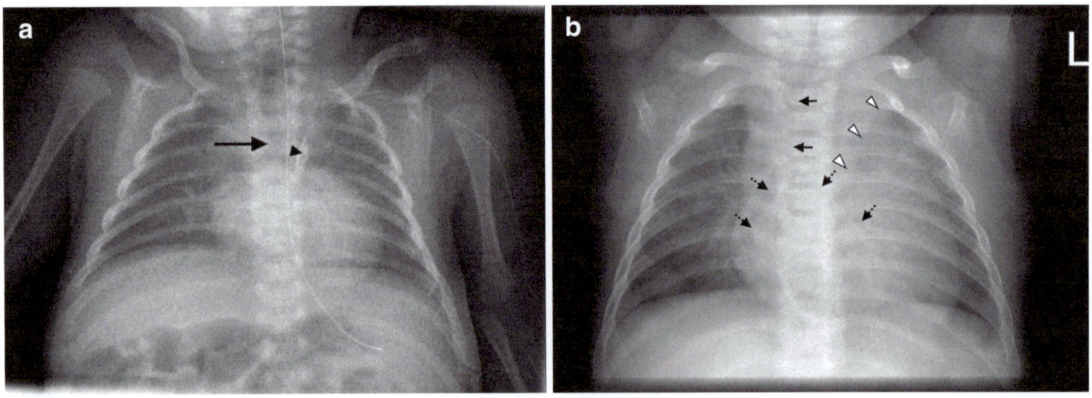

Fig. 5.1 (**a**) Ten-day-old premature baby of 36 gestational weeks. Radiographic technique: 60 kVp, 3 mm Al + 0.1 mm Cu total filtration. Storage phosphor plate, ADC Solo, AGFA Co. Image quality is low; the trachea and main bronchi, the heart border, and the thymus can only be visualized. The vascular pattern of the lungs is not reproduced and the skeleton is barely visualized. However, the position of the central venous line is seen, ending in a persistent left superior vena cava (black arrowhead). Additionally, compression of the lower trachea by a right sided aortic arch can be identified (black arrow). The positioning criteria of European Guidelines are fulfilled, except the fact that the upper arms are unnecessarily exposed (© Springer-Verlag 2018). (**b**) Four-month-old infant with a large thoracic neuroblastoma. Radiographic technique: 63 kVp, 3 mm Al + 0.1 mm Cu total filtration. Needle Image plates, AGFA Co. All image criteria are fulfilled. The lower trachea is compressed along its left side (black arrows) by the mediastinal tumour (white arrowheads). However, the main bronchi are reproduced visually “sharp” (black dashed arrows). The heart, the thymus, the diaphragm, and the vascular pattern of the lungs are also “visually sharp”. Bony structures of the chest are sufficiently reproduced

high noise level the image quality is sufficient to detect anatomical details, like the position of the central venous line, to answer the key question for the referring neonatologist. Although several anatomic structures were not defined as "visually sharp", an additional important pathologic finding can be recognized, i.e. the pathologic impression of the trachea indicative of a right aortic arch. In Fig. 5.1b all positioning and image quality criteria concerning anatomical details are fulfilled.

Image quality criteria of the chest X-ray and skeletal X-rays of infants and toddlers differ significantly from older children, and of course even more from adults [4, 9]. In order to achieve radiographs of adequate image quality for the final correct radiological interpretation, so-called "positioning criteria" were defined, which include the field size, centring and the patient position, and image criteria, referring to important anatomic image details in all paediatric age groups [38]. The great differences can be explained by the changes of the anatomy during the development of the organs. For example, the peripheral lung vessels in newborn babies and young infants are much smaller; the large thymus may overlay the great mediastinal vessels. Further on, the ossification of the thoracic skeleton is immature, i.e. the neurocentric and costovertebral synchondroses still persist up to the first 2 years [48]. Positioning criteria are to a minor degree dose relevant, e.g. inappropriate collimation, rotation of the patient and false centring. However, these simple faults of radiographic technique can sometimes conceal or disclose important pathologic findings, e.g. overinflation of a lung, vascular abnormalities, and mediastinal tumours [35]. These image criteria can be used for image quality assurance measures and auditing [32]. However, there are some limitations to use the criterion "vascular pattern" for chest X-rays if there are extensive changes of the lung parenchyma. The same is true for the criterion "psoas outline/kidney outline" of X-rays of the abdomen in some cases of bowel obstruction or peritonitis [4]. Nevertheless, the adherence to the guidelines guarantees a high image quality level and also provides a reasonably low dose per radiograph, if the recommendations of "good radiographic technique" are followed [25, 32].

The main source of poor image quality is motion artefacts during radiation exposure in paediatric patients less than 5 years of age, because small children are often anxious and do not cooperate. Also mismatched radiographic technique in paediatric radiography is more frequent than in adults [16, 37] and is often associated with a high dose to the patients. Low dose exposures can also lead to non-diagnostic images. As a consequence, all these mistakes can lead to unnecessary retakes or even unneeded imaging, e.g. fluoroscopy, computed tomography, or magnetic resonance imaging. In order to avoid such inadvertent mistakes, radiographers should be well trained and should strictly adhere to the European or national guidelines [4, 9] and the specific operational procedures of the individual radiological departments.

5.3 Image Quality and Radiographic Technique

The radiographic technique differs greatly, especially the beam quality used for standard radiographs of the five paediatric age groups, i.e. newborn babies (including prematures), infants, toddlers, older children, and adolescents, from that used in adults [5, 8, 9]. The introduction of digital X-ray imaging has created the need to recognize some peculiarities of this new modality and implement measures for specific paediatric requirements [40]. Spatial resolution and low contrast details are very important in paediatric imaging. The matrix of the detector and pixel size are the most important characteristic of digital detectors [18]. In order to improve the image processing, the smaller field size, smaller body diameters, higher body water content, and differing absorption characteristics of body organs of paediatric patients influence the radiographic technique [17, 38]. All these factors, often occurring in combination, greatly influence the overall image quality, i.e. the noise level, the visibility of low contrast details, and the spatial resolution of fine anatomical structures, which are significantly smaller in young patients. The computation algorithms of raw data, the use of filters, and post-processing procedures play a central role in CR and DR. A special problem of digital X-ray imaging in small patients is the fact that the grey-scale histograms may vary greatly with the field size and the exposed tissue. The presence of high and low attenuating tissues, e.g. the inclusion of facial bone and overinflated lung of a chest X-ray, can considerably disturb the image computation producing very different grey-scale histograms if the field size is inappropriately small or large [49]. Appropriate field size is a frequent problem in digital paediatric radiography. Thereby, the final image quality may be less in well-collimated chest X-rays compared to X-rays with wider collimation. In addition, the selection and the implementation of the most suitable look up tables are essential for displaying the finest contrast and density structures of the acquired and processed digital data. They must be appropriately adjusted to the different age groups, premature newborns, infants, and small children. They also differ between the air filled lung and bony tissue.

5.3.1 Specific Requirements for Chest Radiography

Chest radiography is the most frequent X-ray examination in all paediatric age groups [26, 43]. The highest numbers are reported in neonates and young infants [26]. Therefore, important imaging details of chest X-rays in this early age group have been comprehensively discussed [38]. But, even at the age between 1 and 5 years some peculiarities must be accepted if high quality chest X-rays have to be obtained. For example, the depth of inspiration—an important positioning criterion—may be very difficult to achieve for the radiographer, in non-cooperative young children [35]. The evaluation of chest X-rays in expiration makes the evaluation of the lung parenchyma for the radiologist more difficult. On the other hand, expiratory chest X-rays are useful in the case of suspected foreign body aspiration. The phenomena of air trapping in cases of pulmonary infection of unintentionally produced expiratory chest images are suitable to recognize the degree of overinflation of the lungs [35]. These peculiarities should be known to radiographers and radiologists, responsible for imaging of paediatric patients and should not lead to a retake in every case.

Medical physicists, radiologists, and radiographers who care for paediatric patients should have at least basic knowledge about anatomical and physiological characteristics of the lungs, the airways, great vessels, the heart, and the paediatric skeleton [34, 38]. In order to identify and evaluate the vascular pattern of the infant lung short exposure times are needed, presumably less than 4 milliseconds (ms) to avoid motion and pulsation artefacts [4]. If kVp values of less than 60 are used, the mAs must be increased to provide enough dose to the detector; this means a longer exposure time. If the voltage is increased to 70 kVp for chest X-rays in small infants, the exposure time could be decreased on the one hand [40], but the visibility of the vascular pattern and tiny pathological changes of the lung parenchyma will substantially be diminished by the loss of contrast [39]. Another often neglected parameter to obtain high quality radiographs in neonates is the focal spot size [12]. The use of the small focus, i.e. 0.6, is recommended in many guidelines. However, mobile units used in neonatal care units often have a focal spot size of 0.8 or even larger. These mobile X-ray equipments are not really suitable for neonatal imaging because of the inacceptable geometrical haziness.

Another characteristic of the infant's lungs is their high density in the first year of life and the fact that large parts of lung tissue are overlaid by water-dense organs, i.e. the thymus and the liver. As sick term babies must be X-rayed within the incubators, some factors must be considered to avoid bad images [38]. The scattered radiation produced by double-walled incubators may degrade the image quality [11]. Moreover, if the radiographer uses the cassette holder, the image quality can be degraded because the baby lies on a mattress and other material which absorb radiation and even produce undesired scatter [38].

Because of the high natural contrast of the air filled lungs and the small sagittal diameter of the thorax, no grid is needed up to the age of 8 years [5, 9]. These age dependent variations of the body size and shape necessitate the use of body weight exposure charts to select the appropriate kVp and mAs values for the different paediatric age groups [5, 17]. As a large European survey has clearly shown, there is no significant difference in image quality of chest X-rays if the radiographic technique with additional filtration was used compared to those without filtration [36]. Later additional filtration with aluminium and/or copper has been recommended in the European and German Guidelines [4, 9]. Thereby, the entrance surface dose and as a consequence the absorbed dose could be lowered by factor 30 without loss of image quality [36, 37] (Fig. 5.2a, b).

The results of animal studies concerning the image quality of chest X-rays cannot be transferred to human imaging, even if the field of view, the body diameters are similar and the same exposure conditions (kVp, additional filtrations) were used [1, 27, 44]. Weighty arguments against these studies are: (1) the tissue absorption characteristics of animal chest organs are greatly different from the human being, especially the fur, the skin, and subcutaneous tissue of the animals' chest, their size and shape, the water content, and vascular pattern of the lungs; (2) the thoracic skeleton of the animals is much more mature; (3) the evaluation of the image was often restricted to the visibility of high contrast material, e.g. tubes and central venous lines; and (4) the exclusion of the evaluation of airways and small pulmonary vessels.

Chest X-rays in patients beyond 8 years of age are usually performed at a chest stand using automatic exposure control, radiographic voltage above 110 kVp, and a grid [9]. As higher kVp is used, more scatter is generated. Thus, heavier additional filtration is needed. The advantage of the high kVp technique is threefold: (1) the absorption of the radiation within the thoracic bones and soft tissues decreases significantly, which improves the detection of tiny vascular and interstitial structures of the lungs; (2) the exposure time decreases significantly, which is partially corrected by the heavier filtration with 0.2 mm copper. Thereby motion artefacts are no longer present; (3) as a consequence the absorbed dose can be significantly reduced.

Chest X-rays performed in all age groups who are severely ill and must be treated in intensive care units profit notably from digital radiography

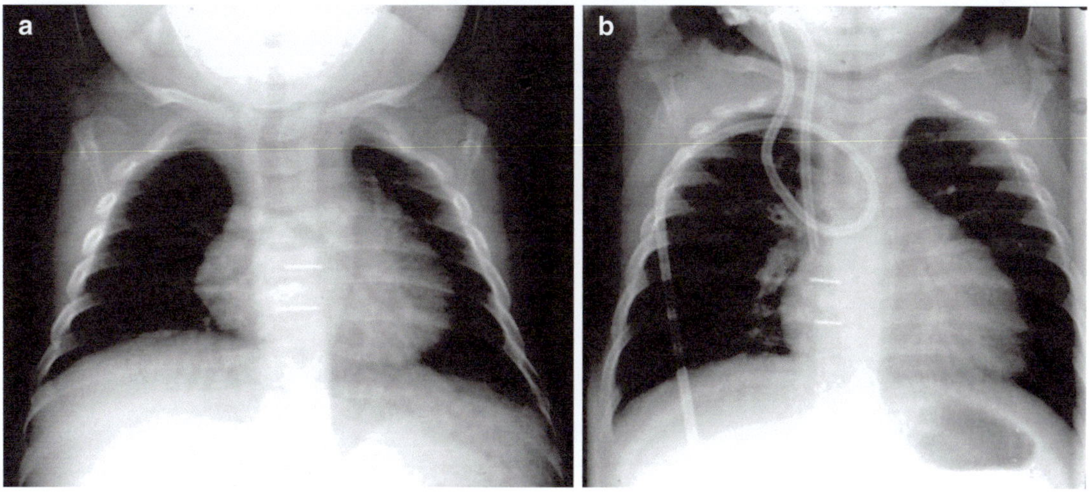

Fig. 5.2 Comparison of two chest X-rays, ap in upright position of 10-month-old infants, which were performed using different radiographic technique. (**a**) kVp was 55, antiscatter grid used, total filtration 3.0 mm Al, screen film sensitivity 100; entrance surface dose 589 μGy. (**b**) kVp was 63, no antiscatter grid, total filtration 3.5 mm Al + 0.1 mm Cu, screen film sensitivity 600; entrance surface dose 21 μGy. Be aware of the very slight different image quality, but notice the great difference of the entrance surface dose, 30:1. (© K. Schneider 2018)

[29]. Retakes are rare because of wide dynamic range and edge enhancement of the images which show much better high contrast details, e.g. the position of tubes, central venous lines, beginning interstitial pulmonary emphysema [3, 28]. These images should be available in addition to regularly processed images which allow a better visualization of low contrast details. The soft tissues algorithms for chest and skeletal X-rays of different paediatric age groups, especially the settings of viewing parameters at the workstations, are often not perfectly elaborated. These parameters should be adapted appropriately by the responsible physicist and radiologist of the department [24, 40].

5.3.2 Specific Requirements for Abdominal Radiographs

Acute abdomen is the most frequent indication to obtain abdominal films in all paediatric age groups. This means that simple radiological signs, i.e. free abdominal air/gas, extent and site of bowel obstruction, the position of tubes or catheters and radio-opaque foreign bodies, have to be detected. The image quality needed to make the correct radiologic diagnosis may be at a relatively low dose level. Even noisy images with suboptimal quality are sufficient in such cases to detect the radiological signs mentioned above. Much more important in this context is the fact that the positioning criterion "field size" is fulfilled. Too narrow collimation of the abdomen with a cropped upper field or false centring causes that the diaphragm is only partially or not at all delineated. In such cases, a small amount of free intraperitoneal air or lower lobe pneumonia can be missed. The same is true concerning the lower field edge. Therefore, incorrect collimation can lead to miss significant pathology in the pelvic region, i.e. congenital anomalies of the rectum and incarcerated inguinal hernia.

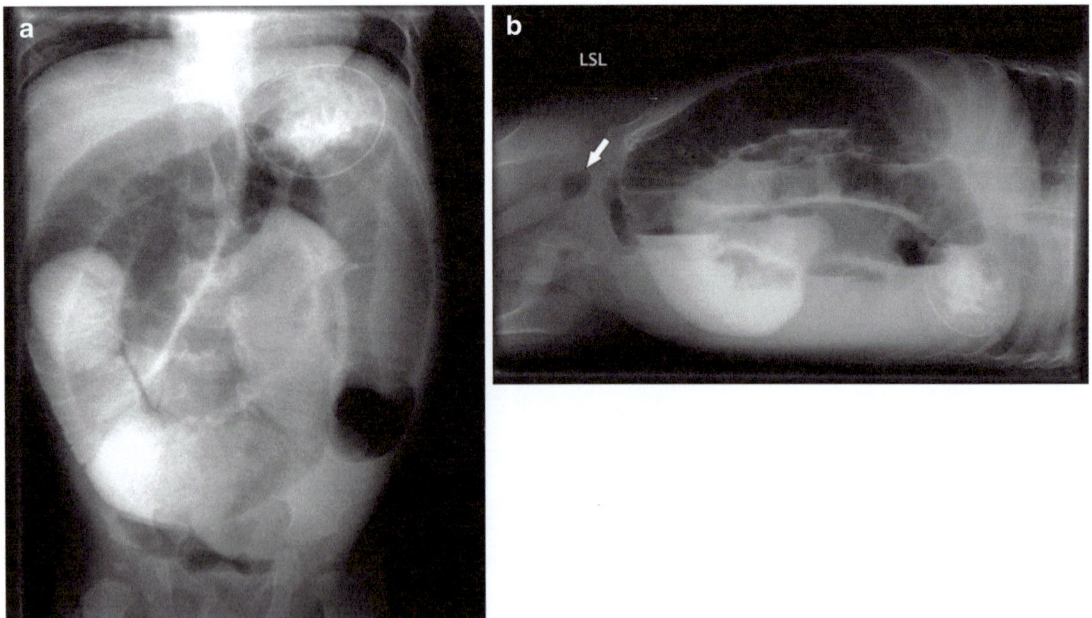

Fig. 5.3 One-year-old infant with severe small bowel obstruction following necrotizing enterocolitis. Follow-up abdominal X-rays. (**a**) ap supine and left lateral decubitus. (**b**) After intragastral administration of iodinated contrast medium shows deep small bowel obstruction. Mind the ideal field size of the abdominal X-rays in both projections. The left lateral decubitus films are slightly rotated which make the interpretation of a gas filled bowel in the region of the right inguinal fossa (white arrow) difficult to interpret. Additional filtration of 1.0 mm Al + 0.1 mm Cu was used, but no grid. (© K. Schneider 2018)

In infants and children up to the age of 5 years, an antiscatter grid can be omitted for abdominal X-rays without loss of relevant image quality if the radiated field is tightly collimated (Fig. 5.3a, b). However, in patients older than 10 years, the use of antiscatter grids is mandatory [4, 5, 9]. There are three patient-related factors that produce more scattering of abdominal radiographs in older patients: the patient diameter, the field size and the density and the physical composition of the irradiated body tissue (fluid filled bowel versus meteorism). The use of high kVp in older children—beyond 90 kVp—amplifies scattering significantly and necessitates at all times the use of an antiscatter grid.

5.3.3 Skeletal System

Radiographic imaging of the skeletal system in paediatrics is the second most frequent X-ray examination [26]. Radiographs of the small and large bones of the extremities predominate [42]. In the last decade, the numbers of skull, spine, and pelvic radiographs have decreased significantly and very often have been replaced by magnetic resonance imaging.

There are some peculiarities of the developing paediatric skeleton which are relevant for the discussion of image quality. Four aspects concerning the "image quality" are relatively important for the detection of bone and soft tissue pathology in young patients: First, the development of the paediatric skeleton is a long-lasting process ending between 18 and 24 years of age, in females and males, respectively [34]. All tubular bones have epiphyseal and some long bones also apophyseal ossification centres which appear at definite time points during growth. Second, correct positioning of patients' extremities and joints is of great importance to obtain the standard projection radiographs [5]. Otherwise, pathologic findings like a fracture line of the traumatized elbow may be superimposed by forearm bones or an ossification centre which may simulate a fracture. Third, the spongiosa and the compacta of infants' bones are particularly different compared to older children or adults. Fourth, the beam quality of X-ray spectrum (kVp, total, filtration, detector dose) should be adapted to the referral criteria. Fifth, the soft tissue layers must always be seen, which means wider collimation.

In order to keep the absorbed dose to the patients as low as possible, some radiologists have defined two different levels of image quality for imaging of the paediatric skeleton [47]. The image quality needed should take into account the referral criteria. The most frequent reason for X-ray imaging in all age groups is accidental trauma of the long and short bones. In most cases, the diagnosis of fractures or luxation is easy even if some noise is in the images. However, some physicists and radiologists recommend for the detection of subtle fractures low kVp technique, between 40 and 50 kVp [10, 14]. As a consequence, the mAs must be increased. However, using such low kVp range, the images may very often be blurred, which is caused by motion artefacts of young uncooperative patients, for whom the exposure time is too long. On the contrary in patients less than 5 years, it is more reasonable to use a range of kilovoltage between 60 and 70 kVp to shorten the exposure time further [3, 11]. If this radiographic technique is used for all types of trauma, the luxation of joints, greenstick and displaced fractures are visible without difficulty. If post-fracture follow-up films within the cast are taken, even kVp values up to 70 for long bones are reasonable to get adequate diagnostic images. The increased noise caused by the use of additional filtration can be tolerated, as the resultant contrast loss has no negative effect on the radiological diagnosis.

However, in patients with battered child syndrome very subtle fractures of the extremities, rib fractures, stress fractures, early bone changes in acute osteomyelitis or some bone tumours are clinically suspected; the additional filtration should be removed because the contrast resolution must be very high [10, 14]. Adequate image contrast is also needed in full spine radiographs in young patients with congenital scoliosis or some rare cases of skeletal dysplasia in order to recognize complex vertebral and rib anomalies

(Fig. 5.4a, b). In all these referrals, the recognition of subtle changes of bone structure is mainly dependent on the beam quality of the X-ray spectrum. But also the type of the detector and the dose at the detector are important factors for high quality radiographs. On the other hand, full spine radiographs in adolescents should be done with high kVp and low detector dose, which allows decreasing the exposure time and so diminishing unwanted motion artefacts [40]. Additional post-processing of the images can increase the image quality considerably at low dose exposures of the full spine [23].

Another example of the difficulty to detect the early findings of bone pathology is acute osteoarthritis (Fig. 5.5a, b). The slight loss of density of the trabecular bone and the loss of the cortical contour of the tarsal bone demonstrate the need of the appropriate radiographic technique and the optimized digital system in order to reach the sufficient image quality. In addition, the importance of optimized post-processing can be impressively demonstrated in the ap projection (Fig. 5.5a). The lateral cortex of the distal fibula and the adjacent soft tissue are not visible. The soft tissue changes especially important for an early radiological diagnosis cannot be visualized in the ap projection in contrast to the lateral projection (Fig. 5.5b).

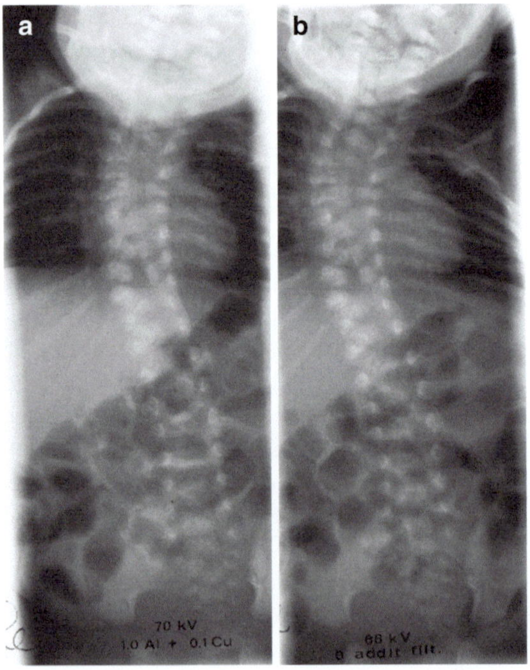

Fig. 5.4 Two full ap spine radiographs of a term baby with congenital scoliosis. (**a**) This radiograph was done with additional filtration of 1.0 mm Al + 0.1 mm Cu. (**b**) This X-ray was repeated without additional filtration using similar kVp and mAs. The severe multiple anomalies of thoracic vertebrae and the rib anomalies are much better seen in the radiograph without additional filtration (**b**). Mind that the cervical spine cannot be sufficiently seen in (**a**). (© K. Schneider 2018)

5.4 Image Detector Systems

Digital radiographic imaging has completely replaced imaging with screen-film-combinations [18]. Actually, different detector systems are available for digital radiography and computed radiography [15, 18, 22, 41]. The innovation process of digital imaging facilitated the development of picture archiving and communication systems. Initially CR was based on digital luminescence detectors released by the Fuji Company using powder image plates within conventional cassettes [20]. Using a special reader the energy

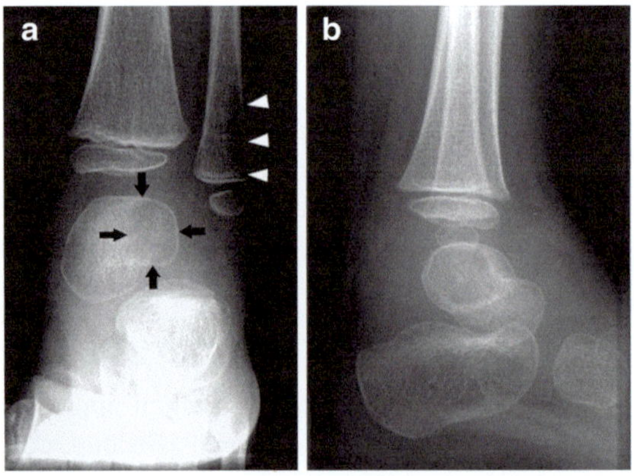

Fig. 5.5 Two year old child with acute osteomyelitis of the talus. CR radiography with needle image plates of Agfa Company. (**a**) The loss of trabecular bone of the talus on the ap radiograph of the ankle (black arrows). Note that the soft tissue swelling and contour of the lateral cortex of the distal fibulas were not displayed in this view, i.e. a typical fault of image processing of CR systems (white arrowheads). (**b**) The lateral view of the ankle was optimally processed showing the soft tissue changes and details of the bony structures. (© K. Schneider 2018)

profile of the stored virtual image was transformed to a histogram with various densities which could be processed to the final image [40]. The image quality of radiographs has been improved in many steps using special algorithms in newer reading systems of computed radiography. First, dual-read-out CR systems have been released [20, 23]. Later the release of needle image plates (NIP) has allowed an image quality similar to the flat detector systems [30]. The higher spatial resolution as well as the contrast resolution can be explained by the unique needle structure of the NIP cassettes based on CsBr and the new wireless detector based on CsI-TFT (Fig. 5.6).

In concurrence to digital luminescence detectors flat panels based on Gd-Oxisulfide (GOS) and CsI/TFT flat detectors came on the market [28, 29]. Direct digital radiography followed next [18, 24]. Now, wireless portable systems are the recent innovation of digital imaging. All the detector systems have profited from newer image readers, using more complex image reconstruction algorithms [24]. The image processing has continuously been improved by software updates, and more important, the speed of computation of the images increased tremendously.

In paediatric patients portable systems are generally preferred, because the vast majority of X-rays are done on the Bucky table or bedside. All these new DR systems stand out due to high DQE [40]. Flat-panel detectors with high DQE have lead to integrated systems using one computer controlled system [24]. Thereby, the image quality and the dose can be more effectively controlled. But there are differences in computing the images between the manufacturers. Different algorithms are used, but all systems are primarily based on the grey-scale histogram [20, 30]. However, the further processing steps using the initial raw data differ considerably. As a consequence, more image details could be displayed at the same detector dose [24].

Figure 5.7 shows the detector sensitivity of commonly used detector material, which is dependent on the voltage. Recommended kVp values used for chest and abdomen X-rays of infants and children are in the range between 60 and 70 kVp. CsI-TFT-detectors respectively CsBr detector of needle image plates (NIP) is superior to GOS detector systems in the low kVp range. However, GOS detectors have the similar performance or have the edge over the CsI-detector if the high kilovoltage technique in older children or adolescents is used.

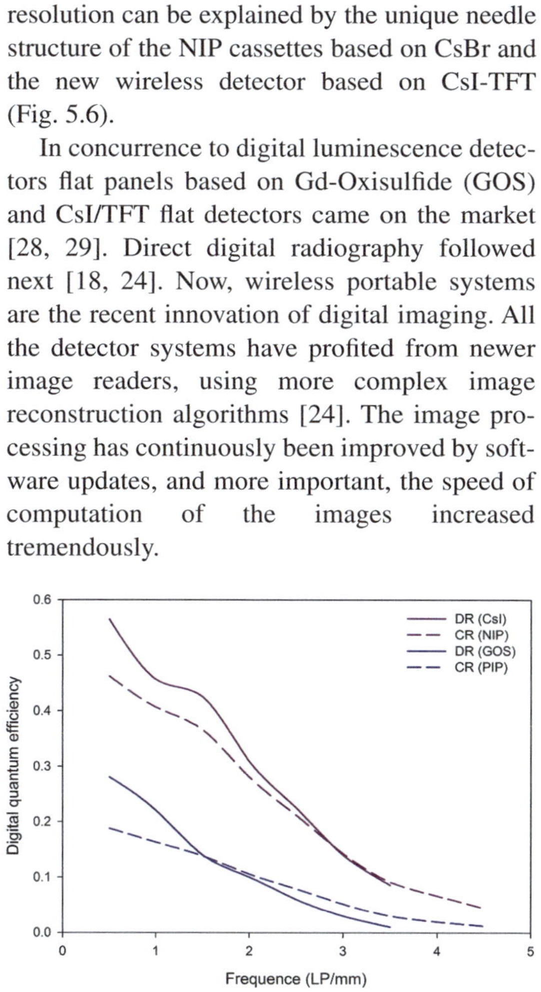

Fig. 5.6 Comparison of different digital detector systems concerning the spatial resolution and digital quantum efficiency for the detector dose of 2.5 μGy. The storage phosphor needle plates (CR-NIP) have the same quantum efficiency as the direct radiography systems with caesium iodine (DR-CsJ). The gadolinium oxisulfide (DR-GOS) and the powder storage phosphor plates (CR-PIP). (Courtesy Agfa Healthcare)

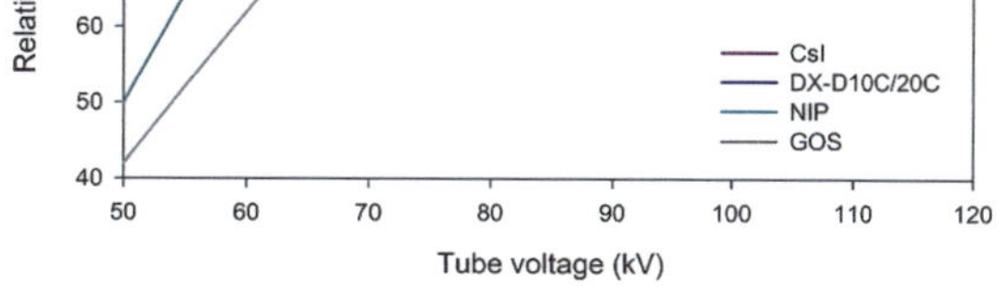

Fig. 5.7 The detector sensitivity is dependent on the voltage. The maximum of the relative sensitivity of caesium iodine (CsJ) and the needle storage phosphor plates (NIP) is around 80 kVp. In contrast the detector sensitivity of gadolinium detectors (GOS) is less in the low kVp range, but higher above 100 kVp than of the caesium based systems. (Courtesy Agfa Healthcare)

The image density and contrast of CR and DR systems can be changed independently from the detector dose. Compared with the film-screen systems the dynamic range of DR systems is increased by a factor of 100 [7, 40]. This wide range can be used in order to save the dose for the patient, if the radiologist can accept some quantity of noise [3]. In the future, noise reduction software could be the solution [24]. However, there is a danger that low contrast details could not be sufficiently displayed [40]. The spatial resolution in digital radiography is not only dependent on the physico-chemical properties of the detector material but also on the numbers of pixel values, which are processed in a predefined matrix [18]. This point is especially important for neonatal imaging, if cassettes with storage phosphor plates are used. The size of storage phosphor cassette is important concerning the image matrix as well as the pixel size. For example, the 35 cm × 43 cm cassette format has a pixel size of 0.2 mm, the 18 cm × 24 cm cassette a pixel size of only 0.1 mm.

Studies with anthropomorphic infant chest phantoms [45] cannot register changes of small anatomic structures, which are blurred by large body movements, respiration, singultus, and pulsation of systemic and pulmonary arteries. Also pulsation artefacts around small lung vessels and larger hilar vessels can significantly degrade the image, especially if the exposure time is too long. As dose reduction in paediatric patients, younger than 5 years, is of utmost importance, the detector dose should be kept low in the range between 0.75 and 2.5 μGy for chest and abdominal X-rays [28, 47]. However, the noise level associated with such low dose exposures may become very disturbing [40]. Finally, the comparison of CR and DR based on CsJ/Gd-Oxisulfide has shown that the image quality with the newer detectors was clearly superior and, in addition, the absorbed dose (DAP values) could be lowered by 40% in the same patient using the same beam quality (kVp, total filtration) but a lower detector dose [28, 29, 33].

5.5 Monitoring Image Quality and Dose

Continuous supervision of applied doses in the routine diagnostic imaging in paediatric radiography is essential in order to cope with the great variation of the dose and image quality. The cause of great differences of dose and image quality in paediatric radiography are the varying patient's diameter, too wide collimation, and the use of weight dependent exposure charts for the different age groups. Dose indicators (DI) can partly assist the radiographers to recognize faulty exposures [2]. However, greater deviations from the normal range must be monitored, whether collimation or centring were done correctly [6, 46]. In addition, the DI values must also be checked with the resultant image quality, e.g. the visualization of certain critical anatomic structures. But this procedure is time consuming and dependent on the individual experience of the radiographers. The great disadvantage of these DI values is the fact that only the relative increase or decrease of the dose can be estimated. Another drawback is that the manufacturer use different dose indicators and different scaling systems [41].

References

1. Ciantar D, Fitzgerald M, Cotterill AD, Pettett A, Cook V, Beluffi G, Pablot SM, Perlmutter N, Schneider K. Correlation between quantitative and subjective assessment of image quality in paediatric radiology. Radiat Prot Dosim. 2000;90:185–8.
2. Cohen MD, Cooper ML, Piersall K, Apgar BK. Quality assurance: using the exposure index and the deviation index to monitor radiation exposure for portable chest radiographs in neonates. Pediatr Radiol. 2011;41:592–601.
3. Cohen MD, Katz BP, Kalasinski LA, White SJ, Smith JA, Long B. Digital imaging with a photostimulable phosphor in the chest of newborns. Radiology. 1991;181:829–32.
4. Commission of the European Communities. European guidelines on quality criteria for diagnostic radiographic images in paediatrics. EUR 16261, SBN 92-827-7843-6; 1996.

5. Cook JV, Shah K, Pablot S, Kyriou J, Pettett A, Fitzgerald M. Guidelines on best practice in the x-ray imaging of children. A manual for all x-rax departments. London: Queen Mary's Hospital for Children, Surrey, and The Radiological Protection Centre; 1998.
6. Dell'Agnolo E, Seidenbusch M, Genzel-Boroviszeny O, Münch HG, Schneider K. Der Neugeborenen-Thorax I—Analyse von Feldgröße, Bildgüte und anderer Qualitätsparameter. Fortschr Röntgenstr. 2013;185:V30.
7. Don S. Radiosensitivity of children: potential for overexposure in CR and DR and magnitude of doses in ordinary radiographic examinations. Pediatr Radiol. 2004;34:S167–72.
8. Fendel H, Schneider K, Kohn MM, Bakowski C. Specific principles for optimization of image quality and patient exposure in paediatric diagnostic imaging. In: Moores BM, Wall BF, Eriskat H, Schibilla H, editors. Optimization of image quality and patient exposure in diagnostic radiology. BIR-Report 20, British Institute of Radiology; 1989. p. 91–101.
9. German Medical Association. Guidelines of the German Medical Association for quality assurance in diagnostic radiology. 2007.
10. Hess R, Neitzel U. Optimizing image quality and dose for digital radiography of distal pediatric extremities using the contrast-to-noise ratio. Fortschr Röntgenstr. 2012;184:643–9.
11. Hinojos-Armendariz VI, Mejía-Rosales SJ, Franco-Cabrera MC. Optimisation of radiation dose and image quality in mobile neonatal chest radiography. Radiography. 2018;24:104–9.
12. Huda W, Abrahams RB. X-ray-based medical imaging and resolution. AJR. 2015;204:W393–7.
13. Illers H, Buhr E, Hoeschen C. Measurement of the detective quantum efficiency (DQE) of digital x-ray detectors according to the novel standard IEC 62220-1. Radiat Prot Dosim. 2005;114:39–44.
14. Jones SA, Ansell C, Jerrom C, Honey D. Optimization of image quality and patient dose in radiographs of paediatric extremities using direct digital radiography. BJR. 2015;88:20140660.
15. Kim HK, Cunningham IA, Yin Z, Cho G. On the development of digital radiography detectors: a review. Int J Precis Eng Manuf. 2008;9:86–100.
16. Klinzmann M. Qualitätsbeurteilung von Röntgenthoraxaufnahmen auf den pädiatrischen Intensivstationen in Abhängigkeit von den ausführenden Berufsgruppen. Dissertation, Justus-Liebig-Universität Gießen; 2015.
17. Knight SP. A paediatric x-ray exposure chart. J Med Radiat Sci. 2014;61:191–201.
18. Körner M, Weber CH, Wirth S, Pfeifer K-J, Reiser MF, Treitl M. Advances in digital radiography: physical principles and system overview. Radiographics. 2007;27:675–86.
19. Liukkonen E, Jartti A, Haapea M, Oikarinen H, Ahvenjärvi L, Mattila S, Neval T, Palosaari K, Perhomaa M, Nieminen MT. Effect of display type and room illuminance in chest radiographs. Eur Radiol. 2016;26:3171–9.
20. MacCutcheon D. Management of pediatric radiation dose using Fuji computed radiography. Pediatr Radiol. 2004;34:S201–6.
21. Marsh DM, Malone JF. Methods and materials for the measurement of subjective and objective measurements of image quality. Radiat Prot Dosim. 2001;94:37–42.
22. Marshall NW, Smet M, Hofmans M, Pauwels H, De Clercq T, Bosmans H. Technical characterization of five x-ray detectors for paediatric radiography applications. Phys Med Biol. 2017;62:N573–86.
23. Neitzel U. Management of pediatric radiation dose using Philips digital radiography. Pediatr Radiol. 2004;34:S221–6.
24. Neitzel U. Status and prospects of digital detector technology for CR and DR. Radiat Prot Dosim. 2005;114:32–8.
25. Perlmutter N, Arthur RJ, Beluffi G, Cook V, Horwitz EA, Kramer P, Montagne JP, Thomas PS, Schneider K. The quality criteria for diagnostic radiographic images in paediatrcs. Radiat Prot Dosim. 1998;80:45–8.
26. Portelli JL, McNulty JP, Bezzina P, Rainford L. Frequency of paediatric medical imaging examinations performed at a European teaching hospital over a 7-year period. Eur Radiol. 2016;26:4221–30.
27. Puig S. Die digitale Thoraxradiographie beim pädiatrischen Patienten. Radiologe. 2003;43:1045–50.
28. Rapp-Bernhardt U, Bernhardt TM, Lenzen H, Esseling R, Roehl FW, Schiborr M, Theobald-Hormann I, Heindel W. Experimental evaluation of a portable indirect flat-panel detector for the pediatric chest: comparison with storage phosphor radiography at different exposures by using a chest phantom. Radiology. 2005;237:485–91.
29. Rapp-Bernhardt U, Roehl F-W, Esseling R, Lenzen H, Schiborr M, Theobald-Hormann I, Heindel W, Bernhardt TM. Portable flat-panel detector for low-dose imaging in a pediatric intensive care unit. Comparison with an asymmetric film-screen System. Invest Radiol. 2005;40:736–41.
30. Schätzing R. Management of pediatric radiation dose using Agfa computed radiography. Pediatr Radiol. 2004;34:S207–14.
31. Schegerer A. Diagnostische Referenzwerte für diagnostische und interventionelle Röntgenanwendungen. Salzgitter: Bundesamt für Strahlenschutz; 2016.
32. Schneider K. Evolution of quality assurance in paediatric radiology. Radiat Prot Dosim. 1995;57: 119–23.
33. Schneider K. Aufnahmetechnik und Strahlenschutz. In: Benz-Bohm G, editor. Kinder-radiologie. Referenz-Reihe radiologische Diagnostik. Stuttgart: Thieme; 2005. p. 1–13.
34. Schneider K. The normal child: growth and development of the infant and child; frequent and important normal variants. In: Daldrup-Link HE, Gooding CA, editors. Essentials of pediatric radiology. A multimo-

dality approach. New York: Cambridge University Press; 2010. p. 1–18.
35. Schneider K. Besonderheiten des Röntgen-Thorax im Kindesalter—was Radiologen wissen sollten. Radiologe. 2018;58:359–76.
36. Schneider K, Fendel H, Bakowski C, Stein E, Kohn M, Kellner M, Schweighofer K, Cartagena G, Padovani R, Panzer W, Scheurer C, Wall B. Results of a dosimetry study in the European Community on frequent x ray examinations in infants. Radiat Prot Dosim. 1992;43:31–6.
37. Schneider K, Kohn MM, Bakowski C, Stein E, Freidhof C, Horwitz AE, Padovani R, Wall B, Panzer W, Fendel H. Impact of radiographic imaging criteria on dose and image quality in infants in an EC-wide survey. Radiat Prot Dosim. 1993;49:73–6.
38. Schneider KO, Seidenbusch MC. Kap. 3. Technische Anforderungen. In: Riccabona M, Beer M, Mentzel H-J, editors. Bildgebung des Thorax bei Neugeborenen und Kleinkindern. Springer; 2019. p. 17–30.
39. Schöfer FH. Optimization of the relation between image quality and dose in pediatric projection radiography (exemplified by the examination of the chest). Dissertation, Physik Department, Technische Universität München; 2008.
40. Seibert JA. Tradeoffs between image quality and dose. Pediatr Radiol. 2004;34:S183–95.
41. Seibert JA, Morin RL. The standardized exposure index for digital radiography: an opportunity for optimization of radiation dose to the pediatric population. Pediatr Radiol. 2004;41:573–81.
42. Seidenbusch M. Rekonstruktion von Organ- und Effektivdosen bei konventionellen Röntgenuntersuchungen am Dr. von Haunerschen Kinderspital der Universität München mit einer Berechnung neuer Konversionsfaktoren für die pädiatrische Radiologie. Dissertation, LMU München, Medizinische Fakultät; 2006.
43. Seidenbusch MC, Schneider K. Strahlenhygienische Aspekte bei der Röntgenuntersuchung des Thorax. Radiologe. 2015;55:580–7.
44. Seifert H, Jesberger H-J, Schneider G, Rein L, Blass G, Limbach H-G, Niewald M, Sitzmann FC, Kramann B. Dose reduction in thorax radiography in simulated neonates with additional filtration and digital luminescence radiography. Acta Radiol 1998;39:514–51.
45. Singh VH, Pradham H. Neonatal chest radiography—comparing image quality and dose for contact-techniques vs. under-tray techniques. Radiogr Open. 2015;2:65–82.
46. Tschauner S, Marterer R, Gübitz M, Kalmar PI, Talakic E, Weissensteiner S, Sorantin E. European guidelines for ap/pa chest x-rays: routinely satisfiable in a paediatric radiology division? Eur Radiol. 2016;26:495–505.
47. Uffmann M, Schaefer-Prokop C. Digital radiography: the balance between image quality and required radiation dose. Eur J Radiol. 2009;72:202–8.
48. Weninger WJ, Meng S, Geyer SH, Weninger SUG. Morphologie und funktionelle Anatomie des wachsenden Thorax. Radiologe. 2003;43:1036–44.
49. Willis CE. Strategies for dose reduction in ordinary radiographic examinations. Pediatr Radiol. 2004;34:S196–200.

6 Some Technical Principles of Conventional Paediatric Radiology

6.1 X-Ray Units

In the recent decades, numerous X-ray equipments for conventional clinical radiology have been developed. However, there is still a lack of child-specific X-ray equipments for paediatric radiology. Thus, most of the X-ray equipment used today in paediatric radiology has been originally adapted to common adult radiology.

6.1.1 Chest Stand

There are several types of chest stand (Fig. 6.1):

- Classic vertical chest stand with cassette holder, antiscatter grid, and automatic exposure control device
- Chest stand with cassette holder and without antiscatter grid, used for chest radiographies in ap projection with the patient positions standing, sitting, and hanging
- Chest stand with digital flat-panel detector system

If staff members or parents hold the patient, a chest stand shielding device (Fig. 6.2) is used for the radiation protection of the staff members or parents.

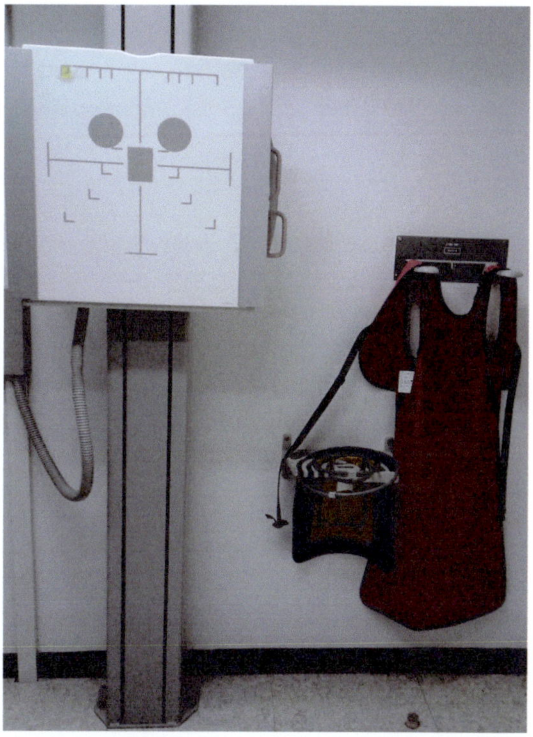

Fig. 6.1 Vertical chest stand. Lead aprons are placed next to the chest stand. (© M. Seidenbusch 2018)

6.1.2 Table

There are several types of tables (Fig. 6.3):

- Classic Bucky table with antiscatter grid implemented into the cassette holder, usually used in patients of more than 5 years of age
- Table with cassette holder, but without antiscatter grid, adjustable in height in order to enable the radiographer to realize large focus-to-detector distances

M. Seidenbusch et al., *Imaging Practice and Radiation Protection in Pediatric Radiology*,
https://doi.org/10.1007/978-3-030-18504-6_6

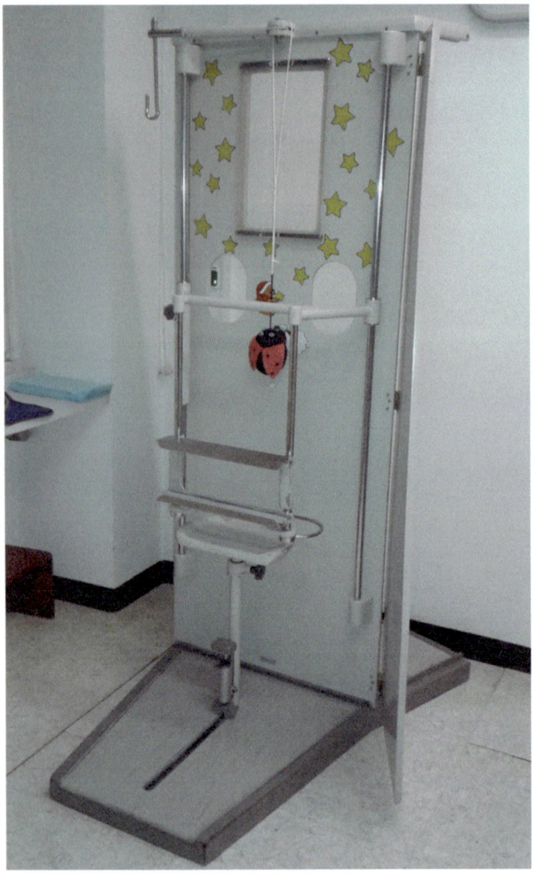

Fig. 6.2 Thoracomat for paediatric vertical chest stands. This shielding device is used for radiation protection of staff members or parents if the patient must be held. (© M. Seidenbusch 2018)

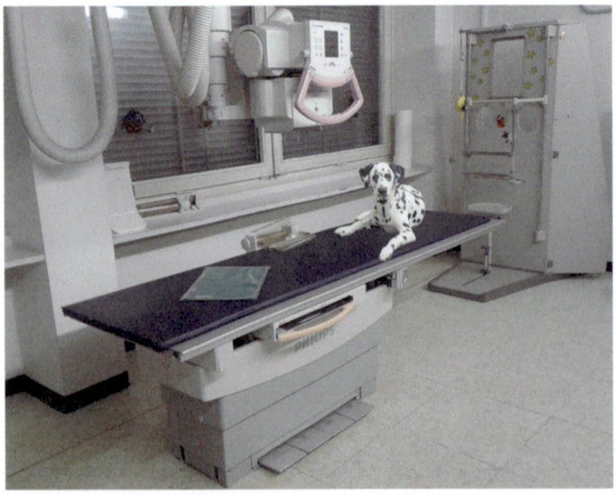

Fig. 6.3 Bucky table. Note that a Bucky table should be adjustable in vertical, horizontal, and longitudinal position. A lead shielding should be always ready to hand. (© M. Seidenbusch 2018)

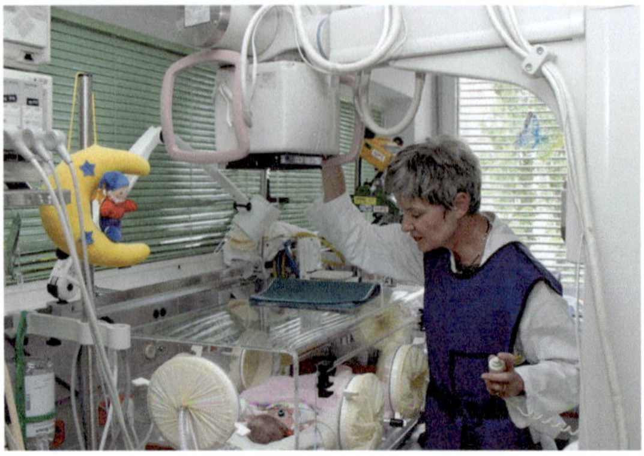

Fig. 6.4 Neonatal chest ap X-ray of a premature baby with a mobile X-ray unit. Note that the chief radiographer (co-author V. Rösenberger) depicted in the image is carefully watching the neonate while simultaneously adjusting the field size and being prepared to make the shot. The patient is placed on the digital detector, the lead shielding is placed on the incubator. (© K. Schneider 2018)

6.1.3 Mobile X-Ray Units

Mobile X-ray units (Fig. 6.4) are used for bedside and incubator radiographs. Till today, most of them are not equipped with automatic exposure control (AEC). Thus, the radiographer should pay attention more than usual that the exposure parameters and field settings are set properly before X-raying. In paediatric radiology, it is favourable if the X-ray tube is equipped with a double focus of 0.6 and 1.3.

6.2 Shielding of Patients

Lead rubber shielding at the entrance side of the patient may reduce radiation exposure due to extrafocal radiation and faulty collimation. Lead shielding at the exit side of the patient is recommended by some authors to reduce backscatter radiation originating from the chest stand device. However, scattered radiation originating within the patient's body is the main source of the radiation exposure beyond the age of 5 years. If shielding is used, the lead rubber should be placed immediately beneath the field edge. If the distance to the field edge is more than 2 cm, shielding is ineffective [4]. Gonadal protection in females for pelvic X-rays is of doubtful value, because the location of the ovaries within the pelvic region varies considerably [1, 3, 5–7, 12]. However, lead rubber shielding should be always done for collimated X-rays of the hip joints in both genders. The male gonads can only be protected safely using testicular capsules. In male infants shielding is preferred, if the testes are not descended because testicular capsules cannot be applied [10, 12]. However, it is questionable whether lead rubber shielding is useful in male infants and young boys [2]. In addition, it goes without saying that shielding of patients and automated exposure control should not be used simultaneously [8].

Nevertheless, gonad shielding should be placed properly in order to achieve maximal radiation protection (Fig. 6.5). Figure 6.5a, b

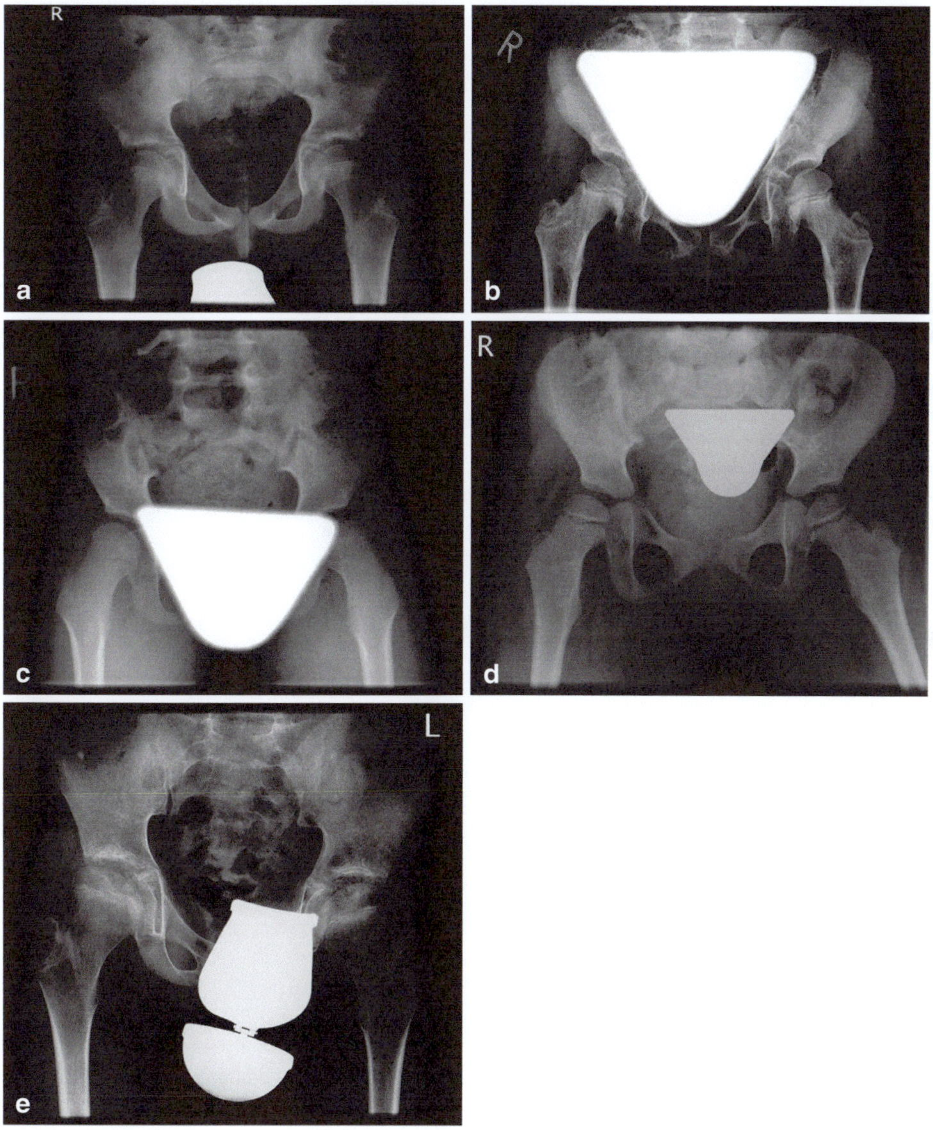

Fig. 6.5 Optimal and suboptimal gonadal shielding in male and female patients during pelvic X-rays. Figure (**a**) shows optimal gonadal shielding using a testicular capsule in a 12-year-old male patient. Figure (**b**) demonstrates nearly optimal gonadal shielding in a 7-year-old female patient. Please note that the left ovary may be not fully protected. Figure (**c**) depicts suboptimal gonadal shielding in a 6-year-old female patient. The gonadal shielding was placed too low and overlies the pubic bones and both acetabula. (**d**) Suboptimal gonadal shielding in a 6-year-old female patient caused by a very small shield. (**e**) Suboptimal gonadal shielding in a 14-year-old male patient by an open testicular capsule. The shielding is useless and hides the left pubic and ischial bone. (© K. Schneider 2018)

shows optimal gonadal shielding. In contrast, Fig. 6.5c–e show several examples of ineffective shielding.

In recent years, the question came up whether patient shielding is useful in diagnostic imaging or not. This discussion still continues today [9]. Nevertheless, in the opinion of the authors of his book, abandoning patient shielding in paediatric radiology would represent an undue simplifying of paediatric imaging. Patient shielding in paediatric radiology should be considered furthermore as an essential prerequisite for paediatric radiation protection and should be performed whenever possible [11].

References

1. Bardo DME, Black M, Schenk K, Zaritzky MF. Location of the ovaries in girls from newborn to 18 years of age: reconsidering ovarian shielding. Pediatr Radiol. 2009;39:253–9.
2. Cook JV, Shah K, Pablot S, Kyriou J, Pettett A, Fitzgerald M. Guidelines on best practice in the x-ray imaging of children. A manual for all x-rax departments. London: Queen Mary's Hospital for Children, Surrey, and The Radiological Protection Centre; 1998.
3. Fawcett SL, Barter SJ. The use of gonad shielding in paediatric hip and pelvis radiographs. Br J Radiol. 2009;82:363–70.
4. Fendel H, Stieve F-E. Strahlenschutz in der Kinderradiologie. NCRP-Bericht Nr. 68; Deutsche Übersetzung. Berlin: H. Hoffmann; 1983.
5. Fochem K, Pape R. Problematik des Ovarialschutzes bei Röntgenaufnahmen des Beckens. Fortschr Röntgenstr. 1962;97:785–93.
6. Forster E. Equipment for diagnostic radiography. Lancaster: MTP Press Ltd; 1985.
7. Frantzen MJ, Robben S, Postma AA, Zoetelief J, Wildberger JE, Kemerink GJ. Gonad shielding in paediatric pelvic radiography: disadvantages prevail over benefit. Insights Imaging. 2012;3:23–32.
8. Kaplan SL, Magill D, Felice MA, Xiao R, Ali S, Zhu X. Female gonadal shielding with automatic exposure control increases radiation risks. Pediatr Radiol. 2018;48:227–34.
9. Marsh RM, Silosky M. Patient shielding in diagnostic imaging: discontinuing a legacy practice. AJR. 2019;212:1–3.
10. Scorer CG. The descent of the testis. Arch Dis Child. 1964;39:605–9.
11. Seidenbusch MC, Schneider K. Strahlenhygienische Aspekte bei der Röntgenuntersuchung des Thorax. Radiologe. 2015;55:580–7.
12. Warlow T, Walker-Birch P, Cosson P. Gonad shielding in paediatric pelvic radiography: effectiveness and practice. Radiography. 2014;20:178–82.

Part III

Standard Operating Procedures (SOP) and Normalized Organ Doses for Conventional Radiographs in Children

"Gut ist der Vorsatz, aber die Erfüllung ist schwer."—"Nothing is more difficult than trying to keep a good resolution."

Johann Wolfgang von Goethe

7 Introduction

For each paediatric X-ray examination listed in Table 7.1 and for each age group (neonates, 1–2 years, 3–7 years, 8–12 years, 13–17 years, young adults) of patients, the following information are provided in this book:

- General referral criteria (indications and contraindications)
- Image criteria
- Field settings
- Patient preparation and radiographic procedure
- Exposure parameters
- Normalized organ doses for optimal and suboptimal field settings

In this way, optimal field settings can be retrieved for most kinds of X-ray examinations in children of any age, and organ doses achieved during these examinations can be calculated from measured dose area products or entrance surface doses by applying normalized organ doses provided for optimal and suboptimal field settings and various exposure parameters.

7.1 Terminology

In order to clear the terminology used in this book, the following definitions of the exposure geometry and of anatomical landmarks are made (Figs. 7.1, 7.2, 7.3, 7.4, and 7.5):

Table 7.1 Overview over the conventional X-ray examinations considered in this book

Chapter	X-ray examination
8	Skull
9	Chest
10	Bony thorax
11	Thoraco-abdominal babygram
12	Abdomen
13	Full spine
14	Cervical spine
15	Thoracic spine
16	Lumbar spine
17	Pelvis
18	Hip joint
19	Thigh
20	Shoulder
21	Clavicle
22	Upper arm
23	Upper extremity
24	Hand
25	Lower extremity
26	Foot
27	Skeletal survey

7.1.1 Patient Position

The *patient position* is defined as the absolute geometrical position of the patient's body in space. In clinical terminology, there are usually used the following terms (Fig. 7.1):

M. Seidenbusch et al., *Imaging Practice and Radiation Protection in Pediatric Radiology*,
https://doi.org/10.1007/978-3-030-18504-6_7

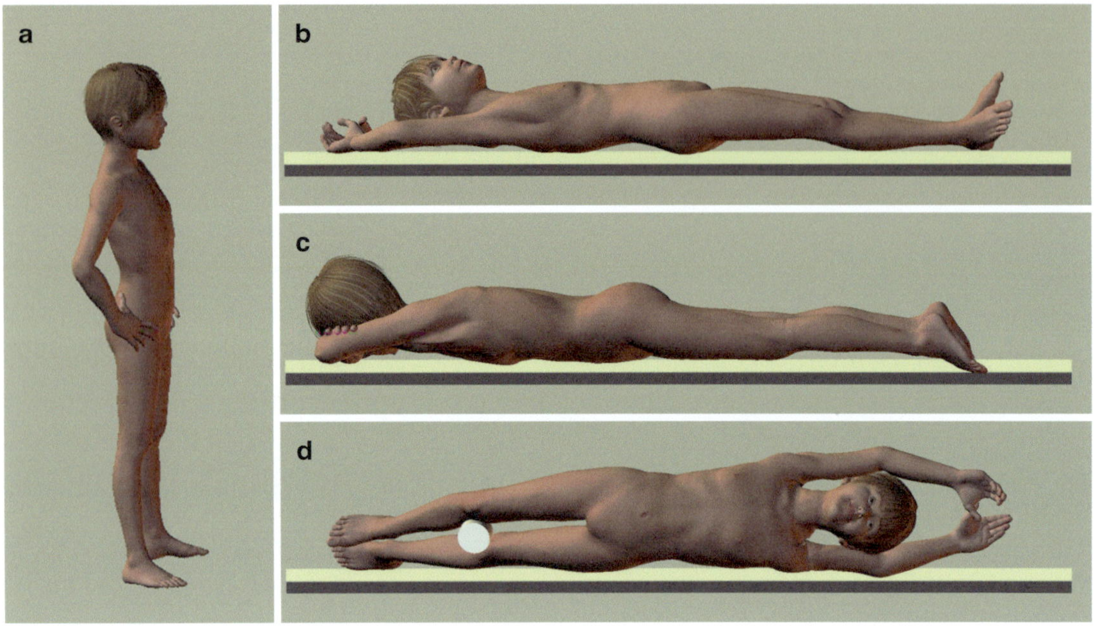

Fig. 7.1 **Patient positions in space**. (**a**) Upright position (standing). (**b**) Supine position. (**c**) Prone position. (**d**) Decubitus position (left). (© V. Rösenberger and M. Seidenbusch 2018)

- Upright position: The patient is hanging, sitting, or standing.
- Supine position: The patient is lying in the bed or on the Bucky table with the face upwards.
- Prone position: The patient is lying on the Bucky table with the face downwards.
- Decubitus position: The patient is lying in the bed or on the Bucky table on the right or left side.

7.1.2 Beam Direction

The *beam direction* is defined as the absolute geometrical direction of the X-ray beam in space. In standard conventional radiology, there are usually applied two beam directions (Fig. 7.2):

- Horizontal beam direction: The X-ray beam is pointed at a wall of the X-ray room.
- Vertical beam direction: The X-ray beam is pointed at the ceiling or the floor of the X-ray room.

7.1.3 Beam Projection

The *beam projection* is defined as the geometrical beam direction relative to the patient (Fig. 7.3). Depending on the localization of the entrance radiation field on the patient's skin surface, there are the following standard beam projections in conventional diagnostic radiology:

- Antero-posterior (ap): X-rays travel from anterior to posterior
- Postero-anterior (pa): X-rays travel from posterior to anterior
- Right lateral (r-lat): X-rays travel from right lateral to left lateral
- Left lateral (l-lat): X-rays travel from left lateral to right lateral
- Right anterior oblique (RAO): X-rays travel from right antero-lateral to left postero-lateral
- Left anterior oblique (LAO): X-rays travel from left antero-lateral to right postero-lateral

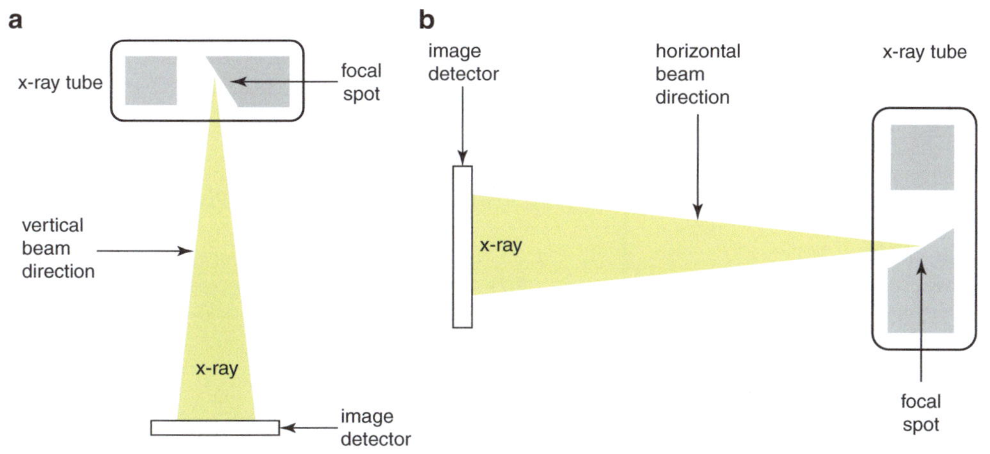

Fig. 7.2 **Beam directions in space**. (**a**) Vertical beam direction. (**b**) Horizontal beam direction. (© M. Seidenbusch 2018)

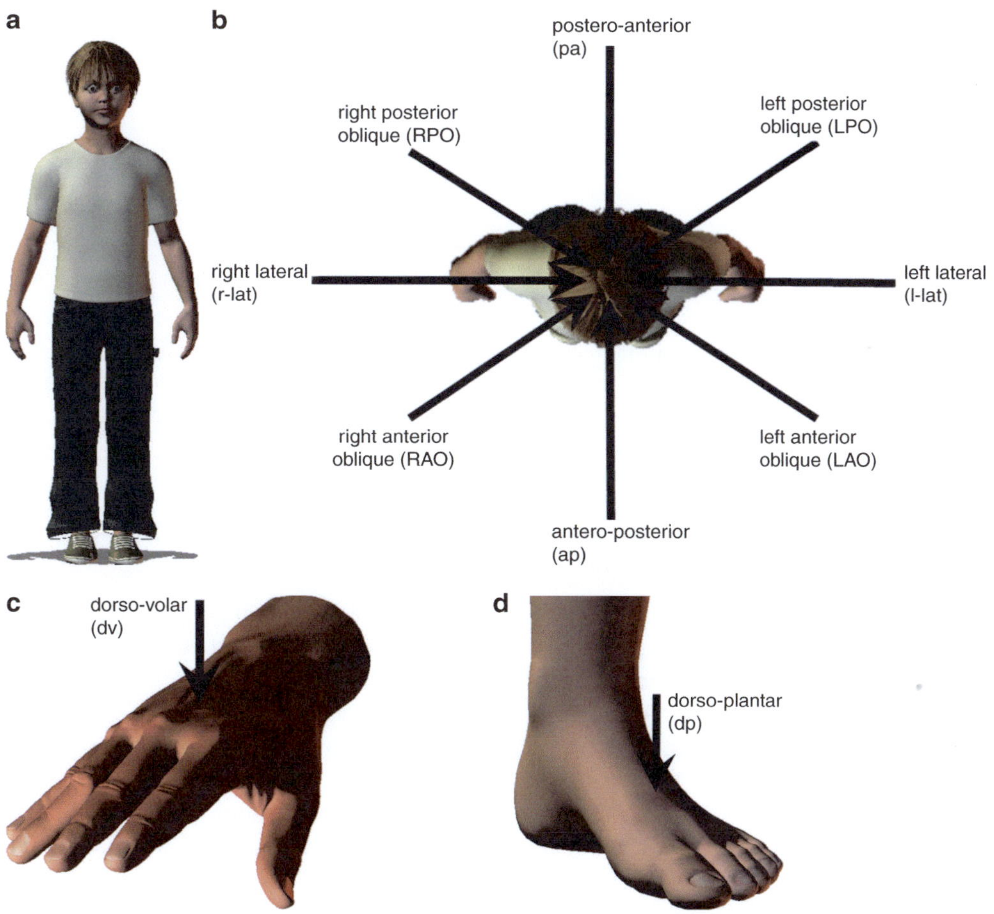

Fig. 7.3 **Beam projections relative to the patient**. (**a**) Patient in ap view. (**b**) Patient in bird's-eye view with standard beam projections drawn in. (**c**) Hand with dorso-volar (dv) beam projection drawn in. (**d**) Foot with dorso-plantar beam projection drawn in. (© V. Rösenberger and M. Seidenbusch 2018)

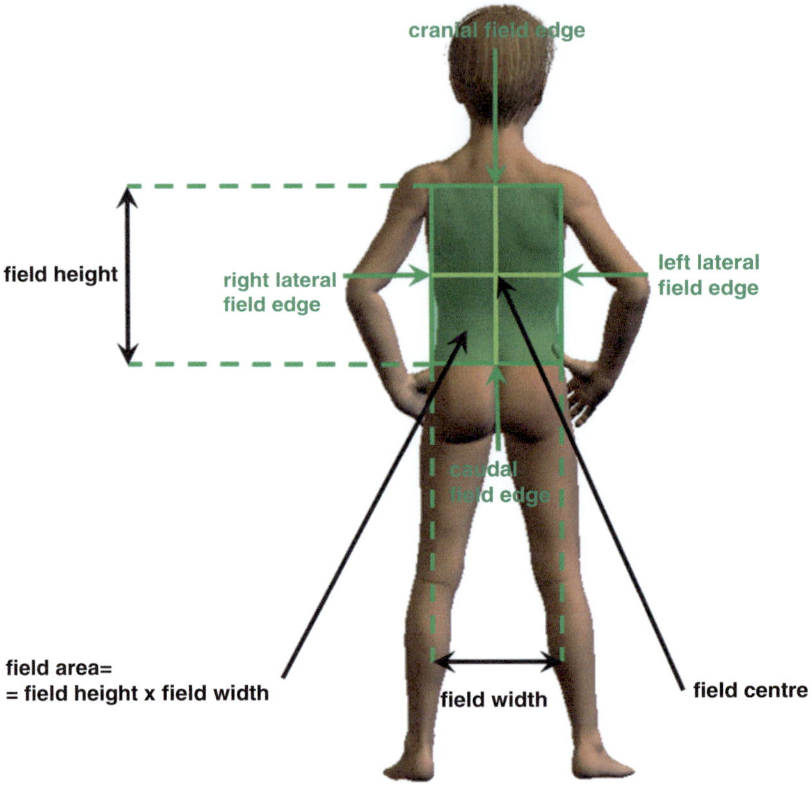

Fig. 7.4 Geometric parameters of the radiation field. The radiation field is bordered by the cranial, caudal, right lateral, and left lateral field edges. The field height is defined as the cranio-caudal length of the field, and the field width is defined by the lateral field lengths. The field area is the product of field height and field width. (© V. Rösenberger and M. Seidenbusch 2018)

- Dorso-volar (dv) in hand radiographs: X-rays travel from dorsal to volar
- Dorso-plantar (dp) in foot radiographs: X-rays travel from dorsal to plantar

Each of these beam projections can be achieved through several combinations of patient position and beam direction (Table 7.2).

7.1.4 Field Size

As a part of the dose area product, the *field size* of a radiograph contributes to the radiation exposure. The geometrical field parameters are defined as follows (Fig. 7.4):

- The *field height* is the field length at patient's skin surface in the longitudinal patient axis.
- The *field width* is the field length at patient's skin surface in a horizontal patient axis.
- The *field size* is the field area at the skin surface of the patient, i.e. the product of field height and field width.

7.1.5 Field Centring

Field centring is defined in this book as positioning the midpoint of the incident radiation field on an anatomical landmark visible on the patient's skin surface.

7.1.6 Field Setting

In this book, the term *field setting* is defined as the combination of field centring and field size. Correct field centring must be considered as an absolute prerequisite for simultaneously depicting the anatomical region of interest (ROI) and reducing the field size and thus the radiation exposure of the patient. Figure 7.5 shows four combinations of field centring and field size. An optimal field setting will be achieved if both field centring and field size are optimal. Suboptimal field settings will be achieved if the field centring or the field size is suboptimal.

It is important to remark that the anatomical region of interest will be depicted correctly in

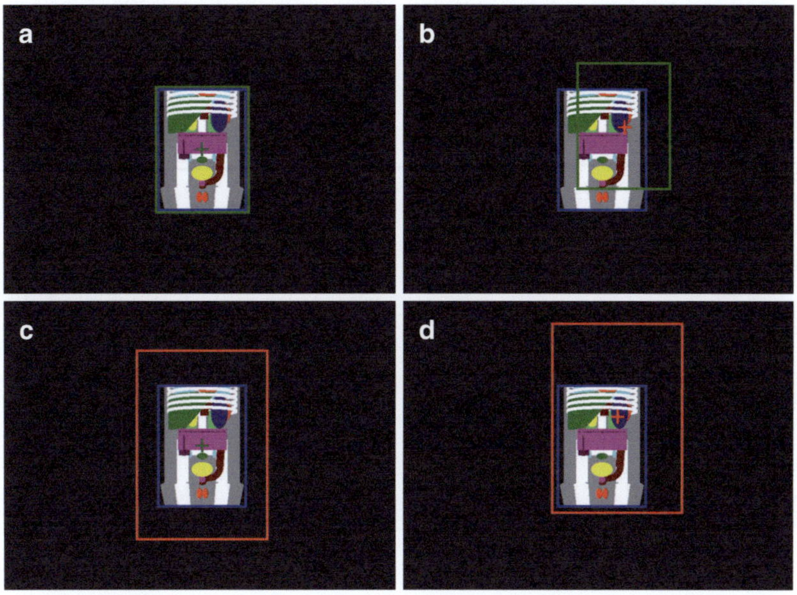

Fig. 7.5 **Optimal and suboptimal field settings**. Blue lines refer to the anatomical region of interest, e.g. full abdomen, which should be depicted in the ap radiograph. Green lines refer to the optimal field size. Red lines refer to the suboptimal field size. Green crosses refer to correct centring point. Red crosses refer to the incorrect centring. (**a**) Optimal field setting = optimal field centring + optimal field size. (**b**) Suboptimal field setting = suboptimal field centring + optimal field size. (**c**) Suboptimal field setting = optimal field centring + suboptimal field size. (**d**) Suboptimal field setting = suboptimal field centring + suboptimal field size. Note that the anatomical region of interest may be depicted in the radiograph through optimal and suboptimal field settings as well. (© M. Seidenbusch 2018)

Table 7.2 Relationship between patient position, beam direction, and beam projection

	Beam direction	Beam projection		Abbreviation
Patient position				
Upright	Horizontal	Sagittal	Antero-posterior	ap
			Postero-anterior	pa
		Lateral	Right lateral	r-lat
			Left lateral	l-lat
		Oblique	Right anterior oblique	RAO
			Left anterior oblique	LAO
Supine	Horizontal	Lateral	Right lateral	r-lat
			Left lateral	l-lat
	Vertical	Sagittal	Antero-posterior	ap
			Postero-anterior	pa
Decubitus	Horizontal	Sagittal	Antero-posterior	ap
			Postero-anterior	pa
	Vertical	Lateral	Right lateral	r-lat
			Left lateral	l-lat
Hand/foot position				
Hand prone	Vertical	–	Dorso-volar	dv
Foot prone	Vertical	–	Dorso-plantar	dp

Annotation*:*
Please note that the beam projection relative to the patient is the result of the combination of the absolute patient position and the absolute beam direction in space

the radiograph not only if an optimal field setting has been achieved but also if a suboptimal field centring goes along with a suboptimal field size which is large enough to encompass the region of interest completely. Note that the latter case may pretend an optimal field setting at first glance, especially if the radiation field boundaries are cut off, e.g. by the imaging processing software. Of course, an adequate radiation protection of the patient will not be achieved in this case.

7.1.7 Anatomical Lines

Anatomical lines which are referred to in this book are described in Fig. 7.6. Please note that the horizontal mammary line and the horizontal navel line have been defined by the authors and are not common in medicine.

7.2 Standard Operating Procedures (SOPs)

The International Council for Harmonization (ICH) defines standard operating procedures (SOPs) for clinical research as "detailed, written instructions to achieve uniformity of the performance of a specific function" [10]. The standard operating procedures presented in this book encompass the following items for each paediatric X-ray examination which were listed in the order of the radiological process:

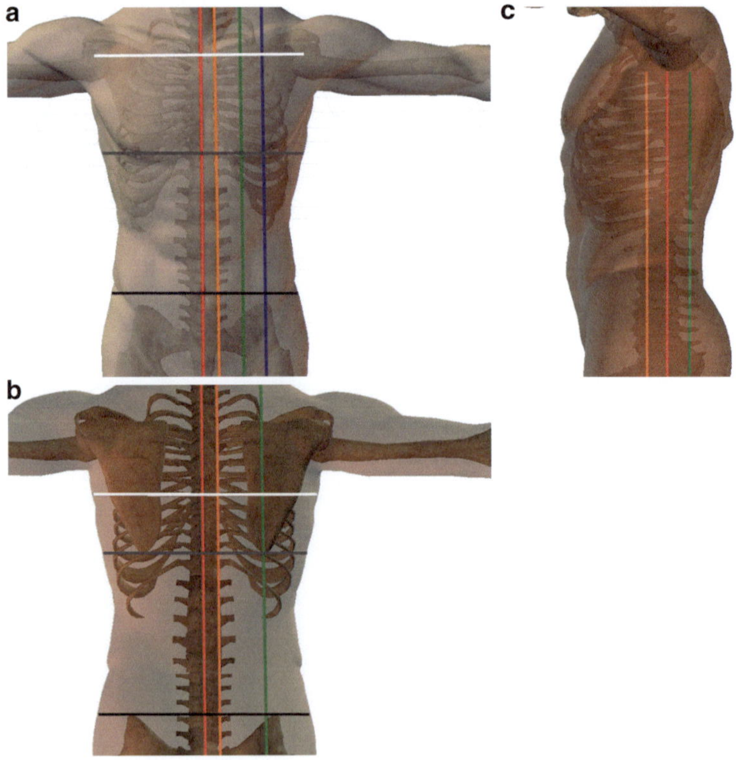

Fig. 7.6 **Anatomical lines**. Anatomical lines are imaginary lines passing through the sagittal, coronal, and horizontal planes of the trunk. The most common lines are described in this figure. (© M. Seidenbusch 2018). (**a**) **Anterior view**. Red = anterior mid line; orange = left parasternal line; green = midclavicular line; blue = mammary line; white = sternoclavicular line; grey = horizontal mammary line; black = horizontal navel line. (**b**) **Posterior view**: Red = posterior mid line; orange = paravertebral line; green = scapular line; white = subaxillary line; grey = subscapular line; black = iliac crest line. (**c**) **Lateral view**. Red = midaxillary line; orange = anterior axillary line; green = posterior axillary line

- Indications and contraindications: First of all, a clinical indication must exist for a specific X-ray examination.
- Image criteria: Depending on the clinical indication, specific anatomical structures must be depicted through the radiographs.
- Field settings: In order to achieve the image criteria required by the clinical indication, correct field settings (i.e. correct field centring on the patient's skin surface and correct field size) must be achieved by the radiographer.
- Patient preparation and radiographic procedure: Depending on the clinical indication and the type of X-ray examination, the patient must be prepared and positioned.
- Exposure parameters: Depending on the type of X-ray examination and the age of the patient, adequate exposure parameters must be chosen in order to achieve an optimal image quality.

7.3 Normalized Organ Doses

All normalized organ doses provided in this book were calculated using the commercially available PCXMC algorithm of Servomaa and Tapiovaara, version 2.0, as described in Chap. 4 [12, 15, 16]. All three-dimensional figures were created with the commercial available Poser® software by Smith Micro Inc., version 10.0, and the open source software MakeHuman, version 1.1.1.

7.3.1 Radiation Quality

As parameters characterizing the beam quality of the X-ray spectrum, the tube voltage, total filtration and anode angle were used [3]. X-ray tube voltages and total filtrations were selected in accordance with the guidelines of the German Medical Association [6] and the European Commission [4] depending on the X-ray examination and on patient's age. The conversion coefficients in this book were calculated for an anode angle of 15°. However, the characteristics of the photon spectrum remain nearly unchanged in the range of the anode angles commonly used in practice (10–20°). According to the European and German guidelines, a total filtration of 3.5 mm aluminium (Al) and 0.1 mm copper (Cu) is strongly recommended in paediatric radiography. Additionally, other radiation qualities which may be in use world-wide were considered in this book.

7.3.2 Beam Direction

Depending on the type of X-ray examination, conversion coefficients are provided in this book for antero-posterior (ap), postero-anterior (pa), and right lateral (r-lat) beam directions (the right lateral beam direction corresponds to the position in which the patient is positioned on the left side) as most of paediatric X-ray examinations are performed in this way.

7.3.3 Field Size

Allowing for the clinical range of radiation field settings, optimal (green field edges in the figures) and suboptimal (red field edges in the figures) standard radiation field settings were defined on the surface of the mathematical phantoms. Suboptimal field settings are achieved during a lot of paediatric X-ray examinations [1, 5, 7, 13, 14, 18], especially in neonates and young infants.

The calculations were performed by neglecting the protective shielding devices that are normally applied to the patient in clinical practice, where possible (lead-rubber covering of non-exposed parts of the trunk, ovarian protection and gonad shields) [6]. The organs and tissues depicted in the field settings are depicted in Fig. 7.7.

7.3.4 Monte Carlo Simulations

In the Monte Carlo simulations, a total of 10,000,000 individual photon histories were taken into consideration for each radiation field. Thus, it was possible to keep the stochastic error of organ doses of organs and tissues hit by the

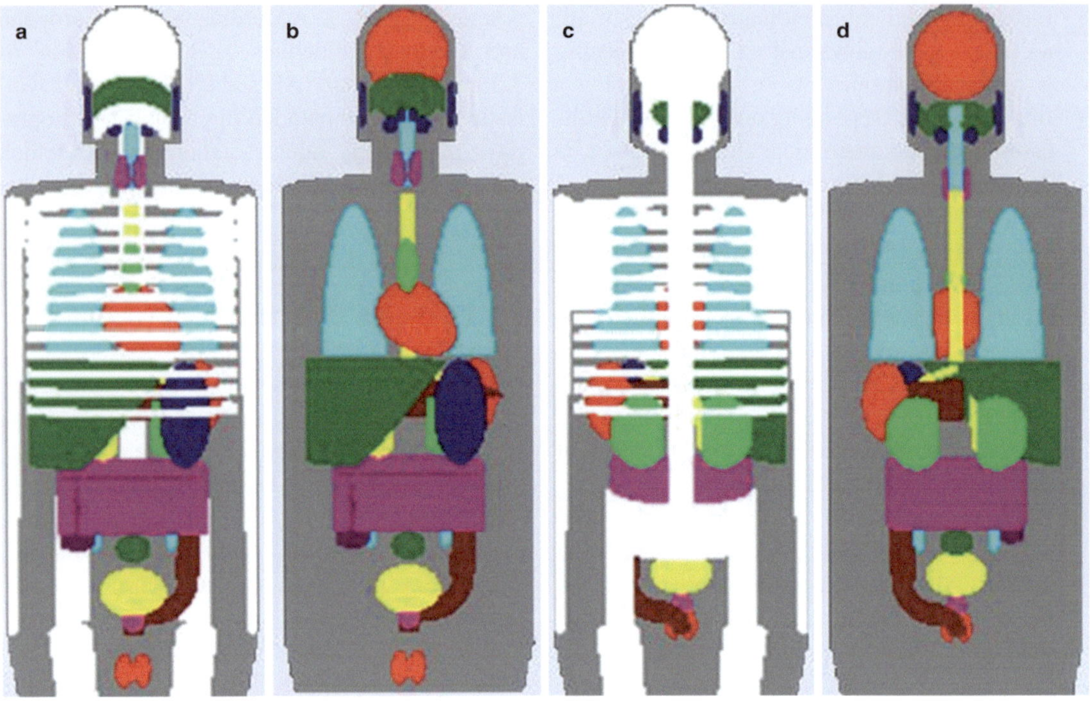

Fig. 7.7 Organs and tissues of the mathematical MIRD phantoms in PCXMC. From left to right: ap view with skeleton, ap view with skeleton removed, pa view with skeleton, pa view with skeleton removed. The pictures of phantom and X-ray image were taken from the user interface of the PCXMC software. (© M. Seidenbusch 2018). Organs and tissues: White = skeleton; grey = soft tissue; red = brain, heart, spleen, testicles; yellow = oseophagus, gall bladder, urinary bladder; dark green = oral mucosa, liver, uterus; light green = thymus, kidneys; dark blue = salivary glands, stomach; light blue = extrapulmonary airways, lungs, ovaries; purple = colon; pink = thyroid gland, small intestine, prostate; brown = pancreas, large intestine

primary radiation field to less than 1% (simple standard deviation). For this reason, conversion coefficients <0.005 mSv/mGy were not listed in the tables.

7.3.5 Effective Dose

First of all, it must be stated that the concept of effective dose has been intended to be a risk estimator for occupational health and has not been defined neither for patients in diagnostic radiology nor for children. Thus, the concept of effective dose ought not be applied for describing the radiation burden to children during paediatric X-ray examinations. On the other hand, effective doses may be suitable in order to compare exposures during paediatric X-ray examinations to each other. Therefore, normalized effective doses are provided in this book, but one should keep in mind that they were calculated considering the tissue weighting factors that have been defined for adults only according to ICRP 60 [8] and ICRP 103 [9] with the algorithm described in [17].

7.3.6 Properties of Normalized Organ Doses

The normalized organ doses listed in this book were calculated for discrete ages and discrete exposure parameters. For clinical applications, an extension to arbitrary ages by linear extrapolation may be desirable. In Fig. 7.4, the dependence of the normalized organ doses of patient age, focus-to-skin distance, and tube voltage is described [11]. As can be seen, a linear extrapolation of the normalized organ dose values tabulated in this book might be possible in most clinical cases without loss of validity (Figs. 7.8, 7.9 and 7.10).

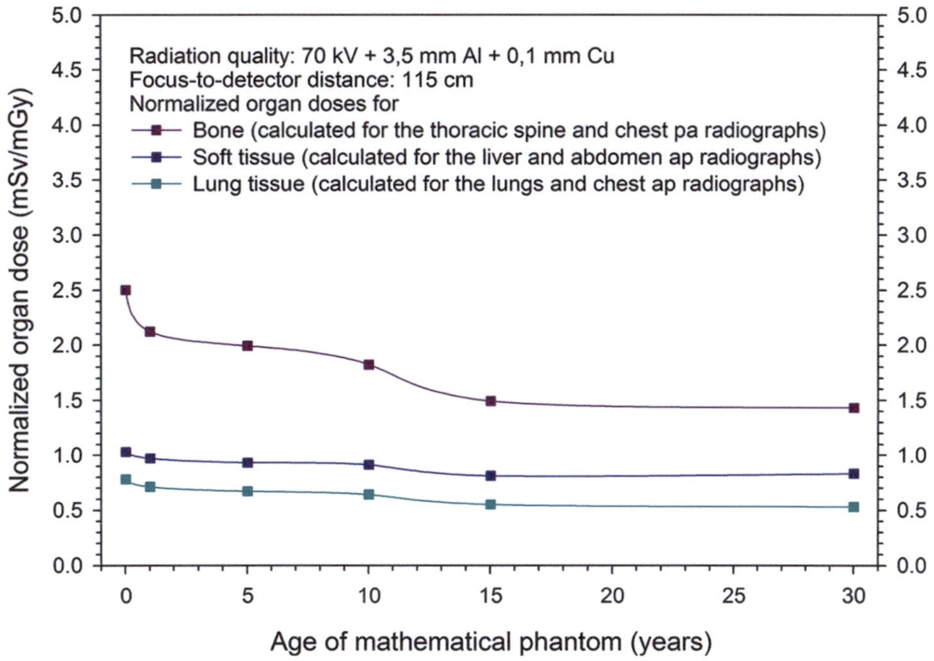

Fig. 7.8 **Normalized organ doses for different tissues as a function of patient's age**. Normalized organ doses were calculated for bone (here: thoracic spine), soft tissue (here: liver), and lung tissue (here: lungs) achieved during chest and abdomen ap radiographs. The percentage of absorbed dose increases with increasing tissue density. Note that the absorbed dose is much higher in bone than in air and that thus normalized organ doses >1 can be achieved. (From [11], © Georg Thieme Verlag 2008, courtesy Georg Thieme Verlag)

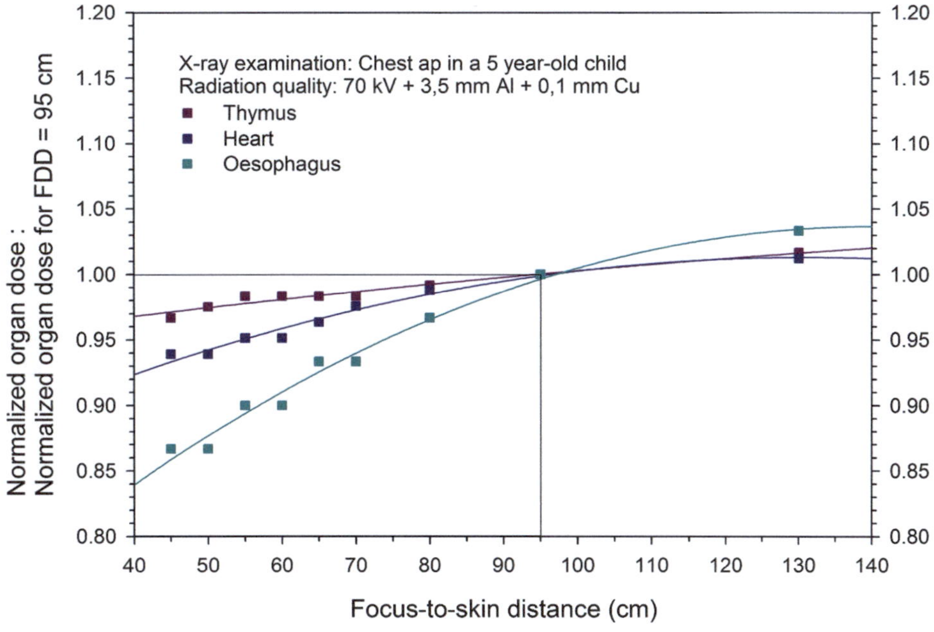

Fig. 7.9 **Normalized organ doses for different organs as a function of focus-to-skin distance**. Normalized organ doses were calculated for the thymus (localized close to the anterior skin surface), the heart (localized in the middle of the thoracic cavity), and the oesophagus (localized far away from the anterior skin surface). The deeper the organ is localized, the more will normalized organ doses depend on the focus-to-skin distance. (From [11], © Georg Thieme Verlag 2008, courtesy Georg Thieme Verlag)

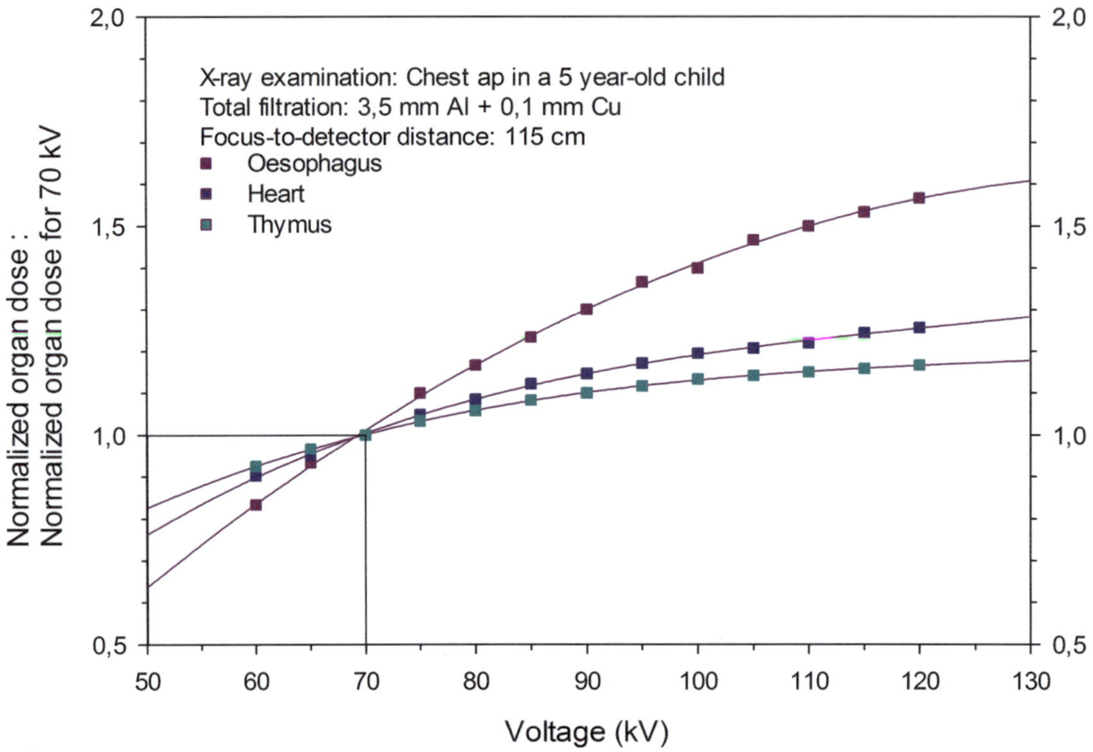

Fig. 7.10 Normalized organ doses for different organs as a function of voltage. Normalized organ doses were calculated for the thymus (localized close to the anterior skin surface), the heart (localized in the middle of the thoracic cavity), and the oesophagus (localized far away from the anterior skin surface). The deeper the organ is localized, the more will normalized organ doses depend on the voltage. (From [11], © Georg Thieme Verlag 2008, courtesy Georg Thieme Verlag)

7.4 How to Calculate Organ and Effective Doses Using This Book

Organ and effective doses can be calculated using the normalized organ doses in this book in various ways. The flow chart in Fig. 7.11 shows some pathways how to calculate organ and effective doses. Green pathways are recommended, and red pathways are not recommended by the authors.

7.4.1 Fastest Way

The fastest way to assess an organ dose is to look up a normalized organ dose in the normalized organ dose tables considering the following issues:

- X-ray examination, e.g. chest → look for the tables "chest"
- Beam projection, e.g. pa → look for the tables "chest pa"
- Patient's age, e.g. 9 years → look for the tables "chest pa; 08–12 years"
- Tube voltage, e.g. 73 kV → look for the table "chest pa; 08–12 years; 70 kV"
- Total filtration, e.g. 3.5 mm Al + 0.1 mm Cu → look for the column "3.5 mm Al; 0.1 mm Cu"
- Requested organ, e.g. thyroid → look for the row "thyroid" and select the normalized organ dose from the table cell "thyroid / 3.5 mm Al + 0.1 mm Cu"

Calculate the requested organ dose like this:

- If the entrance surface dose is known, the organ dose can be calculated by multiplying the normalized organ dose by the entrance surface dose. In this case, the unit of the calculated

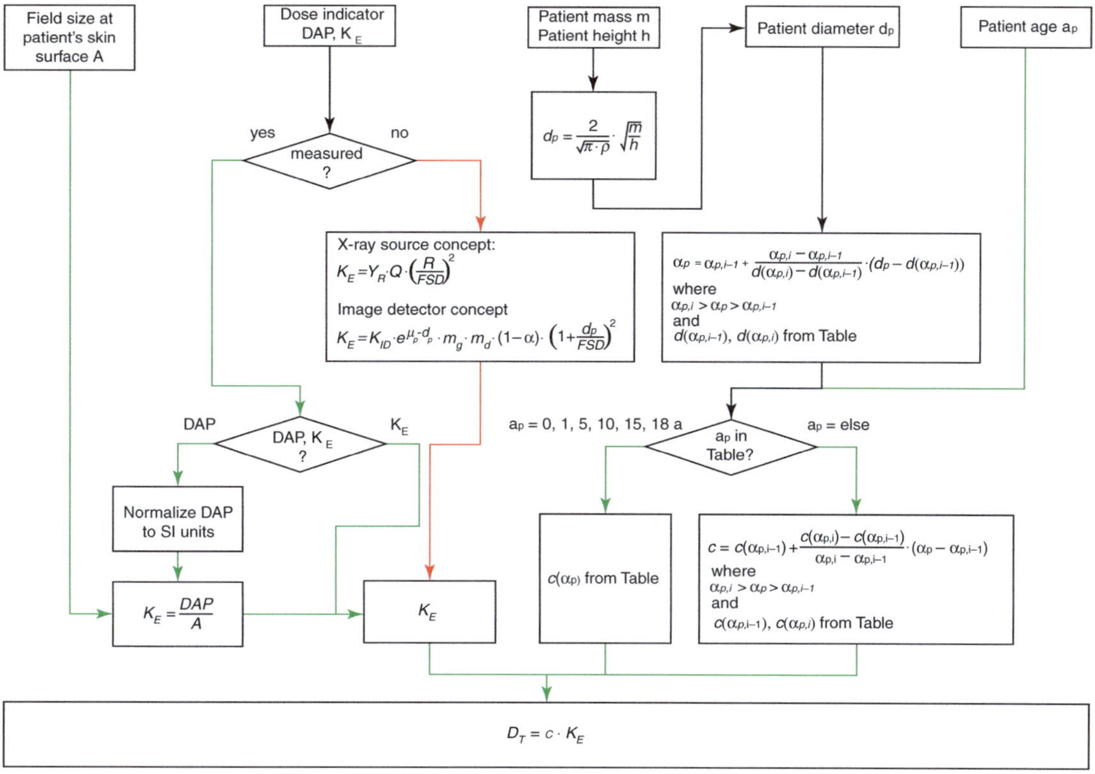

Fig. 7.11 **Flow chart**: Algorithm for how to calculate organ and effective doses from dose area products or entrance surface doses for various age groups or various patient diameters using the normalized organ doses listed in this book. Green pathways are recommended for quick calculations. Red pathways are not recommended

Table 7.3 Conversion coefficients C for various DAP quantities

	mR cm²	cR cm²	dR cm²	R cm²	mGy cm²	cGy cm²	µGy m²	B = A/C
mR cm²	1	0.1	0.01	0.001	0.0084	0.00084	0.00084	
cR cm²	10	1	0.1	0.01	0.084	0.0084	0.0084	
dR cm²	100	10	1	0.1	0.84	0.084	0.084	
R cm²	1000	100	10	1	8.4	0.84	0.84	
mGy cm²	119	11.9	1.19	0.119	1	0.1	0.1	
cGy cm²	1190	119	11.9	1.19	10	1	1	
µGy m²	1190	119	11.9	1.19	10	1	1	
A = C B	**C**							

Note that DAP values in column A can be calculated from DAP values in row B by the Eq. A = C B and vice versa

organ dose is the unit of the entrance surface dose.

- If the entrance surface dose is unknown but the dose area product and the field size at the patient's skin surface is known, the entrance surface dose can be calculated by dividing the dose area product by the field size at the patient's skin surface. Please note that the measuring units are converted correctly. Table 7.3 provides conversion coefficients for converting measuring units of dose area products into each other.
- If neither the entrance dose nor the dose area product is given, one can try to use the X-ray source concept to estimate the entrance surface dose from the mAs product or the image detector concept to assess the entrance surface dose from the image detector dose. However,

these concepts are subjects to an error of at least one magnitude and are not recommended by the authors of this book.

7.4.2 Most Accurate Way

The most accurate way to calculate an organ dose is to look up normalized organ doses in the normalized organ dose tables which correspond to the mathematical intervals of age group and tube voltage considering the following issues:

- X-ray examination, e.g. chest → look for the tables "chest"
- Beam projection, e.g. pa → look for the tables "chest pa"
- Patient's age, e.g. 9 years → look for the tables "chest pa; 03–07 years" (median age: 05 years) and the tables "chest pa; 08–12 years" (median age: 10 years)
- Tube voltage, e.g. 73 kV → look for the tables "chest pa; 03–07 years; 70 kV" and "chest pa; 08–12 years; 70 kV"
- Total filtration, e.g. 3.5 mm Al + 0.1 mm Cu → look for the columns "3.5 mm Al; 0.1 mm Cu" in each table
- Requested organ, e.g. thyroid → look for the row "thyroid" and select the normalized organ dose from the table cell "thyroid / 3.5 mm Al + 0.1 mm Cu" in each table
- Interpolate the normalized organ doses

  Calculate the requested organ dose like this:

- If the entrance surface dose is given, the organ dose can be calculated by multiplying the normalized organ dose by the entrance surface dose. In this case, the unit of the calculated organ dose is the unit of the entrance surface dose.
- If the entrance surface dose is unknown but the dose area product and the field size at the patient's skin surface is known, the entrance surface dose can be calculated by dividing the dose area product by the field size at the patient's skin surface. Please note that the measuring units are converted correctly. Table 7.3 provides conversion coefficients for converting measuring units of dose area products into each other.
- If neither the entrance dose nor the dose area product is known, it is impossible to assess the entrance surface dose in an accurate way.

Table 7.4 Mean body diameters as function of age [2]

	Age (years)				
Anatomical region	0	1	5	10	15
Skull sagittal	11.9	16.2	17.9	18.4	18.9
Skull lateral	9.3	12.4	13.8	14.1	14.6
Chest sagittal	9.3	12.0	13.4	15.3	17.4
Chest lateral	10.6	15.7	18.8	22.1	25.9
Abdomen sagittal	8.8	12.6	13.6	15.9	18.0
Abdomen lateral	10.0	14.7	17.4	21.1	24.5
Pelvis sagittal	7.0	10.6	13.7	16.4	19.6
Pelvis lateral	9.2	15.2	20.2	24.4	30.4
	Diameter (cm)				

7.4.3 Using the Patient's Diameter

Sometimes, it may make sense to estimate organ doses considering the patient's diameter instead of the patient's age. In this case, the virtual age group has to be determined which corresponds to the patient's real diameter. Table 7.4 provides sagittal and lateral diameters of various anatomical regions as a function of the patient's age.

If the patient's body height and body weight are known, the equivalent cylindrical diameter may be calculated from these values.

7.5 Overview Over the Normalized Organ Dose Tables

The following Tables 7.5, 7.6, 7.7, 7.8, 7.9, 7.10, 7.11, 7.12, 7.13, 7.14, 7.15, 7.16, 7.17, 7.18, 7.19, 7.20, and 7.21 provide information about the thematic classification of the organ dose tables in this book.

Table 7.5 Overview over the normalized organ dose tables: Skull

X-ray examination	Focus-to-detector distance (cm)	Projection	Age group (years)	Voltage (kV)	Table
Skull	115	ap	<1	60	8.17
				70	8.18
			1–2	60	8.19
				70	8.20
			3–7	60	8.21
				70	8.22
			8–12	60	8.23
				70	8.24
				80	8.25
			13–17	70	8.26
				80	8.27
			> 17	70	8.28
				80	8.29
		r-lat	< 1	60	8.30
				70	8.31
			1–2	60	8.32
				70	8.33
			3–7	60	8.34
				70	8.35
			8–12	60	8.36
				70	8.37
				80	8.38
			13–17	70	8.39
				80	8.40
			> 17	70	8.41
				80	8.42

Table 7.6 Overview over the normalized organ dose tables: Chest

X-ray examination	Focus-to-detector distance (cm)	Projection	Age group (years)	Voltage (kV)	Table
Chest	100	ap	24 gw	50	9.26
				60	9.27
				70	9.28
				80	9.29
			28 gw	50	9.30
				60	9.31
				70	9.32
				80	9.33
			<1	50	9.34
				60	9.35
				70	9.36
				80	9.37
			1–2	50	9.38
				60	9.39
				70	9.40
				80	9.41
			3–7	60	9.42
				70	9.43
				80	9.44
			8–12	60	9.45
				70	9.46
				80	9.47

(continued)

Table 7.6 (continued)

X-ray examination	Focus-to-detector distance (cm)	Projection	Age group (years)	Voltage (kV)	Table
Chest	100	ap	13–17	60	9.48
				70	9.49
				80	9.50
				110	9.51
				120	9.52
			> 17	60	9.53
				70	9.54
				80	9.55
				110	9.56
				120	9.57
Chest	100	r-lat	24 gw	50	9.58
				60	9.59
				70	9.60
				80	9.61
			28 gw	50	9.62
				60	9.63
				70	9.64
				80	9.65
			<1	50	9.66
				60	9.67
				70	9.68
				80	9.69
			1–2	50	9.70
				60	9.71
				70	9.72
				80	9.73
			3–7	60	9.74
				70	9.75
				80	9.76
			8–12	60	9.77
				70	9.78
				80	9.79
			13–17	60	9.80
				70	9.81
				80	9.82
				110	9.83
				120	9.84
			> 17	60	9.85
				70	9.86
				80	9.87
				110	9.88
				120	9.89
	115	ap	3–7	60	9.90
				70	9.91
				80	9.92
			8–12	60	9.93
				70	9.94
				80	9.95
			13–17	60	9.96
				70	9.97
				80	9.98
				110	9.99
				120	9.100
			>17	60	9.101

Table 7.6 (continued)

X-ray examination	Focus-to-detector distance (cm)	Projection	Age group (years)	Voltage (kV)	Table
Chest	115	ap		70	9.102
				80	9.103
				110	9.104
				120	9.105
Chest		r-lat	3–7	60	9.106
				70	9.107
				80	9.108
			8–12	60	9.109
				70	9.110
				80	9.111
			13–17	60	9.112
				70	9.113
				80	9.114
				110	9.115
				120	9.116
			>17	60	9.117
				70	9.118
				80	9.119
				110	9.120
				120	9.121
Chest	150	ap	<1	50	9.122
				60	9.123
				70	9.124
				80	9.125
			1–2	50	9.126
				60	9.127
				70	9.128
				80	9.129
			3–7	60	9.130
				70	9.131
				80	9.132
		pa	1–2	50	9.133
				60	9.134
				70	9.135
				80	9.136
			3–7	60	9.137
				70	9.138
				80	9.139
			8–12	60	9.140
				70	9.141
				80	9.142
				110	9.143
				120	9.144
			13–17	110	9.145
				120	9.146
			>17	110	9.147
				120	9.148
Chest	150	r-lat	<1	50	9.149
				60	9.150
				70	9.151
				80	9.152
			1–2	50	9.153
				60	9.154

(continued)

Table 7.6 (continued)

X-ray examination	Focus-to-detector distance (cm)	Projection	Age group (years)	Voltage (kV)	Table
Chest	150	r-lat		70	9.155
				80	9.156
			3–7	60	9.157
				70	9.158
				80	9.159
			8–12	60	9.160
				70	9.161
				80	9.162
				110	9.163
				120	9.164
			13–17	110	9.165
				120	9.166
			>17	110	9.167
				120	9.168

Table 7.7 Overview over the normalized organ dose tables: Bony thorax

X-ray examination	Focus-to-detector distance (cm)	Projection	Age group (years)	Voltage (kV)	Table
Bony thorax	100	ap	0	50	10.20
				60	10.21
				70	10.22
		RAO	0	50	10.23
				60	10.24
				70	10.25
		LAO	0	50	10.26
				60	10.27
				70	10.28
	115	ap	1–2	50	10.29
				60	10.30
				70	10.31
			3–7	60	10.32
				70	10.33
			8–12	60	10.34
				70	10.35
			13–17	60	10.36
				70	10.37
			>17	60	10.38
				70	10.39
		RAO	1–2	50	10.40
				60	10.41
				70	10.42
			3–7	60	10.43
				70	10.44
			8–12	60	10.45
				70	10.46
			13–17	60	10.47
				70	10.48
			>17	60	10.49
				70	10.50

Table 7.7 (continued)

X-ray examination	Focus-to-detector distance (cm)	Projection	Age group (years)	Voltage (kV)	Table
Bony thorax	115	LAO	1–2	50	10.51
				60	10.52
				70	10.53
			3–7	60	10.54
				70	10.55
			8–12	60	10.56
				70	10.57
			13–17	60	10.58
				70	10.59
			>17	60	10.60
				70	10.61

Table 7.8 Overview over the normalized organ dose tables: Thoraco-abdominal babygram

X-ray examination	Focus-to-detector distance (cm)	Projection	Age group (years)	Voltage (kV)	Table
Thoraco-abdominal babygram	100	ap	24 gw	50	11.15
				60	11.16
				70	11.17
			28 gw	50	11.18
				60	11.19
				70	11.20
			<1	50	11.21
				60	11.22
				70	11.23
			1–2	60	11.24
				70	11.25
				80	11.26
	115	ap	<1	50	11.27
				60	11.28
				70	11.29
			1–2	60	11.30
				70	11.31
				80	11.32

Table 7.9 Overview over the normalized organ dose tables: Abdomen

X-ray examination	Focus-to-detector distance (cm)	Projection	Age group (years)	Voltage (kV)	Table
Abdomen	100	ap	24 gw	50	12.24
				60	12.25
				70	12.26
			28 gw	50	12.27
				60	12.28
				70	12.29
			<1	60	12.30
				70	12.31
			1–2	60	12.32
				70	12.33
			3–7	60	12.34
				70	12.35
				80	12.36
			8–12	60	12.37
				70	12.38
				80	12.39
			13–17	70	12.40
				80	12.41
				110	12.42
				120	12.43
			>17	70	12.44
				80	12.45
				110	12.46
				120	12.47
Abdomen	100	pa	<1	60	12.48
				70	12.49
			1–2	60	12.50
				70	12.51
			3–7	60	12.52
				70	12.53
				80	12.54
			8–12	60	12.55
				70	12.56
				80	12.57
			13–17	70	12.58
				80	12.59
				110	12.60
				120	12.61
			>17	70	12.62
				80	12.63
				110	12.64
				120	12.65
Abdomen	100	r-lat	<1	60	12.66
				70	12.67
			1–2	60	12.68
				70	12.69
			3–7	60	12.70
				70	12.71
				80	12.72
		l-lat	<1	60	12.73
				70	12.74
			1–2	60	12.75
				70	12.76
			3–7	60	12.77
				70	12.78
				80	12.79

Table 7.9 (continued)

X-ray examination	Focus-to-detector distance (cm)	Projection	Age group (years)	Voltage (kV)	Table
Abdomen	115	ap	24 gw	50	12.80
				60	12.81
				70	12.82
			28 gw	50	12.83
				60	12.84
				70	12.85
			<1	60	12.86
				70	12.87
			1–2	60	12.88
				70	12.89
			3–7	60	12.90
				70	12.91
				80	12.92
			8–12	60	12.93
				70	12.94
				80	12.95
			13–17	70	12.96
				80	12.97
				110	12.98
				120	12.99
			>17	70	12.100
				80	12.101
				110	12.102
				120	12.103
Abdomen	115	pa	<1	60	12.104
				70	12.105
			1–2	60	12.106
				70	12.107
			3–7	60	12.108
				70	12.109
				80	12.110
			8–12	60	12.111
				70	12.112
				80	12.113
			13–17	70	12.114
				80	12.115
				110	12.116
				120	12.117
			>17	70	12.118
				80	12.119
				110	12.120
				120	12.121
		r-lat	<1	60	12.122
				70	12.123
			1–2	60	12.124
				70	12.125
			3–7	60	12.126
				70	12.127
				80	12.128
		l-lat	<1	60	12.129
				70	12.130
			1–2	60	12.131
				70	12.132
			3–7	60	12.133
				70	12.134
				80	12.135

Table 7.10 Overview over the normalized organ dose tables: Full spine

X-ray examination	Focus-to-detector distance (cm)	Projection	Age group (years)	Voltage (kV)	Table
Full spine	100	ap	<1	50	13.19
				60	13.20
				70	13.21
	115	ap	1–2	50	13.22
				60	13.23
				70	13.24
			3–7	60	13.25
				70	13.26
				80	13.27
	200	ap	3–7	60	13.28
				70	13.29
				80	13.30
			8–12	70	13.31
				80	13.32
				90	13.33
			13–17	70	13.34
				80	13.35
				90	13.36
			>17	70	13.37
				80	13.38
				90	13.39
		pa	8–12	70	13.40
				80	13.41
				90	13.42
			13–17	70	13.43
				80	13.44
				90	13.45
			>17	70	13.46
				80	13.47
				90	13.48

Table 7.11 Overview over the normalized organ dose tables: Cervical spine

X-ray examination	Focus-to-detector distance (cm)	Projection	Age group (years)	Voltage (kV)	Table
Cervical spine	100	ap	<1	50	14.19
				60	14.20
				70	14.21
		lat	<1	50	14.22
				60	14.23
				70	14.24
	115	ap	1–2	50	14.25
				60	14.26
				70	14.27
			3–7	60	14.28
				70	14.29
			8–12	60	14.30
				70	14.31
			13–17	60	14.32
				70	14.33
			>17	60	14.34
				70	14.35
		lat	1–2	50	14.36
				60	14.37
				70	14.38
			3–7	60	14.39
				70	14.40
			8–12	60	14.41
				70	14.42
			13–17	60	14.43
				70	14.44
			>17	60	14.45
				70	14.46
Cervical spine	150	ap	8–12	60	14.47
				70	14.48
			13–17	60	14.49
				70	14.50
			>17	60	14.51
				70	14.52
		lat	8–12	60	14.53
				70	14.54
			13–17	60	14.55
				70	14.56
			>17	60	14.57
				70	14.58

Table 7.12 Overview over the normalized organ dose tables: Thoracic spine

X-ray examination	Focus-to-detector distance (cm)	Projection	Age group (years)	Voltage (kV)	Table
Thoracic spine	100	ap	<1	50	15.19
				60	15.20
				70	15.21
		lat	<1	50	15.22
				60	15.23
				70	15.24
	115	ap	1–2	50	15.25
				60	15.26
				70	15.27
			3–7	60	15.28
				70	15.29
			8–12	60	15.30
				70	15.31
				80	15.32
			13–17	60	15.33
				70	15.34
				80	15.35
			>17	60	15.36
				70	15.37
				80	15.38
		lat	1–2	50	15.39
				60	15.40
				70	15.41
			3–7	60	15.42
				70	15.43
			8–12	60	15.44
				70	15.45
				80	15.46
			13–17	60	15.47
				70	15.48
				80	15.49
			>17	60	15.50
				70	15.51
				80	15.52
Thoracic spine	150	ap	8–12	60	15.53
				70	15.54
				80	15.55
			13–17	60	15.56
				70	15.57
				80	15.58
			>17	60	15.59
				70	15.60
				80	15.61
		lat	8–12	60	15.62
				70	15.63
				80	15.64
			13–17	60	15.65
				70	15.66
				80	15.67
			>17	60	15.68
				70	15.69
				80	15.70

Table 7.13 Overview over the normalized organ dose tables: Lumbar spine

X-ray examination	Focus-to-detector distance (cm)	Projection	Age group (years)	Voltage (kV)	Table
Lumbar spine	100	ap	<1	60	16.18
				70	16.19
		lat	<1	60	16.20
				70	16.21
	115	ap	1–2	60	16.22
				70	16.23
			3–7	60	16.24
				70	16.25
			8–12	70	16.26
				80	16.27
			13–17	70	16.28
				80	16.29
			>17	70	16.30
				80	16.31
		lat	1–2	60	16.32
				70	16.33
			3–7	60	16.34
				70	16.35
			8–12	70	16.36
				80	16.37
			13–17	70	16.38
				80	16.39
			>17	70	16.40
				80	16.41
Lumbar spine	150	ap	8–12	70	16.42
				80	16.43
			13–17	70	16.44
				80	16.45
			> 17	70	16.46
				80	16.47
		lat	8–12	70	16.48
				80	16.49
			13–17	70	16.50
				80	16.51
			> 17	70	16.52
				80	16.53

Table 7.14 Overview over the normalized organ dose tables: Pelvis

X-ray examination	Focus-to-detector distance (cm)	Projection	Age group (years)	Voltage (kV)	Table
Pelvis	100	ap	<1	60	17.14
				70	17.15
			1–2	60	17.16
				70	17.17
	115	ap	3–7	60	17.18
				70	17.19
			8–12	60	17.20
				70	17.21
				80	17.22
			13–17	60	17.23
				70	17.24
				80	17.25
			>17	60	17.26
				70	17.27
				80	17.28

Table 7.15 Overview over the normalized organ dose tables: Hip joint

X-ray examination	Focus-to-detector distance (cm)	Projection	Age group (years)	Voltage (kV)	Table
Hip joint	100	ap	<1	60	18.13
				70	18.14
			1–2	60	18.15
				70	18.16
	115	ap	3–7	60	18.17
				70	18.18
			8–12	60	18.19
				70	18.20
				80	18.21
			13–17	60	18.22
				70	18.23
				80	18.24
			>17	60	18.25
				70	18.26
				80	18.27

Table 7.16 Overview over the normalized organ dose tables: Thigh

X-ray examination	Focus-to-detector distance (cm)	Projection	Age group (years)	Voltage (kV)	Table
Thigh	100	ap, lat	< 1	60	19.14
				70	19.15
			1–2	60	19.16
				70	19.17
			3–7	60	19.18
				70	19.19
	115	ap, lat	8–12	60	19.20
				70	19.21
				80	19.22
			13–17	60	19.23
				70	19.24
				80	19.25
			>17	60	19.26
				70	19.27
				80	19.28

Table 7.17 Overview over the normalized organ dose tables: Shoulder

X-ray examination	Focus-to-detector distance (cm)	Projection	Age group (years)	Voltage (kV)	Table
Shoulder	100	ap	<1	60	20.14
				70	20.15
			1–2	60	20.16
				70	20.17
			3–7	60	20.18
				70	20.19
			8–12	60	20.20
				70	20.21
				80	20.22
	115	ap	13–17	60	20.23
				70	20.24
				80	20.25
			>17	60	20.26
				70	20.27
				80	20.28

Table 7.18 Overview over the normalized organ dose tables: Clavicle

X-ray examination	Focus-to-detector distance (cm)	Projection	Age group (years)	Voltage (kV)	Table
Clavicle	100	ap	<1	60	21.14
				70	21.15
			1–2	60	21.16
				70	21.17
	115	ap	3–7	60	21.18
				70	21.19
			8–12	60	21.20
				70	21.21
				80	21.22
			13–17	60	21.23
				70	21.24
				80	21.25
			>17	60	21.26
				70	21.27
				80	21.28

Table 7.19 Overview over the normalized organ dose tables: Upper arm

X-ray examination	Focus-to-detector distance (cm)	Projection	Age group (years)	Voltage (kV)	Table
Upper arm	100	ap, lat	<1	50	22.14
				60	22.15
				70	22.16
			1–2	60	22.17
				70	22.18
			3–7	60	22.19
				70	22.20
			8–12	60	22.21
				70	22.22
				80	22.23
	115	ap, lat	13–17	60	22.24
				70	22.25
				80	22.26
			>17	60	22.27
				70	22.28
				80	22.29

Table 7.20 Overview over the normalized organ dose tables: Upper extremity

X-ray examination	Focus-to-detector distance (cm)	Projection	Age group (years)	Voltage (kV)	Table
Upper extremity	100	ap	<1	50	23.14
				60	23.15
				70	23.16
			1–2	60	23.17
				70	23.18
	115	ap	3–7	60	23.19
				70	23.20
			8–12	60	23.21
				70	23.22
			13–17	60	23.23
				70	23.24
				80	23.25
			>17	60	23.26
				70	23.27
				80	23.28

Table 7.21 Overview over the normalized organ dose tables: Lower extremity

X-ray examination	Focus-to-detector distance (cm)	Projection	Age group (years)	Voltage (kV)	Table
Lower extremity	100	ap	<1	60	25.16
				70	25.17
			1–2	60	25.18
				70	25.19
	115	ap	3–7	60	25.20
				70	25.21
			8–12	60	25.22
				70	25.23
			13–17	60	25.24
				70	25.25
			>17	60	25.26
				70	25.27
	150	ap	8–12	60	25.28
				70	25.29
			13–17	60	25.30
				70	25.31
			>17	60	25.32
				70	25.33

References

1. Bader D, Datz H, Bartal G, Juster AA, Marks K, Smolkin T, Zangen S, Kugelman A, Hoffmann C, Shani G, Ben-Shlomo A, Margaliot M, Sadetzki S. Unintentional exposure of neonates to conventional radiography in the neonatal intensive care units. J Perinatol. 2007;27:579–85.
2. Bohmann I. Ermittlung des Durchstrahlungs-durchmessers bei Säuglingen, Kindern und Jugendlichen zur Aufstellung von Belichtungswerten in der Röntgendiagnostik und Abschätzung der Organdosiswerte bei typischen Röntgenuntersuchungen. GSF-Bericht 16/90; 1990.
3. Boone JM. The three parameters of equivalent spectra as an index of beam quality. Med Phys. 1988;15:304–10.
4. Commission of the European Communities. European guidelines on quality criteria for diagnostic radiographic images in paediatrics. EUR 16261 EN; 1996.
5. Fichtner C, Schneider K, Freidhof C, Endemann B, Horwitz AE, Kohn MM, Fendel H. Critical analysis of field size in chest x-rays of infants—a EC-wide survey in children's clinics. Eur Radiol. 1993;3(Suppl):389.
6. German Medical Association. Guidelines of the German Medical Association for quality assurance in diagnostic radiology; 2007.
7. Griebl G. Variation von Feldgröße und Dosis bei Röntgen-Thoraxaufnahmen beim 5 Jahre alten Kind—

kritische Analyse einer Feldstudie in europäischen Kinderkliniken. Dissertation, Universität München; 2001.
8. International Commission on Radiological Protection. 1990 recommendations of the International Commission on Radiological Protection, ICRP publication 60; 1991.
9. International Commission on Radiological Protection. The 2007 recommendations of the International Commission on Radiological Protection, ICRP Publication 103; 2007.
10. International conference on harmonisation of technical requirements for registration of pharmaceuticals for human use. ICH harmonised tripartite guideline—guideline for good clinical practice E6(R1); 1996.
11. Seidenbusch MC, Regulla D, Schneider K. Radiation exposure of children in pediatric radiology. Part 3: conversion coefficients for reconstruction of organ doses achieved during chest x-ray examinations. Fortschr Röntgenstr. 2008;180:1061–81.
12. Servomaa A, Tapiovaara M. Organ dose calculation in medical x-ray examinations by the program PCXMC. Radiat Prot Dosim. 1998;80:213–9.
13. Soboleski D, Theriault C, Acker A. Unnecessary irradiation to non-thoracic structures during pediatric chest radiography. Pediatr Radiol. 2006;36:22–5.
14. Stollfuss J, Schneider K, Krüger-Stollfuss I. A comparative study of collimation in bedside chest radiography for preterm infants in two teaching hospitals. Eur J Radiol. 2015;2:118–22.
15. Tapiovaara M, Lakkisto M, Servomaa A. PCXMC. A PC-based Monte Carlo program for calculating patient doses in medical x-ray examinations. Finnish Centre for Radiation and Nuclear Safety, Säteilyturvakeskus (STUK), Report STUK A-139; 1997.
16. Tapiovaara M, Siiskonen T. PCXMC. A Monte Carlo program for calculating patient doses in medical x-ray examinations. Säteilyturvakeskus (STUK), Report STUK-A231; 2008.
17. Von Boetticher H, Lachmund J, Looe HK, Hoffmann W, Poppe B. 2007 recommendations of the ICRP change basis for estimation of the effective dose: what is the impact on radiation dose assessment of patient and personnel? Fortschr Röntgenstr. 2008;180:391–5.
18. Zeiss A. Feldgrößen- und Dosisvariation von Röntgen-Thorax-Aufnahmen beim zehnjährigen Kind—kritische Analyse einer Feldstudie in europäischen Kinderkliniken. Dissertation, Universität München; 2002.

Skull

8

8.1 Introduction

There are three different radiographic techniques of skull radiography:

- The standard antero-posterior (ap) and lateral (lat) views serve for imaging of the head, including the cranial vault, the skull base, facial bones, mandible, and the upper cervical spine. This technique is suitable for imaging patients who are suffering from trauma, infection, tumours, and congenital craniofacial anomalies. Note that the pa projection of the skull was not considered here because children very often do not cooperate sufficiently that this position can be correctly obtained.
- Patients with hydrocephalus with the malfunction of the ventriculo-peritoneal shunt need another technique that includes the neck and the upper chest.
- Patients with adjustable flow-control valves after MRI examinations need tightly collimated magnified images of the cranial vault in order to check the indication of the outlet pressure of the valve.

As the first technique is the most common, we only refer to the first technique in this chapter.

The type of skull radiography is determined by the clinical indication. Table 8.1 shows indications and Table 8.2 contraindications for skull radiography. Image criteria as required for radiological findings in skull radiographs are listed in Table 8.3. In order to fulfil these image criteria, the field settings listed in Tables 8.4 and 8.5 should be achieved. These field settings go along with the patient preparation as listed in Table 8.6. Exposure parameter combinations for various age groups and head sizes are listed

Table 8.1 Indications for skull radiographs

Head injury
Abusive head injury
Birth trauma
Accidental trauma
Shunt malfunction in patients with hydrocephalus[a]
Control of shunt valve functions[b]
Skull tumours, e.g. Langerhans cell histiocytosis
Premature closure of cranial sutures
Craniofacial dysplasia (M. Crouzon, Apert syndrome, etc.)
Skeletal dysplasia
Skull infection
Annotations: [a]Wide collimation with lower field edge beneath the clavicles is recommended. [b]Narrow collimation to the valve system is recommended.

Table 8.2 Contraindications for skull radiographs

Open head injury

M. Seidenbusch et al., *Imaging Practice and Radiation Protection in Pediatric Radiology*,
https://doi.org/10.1007/978-3-030-18504-6_8

Table 8.3 Image criteria for skull radiographs

Skull ap
Symmetrical reproduction of the whole cranial vault, the facial bones, and the skull base
Projection of the upper margins of petrous bones into the lower half of the orbits in ap projection
Reproduction of the paranasal sinuses and structures of the temporal bones consistent with age
Visualization of the large fontanelle, the sagittal and lambdoid sutures
Visually sharp reproduction of the outer and inner tables of the entire cranial vault consistent with age
Skull lat
Visually sharp reproduction of the outer and inner tables of the entire cranial vault and the floor of the sella consistent with age
Superimposition of the orbital roofs and the anterior parts of the greater wings of the sphenoid bones
Visualization of the coronal, lambdoid, and squamosal sutures and lateral fontanelles consistent with age
Visually sharp reproduction of the vascular channels and the trabecular structure consistent with age

Table 8.4 Patient positions and beam projections for skull radiographs

Patient position	Beam projections
Sitting	pa[a]
Supine	ap + lat
Decubitus	lat
Prone	pa[a]
Annotations:	
[a]Obsolete.	

Table 8.5 Optimal field settings in skull radiographs

Skull ap	
Upper field edge:	Above the cranial vault
Field centring:	Midline of the skull at the level of the root of the nose
Lower field edge:	Tip of the mandible
Lateral field edges:	Lateral of the cranial vault
Skull lat	
Upper field edge:	Above the cranial vault
Field centring:	Midline of the skull at the level of the cantho-meatal line[a]
Lower field edge:	Tip of the mandible
Anterior field edge:	Tip of the nose
Posterior field edge:	Above the skin surfaces of the occiput
Annotations:	
[a]Depending on skull shape.	

Table 8.6 Patient preparation

General	Remove all clothes, external metallic devices and jewellery. Especially ear rings, piercings, hair clamps, Alice bands, and neck laces should be removed
Positioning	Avoid rotation of the head in the antero-posterior and lateral projection. Uncooperative patients should be fixed with sandbags, rubber devices and/or translucent compressor systems
Attention	On the Bucky table the patient should always be positioned on a foam mat of 10 to 20 cm thickness. So, the lower part of the body will not be cut if lateral projection X-rays (cross table views) in prone or in decubitus position are performed
Shielding	Put lead masks immediately at the adjacent lower field edge chest

in Tables 8.7, 8.8, and 8.9. The age groups in Table 8.8 refer to the standard age groups of the mathematical MIRD phantom family which has been used to calculate the normalized organ doses in this book. Table 8.9 shows radiation qualities (i.e. various combinations of tube voltage and total filtration) which may be applied in clinical practice. Please note that not

Table 8.7 Radiographic procedure

X-ray source/beam quality/grid	Focal spot size between 0.6 and 1.3 dependent on the head size
	Use automatic exposure control (AEC)
	Check the kV
	Check the exposure time
	Check the use of additional filtration
	Check the use of antiscatter grid
	Check the focus-to-detector distance
Field setting	Check the field size
	Check the field centring

Table 8.8 Radiographic techniques of Bucky table skull radiographs

Age group (years)	Beam direction	Patient position	Beam projection	Typical patient diameter (cm)[1]	Optimal field size[1] (cm × cm)	Focus-to-detector distance (cm)	Voltage (kVp)						Antiscatter grid
							50	60	70	80	110	120	
0	Vertical	Supine	ap	11.3	10 × 11	115							
		Decubitus	lat	8.8	12 × 12								
1 – 2		Supine	ap	15.3	14 × 16								
		Decubitus	lat	11.7	16 × 15								
3 – 7		Supine	ap	18.0	15 × 18								
		Decubitus	lat	14.2	18 × 18								
8 – 12		Supine	ap	18.0	16 × 20								
		Decubitus	lat	14.4	19 × 19								
13 – 17		Supine	ap	18.5	17 × 24								
		Decubitus	lat	15.2	20 × 20								
18		Supine	ap	18.5	18 × 25								
		Decubitus	lat	15.5	21 × 21								

Annotations:

[1] Cave: As mathematical phantoms were defined through linear scaling of the patients' bodies, skull sizes have been defined too small in infants and neonates. Thus, ideal field sizes given here may not be valid clinically in the neonate and the 1-year-old child

(black) Parameter combination not recommended, normalized organ doses not provided

(grey) Parameter combination not recommended, normalized organ doses provided

(light blue) Parameter combination recommended, normalized organ doses provided

Table 8.9 Radiation qualities of Bucky table skull radiographs

Voltage (kVp)	Total filtration					
	Without additional filtration		With additional filtration			
	1.5 mm Al	2.5 Al	3.5 Al	4.5 Al	3.5 Al + 0.1 Cu	3.5 + 0.2 Cu
50						
60						
70						
80						
110						
120						

Annotations:

(black) Parameter combination not recommended, normalized organ doses not provided

(grey) Parameter combination not recommended, normalized organ doses provided

(light blue) Parameter combination recommended, normalized organ doses provided

all of these radiation qualities are recommended by the authors. The standard projections of skull radiographies of the patient who is correctly positioned are depicted in Fig. 8.1. Also the optimal field (marked in green), the suboptimal field (marked in red), and the correct shielding are shown. Tables 8.11, 8.12, 8.13, 8.14, 8.15 and 8.16 outline the virtual field settings on the skin surface of mathematical MIRD phantoms of various age groups and the resulting virtual radiographies with and without bone structures, respectively. Normalized organ doses resulting from these virtual field settings have been calculated for all age groups and

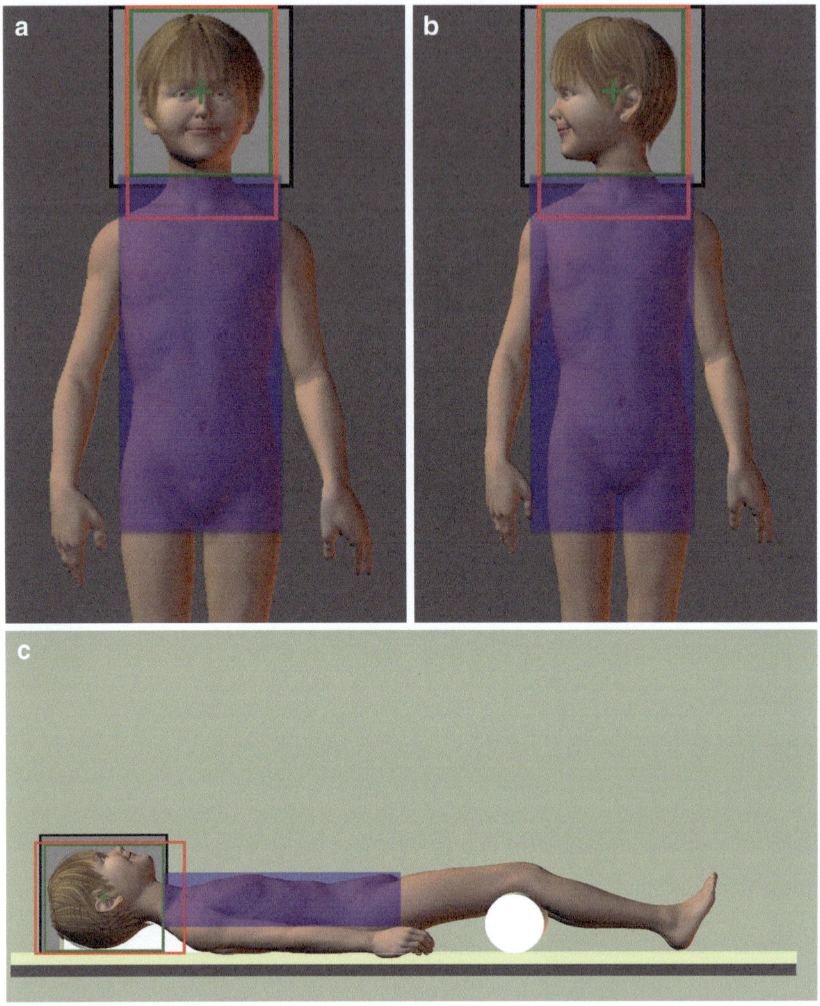

Fig. 8.1 Clinical field settings for skull radiographs. (**a**) Antero-posterior projection. (**b**) Lateral projection, vertical beam direction. (**c**) Lateral projection, horizontal beam direction. ((**c**) V. Rösenberger and M. Seidenbusch 2018) Green lines = optimal field settings; green cross = centring point; red lines = suboptimal field settings; grey shaded area = image detector; blue shaded area = lead shielding of the patient; dark grey shaded area = Bucky table; white cylinder = foam roll; white device behind the head and neck = foam pillow; white foam mat on the top of the Bucky table

have been listed in Tables 8.17, 8.18, 8.19, 8.20, 8.21, 8.22, 8.23, 8.24, 8.25, 8.26, 8.27, 8.28, 8.29, 8.30, 8.31, 8.32, 8.33, 8.34, 8.35, 8.36, 8.37, 8.38, 8.39, 8.40, 8.41 and 8.42. Table cells with normal font refer to optimal field settings and table cells with bold font refer to suboptimal field settings. Please note that normalized organ doses have been provided also for exposure parameter combinations, which are not recommended by the authors.

8.2 Indications and Contraindications

Indications and contraindications for skull radiographs are listed in Tables 8.1 and 8.2. Skull radiographs are very rarely indicated in neonates and infants, as ultrasound examinations (US) of the head, magnetic resonance imaging (MRI), and rarely computed tomographic (CT) examinations of the skull are preferred. In infants and children with premature synostosis, skull radiographs are performed first.

8.3 Image Criteria

The image criteria required for skull radiographs are listed in Table 8.3. See also [2, 3].

8.4 Field Settings

In Table 8.4, patient positions and beam projections used for skull radiographs are listed, and in Table 8.5 the optimal field settings are listed which are required in order to fulfil the image criteria mentioned in Table 8.3. It must be pointed to the fact here that the mathematical phantoms of the different age groups were defined through linear scaling and that therefore the skull sizes have been defined too small in the phantoms of the neonate and 1-year-old infant (see paragraph **4.9.2**). Thus, ideal field sizes drawn in the phantoms may not be valid actually in these age groups. For more details, see e.g. [8–11, 13].

8.5 Patient Preparation and Radiographic Procedure

See Tables 8.6, 8.7 and [1, 5].

8.6 Exposure Parameters

In Tables 8.8 and 8.9, exposure parameter combinations used for skull radiographs in various age groups are listed. Dark grey shaded table cells refer to exposure parameter combinations which are not recommended for paediatric skull radiographs and for which no normalized organ doses have been provided. Light grey and light blue shaded table cells refer to exposure parameter combinations for which normalized organ doses have been provided with light grey shaded table cells referring to exposure parameter combinations being not recommended and light blue shaded table cells referring to exposure parameter combinations following the European Guidelines [2]. Note that due to the fact that the radiation dose is an exponential function of the patient diameter, it may make sense to use the patient diameters instead of the patient's age groups for determining the adequate exposure parameters from Table 8.8. In this case, for determining the corresponding normalized organ doses from Tables 8.17, 8.18, 8.19, 8.20, 8.21, 8.22, 8.23, 8.24, 8.25, 8.26, 8.27, 8.28, 8.29, 8.30, 8.31, 8.32, 8.33, 8.34, 8.35, 8.36, 8.37, 8.38, 8.39, 8.40, 8.41, and 8.42, the age group related to the patient diameter should be used.

8.7 Normalized Organ Doses

For calculating normalized organ doses, the field settings described in Sect. 8.4 were transferred to the mathematical MIRD phantoms. Inconsistencies between the geometrical anatomy of the mathematical phantoms and the real human anatomy have been listed in Table 8.10 whenever they are of relevance for the validity of the normalized organ doses.

In Tables 8.11, 8.12, 8.13, 8.14, 8.15, and 8.16, optimal and suboptimal field settings are listed for the different age groups and the beam projections given in Table 8.8. Green lines refer to optimal field settings, red lines refer to suboptimal field settings, respectively. In the tables, the optimal and suboptimal field settings which were used for performing the calculations of the

Table 8.10 Inconsistencies between the mathematical MIRD phantoms and real human anatomy

Anatomical region	Inconsistencies
Head	No eyes, nose, mouth, and ears have been inserted
	No eye lens has been inserted

For details, see the text

normalized organ doses provided in Tables 8.17, 8.18, 8.19, 8.20, 8.21, 8.22, 8.23, 8.24, 8.25, 8.26, 8.27, 8.28, 8.29, 8.30, 8.31, 8.32, 8.33, 8.34, 8.35, 8.36, 8.37, 8.38, 8.39, 8.40, 8.41, and 8.42 were drawn in on the skin surface of the mathematical phantoms. The virtual mathematical radiographs resulting from these field settings were depicted in the tables with and without bone structures, respectively, in order to enable the reader to visually compare the virtual mathematical radiographs with the real radiographs made under clinical conditions. For comparing the virtual mathematical field settings with the real clinical field settings, the virtual mathematical field edges were also listed in the tables.

As the normalized organ doses provided in Tables 8.17, 8.18, 8.19, 8.20, 8.21, 8.22, 8.23, 8.24, 8.25, 8.26, 8.27, 8.28, 8.29, 8.30, 8.31, 8.32, 8.33, 8.34, 8.35, 8.36, 8.37, 8.38, 8.39, 8.40, 8.41, and 8.42 refer to entrance kerma values, optimal and suboptimal field sizes were provided in the tables in order to enable the reader to calculate entrance air kerma dose values from clinically measured dose area product values. Please note that these optimal and suboptimal field sizes refer to the mathematical phantoms of an age group only and not to real patients with corresponding ages.

Please note also that the anatomy of the mathematical phantoms has been idealized. Therefore, the virtual field settings for calculating the normalized organ doses have been defined with a view to ensuring that the resulting virtual organ doses would be consistent with the real organ doses which would be achieved during real x-ray examinations being performed with the real field settings recommended in Sect. 8.4. Thus, the field settings shown in the following tables may not be consistent with the clinical field settings as recommended in Sect. 8.4.

By the way, suboptimal field sizes were chosen arbitrarily. The range of field sizes has been discussed in literature, e.g. [8–11, 13].

Normalized organ doses for skull radiographs are also listed in e.g. [4, 6, 7, 12].

Table 8.11 Field settings—Skull ap

Age group (years)		Anatomical landmarks			
		Optimal field setting		Suboptimal field setting	
0	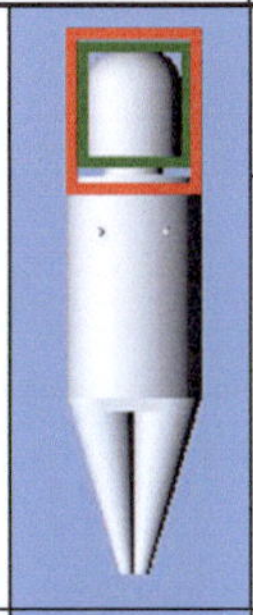	Upper field edge: Above the cranial vault Lower field edge: Tip of the mandible Lateral field edges: Lateral cranial vault Field size = field width × field height: 10 cm × 11 cm = 110 cm²		Upper field edge: Above the cranial vault Lower field edge: Sterno-clavicular joints Lateral field edges: Lateral cranial vault Field size = field width × field height: 10 cm × 14 cm = 140 cm²	

1–2	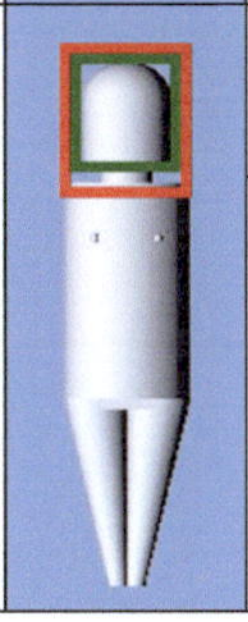	Upper field edge: Above the cranial vault Lower field edge: Tip of the mandible Lateral field edges: Lateral cranial vault Field size = field width × field height: 14 cm × 15 cm = 210 cm²		Upper field edge: Above the cranial vault Lower field edge: Sterno-clavicular joints Lateral field edges: Lateral cranial vault Field size = field width × field height: 14 cm × 19 cm = 266 cm²	

Table 8.12 Field settings—Skull ap

Age group (years)		Anatomical landmarks			
		Optimal field setting		Suboptimal field setting	
3–7	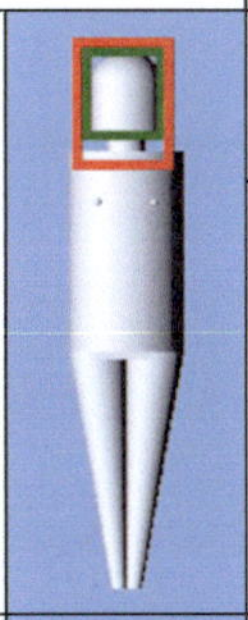	Upper field edge: Above the cranial vault Lower field edge: Tip of the mandible Lateral field edges: Lateral cranial vault Field size = field width × field height: 15 cm × 17 cm = 255 cm²		Upper field edge: Above the cranial vault Lower field edge: Sterno-clavicular joints Lateral field edges: Lateral cranial vault Field size = field width × field height: 15 cm × 23 cm = 345 cm²	

8–12	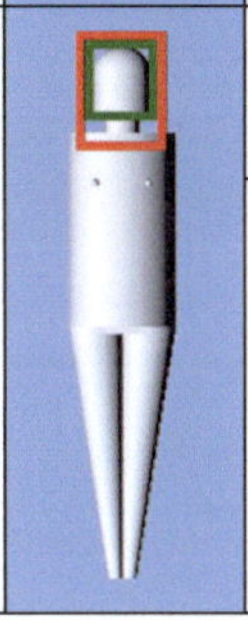	Upper field edge: Above the cranial vault Lower field edge: Tip of the mandible Lateral field edges: Lateral cranial vault Field size = field width × field height: 16 cm × 19 cm = 304 cm²		Upper field edge: Above the cranial vault Lower field edge: Sterno-clavicular joints Lateral field edges: Lateral cranial vault Field size = field width × field height: 16 cm × 26 cm = 416 cm²	

Table 8.13 Field settings—Skull ap

Age group (years)		Anatomical landmarks			
		Optimal field setting		Suboptimal field setting	
13–17	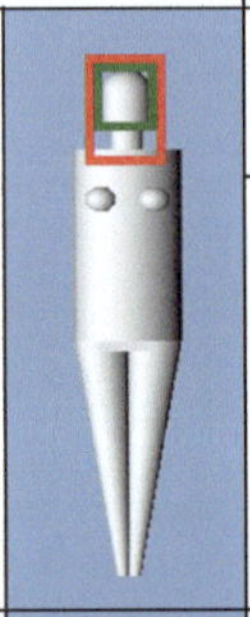	Upper field edge: Above the cranial vault Lower field edge: Tip of the mandible Lateral field edges: Lateral cranial vault Field size = field width × field height: 17 cm × 19 cm = 323 cm²		Upper field edge: Above the cranial vault Lower field edge: Sterno-clavicular joints Lateral field edges: Lateral cranial vault Field size = field width × field height: 17 cm × 30 cm = 510 cm²	

18	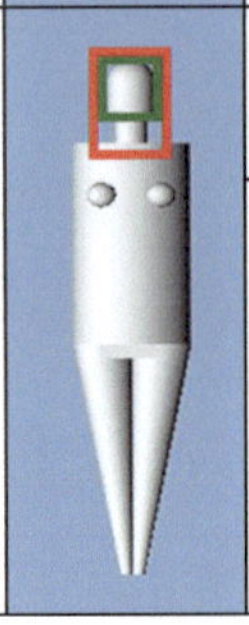	Upper field edge: Above the cranial vault Lower field edge: Tip of the mandible Lateral field edges: Lateral cranial vault Field size = field width × field height: 18 cm × 20 cm = 360 cm²		Upper field edge: Above the cranial vault Lower field edge: Sterno-clavicular joints Lateral field edges: Lateral cranial vault Field size = field width × field height: 18 cm × 32 cm = 576 cm²	

Table 8.14 Field settings—Skull lat

Age group (years)		Anatomical landmarks			
		Optimal field setting		Suboptimal field setting	
0	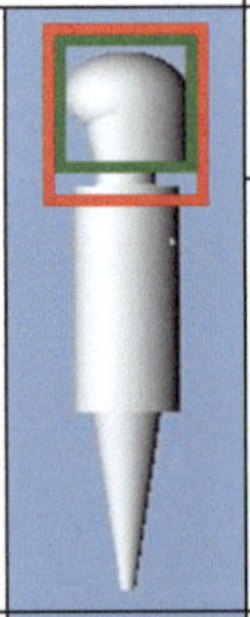	Upper field edge: Above the cranial vault Lower field edge: Tip of the mandible Lateral field edges: Above the skin surface of the forehead / occiput Field size = field width × field height: 12 cm × 12 cm = 144 cm²		Upper field edge: Above the cranial vault Lower field edge: Sterno-clavicular joints Lateral field edges: Above the skin surface of the forehead / occiput Field size = field width × field height: 12 cm × 14 cm = 168 cm²	
				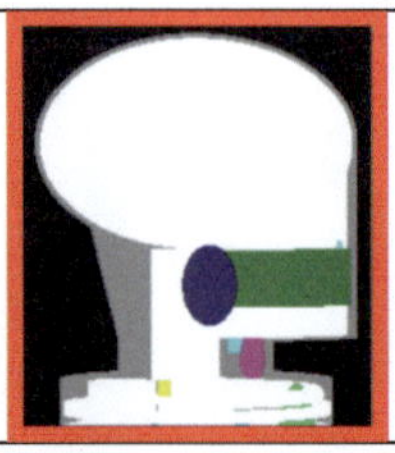	
1–2	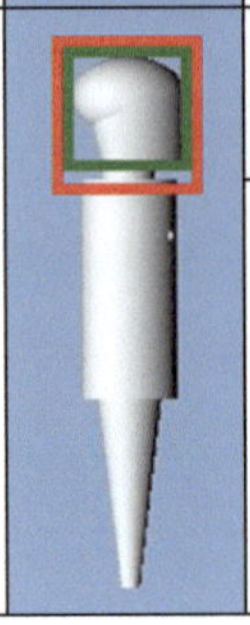	Upper field edge: Above the cranial vault Lower field edge: Tip of the mandible Lateral field edges: Above the skin surface of the forehead / occiput Field size = field width × field height: 16 cm × 15 cm = 240 cm²		Upper field edge: Above the cranial vault Lower field edge: Sterno-clavicular joints Lateral field edges: Above the skin surface of the forehead / occiput Field size = field width × field height: 15 cm × 19 cm = 285 cm²	

Table 8.15 Field settings—Skull lat

Age group (years)		Anatomical landmarks			
		Optimal field setting		Suboptimal field setting	
3–7	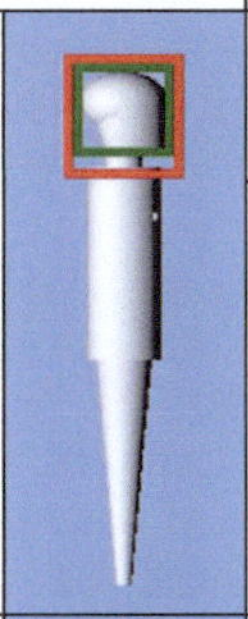	Upper field edge: Above the cranial vault Lower field edge: Tip of the mandible Lateral field edges: Above the skin surface of the forehead / occiput Field size = field width × field height: 18 cm × 18 cm = 324 cm²		Upper field edge: Above the cranial vault Lower field edge: Sterno-clavicular joints Lateral field edges: Above the skin surface of the forehead / occiput Field size = field width × field height: 17 cm × 22 cm = 374 cm²	
				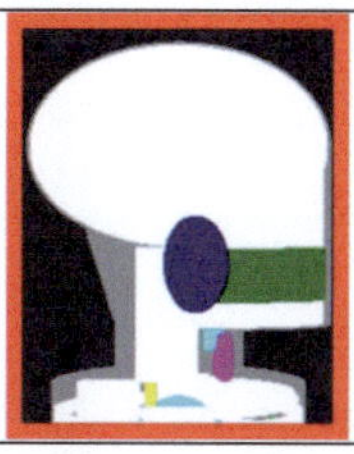	
8–12	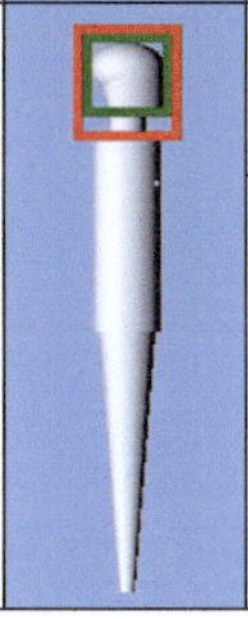	Upper field edge: Above the cranial vault Lower field edge: Tip of the mandible Lateral field edges: Above the skin surface of the forehead / occiput Field size = field width × field height: 19 cm × 19 cm = 361 cm²		Upper field edge: Above the cranial vault Lower field edge: Sterno-clavicular joints Lateral field edges: Above the skin surface of the forehead / occiput Field size = field width × field height: 18 cm × 24 cm = 432 cm²	

Table 8.16 Field settings—Skull lat

Age group (years)		Anatomical landmarks			
		Optimal field setting		Suboptimal field setting	
13–17	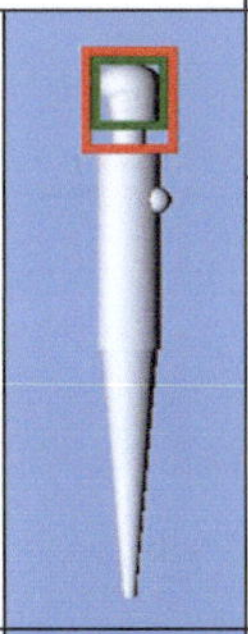	Upper field edge: Above the cranial vault Lower field edge: Tip of the mandible Lateral field edges: Above the skin surface of the forehead / occiput Field size = field width × field height: 20 cm × 20 cm = 400 cm²		Upper field edge: Above the cranial vault Lower field edge: Sterno-clavicular joints Lateral field edges: Above the skin surface of the forehead / occiput Field size = field width × field height: 19 cm × 29 cm = 551 cm²	
			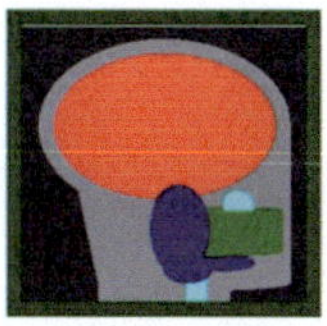		
18	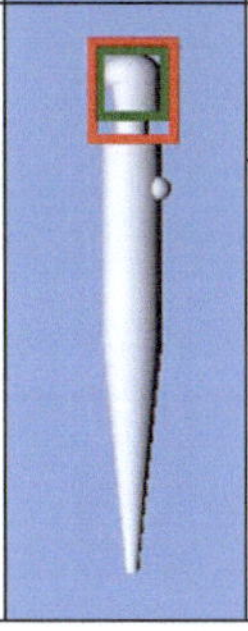	Upper field edge: Above the cranial vault Lower field edge: Tip of the mandible Lateral field edges: Above the skin surface of the forehead / occiput Field size = field width × field height: 21 cm × 21 cm = 441 cm²		Upper field edge: Above the cranial vault Lower field edge: Sterno-clavicular joints Lateral field edges: Above the skin surface of the forehead / occiput Field size = field width × field height: 20 cm × 30 cm = 600 cm²	
		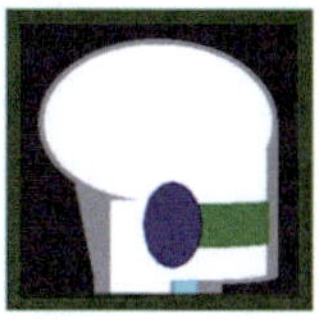	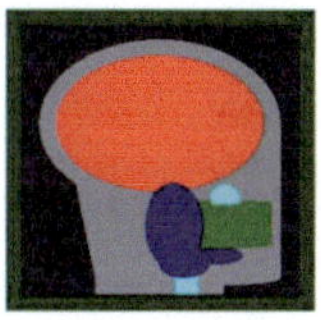		

Table 8.17 Normalized organ doses—Skull ap—<1 year

Beam projection:			ap				**Table cells:**					
Focus-to-detector distance (cm):			115				(+) = normal font = optimal field setting					
Tube voltage (kVp):			60				**(−) = bold font = suboptimal field setting**					
Total filtrations:			See last table row.				Empty table cells correspond to values <0.01					
Organ or tissue	**Organ dose D_T: Entrance dose K_E**							**($mSv * mGy^{-1}$)**				
	(+)	(−)	(+)	(−)	(+)	(−)	(+)	(−)	(+)	(−)	(+)	(−)
Skull	1.21	**1.21**	1.41	**1.41**	1.54	**1.55**	1.64	**1.65**	1.80	**1.82**	1.80	**1.82**
Brain	0.29	**0.29**	0.36	**0.37**	0.42	**0.42**	0.46	**0.46**	0.53	**0.53**	0.51	**0.51**
Oral mucosa	0.60	**0.61**	0.70	**0.72**	0.76	**0.79**	0.81	**0.84**	0.89	**0.93**	0.89	**0.91**
Salivary glands	0.49	**0.51**	0.58	**0.61**	0.64	**0.68**	0.69	**0.73**	0.77	**0.81**	0.76	**0.80**
Cervical spine	0.35	**0.59**	0.46	**0.75**	0.54	**0.88**	0.61	**0.98**	0.73	**1.17**	0.69	**1.12**
Thyroid	0.21	**0.90**	0.25	**1.01**	0.28	**1.08**	0.30	**1.14**	0.35	**1.22**	0.34	**1.21**
Thymus	0.02	**0.21**	0.03	**0.25**	0.03	**0.28**	0.03	**0.30**	0.04	**0.33**	0.04	**0.32**
Thoracic spine	0.01	**0.07**	0.01	**0.09**	0.02	**0.10**	0.02	**0.12**	0.03	**0.15**	0.02	**0.13**
Claviculae	0.10	**1.94**	0.13	**2.23**	0.14	**2.41**	0.16	**2.54**	0.19	**2.74**	0.18	**2.79**
Scapulae	0.02	**0.11**	0.02	**0.14**	0.03	**0.17**	0.03	**0.19**	0.04	**0.23**	0.04	**0.21**
Ribs	0.01	**0.14**	0.02	**0.16**	0.02	**0.18**	0.02	**0.19**	0.03	**0.21**	0.03	**0.21**
Oesophagus		**0.04**		**0.05**	0.01	**0.06**	0.01	**0.07**	0.01	**0.09**	0.01	**0.08**
Trachea	0.54	**0.58**	0.63	**0.68**	0.69	**0.74**	0.73	**0.79**	0.81	**0.87**	0.80	**0.86**
Lung		**0.03**		**0.04**	0.01	**0.05**	0.01	**0.05**	0.01	**0.06**	0.01	**0.06**
Mamma												
Heart		**0.01**		**0.02**		**0.02**		**0.02**		**0.03**		**0.03**
Stomach												
Spleen												
Liver												
Gall bladder												
Pancreas												
Small intestine												
Large intestine												
Sigma/rectum												
Lumbar spine												
Pelvis												
Adrenals												
Kidneys												
Urinary bladder												
Uterus												
Ovaries												
Prostate												
Testes												
Upper arm	0.02	**0.06**	0.02	**0.07**	0.03	**0.09**	0.03	**0.10**	0.04	**0.12**	0.04	**0.11**
Elbow				**0.01**		**0.01**		**0.02**		**0.02**		**0.02**
Forearm												
Upper leg												
Knee												
Lower leg												
Lymph nodes	0.13	**0.17**	0.15	**0.20**	0.16	**0.22**	0.18	**0.24**	0.19	**0.26**	0.19	**0.26**
Skin	0.16	**0.22**	0.17	**0.23**	0.18	**0.24**	0.18	**0.25**	0.19	**0.26**	0.19	**0.26**
Muscle	0.06	**0.09**	0.07	**0.10**	0.08	**0.12**	0.08	**0.12**	0.09	**0.14**	0.09	**0.13**
Skeleton	0.28	**0.34**	0.33	**0.40**	0.36	**0.44**	0.38	**0.47**	0.43	**0.53**	0.43	**0.53**
Red bone marrow	0.14	**0.15**	0.16	**0.18**	0.18	**0.20**	0.19	**0.22**	0.22	**0.24**	0.21	**0.24**
Whole body dose	0.10	**0.13**	0.12	**0.16**	0.14	**0.17**	0.15	**0.19**	0.17	**0.21**	0.16	**0.21**
ED (ICRP 60)	0.03	**0.08**	0.04	**0.09**	0.04	**0.10**	0.05	**0.10**	0.05	**0.11**	0.05	**0.11**
ED (ICRP 103)	0.05	**0.09**	0.06	**0.10**	0.06	**0.11**	0.07	**0.12**	0.08	**0.13**	0.08	**0.13**
Total filtration (mm) Al	**1.5**		**2.5**		**3.5**		**4.5**		**3.5**		**3.5**	
Cu	**0**		**0**		**0**		**0**		**0.1**		**0.2**	

Table 8.18 Normalized organ doses—Skull ap—<1 year

Beam projection:	ap	**Table cells:**	
Focus-to-detector distance (cm):	115	(+) = normal font = optimal field setting	
Tube voltage (kVp):	70	**(−) = bold font = suboptimal field setting**	
Total filtrations:	See last table row.	Empty table cells correspond to values <0.01	

Organ or tissue	**Organ dose D_T: Entrance dose K_E**							**(mSv * mGy^{-1})**				
	(+)	(−)	(+)	(−)	(+)	(−)	(+)	(−)	(+)	(−)	(+)	(−)
Skull	1.30	**1.31**	1.50	**1.51**	1.62	**1.64**	1.72	**1.73**	1.87	**1.89**	1.98	**2.00**
Brain	0.34	**0.34**	0.41	**0.42**	0.47	**0.47**	0.51	**0.51**	0.58	**0.59**	0.65	**0.65**
Oral mucosa	0.65	**0.67**	0.75	**0.77**	0.82	**0.84**	0.86	**0.89**	0.95	**0.98**	1.01	**1.05**
Salivary glands	0.54	**0.57**	0.63	**0.67**	0.69	**0.73**	0.74	**0.78**	0.81	**0.86**	0.88	**0.93**
Cervical spine	0.43	**0.70**	0.54	**0.88**	0.63	**1.01**	0.70	**1.11**	0.83	**1.30**	0.95	**1.47**
Thyroid	0.24	**0.96**	0.28	**1.07**	0.31	**1.14**	0.33	**1.19**	0.37	**1.27**	0.41	**1.34**
Thymus	0.02	**0.24**	0.03	**0.27**	0.03	**0.30**	0.04	**0.31**	0.05	**0.35**	0.05	**0.37**
Thoracic spine	0.01	**0.08**	0.02	**0.11**	0.02	**0.13**	0.03	**0.14**	0.03	**0.17**	0.04	**0.20**
Claviculae	0.12	**2.06**	0.14	**2.33**	0.16	**2.50**	0.18	**2.61**	0.21	**2.78**	0.23	**2.89**
Scapulae	0.02	**0.13**	0.03	**0.17**	0.03	**0.19**	0.04	**0.22**	0.05	**0.26**	0.06	**0.30**
Ribs	0.02	**0.15**	0.02	**0.18**	0.02	**0.19**	0.03	**0.20**	0.03	**0.23**	0.04	**0.24**
Oesophagus		**0.05**	0.01	**0.07**	0.01	**0.08**	0.01	**0.08**	0.02	**0.10**	0.02	**0.12**
Trachea	0.59	**0.64**	0.68	**0.73**	0.74	**0.80**	0.78	**0.84**	0.85	**0.92**	0.91	**0.99**
Lung		**0.04**	0.01	**0.05**	0.01	**0.05**	0.01	**0.06**	0.02	**0.07**	0.02	**0.08**
Mamma												**0.01**
Heart		**0.02**		**0.02**		**0.03**		**0.03**	0.01	**0.03**	0.01	**0.04**
Stomach												
Spleen												**0.01**
Liver												
Gall bladder												
Pancreas												**0.01**
Small intestine												
Large intestine												
Sigma/rectum												
Lumbar spine												
Pelvis												
Adrenals												**0.01**
Kidneys												
Urinary bladder												
Uterus												
Ovaries												
Prostate												
Testes												
Upper arm	0.02	**0.07**	0.03	**0.09**	0.04	**0.10**	0.04	**0.11**	0.05	**0.14**	0.06	**0.16**
Elbow		**0.01**		**0.01**		**0.02**		**0.02**		**0.02**	0.01	**0.03**
Forearm												
Upper leg												
Knee												
Lower leg												
Lymph nodes	0.14	**0.19**	0.16	**0.22**	0.18	**0.24**	0.19	**0.25**	0.21	**0.28**	0.22	**0.30**
Skin	0.17	**0.22**	0.18	**0.24**	0.19	**0.25**	0.19	**0.26**	0.20	**0.27**	0.21	**0.28**
Muscle	0.07	**0.10**	0.08	**0.11**	0.08	**0.13**	0.09	**0.13**	0.10	**0.15**	0.10	**0.16**
Skeleton	0.30	**0.37**	0.35	**0.43**	0.38	**0.47**	0.41	**0.50**	0.45	**0.56**	0.48	**0.60**
Red bone marrow	0.15	**0.17**	0.18	**0.20**	0.20	**0.22**	0.21	**0.24**	0.23	**0.26**	0.26	**0.29**
Whole body dose	0.12	**0.15**	0.14	**0.17**	0.15	**0.19**	0.16	**0.20**	0.18	**0.23**	0.19	**0.25**
ED (ICRP 60)	0.04	**0.08**	0.04	**0.10**	0.05	**0.10**	0.05	**0.11**	0.06	**0.12**	0.06	**0.13**
ED (ICRP 103)	0.05	**0.10**	0.06	**0.11**	0.07	**0.12**	0.07	**0.13**	0.08	**0.14**	0.09	**0.15**
Total filtration (mm) Al	**1.5**		**2.5**		**3.5**		**4.5**		**3.5**		**3.5**	
Cu	**0**		**0**		**0**		**0**		**0.1**		**0.2**	

Table 8.19 Normalized organ doses—Skull ap—1–2 years

Beam projection:			ap			**Table cells:**						
Focus-to-detector distance (cm):			115			(+) = normal font = optimal field setting						
Tube voltage (kVp):			60			**(−) = bold font = suboptimal field setting**						
Total filtrations:			See last table row.			Empty table cells correspond to values <0.01						
Organ or tissue	**Organ dose D_T: Entrance dose K_E**							**(mSv * mGy^{-1})**				
	(+)	(−)	(+)	(−)	(+)	(−)	(+)	(−)	(+)	(−)	(+)	(−)
Skull	0.99	**1.00**	1.18	**1.18**	1.30	**1.31**	1.40	**1.41**	1.56	**1.58**	1.55	**1.56**
Brain	0.20	**0.20**	0.25	**0.25**	0.29	**0.29**	0.32	**0.33**	0.39	**0.39**	0.36	**0.37**
Oral mucosa	0.47	**0.48**	0.55	**0.57**	0.61	**0.63**	0.66	**0.68**	0.74	**0.76**	0.72	**0.74**
Salivary glands	0.35	**0.37**	0.42	**0.45**	0.47	**0.51**	0.51	**0.55**	0.58	**0.63**	0.57	**0.61**
Cervical spine	0.19	**0.40**	0.25	**0.52**	0.31	**0.61**	0.35	**0.69**	0.44	**0.84**	0.39	**0.78**
Thyroid	0.07	**0.88**	0.09	**0.99**	0.10	**1.07**	0.12	**1.12**	0.14	**1.20**	0.13	**1.19**
Thymus	0.01	**0.09**	0.02	**0.11**	0.02	**0.13**	0.02	**0.14**	0.03	**0.16**	0.02	**0.15**
Thoracic spine		**0.04**		**0.05**		**0.06**	0.01	**0.07**	0.01	**0.09**	0.01	**0.08**
Claviculae	0.08	**1.66**	0.10	**1.96**	0.12	**2.16**	0.14	**2.31**	0.17	**2.56**	0.15	**2.58**
Scapulae		**0.05**	0.01	**0.06**	0.01	**0.08**	0.02	**0.09**	0.02	**0.11**	0.02	**0.10**
Ribs		**0.10**	0.01	**0.12**	0.01	**0.13**	0.01	**0.14**	0.02	**0.16**	0.02	**0.16**
Oesophagus		**0.03**		**0.04**		**0.05**		**0.06**	0.01	**0.07**		**0.06**
Trachea	0.44	**0.48**	0.53	**0.58**	0.59	**0.65**	0.64	**0.70**	0.72	**0.79**	0.70	**0.77**
Lung		**0.02**		**0.03**		**0.03**		**0.03**		**0.04**		**0.04**
Mamma												
Heart						**0.01**		**0.01**		**0.02**		**0.01**
Stomach												
Spleen												
Liver												
Gall bladder												
Pancreas												
Small intestine												
Large intestine												
Sigma/rectum												
Lumbar spine												
Pelvis												
Adrenals												
Kidneys												
Urinary bladder												
Uterus												
Ovaries												
Prostate												
Testes												
Upper arm	0.01	**0.04**	0.02	**0.06**	0.02	**0.07**	0.02	**0.08**	0.03	**0.10**	0.02	**0.09**
Elbow												
Forearm												
Upper leg												
Knee												
Lower leg												
Lymph nodes	0.09	**0.14**	0.11	**0.17**	0.13	**0.19**	0.14	**0.20**	0.15	**0.23**	0.15	**0.22**
Skin	0.13	**0.18**	0.14	**0.19**	0.14	**0.20**	0.14	**0.20**	0.15	**0.21**	0.15	**0.21**
Muscle	0.04	**0.06**	0.04	**0.07**	0.05	**0.08**	0.05	**0.08**	0.06	**0.09**	0.06	**0.09**
Skeleton	0.22	**0.27**	0.27	**0.32**	0.30	**0.36**	0.32	**0.39**	0.36	**0.44**	0.36	**0.43**
Red bone marrow	0.09	**0.10**	0.11	**0.12**	0.12	**0.14**	0.13	**0.15**	0.15	**0.17**	0.14	**0.16**
Whole body dose	0.07	**0.10**	0.09	**0.11**	0.10	**0.13**	0.11	**0.14**	0.13	**0.16**	0.12	**0.15**
ED (ICRP 60)	0.02	**0.07**	0.02	**0.08**	0.03	**0.08**	0.03	**0.09**	0.03	**0.10**	0.03	**0.10**
ED (ICRP 103)	0.03	**0.07**	0.04	**0.08**	0.04	**0.09**	0.05	**0.10**	0.05	**0.11**	0.05	**0.11**
Total filtration (mm) Al	**1.5**		**2.5**		**3.5**		**4.5**		**3.5**		**3.5**	
Cu	**0**		**0**		**0**		**0**		**0.1**		**0.2**	

Table 8.20 Normalized organ doses—Skull ap—1–2 years

Beam projection:			ap				**Table cells:**					
Focus-to-detector distance (cm):			115				(+) = normal font = optimal field setting					
Tube voltage (kVp):			70				**(−) = bold font = suboptimal field setting**					
Total filtrations:			See last table row.				Empty table cells correspond to values <0.01					
Organ or tissue	**Organ dose D_T: Entrance dose K_E**							**(mSv * mGy^{-1})**				
	(+)	(−)	(+)	(−)	(+)	(−)	(+)	(−)	(+)	(−)	(+)	(−)
Skull	1.09	**1.10**	1.28	**1.29**	1.40	**1.41**	1.49	**1.51**	1.65	**1.67**	1.78	**1.80**
Brain	0.24	**0.24**	0.30	**0.30**	0.34	**0.34**	0.38	**0.38**	0.45	**0.45**	0.51	**0.51**
Oral mucosa	0.52	**0.54**	0.61	**0.63**	0.67	**0.69**	0.71	**0.74**	0.80	**0.83**	0.87	**0.90**
Salivary glands	0.39	**0.42**	0.47	**0.50**	0.52	**0.56**	0.56	**0.61**	0.64	**0.69**	0.70	**0.75**
Cervical spine	0.25	**0.49**	0.32	**0.63**	0.38	**0.73**	0.43	**0.81**	0.53	**0.98**	0.62	**1.13**
Thyroid	0.09	**0.94**	0.11	**1.05**	0.12	**1.12**	0.14	**1.17**	0.16	**1.26**	0.18	**1.33**
Thymus	0.02	**0.10**	0.02	**0.13**	0.02	**0.14**	0.03	**0.15**	0.03	**0.18**	0.04	**0.20**
Thoracic spine		**0.05**	0.01	**0.07**	0.01	**0.08**	0.01	**0.09**	0.02	**0.11**	0.02	**0.13**
Claviculae	0.10	**1.80**	0.12	**2.09**	0.14	**2.29**	0.16	**2.42**	0.19	**2.65**	0.22	**2.80**
Scapulae	0.01	**0.06**	0.01	**0.08**	0.02	**0.09**	0.02	**0.11**	0.03	**0.13**	0.03	**0.15**
Ribs	0.01	**0.11**	0.01	**0.13**	0.02	**0.15**	0.02	**0.16**	0.02	**0.18**	0.03	**0.19**
Oesophagus		**0.04**		**0.05**		**0.06**	0.01	**0.07**	0.01	**0.08**	0.02	**0.10**
Trachea	0.49	**0.54**	0.58	**0.64**	0.65	**0.71**	0.69	**0.76**	0.78	**0.85**	0.84	**0.93**
Lung		**0.02**		**0.03**		**0.04**		**0.04**	0.01	**0.05**	0.01	**0.06**
Mamma										**0.01**		**0.01**
Heart				**0.01**		**0.01**		**0.02**		**0.02**		**0.03**
Stomach												
Spleen												
Liver												
Gall bladder												
Pancreas												
Small intestine												
Large intestine												
Sigma/rectum												
Lumbar spine												
Pelvis												
Adrenals												
Kidneys												
Urinary bladder												
Uterus												
Ovaries												
Prostate												
Testes												
Upper arm	0.02	**0.06**	0.02	**0.07**	0.02	**0.08**	0.03	**0.09**	0.03	**0.12**	0.04	**0.13**
Elbow								**0.01**		**0.01**		**0.02**
Forearm												
Upper leg												
Knee												
Lower leg												
Lymph nodes	0.11	**0.16**	0.13	**0.19**	0.14	**0.21**	0.15	**0.22**	0.17	**0.25**	0.19	**0.27**
Skin	0.13	**0.19**	0.14	**0.20**	0.15	**0.21**	0.15	**0.21**	0.16	**0.22**	0.17	**0.23**
Muscle	0.04	**0.06**	0.05	**0.08**	0.05	**0.08**	0.06	**0.09**	0.07	**0.10**	0.07	**0.11**
Skeleton	0.25	**0.30**	0.29	**0.35**	0.32	**0.39**	0.35	**0.42**	0.39	**0.47**	0.42	**0.51**
Red bone marrow	0.10	**0.11**	0.12	**0.14**	0.14	**0.15**	0.15	**0.17**	0.17	**0.19**	0.18	**0.21**
Whole body dose	0.09	**0.11**	0.10	**0.13**	0.11	**0.14**	0.12	**0.15**	0.14	**0.18**	0.15	**0.19**
ED (ICRP 60)	0.02	**0.07**	0.03	**0.08**	0.03	**0.09**	0.03	**0.10**	0.04	**0.11**	0.04	**0.11**
ED (ICRP 103)	0.04	**0.08**	0.04	**0.09**	0.05	**0.10**	0.05	**0.11**	0.06	**0.12**	0.07	**0.13**
Total filtration (mm) **Al**	**1.5**		**2.5**		**3.5**		**4.5**		**3.5**		**3.5**	
Cu	**0**		**0**		**0**		**0**		**0.1**		**0.2**	

Table 8.21 Normalized organ doses—Skull ap—3–7 years

Beam projection:				ap				**Table cells:**					
Focus-to-detector distance (cm):				115				(+) = normal font = optimal field setting					
Tube voltage (kVp):				60				**(−) = bold font = suboptimal field setting**					
Total filtrations:				See last table row.				Empty table cells correspond to values <0.01					
Organ or tissue		**Organ dose D_T: Entrance dose K_E**								**(mSv * mGy^{-1})**			
		(+)	(−)	(+)	(−)	(+)	(−)	(+)	(−)	(+)	(−)	(+)	(−)
Skull		0.77	**0.77**	0.93	**0.93**	1.04	**1.05**	1.13	**1.14**	1.29	**1.31**	1.26	**1.27**
Brain		0.13	**0.13**	0.18	**0.18**	0.21	**0.21**	0.24	**0.24**	0.29	**0.29**	0.26	**0.26**
Oral mucosa		0.36	**0.37**	0.43	**0.45**	0.48	**0.50**	0.52	**0.54**	0.59	**0.62**	0.57	**0.60**
Salivary glands		0.25	**0.27**	0.31	**0.34**	0.35	**0.39**	0.39	**0.42**	0.45	**0.49**	0.43	**0.47**
Cervical spine		0.12	**0.33**	0.16	**0.43**	0.19	**0.51**	0.22	**0.58**	0.28	**0.71**	0.24	**0.64**
Thyroid		0.04	**0.83**	0.06	**0.95**	0.07	**1.02**	0.07	**1.08**	0.09	**1.17**	0.08	**1.16**
Thymus			**0.05**		**0.06**		**0.07**		**0.08**	0.01	**0.10**		**0.09**
Thoracic spine			**0.03**		**0.04**		**0.05**		**0.06**		**0.07**		**0.06**
Claviculae		0.05	**1.40**	0.06	**1.68**	0.08	**1.87**	0.09	**2.01**	0.11	**2.25**	0.10	**2.25**
Scapulae			**0.03**		**0.04**		**0.05**		**0.06**	0.01	**0.07**		**0.06**
Ribs			**0.09**		**0.11**		**0.12**		**0.13**	0.01	**0.15**		**0.14**
Oesophagus			**0.02**		**0.03**		**0.03**		**0.04**		**0.04**		**0.04**
Trachea		0.25	**0.30**	0.31	**0.38**	0.36	**0.44**	0.40	**0.48**	0.47	**0.57**	0.45	**0.54**
Lung			**0.02**		**0.02**		**0.03**		**0.03**		**0.04**		**0.04**
Mamma													
Heart											**0.01**		
Stomach													
Spleen													
Liver													
Gall bladder													
Pancreas													
Small intestine													
Large intestine													
Sigma/rectum													
Lumbar spine													
Pelvis													
Adrenals													
Kidneys													
Urinary bladder													
Uterus													
Ovaries													
Prostate													
Testes													
Upper arm			**0.02**		**0.03**		**0.03**	0.01	**0.04**	0.01	**0.05**	0.01	**0.04**
Elbow													
Forearm													
Upper leg													
Knee													
Lower leg													
Lymph nodes		0.06	**0.11**	0.07	**0.13**	0.08	**0.15**	0.09	**0.16**	0.11	**0.18**	0.10	**0.18**
Skin		0.09	**0.13**	0.10	**0.14**	0.10	**0.14**	0.10	**0.15**	0.11	**0.16**	0.10	**0.15**
Muscle		0.02	**0.04**	0.02	**0.04**	0.03	**0.05**	0.03	**0.05**	0.03	**0.06**	0.03	**0.06**
Skeleton		0.20	**0.23**	0.24	**0.28**	0.27	**0.32**	0.30	**0.35**	0.34	**0.40**	0.33	**0.38**
Red bone marrow		0.05	**0.05**	0.06	**0.07**	0.06	**0.08**	0.07	**0.08**	0.08	**0.10**	0.08	**0.09**
Whole body dose		0.05	**0.07**	0.06	**0.09**	0.07	**0.10**	0.08	**0.10**	0.09	**0.12**	0.09	**0.12**
ED (ICRP 60)		0.01	**0.06**	0.01	**0.07**	0.02	**0.07**	0.02	**0.08**	0.02	**0.08**	0.02	**0.08**
ED (ICRP 103)		0.02	**0.06**	0.02	**0.07**	0.03	**0.08**	0.03	**0.08**	0.04	**0.09**	0.03	**0.09**
Total filtration (mm)	**Al**	**1.5**		**2.5**		**3.5**		**4.5**		**3.5**		**3.5**	
	Cu	**0**		**0**		**0**		**0**		**0.1**		**0.2**	

Table 8.22 Normalized organ doses—Skull ap—3–7 years

Beam projection:			ap			**Table cells:**						
Focus-to-detector distance (cm):			115			(+) = normal font = optimal field setting						
Tube voltage (kVp):			70			**(−) = bold font = suboptimal field setting**						
Total filtrations:			See last table row.			Empty table cells correspond to values <0.01						
Organ or tissue	**Organ dose D_T: Entrance dose K_E**							**(mSv * mGy^{-1})**				
	(+)	(−)	(+)	(−)	(+)	(−)	(+)	(−)	(+)	(−)	(+)	(−)
Skull	0.86	**0.87**	1.03	**1.04**	1.15	**1.16**	1.23	**1.25**	1.40	**1.42**	1.53	**1.56**
Brain	0.17	**0.17**	0.22	**0.22**	0.26	**0.26**	0.29	**0.29**	0.35	**0.35**	0.40	**0.41**
Oral mucosa	0.41	**0.43**	0.48	**0.51**	0.54	**0.56**	0.58	**0.61**	0.66	**0.69**	0.73	**0.77**
Salivary glands	0.29	**0.32**	0.35	**0.39**	0.40	**0.44**	0.44	**0.48**	0.50	**0.55**	0.56	**0.62**
Cervical spine	0.16	**0.41**	0.21	**0.53**	0.25	**0.62**	0.28	**0.70**	0.36	**0.85**	0.43	**0.99**
Thyroid	0.05	**0.90**	0.07	**1.01**	0.08	**1.09**	0.09	**1.14**	0.11	**1.23**	0.13	**1.30**
Thymus		**0.06**		**0.07**	0.01	**0.08**	0.01	**0.09**	0.01	**0.11**	0.02	**0.13**
Thoracic spine		**0.04**		**0.05**		**0.06**		**0.07**	0.01	**0.09**	0.01	**0.11**
Claviculae	0.06	**1.54**	0.08	**1.81**	0.09	**2.00**	0.10	**2.13**	0.13	**2.35**	0.15	**2.52**
Scapulae		**0.04**		**0.05**		**0.06**	0.01	**0.07**	0.01	**0.09**	0.02	**0.11**
Ribs		**0.10**		**0.12**		**0.13**	0.01	**0.14**	0.01	**0.16**	0.02	**0.17**
Oesophagus		**0.03**		**0.03**		**0.04**		**0.04**		**0.06**		**0.07**
Trachea	0.29	**0.36**	0.36	**0.44**	0.42	**0.50**	0.46	**0.55**	0.54	**0.64**	0.61	**0.72**
Lung		**0.02**		**0.03**		**0.04**		**0.04**		**0.05**		**0.06**
Mamma												**0.01**
Heart						**0.01**		**0.01**		**0.02**		**0.02**
Stomach												
Spleen												
Liver												
Gall bladder												
Pancreas												
Small intestine												
Large intestine												
Sigma/rectum												
Lumbar spine												
Pelvis												
Adrenals												
Kidneys												
Urinary bladder												
Uterus												
Ovaries												
Prostate												
Testes												
Upper arm		**0.03**	0.01	**0.04**	0.01	**0.04**	0.01	**0.05**	0.02	**0.06**	0.02	**0.07**
Elbow												
Forearm												
Upper leg												
Knee												
Lower leg												
Lymph nodes	0.07	**0.12**	0.08	**0.15**	0.09	**0.16**	0.10	**0.18**	0.12	**0.20**	0.14	**0.22**
Skin	0.09	**0.14**	0.10	**0.14**	0.10	**0.15**	0.11	**0.15**	0.11	**0.16**	0.12	**0.17**
Muscle	0.02	**0.04**	0.03	**0.05**	0.03	**0.06**	0.03	**0.06**	0.04	**0.07**	0.04	**0.08**
Skeleton	0.23	**0.26**	0.27	**0.32**	0.30	**0.35**	0.33	**0.38**	0.37	**0.43**	0.41	**0.48**
Red bone marrow	0.05	**0.06**	0.06	**0.08**	0.07	**0.09**	0.08	**0.10**	0.09	**0.11**	0.10	**0.12**
Whole body dose	0.06	**0.08**	0.07	**0.10**	0.08	**0.11**	0.09	**0.12**	0.10	**0.14**	0.11	**0.15**
ED (ICRP 60)	0.01	**0.06**	0.02	**0.07**	0.02	**0.08**	0.02	**0.08**	0.02	**0.09**	0.03	**0.10**
ED (ICRP 103)	0.02	**0.07**	0.03	**0.08**	0.03	**0.08**	0.04	**0.09**	0.04	**0.10**	0.05	**0.11**
Total filtration (mm) Al	**1.5**		**2.5**		**3.5**		**4.5**		**3.5**		**3.5**	
Cu	**0**		**0**		**0**		**0**		**0.1**		**0.2**	

Table 8.23 Normalized organ doses—Skull ap—8–12 years

Beam projection:			ap		**Table cells:**							
Focus-to-detector distance (cm):			115		(+) = normal font = optimal field setting							
Tube voltage (kVp):			60		**(−) = bold font = suboptimal field setting**							
Total filtrations:			See last table row.		Empty table cells correspond to values <0.01							
Organ or tissue	**Organ dose D_T: Entrance dose K_E**						**(mSv * mGy^{-1})**					
	(+)	(−)	(+)	(−)	(+)	(−)	(+)	(−)	(+)	(−)	(+)	(−)
Skull	0.70	**0.71**	0.86	**0.86**	0.97	**0.98**	1.05	**1.06**	1.21	**1.23**	1.17	**1.18**
Brain	0.12	**0.12**	0.16	**0.16**	0.19	**0.19**	0.21	**0.21**	0.27	**0.26**	0.24	**0.24**
Oral mucosa	0.34	**0.35**	0.40	**0.42**	0.45	**0.47**	0.49	**0.51**	0.56	**0.58**	0.53	**0.55**
Salivary glands	0.23	**0.25**	0.29	**0.31**	0.33	**0.36**	0.36	**0.39**	0.42	**0.46**	0.40	**0.44**
Cervical spine	0.13	**0.33**	0.18	**0.44**	0.22	**0.52**	0.25	**0.59**	0.32	**0.73**	0.27	**0.65**
Thyroid	0.05	**0.78**	0.06	**0.90**	0.07	**0.98**	0.08	**1.04**	0.10	**1.13**	0.09	**1.12**
Thymus		**0.02**		**0.03**		**0.03**		**0.04**		**0.04**		**0.04**
Thoracic spine		**0.02**		**0.03**		**0.03**		**0.04**		**0.05**		**0.04**
Claviculae	0.03	**1.13**	0.04	**1.36**	0.05	**1.52**	0.06	**1.65**	0.07	**1.86**	0.07	**1.85**
Scapulae		**0.01**		**0.02**		**0.02**		**0.03**		**0.04**		**0.03**
Ribs		**0.07**		**0.09**		**0.09**		**0.10**		**0.11**		**0.11**
Oesophagus		**0.01**		**0.02**		**0.02**		**0.03**		**0.03**		**0.03**
Trachea	0.28	**0.42**	0.35	**0.52**	0.41	**0.60**	0.46	**0.66**	0.55	**0.77**	0.52	**0.74**
Lung		**0.01**		**0.02**		**0.02**		**0.02**		**0.03**		**0.02**
Mamma												
Heart												
Stomach												
Spleen												
Liver												
Gall bladder												
Pancreas												
Small intestine												
Large intestine												
Sigma/rectum												
Lumbar spine												
Pelvis												
Adrenals												
Kidneys												
Urinary bladder												
Uterus												
Ovaries												
Prostate												
Testes												
Upper arm		**0.01**		**0.01**		**0.02**		**0.02**		**0.03**		**0.02**
Elbow												
Forearm												
Upper leg												
Knee												
Lower leg												
Lymph nodes	0.06	**0.12**	0.07	**0.14**	0.09	**0.16**	0.10	**0.18**	0.11	**0.20**	0.11	**0.19**
Skin	0.07	**0.10**	0.07	**0.11**	0.07	**0.11**	0.08	**0.11**	0.08	**0.12**	0.08	**0.12**
Muscle	0.01	**0.03**	0.02	**0.03**	0.02	**0.03**	0.02	**0.04**	0.02	**0.04**	0.02	**0.04**
Skeleton	0.14	**0.17**	0.18	**0.21**	0.20	**0.23**	0.22	**0.25**	0.25	**0.30**	0.24	**0.28**
Red bone marrow	0.03	**0.04**	0.04	**0.05**	0.04	**0.05**	0.05	**0.06**	0.06	**0.07**	0.05	**0.07**
Whole body dose	0.04	**0.05**	0.04	**0.06**	0.05	**0.07**	0.05	**0.08**	0.06	**0.09**	0.06	**0.08**
ED (ICRP 60)		**0.05**	0.01	**0.06**	0.01	**0.06**	0.01	**0.07**	0.02	**0.08**	0.02	**0.07**
ED (ICRP 103)	0.02	**0.05**	0.02	**0.06**	0.02	**0.07**	0.03	**0.07**	0.03	**0.08**	0.03	**0.08**
Total filtration (mm) Al	**1.5**		**2.5**		**3.5**		**4.5**		**3.5**		**3.5**	
Cu	**0**		**0**		**0**		**0**		**0.1**		**0.2**	

Table 8.24 Normalized organ doses—Skull ap—8–12 years

Beam projection:				ap			**Table cells:**						
Focus-to-detector distance (cm):				115			(+) = normal font = optimal field setting						
Tube voltage (kVp):				70			**(−) = bold font = suboptimal field setting**						
Total filtrations:				See last table row.			Empty table cells correspond to values <0.01						
Organ or tissue		**Organ dose D_T: Entrance dose K_E**							**($mSv * mGy^{-1}$)**				
		(+)	(−)	(+)	(−)	(+)	(−)	(+)	(−)	(+)	(−)	(+)	(−)
Skull		0.80	**0.81**	0.96	**0.97**	1.07	**1.09**	1.16	**1.18**	1.32	**1.34**	1.46	**1.48**
Brain		0.15	**0.15**	0.20	**0.20**	0.23	**0.23**	0.26	**0.26**	0.32	**0.32**	0.38	**0.38**
Oral mucosa		0.38	**0.40**	0.46	**0.47**	0.51	**0.53**	0.55	**0.57**	0.63	**0.65**	0.70	**0.73**
Salivary glands		0.27	**0.30**	0.33	**0.36**	0.37	**0.41**	0.41	**0.45**	0.47	**0.52**	0.53	**0.58**
Cervical spine		0.18	**0.42**	0.24	**0.55**	0.28	**0.64**	0.32	**0.72**	0.40	**0.88**	0.48	**1.03**
Thyroid		0.06	**0.85**	0.08	**0.97**	0.09	**1.05**	0.10	**1.10**	0.12	**1.20**	0.13	**1.28**
Thymus			**0.03**		**0.03**		**0.04**		**0.04**		**0.05**		**0.07**
Thoracic spine			**0.03**		**0.04**		**0.05**		**0.05**		**0.07**		**0.08**
Claviculae		0.04	**1.25**	0.06	**1.48**	0.07	**1.64**	0.07	**1.76**	0.09	**1.95**	0.11	**2.10**
Scapulae			**0.02**		**0.03**		**0.03**		**0.04**		**0.05**	0.01	**0.06**
Ribs			**0.08**		**0.09**		**0.10**		**0.11**		**0.12**		**0.13**
Oesophagus			**0.02**		**0.03**		**0.03**		**0.03**		**0.04**		**0.05**
Trachea		0.34	**0.49**	0.42	**0.61**	0.49	**0.69**	0.54	**0.75**	0.64	**0.88**	0.73	**0.99**
Lung			**0.02**		**0.02**		**0.02**		**0.03**		**0.03**		**0.04**
Mamma													
Heart													**0.01**
Stomach													
Spleen													
Liver													
Gall bladder													
Pancreas													
Small intestine													
Large intestine													
Sigma/rectum													
Lumbar spine													
Pelvis													
Adrenals													
Kidneys													
Urinary bladder													
Uterus													
Ovaries													
Prostate													
Testes													
Upper arm			**0.01**		**0.02**		**0.02**		**0.03**	0.01	**0.03**	0.01	**0.04**
Elbow													
Forearm													
Upper leg													
Knee													
Lower leg													
Lymph nodes		0.07	**0.14**	0.09	**0.16**	0.10	**0.18**	0.11	**0.20**	0.13	**0.22**	0.15	**0.25**
Skin		0.07	**0.10**	0.07	**0.11**	0.08	**0.12**	0.08	**0.12**	0.08	**0.13**	0.09	**0.13**
Muscle		0.01	**0.03**	0.02	**0.04**	0.02	**0.04**	0.02	**0.04**	0.02	**0.05**	0.03	**0.06**
Skeleton		0.17	**0.19**	0.20	**0.23**	0.22	**0.26**	0.24	**0.28**	0.28	**0.32**	0.31	**0.36**
Red bone marrow		0.04	**0.05**	0.04	**0.06**	0.05	**0.06**	0.06	**0.07**	0.06	**0.08**	0.07	**0.09**
Whole body dose		0.04	**0.06**	0.05	**0.07**	0.06	**0.08**	0.06	**0.09**	0.07	**0.10**	0.08	**0.11**
ED (ICRP 60)		0.01	**0.05**	0.01	**0.06**	0.01	**0.07**	0.02	**0.07**	0.02	**0.08**	0.02	**0.09**
ED (ICRP 103)		0.02	**0.06**	0.02	**0.07**	0.03	**0.08**	0.03	**0.08**	0.04	**0.09**	0.04	**0.10**
Total filtration (mm)	**Al**	**1.5**		**2.5**		**3.5**		**4.5**		**3.5**		**3.5**	
	Cu	**0**		**0**		**0**		**0**		**0.1**		**0.2**	

Table 8.25 Normalized organ doses—Skull ap—8–12 years

Beam projection:			ap			**Table cells:**						
Focus-to-detector distance (cm):			115			(+) = normal font = optimal field setting						
Tube voltage (kVp):			80			**(−) = bold font = suboptimal field setting**						
Total filtrations:			See last table row.			Empty table cells correspond to values <0.01						
Organ or tissue	**Organ dose D_T: Entrance dose K_E**							**(mSv * mGy^{-1})**				
	(+)	(−)	(+)	(−)	(+)	(−)	(+)	(−)	(+)	(−)	(+)	(−)
Skull	0.89	**0.90**	1.05	**1.06**	1.16	**1.17**	1.24	**1.26**	1.40	**1.42**	1.52	**1.54**
Brain	0.19	**0.19**	0.24	**0.24**	0.28	**0.28**	0.31	**0.31**	0.37	**0.37**	0.43	**0.43**
Oral mucosa	0.43	**0.45**	0.51	**0.53**	0.56	**0.59**	0.60	**0.63**	0.68	**0.71**	0.75	**0.79**
Salivary glands	0.31	**0.34**	0.37	**0.41**	0.42	**0.46**	0.45	**0.50**	0.52	**0.57**	0.58	**0.63**
Cervical spine	0.23	**0.52**	0.29	**0.65**	0.34	**0.75**	0.39	**0.83**	0.47	**1.00**	0.56	**1.15**
Thyroid	0.07	**0.91**	0.09	**1.03**	0.10	**1.10**	0.11	**1.16**	0.13	**1.25**	0.15	**1.33**
Thymus		**0.03**		**0.04**		**0.05**		**0.05**		**0.06**	0.01	**0.07**
Thoracic spine		**0.04**		**0.05**		**0.06**		**0.06**		**0.08**	0.01	**0.10**
Claviculae	0.05	**1.36**	0.07	**1.57**	0.08	**1.72**	0.09	**1.82**	0.10	**2.00**	0.12	**2.12**
Scapulae		**0.03**		**0.03**		**0.04**		**0.05**	0.01	**0.06**	0.01	**0.07**
Ribs		**0.09**		**0.10**		**0.11**		**0.12**		**0.13**	0.01	**0.14**
Oesophagus		**0.03**		**0.03**		**0.04**		**0.04**		**0.05**		**0.06**
Trachea	0.40	**0.57**	0.49	**0.68**	0.55	**0.77**	0.61	**0.84**	0.71	**0.96**	0.80	**1.07**
Lung		**0.02**		**0.02**		**0.03**		**0.03**		**0.04**		**0.04**
Mamma												
Heart										**0.01**		**0.01**
Stomach												
Spleen												
Liver												
Gall bladder												
Pancreas												
Small intestine												
Large intestine												
Sigma/rectum												
Lumbar spine												
Pelvis												
Adrenals												
Kidneys												
Urinary bladder												
Uterus												
Ovaries												
Prostate												
Testes												
Upper arm		**0.02**		**0.02**	0.01	**0.03**	0.01	**0.03**	0.01	**0.04**	0.02	**0.05**
Elbow												
Forearm												
Upper leg												
Knee												
Lower leg												
Lymph nodes	0.08	**0.15**	0.10	**0.18**	0.11	**0.20**	0.12	**0.22**	0.14	**0.24**	0.16	**0.27**
Skin	0.07	**0.11**	0.08	**0.12**	0.08	**0.12**	0.08	**0.12**	0.09	**0.13**	0.09	**0.14**
Muscle	0.02	**0.03**	0.02	**0.04**	0.02	**0.04**	0.02	**0.05**	0.03	**0.05**	0.03	**0.06**
Skeleton	0.18	**0.22**	0.22	**0.26**	0.24	**0.28**	0.26	**0.31**	0.29	**0.35**	0.32	**0.38**
Red bone marrow	0.04	**0.05**	0.05	**0.06**	0.06	**0.07**	0.06	**0.08**	0.07	**0.09**	0.08	**0.10**
Whole body dose	0.05	**0.07**	0.06	**0.08**	0.06	**0.09**	0.07	**0.09**	0.08	**0.11**	0.09	**0.12**
ED (ICRP 60)	0.01	**0.06**	0.01	**0.07**	0.02	**0.07**	0.02	**0.08**	0.02	**0.09**	0.02	**0.09**
ED (ICRP 103)	0.02	**0.07**	0.03	**0.08**	0.03	**0.08**	0.03	**0.09**	0.04	**0.10**	0.05	**0.11**
Total filtration (mm) **Al**	**1.5**		**2.5**		**3.5**		**4.5**		**3.5**		**3.5**	
Cu	**0**		**0**		**0**		**0**		**0.1**		**0.2**	

Table 8.26 Normalized organ doses—Skull ap—13–17 years

Beam projection:				ap			**Table cells:**					
Focus-to-detector distance (cm):				115			(+) = normal font = optimal field setting					
Tube voltage (kVp):				70			**(−) = bold font = suboptimal field setting**					
Total filtrations:				See last table row.			Empty table cells correspond to values <0.01					
Organ or tissue	**Organ dose D_T: Entrance dose K_E**							**($mSv * mGy^{-1}$)**				
	(+)	(−)	(+)	(−)	(+)	(−)	(+)	(−)	(+)	(−)	(+)	(−)
Skull	0.70	**0.70**	0.85	**0.85**	0.96	**0.97**	1.04	**1.05**	1.20	**1.22**	1.34	**1.36**
Brain	0.14	**0.14**	0.18	**0.18**	0.21	**0.21**	0.24	**0.24**	0.29	**0.29**	0.35	**0.34**
Oral mucosa	0.34	**0.35**	0.40	**0.42**	0.45	**0.47**	0.49	**0.51**	0.56	**0.59**	0.63	**0.67**
Salivary glands	0.24	**0.26**	0.29	**0.33**	0.33	**0.37**	0.36	**0.41**	0.42	**0.47**	0.48	**0.54**
Cervical spine	0.12	**0.50**	0.16	**0.64**	0.19	**0.76**	0.22	**0.85**	0.28	**1.05**	0.34	**1.24**
Thyroid	0.02	**0.83**	0.02	**0.95**	0.02	**1.04**	0.03	**1.10**	0.03	**1.20**	0.04	**1.28**
Thymus		**0.01**		**0.02**		**0.02**		**0.02**		**0.03**		**0.03**
Thoracic spine		**0.02**		**0.02**		**0.03**		**0.03**		**0.05**		**0.06**
Claviculae	0.02	**0.89**	0.03	**1.06**	0.03	**1.18**	0.04	**1.27**	0.04	**1.43**	0.05	**1.56**
Scapulae				**0.01**		**0.02**		**0.02**		**0.02**		**0.03**
Ribs		**0.06**		**0.07**		**0.08**		**0.08**		**0.09**		**0.10**
Oesophagus		**0.02**		**0.02**		**0.03**		**0.03**		**0.04**		**0.05**
Trachea	0.14	**0.37**	0.18	**0.45**	0.21	**0.51**	0.24	**0.55**	0.29	**0.64**	0.34	**0.72**
Lung				**0.01**		**0.01**		**0.01**		**0.02**		**0.02**
Mamma		**0.02**		**0.02**		**0.03**		**0.03**		**0.04**	0.01	**0.05**
Heart												
Stomach												
Spleen												
Liver												
Gall bladder												
Pancreas												
Small intestine												
Large intestine												
Sigma/rectum												
Lumbar spine												
Pelvis												
Adrenals												
Kidneys												
Urinary bladder												
Uterus												
Ovaries												
Prostate												
Testes												
Upper arm				**0.01**		**0.01**		**0.01**		**0.02**		**0.02**
Elbow												
Forearm												
Upper leg												
Knee												
Lower leg												
Lymph nodes	0.04	**0.11**	0.05	**0.14**	0.06	**0.15**	0.06	**0.16**	0.08	**0.19**	0.09	**0.21**
Skin	0.05	**0.09**	0.05	**0.09**	0.06	**0.10**	0.06	**0.10**	0.06	**0.10**	0.06	**0.11**
Muscle		**0.02**	0.01	**0.03**	0.01	**0.03**	0.01	**0.03**	0.01	**0.04**	0.02	**0.04**
Skeleton	0.11	**0.14**	0.13	**0.17**	0.15	**0.19**	0.16	**0.20**	0.19	**0.24**	0.21	**0.27**
Red bone marrow	0.03	**0.03**	0.03	**0.04**	0.04	**0.05**	0.04	**0.05**	0.05	**0.06**	0.05	**0.07**
Whole body dose	0.03	**0.04**	0.03	**0.05**	0.04	**0.06**	0.04	**0.06**	0.05	**0.07**	0.05	**0.08**
ED (ICRP 60)		**0.05**		**0.06**		**0.07**		**0.07**	0.01	**0.08**	0.01	**0.09**
ED (ICRP 103)	0.01	**0.05**	0.02	**0.06**	0.02	**0.07**	0.02	**0.08**	0.03	**0.09**	0.03	**0.10**
Total filtration (mm) **Al**	**1.5**		**2.5**		**3.5**		**4.5**		**3.5**		**3.5**	
Cu	**0**		**0**		**0**		**0**		**0.1**		**0.2**	

Table 8.27 Normalized organ doses—Skull ap—13–17 years

Beam projection:			ap			**Table cells:**						
Focus-to-detector distance (cm):			115			(+) = normal font = optimal field setting						
Tube voltage (kVp):			80			**(−) = bold font = suboptimal field setting**						
Total filtrations:			See last table row.			Empty table cells correspond to values <0.01						
Organ or tissue	**Organ dose D_T: Entrance dose K_E**							**(mSv * mGy^{-1})**				
	(+)	(−)	(+)	(−)	(+)	(−)	(+)	(−)	(+)	(−)	(+)	(−)
Skull	0.79	**0.79**	0.94	**0.95**	1.05	**1.06**	1.13	**1.15**	1.29	**1.31**	1.42	**1.44**
Brain	0.17	**0.17**	0.22	**0.22**	0.25	**0.25**	0.28	**0.28**	0.34	**0.34**	0.40	**0.40**
Oral mucosa	0.38	**0.40**	0.45	**0.48**	0.50	**0.53**	0.54	**0.57**	0.62	**0.66**	0.69	**0.73**
Salivary glands	0.27	**0.31**	0.33	**0.37**	0.37	**0.42**	0.41	**0.46**	0.47	**0.53**	0.52	**0.59**
Cervical spine	0.16	**0.61**	0.20	**0.77**	0.24	**0.89**	0.27	**1.00**	0.34	**1.20**	0.41	**1.39**
Thyroid	0.02	**0.89**	0.02	**1.02**	0.03	**1.10**	0.03	**1.15**	0.04	**1.26**	0.05	**1.34**
Thymus		**0.02**		**0.02**		**0.02**		**0.03**		**0.03**		**0.04**
Thoracic spine		**0.03**		**0.03**		**0.04**		**0.05**		**0.06**		**0.07**
Claviculae	0.03	**0.97**	0.03	**1.14**	0.04	**1.25**	0.04	**1.34**	0.05	**1.48**	0.06	**1.58**
Scapulae		**0.01**		**0.02**		**0.02**		**0.02**		**0.03**		**0.04**
Ribs		**0.07**		**0.08**		**0.09**		**0.09**		**0.10**		**0.11**
Oesophagus		**0.02**		**0.03**		**0.03**		**0.04**		**0.05**		**0.06**
Trachea	0.17	**0.42**	0.22	**0.51**	0.25	**0.57**	0.28	**0.62**	0.34	**0.71**	0.39	**0.79**
Lung		**0.01**		**0.01**		**0.02**		**0.02**		**0.02**		**0.03**
Mamma		**0.02**		**0.03**		**0.04**		**0.04**	0.01	**0.05**	0.01	**0.06**
Heart												
Stomach												
Spleen												
Liver												
Gall bladder												
Pancreas												
Small intestine												
Large intestine												
Sigma/rectum												
Lumbar spine												
Pelvis												
Adrenals												
Kidneys												
Urinary bladder												
Uterus												
Ovaries												
Prostate												
Testes												
Upper arm		**0.01**		**0.01**		**0.02**		**0.02**		**0.02**		**0.03**
Elbow												
Forearm												
Upper leg												
Knee												
Lower leg												
Lymph nodes	0.05	**0.13**	0.06	**0.15**	0.07	**0.17**	0.07	**0.18**	0.09	**0.21**	0.10	**0.23**
Skin	0.05	**0.09**	0.06	**0.10**	0.06	**0.10**	0.06	**0.10**	0.06	**0.11**	0.07	**0.11**
Muscle	0.01	**0.03**	0.01	**0.03**	0.01	**0.04**	0.01	**0.04**	0.02	**0.04**	0.02	**0.05**
Skeleton	0.12	**0.15**	0.15	**0.18**	0.16	**0.21**	0.18	**0.22**	0.20	**0.26**	0.22	**0.28**
Red bone marrow	0.03	**0.04**	0.04	**0.05**	0.04	**0.06**	0.05	**0.06**	0.05	**0.07**	0.06	**0.08**
Whole body dose	0.03	**0.05**	0.04	**0.06**	0.04	**0.06**	0.04	**0.07**	0.05	**0.08**	0.06	**0.09**
ED (ICRP 60)		**0.06**		**0.07**	0.01	**0.07**	0.01	**0.08**	0.01	**0.08**	0.02	**0.09**
ED (ICRP 103)	0.02	**0.06**	0.02	**0.07**	0.02	**0.08**	0.02	**0.08**	0.03	**0.10**	0.03	**0.11**
Total filtration (mm) Al	**1.5**		**2.5**		**3.5**		**4.5**		**3.5**		**3.5**	
Cu	**0**		**0**		**0**		**0**		**0.1**		**0.2**	

Table 8.28 Normalized organ doses—Skull ap—> 17 years

Beam projection:				ap				**Table cells:**					
Focus-to-detector distance (cm):				115				(+) = normal font = optimal field setting					
Tube voltage (kVp):				70				**(−) = bold font = suboptimal field setting**					
Total filtrations:				See last table row.				Empty table cells correspond to values <0.01					
Organ or tissue		**Organ dose D_T: Entrance dose K_E**							**(mSv * mGy^{-1})**				
		(+)	(−)	(+)	(−)	(+)	(−)	(+)	(−)	(+)	(−)	(+)	(−)
Skull		0.63	**0.63**	0.77	**0.78**	0.87	**0.88**	0.95	**0.97**	1.11	**1.12**	1.24	**1.27**
Brain		0.12	**0.12**	0.16	**0.16**	0.19	**0.19**	0.22	**0.21**	0.27	**0.27**	0.32	**0.32**
Oral mucosa		0.32	**0.34**	0.38	**0.40**	0.43	**0.45**	0.47	**0.49**	0.53	**0.56**	0.60	**0.63**
Salivary glands		0.22	**0.25**	0.27	**0.31**	0.31	**0.35**	0.34	**0.38**	0.39	**0.45**	0.44	**0.51**
Cervical spine		0.09	**0.44**	0.12	**0.57**	0.15	**0.67**	0.17	**0.76**	0.22	**0.94**	0.28	**1.11**
Thyroid		0.01	**0.78**	0.02	**0.90**	0.02	**0.99**	0.02	**1.05**	0.03	**1.16**	0.03	**1.24**
Thymus							**0.01**		**0.01**		**0.02**		**0.02**
Thoracic spine			**0.02**		**0.02**		**0.03**		**0.03**		**0.04**		**0.05**
Claviculae		0.02	**0.82**	0.02	**0.98**	0.03	**1.09**	0.03	**1.17**	0.04	**1.32**	0.04	**1.43**
Scapulae					**0.01**		**0.01**		**0.01**		**0.02**		**0.02**
Ribs			**0.05**		**0.06**		**0.07**		**0.07**		**0.08**		**0.09**
Oesophagus			**0.02**		**0.02**		**0.03**		**0.03**		**0.04**		**0.05**
Trachea		0.10	**0.33**	0.13	**0.41**	0.16	**0.46**	0.18	**0.51**	0.22	**0.59**	0.27	**0.67**
Lung					**0.01**		**0.01**		**0.02**		**0.02**		**0.02**
Mamma													
Heart													
Stomach													
Spleen													
Liver													
Gall bladder													
Pancreas													
Small intestine													
Large intestine													
Sigma/rectum													
Lumbar spine													
Pelvis													
Adrenals													
Kidneys													
Urinary bladder													
Uterus													
Ovaries													
Prostate													
Testes													
Upper arm											**0.01**		**0.01**
Elbow													
Forearm													
Upper leg													
Knee													
Lower leg													
Lymph nodes		0.03	**0.11**	0.04	**0.13**	0.05	**0.14**	0.05	**0.15**	0.06	**0.18**	0.07	**0.20**
Skin		0.04	**0.08**	0.05	**0.08**	0.05	**0.09**	0.05	**0.09**	0.05	**0.09**	0.06	**0.10**
Muscle			**0.02**		**0.02**		**0.03**	0.01	**0.03**	0.01	**0.03**	0.01	**0.04**
Skeleton		0.09	**0.11**	0.11	**0.14**	0.13	**0.16**	0.14	**0.18**	0.16	**0.20**	0.18	**0.23**
Red bone marrow		0.02	**0.03**	0.02	**0.03**	0.03	**0.04**	0.03	**0.04**	0.04	**0.05**	0.04	**0.06**
Whole body dose		0.02	**0.04**	0.03	**0.04**	0.03	**0.05**	0.03	**0.05**	0.04	**0.06**	0.04	**0.07**
ED (ICRP 60)			**0.05**		**0.05**		**0.06**		**0.06**		**0.07**	0.01	**0.08**
ED (ICRP 103)		0.01	**0.05**	0.01	**0.06**	0.02	**0.06**	0.02	**0.07**	0.02	**0.08**	0.02	**0.09**
Total filtration (mm)	**Al**	**1.5**		**2.5**		**3.5**		**4.5**		**3.5**		**3.5**	
	Cu	**0**		**0**		**0**		**0**		**0.1**		**0.2**	

Table 8.29 Normalized organ doses—Skull ap—> 17 years

Beam projection:				ap			**Table cells:**						
Focus-to-detector distance (cm):				115			(+) = normal font = optimal field setting						
Tube voltage (kVp):				80			**(−) = bold font = suboptimal field setting**						
Total filtrations:				See last table row.			Empty table cells correspond to values <0.01						
Organ or tissue		**Organ dose D_T: Entrance dose K_E**							**(mSv * mGy^{-1})**				
		(+)	(−)	(+)	(−)	(+)	(−)	(+)	(−)	(+)	(−)	(+)	(−)
Skull		0.71	**0.72**	0.86	**0.87**	0.96	**0.98**	1.04	**1.06**	1.19	**1.22**	1.32	**1.35**
Brain		0.16	**0.16**	0.20	**0.20**	0.23	**0.23**	0.26	**0.26**	0.32	**0.32**	0.38	**0.37**
Oral mucosa		0.37	**0.38**	0.43	**0.45**	0.48	**0.51**	0.52	**0.55**	0.59	**0.62**	0.66	**0.70**
Salivary glands		0.25	**0.29**	0.31	**0.35**	0.35	**0.40**	0.38	**0.43**	0.44	**0.50**	0.49	**0.56**
Cervical spine		0.13	**0.55**	0.16	**0.69**	0.20	**0.80**	0.22	**0.89**	0.28	**1.08**	0.34	**1.26**
Thyroid		0.02	**0.85**	0.02	**0.97**	0.03	**1.05**	0.03	**1.11**	0.04	**1.21**	0.04	**1.30**
Thymus					**0.01**		**0.02**		**0.02**		**0.02**		**0.03**
Thoracic spine			**0.02**		**0.03**		**0.04**		**0.04**		**0.05**		**0.06**
Claviculae		0.02	**0.89**	0.03	**1.05**	0.03	**1.15**	0.04	**1.23**	0.04	**1.36**	0.05	**1.46**
Scapulae			**0.01**		**0.01**		**0.02**		**0.02**		**0.02**		**0.03**
Ribs			**0.06**		**0.07**		**0.07**		**0.08**		**0.09**		**0.09**
Oesophagus			**0.02**		**0.03**		**0.03**		**0.04**		**0.05**		**0.06**
Trachea		0.13	**0.39**	0.17	**0.47**	0.19	**0.52**	0.22	**0.57**	0.26	**0.66**	0.31	**0.74**
Lung			**0.01**		**0.01**		**0.02**		**0.02**		**0.02**		**0.03**
Mamma													
Heart													
Stomach													
Spleen													
Liver													
Gall bladder													
Pancreas													
Small intestine													
Large intestine													
Sigma/rectum													
Lumbar spine													
Pelvis													
Adrenals													
Kidneys													
Urinary bladder													
Uterus													
Ovaries													
Prostate													
Testes													
Upper arm							**0.01**		**0.01**		**0.02**		**0.02**
Elbow													
Forearm													
Upper leg													
Knee													
Lower leg													
Lymph nodes		0.04	**0.12**	0.05	**0.14**	0.06	**0.16**	0.06	**0.17**	0.07	**0.19**	0.08	**0.21**
Skin		0.05	**0.08**	0.05	**0.09**	0.05	**0.09**	0.05	**0.09**	0.06	**0.10**	0.06	**0.10**
Muscle			**0.02**		**0.03**	0.01	**0.03**	0.01	**0.03**	0.01	**0.04**	0.01	**0.04**
Skeleton		0.10	**0.13**	0.13	**0.16**	0.14	**0.18**	0.15	**0.19**	0.18	**0.22**	0.20	**0.25**
Red bone marrow		0.02	**0.03**	0.03	**0.04**	0.03	**0.05**	0.04	**0.05**	0.04	**0.06**	0.05	**0.07**
Whole body dose		0.03	**0.04**	0.03	**0.05**	0.03	**0.06**	0.04	**0.06**	0.04	**0.07**	0.05	**0.08**
ED (ICRP 60)			**0.05**		**0.06**		**0.07**		**0.07**	0.01	**0.08**	0.01	**0.08**
ED (ICRP 103)		0.01	**0.05**	0.02	**0.06**	0.02	**0.07**	0.02	**0.08**	0.02	**0.09**	0.03	**0.09**
Total filtration (mm)	**Al**	**1.5**		**2.5**		**3.5**		**4.5**		**3.5**		**3.5**	
	Cu	**0**		**0**		**0**		**0**		**0.1**		**0.2**	

Table 8.30 Normalized organ doses—Skull r-lat—< 1 year

Beam projection:			r-lat			**Table cells:**						
Focus-to-detector distance (cm):			115			(+) = normal font = optimal field setting						
Tube voltage (kVp):			60			**(−) = bold font = suboptimal field setting**						
Total filtrations:			See last table row.			Empty table cells correspond to values <0.01						
Organ or tissue	**Organ dose D_T: Entrance dose K_E**							**(mSv * mGy^{-1})**				
	(+)	(−)	(+)	(−)	(+)	(−)	(+)	(−)	(+)	(−)	(+)	(−)
Skull	1.38	**1.42**	1.60	**1.65**	1.74	**1.80**	1.85	**1.91**	2.02	**2.09**	2.03	**2.10**
Brain	0.38	**0.39**	0.47	**0.48**	0.53	**0.55**	0.58	**0.60**	0.66	**0.68**	0.65	**0.67**
Oral mucosa	0.43	**0.46**	0.51	**0.54**	0.57	**0.60**	0.61	**0.65**	0.68	**0.73**	0.67	**0.71**
Salivary glands	0.50	**0.53**	0.56	**0.61**	0.60	**0.65**	0.63	**0.69**	0.69	**0.75**	0.68	**0.73**
Cervical spine	0.62	**0.95**	0.79	**1.19**	0.91	**1.37**	1.00	**1.50**	1.17	**1.75**	1.14	**1.72**
Thyroid	0.14	**0.57**	0.18	**0.68**	0.20	**0.75**	0.22	**0.81**	0.26	**0.90**	0.25	**0.89**
Thymus	0.02	**0.09**	0.02	**0.12**	0.02	**0.13**	0.03	**0.15**	0.03	**0.18**	0.03	**0.17**
Thoracic spine	0.01	**0.11**	0.02	**0.13**	0.02	**0.16**	0.02	**0.18**	0.03	**0.21**	0.03	**0.20**
Claviculae	0.08	**0.92**	0.11	**1.10**	0.12	**1.23**	0.14	**1.33**	0.16	**1.49**	0.16	**1.49**
Scapulae	0.02	**0.23**	0.03	**0.27**	0.03	**0.29**	0.04	**0.31**	0.05	**0.35**	0.04	**0.34**
Ribs	0.01	**0.12**	0.02	**0.14**	0.02	**0.16**	0.02	**0.17**	0.03	**0.19**	0.03	**0.19**
Oesophagus		**0.04**		**0.06**	0.01	**0.06**	0.01	**0.07**	0.02	**0.09**	0.01	**0.08**
Trachea	0.33	**0.37**	0.40	**0.45**	0.45	**0.51**	0.49	**0.56**	0.56	**0.64**	0.55	**0.62**
Lung		**0.02**		**0.03**	0.01	**0.04**	0.01	**0.04**	0.01	**0.05**	0.01	**0.05**
Mamma												
Heart		**0.01**		**0.01**		**0.02**		**0.02**		**0.02**		**0.02**
Stomach												
Spleen												
Liver												
Gall bladder												
Pancreas												
Small intestine												
Large intestine												
Sigma/rectum												
Lumbar spine												
Pelvis												
Adrenals												
Kidneys												
Urinary bladder												
Uterus												
Ovaries												
Prostate												
Testes												
Upper arm	0.02	**0.31**	0.03	**0.36**	0.03	**0.39**	0.04	**0.41**	0.04	**0.45**	0.04	**0.44**
Elbow				**0.01**		**0.01**		**0.01**		**0.02**		**0.02**
Forearm												
Upper leg												
Knee												
Lower leg												
Lymph nodes	0.10	**0.13**	0.11	**0.16**	0.13	**0.17**	0.14	**0.19**	0.15	**0.21**	0.15	**0.20**
Skin	0.16	**0.23**	0.17	**0.24**	0.18	**0.25**	0.18	**0.26**	0.19	**0.27**	0.19	**0.27**
Muscle	0.06	**0.09**	0.07	**0.10**	0.07	**0.11**	0.08	**0.12**	0.09	**0.13**	0.08	**0.13**
Skeleton	0.33	**0.41**	0.38	**0.48**	0.42	**0.53**	0.45	**0.57**	0.50	**0.63**	0.50	**0.63**
Red bone marrow	0.16	**0.18**	0.19	**0.21**	0.20	**0.23**	0.22	**0.25**	0.24	**0.28**	0.24	**0.27**
Whole body dose	0.12	**0.15**	0.14	**0.17**	0.15	**0.19**	0.16	**0.21**	0.18	**0.23**	0.18	**0.23**
ED (ICRP 60)	0.03	**0.06**	0.04	**0.08**	0.04	**0.08**	0.05	**0.09**	0.05	**0.10**	0.05	**0.10**
ED (ICRP 103)	0.05	**0.07**	0.05	**0.09**	0.06	**0.10**	0.07	**0.10**	0.07	**0.12**	0.07	**0.11**
Total filtration (mm) Al	**1.5**		**2.5**		**3.5**		**4.5**		**3.5**		**3.5**	
Cu	**0**		**0**		**0**		**0**		**0.1**		**0.2**	

Table 8.31 Normalized organ doses—Skull r-lat—<1 year

Beam projection:				r-lat			**Table cells:**						
Focus-to-detector distance (cm):				115			(+) = normal font = optimal field setting						
Tube voltage (kVp):				70			**(−) = bold font = suboptimal field setting**						
Total filtrations:				See last table row.			Empty table cells correspond to values <0.01						
Organ or tissue		**Organ dose D_T: Entrance dose K_E**							**(mSv * mGy^{-1})**				
		(+)	(−)	(+)	(−)	(+)	(−)	(+)	(−)	(+)	(−)	(+)	(−)
Skull		1.49	**1.53**	1.70	**1.75**	1.83	**1.89**	1.93	**1.99**	2.08	**2.16**	2.20	**2.28**
Brain		0.44	**0.45**	0.53	**0.54**	0.59	**0.61**	0.64	**0.66**	0.72	**0.75**	0.79	**0.82**
Oral mucosa		0.48	**0.51**	0.56	**0.60**	0.61	**0.66**	0.66	**0.70**	0.73	**0.78**	0.79	**0.85**
Salivary glands		0.53	**0.57**	0.60	**0.65**	0.64	**0.69**	0.67	**0.73**	0.72	**0.79**	0.77	**0.84**
Cervical spine		0.72	**1.10**	0.89	**1.34**	1.01	**1.52**	1.10	**1.65**	1.27	**1.89**	1.41	**2.09**
Thyroid		0.17	**0.63**	0.20	**0.74**	0.23	**0.81**	0.25	**0.87**	0.29	**0.96**	0.32	**1.04**
Thymus		0.02	**0.11**	0.03	**0.13**	0.03	**0.15**	0.03	**0.17**	0.04	**0.20**	0.05	**0.22**
Thoracic spine		0.02	**0.13**	0.02	**0.16**	0.03	**0.18**	0.03	**0.20**	0.04	**0.24**	0.04	**0.27**
Claviculae		0.10	**1.01**	0.12	**1.20**	0.14	**1.32**	0.15	**1.41**	0.18	**1.56**	0.20	**1.67**
Scapulae		0.03	**0.25**	0.03	**0.29**	0.04	**0.32**	0.04	**0.34**	0.05	**0.37**	0.06	**0.40**
Ribs		0.02	**0.13**	0.02	**0.15**	0.02	**0.17**	0.03	**0.19**	0.03	**0.21**	0.04	**0.23**
Oesophagus			**0.05**	0.01	**0.07**	0.01	**0.08**	0.02	**0.08**	0.02	**0.10**	0.02	**0.11**
Trachea		0.37	**0.42**	0.45	**0.51**	0.50	**0.57**	0.54	**0.61**	0.61	**0.69**	0.67	**0.76**
Lung			**0.03**	0.01	**0.04**	0.01	**0.04**	0.01	**0.05**	0.02	**0.06**	0.02	**0.07**
Mamma													
Heart			**0.01**		**0.02**		**0.02**		**0.02**	0.01	**0.03**	0.01	**0.03**
Stomach													
Spleen													
Liver													**0.01**
Gall bladder													
Pancreas													
Small intestine													
Large intestine													
Sigma/rectum													
Lumbar spine													
Pelvis													
Adrenals													**0.01**
Kidneys													
Urinary bladder													
Uterus													
Ovaries													
Prostate													
Testes													
Upper arm		0.03	**0.33**	0.03	**0.38**	0.04	**0.41**	0.04	**0.43**	0.05	**0.47**	0.06	**0.49**
Elbow			**0.01**		**0.01**		**0.02**		**0.02**		**0.02**	0.01	**0.03**
Forearm													
Upper leg													
Knee													
Lower leg													
Lymph nodes		0.11	**0.15**	0.13	**0.17**	0.14	**0.19**	0.15	**0.20**	0.16	**0.22**	0.18	**0.24**
Skin		0.17	**0.24**	0.18	**0.25**	0.18	**0.26**	0.19	**0.27**	0.20	**0.28**	0.21	**0.30**
Muscle		0.06	**0.09**	0.07	**0.11**	0.08	**0.12**	0.08	**0.13**	0.09	**0.14**	0.10	**0.15**
Skeleton		0.36	**0.45**	0.41	**0.52**	0.45	**0.57**	0.47	**0.60**	0.52	**0.66**	0.55	**0.70**
Red bone marrow		0.18	**0.20**	0.20	**0.23**	0.22	**0.25**	0.24	**0.27**	0.26	**0.30**	0.29	**0.33**
Whole body dose		0.13	**0.16**	0.15	**0.19**	0.17	**0.21**	0.18	**0.22**	0.20	**0.25**	0.21	**0.27**
ED (ICRP 60)		0.04	**0.07**	0.04	**0.08**	0.05	**0.09**	0.05	**0.10**	0.06	**0.11**	0.06	**0.12**
ED (ICRP 103)		0.05	**0.08**	0.06	**0.10**	0.07	**0.11**	0.07	**0.11**	0.08	**0.13**	0.09	**0.14**
Total filtration (mm)	**Al**	**1.5**		**2.5**		**3.5**		**4.5**		**3.5**		**3.5**	
	Cu	**0**		**0**		**0**		**0**		**0.1**		**0.2**	

Table 8.32 Normalized organ doses—Skull r-lat—1–2 years

Beam projection:				r-lat			**Table cells:**						
Focus-to-detector distance (cm):				115			(+) = normal font = optimal field setting						
Tube voltage (kVp):				60			**(−) = bold font = suboptimal field setting**						
Total filtrations:				See last table row.			Empty table cells correspond to values <0.01						
Organ or tissue		**Organ dose D_T: Entrance dose K_E**							**(mSv * mGy^{-1})**				
		(+)	(−)	(+)	(−)	(+)	(−)	(+)	(−)	(+)	(−)	(+)	(−)
Skull		1.16	**1.16**	1.37	**1.37**	1.51	**1.52**	1.62	**1.63**	1.80	**1.82**	1.79	**1.80**
Brain		0.28	**0.28**	0.35	**0.35**	0.40	**0.41**	0.45	**0.45**	0.53	**0.53**	0.50	**0.51**
Oral mucosa		0.33	**0.34**	0.40	**0.41**	0.44	**0.46**	0.48	**0.50**	0.55	**0.57**	0.53	**0.55**
Salivary glands		0.41	**0.42**	0.47	**0.49**	0.50	**0.53**	0.53	**0.56**	0.58	**0.61**	0.57	**0.60**
Cervical spine		0.40	**0.61**	0.52	**0.79**	0.61	**0.92**	0.69	**1.03**	0.83	**1.24**	0.78	**1.17**
Thyroid		0.11	**0.50**	0.14	**0.59**	0.16	**0.66**	0.17	**0.71**	0.20	**0.80**	0.19	**0.78**
Thymus		0.01	**0.04**	0.01	**0.05**	0.02	**0.06**	0.02	**0.07**	0.02	**0.09**	0.02	**0.08**
Thoracic spine			**0.07**	0.01	**0.09**	0.02	**0.10**	0.02	**0.12**	0.02	**0.14**	0.02	**0.13**
Claviculae		0.07	**0.70**	0.09	**0.87**	0.11	**0.99**	0.12	**1.08**	0.15	**1.24**	0.13	**1.22**
Scapulae		0.01	**0.07**	0.01	**0.09**	0.02	**0.10**	0.02	**0.11**	0.03	**0.13**	0.02	**0.13**
Ribs			**0.06**	0.01	**0.07**	0.01	**0.09**	0.02	**0.10**	0.02	**0.11**	0.02	**0.11**
Oesophagus			**0.03**		**0.04**		**0.05**	0.01	**0.06**	0.01	**0.07**	0.01	**0.06**
Trachea		0.25	**0.27**	0.32	**0.34**	0.37	**0.39**	0.40	**0.43**	0.47	**0.50**	0.45	**0.48**
Lung			**0.01**		**0.02**		**0.02**		**0.03**	0.01	**0.03**		**0.03**
Mamma											**0.01**		
Heart									**0.01**		**0.01**		**0.01**
Stomach													
Spleen													
Liver													
Gall bladder													
Pancreas													
Small intestine													
Large intestine													
Sigma/rectum													
Lumbar spine													
Pelvis													
Adrenals													
Kidneys													
Urinary bladder													
Uterus													
Ovaries													
Prostate													
Testes													
Upper arm		0.01	**0.17**	0.02	**0.20**	0.02	**0.22**	0.02	**0.23**	0.03	**0.26**	0.03	**0.26**
Elbow													
Forearm													
Upper leg													
Knee													
Lower leg													
Lymph nodes		0.08	**0.10**	0.09	**0.12**	0.10	**0.14**	0.11	**0.15**	0.13	**0.17**	0.12	**0.16**
Skin		0.13	**0.18**	0.14	**0.20**	0.15	**0.20**	0.15	**0.21**	0.16	**0.22**	0.16	**0.22**
Muscle		0.04	**0.05**	0.05	**0.06**	0.05	**0.07**	0.05	**0.08**	0.06	**0.09**	0.06	**0.08**
Skeleton		0.27	**0.31**	0.32	**0.37**	0.36	**0.41**	0.38	**0.45**	0.43	**0.50**	0.43	**0.50**
Red bone marrow		0.11	**0.11**	0.13	**0.14**	0.14	**0.15**	0.15	**0.17**	0.18	**0.19**	0.17	**0.18**
Whole body dose		0.09	**0.11**	0.11	**0.13**	0.12	**0.14**	0.13	**0.15**	0.15	**0.18**	0.14	**0.17**
ED (ICRP 60)		0.02	**0.05**	0.03	**0.06**	0.03	**0.06**	0.04	**0.07**	0.04	**0.08**	0.04	**0.08**
ED (ICRP 103)		0.03	**0.06**	0.04	**0.07**	0.05	**0.07**	0.05	**0.08**	0.06	**0.09**	0.05	**0.09**
Total filtration (mm)	**Al**	**1.5**		**2.5**		**3.5**		**4.5**		**3.5**		**3.5**	
	Cu	**0**		**0**		**0**		**0**		**0.1**		**0.2**	

Table 8.33 Normalized organ doses—Skull r-lat—1–2 years

Beam projection:				r-lat			**Table cells:**						
Focus-to-detector distance (cm):				115			(+) = normal font = optimal field setting						
Tube voltage (kVp):				70			**(−) = bold font = suboptimal field setting**						
Total filtrations:				See last table row.			Empty table cells correspond to values <0.01						
Organ or tissue		**Organ dose D_T: Entrance dose K_E**							**($mSv * mGy^{-1}$)**				
		(+)	(−)	(+)	(−)	(+)	(−)	(+)	(−)	(+)	(−)	(+)	(−)
Skull		1.27	**1.28**	1.48	**1.49**	1.62	**1.63**	1.72	**1.73**	1.90	**1.92**	2.04	**2.06**
Brain		0.33	**0.33**	0.41	**0.41**	0.47	**0.47**	0.51	**0.51**	0.60	**0.60**	0.68	**0.68**
Oral mucosa		0.37	**0.39**	0.44	**0.46**	0.49	**0.51**	0.53	**0.55**	0.60	**0.63**	0.66	**0.69**
Salivary glands		0.44	**0.46**	0.50	**0.53**	0.54	**0.57**	0.57	**0.60**	0.62	**0.66**	0.67	**0.71**
Cervical spine		0.49	**0.74**	0.62	**0.93**	0.72	**1.07**	0.80	**1.18**	0.95	**1.40**	1.09	**1.59**
Thyroid		0.13	**0.55**	0.16	**0.65**	0.18	**0.72**	0.20	**0.77**	0.23	**0.86**	0.25	**0.94**
Thymus		0.01	**0.05**	0.02	**0.06**	0.02	**0.07**	0.02	**0.08**	0.03	**0.10**	0.03	**0.12**
Thoracic spine		0.01	**0.08**	0.02	**0.11**	0.02	**0.12**	0.02	**0.14**	0.03	**0.17**	0.03	**0.19**
Claviculae		0.09	**0.80**	0.11	**0.97**	0.13	**1.09**	0.14	**1.18**	0.17	**1.33**	0.20	**1.46**
Scapulae		0.01	**0.09**	0.02	**0.11**	0.02	**0.12**	0.03	**0.13**	0.03	**0.15**	0.04	**0.17**
Ribs		0.01	**0.07**	0.02	**0.09**	0.02	**0.10**	0.02	**0.11**	0.03	**0.13**	0.03	**0.14**
Oesophagus			**0.04**		**0.05**	0.01	**0.06**	0.01	**0.07**	0.02	**0.08**	0.02	**0.10**
Trachea		0.30	**0.32**	0.37	**0.39**	0.42	**0.44**	0.46	**0.49**	0.53	**0.56**	0.59	**0.63**
Lung			**0.02**		**0.02**		**0.03**	0.01	**0.03**	0.01	**0.04**	0.02	**0.05**
Mamma											**0.01**		**0.02**
Heart							**0.01**		**0.01**		**0.02**		**0.02**
Stomach													
Spleen													
Liver													
Gall bladder													
Pancreas													
Small intestine													
Large intestine													
Sigma/rectum													
Lumbar spine													
Pelvis													
Adrenals													
Kidneys													
Urinary bladder													
Uterus													
Ovaries													
Prostate													
Testes													
Upper arm		0.02	**0.18**	0.02	**0.22**	0.03	**0.24**	0.03	**0.25**	0.04	**0.28**	0.05	**0.31**
Elbow											**0.01**		**0.02**
Forearm													
Upper leg													
Knee													
Lower leg													
Lymph nodes		0.09	**0.12**	0.10	**0.14**	0.11	**0.15**	0.12	**0.16**	0.14	**0.19**	0.15	**0.21**
Skin		0.14	**0.19**	0.15	**0.20**	0.15	**0.21**	0.16	**0.22**	0.17	**0.23**	0.18	**0.24**
Muscle		0.04	**0.06**	0.05	**0.07**	0.06	**0.08**	0.06	**0.09**	0.07	**0.10**	0.07	**0.11**
Skeleton		0.30	**0.35**	0.35	**0.41**	0.39	**0.45**	0.41	**0.48**	0.46	**0.54**	0.50	**0.58**
Red bone marrow		0.12	**0.13**	0.14	**0.16**	0.16	**0.17**	0.17	**0.19**	0.19	**0.21**	0.21	**0.24**
Whole body dose		0.10	**0.12**	0.12	**0.14**	0.13	**0.16**	0.14	**0.17**	0.16	**0.19**	0.18	**0.21**
ED (ICRP 60)		0.03	**0.05**	0.03	**0.06**	0.04	**0.07**	0.04	**0.08**	0.05	**0.09**	0.05	**0.10**
ED (ICRP 103)		0.04	**0.06**	0.05	**0.07**	0.05	**0.08**	0.06	**0.09**	0.06	**0.10**	0.07	**0.11**
Total filtration (mm)	**Al**	**1.5**		**2.5**		**3.5**		**4.5**		**3.5**		**3.5**	
	Cu	**0**		**0**		**0**		**0**		**0.1**		**0.2**	

Table 8.34 Normalized organ doses—Skull r-lat—3–7 years

Beam projection:			r-lat			**Table cells:**						
Focus-to-detector distance (cm):			115			(+) = normal font = optimal field setting						
Tube voltage (kVp):			60			**(−) = bold font = suboptimal field setting**						
Total filtrations:			See last table row.			Empty table cells correspond to values <0.01						
Organ or tissue	**Organ dose D_T: Entrance dose K_E**							**(mSv * mGy^{-1})**				
	(+)	(−)	(+)	(−)	(+)	(−)	(+)	(−)	(+)	(−)	(+)	(−)
Skull	0.85	**0.85**	1.02	**1.02**	1.15	**1.15**	1.24	**1.25**	1.42	**1.42**	1.38	**1.39**
Brain	0.20	**0.20**	0.26	**0.26**	0.30	**0.30**	0.34	**0.34**	0.42	**0.42**	0.38	**0.38**
Oral mucosa	0.25	**0.25**	0.30	**0.31**	0.34	**0.35**	0.37	**0.38**	0.43	**0.45**	0.41	**0.42**
Salivary glands	0.35	**0.37**	0.40	**0.42**	0.44	**0.46**	0.46	**0.48**	0.51	**0.54**	0.49	**0.52**
Cervical spine	0.37	**0.59**	0.48	**0.76**	0.56	**0.89**	0.63	**1.00**	0.77	**1.21**	0.72	**1.13**
Thyroid	0.10	**0.46**	0.12	**0.56**	0.13	**0.62**	0.15	**0.67**	0.17	**0.76**	0.16	**0.74**
Thymus		**0.01**		**0.02**		**0.02**		**0.03**	0.01	**0.03**		**0.03**
Thoracic spine		**0.05**		**0.06**		**0.07**	0.01	**0.08**	0.01	**0.10**	0.01	**0.09**
Claviculae	0.04	**0.51**	0.06	**0.64**	0.07	**0.74**	0.08	**0.82**	0.10	**0.96**	0.09	**0.93**
Scapulae		**0.04**		**0.05**	0.01	**0.05**	0.01	**0.06**	0.01	**0.07**	0.01	**0.07**
Ribs		**0.03**		**0.04**		**0.05**		**0.06**	0.01	**0.07**	0.01	**0.06**
Oesophagus		**0.02**		**0.02**		**0.03**		**0.03**		**0.04**		**0.03**
Trachea	0.15	**0.17**	0.20	**0.22**	0.23	**0.26**	0.26	**0.29**	0.32	**0.36**	0.29	**0.33**
Lung				**0.01**		**0.01**		**0.02**		**0.02**		**0.02**
Mamma												
Heart												
Stomach												
Spleen												
Liver												
Gall bladder												
Pancreas												
Small intestine												
Large intestine												
Sigma/rectum												
Lumbar spine												
Pelvis												
Adrenals												
Kidneys												
Urinary bladder												
Uterus												
Ovaries												
Prostate												
Testes												
Upper arm		**0.11**	0.01	**0.14**	0.01	**0.15**	0.01	**0.16**	0.02	**0.19**	0.01	**0.18**
Elbow												
Forearm												
Upper leg												
Knee												
Lower leg												
Lymph nodes	0.06	**0.08**	0.07	**0.10**	0.08	**0.11**	0.08	**0.12**	0.10	**0.14**	0.09	**0.13**
Skin	0.10	**0.14**	0.10	**0.15**	0.10	**0.15**	0.11	**0.16**	0.11	**0.17**	0.11	**0.16**
Muscle	0.02	**0.03**	0.03	**0.04**	0.03	**0.04**	0.03	**0.05**	0.03	**0.05**	0.03	**0.05**
Skeleton	0.23	**0.25**	0.28	**0.31**	0.31	**0.35**	0.34	**0.38**	0.39	**0.44**	0.38	**0.42**
Red bone marrow	0.05	**0.06**	0.06	**0.07**	0.07	**0.08**	0.08	**0.09**	0.09	**0.10**	0.09	**0.10**
Whole body dose	0.06	**0.07**	0.08	**0.09**	0.09	**0.10**	0.09	**0.11**	0.11	**0.13**	0.10	**0.12**
ED (ICRP 60)	0.02	**0.04**	0.02	**0.04**	0.02	**0.05**	0.02	**0.05**	0.03	**0.06**	0.03	**0.06**
ED (ICRP 103)	0.02	**0.04**	0.03	**0.05**	0.03	**0.06**	0.03	**0.06**	0.04	**0.07**	0.04	**0.07**
Total filtration (mm) Al	**1.5**		**2.5**		**3.5**		**4.5**		**3.5**		**3.5**	
Cu	**0**		**0**		**0**		**0**		**0.1**		**0.2**	

Table 8.35 Normalized organ doses—Skull r-lat—3–7 years

Beam projection:			r-lat			**Table cells:**						
Focus-to-detector distance (cm):			115			(+) = normal font = optimal field setting						
Tube voltage (kVp):			70			**(−) = bold font = suboptimal field setting**						
Total filtrations:			See last table row.			Empty table cells correspond to values <0.01						
Organ or tissue	**Organ dose D_T: Entrance dose K_E**							**(mSv * mGy^{-1})**				
	(+)	(−)	(+)	(−)	(+)	(−)	(+)	(−)	(+)	(−)	(+)	(−)
Skull	0.95	**0.96**	1.13	**1.14**	1.26	**1.26**	1.36	**1.36**	1.53	**1.54**	1.68	**1.69**
Brain	0.25	**0.25**	0.31	**0.31**	0.37	**0.37**	0.41	**0.41**	0.49	**0.49**	0.56	**0.56**
Oral mucosa	0.29	**0.30**	0.34	**0.36**	0.39	**0.40**	0.42	**0.44**	0.49	**0.51**	0.55	**0.57**
Salivary glands	0.39	**0.40**	0.44	**0.46**	0.47	**0.50**	0.50	**0.53**	0.55	**0.58**	0.59	**0.63**
Cervical spine	0.45	**0.72**	0.58	**0.90**	0.67	**1.05**	0.75	**1.16**	0.89	**1.38**	1.03	**1.57**
Thyroid	0.11	**0.52**	0.14	**0.62**	0.15	**0.68**	0.17	**0.73**	0.20	**0.82**	0.22	**0.90**
Thymus		**0.02**		**0.03**		**0.03**	0.01	**0.03**	0.01	**0.04**	0.02	**0.05**
Thoracic spine		**0.06**	0.01	**0.08**	0.01	**0.09**	0.01	**0.10**	0.02	**0.12**	0.02	**0.14**
Claviculae	0.06	**0.60**	0.07	**0.73**	0.09	**0.83**	0.10	**0.91**	0.12	**1.05**	0.14	**1.17**
Scapulae		**0.05**	0.01	**0.06**	0.01	**0.07**	0.02	**0.07**	0.02	**0.09**	0.02	**0.10**
Ribs		**0.04**		**0.05**	0.01	**0.06**	0.01	**0.07**	0.02	**0.08**	0.02	**0.09**
Oesophagus		**0.02**		**0.03**		**0.03**		**0.04**		**0.04**		**0.05**
Trachea	0.19	**0.21**	0.24	**0.27**	0.28	**0.31**	0.31	**0.35**	0.37	**0.42**	0.43	**0.48**
Lung		**0.01**		**0.02**		**0.02**		**0.02**		**0.03**		**0.03**
Mamma												
Heart												**0.01**
Stomach												
Spleen												
Liver												
Gall bladder												
Pancreas												
Small intestine												
Large intestine												
Sigma/rectum												
Lumbar spine												
Pelvis												
Adrenals												
Kidneys												
Urinary bladder												
Uterus												
Ovaries												
Prostate												
Testes												
Upper arm	0.01	**0.13**	0.01	**0.15**	0.02	**0.17**	0.02	**0.18**	0.02	**0.20**	0.03	**0.22**
Elbow												
Forearm												
Upper leg												
Knee												
Lower leg												
Lymph nodes	0.07	**0.09**	0.08	**0.11**	0.09	**0.12**	0.09	**0.13**	0.11	**0.15**	0.12	**0.17**
Skin	0.10	**0.14**	0.11	**0.15**	0.11	**0.16**	0.11	**0.16**	0.12	**0.17**	0.12	**0.18**
Muscle	0.02	**0.04**	0.03	**0.04**	0.03	**0.05**	0.03	**0.05**	0.04	**0.06**	0.04	**0.07**
Skeleton	0.26	**0.29**	0.31	**0.34**	0.35	**0.39**	0.37	**0.42**	0.42	**0.47**	0.47	**0.52**
Red bone marrow	0.06	**0.07**	0.07	**0.08**	0.08	**0.09**	0.09	**0.10**	0.10	**0.12**	0.12	**0.13**
Whole body dose	0.07	**0.09**	0.09	**0.10**	0.10	**0.12**	0.10	**0.12**	0.12	**0.14**	0.13	**0.16**
ED (ICRP 60)	0.02	**0.04**	0.02	**0.05**	0.02	**0.06**	0.03	**0.06**	0.03	**0.07**	0.03	**0.08**
ED (ICRP 103)	0.03	**0.05**	0.03	**0.06**	0.04	**0.06**	0.04	**0.07**	0.05	**0.08**	0.05	**0.09**
Total filtration (mm) Al	**1.5**		**2.5**		**3.5**		**4.5**		**3.5**		**3.5**	
Cu	**0**		**0**		**0**		**0**		**0.1**		**0.2**	

Table 8.36 Normalized organ doses—Skull r-lat—8–12 years

Beam projection:			r-lat			**Table cells:**						
Focus-to-detector distance (cm):			115			(+) = normal font = optimal field setting						
Tube voltage (kVp):			60			**(−) = bold font = suboptimal field setting**						
Total filtrations:			See last table row.			Empty table cells correspond to values <0.01						
Organ or tissue	**Organ dose D_T: Entrance dose K_E**							**($mSv * mGy^{-1}$)**				
	(+)	(−)	(+)	(−)	(+)	(−)	(+)	(−)	(+)	(−)	(+)	(−)
Skull	0.76	**0.77**	0.93	**0.93**	1.05	**1.05**	1.14	**1.15**	1.31	**1.32**	1.27	**1.28**
Brain	0.18	**0.18**	0.23	**0.23**	0.28	**0.28**	0.31	**0.31**	0.38	**0.38**	0.35	**0.35**
Oral mucosa	0.23	**0.23**	0.28	**0.29**	0.31	**0.33**	0.34	**0.36**	0.40	**0.42**	0.37	**0.39**
Salivary glands	0.34	**0.36**	0.39	**0.41**	0.42	**0.45**	0.45	**0.47**	0.49	**0.52**	0.47	**0.50**
Cervical spine	0.30	**0.58**	0.39	**0.76**	0.46	**0.89**	0.52	**1.00**	0.64	**1.22**	0.59	**1.13**
Thyroid	0.03	**0.42**	0.04	**0.51**	0.04	**0.57**	0.05	**0.62**	0.06	**0.71**	0.05	**0.69**
Thymus						**0.01**		**0.01**		**0.02**		**0.01**
Thoracic spine		**0.03**		**0.04**		**0.05**		**0.06**		**0.07**		**0.06**
Claviculae	0.03	**0.26**	0.04	**0.34**	0.04	**0.39**	0.05	**0.44**	0.06	**0.53**	0.05	**0.50**
Scapulae		**0.01**		**0.02**		**0.02**		**0.02**		**0.03**		**0.02**
Ribs		**0.01**		**0.01**		**0.02**		**0.02**		**0.03**		**0.02**
Oesophagus		**0.01**		**0.02**		**0.02**		**0.02**		**0.03**		**0.02**
Trachea	0.16	**0.25**	0.21	**0.32**	0.24	**0.37**	0.28	**0.42**	0.34	**0.51**	0.31	**0.47**
Lung										**0.01**		**0.01**
Mamma												
Heart												
Stomach												
Spleen												
Liver												
Gall bladder												
Pancreas												
Small intestine												
Large intestine												
Sigma/rectum												
Lumbar spine												
Pelvis												
Adrenals												
Kidneys												
Urinary bladder												
Uterus												
Ovaries												
Prostate												
Testes												
Upper arm		**0.01**		**0.02**		**0.02**		**0.03**	0.01	**0.03**		**0.03**
Elbow												
Forearm												
Upper leg												
Knee												
Lower leg												
Lymph nodes	0.05	**0.09**	0.06	**0.10**	0.07	**0.12**	0.08	**0.13**	0.09	**0.15**	0.08	**0.14**
Skin	0.07	**0.11**	0.07	**0.11**	0.08	**0.12**	0.08	**0.12**	0.08	**0.13**	0.08	**0.12**
Muscle	0.01	**0.02**	0.02	**0.03**	0.02	**0.03**	0.02	**0.03**	0.02	**0.04**	0.02	**0.04**
Skeleton	0.16	**0.18**	0.20	**0.22**	0.22	**0.24**	0.24	**0.27**	0.28	**0.31**	0.27	**0.30**
Red bone marrow	0.03	**0.04**	0.04	**0.05**	0.05	**0.05**	0.05	**0.06**	0.06	**0.07**	0.06	**0.07**
Whole body dose	0.04	**0.05**	0.05	**0.06**	0.06	**0.07**	0.06	**0.08**	0.07	**0.09**	0.07	**0.09**
ED (ICRP 60)		**0.03**	0.01	**0.04**	0.01	**0.04**	0.01	**0.05**	0.02	**0.05**	0.01	**0.05**
ED (ICRP 103)	0.02	**0.04**	0.02	**0.04**	0.02	**0.05**	0.03	**0.05**	0.03	**0.06**	0.03	**0.06**
Total filtration (mm) Al	**1.5**		**2.5**		**3.5**		**4.5**		**3.5**		**3.5**	
Cu	**0**		**0**		**0**		**0**		**0.1**		**0.2**	

Table 8.37 Normalized organ doses—Skull r-lat—8–12 years

Beam projection:			r-lat			**Table cells:**						
Focus-to-detector distance (cm):			115			(+) = normal font = optimal field setting						
Tube voltage (kVp):			70			**(−) = bold font = suboptimal field setting**						
Total filtrations:			See last table row.			Empty table cells correspond to values <0.01						
Organ or tissue	**Organ dose D_T: Entrance dose K_E**							**(mSv * mGy^{-1})**				
	(+)	(−)	(+)	(−)	(+)	(−)	(+)	(−)	(+)	(−)	(+)	(−)
Skull	0.87	**0.87**	1.04	**1.05**	1.16	**1.17**	1.25	**1.27**	1.43	**1.44**	1.57	**1.59**
Brain	0.23	**0.23**	0.29	**0.29**	0.34	**0.34**	0.38	**0.38**	0.46	**0.46**	0.53	**0.53**
Oral mucosa	0.26	**0.28**	0.32	**0.33**	0.36	**0.38**	0.39	**0.41**	0.45	**0.48**	0.51	**0.54**
Salivary glands	0.37	**0.39**	0.42	**0.45**	0.46	**0.49**	0.48	**0.52**	0.53	**0.57**	0.58	**0.62**
Cervical spine	0.38	**0.72**	0.48	**0.91**	0.56	**1.06**	0.63	**1.17**	0.76	**1.41**	0.89	**1.62**
Thyroid	0.04	**0.47**	0.05	**0.57**	0.05	**0.63**	0.06	**0.68**	0.08	**0.77**	0.09	**0.85**
Thymus				**0.01**		**0.01**		**0.02**		**0.02**		**0.03**
Thoracic spine		**0.04**		**0.05**		**0.06**		**0.07**		**0.08**	0.01	**0.10**
Claviculae	0.04	**0.32**	0.05	**0.40**	0.05	**0.46**	0.06	**0.50**	0.08	**0.59**	0.09	**0.67**
Scapulae		**0.02**		**0.02**		**0.02**		**0.03**	0.01	**0.04**	0.01	**0.04**
Ribs		**0.01**		**0.02**		**0.02**		**0.03**		**0.03**	0.01	**0.04**
Oesophagus		**0.02**		**0.02**		**0.02**		**0.03**		**0.03**		**0.04**
Trachea	0.20	**0.31**	0.26	**0.39**	0.30	**0.45**	0.34	**0.50**	0.41	**0.59**	0.48	**0.68**
Lung						**0.01**		**0.01**		**0.02**		**0.02**
Mamma												
Heart												
Stomach												
Spleen												
Liver												
Gall bladder												
Pancreas												
Small intestine												
Large intestine												
Sigma/rectum												
Lumbar spine												
Pelvis												
Adrenals												
Kidneys												
Urinary bladder												
Uterus												
Ovaries												
Prostate												
Testes												
Upper arm		**0.02**		**0.03**		**0.03**	0.01	**0.03**	0.01	**0.04**	0.02	**0.05**
Elbow												
Forearm												
Upper leg												
Knee												
Lower leg												
Lymph nodes	0.06	**0.10**	0.07	**0.12**	0.08	**0.13**	0.09	**0.15**	0.10	**0.17**	0.12	**0.19**
Skin	0.07	**0.11**	0.08	**0.12**	0.08	**0.12**	0.08	**0.13**	0.09	**0.13**	0.09	**0.14**
Muscle	0.02	**0.03**	0.02	**0.03**	0.02	**0.03**	0.02	**0.04**	0.02	**0.04**	0.03	**0.05**
Skeleton	0.18	**0.20**	0.22	**0.24**	0.25	**0.27**	0.27	**0.30**	0.31	**0.34**	0.34	**0.38**
Red bone marrow	0.04	**0.05**	0.05	**0.06**	0.06	**0.06**	0.06	**0.07**	0.07	**0.08**	0.08	**0.09**
Whole body dose	0.05	**0.06**	0.06	**0.07**	0.07	**0.08**	0.07	**0.09**	0.08	**0.10**	0.09	**0.11**
ED (ICRP 60)	0.01	**0.03**	0.01	**0.04**	0.01	**0.05**	0.02	**0.05**	0.02	**0.06**	0.02	**0.07**
ED (ICRP 103)	0.02	**0.04**	0.02	**0.05**	0.03	**0.06**	0.03	**0.06**	0.03	**0.07**	0.04	**0.08**
Total filtration (mm) Al	**1.5**		**2.5**		**3.5**		**4.5**		**3.5**		**3.5**	
Cu	**0**		**0**		**0**		**0**		**0.1**		**0.2**	

Table 8.38 Normalized organ doses—Skull r-lat—8–12 years

Beam projection:				r-lat				**Table cells:**					
Focus-to-detector distance (cm):				115				(+) = normal font = optimal field setting					
Tube voltage (kVp):				80				**(−) = bold font = suboptimal field setting**					
Total filtrations:				See last table row.				Empty table cells correspond to values <0.01					
Organ or tissue		**Organ dose D_T: Entrance dose K_E**								**(mSv * mGy^{-1})**			
		(+)	(−)	(+)	(−)	(+)	(−)	(+)	(−)	(+)	(−)	(+)	(−)
Skull		0.96	**0.97**	1.13	**1.14**	1.25	**1.26**	1.34	**1.35**	1.50	**1.52**	1.63	**1.65**
Brain		0.27	**0.27**	0.34	**0.34**	0.39	**0.39**	0.44	**0.44**	0.52	**0.52**	0.60	**0.60**
Oral mucosa		0.30	**0.32**	0.36	**0.38**	0.40	**0.42**	0.44	**0.46**	0.50	**0.53**	0.56	**0.59**
Salivary glands		0.40	**0.43**	0.46	**0.48**	0.49	**0.52**	0.52	**0.55**	0.57	**0.61**	0.61	**0.66**
Cervical spine		0.45	**0.85**	0.56	**1.05**	0.65	**1.20**	0.72	**1.32**	0.85	**1.55**	0.98	**1.75**
Thyroid		0.04	**0.53**	0.06	**0.62**	0.07	**0.69**	0.07	**0.74**	0.09	**0.83**	0.10	**0.90**
Thymus			**0.01**		**0.02**		**0.02**		**0.02**		**0.03**		**0.03**
Thoracic spine			**0.05**		**0.06**		**0.07**		**0.08**	0.01	**0.09**	0.01	**0.11**
Claviculae		0.04	**0.36**	0.06	**0.45**	0.07	**0.51**	0.07	**0.55**	0.09	**0.64**	0.10	**0.72**
Scapulae			**0.02**		**0.03**		**0.03**	0.01	**0.03**	0.01	**0.04**	0.02	**0.05**
Ribs			**0.02**		**0.02**		**0.03**		**0.03**	0.01	**0.04**	0.01	**0.04**
Oesophagus			**0.02**		**0.03**		**0.03**		**0.03**		**0.04**		**0.05**
Trachea		0.25	**0.37**	0.31	**0.45**	0.36	**0.52**	0.40	**0.57**	0.47	**0.67**	0.55	**0.76**
Lung					**0.01**		**0.01**		**0.02**		**0.02**		**0.02**
Mamma													
Heart													
Stomach													
Spleen													
Liver													
Gall bladder													
Pancreas													
Small intestine													
Large intestine													
Sigma/rectum													
Lumbar spine													
Pelvis													
Adrenals													
Kidneys													
Urinary bladder													
Uterus													
Ovaries													
Prostate													
Testes													
Upper arm			**0.02**	0.01	**0.03**	0.01	**0.04**	0.01	**0.04**	0.02	**0.05**	0.02	**0.06**
Elbow													
Forearm													
Upper leg													
Knee													
Lower leg													
Lymph nodes		0.07	**0.11**	0.08	**0.13**	0.09	**0.15**	0.10	**0.16**	0.12	**0.19**	0.13	**0.21**
Skin		0.08	**0.12**	0.08	**0.12**	0.08	**0.13**	0.09	**0.13**	0.09	**0.14**	0.09	**0.14**
Muscle		0.02	**0.03**	0.02	**0.03**	0.02	**0.04**	0.02	**0.04**	0.03	**0.05**	0.03	**0.05**
Skeleton		0.21	**0.23**	0.24	**0.27**	0.27	**0.30**	0.29	**0.32**	0.33	**0.36**	0.36	**0.40**
Red bone marrow		0.05	**0.05**	0.06	**0.06**	0.06	**0.07**	0.07	**0.08**	0.08	**0.09**	0.09	**0.10**
Whole body dose		0.05	**0.07**	0.06	**0.08**	0.07	**0.09**	0.08	**0.10**	0.09	**0.11**	0.10	**0.12**
ED (ICRP 60)		0.01	**0.04**	0.01	**0.05**	0.02	**0.05**	0.02	**0.06**	0.02	**0.06**	0.02	**0.07**
ED (ICRP 103)		0.02	**0.05**	0.03	**0.06**	0.03	**0.06**	0.03	**0.07**	0.04	**0.08**	0.04	**0.08**
Total filtration (mm)	**Al**	**1.5**		**2.5**		**3.5**		**4.5**		**3.5**		**3.5**	
	Cu	**0**		**0**		**0**		**0**		**0.1**		**0.2**	

Table 8.39 Normalized organ doses—Skull r-lat—13–17 years

Beam projection:			r-lat			**Table cells:**						
Focus-to-detector distance (cm):			115			(+) = normal font = optimal field setting						
Tube voltage (kVp):			70			**(−) = bold font = suboptimal field setting**						
Total filtrations:			See last table row.			Empty table cells correspond to values <0.01						
Organ or tissue	**Organ dose D_T: Entrance dose K_E**							**(mSv * mGy^{-1})**				
	(+)	(−)	(+)	(−)	(+)	(−)	(+)	(−)	(+)	(−)	(+)	(−)
Skull	0.75	**0.76**	0.91	**0.92**	1.03	**1.04**	1.12	**1.13**	1.29	**1.30**	1.43	**1.45**
Brain	0.20	**0.20**	0.26	**0.26**	0.31	**0.31**	0.34	**0.34**	0.42	**0.42**	0.49	**0.49**
Oral mucosa	0.23	**0.24**	0.28	**0.30**	0.32	**0.34**	0.35	**0.37**	0.41	**0.43**	0.47	**0.49**
Salivary glands	0.36	**0.38**	0.41	**0.43**	0.44	**0.47**	0.47	**0.50**	0.52	**0.55**	0.56	**0.60**
Cervical spine	0.34	**0.78**	0.44	**0.99**	0.52	**1.16**	0.58	**1.30**	0.71	**1.57**	0.84	**1.82**
Thyroid	0.01	**0.42**	0.02	**0.51**	0.02	**0.57**	0.03	**0.62**	0.03	**0.71**	0.04	**0.78**
Thymus												**0.01**
Thoracic spine		**0.03**		**0.04**		**0.05**		**0.05**		**0.07**		**0.08**
Claviculae	0.02	**0.18**	0.03	**0.22**	0.03	**0.26**	0.03	**0.29**	0.04	**0.35**	0.05	**0.40**
Scapulae				**0.01**		**0.01**		**0.02**		**0.02**		**0.03**
Ribs				**0.01**		**0.01**		**0.02**		**0.02**		**0.02**
Oesophagus		**0.02**		**0.02**		**0.02**		**0.03**		**0.03**		**0.04**
Trachea	0.11	**0.22**	0.14	**0.28**	0.17	**0.33**	0.19	**0.36**	0.23	**0.43**	0.28	**0.50**
Lung												**0.01**
Mamma				**0.01**		**0.01**		**0.01**		**0.02**		**0.02**
Heart												
Stomach												
Spleen												
Liver												
Gall bladder												
Pancreas												
Small intestine												
Large intestine												
Sigma/rectum												
Lumbar spine												
Pelvis												
Adrenals												
Kidneys												
Urinary bladder												
Uterus												
Ovaries												
Prostate												
Testes												
Upper arm				**0.01**		**0.01**		**0.02**		**0.02**		**0.03**
Elbow												
Forearm												
Upper leg												
Knee												
Lower leg												
Lymph nodes	0.05	**0.08**	0.05	**0.10**	0.06	**0.11**	0.07	**0.12**	0.08	**0.14**	0.09	**0.16**
Skin	0.06	**0.09**	0.06	**0.10**	0.06	**0.10**	0.06	**0.11**	0.07	**0.11**	0.07	**0.12**
Muscle	0.01	**0.02**	0.01	**0.02**	0.01	**0.03**	0.02	**0.03**	0.02	**0.03**	0.02	**0.04**
Skeleton	0.12	**0.14**	0.15	**0.17**	0.17	**0.20**	0.18	**0.21**	0.21	**0.25**	0.24	**0.28**
Red bone marrow	0.03	**0.03**	0.04	**0.04**	0.04	**0.05**	0.04	**0.05**	0.05	**0.06**	0.06	**0.07**
Whole body dose	0.03	**0.04**	0.04	**0.05**	0.04	**0.06**	0.05	**0.06**	0.05	**0.07**	0.06	**0.08**
ED (ICRP 60)		**0.03**		**0.04**		**0.04**	0.01	**0.04**	0.01	**0.05**	0.01	**0.06**
ED (ICRP 103)	0.02	**0.04**	0.02	**0.04**	0.02	**0.05**	0.02	**0.05**	0.03	**0.06**	0.03	**0.07**
Total filtration (mm) Al	**1.5**		**2.5**		**3.5**		**4.5**		**3.5**		**3.5**	
Cu	**0**		**0**		**0**		**0**		**0.1**		**0.2**	

Table 8.40 Normalized organ doses—Skull r-lat—13–17 years

Beam projection:			r-lat			**Table cells:**						
Focus-to-detector distance (cm):			115			(+) = normal font = optimal field setting						
Tube voltage (kVp):			80			**(−) = bold font = suboptimal field setting**						
Total filtrations:			See last table row.			Empty table cells correspond to values <0.01						
Organ or tissue	**Organ dose D_T: Entrance dose K_E (mSv * mGy^{-1})**											
	(+)	(−)	(+)	(−)	(+)	(−)	(+)	(−)	(+)	(−)	(+)	(−)
Skull	0.84	**0.85**	1.01	**1.01**	1.12	**1.13**	1.21	**1.22**	1.37	**1.39**	1.51	**1.53**
Brain	0.25	**0.25**	0.31	**0.31**	0.36	**0.36**	0.40	**0.40**	0.48	**0.48**	0.56	**0.56**
Oral mucosa	0.27	**0.28**	0.32	**0.34**	0.37	**0.38**	0.40	**0.42**	0.46	**0.48**	0.52	**0.54**
Salivary glands	0.39	**0.41**	0.44	**0.47**	0.48	**0.51**	0.50	**0.54**	0.56	**0.59**	0.60	**0.65**
Cervical spine	0.42	**0.93**	0.52	**1.16**	0.61	**1.33**	0.68	**1.48**	0.81	**1.75**	0.94	**2.00**
Thyroid	0.02	**0.48**	0.02	**0.57**	0.03	**0.63**	0.03	**0.68**	0.04	**0.76**	0.05	**0.84**
Thymus										**0.01**		**0.02**
Thoracic spine		**0.04**		**0.05**		**0.06**		**0.06**		**0.08**		**0.09**
Claviculae	0.02	**0.21**	0.03	**0.26**	0.04	**0.30**	0.04	**0.33**	0.05	**0.39**	0.06	**0.44**
Scapulae		**0.01**		**0.02**		**0.02**		**0.02**		**0.03**		**0.03**
Ribs		**0.01**		**0.01**		**0.02**		**0.02**		**0.02**		**0.03**
Oesophagus		**0.02**		**0.02**		**0.03**		**0.03**		**0.04**		**0.04**
Trachea	0.14	**0.27**	0.17	**0.33**	0.20	**0.38**	0.23	**0.42**	0.27	**0.49**	0.32	**0.56**
Lung										**0.01**		**0.01**
Mamma		**0.01**		**0.01**		**0.02**		**0.02**		**0.03**		**0.03**
Heart												
Stomach												
Spleen												
Liver												
Gall bladder												
Pancreas												
Small intestine												
Large intestine												
Sigma/rectum												
Lumbar spine												
Pelvis												
Adrenals												
Kidneys												
Urinary bladder												
Uterus												
Ovaries												
Prostate												
Testes												
Upper arm		**0.01**		**0.02**		**0.02**		**0.02**		**0.03**	0.01	**0.03**
Elbow												
Forearm												
Upper leg												
Knee												
Lower leg												
Lymph nodes	0.05	**0.10**	0.06	**0.11**	0.07	**0.13**	0.07	**0.14**	0.09	**0.16**	0.10	**0.17**
Skin	0.06	**0.10**	0.06	**0.10**	0.06	**0.11**	0.07	**0.11**	0.07	**0.12**	0.07	**0.12**
Muscle	0.01	**0.02**	0.01	**0.03**	0.02	**0.03**	0.02	**0.03**	0.02	**0.04**	0.02	**0.04**
Skeleton	0.14	**0.16**	0.17	**0.19**	0.19	**0.22**	0.20	**0.23**	0.23	**0.27**	0.25	**0.30**
Red bone marrow	0.03	**0.04**	0.04	**0.05**	0.05	**0.06**	0.05	**0.06**	0.06	**0.07**	0.07	**0.08**
Whole body dose	0.04	**0.05**	0.04	**0.06**	0.05	**0.06**	0.05	**0.07**	0.06	**0.08**	0.07	**0.09**
ED (ICRP 60)		**0.03**	0.01	**0.04**	0.01	**0.05**	0.01	**0.05**	0.01	**0.06**	0.02	**0.06**
ED (ICRP 103)	0.02	**0.04**	0.02	**0.05**	0.02	**0.06**	0.03	**0.06**	0.03	**0.07**	0.03	**0.08**
Total filtration (mm) Al	**1.5**		**2.5**		**3.5**		**4.5**		**3.5**		**3.5**	
Cu	**0**		**0**		**0**		**0**		**0.1**		**0.2**	

Table 8.41 Normalized organ doses—Skull r-lat—> 17 years

Beam projection:			r-lat			**Table cells:**						
Focus-to-detector distance (cm):			115			(+) = normal font = optimal field setting						
Tube voltage (kVp):			70			**(−) = bold font = suboptimal field setting**						
Total filtrations:			See last table row.			Empty table cells correspond to values <0.01						
Organ or tissue	**Organ dose D_T: Entrance dose K_E**							**(mSv * mGy^{-1})**				
	(+)	(−)	(+)	(−)	(+)	(−)	(+)	(−)	(+)	(−)	(+)	(−)
Skull	0.67	**0.68**	0.82	**0.83**	0.93	**0.94**	1.02	**1.03**	1.18	**1.19**	1.33	**1.34**
Brain	0.18	**0.18**	0.24	**0.24**	0.28	**0.28**	0.32	**0.31**	0.39	**0.39**	0.46	**0.46**
Oral mucosa	0.22	**0.23**	0.27	**0.28**	0.30	**0.32**	0.33	**0.35**	0.39	**0.41**	0.44	**0.46**
Salivary glands	0.34	**0.36**	0.39	**0.41**	0.42	**0.45**	0.45	**0.48**	0.49	**0.53**	0.54	**0.58**
Cervical spine	0.32	**0.75**	0.42	**0.96**	0.49	**1.12**	0.55	**1.26**	0.67	**1.52**	0.79	**1.77**
Thyroid	0.01	**0.41**	0.02	**0.50**	0.02	**0.56**	0.02	**0.61**	0.03	**0.69**	0.04	**0.77**
Thymus												
Thoracic spine		**0.03**		**0.04**		**0.04**		**0.05**		**0.06**		**0.07**
Claviculae	0.02	**0.14**	0.02	**0.18**	0.03	**0.22**	0.03	**0.24**	0.04	**0.29**	0.04	**0.34**
Scapulae						**0.01**		**0.01**		**0.02**		**0.02**
Ribs						**0.01**		**0.01**		**0.02**		**0.02**
Oesophagus		**0.01**		**0.02**		**0.02**		**0.02**		**0.03**		**0.04**
Trachea	0.09	**0.21**	0.12	**0.27**	0.14	**0.31**	0.16	**0.35**	0.19	**0.42**	0.23	**0.48**
Lung												**0.01**
Mamma												
Heart												
Stomach												
Spleen												
Liver												
Gall bladder												
Pancreas												
Small intestine												
Large intestine												
Sigma/rectum												
Lumbar spine												
Pelvis												
Adrenals												
Kidneys												
Urinary bladder												
Uterus												
Ovaries												
Prostate												
Testes												
Upper arm								**0.01**		**0.01**		**0.02**
Elbow												
Forearm												
Upper leg												
Knee												
Lower leg												
Lymph nodes	0.04	**0.08**	0.05	**0.10**	0.05	**0.11**	0.06	**0.12**	0.07	**0.14**	0.08	**0.15**
Skin	0.05	**0.08**	0.05	**0.09**	0.05	**0.09**	0.06	**0.10**	0.06	**0.10**	0.06	**0.11**
Muscle		**0.02**	0.01	**0.02**	0.01	**0.02**	0.01	**0.02**	0.01	**0.03**	0.02	**0.03**
Skeleton	0.10	**0.12**	0.13	**0.15**	0.14	**0.17**	0.16	**0.18**	0.18	**0.22**	0.21	**0.24**
Red bone marrow	0.02	**0.03**	0.03	**0.03**	0.03	**0.04**	0.03	**0.04**	0.04	**0.05**	0.05	**0.06**
Whole body dose	0.03	**0.04**	0.03	**0.04**	0.04	**0.05**	0.04	**0.05**	0.05	**0.06**	0.05	**0.07**
ED (ICRP 60)		**0.03**		**0.03**		**0.04**		**0.04**	0.01	**0.05**	0.01	**0.05**
ED (ICRP 103)	0.01	**0.03**	0.02	**0.04**	0.02	**0.04**	0.02	**0.05**	0.02	**0.06**	0.03	**0.06**
Total filtration (mm) Al	**1.5**		**2.5**		**3.5**		**4.5**		**3.5**		**3.5**	
Cu	**0**		**0**		**0**		**0**		**0.1**		**0.2**	

Table 8.42 Normalized organ doses—Skull r-lat—>17 years

Beam projection:				r-lat			**Table cells:**						
Focus-to-detector distance (cm):				115			(+) = normal font = optimal field setting						
Tube voltage (kVp):				80			**(−) = bold font = suboptimal field setting**						
Total filtrations:				See last table row.			Empty table cells correspond to values <0.01						
Organ or tissue		**Organ dose D_T: Entrance dose K_E**						**(mSv * mGy^{-1})**					
		(+)	(−)	(+)	(−)	(+)	(−)	(+)	(−)	(+)	(−)	(+)	(−)
Skull		0.76	**0.77**	0.92	**0.92**	1.03	**1.04**	1.11	**1.12**	1.27	**1.28**	1.41	**1.42**
Brain		0.23	**0.23**	0.29	**0.29**	0.34	**0.33**	0.37	**0.37**	0.45	**0.45**	0.53	**0.52**
Oral mucosa		0.26	**0.27**	0.31	**0.32**	0.35	**0.36**	0.38	**0.40**	0.44	**0.46**	0.50	**0.52**
Salivary glands		0.37	**0.39**	0.42	**0.45**	0.45	**0.49**	0.48	**0.52**	0.53	**0.57**	0.58	**0.62**
Cervical spine		0.40	**0.90**	0.50	**1.12**	0.58	**1.30**	0.64	**1.43**	0.77	**1.70**	0.89	**1.95**
Thyroid		0.02	**0.46**	0.02	**0.55**	0.03	**0.61**	0.03	**0.66**	0.04	**0.75**	0.04	**0.82**
Thymus													**0.01**
Thoracic spine			**0.03**		**0.04**		**0.05**		**0.06**		**0.07**		**0.08**
Claviculae		0.02	**0.17**	0.03	**0.22**	0.03	**0.25**	0.03	**0.27**	0.04	**0.32**	0.05	**0.37**
Scapulae					**0.01**		**0.01**		**0.02**		**0.02**		**0.02**
Ribs					**0.01**		**0.01**		**0.02**		**0.02**		**0.02**
Oesophagus			**0.02**		**0.02**		**0.03**		**0.03**		**0.04**		**0.04**
Trachea		0.11	**0.26**	0.14	**0.32**	0.17	**0.36**	0.19	**0.40**	0.23	**0.47**	0.27	**0.54**
Lung											**0.01**		**0.01**
Mamma													
Heart													
Stomach													
Spleen													
Liver													
Gall bladder													
Pancreas													
Small intestine													
Large intestine													
Sigma/rectum													
Lumbar spine													
Pelvis													
Adrenals													
Kidneys													
Urinary bladder													
Uterus													
Ovaries													
Prostate													
Testes													
Upper arm					**0.01**		**0.01**		**0.01**		**0.02**		**0.02**
Elbow													
Forearm													
Upper leg													
Knee													
Lower leg													
Lymph nodes		0.05	**0.09**	0.06	**0.11**	0.06	**0.12**	0.07	**0.13**	0.08	**0.15**	0.09	**0.17**
Skin		0.05	**0.09**	0.05	**0.09**	0.06	**0.10**	0.06	**0.10**	0.06	**0.11**	0.06	**0.11**
Muscle		0.01	**0.02**	0.01	**0.02**	0.01	**0.03**	0.01	**0.03**	0.02	**0.03**	0.02	**0.03**
Skeleton		0.12	**0.14**	0.14	**0.17**	0.16	**0.19**	0.17	**0.20**	0.20	**0.23**	0.22	**0.26**
Red bone marrow		0.03	**0.03**	0.03	**0.04**	0.04	**0.05**	0.04	**0.05**	0.05	**0.06**	0.05	**0.07**
Whole body dose		0.03	**0.04**	0.04	**0.05**	0.04	**0.06**	0.04	**0.06**	0.05	**0.07**	0.06	**0.08**
ED (ICRP 60)			**0.03**		**0.04**		**0.04**	0.01	**0.05**	0.01	**0.05**	0.01	**0.06**
ED (ICRP 103)		0.02	**0.04**	0.02	**0.04**	0.02	**0.05**	0.02	**0.05**	0.03	**0.06**	0.03	**0.07**
Total filtration (mm)	**Al**	**1.5**		**2.5**		**3.5**		**4.5**		**3.5**		**3.5**	
	Cu	**0**		**0**		**0**		**0**		**0.1**		**0.2**	

References

1. Bontrager KL. Textbook of radiographic positioning and related anatomy. St. Louis, MO: Mosby Year Book; 1993.
2. Commission of the European Communities. European guidelines on quality criteria for diagnostic radiographic images in paediatrics. EUR 16261, SBN 92-827-7843-6; 1996.
3. Cook JV, Shah K, Pablot S, Kyriou J, Pettett A, Fitzgerald M. Guidelines on best practice in the x-ray imaging of children. A manual for all x-rax departments. London: Queen Mary's Hospital for Children, Surrey, and The Radiological Protection Centre; 1998.
4. Drexler G, Panzer W, Stieve FE. Die Bestimmung von Organdosen in der Röntgendiagnostik. Berlin: Hoffmann Verlag; 1993.
5. Ebel K-D, Willich E. Die Röntgenuntersuchung im Kindesalter—Technik und Indikation. Berlin: Springer; 1968.
6. Hart D, Jones DJ, Wall BF. Estimation of effective dose in diagnostic radiology from entrance surface dose and dose-area-product measurements. Oxon: National Radiological Protection Board NRPB-R262; 1994.
7. Hart D, Jones DJ, Wall BF. Coefficients for estimating effective doses from pediatric x-ray examinations. Oxon: National Radiological Protection Board NRPB-R279; 1996.
8. Hösle M. Untersuchung über die Patientendosis und die Bildqualität bei häufigen Röntgenaufnahmen im Säuglingsalter in Kliniken der alten Bundesländer mit Leitung durch einen Pädiater oder einen Radiologen. Dissertation, Universität München; 1993.
9. Pehe J. Untersuchung über die Patientendosis und die Bildqualität bei häufigen Röntgenaufnahmen in der Kinderradiologie. Dissertation, Universität München; 1993.
10. Schneider K, Fendel H, Bakowski C, Stein E, Kohn M, Kellner M, Schweighofer K, Cartagena G, Padovani R, Panzer W, Scheurer C, Wall B. Results of a dosimetry study in the European Community on frequent x ray examinations in infants. Radiat Prot Dosim. 1992;43:31–6.
11. Schneider K, Kohn MM, Bakowski C, Stein E, Freidhof C, Horwitz AE, Padovani R, Wall B, Panzer W, Fendel H. Impact of radiographic imaging criteria on dose and image quality in infants in an EC-wide survey. Radiat Prot Dosim. 1993;49:73–6.
12. Seidenbusch MC, Regulla D, Schneider K. Radiation exposure of children in pediatric radiology. Part 2: the PAEDOS algorithm for computer-assisted dose reconstruction in pediatric radiology and results for x-ray examinations of the skull. Fortschr Röntgenstr. 2008;180:522–39.
13. Winter W. Untersuchung über die Patientendosis und die Bildqualität bei häufigen Röntgenaufnahmen im Säuglingsalter bei niedergelassenen Ärzten (Pädiater, Orthopäden, Chirurgen). Dissertation, Universität München; 1994.

9 Chest

9.1 Introduction

Conventional chest radiographies are the most frequent X-ray examinations in childhood. In Fig. 9.1, the frequency of chest X-rays in the first five years of life in Dr. von Hauner Children's Hospital between 1976 and 2012 is depicted [20]; it may be interesting that this relation has been remained unchanged for over fifty years [9].

The type of chest radiography is determined by the clinical indication. Table 9.1 shows indications and Table 9.2 contraindications for chest radiography. Image criteria as required for radiological findings in chest radiographs are listed in Table 9.3. In order to fulfil these image criteria, the field settings listed in Tables 9.4 and 9.5 should be achieved. These field settings go along with the patient preparation as listed in Table 9.6. Exposure parameter combinations for various age groups and patient diameters are listed in Tables 9.7, 9.8, 9.9, 9.10, 9.11, 9.12, and 9.13. The age groups in Tables 9.8, 9.10, and 9.12 refer to the standard age groups of the

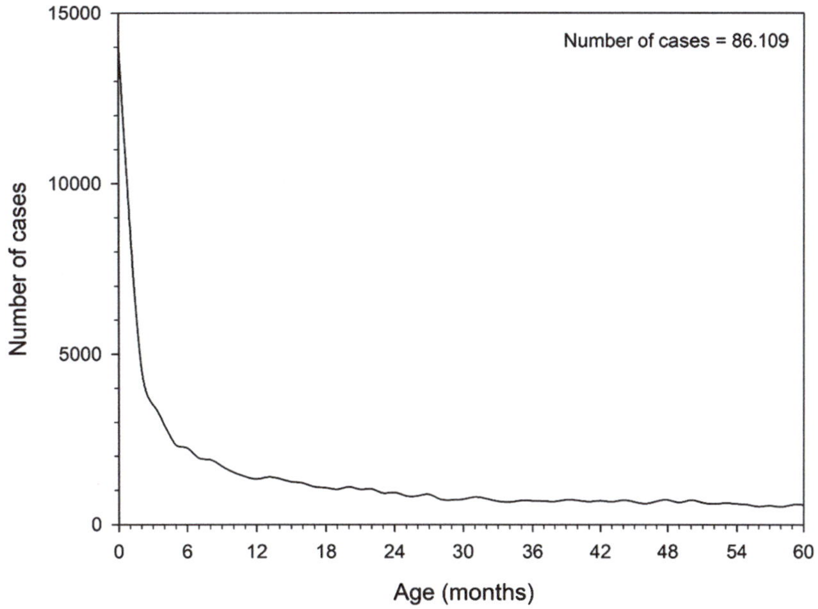

Fig. 9.1 Frequency of chest X-rays in the first five years of life during a 40-year period from 1976 until 2012 at Dr. von Hauner Children's Hospital, University of Munich. About one-third of chest X-rays in children were performed in the first year of age. (From [20], © Springer Verlag 2015, courtesy Springer Verlag)

The original version of this chapter was revised. The correction to this chapter can be found at https://doi.org/10.1007/978-3-030-18504-6_31

M. Seidenbusch et al., *Imaging Practice and Radiation Protection in Pediatric Radiology*,
https://doi.org/10.1007/978-3-030-18504-6_9

Table 9.1 Indications for chest radiographs

Surfactant deficiency syndrome (idiopathic respiratory distress syndrome)
Adult respiratory distress syndrome
Asthma
Pneumothorax
Septicemia
Pneumonia
Pleuritis/pleural empyema
Bronchiectasis
Atelectasis
Tracheal stenosis
Tracheoesophageal fistula
Lung malformation
Congenital heart disease
Endocarditis
Myocarditis
Pericarditis
Cardiomyopathia
Acquired heart disease
Lung edema
Mediastinal tumours
Lung tumours
Lung metastases
Anomalies of the chest wall (e.g. funnel chest, pigeon chest)
Tumours of the chest wall
Congenital diaphragmatic hernia
Chest trauma
Burns/scalding
Chemical aspiration (exhalational pneumonitis)
Aspiration (foreign body aspiration, aspiration pneumonia)
Lung fibrosis
Interstitial lung diseases

Table 9.2 Contraindications for chest radiographs *in upright position*

Neonates < 4 weeks
Severe congenital heart disease (e.g. Fallot's tetralogy, critical aortic stenosis) in neonates
Acute respiratory insufficiency
Septicemia
Acute abdomen
Large abdominal tumours
Severe neurological handicaps
Severe myopathies
Peritoneal dialysis
Mechanical ventilation
Extracorporeal membrane oxygenation (ECMO)

Table 9.3 Image criteria for chest radiographs.

Chest ap/pa
Symmetrical reproduction of the chest without rotation and tilting
Visually sharp reproduction of the trachea and the main bronchi
Reproduction of the vascular pattern of the inner 2/3 of the lungs, except in neonates
Visually sharp reproduction of the heart border and the diaphragm with the costophrenic angles
Reproduction of the spine, the ribs and the paraspinal structures, the retrocardiac lung, and the mediastinum
Chest lat
Visualization of the trachea down to the main bronchi
Visualization of apices of the lung
Visually sharp reproduction of both domes of the diaphragm
Visually sharp reproduction of posterior heart border
Reproduction of the hilar vessels
Reproduction of the sternum, the ribs and the thoracic spine

Table 9.4 Patient positions and beam projections for chest radiographs.

Patient position	Beam projections
Standing	pa + lat
Sitting	ap/pa + lat
Hanging	ap + lat
Supine	ap + lat
Decubitus	ap + lat

mathematical MIRD phantom family which has been used to calculate the normalized organ doses in this book. Tables 9.9, 9.11, and 9.13 show radiation qualities (i.e. various combinations of tube voltage and total filtration) which may be applied in clinical practice. Please note that not all of these radiation qualities are recommended by the authors. The standard projections of chest radiographies of the patient who is correctly positioned are depicted in Figs. 9.3, 9.4, 9.5, and 9.6. Also the optimal field (marked in green), the suboptimal field (marked in red), and the correct shielding are shown. Tables 9.15, 9.16, 9.17, 9.18, 9.19, 9.20, 9.21, 9.22, 9.23, 9.24, and 9.25 outline the virtual field settings on the skin surface of mathematical MIRD phantoms of various age groups and the resulting virtual radiographies with and without

Table 9.5 Optimal field settings for chest radiographs

Chest ap	
Upper field edge:	Tip of the mandible, above the clavicles
Field centring:	Midpoint of the sternum at the level of the mammillae
Lower field edge:	Midpoint between the xiphoid process and the navel
Lateral field edges:	Close to the lateral chest wall
Chest pa	
Upper field edge:	Spinous process of the sixth cervical vertebra
Field centring:	Mid-axillary line at the level of the mamillae[a]
Lower field edge:	Twelfth ribs
Lateral field edges:	Beneath the lateral chest wall
Chest lat	
Upper field edge:	Tip of the mandible/just above the shoulders
Field centring:	Midpoint of the axillae at the level of the mamillae
Lower field edge:	Lateral part of the costal arch
Anterior field edge:	Skin surface of the anterior chest wall
Posterior field edge:	Skin surface of the back
Annotation:	
[a]In daily practice, the subaxillary line will be used.	

Table 9.6 Patient preparation

General	Remove all clothes, external metallic devices, especially piercings, and jewellery from the neck and chest of the patient
	Immobilize infants and uncooperative patients. Alternatively, staff members could hold the patient (in this case also take care for the radiation protection of the staff members)
	For upright chest stand radiographs, infants < 12–24 months of age should be immobilized in a Babix® cradle/sleeve of appropriate size
Positioning	In the supine ap bedside/grid table X-rays the arms should be positioned alongside the chest in slight inward rotation. In chest X-rays ap, pa and chest lateral which are performed in upright position both arms should be held upright
	Rotation of the trunk in the longitudinal and/or horizontal axis should be avoided in the supine, decubitus and upright position
Attention	On the Bucky table the patient should always be positioned on a foam mat of 10 to 20 cm thickness. So, the lower part of the body will not be cut if lateral projection X-rays (cross table views) in prone or in decubitus position are performed
Shielding	Lead mask shielding should be used at the entrance side beneath the lower field edge. Lead shielding at the exit side is recommended by some radiographers

Table 9.7 Radiographic procedure

X-ray source	Use automatic exposure control (AEC) in patients > 10 years of age
	Focal spot size between 0.6 and 1.3 dependent on the age of the patient
	Check the use of additional filtration
	For bedside radiographs, adjust tube angulation dependent on the patient position
	Check the focus-to-detector distance
Field setting	Check the field size
	Check the field centring
Image detector system	Check the use of antiscatter grids
Patient	All chest radiographies should be performed in deep inspiration, only in special cases in exspiration, e.g. search for foreign body aspiration or foreign body ingestion

bone structures, respectively. Normalized organ doses resulting from these virtual field settings have been calculated for all age groups and have been listed in Tables 9.26, 9.27, 9.28, 9.29, 9.30, 9.31, 9.32, 9.33, 9.34, 9.35, 9.36, 9.37, 9.38, 9.39, 9.40, 9.41, 9.42, 9.43, 9.44, 9.45, 9.46, 9.47, 9.48, 9.49, 9.50, 9.51, 9.52, 9.53, 9.54, 9.55, 9.56, 9.57, 9.58, 9.59, 9.60, 9.61, 9.62, 9.63, 9.64, 9.65, 9.66, 9.67, 9.68, 9.69, 9.70, 9.71, 9.72, 9.73, 9.74, 9.75, 9.76, 9.77, 9.78, 9.79, 9.80, 9.81, 9.82, 9.83, 9.84, 9.85, 9.86, 9.87, 9.88, 9.89, 9.90, 9.91, 9.92, 9.93, 9.94, 9.95, 9.96, 9.97, 9.98, 9.99, 9.100, 9.101, 9.102, 9.103, 9.104, 9.105, 9.106, 9.107, 9.108, 9.109, 9.110, 9.111, 9.112, 9.113, 9.114, 9.115, 9.116, 9.117, 9.118, 9.119, 9.120, 9.121, 9.122, 9.123, 9.124, 9.125, 9.126, 9.127, 9.128, 9.129, 9.130, 9.131, 9.132, 9.133, 9.134, 9.135, 9.136, 9.137, 9.138, 9.139, 9.140, 9.141, 9.142, 9.143, 9.144, 9.145, 9.146, 9.147, 9.148, 9.149, 9.150, 9.151, 9.152, 9.153, 9.154, 9.155, 9.156, 9.157, 9.158, 9.159, 9.160, 9.161, 9.162, 9.163, 9.164, 9.165, 9.166, 9.167, and 9.168. Table cells with normal font refer to optimal field settings and table cells with bold font refer to suboptimal field settings. Please note that normalized organ doses have been provided also for exposure parameter combinations, which are not recommended by the authors.

Table 9.8 Radiographic techniques of bedside chest radiographs

Age group (years)	Beam direction	Patient position	Beam projection	Typical patient diameter (cm) With arms	Typical patient diameter (cm) No arms	Optimal field size (cm x cm)	Focus-to-detector distance (cm)	Voltage (kVp) 50	60	70	80	110	120	Antiscatter grid
24 gw[1]	Vertical	Supine	ap	5.2	5.2	6 x 8	100							
	Horizontal		lat	6.9	6.0	6 x 8								
28 gw	Vertical		ap	6.0	6.0	8 x 9								
	Horizontal		lat	8.2	7.1	6 x 9								
0	Vertical		ap	9.8	9.8	11 x 12								
	Horizontal		lat	12.6	10.9	10 x 12								
1–2	Vertical		ap	13.0	13.0	15 x 17								
	Horizontal		lat	17.1	15.2	14 x 17								
3–7	Vertical		ap	15.0	15.0	19 x 22								
	Horizontal		lat	22.6	19.6	14 x 22								
8–12	Vertical		ap	16.8	16.8	23 x 28								
	Horizontal		lat	27.3	23.8	17 x 27								
13–17	Vertical		ap	19.6	19.6	28 x 36								
	Horizontal		lat	34.4	29.6	20 x 34								
18	Vertical		ap	20.0	20.0	32 x 37								
	Horizontal		lat	38.9	34.4	20 x 37								

Annotations:
1 gw = gestational weeks
Parameter combination not recommended, normalized organ doses not provided
Parameter combination not recommended, normalized organ doses provided
Parameter combination recommended, normalized organ doses provided

Table 9.9 Radiation qualities of bedside chest radiographs

Voltage (kVp)	Total filtration: Without additional filtration 1.5 mm Al	2.5 Al	With additional filtration 3.5 Al	4.5 Al	3.5 Al + 0.1 Cu[1]	3.5 Al + 0.2 Cu[1]
50						
60						
70						
80						
110						
120						

Annotations:
1 The total beam filtration is usually 3.5 mm Al + 0.1 mm Cu, in adolescents optionally 3.5 mm Al + 0.2 mm Cu
Parameter combination not recommended, normalized organ doses provided
Parameter combination recommended, normalized organ doses provided

Table 9.10 Radiographic techniques of Bucky table chest radiographs

Age group (years)	Beam direction	Patient position	Beam projection	Typical patient diameter (cm) With arms	Typical patient diameter (cm) No arms	Optimal field size (cm x cm)	Focus-to-detector distance (cm)	Voltage (kVp) 50	60	70	80	110	120	Antiscatter grid
0	Vertical	Supine	ap	9.8	9.8	11 x 12	100[1]							
		Decubitus	lat	12.6	10.9	10 x 12								
	Horizontal		ap	9.8	9.8	11 x 12								
1–2	Vertical	Supine	ap	13.0	13.0	15 x 17								
		Decubitus	lat	17.1	15.2	14 x 17								
	Horizontal		ap	13.0	13.0	15 x 17								
3–7	Vertical	Supine	ap	15.0	15.0	19 x 22	115							
		Decubitus	lat	22.6	19.6	14 x 22								
	Horizontal		ap	15.0	15.0	19 x 22								
8–12	Vertical	Supine	ap	16.8	16.8	23 x 28								
		Decubitus	lat	27.3	23.8	17 x 27								
13–17	Vertical	Supine	ap	19.6	19.6	28 x 36								
		Decubitus	lat	34.4	29.6	20 x 34								
18	Vertical	Supine	ap	20.0	20.0	32 x 37								
		Decubitus	lat	38.9	34.4	20 x 37								

Annotations:
1 The patient is lying on the table top directly on the detector
Parameter combination not recommended, normalized organ doses not provided
Parameter combination not recommended, normalized organ doses provided
Parameter combination recommended, normalized organ doses provided

Table 9.11 Radiation qualities of Bucky table chest radiographs

Voltage (kVp)	Total filtration					
	Without additional filtration		With additional filtration			
	1.5 mm Al	2.5 Al	3.5 Al	4.5 Al	3.5 Al + 0.1 Cu[1]	3.5 Al + 0.2 Cu[1]
50						
60						
70						
80						
110						
120						

Annotations:

1 The total beam filtration is usually 3.5 mm Al + 0.1 mm Cu, in adolescents optionally 3.5 mm Al + 0.2 mm Cu.

(grey) Parameter combination not recommended, but normalized organ doses provided

(light blue) Parameter combination recommended, normalized organ doses provided

Table 9.12 Radiographic techniques of chest stand radiographs.

Age group (years)	Beam direction	Patient position	Beam projection	Typical patient diameter (cm) With arms	Typical patient diameter (cm) No arms	Optimal field size (cm x cm)	Focus-to-detector distance (cm)	Voltage (kVp) 50	60	70	80	110	120	Antiscatter grid
0		Hanging	ap	9.8	9.8	11 x 12								
			lat	12.6	10.9	10 x 12								
1–2		Hanging	ap	13.0	13.0	15 x 17								
			lat	17.1	15.2	14 x 17								
		Sitting	ap, pa	13.0	13.0	15 x 17								
	Horizontal		lat	17.1	15.2	14 x 17	150							
3–7		Sitting	ap, pa	15.0	15.0	19 x 22								
			lat	22.6	19.6	14 x 22								
		Standing	ap, pa	15.0	15.0	19 x 22								
			lat	22.6	19.6	14 x 22								
8–12		Sitting	pa	16.8	16.8	23 x 28								
			lat	27.3	23.8	17 x 27								
		Standing	pa	16.8	16.8	23 x 28								
			lat	27.3	23.8	17 x 27	150							
13–17		Sitting	pa	19.6	19.6	28 x 36								
			lat	34.4	29.6	20 x 34								
		Standing	pa	19.6	19.6	28 x 36								
	Horizontal		lat	34.4	29.6	20 x 34								
18		Sitting	pa	20.0	20.0	32 x 37								
			lat	38.9	34.4	20 x 37								
		Standing	pa	20.0	20.0	32 x 37								
			lat	38.9	34.4	20 x 37								

Annotations:

(white) Not applicable

(black) Parameter combination not recommended, normalized organ doses not provided

(grey) Parameter combination not recommended, but normalized organ doses provided

(light blue) Parameter combination recommended, normalized organ doses provided

Table 9.13 Radiation qualities of chest stand radiographs

Voltage (kVp)	Total filtration					
	Without additional filtration		With additional filtration			
	1.5 mm Al	2.5 Al	3.5 Al	4.5 Al	3.5 Al + 0.1 Cu[1]	3.5 Al + 0.2 Cu[1]
50						
60						
70						
80						
110						
120						

Annotations:

1 The total beam filtration is usually 3.5 mm Al + 0.1 mm Cu, in adolescents optionally 3.5 mm Al + 0.2 mm Cu

(grey) Parameter combination not recommended, normalized organ doses provided

(light blue) Parameter combination recommended, normalized organ doses provided

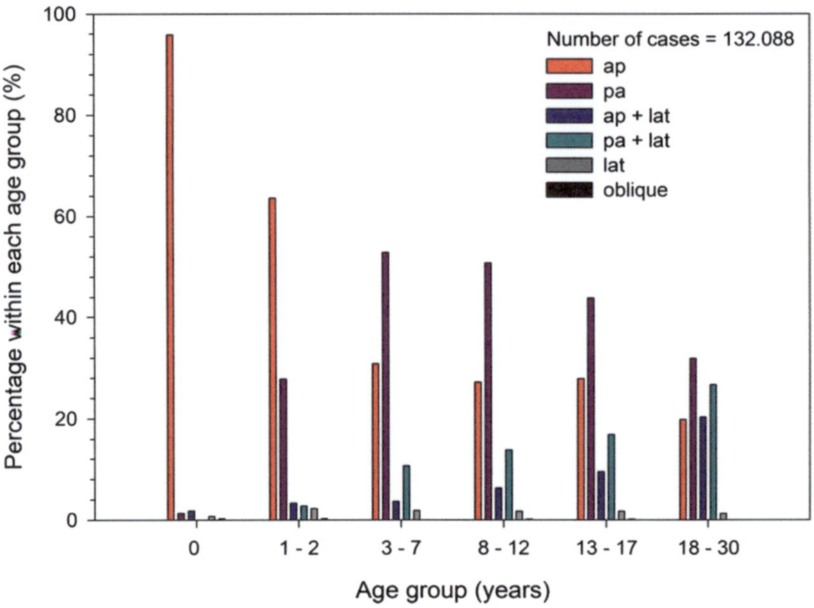

Fig. 9.2 Beam projections used for chest radiographies in various age groups. (From [20], © Springer Verlag 2015, courtesy Springer Verlag)

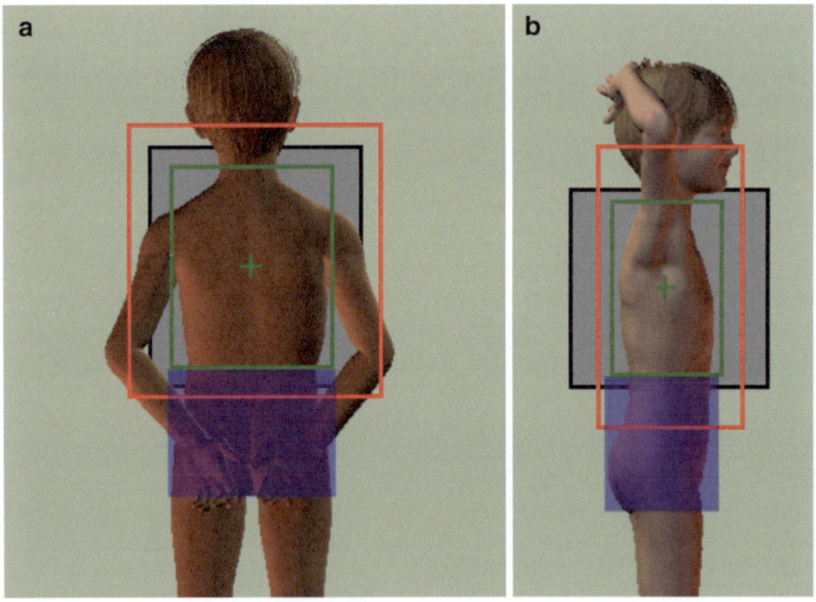

Fig. 9.3 Field settings for chest radiographs in upright position. (**a**) Postero-anterior projection. (**b**) Lateral projection (© V. Rösenberger & M. Seidenbusch 2018). Green lines = optimal field settings; green cross = centring point; red lines = suboptimal field settings; grey shaded area = image detector; blue shaded area = lead mask shielding of the patient

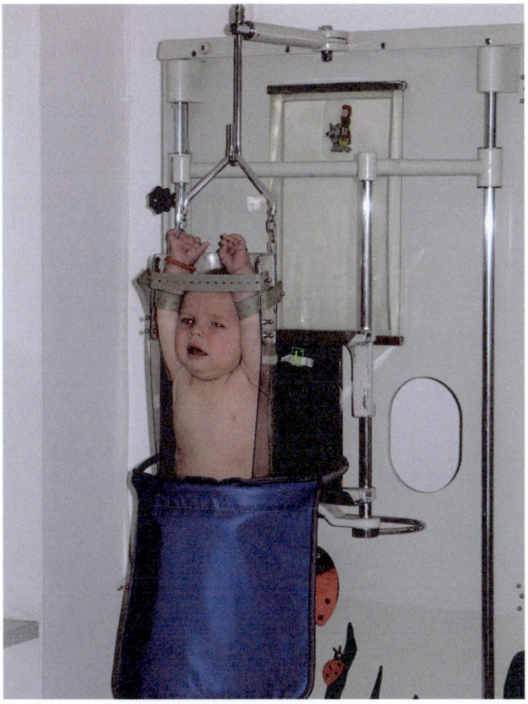

Fig. 9.4 Chest ap X-ray of an infant (hanging). The patient is immobilized in a BABIX® cradle. The head and both extended arms are fixed by rubber bands. The trunk is fixed by a slabstock foam (not visible behind the lead apron). (© V. Rösenberger 2018)

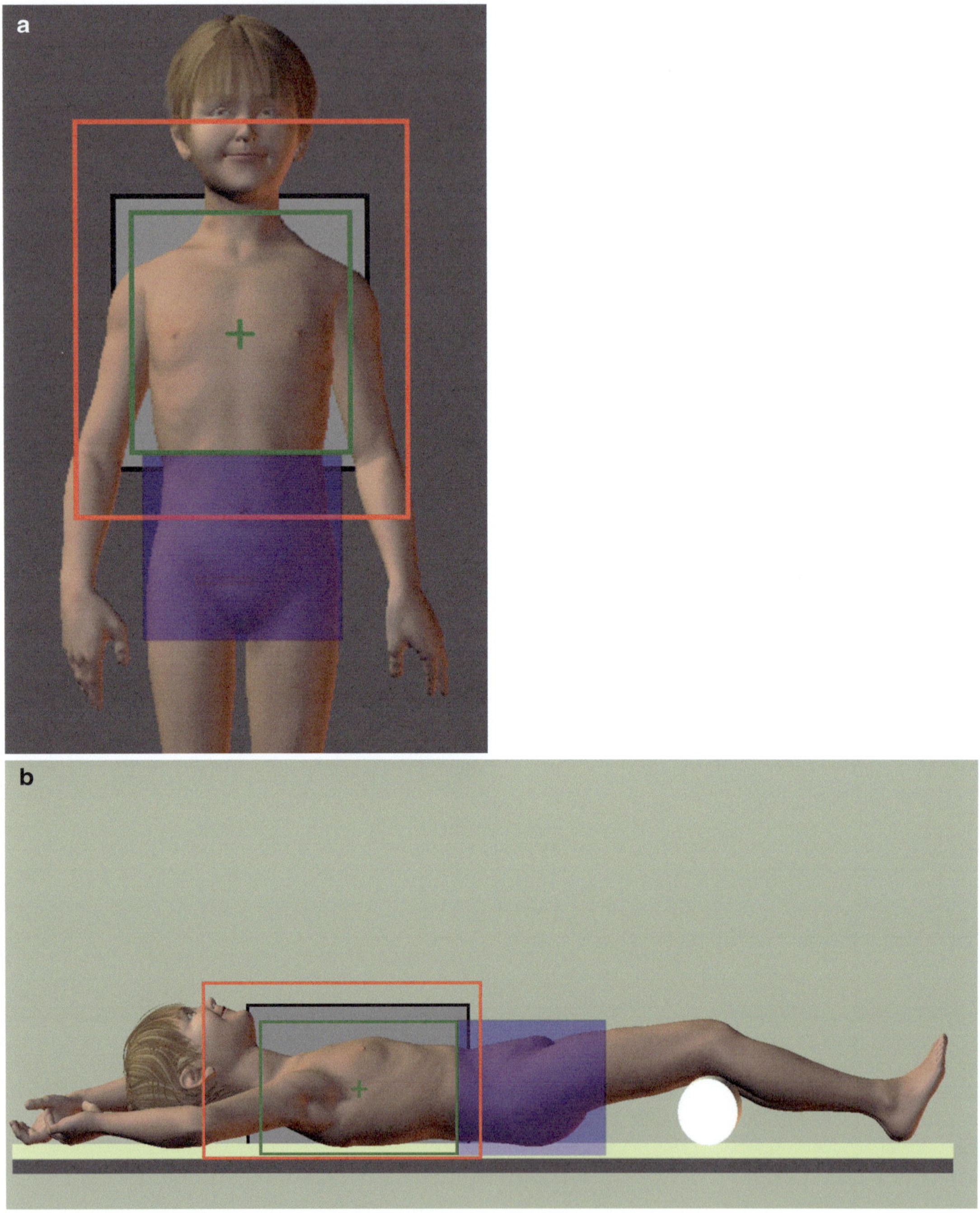

Fig. 9.5 Field settings for chest radiographs in supine position. (**a**) Antero-posterior projection. (**b**) Lateral projection (cross table view) (© V. Rösenberger & M. Seidenbusch 2018). Green lines = optimal field settings; green cross = centring point; red lines = suboptimal field settings; grey shaded area = image detector; blue shaded area = lead mask shielding of the patient; dark grey shaded area = Bucky table; white cylinder = foam roll; white device behind the head and neck = foam pillow; white foam mat on the top of the Bucky table

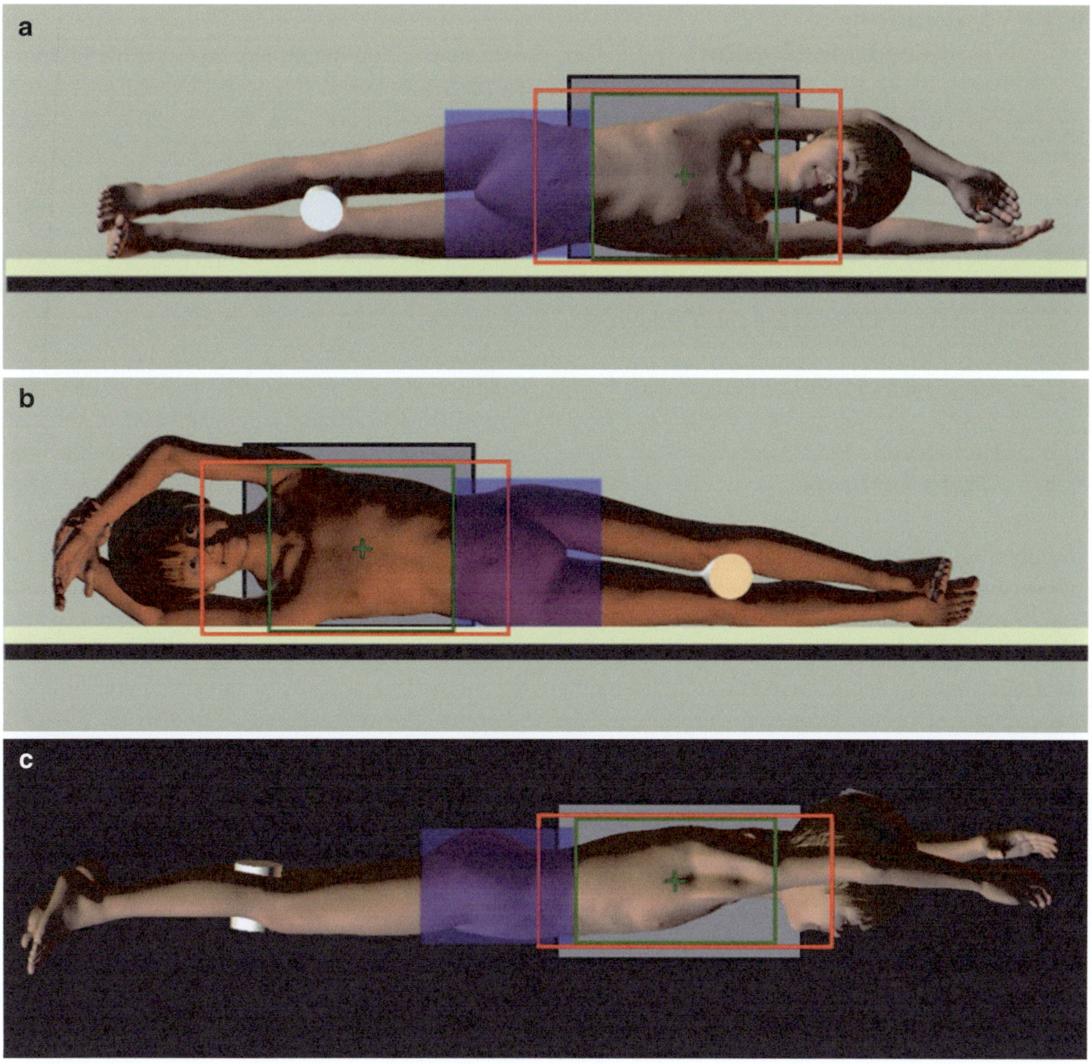

Fig. 9.6 Clinical field settings for chest radiographs in right and left decubitus position. (**a**, **b**) Antero-posterior projection. (**c**) Lateral projection. (© V. Rösenberger & M. Seidenbusch 2018). Green lines = optimal field settings; green cross = centring point; red lines = suboptimal field settings; grey shaded area = image detector; blue shaded area = lead mask shielding of the patient; dark grey shaded area = Bucky table; white cylinder = foam roll; white device behind the head and neck = foam pillow; white foam mat on the top of the Bucky table

9.2 Indications and Contraindications

Indications and contraindications for chest radiographs are given in Tables 9.1 and 9.2.

9.3 Image Criteria

The image criteria required for chest radiographs are listed in Table 9.3.

See also [3, 4, 9, 15, 24].

9.4 Field Settings

In Table 9.4, patient positions and beam projections used for chest radiographs are listed, and in Table 9.5, the optimal field settings are listed which are required in order to fulfil the image criteria mentioned in Table 9.3. As chest radiographs are usually performed in deep inspiration, the ap projection will be preferred in neonates and young infants and uncooperative patients. Thus, the light field and the respiration of the patient can continuously be watched during the X-ray exposure. In cooperative patients older than 5 years of age, the pa projection will be preferred whenever possible in order to reduce the breast dose.

It should be stated here that choosing the beam projection has an impact on the radiation exposure of organs and tissues which are localized close to the patient's skin surface. Thus, from the dosimetric point of view, the radiographer will prefer the pa projection in chest radiographs in order to reduce the organ dose of the mammae which would almost equal the entrance surface dose when the ap projection was taken. As a consequence, the pa projection has been used in the Dr. von Hauner Children's Hospital if the patients were able to sit or stand in front of the vertical chest stand (Fig. 9.2).

Chest ap X-rays in the right and left decubitus projection can be used in uncooperative children to detect foreign body aspiration. In decubitus position, the lower lung is compressed and then smaller than the upper lung. Equal size or increased size of the lower lung is indicative of obstructive emphysema caused by valve close of the bronchus. This technique is no longer be used to detect small pneumothoraces or small pleural effusions.

Lateral view in right or left decubitus position can also be used for detection of foreign body aspiration. The arms are usually fixed behind the back of the patient. Then the suboptimal field size must be used not to miss foreign bodies in pharynx. For more details, see also [2, 5, 6, 10, 11, 14, 16–18, 22, 23 25–28].

9.5 Patient Preparation and Radiographic Procedure

See Tables 9.6, 9.7 and [1, 2].

9.6 Exposure Parameters

In Tables 9.8, 9.9, 9.10, 9.11, 9.12, and 9.13, exposure parameter combinations used for chest

Table 9.14 Inconsistencies between the mathematical MIRD phantoms and real human anatomy. For details, see the text

Anatomical region	Inconsistencies
Head	No eyes, nose, mouth, and ears have been inserted
	No eye lens has been inserted
Chest	No variation of lung density with age
	Mass of glandular tissue is underestimated in newborn infants
Skeleton	The shape of the spine is unrealistic (no lordosis, no kyphosis considered), the vertebral column is not clearly depicted
	Ribs are drawn as complete circles and thicker than in real humans, the course of the ribs is horizontal
	No xiphoid has been inserted

radiographs in various age groups are listed. Dark grey shaded table cells refer to exposure parameter combinations which are not recommended for paediatric chest radiographs and for which no normalized organ doses have been provided. Light grey and light blue shaded table cells refer to exposure parameter combinations for which normalized organ doses have been provided with light grey shaded table cells referring to exposure parameter combinations being not recommended and light blue shaded table cells referring to exposure parameter combinations following the European Guidelines [3]. Note that due to the fact that the radiation dose is an exponential function of the patient diameter, it may make sense to use the patient diameters instead of the patient's age groups for determining the adequate exposure parameters from Tables 9.8, 9.10, and 9.12. In this case, for determining the corresponding normalized organ doses from Tables 9.26, 9.27, 9.28, 9.29, 9.30, 9.31, 9.32, 9.33, 9.34, 9.35, 9.36, 9.37, 9.38, 9.39, 9.40, 9.41, 9.42, 9.43, 9.44, 9.45, 9.46, 9.47, 9.48, 9.49, 9.50, 9.51, 9.52, 9.53, 9.54, 9.55, 9.56, 9.57, 9.58, 9.59, 9.60, 9.61, 9.62, 9.63, 9.64, 9.65, 9.66, 9.67, 9.68, 9.69, 9.70, 9.71, 9.72, 9.73, 9.74, 9.75, 9.76, 9.77, 9.78, 9.79, 9.80, 9.81, 9.82, 9.83, 9.84, 9.85, 9.86, 9.87, 9.88, 9.89, 9.90, 9.91, 9.92, 9.93, 9.94, 9.95, 9.96, 9.97, 9.98, 9.99, 9.100, 9.101, 9.102, 9.103, 9.104, 9.105, 9.106, 9.107, 9.108, 9.109, 9.110, 9.111, 9.112, 9.113, 9.114, 9.115, 9.116, 9.117, 9.118, 9.119, 9.120, 9.121, 9.122, 9.123, 9.124, 9.125, 9.126, 9.127, 9.128, 9.129, 9.130, 9.131, 9.132, 9.133, 9.134, 9.135, 9.136, 9.137, 9.138, 9.139, 9.140, 9.141, 9.142, 9.143, 9.144, 9.145, 9.146, 9.147, 9.148, 9.149, 9.150, 9.151, 9.152, 9.153, 9.154, 9.155, 9.156, 9.157, 9.158, 9.159, 9.160, 9.161, 9.162, 9.163, 9.164, 9.165, 9.166, 9.167, and 9.168, the age group related to the patient diameter should be used.

9.7 Normalized Organ Doses

For calculating normalized organ doses, the field settings described in Sect. 9.4 were transferred to the mathematical MIRD phantoms. Inconsistencies between the geometrical anatomy of the mathematical phantoms and the real human anatomy have been listed in Table 9.14 whenever they are of relevance for the validity of the normalized organ doses.

In Tables 9.15, 9.16, 9.17, 9.18, 9.19, 9.20, 9.21, 9.22, 9.23, 9.24, and 9.25, optimal and suboptimal field settings are listed for the different age groups and the beam projections given in Tables 9.8, 9.9, 9.10, 9.11, 9.12, and 9.13. Green lines refer to optimal field settings, red lines refer to suboptimal field settings, respectively. In the tables, the optimal and suboptimal field settings which were used for performing the calculations of the normalized organ doses provided in Tables 9.26, 9.27, 9.28, 9.29, 9.30, 9.31, 9.32, 9.33, 9.34, 9.35, 9.36, 9.37, 9.38, 9.39, 9.40, 9.41, 9.42, 9.43, 9.44, 9.45, 9.46, 9.47, 9.48, 9.49, 9.50, 9.51, 9.52, 9.53, 9.54, 9.55, 9.56, 9.57, 9.58, 9.59, 9.60, 9.61, 9.62, 9.63, 9.64, 9.65, 9.66, 9.67, 9.68, 9.69, 9.70, 9.71, 9.72, 9.73, 9.74, 9.75, 9.76, 9.77, 9.78, 9.79, 9.80, 9.81, 9.82, 9.83, 9.84, 9.85, 9.86, 9.87, 9.88, 9.89, 9.90, 9.91, 9.92, 9.93, 9.94, 9.95, 9.96, 9.97, 9.98, 9.99, 9.100, 9.101, 9.102, 9.103, 9.104, 9.105, 9.106, 9.107, 9.108, 9.109, 9.110, 9.111, 9.112, 9.113, 9.114, 9.115, 9.116, 9.117, 9.118, 9.119, 9.120, 9.121, 9.122, 9.123, 9.124, 9.125, 9.126, 9.127, 9.128, 9.129, 9.130, 9.131, 9.132, 9.133, 9.134, 9.135, 9.136, 9.137, 9.138, 9.139, 9.140, 9.141, 9.142, 9.143, 9.144, 9.145, 9.146, 9.147, 9.148, 9.149, 9.150, 9.151, 9.152, 9.153, 9.154, 9.155, 9.156, 9.157, 9.158, 9.159, 9.160, 9.161, 9.162, 9.163, 9.164, 9.165, 9.166, 9.167, and 9.168 were drawn in on the skin surface of the mathematical phantoms. The virtual mathematical radiographs resulting from these field settings were depicted in the tables with and without bone structures, respectively, in order to enable the reader to visually compare the virtual mathematical radiographs with the real radiographs made under clinical conditions. For comparing the virtual mathematical field settings with the real clinical field settings, the virtual mathematical field edges were also listed in the tables.

As the normalized organ doses provided in Tables 9.26, 9.27, 9.28, 9.29, 9.30, 9.31, 9.32, 9.33, 9.34, 9.35, 9.36, 9.37, 9.38, 9.39, 9.40, 9.41, 9.42, 9.43, 9.44, 9.45, 9.46, 9.47, 9.48, 9.49, 9.50, 9.51, 9.52, 9.53, 9.54, 9.55, 9.56, 9.57, 9.58, 9.59,

9.60, 9.61, 9.62, 9.63, 9.64, 9.65, 9.66, 9.67, 9.68, 9.69, 9.70, 9.71, 9.72, 9.73, 9.74, 9.75, 9.76, 9.77, 9.78, 9.79, 9.80, 9.81, 9.82, 9.83, 9.84, 9.85, 9.86, 9.87, 9.88, 9.89, 9.90, 9.91, 9.92, 9.93, 9.94, 9.95, 9.96, 9.97, 9.98, 9.99, 9.100, 9.101, 9.102, 9.103, 9.104, 9.105, 9.106, 9.107, 9.108, 9.109, 9.110, 9.111, 9.112, 9.113, 9.114, 9.115, 9.116, 9.117, 9.118, 9.119, 9.120, 9.121, 9.122, 9.123, 9.124, 9.125, 9.126, 9.127, 9.128, 9.129, 9.130, 9.131, 9.132, 9.133, 9.134, 9.135, 9.136, 9.137, 9.138, 9.139, 9.140, 9.141, 9.142, 9.143, 9.144, 9.145, 9.146, 9.147, 9.148, 9.149, 9.150, 9.151, 9.152, 9.153, 9.154, 9.155, 9.156, 9.157, 9.158, 9.159, 9.160, 9.161, 9.162, 9.163, 9.164, 9.165, 9.166, 9.167, and 9.168 refer to entrance kerma values, optimal and suboptimal field sizes were provided in the tables in order to enable the reader to calculate entrance air kerma dose values from clinically measured dose area product values. Please note that these optimal and suboptimal field sizes refer to the mathematical phantoms of an age group only and not to real patients with corresponding ages.

Please note also that the anatomy of the mathematical phantoms has been idealized. Therefore, the virtual field settings for calculating the normalized organ doses have been defined with a view to ensuring that the resulting virtual organ doses would be consistent with the real organ doses which would be achieved during real X-ray examinations being performed with the real field settings recommended in Sect. 9.4. Thus, the field settings shown in the following tables may not be consistent with the clinical field settings as recommended in Sect. 9.4.

For determining the normalized organ doses, in chest radiographs in lateral projection, the arm was removed from the phantom under optimal field settings and was not removed under suboptimal field settings. By the way, suboptimal field settings were chosen arbitrarily. The range of clinical field settings has been discussed in literature [2, 5, 6, 10, 11, 14, 16–18, 22, 23, 25–28].

Normalized organ doses for chest radiographs are also listed in e.g. [7, 12, 13, 19, 21].

Table 9.15 Field settings—Chest ap

Age level (gestational weeks)		Anatomical landmarks: Optimal field setting		Anatomical landmarks: Suboptimal field setting	
24	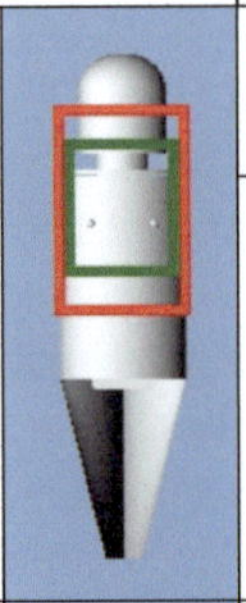	Upper field edge: Tip of the mandible, above the clavicles Lower field edge: 12th rib Lateral field edges: Proximal metaphyses of the humeri Field size = field width x field height: 6 cm x 8 cm = 48 cm^2		Upper field edge: At the upper margins of the parotid glands Lower field edge: At upper margin of the iliac crests Lateral field edges: Both arms included Field size = field width x field height: 7 cm x 12 cm = 84 cm^2	

28	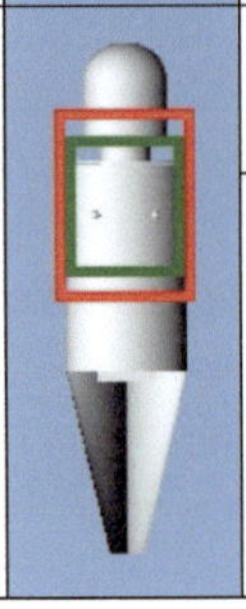	Upper field edge: Tip of the mandible, above the clavicles Lower field edge: 12th rib Lateral field edges: Proximal metaphyses of the humeri Field size = field width x field height: 8 cm x 9 cm = 72 cm^2		Upper field edge: At the upper margins of the parotid glands Lower field edge: At upper margin of the iliac crests Lateral field edges: Both arms included Field size = field width x field height: 9 cm x 13 cm = 117 cm^2	
			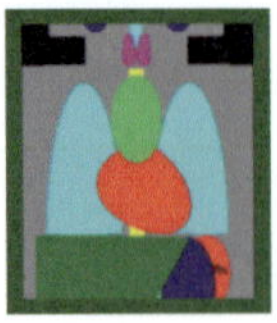	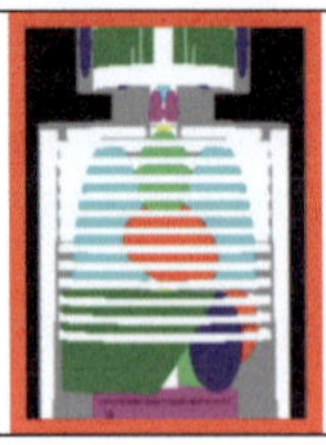	

Table 9.16 Field settings—Chest ap

Age level (gestational weeks)		Anatomical landmarks: Optimal field setting		Anatomical landmarks: Suboptimal field setting	
0	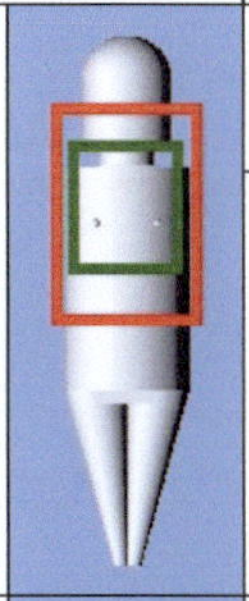	Upper field edge: Tip of the mandible, above the clavicles Lower field edge: 12th rib Lateral field edges: Proximal metaphyses of the humeri Field size = field width x field height: 11 cm x 12 cm = 132 cm^2		Upper field edge: At the upper margins of the parotid glands Lower field edge: At upper margin of the iliac crests Lateral field edges: Both arms included Field size = field width x field height: 14 cm x 20 cm = 280 cm^2	
			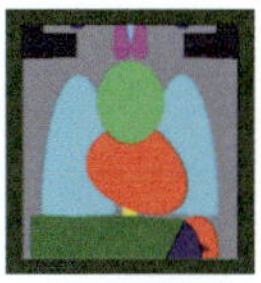		
1–2	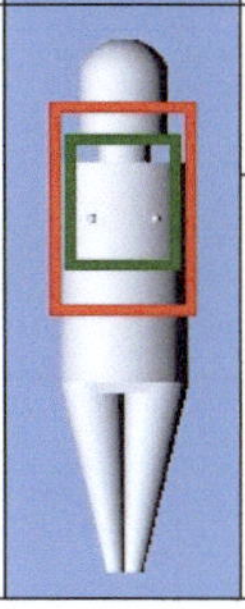	Upper field edge: Tip of the mandible, above the clavicles Lower field edge: 12th rib Lateral field edges: Proximal metaphyses of the humeri Field size = field width x field height: 15 cm x 17 cm = 255 cm^2		Upper field edge: At the upper margins of the parotid glands Lower field edge: At upper margin of the iliac crests Lateral field edges: Both arms included Field size = field width x field height: 18 cm x 28 cm = 504 cm^2	
			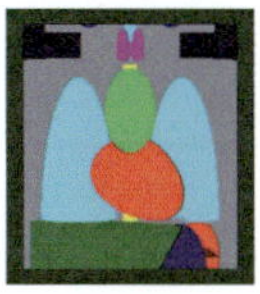		

Table 9.17 Field settings—Chest ap

Age level (gestational weeks)		Anatomical landmarks: Optimal field setting		Anatomical landmarks: Suboptimal field setting	
3–7	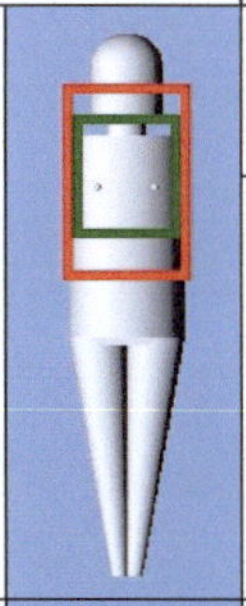	Upper field edge: Tip of the mandible, above the clavicles Lower field edge: 12th rib Lateral field edges: Proximal metaphyses of the humeri Field size = field width x field height: 19 cm x 22 cm = 418 cm^2		Upper field edge: At the upper margins of the parotid glands Lower field edge: At upper margin of the iliac crests Lateral field edges: Both arms included Field size = field width x field height: 21 cm x 37 cm= 777 cm^2	
			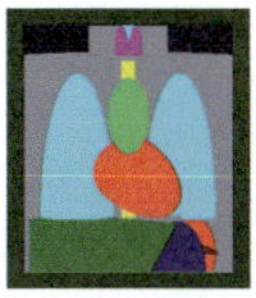		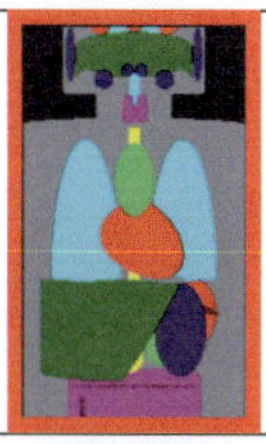
8–12	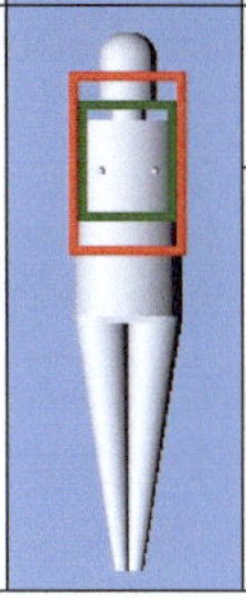	Upper field edge: Tip of the mandible, above the clavicles Lower field edge: 12th rib Lateral field edges: Proximal metaphyses of the humeri Field size = field width x field height: 23 cm x 28 cm = 644 cm^2		Upper field edge: At the upper margins of the parotid glands Lower field edge: At upper margin of the iliac crests Lateral field edges: Both arms included Field size = field width x field height: 26 cm x 44 cm = 1,144 cm^2	
			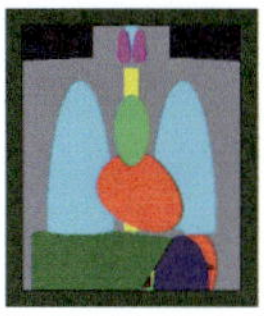		

Table 9.18 Field settings—Chest ap

Age level (gestational weeks)		Anatomical landmarks			
		Optimal field setting		Suboptimal field setting	
13–17	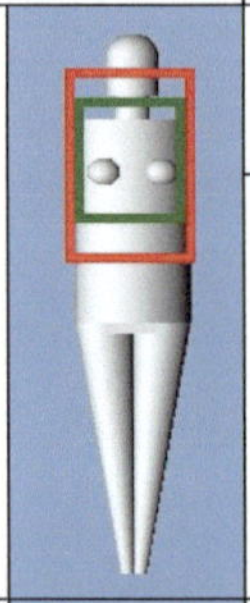	Upper field edge: Tip of the mandible, above the clavicles Lower field edge: 12th rib Lateral field edges: Proximal metaphyses of the humeri Field size = field width x field height: 28 cm x 36 cm = 1,008 cm^2		Upper field edge: At the upper margins of the parotid glands Lower field edge: At upper margin of the iliac crests Lateral field edges: Both arms included Field size = field width x field height: 32 cm x 55 cm = 1,760 cm^2	
			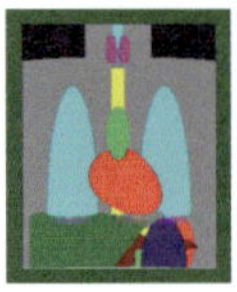	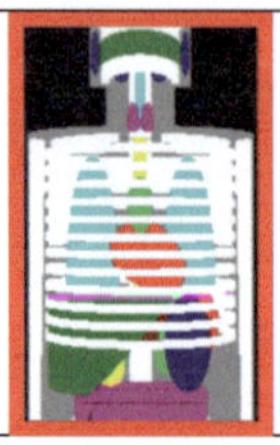	
18	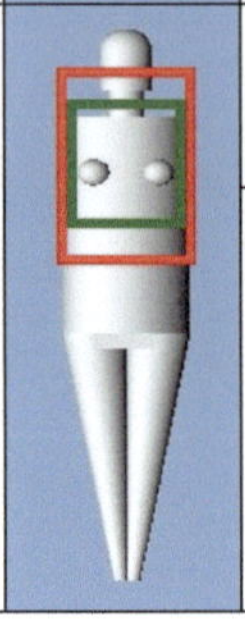	Upper field edge: Tip of the mandible, above the clavicles Lower field edge: 12th rib Lateral field edges: Proximal metaphyses of the humeri Field size = field width x field height: 32 cm x 37 cm = 1,184 cm^2		Upper field edge: At the upper margins of the parotid glands Lower field edge: At upper margin of the iliac crests Lateral field edges: Both arms included Field size = field width x field height: 40 cm x 60 cm = 2,400 cm^2	

Table 9.19 Field settings—Chest pa

Age group (years)		Anatomical landmarks			
		Optimal field setting		Suboptimal field setting	
0	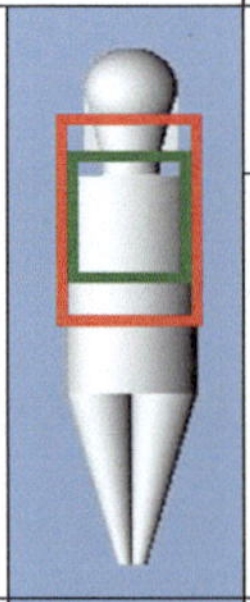	Upper field edge: Above the spinous process of the 6th cervical vertebra Lower field edge: 12th rib Lateral field edges: Proximal metaphyses of the humeri Field size = field width x field height: 11 cm x 12 cm = 132 cm^2		Upper field edge: At the upper margins of the parotid glands Lower field edge: At upper margin of the iliac crests Lateral field edges: Both arms included Field size = field width x field height: 14 cm x 20 cm = 280 cm^2	
		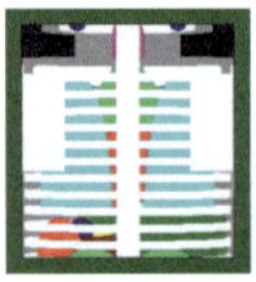			
1–2	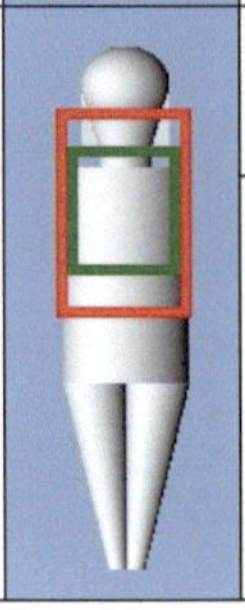	Upper field edge: Above the spinous process of the 6th cervical vertebra Lower field edge: 12th rib Lateral field edges: Proximal metaphyses of the humeri Field size = field width x field height: 15 cm x 17 cm = 255 cm^2		Upper field edge: At the upper margins of the parotid glands Lower field edge: At upper margin of the iliac crests Lateral field edges: Both arms included Field size = field width x field height: 18 cm x 28 cm = 504 cm^2	
			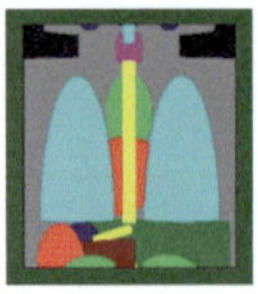	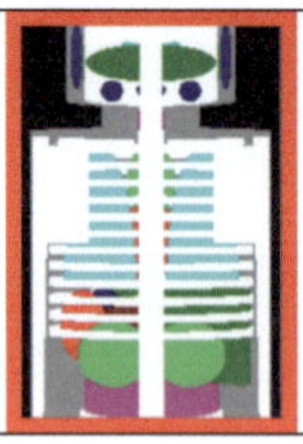	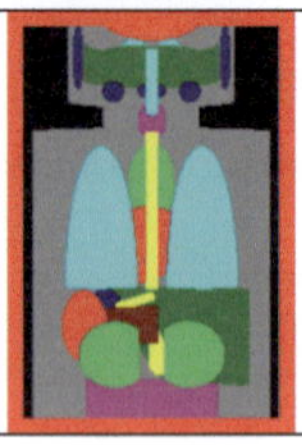

Table 9.20 Field settings—Chest pa

Age group (years)		Anatomical landmarks			
		Optimal field setting		Suboptimal field setting	
3–7	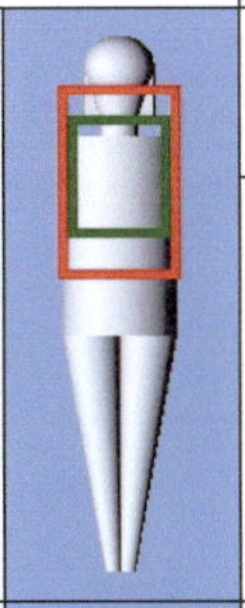	Upper field edge: Above the spinous process of the 6th cervical vertebra Lower field edge: 12th rib Lateral field edges: Proximal metaphyses of the humeri Field size = field width x field height: 19 cm x 22 cm= 418 cm²		Upper field edge: At the upper margins of the parotid glands Lower field edge: At upper margin of the iliac crests Lateral field edges: Both arms included Field size = field width x field height: 21 cm x 37 cm = 777 cm²	
		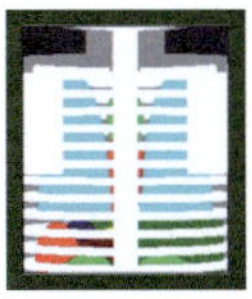		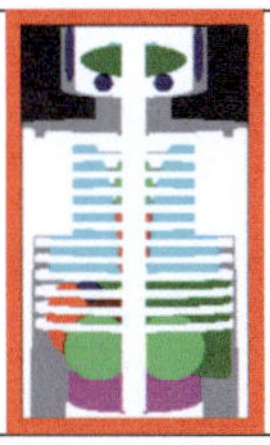	
8–12	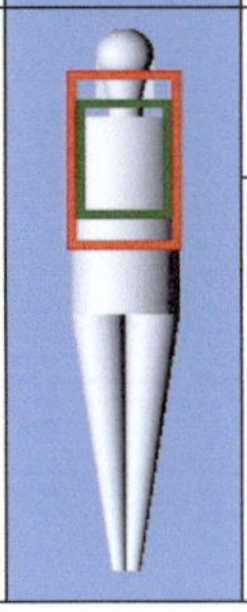	Upper field edge: Above the spinous process of the 6th cervical vertebra Lower field edge: 12th rib Lateral field edges: Proximal metaphyses of the humeri Field size = field width x field height: 23 cm x 28 cm = 644 cm²		Upper field edge: At the upper margins of the parotid glands Lower field edge: At upper margin of the iliac crests Lateral field edges: Both arms included Field size = field width x field height: 26 cm x 44 cm = 1,144 cm²	
		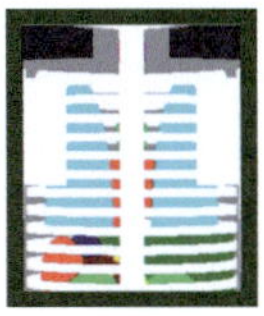	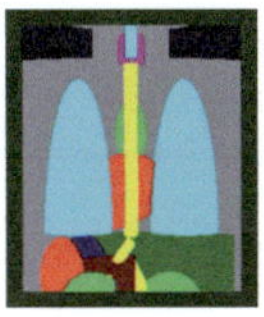	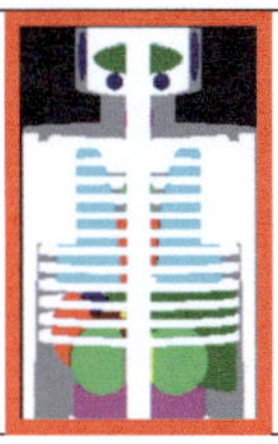	

Table 9.21 Field settings—Chest pa

Age level (gestational weeks)		Anatomical landmarks			
		Optimal field setting		Suboptimal field setting	
13–17	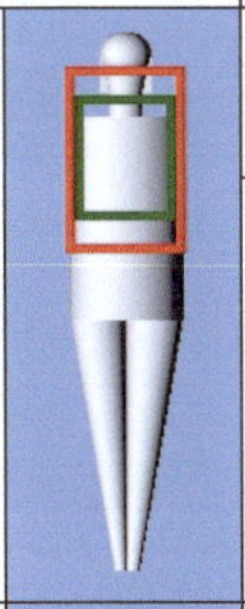	Upper field edge: Above the spinous process of the 6th cervical vertebra Lower field edge: 12th rib Lateral field edges: Proximal metaphyses of the humeri Field size = field width x field height: 28 cm x 36 cm = 1,512 cm²		Upper field edge: At the upper margins of the parotid glands Lower field edge: At upper margin of the iliac crests Lateral field edges: Both arms included Field size = field width x field height: 32 cm x 55 cm = 1,760 cm²	
		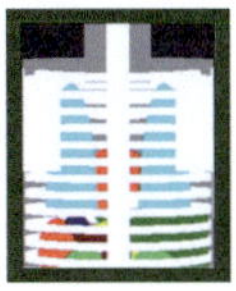		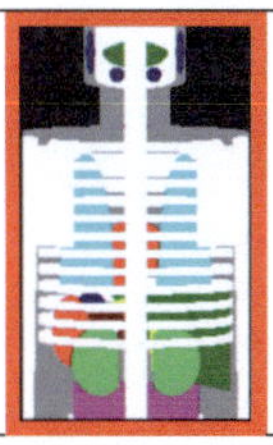	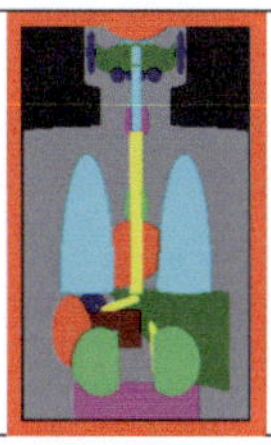
18	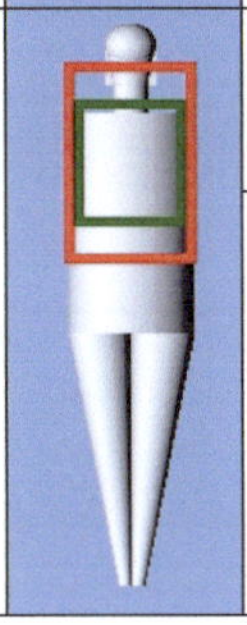	Upper field edge: Above the spinous process of the 6th cervical vertebra Lower field edge: 12th rib Lateral field edges: Proximal metaphyses of the humeri Field size = field width x field height: 32 cm x 37 cm = 1,184 cm²		Upper field edge: At the upper margins of the parotid glands Lower field edge: At upper margin of the iliac crests Lateral field edges: Both arms included Field size = field width x field height: 40 cm x 60 cm = 2,400 cm²	
		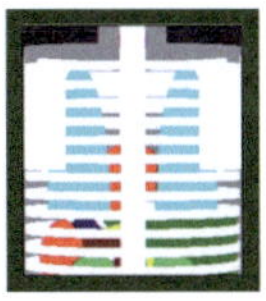			

Table 9.22 Field settings—Chest lat

Age level (gestational weeks)		Anatomical landmarks: Optimal field setting	Anatomical landmarks: Suboptimal field setting
24	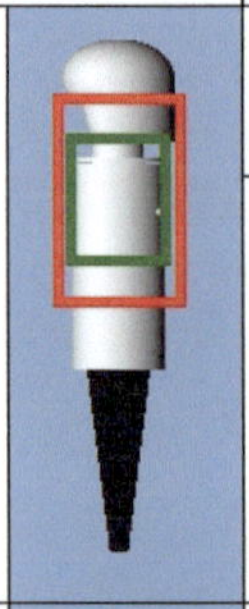	Upper field edge: Above the spinous process of the 6th cervical vertebra Lower field edge: 12th rib included Lateral field edges: Close to the anterior and posterior chest wall Field size = field width x field height: 6 cm x 8 cm = 48 cm^2 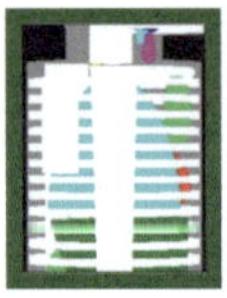 	Upper field edge: At the upper margins of the parotid glands Lower field edge: Upper margin of the iliac crests Lateral field edges: Beneath anterior and posterior chest wall Field size = field width x field height: 7 cm x 12 cm= 84 cm^2 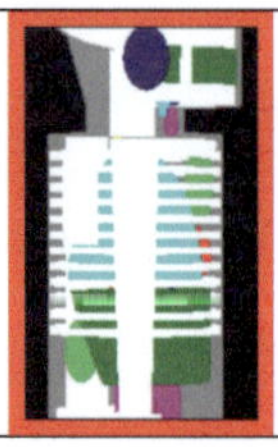 
28	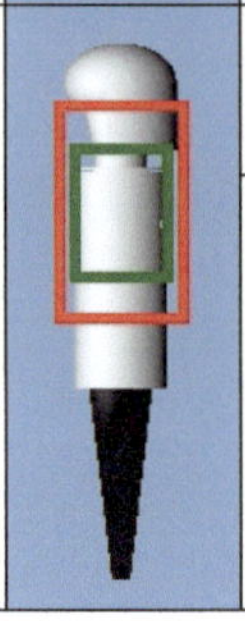	Upper field edge: Above the spinous process of the 6th cervical vertebra Lower field edge: 12th rib included Lateral field edges: Close to the anterior and posterior chest wall Field size = field width x field height: 6 cm x 9 cm = 54 cm^2 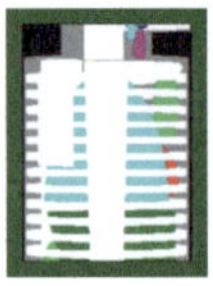 	Upper field edge: At the upper margins of the parotid glands Lower field edge: Upper margin of the iliac crests Lateral field edges: Beneath anterior and posterior chest wall Field size = field width x field height: 8 cm x 14 cm = 112 cm^2  

Table 9.23 Field settings—Chest lat

Age level (gestational weeks)		Anatomical landmarks: Optimal field setting	Anatomical landmarks: Suboptimal field setting
0	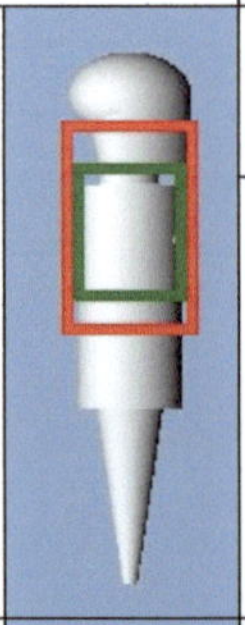	Upper field edge: Above the spinous process of the 6th cervical vertebra Lower field edge: 12th rib included Lateral field edges: Close to the anterior and posterior chest wall Field size = field width x field height: 10 cm x 12 cm = 120 cm^2 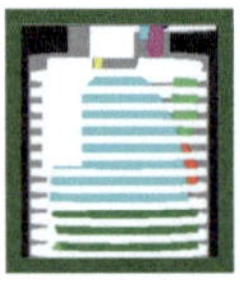 	Upper field edge: At the upper margins of the parotid glands Lower field edge: Upper margin of the iliac crests Lateral field edges: Beneath anterior and posterior chest wall Field size = field width x field height: 12 cm x 20 cm= 240 cm^2  
1–2	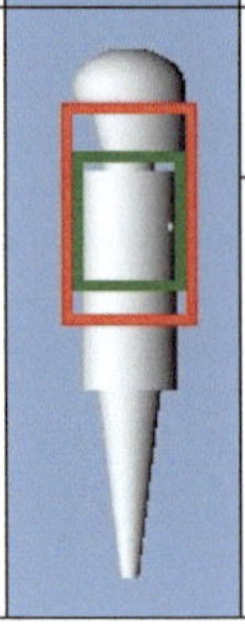	Upper field edge: Above the spinous process of the 6th cervical vertebra Lower field edge: 12th rib included Lateral field edges: Close to the anterior and posterior chest wall Field size = field width x field height: 14 cm x 17 cm = 238 cm^2  	Upper field edge: At the upper margins of the parotid glands Lower field edge: Upper margin of the iliac crests Lateral field edges: Beneath anterior and posterior chest wall Field size = field width x field height: 16 cm x 28 cm = 448 cm^2 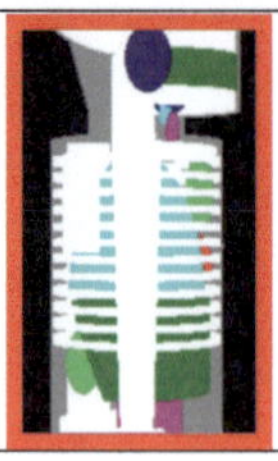 

Table 9.24 Field settings—Chest lat

Age level (gestational weeks)		Anatomical landmarks: Optimal field setting		Anatomical landmarks: Suboptimal field setting	
3–7	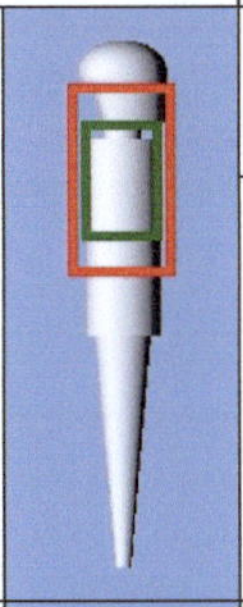	Upper field edge: Above the spinous process of the 6th cervical vertebra Lower field edge: 12th rib included Lateral field edges: Close to the anterior and posterior chest wall Field size = field width x field height: 14 cm x 22 cm= 308 cm^2		Upper field edge: At the upper margins of the parotid glands Lower field edge: Upper margin of the iliac crests Lateral field edges: Beneath anterior and posterior chest wall Field size = field width x field height: 20 cm x 35cm= 700 cm^2	
				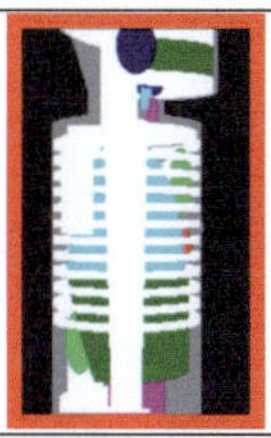	
8–12	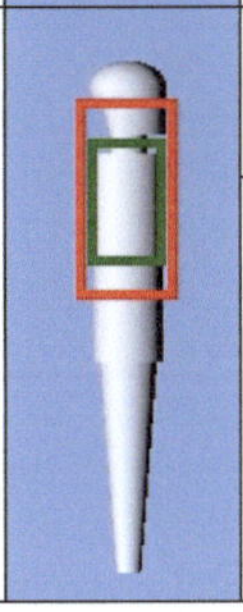	Upper field edge: Above the spinous process of the 6th cervical vertebra Lower field edge: 12th rib included Lateral field edges: Close to the anterior and posterior chest wall Field size = field width x field height: 17 cm x 27 cm = 459 cm^2		Upper field edge: At the upper margins of the parotid glands Lower field edge: Upper margin of the iliac crests Lateral field edges: Beneath anterior and posterior chest wall Field size = field width x field height: 22 cm x 42 cm = 924 cm^2	
			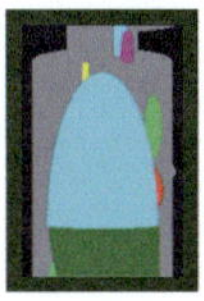		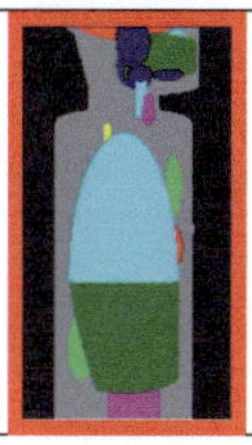

Table 9.25 Field settings—Chest lat

Age level (gestational weeks)		Anatomical landmarks: Optimal field setting		Anatomical landmarks: Suboptimal field setting	
13–17	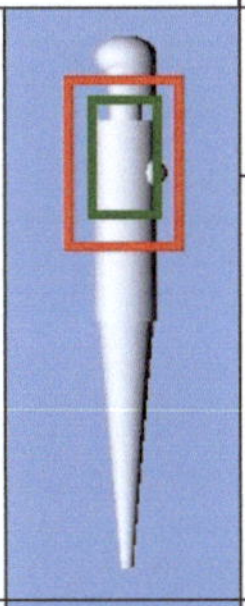	Upper field edge: Above the spinous process of the 6th cervical vertebra Lower field edge: 12th rib included Lateral field edges: Close to the anterior and posterior chest wall Field size = field width x field height: 20 cm x 34 cm= 680 cm^2		Upper field edge: At the upper margins of the parotid glands Lower field edge: Upper margin of the iliac crests Lateral field edges: Beneath anterior and posterior chest wall Field size = field width x field height: 35 cm x 50 cm= 1,750 cm^2	
			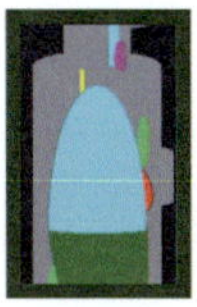		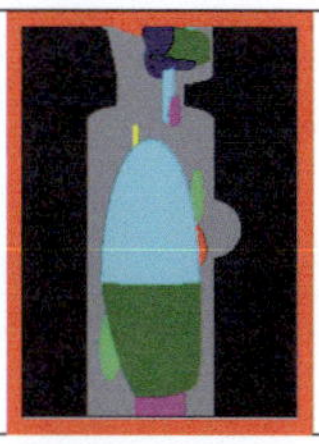
18	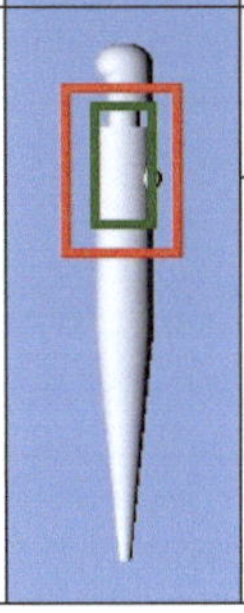	Upper field edge: Above the spinous process of the 6th cervical vertebra Lower field edge: 12th rib included Lateral field edges: Close to the anterior and posterior chest wall Field size = field width x field height: 20 cm x 37 cm = 740 cm^2		Upper field edge: At the upper margins of the parotid glands Lower field edge: Upper margin of the iliac crests Lateral field edges: Beneath anterior and posterior chest wall Field size = field width x field height: 38 cm x 55 cm = 2,090 cm^2	

Table 9.26 Normalized organ doses—Chest ap—24th gestation week

Beam projection:			ap		**Table cells:**							
Focus-to-detector distance (cm):			100		(+) = normal font = optimal field setting							
Tube voltage (kVp):			50		**(−) = bold font = suboptimal field setting**							
Total filtrations:			See last table row.		Empty table cells correspond to values < 0.01							
Organ or tissue	**Organ dose D_T : Entrance dose K_E**							**(mSv × mGy^{-1})**				
	(+)	(−)	(+)	(−)	(+)	(−)	(+)	(−)	(+)	(−)	(+)	(−)
Skull		**0.03**		**0.04**		**0.04**		**0.05**		**0.05**		**0.06**
Brain												
Oral mucosa		**0.05**		**0.06**		**0.06**		**0.06**		**0.07**		**0.07**
Salivary glands		**0.04**		**0.05**		**0.06**	0.01	**0.06**	0.01	**0.06**	0.01	**0.07**
Cervical spine	0.04	**0.07**	0.05	**0.09**	0.06	**0.10**	0.07	**0.11**	0.08	**0.13**	0.08	**0.14**
Thyroid	0.08	**0.08**	0.08	**0.09**	0.09	**0.09**	0.09	**0.09**	0.09	**0.10**	0.09	**0.10**
Thymus	0.07	**0.07**	0.07	**0.08**	0.08	**0.08**	0.08	**0.08**	0.09	**0.09**	0.09	**0.09**
Thoracic spine	0.04	**0.04**	0.05	**0.06**	0.06	**0.07**	0.07	**0.08**	0.08	**0.09**	0.10	**0.10**
Claviculae	0.18	**0.19**	0.21	**0.22**	0.22	**0.23**	0.24	**0.24**	0.25	**0.26**	0.26	**0.27**
Scapulae	0.06	**0.06**	0.07	**0.08**	0.08	**0.09**	0.09	**0.10**	0.10	**0.11**	0.12	**0.12**
Ribs	0.13	**0.15**	0.15	**0.17**	0.16	**0.18**	0.17	**0.19**	0.18	**0.20**	0.19	**0.21**
Oesophagus	0.03	**0.03**	0.04	**0.04**	0.04	**0.04**	0.05	**0.05**	0.05	**0.06**	0.06	**0.06**
Trachea		**0.05**		**0.05**		**0.06**	0.01	**0.06**	0.01	**0.07**	0.01	**0.07**
Lung	0.04	**0.05**	0.05	**0.05**	0.06	**0.06**	0.06	**0.06**	0.06	**0.07**	0.07	**0.07**
Mamma	0.09	**0.09**	0.09	**0.09**	0.09	**0.10**	0.09	**0.10**	0.09	**0.10**	0.10	**0.10**
Heart	0.05	**0.05**	0.06	**0.06**	0.06	**0.07**	0.07	**0.07**	0.07	**0.08**	0.08	**0.08**
Stomach	0.03	**0.05**	0.03	**0.06**	0.03	**0.06**	0.04	**0.07**	0.04	**0.07**	0.04	**0.08**
Spleen	0.02	**0.03**	0.02	**0.03**	0.03	**0.04**	0.03	**0.04**	0.04	**0.05**	0.04	**0.05**
Liver	0.03	**0.05**	0.03	**0.05**	0.04	**0.06**	0.04	**0.06**	0.04	**0.07**	0.05	**0.07**
Gall bladder		**0.04**		**0.05**	0.01	**0.06**	0.01	**0.06**	0.01	**0.07**	0.02	**0.07**
Pancreas	0.03	**0.03**	0.04	**0.04**	0.04	**0.05**	0.04	**0.05**	0.05	**0.06**	0.05	**0.06**
Small intestine		**0.02**		**0.02**		**0.02**		**0.02**		**0.03**		**0.03**
Large intestine		**0.03**		**0.03**		**0.04**		**0.04**		**0.04**		**0.05**
Sigma/Rectum												
Lumbar spine		**0.04**	0.01	**0.06**	0.01	**0.07**	0.02	**0.07**	0.02	**0.09**	0.02	**0.10**
Pelvis								**0.01**		**0.01**		**0.02**
Adrenals	0.02	**0.02**	0.02	**0.02**	0.02	**0.03**	0.03	**0.03**	0.03	**0.03**	0.04	**0.04**
Kidneys		**0.02**		**0.02**		**0.03**	0.01	**0.03**	0.01	**0.03**	0.01	**0.04**
Urinary bladder												
Uterus												
Ovaries												
Prostate												
Testes												
Upper arm	0.01	**0.08**	0.02	**0.10**	0.02	**0.11**	0.02	**0.12**	0.02	**0.13**	0.03	**0.14**
Elbow	0.01	**0.08**	0.01	**0.10**	0.02	**0.11**	0.02	**0.12**	0.02	**0.13**	0.03	**0.14**
Forearm		**0.04**		**0.04**		**0.05**		**0.05**		**0.06**		**0.06**
Upper leg												
Knee												
Lower leg												
Lymph nodes	0.02	**0.04**	0.02	**0.04**	0.02	**0.05**	0.03	**0.05**	0.03	**0.05**	0.03	**0.06**
Skin	0.01	**0.02**	0.01	**0.02**	0.01	**0.03**	0.01	**0.03**	0.02	**0.03**	0.02	**0.03**
Muscle	0.01	**0.02**	0.01	**0.02**	0.01	**0.02**	0.01	**0.03**	0.02	**0.03**	0.02	**0.03**
Skeleton	0.03	**0.05**	0.03	**0.06**	0.04	**0.06**	0.04	**0.07**	0.04	**0.08**	0.05	**0.08**
Red bone marrow		**0.01**		**0.01**		**0.02**		**0.02**	0.01	**0.02**	0.01	**0.02**
Whole body dose	0.01	**0.02**	0.02	**0.03**	0.02	**0.03**	0.02	**0.03**	0.02	**0.03**	0.02	**0.04**
ED (ICRP 60)	0.02	**0.03**	0.02	**0.03**	0.03	**0.04**	0.03	**0.04**	0.03	**0.04**	0.03	**0.04**
ED (ICRP 103)	0.03	**0.04**	0.03	**0.04**	0.03	**0.05**	0.03	**0.05**	0.04	**0.05**	0.04	**0.05**
Total filtration (mm) **Al**	**1.5**		**2.5**		**3.5**		**4.5**		**3.5**		**3.5**	
Cu	**0**		**0**		**0**		**0**		**0.1**		**0.2**	

Table 9.27 Normalized organ doses—Chest ap—24th gestation week

Beam projection:			ap			**Table cells:**						
Focus-to-detector distance (cm):			100			(+) = normal font = optimal field setting						
Tube voltage (kVp):			60			**(−) = bold font = suboptimal field setting**						
Total filtrations:			See last table row.			Empty table cells correspond to values < 0.01						
Organ or tissue	**Organ dose D_T : Entrance dose K_E**						**(mSv × mGy^{-1})**					
	(+)	(−)	(+)	(−)	(+)	(−)	(+)	(−)	(+)	(−)	(+)	(−)
Skull		**0.04**		**0.04**		**0.05**		**0.05**		**0.06**		**0.06**
Brain												
Oral mucosa		**0.05**		**0.06**		**0.06**		**0.07**		**0.07**		**0.07**
Salivary glands		**0.05**		**0.06**	0.01	**0.06**	0.01	**0.06**	0.01	**0.07**	0.01	**0.07**
Cervical spine	0.05	**0.08**	0.06	**0.10**	0.07	**0.11**	0.07	**0.12**	0.08	**0.14**	0.08	**0.14**
Thyroid	0.08	**0.08**	0.09	**0.09**	0.09	**0.09**	0.09	**0.10**	0.09	**0.10**	0.09	**0.10**
Thymus	0.07	**0.07**	0.08	**0.08**	0.08	**0.09**	0.09	**0.09**	0.09	**0.09**	0.09	**0.09**
Thoracic spine	0.05	**0.06**	0.07	**0.07**	0.08	**0.08**	0.08	**0.09**	0.10	**0.11**	0.10	**0.10**
Claviculae	0.20	**0.20**	0.22	**0.23**	0.23	**0.24**	0.24	**0.25**	0.26	**0.27**	0.26	**0.27**
Scapulae	0.07	**0.07**	0.08	**0.09**	0.09	**0.10**	0.10	**0.11**	0.11	**0.12**	0.12	**0.12**
Ribs	0.14	**0.16**	0.16	**0.18**	0.17	**0.19**	0.18	**0.20**	0.19	**0.21**	0.19	**0.21**
Oesophagus	0.04	**0.04**	0.04	**0.05**	0.05	**0.05**	0.05	**0.06**	0.06	**0.06**	0.06	**0.06**
Trachea		**0.05**		**0.06**	0.01	**0.06**	0.01	**0.06**	0.01	**0.07**	0.01	**0.07**
Lung	0.05	**0.05**	0.06	**0.06**	0.06	**0.06**	0.06	**0.07**	0.07	**0.07**	0.07	**0.07**
Mamma	0.09	**0.09**	0.09	**0.10**	0.09	**0.10**	0.09	**0.10**	0.10	**0.10**	0.10	**0.10**
Heart	0.06	**0.06**	0.06	**0.07**	0.07	**0.07**	0.07	**0.08**	0.08	**0.08**	0.08	**0.08**
Stomach	0.03	**0.06**	0.03	**0.06**	0.04	**0.07**	0.04	**0.07**	0.04	**0.08**	0.04	**0.08**
Spleen	0.02	**0.03**	0.03	**0.04**	0.03	**0.04**	0.03	**0.05**	0.04	**0.05**	0.04	**0.05**
Liver	0.03	**0.05**	0.04	**0.06**	0.04	**0.06**	0.04	**0.07**	0.05	**0.07**	0.05	**0.07**
Gall bladder		**0.05**	0.01	**0.06**	0.01	**0.06**	0.01	**0.07**	0.02	**0.07**	0.02	**0.07**
Pancreas	0.03	**0.04**	0.04	**0.05**	0.04	**0.05**	0.05	**0.06**	0.05	**0.06**	0.05	**0.06**
Small intestine		**0.02**		**0.02**		**0.02**		**0.03**		**0.03**		**0.03**
Large intestine		**0.03**		**0.04**		**0.04**		**0.04**		**0.05**		**0.05**
Sigma/Rectum												
Lumbar spine	0.01	**0.05**	0.01	**0.07**	0.02	**0.08**	0.02	**0.09**	0.02	**0.10**	0.02	**0.10**
Pelvis				**0.01**		**0.01**		**0.01**		**0.02**		**0.02**
Adrenals	0.02	**0.02**	0.03	**0.03**	0.03	**0.03**	0.03	**0.03**	0.04	**0.04**	0.04	**0.04**
Kidneys		**0.02**	0.01	**0.03**	0.01	**0.03**	0.01	**0.03**	0.02	**0.04**	0.01	**0.04**
Urinary bladder												
Uterus												
Ovaries												
Prostate												
Testes												
Upper arm	0.01	**0.09**	0.02	**0.11**	0.02	**0.12**	0.02	**0.13**	0.03	**0.14**	0.03	**0.14**
Elbow	0.01	**0.09**	0.02	**0.11**	0.02	**0.12**	0.02	**0.13**	0.03	**0.14**	0.03	**0.14**
Forearm		**0.04**		**0.05**		**0.05**		**0.06**		**0.06**		**0.06**
Upper leg												
Knee												
Lower leg												
Lymph nodes	0.02	**0.04**	0.02	**0.05**	0.03	**0.05**	0.03	**0.05**	0.03	**0.06**	0.03	**0.06**
Skin	0.01	**0.02**	0.01	**0.03**	0.01	**0.03**	0.02	**0.03**	0.02	**0.03**	0.02	**0.03**
Muscle	0.01	**0.02**	0.01	**0.02**	0.02	**0.03**	0.02	**0.03**	0.02	**0.03**	0.02	**0.03**
Skeleton	0.03	**0.06**	0.04	**0.06**	0.04	**0.07**	0.04	**0.08**	0.05	**0.08**	0.05	**0.08**
Red bone marrow		**0.01**		**0.02**	0.01	**0.02**	0.01	**0.02**	0.01	**0.02**	0.01	**0.02**
Whole body dose	0.01	**0.03**	0.02	**0.03**	0.02	**0.03**	0.02	**0.03**	0.02	**0.04**	0.02	**0.04**
ED (ICRP 60)	0.02	**0.03**	0.03	**0.04**	0.03	**0.04**	0.03	**0.04**	0.03	**0.04**	0.03	**0.04**
ED (ICRP 103)	0.03	**0.04**	0.03	**0.05**	0.04	**0.05**	0.04	**0.05**	0.04	**0.05**	0.04	**0.05**
Total filtration (mm) **Al**	**1.5**		**2.5**		**3.5**		**4.5**		**3.5**		**3.5**	
Cu	**0**		**0**		**0**		**0**		**0.1**		**0.2**	

Table 9.28 Normalized organ doses—Chest ap—24th gestation week

Beam projection:			ap		**Table cells:**							
Focus-to-detector distance (cm):			100		(+) = normal font = optimal field setting							
Tube voltage (kVp):			70		**(−) = bold font = suboptimal field setting**							
Total filtrations:			See last table row.		Empty table cells correspond to values < 0.01							
Organ or tissue	**Organ dose D_T : Entrance dose K_E**						**($mSv \times mGy^{-1}$)**					
	(+)	(−)	(+)	(−)	(+)	(−)	(+)	(−)	(+)	(−)	(+)	(−)
Skull		**0.04**		**0.05**		**0.05**		**0.05**		**0.06**		**0.06**
Brain												
Oral mucosa		**0.06**		**0.06**		**0.07**		**0.07**		**0.07**		**0.08**
Salivary glands		**0.05**	0.01	**0.06**	0.01	**0.06**	0.01	**0.07**	0.01	**0.07**	0.01	**0.07**
Cervical spine	0.06	**0.09**	0.07	**0.11**	0.07	**0.12**	0.08	**0.13**	0.09	**0.15**	0.09	**0.16**
Thyroid	0.08	**0.09**	0.09	**0.09**	0.09	**0.10**	0.09	**0.10**	0.10	**0.10**	0.10	**0.10**
Thymus	0.07	**0.08**	0.08	**0.08**	0.09	**0.09**	0.09	**0.09**	0.09	**0.10**	0.10	**0.10**
Thoracic spine	0.06	**0.06**	0.07	**0.08**	0.08	**0.09**	0.09	**0.10**	0.11	**0.11**	0.12	**0.13**
Claviculae	0.21	**0.21**	0.23	**0.24**	0.24	**0.25**	0.24	**0.26**	0.25	**0.27**	0.26	**0.27**
Scapulae	0.07	**0.08**	0.09	**0.10**	0.10	**0.11**	0.11	**0.12**	0.12	**0.13**	0.13	**0.14**
Ribs	0.15	**0.16**	0.16	**0.18**	0.17	**0.19**	0.18	**0.20**	0.19	**0.21**	0.19	**0.22**
Oesophagus	0.04	**0.04**	0.05	**0.05**	0.05	**0.06**	0.06	**0.06**	0.06	**0.07**	0.07	**0.07**
Trachea		**0.05**	0.01	**0.06**	0.01	**0.06**	0.01	**0.07**	0.01	**0.07**	0.01	**0.07**
Lung	0.05	**0.05**	0.06	**0.06**	0.06	**0.07**	0.07	**0.07**	0.07	**0.08**	0.08	**0.08**
Mamma	0.09	**0.09**	0.09	**0.10**	0.09	**0.10**	0.10	**0.10**	0.10	**0.10**	0.10	**0.10**
Heart	0.06	**0.06**	0.07	**0.07**	0.07	**0.08**	0.08	**0.08**	0.08	**0.09**	0.09	**0.09**
Stomach	0.03	**0.06**	0.04	**0.07**	0.04	**0.07**	0.04	**0.08**	0.04	**0.08**	0.05	**0.09**
Spleen	0.03	**0.04**	0.03	**0.04**	0.04	**0.05**	0.04	**0.05**	0.04	**0.06**	0.05	**0.06**
Liver	0.03	**0.06**	0.04	**0.06**	0.04	**0.07**	0.05	**0.07**	0.05	**0.08**	0.05	**0.08**
Gall bladder	0.01	**0.05**	0.01	**0.06**	0.01	**0.07**	0.01	**0.07**	0.02	**0.08**	0.02	**0.08**
Pancreas	0.04	**0.04**	0.04	**0.05**	0.05	**0.06**	0.05	**0.06**	0.06	**0.07**	0.06	**0.07**
Small intestine		**0.02**		**0.02**		**0.03**		**0.03**		**0.03**		**0.03**
Large intestine		**0.03**		**0.04**		**0.04**		**0.04**		**0.05**		**0.05**
Sigma/Rectum												
Lumbar spine	0.01	**0.06**	0.02	**0.08**	0.02	**0.09**	0.02	**0.10**	0.02	**0.11**	0.03	**0.12**
Pelvis				**0.01**		**0.01**		**0.01**		**0.02**		**0.02**
Adrenals	0.02	**0.03**	0.03	**0.03**	0.03	**0.04**	0.03	**0.04**	0.04	**0.04**	0.04	**0.05**
Kidneys		**0.02**	0.01	**0.03**	0.01	**0.03**	0.01	**0.04**	0.02	**0.04**	0.02	**0.05**
Urinary bladder												
Uterus												
Ovaries												
Prostate												
Testes												
Upper arm	0.02	**0.10**	0.02	**0.11**	0.02	**0.12**	0.02	**0.13**	0.03	**0.14**	0.03	**0.15**
Elbow	0.02	**0.10**	0.02	**0.11**	0.02	**0.12**	0.02	**0.13**	0.03	**0.14**	0.03	**0.15**
Forearm		**0.04**		**0.05**		**0.06**		**0.06**		**0.06**		**0.07**
Upper leg												
Knee												
Lower leg												
Lymph nodes	0.02	**0.04**	0.03	**0.05**	0.03	**0.05**	0.03	**0.06**	0.03	**0.06**	0.03	**0.07**
Skin	0.01	**0.03**	0.01	**0.03**	0.02	**0.03**	0.02	**0.03**	0.02	**0.03**	0.02	**0.03**
Muscle	0.01	**0.02**	0.01	**0.03**	0.02	**0.03**	0.02	**0.03**	0.02	**0.03**	0.02	**0.03**
Skeleton	0.03	**0.06**	0.04	**0.07**	0.04	**0.07**	0.04	**0.08**	0.05	**0.09**	0.05	**0.09**
Red bone marrow		**0.02**	0.01	**0.02**	0.01	**0.02**	0.01	**0.02**	0.01	**0.02**	0.01	**0.03**
Whole body dose	0.02	**0.03**	0.02	**0.03**	0.02	**0.03**	0.02	**0.04**	0.02	**0.04**	0.02	**0.04**
ED (ICRP 60)	0.03	**0.03**	0.03	**0.04**	0.03	**0.04**	0.03	**0.04**	0.03	**0.05**	0.04	**0.05**
ED (ICRP 103)	0.03	**0.04**	0.03	**0.05**	0.04	**0.05**	0.04	**0.05**	0.04	**0.06**	0.04	**0.06**
Total filtration (mm) Al	**1.5**		**2.5**		**3.5**		**4.5**		**3.5**		**3.5**	
Cu	**0**		**0**		**0**		**0**		**0.1**		**0.2**	

Table 9.29 Normalized organ doses—Chest ap—24th gestation week

Beam projection:			ap			**Table cells:**						
Focus-to-detector distance (cm):			100			(+) = normal font = optimal field setting						
Tube voltage (kVp):			80			**(−) = bold font = suboptimal field setting**						
Total filtrations:			See last table row.			Empty table cells correspond to values < 0.01						
Organ or tissue	**Organ dose D_T : Entrance dose K_E**							**($mSv \times mGy^{-1}$)**				
	(+)	(−)	(+)	(−)	(+)	(−)	(+)	(−)	(+)	(−)	(+)	(−)
Skull		**0.04**		**0.05**		**0.05**		**0.05**		**0.06**		**0.06**
Brain												
Oral mucosa		**0.06**		**0.07**		**0.07**		**0.07**		**0.08**		**0.08**
Salivary glands		**0.06**	0.01	**0.06**	0.01	**0.06**	0.01	**0.07**	0.01	**0.07**	0.02	**0.07**
Cervical spine	0.06	**0.10**	0.07	**0.12**	0.08	**0.13**	0.08	**0.14**	0.09	**0.15**	0.09	**0.16**
Thyroid	0.08	**0.09**	0.09	**0.09**	0.09	**0.10**	0.09	**0.10**	0.10	**0.10**	0.10	**0.11**
Thymus	0.08	**0.08**	0.08	**0.09**	0.09	**0.09**	0.09	**0.09**	0.10	**0.10**	0.10	**0.10**
Thoracic spine	0.07	**0.07**	0.08	**0.09**	0.09	**0.10**	0.10	**0.11**	0.11	**0.12**	0.12	**0.13**
Claviculae	0.21	**0.22**	0.23	**0.24**	0.24	**0.25**	0.24	**0.25**	0.25	**0.26**	0.25	**0.26**
Scapulae	0.08	**0.09**	0.09	**0.10**	0.10	**0.11**	0.11	**0.12**	0.12	**0.13**	0.13	**0.14**
Ribs	0.15	**0.17**	0.17	**0.19**	0.17	**0.19**	0.18	**0.20**	0.18	**0.21**	0.18	**0.21**
Oesophagus	0.04	**0.05**	0.05	**0.05**	0.06	**0.06**	0.06	**0.06**	0.07	**0.07**	0.07	**0.08**
Trachea		**0.06**	0.01	**0.06**	0.01	**0.07**	0.01	**0.07**	0.01	**0.07**	0.01	**0.08**
Lung	0.06	**0.06**	0.06	**0.07**	0.07	**0.07**	0.07	**0.07**	0.07	**0.08**	0.08	**0.08**
Mamma	0.09	**0.10**	0.09	**0.10**	0.10	**0.10**	0.10	**0.10**	0.10	**0.10**	0.10	**0.10**
Heart	0.06	**0.07**	0.07	**0.07**	0.08	**0.08**	0.08	**0.08**	0.08	**0.09**	0.09	**0.09**
Stomach	0.03	**0.06**	0.04	**0.07**	0.04	**0.08**	0.04	**0.08**	0.05	**0.08**	0.05	**0.09**
Spleen	0.03	**0.04**	0.03	**0.05**	0.04	**0.05**	0.04	**0.05**	0.04	**0.06**	0.05	**0.06**
Liver	0.04	**0.06**	0.04	**0.07**	0.05	**0.07**	0.05	**0.07**	0.05	**0.08**	0.05	**0.08**
Gall bladder	0.01	**0.06**	0.01	**0.07**	0.01	**0.07**	0.02	**0.08**	0.02	**0.08**	0.02	**0.09**
Pancreas	0.04	**0.05**	0.05	**0.06**	0.05	**0.06**	0.05	**0.06**	0.06	**0.07**	0.06	**0.08**
Small intestine		**0.02**		**0.03**		**0.03**		**0.03**		**0.03**		**0.03**
Large intestine		**0.04**		**0.04**		**0.04**		**0.05**		**0.05**		**0.05**
Sigma/Rectum												
Lumbar spine	0.01	**0.07**	0.02	**0.08**	0.02	**0.09**	0.02	**0.10**	0.03	**0.11**	0.03	**0.12**
Pelvis		**0.01**		**0.01**		**0.01**		**0.02**		**0.02**		**0.02**
Adrenals	0.03	**0.03**	0.03	**0.03**	0.03	**0.04**	0.04	**0.04**	0.04	**0.05**	0.05	**0.05**
Kidneys	0.01	**0.03**	0.01	**0.03**	0.01	**0.04**	0.02	**0.04**	0.02	**0.04**	0.02	**0.05**
Urinary bladder												
Uterus												
Ovaries												
Prostate												
Testes												
Upper arm	0.02	**0.10**	0.02	**0.12**	0.02	**0.13**	0.03	**0.13**	0.03	**0.14**	0.03	**0.15**
Elbow	0.02	**0.10**	0.02	**0.12**	0.02	**0.13**	0.02	**0.13**	0.03	**0.14**	0.03	**0.15**
Forearm		**0.05**		**0.05**		**0.06**		**0.06**		**0.06**		**0.07**
Upper leg												
Knee												
Lower leg												
Lymph nodes	0.02	**0.05**	0.03	**0.05**	0.03	**0.06**	0.03	**0.06**	0.03	**0.06**	0.03	**0.07**
Skin	0.01	**0.03**	0.02	**0.03**	0.02	**0.03**	0.02	**0.03**	0.02	**0.03**	0.02	**0.03**
Muscle	0.01	**0.02**	0.02	**0.03**	0.02	**0.03**	0.02	**0.03**	0.02	**0.03**	0.02	**0.03**
Skeleton	0.04	**0.06**	0.04	**0.07**	0.04	**0.08**	0.05	**0.08**	0.05	**0.09**	0.05	**0.09**
Red bone marrow		**0.02**	0.01	**0.02**	0.01	**0.02**	0.01	**0.02**	0.01	**0.03**	0.02	**0.03**
Whole body dose	0.02	**0.03**	0.02	**0.03**	0.02	**0.04**	0.02	**0.04**	0.02	**0.04**	0.02	**0.04**
ED (ICRP 60)	0.03	**0.04**	0.03	**0.04**	0.03	**0.04**	0.03	**0.05**	0.03	**0.05**	0.04	**0.05**
ED (ICRP 103)	0.03	**0.05**	0.04	**0.05**	0.04	**0.05**	0.04	**0.05**	0.04	**0.06**	0.04	**0.06**
Total filtration (mm) **Al**	**1.5**		**2.5**		**3.5**		**4.5**		**3.5**		**3.5**	
Cu	**0**		**0**		**0**		**0**		**0.1**		**0.2**	

Table 9.30 Normalized organ doses—Chest ap—28th gestation week

Beam projection:			ap	**Table cells:**								
Focus-to-detector distance (cm):			100	(+) = normal font = optimal field setting								
Tube voltage (kVp):			50	**(−) = bold font = suboptimal field setting**								
Total filtrations:			See last table row.	Empty table cells correspond to values < 0.01								
Organ or tissue	**Organ dose D_T : Entrance dose K_E (mSv × mGy^{-1})**											
	(+)	(−)	(+)	(−)	(+)	(−)	(+)	(−)	(+)	(−)	(+)	(−)
Skull		**0.02**		**0.03**		**0.03**		**0.03**		**0.04**		**0.04**
Brain												
Oral mucosa		**0.06**		**0.07**		**0.08**		**0.08**		**0.09**		**0.09**
Salivary glands		**0.05**		**0.06**	0.01	**0.07**	0.01	**0.07**	0.01	**0.08**	0.01	**0.08**
Cervical spine	0.06	**0.09**	0.07	**0.11**	0.08	**0.13**	0.09	**0.14**	0.10	**0.16**	0.11	**0.18**
Thyroid	0.11	**0.11**	0.12	**0.12**	0.12	**0.13**	0.13	**0.13**	0.13	**0.14**	0.13	**0.14**
Thymus	0.09	**0.09**	0.10	**0.11**	0.11	**0.11**	0.12	**0.12**	0.13	**0.13**	0.13	**0.14**
Thoracic spine	0.05	**0.06**	0.07	**0.07**	0.08	**0.09**	0.09	**0.10**	0.11	**0.12**	0.13	**0.14**
Claviculae	0.26	**0.26**	0.29	**0.30**	0.32	**0.33**	0.33	**0.34**	0.36	**0.37**	0.37	**0.39**
Scapulae	0.07	**0.08**	0.09	**0.10**	0.11	**0.11**	0.12	**0.12**	0.14	**0.15**	0.16	**0.16**
Ribs	0.20	**0.20**	0.22	**0.23**	0.24	**0.25**	0.25	**0.26**	0.27	**0.28**	0.29	**0.30**
Oesophagus	0.04	**0.04**	0.05	**0.05**	0.06	**0.06**	0.06	**0.07**	0.07	**0.08**	0.08	**0.08**
Trachea	0.01	**0.04**	0.01	**0.04**	0.01	**0.05**	0.01	**0.05**	0.02	**0.05**	0.02	**0.06**
Lung	0.06	**0.06**	0.07	**0.07**	0.08	**0.08**	0.08	**0.09**	0.09	**0.09**	0.10	**0.10**
Mamma	0.13	**0.13**	0.13	**0.13**	0.13	**0.14**	0.14	**0.14**	0.14	**0.14**	0.14	**0.14**
Heart	0.07	**0.07**	0.08	**0.08**	0.09	**0.09**	0.10	**0.10**	0.11	**0.11**	0.11	**0.12**
Stomach	0.04	**0.07**	0.05	**0.08**	0.05	**0.09**	0.06	**0.10**	0.06	**0.10**	0.07	**0.11**
Spleen	0.03	**0.03**	0.04	**0.04**	0.04	**0.05**	0.05	**0.05**	0.05	**0.06**	0.06	**0.07**
Liver	0.04	**0.06**	0.05	**0.07**	0.06	**0.08**	0.06	**0.09**	0.07	**0.10**	0.07	**0.10**
Gall bladder	0.01	**0.06**	0.02	**0.07**	0.02	**0.08**	0.02	**0.09**	0.03	**0.10**	0.03	**0.11**
Pancreas	0.04	**0.04**	0.05	**0.06**	0.05	**0.06**	0.06	**0.07**	0.07	**0.08**	0.08	**0.09**
Small intestine		**0.02**		**0.03**		**0.03**		**0.03**		**0.04**		**0.04**
Large intestine		**0.04**		**0.04**		**0.05**		**0.05**		**0.06**		**0.06**
Sigma/Rectum												**0.01**
Lumbar spine	0.01	**0.05**	0.02	**0.07**	0.02	**0.08**	0.03	**0.09**	0.03	**0.11**	0.04	**0.13**
Pelvis				**0.01**		**0.01**		**0.01**		**0.02**		**0.02**
Adrenals	0.02	**0.02**	0.03	**0.03**	0.03	**0.04**	0.04	**0.04**	0.04	**0.05**	0.05	**0.05**
Kidneys	0.01	**0.02**	0.01	**0.03**	0.02	**0.03**	0.02	**0.04**	0.02	**0.04**	0.02	**0.05**
Urinary bladder												
Uterus												
Ovaries												
Prostate												
Testes												
Upper arm	0.08	**0.15**	0.10	**0.18**	0.11	**0.20**	0.12	**0.22**	0.14	**0.24**	0.15	**0.26**
Elbow	0.08	**0.16**	0.09	**0.19**	0.11	**0.21**	0.12	**0.23**	0.13	**0.26**	0.14	**0.28**
Forearm		**0.08**		**0.09**		**0.10**	0.01	**0.11**	0.01	**0.12**	0.01	**0.13**
Upper leg												
Knee												
Lower leg												
Lymph nodes	0.03	**0.04**	0.03	**0.05**	0.03	**0.06**	0.04	**0.06**	0.04	**0.07**	0.04	**0.08**
Skin	0.02	**0.03**	0.02	**0.04**	0.02	**0.04**	0.02	**0.04**	0.02	**0.04**	0.03	**0.04**
Muscle	0.02	**0.03**	0.02	**0.03**	0.02	**0.03**	0.02	**0.03**	0.02	**0.04**	0.03	**0.04**
Skeleton	0.05	**0.07**	0.06	**0.08**	0.06	**0.09**	0.07	**0.10**	0.07	**0.11**	0.08	**0.12**
Red bone marrow	0.01	**0.02**	0.01	**0.02**	0.01	**0.02**	0.02	**0.02**	0.02	**0.03**	0.02	**0.03**
Whole body dose	0.02	**0.03**	0.02	**0.04**	0.03	**0.04**	0.03	**0.04**	0.03	**0.05**	0.03	**0.05**
ED (ICRP 60)	0.03	**0.04**	0.04	**0.05**	0.04	**0.05**	0.04	**0.05**	0.04	**0.06**	0.05	**0.06**
ED (ICRP 103)	0.04	**0.05**	0.05	**0.06**	0.05	**0.06**	0.05	**0.07**	0.05	**0.07**	0.06	**0.07**
Total filtration (mm) Al	**1.5**		**2.5**		**3.5**		**4.5**		**3.5**		**3.5**	
Cu	**0**		**0**		**0**		**0**		**0.1**		**0.2**	

Table 9.31 Normalized organ doses—Chest ap—28th gestation week

Beam projection:			ap			**Table cells:**						
Focus-to-detector distance (cm):			100			(+) = normal font = optimal field setting						
Tube voltage (kVp):			60			**(–) = bold font = suboptimal field setting**						
Total filtrations:			See last table row.			Empty table cells correspond to values < 0.01						
Organ or tissue	**Organ dose D_T : Entrance dose K_E (mSv x mGy^{-1})**											
	(+)	(–)	(+)	(–)	(+)	(–)	(+)	(–)	(+)	(–)	(+)	(–)
Skull		**0.03**		**0.03**		**0.03**		**0.04**		**0.04**		**0.04**
Brain												
Oral mucosa		**0.07**		**0.08**		**0.08**		**0.09**		**0.09**		**0.09**
Salivary glands		**0.06**	0.01	**0.07**	0.01	**0.07**	0.01	**0.08**	0.01	**0.08**	0.01	**0.08**
Cervical spine	0.07	**0.10**	0.08	**0.13**	0.09	**0.15**	0.10	**0.16**	0.11	**0.18**	0.11	**0.18**
Thyroid	0.11	**0.12**	0.12	**0.13**	0.13	**0.13**	0.13	**0.14**	0.13	**0.14**	0.13	**0.14**
Thymus	0.10	**0.10**	0.11	**0.11**	0.12	**0.12**	0.12	**0.13**	0.13	**0.13**	0.13	**0.14**
Thoracic spine	0.07	**0.07**	0.09	**0.09**	0.10	**0.11**	0.11	**0.12**	0.13	**0.14**	0.13	**0.14**
Claviculae	0.28	**0.28**	0.31	**0.32**	0.33	**0.34**	0.35	**0.36**	0.37	**0.38**	0.37	**0.39**
Scapulae	0.09	**0.09**	0.11	**0.11**	0.13	**0.13**	0.14	**0.14**	0.16	**0.16**	0.16	**0.16**
Ribs	0.21	**0.22**	0.24	**0.25**	0.25	**0.26**	0.27	**0.27**	0.28	**0.29**	0.29	**0.30**
Oesophagus	0.05	**0.05**	0.06	**0.06**	0.07	**0.07**	0.07	**0.08**	0.08	**0.09**	0.08	**0.08**
Trachea	0.01	**0.04**	0.01	**0.05**	0.01	**0.05**	0.02	**0.05**	0.02	**0.06**	0.02	**0.06**
Lung	0.07	**0.07**	0.08	**0.08**	0.09	**0.09**	0.09	**0.09**	0.10	**0.10**	0.10	**0.10**
Mamma	0.13	**0.13**	0.13	**0.14**	0.14	**0.14**	0.14	**0.14**	0.14	**0.14**	0.14	**0.14**
Heart	0.08	**0.08**	0.09	**0.09**	0.10	**0.10**	0.10	**0.11**	0.11	**0.12**	0.11	**0.12**
Stomach	0.05	**0.08**	0.05	**0.09**	0.06	**0.10**	0.06	**0.10**	0.07	**0.11**	0.07	**0.11**
Spleen	0.03	**0.04**	0.04	**0.05**	0.05	**0.06**	0.05	**0.06**	0.06	**0.07**	0.06	**0.07**
Liver	0.05	**0.07**	0.06	**0.08**	0.06	**0.09**	0.07	**0.10**	0.07	**0.10**	0.07	**0.10**
Gall bladder	0.02	**0.07**	0.02	**0.08**	0.02	**0.09**	0.03	**0.10**	0.03	**0.11**	0.03	**0.11**
Pancreas	0.05	**0.05**	0.06	**0.06**	0.06	**0.07**	0.07	**0.08**	0.08	**0.09**	0.08	**0.09**
Small intestine		**0.02**		**0.03**		**0.03**		**0.04**		**0.04**		**0.04**
Large intestine		**0.04**		**0.05**		**0.05**		**0.06**		**0.06**		**0.06**
Sigma/Rectum										**0.01**		**0.01**
Lumbar spine	0.02	**0.07**	0.02	**0.09**	0.03	**0.10**	0.03	**0.11**	0.04	**0.13**	0.04	**0.13**
Pelvis		**0.01**		**0.01**		**0.02**		**0.02**		**0.02**		**0.02**
Adrenals	0.03	**0.03**	0.03	**0.04**	0.04	**0.04**	0.04	**0.05**	0.05	**0.06**	0.05	**0.05**
Kidneys	0.01	**0.03**	0.02	**0.03**	0.02	**0.04**	0.02	**0.04**	0.03	**0.05**	0.02	**0.05**
Urinary bladder												
Uterus												
Ovaries												
Prostate												
Testes												
Upper arm	0.10	**0.17**	0.11	**0.20**	0.13	**0.22**	0.14	**0.24**	0.15	**0.26**	0.15	**0.26**
Elbow	0.09	**0.18**	0.11	**0.21**	0.12	**0.24**	0.13	**0.25**	0.14	**0.28**	0.14	**0.28**
Forearm		**0.09**		**0.10**	0.01	**0.11**	0.01	**0.12**	0.01	**0.13**	0.01	**0.13**
Upper leg												
Knee												
Lower leg												
Lymph nodes	0.03	**0.05**	0.03	**0.06**	0.04	**0.07**	0.04	**0.07**	0.04	**0.08**	0.04	**0.08**
Skin	0.02	**0.04**	0.02	**0.04**	0.02	**0.04**	0.02	**0.04**	0.03	**0.04**	0.03	**0.04**
Muscle	0.02	**0.03**	0.02	**0.03**	0.02	**0.04**	0.02	**0.04**	0.03	**0.04**	0.03	**0.04**
Skeleton	0.05	**0.08**	0.06	**0.09**	0.07	**0.10**	0.07	**0.11**	0.08	**0.12**	0.08	**0.12**
Red bone marrow	0.01	**0.02**	0.01	**0.02**	0.02	**0.02**	0.02	**0.03**	0.02	**0.03**	0.02	**0.03**
Whole body dose	0.02	**0.03**	0.03	**0.04**	0.03	**0.04**	0.03	**0.05**	0.03	**0.05**	0.03	**0.05**
ED (ICRP 60)	0.03	**0.04**	0.04	**0.05**	0.04	**0.05**	0.04	**0.06**	0.05	**0.06**	0.05	**0.06**
ED (ICRP 103)	0.04	**0.06**	0.05	**0.06**	0.05	**0.07**	0.05	**0.07**	0.06	**0.08**	0.06	**0.07**
Total filtration (mm) Al	**1.5**		**2.5**		**3.5**		**4.5**		**3.5**		**3.5**	
Cu	**0**		**0**		**0**		**0**		**0.1**		**0.2**	

Table 9.32 Normalized organ doses—Chest ap—28th gestation week

Beam projection:			ap			**Table cells:**						
Focus-to-detector distance (cm):			100			(+) = normal font = optimal field setting						
Tube voltage (kVp):			70			**(−) = bold font = suboptimal field setting**						
Total filtrations:			See last table row.			Empty table cells correspond to values < 0.01						
Organ or tissue	**Organ dose D_T : Entrance dose K_E**							**(mSv × mGy^{-1})**				
	(+)	(−)	(+)	(−)	(+)	(−)	(+)	(−)	(+)	(−)	(+)	(−)
Skull		**0.03**		**0.03**		**0.04**		**0.04**		**0.04**	0.01	**0.04**
Brain												
Oral mucosa		**0.07**		**0.08**		**0.09**		**0.09**		**0.10**		**0.10**
Salivary glands		**0.06**	0.01	**0.07**	0.01	**0.08**	0.01	**0.08**	0.02	**0.09**	0.02	**0.09**
Cervical spine	0.07	**0.12**	0.09	**0.14**	0.10	**0.16**	0.10	**0.17**	0.12	**0.19**	0.13	**0.21**
Thyroid	0.12	**0.12**	0.12	**0.13**	0.13	**0.14**	0.13	**0.14**	0.14	**0.15**	0.14	**0.15**
Thymus	0.11	**0.11**	0.12	**0.12**	0.12	**0.13**	0.13	**0.13**	0.14	**0.14**	0.14	**0.14**
Thoracic spine	0.08	**0.08**	0.10	**0.10**	0.11	**0.12**	0.12	**0.13**	0.14	**0.15**	0.16	**0.17**
Claviculae	0.29	**0.30**	0.32	**0.33**	0.34	**0.35**	0.35	**0.36**	0.37	**0.38**	0.37	**0.39**
Scapulae	0.10	**0.10**	0.12	**0.13**	0.14	**0.14**	0.15	**0.15**	0.17	**0.17**	0.18	**0.19**
Ribs	0.22	**0.23**	0.25	**0.25**	0.26	**0.27**	0.27	**0.28**	0.29	**0.30**	0.29	**0.31**
Oesophagus	0.05	**0.06**	0.06	**0.07**	0.07	**0.08**	0.08	**0.08**	0.09	**0.09**	0.10	**0.10**
Trachea	0.01	**0.04**	0.01	**0.05**	0.02	**0.05**	0.02	**0.06**	0.02	**0.06**	0.02	**0.06**
Lung	0.07	**0.07**	0.08	**0.09**	0.09	**0.09**	0.09	**0.10**	0.10	**0.11**	0.11	**0.11**
Mamma	0.13	**0.13**	0.14	**0.14**	0.14	**0.14**	0.14	**0.14**	0.14	**0.15**	0.15	**0.15**
Heart	0.08	**0.09**	0.10	**0.10**	0.10	**0.11**	0.11	**0.11**	0.12	**0.12**	0.13	**0.13**
Stomach	0.05	**0.08**	0.06	**0.10**	0.06	**0.10**	0.07	**0.11**	0.07	**0.12**	0.08	**0.12**
Spleen	0.04	**0.05**	0.05	**0.06**	0.05	**0.06**	0.06	**0.07**	0.06	**0.08**	0.07	**0.08**
Liver	0.05	**0.08**	0.06	**0.09**	0.07	**0.10**	0.07	**0.10**	0.08	**0.11**	0.08	**0.12**
Gall bladder	0.02	**0.08**	0.02	**0.09**	0.03	**0.10**	0.03	**0.10**	0.03	**0.11**	0.03	**0.12**
Pancreas	0.05	**0.06**	0.06	**0.07**	0.07	**0.08**	0.07	**0.08**	0.08	**0.10**	0.09	**0.10**
Small intestine		**0.03**		**0.03**		**0.04**		**0.04**		**0.04**		**0.05**
Large intestine		**0.05**		**0.05**		**0.06**		**0.06**		**0.07**		**0.07**
Sigma/Rectum								**0.01**		**0.01**		**0.01**
Lumbar spine	0.02	**0.08**	0.03	**0.10**	0.03	**0.11**	0.04	**0.12**	0.04	**0.14**	0.05	**0.16**
Pelvis		**0.01**		**0.02**		**0.02**		**0.02**		**0.02**		**0.03**
Adrenals	0.03	**0.03**	0.04	**0.04**	0.04	**0.05**	0.05	**0.05**	0.05	**0.06**	0.06	**0.07**
Kidneys	0.02	**0.03**	0.02	**0.04**	0.02	**0.04**	0.02	**0.05**	0.03	**0.06**	0.03	**0.06**
Urinary bladder												
Uterus												**0.01**
Ovaries												**0.01**
Prostate												
Testes												
Upper arm	0.10	**0.18**	0.12	**0.21**	0.14	**0.23**	0.14	**0.25**	0.16	**0.27**	0.17	**0.28**
Elbow	0.10	**0.20**	0.11	**0.23**	0.13	**0.25**	0.13	**0.26**	0.15	**0.28**	0.16	**0.30**
Forearm		**0.09**	0.01	**0.11**	0.01	**0.12**	0.01	**0.12**	0.02	**0.13**	0.02	**0.14**
Upper leg												
Knee												
Lower leg												
Lymph nodes	0.03	**0.06**	0.04	**0.06**	0.04	**0.07**	0.04	**0.07**	0.05	**0.08**	0.05	**0.09**
Skin	0.02	**0.04**	0.02	**0.04**	0.02	**0.04**	0.03	**0.04**	0.03	**0.04**	0.03	**0.05**
Muscle	0.02	**0.03**	0.02	**0.03**	0.02	**0.04**	0.03	**0.04**	0.03	**0.04**	0.03	**0.05**
Skeleton	0.06	**0.08**	0.07	**0.10**	0.07	**0.10**	0.08	**0.11**	0.08	**0.12**	0.09	**0.13**
Red bone marrow	0.01	**0.02**	0.02	**0.02**	0.02	**0.03**	0.02	**0.03**	0.02	**0.03**	0.02	**0.03**
Whole body dose	0.02	**0.04**	0.03	**0.04**	0.03	**0.05**	0.03	**0.05**	0.04	**0.05**	0.04	**0.06**
ED (ICRP 60)	0.04	**0.05**	0.04	**0.05**	0.04	**0.06**	0.05	**0.06**	0.05	**0.07**	0.05	**0.07**
ED (ICRP 103)	0.05	**0.06**	0.05	**0.07**	0.05	**0.07**	0.06	**0.07**	0.06	**0.08**	0.06	**0.08**
Total filtration (mm) **Al**	**1.5**		**2.5**		**3.5**		**4.5**		**3.5**		**3.5**	
Cu	**0**		**0**		**0**		**0**		**0.1**		**0.2**	

Table 9.33 Normalized organ doses—Chest ap—28th gestation week

Beam projection:		ap				**Table cells:**						
Focus-to-detector distance (cm):		100				(+) = normal font = optimal field setting						
Tube voltage (kVp):		80				**(−) = bold font = suboptimal field setting**						
Total filtrations:		See last table row.				Empty table cells correspond to values < 0.01						
Organ or tissue	**Organ dose D_T : Entrance dose K_E**							**(mSv × mGy^{-1})**				
	(+)	(−)	(+)	(−)	(+)	(−)	(+)	(−)	(+)	(−)	(+)	(−)
Skull		**0.03**		**0.03**		**0.04**		**0.04**		**0.04**	0.01	**0.04**
Brain												
Oral mucosa		**0.08**		**0.09**		**0.09**		**0.10**		**0.10**		**0.11**
Salivary glands	0.01	**0.07**	0.01	**0.08**	0.01	**0.08**	0.01	**0.08**	0.02	**0.09**	0.02	**0.09**
Cervical spine	0.08	**0.13**	0.09	**0.15**	0.10	**0.17**	0.11	**0.18**	0.12	**0.20**	0.13	**0.21**
Thyroid	0.12	**0.13**	0.13	**0.13**	0.13	**0.14**	0.13	**0.14**	0.14	**0.15**	0.14	**0.15**
Thymus	0.11	**0.11**	0.12	**0.12**	0.13	**0.13**	0.13	**0.13**	0.14	**0.14**	0.14	**0.15**
Thoracic spine	0.09	**0.09**	0.11	**0.12**	0.12	**0.13**	0.13	**0.14**	0.15	**0.16**	0.16	**0.18**
Claviculae	0.30	**0.31**	0.32	**0.33**	0.34	**0.35**	0.35	**0.36**	0.36	**0.37**	0.36	**0.38**
Scapulae	0.11	**0.11**	0.13	**0.14**	0.14	**0.15**	0.15	**0.16**	0.17	**0.18**	0.18	**0.19**
Ribs	0.23	**0.24**	0.25	**0.26**	0.26	**0.27**	0.27	**0.28**	0.28	**0.29**	0.29	**0.30**
Oesophagus	0.06	**0.06**	0.07	**0.07**	0.08	**0.08**	0.08	**0.09**	0.09	**0.10**	0.10	**0.11**
Trachea	0.01	**0.05**	0.01	**0.05**	0.02	**0.06**	0.02	**0.06**	0.02	**0.06**	0.02	**0.06**
Lung	0.08	**0.08**	0.09	**0.09**	0.09	**0.10**	0.10	**0.10**	0.11	**0.11**	0.11	**0.12**
Mamma	0.13	**0.14**	0.14	**0.14**	0.14	**0.14**	0.14	**0.14**	0.15	**0.15**	0.15	**0.15**
Heart	0.09	**0.09**	0.10	**0.10**	0.11	**0.11**	0.11	**0.12**	0.12	**0.13**	0.13	**0.13**
Stomach	0.05	**0.09**	0.06	**0.10**	0.07	**0.11**	0.07	**0.11**	0.07	**0.12**	0.08	**0.13**
Spleen	0.04	**0.05**	0.05	**0.06**	0.06	**0.07**	0.06	**0.07**	0.07	**0.08**	0.07	**0.09**
Liver	0.06	**0.08**	0.06	**0.09**	0.07	**0.10**	0.07	**0.11**	0.08	**0.11**	0.08	**0.12**
Gall bladder	0.02	**0.08**	0.02	**0.09**	0.03	**0.10**	0.03	**0.11**	0.03	**0.12**	0.03	**0.12**
Pancreas	0.06	**0.07**	0.07	**0.08**	0.07	**0.08**	0.08	**0.09**	0.09	**0.10**	0.09	**0.11**
Small intestine		**0.03**		**0.03**		**0.04**		**0.04**		**0.05**		**0.05**
Large intestine		**0.05**		**0.06**		**0.06**		**0.07**		**0.07**		**0.08**
Sigma/Rectum						**0.01**		**0.01**		**0.01**		**0.01**
Lumbar spine	0.03	**0.09**	0.03	**0.11**	0.04	**0.12**	0.04	**0.13**	0.04	**0.15**	0.05	**0.17**
Pelvis		**0.01**		**0.02**		**0.02**		**0.02**		**0.02**		**0.03**
Adrenals	0.04	**0.04**	0.04	**0.05**	0.05	**0.05**	0.05	**0.06**	0.06	**0.07**	0.06	**0.07**
Kidneys	0.02	**0.04**	0.02	**0.04**	0.02	**0.05**	0.03	**0.05**	0.03	**0.06**	0.03	**0.07**
Urinary bladder												
Uterus												**0.01**
Ovaries										**0.01**		**0.01**
Prostate												
Testes												
Upper arm	0.11	**0.20**	0.13	**0.22**	0.14	**0.24**	0.15	**0.25**	0.16	**0.27**	0.17	**0.28**
Elbow	0.10	**0.21**	0.12	**0.24**	0.13	**0.25**	0.14	**0.27**	0.15	**0.28**	0.16	**0.29**
Forearm	0.01	**0.10**	0.01	**0.11**	0.01	**0.12**	0.01	**0.12**	0.02	**0.13**	0.02	**0.14**
Upper leg												
Knee												
Lower leg												
Lymph nodes	0.03	**0.06**	0.04	**0.07**	0.04	**0.07**	0.04	**0.08**	0.05	**0.08**	0.05	**0.09**
Skin	0.02	**0.04**	0.02	**0.04**	0.03	**0.04**	0.03	**0.04**	0.03	**0.04**	0.03	**0.05**
Muscle	0.02	**0.03**	0.02	**0.04**	0.03	**0.04**	0.03	**0.04**	0.03	**0.04**	0.03	**0.05**
Skeleton	0.06	**0.09**	0.07	**0.10**	0.07	**0.11**	0.08	**0.11**	0.08	**0.12**	0.09	**0.13**
Red bone marrow	0.02	**0.02**	0.02	**0.03**	0.02	**0.03**	0.02	**0.03**	0.02	**0.03**	0.03	**0.04**
Whole body dose	0.03	**0.04**	0.03	**0.04**	0.03	**0.05**	0.03	**0.05**	0.04	**0.05**	0.04	**0.06**
ED (ICRP 60)	0.04	**0.05**	0.04	**0.06**	0.05	**0.06**	0.05	**0.06**	0.05	**0.07**	0.05	**0.07**
ED (ICRP 103)	0.05	**0.06**	0.05	**0.07**	0.06	**0.07**	0.06	**0.08**	0.06	**0.08**	0.06	**0.09**
Total filtration (mm) Al	**1.5**		**2.5**		**3.5**		**4.5**		**3.5**		**3.5**	
Cu	**0**		**0**		**0**		**0**		**0.1**		**0.2**	

Table 9.34 Normalized organ doses—Chest ap—< 1 year

Beam projection:			ap			**Table cells:**						
Focus-to-detector distance (cm):			100			(+) = normal font = optimal field setting						
Tube voltage (kVp):			50			**(−) = bold font = suboptimal field setting**						
Total filtrations:			See last table row.			Empty table cells correspond to values < 0.01						
Organ or tissue	**Organ dose D_T : Entrance dose K_E**								**($mSv \times mGy^{-1}$)**			
	(+)	(−)	(+)	(−)	(+)	(−)	(+)	(−)	(+)	(−)	(+)	(−)
Skull	0.02	**0.20**	0.02	**0.24**	0.03	**0.27**	0.03	**0.30**	0.04	**0.34**	0.04	**0.38**
Brain		**0.02**		**0.02**		**0.03**		**0.03**		**0.04**		**0.04**
Oral mucosa	0.02	**0.53**	0.03	**0.62**	0.04	**0.68**	0.04	**0.73**	0.05	**0.81**	0.06	**0.88**
Salivary glands	0.06	**0.45**	0.07	**0.54**	0.08	**0.61**	0.09	**0.65**	0.11	**0.74**	0.12	**0.81**
Cervical spine	0.30	**0.48**	0.39	**0.63**	0.46	**0.75**	0.52	**0.84**	0.62	**1.02**	0.71	**1.19**
Thyroid	0.83	**0.90**	0.94	**1.03**	1.01	**1.11**	1.06	**1.16**	1.13	**1.26**	1.19	**1.33**
Thymus	0.74	**0.75**	0.86	**0.89**	0.94	**0.97**	1.00	**1.03**	1.10	**1.14**	1.18	**1.22**
Thoracic spine	0.22	**0.24**	0.30	**0.33**	0.36	**0.40**	0.41	**0.46**	0.51	**0.58**	0.62	**0.70**
Claviculae	1.84	**1.91**	2.16	**2.25**	2.37	**2.47**	2.52	**2.63**	2.77	**2.90**	2.96	**3.11**
Scapulae	0.40	**0.42**	0.53	**0.56**	0.63	**0.67**	0.71	**0.76**	0.87	**0.92**	1.01	**1.08**
Ribs	1.26	**1.40**	1.46	**1.62**	1.59	**1.77**	1.69	**1.88**	1.85	**2.06**	1.98	**2.22**
Oesophagus	0.17	**0.19**	0.23	**0.25**	0.27	**0.30**	0.31	**0.34**	0.37	**0.41**	0.43	**0.48**
Trachea	0.08	**0.44**	0.10	**0.52**	0.11	**0.57**	0.12	**0.61**	0.13	**0.67**	0.15	**0.73**
Lung	0.38	**0.40**	0.47	**0.49**	0.53	**0.56**	0.58	**0.61**	0.66	**0.70**	0.73	**0.78**
Mamma	1.18	**1.18**	1.25	**1.25**	1.29	**1.29**	1.31	**1.31**	1.35	**1.34**	1.38	**1.37**
Heart	0.50	**0.52**	0.61	**0.63**	0.69	**0.71**	0.75	**0.77**	0.85	**0.88**	0.94	**0.98**
Stomach	0.21	**0.52**	0.26	**0.63**	0.29	**0.72**	0.32	**0.78**	0.37	**0.89**	0.41	**0.98**
Spleen	0.11	**0.18**	0.15	**0.24**	0.17	**0.28**	0.20	**0.32**	0.24	**0.39**	0.28	**0.45**
Liver	0.23	**0.46**	0.28	**0.56**	0.32	**0.64**	0.35	**0.69**	0.40	**0.79**	0.44	**0.87**
Gall bladder	0.08	**0.36**	0.10	**0.46**	0.12	**0.53**	0.13	**0.59**	0.16	**0.69**	0.18	**0.78**
Pancreas	0.18	**0.25**	0.24	**0.33**	0.28	**0.39**	0.32	**0.44**	0.38	**0.53**	0.44	**0.61**
Small intestine		**0.17**	0.01	**0.22**	0.01	**0.26**	0.02	**0.29**	0.02	**0.34**	0.03	**0.39**
Large intestine	0.01	**0.26**	0.02	**0.33**	0.02	**0.38**	0.02	**0.42**	0.03	**0.49**	0.03	**0.55**
Sigma/Rectum		**0.06**		**0.07**		**0.08**		**0.09**		**0.11**		**0.13**
Lumbar spine	0.04	**0.22**	0.05	**0.30**	0.06	**0.36**	0.07	**0.42**	0.09	**0.52**	0.11	**0.63**
Pelvis		**0.08**		**0.10**	0.01	**0.12**	0.01	**0.14**	0.02	**0.18**	0.02	**0.21**
Adrenals	0.09	**0.11**	0.12	**0.14**	0.14	**0.17**	0.16	**0.20**	0.20	**0.25**	0.24	**0.29**
Kidneys	0.03	**0.10**	0.04	**0.13**	0.05	**0.16**	0.06	**0.18**	0.07	**0.22**	0.09	**0.27**
Urinary bladder		**0.01**		**0.02**		**0.02**		**0.02**		**0.03**		**0.03**
Uterus		**0.03**		**0.04**		**0.05**		**0.05**		**0.06**	0.01	**0.08**
Ovaries		**0.03**		**0.04**		**0.05**		**0.05**		**0.06**		**0.08**
Prostate						**0.01**		**0.01**		**0.02**		**0.02**
Testes												
Upper arm	0.26	**0.97**	0.34	**1.20**	0.40	**1.36**	0.44	**1.49**	0.53	**1.71**	0.61	**1.90**
Elbow	0.20	**1.03**	0.26	**1.28**	0.30	**1.45**	0.34	**1.58**	0.40	**1.81**	0.46	**2.02**
Forearm	0.02	**0.54**	0.02	**0.67**	0.02	**0.76**	0.03	**0.82**	0.04	**0.94**	0.04	**1.04**
Upper leg										**0.01**		**0.02**
Knee												
Lower leg												
Lymph nodes	0.16	**0.34**	0.20	**0.42**	0.22	**0.47**	0.24	**0.51**	0.28	**0.59**	0.31	**0.66**
Skin	0.16	**0.32**	0.17	**0.33**	0.18	**0.35**	0.18	**0.36**	0.19	**0.38**	0.20	**0.39**
Muscle	0.12	**0.21**	0.14	**0.26**	0.16	**0.28**	0.17	**0.31**	0.19	**0.35**	0.21	**0.38**
Skeleton	0.28	**0.49**	0.34	**0.59**	0.38	**0.67**	0.41	**0.73**	0.47	**0.83**	0.52	**0.93**
Red bone marrow	0.07	**0.12**	0.08	**0.15**	0.09	**0.17**	0.10	**0.19**	0.12	**0.22**	0.13	**0.24**
Whole body dose	0.13	**0.23**	0.16	**0.28**	0.18	**0.31**	0.19	**0.34**	0.22	**0.39**	0.24	**0.43**
ED (ICRP 60)	0.21	**0.30**	0.25	**0.36**	0.27	**0.40**	0.29	**0.43**	0.32	**0.48**	0.35	**0.53**
ED (ICRP 103)	0.29	**0.40**	0.33	**0.46**	0.36	**0.50**	0.38	**0.54**	0.42	**0.60**	0.45	**0.65**
Total filtration (mm) Al	**1.5**		**2.5**		**3.5**		**4.5**		**3.5**		**3.5**	
Cu	**0**		**0**		**0**		**0**		**0.1**		**0.2**	

Table 9.35 Normalized organ doses—Chest ap—< 1 year

Beam projection:			ap				**Table cells:**					
Focus-to-detector distance (cm):			100				(+) = normal font = optimal field setting					
Tube voltage (kVp):			60				**(−) = bold font = suboptimal field setting**					
Total filtrations:			See last table row.				Empty table cells correspond to values < 0.01					
Organ or tissue	**Organ dose D_T : Entrance dose K_E**							**(mSv × mGy^{-1})**				
	(+)	(−)	(+)	(−)	(+)	(−)	(+)	(−)	(+)	(−)	(+)	(−)
Skull	0.02	**0.24**	0.03	**0.28**	0.03	**0.31**	0.04	**0.34**	0.05	**0.38**	0.04	**0.38**
Brain		**0.02**		**0.03**		**0.03**		**0.04**		**0.05**		**0.04**
Oral mucosa	0.03	**0.59**	0.04	**0.69**	0.05	**0.75**	0.05	**0.80**	0.06	**0.88**	0.06	**0.88**
Salivary glands	0.07	**0.52**	0.09	**0.61**	0.10	**0.68**	0.11	**0.73**	0.12	**0.81**	0.12	**0.81**
Cervical spine	0.38	**0.62**	0.49	**0.80**	0.56	**0.93**	0.62	**1.04**	0.73	**1.24**	0.71	**1.19**
Thyroid	0.90	**0.98**	1.01	**1.11**	1.07	**1.19**	1.12	**1.25**	1.20	**1.35**	1.19	**1.33**
Thymus	0.82	**0.84**	0.95	**0.98**	1.03	**1.06**	1.09	**1.13**	1.19	**1.23**	1.18	**1.22**
Thoracic spine	0.31	**0.34**	0.40	**0.45**	0.48	**0.54**	0.54	**0.61**	0.67	**0.76**	0.62	**0.70**
Claviculae	2.03	**2.12**	2.35	**2.46**	2.55	**2.67**	2.69	**2.83**	2.91	**3.07**	2.96	**3.11**
Scapulae	0.52	**0.56**	0.67	**0.72**	0.79	**0.84**	0.88	**0.94**	1.05	**1.12**	1.01	**1.08**
Ribs	1.39	**1.55**	1.59	**1.78**	1.72	**1.92**	1.82	**2.03**	1.97	**2.21**	1.98	**2.22**
Oesophagus	0.23	**0.25**	0.29	**0.33**	0.34	**0.38**	0.38	**0.43**	0.46	**0.51**	0.43	**0.48**
Trachea	0.10	**0.49**	0.11	**0.57**	0.13	**0.63**	0.13	**0.67**	0.15	**0.74**	0.15	**0.73**
Lung	0.45	**0.47**	0.54	**0.57**	0.61	**0.64**	0.66	**0.70**	0.74	**0.79**	0.73	**0.78**
Mamma	1.22	**1.22**	1.28	**1.28**	1.32	**1.31**	1.34	**1.34**	1.38	**1.37**	1.38	**1.37**
Heart	0.58	**0.61**	0.70	**0.73**	0.78	**0.82**	0.84	**0.88**	0.95	**1.00**	0.94	**0.98**
Stomach	0.25	**0.61**	0.30	**0.73**	0.34	**0.82**	0.37	**0.88**	0.42	**1.00**	0.41	**0.98**
Spleen	0.15	**0.24**	0.19	**0.30**	0.22	**0.35**	0.25	**0.40**	0.29	**0.48**	0.28	**0.45**
Liver	0.27	**0.54**	0.33	**0.65**	0.37	**0.73**	0.40	**0.79**	0.45	**0.89**	0.44	**0.87**
Gall bladder	0.10	**0.44**	0.12	**0.55**	0.15	**0.63**	0.16	**0.69**	0.19	**0.81**	0.18	**0.78**
Pancreas	0.23	**0.33**	0.30	**0.42**	0.35	**0.48**	0.39	**0.54**	0.46	**0.64**	0.44	**0.61**
Small intestine	0.01	**0.22**	0.02	**0.27**	0.02	**0.31**	0.02	**0.35**	0.03	**0.41**	0.03	**0.39**
Large intestine	0.02	**0.32**	0.02	**0.39**	0.03	**0.45**	0.03	**0.49**	0.04	**0.57**	0.03	**0.55**
Sigma/Rectum		**0.07**		**0.09**		**0.10**		**0.11**	0.01	**0.13**		**0.13**
Lumbar spine	0.05	**0.31**	0.07	**0.41**	0.09	**0.49**	0.10	**0.56**	0.13	**0.69**	0.11	**0.63**
Pelvis	0.01	**0.11**	0.01	**0.14**	0.02	**0.17**	0.02	**0.19**	0.03	**0.23**	0.02	**0.21**
Adrenals	0.12	**0.15**	0.16	**0.19**	0.19	**0.23**	0.21	**0.26**	0.25	**0.31**	0.24	**0.29**
Kidneys	0.04	**0.14**	0.06	**0.18**	0.07	**0.21**	0.08	**0.24**	0.09	**0.29**	0.09	**0.27**
Urinary bladder		**0.02**		**0.02**		**0.03**		**0.03**		**0.04**		**0.03**
Uterus		**0.04**		**0.05**		**0.06**		**0.07**	0.01	**0.09**	0.01	**0.08**
Ovaries		**0.04**		**0.05**		**0.06**		**0.07**	0.01	**0.09**		**0.08**
Prostate				**0.01**		**0.01**		**0.02**		**0.02**		**0.02**
Testes												
Upper arm	0.33	**1.14**	0.42	**1.39**	0.48	**1.56**	0.53	**1.69**	0.62	**1.92**	0.61	**1.90**
Elbow	0.25	**1.22**	0.32	**1.47**	0.37	**1.65**	0.41	**1.79**	0.48	**2.02**	0.46	**2.02**
Forearm	0.02	**0.63**	0.03	**0.77**	0.03	**0.86**	0.04	**0.93**	0.05	**1.04**	0.04	**1.04**
Upper leg				**0.01**		**0.01**		**0.01**		**0.02**		**0.02**
Knee												
Lower leg												
Lymph nodes	0.19	**0.40**	0.23	**0.49**	0.26	**0.55**	0.28	**0.59**	0.32	**0.67**	0.31	**0.66**
Skin	0.17	**0.33**	0.18	**0.35**	0.19	**0.37**	0.19	**0.38**	0.21	**0.40**	0.20	**0.39**
Muscle	0.14	**0.25**	0.16	**0.29**	0.18	**0.32**	0.20	**0.35**	0.22	**0.39**	0.21	**0.38**
Skeleton	0.32	**0.57**	0.39	**0.69**	0.43	**0.77**	0.47	**0.83**	0.53	**0.94**	0.52	**0.93**
Red bone marrow	0.08	**0.15**	0.10	**0.18**	0.11	**0.20**	0.12	**0.22**	0.14	**0.25**	0.13	**0.24**
Whole body dose	0.15	**0.27**	0.18	**0.32**	0.20	**0.36**	0.22	**0.39**	0.24	**0.44**	0.24	**0.43**
ED (ICRP 60)	0.24	**0.35**	0.28	**0.41**	0.30	**0.45**	0.32	**0.48**	0.35	**0.54**	0.35	**0.53**
ED (ICRP 103)	0.32	**0.45**	0.37	**0.51**	0.39	**0.56**	0.42	**0.60**	0.45	**0.66**	0.45	**0.65**
Total filtration (mm) Al	**1.5**		**2.5**		**3.5**		**4.5**		**3.5**		**3.5**	
Cu	**0**		**0**		**0**		**0**		**0.1**		**0.2**	

Table 9.36 Normalized organ doses—Chest ap—< 1 year

Beam projection:			ap		**Table cells:**							
Focus-to-detector distance (cm):			100		(+) = normal font = optimal field setting							
Tube voltage (kVp):			70		**(–) = bold font = suboptimal field setting**							
Total filtrations:			See last table row.		Empty table cells correspond to values < 0.01							
Organ or tissue	**Organ dose D_T : Entrance dose K_E**							**(mSv × mGy^{-1})**				
	(+)	(–)	(+)	(–)	(+)	(–)	(+)	(–)	(+)	(–)	(+)	(–)
Skull	0.03	**0.26**	0.04	**0.31**	0.04	**0.34**	0.05	**0.36**	0.05	**0.41**	0.06	**0.44**
Brain		**0.03**		**0.04**		**0.04**		**0.05**	0.01	**0.06**	0.01	**0.07**
Oral mucosa	0.04	**0.65**	0.05	**0.74**	0.06	**0.81**	0.06	**0.85**	0.07	**0.94**	0.08	**1.00**
Salivary glands	0.08	**0.57**	0.10	**0.67**	0.11	**0.73**	0.12	**0.78**	0.14	**0.86**	0.15	**0.93**
Cervical spine	0.45	**0.74**	0.56	**0.93**	0.63	**1.06**	0.70	**1.17**	0.81	**1.37**	0.90	**1.55**
Thyroid	0.95	**1.05**	1.06	**1.17**	1.12	**1.25**	1.17	**1.31**	1.24	**1.40**	1.29	**1.48**
Thymus	0.89	**0.91**	1.01	**1.04**	1.09	**1.13**	1.15	**1.19**	1.24	**1.29**	1.31	**1.37**
Thoracic spine	0.38	**0.43**	0.49	**0.55**	0.57	**0.65**	0.64	**0.73**	0.77	**0.88**	0.89	**1.02**
Claviculae	2.17	**2.27**	2.46	**2.58**	2.64	**2.78**	2.77	**2.92**	2.96	**3.14**	3.08	**3.28**
Scapulae	0.62	**0.67**	0.78	**0.84**	0.90	**0.96**	0.99	**1.06**	1.16	**1.25**	1.31	**1.41**
Ribs	1.48	**1.66**	1.68	**1.87**	1.80	**2.02**	1.88	**2.12**	2.02	**2.28**	2.12	**2.40**
Oesophagus	0.28	**0.31**	0.35	**0.39**	0.40	**0.45**	0.44	**0.49**	0.52	**0.58**	0.59	**0.66**
Trachea	0.11	**0.54**	0.12	**0.62**	0.14	**0.67**	0.15	**0.71**	0.16	**0.78**	0.18	**0.83**
Lung	0.51	**0.53**	0.60	**0.64**	0.67	**0.71**	0.71	**0.76**	0.80	**0.86**	0.87	**0.94**
Mamma	1.25	**1.24**	1.31	**1.30**	1.34	**1.33**	1.37	**1.35**	1.40	**1.38**	1.43	**1.40**
Heart	0.65	**0.68**	0.77	**0.81**	0.85	**0.89**	0.91	**0.96**	1.02	**1.07**	1.10	**1.17**
Stomach	0.28	**0.68**	0.33	**0.81**	0.37	**0.89**	0.40	**0.96**	0.45	**1.07**	0.49	**1.17**
Spleen	0.18	**0.29**	0.22	**0.36**	0.26	**0.41**	0.28	**0.46**	0.33	**0.54**	0.38	**0.61**
Liver	0.31	**0.61**	0.36	**0.72**	0.41	**0.80**	0.44	**0.86**	0.49	**0.96**	0.54	**1.05**
Gall bladder	0.12	**0.51**	0.15	**0.63**	0.17	**0.71**	0.18	**0.77**	0.22	**0.89**	0.24	**0.98**
Pancreas	0.28	**0.39**	0.35	**0.49**	0.40	**0.56**	0.44	**0.62**	0.51	**0.72**	0.57	**0.82**
Small intestine	0.02	**0.26**	0.02	**0.32**	0.03	**0.36**	0.03	**0.39**	0.04	**0.45**	0.04	**0.51**
Large intestine	0.02	**0.37**	0.03	**0.45**	0.03	**0.50**	0.04	**0.54**	0.04	**0.62**	0.05	**0.69**
Sigma/Rectum		**0.08**		**0.10**		**0.12**	0.01	**0.13**	0.01	**0.15**	0.02	**0.17**
Lumbar spine	0.07	**0.39**	0.09	**0.50**	0.11	**0.59**	0.13	**0.67**	0.16	**0.81**	0.18	**0.94**
Pelvis	0.01	**0.13**	0.02	**0.17**	0.02	**0.20**	0.03	**0.22**	0.03	**0.27**	0.04	**0.31**
Adrenals	0.15	**0.18**	0.19	**0.24**	0.22	**0.27**	0.25	**0.31**	0.30	**0.37**	0.34	**0.43**
Kidneys	0.05	**0.17**	0.07	**0.22**	0.08	**0.25**	0.09	**0.28**	0.11	**0.34**	0.13	**0.40**
Urinary bladder		**0.02**		**0.03**		**0.03**		**0.04**		**0.05**		**0.06**
Uterus		**0.05**		**0.06**	0.01	**0.08**	0.01	**0.08**	0.02	**0.10**	0.02	**0.12**
Ovaries		**0.05**		**0.06**		**0.08**	0.01	**0.09**	0.01	**0.10**	0.02	**0.12**
Prostate		**0.01**		**0.02**		**0.02**		**0.02**		**0.03**		**0.03**
Testes										**0.01**		**0.02**
Upper arm	0.38	**1.27**	0.47	**1.52**	0.54	**1.68**	0.59	**1.81**	0.68	**2.02**	0.75	**2.19**
Elbow	0.29	**1.35**	0.36	**1.61**	0.41	**1.78**	0.45	**1.91**	0.52	**2.13**	0.58	**2.30**
Forearm	0.03	**0.70**	0.04	**0.83**	0.04	**0.92**	0.05	**0.99**	0.06	**1.10**	0.07	**1.18**
Upper leg		**0.01**		**0.01**		**0.02**		**0.02**		**0.03**		**0.03**
Knee												
Lower leg												
Lymph nodes	0.22	**0.45**	0.26	**0.54**	0.28	**0.60**	0.31	**0.65**	0.35	**0.73**	0.38	**0.80**
Skin	0.18	**0.35**	0.19	**0.37**	0.20	**0.38**	0.20	**0.39**	0.22	**0.42**	0.23	**0.44**
Muscle	0.15	**0.28**	0.18	**0.32**	0.20	**0.35**	0.21	**0.38**	0.24	**0.42**	0.26	**0.46**
Skeleton	0.36	**0.64**	0.42	**0.75**	0.47	**0.83**	0.50	**0.90**	0.56	**1.00**	0.61	**1.09**
Red bone marrow	0.09	**0.17**	0.11	**0.20**	0.12	**0.23**	0.13	**0.25**	0.15	**0.28**	0.17	**0.31**
Whole body dose	0.17	**0.30**	0.20	**0.36**	0.22	**0.39**	0.24	**0.42**	0.26	**0.47**	0.29	**0.51**
ED (ICRP 60)	0.26	**0.38**	0.30	**0.45**	0.32	**0.49**	0.34	**0.52**	0.38	**0.58**	0.41	**0.63**
ED (ICRP 103)	0.35	**0.49**	0.39	**0.56**	0.42	**0.60**	0.44	**0.64**	0.48	**0.70**	0.51	**0.75**
Total filtration (mm) **Al**	**1.5**		**2.5**		**3.5**		**4.5**		**3.5**		**3.5**	
Cu	**0**		**0**		**0**		**0**		**0.1**		**0.2**	

Table 9.37 Normalized organ doses—Chest ap—< 1 year

Beam projection:			ap				**Table cells:**					
Focus-to-detector distance (cm):			100				(+) = normal font = optimal field setting					
Tube voltage (kVp):			80				**(−) = bold font = suboptimal field setting**					
Total filtrations:			See last table row.				Empty table cells correspond to values < 0.01					
Organ or tissue	**Organ dose D_T : Entrance dose K_E**							**(mSv × mGy^{-1})**				
	(+)	(−)	(+)	(−)	(+)	(−)	(+)	(−)	(+)	(−)	(+)	(−)
Skull	0.03	**0.29**	0.04	**0.33**	0.05	**0.36**	0.05	**0.38**	0.06	**0.42**	0.07	**0.45**
Brain		**0.03**		**0.04**	0.01	**0.05**	0.01	**0.06**	0.01	**0.07**	0.02	**0.08**
Oral mucosa	0.04	**0.70**	0.05	**0.79**	0.06	**0.85**	0.07	**0.90**	0.08	**0.97**	0.09	**1.03**
Salivary glands	0.09	**0.62**	0.11	**0.72**	0.12	**0.78**	0.13	**0.83**	0.14	**0.90**	0.16	**0.97**
Cervical spine	0.51	**0.85**	0.61	**1.03**	0.69	**1.17**	0.75	**1.27**	0.85	**1.46**	0.93	**1.62**
Thyroid	1.00	**1.11**	1.10	**1.23**	1.16	**1.30**	1.20	**1.36**	1.27	**1.45**	1.32	**1.51**
Thymus	0.94	**0.98**	1.06	**1.10**	1.14	**1.18**	1.19	**1.24**	1.28	**1.34**	1.35	**1.41**
Thoracic spine	0.45	**0.51**	0.56	**0.64**	0.65	**0.74**	0.72	**0.82**	0.85	**0.97**	0.96	**1.10**
Claviculae	2.27	**2.38**	2.53	**2.67**	2.69	**2.84**	2.80	**2.96**	2.95	**3.14**	3.03	**3.23**
Scapulae	0.71	**0.76**	0.87	**0.93**	0.99	**1.06**	1.07	**1.16**	1.24	**1.33**	1.36	**1.48**
Ribs	1.56	**1.74**	1.73	**1.94**	1.84	**2.07**	1.92	**2.16**	2.03	**2.30**	2.10	**2.39**
Oesophagus	0.32	**0.36**	0.39	**0.44**	0.45	**0.50**	0.49	**0.55**	0.57	**0.64**	0.64	**0.72**
Trachea	0.12	**0.58**	0.13	**0.66**	0.15	**0.71**	0.16	**0.75**	0.17	**0.81**	0.19	**0.86**
Lung	0.56	**0.59**	0.65	**0.69**	0.71	**0.76**	0.76	**0.81**	0.84	**0.91**	0.91	**0.98**
Mamma	1.27	**1.27**	1.33	**1.32**	1.36	**1.35**	1.38	**1.37**	1.41	**1.39**	1.44	**1.41**
Heart	0.71	**0.75**	0.83	**0.87**	0.91	**0.96**	0.97	**1.02**	1.07	**1.13**	1.15	**1.22**
Stomach	0.31	**0.75**	0.36	**0.87**	0.40	**0.96**	0.43	**1.02**	0.48	**1.13**	0.52	**1.22**
Spleen	0.21	**0.33**	0.25	**0.41**	0.29	**0.47**	0.32	**0.51**	0.37	**0.59**	0.41	**0.67**
Liver	0.34	**0.67**	0.40	**0.78**	0.44	**0.86**	0.47	**0.92**	0.52	**1.02**	0.57	**1.10**
Gall bladder	0.14	**0.58**	0.16	**0.69**	0.19	**0.78**	0.20	**0.84**	0.24	**0.95**	0.26	**1.04**
Pancreas	0.32	**0.45**	0.39	**0.55**	0.44	**0.63**	0.48	**0.68**	0.55	**0.79**	0.62	**0.88**
Small intestine	0.02	**0.29**	0.03	**0.35**	0.03	**0.40**	0.04	**0.43**	0.04	**0.49**	0.05	**0.54**
Large intestine	0.03	**0.41**	0.03	**0.49**	0.04	**0.55**	0.04	**0.59**	0.05	**0.67**	0.06	**0.73**
Sigma/Rectum		**0.10**	0.01	**0.12**	0.01	**0.13**	0.01	**0.14**	0.02	**0.16**	0.02	**0.18**
Lumbar spine	0.09	**0.47**	0.11	**0.59**	0.13	**0.68**	0.15	**0.75**	0.18	**0.89**	0.21	**1.02**
Pelvis	0.02	**0.16**	0.02	**0.20**	0.03	**0.23**	0.03	**0.25**	0.04	**0.30**	0.05	**0.34**
Adrenals	0.18	**0.22**	0.22	**0.28**	0.25	**0.32**	0.28	**0.35**	0.33	**0.41**	0.38	**0.47**
Kidneys	0.07	**0.21**	0.08	**0.26**	0.10	**0.30**	0.11	**0.33**	0.13	**0.39**	0.15	**0.44**
Urinary bladder		**0.03**		**0.03**		**0.04**		**0.04**		**0.05**	0.01	**0.06**
Uterus		**0.06**	0.01	**0.08**	0.01	**0.09**	0.02	**0.10**	0.02	**0.12**	0.02	**0.13**
Ovaries		**0.06**	0.01	**0.08**	0.01	**0.09**	0.01	**0.10**	0.02	**0.12**	0.02	**0.14**
Prostate		**0.02**		**0.02**		**0.02**		**0.03**		**0.03**		**0.04**
Testes						**0.01**		**0.01**		**0.02**		**0.02**
Upper arm	0.43	**1.38**	0.52	**1.62**	0.58	**1.77**	0.63	**1.89**	0.71	**2.08**	0.78	**2.21**
Elbow	0.33	**1.47**	0.40	**1.71**	0.45	**1.87**	0.48	**1.99**	0.55	**2.18**	0.60	**2.32**
Forearm	0.03	**0.76**	0.04	**0.88**	0.05	**0.97**	0.06	**1.03**	0.07	**1.12**	0.08	**1.19**
Upper leg		**0.01**		**0.02**		**0.02**		**0.02**		**0.03**		**0.04**
Knee												**0.01**
Lower leg												
Lymph nodes	0.24	**0.50**	0.28	**0.59**	0.31	**0.65**	0.33	**0.70**	0.37	**0.78**	0.40	**0.85**
Skin	0.18	**0.36**	0.20	**0.38**	0.20	**0.40**	0.21	**0.41**	0.22	**0.43**	0.23	**0.45**
Muscle	0.17	**0.30**	0.20	**0.35**	0.21	**0.38**	0.23	**0.41**	0.25	**0.45**	0.27	**0.49**
Skeleton	0.39	**0.69**	0.45	**0.81**	0.50	**0.88**	0.53	**0.94**	0.58	**1.04**	0.62	**1.11**
Red bone marrow	0.10	**0.19**	0.12	**0.23**	0.14	**0.25**	0.15	**0.27**	0.17	**0.31**	0.18	**0.34**
Whole body dose	0.19	**0.33**	0.22	**0.39**	0.24	**0.42**	0.25	**0.45**	0.28	**0.50**	0.30	**0.54**
ED (ICRP 60)	0.28	**0.42**	0.32	**0.48**	0.34	**0.52**	0.36	**0.55**	0.40	**0.61**	0.42	**0.66**
ED (ICRP 103)	0.37	**0.52**	0.41	**0.59**	0.44	**0.64**	0.46	**0.67**	0.50	**0.73**	0.53	**0.78**
Total filtration (mm) Al	**1.5**		**2.5**		**3.5**		**4.5**		**3.5**		**3.5**	
Cu	**0**		**0**		**0**		**0**		**0.1**		**0.2**	

Table 9.38 Normalized organ doses—Chest ap—1–2 years

Beam projection:			ap			**Table cells:**						
Focus-to-detector distance (cm):			100			(+) = normal font = optimal field setting						
Tube voltage (kVp):			50			**(−) = bold font = suboptimal field setting**						
Total filtrations:			See last table row.			Empty table cells correspond to values < 0.01						
Organ or tissue	**Organ dose D_T : Entrance dose K_E**							**(mSv × mGy^{-1})**				
	(+)	(−)	(+)	(−)	(+)	(−)	(+)	(−)	(+)	(−)	(+)	(−)
Skull	0.01	**0.21**	0.02	**0.25**	0.02	**0.28**	0.02	**0.31**	0.03	**0.35**	0.03	**0.39**
Brain		**0.01**		**0.02**		**0.02**		**0.02**		**0.03**		**0.03**
Oral mucosa	0.02	**0.41**	0.02	**0.49**	0.03	**0.54**	0.03	**0.58**	0.04	**0.66**	0.05	**0.72**
Salivary glands	0.04	**0.31**	0.05	**0.39**	0.06	**0.44**	0.07	**0.48**	0.08	**0.55**	0.09	**0.62**
Cervical spine	0.22	**0.32**	0.30	**0.43**	0.35	**0.52**	0.40	**0.59**	0.49	**0.73**	0.57	**0.87**
Thyroid	0.79	**0.83**	0.90	**0.95**	0.97	**1.03**	1.02	**1.08**	1.10	**1.17**	1.17	**1.25**
Thymus	0.63	**0.65**	0.76	**0.78**	0.85	**0.87**	0.92	**0.94**	1.03	**1.06**	1.13	**1.16**
Thoracic spine	0.12	**0.14**	0.17	**0.19**	0.21	**0.23**	0.24	**0.27**	0.31	**0.35**	0.39	**0.44**
Claviculae	1.55	**1.62**	1.88	**1.97**	2.10	**2.21**	2.27	**2.39**	2.56	**2.71**	2.80	**2.97**
Scapulae	0.20	**0.21**	0.27	**0.28**	0.32	**0.34**	0.37	**0.40**	0.46	**0.50**	0.56	**0.60**
Ribs	1.16	**1.26**	1.36	**1.49**	1.50	**1.64**	1.61	**1.76**	1.78	**1.96**	1.94	**2.13**
Oesophagus	0.14	**0.15**	0.19	**0.21**	0.23	**0.25**	0.26	**0.29**	0.32	**0.36**	0.38	**0.43**
Trachea	0.07	**0.37**	0.08	**0.45**	0.09	**0.51**	0.10	**0.55**	0.11	**0.62**	0.13	**0.68**
Lung	0.30	**0.31**	0.37	**0.39**	0.43	**0.45**	0.47	**0.50**	0.55	**0.58**	0.62	**0.66**
Mamma	1.09	**1.10**	1.18	**1.18**	1.22	**1.23**	1.25	**1.25**	1.29	**1.30**	1.33	**1.34**
Heart	0.41	**0.42**	0.51	**0.53**	0.58	**0.60**	0.64	**0.67**	0.74	**0.77**	0.84	**0.87**
Stomach	0.18	**0.41**	0.22	**0.51**	0.26	**0.58**	0.28	**0.64**	0.33	**0.74**	0.37	**0.83**
Spleen	0.08	**0.12**	0.11	**0.16**	0.13	**0.19**	0.15	**0.22**	0.18	**0.27**	0.22	**0.32**
Liver	0.18	**0.35**	0.23	**0.44**	0.27	**0.50**	0.30	**0.55**	0.35	**0.64**	0.40	**0.73**
Gall bladder	0.04	**0.30**	0.06	**0.39**	0.07	**0.45**	0.08	**0.51**	0.10	**0.60**	0.12	**0.70**
Pancreas	0.13	**0.18**	0.18	**0.24**	0.21	**0.28**	0.24	**0.32**	0.29	**0.40**	0.35	**0.47**
Small intestine		**0.13**		**0.17**		**0.20**	0.01	**0.22**	0.02	**0.27**	0.02	**0.32**
Large intestine		**0.20**	0.01	**0.25**	0.01	**0.30**	0.02	**0.33**	0.02	**0.40**	0.03	**0.46**
Sigma/Rectum		**0.04**		**0.05**		**0.06**		**0.07**		**0.09**		**0.10**
Lumbar spine	0.02	**0.12**	0.03	**0.16**	0.04	**0.20**	0.05	**0.24**	0.07	**0.31**	0.08	**0.39**
Pelvis		**0.05**		**0.07**		**0.08**		**0.10**		**0.12**	0.01	**0.15**
Adrenals	0.05	**0.07**	0.07	**0.09**	0.09	**0.11**	0.11	**0.13**	0.13	**0.17**	0.16	**0.21**
Kidneys	0.02	**0.06**	0.03	**0.08**	0.04	**0.10**	0.04	**0.11**	0.05	**0.14**	0.07	**0.17**
Urinary bladder				**0.01**		**0.01**		**0.01**		**0.02**		**0.02**
Uterus		**0.02**		**0.03**		**0.03**		**0.04**		**0.05**		**0.06**
Ovaries		**0.02**		**0.03**		**0.03**		**0.04**		**0.05**		**0.06**
Prostate												**0.01**
Testes												
Upper arm	0.17	**0.65**	0.22	**0.82**	0.27	**0.95**	0.30	**1.06**	0.37	**1.25**	0.44	**1.43**
Elbow	0.13	**0.70**	0.18	**0.89**	0.22	**1.03**	0.24	**1.14**	0.30	**1.34**	0.36	**1.53**
Forearm		**0.38**	0.01	**0.48**	0.02	**0.55**	0.02	**0.61**	0.03	**0.72**	0.03	**0.82**
Upper leg												
Knee												
Lower leg												
Lymph nodes	0.13	**0.27**	0.16	**0.34**	0.18	**0.38**	0.20	**0.42**	0.23	**0.49**	0.26	**0.56**
Skin	0.14	**0.28**	0.15	**0.29**	0.15	**0.30**	0.16	**0.31**	0.17	**0.32**	0.17	**0.34**
Muscle	0.08	**0.15**	0.10	**0.18**	0.12	**0.21**	0.13	**0.22**	0.14	**0.25**	0.16	**0.28**
Skeleton	0.20	**0.35**	0.25	**0.43**	0.28	**0.49**	0.30	**0.54**	0.35	**0.62**	0.40	**0.70**
Red bone marrow	0.05	**0.08**	0.06	**0.11**	0.07	**0.12**	0.07	**0.13**	0.09	**0.15**	0.10	**0.18**
Whole body dose	0.10	**0.18**	0.12	**0.21**	0.14	**0.24**	0.15	**0.26**	0.17	**0.30**	0.19	**0.34**
ED (ICRP 60)	0.18	**0.25**	0.21	**0.30**	0.24	**0.33**	0.25	**0.36**	0.28	**0.41**	0.31	**0.45**
ED (ICRP 103)	0.26	**0.33**	0.30	**0.39**	0.32	**0.43**	0.34	**0.46**	0.38	**0.52**	0.41	**0.57**
Total filtration (mm) **Al**	**1.5**		**2.5**		**3.5**		**4.5**		**3.5**		**3.5**	
Cu	**0**		**0**		**0**		**0**		**0.1**		**0.2**	

Table 9.39 Normalized organ doses—Chest ap—1–2 years

Beam projection:			ap				**Table cells:**					
Focus-to-detector distance (cm):			100				(+) = normal font = optimal field setting					
Tube voltage (kVp):			60				**(−) = bold font = suboptimal field setting**					
Total filtrations:			See last table row.				Empty table cells correspond to values < 0.01					
Organ or tissue	**Organ dose D_T : Entrance dose K_E**						**(mSv × mGy^{-1})**					
	(+)	(−)	(+)	(−)	(+)	(−)	(+)	(−)	(+)	(−)	(+)	(−)
Skull	0.02	**0.24**	0.02	**0.29**	0.03	**0.33**	0.03	**0.36**	0.04	**0.41**	0.03	**0.39**
Brain		**0.02**		**0.02**		**0.03**		**0.03**		**0.04**		**0.03**
Oral mucosa	0.02	**0.47**	0.03	**0.55**	0.04	**0.61**	0.04	**0.66**	0.05	**0.74**	0.05	**0.72**
Salivary glands	0.05	**0.37**	0.06	**0.46**	0.07	**0.51**	0.08	**0.56**	0.09	**0.63**	0.09	**0.62**
Cervical spine	0.30	**0.44**	0.38	**0.58**	0.45	**0.68**	0.50	**0.77**	0.61	**0.94**	0.57	**0.87**
Thyroid	0.86	**0.91**	0.97	**1.04**	1.04	**1.12**	1.09	**1.17**	1.18	**1.27**	1.17	**1.25**
Thymus	0.73	**0.75**	0.86	**0.89**	0.96	**0.99**	1.03	**1.06**	1.15	**1.18**	1.13	**1.16**
Thoracic spine	0.19	**0.21**	0.25	**0.29**	0.31	**0.35**	0.35	**0.40**	0.45	**0.51**	0.39	**0.44**
Claviculae	1.77	**1.86**	2.11	**2.23**	2.34	**2.47**	2.51	**2.66**	2.79	**2.97**	2.80	**2.97**
Scapulae	0.28	**0.30**	0.36	**0.39**	0.43	**0.46**	0.49	**0.52**	0.60	**0.65**	0.56	**0.60**
Ribs	1.30	**1.43**	1.52	**1.66**	1.66	**1.82**	1.76	**1.94**	1.94	**2.15**	1.94	**2.13**
Oesophagus	0.19	**0.22**	0.25	**0.28**	0.30	**0.34**	0.34	**0.38**	0.42	**0.46**	0.38	**0.43**
Trachea	0.08	**0.43**	0.10	**0.52**	0.11	**0.57**	0.12	**0.62**	0.13	**0.69**	0.13	**0.68**
Lung	0.36	**0.38**	0.44	**0.47**	0.51	**0.54**	0.55	**0.59**	0.64	**0.68**	0.62	**0.66**
Mamma	1.14	**1.15**	1.22	**1.23**	1.26	**1.27**	1.29	**1.30**	1.34	**1.35**	1.33	**1.34**
Heart	0.49	**0.51**	0.61	**0.63**	0.69	**0.72**	0.75	**0.78**	0.86	**0.90**	0.84	**0.87**
Stomach	0.22	**0.49**	0.27	**0.61**	0.30	**0.68**	0.33	**0.75**	0.39	**0.86**	0.37	**0.83**
Spleen	0.11	**0.16**	0.14	**0.21**	0.17	**0.25**	0.19	**0.29**	0.23	**0.35**	0.22	**0.32**
Liver	0.23	**0.43**	0.28	**0.52**	0.32	**0.60**	0.35	**0.65**	0.41	**0.76**	0.40	**0.73**
Gall bladder	0.06	**0.38**	0.08	**0.48**	0.09	**0.56**	0.11	**0.62**	0.13	**0.73**	0.12	**0.70**
Pancreas	0.18	**0.24**	0.23	**0.31**	0.27	**0.37**	0.31	**0.42**	0.37	**0.51**	0.35	**0.47**
Small intestine		**0.17**	0.01	**0.22**	0.02	**0.25**	0.02	**0.28**	0.02	**0.34**	0.02	**0.32**
Large intestine	0.01	**0.25**	0.02	**0.32**	0.02	**0.37**	0.02	**0.41**	0.03	**0.48**	0.03	**0.46**
Sigma/Rectum		**0.05**		**0.07**		**0.08**		**0.09**		**0.11**		**0.10**
Lumbar spine	0.04	**0.19**	0.05	**0.25**	0.07	**0.30**	0.08	**0.35**	0.10	**0.45**	0.08	**0.39**
Pelvis		**0.07**		**0.10**		**0.12**	0.01	**0.14**	0.02	**0.17**	0.01	**0.15**
Adrenals	0.08	**0.10**	0.11	**0.13**	0.13	**0.16**	0.15	**0.18**	0.18	**0.23**	0.16	**0.21**
Kidneys	0.03	**0.08**	0.04	**0.11**	0.05	**0.14**	0.06	**0.16**	0.08	**0.20**	0.07	**0.17**
Urinary bladder		**0.01**		**0.02**		**0.02**		**0.02**		**0.03**		**0.02**
Uterus		**0.03**		**0.04**		**0.05**		**0.06**		**0.07**		**0.06**
Ovaries		**0.03**		**0.04**		**0.05**		**0.06**		**0.07**		**0.06**
Prostate										**0.01**		**0.01**
Testes												
Upper arm	0.23	**0.81**	0.30	**1.00**	0.35	**1.15**	0.39	**1.26**	0.47	**1.48**	0.44	**1.43**
Elbow	0.18	**0.86**	0.24	**1.08**	0.28	**1.23**	0.32	**1.35**	0.39	**1.58**	0.36	**1.53**
Forearm	0.02	**0.46**	0.02	**0.58**	0.03	**0.66**	0.03	**0.72**	0.04	**0.84**	0.03	**0.82**
Upper leg												
Knee												
Lower leg												
Lymph nodes	0.16	**0.33**	0.19	**0.41**	0.22	**0.46**	0.24	**0.50**	0.28	**0.58**	0.26	**0.56**
Skin	0.15	**0.29**	0.16	**0.31**	0.16	**0.32**	0.17	**0.33**	0.18	**0.35**	0.17	**0.34**
Muscle	0.10	**0.18**	0.12	**0.21**	0.14	**0.24**	0.15	**0.26**	0.17	**0.30**	0.16	**0.28**
Skeleton	0.24	**0.42**	0.29	**0.51**	0.33	**0.58**	0.36	**0.63**	0.41	**0.73**	0.40	**0.70**
Red bone marrow	0.06	**0.11**	0.07	**0.13**	0.08	**0.15**	0.09	**0.16**	0.10	**0.19**	0.10	**0.18**
Whole body dose	0.12	**0.21**	0.14	**0.25**	0.16	**0.28**	0.17	**0.31**	0.20	**0.35**	0.19	**0.34**
ED (ICRP 60)	0.21	**0.29**	0.24	**0.35**	0.27	**0.39**	0.29	**0.42**	0.32	**0.47**	0.31	**0.45**
ED (ICRP 103)	0.29	**0.38**	0.33	**0.45**	0.36	**0.49**	0.38	**0.52**	0.42	**0.59**	0.41	**0.57**
Total filtration (mm) Al	**1.5**		**2.5**		**3.5**		**4.5**		**3.5**		**3.5**	
Cu	**0**		**0**		**0**		**0**		**0.1**		**0.2**	

Table 9.40 Normalized organ doses—Chest ap—1–2 years

Beam projection:			ap		**Table cells:**							
Focus-to-detector distance (cm):			100		(+) = normal font = optimal field setting							
Tube voltage (kVp):			70		**(−) = bold font = suboptimal field setting**							
Total filtrations:			See last table row.		Empty table cells correspond to values < 0.01							
Organ or tissue	**Organ dose D_T : Entrance dose K_E**						**(mSv × mGy^{-1})**					
	(+)	(−)	(+)	(−)	(+)	(−)	(+)	(−)	(+)	(−)	(+)	(−)
Skull	0.02	**0.27**	0.03	**0.33**	0.03	**0.36**	0.04	**0.39**	0.05	**0.44**	0.05	**0.48**
Brain		**0.02**		**0.03**		**0.04**		**0.04**		**0.05**		**0.06**
Oral mucosa	0.03	**0.52**	0.04	**0.61**	0.05	**0.67**	0.05	**0.72**	0.06	**0.80**	0.07	**0.87**
Salivary glands	0.06	**0.43**	0.07	**0.51**	0.08	**0.57**	0.09	**0.61**	0.11	**0.69**	0.12	**0.76**
Cervical spine	0.36	**0.55**	0.46	**0.70**	0.53	**0.81**	0.58	**0.90**	0.69	**1.08**	0.79	**1.25**
Thyroid	0.92	**0.98**	1.03	**1.11**	1.10	**1.19**	1.15	**1.24**	1.23	**1.34**	1.29	**1.43**
Thymus	0.81	**0.83**	0.94	**0.97**	1.04	**1.07**	1.10	**1.14**	1.22	**1.27**	1.32	**1.37**
Thoracic spine	0.25	**0.28**	0.33	**0.37**	0.39	**0.44**	0.44	**0.50**	0.55	**0.63**	0.66	**0.75**
Claviculae	1.94	**2.05**	2.27	**2.40**	2.48	**2.64**	2.64	**2.81**	2.90	**3.09**	3.09	**3.31**
Scapulae	0.35	**0.37**	0.44	**0.47**	0.52	**0.56**	0.58	**0.62**	0.70	**0.75**	0.81	**0.87**
Ribs	1.41	**1.55**	1.62	**1.78**	1.75	**1.94**	1.85	**2.05**	2.02	**2.24**	2.14	**2.39**
Oesophagus	0.24	**0.27**	0.31	**0.35**	0.36	**0.41**	0.41	**0.45**	0.49	**0.55**	0.57	**0.64**
Trachea	0.09	**0.48**	0.11	**0.57**	0.12	**0.63**	0.13	**0.67**	0.14	**0.75**	0.16	**0.81**
Lung	0.42	**0.44**	0.50	**0.53**	0.57	**0.60**	0.62	**0.66**	0.70	**0.75**	0.78	**0.84**
Mamma	1.18	**1.19**	1.25	**1.26**	1.30	**1.31**	1.32	**1.34**	1.37	**1.39**	1.41	**1.43**
Heart	0.57	**0.59**	0.68	**0.71**	0.77	**0.80**	0.83	**0.87**	0.95	**1.00**	1.04	**1.10**
Stomach	0.25	**0.57**	0.30	**0.68**	0.34	**0.76**	0.37	**0.83**	0.43	**0.94**	0.47	**1.04**
Spleen	0.14	**0.21**	0.18	**0.26**	0.21	**0.31**	0.23	**0.35**	0.28	**0.42**	0.32	**0.49**
Liver	0.27	**0.49**	0.32	**0.60**	0.37	**0.67**	0.40	**0.73**	0.46	**0.84**	0.51	**0.93**
Gall bladder	0.08	**0.45**	0.10	**0.56**	0.11	**0.64**	0.13	**0.71**	0.16	**0.83**	0.18	**0.93**
Pancreas	0.22	**0.30**	0.28	**0.38**	0.33	**0.45**	0.36	**0.50**	0.43	**0.60**	0.50	**0.69**
Small intestine	0.01	**0.21**	0.02	**0.26**	0.02	**0.30**	0.02	**0.33**	0.03	**0.39**	0.04	**0.45**
Large intestine	0.02	**0.30**	0.02	**0.37**	0.03	**0.42**	0.03	**0.46**	0.04	**0.54**	0.05	**0.61**
Sigma/Rectum		**0.07**		**0.08**		**0.10**		**0.11**		**0.13**	0.01	**0.15**
Lumbar spine	0.06	**0.25**	0.08	**0.33**	0.09	**0.39**	0.10	**0.45**	0.13	**0.56**	0.16	**0.67**
Pelvis		**0.10**	0.01	**0.13**	0.01	**0.15**	0.02	**0.17**	0.02	**0.21**	0.03	**0.25**
Adrenals	0.11	**0.13**	0.14	**0.17**	0.16	**0.20**	0.18	**0.23**	0.22	**0.29**	0.27	**0.34**
Kidneys	0.04	**0.11**	0.06	**0.15**	0.07	**0.17**	0.08	**0.20**	0.10	**0.24**	0.11	**0.29**
Urinary bladder		**0.02**		**0.02**		**0.03**		**0.03**		**0.04**		**0.04**
Uterus		**0.04**		**0.05**		**0.06**		**0.07**	0.01	**0.09**	0.01	**0.11**
Ovaries		**0.04**		**0.05**		**0.06**	0.01	**0.07**	0.01	**0.09**	0.02	**0.11**
Prostate						**0.01**		**0.01**		**0.02**		**0.02**
Testes												**0.01**
Upper arm	0.28	**0.93**	0.35	**1.14**	0.41	**1.29**	0.45	**1.40**	0.54	**1.62**	0.62	**1.80**
Elbow	0.23	**1.00**	0.29	**1.22**	0.33	**1.38**	0.37	**1.50**	0.45	**1.72**	0.51	**1.91**
Forearm	0.02	**0.53**	0.03	**0.65**	0.03	**0.73**	0.04	**0.80**	0.05	**0.91**	0.06	**1.01**
Upper leg										**0.01**		**0.02**
Knee												
Lower leg												
Lymph nodes	0.18	**0.38**	0.22	**0.46**	0.25	**0.52**	0.27	**0.57**	0.31	**0.65**	0.34	**0.73**
Skin	0.15	**0.30**	0.16	**0.32**	0.17	**0.33**	0.18	**0.34**	0.19	**0.36**	0.20	**0.38**
Muscle	0.11	**0.20**	0.14	**0.24**	0.15	**0.27**	0.16	**0.29**	0.18	**0.33**	0.20	**0.36**
Skeleton	0.27	**0.48**	0.33	**0.58**	0.36	**0.64**	0.39	**0.70**	0.45	**0.80**	0.49	**0.88**
Red bone marrow	0.07	**0.12**	0.08	**0.15**	0.09	**0.17**	0.10	**0.19**	0.12	**0.22**	0.13	**0.25**
Whole body dose	0.13	**0.24**	0.16	**0.28**	0.18	**0.32**	0.19	**0.34**	0.22	**0.39**	0.24	**0.43**
ED (ICRP 60)	0.23	**0.33**	0.27	**0.39**	0.29	**0.43**	0.31	**0.46**	0.35	**0.52**	0.38	**0.57**
ED (ICRP 103)	0.31	**0.42**	0.36	**0.49**	0.38	**0.54**	0.41	**0.57**	0.45	**0.64**	0.48	**0.69**
Total filtration (mm) Al	**1.5**		**2.5**		**3.5**		**4.5**		**3.5**		**3.5**	
Cu	**0**		**0**		**0**		**0**		**0.1**		**0.2**	

Table 9.41 Normalized organ doses—Chest ap—1–2 years

Beam projection:			ap			**Table cells:**						
Focus-to-detector distance (cm):			100			(+) = normal font = optimal field setting						
Tube voltage (kVp):			80			**(−) = bold font = suboptimal field setting**						
Total filtrations:			See last table row.			Empty table cells correspond to values < 0.01						
Organ or tissue	**Organ dose D_T : Entrance dose K_E**							**(mSv × mGy^{-1})**				
	(+)	(−)	(+)	(−)	(+)	(−)	(+)	(−)	(+)	(−)	(+)	(−)
Skull	0.03	**0.30**	0.03	**0.35**	0.04	**0.39**	0.04	**0.41**	0.05	**0.46**	0.06	**0.50**
Brain		**0.03**		**0.04**		**0.04**		**0.05**	0.01	**0.06**	0.01	**0.07**
Oral mucosa	0.04	**0.57**	0.05	**0.66**	0.05	**0.72**	0.06	**0.77**	0.07	**0.85**	0.08	**0.91**
Salivary glands	0.07	**0.47**	0.08	**0.56**	0.09	**0.61**	0.10	**0.66**	0.12	**0.74**	0.13	**0.80**
Cervical spine	0.42	**0.65**	0.52	**0.80**	0.59	**0.92**	0.64	**1.01**	0.75	**1.19**	0.84	**1.35**
Thyroid	0.97	**1.05**	1.08	**1.17**	1.14	**1.25**	1.19	**1.30**	1.27	**1.40**	1.33	**1.48**
Thymus	0.88	**0.91**	1.01	**1.05**	1.10	**1.14**	1.17	**1.21**	1.28	**1.34**	1.37	**1.43**
Thoracic spine	0.31	**0.36**	0.40	**0.45**	0.47	**0.53**	0.52	**0.59**	0.63	**0.73**	0.74	**0.85**
Claviculae	2.07	**2.20**	2.38	**2.53**	2.57	**2.74**	2.71	**2.90**	2.93	**3.14**	3.08	**3.31**
Scapulae	0.41	**0.44**	0.51	**0.55**	0.59	**0.63**	0.65	**0.70**	0.77	**0.83**	0.87	**0.95**
Ribs	1.50	**1.65**	1.69	**1.87**	1.82	**2.02**	1.91	**2.12**	2.05	**2.29**	2.15	**2.41**
Oesophagus	0.29	**0.33**	0.37	**0.41**	0.42	**0.47**	0.47	**0.52**	0.55	**0.62**	0.63	**0.71**
Trachea	0.10	**0.53**	0.12	**0.62**	0.13	**0.67**	0.14	**0.72**	0.15	**0.79**	0.17	**0.86**
Lung	0.47	**0.50**	0.56	**0.59**	0.62	**0.66**	0.67	**0.72**	0.75	**0.81**	0.83	**0.89**
Mamma	1.21	**1.22**	1.28	**1.30**	1.32	**1.34**	1.35	**1.37**	1.40	**1.42**	1.43	**1.46**
Heart	0.63	**0.66**	0.75	**0.79**	0.84	**0.88**	0.90	**0.95**	1.01	**1.07**	1.11	**1.17**
Stomach	0.28	**0.63**	0.34	**0.75**	0.38	**0.83**	0.41	**0.90**	0.46	**1.01**	0.51	**1.10**
Spleen	0.17	**0.25**	0.21	**0.31**	0.24	**0.36**	0.26	**0.40**	0.31	**0.48**	0.36	**0.55**
Liver	0.30	**0.56**	0.36	**0.66**	0.40	**0.74**	0.44	**0.80**	0.50	**0.90**	0.55	**0.99**
Gall bladder	0.09	**0.52**	0.12	**0.64**	0.13	**0.72**	0.15	**0.78**	0.18	**0.90**	0.20	**1.01**
Pancreas	0.26	**0.36**	0.33	**0.45**	0.37	**0.51**	0.41	**0.57**	0.48	**0.67**	0.55	**0.76**
Small intestine	0.02	**0.24**	0.02	**0.30**	0.03	**0.34**	0.03	**0.37**	0.04	**0.43**	0.04	**0.49**
Large intestine	0.02	**0.34**	0.03	**0.42**	0.03	**0.47**	0.04	**0.51**	0.05	**0.59**	0.05	**0.66**
Sigma/Rectum		**0.08**		**0.10**		**0.11**		**0.12**	0.01	**0.14**	0.01	**0.16**
Lumbar spine	0.07	**0.32**	0.10	**0.40**	0.11	**0.47**	0.13	**0.53**	0.16	**0.65**	0.19	**0.76**
Pelvis	0.01	**0.12**	0.02	**0.15**	0.02	**0.18**	0.02	**0.20**	0.03	**0.24**	0.03	**0.28**
Adrenals	0.13	**0.17**	0.17	**0.21**	0.19	**0.25**	0.22	**0.28**	0.26	**0.33**	0.30	**0.39**
Kidneys	0.05	**0.14**	0.07	**0.18**	0.08	**0.21**	0.09	**0.23**	0.11	**0.28**	0.13	**0.33**
Urinary bladder		**0.02**		**0.03**		**0.03**		**0.04**		**0.04**		**0.05**
Uterus		**0.05**		**0.07**		**0.08**	0.01	**0.09**	0.01	**0.10**	0.02	**0.12**
Ovaries		**0.05**	0.01	**0.07**	0.01	**0.08**	0.01	**0.09**	0.02	**0.11**	0.02	**0.13**
Prostate				**0.01**		**0.01**		**0.02**		**0.02**		**0.02**
Testes										**0.01**		**0.01**
Upper arm	0.33	**1.04**	0.40	**1.25**	0.46	**1.40**	0.50	**1.51**	0.59	**1.71**	0.66	**1.87**
Elbow	0.27	**1.11**	0.33	**1.33**	0.38	**1.49**	0.41	**1.60**	0.49	**1.81**	0.55	**1.98**
Forearm	0.03	**0.59**	0.03	**0.71**	0.04	**0.79**	0.05	**0.85**	0.06	**0.96**	0.07	**1.04**
Upper leg						**0.01**		**0.01**		**0.02**		**0.02**
Knee												
Lower leg												
Lymph nodes	0.21	**0.43**	0.24	**0.52**	0.27	**0.58**	0.29	**0.62**	0.33	**0.71**	0.37	**0.78**
Skin	0.16	**0.31**	0.17	**0.33**	0.18	**0.35**	0.19	**0.36**	0.20	**0.38**	0.21	**0.40**
Muscle	0.13	**0.23**	0.15	**0.27**	0.17	**0.29**	0.18	**0.31**	0.20	**0.35**	0.22	**0.39**
Skeleton	0.30	**0.53**	0.35	**0.63**	0.39	**0.70**	0.42	**0.75**	0.47	**0.84**	0.51	**0.92**
Red bone marrow	0.08	**0.14**	0.09	**0.17**	0.11	**0.19**	0.11	**0.21**	0.13	**0.24**	0.15	**0.27**
Whole body dose	0.15	**0.27**	0.18	**0.31**	0.19	**0.34**	0.21	**0.37**	0.23	**0.42**	0.26	**0.46**
ED (ICRP 60)	0.25	**0.36**	0.29	**0.42**	0.32	**0.46**	0.34	**0.50**	0.37	**0.56**	0.40	**0.61**
ED (ICRP 103)	0.34	**0.46**	0.38	**0.53**	0.41	**0.58**	0.43	**0.61**	0.47	**0.68**	0.50	**0.73**
Total filtration (mm) **Al**	**1.5**		**2.5**		**3.5**		**4.5**		**3.5**		**3.5**	
Cu	**0**		**0**		**0**		**0**		**0.1**		**0.2**	

Table 9.42 Normalized organ doses—Chest ap—3–7 years

Beam projection:			ap			**Table cells:**						
Focus-to-detector distance (cm):			100			(+) = normal font = optimal field setting						
Tube voltage (kVp):			60			**(−) = bold font = suboptimal field setting**						
Total filtrations:			See last table row.			Empty table cells correspond to values < 0.01						
Organ or tissue	**Organ dose D_T : Entrance dose K_E**							**(mSv × mGy^{-1})**				
	(+)	(−)	(+)	(−)	(+)	(−)	(+)	(−)	(+)	(−)	(+)	(−)
Skull	0.01	**0.25**	0.02	**0.31**	0.02	**0.35**	0.02	**0.38**	0.03	**0.45**	0.03	**0.43**
Brain		**0.01**		**0.01**		**0.02**		**0.02**		**0.03**		**0.02**
Oral mucosa	0.02	**0.37**	0.02	**0.44**	0.03	**0.49**	0.03	**0.53**	0.04	**0.61**	0.04	**0.58**
Salivary glands	0.03	**0.29**	0.04	**0.36**	0.04	**0.41**	0.05	**0.45**	0.06	**0.52**	0.05	**0.50**
Cervical spine	0.25	**0.38**	0.32	**0.50**	0.38	**0.59**	0.43	**0.67**	0.52	**0.82**	0.48	**0.75**
Thyroid	0.76	**0.80**	0.86	**0.93**	0.93	**1.01**	0.98	**1.07**	1.07	**1.17**	1.05	**1.15**
Thymus	0.69	**0.70**	0.83	**0.85**	0.93	**0.95**	1.00	**1.03**	1.14	**1.17**	1.11	**1.14**
Thoracic spine	0.14	**0.16**	0.19	**0.21**	0.24	**0.26**	0.27	**0.30**	0.35	**0.39**	0.29	**0.33**
Claviculae	1.59	**1.67**	1.91	**2.01**	2.14	**2.26**	2.31	**2.44**	2.60	**2.76**	2.59	**2.74**
Scapulae	0.21	**0.23**	0.28	**0.31**	0.34	**0.37**	0.39	**0.42**	0.48	**0.52**	0.44	**0.47**
Ribs	1.27	**1.35**	1.48	**1.58**	1.63	**1.74**	1.74	**1.86**	1.92	**2.06**	1.91	**2.04**
Oesophagus	0.12	**0.13**	0.16	**0.17**	0.19	**0.20**	0.21	**0.23**	0.26	**0.29**	0.24	**0.26**
Trachea	0.08	**0.30**	0.09	**0.37**	0.10	**0.42**	0.11	**0.47**	0.12	**0.54**	0.12	**0.52**
Lung	0.32	**0.34**	0.41	**0.43**	0.47	**0.49**	0.51	**0.54**	0.60	**0.63**	0.57	**0.61**
Mamma	1.12	**1.16**	1.21	**1.25**	1.26	**1.30**	1.29	**1.33**	1.35	**1.39**	1.34	**1.38**
Heart	0.44	**0.45**	0.54	**0.56**	0.62	**0.65**	0.69	**0.71**	0.80	**0.84**	0.77	**0.80**
Stomach	0.23	**0.48**	0.29	**0.60**	0.34	**0.68**	0.37	**0.75**	0.43	**0.87**	0.41	**0.84**
Spleen	0.09	**0.13**	0.12	**0.17**	0.15	**0.21**	0.17	**0.24**	0.21	**0.30**	0.19	**0.26**
Liver	0.22	**0.38**	0.28	**0.48**	0.32	**0.55**	0.35	**0.60**	0.42	**0.71**	0.40	**0.67**
Gall bladder	0.07	**0.34**	0.10	**0.43**	0.12	**0.51**	0.13	**0.57**	0.16	**0.68**	0.14	**0.64**
Pancreas	0.16	**0.21**	0.21	**0.27**	0.25	**0.32**	0.28	**0.37**	0.35	**0.45**	0.32	**0.41**
Small intestine		**0.18**		**0.23**	0.01	**0.27**	0.01	**0.31**	0.02	**0.37**	0.01	**0.34**
Large intestine		**0.24**	0.01	**0.31**	0.02	**0.36**	0.02	**0.40**	0.02	**0.48**	0.02	**0.45**
Sigma/Rectum		**0.06**		**0.08**		**0.09**		**0.10**		**0.12**		**0.11**
Lumbar spine	0.03	**0.14**	0.05	**0.19**	0.06	**0.23**	0.07	**0.26**	0.09	**0.34**	0.07	**0.28**
Pelvis		**0.07**		**0.10**		**0.12**		**0.14**		**0.18**		**0.15**
Adrenals	0.07	**0.08**	0.09	**0.11**	0.11	**0.13**	0.12	**0.15**	0.16	**0.19**	0.14	**0.16**
Kidneys	0.03	**0.07**	0.04	**0.09**	0.05	**0.11**	0.06	**0.12**	0.07	**0.16**	0.06	**0.13**
Urinary bladder				**0.01**		**0.02**		**0.02**		**0.02**		**0.02**
Uterus		**0.03**		**0.04**		**0.05**		**0.06**		**0.07**		**0.06**
Ovaries		**0.03**		**0.04**		**0.05**		**0.06**		**0.08**		**0.07**
Prostate												
Testes												
Upper arm	0.17	**0.67**	0.22	**0.85**	0.26	**0.98**	0.30	**1.08**	0.37	**1.28**	0.33	**1.22**
Elbow	0.14	**0.72**	0.19	**0.91**	0.22	**1.05**	0.25	**1.16**	0.31	**1.37**	0.28	**1.31**
Forearm	0.01	**0.42**	0.02	**0.53**	0.02	**0.61**	0.02	**0.67**	0.03	**0.79**	0.03	**0.76**
Upper leg												
Knee												
Lower leg												
Lymph nodes	0.14	**0.28**	0.17	**0.35**	0.20	**0.41**	0.22	**0.45**	0.25	**0.53**	0.24	**0.50**
Skin	0.12	**0.26**	0.13	**0.27**	0.14	**0.28**	0.14	**0.29**	0.15	**0.31**	0.14	**0.30**
Muscle	0.09	**0.15**	0.10	**0.18**	0.12	**0.21**	0.13	**0.22**	0.15	**0.26**	0.14	**0.24**
Skeleton	0.18	**0.33**	0.22	**0.40**	0.24	**0.46**	0.27	**0.50**	0.31	**0.58**	0.29	**0.56**
Red bone marrow	0.06	**0.09**	0.07	**0.12**	0.08	**0.13**	0.09	**0.15**	0.10	**0.17**	0.09	**0.16**
Whole body dose	0.10	**0.18**	0.12	**0.22**	0.14	**0.25**	0.15	**0.27**	0.18	**0.31**	0.17	**0.30**
ED (ICRP 60)	0.19	**0.27**	0.23	**0.32**	0.25	**0.36**	0.27	**0.39**	0.31	**0.45**	0.30	**0.43**
ED (ICRP 103)	0.27	**0.36**	0.31	**0.42**	0.34	**0.47**	0.36	**0.50**	0.40	**0.56**	0.39	**0.54**
Total filtration (mm) Al	**1.5**		**2.5**		**3.5**		**4.5**		**3.5**		**3.5**	
Cu	**0**		**0**		**0**		**0**		**0.1**		**0.2**	

Table 9.43 Normalized organ doses—Chest ap—3–7 years

Beam projection:			ap				**Table cells:**					
Focus-to-detector distance (cm):			100				(+) = normal font = optimal field setting					
Tube voltage (kVp):			70				**(−) = bold font = suboptimal field setting**					
Total filtrations:			See last table row.				Empty table cells correspond to values < 0.01					
Organ or tissue	**Organ dose D_T : Entrance dose K_E**								**(mSv × mGy^{-1})**			
	(+)	(−)	(+)	(−)	(+)	(−)	(+)	(−)	(+)	(−)	(+)	(−)
Skull	0.02	**0.29**	0.02	**0.35**	0.03	**0.39**	0.03	**0.43**	0.04	**0.49**	0.05	**0.54**
Brain		**0.02**		**0.02**		**0.03**		**0.03**		**0.04**		**0.05**
Oral mucosa	0.02	**0.42**	0.03	**0.49**	0.04	**0.55**	0.04	**0.59**	0.05	**0.67**	0.06	**0.74**
Salivary glands	0.04	**0.34**	0.05	**0.41**	0.05	**0.46**	0.06	**0.51**	0.07	**0.58**	0.08	**0.65**
Cervical spine	0.31	**0.48**	0.39	**0.61**	0.45	**0.71**	0.50	**0.80**	0.61	**0.97**	0.70	**1.13**
Thyroid	0.82	**0.88**	0.92	**1.00**	0.99	**1.08**	1.04	**1.15**	1.13	**1.25**	1.20	**1.35**
Thymus	0.78	**0.80**	0.92	**0.94**	1.02	**1.05**	1.10	**1.13**	1.23	**1.27**	1.34	**1.39**
Thoracic spine	0.20	**0.22**	0.26	**0.29**	0.31	**0.35**	0.36	**0.40**	0.45	**0.51**	0.55	**0.62**
Claviculae	1.76	**1.86**	2.08	**2.20**	2.30	**2.44**	2.46	**2.62**	2.73	**2.92**	2.94	**3.15**
Scapulae	0.27	**0.30**	0.35	**0.38**	0.42	**0.45**	0.47	**0.51**	0.58	**0.63**	0.68	**0.74**
Ribs	1.38	**1.47**	1.59	**1.71**	1.73	**1.86**	1.84	**1.98**	2.01	**2.18**	2.15	**2.33**
Oesophagus	0.15	**0.17**	0.20	**0.22**	0.23	**0.26**	0.26	**0.29**	0.32	**0.35**	0.38	**0.42**
Trachea	0.09	**0.35**	0.10	**0.43**	0.11	**0.48**	0.12	**0.53**	0.14	**0.61**	0.15	**0.68**
Lung	0.38	**0.40**	0.47	**0.49**	0.53	**0.56**	0.58	**0.61**	0.67	**0.71**	0.75	**0.80**
Mamma	1.17	**1.20**	1.25	**1.29**	1.30	**1.33**	1.33	**1.37**	1.39	**1.42**	1.44	**1.47**
Heart	0.51	**0.53**	0.63	**0.65**	0.71	**0.74**	0.78	**0.81**	0.90	**0.94**	1.01	**1.06**
Stomach	0.28	**0.56**	0.34	**0.68**	0.38	**0.77**	0.42	**0.84**	0.49	**0.97**	0.55	**1.08**
Spleen	0.12	**0.17**	0.16	**0.22**	0.19	**0.26**	0.21	**0.30**	0.26	**0.36**	0.30	**0.43**
Liver	0.26	**0.45**	0.32	**0.55**	0.37	**0.63**	0.41	**0.69**	0.47	**0.80**	0.53	**0.90**
Gall bladder	0.09	**0.41**	0.12	**0.52**	0.14	**0.60**	0.16	**0.66**	0.20	**0.79**	0.23	**0.90**
Pancreas	0.20	**0.26**	0.26	**0.34**	0.31	**0.40**	0.34	**0.45**	0.42	**0.55**	0.49	**0.64**
Small intestine	0.01	**0.22**	0.01	**0.28**	0.02	**0.33**	0.02	**0.36**	0.02	**0.43**	0.03	**0.50**
Large intestine	0.01	**0.30**	0.02	**0.37**	0.02	**0.43**	0.03	**0.47**	0.03	**0.56**	0.04	**0.63**
Sigma/Rectum		**0.08**		**0.09**		**0.11**		**0.12**		**0.15**		**0.17**
Lumbar spine	0.05	**0.19**	0.07	**0.26**	0.08	**0.31**	0.09	**0.35**	0.12	**0.45**	0.15	**0.55**
Pelvis		**0.10**		**0.13**		**0.16**	0.01	**0.18**	0.01	**0.22**	0.02	**0.27**
Adrenals	0.09	**0.11**	0.12	**0.14**	0.14	**0.17**	0.16	**0.20**	0.20	**0.25**	0.24	**0.30**
Kidneys	0.04	**0.09**	0.05	**0.12**	0.06	**0.14**	0.07	**0.16**	0.09	**0.20**	0.11	**0.25**
Urinary bladder		**0.01**		**0.02**		**0.02**		**0.03**		**0.03**		**0.04**
Uterus		**0.04**		**0.05**		**0.06**		**0.07**		**0.09**	0.01	**0.11**
Ovaries		**0.05**		**0.06**		**0.07**		**0.08**		**0.10**		**0.12**
Prostate										**0.01**		**0.02**
Testes												
Upper arm	0.21	**0.79**	0.27	**0.98**	0.32	**1.11**	0.36	**1.22**	0.43	**1.42**	0.51	**1.60**
Elbow	0.18	**0.85**	0.23	**1.04**	0.27	**1.19**	0.31	**1.31**	0.37	**1.52**	0.44	**1.71**
Forearm	0.02	**0.49**	0.02	**0.60**	0.03	**0.69**	0.03	**0.75**	0.04	**0.87**	0.05	**0.97**
Upper leg												
Knee												
Lower leg												
Lymph nodes	0.16	**0.34**	0.20	**0.41**	0.23	**0.47**	0.25	**0.52**	0.29	**0.61**	0.33	**0.69**
Skin	0.13	**0.27**	0.14	**0.28**	0.14	**0.29**	0.15	**0.30**	0.16	**0.32**	0.17	**0.34**
Muscle	0.10	**0.17**	0.12	**0.21**	0.13	**0.23**	0.14	**0.25**	0.17	**0.29**	0.18	**0.32**
Skeleton	0.20	**0.38**	0.24	**0.46**	0.27	**0.52**	0.30	**0.56**	0.34	**0.65**	0.37	**0.72**
Red bone marrow	0.07	**0.11**	0.08	**0.14**	0.09	**0.16**	0.10	**0.17**	0.11	**0.20**	0.13	**0.23**
Whole body dose	0.12	**0.21**	0.14	**0.25**	0.16	**0.28**	0.17	**0.31**	0.20	**0.35**	0.22	**0.39**
ED (ICRP 60)	0.22	**0.31**	0.25	**0.37**	0.28	**0.41**	0.30	**0.44**	0.34	**0.50**	0.37	**0.56**
ED (ICRP 103)	0.30	**0.40**	0.34	**0.47**	0.37	**0.52**	0.40	**0.55**	0.44	**0.62**	0.48	**0.68**
Total filtration (mm) **Al**	**1.5**		**2.5**		**3.5**		**4.5**		**3.5**		**3.5**	
Cu	**0**		**0**		**0**		**0**		**0.1**		**0.2**	

Table 9.44 Normalized organ doses—Chest ap—3–7 years

Beam projection:		ap			**Table cells:**							
Focus-to-detector distance (cm):		100			(+) = normal font = optimal field setting							
Tube voltage (kVp):		80			**(−) = bold font = suboptimal field setting**							
Total filtrations:		See last table row.			Empty table cells correspond to values < 0.01							
Organ or tissue	**Organ dose D_T : Entrance dose K_E**						**(mSv × mGy^{-1})**					
	(+)	(−)	(+)	(−)	(+)	(−)	(+)	(−)	(+)	(−)	(+)	(−)
Skull	0.02	**0.33**	0.03	**0.39**	0.03	**0.43**	0.04	**0.46**	0.05	**0.52**	0.05	**0.57**
Brain		**0.02**		**0.03**		**0.03**		**0.04**		**0.05**		**0.06**
Oral mucosa	0.03	**0.47**	0.04	**0.55**	0.04	**0.60**	0.05	**0.65**	0.06	**0.73**	0.07	**0.80**
Salivary glands	0.04	**0.39**	0.05	**0.46**	0.06	**0.51**	0.07	**0.56**	0.08	**0.63**	0.09	**0.70**
Cervical spine	0.36	**0.57**	0.45	**0.72**	0.52	**0.83**	0.57	**0.91**	0.67	**1.09**	0.76	**1.25**
Thyroid	0.87	**0.95**	0.98	**1.07**	1.05	**1.15**	1.09	**1.21**	1.18	**1.32**	1.25	**1.41**
Thymus	0.86	**0.88**	1.00	**1.03**	1.10	**1.13**	1.17	**1.21**	1.30	**1.35**	1.41	**1.46**
Thoracic spine	0.25	**0.29**	0.33	**0.37**	0.38	**0.43**	0.43	**0.49**	0.54	**0.61**	0.64	**0.72**
Claviculae	1.90	**2.01**	2.20	**2.34**	2.41	**2.57**	2.56	**2.73**	2.80	**3.00**	2.96	**3.19**
Scapulae	0.33	**0.36**	0.42	**0.46**	0.49	**0.53**	0.54	**0.59**	0.65	**0.71**	0.75	**0.82**
Ribs	1.48	**1.58**	1.68	**1.80**	1.81	**1.95**	1.90	**2.06**	2.06	**2.24**	2.17	**2.37**
Oesophagus	0.19	**0.21**	0.24	**0.26**	0.28	**0.31**	0.31	**0.34**	0.37	**0.41**	0.43	**0.48**
Trachea	0.10	**0.40**	0.11	**0.48**	0.12	**0.53**	0.13	**0.58**	0.15	**0.66**	0.16	**0.73**
Lung	0.43	**0.46**	0.52	**0.56**	0.59	**0.62**	0.64	**0.68**	0.73	**0.78**	0.81	**0.86**
Mamma	1.21	**1.24**	1.29	**1.32**	1.33	**1.37**	1.37	**1.40**	1.42	**1.46**	1.47	**1.50**
Heart	0.58	**0.61**	0.70	**0.73**	0.79	**0.82**	0.85	**0.90**	0.98	**1.03**	1.08	**1.14**
Stomach	0.32	**0.64**	0.38	**0.76**	0.43	**0.85**	0.46	**0.92**	0.53	**1.05**	0.59	**1.16**
Spleen	0.15	**0.21**	0.19	**0.27**	0.22	**0.31**	0.25	**0.35**	0.30	**0.42**	0.35	**0.49**
Liver	0.30	**0.52**	0.37	**0.62**	0.41	**0.70**	0.45	**0.76**	0.52	**0.88**	0.58	**0.97**
Gall bladder	0.12	**0.49**	0.15	**0.60**	0.17	**0.68**	0.19	**0.75**	0.22	**0.88**	0.26	**0.99**
Pancreas	0.25	**0.32**	0.31	**0.41**	0.36	**0.47**	0.40	**0.52**	0.48	**0.63**	0.55	**0.72**
Small intestine	0.01	**0.27**	0.02	**0.33**	0.02	**0.38**	0.02	**0.41**	0.03	**0.49**	0.04	**0.56**
Large intestine	0.02	**0.35**	0.02	**0.42**	0.03	**0.48**	0.03	**0.53**	0.04	**0.62**	0.05	**0.70**
Sigma/Rectum		**0.09**		**0.11**		**0.13**		**0.14**		**0.17**	0.01	**0.19**
Lumbar spine	0.07	**0.25**	0.09	**0.32**	0.11	**0.38**	0.12	**0.44**	0.15	**0.54**	0.18	**0.65**
Pelvis		**0.13**	0.01	**0.16**	0.01	**0.19**	0.01	**0.21**	0.02	**0.26**	0.02	**0.31**
Adrenals	0.12	**0.14**	0.15	**0.18**	0.17	**0.21**	0.19	**0.24**	0.24	**0.29**	0.28	**0.35**
Kidneys	0.05	**0.12**	0.07	**0.15**	0.08	**0.18**	0.09	**0.20**	0.11	**0.25**	0.13	**0.29**
Urinary bladder		**0.02**		**0.02**		**0.03**		**0.03**		**0.04**		**0.05**
Uterus		**0.05**		**0.07**		**0.08**		**0.09**	0.01	**0.11**	0.01	**0.13**
Ovaries		**0.06**		**0.07**		**0.09**		**0.10**		**0.12**	0.01	**0.14**
Prostate						**0.01**		**0.01**		**0.02**		**0.02**
Testes												
Upper arm	0.25	**0.90**	0.32	**1.09**	0.37	**1.22**	0.41	**1.33**	0.48	**1.52**	0.55	**1.68**
Elbow	0.22	**0.96**	0.27	**1.16**	0.32	**1.31**	0.35	**1.42**	0.42	**1.62**	0.48	**1.79**
Forearm	0.02	**0.55**	0.03	**0.67**	0.04	**0.75**	0.04	**0.81**	0.05	**0.92**	0.06	**1.02**
Upper leg												
Knee												
Lower leg												
Lymph nodes	0.19	**0.39**	0.23	**0.47**	0.25	**0.53**	0.28	**0.58**	0.32	**0.67**	0.35	**0.75**
Skin	0.13	**0.28**	0.14	**0.29**	0.15	**0.31**	0.15	**0.32**	0.16	**0.33**	0.17	**0.35**
Muscle	0.11	**0.20**	0.13	**0.23**	0.15	**0.26**	0.16	**0.28**	0.18	**0.31**	0.20	**0.35**
Skeleton	0.23	**0.43**	0.27	**0.51**	0.30	**0.56**	0.32	**0.61**	0.36	**0.69**	0.39	**0.76**
Red bone marrow	0.08	**0.13**	0.09	**0.16**	0.10	**0.18**	0.11	**0.19**	0.13	**0.23**	0.14	**0.26**
Whole body dose	0.13	**0.24**	0.16	**0.28**	0.18	**0.31**	0.19	**0.34**	0.21	**0.38**	0.24	**0.42**
ED (ICRP 60)	0.24	**0.35**	0.28	**0.40**	0.30	**0.45**	0.33	**0.48**	0.36	**0.54**	0.40	**0.60**
ED (ICRP 103)	0.33	**0.44**	0.37	**0.51**	0.40	**0.56**	0.42	**0.60**	0.47	**0.67**	0.50	**0.73**
Total filtration (mm) **Al**	**1.5**		**2.5**		**3.5**		**4.5**		**3.5**		**3.5**	
Cu	**0**		**0**		**0**		**0**		**0.1**		**0.2**	

Table 9.45 Normalized organ doses—Chest ap—8–12 years

Beam projection:			ap				**Table cells:**					
Focus-to-detector distance (cm):			100				(+) = normal font = optimal field setting					
Tube voltage (kVp):			60				**(−) = bold font = suboptimal field setting**					
Total filtrations:			See last table row.				Empty table cells correspond to values < 0.01					
Organ or tissue	**Organ dose D_T : Entrance dose K_E**								**(mSv × mGy^{-1})**			
	(+)	(−)	(+)	(−)	(+)	(−)	(+)	(−)	(+)	(−)	(+)	(−)
Skull	0.01	**0.25**	0.02	**0.31**	0.02	**0.35**	0.02	**0.39**	0.03	**0.45**	0.02	**0.43**
Brain		**0.01**		**0.01**		**0.02**		**0.02**		**0.03**		**0.02**
Oral mucosa	0.01	**0.33**	0.02	**0.40**	0.02	**0.45**	0.03	**0.49**	0.03	**0.56**	0.03	**0.53**
Salivary glands	0.03	**0.28**	0.04	**0.34**	0.04	**0.39**	0.05	**0.43**	0.06	**0.50**	0.06	**0.48**
Cervical spine	0.25	**0.37**	0.33	**0.49**	0.39	**0.58**	0.44	**0.66**	0.54	**0.82**	0.49	**0.73**
Thyroid	0.69	**0.74**	0.80	**0.86**	0.88	**0.94**	0.93	**1.00**	1.02	**1.10**	1.00	**1.08**
Thymus	0.66	**0.67**	0.80	**0.82**	0.90	**0.92**	0.98	**1.01**	1.13	**1.16**	1.10	**1.12**
Thoracic spine	0.11	**0.12**	0.15	**0.16**	0.18	**0.20**	0.21	**0.23**	0.28	**0.30**	0.22	**0.24**
Claviculae	1.44	**1.51**	1.75	**1.85**	1.97	**2.08**	2.15	**2.26**	2.45	**2.59**	2.42	**2.55**
Scapulae	0.16	**0.17**	0.22	**0.23**	0.26	**0.28**	0.30	**0.32**	0.38	**0.41**	0.34	**0.36**
Ribs	1.21	**1.27**	1.42	**1.50**	1.56	**1.65**	1.67	**1.77**	1.86	**1.98**	1.84	**1.95**
Oesophagus	0.08	**0.09**	0.11	**0.12**	0.13	**0.14**	0.15	**0.16**	0.19	**0.20**	0.16	**0.17**
Trachea	0.19	**0.42**	0.22	**0.53**	0.25	**0.60**	0.27	**0.66**	0.30	**0.77**	0.29	**0.73**
Lung	0.30	**0.31**	0.38	**0.39**	0.43	**0.45**	0.48	**0.50**	0.57	**0.60**	0.54	**0.56**
Mamma	1.11	**1.15**	1.20	**1.24**	1.26	**1.30**	1.30	**1.34**	1.36	**1.40**	1.34	**1.39**
Heart	0.39	**0.40**	0.49	**0.51**	0.57	**0.59**	0.63	**0.65**	0.75	**0.78**	0.70	**0.73**
Stomach	0.22	**0.44**	0.28	**0.54**	0.32	**0.62**	0.36	**0.69**	0.42	**0.81**	0.40	**0.77**
Spleen	0.08	**0.11**	0.11	**0.14**	0.13	**0.17**	0.15	**0.20**	0.19	**0.25**	0.16	**0.21**
Liver	0.20	**0.32**	0.26	**0.41**	0.30	**0.47**	0.33	**0.52**	0.40	**0.62**	0.37	**0.58**
Gall bladder	0.07	**0.29**	0.09	**0.38**	0.11	**0.45**	0.12	**0.50**	0.15	**0.61**	0.13	**0.56**
Pancreas	0.14	**0.17**	0.18	**0.23**	0.22	**0.28**	0.25	**0.32**	0.31	**0.40**	0.28	**0.35**
Small intestine		**0.10**		**0.13**		**0.16**		**0.18**	0.01	**0.22**		**0.20**
Large intestine		**0.17**		**0.22**	0.01	**0.26**	0.01	**0.29**	0.02	**0.36**	0.01	**0.33**
Sigma/Rectum		**0.03**		**0.04**		**0.05**		**0.05**		**0.06**		**0.06**
Lumbar spine	0.03	**0.09**	0.04	**0.13**	0.04	**0.16**	0.05	**0.18**	0.07	**0.24**	0.05	**0.19**
Pelvis		**0.03**		**0.04**		**0.05**		**0.06**		**0.08**		**0.06**
Adrenals	0.05	**0.06**	0.07	**0.09**	0.09	**0.10**	0.10	**0.12**	0.13	**0.16**	0.11	**0.13**
Kidneys	0.02	**0.05**	0.03	**0.07**	0.04	**0.08**	0.05	**0.10**	0.06	**0.12**	0.05	**0.10**
Urinary bladder												
Uterus		**0.01**		**0.02**		**0.02**		**0.02**		**0.03**		**0.02**
Ovaries		**0.01**		**0.02**		**0.02**		**0.02**		**0.03**		**0.02**
Prostate												
Testes												
Upper arm	0.23	**0.57**	0.30	**0.73**	0.36	**0.84**	0.40	**0.94**	0.49	**1.12**	0.45	**1.06**
Elbow	0.21	**0.61**	0.28	**0.78**	0.33	**0.90**	0.38	**1.00**	0.46	**1.20**	0.42	**1.13**
Forearm	0.01	**0.30**	0.02	**0.38**	0.02	**0.44**	0.03	**0.49**	0.03	**0.58**	0.03	**0.56**
Upper leg												
Knee												
Lower leg												
Lymph nodes	0.14	**0.26**	0.17	**0.32**	0.20	**0.37**	0.22	**0.41**	0.26	**0.49**	0.24	**0.46**
Skin	0.12	**0.23**	0.13	**0.24**	0.13	**0.25**	0.14	**0.25**	0.15	**0.27**	0.14	**0.26**
Muscle	0.08	**0.13**	0.10	**0.16**	0.11	**0.18**	0.12	**0.19**	0.14	**0.22**	0.14	**0.21**
Skeleton	0.17	**0.28**	0.21	**0.34**	0.23	**0.39**	0.26	**0.43**	0.30	**0.50**	0.28	**0.47**
Red bone marrow	0.06	**0.09**	0.08	**0.11**	0.09	**0.12**	0.10	**0.14**	0.11	**0.16**	0.10	**0.15**
Whole body dose	0.10	**0.16**	0.12	**0.19**	0.14	**0.22**	0.15	**0.24**	0.17	**0.27**	0.16	**0.26**
ED (ICRP 60)	0.18	**0.24**	0.22	**0.29**	0.24	**0.32**	0.26	**0.35**	0.30	**0.40**	0.28	**0.38**
ED (ICRP 103)	0.26	**0.33**	0.30	**0.39**	0.33	**0.43**	0.35	**0.46**	0.40	**0.52**	0.38	**0.50**
Total filtration (mm) Al	**1.5**		**2.5**		**3.5**		**4.5**		**3.5**		**3.5**	
Cu	**0**		**0**		**0**		**0**		**0.1**		**0.2**	

Table 9.46 Normalized organ doses—Chest ap—8–12 years

Beam projection:	ap						**Table cells:**					
Focus-to-detector distance (cm):	100						(+) = normal font = optimal field setting					
Tube voltage (kVp):	70						**(−) = bold font = suboptimal field setting**					
Total filtrations:	See last table row.						Empty table cells correspond to values < 0.01					
Organ or tissue	**Organ dose D_T : Entrance dose K_E**						**(mSv × mGy^{-1})**					
	(+)	(−)	(+)	(−)	(+)	(−)	(+)	(−)	(+)	(−)	(+)	(−)
Skull	0.02	**0.29**	0.02	**0.35**	0.03	**0.40**	0.03	**0.43**	0.04	**0.50**	0.04	**0.56**
Brain		**0.02**		**0.02**		**0.02**		**0.03**		**0.04**		**0.05**
Oral mucosa	0.02	**0.38**	0.02	**0.45**	0.03	**0.51**	0.03	**0.55**	0.04	**0.62**	0.05	**0.69**
Salivary glands	0.04	**0.32**	0.05	**0.39**	0.05	**0.45**	0.06	**0.49**	0.07	**0.56**	0.08	**0.63**
Cervical spine	0.31	**0.47**	0.40	**0.61**	0.47	**0.71**	0.52	**0.80**	0.64	**0.98**	0.74	**1.15**
Thyroid	0.76	**0.81**	0.87	**0.93**	0.94	**1.01**	0.99	**1.07**	1.09	**1.18**	1.16	**1.28**
Thymus	0.75	**0.77**	0.90	**0.92**	1.01	**1.03**	1.09	**1.11**	1.24	**1.27**	1.36	**1.40**
Thoracic spine	0.16	**0.17**	0.21	**0.23**	0.25	**0.27**	0.29	**0.32**	0.37	**0.40**	0.45	**0.50**
Claviculae	1.62	**1.70**	1.93	**2.04**	2.15	**2.28**	2.32	**2.45**	2.61	**2.77**	2.83	**3.02**
Scapulae	0.22	**0.23**	0.28	**0.30**	0.34	**0.36**	0.38	**0.41**	0.47	**0.51**	0.56	**0.60**
Ribs	1.32	**1.40**	1.53	**1.63**	1.67	**1.78**	1.78	**1.90**	1.96	**2.10**	2.11	**2.26**
Oesophagus	0.11	**0.12**	0.14	**0.15**	0.17	**0.18**	0.19	**0.21**	0.24	**0.26**	0.29	**0.31**
Trachea	0.21	**0.50**	0.25	**0.60**	0.28	**0.68**	0.29	**0.75**	0.33	**0.86**	0.36	**0.96**
Lung	0.35	**0.37**	0.44	**0.46**	0.50	**0.53**	0.55	**0.58**	0.65	**0.68**	0.73	**0.77**
Mamma	1.16	**1.20**	1.25	**1.29**	1.30	**1.34**	1.34	**1.38**	1.40	**1.45**	1.45	**1.50**
Heart	0.47	**0.48**	0.58	**0.60**	0.66	**0.69**	0.72	**0.75**	0.85	**0.89**	0.96	**1.01**
Stomach	0.26	**0.51**	0.33	**0.63**	0.37	**0.72**	0.41	**0.79**	0.48	**0.91**	0.54	**1.03**
Spleen	0.11	**0.14**	0.14	**0.19**	0.17	**0.22**	0.19	**0.25**	0.23	**0.31**	0.28	**0.37**
Liver	0.24	**0.39**	0.30	**0.48**	0.35	**0.55**	0.39	**0.61**	0.46	**0.72**	0.52	**0.81**
Gall bladder	0.09	**0.36**	0.11	**0.46**	0.13	**0.53**	0.15	**0.60**	0.19	**0.71**	0.22	**0.82**
Pancreas	0.18	**0.23**	0.23	**0.29**	0.28	**0.35**	0.31	**0.39**	0.38	**0.49**	0.45	**0.58**
Small intestine		**0.13**		**0.17**	0.01	**0.20**	0.01	**0.22**	0.02	**0.27**	0.02	**0.31**
Large intestine	0.01	**0.21**	0.01	**0.27**	0.02	**0.31**	0.02	**0.35**	0.03	**0.42**	0.03	**0.48**
Sigma/Rectum		**0.04**		**0.05**		**0.06**		**0.06**		**0.08**		**0.09**
Lumbar spine	0.04	**0.14**	0.05	**0.18**	0.07	**0.22**	0.08	**0.26**	0.10	**0.33**	0.12	**0.41**
Pelvis		**0.05**		**0.06**		**0.07**		**0.08**		**0.11**	0.01	**0.13**
Adrenals	0.08	**0.09**	0.10	**0.12**	0.12	**0.14**	0.14	**0.16**	0.18	**0.21**	0.21	**0.25**
Kidneys	0.03	**0.07**	0.05	**0.09**	0.06	**0.11**	0.06	**0.13**	0.08	**0.16**	0.10	**0.20**
Urinary bladder								**0.01**		**0.01**		**0.02**
Uterus		**0.02**		**0.02**		**0.03**		**0.03**		**0.04**		**0.05**
Ovaries		**0.02**		**0.03**		**0.03**		**0.04**		**0.05**		**0.06**
Prostate												
Testes												
Upper arm	0.29	**0.68**	0.37	**0.85**	0.43	**0.97**	0.48	**1.07**	0.58	**1.26**	0.68	**1.43**
Elbow	0.27	**0.73**	0.34	**0.91**	0.40	**1.04**	0.45	**1.15**	0.54	**1.35**	0.63	**1.52**
Forearm	0.02	**0.36**	0.03	**0.44**	0.03	**0.51**	0.04	**0.56**	0.04	**0.65**	0.05	**0.74**
Upper leg												
Knee												
Lower leg												
Lymph nodes	0.17	**0.31**	0.20	**0.38**	0.23	**0.44**	0.25	**0.48**	0.30	**0.56**	0.34	**0.64**
Skin	0.13	**0.24**	0.14	**0.25**	0.14	**0.26**	0.15	**0.27**	0.15	**0.28**	0.16	**0.30**
Muscle	0.10	**0.15**	0.12	**0.18**	0.13	**0.20**	0.14	**0.22**	0.16	**0.25**	0.18	**0.28**
Skeleton	0.20	**0.32**	0.24	**0.39**	0.26	**0.44**	0.29	**0.48**	0.33	**0.56**	0.37	**0.62**
Red bone marrow	0.08	**0.11**	0.09	**0.13**	0.10	**0.15**	0.11	**0.16**	0.13	**0.19**	0.15	**0.22**
Whole body dose	0.11	**0.18**	0.14	**0.22**	0.16	**0.25**	0.17	**0.27**	0.19	**0.31**	0.22	**0.35**
ED (ICRP 60)	0.21	**0.28**	0.24	**0.33**	0.27	**0.37**	0.29	**0.40**	0.33	**0.45**	0.36	**0.51**
ED (ICRP 103)	0.29	**0.37**	0.33	**0.43**	0.36	**0.48**	0.39	**0.51**	0.43	**0.58**	0.47	**0.64**
Total filtration (mm) Al	**1.5**		**2.5**		**3.5**		**4.5**		**3.5**		**3.5**	
Cu	**0**		**0**		**0**		**0**		**0.1**		**0.2**	

Table 9.47 Normalized organ doses—Chest ap—8–12 years

Beam projection:			ap				**Table cells:**					
Focus-to-detector distance (cm):			100				(+) = normal font = optimal field setting					
Tube voltage (kVp):			80				**(−) = bold font = suboptimal field setting**					
Total filtrations:			See last table row.				Empty table cells correspond to values < 0.01					
Organ or tissue	**Organ dose D_T : Entrance dose K_E**								**(mSv × mGy^{-1})**			
	(+)	(−)	(+)	(−)	(+)	(−)	(+)	(−)	(+)	(−)	(+)	(−)
Skull	0.02	**0.33**	0.03	**0.39**	0.03	**0.43**	0.04	**0.47**	0.04	**0.53**	0.05	**0.59**
Brain		**0.02**		**0.03**		**0.03**		**0.04**		**0.05**		**0.06**
Oral mucosa	0.02	**0.43**	0.03	**0.50**	0.04	**0.56**	0.04	**0.60**	0.05	**0.68**	0.06	**0.75**
Salivary glands	0.04	**0.37**	0.05	**0.44**	0.06	**0.49**	0.07	**0.54**	0.08	**0.61**	0.09	**0.68**
Cervical spine	0.38	**0.57**	0.47	**0.72**	0.54	**0.83**	0.60	**0.92**	0.71	**1.11**	0.81	**1.28**
Thyroid	0.82	**0.88**	0.92	**1.00**	1.00	**1.08**	1.05	**1.14**	1.14	**1.25**	1.21	**1.34**
Thymus	0.84	**0.86**	0.99	**1.01**	1.10	**1.12**	1.18	**1.21**	1.32	**1.36**	1.44	**1.49**
Thoracic spine	0.21	**0.23**	0.27	**0.29**	0.32	**0.35**	0.36	**0.40**	0.45	**0.50**	0.54	**0.60**
Claviculae	1.77	**1.87**	2.07	**2.19**	2.28	**2.42**	2.43	**2.58**	2.69	**2.87**	2.89	**3.08**
Scapulae	0.27	**0.29**	0.34	**0.37**	0.40	**0.43**	0.45	**0.48**	0.55	**0.59**	0.64	**0.68**
Ribs	1.42	**1.51**	1.62	**1.73**	1.76	**1.87**	1.85	**1.98**	2.02	**2.16**	2.14	**2.31**
Oesophagus	0.14	**0.15**	0.18	**0.19**	0.21	**0.22**	0.23	**0.25**	0.28	**0.31**	0.33	**0.37**
Trachea	0.23	**0.57**	0.27	**0.68**	0.30	**0.76**	0.32	**0.82**	0.35	**0.94**	0.38	**1.05**
Lung	0.41	**0.43**	0.50	**0.52**	0.56	**0.59**	0.61	**0.65**	0.71	**0.75**	0.79	**0.84**
Mamma	1.20	**1.24**	1.29	**1.33**	1.34	**1.38**	1.37	**1.42**	1.44	**1.48**	1.48	**1.53**
Heart	0.54	**0.56**	0.66	**0.68**	0.74	**0.77**	0.81	**0.84**	0.94	**0.98**	1.05	**1.10**
Stomach	0.31	**0.59**	0.37	**0.71**	0.42	**0.80**	0.46	**0.87**	0.53	**1.00**	0.59	**1.11**
Spleen	0.14	**0.18**	0.17	**0.23**	0.20	**0.27**	0.23	**0.30**	0.28	**0.37**	0.32	**0.43**
Liver	0.29	**0.45**	0.35	**0.55**	0.40	**0.62**	0.44	**0.68**	0.51	**0.79**	0.58	**0.89**
Gall bladder	0.11	**0.43**	0.14	**0.54**	0.16	**0.62**	0.18	**0.68**	0.22	**0.80**	0.25	**0.91**
Pancreas	0.23	**0.28**	0.28	**0.36**	0.33	**0.42**	0.37	**0.47**	0.45	**0.57**	0.52	**0.66**
Small intestine	0.01	**0.16**	0.01	**0.20**	0.02	**0.23**	0.02	**0.26**	0.02	**0.31**	0.03	**0.36**
Large intestine	0.01	**0.26**	0.02	**0.32**	0.02	**0.36**	0.03	**0.40**	0.03	**0.47**	0.04	**0.54**
Sigma/Rectum		**0.05**		**0.06**		**0.07**		**0.07**		**0.09**		**0.10**
Lumbar spine	0.06	**0.19**	0.07	**0.24**	0.09	**0.29**	0.10	**0.33**	0.13	**0.41**	0.15	**0.49**
Pelvis		**0.06**		**0.08**		**0.09**		**0.10**	0.01	**0.13**	0.01	**0.16**
Adrenals	0.10	**0.12**	0.13	**0.15**	0.15	**0.18**	0.17	**0.20**	0.22	**0.26**	0.26	**0.31**
Kidneys	0.05	**0.09**	0.06	**0.12**	0.07	**0.14**	0.08	**0.16**	0.10	**0.20**	0.12	**0.24**
Urinary bladder						**0.01**		**0.01**		**0.02**		**0.02**
Uterus		**0.03**		**0.03**		**0.04**		**0.04**		**0.06**		**0.07**
Ovaries		**0.03**		**0.03**		**0.04**		**0.05**		**0.06**		**0.07**
Prostate												
Testes												
Upper arm	0.35	**0.79**	0.43	**0.96**	0.50	**1.08**	0.55	**1.18**	0.65	**1.37**	0.74	**1.52**
Elbow	0.32	**0.84**	0.40	**1.02**	0.46	**1.16**	0.51	**1.26**	0.61	**1.45**	0.69	**1.62**
Forearm	0.03	**0.41**	0.03	**0.50**	0.04	**0.56**	0.04	**0.61**	0.05	**0.70**	0.06	**0.78**
Upper leg												
Knee												
Lower leg												
Lymph nodes	0.19	**0.36**	0.23	**0.44**	0.26	**0.49**	0.28	**0.54**	0.33	**0.63**	0.37	**0.71**
Skin	0.13	**0.25**	0.14	**0.26**	0.15	**0.27**	0.15	**0.28**	0.16	**0.29**	0.17	**0.31**
Muscle	0.11	**0.17**	0.13	**0.20**	0.14	**0.22**	0.16	**0.24**	0.18	**0.28**	0.20	**0.31**
Skeleton	0.22	**0.37**	0.26	**0.44**	0.29	**0.49**	0.31	**0.53**	0.35	**0.60**	0.39	**0.66**
Red bone marrow	0.09	**0.12**	0.10	**0.15**	0.12	**0.17**	0.13	**0.18**	0.15	**0.22**	0.17	**0.25**
Whole body dose	0.13	**0.21**	0.15	**0.25**	0.17	**0.27**	0.19	**0.30**	0.21	**0.34**	0.24	**0.38**
ED (ICRP 60)	0.23	**0.31**	0.27	**0.37**	0.29	**0.41**	0.32	**0.44**	0.35	**0.50**	0.39	**0.55**
ED (ICRP 103)	0.32	**0.41**	0.36	**0.48**	0.39	**0.52**	0.42	**0.56**	0.46	**0.62**	0.50	**0.68**
Total filtration (mm) **Al**	**1.5**		**2.5**		**3.5**		**4.5**		**3.5**		**3.5**	
Cu	**0**		**0**		**0**		**0**		**0.1**		**0.2**	

Table 9.48 Normalized organ doses—Chest ap—13–17 years

Beam projection:			ap				**Table cells:**					
Focus-to-detector distance (cm):			100				(+) = normal font = optimal field setting					
Tube voltage (kVp):			60				**(−) = bold font = suboptimal field setting**					
Total filtrations:			See last table row.				Empty table cells correspond to values < 0.01					
Organ or tissue	**Organ dose D_T : Entrance dose K_E**								**(mSv × mGy^{-1})**			
	(+)	(−)	(+)	(−)	(+)	(−)	(+)	(−)	(+)	(−)	(+)	(−)
Skull	0.01	**0.24**	0.01	**0.30**	0.02	**0.34**	0.02	**0.38**	0.02	**0.45**	0.02	**0.42**
Brain				**0.01**		**0.02**		**0.02**		**0.03**		**0.02**
Oral mucosa		**0.28**	0.01	**0.33**	0.01	**0.38**	0.02	**0.41**	0.02	**0.47**	0.02	**0.45**
Salivary glands	0.02	**0.25**	0.02	**0.31**	0.03	**0.35**	0.03	**0.39**	0.04	**0.45**	0.04	**0.43**
Cervical spine	0.25	**0.39**	0.33	**0.52**	0.40	**0.62**	0.45	**0.70**	0.56	**0.88**	0.50	**0.77**
Thyroid	0.62	**0.66**	0.73	**0.78**	0.81	**0.86**	0.86	**0.92**	0.96	**1.02**	0.94	**1.00**
Thymus	0.60	**0.63**	0.75	**0.78**	0.85	**0.89**	0.93	**0.98**	1.08	**1.14**	1.04	**1.09**
Thoracic spine	0.06	**0.07**	0.09	**0.10**	0.11	**0.12**	0.13	**0.14**	0.17	**0.19**	0.13	**0.14**
Claviculae	1.19	**1.25**	1.47	**1.55**	1.67	**1.77**	1.83	**1.93**	2.12	**2.24**	2.07	**2.19**
Scapulae	0.08	**0.08**	0.11	**0.12**	0.13	**0.14**	0.15	**0.17**	0.20	**0.22**	0.16	**0.17**
Ribs	1.01	**1.08**	1.20	**1.28**	1.33	**1.43**	1.43	**1.53**	1.61	**1.73**	1.58	**1.69**
Oesophagus	0.08	**0.09**	0.11	**0.12**	0.14	**0.15**	0.16	**0.17**	0.20	**0.22**	0.17	**0.18**
Trachea	0.17	**0.31**	0.20	**0.38**	0.22	**0.44**	0.24	**0.48**	0.28	**0.56**	0.27	**0.53**
Lung	0.19	**0.20**	0.25	**0.26**	0.29	**0.31**	0.33	**0.35**	0.40	**0.43**	0.37	**0.39**
Mamma	7.94	**8.29**	9.19	**9.61**	10.0	**10.5**	10.6	**11.1**	11.7	**12.2**	11.5	**12.0**
Heart	0.34	**0.35**	0.43	**0.45**	0.50	**0.53**	0.56	**0.59**	0.67	**0.71**	0.62	**0.65**
Stomach	0.18	**0.35**	0.23	**0.44**	0.27	**0.52**	0.30	**0.58**	0.37	**0.69**	0.34	**0.64**
Spleen	0.06	**0.07**	0.08	**0.10**	0.09	**0.12**	0.11	**0.14**	0.14	**0.18**	0.12	**0.15**
Liver	0.17	**0.26**	0.22	**0.33**	0.26	**0.39**	0.29	**0.43**	0.36	**0.52**	0.33	**0.48**
Gall bladder	0.05	**0.24**	0.06	**0.31**	0.08	**0.37**	0.09	**0.42**	0.11	**0.52**	0.10	**0.47**
Pancreas	0.10	**0.13**	0.14	**0.17**	0.17	**0.21**	0.19	**0.24**	0.24	**0.31**	0.21	**0.26**
Small intestine		**0.08**		**0.10**		**0.12**		**0.14**		**0.18**		**0.15**
Large intestine		**0.13**		**0.18**		**0.21**		**0.24**	0.01	**0.29**		**0.26**
Sigma/Rectum		**0.02**		**0.03**		**0.04**		**0.04**		**0.05**		**0.04**
Lumbar spine	0.02	**0.06**	0.02	**0.08**	0.03	**0.09**	0.03	**0.11**	0.04	**0.15**	0.03	**0.11**
Pelvis		**0.02**		**0.03**		**0.04**		**0.04**		**0.05**		**0.04**
Adrenals	0.03	**0.04**	0.05	**0.06**	0.06	**0.07**	0.07	**0.08**	0.09	**0.11**	0.07	**0.08**
Kidneys	0.02	**0.03**	0.02	**0.04**	0.03	**0.05**	0.03	**0.06**	0.04	**0.08**	0.03	**0.06**
Urinary bladder												
Uterus		**0.02**		**0.03**		**0.04**		**0.05**		**0.06**		**0.05**
Ovaries				**0.01**		**0.01**		**0.02**		**0.02**		**0.02**
Prostate												
Testes												
Upper arm	0.16	**0.44**	0.22	**0.56**	0.26	**0.66**	0.30	**0.74**	0.37	**0.89**	0.33	**0.83**
Elbow	0.15	**0.46**	0.20	**0.59**	0.24	**0.69**	0.27	**0.78**	0.34	**0.93**	0.30	**0.87**
Forearm	0.01	**0.23**	0.02	**0.29**	0.02	**0.34**	0.02	**0.38**	0.03	**0.46**	0.02	**0.43**
Upper leg												
Knee												
Lower leg												
Lymph nodes	0.12	**0.21**	0.14	**0.26**	0.17	**0.30**	0.18	**0.34**	0.22	**0.40**	0.20	**0.37**
Skin	0.14	**0.24**	0.15	**0.26**	0.16	**0.27**	0.16	**0.28**	0.17	**0.29**	0.17	**0.28**
Muscle	0.07	**0.11**	0.09	**0.14**	0.10	**0.16**	0.11	**0.17**	0.13	**0.20**	0.12	**0.19**
Skeleton	0.15	**0.24**	0.18	**0.29**	0.20	**0.33**	0.22	**0.37**	0.26	**0.43**	0.25	**0.41**
Red bone marrow	0.06	**0.08**	0.08	**0.10**	0.09	**0.12**	0.10	**0.13**	0.11	**0.15**	0.10	**0.14**
Whole body dose	0.09	**0.14**	0.11	**0.17**	0.13	**0.20**	0.14	**0.22**	0.16	**0.25**	0.15	**0.24**
ED (ICRP 60)	0.50	**0.56**	0.59	**0.67**	0.65	**0.74**	0.70	**0.79**	0.78	**0.90**	0.76	**0.86**
ED (ICRP 103)	1.06	**1.15**	1.23	**1.35**	1.35	**1.48**	1.44	**1.59**	1.60	**1.77**	1.56	**1.72**
Total filtration (mm) Al	**1.5**		**2.5**		**3.5**		**4.5**		**3.5**		**3.5**	
Cu	**0**		**0**		**0**		**0**		**0.1**		**0.2**	

Table 9.49 Normalized organ doses—Chest ap—13–17 years

Beam projection:			ap			**Table cells:**						
Focus-to-detector distance (cm):			100			(+) = normal font = optimal field setting						
Tube voltage (kVp):			70			**(−) = bold font = suboptimal field setting**						
Total filtrations:			See last table row.			Empty table cells correspond to values < 0.01						
Organ or tissue	**Organ dose D_T : Entrance dose K_E**							**(mSv × mGy^{-1})**				
	(+)	(−)	(+)	(−)	(+)	(−)	(+)	(−)	(+)	(−)	(+)	(−)
Skull	0.01	**0.28**	0.02	**0.34**	0.02	**0.39**	0.02	**0.43**	0.03	**0.51**	0.04	**0.57**
Brain		**0.01**		**0.02**		**0.02**		**0.03**		**0.04**		**0.04**
Oral mucosa	0.01	**0.32**	0.02	**0.38**	0.02	**0.43**	0.02	**0.46**	0.03	**0.53**	0.03	**0.60**
Salivary glands	0.02	**0.29**	0.03	**0.36**	0.04	**0.40**	0.04	**0.44**	0.05	**0.51**	0.06	**0.58**
Cervical spine	0.32	**0.51**	0.42	**0.65**	0.49	**0.77**	0.55	**0.87**	0.67	**1.07**	0.79	**1.27**
Thyroid	0.69	**0.73**	0.80	**0.85**	0.88	**0.93**	0.93	**1.00**	1.03	**1.10**	1.12	**1.20**
Thymus	0.70	**0.73**	0.85	**0.89**	0.96	**1.01**	1.05	**1.10**	1.21	**1.27**	1.35	**1.42**
Thoracic spine	0.10	**0.11**	0.13	**0.15**	0.16	**0.18**	0.19	**0.21**	0.24	**0.27**	0.30	**0.34**
Claviculae	1.36	**1.44**	1.65	**1.74**	1.85	**1.96**	2.01	**2.13**	2.29	**2.43**	2.52	**2.68**
Scapulae	0.11	**0.12**	0.15	**0.16**	0.18	**0.19**	0.21	**0.22**	0.26	**0.29**	0.32	**0.35**
Ribs	1.12	**1.20**	1.31	**1.41**	1.44	**1.55**	1.54	**1.66**	1.72	**1.86**	1.87	**2.02**
Oesophagus	0.11	**0.13**	0.15	**0.17**	0.18	**0.20**	0.21	**0.23**	0.26	**0.29**	0.32	**0.35**
Trachea	0.19	**0.36**	0.22	**0.44**	0.25	**0.50**	0.27	**0.54**	0.30	**0.63**	0.33	**0.71**
Lung	0.24	**0.26**	0.31	**0.32**	0.36	**0.38**	0.40	**0.42**	0.48	**0.51**	0.56	**0.60**
Mamma	8.66	**9.06**	9.92	**10.4**	10.8	**11.3**	11.4	**11.9**	12.5	**13.0**	13.3	**14.0**
Heart	0.41	**0.43**	0.52	**0.54**	0.59	**0.62**	0.66	**0.69**	0.78	**0.82**	0.90	**0.95**
Stomach	0.22	**0.42**	0.28	**0.53**	0.32	**0.61**	0.36	**0.67**	0.43	**0.80**	0.49	**0.91**
Spleen	0.08	**0.10**	0.10	**0.14**	0.12	**0.17**	0.14	**0.19**	0.18	**0.24**	0.22	**0.29**
Liver	0.22	**0.32**	0.27	**0.40**	0.32	**0.46**	0.35	**0.51**	0.42	**0.62**	0.49	**0.71**
Gall bladder	0.07	**0.30**	0.09	**0.39**	0.10	**0.46**	0.12	**0.51**	0.15	**0.62**	0.18	**0.73**
Pancreas	0.14	**0.18**	0.18	**0.23**	0.22	**0.27**	0.25	**0.31**	0.31	**0.39**	0.37	**0.47**
Small intestine		**0.10**		**0.13**		**0.16**		**0.18**	0.01	**0.22**	0.02	**0.26**
Large intestine		**0.17**		**0.22**	0.01	**0.26**	0.01	**0.29**	0.02	**0.35**	0.02	**0.41**
Sigma/Rectum		**0.03**		**0.04**		**0.05**		**0.05**		**0.06**		**0.08**
Lumbar spine	0.03	**0.09**	0.03	**0.12**	0.04	**0.14**	0.05	**0.16**	0.07	**0.21**	0.08	**0.27**
Pelvis		**0.03**		**0.04**		**0.05**		**0.06**		**0.07**		**0.09**
Adrenals	0.05	**0.06**	0.07	**0.08**	0.08	**0.10**	0.10	**0.12**	0.13	**0.15**	0.16	**0.19**
Kidneys	0.02	**0.05**	0.03	**0.06**	0.04	**0.08**	0.05	**0.09**	0.06	**0.12**	0.07	**0.14**
Urinary bladder												
Uterus		**0.04**		**0.05**		**0.06**		**0.07**		**0.09**		**0.11**
Ovaries		**0.01**		**0.02**		**0.02**		**0.03**		**0.03**		**0.04**
Prostate												
Testes												
Upper arm	0.21	**0.54**	0.27	**0.67**	0.32	**0.78**	0.36	**0.86**	0.45	**1.02**	0.53	**1.17**
Elbow	0.20	**0.56**	0.25	**0.71**	0.30	**0.82**	0.34	**0.90**	0.41	**1.07**	0.49	**1.23**
Forearm	0.02	**0.27**	0.02	**0.34**	0.03	**0.40**	0.03	**0.44**	0.04	**0.52**	0.04	**0.60**
Upper leg												
Knee												
Lower leg												
Lymph nodes	0.14	**0.25**	0.17	**0.31**	0.20	**0.36**	0.22	**0.40**	0.26	**0.47**	0.29	**0.54**
Skin	0.15	**0.25**	0.16	**0.27**	0.17	**0.28**	0.17	**0.29**	0.18	**0.31**	0.19	**0.32**
Muscle	0.09	**0.13**	0.11	**0.16**	0.12	**0.18**	0.13	**0.20**	0.15	**0.23**	0.17	**0.26**
Skeleton	0.17	**0.28**	0.21	**0.34**	0.23	**0.38**	0.25	**0.42**	0.29	**0.49**	0.33	**0.55**
Red bone marrow	0.07	**0.10**	0.09	**0.12**	0.10	**0.14**	0.11	**0.15**	0.13	**0.18**	0.15	**0.21**
Whole body dose	0.11	**0.17**	0.13	**0.20**	0.15	**0.23**	0.16	**0.25**	0.19	**0.29**	0.21	**0.33**
ED (ICRP 60)	0.56	**0.63**	0.65	**0.74**	0.71	**0.82**	0.76	**0.87**	0.85	**0.98**	0.93	**1.08**
ED (ICRP 103)	1.16	**1.28**	1.34	**1.48**	1.47	**1.62**	1.56	**1.72**	1.72	**1.92**	1.86	**2.08**
Total filtration (mm) **Al**	**1.5**		**2.5**		**3.5**		**4.5**		**3.5**		**3.5**	
Cu	**0**		**0**		**0**		**0**		**0.1**		**0.2**	

Table 9.50 Normalized organ doses—Chest ap—13–17 years

Beam projection:			ap			**Table cells:**						
Focus-to-detector distance (cm):			100			(+) = normal font = optimal field setting						
Tube voltage (kVp):			80			**(−) = bold font = suboptimal field setting**						
Total filtrations:			See last table row.			Empty table cells correspond to values < 0.01						
Organ or tissue	**Organ dose D_T : Entrance dose K_E**						**(mSv × mGy^{-1})**					
	(+)	(−)	(+)	(−)	(+)	(−)	(+)	(−)	(+)	(−)	(+)	(−)
Skull	0.02	**0.32**	0.02	**0.39**	0.03	**0.44**	0.03	**0.48**	0.04	**0.55**	0.04	**0.61**
Brain		**0.02**		**0.03**		**0.03**		**0.04**		**0.05**		**0.06**
Oral mucosa	0.02	**0.36**	0.02	**0.43**	0.02	**0.48**	0.03	**0.52**	0.03	**0.59**	0.04	**0.65**
Salivary glands	0.03	**0.33**	0.04	**0.40**	0.04	**0.45**	0.05	**0.49**	0.06	**0.56**	0.07	**0.63**
Cervical spine	0.40	**0.63**	0.50	**0.79**	0.58	**0.91**	0.64	**1.02**	0.77	**1.23**	0.89	**1.43**
Thyroid	0.75	**0.80**	0.86	**0.92**	0.94	**1.00**	0.99	**1.06**	1.09	**1.17**	1.17	**1.26**
Thymus	0.80	**0.83**	0.95	**1.00**	1.07	**1.12**	1.15	**1.21**	1.31	**1.38**	1.45	**1.52**
Thoracic spine	0.14	**0.15**	0.18	**0.20**	0.21	**0.24**	0.25	**0.27**	0.31	**0.35**	0.38	**0.42**
Claviculae	1.51	**1.59**	1.79	**1.90**	1.99	**2.11**	2.14	**2.27**	2.40	**2.55**	2.60	**2.77**
Scapulae	0.15	**0.16**	0.19	**0.21**	0.23	**0.25**	0.26	**0.28**	0.32	**0.35**	0.38	**0.42**
Ribs	1.22	**1.31**	1.41	**1.51**	1.53	**1.65**	1.63	**1.76**	1.79	**1.94**	1.92	**2.09**
Oesophagus	0.15	**0.17**	0.19	**0.21**	0.23	**0.25**	0.26	**0.28**	0.32	**0.35**	0.38	**0.42**
Trachea	0.21	**0.42**	0.25	**0.50**	0.27	**0.56**	0.29	**0.60**	0.33	**0.69**	0.35	**0.77**
Lung	0.29	**0.31**	0.36	**0.38**	0.42	**0.44**	0.46	**0.49**	0.55	**0.58**	0.63	**0.67**
Mamma	9.34	**9.77**	10.6	**11.1**	11.4	**11.9**	12.0	**12.6**	13.1	**13.7**	13.9	**14.6**
Heart	0.49	**0.51**	0.60	**0.63**	0.68	**0.72**	0.75	**0.79**	0.88	**0.92**	1.00	**1.05**
Stomach	0.26	**0.50**	0.33	**0.61**	0.37	**0.70**	0.41	**0.76**	0.48	**0.89**	0.55	**1.01**
Spleen	0.10	**0.14**	0.13	**0.18**	0.16	**0.21**	0.18	**0.24**	0.22	**0.29**	0.26	**0.35**
Liver	0.26	**0.38**	0.32	**0.47**	0.37	**0.54**	0.41	**0.59**	0.48	**0.70**	0.55	**0.79**
Gall bladder	0.08	**0.37**	0.11	**0.46**	0.13	**0.54**	0.14	**0.60**	0.17	**0.71**	0.21	**0.82**
Pancreas	0.18	**0.23**	0.23	**0.29**	0.27	**0.34**	0.30	**0.38**	0.37	**0.47**	0.44	**0.56**
Small intestine		**0.13**		**0.16**	0.01	**0.19**	0.01	**0.21**	0.02	**0.26**	0.02	**0.30**
Large intestine	0.01	**0.21**	0.01	**0.26**	0.02	**0.30**	0.02	**0.34**	0.02	**0.40**	0.03	**0.46**
Sigma/Rectum		**0.04**		**0.05**		**0.06**		**0.06**		**0.08**		**0.09**
Lumbar spine	0.04	**0.12**	0.05	**0.16**	0.06	**0.19**	0.07	**0.22**	0.09	**0.28**	0.11	**0.34**
Pelvis		**0.04**		**0.05**		**0.07**		**0.07**		**0.09**		**0.11**
Adrenals	0.07	**0.09**	0.09	**0.11**	0.11	**0.13**	0.13	**0.15**	0.16	**0.19**	0.19	**0.23**
Kidneys	0.03	**0.07**	0.04	**0.09**	0.05	**0.10**	0.06	**0.12**	0.08	**0.15**	0.09	**0.18**
Urinary bladder										**0.01**		**0.01**
Uterus		**0.05**		**0.07**		**0.08**		**0.09**		**0.12**	0.01	**0.15**
Ovaries		**0.02**		**0.03**		**0.03**		**0.04**		**0.05**		**0.06**
Prostate												
Testes												
Upper arm	0.26	**0.63**	0.33	**0.77**	0.38	**0.88**	0.42	**0.96**	0.51	**1.12**	0.59	**1.27**
Elbow	0.24	**0.66**	0.30	**0.81**	0.35	**0.92**	0.39	**1.01**	0.47	**1.18**	0.55	**1.33**
Forearm	0.02	**0.32**	0.03	**0.39**	0.03	**0.45**	0.04	**0.49**	0.04	**0.57**	0.05	**0.64**
Upper leg												
Knee												
Lower leg												
Lymph nodes	0.16	**0.30**	0.20	**0.36**	0.23	**0.41**	0.25	**0.46**	0.29	**0.53**	0.33	**0.61**
Skin	0.16	**0.26**	0.17	**0.28**	0.17	**0.29**	0.18	**0.30**	0.19	**0.32**	0.20	**0.33**
Muscle	0.10	**0.16**	0.12	**0.19**	0.14	**0.21**	0.15	**0.23**	0.17	**0.26**	0.19	**0.29**
Skeleton	0.19	**0.32**	0.23	**0.38**	0.26	**0.43**	0.28	**0.46**	0.32	**0.53**	0.35	**0.59**
Red bone marrow	0.09	**0.12**	0.10	**0.14**	0.12	**0.16**	0.13	**0.18**	0.15	**0.21**	0.17	**0.24**
Whole body dose	0.12	**0.19**	0.15	**0.23**	0.17	**0.26**	0.18	**0.28**	0.21	**0.32**	0.23	**0.36**
ED (ICRP 60)	0.61	**0.70**	0.70	**0.81**	0.77	**0.89**	0.82	**0.95**	0.91	**1.06**	0.98	**1.15**
ED (ICRP 103)	1.27	**1.39**	1.45	**1.60**	1.57	**1.74**	1.66	**1.85**	1.82	**2.04**	1.96	**2.20**
Total filtration (mm) **Al**	**1.5**		**2.5**		**3.5**		**4.5**		**3.5**		**3.5**	
Cu	**0**		**0**		**0**		**0**		**0.1**		**0.2**	

Table 9.51 Normalized organ doses—Chest ap—13–17 years

Beam projection:			ap				**Table cells:**					
Focus-to-detector distance (cm):			100				(+) = normal font = optimal field setting					
Tube voltage (kVp):			110				**(−) = bold font = suboptimal field setting**					
Total filtrations:			See last table row.				Empty table cells correspond to values < 0.01					
Organ or tissue	**Organ dose D_T : Entrance dose K_E**								**(mSv × mGy⁻¹)**			
	(+)	(−)	(+)	(−)	(+)	(−)	(+)	(−)	(+)	(−)	(+)	(−)
Skull	0.03	**0.42**	0.03	**0.48**	0.04	**0.52**	0.04	**0.55**	0.05	**0.61**	0.05	**0.65**
Brain		**0.04**		**0.05**		**0.05**		**0.06**		**0.07**		**0.08**
Oral mucosa	0.03	**0.48**	0.03	**0.55**	0.04	**0.59**	0.04	**0.63**	0.04	**0.70**	0.05	**0.75**
Salivary glands	0.04	**0.45**	0.05	**0.51**	0.06	**0.56**	0.06	**0.60**	0.07	**0.66**	0.08	**0.71**
Cervical spine	0.57	**0.92**	0.67	**1.08**	0.75	**1.20**	0.81	**1.30**	0.92	**1.49**	1.01	**1.64**
Thyroid	0.90	**0.97**	1.00	**1.08**	1.07	**1.15**	1.11	**1.20**	1.19	**1.29**	1.26	**1.36**
Thymus	1.03	**1.08**	1.18	**1.24**	1.28	**1.34**	1.35	**1.42**	1.49	**1.57**	1.60	**1.68**
Thoracic spine	0.25	**0.27**	0.30	**0.33**	0.34	**0.38**	0.38	**0.42**	0.45	**0.50**	0.51	**0.57**
Claviculae	1.81	**1.93**	2.05	**2.18**	2.20	**2.34**	2.30	**2.46**	2.48	**2.65**	2.59	**2.77**
Scapulae	0.24	**0.26**	0.29	**0.32**	0.33	**0.36**	0.36	**0.39**	0.42	**0.46**	0.48	**0.52**
Ribs	1.42	**1.53**	1.57	**1.70**	1.67	**1.81**	1.73	**1.89**	1.84	**2.01**	1.91	**2.09**
Oesophagus	0.25	**0.28**	0.30	**0.33**	0.34	**0.37**	0.37	**0.41**	0.44	**0.48**	0.50	**0.54**
Trachea	0.26	**0.55**	0.30	**0.63**	0.32	**0.68**	0.34	**0.73**	0.37	**0.80**	0.39	**0.87**
Lung	0.42	**0.45**	0.49	**0.53**	0.55	**0.59**	0.59	**0.63**	0.67	**0.72**	0.74	**0.79**
Mamma	11.0	**11.5**	12.0	**12.6**	12.7	**13.4**	13.2	**13.9**	14.1	**14.8**	14.7	**15.5**
Heart	0.68	**0.72**	0.79	**0.84**	0.87	**0.92**	0.94	**0.99**	1.05	**1.11**	1.15	**1.22**
Stomach	0.37	**0.69**	0.44	**0.80**	0.48	**0.89**	0.52	**0.95**	0.58	**1.06**	0.64	**1.16**
Spleen	0.17	**0.23**	0.21	**0.28**	0.23	**0.31**	0.26	**0.34**	0.30	**0.40**	0.34	**0.46**
Liver	0.37	**0.54**	0.44	**0.63**	0.48	**0.70**	0.52	**0.75**	0.59	**0.85**	0.65	**0.93**
Gall bladder	0.14	**0.55**	0.16	**0.65**	0.18	**0.72**	0.20	**0.77**	0.23	**0.88**	0.26	**0.98**
Pancreas	0.29	**0.37**	0.34	**0.44**	0.39	**0.49**	0.42	**0.54**	0.49	**0.63**	0.55	**0.71**
Small intestine	0.01	**0.20**	0.02	**0.24**	0.02	**0.27**	0.02	**0.29**	0.03	**0.34**	0.03	**0.38**
Large intestine	0.02	**0.31**	0.02	**0.37**	0.03	**0.41**	0.03	**0.44**	0.04	**0.50**	0.04	**0.56**
Sigma/Rectum		**0.06**		**0.07**		**0.08**		**0.09**		**0.10**		**0.11**
Lumbar spine	0.07	**0.22**	0.09	**0.27**	0.10	**0.31**	0.11	**0.34**	0.13	**0.40**	0.16	**0.46**
Pelvis		**0.07**		**0.09**		**0.10**		**0.11**	0.01	**0.13**	0.01	**0.15**
Adrenals	0.13	**0.16**	0.16	**0.19**	0.18	**0.22**	0.20	**0.24**	0.23	**0.28**	0.27	**0.33**
Kidneys	0.06	**0.12**	0.08	**0.15**	0.09	**0.17**	0.10	**0.18**	0.12	**0.22**	0.13	**0.25**
Urinary bladder				**0.01**		**0.01**		**0.01**		**0.02**		**0.02**
Uterus		**0.10**	0.01	**0.12**	0.01	**0.14**	0.01	**0.15**	0.02	**0.18**	0.02	**0.21**
Ovaries		**0.04**		**0.05**		**0.05**		**0.06**		**0.07**		**0.08**
Prostate												
Testes												
Upper arm	0.38	**0.84**	0.45	**0.97**	0.50	**1.07**	0.54	**1.14**	0.61	**1.27**	0.67	**1.37**
Elbow	0.35	**0.88**	0.41	**1.02**	0.46	**1.12**	0.50	**1.19**	0.57	**1.33**	0.62	**1.44**
Forearm	0.03	**0.42**	0.04	**0.49**	0.05	**0.54**	0.05	**0.58**	0.06	**0.64**	0.06	**0.69**
Upper leg												
Knee												
Lower leg												
Lymph nodes	0.23	**0.42**	0.27	**0.49**	0.29	**0.54**	0.31	**0.58**	0.35	**0.65**	0.39	**0.72**
Skin	0.17	**0.29**	0.18	**0.31**	0.19	**0.32**	0.20	**0.33**	0.21	**0.34**	0.22	**0.36**
Muscle	0.14	**0.21**	0.16	**0.24**	0.17	**0.26**	0.18	**0.28**	0.21	**0.31**	0.22	**0.34**
Skeleton	0.25	**0.41**	0.28	**0.47**	0.30	**0.51**	0.32	**0.54**	0.35	**0.59**	0.38	**0.64**
Red bone marrow	0.12	**0.17**	0.14	**0.20**	0.15	**0.22**	0.16	**0.23**	0.18	**0.26**	0.20	**0.29**
Whole body dose	0.17	**0.26**	0.19	**0.29**	0.21	**0.32**	0.22	**0.34**	0.24	**0.38**	0.26	**0.41**
ED (ICRP 60)	0.75	**0.87**	0.83	**0.97**	0.89	**1.04**	0.93	**1.09**	1.01	**1.19**	1.07	**1.27**
ED (ICRP 103)	1.52	**1.69**	1.68	**1.87**	1.79	**2.00**	1.86	**2.09**	2.00	**2.25**	2.11	**2.38**
Total filtration (mm) **Al**	**1.5**		**2.5**		**3.5**		**4.5**		**3.5**		**3.5**	
Cu	**0**		**0**		**0**		**0**		**0.1**		**0.2**	

Table 9.52 Normalized organ doses—Chest ap—13–17 years

Beam projection:			ap			**Table cells:**						
Focus-to-detector distance (cm):			100			(+) = normal font = optimal field setting						
Tube voltage (kVp):			120			**(−) = bold font = suboptimal field setting**						
Total filtrations:			See last table row.			Empty table cells correspond to values < 0.01						
Organ or tissue	**Organ dose D_T : Entrance dose K_E**							**(mSv × mGy^{-1})**				
	(+)	(−)	(+)	(−)	(+)	(−)	(+)	(−)	(+)	(−)	(+)	(−)
Skull	0.03	**0.44**	0.04	**0.50**	0.04	**0.53**	0.04	**0.56**	0.05	**0.61**	0.06	**0.65**
Brain		**0.04**		**0.05**		**0.06**		**0.06**		**0.08**		**0.09**
Oral mucosa	0.03	**0.51**	0.03	**0.57**	0.04	**0.62**	0.04	**0.65**	0.05	**0.72**	0.05	**0.77**
Salivary glands	0.05	**0.48**	0.06	**0.54**	0.06	**0.59**	0.07	**0.62**	0.07	**0.68**	0.08	**0.73**
Cervical spine	0.61	**0.99**	0.71	**1.15**	0.78	**1.26**	0.84	**1.36**	0.94	**1.52**	1.02	**1.66**
Thyroid	0.94	**1.01**	1.03	**1.11**	1.09	**1.18**	1.14	**1.23**	1.21	**1.31**	1.27	**1.38**
Thymus	1.09	**1.15**	1.23	**1.29**	1.33	**1.39**	1.40	**1.47**	1.52	**1.60**	1.62	**1.70**
Thoracic spine	0.28	**0.31**	0.33	**0.37**	0.37	**0.41**	0.40	**0.45**	0.47	**0.53**	0.53	**0.60**
Claviculae	1.88	**2.00**	2.09	**2.23**	2.22	**2.37**	2.32	**2.48**	2.46	**2.63**	2.55	**2.73**
Scapulae	0.27	**0.29**	0.32	**0.34**	0.35	**0.38**	0.38	**0.42**	0.44	**0.48**	0.49	**0.54**
Ribs	1.46	**1.58**	1.60	**1.73**	1.68	**1.83**	1.74	**1.90**	1.83	**2.00**	1.88	**2.07**
Oesophagus	0.28	**0.31**	0.33	**0.36**	0.37	**0.41**	0.40	**0.44**	0.46	**0.51**	0.52	**0.57**
Trachea	0.28	**0.59**	0.31	**0.66**	0.33	**0.72**	0.35	**0.76**	0.37	**0.83**	0.39	**0.89**
Lung	0.46	**0.49**	0.53	**0.57**	0.58	**0.62**	0.62	**0.66**	0.69	**0.75**	0.76	**0.81**
Mamma	11.4	**11.9**	12.4	**13.0**	13.0	**13.7**	13.5	**14.2**	14.3	**15.0**	14.8	**15.6**
Heart	0.73	**0.78**	0.84	**0.89**	0.92	**0.97**	0.98	**1.03**	1.09	**1.15**	1.18	**1.25**
Stomach	0.40	**0.74**	0.46	**0.85**	0.51	**0.93**	0.54	**0.99**	0.60	**1.10**	0.65	**1.18**
Spleen	0.19	**0.26**	0.23	**0.30**	0.25	**0.34**	0.27	**0.37**	0.32	**0.43**	0.35	**0.48**
Liver	0.40	**0.58**	0.47	**0.67**	0.51	**0.74**	0.54	**0.79**	0.61	**0.88**	0.66	**0.96**
Gall bladder	0.15	**0.60**	0.18	**0.69**	0.20	**0.76**	0.21	**0.82**	0.24	**0.92**	0.27	**1.00**
Pancreas	0.32	**0.41**	0.37	**0.48**	0.42	**0.53**	0.45	**0.57**	0.51	**0.66**	0.57	**0.73**
Small intestine	0.02	**0.22**	0.02	**0.26**	0.02	**0.29**	0.02	**0.31**	0.03	**0.35**	0.03	**0.39**
Large intestine	0.02	**0.34**	0.03	**0.39**	0.03	**0.43**	0.03	**0.46**	0.04	**0.52**	0.05	**0.57**
Sigma/Rectum		**0.07**		**0.08**		**0.09**		**0.09**		**0.10**		**0.12**
Lumbar spine	0.08	**0.25**	0.10	**0.30**	0.11	**0.33**	0.12	**0.37**	0.14	**0.43**	0.16	**0.49**
Pelvis		**0.08**		**0.10**		**0.11**	0.01	**0.12**	0.01	**0.14**	0.01	**0.16**
Adrenals	0.15	**0.18**	0.18	**0.21**	0.20	**0.24**	0.21	**0.26**	0.25	**0.30**	0.28	**0.34**
Kidneys	0.07	**0.14**	0.09	**0.16**	0.10	**0.18**	0.11	**0.20**	0.12	**0.23**	0.14	**0.27**
Urinary bladder		**0.01**		**0.01**		**0.02**		**0.02**		**0.02**		**0.02**
Uterus	0.01	**0.11**	0.01	**0.13**	0.01	**0.15**	0.02	**0.16**	0.02	**0.19**	0.02	**0.22**
Ovaries		**0.04**		**0.05**		**0.06**		**0.06**		**0.07**		**0.08**
Prostate												
Testes												
Upper arm	0.41	**0.89**	0.47	**1.02**	0.52	**1.10**	0.56	**1.17**	0.63	**1.29**	0.68	**1.37**
Elbow	0.38	**0.93**	0.44	**1.06**	0.48	**1.16**	0.52	**1.23**	0.58	**1.35**	0.63	**1.44**
Forearm	0.04	**0.45**	0.04	**0.51**	0.05	**0.56**	0.05	**0.59**	0.06	**0.65**	0.07	**0.69**
Upper leg												
Knee												
Lower leg												
Lymph nodes	0.25	**0.46**	0.28	**0.52**	0.31	**0.57**	0.33	**0.61**	0.36	**0.68**	0.40	**0.74**
Skin	0.18	**0.30**	0.19	**0.32**	0.20	**0.33**	0.20	**0.33**	0.21	**0.35**	0.22	**0.36**
Muscle	0.15	**0.22**	0.17	**0.25**	0.18	**0.28**	0.19	**0.29**	0.21	**0.32**	0.23	**0.35**
Skeleton	0.26	**0.43**	0.29	**0.49**	0.31	**0.52**	0.33	**0.55**	0.36	**0.60**	0.38	**0.64**
Red bone marrow	0.13	**0.18**	0.15	**0.21**	0.16	**0.23**	0.17	**0.25**	0.19	**0.28**	0.21	**0.30**
Whole body dose	0.18	**0.27**	0.20	**0.31**	0.22	**0.33**	0.23	**0.35**	0.25	**0.39**	0.27	**0.42**
ED (ICRP 60)	0.78	**0.91**	0.86	**1.01**	0.92	**1.08**	0.96	**1.13**	1.03	**1.22**	1.08	**1.29**
ED (ICRP 103)	1.58	**1.77**	1.74	**1.94**	1.84	**2.06**	1.91	**2.14**	2.03	**2.29**	2.13	**2.41**
Total filtration (mm) **Al**	**1.5**		**2.5**		**3.5**		**4.5**		**3.5**		**3.5**	
Cu	**0**		**0**		**0**		**0**		**0.1**		**0.2**	

Table 9.53 Normalized organ doses—Chest ap—> 17 years

Beam projection:			ap				**Table cells:**					
Focus-to-detector distance (cm):			100				(+) = normal font = optimal field setting					
Tube voltage (kVp):			60				**(−) = bold font = suboptimal field setting**					
Total filtrations:			See last table row.				Empty table cells correspond to values < 0.01					
Organ or tissue	**Organ dose D_T : Entrance dose K_E**								**($mSv \times mGy^{-1}$)**			
	(+)	(−)	(+)	(−)	(+)	(−)	(+)	(−)	(+)	(−)	(+)	(−)
Skull		**0.19**		**0.24**	0.01	**0.28**	0.01	**0.31**	0.02	**0.37**	0.01	**0.35**
Brain								**0.01**		**0.02**		**0.01**
Oral mucosa		**0.24**		**0.30**		**0.33**		**0.36**	0.01	**0.42**		**0.40**
Salivary glands		**0.23**	0.01	**0.29**	0.01	**0.33**	0.02	**0.37**	0.02	**0.43**	0.02	**0.41**
Cervical spine	0.13	**0.34**	0.17	**0.45**	0.21	**0.54**	0.24	**0.62**	0.30	**0.77**	0.26	**0.68**
Thyroid	0.49	**0.61**	0.58	**0.73**	0.64	**0.81**	0.69	**0.87**	0.78	**0.98**	0.75	**0.95**
Thymus	0.60	**0.64**	0.74	**0.79**	0.85	**0.91**	0.93	**1.00**	1.09	**1.16**	1.04	**1.11**
Thoracic spine	0.06	**0.07**	0.08	**0.09**	0.10	**0.11**	0.12	**0.14**	0.16	**0.18**	0.12	**0.13**
Claviculae	1.17	**1.25**	1.44	**1.55**	1.65	**1.77**	1.81	**1.94**	2.09	**2.26**	2.04	**2.20**
Scapulae	0.08	**0.09**	0.11	**0.12**	0.14	**0.15**	0.16	**0.18**	0.21	**0.23**	0.17	**0.19**
Ribs	1.03	**1.12**	1.23	**1.34**	1.36	**1.48**	1.47	**1.60**	1.65	**1.80**	1.62	**1.76**
Oesophagus	0.08	**0.09**	0.11	**0.13**	0.14	**0.15**	0.16	**0.18**	0.21	**0.23**	0.17	**0.19**
Trachea	0.09	**0.27**	0.11	**0.34**	0.13	**0.39**	0.14	**0.42**	0.16	**0.49**	0.15	**0.47**
Lung	0.21	**0.22**	0.27	**0.29**	0.32	**0.34**	0.36	**0.38**	0.43	**0.47**	0.40	**0.43**
Mamma	0.86	**0.92**	1.00	**1.06**	1.09	**1.16**	1.16	**1.23**	1.28	**1.36**	1.26	**1.34**
Heart	0.32	**0.34**	0.41	**0.44**	0.48	**0.51**	0.54	**0.57**	0.65	**0.69**	0.60	**0.64**
Stomach	0.18	**0.35**	0.23	**0.45**	0.27	**0.52**	0.30	**0.58**	0.36	**0.70**	0.33	**0.65**
Spleen	0.06	**0.07**	0.07	**0.10**	0.09	**0.12**	0.11	**0.14**	0.14	**0.18**	0.11	**0.15**
Liver	0.17	**0.26**	0.21	**0.33**	0.25	**0.39**	0.28	**0.43**	0.34	**0.52**	0.31	**0.48**
Gall bladder	0.04	**0.23**	0.06	**0.31**	0.07	**0.36**	0.08	**0.41**	0.10	**0.51**	0.08	**0.46**
Pancreas	0.10	**0.13**	0.14	**0.17**	0.16	**0.21**	0.19	**0.24**	0.24	**0.30**	0.20	**0.26**
Small intestine		**0.08**		**0.11**		**0.13**		**0.15**		**0.18**		**0.16**
Large intestine		**0.14**		**0.18**		**0.21**		**0.24**		**0.30**		**0.27**
Sigma/Rectum		**0.02**		**0.03**		**0.04**		**0.04**		**0.05**		**0.05**
Lumbar spine	0.01	**0.05**	0.02	**0.07**	0.02	**0.09**	0.03	**0.10**	0.04	**0.14**	0.03	**0.10**
Pelvis		**0.02**		**0.03**		**0.03**		**0.04**		**0.05**		**0.04**
Adrenals	0.04	**0.04**	0.05	**0.06**	0.06	**0.07**	0.07	**0.08**	0.09	**0.11**	0.07	**0.08**
Kidneys	0.01	**0.03**	0.02	**0.04**	0.03	**0.05**	0.03	**0.06**	0.04	**0.08**	0.03	**0.06**
Urinary bladder												
Uterus				**0.01**		**0.01**		**0.02**		**0.02**		**0.02**
Ovaries				**0.01**		**0.01**		**0.02**		**0.02**		**0.02**
Prostate												
Testes												
Upper arm	0.13	**0.43**	0.17	**0.55**	0.21	**0.64**	0.24	**0.72**	0.30	**0.87**	0.26	**0.81**
Elbow	0.12	**0.45**	0.16	**0.58**	0.19	**0.68**	0.21	**0.76**	0.27	**0.92**	0.24	**0.86**
Forearm		**0.22**	0.01	**0.29**	0.01	**0.34**	0.02	**0.38**	0.02	**0.45**	0.02	**0.42**
Upper leg												
Knee												
Lower leg												
Lymph nodes	0.10	**0.20**	0.12	**0.25**	0.14	**0.29**	0.16	**0.33**	0.19	**0.39**	0.17	**0.36**
Skin	0.14	**0.24**	0.14	**0.26**	0.15	**0.27**	0.16	**0.28**	0.17	**0.29**	0.16	**0.28**
Muscle	0.07	**0.11**	0.09	**0.14**	0.10	**0.16**	0.11	**0.17**	0.13	**0.20**	0.12	**0.19**
Skeleton	0.14	**0.22**	0.17	**0.28**	0.19	**0.32**	0.21	**0.35**	0.24	**0.41**	0.23	**0.39**
Red bone marrow	0.07	**0.09**	0.09	**0.11**	0.10	**0.13**	0.11	**0.14**	0.13	**0.16**	0.12	**0.15**
Whole body dose	0.09	**0.14**	0.11	**0.17**	0.12	**0.20**	0.13	**0.21**	0.16	**0.25**	0.15	**0.23**
ED (ICRP 60)	0.14	**0.19**	0.17	**0.24**	0.20	**0.27**	0.22	**0.30**	0.25	**0.36**	0.24	**0.33**
ED (ICRP 103)	0.20	**0.27**	0.25	**0.32**	0.28	**0.37**	0.30	**0.40**	0.35	**0.47**	0.33	**0.44**
Total filtration (mm) Al	**1.5**		**2.5**		**3.5**		**4.5**		**3.5**		**3.5**	
Cu	**0**		**0**		**0**		**0**		**0.1**		**0.2**	

Table 9.54 Normalized organ doses—Chest ap—> 17 years

Beam projection:			ap			**Table cells:**						
Focus-to-detector distance (cm):			100			(+) = normal font = optimal field setting						
Tube voltage (kVp):			70			**(−) = bold font = suboptimal field setting**						
Total filtrations:			See last table row.			Empty table cells correspond to values < 0.01						
Organ or tissue	**Organ dose D_T : Entrance dose K_E**							**(mSv × mGy^{-1})**				
	(+)	(−)	(+)	(−)	(+)	(−)	(+)	(−)	(+)	(−)	(+)	(−)
Skull		**0.23**	0.01	**0.28**	0.01	**0.33**	0.02	**0.36**	0.02	**0.43**	0.03	**0.48**
Brain				**0.01**		**0.01**		**0.02**		**0.02**		**0.03**
Oral mucosa		**0.28**		**0.34**	0.01	**0.38**	0.01	**0.42**	0.02	**0.48**	0.02	**0.54**
Salivary glands	0.01	**0.28**	0.02	**0.34**	0.02	**0.38**	0.02	**0.42**	0.03	**0.49**	0.03	**0.55**
Cervical spine	0.17	**0.45**	0.22	**0.58**	0.26	**0.68**	0.30	**0.77**	0.37	**0.95**	0.44	**1.13**
Thyroid	0.55	**0.69**	0.64	**0.80**	0.70	**0.89**	0.75	**0.95**	0.84	**1.06**	0.91	**1.15**
Thymus	0.70	**0.75**	0.85	**0.91**	0.97	**1.03**	1.05	**1.12**	1.22	**1.30**	1.36	**1.45**
Thoracic spine	0.09	**0.10**	0.13	**0.14**	0.15	**0.17**	0.18	**0.20**	0.23	**0.26**	0.29	**0.33**
Claviculae	1.34	**1.44**	1.62	**1.75**	1.83	**1.98**	1.99	**2.15**	2.27	**2.46**	2.51	**2.72**
Scapulae	0.12	**0.13**	0.16	**0.17**	0.19	**0.21**	0.22	**0.24**	0.28	**0.30**	0.34	**0.37**
Ribs	1.15	**1.25**	1.35	**1.47**	1.48	**1.62**	1.58	**1.73**	1.76	**1.94**	1.91	**2.11**
Oesophagus	0.12	**0.13**	0.16	**0.17**	0.19	**0.21**	0.22	**0.24**	0.27	**0.30**	0.33	**0.37**
Trachea	0.10	**0.32**	0.13	**0.39**	0.14	**0.44**	0.16	**0.48**	0.18	**0.56**	0.20	**0.63**
Lung	0.26	**0.28**	0.33	**0.35**	0.38	**0.41**	0.43	**0.46**	0.51	**0.55**	0.59	**0.64**
Mamma	0.94	**1.00**	1.08	**1.15**	1.17	**1.25**	1.24	**1.32**	1.36	**1.45**	1.46	**1.55**
Heart	0.40	**0.42**	0.50	**0.53**	0.58	**0.61**	0.64	**0.68**	0.76	**0.81**	0.88	**0.94**
Stomach	0.22	**0.43**	0.28	**0.53**	0.32	**0.62**	0.36	**0.68**	0.42	**0.81**	0.49	**0.93**
Spleen	0.08	**0.10**	0.10	**0.14**	0.12	**0.17**	0.14	**0.19**	0.18	**0.24**	0.22	**0.29**
Liver	0.21	**0.32**	0.26	**0.40**	0.30	**0.46**	0.34	**0.52**	0.41	**0.62**	0.47	**0.72**
Gall bladder	0.06	**0.30**	0.08	**0.38**	0.09	**0.45**	0.10	**0.50**	0.13	**0.62**	0.16	**0.72**
Pancreas	0.14	**0.17**	0.18	**0.23**	0.21	**0.27**	0.24	**0.31**	0.31	**0.39**	0.37	**0.47**
Small intestine		**0.11**		**0.14**		**0.16**		**0.18**		**0.23**	0.01	**0.27**
Large intestine		**0.18**		**0.23**		**0.26**	0.01	**0.30**	0.01	**0.36**	0.02	**0.42**
Sigma/Rectum		**0.03**		**0.04**		**0.05**		**0.05**		**0.07**		**0.08**
Lumbar spine	0.02	**0.08**	0.03	**0.11**	0.04	**0.13**	0.04	**0.15**	0.06	**0.20**	0.07	**0.25**
Pelvis		**0.03**		**0.04**		**0.05**		**0.06**		**0.07**		**0.09**
Adrenals	0.05	**0.06**	0.07	**0.08**	0.09	**0.10**	0.10	**0.12**	0.13	**0.15**	0.16	**0.19**
Kidneys	0.02	**0.05**	0.03	**0.06**	0.04	**0.08**	0.04	**0.09**	0.06	**0.11**	0.07	**0.14**
Urinary bladder												
Uterus		**0.01**		**0.02**		**0.02**		**0.02**		**0.03**		**0.04**
Ovaries		**0.01**		**0.02**		**0.02**		**0.02**		**0.03**		**0.04**
Prostate												
Testes												
Upper arm	0.17	**0.53**	0.22	**0.66**	0.26	**0.76**	0.30	**0.85**	0.37	**1.01**	0.43	**1.16**
Elbow	0.15	**0.55**	0.20	**0.70**	0.24	**0.81**	0.27	**0.89**	0.33	**1.07**	0.40	**1.22**
Forearm	0.01	**0.27**	0.02	**0.34**	0.02	**0.39**	0.02	**0.44**	0.03	**0.52**	0.03	**0.60**
Upper leg												
Knee												
Lower leg												
Lymph nodes	0.12	**0.24**	0.15	**0.30**	0.17	**0.35**	0.19	**0.39**	0.23	**0.46**	0.26	**0.54**
Skin	0.14	**0.25**	0.15	**0.27**	0.16	**0.28**	0.16	**0.29**	0.18	**0.31**	0.19	**0.32**
Muscle	0.08	**0.13**	0.10	**0.16**	0.11	**0.18**	0.12	**0.20**	0.15	**0.23**	0.17	**0.26**
Skeleton	0.16	**0.26**	0.19	**0.32**	0.22	**0.37**	0.24	**0.40**	0.27	**0.47**	0.31	**0.53**
Red bone marrow	0.08	**0.11**	0.10	**0.13**	0.11	**0.15**	0.13	**0.16**	0.15	**0.19**	0.17	**0.22**
Whole body dose	0.10	**0.17**	0.12	**0.20**	0.14	**0.23**	0.15	**0.25**	0.18	**0.29**	0.20	**0.32**
ED (ICRP 60)	0.17	**0.23**	0.20	**0.28**	0.23	**0.32**	0.25	**0.35**	0.29	**0.41**	0.33	**0.46**
ED (ICRP 103)	0.23	**0.31**	0.28	**0.37**	0.31	**0.42**	0.34	**0.46**	0.39	**0.53**	0.43	**0.59**
Total filtration (mm) **Al**	**1.5**		**2.5**		**3.5**		**4.5**		**3.5**		**3.5**	
Cu	**0**		**0**		**0**		**0**		**0.1**		**0.2**	

Table 9.55 Normalized organ doses—Chest ap—> 17 years

Beam projection:	ap				**Table cells:**							
Focus-to-detector distance (cm):	100				(+) = normal font = optimal field setting							
Tube voltage (kVp):	80				**(−) = bold font = suboptimal field setting**							
Total filtrations:	See last table row.				Empty table cells correspond to values < 0.01							
Organ or tissue	**Organ dose D_T : Entrance dose K_E**								**(mSv × mGy^{-1})**			
	(+)	(−)	(+)	(−)	(+)	(−)	(+)	(−)	(+)	(−)	(+)	(−)
Skull	0.01	**0.27**	0.02	**0.32**	0.02	**0.37**	0.02	**0.40**	0.03	**0.47**	0.03	**0.52**
Brain		**0.01**		**0.02**		**0.02**		**0.02**		**0.03**		**0.04**
Oral mucosa		**0.33**	0.01	**0.39**	0.01	**0.43**	0.02	**0.46**	0.02	**0.53**	0.02	**0.59**
Salivary glands	0.02	**0.32**	0.02	**0.38**	0.02	**0.43**	0.03	**0.47**	0.03	**0.54**	0.04	**0.60**
Cervical spine	0.21	**0.55**	0.27	**0.70**	0.32	**0.82**	0.36	**0.91**	0.43	**1.11**	0.51	**1.29**
Thyroid	0.60	**0.75**	0.69	**0.87**	0.76	**0.95**	0.81	**1.02**	0.89	**1.13**	0.97	**1.22**
Thymus	0.80	**0.85**	0.96	**1.02**	1.07	**1.14**	1.16	**1.24**	1.32	**1.41**	1.47	**1.56**
Thoracic spine	0.13	**0.15**	0.17	**0.19**	0.21	**0.23**	0.24	**0.26**	0.30	**0.34**	0.37	**0.41**
Claviculae	1.49	**1.61**	1.77	**1.92**	1.97	**2.13**	2.12	**2.30**	2.39	**2.59**	2.60	**2.83**
Scapulae	0.16	**0.17**	0.20	**0.22**	0.24	**0.26**	0.27	**0.30**	0.34	**0.37**	0.40	**0.45**
Ribs	1.25	**1.36**	1.44	**1.58**	1.57	**1.72**	1.67	**1.83**	1.84	**2.02**	1.97	**2.18**
Oesophagus	0.16	**0.17**	0.20	**0.22**	0.24	**0.27**	0.27	**0.30**	0.34	**0.37**	0.40	**0.45**
Trachea	0.12	**0.37**	0.14	**0.44**	0.16	**0.50**	0.17	**0.54**	0.20	**0.62**	0.22	**0.69**
Lung	0.31	**0.34**	0.39	**0.42**	0.44	**0.48**	0.49	**0.53**	0.58	**0.63**	0.66	**0.72**
Mamma	1.02	**1.08**	1.15	**1.23**	1.24	**1.32**	1.31	**1.40**	1.43	**1.52**	1.52	**1.62**
Heart	0.47	**0.50**	0.58	**0.62**	0.66	**0.71**	0.73	**0.78**	0.86	**0.92**	0.98	**1.05**
Stomach	0.26	**0.50**	0.32	**0.62**	0.37	**0.70**	0.41	**0.77**	0.48	**0.91**	0.55	**1.03**
Spleen	0.10	**0.14**	0.13	**0.18**	0.15	**0.21**	0.17	**0.24**	0.22	**0.30**	0.26	**0.35**
Liver	0.25	**0.38**	0.31	**0.47**	0.36	**0.54**	0.39	**0.60**	0.47	**0.70**	0.53	**0.81**
Gall bladder	0.08	**0.37**	0.10	**0.46**	0.11	**0.53**	0.13	**0.59**	0.16	**0.71**	0.19	**0.82**
Pancreas	0.18	**0.23**	0.23	**0.29**	0.27	**0.34**	0.30	**0.38**	0.37	**0.47**	0.44	**0.56**
Small intestine		**0.13**		**0.17**		**0.20**	0.01	**0.22**	0.01	**0.27**	0.02	**0.32**
Large intestine		**0.22**	0.01	**0.27**	0.01	**0.31**	0.02	**0.35**	0.02	**0.42**	0.02	**0.48**
Sigma/Rectum		**0.04**		**0.05**		**0.06**		**0.07**		**0.08**		**0.09**
Lumbar spine	0.03	**0.11**	0.04	**0.15**	0.05	**0.18**	0.06	**0.21**	0.08	**0.26**	0.10	**0.32**
Pelvis		**0.04**		**0.05**		**0.07**		**0.07**		**0.09**		**0.11**
Adrenals	0.07	**0.09**	0.10	**0.11**	0.11	**0.14**	0.13	**0.16**	0.16	**0.20**	0.20	**0.24**
Kidneys	0.03	**0.07**	0.04	**0.09**	0.05	**0.10**	0.06	**0.12**	0.07	**0.15**	0.09	**0.18**
Urinary bladder												**0.01**
Uterus		**0.02**		**0.02**		**0.03**		**0.03**		**0.04**		**0.05**
Ovaries		**0.02**		**0.02**		**0.03**		**0.03**		**0.04**		**0.05**
Prostate												
Testes												
Upper arm	0.21	**0.62**	0.27	**0.76**	0.31	**0.87**	0.35	**0.95**	0.42	**1.11**	0.49	**1.26**
Elbow	0.19	**0.65**	0.24	**0.80**	0.29	**0.91**	0.32	**1.00**	0.39	**1.18**	0.45	**1.33**
Forearm	0.02	**0.32**	0.02	**0.39**	0.02	**0.45**	0.03	**0.49**	0.03	**0.57**	0.04	**0.65**
Upper leg												
Knee												
Lower leg												
Lymph nodes	0.14	**0.29**	0.18	**0.36**	0.20	**0.41**	0.22	**0.45**	0.26	**0.53**	0.30	**0.60**
Skin	0.15	**0.26**	0.16	**0.28**	0.17	**0.29**	0.17	**0.30**	0.18	**0.32**	0.19	**0.34**
Muscle	0.10	**0.15**	0.11	**0.18**	0.13	**0.21**	0.14	**0.22**	0.16	**0.26**	0.18	**0.29**
Skeleton	0.18	**0.30**	0.22	**0.36**	0.24	**0.41**	0.26	**0.44**	0.30	**0.51**	0.33	**0.57**
Red bone marrow	0.10	**0.13**	0.12	**0.15**	0.13	**0.17**	0.14	**0.19**	0.16	**0.22**	0.19	**0.25**
Whole body dose	0.12	**0.19**	0.14	**0.23**	0.16	**0.25**	0.17	**0.28**	0.20	**0.32**	0.22	**0.36**
ED (ICRP 60)	0.19	**0.27**	0.23	**0.32**	0.26	**0.36**	0.28	**0.39**	0.32	**0.46**	0.36	**0.51**
ED (ICRP 103)	0.26	**0.35**	0.31	**0.42**	0.35	**0.47**	0.37	**0.51**	0.42	**0.58**	0.47	**0.65**
Total filtration (mm) Al	**1.5**		**2.5**		**3.5**		**4.5**		**3.5**		**3.5**	
Cu	**0**		**0**		**0**		**0**		**0.1**		**0.2**	

Table 9.56 Normalized organ doses—Chest ap—> 17 years

Beam projection:			ap		**Table cells:**							
Focus-to-detector distance (cm):			100		(+) = normal font = optimal field setting							
Tube voltage (kVp):			110		**(–) = bold font = suboptimal field setting**							
Total filtrations:			See last table row.		Empty table cells correspond to values < 0.01							
Organ or tissue	**Organ dose D_T : Entrance dose K_E**						**(mSv × mGy^{-1})**					
	(+)	(–)	(+)	(–)	(+)	(–)	(+)	(–)	(+)	(–)	(+)	(–)
Skull	0.02	**0.35**	0.02	**0.41**	0.03	**0.44**	0.03	**0.47**	0.04	**0.53**	0.04	**0.57**
Brain		**0.03**		**0.03**		**0.04**		**0.04**		**0.05**		**0.06**
Oral mucosa	0.02	**0.43**	0.02	**0.49**	0.02	**0.54**	0.02	**0.57**	0.03	**0.63**	0.03	**0.69**
Salivary glands	0.03	**0.43**	0.03	**0.49**	0.04	**0.54**	0.04	**0.57**	0.04	**0.64**	0.05	**0.69**
Cervical spine	0.33	**0.83**	0.39	**0.98**	0.43	**1.09**	0.47	**1.18**	0.54	**1.36**	0.60	**1.50**
Thyroid	0.73	**0.92**	0.82	**1.03**	0.87	**1.10**	0.92	**1.16**	0.99	**1.25**	1.04	**1.32**
Thymus	1.04	**1.11**	1.19	**1.27**	1.29	**1.38**	1.37	**1.46**	1.51	**1.61**	1.62	**1.73**
Thoracic spine	0.24	**0.27**	0.29	**0.33**	0.33	**0.37**	0.37	**0.41**	0.44	**0.49**	0.50	**0.57**
Claviculae	1.80	**1.96**	2.04	**2.22**	2.19	**2.39**	2.30	**2.51**	2.48	**2.71**	2.60	**2.84**
Scapulae	0.25	**0.28**	0.31	**0.34**	0.35	**0.38**	0.38	**0.42**	0.44	**0.49**	0.50	**0.56**
Ribs	1.45	**1.60**	1.61	**1.77**	1.71	**1.89**	1.78	**1.97**	1.89	**2.10**	1.96	**2.18**
Oesophagus	0.27	**0.30**	0.32	**0.36**	0.36	**0.40**	0.40	**0.44**	0.47	**0.52**	0.53	**0.59**
Trachea	0.16	**0.49**	0.18	**0.56**	0.20	**0.62**	0.21	**0.65**	0.23	**0.73**	0.25	**0.79**
Lung	0.45	**0.49**	0.52	**0.57**	0.58	**0.63**	0.62	**0.68**	0.71	**0.77**	0.78	**0.85**
Mamma	1.20	**1.27**	1.32	**1.40**	1.39	**1.48**	1.45	**1.54**	1.54	**1.64**	1.61	**1.72**
Heart	0.67	**0.71**	0.78	**0.83**	0.86	**0.92**	0.92	**0.99**	1.04	**1.11**	1.14	**1.22**
Stomach	0.37	**0.70**	0.43	**0.82**	0.48	**0.90**	0.51	**0.97**	0.58	**1.09**	0.64	**1.19**
Spleen	0.17	**0.23**	0.21	**0.28**	0.23	**0.32**	0.25	**0.35**	0.30	**0.41**	0.34	**0.46**
Liver	0.36	**0.55**	0.42	**0.64**	0.47	**0.71**	0.51	**0.76**	0.57	**0.86**	0.63	**0.95**
Gall bladder	0.12	**0.55**	0.15	**0.65**	0.17	**0.72**	0.18	**0.78**	0.21	**0.89**	0.24	**0.99**
Pancreas	0.29	**0.37**	0.34	**0.44**	0.39	**0.50**	0.42	**0.54**	0.49	**0.63**	0.55	**0.72**
Small intestine	0.01	**0.21**	0.01	**0.25**	0.02	**0.28**	0.02	**0.31**	0.02	**0.35**	0.03	**0.40**
Large intestine	0.02	**0.32**	0.02	**0.38**	0.02	**0.42**	0.03	**0.46**	0.03	**0.52**	0.04	**0.58**
Sigma/Rectum		**0.06**		**0.07**		**0.08**		**0.09**		**0.10**		**0.12**
Lumbar spine	0.07	**0.21**	0.08	**0.26**	0.09	**0.30**	0.10	**0.33**	0.12	**0.39**	0.14	**0.45**
Pelvis		**0.07**		**0.09**		**0.10**		**0.11**		**0.13**		**0.15**
Adrenals	0.13	**0.16**	0.16	**0.20**	0.18	**0.22**	0.20	**0.24**	0.24	**0.29**	0.28	**0.34**
Kidneys	0.06	**0.12**	0.07	**0.15**	0.08	**0.17**	0.09	**0.19**	0.11	**0.22**	0.13	**0.25**
Urinary bladder				**0.01**		**0.01**		**0.01**		**0.02**		**0.02**
Uterus		**0.04**		**0.04**		**0.05**		**0.06**		**0.07**		**0.08**
Ovaries		**0.04**		**0.04**		**0.05**		**0.06**		**0.07**		**0.08**
Prostate												
Testes												
Upper arm	0.32	**0.83**	0.37	**0.97**	0.42	**1.06**	0.45	**1.13**	0.52	**1.27**	0.57	**1.37**
Elbow	0.29	**0.88**	0.34	**1.02**	0.38	**1.12**	0.41	**1.19**	0.48	**1.33**	0.53	**1.44**
Forearm	0.03	**0.42**	0.03	**0.49**	0.04	**0.54**	0.04	**0.58**	0.04	**0.64**	0.05	**0.70**
Upper leg												
Knee												
Lower leg												
Lymph nodes	0.21	**0.42**	0.24	**0.48**	0.26	**0.54**	0.28	**0.58**	0.32	**0.65**	0.35	**0.72**
Skin	0.17	**0.29**	0.18	**0.31**	0.18	**0.32**	0.19	**0.33**	0.20	**0.35**	0.21	**0.36**
Muscle	0.13	**0.21**	0.15	**0.24**	0.16	**0.26**	0.18	**0.28**	0.20	**0.31**	0.22	**0.34**
Skeleton	0.23	**0.39**	0.26	**0.45**	0.28	**0.49**	0.30	**0.52**	0.33	**0.57**	0.35	**0.61**
Red bone marrow	0.14	**0.18**	0.16	**0.21**	0.17	**0.23**	0.18	**0.25**	0.20	**0.28**	0.22	**0.31**
Whole body dose	0.16	**0.25**	0.18	**0.29**	0.20	**0.32**	0.21	**0.34**	0.24	**0.38**	0.26	**0.41**
ED (ICRP 60)	0.26	**0.36**	0.29	**0.42**	0.32	**0.46**	0.34	**0.49**	0.38	**0.55**	0.41	**0.59**
ED (ICRP 103)	0.34	**0.47**	0.39	**0.53**	0.42	**0.58**	0.45	**0.62**	0.49	**0.68**	0.53	**0.74**
Total filtration (mm) Al	**1.5**		**2.5**		**3.5**		**4.5**		**3.5**		**3.5**	
Cu	**0**		**0**		**0**		**0**		**0.1**		**0.2**	

Table 9.57 Normalized organ doses—Chest ap—> 17 years

Beam projection:			ap				**Table cells:**					
Focus-to-detector distance (cm):			100				(+) = normal font = optimal field setting					
Tube voltage (kVp):			120				**(−) = bold font = suboptimal field setting**					
Total filtrations:			See last table row.				Empty table cells correspond to values < 0.01					
Organ or tissue	**Organ dose D_T : Entrance dose K_E**								**($mSv \times mGy^{-1}$)**			
	(+)	(−)	(+)	(−)	(+)	(−)	(+)	(−)	(+)	(−)	(+)	(−)
Skull	0.02	**0.37**	0.03	**0.42**	0.03	**0.46**	0.03	**0.49**	0.04	**0.53**	0.04	**0.57**
Brain		**0.03**		**0.04**		**0.04**		**0.05**		**0.05**		**0.06**
Oral mucosa	0.02	**0.46**	0.02	**0.52**	0.02	**0.56**	0.03	**0.60**	0.03	**0.66**	0.03	**0.71**
Salivary glands	0.03	**0.46**	0.03	**0.52**	0.04	**0.57**	0.04	**0.60**	0.05	**0.66**	0.05	**0.71**
Cervical spine	0.35	**0.90**	0.41	**1.04**	0.46	**1.15**	0.49	**1.24**	0.56	**1.39**	0.61	**1.53**
Thyroid	0.77	**0.97**	0.85	**1.07**	0.90	**1.14**	0.94	**1.19**	1.01	**1.27**	1.06	**1.34**
Thymus	1.10	**1.18**	1.25	**1.33**	1.34	**1.43**	1.42	**1.51**	1.54	**1.64**	1.65	**1.75**
Thoracic spine	0.27	**0.31**	0.32	**0.36**	0.36	**0.41**	0.39	**0.45**	0.46	**0.52**	0.52	**0.59**
Claviculae	1.87	**2.03**	2.09	**2.27**	2.22	**2.42**	2.32	**2.53**	2.47	**2.70**	2.56	**2.80**
Scapulae	0.28	**0.31**	0.33	**0.37**	0.37	**0.41**	0.40	**0.45**	0.46	**0.51**	0.51	**0.57**
Ribs	1.50	**1.65**	1.64	**1.81**	1.73	**1.91**	1.79	**1.98**	1.88	**2.09**	1.94	**2.16**
Oesophagus	0.30	**0.33**	0.35	**0.39**	0.40	**0.44**	0.43	**0.48**	0.50	**0.55**	0.56	**0.62**
Trachea	0.17	**0.53**	0.19	**0.60**	0.20	**0.64**	0.22	**0.68**	0.24	**0.75**	0.25	**0.80**
Lung	0.48	**0.53**	0.56	**0.61**	0.61	**0.67**	0.65	**0.71**	0.73	**0.80**	0.80	**0.87**
Mamma	1.24	**1.32**	1.35	**1.44**	1.43	**1.52**	1.48	**1.57**	1.56	**1.67**	1.63	**1.73**
Heart	0.72	**0.77**	0.83	**0.89**	0.90	**0.97**	0.96	**1.03**	1.07	**1.15**	1.16	**1.25**
Stomach	0.40	**0.76**	0.46	**0.87**	0.51	**0.95**	0.54	**1.01**	0.60	**1.12**	0.65	**1.21**
Spleen	0.19	**0.26**	0.23	**0.31**	0.25	**0.35**	0.27	**0.37**	0.32	**0.43**	0.35	**0.49**
Liver	0.39	**0.59**	0.45	**0.68**	0.50	**0.75**	0.53	**0.80**	0.59	**0.90**	0.65	**0.98**
Gall bladder	0.14	**0.60**	0.16	**0.69**	0.18	**0.76**	0.19	**0.82**	0.22	**0.92**	0.25	**1.01**
Pancreas	0.32	**0.41**	0.37	**0.48**	0.42	**0.54**	0.45	**0.58**	0.52	**0.67**	0.58	**0.75**
Small intestine	0.01	**0.23**	0.02	**0.27**	0.02	**0.30**	0.02	**0.33**	0.02	**0.37**	0.03	**0.41**
Large intestine	0.02	**0.35**	0.02	**0.41**	0.03	**0.45**	0.03	**0.48**	0.03	**0.54**	0.04	**0.60**
Sigma/Rectum		**0.07**		**0.08**		**0.09**		**0.10**		**0.11**		**0.12**
Lumbar spine	0.07	**0.24**	0.09	**0.29**	0.10	**0.33**	0.11	**0.36**	0.13	**0.42**	0.15	**0.48**
Pelvis		**0.08**		**0.10**		**0.11**		**0.12**		**0.14**	0.01	**0.16**
Adrenals	0.15	**0.18**	0.18	**0.22**	0.20	**0.24**	0.22	**0.27**	0.26	**0.31**	0.29	**0.36**
Kidneys	0.07	**0.14**	0.08	**0.17**	0.09	**0.19**	0.10	**0.20**	0.12	**0.24**	0.14	**0.27**
Urinary bladder		**0.01**		**0.01**		**0.01**		**0.01**		**0.02**		**0.02**
Uterus		**0.04**		**0.05**		**0.06**		**0.06**		**0.07**		**0.08**
Ovaries		**0.04**		**0.05**		**0.06**		**0.06**		**0.07**		**0.08**
Prostate												
Testes												
Upper arm	0.34	**0.88**	0.40	**1.01**	0.44	**1.10**	0.47	**1.17**	0.53	**1.29**	0.58	**1.38**
Elbow	0.31	**0.93**	0.37	**1.06**	0.40	**1.16**	0.43	**1.23**	0.49	**1.35**	0.54	**1.44**
Forearm	0.03	**0.45**	0.03	**0.51**	0.04	**0.56**	0.04	**0.59**	0.05	**0.65**	0.05	**0.70**
Upper leg												
Knee												
Lower leg												
Lymph nodes	0.22	**0.45**	0.26	**0.52**	0.28	**0.57**	0.30	**0.60**	0.33	**0.68**	0.37	**0.74**
Skin	0.17	**0.30**	0.18	**0.32**	0.19	**0.33**	0.19	**0.34**	0.20	**0.35**	0.21	**0.36**
Muscle	0.14	**0.22**	0.16	**0.25**	0.17	**0.27**	0.18	**0.29**	0.20	**0.32**	0.22	**0.35**
Skeleton	0.24	**0.41**	0.27	**0.47**	0.29	**0.50**	0.31	**0.53**	0.33	**0.58**	0.35	**0.61**
Red bone marrow	0.15	**0.20**	0.17	**0.23**	0.18	**0.25**	0.19	**0.26**	0.21	**0.30**	0.23	**0.32**
Whole body dose	0.17	**0.27**	0.19	**0.31**	0.21	**0.33**	0.22	**0.35**	0.24	**0.39**	0.26	**0.42**
ED (ICRP 60)	0.27	**0.39**	0.31	**0.44**	0.33	**0.48**	0.35	**0.51**	0.39	**0.56**	0.42	**0.61**
ED (ICRP 103)	0.36	**0.50**	0.41	**0.56**	0.44	**0.61**	0.46	**0.64**	0.51	**0.70**	0.54	**0.76**
Total filtration (mm) **Al**	**1.5**		**2.5**		**3.5**		**4.5**		**3.5**		**3.5**	
Cu	**0**		**0**		**0**		**0**		**0.1**		**0.2**	

Table 9.58 Normalized organ doses—Chest r-lat—24th gestation week

Beam projection:			r-lat				**Table cells:**					
Focus-to-detector distance (cm):			100				(+) = normal font = optimal field setting					
Tube voltage (kVp):			50				**(−) = bold font = suboptimal field setting**					
Total filtrations:			See last table row.				Empty table cells correspond to values < 0.01					
Organ or tissue	**Organ dose D_T : Entrance dose K_E**							**(mSv × mGy^{-1})**				
	(+)	(−)	(+)	(−)	(+)	(−)	(+)	(−)	(+)	(−)	(+)	(−)
Skull		**0.03**		**0.03**		**0.04**		**0.04**		**0.04**		**0.05**
Brain												
Oral mucosa		**0.04**		**0.04**		**0.04**		**0.05**		**0.05**		**0.06**
Salivary glands		**0.04**		**0.04**		**0.05**		**0.05**		**0.05**		**0.05**
Cervical spine	0.04	**0.08**	0.05	**0.10**	0.06	**0.11**	0.07	**0.12**	0.07	**0.14**	0.08	**0.15**
Thyroid	0.05	**0.05**	0.06	**0.06**	0.06	**0.07**	0.06	**0.07**	0.07	**0.07**	0.07	**0.08**
Thymus	0.03	**0.03**	0.03	**0.03**	0.04	**0.04**	0.04	**0.04**	0.05	**0.05**	0.05	**0.05**
Thoracic spine	0.07	**0.07**	0.08	**0.09**	0.10	**0.10**	0.11	**0.11**	0.12	**0.13**	0.14	**0.15**
Claviculae	0.09	**0.09**	0.11	**0.11**	0.12	**0.12**	0.13	**0.13**	0.14	**0.15**	0.15	**0.16**
Scapulae	0.12	**0.12**	0.13	**0.13**	0.14	**0.14**	0.15	**0.15**	0.16	**0.16**	0.17	**0.17**
Ribs	0.11	**0.11**	0.13	**0.13**	0.14	**0.14**	0.15	**0.15**	0.16	**0.17**	0.17	**0.18**
Oesophagus	0.03	**0.03**	0.03	**0.03**	0.04	**0.04**	0.04	**0.04**	0.05	**0.05**	0.05	**0.06**
Trachea		**0.03**		**0.04**		**0.04**		**0.04**		**0.05**		**0.05**
Lung	0.03	**0.03**	0.04	**0.04**	0.04	**0.04**	0.04	**0.05**	0.05	**0.05**	0.05	**0.06**
Mamma	0.05	**0.05**	0.05	**0.05**	0.05	**0.06**	0.05	**0.06**	0.06	**0.06**	0.06	**0.06**
Heart	0.03	**0.03**	0.03	**0.03**	0.04	**0.04**	0.04	**0.04**	0.05	**0.05**	0.05	**0.05**
Stomach		**0.01**	0.01	**0.02**	0.01	**0.02**	0.01	**0.02**	0.02	**0.03**	0.02	**0.03**
Spleen			0.01	**0.01**	0.01	**0.02**	0.01	**0.02**	0.02	**0.02**	0.02	**0.02**
Liver	0.02	**0.04**	0.03	**0.05**	0.03	**0.05**	0.04	**0.06**	0.04	**0.06**	0.04	**0.07**
Gall bladder		**0.03**		**0.03**		**0.04**		**0.04**	0.01	**0.05**	0.01	**0.05**
Pancreas	0.01	**0.02**	0.02	**0.02**	0.02	**0.02**	0.02	**0.03**	0.03	**0.03**	0.03	**0.04**
Small intestine		**0.01**		**0.02**		**0.02**		**0.02**		**0.02**		**0.03**
Large intestine		**0.02**		**0.02**		**0.03**		**0.03**		**0.03**		**0.04**
Sigma/Rectum												
Lumbar spine	0.01	**0.07**	0.01	**0.08**	0.02	**0.10**	0.02	**0.11**	0.02	**0.12**	0.02	**0.14**
Pelvis		**0.01**		**0.01**		**0.02**		**0.02**		**0.02**		**0.02**
Adrenals	0.02	**0.02**	0.02	**0.03**	0.03	**0.03**	0.03	**0.03**	0.04	**0.04**	0.04	**0.04**
Kidneys		**0.03**	0.01	**0.03**	0.01	**0.04**	0.01	**0.04**	0.02	**0.04**	0.02	**0.05**
Urinary bladder												
Uterus												
Ovaries												
Prostate												
Testes												
Upper arm	0.12	**0.12**	0.13	**0.13**	0.14	**0.14**	0.15	**0.15**	0.16	**0.16**	0.17	**0.17**
Elbow	0.12	**0.12**	0.13	**0.14**	0.14	**0.15**	0.15	**0.15**	0.16	**0.17**	0.17	**0.18**
Forearm		**0.06**		**0.07**		**0.07**		**0.07**		**0.08**		**0.08**
Upper leg												
Knee												
Lower leg												
Lymph nodes	0.01	**0.02**	0.01	**0.03**	0.02	**0.03**	0.02	**0.03**	0.02	**0.04**	0.02	**0.04**
Skin	0.02	**0.02**	0.02	**0.02**	0.02	**0.03**	0.02	**0.03**	0.02	**0.03**	0.02	**0.03**
Muscle		**0.02**	0.01	**0.02**	0.01	**0.02**	0.01	**0.02**	0.01	**0.02**	0.02	**0.03**
Skeleton	0.04	**0.05**	0.05	**0.06**	0.05	**0.07**	0.05	**0.08**	0.06	**0.08**	0.07	**0.09**
Red bone marrow		**0.01**	0.01	**0.02**	0.01	**0.02**	0.01	**0.02**	0.01	**0.02**	0.02	**0.02**
Whole body dose	0.01	**0.02**	0.01	**0.02**	0.02	**0.03**	0.02	**0.03**	0.02	**0.03**	0.02	**0.03**
ED (ICRP 60)	0.01	**0.02**	0.02	**0.02**	0.02	**0.02**	0.02	**0.03**	0.02	**0.03**	0.02	**0.03**
ED (ICRP 103)	0.02	**0.02**	0.02	**0.03**	0.02	**0.03**	0.02	**0.03**	0.03	**0.03**	0.03	**0.04**
Total filtration (mm) **Al**	**1.5**		**2.5**		**3.5**		**4.5**		**3.5**		**3.5**	
Cu	**0**		**0**		**0**		**0**		**0.1**		**0.2**	

Table 9.59 Normalized organ doses—Chest r-lat—24th gestation week

Beam projection:			r-lat				**Table cells:**					
Focus-to-detector distance (cm):			100				(+) = normal font = optimal field setting					
Tube voltage (kVp):			60				**(−) = bold font = suboptimal field setting**					
Total filtrations:			See last table row.				Empty table cells correspond to values < 0.01					
Organ or tissue	**Organ dose D_T : Entrance dose K_E**								**(mSv × mGy^{-1})**			
	(+)	(−)	(+)	(−)	(+)	(−)	(+)	(−)	(+)	(−)	(+)	(−)
Skull		**0.03**		**0.04**		**0.04**		**0.04**		**0.05**		**0.05**
Brain												
Oral mucosa		**0.04**		**0.05**		**0.05**		**0.05**		**0.06**		**0.06**
Salivary glands		**0.04**		**0.05**		**0.05**		**0.05**		**0.05**		**0.05**
Cervical spine	0.05	**0.09**	0.06	**0.11**	0.07	**0.13**	0.07	**0.14**	0.08	**0.15**	0.08	**0.15**
Thyroid	0.05	**0.06**	0.06	**0.07**	0.07	**0.07**	0.07	**0.07**	0.07	**0.08**	0.07	**0.08**
Thymus	0.03	**0.03**	0.04	**0.04**	0.04	**0.04**	0.05	**0.05**	0.05	**0.05**	0.05	**0.05**
Thoracic spine	0.08	**0.08**	0.10	**0.10**	0.11	**0.12**	0.12	**0.13**	0.14	**0.15**	0.14	**0.15**
Claviculae	0.10	**0.10**	0.12	**0.12**	0.13	**0.13**	0.14	**0.14**	0.15	**0.16**	0.15	**0.16**
Scapulae	0.13	**0.13**	0.14	**0.14**	0.15	**0.15**	0.16	**0.16**	0.17	**0.17**	0.17	**0.17**
Ribs	0.12	**0.12**	0.14	**0.14**	0.15	**0.15**	0.16	**0.16**	0.17	**0.18**	0.17	**0.18**
Oesophagus	0.03	**0.03**	0.04	**0.04**	0.04	**0.04**	0.05	**0.05**	0.05	**0.06**	0.05	**0.06**
Trachea		**0.04**		**0.04**		**0.05**		**0.05**		**0.05**		**0.05**
Lung	0.03	**0.04**	0.04	**0.04**	0.05	**0.05**	0.05	**0.05**	0.05	**0.06**	0.05	**0.06**
Mamma	0.05	**0.05**	0.05	**0.06**	0.05	**0.06**	0.06	**0.06**	0.06	**0.06**	0.06	**0.06**
Heart	0.03	**0.03**	0.04	**0.04**	0.04	**0.04**	0.05	**0.05**	0.05	**0.05**	0.05	**0.05**
Stomach	0.01	**0.02**	0.01	**0.02**	0.01	**0.02**	0.02	**0.03**	0.02	**0.03**	0.02	**0.03**
Spleen	0.01	**0.01**	0.01	**0.02**	0.02	**0.02**	0.02	**0.02**	0.02	**0.03**	0.02	**0.02**
Liver	0.03	**0.04**	0.03	**0.05**	0.04	**0.06**	0.04	**0.06**	0.04	**0.07**	0.04	**0.07**
Gall bladder		**0.03**		**0.04**		**0.04**	0.01	**0.05**	0.01	**0.05**	0.01	**0.05**
Pancreas	0.02	**0.02**	0.02	**0.02**	0.02	**0.03**	0.03	**0.03**	0.03	**0.04**	0.03	**0.04**
Small intestine		**0.02**		**0.02**		**0.02**		**0.02**		**0.03**		**0.03**
Large intestine		**0.02**		**0.03**		**0.03**		**0.03**		**0.04**		**0.04**
Sigma/Rectum												
Lumbar spine	0.01	**0.08**	0.02	**0.10**	0.02	**0.11**	0.02	**0.12**	0.02	**0.14**	0.02	**0.14**
Pelvis		**0.01**		**0.02**		**0.02**		**0.02**		**0.02**		**0.02**
Adrenals	0.02	**0.03**	0.03	**0.03**	0.03	**0.04**	0.04	**0.04**	0.04	**0.04**	0.04	**0.04**
Kidneys	0.01	**0.03**	0.01	**0.04**	0.01	**0.04**	0.02	**0.04**	0.02	**0.05**	0.02	**0.05**
Urinary bladder												
Uterus												
Ovaries												
Prostate												
Testes												
Upper arm	0.13	**0.13**	0.14	**0.14**	0.15	**0.15**	0.16	**0.16**	0.17	**0.17**	0.17	**0.17**
Elbow	0.13	**0.13**	0.14	**0.15**	0.15	**0.16**	0.16	**0.16**	0.17	**0.17**	0.17	**0.18**
Forearm		**0.06**		**0.07**		**0.07**		**0.08**		**0.08**		**0.08**
Upper leg												
Knee												
Lower leg												
Lymph nodes	0.01	**0.03**	0.02	**0.03**	0.02	**0.04**	0.02	**0.04**	0.02	**0.04**	0.02	**0.04**
Skin	0.02	**0.02**	0.02	**0.03**	0.02	**0.03**	0.02	**0.03**	0.02	**0.03**	0.02	**0.03**
Muscle	0.01	**0.02**	0.01	**0.02**	0.01	**0.02**	0.01	**0.02**	0.02	**0.03**	0.02	**0.03**
Skeleton	0.04	**0.06**	0.05	**0.07**	0.06	**0.08**	0.06	**0.08**	0.06	**0.09**	0.07	**0.09**
Red bone marrow		**0.02**	0.01	**0.02**	0.01	**0.02**	0.01	**0.02**	0.02	**0.02**	0.02	**0.02**
Whole body dose	0.01	**0.02**	0.02	**0.03**	0.02	**0.03**	0.02	**0.03**	0.02	**0.03**	0.02	**0.03**
ED (ICRP 60)	0.02	**0.02**	0.02	**0.02**	0.02	**0.03**	0.02	**0.03**	0.02	**0.03**	0.02	**0.03**
ED (ICRP 103)	0.02	**0.03**	0.02	**0.03**	0.02	**0.03**	0.03	**0.03**	0.03	**0.04**	0.03	**0.04**
Total filtration (mm) Al	**1.5**		**2.5**		**3.5**		**4.5**		**3.5**		**3.5**	
Cu	**0**		**0**		**0**		**0**		**0.1**		**0.2**	

Table 9.60 Normalized organ doses—Chest r-lat—24th gestation week

Beam projection:			r-lat		**Table cells:**							
Focus-to-detector distance (cm):			100		(+) = normal font = optimal field setting							
Tube voltage (kVp):			70		**(−) = bold font = suboptimal field setting**							
Total filtrations:			See last table row.		Empty table cells correspond to values < 0.01							
Organ or tissue	**Organ dose D_T : Entrance dose K_E**							**(mSv × mGy⁻¹)**				
	(+)	(−)	(+)	(−)	(+)	(−)	(+)	(−)	(+)	(−)	(+)	(−)
Skull		**0.03**		**0.04**		**0.04**		**0.05**		**0.05**		**0.05**
Brain												
Oral mucosa		**0.04**		**0.05**		**0.05**		**0.05**		**0.06**		**0.06**
Salivary glands		**0.04**		**0.05**		**0.05**		**0.05**		**0.06**		**0.06**
Cervical spine	0.06	**0.10**	0.06	**0.12**	0.07	**0.13**	0.08	**0.14**	0.08	**0.16**	0.09	**0.17**
Thyroid	0.06	**0.06**	0.06	**0.07**	0.07	**0.07**	0.07	**0.08**	0.07	**0.08**	0.08	**0.09**
Thymus	0.04	**0.04**	0.04	**0.04**	0.05	**0.05**	0.05	**0.05**	0.06	**0.06**	0.06	**0.06**
Thoracic spine	0.09	**0.09**	0.11	**0.11**	0.12	**0.13**	0.13	**0.14**	0.15	**0.15**	0.16	**0.17**
Claviculae	0.11	**0.11**	0.12	**0.13**	0.13	**0.14**	0.14	**0.15**	0.15	**0.16**	0.16	**0.17**
Scapulae	0.13	**0.13**	0.15	**0.15**	0.16	**0.16**	0.16	**0.16**	0.17	**0.17**	0.17	**0.18**
Ribs	0.13	**0.13**	0.14	**0.15**	0.15	**0.16**	0.16	**0.17**	0.17	**0.18**	0.18	**0.19**
Oesophagus	0.03	**0.04**	0.04	**0.04**	0.05	**0.05**	0.05	**0.05**	0.06	**0.06**	0.06	**0.07**
Trachea		**0.04**		**0.05**		**0.05**		**0.05**		**0.06**	0.01	**0.06**
Lung	0.04	**0.04**	0.04	**0.05**	0.05	**0.05**	0.05	**0.05**	0.06	**0.06**	0.06	**0.06**
Mamma	0.05	**0.05**	0.05	**0.06**	0.06	**0.06**	0.06	**0.06**	0.06	**0.07**	0.06	**0.07**
Heart	0.03	**0.04**	0.04	**0.04**	0.05	**0.05**	0.05	**0.05**	0.05	**0.06**	0.06	**0.06**
Stomach	0.01	**0.02**	0.01	**0.02**	0.02	**0.03**	0.02	**0.03**	0.02	**0.03**	0.02	**0.04**
Spleen	0.01	**0.02**	0.02	**0.02**	0.02	**0.02**	0.02	**0.02**	0.02	**0.03**	0.03	**0.03**
Liver	0.03	**0.05**	0.04	**0.06**	0.04	**0.06**	0.04	**0.07**	0.05	**0.07**	0.05	**0.08**
Gall bladder		**0.03**		**0.04**	0.01	**0.05**	0.01	**0.05**	0.01	**0.06**	0.02	**0.06**
Pancreas	0.02	**0.02**	0.02	**0.03**	0.03	**0.03**	0.03	**0.04**	0.03	**0.04**	0.04	**0.05**
Small intestine		**0.02**		**0.02**		**0.03**		**0.03**		**0.03**		**0.03**
Large intestine		**0.02**		**0.03**		**0.03**		**0.04**		**0.04**		**0.04**
Sigma/Rectum												
Lumbar spine	0.01	**0.09**	0.02	**0.11**	0.02	**0.12**	0.02	**0.13**	0.03	**0.15**	0.03	**0.16**
Pelvis		**0.02**		**0.02**		**0.02**		**0.02**		**0.03**		**0.03**
Adrenals	0.03	**0.03**	0.03	**0.04**	0.04	**0.04**	0.04	**0.04**	0.04	**0.05**	0.05	**0.05**
Kidneys	0.01	**0.04**	0.01	**0.04**	0.02	**0.05**	0.02	**0.05**	0.02	**0.05**	0.02	**0.06**
Urinary bladder												
Uterus												
Ovaries												
Prostate												
Testes												
Upper arm	0.13	**0.13**	0.15	**0.15**	0.15	**0.16**	0.16	**0.16**	0.17	**0.17**	0.17	**0.18**
Elbow	0.13	**0.14**	0.15	**0.15**	0.15	**0.16**	0.16	**0.17**	0.17	**0.18**	0.17	**0.18**
Forearm		**0.07**		**0.07**		**0.08**		**0.08**		**0.08**		**0.09**
Upper leg												
Knee												
Lower leg												
Lymph nodes	0.01	**0.03**	0.02	**0.04**	0.02	**0.04**	0.02	**0.04**	0.02	**0.05**	0.03	**0.05**
Skin	0.02	**0.03**	0.02	**0.03**	0.02	**0.03**	0.02	**0.03**	0.02	**0.03**	0.02	**0.03**
Muscle	0.01	**0.02**	0.01	**0.02**	0.01	**0.02**	0.01	**0.02**	0.02	**0.03**	0.02	**0.03**
Skeleton	0.05	**0.07**	0.05	**0.07**	0.06	**0.08**	0.06	**0.09**	0.07	**0.09**	0.07	**0.10**
Red bone marrow	0.01	**0.02**	0.01	**0.02**	0.01	**0.02**	0.01	**0.02**	0.02	**0.03**	0.02	**0.03**
Whole body dose	0.02	**0.02**	0.02	**0.03**	0.02	**0.03**	0.02	**0.03**	0.02	**0.04**	0.02	**0.04**
ED (ICRP 60)	0.02	**0.02**	0.02	**0.03**	0.02	**0.03**	0.02	**0.03**	0.03	**0.04**	0.03	**0.04**
ED (ICRP 103)	0.02	**0.03**	0.02	**0.03**	0.03	**0.04**	0.03	**0.04**	0.03	**0.04**	0.03	**0.04**
Total filtration (mm) **Al**	**1.5**		**2.5**		**3.5**		**4.5**		**3.5**		**3.5**	
Cu	**0**		**0**		**0**		**0**		**0.1**		**0.2**	

Table 9.61 Normalized organ doses—Chest r-lat—24th gestation week

Beam projection:			r-lat			**Table cells:**						
Focus-to-detector distance (cm):			100			(+) = normal font = optimal field setting						
Tube voltage (kVp):			80			**(–) = bold font = suboptimal field setting**						
Total filtrations:			See last table row.			Empty table cells correspond to values < 0.01						
Organ or tissue	**Organ dose D_T : Entrance dose K_E**							**(mSv × mGy^{-1})**				
	(+)	(–)	(+)	(–)	(+)	(–)	(+)	(–)	(+)	(–)	(+)	(–)
Skull		**0.04**		**0.04**		**0.04**		**0.05**		**0.05**		**0.05**
Brain												
Oral mucosa		**0.05**		**0.05**		**0.05**		**0.06**		**0.06**		**0.06**
Salivary glands		**0.05**		**0.05**		**0.05**		**0.06**		**0.06**		**0.06**
Cervical spine	0.06	**0.11**	0.07	**0.13**	0.07	**0.14**	0.08	**0.15**	0.08	**0.16**	0.09	**0.17**
Thyroid	0.06	**0.07**	0.07	**0.07**	0.07	**0.08**	0.07	**0.08**	0.08	**0.08**	0.08	**0.09**
Thymus	0.04	**0.04**	0.05	**0.05**	0.05	**0.05**	0.05	**0.05**	0.06	**0.06**	0.06	**0.06**
Thoracic spine	0.10	**0.10**	0.11	**0.12**	0.13	**0.13**	0.14	**0.14**	0.15	**0.16**	0.16	**0.17**
Claviculae	0.11	**0.12**	0.13	**0.13**	0.14	**0.14**	0.14	**0.15**	0.15	**0.16**	0.16	**0.17**
Scapulae	0.14	**0.14**	0.15	**0.15**	0.16	**0.16**	0.16	**0.16**	0.17	**0.17**	0.17	**0.17**
Ribs	0.13	**0.14**	0.15	**0.15**	0.16	**0.16**	0.16	**0.17**	0.17	**0.18**	0.18	**0.18**
Oesophagus	0.04	**0.04**	0.05	**0.05**	0.05	**0.05**	0.06	**0.06**	0.06	**0.07**	0.07	**0.07**
Trachea		**0.04**		**0.05**		**0.05**		**0.05**	0.01	**0.06**	0.01	**0.06**
Lung	0.04	**0.04**	0.05	**0.05**	0.05	**0.05**	0.06	**0.06**	0.06	**0.06**	0.06	**0.07**
Mamma	0.05	**0.06**	0.06	**0.06**	0.06	**0.06**	0.06	**0.06**	0.06	**0.07**	0.07	**0.07**
Heart	0.04	**0.04**	0.04	**0.05**	0.05	**0.05**	0.05	**0.05**	0.06	**0.06**	0.06	**0.07**
Stomach	0.01	**0.02**	0.02	**0.03**	0.02	**0.03**	0.02	**0.03**	0.02	**0.04**	0.03	**0.04**
Spleen	0.01	**0.02**	0.02	**0.02**	0.02	**0.03**	0.02	**0.03**	0.03	**0.03**	0.03	**0.04**
Liver	0.03	**0.05**	0.04	**0.06**	0.04	**0.07**	0.04	**0.07**	0.05	**0.07**	0.05	**0.08**
Gall bladder		**0.04**	0.01	**0.05**	0.01	**0.05**	0.01	**0.05**	0.01	**0.06**	0.02	**0.06**
Pancreas	0.02	**0.03**	0.03	**0.03**	0.03	**0.04**	0.03	**0.04**	0.04	**0.04**	0.04	**0.05**
Small intestine		**0.02**		**0.02**		**0.03**		**0.03**		**0.03**		**0.04**
Large intestine		**0.03**		**0.03**		**0.04**		**0.04**		**0.04**		**0.04**
Sigma/Rectum												
Lumbar spine	0.02	**0.10**	0.02	**0.11**	0.02	**0.13**	0.02	**0.13**	0.03	**0.15**	0.03	**0.16**
Pelvis		**0.02**		**0.02**		**0.02**		**0.02**		**0.03**		**0.03**
Adrenals	0.03	**0.03**	0.04	**0.04**	0.04	**0.04**	0.04	**0.05**	0.05	**0.05**	0.05	**0.06**
Kidneys	0.01	**0.04**	0.02	**0.04**	0.02	**0.05**	0.02	**0.05**	0.02	**0.06**	0.02	**0.06**
Urinary bladder												
Uterus												
Ovaries												**0.01**
Prostate												
Testes												
Upper arm	0.14	**0.14**	0.15	**0.15**	0.16	**0.16**	0.16	**0.16**	0.17	**0.17**	0.17	**0.17**
Elbow	0.14	**0.14**	0.15	**0.15**	0.16	**0.16**	0.16	**0.17**	0.17	**0.17**	0.17	**0.18**
Forearm		**0.07**		**0.07**		**0.08**		**0.08**		**0.08**		**0.08**
Upper leg												
Knee												
Lower leg												
Lymph nodes	0.02	**0.03**	0.02	**0.04**	0.02	**0.04**	0.02	**0.05**	0.02	**0.05**	0.03	**0.05**
Skin	0.02	**0.03**	0.02	**0.03**	0.02	**0.03**	0.02	**0.03**	0.02	**0.03**	0.02	**0.03**
Muscle	0.01	**0.02**	0.01	**0.02**	0.01	**0.02**	0.02	**0.03**	0.02	**0.03**	0.02	**0.03**
Skeleton	0.05	**0.07**	0.06	**0.08**	0.06	**0.08**	0.06	**0.09**	0.07	**0.09**	0.07	**0.10**
Red bone marrow	0.01	**0.02**	0.01	**0.02**	0.02	**0.02**	0.02	**0.02**	0.02	**0.03**	0.02	**0.03**
Whole body dose	0.02	**0.03**	0.02	**0.03**	0.02	**0.03**	0.02	**0.03**	0.02	**0.04**	0.02	**0.04**
ED (ICRP 60)	0.02	**0.03**	0.02	**0.03**	0.02	**0.03**	0.02	**0.03**	0.03	**0.04**	0.03	**0.04**
ED (ICRP 103)	0.02	**0.03**	0.03	**0.03**	0.03	**0.04**	0.03	**0.04**	0.03	**0.04**	0.03	**0.05**
Total filtration (mm) **Al**	**1.5**		**2.5**		**3.5**		**4.5**		**3.5**		**3.5**	
Cu	**0**		**0**		**0**		**0**		**0.1**		**0.2**	

Table 9.62 Normalized organ doses—Chest r-lat—28th gestation week

Beam projection:			r-lat			**Table cells:**						
Focus-to-detector distance (cm):			100			(+) = normal font = optimal field setting						
Tube voltage (kVp):			50			**(−) = bold font = suboptimal field setting**						
Total filtrations:			See last table row.			Empty table cells correspond to values < 0.01						
Organ or tissue	**Organ dose D_T : Entrance dose K_E**							**(mSv × mGy^{-1})**				
	(+)	(−)	(+)	(−)	(+)	(−)	(+)	(−)	(+)	(−)	(+)	(−)
Skull		**0.04**		**0.05**		**0.06**		**0.06**		**0.07**		**0.07**
Brain												**0.01**
Oral mucosa		**0.05**		**0.06**		**0.06**		**0.07**		**0.07**		**0.08**
Salivary glands		**0.05**		**0.06**		**0.06**		**0.07**		**0.07**		**0.08**
Cervical spine	0.05	**0.10**	0.07	**0.13**	0.08	**0.15**	0.08	**0.16**	0.09	**0.19**	0.10	**0.21**
Thyroid	0.07	**0.07**	0.08	**0.08**	0.08	**0.09**	0.09	**0.09**	0.09	**0.10**	0.10	**0.11**
Thymus	0.04	**0.04**	0.04	**0.05**	0.05	**0.05**	0.05	**0.06**	0.06	**0.07**	0.07	**0.07**
Thoracic spine	0.08	**0.09**	0.11	**0.11**	0.12	**0.13**	0.14	**0.14**	0.16	**0.17**	0.18	**0.19**
Claviculae	0.12	**0.12**	0.14	**0.15**	0.16	**0.16**	0.17	**0.18**	0.19	**0.20**	0.20	**0.22**
Scapulae	0.16	**0.16**	0.18	**0.19**	0.20	**0.20**	0.21	**0.21**	0.22	**0.23**	0.23	**0.24**
Ribs	0.14	**0.15**	0.17	**0.18**	0.19	**0.19**	0.20	**0.21**	0.22	**0.23**	0.24	**0.25**
Oesophagus	0.03	**0.03**	0.04	**0.04**	0.05	**0.05**	0.05	**0.06**	0.06	**0.07**	0.07	**0.07**
Trachea		**0.04**		**0.05**		**0.06**		**0.06**	0.01	**0.07**	0.01	**0.07**
Lung	0.04	**0.04**	0.05	**0.05**	0.05	**0.06**	0.06	**0.06**	0.07	**0.07**	0.07	**0.08**
Mamma	0.07	**0.07**	0.07	**0.07**	0.07	**0.07**	0.08	**0.08**	0.08	**0.08**	0.08	**0.08**
Heart	0.03	**0.03**	0.04	**0.04**	0.05	**0.05**	0.05	**0.05**	0.06	**0.06**	0.07	**0.07**
Stomach		**0.02**	0.01	**0.02**	0.02	**0.03**	0.02	**0.03**	0.02	**0.03**	0.02	**0.04**
Spleen		**0.01**	0.01	**0.02**	0.01	**0.02**	0.02	**0.02**	0.02	**0.03**	0.02	**0.03**
Liver	0.03	**0.05**	0.04	**0.06**	0.04	**0.07**	0.05	**0.08**	0.05	**0.09**	0.06	**0.09**
Gall bladder		**0.03**		**0.04**	0.01	**0.05**	0.01	**0.05**	0.01	**0.06**	0.02	**0.07**
Pancreas	0.02	**0.02**	0.02	**0.03**	0.02	**0.03**	0.03	**0.03**	0.03	**0.04**	0.04	**0.05**
Small intestine		**0.02**		**0.02**		**0.02**		**0.03**		**0.03**		**0.04**
Large intestine		**0.02**		**0.03**		**0.03**		**0.04**		**0.04**		**0.05**
Sigma/Rectum												
Lumbar spine	0.01	**0.08**	0.01	**0.11**	0.02	**0.12**	0.02	**0.14**	0.02	**0.16**	0.03	**0.18**
Pelvis		**0.01**		**0.02**		**0.02**		**0.02**		**0.03**		**0.03**
Adrenals	0.02	**0.03**	0.03	**0.03**	0.04	**0.04**	0.04	**0.04**	0.05	**0.05**	0.05	**0.06**
Kidneys	0.01	**0.04**	0.01	**0.04**	0.02	**0.05**	0.02	**0.05**	0.02	**0.06**	0.02	**0.07**
Urinary bladder												
Uterus												
Ovaries												
Prostate												
Testes												
Upper arm	0.16	**0.16**	0.18	**0.18**	0.20	**0.20**	0.21	**0.21**	0.22	**0.23**	0.24	**0.24**
Elbow	0.16	**0.17**	0.18	**0.19**	0.19	**0.20**	0.20	**0.21**	0.22	**0.23**	0.23	**0.24**
Forearm		**0.08**		**0.09**		**0.10**		**0.10**		**0.11**		**0.11**
Upper leg												
Knee												
Lower leg												
Lymph nodes	0.01	**0.03**	0.02	**0.04**	0.02	**0.04**	0.02	**0.05**	0.03	**0.05**	0.03	**0.06**
Skin	0.02	**0.03**	0.02	**0.03**	0.02	**0.04**	0.02	**0.04**	0.02	**0.04**	0.03	**0.04**
Muscle	0.01	**0.02**	0.01	**0.02**	0.02	**0.03**	0.02	**0.03**	0.02	**0.03**	0.02	**0.03**
Skeleton	0.05	**0.07**	0.06	**0.09**	0.07	**0.10**	0.07	**0.10**	0.08	**0.12**	0.09	**0.13**
Red bone marrow	0.01	**0.02**	0.01	**0.02**	0.02	**0.02**	0.02	**0.03**	0.02	**0.03**	0.02	**0.03**
Whole body dose	0.02	**0.03**	0.02	**0.03**	0.02	**0.04**	0.02	**0.04**	0.03	**0.04**	0.03	**0.05**
ED (ICRP 60)	0.02	**0.02**	0.02	**0.03**	0.02	**0.03**	0.03	**0.04**	0.03	**0.04**	0.03	**0.04**
ED (ICRP 103)	0.02	**0.03**	0.03	**0.04**	0.03	**0.04**	0.03	**0.04**	0.03	**0.05**	0.04	**0.05**
Total filtration (mm) **Al**	**1.5**		**2.5**		**3.5**		**4.5**		**3.5**		**3.5**	
Cu	**0**		**0**		**0**		**0**		**0.1**		**0.2**	

Table 9.63 Normalized organ doses—Chest r-lat—28th gestation week

Beam projection:			r-lat			**Table cells:**						
Focus-to-detector distance (cm):			100			(+) = normal font = optimal field setting						
Tube voltage (kVp):			60			**(−) = bold font = suboptimal field setting**						
Total filtrations:			See last table row.			Empty table cells correspond to values < 0.01						
Organ or tissue	**Organ dose D_T : Entrance dose K_E**							**($mSv \times mGy^{-1}$)**				
	(+)	(−)	(+)	(−)	(+)	(−)	(+)	(−)	(+)	(−)	(+)	(−)
Skull		**0.05**		**0.06**		**0.06**		**0.07**		**0.07**		**0.07**
Brain										**0.01**		**0.01**
Oral mucosa		**0.05**		**0.06**		**0.07**		**0.07**		**0.08**		**0.08**
Salivary glands		**0.06**		**0.06**		**0.07**		**0.07**		**0.08**		**0.08**
Cervical spine	0.06	**0.12**	0.08	**0.15**	0.09	**0.17**	0.09	**0.18**	0.10	**0.21**	0.10	**0.21**
Thyroid	0.07	**0.08**	0.08	**0.09**	0.09	**0.10**	0.09	**0.10**	0.10	**0.11**	0.10	**0.11**
Thymus	0.04	**0.04**	0.05	**0.05**	0.06	**0.06**	0.06	**0.06**	0.07	**0.07**	0.07	**0.07**
Thoracic spine	0.10	**0.11**	0.13	**0.13**	0.15	**0.15**	0.16	**0.17**	0.18	**0.19**	0.18	**0.19**
Claviculae	0.13	**0.14**	0.16	**0.16**	0.17	**0.18**	0.18	**0.19**	0.20	**0.21**	0.20	**0.22**
Scapulae	0.17	**0.18**	0.19	**0.20**	0.21	**0.21**	0.22	**0.22**	0.23	**0.24**	0.23	**0.24**
Ribs	0.16	**0.17**	0.19	**0.19**	0.20	**0.21**	0.21	**0.22**	0.23	**0.24**	0.24	**0.25**
Oesophagus	0.04	**0.04**	0.05	**0.05**	0.06	**0.06**	0.06	**0.07**	0.07	**0.08**	0.07	**0.07**
Trachea		**0.05**		**0.06**	0.01	**0.06**	0.01	**0.07**	0.01	**0.07**	0.01	**0.07**
Lung	0.05	**0.05**	0.06	**0.06**	0.06	**0.06**	0.07	**0.07**	0.07	**0.08**	0.07	**0.08**
Mamma	0.07	**0.07**	0.07	**0.07**	0.08	**0.08**	0.08	**0.08**	0.08	**0.08**	0.08	**0.08**
Heart	0.04	**0.04**	0.05	**0.05**	0.06	**0.06**	0.06	**0.06**	0.07	**0.07**	0.07	**0.07**
Stomach	0.01	**0.02**	0.02	**0.03**	0.02	**0.03**	0.02	**0.03**	0.02	**0.04**	0.02	**0.04**
Spleen	0.01	**0.02**	0.02	**0.02**	0.02	**0.02**	0.02	**0.03**	0.03	**0.03**	0.02	**0.03**
Liver	0.04	**0.06**	0.04	**0.07**	0.05	**0.08**	0.05	**0.09**	0.06	**0.09**	0.06	**0.09**
Gall bladder		**0.04**	0.01	**0.05**	0.01	**0.06**	0.01	**0.06**	0.02	**0.07**	0.02	**0.07**
Pancreas	0.02	**0.03**	0.03	**0.03**	0.03	**0.04**	0.03	**0.04**	0.04	**0.05**	0.04	**0.05**
Small intestine		**0.02**		**0.03**		**0.03**		**0.03**		**0.04**		**0.04**
Large intestine		**0.03**		**0.04**		**0.04**		**0.04**		**0.05**		**0.05**
Sigma/Rectum												
Lumbar spine	0.01	**0.10**	0.02	**0.13**	0.02	**0.14**	0.02	**0.16**	0.03	**0.18**	0.03	**0.18**
Pelvis		**0.02**		**0.02**		**0.02**		**0.03**		**0.03**		**0.03**
Adrenals	0.03	**0.03**	0.04	**0.04**	0.04	**0.05**	0.05	**0.05**	0.05	**0.06**	0.05	**0.06**
Kidneys	0.01	**0.04**	0.02	**0.05**	0.02	**0.06**	0.02	**0.06**	0.02	**0.07**	0.02	**0.07**
Urinary bladder												
Uterus												
Ovaries												
Prostate												
Testes												
Upper arm	0.17	**0.18**	0.19	**0.20**	0.21	**0.21**	0.22	**0.22**	0.23	**0.24**	0.24	**0.24**
Elbow	0.17	**0.18**	0.19	**0.20**	0.21	**0.22**	0.21	**0.23**	0.23	**0.24**	0.23	**0.24**
Forearm		**0.09**		**0.10**		**0.10**		**0.11**		**0.11**		**0.11**
Upper leg												
Knee												
Lower leg												
Lymph nodes	0.02	**0.04**	0.02	**0.04**	0.02	**0.05**	0.03	**0.05**	0.03	**0.06**	0.03	**0.06**
Skin	0.02	**0.03**	0.02	**0.04**	0.02	**0.04**	0.02	**0.04**	0.03	**0.04**	0.03	**0.04**
Muscle	0.01	**0.02**	0.02	**0.03**	0.02	**0.03**	0.02	**0.03**	0.02	**0.04**	0.02	**0.03**
Skeleton	0.06	**0.08**	0.07	**0.10**	0.08	**0.11**	0.08	**0.11**	0.09	**0.12**	0.09	**0.13**
Red bone marrow	0.01	**0.02**	0.02	**0.02**	0.02	**0.03**	0.02	**0.03**	0.02	**0.03**	0.02	**0.03**
Whole body dose	0.02	**0.03**	0.02	**0.04**	0.02	**0.04**	0.03	**0.04**	0.03	**0.05**	0.03	**0.05**
ED (ICRP 60)	0.02	**0.03**	0.02	**0.03**	0.03	**0.04**	0.03	**0.04**	0.03	**0.04**	0.03	**0.04**
ED (ICRP 103)	0.03	**0.03**	0.03	**0.04**	0.03	**0.04**	0.03	**0.05**	0.04	**0.05**	0.04	**0.05**
Total filtration (mm) **Al**	**1.5**		**2.5**		**3.5**		**4.5**		**3.5**		**3.5**	
Cu	**0**		**0**		**0**		**0**		**0.1**		**0.2**	

Table 9.64 Normalized organ doses—Chest r-lat—28th gestation week

Beam projection:			r-lat		**Table cells:**							
Focus-to-detector distance (cm):			100		(+) = normal font = optimal field setting							
Tube voltage (kVp):			70		**(−) = bold font = suboptimal field setting**							
Total filtrations:			See last table row.		Empty table cells correspond to values < 0.01							
Organ or tissue	**Organ dose D_T : Entrance dose K_E**						**(mSv × mGy⁻¹)**					
	(+)	(−)	(+)	(−)	(+)	(−)	(+)	(−)	(+)	(−)	(+)	(−)
Skull		**0.05**		**0.06**		**0.07**		**0.07**		**0.08**		**0.08**
Brain										**0.01**		**0.01**
Oral mucosa		**0.06**		**0.07**		**0.07**		**0.08**		**0.08**		**0.09**
Salivary glands		**0.06**		**0.07**		**0.07**		**0.08**		**0.08**		**0.08**
Cervical spine	0.07	**0.14**	0.08	**0.16**	0.09	**0.18**	0.10	**0.19**	0.11	**0.22**	0.11	**0.23**
Thyroid	0.08	**0.08**	0.09	**0.09**	0.09	**0.10**	0.10	**0.11**	0.10	**0.11**	0.11	**0.12**
Thymus	0.05	**0.05**	0.06	**0.06**	0.06	**0.07**	0.07	**0.07**	0.08	**0.08**	0.08	**0.09**
Thoracic spine	0.12	**0.12**	0.14	**0.15**	0.16	**0.17**	0.17	**0.18**	0.20	**0.21**	0.21	**0.23**
Claviculae	0.14	**0.15**	0.17	**0.18**	0.18	**0.19**	0.19	**0.20**	0.21	**0.22**	0.22	**0.23**
Scapulae	0.18	**0.19**	0.20	**0.21**	0.21	**0.22**	0.22	**0.23**	0.24	**0.24**	0.24	**0.25**
Ribs	0.17	**0.18**	0.20	**0.20**	0.21	**0.22**	0.22	**0.23**	0.24	**0.25**	0.25	**0.26**
Oesophagus	0.05	**0.05**	0.06	**0.06**	0.06	**0.07**	0.07	**0.07**	0.08	**0.08**	0.09	**0.09**
Trachea		**0.05**		**0.06**	0.01	**0.07**	0.01	**0.07**	0.01	**0.08**	0.01	**0.08**
Lung	0.05	**0.05**	0.06	**0.06**	0.07	**0.07**	0.07	**0.07**	0.08	**0.08**	0.09	**0.09**
Mamma	0.07	**0.07**	0.08	**0.08**	0.08	**0.08**	0.08	**0.08**	0.09	**0.09**	0.09	**0.09**
Heart	0.05	**0.05**	0.06	**0.06**	0.06	**0.06**	0.07	**0.07**	0.07	**0.08**	0.08	**0.09**
Stomach	0.01	**0.03**	0.02	**0.03**	0.02	**0.04**	0.02	**0.04**	0.03	**0.05**	0.03	**0.05**
Spleen	0.01	**0.02**	0.02	**0.02**	0.02	**0.03**	0.02	**0.03**	0.03	**0.04**	0.03	**0.04**
Liver	0.04	**0.07**	0.05	**0.08**	0.05	**0.09**	0.06	**0.09**	0.06	**0.10**	0.07	**0.11**
Gall bladder	0.01	**0.05**	0.01	**0.06**	0.01	**0.06**	0.02	**0.07**	0.02	**0.08**	0.02	**0.08**
Pancreas	0.02	**0.03**	0.03	**0.04**	0.04	**0.04**	0.04	**0.05**	0.05	**0.06**	0.05	**0.06**
Small intestine		**0.02**		**0.03**		**0.03**		**0.04**		**0.04**		**0.05**
Large intestine		**0.03**		**0.04**		**0.04**		**0.05**		**0.05**		**0.06**
Sigma/Rectum												**0.01**
Lumbar spine	0.02	**0.12**	0.02	**0.14**	0.02	**0.16**	0.03	**0.17**	0.03	**0.19**	0.04	**0.21**
Pelvis		**0.02**		**0.02**		**0.03**		**0.03**		**0.03**		**0.04**
Adrenals	0.04	**0.04**	0.04	**0.05**	0.05	**0.05**	0.05	**0.06**	0.06	**0.07**	0.07	**0.07**
Kidneys	0.01	**0.05**	0.02	**0.06**	0.02	**0.06**	0.02	**0.07**	0.02	**0.07**	0.03	**0.08**
Urinary bladder												
Uterus												
Ovaries										**0.01**		**0.01**
Prostate												
Testes												
Upper arm	0.18	**0.18**	0.20	**0.21**	0.21	**0.22**	0.22	**0.23**	0.24	**0.24**	0.24	**0.25**
Elbow	0.18	**0.19**	0.20	**0.21**	0.21	**0.22**	0.22	**0.23**	0.23	**0.25**	0.24	**0.25**
Forearm		**0.09**		**0.10**		**0.10**		**0.11**		**0.11**	0.01	**0.12**
Upper leg												
Knee												
Lower leg												
Lymph nodes	0.02	**0.04**	0.02	**0.05**	0.03	**0.05**	0.03	**0.06**	0.03	**0.06**	0.03	**0.07**
Skin	0.02	**0.04**	0.02	**0.04**	0.02	**0.04**	0.03	**0.04**	0.03	**0.04**	0.03	**0.04**
Muscle	0.01	**0.03**	0.02	**0.03**	0.02	**0.03**	0.02	**0.03**	0.02	**0.04**	0.02	**0.04**
Skeleton	0.06	**0.09**	0.07	**0.10**	0.08	**0.11**	0.08	**0.12**	0.09	**0.13**	0.10	**0.14**
Red bone marrow	0.01	**0.02**	0.02	**0.03**	0.02	**0.03**	0.02	**0.03**	0.02	**0.04**	0.02	**0.04**
Whole body dose	0.02	**0.03**	0.02	**0.04**	0.03	**0.04**	0.03	**0.04**	0.03	**0.05**	0.03	**0.05**
ED (ICRP 60)	0.02	**0.03**	0.03	**0.04**	0.03	**0.04**	0.03	**0.04**	0.04	**0.05**	0.04	**0.05**
ED (ICRP 103)	0.03	**0.04**	0.03	**0.04**	0.03	**0.05**	0.04	**0.05**	0.04	**0.06**	0.04	**0.06**
Total filtration (mm) Al	**1.5**		**2.5**		**3.5**		**4.5**		**3.5**		**3.5**	
Cu	**0**		**0**		**0**		**0**		**0.1**		**0.2**	

Table 9.65 Normalized organ doses—Chest r-lat—28th gestation week

Beam projection:			r-lat				**Table cells:**					
Focus-to-detector distance (cm):			100				(+) = normal font = optimal field setting					
Tube voltage (kVp):			80				**(−) = bold font = suboptimal field setting**					
Total filtrations:			See last table row.				Empty table cells correspond to values < 0.01					
Organ or tissue	**Organ dose D_T : Entrance dose K_E**							**($mSv \times mGy^{-1}$)**				
	(+)	(−)	(+)	(−)	(+)	(−)	(+)	(−)	(+)	(−)	(+)	(−)
Skull		**0.06**		**0.06**		**0.07**		**0.07**		**0.08**		**0.08**
Brain								**0.01**		**0.01**		**0.01**
Oral mucosa		**0.06**		**0.07**		**0.08**		**0.08**		**0.09**		**0.09**
Salivary glands		**0.06**		**0.07**		**0.08**		**0.08**		**0.08**		**0.09**
Cervical spine	0.08	**0.15**	0.09	**0.17**	0.09	**0.19**	0.10	**0.20**	0.11	**0.22**	0.11	**0.23**
Thyroid	0.08	**0.09**	0.09	**0.10**	0.10	**0.10**	0.10	**0.11**	0.11	**0.12**	0.11	**0.12**
Thymus	0.05	**0.05**	0.06	**0.06**	0.07	**0.07**	0.07	**0.08**	0.08	**0.08**	0.09	**0.09**
Thoracic spine	0.13	**0.13**	0.15	**0.16**	0.17	**0.18**	0.18	**0.19**	0.20	**0.21**	0.22	**0.23**
Claviculae	0.15	**0.16**	0.17	**0.18**	0.19	**0.20**	0.20	**0.21**	0.21	**0.22**	0.22	**0.23**
Scapulae	0.19	**0.19**	0.21	**0.21**	0.22	**0.22**	0.22	**0.23**	0.23	**0.24**	0.24	**0.25**
Ribs	0.18	**0.19**	0.20	**0.21**	0.22	**0.23**	0.23	**0.23**	0.24	**0.25**	0.25	**0.26**
Oesophagus	0.05	**0.06**	0.06	**0.07**	0.07	**0.07**	0.08	**0.08**	0.08	**0.09**	0.09	**0.10**
Trachea		**0.06**	0.01	**0.07**	0.01	**0.07**	0.01	**0.08**	0.01	**0.08**	0.01	**0.09**
Lung	0.06	**0.06**	0.07	**0.07**	0.07	**0.07**	0.08	**0.08**	0.08	**0.09**	0.09	**0.09**
Mamma	0.07	**0.07**	0.08	**0.08**	0.08	**0.08**	0.09	**0.09**	0.09	**0.09**	0.09	**0.09**
Heart	0.05	**0.05**	0.06	**0.06**	0.07	**0.07**	0.07	**0.07**	0.08	**0.08**	0.09	**0.09**
Stomach	0.02	**0.03**	0.02	**0.04**	0.02	**0.04**	0.03	**0.04**	0.03	**0.05**	0.03	**0.06**
Spleen	0.02	**0.02**	0.02	**0.03**	0.02	**0.03**	0.03	**0.04**	0.03	**0.04**	0.04	**0.05**
Liver	0.04	**0.07**	0.05	**0.08**	0.06	**0.09**	0.06	**0.10**	0.07	**0.11**	0.07	**0.11**
Gall bladder	0.01	**0.05**	0.01	**0.06**	0.02	**0.07**	0.02	**0.07**	0.02	**0.08**	0.02	**0.09**
Pancreas	0.03	**0.04**	0.04	**0.04**	0.04	**0.05**	0.04	**0.05**	0.05	**0.06**	0.06	**0.07**
Small intestine		**0.03**		**0.03**		**0.04**		**0.04**		**0.04**		**0.05**
Large intestine		**0.04**		**0.04**		**0.05**		**0.05**		**0.06**		**0.06**
Sigma/Rectum										**0.01**		**0.01**
Lumbar spine	0.02	**0.13**	0.02	**0.15**	0.03	**0.17**	0.03	**0.18**	0.03	**0.20**	0.04	**0.22**
Pelvis		**0.02**		**0.03**		**0.03**		**0.03**		**0.04**		**0.04**
Adrenals	0.04	**0.04**	0.05	**0.05**	0.05	**0.06**	0.06	**0.06**	0.06	**0.07**	0.07	**0.08**
Kidneys	0.02	**0.05**	0.02	**0.06**	0.02	**0.07**	0.02	**0.07**	0.03	**0.08**	0.03	**0.08**
Urinary bladder												
Uterus												**0.01**
Ovaries								**0.01**		**0.01**		**0.01**
Prostate												
Testes												
Upper arm	0.19	**0.19**	0.21	**0.21**	0.22	**0.22**	0.22	**0.23**	0.23	**0.24**	0.24	**0.25**
Elbow	0.19	**0.20**	0.20	**0.22**	0.21	**0.23**	0.22	**0.23**	0.23	**0.24**	0.23	**0.25**
Forearm		**0.09**		**0.10**		**0.11**		**0.11**		**0.11**	0.01	**0.12**
Upper leg												
Knee												
Lower leg												
Lymph nodes	0.02	**0.05**	0.03	**0.05**	0.03	**0.06**	0.03	**0.06**	0.03	**0.07**	0.04	**0.07**
Skin	0.02	**0.04**	0.02	**0.04**	0.03	**0.04**	0.03	**0.04**	0.03	**0.04**	0.03	**0.05**
Muscle	0.02	**0.03**	0.02	**0.03**	0.02	**0.03**	0.02	**0.04**	0.02	**0.04**	0.02	**0.04**
Skeleton	0.07	**0.09**	0.08	**0.11**	0.08	**0.12**	0.09	**0.12**	0.09	**0.13**	0.09	**0.13**
Red bone marrow	0.02	**0.03**	0.02	**0.03**	0.02	**0.03**	0.02	**0.03**	0.02	**0.04**	0.03	**0.04**
Whole body dose	0.02	**0.04**	0.03	**0.04**	0.03	**0.04**	0.03	**0.05**	0.03	**0.05**	0.03	**0.05**
ED (ICRP 60)	0.03	**0.03**	0.03	**0.04**	0.03	**0.04**	0.03	**0.05**	0.04	**0.05**	0.04	**0.05**
ED (ICRP 103)	0.03	**0.04**	0.03	**0.05**	0.04	**0.05**	0.04	**0.05**	0.04	**0.06**	0.05	**0.06**
Total filtration (mm) Al	**1.5**		**2.5**		**3.5**		**4.5**		**3.5**		**3.5**	
Cu	**0**		**0**		**0**		**0**		**0.1**		**0.2**	

Table 9.66 Normalized organ doses—Chest r-lat—< 1 year

Beam projection:			r-lat			**Table cells:**						
Focus-to-detector distance (cm):			100			(+) = normal font = optimal field setting						
Tube voltage (kVp):			50			**(−) = bold font = suboptimal field setting**						
Total filtrations:			See last table row.			Empty table cells correspond to values < 0.01						
Organ or tissue	**Organ dose D_T : Entrance dose K_E**							**(mSv × mGy^{-1})**				
	(+)	(−)	(+)	(−)	(+)	(−)	(+)	(−)	(+)	(−)	(+)	(−)
Skull	0.01	**0.20**	0.02	**0.24**	0.02	**0.28**	0.02	**0.30**	0.03	**0.35**	0.03	**0.39**
Brain		**0.02**		**0.03**		**0.03**		**0.04**		**0.05**		**0.06**
Oral mucosa	0.02	**0.36**	0.02	**0.44**	0.03	**0.49**	0.03	**0.52**	0.04	**0.59**	0.04	**0.65**
Salivary glands	0.03	**0.47**	0.04	**0.54**	0.05	**0.58**	0.06	**0.61**	0.07	**0.66**	0.08	**0.71**
Cervical spine	0.33	**0.73**	0.42	**0.93**	0.49	**1.09**	0.54	**1.21**	0.64	**1.43**	0.73	**1.63**
Thyroid	0.44	**0.50**	0.52	**0.60**	0.58	**0.67**	0.62	**0.72**	0.69	**0.82**	0.75	**0.90**
Thymus	0.21	**0.22**	0.27	**0.29**	0.31	**0.33**	0.35	**0.37**	0.41	**0.44**	0.47	**0.51**
Thoracic spine	0.48	**0.51**	0.63	**0.67**	0.75	**0.80**	0.84	**0.90**	1.01	**1.09**	1.18	**1.28**
Claviculae	0.80	**0.87**	0.97	**1.06**	1.09	**1.19**	1.18	**1.30**	1.33	**1.47**	1.47	**1.63**
Scapulae	1.53	**1.60**	1.76	**1.84**	1.90	**1.99**	2.01	**2.10**	2.17	**2.28**	2.31	**2.43**
Ribs	1.11	**1.18**	1.30	**1.39**	1.43	**1.52**	1.52	**1.62**	1.67	**1.79**	1.80	**1.93**
Oesophagus	0.25	**0.26**	0.32	**0.34**	0.37	**0.40**	0.41	**0.44**	0.49	**0.52**	0.56	**0.60**
Trachea	0.05	**0.29**	0.06	**0.36**	0.07	**0.40**	0.08	**0.44**	0.09	**0.51**	0.10	**0.56**
Lung	0.37	**0.39**	0.45	**0.47**	0.50	**0.53**	0.55	**0.58**	0.62	**0.66**	0.68	**0.73**
Mamma	0.55	**0.56**	0.59	**0.61**	0.62	**0.63**	0.64	**0.65**	0.67	**0.68**	0.70	**0.71**
Heart	0.23	**0.24**	0.30	**0.31**	0.34	**0.36**	0.38	**0.41**	0.46	**0.48**	0.52	**0.56**
Stomach	0.05	**0.09**	0.07	**0.12**	0.09	**0.15**	0.10	**0.17**	0.12	**0.21**	0.15	**0.25**
Spleen	0.05	**0.06**	0.06	**0.09**	0.08	**0.10**	0.09	**0.12**	0.11	**0.15**	0.14	**0.19**
Liver	0.26	**0.49**	0.32	**0.60**	0.36	**0.67**	0.39	**0.73**	0.45	**0.82**	0.50	**0.90**
Gall bladder	0.07	**0.20**	0.09	**0.26**	0.11	**0.31**	0.12	**0.35**	0.15	**0.42**	0.17	**0.49**
Pancreas	0.12	**0.14**	0.15	**0.19**	0.18	**0.23**	0.21	**0.26**	0.26	**0.32**	0.30	**0.38**
Small intestine		**0.13**	0.01	**0.17**	0.01	**0.20**	0.02	**0.22**	0.02	**0.26**	0.03	**0.31**
Large intestine	0.01	**0.19**	0.01	**0.24**	0.02	**0.28**	0.02	**0.31**	0.03	**0.36**	0.03	**0.41**
Sigma/Rectum		**0.02**		**0.03**		**0.03**		**0.04**		**0.05**		**0.06**
Lumbar spine	0.06	**0.43**	0.08	**0.57**	0.10	**0.68**	0.11	**0.77**	0.14	**0.93**	0.17	**1.09**
Pelvis		**0.09**		**0.11**	0.01	**0.13**	0.01	**0.15**	0.02	**0.19**	0.02	**0.22**
Adrenals	0.16	**0.18**	0.21	**0.24**	0.24	**0.28**	0.27	**0.31**	0.33	**0.38**	0.38	**0.43**
Kidneys	0.06	**0.24**	0.08	**0.30**	0.09	**0.35**	0.11	**0.38**	0.13	**0.45**	0.15	**0.51**
Urinary bladder				**0.01**		**0.01**		**0.02**		**0.02**		**0.03**
Uterus		**0.02**		**0.03**		**0.04**		**0.04**		**0.05**	0.01	**0.06**
Ovaries		**0.02**		**0.03**		**0.04**		**0.04**		**0.06**		**0.07**
Prostate										**0.01**		**0.01**
Testes												
Upper arm												
Elbow												
Forearm												
Upper leg												**0.01**
Knee												
Lower leg												
Lymph nodes	0.11	**0.23**	0.14	**0.29**	0.16	**0.33**	0.17	**0.37**	0.20	**0.42**	0.23	**0.48**
Skin	0.18	**0.29**	0.19	**0.31**	0.19	**0.32**	0.20	**0.33**	0.21	**0.34**	0.22	**0.36**
Muscle	0.10	**0.17**	0.12	**0.20**	0.13	**0.23**	0.14	**0.25**	0.16	**0.28**	0.18	**0.31**
Skeleton	0.28	**0.38**	0.34	**0.46**	0.38	**0.52**	0.41	**0.57**	0.46	**0.65**	0.51	**0.72**
Red bone marrow	0.07	**0.11**	0.09	**0.13**	0.10	**0.15**	0.11	**0.16**	0.13	**0.19**	0.14	**0.21**
Whole body dose	0.11	**0.19**	0.14	**0.23**	0.15	**0.26**	0.17	**0.28**	0.19	**0.32**	0.21	**0.35**
ED (ICRP 60)	0.14	**0.19**	0.17	**0.23**	0.19	**0.26**	0.21	**0.28**	0.23	**0.32**	0.26	**0.36**
ED (ICRP 103)	0.18	**0.24**	0.21	**0.28**	0.23	**0.31**	0.25	**0.34**	0.28	**0.38**	0.31	**0.42**
Total filtration (mm) Al	**1.5**		**2.5**		**3.5**		**4.5**		**3.5**		**3.5**	
Cu	**0**		**0**		**0**		**0**		**0.1**		**0.2**	

Table 9.67 Normalized organ doses—Chest r-lat—< 1 year

Beam projection:			r-lat				**Table cells:**					
Focus-to-detector distance (cm):			100				(+) = normal font = optimal field setting					
Tube voltage (kVp):			60				**(−) = bold font = suboptimal field setting**					
Total filtrations:			See last table row.				Empty table cells correspond to values < 0.01					
Organ or tissue	**Organ dose D_T : Entrance dose K_E**								**(mSv × mGy^{-1})**			
	(+)	(−)	(+)	(−)	(+)	(−)	(+)	(−)	(+)	(−)	(+)	(−)
Skull	0.02	**0.23**	0.02	**0.28**	0.03	**0.32**	0.03	**0.35**	0.04	**0.39**	0.03	**0.39**
Brain		**0.03**		**0.04**		**0.04**		**0.05**		**0.06**		**0.06**
Oral mucosa	0.02	**0.42**	0.03	**0.50**	0.04	**0.55**	0.04	**0.59**	0.05	**0.66**	0.04	**0.65**
Salivary glands	0.04	**0.52**	0.05	**0.59**	0.06	**0.63**	0.07	**0.66**	0.08	**0.72**	0.08	**0.71**
Cervical spine	0.41	**0.90**	0.51	**1.13**	0.58	**1.30**	0.64	**1.43**	0.74	**1.66**	0.73	**1.63**
Thyroid	0.50	**0.57**	0.58	**0.68**	0.64	**0.76**	0.69	**0.81**	0.76	**0.91**	0.75	**0.90**
Thymus	0.26	**0.28**	0.33	**0.35**	0.38	**0.41**	0.42	**0.45**	0.49	**0.53**	0.47	**0.51**
Thoracic spine	0.62	**0.66**	0.79	**0.85**	0.92	**1.00**	1.03	**1.11**	1.22	**1.33**	1.18	**1.28**
Claviculae	0.92	**1.01**	1.10	**1.21**	1.22	**1.35**	1.31	**1.45**	1.47	**1.63**	1.47	**1.63**
Scapulae	1.67	**1.75**	1.90	**1.99**	2.04	**2.14**	2.14	**2.25**	2.30	**2.42**	2.31	**2.43**
Ribs	1.24	**1.32**	1.43	**1.53**	1.55	**1.66**	1.64	**1.76**	1.79	**1.93**	1.80	**1.93**
Oesophagus	0.31	**0.33**	0.39	**0.42**	0.45	**0.48**	0.49	**0.53**	0.58	**0.62**	0.56	**0.60**
Trachea	0.06	**0.34**	0.08	**0.42**	0.09	**0.47**	0.09	**0.51**	0.11	**0.58**	0.10	**0.56**
Lung	0.43	**0.46**	0.52	**0.55**	0.57	**0.61**	0.62	**0.66**	0.69	**0.74**	0.68	**0.73**
Mamma	0.58	**0.59**	0.62	**0.64**	0.65	**0.67**	0.67	**0.69**	0.71	**0.73**	0.70	**0.71**
Heart	0.29	**0.31**	0.36	**0.39**	0.42	**0.44**	0.46	**0.49**	0.54	**0.58**	0.52	**0.56**
Stomach	0.07	**0.13**	0.10	**0.17**	0.12	**0.20**	0.13	**0.23**	0.16	**0.28**	0.15	**0.25**
Spleen	0.07	**0.09**	0.09	**0.12**	0.11	**0.15**	0.12	**0.17**	0.15	**0.21**	0.14	**0.19**
Liver	0.31	**0.57**	0.37	**0.68**	0.41	**0.76**	0.45	**0.82**	0.51	**0.92**	0.50	**0.90**
Gall bladder	0.09	**0.26**	0.12	**0.33**	0.14	**0.39**	0.15	**0.43**	0.18	**0.52**	0.17	**0.49**
Pancreas	0.16	**0.20**	0.20	**0.25**	0.24	**0.30**	0.26	**0.34**	0.32	**0.41**	0.30	**0.38**
Small intestine	0.01	**0.17**	0.02	**0.21**	0.02	**0.24**	0.02	**0.27**	0.03	**0.32**	0.03	**0.31**
Large intestine	0.02	**0.23**	0.02	**0.29**	0.03	**0.33**	0.03	**0.37**	0.04	**0.43**	0.03	**0.41**
Sigma/Rectum		**0.03**		**0.04**		**0.04**		**0.05**		**0.06**		**0.06**
Lumbar spine	0.09	**0.56**	0.11	**0.73**	0.14	**0.85**	0.15	**0.95**	0.19	**1.14**	0.17	**1.09**
Pelvis	0.01	**0.11**	0.01	**0.15**	0.02	**0.17**	0.02	**0.19**	0.03	**0.24**	0.02	**0.22**
Adrenals	0.21	**0.24**	0.26	**0.30**	0.30	**0.35**	0.33	**0.39**	0.39	**0.46**	0.38	**0.43**
Kidneys	0.08	**0.30**	0.10	**0.37**	0.12	**0.42**	0.13	**0.46**	0.16	**0.53**	0.15	**0.51**
Urinary bladder		**0.01**		**0.02**		**0.02**		**0.02**		**0.03**		**0.03**
Uterus		**0.03**		**0.04**		**0.05**		**0.06**	0.01	**0.07**	0.01	**0.06**
Ovaries		**0.03**		**0.04**		**0.05**		**0.06**	0.01	**0.07**		**0.07**
Prostate								**0.01**		**0.01**		**0.01**
Testes												
Upper arm												
Elbow												
Forearm												
Upper leg						**0.01**		**0.01**		**0.02**		**0.01**
Knee												
Lower leg												
Lymph nodes	0.13	**0.28**	0.17	**0.35**	0.19	**0.39**	0.21	**0.43**	0.24	**0.50**	0.23	**0.48**
Skin	0.18	**0.30**	0.20	**0.32**	0.20	**0.34**	0.21	**0.35**	0.22	**0.37**	0.22	**0.36**
Muscle	0.11	**0.20**	0.13	**0.24**	0.15	**0.26**	0.16	**0.28**	0.18	**0.32**	0.18	**0.31**
Skeleton	0.32	**0.45**	0.39	**0.54**	0.43	**0.60**	0.46	**0.65**	0.51	**0.73**	0.51	**0.72**
Red bone marrow	0.09	**0.13**	0.10	**0.16**	0.12	**0.18**	0.13	**0.19**	0.15	**0.22**	0.14	**0.21**
Whole body dose	0.13	**0.22**	0.16	**0.27**	0.18	**0.30**	0.19	**0.32**	0.21	**0.36**	0.21	**0.35**
ED (ICRP 60)	0.17	**0.22**	0.20	**0.27**	0.22	**0.30**	0.24	**0.33**	0.27	**0.37**	0.26	**0.36**
ED (ICRP 103)	0.20	**0.27**	0.24	**0.32**	0.26	**0.36**	0.28	**0.39**	0.31	**0.44**	0.31	**0.42**
Total filtration (mm) **Al**	**1.5**		**2.5**		**3.5**		**4.5**		**3.5**		**3.5**	
Cu	**0**		**0**		**0**		**0**		**0.1**		**0.2**	

Table 9.68 Normalized organ doses—Chest r-lat—< 1 year

Beam projection:		r-lat			**Table cells:**							
Focus-to-detector distance (cm):		100			(+) = normal font = optimal field setting							
Tube voltage (kVp):		70			**(−) = bold font = suboptimal field setting**							
Total filtrations:		See last table row.			Empty table cells correspond to values < 0.01							
Organ or tissue	**Organ dose D_T : Entrance dose K_E**							**(mSv × mGy^{-1})**				
	(+)	(−)	(+)	(−)	(+)	(−)	(+)	(−)	(+)	(−)	(+)	(−)
Skull	0.02	**0.26**	0.03	**0.31**	0.03	**0.35**	0.04	**0.38**	0.04	**0.42**	0.05	**0.46**
Brain		**0.04**		**0.04**		**0.05**		**0.06**	0.01	**0.07**	0.01	**0.08**
Oral mucosa	0.03	**0.47**	0.04	**0.54**	0.04	**0.60**	0.05	**0.64**	0.06	**0.71**	0.06	**0.76**
Salivary glands	0.05	**0.56**	0.06	**0.63**	0.07	**0.67**	0.08	**0.70**	0.09	**0.76**	0.10	**0.81**
Cervical spine	0.47	**1.04**	0.57	**1.27**	0.64	**1.44**	0.70	**1.57**	0.80	**1.79**	0.88	**1.98**
Thyroid	0.55	**0.64**	0.63	**0.75**	0.69	**0.82**	0.74	**0.88**	0.81	**0.97**	0.87	**1.05**
Thymus	0.31	**0.33**	0.38	**0.41**	0.43	**0.46**	0.47	**0.51**	0.54	**0.59**	0.61	**0.66**
Thoracic spine	0.73	**0.79**	0.91	**0.99**	1.05	**1.14**	1.15	**1.26**	1.35	**1.47**	1.51	**1.66**
Claviculae	1.01	**1.11**	1.19	**1.31**	1.30	**1.45**	1.39	**1.55**	1.53	**1.71**	1.64	**1.84**
Scapulae	1.77	**1.86**	1.98	**2.09**	2.11	**2.23**	2.20	**2.32**	2.35	**2.48**	2.44	**2.59**
Ribs	1.33	**1.42**	1.51	**1.62**	1.62	**1.75**	1.71	**1.84**	1.84	**1.99**	1.94	**2.11**
Oesophagus	0.36	**0.39**	0.45	**0.48**	0.51	**0.55**	0.55	**0.60**	0.64	**0.69**	0.71	**0.78**
Trachea	0.07	**0.39**	0.09	**0.46**	0.10	**0.51**	0.10	**0.55**	0.12	**0.62**	0.13	**0.68**
Lung	0.48	**0.51**	0.57	**0.60**	0.62	**0.67**	0.67	**0.71**	0.74	**0.80**	0.81	**0.87**
Mamma	0.60	**0.62**	0.65	**0.67**	0.68	**0.70**	0.70	**0.72**	0.73	**0.76**	0.76	**0.80**
Heart	0.34	**0.36**	0.42	**0.44**	0.47	**0.51**	0.52	**0.56**	0.60	**0.64**	0.67	**0.72**
Stomach	0.09	**0.16**	0.12	**0.21**	0.14	**0.24**	0.16	**0.27**	0.19	**0.33**	0.22	**0.38**
Spleen	0.09	**0.12**	0.11	**0.15**	0.13	**0.18**	0.15	**0.20**	0.18	**0.25**	0.22	**0.30**
Liver	0.35	**0.64**	0.41	**0.75**	0.45	**0.83**	0.49	**0.88**	0.55	**0.98**	0.59	**1.07**
Gall bladder	0.11	**0.31**	0.14	**0.39**	0.16	**0.45**	0.18	**0.49**	0.21	**0.58**	0.24	**0.66**
Pancreas	0.19	**0.24**	0.24	**0.31**	0.28	**0.36**	0.31	**0.40**	0.37	**0.47**	0.42	**0.54**
Small intestine	0.02	**0.20**	0.02	**0.25**	0.03	**0.28**	0.03	**0.31**	0.04	**0.36**	0.04	**0.41**
Large intestine	0.02	**0.27**	0.03	**0.33**	0.03	**0.38**	0.04	**0.41**	0.04	**0.47**	0.05	**0.53**
Sigma/Rectum		**0.04**		**0.05**		**0.05**		**0.06**	0.01	**0.08**	0.01	**0.09**
Lumbar spine	0.11	**0.68**	0.14	**0.85**	0.16	**0.98**	0.18	**1.08**	0.22	**1.28**	0.26	**1.44**
Pelvis	0.01	**0.14**	0.02	**0.18**	0.02	**0.20**	0.03	**0.23**	0.03	**0.27**	0.04	**0.31**
Adrenals	0.24	**0.28**	0.30	**0.35**	0.34	**0.40**	0.38	**0.44**	0.44	**0.52**	0.50	**0.58**
Kidneys	0.10	**0.34**	0.12	**0.42**	0.14	**0.47**	0.15	**0.51**	0.18	**0.59**	0.20	**0.66**
Urinary bladder		**0.02**		**0.02**		**0.03**		**0.03**		**0.04**		**0.04**
Uterus		**0.04**		**0.05**	0.01	**0.06**	0.01	**0.07**	0.02	**0.09**	0.02	**0.10**
Ovaries		**0.04**		**0.06**	0.01	**0.07**	0.01	**0.07**	0.02	**0.09**	0.02	**0.10**
Prostate				**0.01**		**0.01**		**0.01**		**0.02**		**0.02**
Testes										**0.01**		**0.01**
Upper arm												
Elbow												
Forearm												
Upper leg				**0.01**		**0.01**		**0.02**		**0.02**		**0.03**
Knee												
Lower leg												
Lymph nodes	0.16	**0.33**	0.19	**0.39**	0.21	**0.44**	0.23	**0.48**	0.27	**0.55**	0.30	**0.61**
Skin	0.19	**0.32**	0.20	**0.34**	0.21	**0.35**	0.22	**0.36**	0.23	**0.38**	0.24	**0.40**
Muscle	0.13	**0.22**	0.15	**0.26**	0.16	**0.29**	0.18	**0.31**	0.20	**0.35**	0.22	**0.38**
Skeleton	0.36	**0.50**	0.42	**0.59**	0.46	**0.65**	0.49	**0.70**	0.54	**0.78**	0.58	**0.85**
Red bone marrow	0.10	**0.15**	0.12	**0.18**	0.13	**0.20**	0.14	**0.22**	0.16	**0.25**	0.18	**0.27**
Whole body dose	0.15	**0.25**	0.17	**0.30**	0.19	**0.33**	0.21	**0.35**	0.23	**0.39**	0.25	**0.43**
ED (ICRP 60)	0.19	**0.25**	0.22	**0.30**	0.24	**0.34**	0.26	**0.36**	0.29	**0.41**	0.32	**0.45**
ED (ICRP 103)	0.23	**0.31**	0.26	**0.36**	0.29	**0.40**	0.30	**0.42**	0.34	**0.48**	0.37	**0.52**
Total filtration (mm) **Al**	**1.5**		**2.5**		**3.5**		**4.5**		**3.5**		**3.5**	
Cu	**0**		**0**		**0**		**0**		**0.1**		**0.2**	

Table 9.69 Normalized organ doses—Chest r-lat—< 1 year

Beam projection:			r-lat				**Table cells:**					
Focus-to-detector distance (cm):			100				(+) = normal font = optimal field setting					
Tube voltage (kVp):			80				**(−) = bold font = suboptimal field setting**					
Total filtrations:			See last table row.				Empty table cells correspond to values < 0.01					
Organ or tissue	**Organ dose D_T : Entrance dose K_E**								**(mSv × mGy^{-1})**			
	(+)	(−)	(+)	(−)	(+)	(−)	(+)	(−)	(+)	(−)	(+)	(−)
Skull	0.03	**0.29**	0.03	**0.34**	0.04	**0.37**	0.04	**0.40**	0.05	**0.44**	0.06	**0.47**
Brain		**0.04**		**0.05**		**0.06**	0.01	**0.07**	0.01	**0.08**	0.01	**0.09**
Oral mucosa	0.03	**0.51**	0.04	**0.59**	0.05	**0.64**	0.05	**0.68**	0.06	**0.74**	0.07	**0.80**
Salivary glands	0.06	**0.59**	0.07	**0.66**	0.08	**0.70**	0.09	**0.73**	0.10	**0.79**	0.11	**0.83**
Cervical spine	0.52	**1.15**	0.62	**1.38**	0.69	**1.54**	0.74	**1.66**	0.83	**1.87**	0.90	**2.03**
Thyroid	0.59	**0.69**	0.68	**0.80**	0.73	**0.87**	0.78	**0.93**	0.85	**1.02**	0.90	**1.10**
Thymus	0.35	**0.38**	0.42	**0.46**	0.47	**0.51**	0.51	**0.56**	0.59	**0.64**	0.65	**0.71**
Thoracic spine	0.83	**0.90**	1.01	**1.10**	1.15	**1.25**	1.25	**1.36**	1.43	**1.57**	1.57	**1.73**
Claviculae	1.09	**1.20**	1.25	**1.39**	1.36	**1.51**	1.44	**1.60**	1.56	**1.75**	1.65	**1.85**
Scapulae	1.85	**1.94**	2.04	**2.15**	2.16	**2.28**	2.23	**2.36**	2.35	**2.49**	2.42	**2.57**
Ribs	1.40	**1.50**	1.56	**1.69**	1.67	**1.81**	1.74	**1.89**	1.85	**2.02**	1.92	**2.10**
Oesophagus	0.41	**0.45**	0.50	**0.54**	0.56	**0.61**	0.60	**0.66**	0.69	**0.75**	0.76	**0.83**
Trachea	0.08	**0.43**	0.09	**0.50**	0.11	**0.55**	0.11	**0.59**	0.13	**0.66**	0.14	**0.71**
Lung	0.53	**0.56**	0.61	**0.65**	0.67	**0.72**	0.71	**0.76**	0.78	**0.84**	0.84	**0.91**
Mamma	0.63	**0.65**	0.67	**0.69**	0.70	**0.72**	0.72	**0.75**	0.75	**0.79**	0.78	**0.82**
Heart	0.39	**0.41**	0.47	**0.50**	0.52	**0.56**	0.57	**0.61**	0.65	**0.70**	0.71	**0.77**
Stomach	0.11	**0.19**	0.14	**0.24**	0.16	**0.28**	0.18	**0.31**	0.21	**0.37**	0.24	**0.42**
Spleen	0.11	**0.15**	0.14	**0.19**	0.16	**0.22**	0.18	**0.24**	0.21	**0.29**	0.24	**0.33**
Liver	0.38	**0.70**	0.44	**0.81**	0.49	**0.88**	0.52	**0.94**	0.58	**1.04**	0.62	**1.11**
Gall bladder	0.13	**0.36**	0.16	**0.44**	0.18	**0.50**	0.20	**0.55**	0.23	**0.64**	0.26	**0.71**
Pancreas	0.22	**0.29**	0.28	**0.35**	0.32	**0.41**	0.35	**0.45**	0.41	**0.53**	0.46	**0.60**
Small intestine	0.02	**0.23**	0.03	**0.28**	0.03	**0.32**	0.03	**0.34**	0.04	**0.40**	0.05	**0.44**
Large intestine	0.03	**0.31**	0.03	**0.37**	0.04	**0.41**	0.04	**0.45**	0.05	**0.51**	0.06	**0.56**
Sigma/Rectum		**0.04**		**0.06**	0.01	**0.06**	0.01	**0.07**	0.01	**0.09**	0.02	**0.10**
Lumbar spine	0.13	**0.78**	0.16	**0.95**	0.19	**1.08**	0.21	**1.18**	0.25	**1.36**	0.28	**1.52**
Pelvis	0.02	**0.16**	0.02	**0.20**	0.03	**0.23**	0.03	**0.25**	0.04	**0.29**	0.05	**0.33**
Adrenals	0.28	**0.33**	0.34	**0.40**	0.38	**0.45**	0.42	**0.49**	0.48	**0.57**	0.53	**0.63**
Kidneys	0.11	**0.39**	0.14	**0.46**	0.16	**0.52**	0.17	**0.56**	0.20	**0.64**	0.22	**0.70**
Urinary bladder		**0.02**		**0.03**		**0.03**		**0.04**		**0.04**	0.01	**0.05**
Uterus		**0.05**	0.01	**0.06**	0.01	**0.07**	0.01	**0.08**	0.02	**0.10**	0.02	**0.11**
Ovaries		**0.05**	0.01	**0.07**	0.01	**0.08**	0.02	**0.08**	0.02	**0.10**	0.03	**0.12**
Prostate		**0.01**		**0.01**		**0.02**		**0.02**		**0.02**		**0.03**
Testes								**0.01**		**0.01**		**0.02**
Upper arm												
Elbow												
Forearm												
Upper leg		**0.01**		**0.01**		**0.02**		**0.02**		**0.03**		**0.03**
Knee												
Lower leg												
Lymph nodes	0.18	**0.37**	0.21	**0.44**	0.24	**0.48**	0.25	**0.52**	0.29	**0.59**	0.32	**0.65**
Skin	0.20	**0.33**	0.21	**0.35**	0.22	**0.36**	0.23	**0.37**	0.24	**0.39**	0.25	**0.41**
Muscle	0.14	**0.25**	0.16	**0.28**	0.18	**0.31**	0.19	**0.33**	0.21	**0.37**	0.23	**0.40**
Skeleton	0.39	**0.54**	0.44	**0.63**	0.48	**0.69**	0.51	**0.73**	0.56	**0.81**	0.59	**0.86**
Red bone marrow	0.11	**0.17**	0.13	**0.20**	0.14	**0.22**	0.15	**0.24**	0.17	**0.27**	0.19	**0.29**
Whole body dose	0.16	**0.28**	0.19	**0.32**	0.21	**0.35**	0.22	**0.37**	0.24	**0.42**	0.26	**0.45**
ED (ICRP 60)	0.21	**0.28**	0.24	**0.33**	0.26	**0.36**	0.28	**0.39**	0.31	**0.44**	0.33	**0.48**
ED (ICRP 103)	0.25	**0.34**	0.28	**0.39**	0.31	**0.43**	0.33	**0.46**	0.36	**0.51**	0.39	**0.55**
Total filtration (mm) Al	**1.5**		**2.5**		**3.5**		**4.5**		**3.5**		**3.5**	
Cu	**0**		**0**		**0**		**0**		**0.1**		**0.2**	

Table 9.70 Normalized organ doses—Chest r-lat—1–2 years

Beam projection:		r-lat				**Table cells:**						
Focus-to-detector distance (cm):		100				(+) = normal font = optimal field setting						
Tube voltage (kVp):		50				**(−) = bold font = suboptimal field setting**						
Total filtrations:		See last table row.				Empty table cells correspond to values < 0.01						
Organ or tissue	**Organ dose D_T : Entrance dose K_E**							**($mSv \times mGy^{-1}$)**				
	(+)	(−)	(+)	(−)	(+)	(−)	(+)	(−)	(+)	(−)	(+)	(−)
Skull		**0.20**	0.01	**0.25**	0.01	**0.28**	0.02	**0.31**	0.02	**0.36**	0.02	**0.41**
Brain		**0.02**		**0.03**		**0.03**		**0.04**		**0.04**		**0.05**
Oral mucosa	0.01	**0.26**	0.01	**0.32**	0.02	**0.36**	0.02	**0.39**	0.02	**0.44**	0.03	**0.50**
Salivary glands	0.02	**0.36**	0.02	**0.41**	0.03	**0.45**	0.03	**0.47**	0.04	**0.52**	0.05	**0.56**
Cervical spine	0.19	**0.43**	0.25	**0.57**	0.30	**0.68**	0.34	**0.77**	0.41	**0.94**	0.48	**1.10**
Thyroid	0.31	**0.36**	0.38	**0.44**	0.43	**0.49**	0.46	**0.54**	0.53	**0.62**	0.59	**0.69**
Thymus	0.13	**0.14**	0.18	**0.18**	0.21	**0.22**	0.24	**0.25**	0.29	**0.31**	0.34	**0.36**
Thoracic spine	0.28	**0.29**	0.38	**0.40**	0.46	**0.48**	0.53	**0.55**	0.66	**0.69**	0.79	**0.83**
Claviculae	0.60	**0.65**	0.75	**0.81**	0.86	**0.93**	0.94	**1.03**	1.09	**1.20**	1.23	**1.35**
Scapulae	1.05	**1.07**	1.24	**1.26**	1.37	**1.39**	1.46	**1.48**	1.61	**1.64**	1.75	**1.78**
Ribs	0.96	**0.98**	1.14	**1.17**	1.27	**1.30**	1.36	**1.40**	1.53	**1.57**	1.67	**1.72**
Oesophagus	0.19	**0.20**	0.25	**0.26**	0.30	**0.31**	0.34	**0.35**	0.41	**0.43**	0.48	**0.50**
Trachea	0.03	**0.20**	0.04	**0.25**	0.04	**0.29**	0.05	**0.32**	0.06	**0.38**	0.07	**0.43**
Lung	0.28	**0.29**	0.35	**0.36**	0.40	**0.41**	0.44	**0.45**	0.51	**0.53**	0.57	**0.59**
Mamma	0.50	**0.50**	0.54	**0.55**	0.57	**0.57**	0.59	**0.59**	0.62	**0.63**	0.65	**0.65**
Heart	0.16	**0.16**	0.21	**0.21**	0.24	**0.25**	0.28	**0.28**	0.34	**0.34**	0.39	**0.40**
Stomach	0.03	**0.05**	0.05	**0.06**	0.06	**0.08**	0.06	**0.09**	0.08	**0.11**	0.10	**0.14**
Spleen	0.02	**0.03**	0.03	**0.04**	0.04	**0.05**	0.05	**0.06**	0.06	**0.08**	0.08	**0.10**
Liver	0.26	**0.39**	0.32	**0.48**	0.37	**0.54**	0.40	**0.59**	0.47	**0.69**	0.53	**0.77**
Gall bladder	0.06	**0.14**	0.08	**0.19**	0.09	**0.23**	0.11	**0.26**	0.13	**0.33**	0.16	**0.39**
Pancreas	0.07	**0.09**	0.10	**0.12**	0.12	**0.14**	0.14	**0.16**	0.18	**0.21**	0.21	**0.25**
Small intestine		**0.09**	0.01	**0.11**	0.01	**0.14**	0.01	**0.15**	0.02	**0.19**	0.02	**0.22**
Large intestine	0.01	**0.13**	0.01	**0.17**	0.02	**0.20**	0.02	**0.22**	0.03	**0.27**	0.03	**0.31**
Sigma/Rectum				**0.01**		**0.02**		**0.02**		**0.02**		**0.03**
Lumbar spine	0.07	**0.24**	0.10	**0.32**	0.12	**0.39**	0.14	**0.45**	0.18	**0.57**	0.22	**0.69**
Pelvis		**0.06**		**0.08**		**0.09**		**0.10**	0.01	**0.13**	0.02	**0.16**
Adrenals	0.10	**0.11**	0.13	**0.14**	0.16	**0.17**	0.18	**0.20**	0.23	**0.24**	0.27	**0.29**
Kidneys	0.08	**0.16**	0.10	**0.21**	0.12	**0.24**	0.14	**0.27**	0.16	**0.32**	0.19	**0.37**
Urinary bladder								**0.01**		**0.01**		**0.02**
Uterus		**0.01**		**0.02**		**0.02**		**0.03**		**0.04**		**0.05**
Ovaries		**0.02**		**0.02**		**0.03**		**0.03**		**0.04**		**0.05**
Prostate												**0.01**
Testes												
Upper arm												
Elbow												
Forearm												
Upper leg												
Knee												
Lower leg												
Lymph nodes	0.08	**0.16**	0.10	**0.21**	0.12	**0.24**	0.13	**0.27**	0.16	**0.32**	0.18	**0.36**
Skin	0.15	**0.25**	0.16	**0.26**	0.17	**0.27**	0.17	**0.28**	0.18	**0.29**	0.19	**0.30**
Muscle	0.07	**0.11**	0.08	**0.14**	0.09	**0.15**	0.10	**0.17**	0.11	**0.19**	0.13	**0.21**
Skeleton	0.20	**0.26**	0.24	**0.32**	0.27	**0.37**	0.29	**0.41**	0.34	**0.47**	0.38	**0.53**
Red bone marrow	0.05	**0.07**	0.06	**0.09**	0.07	**0.10**	0.08	**0.12**	0.09	**0.14**	0.10	**0.16**
Whole body dose	0.09	**0.14**	0.10	**0.17**	0.12	**0.19**	0.13	**0.20**	0.15	**0.24**	0.17	**0.26**
ED (ICRP 60)	0.11	**0.14**	0.14	**0.17**	0.16	**0.20**	0.17	**0.22**	0.20	**0.25**	0.22	**0.28**
ED (ICRP 103)	0.15	**0.18**	0.17	**0.22**	0.19	**0.24**	0.21	**0.26**	0.24	**0.30**	0.26	**0.34**
Total filtration (mm) Al	**1.5**		**2.5**		**3.5**		**4.5**		**3.5**		**3.5**	
Cu	**0**		**0**		**0**		**0**		**0.1**		**0.2**	

Table 9.71 Normalized organ doses—Chest r-lat—1–2 years

Beam projection:			r-lat				**Table cells:**					
Focus-to-detector distance (cm):			100				(+) = normal font = optimal field setting					
Tube voltage (kVp):			60				**(–) = bold font = suboptimal field setting**					
Total filtrations:			See last table row.				Empty table cells correspond to values < 0.01					
Organ or tissue	**Organ dose D_T : Entrance dose K_E**								**(mSv × mGy^{-1})**			
	(+)	(–)	(+)	(–)	(+)	(–)	(+)	(–)	(+)	(–)	(+)	(–)
Skull	0.01	**0.24**	0.02	**0.30**	0.02	**0.34**	0.02	**0.37**	0.03	**0.42**	0.02	**0.41**
Brain		**0.03**		**0.04**		**0.04**		**0.05**		**0.06**		**0.05**
Oral mucosa	0.01	**0.31**	0.02	**0.37**	0.02	**0.42**	0.03	**0.45**	0.03	**0.51**	0.03	**0.50**
Salivary glands	0.02	**0.40**	0.03	**0.46**	0.04	**0.50**	0.04	**0.52**	0.05	**0.58**	0.05	**0.56**
Cervical spine	0.25	**0.57**	0.33	**0.74**	0.38	**0.87**	0.42	**0.97**	0.51	**1.16**	0.48	**1.10**
Thyroid	0.37	**0.42**	0.44	**0.51**	0.49	**0.57**	0.53	**0.62**	0.60	**0.71**	0.59	**0.69**
Thymus	0.18	**0.19**	0.23	**0.24**	0.27	**0.29**	0.30	**0.32**	0.37	**0.39**	0.34	**0.36**
Thoracic spine	0.40	**0.41**	0.52	**0.54**	0.62	**0.65**	0.70	**0.73**	0.87	**0.91**	0.79	**0.83**
Claviculae	0.72	**0.78**	0.88	**0.96**	1.00	**1.09**	1.09	**1.20**	1.25	**1.38**	1.23	**1.35**
Scapulae	1.18	**1.20**	1.37	**1.40**	1.50	**1.53**	1.59	**1.62**	1.74	**1.78**	1.75	**1.78**
Ribs	1.09	**1.12**	1.28	**1.32**	1.41	**1.45**	1.51	**1.56**	1.68	**1.73**	1.67	**1.72**
Oesophagus	0.26	**0.27**	0.33	**0.34**	0.38	**0.40**	0.43	**0.45**	0.51	**0.54**	0.48	**0.50**
Trachea	0.04	**0.25**	0.05	**0.31**	0.05	**0.35**	0.06	**0.39**	0.07	**0.45**	0.07	**0.43**
Lung	0.34	**0.35**	0.42	**0.43**	0.47	**0.49**	0.52	**0.53**	0.59	**0.61**	0.57	**0.59**
Mamma	0.53	**0.53**	0.58	**0.58**	0.60	**0.61**	0.62	**0.63**	0.66	**0.66**	0.65	**0.65**
Heart	0.21	**0.21**	0.27	**0.27**	0.31	**0.32**	0.35	**0.36**	0.42	**0.43**	0.39	**0.40**
Stomach	0.05	**0.07**	0.07	**0.09**	0.08	**0.11**	0.09	**0.13**	0.12	**0.16**	0.10	**0.14**
Spleen	0.04	**0.05**	0.05	**0.06**	0.06	**0.08**	0.07	**0.09**	0.09	**0.11**	0.08	**0.10**
Liver	0.31	**0.46**	0.38	**0.56**	0.43	**0.64**	0.47	**0.69**	0.55	**0.79**	0.53	**0.77**
Gall bladder	0.08	**0.20**	0.11	**0.26**	0.13	**0.30**	0.14	**0.34**	0.18	**0.42**	0.16	**0.39**
Pancreas	0.11	**0.13**	0.14	**0.17**	0.17	**0.20**	0.19	**0.23**	0.24	**0.28**	0.21	**0.25**
Small intestine	0.01	**0.12**	0.02	**0.15**	0.02	**0.18**	0.02	**0.20**	0.03	**0.24**	0.02	**0.22**
Large intestine	0.02	**0.17**	0.02	**0.21**	0.03	**0.25**	0.03	**0.28**	0.04	**0.33**	0.03	**0.31**
Sigma/Rectum		**0.02**		**0.02**		**0.03**		**0.03**		**0.04**		**0.03**
Lumbar spine	0.11	**0.34**	0.14	**0.45**	0.17	**0.54**	0.19	**0.61**	0.24	**0.75**	0.22	**0.69**
Pelvis		**0.08**	0.01	**0.10**	0.01	**0.12**	0.01	**0.14**	0.02	**0.17**	0.02	**0.16**
Adrenals	0.14	**0.15**	0.18	**0.20**	0.21	**0.23**	0.24	**0.26**	0.29	**0.32**	0.27	**0.29**
Kidneys	0.10	**0.21**	0.13	**0.26**	0.15	**0.30**	0.17	**0.33**	0.20	**0.39**	0.19	**0.37**
Urinary bladder				**0.01**		**0.01**		**0.02**		**0.02**		**0.02**
Uterus		**0.02**		**0.03**		**0.04**		**0.04**	0.01	**0.05**		**0.05**
Ovaries		**0.03**		**0.03**		**0.04**		**0.05**		**0.06**		**0.05**
Prostate										**0.01**		**0.01**
Testes												
Upper arm												
Elbow												
Forearm												
Upper leg												
Knee												
Lower leg												
Lymph nodes	0.10	**0.21**	0.13	**0.26**	0.15	**0.30**	0.16	**0.33**	0.19	**0.38**	0.18	**0.36**
Skin	0.16	**0.26**	0.17	**0.27**	0.18	**0.29**	0.18	**0.29**	0.19	**0.31**	0.19	**0.30**
Muscle	0.08	**0.13**	0.10	**0.16**	0.11	**0.18**	0.12	**0.20**	0.14	**0.22**	0.13	**0.21**
Skeleton	0.23	**0.32**	0.28	**0.39**	0.31	**0.44**	0.34	**0.48**	0.39	**0.55**	0.38	**0.53**
Red bone marrow	0.06	**0.09**	0.07	**0.11**	0.08	**0.13**	0.09	**0.14**	0.11	**0.17**	0.10	**0.16**
Whole body dose	0.10	**0.16**	0.12	**0.20**	0.14	**0.22**	0.15	**0.24**	0.17	**0.28**	0.17	**0.26**
ED (ICRP 60)	0.14	**0.17**	0.16	**0.21**	0.18	**0.24**	0.20	**0.26**	0.23	**0.30**	0.22	**0.28**
ED (ICRP 103)	0.17	**0.21**	0.20	**0.25**	0.22	**0.28**	0.24	**0.31**	0.27	**0.35**	0.26	**0.34**
Total filtration (mm) **Al**	**1.5**		**2.5**		**3.5**		**4.5**		**3.5**		**3.5**	
Cu	**0**		**0**		**0**		**0**		**0.1**		**0.2**	

Table 9.72 Normalized organ doses—Chest r-lat—1–2 years

Beam projection:			r-lat				**Table cells:**					
Focus-to-detector distance (cm):			100				(+) = normal font = optimal field setting					
Tube voltage (kVp):			70				**(−) = bold font = suboptimal field setting**					
Total filtrations:			See last table row.				Empty table cells correspond to values < 0.01					
Organ or tissue	**Organ dose D_T : Entrance dose K_E**								**(mSv × mGy^{-1})**			
	(+)	(−)	(+)	(−)	(+)	(−)	(+)	(−)	(+)	(−)	(+)	(−)
Skull	0.02	**0.28**	0.02	**0.33**	0.02	**0.37**	0.03	**0.40**	0.03	**0.46**	0.04	**0.50**
Brain		**0.03**		**0.04**		**0.05**		**0.06**		**0.07**		**0.08**
Oral mucosa	0.02	**0.35**	0.02	**0.42**	0.03	**0.46**	0.03	**0.50**	0.04	**0.56**	0.05	**0.62**
Salivary glands	0.03	**0.43**	0.04	**0.49**	0.04	**0.53**	0.05	**0.56**	0.06	**0.62**	0.07	**0.66**
Cervical spine	0.30	**0.69**	0.38	**0.87**	0.44	**1.00**	0.48	**1.11**	0.57	**1.31**	0.65	**1.49**
Thyroid	0.41	**0.48**	0.49	**0.57**	0.54	**0.64**	0.58	**0.69**	0.66	**0.78**	0.72	**0.85**
Thymus	0.22	**0.23**	0.28	**0.29**	0.32	**0.34**	0.36	**0.38**	0.42	**0.45**	0.49	**0.52**
Thoracic spine	0.50	**0.52**	0.64	**0.67**	0.75	**0.78**	0.84	**0.88**	1.01	**1.06**	1.17	**1.24**
Claviculae	0.81	**0.89**	0.98	**1.08**	1.10	**1.21**	1.18	**1.31**	1.34	**1.48**	1.47	**1.63**
Scapulae	1.27	**1.30**	1.46	**1.49**	1.58	**1.61**	1.67	**1.70**	1.81	**1.85**	1.91	**1.97**
Ribs	1.19	**1.23**	1.38	**1.42**	1.51	**1.56**	1.60	**1.65**	1.76	**1.82**	1.88	**1.95**
Oesophagus	0.31	**0.33**	0.39	**0.41**	0.45	**0.48**	0.50	**0.53**	0.59	**0.63**	0.68	**0.72**
Trachea	0.04	**0.29**	0.05	**0.35**	0.06	**0.40**	0.07	**0.44**	0.08	**0.50**	0.09	**0.56**
Lung	0.39	**0.40**	0.47	**0.49**	0.53	**0.54**	0.57	**0.59**	0.65	**0.67**	0.72	**0.75**
Mamma	0.56	**0.56**	0.60	**0.61**	0.63	**0.64**	0.65	**0.66**	0.69	**0.70**	0.72	**0.73**
Heart	0.25	**0.26**	0.32	**0.33**	0.37	**0.38**	0.41	**0.42**	0.48	**0.50**	0.55	**0.57**
Stomach	0.07	**0.09**	0.09	**0.12**	0.10	**0.14**	0.12	**0.16**	0.14	**0.20**	0.17	**0.24**
Spleen	0.05	**0.06**	0.07	**0.09**	0.08	**0.10**	0.10	**0.12**	0.12	**0.15**	0.15	**0.18**
Liver	0.36	**0.53**	0.43	**0.63**	0.49	**0.71**	0.53	**0.77**	0.60	**0.87**	0.67	**0.96**
Gall bladder	0.10	**0.25**	0.13	**0.31**	0.16	**0.36**	0.17	**0.41**	0.21	**0.49**	0.25	**0.56**
Pancreas	0.14	**0.16**	0.18	**0.21**	0.21	**0.25**	0.23	**0.28**	0.29	**0.34**	0.34	**0.40**
Small intestine	0.02	**0.14**	0.02	**0.18**	0.03	**0.21**	0.03	**0.24**	0.04	**0.28**	0.04	**0.33**
Large intestine	0.02	**0.20**	0.03	**0.25**	0.03	**0.29**	0.04	**0.32**	0.05	**0.38**	0.06	**0.43**
Sigma/Rectum		**0.02**		**0.03**		**0.03**		**0.04**		**0.05**	0.01	**0.06**
Lumbar spine	0.14	**0.43**	0.18	**0.56**	0.21	**0.65**	0.24	**0.73**	0.30	**0.89**	0.35	**1.04**
Pelvis	0.01	**0.10**	0.01	**0.13**	0.02	**0.15**	0.02	**0.17**	0.03	**0.21**	0.03	**0.25**
Adrenals	0.17	**0.19**	0.22	**0.24**	0.26	**0.28**	0.29	**0.31**	0.35	**0.38**	0.40	**0.44**
Kidneys	0.13	**0.25**	0.16	**0.31**	0.18	**0.35**	0.20	**0.38**	0.24	**0.45**	0.27	**0.51**
Urinary bladder		**0.01**		**0.02**		**0.02**		**0.02**		**0.03**		**0.03**
Uterus		**0.03**		**0.04**	0.01	**0.05**	0.01	**0.05**	0.02	**0.07**	0.02	**0.08**
Ovaries		**0.03**		**0.04**		**0.05**	0.01	**0.06**	0.01	**0.08**	0.02	**0.09**
Prostate						**0.01**		**0.01**		**0.02**		**0.02**
Testes												
Upper arm												
Elbow												
Forearm												
Upper leg												**0.01**
Knee												
Lower leg												
Lymph nodes	0.12	**0.25**	0.15	**0.30**	0.17	**0.34**	0.19	**0.37**	0.22	**0.44**	0.25	**0.50**
Skin	0.17	**0.27**	0.18	**0.29**	0.19	**0.30**	0.19	**0.31**	0.20	**0.32**	0.21	**0.34**
Muscle	0.09	**0.15**	0.11	**0.18**	0.12	**0.20**	0.13	**0.22**	0.15	**0.25**	0.17	**0.28**
Skeleton	0.26	**0.36**	0.31	**0.43**	0.35	**0.49**	0.37	**0.53**	0.42	**0.60**	0.46	**0.66**
Red bone marrow	0.07	**0.11**	0.09	**0.13**	0.10	**0.15**	0.11	**0.16**	0.12	**0.19**	0.14	**0.22**
Whole body dose	0.12	**0.19**	0.14	**0.22**	0.16	**0.25**	0.17	**0.27**	0.19	**0.31**	0.21	**0.34**
ED (ICRP 60)	0.16	**0.20**	0.19	**0.24**	0.21	**0.27**	0.22	**0.29**	0.26	**0.33**	0.28	**0.37**
ED (ICRP 103)	0.19	**0.24**	0.22	**0.29**	0.25	**0.32**	0.27	**0.34**	0.30	**0.39**	0.33	**0.43**
Total filtration (mm) **Al**	**1.5**		**2.5**		**3.5**		**4.5**		**3.5**		**3.5**	
Cu	**0**		**0**		**0**		**0**		**0.1**		**0.2**	

Table 9.73 Normalized organ doses—Chest r-lat—1–2 years

Beam projection:			r-lat			**Table cells:**						
Focus-to-detector distance (cm):			100			(+) = normal font = optimal field setting						
Tube voltage (kVp):			80			**(−) = bold font = suboptimal field setting**						
Total filtrations:			See last table row.			Empty table cells correspond to values < 0.01						
Organ or tissue	**Organ dose D_T : Entrance dose K_E**								**(mSv × mGy^{-1})**			
	(+)	(−)	(+)	(−)	(+)	(−)	(+)	(−)	(+)	(−)	(+)	(−)
Skull	0.02	**0.31**	0.03	**0.36**	0.03	**0.40**	0.03	**0.43**	0.04	**0.48**	0.05	**0.52**
Brain		**0.04**		**0.05**		**0.06**		**0.07**		**0.08**	0.01	**0.10**
Oral mucosa	0.02	**0.39**	0.03	**0.46**	0.03	**0.50**	0.04	**0.54**	0.05	**0.61**	0.05	**0.66**
Salivary glands	0.04	**0.47**	0.04	**0.53**	0.05	**0.57**	0.06	**0.60**	0.07	**0.65**	0.08	**0.70**
Cervical spine	0.35	**0.80**	0.43	**0.98**	0.49	**1.12**	0.53	**1.22**	0.62	**1.42**	0.69	**1.58**
Thyroid	0.46	**0.53**	0.54	**0.63**	0.59	**0.69**	0.63	**0.74**	0.70	**0.83**	0.76	**0.91**
Thymus	0.26	**0.27**	0.32	**0.34**	0.37	**0.39**	0.40	**0.43**	0.47	**0.50**	0.54	**0.57**
Thoracic spine	0.59	**0.62**	0.74	**0.78**	0.85	**0.90**	0.94	**1.00**	1.12	**1.18**	1.27	**1.35**
Claviculae	0.89	**0.98**	1.06	**1.16**	1.17	**1.29**	1.25	**1.38**	1.39	**1.55**	1.50	**1.67**
Scapulae	1.35	**1.38**	1.52	**1.56**	1.63	**1.67**	1.71	**1.75**	1.83	**1.88**	1.91	**1.97**
Ribs	1.27	**1.31**	1.45	**1.50**	1.57	**1.63**	1.66	**1.72**	1.79	**1.86**	1.89	**1.97**
Oesophagus	0.37	**0.39**	0.45	**0.48**	0.51	**0.54**	0.56	**0.60**	0.66	**0.70**	0.74	**0.79**
Trachea	0.05	**0.33**	0.06	**0.39**	0.07	**0.44**	0.08	**0.48**	0.09	**0.55**	0.10	**0.61**
Lung	0.44	**0.45**	0.52	**0.54**	0.57	**0.60**	0.62	**0.64**	0.70	**0.72**	0.76	**0.79**
Mamma	0.58	**0.59**	0.62	**0.63**	0.65	**0.66**	0.67	**0.68**	0.71	**0.72**	0.74	**0.75**
Heart	0.30	**0.31**	0.37	**0.38**	0.42	**0.43**	0.46	**0.48**	0.54	**0.56**	0.61	**0.63**
Stomach	0.08	**0.12**	0.11	**0.15**	0.12	**0.17**	0.14	**0.19**	0.17	**0.23**	0.20	**0.27**
Spleen	0.07	**0.08**	0.09	**0.11**	0.11	**0.13**	0.12	**0.14**	0.15	**0.18**	0.17	**0.21**
Liver	0.40	**0.59**	0.48	**0.70**	0.53	**0.77**	0.57	**0.83**	0.65	**0.93**	0.71	**1.02**
Gall bladder	0.13	**0.29**	0.16	**0.37**	0.18	**0.42**	0.20	**0.46**	0.24	**0.55**	0.28	**0.62**
Pancreas	0.17	**0.20**	0.21	**0.25**	0.25	**0.29**	0.27	**0.33**	0.33	**0.39**	0.38	**0.46**
Small intestine	0.02	**0.17**	0.03	**0.21**	0.03	**0.25**	0.04	**0.27**	0.04	**0.32**	0.05	**0.36**
Large intestine	0.03	**0.24**	0.03	**0.29**	0.04	**0.33**	0.05	**0.36**	0.06	**0.42**	0.07	**0.47**
Sigma/Rectum		**0.03**		**0.04**		**0.04**		**0.05**	0.01	**0.06**	0.01	**0.07**
Lumbar spine	0.17	**0.52**	0.22	**0.65**	0.25	**0.75**	0.28	**0.83**	0.34	**0.99**	0.39	**1.14**
Pelvis	0.02	**0.12**	0.02	**0.15**	0.02	**0.18**	0.03	**0.20**	0.03	**0.24**	0.04	**0.27**
Adrenals	0.21	**0.23**	0.26	**0.29**	0.30	**0.33**	0.33	**0.36**	0.40	**0.43**	0.45	**0.50**
Kidneys	0.15	**0.29**	0.18	**0.35**	0.21	**0.39**	0.23	**0.43**	0.26	**0.50**	0.30	**0.56**
Urinary bladder		**0.02**		**0.02**		**0.02**		**0.03**		**0.03**		**0.04**
Uterus		**0.04**	0.01	**0.05**	0.01	**0.06**	0.01	**0.07**	0.02	**0.08**	0.02	**0.10**
Ovaries		**0.04**		**0.05**	0.01	**0.06**	0.01	**0.07**	0.02	**0.09**	0.02	**0.10**
Prostate				**0.01**		**0.01**		**0.02**		**0.02**		**0.02**
Testes												**0.01**
Upper arm												
Elbow												
Forearm												
Upper leg										**0.01**		**0.01**
Knee												
Lower leg												
Lymph nodes	0.14	**0.28**	0.17	**0.34**	0.20	**0.38**	0.21	**0.42**	0.25	**0.48**	0.28	**0.54**
Skin	0.18	**0.28**	0.19	**0.30**	0.19	**0.31**	0.20	**0.32**	0.21	**0.34**	0.22	**0.35**
Muscle	0.10	**0.17**	0.12	**0.20**	0.13	**0.22**	0.15	**0.24**	0.16	**0.27**	0.18	**0.30**
Skeleton	0.29	**0.40**	0.34	**0.47**	0.37	**0.52**	0.40	**0.56**	0.44	**0.63**	0.48	**0.69**
Red bone marrow	0.08	**0.13**	0.10	**0.15**	0.11	**0.17**	0.12	**0.18**	0.14	**0.21**	0.15	**0.24**
Whole body dose	0.13	**0.21**	0.15	**0.25**	0.17	**0.27**	0.18	**0.29**	0.21	**0.33**	0.23	**0.36**
ED (ICRP 60)	0.17	**0.22**	0.21	**0.27**	0.23	**0.30**	0.24	**0.32**	0.28	**0.36**	0.30	**0.40**
ED (ICRP 103)	0.21	**0.27**	0.25	**0.32**	0.27	**0.35**	0.29	**0.38**	0.32	**0.42**	0.35	**0.47**
Total filtration (mm) **Al**	**1.5**		**2.5**		**3.5**		**4.5**		**3.5**		**3.5**	
Cu	**0**		**0**		**0**		**0**		**0.1**		**0.2**	

Table 9.74 Normalized organ doses—Chest r-lat—3–7 years

Beam projection:			r-lat			**Table cells:**						
Focus-to-detector distance (cm):			100			(+) = normal font = optimal field setting						
Tube voltage (kVp):			60			**(−) = bold font = suboptimal field setting**						
Total filtrations:			See last table row.			Empty table cells correspond to values < 0.01						
Organ or tissue	**Organ dose D_T : Entrance dose K_E**						**(mSv × mGy^{-1})**					
	(+)	(−)	(+)	(−)	(+)	(−)	(+)	(−)	(+)	(−)	(+)	(−)
Skull	0.01	**0.19**	0.02	**0.23**	0.02	**0.27**	0.02	**0.30**	0.03	**0.35**	0.03	**0.33**
Brain		**0.01**		**0.02**		**0.02**		**0.03**		**0.03**		**0.03**
Oral mucosa	0.01	**0.22**	0.02	**0.27**	0.02	**0.30**	0.02	**0.33**	0.03	**0.38**	0.02	**0.37**
Salivary glands	0.02	**0.31**	0.03	**0.36**	0.03	**0.39**	0.04	**0.41**	0.05	**0.46**	0.04	**0.44**
Cervical spine	0.25	**0.50**	0.32	**0.65**	0.37	**0.76**	0.41	**0.86**	0.50	**1.03**	0.47	**0.97**
Thyroid	0.28	**0.30**	0.34	**0.37**	0.38	**0.41**	0.41	**0.45**	0.47	**0.52**	0.46	**0.50**
Thymus	0.11	**0.12**	0.15	**0.16**	0.18	**0.19**	0.20	**0.21**	0.25	**0.26**	0.23	**0.24**
Thoracic spine	0.24	**0.25**	0.32	**0.34**	0.39	**0.41**	0.45	**0.47**	0.56	**0.59**	0.49	**0.52**
Claviculae	0.52	**0.57**	0.65	**0.71**	0.75	**0.82**	0.83	**0.90**	0.96	**1.06**	0.94	**1.02**
Scapulae	1.01	**1.03**	1.19	**1.22**	1.31	**1.34**	1.40	**1.44**	1.56	**1.60**	1.55	**1.59**
Ribs	0.88	**0.91**	1.04	**1.08**	1.16	**1.20**	1.25	**1.29**	1.40	**1.45**	1.38	**1.43**
Oesophagus	0.12	**0.13**	0.16	**0.17**	0.19	**0.20**	0.22	**0.23**	0.27	**0.28**	0.24	**0.25**
Trachea	0.04	**0.16**	0.06	**0.20**	0.06	**0.23**	0.07	**0.26**	0.08	**0.31**	0.08	**0.29**
Lung	0.27	**0.28**	0.33	**0.34**	0.38	**0.39**	0.42	**0.43**	0.49	**0.50**	0.47	**0.48**
Mamma	0.49	**0.49**	0.53	**0.54**	0.56	**0.56**	0.58	**0.58**	0.61	**0.61**	0.60	**0.60**
Heart	0.14	**0.14**	0.18	**0.19**	0.22	**0.22**	0.24	**0.25**	0.30	**0.31**	0.27	**0.28**
Stomach	0.03	**0.04**	0.04	**0.05**	0.04	**0.06**	0.05	**0.07**	0.07	**0.09**	0.05	**0.08**
Spleen	0.02	**0.02**	0.03	**0.03**	0.03	**0.04**	0.04	**0.04**	0.05	**0.06**	0.04	**0.04**
Liver	0.23	**0.39**	0.29	**0.48**	0.33	**0.54**	0.37	**0.60**	0.43	**0.70**	0.41	**0.67**
Gall bladder	0.04	**0.11**	0.06	**0.15**	0.07	**0.18**	0.08	**0.21**	0.11	**0.26**	0.09	**0.23**
Pancreas	0.06	**0.07**	0.08	**0.10**	0.10	**0.12**	0.11	**0.14**	0.14	**0.17**	0.12	**0.15**
Small intestine		**0.08**		**0.10**		**0.12**	0.01	**0.14**	0.01	**0.17**	0.01	**0.15**
Large intestine		**0.12**	0.01	**0.15**	0.01	**0.18**	0.02	**0.20**	0.02	**0.25**	0.02	**0.23**
Sigma/Rectum				**0.01**		**0.01**		**0.01**		**0.02**		**0.02**
Lumbar spine	0.05	**0.19**	0.07	**0.25**	0.08	**0.31**	0.10	**0.35**	0.13	**0.45**	0.10	**0.39**
Pelvis		**0.05**		**0.07**		**0.08**		**0.09**		**0.12**		**0.10**
Adrenals	0.09	**0.10**	0.12	**0.13**	0.14	**0.16**	0.16	**0.18**	0.20	**0.23**	0.18	**0.20**
Kidneys	0.06	**0.14**	0.08	**0.18**	0.09	**0.22**	0.10	**0.24**	0.12	**0.29**	0.11	**0.27**
Urinary bladder										**0.01**		
Uterus		**0.01**		**0.02**		**0.02**		**0.03**		**0.04**		**0.03**
Ovaries		**0.02**		**0.02**		**0.03**		**0.03**		**0.04**		**0.03**
Prostate												
Testes												
Upper arm												
Elbow												
Forearm												
Upper leg												
Knee												
Lower leg												
Lymph nodes	0.07	**0.14**	0.09	**0.17**	0.10	**0.20**	0.12	**0.23**	0.14	**0.27**	0.13	**0.25**
Skin	0.11	**0.17**	0.11	**0.18**	0.12	**0.19**	0.12	**0.20**	0.13	**0.21**	0.13	**0.20**
Muscle	0.06	**0.09**	0.07	**0.11**	0.08	**0.13**	0.09	**0.14**	0.10	**0.16**	0.10	**0.15**
Skeleton	0.15	**0.21**	0.18	**0.26**	0.20	**0.30**	0.22	**0.33**	0.26	**0.38**	0.25	**0.36**
Red bone marrow	0.05	**0.07**	0.06	**0.08**	0.07	**0.10**	0.07	**0.11**	0.09	**0.13**	0.08	**0.12**
Whole body dose	0.08	**0.12**	0.09	**0.14**	0.10	**0.16**	0.11	**0.18**	0.13	**0.21**	0.12	**0.19**
ED (ICRP 60)	0.10	**0.13**	0.12	**0.16**	0.14	**0.18**	0.15	**0.19**	0.18	**0.23**	0.17	**0.21**
ED (ICRP 103)	0.13	**0.16**	0.16	**0.20**	0.18	**0.22**	0.19	**0.24**	0.22	**0.28**	0.21	**0.26**
Total filtration (mm) **Al**	**1.5**		**2.5**		**3.5**		**4.5**		**3.5**		**3.5**	
Cu	**0**		**0**		**0**		**0**		**0.1**		**0.2**	

Table 9.75 Normalized organ doses—Chest r-lat—3–7 years

Beam projection:			r-lat			**Table cells:**						
Focus-to-detector distance (cm):			100			(+) = normal font = optimal field setting						
Tube voltage (kVp):			70			**(−) = bold font = suboptimal field setting**						
Total filtrations:			See last table row.			Empty table cells correspond to values < 0.01						
Organ or tissue	**Organ dose D_T : Entrance dose K_E**							**(mSv × mGy^{-1})**				
	(+)	(−)	(+)	(−)	(+)	(−)	(+)	(−)	(+)	(−)	(+)	(−)
Skull	0.02	**0.22**	0.02	**0.27**	0.03	**0.30**	0.03	**0.33**	0.04	**0.38**	0.04	**0.43**
Brain		**0.02**		**0.02**		**0.03**		**0.03**		**0.04**		**0.05**
Oral mucosa	0.02	**0.26**	0.02	**0.31**	0.03	**0.35**	0.03	**0.38**	0.04	**0.43**	0.04	**0.49**
Salivary glands	0.03	**0.34**	0.03	**0.39**	0.04	**0.42**	0.04	**0.45**	0.05	**0.50**	0.06	**0.54**
Cervical spine	0.30	**0.61**	0.38	**0.77**	0.43	**0.90**	0.48	**0.99**	0.57	**1.18**	0.65	**1.35**
Thyroid	0.32	**0.34**	0.38	**0.42**	0.42	**0.47**	0.46	**0.51**	0.52	**0.59**	0.58	**0.65**
Thymus	0.15	**0.15**	0.19	**0.20**	0.22	**0.23**	0.25	**0.26**	0.30	**0.32**	0.36	**0.37**
Thoracic spine	0.32	**0.33**	0.41	**0.44**	0.49	**0.52**	0.55	**0.59**	0.69	**0.73**	0.81	**0.86**
Claviculae	0.61	**0.66**	0.74	**0.81**	0.84	**0.92**	0.92	**1.00**	1.05	**1.16**	1.17	**1.29**
Scapulae	1.10	**1.13**	1.28	**1.31**	1.40	**1.44**	1.48	**1.53**	1.63	**1.68**	1.74	**1.80**
Ribs	0.97	**1.00**	1.14	**1.18**	1.25	**1.30**	1.33	**1.39**	1.48	**1.54**	1.60	**1.67**
Oesophagus	0.16	**0.17**	0.20	**0.21**	0.24	**0.25**	0.26	**0.28**	0.32	**0.34**	0.37	**0.39**
Trachea	0.05	**0.19**	0.06	**0.24**	0.07	**0.27**	0.08	**0.30**	0.09	**0.36**	0.10	**0.41**
Lung	0.31	**0.32**	0.38	**0.39**	0.43	**0.45**	0.47	**0.49**	0.54	**0.57**	0.61	**0.63**
Mamma	0.51	**0.52**	0.56	**0.56**	0.58	**0.59**	0.60	**0.61**	0.64	**0.64**	0.67	**0.67**
Heart	0.18	**0.18**	0.22	**0.23**	0.26	**0.27**	0.29	**0.31**	0.36	**0.37**	0.42	**0.43**
Stomach	0.04	**0.05**	0.05	**0.07**	0.06	**0.08**	0.07	**0.10**	0.09	**0.12**	0.11	**0.15**
Spleen	0.03	**0.03**	0.04	**0.04**	0.04	**0.05**	0.05	**0.06**	0.07	**0.08**	0.08	**0.10**
Liver	0.28	**0.45**	0.34	**0.55**	0.38	**0.62**	0.42	**0.67**	0.49	**0.78**	0.55	**0.87**
Gall bladder	0.06	**0.15**	0.08	**0.19**	0.09	**0.23**	0.11	**0.26**	0.13	**0.32**	0.16	**0.38**
Pancreas	0.08	**0.10**	0.11	**0.13**	0.13	**0.16**	0.15	**0.18**	0.18	**0.22**	0.22	**0.27**
Small intestine		**0.10**	0.01	**0.13**	0.01	**0.15**	0.02	**0.17**	0.02	**0.20**	0.02	**0.24**
Large intestine	0.01	**0.15**	0.02	**0.19**	0.02	**0.22**	0.02	**0.24**	0.03	**0.29**	0.03	**0.34**
Sigma/Rectum		**0.01**		**0.01**		**0.02**		**0.02**		**0.03**		**0.03**
Lumbar spine	0.07	**0.25**	0.09	**0.33**	0.11	**0.39**	0.13	**0.45**	0.16	**0.56**	0.20	**0.67**
Pelvis		**0.07**		**0.09**		**0.10**		**0.12**	0.01	**0.14**	0.01	**0.17**
Adrenals	0.12	**0.13**	0.15	**0.17**	0.18	**0.20**	0.20	**0.23**	0.25	**0.28**	0.30	**0.33**
Kidneys	0.07	**0.18**	0.09	**0.22**	0.11	**0.26**	0.12	**0.29**	0.15	**0.34**	0.17	**0.39**
Urinary bladder								**0.01**		**0.01**		**0.02**
Uterus		**0.02**		**0.03**		**0.03**		**0.04**		**0.05**		**0.06**
Ovaries		**0.02**		**0.03**		**0.04**		**0.04**		**0.05**		**0.06**
Prostate												
Testes												
Upper arm												
Elbow												
Forearm												
Upper leg												
Knee												
Lower leg												
Lymph nodes	0.09	**0.17**	0.11	**0.21**	0.12	**0.24**	0.14	**0.27**	0.16	**0.32**	0.19	**0.36**
Skin	0.11	**0.18**	0.12	**0.19**	0.13	**0.20**	0.13	**0.21**	0.14	**0.22**	0.15	**0.23**
Muscle	0.07	**0.11**	0.08	**0.13**	0.09	**0.15**	0.10	**0.16**	0.11	**0.18**	0.13	**0.20**
Skeleton	0.17	**0.25**	0.20	**0.30**	0.23	**0.34**	0.25	**0.37**	0.28	**0.42**	0.31	**0.47**
Red bone marrow	0.06	**0.08**	0.07	**0.10**	0.08	**0.11**	0.09	**0.13**	0.10	**0.15**	0.11	**0.17**
Whole body dose	0.09	**0.14**	0.11	**0.16**	0.12	**0.19**	0.13	**0.20**	0.15	**0.23**	0.17	**0.26**
ED (ICRP 60)	0.12	**0.15**	0.14	**0.18**	0.16	**0.20**	0.17	**0.22**	0.20	**0.26**	0.22	**0.29**
ED (ICRP 103)	0.15	**0.19**	0.18	**0.22**	0.20	**0.25**	0.21	**0.27**	0.24	**0.31**	0.27	**0.35**
Total filtration (mm) **Al**	**1.5**		**2.5**		**3.5**		**4.5**		**3.5**		**3.5**	
Cu	**0**		**0**		**0**		**0**		**0.1**		**0.2**	

Table 9.76 Normalized organ doses—Chest r-lat—3–7 years

Beam projection:			r-lat			**Table cells:**						
Focus-to-detector distance (cm):			100			(+) = normal font = optimal field setting						
Tube voltage (kVp):			80			**(−) = bold font = suboptimal field setting**						
Total filtrations:			See last table row.			Empty table cells correspond to values < 0.01						
Organ or tissue	**Organ dose D_T : Entrance dose K_E**							**($mSv \times mGy^{-1}$)**				
	(+)	(−)	(+)	(−)	(+)	(−)	(+)	(−)	(+)	(−)	(+)	(−)
Skull	0.02	**0.25**	0.03	**0.30**	0.03	**0.33**	0.03	**0.36**	0.04	**0.41**	0.05	**0.46**
Brain		**0.02**		**0.03**		**0.04**		**0.04**		**0.05**		**0.06**
Oral mucosa	0.02	**0.29**	0.03	**0.35**	0.03	**0.39**	0.03	**0.42**	0.04	**0.48**	0.05	**0.53**
Salivary glands	0.03	**0.37**	0.04	**0.42**	0.05	**0.46**	0.05	**0.48**	0.06	**0.53**	0.07	**0.57**
Cervical spine	0.35	**0.72**	0.43	**0.88**	0.49	**1.01**	0.53	**1.11**	0.62	**1.29**	0.70	**1.45**
Thyroid	0.36	**0.39**	0.42	**0.47**	0.46	**0.52**	0.50	**0.56**	0.56	**0.64**	0.62	**0.70**
Thymus	0.18	**0.19**	0.23	**0.24**	0.26	**0.27**	0.29	**0.30**	0.35	**0.37**	0.40	**0.42**
Thoracic spine	0.39	**0.42**	0.50	**0.53**	0.58	**0.62**	0.65	**0.69**	0.79	**0.84**	0.92	**0.97**
Claviculae	0.68	**0.74**	0.81	**0.89**	0.91	**1.00**	0.98	**1.08**	1.11	**1.23**	1.22	**1.35**
Scapulae	1.18	**1.21**	1.35	**1.39**	1.46	**1.50**	1.54	**1.58**	1.66	**1.72**	1.75	**1.81**
Ribs	1.05	**1.09**	1.21	**1.26**	1.31	**1.37**	1.39	**1.45**	1.53	**1.59**	1.62	**1.70**
Oesophagus	0.19	**0.20**	0.24	**0.25**	0.28	**0.29**	0.31	**0.32**	0.37	**0.39**	0.42	**0.45**
Trachea	0.06	**0.22**	0.07	**0.27**	0.08	**0.31**	0.09	**0.34**	0.10	**0.40**	0.11	**0.45**
Lung	0.36	**0.37**	0.43	**0.44**	0.48	**0.50**	0.52	**0.54**	0.59	**0.62**	0.65	**0.68**
Mamma	0.53	**0.54**	0.58	**0.58**	0.61	**0.61**	0.63	**0.63**	0.66	**0.66**	0.69	**0.69**
Heart	0.21	**0.22**	0.27	**0.28**	0.31	**0.32**	0.34	**0.35**	0.41	**0.42**	0.47	**0.49**
Stomach	0.05	**0.07**	0.06	**0.09**	0.07	**0.11**	0.08	**0.12**	0.11	**0.15**	0.13	**0.18**
Spleen	0.04	**0.05**	0.05	**0.06**	0.06	**0.07**	0.07	**0.08**	0.08	**0.10**	0.10	**0.12**
Liver	0.32	**0.51**	0.38	**0.61**	0.43	**0.68**	0.47	**0.74**	0.53	**0.84**	0.59	**0.93**
Gall bladder	0.08	**0.19**	0.10	**0.23**	0.11	**0.27**	0.13	**0.31**	0.16	**0.37**	0.18	**0.43**
Pancreas	0.10	**0.13**	0.13	**0.16**	0.16	**0.19**	0.18	**0.22**	0.22	**0.27**	0.26	**0.31**
Small intestine	0.01	**0.12**	0.01	**0.15**	0.02	**0.18**	0.02	**0.20**	0.03	**0.24**	0.03	**0.28**
Large intestine	0.02	**0.18**	0.02	**0.22**	0.02	**0.25**	0.03	**0.28**	0.03	**0.33**	0.04	**0.37**
Sigma/Rectum		**0.02**		**0.02**		**0.02**		**0.03**		**0.03**		**0.04**
Lumbar spine	0.09	**0.32**	0.12	**0.40**	0.14	**0.47**	0.16	**0.53**	0.19	**0.65**	0.23	**0.76**
Pelvis		**0.08**		**0.11**	0.01	**0.12**	0.01	**0.14**	0.02	**0.17**	0.02	**0.20**
Adrenals	0.15	**0.17**	0.19	**0.21**	0.22	**0.24**	0.24	**0.27**	0.29	**0.33**	0.34	**0.39**
Kidneys	0.09	**0.21**	0.11	**0.26**	0.13	**0.30**	0.14	**0.33**	0.17	**0.38**	0.20	**0.44**
Urinary bladder				**0.01**		**0.01**		**0.02**		**0.02**		**0.02**
Uterus		**0.03**		**0.03**		**0.04**		**0.05**		**0.06**		**0.07**
Ovaries		**0.03**		**0.04**		**0.05**		**0.05**		**0.06**	0.01	**0.07**
Prostate												**0.01**
Testes												
Upper arm												
Elbow												
Forearm												
Upper leg												
Knee												
Lower leg												
Lymph nodes	0.10	**0.20**	0.12	**0.24**	0.14	**0.28**	0.16	**0.30**	0.18	**0.36**	0.21	**0.40**
Skin	0.12	**0.19**	0.13	**0.20**	0.13	**0.21**	0.14	**0.22**	0.14	**0.23**	0.15	**0.24**
Muscle	0.08	**0.12**	0.09	**0.15**	0.10	**0.16**	0.11	**0.17**	0.13	**0.20**	0.14	**0.22**
Skeleton	0.19	**0.28**	0.22	**0.33**	0.25	**0.37**	0.26	**0.40**	0.30	**0.45**	0.33	**0.50**
Red bone marrow	0.07	**0.10**	0.08	**0.12**	0.09	**0.13**	0.10	**0.14**	0.11	**0.17**	0.13	**0.19**
Whole body dose	0.10	**0.16**	0.12	**0.18**	0.13	**0.21**	0.14	**0.22**	0.16	**0.25**	0.18	**0.28**
ED (ICRP 60)	0.13	**0.17**	0.16	**0.20**	0.18	**0.23**	0.19	**0.25**	0.22	**0.29**	0.24	**0.32**
ED (ICRP 103)	0.17	**0.21**	0.20	**0.25**	0.22	**0.28**	0.23	**0.30**	0.26	**0.34**	0.29	**0.38**
Total filtration (mm) **Al**	**1.5**		**2.5**		**3.5**		**4.5**		**3.5**		**3.5**	
Cu	**0**		**0**		**0**		**0**		**0.1**		**0.2**	

Table 9.77 Normalized organ doses—Chest r-lat—8–12 years

Beam projection:		r-lat		**Table cells:**								
Focus-to-detector distance (cm):		100		(+) = normal font = optimal field setting								
Tube voltage (kVp):		60		**(−) = bold font = suboptimal field setting**								
Total filtrations:		See last table row.		Empty table cells correspond to values < 0.01								
Organ or tissue	**Organ dose D_T : Entrance dose K_E**							**(mSv × mGy^{-1})**				
	(+)	(−)	(+)	(−)	(+)	(−)	(+)	(−)	(+)	(−)	(+)	(−)
Skull	0.01	**0.17**	0.02	**0.21**	0.02	**0.24**	0.02	**0.26**	0.03	**0.31**	0.03	**0.30**
Brain		**0.01**		**0.01**		**0.02**		**0.02**		**0.03**		**0.02**
Oral mucosa		**0.19**	0.01	**0.23**	0.02	**0.26**	0.02	**0.28**	0.02	**0.33**	0.02	**0.31**
Salivary glands	0.02	**0.29**	0.03	**0.33**	0.03	**0.36**	0.04	**0.39**	0.05	**0.43**	0.04	**0.42**
Cervical spine	0.23	**0.43**	0.30	**0.56**	0.36	**0.67**	0.40	**0.75**	0.49	**0.92**	0.45	**0.84**
Thyroid	0.21	**0.21**	0.26	**0.26**	0.30	**0.29**	0.33	**0.32**	0.38	**0.38**	0.36	**0.36**
Thymus	0.08	**0.08**	0.10	**0.10**	0.12	**0.13**	0.14	**0.14**	0.18	**0.18**	0.15	**0.16**
Thoracic spine	0.15	**0.16**	0.20	**0.21**	0.24	**0.26**	0.28	**0.30**	0.36	**0.39**	0.31	**0.32**
Claviculae	0.40	**0.42**	0.50	**0.53**	0.58	**0.62**	0.65	**0.69**	0.77	**0.82**	0.73	**0.78**
Scapulae	0.89	**0.92**	1.06	**1.09**	1.18	**1.22**	1.27	**1.31**	1.42	**1.47**	1.41	**1.46**
Ribs	0.70	**0.76**	0.84	**0.91**	0.94	**1.02**	1.01	**1.10**	1.15	**1.25**	1.13	**1.22**
Oesophagus	0.08	**0.08**	0.10	**0.11**	0.12	**0.13**	0.14	**0.15**	0.18	**0.19**	0.16	**0.16**
Trachea	0.09	**0.20**	0.11	**0.26**	0.13	**0.30**	0.14	**0.33**	0.17	**0.40**	0.16	**0.37**
Lung	0.21	**0.21**	0.26	**0.27**	0.30	**0.31**	0.33	**0.35**	0.40	**0.41**	0.38	**0.39**
Mamma	0.45	**0.46**	0.49	**0.50**	0.52	**0.53**	0.54	**0.54**	0.57	**0.57**	0.56	**0.57**
Heart	0.09	**0.10**	0.12	**0.13**	0.15	**0.16**	0.17	**0.18**	0.21	**0.22**	0.19	**0.20**
Stomach	0.01	**0.02**	0.02	**0.03**	0.02	**0.03**	0.03	**0.04**	0.03	**0.05**	0.03	**0.04**
Spleen		**0.01**	0.01	**0.02**	0.02	**0.02**	0.02	**0.02**	0.02	**0.03**	0.02	**0.02**
Liver	0.18	**0.34**	0.22	**0.42**	0.26	**0.48**	0.29	**0.53**	0.34	**0.63**	0.32	**0.59**
Gall bladder	0.03	**0.09**	0.03	**0.12**	0.04	**0.15**	0.05	**0.17**	0.06	**0.21**	0.05	**0.18**
Pancreas	0.03	**0.04**	0.05	**0.06**	0.06	**0.07**	0.06	**0.08**	0.08	**0.11**	0.07	**0.09**
Small intestine		**0.05**		**0.06**		**0.08**		**0.09**		**0.11**		**0.10**
Large intestine		**0.08**		**0.10**		**0.12**		**0.14**		**0.17**		**0.15**
Sigma/Rectum												
Lumbar spine	0.02	**0.11**	0.03	**0.15**	0.04	**0.18**	0.04	**0.21**	0.05	**0.27**	0.04	**0.22**
Pelvis		**0.03**		**0.04**		**0.04**		**0.05**		**0.07**		**0.06**
Adrenals	0.06	**0.07**	0.08	**0.09**	0.10	**0.11**	0.11	**0.13**	0.14	**0.17**	0.12	**0.14**
Kidneys	0.03	**0.10**	0.04	**0.13**	0.04	**0.16**	0.05	**0.18**	0.06	**0.22**	0.05	**0.20**
Urinary bladder												
Uterus				**0.01**		**0.01**		**0.01**		**0.02**		**0.01**
Ovaries				**0.01**		**0.01**		**0.01**		**0.02**		**0.02**
Prostate												
Testes												
Upper arm												
Elbow												
Forearm												
Upper leg												
Knee												
Lower leg												
Lymph nodes	0.06	**0.11**	0.07	**0.14**	0.08	**0.17**	0.09	**0.19**	0.11	**0.23**	0.10	**0.21**
Skin	0.09	**0.15**	0.10	**0.16**	0.10	**0.16**	0.11	**0.17**	0.11	**0.18**	0.11	**0.17**
Muscle	0.05	**0.07**	0.06	**0.09**	0.06	**0.10**	0.07	**0.11**	0.08	**0.13**	0.08	**0.12**
Skeleton	0.12	**0.16**	0.14	**0.20**	0.16	**0.23**	0.18	**0.25**	0.20	**0.30**	0.20	**0.28**
Red bone marrow	0.05	**0.06**	0.06	**0.08**	0.06	**0.09**	0.07	**0.10**	0.08	**0.12**	0.08	**0.10**
Whole body dose	0.06	**0.09**	0.07	**0.11**	0.08	**0.13**	0.09	**0.14**	0.11	**0.17**	0.10	**0.16**
ED (ICRP 60)	0.08	**0.10**	0.10	**0.12**	0.11	**0.14**	0.12	**0.15**	0.14	**0.18**	0.13	**0.17**
ED (ICRP 103)	0.11	**0.14**	0.13	**0.16**	0.15	**0.18**	0.16	**0.20**	0.18	**0.23**	0.17	**0.21**
Total filtration (mm) Al	**1.5**		**2.5**		**3.5**		**4.5**		**3.5**		**3.5**	
Cu	**0**		**0**		**0**		**0**		**0.1**		**0.2**	

Table 9.78 Normalized organ doses—Chest r-lat—8–12 years

Beam projection:			r-lat				**Table cells:**					
Focus-to-detector distance (cm):			100				(+) = normal font = optimal field setting					
Tube voltage (kVp):			70				**(−) = bold font = suboptimal field setting**					
Total filtrations:			See last table row.				Empty table cells correspond to values < 0.01					
Organ or tissue	**Organ dose D_T : Entrance dose K_E**							**($mSv \times mGy^{-1}$)**				
	(+)	(−)	(+)	(−)	(+)	(−)	(+)	(−)	(+)	(−)	(+)	(−)
Skull	0.02	**0.20**	0.02	**0.24**	0.03	**0.27**	0.03	**0.30**	0.03	**0.35**	0.04	**0.39**
Brain		**0.02**		**0.02**		**0.02**		**0.03**		**0.04**		**0.04**
Oral mucosa	0.01	**0.22**	0.02	**0.26**	0.02	**0.30**	0.02	**0.32**	0.03	**0.37**	0.03	**0.42**
Salivary glands	0.03	**0.32**	0.04	**0.37**	0.04	**0.40**	0.05	**0.42**	0.06	**0.47**	0.06	**0.51**
Cervical spine	0.29	**0.54**	0.37	**0.68**	0.42	**0.80**	0.47	**0.89**	0.57	**1.07**	0.65	**1.23**
Thyroid	0.25	**0.24**	0.30	**0.30**	0.34	**0.34**	0.37	**0.37**	0.43	**0.43**	0.47	**0.49**
Thymus	0.10	**0.10**	0.13	**0.14**	0.16	**0.16**	0.18	**0.18**	0.22	**0.23**	0.26	**0.27**
Thoracic spine	0.20	**0.22**	0.27	**0.29**	0.32	**0.34**	0.37	**0.39**	0.46	**0.49**	0.56	**0.60**
Claviculae	0.47	**0.50**	0.58	**0.62**	0.67	**0.71**	0.73	**0.78**	0.86	**0.92**	0.96	**1.03**
Scapulae	0.98	**1.01**	1.15	**1.19**	1.27	**1.31**	1.36	**1.40**	1.50	**1.56**	1.62	**1.68**
Ribs	0.78	**0.85**	0.92	**1.01**	1.02	**1.11**	1.10	**1.19**	1.23	**1.34**	1.34	**1.46**
Oesophagus	0.10	**0.11**	0.13	**0.14**	0.16	**0.17**	0.18	**0.19**	0.22	**0.23**	0.26	**0.28**
Trachea	0.11	**0.24**	0.13	**0.31**	0.15	**0.35**	0.16	**0.39**	0.19	**0.47**	0.21	**0.54**
Lung	0.25	**0.26**	0.31	**0.32**	0.35	**0.36**	0.39	**0.40**	0.45	**0.47**	0.51	**0.53**
Mamma	0.47	**0.48**	0.51	**0.52**	0.54	**0.55**	0.56	**0.57**	0.59	**0.60**	0.62	**0.63**
Heart	0.12	**0.13**	0.16	**0.17**	0.19	**0.20**	0.21	**0.22**	0.26	**0.27**	0.31	**0.33**
Stomach	0.02	**0.03**	0.03	**0.04**	0.03	**0.05**	0.04	**0.06**	0.05	**0.07**	0.06	**0.09**
Spleen	0.01	**0.02**	0.02	**0.02**	0.02	**0.03**	0.03	**0.04**	0.04	**0.05**	0.04	**0.06**
Liver	0.21	**0.40**	0.26	**0.49**	0.30	**0.56**	0.33	**0.61**	0.39	**0.71**	0.44	**0.80**
Gall bladder	0.04	**0.12**	0.05	**0.16**	0.06	**0.19**	0.06	**0.21**	0.08	**0.27**	0.10	**0.32**
Pancreas	0.05	**0.06**	0.06	**0.08**	0.08	**0.10**	0.09	**0.11**	0.11	**0.15**	0.14	**0.18**
Small intestine		**0.07**		**0.08**		**0.10**		**0.11**		**0.14**	0.01	**0.17**
Large intestine		**0.10**		**0.13**		**0.15**	0.01	**0.17**	0.01	**0.21**	0.02	**0.24**
Sigma/Rectum								**0.01**		**0.01**		**0.02**
Lumbar spine	0.03	**0.15**	0.04	**0.20**	0.05	**0.24**	0.06	**0.28**	0.08	**0.35**	0.09	**0.43**
Pelvis		**0.04**		**0.05**		**0.06**		**0.07**		**0.09**		**0.10**
Adrenals	0.08	**0.10**	0.11	**0.13**	0.13	**0.15**	0.14	**0.17**	0.18	**0.21**	0.22	**0.26**
Kidneys	0.04	**0.13**	0.05	**0.16**	0.06	**0.19**	0.06	**0.21**	0.08	**0.26**	0.09	**0.30**
Urinary bladder												**0.01**
Uterus		**0.01**		**0.02**		**0.02**		**0.02**		**0.03**		**0.03**
Ovaries		**0.01**		**0.02**		**0.02**		**0.02**		**0.03**		**0.04**
Prostate												
Testes												
Upper arm												
Elbow												
Forearm												
Upper leg												
Knee												
Lower leg												
Lymph nodes	0.07	**0.14**	0.09	**0.18**	0.10	**0.20**	0.11	**0.22**	0.14	**0.27**	0.16	**0.31**
Skin	0.10	**0.15**	0.10	**0.16**	0.11	**0.17**	0.11	**0.17**	0.12	**0.18**	0.12	**0.19**
Muscle	0.06	**0.09**	0.07	**0.10**	0.07	**0.12**	0.08	**0.13**	0.09	**0.14**	0.11	**0.16**
Skeleton	0.13	**0.19**	0.16	**0.23**	0.18	**0.26**	0.20	**0.29**	0.23	**0.33**	0.25	**0.37**
Red bone marrow	0.05	**0.07**	0.07	**0.09**	0.08	**0.10**	0.08	**0.11**	0.10	**0.14**	0.11	**0.16**
Whole body dose	0.07	**0.11**	0.09	**0.13**	0.10	**0.15**	0.10	**0.16**	0.12	**0.19**	0.14	**0.21**
ED (ICRP 60)	0.09	**0.12**	0.11	**0.14**	0.13	**0.16**	0.14	**0.18**	0.16	**0.21**	0.18	**0.24**
ED (ICRP 103)	0.13	**0.16**	0.15	**0.19**	0.16	**0.21**	0.18	**0.23**	0.20	**0.26**	0.22	**0.29**
Total filtration (mm) **Al**	**1.5**		**2.5**		**3.5**		**4.5**		**3.5**		**3.5**	
Cu	**0**		**0**		**0**		**0**		**0.1**		**0.2**	

Table 9.79 Normalized organ doses—Chest r-lat—8–12 years

Beam projection:			r-lat			**Table cells:**						
Focus-to-detector distance (cm):			100			(+) = normal font = optimal field setting						
Tube voltage (kVp):			80			**(−) = bold font = suboptimal field setting**						
Total filtrations:			See last table row.			Empty table cells correspond to values < 0.01						
Organ or tissue	**Organ dose D_T : Entrance dose K_E (mSv × mGy^{-1})**											
	(+)	(−)	(+)	(−)	(+)	(−)	(+)	(−)	(+)	(−)	(+)	(−)
Skull	0.02	**0.22**	0.03	**0.27**	0.03	**0.30**	0.03	**0.33**	0.04	**0.38**	0.05	**0.42**
Brain		**0.02**		**0.03**		**0.03**		**0.04**		**0.04**		**0.05**
Oral mucosa	0.02	**0.25**	0.02	**0.30**	0.02	**0.33**	0.03	**0.36**	0.03	**0.41**	0.04	**0.46**
Salivary glands	0.03	**0.35**	0.04	**0.40**	0.05	**0.43**	0.05	**0.45**	0.06	**0.50**	0.07	**0.54**
Cervical spine	0.34	**0.64**	0.42	**0.79**	0.48	**0.91**	0.53	**1.00**	0.62	**1.18**	0.71	**1.34**
Thyroid	0.28	**0.28**	0.34	**0.34**	0.38	**0.38**	0.41	**0.42**	0.46	**0.48**	0.51	**0.54**
Thymus	0.13	**0.13**	0.16	**0.17**	0.19	**0.20**	0.21	**0.22**	0.26	**0.27**	0.31	**0.32**
Thoracic spine	0.26	**0.28**	0.34	**0.36**	0.39	**0.42**	0.45	**0.48**	0.55	**0.59**	0.65	**0.70**
Claviculae	0.54	**0.57**	0.65	**0.70**	0.73	**0.79**	0.80	**0.86**	0.92	**0.99**	1.01	**1.10**
Scapulae	1.06	**1.10**	1.23	**1.27**	1.33	**1.38**	1.41	**1.46**	1.55	**1.60**	1.64	**1.70**
Ribs	0.85	**0.93**	0.99	**1.08**	1.09	**1.18**	1.16	**1.26**	1.28	**1.39**	1.37	**1.50**
Oesophagus	0.13	**0.14**	0.16	**0.17**	0.19	**0.20**	0.21	**0.23**	0.26	**0.27**	0.30	**0.32**
Trachea	0.12	**0.29**	0.15	**0.36**	0.17	**0.41**	0.18	**0.45**	0.21	**0.52**	0.23	**0.59**
Lung	0.29	**0.30**	0.35	**0.36**	0.39	**0.41**	0.43	**0.45**	0.50	**0.52**	0.56	**0.58**
Mamma	0.49	**0.51**	0.54	**0.55**	0.56	**0.57**	0.58	**0.59**	0.61	**0.62**	0.64	**0.65**
Heart	0.15	**0.16**	0.20	**0.20**	0.23	**0.24**	0.25	**0.27**	0.31	**0.32**	0.36	**0.38**
Stomach	0.03	**0.04**	0.04	**0.05**	0.04	**0.06**	0.05	**0.07**	0.06	**0.09**	0.07	**0.11**
Spleen	0.02	**0.03**	0.03	**0.03**	0.03	**0.04**	0.04	**0.05**	0.05	**0.06**	0.06	**0.08**
Liver	0.25	**0.46**	0.30	**0.55**	0.34	**0.62**	0.37	**0.68**	0.43	**0.78**	0.49	**0.87**
Gall bladder	0.05	**0.15**	0.06	**0.20**	0.07	**0.23**	0.08	**0.26**	0.10	**0.32**	0.12	**0.37**
Pancreas	0.06	**0.08**	0.08	**0.11**	0.10	**0.13**	0.11	**0.15**	0.14	**0.18**	0.17	**0.22**
Small intestine		**0.08**		**0.10**		**0.12**		**0.14**	0.01	**0.17**	0.02	**0.19**
Large intestine		**0.12**		**0.16**	0.01	**0.18**	0.01	**0.20**	0.02	**0.24**	0.02	**0.28**
Sigma/Rectum				**0.01**		**0.01**		**0.01**		**0.02**		**0.02**
Lumbar spine	0.04	**0.20**	0.06	**0.25**	0.07	**0.30**	0.08	**0.34**	0.10	**0.42**	0.12	**0.50**
Pelvis		**0.05**		**0.06**		**0.07**		**0.08**		**0.10**		**0.12**
Adrenals	0.11	**0.12**	0.13	**0.16**	0.16	**0.18**	0.18	**0.21**	0.22	**0.26**	0.25	**0.30**
Kidneys	0.05	**0.16**	0.06	**0.20**	0.07	**0.23**	0.08	**0.25**	0.09	**0.30**	0.11	**0.34**
Urinary bladder										**0.01**		**0.01**
Uterus		**0.02**		**0.02**		**0.02**		**0.03**		**0.03**		**0.04**
Ovaries		**0.02**		**0.02**		**0.03**		**0.03**		**0.04**		**0.05**
Prostate												
Testes												
Upper arm												
Elbow												
Forearm												
Upper leg												
Knee												
Lower leg												
Lymph nodes	0.08	**0.17**	0.10	**0.21**	0.12	**0.24**	0.13	**0.26**	0.15	**0.31**	0.18	**0.35**
Skin	0.10	**0.16**	0.11	**0.17**	0.11	**0.18**	0.12	**0.18**	0.12	**0.19**	0.13	**0.20**
Muscle	0.06	**0.10**	0.07	**0.12**	0.08	**0.13**	0.09	**0.14**	0.10	**0.16**	0.12	**0.18**
Skeleton	0.15	**0.22**	0.18	**0.26**	0.20	**0.29**	0.21	**0.31**	0.24	**0.36**	0.26	**0.39**
Red bone marrow	0.06	**0.09**	0.08	**0.11**	0.09	**0.12**	0.10	**0.13**	0.11	**0.16**	0.13	**0.18**
Whole body dose	0.08	**0.13**	0.10	**0.15**	0.11	**0.17**	0.12	**0.18**	0.13	**0.21**	0.15	**0.23**
ED (ICRP 60)	0.11	**0.14**	0.13	**0.16**	0.14	**0.19**	0.15	**0.20**	0.18	**0.23**	0.20	**0.26**
ED (ICRP 103)	0.14	**0.18**	0.16	**0.21**	0.18	**0.23**	0.19	**0.25**	0.22	**0.29**	0.24	**0.32**
Total filtration (mm) Al	**1.5**		**2.5**		**3.5**		**4.5**		**3.5**		**3.5**	
Cu	**0**		**0**		**0**		**0**		**0.1**		**0.2**	

Table 9.80 Normalized organ doses—Chest r-lat—13–17 years

Beam projection:			r-lat				**Table cells:**					
Focus-to-detector distance (cm):			100				(+) = normal font = optimal field setting					
Tube voltage (kVp):			60				**(–) = bold font = suboptimal field setting**					
Total filtrations:			See last table row.				Empty table cells correspond to values < 0.01					
Organ or tissue	**Organ dose D_T : Entrance dose K_E**								**(mSv × mGy^{-1})**			
	(+)	(–)	(+)	(–)	(+)	(–)	(+)	(–)	(+)	(–)	(+)	(–)
Skull	0.01	**0.13**	0.01	**0.17**	0.02	**0.20**	0.02	**0.22**	0.02	**0.26**	0.02	**0.24**
Brain				**0.01**		**0.01**		**0.02**		**0.02**		**0.02**
Oral mucosa		**0.14**		**0.17**	0.01	**0.20**	0.01	**0.22**	0.02	**0.26**	0.01	**0.24**
Salivary glands	0.02	**0.22**	0.02	**0.26**	0.02	**0.28**	0.03	**0.30**	0.03	**0.34**	0.03	**0.33**
Cervical spine	0.23	**0.38**	0.30	**0.50**	0.35	**0.59**	0.40	**0.67**	0.49	**0.83**	0.44	**0.75**
Thyroid	0.12	**0.10**	0.15	**0.12**	0.18	**0.14**	0.20	**0.16**	0.23	**0.20**	0.22	**0.18**
Thymus	0.04	**0.04**	0.06	**0.06**	0.07	**0.07**	0.08	**0.08**	0.10	**0.11**	0.08	**0.09**
Thoracic spine	0.07	**0.08**	0.10	**0.11**	0.12	**0.13**	0.14	**0.16**	0.19	**0.21**	0.15	**0.16**
Claviculae	0.28	**0.30**	0.36	**0.38**	0.42	**0.45**	0.47	**0.50**	0.57	**0.60**	0.53	**0.57**
Scapulae	0.75	**0.79**	0.90	**0.95**	1.02	**1.07**	1.10	**1.16**	1.25	**1.32**	1.23	**1.30**
Ribs	0.56	**0.62**	0.68	**0.75**	0.76	**0.85**	0.83	**0.92**	0.95	**1.05**	0.92	**1.03**
Oesophagus	0.05	**0.05**	0.07	**0.07**	0.08	**0.09**	0.10	**0.10**	0.12	**0.13**	0.10	**0.11**
Trachea	0.07	**0.12**	0.08	**0.15**	0.10	**0.18**	0.11	**0.20**	0.13	**0.24**	0.12	**0.22**
Lung	0.15	**0.16**	0.19	**0.20**	0.22	**0.24**	0.25	**0.27**	0.30	**0.32**	0.28	**0.30**
Mamma	1.99	**2.65**	2.36	**3.13**	2.61	**3.46**	2.81	**3.71**	3.16	**4.16**	3.06	**4.04**
Heart	0.05	**0.06**	0.07	**0.08**	0.09	**0.09**	0.10	**0.11**	0.13	**0.14**	0.11	**0.12**
Stomach				**0.01**	0.01	**0.01**	0.01	**0.02**	0.02	**0.02**	0.01	**0.02**
Spleen									0.01	**0.01**		
Liver	0.16	**0.29**	0.20	**0.37**	0.23	**0.42**	0.26	**0.47**	0.31	**0.56**	0.29	**0.52**
Gall bladder	0.02	**0.08**	0.03	**0.11**	0.03	**0.14**	0.04	**0.16**	0.05	**0.20**	0.04	**0.17**
Pancreas	0.02	**0.02**	0.02	**0.03**	0.03	**0.04**	0.03	**0.04**	0.05	**0.06**	0.03	**0.04**
Small intestine		**0.03**		**0.03**		**0.04**		**0.05**		**0.06**		**0.05**
Large intestine		**0.04**		**0.06**		**0.07**		**0.08**		**0.10**		**0.09**
Sigma/Rectum												
Lumbar spine	0.01	**0.05**	0.02	**0.07**	0.02	**0.08**	0.02	**0.10**	0.03	**0.13**	0.02	**0.10**
Pelvis		**0.01**		**0.02**		**0.02**		**0.02**		**0.03**		**0.02**
Adrenals	0.04	**0.04**	0.05	**0.06**	0.06	**0.07**	0.07	**0.08**	0.09	**0.10**	0.07	**0.08**
Kidneys	0.02	**0.06**	0.03	**0.08**	0.03	**0.10**	0.04	**0.11**	0.05	**0.14**	0.04	**0.13**
Urinary bladder												
Uterus				**0.01**		**0.02**		**0.02**		**0.03**		**0.02**
Ovaries												
Prostate												
Testes												
Upper arm												
Elbow												
Forearm												
Upper leg												
Knee												
Lower leg												
Lymph nodes	0.04	**0.08**	0.05	**0.10**	0.06	**0.11**	0.07	**0.13**	0.08	**0.15**	0.07	**0.14**
Skin	0.10	**0.16**	0.11	**0.17**	0.12	**0.17**	0.12	**0.18**	0.13	**0.19**	0.12	**0.19**
Muscle	0.04	**0.06**	0.05	**0.07**	0.05	**0.08**	0.06	**0.09**	0.07	**0.10**	0.07	**0.10**
Skeleton	0.10	**0.13**	0.12	**0.16**	0.14	**0.19**	0.15	**0.20**	0.17	**0.24**	0.17	**0.23**
Red bone marrow	0.04	**0.05**	0.05	**0.07**	0.06	**0.08**	0.07	**0.09**	0.08	**0.10**	0.07	**0.09**
Whole body dose	0.05	**0.08**	0.06	**0.10**	0.07	**0.11**	0.08	**0.12**	0.09	**0.14**	0.09	**0.13**
ED (ICRP 60)	0.14	**0.19**	0.17	**0.23**	0.19	**0.26**	0.21	**0.28**	0.24	**0.32**	0.23	**0.30**
ED (ICRP 103)	0.28	**0.38**	0.34	**0.45**	0.38	**0.50**	0.41	**0.54**	0.46	**0.62**	0.44	**0.59**
Total filtration (mm) Al	**1.5**		**2.5**		**3.5**		**4.5**		**3.5**		**3.5**	
Cu	**0**		**0**		**0**		**0**		**0.1**		**0.2**	

Table 9.81 Normalized organ doses—Chest r-lat—13–17 years

Beam projection:			r-lat			**Table cells:**						
Focus-to-detector distance (cm):			100			(+) = normal font = optimal field setting						
Tube voltage (kVp):			70			**(−) = bold font = suboptimal field setting**						
Total filtrations:			See last table row.			Empty table cells correspond to values < 0.01						
Organ or tissue	**Organ dose D_T : Entrance dose K_E**								**(mSv × mGy^{-1})**			
	(+)	(−)	(+)	(−)	(+)	(−)	(+)	(−)	(+)	(−)	(+)	(−)
Skull	0.01	**0.16**	0.02	**0.20**	0.02	**0.23**	0.02	**0.25**	0.03	**0.30**	0.03	**0.34**
Brain		**0.01**		**0.02**		**0.02**		**0.02**		**0.03**		**0.03**
Oral mucosa		**0.17**	0.01	**0.20**	0.01	**0.23**	0.02	**0.25**	0.02	**0.29**	0.02	**0.33**
Salivary glands	0.02	**0.25**	0.03	**0.29**	0.03	**0.31**	0.03	**0.34**	0.04	**0.38**	0.05	**0.41**
Cervical spine	0.29	**0.48**	0.36	**0.62**	0.43	**0.72**	0.48	**0.81**	0.58	**0.99**	0.67	**1.15**
Thyroid	0.15	**0.12**	0.18	**0.15**	0.21	**0.17**	0.23	**0.19**	0.27	**0.23**	0.30	**0.27**
Thymus	0.06	**0.06**	0.08	**0.08**	0.09	**0.10**	0.11	**0.11**	0.14	**0.14**	0.17	**0.18**
Thoracic spine	0.11	**0.12**	0.14	**0.16**	0.17	**0.19**	0.20	**0.22**	0.26	**0.28**	0.32	**0.35**
Claviculae	0.34	**0.36**	0.43	**0.45**	0.49	**0.52**	0.54	**0.58**	0.64	**0.69**	0.73	**0.79**
Scapulae	0.84	**0.89**	1.00	**1.05**	1.11	**1.17**	1.19	**1.26**	1.34	**1.42**	1.46	**1.54**
Ribs	0.63	**0.71**	0.75	**0.84**	0.84	**0.94**	0.91	**1.01**	1.02	**1.14**	1.13	**1.25**
Oesophagus	0.07	**0.08**	0.09	**0.10**	0.11	**0.12**	0.13	**0.14**	0.16	**0.17**	0.20	**0.21**
Trachea	0.08	**0.14**	0.10	**0.18**	0.11	**0.21**	0.13	**0.23**	0.15	**0.28**	0.17	**0.32**
Lung	0.18	**0.19**	0.23	**0.24**	0.27	**0.28**	0.30	**0.32**	0.35	**0.38**	0.41	**0.44**
Mamma	2.23	**2.96**	2.61	**3.45**	2.88	**3.79**	3.08	**4.05**	3.46	**4.52**	3.78	**4.92**
Heart	0.07	**0.08**	0.10	**0.10**	0.12	**0.12**	0.13	**0.14**	0.17	**0.18**	0.20	**0.22**
Stomach		**0.01**	0.01	**0.02**	0.02	**0.02**	0.02	**0.03**	0.02	**0.03**	0.03	**0.04**
Spleen				**0.01**	0.01	**0.01**	0.01	**0.02**	0.02	**0.02**	0.02	**0.03**
Liver	0.19	**0.35**	0.24	**0.43**	0.28	**0.50**	0.31	**0.55**	0.37	**0.65**	0.42	**0.74**
Gall bladder	0.03	**0.11**	0.04	**0.15**	0.05	**0.18**	0.05	**0.20**	0.07	**0.26**	0.08	**0.31**
Pancreas	0.03	**0.03**	0.04	**0.05**	0.04	**0.06**	0.05	**0.06**	0.06	**0.08**	0.08	**0.10**
Small intestine		**0.04**		**0.05**		**0.06**		**0.06**		**0.08**		**0.10**
Large intestine		**0.06**		**0.08**		**0.09**		**0.10**		**0.13**	0.01	**0.15**
Sigma/Rectum												
Lumbar spine	0.02	**0.07**	0.02	**0.10**	0.03	**0.12**	0.04	**0.14**	0.05	**0.18**	0.06	**0.22**
Pelvis		**0.02**		**0.02**		**0.03**		**0.03**		**0.04**		**0.05**
Adrenals	0.05	**0.06**	0.07	**0.08**	0.08	**0.09**	0.09	**0.11**	0.12	**0.14**	0.14	**0.17**
Kidneys	0.03	**0.08**	0.03	**0.11**	0.04	**0.13**	0.05	**0.14**	0.06	**0.18**	0.07	**0.21**
Urinary bladder												
Uterus		**0.02**		**0.02**		**0.03**		**0.03**		**0.04**		**0.05**
Ovaries								**0.01**		**0.01**		**0.02**
Prostate												
Testes												
Upper arm												
Elbow												
Forearm												
Upper leg												
Knee												
Lower leg												
Lymph nodes	0.05	**0.09**	0.06	**0.12**	0.07	**0.14**	0.08	**0.15**	0.10	**0.19**	0.11	**0.22**
Skin	0.11	**0.16**	0.12	**0.17**	0.12	**0.18**	0.13	**0.19**	0.13	**0.20**	0.14	**0.21**
Muscle	0.05	**0.07**	0.06	**0.08**	0.06	**0.09**	0.07	**0.10**	0.08	**0.12**	0.09	**0.14**
Skeleton	0.11	**0.15**	0.14	**0.19**	0.15	**0.21**	0.17	**0.23**	0.19	**0.27**	0.22	**0.30**
Red bone marrow	0.05	**0.07**	0.06	**0.08**	0.07	**0.09**	0.08	**0.10**	0.09	**0.12**	0.11	**0.14**
Whole body dose	0.06	**0.09**	0.07	**0.11**	0.08	**0.12**	0.09	**0.14**	0.11	**0.16**	0.12	**0.18**
ED (ICRP 60)	0.16	**0.22**	0.20	**0.26**	0.22	**0.29**	0.24	**0.31**	0.27	**0.36**	0.31	**0.40**
ED (ICRP 103)	0.32	**0.43**	0.38	**0.50**	0.42	**0.56**	0.45	**0.60**	0.51	**0.68**	0.57	**0.75**
Total filtration (mm) **Al**	**1.5**		**2.5**		**3.5**		**4.5**		**3.5**		**3.5**	
Cu	**0**		**0**		**0**		**0**		**0.1**		**0.2**	

Table 9.82 Normalized organ doses—Chest r-lat—13–17 years

Beam projection:			r-lat		**Table cells:**							
Focus-to-detector distance (cm):			100		(+) = normal font = optimal field setting							
Tube voltage (kVp):			80		**(−) = bold font = suboptimal field setting**							
Total filtrations:			See last table row.		Empty table cells correspond to values < 0.01							
Organ or tissue	**Organ dose D_T : Entrance dose K_E**						**(mSv × mGy^{-1})**					
	(+)	(−)	(+)	(−)	(+)	(−)	(+)	(−)	(+)	(−)	(+)	(−)
Skull	0.02	**0.19**	0.02	**0.23**	0.03	**0.26**	0.03	**0.28**	0.03	**0.33**	0.04	**0.37**
Brain		**0.02**		**0.02**		**0.02**		**0.03**		**0.03**		**0.04**
Oral mucosa	0.01	**0.19**	0.01	**0.23**	0.02	**0.26**	0.02	**0.28**	0.02	**0.33**	0.03	**0.37**
Salivary glands	0.02	**0.27**	0.03	**0.31**	0.04	**0.34**	0.04	**0.37**	0.05	**0.41**	0.05	**0.44**
Cervical spine	0.34	**0.58**	0.43	**0.73**	0.49	**0.84**	0.55	**0.93**	0.65	**1.11**	0.74	**1.28**
Thyroid	0.17	**0.14**	0.21	**0.18**	0.24	**0.20**	0.26	**0.22**	0.30	**0.27**	0.34	**0.30**
Thymus	0.08	**0.08**	0.10	**0.11**	0.12	**0.12**	0.14	**0.14**	0.17	**0.18**	0.20	**0.21**
Thoracic spine	0.15	**0.16**	0.19	**0.20**	0.22	**0.24**	0.25	**0.28**	0.32	**0.35**	0.39	**0.42**
Claviculae	0.39	**0.42**	0.48	**0.52**	0.55	**0.59**	0.60	**0.64**	0.70	**0.75**	0.78	**0.84**
Scapulae	0.92	**0.97**	1.07	**1.13**	1.18	**1.25**	1.26	**1.33**	1.39	**1.47**	1.50	**1.59**
Ribs	0.70	**0.78**	0.82	**0.91**	0.90	**1.01**	0.96	**1.08**	1.08	**1.20**	1.17	**1.30**
Oesophagus	0.09	**0.10**	0.12	**0.13**	0.14	**0.15**	0.16	**0.17**	0.20	**0.22**	0.24	**0.26**
Trachea	0.09	**0.17**	0.11	**0.21**	0.13	**0.24**	0.14	**0.27**	0.16	**0.32**	0.18	**0.36**
Lung	0.22	**0.23**	0.27	**0.29**	0.31	**0.33**	0.34	**0.36**	0.40	**0.43**	0.45	**0.49**
Mamma	2.47	**3.25**	2.85	**3.75**	3.12	**4.09**	3.33	**4.35**	3.71	**4.83**	4.03	**5.23**
Heart	0.10	**0.10**	0.12	**0.13**	0.14	**0.16**	0.16	**0.18**	0.20	**0.22**	0.24	**0.26**
Stomach	0.01	**0.02**	0.02	**0.03**	0.02	**0.03**	0.03	**0.04**	0.03	**0.05**	0.04	**0.06**
Spleen		**0.01**	0.01	**0.02**	0.02	**0.02**	0.02	**0.02**	0.02	**0.03**	0.03	**0.04**
Liver	0.23	**0.41**	0.28	**0.50**	0.32	**0.56**	0.35	**0.62**	0.41	**0.72**	0.47	**0.81**
Gall bladder	0.04	**0.15**	0.05	**0.19**	0.06	**0.22**	0.07	**0.25**	0.08	**0.31**	0.10	**0.36**
Pancreas	0.04	**0.05**	0.05	**0.06**	0.06	**0.07**	0.07	**0.08**	0.08	**0.11**	0.10	**0.13**
Small intestine		**0.05**		**0.06**		**0.07**		**0.08**		**0.10**		**0.12**
Large intestine		**0.07**		**0.09**		**0.11**		**0.12**	0.01	**0.15**	0.01	**0.18**
Sigma/Rectum												**0.01**
Lumbar spine	0.03	**0.10**	0.03	**0.13**	0.04	**0.16**	0.05	**0.18**	0.06	**0.23**	0.08	**0.28**
Pelvis		**0.02**		**0.03**		**0.04**		**0.04**		**0.05**		**0.06**
Adrenals	0.07	**0.08**	0.09	**0.10**	0.10	**0.12**	0.12	**0.14**	0.15	**0.17**	0.17	**0.20**
Kidneys	0.03	**0.10**	0.04	**0.13**	0.05	**0.15**	0.06	**0.17**	0.07	**0.21**	0.09	**0.24**
Urinary bladder												
Uterus		**0.02**		**0.03**		**0.04**		**0.04**		**0.05**		**0.07**
Ovaries				**0.01**		**0.01**		**0.01**		**0.02**		**0.02**
Prostate												
Testes												
Upper arm												
Elbow												
Forearm												
Upper leg												
Knee												
Lower leg												
Lymph nodes	0.06	**0.12**	0.07	**0.14**	0.09	**0.16**	0.09	**0.18**	0.11	**0.22**	0.13	**0.25**
Skin	0.11	**0.17**	0.12	**0.18**	0.13	**0.19**	0.13	**0.20**	0.14	**0.21**	0.15	**0.22**
Muscle	0.05	**0.08**	0.06	**0.10**	0.07	**0.11**	0.08	**0.12**	0.09	**0.13**	0.10	**0.15**
Skeleton	0.13	**0.17**	0.15	**0.21**	0.17	**0.23**	0.18	**0.25**	0.21	**0.29**	0.23	**0.32**
Red bone marrow	0.06	**0.08**	0.07	**0.10**	0.08	**0.11**	0.09	**0.12**	0.11	**0.14**	0.12	**0.16**
Whole body dose	0.07	**0.11**	0.08	**0.13**	0.09	**0.14**	0.10	**0.15**	0.12	**0.18**	0.13	**0.20**
ED (ICRP 60)	0.19	**0.25**	0.22	**0.29**	0.25	**0.32**	0.26	**0.35**	0.30	**0.40**	0.33	**0.44**
ED (ICRP 103)	0.36	**0.48**	0.42	**0.56**	0.46	**0.61**	0.50	**0.66**	0.56	**0.74**	0.62	**0.81**
Total filtration (mm) Al	**1.5**		**2.5**		**3.5**		**4.5**		**3.5**		**3.5**	
Cu	**0**		**0**		**0**		**0**		**0.1**		**0.2**	

Table 9.83 Normalized organ doses—Chest r-lat—13–17 years

Beam projection:			r-lat			**Table cells:**						
Focus-to-detector distance (cm):			100			(+) = normal font = optimal field setting						
Tube voltage (kVp):			110			**(−) = bold font = suboptimal field setting**						
Total filtrations:			See last table row.			Empty table cells correspond to values < 0.01						
Organ or tissue	**Organ dose D_T : Entrance dose K_E**							**(mSv × mGy^{-1})**				
	(+)	(−)	(+)	(−)	(+)	(−)	(+)	(−)	(+)	(−)	(+)	(−)
Skull	0.03	**0.25**	0.03	**0.28**	0.03	**0.31**	0.04	**0.33**	0.04	**0.37**	0.05	**0.40**
Brain		**0.03**		**0.03**		**0.04**		**0.04**		**0.05**		**0.06**
Oral mucosa	0.02	**0.26**	0.02	**0.30**	0.03	**0.33**	0.03	**0.36**	0.03	**0.40**	0.04	**0.43**
Salivary glands	0.04	**0.34**	0.04	**0.38**	0.05	**0.41**	0.05	**0.43**	0.06	**0.46**	0.07	**0.50**
Cervical spine	0.48	**0.82**	0.56	**0.97**	0.62	**1.07**	0.67	**1.15**	0.76	**1.30**	0.83	**1.43**
Thyroid	0.24	**0.21**	0.27	**0.24**	0.30	**0.27**	0.32	**0.29**	0.36	**0.33**	0.39	**0.37**
Thymus	0.13	**0.14**	0.16	**0.17**	0.18	**0.19**	0.20	**0.21**	0.24	**0.25**	0.27	**0.28**
Thoracic spine	0.25	**0.27**	0.30	**0.33**	0.34	**0.37**	0.37	**0.41**	0.44	**0.48**	0.50	**0.55**
Claviculae	0.52	**0.55**	0.60	**0.64**	0.65	**0.70**	0.70	**0.75**	0.77	**0.83**	0.83	**0.89**
Scapulae	1.08	**1.14**	1.20	**1.28**	1.28	**1.36**	1.34	**1.42**	1.43	**1.51**	1.48	**1.57**
Ribs	0.83	**0.93**	0.93	**1.04**	1.00	**1.12**	1.05	**1.17**	1.12	**1.26**	1.18	**1.32**
Oesophagus	0.16	**0.17**	0.19	**0.21**	0.22	**0.23**	0.24	**0.26**	0.28	**0.30**	0.31	**0.34**
Trachea	0.13	**0.25**	0.15	**0.29**	0.16	**0.32**	0.17	**0.35**	0.19	**0.39**	0.21	**0.43**
Lung	0.31	**0.33**	0.36	**0.39**	0.40	**0.43**	0.43	**0.46**	0.48	**0.52**	0.53	**0.57**
Mamma	3.06	**3.99**	3.42	**4.45**	3.67	**4.76**	3.85	**4.99**	4.18	**5.40**	4.45	**5.72**
Heart	0.16	**0.17**	0.19	**0.21**	0.22	**0.23**	0.24	**0.26**	0.28	**0.30**	0.31	**0.34**
Stomach	0.03	**0.04**	0.03	**0.05**	0.04	**0.06**	0.04	**0.06**	0.05	**0.08**	0.06	**0.09**
Spleen	0.02	**0.03**	0.03	**0.03**	0.03	**0.04**	0.03	**0.04**	0.04	**0.05**	0.05	**0.06**
Liver	0.32	**0.56**	0.38	**0.65**	0.41	**0.71**	0.44	**0.76**	0.50	**0.85**	0.55	**0.93**
Gall bladder	0.07	**0.24**	0.08	**0.29**	0.09	**0.32**	0.10	**0.35**	0.12	**0.41**	0.14	**0.46**
Pancreas	0.07	**0.09**	0.08	**0.11**	0.10	**0.12**	0.11	**0.14**	0.13	**0.16**	0.15	**0.19**
Small intestine		**0.08**		**0.09**		**0.11**	0.01	**0.12**	0.01	**0.14**	0.01	**0.16**
Large intestine		**0.12**	0.01	**0.14**	0.01	**0.16**	0.01	**0.17**	0.02	**0.20**	0.02	**0.23**
Sigma/Rectum						**0.01**		**0.01**		**0.01**		**0.02**
Lumbar spine	0.05	**0.18**	0.06	**0.22**	0.07	**0.25**	0.08	**0.27**	0.09	**0.32**	0.11	**0.37**
Pelvis		**0.04**		**0.05**		**0.06**		**0.06**		**0.07**		**0.09**
Adrenals	0.12	**0.14**	0.14	**0.16**	0.16	**0.19**	0.17	**0.20**	0.20	**0.24**	0.23	**0.27**
Kidneys	0.06	**0.16**	0.07	**0.19**	0.08	**0.22**	0.09	**0.24**	0.10	**0.27**	0.11	**0.31**
Urinary bladder												**0.01**
Uterus		**0.05**		**0.06**		**0.06**		**0.07**		**0.09**	0.01	**0.10**
Ovaries		**0.02**		**0.02**		**0.02**		**0.03**		**0.03**		**0.04**
Prostate												
Testes												
Upper arm												
Elbow												
Forearm												
Upper leg												
Knee												
Lower leg												
Lymph nodes	0.09	**0.17**	0.11	**0.20**	0.12	**0.22**	0.13	**0.24**	0.15	**0.28**	0.16	**0.31**
Skin	0.13	**0.19**	0.13	**0.20**	0.14	**0.21**	0.14	**0.21**	0.15	**0.22**	0.16	**0.23**
Muscle	0.07	**0.11**	0.09	**0.13**	0.09	**0.14**	0.10	**0.15**	0.11	**0.16**	0.12	**0.18**
Skeleton	0.16	**0.22**	0.18	**0.26**	0.20	**0.28**	0.21	**0.29**	0.23	**0.32**	0.24	**0.35**
Red bone marrow	0.09	**0.11**	0.10	**0.13**	0.11	**0.15**	0.12	**0.16**	0.13	**0.18**	0.15	**0.20**
Whole body dose	0.09	**0.14**	0.11	**0.16**	0.12	**0.18**	0.13	**0.19**	0.14	**0.21**	0.15	**0.23**
ED (ICRP 60)	0.24	**0.32**	0.28	**0.37**	0.30	**0.40**	0.32	**0.42**	0.35	**0.47**	0.38	**0.50**
ED (ICRP 103)	0.46	**0.61**	0.52	**0.68**	0.56	**0.74**	0.59	**0.78**	0.65	**0.85**	0.69	**0.91**
Total filtration (mm) **Al**	**1.5**		**2.5**		**3.5**		**4.5**		**3.5**		**3.5**	
Cu	**0**		**0**		**0**		**0**		**0.1**		**0.2**	

Table 9.84 Normalized organ doses—Chest r-lat—13–17 years

Beam projection:		r-lat		**Table cells:**								
Focus-to-detector distance (cm):		100		(+) = normal font = optimal field setting								
Tube voltage (kVp):		120		**(–) = bold font = suboptimal field setting**								
Total filtrations:		See last table row.		Empty table cells correspond to values < 0.01								
Organ or tissue	**Organ dose D_T : Entrance dose K_E**						**(mSv × mGy^{-1})**					
	(+)	(–)	(+)	(–)	(+)	(–)	(+)	(–)	(+)	(–)	(+)	(–)
Skull	0.03	**0.26**	0.03	**0.30**	0.04	**0.32**	0.04	**0.34**	0.04	**0.37**	0.05	**0.40**
Brain		**0.03**		**0.04**		**0.04**		**0.05**		**0.05**		**0.06**
Oral mucosa	0.02	**0.28**	0.03	**0.32**	0.03	**0.35**	0.03	**0.37**	0.03	**0.41**	0.04	**0.45**
Salivary glands	0.04	**0.36**	0.05	**0.40**	0.05	**0.42**	0.06	**0.44**	0.06	**0.48**	0.07	**0.51**
Cervical spine	0.52	**0.88**	0.59	**1.02**	0.65	**1.12**	0.69	**1.19**	0.77	**1.33**	0.83	**1.44**
Thyroid	0.26	**0.23**	0.29	**0.26**	0.32	**0.29**	0.34	**0.31**	0.37	**0.35**	0.40	**0.38**
Thymus	0.15	**0.16**	0.18	**0.19**	0.20	**0.21**	0.22	**0.23**	0.25	**0.26**	0.28	**0.29**
Thoracic spine	0.28	**0.30**	0.33	**0.36**	0.37	**0.40**	0.40	**0.44**	0.46	**0.50**	0.52	**0.57**
Claviculae	0.54	**0.59**	0.62	**0.67**	0.67	**0.72**	0.71	**0.77**	0.78	**0.84**	0.82	**0.89**
Scapulae	1.11	**1.18**	1.23	**1.30**	1.30	**1.37**	1.35	**1.43**	1.42	**1.50**	1.46	**1.55**
Ribs	0.86	**0.97**	0.96	**1.07**	1.01	**1.13**	1.06	**1.18**	1.12	**1.25**	1.16	**1.30**
Oesophagus	0.18	**0.19**	0.21	**0.23**	0.23	**0.25**	0.25	**0.28**	0.29	**0.32**	0.33	**0.36**
Trachea	0.14	**0.27**	0.16	**0.31**	0.17	**0.34**	0.18	**0.36**	0.20	**0.41**	0.22	**0.45**
Lung	0.33	**0.36**	0.38	**0.41**	0.42	**0.45**	0.45	**0.48**	0.50	**0.54**	0.54	**0.59**
Mamma	3.22	**4.19**	3.56	**4.63**	3.80	**4.92**	3.97	**5.13**	4.28	**5.51**	4.52	**5.81**
Heart	0.18	**0.19**	0.21	**0.23**	0.23	**0.25**	0.25	**0.28**	0.29	**0.32**	0.33	**0.36**
Stomach	0.03	**0.05**	0.04	**0.06**	0.04	**0.06**	0.05	**0.07**	0.06	**0.08**	0.07	**0.10**
Spleen	0.02	**0.03**	0.03	**0.04**	0.03	**0.04**	0.04	**0.05**	0.04	**0.05**	0.05	**0.06**
Liver	0.35	**0.60**	0.40	**0.69**	0.44	**0.75**	0.46	**0.79**	0.52	**0.88**	0.56	**0.95**
Gall bladder	0.08	**0.27**	0.09	**0.31**	0.10	**0.35**	0.11	**0.38**	0.13	**0.43**	0.14	**0.48**
Pancreas	0.08	**0.10**	0.09	**0.12**	0.11	**0.14**	0.12	**0.15**	0.14	**0.18**	0.15	**0.20**
Small intestine		**0.09**		**0.10**	0.01	**0.12**	0.01	**0.13**	0.01	**0.15**	0.02	**0.16**
Large intestine	0.01	**0.13**	0.01	**0.16**	0.02	**0.17**	0.02	**0.19**	0.02	**0.21**	0.02	**0.24**
Sigma/Rectum				**0.01**		**0.01**		**0.01**		**0.02**		**0.02**
Lumbar spine	0.06	**0.20**	0.07	**0.24**	0.08	**0.27**	0.08	**0.29**	0.10	**0.34**	0.11	**0.38**
Pelvis		**0.05**		**0.06**		**0.06**		**0.07**		**0.08**		**0.09**
Adrenals	0.13	**0.15**	0.16	**0.18**	0.17	**0.20**	0.19	**0.22**	0.22	**0.26**	0.25	**0.29**
Kidneys	0.06	**0.18**	0.08	**0.21**	0.09	**0.23**	0.09	**0.25**	0.11	**0.29**	0.12	**0.32**
Urinary bladder												**0.01**
Uterus		**0.05**		**0.06**		**0.07**		**0.08**		**0.09**	0.01	**0.11**
Ovaries		**0.02**		**0.02**		**0.03**		**0.03**		**0.03**		**0.04**
Prostate												
Testes												
Upper arm												
Elbow												
Forearm												
Upper leg												
Knee												
Lower leg												
Lymph nodes	0.10	**0.19**	0.11	**0.22**	0.13	**0.24**	0.14	**0.26**	0.15	**0.29**	0.17	**0.32**
Skin	0.13	**0.19**	0.14	**0.21**	0.14	**0.21**	0.15	**0.22**	0.15	**0.23**	0.16	**0.24**
Muscle	0.08	**0.12**	0.09	**0.13**	0.10	**0.14**	0.10	**0.15**	0.12	**0.17**	0.13	**0.18**
Skeleton	0.17	**0.23**	0.19	**0.26**	0.20	**0.28**	0.21	**0.30**	0.23	**0.33**	0.24	**0.34**
Red bone marrow	0.09	**0.13**	0.11	**0.14**	0.12	**0.16**	0.12	**0.17**	0.14	**0.19**	0.15	**0.21**
Whole body dose	0.10	**0.15**	0.11	**0.17**	0.12	**0.18**	0.13	**0.19**	0.14	**0.21**	0.16	**0.23**
ED (ICRP 60)	0.26	**0.34**	0.29	**0.39**	0.32	**0.42**	0.33	**0.44**	0.37	**0.48**	0.39	**0.52**
ED (ICRP 103)	0.48	**0.64**	0.54	**0.72**	0.58	**0.77**	0.61	**0.81**	0.66	**0.88**	0.71	**0.93**
Total filtration (mm) Al	**1.5**		**2.5**		**3.5**		**4.5**		**3.5**		**3.5**	
Cu	**0**		**0**		**0**		**0**		**0.1**		**0.2**	

Table 9.85 Normalized organ doses—Chest r-lat—> 17 years

Beam projection:			r-lat		**Table cells:**							
Focus-to-detector distance (cm):			100		(+) = normal font = optimal field setting							
Tube voltage (kVp):			60		**(−) = bold font = suboptimal field setting**							
Total filtrations:			See last table row.		Empty table cells correspond to values < 0.01							
Organ or tissue	**Organ dose D_T : Entrance dose K_E**							**(mSv × mGy^{-1})**				
	(+)	(−)	(+)	(−)	(+)	(−)	(+)	(−)	(+)	(−)	(+)	(−)
Skull	0.02	**0.14**	0.02	**0.18**	0.03	**0.21**	0.03	**0.23**	0.04	**0.28**	0.03	**0.26**
Brain		**0.01**		**0.02**		**0.02**		**0.02**		**0.03**		**0.03**
Oral mucosa		**0.12**	0.01	**0.15**	0.01	**0.17**	0.02	**0.19**	0.02	**0.22**	0.02	**0.21**
Salivary glands	0.03	**0.23**	0.04	**0.27**	0.05	**0.30**	0.05	**0.32**	0.06	**0.36**	0.06	**0.34**
Cervical spine	0.21	**0.30**	0.28	**0.39**	0.33	**0.47**	0.37	**0.53**	0.46	**0.66**	0.42	**0.59**
Thyroid	0.09	**0.06**	0.11	**0.08**	0.13	**0.09**	0.14	**0.10**	0.17	**0.13**	0.16	**0.11**
Thymus	0.03	**0.03**	0.04	**0.04**	0.05	**0.05**	0.05	**0.06**	0.07	**0.08**	0.05	**0.06**
Thoracic spine	0.05	**0.05**	0.07	**0.07**	0.08	**0.09**	0.10	**0.11**	0.13	**0.14**	0.10	**0.11**
Claviculae	0.21	**0.23**	0.28	**0.29**	0.33	**0.35**	0.37	**0.39**	0.45	**0.47**	0.41	**0.44**
Scapulae	0.65	**0.70**	0.79	**0.85**	0.90	**0.97**	0.98	**1.05**	1.12	**1.21**	1.10	**1.18**
Ribs	0.46	**0.54**	0.56	**0.65**	0.63	**0.73**	0.68	**0.80**	0.79	**0.92**	0.76	**0.89**
Oesophagus	0.04	**0.04**	0.05	**0.06**	0.06	**0.07**	0.07	**0.08**	0.10	**0.10**	0.08	**0.08**
Trachea	0.07	**0.09**	0.09	**0.12**	0.10	**0.14**	0.11	**0.16**	0.13	**0.19**	0.13	**0.17**
Lung	0.12	**0.13**	0.15	**0.17**	0.18	**0.20**	0.20	**0.22**	0.25	**0.27**	0.23	**0.25**
Mamma	0.20	**0.27**	0.24	**0.32**	0.27	**0.36**	0.29	**0.39**	0.32	**0.43**	0.31	**0.42**
Heart	0.04	**0.04**	0.05	**0.05**	0.06	**0.07**	0.07	**0.08**	0.09	**0.10**	0.07	**0.08**
Stomach										**0.01**		
Spleen												
Liver	0.11	**0.24**	0.14	**0.30**	0.16	**0.35**	0.18	**0.39**	0.22	**0.47**	0.20	**0.43**
Gall bladder	0.01	**0.05**	0.01	**0.06**	0.02	**0.08**	0.02	**0.09**	0.03	**0.12**	0.02	**0.10**
Pancreas		**0.01**	0.01	**0.02**	0.02	**0.02**	0.02	**0.03**	0.03	**0.04**	0.02	**0.03**
Small intestine		**0.02**		**0.02**		**0.03**		**0.03**		**0.04**		**0.04**
Large intestine		**0.03**		**0.04**		**0.05**		**0.06**		**0.08**		**0.06**
Sigma/Rectum												
Lumbar spine		**0.03**		**0.04**	0.01	**0.05**	0.01	**0.05**	0.02	**0.07**	0.01	**0.05**
Pelvis				**0.01**		**0.02**		**0.02**		**0.02**		**0.02**
Adrenals	0.02	**0.03**	0.03	**0.04**	0.04	**0.05**	0.05	**0.05**	0.06	**0.07**	0.05	**0.06**
Kidneys	0.01	**0.04**	0.02	**0.06**	0.02	**0.07**	0.02	**0.08**	0.03	**0.10**	0.02	**0.09**
Urinary bladder												
Uterus												
Ovaries												
Prostate												
Testes												
Upper arm												
Elbow												
Forearm												
Upper leg												
Knee												
Lower leg												
Lymph nodes	0.03	**0.06**	0.04	**0.08**	0.05	**0.09**	0.05	**0.10**	0.07	**0.12**	0.06	**0.11**
Skin	0.08	**0.12**	0.09	**0.13**	0.09	**0.14**	0.09	**0.14**	0.10	**0.15**	0.10	**0.15**
Muscle	0.03	**0.05**	0.04	**0.06**	0.05	**0.07**	0.05	**0.07**	0.06	**0.09**	0.05	**0.08**
Skeleton	0.08	**0.11**	0.10	**0.14**	0.11	**0.16**	0.12	**0.17**	0.14	**0.20**	0.14	**0.19**
Red bone marrow	0.04	**0.05**	0.05	**0.06**	0.06	**0.07**	0.06	**0.08**	0.07	**0.10**	0.07	**0.09**
Whole body dose	0.04	**0.07**	0.05	**0.08**	0.06	**0.09**	0.07	**0.10**	0.08	**0.12**	0.07	**0.11**
ED (ICRP 60)	0.04	**0.06**	0.05	**0.07**	0.06	**0.08**	0.07	**0.09**	0.08	**0.11**	0.08	**0.10**
ED (ICRP 103)	0.06	**0.08**	0.07	**0.10**	0.08	**0.11**	0.09	**0.13**	0.11	**0.15**	0.10	**0.14**
Total filtration (mm) **Al**	**1.5**		**2.5**		**3.5**		**4.5**		**3.5**		**3.5**	
Cu	**0**		**0**		**0**		**0**		**0.1**		**0.2**	

Table 9.86 Normalized organ doses—Chest r-lat—> 17 years

Beam projection:		r-lat		**Table cells:**								
Focus-to-detector distance (cm):		100		(+) = normal font = optimal field setting								
Tube voltage (kVp):		70		**(−) = bold font = suboptimal field setting**								
Total filtrations:		See last table row.		Empty table cells correspond to values < 0.01								
Organ or tissue	**Organ dose D_T : Entrance dose K_E**						**(mSv × mGy^{-1})**					
	(+)	(−)	(+)	(−)	(+)	(−)	(+)	(−)	(+)	(−)	(+)	(−)
Skull	0.02	**0.17**	0.03	**0.21**	0.03	**0.25**	0.04	**0.27**	0.04	**0.32**	0.05	**0.37**
Brain		**0.02**		**0.02**		**0.03**		**0.03**		**0.04**		**0.05**
Oral mucosa	0.01	**0.14**	0.01	**0.18**	0.02	**0.20**	0.02	**0.22**	0.02	**0.26**	0.03	**0.29**
Salivary glands	0.04	**0.26**	0.05	**0.30**	0.06	**0.33**	0.06	**0.35**	0.08	**0.39**	0.09	**0.43**
Cervical spine	0.27	**0.38**	0.34	**0.49**	0.40	**0.58**	0.45	**0.65**	0.55	**0.79**	0.64	**0.93**
Thyroid	0.11	**0.08**	0.13	**0.10**	0.15	**0.11**	0.17	**0.13**	0.20	**0.15**	0.23	**0.18**
Thymus	0.04	**0.04**	0.05	**0.06**	0.06	**0.07**	0.07	**0.08**	0.10	**0.10**	0.12	**0.13**
Thoracic spine	0.07	**0.08**	0.10	**0.11**	0.12	**0.13**	0.14	**0.15**	0.18	**0.20**	0.22	**0.24**
Claviculae	0.26	**0.28**	0.33	**0.35**	0.39	**0.41**	0.43	**0.46**	0.51	**0.55**	0.59	**0.63**
Scapulae	0.74	**0.79**	0.88	**0.95**	0.99	**1.07**	1.07	**1.15**	1.21	**1.31**	1.33	**1.44**
Ribs	0.52	**0.61**	0.62	**0.73**	0.70	**0.81**	0.75	**0.88**	0.86	**1.00**	0.95	**1.10**
Oesophagus	0.05	**0.06**	0.07	**0.08**	0.09	**0.10**	0.10	**0.11**	0.13	**0.14**	0.16	**0.17**
Trachea	0.08	**0.11**	0.10	**0.14**	0.12	**0.17**	0.13	**0.19**	0.15	**0.23**	0.17	**0.26**
Lung	0.15	**0.16**	0.19	**0.21**	0.22	**0.24**	0.24	**0.27**	0.29	**0.32**	0.34	**0.38**
Mamma	0.23	**0.31**	0.27	**0.36**	0.29	**0.39**	0.32	**0.42**	0.35	**0.47**	0.39	**0.51**
Heart	0.05	**0.06**	0.07	**0.08**	0.08	**0.09**	0.09	**0.10**	0.12	**0.13**	0.15	**0.16**
Stomach						**0.01**		**0.01**	0.01	**0.02**	0.02	**0.02**
Spleen										**0.01**	0.01	**0.01**
Liver	0.14	**0.29**	0.17	**0.36**	0.20	**0.41**	0.22	**0.46**	0.27	**0.54**	0.31	**0.62**
Gall bladder	0.02	**0.07**	0.02	**0.09**	0.03	**0.11**	0.03	**0.13**	0.04	**0.16**	0.05	**0.20**
Pancreas	0.02	**0.02**	0.02	**0.03**	0.03	**0.03**	0.03	**0.04**	0.04	**0.05**	0.05	**0.07**
Small intestine		**0.03**		**0.03**		**0.04**		**0.05**		**0.06**		**0.07**
Large intestine		**0.04**		**0.06**		**0.07**		**0.08**		**0.10**		**0.12**
Sigma/Rectum												
Lumbar spine		**0.04**	0.01	**0.06**	0.02	**0.07**	0.02	**0.08**	0.03	**0.10**	0.03	**0.13**
Pelvis		**0.01**		**0.02**		**0.02**		**0.03**		**0.03**		**0.04**
Adrenals	0.03	**0.04**	0.05	**0.06**	0.05	**0.07**	0.06	**0.08**	0.08	**0.10**	0.10	**0.12**
Kidneys	0.02	**0.06**	0.02	**0.08**	0.03	**0.09**	0.03	**0.10**	0.04	**0.13**	0.05	**0.15**
Urinary bladder												
Uterus										**0.01**		**0.01**
Ovaries										**0.01**		**0.02**
Prostate												
Testes												
Upper arm												
Elbow												
Forearm												
Upper leg												
Knee												
Lower leg												
Lymph nodes	0.04	**0.08**	0.05	**0.10**	0.06	**0.11**	0.07	**0.12**	0.08	**0.15**	0.10	**0.17**
Skin	0.08	**0.13**	0.09	**0.14**	0.09	**0.14**	0.10	**0.15**	0.10	**0.16**	0.11	**0.17**
Muscle	0.04	**0.06**	0.05	**0.07**	0.05	**0.08**	0.06	**0.09**	0.07	**0.10**	0.08	**0.11**
Skeleton	0.09	**0.13**	0.11	**0.16**	0.13	**0.18**	0.14	**0.20**	0.16	**0.23**	0.18	**0.26**
Red bone marrow	0.05	**0.06**	0.06	**0.08**	0.07	**0.09**	0.07	**0.10**	0.09	**0.11**	0.10	**0.13**
Whole body dose	0.05	**0.08**	0.06	**0.09**	0.07	**0.11**	0.08	**0.12**	0.09	**0.14**	0.10	**0.15**
ED (ICRP 60)	0.05	**0.07**	0.06	**0.09**	0.07	**0.10**	0.08	**0.11**	0.10	**0.13**	0.11	**0.16**
ED (ICRP 103)	0.07	**0.10**	0.08	**0.12**	0.09	**0.13**	0.10	**0.15**	0.12	**0.17**	0.14	**0.20**
Total filtration (mm) Al	**1.5**		**2.5**		**3.5**		**4.5**		**3.5**		**3.5**	
Cu	**0**		**0**		**0**		**0**		**0.1**		**0.2**	

Table 9.87 Normalized organ doses—Chest r-lat—> 17 years

Beam projection:			r-lat			**Table cells:**						
Focus-to-detector distance (cm):			100			(+) = normal font = optimal field setting						
Tube voltage (kVp):			80			**(−) = bold font = suboptimal field setting**						
Total filtrations:			See last table row.			Empty table cells correspond to values < 0.01						
Organ or tissue	**Organ dose D_T : Entrance dose K_E**								**($mSv \times mGy^{-1}$)**			
	(+)	(−)	(+)	(−)	(+)	(−)	(+)	(−)	(+)	(−)	(+)	(−)
Skull	0.03	**0.20**	0.03	**0.25**	0.04	**0.28**	0.04	**0.31**	0.05	**0.36**	0.06	**0.40**
Brain		**0.02**		**0.03**		**0.04**		**0.04**		**0.05**		**0.06**
Oral mucosa	0.01	**0.17**	0.02	**0.20**	0.02	**0.23**	0.02	**0.25**	0.03	**0.29**	0.04	**0.33**
Salivary glands	0.05	**0.28**	0.06	**0.33**	0.07	**0.36**	0.07	**0.38**	0.09	**0.42**	0.10	**0.46**
Cervical spine	0.32	**0.47**	0.41	**0.59**	0.47	**0.68**	0.52	**0.75**	0.62	**0.90**	0.71	**1.04**
Thyroid	0.13	**0.09**	0.15	**0.12**	0.18	**0.13**	0.19	**0.15**	0.23	**0.18**	0.26	**0.21**
Thymus	0.05	**0.06**	0.07	**0.08**	0.08	**0.09**	0.10	**0.10**	0.12	**0.13**	0.15	**0.16**
Thoracic spine	0.10	**0.11**	0.13	**0.14**	0.16	**0.17**	0.18	**0.20**	0.22	**0.25**	0.27	**0.31**
Claviculae	0.31	**0.33**	0.38	**0.41**	0.44	**0.47**	0.48	**0.52**	0.57	**0.61**	0.64	**0.69**
Scapulae	0.81	**0.88**	0.96	**1.03**	1.06	**1.14**	1.13	**1.23**	1.27	**1.37**	1.37	**1.48**
Ribs	0.58	**0.67**	0.68	**0.79**	0.75	**0.88**	0.81	**0.94**	0.91	**1.05**	0.98	**1.14**
Oesophagus	0.07	**0.08**	0.09	**0.10**	0.11	**0.12**	0.13	**0.14**	0.16	**0.18**	0.19	**0.21**
Trachea	0.10	**0.14**	0.12	**0.17**	0.13	**0.20**	0.15	**0.22**	0.17	**0.26**	0.19	**0.30**
Lung	0.18	**0.20**	0.22	**0.24**	0.25	**0.28**	0.28	**0.31**	0.33	**0.37**	0.38	**0.42**
Mamma	0.25	**0.34**	0.29	**0.39**	0.32	**0.43**	0.34	**0.45**	0.38	**0.50**	0.41	**0.54**
Heart	0.07	**0.08**	0.09	**0.10**	0.10	**0.12**	0.12	**0.13**	0.15	**0.16**	0.18	**0.20**
Stomach		**0.01**	0.01	**0.01**	0.01	**0.02**	0.01	**0.02**	0.02	**0.03**	0.02	**0.03**
Spleen						**0.01**		**0.01**	0.01	**0.02**	0.02	**0.02**
Liver	0.16	**0.34**	0.20	**0.42**	0.23	**0.47**	0.26	**0.52**	0.31	**0.61**	0.35	**0.69**
Gall bladder	0.02	**0.09**	0.03	**0.12**	0.04	**0.14**	0.04	**0.16**	0.05	**0.20**	0.06	**0.24**
Pancreas	0.02	**0.03**	0.03	**0.04**	0.04	**0.05**	0.04	**0.06**	0.05	**0.07**	0.06	**0.09**
Small intestine		**0.03**		**0.04**		**0.05**		**0.06**		**0.07**		**0.09**
Large intestine		**0.06**		**0.07**		**0.09**		**0.10**		**0.12**		**0.14**
Sigma/Rectum												
Lumbar spine	0.01	**0.06**	0.02	**0.08**	0.02	**0.09**	0.03	**0.11**	0.03	**0.14**	0.04	**0.17**
Pelvis		**0.02**		**0.03**		**0.03**		**0.03**		**0.04**		**0.05**
Adrenals	0.05	**0.06**	0.06	**0.07**	0.07	**0.09**	0.08	**0.10**	0.10	**0.13**	0.12	**0.15**
Kidneys	0.02	**0.07**	0.03	**0.09**	0.03	**0.11**	0.04	**0.12**	0.05	**0.15**	0.06	**0.18**
Urinary bladder												
Uterus						**0.01**		**0.01**		**0.02**		**0.02**
Ovaries						**0.01**		**0.01**		**0.02**		**0.02**
Prostate												
Testes												
Upper arm												
Elbow												
Forearm												
Upper leg												
Knee												
Lower leg												
Lymph nodes	0.05	**0.09**	0.06	**0.11**	0.07	**0.13**	0.08	**0.14**	0.10	**0.17**	0.11	**0.20**
Skin	0.09	**0.13**	0.09	**0.14**	0.10	**0.15**	0.10	**0.16**	0.11	**0.17**	0.11	**0.18**
Muscle	0.04	**0.07**	0.05	**0.08**	0.06	**0.09**	0.07	**0.10**	0.08	**0.11**	0.08	**0.13**
Skeleton	0.11	**0.15**	0.13	**0.18**	0.14	**0.20**	0.15	**0.22**	0.17	**0.25**	0.19	**0.28**
Red bone marrow	0.06	**0.07**	0.07	**0.09**	0.08	**0.10**	0.08	**0.11**	0.10	**0.13**	0.11	**0.15**
Whole body dose	0.06	**0.09**	0.07	**0.11**	0.08	**0.12**	0.09	**0.13**	0.10	**0.15**	0.11	**0.17**
ED (ICRP 60)	0.06	**0.08**	0.08	**0.10**	0.09	**0.12**	0.10	**0.13**	0.11	**0.15**	0.13	**0.18**
ED (ICRP 103)	0.08	**0.11**	0.10	**0.14**	0.11	**0.15**	0.12	**0.17**	0.14	**0.19**	0.16	**0.22**
Total filtration (mm) **Al**	**1.5**		**2.5**		**3.5**		**4.5**		**3.5**		**3.5**	
Cu	**0**		**0**		**0**		**0**		**0.1**		**0.2**	

Table 9.88 Normalized organ doses—Chest r-lat—> 17 years

Beam projection:			r-lat			**Table cells:**						
Focus-to-detector distance (cm):			100			(+) = normal font = optimal field setting						
Tube voltage (kVp):			110			**(−) = bold font = suboptimal field setting**						
Total filtrations:			See last table row.			Empty table cells correspond to values < 0.01						
Organ or tissue	**Organ dose D_T : Entrance dose K_E**							**(mSv × mGy^{-1})**				
	(+)	(−)	(+)	(−)	(+)	(−)	(+)	(−)	(+)	(−)	(+)	(−)
Skull	0.04	**0.27**	0.05	**0.31**	0.05	**0.34**	0.05	**0.37**	0.06	**0.41**	0.07	**0.44**
Brain		**0.04**		**0.05**		**0.06**		**0.06**		**0.07**		**0.08**
Oral mucosa	0.02	**0.23**	0.03	**0.27**	0.03	**0.30**	0.03	**0.32**	0.04	**0.36**	0.05	**0.39**
Salivary glands	0.07	**0.35**	0.08	**0.39**	0.09	**0.42**	0.09	**0.44**	0.11	**0.48**	0.12	**0.51**
Cervical spine	0.46	**0.67**	0.54	**0.79**	0.60	**0.88**	0.64	**0.95**	0.73	**1.07**	0.79	**1.18**
Thyroid	0.18	**0.14**	0.21	**0.17**	0.23	**0.18**	0.25	**0.20**	0.28	**0.23**	0.30	**0.25**
Thymus	0.10	**0.11**	0.12	**0.13**	0.13	**0.15**	0.15	**0.16**	0.17	**0.19**	0.20	**0.22**
Thoracic spine	0.18	**0.20**	0.21	**0.24**	0.24	**0.27**	0.27	**0.30**	0.32	**0.36**	0.36	**0.41**
Claviculae	0.42	**0.45**	0.49	**0.52**	0.53	**0.58**	0.57	**0.62**	0.64	**0.69**	0.69	**0.75**
Scapulae	0.97	**1.05**	1.09	**1.18**	1.17	**1.27**	1.22	**1.33**	1.31	**1.42**	1.37	**1.49**
Ribs	0.70	**0.81**	0.79	**0.91**	0.84	**0.98**	0.88	**1.03**	0.95	**1.11**	1.00	**1.16**
Oesophagus	0.13	**0.14**	0.16	**0.17**	0.18	**0.19**	0.19	**0.21**	0.23	**0.25**	0.26	**0.29**
Trachea	0.13	**0.20**	0.15	**0.24**	0.17	**0.27**	0.18	**0.29**	0.20	**0.33**	0.22	**0.36**
Lung	0.26	**0.28**	0.30	**0.33**	0.33	**0.37**	0.36	**0.40**	0.41	**0.45**	0.45	**0.50**
Mamma	0.31	**0.41**	0.35	**0.46**	0.37	**0.49**	0.39	**0.52**	0.43	**0.56**	0.45	**0.60**
Heart	0.12	**0.13**	0.14	**0.16**	0.16	**0.18**	0.18	**0.20**	0.21	**0.23**	0.24	**0.26**
Stomach	0.02	**0.02**	0.02	**0.03**	0.02	**0.03**	0.03	**0.04**	0.03	**0.04**	0.04	**0.05**
Spleen	0.01	**0.01**	0.01	**0.02**	0.02	**0.02**	0.02	**0.02**	0.02	**0.03**	0.03	**0.03**
Liver	0.24	**0.48**	0.28	**0.56**	0.31	**0.61**	0.33	**0.66**	0.38	**0.74**	0.42	**0.81**
Gall bladder	0.04	**0.16**	0.05	**0.19**	0.06	**0.22**	0.06	**0.24**	0.08	**0.28**	0.09	**0.32**
Pancreas	0.04	**0.06**	0.05	**0.07**	0.06	**0.08**	0.07	**0.09**	0.08	**0.11**	0.10	**0.13**
Small intestine		**0.06**		**0.07**		**0.08**		**0.09**		**0.10**		**0.12**
Large intestine		**0.09**		**0.11**		**0.13**		**0.14**	0.01	**0.16**	0.01	**0.18**
Sigma/Rectum												
Lumbar spine	0.03	**0.11**	0.04	**0.13**	0.04	**0.15**	0.05	**0.17**	0.06	**0.20**	0.06	**0.23**
Pelvis		**0.03**		**0.04**		**0.05**		**0.05**		**0.06**		**0.07**
Adrenals	0.08	**0.10**	0.10	**0.12**	0.11	**0.14**	0.12	**0.15**	0.15	**0.18**	0.17	**0.21**
Kidneys	0.04	**0.12**	0.05	**0.14**	0.05	**0.16**	0.06	**0.18**	0.07	**0.20**	0.08	**0.23**
Urinary bladder												
Uterus		**0.01**		**0.02**		**0.02**		**0.02**		**0.03**		**0.03**
Ovaries		**0.01**		**0.02**		**0.02**		**0.02**		**0.03**		**0.03**
Prostate												
Testes												
Upper arm												
Elbow												
Forearm												
Upper leg												
Knee												
Lower leg												
Lymph nodes	0.08	**0.14**	0.09	**0.16**	0.10	**0.18**	0.11	**0.20**	0.12	**0.22**	0.14	**0.25**
Skin	0.10	**0.15**	0.10	**0.16**	0.11	**0.17**	0.11	**0.17**	0.12	**0.18**	0.12	**0.19**
Muscle	0.06	**0.09**	0.07	**0.10**	0.08	**0.11**	0.08	**0.12**	0.09	**0.14**	0.10	**0.15**
Skeleton	0.13	**0.19**	0.15	**0.22**	0.17	**0.24**	0.18	**0.25**	0.19	**0.28**	0.21	**0.30**
Red bone marrow	0.08	**0.11**	0.09	**0.13**	0.10	**0.14**	0.11	**0.15**	0.12	**0.17**	0.14	**0.19**
Whole body dose	0.08	**0.12**	0.09	**0.14**	0.10	**0.15**	0.11	**0.16**	0.12	**0.18**	0.13	**0.19**
ED (ICRP 60)	0.09	**0.12**	0.10	**0.14**	0.11	**0.16**	0.12	**0.17**	0.14	**0.19**	0.15	**0.21**
ED (ICRP 103)	0.11	**0.16**	0.13	**0.18**	0.14	**0.20**	0.15	**0.21**	0.17	**0.24**	0.18	**0.26**
Total filtration (mm) Al	**1.5**		**2.5**		**3.5**		**4.5**		**3.5**		**3.5**	
Cu	**0**		**0**		**0**		**0**		**0.1**		**0.2**	

Table 9.89 Normalized organ doses—Chest r-lat—> 17 years

Beam projection:			r-lat				**Table cells:**					
Focus-to-detector distance (cm):			100				(+) = normal font = optimal field setting					
Tube voltage (kVp):			120				**(−) = bold font = suboptimal field setting**					
Total filtrations:			See last table row.				Empty table cells correspond to values < 0.01					
Organ or tissue	**Organ dose D_T : Entrance dose K_E**						**(mSv × mGy^{-1})**					
	(+)	(−)	(+)	(−)	(+)	(−)	(+)	(−)	(+)	(−)	(+)	(−)
Skull	0.04	**0.29**	0.05	**0.33**	0.05	**0.35**	0.06	**0.38**	0.06	**0.41**	0.07	**0.44**
Brain		**0.05**		**0.06**		**0.06**		**0.07**		**0.08**		**0.09**
Oral mucosa	0.03	**0.25**	0.03	**0.29**	0.03	**0.32**	0.04	**0.34**	0.04	**0.37**	0.05	**0.41**
Salivary glands	0.07	**0.37**	0.08	**0.41**	0.09	**0.43**	0.10	**0.45**	0.11	**0.49**	0.12	**0.52**
Cervical spine	0.49	**0.72**	0.57	**0.84**	0.62	**0.92**	0.66	**0.98**	0.74	**1.10**	0.80	**1.19**
Thyroid	0.19	**0.15**	0.22	**0.18**	0.24	**0.20**	0.26	**0.21**	0.29	**0.24**	0.31	**0.26**
Thymus	0.11	**0.12**	0.13	**0.14**	0.15	**0.16**	0.16	**0.17**	0.19	**0.20**	0.21	**0.23**
Thoracic spine	0.20	**0.22**	0.24	**0.26**	0.26	**0.30**	0.29	**0.32**	0.34	**0.38**	0.38	**0.43**
Claviculae	0.44	**0.48**	0.51	**0.55**	0.55	**0.60**	0.59	**0.63**	0.64	**0.70**	0.69	**0.75**
Scapulae	1.01	**1.09**	1.11	**1.21**	1.18	**1.28**	1.23	**1.34**	1.31	**1.42**	1.35	**1.47**
Ribs	0.73	**0.84**	0.80	**0.94**	0.86	**1.00**	0.89	**1.04**	0.95	**1.11**	0.99	**1.15**
Oesophagus	0.15	**0.16**	0.17	**0.19**	0.19	**0.21**	0.21	**0.23**	0.24	**0.27**	0.28	**0.31**
Trachea	0.14	**0.22**	0.16	**0.26**	0.18	**0.28**	0.19	**0.30**	0.21	**0.34**	0.23	**0.38**
Lung	0.28	**0.31**	0.32	**0.36**	0.35	**0.39**	0.38	**0.42**	0.42	**0.47**	0.46	**0.51**
Mamma	0.33	**0.43**	0.36	**0.48**	0.39	**0.51**	0.41	**0.53**	0.44	**0.57**	0.46	**0.60**
Heart	0.13	**0.15**	0.16	**0.17**	0.18	**0.19**	0.19	**0.21**	0.22	**0.25**	0.25	**0.28**
Stomach	0.02	**0.03**	0.02	**0.03**	0.03	**0.04**	0.03	**0.04**	0.03	**0.05**	0.04	**0.06**
Spleen	0.01	**0.02**	0.02	**0.02**	0.02	**0.02**	0.02	**0.03**	0.02	**0.03**	0.03	**0.04**
Liver	0.26	**0.51**	0.30	**0.59**	0.33	**0.64**	0.35	**0.69**	0.39	**0.76**	0.43	**0.83**
Gall bladder	0.05	**0.18**	0.06	**0.21**	0.06	**0.23**	0.07	**0.25**	0.08	**0.30**	0.09	**0.33**
Pancreas	0.05	**0.07**	0.06	**0.08**	0.07	**0.09**	0.08	**0.10**	0.09	**0.12**	0.10	**0.14**
Small intestine		**0.07**		**0.08**		**0.09**		**0.10**		**0.11**		**0.13**
Large intestine		**0.11**		**0.12**		**0.14**	0.01	**0.15**	0.01	**0.17**	0.01	**0.19**
Sigma/Rectum												**0.01**
Lumbar spine	0.03	**0.13**	0.04	**0.15**	0.05	**0.17**	0.05	**0.18**	0.06	**0.22**	0.07	**0.25**
Pelvis		**0.04**		**0.05**		**0.05**		**0.06**		**0.06**		**0.07**
Adrenals	0.09	**0.12**	0.11	**0.14**	0.12	**0.15**	0.14	**0.17**	0.16	**0.20**	0.18	**0.22**
Kidneys	0.04	**0.13**	0.05	**0.16**	0.06	**0.17**	0.06	**0.19**	0.07	**0.22**	0.08	**0.24**
Urinary bladder												
Uterus		**0.02**		**0.02**		**0.02**		**0.02**		**0.03**		**0.03**
Ovaries		**0.02**		**0.02**		**0.02**		**0.02**		**0.03**		**0.03**
Prostate												
Testes												
Upper arm												
Elbow												
Forearm												
Upper leg												
Knee												
Lower leg												
Lymph nodes	0.08	**0.15**	0.10	**0.18**	0.11	**0.19**	0.11	**0.21**	0.13	**0.24**	0.14	**0.26**
Skin	0.10	**0.16**	0.11	**0.16**	0.11	**0.17**	0.12	**0.18**	0.12	**0.18**	0.13	**0.19**
Muscle	0.07	**0.10**	0.07	**0.11**	0.08	**0.12**	0.09	**0.13**	0.10	**0.14**	0.10	**0.15**
Skeleton	0.14	**0.20**	0.16	**0.23**	0.17	**0.25**	0.18	**0.26**	0.19	**0.28**	0.20	**0.30**
Red bone marrow	0.09	**0.12**	0.10	**0.13**	0.11	**0.15**	0.12	**0.16**	0.13	**0.18**	0.14	**0.20**
Whole body dose	0.08	**0.13**	0.10	**0.15**	0.10	**0.16**	0.11	**0.17**	0.12	**0.18**	0.13	**0.20**
ED (ICRP 60)	0.10	**0.13**	0.11	**0.15**	0.12	**0.17**	0.13	**0.18**	0.14	**0.20**	0.16	**0.22**
ED (ICRP 103)	0.12	**0.17**	0.14	**0.19**	0.15	**0.21**	0.16	**0.22**	0.18	**0.25**	0.19	**0.27**
Total filtration (mm) **Al**	**1.5**		**2.5**		**3.5**		**4.5**		**3.5**		**3.5**	
Cu	**0**		**0**		**0**		**0**		**0.1**		**0.2**	

Table 9.90 Normalized organ doses—Chest ap—3–7 years

Beam projection:			ap			**Table cells:**						
Focus-to-detector distance (cm):			115			(+) = normal font = optimal field setting						
Tube voltage (kVp):			60			**(−) = bold font = suboptimal field setting**						
Total filtrations:			See last table row.			Empty table cells correspond to values < 0.01						
Organ or tissue	**Organ dose D_T : Entrance dose K_E**							**(mSv × mGy^{-1})**				
	(+)	(−)	(+)	(−)	(+)	(−)	(+)	(−)	(+)	(−)	(+)	(−)
Skull	0.01	**0.25**	0.02	**0.31**	0.02	**0.35**	0.02	**0.38**	0.03	**0.45**	0.03	**0.43**
Brain		**0.01**		**0.01**		**0.02**		**0.02**		**0.03**		**0.02**
Oral mucosa	0.02	**0.37**	0.02	**0.44**	0.03	**0.49**	0.03	**0.53**	0.04	**0.61**	0.03	**0.58**
Salivary glands	0.03	**0.30**	0.04	**0.37**	0.04	**0.42**	0.05	**0.46**	0.06	**0.53**	0.05	**0.51**
Cervical spine	0.25	**0.38**	0.32	**0.50**	0.38	**0.60**	0.42	**0.67**	0.52	**0.83**	0.48	**0.75**
Thyroid	0.78	**0.83**	0.90	**0.95**	0.97	**1.03**	1.02	**1.09**	1.10	**1.20**	1.09	**1.17**
Thymus	0.69	**0.71**	0.83	**0.85**	0.93	**0.96**	1.01	**1.03**	1.14	**1.17**	1.12	**1.15**
Thoracic spine	0.15	**0.16**	0.20	**0.22**	0.24	**0.27**	0.28	**0.31**	0.36	**0.40**	0.30	**0.33**
Claviculae	1.60	**1.68**	1.93	**2.03**	2.16	**2.27**	2.33	**2.46**	2.63	**2.78**	2.62	**2.76**
Scapulae	0.22	**0.24**	0.29	**0.31**	0.35	**0.38**	0.40	**0.43**	0.50	**0.54**	0.45	**0.48**
Ribs	1.28	**1.36**	1.50	**1.59**	1.64	**1.75**	1.75	**1.87**	1.94	**2.08**	1.92	**2.06**
Oesophagus	0.12	**0.13**	0.16	**0.18**	0.19	**0.21**	0.22	**0.24**	0.27	**0.30**	0.24	**0.27**
Trachea	0.07	**0.30**	0.09	**0.37**	0.10	**0.43**	0.10	**0.47**	0.12	**0.54**	0.11	**0.52**
Lung	0.33	**0.35**	0.41	**0.43**	0.48	**0.50**	0.52	**0.55**	0.61	**0.65**	0.59	**0.62**
Mamma	1.12	**1.16**	1.21	**1.24**	1.26	**1.30**	1.29	**1.33**	1.35	**1.39**	1.34	**1.38**
Heart	0.44	**0.46**	0.55	**0.57**	0.63	**0.65**	0.69	**0.72**	0.81	**0.84**	0.78	**0.80**
Stomach	0.23	**0.49**	0.29	**0.61**	0.33	**0.69**	0.37	**0.76**	0.43	**0.88**	0.41	**0.85**
Spleen	0.10	**0.14**	0.13	**0.18**	0.15	**0.22**	0.17	**0.25**	0.21	**0.31**	0.19	**0.27**
Liver	0.22	**0.39**	0.28	**0.48**	0.32	**0.56**	0.35	**0.61**	0.42	**0.72**	0.40	**0.68**
Gall bladder	0.07	**0.34**	0.09	**0.44**	0.11	**0.52**	0.13	**0.58**	0.16	**0.69**	0.14	**0.65**
Pancreas	0.16	**0.21**	0.21	**0.28**	0.25	**0.33**	0.29	**0.37**	0.35	**0.46**	0.32	**0.42**
Small intestine		**0.18**		**0.23**	0.01	**0.27**	0.01	**0.30**	0.02	**0.37**	0.01	**0.34**
Large intestine		**0.25**	0.01	**0.31**	0.02	**0.37**	0.02	**0.41**	0.02	**0.49**	0.02	**0.46**
Sigma/Rectum		**0.06**		**0.08**		**0.09**		**0.10**		**0.12**		**0.11**
Lumbar spine	0.03	**0.14**	0.05	**0.19**	0.06	**0.24**	0.07	**0.27**	0.09	**0.36**	0.07	**0.29**
Pelvis		**0.07**		**0.10**		**0.12**		**0.13**		**0.17**		**0.15**
Adrenals	0.07	**0.08**	0.09	**0.11**	0.11	**0.13**	0.13	**0.15**	0.16	**0.20**	0.14	**0.17**
Kidneys	0.03	**0.07**	0.04	**0.09**	0.05	**0.11**	0.05	**0.13**	0.07	**0.16**	0.06	**0.14**
Urinary bladder				**0.01**		**0.02**		**0.02**		**0.02**		**0.02**
Uterus		**0.03**		**0.04**		**0.05**		**0.06**		**0.07**		**0.06**
Ovaries		**0.03**		**0.04**		**0.05**		**0.06**		**0.08**		**0.07**
Prostate												
Testes												
Upper arm	0.14	**0.68**	0.19	**0.86**	0.22	**0.99**	0.25	**1.10**	0.32	**1.30**	0.28	**1.24**
Elbow	0.12	**0.74**	0.16	**0.93**	0.19	**1.07**	0.22	**1.18**	0.28	**1.39**	0.25	**1.34**
Forearm	0.01	**0.42**	0.02	**0.53**	0.02	**0.61**	0.02	**0.68**	0.03	**0.80**	0.03	**0.77**
Upper leg												
Knee												
Lower leg												
Lymph nodes	0.14	**0.29**	0.18	**0.36**	0.20	**0.41**	0.22	**0.45**	0.26	**0.54**	0.24	**0.51**
Skin	0.12	**0.25**	0.13	**0.26**	0.13	**0.27**	0.14	**0.28**	0.15	**0.29**	0.14	**0.29**
Muscle	0.09	**0.15**	0.11	**0.18**	0.12	**0.21**	0.13	**0.23**	0.15	**0.26**	0.14	**0.25**
Skeleton	0.18	**0.33**	0.22	**0.41**	0.24	**0.46**	0.26	**0.51**	0.30	**0.59**	0.29	**0.56**
Red bone marrow	0.06	**0.09**	0.07	**0.12**	0.08	**0.13**	0.09	**0.15**	0.10	**0.17**	0.09	**0.16**
Whole body dose	0.10	**0.18**	0.12	**0.22**	0.14	**0.25**	0.15	**0.27**	0.18	**0.32**	0.17	**0.30**
ED (ICRP 60)	0.20	**0.27**	0.23	**0.33**	0.26	**0.37**	0.28	**0.40**	0.31	**0.46**	0.30	**0.44**
ED (ICRP 103)	0.27	**0.36**	0.32	**0.43**	0.34	**0.47**	0.37	**0.51**	0.41	**0.57**	0.40	**0.55**
Total filtration (mm) Al	**1.5**		**2.5**		**3.5**		**4.5**		**3.5**		**3.5**	
Cu	**0**		**0**		**0**		**0**		**0.1**		**0.2**	

Table 9.91 Normalized organ doses—Chest ap—3–7 years

Beam projection:			ap				**Table cells:**					
Focus-to-detector distance (cm):			115				(+) = normal font = optimal field setting					
Tube voltage (kVp):			70				**(−) = bold font = suboptimal field setting**					
Total filtrations:			See last table row.				Empty table cells correspond to values < 0.01					
Organ or tissue	**Organ dose D_T : Entrance dose K_E**								**(mSv × mGy^{-1})**			
	(+)	(−)	(+)	(−)	(+)	(−)	(+)	(−)	(+)	(−)	(+)	(−)
Skull	0.02	**0.29**	0.02	**0.35**	0.03	**0.39**	0.03	**0.43**	0.04	**0.49**	0.05	**0.54**
Brain		**0.02**		**0.02**		**0.02**		**0.03**		**0.04**		**0.04**
Oral mucosa	0.02	**0.42**	0.03	**0.50**	0.03	**0.55**	0.04	**0.60**	0.05	**0.68**	0.06	**0.75**
Salivary glands	0.04	**0.35**	0.04	**0.42**	0.05	**0.47**	0.06	**0.51**	0.07	**0.59**	0.08	**0.66**
Cervical spine	0.31	**0.48**	0.39	**0.62**	0.45	**0.72**	0.50	**0.81**	0.60	**0.98**	0.70	**1.15**
Thyroid	0.85	**0.90**	0.96	**1.03**	1.03	**1.11**	1.08	**1.17**	1.16	**1.27**	1.23	**1.36**
Thymus	0.78	**0.80**	0.92	**0.95**	1.03	**1.05**	1.10	**1.13**	1.24	**1.27**	1.35	**1.39**
Thoracic spine	0.20	**0.23**	0.27	**0.30**	0.32	**0.36**	0.37	**0.41**	0.46	**0.52**	0.56	**0.63**
Claviculae	1.78	**1.87**	2.10	**2.22**	2.32	**2.46**	2.49	**2.64**	2.76	**2.94**	2.97	**3.18**
Scapulae	0.28	**0.30**	0.36	**0.39**	0.43	**0.46**	0.48	**0.52**	0.59	**0.64**	0.69	**0.75**
Ribs	1.39	**1.48**	1.61	**1.72**	1.75	**1.88**	1.85	**1.99**	2.03	**2.20**	2.17	**2.36**
Oesophagus	0.15	**0.17**	0.20	**0.22**	0.24	**0.26**	0.27	**0.30**	0.33	**0.37**	0.39	**0.43**
Trachea	0.08	**0.35**	0.10	**0.43**	0.11	**0.48**	0.12	**0.53**	0.13	**0.60**	0.14	**0.67**
Lung	0.39	**0.41**	0.48	**0.50**	0.54	**0.57**	0.59	**0.63**	0.68	**0.73**	0.77	**0.82**
Mamma	1.17	**1.20**	1.25	**1.29**	1.30	**1.34**	1.33	**1.37**	1.39	**1.43**	1.44	**1.47**
Heart	0.52	**0.54**	0.63	**0.66**	0.72	**0.75**	0.78	**0.82**	0.91	**0.95**	1.02	**1.07**
Stomach	0.27	**0.57**	0.34	**0.69**	0.38	**0.78**	0.42	**0.85**	0.49	**0.99**	0.55	**1.10**
Spleen	0.12	**0.18**	0.16	**0.23**	0.19	**0.27**	0.21	**0.31**	0.26	**0.38**	0.31	**0.44**
Liver	0.26	**0.46**	0.32	**0.56**	0.37	**0.64**	0.41	**0.70**	0.47	**0.81**	0.54	**0.91**
Gall bladder	0.09	**0.42**	0.12	**0.53**	0.14	**0.61**	0.16	**0.68**	0.19	**0.80**	0.22	**0.92**
Pancreas	0.21	**0.27**	0.26	**0.35**	0.31	**0.41**	0.35	**0.46**	0.42	**0.56**	0.49	**0.65**
Small intestine	0.01	**0.22**	0.01	**0.28**	0.02	**0.32**	0.02	**0.36**	0.02	**0.43**	0.03	**0.49**
Large intestine	0.01	**0.30**	0.02	**0.37**	0.02	**0.43**	0.03	**0.47**	0.03	**0.56**	0.04	**0.64**
Sigma/Rectum		**0.07**		**0.09**		**0.11**		**0.12**		**0.14**		**0.17**
Lumbar spine	0.05	**0.20**	0.06	**0.26**	0.08	**0.32**	0.09	**0.36**	0.12	**0.46**	0.15	**0.56**
Pelvis		**0.10**		**0.13**		**0.15**	0.01	**0.17**	0.01	**0.21**	0.02	**0.25**
Adrenals	0.09	**0.11**	0.12	**0.15**	0.14	**0.18**	0.16	**0.20**	0.21	**0.25**	0.25	**0.30**
Kidneys	0.04	**0.09**	0.05	**0.12**	0.06	**0.15**	0.07	**0.17**	0.09	**0.21**	0.11	**0.25**
Urinary bladder		**0.01**		**0.02**		**0.02**		**0.02**		**0.03**		**0.04**
Uterus		**0.04**		**0.05**		**0.06**		**0.07**		**0.09**		**0.11**
Ovaries		**0.04**		**0.06**		**0.07**		**0.08**		**0.10**	0.01	**0.12**
Prostate										**0.01**		**0.02**
Testes												
Upper arm	0.18	**0.81**	0.23	**0.99**	0.28	**1.13**	0.31	**1.24**	0.38	**1.45**	0.44	**1.63**
Elbow	0.16	**0.87**	0.20	**1.07**	0.24	**1.21**	0.27	**1.33**	0.33	**1.55**	0.39	**1.74**
Forearm	0.02	**0.50**	0.02	**0.61**	0.03	**0.69**	0.03	**0.76**	0.04	**0.88**	0.05	**0.98**
Upper leg												
Knee												
Lower leg												
Lymph nodes	0.17	**0.34**	0.20	**0.42**	0.23	**0.48**	0.25	**0.52**	0.29	**0.61**	0.33	**0.69**
Skin	0.13	**0.26**	0.14	**0.27**	0.14	**0.28**	0.15	**0.29**	0.16	**0.31**	0.16	**0.32**
Muscle	0.10	**0.18**	0.12	**0.21**	0.13	**0.23**	0.15	**0.25**	0.17	**0.29**	0.19	**0.32**
Skeleton	0.20	**0.38**	0.24	**0.46**	0.27	**0.52**	0.29	**0.57**	0.34	**0.65**	0.37	**0.73**
Red bone marrow	0.07	**0.11**	0.08	**0.14**	0.09	**0.16**	0.10	**0.17**	0.11	**0.20**	0.13	**0.23**
Whole body dose	0.12	**0.21**	0.14	**0.25**	0.16	**0.28**	0.17	**0.31**	0.20	**0.35**	0.22	**0.40**
ED (ICRP 60)	0.22	**0.31**	0.26	**0.37**	0.28	**0.41**	0.30	**0.44**	0.34	**0.51**	0.37	**0.56**
ED (ICRP 103)	0.30	**0.41**	0.35	**0.47**	0.38	**0.52**	0.40	**0.56**	0.44	**0.62**	0.48	**0.69**
Total filtration (mm) **Al**	**1.5**		**2.5**		**3.5**		**4.5**		**3.5**		**3.5**	
Cu	**0**		**0**		**0**		**0**		**0.1**		**0.2**	

Table 9.92 Normalized organ doses—Chest ap—3–7 years

Beam projection:			ap		**Table cells:**							
Focus-to-detector distance (cm):			115		(+) = normal font = optimal field setting							
Tube voltage (kVp):			80		**(−) = bold font = suboptimal field setting**							
Total filtrations:			See last table row.		Empty table cells correspond to values < 0.01							
Organ or tissue	**Organ dose D_T : Entrance dose K_E**						**(mSv × mGy^{-1})**					
	(+)	(−)	(+)	(−)	(+)	(−)	(+)	(−)	(+)	(−)	(+)	(−)
Skull	0.02	**0.33**	0.03	**0.39**	0.03	**0.43**	0.04	**0.46**	0.05	**0.52**	0.05	**0.57**
Brain		**0.02**		**0.03**		**0.03**		**0.04**		**0.04**		**0.05**
Oral mucosa	0.03	**0.47**	0.04	**0.55**	0.04	**0.61**	0.05	**0.65**	0.06	**0.73**	0.07	**0.81**
Salivary glands	0.04	**0.39**	0.05	**0.47**	0.06	**0.52**	0.07	**0.56**	0.08	**0.64**	0.09	**0.71**
Cervical spine	0.36	**0.58**	0.45	**0.73**	0.51	**0.84**	0.57	**0.93**	0.67	**1.10**	0.76	**1.26**
Thyroid	0.90	**0.97**	1.01	**1.09**	1.07	**1.17**	1.12	**1.23**	1.21	**1.33**	1.27	**1.42**
Thymus	0.86	**0.88**	1.01	**1.03**	1.11	**1.14**	1.18	**1.21**	1.31	**1.35**	1.42	**1.47**
Thoracic spine	0.26	**0.29**	0.33	**0.38**	0.39	**0.44**	0.44	**0.50**	0.55	**0.62**	0.65	**0.74**
Claviculae	1.92	**2.03**	2.23	**2.36**	2.43	**2.59**	2.58	**2.75**	2.82	**3.02**	3.00	**3.22**
Scapulae	0.34	**0.37**	0.43	**0.47**	0.50	**0.54**	0.55	**0.60**	0.66	**0.72**	0.76	**0.83**
Ribs	1.49	**1.59**	1.69	**1.82**	1.82	**1.97**	1.92	**2.07**	2.08	**2.26**	2.19	**2.39**
Oesophagus	0.19	**0.21**	0.24	**0.27**	0.28	**0.31**	0.31	**0.35**	0.38	**0.42**	0.44	**0.49**
Trachea	0.09	**0.40**	0.11	**0.48**	0.12	**0.53**	0.13	**0.58**	0.14	**0.66**	0.16	**0.73**
Lung	0.44	**0.47**	0.53	**0.57**	0.60	**0.64**	0.65	**0.69**	0.74	**0.79**	0.82	**0.88**
Mamma	1.21	**1.24**	1.29	**1.32**	1.34	**1.37**	1.37	**1.41**	1.43	**1.46**	1.47	**1.51**
Heart	0.59	**0.61**	0.71	**0.74**	0.80	**0.83**	0.86	**0.90**	0.99	**1.03**	1.09	**1.15**
Stomach	0.31	**0.65**	0.38	**0.78**	0.43	**0.87**	0.46	**0.94**	0.53	**1.07**	0.59	**1.18**
Spleen	0.15	**0.22**	0.19	**0.28**	0.22	**0.32**	0.25	**0.36**	0.30	**0.44**	0.35	**0.51**
Liver	0.30	**0.53**	0.37	**0.63**	0.41	**0.71**	0.45	**0.77**	0.52	**0.89**	0.58	**0.99**
Gall bladder	0.11	**0.49**	0.14	**0.61**	0.16	**0.69**	0.18	**0.76**	0.22	**0.89**	0.25	**1.00**
Pancreas	0.25	**0.33**	0.31	**0.41**	0.36	**0.48**	0.40	**0.53**	0.48	**0.64**	0.55	**0.74**
Small intestine	0.01	**0.26**	0.02	**0.32**	0.02	**0.37**	0.02	**0.41**	0.03	**0.48**	0.04	**0.55**
Large intestine	0.02	**0.35**	0.02	**0.43**	0.03	**0.49**	0.03	**0.53**	0.04	**0.62**	0.05	**0.70**
Sigma/Rectum		**0.09**		**0.11**		**0.13**		**0.14**		**0.16**	0.01	**0.19**
Lumbar spine	0.07	**0.26**	0.09	**0.34**	0.10	**0.40**	0.12	**0.45**	0.15	**0.56**	0.18	**0.67**
Pelvis		**0.12**		**0.16**	0.01	**0.18**	0.01	**0.20**	0.02	**0.25**	0.02	**0.29**
Adrenals	0.12	**0.15**	0.15	**0.19**	0.18	**0.22**	0.20	**0.25**	0.25	**0.30**	0.29	**0.36**
Kidneys	0.05	**0.12**	0.07	**0.15**	0.08	**0.18**	0.09	**0.21**	0.11	**0.25**	0.13	**0.30**
Urinary bladder		**0.02**		**0.02**		**0.03**		**0.03**		**0.04**		**0.05**
Uterus		**0.05**		**0.07**		**0.08**		**0.09**		**0.11**	0.01	**0.13**
Ovaries		**0.06**		**0.07**		**0.08**		**0.09**	0.01	**0.12**	0.02	**0.14**
Prostate						**0.01**		**0.01**		**0.02**		**0.02**
Testes												
Upper arm	0.22	**0.92**	0.28	**1.11**	0.32	**1.25**	0.36	**1.35**	0.43	**1.55**	0.49	**1.72**
Elbow	0.19	**0.98**	0.24	**1.18**	0.28	**1.33**	0.31	**1.45**	0.38	**1.65**	0.44	**1.82**
Forearm	0.02	**0.56**	0.03	**0.67**	0.03	**0.76**	0.04	**0.82**	0.05	**0.93**	0.06	**1.03**
Upper leg												
Knee												
Lower leg												
Lymph nodes	0.19	**0.39**	0.23	**0.48**	0.26	**0.54**	0.28	**0.58**	0.32	**0.67**	0.36	**0.75**
Skin	0.13	**0.27**	0.14	**0.28**	0.15	**0.29**	0.15	**0.30**	0.16	**0.32**	0.17	**0.34**
Muscle	0.11	**0.20**	0.13	**0.23**	0.15	**0.26**	0.16	**0.28**	0.18	**0.32**	0.20	**0.35**
Skeleton	0.23	**0.43**	0.27	**0.51**	0.30	**0.57**	0.32	**0.61**	0.36	**0.70**	0.39	**0.77**
Red bone marrow	0.08	**0.13**	0.09	**0.16**	0.10	**0.18**	0.11	**0.20**	0.13	**0.23**	0.14	**0.26**
Whole body dose	0.13	**0.24**	0.16	**0.28**	0.18	**0.31**	0.19	**0.34**	0.22	**0.38**	0.24	**0.43**
ED (ICRP 60)	0.24	**0.35**	0.28	**0.41**	0.31	**0.45**	0.33	**0.49**	0.37	**0.55**	0.40	**0.60**
ED (ICRP 103)	0.33	**0.45**	0.37	**0.52**	0.40	**0.56**	0.43	**0.60**	0.47	**0.67**	0.51	**0.73**
Total filtration (mm) **Al**	**1.5**		**2.5**		**3.5**		**4.5**		**3.5**		**3.5**	
Cu	**0**		**0**		**0**		**0**		**0.1**		**0.2**	

Table 9.93 Normalized organ doses—Chest ap—8–12 years

Beam projection:			ap				**Table cells:**					
Focus-to-detector distance (cm):			115				(+) = normal font = optimal field setting					
Tube voltage (kVp):			60				**(−) = bold font = suboptimal field setting**					
Total filtrations:			See last table row.				Empty table cells correspond to values < 0.01					
Organ or tissue	**Organ dose D_T : Entrance dose K_E**								**(mSv × mGy^{-1})**			
	(+)	(−)	(+)	(−)	(+)	(−)	(+)	(−)	(+)	(−)	(+)	(−)
Skull	0.01	**0.25**	0.02	**0.31**	0.02	**0.35**	0.02	**0.38**	0.03	**0.45**	0.02	**0.43**
Brain				**0.01**		**0.02**		**0.02**		**0.02**		**0.02**
Oral mucosa	0.01	**0.34**	0.02	**0.41**	0.02	**0.46**	0.03	**0.49**	0.03	**0.57**	0.03	**0.54**
Salivary glands	0.03	**0.28**	0.04	**0.35**	0.04	**0.40**	0.05	**0.43**	0.06	**0.50**	0.05	**0.48**
Cervical spine	0.25	**0.37**	0.33	**0.49**	0.39	**0.59**	0.44	**0.67**	0.54	**0.83**	0.49	**0.74**
Thyroid	0.72	**0.77**	0.83	**0.89**	0.91	**0.97**	0.96	**1.03**	1.05	**1.14**	1.04	**1.12**
Thymus	0.66	**0.67**	0.80	**0.82**	0.90	**0.92**	0.98	**1.01**	1.13	**1.15**	1.09	**1.12**
Thoracic spine	0.11	**0.12**	0.15	**0.16**	0.19	**0.20**	0.22	**0.24**	0.28	**0.31**	0.23	**0.25**
Claviculae	1.46	**1.52**	1.78	**1.86**	2.01	**2.10**	2.18	**2.28**	2.49	**2.61**	2.46	**2.57**
Scapulae	0.17	**0.18**	0.23	**0.24**	0.27	**0.29**	0.31	**0.33**	0.40	**0.42**	0.35	**0.37**
Ribs	1.22	**1.28**	1.43	**1.51**	1.58	**1.67**	1.68	**1.79**	1.88	**2.00**	1.85	**1.96**
Oesophagus	0.08	**0.09**	0.11	**0.12**	0.13	**0.14**	0.15	**0.17**	0.19	**0.21**	0.17	**0.18**
Trachea	0.18	**0.42**	0.22	**0.52**	0.24	**0.60**	0.26	**0.66**	0.30	**0.76**	0.29	**0.73**
Lung	0.30	**0.32**	0.38	**0.40**	0.44	**0.46**	0.49	**0.51**	0.58	**0.61**	0.55	**0.57**
Mamma	1.10	**1.13**	1.19	**1.22**	1.25	**1.28**	1.29	**1.32**	1.35	**1.38**	1.34	**1.36**
Heart	0.39	**0.41**	0.50	**0.51**	0.57	**0.59**	0.63	**0.66**	0.75	**0.78**	0.71	**0.74**
Stomach	0.22	**0.44**	0.28	**0.55**	0.32	**0.64**	0.36	**0.70**	0.42	**0.82**	0.40	**0.78**
Spleen	0.08	**0.11**	0.11	**0.15**	0.13	**0.18**	0.15	**0.20**	0.19	**0.26**	0.16	**0.22**
Liver	0.20	**0.33**	0.26	**0.41**	0.30	**0.48**	0.33	**0.53**	0.40	**0.63**	0.37	**0.60**
Gall bladder	0.06	**0.30**	0.09	**0.38**	0.10	**0.45**	0.12	**0.51**	0.15	**0.62**	0.13	**0.57**
Pancreas	0.14	**0.18**	0.19	**0.24**	0.22	**0.28**	0.26	**0.32**	0.32	**0.41**	0.28	**0.36**
Small intestine		**0.10**		**0.13**		**0.15**		**0.18**	0.01	**0.22**		**0.19**
Large intestine		**0.17**		**0.22**	0.01	**0.26**	0.01	**0.30**	0.02	**0.36**	0.01	**0.33**
Sigma/Rectum		**0.03**		**0.04**		**0.04**		**0.05**		**0.06**		**0.05**
Lumbar spine	0.03	**0.10**	0.03	**0.13**	0.04	**0.16**	0.05	**0.19**	0.07	**0.25**	0.05	**0.20**
Pelvis		**0.03**		**0.04**		**0.05**		**0.06**		**0.07**		**0.06**
Adrenals	0.06	**0.07**	0.08	**0.09**	0.09	**0.11**	0.11	**0.13**	0.14	**0.16**	0.11	**0.14**
Kidneys	0.02	**0.05**	0.03	**0.07**	0.04	**0.09**	0.05	**0.10**	0.06	**0.13**	0.05	**0.10**
Urinary bladder												
Uterus		**0.01**		**0.02**		**0.02**		**0.02**		**0.03**		**0.02**
Ovaries		**0.01**		**0.02**		**0.02**		**0.02**		**0.03**		**0.02**
Prostate												
Testes												
Upper arm	0.20	**0.60**	0.26	**0.76**	0.31	**0.88**	0.35	**0.98**	0.43	**1.17**	0.39	**1.10**
Elbow	0.17	**0.65**	0.23	**0.82**	0.27	**0.95**	0.31	**1.06**	0.38	**1.26**	0.35	**1.19**
Forearm	0.01	**0.32**	0.02	**0.40**	0.02	**0.46**	0.02	**0.51**	0.03	**0.61**	0.02	**0.58**
Upper leg												
Knee												
Lower leg												
Lymph nodes	0.14	**0.26**	0.18	**0.33**	0.20	**0.38**	0.22	**0.42**	0.26	**0.49**	0.24	**0.46**
Skin	0.12	**0.23**	0.13	**0.24**	0.13	**0.25**	0.14	**0.26**	0.14	**0.27**	0.14	**0.26**
Muscle	0.08	**0.13**	0.10	**0.16**	0.11	**0.18**	0.12	**0.19**	0.14	**0.22**	0.14	**0.21**
Skeleton	0.17	**0.28**	0.20	**0.35**	0.23	**0.40**	0.25	**0.44**	0.29	**0.51**	0.28	**0.48**
Red bone marrow	0.06	**0.09**	0.08	**0.11**	0.09	**0.13**	0.10	**0.14**	0.11	**0.16**	0.10	**0.15**
Whole body dose	0.10	**0.16**	0.12	**0.19**	0.14	**0.22**	0.15	**0.24**	0.17	**0.28**	0.16	**0.26**
ED (ICRP 60)	0.18	**0.24**	0.22	**0.29**	0.24	**0.33**	0.26	**0.36**	0.30	**0.41**	0.29	**0.39**
ED (ICRP 103)	0.26	**0.33**	0.30	**0.39**	0.33	**0.43**	0.36	**0.47**	0.40	**0.53**	0.38	**0.50**
Total filtration (mm) **Al**	**1.5**		**2.5**		**3.5**		**4.5**		**3.5**		**3.5**	
Cu	**0**		**0**		**0**		**0**		**0.1**		**0.2**	

Table 9.94 Normalized organ doses—Chest ap—8–12 years

Beam projection:		ap				**Table cells:**						
Focus-to-detector distance (cm):		115				(+) = normal font = optimal field setting						
Tube voltage (kVp):		70				**(−) = bold font = suboptimal field setting**						
Total filtrations:		See last table row.				Empty table cells correspond to values < 0.01						
Organ or tissue	**Organ dose D_T : Entrance dose K_E**						**(mSv × mGy⁻¹)**					
	(+)	(−)	(+)	(−)	(+)	(−)	(+)	(−)	(+)	(−)	(+)	(−)
Skull	0.02	**0.29**	0.02	**0.35**	0.03	**0.40**	0.03	**0.43**	0.04	**0.50**	0.04	**0.56**
Brain		**0.01**		**0.02**		**0.02**		**0.03**		**0.03**		**0.04**
Oral mucosa	0.02	**0.39**	0.02	**0.46**	0.03	**0.51**	0.03	**0.55**	0.04	**0.63**	0.05	**0.70**
Salivary glands	0.03	**0.33**	0.04	**0.40**	0.05	**0.45**	0.06	**0.49**	0.07	**0.57**	0.08	**0.63**
Cervical spine	0.31	**0.48**	0.40	**0.62**	0.47	**0.72**	0.52	**0.81**	0.63	**0.99**	0.74	**1.16**
Thyroid	0.78	**0.84**	0.90	**0.97**	0.97	**1.05**	1.02	**1.11**	1.12	**1.22**	1.20	**1.31**
Thymus	0.75	**0.77**	0.90	**0.92**	1.01	**1.03**	1.09	**1.11**	1.24	**1.27**	1.36	**1.40**
Thoracic spine	0.16	**0.17**	0.21	**0.23**	0.26	**0.28**	0.30	**0.32**	0.38	**0.41**	0.46	**0.51**
Claviculae	1.64	**1.72**	1.96	**2.06**	2.19	**2.29**	2.36	**2.47**	2.65	**2.79**	2.88	**3.04**
Scapulae	0.22	**0.24**	0.29	**0.31**	0.35	**0.37**	0.39	**0.42**	0.49	**0.52**	0.58	**0.62**
Ribs	1.34	**1.41**	1.55	**1.64**	1.69	**1.80**	1.80	**1.91**	1.98	**2.12**	2.13	**2.28**
Oesophagus	0.11	**0.12**	0.15	**0.16**	0.17	**0.19**	0.20	**0.21**	0.25	**0.27**	0.29	**0.32**
Trachea	0.21	**0.50**	0.24	**0.60**	0.27	**0.68**	0.29	**0.74**	0.32	**0.86**	0.35	**0.96**
Lung	0.36	**0.38**	0.45	**0.47**	0.51	**0.54**	0.56	**0.59**	0.66	**0.69**	0.75	**0.78**
Mamma	1.15	**1.18**	1.24	**1.27**	1.29	**1.32**	1.33	**1.36**	1.40	**1.42**	1.45	**1.48**
Heart	0.47	**0.49**	0.58	**0.60**	0.66	**0.69**	0.73	**0.76**	0.86	**0.89**	0.97	**1.01**
Stomach	0.26	**0.52**	0.33	**0.64**	0.37	**0.73**	0.41	**0.80**	0.48	**0.93**	0.54	**1.05**
Spleen	0.11	**0.15**	0.14	**0.19**	0.17	**0.23**	0.19	**0.26**	0.24	**0.32**	0.28	**0.39**
Liver	0.25	**0.39**	0.31	**0.49**	0.35	**0.56**	0.39	**0.62**	0.46	**0.73**	0.53	**0.83**
Gall bladder	0.08	**0.37**	0.11	**0.47**	0.13	**0.54**	0.15	**0.60**	0.18	**0.72**	0.21	**0.84**
Pancreas	0.18	**0.23**	0.24	**0.30**	0.28	**0.36**	0.32	**0.41**	0.39	**0.50**	0.46	**0.59**
Small intestine		**0.13**		**0.16**	0.01	**0.19**	0.01	**0.21**	0.02	**0.26**	0.02	**0.31**
Large intestine	0.01	**0.21**	0.01	**0.27**	0.02	**0.32**	0.02	**0.35**	0.03	**0.42**	0.03	**0.48**
Sigma/Rectum		**0.04**		**0.05**		**0.05**		**0.06**		**0.07**		**0.09**
Lumbar spine	0.04	**0.14**	0.05	**0.19**	0.06	**0.23**	0.07	**0.26**	0.10	**0.34**	0.12	**0.42**
Pelvis		**0.04**		**0.05**		**0.07**		**0.08**		**0.10**	0.01	**0.12**
Adrenals	0.08	**0.09**	0.10	**0.12**	0.12	**0.15**	0.14	**0.17**	0.18	**0.22**	0.22	**0.26**
Kidneys	0.03	**0.07**	0.05	**0.10**	0.05	**0.12**	0.06	**0.13**	0.08	**0.17**	0.10	**0.21**
Urinary bladder								**0.01**		**0.01**		**0.02**
Uterus		**0.02**		**0.02**		**0.03**		**0.03**		**0.04**		**0.05**
Ovaries		**0.02**		**0.02**		**0.03**		**0.03**		**0.04**		**0.05**
Prostate												
Testes												
Upper arm	0.25	**0.72**	0.32	**0.89**	0.37	**1.02**	0.42	**1.12**	0.51	**1.32**	0.59	**1.49**
Elbow	0.22	**0.77**	0.28	**0.96**	0.33	**1.09**	0.37	**1.20**	0.46	**1.41**	0.53	**1.60**
Forearm	0.02	**0.37**	0.02	**0.46**	0.03	**0.53**	0.03	**0.58**	0.04	**0.68**	0.05	**0.76**
Upper leg												
Knee												
Lower leg												
Lymph nodes	0.17	**0.31**	0.21	**0.38**	0.23	**0.44**	0.26	**0.48**	0.30	**0.57**	0.34	**0.65**
Skin	0.13	**0.24**	0.13	**0.25**	0.14	**0.26**	0.14	**0.27**	0.15	**0.28**	0.16	**0.30**
Muscle	0.10	**0.15**	0.12	**0.18**	0.13	**0.20**	0.14	**0.22**	0.16	**0.25**	0.18	**0.28**
Skeleton	0.19	**0.33**	0.23	**0.40**	0.26	**0.45**	0.28	**0.49**	0.32	**0.57**	0.36	**0.63**
Red bone marrow	0.08	**0.11**	0.09	**0.13**	0.10	**0.15**	0.11	**0.16**	0.13	**0.19**	0.15	**0.22**
Whole body dose	0.12	**0.18**	0.14	**0.22**	0.16	**0.25**	0.17	**0.27**	0.20	**0.31**	0.22	**0.35**
ED (ICRP 60)	0.21	**0.28**	0.25	**0.33**	0.27	**0.37**	0.29	**0.40**	0.33	**0.46**	0.37	**0.51**
ED (ICRP 103)	0.29	**0.37**	0.33	**0.44**	0.37	**0.48**	0.39	**0.52**	0.43	**0.58**	0.47	**0.64**
Total filtration (mm) **Al**	**1.5**		**2.5**		**3.5**		**4.5**		**3.5**		**3.5**	
Cu	**0**		**0**		**0**		**0**		**0.1**		**0.2**	

Table 9.95 Normalized organ doses—Chest ap—8–12 years

Beam projection:		ap					**Table cells:**						
Focus-to-detector distance (cm):		115					(+) = normal font = optimal field setting						
Tube voltage (kVp):		80					**(−) = bold font = suboptimal field setting**						
Total filtrations:		See last table row.					Empty table cells correspond to values < 0.01						
Organ or tissue		**Organ dose D_T : Entrance dose K_E**								**(mSv x mGy^{-1})**			
		(+)	(−)	(+)	(−)	(+)	(−)	(+)	(−)	(+)	(−)	(+)	(−)
Skull		0.02	**0.33**	0.03	**0.39**	0.03	**0.43**	0.03	**0.47**	0.04	**0.53**	0.05	**0.59**
Brain			**0.02**		**0.03**		**0.03**		**0.03**		**0.04**		**0.05**
Oral mucosa		0.02	**0.44**	0.03	**0.51**	0.03	**0.57**	0.04	**0.61**	0.05	**0.69**	0.06	**0.76**
Salivary glands		0.04	**0.37**	0.05	**0.45**	0.06	**0.50**	0.07	**0.54**	0.08	**0.62**	0.09	**0.68**
Cervical spine		0.38	**0.58**	0.47	**0.73**	0.54	**0.84**	0.60	**0.94**	0.71	**1.12**	0.81	**1.30**
Thyroid		0.84	**0.91**	0.95	**1.03**	1.03	**1.12**	1.08	**1.18**	1.17	**1.28**	1.25	**1.37**
Thymus		0.84	**0.86**	0.99	**1.01**	1.10	**1.12**	1.18	**1.21**	1.32	**1.36**	1.44	**1.48**
Thoracic spine		0.21	**0.23**	0.27	**0.30**	0.33	**0.36**	0.37	**0.41**	0.46	**0.51**	0.56	**0.61**
Claviculae		1.80	**1.88**	2.10	**2.21**	2.32	**2.44**	2.47	**2.60**	2.73	**2.89**	2.93	**3.10**
Scapulae		0.28	**0.30**	0.35	**0.38**	0.41	**0.44**	0.46	**0.49**	0.56	**0.60**	0.65	**0.70**
Ribs		1.43	**1.52**	1.64	**1.74**	1.77	**1.89**	1.87	**2.00**	2.04	**2.19**	2.16	**2.33**
Oesophagus		0.14	**0.15**	0.18	**0.20**	0.21	**0.23**	0.24	**0.26**	0.29	**0.32**	0.34	**0.37**
Trachea		0.23	**0.57**	0.27	**0.68**	0.29	**0.76**	0.31	**0.82**	0.35	**0.94**	0.38	**1.04**
Lung		0.42	**0.44**	0.51	**0.53**	0.58	**0.60**	0.63	**0.66**	0.73	**0.76**	0.81	**0.85**
Mamma		1.19	**1.22**	1.28	**1.31**	1.33	**1.36**	1.37	**1.40**	1.43	**1.46**	1.49	**1.51**
Heart		0.54	**0.57**	0.66	**0.69**	0.75	**0.78**	0.82	**0.85**	0.94	**0.99**	1.06	**1.11**
Stomach		0.30	**0.60**	0.37	**0.72**	0.42	**0.82**	0.46	**0.89**	0.53	**1.02**	0.59	**1.13**
Spleen		0.14	**0.19**	0.17	**0.24**	0.20	**0.28**	0.23	**0.31**	0.28	**0.38**	0.33	**0.45**
Liver		0.29	**0.46**	0.35	**0.56**	0.40	**0.64**	0.44	**0.70**	0.51	**0.81**	0.58	**0.91**
Gall bladder		0.11	**0.44**	0.13	**0.55**	0.16	**0.63**	0.17	**0.69**	0.21	**0.82**	0.25	**0.93**
Pancreas		0.23	**0.29**	0.29	**0.37**	0.34	**0.43**	0.38	**0.48**	0.45	**0.58**	0.53	**0.68**
Small intestine		0.01	**0.16**	0.01	**0.20**	0.02	**0.23**	0.02	**0.25**	0.02	**0.30**	0.03	**0.35**
Large intestine		0.01	**0.26**	0.02	**0.32**	0.02	**0.36**	0.03	**0.40**	0.03	**0.47**	0.04	**0.54**
Sigma/Rectum			**0.04**		**0.06**		**0.06**		**0.07**		**0.09**		**0.10**
Lumbar spine		0.05	**0.19**	0.07	**0.25**	0.08	**0.29**	0.10	**0.34**	0.12	**0.42**	0.15	**0.51**
Pelvis			**0.05**		**0.07**		**0.08**		**0.10**	0.01	**0.12**	0.01	**0.14**
Adrenals		0.10	**0.12**	0.13	**0.16**	0.16	**0.19**	0.18	**0.21**	0.22	**0.27**	0.26	**0.32**
Kidneys		0.05	**0.10**	0.06	**0.12**	0.07	**0.15**	0.08	**0.17**	0.10	**0.21**	0.12	**0.25**
Urinary bladder					**0.01**		**0.01**		**0.01**		**0.02**		**0.02**
Uterus			**0.03**		**0.03**		**0.04**		**0.04**		**0.05**		**0.07**
Ovaries			**0.03**		**0.03**		**0.04**		**0.04**		**0.06**		**0.07**
Prostate													
Testes													
Upper arm		0.30	**0.82**	0.37	**1.00**	0.43	**1.13**	0.48	**1.23**	0.57	**1.42**	0.65	**1.59**
Elbow		0.27	**0.88**	0.34	**1.07**	0.39	**1.21**	0.43	**1.32**	0.51	**1.52**	0.59	**1.70**
Forearm		0.02	**0.43**	0.03	**0.52**	0.03	**0.58**	0.04	**0.63**	0.04	**0.73**	0.05	**0.81**
Upper leg													
Knee													
Lower leg													
Lymph nodes		0.20	**0.36**	0.23	**0.44**	0.26	**0.50**	0.29	**0.54**	0.33	**0.63**	0.37	**0.71**
Skin		0.13	**0.25**	0.14	**0.26**	0.15	**0.27**	0.15	**0.28**	0.16	**0.30**	0.17	**0.31**
Muscle		0.11	**0.17**	0.13	**0.20**	0.15	**0.23**	0.16	**0.24**	0.18	**0.28**	0.20	**0.31**
Skeleton		0.22	**0.37**	0.26	**0.44**	0.29	**0.49**	0.31	**0.53**	0.35	**0.61**	0.38	**0.67**
Red bone marrow		0.09	**0.13**	0.10	**0.15**	0.12	**0.17**	0.13	**0.19**	0.15	**0.22**	0.17	**0.25**
Whole body dose		0.13	**0.21**	0.16	**0.25**	0.17	**0.28**	0.19	**0.30**	0.21	**0.34**	0.24	**0.38**
ED (ICRP 60)		0.23	**0.31**	0.27	**0.37**	0.30	**0.41**	0.32	**0.44**	0.36	**0.50**	0.39	**0.55**
ED (ICRP 103)		0.32	**0.41**	0.36	**0.48**	0.39	**0.53**	0.42	**0.56**	0.46	**0.63**	0.50	**0.69**
Total filtration (mm)	**Al**	1.5		2.5		3.5		4.5		3.5		3.5	
	Cu	**0**		**0**		**0**		**0**		**0.1**		**0.2**	

Table 9.96 Normalized organ doses—Chest ap—13–17 years

Beam projection:			ap			**Table cells:**						
Focus-to-detector distance (cm):			115			(+) = normal font = optimal field setting						
Tube voltage (kVp):			60			**(−) = bold font = suboptimal field setting**						
Total filtrations:			See last table row.			Empty table cells correspond to values < 0.01						
Organ or tissue	**Organ dose D_T : Entrance dose K_E**							**(mSv × mGy^{-1})**				
	(+)	(−)	(+)	(−)	(+)	(−)	(+)	(−)	(+)	(−)	(+)	(−)
Skull		**0.23**	0.01	**0.29**	0.02	**0.34**	0.02	**0.38**	0.02	**0.45**	0.02	**0.42**
Brain				**0.01**		**0.01**		**0.02**		**0.02**		**0.02**
Oral mucosa		**0.28**	0.01	**0.34**	0.01	**0.39**	0.02	**0.42**	0.02	**0.49**	0.02	**0.46**
Salivary glands	0.02	**0.25**	0.02	**0.31**	0.03	**0.35**	0.03	**0.39**	0.04	**0.46**	0.03	**0.43**
Cervical spine	0.25	**0.40**	0.34	**0.53**	0.40	**0.64**	0.45	**0.73**	0.56	**0.91**	0.50	**0.80**
Thyroid	0.66	**0.69**	0.77	**0.81**	0.85	**0.89**	0.91	**0.95**	1.01	**1.06**	0.99	**1.04**
Thymus	0.61	**0.62**	0.75	**0.77**	0.86	**0.88**	0.94	**0.96**	1.09	**1.12**	1.05	**1.07**
Thoracic spine	0.07	**0.07**	0.09	**0.10**	0.11	**0.12**	0.13	**0.15**	0.18	**0.19**	0.13	**0.15**
Claviculae	1.21	**1.26**	1.50	**1.56**	1.70	**1.78**	1.86	**1.95**	2.16	**2.25**	2.11	**2.20**
Scapulae	0.08	**0.09**	0.11	**0.12**	0.14	**0.15**	0.16	**0.17**	0.21	**0.22**	0.17	**0.18**
Ribs	1.02	**1.08**	1.21	**1.29**	1.34	**1.43**	1.45	**1.54**	1.63	**1.74**	1.59	**1.70**
Oesophagus	0.08	**0.09**	0.11	**0.12**	0.14	**0.15**	0.16	**0.17**	0.21	**0.22**	0.17	**0.18**
Trachea	0.17	**0.32**	0.20	**0.39**	0.22	**0.44**	0.24	**0.49**	0.28	**0.56**	0.27	**0.54**
Lung	0.20	**0.21**	0.26	**0.27**	0.30	**0.32**	0.34	**0.36**	0.42	**0.44**	0.38	**0.40**
Mamma	7.91	**8.17**	9.16	**9.46**	9.99	**10.3**	10.6	**11.0**	11.7	**12.1**	11.5	**11.9**
Heart	0.34	**0.35**	0.43	**0.45**	0.51	**0.53**	0.56	**0.59**	0.68	**0.71**	0.63	**0.65**
Stomach	0.18	**0.36**	0.23	**0.46**	0.27	**0.53**	0.30	**0.59**	0.37	**0.71**	0.34	**0.66**
Spleen	0.06	**0.08**	0.08	**0.11**	0.10	**0.13**	0.11	**0.15**	0.14	**0.19**	0.12	**0.16**
Liver	0.18	**0.26**	0.23	**0.34**	0.27	**0.40**	0.30	**0.44**	0.36	**0.54**	0.33	**0.49**
Gall bladder	0.05	**0.25**	0.06	**0.32**	0.07	**0.38**	0.09	**0.43**	0.11	**0.53**	0.09	**0.48**
Pancreas	0.11	**0.13**	0.14	**0.18**	0.17	**0.22**	0.20	**0.25**	0.25	**0.32**	0.21	**0.27**
Small intestine		**0.08**		**0.10**		**0.12**		**0.14**		**0.17**		**0.15**
Large intestine		**0.14**		**0.18**		**0.21**		**0.24**	0.01	**0.30**		**0.27**
Sigma/Rectum		**0.02**		**0.03**		**0.03**		**0.04**		**0.05**		**0.04**
Lumbar spine	0.02	**0.06**	0.02	**0.08**	0.03	**0.10**	0.03	**0.12**	0.04	**0.16**	0.03	**0.12**
Pelvis		**0.02**		**0.03**		**0.03**		**0.04**		**0.05**		**0.04**
Adrenals	0.04	**0.04**	0.05	**0.06**	0.06	**0.07**	0.07	**0.09**	0.10	**0.11**	0.08	**0.09**
Kidneys	0.02	**0.03**	0.02	**0.05**	0.03	**0.06**	0.03	**0.07**	0.04	**0.09**	0.03	**0.07**
Urinary bladder												
Uterus		**0.02**		**0.03**		**0.04**		**0.04**		**0.06**		**0.04**
Ovaries				**0.01**		**0.01**		**0.02**		**0.02**		**0.02**
Prostate												
Testes												
Upper arm	0.14	**0.47**	0.18	**0.60**	0.21	**0.70**	0.25	**0.78**	0.31	**0.94**	0.27	**0.88**
Elbow	0.12	**0.50**	0.16	**0.64**	0.19	**0.75**	0.22	**0.83**	0.27	**1.00**	0.24	**0.94**
Forearm		**0.24**	0.01	**0.31**	0.01	**0.36**	0.02	**0.40**	0.02	**0.49**	0.02	**0.46**
Upper leg												
Knee												
Lower leg												
Lymph nodes	0.12	**0.21**	0.15	**0.26**	0.17	**0.31**	0.19	**0.34**	0.22	**0.41**	0.21	**0.38**
Skin	0.14	**0.24**	0.15	**0.26**	0.16	**0.27**	0.16	**0.28**	0.17	**0.29**	0.17	**0.29**
Muscle	0.07	**0.12**	0.09	**0.14**	0.10	**0.16**	0.11	**0.18**	0.13	**0.21**	0.12	**0.19**
Skeleton	0.15	**0.24**	0.18	**0.30**	0.20	**0.34**	0.22	**0.38**	0.26	**0.44**	0.24	**0.42**
Red bone marrow	0.06	**0.08**	0.08	**0.10**	0.09	**0.12**	0.10	**0.13**	0.11	**0.15**	0.10	**0.14**
Whole body dose	0.09	**0.14**	0.11	**0.18**	0.13	**0.20**	0.14	**0.22**	0.16	**0.26**	0.15	**0.24**
ED (ICRP 60)	0.50	**0.56**	0.59	**0.66**	0.65	**0.74**	0.70	**0.79**	0.78	**0.89**	0.76	**0.86**
ED (ICRP 103)	1.06	**1.14**	1.23	**1.34**	1.35	**1.47**	1.44	**1.57**	1.60	**1.75**	1.56	**1.71**
Total filtration (mm) **Al**	**1.5**		**2.5**		**3.5**		**4.5**		**3.5**		**3.5**	
Cu	**0**		**0**		**0**		**0**		**0.1**		**0.2**	

Table 9.97 Normalized organ doses—Chest ap—13–17 years

Beam projection:			ap			**Table cells:**						
Focus-to-detector distance (cm):			115			(+) = normal font = optimal field setting						
Tube voltage (kVp):			70			**(−) = bold font = suboptimal field setting**						
Total filtrations:			See last table row.			Empty table cells correspond to values < 0.01						
Organ or tissue	**Organ dose D_T : Entrance dose K_E**								**(mSv × mGy^{-1})**			
	(+)	(−)	(+)	(−)	(+)	(−)	(+)	(−)	(+)	(−)	(+)	(−)
Skull	0.01	**0.28**	0.02	**0.34**	0.02	**0.39**	0.02	**0.43**	0.03	**0.50**	0.04	**0.57**
Brain		**0.01**		**0.02**		**0.02**		**0.02**		**0.03**		**0.04**
Oral mucosa	0.01	**0.33**	0.02	**0.39**	0.02	**0.44**	0.02	**0.48**	0.03	**0.55**	0.03	**0.61**
Salivary glands	0.02	**0.29**	0.03	**0.36**	0.03	**0.41**	0.04	**0.44**	0.05	**0.52**	0.05	**0.58**
Cervical spine	0.33	**0.52**	0.42	**0.68**	0.49	**0.80**	0.56	**0.90**	0.68	**1.10**	0.80	**1.30**
Thyroid	0.73	**0.76**	0.84	**0.89**	0.92	**0.97**	0.98	**1.03**	1.08	**1.14**	1.16	**1.24**
Thymus	0.70	**0.72**	0.86	**0.88**	0.97	**1.00**	1.06	**1.08**	1.22	**1.25**	1.36	**1.40**
Thoracic spine	0.10	**0.11**	0.14	**0.15**	0.17	**0.18**	0.19	**0.21**	0.25	**0.28**	0.31	**0.35**
Claviculae	1.39	**1.44**	1.68	**1.75**	1.89	**1.97**	2.05	**2.14**	2.33	**2.45**	2.57	**2.70**
Scapulae	0.12	**0.13**	0.15	**0.17**	0.19	**0.20**	0.21	**0.23**	0.27	**0.30**	0.34	**0.36**
Ribs	1.13	**1.21**	1.33	**1.42**	1.46	**1.56**	1.56	**1.67**	1.74	**1.87**	1.89	**2.04**
Oesophagus	0.12	**0.13**	0.16	**0.17**	0.19	**0.20**	0.21	**0.23**	0.27	**0.29**	0.33	**0.35**
Trachea	0.19	**0.37**	0.22	**0.45**	0.25	**0.51**	0.27	**0.55**	0.30	**0.64**	0.33	**0.71**
Lung	0.25	**0.26**	0.31	**0.33**	0.37	**0.39**	0.41	**0.43**	0.49	**0.52**	0.57	**0.61**
Mamma	8.64	**8.92**	9.89	**10.2**	10.7	**11.1**	11.3	**11.7**	12.4	**12.9**	13.3	**13.8**
Heart	0.42	**0.43**	0.52	**0.54**	0.60	**0.62**	0.66	**0.69**	0.79	**0.82**	0.91	**0.95**
Stomach	0.22	**0.44**	0.28	**0.54**	0.32	**0.63**	0.36	**0.69**	0.43	**0.82**	0.49	**0.94**
Spleen	0.08	**0.11**	0.11	**0.14**	0.13	**0.17**	0.15	**0.20**	0.18	**0.25**	0.22	**0.30**
Liver	0.22	**0.32**	0.28	**0.41**	0.32	**0.47**	0.36	**0.53**	0.43	**0.63**	0.50	**0.73**
Gall bladder	0.06	**0.31**	0.08	**0.40**	0.10	**0.47**	0.11	**0.53**	0.14	**0.64**	0.17	**0.75**
Pancreas	0.14	**0.18**	0.19	**0.24**	0.22	**0.28**	0.25	**0.32**	0.32	**0.41**	0.38	**0.49**
Small intestine		**0.10**		**0.13**		**0.15**		**0.17**	0.01	**0.21**	0.01	**0.26**
Large intestine		**0.18**		**0.22**	0.01	**0.26**	0.01	**0.29**	0.02	**0.36**	0.02	**0.42**
Sigma/Rectum		**0.03**		**0.04**		**0.04**		**0.05**		**0.06**		**0.07**
Lumbar spine	0.02	**0.09**	0.03	**0.12**	0.04	**0.15**	0.05	**0.17**	0.06	**0.22**	0.08	**0.28**
Pelvis		**0.03**		**0.04**		**0.05**		**0.05**		**0.07**		**0.08**
Adrenals	0.06	**0.07**	0.07	**0.09**	0.09	**0.11**	0.10	**0.12**	0.13	**0.16**	0.16	**0.20**
Kidneys	0.02	**0.05**	0.03	**0.07**	0.04	**0.08**	0.04	**0.09**	0.06	**0.12**	0.07	**0.15**
Urinary bladder												
Uterus		**0.03**		**0.05**		**0.06**		**0.07**		**0.09**		**0.11**
Ovaries		**0.01**		**0.02**		**0.02**		**0.03**		**0.03**		**0.04**
Prostate												
Testes												
Upper arm	0.18	**0.57**	0.23	**0.71**	0.27	**0.82**	0.31	**0.91**	0.38	**1.08**	0.45	**1.24**
Elbow	0.16	**0.61**	0.20	**0.76**	0.24	**0.87**	0.27	**0.97**	0.33	**1.15**	0.40	**1.31**
Forearm	0.01	**0.29**	0.02	**0.37**	0.02	**0.42**	0.02	**0.47**	0.03	**0.55**	0.03	**0.63**
Upper leg												
Knee												
Lower leg												
Lymph nodes	0.14	**0.26**	0.18	**0.32**	0.20	**0.37**	0.22	**0.40**	0.26	**0.48**	0.30	**0.55**
Skin	0.15	**0.25**	0.16	**0.27**	0.16	**0.28**	0.17	**0.29**	0.18	**0.31**	0.19	**0.32**
Muscle	0.09	**0.14**	0.11	**0.17**	0.12	**0.19**	0.13	**0.20**	0.15	**0.24**	0.18	**0.27**
Skeleton	0.17	**0.28**	0.20	**0.35**	0.23	**0.39**	0.25	**0.43**	0.29	**0.50**	0.33	**0.57**
Red bone marrow	0.08	**0.10**	0.09	**0.12**	0.10	**0.14**	0.11	**0.15**	0.13	**0.18**	0.15	**0.21**
Whole body dose	0.11	**0.17**	0.13	**0.20**	0.15	**0.23**	0.16	**0.25**	0.19	**0.29**	0.21	**0.33**
ED (ICRP 60)	0.56	**0.63**	0.65	**0.74**	0.71	**0.81**	0.76	**0.87**	0.85	**0.98**	0.93	**1.08**
ED (ICRP 103)	1.16	**1.26**	1.34	**1.47**	1.47	**1.60**	1.56	**1.71**	1.72	**1.90**	1.86	**2.06**
Total filtration (mm) **Al**	**1.5**		**2.5**		**3.5**		**4.5**		**3.5**		**3.5**	
Cu	**0**		**0**		**0**		**0**		**0.1**		**0.2**	

Table 9.98 Normalized organ doses—Chest ap—13–17 years

Beam projection:			ap			**Table cells:**						
Focus-to-detector distance (cm):			115			(+) = normal font = optimal field setting						
Tube voltage (kVp):			80			**(−) = bold font = suboptimal field setting**						
Total filtrations:			See last table row.			Empty table cells correspond to values < 0.01						
Organ or tissue	**Organ dose D_T : Entrance dose K_E**							**(mSv × mGy^{-1})**				
	(+)	(−)	(+)	(−)	(+)	(−)	(+)	(−)	(+)	(−)	(+)	(−)
Skull	0.02	**0.32**	0.02	**0.39**	0.03	**0.44**	0.03	**0.47**	0.04	**0.55**	0.04	**0.61**
Brain		**0.02**		**0.02**		**0.03**		**0.03**		**0.04**		**0.05**
Oral mucosa	0.02	**0.37**	0.02	**0.44**	0.02	**0.49**	0.03	**0.53**	0.03	**0.61**	0.04	**0.67**
Salivary glands	0.03	**0.34**	0.03	**0.41**	0.04	**0.46**	0.04	**0.49**	0.05	**0.57**	0.06	**0.63**
Cervical spine	0.40	**0.64**	0.50	**0.81**	0.58	**0.94**	0.64	**1.05**	0.77	**1.27**	0.89	**1.47**
Thyroid	0.79	**0.83**	0.91	**0.96**	0.98	**1.04**	1.04	**1.10**	1.14	**1.21**	1.22	**1.30**
Thymus	0.80	**0.82**	0.96	**0.98**	1.07	**1.10**	1.16	**1.19**	1.32	**1.36**	1.46	**1.50**
Thoracic spine	0.14	**0.16**	0.19	**0.20**	0.22	**0.24**	0.25	**0.28**	0.32	**0.36**	0.39	**0.43**
Claviculae	1.54	**1.61**	1.82	**1.91**	2.02	**2.12**	2.18	**2.28**	2.44	**2.57**	2.65	**2.79**
Scapulae	0.15	**0.17**	0.20	**0.22**	0.24	**0.26**	0.27	**0.29**	0.33	**0.36**	0.40	**0.43**
Ribs	1.23	**1.32**	1.42	**1.52**	1.55	**1.67**	1.65	**1.77**	1.81	**1.96**	1.94	**2.10**
Oesophagus	0.16	**0.17**	0.20	**0.21**	0.24	**0.25**	0.27	**0.29**	0.33	**0.35**	0.39	**0.42**
Trachea	0.21	**0.42**	0.25	**0.50**	0.27	**0.56**	0.29	**0.61**	0.32	**0.70**	0.35	**0.78**
Lung	0.30	**0.32**	0.37	**0.39**	0.43	**0.45**	0.47	**0.50**	0.56	**0.59**	0.64	**0.68**
Mamma	9.31	**9.63**	10.5	**10.9**	11.4	**11.8**	12.0	**12.4**	13.0	**13.5**	13.9	**14.4**
Heart	0.49	**0.51**	0.60	**0.63**	0.69	**0.72**	0.76	**0.79**	0.89	**0.93**	1.00	**1.05**
Stomach	0.26	**0.51**	0.33	**0.63**	0.37	**0.71**	0.41	**0.78**	0.48	**0.91**	0.55	**1.03**
Spleen	0.11	**0.14**	0.14	**0.18**	0.16	**0.22**	0.18	**0.25**	0.22	**0.31**	0.27	**0.36**
Liver	0.26	**0.39**	0.32	**0.48**	0.37	**0.55**	0.41	**0.60**	0.49	**0.71**	0.55	**0.81**
Gall bladder	0.08	**0.38**	0.10	**0.48**	0.12	**0.55**	0.14	**0.61**	0.17	**0.73**	0.20	**0.84**
Pancreas	0.18	**0.23**	0.23	**0.30**	0.27	**0.35**	0.31	**0.40**	0.38	**0.49**	0.44	**0.58**
Small intestine		**0.13**		**0.16**	0.01	**0.19**	0.01	**0.21**	0.02	**0.25**	0.02	**0.30**
Large intestine	0.01	**0.21**	0.01	**0.27**	0.02	**0.31**	0.02	**0.34**	0.02	**0.41**	0.03	**0.47**
Sigma/Rectum		**0.04**		**0.04**		**0.05**		**0.06**		**0.07**		**0.08**
Lumbar spine	0.04	**0.13**	0.05	**0.17**	0.06	**0.20**	0.07	**0.23**	0.08	**0.29**	0.10	**0.35**
Pelvis		**0.04**		**0.05**		**0.06**		**0.07**		**0.09**		**0.11**
Adrenals	0.08	**0.09**	0.10	**0.12**	0.12	**0.14**	0.13	**0.16**	0.17	**0.20**	0.20	**0.25**
Kidneys	0.03	**0.07**	0.04	**0.09**	0.05	**0.11**	0.06	**0.12**	0.08	**0.15**	0.09	**0.19**
Urinary bladder										**0.01**		**0.01**
Uterus		**0.05**		**0.06**		**0.08**		**0.09**		**0.11**	0.01	**0.14**
Ovaries		**0.02**		**0.03**		**0.03**		**0.04**		**0.05**		**0.06**
Prostate												
Testes												
Upper arm	0.22	**0.67**	0.28	**0.82**	0.32	**0.93**	0.36	**1.02**	0.43	**1.19**	0.51	**1.34**
Elbow	0.19	**0.71**	0.25	**0.87**	0.29	**0.98**	0.32	**1.08**	0.39	**1.26**	0.45	**1.42**
Forearm	0.02	**0.34**	0.02	**0.42**	0.02	**0.47**	0.03	**0.52**	0.03	**0.60**	0.04	**0.68**
Upper leg												
Knee												
Lower leg												
Lymph nodes	0.17	**0.30**	0.20	**0.37**	0.23	**0.42**	0.25	**0.46**	0.29	**0.54**	0.33	**0.62**
Skin	0.15	**0.26**	0.16	**0.28**	0.17	**0.29**	0.18	**0.30**	0.19	**0.32**	0.20	**0.34**
Muscle	0.10	**0.16**	0.12	**0.19**	0.14	**0.21**	0.15	**0.23**	0.17	**0.26**	0.19	**0.30**
Skeleton	0.19	**0.32**	0.23	**0.39**	0.25	**0.44**	0.28	**0.47**	0.31	**0.55**	0.35	**0.61**
Red bone marrow	0.09	**0.12**	0.11	**0.14**	0.12	**0.16**	0.13	**0.18**	0.15	**0.21**	0.17	**0.24**
Whole body dose	0.12	**0.19**	0.15	**0.23**	0.17	**0.26**	0.18	**0.28**	0.21	**0.32**	0.23	**0.36**
ED (ICRP 60)	0.61	**0.70**	0.71	**0.81**	0.77	**0.88**	0.82	**0.94**	0.91	**1.05**	0.98	**1.15**
ED (ICRP 103)	1.27	**1.38**	1.45	**1.59**	1.57	**1.72**	1.66	**1.83**	1.82	**2.02**	1.96	**2.18**
Total filtration (mm) **Al**	**1.5**		**2.5**		**3.5**		**4.5**		**3.5**		**3.5**	
Cu	**0**		**0**		**0**		**0**		**0.1**		**0.2**	

Table 9.99 Normalized organ doses—Chest ap—13–17 years

Beam projection:			ap			**Table cells:**						
Focus-to-detector distance (cm):			115			(+) = normal font = optimal field setting						
Tube voltage (kVp):			110			**(–) = bold font = suboptimal field setting**						
Total filtrations:			See last table row.			Empty table cells correspond to values < 0.01						
Organ or tissue	**Organ dose D_T : Entrance dose K_E**							**(mSv × mGy^{-1})**				
	(+)	(–)	(+)	(–)	(+)	(–)	(+)	(–)	(+)	(–)	(+)	(–)
Skull	0.03	**0.41**	0.03	**0.47**	0.04	**0.52**	0.04	**0.55**	0.05	**0.60**	0.05	**0.65**
Brain		**0.03**		**0.04**		**0.05**		**0.05**		**0.06**		**0.07**
Oral mucosa	0.03	**0.49**	0.03	**0.56**	0.03	**0.61**	0.04	**0.65**	0.04	**0.71**	0.05	**0.77**
Salivary glands	0.04	**0.45**	0.05	**0.52**	0.06	**0.57**	0.06	**0.60**	0.07	**0.67**	0.08	**0.72**
Cervical spine	0.57	**0.94**	0.68	**1.11**	0.75	**1.24**	0.81	**1.34**	0.92	**1.53**	1.01	**1.69**
Thyroid	0.95	**1.00**	1.05	**1.11**	1.11	**1.19**	1.16	**1.24**	1.24	**1.33**	1.30	**1.40**
Thymus	1.04	**1.07**	1.18	**1.22**	1.28	**1.33**	1.36	**1.40**	1.50	**1.54**	1.60	**1.66**
Thoracic spine	0.25	**0.28**	0.31	**0.34**	0.35	**0.39**	0.39	**0.43**	0.46	**0.51**	0.53	**0.59**
Claviculae	1.85	**1.94**	2.08	**2.19**	2.24	**2.36**	2.35	**2.47**	2.52	**2.66**	2.63	**2.79**
Scapulae	0.25	**0.27**	0.30	**0.33**	0.34	**0.37**	0.37	**0.41**	0.44	**0.48**	0.49	**0.54**
Ribs	1.43	**1.55**	1.59	**1.72**	1.69	**1.82**	1.75	**1.90**	1.86	**2.03**	1.93	**2.11**
Oesophagus	0.26	**0.28**	0.31	**0.34**	0.35	**0.38**	0.38	**0.42**	0.45	**0.49**	0.51	**0.55**
Trachea	0.26	**0.56**	0.30	**0.64**	0.32	**0.69**	0.33	**0.74**	0.36	**0.81**	0.39	**0.88**
Lung	0.43	**0.46**	0.51	**0.54**	0.56	**0.60**	0.60	**0.64**	0.69	**0.73**	0.75	**0.81**
Mamma	10.9	**11.3**	12.0	**12.4**	12.7	**13.2**	13.2	**13.7**	14.0	**14.6**	14.7	**15.3**
Heart	0.69	**0.72**	0.80	**0.84**	0.88	**0.92**	0.94	**0.99**	1.06	**1.12**	1.16	**1.22**
Stomach	0.37	**0.71**	0.44	**0.82**	0.48	**0.91**	0.52	**0.97**	0.58	**1.09**	0.64	**1.19**
Spleen	0.18	**0.24**	0.21	**0.29**	0.24	**0.33**	0.26	**0.36**	0.31	**0.42**	0.35	**0.48**
Liver	0.38	**0.55**	0.44	**0.65**	0.49	**0.72**	0.52	**0.77**	0.59	**0.87**	0.65	**0.95**
Gall bladder	0.13	**0.56**	0.16	**0.66**	0.18	**0.74**	0.19	**0.79**	0.22	**0.91**	0.25	**1.00**
Pancreas	0.29	**0.38**	0.35	**0.45**	0.39	**0.51**	0.43	**0.55**	0.50	**0.64**	0.56	**0.73**
Small intestine	0.01	**0.20**	0.02	**0.23**	0.02	**0.26**	0.02	**0.29**	0.03	**0.33**	0.03	**0.37**
Large intestine	0.02	**0.32**	0.02	**0.37**	0.03	**0.41**	0.03	**0.45**	0.04	**0.51**	0.04	**0.56**
Sigma/Rectum		**0.06**		**0.07**		**0.07**		**0.08**		**0.09**		**0.11**
Lumbar spine	0.07	**0.23**	0.09	**0.28**	0.10	**0.32**	0.11	**0.35**	0.13	**0.42**	0.15	**0.48**
Pelvis		**0.07**		**0.08**		**0.09**		**0.10**	0.01	**0.12**	0.01	**0.14**
Adrenals	0.14	**0.16**	0.16	**0.20**	0.19	**0.23**	0.20	**0.25**	0.24	**0.30**	0.28	**0.34**
Kidneys	0.06	**0.13**	0.08	**0.15**	0.09	**0.17**	0.09	**0.19**	0.11	**0.23**	0.13	**0.26**
Urinary bladder				**0.01**		**0.01**		**0.01**		**0.02**		**0.02**
Uterus		**0.09**	0.01	**0.11**	0.01	**0.13**	0.01	**0.14**	0.02	**0.17**	0.02	**0.20**
Ovaries		**0.04**		**0.04**		**0.05**		**0.06**		**0.07**		**0.08**
Prostate												
Testes												
Upper arm	0.32	**0.89**	0.38	**1.03**	0.43	**1.13**	0.46	**1.20**	0.53	**1.34**	0.59	**1.44**
Elbow	0.29	**0.94**	0.34	**1.09**	0.38	**1.19**	0.41	**1.27**	0.48	**1.41**	0.53	**1.52**
Forearm	0.03	**0.45**	0.03	**0.52**	0.04	**0.57**	0.04	**0.61**	0.05	**0.67**	0.05	**0.72**
Upper leg												
Knee												
Lower leg												
Lymph nodes	0.23	**0.43**	0.27	**0.50**	0.30	**0.55**	0.32	**0.59**	0.36	**0.66**	0.39	**0.73**
Skin	0.17	**0.29**	0.18	**0.31**	0.19	**0.32**	0.20	**0.33**	0.21	**0.35**	0.22	**0.36**
Muscle	0.14	**0.21**	0.16	**0.24**	0.17	**0.27**	0.19	**0.28**	0.21	**0.32**	0.23	**0.34**
Skeleton	0.24	**0.42**	0.28	**0.48**	0.30	**0.52**	0.32	**0.55**	0.35	**0.61**	0.37	**0.65**
Red bone marrow	0.12	**0.17**	0.14	**0.20**	0.15	**0.22**	0.16	**0.24**	0.19	**0.27**	0.20	**0.29**
Whole body dose	0.17	**0.26**	0.19	**0.30**	0.21	**0.32**	0.22	**0.34**	0.25	**0.38**	0.27	**0.41**
ED (ICRP 60)	0.75	**0.87**	0.84	**0.97**	0.89	**1.04**	0.94	**1.09**	1.01	**1.19**	1.07	**1.27**
ED (ICRP 103)	1.52	**1.68**	1.68	**1.86**	1.79	**1.98**	1.86	**2.07**	2.00	**2.24**	2.11	**2.36**
Total filtration (mm) **Al**	**1.5**		**2.5**		**3.5**		**4.5**		**3.5**		**3.5**	
Cu	**0**		**0**		**0**		**0**		**0.1**		**0.2**	

Table 9.100 Normalized organ doses—Chest ap—13–17 years

Beam projection:			ap			**Table cells:**						
Focus-to-detector distance (cm):			115			(+) = normal font = optimal field setting						
Tube voltage (kVp):			120			**(−) = bold font = suboptimal field setting**						
Total filtrations:			See last table row.			Empty table cells correspond to values < 0.01						
Organ or tissue	**Organ dose D_T : Entrance dose K_E**							**(mSv × mGy^{-1})**				
	(+)	(−)	(+)	(−)	(+)	(−)	(+)	(−)	(+)	(−)	(+)	(−)
Skull	0.03	**0.43**	0.04	**0.49**	0.04	**0.53**	0.04	**0.56**	0.05	**0.61**	0.06	**0.65**
Brain		**0.04**		**0.05**		**0.05**		**0.06**		**0.07**		**0.08**
Oral mucosa	0.03	**0.52**	0.03	**0.59**	0.04	**0.64**	0.04	**0.67**	0.05	**0.74**	0.05	**0.79**
Salivary glands	0.05	**0.48**	0.05	**0.55**	0.06	**0.59**	0.06	**0.63**	0.07	**0.69**	0.08	**0.74**
Cervical spine	0.62	**1.02**	0.72	**1.18**	0.79	**1.30**	0.84	**1.39**	0.94	**1.57**	1.02	**1.71**
Thyroid	0.99	**1.05**	1.08	**1.15**	1.14	**1.22**	1.18	**1.27**	1.26	**1.35**	1.32	**1.42**
Thymus	1.10	**1.13**	1.24	**1.28**	1.33	**1.38**	1.40	**1.45**	1.53	**1.58**	1.62	**1.68**
Thoracic spine	0.29	**0.32**	0.34	**0.38**	0.38	**0.42**	0.42	**0.46**	0.49	**0.54**	0.55	**0.61**
Claviculae	1.91	**2.01**	2.13	**2.24**	2.26	**2.39**	2.36	**2.49**	2.51	**2.65**	2.59	**2.75**
Scapulae	0.28	**0.30**	0.33	**0.36**	0.36	**0.40**	0.39	**0.43**	0.45	**0.50**	0.51	**0.55**
Ribs	1.48	**1.59**	1.62	**1.75**	1.70	**1.85**	1.76	**1.91**	1.85	**2.02**	1.91	**2.08**
Oesophagus	0.29	**0.31**	0.34	**0.37**	0.38	**0.41**	0.41	**0.45**	0.47	**0.52**	0.53	**0.58**
Trachea	0.28	**0.59**	0.31	**0.67**	0.33	**0.72**	0.34	**0.76**	0.37	**0.84**	0.39	**0.90**
Lung	0.47	**0.50**	0.54	**0.58**	0.59	**0.63**	0.63	**0.68**	0.71	**0.76**	0.77	**0.83**
Mamma	11.3	**11.8**	12.3	**12.8**	13.0	**13.5**	13.4	**14.0**	14.2	**14.8**	14.8	**15.4**
Heart	0.74	**0.78**	0.85	**0.89**	0.93	**0.97**	0.99	**1.04**	1.10	**1.15**	1.19	**1.25**
Stomach	0.40	**0.76**	0.46	**0.87**	0.51	**0.95**	0.54	**1.01**	0.60	**1.12**	0.65	**1.21**
Spleen	0.20	**0.27**	0.23	**0.32**	0.26	**0.35**	0.28	**0.38**	0.32	**0.44**	0.36	**0.50**
Liver	0.41	**0.60**	0.47	**0.69**	0.52	**0.75**	0.55	**0.81**	0.61	**0.90**	0.67	**0.98**
Gall bladder	0.14	**0.61**	0.17	**0.71**	0.19	**0.78**	0.20	**0.84**	0.23	**0.94**	0.26	**1.03**
Pancreas	0.32	**0.42**	0.38	**0.49**	0.42	**0.55**	0.46	**0.59**	0.52	**0.68**	0.58	**0.75**
Small intestine	0.02	**0.22**	0.02	**0.25**	0.02	**0.28**	0.02	**0.30**	0.03	**0.35**	0.03	**0.38**
Large intestine	0.02	**0.34**	0.03	**0.40**	0.03	**0.44**	0.03	**0.47**	0.04	**0.53**	0.04	**0.58**
Sigma/Rectum		**0.06**		**0.07**		**0.08**		**0.09**		**0.10**		**0.11**
Lumbar spine	0.08	**0.26**	0.10	**0.31**	0.11	**0.35**	0.12	**0.38**	0.14	**0.44**	0.16	**0.50**
Pelvis		**0.08**		**0.09**		**0.10**		**0.11**	0.01	**0.13**	0.01	**0.15**
Adrenals	0.15	**0.19**	0.18	**0.22**	0.20	**0.25**	0.22	**0.27**	0.26	**0.32**	0.29	**0.36**
Kidneys	0.07	**0.14**	0.08	**0.17**	0.10	**0.19**	0.10	**0.21**	0.12	**0.24**	0.14	**0.28**
Urinary bladder		**0.01**		**0.01**		**0.01**		**0.02**		**0.02**		**0.02**
Uterus	0.01	**0.11**	0.01	**0.13**	0.01	**0.14**	0.02	**0.16**	0.02	**0.18**	0.02	**0.21**
Ovaries		**0.04**		**0.05**		**0.06**		**0.06**		**0.07**		**0.08**
Prostate												
Testes												
Upper arm	0.35	**0.94**	0.41	**1.07**	0.45	**1.17**	0.48	**1.23**	0.54	**1.36**	0.59	**1.45**
Elbow	0.31	**0.99**	0.36	**1.13**	0.40	**1.23**	0.43	**1.30**	0.49	**1.43**	0.53	**1.52**
Forearm	0.03	**0.47**	0.04	**0.54**	0.04	**0.59**	0.04	**0.62**	0.05	**0.68**	0.05	**0.72**
Upper leg												
Knee												
Lower leg												
Lymph nodes	0.25	**0.46**	0.29	**0.53**	0.31	**0.58**	0.33	**0.62**	0.37	**0.69**	0.40	**0.75**
Skin	0.18	**0.30**	0.19	**0.32**	0.19	**0.33**	0.20	**0.34**	0.21	**0.35**	0.22	**0.37**
Muscle	0.15	**0.23**	0.17	**0.26**	0.18	**0.28**	0.19	**0.30**	0.21	**0.33**	0.23	**0.35**
Skeleton	0.26	**0.44**	0.29	**0.50**	0.31	**0.53**	0.32	**0.56**	0.35	**0.61**	0.37	**0.65**
Red bone marrow	0.13	**0.19**	0.15	**0.21**	0.16	**0.23**	0.17	**0.25**	0.19	**0.28**	0.21	**0.31**
Whole body dose	0.18	**0.28**	0.20	**0.31**	0.22	**0.34**	0.23	**0.36**	0.25	**0.39**	0.27	**0.42**
ED (ICRP 60)	0.79	**0.91**	0.87	**1.01**	0.92	**1.08**	0.96	**1.13**	1.03	**1.21**	1.09	**1.28**
ED (ICRP 103)	1.58	**1.75**	1.74	**1.93**	1.84	**2.04**	1.91	**2.13**	2.03	**2.28**	2.13	**2.39**
Total filtration (mm) **Al**	**1.5**		**2.5**		**3.5**		**4.5**		**3.5**		**3.5**	
Cu	**0**		**0**		**0**		**0**		**0.1**		**0.2**	

Table 9.101 Normalized organ doses—Chest ap—> 17 years

Beam projection:		ap				**Table cells:**						
Focus-to-detector distance (cm):		115				(+) = normal font = optimal field setting						
Tube voltage (kVp):		60				**(−) = bold font = suboptimal field setting**						
Total filtrations:		See last table row.				Empty table cells correspond to values < 0.01						
Organ or tissue	**Organ dose D_T : Entrance dose K_E**							**($mSv \times mGy^{-1}$)**				
	(+)	(−)	(+)	(−)	(+)	(−)	(+)	(−)	(+)	(−)	(+)	(−)
Skull		**0.19**		**0.24**	0.01	**0.28**	0.01	**0.31**	0.02	**0.37**	0.01	**0.34**
Brain								**0.01**		**0.01**		
Oral mucosa		**0.25**		**0.30**		**0.34**		**0.37**	0.01	**0.43**		**0.41**
Salivary glands		**0.23**	0.01	**0.29**	0.01	**0.34**	0.02	**0.37**	0.02	**0.43**	0.02	**0.41**
Cervical spine	0.13	**0.35**	0.17	**0.47**	0.21	**0.56**	0.24	**0.64**	0.30	**0.80**	0.26	**0.70**
Thyroid	0.50	**0.64**	0.59	**0.76**	0.65	**0.85**	0.70	**0.91**	0.78	**1.02**	0.76	**0.99**
Thymus	0.60	**0.62**	0.74	**0.78**	0.85	**0.89**	0.93	**0.97**	1.09	**1.14**	1.04	**1.09**
Thoracic spine	0.06	**0.07**	0.09	**0.09**	0.11	**0.12**	0.13	**0.14**	0.17	**0.18**	0.12	**0.14**
Claviculae	1.20	**1.27**	1.48	**1.57**	1.69	**1.79**	1.85	**1.96**	2.14	**2.28**	2.09	**2.22**
Scapulae	0.09	**0.09**	0.12	**0.13**	0.14	**0.16**	0.17	**0.18**	0.22	**0.24**	0.18	**0.19**
Ribs	1.04	**1.12**	1.24	**1.34**	1.38	**1.49**	1.48	**1.60**	1.67	**1.81**	1.63	**1.76**
Oesophagus	0.09	**0.09**	0.12	**0.13**	0.14	**0.16**	0.17	**0.18**	0.21	**0.24**	0.18	**0.19**
Trachea	0.09	**0.28**	0.11	**0.34**	0.12	**0.39**	0.13	**0.43**	0.16	**0.50**	0.15	**0.47**
Lung	0.21	**0.23**	0.28	**0.29**	0.33	**0.35**	0.37	**0.39**	0.45	**0.47**	0.41	**0.43**
Mamma	0.86	**0.90**	1.00	**1.04**	1.09	**1.14**	1.16	**1.21**	1.27	**1.33**	1.25	**1.31**
Heart	0.33	**0.34**	0.42	**0.44**	0.49	**0.51**	0.54	**0.57**	0.66	**0.69**	0.60	**0.63**
Stomach	0.18	**0.36**	0.23	**0.46**	0.27	**0.53**	0.30	**0.60**	0.36	**0.71**	0.33	**0.66**
Spleen	0.06	**0.08**	0.08	**0.10**	0.09	**0.13**	0.11	**0.15**	0.14	**0.19**	0.12	**0.16**
Liver	0.17	**0.26**	0.22	**0.34**	0.26	**0.40**	0.29	**0.44**	0.35	**0.54**	0.32	**0.49**
Gall bladder	0.04	**0.24**	0.05	**0.32**	0.06	**0.38**	0.07	**0.43**	0.10	**0.53**	0.08	**0.47**
Pancreas	0.10	**0.13**	0.14	**0.18**	0.17	**0.21**	0.19	**0.25**	0.25	**0.31**	0.21	**0.26**
Small intestine		**0.08**		**0.11**		**0.13**		**0.14**		**0.18**		**0.16**
Large intestine		**0.14**		**0.18**		**0.22**		**0.25**		**0.30**		**0.27**
Sigma/Rectum		**0.02**		**0.03**		**0.04**		**0.04**		**0.05**		**0.04**
Lumbar spine	0.01	**0.05**	0.02	**0.07**	0.02	**0.09**	0.03	**0.11**	0.04	**0.15**	0.03	**0.11**
Pelvis		**0.02**		**0.03**		**0.03**		**0.04**		**0.05**		**0.04**
Adrenals	0.04	**0.04**	0.05	**0.06**	0.06	**0.07**	0.07	**0.08**	0.10	**0.11**	0.07	**0.09**
Kidneys	0.01	**0.03**	0.02	**0.04**	0.03	**0.06**	0.03	**0.06**	0.04	**0.09**	0.03	**0.07**
Urinary bladder												
Uterus				**0.01**		**0.01**		**0.02**		**0.02**		**0.02**
Ovaries				**0.01**		**0.01**		**0.02**		**0.02**		**0.02**
Prostate												
Testes												
Upper arm	0.10	**0.46**	0.14	**0.59**	0.17	**0.69**	0.19	**0.77**	0.24	**0.93**	0.21	**0.86**
Elbow	0.09	**0.49**	0.12	**0.63**	0.14	**0.73**	0.17	**0.82**	0.21	**0.99**	0.18	**0.92**
Forearm		**0.24**		**0.31**	0.01	**0.36**	0.01	**0.40**	0.02	**0.49**	0.01	**0.45**
Upper leg												
Knee												
Lower leg												
Lymph nodes	0.10	**0.20**	0.12	**0.26**	0.14	**0.30**	0.16	**0.33**	0.19	**0.40**	0.18	**0.37**
Skin	0.13	**0.24**	0.14	**0.26**	0.15	**0.27**	0.16	**0.28**	0.16	**0.29**	0.16	**0.29**
Muscle	0.07	**0.11**	0.09	**0.14**	0.10	**0.16**	0.11	**0.17**	0.13	**0.20**	0.12	**0.19**
Skeleton	0.14	**0.23**	0.17	**0.28**	0.19	**0.32**	0.21	**0.36**	0.24	**0.42**	0.23	**0.40**
Red bone marrow	0.07	**0.09**	0.09	**0.11**	0.10	**0.13**	0.11	**0.14**	0.13	**0.16**	0.12	**0.15**
Whole body dose	0.09	**0.14**	0.11	**0.17**	0.12	**0.20**	0.13	**0.22**	0.16	**0.25**	0.15	**0.24**
ED (ICRP 60)	0.14	**0.20**	0.18	**0.24**	0.20	**0.28**	0.22	**0.31**	0.26	**0.36**	0.24	**0.34**
ED (ICRP 103)	0.20	**0.27**	0.25	**0.33**	0.28	**0.37**	0.30	**0.40**	0.35	**0.47**	0.33	**0.44**
Total filtration (mm) Al	**1.5**		**2.5**		**3.5**		**4.5**		**3.5**		**3.5**	
Cu	**0**		**0**		**0**		**0**		**0.1**		**0.2**	

Table 9.102 Normalized organ doses—Chest ap—> 17 years

Beam projection:			ap				**Table cells:**					
Focus-to-detector distance (cm):			115				(+) = normal font = optimal field setting					
Tube voltage (kVp):			70				**(−) = bold font = suboptimal field setting**					
Total filtrations:			See last table row.				Empty table cells correspond to values < 0.01					
Organ or tissue	**Organ dose D_T : Entrance dose K_E**							**(mSv × mGy^{-1})**				
	(+)	(−)	(+)	(−)	(+)	(−)	(+)	(−)	(+)	(−)	(+)	(−)
Skull		**0.23**	0.01	**0.28**	0.01	**0.32**	0.02	**0.36**	0.02	**0.42**	0.03	**0.48**
Brain				**0.01**		**0.01**		**0.02**		**0.02**		**0.03**
Oral mucosa		**0.29**		**0.35**	0.01	**0.39**	0.01	**0.43**	0.02	**0.49**	0.02	**0.55**
Salivary glands	0.01	**0.28**	0.02	**0.34**	0.02	**0.39**	0.02	**0.42**	0.03	**0.49**	0.03	**0.56**
Cervical spine	0.17	**0.46**	0.22	**0.60**	0.26	**0.70**	0.30	**0.80**	0.37	**0.98**	0.44	**1.17**
Thyroid	0.55	**0.72**	0.65	**0.84**	0.71	**0.92**	0.76	**0.99**	0.84	**1.10**	0.91	**1.20**
Thymus	0.70	**0.73**	0.86	**0.89**	0.97	**1.01**	1.06	**1.10**	1.22	**1.27**	1.36	**1.43**
Thoracic spine	0.10	**0.11**	0.13	**0.14**	0.16	**0.17**	0.18	**0.20**	0.24	**0.27**	0.30	**0.33**
Claviculae	1.37	**1.46**	1.66	**1.77**	1.87	**1.99**	2.04	**2.17**	2.33	**2.48**	2.57	**2.74**
Scapulae	0.12	**0.13**	0.16	**0.18**	0.20	**0.22**	0.23	**0.25**	0.29	**0.32**	0.35	**0.39**
Ribs	1.16	**1.25**	1.36	**1.47**	1.50	**1.62**	1.60	**1.74**	1.78	**1.94**	1.93	**2.11**
Oesophagus	0.12	**0.13**	0.16	**0.18**	0.19	**0.21**	0.22	**0.24**	0.28	**0.31**	0.34	**0.38**
Trachea	0.10	**0.32**	0.12	**0.39**	0.14	**0.45**	0.15	**0.49**	0.17	**0.56**	0.20	**0.63**
Lung	0.27	**0.28**	0.34	**0.36**	0.39	**0.42**	0.44	**0.47**	0.53	**0.56**	0.61	**0.65**
Mamma	0.94	**0.98**	1.08	**1.13**	1.17	**1.22**	1.24	**1.30**	1.36	**1.42**	1.45	**1.53**
Heart	0.40	**0.42**	0.50	**0.53**	0.58	**0.61**	0.64	**0.68**	0.77	**0.81**	0.88	**0.93**
Stomach	0.22	**0.44**	0.28	**0.55**	0.32	**0.63**	0.36	**0.70**	0.43	**0.83**	0.49	**0.95**
Spleen	0.08	**0.11**	0.10	**0.14**	0.13	**0.17**	0.14	**0.20**	0.18	**0.25**	0.22	**0.31**
Liver	0.21	**0.33**	0.27	**0.41**	0.31	**0.48**	0.34	**0.53**	0.41	**0.63**	0.48	**0.73**
Gall bladder	0.06	**0.31**	0.07	**0.40**	0.09	**0.46**	0.10	**0.52**	0.12	**0.64**	0.15	**0.75**
Pancreas	0.14	**0.18**	0.18	**0.24**	0.22	**0.28**	0.25	**0.32**	0.31	**0.40**	0.38	**0.49**
Small intestine		**0.11**		**0.14**		**0.16**		**0.18**		**0.22**	0.01	**0.27**
Large intestine		**0.18**		**0.23**		**0.27**	0.01	**0.30**	0.01	**0.37**	0.02	**0.43**
Sigma/Rectum		**0.03**		**0.04**		**0.05**		**0.05**		**0.06**		**0.08**
Lumbar spine	0.02	**0.08**	0.03	**0.11**	0.04	**0.14**	0.04	**0.16**	0.06	**0.21**	0.07	**0.26**
Pelvis		**0.03**		**0.04**		**0.05**		**0.05**		**0.07**		**0.09**
Adrenals	0.05	**0.06**	0.07	**0.09**	0.09	**0.10**	0.10	**0.12**	0.13	**0.16**	0.16	**0.19**
Kidneys	0.02	**0.05**	0.03	**0.07**	0.04	**0.08**	0.04	**0.09**	0.06	**0.12**	0.07	**0.15**
Urinary bladder												
Uterus		**0.01**		**0.02**		**0.02**		**0.02**		**0.03**		**0.04**
Ovaries		**0.01**		**0.02**		**0.02**		**0.02**		**0.03**		**0.04**
Prostate												
Testes												
Upper arm	0.14	**0.56**	0.18	**0.70**	0.21	**0.81**	0.24	**0.90**	0.30	**1.07**	0.36	**1.23**
Elbow	0.12	**0.60**	0.16	**0.75**	0.19	**0.86**	0.21	**0.96**	0.27	**1.14**	0.32	**1.30**
Forearm		**0.29**	0.01	**0.37**	0.01	**0.42**	0.02	**0.47**	0.02	**0.55**	0.03	**0.63**
Upper leg												
Knee												
Lower leg												
Lymph nodes	0.12	**0.25**	0.15	**0.31**	0.17	**0.36**	0.19	**0.40**	0.23	**0.47**	0.27	**0.54**
Skin	0.14	**0.25**	0.15	**0.27**	0.16	**0.28**	0.16	**0.29**	0.17	**0.31**	0.18	**0.32**
Muscle	0.08	**0.13**	0.10	**0.16**	0.11	**0.18**	0.13	**0.20**	0.15	**0.23**	0.17	**0.27**
Skeleton	0.16	**0.27**	0.19	**0.33**	0.22	**0.37**	0.23	**0.41**	0.27	**0.48**	0.30	**0.54**
Red bone marrow	0.09	**0.11**	0.10	**0.13**	0.12	**0.15**	0.13	**0.17**	0.15	**0.19**	0.17	**0.22**
Whole body dose	0.10	**0.17**	0.13	**0.20**	0.14	**0.23**	0.15	**0.25**	0.18	**0.29**	0.20	**0.33**
ED (ICRP 60)	0.17	**0.23**	0.20	**0.29**	0.23	**0.32**	0.25	**0.35**	0.29	**0.41**	0.33	**0.47**
ED (ICRP 103)	0.23	**0.31**	0.28	**0.38**	0.31	**0.42**	0.34	**0.46**	0.39	**0.53**	0.43	**0.60**
Total filtration (mm) **Al**	**1.5**		**2.5**		**3.5**		**4.5**		**3.5**		**3.5**	
Cu	**0**		**0**		**0**		**0**		**0.1**		**0.2**	

Table 9.103 Normalized organ doses—Chest ap—> 17 years

Beam projection:			ap		**Table cells:**							
Focus-to-detector distance (cm):			115		(+) = normal font = optimal field setting							
Tube voltage (kVp):			80		**(−) = bold font = suboptimal field setting**							
Total filtrations:			See last table row.		Empty table cells correspond to values < 0.01							
Organ or tissue	**Organ dose D_T : Entrance dose K_E**							**(mSv × mGy^{-1})**				
	(+)	(−)	(+)	(−)	(+)	(−)	(+)	(−)	(+)	(−)	(+)	(−)
Skull	0.01	**0.26**	0.02	**0.32**	0.02	**0.36**	0.02	**0.40**	0.03	**0.46**	0.03	**0.52**
Brain		**0.01**		**0.02**		**0.02**		**0.02**		**0.03**		**0.03**
Oral mucosa		**0.33**	0.01	**0.40**	0.01	**0.44**	0.02	**0.48**	0.02	**0.55**	0.02	**0.61**
Salivary glands	0.02	**0.32**	0.02	**0.39**	0.02	**0.43**	0.03	**0.47**	0.03	**0.54**	0.04	**0.61**
Cervical spine	0.21	**0.57**	0.27	**0.72**	0.32	**0.84**	0.35	**0.94**	0.43	**1.14**	0.50	**1.33**
Thyroid	0.61	**0.79**	0.70	**0.91**	0.76	**1.00**	0.81	**1.06**	0.90	**1.17**	0.97	**1.27**
Thymus	0.80	**0.83**	0.96	**1.00**	1.07	**1.12**	1.16	**1.21**	1.33	**1.39**	1.47	**1.54**
Thoracic spine	0.14	**0.15**	0.18	**0.20**	0.21	**0.24**	0.24	**0.27**	0.31	**0.34**	0.38	**0.42**
Claviculae	1.52	**1.62**	1.81	**1.93**	2.02	**2.15**	2.17	**2.32**	2.44	**2.61**	2.65	**2.85**
Scapulae	0.16	**0.18**	0.21	**0.23**	0.25	**0.27**	0.28	**0.31**	0.35	**0.39**	0.42	**0.46**
Ribs	1.26	**1.37**	1.46	**1.58**	1.59	**1.73**	1.69	**1.84**	1.86	**2.03**	1.99	**2.18**
Oesophagus	0.16	**0.18**	0.21	**0.23**	0.25	**0.27**	0.28	**0.31**	0.35	**0.38**	0.42	**0.46**
Trachea	0.12	**0.37**	0.14	**0.45**	0.16	**0.50**	0.17	**0.54**	0.19	**0.62**	0.21	**0.69**
Lung	0.32	**0.34**	0.40	**0.42**	0.46	**0.49**	0.50	**0.54**	0.59	**0.64**	0.68	**0.73**
Mamma	1.01	**1.06**	1.15	**1.20**	1.24	**1.30**	1.31	**1.37**	1.42	**1.49**	1.52	**1.59**
Heart	0.48	**0.50**	0.59	**0.62**	0.67	**0.71**	0.74	**0.78**	0.87	**0.92**	0.98	**1.04**
Stomach	0.26	**0.52**	0.32	**0.63**	0.37	**0.72**	0.41	**0.79**	0.48	**0.93**	0.55	**1.05**
Spleen	0.10	**0.14**	0.13	**0.19**	0.16	**0.22**	0.18	**0.25**	0.22	**0.31**	0.27	**0.37**
Liver	0.25	**0.39**	0.31	**0.48**	0.36	**0.55**	0.40	**0.61**	0.47	**0.72**	0.54	**0.82**
Gall bladder	0.07	**0.38**	0.09	**0.47**	0.11	**0.55**	0.12	**0.61**	0.15	**0.73**	0.18	**0.85**
Pancreas	0.18	**0.23**	0.23	**0.30**	0.27	**0.35**	0.31	**0.39**	0.38	**0.49**	0.45	**0.58**
Small intestine		**0.13**		**0.17**		**0.20**		**0.22**	0.01	**0.27**	0.02	**0.31**
Large intestine		**0.22**	0.01	**0.27**	0.01	**0.32**	0.01	**0.35**	0.02	**0.42**	0.02	**0.49**
Sigma/Rectum		**0.04**		**0.05**		**0.06**		**0.06**		**0.08**		**0.09**
Lumbar spine	0.03	**0.12**	0.04	**0.16**	0.05	**0.19**	0.06	**0.22**	0.07	**0.28**	0.09	**0.34**
Pelvis		**0.04**		**0.05**		**0.06**		**0.07**		**0.09**		**0.11**
Adrenals	0.08	**0.09**	0.10	**0.12**	0.12	**0.14**	0.13	**0.16**	0.17	**0.20**	0.20	**0.24**
Kidneys	0.03	**0.07**	0.04	**0.09**	0.05	**0.11**	0.06	**0.12**	0.07	**0.15**	0.09	**0.19**
Urinary bladder												**0.01**
Uterus		**0.02**		**0.02**		**0.03**		**0.03**		**0.04**		**0.05**
Ovaries		**0.02**		**0.02**		**0.03**		**0.03**		**0.04**		**0.05**
Prostate												
Testes												
Upper arm	0.17	**0.66**	0.22	**0.81**	0.26	**0.92**	0.29	**1.01**	0.35	**1.18**	0.42	**1.33**
Elbow	0.15	**0.70**	0.19	**0.86**	0.23	**0.97**	0.26	**1.07**	0.31	**1.25**	0.37	**1.41**
Forearm	0.01	**0.34**	0.02	**0.42**	0.02	**0.47**	0.02	**0.52**	0.03	**0.61**	0.03	**0.68**
Upper leg												
Knee												
Lower leg												
Lymph nodes	0.14	**0.30**	0.18	**0.36**	0.20	**0.41**	0.22	**0.45**	0.26	**0.54**	0.30	**0.61**
Skin	0.15	**0.26**	0.16	**0.28**	0.17	**0.29**	0.17	**0.30**	0.18	**0.32**	0.19	**0.34**
Muscle	0.10	**0.16**	0.12	**0.19**	0.13	**0.21**	0.14	**0.23**	0.16	**0.26**	0.19	**0.29**
Skeleton	0.18	**0.31**	0.21	**0.37**	0.24	**0.42**	0.26	**0.45**	0.29	**0.52**	0.32	**0.58**
Red bone marrow	0.10	**0.13**	0.12	**0.15**	0.13	**0.17**	0.14	**0.19**	0.17	**0.22**	0.19	**0.26**
Whole body dose	0.12	**0.19**	0.14	**0.23**	0.16	**0.26**	0.17	**0.28**	0.20	**0.32**	0.22	**0.36**
ED (ICRP 60)	0.19	**0.27**	0.23	**0.33**	0.26	**0.37**	0.28	**0.40**	0.32	**0.46**	0.36	**0.52**
ED (ICRP 103)	0.27	**0.36**	0.31	**0.42**	0.35	**0.47**	0.38	**0.51**	0.43	**0.59**	0.47	**0.65**
Total filtration (mm) **Al**	**1.5**		**2.5**		**3.5**		**4.5**		**3.5**		**3.5**	
Cu	**0**		**0**		**0**		**0**		**0.1**		**0.2**	

Table 9.104 Normalized organ doses—Chest ap—> 17 years

Beam projection:			ap			**Table cells:**						
Focus-to-detector distance (cm):			115			(+) = normal font = optimal field setting						
Tube voltage (kVp):			110			**(−) = bold font = suboptimal field setting**						
Total filtrations:			See last table row.			Empty table cells correspond to values < 0.01						
Organ or tissue	**Organ dose D_T : Entrance dose K_E**								**(mSv × mGy^{-1})**			
	(+)	(−)	(+)	(−)	(+)	(−)	(+)	(−)	(+)	(−)	(+)	(−)
Skull	0.02	**0.35**	0.02	**0.40**	0.03	**0.44**	0.03	**0.47**	0.04	**0.52**	0.04	**0.56**
Brain		**0.02**		**0.03**		**0.03**		**0.04**		**0.04**		**0.05**
Oral mucosa	0.02	**0.44**	0.02	**0.51**	0.02	**0.55**	0.02	**0.58**	0.03	**0.65**	0.03	**0.71**
Salivary glands	0.03	**0.43**	0.03	**0.50**	0.03	**0.54**	0.04	**0.58**	0.04	**0.64**	0.05	**0.70**
Cervical spine	0.32	**0.85**	0.38	**1.01**	0.43	**1.12**	0.47	**1.22**	0.54	**1.39**	0.60	**1.54**
Thyroid	0.74	**0.96**	0.82	**1.07**	0.88	**1.15**	0.92	**1.20**	0.99	**1.29**	1.05	**1.37**
Thymus	1.04	**1.09**	1.19	**1.25**	1.30	**1.35**	1.37	**1.44**	1.51	**1.58**	1.62	**1.70**
Thoracic spine	0.25	**0.28**	0.30	**0.34**	0.34	**0.38**	0.38	**0.42**	0.45	**0.50**	0.52	**0.58**
Claviculae	1.84	**1.97**	2.08	**2.23**	2.24	**2.40**	2.35	**2.53**	2.53	**2.72**	2.65	**2.86**
Scapulae	0.26	**0.29**	0.32	**0.35**	0.36	**0.40**	0.39	**0.43**	0.46	**0.51**	0.52	**0.58**
Ribs	1.47	**1.60**	1.63	**1.78**	1.73	**1.89**	1.80	**1.98**	1.91	**2.11**	1.98	**2.19**
Oesophagus	0.27	**0.30**	0.33	**0.37**	0.37	**0.41**	0.41	**0.45**	0.48	**0.53**	0.55	**0.61**
Trachea	0.15	**0.50**	0.18	**0.57**	0.19	**0.62**	0.20	**0.66**	0.22	**0.73**	0.24	**0.79**
Lung	0.46	**0.49**	0.54	**0.58**	0.59	**0.64**	0.64	**0.69**	0.72	**0.78**	0.80	**0.86**
Mamma	1.19	**1.25**	1.31	**1.38**	1.39	**1.46**	1.44	**1.51**	1.54	**1.61**	1.61	**1.69**
Heart	0.67	**0.71**	0.78	**0.83**	0.86	**0.92**	0.93	**0.99**	1.05	**1.11**	1.15	**1.22**
Stomach	0.37	**0.72**	0.43	**0.84**	0.48	**0.92**	0.52	**0.99**	0.58	**1.11**	0.64	**1.21**
Spleen	0.18	**0.24**	0.21	**0.29**	0.24	**0.33**	0.26	**0.36**	0.31	**0.42**	0.35	**0.48**
Liver	0.37	**0.56**	0.43	**0.66**	0.47	**0.72**	0.51	**0.78**	0.58	**0.88**	0.64	**0.97**
Gall bladder	0.12	**0.56**	0.14	**0.67**	0.16	**0.74**	0.18	**0.80**	0.20	**0.92**	0.23	**1.01**
Pancreas	0.29	**0.38**	0.35	**0.45**	0.40	**0.51**	0.43	**0.56**	0.50	**0.65**	0.57	**0.74**
Small intestine	0.01	**0.21**	0.01	**0.25**	0.02	**0.28**	0.02	**0.30**	0.02	**0.35**	0.02	**0.39**
Large intestine	0.02	**0.33**	0.02	**0.38**	0.02	**0.43**	0.03	**0.46**	0.03	**0.53**	0.04	**0.59**
Sigma/Rectum		**0.06**		**0.07**		**0.08**		**0.09**		**0.10**		**0.11**
Lumbar spine	0.06	**0.22**	0.08	**0.27**	0.09	**0.31**	0.10	**0.34**	0.12	**0.41**	0.14	**0.47**
Pelvis		**0.07**		**0.08**		**0.10**		**0.11**		**0.13**		**0.14**
Adrenals	0.14	**0.17**	0.17	**0.20**	0.19	**0.23**	0.21	**0.25**	0.24	**0.30**	0.28	**0.34**
Kidneys	0.06	**0.13**	0.07	**0.15**	0.08	**0.18**	0.09	**0.19**	0.11	**0.23**	0.13	**0.27**
Urinary bladder						**0.01**		**0.01**		**0.02**		**0.02**
Uterus		**0.03**		**0.04**		**0.05**		**0.05**		**0.06**		**0.07**
Ovaries		**0.03**		**0.04**		**0.05**		**0.05**		**0.06**		**0.07**
Prostate												
Testes												
Upper arm	0.27	**0.88**	0.32	**1.02**	0.35	**1.12**	0.38	**1.20**	0.44	**1.34**	0.49	**1.44**
Elbow	0.23	**0.93**	0.28	**1.08**	0.31	**1.19**	0.34	**1.27**	0.39	**1.41**	0.44	**1.52**
Forearm	0.02	**0.45**	0.03	**0.52**	0.03	**0.57**	0.03	**0.61**	0.04	**0.68**	0.04	**0.73**
Upper leg												
Knee												
Lower leg												
Lymph nodes	0.21	**0.42**	0.24	**0.49**	0.27	**0.54**	0.29	**0.58**	0.32	**0.66**	0.36	**0.73**
Skin	0.17	**0.29**	0.18	**0.31**	0.18	**0.32**	0.19	**0.33**	0.20	**0.35**	0.21	**0.36**
Muscle	0.13	**0.21**	0.15	**0.24**	0.17	**0.26**	0.18	**0.28**	0.20	**0.31**	0.22	**0.34**
Skeleton	0.23	**0.40**	0.26	**0.46**	0.28	**0.50**	0.30	**0.53**	0.32	**0.58**	0.35	**0.62**
Red bone marrow	0.14	**0.18**	0.16	**0.21**	0.17	**0.23**	0.18	**0.25**	0.21	**0.28**	0.23	**0.32**
Whole body dose	0.16	**0.26**	0.18	**0.29**	0.20	**0.32**	0.21	**0.34**	0.24	**0.38**	0.26	**0.41**
ED (ICRP 60)	0.26	**0.37**	0.30	**0.42**	0.32	**0.46**	0.34	**0.49**	0.38	**0.55**	0.41	**0.60**
ED (ICRP 103)	0.34	**0.47**	0.39	**0.54**	0.42	**0.59**	0.45	**0.62**	0.49	**0.69**	0.53	**0.75**
Total filtration (mm) **Al**	**1.5**		**2.5**		**3.5**		**4.5**		**3.5**		**3.5**	
Cu	**0**		**0**		**0**		**0**		**0.1**		**0.2**	

Table 9.105 Normalized organ doses—Chest ap—> 17 years

Beam projection:			ap		**Table cells:**							
Focus-to-detector distance (cm):			115		(+) = normal font = optimal field setting							
Tube voltage (kVp):			120		**(−) = bold font = suboptimal field setting**							
Total filtrations:			See last table row.		Empty table cells correspond to values < 0.01							
Organ or tissue	**Organ dose D_T : Entrance dose K_E**							**(mSv × mGy^{-1})**				
	(+)	(−)	(+)	(−)	(+)	(−)	(+)	(−)	(+)	(−)	(+)	(−)
Skull	0.02	**0.37**	0.03	**0.42**	0.03	**0.45**	0.03	**0.48**	0.04	**0.52**	0.04	**0.56**
Brain		**0.03**		**0.03**		**0.04**		**0.04**		**0.05**		**0.05**
Oral mucosa	0.02	**0.47**	0.02	**0.54**	0.02	**0.58**	0.03	**0.61**	0.03	**0.67**	0.03	**0.72**
Salivary glands	0.03	**0.46**	0.03	**0.53**	0.04	**0.57**	0.04	**0.60**	0.05	**0.66**	0.05	**0.72**
Cervical spine	0.35	**0.92**	0.41	**1.07**	0.45	**1.18**	0.49	**1.27**	0.55	**1.43**	0.61	**1.57**
Thyroid	0.77	**1.01**	0.85	**1.11**	0.90	**1.18**	0.94	**1.23**	1.01	**1.32**	1.06	**1.38**
Thymus	1.11	**1.16**	1.25	**1.31**	1.35	**1.41**	1.42	**1.48**	1.55	**1.62**	1.65	**1.73**
Thoracic spine	0.28	**0.31**	0.33	**0.37**	0.37	**0.42**	0.41	**0.46**	0.48	**0.53**	0.54	**0.61**
Claviculae	1.91	**2.05**	2.13	**2.29**	2.27	**2.44**	2.37	**2.55**	2.52	**2.71**	2.61	**2.82**
Scapulae	0.29	**0.32**	0.34	**0.38**	0.38	**0.43**	0.41	**0.46**	0.47	**0.53**	0.53	**0.59**
Ribs	1.51	**1.65**	1.66	**1.82**	1.75	**1.92**	1.81	**1.99**	1.90	**2.10**	1.96	**2.17**
Oesophagus	0.31	**0.34**	0.36	**0.40**	0.41	**0.45**	0.44	**0.49**	0.51	**0.56**	0.57	**0.63**
Trachea	0.16	**0.53**	0.19	**0.60**	0.20	**0.65**	0.21	**0.68**	0.23	**0.75**	0.25	**0.81**
Lung	0.50	**0.53**	0.57	**0.62**	0.63	**0.67**	0.67	**0.72**	0.75	**0.81**	0.82	**0.88**
Mamma	1.24	**1.30**	1.35	**1.42**	1.42	**1.49**	1.47	**1.54**	1.56	**1.64**	1.62	**1.70**
Heart	0.73	**0.77**	0.83	**0.88**	0.91	**0.97**	0.97	**1.03**	1.08	**1.15**	1.17	**1.25**
Stomach	0.40	**0.78**	0.46	**0.89**	0.51	**0.97**	0.54	**1.03**	0.60	**1.15**	0.65	**1.24**
Spleen	0.20	**0.27**	0.23	**0.32**	0.26	**0.36**	0.28	**0.39**	0.32	**0.45**	0.36	**0.50**
Liver	0.40	**0.61**	0.46	**0.70**	0.50	**0.77**	0.54	**0.82**	0.60	**0.91**	0.66	**1.00**
Gall bladder	0.13	**0.62**	0.16	**0.71**	0.17	**0.79**	0.19	**0.84**	0.22	**0.95**	0.24	**1.04**
Pancreas	0.33	**0.42**	0.38	**0.50**	0.43	**0.55**	0.46	**0.60**	0.53	**0.69**	0.59	**0.77**
Small intestine	0.01	**0.23**	0.02	**0.27**	0.02	**0.30**	0.02	**0.32**	0.02	**0.37**	0.03	**0.41**
Large intestine	0.02	**0.36**	0.02	**0.41**	0.03	**0.46**	0.03	**0.49**	0.03	**0.55**	0.04	**0.60**
Sigma/Rectum		**0.07**		**0.08**		**0.09**		**0.09**		**0.10**		**0.12**
Lumbar spine	0.07	**0.25**	0.09	**0.30**	0.10	**0.34**	0.11	**0.37**	0.13	**0.44**	0.14	**0.50**
Pelvis		**0.08**		**0.09**		**0.10**		**0.11**		**0.13**	0.01	**0.15**
Adrenals	0.15	**0.19**	0.18	**0.22**	0.21	**0.25**	0.23	**0.27**	0.26	**0.32**	0.30	**0.36**
Kidneys	0.07	**0.15**	0.08	**0.17**	0.09	**0.19**	0.10	**0.21**	0.12	**0.25**	0.13	**0.28**
Urinary bladder				**0.01**		**0.01**		**0.01**		**0.02**		**0.02**
Uterus		**0.04**		**0.05**		**0.05**		**0.06**		**0.07**		**0.08**
Ovaries		**0.04**		**0.05**		**0.05**		**0.06**		**0.07**		**0.08**
Prostate												
Testes												
Upper arm	0.29	**0.94**	0.34	**1.07**	0.37	**1.16**	0.40	**1.23**	0.45	**1.35**	0.50	**1.45**
Elbow	0.26	**0.99**	0.30	**1.13**	0.33	**1.23**	0.36	**1.30**	0.41	**1.43**	0.45	**1.53**
Forearm	0.02	**0.48**	0.03	**0.54**	0.03	**0.59**	0.03	**0.63**	0.04	**0.69**	0.04	**0.73**
Upper leg												
Knee												
Lower leg												
Lymph nodes	0.22	**0.46**	0.26	**0.53**	0.28	**0.57**	0.30	**0.61**	0.34	**0.69**	0.37	**0.75**
Skin	0.17	**0.30**	0.18	**0.32**	0.19	**0.33**	0.19	**0.34**	0.20	**0.35**	0.21	**0.37**
Muscle	0.14	**0.23**	0.16	**0.26**	0.17	**0.28**	0.19	**0.29**	0.21	**0.32**	0.22	**0.35**
Skeleton	0.24	**0.42**	0.27	**0.48**	0.29	**0.51**	0.30	**0.54**	0.33	**0.59**	0.35	**0.63**
Red bone marrow	0.15	**0.20**	0.17	**0.23**	0.18	**0.25**	0.19	**0.27**	0.21	**0.30**	0.23	**0.33**
Whole body dose	0.17	**0.27**	0.19	**0.31**	0.21	**0.34**	0.22	**0.36**	0.24	**0.39**	0.26	**0.42**
ED (ICRP 60)	0.28	**0.39**	0.31	**0.45**	0.34	**0.49**	0.36	**0.52**	0.39	**0.57**	0.42	**0.61**
ED (ICRP 103)	0.37	**0.50**	0.41	**0.57**	0.44	**0.61**	0.47	**0.65**	0.51	**0.71**	0.54	**0.76**
Total filtration (mm) Al	**1.5**		**2.5**		**3.5**		**4.5**		**3.5**		**3.5**	
Cu	**0**		**0**		**0**		**0**		**0.1**		**0.2**	

Table 9.106 Normalized organ doses—Chest r-lat—3–7 years

Beam projection:			r-lat				**Table cells:**					
Focus-to-detector distance (cm):			115				(+) = normal font = optimal field setting					
Tube voltage (kVp):			60				**(–) = bold font = suboptimal field setting**					
Total filtrations:			See last table row.				Empty table cells correspond to values < 0.01					
Organ or tissue		**Organ dose D_T : Entrance dose K_E**						**(mSv × mGy^{-1})**				
	(+)	(–)	(+)	(–)	(+)	(–)	(+)	(–)	(+)	(–)	(+)	(–)
Skull	0.01	**0.19**	0.01	**0.24**	0.02	**0.27**	0.02	**0.30**	0.03	**0.35**	0.02	**0.33**
Brain		**0.01**		**0.02**		**0.02**		**0.02**		**0.03**		**0.03**
Oral mucosa	0.01	**0.22**	0.02	**0.27**	0.02	**0.31**	0.02	**0.34**	0.03	**0.39**	0.02	**0.37**
Salivary glands	0.02	**0.31**	0.02	**0.36**	0.03	**0.39**	0.03	**0.42**	0.04	**0.46**	0.04	**0.44**
Cervical spine	0.24	**0.49**	0.31	**0.63**	0.37	**0.74**	0.41	**0.83**	0.50	**1.01**	0.47	**0.94**
Thyroid	0.31	**0.30**	0.37	**0.37**	0.42	**0.42**	0.45	**0.46**	0.52	**0.52**	0.50	**0.50**
Thymus	0.12	**0.11**	0.16	**0.15**	0.19	**0.18**	0.21	**0.20**	0.26	**0.25**	0.24	**0.22**
Thoracic spine	0.25	**0.22**	0.33	**0.29**	0.40	**0.35**	0.46	**0.41**	0.58	**0.52**	0.51	**0.45**
Claviculae	0.54	**0.49**	0.67	**0.62**	0.77	**0.72**	0.85	**0.80**	0.99	**0.94**	0.96	**0.90**
Scapulae	1.02	**0.96**	1.20	**1.14**	1.32	**1.26**	1.42	**1.35**	1.57	**1.50**	1.57	**1.49**
Ribs	0.90	**0.70**	1.07	**0.86**	1.19	**0.98**	1.27	**1.07**	1.43	**1.24**	1.41	**1.20**
Oesophagus	0.13	**0.08**	0.17	**0.11**	0.20	**0.13**	0.23	**0.15**	0.28	**0.19**	0.25	**0.16**
Trachea	0.04	**0.16**	0.05	**0.20**	0.06	**0.24**	0.06	**0.26**	0.07	**0.31**	0.07	**0.29**
Lung	0.27	**0.18**	0.34	**0.23**	0.39	**0.27**	0.43	**0.30**	0.50	**0.36**	0.48	**0.33**
Mamma	0.49	**0.48**	0.54	**0.52**	0.56	**0.55**	0.58	**0.57**	0.61	**0.60**	0.61	**0.59**
Heart	0.14	**0.11**	0.19	**0.14**	0.22	**0.17**	0.25	**0.19**	0.31	**0.24**	0.28	**0.21**
Stomach	0.03	**0.03**	0.04	**0.04**	0.04	**0.05**	0.05	**0.06**	0.07	**0.08**	0.05	**0.06**
Spleen	0.02	**0.02**	0.03	**0.02**	0.03	**0.03**	0.04	**0.03**	0.05	**0.05**	0.04	**0.03**
Liver	0.23	**0.26**	0.29	**0.33**	0.34	**0.39**	0.37	**0.43**	0.44	**0.52**	0.41	**0.48**
Gall bladder	0.04	**0.09**	0.06	**0.12**	0.07	**0.14**	0.08	**0.16**	0.10	**0.20**	0.09	**0.18**
Pancreas	0.06	**0.05**	0.08	**0.07**	0.10	**0.08**	0.12	**0.09**	0.15	**0.12**	0.13	**0.10**
Small intestine		**0.05**		**0.07**		**0.09**	0.01	**0.10**	0.01	**0.12**	0.01	**0.11**
Large intestine		**0.09**	0.01	**0.11**	0.01	**0.13**	0.02	**0.15**	0.02	**0.19**	0.02	**0.17**
Sigma/Rectum								**0.01**		**0.02**		**0.01**
Lumbar spine	0.05	**0.17**	0.07	**0.23**	0.08	**0.28**	0.09	**0.32**	0.12	**0.41**	0.10	**0.35**
Pelvis		**0.05**		**0.06**		**0.08**		**0.09**		**0.11**		**0.09**
Adrenals	0.09	**0.08**	0.12	**0.11**	0.15	**0.14**	0.17	**0.16**	0.21	**0.20**	0.18	**0.17**
Kidneys	0.06	**0.14**	0.08	**0.17**	0.09	**0.20**	0.10	**0.23**	0.12	**0.27**	0.11	**0.25**
Urinary bladder												
Uterus		**0.01**		**0.02**		**0.02**		**0.02**		**0.03**		**0.02**
Ovaries		**0.01**		**0.02**		**0.02**		**0.02**		**0.03**		**0.03**
Prostate												
Testes												
Upper arm		**1.04**		**1.22**		**1.35**		**1.44**		**1.60**		**1.59**
Elbow		**1.11**		**1.30**		**1.43**		**1.52**		**1.69**		**1.68**
Forearm		**0.49**		**0.58**		**0.63**		**0.67**		**0.74**		**0.74**
Upper leg												
Knee												
Lower leg												
Lymph nodes	0.07	**0.12**	0.09	**0.15**	0.11	**0.17**	0.12	**0.19**	0.14	**0.23**	0.13	**0.21**
Skin	0.11	**0.21**	0.12	**0.22**	0.12	**0.23**	0.13	**0.23**	0.13	**0.25**	0.13	**0.24**
Muscle	0.06	**0.09**	0.07	**0.11**	0.08	**0.12**	0.09	**0.13**	0.10	**0.15**	0.10	**0.14**
Skeleton	0.15	**0.31**	0.18	**0.37**	0.21	**0.42**	0.23	**0.46**	0.26	**0.53**	0.25	**0.51**
Red bone marrow	0.05	**0.08**	0.06	**0.10**	0.07	**0.11**	0.08	**0.12**	0.09	**0.15**	0.08	**0.13**
Whole body dose	0.08	**0.12**	0.09	**0.15**	0.11	**0.17**	0.12	**0.18**	0.13	**0.21**	0.13	**0.20**
ED (ICRP 60)	0.10	**0.10**	0.13	**0.13**	0.14	**0.15**	0.16	**0.16**	0.18	**0.19**	0.17	**0.18**
ED (ICRP 103)	0.14	**0.14**	0.16	**0.17**	0.18	**0.19**	0.19	**0.21**	0.22	**0.24**	0.21	**0.22**
Total filtration (mm) **Al**	**1.5**		**2.5**		**3.5**		**4.5**		**3.5**		**3.5**	
Cu	**0**		**0**		**0**		**0**		**0.1**		**0.2**	

Table 9.107 Normalized organ doses—Chest r-lat—3–7 years

Beam projection:		r-lat				**Table cells:**							
Focus-to-detector distance (cm):		115				(+) = normal font = optimal field setting							
Tube voltage (kVp):		70				**(−) = bold font = suboptimal field setting**							
Total filtrations:		See last table row.				Empty table cells correspond to values < 0.01							
Organ or tissue		**Organ dose D_T : Entrance dose K_E**								**(mSv × mGy^{-1})**			
		(+)	(−)	(+)	(−)	(+)	(−)	(+)	(−)	(+)	(−)	(+)	(−)
Skull		0.01	**0.22**	0.02	**0.27**	0.02	**0.30**	0.03	**0.33**	0.03	**0.38**	0.04	**0.43**
Brain			**0.02**		**0.02**		**0.03**		**0.03**		**0.04**		**0.05**
Oral mucosa		0.02	**0.26**	0.02	**0.31**	0.02	**0.35**	0.03	**0.38**	0.03	**0.44**	0.04	**0.49**
Salivary glands		0.02	**0.34**	0.03	**0.39**	0.04	**0.43**	0.04	**0.45**	0.05	**0.50**	0.06	**0.54**
Cervical spine		0.30	**0.60**	0.37	**0.75**	0.43	**0.87**	0.48	**0.97**	0.56	**1.15**	0.65	**1.32**
Thyroid		0.35	**0.35**	0.42	**0.42**	0.47	**0.47**	0.50	**0.51**	0.57	**0.58**	0.63	**0.65**
Thymus		0.15	**0.14**	0.20	**0.19**	0.23	**0.22**	0.26	**0.25**	0.32	**0.30**	0.37	**0.36**
Thoracic spine		0.33	**0.29**	0.43	**0.38**	0.51	**0.45**	0.57	**0.51**	0.71	**0.64**	0.84	**0.77**
Claviculae		0.62	**0.58**	0.76	**0.72**	0.86	**0.82**	0.94	**0.90**	1.08	**1.05**	1.20	**1.18**
Scapulae		1.11	**1.05**	1.29	**1.23**	1.41	**1.35**	1.50	**1.43**	1.65	**1.58**	1.76	**1.70**
Ribs		0.99	**0.80**	1.16	**0.96**	1.28	**1.08**	1.37	**1.18**	1.52	**1.34**	1.64	**1.48**
Oesophagus		0.16	**0.11**	0.21	**0.14**	0.24	**0.17**	0.27	**0.19**	0.33	**0.24**	0.39	**0.28**
Trachea		0.05	**0.19**	0.06	**0.24**	0.07	**0.28**	0.07	**0.31**	0.08	**0.36**	0.09	**0.42**
Lung		0.32	**0.22**	0.39	**0.27**	0.44	**0.32**	0.48	**0.35**	0.55	**0.42**	0.62	**0.48**
Mamma		0.52	**0.50**	0.56	**0.55**	0.59	**0.57**	0.61	**0.59**	0.64	**0.63**	0.67	**0.66**
Heart		0.18	**0.14**	0.23	**0.18**	0.27	**0.21**	0.30	**0.24**	0.37	**0.30**	0.43	**0.35**
Stomach		0.04	**0.04**	0.05	**0.06**	0.06	**0.07**	0.07	**0.08**	0.09	**0.10**	0.11	**0.13**
Spleen		0.03	**0.03**	0.04	**0.04**	0.05	**0.04**	0.05	**0.05**	0.07	**0.07**	0.08	**0.08**
Liver		0.28	**0.32**	0.34	**0.40**	0.39	**0.46**	0.42	**0.51**	0.49	**0.60**	0.55	**0.69**
Gall bladder		0.06	**0.12**	0.08	**0.15**	0.09	**0.18**	0.10	**0.21**	0.13	**0.26**	0.15	**0.31**
Pancreas		0.08	**0.07**	0.11	**0.09**	0.13	**0.11**	0.15	**0.13**	0.19	**0.16**	0.22	**0.20**
Small intestine			**0.07**	0.01	**0.09**	0.01	**0.11**	0.02	**0.13**	0.02	**0.16**	0.02	**0.19**
Large intestine		0.01	**0.11**	0.02	**0.14**	0.02	**0.17**	0.02	**0.19**	0.03	**0.23**	0.03	**0.27**
Sigma/Rectum					**0.01**		**0.01**		**0.02**		**0.02**		**0.03**
Lumbar spine		0.07	**0.23**	0.09	**0.30**	0.11	**0.36**	0.12	**0.41**	0.16	**0.51**	0.19	**0.61**
Pelvis			**0.06**		**0.08**		**0.10**		**0.11**	0.01	**0.14**	0.01	**0.16**
Adrenals		0.12	**0.11**	0.16	**0.15**	0.18	**0.18**	0.21	**0.20**	0.26	**0.25**	0.30	**0.29**
Kidneys		0.07	**0.17**	0.09	**0.21**	0.11	**0.24**	0.12	**0.27**	0.15	**0.32**	0.17	**0.37**
Urinary bladder									**0.01**		**0.01**		**0.02**
Uterus			**0.02**		**0.02**		**0.03**		**0.03**		**0.04**		**0.05**
Ovaries			**0.02**		**0.02**		**0.03**		**0.03**		**0.04**		**0.05**
Prostate													
Testes													
Upper arm			**1.14**		**1.32**		**1.44**		**1.53**		**1.68**		**1.80**
Elbow			**1.21**		**1.39**		**1.52**		**1.61**		**1.76**		**1.88**
Forearm			**0.54**		**0.62**		**0.67**		**0.71**		**0.77**		**0.82**
Upper leg													
Knee													
Lower leg													
Lymph nodes		0.09	**0.14**	0.11	**0.18**	0.13	**0.21**	0.14	**0.23**	0.17	**0.27**	0.19	**0.31**
Skin		0.11	**0.22**	0.12	**0.23**	0.13	**0.24**	0.13	**0.24**	0.14	**0.26**	0.15	**0.27**
Muscle		0.07	**0.10**	0.08	**0.12**	0.09	**0.14**	0.10	**0.15**	0.12	**0.17**	0.13	**0.19**
Skeleton		0.17	**0.35**	0.21	**0.42**	0.23	**0.47**	0.25	**0.51**	0.29	**0.58**	0.32	**0.64**
Red bone marrow		0.06	**0.09**	0.07	**0.12**	0.08	**0.13**	0.09	**0.14**	0.10	**0.17**	0.12	**0.19**
Whole body dose		0.09	**0.14**	0.11	**0.17**	0.12	**0.19**	0.13	**0.21**	0.15	**0.24**	0.17	**0.27**
ED (ICRP 60)		0.12	**0.12**	0.15	**0.15**	0.16	**0.17**	0.18	**0.19**	0.20	**0.22**	0.23	**0.25**
ED (ICRP 103)		0.15	**0.16**	0.18	**0.20**	0.20	**0.22**	0.22	**0.24**	0.24	**0.27**	0.27	**0.31**
Total filtration (mm)	**Al**	**1.5**		**2.5**		**3.5**		**4.5**		**3.5**		**3.5**	
	Cu	**0**		**0**		**0**		**0**		**0.1**		**0.2**	

Table 9.108 Normalized organ doses—Chest r-lat—3–7 years

Beam projection:			r-lat				**Table cells:**					
Focus-to-detector distance (cm):			115				(+) = normal font = optimal field setting					
Tube voltage (kVp):			80				**(−) = bold font = suboptimal field setting**					
Total filtrations:			See last table row.				Empty table cells correspond to values < 0.01					
Organ or tissue	**Organ dose D_T : Entrance dose K_E**								**(mSv × mGy⁻¹)**			
	(+)	(−)	(+)	(−)	(+)	(−)	(+)	(−)	(+)	(−)	(+)	(−)
Skull	0.02	**0.25**	0.02	**0.30**	0.03	**0.33**	0.03	**0.36**	0.04	**0.41**	0.04	**0.45**
Brain		**0.02**		**0.03**		**0.04**		**0.04**		**0.05**		**0.06**
Oral mucosa	0.02	**0.29**	0.02	**0.35**	0.03	**0.39**	0.03	**0.42**	0.04	**0.48**	0.05	**0.53**
Salivary glands	0.03	**0.37**	0.04	**0.42**	0.04	**0.46**	0.05	**0.48**	0.05	**0.53**	0.06	**0.57**
Cervical spine	0.35	**0.70**	0.42	**0.86**	0.48	**0.98**	0.53	**1.08**	0.62	**1.26**	0.69	**1.42**
Thyroid	0.39	**0.39**	0.46	**0.47**	0.51	**0.52**	0.55	**0.56**	0.61	**0.63**	0.67	**0.70**
Thymus	0.19	**0.18**	0.24	**0.22**	0.27	**0.26**	0.30	**0.29**	0.36	**0.35**	0.42	**0.41**
Thoracic spine	0.41	**0.37**	0.51	**0.46**	0.60	**0.54**	0.67	**0.61**	0.81	**0.74**	0.95	**0.87**
Claviculae	0.70	**0.66**	0.83	**0.80**	0.93	**0.90**	1.01	**0.98**	1.14	**1.12**	1.24	**1.24**
Scapulae	1.19	**1.13**	1.36	**1.30**	1.47	**1.41**	1.55	**1.49**	1.68	**1.62**	1.77	**1.72**
Ribs	1.07	**0.89**	1.24	**1.05**	1.35	**1.17**	1.43	**1.26**	1.56	**1.41**	1.66	**1.54**
Oesophagus	0.20	**0.14**	0.25	**0.17**	0.29	**0.20**	0.32	**0.23**	0.38	**0.28**	0.43	**0.33**
Trachea	0.06	**0.23**	0.07	**0.28**	0.07	**0.32**	0.08	**0.35**	0.09	**0.41**	0.10	**0.46**
Lung	0.36	**0.26**	0.43	**0.32**	0.49	**0.36**	0.53	**0.40**	0.60	**0.47**	0.67	**0.53**
Mamma	0.54	**0.53**	0.59	**0.57**	0.61	**0.60**	0.63	**0.62**	0.67	**0.65**	0.70	**0.68**
Heart	0.22	**0.17**	0.28	**0.22**	0.32	**0.26**	0.35	**0.29**	0.42	**0.35**	0.48	**0.40**
Stomach	0.05	**0.06**	0.06	**0.08**	0.08	**0.09**	0.09	**0.10**	0.11	**0.13**	0.13	**0.16**
Spleen	0.04	**0.04**	0.05	**0.05**	0.06	**0.06**	0.07	**0.07**	0.08	**0.09**	0.10	**0.10**
Liver	0.32	**0.37**	0.38	**0.46**	0.43	**0.52**	0.47	**0.57**	0.54	**0.67**	0.60	**0.76**
Gall bladder	0.07	**0.15**	0.09	**0.19**	0.11	**0.22**	0.12	**0.25**	0.15	**0.31**	0.18	**0.36**
Pancreas	0.11	**0.09**	0.14	**0.12**	0.16	**0.14**	0.18	**0.16**	0.22	**0.20**	0.26	**0.24**
Small intestine	0.01	**0.09**	0.01	**0.12**	0.02	**0.14**	0.02	**0.15**	0.03	**0.19**	0.03	**0.22**
Large intestine	0.02	**0.14**	0.02	**0.17**	0.02	**0.20**	0.03	**0.22**	0.03	**0.26**	0.04	**0.31**
Sigma/Rectum		**0.01**		**0.02**		**0.02**		**0.02**		**0.03**		**0.03**
Lumbar spine	0.09	**0.29**	0.11	**0.37**	0.13	**0.44**	0.15	**0.49**	0.19	**0.60**	0.22	**0.70**
Pelvis		**0.08**		**0.10**	0.01	**0.12**	0.01	**0.13**	0.02	**0.16**	0.02	**0.19**
Adrenals	0.15	**0.14**	0.19	**0.18**	0.22	**0.21**	0.25	**0.24**	0.30	**0.29**	0.35	**0.34**
Kidneys	0.09	**0.20**	0.11	**0.25**	0.13	**0.28**	0.14	**0.31**	0.17	**0.36**	0.20	**0.41**
Urinary bladder				**0.01**		**0.01**		**0.01**		**0.02**		**0.02**
Uterus		**0.02**		**0.03**		**0.04**		**0.04**		**0.05**		**0.06**
Ovaries		**0.03**		**0.03**		**0.04**		**0.04**		**0.05**	0.01	**0.06**
Prostate												**0.01**
Testes												
Upper arm		**1.22**		**1.39**		**1.51**		**1.59**		**1.73**		**1.82**
Elbow		**1.29**		**1.47**		**1.58**		**1.67**		**1.80**		**1.90**
Forearm		**0.57**		**0.65**		**0.70**		**0.73**		**0.79**		**0.83**
Upper leg												
Knee												
Lower leg												
Lymph nodes	0.10	**0.17**	0.13	**0.21**	0.15	**0.24**	0.16	**0.26**	0.19	**0.31**	0.21	**0.35**
Skin	0.12	**0.22**	0.13	**0.24**	0.13	**0.25**	0.14	**0.25**	0.15	**0.27**	0.15	**0.28**
Muscle	0.08	**0.12**	0.09	**0.14**	0.10	**0.15**	0.11	**0.17**	0.13	**0.19**	0.14	**0.21**
Skeleton	0.19	**0.39**	0.23	**0.46**	0.25	**0.51**	0.27	**0.55**	0.30	**0.61**	0.33	**0.67**
Red bone marrow	0.07	**0.11**	0.08	**0.13**	0.09	**0.15**	0.10	**0.16**	0.11	**0.19**	0.13	**0.22**
Whole body dose	0.10	**0.16**	0.12	**0.19**	0.13	**0.21**	0.14	**0.23**	0.16	**0.26**	0.18	**0.29**
ED (ICRP 60)	0.14	**0.15**	0.16	**0.17**	0.18	**0.20**	0.20	**0.21**	0.22	**0.25**	0.25	**0.28**
ED (ICRP 103)	0.17	**0.19**	0.20	**0.22**	0.22	**0.24**	0.24	**0.26**	0.27	**0.30**	0.29	**0.34**
Total filtration (mm) Al	**1.5**		**2.5**		**3.5**		**4.5**		**3.5**		**3.5**	
Cu	**0**		**0**		**0**		**0**		**0.1**		**0.2**	

Table 9.109 Normalized organ doses—Chest r-lat—8–12 years

Beam projection:			r-lat			**Table cells:**						
Focus-to-detector distance (cm):			115			(+) = normal font = optimal field setting						
Tube voltage (kVp):			60			**(−) = bold font = suboptimal field setting**						
Total filtrations:			See last table row.			Empty table cells correspond to values < 0.01						
Organ or tissue	**Organ dose D_T : Entrance dose K_E**							**(mSv × mGy^{-1})**				
	(+)	(−)	(+)	(−)	(+)	(−)	(+)	(−)	(+)	(−)	(+)	(−)
Skull	0.01	**0.17**	0.01	**0.21**	0.02	**0.24**	0.02	**0.26**	0.02	**0.31**	0.02	**0.29**
Brain		**0.01**		**0.01**		**0.02**		**0.02**		**0.03**		**0.02**
Oral mucosa		**0.19**	0.01	**0.23**	0.01	**0.26**	0.02	**0.29**	0.02	**0.33**	0.02	**0.31**
Salivary glands	0.02	**0.29**	0.02	**0.33**	0.03	**0.36**	0.03	**0.38**	0.04	**0.43**	0.03	**0.41**
Cervical spine	0.23	**0.43**	0.30	**0.56**	0.36	**0.66**	0.40	**0.74**	0.49	**0.91**	0.45	**0.84**
Thyroid	0.25	**0.21**	0.30	**0.26**	0.34	**0.30**	0.37	**0.33**	0.43	**0.39**	0.41	**0.37**
Thymus	0.08	**0.07**	0.11	**0.09**	0.13	**0.11**	0.15	**0.13**	0.19	**0.17**	0.16	**0.14**
Thoracic spine	0.16	**0.13**	0.21	**0.18**	0.26	**0.21**	0.30	**0.25**	0.38	**0.32**	0.32	**0.27**
Claviculae	0.41	**0.35**	0.52	**0.45**	0.61	**0.53**	0.67	**0.59**	0.79	**0.71**	0.76	**0.67**
Scapulae	0.90	**0.83**	1.08	**1.00**	1.20	**1.11**	1.29	**1.20**	1.45	**1.35**	1.43	**1.34**
Ribs	0.72	**0.55**	0.86	**0.68**	0.97	**0.78**	1.04	**0.86**	1.18	**1.01**	1.16	**0.98**
Oesophagus	0.08	**0.05**	0.11	**0.06**	0.13	**0.08**	0.15	**0.09**	0.18	**0.12**	0.16	**0.10**
Trachea	0.09	**0.20**	0.11	**0.25**	0.13	**0.30**	0.14	**0.33**	0.16	**0.40**	0.15	**0.37**
Lung	0.21	**0.13**	0.27	**0.17**	0.31	**0.20**	0.34	**0.23**	0.41	**0.28**	0.38	**0.25**
Mamma	0.46	**0.45**	0.50	**0.50**	0.53	**0.52**	0.55	**0.54**	0.58	**0.57**	0.58	**0.56**
Heart	0.10	**0.07**	0.13	**0.09**	0.16	**0.11**	0.18	**0.13**	0.22	**0.16**	0.20	**0.14**
Stomach	0.01	**0.02**	0.02	**0.02**	0.02	**0.03**	0.03	**0.03**	0.03	**0.04**	0.03	**0.03**
Spleen			0.01	**0.01**	0.02	**0.01**	0.02	**0.02**	0.03	**0.02**	0.02	**0.02**
Liver	0.18	**0.21**	0.22	**0.27**	0.26	**0.32**	0.29	**0.36**	0.34	**0.43**	0.32	**0.40**
Gall bladder	0.02	**0.07**	0.03	**0.09**	0.04	**0.11**	0.05	**0.12**	0.06	**0.16**	0.05	**0.13**
Pancreas	0.03	**0.03**	0.05	**0.04**	0.06	**0.05**	0.07	**0.06**	0.09	**0.07**	0.07	**0.06**
Small intestine		**0.03**		**0.04**		**0.05**		**0.06**		**0.08**		**0.07**
Large intestine		**0.06**		**0.07**		**0.09**		**0.10**		**0.13**		**0.11**
Sigma/Rectum												
Lumbar spine	0.02	**0.10**	0.03	**0.13**	0.03	**0.16**	0.04	**0.19**	0.05	**0.24**	0.04	**0.20**
Pelvis		**0.03**		**0.03**		**0.04**		**0.05**		**0.06**		**0.05**
Adrenals	0.06	**0.06**	0.08	**0.08**	0.10	**0.09**	0.12	**0.11**	0.15	**0.14**	0.13	**0.12**
Kidneys	0.03	**0.09**	0.04	**0.12**	0.04	**0.15**	0.05	**0.16**	0.06	**0.20**	0.05	**0.18**
Urinary bladder												
Uterus						**0.01**		**0.01**		**0.02**		**0.01**
Ovaries						**0.01**		**0.01**		**0.02**		**0.02**
Prostate												
Testes												
Upper arm		**0.94**		**1.11**		**1.23**		**1.32**		**1.49**		**1.47**
Elbow		**1.01**		**1.20**		**1.32**		**1.42**		**1.58**		**1.57**
Forearm		**0.41**		**0.49**		**0.54**		**0.58**		**0.64**		**0.64**
Upper leg												
Knee												
Lower leg												
Lymph nodes	0.06	**0.10**	0.07	**0.12**	0.09	**0.14**	0.10	**0.16**	0.12	**0.19**	0.11	**0.17**
Skin	0.09	**0.18**	0.10	**0.19**	0.10	**0.19**	0.11	**0.20**	0.11	**0.21**	0.11	**0.21**
Muscle	0.05	**0.07**	0.06	**0.08**	0.07	**0.09**	0.07	**0.10**	0.08	**0.12**	0.08	**0.11**
Skeleton	0.12	**0.24**	0.14	**0.30**	0.16	**0.34**	0.18	**0.37**	0.21	**0.43**	0.20	**0.41**
Red bone marrow	0.05	**0.06**	0.06	**0.08**	0.07	**0.09**	0.07	**0.10**	0.09	**0.12**	0.08	**0.11**
Whole body dose	0.06	**0.10**	0.08	**0.12**	0.09	**0.13**	0.09	**0.14**	0.11	**0.17**	0.10	**0.16**
ED (ICRP 60)	0.08	**0.08**	0.10	**0.10**	0.12	**0.11**	0.13	**0.13**	0.15	**0.15**	0.14	**0.14**
ED (ICRP 103)	0.11	**0.12**	0.14	**0.14**	0.15	**0.16**	0.16	**0.17**	0.19	**0.20**	0.18	**0.18**
Total filtration (mm) **Al**	**1.5**		**2.5**		**3.5**		**4.5**		**3.5**		**3.5**	
Cu	**0**		**0**		**0**		**0**		**0.1**		**0.2**	

Table 9.110 Normalized organ doses—Chest r-lat—8–12 years

Beam projection:			r-lat				**Table cells:**					
Focus-to-detector distance (cm):			115				(+) = normal font = optimal field setting					
Tube voltage (kVp):			70				**(−) = bold font = suboptimal field setting**					
Total filtrations:			See last table row.				Empty table cells correspond to values < 0.01					
Organ or tissue	**Organ dose D_T : Entrance dose K_E**						**(mSv × mGy^{-1})**					
	(+)	(−)	(+)	(−)	(+)	(−)	(+)	(−)	(+)	(−)	(+)	(−)
Skull	0.01	**0.20**	0.02	**0.24**	0.02	**0.27**	0.02	**0.30**	0.03	**0.35**	0.03	**0.39**
Brain		**0.01**		**0.02**		**0.02**		**0.03**		**0.03**		**0.04**
Oral mucosa	0.01	**0.22**	0.02	**0.27**	0.02	**0.30**	0.02	**0.33**	0.03	**0.38**	0.03	**0.42**
Salivary glands	0.02	**0.32**	0.03	**0.36**	0.03	**0.39**	0.04	**0.42**	0.05	**0.46**	0.05	**0.50**
Cervical spine	0.29	**0.53**	0.37	**0.68**	0.42	**0.79**	0.47	**0.88**	0.56	**1.05**	0.65	**1.22**
Thyroid	0.28	**0.25**	0.34	**0.31**	0.38	**0.35**	0.42	**0.38**	0.48	**0.44**	0.53	**0.49**
Thymus	0.11	**0.10**	0.14	**0.12**	0.17	**0.15**	0.19	**0.17**	0.23	**0.21**	0.28	**0.25**
Thoracic spine	0.21	**0.18**	0.28	**0.24**	0.34	**0.29**	0.38	**0.33**	0.48	**0.42**	0.58	**0.51**
Claviculae	0.49	**0.42**	0.60	**0.53**	0.69	**0.61**	0.76	**0.68**	0.88	**0.80**	1.00	**0.92**
Scapulae	1.00	**0.92**	1.17	**1.09**	1.29	**1.20**	1.38	**1.29**	1.52	**1.44**	1.64	**1.55**
Ribs	0.80	**0.64**	0.95	**0.78**	1.05	**0.88**	1.13	**0.97**	1.26	**1.11**	1.38	**1.24**
Oesophagus	0.11	**0.07**	0.14	**0.09**	0.16	**0.11**	0.18	**0.12**	0.23	**0.15**	0.27	**0.19**
Trachea	0.10	**0.24**	0.13	**0.31**	0.15	**0.35**	0.16	**0.39**	0.18	**0.46**	0.21	**0.53**
Lung	0.25	**0.16**	0.31	**0.21**	0.36	**0.24**	0.39	**0.27**	0.46	**0.33**	0.52	**0.39**
Mamma	0.48	**0.48**	0.53	**0.52**	0.56	**0.54**	0.58	**0.56**	0.61	**0.59**	0.64	**0.61**
Heart	0.13	**0.09**	0.17	**0.12**	0.20	**0.15**	0.22	**0.17**	0.27	**0.21**	0.32	**0.25**
Stomach	0.02	**0.02**	0.03	**0.03**	0.03	**0.04**	0.04	**0.05**	0.05	**0.06**	0.06	**0.07**
Spleen	0.01	**0.01**	0.02	**0.02**	0.02	**0.02**	0.03	**0.03**	0.04	**0.04**	0.05	**0.04**
Liver	0.21	**0.26**	0.26	**0.33**	0.30	**0.38**	0.33	**0.43**	0.39	**0.51**	0.45	**0.60**
Gall bladder	0.03	**0.09**	0.05	**0.12**	0.05	**0.14**	0.06	**0.16**	0.08	**0.21**	0.10	**0.25**
Pancreas	0.05	**0.04**	0.07	**0.06**	0.08	**0.07**	0.09	**0.08**	0.12	**0.10**	0.14	**0.13**
Small intestine		**0.05**		**0.06**		**0.07**		**0.08**		**0.10**	0.01	**0.13**
Large intestine		**0.07**		**0.10**		**0.11**	0.01	**0.13**	0.01	**0.16**	0.02	**0.19**
Sigma/Rectum										**0.01**		**0.01**
Lumbar spine	0.03	**0.14**	0.04	**0.18**	0.05	**0.22**	0.05	**0.25**	0.07	**0.32**	0.09	**0.39**
Pelvis		**0.04**		**0.05**		**0.06**		**0.06**		**0.08**		**0.10**
Adrenals	0.08	**0.08**	0.11	**0.10**	0.13	**0.13**	0.15	**0.14**	0.19	**0.18**	0.23	**0.22**
Kidneys	0.04	**0.12**	0.05	**0.15**	0.05	**0.18**	0.06	**0.20**	0.08	**0.24**	0.09	**0.28**
Urinary bladder												
Uterus				**0.01**		**0.02**		**0.02**		**0.02**		**0.03**
Ovaries		**0.01**		**0.01**		**0.02**		**0.02**		**0.03**		**0.03**
Prostate												
Testes												
Upper arm		**1.03**		**1.21**		**1.33**		**1.42**		**1.57**		**1.70**
Elbow		**1.11**		**1.30**		**1.42**		**1.51**		**1.67**		**1.79**
Forearm		**0.45**		**0.53**		**0.58**		**0.61**		**0.68**		**0.72**
Upper leg												
Knee												
Lower leg												
Lymph nodes	0.07	**0.12**	0.09	**0.15**	0.10	**0.17**	0.12	**0.19**	0.14	**0.23**	0.16	**0.27**
Skin	0.10	**0.18**	0.11	**0.20**	0.11	**0.20**	0.11	**0.21**	0.12	**0.22**	0.13	**0.23**
Muscle	0.06	**0.08**	0.07	**0.09**	0.08	**0.11**	0.08	**0.12**	0.10	**0.13**	0.11	**0.15**
Skeleton	0.14	**0.28**	0.16	**0.34**	0.18	**0.38**	0.20	**0.41**	0.23	**0.47**	0.26	**0.53**
Red bone marrow	0.06	**0.08**	0.07	**0.10**	0.08	**0.11**	0.09	**0.12**	0.10	**0.15**	0.11	**0.17**
Whole body dose	0.07	**0.11**	0.09	**0.13**	0.10	**0.15**	0.11	**0.17**	0.12	**0.19**	0.14	**0.22**
ED (ICRP 60)	0.10	**0.10**	0.12	**0.12**	0.13	**0.13**	0.14	**0.15**	0.17	**0.17**	0.19	**0.20**
ED (ICRP 103)	0.13	**0.14**	0.15	**0.16**	0.17	**0.18**	0.18	**0.19**	0.21	**0.22**	0.23	**0.25**
Total filtration (mm) Al	**1.5**		**2.5**		**3.5**		**4.5**		**3.5**		**3.5**	
Cu	**0**		**0**		**0**		**0**		**0.1**		**0.2**	

Table 9.111 Normalized organ doses—Chest r-lat—8–12 years

Beam projection:			r-lat				**Table cells:**					
Focus-to-detector distance (cm):			115				(+) = normal font = optimal field setting					
Tube voltage (kVp):			80				**(−) = bold font = suboptimal field setting**					
Total filtrations:			See last table row.				Empty table cells correspond to values < 0.01					
Organ or tissue	**Organ dose D_T : Entrance dose K_E**							**(mSv × mGy^{-1})**				
	(+)	(−)	(+)	(−)	(+)	(−)	(+)	(−)	(+)	(−)	(+)	(−)
Skull	0.02	**0.22**	0.02	**0.27**	0.02	**0.30**	0.03	**0.33**	0.03	**0.38**	0.04	**0.42**
Brain		**0.02**		**0.03**		**0.03**		**0.03**		**0.04**		**0.05**
Oral mucosa	0.02	**0.25**	0.02	**0.30**	0.02	**0.34**	0.03	**0.36**	0.03	**0.42**	0.04	**0.46**
Salivary glands	0.03	**0.35**	0.03	**0.39**	0.04	**0.42**	0.04	**0.45**	0.05	**0.50**	0.06	**0.54**
Cervical spine	0.34	**0.63**	0.42	**0.78**	0.48	**0.90**	0.53	**0.99**	0.62	**1.17**	0.71	**1.32**
Thyroid	0.32	**0.29**	0.38	**0.34**	0.42	**0.39**	0.46	**0.42**	0.52	**0.48**	0.57	**0.53**
Thymus	0.14	**0.12**	0.17	**0.16**	0.20	**0.18**	0.23	**0.20**	0.27	**0.25**	0.32	**0.30**
Thoracic spine	0.27	**0.24**	0.35	**0.30**	0.41	**0.36**	0.46	**0.41**	0.57	**0.50**	0.68	**0.60**
Claviculae	0.56	**0.49**	0.67	**0.60**	0.76	**0.69**	0.83	**0.75**	0.95	**0.87**	1.05	**0.98**
Scapulae	1.08	**1.00**	1.24	**1.16**	1.35	**1.27**	1.43	**1.35**	1.57	**1.49**	1.67	**1.59**
Ribs	0.88	**0.72**	1.02	**0.86**	1.12	**0.96**	1.19	**1.04**	1.32	**1.19**	1.41	**1.30**
Oesophagus	0.13	**0.09**	0.17	**0.11**	0.20	**0.13**	0.22	**0.15**	0.27	**0.19**	0.31	**0.23**
Trachea	0.12	**0.29**	0.14	**0.36**	0.16	**0.41**	0.18	**0.45**	0.20	**0.52**	0.22	**0.59**
Lung	0.29	**0.20**	0.36	**0.25**	0.40	**0.29**	0.44	**0.32**	0.51	**0.38**	0.57	**0.43**
Mamma	0.51	**0.50**	0.55	**0.54**	0.58	**0.56**	0.60	**0.58**	0.63	**0.61**	0.66	**0.63**
Heart	0.16	**0.12**	0.20	**0.15**	0.24	**0.18**	0.26	**0.20**	0.32	**0.25**	0.37	**0.30**
Stomach	0.03	**0.03**	0.04	**0.04**	0.04	**0.05**	0.05	**0.06**	0.06	**0.08**	0.08	**0.10**
Spleen	0.02	**0.02**	0.03	**0.03**	0.03	**0.03**	0.04	**0.04**	0.05	**0.05**	0.06	**0.06**
Liver	0.25	**0.31**	0.30	**0.39**	0.34	**0.45**	0.37	**0.49**	0.44	**0.58**	0.49	**0.67**
Gall bladder	0.04	**0.12**	0.06	**0.15**	0.07	**0.18**	0.08	**0.20**	0.10	**0.25**	0.12	**0.30**
Pancreas	0.07	**0.06**	0.09	**0.08**	0.10	**0.09**	0.12	**0.11**	0.14	**0.13**	0.17	**0.16**
Small intestine		**0.06**		**0.08**		**0.09**		**0.10**	0.01	**0.13**	0.02	**0.15**
Large intestine		**0.09**		**0.12**	0.01	**0.14**	0.01	**0.16**	0.02	**0.19**	0.02	**0.22**
Sigma/Rectum						**0.01**		**0.01**		**0.01**		**0.02**
Lumbar spine	0.04	**0.18**	0.05	**0.23**	0.06	**0.28**	0.07	**0.31**	0.09	**0.39**	0.11	**0.46**
Pelvis		**0.05**		**0.06**		**0.07**		**0.08**		**0.10**		**0.12**
Adrenals	0.11	**0.10**	0.14	**0.13**	0.16	**0.16**	0.18	**0.18**	0.22	**0.22**	0.27	**0.26**
Kidneys	0.05	**0.15**	0.06	**0.18**	0.07	**0.21**	0.08	**0.23**	0.09	**0.28**	0.11	**0.32**
Urinary bladder										**0.01**		**0.01**
Uterus		**0.01**		**0.02**		**0.02**		**0.02**		**0.03**		**0.04**
Ovaries		**0.02**		**0.02**		**0.02**		**0.03**		**0.03**		**0.04**
Prostate												
Testes												
Upper arm		**1.12**		**1.28**		**1.40**		**1.48**		**1.62**		**1.73**
Elbow		**1.20**		**1.37**		**1.49**		**1.57**		**1.71**		**1.82**
Forearm		**0.49**		**0.56**		**0.60**		**0.64**		**0.69**		**0.73**
Upper leg												
Knee												
Lower leg												
Lymph nodes	0.09	**0.15**	0.11	**0.18**	0.12	**0.20**	0.13	**0.22**	0.16	**0.27**	0.18	**0.30**
Skin	0.10	**0.19**	0.11	**0.20**	0.11	**0.21**	0.12	**0.22**	0.13	**0.23**	0.13	**0.24**
Muscle	0.06	**0.09**	0.08	**0.11**	0.09	**0.12**	0.09	**0.13**	0.11	**0.15**	0.12	**0.17**
Skeleton	0.15	**0.31**	0.18	**0.37**	0.20	**0.41**	0.22	**0.45**	0.25	**0.51**	0.27	**0.55**
Red bone marrow	0.07	**0.09**	0.08	**0.11**	0.09	**0.13**	0.10	**0.14**	0.11	**0.17**	0.13	**0.19**
Whole body dose	0.08	**0.13**	0.10	**0.15**	0.11	**0.17**	0.12	**0.18**	0.13	**0.21**	0.15	**0.24**
ED (ICRP 60)	0.11	**0.11**	0.13	**0.14**	0.15	**0.16**	0.16	**0.17**	0.18	**0.20**	0.20	**0.22**
ED (ICRP 103)	0.15	**0.15**	0.17	**0.18**	0.19	**0.20**	0.20	**0.22**	0.23	**0.25**	0.25	**0.28**
Total filtration (mm) Al	**1.5**		**2.5**		**3.5**		**4.5**		**3.5**		**3.5**	
Cu	**0**		**0**		**0**		**0**		**0.1**		**0.2**	

Table 9.112 Normalized organ doses—Chest r-lat—13–17 years

Beam projection:			r-lat			**Table cells:**						
Focus-to-detector distance (cm):			115			(+) = normal font = optimal field setting						
Tube voltage (kVp):			60			**(−) = bold font = suboptimal field setting**						
Total filtrations:			See last table row.			Empty table cells correspond to values < 0.01						
Organ or tissue	**Organ dose D_T : Entrance dose K_E**							**(mSv × mGy^{-1})**				
	(+)	(−)	(+)	(−)	(+)	(−)	(+)	(−)	(+)	(−)	(+)	(−)
Skull		0.13		0.16	0.01	0.19	0.01	0.21	0.02	0.25	0.01	0.24
Brain						0.01		0.01		0.02		0.01
Oral mucosa		0.14		0.18		0.20		0.22	0.01	0.26	0.01	0.24
Salivary glands	0.01	0.22	0.02	0.26	0.02	0.28	0.02	0.30	0.03	0.34	0.02	0.32
Cervical spine	0.23	0.39	0.30	0.51	0.36	0.60	0.41	0.68	0.50	0.84	0.45	0.76
Thyroid	0.16	0.11	0.20	0.14	0.23	0.16	0.25	0.18	0.29	0.22	0.28	0.20
Thymus	0.04	0.04	0.06	0.05	0.07	0.06	0.09	0.07	0.11	0.09	0.09	0.08
Thoracic spine	0.08	0.06	0.11	0.08	0.13	0.10	0.15	0.12	0.20	0.16	0.16	0.12
Claviculae	0.29	0.23	0.38	0.30	0.44	0.36	0.49	0.40	0.59	0.49	0.56	0.45
Scapulae	0.76	0.69	0.92	0.84	1.03	0.95	1.12	1.03	1.27	1.18	1.25	1.16
Ribs	0.58	0.41	0.70	0.51	0.79	0.60	0.85	0.66	0.98	0.79	0.95	0.75
Oesophagus	0.05	0.03	0.07	0.04	0.09	0.05	0.10	0.06	0.13	0.08	0.11	0.06
Trachea	0.07	0.12	0.09	0.16	0.10	0.18	0.11	0.20	0.13	0.25	0.12	0.23
Lung	0.15	0.09	0.19	0.12	0.23	0.14	0.26	0.16	0.31	0.20	0.29	0.17
Mamma	1.98	2.53	2.34	2.98	2.60	3.29	2.79	3.52	3.15	3.95	3.04	3.84
Heart	0.06	0.04	0.08	0.05	0.09	0.06	0.11	0.07	0.13	0.09	0.11	0.07
Stomach					0.01	0.01	0.01	0.01	0.02	0.02	0.01	0.01
Spleen									0.01			
Liver	0.16	0.16	0.20	0.21	0.24	0.25	0.26	0.28	0.32	0.35	0.29	0.31
Gall bladder	0.02	0.06	0.03	0.08	0.03	0.10	0.04	0.11	0.05	0.14	0.04	0.12
Pancreas	0.02	0.01	0.03	0.02	0.03	0.02	0.04	0.03	0.05	0.04	0.04	0.03
Small intestine		0.02		0.02		0.03		0.03		0.04		0.03
Large intestine		0.03		0.04		0.05		0.06		0.07		0.06
Sigma/Rectum												
Lumbar spine	0.01	0.04	0.02	0.06	0.02	0.08	0.02	0.09	0.03	0.12	0.02	0.09
Pelvis		0.01		0.02		0.02		0.02		0.03		0.02
Adrenals	0.04	0.03	0.05	0.04	0.06	0.05	0.07	0.06	0.09	0.08	0.08	0.06
Kidneys	0.02	0.06	0.03	0.08	0.03	0.09	0.04	0.11	0.04	0.13	0.04	0.12
Urinary bladder												
Uterus						0.01		0.01		0.02		0.01
Ovaries												
Prostate												
Testes												
Upper arm		0.84		1.01		1.13		1.22		1.38		1.36
Elbow		0.91		1.09		1.22		1.31		1.48		1.46
Forearm		0.32		0.39		0.43		0.46		0.52		0.52
Upper leg												
Knee												
Lower leg												
Lymph nodes	0.04	0.06	0.05	0.08	0.06	0.09	0.07	0.11	0.08	0.13	0.07	0.11
Skin	0.11	0.16	0.11	0.17	0.12	0.18	0.12	0.19	0.13	0.20	0.13	0.19
Muscle	0.04	0.05	0.05	0.06	0.06	0.07	0.06	0.08	0.07	0.09	0.07	0.09
Skeleton	0.10	0.21	0.12	0.25	0.14	0.29	0.15	0.32	0.18	0.37	0.17	0.35
Red bone marrow	0.04	0.05	0.05	0.06	0.06	0.08	0.07	0.08	0.08	0.10	0.07	0.09
Whole body dose	0.05	0.08	0.07	0.10	0.07	0.11	0.08	0.12	0.10	0.14	0.09	0.13
ED (ICRP 60)	0.14	0.17	0.17	0.20	0.20	0.22	0.21	0.24	0.25	0.28	0.23	0.27
ED (ICRP 103)	0.28	0.35	0.34	0.41	0.38	0.46	0.41	0.50	0.47	0.56	0.45	0.54
Total filtration (mm) **Al**	**1.5**		**2.5**		**3.5**		**4.5**		**3.5**		**3.5**	
Cu	**0**		**0**		**0**		**0**		**0.1**		**0.2**	

Table 9.113 Normalized organ doses—Chest r-lat—13–17 years

Beam projection:			r-lat			**Table cells:**						
Focus-to-detector distance (cm):			115			(+) = normal font = optimal field setting						
Tube voltage (kVp):			70			**(−) = bold font = suboptimal field setting**						
Total filtrations:			See last table row.			Empty table cells correspond to values < 0.01						
Organ or tissue	**Organ dose D_T : Entrance dose K_E**							**(mSv x mGy^{-1})**				
	(+)	(−)	(+)	(−)	(+)	(−)	(+)	(−)	(+)	(−)	(+)	(−)
Skull		**0.16**	0.01	**0.19**	0.01	**0.22**	0.02	**0.25**	0.02	**0.29**	0.03	**0.33**
Brain		**0.01**		**0.01**		**0.02**		**0.02**		**0.02**		**0.03**
Oral mucosa		**0.17**		**0.21**	0.01	**0.23**	0.01	**0.26**	0.02	**0.30**	0.02	**0.34**
Salivary glands	0.02	**0.24**	0.02	**0.28**	0.02	**0.31**	0.03	**0.33**	0.03	**0.37**	0.04	**0.41**
Cervical spine	0.29	**0.49**	0.37	**0.63**	0.43	**0.73**	0.49	**0.82**	0.59	**1.00**	0.68	**1.17**
Thyroid	0.19	**0.14**	0.23	**0.17**	0.26	**0.19**	0.28	**0.21**	0.33	**0.25**	0.37	**0.29**
Thymus	0.06	**0.05**	0.08	**0.07**	0.10	**0.09**	0.12	**0.10**	0.15	**0.13**	0.18	**0.16**
Thoracic spine	0.11	**0.09**	0.15	**0.12**	0.18	**0.15**	0.21	**0.17**	0.27	**0.22**	0.34	**0.28**
Claviculae	0.36	**0.29**	0.45	**0.37**	0.51	**0.43**	0.57	**0.47**	0.67	**0.57**	0.77	**0.66**
Scapulae	0.85	**0.78**	1.02	**0.94**	1.13	**1.04**	1.21	**1.13**	1.36	**1.27**	1.48	**1.39**
Ribs	0.65	**0.48**	0.78	**0.60**	0.87	**0.69**	0.94	**0.76**	1.06	**0.89**	1.16	**1.00**
Oesophagus	0.07	**0.04**	0.10	**0.06**	0.12	**0.07**	0.13	**0.08**	0.17	**0.11**	0.20	**0.13**
Trachea	0.08	**0.15**	0.10	**0.19**	0.12	**0.22**	0.13	**0.24**	0.15	**0.29**	0.17	**0.33**
Lung	0.19	**0.12**	0.23	**0.15**	0.27	**0.18**	0.30	**0.20**	0.36	**0.25**	0.42	**0.29**
Mamma	2.22	**2.81**	2.60	**3.28**	2.87	**3.60**	3.07	**3.84**	3.45	**4.29**	3.78	**4.67**
Heart	0.08	**0.05**	0.10	**0.07**	0.12	**0.09**	0.14	**0.10**	0.17	**0.13**	0.21	**0.16**
Stomach		**0.01**	0.01	**0.01**	0.02	**0.02**	0.02	**0.02**	0.02	**0.03**	0.03	**0.04**
Spleen					0.01		0.01	**0.01**	0.02	**0.02**	0.02	**0.02**
Liver	0.19	**0.20**	0.24	**0.26**	0.28	**0.31**	0.31	**0.35**	0.37	**0.42**	0.43	**0.50**
Gall bladder	0.03	**0.08**	0.04	**0.11**	0.04	**0.13**	0.05	**0.15**	0.06	**0.19**	0.08	**0.23**
Pancreas	0.03	**0.02**	0.04	**0.03**	0.05	**0.04**	0.05	**0.04**	0.07	**0.06**	0.08	**0.07**
Small intestine		**0.02**		**0.03**		**0.04**		**0.05**		**0.06**		**0.07**
Large intestine		**0.04**		**0.06**		**0.07**		**0.08**		**0.10**	0.01	**0.12**
Sigma/Rectum												
Lumbar spine	0.02	**0.07**	0.02	**0.09**	0.03	**0.11**	0.03	**0.13**	0.04	**0.16**	0.06	**0.20**
Pelvis		**0.02**		**0.02**		**0.03**		**0.03**		**0.04**		**0.05**
Adrenals	0.05	**0.05**	0.07	**0.06**	0.08	**0.07**	0.10	**0.08**	0.12	**0.11**	0.15	**0.13**
Kidneys	0.03	**0.08**	0.03	**0.10**	0.04	**0.12**	0.05	**0.13**	0.06	**0.16**	0.07	**0.19**
Urinary bladder												
Uterus		**0.01**		**0.02**		**0.02**		**0.02**		**0.03**		**0.04**
Ovaries								**0.01**		**0.01**		**0.02**
Prostate												
Testes												
Upper arm		**0.94**		**1.11**		**1.23**		**1.32**		**1.48**		**1.61**
Elbow		**1.02**		**1.20**		**1.32**		**1.41**		**1.58**		**1.71**
Forearm		**0.36**		**0.42**		**0.47**		**0.50**		**0.56**		**0.60**
Upper leg												
Knee												
Lower leg												
Lymph nodes	0.05	**0.08**	0.06	**0.10**	0.08	**0.12**	0.08	**0.13**	0.10	**0.16**	0.12	**0.18**
Skin	0.11	**0.17**	0.12	**0.18**	0.12	**0.19**	0.13	**0.20**	0.14	**0.21**	0.14	**0.22**
Muscle	0.05	**0.06**	0.06	**0.07**	0.07	**0.08**	0.07	**0.09**	0.08	**0.11**	0.10	**0.12**
Skeleton	0.12	**0.24**	0.14	**0.29**	0.16	**0.33**	0.17	**0.36**	0.20	**0.41**	0.22	**0.46**
Red bone marrow	0.05	**0.06**	0.06	**0.08**	0.07	**0.09**	0.08	**0.10**	0.10	**0.12**	0.11	**0.14**
Whole body dose	0.06	**0.09**	0.08	**0.11**	0.09	**0.13**	0.09	**0.14**	0.11	**0.16**	0.12	**0.18**
ED (ICRP 60)	0.17	**0.19**	0.20	**0.23**	0.22	**0.26**	0.24	**0.28**	0.28	**0.32**	0.31	**0.36**
ED (ICRP 103)	0.32	**0.39**	0.38	**0.46**	0.42	**0.51**	0.46	**0.55**	0.52	**0.62**	0.57	**0.69**
Total filtration (mm) Al	**1.5**		**2.5**		**3.5**		**4.5**		**3.5**		**3.5**	
Cu	**0**		**0**		**0**		**0**		**0.1**		**0.2**	

Table 9.114 Normalized organ doses—Chest r-lat—13–17 years

Beam projection:		r-lat				**Table cells:**						
Focus-to-detector distance (cm):		115				(+) = normal font = optimal field setting						
Tube voltage (kVp):		80				**(−) = bold font = suboptimal field setting**						
Total filtrations:		See last table row.				Empty table cells correspond to values < 0.01						
Organ or tissue	**Organ dose D_T : Entrance dose K_E**							**(mSv × mGy^{-1})**				
	(+)	(−)	(+)	(−)	(+)	(−)	(+)	(−)	(+)	(−)	(+)	(−)
Skull	0.01	**0.18**	0.02	**0.22**	0.02	**0.25**	0.02	**0.27**	0.03	**0.32**	0.03	**0.36**
Brain		**0.01**		**0.02**		**0.02**		**0.02**		**0.03**		**0.04**
Oral mucosa		**0.20**	0.01	**0.24**	0.01	**0.27**	0.02	**0.29**	0.02	**0.33**	0.02	**0.38**
Salivary glands	0.02	**0.27**	0.02	**0.31**	0.03	**0.34**	0.03	**0.36**	0.04	**0.40**	0.04	**0.44**
Cervical spine	0.35	**0.59**	0.44	**0.74**	0.50	**0.85**	0.56	**0.94**	0.66	**1.12**	0.75	**1.29**
Thyroid	0.22	**0.16**	0.26	**0.20**	0.29	**0.22**	0.32	**0.24**	0.37	**0.28**	0.41	**0.32**
Thymus	0.08	**0.07**	0.11	**0.09**	0.13	**0.11**	0.14	**0.13**	0.18	**0.16**	0.22	**0.19**
Thoracic spine	0.15	**0.13**	0.20	**0.17**	0.24	**0.20**	0.27	**0.22**	0.34	**0.28**	0.41	**0.35**
Claviculae	0.41	**0.34**	0.51	**0.42**	0.57	**0.48**	0.63	**0.53**	0.73	**0.63**	0.82	**0.72**
Scapulae	0.93	**0.86**	1.09	**1.01**	1.20	**1.12**	1.28	**1.19**	1.41	**1.33**	1.52	**1.43**
Ribs	0.72	**0.55**	0.84	**0.67**	0.93	**0.76**	1.00	**0.83**	1.11	**0.96**	1.21	**1.07**
Oesophagus	0.10	**0.06**	0.12	**0.08**	0.15	**0.10**	0.17	**0.11**	0.21	**0.14**	0.25	**0.17**
Trachea	0.09	**0.18**	0.12	**0.22**	0.13	**0.25**	0.14	**0.28**	0.16	**0.33**	0.18	**0.37**
Lung	0.22	**0.14**	0.27	**0.18**	0.31	**0.21**	0.35	**0.24**	0.41	**0.29**	0.47	**0.34**
Mamma	2.46	**3.09**	2.85	**3.56**	3.12	**3.89**	3.32	**4.13**	3.70	**4.57**	4.03	**4.95**
Heart	0.10	**0.07**	0.13	**0.09**	0.15	**0.11**	0.17	**0.13**	0.21	**0.16**	0.25	**0.19**
Stomach	0.01	**0.02**	0.02	**0.02**	0.02	**0.03**	0.03	**0.03**	0.03	**0.04**	0.04	**0.05**
Spleen			0.01	**0.01**	0.02	**0.01**	0.02	**0.02**	0.02	**0.02**	0.03	**0.03**
Liver	0.23	**0.25**	0.28	**0.32**	0.32	**0.37**	0.36	**0.41**	0.42	**0.49**	0.48	**0.57**
Gall bladder	0.04	**0.11**	0.05	**0.14**	0.06	**0.17**	0.06	**0.19**	0.08	**0.23**	0.10	**0.28**
Pancreas	0.04	**0.03**	0.05	**0.04**	0.06	**0.05**	0.07	**0.06**	0.09	**0.08**	0.11	**0.10**
Small intestine		**0.03**		**0.04**		**0.05**		**0.06**		**0.07**		**0.09**
Large intestine		**0.06**		**0.07**		**0.08**		**0.09**	0.01	**0.12**	0.01	**0.14**
Sigma/Rectum												
Lumbar spine	0.03	**0.09**	0.03	**0.12**	0.04	**0.14**	0.05	**0.16**	0.06	**0.21**	0.07	**0.25**
Pelvis		**0.02**		**0.03**		**0.03**		**0.04**		**0.05**		**0.06**
Adrenals	0.07	**0.06**	0.09	**0.08**	0.11	**0.10**	0.12	**0.11**	0.15	**0.14**	0.18	**0.17**
Kidneys	0.03	**0.10**	0.04	**0.12**	0.05	**0.14**	0.06	**0.16**	0.07	**0.19**	0.08	**0.23**
Urinary bladder												
Uterus		**0.02**		**0.02**		**0.03**		**0.03**		**0.04**		**0.05**
Ovaries				**0.01**		**0.01**		**0.01**		**0.02**		**0.02**
Prostate												
Testes												
Upper arm		**1.03**		**1.19**		**1.31**		**1.39**		**1.54**		**1.65**
Elbow		**1.10**		**1.28**		**1.40**		**1.49**		**1.64**		**1.75**
Forearm		**0.39**		**0.45**		**0.49**		**0.52**		**0.58**		**0.62**
Upper leg												
Knee												
Lower leg												
Lymph nodes	0.06	**0.10**	0.08	**0.12**	0.09	**0.14**	0.10	**0.15**	0.12	**0.18**	0.14	**0.21**
Skin	0.12	**0.18**	0.12	**0.19**	0.13	**0.20**	0.13	**0.20**	0.14	**0.21**	0.15	**0.22**
Muscle	0.06	**0.07**	0.07	**0.09**	0.07	**0.10**	0.08	**0.10**	0.09	**0.12**	0.11	**0.14**
Skeleton	0.13	**0.27**	0.16	**0.32**	0.17	**0.36**	0.19	**0.39**	0.21	**0.44**	0.24	**0.49**
Red bone marrow	0.06	**0.08**	0.08	**0.09**	0.09	**0.11**	0.09	**0.12**	0.11	**0.14**	0.12	**0.16**
Whole body dose	0.07	**0.11**	0.09	**0.13**	0.10	**0.14**	0.11	**0.15**	0.12	**0.18**	0.14	**0.20**
ED (ICRP 60)	0.19	**0.22**	0.22	**0.26**	0.25	**0.29**	0.27	**0.31**	0.31	**0.35**	0.34	**0.39**
ED (ICRP 103)	0.36	**0.44**	0.42	**0.51**	0.47	**0.56**	0.50	**0.60**	0.56	**0.68**	0.62	**0.74**
Total filtration (mm) Al	**1.5**		**2.5**		**3.5**		**4.5**		**3.5**		**3.5**	
Cu	**0**		**0**		**0**		**0**		**0.1**		**0.2**	

Table 9.115 Normalized organ doses—Chest r-lat—13–17 years

Beam projection:			r-lat			**Table cells:**						
Focus-to-detector distance (cm):			115			(+) = normal font = optimal field setting						
Tube voltage (kVp):			110			**(−) = bold font = suboptimal field setting**						
Total filtrations:			See last table row.			Empty table cells correspond to values < 0.01						
Organ or tissue	**Organ dose D_T : Entrance dose K_E**							**(mSv × mGy^{-1})**				
	(+)	(−)	(+)	(−)	(+)	(−)	(+)	(−)	(+)	(−)	(+)	(−)
Skull	0.02	**0.24**	0.02	**0.28**	0.03	**0.30**	0.03	**0.32**	0.03	**0.36**	0.04	**0.39**
Brain		**0.02**		**0.03**		**0.03**		**0.04**		**0.04**		**0.05**
Oral mucosa	0.02	**0.27**	0.02	**0.31**	0.02	**0.34**	0.02	**0.36**	0.03	**0.41**	0.03	**0.44**
Salivary glands	0.03	**0.34**	0.03	**0.37**	0.04	**0.40**	0.04	**0.42**	0.05	**0.46**	0.05	**0.49**
Cervical spine	0.49	**0.83**	0.57	**0.97**	0.63	**1.08**	0.68	**1.16**	0.77	**1.31**	0.84	**1.43**
Thyroid	0.29	**0.23**	0.34	**0.26**	0.37	**0.29**	0.39	**0.31**	0.43	**0.35**	0.47	**0.38**
Thymus	0.14	**0.13**	0.17	**0.15**	0.20	**0.17**	0.21	**0.19**	0.25	**0.22**	0.29	**0.26**
Thoracic spine	0.26	**0.22**	0.32	**0.27**	0.36	**0.31**	0.39	**0.34**	0.46	**0.40**	0.53	**0.46**
Claviculae	0.54	**0.46**	0.62	**0.54**	0.68	**0.60**	0.73	**0.64**	0.80	**0.72**	0.86	**0.78**
Scapulae	1.10	**1.03**	1.22	**1.15**	1.30	**1.23**	1.36	**1.29**	1.45	**1.38**	1.50	**1.43**
Ribs	0.86	**0.72**	0.97	**0.82**	1.03	**0.90**	1.08	**0.95**	1.16	**1.05**	1.22	**1.12**
Oesophagus	0.16	**0.11**	0.20	**0.14**	0.22	**0.16**	0.24	**0.17**	0.29	**0.21**	0.33	**0.24**
Trachea	0.13	**0.26**	0.15	**0.30**	0.16	**0.33**	0.17	**0.35**	0.19	**0.40**	0.21	**0.44**
Lung	0.32	**0.22**	0.37	**0.27**	0.41	**0.30**	0.44	**0.32**	0.50	**0.37**	0.54	**0.42**
Mamma	3.06	**3.79**	3.42	**4.23**	3.67	**4.52**	3.85	**4.74**	4.19	**5.12**	4.46	**5.43**
Heart	0.17	**0.13**	0.20	**0.15**	0.23	**0.18**	0.25	**0.19**	0.29	**0.23**	0.33	**0.26**
Stomach	0.03	**0.03**	0.04	**0.04**	0.04	**0.05**	0.04	**0.05**	0.05	**0.06**	0.06	**0.07**
Spleen	0.02	**0.02**	0.03	**0.03**	0.03	**0.03**	0.03	**0.03**	0.04	**0.04**	0.05	**0.05**
Liver	0.33	**0.38**	0.38	**0.45**	0.42	**0.50**	0.45	**0.54**	0.51	**0.62**	0.56	**0.69**
Gall bladder	0.06	**0.19**	0.08	**0.22**	0.09	**0.25**	0.10	**0.28**	0.11	**0.32**	0.13	**0.37**
Pancreas	0.07	**0.07**	0.09	**0.08**	0.10	**0.09**	0.11	**0.10**	0.13	**0.12**	0.15	**0.14**
Small intestine		**0.06**		**0.07**		**0.08**		**0.09**	0.01	**0.11**	0.01	**0.12**
Large intestine		**0.09**	0.01	**0.11**	0.01	**0.13**	0.01	**0.14**	0.02	**0.16**	0.02	**0.18**
Sigma/Rectum										**0.01**		**0.01**
Lumbar spine	0.05	**0.16**	0.06	**0.20**	0.07	**0.22**	0.07	**0.25**	0.09	**0.29**	0.10	**0.33**
Pelvis		**0.04**		**0.05**		**0.05**		**0.06**		**0.07**		**0.08**
Adrenals	0.12	**0.11**	0.15	**0.14**	0.16	**0.15**	0.18	**0.17**	0.21	**0.20**	0.24	**0.23**
Kidneys	0.06	**0.15**	0.07	**0.18**	0.08	**0.20**	0.08	**0.22**	0.10	**0.25**	0.11	**0.28**
Urinary bladder												
Uterus		**0.04**		**0.05**		**0.05**		**0.06**		**0.07**		**0.09**
Ovaries		**0.02**		**0.02**		**0.02**		**0.03**		**0.03**		**0.03**
Prostate												
Testes												
Upper arm		**1.20**		**1.34**		**1.42**		**1.49**		**1.58**		**1.64**
Elbow		**1.28**		**1.42**		**1.51**		**1.57**		**1.67**		**1.73**
Forearm		**0.45**		**0.50**		**0.53**		**0.55**		**0.59**		**0.61**
Upper leg												
Knee												
Lower leg												
Lymph nodes	0.09	**0.15**	0.11	**0.17**	0.12	**0.19**	0.13	**0.21**	0.15	**0.24**	0.17	**0.27**
Skin	0.13	**0.20**	0.14	**0.21**	0.14	**0.22**	0.15	**0.22**	0.15	**0.23**	0.16	**0.24**
Muscle	0.08	**0.10**	0.09	**0.11**	0.10	**0.13**	0.10	**0.13**	0.11	**0.15**	0.13	**0.17**
Skeleton	0.16	**0.34**	0.19	**0.38**	0.20	**0.42**	0.21	**0.44**	0.23	**0.48**	0.25	**0.51**
Red bone marrow	0.09	**0.11**	0.10	**0.13**	0.11	**0.15**	0.12	**0.16**	0.14	**0.18**	0.15	**0.20**
Whole body dose	0.10	**0.14**	0.11	**0.16**	0.12	**0.18**	0.13	**0.19**	0.14	**0.21**	0.16	**0.23**
ED (ICRP 60)	0.25	**0.29**	0.28	**0.33**	0.31	**0.35**	0.33	**0.38**	0.36	**0.42**	0.39	**0.45**
ED (ICRP 103)	0.46	**0.56**	0.52	**0.63**	0.56	**0.68**	0.59	**0.72**	0.65	**0.78**	0.70	**0.84**
Total filtration (mm) Al	**1.5**		**2.5**		**3.5**		**4.5**		**3.5**		**3.5**	
Cu	**0**		**0**		**0**		**0**		**0.1**		**0.2**	

Table 9.116 Normalized organ doses—Chest r-lat—13–17 years

Beam projection:			r-lat		**Table cells:**							
Focus-to-detector distance (cm):			115		(+) = normal font = optimal field setting							
Tube voltage (kVp):			120		**(−) = bold font = suboptimal field setting**							
Total filtrations:			See last table row.		Empty table cells correspond to values < 0.01							
Organ or tissue	**Organ dose D_T : Entrance dose K_E**						**(mSv × mGy⁻¹)**					
	(+)	(−)	(+)	(−)	(+)	(−)	(+)	(−)	(+)	(−)	(+)	(−)
Skull	0.02	**0.25**	0.02	**0.29**	0.03	**0.31**	0.03	**0.33**	0.03	**0.36**	0.04	**0.39**
Brain		**0.03**		**0.03**		**0.04**		**0.04**		**0.05**		**0.05**
Oral mucosa	0.02	**0.29**	0.02	**0.33**	0.02	**0.36**	0.03	**0.38**	0.03	**0.42**	0.03	**0.46**
Salivary glands	0.03	**0.35**	0.04	**0.39**	0.04	**0.42**	0.04	**0.44**	0.05	**0.47**	0.06	**0.50**
Cervical spine	0.52	**0.89**	0.60	**1.03**	0.66	**1.12**	0.70	**1.20**	0.78	**1.33**	0.84	**1.44**
Thyroid	0.31	**0.24**	0.35	**0.28**	0.38	**0.30**	0.41	**0.32**	0.45	**0.36**	0.48	**0.39**
Thymus	0.16	**0.14**	0.19	**0.17**	0.21	**0.19**	0.23	**0.21**	0.27	**0.24**	0.30	**0.27**
Thoracic spine	0.29	**0.25**	0.35	**0.30**	0.39	**0.33**	0.42	**0.36**	0.49	**0.42**	0.55	**0.48**
Claviculae	0.57	**0.49**	0.65	**0.57**	0.70	**0.62**	0.74	**0.66**	0.81	**0.73**	0.86	**0.78**
Scapulae	1.13	**1.06**	1.25	**1.17**	1.32	**1.24**	1.37	**1.29**	1.44	**1.37**	1.48	**1.41**
Ribs	0.89	**0.75**	0.99	**0.85**	1.05	**0.92**	1.09	**0.97**	1.16	**1.06**	1.20	**1.12**
Oesophagus	0.18	**0.13**	0.22	**0.15**	0.24	**0.17**	0.26	**0.19**	0.30	**0.22**	0.34	**0.25**
Trachea	0.14	**0.28**	0.16	**0.32**	0.17	**0.35**	0.18	**0.37**	0.20	**0.42**	0.22	**0.45**
Lung	0.34	**0.25**	0.39	**0.29**	0.43	**0.32**	0.46	**0.34**	0.51	**0.39**	0.56	**0.43**
Mamma	3.22	**3.98**	3.56	**4.40**	3.80	**4.67**	3.97	**4.87**	4.29	**5.23**	4.54	**5.52**
Heart	0.19	**0.14**	0.22	**0.17**	0.25	**0.19**	0.27	**0.21**	0.31	**0.24**	0.34	**0.28**
Stomach	0.03	**0.04**	0.04	**0.05**	0.05	**0.05**	0.05	**0.06**	0.06	**0.07**	0.07	**0.08**
Spleen	0.02	**0.02**	0.03	**0.03**	0.03	**0.03**	0.04	**0.04**	0.04	**0.04**	0.05	**0.05**
Liver	0.35	**0.42**	0.40	**0.49**	0.44	**0.54**	0.47	**0.58**	0.52	**0.65**	0.57	**0.72**
Gall bladder	0.07	**0.21**	0.09	**0.25**	0.10	**0.27**	0.10	**0.30**	0.12	**0.34**	0.14	**0.39**
Pancreas	0.08	**0.08**	0.10	**0.09**	0.11	**0.10**	0.12	**0.11**	0.14	**0.13**	0.16	**0.15**
Small intestine		**0.07**		**0.08**	0.01	**0.09**	0.01	**0.10**	0.01	**0.11**	0.02	**0.13**
Large intestine	0.01	**0.10**	0.01	**0.12**	0.01	**0.14**	0.02	**0.15**	0.02	**0.17**	0.02	**0.19**
Sigma/Rectum								**0.01**		**0.01**		**0.01**
Lumbar spine	0.05	**0.18**	0.06	**0.22**	0.07	**0.24**	0.08	**0.27**	0.09	**0.31**	0.11	**0.35**
Pelvis		**0.04**		**0.05**		**0.06**		**0.06**		**0.07**		**0.08**
Adrenals	0.14	**0.13**	0.16	**0.15**	0.18	**0.17**	0.20	**0.19**	0.23	**0.22**	0.25	**0.25**
Kidneys	0.06	**0.17**	0.07	**0.20**	0.08	**0.22**	0.09	**0.23**	0.10	**0.27**	0.12	**0.30**
Urinary bladder												**0.01**
Uterus		**0.05**		**0.05**		**0.06**		**0.07**		**0.08**	0.01	**0.09**
Ovaries		**0.02**		**0.02**		**0.03**		**0.03**		**0.03**		**0.04**
Prostate												
Testes												
Upper arm		**1.24**		**1.36**		**1.44**		**1.49**		**1.57**		**1.62**
Elbow		**1.32**		**1.45**		**1.52**		**1.58**		**1.66**		**1.70**
Forearm		**0.46**		**0.51**		**0.54**		**0.55**		**0.58**		**0.60**
Upper leg												
Knee												
Lower leg												
Lymph nodes	0.10	**0.16**	0.12	**0.19**	0.13	**0.21**	0.14	**0.22**	0.16	**0.25**	0.17	**0.28**
Skin	0.13	**0.20**	0.14	**0.21**	0.15	**0.22**	0.15	**0.23**	0.16	**0.24**	0.16	**0.24**
Muscle	0.08	**0.11**	0.09	**0.12**	0.10	**0.13**	0.11	**0.14**	0.12	**0.16**	0.13	**0.17**
Skeleton	0.17	**0.35**	0.19	**0.40**	0.21	**0.43**	0.22	**0.45**	0.24	**0.48**	0.25	**0.51**
Red bone marrow	0.10	**0.13**	0.11	**0.15**	0.12	**0.16**	0.13	**0.17**	0.14	**0.19**	0.16	**0.21**
Whole body dose	0.10	**0.15**	0.12	**0.17**	0.13	**0.19**	0.13	**0.20**	0.15	**0.22**	0.16	**0.23**
ED (ICRP 60)	0.27	**0.31**	0.30	**0.34**	0.32	**0.37**	0.34	**0.39**	0.37	**0.43**	0.40	**0.46**
ED (ICRP 103)	0.49	**0.59**	0.55	**0.66**	0.59	**0.71**	0.62	**0.74**	0.67	**0.81**	0.72	**0.86**
Total filtration (mm) Al	**1.5**		**2.5**		**3.5**		**4.5**		**3.5**		**3.5**	
Cu	**0**		**0**		**0**		**0**		**0.1**		**0.2**	

Table 9.117 Normalized organ doses—Chest r-lat—> 17 years

Beam projection:			r-lat			**Table cells:**						
Focus-to-detector distance (cm):			115			(+) = normal font = optimal field setting						
Tube voltage (kVp):			60			**(−) = bold font = suboptimal field setting**						
Total filtrations:			See last table row.			Empty table cells correspond to values < 0.01						
Organ or tissue	**Organ dose D_T : Entrance dose K_E**							**(mSv × mGy^{-1})**				
	(+)	(−)	(+)	(−)	(+)	(−)	(+)	(−)	(+)	(−)	(+)	(−)
Skull	0.01	**0.14**	0.01	**0.18**	0.02	**0.21**	0.02	**0.23**	0.02	**0.28**	0.02	**0.26**
Brain		**0.01**		**0.01**		**0.02**		**0.02**		**0.03**		**0.02**
Oral mucosa		**0.13**		**0.15**		**0.18**	0.01	**0.20**	0.01	**0.23**	0.01	**0.21**
Salivary glands	0.02	**0.23**	0.02	**0.27**	0.03	**0.30**	0.03	**0.32**	0.04	**0.35**	0.03	**0.34**
Cervical spine	0.22	**0.31**	0.29	**0.41**	0.34	**0.49**	0.39	**0.56**	0.48	**0.69**	0.43	**0.62**
Thyroid	0.13	**0.07**	0.16	**0.09**	0.18	**0.10**	0.20	**0.12**	0.24	**0.14**	0.22	**0.13**
Thymus	0.03	**0.02**	0.04	**0.03**	0.05	**0.04**	0.06	**0.05**	0.08	**0.06**	0.06	**0.05**
Thoracic spine	0.05	**0.04**	0.07	**0.05**	0.09	**0.07**	0.10	**0.08**	0.14	**0.10**	0.10	**0.08**
Claviculae	0.23	**0.17**	0.30	**0.22**	0.35	**0.26**	0.39	**0.30**	0.47	**0.36**	0.44	**0.33**
Scapulae	0.67	**0.60**	0.81	**0.73**	0.92	**0.83**	1.00	**0.91**	1.15	**1.05**	1.12	**1.02**
Ribs	0.47	**0.32**	0.58	**0.41**	0.65	**0.47**	0.71	**0.53**	0.82	**0.64**	0.79	**0.60**
Oesophagus	0.04	**0.02**	0.05	**0.03**	0.07	**0.04**	0.08	**0.04**	0.10	**0.06**	0.08	**0.04**
Trachea	0.07	**0.09**	0.09	**0.12**	0.10	**0.14**	0.11	**0.16**	0.14	**0.20**	0.13	**0.18**
Lung	0.12	**0.07**	0.16	**0.09**	0.19	**0.11**	0.21	**0.13**	0.26	**0.16**	0.24	**0.14**
Mamma	0.20	**0.26**	0.24	**0.30**	0.26	**0.33**	0.28	**0.36**	0.32	**0.40**	0.31	**0.39**
Heart	0.04	**0.02**	0.05	**0.03**	0.06	**0.04**	0.07	**0.05**	0.10	**0.06**	0.08	**0.05**
Stomach												
Spleen												
Liver	0.11	**0.12**	0.14	**0.16**	0.17	**0.19**	0.19	**0.22**	0.23	**0.27**	0.21	**0.24**
Gall bladder		**0.03**	0.01	**0.05**	0.02	**0.06**	0.02	**0.06**	0.03	**0.09**	0.02	**0.07**
Pancreas	0.01		0.01	**0.01**	0.02	**0.01**	0.02	**0.02**	0.03	**0.02**	0.02	**0.02**
Small intestine		**0.01**		**0.02**		**0.02**		**0.02**		**0.03**		**0.02**
Large intestine		**0.02**		**0.03**		**0.04**		**0.04**		**0.05**		**0.04**
Sigma/Rectum												
Lumbar spine		**0.02**		**0.03**		**0.04**	0.01	**0.05**	0.02	**0.06**	0.01	**0.05**
Pelvis				**0.01**		**0.02**		**0.02**		**0.02**		**0.02**
Adrenals	0.02	**0.02**	0.03	**0.03**	0.04	**0.04**	0.05	**0.04**	0.06	**0.05**	0.05	**0.04**
Kidneys	0.01	**0.04**	0.01	**0.05**	0.02	**0.06**	0.02	**0.07**	0.03	**0.09**	0.02	**0.08**
Urinary bladder												
Uterus												
Ovaries												
Prostate												
Testes												
Upper arm		**0.78**		**0.94**		**1.06**		**1.15**		**1.31**		**1.28**
Elbow		**0.84**		**1.02**		**1.14**		**1.23**		**1.40**		**1.38**
Forearm		**0.29**		**0.35**		**0.40**		**0.43**		**0.49**		**0.48**
Upper leg												
Knee												
Lower leg												
Lymph nodes	0.03	**0.05**	0.04	**0.07**	0.05	**0.08**	0.06	**0.08**	0.07	**0.10**	0.06	**0.09**
Skin	0.08	**0.15**	0.09	**0.16**	0.09	**0.17**	0.10	**0.17**	0.10	**0.18**	0.10	**0.18**
Muscle	0.03	**0.04**	0.04	**0.05**	0.05	**0.06**	0.05	**0.06**	0.06	**0.07**	0.06	**0.07**
Skeleton	0.08	**0.18**	0.10	**0.22**	0.12	**0.25**	0.13	**0.28**	0.15	**0.32**	0.14	**0.31**
Red bone marrow	0.04	**0.04**	0.05	**0.05**	0.06	**0.06**	0.06	**0.07**	0.08	**0.08**	0.07	**0.07**
Whole body dose	0.04	**0.07**	0.05	**0.08**	0.06	**0.09**	0.07	**0.10**	0.08	**0.12**	0.07	**0.11**
ED (ICRP 60)	0.05	**0.04**	0.06	**0.05**	0.07	**0.06**	0.07	**0.07**	0.09	**0.08**	0.08	**0.08**
ED (ICRP 103)	0.06	**0.06**	0.07	**0.08**	0.08	**0.09**	0.09	**0.10**	0.11	**0.12**	0.10	**0.11**
Total filtration (mm) **Al**	**1.5**		**2.5**		**3.5**		**4.5**		**3.5**		**3.5**	
Cu	**0**		**0**		**0**		**0**		**0.1**		**0.2**	

Table 9.118 Normalized organ doses—Chest r-lat—> 17 years

Beam projection:			r-lat			**Table cells:**						
Focus-to-detector distance (cm):			115			(+) = normal font = optimal field setting						
Tube voltage (kVp):			70			**(−) = bold font = suboptimal field setting**						
Total filtrations:			See last table row.			Empty table cells correspond to values < 0.01						
Organ or tissue	**Organ dose D_T : Entrance dose K_E**							**(mSv × mGy^{-1})**				
	(+)	(−)	(+)	(−)	(+)	(−)	(+)	(−)	(+)	(−)	(+)	(−)
Skull	0.01	**0.17**	0.02	**0.21**	0.02	**0.24**	0.02	**0.27**	0.03	**0.32**	0.04	**0.36**
Brain		**0.01**		**0.02**		**0.02**		**0.03**		**0.04**		**0.04**
Oral mucosa		**0.15**	0.01	**0.18**	0.01	**0.21**	0.01	**0.23**	0.02	**0.27**	0.02	**0.30**
Salivary glands	0.02	**0.26**	0.03	**0.30**	0.03	**0.33**	0.04	**0.35**	0.04	**0.39**	0.05	**0.42**
Cervical spine	0.28	**0.40**	0.36	**0.51**	0.42	**0.60**	0.46	**0.67**	0.56	**0.82**	0.66	**0.96**
Thyroid	0.15	**0.08**	0.19	**0.11**	0.21	**0.12**	0.23	**0.14**	0.27	**0.17**	0.31	**0.19**
Thymus	0.04	**0.04**	0.06	**0.05**	0.07	**0.06**	0.08	**0.07**	0.10	**0.09**	0.13	**0.11**
Thoracic spine	0.08	**0.06**	0.10	**0.08**	0.13	**0.10**	0.15	**0.11**	0.19	**0.15**	0.24	**0.19**
Claviculae	0.28	**0.21**	0.36	**0.27**	0.41	**0.32**	0.46	**0.36**	0.55	**0.43**	0.63	**0.50**
Scapulae	0.75	**0.68**	0.91	**0.82**	1.01	**0.92**	1.09	**1.00**	1.24	**1.14**	1.36	**1.26**
Ribs	0.54	**0.38**	0.65	**0.48**	0.73	**0.55**	0.78	**0.61**	0.89	**0.72**	0.98	**0.83**
Oesophagus	0.06	**0.03**	0.08	**0.04**	0.09	**0.05**	0.11	**0.06**	0.13	**0.08**	0.17	**0.10**
Trachea	0.08	**0.12**	0.10	**0.15**	0.12	**0.17**	0.13	**0.19**	0.15	**0.23**	0.18	**0.27**
Lung	0.15	**0.09**	0.19	**0.12**	0.23	**0.14**	0.25	**0.16**	0.30	**0.20**	0.35	**0.24**
Mamma	0.23	**0.29**	0.26	**0.33**	0.29	**0.37**	0.31	**0.39**	0.35	**0.44**	0.38	**0.48**
Heart	0.05	**0.04**	0.07	**0.05**	0.09	**0.06**	0.10	**0.07**	0.13	**0.09**	0.16	**0.11**
Stomach							0.01	**0.01**	0.01	**0.01**	0.02	**0.02**
Spleen											0.01	
Liver	0.14	**0.16**	0.17	**0.20**	0.20	**0.24**	0.23	**0.27**	0.27	**0.34**	0.32	**0.40**
Gall bladder	0.02	**0.05**	0.02	**0.06**	0.02	**0.08**	0.03	**0.09**	0.04	**0.12**	0.05	**0.14**
Pancreas	0.02	**0.01**	0.02	**0.02**	0.03	**0.02**	0.03	**0.03**	0.04	**0.03**	0.05	**0.04**
Small intestine		**0.02**		**0.02**		**0.03**		**0.03**		**0.04**		**0.05**
Large intestine		**0.03**		**0.04**		**0.05**		**0.06**		**0.07**		**0.09**
Sigma/Rectum												
Lumbar spine		**0.04**	0.01	**0.05**	0.02	**0.06**	0.02	**0.07**	0.02	**0.09**	0.03	**0.12**
Pelvis		**0.01**		**0.02**		**0.02**		**0.03**		**0.03**		**0.04**
Adrenals	0.04	**0.03**	0.05	**0.04**	0.06	**0.05**	0.07	**0.06**	0.09	**0.08**	0.11	**0.09**
Kidneys	0.02	**0.05**	0.02	**0.07**	0.02	**0.08**	0.03	**0.09**	0.04	**0.12**	0.04	**0.14**
Urinary bladder												
Uterus												**0.01**
Ovaries												**0.01**
Prostate												
Testes												
Upper arm		**0.88**		**1.04**		**1.16**		**1.25**		**1.41**		**1.54**
Elbow		**0.95**		**1.12**		**1.25**		**1.34**		**1.51**		**1.64**
Forearm		**0.33**		**0.39**		**0.43**		**0.47**		**0.52**		**0.57**
Upper leg												
Knee												
Lower leg												
Lymph nodes	0.04	**0.06**	0.05	**0.08**	0.06	**0.09**	0.07	**0.10**	0.08	**0.12**	0.10	**0.15**
Skin	0.09	**0.16**	0.09	**0.17**	0.10	**0.18**	0.10	**0.18**	0.11	**0.19**	0.11	**0.20**
Muscle	0.04	**0.05**	0.05	**0.06**	0.05	**0.07**	0.06	**0.07**	0.07	**0.09**	0.08	**0.10**
Skeleton	0.10	**0.21**	0.12	**0.25**	0.13	**0.29**	0.14	**0.31**	0.17	**0.36**	0.19	**0.41**
Red bone marrow	0.05	**0.05**	0.06	**0.07**	0.07	**0.08**	0.08	**0.08**	0.09	**0.10**	0.10	**0.12**
Whole body dose	0.05	**0.08**	0.06	**0.09**	0.07	**0.11**	0.08	**0.12**	0.09	**0.14**	0.10	**0.15**
ED (ICRP 60)	0.06	**0.05**	0.07	**0.07**	0.08	**0.08**	0.09	**0.08**	0.10	**0.10**	0.12	**0.12**
ED (ICRP 103)	0.07	**0.08**	0.09	**0.09**	0.10	**0.11**	0.11	**0.12**	0.13	**0.14**	0.15	**0.16**
Total filtration (mm) **Al**	**1.5**		**2.5**		**3.5**		**4.5**		**3.5**		**3.5**	
Cu	**0**		**0**		**0**		**0**		**0.1**		**0.2**	

Table 9.119 Normalized organ doses—Chest r-lat—> 17 years

Beam projection:			r-lat			**Table cells:**						
Focus-to-detector distance (cm):			115			(+) = normal font = optimal field setting						
Tube voltage (kVp):			80			**(−) = bold font = suboptimal field setting**						
Total filtrations:			See last table row.			Empty table cells correspond to values < 0.01						
Organ or tissue	**Organ dose D_T : Entrance dose K_E**							**(mSv x mGy^{-1})**				
	(+)	(−)	(+)	(−)	(+)	(−)	(+)	(−)	(+)	(−)	(+)	(−)
Skull	0.02	**0.20**	0.02	**0.24**	0.03	**0.27**	0.03	**0.30**	0.03	**0.35**	0.04	**0.40**
Brain		**0.02**		**0.03**		**0.03**		**0.04**		**0.04**		**0.05**
Oral mucosa	0.01	**0.17**	0.01	**0.21**	0.02	**0.24**	0.02	**0.26**	0.02	**0.30**	0.03	**0.34**
Salivary glands	0.03	**0.28**	0.03	**0.32**	0.04	**0.35**	0.04	**0.37**	0.05	**0.42**	0.06	**0.45**
Cervical spine	0.33	**0.48**	0.42	**0.61**	0.48	**0.70**	0.53	**0.78**	0.63	**0.93**	0.73	**1.07**
Thyroid	0.17	**0.10**	0.21	**0.13**	0.24	**0.15**	0.26	**0.16**	0.30	**0.19**	0.34	**0.22**
Thymus	0.06	**0.05**	0.08	**0.06**	0.09	**0.08**	0.10	**0.09**	0.13	**0.11**	0.16	**0.14**
Thoracic spine	0.11	**0.08**	0.14	**0.11**	0.17	**0.13**	0.19	**0.15**	0.24	**0.19**	0.29	**0.24**
Claviculae	0.33	**0.25**	0.41	**0.32**	0.47	**0.37**	0.51	**0.41**	0.60	**0.49**	0.68	**0.56**
Scapulae	0.83	**0.76**	0.98	**0.90**	1.08	**0.99**	1.16	**1.07**	1.29	**1.20**	1.40	**1.30**
Ribs	0.60	**0.45**	0.71	**0.55**	0.78	**0.62**	0.84	**0.68**	0.94	**0.79**	1.03	**0.89**
Oesophagus	0.08	**0.05**	0.10	**0.06**	0.12	**0.07**	0.13	**0.08**	0.17	**0.11**	0.20	**0.13**
Trachea	0.10	**0.14**	0.12	**0.18**	0.14	**0.20**	0.15	**0.22**	0.17	**0.27**	0.19	**0.31**
Lung	0.18	**0.12**	0.23	**0.15**	0.26	**0.17**	0.29	**0.19**	0.35	**0.23**	0.40	**0.28**
Mamma	0.25	**0.31**	0.29	**0.36**	0.32	**0.40**	0.34	**0.42**	0.38	**0.47**	0.41	**0.50**
Heart	0.07	**0.05**	0.09	**0.06**	0.11	**0.08**	0.13	**0.09**	0.16	**0.11**	0.19	**0.13**
Stomach			0.01	**0.01**	0.01	**0.01**	0.01	**0.02**	0.02	**0.02**	0.02	**0.03**
Spleen									0.01	**0.01**	0.02	**0.01**
Liver	0.17	**0.20**	0.21	**0.25**	0.24	**0.29**	0.26	**0.33**	0.31	**0.40**	0.36	**0.46**
Gall bladder	0.02	**0.07**	0.03	**0.09**	0.03	**0.10**	0.04	**0.12**	0.05	**0.15**	0.06	**0.18**
Pancreas	0.02	**0.02**	0.03	**0.03**	0.04	**0.03**	0.04	**0.04**	0.05	**0.05**	0.07	**0.06**
Small intestine		**0.02**		**0.03**		**0.04**		**0.04**		**0.05**		**0.06**
Large intestine		**0.04**		**0.05**		**0.06**		**0.07**		**0.09**		**0.11**
Sigma/Rectum												
Lumbar spine	0.01	**0.05**	0.02	**0.07**	0.02	**0.08**	0.03	**0.10**	0.03	**0.12**	0.04	**0.15**
Pelvis		**0.02**		**0.02**		**0.03**		**0.03**		**0.04**		**0.05**
Adrenals	0.05	**0.04**	0.06	**0.06**	0.08	**0.07**	0.09	**0.08**	0.11	**0.10**	0.13	**0.12**
Kidneys	0.02	**0.07**	0.03	**0.09**	0.03	**0.10**	0.04	**0.11**	0.05	**0.14**	0.05	**0.16**
Urinary bladder												
Uterus										**0.01**		**0.02**
Ovaries								**0.01**		**0.01**		**0.02**
Prostate												
Testes												
Upper arm		**0.96**		**1.13**		**1.24**		**1.33**		**1.47**		**1.59**
Elbow		**1.04**		**1.21**		**1.33**		**1.42**		**1.57**		**1.69**
Forearm		**0.36**		**0.42**		**0.46**		**0.49**		**0.55**		**0.59**
Upper leg												
Knee												
Lower leg												
Lymph nodes	0.05	**0.08**	0.06	**0.10**	0.07	**0.11**	0.08	**0.12**	0.10	**0.15**	0.11	**0.17**
Skin	0.09	**0.16**	0.10	**0.17**	0.10	**0.18**	0.11	**0.19**	0.11	**0.20**	0.12	**0.21**
Muscle	0.05	**0.06**	0.06	**0.07**	0.06	**0.08**	0.07	**0.08**	0.08	**0.10**	0.09	**0.11**
Skeleton	0.11	**0.23**	0.13	**0.28**	0.14	**0.31**	0.16	**0.34**	0.18	**0.39**	0.20	**0.43**
Red bone marrow	0.06	**0.06**	0.07	**0.08**	0.08	**0.09**	0.09	**0.10**	0.10	**0.12**	0.12	**0.14**
Whole body dose	0.06	**0.09**	0.07	**0.11**	0.08	**0.12**	0.09	**0.13**	0.10	**0.15**	0.11	**0.17**
ED (ICRP 60)	0.07	**0.06**	0.08	**0.08**	0.09	**0.09**	0.10	**0.10**	0.12	**0.12**	0.13	**0.14**
ED (ICRP 103)	0.08	**0.09**	0.10	**0.11**	0.11	**0.12**	0.12	**0.14**	0.14	**0.16**	0.16	**0.18**
Total filtration (mm) **Al**	**1.5**		**2.5**		**3.5**		**4.5**		**3.5**		**3.5**	
Cu	**0**		**0**		**0**		**0**		**0.1**		**0.2**	

Table 9.120 Normalized organ doses—Chest r-lat—> 17 years

Beam projection:			r-lat			**Table cells:**						
Focus-to-detector distance (cm):			115			(+) = normal font = optimal field setting						
Tube voltage (kVp):			110			**(−) = bold font = suboptimal field setting**						
Total filtrations:			See last table row.			Empty table cells correspond to values < 0.01						
Organ or tissue	**Organ dose D_T : Entrance dose K_E**						**(mSv × mGy^{-1})**					
	(+)	(−)	(+)	(−)	(+)	(−)	(+)	(−)	(+)	(−)	(+)	(−)
Skull	0.03	**0.26**	0.03	**0.31**	0.03	**0.33**	0.04	**0.36**	0.04	**0.40**	0.05	**0.43**
Brain		**0.04**		**0.04**		**0.05**		**0.05**		**0.06**		**0.07**
Oral mucosa	0.02	**0.24**	0.02	**0.28**	0.02	**0.31**	0.03	**0.33**	0.03	**0.37**	0.04	**0.40**
Salivary glands	0.04	**0.35**	0.05	**0.39**	0.05	**0.42**	0.06	**0.44**	0.06	**0.47**	0.07	**0.50**
Cervical spine	0.47	**0.69**	0.55	**0.81**	0.61	**0.90**	0.66	**0.97**	0.74	**1.10**	0.81	**1.21**
Thyroid	0.24	**0.15**	0.28	**0.18**	0.30	**0.20**	0.32	**0.21**	0.36	**0.24**	0.39	**0.26**
Thymus	0.11	**0.09**	0.13	**0.11**	0.14	**0.13**	0.16	**0.14**	0.19	**0.16**	0.22	**0.19**
Thoracic spine	0.19	**0.15**	0.23	**0.19**	0.26	**0.21**	0.29	**0.24**	0.34	**0.28**	0.39	**0.32**
Claviculae	0.44	**0.36**	0.52	**0.42**	0.57	**0.47**	0.61	**0.50**	0.68	**0.57**	0.73	**0.62**
Scapulae	0.99	**0.92**	1.12	**1.03**	1.19	**1.11**	1.25	**1.17**	1.34	**1.26**	1.40	**1.32**
Ribs	0.73	**0.59**	0.82	**0.68**	0.88	**0.74**	0.92	**0.79**	0.99	**0.88**	1.04	**0.94**
Oesophagus	0.14	**0.09**	0.16	**0.11**	0.19	**0.13**	0.20	**0.14**	0.24	**0.17**	0.28	**0.19**
Trachea	0.14	**0.21**	0.16	**0.25**	0.17	**0.27**	0.18	**0.29**	0.20	**0.33**	0.22	**0.37**
Lung	0.27	**0.18**	0.31	**0.22**	0.35	**0.24**	0.37	**0.27**	0.42	**0.31**	0.47	**0.34**
Mamma	0.31	**0.39**	0.35	**0.43**	0.37	**0.46**	0.39	**0.48**	0.43	**0.52**	0.45	**0.55**
Heart	0.13	**0.09**	0.15	**0.11**	0.17	**0.12**	0.19	**0.14**	0.22	**0.16**	0.25	**0.19**
Stomach	0.02	**0.02**	0.02	**0.02**	0.02	**0.03**	0.03	**0.03**	0.03	**0.04**	0.04	**0.04**
Spleen	0.01	**0.01**	0.01	**0.01**	0.02	**0.02**	0.02	**0.02**	0.02	**0.02**	0.03	**0.03**
Liver	0.24	**0.31**	0.28	**0.37**	0.31	**0.41**	0.34	**0.44**	0.38	**0.51**	0.42	**0.57**
Gall bladder	0.04	**0.12**	0.05	**0.14**	0.05	**0.16**	0.06	**0.18**	0.07	**0.21**	0.08	**0.24**
Pancreas	0.05	**0.04**	0.06	**0.05**	0.06	**0.06**	0.07	**0.07**	0.09	**0.08**	0.10	**0.09**
Small intestine		**0.04**		**0.05**		**0.06**		**0.07**		**0.08**		**0.09**
Large intestine		**0.07**		**0.09**		**0.10**		**0.11**	0.01	**0.13**	0.01	**0.15**
Sigma/Rectum												
Lumbar spine	0.03	**0.10**	0.03	**0.12**	0.04	**0.14**	0.04	**0.15**	0.05	**0.18**	0.06	**0.21**
Pelvis		**0.03**		**0.04**		**0.05**		**0.05**		**0.06**		**0.07**
Adrenals	0.09	**0.08**	0.11	**0.10**	0.12	**0.11**	0.13	**0.12**	0.16	**0.15**	0.18	**0.17**
Kidneys	0.04	**0.11**	0.04	**0.13**	0.05	**0.15**	0.06	**0.16**	0.06	**0.19**	0.07	**0.21**
Urinary bladder												
Uterus		**0.01**		**0.01**		**0.02**		**0.02**		**0.02**		**0.03**
Ovaries		**0.01**		**0.01**		**0.02**		**0.02**		**0.02**		**0.03**
Prostate												
Testes												
Upper arm		**1.14**		**1.28**		**1.36**		**1.43**		**1.53**		**1.59**
Elbow		**1.22**		**1.36**		**1.45**		**1.51**		**1.61**		**1.68**
Forearm		**0.42**		**0.47**		**0.50**		**0.53**		**0.56**		**0.58**
Upper leg												
Knee												
Lower leg												
Lymph nodes	0.08	**0.12**	0.09	**0.14**	0.10	**0.15**	0.11	**0.17**	0.13	**0.19**	0.14	**0.21**
Skin	0.10	**0.18**	0.11	**0.19**	0.11	**0.20**	0.12	**0.20**	0.12	**0.21**	0.13	**0.22**
Muscle	0.06	**0.08**	0.07	**0.09**	0.08	**0.10**	0.08	**0.11**	0.09	**0.12**	0.10	**0.13**
Skeleton	0.14	**0.30**	0.16	**0.34**	0.17	**0.37**	0.18	**0.39**	0.20	**0.43**	0.21	**0.45**
Red bone marrow	0.08	**0.10**	0.10	**0.11**	0.11	**0.13**	0.11	**0.14**	0.13	**0.16**	0.14	**0.18**
Whole body dose	0.08	**0.12**	0.09	**0.14**	0.10	**0.15**	0.11	**0.16**	0.12	**0.18**	0.13	**0.20**
ED (ICRP 60)	0.09	**0.10**	0.11	**0.11**	0.12	**0.12**	0.13	**0.13**	0.15	**0.15**	0.16	**0.17**
ED (ICRP 103)	0.11	**0.13**	0.13	**0.15**	0.15	**0.16**	0.16	**0.18**	0.17	**0.20**	0.19	**0.22**
Total filtration (mm) **Al**	**1.5**		**2.5**		**3.5**		**4.5**		**3.5**		**3.5**	
Cu	**0**		**0**		**0**		**0**		**0.1**		**0.2**	

Table 9.121 Normalized organ doses—Chest r-lat—> 17 years

Beam projection:			r-lat			**Table cells:**						
Focus-to-detector distance (cm):			115			(+) = normal font = optimal field setting						
Tube voltage (kVp):			120			**(−) = bold font = suboptimal field setting**						
Total filtrations:			See last table row.			Empty table cells correspond to values < 0.01						
Organ or tissue	**Organ dose D_T : Entrance dose K_E**							**(mSv × mGy^{-1})**				
	(+)	(−)	(+)	(−)	(+)	(−)	(+)	(−)	(+)	(−)	(+)	(−)
Skull	0.03	**0.28**	0.03	**0.32**	0.04	**0.35**	0.04	**0.37**	0.04	**0.40**	0.05	**0.43**
Brain		**0.04**		**0.05**		**0.05**		**0.06**		**0.07**		**0.08**
Oral mucosa	0.02	**0.26**	0.02	**0.30**	0.03	**0.32**	0.03	**0.35**	0.03	**0.38**	0.04	**0.42**
Salivary glands	0.04	**0.37**	0.05	**0.40**	0.06	**0.43**	0.06	**0.45**	0.07	**0.48**	0.07	**0.51**
Cervical spine	0.50	**0.74**	0.58	**0.86**	0.64	**0.94**	0.68	**1.00**	0.76	**1.12**	0.82	**1.22**
Thyroid	0.26	**0.16**	0.29	**0.19**	0.32	**0.21**	0.34	**0.22**	0.37	**0.25**	0.40	**0.27**
Thymus	0.12	**0.10**	0.14	**0.12**	0.16	**0.14**	0.17	**0.15**	0.20	**0.18**	0.23	**0.20**
Thoracic spine	0.21	**0.17**	0.25	**0.21**	0.28	**0.23**	0.31	**0.26**	0.36	**0.30**	0.41	**0.34**
Claviculae	0.47	**0.38**	0.54	**0.44**	0.59	**0.49**	0.62	**0.52**	0.68	**0.58**	0.73	**0.62**
Scapulae	1.03	**0.95**	1.14	**1.06**	1.21	**1.13**	1.26	**1.18**	1.33	**1.25**	1.38	**1.30**
Ribs	0.75	**0.62**	0.84	**0.71**	0.89	**0.77**	0.93	**0.81**	0.99	**0.89**	1.03	**0.94**
Oesophagus	0.15	**0.10**	0.18	**0.12**	0.20	**0.14**	0.22	**0.15**	0.26	**0.18**	0.29	**0.21**
Trachea	0.15	**0.23**	0.17	**0.26**	0.18	**0.29**	0.19	**0.31**	0.21	**0.35**	0.23	**0.38**
Lung	0.29	**0.20**	0.33	**0.24**	0.37	**0.26**	0.39	**0.28**	0.44	**0.32**	0.48	**0.36**
Mamma	0.33	**0.41**	0.36	**0.45**	0.39	**0.48**	0.40	**0.50**	0.44	**0.53**	0.46	**0.56**
Heart	0.14	**0.10**	0.17	**0.12**	0.19	**0.14**	0.20	**0.15**	0.23	**0.18**	0.26	**0.20**
Stomach	0.02	**0.02**	0.02	**0.03**	0.03	**0.03**	0.03	**0.03**	0.04	**0.04**	0.04	**0.05**
Spleen	0.01	**0.01**	0.02	**0.02**	0.02	**0.02**	0.02	**0.02**	0.03	**0.02**	0.03	**0.03**
Liver	0.26	**0.34**	0.30	**0.40**	0.33	**0.44**	0.36	**0.47**	0.40	**0.54**	0.44	**0.59**
Gall bladder	0.05	**0.14**	0.05	**0.16**	0.06	**0.18**	0.07	**0.20**	0.08	**0.23**	0.09	**0.26**
Pancreas	0.05	**0.05**	0.06	**0.06**	0.07	**0.07**	0.08	**0.07**	0.09	**0.09**	0.11	**0.10**
Small intestine		**0.05**		**0.06**		**0.07**		**0.07**		**0.09**		**0.10**
Large intestine		**0.08**		**0.10**		**0.11**		**0.12**	0.01	**0.14**	0.01	**0.15**
Sigma/Rectum												
Lumbar spine	0.03	**0.11**	0.04	**0.13**	0.04	**0.15**	0.05	**0.17**	0.06	**0.19**	0.07	**0.22**
Pelvis		**0.04**		**0.04**		**0.05**		**0.05**		**0.06**		**0.07**
Adrenals	0.10	**0.09**	0.12	**0.11**	0.13	**0.12**	0.14	**0.14**	0.17	**0.16**	0.19	**0.18**
Kidneys	0.04	**0.12**	0.05	**0.14**	0.06	**0.16**	0.06	**0.17**	0.07	**0.20**	0.08	**0.22**
Urinary bladder												
Uterus		**0.01**		**0.02**		**0.02**		**0.02**		**0.02**		**0.03**
Ovaries		**0.01**		**0.02**		**0.02**		**0.02**		**0.03**		**0.03**
Prostate												
Testes												
Upper arm		**1.18**		**1.30**		**1.38**		**1.44**		**1.52**		**1.57**
Elbow		**1.26**		**1.38**		**1.46**		**1.52**		**1.60**		**1.65**
Forearm		**0.44**		**0.48**		**0.51**		**0.53**		**0.56**		**0.57**
Upper leg												
Knee												
Lower leg												
Lymph nodes	0.09	**0.13**	0.10	**0.15**	0.11	**0.17**	0.12	**0.18**	0.13	**0.20**	0.15	**0.22**
Skin	0.11	**0.19**	0.11	**0.20**	0.12	**0.20**	0.12	**0.21**	0.13	**0.22**	0.13	**0.22**
Muscle	0.07	**0.09**	0.08	**0.10**	0.08	**0.11**	0.09	**0.11**	0.10	**0.13**	0.11	**0.14**
Skeleton	0.14	**0.31**	0.16	**0.35**	0.17	**0.38**	0.18	**0.40**	0.20	**0.43**	0.21	**0.45**
Red bone marrow	0.09	**0.11**	0.10	**0.12**	0.11	**0.14**	0.12	**0.15**	0.14	**0.17**	0.15	**0.19**
Whole body dose	0.09	**0.13**	0.10	**0.15**	0.11	**0.16**	0.11	**0.17**	0.12	**0.19**	0.13	**0.20**
ED (ICRP 60)	0.10	**0.10**	0.12	**0.12**	0.13	**0.13**	0.14	**0.14**	0.15	**0.16**	0.17	**0.18**
ED (ICRP 103)	0.12	**0.14**	0.14	**0.16**	0.15	**0.17**	0.16	**0.18**	0.18	**0.21**	0.20	**0.23**
Total filtration (mm) **Al**	**1.5**		**2.5**		**3.5**		**4.5**		**3.5**		**3.5**	
Cu	**0**		**0**		**0**		**0**		**0.1**		**0.2**	

Table 9.122 Normalized organ doses—Chest ap—< 1 year

Beam projection:			ap		**Table cells:**							
Focus-to-detector distance (cm):			150		(+) = normal font = optimal field setting							
Tube voltage (kVp):			50		**(−) = bold font = suboptimal field setting**							
Total filtrations:			See last table row.		Empty table cells correspond to values < 0.01							
Organ or tissue	**Organ dose D_T : Entrance dose K_E**						**($mSv \times mGy^{-1}$)**					
	(+)	(−)	(+)	(−)	(+)	(−)	(+)	(−)	(+)	(−)	(+)	(−)
Skull	0.02	**0.20**	0.02	**0.24**	0.03	**0.27**	0.03	**0.29**	0.03	**0.33**	0.04	**0.36**
Brain		**0.01**		**0.02**		**0.02**		**0.03**		**0.03**		**0.04**
Oral mucosa	0.02	**0.54**	0.03	**0.63**	0.04	**0.70**	0.04	**0.74**	0.05	**0.82**	0.06	**0.89**
Salivary glands	0.05	**0.47**	0.06	**0.56**	0.07	**0.63**	0.08	**0.68**	0.09	**0.76**	0.10	**0.83**
Cervical spine	0.30	**0.49**	0.39	**0.65**	0.46	**0.76**	0.52	**0.86**	0.62	**1.04**	0.71	**1.21**
Thyroid	0.85	**0.91**	0.96	**1.04**	1.02	**1.12**	1.07	**1.17**	1.14	**1.27**	1.20	**1.35**
Thymus	0.74	**0.76**	0.87	**0.89**	0.95	**0.98**	1.01	**1.04**	1.11	**1.15**	1.20	**1.23**
Thoracic spine	0.23	**0.26**	0.31	**0.35**	0.37	**0.42**	0.43	**0.48**	0.53	**0.61**	0.64	**0.73**
Claviculae	1.87	**1.94**	2.19	**2.29**	2.40	**2.51**	2.56	**2.67**	2.80	**2.94**	3.00	**3.16**
Scapulae	0.42	**0.45**	0.55	**0.59**	0.66	**0.70**	0.74	**0.80**	0.91	**0.97**	1.06	**1.14**
Ribs	1.28	**1.43**	1.48	**1.66**	1.61	**1.81**	1.71	**1.92**	1.87	**2.11**	2.01	**2.27**
Oesophagus	0.18	**0.20**	0.24	**0.26**	0.28	**0.31**	0.32	**0.35**	0.39	**0.43**	0.45	**0.50**
Trachea	0.08	**0.44**	0.10	**0.52**	0.11	**0.57**	0.11	**0.61**	0.13	**0.67**	0.14	**0.73**
Lung	0.40	**0.42**	0.49	**0.51**	0.55	**0.58**	0.60	**0.63**	0.69	**0.72**	0.76	**0.81**
Mamma	1.17	**1.20**	1.23	**1.26**	1.26	**1.29**	1.29	**1.32**	1.32	**1.35**	1.36	**1.37**
Heart	0.51	**0.53**	0.62	**0.64**	0.70	**0.73**	0.76	**0.79**	0.86	**0.90**	0.96	**1.00**
Stomach	0.21	**0.53**	0.26	**0.65**	0.29	**0.74**	0.32	**0.80**	0.36	**0.91**	0.40	**1.01**
Spleen	0.11	**0.19**	0.15	**0.25**	0.17	**0.30**	0.19	**0.33**	0.23	**0.40**	0.27	**0.47**
Liver	0.23	**0.47**	0.28	**0.58**	0.32	**0.65**	0.35	**0.71**	0.40	**0.81**	0.44	**0.90**
Gall bladder	0.07	**0.37**	0.09	**0.47**	0.11	**0.55**	0.12	**0.60**	0.15	**0.71**	0.17	**0.81**
Pancreas	0.18	**0.26**	0.24	**0.34**	0.28	**0.40**	0.31	**0.45**	0.38	**0.55**	0.43	**0.63**
Small intestine		**0.17**	0.01	**0.21**	0.01	**0.25**	0.02	**0.27**	0.02	**0.33**	0.03	**0.37**
Large intestine	0.01	**0.26**	0.02	**0.33**	0.02	**0.38**	0.02	**0.42**	0.03	**0.49**	0.03	**0.55**
Sigma/Rectum		**0.05**		**0.07**		**0.08**		**0.09**		**0.10**		**0.12**
Lumbar spine	0.03	**0.23**	0.05	**0.31**	0.06	**0.38**	0.07	**0.43**	0.09	**0.55**	0.11	**0.66**
Pelvis		**0.06**		**0.09**	0.01	**0.11**	0.01	**0.12**	0.02	**0.15**	0.02	**0.18**
Adrenals	0.09	**0.11**	0.12	**0.15**	0.15	**0.18**	0.17	**0.21**	0.21	**0.26**	0.24	**0.30**
Kidneys	0.03	**0.10**	0.04	**0.14**	0.04	**0.17**	0.05	**0.19**	0.06	**0.24**	0.08	**0.28**
Urinary bladder		**0.01**		**0.01**		**0.02**		**0.02**		**0.03**		**0.03**
Uterus		**0.03**		**0.04**		**0.04**		**0.05**		**0.06**	0.01	**0.08**
Ovaries		**0.03**		**0.04**		**0.05**		**0.05**		**0.06**		**0.08**
Prostate						**0.01**		**0.01**		**0.02**		**0.02**
Testes												
Upper arm	0.19	**1.03**	0.25	**1.27**	0.29	**1.44**	0.33	**1.57**	0.39	**1.81**	0.46	**2.01**
Elbow	0.14	**1.10**	0.19	**1.36**	0.22	**1.54**	0.25	**1.68**	0.31	**1.92**	0.36	**2.13**
Forearm	0.01	**0.56**	0.02	**0.69**	0.02	**0.78**	0.03	**0.85**	0.03	**0.97**	0.04	**1.07**
Upper leg										**0.01**		**0.01**
Knee												
Lower leg												
Lymph nodes	0.16	**0.34**	0.20	**0.42**	0.22	**0.48**	0.24	**0.52**	0.28	**0.60**	0.31	**0.67**
Skin	0.16	**0.32**	0.17	**0.34**	0.17	**0.35**	0.18	**0.36**	0.19	**0.38**	0.20	**0.40**
Muscle	0.12	**0.22**	0.14	**0.26**	0.16	**0.29**	0.17	**0.31**	0.20	**0.35**	0.22	**0.39**
Skeleton	0.27	**0.50**	0.33	**0.61**	0.37	**0.69**	0.40	**0.75**	0.46	**0.85**	0.51	**0.95**
Red bone marrow	0.07	**0.12**	0.08	**0.15**	0.09	**0.17**	0.10	**0.19**	0.12	**0.22**	0.13	**0.25**
Whole body dose	0.13	**0.24**	0.16	**0.29**	0.18	**0.32**	0.19	**0.35**	0.22	**0.39**	0.24	**0.43**
ED (ICRP 60)	0.21	**0.31**	0.25	**0.37**	0.27	**0.40**	0.29	**0.44**	0.32	**0.49**	0.35	**0.54**
ED (ICRP 103)	0.29	**0.40**	0.33	**0.47**	0.36	**0.51**	0.38	**0.55**	0.42	**0.61**	0.45	**0.66**
Total filtration (mm) **Al**	**1.5**		**2.5**		**3.5**		**4.5**		**3.5**		**3.5**	
Cu	**0**		**0**		**0**		**0**		**0.1**		**0.2**	

Table 9.123 Normalized organ doses—Chest ap—< 1 year

Beam projection:			ap				**Table cells:**					
Focus-to-detector distance (cm):			150				(+) = normal font = optimal field setting					
Tube voltage (kVp):			60				**(−) = bold font = suboptimal field setting**					
Total filtrations:			See last table row.				Empty table cells correspond to values < 0.01					
Organ or tissue	**Organ dose D_T : Entrance dose K_E**								**(mSv × mGy^{-1})**			
	(+)	(−)	(+)	(−)	(+)	(−)	(+)	(−)	(+)	(−)	(+)	(−)
Skull	0.02	**0.23**	0.03	**0.27**	0.03	**0.30**	0.04	**0.33**	0.04	**0.37**	0.04	**0.36**
Brain		**0.02**		**0.03**		**0.03**		**0.04**		**0.04**		**0.04**
Oral mucosa	0.03	**0.60**	0.04	**0.70**	0.05	**0.77**	0.05	**0.82**	0.06	**0.90**	0.06	**0.89**
Salivary glands	0.06	**0.54**	0.07	**0.64**	0.08	**0.70**	0.09	**0.75**	0.11	**0.84**	0.10	**0.83**
Cervical spine	0.38	**0.64**	0.49	**0.82**	0.56	**0.95**	0.62	**1.06**	0.73	**1.27**	0.71	**1.21**
Thyroid	0.91	**0.99**	1.02	**1.12**	1.09	**1.21**	1.13	**1.26**	1.21	**1.36**	1.20	**1.35**
Thymus	0.83	**0.85**	0.96	**0.98**	1.04	**1.07**	1.10	**1.13**	1.20	**1.24**	1.20	**1.23**
Thoracic spine	0.32	**0.36**	0.42	**0.48**	0.50	**0.57**	0.56	**0.64**	0.69	**0.79**	0.64	**0.73**
Claviculae	2.06	**2.15**	2.38	**2.49**	2.58	**2.71**	2.73	**2.87**	2.95	**3.12**	3.00	**3.16**
Scapulae	0.55	**0.59**	0.70	**0.76**	0.82	**0.89**	0.92	**0.99**	1.10	**1.18**	1.06	**1.14**
Ribs	1.41	**1.58**	1.61	**1.81**	1.74	**1.97**	1.84	**2.08**	2.00	**2.27**	2.01	**2.27**
Oesophagus	0.24	**0.26**	0.30	**0.34**	0.36	**0.40**	0.40	**0.44**	0.47	**0.53**	0.45	**0.50**
Trachea	0.09	**0.49**	0.11	**0.57**	0.12	**0.63**	0.13	**0.67**	0.15	**0.73**	0.14	**0.73**
Lung	0.47	**0.49**	0.56	**0.59**	0.63	**0.67**	0.68	**0.72**	0.77	**0.82**	0.76	**0.81**
Mamma	1.20	**1.23**	1.26	**1.29**	1.30	**1.32**	1.32	**1.34**	1.36	**1.37**	1.36	**1.37**
Heart	0.59	**0.62**	0.71	**0.74**	0.80	**0.83**	0.86	**0.90**	0.97	**1.02**	0.96	**1.00**
Stomach	0.25	**0.62**	0.30	**0.75**	0.33	**0.84**	0.36	**0.91**	0.41	**1.02**	0.40	**1.01**
Spleen	0.14	**0.25**	0.19	**0.32**	0.22	**0.37**	0.24	**0.42**	0.29	**0.50**	0.27	**0.47**
Liver	0.27	**0.56**	0.33	**0.67**	0.37	**0.75**	0.40	**0.81**	0.45	**0.92**	0.44	**0.90**
Gall bladder	0.09	**0.46**	0.12	**0.57**	0.13	**0.65**	0.15	**0.71**	0.18	**0.83**	0.17	**0.81**
Pancreas	0.23	**0.34**	0.30	**0.43**	0.34	**0.50**	0.38	**0.56**	0.45	**0.66**	0.43	**0.63**
Small intestine	0.01	**0.21**	0.02	**0.26**	0.02	**0.30**	0.02	**0.33**	0.03	**0.39**	0.03	**0.37**
Large intestine	0.02	**0.32**	0.02	**0.39**	0.03	**0.45**	0.03	**0.49**	0.04	**0.57**	0.03	**0.55**
Sigma/Rectum		**0.07**		**0.08**		**0.09**		**0.10**	0.01	**0.12**		**0.12**
Lumbar spine	0.05	**0.32**	0.07	**0.43**	0.08	**0.51**	0.10	**0.58**	0.12	**0.72**	0.11	**0.66**
Pelvis		**0.09**	0.01	**0.12**	0.02	**0.14**	0.02	**0.16**	0.02	**0.20**	0.02	**0.18**
Adrenals	0.12	**0.15**	0.16	**0.20**	0.19	**0.24**	0.22	**0.27**	0.26	**0.33**	0.24	**0.30**
Kidneys	0.04	**0.14**	0.05	**0.19**	0.06	**0.22**	0.07	**0.25**	0.09	**0.31**	0.08	**0.28**
Urinary bladder		**0.02**		**0.02**		**0.03**		**0.03**		**0.04**		**0.03**
Uterus		**0.04**		**0.05**		**0.06**	0.01	**0.07**	0.01	**0.08**	0.01	**0.08**
Ovaries		**0.04**		**0.05**		**0.06**		**0.07**	0.01	**0.09**		**0.08**
Prostate				**0.01**		**0.01**		**0.02**		**0.02**		**0.02**
Testes												
Upper arm	0.24	**1.21**	0.31	**1.47**	0.36	**1.65**	0.40	**1.78**	0.47	**2.02**	0.46	**2.01**
Elbow	0.19	**1.29**	0.24	**1.56**	0.28	**1.75**	0.31	**1.89**	0.37	**2.14**	0.36	**2.13**
Forearm	0.02	**0.65**	0.03	**0.79**	0.03	**0.88**	0.04	**0.95**	0.05	**1.07**	0.04	**1.07**
Upper leg						**0.01**		**0.01**		**0.02**		**0.01**
Knee												
Lower leg												
Lymph nodes	0.19	**0.41**	0.23	**0.49**	0.26	**0.55**	0.28	**0.60**	0.32	**0.68**	0.31	**0.67**
Skin	0.17	**0.34**	0.18	**0.36**	0.19	**0.37**	0.19	**0.38**	0.20	**0.41**	0.20	**0.40**
Muscle	0.14	**0.25**	0.16	**0.30**	0.18	**0.33**	0.20	**0.35**	0.22	**0.40**	0.22	**0.39**
Skeleton	0.32	**0.59**	0.38	**0.70**	0.42	**0.79**	0.46	**0.85**	0.52	**0.97**	0.51	**0.95**
Red bone marrow	0.08	**0.15**	0.10	**0.18**	0.11	**0.21**	0.12	**0.22**	0.14	**0.26**	0.13	**0.25**
Whole body dose	0.15	**0.28**	0.18	**0.33**	0.20	**0.37**	0.22	**0.39**	0.24	**0.45**	0.24	**0.43**
ED (ICRP 60)	0.24	**0.35**	0.28	**0.41**	0.30	**0.46**	0.32	**0.49**	0.36	**0.55**	0.35	**0.54**
ED (ICRP 103)	0.32	**0.45**	0.37	**0.52**	0.40	**0.57**	0.42	**0.61**	0.45	**0.67**	0.45	**0.66**
Total filtration (mm) **Al**	**1.5**		**2.5**		**3.5**		**4.5**		**3.5**		**3.5**	
Cu	**0**		**0**		**0**		**0**		**0.1**		**0.2**	

Table 9.124 Normalized organ doses—Chest ap—< 1 year

Beam projection:			ap			**Table cells:**						
Focus-to-detector distance (cm):			150			(+) = normal font = optimal field setting						
Tube voltage (kVp):			70			**(−) = bold font = suboptimal field setting**						
Total filtrations:			See last table row.			Empty table cells correspond to values < 0.01						
Organ or tissue	**Organ dose D_T : Entrance dose K_E**							**(mSv × mGy^{-1})**				
	(+)	(−)	(+)	(−)	(+)	(−)	(+)	(−)	(+)	(−)	(+)	(−)
Skull	0.03	**0.25**	0.03	**0.30**	0.04	**0.33**	0.04	**0.35**	0.05	**0.39**	0.06	**0.43**
Brain		**0.03**		**0.03**		**0.04**		**0.04**	0.01	**0.05**	0.01	**0.06**
Oral mucosa	0.04	**0.66**	0.05	**0.76**	0.05	**0.82**	0.06	**0.87**	0.07	**0.95**	0.08	**1.02**
Salivary glands	0.07	**0.59**	0.08	**0.69**	0.09	**0.76**	0.10	**0.81**	0.12	**0.89**	0.13	**0.96**
Cervical spine	0.45	**0.76**	0.56	**0.95**	0.63	**1.09**	0.69	**1.20**	0.80	**1.41**	0.90	**1.58**
Thyroid	0.97	**1.06**	1.07	**1.19**	1.13	**1.27**	1.18	**1.33**	1.25	**1.42**	1.30	**1.50**
Thymus	0.89	**0.92**	1.02	**1.05**	1.10	**1.14**	1.16	**1.20**	1.26	**1.30**	1.33	**1.38**
Thoracic spine	0.40	**0.45**	0.51	**0.58**	0.59	**0.68**	0.66	**0.76**	0.80	**0.92**	0.93	**1.07**
Claviculae	2.20	**2.30**	2.50	**2.62**	2.68	**2.83**	2.81	**2.97**	3.00	**3.19**	3.12	**3.33**
Scapulae	0.65	**0.70**	0.82	**0.88**	0.94	**1.01**	1.04	**1.12**	1.21	**1.32**	1.37	**1.49**
Ribs	1.50	**1.69**	1.70	**1.92**	1.82	**2.06**	1.91	**2.17**	2.05	**2.34**	2.15	**2.46**
Oesophagus	0.29	**0.32**	0.36	**0.40**	0.41	**0.46**	0.46	**0.51**	0.54	**0.61**	0.61	**0.69**
Trachea	0.10	**0.54**	0.12	**0.62**	0.13	**0.67**	0.14	**0.71**	0.16	**0.78**	0.17	**0.83**
Lung	0.52	**0.55**	0.62	**0.66**	0.69	**0.73**	0.74	**0.79**	0.83	**0.89**	0.90	**0.97**
Mamma	1.23	**1.26**	1.29	**1.31**	1.32	**1.34**	1.34	**1.36**	1.38	**1.39**	1.41	**1.41**
Heart	0.66	**0.69**	0.78	**0.82**	0.87	**0.91**	0.93	**0.98**	1.04	**1.09**	1.13	**1.19**
Stomach	0.28	**0.70**	0.33	**0.83**	0.37	**0.92**	0.40	**0.98**	0.45	**1.10**	0.49	**1.19**
Spleen	0.17	**0.30**	0.22	**0.38**	0.25	**0.43**	0.28	**0.48**	0.33	**0.57**	0.37	**0.64**
Liver	0.31	**0.62**	0.36	**0.74**	0.40	**0.82**	0.44	**0.88**	0.49	**0.99**	0.54	**1.08**
Gall bladder	0.11	**0.53**	0.14	**0.64**	0.16	**0.73**	0.17	**0.79**	0.20	**0.91**	0.23	**1.00**
Pancreas	0.28	**0.40**	0.34	**0.50**	0.39	**0.58**	0.43	**0.64**	0.51	**0.75**	0.57	**0.84**
Small intestine	0.02	**0.24**	0.02	**0.30**	0.03	**0.34**	0.03	**0.37**	0.04	**0.43**	0.04	**0.49**
Large intestine	0.02	**0.37**	0.03	**0.44**	0.03	**0.50**	0.04	**0.54**	0.04	**0.62**	0.05	**0.69**
Sigma/Rectum		**0.08**		**0.10**		**0.11**	0.01	**0.12**	0.01	**0.14**	0.02	**0.16**
Lumbar spine	0.07	**0.41**	0.09	**0.53**	0.11	**0.62**	0.12	**0.69**	0.15	**0.84**	0.18	**0.98**
Pelvis	0.01	**0.12**	0.02	**0.15**	0.02	**0.17**	0.03	**0.19**	0.03	**0.24**	0.04	**0.28**
Adrenals	0.15	**0.19**	0.20	**0.24**	0.23	**0.29**	0.25	**0.32**	0.31	**0.38**	0.35	**0.44**
Kidneys	0.05	**0.18**	0.06	**0.23**	0.08	**0.27**	0.08	**0.30**	0.10	**0.36**	0.12	**0.42**
Urinary bladder		**0.02**		**0.03**		**0.03**		**0.04**		**0.05**		**0.05**
Uterus		**0.05**		**0.06**	0.01	**0.07**	0.01	**0.08**	0.02	**0.10**	0.02	**0.12**
Ovaries		**0.05**		**0.06**		**0.08**	0.01	**0.09**	0.01	**0.10**	0.02	**0.12**
Prostate		**0.01**		**0.02**		**0.02**		**0.02**		**0.03**		**0.03**
Testes										**0.01**		**0.01**
Upper arm	0.29	**1.35**	0.36	**1.60**	0.41	**1.78**	0.45	**1.91**	0.52	**2.13**	0.59	**2.30**
Elbow	0.22	**1.44**	0.28	**1.70**	0.32	**1.89**	0.35	**2.02**	0.41	**2.25**	0.47	**2.43**
Forearm	0.03	**0.72**	0.03	**0.86**	0.04	**0.95**	0.05	**1.01**	0.06	**1.12**	0.07	**1.21**
Upper leg		**0.01**		**0.01**		**0.02**		**0.02**		**0.02**		**0.03**
Knee												
Lower leg												
Lymph nodes	0.22	**0.46**	0.26	**0.55**	0.28	**0.61**	0.31	**0.66**	0.35	**0.74**	0.38	**0.81**
Skin	0.17	**0.35**	0.19	**0.37**	0.19	**0.39**	0.20	**0.40**	0.21	**0.42**	0.22	**0.44**
Muscle	0.16	**0.28**	0.18	**0.33**	0.20	**0.36**	0.21	**0.39**	0.24	**0.43**	0.26	**0.47**
Skeleton	0.36	**0.65**	0.42	**0.77**	0.46	**0.86**	0.49	**0.92**	0.55	**1.03**	0.60	**1.12**
Red bone marrow	0.09	**0.17**	0.11	**0.21**	0.12	**0.23**	0.13	**0.25**	0.15	**0.29**	0.17	**0.32**
Whole body dose	0.17	**0.31**	0.20	**0.36**	0.22	**0.40**	0.24	**0.43**	0.26	**0.48**	0.29	**0.52**
ED (ICRP 60)	0.26	**0.39**	0.30	**0.45**	0.33	**0.50**	0.35	**0.53**	0.38	**0.59**	0.41	**0.64**
ED (ICRP 103)	0.35	**0.50**	0.39	**0.57**	0.42	**0.61**	0.44	**0.65**	0.48	**0.71**	0.51	**0.77**
Total filtration (mm) **Al**	**1.5**		**2.5**		**3.5**		**4.5**		**3.5**		**3.5**	
Cu	**0**		**0**		**0**		**0**		**0.1**		**0.2**	

Table 9.125 Normalized organ doses—Chest ap—< 1 year

Beam projection:		ap				**Table cells:**						
Focus-to-detector distance (cm):		150				(+) = normal font = optimal field setting						
Tube voltage (kVp):		80				**(−) = bold font = suboptimal field setting**						
Total filtrations:		See last table row.				Empty table cells correspond to values < 0.01						
Organ or tissue	**Organ dose D_T : Entrance dose K_E**							**(mSv x mGy^{-1})**				
	(+)	(−)	(+)	(−)	(+)	(−)	(+)	(−)	(+)	(−)	(+)	(−)
Skull	0.03	**0.28**	0.04	**0.32**	0.04	**0.35**	0.05	**0.37**	0.06	**0.41**	0.07	**0.43**
Brain		**0.03**		**0.04**	0.01	**0.05**	0.01	**0.05**	0.01	**0.06**	0.02	**0.07**
Oral mucosa	0.04	**0.71**	0.05	**0.81**	0.06	**0.87**	0.07	**0.91**	0.08	**0.99**	0.09	**1.05**
Salivary glands	0.08	**0.64**	0.09	**0.74**	0.10	**0.81**	0.11	**0.86**	0.13	**0.94**	0.14	**1.00**
Cervical spine	0.51	**0.87**	0.61	**1.06**	0.69	**1.20**	0.75	**1.30**	0.85	**1.50**	0.93	**1.66**
Thyroid	1.01	**1.12**	1.11	**1.24**	1.17	**1.32**	1.21	**1.38**	1.28	**1.47**	1.33	**1.54**
Thymus	0.95	**0.98**	1.07	**1.11**	1.15	**1.19**	1.21	**1.25**	1.29	**1.35**	1.36	**1.42**
Thoracic spine	0.47	**0.54**	0.58	**0.67**	0.67	**0.78**	0.74	**0.86**	0.88	**1.02**	1.00	**1.16**
Claviculae	2.30	**2.42**	2.57	**2.71**	2.73	**2.89**	2.84	**3.01**	2.99	**3.18**	3.07	**3.28**
Scapulae	0.74	**0.80**	0.91	**0.98**	1.03	**1.12**	1.12	**1.22**	1.29	**1.41**	1.42	**1.56**
Ribs	1.58	**1.78**	1.76	**1.99**	1.87	**2.12**	1.94	**2.21**	2.06	**2.36**	2.13	**2.45**
Oesophagus	0.33	**0.37**	0.41	**0.46**	0.46	**0.52**	0.51	**0.57**	0.59	**0.67**	0.66	**0.75**
Trachea	0.11	**0.58**	0.13	**0.66**	0.14	**0.71**	0.15	**0.75**	0.17	**0.81**	0.18	**0.86**
Lung	0.58	**0.61**	0.67	**0.72**	0.74	**0.79**	0.79	**0.84**	0.87	**0.94**	0.94	**1.02**
Mamma	1.26	**1.28**	1.31	**1.33**	1.34	**1.36**	1.36	**1.38**	1.40	**1.41**	1.42	**1.43**
Heart	0.73	**0.76**	0.85	**0.89**	0.93	**0.98**	0.99	**1.04**	1.09	**1.15**	1.17	**1.24**
Stomach	0.31	**0.77**	0.36	**0.89**	0.40	**0.98**	0.42	**1.04**	0.47	**1.15**	0.51	**1.24**
Spleen	0.20	**0.35**	0.25	**0.43**	0.28	**0.49**	0.31	**0.54**	0.36	**0.62**	0.40	**0.70**
Liver	0.34	**0.69**	0.40	**0.80**	0.44	**0.88**	0.47	**0.94**	0.52	**1.04**	0.56	**1.13**
Gall bladder	0.13	**0.59**	0.15	**0.71**	0.17	**0.79**	0.19	**0.86**	0.22	**0.97**	0.25	**1.06**
Pancreas	0.32	**0.47**	0.39	**0.57**	0.44	**0.65**	0.48	**0.70**	0.55	**0.81**	0.61	**0.91**
Small intestine	0.02	**0.28**	0.03	**0.34**	0.03	**0.38**	0.04	**0.41**	0.04	**0.47**	0.05	**0.52**
Large intestine	0.03	**0.41**	0.03	**0.49**	0.04	**0.54**	0.04	**0.59**	0.05	**0.66**	0.06	**0.72**
Sigma/Rectum		**0.09**		**0.11**	0.01	**0.12**	0.01	**0.13**	0.02	**0.15**	0.02	**0.17**
Lumbar spine	0.09	**0.49**	0.11	**0.61**	0.13	**0.71**	0.14	**0.78**	0.17	**0.93**	0.20	**1.06**
Pelvis	0.02	**0.14**	0.02	**0.17**	0.03	**0.20**	0.03	**0.22**	0.04	**0.26**	0.05	**0.30**
Adrenals	0.18	**0.23**	0.23	**0.29**	0.26	**0.33**	0.29	**0.36**	0.34	**0.43**	0.39	**0.49**
Kidneys	0.06	**0.22**	0.08	**0.27**	0.09	**0.31**	0.10	**0.34**	0.12	**0.41**	0.14	**0.47**
Urinary bladder		**0.03**		**0.03**		**0.04**		**0.04**		**0.05**	0.01	**0.06**
Uterus		**0.06**	0.01	**0.07**	0.01	**0.09**	0.02	**0.10**	0.02	**0.11**	0.02	**0.13**
Ovaries		**0.06**	0.01	**0.08**	0.01	**0.09**	0.02	**0.10**	0.02	**0.12**	0.02	**0.14**
Prostate		**0.02**		**0.02**		**0.02**		**0.03**		**0.03**		**0.04**
Testes										**0.01**		**0.01**
Upper arm	0.33	**1.46**	0.39	**1.71**	0.44	**1.87**	0.48	**1.99**	0.55	**2.18**	0.61	**2.32**
Elbow	0.25	**1.56**	0.31	**1.81**	0.35	**1.98**	0.38	**2.10**	0.44	**2.30**	0.49	**2.45**
Forearm	0.03	**0.78**	0.04	**0.91**	0.05	**0.99**	0.05	**1.05**	0.06	**1.15**	0.07	**1.22**
Upper leg		**0.01**		**0.02**		**0.02**		**0.02**		**0.03**		**0.04**
Knee												**0.01**
Lower leg												
Lymph nodes	0.24	**0.51**	0.28	**0.60**	0.31	**0.66**	0.33	**0.71**	0.37	**0.79**	0.40	**0.86**
Skin	0.18	**0.36**	0.19	**0.39**	0.20	**0.40**	0.21	**0.42**	0.22	**0.44**	0.23	**0.46**
Muscle	0.17	**0.31**	0.20	**0.35**	0.22	**0.39**	0.23	**0.41**	0.25	**0.46**	0.28	**0.49**
Skeleton	0.39	**0.71**	0.45	**0.83**	0.49	**0.91**	0.52	**0.97**	0.57	**1.07**	0.61	**1.14**
Red bone marrow	0.10	**0.19**	0.12	**0.23**	0.14	**0.26**	0.15	**0.27**	0.17	**0.31**	0.18	**0.34**
Whole body dose	0.19	**0.34**	0.22	**0.39**	0.24	**0.43**	0.25	**0.46**	0.28	**0.51**	0.30	**0.55**
ED (ICRP 60)	0.28	**0.43**	0.32	**0.49**	0.35	**0.53**	0.37	**0.56**	0.40	**0.62**	0.43	**0.67**
ED (ICRP 103)	0.37	**0.53**	0.41	**0.60**	0.44	**0.65**	0.46	**0.69**	0.50	**0.75**	0.53	**0.80**
Total filtration (mm) **Al**	**1.5**		**2.5**		**3.5**		**4.5**		**3.5**		**3.5**	
Cu	**0**		**0**		**0**		**0**		**0.1**		**0.2**	

Table 9.126 Normalized organ doses—Chest ap—1–2 years

Beam projection:			ap			**Table cells:**						
Focus-to-detector distance (cm):			150			(+) = normal font = optimal field setting						
Tube voltage (kVp):			50			**(−) = bold font = suboptimal field setting**						
Total filtrations:			See last table row.			Empty table cells correspond to values < 0.01						
Organ or tissue	**Organ dose D_T : Entrance dose K_E**						**(mSv × mGy^{-1})**					
	(+)	(−)	(+)	(−)	(+)	(−)	(+)	(−)	(+)	(−)	(+)	(−)
Skull	0.01	**0.20**	0.01	**0.25**	0.02	**0.28**	0.02	**0.30**	0.03	**0.35**	0.03	**0.39**
Brain				**0.01**		**0.02**		**0.02**		**0.03**		**0.03**
Oral mucosa	0.02	**0.42**	0.02	**0.50**	0.03	**0.56**	0.03	**0.60**	0.04	**0.67**	0.04	**0.74**
Salivary glands	0.03	**0.33**	0.04	**0.41**	0.05	**0.46**	0.05	**0.51**	0.07	**0.58**	0.08	**0.65**
Cervical spine	0.23	**0.33**	0.30	**0.44**	0.35	**0.53**	0.40	**0.61**	0.49	**0.75**	0.57	**0.89**
Thyroid	0.82	**0.87**	0.94	**0.99**	1.01	**1.07**	1.06	**1.13**	1.14	**1.22**	1.20	**1.30**
Thymus	0.64	**0.65**	0.77	**0.79**	0.86	**0.88**	0.93	**0.95**	1.04	**1.07**	1.14	**1.18**
Thoracic spine	0.13	**0.15**	0.18	**0.20**	0.22	**0.25**	0.26	**0.29**	0.33	**0.38**	0.41	**0.47**
Claviculae	1.58	**1.65**	1.92	**2.00**	2.15	**2.25**	2.32	**2.43**	2.61	**2.75**	2.86	**3.03**
Scapulae	0.21	**0.22**	0.28	**0.30**	0.34	**0.37**	0.39	**0.42**	0.49	**0.53**	0.59	**0.63**
Ribs	1.17	**1.29**	1.38	**1.52**	1.52	**1.68**	1.63	**1.80**	1.81	**2.01**	1.96	**2.19**
Oesophagus	0.15	**0.16**	0.20	**0.22**	0.24	**0.26**	0.27	**0.30**	0.33	**0.37**	0.40	**0.44**
Trachea	0.07	**0.37**	0.08	**0.45**	0.09	**0.50**	0.10	**0.55**	0.11	**0.62**	0.12	**0.68**
Lung	0.31	**0.32**	0.39	**0.41**	0.45	**0.47**	0.49	**0.52**	0.57	**0.61**	0.65	**0.69**
Mamma	1.09	**1.11**	1.17	**1.19**	1.21	**1.24**	1.24	**1.27**	1.29	**1.31**	1.33	**1.35**
Heart	0.42	**0.43**	0.52	**0.54**	0.60	**0.62**	0.65	**0.68**	0.76	**0.79**	0.86	**0.89**
Stomach	0.17	**0.43**	0.22	**0.53**	0.25	**0.61**	0.28	**0.66**	0.33	**0.77**	0.37	**0.86**
Spleen	0.08	**0.13**	0.11	**0.17**	0.13	**0.21**	0.14	**0.23**	0.18	**0.29**	0.21	**0.35**
Liver	0.19	**0.36**	0.23	**0.45**	0.27	**0.52**	0.30	**0.57**	0.35	**0.67**	0.40	**0.75**
Gall bladder	0.04	**0.31**	0.06	**0.40**	0.07	**0.47**	0.08	**0.53**	0.10	**0.63**	0.11	**0.73**
Pancreas	0.13	**0.19**	0.18	**0.25**	0.21	**0.30**	0.24	**0.34**	0.30	**0.42**	0.35	**0.50**
Small intestine		**0.12**		**0.16**		**0.19**	0.01	**0.21**	0.01	**0.26**	0.02	**0.30**
Large intestine		**0.20**	0.01	**0.25**	0.01	**0.30**	0.02	**0.33**	0.02	**0.40**	0.02	**0.46**
Sigma/Rectum		**0.04**		**0.05**		**0.06**		**0.06**		**0.08**		**0.09**
Lumbar spine	0.02	**0.13**	0.03	**0.17**	0.04	**0.21**	0.04	**0.25**	0.06	**0.33**	0.07	**0.41**
Pelvis		**0.04**		**0.06**		**0.07**		**0.08**		**0.10**	0.01	**0.13**
Adrenals	0.06	**0.07**	0.08	**0.10**	0.10	**0.12**	0.11	**0.14**	0.14	**0.18**	0.17	**0.21**
Kidneys	0.02	**0.06**	0.03	**0.08**	0.03	**0.10**	0.04	**0.12**	0.05	**0.15**	0.06	**0.18**
Urinary bladder						**0.01**		**0.01**		**0.02**		**0.02**
Uterus		**0.02**		**0.03**		**0.03**		**0.04**		**0.05**		**0.06**
Ovaries		**0.02**		**0.03**		**0.03**		**0.04**		**0.05**		**0.06**
Prostate												**0.01**
Testes												
Upper arm	0.11	**0.71**	0.15	**0.90**	0.18	**1.04**	0.21	**1.15**	0.26	**1.36**	0.32	**1.55**
Elbow	0.10	**0.77**	0.13	**0.97**	0.16	**1.12**	0.18	**1.24**	0.23	**1.46**	0.28	**1.67**
Forearm		**0.40**	0.01	**0.50**	0.02	**0.58**	0.02	**0.64**	0.02	**0.75**	0.03	**0.85**
Upper leg												
Knee												
Lower leg												
Lymph nodes	0.13	**0.28**	0.16	**0.34**	0.19	**0.39**	0.20	**0.43**	0.24	**0.51**	0.27	**0.57**
Skin	0.14	**0.28**	0.15	**0.30**	0.15	**0.31**	0.16	**0.32**	0.16	**0.33**	0.17	**0.35**
Muscle	0.08	**0.15**	0.10	**0.19**	0.12	**0.21**	0.13	**0.23**	0.14	**0.26**	0.16	**0.29**
Skeleton	0.20	**0.36**	0.24	**0.44**	0.28	**0.51**	0.30	**0.55**	0.35	**0.64**	0.39	**0.73**
Red bone marrow	0.05	**0.09**	0.06	**0.11**	0.07	**0.12**	0.07	**0.13**	0.09	**0.16**	0.10	**0.18**
Whole body dose	0.10	**0.18**	0.12	**0.22**	0.14	**0.25**	0.15	**0.27**	0.17	**0.31**	0.19	**0.35**
ED (ICRP 60)	0.18	**0.26**	0.22	**0.31**	0.24	**0.34**	0.26	**0.37**	0.29	**0.42**	0.32	**0.47**
ED (ICRP 103)	0.26	**0.34**	0.30	**0.40**	0.32	**0.44**	0.34	**0.47**	0.38	**0.53**	0.41	**0.58**
Total filtration (mm) Al	**1.5**		**2.5**		**3.5**		**4.5**		**3.5**		**3.5**	
Cu	**0**		**0**		**0**		**0**		**0.1**		**0.2**	

Table 9.127 Normalized organ doses—Chest ap—1–2 years

Beam projection:			ap				**Table cells:**					
Focus-to-detector distance (cm):			150				(+) = normal font = optimal field setting					
Tube voltage (kVp):			60				**(−) = bold font = suboptimal field setting**					
Total filtrations:			See last table row.				Empty table cells correspond to values < 0.01					
Organ or tissue	**Organ dose D_T : Entrance dose K_E**								**(mSv × mGy^{-1})**			
	(+)	(−)	(+)	(−)	(+)	(−)	(+)	(−)	(+)	(−)	(+)	(−)
Skull	0.02	**0.24**	0.02	**0.29**	0.03	**0.32**	0.03	**0.35**	0.04	**0.40**	0.03	**0.39**
Brain		**0.02**		**0.02**		**0.02**		**0.03**		**0.04**		**0.03**
Oral mucosa	0.02	**0.48**	0.03	**0.57**	0.04	**0.63**	0.04	**0.68**	0.05	**0.76**	0.04	**0.74**
Salivary glands	0.04	**0.40**	0.05	**0.48**	0.06	**0.54**	0.07	**0.58**	0.08	**0.66**	0.08	**0.65**
Cervical spine	0.30	**0.46**	0.39	**0.59**	0.45	**0.70**	0.50	**0.79**	0.61	**0.97**	0.57	**0.89**
Thyroid	0.90	**0.95**	1.01	**1.08**	1.08	**1.17**	1.13	**1.22**	1.21	**1.32**	1.20	**1.30**
Thymus	0.74	**0.75**	0.87	**0.90**	0.97	**1.00**	1.04	**1.07**	1.16	**1.20**	1.14	**1.18**
Thoracic spine	0.20	**0.23**	0.27	**0.30**	0.32	**0.37**	0.37	**0.42**	0.47	**0.54**	0.41	**0.47**
Claviculae	1.81	**1.90**	2.16	**2.27**	2.39	**2.52**	2.56	**2.70**	2.85	**3.02**	2.86	**3.03**
Scapulae	0.29	**0.31**	0.38	**0.41**	0.46	**0.49**	0.52	**0.56**	0.64	**0.69**	0.59	**0.63**
Ribs	1.32	**1.46**	1.54	**1.70**	1.68	**1.87**	1.79	**1.99**	1.97	**2.20**	1.96	**2.19**
Oesophagus	0.20	**0.22**	0.27	**0.29**	0.31	**0.35**	0.35	**0.40**	0.43	**0.49**	0.40	**0.44**
Trachea	0.08	**0.43**	0.09	**0.52**	0.10	**0.57**	0.11	**0.62**	0.12	**0.69**	0.12	**0.68**
Lung	0.38	**0.40**	0.47	**0.49**	0.53	**0.56**	0.58	**0.61**	0.67	**0.71**	0.65	**0.69**
Mamma	1.14	**1.16**	1.22	**1.24**	1.26	**1.28**	1.29	**1.32**	1.34	**1.36**	1.33	**1.35**
Heart	0.50	**0.52**	0.62	**0.64**	0.70	**0.73**	0.77	**0.80**	0.88	**0.92**	0.86	**0.89**
Stomach	0.21	**0.51**	0.26	**0.63**	0.30	**0.71**	0.33	**0.77**	0.38	**0.89**	0.37	**0.86**
Spleen	0.11	**0.18**	0.14	**0.23**	0.17	**0.27**	0.19	**0.31**	0.23	**0.38**	0.21	**0.35**
Liver	0.23	**0.44**	0.28	**0.54**	0.32	**0.62**	0.35	**0.68**	0.41	**0.78**	0.40	**0.75**
Gall bladder	0.06	**0.40**	0.08	**0.50**	0.09	**0.58**	0.10	**0.64**	0.13	**0.76**	0.11	**0.73**
Pancreas	0.18	**0.25**	0.23	**0.33**	0.28	**0.39**	0.31	**0.44**	0.38	**0.54**	0.35	**0.50**
Small intestine		**0.16**	0.01	**0.21**	0.02	**0.24**	0.02	**0.27**	0.02	**0.32**	0.02	**0.30**
Large intestine	0.01	**0.25**	0.02	**0.32**	0.02	**0.37**	0.02	**0.41**	0.03	**0.48**	0.02	**0.46**
Sigma/Rectum		**0.05**		**0.06**		**0.07**		**0.08**		**0.10**		**0.09**
Lumbar spine	0.04	**0.20**	0.05	**0.26**	0.06	**0.32**	0.07	**0.37**	0.09	**0.47**	0.07	**0.41**
Pelvis		**0.06**		**0.08**		**0.10**	0.01	**0.12**	0.01	**0.15**	0.01	**0.13**
Adrenals	0.09	**0.11**	0.11	**0.14**	0.14	**0.17**	0.15	**0.19**	0.19	**0.24**	0.17	**0.21**
Kidneys	0.03	**0.09**	0.04	**0.12**	0.05	**0.15**	0.05	**0.17**	0.07	**0.21**	0.06	**0.18**
Urinary bladder		**0.01**		**0.01**		**0.02**		**0.02**		**0.03**		**0.02**
Uterus		**0.03**		**0.04**		**0.05**		**0.06**		**0.07**		**0.06**
Ovaries		**0.03**		**0.04**		**0.05**		**0.05**		**0.07**		**0.06**
Prostate										**0.01**		**0.01**
Testes												
Upper arm	0.16	**0.88**	0.21	**1.09**	0.25	**1.25**	0.28	**1.37**	0.35	**1.60**	0.32	**1.55**
Elbow	0.14	**0.95**	0.18	**1.17**	0.22	**1.34**	0.25	**1.47**	0.30	**1.71**	0.28	**1.67**
Forearm	0.01	**0.48**	0.02	**0.60**	0.02	**0.68**	0.03	**0.75**	0.03	**0.87**	0.03	**0.85**
Upper leg												
Knee												
Lower leg												
Lymph nodes	0.16	**0.34**	0.20	**0.41**	0.22	**0.47**	0.24	**0.52**	0.28	**0.60**	0.27	**0.57**
Skin	0.14	**0.29**	0.15	**0.31**	0.16	**0.33**	0.17	**0.34**	0.18	**0.35**	0.17	**0.35**
Muscle	0.10	**0.18**	0.12	**0.22**	0.14	**0.24**	0.15	**0.26**	0.17	**0.30**	0.16	**0.29**
Skeleton	0.24	**0.43**	0.29	**0.53**	0.32	**0.60**	0.35	**0.65**	0.40	**0.75**	0.39	**0.73**
Red bone marrow	0.06	**0.11**	0.07	**0.13**	0.08	**0.15**	0.09	**0.17**	0.10	**0.19**	0.10	**0.18**
Whole body dose	0.12	**0.21**	0.14	**0.26**	0.16	**0.29**	0.17	**0.31**	0.20	**0.36**	0.19	**0.35**
ED (ICRP 60)	0.21	**0.30**	0.25	**0.36**	0.27	**0.40**	0.29	**0.43**	0.33	**0.49**	0.32	**0.47**
ED (ICRP 103)	0.29	**0.39**	0.33	**0.46**	0.36	**0.50**	0.38	**0.54**	0.42	**0.60**	0.41	**0.58**
Total filtration (mm) Al	**1.5**		**2.5**		**3.5**		**4.5**		**3.5**		**3.5**	
Cu	**0**		**0**		**0**		**0**		**0.1**		**0.2**	

Table 9.128 Normalized organ doses—Chest ap—1–2 years

Beam projection:			ap		**Table cells:**							
Focus-to-detector distance (cm):			150		(+) = normal font = optimal field setting							
Tube voltage (kVp):			70		**(–) = bold font = suboptimal field setting**							
Total filtrations:			See last table row.		Empty table cells correspond to values < 0.01							
Organ or tissue	**Organ dose D_T : Entrance dose K_E**								**(mSv × mGy^{-1})**			
	(+)	(–)	(+)	(–)	(+)	(–)	(+)	(–)	(+)	(–)	(+)	(–)
Skull	0.02	**0.27**	0.03	**0.32**	0.03	**0.35**	0.04	**0.38**	0.04	**0.43**	0.05	**0.47**
Brain		**0.02**		**0.03**		**0.03**		**0.04**		**0.05**		**0.06**
Oral mucosa	0.03	**0.54**	0.04	**0.63**	0.04	**0.69**	0.05	**0.73**	0.06	**0.82**	0.07	**0.89**
Salivary glands	0.05	**0.45**	0.06	**0.53**	0.07	**0.59**	0.08	**0.64**	0.09	**0.72**	0.10	**0.79**
Cervical spine	0.36	**0.56**	0.46	**0.72**	0.53	**0.83**	0.58	**0.93**	0.69	**1.12**	0.79	**1.29**
Thyroid	0.95	**1.03**	1.07	**1.15**	1.13	**1.23**	1.18	**1.29**	1.26	**1.39**	1.33	**1.48**
Thymus	0.82	**0.84**	0.95	**0.98**	1.05	**1.08**	1.12	**1.16**	1.24	**1.28**	1.34	**1.39**
Thoracic spine	0.27	**0.30**	0.35	**0.39**	0.41	**0.47**	0.47	**0.53**	0.58	**0.67**	0.69	**0.80**
Claviculae	1.98	**2.09**	2.31	**2.44**	2.53	**2.68**	2.69	**2.86**	2.95	**3.15**	3.14	**3.37**
Scapulae	0.36	**0.39**	0.47	**0.50**	0.55	**0.59**	0.61	**0.66**	0.73	**0.80**	0.85	**0.93**
Ribs	1.43	**1.59**	1.64	**1.83**	1.78	**1.99**	1.88	**2.10**	2.05	**2.30**	2.18	**2.46**
Oesophagus	0.25	**0.28**	0.32	**0.36**	0.38	**0.43**	0.42	**0.48**	0.51	**0.58**	0.59	**0.67**
Trachea	0.09	**0.48**	0.10	**0.57**	0.11	**0.63**	0.12	**0.67**	0.14	**0.75**	0.15	**0.81**
Lung	0.44	**0.46**	0.53	**0.56**	0.59	**0.63**	0.64	**0.69**	0.74	**0.79**	0.81	**0.87**
Mamma	1.18	**1.20**	1.25	**1.28**	1.30	**1.32**	1.33	**1.35**	1.38	**1.40**	1.42	**1.45**
Heart	0.58	**0.60**	0.70	**0.73**	0.78	**0.82**	0.85	**0.89**	0.97	**1.01**	1.07	**1.13**
Stomach	0.25	**0.59**	0.30	**0.71**	0.34	**0.79**	0.37	**0.86**	0.42	**0.98**	0.47	**1.08**
Spleen	0.14	**0.22**	0.18	**0.28**	0.20	**0.33**	0.23	**0.37**	0.28	**0.45**	0.32	**0.52**
Liver	0.27	**0.51**	0.32	**0.62**	0.37	**0.70**	0.40	**0.76**	0.46	**0.87**	0.51	**0.96**
Gall bladder	0.07	**0.47**	0.09	**0.58**	0.11	**0.67**	0.12	**0.73**	0.15	**0.86**	0.17	**0.96**
Pancreas	0.22	**0.32**	0.28	**0.40**	0.33	**0.47**	0.36	**0.52**	0.44	**0.63**	0.50	**0.73**
Small intestine	0.01	**0.19**	0.02	**0.24**	0.02	**0.28**	0.02	**0.31**	0.03	**0.37**	0.04	**0.42**
Large intestine	0.02	**0.30**	0.02	**0.37**	0.03	**0.42**	0.03	**0.46**	0.04	**0.54**	0.04	**0.61**
Sigma/Rectum		**0.06**		**0.08**		**0.09**		**0.10**		**0.12**	0.01	**0.13**
Lumbar spine	0.05	**0.26**	0.07	**0.35**	0.08	**0.41**	0.09	**0.47**	0.12	**0.59**	0.15	**0.71**
Pelvis		**0.08**	0.01	**0.11**	0.01	**0.13**	0.02	**0.15**	0.02	**0.18**	0.03	**0.22**
Adrenals	0.11	**0.14**	0.14	**0.18**	0.17	**0.22**	0.19	**0.24**	0.24	**0.30**	0.28	**0.36**
Kidneys	0.04	**0.12**	0.05	**0.16**	0.06	**0.19**	0.07	**0.21**	0.09	**0.26**	0.10	**0.31**
Urinary bladder		**0.02**		**0.02**		**0.02**		**0.03**		**0.03**		**0.04**
Uterus		**0.04**		**0.05**		**0.06**		**0.07**	0.01	**0.09**	0.01	**0.10**
Ovaries		**0.04**		**0.05**		**0.06**		**0.07**		**0.09**	0.01	**0.10**
Prostate						**0.01**		**0.01**		**0.02**		**0.02**
Testes												**0.01**
Upper arm	0.20	**1.01**	0.26	**1.23**	0.30	**1.39**	0.33	**1.52**	0.40	**1.74**	0.47	**1.94**
Elbow	0.18	**1.09**	0.22	**1.32**	0.26	**1.49**	0.29	**1.62**	0.36	**1.86**	0.41	**2.06**
Forearm	0.02	**0.55**	0.03	**0.67**	0.03	**0.76**	0.04	**0.82**	0.04	**0.94**	0.05	**1.04**
Upper leg										**0.01**		**0.01**
Knee												
Lower leg												
Lymph nodes	0.18	**0.39**	0.22	**0.47**	0.25	**0.53**	0.27	**0.58**	0.31	**0.66**	0.35	**0.74**
Skin	0.15	**0.31**	0.16	**0.33**	0.17	**0.34**	0.18	**0.35**	0.19	**0.37**	0.20	**0.39**
Muscle	0.11	**0.21**	0.14	**0.25**	0.15	**0.27**	0.16	**0.29**	0.19	**0.33**	0.21	**0.37**
Skeleton	0.27	**0.50**	0.32	**0.59**	0.36	**0.67**	0.39	**0.72**	0.44	**0.82**	0.49	**0.91**
Red bone marrow	0.07	**0.13**	0.08	**0.15**	0.09	**0.17**	0.10	**0.19**	0.12	**0.22**	0.13	**0.25**
Whole body dose	0.13	**0.24**	0.16	**0.29**	0.18	**0.32**	0.19	**0.35**	0.22	**0.40**	0.24	**0.44**
ED (ICRP 60)	0.24	**0.34**	0.27	**0.40**	0.30	**0.44**	0.32	**0.47**	0.36	**0.53**	0.39	**0.59**
ED (ICRP 103)	0.32	**0.43**	0.36	**0.50**	0.39	**0.55**	0.41	**0.59**	0.45	**0.65**	0.49	**0.71**
Total filtration (mm) Al	**1.5**		**2.5**		**3.5**		**4.5**		**3.5**		**3.5**	
Cu	**0**		**0**		**0**		**0**		**0.1**		**0.2**	

Table 9.129 Normalized organ doses—Chest ap—1–2 years

Beam projection:		ap				**Table cells:**						
Focus-to-detector distance (cm):		150				(+) = normal font = optimal field setting						
Tube voltage (kVp):		80				**(−) = bold font = suboptimal field setting**						
Total filtrations:		See last table row.				Empty table cells correspond to values < 0.01						
Organ or tissue	**Organ dose D_T : Entrance dose K_E**							**(mSv × mGy^{-1})**				
	(+)	(−)	(+)	(−)	(+)	(−)	(+)	(−)	(+)	(−)	(+)	(−)
Skull	0.03	**0.30**	0.03	**0.34**	0.04	**0.38**	0.04	**0.40**	0.05	**0.45**	0.06	**0.48**
Brain		**0.03**		**0.03**		**0.04**		**0.04**		**0.05**	0.01	**0.06**
Oral mucosa	0.04	**0.59**	0.04	**0.68**	0.05	**0.74**	0.06	**0.78**	0.07	**0.87**	0.08	**0.94**
Salivary glands	0.06	**0.50**	0.07	**0.58**	0.08	**0.64**	0.09	**0.69**	0.10	**0.77**	0.11	**0.84**
Cervical spine	0.42	**0.67**	0.52	**0.83**	0.59	**0.95**	0.64	**1.04**	0.75	**1.23**	0.84	**1.39**
Thyroid	1.01	**1.09**	1.11	**1.22**	1.18	**1.29**	1.23	**1.35**	1.30	**1.45**	1.36	**1.53**
Thymus	0.89	**0.92**	1.02	**1.06**	1.12	**1.15**	1.18	**1.23**	1.30	**1.35**	1.39	**1.45**
Thoracic spine	0.33	**0.38**	0.42	**0.48**	0.49	**0.56**	0.55	**0.63**	0.67	**0.77**	0.78	**0.90**
Claviculae	2.12	**2.24**	2.43	**2.57**	2.63	**2.79**	2.77	**2.95**	2.99	**3.20**	3.13	**3.37**
Scapulae	0.43	**0.47**	0.54	**0.58**	0.62	**0.67**	0.68	**0.75**	0.81	**0.88**	0.91	**1.00**
Ribs	1.52	**1.69**	1.72	**1.92**	1.85	**2.07**	1.94	**2.18**	2.08	**2.35**	2.19	**2.48**
Oesophagus	0.31	**0.34**	0.38	**0.43**	0.44	**0.50**	0.48	**0.55**	0.57	**0.65**	0.66	**0.75**
Trachea	0.10	**0.53**	0.11	**0.62**	0.12	**0.67**	0.13	**0.72**	0.15	**0.79**	0.16	**0.85**
Lung	0.49	**0.52**	0.58	**0.62**	0.65	**0.69**	0.70	**0.75**	0.79	**0.85**	0.86	**0.93**
Mamma	1.21	**1.24**	1.28	**1.31**	1.33	**1.35**	1.36	**1.38**	1.40	**1.43**	1.44	**1.48**
Heart	0.65	**0.68**	0.77	**0.80**	0.85	**0.90**	0.92	**0.96**	1.04	**1.09**	1.13	**1.20**
Stomach	0.28	**0.66**	0.33	**0.78**	0.37	**0.86**	0.40	**0.93**	0.46	**1.04**	0.50	**1.14**
Spleen	0.17	**0.27**	0.21	**0.34**	0.24	**0.39**	0.26	**0.43**	0.31	**0.51**	0.36	**0.58**
Liver	0.30	**0.58**	0.36	**0.69**	0.40	**0.76**	0.44	**0.82**	0.50	**0.94**	0.55	**1.03**
Gall bladder	0.09	**0.54**	0.11	**0.66**	0.13	**0.74**	0.14	**0.81**	0.17	**0.93**	0.19	**1.04**
Pancreas	0.26	**0.38**	0.33	**0.47**	0.38	**0.54**	0.41	**0.60**	0.49	**0.70**	0.55	**0.80**
Small intestine	0.02	**0.23**	0.02	**0.28**	0.03	**0.32**	0.03	**0.35**	0.04	**0.41**	0.04	**0.46**
Large intestine	0.02	**0.34**	0.03	**0.42**	0.03	**0.47**	0.04	**0.51**	0.04	**0.59**	0.05	**0.65**
Sigma/Rectum		**0.07**		**0.09**		**0.10**		**0.11**	0.01	**0.13**	0.01	**0.15**
Lumbar spine	0.07	**0.33**	0.09	**0.43**	0.10	**0.50**	0.12	**0.56**	0.14	**0.69**	0.17	**0.80**
Pelvis	0.01	**0.10**	0.02	**0.13**	0.02	**0.15**	0.02	**0.17**	0.03	**0.21**	0.03	**0.25**
Adrenals	0.14	**0.18**	0.17	**0.22**	0.20	**0.26**	0.23	**0.29**	0.27	**0.35**	0.32	**0.41**
Kidneys	0.05	**0.15**	0.06	**0.19**	0.08	**0.22**	0.08	**0.25**	0.10	**0.30**	0.12	**0.35**
Urinary bladder		**0.02**		**0.03**		**0.03**		**0.03**		**0.04**		**0.05**
Uterus		**0.05**		**0.06**		**0.07**	0.01	**0.08**	0.01	**0.10**	0.02	**0.12**
Ovaries		**0.05**		**0.06**		**0.07**	0.01	**0.08**	0.01	**0.10**	0.02	**0.12**
Prostate				**0.01**		**0.01**		**0.02**		**0.02**		**0.02**
Testes										**0.01**		**0.01**
Upper arm	0.24	**1.13**	0.30	**1.35**	0.34	**1.51**	0.38	**1.63**	0.45	**1.84**	0.51	**2.01**
Elbow	0.21	**1.21**	0.26	**1.44**	0.30	**1.61**	0.33	**1.73**	0.40	**1.95**	0.45	**2.13**
Forearm	0.03	**0.62**	0.03	**0.73**	0.04	**0.81**	0.04	**0.88**	0.05	**0.98**	0.06	**1.07**
Upper leg						**0.01**		**0.01**		**0.02**		**0.02**
Knee												
Lower leg												
Lymph nodes	0.21	**0.44**	0.25	**0.53**	0.27	**0.59**	0.30	**0.63**	0.34	**0.72**	0.37	**0.80**
Skin	0.16	**0.32**	0.17	**0.34**	0.18	**0.35**	0.18	**0.37**	0.19	**0.39**	0.20	**0.41**
Muscle	0.13	**0.23**	0.15	**0.27**	0.17	**0.30**	0.18	**0.32**	0.20	**0.36**	0.22	**0.40**
Skeleton	0.30	**0.55**	0.35	**0.65**	0.39	**0.72**	0.41	**0.77**	0.46	**0.87**	0.51	**0.95**
Red bone marrow	0.08	**0.15**	0.09	**0.18**	0.11	**0.20**	0.11	**0.21**	0.13	**0.25**	0.15	**0.28**
Whole body dose	0.15	**0.27**	0.18	**0.32**	0.19	**0.35**	0.21	**0.38**	0.24	**0.43**	0.26	**0.47**
ED (ICRP 60)	0.26	**0.37**	0.30	**0.43**	0.32	**0.48**	0.34	**0.51**	0.38	**0.57**	0.41	**0.62**
ED (ICRP 103)	0.34	**0.48**	0.39	**0.54**	0.42	**0.59**	0.44	**0.63**	0.48	**0.70**	0.51	**0.75**
Total filtration (mm) Al	**1.5**		**2.5**		**3.5**		**4.5**		**3.5**		**3.5**	
Cu	**0**		**0**		**0**		**0**		**0.1**		**0.2**	

Table 9.130 Normalized organ doses—Chest ap—3–7 years

Beam projection:		ap			**Table cells:**							
Focus-to-detector distance (cm):		150			(+) = normal font = optimal field setting							
Tube voltage (kVp):		60			**(−) = bold font = suboptimal field setting**							
Total filtrations:		See last table row.			Empty table cells correspond to values < 0.01							
Organ or tissue	**Organ dose D_T : Entrance dose K_E**							**(mSv × mGy⁻¹)**				
	(+)	(−)	(+)	(−)	(+)	(−)	(+)	(−)	(+)	(−)	(+)	(−)
Skull	0.01	**0.25**	0.02	**0.31**	0.02	**0.35**	0.02	**0.38**	0.03	**0.44**	0.03	**0.43**
Brain				**0.01**		**0.02**		**0.02**		**0.02**		**0.02**
Oral mucosa	0.02	**0.38**	0.02	**0.45**	0.03	**0.51**	0.03	**0.55**	0.04	**0.63**	0.03	**0.60**
Salivary glands	0.03	**0.30**	0.04	**0.37**	0.04	**0.43**	0.05	**0.47**	0.06	**0.54**	0.05	**0.52**
Cervical spine	0.25	**0.39**	0.32	**0.51**	0.38	**0.61**	0.42	**0.68**	0.52	**0.84**	0.48	**0.77**
Thyroid	0.82	**0.86**	0.93	**0.99**	1.00	**1.07**	1.06	**1.13**	1.15	**1.24**	1.13	**1.22**
Thymus	0.70	**0.71**	0.84	**0.85**	0.94	**0.96**	1.02	**1.03**	1.15	**1.17**	1.13	**1.15**
Thoracic spine	0.15	**0.17**	0.21	**0.23**	0.25	**0.28**	0.29	**0.32**	0.37	**0.42**	0.31	**0.35**
Claviculae	1.63	**1.70**	1.97	**2.06**	2.20	**2.31**	2.38	**2.50**	2.68	**2.82**	2.67	**2.81**
Scapulae	0.23	**0.24**	0.30	**0.32**	0.36	**0.39**	0.41	**0.44**	0.51	**0.55**	0.46	**0.50**
Ribs	1.29	**1.37**	1.51	**1.61**	1.66	**1.78**	1.77	**1.90**	1.96	**2.11**	1.95	**2.09**
Oesophagus	0.12	**0.14**	0.16	**0.18**	0.20	**0.22**	0.22	**0.25**	0.28	**0.31**	0.25	**0.27**
Trachea	0.07	**0.30**	0.08	**0.37**	0.09	**0.42**	0.10	**0.46**	0.11	**0.53**	0.11	**0.51**
Lung	0.34	**0.36**	0.43	**0.45**	0.49	**0.51**	0.54	**0.57**	0.63	**0.66**	0.61	**0.63**
Mamma	1.13	**1.14**	1.22	**1.23**	1.27	**1.28**	1.30	**1.32**	1.36	**1.38**	1.34	**1.37**
Heart	0.45	**0.46**	0.56	**0.57**	0.64	**0.66**	0.70	**0.73**	0.82	**0.85**	0.79	**0.81**
Stomach	0.23	**0.50**	0.29	**0.62**	0.33	**0.71**	0.37	**0.78**	0.43	**0.90**	0.41	**0.87**
Spleen	0.10	**0.14**	0.13	**0.19**	0.15	**0.23**	0.17	**0.26**	0.21	**0.32**	0.19	**0.29**
Liver	0.22	**0.40**	0.28	**0.50**	0.32	**0.57**	0.36	**0.63**	0.42	**0.74**	0.40	**0.70**
Gall bladder	0.07	**0.35**	0.09	**0.46**	0.11	**0.53**	0.12	**0.59**	0.15	**0.71**	0.14	**0.67**
Pancreas	0.16	**0.22**	0.22	**0.29**	0.26	**0.34**	0.29	**0.39**	0.36	**0.48**	0.32	**0.43**
Small intestine		**0.17**		**0.23**	0.01	**0.26**	0.01	**0.30**	0.02	**0.36**	0.01	**0.33**
Large intestine		**0.25**	0.01	**0.32**	0.02	**0.37**	0.02	**0.41**	0.02	**0.49**	0.02	**0.46**
Sigma/Rectum		**0.06**		**0.07**		**0.09**		**0.10**		**0.12**		**0.11**
Lumbar spine	0.03	**0.15**	0.04	**0.20**	0.05	**0.24**	0.06	**0.28**	0.08	**0.37**	0.06	**0.30**
Pelvis		**0.07**		**0.09**		**0.11**		**0.12**		**0.16**		**0.13**
Adrenals	0.07	**0.09**	0.10	**0.12**	0.12	**0.14**	0.13	**0.16**	0.17	**0.21**	0.15	**0.18**
Kidneys	0.03	**0.07**	0.04	**0.10**	0.04	**0.12**	0.05	**0.13**	0.07	**0.17**	0.06	**0.15**
Urinary bladder				**0.01**		**0.01**		**0.02**		**0.02**		**0.02**
Uterus		**0.03**		**0.04**		**0.04**		**0.05**		**0.07**		**0.06**
Ovaries		**0.03**		**0.04**		**0.05**		**0.05**		**0.07**		**0.06**
Prostate												
Testes												
Upper arm	0.12	**0.63**	0.15	**0.80**	0.19	**0.92**	0.21	**1.03**	0.27	**1.22**	0.23	**1.16**
Elbow	0.11	**0.67**	0.15	**0.84**	0.18	**0.98**	0.21	**1.08**	0.26	**1.29**	0.23	**1.23**
Forearm	0.01	**0.37**	0.02	**0.46**	0.02	**0.54**	0.02	**0.59**	0.03	**0.70**	0.02	**0.68**
Upper leg												
Knee												
Lower leg												
Lymph nodes	0.14	**0.29**	0.18	**0.36**	0.20	**0.42**	0.22	**0.46**	0.26	**0.54**	0.25	**0.51**
Skin	0.12	**0.24**	0.13	**0.25**	0.13	**0.26**	0.14	**0.27**	0.15	**0.29**	0.14	**0.28**
Muscle	0.09	**0.16**	0.11	**0.19**	0.12	**0.21**	0.13	**0.23**	0.15	**0.26**	0.14	**0.25**
Skeleton	0.18	**0.33**	0.22	**0.40**	0.24	**0.46**	0.26	**0.50**	0.30	**0.58**	0.29	**0.55**
Red bone marrow	0.06	**0.09**	0.07	**0.12**	0.08	**0.13**	0.09	**0.15**	0.10	**0.17**	0.09	**0.16**
Whole body dose	0.10	**0.18**	0.13	**0.22**	0.14	**0.25**	0.15	**0.28**	0.18	**0.32**	0.17	**0.30**
ED (ICRP 60)	0.20	**0.28**	0.24	**0.33**	0.26	**0.37**	0.28	**0.40**	0.32	**0.46**	0.30	**0.44**
ED (ICRP 103)	0.28	**0.37**	0.32	**0.43**	0.35	**0.47**	0.37	**0.51**	0.41	**0.58**	0.40	**0.55**
Total filtration (mm) Al	**1.5**		**2.5**		**3.5**		**4.5**		**3.5**		**3.5**	
Cu	**0**		**0**		**0**		**0**		**0.1**		**0.2**	

Table 9.131 Normalized organ doses—Chest ap—3–7 years

Beam projection:		ap				**Table cells:**						
Focus-to-detector distance (cm):		150				(+) = normal font = optimal field setting						
Tube voltage (kVp):		70				**(−) = bold font = suboptimal field setting**						
Total filtrations:		See last table row.				Empty table cells correspond to values < 0.01						
Organ or tissue	**Organ dose D_T : Entrance dose K_E**							**(mSv × mGy^{-1})**				
	(+)	(−)	(+)	(−)	(+)	(−)	(+)	(−)	(+)	(−)	(+)	(−)
Skull	0.02	**0.29**	0.02	**0.35**	0.03	**0.39**	0.03	**0.42**	0.04	**0.49**	0.05	**0.54**
Brain		**0.01**		**0.02**		**0.02**		**0.03**		**0.03**		**0.04**
Oral mucosa	0.02	**0.43**	0.03	**0.51**	0.03	**0.57**	0.04	**0.61**	0.05	**0.69**	0.06	**0.77**
Salivary glands	0.03	**0.35**	0.04	**0.43**	0.05	**0.48**	0.06	**0.52**	0.07	**0.60**	0.08	**0.67**
Cervical spine	0.30	**0.49**	0.38	**0.63**	0.45	**0.74**	0.50	**0.82**	0.60	**1.00**	0.69	**1.17**
Thyroid	0.88	**0.94**	0.99	**1.06**	1.06	**1.15**	1.12	**1.21**	1.20	**1.32**	1.27	**1.41**
Thymus	0.79	**0.80**	0.93	**0.95**	1.03	**1.05**	1.11	**1.13**	1.25	**1.28**	1.36	**1.40**
Thoracic spine	0.21	**0.23**	0.28	**0.31**	0.33	**0.37**	0.38	**0.43**	0.48	**0.54**	0.58	**0.66**
Claviculae	1.81	**1.90**	2.14	**2.25**	2.37	**2.49**	2.53	**2.67**	2.81	**2.98**	3.03	**3.22**
Scapulae	0.29	**0.31**	0.38	**0.41**	0.44	**0.48**	0.50	**0.54**	0.61	**0.66**	0.72	**0.78**
Ribs	1.41	**1.51**	1.63	**1.74**	1.77	**1.90**	1.87	**2.02**	2.05	**2.23**	2.20	**2.39**
Oesophagus	0.16	**0.18**	0.21	**0.23**	0.24	**0.27**	0.28	**0.30**	0.34	**0.37**	0.40	**0.44**
Trachea	0.08	**0.35**	0.09	**0.42**	0.10	**0.48**	0.11	**0.52**	0.12	**0.60**	0.14	**0.67**
Lung	0.40	**0.42**	0.49	**0.52**	0.56	**0.59**	0.61	**0.64**	0.70	**0.75**	0.79	**0.84**
Mamma	1.18	**1.19**	1.26	**1.28**	1.31	**1.33**	1.34	**1.36**	1.39	**1.42**	1.44	**1.47**
Heart	0.52	**0.54**	0.64	**0.66**	0.73	**0.75**	0.80	**0.83**	0.92	**0.96**	1.03	**1.08**
Stomach	0.27	**0.59**	0.34	**0.71**	0.38	**0.80**	0.42	**0.88**	0.48	**1.01**	0.54	**1.12**
Spleen	0.12	**0.19**	0.16	**0.24**	0.19	**0.28**	0.21	**0.32**	0.26	**0.39**	0.31	**0.46**
Liver	0.26	**0.47**	0.33	**0.57**	0.37	**0.65**	0.41	**0.71**	0.48	**0.83**	0.54	**0.94**
Gall bladder	0.09	**0.43**	0.11	**0.54**	0.13	**0.63**	0.15	**0.69**	0.18	**0.82**	0.22	**0.94**
Pancreas	0.21	**0.28**	0.27	**0.36**	0.31	**0.42**	0.35	**0.47**	0.43	**0.57**	0.50	**0.67**
Small intestine		**0.22**	0.01	**0.27**	0.02	**0.32**	0.02	**0.35**	0.02	**0.42**	0.03	**0.49**
Large intestine	0.01	**0.30**	0.02	**0.38**	0.02	**0.43**	0.02	**0.48**	0.03	**0.56**	0.04	**0.64**
Sigma/Rectum		**0.07**		**0.09**		**0.10**		**0.11**		**0.14**		**0.16**
Lumbar spine	0.05	**0.21**	0.06	**0.27**	0.07	**0.33**	0.08	**0.38**	0.11	**0.48**	0.14	**0.59**
Pelvis		**0.09**		**0.12**		**0.14**		**0.16**	0.01	**0.20**	0.02	**0.24**
Adrenals	0.10	**0.12**	0.13	**0.15**	0.15	**0.18**	0.17	**0.21**	0.21	**0.26**	0.25	**0.32**
Kidneys	0.04	**0.10**	0.05	**0.13**	0.06	**0.15**	0.07	**0.17**	0.09	**0.22**	0.11	**0.27**
Urinary bladder		**0.01**		**0.02**		**0.02**		**0.02**		**0.03**		**0.04**
Uterus		**0.04**		**0.05**		**0.06**		**0.07**		**0.08**		**0.10**
Ovaries		**0.04**		**0.05**		**0.06**		**0.07**		**0.09**		**0.11**
Prostate										**0.01**		**0.02**
Testes												
Upper arm	0.15	**0.75**	0.20	**0.93**	0.23	**1.06**	0.26	**1.17**	0.32	**1.36**	0.38	**1.54**
Elbow	0.15	**0.79**	0.19	**0.98**	0.22	**1.12**	0.25	**1.23**	0.31	**1.44**	0.37	**1.62**
Forearm	0.02	**0.43**	0.02	**0.53**	0.03	**0.61**	0.03	**0.67**	0.04	**0.78**	0.04	**0.87**
Upper leg												
Knee												
Lower leg												
Lymph nodes	0.17	**0.35**	0.21	**0.42**	0.23	**0.48**	0.25	**0.53**	0.29	**0.62**	0.33	**0.70**
Skin	0.13	**0.25**	0.13	**0.26**	0.14	**0.28**	0.14	**0.28**	0.15	**0.30**	0.16	**0.32**
Muscle	0.10	**0.18**	0.12	**0.21**	0.13	**0.24**	0.15	**0.26**	0.17	**0.29**	0.19	**0.33**
Skeleton	0.20	**0.38**	0.24	**0.46**	0.27	**0.51**	0.29	**0.56**	0.34	**0.64**	0.37	**0.72**
Red bone marrow	0.07	**0.11**	0.08	**0.14**	0.09	**0.16**	0.10	**0.17**	0.12	**0.20**	0.13	**0.23**
Whole body dose	0.12	**0.21**	0.14	**0.25**	0.16	**0.29**	0.17	**0.31**	0.20	**0.36**	0.22	**0.40**
ED (ICRP 60)	0.22	**0.32**	0.26	**0.37**	0.29	**0.42**	0.31	**0.45**	0.35	**0.51**	0.38	**0.57**
ED (ICRP 103)	0.31	**0.41**	0.35	**0.48**	0.38	**0.53**	0.40	**0.56**	0.45	**0.63**	0.48	**0.70**
Total filtration (mm) Al	**1.5**		**2.5**		**3.5**		**4.5**		**3.5**		**3.5**	
Cu	**0**		**0**		**0**		**0**		**0.1**		**0.2**	

Table 9.132 Normalized organ doses—Chest ap—3–7 years

Beam projection:			ap				**Table cells:**					
Focus-to-detector distance (cm):			150				(+) = normal font = optimal field setting					
Tube voltage (kVp):			80				**(−) = bold font = suboptimal field setting**					
Total filtrations:			See last table row.				Empty table cells correspond to values < 0.01					
Organ or tissue	**Organ dose D_T : Entrance dose K_E**							**(mSv × mGy^{-1})**				
	(+)	(−)	(+)	(−)	(+)	(−)	(+)	(−)	(+)	(−)	(+)	(−)
Skull	0.02	**0.32**	0.03	**0.38**	0.03	**0.42**	0.04	**0.46**	0.04	**0.52**	0.05	**0.56**
Brain		**0.02**		**0.02**		**0.03**		**0.03**		**0.04**		**0.05**
Oral mucosa	0.03	**0.48**	0.04	**0.56**	0.04	**0.62**	0.05	**0.67**	0.06	**0.75**	0.06	**0.82**
Salivary glands	0.04	**0.40**	0.05	**0.48**	0.06	**0.53**	0.06	**0.58**	0.08	**0.65**	0.09	**0.72**
Cervical spine	0.36	**0.59**	0.44	**0.74**	0.51	**0.85**	0.56	**0.94**	0.66	**1.12**	0.75	**1.28**
Thyroid	0.94	**1.00**	1.05	**1.13**	1.11	**1.21**	1.16	**1.28**	1.25	**1.38**	1.32	**1.47**
Thymus	0.87	**0.88**	1.01	**1.03**	1.11	**1.14**	1.19	**1.22**	1.32	**1.36**	1.43	**1.47**
Thoracic spine	0.27	**0.30**	0.35	**0.39**	0.41	**0.46**	0.46	**0.52**	0.57	**0.64**	0.67	**0.77**
Claviculae	1.95	**2.06**	2.27	**2.40**	2.48	**2.62**	2.63	**2.79**	2.87	**3.06**	3.05	**3.26**
Scapulae	0.35	**0.38**	0.44	**0.48**	0.51	**0.56**	0.57	**0.62**	0.69	**0.75**	0.79	**0.86**
Ribs	1.51	**1.62**	1.71	**1.84**	1.84	**2.00**	1.94	**2.11**	2.10	**2.29**	2.22	**2.43**
Oesophagus	0.20	**0.22**	0.25	**0.28**	0.29	**0.32**	0.32	**0.36**	0.39	**0.43**	0.45	**0.50**
Trachea	0.09	**0.39**	0.10	**0.47**	0.11	**0.53**	0.12	**0.57**	0.13	**0.65**	0.15	**0.72**
Lung	0.46	**0.48**	0.55	**0.58**	0.62	**0.65**	0.67	**0.71**	0.76	**0.81**	0.85	**0.90**
Mamma	1.22	**1.23**	1.29	**1.32**	1.34	**1.37**	1.37	**1.40**	1.42	**1.46**	1.47	**1.51**
Heart	0.60	**0.62**	0.72	**0.75**	0.81	**0.84**	0.88	**0.91**	1.00	**1.05**	1.11	**1.16**
Stomach	0.31	**0.66**	0.38	**0.79**	0.42	**0.89**	0.46	**0.96**	0.53	**1.09**	0.59	**1.21**
Spleen	0.15	**0.23**	0.19	**0.29**	0.23	**0.34**	0.25	**0.38**	0.30	**0.45**	0.35	**0.53**
Liver	0.30	**0.54**	0.37	**0.65**	0.42	**0.73**	0.45	**0.79**	0.52	**0.91**	0.58	**1.01**
Gall bladder	0.11	**0.51**	0.14	**0.62**	0.16	**0.71**	0.18	**0.78**	0.21	**0.91**	0.25	**1.03**
Pancreas	0.25	**0.34**	0.32	**0.43**	0.37	**0.49**	0.41	**0.55**	0.49	**0.66**	0.56	**0.76**
Small intestine	0.01	**0.26**	0.02	**0.32**	0.02	**0.37**	0.02	**0.40**	0.03	**0.47**	0.04	**0.54**
Large intestine	0.02	**0.35**	0.02	**0.43**	0.03	**0.49**	0.03	**0.54**	0.04	**0.63**	0.05	**0.70**
Sigma/Rectum		**0.08**		**0.10**		**0.12**		**0.13**		**0.16**	0.01	**0.18**
Lumbar spine	0.06	**0.27**	0.08	**0.35**	0.10	**0.41**	0.11	**0.46**	0.14	**0.58**	0.17	**0.69**
Pelvis		**0.11**		**0.14**	0.01	**0.17**	0.01	**0.19**	0.02	**0.23**	0.02	**0.27**
Adrenals	0.12	**0.15**	0.16	**0.19**	0.18	**0.23**	0.21	**0.26**	0.25	**0.32**	0.30	**0.37**
Kidneys	0.05	**0.13**	0.06	**0.16**	0.08	**0.19**	0.08	**0.21**	0.11	**0.26**	0.13	**0.31**
Urinary bladder		**0.02**		**0.02**		**0.03**		**0.03**		**0.04**		**0.04**
Uterus		**0.05**		**0.06**		**0.07**		**0.08**		**0.10**	0.01	**0.12**
Ovaries		**0.05**		**0.07**		**0.08**		**0.09**		**0.11**	0.01	**0.13**
Prostate						**0.01**		**0.01**		**0.02**		**0.02**
Testes												
Upper arm	0.19	**0.86**	0.23	**1.04**	0.27	**1.17**	0.30	**1.27**	0.37	**1.47**	0.43	**1.63**
Elbow	0.18	**0.90**	0.23	**1.09**	0.26	**1.23**	0.30	**1.34**	0.36	**1.54**	0.41	**1.71**
Forearm	0.02	**0.49**	0.03	**0.59**	0.03	**0.67**	0.04	**0.73**	0.04	**0.83**	0.05	**0.92**
Upper leg												
Knee												
Lower leg												
Lymph nodes	0.19	**0.40**	0.23	**0.48**	0.26	**0.54**	0.28	**0.59**	0.32	**0.68**	0.36	**0.76**
Skin	0.13	**0.26**	0.14	**0.28**	0.15	**0.29**	0.15	**0.30**	0.16	**0.31**	0.17	**0.33**
Muscle	0.11	**0.20**	0.13	**0.24**	0.15	**0.26**	0.16	**0.28**	0.18	**0.32**	0.20	**0.35**
Skeleton	0.23	**0.42**	0.27	**0.50**	0.30	**0.56**	0.32	**0.61**	0.36	**0.69**	0.39	**0.76**
Red bone marrow	0.08	**0.13**	0.09	**0.16**	0.10	**0.18**	0.11	**0.19**	0.13	**0.23**	0.14	**0.26**
Whole body dose	0.13	**0.24**	0.16	**0.28**	0.18	**0.32**	0.19	**0.34**	0.22	**0.39**	0.24	**0.43**
ED (ICRP 60)	0.25	**0.35**	0.29	**0.42**	0.31	**0.46**	0.33	**0.49**	0.37	**0.56**	0.40	**0.61**
ED (ICRP 103)	0.33	**0.45**	0.38	**0.52**	0.41	**0.57**	0.43	**0.61**	0.47	**0.68**	0.51	**0.74**
Total filtration (mm) **Al**	**1.5**		**2.5**		**3.5**		**4.5**		**3.5**		**3.5**	
Cu	**0**		**0**		**0**		**0**		**0.1**		**0.2**	

Table 9.133 Normalized organ doses—Chest pa—1–2 years

Beam projection:			pa				**Table cells:**					
Focus-to-detector distance (cm):			150				(+) = normal font = optimal field setting					
Tube voltage (kVp):			50				**(−) = bold font = suboptimal field setting**					
Total filtrations:			See last table row.				Empty table cells correspond to values < 0.01					
Organ or tissue	**Organ dose D_T : Entrance dose K_E**								**(mSv x mGy^{-1})**			
	(+)	(−)	(+)	(−)	(+)	(−)	(+)	(−)	(+)	(−)	(+)	(−)
Skull	0.01	**0.13**	0.02	**0.17**	0.02	**0.20**	0.02	**0.22**	0.03	**0.26**	0.04	**0.30**
Brain		**0.01**		**0.02**		**0.02**		**0.03**		**0.04**		**0.04**
Oral mucosa	0.01	**0.15**	0.02	**0.20**	0.02	**0.23**	0.02	**0.25**	0.03	**0.30**	0.03	**0.34**
Salivary glands	0.03	**0.38**	0.04	**0.45**	0.04	**0.49**	0.05	**0.53**	0.06	**0.59**	0.07	**0.64**
Cervical spine	0.25	**0.56**	0.32	**0.74**	0.38	**0.87**	0.43	**0.98**	0.52	**1.19**	0.61	**1.39**
Thyroid	0.07	**0.08**	0.09	**0.11**	0.11	**0.14**	0.13	**0.16**	0.16	**0.20**	0.19	**0.24**
Thymus	0.04	**0.05**	0.06	**0.06**	0.07	**0.08**	0.08	**0.09**	0.11	**0.12**	0.13	**0.15**
Thoracic spine	0.74	**0.78**	0.95	**1.02**	1.12	**1.20**	1.26	**1.34**	1.51	**1.62**	1.76	**1.89**
Claviculae	0.16	**0.19**	0.21	**0.25**	0.26	**0.31**	0.30	**0.35**	0.37	**0.44**	0.45	**0.54**
Scapulae	2.08	**2.13**	2.43	**2.49**	2.65	**2.72**	2.81	**2.89**	3.08	**3.17**	3.30	**3.41**
Ribs	1.07	**1.19**	1.27	**1.41**	1.41	**1.56**	1.51	**1.68**	1.68	**1.88**	1.84	**2.06**
Oesophagus	0.14	**0.15**	0.18	**0.20**	0.22	**0.25**	0.25	**0.28**	0.32	**0.35**	0.39	**0.43**
Trachea	0.01	**0.05**	0.02	**0.07**	0.02	**0.08**	0.03	**0.09**	0.03	**0.12**	0.04	**0.15**
Lung	0.34	**0.35**	0.42	**0.44**	0.48	**0.50**	0.53	**0.55**	0.62	**0.65**	0.69	**0.73**
Mamma	0.10	**0.10**	0.13	**0.14**	0.16	**0.16**	0.18	**0.18**	0.22	**0.23**	0.26	**0.26**
Heart	0.10	**0.10**	0.13	**0.14**	0.16	**0.17**	0.18	**0.19**	0.22	**0.24**	0.27	**0.29**
Stomach	0.04	**0.09**	0.06	**0.12**	0.07	**0.15**	0.08	**0.17**	0.10	**0.21**	0.13	**0.26**
Spleen	0.21	**0.36**	0.27	**0.46**	0.31	**0.53**	0.35	**0.59**	0.41	**0.69**	0.46	**0.78**
Liver	0.09	**0.16**	0.12	**0.21**	0.14	**0.25**	0.16	**0.28**	0.19	**0.34**	0.23	**0.40**
Gall bladder	0.02	**0.08**	0.03	**0.11**	0.03	**0.13**	0.04	**0.15**	0.05	**0.19**	0.06	**0.24**
Pancreas	0.11	**0.15**	0.14	**0.20**	0.17	**0.25**	0.20	**0.28**	0.25	**0.35**	0.29	**0.42**
Small intestine		**0.08**		**0.11**		**0.13**		**0.15**	0.01	**0.18**	0.02	**0.22**
Large intestine		**0.08**		**0.11**		**0.13**	0.01	**0.15**	0.01	**0.18**	0.02	**0.22**
Sigma/Rectum		**0.02**		**0.03**		**0.03**		**0.04**		**0.05**		**0.06**
Lumbar spine	0.04	**0.80**	0.05	**1.03**	0.06	**1.21**	0.07	**1.35**	0.09	**1.62**	0.12	**1.87**
Pelvis		**0.06**		**0.08**		**0.10**		**0.11**		**0.14**	0.01	**0.17**
Adrenals	0.33	**0.37**	0.42	**0.47**	0.48	**0.54**	0.53	**0.60**	0.62	**0.70**	0.70	**0.80**
Kidneys	0.08	**0.58**	0.09	**0.70**	0.11	**0.78**	0.12	**0.84**	0.14	**0.95**	0.16	**1.05**
Urinary bladder										**0.01**		**0.02**
Uterus		**0.01**		**0.02**		**0.02**		**0.03**		**0.04**		**0.05**
Ovaries		**0.01**		**0.02**		**0.02**		**0.03**		**0.04**		**0.04**
Prostate												**0.01**
Testes												
Upper arm	0.09	**0.69**	0.13	**0.87**	0.16	**1.01**	0.18	**1.12**	0.23	**1.32**	0.28	**1.51**
Elbow	0.09	**0.76**	0.13	**0.97**	0.15	**1.12**	0.18	**1.24**	0.22	**1.45**	0.27	**1.66**
Forearm		**0.40**	0.01	**0.50**	0.01	**0.58**	0.02	**0.64**	0.02	**0.75**	0.03	**0.85**
Upper leg												
Knee												
Lower leg												
Lymph nodes	0.07	**0.14**	0.09	**0.18**	0.10	**0.21**	0.11	**0.23**	0.14	**0.28**	0.16	**0.33**
Skin	0.13	**0.27**	0.14	**0.28**	0.14	**0.29**	0.15	**0.30**	0.16	**0.31**	0.16	**0.33**
Muscle	0.08	**0.15**	0.10	**0.18**	0.11	**0.20**	0.12	**0.22**	0.13	**0.25**	0.15	**0.28**
Skeleton	0.30	**0.48**	0.36	**0.59**	0.41	**0.68**	0.44	**0.74**	0.51	**0.86**	0.57	**0.97**
Red bone marrow	0.07	**0.12**	0.09	**0.15**	0.10	**0.17**	0.11	**0.19**	0.13	**0.22**	0.15	**0.25**
Whole body dose	0.10	**0.18**	0.12	**0.22**	0.14	**0.25**	0.15	**0.27**	0.17	**0.31**	0.20	**0.35**
ED (ICRP 60)	0.08	**0.11**	0.11	**0.15**	0.12	**0.17**	0.14	**0.19**	0.16	**0.23**	0.19	**0.27**
ED (ICRP 103)	0.09	**0.13**	0.12	**0.17**	0.14	**0.20**	0.15	**0.22**	0.18	**0.26**	0.21	**0.30**
Total filtration (mm) **Al**	**1.5**		**2.5**		**3.5**		**4.5**		**3.5**		**3.5**	
Cu	**0**		**0**		**0**		**0**		**0.1**		**0.2**	

Table 9.134 Normalized organ doses—Chest pa—1–2 years

Beam projection:			pa			**Table cells:**						
Focus-to-detector distance (cm):			150			(+) = normal font = optimal field setting						
Tube voltage (kVp):			60			**(−) = bold font = suboptimal field setting**						
Total filtrations:			See last table row.			Empty table cells correspond to values < 0.01						
Organ or tissue	**Organ dose D_T : Entrance dose K_E**							**(mSv × mGy^{-1})**				
	(+)	(−)	(+)	(−)	(+)	(−)	(+)	(−)	(+)	(−)	(+)	(−)
Skull	0.02	**0.17**	0.02	**0.21**	0.03	**0.24**	0.03	**0.27**	0.04	**0.32**	0.04	**0.30**
Brain		**0.02**		**0.03**		**0.03**		**0.04**		**0.05**		**0.04**
Oral mucosa	0.02	**0.19**	0.02	**0.24**	0.03	**0.28**	0.03	**0.31**	0.04	**0.36**	0.03	**0.34**
Salivary glands	0.04	**0.43**	0.05	**0.50**	0.05	**0.55**	0.06	**0.59**	0.07	**0.66**	0.07	**0.64**
Cervical spine	0.32	**0.74**	0.41	**0.94**	0.48	**1.10**	0.54	**1.22**	0.65	**1.46**	0.61	**1.39**
Thyroid	0.10	**0.12**	0.13	**0.16**	0.15	**0.19**	0.18	**0.22**	0.22	**0.28**	0.19	**0.24**
Thymus	0.06	**0.07**	0.09	**0.10**	0.10	**0.12**	0.12	**0.13**	0.15	**0.17**	0.13	**0.15**
Thoracic spine	0.95	**1.01**	1.20	**1.29**	1.39	**1.50**	1.55	**1.67**	1.84	**1.98**	1.76	**1.89**
Claviculae	0.22	**0.27**	0.29	**0.35**	0.35	**0.42**	0.40	**0.48**	0.49	**0.59**	0.45	**0.54**
Scapulae	2.30	**2.36**	2.64	**2.72**	2.86	**2.95**	3.02	**3.12**	3.28	**3.39**	3.30	**3.41**
Ribs	1.22	**1.35**	1.43	**1.59**	1.57	**1.75**	1.67	**1.87**	1.85	**2.08**	1.84	**2.06**
Oesophagus	0.20	**0.22**	0.26	**0.29**	0.31	**0.34**	0.35	**0.39**	0.43	**0.48**	0.39	**0.43**
Trachea	0.02	**0.07**	0.03	**0.10**	0.03	**0.12**	0.04	**0.13**	0.05	**0.17**	0.04	**0.15**
Lung	0.41	**0.43**	0.50	**0.53**	0.57	**0.60**	0.62	**0.65**	0.71	**0.76**	0.69	**0.73**
Mamma	0.13	**0.14**	0.17	**0.18**	0.21	**0.21**	0.23	**0.24**	0.28	**0.28**	0.26	**0.26**
Heart	0.14	**0.15**	0.18	**0.19**	0.21	**0.23**	0.24	**0.26**	0.30	**0.32**	0.27	**0.29**
Stomach	0.06	**0.13**	0.08	**0.17**	0.10	**0.20**	0.11	**0.23**	0.14	**0.29**	0.13	**0.26**
Spleen	0.26	**0.45**	0.33	**0.55**	0.38	**0.64**	0.41	**0.70**	0.48	**0.81**	0.46	**0.78**
Liver	0.12	**0.21**	0.16	**0.27**	0.18	**0.32**	0.20	**0.35**	0.24	**0.43**	0.23	**0.40**
Gall bladder	0.03	**0.12**	0.04	**0.16**	0.05	**0.19**	0.06	**0.22**	0.07	**0.27**	0.06	**0.24**
Pancreas	0.15	**0.21**	0.20	**0.28**	0.23	**0.33**	0.26	**0.37**	0.32	**0.46**	0.29	**0.42**
Small intestine		**0.11**	0.01	**0.15**	0.01	**0.17**	0.02	**0.20**	0.02	**0.24**	0.02	**0.22**
Large intestine		**0.11**	0.01	**0.15**	0.02	**0.17**	0.02	**0.20**	0.02	**0.24**	0.02	**0.22**
Sigma/Rectum		**0.03**		**0.04**		**0.05**		**0.05**		**0.07**		**0.06**
Lumbar spine	0.06	**1.02**	0.08	**1.29**	0.09	**1.49**	0.11	**1.65**	0.14	**1.95**	0.12	**1.87**
Pelvis		**0.08**		**0.11**		**0.13**	0.01	**0.15**	0.01	**0.19**	0.01	**0.17**
Adrenals	0.40	**0.46**	0.50	**0.57**	0.57	**0.65**	0.62	**0.71**	0.72	**0.83**	0.70	**0.80**
Kidneys	0.09	**0.67**	0.12	**0.80**	0.13	**0.89**	0.15	**0.95**	0.17	**1.07**	0.16	**1.05**
Urinary bladder				**0.01**		**0.01**		**0.01**		**0.02**		**0.02**
Uterus		**0.02**		**0.03**		**0.04**		**0.04**		**0.06**		**0.05**
Ovaries		**0.02**		**0.03**		**0.04**		**0.04**		**0.05**		**0.04**
Prostate										**0.01**		**0.01**
Testes												
Upper arm	0.14	**0.85**	0.18	**1.06**	0.22	**1.21**	0.25	**1.33**	0.31	**1.56**	0.28	**1.51**
Elbow	0.13	**0.94**	0.18	**1.17**	0.21	**1.33**	0.24	**1.46**	0.30	**1.70**	0.27	**1.66**
Forearm	0.01	**0.49**	0.02	**0.60**	0.02	**0.69**	0.03	**0.75**	0.03	**0.87**	0.03	**0.85**
Upper leg												
Knee												
Lower leg												
Lymph nodes	0.09	**0.18**	0.11	**0.23**	0.13	**0.27**	0.15	**0.30**	0.18	**0.36**	0.16	**0.33**
Skin	0.14	**0.28**	0.15	**0.30**	0.15	**0.31**	0.16	**0.32**	0.17	**0.34**	0.16	**0.33**
Muscle	0.09	**0.18**	0.11	**0.21**	0.13	**0.23**	0.14	**0.25**	0.16	**0.29**	0.15	**0.28**
Skeleton	0.35	**0.58**	0.42	**0.70**	0.47	**0.80**	0.51	**0.87**	0.59	**1.00**	0.57	**0.97**
Red bone marrow	0.09	**0.15**	0.11	**0.18**	0.13	**0.21**	0.14	**0.23**	0.16	**0.27**	0.15	**0.25**
Whole body dose	0.12	**0.22**	0.15	**0.26**	0.16	**0.29**	0.18	**0.32**	0.20	**0.37**	0.20	**0.35**
ED (ICRP 60)	0.11	**0.15**	0.13	**0.19**	0.15	**0.22**	0.17	**0.24**	0.20	**0.29**	0.19	**0.27**
ED (ICRP 103)	0.12	**0.17**	0.15	**0.21**	0.17	**0.25**	0.19	**0.27**	0.22	**0.33**	0.21	**0.30**
Total filtration (mm) **Al**	**1.5**		**2.5**		**3.5**		**4.5**		**3.5**		**3.5**	
Cu	**0**		**0**		**0**		**0**		**0.1**		**0.2**	

Table 9.135 Normalized organ doses—Chest pa—1–2 years

Beam projection:			pa				**Table cells:**					
Focus-to-detector distance (cm):			150				(+) = normal font = optimal field setting					
Tube voltage (kVp):			70				**(−) = bold font = suboptimal field setting**					
Total filtrations:			See last table row.				Empty table cells correspond to values < 0.01					
Organ or tissue	**Organ dose D_T : Entrance dose K_E (mSv × mGy^{-1})**											
	(+)	(−)	(+)	(−)	(+)	(−)	(+)	(−)	(+)	(−)	(+)	(−)
Skull	0.02	**0.20**	0.03	**0.24**	0.03	**0.28**	0.04	**0.31**	0.05	**0.36**	0.06	**0.40**
Brain		**0.03**		**0.04**		**0.04**		**0.05**		**0.06**	0.01	**0.07**
Oral mucosa	0.02	**0.23**	0.03	**0.28**	0.03	**0.32**	0.04	**0.35**	0.05	**0.41**	0.05	**0.47**
Salivary glands	0.04	**0.47**	0.06	**0.55**	0.06	**0.60**	0.07	**0.64**	0.08	**0.71**	0.10	**0.76**
Cervical spine	0.39	**0.88**	0.49	**1.10**	0.56	**1.26**	0.62	**1.39**	0.73	**1.64**	0.83	**1.86**
Thyroid	0.13	**0.16**	0.16	**0.21**	0.19	**0.25**	0.22	**0.28**	0.27	**0.34**	0.32	**0.40**
Thymus	0.09	**0.10**	0.11	**0.13**	0.13	**0.15**	0.15	**0.17**	0.19	**0.21**	0.23	**0.26**
Thoracic spine	1.12	**1.21**	1.39	**1.50**	1.59	**1.72**	1.75	**1.89**	2.04	**2.22**	2.30	**2.51**
Claviculae	0.28	**0.34**	0.36	**0.43**	0.42	**0.51**	0.48	**0.57**	0.58	**0.70**	0.67	**0.81**
Scapulae	2.45	**2.52**	2.78	**2.87**	2.98	**3.08**	3.12	**3.23**	3.35	**3.48**	3.50	**3.65**
Ribs	1.33	**1.48**	1.53	**1.71**	1.67	**1.87**	1.77	**1.99**	1.94	**2.19**	2.07	**2.35**
Oesophagus	0.25	**0.28**	0.32	**0.36**	0.38	**0.42**	0.43	**0.48**	0.52	**0.58**	0.62	**0.69**
Trachea	0.03	**0.10**	0.03	**0.13**	0.04	**0.15**	0.05	**0.17**	0.06	**0.21**	0.07	**0.25**
Lung	0.47	**0.49**	0.56	**0.60**	0.63	**0.67**	0.69	**0.73**	0.78	**0.84**	0.87	**0.93**
Mamma	0.17	**0.17**	0.21	**0.22**	0.25	**0.25**	0.27	**0.28**	0.33	**0.33**	0.37	**0.38**
Heart	0.17	**0.19**	0.22	**0.24**	0.26	**0.28**	0.29	**0.32**	0.35	**0.39**	0.41	**0.46**
Stomach	0.08	**0.17**	0.11	**0.21**	0.12	**0.25**	0.14	**0.28**	0.17	**0.35**	0.20	**0.41**
Spleen	0.31	**0.52**	0.37	**0.64**	0.42	**0.72**	0.46	**0.78**	0.53	**0.90**	0.59	**1.01**
Liver	0.15	**0.26**	0.19	**0.33**	0.22	**0.38**	0.24	**0.42**	0.28	**0.49**	0.32	**0.57**
Gall bladder	0.04	**0.16**	0.06	**0.20**	0.07	**0.24**	0.08	**0.27**	0.09	**0.34**	0.11	**0.40**
Pancreas	0.19	**0.27**	0.24	**0.34**	0.28	**0.40**	0.31	**0.45**	0.38	**0.55**	0.44	**0.64**
Small intestine	0.01	**0.14**	0.01	**0.18**	0.02	**0.21**	0.02	**0.24**	0.03	**0.29**	0.03	**0.33**
Large intestine	0.01	**0.14**	0.02	**0.18**	0.02	**0.21**	0.02	**0.24**	0.03	**0.29**	0.04	**0.34**
Sigma/Rectum		**0.04**		**0.05**		**0.06**		**0.07**		**0.08**	0.01	**0.10**
Lumbar spine	0.08	**1.20**	0.10	**1.49**	0.12	**1.70**	0.14	**1.86**	0.17	**2.17**	0.21	**2.44**
Pelvis		**0.11**	0.01	**0.14**	0.01	**0.16**	0.02	**0.18**	0.02	**0.22**	0.03	**0.26**
Adrenals	0.47	**0.53**	0.57	**0.65**	0.64	**0.74**	0.70	**0.80**	0.80	**0.93**	0.89	**1.04**
Kidneys	0.11	**0.75**	0.13	**0.88**	0.15	**0.97**	0.17	**1.04**	0.19	**1.15**	0.22	**1.25**
Urinary bladder		**0.01**		**0.01**		**0.02**		**0.02**		**0.03**		**0.03**
Uterus		**0.03**		**0.04**		**0.05**		**0.06**	0.01	**0.07**	0.01	**0.09**
Ovaries		**0.03**		**0.04**		**0.05**		**0.06**	0.01	**0.07**	0.01	**0.08**
Prostate						**0.01**		**0.01**		**0.02**		**0.02**
Testes												**0.01**
Upper arm	0.18	**0.99**	0.23	**1.20**	0.27	**1.36**	0.30	**1.48**	0.37	**1.70**	0.43	**1.89**
Elbow	0.17	**1.08**	0.22	**1.32**	0.26	**1.49**	0.29	**1.62**	0.35	**1.85**	0.41	**2.05**
Forearm	0.02	**0.56**	0.02	**0.67**	0.03	**0.76**	0.03	**0.82**	0.04	**0.94**	0.05	**1.04**
Upper leg										**0.01**		**0.01**
Knee												
Lower leg												
Lymph nodes	0.11	**0.22**	0.14	**0.27**	0.16	**0.32**	0.17	**0.35**	0.21	**0.42**	0.24	**0.48**
Skin	0.14	**0.29**	0.15	**0.31**	0.16	**0.32**	0.17	**0.33**	0.18	**0.35**	0.19	**0.37**
Muscle	0.11	**0.20**	0.13	**0.24**	0.14	**0.26**	0.15	**0.28**	0.18	**0.32**	0.20	**0.35**
Skeleton	0.39	**0.66**	0.47	**0.79**	0.52	**0.88**	0.56	**0.95**	0.63	**1.09**	0.69	**1.20**
Red bone marrow	0.11	**0.17**	0.13	**0.21**	0.14	**0.24**	0.16	**0.26**	0.18	**0.30**	0.20	**0.34**
Whole body dose	0.14	**0.25**	0.16	**0.30**	0.18	**0.33**	0.20	**0.36**	0.23	**0.41**	0.25	**0.45**
ED (ICRP 60)	0.13	**0.18**	0.16	**0.22**	0.18	**0.26**	0.19	**0.28**	0.23	**0.33**	0.26	**0.38**
ED (ICRP 103)	0.14	**0.20**	0.17	**0.25**	0.20	**0.29**	0.22	**0.32**	0.26	**0.38**	0.29	**0.43**
Total filtration (mm) Al	**1.5**		**2.5**		**3.5**		**4.5**		**3.5**		**3.5**	
Cu	**0**		**0**		**0**		**0**		**0.1**		**0.2**	

Table 9.136 Normalized organ doses—Chest pa—1–2 years

Beam projection:			pa			**Table cells:**						
Focus-to-detector distance (cm):			150			(+) = normal font = optimal field setting						
Tube voltage (kVp):			80			**(−) = bold font = suboptimal field setting**						
Total filtrations:			See last table row.			Empty table cells correspond to values < 0.01						
Organ or tissue	**Organ dose D_T : Entrance dose K_E**							**($mSv \times mGy^{-1}$)**				
	(+)	(−)	(+)	(−)	(+)	(−)	(+)	(−)	(+)	(−)	(+)	(−)
Skull	0.03	**0.23**	0.04	**0.27**	0.04	**0.31**	0.04	**0.33**	0.05	**0.38**	0.06	**0.42**
Brain		**0.04**		**0.04**		**0.05**		**0.06**	0.01	**0.07**	0.01	**0.08**
Oral mucosa	0.03	**0.27**	0.03	**0.32**	0.04	**0.36**	0.04	**0.39**	0.05	**0.46**	0.06	**0.51**
Salivary glands	0.05	**0.52**	0.06	**0.59**	0.07	**0.64**	0.08	**0.68**	0.10	**0.75**	0.11	**0.80**
Cervical spine	0.45	**1.01**	0.55	**1.24**	0.62	**1.40**	0.68	**1.53**	0.79	**1.77**	0.88	**1.97**
Thyroid	0.16	**0.20**	0.20	**0.25**	0.23	**0.29**	0.26	**0.33**	0.32	**0.40**	0.37	**0.46**
Thymus	0.11	**0.12**	0.14	**0.16**	0.16	**0.18**	0.18	**0.21**	0.22	**0.25**	0.26	**0.30**
Thoracic spine	1.27	**1.38**	1.55	**1.68**	1.75	**1.90**	1.90	**2.07**	2.18	**2.38**	2.42	**2.65**
Claviculae	0.34	**0.40**	0.42	**0.51**	0.49	**0.59**	0.54	**0.65**	0.64	**0.77**	0.73	**0.88**
Scapulae	2.57	**2.65**	2.87	**2.97**	3.05	**3.16**	3.17	**3.29**	3.35	**3.49**	3.46	**3.62**
Ribs	1.42	**1.59**	1.61	**1.81**	1.74	**1.96**	1.83	**2.06**	1.98	**2.24**	2.09	**2.38**
Oesophagus	0.31	**0.34**	0.39	**0.43**	0.45	**0.50**	0.50	**0.56**	0.60	**0.67**	0.70	**0.78**
Trachea	0.03	**0.12**	0.04	**0.16**	0.05	**0.18**	0.06	**0.20**	0.07	**0.25**	0.08	**0.29**
Lung	0.52	**0.55**	0.62	**0.66**	0.69	**0.74**	0.75	**0.79**	0.84	**0.90**	0.92	**0.99**
Mamma	0.20	**0.20**	0.25	**0.25**	0.28	**0.29**	0.31	**0.32**	0.37	**0.38**	0.42	**0.43**
Heart	0.21	**0.23**	0.26	**0.29**	0.31	**0.34**	0.34	**0.37**	0.41	**0.45**	0.47	**0.52**
Stomach	0.10	**0.20**	0.13	**0.26**	0.15	**0.30**	0.16	**0.33**	0.20	**0.40**	0.23	**0.46**
Spleen	0.35	**0.59**	0.42	**0.71**	0.47	**0.79**	0.51	**0.86**	0.58	**0.98**	0.64	**1.08**
Liver	0.18	**0.31**	0.22	**0.38**	0.25	**0.43**	0.27	**0.47**	0.32	**0.55**	0.36	**0.63**
Gall bladder	0.05	**0.20**	0.07	**0.25**	0.08	**0.29**	0.09	**0.33**	0.11	**0.40**	0.13	**0.47**
Pancreas	0.23	**0.33**	0.28	**0.41**	0.33	**0.47**	0.36	**0.52**	0.43	**0.62**	0.50	**0.72**
Small intestine	0.01	**0.17**	0.02	**0.21**	0.02	**0.25**	0.03	**0.27**	0.03	**0.33**	0.04	**0.37**
Large intestine	0.02	**0.17**	0.02	**0.22**	0.03	**0.25**	0.03	**0.28**	0.04	**0.34**	0.04	**0.39**
Sigma/Rectum		**0.05**		**0.06**		**0.07**		**0.08**	0.01	**0.10**	0.01	**0.11**
Lumbar spine	0.10	**1.37**	0.13	**1.65**	0.15	**1.86**	0.17	**2.02**	0.20	**2.32**	0.24	**2.56**
Pelvis	0.01	**0.13**	0.02	**0.16**	0.02	**0.19**	0.02	**0.21**	0.03	**0.25**	0.03	**0.29**
Adrenals	0.53	**0.61**	0.63	**0.73**	0.71	**0.82**	0.77	**0.88**	0.87	**1.01**	0.96	**1.11**
Kidneys	0.13	**0.82**	0.15	**0.95**	0.17	**1.04**	0.18	**1.11**	0.21	**1.22**	0.24	**1.32**
Urinary bladder		**0.01**		**0.02**		**0.02**		**0.03**		**0.03**		**0.04**
Uterus		**0.04**		**0.05**		**0.06**	0.01	**0.07**	0.01	**0.08**	0.02	**0.10**
Ovaries		**0.04**		**0.05**	0.01	**0.06**	0.01	**0.07**	0.01	**0.09**	0.02	**0.10**
Prostate				**0.01**		**0.01**		**0.01**		**0.02**		**0.02**
Testes										**0.01**		**0.01**
Upper arm	0.21	**1.10**	0.27	**1.32**	0.31	**1.47**	0.34	**1.59**	0.41	**1.80**	0.47	**1.97**
Elbow	0.20	**1.21**	0.25	**1.44**	0.29	**1.60**	0.32	**1.73**	0.39	**1.95**	0.44	**2.12**
Forearm	0.02	**0.62**	0.03	**0.73**	0.04	**0.82**	0.04	**0.88**	0.05	**0.98**	0.06	**1.07**
Upper leg								**0.01**		**0.01**		**0.02**
Knee												
Lower leg												
Lymph nodes	0.13	**0.26**	0.16	**0.32**	0.18	**0.36**	0.20	**0.40**	0.23	**0.47**	0.27	**0.53**
Skin	0.15	**0.30**	0.16	**0.32**	0.17	**0.34**	0.18	**0.35**	0.19	**0.37**	0.20	**0.39**
Muscle	0.12	**0.22**	0.14	**0.26**	0.16	**0.29**	0.17	**0.31**	0.19	**0.35**	0.21	**0.38**
Skeleton	0.43	**0.73**	0.51	**0.86**	0.56	**0.95**	0.59	**1.02**	0.66	**1.14**	0.71	**1.24**
Red bone marrow	0.12	**0.20**	0.14	**0.24**	0.16	**0.27**	0.17	**0.29**	0.20	**0.33**	0.22	**0.37**
Whole body dose	0.15	**0.28**	0.18	**0.33**	0.20	**0.36**	0.21	**0.39**	0.24	**0.44**	0.27	**0.48**
ED (ICRP 60)	0.15	**0.21**	0.18	**0.26**	0.20	**0.29**	0.22	**0.32**	0.25	**0.37**	0.28	**0.42**
ED (ICRP 103)	0.16	**0.24**	0.20	**0.29**	0.22	**0.33**	0.24	**0.36**	0.28	**0.42**	0.32	**0.47**
Total filtration (mm) **Al**	**1.5**		**2.5**		**3.5**		**4.5**		**3.5**		**3.5**	
Cu	**0**		**0**		**0**		**0**		**0.1**		**0.2**	

Table 9.137 Normalized organ doses—Chest pa—3–7 years

Beam projection:			pa			**Table cells:**						
Focus-to-detector distance (cm):			150			(+) = normal font = optimal field setting						
Tube voltage (kVp):			60			**(−) = bold font = suboptimal field setting**						
Total filtrations:			See last table row.			Empty table cells correspond to values < 0.01						
Organ or tissue	**Organ dose D_T : Entrance dose K_E**							**(mSv × mGy^{-1})**				
	(+)	(−)	(+)	(−)	(+)	(−)	(+)	(−)	(+)	(−)	(+)	(−)
Skull	0.01	**0.16**	0.02	**0.20**	0.02	**0.23**	0.02	**0.26**	0.03	**0.31**	0.02	**0.29**
Brain		**0.02**		**0.02**		**0.03**		**0.03**		**0.04**		**0.03**
Oral mucosa	0.01	**0.16**	0.01	**0.20**	0.02	**0.23**	0.02	**0.26**	0.03	**0.31**	0.02	**0.29**
Salivary glands	0.02	**0.40**	0.02	**0.47**	0.03	**0.52**	0.03	**0.56**	0.04	**0.62**	0.03	**0.61**
Cervical spine	0.32	**0.80**	0.41	**1.02**	0.48	**1.18**	0.53	**1.32**	0.64	**1.57**	0.60	**1.49**
Thyroid	0.08	**0.11**	0.11	**0.14**	0.13	**0.17**	0.15	**0.20**	0.19	**0.25**	0.16	**0.21**
Thymus	0.05	**0.06**	0.07	**0.07**	0.08	**0.09**	0.10	**0.11**	0.12	**0.14**	0.10	**0.11**
Thoracic spine	0.84	**0.88**	1.08	**1.13**	1.26	**1.32**	1.41	**1.48**	1.69	**1.78**	1.59	**1.67**
Claviculae	0.16	**0.20**	0.21	**0.26**	0.25	**0.31**	0.29	**0.36**	0.36	**0.45**	0.32	**0.40**
Scapulae	2.11	**2.21**	2.45	**2.57**	2.68	**2.81**	2.84	**2.98**	3.11	**3.27**	3.12	**3.28**
Ribs	1.20	**1.27**	1.40	**1.50**	1.55	**1.66**	1.65	**1.77**	1.84	**1.99**	1.82	**1.95**
Oesophagus	0.12	**0.13**	0.17	**0.18**	0.20	**0.21**	0.23	**0.25**	0.29	**0.31**	0.25	**0.27**
Trachea	0.02	**0.06**	0.02	**0.09**	0.03	**0.10**	0.03	**0.12**	0.04	**0.15**	0.03	**0.13**
Lung	0.36	**0.38**	0.45	**0.47**	0.52	**0.54**	0.57	**0.60**	0.66	**0.70**	0.64	**0.67**
Mamma	0.11	**0.11**	0.14	**0.15**	0.17	**0.18**	0.19	**0.20**	0.23	**0.24**	0.21	**0.22**
Heart	0.12	**0.13**	0.16	**0.17**	0.19	**0.20**	0.21	**0.23**	0.27	**0.29**	0.24	**0.25**
Stomach	0.06	**0.11**	0.08	**0.15**	0.09	**0.18**	0.11	**0.20**	0.14	**0.26**	0.12	**0.22**
Spleen	0.27	**0.41**	0.35	**0.52**	0.40	**0.60**	0.44	**0.66**	0.52	**0.78**	0.50	**0.74**
Liver	0.12	**0.19**	0.15	**0.25**	0.18	**0.29**	0.20	**0.33**	0.24	**0.40**	0.22	**0.36**
Gall bladder	0.03	**0.09**	0.04	**0.12**	0.05	**0.15**	0.06	**0.17**	0.07	**0.23**	0.06	**0.19**
Pancreas	0.14	**0.18**	0.18	**0.24**	0.21	**0.28**	0.24	**0.32**	0.30	**0.40**	0.27	**0.35**
Small intestine		**0.11**		**0.14**		**0.17**	0.01	**0.19**	0.01	**0.24**	0.01	**0.21**
Large intestine		**0.10**		**0.13**	0.01	**0.15**	0.01	**0.18**	0.02	**0.22**	0.01	**0.19**
Sigma/Rectum		**0.03**		**0.04**		**0.04**		**0.05**		**0.06**		**0.06**
Lumbar spine	0.05	**0.91**	0.07	**1.17**	0.09	**1.36**	0.10	**1.51**	0.13	**1.82**	0.11	**1.71**
Pelvis		**0.15**		**0.19**		**0.23**		**0.25**		**0.31**		**0.28**
Adrenals	0.38	**0.41**	0.48	**0.52**	0.55	**0.60**	0.61	**0.67**	0.71	**0.79**	0.68	**0.75**
Kidneys	0.12	**0.63**	0.15	**0.76**	0.18	**0.85**	0.19	**0.93**	0.23	**1.05**	0.22	**1.02**
Urinary bladder								**0.01**		**0.02**		**0.01**
Uterus		**0.02**		**0.02**		**0.03**		**0.03**		**0.04**		**0.03**
Ovaries		**0.02**		**0.03**		**0.04**		**0.05**		**0.06**		**0.05**
Prostate												
Testes												
Upper arm	0.10	**0.61**	0.13	**0.77**	0.16	**0.89**	0.18	**0.99**	0.23	**1.18**	0.20	**1.12**
Elbow	0.11	**0.66**	0.14	**0.84**	0.17	**0.97**	0.20	**1.08**	0.25	**1.28**	0.22	**1.22**
Forearm	0.01	**0.37**	0.01	**0.46**	0.02	**0.53**	0.02	**0.59**	0.03	**0.70**	0.02	**0.67**
Upper leg												
Knee												
Lower leg												
Lymph nodes	0.07	**0.16**	0.10	**0.20**	0.11	**0.24**	0.13	**0.26**	0.16	**0.32**	0.14	**0.29**
Skin	0.12	**0.23**	0.13	**0.24**	0.13	**0.25**	0.14	**0.26**	0.14	**0.27**	0.14	**0.27**
Muscle	0.08	**0.15**	0.10	**0.18**	0.11	**0.20**	0.12	**0.22**	0.14	**0.25**	0.13	**0.24**
Skeleton	0.27	**0.43**	0.32	**0.53**	0.36	**0.60**	0.39	**0.66**	0.45	**0.77**	0.44	**0.74**
Red bone marrow	0.09	**0.14**	0.10	**0.17**	0.12	**0.19**	0.13	**0.22**	0.15	**0.25**	0.14	**0.24**
Whole body dose	0.11	**0.19**	0.13	**0.23**	0.14	**0.26**	0.16	**0.28**	0.18	**0.32**	0.17	**0.31**
ED (ICRP 60)	0.09	**0.13**	0.12	**0.16**	0.13	**0.19**	0.15	**0.21**	0.18	**0.26**	0.17	**0.24**
ED (ICRP 103)	0.10	**0.15**	0.13	**0.19**	0.15	**0.22**	0.17	**0.24**	0.20	**0.29**	0.19	**0.27**
Total filtration (mm) **Al**	**1.5**		**2.5**		**3.5**		**4.5**		**3.5**		**3.5**	
Cu	**0**		**0**		**0**		**0**		**0.1**		**0.2**	

Table 9.138 Normalized organ doses—Chest pa—3–7 years

Beam projection:	pa	**Table cells:**	
Focus-to-detector distance (cm):	150		(+) = normal font = optimal field setting
Tube voltage (kVp):	70		(−) = **bold font = suboptimal field setting**
Total filtrations:	See last table row.		Empty table cells correspond to values < 0.01

Organ or tissue	**Organ dose D_T : Entrance dose K_E**								**(mSv × mGy⁻¹)**			
	(+)	(−)	(+)	(−)	(+)	(−)	(+)	(−)	(+)	(−)	(+)	(−)
Skull	0.02	**0.19**	0.02	**0.24**	0.03	**0.27**	0.03	**0.30**	0.04	**0.35**	0.04	**0.40**
Brain		**0.02**		**0.03**		**0.03**		**0.04**		**0.05**		**0.06**
Oral mucosa	0.01	**0.19**	0.02	**0.24**	0.02	**0.27**	0.03	**0.30**	0.03	**0.36**	0.04	**0.41**
Salivary glands	0.02	**0.44**	0.03	**0.52**	0.04	**0.57**	0.04	**0.61**	0.05	**0.67**	0.06	**0.73**
Cervical spine	0.39	**0.96**	0.48	**1.19**	0.56	**1.37**	0.61	**1.51**	0.72	**1.78**	0.83	**2.02**
Thyroid	0.11	**0.14**	0.14	**0.19**	0.17	**0.22**	0.19	**0.25**	0.24	**0.31**	0.28	**0.37**
Thymus	0.07	**0.08**	0.09	**0.10**	0.11	**0.12**	0.13	**0.14**	0.16	**0.18**	0.19	**0.22**
Thoracic spine	1.01	**1.07**	1.27	**1.34**	1.47	**1.55**	1.63	**1.72**	1.93	**2.04**	2.20	**2.33**
Claviculae	0.20	**0.26**	0.27	**0.33**	0.32	**0.39**	0.36	**0.44**	0.44	**0.55**	0.53	**0.65**
Scapulae	2.28	**2.39**	2.60	**2.73**	2.81	**2.96**	2.97	**3.12**	3.21	**3.38**	3.39	**3.58**
Ribs	1.31	**1.40**	1.52	**1.63**	1.66	**1.79**	1.76	**1.90**	1.94	**2.11**	2.08	**2.28**
Oesophagus	0.17	**0.18**	0.22	**0.23**	0.26	**0.28**	0.29	**0.31**	0.36	**0.39**	0.43	**0.47**
Trachea	0.02	**0.09**	0.03	**0.11**	0.03	**0.13**	0.04	**0.15**	0.05	**0.19**	0.06	**0.23**
Lung	0.42	**0.44**	0.52	**0.55**	0.59	**0.62**	0.64	**0.68**	0.74	**0.78**	0.83	**0.88**
Mamma	0.14	**0.15**	0.18	**0.19**	0.21	**0.22**	0.23	**0.24**	0.28	**0.29**	0.32	**0.34**
Heart	0.15	**0.17**	0.20	**0.21**	0.24	**0.25**	0.27	**0.29**	0.33	**0.35**	0.39	**0.42**
Stomach	0.08	**0.15**	0.10	**0.19**	0.12	**0.23**	0.14	**0.26**	0.17	**0.32**	0.20	**0.38**
Spleen	0.33	**0.49**	0.40	**0.60**	0.46	**0.69**	0.50	**0.76**	0.59	**0.88**	0.66	**0.99**
Liver	0.15	**0.24**	0.18	**0.30**	0.21	**0.35**	0.24	**0.39**	0.29	**0.47**	0.33	**0.54**
Gall bladder	0.04	**0.13**	0.06	**0.17**	0.07	**0.20**	0.08	**0.23**	0.10	**0.29**	0.12	**0.36**
Pancreas	0.18	**0.23**	0.23	**0.30**	0.27	**0.36**	0.30	**0.40**	0.37	**0.49**	0.44	**0.59**
Small intestine		**0.14**	0.01	**0.18**	0.01	**0.21**	0.02	**0.24**	0.02	**0.30**	0.03	**0.35**
Large intestine	0.01	**0.13**	0.01	**0.17**	0.02	**0.20**	0.02	**0.22**	0.02	**0.28**	0.03	**0.33**
Sigma/Rectum		**0.04**		**0.05**		**0.06**		**0.07**		**0.08**		**0.10**
Lumbar spine	0.08	**1.10**	0.10	**1.37**	0.12	**1.58**	0.14	**1.75**	0.17	**2.06**	0.21	**2.35**
Pelvis		**0.19**		**0.23**		**0.27**		**0.30**	0.01	**0.35**	0.02	**0.41**
Adrenals	0.45	**0.49**	0.55	**0.61**	0.63	**0.70**	0.69	**0.77**	0.80	**0.90**	0.90	**1.02**
Kidneys	0.14	**0.71**	0.18	**0.85**	0.20	**0.94**	0.22	**1.02**	0.26	**1.15**	0.29	**1.26**
Urinary bladder				**0.01**		**0.01**		**0.02**		**0.02**		**0.03**
Uterus		**0.03**		**0.03**		**0.04**		**0.05**		**0.06**		**0.08**
Ovaries		**0.03**		**0.05**		**0.06**		**0.06**		**0.08**		**0.10**
Prostate										**0.01**		**0.02**
Testes												
Upper arm	0.13	**0.72**	0.17	**0.90**	0.20	**1.03**	0.23	**1.13**	0.29	**1.32**	0.34	**1.50**
Elbow	0.14	**0.78**	0.18	**0.97**	0.22	**1.11**	0.25	**1.22**	0.30	**1.43**	0.36	**1.61**
Forearm	0.02	**0.43**	0.02	**0.53**	0.02	**0.61**	0.03	**0.67**	0.04	**0.78**	0.04	**0.87**
Upper leg												
Knee												
Lower leg												
Lymph nodes	0.09	**0.20**	0.12	**0.25**	0.14	**0.29**	0.15	**0.32**	0.19	**0.38**	0.22	**0.45**
Skin	0.13	**0.24**	0.13	**0.25**	0.14	**0.26**	0.14	**0.27**	0.15	**0.29**	0.16	**0.30**
Muscle	0.09	**0.17**	0.11	**0.20**	0.12	**0.23**	0.13	**0.24**	0.15	**0.28**	0.17	**0.31**
Skeleton	0.30	**0.50**	0.36	**0.60**	0.40	**0.68**	0.43	**0.74**	0.49	**0.85**	0.55	**0.95**
Red bone marrow	0.10	**0.16**	0.12	**0.20**	0.14	**0.23**	0.15	**0.25**	0.17	**0.29**	0.19	**0.34**
Whole body dose	0.12	**0.22**	0.15	**0.26**	0.16	**0.29**	0.18	**0.32**	0.20	**0.36**	0.23	**0.41**
ED (ICRP 60)	0.11	**0.16**	0.14	**0.20**	0.16	**0.23**	0.17	**0.25**	0.21	**0.30**	0.24	**0.35**
ED (ICRP 103)	0.12	**0.18**	0.15	**0.23**	0.18	**0.26**	0.20	**0.29**	0.23	**0.34**	0.27	**0.40**
Total filtration (mm) **Al**	**1.5**		**2.5**		**3.5**		**4.5**		**3.5**		**3.5**	
Cu	**0**		**0**		**0**		**0**		**0.1**		**0.2**	

Table 9.139 Normalized organ doses—Chest pa—3–7 years

Beam projection:		pa				**Table cells:**						
Focus-to-detector distance (cm):		150				(+) = normal font = optimal field setting						
Tube voltage (kVp):		80				**(−) = bold font = suboptimal field setting**						
Total filtrations:		See last table row.				Empty table cells correspond to values < 0.01						
Organ or tissue	**Organ dose D_T : Entrance dose K_E**							**(mSv × mGy^{-1})**				
	(+)	(−)	(+)	(−)	(+)	(−)	(+)	(−)	(+)	(−)	(+)	(−)
Skull	0.02	**0.22**	0.03	**0.27**	0.03	**0.30**	0.04	**0.33**	0.04	**0.39**	0.05	**0.43**
Brain		**0.03**		**0.04**		**0.04**		**0.05**		**0.06**	0.01	**0.07**
Oral mucosa	0.02	**0.23**	0.02	**0.28**	0.03	**0.31**	0.03	**0.34**	0.04	**0.40**	0.05	**0.46**
Salivary glands	0.03	**0.49**	0.04	**0.56**	0.04	**0.61**	0.05	**0.65**	0.06	**0.72**	0.07	**0.78**
Cervical spine	0.45	**1.10**	0.55	**1.34**	0.62	**1.52**	0.68	**1.66**	0.79	**1.92**	0.89	**2.15**
Thyroid	0.14	**0.18**	0.17	**0.23**	0.20	**0.27**	0.23	**0.30**	0.28	**0.37**	0.33	**0.44**
Thymus	0.09	**0.10**	0.12	**0.13**	0.14	**0.15**	0.16	**0.17**	0.19	**0.22**	0.23	**0.26**
Thoracic spine	1.18	**1.24**	1.44	**1.53**	1.64	**1.74**	1.80	**1.91**	2.10	**2.23**	2.35	**2.51**
Claviculae	0.25	**0.31**	0.32	**0.40**	0.37	**0.46**	0.42	**0.52**	0.51	**0.63**	0.59	**0.73**
Scapulae	2.40	**2.53**	2.71	**2.85**	2.90	**3.06**	3.03	**3.20**	3.24	**3.42**	3.37	**3.57**
Ribs	1.41	**1.51**	1.60	**1.73**	1.74	**1.88**	1.83	**1.99**	1.99	**2.18**	2.12	**2.32**
Oesophagus	0.21	**0.23**	0.27	**0.29**	0.31	**0.34**	0.35	**0.38**	0.42	**0.46**	0.50	**0.55**
Trachea	0.03	**0.11**	0.04	**0.14**	0.04	**0.16**	0.05	**0.19**	0.06	**0.23**	0.07	**0.27**
Lung	0.48	**0.51**	0.58	**0.61**	0.65	**0.69**	0.70	**0.75**	0.80	**0.85**	0.89	**0.95**
Mamma	0.17	**0.18**	0.21	**0.22**	0.24	**0.26**	0.27	**0.29**	0.32	**0.34**	0.36	**0.39**
Heart	0.19	**0.21**	0.24	**0.26**	0.28	**0.31**	0.32	**0.34**	0.38	**0.41**	0.45	**0.48**
Stomach	0.10	**0.18**	0.13	**0.23**	0.15	**0.27**	0.16	**0.31**	0.20	**0.37**	0.24	**0.44**
Spleen	0.37	**0.56**	0.45	**0.68**	0.51	**0.77**	0.56	**0.84**	0.64	**0.97**	0.72	**1.08**
Liver	0.18	**0.29**	0.22	**0.36**	0.25	**0.41**	0.28	**0.45**	0.33	**0.53**	0.37	**0.61**
Gall bladder	0.06	**0.17**	0.07	**0.22**	0.09	**0.25**	0.10	**0.29**	0.12	**0.36**	0.14	**0.42**
Pancreas	0.22	**0.29**	0.27	**0.37**	0.32	**0.43**	0.36	**0.48**	0.43	**0.58**	0.50	**0.67**
Small intestine	0.01	**0.18**	0.02	**0.22**	0.02	**0.26**	0.02	**0.29**	0.03	**0.35**	0.03	**0.40**
Large intestine	0.01	**0.16**	0.02	**0.20**	0.02	**0.24**	0.03	**0.27**	0.03	**0.32**	0.04	**0.38**
Sigma/Rectum		**0.05**		**0.06**		**0.07**		**0.08**		**0.10**		**0.12**
Lumbar spine	0.10	**1.27**	0.13	**1.56**	0.15	**1.77**	0.17	**1.93**	0.21	**2.24**	0.25	**2.51**
Pelvis		**0.22**		**0.27**	0.01	**0.31**	0.01	**0.34**	0.02	**0.39**	0.02	**0.45**
Adrenals	0.52	**0.57**	0.62	**0.69**	0.70	**0.79**	0.76	**0.86**	0.88	**0.99**	0.98	**1.11**
Kidneys	0.17	**0.79**	0.20	**0.93**	0.23	**1.03**	0.25	**1.10**	0.28	**1.23**	0.32	**1.33**
Urinary bladder		**0.01**		**0.02**		**0.02**		**0.02**		**0.03**		**0.03**
Uterus		**0.04**		**0.05**		**0.05**		**0.06**		**0.08**	0.01	**0.10**
Ovaries		**0.05**		**0.06**		**0.07**		**0.08**		**0.10**		**0.12**
Prostate						**0.01**		**0.01**		**0.02**		**0.02**
Testes												
Upper arm	0.16	**0.83**	0.21	**1.01**	0.24	**1.14**	0.27	**1.24**	0.33	**1.43**	0.39	**1.59**
Elbow	0.17	**0.89**	0.22	**1.08**	0.26	**1.22**	0.29	**1.33**	0.35	**1.53**	0.40	**1.70**
Forearm	0.02	**0.49**	0.03	**0.59**	0.03	**0.67**	0.03	**0.72**	0.04	**0.83**	0.05	**0.91**
Upper leg												
Knee												
Lower leg												
Lymph nodes	0.11	**0.24**	0.14	**0.29**	0.16	**0.34**	0.18	**0.37**	0.21	**0.44**	0.25	**0.51**
Skin	0.13	**0.25**	0.14	**0.26**	0.15	**0.27**	0.15	**0.28**	0.16	**0.30**	0.17	**0.32**
Muscle	0.10	**0.19**	0.12	**0.23**	0.14	**0.25**	0.15	**0.27**	0.17	**0.31**	0.19	**0.34**
Skeleton	0.33	**0.56**	0.39	**0.66**	0.43	**0.74**	0.47	**0.80**	0.52	**0.90**	0.57	**0.99**
Red bone marrow	0.11	**0.19**	0.14	**0.23**	0.15	**0.26**	0.17	**0.28**	0.19	**0.33**	0.21	**0.37**
Whole body dose	0.14	**0.24**	0.16	**0.29**	0.18	**0.32**	0.19	**0.35**	0.22	**0.40**	0.25	**0.44**
ED (ICRP 60)	0.13	**0.19**	0.16	**0.23**	0.18	**0.26**	0.20	**0.29**	0.23	**0.34**	0.26	**0.39**
ED (ICRP 103)	0.15	**0.21**	0.18	**0.26**	0.20	**0.30**	0.22	**0.33**	0.26	**0.39**	0.30	**0.44**
Total filtration (mm) **Al**	**1.5**		**2.5**		**3.5**		**4.5**		**3.5**		**3.5**	
Cu	**0**		**0**		**0**		**0**		**0.1**		**0.2**	

Table 9.140 Normalized organ doses—Chest pa—8–12 years

Beam projection:		pa				**Table cells:**						
Focus-to-detector distance (cm):		150				(+) = normal font = optimal field setting						
Tube voltage (kVp):		60				**(−) = bold font = suboptimal field setting**						
Total filtrations:		See last table row.				Empty table cells correspond to values < 0.01						
Organ or tissue	**Organ dose D_T : Entrance dose K_E**							**(mSv × mGy^{-1})**				
	(+)	(−)	(+)	(−)	(+)	(−)	(+)	(−)	(+)	(−)	(+)	(−)
Skull	0.01	**0.14**	0.01	**0.18**	0.02	**0.21**	0.02	**0.24**	0.02	**0.29**	0.02	**0.26**
Brain		**0.01**		**0.02**		**0.02**		**0.03**		**0.04**		**0.03**
Oral mucosa		**0.13**	0.01	**0.16**	0.01	**0.19**	0.02	**0.21**	0.02	**0.25**	0.02	**0.23**
Salivary glands	0.02	**0.37**	0.02	**0.44**	0.03	**0.49**	0.03	**0.53**	0.04	**0.59**	0.03	**0.58**
Cervical spine	0.34	**0.79**	0.44	**1.01**	0.51	**1.17**	0.58	**1.31**	0.69	**1.58**	0.64	**1.48**
Thyroid	0.07	**0.09**	0.09	**0.12**	0.11	**0.14**	0.13	**0.16**	0.17	**0.21**	0.14	**0.18**
Thymus	0.04	**0.04**	0.05	**0.06**	0.07	**0.07**	0.08	**0.08**	0.10	**0.11**	0.08	**0.09**
Thoracic spine	0.74	**0.77**	0.96	**1.00**	1.13	**1.17**	1.27	**1.32**	1.54	**1.61**	1.43	**1.48**
Claviculae	0.12	**0.14**	0.16	**0.19**	0.19	**0.23**	0.22	**0.27**	0.29	**0.34**	0.24	**0.29**
Scapulae	2.05	**2.10**	2.39	**2.46**	2.63	**2.70**	2.80	**2.88**	3.09	**3.18**	3.08	**3.17**
Ribs	1.14	**1.20**	1.34	**1.42**	1.49	**1.57**	1.59	**1.69**	1.78	**1.90**	1.75	**1.86**
Oesophagus	0.10	**0.10**	0.13	**0.14**	0.16	**0.17**	0.18	**0.19**	0.23	**0.25**	0.20	**0.21**
Trachea	0.03	**0.09**	0.04	**0.11**	0.05	**0.14**	0.06	**0.16**	0.07	**0.20**	0.06	**0.17**
Lung	0.33	**0.34**	0.42	**0.43**	0.48	**0.50**	0.53	**0.55**	0.63	**0.65**	0.60	**0.62**
Mamma	0.09	**0.09**	0.12	**0.12**	0.14	**0.14**	0.16	**0.16**	0.20	**0.20**	0.18	**0.18**
Heart	0.10	**0.10**	0.13	**0.14**	0.16	**0.16**	0.18	**0.19**	0.23	**0.24**	0.20	**0.20**
Stomach	0.05	**0.09**	0.07	**0.12**	0.08	**0.14**	0.09	**0.17**	0.12	**0.21**	0.10	**0.18**
Spleen	0.26	**0.37**	0.33	**0.47**	0.38	**0.54**	0.42	**0.61**	0.50	**0.72**	0.48	**0.68**
Liver	0.12	**0.18**	0.15	**0.23**	0.18	**0.27**	0.20	**0.30**	0.24	**0.37**	0.22	**0.33**
Gall bladder	0.03	**0.08**	0.03	**0.11**	0.04	**0.13**	0.05	**0.15**	0.06	**0.20**	0.05	**0.16**
Pancreas	0.12	**0.15**	0.16	**0.20**	0.19	**0.24**	0.22	**0.28**	0.27	**0.35**	0.24	**0.30**
Small intestine		**0.06**		**0.09**		**0.10**		**0.12**	0.01	**0.15**		**0.13**
Large intestine		**0.07**		**0.09**		**0.11**		**0.12**	0.01	**0.16**		**0.13**
Sigma/Rectum		**0.01**		**0.02**		**0.02**		**0.03**		**0.03**		**0.03**
Lumbar spine	0.05	**0.71**	0.06	**0.92**	0.08	**1.08**	0.09	**1.21**	0.12	**1.46**	0.10	**1.35**
Pelvis		**0.03**		**0.05**		**0.06**		**0.07**		**0.08**		**0.07**
Adrenals	0.36	**0.39**	0.46	**0.49**	0.53	**0.57**	0.59	**0.64**	0.70	**0.76**	0.66	**0.71**
Kidneys	0.12	**0.58**	0.15	**0.71**	0.18	**0.80**	0.20	**0.88**	0.23	**1.01**	0.22	**0.97**
Urinary bladder												
Uterus				**0.01**		**0.01**		**0.02**		**0.02**		**0.02**
Ovaries				**0.01**		**0.02**		**0.02**		**0.03**		**0.02**
Prostate												
Testes												
Upper arm	0.13	**0.63**	0.17	**0.79**	0.20	**0.91**	0.23	**1.01**	0.29	**1.21**	0.26	**1.14**
Elbow	0.12	**0.70**	0.16	**0.88**	0.19	**1.02**	0.22	**1.13**	0.28	**1.34**	0.24	**1.27**
Forearm		**0.33**	0.01	**0.42**	0.02	**0.48**	0.02	**0.53**	0.02	**0.63**	0.02	**0.60**
Upper leg												
Knee												
Lower leg												
Lymph nodes	0.07	**0.14**	0.09	**0.17**	0.10	**0.20**	0.12	**0.23**	0.15	**0.28**	0.13	**0.25**
Skin	0.12	**0.22**	0.13	**0.23**	0.13	**0.24**	0.13	**0.25**	0.14	**0.26**	0.14	**0.25**
Muscle	0.08	**0.13**	0.09	**0.15**	0.10	**0.17**	0.11	**0.19**	0.13	**0.21**	0.12	**0.20**
Skeleton	0.25	**0.38**	0.30	**0.47**	0.34	**0.53**	0.37	**0.59**	0.43	**0.68**	0.41	**0.65**
Red bone marrow	0.10	**0.13**	0.12	**0.17**	0.14	**0.19**	0.15	**0.21**	0.18	**0.25**	0.16	**0.23**
Whole body dose	0.10	**0.16**	0.12	**0.20**	0.14	**0.23**	0.15	**0.25**	0.18	**0.29**	0.17	**0.27**
ED (ICRP 60)	0.08	**0.11**	0.11	**0.14**	0.12	**0.17**	0.14	**0.19**	0.17	**0.22**	0.15	**0.20**
ED (ICRP 103)	0.09	**0.13**	0.12	**0.16**	0.14	**0.19**	0.16	**0.21**	0.19	**0.26**	0.17	**0.23**
Total filtration (mm) **Al**	**1.5**		**2.5**		**3.5**		**4.5**		**3.5**		**3.5**	
Cu	**0**		**0**		**0**		**0**		**0.1**		**0.2**	

Table 9.141 Normalized organ doses—Chest pa—8–12 years

Beam projection:			pa				**Table cells:**					
Focus-to-detector distance (cm):			150				(+) = normal font = optimal field setting					
Tube voltage (kVp):			70				**(−) = bold font = suboptimal field setting**					
Total filtrations:			See last table row.				Empty table cells correspond to values < 0.01					
Organ or tissue	**Organ dose D_T : Entrance dose K_E**							**(mSv × mGy^{-1})**				
	(+)	(−)	(+)	(−)	(+)	(−)	(+)	(−)	(+)	(−)	(+)	(−)
Skull	0.01	**0.17**	0.02	**0.22**	0.02	**0.25**	0.03	**0.28**	0.03	**0.33**	0.04	**0.38**
Brain		**0.02**		**0.03**		**0.03**		**0.04**		**0.05**		**0.06**
Oral mucosa	0.01	**0.16**	0.01	**0.20**	0.02	**0.23**	0.02	**0.25**	0.03	**0.30**	0.03	**0.35**
Salivary glands	0.02	**0.42**	0.03	**0.49**	0.03	**0.54**	0.04	**0.58**	0.05	**0.65**	0.06	**0.70**
Cervical spine	0.42	**0.95**	0.53	**1.19**	0.61	**1.38**	0.67	**1.52**	0.80	**1.81**	0.91	**2.07**
Thyroid	0.10	**0.12**	0.12	**0.16**	0.15	**0.19**	0.17	**0.21**	0.21	**0.26**	0.25	**0.32**
Thymus	0.06	**0.06**	0.08	**0.08**	0.09	**0.10**	0.11	**0.12**	0.14	**0.15**	0.17	**0.18**
Thoracic spine	0.91	**0.95**	1.16	**1.21**	1.34	**1.40**	1.49	**1.56**	1.79	**1.87**	2.06	**2.16**
Claviculae	0.16	**0.19**	0.21	**0.25**	0.25	**0.30**	0.29	**0.34**	0.36	**0.43**	0.43	**0.51**
Scapulae	2.22	**2.28**	2.56	**2.63**	2.78	**2.86**	2.94	**3.03**	3.21	**3.31**	3.41	**3.53**
Ribs	1.26	**1.33**	1.46	**1.55**	1.60	**1.71**	1.71	**1.82**	1.89	**2.03**	2.04	**2.20**
Oesophagus	0.13	**0.14**	0.18	**0.19**	0.21	**0.22**	0.24	**0.26**	0.30	**0.32**	0.37	**0.39**
Trachea	0.04	**0.12**	0.06	**0.15**	0.07	**0.18**	0.08	**0.21**	0.10	**0.26**	0.12	**0.32**
Lung	0.39	**0.41**	0.49	**0.50**	0.55	**0.58**	0.61	**0.63**	0.71	**0.74**	0.80	**0.84**
Mamma	0.12	**0.12**	0.15	**0.15**	0.18	**0.17**	0.20	**0.20**	0.24	**0.24**	0.29	**0.28**
Heart	0.13	**0.14**	0.17	**0.18**	0.20	**0.22**	0.23	**0.25**	0.29	**0.31**	0.35	**0.37**
Stomach	0.07	**0.12**	0.09	**0.16**	0.11	**0.19**	0.12	**0.22**	0.15	**0.27**	0.18	**0.33**
Spleen	0.31	**0.44**	0.39	**0.55**	0.44	**0.64**	0.49	**0.70**	0.58	**0.83**	0.66	**0.95**
Liver	0.15	**0.22**	0.19	**0.28**	0.22	**0.33**	0.24	**0.37**	0.29	**0.44**	0.34	**0.52**
Gall bladder	0.04	**0.11**	0.05	**0.15**	0.06	**0.18**	0.07	**0.20**	0.09	**0.26**	0.11	**0.32**
Pancreas	0.16	**0.20**	0.20	**0.27**	0.24	**0.32**	0.27	**0.36**	0.34	**0.45**	0.41	**0.53**
Small intestine		**0.09**		**0.11**		**0.13**	0.01	**0.15**	0.01	**0.19**	0.02	**0.22**
Large intestine		**0.09**	0.01	**0.12**	0.01	**0.14**	0.01	**0.16**	0.02	**0.20**	0.02	**0.24**
Sigma/Rectum		**0.02**		**0.03**		**0.03**		**0.04**		**0.04**		**0.05**
Lumbar spine	0.07	**0.87**	0.09	**1.10**	0.11	**1.27**	0.12	**1.41**	0.16	**1.69**	0.19	**1.94**
Pelvis		**0.05**		**0.06**		**0.08**		**0.09**		**0.11**	0.01	**0.13**
Adrenals	0.43	**0.47**	0.54	**0.59**	0.62	**0.67**	0.68	**0.74**	0.80	**0.88**	0.91	**1.00**
Kidneys	0.15	**0.67**	0.18	**0.80**	0.21	**0.90**	0.23	**0.97**	0.27	**1.11**	0.30	**1.23**
Urinary bladder												**0.01**
Uterus		**0.01**		**0.02**		**0.02**		**0.02**		**0.03**		**0.04**
Ovaries		**0.02**		**0.02**		**0.03**		**0.03**		**0.04**		**0.05**
Prostate												
Testes												
Upper arm	0.17	**0.74**	0.22	**0.92**	0.26	**1.05**	0.29	**1.16**	0.36	**1.36**	0.43	**1.54**
Elbow	0.16	**0.82**	0.21	**1.02**	0.24	**1.17**	0.27	**1.28**	0.34	**1.50**	0.40	**1.69**
Forearm	0.01	**0.39**	0.02	**0.48**	0.02	**0.55**	0.02	**0.60**	0.03	**0.70**	0.04	**0.79**
Upper leg												
Knee												
Lower leg												
Lymph nodes	0.09	**0.17**	0.11	**0.22**	0.13	**0.25**	0.15	**0.28**	0.18	**0.34**	0.21	**0.40**
Skin	0.12	**0.23**	0.13	**0.24**	0.14	**0.25**	0.14	**0.26**	0.15	**0.27**	0.16	**0.29**
Muscle	0.09	**0.14**	0.11	**0.17**	0.12	**0.19**	0.13	**0.21**	0.15	**0.24**	0.17	**0.27**
Skeleton	0.28	**0.44**	0.34	**0.53**	0.38	**0.60**	0.41	**0.66**	0.47	**0.76**	0.53	**0.85**
Red bone marrow	0.12	**0.16**	0.14	**0.20**	0.16	**0.23**	0.18	**0.25**	0.20	**0.30**	0.23	**0.34**
Whole body dose	0.12	**0.19**	0.14	**0.23**	0.16	**0.26**	0.17	**0.28**	0.20	**0.32**	0.22	**0.36**
ED (ICRP 60)	0.10	**0.14**	0.13	**0.17**	0.15	**0.20**	0.17	**0.22**	0.20	**0.27**	0.23	**0.31**
ED (ICRP 103)	0.12	**0.16**	0.14	**0.20**	0.17	**0.23**	0.19	**0.25**	0.22	**0.31**	0.26	**0.35**
Total filtration (mm) Al	**1.5**		**2.5**		**3.5**		**4.5**		**3.5**		**3.5**	
Cu	**0**		**0**		**0**		**0**		**0.1**		**0.2**	

Table 9.142 Normalized organ doses—Chest pa—8–12 years

Beam projection:			pa			**Table cells:**						
Focus-to-detector distance (cm):			150			(+) = normal font = optimal field setting						
Tube voltage (kVp):			80			**(−) = bold font = suboptimal field setting**						
Total filtrations:			See last table row.			Empty table cells correspond to values < 0.01						
Organ or tissue	**Organ dose D_T : Entrance dose K_E**						**($mSv \times mGy^{-1}$)**					
	(+)	(−)	(+)	(−)	(+)	(−)	(+)	(−)	(+)	(−)	(+)	(−)
Skull	0.02	**0.20**	0.02	**0.25**	0.03	**0.28**	0.03	**0.31**	0.04	**0.37**	0.05	**0.42**
Brain		**0.03**		**0.03**		**0.04**		**0.05**		**0.06**		**0.07**
Oral mucosa	0.02	**0.19**	0.02	**0.23**	0.02	**0.26**	0.03	**0.29**	0.03	**0.34**	0.04	**0.39**
Salivary glands	0.03	**0.46**	0.04	**0.53**	0.04	**0.58**	0.05	**0.62**	0.06	**0.69**	0.07	**0.75**
Cervical spine	0.49	**1.11**	0.60	**1.36**	0.68	**1.55**	0.75	**1.70**	0.88	**1.98**	0.99	**2.22**
Thyroid	0.12	**0.15**	0.16	**0.20**	0.18	**0.23**	0.21	**0.26**	0.26	**0.32**	0.30	**0.38**
Thymus	0.08	**0.08**	0.10	**0.11**	0.12	**0.13**	0.14	**0.15**	0.17	**0.19**	0.21	**0.22**
Thoracic spine	1.08	**1.13**	1.33	**1.40**	1.53	**1.60**	1.68	**1.76**	1.98	**2.08**	2.24	**2.36**
Claviculae	0.20	**0.24**	0.26	**0.31**	0.30	**0.36**	0.34	**0.41**	0.42	**0.50**	0.50	**0.59**
Scapulae	2.36	**2.43**	2.68	**2.76**	2.89	**2.98**	3.03	**3.13**	3.26	**3.38**	3.42	**3.55**
Ribs	1.36	**1.44**	1.56	**1.66**	1.69	**1.81**	1.79	**1.92**	1.96	**2.11**	2.09	**2.26**
Oesophagus	0.17	**0.19**	0.22	**0.24**	0.26	**0.28**	0.30	**0.32**	0.37	**0.39**	0.43	**0.47**
Trachea	0.06	**0.15**	0.07	**0.20**	0.08	**0.23**	0.09	**0.26**	0.12	**0.32**	0.14	**0.38**
Lung	0.45	**0.47**	0.55	**0.57**	0.62	**0.65**	0.68	**0.71**	0.78	**0.82**	0.87	**0.91**
Mamma	0.14	**0.14**	0.18	**0.18**	0.21	**0.21**	0.23	**0.23**	0.28	**0.28**	0.33	**0.32**
Heart	0.17	**0.18**	0.21	**0.23**	0.25	**0.27**	0.28	**0.30**	0.35	**0.37**	0.41	**0.44**
Stomach	0.09	**0.16**	0.11	**0.20**	0.13	**0.23**	0.15	**0.26**	0.18	**0.33**	0.22	**0.39**
Spleen	0.36	**0.52**	0.44	**0.64**	0.50	**0.72**	0.55	**0.79**	0.64	**0.92**	0.72	**1.03**
Liver	0.18	**0.27**	0.22	**0.34**	0.25	**0.39**	0.28	**0.43**	0.33	**0.51**	0.39	**0.59**
Gall bladder	0.05	**0.15**	0.07	**0.19**	0.08	**0.23**	0.09	**0.26**	0.11	**0.32**	0.13	**0.38**
Pancreas	0.20	**0.26**	0.25	**0.33**	0.29	**0.39**	0.33	**0.43**	0.40	**0.53**	0.47	**0.62**
Small intestine		**0.11**	0.01	**0.14**	0.01	**0.16**	0.02	**0.18**	0.02	**0.22**	0.02	**0.26**
Large intestine	0.01	**0.12**	0.01	**0.15**	0.02	**0.17**	0.02	**0.19**	0.02	**0.24**	0.03	**0.28**
Sigma/Rectum		**0.03**		**0.03**		**0.04**		**0.04**		**0.06**		**0.07**
Lumbar spine	0.09	**1.03**	0.11	**1.27**	0.13	**1.44**	0.15	**1.59**	0.19	**1.86**	0.23	**2.11**
Pelvis		**0.06**		**0.08**		**0.10**		**0.11**	0.01	**0.13**	0.01	**0.16**
Adrenals	0.51	**0.55**	0.62	**0.67**	0.70	**0.77**	0.76	**0.84**	0.89	**0.98**	0.99	**1.10**
Kidneys	0.17	**0.75**	0.21	**0.89**	0.23	**0.99**	0.26	**1.06**	0.30	**1.20**	0.33	**1.31**
Urinary bladder										**0.01**		**0.02**
Uterus		**0.02**		**0.02**		**0.03**		**0.03**		**0.04**		**0.05**
Ovaries		**0.02**		**0.03**		**0.04**		**0.04**		**0.05**		**0.06**
Prostate												
Testes												
Upper arm	0.21	**0.85**	0.26	**1.04**	0.31	**1.17**	0.34	**1.27**	0.42	**1.47**	0.49	**1.64**
Elbow	0.20	**0.94**	0.25	**1.14**	0.29	**1.29**	0.32	**1.40**	0.39	**1.61**	0.45	**1.79**
Forearm	0.02	**0.44**	0.02	**0.53**	0.03	**0.60**	0.03	**0.65**	0.04	**0.75**	0.04	**0.83**
Upper leg												
Knee												
Lower leg												
Lymph nodes	0.11	**0.21**	0.13	**0.26**	0.16	**0.30**	0.17	**0.33**	0.21	**0.40**	0.24	**0.46**
Skin	0.13	**0.24**	0.14	**0.25**	0.14	**0.26**	0.15	**0.27**	0.16	**0.29**	0.17	**0.30**
Muscle	0.10	**0.16**	0.12	**0.19**	0.13	**0.22**	0.14	**0.23**	0.17	**0.27**	0.19	**0.30**
Skeleton	0.32	**0.50**	0.37	**0.59**	0.41	**0.66**	0.45	**0.71**	0.50	**0.81**	0.55	**0.90**
Red bone marrow	0.13	**0.19**	0.16	**0.23**	0.18	**0.26**	0.20	**0.29**	0.23	**0.34**	0.26	**0.38**
Whole body dose	0.13	**0.22**	0.16	**0.26**	0.18	**0.29**	0.19	**0.31**	0.22	**0.36**	0.24	**0.40**
ED (ICRP 60)	0.12	**0.17**	0.15	**0.20**	0.17	**0.23**	0.19	**0.26**	0.22	**0.31**	0.26	**0.35**
ED (ICRP 103)	0.14	**0.19**	0.17	**0.23**	0.19	**0.27**	0.21	**0.29**	0.25	**0.35**	0.29	**0.40**
Total filtration (mm) **Al**	**1.5**		**2.5**		**3.5**		**4.5**		**3.5**		**3.5**	
Cu	**0**		**0**		**0**		**0**		**0.1**		**0.2**	

Table 9.143 Normalized organ doses—Chest pa—8–12 years

Beam projection:		pa		**Table cells:**								
Focus-to-detector distance (cm):		150		(+) = normal font = optimal field setting								
Tube voltage (kVp):		110		**(−) = bold font = suboptimal field setting**								
Total filtrations:		See last table row.		Empty table cells correspond to values < 0.01								
Organ or tissue	**Organ dose D_T : Entrance dose K_E**						**(mSv × mGy^{-1})**					
	(+)	(−)	(+)	(−)	(+)	(−)	(+)	(−)	(+)	(−)	(+)	(−)
Skull	0.03	**0.27**	0.04	**0.32**	0.04	**0.35**	0.04	**0.37**	0.05	**0.42**	0.06	**0.45**
Brain		**0.05**		**0.05**		**0.06**		**0.07**	0.01	**0.08**	0.01	**0.09**
Oral mucosa	0.03	**0.27**	0.03	**0.32**	0.04	**0.35**	0.04	**0.38**	0.05	**0.43**	0.05	**0.47**
Salivary glands	0.04	**0.57**	0.05	**0.64**	0.06	**0.68**	0.06	**0.72**	0.07	**0.78**	0.08	**0.83**
Cervical spine	0.65	**1.47**	0.76	**1.70**	0.83	**1.86**	0.89	**1.98**	0.99	**2.20**	1.07	**2.37**
Thyroid	0.20	**0.25**	0.24	**0.30**	0.27	**0.34**	0.30	**0.37**	0.35	**0.43**	0.39	**0.48**
Thymus	0.14	**0.15**	0.17	**0.18**	0.19	**0.20**	0.21	**0.22**	0.24	**0.26**	0.28	**0.30**
Thoracic spine	1.46	**1.54**	1.70	**1.79**	1.87	**1.97**	2.00	**2.11**	2.24	**2.37**	2.42	**2.57**
Claviculae	0.31	**0.37**	0.37	**0.44**	0.42	**0.49**	0.45	**0.54**	0.52	**0.62**	0.58	**0.69**
Scapulae	2.62	**2.71**	2.86	**2.96**	3.00	**3.10**	3.09	**3.20**	3.21	**3.33**	3.26	**3.40**
Ribs	1.56	**1.67**	1.72	**1.85**	1.81	**1.96**	1.88	**2.04**	1.99	**2.16**	2.06	**2.24**
Oesophagus	0.29	**0.31**	0.34	**0.37**	0.39	**0.42**	0.42	**0.45**	0.49	**0.53**	0.56	**0.60**
Trachea	0.09	**0.25**	0.11	**0.30**	0.13	**0.34**	0.14	**0.38**	0.16	**0.44**	0.18	**0.50**
Lung	0.61	**0.64**	0.70	**0.73**	0.77	**0.80**	0.82	**0.86**	0.91	**0.95**	0.98	**1.03**
Mamma	0.22	**0.22**	0.26	**0.26**	0.29	**0.29**	0.32	**0.32**	0.37	**0.37**	0.41	**0.41**
Heart	0.27	**0.29**	0.32	**0.34**	0.36	**0.39**	0.39	**0.42**	0.46	**0.49**	0.51	**0.55**
Stomach	0.14	**0.25**	0.17	**0.30**	0.19	**0.34**	0.21	**0.37**	0.25	**0.43**	0.28	**0.49**
Spleen	0.49	**0.71**	0.57	**0.83**	0.63	**0.91**	0.67	**0.97**	0.75	**1.08**	0.81	**1.18**
Liver	0.26	**0.40**	0.31	**0.47**	0.34	**0.52**	0.37	**0.56**	0.42	**0.64**	0.46	**0.70**
Gall bladder	0.09	**0.25**	0.11	**0.30**	0.12	**0.34**	0.13	**0.38**	0.16	**0.44**	0.18	**0.50**
Pancreas	0.31	**0.41**	0.37	**0.49**	0.42	**0.55**	0.45	**0.60**	0.53	**0.69**	0.59	**0.78**
Small intestine	0.02	**0.18**	0.02	**0.21**	0.02	**0.23**	0.03	**0.26**	0.03	**0.30**	0.04	**0.33**
Large intestine	0.02	**0.19**	0.03	**0.22**	0.03	**0.25**	0.03	**0.27**	0.04	**0.32**	0.04	**0.36**
Sigma/Rectum		**0.04**		**0.05**		**0.06**		**0.07**		**0.08**		**0.09**
Lumbar spine	0.15	**1.38**	0.17	**1.61**	0.20	**1.77**	0.21	**1.89**	0.25	**2.11**	0.28	**2.28**
Pelvis		**0.10**	0.01	**0.12**	0.01	**0.14**	0.01	**0.15**	0.02	**0.17**	0.02	**0.20**
Adrenals	0.69	**0.76**	0.80	**0.88**	0.88	**0.97**	0.93	**1.04**	1.04	**1.17**	1.13	**1.27**
Kidneys	0.23	**0.95**	0.27	**1.08**	0.29	**1.17**	0.31	**1.23**	0.35	**1.35**	0.38	**1.44**
Urinary bladder		**0.01**		**0.01**		**0.02**		**0.02**		**0.02**		**0.02**
Uterus		**0.03**		**0.04**		**0.05**		**0.05**		**0.06**	0.01	**0.07**
Ovaries		**0.04**		**0.05**		**0.06**		**0.06**		**0.08**	0.01	**0.09**
Prostate												
Testes												
Upper arm	0.31	**1.10**	0.36	**1.26**	0.41	**1.38**	0.44	**1.46**	0.50	**1.61**	0.56	**1.72**
Elbow	0.29	**1.21**	0.34	**1.38**	0.38	**1.51**	0.41	**1.60**	0.47	**1.75**	0.52	**1.87**
Forearm	0.03	**0.56**	0.03	**0.64**	0.04	**0.70**	0.04	**0.74**	0.05	**0.81**	0.06	**0.86**
Upper leg												
Knee												
Lower leg												
Lymph nodes	0.16	**0.31**	0.19	**0.37**	0.22	**0.41**	0.23	**0.44**	0.27	**0.50**	0.30	**0.56**
Skin	0.15	**0.26**	0.15	**0.28**	0.16	**0.29**	0.16	**0.30**	0.17	**0.31**	0.18	**0.32**
Muscle	0.13	**0.22**	0.15	**0.25**	0.17	**0.27**	0.18	**0.28**	0.20	**0.32**	0.21	**0.34**
Skeleton	0.39	**0.62**	0.44	**0.70**	0.47	**0.76**	0.49	**0.80**	0.54	**0.87**	0.57	**0.93**
Red bone marrow	0.18	**0.27**	0.21	**0.31**	0.23	**0.34**	0.25	**0.37**	0.28	**0.41**	0.30	**0.45**
Whole body dose	0.17	**0.28**	0.20	**0.32**	0.22	**0.35**	0.23	**0.37**	0.25	**0.41**	0.27	**0.44**
ED (ICRP 60)	0.18	**0.24**	0.21	**0.28**	0.23	**0.31**	0.24	**0.34**	0.28	**0.38**	0.30	**0.42**
ED (ICRP 103)	0.20	**0.27**	0.23	**0.32**	0.26	**0.36**	0.27	**0.38**	0.31	**0.43**	0.34	**0.48**
Total filtration (mm) **Al**	**1.5**		**2.5**		**3.5**		**4.5**		**3.5**		**3.5**	
Cu	**0**		**0**		**0**		**0**		**0.1**		**0.2**	

Table 9.144 Normalized organ doses—Chest pa—8–12 years

Beam projection:			pa				**Table cells:**					
Focus-to-detector distance (cm):			150				(+) = normal font = optimal field setting					
Tube voltage (kVp):			120				**(−) = bold font = suboptimal field setting**					
Total filtrations:			See last table row.				Empty table cells correspond to values < 0.01					
Organ or tissue	**Organ dose D_T : Entrance dose K_E**							**(mSv × mGy^{-1})**				
	(+)	(−)	(+)	(−)	(+)	(−)	(+)	(−)	(+)	(−)	(+)	(−)
Skull	0.03	**0.29**	0.04	**0.33**	0.04	**0.36**	0.05	**0.38**	0.05	**0.42**	0.06	**0.45**
Brain		**0.05**		**0.06**		**0.07**		**0.07**	0.01	**0.08**	0.01	**0.09**
Oral mucosa	0.03	**0.30**	0.03	**0.34**	0.04	**0.37**	0.04	**0.40**	0.05	**0.45**	0.05	**0.49**
Salivary glands	0.05	**0.60**	0.06	**0.66**	0.06	**0.71**	0.07	**0.74**	0.08	**0.80**	0.08	**0.84**
Cervical spine	0.69	**1.55**	0.79	**1.77**	0.86	**1.91**	0.91	**2.03**	1.00	**2.22**	1.07	**2.36**
Thyroid	0.23	**0.28**	0.27	**0.33**	0.30	**0.37**	0.32	**0.40**	0.37	**0.45**	0.41	**0.51**
Thymus	0.16	**0.17**	0.18	**0.20**	0.21	**0.22**	0.22	**0.24**	0.26	**0.28**	0.29	**0.32**
Thoracic spine	1.55	**1.63**	1.77	**1.87**	1.93	**2.04**	2.05	**2.17**	2.26	**2.40**	2.42	**2.57**
Claviculae	0.34	**0.40**	0.40	**0.47**	0.44	**0.52**	0.47	**0.56**	0.54	**0.64**	0.59	**0.70**
Scapulae	2.67	**2.76**	2.88	**2.98**	3.00	**3.10**	3.07	**3.18**	3.16	**3.29**	3.19	**3.32**
Ribs	1.60	**1.72**	1.74	**1.88**	1.83	**1.98**	1.89	**2.04**	1.97	**2.15**	2.02	**2.21**
Oesophagus	0.32	**0.34**	0.38	**0.40**	0.42	**0.45**	0.45	**0.49**	0.52	**0.56**	0.58	**0.63**
Trachea	0.10	**0.28**	0.12	**0.33**	0.14	**0.37**	0.15	**0.40**	0.17	**0.46**	0.19	**0.52**
Lung	0.65	**0.68**	0.74	**0.77**	0.80	**0.84**	0.85	**0.89**	0.93	**0.98**	1.00	**1.05**
Mamma	0.24	**0.25**	0.29	**0.29**	0.32	**0.32**	0.34	**0.34**	0.38	**0.39**	0.42	**0.43**
Heart	0.30	**0.32**	0.35	**0.37**	0.39	**0.42**	0.42	**0.45**	0.48	**0.52**	0.53	**0.57**
Stomach	0.16	**0.28**	0.19	**0.33**	0.21	**0.37**	0.23	**0.40**	0.26	**0.46**	0.29	**0.51**
Spleen	0.53	**0.76**	0.61	**0.87**	0.66	**0.95**	0.70	**1.01**	0.77	**1.11**	0.83	**1.20**
Liver	0.28	**0.43**	0.33	**0.50**	0.36	**0.55**	0.39	**0.59**	0.44	**0.66**	0.48	**0.73**
Gall bladder	0.10	**0.28**	0.12	**0.33**	0.13	**0.37**	0.14	**0.40**	0.17	**0.47**	0.19	**0.52**
Pancreas	0.35	**0.45**	0.41	**0.53**	0.45	**0.59**	0.48	**0.64**	0.55	**0.73**	0.61	**0.81**
Small intestine	0.02	**0.19**	0.02	**0.23**	0.03	**0.25**	0.03	**0.27**	0.03	**0.31**	0.04	**0.35**
Large intestine	0.02	**0.21**	0.03	**0.24**	0.03	**0.27**	0.03	**0.29**	0.04	**0.34**	0.05	**0.37**
Sigma/Rectum		**0.05**		**0.06**		**0.06**		**0.07**		**0.08**	0.01	**0.09**
Lumbar spine	0.16	**1.47**	0.19	**1.68**	0.21	**1.83**	0.23	**1.94**	0.26	**2.13**	0.29	**2.28**
Pelvis	0.01	**0.11**	0.01	**0.13**	0.01	**0.15**	0.02	**0.16**	0.02	**0.18**	0.02	**0.20**
Adrenals	0.74	**0.82**	0.84	**0.94**	0.92	**1.02**	0.97	**1.09**	1.07	**1.20**	1.15	**1.30**
Kidneys	0.25	**1.01**	0.28	**1.13**	0.31	**1.21**	0.33	**1.27**	0.36	**1.37**	0.39	**1.45**
Urinary bladder		**0.01**		**0.02**		**0.02**		**0.02**		**0.02**		**0.03**
Uterus		**0.04**		**0.05**		**0.05**		**0.06**		**0.07**	0.01	**0.08**
Ovaries		**0.05**		**0.06**		**0.06**		**0.07**	0.01	**0.08**	0.01	**0.09**
Prostate												
Testes												
Upper arm	0.33	**1.16**	0.39	**1.31**	0.43	**1.41**	0.46	**1.49**	0.52	**1.62**	0.56	**1.71**
Elbow	0.31	**1.27**	0.36	**1.43**	0.40	**1.54**	0.42	**1.62**	0.48	**1.76**	0.52	**1.86**
Forearm	0.03	**0.59**	0.04	**0.66**	0.04	**0.71**	0.05	**0.75**	0.05	**0.81**	0.06	**0.85**
Upper leg												
Knee												
Lower leg												
Lymph nodes	0.18	**0.34**	0.21	**0.40**	0.23	**0.43**	0.25	**0.47**	0.28	**0.53**	0.31	**0.58**
Skin	0.15	**0.27**	0.16	**0.28**	0.16	**0.29**	0.17	**0.30**	0.18	**0.32**	0.18	**0.33**
Muscle	0.14	**0.23**	0.16	**0.26**	0.18	**0.28**	0.19	**0.30**	0.20	**0.33**	0.22	**0.35**
Skeleton	0.40	**0.65**	0.45	**0.72**	0.48	**0.78**	0.50	**0.81**	0.54	**0.88**	0.56	**0.92**
Red bone marrow	0.20	**0.29**	0.22	**0.33**	0.24	**0.36**	0.26	**0.38**	0.29	**0.43**	0.31	**0.47**
Whole body dose	0.19	**0.30**	0.21	**0.34**	0.22	**0.36**	0.24	**0.38**	0.26	**0.42**	0.28	**0.45**
ED (ICRP 60)	0.19	**0.26**	0.22	**0.30**	0.24	**0.33**	0.26	**0.35**	0.29	**0.40**	0.31	**0.44**
ED (ICRP 103)	0.21	**0.30**	0.25	**0.34**	0.27	**0.38**	0.29	**0.40**	0.32	**0.45**	0.35	**0.49**
Total filtration (mm) **Al**	**1.5**		**2.5**		**3.5**		**4.5**		**3.5**		**3.5**	
Cu	**0**		**0**		**0**		**0**		**0.1**		**0.2**	

Table 9.145 Normalized organ doses—Chest pa—13–17 years

Beam projection:			pa			**Table cells:**						
Focus-to-detector distance (cm):			150			(+) = normal font = optimal field setting						
Tube voltage (kVp):			110			**(−) = bold font = suboptimal field setting**						
Total filtrations:			See last table row.			Empty table cells correspond to values < 0.01						
Organ or tissue	**Organ dose D_T : Entrance dose K_E**							**(mSv × mGy^{-1})**				
	(+)	(−)	(+)	(−)	(+)	(−)	(+)	(−)	(+)	(−)	(+)	(−)
Skull	0.02	**0.26**	0.03	**0.30**	0.03	**0.33**	0.03	**0.35**	0.04	**0.40**	0.04	**0.44**
Brain		**0.05**		**0.06**		**0.06**		**0.07**		**0.08**		**0.09**
Oral mucosa	0.02	**0.22**	0.02	**0.26**	0.02	**0.29**	0.02	**0.31**	0.03	**0.36**	0.03	**0.40**
Salivary glands	0.03	**0.52**	0.04	**0.59**	0.04	**0.63**	0.04	**0.67**	0.05	**0.73**	0.06	**0.77**
Cervical spine	0.73	**1.59**	0.86	**1.85**	0.95	**2.04**	1.01	**2.18**	1.14	**2.44**	1.24	**2.65**
Thyroid	0.16	**0.19**	0.19	**0.23**	0.22	**0.26**	0.24	**0.28**	0.28	**0.33**	0.32	**0.38**
Thymus	0.11	**0.12**	0.13	**0.14**	0.15	**0.16**	0.17	**0.18**	0.20	**0.22**	0.23	**0.25**
Thoracic spine	1.30	**1.37**	1.53	**1.61**	1.69	**1.78**	1.82	**1.92**	2.06	**2.17**	2.25	**2.38**
Claviculae	0.22	**0.26**	0.27	**0.31**	0.30	**0.35**	0.33	**0.38**	0.39	**0.45**	0.44	**0.51**
Scapulae	2.47	**2.57**	2.71	**2.83**	2.86	**2.99**	2.96	**3.10**	3.11	**3.26**	3.19	**3.35**
Ribs	1.46	**1.57**	1.62	**1.75**	1.72	**1.86**	1.79	**1.94**	1.91	**2.07**	1.98	**2.16**
Oesophagus	0.32	**0.34**	0.38	**0.41**	0.43	**0.46**	0.47	**0.51**	0.56	**0.60**	0.64	**0.68**
Trachea	0.07	**0.17**	0.08	**0.20**	0.09	**0.23**	0.10	**0.25**	0.12	**0.30**	0.14	**0.34**
Lung	0.54	**0.57**	0.63	**0.66**	0.70	**0.73**	0.75	**0.79**	0.84	**0.89**	0.92	**0.97**
Mamma	1.18	**1.24**	1.42	**1.49**	1.60	**1.69**	1.75	**1.85**	2.06	**2.17**	2.34	**2.47**
Heart	0.22	**0.23**	0.26	**0.28**	0.29	**0.32**	0.32	**0.35**	0.38	**0.41**	0.43	**0.47**
Stomach	0.12	**0.21**	0.15	**0.25**	0.17	**0.29**	0.18	**0.31**	0.21	**0.37**	0.25	**0.42**
Spleen	0.46	**0.65**	0.54	**0.76**	0.60	**0.84**	0.64	**0.91**	0.72	**1.02**	0.79	**1.12**
Liver	0.25	**0.35**	0.29	**0.42**	0.33	**0.47**	0.35	**0.51**	0.41	**0.58**	0.45	**0.65**
Gall bladder	0.07	**0.22**	0.09	**0.27**	0.10	**0.30**	0.11	**0.33**	0.13	**0.39**	0.15	**0.45**
Pancreas	0.27	**0.34**	0.32	**0.41**	0.36	**0.47**	0.39	**0.51**	0.46	**0.60**	0.52	**0.68**
Small intestine	0.01	**0.14**	0.01	**0.17**	0.02	**0.19**	0.02	**0.21**	0.02	**0.25**	0.03	**0.28**
Large intestine	0.01	**0.15**	0.02	**0.18**	0.02	**0.20**	0.02	**0.22**	0.03	**0.26**	0.03	**0.30**
Sigma/Rectum		**0.03**		**0.04**		**0.05**		**0.05**		**0.06**		**0.07**
Lumbar spine	0.12	**1.20**	0.15	**1.41**	0.17	**1.56**	0.18	**1.67**	0.22	**1.88**	0.25	**2.06**
Pelvis		**0.08**		**0.10**		**0.11**		**0.12**	0.01	**0.14**	0.01	**0.16**
Adrenals	0.64	**0.71**	0.75	**0.83**	0.83	**0.91**	0.89	**0.98**	1.00	**1.11**	1.09	**1.22**
Kidneys	0.22	**0.90**	0.26	**1.03**	0.29	**1.12**	0.31	**1.19**	0.34	**1.31**	0.38	**1.41**
Urinary bladder								**0.01**		**0.01**		**0.02**
Uterus		**0.07**		**0.09**	0.01	**0.10**	0.01	**0.11**	0.01	**0.13**	0.02	**0.16**
Ovaries		**0.03**		**0.04**		**0.04**		**0.05**		**0.06**		**0.07**
Prostate												
Testes												
Upper arm	0.23	**0.94**	0.27	**1.09**	0.31	**1.20**	0.34	**1.28**	0.39	**1.42**	0.44	**1.53**
Elbow	0.23	**1.03**	0.27	**1.19**	0.30	**1.31**	0.33	**1.39**	0.38	**1.55**	0.43	**1.66**
Forearm	0.02	**0.48**	0.03	**0.55**	0.03	**0.60**	0.03	**0.64**	0.04	**0.71**	0.04	**0.76**
Upper leg												
Knee												
Lower leg												
Lymph nodes	0.14	**0.27**	0.17	**0.31**	0.19	**0.35**	0.20	**0.38**	0.24	**0.44**	0.27	**0.49**
Skin	0.15	**0.27**	0.16	**0.28**	0.16	**0.29**	0.17	**0.30**	0.18	**0.32**	0.19	**0.33**
Muscle	0.13	**0.21**	0.15	**0.24**	0.17	**0.26**	0.18	**0.28**	0.20	**0.31**	0.22	**0.34**
Skeleton	0.38	**0.59**	0.43	**0.68**	0.47	**0.74**	0.49	**0.78**	0.54	**0.86**	0.57	**0.91**
Red bone marrow	0.21	**0.29**	0.24	**0.33**	0.26	**0.37**	0.28	**0.40**	0.31	**0.45**	0.35	**0.49**
Whole body dose	0.17	**0.27**	0.20	**0.31**	0.21	**0.34**	0.23	**0.36**	0.25	**0.40**	0.27	**0.43**
ED (ICRP 60)	0.21	**0.27**	0.25	**0.32**	0.28	**0.36**	0.31	**0.39**	0.35	**0.45**	0.39	**0.50**
ED (ICRP 103)	0.30	**0.37**	0.36	**0.44**	0.40	**0.50**	0.43	**0.54**	0.50	**0.62**	0.56	**0.70**
Total filtration (mm) **Al**	**1.5**		**2.5**		**3.5**		**4.5**		**3.5**		**3.5**	
Cu	**0**		**0**		**0**		**0**		**0.1**		**0.2**	

Table 9.146 Normalized organ doses—Chest pa—13–17 years

Beam projection:			pa					**Table cells:**				
Focus-to-detector distance (cm):			150					(+) = normal font = optimal field setting				
Tube voltage (kVp):			120					**(−) = bold font = suboptimal field setting**				
Total filtrations:			See last table row.					Empty table cells correspond to values < 0.01				
Organ or tissue	**Organ dose D_T : Entrance dose K_E**							**(mSv × mGy^{-1})**				
	(+)	(−)	(+)	(−)	(+)	(−)	(+)	(−)	(+)	(−)	(+)	(−)
Skull	0.02	**0.27**	0.03	**0.32**	0.03	**0.34**	0.03	**0.37**	0.04	**0.41**	0.04	**0.44**
Brain		**0.05**		**0.06**		**0.07**		**0.07**		**0.09**		**0.10**
Oral mucosa	0.02	**0.24**	0.02	**0.28**	0.02	**0.31**	0.03	**0.33**	0.03	**0.37**	0.04	**0.41**
Salivary glands	0.03	**0.55**	0.04	**0.62**	0.04	**0.66**	0.05	**0.69**	0.05	**0.74**	0.06	**0.79**
Cervical spine	0.78	**1.69**	0.90	**1.94**	0.98	**2.11**	1.04	**2.25**	1.16	**2.48**	1.25	**2.66**
Thyroid	0.18	**0.21**	0.21	**0.25**	0.24	**0.28**	0.26	**0.30**	0.30	**0.35**	0.34	**0.40**
Thymus	0.13	**0.13**	0.15	**0.16**	0.17	**0.18**	0.18	**0.20**	0.22	**0.23**	0.25	**0.27**
Thoracic spine	1.40	**1.47**	1.61	**1.70**	1.77	**1.86**	1.88	**1.99**	2.10	**2.22**	2.27	**2.40**
Claviculae	0.25	**0.29**	0.29	**0.34**	0.32	**0.38**	0.35	**0.41**	0.40	**0.47**	0.45	**0.52**
Scapulae	2.52	**2.63**	2.74	**2.87**	2.87	**3.00**	2.96	**3.10**	3.07	**3.22**	3.13	**3.28**
Ribs	1.51	**1.62**	1.65	**1.78**	1.74	**1.88**	1.80	**1.95**	1.90	**2.06**	1.96	**2.13**
Oesophagus	0.36	**0.38**	0.42	**0.45**	0.47	**0.51**	0.51	**0.55**	0.59	**0.64**	0.67	**0.72**
Trachea	0.08	**0.19**	0.09	**0.23**	0.10	**0.25**	0.11	**0.27**	0.13	**0.32**	0.14	**0.36**
Lung	0.59	**0.62**	0.67	**0.71**	0.73	**0.77**	0.78	**0.82**	0.87	**0.92**	0.94	**0.99**
Mamma	1.33	**1.40**	1.57	**1.65**	1.75	**1.84**	1.90	**2.00**	2.20	**2.32**	2.47	**2.61**
Heart	0.24	**0.26**	0.29	**0.31**	0.32	**0.34**	0.35	**0.37**	0.40	**0.43**	0.45	**0.49**
Stomach	0.14	**0.24**	0.16	**0.28**	0.18	**0.31**	0.20	**0.34**	0.23	**0.39**	0.26	**0.44**
Spleen	0.50	**0.71**	0.57	**0.81**	0.63	**0.89**	0.67	**0.95**	0.75	**1.06**	0.81	**1.15**
Liver	0.27	**0.39**	0.32	**0.45**	0.35	**0.50**	0.38	**0.54**	0.43	**0.61**	0.47	**0.67**
Gall bladder	0.08	**0.25**	0.10	**0.30**	0.11	**0.33**	0.12	**0.36**	0.14	**0.42**	0.16	**0.47**
Pancreas	0.30	**0.38**	0.35	**0.45**	0.39	**0.51**	0.42	**0.55**	0.49	**0.63**	0.54	**0.71**
Small intestine	0.01	**0.16**	0.02	**0.19**	0.02	**0.21**	0.02	**0.23**	0.02	**0.26**	0.03	**0.30**
Large intestine	0.02	**0.17**	0.02	**0.20**	0.02	**0.22**	0.03	**0.24**	0.03	**0.28**	0.03	**0.31**
Sigma/Rectum		**0.04**		**0.05**		**0.05**		**0.06**		**0.06**		**0.07**
Lumbar spine	0.14	**1.29**	0.16	**1.48**	0.18	**1.62**	0.20	**1.73**	0.23	**1.92**	0.26	**2.07**
Pelvis		**0.09**		**0.10**		**0.12**		**0.13**	0.01	**0.15**	0.01	**0.16**
Adrenals	0.69	**0.77**	0.80	**0.88**	0.87	**0.97**	0.93	**1.03**	1.03	**1.15**	1.12	**1.25**
Kidneys	0.24	**0.96**	0.28	**1.08**	0.30	**1.17**	0.32	**1.23**	0.36	**1.34**	0.39	**1.43**
Urinary bladder						**0.01**		**0.01**		**0.01**		**0.02**
Uterus		**0.08**	0.01	**0.10**	0.01	**0.11**	0.01	**0.12**	0.02	**0.15**	0.02	**0.17**
Ovaries		**0.04**		**0.04**		**0.05**		**0.05**		**0.06**		**0.07**
Prostate												
Testes												
Upper arm	0.25	**1.00**	0.30	**1.14**	0.33	**1.24**	0.36	**1.31**	0.41	**1.44**	0.45	**1.53**
Elbow	0.25	**1.09**	0.29	**1.24**	0.32	**1.35**	0.35	**1.43**	0.39	**1.56**	0.44	**1.66**
Forearm	0.02	**0.50**	0.03	**0.57**	0.03	**0.62**	0.04	**0.65**	0.04	**0.71**	0.05	**0.76**
Upper leg												
Knee												
Lower leg												
Lymph nodes	0.16	**0.29**	0.18	**0.34**	0.20	**0.38**	0.22	**0.40**	0.25	**0.46**	0.28	**0.51**
Skin	0.15	**0.28**	0.16	**0.29**	0.17	**0.30**	0.17	**0.31**	0.18	**0.32**	0.19	**0.34**
Muscle	0.14	**0.22**	0.16	**0.25**	0.18	**0.27**	0.19	**0.29**	0.21	**0.32**	0.23	**0.35**
Skeleton	0.40	**0.62**	0.45	**0.70**	0.48	**0.76**	0.50	**0.79**	0.54	**0.86**	0.57	**0.91**
Red bone marrow	0.22	**0.31**	0.25	**0.36**	0.28	**0.39**	0.29	**0.42**	0.33	**0.47**	0.36	**0.51**
Whole body dose	0.18	**0.29**	0.21	**0.33**	0.22	**0.35**	0.24	**0.37**	0.26	**0.41**	0.28	**0.44**
ED (ICRP 60)	0.24	**0.30**	0.27	**0.35**	0.30	**0.39**	0.33	**0.41**	0.37	**0.47**	0.41	**0.52**
ED (ICRP 103)	0.33	**0.41**	0.39	**0.48**	0.43	**0.53**	0.46	**0.58**	0.53	**0.66**	0.59	**0.73**
Total filtration (mm) Al	**1.5**		**2.5**		**3.5**		**4.5**		**3.5**		**3.5**	
Cu	**0**		**0**		**0**		**0**		**0.1**		**0.2**	

Table 9.147 Normalized organ doses—Chest pa—> 17 years

Beam projection:			pa			**Table cells:**						
Focus-to-detector distance (cm):			150			(+) = normal font = optimal field setting						
Tube voltage (kVp):			110			**(−) = bold font = suboptimal field setting**						
Total filtrations:			See last table row.			Empty table cells correspond to values < 0.01						
Organ or tissue	**Organ dose D_T : Entrance dose K_E**							**(mSv × mGy^{-1})**				
	(+)	(−)	(+)	(−)	(+)	(−)	(+)	(−)	(+)	(−)	(+)	(−)
Skull	0.02	**0.20**	0.02	**0.24**	0.02	**0.26**	0.02	**0.28**	0.03	**0.32**	0.03	**0.36**
Brain		**0.03**		**0.04**		**0.04**		**0.04**		**0.05**		**0.06**
Oral mucosa	0.01	**0.20**	0.01	**0.24**	0.01	**0.27**	0.02	**0.29**	0.02	**0.33**	0.02	**0.37**
Salivary glands	0.02	**0.49**	0.02	**0.55**	0.03	**0.59**	0.03	**0.63**	0.03	**0.68**	0.04	**0.73**
Cervical spine	0.44	**1.64**	0.52	**1.90**	0.57	**2.09**	0.61	**2.24**	0.69	**2.50**	0.76	**2.71**
Thyroid	0.13	**0.19**	0.16	**0.22**	0.18	**0.25**	0.20	**0.28**	0.24	**0.33**	0.27	**0.37**
Thymus	0.11	**0.12**	0.13	**0.14**	0.15	**0.16**	0.17	**0.18**	0.20	**0.21**	0.23	**0.25**
Thoracic spine	1.27	**1.34**	1.49	**1.58**	1.65	**1.75**	1.78	**1.89**	2.02	**2.14**	2.21	**2.35**
Claviculae	0.22	**0.26**	0.26	**0.31**	0.29	**0.35**	0.32	**0.39**	0.38	**0.45**	0.43	**0.51**
Scapulae	2.42	**2.58**	2.67	**2.85**	2.82	**3.01**	2.92	**3.12**	3.07	**3.28**	3.15	**3.38**
Ribs	1.48	**1.60**	1.63	**1.78**	1.74	**1.89**	1.81	**1.98**	1.93	**2.11**	2.00	**2.20**
Oesophagus	0.35	**0.37**	0.42	**0.45**	0.47	**0.51**	0.52	**0.56**	0.61	**0.66**	0.70	**0.76**
Trachea	0.05	**0.17**	0.06	**0.20**	0.07	**0.23**	0.07	**0.25**	0.08	**0.30**	0.10	**0.34**
Lung	0.56	**0.59**	0.65	**0.69**	0.72	**0.76**	0.77	**0.81**	0.86	**0.91**	0.94	**1.00**
Mamma	0.15	**0.16**	0.18	**0.19**	0.20	**0.21**	0.22	**0.23**	0.26	**0.27**	0.29	**0.31**
Heart	0.22	**0.24**	0.27	**0.29**	0.30	**0.33**	0.33	**0.36**	0.39	**0.42**	0.45	**0.48**
Stomach	0.12	**0.21**	0.15	**0.26**	0.16	**0.29**	0.18	**0.32**	0.21	**0.38**	0.24	**0.43**
Spleen	0.46	**0.66**	0.53	**0.77**	0.59	**0.85**	0.63	**0.92**	0.72	**1.04**	0.79	**1.14**
Liver	0.24	**0.36**	0.29	**0.43**	0.32	**0.48**	0.35	**0.52**	0.40	**0.60**	0.45	**0.67**
Gall bladder	0.07	**0.22**	0.08	**0.26**	0.09	**0.30**	0.10	**0.33**	0.12	**0.39**	0.14	**0.45**
Pancreas	0.26	**0.34**	0.32	**0.41**	0.36	**0.47**	0.39	**0.51**	0.46	**0.60**	0.52	**0.68**
Small intestine		**0.15**	0.01	**0.18**	0.01	**0.20**	0.01	**0.22**	0.02	**0.26**	0.02	**0.30**
Large intestine	0.01	**0.15**	0.02	**0.19**	0.02	**0.21**	0.02	**0.23**	0.02	**0.27**	0.03	**0.31**
Sigma/Rectum		**0.04**		**0.04**		**0.05**		**0.05**		**0.06**		**0.07**
Lumbar spine	0.11	**1.21**	0.13	**1.42**	0.15	**1.57**	0.16	**1.68**	0.19	**1.90**	0.22	**2.08**
Pelvis		**0.08**		**0.10**		**0.11**		**0.12**		**0.14**		**0.16**
Adrenals	0.65	**0.72**	0.76	**0.84**	0.84	**0.93**	0.90	**1.00**	1.01	**1.13**	1.11	**1.25**
Kidneys	0.22	**0.92**	0.25	**1.05**	0.28	**1.14**	0.30	**1.21**	0.34	**1.34**	0.37	**1.44**
Urinary bladder										**0.01**		**0.01**
Uterus		**0.03**		**0.03**		**0.04**		**0.04**		**0.05**		**0.06**
Ovaries		**0.03**		**0.03**		**0.04**		**0.04**		**0.05**		**0.06**
Prostate												
Testes												
Upper arm	0.18	**0.93**	0.22	**1.08**	0.25	**1.19**	0.27	**1.27**	0.32	**1.41**	0.36	**1.53**
Elbow	0.19	**1.03**	0.23	**1.19**	0.26	**1.30**	0.28	**1.39**	0.33	**1.54**	0.37	**1.66**
Forearm	0.02	**0.49**	0.02	**0.56**	0.02	**0.61**	0.03	**0.65**	0.03	**0.72**	0.04	**0.78**
Upper leg												
Knee												
Lower leg												
Lymph nodes	0.14	**0.27**	0.16	**0.32**	0.18	**0.35**	0.20	**0.38**	0.23	**0.44**	0.26	**0.50**
Skin	0.15	**0.27**	0.15	**0.29**	0.16	**0.30**	0.17	**0.31**	0.17	**0.32**	0.18	**0.34**
Muscle	0.12	**0.20**	0.14	**0.23**	0.16	**0.25**	0.17	**0.27**	0.19	**0.30**	0.21	**0.33**
Skeleton	0.36	**0.57**	0.40	**0.65**	0.44	**0.71**	0.46	**0.75**	0.50	**0.83**	0.53	**0.89**
Red bone marrow	0.23	**0.31**	0.26	**0.37**	0.29	**0.40**	0.31	**0.43**	0.35	**0.49**	0.38	**0.54**
Whole body dose	0.16	**0.27**	0.19	**0.31**	0.21	**0.33**	0.22	**0.36**	0.24	**0.40**	0.27	**0.43**
ED (ICRP 60)	0.17	**0.22**	0.19	**0.26**	0.22	**0.29**	0.23	**0.32**	0.27	**0.37**	0.30	**0.41**
ED (ICRP 103)	0.18	**0.25**	0.21	**0.29**	0.23	**0.33**	0.25	**0.35**	0.29	**0.41**	0.32	**0.45**
Total filtration (mm) **Al**	**1.5**		**2.5**		**3.5**		**4.5**		**3.5**		**3.5**	
Cu	**0**		**0**		**0**		**0**		**0.1**		**0.2**	

Table 9.148 Normalized organ doses—Chest pa—> 17 years

Beam projection:		pa			Table cells:							
Focus-to-detector distance (cm):		150			(+) = normal font = optimal field setting							
Tube voltage (kVp):		120			**(−) = bold font = suboptimal field setting**							
Total filtrations:		See last table row.			Empty table cells correspond to values < 0.01							
Organ or tissue	**Organ dose D_T : Entrance dose K_E**						**($mSv \times mGy^{-1}$)**					
	(+)	(−)	(+)	(−)	(+)	(−)	(+)	(−)	(+)	(−)	(+)	(−)
Skull	0.02	**0.22**	0.02	**0.25**	0.02	**0.27**	0.03	**0.29**	0.03	**0.33**	0.03	**0.36**
Brain		**0.03**		**0.04**		**0.04**		**0.05**		**0.06**		**0.06**
Oral mucosa	0.01	**0.22**	0.01	**0.26**	0.02	**0.29**	0.02	**0.31**	0.02	**0.35**	0.02	**0.38**
Salivary glands	0.02	**0.52**	0.02	**0.58**	0.03	**0.62**	0.03	**0.65**	0.03	**0.70**	0.04	**0.75**
Cervical spine	0.47	**1.74**	0.54	**1.99**	0.60	**2.17**	0.64	**2.30**	0.71	**2.54**	0.77	**2.72**
Thyroid	0.15	**0.21**	0.18	**0.25**	0.20	**0.28**	0.22	**0.30**	0.26	**0.35**	0.29	**0.40**
Thymus	0.12	**0.13**	0.15	**0.16**	0.17	**0.18**	0.18	**0.20**	0.22	**0.23**	0.25	**0.26**
Thoracic spine	1.36	**1.44**	1.58	**1.67**	1.73	**1.83**	1.85	**1.96**	2.06	**2.19**	2.23	**2.38**
Claviculae	0.24	**0.29**	0.28	**0.34**	0.31	**0.38**	0.34	**0.41**	0.39	**0.47**	0.44	**0.53**
Scapulae	2.48	**2.65**	2.70	**2.88**	2.83	**3.02**	2.91	**3.12**	3.03	**3.25**	3.09	**3.32**
Ribs	1.52	**1.65**	1.67	**1.81**	1.76	**1.92**	1.82	**1.99**	1.92	**2.11**	1.98	**2.18**
Oesophagus	0.39	**0.42**	0.46	**0.50**	0.52	**0.56**	0.56	**0.61**	0.66	**0.71**	0.74	**0.80**
Trachea	0.05	**0.19**	0.06	**0.22**	0.07	**0.25**	0.08	**0.27**	0.09	**0.32**	0.10	**0.36**
Lung	0.60	**0.64**	0.69	**0.73**	0.75	**0.80**	0.80	**0.85**	0.89	**0.94**	0.96	**1.02**
Mamma	0.17	**0.18**	0.20	**0.21**	0.22	**0.23**	0.24	**0.25**	0.28	**0.29**	0.31	**0.33**
Heart	0.25	**0.27**	0.30	**0.32**	0.33	**0.36**	0.36	**0.39**	0.42	**0.45**	0.47	**0.51**
Stomach	0.14	**0.24**	0.16	**0.28**	0.18	**0.32**	0.20	**0.35**	0.23	**0.40**	0.26	**0.45**
Spleen	0.49	**0.71**	0.57	**0.82**	0.62	**0.90**	0.67	**0.96**	0.74	**1.08**	0.81	**1.17**
Liver	0.27	**0.40**	0.31	**0.46**	0.35	**0.51**	0.37	**0.55**	0.42	**0.62**	0.46	**0.69**
Gall bladder	0.08	**0.25**	0.09	**0.29**	0.10	**0.33**	0.11	**0.36**	0.13	**0.42**	0.15	**0.47**
Pancreas	0.29	**0.38**	0.35	**0.45**	0.39	**0.51**	0.42	**0.55**	0.48	**0.64**	0.54	**0.71**
Small intestine	0.01	**0.17**	0.01	**0.20**	0.02	**0.22**	0.02	**0.24**	0.02	**0.28**	0.02	**0.31**
Large intestine	0.01	**0.17**	0.02	**0.21**	0.02	**0.23**	0.02	**0.25**	0.03	**0.29**	0.03	**0.33**
Sigma/Rectum		**0.04**		**0.05**		**0.05**		**0.06**		**0.07**		**0.08**
Lumbar spine	0.12	**1.30**	0.14	**1.49**	0.16	**1.64**	0.18	**1.74**	0.20	**1.94**	0.23	**2.10**
Pelvis		**0.09**		**0.11**		**0.12**		**0.13**		**0.15**	0.01	**0.17**
Adrenals	0.70	**0.78**	0.81	**0.90**	0.88	**0.99**	0.94	**1.05**	1.05	**1.18**	1.14	**1.28**
Kidneys	0.24	**0.97**	0.27	**1.10**	0.30	**1.19**	0.31	**1.26**	0.35	**1.37**	0.38	**1.47**
Urinary bladder								**0.01**		**0.01**		**0.02**
Uterus		**0.03**		**0.04**		**0.04**		**0.04**		**0.05**		**0.06**
Ovaries		**0.03**		**0.04**		**0.05**		**0.05**		**0.06**		**0.07**
Prostate												
Testes												
Upper arm	0.20	**0.99**	0.24	**1.13**	0.27	**1.23**	0.29	**1.30**	0.33	**1.43**	0.37	**1.53**
Elbow	0.21	**1.09**	0.25	**1.24**	0.28	**1.34**	0.30	**1.42**	0.34	**1.56**	0.38	**1.66**
Forearm	0.02	**0.51**	0.02	**0.58**	0.03	**0.63**	0.03	**0.67**	0.03	**0.73**	0.04	**0.78**
Upper leg												
Knee												
Lower leg												
Lymph nodes	0.15	**0.29**	0.18	**0.34**	0.20	**0.38**	0.21	**0.41**	0.24	**0.47**	0.27	**0.52**
Skin	0.15	**0.28**	0.16	**0.30**	0.16	**0.31**	0.17	**0.31**	0.18	**0.33**	0.19	**0.34**
Muscle	0.13	**0.22**	0.15	**0.25**	0.17	**0.27**	0.18	**0.28**	0.19	**0.31**	0.21	**0.34**
Skeleton	0.37	**0.60**	0.42	**0.68**	0.45	**0.73**	0.47	**0.77**	0.51	**0.84**	0.53	**0.88**
Red bone marrow	0.24	**0.34**	0.28	**0.39**	0.30	**0.43**	0.32	**0.46**	0.36	**0.51**	0.39	**0.56**
Whole body dose	0.18	**0.29**	0.20	**0.32**	0.22	**0.35**	0.23	**0.37**	0.25	**0.41**	0.27	**0.44**
ED (ICRP 60)	0.18	**0.25**	0.21	**0.29**	0.23	**0.32**	0.25	**0.34**	0.28	**0.38**	0.31	**0.42**
ED (ICRP 103)	0.20	**0.27**	0.23	**0.32**	0.25	**0.35**	0.27	**0.38**	0.30	**0.43**	0.33	**0.47**
Total filtration (mm) **Al**	**1.5**		**2.5**		**3.5**		**4.5**		**3.5**		**3.5**	
Cu	**0**		**0**		**0**		**0**		**0.1**		**0.2**	

Table 9.149 Normalized organ doses—Chest r-lat—< 1 year

Beam projection:		r-lat				**Table cells:**						
Focus-to-detector distance (cm):		150				(+) = normal font = optimal field setting						
Tube voltage (kVp):		50				**(−) = bold font = suboptimal field setting**						
Total filtrations:		See last table row.				Empty table cells correspond to values < 0.01						
Organ or tissue	**Organ dose D_T : Entrance dose K_E**								**(mSv × mGy^{-1})**			
	(+)	(−)	(+)	(−)	(+)	(−)	(+)	(−)	(+)	(−)	(+)	(−)
Skull	0.01	**0.19**	0.02	**0.23**	0.02	**0.26**	0.02	**0.28**	0.03	**0.33**	0.03	**0.37**
Brain		**0.02**		**0.02**		**0.03**		**0.03**		**0.04**		**0.05**
Oral mucosa	0.02	**0.37**	0.02	**0.45**	0.03	**0.50**	0.03	**0.54**	0.04	**0.60**	0.04	**0.66**
Salivary glands	0.03	**0.47**	0.04	**0.54**	0.04	**0.58**	0.05	**0.61**	0.06	**0.67**	0.07	**0.71**
Cervical spine	0.33	**0.74**	0.42	**0.95**	0.49	**1.11**	0.54	**1.23**	0.64	**1.46**	0.72	**1.66**
Thyroid	0.45	**0.51**	0.54	**0.62**	0.60	**0.69**	0.64	**0.75**	0.71	**0.84**	0.78	**0.92**
Thymus	0.22	**0.23**	0.28	**0.29**	0.33	**0.34**	0.37	**0.38**	0.44	**0.46**	0.50	**0.52**
Thoracic spine	0.50	**0.49**	0.66	**0.65**	0.78	**0.78**	0.88	**0.88**	1.06	**1.07**	1.23	**1.25**
Claviculae	0.82	**0.82**	1.00	**1.01**	1.12	**1.15**	1.21	**1.25**	1.37	**1.44**	1.51	**1.60**
Scapulae	1.56	**1.56**	1.79	**1.80**	1.94	**1.96**	2.04	**2.07**	2.21	**2.25**	2.35	**2.41**
Ribs	1.15	**1.04**	1.34	**1.25**	1.47	**1.40**	1.56	**1.51**	1.72	**1.69**	1.85	**1.86**
Oesophagus	0.26	**0.20**	0.33	**0.26**	0.38	**0.31**	0.43	**0.35**	0.50	**0.42**	0.58	**0.50**
Trachea	0.05	**0.29**	0.06	**0.36**	0.07	**0.41**	0.07	**0.45**	0.09	**0.52**	0.10	**0.58**
Lung	0.38	**0.30**	0.46	**0.37**	0.52	**0.43**	0.56	**0.47**	0.64	**0.55**	0.70	**0.62**
Mamma	0.57	**0.55**	0.61	**0.59**	0.64	**0.61**	0.66	**0.63**	0.69	**0.66**	0.72	**0.69**
Heart	0.24	**0.21**	0.31	**0.27**	0.36	**0.32**	0.40	**0.36**	0.47	**0.43**	0.54	**0.50**
Stomach	0.05	**0.08**	0.07	**0.12**	0.09	**0.14**	0.10	**0.16**	0.12	**0.20**	0.15	**0.24**
Spleen	0.05	**0.06**	0.06	**0.08**	0.08	**0.09**	0.09	**0.11**	0.11	**0.14**	0.14	**0.17**
Liver	0.26	**0.39**	0.32	**0.49**	0.36	**0.56**	0.40	**0.61**	0.45	**0.71**	0.50	**0.79**
Gall bladder	0.06	**0.17**	0.08	**0.22**	0.10	**0.27**	0.11	**0.30**	0.14	**0.37**	0.16	**0.43**
Pancreas	0.12	**0.11**	0.15	**0.15**	0.19	**0.18**	0.21	**0.20**	0.26	**0.26**	0.30	**0.31**
Small intestine		**0.10**	0.01	**0.13**	0.01	**0.15**	0.02	**0.17**	0.02	**0.21**	0.03	**0.24**
Large intestine	0.01	**0.15**	0.01	**0.19**	0.02	**0.22**	0.02	**0.25**	0.03	**0.30**	0.03	**0.35**
Sigma/Rectum		**0.01**		**0.02**		**0.02**		**0.03**		**0.04**		**0.05**
Lumbar spine	0.05	**0.42**	0.07	**0.56**	0.09	**0.67**	0.10	**0.76**	0.13	**0.92**	0.15	**1.08**
Pelvis		**0.07**		**0.09**	0.01	**0.11**	0.01	**0.13**	0.02	**0.16**	0.02	**0.19**
Adrenals	0.17	**0.18**	0.22	**0.23**	0.25	**0.27**	0.28	**0.30**	0.34	**0.36**	0.39	**0.42**
Kidneys	0.06	**0.24**	0.08	**0.30**	0.09	**0.35**	0.10	**0.38**	0.12	**0.45**	0.14	**0.51**
Urinary bladder				**0.01**		**0.01**		**0.01**		**0.02**		**0.02**
Uterus		**0.02**		**0.03**		**0.03**		**0.04**		**0.05**	0.01	**0.06**
Ovaries		**0.02**		**0.03**		**0.03**		**0.04**		**0.05**		**0.06**
Prostate										**0.01**		**0.01**
Testes												
Upper arm		**1.29**		**1.49**		**1.61**		**1.70**		**1.84**		**1.97**
Elbow		**1.35**		**1.55**		**1.67**		**1.76**		**1.90**		**2.02**
Forearm		**0.63**		**0.71**		**0.76**		**0.80**		**0.86**		**0.91**
Upper leg												**0.01**
Knee												
Lower leg												
Lymph nodes	0.11	**0.21**	0.14	**0.26**	0.16	**0.30**	0.18	**0.33**	0.21	**0.39**	0.23	**0.44**
Skin	0.18	**0.29**	0.19	**0.31**	0.20	**0.32**	0.20	**0.33**	0.21	**0.35**	0.22	**0.37**
Muscle	0.10	**0.17**	0.12	**0.20**	0.13	**0.23**	0.14	**0.24**	0.16	**0.28**	0.18	**0.31**
Skeleton	0.29	**0.55**	0.35	**0.66**	0.39	**0.74**	0.42	**0.79**	0.47	**0.90**	0.52	**0.99**
Red bone marrow	0.07	**0.13**	0.09	**0.16**	0.10	**0.18**	0.11	**0.20**	0.13	**0.23**	0.14	**0.26**
Whole body dose	0.12	**0.20**	0.14	**0.24**	0.16	**0.27**	0.17	**0.29**	0.19	**0.33**	0.21	**0.37**
ED (ICRP 60)	0.15	**0.17**	0.18	**0.21**	0.20	**0.24**	0.21	**0.26**	0.24	**0.30**	0.27	**0.34**
ED (ICRP 103)	0.18	**0.22**	0.22	**0.26**	0.24	**0.29**	0.26	**0.31**	0.29	**0.36**	0.31	**0.40**
Total filtration (mm) **Al**	**1.5**		**2.5**		**3.5**		**4.5**		**3.5**		**3.5**	
Cu	**0**		**0**		**0**		**0**		**0.1**		**0.2**	

Table 9.150 Normalized organ doses—Chest r-lat—< 1 year

Beam projection:			r-lat			**Table cells:**						
Focus-to-detector distance (cm):			150			(+) = normal font = optimal field setting						
Tube voltage (kVp):			60			**(−) = bold font = suboptimal field setting**						
Total filtrations:			See last table row.			Empty table cells correspond to values < 0.01						
Organ or tissue	**Organ dose D_T : Entrance dose K_E**							**(mSv × mGy^{-1})**				
	(+)	(−)	(+)	(−)	(+)	(−)	(+)	(−)	(+)	(−)	(+)	(−)
Skull	0.02	**0.22**	0.02	**0.27**	0.03	**0.30**	0.03	**0.33**	0.04	**0.37**	0.03	**0.37**
Brain		**0.03**		**0.03**		**0.04**		**0.04**		**0.05**		**0.05**
Oral mucosa	0.02	**0.43**	0.03	**0.51**	0.03	**0.56**	0.04	**0.60**	0.05	**0.68**	0.04	**0.66**
Salivary glands	0.04	**0.52**	0.05	**0.59**	0.05	**0.63**	0.06	**0.67**	0.07	**0.72**	0.07	**0.71**
Cervical spine	0.41	**0.92**	0.51	**1.15**	0.58	**1.32**	0.63	**1.45**	0.74	**1.69**	0.72	**1.66**
Thyroid	0.51	**0.59**	0.60	**0.70**	0.66	**0.78**	0.71	**0.84**	0.79	**0.94**	0.78	**0.92**
Thymus	0.28	**0.29**	0.35	**0.36**	0.40	**0.42**	0.44	**0.46**	0.52	**0.54**	0.50	**0.52**
Thoracic spine	0.65	**0.65**	0.83	**0.83**	0.96	**0.97**	1.07	**1.09**	1.27	**1.30**	1.23	**1.25**
Claviculae	0.95	**0.97**	1.13	**1.17**	1.26	**1.31**	1.35	**1.42**	1.51	**1.60**	1.51	**1.60**
Scapulae	1.70	**1.72**	1.93	**1.96**	2.07	**2.11**	2.17	**2.22**	2.34	**2.40**	2.35	**2.41**
Ribs	1.28	**1.19**	1.47	**1.41**	1.60	**1.56**	1.69	**1.67**	1.85	**1.86**	1.85	**1.86**
Oesophagus	0.32	**0.26**	0.40	**0.34**	0.46	**0.39**	0.51	**0.44**	0.60	**0.52**	0.58	**0.50**
Trachea	0.06	**0.35**	0.07	**0.42**	0.08	**0.48**	0.09	**0.52**	0.10	**0.59**	0.10	**0.58**
Lung	0.44	**0.36**	0.53	**0.44**	0.59	**0.51**	0.63	**0.55**	0.71	**0.64**	0.70	**0.62**
Mamma	0.60	**0.58**	0.65	**0.62**	0.67	**0.65**	0.69	**0.67**	0.73	**0.71**	0.72	**0.69**
Heart	0.30	**0.27**	0.38	**0.34**	0.44	**0.40**	0.48	**0.44**	0.56	**0.53**	0.54	**0.50**
Stomach	0.07	**0.12**	0.10	**0.16**	0.12	**0.19**	0.13	**0.21**	0.16	**0.26**	0.15	**0.24**
Spleen	0.07	**0.08**	0.09	**0.11**	0.11	**0.13**	0.12	**0.15**	0.15	**0.19**	0.14	**0.17**
Liver	0.31	**0.47**	0.37	**0.58**	0.42	**0.65**	0.45	**0.71**	0.51	**0.81**	0.50	**0.79**
Gall bladder	0.08	**0.22**	0.11	**0.29**	0.13	**0.34**	0.14	**0.38**	0.17	**0.46**	0.16	**0.43**
Pancreas	0.16	**0.15**	0.20	**0.20**	0.24	**0.24**	0.27	**0.27**	0.32	**0.34**	0.30	**0.31**
Small intestine	0.01	**0.13**	0.02	**0.17**	0.02	**0.19**	0.02	**0.22**	0.03	**0.26**	0.03	**0.24**
Large intestine	0.02	**0.19**	0.02	**0.24**	0.03	**0.28**	0.03	**0.31**	0.04	**0.37**	0.03	**0.35**
Sigma/Rectum		**0.02**		**0.03**		**0.04**		**0.04**		**0.05**		**0.05**
Lumbar spine	0.08	**0.56**	0.10	**0.72**	0.12	**0.84**	0.14	**0.94**	0.17	**1.13**	0.15	**1.08**
Pelvis	0.01	**0.10**	0.01	**0.13**	0.02	**0.15**	0.02	**0.17**	0.03	**0.20**	0.02	**0.19**
Adrenals	0.21	**0.23**	0.27	**0.29**	0.31	**0.34**	0.34	**0.37**	0.41	**0.44**	0.39	**0.42**
Kidneys	0.08	**0.30**	0.10	**0.37**	0.11	**0.42**	0.13	**0.46**	0.15	**0.53**	0.14	**0.51**
Urinary bladder		**0.01**		**0.02**		**0.02**		**0.02**		**0.03**		**0.02**
Uterus		**0.03**		**0.04**		**0.05**		**0.05**	0.01	**0.06**	0.01	**0.06**
Ovaries		**0.03**		**0.04**		**0.05**		**0.05**		**0.07**		**0.06**
Prostate								**0.01**		**0.02**		**0.01**
Testes												
Upper arm		**1.42**		**1.61**		**1.73**		**1.82**		**1.96**		**1.97**
Elbow		**1.48**		**1.67**		**1.79**		**1.87**		**2.01**		**2.02**
Forearm		**0.68**		**0.76**		**0.81**		**0.85**		**0.91**		**0.91**
Upper leg								**0.01**		**0.01**		**0.01**
Knee												
Lower leg												
Lymph nodes	0.14	**0.26**	0.17	**0.31**	0.19	**0.36**	0.21	**0.39**	0.24	**0.46**	0.23	**0.44**
Skin	0.19	**0.31**	0.20	**0.33**	0.21	**0.34**	0.21	**0.35**	0.23	**0.37**	0.22	**0.37**
Muscle	0.11	**0.20**	0.14	**0.23**	0.15	**0.26**	0.16	**0.28**	0.19	**0.32**	0.18	**0.31**
Skeleton	0.33	**0.63**	0.40	**0.75**	0.44	**0.83**	0.47	**0.89**	0.53	**1.00**	0.52	**0.99**
Red bone marrow	0.09	**0.16**	0.11	**0.19**	0.12	**0.22**	0.13	**0.24**	0.15	**0.27**	0.14	**0.26**
Whole body dose	0.14	**0.23**	0.16	**0.28**	0.18	**0.31**	0.19	**0.34**	0.22	**0.38**	0.21	**0.37**
ED (ICRP 60)	0.17	**0.20**	0.20	**0.25**	0.22	**0.28**	0.24	**0.30**	0.27	**0.35**	0.27	**0.34**
ED (ICRP 103)	0.21	**0.25**	0.24	**0.30**	0.27	**0.34**	0.29	**0.36**	0.32	**0.41**	0.31	**0.40**
Total filtration (mm) Al	**1.5**		**2.5**		**3.5**		**4.5**		**3.5**		**3.5**	
Cu	**0**		**0**		**0**		**0**		**0.1**		**0.2**	

Table 9.151 Normalized organ doses—Chest r-lat—< 1 year

Beam projection:			r-lat				**Table cells:**					
Focus-to-detector distance (cm):			150				(+) = normal font = optimal field setting					
Tube voltage (kVp):			70				**(−) = bold font = suboptimal field setting**					
Total filtrations:			See last table row.				Empty table cells correspond to values < 0.01					
Organ or tissue	**Organ dose D_T : Entrance dose K_E**								**(mSv × mGy^{-1})**			
	(+)	(−)	(+)	(−)	(+)	(−)	(+)	(−)	(+)	(−)	(+)	(−)
Skull	0.02	**0.25**	0.03	**0.30**	0.03	**0.33**	0.03	**0.35**	0.04	**0.40**	0.05	**0.43**
Brain		**0.03**		**0.04**		**0.05**		**0.05**	0.01	**0.06**	0.01	**0.07**
Oral mucosa	0.03	**0.48**	0.03	**0.56**	0.04	**0.61**	0.05	**0.65**	0.05	**0.72**	0.06	**0.78**
Salivary glands	0.04	**0.56**	0.05	**0.63**	0.06	**0.67**	0.07	**0.71**	0.08	**0.76**	0.09	**0.81**
Cervical spine	0.46	**1.06**	0.57	**1.30**	0.64	**1.47**	0.69	**1.60**	0.79	**1.83**	0.87	**2.02**
Thyroid	0.56	**0.66**	0.65	**0.77**	0.71	**0.85**	0.76	**0.90**	0.83	**1.00**	0.89	**1.08**
Thymus	0.32	**0.34**	0.40	**0.42**	0.45	**0.48**	0.50	**0.52**	0.57	**0.61**	0.64	**0.68**
Thoracic spine	0.76	**0.77**	0.95	**0.97**	1.09	**1.12**	1.20	**1.24**	1.41	**1.45**	1.58	**1.64**
Claviculae	1.04	**1.08**	1.22	**1.28**	1.34	**1.41**	1.43	**1.52**	1.58	**1.69**	1.69	**1.82**
Scapulae	1.80	**1.83**	2.02	**2.06**	2.15	**2.21**	2.24	**2.31**	2.39	**2.47**	2.49	**2.59**
Ribs	1.37	**1.31**	1.56	**1.52**	1.68	**1.66**	1.76	**1.77**	1.90	**1.94**	2.00	**2.07**
Oesophagus	0.38	**0.32**	0.46	**0.40**	0.52	**0.46**	0.57	**0.50**	0.66	**0.60**	0.74	**0.68**
Trachea	0.07	**0.39**	0.08	**0.47**	0.09	**0.52**	0.10	**0.56**	0.11	**0.64**	0.12	**0.69**
Lung	0.49	**0.42**	0.58	**0.50**	0.64	**0.57**	0.69	**0.61**	0.76	**0.70**	0.83	**0.78**
Mamma	0.63	**0.61**	0.67	**0.66**	0.70	**0.69**	0.72	**0.71**	0.75	**0.76**	0.78	**0.80**
Heart	0.35	**0.32**	0.43	**0.40**	0.49	**0.46**	0.54	**0.51**	0.62	**0.60**	0.70	**0.68**
Stomach	0.09	**0.15**	0.12	**0.20**	0.14	**0.23**	0.16	**0.26**	0.19	**0.31**	0.22	**0.37**
Spleen	0.09	**0.11**	0.11	**0.14**	0.13	**0.17**	0.15	**0.19**	0.18	**0.24**	0.22	**0.28**
Liver	0.35	**0.54**	0.41	**0.65**	0.46	**0.72**	0.49	**0.78**	0.55	**0.89**	0.60	**0.98**
Gall bladder	0.10	**0.27**	0.13	**0.35**	0.15	**0.40**	0.17	**0.44**	0.20	**0.53**	0.23	**0.60**
Pancreas	0.19	**0.20**	0.24	**0.25**	0.28	**0.29**	0.31	**0.33**	0.37	**0.40**	0.43	**0.47**
Small intestine	0.02	**0.16**	0.02	**0.20**	0.03	**0.23**	0.03	**0.25**	0.04	**0.30**	0.04	**0.35**
Large intestine	0.02	**0.23**	0.03	**0.28**	0.03	**0.32**	0.04	**0.35**	0.04	**0.41**	0.05	**0.47**
Sigma/Rectum		**0.03**		**0.04**		**0.05**		**0.05**	0.01	**0.06**	0.01	**0.08**
Lumbar spine	0.10	**0.67**	0.12	**0.84**	0.15	**0.97**	0.16	**1.07**	0.20	**1.26**	0.23	**1.43**
Pelvis	0.01	**0.12**	0.02	**0.15**	0.02	**0.18**	0.03	**0.20**	0.03	**0.24**	0.04	**0.27**
Adrenals	0.25	**0.27**	0.31	**0.34**	0.36	**0.39**	0.39	**0.43**	0.46	**0.51**	0.51	**0.57**
Kidneys	0.09	**0.34**	0.11	**0.42**	0.13	**0.47**	0.14	**0.52**	0.17	**0.59**	0.19	**0.66**
Urinary bladder		**0.02**		**0.02**		**0.02**		**0.03**		**0.03**		**0.04**
Uterus		**0.04**		**0.05**	0.01	**0.06**	0.01	**0.06**	0.02	**0.08**	0.02	**0.09**
Ovaries		**0.04**		**0.05**		**0.06**	0.01	**0.07**	0.01	**0.08**	0.02	**0.10**
Prostate				**0.01**		**0.01**		**0.02**		**0.02**		**0.03**
Testes												**0.01**
Upper arm		**1.51**		**1.69**		**1.80**		**1.88**		**2.01**		**2.10**
Elbow		**1.56**		**1.74**		**1.86**		**1.93**		**2.06**		**2.15**
Forearm		**0.72**		**0.80**		**0.84**		**0.88**		**0.93**		**0.97**
Upper leg				**0.01**		**0.01**		**0.02**		**0.02**		**0.03**
Knee												
Lower leg												
Lymph nodes	0.16	**0.30**	0.19	**0.36**	0.22	**0.41**	0.23	**0.44**	0.27	**0.51**	0.30	**0.57**
Skin	0.19	**0.32**	0.21	**0.34**	0.22	**0.36**	0.22	**0.37**	0.23	**0.39**	0.25	**0.41**
Muscle	0.13	**0.22**	0.15	**0.26**	0.17	**0.29**	0.18	**0.31**	0.20	**0.35**	0.22	**0.39**
Skeleton	0.37	**0.70**	0.43	**0.82**	0.47	**0.90**	0.50	**0.96**	0.56	**1.06**	0.60	**1.14**
Red bone marrow	0.10	**0.18**	0.12	**0.22**	0.13	**0.24**	0.15	**0.26**	0.17	**0.30**	0.18	**0.33**
Whole body dose	0.15	**0.26**	0.18	**0.31**	0.20	**0.34**	0.21	**0.37**	0.24	**0.41**	0.26	**0.45**
ED (ICRP 60)	0.19	**0.23**	0.22	**0.28**	0.25	**0.31**	0.26	**0.34**	0.30	**0.39**	0.32	**0.43**
ED (ICRP 103)	0.23	**0.29**	0.27	**0.34**	0.29	**0.37**	0.31	**0.40**	0.35	**0.46**	0.38	**0.51**
Total filtration (mm) Al	**1.5**		**2.5**		**3.5**		**4.5**		**3.5**		**3.5**	
Cu	**0**		**0**		**0**		**0**		**0.1**		**0.2**	

Table 9.152 Normalized organ doses—Chest r-lat—< 1 year

Beam projection:			r-lat			**Table cells:**						
Focus-to-detector distance (cm):			150			(+) = normal font = optimal field setting						
Tube voltage (kVp):			80			(−) = **bold font = suboptimal field setting**						
Total filtrations:			See last table row.			Empty table cells correspond to values < 0.01						
Organ or tissue	**Organ dose D_T : Entrance dose K_E**							**(mSv × mGy^{-1})**				
	(+)	(−)	(+)	(−)	(+)	(−)	(+)	(−)	(+)	(−)	(+)	(−)
Skull	0.02	**0.27**	0.03	**0.32**	0.04	**0.35**	0.04	**0.37**	0.05	**0.41**	0.05	**0.45**
Brain		**0.04**		**0.05**		**0.05**	0.01	**0.06**	0.01	**0.07**	0.01	**0.08**
Oral mucosa	0.03	**0.52**	0.04	**0.60**	0.05	**0.65**	0.05	**0.69**	0.06	**0.76**	0.07	**0.82**
Salivary glands	0.05	**0.59**	0.06	**0.66**	0.07	**0.71**	0.08	**0.74**	0.09	**0.80**	0.10	**0.84**
Cervical spine	0.52	**1.18**	0.62	**1.41**	0.68	**1.58**	0.74	**1.70**	0.83	**1.91**	0.90	**2.08**
Thyroid	0.61	**0.72**	0.70	**0.83**	0.76	**0.90**	0.80	**0.96**	0.87	**1.05**	0.93	**1.13**
Thymus	0.37	**0.39**	0.45	**0.47**	0.50	**0.53**	0.54	**0.57**	0.62	**0.66**	0.68	**0.73**
Thoracic spine	0.87	**0.89**	1.06	**1.09**	1.20	**1.23**	1.30	**1.35**	1.49	**1.55**	1.64	**1.72**
Claviculae	1.12	**1.17**	1.29	**1.36**	1.40	**1.49**	1.48	**1.58**	1.61	**1.73**	1.70	**1.84**
Scapulae	1.88	**1.92**	2.08	**2.13**	2.20	**2.26**	2.28	**2.35**	2.39	**2.49**	2.47	**2.58**
Ribs	1.44	**1.40**	1.61	**1.60**	1.72	**1.73**	1.80	**1.83**	1.91	**1.98**	1.99	**2.08**
Oesophagus	0.43	**0.37**	0.51	**0.45**	0.58	**0.51**	0.63	**0.56**	0.71	**0.66**	0.79	**0.74**
Trachea	0.07	**0.44**	0.09	**0.51**	0.10	**0.57**	0.11	**0.60**	0.12	**0.67**	0.13	**0.73**
Lung	0.54	**0.47**	0.63	**0.56**	0.69	**0.62**	0.73	**0.67**	0.81	**0.76**	0.87	**0.83**
Mamma	0.65	**0.64**	0.69	**0.69**	0.72	**0.72**	0.74	**0.75**	0.78	**0.80**	0.81	**0.84**
Heart	0.40	**0.37**	0.49	**0.46**	0.54	**0.52**	0.59	**0.57**	0.67	**0.66**	0.74	**0.73**
Stomach	0.11	**0.19**	0.14	**0.23**	0.16	**0.27**	0.18	**0.30**	0.21	**0.36**	0.25	**0.41**
Spleen	0.11	**0.14**	0.14	**0.17**	0.16	**0.20**	0.18	**0.23**	0.21	**0.27**	0.25	**0.32**
Liver	0.38	**0.60**	0.45	**0.71**	0.49	**0.78**	0.52	**0.84**	0.58	**0.95**	0.63	**1.03**
Gall bladder	0.12	**0.32**	0.15	**0.40**	0.17	**0.45**	0.19	**0.50**	0.22	**0.58**	0.25	**0.66**
Pancreas	0.23	**0.24**	0.28	**0.30**	0.32	**0.34**	0.35	**0.38**	0.41	**0.46**	0.47	**0.52**
Small intestine	0.02	**0.19**	0.03	**0.23**	0.03	**0.26**	0.03	**0.29**	0.04	**0.34**	0.05	**0.38**
Large intestine	0.03	**0.26**	0.03	**0.32**	0.04	**0.36**	0.04	**0.39**	0.05	**0.45**	0.06	**0.50**
Sigma/Rectum		**0.04**		**0.05**	0.01	**0.05**	0.01	**0.06**	0.01	**0.07**	0.02	**0.09**
Lumbar spine	0.12	**0.77**	0.15	**0.94**	0.17	**1.07**	0.19	**1.17**	0.22	**1.35**	0.26	**1.50**
Pelvis	0.02	**0.14**	0.02	**0.18**	0.03	**0.20**	0.03	**0.22**	0.04	**0.26**	0.05	**0.30**
Adrenals	0.29	**0.32**	0.35	**0.39**	0.40	**0.44**	0.43	**0.48**	0.50	**0.56**	0.55	**0.62**
Kidneys	0.11	**0.39**	0.13	**0.47**	0.15	**0.52**	0.16	**0.56**	0.19	**0.64**	0.21	**0.71**
Urinary bladder		**0.02**		**0.03**		**0.03**		**0.03**		**0.04**	0.01	**0.05**
Uterus		**0.05**	0.01	**0.06**	0.01	**0.07**	0.02	**0.07**	0.02	**0.09**	0.02	**0.10**
Ovaries		**0.05**	0.01	**0.06**	0.01	**0.07**	0.01	**0.08**	0.02	**0.10**	0.02	**0.11**
Prostate		**0.01**		**0.01**		**0.02**		**0.02**		**0.02**		**0.03**
Testes										**0.01**		**0.01**
Upper arm		**1.57**		**1.74**		**1.84**		**1.91**		**2.02**		**2.09**
Elbow		**1.63**		**1.80**		**1.90**		**1.97**		**2.07**		**2.13**
Forearm		**0.75**		**0.82**		**0.86**		**0.89**		**0.94**		**0.97**
Upper leg		**0.01**		**0.01**		**0.02**		**0.02**		**0.03**		**0.03**
Knee												
Lower leg												
Lymph nodes	0.18	**0.34**	0.21	**0.40**	0.24	**0.45**	0.26	**0.49**	0.29	**0.55**	0.32	**0.61**
Skin	0.20	**0.33**	0.21	**0.36**	0.22	**0.37**	0.23	**0.38**	0.24	**0.40**	0.25	**0.42**
Muscle	0.14	**0.25**	0.17	**0.29**	0.18	**0.31**	0.19	**0.34**	0.22	**0.38**	0.24	**0.41**
Skeleton	0.40	**0.75**	0.45	**0.87**	0.49	**0.94**	0.52	**1.00**	0.57	**1.09**	0.61	**1.16**
Red bone marrow	0.11	**0.20**	0.13	**0.24**	0.15	**0.27**	0.16	**0.29**	0.18	**0.32**	0.19	**0.35**
Whole body dose	0.17	**0.29**	0.19	**0.34**	0.21	**0.37**	0.23	**0.39**	0.25	**0.44**	0.27	**0.48**
ED (ICRP 60)	0.21	**0.26**	0.24	**0.31**	0.27	**0.35**	0.28	**0.37**	0.32	**0.42**	0.34	**0.46**
ED (ICRP 103)	0.25	**0.32**	0.29	**0.37**	0.31	**0.41**	0.33	**0.44**	0.37	**0.49**	0.40	**0.54**
Total filtration (mm) Al	**1.5**		**2.5**		**3.5**		**4.5**		**3.5**		**3.5**	
Cu	**0**		**0**		**0**		**0**		**0.1**		**0.2**	

Table 9.153 Normalized organ doses—Chest r-lat—1–2 years

Beam projection:			r-lat			**Table cells:**						
Focus-to-detector distance (cm):			150			(+) = normal font = optimal field setting						
Tube voltage (kVp):			50			**(−) = bold font = suboptimal field setting**						
Total filtrations:			See last table row.			Empty table cells correspond to values < 0.01						
Organ or tissue	**Organ dose D_T : Entrance dose K_E**							**(mSv × mGy^{-1})**				
	(+)	(−)	(+)	(−)	(+)	(−)	(+)	(−)	(+)	(−)	(+)	(−)
Skull		**0.19**		**0.23**	0.01	**0.27**	0.01	**0.29**	0.02	**0.34**	0.02	**0.38**
Brain		**0.02**		**0.02**		**0.03**		**0.03**		**0.04**		**0.04**
Oral mucosa		**0.27**	0.01	**0.33**	0.02	**0.37**	0.02	**0.40**	0.02	**0.46**	0.03	**0.52**
Salivary glands	0.02	**0.36**	0.02	**0.41**	0.02	**0.45**	0.03	**0.48**	0.03	**0.52**	0.04	**0.56**
Cervical spine	0.19	**0.45**	0.25	**0.59**	0.29	**0.70**	0.33	**0.79**	0.40	**0.96**	0.47	**1.13**
Thyroid	0.36	**0.40**	0.43	**0.48**	0.49	**0.54**	0.53	**0.59**	0.60	**0.67**	0.66	**0.75**
Thymus	0.14	**0.14**	0.19	**0.19**	0.23	**0.23**	0.26	**0.26**	0.32	**0.32**	0.37	**0.37**
Thoracic spine	0.30	**0.27**	0.41	**0.37**	0.49	**0.45**	0.56	**0.52**	0.70	**0.65**	0.85	**0.79**
Claviculae	0.62	**0.59**	0.78	**0.75**	0.89	**0.87**	0.98	**0.96**	1.14	**1.13**	1.28	**1.29**
Scapulae	1.08	**1.03**	1.27	**1.22**	1.39	**1.34**	1.49	**1.44**	1.65	**1.60**	1.78	**1.73**
Ribs	1.00	**0.80**	1.19	**0.99**	1.33	**1.12**	1.42	**1.22**	1.60	**1.41**	1.75	**1.57**
Oesophagus	0.20	**0.12**	0.27	**0.17**	0.32	**0.20**	0.36	**0.23**	0.43	**0.29**	0.51	**0.36**
Trachea	0.02	**0.21**	0.03	**0.27**	0.04	**0.31**	0.04	**0.34**	0.05	**0.40**	0.06	**0.45**
Lung	0.29	**0.19**	0.36	**0.25**	0.42	**0.29**	0.46	**0.33**	0.53	**0.39**	0.60	**0.46**
Mamma	0.52	**0.50**	0.57	**0.55**	0.59	**0.58**	0.61	**0.60**	0.64	**0.63**	0.67	**0.66**
Heart	0.17	**0.13**	0.22	**0.18**	0.26	**0.21**	0.29	**0.24**	0.36	**0.30**	0.42	**0.35**
Stomach	0.03	**0.04**	0.05	**0.06**	0.06	**0.07**	0.07	**0.08**	0.09	**0.10**	0.10	**0.13**
Spleen	0.03	**0.02**	0.04	**0.03**	0.04	**0.04**	0.05	**0.05**	0.07	**0.06**	0.08	**0.08**
Liver	0.26	**0.27**	0.33	**0.35**	0.37	**0.41**	0.41	**0.46**	0.48	**0.54**	0.54	**0.62**
Gall bladder	0.05	**0.12**	0.07	**0.16**	0.09	**0.19**	0.10	**0.22**	0.13	**0.27**	0.15	**0.32**
Pancreas	0.08	**0.06**	0.11	**0.08**	0.13	**0.10**	0.15	**0.11**	0.18	**0.15**	0.22	**0.18**
Small intestine		**0.06**		**0.08**	0.01	**0.10**	0.01	**0.11**	0.02	**0.14**	0.02	**0.16**
Large intestine		**0.09**	0.01	**0.12**	0.02	**0.15**	0.02	**0.17**	0.03	**0.21**	0.03	**0.24**
Sigma/Rectum				**0.01**		**0.01**		**0.01**		**0.02**		**0.02**
Lumbar spine	0.07	**0.24**	0.09	**0.32**	0.11	**0.39**	0.13	**0.45**	0.17	**0.56**	0.20	**0.68**
Pelvis		**0.04**		**0.06**		**0.07**		**0.08**	0.01	**0.10**	0.01	**0.13**
Adrenals	0.11	**0.10**	0.14	**0.13**	0.17	**0.16**	0.19	**0.18**	0.24	**0.23**	0.29	**0.27**
Kidneys	0.08	**0.16**	0.10	**0.21**	0.12	**0.24**	0.13	**0.27**	0.16	**0.32**	0.19	**0.37**
Urinary bladder										**0.01**		**0.01**
Uterus				**0.01**		**0.02**		**0.02**		**0.03**		**0.04**
Ovaries		**0.01**		**0.02**		**0.02**		**0.02**		**0.03**		**0.04**
Prostate												
Testes												
Upper arm		**1.05**		**1.23**		**1.36**		**1.45**		**1.62**		**1.75**
Elbow		**1.12**		**1.31**		**1.44**		**1.53**		**1.69**		**1.83**
Forearm		**0.52**		**0.60**		**0.66**		**0.70**		**0.77**		**0.83**
Upper leg												
Knee												
Lower leg												
Lymph nodes	0.08	**0.14**	0.10	**0.18**	0.12	**0.21**	0.14	**0.23**	0.16	**0.28**	0.19	**0.32**
Skin	0.16	**0.25**	0.16	**0.27**	0.17	**0.28**	0.18	**0.28**	0.18	**0.30**	0.19	**0.31**
Muscle	0.07	**0.11**	0.08	**0.13**	0.09	**0.15**	0.10	**0.16**	0.12	**0.19**	0.13	**0.21**
Skeleton	0.20	**0.38**	0.25	**0.47**	0.28	**0.53**	0.30	**0.58**	0.35	**0.66**	0.39	**0.75**
Red bone marrow	0.05	**0.09**	0.06	**0.11**	0.07	**0.12**	0.08	**0.14**	0.09	**0.16**	0.10	**0.18**
Whole body dose	0.09	**0.14**	0.11	**0.17**	0.12	**0.20**	0.13	**0.21**	0.15	**0.25**	0.17	**0.28**
ED (ICRP 60)	0.12	**0.12**	0.14	**0.15**	0.16	**0.17**	0.18	**0.19**	0.21	**0.22**	0.23	**0.25**
ED (ICRP 103)	0.15	**0.16**	0.18	**0.19**	0.20	**0.22**	0.22	**0.24**	0.25	**0.27**	0.27	**0.31**
Total filtration (mm) **Al**	**1.5**		**2.5**		**3.5**		**4.5**		**3.5**		**3.5**	
Cu	**0**		**0**		**0**		**0**		**0.1**		**0.2**	

Table 9.154 Normalized organ doses—Chest r-lat—1–2 years

Beam projection:		r-lat			**Table cells:**							
Focus-to-detector distance (cm):		150			(+) = normal font = optimal field setting							
Tube voltage (kVp):		60			**(−) = bold font = suboptimal field setting**							
Total filtrations:		See last table row.			Empty table cells correspond to values < 0.01							
Organ or tissue	**Organ dose D_T : Entrance dose K_E**						**($mSv \times mGy^{-1}$)**					
	(+)	(−)	(+)	(−)	(+)	(−)	(+)	(−)	(+)	(−)	(+)	(−)
Skull	0.01	**0.23**	0.01	**0.28**	0.02	**0.32**	0.02	**0.34**	0.02	**0.40**	0.02	**0.38**
Brain		**0.02**		**0.03**		**0.04**		**0.04**		**0.05**		**0.04**
Oral mucosa	0.01	**0.32**	0.02	**0.39**	0.02	**0.43**	0.02	**0.47**	0.03	**0.53**	0.03	**0.52**
Salivary glands	0.02	**0.40**	0.03	**0.46**	0.03	**0.50**	0.04	**0.53**	0.04	**0.58**	0.04	**0.56**
Cervical spine	0.25	**0.59**	0.32	**0.76**	0.37	**0.89**	0.41	**0.99**	0.50	**1.19**	0.47	**1.13**
Thyroid	0.42	**0.47**	0.50	**0.56**	0.55	**0.63**	0.60	**0.68**	0.67	**0.77**	0.66	**0.75**
Thymus	0.19	**0.19**	0.25	**0.25**	0.29	**0.29**	0.33	**0.33**	0.40	**0.40**	0.37	**0.37**
Thoracic spine	0.42	**0.39**	0.55	**0.52**	0.66	**0.62**	0.75	**0.70**	0.92	**0.87**	0.85	**0.79**
Claviculae	0.75	**0.73**	0.92	**0.90**	1.04	**1.03**	1.13	**1.14**	1.30	**1.32**	1.28	**1.29**
Scapulae	1.21	**1.16**	1.40	**1.36**	1.53	**1.48**	1.62	**1.58**	1.78	**1.74**	1.78	**1.73**
Ribs	1.14	**0.95**	1.34	**1.15**	1.48	**1.29**	1.58	**1.40**	1.76	**1.60**	1.75	**1.57**
Oesophagus	0.27	**0.18**	0.35	**0.24**	0.40	**0.28**	0.45	**0.32**	0.54	**0.40**	0.51	**0.36**
Trachea	0.03	**0.26**	0.04	**0.32**	0.05	**0.37**	0.05	**0.40**	0.06	**0.47**	0.06	**0.45**
Lung	0.35	**0.25**	0.43	**0.32**	0.49	**0.37**	0.53	**0.41**	0.61	**0.48**	0.60	**0.46**
Mamma	0.55	**0.54**	0.60	**0.58**	0.63	**0.61**	0.65	**0.63**	0.68	**0.67**	0.67	**0.66**
Heart	0.22	**0.18**	0.28	**0.24**	0.33	**0.28**	0.37	**0.31**	0.45	**0.38**	0.42	**0.35**
Stomach	0.05	**0.06**	0.07	**0.09**	0.08	**0.10**	0.10	**0.12**	0.12	**0.15**	0.10	**0.13**
Spleen	0.04	**0.04**	0.06	**0.05**	0.07	**0.07**	0.08	**0.08**	0.10	**0.10**	0.08	**0.08**
Liver	0.32	**0.35**	0.39	**0.44**	0.44	**0.50**	0.48	**0.55**	0.55	**0.65**	0.54	**0.62**
Gall bladder	0.08	**0.16**	0.10	**0.21**	0.12	**0.26**	0.14	**0.29**	0.17	**0.36**	0.15	**0.32**
Pancreas	0.11	**0.09**	0.15	**0.12**	0.18	**0.15**	0.20	**0.17**	0.25	**0.21**	0.22	**0.18**
Small intestine	0.01	**0.08**	0.02	**0.11**	0.02	**0.13**	0.02	**0.15**	0.03	**0.18**	0.02	**0.16**
Large intestine	0.02	**0.13**	0.02	**0.17**	0.03	**0.19**	0.03	**0.22**	0.04	**0.27**	0.03	**0.24**
Sigma/Rectum		**0.01**		**0.02**		**0.02**		**0.02**		**0.03**		**0.02**
Lumbar spine	0.10	**0.34**	0.13	**0.44**	0.16	**0.53**	0.18	**0.60**	0.23	**0.75**	0.20	**0.68**
Pelvis		**0.06**		**0.08**	0.01	**0.10**	0.01	**0.12**	0.02	**0.14**	0.01	**0.13**
Adrenals	0.15	**0.14**	0.19	**0.18**	0.23	**0.22**	0.25	**0.24**	0.31	**0.30**	0.29	**0.27**
Kidneys	0.10	**0.21**	0.13	**0.26**	0.15	**0.30**	0.17	**0.33**	0.20	**0.40**	0.19	**0.37**
Urinary bladder						**0.01**		**0.01**		**0.02**		**0.01**
Uterus		**0.02**		**0.02**		**0.03**		**0.03**	0.01	**0.04**		**0.04**
Ovaries		**0.02**		**0.03**		**0.03**		**0.04**		**0.05**		**0.04**
Prostate												
Testes												
Upper arm		**1.18**		**1.37**		**1.50**		**1.60**		**1.76**		**1.75**
Elbow		**1.25**		**1.45**		**1.58**		**1.67**		**1.84**		**1.83**
Forearm		**0.57**		**0.66**		**0.72**		**0.76**		**0.83**		**0.83**
Upper leg												
Knee												
Lower leg												
Lymph nodes	0.10	**0.18**	0.13	**0.23**	0.15	**0.26**	0.17	**0.29**	0.20	**0.34**	0.19	**0.32**
Skin	0.16	**0.26**	0.17	**0.28**	0.18	**0.29**	0.19	**0.30**	0.20	**0.32**	0.19	**0.31**
Muscle	0.08	**0.13**	0.10	**0.16**	0.11	**0.18**	0.12	**0.19**	0.14	**0.22**	0.13	**0.21**
Skeleton	0.24	**0.46**	0.29	**0.55**	0.33	**0.62**	0.35	**0.67**	0.40	**0.77**	0.39	**0.75**
Red bone marrow	0.06	**0.11**	0.08	**0.13**	0.09	**0.15**	0.10	**0.17**	0.11	**0.20**	0.10	**0.18**
Whole body dose	0.11	**0.17**	0.13	**0.21**	0.14	**0.23**	0.16	**0.25**	0.18	**0.29**	0.17	**0.28**
ED (ICRP 60)	0.14	**0.15**	0.17	**0.18**	0.19	**0.21**	0.21	**0.23**	0.24	**0.27**	0.23	**0.25**
ED (ICRP 103)	0.18	**0.19**	0.21	**0.23**	0.23	**0.26**	0.25	**0.28**	0.28	**0.32**	0.27	**0.31**
Total filtration (mm) **Al**	**1.5**		**2.5**		**3.5**		**4.5**		**3.5**		**3.5**	
Cu	**0**		**0**		**0**		**0**		**0.1**		**0.2**	

Table 9.155 Normalized organ doses—Chest r-lat—1–2 years

Beam projection:			r-lat				**Table cells:**					
Focus-to-detector distance (cm):			150				(+) = normal font = optimal field setting					
Tube voltage (kVp):			70				**(−) = bold font = suboptimal field setting**					
Total filtrations:			See last table row.				Empty table cells correspond to values < 0.01					
Organ or tissue	**Organ dose D_T : Entrance dose K_E**							**(mSv × mGy^{-1})**				
	(+)	(−)	(+)	(−)	(+)	(−)	(+)	(−)	(+)	(−)	(+)	(−)
Skull	0.01	**0.26**	0.02	**0.31**	0.02	**0.35**	0.02	**0.38**	0.03	**0.43**	0.04	**0.48**
Brain		**0.03**		**0.04**		**0.04**		**0.05**		**0.06**		**0.07**
Oral mucosa	0.02	**0.36**	0.02	**0.43**	0.03	**0.48**	0.03	**0.52**	0.04	**0.58**	0.04	**0.64**
Salivary glands	0.03	**0.44**	0.03	**0.50**	0.04	**0.54**	0.04	**0.57**	0.05	**0.62**	0.06	**0.67**
Cervical spine	0.30	**0.71**	0.37	**0.89**	0.43	**1.03**	0.47	**1.14**	0.56	**1.35**	0.63	**1.53**
Thyroid	0.47	**0.53**	0.55	**0.62**	0.60	**0.69**	0.65	**0.74**	0.72	**0.84**	0.79	**0.92**
Thymus	0.24	**0.24**	0.30	**0.30**	0.35	**0.35**	0.39	**0.39**	0.46	**0.46**	0.53	**0.53**
Thoracic spine	0.53	**0.50**	0.68	**0.64**	0.79	**0.75**	0.89	**0.85**	1.08	**1.03**	1.25	**1.20**
Claviculae	0.85	**0.84**	1.02	**1.02**	1.14	**1.15**	1.23	**1.25**	1.39	**1.44**	1.52	**1.59**
Scapulae	1.30	**1.26**	1.49	**1.45**	1.61	**1.58**	1.70	**1.67**	1.85	**1.82**	1.96	**1.93**
Ribs	1.25	**1.07**	1.44	**1.27**	1.58	**1.41**	1.67	**1.52**	1.84	**1.71**	1.97	**1.86**
Oesophagus	0.33	**0.23**	0.41	**0.30**	0.47	**0.35**	0.53	**0.39**	0.62	**0.48**	0.71	**0.56**
Trachea	0.04	**0.30**	0.05	**0.37**	0.05	**0.42**	0.06	**0.45**	0.07	**0.52**	0.08	**0.58**
Lung	0.41	**0.30**	0.49	**0.37**	0.55	**0.43**	0.59	**0.47**	0.67	**0.55**	0.74	**0.62**
Mamma	0.58	**0.56**	0.62	**0.61**	0.65	**0.64**	0.67	**0.66**	0.71	**0.70**	0.74	**0.74**
Heart	0.27	**0.23**	0.34	**0.29**	0.39	**0.34**	0.43	**0.37**	0.51	**0.45**	0.59	**0.52**
Stomach	0.07	**0.09**	0.09	**0.11**	0.11	**0.13**	0.12	**0.15**	0.15	**0.19**	0.18	**0.23**
Spleen	0.06	**0.06**	0.07	**0.07**	0.09	**0.09**	0.10	**0.10**	0.13	**0.13**	0.16	**0.16**
Liver	0.36	**0.41**	0.44	**0.51**	0.49	**0.58**	0.54	**0.63**	0.61	**0.74**	0.68	**0.83**
Gall bladder	0.10	**0.21**	0.12	**0.27**	0.15	**0.31**	0.16	**0.35**	0.20	**0.42**	0.23	**0.50**
Pancreas	0.14	**0.12**	0.19	**0.16**	0.22	**0.19**	0.25	**0.22**	0.30	**0.27**	0.35	**0.32**
Small intestine	0.02	**0.11**	0.02	**0.14**	0.02	**0.16**	0.03	**0.18**	0.04	**0.22**	0.04	**0.26**
Large intestine	0.02	**0.16**	0.03	**0.20**	0.03	**0.23**	0.04	**0.26**	0.05	**0.31**	0.06	**0.36**
Sigma/Rectum		**0.02**		**0.02**		**0.03**		**0.03**		**0.04**	0.01	**0.05**
Lumbar spine	0.13	**0.43**	0.17	**0.55**	0.20	**0.65**	0.23	**0.73**	0.28	**0.88**	0.33	**1.03**
Pelvis	0.01	**0.08**	0.01	**0.11**	0.02	**0.13**	0.02	**0.14**	0.03	**0.18**	0.03	**0.21**
Adrenals	0.18	**0.18**	0.23	**0.23**	0.27	**0.27**	0.31	**0.30**	0.37	**0.36**	0.43	**0.43**
Kidneys	0.12	**0.25**	0.15	**0.31**	0.18	**0.35**	0.20	**0.39**	0.23	**0.45**	0.27	**0.51**
Urinary bladder		**0.01**		**0.01**		**0.02**		**0.02**		**0.02**		**0.03**
Uterus		**0.02**		**0.03**		**0.04**	0.01	**0.04**	0.01	**0.06**	0.02	**0.07**
Ovaries		**0.03**		**0.04**		**0.04**	0.01	**0.05**	0.01	**0.06**	0.02	**0.07**
Prostate												**0.01**
Testes												
Upper arm		**1.28**		**1.47**		**1.59**		**1.69**		**1.84**		**1.96**
Elbow		**1.35**		**1.54**		**1.67**		**1.76**		**1.91**		**2.03**
Forearm		**0.62**		**0.70**		**0.76**		**0.80**		**0.86**		**0.91**
Upper leg												**0.01**
Knee												
Lower leg												
Lymph nodes	0.13	**0.22**	0.16	**0.27**	0.18	**0.31**	0.20	**0.34**	0.23	**0.39**	0.26	**0.45**
Skin	0.17	**0.28**	0.18	**0.29**	0.19	**0.30**	0.20	**0.31**	0.21	**0.33**	0.22	**0.35**
Muscle	0.09	**0.15**	0.11	**0.18**	0.13	**0.20**	0.14	**0.22**	0.16	**0.25**	0.17	**0.28**
Skeleton	0.27	**0.51**	0.32	**0.61**	0.36	**0.68**	0.39	**0.73**	0.44	**0.83**	0.48	**0.91**
Red bone marrow	0.07	**0.13**	0.09	**0.16**	0.10	**0.18**	0.11	**0.19**	0.13	**0.22**	0.14	**0.25**
Whole body dose	0.12	**0.20**	0.14	**0.23**	0.16	**0.26**	0.17	**0.28**	0.20	**0.32**	0.22	**0.36**
ED (ICRP 60)	0.16	**0.18**	0.19	**0.21**	0.22	**0.24**	0.23	**0.26**	0.27	**0.30**	0.30	**0.34**
ED (ICRP 103)	0.20	**0.22**	0.23	**0.26**	0.26	**0.29**	0.28	**0.32**	0.31	**0.36**	0.34	**0.41**
Total filtration (mm) **Al**	**1.5**		**2.5**		**3.5**		**4.5**		**3.5**		**3.5**	
Cu	**0**		**0**		**0**		**0**		**0.1**		**0.2**	

Table 9.156 Normalized organ doses—Chest r-lat—1–2 years

Beam projection:			r-lat		**Table cells:**							
Focus-to-detector distance (cm):			150		(+) = normal font = optimal field setting							
Tube voltage (kVp):			80		**(−) = bold font = suboptimal field setting**							
Total filtrations:			See last table row.		Empty table cells correspond to values < 0.01							
Organ or tissue	**Organ dose D_T : Entrance dose K_E**						**(mSv × mGy⁻¹)**					
	(+)	(−)	(+)	(−)	(+)	(−)	(+)	(−)	(+)	(−)	(+)	(−)
Skull	0.02	**0.29**	0.02	**0.34**	0.03	**0.38**	0.03	**0.41**	0.04	**0.46**	0.04	**0.50**
Brain		**0.04**		**0.05**		**0.05**		**0.06**		**0.07**		**0.08**
Oral mucosa	0.02	**0.40**	0.03	**0.47**	0.03	**0.52**	0.04	**0.56**	0.04	**0.63**	0.05	**0.68**
Salivary glands	0.03	**0.47**	0.04	**0.53**	0.05	**0.57**	0.05	**0.60**	0.06	**0.66**	0.07	**0.70**
Cervical spine	0.34	**0.82**	0.42	**1.01**	0.48	**1.15**	0.52	**1.25**	0.60	**1.46**	0.67	**1.63**
Thyroid	0.51	**0.58**	0.59	**0.68**	0.65	**0.75**	0.69	**0.80**	0.77	**0.89**	0.83	**0.97**
Thymus	0.28	**0.28**	0.35	**0.35**	0.40	**0.40**	0.44	**0.44**	0.51	**0.52**	0.58	**0.59**
Thoracic spine	0.63	**0.60**	0.79	**0.75**	0.91	**0.87**	1.00	**0.96**	1.19	**1.15**	1.36	**1.32**
Claviculae	0.93	**0.93**	1.10	**1.11**	1.21	**1.24**	1.30	**1.34**	1.44	**1.51**	1.56	**1.64**
Scapulae	1.38	**1.34**	1.56	**1.52**	1.67	**1.64**	1.75	**1.72**	1.87	**1.85**	1.96	**1.95**
Ribs	1.33	**1.17**	1.52	**1.36**	1.64	**1.50**	1.73	**1.60**	1.88	**1.77**	1.98	**1.91**
Oesophagus	0.39	**0.28**	0.47	**0.36**	0.54	**0.41**	0.59	**0.46**	0.69	**0.55**	0.78	**0.64**
Trachea	0.04	**0.34**	0.05	**0.41**	0.06	**0.46**	0.06	**0.50**	0.07	**0.57**	0.08	**0.63**
Lung	0.45	**0.35**	0.54	**0.42**	0.60	**0.48**	0.64	**0.52**	0.72	**0.60**	0.79	**0.68**
Mamma	0.60	**0.59**	0.65	**0.64**	0.68	**0.67**	0.70	**0.69**	0.73	**0.73**	0.76	**0.77**
Heart	0.32	**0.27**	0.39	**0.34**	0.44	**0.39**	0.49	**0.43**	0.57	**0.51**	0.64	**0.58**
Stomach	0.09	**0.11**	0.11	**0.14**	0.13	**0.16**	0.14	**0.18**	0.18	**0.22**	0.21	**0.26**
Spleen	0.07	**0.07**	0.09	**0.10**	0.11	**0.11**	0.13	**0.13**	0.16	**0.16**	0.18	**0.19**
Liver	0.41	**0.47**	0.49	**0.57**	0.54	**0.64**	0.58	**0.70**	0.66	**0.80**	0.72	**0.90**
Gall bladder	0.12	**0.25**	0.15	**0.32**	0.17	**0.36**	0.19	**0.40**	0.23	**0.48**	0.26	**0.56**
Pancreas	0.18	**0.16**	0.22	**0.20**	0.26	**0.23**	0.29	**0.26**	0.35	**0.32**	0.40	**0.38**
Small intestine	0.02	**0.13**	0.03	**0.16**	0.03	**0.19**	0.04	**0.21**	0.04	**0.25**	0.05	**0.29**
Large intestine	0.03	**0.19**	0.03	**0.24**	0.04	**0.27**	0.05	**0.30**	0.06	**0.35**	0.07	**0.40**
Sigma/Rectum		**0.02**		**0.03**		**0.03**		**0.04**	0.01	**0.05**	0.01	**0.06**
Lumbar spine	0.16	**0.51**	0.20	**0.65**	0.24	**0.75**	0.27	**0.83**	0.32	**0.99**	0.37	**1.13**
Pelvis	0.01	**0.10**	0.02	**0.13**	0.02	**0.15**	0.03	**0.17**	0.03	**0.20**	0.04	**0.24**
Adrenals	0.22	**0.22**	0.28	**0.27**	0.32	**0.31**	0.35	**0.35**	0.42	**0.42**	0.48	**0.48**
Kidneys	0.15	**0.29**	0.18	**0.35**	0.20	**0.40**	0.22	**0.43**	0.26	**0.50**	0.29	**0.56**
Urinary bladder		**0.01**		**0.02**		**0.02**		**0.02**		**0.03**		**0.04**
Uterus		**0.03**	0.01	**0.04**	0.01	**0.05**	0.01	**0.05**	0.02	**0.07**	0.02	**0.08**
Ovaries		**0.04**		**0.05**	0.01	**0.05**	0.01	**0.06**	0.02	**0.07**	0.02	**0.09**
Prostate										**0.01**		**0.02**
Testes												
Upper arm		**1.36**		**1.54**		**1.66**		**1.74**		**1.87**		**1.97**
Elbow		**1.43**		**1.61**		**1.73**		**1.81**		**1.94**		**2.03**
Forearm		**0.65**		**0.73**		**0.78**		**0.82**		**0.87**		**0.91**
Upper leg										**0.01**		**0.01**
Knee												
Lower leg												
Lymph nodes	0.15	**0.25**	0.18	**0.31**	0.20	**0.35**	0.22	**0.38**	0.26	**0.44**	0.29	**0.50**
Skin	0.18	**0.29**	0.19	**0.30**	0.20	**0.32**	0.20	**0.33**	0.22	**0.34**	0.23	**0.36**
Muscle	0.11	**0.17**	0.13	**0.20**	0.14	**0.22**	0.15	**0.24**	0.17	**0.27**	0.19	**0.30**
Skeleton	0.30	**0.57**	0.35	**0.66**	0.38	**0.73**	0.41	**0.78**	0.46	**0.87**	0.50	**0.94**
Red bone marrow	0.08	**0.15**	0.10	**0.18**	0.11	**0.20**	0.12	**0.22**	0.14	**0.25**	0.16	**0.28**
Whole body dose	0.14	**0.22**	0.16	**0.26**	0.18	**0.29**	0.19	**0.31**	0.21	**0.35**	0.23	**0.38**
ED (ICRP 60)	0.18	**0.20**	0.21	**0.24**	0.24	**0.27**	0.26	**0.29**	0.29	**0.34**	0.32	**0.38**
ED (ICRP 103)	0.22	**0.25**	0.26	**0.29**	0.28	**0.33**	0.30	**0.35**	0.34	**0.40**	0.37	**0.44**
Total filtration (mm) Al	**1.5**		**2.5**		**3.5**		**4.5**		**3.5**		**3.5**	
Cu	**0**		**0**		**0**		**0**		**0.1**		**0.2**	

Table 9.157 Normalized organ doses—Chest r-lat—3–7 years

Beam projection:			r-lat			**Table cells:**						
Focus-to-detector distance (cm):			150					(+) = normal font = optimal field setting				
Tube voltage (kVp):			60					**(–) = bold font = suboptimal field setting**				
Total filtrations:			See last table row.					Empty table cells correspond to values < 0.01				
Organ or tissue	**Organ dose D_T : Entrance dose K_E**							**(mSv × mGy^{-1})**				
	(+)	(–)	(+)	(–)	(+)	(–)	(+)	(–)	(+)	(–)	(+)	(–)
Skull		**0.17**	0.01	**0.21**	0.01	**0.25**	0.02	**0.27**	0.02	**0.32**	0.02	**0.30**
Brain				**0.01**		**0.02**		**0.02**		**0.02**		**0.02**
Oral mucosa	0.01	**0.23**	0.01	**0.28**	0.02	**0.32**	0.02	**0.35**	0.03	**0.40**	0.02	**0.38**
Salivary glands	0.02	**0.30**	0.02	**0.35**	0.03	**0.38**	0.03	**0.41**	0.04	**0.45**	0.03	**0.44**
Cervical spine	0.24	**0.54**	0.31	**0.69**	0.37	**0.81**	0.41	**0.91**	0.49	**1.09**	0.46	**1.03**
Thyroid	0.36	**0.36**	0.43	**0.44**	0.48	**0.49**	0.52	**0.54**	0.59	**0.61**	0.57	**0.59**
Thymus	0.13	**0.12**	0.17	**0.16**	0.20	**0.19**	0.23	**0.22**	0.28	**0.27**	0.26	**0.24**
Thoracic spine	0.26	**0.23**	0.35	**0.31**	0.42	**0.38**	0.48	**0.43**	0.61	**0.55**	0.54	**0.48**
Claviculae	0.56	**0.52**	0.70	**0.66**	0.80	**0.76**	0.88	**0.84**	1.02	**0.99**	0.99	**0.95**
Scapulae	1.04	**0.98**	1.22	**1.16**	1.35	**1.28**	1.44	**1.38**	1.60	**1.54**	1.60	**1.52**
Ribs	0.93	**0.73**	1.11	**0.90**	1.23	**1.02**	1.32	**1.12**	1.49	**1.29**	1.47	**1.26**
Oesophagus	0.13	**0.09**	0.17	**0.11**	0.21	**0.14**	0.23	**0.16**	0.29	**0.20**	0.26	**0.17**
Trachea	0.04	**0.16**	0.04	**0.21**	0.05	**0.24**	0.06	**0.27**	0.07	**0.32**	0.06	**0.30**
Lung	0.28	**0.18**	0.35	**0.24**	0.40	**0.28**	0.44	**0.31**	0.51	**0.37**	0.49	**0.35**
Mamma	0.50	**0.49**	0.55	**0.54**	0.58	**0.57**	0.60	**0.59**	0.64	**0.62**	0.63	**0.61**
Heart	0.15	**0.11**	0.20	**0.15**	0.23	**0.18**	0.26	**0.20**	0.32	**0.26**	0.30	**0.23**
Stomach	0.03	**0.03**	0.04	**0.05**	0.04	**0.06**	0.05	**0.07**	0.07	**0.08**	0.06	**0.07**
Spleen	0.02	**0.02**	0.03	**0.03**	0.03	**0.03**	0.04	**0.04**	0.05	**0.05**	0.04	**0.04**
Liver	0.24	**0.27**	0.30	**0.34**	0.34	**0.40**	0.38	**0.45**	0.44	**0.54**	0.42	**0.50**
Gall bladder	0.04	**0.09**	0.06	**0.13**	0.07	**0.15**	0.08	**0.17**	0.10	**0.22**	0.08	**0.19**
Pancreas	0.06	**0.05**	0.09	**0.07**	0.10	**0.09**	0.12	**0.10**	0.15	**0.13**	0.13	**0.10**
Small intestine		**0.05**		**0.07**		**0.08**	0.01	**0.10**	0.01	**0.12**	0.01	**0.11**
Large intestine		**0.09**	0.01	**0.11**	0.01	**0.13**	0.01	**0.15**	0.02	**0.19**	0.02	**0.17**
Sigma/Rectum								**0.01**		**0.01**		**0.01**
Lumbar spine	0.05	**0.19**	0.06	**0.25**	0.08	**0.31**	0.09	**0.35**	0.11	**0.45**	0.09	**0.39**
Pelvis		**0.04**		**0.05**		**0.06**		**0.07**		**0.09**		**0.08**
Adrenals	0.10	**0.09**	0.13	**0.12**	0.15	**0.15**	0.17	**0.17**	0.22	**0.21**	0.19	**0.18**
Kidneys	0.06	**0.15**	0.07	**0.19**	0.09	**0.22**	0.10	**0.24**	0.12	**0.29**	0.11	**0.27**
Urinary bladder												
Uterus		**0.01**		**0.02**		**0.02**		**0.02**		**0.03**		**0.02**
Ovaries				**0.01**		**0.02**		**0.02**		**0.03**		**0.02**
Prostate												
Testes												
Upper arm		**1.03**		**1.22**		**1.35**		**1.44**		**1.60**		**1.59**
Elbow		**1.10**		**1.29**		**1.42**		**1.52**		**1.68**		**1.67**
Forearm		**0.50**		**0.58**		**0.63**		**0.68**		**0.75**		**0.74**
Upper leg												
Knee												
Lower leg												
Lymph nodes	0.07	**0.12**	0.09	**0.15**	0.11	**0.18**	0.12	**0.20**	0.15	**0.24**	0.13	**0.22**
Skin	0.11	**0.21**	0.12	**0.23**	0.12	**0.23**	0.13	**0.24**	0.14	**0.25**	0.13	**0.25**
Muscle	0.06	**0.09**	0.07	**0.11**	0.08	**0.12**	0.09	**0.13**	0.11	**0.16**	0.10	**0.15**
Skeleton	0.16	**0.31**	0.19	**0.38**	0.21	**0.43**	0.23	**0.46**	0.27	**0.53**	0.26	**0.52**
Red bone marrow	0.05	**0.08**	0.06	**0.10**	0.07	**0.11**	0.08	**0.12**	0.09	**0.15**	0.09	**0.14**
Whole body dose	0.08	**0.12**	0.10	**0.15**	0.11	**0.17**	0.12	**0.19**	0.14	**0.22**	0.13	**0.20**
ED (ICRP 60)	0.11	**0.11**	0.13	**0.14**	0.15	**0.16**	0.16	**0.17**	0.19	**0.20**	0.18	**0.19**
ED (ICRP 103)	0.14	**0.15**	0.17	**0.18**	0.19	**0.20**	0.20	**0.22**	0.23	**0.25**	0.22	**0.24**
Total filtration (mm) Al	**1.5**		**2.5**		**3.5**		**4.5**		**3.5**		**3.5**	
Cu	**0**		**0**		**0**		**0**		**0.1**		**0.2**	

Table 9.158 Normalized organ doses—Chest r-lat—3–7 years

Beam projection:			r-lat				**Table cells:**					
Focus-to-detector distance (cm):			150				(+) = normal font = optimal field setting					
Tube voltage (kVp):			70				**(−) = bold font = suboptimal field setting**					
Total filtrations:			See last table row.				Empty table cells correspond to values < 0.01					
Organ or tissue	**Organ dose D_T : Entrance dose K_E**						**(mSv × mGy^{-1})**					
	(+)	(−)	(+)	(−)	(+)	(−)	(+)	(−)	(+)	(−)	(+)	(−)
Skull	0.01	**0.20**	0.02	**0.25**	0.02	**0.28**	0.02	**0.30**	0.03	**0.35**	0.03	**0.39**
Brain		**0.01**		**0.02**		**0.02**		**0.02**		**0.03**		**0.04**
Oral mucosa	0.01	**0.27**	0.02	**0.32**	0.02	**0.36**	0.03	**0.40**	0.03	**0.45**	0.04	**0.51**
Salivary glands	0.02	**0.33**	0.03	**0.38**	0.03	**0.42**	0.04	**0.44**	0.04	**0.49**	0.05	**0.53**
Cervical spine	0.30	**0.65**	0.37	**0.82**	0.43	**0.95**	0.47	**1.05**	0.56	**1.25**	0.64	**1.43**
Thyroid	0.40	**0.41**	0.48	**0.49**	0.53	**0.55**	0.57	**0.60**	0.64	**0.68**	0.70	**0.75**
Thymus	0.17	**0.16**	0.21	**0.20**	0.25	**0.24**	0.28	**0.27**	0.34	**0.33**	0.40	**0.38**
Thoracic spine	0.35	**0.31**	0.45	**0.41**	0.53	**0.48**	0.60	**0.55**	0.75	**0.68**	0.88	**0.82**
Claviculae	0.65	**0.61**	0.79	**0.76**	0.89	**0.86**	0.97	**0.95**	1.12	**1.10**	1.24	**1.24**
Scapulae	1.13	**1.08**	1.32	**1.26**	1.44	**1.38**	1.53	**1.47**	1.68	**1.62**	1.79	**1.74**
Ribs	1.03	**0.84**	1.21	**1.01**	1.33	**1.14**	1.42	**1.23**	1.57	**1.41**	1.70	**1.55**
Oesophagus	0.17	**0.11**	0.22	**0.15**	0.25	**0.18**	0.28	**0.20**	0.34	**0.25**	0.40	**0.30**
Trachea	0.04	**0.20**	0.05	**0.25**	0.06	**0.28**	0.07	**0.31**	0.08	**0.37**	0.09	**0.43**
Lung	0.33	**0.23**	0.40	**0.28**	0.45	**0.33**	0.49	**0.37**	0.57	**0.44**	0.64	**0.50**
Mamma	0.53	**0.52**	0.58	**0.57**	0.61	**0.59**	0.63	**0.61**	0.67	**0.65**	0.70	**0.68**
Heart	0.19	**0.15**	0.24	**0.19**	0.29	**0.23**	0.32	**0.26**	0.39	**0.31**	0.45	**0.37**
Stomach	0.04	**0.05**	0.05	**0.06**	0.06	**0.08**	0.07	**0.09**	0.09	**0.11**	0.11	**0.14**
Spleen	0.03	**0.03**	0.04	**0.04**	0.05	**0.05**	0.05	**0.05**	0.07	**0.07**	0.09	**0.09**
Liver	0.28	**0.33**	0.34	**0.41**	0.39	**0.48**	0.43	**0.53**	0.50	**0.62**	0.56	**0.72**
Gall bladder	0.06	**0.13**	0.07	**0.16**	0.09	**0.20**	0.10	**0.22**	0.13	**0.28**	0.15	**0.33**
Pancreas	0.09	**0.07**	0.11	**0.10**	0.13	**0.12**	0.15	**0.13**	0.19	**0.17**	0.23	**0.21**
Small intestine		**0.07**	0.01	**0.09**	0.01	**0.11**	0.02	**0.12**	0.02	**0.15**	0.02	**0.18**
Large intestine	0.01	**0.11**	0.01	**0.14**	0.02	**0.17**	0.02	**0.19**	0.03	**0.23**	0.03	**0.27**
Sigma/Rectum				**0.01**		**0.01**		**0.02**		**0.02**		**0.03**
Lumbar spine	0.06	**0.25**	0.08	**0.33**	0.10	**0.39**	0.12	**0.45**	0.15	**0.55**	0.18	**0.66**
Pelvis		**0.05**		**0.07**		**0.08**		**0.09**	0.01	**0.12**	0.01	**0.14**
Adrenals	0.13	**0.12**	0.16	**0.16**	0.19	**0.19**	0.22	**0.21**	0.27	**0.26**	0.31	**0.31**
Kidneys	0.07	**0.18**	0.09	**0.22**	0.11	**0.26**	0.12	**0.29**	0.14	**0.34**	0.17	**0.39**
Urinary bladder								**0.01**		**0.01**		**0.02**
Uterus		**0.02**		**0.02**		**0.03**		**0.03**		**0.04**		**0.05**
Ovaries		**0.01**		**0.02**		**0.02**		**0.03**		**0.04**		**0.05**
Prostate												
Testes												
Upper arm		**1.13**		**1.32**		**1.44**		**1.53**		**1.69**		**1.81**
Elbow		**1.20**		**1.39**		**1.52**		**1.61**		**1.76**		**1.88**
Forearm		**0.54**		**0.62**		**0.67**		**0.71**		**0.78**		**0.83**
Upper leg												
Knee												
Lower leg												
Lymph nodes	0.09	**0.15**	0.11	**0.19**	0.13	**0.21**	0.14	**0.24**	0.17	**0.28**	0.20	**0.33**
Skin	0.12	**0.22**	0.12	**0.23**	0.13	**0.24**	0.13	**0.25**	0.14	**0.26**	0.15	**0.28**
Muscle	0.07	**0.10**	0.09	**0.13**	0.10	**0.14**	0.10	**0.15**	0.12	**0.18**	0.13	**0.20**
Skeleton	0.18	**0.35**	0.21	**0.42**	0.24	**0.47**	0.26	**0.51**	0.29	**0.59**	0.33	**0.65**
Red bone marrow	0.06	**0.09**	0.07	**0.12**	0.08	**0.13**	0.09	**0.15**	0.11	**0.17**	0.12	**0.20**
Whole body dose	0.09	**0.14**	0.11	**0.17**	0.12	**0.19**	0.13	**0.21**	0.15	**0.24**	0.17	**0.27**
ED (ICRP 60)	0.13	**0.13**	0.15	**0.16**	0.17	**0.18**	0.18	**0.20**	0.21	**0.23**	0.24	**0.26**
ED (ICRP 103)	0.16	**0.17**	0.19	**0.20**	0.21	**0.23**	0.22	**0.25**	0.25	**0.29**	0.28	**0.32**
Total filtration (mm) Al	**1.5**		**2.5**		**3.5**		**4.5**		**3.5**		**3.5**	
Cu	**0**		**0**		**0**		**0**		**0.1**		**0.2**	

Table 9.159 Normalized organ doses—Chest r-lat—3–7 years

Beam projection:			r-lat			**Table cells:**						
Focus-to-detector distance (cm):			150			(+) = normal font = optimal field setting						
Tube voltage (kVp):			80			**(−) = bold font = suboptimal field setting**						
Total filtrations:			See last table row.			Empty table cells correspond to values < 0.01						
Organ or tissue	**Organ dose D_T : Entrance dose K_E**							**(mSv × mGy^{-1})**				
	(+)	(−)	(+)	(−)	(+)	(−)	(+)	(−)	(+)	(−)	(+)	(−)
Skull	0.02	**0.23**	0.02	**0.27**	0.02	**0.31**	0.03	**0.33**	0.03	**0.38**	0.04	**0.42**
Brain		**0.02**		**0.02**		**0.03**		**0.03**		**0.04**		**0.05**
Oral mucosa	0.02	**0.31**	0.02	**0.36**	0.03	**0.40**	0.03	**0.44**	0.04	**0.50**	0.04	**0.55**
Salivary glands	0.03	**0.36**	0.03	**0.41**	0.04	**0.45**	0.04	**0.48**	0.05	**0.52**	0.06	**0.57**
Cervical spine	0.34	**0.76**	0.42	**0.93**	0.48	**1.06**	0.53	**1.17**	0.61	**1.36**	0.69	**1.53**
Thyroid	0.45	**0.46**	0.52	**0.55**	0.58	**0.61**	0.62	**0.65**	0.69	**0.73**	0.75	**0.80**
Thymus	0.20	**0.19**	0.25	**0.24**	0.29	**0.28**	0.33	**0.31**	0.39	**0.38**	0.45	**0.44**
Thoracic spine	0.43	**0.39**	0.54	**0.50**	0.63	**0.58**	0.70	**0.65**	0.85	**0.79**	0.99	**0.93**
Claviculae	0.72	**0.69**	0.86	**0.84**	0.97	**0.95**	1.04	**1.03**	1.18	**1.18**	1.29	**1.30**
Scapulae	1.21	**1.16**	1.39	**1.33**	1.50	**1.45**	1.58	**1.53**	1.71	**1.66**	1.81	**1.76**
Ribs	1.11	**0.93**	1.28	**1.10**	1.40	**1.22**	1.48	**1.32**	1.62	**1.48**	1.73	**1.61**
Oesophagus	0.21	**0.15**	0.26	**0.19**	0.30	**0.22**	0.33	**0.24**	0.39	**0.30**	0.45	**0.35**
Trachea	0.05	**0.23**	0.06	**0.29**	0.07	**0.32**	0.07	**0.36**	0.08	**0.42**	0.09	**0.47**
Lung	0.37	**0.27**	0.45	**0.33**	0.50	**0.38**	0.54	**0.42**	0.62	**0.49**	0.69	**0.56**
Mamma	0.56	**0.55**	0.60	**0.59**	0.63	**0.62**	0.65	**0.64**	0.69	**0.68**	0.72	**0.71**
Heart	0.23	**0.18**	0.29	**0.23**	0.33	**0.27**	0.37	**0.30**	0.44	**0.37**	0.51	**0.43**
Stomach	0.05	**0.06**	0.07	**0.08**	0.08	**0.10**	0.09	**0.11**	0.11	**0.14**	0.13	**0.17**
Spleen	0.04	**0.04**	0.05	**0.05**	0.06	**0.06**	0.07	**0.07**	0.09	**0.09**	0.11	**0.11**
Liver	0.32	**0.39**	0.39	**0.48**	0.44	**0.54**	0.47	**0.60**	0.54	**0.70**	0.61	**0.79**
Gall bladder	0.07	**0.16**	0.09	**0.20**	0.11	**0.24**	0.12	**0.27**	0.15	**0.33**	0.18	**0.38**
Pancreas	0.11	**0.10**	0.14	**0.13**	0.17	**0.15**	0.19	**0.17**	0.23	**0.21**	0.27	**0.26**
Small intestine	0.01	**0.09**	0.01	**0.11**	0.02	**0.13**	0.02	**0.15**	0.02	**0.18**	0.03	**0.21**
Large intestine	0.02	**0.14**	0.02	**0.17**	0.02	**0.20**	0.03	**0.22**	0.03	**0.26**	0.04	**0.31**
Sigma/Rectum		**0.01**		**0.02**		**0.02**		**0.02**		**0.03**		**0.03**
Lumbar spine	0.08	**0.32**	0.11	**0.40**	0.13	**0.47**	0.14	**0.53**	0.18	**0.65**	0.21	**0.76**
Pelvis		**0.07**		**0.09**	0.01	**0.10**	0.01	**0.11**	0.02	**0.14**	0.02	**0.16**
Adrenals	0.16	**0.15**	0.20	**0.19**	0.23	**0.23**	0.26	**0.25**	0.31	**0.31**	0.36	**0.36**
Kidneys	0.09	**0.21**	0.11	**0.26**	0.12	**0.30**	0.14	**0.33**	0.16	**0.39**	0.19	**0.44**
Urinary bladder				**0.01**		**0.01**		**0.01**		**0.02**		**0.02**
Uterus		**0.02**		**0.03**		**0.03**		**0.04**		**0.05**	0.01	**0.06**
Ovaries		**0.02**		**0.03**		**0.03**		**0.04**		**0.05**		**0.06**
Prostate												**0.01**
Testes												
Upper arm		**1.22**		**1.39**		**1.51**		**1.59**		**1.73**		**1.83**
Elbow		**1.29**		**1.47**		**1.58**		**1.67**		**1.80**		**1.90**
Forearm		**0.57**		**0.65**		**0.70**		**0.74**		**0.79**		**0.83**
Upper leg												
Knee												
Lower leg												
Lymph nodes	0.11	**0.18**	0.13	**0.22**	0.15	**0.25**	0.16	**0.27**	0.19	**0.32**	0.22	**0.37**
Skin	0.12	**0.23**	0.13	**0.24**	0.14	**0.25**	0.14	**0.26**	0.15	**0.28**	0.16	**0.29**
Muscle	0.08	**0.12**	0.10	**0.14**	0.11	**0.16**	0.12	**0.17**	0.13	**0.20**	0.15	**0.22**
Skeleton	0.20	**0.39**	0.23	**0.46**	0.26	**0.51**	0.28	**0.55**	0.31	**0.62**	0.34	**0.68**
Red bone marrow	0.07	**0.11**	0.08	**0.13**	0.09	**0.15**	0.10	**0.17**	0.12	**0.19**	0.13	**0.22**
Whole body dose	0.10	**0.16**	0.12	**0.19**	0.14	**0.22**	0.15	**0.23**	0.17	**0.27**	0.19	**0.30**
ED (ICRP 60)	0.14	**0.15**	0.17	**0.18**	0.19	**0.21**	0.20	**0.22**	0.23	**0.26**	0.26	**0.29**
ED (ICRP 103)	0.18	**0.20**	0.21	**0.23**	0.23	**0.26**	0.25	**0.28**	0.28	**0.32**	0.30	**0.35**
Total filtration (mm) Al	**1.5**		**2.5**		**3.5**		**4.5**		**3.5**		**3.5**	
Cu	**0**		**0**		**0**		**0**		**0.1**		**0.2**	

Table 9.160 Normalized organ doses—Chest r-lat—8–12 years

Beam projection:			r-lat			**Table cells:**						
Focus-to-detector distance (cm):			150			(+) = normal font = optimal field setting						
Tube voltage (kVp):			60			**(−) = bold font = suboptimal field setting**						
Total filtrations:			See last table row.			Empty table cells correspond to values < 0.01						
Organ or tissue	**Organ dose D_T : Entrance dose K_E**						**(mSv × mGy^{-1})**					
	(+)	(−)	(+)	(−)	(+)	(−)	(+)	(−)	(+)	(−)	(+)	(−)
Skull		**0.15**	0.01	**0.18**	0.01	**0.21**	0.01	**0.23**	0.02	**0.27**	0.02	**0.26**
Brain						**0.01**		**0.01**		**0.02**		**0.01**
Oral mucosa		**0.20**	0.01	**0.25**	0.01	**0.28**	0.01	**0.31**	0.02	**0.35**	0.01	**0.34**
Salivary glands	0.02	**0.28**	0.02	**0.32**	0.02	**0.35**	0.03	**0.37**	0.03	**0.41**	0.03	**0.40**
Cervical spine	0.24	**0.49**	0.30	**0.63**	0.36	**0.75**	0.40	**0.84**	0.49	**1.02**	0.45	**0.94**
Thyroid	0.30	**0.29**	0.37	**0.35**	0.41	**0.40**	0.45	**0.44**	0.51	**0.50**	0.50	**0.49**
Thymus	0.09	**0.08**	0.12	**0.11**	0.14	**0.13**	0.16	**0.15**	0.21	**0.19**	0.18	**0.16**
Thoracic spine	0.17	**0.14**	0.22	**0.19**	0.27	**0.24**	0.31	**0.27**	0.40	**0.35**	0.34	**0.29**
Claviculae	0.44	**0.38**	0.55	**0.49**	0.64	**0.57**	0.71	**0.63**	0.84	**0.76**	0.80	**0.72**
Scapulae	0.92	**0.86**	1.10	**1.03**	1.22	**1.14**	1.31	**1.23**	1.47	**1.39**	1.46	**1.37**
Ribs	0.75	**0.59**	0.90	**0.73**	1.00	**0.84**	1.09	**0.92**	1.23	**1.08**	1.21	**1.04**
Oesophagus	0.09	**0.05**	0.11	**0.07**	0.14	**0.09**	0.16	**0.10**	0.20	**0.13**	0.17	**0.11**
Trachea	0.09	**0.21**	0.11	**0.27**	0.12	**0.32**	0.14	**0.35**	0.16	**0.42**	0.15	**0.39**
Lung	0.22	**0.14**	0.27	**0.18**	0.32	**0.21**	0.35	**0.24**	0.42	**0.29**	0.40	**0.26**
Mamma	0.47	**0.46**	0.52	**0.51**	0.55	**0.53**	0.57	**0.55**	0.60	**0.58**	0.59	**0.57**
Heart	0.10	**0.07**	0.14	**0.10**	0.17	**0.12**	0.19	**0.14**	0.24	**0.18**	0.21	**0.15**
Stomach	0.01	**0.02**	0.02	**0.02**	0.02	**0.03**	0.03	**0.04**	0.04	**0.05**	0.03	**0.04**
Spleen			0.01	**0.01**	0.02	**0.02**	0.02	**0.02**	0.03	**0.03**	0.02	**0.02**
Liver	0.18	**0.22**	0.23	**0.28**	0.26	**0.33**	0.29	**0.38**	0.35	**0.46**	0.32	**0.42**
Gall bladder	0.02	**0.07**	0.03	**0.10**	0.04	**0.12**	0.04	**0.14**	0.06	**0.17**	0.05	**0.15**
Pancreas	0.04	**0.03**	0.05	**0.04**	0.06	**0.05**	0.07	**0.06**	0.09	**0.08**	0.07	**0.06**
Small intestine		**0.03**		**0.04**		**0.05**		**0.06**		**0.08**		**0.06**
Large intestine		**0.06**		**0.07**		**0.09**		**0.10**		**0.13**		**0.11**
Sigma/Rectum												
Lumbar spine	0.02	**0.11**	0.02	**0.15**	0.03	**0.18**	0.03	**0.21**	0.05	**0.27**	0.03	**0.23**
Pelvis		**0.02**		**0.03**		**0.03**		**0.04**		**0.05**		**0.04**
Adrenals	0.07	**0.06**	0.09	**0.08**	0.11	**0.10**	0.12	**0.12**	0.15	**0.15**	0.13	**0.13**
Kidneys	0.03	**0.10**	0.03	**0.14**	0.04	**0.16**	0.05	**0.18**	0.06	**0.22**	0.05	**0.20**
Urinary bladder												
Uterus								**0.01**		**0.01**		**0.01**
Ovaries								**0.01**		**0.02**		
Prostate												
Testes												
Upper arm		**0.93**		**1.10**		**1.23**		**1.32**		**1.48**		**1.46**
Elbow		**1.01**		**1.19**		**1.32**		**1.41**		**1.58**		**1.56**
Forearm		**0.42**		**0.49**		**0.54**		**0.58**		**0.65**		**0.64**
Upper leg												
Knee												
Lower leg												
Lymph nodes	0.06	**0.10**	0.08	**0.13**	0.09	**0.15**	0.10	**0.17**	0.12	**0.20**	0.11	**0.19**
Skin	0.10	**0.18**	0.10	**0.19**	0.11	**0.20**	0.11	**0.21**	0.12	**0.22**	0.12	**0.21**
Muscle	0.05	**0.07**	0.06	**0.08**	0.07	**0.10**	0.07	**0.10**	0.09	**0.12**	0.08	**0.11**
Skeleton	0.12	**0.25**	0.15	**0.30**	0.17	**0.34**	0.18	**0.38**	0.21	**0.44**	0.21	**0.42**
Red bone marrow	0.05	**0.07**	0.06	**0.08**	0.07	**0.09**	0.08	**0.11**	0.09	**0.13**	0.08	**0.11**
Whole body dose	0.06	**0.10**	0.08	**0.12**	0.09	**0.14**	0.10	**0.15**	0.11	**0.17**	0.11	**0.16**
ED (ICRP 60)	0.09	**0.09**	0.11	**0.11**	0.12	**0.12**	0.13	**0.14**	0.15	**0.16**	0.15	**0.15**
ED (ICRP 103)	0.12	**0.12**	0.14	**0.15**	0.16	**0.17**	0.17	**0.18**	0.19	**0.21**	0.18	**0.19**
Total filtration (mm) Al	**1.5**		**2.5**		**3.5**		**4.5**		**3.5**		**3.5**	
Cu	**0**		**0**		**0**		**0**		**0.1**		**0.2**	

Table 9.161 Normalized organ doses—Chest r-lat—8–12 years

Beam projection:		r-lat			**Table cells:**							
Focus-to-detector distance (cm):		150			(+) = normal font = optimal field setting							
Tube voltage (kVp):		70			**(−) = bold font = suboptimal field setting**							
Total filtrations:		See last table row.			Empty table cells correspond to values < 0.01							
Organ or tissue	**Organ dose D_T : Entrance dose K_E**							**(mSv × mGy^{-1})**				
	(+)	(−)	(+)	(−)	(+)	(−)	(+)	(−)	(+)	(−)	(+)	(−)
Skull	0.01	**0.17**	0.01	**0.21**	0.02	**0.24**	0.02	**0.26**	0.02	**0.30**	0.03	**0.34**
Brain				**0.01**		**0.02**		**0.02**		**0.02**		**0.03**
Oral mucosa	0.01	**0.24**	0.01	**0.28**	0.02	**0.32**	0.02	**0.35**	0.02	**0.40**	0.03	**0.45**
Salivary glands	0.02	**0.31**	0.02	**0.35**	0.03	**0.38**	0.03	**0.41**	0.04	**0.45**	0.05	**0.49**
Cervical spine	0.29	**0.60**	0.37	**0.76**	0.42	**0.89**	0.47	**0.99**	0.56	**1.18**	0.65	**1.36**
Thyroid	0.34	**0.33**	0.41	**0.40**	0.46	**0.45**	0.50	**0.49**	0.56	**0.56**	0.62	**0.62**
Thymus	0.12	**0.11**	0.15	**0.14**	0.18	**0.17**	0.21	**0.19**	0.26	**0.23**	0.31	**0.28**
Thoracic spine	0.23	**0.20**	0.30	**0.26**	0.36	**0.32**	0.41	**0.36**	0.51	**0.46**	0.62	**0.56**
Claviculae	0.51	**0.46**	0.64	**0.57**	0.73	**0.66**	0.80	**0.73**	0.93	**0.86**	1.05	**0.98**
Scapulae	1.02	**0.95**	1.20	**1.12**	1.32	**1.24**	1.40	**1.33**	1.56	**1.48**	1.67	**1.60**
Ribs	0.84	**0.68**	0.99	**0.83**	1.09	**0.94**	1.17	**1.03**	1.32	**1.19**	1.43	**1.32**
Oesophagus	0.11	**0.07**	0.15	**0.10**	0.17	**0.12**	0.20	**0.13**	0.24	**0.17**	0.29	**0.20**
Trachea	0.10	**0.26**	0.13	**0.32**	0.14	**0.37**	0.15	**0.41**	0.18	**0.49**	0.20	**0.56**
Lung	0.26	**0.17**	0.32	**0.22**	0.37	**0.26**	0.41	**0.29**	0.48	**0.35**	0.54	**0.40**
Mamma	0.50	**0.49**	0.54	**0.53**	0.57	**0.56**	0.59	**0.57**	0.62	**0.61**	0.65	**0.63**
Heart	0.14	**0.10**	0.18	**0.13**	0.21	**0.16**	0.24	**0.18**	0.29	**0.22**	0.35	**0.27**
Stomach	0.02	**0.03**	0.03	**0.04**	0.03	**0.04**	0.04	**0.05**	0.05	**0.07**	0.06	**0.08**
Spleen	0.01	**0.02**	0.02	**0.02**	0.02	**0.03**	0.03	**0.03**	0.04	**0.04**	0.05	**0.05**
Liver	0.22	**0.27**	0.27	**0.35**	0.31	**0.40**	0.34	**0.45**	0.40	**0.54**	0.45	**0.63**
Gall bladder	0.03	**0.10**	0.04	**0.13**	0.05	**0.16**	0.06	**0.18**	0.07	**0.23**	0.09	**0.27**
Pancreas	0.05	**0.05**	0.07	**0.06**	0.08	**0.08**	0.09	**0.09**	0.12	**0.11**	0.15	**0.14**
Small intestine		**0.04**		**0.06**		**0.07**		**0.08**		**0.10**	0.01	**0.12**
Large intestine		**0.07**		**0.10**		**0.11**		**0.13**	0.01	**0.16**	0.02	**0.19**
Sigma/Rectum										**0.01**		**0.01**
Lumbar spine	0.03	**0.15**	0.03	**0.20**	0.04	**0.24**	0.05	**0.28**	0.06	**0.35**	0.08	**0.43**
Pelvis		**0.03**		**0.04**		**0.05**		**0.05**		**0.07**		**0.08**
Adrenals	0.09	**0.09**	0.12	**0.11**	0.14	**0.13**	0.16	**0.15**	0.19	**0.19**	0.23	**0.23**
Kidneys	0.03	**0.13**	0.04	**0.17**	0.05	**0.20**	0.06	**0.22**	0.07	**0.26**	0.09	**0.31**
Urinary bladder												
Uterus				**0.01**		**0.01**		**0.02**		**0.02**		**0.03**
Ovaries				**0.01**		**0.01**		**0.02**		**0.02**		**0.03**
Prostate												
Testes												
Upper arm		**1.03**		**1.20**		**1.32**		**1.41**		**1.57**		**1.69**
Elbow		**1.11**		**1.29**		**1.41**		**1.51**		**1.66**		**1.79**
Forearm		**0.46**		**0.53**		**0.58**		**0.62**		**0.68**		**0.73**
Upper leg												
Knee												
Lower leg												
Lymph nodes	0.08	**0.13**	0.10	**0.16**	0.11	**0.18**	0.12	**0.20**	0.14	**0.24**	0.17	**0.28**
Skin	0.10	**0.19**	0.11	**0.20**	0.11	**0.21**	0.12	**0.22**	0.12	**0.23**	0.13	**0.24**
Muscle	0.06	**0.08**	0.07	**0.10**	0.08	**0.11**	0.09	**0.12**	0.10	**0.14**	0.11	**0.16**
Skeleton	0.14	**0.28**	0.17	**0.34**	0.19	**0.39**	0.21	**0.42**	0.24	**0.48**	0.26	**0.53**
Red bone marrow	0.06	**0.08**	0.07	**0.10**	0.08	**0.11**	0.09	**0.13**	0.10	**0.15**	0.12	**0.17**
Whole body dose	0.07	**0.12**	0.09	**0.14**	0.10	**0.16**	0.11	**0.17**	0.13	**0.20**	0.14	**0.22**
ED (ICRP 60)	0.10	**0.10**	0.12	**0.13**	0.14	**0.14**	0.15	**0.16**	0.17	**0.19**	0.20	**0.21**
ED (ICRP 103)	0.14	**0.14**	0.16	**0.17**	0.18	**0.19**	0.19	**0.21**	0.22	**0.24**	0.24	**0.27**
Total filtration (mm) **Al**	**1.5**		**2.5**		**3.5**		**4.5**		**3.5**		**3.5**	
Cu	**0**		**0**		**0**		**0**		**0.1**		**0.2**	

Table 9.162 Normalized organ doses—Chest r-lat—8–12 years

Beam projection:		r-lat			**Table cells:**							
Focus-to-detector distance (cm):		150					(+) = normal font = optimal field setting					
Tube voltage (kVp):		80					**(−) = bold font = suboptimal field setting**					
Total filtrations:		See last table row.					Empty table cells correspond to values < 0.01					
Organ or tissue	**Organ dose D_T : Entrance dose K_E**							**(mSv × mGy^{-1})**				
	(+)	(−)	(+)	(−)	(+)	(−)	(+)	(−)	(+)	(−)	(+)	(−)
Skull	0.01	**0.20**	0.02	**0.24**	0.02	**0.26**	0.02	**0.29**	0.03	**0.33**	0.03	**0.37**
Brain		**0.01**		**0.02**		**0.02**		**0.02**		**0.03**		**0.04**
Oral mucosa	0.01	**0.27**	0.02	**0.32**	0.02	**0.36**	0.02	**0.39**	0.03	**0.44**	0.03	**0.49**
Salivary glands	0.02	**0.33**	0.03	**0.38**	0.03	**0.41**	0.04	**0.44**	0.05	**0.49**	0.05	**0.53**
Cervical spine	0.34	**0.71**	0.42	**0.88**	0.48	**1.00**	0.53	**1.11**	0.62	**1.30**	0.70	**1.47**
Thyroid	0.38	**0.38**	0.45	**0.45**	0.50	**0.50**	0.54	**0.54**	0.61	**0.61**	0.67	**0.67**
Thymus	0.15	**0.14**	0.19	**0.17**	0.22	**0.20**	0.25	**0.23**	0.30	**0.28**	0.35	**0.33**
Thoracic spine	0.29	**0.26**	0.37	**0.33**	0.44	**0.39**	0.49	**0.44**	0.61	**0.55**	0.72	**0.65**
Claviculae	0.58	**0.53**	0.71	**0.65**	0.80	**0.74**	0.87	**0.81**	0.99	**0.94**	1.10	**1.05**
Scapulae	1.10	**1.03**	1.27	**1.20**	1.38	**1.31**	1.46	**1.39**	1.60	**1.53**	1.70	**1.63**
Ribs	0.91	**0.77**	1.06	**0.92**	1.16	**1.03**	1.24	**1.11**	1.37	**1.26**	1.47	**1.38**
Oesophagus	0.14	**0.10**	0.18	**0.12**	0.21	**0.15**	0.23	**0.16**	0.28	**0.20**	0.33	**0.24**
Trachea	0.12	**0.31**	0.14	**0.38**	0.16	**0.43**	0.17	**0.47**	0.20	**0.55**	0.22	**0.62**
Lung	0.30	**0.21**	0.37	**0.26**	0.42	**0.30**	0.46	**0.33**	0.53	**0.40**	0.59	**0.46**
Mamma	0.52	**0.51**	0.57	**0.55**	0.59	**0.58**	0.61	**0.60**	0.65	**0.63**	0.68	**0.66**
Heart	0.17	**0.13**	0.22	**0.17**	0.25	**0.20**	0.28	**0.22**	0.34	**0.27**	0.40	**0.32**
Stomach	0.03	**0.04**	0.04	**0.05**	0.04	**0.06**	0.05	**0.07**	0.06	**0.09**	0.08	**0.11**
Spleen	0.02	**0.02**	0.03	**0.03**	0.03	**0.04**	0.04	**0.04**	0.05	**0.05**	0.06	**0.07**
Liver	0.25	**0.33**	0.31	**0.41**	0.35	**0.47**	0.38	**0.52**	0.44	**0.61**	0.49	**0.70**
Gall bladder	0.04	**0.13**	0.06	**0.17**	0.07	**0.20**	0.07	**0.22**	0.09	**0.27**	0.11	**0.32**
Pancreas	0.07	**0.06**	0.09	**0.08**	0.10	**0.10**	0.12	**0.11**	0.15	**0.14**	0.18	**0.18**
Small intestine		**0.06**		**0.07**		**0.09**		**0.10**	0.01	**0.12**	0.01	**0.15**
Large intestine		**0.09**		**0.12**	0.01	**0.14**	0.01	**0.15**	0.02	**0.19**	0.02	**0.22**
Sigma/Rectum								**0.01**		**0.01**		**0.02**
Lumbar spine	0.04	**0.20**	0.05	**0.26**	0.06	**0.31**	0.06	**0.34**	0.08	**0.43**	0.10	**0.51**
Pelvis		**0.04**		**0.05**		**0.06**		**0.07**		**0.08**		**0.10**
Adrenals	0.11	**0.11**	0.14	**0.14**	0.17	**0.17**	0.19	**0.19**	0.23	**0.23**	0.28	**0.28**
Kidneys	0.04	**0.16**	0.05	**0.20**	0.06	**0.23**	0.07	**0.26**	0.09	**0.30**	0.10	**0.35**
Urinary bladder												**0.01**
Uterus		**0.01**		**0.02**		**0.02**		**0.02**		**0.03**		**0.03**
Ovaries		**0.01**		**0.02**		**0.02**		**0.02**		**0.03**		**0.04**
Prostate												
Testes												
Upper arm		**1.11**		**1.28**		**1.39**		**1.48**		**1.62**		**1.72**
Elbow		**1.19**		**1.37**		**1.48**		**1.57**		**1.71**		**1.82**
Forearm		**0.49**		**0.56**		**0.61**		**0.64**		**0.70**		**0.74**
Upper leg												
Knee												
Lower leg												
Lymph nodes	0.09	**0.15**	0.11	**0.19**	0.13	**0.22**	0.14	**0.24**	0.16	**0.28**	0.19	**0.32**
Skin	0.11	**0.20**	0.11	**0.21**	0.12	**0.22**	0.12	**0.22**	0.13	**0.24**	0.14	**0.25**
Muscle	0.07	**0.09**	0.08	**0.11**	0.09	**0.13**	0.10	**0.14**	0.11	**0.16**	0.12	**0.18**
Skeleton	0.16	**0.32**	0.19	**0.38**	0.21	**0.42**	0.22	**0.45**	0.25	**0.51**	0.28	**0.56**
Red bone marrow	0.07	**0.10**	0.08	**0.12**	0.09	**0.13**	0.10	**0.15**	0.12	**0.17**	0.13	**0.20**
Whole body dose	0.08	**0.13**	0.10	**0.16**	0.11	**0.18**	0.12	**0.19**	0.14	**0.22**	0.16	**0.24**
ED (ICRP 60)	0.12	**0.12**	0.14	**0.15**	0.16	**0.17**	0.17	**0.18**	0.19	**0.21**	0.22	**0.24**
ED (ICRP 103)	0.15	**0.16**	0.18	**0.19**	0.19	**0.21**	0.21	**0.23**	0.24	**0.26**	0.26	**0.30**
Total filtration (mm) Al	**1.5**		**2.5**		**3.5**		**4.5**		**3.5**		**3.5**	
Cu	**0**		**0**		**0**		**0**		**0.1**		**0.2**	

Table 9.163 Normalized organ doses—Chest r-lat—8–12 years

Beam projection:			r-lat			**Table cells:**						
Focus-to-detector distance (cm):			150			(+) = normal font = optimal field setting						
Tube voltage (kVp):			110			**(−) = bold font = suboptimal field setting**						
Total filtrations:			See last table row.			Empty table cells correspond to values < 0.01						
Organ or tissue	**Organ dose D_T : Entrance dose K_E**							**(mSv × mGy^{-1})**				
	(+)	(−)	(+)	(−)	(+)	(−)	(+)	(−)	(+)	(−)	(+)	(−)
Skull	0.02	**0.25**	0.03	**0.29**	0.03	**0.31**	0.03	**0.33**	0.04	**0.36**	0.04	**0.38**
Brain		**0.02**		**0.03**		**0.03**		**0.04**		**0.04**		**0.05**
Oral mucosa	0.02	**0.36**	0.03	**0.41**	0.03	**0.44**	0.03	**0.47**	0.04	**0.52**	0.04	**0.57**
Salivary glands	0.04	**0.41**	0.04	**0.45**	0.05	**0.48**	0.05	**0.51**	0.06	**0.55**	0.07	**0.58**
Cervical spine	0.46	**0.96**	0.53	**1.12**	0.59	**1.23**	0.63	**1.31**	0.70	**1.47**	0.76	**1.59**
Thyroid	0.49	**0.49**	0.56	**0.56**	0.60	**0.60**	0.63	**0.64**	0.69	**0.70**	0.74	**0.75**
Thymus	0.23	**0.22**	0.28	**0.26**	0.31	**0.29**	0.34	**0.32**	0.39	**0.37**	0.44	**0.41**
Thoracic spine	0.45	**0.41**	0.54	**0.50**	0.61	**0.56**	0.66	**0.61**	0.77	**0.71**	0.86	**0.80**
Claviculae	0.74	**0.69**	0.85	**0.80**	0.92	**0.87**	0.97	**0.93**	1.07	**1.03**	1.13	**1.10**
Scapulae	1.26	**1.20**	1.39	**1.33**	1.47	**1.41**	1.53	**1.47**	1.61	**1.55**	1.65	**1.60**
Ribs	1.07	**0.95**	1.19	**1.08**	1.27	**1.17**	1.32	**1.23**	1.40	**1.34**	1.46	**1.41**
Oesophagus	0.22	**0.16**	0.26	**0.19**	0.29	**0.22**	0.32	**0.24**	0.37	**0.28**	0.41	**0.32**
Trachea	0.16	**0.43**	0.18	**0.50**	0.19	**0.55**	0.21	**0.59**	0.23	**0.66**	0.25	**0.73**
Lung	0.41	**0.31**	0.47	**0.36**	0.52	**0.40**	0.55	**0.43**	0.62	**0.49**	0.67	**0.54**
Mamma	0.58	**0.57**	0.62	**0.61**	0.65	**0.63**	0.66	**0.65**	0.70	**0.68**	0.72	**0.70**
Heart	0.26	**0.21**	0.31	**0.25**	0.35	**0.29**	0.38	**0.31**	0.44	**0.36**	0.49	**0.41**
Stomach	0.05	**0.07**	0.06	**0.09**	0.07	**0.10**	0.08	**0.11**	0.09	**0.13**	0.11	**0.15**
Spleen	0.04	**0.05**	0.05	**0.06**	0.06	**0.06**	0.06	**0.07**	0.08	**0.09**	0.09	**0.10**
Liver	0.34	**0.47**	0.40	**0.55**	0.43	**0.61**	0.46	**0.66**	0.52	**0.75**	0.56	**0.82**
Gall bladder	0.07	**0.21**	0.09	**0.26**	0.10	**0.29**	0.11	**0.31**	0.13	**0.37**	0.14	**0.42**
Pancreas	0.12	**0.12**	0.14	**0.14**	0.16	**0.16**	0.18	**0.18**	0.21	**0.21**	0.24	**0.25**
Small intestine	0.01	**0.10**	0.01	**0.12**	0.01	**0.13**	0.02	**0.14**	0.02	**0.17**	0.02	**0.19**
Large intestine	0.01	**0.15**	0.02	**0.18**	0.02	**0.20**	0.02	**0.21**	0.03	**0.25**	0.03	**0.28**
Sigma/Rectum		**0.01**		**0.02**		**0.02**		**0.02**		**0.02**		**0.03**
Lumbar spine	0.06	**0.32**	0.08	**0.39**	0.09	**0.44**	0.10	**0.48**	0.12	**0.55**	0.13	**0.63**
Pelvis		**0.06**		**0.08**		**0.09**		**0.09**	0.01	**0.11**	0.01	**0.12**
Adrenals	0.18	**0.19**	0.22	**0.22**	0.25	**0.25**	0.27	**0.27**	0.31	**0.32**	0.35	**0.36**
Kidneys	0.07	**0.24**	0.08	**0.28**	0.09	**0.31**	0.10	**0.34**	0.12	**0.38**	0.13	**0.42**
Urinary bladder						**0.01**		**0.01**		**0.02**		**0.02**
Uterus		**0.02**		**0.03**		**0.03**		**0.03**		**0.04**		**0.05**
Ovaries		**0.03**		**0.03**		**0.04**		**0.04**		**0.05**		**0.06**
Prostate												
Testes												
Upper arm		**1.28**		**1.41**		**1.49**		**1.55**		**1.64**		**1.69**
Elbow		**1.36**		**1.49**		**1.58**		**1.64**		**1.72**		**1.77**
Forearm		**0.56**		**0.61**		**0.64**		**0.67**		**0.70**		**0.72**
Upper leg												
Knee												
Lower leg												
Lymph nodes	0.13	**0.22**	0.15	**0.26**	0.17	**0.29**	0.18	**0.31**	0.20	**0.35**	0.22	**0.39**
Skin	0.12	**0.22**	0.12	**0.23**	0.13	**0.24**	0.13	**0.24**	0.14	**0.25**	0.15	**0.26**
Muscle	0.09	**0.13**	0.10	**0.15**	0.11	**0.16**	0.12	**0.17**	0.13	**0.19**	0.14	**0.21**
Skeleton	0.19	**0.39**	0.22	**0.44**	0.24	**0.48**	0.25	**0.50**	0.27	**0.55**	0.29	**0.58**
Red bone marrow	0.09	**0.14**	0.11	**0.16**	0.12	**0.18**	0.13	**0.19**	0.14	**0.22**	0.16	**0.24**
Whole body dose	0.11	**0.17**	0.13	**0.20**	0.14	**0.22**	0.15	**0.23**	0.16	**0.25**	0.17	**0.28**
ED (ICRP 60)	0.16	**0.17**	0.18	**0.20**	0.19	**0.22**	0.21	**0.23**	0.23	**0.26**	0.25	**0.28**
ED (ICRP 103)	0.19	**0.22**	0.22	**0.25**	0.24	**0.27**	0.25	**0.29**	0.27	**0.32**	0.30	**0.35**
Total filtration (mm) **Al**	**1.5**		**2.5**		**3.5**		**4.5**		**3.5**		**3.5**	
Cu	**0**		**0**		**0**		**0**		**0.1**		**0.2**	

Table 9.164 Normalized organ doses—Chest r-lat—8–12 years

Beam projection:			r-lat				**Table cells:**					
Focus-to-detector distance (cm):			150				(+) = normal font = optimal field setting					
Tube voltage (kVp):			120				**(−) = bold font = suboptimal field setting**					
Total filtrations:			See last table row.				Empty table cells correspond to values < 0.01					
Organ or tissue	**Organ dose D_T : Entrance dose K_E**						**(mSv × mGy^{-1})**					
	(+)	(−)	(+)	(−)	(+)	(−)	(+)	(−)	(+)	(−)	(+)	(−)
Skull	0.02	**0.26**	0.03	**0.30**	0.03	**0.32**	0.03	**0.33**	0.04	**0.36**	0.04	**0.38**
Brain		**0.03**		**0.03**		**0.04**		**0.04**		**0.04**		**0.05**
Oral mucosa	0.02	**0.38**	0.03	**0.43**	0.03	**0.47**	0.03	**0.49**	0.04	**0.54**	0.04	**0.58**
Salivary glands	0.04	**0.43**	0.05	**0.47**	0.05	**0.50**	0.05	**0.52**	0.06	**0.56**	0.07	**0.59**
Cervical spine	0.49	**1.02**	0.56	**1.17**	0.61	**1.27**	0.64	**1.35**	0.71	**1.49**	0.76	**1.59**
Thyroid	0.52	**0.52**	0.58	**0.58**	0.62	**0.63**	0.66	**0.66**	0.71	**0.72**	0.75	**0.77**
Thymus	0.26	**0.24**	0.30	**0.28**	0.33	**0.31**	0.36	**0.34**	0.41	**0.38**	0.45	**0.43**
Thoracic spine	0.49	**0.45**	0.58	**0.53**	0.64	**0.59**	0.69	**0.64**	0.79	**0.74**	0.87	**0.82**
Claviculae	0.78	**0.73**	0.87	**0.83**	0.94	**0.90**	0.99	**0.95**	1.07	**1.04**	1.12	**1.10**
Scapulae	1.29	**1.23**	1.41	**1.35**	1.48	**1.42**	1.53	**1.47**	1.59	**1.54**	1.62	**1.58**
Ribs	1.10	**0.99**	1.21	**1.11**	1.28	**1.19**	1.32	**1.25**	1.39	**1.34**	1.44	**1.40**
Oesophagus	0.24	**0.18**	0.28	**0.21**	0.31	**0.24**	0.34	**0.26**	0.38	**0.30**	0.42	**0.34**
Trachea	0.17	**0.46**	0.19	**0.53**	0.20	**0.58**	0.22	**0.62**	0.24	**0.69**	0.25	**0.75**
Lung	0.44	**0.33**	0.50	**0.39**	0.54	**0.42**	0.58	**0.45**	0.63	**0.51**	0.68	**0.56**
Mamma	0.60	**0.59**	0.64	**0.62**	0.66	**0.64**	0.68	**0.66**	0.71	**0.69**	0.73	**0.71**
Heart	0.29	**0.24**	0.34	**0.28**	0.38	**0.31**	0.40	**0.33**	0.46	**0.38**	0.51	**0.43**
Stomach	0.06	**0.08**	0.07	**0.10**	0.08	**0.11**	0.08	**0.12**	0.10	**0.14**	0.11	**0.16**
Spleen	0.05	**0.05**	0.06	**0.06**	0.06	**0.07**	0.07	**0.08**	0.08	**0.09**	0.10	**0.11**
Liver	0.37	**0.51**	0.42	**0.59**	0.46	**0.65**	0.48	**0.69**	0.53	**0.78**	0.58	**0.84**
Gall bladder	0.08	**0.24**	0.10	**0.28**	0.11	**0.31**	0.12	**0.34**	0.13	**0.39**	0.15	**0.43**
Pancreas	0.13	**0.13**	0.16	**0.16**	0.17	**0.18**	0.19	**0.20**	0.22	**0.23**	0.25	**0.26**
Small intestine	0.01	**0.11**	0.01	**0.13**	0.02	**0.14**	0.02	**0.16**	0.02	**0.18**	0.02	**0.20**
Large intestine	0.02	**0.16**	0.02	**0.19**	0.02	**0.21**	0.02	**0.23**	0.03	**0.26**	0.03	**0.29**
Sigma/Rectum		**0.01**		**0.02**		**0.02**		**0.02**		**0.03**		**0.03**
Lumbar spine	0.07	**0.36**	0.09	**0.42**	0.10	**0.47**	0.11	**0.50**	0.12	**0.58**	0.14	**0.64**
Pelvis		**0.07**		**0.08**		**0.09**		**0.10**	0.01	**0.12**	0.01	**0.13**
Adrenals	0.20	**0.21**	0.24	**0.24**	0.27	**0.27**	0.29	**0.30**	0.33	**0.34**	0.37	**0.38**
Kidneys	0.08	**0.26**	0.09	**0.30**	0.10	**0.33**	0.11	**0.36**	0.12	**0.40**	0.14	**0.44**
Urinary bladder				**0.01**		**0.01**		**0.01**		**0.02**		**0.02**
Uterus		**0.03**		**0.03**		**0.03**		**0.04**		**0.04**		**0.05**
Ovaries		**0.03**		**0.03**		**0.04**		**0.04**		**0.05**		**0.06**
Prostate												
Testes												
Upper arm		**1.31**		**1.43**		**1.50**		**1.55**		**1.62**		**1.66**
Elbow		**1.39**		**1.51**		**1.59**		**1.64**		**1.70**		**1.74**
Forearm		**0.57**		**0.62**		**0.65**		**0.66**		**0.69**		**0.70**
Upper leg												
Knee												
Lower leg												
Lymph nodes	0.14	**0.24**	0.16	**0.28**	0.18	**0.31**	0.19	**0.33**	0.21	**0.37**	0.23	**0.40**
Skin	0.12	**0.22**	0.13	**0.23**	0.13	**0.24**	0.14	**0.25**	0.14	**0.26**	0.15	**0.27**
Muscle	0.10	**0.14**	0.11	**0.15**	0.12	**0.17**	0.12	**0.18**	0.14	**0.20**	0.15	**0.21**
Skeleton	0.20	**0.41**	0.23	**0.46**	0.24	**0.49**	0.25	**0.51**	0.27	**0.55**	0.29	**0.57**
Red bone marrow	0.10	**0.15**	0.12	**0.17**	0.13	**0.19**	0.14	**0.20**	0.15	**0.23**	0.16	**0.25**
Whole body dose	0.12	**0.19**	0.13	**0.21**	0.14	**0.23**	0.15	**0.24**	0.17	**0.26**	0.18	**0.28**
ED (ICRP 60)	0.17	**0.18**	0.19	**0.21**	0.20	**0.23**	0.21	**0.24**	0.23	**0.27**	0.25	**0.29**
ED (ICRP 103)	0.21	**0.23**	0.23	**0.26**	0.25	**0.28**	0.26	**0.30**	0.28	**0.33**	0.30	**0.36**
Total filtration (mm) **Al**	**1.5**		**2.5**		**3.5**		**4.5**		**3.5**		**3.5**	
Cu	**0**		**0**		**0**		**0**		**0.1**		**0.2**	

Table 9.165 Normalized organ doses—Chest r-lat—13–17 years

Beam projection:			r-lat		**Table cells:**							
Focus-to-detector distance (cm):			150		(+) = normal font = optimal field setting							
Tube voltage (kVp):			110		**(–) = bold font = suboptimal field setting**							
Total filtrations:			See last table row.		Empty table cells correspond to values < 0.01							
Organ or tissue	**Organ dose D_T : Entrance dose K_E**						**(mSv × mGy^{-1})**					
	(+)	(–)	(+)	(–)	(+)	(–)	(+)	(–)	(+)	(–)	(+)	(–)
Skull	0.02	**0.19**	0.02	**0.22**	0.02	**0.24**	0.03	**0.26**	0.03	**0.29**	0.03	**0.31**
Brain		**0.02**		**0.02**		**0.02**		**0.02**		**0.03**		**0.03**
Oral mucosa	0.01	**0.26**	0.02	**0.30**	0.02	**0.33**	0.02	**0.35**	0.03	**0.40**	0.03	**0.43**
Salivary glands	0.03	**0.29**	0.03	**0.33**	0.03	**0.36**	0.04	**0.38**	0.04	**0.41**	0.05	**0.44**
Cervical spine	0.50	**0.96**	0.58	**1.12**	0.64	**1.24**	0.69	**1.34**	0.78	**1.50**	0.85	**1.64**
Thyroid	0.38	**0.34**	0.44	**0.39**	0.48	**0.42**	0.51	**0.45**	0.56	**0.50**	0.60	**0.54**
Thymus	0.16	**0.14**	0.19	**0.17**	0.22	**0.19**	0.24	**0.21**	0.28	**0.25**	0.32	**0.29**
Thoracic spine	0.29	**0.25**	0.34	**0.30**	0.39	**0.34**	0.43	**0.38**	0.50	**0.44**	0.57	**0.51**
Claviculae	0.58	**0.51**	0.67	**0.59**	0.73	**0.65**	0.78	**0.70**	0.86	**0.78**	0.92	**0.84**
Scapulae	1.13	**1.06**	1.26	**1.18**	1.34	**1.26**	1.40	**1.32**	1.49	**1.41**	1.54	**1.47**
Ribs	0.90	**0.77**	1.01	**0.88**	1.09	**0.96**	1.14	**1.02**	1.22	**1.12**	1.28	**1.19**
Oesophagus	0.18	**0.12**	0.21	**0.15**	0.24	**0.17**	0.26	**0.19**	0.31	**0.22**	0.35	**0.26**
Trachea	0.13	**0.26**	0.15	**0.31**	0.16	**0.34**	0.18	**0.36**	0.20	**0.41**	0.21	**0.45**
Lung	0.33	**0.24**	0.39	**0.28**	0.43	**0.31**	0.46	**0.34**	0.52	**0.39**	0.57	**0.44**
Mamma	3.05	**3.96**	3.41	**4.42**	3.66	**4.72**	3.85	**4.95**	4.19	**5.36**	4.46	**5.69**
Heart	0.18	**0.14**	0.22	**0.17**	0.24	**0.19**	0.27	**0.21**	0.31	**0.25**	0.35	**0.28**
Stomach	0.03	**0.04**	0.04	**0.05**	0.04	**0.05**	0.05	**0.06**	0.06	**0.07**	0.06	**0.08**
Spleen	0.02	**0.02**	0.03	**0.03**	0.03	**0.03**	0.04	**0.04**	0.04	**0.04**	0.05	**0.05**
Liver	0.33	**0.40**	0.39	**0.48**	0.43	**0.53**	0.46	**0.58**	0.52	**0.66**	0.56	**0.73**
Gall bladder	0.06	**0.20**	0.07	**0.24**	0.08	**0.28**	0.09	**0.30**	0.11	**0.35**	0.12	**0.40**
Pancreas	0.08	**0.07**	0.09	**0.09**	0.10	**0.10**	0.11	**0.11**	0.14	**0.13**	0.16	**0.16**
Small intestine		**0.06**		**0.07**		**0.08**		**0.09**	0.01	**0.10**	0.01	**0.12**
Large intestine		**0.09**	0.01	**0.11**	0.01	**0.13**	0.01	**0.14**	0.02	**0.16**	0.02	**0.18**
Sigma/Rectum										**0.01**		**0.01**
Lumbar spine	0.04	**0.19**	0.05	**0.23**	0.06	**0.26**	0.07	**0.28**	0.08	**0.33**	0.09	**0.38**
Pelvis		**0.03**		**0.04**		**0.04**		**0.05**		**0.06**		**0.07**
Adrenals	0.13	**0.12**	0.16	**0.15**	0.18	**0.17**	0.19	**0.19**	0.23	**0.22**	0.26	**0.26**
Kidneys	0.05	**0.17**	0.06	**0.20**	0.07	**0.23**	0.08	**0.25**	0.09	**0.28**	0.11	**0.32**
Urinary bladder												
Uterus		**0.04**		**0.04**		**0.05**		**0.05**		**0.07**		**0.08**
Ovaries		**0.01**		**0.02**		**0.02**		**0.02**		**0.02**		**0.03**
Prostate												
Testes												
Upper arm		**1.18**		**1.32**		**1.40**		**1.46**		**1.56**		**1.62**
Elbow		**1.26**		**1.40**		**1.49**		**1.55**		**1.65**		**1.70**
Forearm		**0.45**		**0.50**		**0.53**		**0.55**		**0.59**		**0.61**
Upper leg												
Knee												
Lower leg												
Lymph nodes	0.10	**0.16**	0.12	**0.18**	0.13	**0.20**	0.14	**0.22**	0.16	**0.25**	0.18	**0.28**
Skin	0.13	**0.21**	0.14	**0.22**	0.15	**0.22**	0.15	**0.23**	0.16	**0.24**	0.17	**0.25**
Muscle	0.08	**0.10**	0.09	**0.12**	0.10	**0.13**	0.11	**0.14**	0.12	**0.16**	0.13	**0.17**
Skeleton	0.17	**0.34**	0.20	**0.39**	0.21	**0.42**	0.22	**0.44**	0.24	**0.48**	0.26	**0.51**
Red bone marrow	0.09	**0.12**	0.11	**0.14**	0.12	**0.15**	0.13	**0.17**	0.14	**0.19**	0.16	**0.21**
Whole body dose	0.10	**0.15**	0.12	**0.17**	0.13	**0.18**	0.13	**0.20**	0.15	**0.22**	0.16	**0.24**
ED (ICRP 60)	0.26	**0.30**	0.29	**0.35**	0.32	**0.38**	0.34	**0.40**	0.37	**0.44**	0.40	**0.48**
ED (ICRP 103)	0.47	**0.59**	0.53	**0.66**	0.57	**0.71**	0.60	**0.75**	0.66	**0.82**	0.71	**0.89**
Total filtration (mm) Al	**1.5**		**2.5**		**3.5**		**4.5**		**3.5**		**3.5**	
Cu	**0**		**0**		**0**		**0**		**0.1**		**0.2**	

Table 9.166 Normalized organ doses—Chest r-lat—13–17 years

Beam projection:		r-lat				**Table cells:**						
Focus-to-detector distance (cm):		150				(+) = normal font = optimal field setting						
Tube voltage (kVp):		120				**(−) = bold font = suboptimal field setting**						
Total filtrations:		See last table row.				Empty table cells correspond to values < 0.01						
Organ or tissue	**Organ dose D_T : Entrance dose K_E**							**(mSv × mGy⁻¹)**				
	(+)	(−)	(+)	(−)	(+)	(−)	(+)	(−)	(+)	(−)	(+)	(−)
Skull	0.02	**0.21**	0.02	**0.23**	0.03	**0.25**	0.03	**0.27**	0.03	**0.29**	0.04	**0.31**
Brain		**0.02**		**0.02**		**0.02**		**0.03**		**0.03**		**0.03**
Oral mucosa	0.02	**0.28**	0.02	**0.32**	0.02	**0.35**	0.02	**0.37**	0.03	**0.41**	0.03	**0.45**
Salivary glands	0.03	**0.31**	0.03	**0.35**	0.04	**0.37**	0.04	**0.39**	0.04	**0.43**	0.05	**0.45**
Cervical spine	0.53	**1.03**	0.61	**1.18**	0.67	**1.29**	0.71	**1.38**	0.79	**1.53**	0.85	**1.65**
Thyroid	0.41	**0.36**	0.46	**0.41**	0.50	**0.45**	0.53	**0.47**	0.58	**0.52**	0.62	**0.56**
Thymus	0.18	**0.16**	0.21	**0.19**	0.24	**0.21**	0.26	**0.23**	0.29	**0.27**	0.33	**0.30**
Thoracic spine	0.32	**0.28**	0.38	**0.33**	0.42	**0.37**	0.46	**0.40**	0.53	**0.47**	0.59	**0.53**
Claviculae	0.61	**0.54**	0.69	**0.62**	0.75	**0.67**	0.79	**0.72**	0.87	**0.79**	0.92	**0.84**
Scapulae	1.16	**1.09**	1.28	**1.21**	1.35	**1.28**	1.40	**1.33**	1.48	**1.41**	1.52	**1.45**
Ribs	0.94	**0.81**	1.04	**0.91**	1.10	**0.98**	1.15	**1.04**	1.22	**1.13**	1.27	**1.19**
Oesophagus	0.20	**0.14**	0.23	**0.17**	0.26	**0.19**	0.28	**0.21**	0.33	**0.24**	0.37	**0.27**
Trachea	0.14	**0.29**	0.16	**0.33**	0.17	**0.36**	0.18	**0.38**	0.20	**0.42**	0.22	**0.46**
Lung	0.36	**0.26**	0.41	**0.30**	0.45	**0.34**	0.48	**0.36**	0.54	**0.41**	0.58	**0.46**
Mamma	3.21	**4.16**	3.56	**4.59**	3.79	**4.88**	3.97	**5.10**	4.29	**5.48**	4.54	**5.78**
Heart	0.20	**0.16**	0.24	**0.19**	0.26	**0.21**	0.29	**0.23**	0.33	**0.27**	0.37	**0.30**
Stomach	0.03	**0.04**	0.04	**0.05**	0.05	**0.06**	0.05	**0.07**	0.06	**0.08**	0.07	**0.09**
Spleen	0.03	**0.03**	0.03	**0.03**	0.04	**0.04**	0.04	**0.04**	0.05	**0.05**	0.06	**0.06**
Liver	0.36	**0.44**	0.41	**0.51**	0.45	**0.57**	0.48	**0.61**	0.53	**0.69**	0.58	**0.76**
Gall bladder	0.07	**0.23**	0.08	**0.27**	0.09	**0.30**	0.10	**0.32**	0.11	**0.37**	0.13	**0.42**
Pancreas	0.09	**0.08**	0.10	**0.10**	0.12	**0.11**	0.13	**0.12**	0.15	**0.15**	0.17	**0.17**
Small intestine		**0.07**		**0.08**		**0.09**	0.01	**0.09**	0.01	**0.11**	0.01	**0.13**
Large intestine	0.01	**0.10**	0.01	**0.12**	0.01	**0.14**	0.02	**0.15**	0.02	**0.17**	0.02	**0.19**
Sigma/Rectum								**0.01**		**0.01**		**0.01**
Lumbar spine	0.05	**0.21**	0.06	**0.25**	0.07	**0.28**	0.07	**0.30**	0.09	**0.35**	0.10	**0.40**
Pelvis		**0.04**		**0.04**		**0.05**		**0.05**		**0.06**		**0.07**
Adrenals	0.15	**0.14**	0.17	**0.17**	0.19	**0.19**	0.21	**0.20**	0.24	**0.24**	0.27	**0.27**
Kidneys	0.06	**0.19**	0.07	**0.22**	0.08	**0.24**	0.09	**0.26**	0.10	**0.30**	0.11	**0.33**
Urinary bladder												
Uterus		**0.04**		**0.05**		**0.06**		**0.06**		**0.07**	0.01	**0.08**
Ovaries		**0.01**		**0.02**		**0.02**		**0.02**		**0.03**		**0.03**
Prostate												
Testes												
Upper arm		**1.22**		**1.34**		**1.42**		**1.47**		**1.55**		**1.59**
Elbow		**1.30**		**1.42**		**1.50**		**1.55**		**1.63**		**1.68**
Forearm		**0.47**		**0.51**		**0.54**		**0.56**		**0.58**		**0.60**
Upper leg												
Knee												
Lower leg												
Lymph nodes	0.11	**0.17**	0.13	**0.20**	0.14	**0.22**	0.15	**0.23**	0.17	**0.26**	0.19	**0.29**
Skin	0.14	**0.21**	0.15	**0.22**	0.15	**0.23**	0.16	**0.23**	0.16	**0.25**	0.17	**0.25**
Muscle	0.09	**0.11**	0.10	**0.13**	0.11	**0.14**	0.11	**0.15**	0.12	**0.16**	0.13	**0.18**
Skeleton	0.18	**0.36**	0.20	**0.40**	0.22	**0.43**	0.23	**0.45**	0.25	**0.49**	0.26	**0.51**
Red bone marrow	0.10	**0.13**	0.12	**0.15**	0.13	**0.16**	0.13	**0.18**	0.15	**0.20**	0.16	**0.22**
Whole body dose	0.11	**0.16**	0.12	**0.18**	0.13	**0.19**	0.14	**0.20**	0.15	**0.22**	0.17	**0.24**
ED (ICRP 60)	0.27	**0.33**	0.31	**0.37**	0.33	**0.39**	0.35	**0.42**	0.38	**0.46**	0.41	**0.49**
ED (ICRP 103)	0.50	**0.62**	0.55	**0.69**	0.60	**0.74**	0.63	**0.78**	0.68	**0.85**	0.73	**0.90**
Total filtration (mm) Al	**1.5**		**2.5**		**3.5**		**4.5**		**3.5**		**3.5**	
Cu	**0**		**0**		**0**		**0**		**0.1**		**0.2**	

Table 9.167 Normalized organ doses—Chest r-lat—> 17 years

Beam projection:			r-lat			**Table cells:**						
Focus-to-detector distance (cm):			150			(+) = normal font = optimal field setting						
Tube voltage (kVp):			110			**(−) = bold font = suboptimal field setting**						
Total filtrations:			See last table row.			Empty table cells correspond to values < 0.01						
Organ or tissue	**Organ dose D_T : Entrance dose K_E**						**(mSv × mGy^{-1})**					
	(+)	(−)	(+)	(−)	(+)	(−)	(+)	(−)	(+)	(−)	(+)	(−)
Skull	0.02	**0.20**	0.02	**0.24**	0.02	**0.26**	0.03	**0.28**	0.03	**0.31**	0.03	**0.33**
Brain		**0.02**		**0.02**		**0.02**		**0.02**		**0.03**		**0.03**
Oral mucosa	0.01	**0.25**	0.02	**0.29**	0.02	**0.32**	0.02	**0.34**	0.02	**0.39**	0.03	**0.42**
Salivary glands	0.03	**0.31**	0.03	**0.35**	0.04	**0.37**	0.04	**0.39**	0.04	**0.43**	0.05	**0.46**
Cervical spine	0.48	**0.85**	0.57	**1.00**	0.63	**1.11**	0.67	**1.19**	0.76	**1.34**	0.83	**1.47**
Thyroid	0.34	**0.27**	0.38	**0.31**	0.42	**0.34**	0.44	**0.36**	0.49	**0.40**	0.53	**0.44**
Thymus	0.12	**0.10**	0.14	**0.12**	0.16	**0.14**	0.18	**0.16**	0.21	**0.18**	0.24	**0.21**
Thoracic spine	0.21	**0.18**	0.25	**0.21**	0.29	**0.24**	0.32	**0.27**	0.38	**0.32**	0.43	**0.37**
Claviculae	0.48	**0.41**	0.56	**0.48**	0.61	**0.53**	0.66	**0.57**	0.73	**0.64**	0.79	**0.70**
Scapulae	1.03	**0.95**	1.15	**1.07**	1.23	**1.15**	1.29	**1.20**	1.38	**1.30**	1.44	**1.36**
Ribs	0.77	**0.64**	0.87	**0.74**	0.93	**0.81**	0.98	**0.86**	1.05	**0.95**	1.11	**1.02**
Oesophagus	0.15	**0.10**	0.18	**0.12**	0.20	**0.14**	0.22	**0.15**	0.26	**0.18**	0.30	**0.21**
Trachea	0.14	**0.25**	0.16	**0.29**	0.17	**0.32**	0.19	**0.34**	0.21	**0.39**	0.23	**0.43**
Lung	0.28	**0.19**	0.33	**0.23**	0.36	**0.26**	0.39	**0.28**	0.45	**0.33**	0.49	**0.37**
Mamma	0.31	**0.41**	0.34	**0.45**	0.37	**0.49**	0.39	**0.51**	0.42	**0.55**	0.45	**0.58**
Heart	0.14	**0.10**	0.17	**0.12**	0.19	**0.14**	0.21	**0.15**	0.24	**0.18**	0.27	**0.21**
Stomach	0.02	**0.02**	0.02	**0.03**	0.02	**0.03**	0.03	**0.03**	0.03	**0.04**	0.04	**0.05**
Spleen	0.01	**0.01**	0.02	**0.02**	0.02	**0.02**	0.02	**0.02**	0.02	**0.02**	0.03	**0.03**
Liver	0.25	**0.33**	0.29	**0.39**	0.32	**0.44**	0.34	**0.48**	0.39	**0.55**	0.43	**0.62**
Gall bladder	0.04	**0.14**	0.05	**0.16**	0.05	**0.19**	0.06	**0.21**	0.07	**0.24**	0.08	**0.28**
Pancreas	0.05	**0.05**	0.06	**0.06**	0.07	**0.07**	0.08	**0.07**	0.09	**0.09**	0.10	**0.11**
Small intestine		**0.04**		**0.05**		**0.06**		**0.07**		**0.08**		**0.09**
Large intestine		**0.07**		**0.09**		**0.10**		**0.11**	0.01	**0.13**	0.01	**0.15**
Sigma/Rectum												
Lumbar spine	0.03	**0.12**	0.03	**0.15**	0.04	**0.17**	0.04	**0.18**	0.05	**0.22**	0.06	**0.25**
Pelvis		**0.03**		**0.03**		**0.04**		**0.04**		**0.05**		**0.05**
Adrenals	0.10	**0.09**	0.12	**0.11**	0.13	**0.13**	0.14	**0.14**	0.17	**0.17**	0.20	**0.19**
Kidneys	0.03	**0.13**	0.04	**0.15**	0.05	**0.17**	0.05	**0.19**	0.06	**0.22**	0.07	**0.24**
Urinary bladder												
Uterus				**0.01**		**0.01**		**0.02**		**0.02**		**0.02**
Ovaries				**0.01**		**0.01**		**0.01**		**0.02**		**0.02**
Prostate												
Testes												
Upper arm		**1.11**		**1.24**		**1.33**		**1.39**		**1.49**		**1.55**
Elbow		**1.19**		**1.33**		**1.42**		**1.48**		**1.58**		**1.64**
Forearm		**0.43**		**0.48**		**0.51**		**0.53**		**0.56**		**0.58**
Upper leg												
Knee												
Lower leg												
Lymph nodes	0.08	**0.13**	0.10	**0.15**	0.11	**0.17**	0.12	**0.18**	0.14	**0.21**	0.15	**0.24**
Skin	0.11	**0.19**	0.11	**0.20**	0.12	**0.21**	0.12	**0.21**	0.13	**0.22**	0.14	**0.23**
Muscle	0.07	**0.09**	0.08	**0.10**	0.08	**0.11**	0.09	**0.12**	0.10	**0.13**	0.11	**0.14**
Skeleton	0.14	**0.30**	0.16	**0.34**	0.18	**0.37**	0.19	**0.39**	0.21	**0.43**	0.22	**0.45**
Red bone marrow	0.09	**0.10**	0.10	**0.12**	0.11	**0.13**	0.12	**0.14**	0.14	**0.17**	0.15	**0.19**
Whole body dose	0.08	**0.13**	0.10	**0.14**	0.11	**0.16**	0.11	**0.17**	0.13	**0.19**	0.14	**0.20**
ED (ICRP 60)	0.10	**0.11**	0.12	**0.12**	0.13	**0.14**	0.14	**0.15**	0.16	**0.17**	0.17	**0.19**
ED (ICRP 103)	0.12	**0.14**	0.14	**0.16**	0.15	**0.18**	0.16	**0.19**	0.19	**0.22**	0.20	**0.24**
Total filtration (mm) Al	**1.5**		**2.5**		**3.5**		**4.5**		**3.5**		**3.5**	
Cu	**0**		**0**		**0**		**0**		**0.1**		**0.2**	

Table 9.168 Normalized organ doses—Chest r-lat— >17 years

Beam projection:			r-lat		**Table cells:**							
Focus-to-detector distance (cm):			150		(+) = normal font = optimal field setting							
Tube voltage (kVp):			120		**(−) = bold font = suboptimal field setting**							
Total filtrations:			See last table row.		Empty table cells correspond to values < 0.01							
Organ or tissue	**Organ dose D_T : Entrance dose K_E**						**(mSv × mGy^{-1})**					
	(+)	(−)	(+)	(−)	(+)	(−)	(+)	(−)	(+)	(−)	(+)	(−)
Skull	0.02	**0.22**	0.02	**0.25**	0.02	**0.27**	0.03	**0.28**	0.03	**0.31**	0.03	**0.34**
Brain		**0.02**		**0.02**		**0.02**		**0.03**		**0.03**		**0.04**
Oral mucosa	0.02	**0.28**	0.02	**0.31**	0.02	**0.34**	0.02	**0.36**	0.03	**0.40**	0.03	**0.44**
Salivary glands	0.03	**0.33**	0.03	**0.36**	0.04	**0.39**	0.04	**0.41**	0.05	**0.44**	0.05	**0.47**
Cervical spine	0.52	**0.91**	0.60	**1.05**	0.65	**1.15**	0.70	**1.23**	0.77	**1.37**	0.84	**1.48**
Thyroid	0.36	**0.29**	0.41	**0.33**	0.44	**0.36**	0.46	**0.38**	0.51	**0.42**	0.54	**0.45**
Thymus	0.14	**0.12**	0.16	**0.14**	0.18	**0.16**	0.20	**0.17**	0.23	**0.20**	0.26	**0.23**
Thoracic spine	0.24	**0.20**	0.28	**0.24**	0.31	**0.27**	0.34	**0.29**	0.40	**0.34**	0.45	**0.39**
Claviculae	0.51	**0.43**	0.58	**0.50**	0.63	**0.55**	0.67	**0.58**	0.74	**0.65**	0.79	**0.70**
Scapulae	1.06	**0.99**	1.18	**1.10**	1.25	**1.17**	1.30	**1.22**	1.38	**1.29**	1.42	**1.34**
Ribs	0.80	**0.68**	0.89	**0.77**	0.95	**0.83**	0.99	**0.88**	1.05	**0.96**	1.10	**1.02**
Oesophagus	0.17	**0.11**	0.20	**0.14**	0.22	**0.15**	0.24	**0.17**	0.28	**0.20**	0.31	**0.23**
Trachea	0.15	**0.27**	0.17	**0.31**	0.18	**0.34**	0.19	**0.36**	0.21	**0.40**	0.23	**0.44**
Lung	0.30	**0.22**	0.35	**0.25**	0.39	**0.28**	0.41	**0.30**	0.46	**0.34**	0.50	**0.38**
Mamma	0.32	**0.43**	0.36	**0.47**	0.38	**0.50**	0.40	**0.52**	0.43	**0.56**	0.46	**0.59**
Heart	0.15	**0.12**	0.18	**0.14**	0.20	**0.15**	0.22	**0.17**	0.26	**0.20**	0.29	**0.22**
Stomach	0.02	**0.03**	0.02	**0.03**	0.03	**0.04**	0.03	**0.04**	0.04	**0.05**	0.04	**0.05**
Spleen	0.02	**0.02**	0.02	**0.02**	0.02	**0.02**	0.02	**0.02**	0.03	**0.03**	0.03	**0.03**
Liver	0.27	**0.36**	0.31	**0.43**	0.34	**0.47**	0.36	**0.51**	0.41	**0.58**	0.44	**0.64**
Gall bladder	0.04	**0.15**	0.05	**0.18**	0.06	**0.21**	0.06	**0.22**	0.07	**0.26**	0.08	**0.29**
Pancreas	0.06	**0.06**	0.07	**0.07**	0.08	**0.08**	0.08	**0.08**	0.10	**0.10**	0.11	**0.11**
Small intestine		**0.05**		**0.06**		**0.07**		**0.07**		**0.09**		**0.10**
Large intestine		**0.08**		**0.10**		**0.11**		**0.12**	0.01	**0.14**	0.01	**0.15**
Sigma/Rectum												
Lumbar spine	0.03	**0.14**	0.04	**0.16**	0.04	**0.18**	0.04	**0.20**	0.05	**0.23**	0.06	**0.27**
Pelvis		**0.03**		**0.04**		**0.04**		**0.04**		**0.05**		**0.06**
Adrenals	0.11	**0.10**	0.13	**0.12**	0.14	**0.14**	0.16	**0.15**	0.18	**0.18**	0.21	**0.20**
Kidneys	0.04	**0.14**	0.05	**0.17**	0.05	**0.19**	0.06	**0.20**	0.07	**0.23**	0.07	**0.25**
Urinary bladder												
Uterus		**0.01**		**0.01**		**0.02**		**0.02**		**0.02**		**0.02**
Ovaries		**0.01**		**0.01**		**0.01**		**0.02**		**0.02**		**0.02**
Prostate												
Testes												
Upper arm		**1.15**		**1.27**		**1.34**		**1.40**		**1.48**		**1.53**
Elbow		**1.23**		**1.35**		**1.43**		**1.49**		**1.57**		**1.62**
Forearm		**0.44**		**0.48**		**0.51**		**0.53**		**0.56**		**0.57**
Upper leg												
Knee												
Lower leg												
Lymph nodes	0.09	**0.14**	0.11	**0.17**	0.12	**0.18**	0.13	**0.20**	0.14	**0.22**	0.16	**0.25**
Skin	0.11	**0.20**	0.12	**0.21**	0.12	**0.21**	0.13	**0.22**	0.13	**0.23**	0.14	**0.24**
Muscle	0.07	**0.09**	0.08	**0.11**	0.09	**0.11**	0.09	**0.12**	0.10	**0.14**	0.11	**0.15**
Skeleton	0.15	**0.31**	0.17	**0.35**	0.18	**0.38**	0.19	**0.40**	0.21	**0.43**	0.22	**0.45**
Red bone marrow	0.10	**0.11**	0.11	**0.13**	0.12	**0.14**	0.13	**0.15**	0.14	**0.17**	0.16	**0.19**
Whole body dose	0.09	**0.13**	0.10	**0.15**	0.11	**0.17**	0.12	**0.18**	0.13	**0.19**	0.14	**0.21**
ED (ICRP 60)	0.11	**0.12**	0.13	**0.13**	0.14	**0.15**	0.15	**0.16**	0.16	**0.18**	0.18	**0.20**
ED (ICRP 103)	0.13	**0.15**	0.15	**0.17**	0.16	**0.19**	0.17	**0.20**	0.19	**0.22**	0.21	**0.25**
Total filtration (mm) Al	**1.5**		**2.5**		**3.5**		**4.5**		**3.5**		**3.5**	
Cu	**0**		**0**		**0**		**0**		**0.1**		**0.2**	

References

1. Bontrager KL. Textbook of radiographic positioning and related anatomy: Mosby; 1993.
2. Bader D, Datz H, Bartal G, Juster AA, Marks K, Smolkin T, et al. Unintentional exposure of neonates to conventional radiography in the neonatal intensive care units. J Perinat. 2007;27:579–85.
3. Commission of the European Communities. European guidelines on quality criteria for diagnostic radiographic images in paediatrics. EUR 16261, SBN 92-827-7843-6; 1996
4. Cook JV, Shah K, Pablot S, Kyriou J, Pettett A, Fitzgerald M. Guidelines on best practice in the X-ray imaging of children. A manual for all X-ray departments. London: Queen Mary's Hospital for Children, Surrey, and The Radiological Protection Centre; 1998.
5. Datz H, Ben-Shlomo A, Bader D, Sadetzki S, Juster-Reicher A, Marks K, et al. The additional dose to radiosensitive organs caused by using under-collimated X-ray beams in neonatal intensive care radiography. Radiat Prot Dosim. 2008;130:518–24.
6. Dell'Agnolo E, Seidenbusch M, Genzel-Boroviszeny O, Münch HG, Schneider K. Der Neugeborenen-Thorax I - Analyse von Feldgröße, Bildgüte und anderer Qualitätsparameter. Fortschr Röntgenstr. 2013;185:V30.
7. Drexler G, Panzer W, Stieve FE. Die Bestimmung von Organdosen in der Röntgendiagnostik. Berlin: Hoffmann Verlag; 1993.
8. Ebel KD, Willich E. Die Röntgenuntersuchung im Kindesalter – Technik und Indikation: Berlin, Springer Verlag; 1968.
9. Fendel H. Radiation problems in Roentgen examinations of the chest. Progr Pediatr Radiol. 1967;1:18–32.
10. Fichtner C, Schneider K, Freidhof C, Endemann B, Horwitz AE, Kohn MM, et al. Critical analysis of field size in chest X-rays of infants - a EC-wide survey in children's clinics. Eur Radiol. 1993;3:389.
11. Griebl G. Variation von Feldgröße und Dosis bei Röntgen-Thoraxaufnahmen beim 5 Jahre alten Kind – kritische Analyse einer Feldstudie in europäischen Kinderkliniken. Dissertation, Universität München; 2001.
12. Hart D, Jones DJ, Wall BF. Estimation of effective dose in diagnostic radiology from entrance surface dose and dose-area-product measurements. National Radiological Protection Board NRPB-R262; 1994.
13. Hart D, Jones DJ, Wall BF. Coefficients for estimating effective doses from pediatric X-ray examinations. National Radiological Protection Board NRPB-R279; 1996
14. Klinzmann M. Qualitätsbeurteilung von Röntgenthoraxaufnahmen auf den pädiatrischen Intensivstationen in Abhängigkeit von den ausführenden Berufsgruppen. Dissertation, Justus-Liebig-Universität Gießen; 2015.
15. Riccabona M, Beer M, Mentzel H-J (Hrsg.). Bildgebung des Thorax bei Neugeborenen und Kleinkindern. Springer Verlag; 2019
16. Schneider K, Fendel H, Bakowski C, Stein E, Kohn M, Kellner M, et al. Results of a dosimetry study in the European Community on frequent x ray examinations in infants. Radiat Prot Dosim. 1992;43:31–6.
17. Schneider K, Kohn MM, Bakowski C, Stein E, Freidhof C, Horwitz AE, et al. Impact of radiographic imaging criteria on dose and image quality in infants in an EC-wide survey. Radiat Prot Dosim. 1993;49:73–6.
18. Schneider K, Seidenbusch M. Technische Anforderungen. In: Riccabona M, Beer M, Mentzel H-J (Hrsg.). Bildgebung des Thorax bei Neugeborenen und Kleinkindern. Springer Verlag; 2019 p. 17-29
19. Seidenbusch MC, Regulla D, Schneider K. Radiation exposure of children in pediatric radiology. Part 3: Conversion coefficients for reconstruction of organ doses achieved during chest X-ray examinations. Fortschr Röntgenstr. 2008;180:1061–81.
20. Seidenbusch MC, Schneider K. Strahlenhygienische Aspekte bei der Röntgenuntersuchung des Thorax. Radiologe 2015;55:580–587
21. Smans K, Tapiovaara M, Cannie M, Struelens L, Vanhavere F, Smet M, et al. Calculation of organ doses in X-ray examinations of premature babies. Med Phys. 2008;35:556–68.
22. Soboleski D, Theriault C, Acker A. Unnecessary irradiation to non-thoracic structures during pediatric chest radiography. Pediatr Radiol. 2006;36:22–5.
23. Stollfuss J, Schneider K, Krüger-Stollfuss I. A comparative study of collimation in bedside chest radiography for preterm infants in two teaching hospitals. Eur J Radiol Open. 2015;2:118–22.
24. Tschauner S, Marterer R, Gübitz M, Kalmar PI, Talakic E, Weissensteiner S, et al. European guidelines for ap/pa chest X-rays: routinely satisfiable in a paediatric radiology division? Eur Radiol. 2016;26:495–505.
25. Weigl A. Untersuchung über Patientendosis und Bildqualität bei Röntgen-Thoraxaufnahmen im Säuglingsalter in Kliniken der neuen Bundesländer. Dissertation, Universität München; 1995.
26. Weisbach M. Untersuchung über die Patientendosis bei Röntgenaufnahmen des Thorax in der Kinderradiologie. Dissertation, Universität München; 1991.
27. Winter W. Untersuchung über die Patientendosis und die Bildqualität bei häufigen Röntgenaufnahmen im Säuglingsalter bei niedergelassenen Ärzten (Pädiater, Orthopäden, Chirurgen). Dissertation, Universität München; 1994.
28. Zeiss A. Feldgrößen- und Dosisvariation von Röntgen-Thorax-Aufnahmen beim zehnjährigen Kind – kritische Analyse einer Feldstudie in europäischen Kinderkliniken. Dissertation, Universität München; 2002.

10 Bony Thorax

10.1 Introduction

The type of bony thorax radiography is determined by the clinical indication. Table 10.1 shows indications and Table 10.2 contraindications for bony thorax. Image criteria as required for radiological findings in bony thorax radiographs are listed in Table 10.3. In order to fulfil these image criteria, the field settings listed in Tables 10.4 and 10.5 should be achieved. These field settings go along with the patient preparation as listed in Table 10.6. Exposure parameter combinations for various age groups and patient diameters are listed in Tables 10.7, 10.8, and 10.9. The age groups in Table 10.8 refer to the standard age groups of the mathematical MIRD phantom family which has been used to calculate the normalized organ doses in this book.

Table 10.1 Indications for the bony thorax

Chest trauma
Battered child syndrome
Skeletal dysplasia
Inflammatory diseases of the chest wall, pleura, and the lung, e.g. aspergillosis, actinomycosis
Osteomyelitis of the clavicle, ribs, and sternum[a]
Benign and malignant tumours of the bony thorax[a]
Soft tissue tumours[a]
Annotation:
[a]Today MRI is the preferred imaging.

Table 10.2 Contraindications for the bony thorax

None

Table 10.3 Image criteria for the bony thorax

Bony thorax ap
Symmetrical reproduction of the chest without rotation and tilting
Reproduction of the trachea, the main bronchi, and the vascular pattern of the lungs
Reproduction of the heart border and the diaphragm with the costophrenic angles
Visually sharp reproduction of the thoracic and upper lumber spine, the paraspinal structures, the ribs, the scapulae, the clavicles, and the proximal metaphyses of the humeri
Bony thorax of oblique views
Visually sharp reproduction of the clavicles, the scapulae, the sternum, and the thoracic spine
Visually sharp reproduction of the sternum and the thoracic spine

Table 10.4 Patient positions and beam projections for the bony thorax

Patient position	Beam projections
Supine	ap
Supine elevated (30°–45°)	RAO + LAO (30°–45°)

Table 10.9 shows radiation qualities (i.e. various combinations of tube voltage and total filtration) which may be applied in clinical practice. Please note that not all of these radiation qualities are recommended by the authors. The standard

The original version of this chapter was revised. The correction to this chapter can be found at https://doi.org/10.1007/978-3-030-18504-6_31

M. Seidenbusch et al., *Imaging Practice and Radiation Protection in Pediatric Radiology*, https://doi.org/10.1007/978-3-030-18504-6_10

projections of bony thorax of the patient who is correctly positioned are depicted in Fig. 10.1. Also the optimal field (marked in green), the suboptimal field (marked in red), and the correct shielding are shown. Tables 10.11, 10.12, 10.13, 10.14, 10.15, 10.16, 10.17, 10.18, and 10.19 outline the virtual field settings on the skin sur-

Table 10.5 Optimal field settings for the bony thorax

Bony thorax ap	
Upper field edge:	Tip of the mandible, above the clavicles
Field centring:	Midpoint of the sternum at the level of the mamillae
Lower field edge:	Midpoint at the lower third between the xiphoid process and the navel
Lateral field edges:	Close to the skin of lateral chest wall including the proximal humeri
Bony thorax RAO	
Upper field edge:	Tip of the mandible/just above the shoulders
Field centring:	Midpoint of the mammillary line
Lower field edge:	Below the lateral part of the costal arch
Lateral field edges:	Skin surface of the chest wall
Bony thorax LAO	
Upper field edge:	Tip of the mandible/just above the shoulders
Field centring:	Midpoint of the mamillary line
Lower field edge:	Below the lateral part of the costal arch
Lateral field edges:	Skin surface of the chest wall

Table 10.6 Patient preparation

General	Remove clothes, neck laces, and all metallic devices, especially piercings, from the neck, chest, and abdomen of the patient
	Immobilize infants in the supine position. Put sandbags over the arms and legs. Alternatively, accompanying persons or staff members could hold the patient (in this case also take care for the radiation protection of the accompanying persons or staff members)
	Place eventually small X-ray positive markers at palpable tumours or on pain points
Positioning	Supine; using wedge shaped foam pads for RAO and LAO projection
Attention	On the Bucky table the patient should always be positioned on a foam mat of 10 to 20 cm thickness. So, the lower part of the body will not be cut if lateral projection X-rays (cross table views) in prone or in decubitus position are performed
Shielding	Lead masking at lower field border

Table 10.7 Radiographic procedure

X-ray source	The mAs product used in free-hand exposure is chosen according to age and body weight
	Focal spot size between 0.6 and 1.0
	Check the use of additional filtration
	Check the focus-to-detector distance
Field setting	Check the field size
	Check the field centring
Image detector system	Check the use of antiscatter grids
Patient	The arms should be positioned alongside the chest
	The exposure should be performed in deep inspiration

Table 10.8 Radiographic techniques of Bucky table bony thorax radiographs

Age group (years)	Beam direction	Patient position	Beam projection	Typical patient diameter (cm)		Optimal field size (cm x cm)	Focus-to-detector distance (cm)	Voltage (kVp)						Antiscatter grid
				With arms	No arms			50	60	70	80	110	120	
0	Vertical	Supine	ap	9.8	9.8	11 x 13	100[1]							
			RAO	-	-	10 x 12								
			LAO	-	-									
1 – 2			ap	13.0	13.0	15 x 18	115							
			RAO	-	-	14 x 18								
			LAO	-	-									
3 – 7			ap	15.0	15.0	19 x 24								
			RAO	-	-	17 x 24								
			LAO	-	-									
8 – 12			ap	16.8	16.8	23 x 30								
			RAO	-	-	20 x 31								
			LAO	-	-									
13 – 17			ap	19.6	19.6	28 x 39								
			RAO	-	-	24 x 39								
			LAO	-	-									
18			ap	20.0	20.0	33 x 44								
			LAO	-	-	28 x 43								
			RAO	-	-									

Annotations:
[1] The patient is lying on the table top directly on the detector.
(dark grey) Parameter combination not recommended, normalized organ doses not provided
(light grey) Parameter combination not recommended, normalized organ doses provided
(light blue) Parameter combination recommended, normalized organ doses provided

Table 10.9 Radiation qualities of Bucky table bony thorax radiographs

Voltage (kVp)	Total filtration					
	Without additional filtration		With additional filtration			
	1.5 mm Al	2.5 Al	3.5 Al	4.5 Al	3.5 Al + 0.1 Cu	3.5 + 0.2 Cu
50						
60						
70						
80						
110						
120						

Annotations:

- Parameter combination not recommended, normalized organ doses not provided
- Parameter combination not recommended, normalized organ doses provided
- Parameter combination recommended, normalized organ doses provided

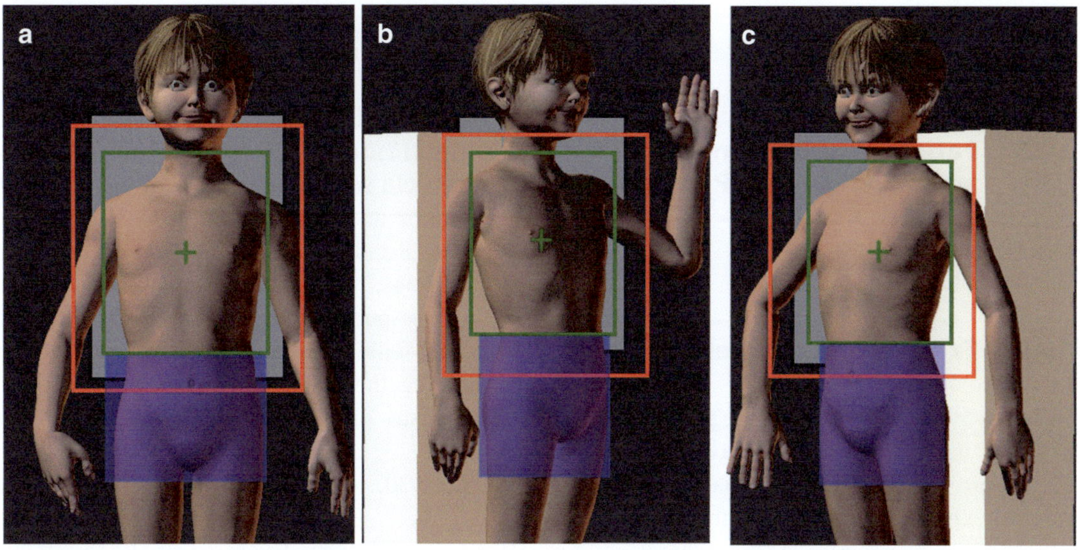

Fig. 10.1 **Field settings for the bony thorax.** (**a**) Anteroposterior projection. (**b**) Right anterior oblique projection. (**c**) Left anterior oblique projection. (© V. Rösenberger and M. Seidenbusch 2018). Green lines = optimal field settings; green cross = centring point; red lines = suboptimal field settings; grey shaded area = image detector; blue shaded area = lead mask shielding of the patient; dark grey shaded area = Bucky table; white triangle = triangular foam wedge

face of mathematical MIRD phantoms of various age groups and the resulting virtual radiographies with and without bone structures, respectively. Normalized organ doses resulting from these virtual field settings have been calculated for all age groups and have been listed in Tables 10.20, 10.21, 10.22, 10.23, 10.24, 10.25, 10.26, 10.27, 10.28, 10.29, 10.30, 10.31, 10.32, 10.33, 10.34, 10.35, 10.36, 10.37, 10.38, 10.39, 10.40, 10.41, 10.42, 10.43, 10.44, 10.45, 10.46, 10.47, 10.48, 10.49, 10.50, 10.51, 10.52, 10.53, 10.54, 10.55, 10.56, 10.57, 10.58, 10.59, 10.60, and 10.61. Table cells with normal font refer to optimal field settings and table cells with bold font refer to suboptimal field settings. Please note that normalized organ doses have been provided also for exposure parameter combinations, which are not recommended by the authors.

10.2 Indications and Contraindications

Indications and contraindications for bony thorax are given in Tables 10.1 and 10.2.

10.3 Image Criteria

The image criteria required for the bony thorax are listed in Table 10.3.

10.4 Field Settings

In Table 10.4, patient positions and beam projections used for bony thorax radiographs are listed, and in Table 10.5 the optimal field settings are listed which are required in order to fulfil the image criteria mentioned in Table 10.3.

10.5 Patient Preparation and Radiographic Procedure

See Tables 10.6, 10.7 and [1].

10.6 Exposure Parameters

In Tables 10.8 and 10.9, exposure parameter combinations used for the bony thorax in various age groups are listed. Dark grey shaded table cells refer to exposure parameter combinations which are not recommended for the bony thorax and for which no normalized organ doses have been provided. Light grey and light blue shaded table cells refer to exposure parameter combinations for which normalized organ doses have been provided with light grey shaded table cells referring to exposure parameter combinations being not recommended and light blue shaded table cells referring to exposure parameter combinations being recommended for the bony thorax. Note that due to the fact that the radiation dose is an exponential function of the patient diameter, it may make sense to use the patient diameters instead of the patient's age groups for determining the adequate exposure parameters from Table 10.8. In this case, for determining the corresponding normalized organ doses from Tables 10.20, 10.21, 10.22, 10.23, 10.24, 10.25, 10.26, 10.27, 10.28, 10.29, 10.30, 10.31, 10.32, 10.33, 10.34, 10.35, 10.36, 10.37, 10.38, 10.39, 10.40, 10.41, 10.42, 10.43, 10.44, 10.45, 10.46, 10.47, 10.48, 10.49, 10.50, 10.51, 10.52, 10.53, 10.54, 10.55, 10.56, 10.57, 10.58, 10.59, 10.60, and 10.61, the age group related to the patient diameter should be used.

10.7 Normalized Organ Doses

For calculating normalized organ doses, the field settings described in Sect. 10.4 were transferred to the mathematical MIRD phantoms. Inconsistencies between the geometrical anatomy of the mathematical phantoms and the real human anatomy have been listed in Table 10.10 whenever they are of relevance for the validity of the normalized organ doses.

In Tables 10.11, 10.12, 10.13, 10.14, 10.15, 10.16, 10.17, 10.18, and 10.19, optimal and suboptimal field settings are listed for the different age groups and the beam projections given in Tables 10.8 and 10.9. Green lines refer to optimal field settings, red lines refer to suboptimal field settings, respectively. In the tables, the optimal and suboptimal field settings which were used for performing the calculations of the normalized organ doses provided in Tables 10.20, 10.21, 10.22, 10.23, 10.24, 10.25, 10.26, 10.27, 10.28, 10.29, 10.30, 10.31, 10.32, 10.33, 10.34, 10.35, 10.36, 10.37, 10.38, 10.39, 10.40, 10.41, 10.42, 10.43, 10.44, 10.45, 10.46, 10.47, 10.48, 10.49, 10.50, 10.51, 10.52, 10.53, 10.54, 10.55, 10.56, 10.57, 10.58, 10.59, 10.60, and 10.61 were drawn in on the skin surface of the mathematical phan-

Table 10.10 Inconsistencies between the mathematical MIRD phantoms and real human anatomy

Anatomical region	Inconsistencies
Chest	No variation of lung density with age
	Mass of glandular tissue is underestimated in newborn infants
Skeleton	The shape of the spine is unrealistic (no lordosis, no kyphosis considered), the vertebral column is not clearly depicted
	Ribs are drawn as complete circles and thicker than in real humans, the course of the ribs is horizontal
	No xiphoid has been inserted
Extremties	Arms are constructed too long in the ages 0, 1, and 5 years
	Position of the extremities cannot be changed

For details, see the text

toms. The virtual mathematical radiographs resulting from these field settings were depicted in the tables with and without bone structures, respectively, in order to enable the reader to visually compare the virtual mathematical radiographs with the real radiographs made under clinical conditions. For comparing the virtual mathematical field settings with the real clinical field settings, the virtual mathematical field edges were also listed in the tables.

As the normalized organ doses provided in Tables 10.20, 10.21, 10.22, 10.23, 10.24, 10.25, 10.26, 10.27, 10.28, 10.29, 10.30, 10.31, 10.32, 10.33, 10.34, 10.35, 10.36, 10.37, 10.38, 10.39, 10.40, 10.41, 10.42, 10.43, 10.44, 10.45, 10.46, 10.47, 10.48, 10.49, 10.50, 10.51, 10.52, 10.53, 10.54, 10.55, 10.56, 10.57, 10.58, 10.59, 10.60, and 10.61 refer to entrance air kerma values, optimal and suboptimal field sizes were provided in the tables in order tom enable the reader to calculate entrance air kerma dose values from clinically measured dose area product values. Please note that these optimal and suboptimal field sizes refer to the mathematical phantoms of an age group only and not to real patients with corresponding ages.

Please note also that the anatomy of the mathematical phantoms has been idealized. Therefore, the virtual field settings for calculating the normalized organ doses have been defined with a view to ensuring that the resulting virtual organ doses would be consistent with the real organ doses which would be achieved during real X-ray examinations being performed with the real field settings recommended in Sect. 10.4. Thus, the field settings shown in the following tables may bot be consistent with the clinical field settings as recommended in Sect. 10.4.

Table 10.11 Field settings—Bony thorax ap

Age group (years)		Anatomical landmarks			
		Optimal field setting		**Suboptimal field setting**	
0	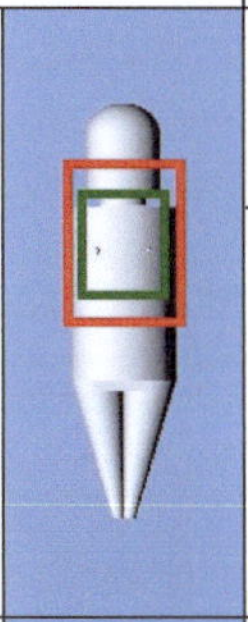	Upper field edge: Tip of the mandible Lower field edge: Just below the 12th rib Lateral field edges: Close to the chest and abdominal wall. Proximal metaphyses of the humeri and femora included Field size = field width x field height: 11 cm x 13 cm = 143 cm^2		Upper field edge: Lower third of the skull Lower field edge: At the upper margins of the iliac crests Lateral field edges: Lateral to the chest and abdominal wall. Both arms included Field size = field width x field height: 14 cm x 20 cm = 280 cm^2	
		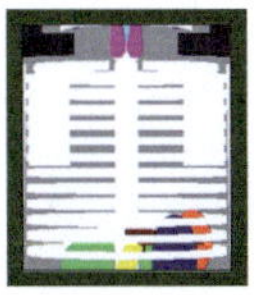	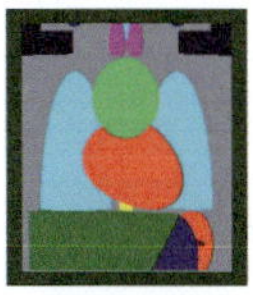	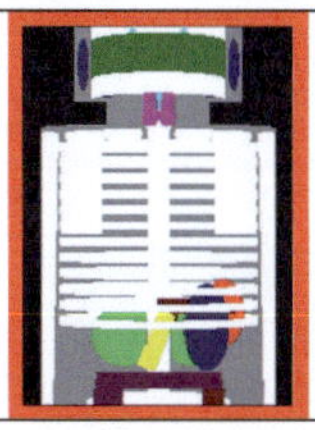	
1 - 2	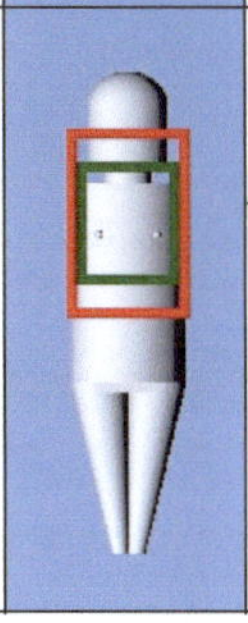	Upper field edge: Tip of the mandible Lower field edge: Just below the 12th rib Lateral field edges: Close to the chest and abdominal wall. Proximal metaphyses of the humeri and femora included Field size = field width x field height: 15 cm x 18 cm = 270 cm^2		Upper field edge: Lower third of the skull Lower field edge: At the upper margins of the iliac crests Lateral field edges: Lateral to the chest and abdominal wall. Both arms included Field size = field width x field height: 18 cm x 28 cm = 504 cm^2	
		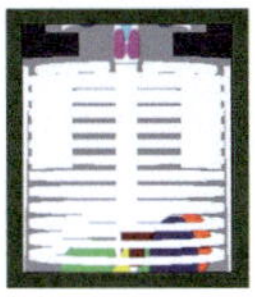		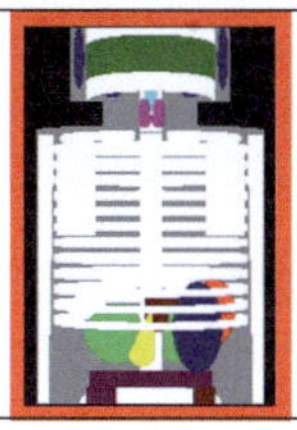	

Table 10.12 Field settings—Bony thorax ap

Age group (years)		Anatomical landmarks			
		Optimal field setting		Suboptimal field setting	
3 - 7	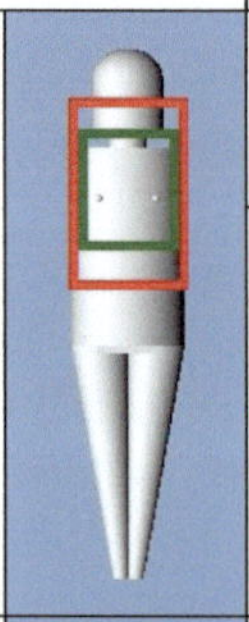	Upper field edge: Tip of the mandible Lower field edge: Just below the 12th rib Lateral field edges: Close to the chest and abdominal wall. Proximal metaphyses of the humeri and femora included Field size = field width x field height: 19 cm x 24 cm= 456 cm^2		Upper field edge: Lower third of the skull Lower field edge: At the upper margins of the iliac crests Lateral field edges: Lateral to the chest and abdominal wall. Both arms included Field size = field width x field height: 21 cm x 37 cm= 777 cm^2	
		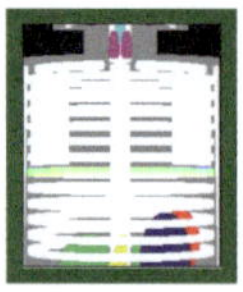	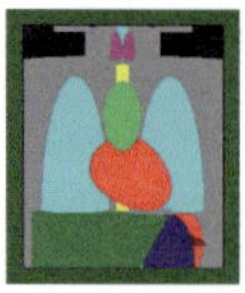	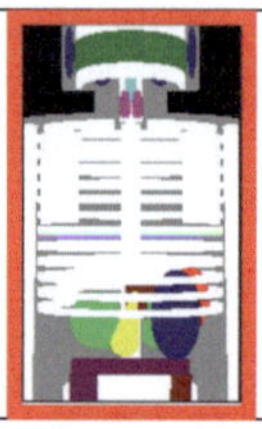	
8 - 12	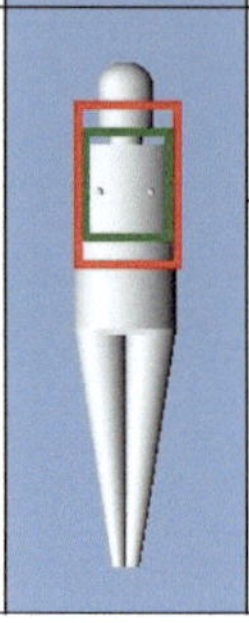	Upper field edge: Tip of the mandible Lower field edge: Just below the 12th rib Lateral field edges: Close to the chest and abdominal wall. Proximal metaphyses of the humeri and femora included Field size = field width x field height: 23 cm x 30 cm = 690 cm^2		Upper field edge: Lower third of the skull Lower field edge: At the upper margins of the iliac crests Lateral field edges: Lateral to the chest and abdominal wall. Both arms included Field size = field width x field height: 26 cm x 44 cm = 1,144 cm^2	
			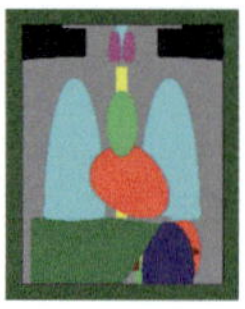		

Table 10.13 Field settings—Bony thorax ap

Age group (years)		Anatomical landmarks			
		Optimal field setting		Suboptimal field setting	
13 - 17	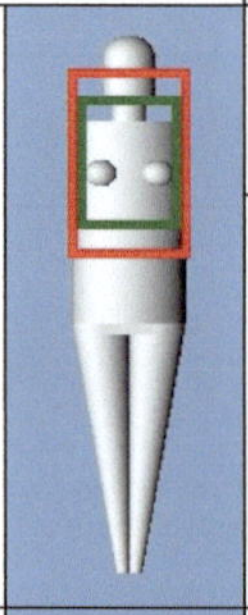	Upper field edge: Tip of the mandible Lower field edge: Just below the 12th rib Lateral field edges: Close to the chest and abdominal wall. Proximal metaphyses of the humeri and femora included Field size = field width x field height: 28 cm x 39 cm= 1,092 cm^2		Upper field edge: Lower third of the skull Lower field edge: At the upper margins of the iliac crests Lateral field edges: Lateral to the chest and abdominal wall. Both arms included Field size = field width x field height: 32 cm x 55 cm= 1,760 cm^2	
		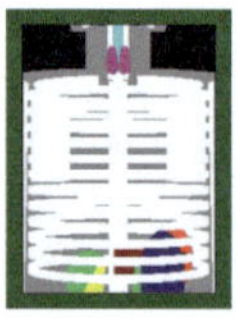			
18	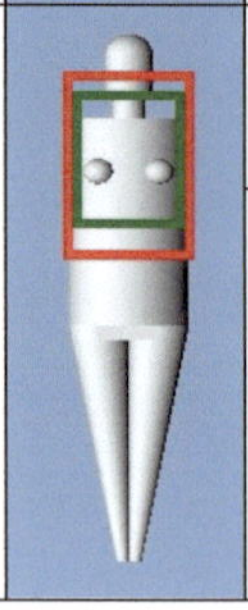	Upper field edge: Tip of the mandible Lower field edge: Just below the 12th rib Lateral field edges: Close to the chest and abdominal wall. Proximal metaphyses of the humeri and femora included Field size = field width x field height: 33 cm x 44 cm = 1,452 cm^2		Upper field edge: Lower third of the skull Lower field edge: At the upper margins of the iliac crests Lateral field edges: Lateral to the chest and abdominal wall. Both arms included Field size = field width x field height: 40 cm x 60 cm = 2,400 cm^2	
				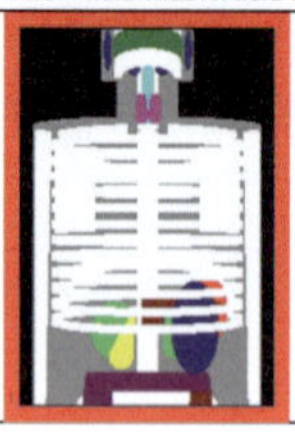	

Table 10.14 Field settings—Bony thorax RAO

Age group (years)		Anatomical landmarks			
		Optimal field setting		Suboptimal field setting	
0	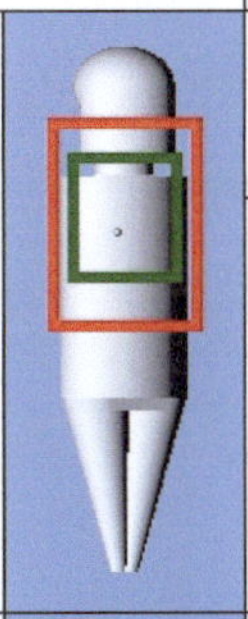	Upper field edge: Tip of the mandible Lower field edge: Just below the 12th rib Lateral field edges: Close to the chest and abdominal wall. Proximal metaphyses of the humeri and femora included Field size = field width x field height: 10 cm x 12 cm= 120 cm^2		Upper field edge: Lower third of the skull Lower field edge: At the upper margins of the iliac crests Lateral field edges: Lateral to the chest and abdominal wall. Both arms included Field size = field width x field height: 14 cm x 20 cm= 280 cm^2	
		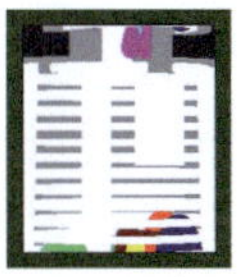		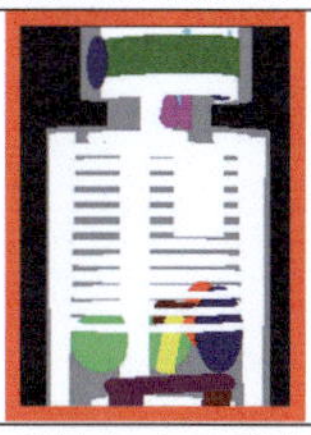	
1 - 2	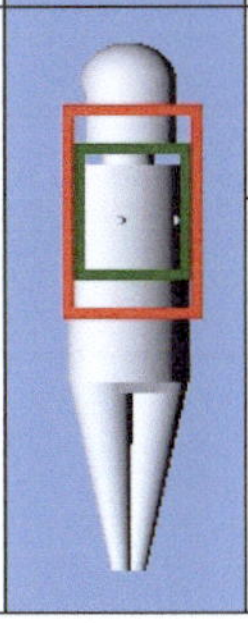	Upper field edge: Tip of the mandible Lower field edge: Just below the 12th rib Lateral field edges: Close to the chest and abdominal wall. Proximal metaphyses of the humeri and femora included Field size = field width x field height: 14 cm x 18 cm = 252 cm^2		Upper field edge: Lower third of the skull Lower field edge: At the upper margins of the iliac crests Lateral field edges: Lateral to the chest and abdominal wall. Both arms included Field size = field width x field height: 18 cm x 28 cm = 504 cm^2	
		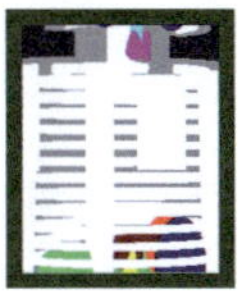	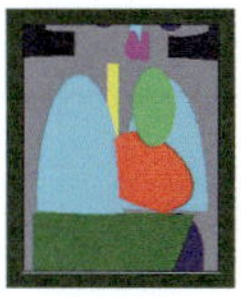		

Table 10.15 Field settings—Bony thorax RAO

Age group (years)		Anatomical landmarks			
		Optimal field setting		Suboptimal field setting	
3 - 7	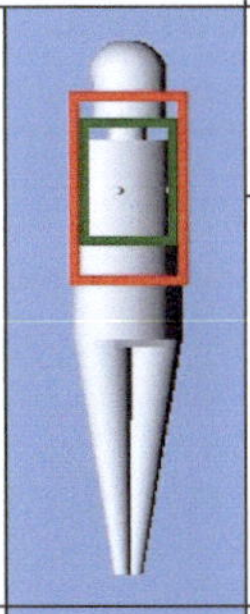	Upper field edge: Tip of the mandible Lower field edge: Just below the 12th rib Lateral field edges: Close to the chest and abdominal wall. Proximal metaphyses of the humeri and femora included Field size = field width x field height: 17 cm x 24 cm= 408 cm^2		Upper field edge: Lower third of the skull Lower field edge: At the upper margins of the iliac crests Lateral field edges: Lateral to the chest and abdominal wall. Both arms included Field size = field width x field height: 22 cm x 37 cm= 814 cm^2	
		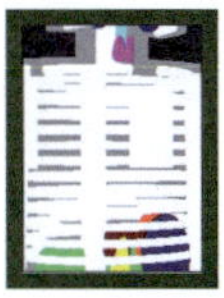	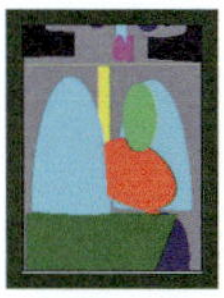		
8 - 12	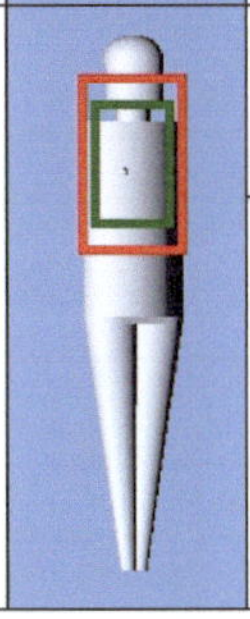	Upper field edge: Tip of the mandible Lower field edge: Just below the 12th rib Lateral field edges: Close to the chest and abdominal wall. Proximal metaphyses of the humeri and femora included Field size = field width x field height: 20 cm x 31 cm = 620 cm^2		Upper field edge: Lower third of the skull Lower field edge: At the upper margins of the iliac crests Lateral field edges: Lateral to the chest and abdominal wall. Both arms included Field size = field width x field height: 26 cm x 44 cm = 1,144 cm^2	
		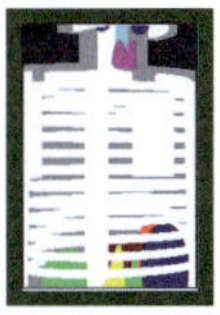			

Table 10.16 Field settings—Bony thorax RAO

Age group (years)		Anatomical landmarks			
		Optimal field setting		Suboptimal field setting	
13 - 17	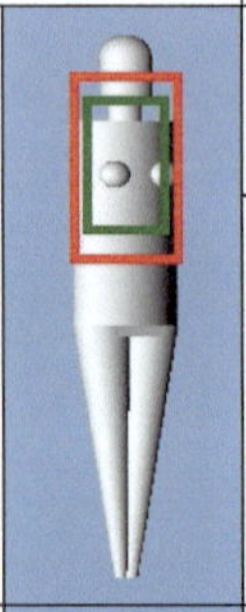	Upper field edge: Tip of the mandible Lower field edge: Just below the 12th rib Lateral field edges: Close to the chest and abdominal wall. Proximal metaphyses of the humeri and femora included Field size = field width x field height: 24 cm x 39 cm= 936 cm^2		Upper field edge: Lower third of the skull Lower field edge: At the upper margins of the iliac crests Lateral field edges: Lateral to the chest and abdominal wall. Both arms included Field size = field width x field height: 32 cm x 55 cm= 1,760 cm^2	

18	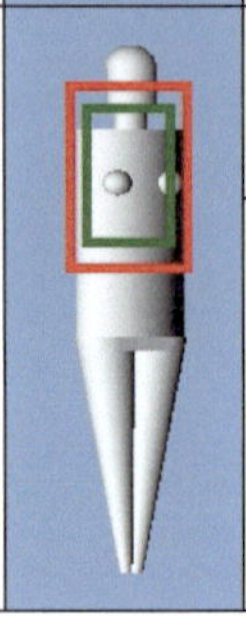	Upper field edge: Tip of the mandible Lower field edge: Just below the 12th rib Lateral field edges: Close to the chest and abdominal wall. Proximal metaphyses of the humeri and femora included Field size = field width x field height: 28 cm x 43 cm = 1,204 cm^2		Upper field edge: Lower third of the skull Lower field edge: At the upper margins of the iliac crests Lateral field edges: Lateral to the chest and abdominal wall. Both arms included Field size = field width x field height: 40 cm x 60 cm = 2,400 cm^2	
		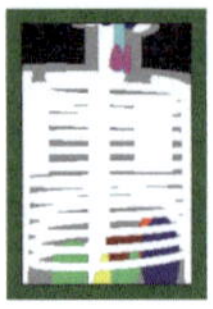			

Table 10.17 Field settings—Bony thorax LAO

Age group (years)		Anatomical landmarks			
		Optimal field setting		Suboptimal field setting	
0	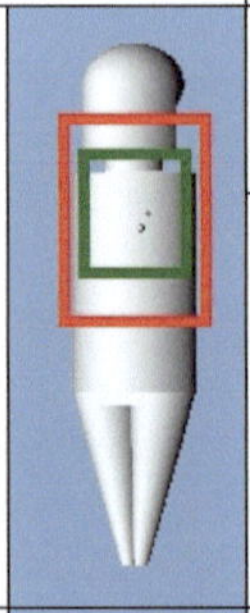	Upper field edge: Tip of the mandible Lower field edge: Just below the 12th rib Lateral field edges: Close to the chest and abdominal wall. Proximal metaphyses of the humeri and femora included Field size = field width x field height: 10 cm x 12 cm= 120 cm^2		Upper field edge: Lower third of the skull Lower field edge: At the upper margins of the iliac crests Lateral field edges: Lateral to the chest and abdominal wall. Both arms included Field size = field width x field height: 14 cm x 20 cm = 280 cm^2	
		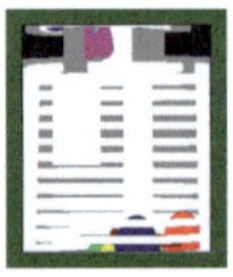		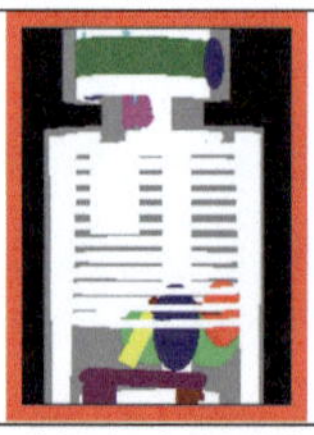	
1 - 2	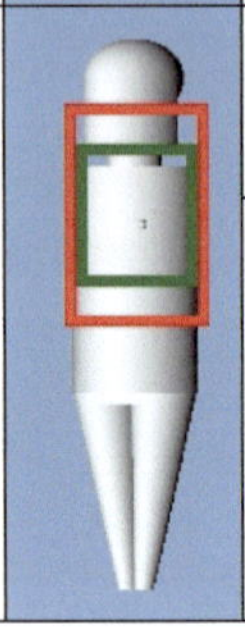	Upper field edge: Tip of the mandible Lower field edge: Just below the 12th rib Lateral field edges: Close to the chest and abdominal wall. Proximal metaphyses of the humeri and femora included Field size = field width x field height: 14 cm x 18 cm = 252 cm^2		Upper field edge: Lower third of the skull Lower field edge: At the upper margins of the iliac crests Lateral field edges: Lateral to the chest and abdominal wall. Both arms included Field size = field width x field height: 18 cm x 28 cm = 504 cm^2	
		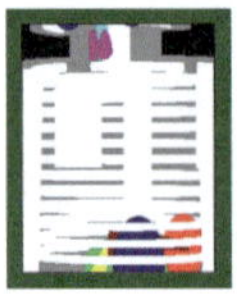		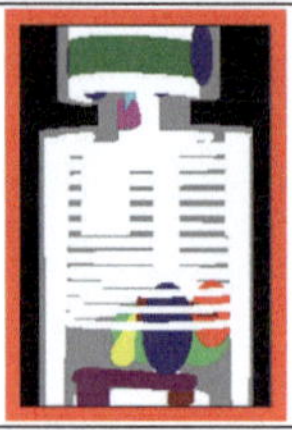	

Table 10.18 Field settings—Bony thorax LAO

Age group (years)		Anatomical landmarks			
		Optimal field setting		**Suboptimal field setting**	
3 - 7	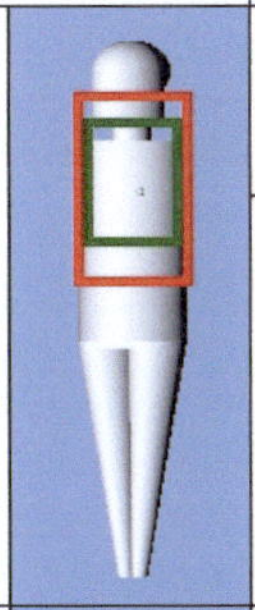	Upper field edge: Tip of the mandible Lower field edge: Just below the 12th rib Lateral field edges: Close to the chest and abdominal wall. Proximal metaphyses of the humeri and femora included Field size = field width x field height: 17 cm x 24 cm= 408 cm^2		Upper field edge: Lower third of the skull Lower field edge: At the upper margins of the iliac crests Lateral field edges: Lateral to the chest and abdominal wall. Both arms included Field size = field width x field height: 22 cm x 37 cm= 814 cm^2	
			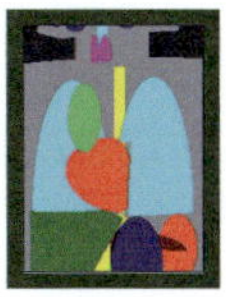		
8 - 12	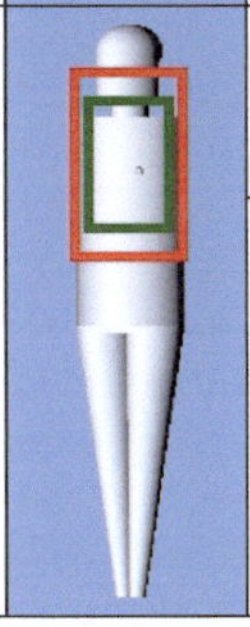	Upper field edge: Tip of the mandible Lower field edge: Just below the 12th rib Lateral field edges: Close to the chest and abdominal wall. Proximal metaphyses of the humeri and femora included Field size = field width x field height: 20 cm x 31 cm = 620 cm^2		Upper field edge: Lower third of the skull Lower field edge: At the upper margins of the iliac crests Lateral field edges: Lateral to the chest and abdominal wall. Both arms included Field size = field width x field height: 26 cm x 44 cm = 1,144 cm^2	
		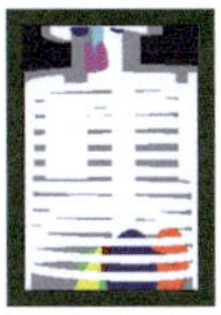			

Table 10.19 Field settings—Bony thorax LAO

Age group (years)		Anatomical landmarks			
		Optimal field setting		**Suboptimal field setting**	
13 - 17	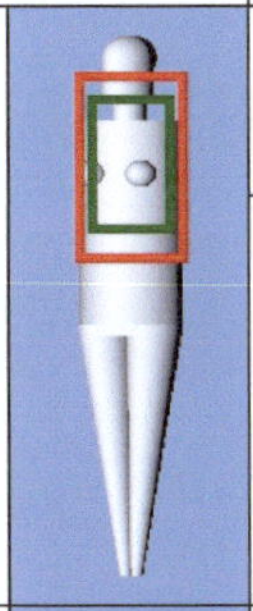	Upper field edge: Tip of the mandible Lower field edge: Just below the 12th rib Lateral field edges: Close to the chest and abdominal wall. Proximal metaphyses of the humeri and femora included Field size = field width x field height: 24 cm x 39 cm= 936 cm^2		Upper field edge: Lower third of the skull Lower field edge: At the upper margins of the iliac crests Lateral field edges: Lateral to the chest and abdominal wall. Both arms included Field size = field width x field height: 32 cm x 55 cm= 1,760 cm^2	
		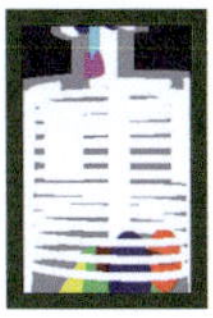			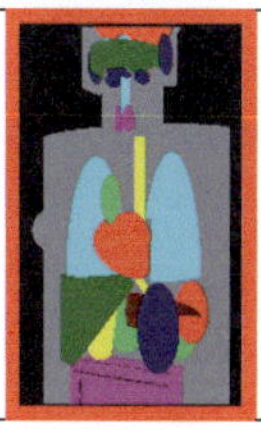
18	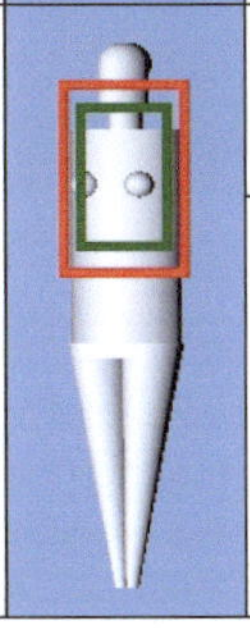	Upper field edge: Tip of the mandible Lower field edge: Just below the 12th rib Lateral field edges: Close to the chest and abdominal wall. Proximal metaphyses of the humeri and femora included Field size = field width x field height: 28 cm x 43 cm = 1,204 cm^2		Upper field edge: Lower third of the skull Lower field edge: At the upper margins of the iliac crests Lateral field edges: Lateral to the chest and abdominal wall. Both arms included Field size = field width x field height: 40 cm x 60 cm = 2,400 cm^2	
			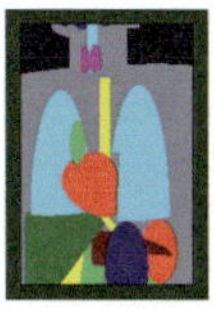		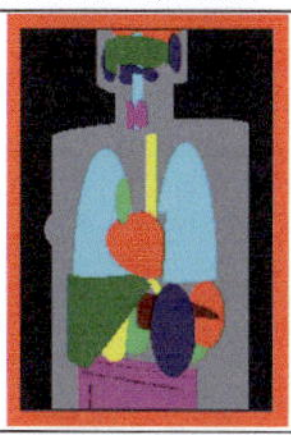

Table 10.20 Normalized organ doses—Bony thorax ap— <1 year

Beam projection:		ap			**Table cells:**							
Focus-to-detector distance (cm):		100			(+) = normal font = optimal field setting							
Tube voltage (kVp):		50			**(−) = bold font = suboptimal field setting**							
Total filtrations:		See last table row			Empty table cells correspond to values <0.01							
Organ or tissue	**Organ dose D_T: Entrance dose K_E**						**(mSv * mGy^{-1})**					
	(+)	(−)	(+)	(−)	(+)	(−)	(+)	(−)	(+)	(−)	(+)	(−)
Skull	0.02	**0.20**	0.02	**0.24**	0.03	**0.27**	0.03	**0.30**	0.04	**0.34**	0.04	**0.38**
Brain		**0.02**		**0.02**		**0.03**		**0.03**		**0.04**		**0.04**
Oral mucosa	0.02	**0.52**	0.03	**0.62**	0.04	**0.68**	0.04	**0.73**	0.05	**0.81**	0.06	**0.87**
Salivary glands	0.06	**0.45**	0.07	**0.54**	0.09	**0.60**	0.09	**0.65**	0.11	**0.73**	0.12	**0.80**
Cervical spine	0.31	**0.48**	0.40	**0.63**	0.47	**0.74**	0.52	**0.84**	0.62	**1.02**	0.72	**1.18**
Thyroid	0.84	**0.89**	0.95	**1.02**	1.02	**1.10**	1.06	**1.16**	1.14	**1.26**	1.20	**1.34**
Thymus	0.74	**0.75**	0.86	**0.88**	0.95	**0.97**	1.01	**1.03**	1.11	**1.14**	1.19	**1.22**
Thoracic spine	0.22	**0.24**	0.31	**0.33**	0.37	**0.40**	0.42	**0.46**	0.53	**0.58**	0.64	**0.70**
Claviculae	1.83	**1.91**	2.16	**2.25**	2.36	**2.47**	2.51	**2.63**	2.75	**2.90**	2.95	**3.11**
Scapulae	0.41	**0.42**	0.54	**0.56**	0.64	**0.67**	0.72	**0.76**	0.88	**0.92**	1.02	**1.08**
Ribs	1.35	**1.40**	1.56	**1.62**	1.70	**1.77**	1.80	**1.88**	1.97	**2.06**	2.11	**2.22**
Oesophagus	0.18	**0.19**	0.24	**0.26**	0.28	**0.30**	0.32	**0.34**	0.39	**0.42**	0.45	**0.49**
Trachea	0.09	**0.44**	0.10	**0.52**	0.11	**0.57**	0.12	**0.61**	0.14	**0.67**	0.15	**0.73**
Lung	0.39	**0.40**	0.47	**0.49**	0.54	**0.56**	0.58	**0.61**	0.67	**0.70**	0.74	**0.78**
Mamma	1.16	**1.18**	1.23	**1.25**	1.27	**1.29**	1.29	**1.31**	1.33	**1.34**	1.36	**1.37**
Heart	0.50	**0.52**	0.62	**0.63**	0.69	**0.71**	0.75	**0.78**	0.86	**0.88**	0.95	**0.98**
Stomach	0.33	**0.52**	0.41	**0.64**	0.46	**0.72**	0.50	**0.78**	0.57	**0.89**	0.63	**0.98**
Spleen	0.15	**0.18**	0.20	**0.24**	0.23	**0.28**	0.26	**0.32**	0.32	**0.38**	0.37	**0.45**
Liver	0.31	**0.46**	0.38	**0.56**	0.43	**0.64**	0.47	**0.69**	0.54	**0.79**	0.60	**0.87**
Gall bladder	0.17	**0.36**	0.21	**0.46**	0.25	**0.53**	0.28	**0.59**	0.33	**0.69**	0.37	**0.79**
Pancreas	0.22	**0.25**	0.28	**0.33**	0.33	**0.39**	0.37	**0.44**	0.45	**0.53**	0.52	**0.61**
Small intestine	0.01	**0.17**	0.02	**0.22**	0.02	**0.26**	0.03	**0.29**	0.03	**0.34**	0.04	**0.39**
Large intestine	0.02	**0.26**	0.03	**0.33**	0.03	**0.38**	0.04	**0.42**	0.04	**0.49**	0.05	**0.55**
Sigma/rectum		**0.06**		**0.07**		**0.08**		**0.09**	0.01	**0.11**	0.01	**0.13**
Lumbar spine	0.07	**0.22**	0.10	**0.30**	0.13	**0.36**	0.14	**0.41**	0.18	**0.52**	0.23	**0.63**
Pelvis		**0.08**	0.01	**0.10**	0.02	**0.12**	0.02	**0.14**	0.03	**0.18**	0.03	**0.21**
Adrenals	0.09	**0.11**	0.13	**0.14**	0.15	**0.17**	0.17	**0.20**	0.21	**0.25**	0.25	**0.29**
Kidneys	0.06	**0.10**	0.08	**0.13**	0.09	**0.16**	0.10	**0.18**	0.13	**0.22**	0.15	**0.27**
Urinary bladder		**0.01**		**0.02**		**0.02**		**0.02**		**0.03**		**0.03**
Uterus		**0.03**		**0.04**		**0.04**		**0.05**	0.01	**0.06**	0.02	**0.08**
Ovaries		**0.03**		**0.04**		**0.05**		**0.05**		**0.06**	0.01	**0.08**
Prostate						**0.01**		**0.01**		**0.02**		**0.02**
Testes												
Upper arm	0.27	**0.97**	0.34	**1.20**	0.40	**1.36**	0.45	**1.49**	0.53	**1.71**	0.61	**1.91**
Elbow	0.21	**1.03**	0.28	**1.28**	0.33	**1.45**	0.36	**1.58**	0.44	**1.81**	0.51	**2.01**
Forearm	0.03	**0.54**	0.04	**0.67**	0.05	**0.76**	0.05	**0.82**	0.07	**0.94**	0.08	**1.04**
Upper leg										**0.01**		**0.01**
Knee												
Lower leg												
Lymph nodes	0.18	**0.34**	0.22	**0.42**	0.25	**0.47**	0.28	**0.51**	0.32	**0.59**	0.35	**0.66**
Skin	0.17	**0.32**	0.18	**0.33**	0.19	**0.35**	0.20	**0.36**	0.21	**0.38**	0.22	**0.39**
Muscle	0.13	**0.21**	0.15	**0.26**	0.17	**0.28**	0.19	**0.31**	0.21	**0.35**	0.23	**0.38**
Skeleton	0.29	**0.49**	0.36	**0.59**	0.40	**0.67**	0.43	**0.73**	0.49	**0.83**	0.55	**0.93**
Red bone marrow	0.07	**0.12**	0.09	**0.15**	0.10	**0.17**	0.11	**0.19**	0.12	**0.22**	0.14	**0.24**
Whole body dose	0.14	**0.23**	0.17	**0.28**	0.19	**0.31**	0.21	**0.34**	0.24	**0.39**	0.26	**0.43**
ED (ICRP 60)	0.23	**0.30**	0.27	**0.36**	0.30	**0.40**	0.32	**0.43**	0.36	**0.48**	0.39	**0.53**
ED (ICRP 103)	0.31	**0.40**	0.36	**0.46**	0.39	**0.50**	0.41	**0.54**	0.45	**0.60**	0.49	**0.65**
Total filtration (mm) Al	**1.5**		**2.5**		**3.5**		**4.5**		**3.5**		**3.5**	
Cu	**0**		**0**		**0**		**0**		**0.1**		**0.2**	

Table 10.21 Normalized organ doses—Bony thorax ap— <1 year

Beam projection:			ap			Table cells:						
Focus-to-detector distance (cm):			100			(+) = normal font = optimal field setting						
Tube voltage (kVp):			60			**(−) = bold font = suboptimal field setting**						
Total filtrations:			See last table row			Empty table cells correspond to values <0.01						
Organ or tissue	**Organ dose D_T: Entrance dose K_E**						**($mSv * mGy^{-1}$)**					
	(+)	(−)	(+)	(−)	(+)	(−)	(+)	(−)	(+)	(−)	(+)	(−)
Skull	0.02	**0.24**	0.03	**0.28**	0.04	**0.31**	0.04	**0.34**	0.05	**0.38**	0.04	**0.38**
Brain		**0.02**		**0.03**		**0.03**		**0.04**		**0.05**		**0.04**
Oral mucosa	0.03	**0.59**	0.04	**0.69**	0.05	**0.75**	0.05	**0.80**	0.06	**0.88**	0.06	**0.87**
Salivary glands	0.07	**0.52**	0.09	**0.61**	0.10	**0.68**	0.11	**0.73**	0.13	**0.81**	0.12	**0.80**
Cervical spine	0.39	**0.62**	0.49	**0.80**	0.57	**0.93**	0.63	**1.03**	0.74	**1.23**	0.72	**1.18**
Thyroid	0.91	**0.98**	1.02	**1.11**	1.08	**1.19**	1.13	**1.25**	1.20	**1.35**	1.20	**1.34**
Thymus	0.82	**0.84**	0.95	**0.97**	1.03	**1.06**	1.09	**1.12**	1.19	**1.23**	1.19	**1.22**
Thoracic spine	0.32	**0.34**	0.42	**0.45**	0.50	**0.54**	0.56	**0.61**	0.69	**0.76**	0.64	**0.70**
Claviculae	2.03	**2.12**	2.34	**2.45**	2.54	**2.67**	2.68	**2.82**	2.90	**3.07**	2.95	**3.11**
Scapulae	0.53	**0.56**	0.68	**0.72**	0.80	**0.84**	0.89	**0.94**	1.06	**1.12**	1.02	**1.08**
Ribs	1.49	**1.55**	1.70	**1.78**	1.84	**1.92**	1.94	**2.03**	2.10	**2.21**	2.11	**2.22**
Oesophagus	0.24	**0.25**	0.30	**0.33**	0.36	**0.38**	0.40	**0.43**	0.47	**0.51**	0.45	**0.49**
Trachea	0.10	**0.49**	0.12	**0.58**	0.13	**0.63**	0.14	**0.67**	0.15	**0.74**	0.15	**0.73**
Lung	0.45	**0.47**	0.55	**0.57**	0.61	**0.64**	0.66	**0.70**	0.75	**0.79**	0.74	**0.78**
Mamma	1.20	**1.22**	1.27	**1.28**	1.30	**1.32**	1.33	**1.34**	1.37	**1.37**	1.36	**1.37**
Heart	0.59	**0.61**	0.71	**0.73**	0.79	**0.82**	0.85	**0.88**	0.96	**1.00**	0.95	**0.98**
Stomach	0.39	**0.61**	0.47	**0.73**	0.52	**0.82**	0.57	**0.88**	0.64	**1.00**	0.63	**0.98**
Spleen	0.19	**0.24**	0.25	**0.30**	0.29	**0.35**	0.32	**0.40**	0.39	**0.47**	0.37	**0.45**
Liver	0.37	**0.54**	0.44	**0.65**	0.50	**0.73**	0.54	**0.79**	0.61	**0.89**	0.60	**0.87**
Gall bladder	0.21	**0.44**	0.26	**0.55**	0.30	**0.63**	0.33	**0.70**	0.39	**0.81**	0.37	**0.79**
Pancreas	0.28	**0.33**	0.35	**0.42**	0.41	**0.48**	0.46	**0.54**	0.54	**0.64**	0.52	**0.61**
Small intestine	0.02	**0.22**	0.03	**0.27**	0.03	**0.31**	0.04	**0.35**	0.05	**0.41**	0.04	**0.39**
Large intestine	0.03	**0.32**	0.04	**0.40**	0.04	**0.45**	0.05	**0.49**	0.06	**0.57**	0.05	**0.55**
Sigma/rectum		**0.07**		**0.09**	0.01	**0.10**	0.01	**0.11**	0.02	**0.13**	0.01	**0.13**
Lumbar spine	0.11	**0.31**	0.15	**0.41**	0.18	**0.49**	0.20	**0.56**	0.25	**0.69**	0.23	**0.63**
Pelvis	0.02	**0.11**	0.02	**0.14**	0.03	**0.17**	0.03	**0.19**	0.04	**0.23**	0.03	**0.21**
Adrenals	0.13	**0.15**	0.17	**0.19**	0.20	**0.23**	0.22	**0.26**	0.27	**0.32**	0.25	**0.29**
Kidneys	0.08	**0.14**	0.10	**0.18**	0.12	**0.21**	0.14	**0.24**	0.17	**0.29**	0.15	**0.27**
Urinary bladder		**0.02**		**0.02**		**0.03**		**0.03**		**0.04**		**0.03**
Uterus		**0.04**	0.01	**0.05**	0.01	**0.06**	0.01	**0.07**	0.02	**0.09**	0.02	**0.08**
Ovaries		**0.04**		**0.05**	0.01	**0.06**	0.01	**0.07**	0.02	**0.09**	0.01	**0.08**
Prostate				**0.01**		**0.01**		**0.02**		**0.02**		**0.02**
Testes												
Upper arm	0.33	**1.14**	0.42	**1.39**	0.48	**1.56**	0.53	**1.69**	0.63	**1.91**	0.61	**1.91**
Elbow	0.27	**1.21**	0.34	**1.47**	0.40	**1.65**	0.44	**1.78**	0.52	**2.02**	0.51	**2.01**
Forearm	0.04	**0.63**	0.05	**0.77**	0.06	**0.86**	0.07	**0.93**	0.08	**1.04**	0.08	**1.04**
Upper leg						**0.01**		**0.01**		**0.02**		**0.01**
Knee												
Lower leg												
Lymph nodes	0.22	**0.40**	0.26	**0.49**	0.29	**0.55**	0.32	**0.59**	0.36	**0.67**	0.35	**0.66**
Skin	0.18	**0.33**	0.19	**0.35**	0.20	**0.37**	0.21	**0.38**	0.22	**0.40**	0.22	**0.39**
Muscle	0.15	**0.25**	0.18	**0.29**	0.20	**0.32**	0.21	**0.35**	0.24	**0.39**	0.23	**0.38**
Skeleton	0.34	**0.57**	0.41	**0.69**	0.46	**0.77**	0.49	**0.83**	0.56	**0.94**	0.55	**0.93**
Red bone marrow	0.09	**0.15**	0.10	**0.18**	0.12	**0.20**	0.13	**0.22**	0.15	**0.25**	0.14	**0.24**
Whole body dose	0.17	**0.27**	0.20	**0.32**	0.22	**0.36**	0.24	**0.39**	0.27	**0.44**	0.26	**0.43**
ED (ICRP 60)	0.26	**0.35**	0.31	**0.41**	0.33	**0.45**	0.36	**0.48**	0.40	**0.54**	0.39	**0.53**
ED (ICRP 103)	0.35	**0.45**	0.40	**0.51**	0.43	**0.56**	0.45	**0.60**	0.50	**0.66**	0.49	**0.65**
Total filtration (mm) Al	**1.5**		**2.5**		**3.5**		**4.5**		**3.5**		**3.5**	
Cu	**0**		**0**		**0**		**0**		**0.1**		**0.2**	

Table 10.22 Normalized organ doses—Bony thorax ap— <1 year

Beam projection:			ap			Table cells:						
Focus-to-detector distance (cm):			100			(+) = normal font = optimal field setting						
Tube voltage (kVp):			70			**(−) = bold font = suboptimal field setting**						
Total filtrations:			See last table row			Empty table cells correspond to values <0.01						
Organ or tissue	**Organ dose D_T: Entrance dose K_E**							**(mSv * mGy^{-1})**				
	(+)	(−)	(+)	(−)	(+)	(−)	(+)	(−)	(+)	(−)	(+)	(−)
Skull	0.03	**0.26**	0.04	**0.31**	0.04	**0.34**	0.05	**0.36**	0.06	**0.41**	0.06	**0.44**
Brain		**0.03**		**0.04**		**0.04**		**0.05**	0.01	**0.06**	0.01	**0.07**
Oral mucosa	0.04	**0.65**	0.05	**0.74**	0.06	**0.81**	0.06	**0.85**	0.07	**0.93**	0.08	**1.00**
Salivary glands	0.08	**0.57**	0.10	**0.67**	0.11	**0.73**	0.12	**0.78**	0.14	**0.86**	0.16	**0.93**
Cervical spine	0.45	**0.74**	0.56	**0.92**	0.64	**1.06**	0.70	**1.17**	0.81	**1.37**	0.91	**1.54**
Thyroid	0.96	**1.05**	1.06	**1.17**	1.13	**1.25**	1.17	**1.31**	1.24	**1.40**	1.30	**1.48**
Thymus	0.89	**0.91**	1.01	**1.04**	1.09	**1.13**	1.15	**1.19**	1.25	**1.29**	1.32	**1.37**
Thoracic spine	0.39	**0.43**	0.51	**0.55**	0.59	**0.65**	0.66	**0.73**	0.80	**0.88**	0.93	**1.02**
Claviculae	2.16	**2.26**	2.45	**2.58**	2.63	**2.78**	2.76	**2.92**	2.95	**3.13**	3.07	**3.28**
Scapulae	0.63	**0.67**	0.79	**0.84**	0.91	**0.96**	1.00	**1.06**	1.17	**1.25**	1.32	**1.41**
Ribs	1.59	**1.66**	1.79	**1.88**	1.92	**2.02**	2.01	**2.12**	2.15	**2.28**	2.26	**2.40**
Oesophagus	0.29	**0.31**	0.36	**0.39**	0.41	**0.45**	0.46	**0.50**	0.54	**0.58**	0.61	**0.67**
Trachea	0.11	**0.54**	0.13	**0.62**	0.14	**0.67**	0.15	**0.71**	0.17	**0.78**	0.18	**0.83**
Lung	0.51	**0.53**	0.61	**0.64**	0.67	**0.71**	0.72	**0.76**	0.81	**0.86**	0.88	**0.93**
Mamma	1.23	**1.25**	1.29	**1.31**	1.33	**1.34**	1.35	**1.36**	1.39	**1.40**	1.42	**1.43**
Heart	0.66	**0.68**	0.78	**0.81**	0.86	**0.89**	0.92	**0.96**	1.03	**1.07**	1.12	**1.17**
Stomach	0.44	**0.68**	0.52	**0.81**	0.57	**0.89**	0.61	**0.96**	0.69	**1.07**	0.75	**1.17**
Spleen	0.23	**0.28**	0.29	**0.36**	0.34	**0.41**	0.37	**0.46**	0.44	**0.54**	0.50	**0.61**
Liver	0.41	**0.61**	0.49	**0.72**	0.55	**0.80**	0.59	**0.86**	0.66	**0.96**	0.72	**1.05**
Gall bladder	0.24	**0.51**	0.30	**0.63**	0.34	**0.71**	0.37	**0.77**	0.43	**0.89**	0.48	**0.99**
Pancreas	0.33	**0.39**	0.41	**0.49**	0.47	**0.56**	0.52	**0.61**	0.61	**0.72**	0.68	**0.81**
Small intestine	0.03	**0.26**	0.03	**0.32**	0.04	**0.36**	0.05	**0.39**	0.06	**0.45**	0.07	**0.51**
Large intestine	0.03	**0.37**	0.04	**0.45**	0.05	**0.50**	0.06	**0.55**	0.07	**0.62**	0.08	**0.69**
Sigma/rectum		**0.08**	0.01	**0.10**	0.01	**0.12**	0.02	**0.13**	0.02	**0.15**	0.02	**0.17**
Lumbar spine	0.14	**0.39**	0.18	**0.50**	0.22	**0.59**	0.24	**0.67**	0.30	**0.81**	0.35	**0.94**
Pelvis	0.02	**0.13**	0.03	**0.17**	0.03	**0.20**	0.04	**0.22**	0.05	**0.27**	0.06	**0.31**
Adrenals	0.16	**0.18**	0.20	**0.24**	0.24	**0.27**	0.27	**0.31**	0.32	**0.37**	0.37	**0.43**
Kidneys	0.10	**0.17**	0.13	**0.22**	0.15	**0.25**	0.16	**0.28**	0.20	**0.34**	0.23	**0.40**
Urinary bladder		**0.02**		**0.03**		**0.03**		**0.04**	0.01	**0.05**	0.01	**0.05**
Uterus	0.01	**0.05**	0.01	**0.06**	0.02	**0.07**	0.02	**0.08**	0.02	**0.10**	0.03	**0.12**
Ovaries		**0.05**	0.01	**0.06**	0.02	**0.08**	0.02	**0.09**	0.02	**0.10**	0.03	**0.12**
Prostate		**0.01**		**0.02**		**0.02**		**0.02**		**0.03**		**0.03**
Testes										**0.01**		**0.02**
Upper arm	0.39	**1.27**	0.48	**1.52**	0.54	**1.68**	0.59	**1.81**	0.68	**2.02**	0.76	**2.19**
Elbow	0.32	**1.35**	0.39	**1.60**	0.45	**1.78**	0.49	**1.91**	0.57	**2.13**	0.64	**2.30**
Forearm	0.05	**0.70**	0.06	**0.83**	0.07	**0.92**	0.08	**0.99**	0.10	**1.10**	0.11	**1.18**
Upper leg		**0.01**		**0.01**		**0.02**		**0.02**		**0.02**		**0.03**
Knee												
Lower leg												
Lymph nodes	0.25	**0.45**	0.29	**0.54**	0.33	**0.60**	0.35	**0.65**	0.40	**0.73**	0.44	**0.80**
Skin	0.19	**0.35**	0.20	**0.37**	0.21	**0.38**	0.22	**0.39**	0.23	**0.42**	0.24	**0.44**
Muscle	0.17	**0.28**	0.20	**0.32**	0.22	**0.35**	0.23	**0.38**	0.26	**0.42**	0.28	**0.46**
Skeleton	0.38	**0.64**	0.45	**0.75**	0.50	**0.83**	0.53	**0.90**	0.60	**1.00**	0.65	**1.09**
Red bone marrow	0.10	**0.17**	0.12	**0.20**	0.13	**0.23**	0.14	**0.25**	0.16	**0.28**	0.18	**0.32**
Whole body dose	0.19	**0.30**	0.22	**0.36**	0.24	**0.39**	0.26	**0.42**	0.29	**0.47**	0.32	**0.51**
ED (ICRP 60)	0.29	**0.38**	0.33	**0.45**	0.36	**0.49**	0.38	**0.52**	0.42	**0.58**	0.46	**0.63**
ED (ICRP 103)	0.38	**0.49**	0.42	**0.56**	0.46	**0.60**	0.48	**0.64**	0.52	**0.70**	0.56	**0.76**
Total filtration (mm) Al	**1.5**		**2.5**		**3.5**		**4.5**		**3.5**		**3.5**	
Cu	**0**		**0**		**0**		**0**		**0.1**		**0.2**	

Table 10.23 Normalized organ doses—Bony thorax RAO— <1 year

Beam projection:			RAO			Table cells:						
Focus-to-detector distance (cm):			100			(+) = normal font = optimal field setting						
Tube voltage (kVp):			50			**(−) = bold font = suboptimal field setting**						
Total filtrations:			See last table row			Empty table cells correspond to values <0.01						
Organ or tissue	**Organ dose D_T: Entrance dose K_E**							**(mSv * mGy^{-1})**				
	(+)	(−)	(+)	(−)	(+)	(−)	(+)	(−)	(+)	(−)	(+)	(−)
Skull	0.02	**0.20**	0.02	**0.25**	0.03	**0.28**	0.03	**0.30**	0.04	**0.34**	0.04	**0.38**
Brain		**0.02**		**0.02**		**0.03**		**0.03**		**0.04**		**0.05**
Oral mucosa	0.02	**0.50**	0.03	**0.59**	0.04	**0.65**	0.04	**0.70**	0.05	**0.77**	0.06	**0.84**
Salivary glands	0.06	**0.48**	0.07	**0.56**	0.08	**0.61**	0.09	**0.65**	0.11	**0.72**	0.12	**0.77**
Cervical spine	0.31	**0.51**	0.40	**0.67**	0.47	**0.79**	0.52	**0.89**	0.63	**1.07**	0.72	**1.24**
Thyroid	0.79	**0.85**	0.90	**0.98**	0.97	**1.06**	1.02	**1.11**	1.09	**1.21**	1.16	**1.29**
Thymus	0.67	**0.68**	0.79	**0.81**	0.87	**0.89**	0.92	**0.95**	1.02	**1.06**	1.10	**1.14**
Thoracic spine	0.29	**0.32**	0.39	**0.43**	0.47	**0.51**	0.53	**0.58**	0.66	**0.72**	0.78	**0.86**
Claviculae	1.64	**1.72**	1.94	**2.03**	2.13	**2.24**	2.27	**2.39**	2.50	**2.65**	2.69	**2.86**
Scapulae	0.25	**0.37**	0.34	**0.50**	0.40	**0.60**	0.46	**0.68**	0.57	**0.84**	0.68	**0.99**
Ribs	1.15	**1.29**	1.33	**1.51**	1.46	**1.65**	1.55	**1.76**	1.70	**1.94**	1.83	**2.09**
Oesophagus	0.18	**0.19**	0.23	**0.26**	0.27	**0.30**	0.31	**0.34**	0.38	**0.41**	0.44	**0.48**
Trachea	0.08	**0.43**	0.10	**0.51**	0.11	**0.57**	0.12	**0.61**	0.13	**0.67**	0.15	**0.73**
Lung	0.34	**0.36**	0.42	**0.45**	0.48	**0.51**	0.52	**0.55**	0.60	**0.64**	0.67	**0.71**
Mamma	1.03	**1.18**	1.10	**1.25**	1.13	**1.29**	1.15	**1.31**	1.18	**1.34**	1.21	**1.37**
Heart	0.43	**0.45**	0.53	**0.56**	0.61	**0.63**	0.66	**0.69**	0.76	**0.80**	0.84	**0.89**
Stomach	0.16	**0.37**	0.20	**0.47**	0.23	**0.53**	0.25	**0.59**	0.29	**0.68**	0.33	**0.77**
Spleen	0.07	**0.11**	0.09	**0.15**	0.11	**0.18**	0.13	**0.20**	0.16	**0.25**	0.19	**0.30**
Liver	0.24	**0.51**	0.30	**0.62**	0.34	**0.70**	0.37	**0.75**	0.42	**0.86**	0.47	**0.95**
Gall bladder	0.07	**0.33**	0.09	**0.42**	0.11	**0.49**	0.12	**0.54**	0.15	**0.64**	0.17	**0.74**
Pancreas	0.14	**0.20**	0.19	**0.27**	0.23	**0.32**	0.25	**0.36**	0.31	**0.44**	0.36	**0.52**
Small intestine		**0.16**	0.01	**0.21**	0.01	**0.24**	0.02	**0.27**	0.02	**0.32**	0.03	**0.37**
Large intestine	0.01	**0.25**	0.01	**0.31**	0.02	**0.36**	0.02	**0.40**	0.03	**0.46**	0.03	**0.52**
Sigma/rectum		**0.04**		**0.06**		**0.07**		**0.08**		**0.09**		**0.11**
Lumbar spine	0.03	**0.22**	0.05	**0.30**	0.06	**0.36**	0.07	**0.41**	0.09	**0.52**	0.11	**0.63**
Pelvis		**0.07**		**0.10**	0.01	**0.12**	0.01	**0.13**	0.02	**0.17**	0.02	**0.20**
Adrenals	0.09	**0.11**	0.12	**0.14**	0.14	**0.17**	0.16	**0.20**	0.20	**0.25**	0.24	**0.29**
Kidneys	0.03	**0.10**	0.04	**0.13**	0.05	**0.16**	0.05	**0.18**	0.07	**0.23**	0.08	**0.27**
Urinary bladder		**0.01**		**0.01**		**0.02**		**0.02**		**0.03**		**0.03**
Uterus		**0.03**		**0.04**		**0.04**		**0.05**		**0.06**	0.01	**0.08**
Ovaries		**0.03**		**0.04**		**0.05**		**0.05**		**0.06**		**0.08**
Prostate						**0.01**		**0.01**		**0.02**		**0.02**
Testes												
Upper arm	0.64	**1.06**	0.77	**1.28**	0.87	**1.43**	0.94	**1.54**	1.07	**1.73**	1.18	**1.90**
Elbow	0.59	**1.14**	0.71	**1.37**	0.80	**1.52**	0.87	**1.63**	0.99	**1.84**	1.09	**2.01**
Forearm	0.01	**0.58**	0.02	**0.69**	0.02	**0.76**	0.03	**0.82**	0.03	**0.92**	0.04	**1.00**
Upper leg										**0.01**		**0.01**
Knee												
Lower leg												
Lymph nodes	0.14	**0.31**	0.18	**0.39**	0.20	**0.44**	0.22	**0.48**	0.25	**0.55**	0.28	**0.61**
Skin	0.15	**0.31**	0.16	**0.33**	0.17	**0.34**	0.17	**0.35**	0.18	**0.37**	0.19	**0.39**
Muscle	0.11	**0.20**	0.13	**0.24**	0.15	**0.27**	0.16	**0.29**	0.18	**0.33**	0.20	**0.36**
Skeleton	0.31	**0.49**	0.37	**0.59**	0.42	**0.67**	0.45	**0.72**	0.52	**0.83**	0.57	**0.92**
Red bone marrow	0.07	**0.12**	0.09	**0.15**	0.10	**0.17**	0.11	**0.19**	0.12	**0.21**	0.14	**0.24**
Whole body dose	0.13	**0.23**	0.15	**0.27**	0.17	**0.30**	0.18	**0.33**	0.21	**0.37**	0.23	**0.41**
ED (ICRP 60)	0.19	**0.28**	0.22	**0.33**	0.25	**0.37**	0.26	**0.39**	0.29	**0.44**	0.32	**0.49**
ED (ICRP 103)	0.26	**0.37**	0.30	**0.43**	0.33	**0.47**	0.34	**0.50**	0.38	**0.56**	0.41	**0.61**
Total filtration (mm) **Al**	**1.5**		**2.5**		**3.5**		**4.5**		**3.5**		**3.5**	
Cu	**0**		**0**		**0**		**0**		**0.1**		**0.2**	

Table 10.24 Normalized organ doses—Bony thorax RAO— <1 year

Beam projection:			RAO			**Table cells:**						
Focus-to-detector distance (cm):			100			(+) = normal font = optimal field setting						
Tube voltage (kVp):			60			**(−) = bold font = suboptimal field setting**						
Total filtrations:			See last table row			Empty table cells correspond to values <0.01						
Organ or tissue	**Organ dose D_T: Entrance dose K_E**							**(mSv * mGy^{-1})**				
	(+)	(−)	(+)	(−)	(+)	(−)	(+)	(−)	(+)	(−)	(+)	(−)
Skull	0.02	**0.24**	0.03	**0.28**	0.04	**0.32**	0.04	**0.34**	0.05	**0.38**	0.04	**0.38**
Brain		**0.02**		**0.03**		**0.04**		**0.04**		**0.05**		**0.05**
Oral mucosa	0.03	**0.57**	0.04	**0.66**	0.05	**0.72**	0.05	**0.77**	0.06	**0.85**	0.06	**0.84**
Salivary glands	0.07	**0.53**	0.09	**0.62**	0.10	**0.67**	0.11	**0.72**	0.13	**0.79**	0.12	**0.77**
Cervical spine	0.39	**0.66**	0.49	**0.84**	0.57	**0.97**	0.63	**1.08**	0.74	**1.29**	0.72	**1.24**
Thyroid	0.86	**0.93**	0.97	**1.06**	1.04	**1.14**	1.08	**1.20**	1.16	**1.30**	1.16	**1.29**
Thymus	0.75	**0.77**	0.87	**0.90**	0.95	**0.98**	1.01	**1.05**	1.11	**1.15**	1.10	**1.14**
Thoracic spine	0.39	**0.43**	0.51	**0.57**	0.61	**0.67**	0.68	**0.75**	0.83	**0.92**	0.78	**0.86**
Claviculae	1.82	**1.92**	2.11	**2.24**	2.30	**2.44**	2.44	**2.59**	2.65	**2.83**	2.69	**2.86**
Scapulae	0.34	**0.50**	0.44	**0.65**	0.52	**0.76**	0.59	**0.86**	0.72	**1.04**	0.68	**0.99**
Ribs	1.27	**1.44**	1.46	**1.66**	1.58	**1.80**	1.67	**1.91**	1.82	**2.09**	1.83	**2.09**
Oesophagus	0.23	**0.25**	0.30	**0.33**	0.35	**0.38**	0.39	**0.43**	0.46	**0.51**	0.44	**0.48**
Trachea	0.09	**0.49**	0.11	**0.57**	0.12	**0.63**	0.13	**0.67**	0.15	**0.74**	0.15	**0.73**
Lung	0.41	**0.43**	0.49	**0.52**	0.55	**0.59**	0.60	**0.64**	0.68	**0.73**	0.67	**0.71**
Mamma	1.07	**1.21**	1.12	**1.28**	1.16	**1.31**	1.18	**1.34**	1.21	**1.37**	1.21	**1.37**
Heart	0.51	**0.54**	0.62	**0.65**	0.70	**0.73**	0.76	**0.80**	0.86	**0.91**	0.84	**0.89**
Stomach	0.19	**0.45**	0.24	**0.55**	0.27	**0.63**	0.30	**0.68**	0.34	**0.79**	0.33	**0.77**
Spleen	0.09	**0.15**	0.12	**0.20**	0.15	**0.23**	0.16	**0.26**	0.20	**0.32**	0.19	**0.30**
Liver	0.29	**0.59**	0.35	**0.71**	0.39	**0.79**	0.42	**0.85**	0.48	**0.96**	0.47	**0.95**
Gall bladder	0.09	**0.41**	0.12	**0.51**	0.13	**0.59**	0.15	**0.65**	0.18	**0.76**	0.17	**0.74**
Pancreas	0.19	**0.27**	0.24	**0.35**	0.28	**0.41**	0.32	**0.46**	0.38	**0.55**	0.36	**0.52**
Small intestine	0.01	**0.20**	0.02	**0.26**	0.02	**0.30**	0.02	**0.33**	0.03	**0.39**	0.03	**0.37**
Large intestine	0.02	**0.30**	0.02	**0.37**	0.02	**0.42**	0.03	**0.47**	0.04	**0.54**	0.03	**0.52**
Sigma/rectum		**0.06**		**0.07**		**0.09**		**0.10**	0.01	**0.11**		**0.11**
Lumbar spine	0.05	**0.31**	0.07	**0.41**	0.08	**0.49**	0.10	**0.56**	0.12	**0.69**	0.11	**0.63**
Pelvis		**0.10**	0.01	**0.13**	0.02	**0.16**	0.02	**0.18**	0.02	**0.22**	0.02	**0.20**
Adrenals	0.12	**0.15**	0.16	**0.19**	0.18	**0.23**	0.21	**0.26**	0.25	**0.31**	0.24	**0.29**
Kidneys	0.04	**0.14**	0.05	**0.18**	0.06	**0.21**	0.07	**0.24**	0.09	**0.29**	0.08	**0.27**
Urinary bladder		**0.02**		**0.02**		**0.03**		**0.03**		**0.04**		**0.03**
Uterus		**0.04**		**0.05**		**0.06**		**0.07**	0.01	**0.08**	0.01	**0.08**
Ovaries		**0.04**		**0.05**		**0.06**		**0.07**	0.01	**0.09**		**0.08**
Prostate				**0.01**		**0.01**		**0.02**		**0.02**		**0.02**
Testes												
Upper arm	0.74	**1.22**	0.89	**1.44**	0.99	**1.59**	1.06	**1.71**	1.19	**1.91**	1.18	**1.90**
Elbow	0.68	**1.30**	0.82	**1.53**	0.91	**1.69**	0.98	**1.81**	1.10	**2.01**	1.09	**2.01**
Forearm	0.02	**0.65**	0.03	**0.77**	0.03	**0.85**	0.04	**0.91**	0.05	**1.01**	0.04	**1.00**
Upper leg						**0.01**		**0.01**		**0.02**		**0.01**
Knee												
Lower leg												
Lymph nodes	0.17	**0.37**	0.21	**0.45**	0.24	**0.51**	0.26	**0.55**	0.29	**0.63**	0.28	**0.61**
Skin	0.16	**0.32**	0.17	**0.34**	0.18	**0.36**	0.18	**0.37**	0.19	**0.39**	0.19	**0.39**
Muscle	0.13	**0.23**	0.15	**0.28**	0.17	**0.31**	0.18	**0.33**	0.20	**0.37**	0.20	**0.36**
Skeleton	0.36	**0.57**	0.43	**0.69**	0.48	**0.77**	0.51	**0.83**	0.58	**0.93**	0.57	**0.92**
Red bone marrow	0.08	**0.15**	0.10	**0.18**	0.12	**0.20**	0.13	**0.22**	0.14	**0.25**	0.14	**0.24**
Whole body dose	0.15	**0.26**	0.17	**0.31**	0.19	**0.35**	0.21	**0.37**	0.24	**0.42**	0.23	**0.41**
ED (ICRP 60)	0.22	**0.32**	0.25	**0.38**	0.28	**0.42**	0.29	**0.45**	0.33	**0.50**	0.32	**0.49**
ED (ICRP 103)	0.29	**0.42**	0.33	**0.48**	0.36	**0.53**	0.38	**0.56**	0.41	**0.62**	0.41	**0.61**
Total filtration (mm) Al	**1.5**		**2.5**		**3.5**		**4.5**		**3.5**		**3.5**	
Cu	**0**		**0**		**0**		**0**		**0.1**		**0.2**	

Table 10.25 Normalized organ doses—Bony thorax RAO— <1 year

Beam projection:			RAO			**Table cells:**						
Focus-to-detector distance (cm):			100			(+) = normal font = optimal field setting						
Tube voltage (kVp):			70			**(−) = bold font = suboptimal field setting**						
Total filtrations:			See last table row			Empty table cells correspond to values <0.01						
Organ or tissue	**Organ dose D_T: Entrance dose K_E**							**(mSv * mGy^{-1})**				
	(+)	(−)	(+)	(−)	(+)	(−)	(+)	(−)	(+)	(−)	(+)	(−)
Skull	0.03	0.26	0.04	0.31	0.04	0.34	0.05	0.37	0.05	0.41	0.06	0.45
Brain		0.03		0.04		0.04		0.05	0.01	0.06	0.01	0.07
Oral mucosa	0.04	0.62	0.05	0.71	0.05	0.78	0.06	0.82	0.07	0.90	0.08	0.96
Salivary glands	0.08	0.58	0.10	0.67	0.11	0.72	0.12	0.76	0.14	0.83	0.15	0.89
Cervical spine	0.46	0.78	0.56	0.97	0.64	1.11	0.70	1.22	0.81	1.42	0.90	1.60
Thyroid	0.91	1.00	1.02	1.13	1.08	1.21	1.13	1.26	1.20	1.36	1.26	1.43
Thymus	0.81	0.84	0.93	0.97	1.01	1.05	1.07	1.11	1.17	1.21	1.24	1.29
Thoracic spine	0.48	0.53	0.61	0.68	0.71	0.79	0.79	0.88	0.94	1.05	1.07	1.21
Claviculae	1.95	2.06	2.23	2.37	2.40	2.56	2.52	2.69	2.71	2.91	2.83	3.05
Scapulae	0.41	0.61	0.53	0.77	0.61	0.89	0.68	0.99	0.81	1.17	0.93	1.34
Ribs	1.36	1.55	1.54	1.76	1.66	1.90	1.74	2.00	1.88	2.16	1.98	2.28
Oesophagus	0.28	0.31	0.35	0.39	0.40	0.45	0.44	0.49	0.52	0.58	0.59	0.66
Trachea	0.11	0.54	0.12	0.62	0.14	0.67	0.15	0.71	0.16	0.78	0.18	0.84
Lung	0.46	0.49	0.55	0.58	0.61	0.65	0.65	0.70	0.73	0.79	0.80	0.87
Mamma	1.09	1.24	1.15	1.30	1.18	1.33	1.20	1.35	1.23	1.38	1.25	1.40
Heart	0.58	0.61	0.69	0.72	0.76	0.81	0.82	0.87	0.92	0.98	1.01	1.08
Stomach	0.22	0.51	0.27	0.62	0.30	0.70	0.33	0.76	0.38	0.86	0.42	0.95
Spleen	0.12	0.19	0.15	0.24	0.18	0.28	0.20	0.31	0.24	0.37	0.27	0.43
Liver	0.32	0.66	0.38	0.78	0.43	0.86	0.46	0.93	0.51	1.04	0.56	1.12
Gall bladder	0.11	0.48	0.14	0.59	0.16	0.66	0.17	0.72	0.20	0.84	0.23	0.93
Pancreas	0.23	0.33	0.29	0.41	0.33	0.48	0.37	0.53	0.43	0.62	0.49	0.71
Small intestine	0.02	0.24	0.02	0.30	0.03	0.34	0.03	0.37	0.04	0.43	0.04	0.49
Large intestine	0.02	0.35	0.03	0.42	0.03	0.48	0.04	0.52	0.04	0.59	0.05	0.65
Sigma/rectum		0.07		0.09		0.10	0.01	0.11	0.01	0.13	0.02	0.15
Lumbar spine	0.07	0.39	0.09	0.50	0.11	0.59	0.12	0.66	0.15	0.80	0.18	0.93
Pelvis	0.01	0.13	0.02	0.16	0.02	0.19	0.02	0.21	0.03	0.26	0.04	0.30
Adrenals	0.15	0.18	0.19	0.24	0.22	0.27	0.24	0.31	0.29	0.37	0.34	0.43
Kidneys	0.05	0.17	0.07	0.22	0.08	0.26	0.09	0.29	0.11	0.35	0.13	0.40
Urinary bladder		0.02		0.03		0.03		0.04		0.05		0.05
Uterus		0.05		0.06	0.01	0.07	0.01	0.08	0.02	0.10	0.02	0.12
Ovaries		0.05		0.07	0.01	0.08	0.01	0.09	0.02	0.11	0.02	0.12
Prostate		0.01		0.02		0.02		0.02		0.03		0.03
Testes										0.01		0.01
Upper arm	0.82	1.33	0.96	1.55	1.06	1.70	1.13	1.81	1.26	1.99	1.36	2.13
Elbow	0.76	1.42	0.89	1.65	0.98	1.80	1.04	1.91	1.16	2.10	1.25	2.24
Forearm	0.03	0.71	0.03	0.83	0.04	0.90	0.05	0.96	0.06	1.05	0.07	1.12
Upper leg		0.01		0.01		0.02		0.02		0.02		0.03
Knee												
Lower leg												
Lymph nodes	0.20	0.42	0.23	0.51	0.26	0.56	0.28	0.61	0.32	0.69	0.35	0.76
Skin	0.17	0.34	0.18	0.36	0.19	0.37	0.19	0.39	0.20	0.41	0.21	0.43
Muscle	0.14	0.26	0.17	0.31	0.18	0.34	0.20	0.36	0.22	0.40	0.24	0.44
Skeleton	0.40	0.64	0.47	0.75	0.52	0.83	0.55	0.89	0.62	1.00	0.67	1.08
Red bone marrow	0.10	0.17	0.12	0.20	0.13	0.23	0.14	0.25	0.16	0.28	0.18	0.31
Whole body dose	0.16	0.29	0.19	0.35	0.21	0.38	0.23	0.41	0.26	0.46	0.28	0.50
ED (ICRP 60)	0.24	0.35	0.27	0.41	0.30	0.45	0.32	0.49	0.35	0.54	0.38	0.59
ED (ICRP 103)	0.31	0.46	0.35	0.52	0.38	0.57	0.40	0.60	0.44	0.66	0.47	0.71
Total filtration (mm) Al	**1.5**		**2.5**		**3.5**		**4.5**		**3.5**		**3.5**	
Cu	**0**		**0**		**0**		**0**		**0.1**		**0.2**	

Table 10.26 Normalized organ doses—Bony thorax LAO— <1 year

Beam projection:			LAO			Table cells:						
Focus-to-detector distance (cm):			100			(+) = normal font = optimal field setting						
Tube voltage (kVp):			50			**(−) = bold font = suboptimal field setting**						
Total filtrations:			See last table row			Empty table cells correspond to values <0.01						
Organ or tissue	**Organ dose D_T: Entrance dose K_E**						**(mSv * mGy^{-1})**					
	(+)	(−)	(+)	(−)	(+)	(−)	(+)	(−)	(+)	(−)	(+)	(−)
Skull	0.02	**0.20**	0.02	**0.25**	0.03	**0.28**	0.03	**0.30**	0.04	**0.34**	0.04	**0.38**
Brain		**0.02**		**0.02**		**0.03**		**0.03**		**0.04**		**0.05**
Oral mucosa	0.02	**0.51**	0.03	**0.60**	0.04	**0.65**	0.04	**0.70**	0.05	**0.77**	0.06	**0.84**
Salivary glands	0.06	**0.47**	0.07	**0.56**	0.08	**0.61**	0.09	**0.65**	0.11	**0.72**	0.12	**0.78**
Cervical spine	0.31	**0.51**	0.40	**0.67**	0.47	**0.79**	0.53	**0.88**	0.63	**1.07**	0.72	**1.24**
Thyroid	0.79	**0.84**	0.90	**0.97**	0.97	**1.05**	1.02	**1.11**	1.09	**1.20**	1.15	**1.28**
Thymus	0.67	**0.68**	0.79	**0.81**	0.87	**0.89**	0.92	**0.95**	1.02	**1.06**	1.10	**1.15**
Thoracic spine	0.26	**0.28**	0.35	**0.38**	0.42	**0.46**	0.48	**0.53**	0.60	**0.66**	0.71	**0.79**
Claviculae	1.64	**1.71**	1.93	**2.03**	2.13	**2.24**	2.27	**2.39**	2.50	**2.64**	2.68	**2.85**
Scapulae	0.26	**0.38**	0.34	**0.50**	0.41	**0.60**	0.47	**0.69**	0.58	**0.85**	0.69	**1.00**
Ribs	1.15	**1.30**	1.34	**1.51**	1.46	**1.66**	1.55	**1.76**	1.70	**1.94**	1.83	**2.09**
Oesophagus	0.17	**0.19**	0.23	**0.25**	0.27	**0.30**	0.30	**0.34**	0.37	**0.41**	0.43	**0.48**
Trachea	0.08	**0.44**	0.10	**0.52**	0.11	**0.57**	0.12	**0.61**	0.13	**0.68**	0.14	**0.74**
Lung	0.32	**0.34**	0.40	**0.42**	0.45	**0.48**	0.50	**0.53**	0.57	**0.61**	0.64	**0.69**
Mamma	1.04	**1.17**	1.10	**1.23**	1.13	**1.27**	1.15	**1.29**	1.18	**1.32**	1.20	**1.34**
Heart	0.48	**0.50**	0.59	**0.61**	0.66	**0.69**	0.72	**0.75**	0.82	**0.86**	0.91	**0.95**
Stomach	0.22	**0.57**	0.27	**0.69**	0.31	**0.78**	0.33	**0.84**	0.38	**0.95**	0.42	**1.05**
Spleen	0.15	**0.26**	0.20	**0.34**	0.23	**0.40**	0.26	**0.44**	0.31	**0.53**	0.35	**0.61**
Liver	0.18	**0.35**	0.22	**0.43**	0.25	**0.49**	0.28	**0.54**	0.32	**0.62**	0.36	**0.70**
Gall bladder	0.07	**0.32**	0.09	**0.41**	0.11	**0.48**	0.12	**0.53**	0.15	**0.63**	0.17	**0.72**
Pancreas	0.19	**0.27**	0.25	**0.35**	0.30	**0.41**	0.33	**0.46**	0.40	**0.55**	0.46	**0.64**
Small intestine		**0.16**	0.01	**0.21**	0.01	**0.24**	0.02	**0.27**	0.02	**0.32**	0.03	**0.37**
Large intestine	0.01	**0.23**	0.01	**0.30**	0.02	**0.34**	0.02	**0.38**	0.03	**0.44**	0.03	**0.50**
Sigma/rectum		**0.06**		**0.07**		**0.09**		**0.09**		**0.11**		**0.13**
Lumbar spine	0.03	**0.22**	0.05	**0.30**	0.06	**0.36**	0.07	**0.41**	0.09	**0.52**	0.11	**0.63**
Pelvis		**0.07**		**0.10**	0.01	**0.12**	0.01	**0.13**	0.02	**0.17**	0.02	**0.20**
Adrenals	0.09	**0.11**	0.12	**0.14**	0.14	**0.17**	0.16	**0.19**	0.19	**0.24**	0.23	**0.29**
Kidneys	0.03	**0.10**	0.04	**0.13**	0.05	**0.16**	0.05	**0.18**	0.07	**0.23**	0.08	**0.27**
Urinary bladder		**0.01**		**0.01**		**0.02**		**0.02**		**0.03**		**0.03**
Uterus		**0.03**		**0.04**		**0.04**		**0.05**		**0.06**		**0.08**
Ovaries		**0.03**		**0.04**		**0.05**		**0.05**		**0.06**		**0.08**
Prostate						**0.01**		**0.01**		**0.02**		**0.02**
Testes												
Upper arm	0.64	**1.07**	0.78	**1.28**	0.87	**1.43**	0.94	**1.54**	1.07	**1.74**	1.19	**1.91**
Elbow	0.59	**1.14**	0.72	**1.36**	0.81	**1.52**	0.87	**1.63**	0.99	**1.83**	1.10	**2.01**
Forearm	0.01	**0.57**	0.02	**0.69**	0.02	**0.76**	0.03	**0.82**	0.03	**0.92**	0.04	**1.00**
Upper leg										**0.01**		**0.01**
Knee												
Lower leg												
Lymph nodes	0.15	**0.33**	0.19	**0.40**	0.21	**0.46**	0.23	**0.50**	0.27	**0.57**	0.30	**0.64**
Skin	0.15	**0.31**	0.16	**0.33**	0.17	**0.34**	0.17	**0.35**	0.18	**0.37**	0.19	**0.39**
Muscle	0.11	**0.21**	0.13	**0.25**	0.15	**0.28**	0.16	**0.30**	0.18	**0.34**	0.20	**0.37**
Skeleton	0.31	**0.49**	0.37	**0.59**	0.41	**0.66**	0.45	**0.72**	0.51	**0.82**	0.57	**0.91**
Red bone marrow	0.07	**0.12**	0.08	**0.15**	0.10	**0.17**	0.10	**0.18**	0.12	**0.21**	0.14	**0.24**
Whole body dose	0.13	**0.23**	0.15	**0.27**	0.17	**0.30**	0.18	**0.33**	0.21	**0.37**	0.23	**0.41**
ED (ICRP 60)	0.19	**0.29**	0.23	**0.34**	0.25	**0.38**	0.27	**0.41**	0.30	**0.46**	0.32	**0.51**
ED (ICRP 103)	0.27	**0.39**	0.30	**0.45**	0.33	**0.49**	0.35	**0.52**	0.38	**0.58**	0.41	**0.63**
Total filtration (mm) Al	**1.5**		**2.5**		**3.5**		**4.5**		**3.5**		**3.5**	
Cu	**0**		**0**		**0**		**0**		**0.1**		**0.2**	

Table 10.27 Normalized organ doses—Bony thorax LAO— <1 year

Beam projection:			LAO			**Table cells:**						
Focus-to-detector distance (cm):			100			(+) = normal font = optimal field setting						
Tube voltage (kVp):			60			**(−) = bold font = suboptimal field setting**						
Total filtrations:			See last table row			Empty table cells correspond to values <0.01						
Organ or tissue	**Organ dose D_T: Entrance dose K_E**							**(mSv * mGy^{-1})**				
	(+)	(−)	(+)	(−)	(+)	(−)	(+)	(−)	(+)	(−)	(+)	(−)
Skull	0.02	**0.24**	0.03	**0.28**	0.04	**0.32**	0.04	**0.34**	0.05	**0.38**	0.04	**0.38**
Brain		**0.02**		**0.03**		**0.04**		**0.04**		**0.05**		**0.05**
Oral mucosa	0.03	**0.57**	0.04	**0.66**	0.05	**0.72**	0.05	**0.77**	0.06	**0.85**	0.06	**0.84**
Salivary glands	0.07	**0.53**	0.09	**0.62**	0.10	**0.67**	0.11	**0.71**	0.13	**0.78**	0.12	**0.78**
Cervical spine	0.39	**0.66**	0.49	**0.84**	0.57	**0.97**	0.63	**1.08**	0.74	**1.29**	0.72	**1.24**
Thyroid	0.86	**0.92**	0.97	**1.05**	1.04	**1.13**	1.08	**1.19**	1.16	**1.29**	1.15	**1.28**
Thymus	0.75	**0.77**	0.87	**0.90**	0.95	**0.99**	1.01	**1.05**	1.11	**1.15**	1.10	**1.15**
Thoracic spine	0.35	**0.39**	0.46	**0.52**	0.55	**0.61**	0.62	**0.69**	0.76	**0.85**	0.71	**0.79**
Claviculae	1.82	**1.91**	2.11	**2.23**	2.30	**2.43**	2.43	**2.58**	2.65	**2.82**	2.68	**2.85**
Scapulae	0.35	**0.51**	0.45	**0.66**	0.53	**0.78**	0.60	**0.87**	0.73	**1.06**	0.69	**1.00**
Ribs	1.27	**1.44**	1.46	**1.66**	1.59	**1.81**	1.68	**1.91**	1.83	**2.09**	1.83	**2.09**
Oesophagus	0.23	**0.25**	0.29	**0.32**	0.34	**0.38**	0.38	**0.42**	0.45	**0.51**	0.43	**0.48**
Trachea	0.09	**0.49**	0.11	**0.58**	0.12	**0.63**	0.13	**0.67**	0.15	**0.75**	0.14	**0.74**
Lung	0.38	**0.41**	0.47	**0.50**	0.53	**0.56**	0.57	**0.61**	0.65	**0.70**	0.64	**0.69**
Mamma	1.07	**1.21**	1.12	**1.27**	1.15	**1.30**	1.17	**1.32**	1.20	**1.35**	1.20	**1.34**
Heart	0.56	**0.59**	0.68	**0.71**	0.76	**0.79**	0.82	**0.86**	0.92	**0.97**	0.91	**0.95**
Stomach	0.26	**0.66**	0.32	**0.79**	0.35	**0.88**	0.38	**0.95**	0.43	**1.07**	0.42	**1.05**
Spleen	0.19	**0.33**	0.24	**0.42**	0.28	**0.48**	0.31	**0.54**	0.37	**0.63**	0.35	**0.61**
Liver	0.22	**0.42**	0.26	**0.51**	0.30	**0.57**	0.33	**0.62**	0.37	**0.71**	0.36	**0.70**
Gall bladder	0.09	**0.40**	0.12	**0.50**	0.14	**0.58**	0.15	**0.64**	0.18	**0.74**	0.17	**0.72**
Pancreas	0.25	**0.34**	0.31	**0.44**	0.36	**0.51**	0.40	**0.56**	0.47	**0.67**	0.46	**0.64**
Small intestine	0.01	**0.21**	0.02	**0.26**	0.02	**0.30**	0.02	**0.33**	0.03	**0.39**	0.03	**0.37**
Large intestine	0.02	**0.29**	0.02	**0.36**	0.02	**0.41**	0.03	**0.45**	0.03	**0.52**	0.03	**0.50**
Sigma/rectum		**0.07**		**0.09**		**0.10**		**0.11**	0.01	**0.13**		**0.13**
Lumbar spine	0.05	**0.31**	0.07	**0.41**	0.08	**0.49**	0.10	**0.56**	0.12	**0.69**	0.11	**0.63**
Pelvis		**0.10**	0.01	**0.13**	0.02	**0.16**	0.02	**0.18**	0.02	**0.22**	0.02	**0.20**
Adrenals	0.12	**0.15**	0.15	**0.19**	0.18	**0.22**	0.20	**0.25**	0.25	**0.31**	0.23	**0.29**
Kidneys	0.04	**0.14**	0.05	**0.18**	0.06	**0.21**	0.07	**0.24**	0.09	**0.29**	0.08	**0.27**
Urinary bladder		**0.02**		**0.02**		**0.03**		**0.03**		**0.04**		**0.03**
Uterus		**0.04**		**0.05**		**0.06**		**0.07**	0.01	**0.08**		**0.08**
Ovaries		**0.04**		**0.05**		**0.06**		**0.07**		**0.09**		**0.08**
Prostate				**0.01**		**0.01**		**0.02**		**0.02**		**0.02**
Testes												
Upper arm	0.74	**1.22**	0.89	**1.45**	0.99	**1.60**	1.06	**1.71**	1.20	**1.91**	1.19	**1.91**
Elbow	0.69	**1.30**	0.82	**1.53**	0.91	**1.69**	0.98	**1.81**	1.10	**2.01**	1.10	**2.01**
Forearm	0.02	**0.65**	0.03	**0.77**	0.03	**0.85**	0.04	**0.90**	0.05	**1.00**	0.04	**1.00**
Upper leg						**0.01**		**0.01**		**0.02**		**0.01**
Knee												
Lower leg												
Lymph nodes	0.18	**0.39**	0.22	**0.47**	0.25	**0.53**	0.27	**0.58**	0.31	**0.66**	0.30	**0.64**
Skin	0.16	**0.32**	0.17	**0.34**	0.18	**0.36**	0.18	**0.37**	0.19	**0.39**	0.19	**0.39**
Muscle	0.13	**0.24**	0.15	**0.29**	0.17	**0.32**	0.18	**0.34**	0.21	**0.38**	0.20	**0.37**
Skeleton	0.36	**0.57**	0.42	**0.68**	0.47	**0.76**	0.51	**0.82**	0.58	**0.93**	0.57	**0.91**
Red bone marrow	0.08	**0.15**	0.10	**0.18**	0.11	**0.20**	0.12	**0.22**	0.14	**0.25**	0.14	**0.24**
Whole body dose	0.15	**0.26**	0.17	**0.31**	0.19	**0.35**	0.21	**0.37**	0.24	**0.42**	0.23	**0.41**
ED (ICRP 60)	0.22	**0.33**	0.25	**0.39**	0.28	**0.43**	0.30	**0.46**	0.33	**0.52**	0.32	**0.51**
ED (ICRP 103)	0.30	**0.43**	0.34	**0.50**	0.36	**0.54**	0.38	**0.58**	0.42	**0.64**	0.41	**0.63**
Total filtration (mm) **Al**	**1.5**		**2.5**		**3.5**		**4.5**		**3.5**		**3.5**	
Cu	**0**		**0**		**0**		**0**		**0.1**		**0.2**	

Table 10.28 Normalized organ doses—Bony thorax LAO— <1 year

Beam projection:			LAO			Table cells:						
Focus-to-detector distance (cm):			100			(+) = normal font = optimal field setting						
Tube voltage (kVp):			70			**(−) = bold font = suboptimal field setting**						
Total filtrations:			See last table row			Empty table cells correspond to values <0.01						
Organ or tissue	**Organ dose D_T: Entrance dose K_E**							**(mSv * mGy^{-1})**				
	(+)	(−)	(+)	(−)	(+)	(−)	(+)	(−)	(+)	(−)	(+)	(−)
Skull	0.03	**0.26**	0.04	**0.31**	0.04	**0.34**	0.05	**0.37**	0.05	**0.41**	0.06	**0.45**
Brain		**0.03**		**0.04**		**0.04**		**0.05**	0.01	**0.06**	0.01	**0.07**
Oral mucosa	0.04	**0.62**	0.05	**0.71**	0.05	**0.78**	0.06	**0.82**	0.07	**0.90**	0.08	**0.96**
Salivary glands	0.08	**0.58**	0.10	**0.66**	0.11	**0.72**	0.12	**0.76**	0.14	**0.83**	0.15	**0.89**
Cervical spine	0.46	**0.78**	0.56	**0.97**	0.64	**1.11**	0.70	**1.22**	0.81	**1.42**	0.90	**1.60**
Thyroid	0.91	**0.99**	1.02	**1.12**	1.08	**1.20**	1.13	**1.25**	1.20	**1.35**	1.26	**1.42**
Thymus	0.81	**0.84**	0.93	**0.97**	1.01	**1.05**	1.07	**1.11**	1.16	**1.22**	1.24	**1.30**
Thoracic spine	0.44	**0.49**	0.56	**0.62**	0.65	**0.73**	0.72	**0.81**	0.87	**0.98**	1.00	**1.13**
Claviculae	1.95	**2.06**	2.22	**2.36**	2.40	**2.55**	2.52	**2.68**	2.71	**2.89**	2.83	**3.03**
Scapulae	0.42	**0.61**	0.54	**0.78**	0.62	**0.90**	0.69	**1.00**	0.83	**1.19**	0.95	**1.36**
Ribs	1.36	**1.55**	1.55	**1.76**	1.66	**1.90**	1.74	**2.00**	1.88	**2.16**	1.98	**2.29**
Oesophagus	0.27	**0.31**	0.34	**0.38**	0.39	**0.44**	0.43	**0.49**	0.51	**0.58**	0.58	**0.66**
Trachea	0.10	**0.54**	0.12	**0.62**	0.13	**0.68**	0.14	**0.72**	0.16	**0.79**	0.18	**0.85**
Lung	0.43	**0.46**	0.52	**0.56**	0.58	**0.63**	0.63	**0.68**	0.71	**0.77**	0.78	**0.84**
Mamma	1.09	**1.23**	1.14	**1.29**	1.17	**1.32**	1.19	**1.34**	1.22	**1.37**	1.24	**1.39**
Heart	0.63	**0.66**	0.75	**0.78**	0.83	**0.87**	0.89	**0.93**	0.99	**1.05**	1.07	**1.14**
Stomach	0.29	**0.74**	0.35	**0.87**	0.39	**0.96**	0.42	**1.03**	0.47	**1.14**	0.51	**1.24**
Spleen	0.23	**0.39**	0.28	**0.48**	0.32	**0.55**	0.35	**0.60**	0.41	**0.70**	0.46	**0.79**
Liver	0.25	**0.47**	0.30	**0.57**	0.33	**0.64**	0.36	**0.69**	0.41	**0.78**	0.45	**0.86**
Gall bladder	0.11	**0.47**	0.14	**0.58**	0.16	**0.65**	0.17	**0.71**	0.20	**0.82**	0.23	**0.92**
Pancreas	0.29	**0.41**	0.36	**0.51**	0.41	**0.58**	0.45	**0.64**	0.53	**0.75**	0.59	**0.84**
Small intestine	0.02	**0.24**	0.02	**0.30**	0.03	**0.34**	0.03	**0.37**	0.04	**0.43**	0.04	**0.49**
Large intestine	0.02	**0.33**	0.03	**0.41**	0.03	**0.46**	0.03	**0.50**	0.04	**0.57**	0.05	**0.64**
Sigma/rectum		**0.09**		**0.10**		**0.12**	0.01	**0.13**	0.01	**0.15**	0.02	**0.17**
Lumbar spine	0.07	**0.39**	0.09	**0.50**	0.11	**0.59**	0.12	**0.66**	0.15	**0.80**	0.18	**0.94**
Pelvis	0.01	**0.13**	0.02	**0.16**	0.02	**0.19**	0.02	**0.21**	0.03	**0.26**	0.04	**0.30**
Adrenals	0.15	**0.18**	0.19	**0.23**	0.22	**0.27**	0.24	**0.30**	0.29	**0.36**	0.33	**0.42**
Kidneys	0.05	**0.17**	0.07	**0.22**	0.08	**0.26**	0.09	**0.29**	0.11	**0.35**	0.12	**0.40**
Urinary bladder		**0.02**		**0.03**		**0.03**		**0.04**		**0.05**		**0.05**
Uterus		**0.05**		**0.06**	0.01	**0.07**	0.01	**0.08**	0.02	**0.10**	0.02	**0.12**
Ovaries		**0.05**		**0.06**		**0.08**	0.01	**0.09**	0.01	**0.10**	0.02	**0.12**
Prostate		**0.01**		**0.02**		**0.02**		**0.02**		**0.03**		**0.03**
Testes										**0.01**		**0.02**
Upper arm	0.82	**1.34**	0.96	**1.56**	1.06	**1.71**	1.14	**1.82**	1.26	**2.00**	1.36	**2.14**
Elbow	0.76	**1.42**	0.89	**1.65**	0.98	**1.80**	1.05	**1.91**	1.16	**2.10**	1.25	**2.24**
Forearm	0.03	**0.71**	0.03	**0.83**	0.04	**0.90**	0.05	**0.96**	0.06	**1.05**	0.07	**1.12**
Upper leg		**0.01**		**0.01**		**0.02**		**0.02**		**0.02**		**0.03**
Knee												
Lower leg												
Lymph nodes	0.21	**0.44**	0.25	**0.53**	0.28	**0.59**	0.30	**0.63**	0.34	**0.72**	0.37	**0.78**
Skin	0.17	**0.34**	0.18	**0.36**	0.19	**0.37**	0.19	**0.39**	0.20	**0.41**	0.21	**0.43**
Muscle	0.15	**0.27**	0.17	**0.31**	0.19	**0.35**	0.20	**0.37**	0.23	**0.42**	0.25	**0.45**
Skeleton	0.40	**0.63**	0.46	**0.75**	0.51	**0.83**	0.55	**0.88**	0.61	**0.99**	0.66	**1.08**
Red bone marrow	0.10	**0.17**	0.11	**0.20**	0.13	**0.23**	0.14	**0.24**	0.16	**0.28**	0.18	**0.31**
Whole body dose	0.16	**0.29**	0.19	**0.34**	0.21	**0.38**	0.23	**0.41**	0.26	**0.46**	0.28	**0.50**
ED (ICRP 60)	0.24	**0.37**	0.28	**0.43**	0.30	**0.47**	0.32	**0.50**	0.35	**0.56**	0.38	**0.61**
ED (ICRP 103)	0.32	**0.47**	0.36	**0.54**	0.39	**0.59**	0.41	**0.62**	0.44	**0.68**	0.47	**0.73**
Total filtration (mm) Al	**1.5**		**2.5**		**3.5**		**4.5**		**3.5**		**3.5**	
Cu	**0**		**0**		**0**		**0**		**0.1**		**0.2**	

Table 10.29 Normalized organ doses—Bony thorax ap—1–2 years

Beam projection:			ap				**Table cells:**					
Focus-to-detector distance (cm):			115				(+) = normal font = optimal field setting					
Tube voltage (kVp):			50				**(−) = bold font = suboptimal field setting**					
Total filtrations:			See last table row				Empty table cells correspond to values <0.01					
Organ or tissue	**Organ dose D_T: Entrance dose K_E**							**(mSv * mGy^{-1})**				
	(+)	(−)	(+)	(−)	(+)	(−)	(+)	(−)	(+)	(−)	(+)	(−)
Skull	0.01	**0.21**	0.02	**0.25**	0.02	**0.28**	0.02	**0.31**	0.03	**0.35**	0.03	**0.39**
Brain		**0.01**		**0.01**		**0.02**		**0.02**		**0.03**		**0.03**
Oral mucosa	0.02	**0.41**	0.02	**0.49**	0.03	**0.55**	0.03	**0.59**	0.04	**0.66**	0.05	**0.73**
Salivary glands	0.04	**0.32**	0.05	**0.40**	0.06	**0.45**	0.06	**0.49**	0.07	**0.56**	0.09	**0.63**
Cervical spine	0.23	**0.33**	0.30	**0.44**	0.36	**0.52**	0.40	**0.59**	0.49	**0.74**	0.58	**0.88**
Thyroid	0.80	**0.84**	0.91	**0.97**	0.98	**1.05**	1.03	**1.10**	1.11	**1.19**	1.18	**1.27**
Thymus	0.64	**0.65**	0.77	**0.78**	0.86	**0.88**	0.92	**0.95**	1.04	**1.07**	1.14	**1.17**
Thoracic spine	0.13	**0.14**	0.18	**0.20**	0.22	**0.24**	0.26	**0.28**	0.33	**0.36**	0.41	**0.45**
Claviculae	1.57	**1.63**	1.90	**1.98**	2.12	**2.22**	2.29	**2.40**	2.59	**2.72**	2.84	**2.99**
Scapulae	0.21	**0.22**	0.28	**0.29**	0.34	**0.35**	0.39	**0.41**	0.48	**0.51**	0.58	**0.61**
Ribs	1.24	**1.27**	1.46	**1.50**	1.60	**1.66**	1.71	**1.78**	1.90	**1.98**	2.06	**2.15**
Oesophagus	0.15	**0.16**	0.20	**0.21**	0.24	**0.26**	0.27	**0.29**	0.33	**0.36**	0.40	**0.44**
Trachea	0.07	**0.38**	0.08	**0.45**	0.09	**0.51**	0.10	**0.55**	0.11	**0.62**	0.12	**0.69**
Lung	0.30	**0.32**	0.38	**0.40**	0.44	**0.46**	0.49	**0.50**	0.57	**0.59**	0.64	**0.67**
Mamma	1.10	**1.10**	1.18	**1.19**	1.22	**1.23**	1.25	**1.26**	1.30	**1.31**	1.34	**1.35**
Heart	0.42	**0.43**	0.52	**0.53**	0.59	**0.61**	0.65	**0.67**	0.76	**0.78**	0.85	**0.88**
Stomach	0.24	**0.42**	0.30	**0.52**	0.35	**0.59**	0.38	**0.65**	0.44	**0.75**	0.50	**0.85**
Spleen	0.10	**0.12**	0.13	**0.16**	0.16	**0.20**	0.18	**0.22**	0.22	**0.28**	0.27	**0.34**
Liver	0.23	**0.35**	0.29	**0.44**	0.33	**0.51**	0.37	**0.56**	0.43	**0.65**	0.49	**0.74**
Gall bladder	0.08	**0.30**	0.11	**0.39**	0.13	**0.46**	0.14	**0.51**	0.18	**0.61**	0.21	**0.71**
Pancreas	0.15	**0.18**	0.20	**0.24**	0.24	**0.29**	0.28	**0.33**	0.34	**0.40**	0.40	**0.48**
Small intestine		**0.13**	0.01	**0.17**	0.01	**0.20**	0.02	**0.22**	0.02	**0.27**	0.03	**0.31**
Large intestine	0.01	**0.20**	0.02	**0.26**	0.02	**0.30**	0.02	**0.33**	0.03	**0.40**	0.04	**0.46**
Sigma/rectum		**0.04**		**0.05**		**0.06**		**0.07**		**0.08**		**0.10**
Lumbar spine	0.04	**0.12**	0.05	**0.17**	0.07	**0.21**	0.08	**0.24**	0.10	**0.32**	0.13	**0.39**
Pelvis		**0.05**		**0.06**		**0.08**		**0.09**	0.01	**0.11**	0.02	**0.14**
Adrenals	0.06	**0.07**	0.08	**0.09**	0.10	**0.12**	0.11	**0.13**	0.15	**0.17**	0.18	**0.21**
Kidneys	0.03	**0.06**	0.04	**0.08**	0.05	**0.10**	0.06	**0.11**	0.08	**0.14**	0.10	**0.18**
Urinary bladder						**0.01**		**0.01**		**0.02**		**0.02**
Uterus		**0.02**		**0.03**		**0.03**		**0.04**		**0.05**		**0.06**
Ovaries		**0.02**		**0.03**		**0.03**		**0.04**		**0.05**		**0.06**
Prostate												**0.01**
Testes												
Upper arm	0.25	**0.67**	0.33	**0.85**	0.39	**0.99**	0.44	**1.09**	0.53	**1.29**	0.62	**1.48**
Elbow	0.22	**0.73**	0.30	**0.92**	0.35	**1.06**	0.39	**1.18**	0.48	**1.39**	0.56	**1.59**
Forearm	0.02	**0.39**	0.03	**0.49**	0.04	**0.56**	0.04	**0.62**	0.05	**0.73**	0.06	**0.83**
Upper leg												
Knee												
Lower leg												
Lymph nodes	0.14	**0.27**	0.18	**0.34**	0.20	**0.39**	0.22	**0.43**	0.26	**0.50**	0.29	**0.57**
Skin	0.15	**0.28**	0.16	**0.29**	0.17	**0.30**	0.17	**0.31**	0.18	**0.33**	0.19	**0.34**
Muscle	0.09	**0.15**	0.11	**0.18**	0.13	**0.21**	0.14	**0.22**	0.16	**0.26**	0.17	**0.29**
Skeleton	0.22	**0.35**	0.27	**0.44**	0.31	**0.50**	0.34	**0.54**	0.39	**0.63**	0.44	**0.71**
Red bone marrow	0.05	**0.09**	0.06	**0.11**	0.07	**0.12**	0.08	**0.13**	0.09	**0.16**	0.11	**0.18**
Whole body dose	0.11	**0.18**	0.13	**0.22**	0.15	**0.24**	0.16	**0.26**	0.19	**0.30**	0.21	**0.34**
ED (ICRP 60)	0.20	**0.25**	0.23	**0.30**	0.26	**0.34**	0.28	**0.36**	0.31	**0.41**	0.34	**0.46**
ED (ICRP 103)	0.27	**0.34**	0.31	**0.40**	0.34	**0.44**	0.36	**0.47**	0.40	**0.52**	0.44	**0.57**
Total filtration (mm) **Al**	**1.5**		**2.5**		**3.5**		**4.5**		**3.5**		**3.5**	
Cu	**0**		**0**		**0**		**0**		**0.1**		**0.2**	

Table 10.30 Normalized organ doses—Bony thorax ap—1–2 years

Beam projection:			ap			**Table cells:**						
Focus-to-detector distance (cm):			115			(+) = normal font = optimal field setting						
Tube voltage (kVp):			60			**(−) = bold font = suboptimal field setting**						
Total filtrations:			See last table row			Empty table cells correspond to values <0.01						
Organ or tissue	**Organ dose D_T: Entrance dose K_E**							**(mSv * mGy^{-1})**				
	(+)	(−)	(+)	(−)	(+)	(−)	(+)	(−)	(+)	(−)	(+)	(−)
Skull	0.02	**0.24**	0.02	**0.29**	0.03	**0.33**	0.03	**0.35**	0.04	**0.40**	0.03	**0.39**
Brain		**0.02**		**0.02**		**0.03**		**0.03**		**0.04**		**0.03**
Oral mucosa	0.02	**0.47**	0.03	**0.56**	0.04	**0.62**	0.04	**0.66**	0.05	**0.75**	0.05	**0.73**
Salivary glands	0.05	**0.38**	0.06	**0.46**	0.07	**0.52**	0.08	**0.57**	0.09	**0.65**	0.09	**0.63**
Cervical spine	0.30	**0.45**	0.39	**0.58**	0.45	**0.69**	0.51	**0.78**	0.61	**0.95**	0.58	**0.88**
Thyroid	0.87	**0.93**	0.99	**1.06**	1.06	**1.14**	1.11	**1.19**	1.19	**1.29**	1.18	**1.27**
Thymus	0.73	**0.75**	0.87	**0.89**	0.96	**0.99**	1.03	**1.06**	1.16	**1.19**	1.14	**1.17**
Thoracic spine	0.20	**0.22**	0.27	**0.29**	0.32	**0.36**	0.37	**0.41**	0.47	**0.52**	0.41	**0.45**
Claviculae	1.79	**1.88**	2.13	**2.24**	2.36	**2.49**	2.53	**2.67**	2.82	**2.98**	2.84	**2.99**
Scapulae	0.29	**0.30**	0.38	**0.40**	0.45	**0.47**	0.51	**0.54**	0.63	**0.66**	0.58	**0.61**
Ribs	1.39	**1.44**	1.62	**1.68**	1.77	**1.84**	1.88	**1.96**	2.07	**2.17**	2.06	**2.15**
Oesophagus	0.20	**0.22**	0.26	**0.29**	0.31	**0.34**	0.35	**0.39**	0.43	**0.48**	0.40	**0.44**
Trachea	0.08	**0.44**	0.10	**0.52**	0.11	**0.58**	0.11	**0.62**	0.13	**0.70**	0.12	**0.69**
Lung	0.37	**0.39**	0.46	**0.48**	0.52	**0.54**	0.57	**0.60**	0.66	**0.69**	0.64	**0.67**
Mamma	1.14	**1.16**	1.22	**1.23**	1.27	**1.28**	1.30	**1.31**	1.34	**1.36**	1.34	**1.35**
Heart	0.50	**0.52**	0.62	**0.64**	0.70	**0.72**	0.76	**0.79**	0.88	**0.91**	0.85	**0.88**
Stomach	0.29	**0.50**	0.36	**0.62**	0.41	**0.70**	0.45	**0.76**	0.52	**0.87**	0.50	**0.85**
Spleen	0.13	**0.17**	0.18	**0.22**	0.21	**0.26**	0.24	**0.30**	0.29	**0.37**	0.27	**0.34**
Liver	0.28	**0.43**	0.35	**0.53**	0.40	**0.60**	0.44	**0.66**	0.51	**0.77**	0.49	**0.74**
Gall bladder	0.11	**0.38**	0.14	**0.49**	0.17	**0.56**	0.19	**0.63**	0.22	**0.74**	0.21	**0.71**
Pancreas	0.21	**0.25**	0.27	**0.32**	0.32	**0.38**	0.36	**0.43**	0.43	**0.52**	0.40	**0.48**
Small intestine	0.01	**0.17**	0.02	**0.21**	0.02	**0.25**	0.03	**0.28**	0.03	**0.33**	0.03	**0.31**
Large intestine	0.02	**0.25**	0.02	**0.32**	0.03	**0.37**	0.03	**0.41**	0.04	**0.48**	0.04	**0.46**
Sigma/rectum		**0.05**		**0.07**		**0.08**		**0.09**	0.01	**0.10**		**0.10**
Lumbar spine	0.06	**0.19**	0.08	**0.26**	0.10	**0.31**	0.12	**0.36**	0.15	**0.46**	0.13	**0.39**
Pelvis		**0.07**	0.01	**0.09**	0.01	**0.11**	0.02	**0.13**	0.02	**0.16**	0.02	**0.14**
Adrenals	0.09	**0.10**	0.12	**0.14**	0.14	**0.17**	0.16	**0.19**	0.20	**0.24**	0.18	**0.21**
Kidneys	0.05	**0.09**	0.06	**0.12**	0.08	**0.14**	0.09	**0.16**	0.11	**0.20**	0.10	**0.18**
Urinary bladder		**0.01**		**0.02**		**0.02**		**0.02**		**0.03**		**0.02**
Uterus		**0.03**		**0.04**		**0.05**		**0.06**	0.01	**0.07**		**0.06**
Ovaries		**0.03**		**0.04**		**0.05**		**0.06**	0.01	**0.07**		**0.06**
Prostate										**0.01**		**0.01**
Testes												
Upper arm	0.33	**0.83**	0.42	**1.04**	0.49	**1.19**	0.55	**1.31**	0.66	**1.53**	0.62	**1.48**
Elbow	0.30	**0.90**	0.38	**1.11**	0.44	**1.27**	0.50	**1.40**	0.60	**1.63**	0.56	**1.59**
Forearm	0.03	**0.47**	0.04	**0.59**	0.05	**0.67**	0.06	**0.73**	0.07	**0.85**	0.06	**0.83**
Upper leg												
Knee												
Lower leg												
Lymph nodes	0.17	**0.33**	0.21	**0.41**	0.24	**0.47**	0.26	**0.51**	0.30	**0.59**	0.29	**0.57**
Skin	0.16	**0.29**	0.17	**0.31**	0.18	**0.32**	0.18	**0.33**	0.19	**0.35**	0.19	**0.34**
Muscle	0.11	**0.18**	0.13	**0.22**	0.15	**0.24**	0.16	**0.26**	0.18	**0.30**	0.17	**0.29**
Skeleton	0.27	**0.43**	0.32	**0.52**	0.36	**0.59**	0.40	**0.64**	0.45	**0.74**	0.44	**0.71**
Red bone marrow	0.06	**0.11**	0.08	**0.13**	0.09	**0.15**	0.10	**0.16**	0.11	**0.19**	0.11	**0.18**
Whole body dose	0.13	**0.21**	0.16	**0.25**	0.18	**0.29**	0.19	**0.31**	0.22	**0.35**	0.21	**0.34**
ED (ICRP 60)	0.23	**0.29**	0.26	**0.35**	0.29	**0.39**	0.31	**0.42**	0.35	**0.48**	0.34	**0.46**
ED (ICRP 103)	0.30	**0.39**	0.35	**0.45**	0.38	**0.50**	0.40	**0.53**	0.45	**0.59**	0.44	**0.57**
Total filtration (mm) Al	**1.5**		**2.5**		**3.5**		**4.5**		**3.5**		**3.5**	
Cu	**0**		**0**		**0**		**0**		**0.1**		**0.2**	

Table 10.31 Normalized organ doses—Bony thorax ap—1–2 years

Beam projection:				ap				**Table cells:**				
Focus-to-detector distance (cm):				115				(+) = normal font = optimal field setting				
Tube voltage (kVp):				70				**(−) = bold font = suboptimal field setting**				
Total filtrations:				See last table row				Empty table cells correspond to values <0.01				
Organ or tissue	**Organ dose D_T: Entrance dose K_E**							**(mSv * mGy^{-1})**				
	(+)	(−)	(+)	(−)	(+)	(−)	(+)	(−)	(+)	(−)	(+)	(−)
Skull	0.02	**0.27**	0.03	**0.32**	0.03	**0.36**	0.04	**0.39**	0.05	**0.43**	0.05	**0.48**
Brain		**0.02**		**0.03**		**0.03**		**0.04**		**0.05**	0.01	**0.06**
Oral mucosa	0.03	**0.53**	0.04	**0.61**	0.05	**0.68**	0.05	**0.72**	0.06	**0.81**	0.07	**0.88**
Salivary glands	0.06	**0.43**	0.07	**0.52**	0.08	**0.58**	0.09	**0.62**	0.10	**0.70**	0.12	**0.77**
Cervical spine	0.37	**0.55**	0.46	**0.70**	0.53	**0.82**	0.59	**0.91**	0.70	**1.10**	0.80	**1.26**
Thyroid	0.93	**1.00**	1.04	**1.12**	1.11	**1.20**	1.16	**1.26**	1.24	**1.36**	1.30	**1.44**
Thymus	0.81	**0.83**	0.95	**0.98**	1.04	**1.08**	1.11	**1.15**	1.24	**1.28**	1.33	**1.39**
Thoracic spine	0.27	**0.29**	0.35	**0.38**	0.41	**0.45**	0.47	**0.52**	0.58	**0.64**	0.70	**0.77**
Claviculae	1.96	**2.06**	2.29	**2.41**	2.51	**2.65**	2.67	**2.82**	2.93	**3.11**	3.12	**3.33**
Scapulae	0.36	**0.38**	0.46	**0.49**	0.54	**0.57**	0.60	**0.64**	0.73	**0.77**	0.84	**0.89**
Ribs	1.50	**1.57**	1.72	**1.80**	1.87	**1.96**	1.97	**2.07**	2.15	**2.27**	2.28	**2.42**
Oesophagus	0.25	**0.28**	0.33	**0.36**	0.38	**0.42**	0.42	**0.47**	0.51	**0.56**	0.60	**0.66**
Trachea	0.09	**0.49**	0.11	**0.57**	0.12	**0.63**	0.13	**0.68**	0.14	**0.75**	0.15	**0.82**
Lung	0.43	**0.45**	0.52	**0.54**	0.59	**0.61**	0.64	**0.67**	0.73	**0.77**	0.81	**0.85**
Mamma	1.18	**1.20**	1.26	**1.27**	1.30	**1.32**	1.33	**1.35**	1.37	**1.40**	1.41	**1.45**
Heart	0.58	**0.60**	0.70	**0.72**	0.78	**0.81**	0.85	**0.88**	0.97	**1.00**	1.07	**1.11**
Stomach	0.34	**0.58**	0.41	**0.69**	0.46	**0.78**	0.50	**0.84**	0.57	**0.96**	0.63	**1.06**
Spleen	0.17	**0.21**	0.22	**0.27**	0.25	**0.32**	0.28	**0.36**	0.34	**0.43**	0.40	**0.50**
Liver	0.33	**0.50**	0.40	**0.60**	0.45	**0.68**	0.49	**0.74**	0.57	**0.85**	0.63	**0.94**
Gall bladder	0.13	**0.46**	0.17	**0.57**	0.20	**0.65**	0.22	**0.71**	0.26	**0.83**	0.30	**0.94**
Pancreas	0.26	**0.31**	0.32	**0.39**	0.38	**0.46**	0.42	**0.51**	0.50	**0.61**	0.58	**0.71**
Small intestine	0.02	**0.20**	0.02	**0.25**	0.03	**0.29**	0.03	**0.32**	0.04	**0.38**	0.05	**0.44**
Large intestine	0.02	**0.30**	0.03	**0.37**	0.04	**0.42**	0.04	**0.46**	0.05	**0.54**	0.07	**0.61**
Sigma/rectum		**0.06**		**0.08**		**0.09**	0.01	**0.10**	0.01	**0.12**	0.02	**0.14**
Lumbar spine	0.09	**0.26**	0.11	**0.34**	0.14	**0.40**	0.16	**0.46**	0.20	**0.57**	0.24	**0.69**
Pelvis	0.01	**0.09**	0.02	**0.12**	0.02	**0.14**	0.02	**0.16**	0.03	**0.20**	0.04	**0.24**
Adrenals	0.12	**0.14**	0.15	**0.18**	0.18	**0.21**	0.20	**0.24**	0.25	**0.29**	0.29	**0.35**
Kidneys	0.06	**0.12**	0.08	**0.15**	0.10	**0.18**	0.11	**0.20**	0.14	**0.25**	0.16	**0.30**
Urinary bladder		**0.02**		**0.02**		**0.02**		**0.03**		**0.04**		**0.04**
Uterus		**0.04**		**0.05**	0.01	**0.06**	0.01	**0.07**	0.02	**0.09**	0.02	**0.11**
Ovaries		**0.04**		**0.05**		**0.06**	0.01	**0.07**	0.02	**0.09**	0.02	**0.11**
Prostate						**0.01**		**0.01**		**0.02**		**0.02**
Testes												**0.01**
Upper arm	0.40	**0.96**	0.49	**1.18**	0.57	**1.33**	0.63	**1.45**	0.74	**1.67**	0.85	**1.85**
Elbow	0.36	**1.03**	0.45	**1.26**	0.51	**1.42**	0.57	**1.55**	0.67	**1.78**	0.76	**1.97**
Forearm	0.04	**0.54**	0.05	**0.66**	0.06	**0.74**	0.07	**0.81**	0.09	**0.92**	0.10	**1.02**
Upper leg										**0.01**		**0.01**
Knee												
Lower leg												
Lymph nodes	0.20	**0.39**	0.24	**0.47**	0.27	**0.53**	0.30	**0.57**	0.34	**0.66**	0.38	**0.73**
Skin	0.17	**0.30**	0.18	**0.32**	0.19	**0.34**	0.19	**0.35**	0.20	**0.37**	0.22	**0.39**
Muscle	0.12	**0.21**	0.15	**0.24**	0.16	**0.27**	0.18	**0.29**	0.20	**0.33**	0.22	**0.36**
Skeleton	0.30	**0.49**	0.36	**0.58**	0.40	**0.65**	0.44	**0.71**	0.49	**0.81**	0.55	**0.89**
Red bone marrow	0.08	**0.13**	0.09	**0.15**	0.10	**0.17**	0.11	**0.19**	0.13	**0.22**	0.15	**0.25**
Whole body dose	0.15	**0.24**	0.18	**0.29**	0.20	**0.32**	0.21	**0.34**	0.24	**0.39**	0.27	**0.43**
ED (ICRP 60)	0.25	**0.33**	0.29	**0.39**	0.32	**0.43**	0.34	**0.47**	0.38	**0.52**	0.42	**0.58**
ED (ICRP 103)	0.33	**0.43**	0.38	**0.50**	0.41	**0.54**	0.44	**0.58**	0.48	**0.64**	0.52	**0.70**
Total filtration (mm) Al	**1.5**		**2.5**		**3.5**		**4.5**		**3.5**		**3.5**	
Cu	**0**		**0**		**0**		**0**		**0.1**		**0.2**	

Table 10.32 Normalized organ doses—Bony thorax ap—3–7 years

Beam projection:	ap	**Table cells:**
Focus-to-detector distance (cm):	115	(+) = normal font = optimal field setting
Tube voltage (kVp):	60	**(−) = bold font = suboptimal field setting**
Total filtrations:	See last table row	Empty table cells correspond to values <0.01

Organ or tissue	**Organ dose D_T: Entrance dose K_E**						**(mSv * mGy^{-1})**					
	(+)	(−)	(+)	(−)	(+)	(−)	(+)	(−)	(+)	(−)	(+)	(−)
Skull	0.02	**0.25**	0.02	**0.31**	0.03	**0.35**	0.03	**0.38**	0.04	**0.45**	0.03	**0.43**
Brain		**0.01**		**0.01**		**0.02**		**0.02**		**0.03**		**0.02**
Oral mucosa	0.02	**0.37**	0.03	**0.44**	0.03	**0.49**	0.04	**0.54**	0.05	**0.61**	0.04	**0.59**
Salivary glands	0.04	**0.30**	0.05	**0.37**	0.06	**0.42**	0.07	**0.46**	0.08	**0.53**	0.07	**0.51**
Cervical spine	0.29	**0.38**	0.37	**0.50**	0.44	**0.60**	0.49	**0.67**	0.60	**0.83**	0.56	**0.75**
Thyroid	0.79	**0.83**	0.90	**0.95**	0.97	**1.03**	1.03	**1.09**	1.12	**1.20**	1.10	**1.17**
Thymus	0.69	**0.71**	0.84	**0.85**	0.94	**0.96**	1.01	**1.03**	1.15	**1.17**	1.12	**1.15**
Thoracic spine	0.15	**0.16**	0.20	**0.22**	0.25	**0.27**	0.28	**0.31**	0.37	**0.40**	0.31	**0.33**
Claviculae	1.62	**1.68**	1.95	**2.03**	2.18	**2.27**	2.35	**2.46**	2.66	**2.78**	2.65	**2.76**
Scapulae	0.22	**0.24**	0.30	**0.31**	0.35	**0.38**	0.40	**0.43**	0.50	**0.54**	0.45	**0.48**
Ribs	1.30	**1.36**	1.52	**1.59**	1.67	**1.75**	1.78	**1.87**	1.98	**2.08**	1.96	**2.06**
Oesophagus	0.12	**0.13**	0.16	**0.18**	0.19	**0.21**	0.22	**0.24**	0.27	**0.30**	0.24	**0.27**
Trachea	0.10	**0.30**	0.11	**0.37**	0.13	**0.43**	0.14	**0.47**	0.15	**0.54**	0.15	**0.52**
Lung	0.33	**0.35**	0.42	**0.43**	0.48	**0.50**	0.53	**0.55**	0.62	**0.65**	0.59	**0.62**
Mamma	1.13	**1.16**	1.21	**1.25**	1.27	**1.30**	1.30	**1.33**	1.36	**1.39**	1.35	**1.38**
Heart	0.44	**0.46**	0.55	**0.57**	0.63	**0.65**	0.70	**0.72**	0.82	**0.84**	0.78	**0.80**
Stomach	0.27	**0.49**	0.33	**0.61**	0.38	**0.69**	0.42	**0.76**	0.49	**0.88**	0.47	**0.85**
Spleen	0.10	**0.14**	0.14	**0.18**	0.16	**0.22**	0.19	**0.25**	0.23	**0.31**	0.21	**0.27**
Liver	0.24	**0.39**	0.30	**0.48**	0.35	**0.56**	0.39	**0.61**	0.46	**0.72**	0.43	**0.68**
Gall bladder	0.09	**0.34**	0.12	**0.44**	0.14	**0.52**	0.16	**0.58**	0.20	**0.70**	0.18	**0.65**
Pancreas	0.17	**0.21**	0.22	**0.28**	0.27	**0.33**	0.30	**0.37**	0.37	**0.46**	0.34	**0.42**
Small intestine		**0.18**	0.01	**0.23**	0.01	**0.27**	0.02	**0.30**	0.02	**0.37**	0.02	**0.34**
Large intestine	0.01	**0.25**	0.02	**0.31**	0.02	**0.37**	0.02	**0.41**	0.03	**0.49**	0.02	**0.46**
Sigma/rectum		**0.06**		**0.08**		**0.09**		**0.10**		**0.12**		**0.11**
Lumbar spine	0.04	**0.14**	0.05	**0.19**	0.07	**0.24**	0.08	**0.27**	0.10	**0.35**	0.08	**0.29**
Pelvis		**0.07**		**0.10**		**0.12**		**0.13**	0.01	**0.17**		**0.14**
Adrenals	0.07	**0.08**	0.09	**0.11**	0.11	**0.13**	0.13	**0.15**	0.17	**0.20**	0.14	**0.17**
Kidneys	0.03	**0.07**	0.05	**0.09**	0.06	**0.11**	0.06	**0.13**	0.08	**0.16**	0.07	**0.14**
Urinary bladder				**0.01**		**0.02**		**0.02**		**0.02**		**0.02**
Uterus		**0.03**		**0.04**		**0.05**		**0.06**		**0.07**		**0.06**
Ovaries		**0.03**		**0.04**		**0.05**		**0.06**		**0.08**		**0.07**
Prostate												
Testes												
Upper arm	0.22	**0.69**	0.28	**0.86**	0.33	**1.00**	0.38	**1.10**	0.46	**1.30**	0.42	**1.24**
Elbow	0.19	**0.74**	0.24	**0.93**	0.29	**1.07**	0.33	**1.18**	0.40	**1.39**	0.37	**1.34**
Forearm	0.02	**0.42**	0.02	**0.53**	0.03	**0.61**	0.03	**0.68**	0.04	**0.80**	0.04	**0.77**
Upper leg												
Knee												
Lower leg												
Lymph nodes	0.15	**0.29**	0.19	**0.36**	0.21	**0.41**	0.23	**0.46**	0.27	**0.54**	0.26	**0.51**
Skin	0.13	**0.25**	0.14	**0.26**	0.15	**0.27**	0.15	**0.28**	0.16	**0.29**	0.16	**0.29**
Muscle	0.09	**0.15**	0.11	**0.18**	0.13	**0.21**	0.14	**0.23**	0.16	**0.26**	0.15	**0.25**
Skeleton	0.19	**0.33**	0.23	**0.41**	0.26	**0.46**	0.28	**0.51**	0.33	**0.59**	0.31	**0.56**
Red bone marrow	0.06	**0.09**	0.07	**0.12**	0.08	**0.13**	0.09	**0.15**	0.10	**0.17**	0.10	**0.16**
Whole body dose	0.11	**0.18**	0.13	**0.22**	0.15	**0.25**	0.16	**0.27**	0.19	**0.32**	0.18	**0.30**
ED (ICRP 60)	0.20	**0.27**	0.24	**0.33**	0.27	**0.37**	0.29	**0.40**	0.32	**0.46**	0.31	**0.44**
ED (ICRP 103)	0.28	**0.36**	0.33	**0.43**	0.36	**0.47**	0.38	**0.51**	0.42	**0.57**	0.41	**0.55**
Total filtration (mm) **Al**	**1.5**		**2.5**		**3.5**		**4.5**		**3.5**		**3.5**	
Cu	**0**		**0**		**0**		**0**		**0.1**		**0.2**	

Table 10.33 Normalized organ doses—Bony thorax ap—3–7 years

Beam projection:			ap			**Table cells:**						
Focus-to-detector distance (cm):			115			(+) = normal font = optimal field setting						
Tube voltage (kVp):			70			**(−) = bold font = suboptimal field setting**						
Total filtrations:			See last table row			Empty table cells correspond to values <0.01						
Organ or tissue	**Organ dose D_T: Entrance dose K_E**							**(mSv * mGy^{-1})**				
	(+)	(−)	(+)	(−)	(+)	(−)	(+)	(−)	(+)	(−)	(+)	(−)
Skull	0.02	**0.29**	0.03	**0.35**	0.03	**0.39**	0.04	**0.43**	0.05	**0.49**	0.06	**0.54**
Brain		**0.02**		**0.02**		**0.02**		**0.03**		**0.04**		**0.04**
Oral mucosa	0.03	**0.42**	0.04	**0.50**	0.04	**0.55**	0.05	**0.60**	0.06	**0.68**	0.07	**0.75**
Salivary glands	0.05	**0.35**	0.06	**0.42**	0.07	**0.47**	0.08	**0.52**	0.09	**0.59**	0.10	**0.66**
Cervical spine	0.35	**0.48**	0.45	**0.62**	0.52	**0.72**	0.58	**0.81**	0.70	**0.98**	0.80	**1.15**
Thyroid	0.85	**0.90**	0.96	**1.03**	1.04	**1.11**	1.09	**1.17**	1.18	**1.27**	1.25	**1.36**
Thymus	0.78	**0.80**	0.93	**0.95**	1.03	**1.05**	1.11	**1.13**	1.24	**1.27**	1.36	**1.39**
Thoracic spine	0.21	**0.23**	0.27	**0.30**	0.33	**0.36**	0.37	**0.41**	0.47	**0.52**	0.57	**0.63**
Claviculae	1.79	**1.87**	2.12	**2.22**	2.35	**2.46**	2.51	**2.64**	2.79	**2.94**	3.01	**3.18**
Scapulae	0.29	**0.30**	0.37	**0.39**	0.44	**0.46**	0.49	**0.52**	0.60	**0.64**	0.71	**0.75**
Ribs	1.42	**1.48**	1.64	**1.72**	1.78	**1.88**	1.89	**1.99**	2.07	**2.20**	2.22	**2.36**
Oesophagus	0.16	**0.17**	0.20	**0.22**	0.24	**0.26**	0.27	**0.30**	0.33	**0.37**	0.39	**0.43**
Trachea	0.11	**0.35**	0.13	**0.43**	0.14	**0.48**	0.15	**0.52**	0.17	**0.60**	0.18	**0.67**
Lung	0.39	**0.41**	0.48	**0.50**	0.55	**0.57**	0.60	**0.63**	0.69	**0.73**	0.77	**0.82**
Mamma	1.17	**1.20**	1.26	**1.29**	1.31	**1.34**	1.34	**1.37**	1.40	**1.43**	1.44	**1.47**
Heart	0.52	**0.54**	0.64	**0.66**	0.72	**0.75**	0.79	**0.82**	0.91	**0.95**	1.02	**1.07**
Stomach	0.31	**0.57**	0.38	**0.69**	0.43	**0.78**	0.47	**0.86**	0.55	**0.99**	0.62	**1.10**
Spleen	0.13	**0.18**	0.17	**0.23**	0.20	**0.27**	0.23	**0.31**	0.28	**0.38**	0.33	**0.44**
Liver	0.29	**0.46**	0.35	**0.56**	0.40	**0.64**	0.44	**0.70**	0.52	**0.81**	0.58	**0.91**
Gall bladder	0.12	**0.42**	0.15	**0.53**	0.17	**0.61**	0.19	**0.68**	0.24	**0.80**	0.27	**0.92**
Pancreas	0.22	**0.27**	0.28	**0.35**	0.33	**0.41**	0.36	**0.46**	0.44	**0.56**	0.52	**0.65**
Small intestine	0.01	**0.22**	0.02	**0.28**	0.02	**0.32**	0.02	**0.36**	0.03	**0.43**	0.04	**0.49**
Large intestine	0.02	**0.30**	0.02	**0.37**	0.03	**0.43**	0.03	**0.47**	0.04	**0.56**	0.05	**0.64**
Sigma/rectum		**0.07**		**0.09**		**0.11**		**0.12**		**0.14**		**0.17**
Lumbar spine	0.06	**0.20**	0.08	**0.26**	0.09	**0.32**	0.11	**0.36**	0.14	**0.46**	0.17	**0.56**
Pelvis		**0.10**		**0.13**	0.01	**0.15**	0.01	**0.17**	0.02	**0.21**	0.02	**0.25**
Adrenals	0.09	**0.11**	0.12	**0.15**	0.15	**0.18**	0.17	**0.20**	0.21	**0.25**	0.25	**0.31**
Kidneys	0.05	**0.09**	0.06	**0.12**	0.07	**0.15**	0.08	**0.17**	0.11	**0.21**	0.13	**0.25**
Urinary bladder		**0.01**		**0.02**		**0.02**		**0.02**		**0.03**		**0.04**
Uterus		**0.04**		**0.05**		**0.06**		**0.07**		**0.09**	0.01	**0.11**
Ovaries		**0.04**		**0.06**		**0.07**		**0.08**	0.01	**0.10**	0.01	**0.12**
Prostate										**0.01**		**0.02**
Testes												
Upper arm	0.27	**0.81**	0.34	**1.00**	0.40	**1.13**	0.45	**1.24**	0.54	**1.45**	0.62	**1.63**
Elbow	0.23	**0.87**	0.30	**1.07**	0.35	**1.21**	0.39	**1.33**	0.47	**1.55**	0.55	**1.74**
Forearm	0.02	**0.50**	0.03	**0.61**	0.04	**0.69**	0.04	**0.76**	0.05	**0.88**	0.06	**0.98**
Upper leg												
Knee												
Lower leg												
Lymph nodes	0.18	**0.34**	0.22	**0.42**	0.24	**0.48**	0.27	**0.52**	0.31	**0.61**	0.35	**0.69**
Skin	0.14	**0.26**	0.15	**0.27**	0.16	**0.28**	0.16	**0.29**	0.17	**0.31**	0.18	**0.32**
Muscle	0.11	**0.18**	0.13	**0.21**	0.14	**0.23**	0.15	**0.25**	0.18	**0.29**	0.20	**0.32**
Skeleton	0.22	**0.38**	0.26	**0.46**	0.29	**0.52**	0.32	**0.57**	0.36	**0.65**	0.40	**0.73**
Red bone marrow	0.07	**0.11**	0.08	**0.14**	0.09	**0.16**	0.10	**0.17**	0.12	**0.20**	0.14	**0.23**
Whole body dose	0.13	**0.21**	0.15	**0.25**	0.17	**0.28**	0.18	**0.31**	0.21	**0.35**	0.23	**0.40**
ED (ICRP 60)	0.23	**0.31**	0.27	**0.37**	0.29	**0.41**	0.32	**0.44**	0.36	**0.51**	0.39	**0.56**
ED (ICRP 103)	0.31	**0.41**	0.36	**0.47**	0.39	**0.52**	0.41	**0.56**	0.46	**0.63**	0.50	**0.69**
Total filtration (mm) **Al**	**1.5**		**2.5**		**3.5**		**4.5**		**3.5**		**3.5**	
Cu	**0**		**0**		**0**		**0**		**0.1**		**0.2**	

Table 10.34 Normalized organ doses—Bony thorax ap—8–12 years

Beam projection:			ap			**Table cells:**						
Focus-to-detector distance (cm):			115			(+) = normal font = optimal field setting						
Tube voltage (kVp):			60			**(−) = bold font = suboptimal field setting**						
Total filtrations:			See last table row			Empty table cells correspond to values <0.01						
Organ or tissue	**Organ dose D_T: Entrance dose K_E**							**(mSv * mGy^{-1})**				
	(+)	(−)	(+)	(−)	(+)	(−)	(+)	(−)	(+)	(−)	(+)	(−)
Skull	0.01	**0.25**	0.02	**0.31**	0.02	**0.35**	0.02	**0.38**	0.03	**0.45**	0.03	**0.43**
Brain				**0.01**		**0.02**		**0.02**		**0.02**		**0.02**
Oral mucosa	0.01	**0.34**	0.02	**0.41**	0.02	**0.46**	0.03	**0.49**	0.03	**0.57**	0.03	**0.54**
Salivary glands	0.03	**0.28**	0.04	**0.35**	0.05	**0.40**	0.05	**0.43**	0.06	**0.50**	0.06	**0.48**
Cervical spine	0.27	**0.38**	0.35	**0.49**	0.41	**0.59**	0.46	**0.67**	0.57	**0.83**	0.52	**0.74**
Thyroid	0.72	**0.76**	0.83	**0.89**	0.91	**0.97**	0.96	**1.03**	1.06	**1.14**	1.04	**1.11**
Thymus	0.66	**0.67**	0.80	**0.82**	0.90	**0.92**	0.98	**1.01**	1.13	**1.15**	1.09	**1.12**
Thoracic spine	0.11	**0.12**	0.15	**0.17**	0.19	**0.20**	0.22	**0.24**	0.29	**0.31**	0.23	**0.25**
Claviculae	1.47	**1.53**	1.79	**1.86**	2.01	**2.10**	2.19	**2.29**	2.49	**2.61**	2.46	**2.58**
Scapulae	0.17	**0.18**	0.23	**0.24**	0.27	**0.29**	0.31	**0.33**	0.40	**0.42**	0.35	**0.37**
Ribs	1.24	**1.28**	1.45	**1.51**	1.60	**1.67**	1.71	**1.78**	1.91	**1.99**	1.88	**1.96**
Oesophagus	0.08	**0.09**	0.11	**0.12**	0.14	**0.14**	0.16	**0.17**	0.20	**0.21**	0.17	**0.18**
Trachea	0.20	**0.42**	0.24	**0.52**	0.26	**0.60**	0.28	**0.65**	0.32	**0.76**	0.31	**0.73**
Lung	0.31	**0.32**	0.39	**0.40**	0.45	**0.46**	0.49	**0.51**	0.58	**0.61**	0.55	**0.57**
Mamma	1.11	**1.13**	1.21	**1.23**	1.26	**1.29**	1.30	**1.32**	1.36	**1.38**	1.34	**1.37**
Heart	0.40	**0.40**	0.50	**0.51**	0.58	**0.59**	0.64	**0.66**	0.76	**0.78**	0.71	**0.73**
Stomach	0.28	**0.44**	0.35	**0.55**	0.41	**0.64**	0.45	**0.70**	0.53	**0.82**	0.50	**0.78**
Spleen	0.09	**0.11**	0.12	**0.15**	0.15	**0.18**	0.17	**0.20**	0.21	**0.26**	0.18	**0.22**
Liver	0.24	**0.33**	0.30	**0.41**	0.35	**0.48**	0.39	**0.53**	0.46	**0.64**	0.43	**0.60**
Gall bladder	0.10	**0.30**	0.14	**0.39**	0.16	**0.45**	0.18	**0.51**	0.23	**0.62**	0.20	**0.57**
Pancreas	0.15	**0.18**	0.20	**0.23**	0.24	**0.28**	0.28	**0.32**	0.34	**0.40**	0.31	**0.36**
Small intestine		**0.10**		**0.13**	0.01	**0.15**	0.01	**0.18**	0.02	**0.22**	0.01	**0.19**
Large intestine	0.01	**0.17**	0.01	**0.23**	0.02	**0.26**	0.02	**0.30**	0.03	**0.36**	0.02	**0.33**
Sigma/rectum		**0.03**		**0.04**		**0.04**		**0.05**		**0.06**		**0.05**
Lumbar spine	0.04	**0.10**	0.05	**0.13**	0.06	**0.16**	0.07	**0.19**	0.09	**0.25**	0.07	**0.20**
Pelvis		**0.03**		**0.04**		**0.05**		**0.06**		**0.07**		**0.06**
Adrenals	0.06	**0.06**	0.08	**0.09**	0.10	**0.11**	0.11	**0.12**	0.15	**0.16**	0.12	**0.13**
Kidneys	0.03	**0.05**	0.04	**0.07**	0.05	**0.09**	0.06	**0.10**	0.08	**0.13**	0.06	**0.11**
Urinary bladder												
Uterus		**0.01**		**0.02**		**0.02**		**0.02**		**0.03**		**0.02**
Ovaries		**0.02**		**0.02**		**0.03**		**0.03**		**0.04**		**0.03**
Prostate												
Testes												
Upper arm	0.20	**0.60**	0.26	**0.76**	0.31	**0.88**	0.35	**0.98**	0.43	**1.17**	0.39	**1.10**
Elbow	0.18	**0.65**	0.24	**0.82**	0.28	**0.95**	0.32	**1.06**	0.40	**1.26**	0.36	**1.19**
Forearm	0.02	**0.31**	0.03	**0.40**	0.03	**0.46**	0.04	**0.51**	0.05	**0.61**	0.04	**0.58**
Upper leg												
Knee												
Lower leg												
Lymph nodes	0.15	**0.26**	0.19	**0.33**	0.22	**0.38**	0.24	**0.42**	0.28	**0.49**	0.26	**0.46**
Skin	0.13	**0.23**	0.13	**0.24**	0.14	**0.25**	0.14	**0.26**	0.15	**0.27**	0.15	**0.26**
Muscle	0.09	**0.13**	0.11	**0.16**	0.12	**0.18**	0.13	**0.19**	0.15	**0.22**	0.14	**0.21**
Skeleton	0.17	**0.28**	0.21	**0.35**	0.24	**0.40**	0.26	**0.44**	0.30	**0.51**	0.28	**0.48**
Red bone marrow	0.07	**0.09**	0.08	**0.11**	0.09	**0.13**	0.10	**0.14**	0.12	**0.16**	0.11	**0.15**
Whole body dose	0.10	**0.16**	0.13	**0.19**	0.14	**0.22**	0.16	**0.24**	0.18	**0.28**	0.17	**0.26**
ED (ICRP 60)	0.19	**0.24**	0.23	**0.29**	0.26	**0.33**	0.28	**0.36**	0.32	**0.41**	0.30	**0.39**
ED (ICRP 103)	0.27	**0.33**	0.32	**0.39**	0.35	**0.43**	0.37	**0.47**	0.42	**0.53**	0.40	**0.51**
Total filtration (mm) **Al**	**1.5**		**2.5**		**3.5**		**4.5**		**3.5**		**3.5**	
Cu	**0**		**0**		**0**		**0**		**0.1**		**0.2**	

Table 10.35 Normalized organ doses—Bony thorax ap—8–12 years

Beam projection:			ap				**Table cells:**					
Focus-to-detector distance (cm):			115				(+) = normal font = optimal field setting					
Tube voltage (kVp):			70				**(−) = bold font = suboptimal field setting**					
Total filtrations:			See last table row				Empty table cells correspond to values <0.01					
Organ or tissue	**Organ dose D_T: Entrance dose K_E**							**(mSv * mGy^{-1})**				
	(+)	(−)	(+)	(−)	(+)	(−)	(+)	(−)	(+)	(−)	(+)	(−)
Skull	0.02	**0.29**	0.02	**0.35**	0.03	**0.40**	0.03	**0.43**	0.04	**0.50**	0.05	**0.55**
Brain		**0.01**		**0.02**		**0.02**		**0.03**		**0.03**		**0.04**
Oral mucosa	0.02	**0.39**	0.03	**0.46**	0.03	**0.51**	0.03	**0.55**	0.04	**0.63**	0.05	**0.70**
Salivary glands	0.04	**0.33**	0.05	**0.40**	0.06	**0.45**	0.06	**0.49**	0.08	**0.57**	0.09	**0.63**
Cervical spine	0.33	**0.48**	0.43	**0.62**	0.50	**0.72**	0.56	**0.81**	0.67	**0.99**	0.78	**1.17**
Thyroid	0.78	**0.84**	0.90	**0.96**	0.97	**1.05**	1.03	**1.11**	1.12	**1.22**	1.20	**1.31**
Thymus	0.75	**0.77**	0.90	**0.92**	1.01	**1.03**	1.09	**1.11**	1.24	**1.27**	1.36	**1.40**
Thoracic spine	0.16	**0.17**	0.22	**0.23**	0.26	**0.28**	0.30	**0.32**	0.39	**0.42**	0.47	**0.51**
Claviculae	1.65	**1.72**	1.97	**2.06**	2.19	**2.30**	2.36	**2.48**	2.65	**2.79**	2.88	**3.05**
Scapulae	0.22	**0.24**	0.29	**0.31**	0.35	**0.37**	0.39	**0.42**	0.49	**0.52**	0.58	**0.62**
Ribs	1.36	**1.41**	1.57	**1.64**	1.72	**1.80**	1.83	**1.91**	2.01	**2.11**	2.16	**2.28**
Oesophagus	0.11	**0.12**	0.15	**0.16**	0.18	**0.19**	0.20	**0.21**	0.25	**0.27**	0.30	**0.32**
Trachea	0.22	**0.49**	0.26	**0.60**	0.29	**0.68**	0.31	**0.74**	0.35	**0.86**	0.38	**0.96**
Lung	0.36	**0.38**	0.45	**0.47**	0.52	**0.54**	0.57	**0.59**	0.66	**0.69**	0.75	**0.78**
Mamma	1.16	**1.18**	1.25	**1.28**	1.31	**1.33**	1.34	**1.37**	1.41	**1.43**	1.46	**1.48**
Heart	0.47	**0.49**	0.58	**0.60**	0.67	**0.69**	0.74	**0.76**	0.86	**0.89**	0.98	**1.01**
Stomach	0.33	**0.52**	0.41	**0.64**	0.47	**0.73**	0.51	**0.80**	0.60	**0.93**	0.67	**1.05**
Spleen	0.12	**0.15**	0.16	**0.19**	0.19	**0.23**	0.21	**0.26**	0.26	**0.32**	0.31	**0.39**
Liver	0.28	**0.39**	0.35	**0.49**	0.41	**0.56**	0.45	**0.62**	0.53	**0.73**	0.60	**0.83**
Gall bladder	0.13	**0.37**	0.17	**0.47**	0.20	**0.55**	0.23	**0.61**	0.27	**0.73**	0.32	**0.84**
Pancreas	0.20	**0.23**	0.26	**0.30**	0.30	**0.36**	0.34	**0.40**	0.42	**0.50**	0.50	**0.59**
Small intestine	0.01	**0.13**	0.01	**0.16**	0.02	**0.19**	0.02	**0.21**	0.03	**0.26**	0.03	**0.31**
Large intestine	0.02	**0.22**	0.02	**0.27**	0.02	**0.32**	0.03	**0.35**	0.04	**0.42**	0.05	**0.48**
Sigma/rectum		**0.04**		**0.05**		**0.05**		**0.06**		**0.07**		**0.09**
Lumbar spine	0.05	**0.14**	0.07	**0.19**	0.09	**0.23**	0.10	**0.26**	0.13	**0.34**	0.16	**0.42**
Pelvis		**0.04**		**0.05**		**0.07**		**0.08**	0.01	**0.10**	0.01	**0.12**
Adrenals	0.08	**0.09**	0.11	**0.12**	0.13	**0.15**	0.15	**0.17**	0.19	**0.21**	0.23	**0.26**
Kidneys	0.04	**0.07**	0.06	**0.10**	0.07	**0.12**	0.08	**0.13**	0.10	**0.17**	0.13	**0.21**
Urinary bladder								**0.01**		**0.01**		**0.02**
Uterus		**0.02**		**0.02**		**0.03**		**0.03**		**0.04**		**0.05**
Ovaries		**0.02**		**0.03**		**0.03**		**0.04**		**0.05**		**0.06**
Prostate												
Testes												
Upper arm	0.25	**0.72**	0.32	**0.89**	0.37	**1.02**	0.42	**1.12**	0.51	**1.32**	0.59	**1.49**
Elbow	0.23	**0.77**	0.29	**0.96**	0.34	**1.09**	0.39	**1.20**	0.47	**1.41**	0.55	**1.60**
Forearm	0.03	**0.37**	0.04	**0.46**	0.04	**0.53**	0.05	**0.58**	0.06	**0.68**	0.07	**0.76**
Upper leg												
Knee												
Lower leg												
Lymph nodes	0.18	**0.31**	0.22	**0.38**	0.25	**0.44**	0.28	**0.48**	0.32	**0.57**	0.37	**0.65**
Skin	0.13	**0.24**	0.14	**0.25**	0.15	**0.26**	0.15	**0.27**	0.16	**0.28**	0.17	**0.30**
Muscle	0.10	**0.15**	0.12	**0.18**	0.14	**0.20**	0.15	**0.22**	0.17	**0.25**	0.19	**0.28**
Skeleton	0.20	**0.33**	0.24	**0.40**	0.27	**0.45**	0.29	**0.49**	0.33	**0.57**	0.37	**0.63**
Red bone marrow	0.08	**0.11**	0.09	**0.13**	0.11	**0.15**	0.12	**0.16**	0.13	**0.19**	0.15	**0.22**
Whole body dose	0.12	**0.18**	0.15	**0.22**	0.16	**0.25**	0.18	**0.27**	0.21	**0.31**	0.23	**0.35**
ED (ICRP 60)	0.22	**0.28**	0.26	**0.33**	0.29	**0.37**	0.31	**0.40**	0.35	**0.46**	0.39	**0.51**
ED (ICRP 103)	0.30	**0.37**	0.35	**0.44**	0.38	**0.48**	0.41	**0.52**	0.46	**0.58**	0.50	**0.64**
Total filtration (mm) Al	**1.5**		**2.5**		**3.5**		**4.5**		**3.5**		**3.5**	
Cu	**0**		**0**		**0**		**0**		**0.1**		**0.2**	

Table 10.36 Normalized organ doses—Bony thorax ap—13–17 years

Beam projection:			ap			**Table cells:**						
Focus-to-detector distance (cm):			115			(+) = normal font = optimal field setting						
Tube voltage (kVp):			60			**(−) = bold font = suboptimal field setting**						
Total filtrations:			See last table row			Empty table cells correspond to values <0.01						
Organ or tissue	**Organ dose D_T: Entrance dose K_E**						**(mSv * mGy^{-1})**					
	(+)	(−)	(+)	(−)	(+)	(−)	(+)	(−)	(+)	(−)	(+)	(−)
Skull	0.02	**0.23**	0.02	**0.29**	0.03	**0.34**	0.03	**0.38**	0.04	**0.45**	0.03	**0.42**
Brain				**0.01**		**0.01**		**0.02**		**0.02**		**0.02**
Oral mucosa	0.01	**0.29**	0.02	**0.34**	0.02	**0.39**	0.02	**0.42**	0.03	**0.49**	0.03	**0.46**
Salivary glands	0.04	**0.25**	0.05	**0.31**	0.05	**0.36**	0.06	**0.39**	0.07	**0.46**	0.07	**0.43**
Cervical spine	0.32	**0.40**	0.42	**0.53**	0.50	**0.63**	0.57	**0.72**	0.71	**0.90**	0.63	**0.80**
Thyroid	0.67	**0.69**	0.78	**0.81**	0.86	**0.89**	0.92	**0.95**	1.02	**1.06**	1.00	**1.03**
Thymus	0.60	**0.62**	0.75	**0.77**	0.86	**0.88**	0.94	**0.96**	1.09	**1.12**	1.05	**1.08**
Thoracic spine	0.07	**0.07**	0.09	**0.10**	0.12	**0.12**	0.14	**0.15**	0.18	**0.19**	0.14	**0.15**
Claviculae	1.22	**1.26**	1.51	**1.56**	1.71	**1.78**	1.87	**1.95**	2.17	**2.26**	2.12	**2.20**
Scapulae	0.08	**0.09**	0.11	**0.12**	0.14	**0.15**	0.16	**0.17**	0.21	**0.22**	0.17	**0.18**
Ribs	1.04	**1.08**	1.23	**1.29**	1.37	**1.43**	1.47	**1.54**	1.65	**1.74**	1.62	**1.70**
Oesophagus	0.09	**0.09**	0.12	**0.12**	0.14	**0.15**	0.16	**0.17**	0.21	**0.22**	0.17	**0.19**
Trachea	0.21	**0.32**	0.26	**0.39**	0.29	**0.45**	0.31	**0.49**	0.35	**0.57**	0.34	**0.54**
Lung	0.20	**0.21**	0.26	**0.27**	0.30	**0.32**	0.34	**0.36**	0.42	**0.44**	0.38	**0.40**
Mamma	7.94	**8.17**	9.20	**9.47**	10.0	**10.3**	10.6	**11.0**	11.7	**12.1**	11.5	**11.9**
Heart	0.34	**0.35**	0.44	**0.45**	0.51	**0.52**	0.57	**0.59**	0.68	**0.70**	0.63	**0.65**
Stomach	0.23	**0.36**	0.30	**0.46**	0.34	**0.53**	0.38	**0.59**	0.46	**0.71**	0.43	**0.66**
Spleen	0.06	**0.08**	0.09	**0.11**	0.11	**0.13**	0.12	**0.15**	0.16	**0.19**	0.13	**0.16**
Liver	0.20	**0.26**	0.26	**0.34**	0.30	**0.40**	0.34	**0.44**	0.41	**0.54**	0.38	**0.49**
Gall bladder	0.08	**0.25**	0.11	**0.32**	0.13	**0.38**	0.15	**0.44**	0.18	**0.54**	0.16	**0.49**
Pancreas	0.11	**0.13**	0.15	**0.18**	0.18	**0.22**	0.21	**0.25**	0.27	**0.32**	0.23	**0.27**
Small intestine		**0.08**		**0.10**		**0.12**		**0.14**	0.01	**0.17**		**0.15**
Large intestine		**0.14**	0.01	**0.18**	0.01	**0.21**	0.01	**0.24**	0.02	**0.30**	0.01	**0.27**
Sigma/rectum		**0.02**		**0.03**		**0.03**		**0.04**		**0.05**		**0.04**
Lumbar spine	0.02	**0.06**	0.03	**0.08**	0.04	**0.10**	0.05	**0.12**	0.06	**0.16**	0.04	**0.12**
Pelvis		**0.02**		**0.03**		**0.03**		**0.04**		**0.05**		**0.04**
Adrenals	0.04	**0.04**	0.05	**0.06**	0.07	**0.07**	0.08	**0.08**	0.10	**0.11**	0.08	**0.09**
Kidneys	0.02	**0.03**	0.03	**0.05**	0.03	**0.06**	0.04	**0.07**	0.05	**0.09**	0.04	**0.07**
Urinary bladder												
Uterus		**0.02**		**0.03**		**0.04**		**0.04**		**0.06**		**0.04**
Ovaries				**0.01**		**0.01**		**0.02**		**0.02**		**0.02**
Prostate												
Testes												
Upper arm	0.14	**0.47**	0.18	**0.60**	0.22	**0.70**	0.25	**0.78**	0.31	**0.94**	0.27	**0.88**
Elbow	0.12	**0.50**	0.16	**0.64**	0.19	**0.75**	0.22	**0.83**	0.28	**1.00**	0.24	**0.94**
Forearm	0.02	**0.24**	0.02	**0.31**	0.03	**0.36**	0.03	**0.40**	0.04	**0.49**	0.03	**0.46**
Upper leg												
Knee												
Lower leg												
Lymph nodes	0.13	**0.21**	0.17	**0.26**	0.19	**0.31**	0.21	**0.34**	0.25	**0.41**	0.23	**0.38**
Skin	0.15	**0.24**	0.16	**0.26**	0.17	**0.27**	0.17	**0.28**	0.18	**0.29**	0.18	**0.29**
Muscle	0.08	**0.12**	0.10	**0.14**	0.11	**0.16**	0.12	**0.18**	0.14	**0.21**	0.13	**0.19**
Skeleton	0.15	**0.24**	0.19	**0.30**	0.21	**0.34**	0.23	**0.38**	0.27	**0.44**	0.25	**0.42**
Red bone marrow	0.06	**0.08**	0.08	**0.10**	0.09	**0.12**	0.10	**0.13**	0.12	**0.15**	0.11	**0.14**
Whole body dose	0.10	**0.14**	0.12	**0.18**	0.14	**0.20**	0.15	**0.22**	0.17	**0.26**	0.16	**0.24**
ED (ICRP 60)	0.51	**0.56**	0.60	**0.66**	0.67	**0.74**	0.71	**0.79**	0.80	**0.89**	0.78	**0.86**
ED (ICRP 103)	1.07	**1.14**	1.25	**1.34**	1.37	**1.47**	1.46	**1.57**	1.63	**1.75**	1.59	**1.71**
Total filtration (mm) **Al**	**1.5**		**2.5**		**3.5**		**4.5**		**3.5**		**3.5**	
Cu	**0**		**0**		**0**		**0**		**0.1**		**0.2**	

Table 10.37 Normalized organ doses—Bony thorax ap—13–17 years

Beam projection:			ap			**Table cells:**						
Focus-to-detector distance (cm):			115			(+) = normal font = optimal field setting						
Tube voltage (kVp):			70			**(−) = bold font = suboptimal field setting**						
Total filtrations:			See last table row			Empty table cells correspond to values <0.01						
Organ or tissue	**Organ dose D_T: Entrance dose K_E**						**(mSv * mGy^{-1})**					
	(+)	(−)	(+)	(−)	(+)	(−)	(+)	(−)	(+)	(−)	(+)	(−)
Skull	0.02	**0.28**	0.03	**0.34**	0.03	**0.39**	0.04	**0.43**	0.05	**0.50**	0.05	**0.57**
Brain		**0.01**		**0.02**		**0.02**		**0.02**		**0.03**		**0.04**
Oral mucosa	0.02	**0.33**	0.02	**0.39**	0.03	**0.44**	0.03	**0.48**	0.04	**0.55**	0.04	**0.61**
Salivary glands	0.05	**0.29**	0.06	**0.36**	0.06	**0.41**	0.07	**0.45**	0.09	**0.52**	0.10	**0.58**
Cervical spine	0.41	**0.52**	0.53	**0.67**	0.62	**0.79**	0.69	**0.89**	0.85	**1.10**	0.99	**1.30**
Thyroid	0.74	**0.76**	0.85	**0.88**	0.93	**0.97**	0.99	**1.03**	1.09	**1.14**	1.18	**1.23**
Thymus	0.70	**0.72**	0.86	**0.88**	0.97	**1.00**	1.06	**1.09**	1.22	**1.25**	1.36	**1.40**
Thoracic spine	0.10	**0.11**	0.14	**0.15**	0.17	**0.18**	0.20	**0.21**	0.26	**0.28**	0.32	**0.35**
Claviculae	1.39	**1.45**	1.69	**1.76**	1.90	**1.98**	2.06	**2.14**	2.34	**2.45**	2.58	**2.70**
Scapulae	0.12	**0.13**	0.16	**0.17**	0.19	**0.20**	0.22	**0.23**	0.28	**0.30**	0.34	**0.36**
Ribs	1.15	**1.21**	1.35	**1.42**	1.48	**1.56**	1.59	**1.67**	1.77	**1.87**	1.92	**2.04**
Oesophagus	0.12	**0.13**	0.16	**0.17**	0.19	**0.20**	0.22	**0.23**	0.27	**0.29**	0.33	**0.36**
Trachea	0.24	**0.37**	0.29	**0.45**	0.32	**0.51**	0.34	**0.56**	0.38	**0.64**	0.42	**0.72**
Lung	0.25	**0.26**	0.32	**0.33**	0.37	**0.39**	0.41	**0.43**	0.50	**0.52**	0.58	**0.61**
Mamma	8.67	**8.93**	9.93	**10.2**	10.8	**11.1**	11.4	**11.7**	12.5	**12.9**	13.4	**13.8**
Heart	0.42	**0.43**	0.52	**0.54**	0.60	**0.62**	0.67	**0.69**	0.80	**0.82**	0.91	**0.94**
Stomach	0.28	**0.44**	0.35	**0.54**	0.41	**0.63**	0.45	**0.69**	0.53	**0.82**	0.61	**0.94**
Spleen	0.09	**0.11**	0.12	**0.14**	0.14	**0.17**	0.16	**0.20**	0.20	**0.25**	0.25	**0.31**
Liver	0.25	**0.32**	0.31	**0.41**	0.36	**0.47**	0.40	**0.53**	0.48	**0.63**	0.56	**0.73**
Gall bladder	0.11	**0.31**	0.14	**0.40**	0.16	**0.47**	0.18	**0.53**	0.23	**0.64**	0.27	**0.75**
Pancreas	0.15	**0.18**	0.20	**0.24**	0.24	**0.28**	0.27	**0.32**	0.34	**0.40**	0.41	**0.49**
Small intestine		**0.10**		**0.13**	0.01	**0.15**	0.01	**0.17**	0.02	**0.21**	0.02	**0.25**
Large intestine	0.01	**0.18**	0.02	**0.22**	0.02	**0.26**	0.02	**0.29**	0.03	**0.36**	0.03	**0.42**
Sigma/rectum		**0.03**		**0.04**		**0.04**		**0.05**		**0.06**		**0.07**
Lumbar spine	0.04	**0.09**	0.05	**0.12**	0.06	**0.15**	0.07	**0.17**	0.09	**0.22**	0.11	**0.28**
Pelvis		**0.03**		**0.04**		**0.05**		**0.05**		**0.07**		**0.08**
Adrenals	0.06	**0.06**	0.08	**0.08**	0.09	**0.10**	0.11	**0.12**	0.14	**0.15**	0.17	**0.19**
Kidneys	0.03	**0.05**	0.04	**0.07**	0.05	**0.08**	0.06	**0.09**	0.07	**0.12**	0.09	**0.15**
Urinary bladder												
Uterus		**0.03**		**0.05**		**0.06**		**0.06**		**0.09**	0.01	**0.11**
Ovaries		**0.01**		**0.02**		**0.02**		**0.03**		**0.03**		**0.04**
Prostate												
Testes												
Upper arm	0.18	**0.57**	0.23	**0.71**	0.27	**0.82**	0.31	**0.91**	0.38	**1.08**	0.45	**1.24**
Elbow	0.16	**0.61**	0.21	**0.76**	0.24	**0.88**	0.28	**0.97**	0.34	**1.15**	0.41	**1.31**
Forearm	0.02	**0.29**	0.03	**0.37**	0.03	**0.42**	0.04	**0.47**	0.05	**0.55**	0.06	**0.63**
Upper leg												
Knee												
Lower leg												
Lymph nodes	0.16	**0.26**	0.20	**0.32**	0.23	**0.37**	0.25	**0.40**	0.29	**0.48**	0.34	**0.55**
Skin	0.16	**0.25**	0.17	**0.27**	0.18	**0.28**	0.18	**0.29**	0.19	**0.31**	0.20	**0.32**
Muscle	0.09	**0.14**	0.11	**0.17**	0.13	**0.19**	0.14	**0.20**	0.17	**0.24**	0.19	**0.27**
Skeleton	0.18	**0.28**	0.21	**0.35**	0.24	**0.39**	0.26	**0.43**	0.30	**0.50**	0.34	**0.57**
Red bone marrow	0.08	**0.10**	0.09	**0.12**	0.11	**0.14**	0.12	**0.15**	0.14	**0.18**	0.16	**0.21**
Whole body dose	0.12	**0.17**	0.14	**0.20**	0.16	**0.23**	0.17	**0.25**	0.20	**0.29**	0.23	**0.33**
ED (ICRP 60)	0.57	**0.63**	0.67	**0.74**	0.73	**0.81**	0.78	**0.87**	0.87	**0.98**	0.95	**1.08**
ED (ICRP 103)	1.18	**1.26**	1.36	**1.47**	1.49	**1.61**	1.58	**1.71**	1.75	**1.90**	1.89	**2.06**
Total filtration (mm) Al	**1.5**		**2.5**		**3.5**		**4.5**		**3.5**		**3.5**	
Cu	**0**		**0**		**0**		**0**		**0.1**		**0.2**	

Table 10.38 Normalized organ doses—Bony thorax ap— >17 years

Beam projection:			ap			**Table cells:**						
Focus-to-detector distance (cm):			115			(+) = normal font = optimal field setting						
Tube voltage (kVp):			60			**(−) = bold font = suboptimal field setting**						
Total filtrations:			See last table row			Empty table cells correspond to values <0.01						
Organ or tissue	**Organ dose D_T: Entrance dose K_E**						**($mSv * mGy^{-1}$)**					
	(+)	(−)	(+)	(−)	(+)	(−)	(+)	(−)	(+)	(−)	(+)	(−)
Skull	0.03	**0.19**	0.04	**0.24**	0.05	**0.28**	0.05	**0.31**	0.06	**0.37**	0.06	**0.34**
Brain								**0.01**		**0.01**		
Oral mucosa	0.02	**0.25**	0.02	**0.31**	0.03	**0.35**	0.03	**0.38**	0.04	**0.44**	0.03	**0.41**
Salivary glands	0.07	**0.24**	0.09	**0.29**	0.11	**0.34**	0.12	**0.37**	0.14	**0.44**	0.13	**0.41**
Cervical spine	0.31	**0.35**	0.41	**0.46**	0.49	**0.56**	0.56	**0.63**	0.70	**0.80**	0.62	**0.70**
Thyroid	0.64	**0.65**	0.75	**0.77**	0.83	**0.85**	0.89	**0.92**	1.00	**1.03**	0.98	**1.00**
Thymus	0.61	**0.62**	0.75	**0.77**	0.86	**0.88**	0.94	**0.97**	1.10	**1.13**	1.05	**1.08**
Thoracic spine	0.06	**0.07**	0.09	**0.09**	0.11	**0.12**	0.13	**0.14**	0.17	**0.18**	0.13	**0.14**
Claviculae	1.23	**1.27**	1.52	**1.57**	1.73	**1.79**	1.90	**1.97**	2.20	**2.28**	2.14	**2.22**
Scapulae	0.09	**0.09**	0.12	**0.13**	0.15	**0.16**	0.18	**0.18**	0.23	**0.24**	0.19	**0.19**
Ribs	1.07	**1.12**	1.28	**1.34**	1.42	**1.49**	1.52	**1.60**	1.72	**1.80**	1.68	**1.76**
Oesophagus	0.09	**0.09**	0.12	**0.13**	0.15	**0.16**	0.17	**0.18**	0.22	**0.24**	0.18	**0.19**
Trachea	0.23	**0.28**	0.28	**0.34**	0.31	**0.39**	0.34	**0.43**	0.39	**0.50**	0.38	**0.47**
Lung	0.22	**0.23**	0.28	**0.29**	0.33	**0.35**	0.37	**0.39**	0.45	**0.47**	0.41	**0.43**
Mamma	0.87	**0.90**	1.01	**1.04**	1.10	**1.14**	1.17	**1.21**	1.29	**1.34**	1.26	**1.31**
Heart	0.33	**0.34**	0.42	**0.44**	0.49	**0.51**	0.55	**0.57**	0.66	**0.69**	0.61	**0.63**
Stomach	0.22	**0.36**	0.29	**0.46**	0.34	**0.53**	0.37	**0.60**	0.45	**0.71**	0.42	**0.66**
Spleen	0.06	**0.08**	0.09	**0.10**	0.10	**0.13**	0.12	**0.15**	0.16	**0.19**	0.13	**0.16**
Liver	0.19	**0.26**	0.25	**0.34**	0.29	**0.40**	0.33	**0.44**	0.40	**0.54**	0.36	**0.49**
Gall bladder	0.07	**0.24**	0.09	**0.32**	0.11	**0.38**	0.13	**0.43**	0.16	**0.53**	0.14	**0.47**
Pancreas	0.11	**0.13**	0.15	**0.18**	0.18	**0.21**	0.21	**0.25**	0.26	**0.31**	0.22	**0.26**
Small intestine		**0.08**		**0.10**		**0.13**		**0.14**		**0.18**		**0.16**
Large intestine		**0.14**		**0.18**	0.01	**0.22**	0.01	**0.25**	0.02	**0.30**	0.01	**0.27**
Sigma/rectum		**0.02**		**0.03**		**0.04**		**0.04**		**0.05**		**0.04**
Lumbar spine	0.02	**0.05**	0.03	**0.07**	0.03	**0.09**	0.04	**0.11**	0.05	**0.15**	0.04	**0.11**
Pelvis		**0.02**		**0.03**		**0.03**		**0.04**		**0.05**		**0.04**
Adrenals	0.04	**0.04**	0.05	**0.06**	0.07	**0.07**	0.08	**0.08**	0.10	**0.11**	0.08	**0.09**
Kidneys	0.02	**0.03**	0.03	**0.04**	0.03	**0.06**	0.04	**0.06**	0.05	**0.09**	0.04	**0.07**
Urinary bladder												
Uterus				**0.01**		**0.01**		**0.02**		**0.02**		**0.02**
Ovaries				**0.01**		**0.01**		**0.02**		**0.02**		**0.02**
Prostate												
Testes												
Upper arm	0.17	**0.46**	0.23	**0.59**	0.27	**0.69**	0.31	**0.77**	0.38	**0.93**	0.34	**0.86**
Elbow	0.16	**0.49**	0.21	**0.63**	0.25	**0.73**	0.29	**0.82**	0.36	**0.99**	0.32	**0.92**
Forearm	0.02	**0.24**	0.02	**0.31**	0.03	**0.36**	0.03	**0.40**	0.04	**0.48**	0.04	**0.45**
Upper leg												
Knee												
Lower leg												
Lymph nodes	0.14	**0.20**	0.17	**0.26**	0.20	**0.30**	0.22	**0.33**	0.26	**0.40**	0.24	**0.37**
Skin	0.16	**0.24**	0.17	**0.26**	0.18	**0.27**	0.18	**0.28**	0.19	**0.29**	0.19	**0.29**
Muscle	0.08	**0.11**	0.10	**0.14**	0.11	**0.16**	0.12	**0.17**	0.14	**0.20**	0.13	**0.19**
Skeleton	0.16	**0.23**	0.19	**0.28**	0.22	**0.32**	0.24	**0.36**	0.28	**0.42**	0.26	**0.40**
Red bone marrow	0.08	**0.09**	0.09	**0.11**	0.11	**0.13**	0.12	**0.14**	0.14	**0.16**	0.13	**0.15**
Whole body dose	0.10	**0.14**	0.12	**0.17**	0.14	**0.20**	0.15	**0.22**	0.18	**0.25**	0.17	**0.24**
ED (ICRP 60)	0.16	**0.20**	0.20	**0.24**	0.22	**0.28**	0.24	**0.31**	0.28	**0.36**	0.27	**0.34**
ED (ICRP 103)	0.22	**0.27**	0.27	**0.33**	0.30	**0.37**	0.33	**0.40**	0.38	**0.47**	0.36	**0.44**
Total filtration (mm) **Al**	**1.5**		**2.5**		**3.5**		**4.5**		**3.5**		**3.5**	
Cu	**0**		**0**		**0**		**0**		**0.1**		**0.2**	

Table 10.39 Normalized organ doses—Bony thorax ap— >17 years

Beam projection:			ap		**Table cells:**							
Focus-to-detector distance (cm):			115		(+) = normal font = optimal field setting							
Tube voltage (kVp):			70		**(−) = bold font = suboptimal field setting**							
Total filtrations:			See last table row		Empty table cells correspond to values <0.01							
Organ or tissue	**Organ dose D_T: Entrance dose K_E**							**(mSv * mGy^{-1})**				
	(+)	(−)	(+)	(−)	(+)	(−)	(+)	(−)	(+)	(−)	(+)	(−)
Skull	0.04	**0.23**	0.05	**0.28**	0.06	**0.32**	0.06	**0.36**	0.08	**0.42**	0.09	**0.48**
Brain				**0.01**		**0.01**		**0.02**		**0.02**		**0.03**
Oral mucosa	0.02	**0.29**	0.03	**0.35**	0.03	**0.39**	0.04	**0.43**	0.05	**0.49**	0.05	**0.55**
Salivary glands	0.09	**0.28**	0.11	**0.34**	0.13	**0.39**	0.14	**0.43**	0.16	**0.49**	0.18	**0.56**
Cervical spine	0.40	**0.46**	0.52	**0.59**	0.61	**0.70**	0.69	**0.79**	0.85	**0.98**	1.00	**1.16**
Thyroid	0.71	**0.72**	0.83	**0.85**	0.91	**0.93**	0.97	**1.00**	1.08	**1.11**	1.17	**1.20**
Thymus	0.71	**0.73**	0.86	**0.89**	0.98	**1.01**	1.07	**1.10**	1.23	**1.27**	1.38	**1.42**
Thoracic spine	0.10	**0.11**	0.13	**0.14**	0.16	**0.17**	0.19	**0.20**	0.25	**0.26**	0.31	**0.33**
Claviculae	1.41	**1.46**	1.71	**1.77**	1.92	**2.00**	2.09	**2.17**	2.39	**2.48**	2.64	**2.75**
Scapulae	0.13	**0.13**	0.17	**0.18**	0.21	**0.22**	0.24	**0.25**	0.30	**0.32**	0.37	**0.39**
Ribs	1.19	**1.25**	1.40	**1.47**	1.54	**1.62**	1.65	**1.74**	1.84	**1.94**	2.00	**2.11**
Oesophagus	0.13	**0.13**	0.17	**0.18**	0.20	**0.21**	0.23	**0.25**	0.29	**0.31**	0.35	**0.38**
Trachea	0.26	**0.32**	0.31	**0.39**	0.35	**0.45**	0.38	**0.49**	0.43	**0.56**	0.47	**0.64**
Lung	0.27	**0.28**	0.34	**0.36**	0.40	**0.42**	0.44	**0.47**	0.53	**0.56**	0.62	**0.65**
Mamma	0.95	**0.98**	1.09	**1.13**	1.18	**1.23**	1.25	**1.30**	1.37	**1.42**	1.47	**1.53**
Heart	0.40	**0.42**	0.51	**0.53**	0.58	**0.61**	0.65	**0.68**	0.77	**0.81**	0.89	**0.93**
Stomach	0.28	**0.44**	0.35	**0.55**	0.40	**0.63**	0.44	**0.70**	0.53	**0.83**	0.60	**0.95**
Spleen	0.09	**0.11**	0.12	**0.14**	0.14	**0.17**	0.16	**0.20**	0.20	**0.25**	0.25	**0.31**
Liver	0.24	**0.33**	0.30	**0.41**	0.35	**0.48**	0.39	**0.53**	0.47	**0.63**	0.55	**0.73**
Gall bladder	0.09	**0.31**	0.12	**0.40**	0.15	**0.47**	0.16	**0.52**	0.20	**0.64**	0.24	**0.75**
Pancreas	0.15	**0.18**	0.20	**0.24**	0.24	**0.28**	0.27	**0.32**	0.34	**0.40**	0.41	**0.49**
Small intestine		**0.10**		**0.14**		**0.16**	0.01	**0.18**	0.01	**0.22**	0.02	**0.27**
Large intestine		**0.18**	0.01	**0.23**	0.02	**0.27**	0.02	**0.30**	0.02	**0.37**	0.03	**0.43**
Sigma/rectum		**0.03**		**0.04**		**0.05**		**0.05**		**0.06**		**0.08**
Lumbar spine	0.03	**0.08**	0.04	**0.11**	0.05	**0.14**	0.06	**0.16**	0.08	**0.21**	0.10	**0.27**
Pelvis		**0.03**		**0.04**		**0.05**		**0.05**		**0.07**		**0.09**
Adrenals	0.06	**0.06**	0.08	**0.09**	0.09	**0.10**	0.11	**0.12**	0.14	**0.16**	0.17	**0.19**
Kidneys	0.03	**0.05**	0.04	**0.07**	0.05	**0.08**	0.06	**0.09**	0.07	**0.12**	0.09	**0.15**
Urinary bladder												
Uterus		**0.01**		**0.02**		**0.02**		**0.02**		**0.03**		**0.04**
Ovaries		**0.01**		**0.02**		**0.02**		**0.02**		**0.03**		**0.04**
Prostate												
Testes												
Upper arm	0.22	**0.56**	0.29	**0.70**	0.34	**0.81**	0.38	**0.90**	0.47	**1.07**	0.55	**1.23**
Elbow	0.21	**0.60**	0.27	**0.75**	0.31	**0.86**	0.35	**0.96**	0.43	**1.14**	0.51	**1.31**
Forearm	0.02	**0.29**	0.03	**0.37**	0.04	**0.42**	0.04	**0.47**	0.05	**0.55**	0.06	**0.63**
Upper leg												
Knee												
Lower leg												
Lymph nodes	0.16	**0.25**	0.20	**0.31**	0.23	**0.36**	0.26	**0.40**	0.30	**0.47**	0.35	**0.54**
Skin	0.17	**0.25**	0.18	**0.27**	0.19	**0.28**	0.19	**0.29**	0.20	**0.31**	0.21	**0.32**
Muscle	0.09	**0.13**	0.11	**0.16**	0.13	**0.18**	0.14	**0.20**	0.16	**0.23**	0.19	**0.27**
Skeleton	0.18	**0.27**	0.22	**0.33**	0.25	**0.37**	0.27	**0.41**	0.31	**0.48**	0.35	**0.54**
Red bone marrow	0.09	**0.11**	0.11	**0.13**	0.12	**0.15**	0.14	**0.17**	0.16	**0.19**	0.18	**0.22**
Whole body dose	0.12	**0.17**	0.14	**0.20**	0.16	**0.23**	0.17	**0.25**	0.20	**0.29**	0.23	**0.33**
ED (ICRP 60)	0.19	**0.23**	0.23	**0.29**	0.26	**0.32**	0.28	**0.36**	0.32	**0.42**	0.36	**0.47**
ED (ICRP 103)	0.26	**0.31**	0.31	**0.38**	0.34	**0.42**	0.37	**0.46**	0.42	**0.53**	0.47	**0.60**
Total filtration (mm) **Al**	**1.5**		**2.5**		**3.5**		**4.5**		**3.5**		**3.5**	
Cu	**0**		**0**		**0**		**0**		**0.1**		**0.2**	

Table 10.40 Normalized organ doses—Bony thorax RAO—1–2 years

Beam projection:				RAO			**Table cells:**					
Focus-to-detector distance (cm):				115			(+) = normal font = optimal field setting					
Tube voltage (kVp):				50			**(−) = bold font = suboptimal field setting**					
Total filtrations:				See last table row			Empty table cells correspond to values <0.01					
Organ or tissue	**Organ dose D_T: Entrance dose K_E**							**(mSv * mGy^{-1})**				
	(+)	(−)	(+)	(−)	(+)	(−)	(+)	(−)	(+)	(−)	(+)	(−)
Skull	0.01	**0.20**	0.02	**0.25**	0.02	**0.28**	0.02	**0.31**	0.03	**0.35**	0.03	**0.39**
Brain		**0.01**		**0.02**		**0.02**		**0.02**		**0.03**		**0.04**
Oral mucosa	0.02	**0.39**	0.02	**0.47**	0.03	**0.52**	0.03	**0.56**	0.04	**0.63**	0.05	**0.69**
Salivary glands	0.04	**0.36**	0.05	**0.42**	0.06	**0.47**	0.06	**0.50**	0.08	**0.57**	0.09	**0.62**
Cervical spine	0.23	**0.34**	0.30	**0.46**	0.36	**0.55**	0.41	**0.62**	0.49	**0.77**	0.58	**0.91**
Thyroid	0.74	**0.78**	0.85	**0.90**	0.92	**0.98**	0.97	**1.03**	1.05	**1.13**	1.11	**1.21**
Thymus	0.56	**0.58**	0.68	**0.70**	0.77	**0.79**	0.83	**0.86**	0.95	**0.97**	1.04	**1.08**
Thoracic spine	0.18	**0.19**	0.25	**0.26**	0.30	**0.32**	0.35	**0.37**	0.44	**0.47**	0.54	**0.58**
Claviculae	1.39	**1.45**	1.69	**1.77**	1.89	**1.99**	2.05	**2.16**	2.32	**2.45**	2.55	**2.70**
Scapulae	0.13	**0.16**	0.19	**0.23**	0.23	**0.28**	0.26	**0.32**	0.34	**0.41**	0.41	**0.50**
Ribs	1.12	**1.16**	1.32	**1.37**	1.46	**1.52**	1.56	**1.63**	1.73	**1.82**	1.89	**1.99**
Oesophagus	0.14	**0.15**	0.19	**0.21**	0.23	**0.25**	0.27	**0.28**	0.33	**0.35**	0.39	**0.42**
Trachea	0.07	**0.37**	0.08	**0.44**	0.09	**0.50**	0.10	**0.54**	0.11	**0.61**	0.12	**0.68**
Lung	0.27	**0.28**	0.34	**0.36**	0.40	**0.41**	0.44	**0.45**	0.51	**0.53**	0.58	**0.61**
Mamma	1.05	**1.07**	1.13	**1.15**	1.18	**1.20**	1.20	**1.23**	1.25	**1.28**	1.29	**1.33**
Heart	0.35	**0.36**	0.44	**0.45**	0.51	**0.52**	0.56	**0.58**	0.66	**0.68**	0.75	**0.77**
Stomach	0.17	**0.28**	0.22	**0.36**	0.26	**0.42**	0.28	**0.46**	0.34	**0.55**	0.39	**0.63**
Spleen	0.05	**0.07**	0.07	**0.09**	0.09	**0.11**	0.10	**0.13**	0.13	**0.16**	0.16	**0.20**
Liver	0.25	**0.39**	0.31	**0.49**	0.36	**0.56**	0.40	**0.62**	0.47	**0.72**	0.53	**0.81**
Gall bladder	0.08	**0.28**	0.10	**0.36**	0.12	**0.43**	0.14	**0.48**	0.17	**0.58**	0.20	**0.67**
Pancreas	0.12	**0.14**	0.16	**0.19**	0.19	**0.23**	0.22	**0.26**	0.27	**0.33**	0.32	**0.39**
Small intestine		**0.12**	0.01	**0.15**	0.01	**0.18**	0.02	**0.20**	0.02	**0.25**	0.03	**0.29**
Large intestine	0.01	**0.18**	0.02	**0.24**	0.02	**0.28**	0.02	**0.31**	0.03	**0.37**	0.04	**0.43**
Sigma/rectum		**0.03**		**0.04**		**0.05**		**0.05**		**0.07**		**0.08**
Lumbar spine	0.04	**0.12**	0.05	**0.16**	0.06	**0.20**	0.07	**0.24**	0.10	**0.31**	0.12	**0.38**
Pelvis		**0.04**		**0.06**		**0.07**		**0.08**	0.01	**0.11**	0.02	**0.13**
Adrenals	0.06	**0.07**	0.08	**0.09**	0.10	**0.11**	0.11	**0.13**	0.14	**0.17**	0.17	**0.20**
Kidneys	0.03	**0.06**	0.04	**0.08**	0.05	**0.10**	0.06	**0.11**	0.07	**0.14**	0.09	**0.17**
Urinary bladder						**0.01**		**0.01**		**0.02**		**0.02**
Uterus		**0.02**		**0.02**		**0.03**		**0.04**		**0.05**		**0.06**
Ovaries		**0.02**		**0.03**		**0.03**		**0.04**		**0.05**		**0.06**
Prostate												**0.01**
Testes												
Upper arm	0.54	**0.78**	0.67	**0.96**	0.76	**1.09**	0.84	**1.19**	0.98	**1.38**	1.10	**1.55**
Elbow	0.57	**0.85**	0.70	**1.04**	0.80	**1.18**	0.88	**1.29**	1.02	**1.48**	1.15	**1.66**
Forearm	0.03	**0.43**	0.04	**0.53**	0.05	**0.59**	0.06	**0.65**	0.07	**0.74**	0.09	**0.83**
Upper leg												
Knee												
Lower leg												
Lymph nodes	0.13	**0.25**	0.16	**0.31**	0.18	**0.36**	0.20	**0.39**	0.23	**0.46**	0.26	**0.52**
Skin	0.15	**0.27**	0.15	**0.29**	0.16	**0.30**	0.16	**0.30**	0.17	**0.32**	0.18	**0.33**
Muscle	0.08	**0.14**	0.10	**0.17**	0.11	**0.19**	0.12	**0.21**	0.14	**0.24**	0.16	**0.27**
Skeleton	0.24	**0.36**	0.30	**0.44**	0.34	**0.50**	0.37	**0.54**	0.43	**0.63**	0.48	**0.71**
Red bone marrow	0.05	**0.08**	0.07	**0.10**	0.08	**0.12**	0.08	**0.13**	0.10	**0.15**	0.11	**0.18**
Whole body dose	0.10	**0.17**	0.13	**0.21**	0.14	**0.23**	0.16	**0.25**	0.18	**0.29**	0.20	**0.33**
ED (ICRP 60)	0.18	**0.23**	0.21	**0.27**	0.23	**0.30**	0.25	**0.33**	0.28	**0.38**	0.31	**0.42**
ED (ICRP 103)	0.25	**0.31**	0.29	**0.36**	0.31	**0.40**	0.33	**0.43**	0.37	**0.48**	0.40	**0.53**
Total filtration (mm) Al	**1.5**		**2.5**		**3.5**		**4.5**		**3.5**		**3.5**	
Cu	**0**		**0**		**0**		**0**		**0.1**		**0.2**	

Table 10.41 Normalized organ doses—Bony thorax RAO—1–2 years

Beam projection:				RAO			**Table cells:**					
Focus-to-detector distance (cm):				115			(+) = normal font = optimal field setting					
Tube voltage (kVp):				60			**(−) = bold font = suboptimal field setting**					
Total filtrations:				See last table row			Empty table cells correspond to values <0.01					
Organ or tissue	**Organ dose D_T: Entrance dose K_E**							**(mSv * mGy^{-1})**				
	(+)	(−)	(+)	(−)	(+)	(−)	(+)	(−)	(+)	(−)	(+)	(−)
Skull	0.02	**0.24**	0.02	**0.29**	0.03	**0.33**	0.03	**0.35**	0.04	**0.40**	0.03	**0.39**
Brain		**0.02**		**0.02**		**0.03**		**0.03**		**0.04**		**0.04**
Oral mucosa	0.02	**0.45**	0.03	**0.53**	0.04	**0.59**	0.04	**0.63**	0.05	**0.71**	0.05	**0.69**
Salivary glands	0.05	**0.41**	0.06	**0.48**	0.07	**0.53**	0.08	**0.57**	0.09	**0.64**	0.09	**0.62**
Cervical spine	0.30	**0.47**	0.39	**0.61**	0.46	**0.72**	0.51	**0.81**	0.62	**0.98**	0.58	**0.91**
Thyroid	0.81	**0.86**	0.92	**0.99**	0.99	**1.07**	1.04	**1.13**	1.13	**1.23**	1.11	**1.21**
Thymus	0.66	**0.67**	0.78	**0.81**	0.87	**0.90**	0.94	**0.97**	1.06	**1.10**	1.04	**1.08**
Thoracic spine	0.26	**0.28**	0.35	**0.38**	0.42	**0.46**	0.48	**0.52**	0.61	**0.65**	0.54	**0.58**
Claviculae	1.60	**1.68**	1.91	**2.01**	2.12	**2.24**	2.28	**2.41**	2.54	**2.71**	2.55	**2.70**
Scapulae	0.20	**0.24**	0.27	**0.32**	0.32	**0.39**	0.37	**0.45**	0.46	**0.56**	0.41	**0.50**
Ribs	1.26	**1.32**	1.47	**1.54**	1.61	**1.69**	1.72	**1.81**	1.90	**2.01**	1.89	**1.99**
Oesophagus	0.20	**0.21**	0.26	**0.28**	0.31	**0.33**	0.35	**0.38**	0.43	**0.46**	0.39	**0.42**
Trachea	0.08	**0.43**	0.09	**0.51**	0.10	**0.57**	0.11	**0.62**	0.13	**0.69**	0.12	**0.68**
Lung	0.33	**0.35**	0.41	**0.43**	0.47	**0.49**	0.52	**0.54**	0.60	**0.63**	0.58	**0.61**
Mamma	1.10	**1.12**	1.18	**1.20**	1.22	**1.25**	1.25	**1.29**	1.29	**1.34**	1.29	**1.33**
Heart	0.43	**0.44**	0.53	**0.55**	0.61	**0.63**	0.67	**0.69**	0.78	**0.80**	0.75	**0.77**
Stomach	0.22	**0.35**	0.27	**0.44**	0.31	**0.51**	0.34	**0.56**	0.40	**0.65**	0.39	**0.63**
Spleen	0.08	**0.10**	0.10	**0.13**	0.12	**0.16**	0.14	**0.18**	0.18	**0.23**	0.16	**0.20**
Liver	0.31	**0.48**	0.38	**0.59**	0.43	**0.67**	0.47	**0.73**	0.55	**0.84**	0.53	**0.81**
Gall bladder	0.10	**0.36**	0.13	**0.46**	0.16	**0.53**	0.18	**0.59**	0.21	**0.70**	0.20	**0.67**
Pancreas	0.16	**0.20**	0.21	**0.26**	0.25	**0.31**	0.28	**0.35**	0.35	**0.43**	0.32	**0.39**
Small intestine	0.01	**0.15**	0.02	**0.20**	0.02	**0.23**	0.02	**0.26**	0.03	**0.31**	0.03	**0.29**
Large intestine	0.02	**0.23**	0.02	**0.29**	0.03	**0.34**	0.03	**0.38**	0.04	**0.45**	0.04	**0.43**
Sigma/rectum		**0.04**		**0.05**		**0.06**		**0.07**		**0.09**		**0.08**
Lumbar spine	0.06	**0.18**	0.08	**0.25**	0.10	**0.30**	0.12	**0.35**	0.15	**0.45**	0.12	**0.38**
Pelvis		**0.06**	0.01	**0.09**	0.01	**0.10**	0.02	**0.12**	0.02	**0.15**	0.02	**0.13**
Adrenals	0.09	**0.10**	0.11	**0.13**	0.14	**0.16**	0.16	**0.18**	0.20	**0.23**	0.17	**0.20**
Kidneys	0.05	**0.09**	0.06	**0.12**	0.07	**0.14**	0.08	**0.16**	0.11	**0.20**	0.09	**0.17**
Urinary bladder		**0.01**		**0.01**		**0.02**		**0.02**		**0.03**		**0.02**
Uterus		**0.03**		**0.04**		**0.05**		**0.05**	0.01	**0.07**		**0.06**
Ovaries		**0.03**		**0.04**		**0.05**		**0.05**		**0.07**		**0.06**
Prostate								**0.01**		**0.01**		**0.01**
Testes												
Upper arm	0.65	**0.93**	0.80	**1.13**	0.90	**1.27**	0.98	**1.38**	1.14	**1.58**	1.10	**1.55**
Elbow	0.68	**1.01**	0.83	**1.22**	0.94	**1.37**	1.03	**1.48**	1.18	**1.69**	1.15	**1.66**
Forearm	0.05	**0.51**	0.06	**0.61**	0.07	**0.69**	0.08	**0.74**	0.09	**0.85**	0.09	**0.83**
Upper leg												
Knee												
Lower leg												
Lymph nodes	0.15	**0.31**	0.19	**0.38**	0.22	**0.43**	0.24	**0.47**	0.27	**0.55**	0.26	**0.52**
Skin	0.15	**0.28**	0.16	**0.30**	0.17	**0.31**	0.18	**0.32**	0.19	**0.34**	0.18	**0.33**
Muscle	0.10	**0.17**	0.12	**0.20**	0.13	**0.22**	0.15	**0.24**	0.17	**0.28**	0.16	**0.27**
Skeleton	0.29	**0.43**	0.35	**0.52**	0.40	**0.58**	0.43	**0.64**	0.50	**0.73**	0.48	**0.71**
Red bone marrow	0.07	**0.11**	0.08	**0.13**	0.09	**0.15**	0.10	**0.16**	0.12	**0.19**	0.11	**0.18**
Whole body dose	0.12	**0.20**	0.15	**0.24**	0.17	**0.27**	0.18	**0.30**	0.21	**0.34**	0.20	**0.33**
ED (ICRP 60)	0.21	**0.27**	0.24	**0.32**	0.27	**0.35**	0.29	**0.38**	0.33	**0.44**	0.31	**0.42**
ED (ICRP 103)	0.28	**0.35**	0.32	**0.41**	0.35	**0.46**	0.37	**0.49**	0.41	**0.55**	0.40	**0.53**
Total filtration (mm) **Al**	**1.5**		**2.5**		**3.5**		**4.5**		**3.5**		**3.5**	
Cu	**0**		**0**		**0**		**0**		**0.1**		**0.2**	

Table 10.42 Normalized organ doses—Bony thorax RAO—1–2 years

Beam projection:			RAO			**Table cells:**						
Focus-to-detector distance (cm):			115			(+) = normal font = optimal field setting						
Tube voltage (kVp):			70			**(−) = bold font = suboptimal field setting**						
Total filtrations:			See last table row			Empty table cells correspond to values <0.01						
Organ or tissue	**Organ dose D_T: Entrance dose K_E**						**(mSv * mGy^{-1})**					
	(+)	(−)	(+)	(−)	(+)	(−)	(+)	(−)	(+)	(−)	(+)	(−)
Skull	0.02	**0.27**	0.03	**0.32**	0.03	**0.36**	0.04	**0.39**	0.05	**0.44**	0.05	**0.48**
Brain		**0.02**		**0.03**		**0.04**		**0.04**		**0.05**		**0.06**
Oral mucosa	0.03	**0.50**	0.04	**0.59**	0.05	**0.64**	0.05	**0.69**	0.06	**0.77**	0.07	**0.84**
Salivary glands	0.06	**0.45**	0.07	**0.53**	0.08	**0.58**	0.09	**0.62**	0.10	**0.69**	0.12	**0.75**
Cervical spine	0.37	**0.57**	0.46	**0.73**	0.53	**0.85**	0.59	**0.94**	0.70	**1.13**	0.80	**1.30**
Thyroid	0.87	**0.93**	0.98	**1.06**	1.05	**1.14**	1.10	**1.19**	1.18	**1.29**	1.25	**1.38**
Thymus	0.73	**0.75**	0.86	**0.89**	0.95	**0.99**	1.02	**1.06**	1.14	**1.18**	1.23	**1.29**
Thoracic spine	0.34	**0.37**	0.44	**0.48**	0.52	**0.57**	0.59	**0.64**	0.73	**0.79**	0.86	**0.94**
Claviculae	1.76	**1.86**	2.06	**2.18**	2.26	**2.40**	2.41	**2.56**	2.66	**2.84**	2.84	**3.04**
Scapulae	0.26	**0.31**	0.34	**0.41**	0.40	**0.48**	0.45	**0.54**	0.55	**0.67**	0.65	**0.79**
Ribs	1.37	**1.44**	1.58	**1.66**	1.71	**1.81**	1.81	**1.92**	1.98	**2.11**	2.11	**2.26**
Oesophagus	0.25	**0.27**	0.32	**0.35**	0.37	**0.40**	0.42	**0.45**	0.50	**0.55**	0.59	**0.64**
Trachea	0.09	**0.48**	0.10	**0.57**	0.11	**0.63**	0.12	**0.67**	0.14	**0.75**	0.15	**0.82**
Lung	0.39	**0.41**	0.47	**0.49**	0.53	**0.56**	0.58	**0.61**	0.66	**0.70**	0.74	**0.78**
Mamma	1.14	**1.16**	1.21	**1.24**	1.25	**1.29**	1.28	**1.32**	1.32	**1.38**	1.36	**1.42**
Heart	0.50	**0.52**	0.61	**0.63**	0.69	**0.71**	0.75	**0.78**	0.86	**0.89**	0.96	**1.00**
Stomach	0.26	**0.41**	0.31	**0.51**	0.36	**0.58**	0.39	**0.63**	0.45	**0.73**	0.51	**0.82**
Spleen	0.10	**0.13**	0.13	**0.17**	0.16	**0.20**	0.18	**0.23**	0.22	**0.28**	0.26	**0.33**
Liver	0.36	**0.55**	0.43	**0.66**	0.49	**0.75**	0.53	**0.81**	0.61	**0.93**	0.67	**1.03**
Gall bladder	0.13	**0.43**	0.16	**0.54**	0.19	**0.61**	0.21	**0.68**	0.25	**0.79**	0.29	**0.90**
Pancreas	0.20	**0.25**	0.26	**0.32**	0.30	**0.37**	0.34	**0.42**	0.41	**0.51**	0.48	**0.59**
Small intestine	0.02	**0.19**	0.02	**0.24**	0.03	**0.27**	0.03	**0.30**	0.04	**0.36**	0.05	**0.41**
Large intestine	0.02	**0.28**	0.03	**0.34**	0.04	**0.39**	0.04	**0.43**	0.05	**0.51**	0.06	**0.57**
Sigma/rectum		**0.05**		**0.07**		**0.08**		**0.09**	0.01	**0.10**	0.01	**0.12**
Lumbar spine	0.08	**0.25**	0.11	**0.33**	0.13	**0.39**	0.15	**0.44**	0.19	**0.56**	0.23	**0.67**
Pelvis	0.01	**0.09**	0.02	**0.11**	0.02	**0.13**	0.02	**0.15**	0.03	**0.19**	0.03	**0.22**
Adrenals	0.11	**0.13**	0.15	**0.17**	0.17	**0.20**	0.20	**0.23**	0.24	**0.29**	0.29	**0.34**
Kidneys	0.06	**0.11**	0.08	**0.15**	0.09	**0.18**	0.11	**0.20**	0.13	**0.25**	0.16	**0.29**
Urinary bladder		**0.01**		**0.02**		**0.02**		**0.03**		**0.03**		**0.04**
Uterus		**0.04**		**0.05**	0.01	**0.06**	0.01	**0.07**	0.02	**0.08**	0.02	**0.10**
Ovaries		**0.04**		**0.05**		**0.06**	0.01	**0.07**	0.01	**0.09**	0.02	**0.10**
Prostate				**0.01**		**0.01**		**0.01**		**0.02**		**0.02**
Testes												**0.01**
Upper arm	0.74	**1.05**	0.89	**1.25**	1.00	**1.40**	1.08	**1.51**	1.23	**1.70**	1.36	**1.86**
Elbow	0.78	**1.13**	0.93	**1.35**	1.04	**1.50**	1.12	**1.61**	1.28	**1.81**	1.40	**1.98**
Forearm	0.06	**0.57**	0.07	**0.67**	0.08	**0.75**	0.09	**0.80**	0.11	**0.90**	0.13	**0.98**
Upper leg										**0.01**		**0.01**
Knee												
Lower leg												
Lymph nodes	0.18	**0.36**	0.22	**0.43**	0.25	**0.49**	0.27	**0.53**	0.31	**0.61**	0.35	**0.68**
Skin	0.16	**0.30**	0.17	**0.32**	0.18	**0.33**	0.18	**0.34**	0.20	**0.36**	0.21	**0.38**
Muscle	0.11	**0.19**	0.13	**0.23**	0.15	**0.25**	0.16	**0.27**	0.19	**0.31**	0.21	**0.34**
Skeleton	0.33	**0.49**	0.40	**0.58**	0.44	**0.65**	0.48	**0.70**	0.54	**0.80**	0.60	**0.88**
Red bone marrow	0.08	**0.12**	0.09	**0.15**	0.11	**0.17**	0.12	**0.19**	0.14	**0.22**	0.15	**0.25**
Whole body dose	0.14	**0.23**	0.17	**0.27**	0.19	**0.31**	0.20	**0.33**	0.23	**0.38**	0.26	**0.42**
ED (ICRP 60)	0.23	**0.30**	0.27	**0.36**	0.30	**0.40**	0.32	**0.43**	0.36	**0.48**	0.39	**0.53**
ED (ICRP 103)	0.31	**0.39**	0.35	**0.46**	0.38	**0.50**	0.41	**0.53**	0.45	**0.60**	0.48	**0.65**
Total filtration (mm) **Al**	**1.5**		**2.5**		**3.5**		**4.5**		**3.5**		**3.5**	
Cu	**0**		**0**		**0**		**0**		**0.1**		**0.2**	

Table 10.43 Normalized organ doses—Bony thorax RAO—3–7 years

Beam projection:			RAO				**Table cells:**					
Focus-to-detector distance (cm):			115				(+) = normal font = optimal field setting					
Tube voltage (kVp):			60				**(−) = bold font = suboptimal field setting**					
Total filtrations:			See last table row				Empty table cells correspond to values <0.01					
Organ or tissue	**Organ dose D_T: Entrance dose K_E**							**(mSv * mGy^{-1})**				
	(+)	(−)	(+)	(−)	(+)	(−)	(+)	(−)	(+)	(−)	(+)	(−)
Skull	0.02	**0.25**	0.03	**0.31**	0.03	**0.35**	0.04	**0.39**	0.05	**0.45**	0.04	**0.43**
Brain		**0.01**		**0.02**		**0.02**		**0.02**		**0.03**		**0.02**
Oral mucosa	0.02	**0.35**	0.03	**0.42**	0.04	**0.47**	0.04	**0.51**	0.05	**0.58**	0.05	**0.56**
Salivary glands	0.05	**0.34**	0.07	**0.40**	0.08	**0.44**	0.09	**0.47**	0.10	**0.54**	0.09	**0.52**
Cervical spine	0.30	**0.39**	0.39	**0.51**	0.46	**0.60**	0.51	**0.68**	0.62	**0.84**	0.58	**0.77**
Thyroid	0.73	**0.76**	0.84	**0.87**	0.91	**0.95**	0.97	**1.01**	1.06	**1.11**	1.05	**1.08**
Thymus	0.61	**0.62**	0.74	**0.76**	0.83	**0.85**	0.90	**0.93**	1.03	**1.06**	1.01	**1.03**
Thoracic spine	0.18	**0.20**	0.25	**0.27**	0.30	**0.33**	0.35	**0.38**	0.44	**0.48**	0.38	**0.41**
Claviculae	1.43	**1.50**	1.73	**1.82**	1.94	**2.04**	2.10	**2.21**	2.37	**2.51**	2.36	**2.49**
Scapulae	0.14	**0.18**	0.18	**0.24**	0.22	**0.29**	0.26	**0.34**	0.33	**0.43**	0.28	**0.37**
Ribs	1.17	**1.23**	1.38	**1.45**	1.51	**1.60**	1.61	**1.71**	1.79	**1.91**	1.78	**1.88**
Oesophagus	0.11	**0.12**	0.15	**0.16**	0.18	**0.20**	0.21	**0.22**	0.26	**0.28**	0.23	**0.25**
Trachea	0.10	**0.29**	0.12	**0.36**	0.13	**0.41**	0.14	**0.45**	0.16	**0.53**	0.15	**0.50**
Lung	0.30	**0.31**	0.37	**0.39**	0.43	**0.45**	0.47	**0.50**	0.55	**0.58**	0.53	**0.55**
Mamma	1.07	**1.09**	1.16	**1.18**	1.20	**1.23**	1.24	**1.27**	1.29	**1.32**	1.28	**1.31**
Heart	0.36	**0.38**	0.46	**0.48**	0.53	**0.55**	0.59	**0.61**	0.69	**0.72**	0.66	**0.68**
Stomach	0.20	**0.33**	0.26	**0.42**	0.30	**0.48**	0.33	**0.54**	0.39	**0.63**	0.37	**0.60**
Spleen	0.06	**0.07**	0.08	**0.10**	0.09	**0.12**	0.11	**0.14**	0.14	**0.18**	0.12	**0.15**
Liver	0.27	**0.43**	0.33	**0.53**	0.38	**0.61**	0.42	**0.67**	0.50	**0.79**	0.48	**0.75**
Gall bladder	0.09	**0.30**	0.12	**0.39**	0.15	**0.46**	0.16	**0.52**	0.20	**0.62**	0.18	**0.58**
Pancreas	0.13	**0.16**	0.17	**0.21**	0.20	**0.25**	0.23	**0.29**	0.29	**0.36**	0.26	**0.32**
Small intestine		**0.16**	0.01	**0.20**	0.01	**0.24**	0.02	**0.27**	0.02	**0.33**	0.02	**0.30**
Large intestine	0.01	**0.23**	0.02	**0.29**	0.02	**0.34**	0.02	**0.38**	0.03	**0.45**	0.02	**0.42**
Sigma/rectum		**0.04**		**0.06**		**0.07**		**0.08**		**0.09**		**0.09**
Lumbar spine	0.04	**0.13**	0.05	**0.17**	0.07	**0.21**	0.08	**0.25**	0.10	**0.32**	0.08	**0.26**
Pelvis		**0.06**		**0.09**		**0.10**		**0.12**	0.01	**0.15**		**0.13**
Adrenals	0.07	**0.08**	0.09	**0.10**	0.11	**0.13**	0.13	**0.15**	0.16	**0.19**	0.14	**0.16**
Kidneys	0.03	**0.06**	0.04	**0.09**	0.05	**0.11**	0.06	**0.12**	0.08	**0.15**	0.07	**0.13**
Urinary bladder				**0.01**		**0.01**		**0.02**		**0.02**		**0.02**
Uterus		**0.03**		**0.04**		**0.05**		**0.05**		**0.07**		**0.06**
Ovaries		**0.03**		**0.04**		**0.05**		**0.06**		**0.08**		**0.07**
Prostate												
Testes												
Upper arm	0.30	**0.79**	0.38	**0.97**	0.44	**1.10**	0.49	**1.20**	0.58	**1.39**	0.55	**1.34**
Elbow	0.27	**0.85**	0.34	**1.04**	0.40	**1.18**	0.45	**1.29**	0.53	**1.49**	0.50	**1.45**
Forearm	0.02	**0.47**	0.03	**0.57**	0.04	**0.64**	0.04	**0.70**	0.05	**0.81**	0.05	**0.79**
Upper leg												
Knee												
Lower leg												
Lymph nodes	0.14	**0.26**	0.17	**0.32**	0.19	**0.37**	0.21	**0.41**	0.25	**0.48**	0.23	**0.45**
Skin	0.13	**0.25**	0.14	**0.27**	0.15	**0.28**	0.15	**0.28**	0.16	**0.30**	0.15	**0.29**
Muscle	0.08	**0.14**	0.10	**0.17**	0.11	**0.19**	0.12	**0.21**	0.14	**0.24**	0.14	**0.22**
Skeleton	0.19	**0.33**	0.23	**0.40**	0.26	**0.46**	0.28	**0.50**	0.32	**0.58**	0.31	**0.56**
Red bone marrow	0.06	**0.09**	0.07	**0.11**	0.08	**0.13**	0.09	**0.14**	0.10	**0.17**	0.09	**0.15**
Whole body dose	0.10	**0.17**	0.12	**0.21**	0.14	**0.24**	0.15	**0.26**	0.18	**0.30**	0.17	**0.28**
ED (ICRP 60)	0.18	**0.24**	0.22	**0.29**	0.24	**0.33**	0.26	**0.35**	0.30	**0.41**	0.29	**0.39**
ED (ICRP 103)	0.26	**0.32**	0.30	**0.38**	0.33	**0.42**	0.35	**0.45**	0.39	**0.51**	0.38	**0.49**
Total filtration (mm) **Al**	**1.5**		**2.5**		**3.5**		**4.5**		**3.5**		**3.5**	
Cu	**0**		**0**		**0**		**0**		**0.1**		**0.2**	

Table 10.44 Normalized organ doses—Bony thorax RAO—3–7 years

Beam projection:	RAO	**Table cells:**
Focus-to-detector distance (cm):	115	(+) = normal font = optimal field setting
Tube voltage (kVp):	70	**(−) = bold font = suboptimal field setting**
Total filtrations:	See last table row	Empty table cells correspond to values <0.01

Organ or tissue	**Organ dose D_T: Entrance dose K_E**							**(mSv * mGy^{-1})**				
	(+)	(−)	(+)	(−)	(+)	(−)	(+)	(−)	(+)	(−)	(+)	(−)
Skull	0.03	**0.29**	0.04	**0.35**	0.04	**0.39**	0.05	**0.43**	0.06	**0.49**	0.07	**0.54**
Brain		**0.02**		**0.02**		**0.03**		**0.03**		**0.04**		**0.05**
Oral mucosa	0.03	**0.40**	0.04	**0.48**	0.04	**0.53**	0.05	**0.57**	0.06	**0.65**	0.07	**0.72**
Salivary glands	0.06	**0.38**	0.08	**0.44**	0.09	**0.49**	0.10	**0.52**	0.12	**0.59**	0.13	**0.65**
Cervical spine	0.37	**0.49**	0.47	**0.63**	0.54	**0.73**	0.60	**0.82**	0.72	**0.99**	0.83	**1.16**
Thyroid	0.79	**0.83**	0.90	**0.94**	0.98	**1.02**	1.03	**1.08**	1.12	**1.18**	1.19	**1.27**
Thymus	0.69	**0.71**	0.83	**0.85**	0.92	**0.95**	1.00	**1.02**	1.13	**1.16**	1.24	**1.27**
Thoracic spine	0.25	**0.27**	0.33	**0.36**	0.39	**0.42**	0.44	**0.48**	0.56	**0.61**	0.67	**0.73**
Claviculae	1.59	**1.68**	1.89	**1.99**	2.09	**2.21**	2.25	**2.38**	2.51	**2.66**	2.71	**2.89**
Scapulae	0.18	**0.24**	0.24	**0.31**	0.28	**0.37**	0.32	**0.42**	0.40	**0.53**	0.48	**0.63**
Ribs	1.28	**1.35**	1.48	**1.57**	1.62	**1.72**	1.72	**1.83**	1.89	**2.02**	2.02	**2.18**
Oesophagus	0.15	**0.16**	0.19	**0.21**	0.23	**0.25**	0.26	**0.28**	0.31	**0.34**	0.37	**0.40**
Trachea	0.11	**0.34**	0.13	**0.41**	0.14	**0.47**	0.16	**0.51**	0.17	**0.59**	0.19	**0.66**
Lung	0.35	**0.37**	0.43	**0.45**	0.49	**0.52**	0.53	**0.56**	0.62	**0.66**	0.70	**0.74**
Mamma	1.11	**1.14**	1.20	**1.22**	1.24	**1.27**	1.27	**1.31**	1.33	**1.37**	1.37	**1.42**
Heart	0.43	**0.45**	0.53	**0.56**	0.61	**0.64**	0.67	**0.70**	0.78	**0.82**	0.88	**0.93**
Stomach	0.24	**0.40**	0.30	**0.49**	0.35	**0.56**	0.38	**0.62**	0.45	**0.72**	0.51	**0.82**
Spleen	0.08	**0.10**	0.10	**0.13**	0.12	**0.16**	0.14	**0.18**	0.17	**0.22**	0.21	**0.27**
Liver	0.32	**0.50**	0.39	**0.61**	0.44	**0.70**	0.48	**0.76**	0.56	**0.88**	0.63	**0.99**
Gall bladder	0.12	**0.37**	0.15	**0.47**	0.18	**0.55**	0.20	**0.61**	0.24	**0.73**	0.28	**0.83**
Pancreas	0.17	**0.21**	0.22	**0.27**	0.25	**0.32**	0.29	**0.36**	0.35	**0.44**	0.41	**0.52**
Small intestine	0.01	**0.20**	0.02	**0.25**	0.02	**0.29**	0.02	**0.32**	0.03	**0.39**	0.04	**0.45**
Large intestine	0.02	**0.27**	0.02	**0.34**	0.03	**0.40**	0.03	**0.44**	0.04	**0.52**	0.05	**0.59**
Sigma/rectum		**0.06**		**0.07**		**0.08**		**0.09**		**0.11**		**0.13**
Lumbar spine	0.06	**0.18**	0.08	**0.24**	0.09	**0.29**	0.11	**0.33**	0.14	**0.42**	0.17	**0.52**
Pelvis		**0.09**		**0.11**	0.01	**0.13**	0.01	**0.15**	0.02	**0.19**	0.02	**0.23**
Adrenals	0.09	**0.11**	0.12	**0.14**	0.14	**0.17**	0.16	**0.19**	0.20	**0.24**	0.24	**0.29**
Kidneys	0.05	**0.09**	0.06	**0.12**	0.07	**0.14**	0.08	**0.16**	0.10	**0.20**	0.13	**0.24**
Urinary bladder		**0.01**		**0.02**		**0.02**		**0.02**		**0.03**		**0.04**
Uterus		**0.04**		**0.05**		**0.06**		**0.07**		**0.09**		**0.11**
Ovaries		**0.04**		**0.06**		**0.07**		**0.08**		**0.10**	0.01	**0.12**
Prostate										**0.01**		**0.02**
Testes												
Upper arm	0.36	**0.90**	0.45	**1.09**	0.51	**1.22**	0.56	**1.33**	0.66	**1.52**	0.75	**1.68**
Elbow	0.33	**0.97**	0.41	**1.17**	0.47	**1.31**	0.52	**1.42**	0.61	**1.62**	0.70	**1.80**
Forearm	0.03	**0.53**	0.04	**0.64**	0.05	**0.71**	0.05	**0.77**	0.06	**0.88**	0.08	**0.97**
Upper leg												
Knee												
Lower leg												
Lymph nodes	0.16	**0.31**	0.20	**0.38**	0.22	**0.43**	0.24	**0.47**	0.28	**0.55**	0.32	**0.62**
Skin	0.14	**0.26**	0.15	**0.28**	0.15	**0.29**	0.16	**0.30**	0.17	**0.31**	0.18	**0.33**
Muscle	0.10	**0.16**	0.12	**0.19**	0.13	**0.21**	0.14	**0.23**	0.16	**0.27**	0.18	**0.30**
Skeleton	0.21	**0.38**	0.26	**0.46**	0.29	**0.51**	0.31	**0.56**	0.36	**0.64**	0.40	**0.71**
Red bone marrow	0.07	**0.11**	0.08	**0.13**	0.09	**0.15**	0.10	**0.17**	0.12	**0.20**	0.13	**0.22**
Whole body dose	0.12	**0.20**	0.14	**0.24**	0.16	**0.27**	0.17	**0.29**	0.20	**0.34**	0.22	**0.37**
ED (ICRP 60)	0.21	**0.28**	0.24	**0.33**	0.27	**0.37**	0.29	**0.40**	0.33	**0.45**	0.36	**0.50**
ED (ICRP 103)	0.29	**0.36**	0.33	**0.42**	0.36	**0.47**	0.38	**0.50**	0.42	**0.56**	0.46	**0.62**
Total filtration (mm) Al	**1.5**		**2.5**		**3.5**		**4.5**		**3.5**		**3.5**	
Cu	**0**		**0**		**0**		**0**		**0.1**		**0.2**	

Table 10.45 Normalized organ doses—Bony thorax RAO—8–12 years

Beam projection:			RAO				**Table cells:**					
Focus-to-detector distance (cm):			115				(+) = normal font = optimal field setting					
Tube voltage (kVp):			60				**(−) = bold font = suboptimal field setting**					
Total filtrations:			See last table row				Empty table cells correspond to values <0.01					
Organ or tissue	**Organ dose D_T: Entrance dose K_E**							**($mSv * mGy^{-1}$)**				
	(+)	(−)	(+)	(−)	(+)	(−)	(+)	(−)	(+)	(−)	(+)	(−)
Skull	0.03	**0.25**	0.04	**0.31**	0.05	**0.35**	0.05	**0.39**	0.06	**0.45**	0.06	**0.43**
Brain		**0.01**		**0.01**		**0.02**		**0.02**		**0.03**		**0.02**
Oral mucosa	0.02	**0.32**	0.03	**0.39**	0.03	**0.43**	0.04	**0.47**	0.05	**0.54**	0.04	**0.51**
Salivary glands	0.07	**0.32**	0.09	**0.38**	0.10	**0.42**	0.11	**0.45**	0.13	**0.51**	0.13	**0.49**
Cervical spine	0.30	**0.37**	0.39	**0.49**	0.47	**0.58**	0.53	**0.66**	0.65	**0.82**	0.59	**0.74**
Thyroid	0.66	**0.69**	0.77	**0.81**	0.85	**0.89**	0.90	**0.95**	0.99	**1.05**	0.98	**1.03**
Thymus	0.56	**0.58**	0.69	**0.71**	0.79	**0.81**	0.86	**0.88**	0.99	**1.02**	0.96	**0.98**
Thoracic spine	0.13	**0.14**	0.18	**0.19**	0.22	**0.23**	0.25	**0.27**	0.33	**0.35**	0.27	**0.28**
Claviculae	1.28	**1.34**	1.57	**1.65**	1.77	**1.86**	1.93	**2.03**	2.21	**2.33**	2.18	**2.29**
Scapulae	0.11	**0.14**	0.15	**0.19**	0.19	**0.23**	0.22	**0.26**	0.28	**0.34**	0.24	**0.29**
Ribs	1.08	**1.16**	1.27	**1.37**	1.41	**1.51**	1.50	**1.62**	1.68	**1.82**	1.66	**1.79**
Oesophagus	0.08	**0.08**	0.10	**0.11**	0.13	**0.13**	0.14	**0.15**	0.18	**0.20**	0.16	**0.17**
Trachea	0.22	**0.40**	0.26	**0.50**	0.30	**0.57**	0.32	**0.63**	0.36	**0.73**	0.35	**0.70**
Lung	0.26	**0.28**	0.33	**0.35**	0.38	**0.41**	0.43	**0.45**	0.51	**0.54**	0.48	**0.50**
Mamma	0.96	**1.08**	1.05	**1.17**	1.10	**1.23**	1.13	**1.27**	1.18	**1.33**	1.17	**1.32**
Heart	0.31	**0.33**	0.40	**0.42**	0.47	**0.49**	0.52	**0.54**	0.62	**0.65**	0.58	**0.61**
Stomach	0.19	**0.29**	0.24	**0.37**	0.28	**0.43**	0.32	**0.47**	0.38	**0.57**	0.35	**0.53**
Spleen	0.04	**0.06**	0.06	**0.08**	0.07	**0.09**	0.08	**0.11**	0.11	**0.14**	0.09	**0.11**
Liver	0.25	**0.37**	0.32	**0.46**	0.37	**0.54**	0.41	**0.59**	0.49	**0.71**	0.46	**0.66**
Gall bladder	0.10	**0.27**	0.13	**0.35**	0.15	**0.41**	0.17	**0.47**	0.21	**0.57**	0.19	**0.52**
Pancreas	0.11	**0.13**	0.14	**0.17**	0.17	**0.21**	0.20	**0.24**	0.25	**0.30**	0.22	**0.26**
Small intestine		**0.09**		**0.12**	0.01	**0.14**	0.01	**0.16**	0.02	**0.19**	0.01	**0.17**
Large intestine		**0.15**	0.01	**0.19**	0.02	**0.23**	0.02	**0.26**	0.03	**0.31**	0.02	**0.29**
Sigma/rectum		**0.02**		**0.03**		**0.04**		**0.04**		**0.05**		**0.05**
Lumbar spine	0.03	**0.09**	0.04	**0.12**	0.05	**0.14**	0.06	**0.17**	0.08	**0.22**	0.06	**0.17**
Pelvis		**0.03**		**0.03**		**0.04**		**0.05**		**0.06**		**0.05**
Adrenals	0.05	**0.06**	0.07	**0.08**	0.09	**0.10**	0.10	**0.12**	0.13	**0.15**	0.11	**0.12**
Kidneys	0.03	**0.05**	0.04	**0.06**	0.05	**0.08**	0.05	**0.09**	0.07	**0.12**	0.06	**0.10**
Urinary bladder												
Uterus		**0.01**		**0.02**		**0.02**		**0.02**		**0.03**		**0.02**
Ovaries		**0.01**		**0.02**		**0.02**		**0.02**		**0.03**		**0.02**
Prostate												
Testes												
Upper arm	0.29	**0.68**	0.36	**0.85**	0.42	**0.96**	0.46	**1.06**	0.55	**1.24**	0.52	**1.19**
Elbow	0.27	**0.74**	0.34	**0.92**	0.39	**1.04**	0.43	**1.14**	0.52	**1.33**	0.48	**1.28**
Forearm	0.02	**0.33**	0.03	**0.41**	0.03	**0.47**	0.04	**0.51**	0.05	**0.60**	0.04	**0.58**
Upper leg												
Knee												
Lower leg												
Lymph nodes	0.14	**0.23**	0.17	**0.29**	0.20	**0.34**	0.22	**0.37**	0.26	**0.44**	0.24	**0.41**
Skin	0.12	**0.22**	0.13	**0.23**	0.14	**0.24**	0.14	**0.25**	0.15	**0.26**	0.14	**0.26**
Muscle	0.08	**0.12**	0.09	**0.14**	0.11	**0.16**	0.12	**0.17**	0.13	**0.20**	0.13	**0.19**
Skeleton	0.17	**0.28**	0.21	**0.34**	0.23	**0.39**	0.25	**0.43**	0.29	**0.49**	0.28	**0.47**
Red bone marrow	0.06	**0.08**	0.07	**0.10**	0.08	**0.12**	0.09	**0.13**	0.11	**0.16**	0.10	**0.14**
Whole body dose	0.09	**0.15**	0.12	**0.18**	0.13	**0.20**	0.14	**0.22**	0.17	**0.26**	0.16	**0.24**
ED (ICRP 60)	0.17	**0.21**	0.20	**0.26**	0.22	**0.29**	0.24	**0.31**	0.28	**0.36**	0.26	**0.34**
ED (ICRP 103)	0.24	**0.30**	0.27	**0.35**	0.30	**0.39**	0.32	**0.41**	0.36	**0.47**	0.35	**0.45**
Total filtration (mm) Al	**1.5**		**2.5**		**3.5**		**4.5**		**3.5**		**3.5**	
Cu	**0**		**0**		**0**		**0**		**0.1**		**0.2**	

Table 10.46 Normalized organ doses—Bony thorax RAO—8–12 years

Beam projection:			RAO			**Table cells:**						
Focus-to-detector distance (cm):			115			(+) = normal font = optimal field setting						
Tube voltage (kVp):			70			**(−) = bold font = suboptimal field setting**						
Total filtrations:			See last table row			Empty table cells correspond to values <0.01						
Organ or tissue	**Organ dose D_T: Entrance dose K_E**							**(mSv * mGy^{-1})**				
	(+)	(−)	(+)	(−)	(+)	(−)	(+)	(−)	(+)	(−)	(+)	(−)
Skull	0.04	**0.29**	0.05	**0.35**	0.05	**0.40**	0.06	**0.44**	0.07	**0.50**	0.08	**0.56**
Brain		**0.02**		**0.02**		**0.03**		**0.03**		**0.04**		**0.05**
Oral mucosa	0.03	**0.37**	0.04	**0.44**	0.04	**0.49**	0.05	**0.53**	0.06	**0.60**	0.07	**0.67**
Salivary glands	0.08	**0.36**	0.10	**0.42**	0.12	**0.47**	0.13	**0.50**	0.15	**0.57**	0.17	**0.62**
Cervical spine	0.38	**0.47**	0.48	**0.61**	0.56	**0.72**	0.63	**0.81**	0.76	**0.98**	0.88	**1.15**
Thyroid	0.73	**0.77**	0.84	**0.89**	0.91	**0.97**	0.97	**1.03**	1.06	**1.13**	1.14	**1.22**
Thymus	0.65	**0.67**	0.78	**0.81**	0.88	**0.91**	0.96	**0.99**	1.10	**1.13**	1.22	**1.25**
Thoracic spine	0.18	**0.20**	0.24	**0.26**	0.29	**0.31**	0.34	**0.36**	0.43	**0.46**	0.52	**0.56**
Claviculae	1.45	**1.52**	1.74	**1.83**	1.94	**2.04**	2.09	**2.21**	2.36	**2.50**	2.58	**2.74**
Scapulae	0.16	**0.19**	0.21	**0.25**	0.25	**0.30**	0.28	**0.34**	0.35	**0.43**	0.43	**0.52**
Ribs	1.19	**1.28**	1.38	**1.49**	1.51	**1.64**	1.61	**1.74**	1.78	**1.94**	1.91	**2.09**
Oesophagus	0.10	**0.11**	0.14	**0.15**	0.16	**0.18**	0.18	**0.20**	0.23	**0.25**	0.28	**0.30**
Trachea	0.25	**0.47**	0.29	**0.58**	0.33	**0.65**	0.35	**0.71**	0.40	**0.83**	0.44	**0.93**
Lung	0.31	**0.33**	0.39	**0.41**	0.45	**0.47**	0.49	**0.52**	0.58	**0.61**	0.66	**0.70**
Mamma	1.01	**1.13**	1.09	**1.22**	1.14	**1.27**	1.17	**1.31**	1.23	**1.38**	1.27	**1.43**
Heart	0.38	**0.40**	0.48	**0.50**	0.55	**0.57**	0.61	**0.64**	0.72	**0.75**	0.82	**0.86**
Stomach	0.23	**0.35**	0.29	**0.44**	0.33	**0.50**	0.37	**0.55**	0.44	**0.65**	0.50	**0.75**
Spleen	0.06	**0.08**	0.08	**0.10**	0.10	**0.13**	0.11	**0.14**	0.14	**0.18**	0.17	**0.22**
Liver	0.30	**0.44**	0.38	**0.55**	0.43	**0.62**	0.48	**0.69**	0.56	**0.81**	0.64	**0.92**
Gall bladder	0.12	**0.34**	0.16	**0.43**	0.19	**0.50**	0.21	**0.56**	0.26	**0.67**	0.30	**0.78**
Pancreas	0.14	**0.17**	0.19	**0.22**	0.22	**0.27**	0.25	**0.30**	0.31	**0.38**	0.38	**0.45**
Small intestine		**0.11**	0.01	**0.15**	0.02	**0.17**	0.02	**0.19**	0.02	**0.24**	0.03	**0.28**
Large intestine	0.01	**0.19**	0.02	**0.24**	0.02	**0.28**	0.03	**0.31**	0.03	**0.37**	0.04	**0.43**
Sigma/rectum		**0.03**		**0.04**		**0.05**		**0.05**		**0.06**		**0.08**
Lumbar spine	0.05	**0.13**	0.06	**0.17**	0.08	**0.20**	0.09	**0.23**	0.12	**0.30**	0.15	**0.37**
Pelvis		**0.04**		**0.05**		**0.06**		**0.07**	0.01	**0.09**	0.01	**0.11**
Adrenals	0.07	**0.09**	0.10	**0.11**	0.12	**0.14**	0.14	**0.16**	0.17	**0.20**	0.21	**0.25**
Kidneys	0.04	**0.07**	0.05	**0.09**	0.06	**0.11**	0.07	**0.12**	0.09	**0.16**	0.11	**0.19**
Urinary bladder										**0.01**		**0.02**
Uterus		**0.02**		**0.02**		**0.03**		**0.03**		**0.04**		**0.05**
Ovaries		**0.02**		**0.02**		**0.03**		**0.03**		**0.04**		**0.05**
Prostate												
Testes												
Upper arm	0.34	**0.79**	0.42	**0.96**	0.49	**1.09**	0.53	**1.19**	0.63	**1.37**	0.71	**1.53**
Elbow	0.32	**0.86**	0.40	**1.04**	0.45	**1.17**	0.50	**1.28**	0.59	**1.47**	0.67	**1.63**
Forearm	0.03	**0.38**	0.04	**0.47**	0.04	**0.53**	0.05	**0.57**	0.06	**0.66**	0.07	**0.73**
Upper leg												
Knee												
Lower leg												
Lymph nodes	0.17	**0.28**	0.20	**0.34**	0.23	**0.39**	0.25	**0.43**	0.30	**0.51**	0.34	**0.58**
Skin	0.13	**0.23**	0.14	**0.24**	0.14	**0.25**	0.15	**0.26**	0.16	**0.27**	0.16	**0.29**
Muscle	0.09	**0.14**	0.11	**0.16**	0.12	**0.18**	0.13	**0.20**	0.15	**0.23**	0.17	**0.26**
Skeleton	0.19	**0.32**	0.23	**0.39**	0.26	**0.44**	0.28	**0.48**	0.33	**0.55**	0.36	**0.61**
Red bone marrow	0.07	**0.10**	0.09	**0.12**	0.10	**0.14**	0.11	**0.15**	0.13	**0.18**	0.14	**0.21**
Whole body dose	0.11	**0.17**	0.13	**0.21**	0.15	**0.23**	0.16	**0.25**	0.19	**0.29**	0.21	**0.33**
ED (ICRP 60)	0.19	**0.24**	0.23	**0.29**	0.25	**0.33**	0.27	**0.35**	0.31	**0.41**	0.34	**0.45**
ED (ICRP 103)	0.26	**0.33**	0.30	**0.39**	0.33	**0.43**	0.36	**0.46**	0.40	**0.52**	0.44	**0.58**
Total filtration (mm) Al	**1.5**		**2.5**		**3.5**		**4.5**		**3.5**		**3.5**	
Cu	**0**		**0**		**0**		**0**		**0.1**		**0.2**	

Table 10.47 Normalized organ doses—Bony thorax RAO—13–17 years

Beam projection:	RAO	**Table cells:**
Focus-to-detector distance (cm):	115	(+) = normal font = optimal field setting
Tube voltage (kVp):	60	**(−) = bold font = suboptimal field setting**
Total filtrations:	See last table row	Empty table cells correspond to values <0.01

Organ or tissue	**Organ dose D_T: Entrance dose K_E**						**(mSv * mGy^{-1})**					
	(+)	(−)	(+)	(−)	(+)	(−)	(+)	(−)	(+)	(−)	(+)	(−)
Skull	0.02	**0.26**	0.03	**0.33**	0.04	**0.38**	0.04	**0.42**	0.05	**0.49**	0.04	**0.46**
Brain		**0.02**		**0.02**		**0.03**		**0.03**		**0.04**		**0.03**
Oral mucosa	0.02	**0.25**	0.02	**0.31**	0.02	**0.35**	0.03	**0.38**	0.03	**0.44**	0.03	**0.41**
Salivary glands	0.06	**0.27**	0.07	**0.33**	0.08	**0.36**	0.09	**0.39**	0.11	**0.45**	0.10	**0.43**
Cervical spine	0.31	**0.37**	0.41	**0.48**	0.49	**0.58**	0.56	**0.66**	0.69	**0.82**	0.62	**0.73**
Thyroid	0.57	**0.58**	0.67	**0.68**	0.74	**0.75**	0.79	**0.81**	0.88	**0.91**	0.86	**0.88**
Thymus	0.48	**0.48**	0.60	**0.61**	0.69	**0.70**	0.76	**0.77**	0.89	**0.90**	0.85	**0.86**
Thoracic spine	0.06	**0.06**	0.09	**0.09**	0.11	**0.11**	0.12	**0.13**	0.17	**0.17**	0.13	**0.13**
Claviculae	1.01	**1.03**	1.25	**1.28**	1.42	**1.46**	1.56	**1.60**	1.81	**1.86**	1.76	**1.81**
Scapulae	0.07	**0.08**	0.09	**0.11**	0.11	**0.13**	0.13	**0.15**	0.17	**0.20**	0.14	**0.16**
Ribs	0.90	**0.93**	1.07	**1.11**	1.19	**1.24**	1.28	**1.34**	1.45	**1.51**	1.42	**1.47**
Oesophagus	0.06	**0.07**	0.09	**0.09**	0.11	**0.11**	0.12	**0.13**	0.16	**0.16**	0.13	**0.13**
Trachea	0.20	**0.28**	0.24	**0.34**	0.27	**0.39**	0.29	**0.43**	0.33	**0.50**	0.33	**0.48**
Lung	0.17	**0.18**	0.22	**0.23**	0.26	**0.27**	0.30	**0.31**	0.36	**0.38**	0.33	**0.34**
Mamma	5.61	**7.02**	6.51	**8.15**	7.11	**8.91**	7.56	**9.47**	8.34	**10.4**	8.18	**10.3**
Heart	0.24	**0.25**	0.31	**0.32**	0.37	**0.38**	0.41	**0.42**	0.50	**0.51**	0.46	**0.47**
Stomach	0.15	**0.20**	0.20	**0.26**	0.23	**0.31**	0.26	**0.35**	0.32	**0.42**	0.29	**0.38**
Spleen	0.03	**0.03**	0.04	**0.05**	0.05	**0.06**	0.06	**0.07**	0.07	**0.09**	0.06	**0.07**
Liver	0.22	**0.29**	0.28	**0.37**	0.33	**0.43**	0.37	**0.48**	0.44	**0.58**	0.41	**0.53**
Gall bladder	0.08	**0.23**	0.11	**0.30**	0.13	**0.36**	0.15	**0.41**	0.19	**0.50**	0.16	**0.45**
Pancreas	0.07	**0.08**	0.10	**0.11**	0.12	**0.14**	0.14	**0.16**	0.18	**0.21**	0.15	**0.17**
Small intestine		**0.08**		**0.10**		**0.12**		**0.14**	0.01	**0.17**		**0.15**
Large intestine		**0.12**	0.01	**0.16**	0.01	**0.19**	0.01	**0.21**	0.02	**0.27**	0.02	**0.24**
Sigma/rectum		**0.02**		**0.03**		**0.03**		**0.04**		**0.05**		**0.04**
Lumbar spine	0.02	**0.05**	0.03	**0.06**	0.03	**0.08**	0.04	**0.10**	0.05	**0.13**	0.04	**0.09**
Pelvis		**0.02**		**0.03**		**0.03**		**0.04**		**0.05**		**0.04**
Adrenals	0.03	**0.04**	0.05	**0.05**	0.06	**0.06**	0.07	**0.07**	0.09	**0.10**	0.07	**0.08**
Kidneys	0.02	**0.03**	0.02	**0.04**	0.03	**0.05**	0.04	**0.06**	0.05	**0.07**	0.04	**0.06**
Urinary bladder												
Uterus		**0.03**		**0.04**		**0.04**		**0.05**		**0.07**		**0.05**
Ovaries				**0.01**		**0.02**		**0.02**		**0.03**		**0.02**
Prostate												
Testes												
Upper arm	0.18	**0.52**	0.23	**0.65**	0.27	**0.75**	0.30	**0.83**	0.36	**0.97**	0.33	**0.92**
Elbow	0.15	**0.56**	0.20	**0.70**	0.23	**0.80**	0.26	**0.88**	0.32	**1.04**	0.29	**0.98**
Forearm	0.02	**0.27**	0.02	**0.33**	0.03	**0.38**	0.03	**0.42**	0.04	**0.49**	0.03	**0.47**
Upper leg												
Knee												
Lower leg												
Lymph nodes	0.11	**0.18**	0.14	**0.23**	0.16	**0.26**	0.18	**0.29**	0.21	**0.35**	0.20	**0.32**
Skin	0.13	**0.22**	0.14	**0.24**	0.15	**0.25**	0.15	**0.26**	0.16	**0.27**	0.16	**0.26**
Muscle	0.07	**0.10**	0.08	**0.12**	0.09	**0.14**	0.10	**0.15**	0.12	**0.18**	0.11	**0.16**
Skeleton	0.14	**0.23**	0.17	**0.29**	0.19	**0.33**	0.21	**0.36**	0.25	**0.42**	0.23	**0.40**
Red bone marrow	0.06	**0.07**	0.07	**0.09**	0.08	**0.11**	0.09	**0.12**	0.10	**0.14**	0.10	**0.13**
Whole body dose	0.09	**0.13**	0.10	**0.16**	0.12	**0.18**	0.13	**0.19**	0.15	**0.23**	0.14	**0.21**
ED (ICRP 60)	0.38	**0.47**	0.44	**0.56**	0.49	**0.62**	0.53	**0.67**	0.60	**0.76**	0.58	**0.73**
ED (ICRP 103)	0.77	**0.97**	0.90	**1.14**	0.99	**1.25**	1.06	**1.34**	1.18	**1.50**	1.15	**1.46**
Total filtration (mm) Al	**1.5**		**2.5**		**3.5**		**4.5**		**3.5**		**3.5**	
Cu	**0**		**0**		**0**		**0**		**0.1**		**0.2**	

Table 10.48 Normalized organ doses—Bony thorax RAO—13–17 years

Beam projection:			RAO			**Table cells:**						
Focus-to-detector distance (cm):			115			(+) = normal font = optimal field setting						
Tube voltage (kVp):			70			**(−) = bold font = suboptimal field setting**						
Total filtrations:			See last table row			Empty table cells correspond to values <0.01						
Organ or tissue	**Organ dose D_T: Entrance dose K_E**						**(mSv * mGy^{-1})**					
	(+)	(−)	(+)	(−)	(+)	(−)	(+)	(−)	(+)	(−)	(+)	(−)
Skull	0.03	**0.31**	0.04	**0.38**	0.04	**0.43**	0.05	**0.47**	0.06	**0.55**	0.07	**0.63**
Brain		**0.02**		**0.03**		**0.04**		**0.04**		**0.05**		**0.07**
Oral mucosa	0.02	**0.30**	0.03	**0.35**	0.03	**0.40**	0.03	**0.43**	0.04	**0.49**	0.05	**0.55**
Salivary glands	0.07	**0.31**	0.08	**0.37**	0.10	**0.41**	0.11	**0.44**	0.12	**0.50**	0.14	**0.55**
Cervical spine	0.40	**0.48**	0.52	**0.61**	0.60	**0.72**	0.68	**0.82**	0.83	**1.01**	0.97	**1.19**
Thyroid	0.63	**0.64**	0.73	**0.75**	0.80	**0.82**	0.86	**0.88**	0.95	**0.98**	1.03	**1.06**
Thymus	0.56	**0.57**	0.69	**0.70**	0.79	**0.80**	0.86	**0.87**	1.00	**1.02**	1.12	**1.14**
Thoracic spine	0.09	**0.10**	0.13	**0.13**	0.15	**0.16**	0.18	**0.19**	0.23	**0.25**	0.29	**0.31**
Claviculae	1.16	**1.19**	1.41	**1.44**	1.58	**1.63**	1.72	**1.77**	1.97	**2.03**	2.18	**2.25**
Scapulae	0.10	**0.11**	0.13	**0.15**	0.15	**0.18**	0.18	**0.21**	0.23	**0.27**	0.28	**0.33**
Ribs	1.00	**1.04**	1.17	**1.22**	1.30	**1.35**	1.39	**1.45**	1.55	**1.63**	1.69	**1.78**
Oesophagus	0.09	**0.09**	0.12	**0.12**	0.14	**0.15**	0.17	**0.17**	0.21	**0.22**	0.26	**0.27**
Trachea	0.23	**0.33**	0.27	**0.40**	0.30	**0.45**	0.32	**0.49**	0.37	**0.57**	0.40	**0.64**
Lung	0.22	**0.22**	0.27	**0.28**	0.32	**0.33**	0.36	**0.37**	0.43	**0.45**	0.50	**0.52**
Mamma	6.15	**7.69**	7.06	**8.83**	7.66	**9.59**	8.12	**10.2**	8.91	**11.2**	9.58	**12.0**
Heart	0.30	**0.31**	0.38	**0.39**	0.45	**0.46**	0.50	**0.51**	0.60	**0.61**	0.69	**0.71**
Stomach	0.19	**0.25**	0.24	**0.32**	0.28	**0.37**	0.31	**0.42**	0.38	**0.50**	0.44	**0.58**
Spleen	0.04	**0.05**	0.06	**0.07**	0.07	**0.08**	0.08	**0.09**	0.10	**0.12**	0.12	**0.15**
Liver	0.27	**0.35**	0.34	**0.44**	0.39	**0.51**	0.43	**0.57**	0.52	**0.68**	0.60	**0.78**
Gall bladder	0.11	**0.29**	0.14	**0.38**	0.17	**0.44**	0.19	**0.49**	0.23	**0.60**	0.27	**0.70**
Pancreas	0.10	**0.12**	0.14	**0.16**	0.16	**0.19**	0.19	**0.21**	0.24	**0.27**	0.29	**0.33**
Small intestine		**0.10**		**0.13**	0.01	**0.15**	0.01	**0.17**	0.02	**0.22**	0.02	**0.26**
Large intestine	0.01	**0.16**	0.02	**0.20**	0.02	**0.23**	0.02	**0.26**	0.03	**0.32**	0.03	**0.38**
Sigma/rectum		**0.03**		**0.04**		**0.04**		**0.05**		**0.06**		**0.07**
Lumbar spine	0.03	**0.07**	0.04	**0.10**	0.05	**0.12**	0.06	**0.14**	0.08	**0.18**	0.10	**0.23**
Pelvis		**0.03**		**0.04**		**0.05**		**0.06**		**0.07**		**0.09**
Adrenals	0.05	**0.06**	0.07	**0.07**	0.08	**0.09**	0.09	**0.10**	0.12	**0.13**	0.15	**0.17**
Kidneys	0.03	**0.04**	0.04	**0.06**	0.04	**0.07**	0.05	**0.08**	0.07	**0.10**	0.08	**0.13**
Urinary bladder												**0.01**
Uterus		**0.04**		**0.05**		**0.07**		**0.08**		**0.10**	0.01	**0.12**
Ovaries		**0.01**		**0.02**		**0.02**		**0.03**		**0.04**		**0.05**
Prostate												
Testes												
Upper arm	0.22	**0.62**	0.28	**0.75**	0.32	**0.86**	0.35	**0.94**	0.42	**1.09**	0.49	**1.23**
Elbow	0.19	**0.66**	0.24	**0.81**	0.28	**0.91**	0.31	**1.00**	0.38	**1.16**	0.44	**1.30**
Forearm	0.02	**0.31**	0.03	**0.38**	0.03	**0.43**	0.04	**0.47**	0.05	**0.55**	0.06	**0.62**
Upper leg												
Knee												
Lower leg												
Lymph nodes	0.14	**0.22**	0.17	**0.27**	0.19	**0.31**	0.21	**0.34**	0.25	**0.41**	0.29	**0.47**
Skin	0.14	**0.23**	0.15	**0.25**	0.16	**0.26**	0.16	**0.27**	0.17	**0.28**	0.18	**0.30**
Muscle	0.08	**0.12**	0.10	**0.14**	0.11	**0.16**	0.12	**0.17**	0.14	**0.20**	0.16	**0.23**
Skeleton	0.16	**0.27**	0.20	**0.33**	0.22	**0.37**	0.24	**0.41**	0.28	**0.47**	0.31	**0.53**
Red bone marrow	0.07	**0.09**	0.08	**0.11**	0.09	**0.13**	0.10	**0.14**	0.12	**0.17**	0.14	**0.19**
Whole body dose	0.10	**0.15**	0.12	**0.18**	0.14	**0.21**	0.15	**0.22**	0.17	**0.26**	0.20	**0.30**
ED (ICRP 60)	0.42	**0.53**	0.49	**0.62**	0.54	**0.69**	0.58	**0.74**	0.65	**0.83**	0.72	**0.91**
ED (ICRP 103)	0.85	**1.08**	0.99	**1.25**	1.08	**1.37**	1.15	**1.46**	1.28	**1.63**	1.39	**1.77**
Total filtration (mm) Al	**1.5**		**2.5**		**3.5**		**4.5**		**3.5**		**3.5**	
Cu	**0**		**0**		**0**		**0**		**0.1**		**0.2**	

Table 10.49 Normalized organ doses—Bony thorax RAO— >17 years

Beam projection:			RAO			**Table cells:**						
Focus-to-detector distance (cm):			115			(+) = normal font = optimal field setting						
Tube voltage (kVp):			60			**(−) = bold font = suboptimal field setting**						
Total filtrations:			See last table row			Empty table cells correspond to values <0.01						
Organ or tissue	**Organ dose D_T: Entrance dose K_E**							**(mSv * mGy^{-1})**				
	(+)	(−)	(+)	(−)	(+)	(−)	(+)	(−)	(+)	(−)	(+)	(−)
Skull	0.02	**0.21**	0.03	**0.27**	0.03	**0.31**	0.03	**0.34**	0.04	**0.41**	0.04	**0.39**
Brain				**0.01**		**0.02**		**0.02**		**0.02**		**0.02**
Oral mucosa	0.01	**0.23**	0.02	**0.28**	0.02	**0.32**	0.02	**0.35**	0.03	**0.40**	0.02	**0.38**
Salivary glands	0.05	**0.26**	0.06	**0.31**	0.07	**0.35**	0.08	**0.38**	0.09	**0.43**	0.09	**0.41**
Cervical spine	0.27	**0.31**	0.35	**0.42**	0.42	**0.50**	0.48	**0.57**	0.59	**0.72**	0.53	**0.63**
Thyroid	0.52	**0.52**	0.62	**0.63**	0.69	**0.70**	0.74	**0.75**	0.83	**0.85**	0.81	**0.82**
Thymus	0.47	**0.48**	0.59	**0.60**	0.68	**0.70**	0.75	**0.77**	0.88	**0.90**	0.84	**0.86**
Thoracic spine	0.06	**0.06**	0.08	**0.08**	0.10	**0.10**	0.12	**0.12**	0.15	**0.16**	0.11	**0.12**
Claviculae	0.99	**1.03**	1.23	**1.28**	1.41	**1.47**	1.54	**1.61**	1.80	**1.88**	1.75	**1.82**
Scapulae	0.07	**0.08**	0.10	**0.11**	0.12	**0.14**	0.15	**0.16**	0.19	**0.21**	0.15	**0.17**
Ribs	0.90	**0.96**	1.08	**1.14**	1.20	**1.27**	1.30	**1.38**	1.47	**1.56**	1.43	**1.52**
Oesophagus	0.06	**0.07**	0.09	**0.09**	0.11	**0.11**	0.12	**0.13**	0.16	**0.17**	0.13	**0.14**
Trachea	0.19	**0.25**	0.23	**0.31**	0.26	**0.35**	0.28	**0.39**	0.32	**0.46**	0.31	**0.43**
Lung	0.18	**0.19**	0.23	**0.25**	0.28	**0.29**	0.31	**0.33**	0.38	**0.40**	0.34	**0.36**
Mamma	0.62	**0.78**	0.73	**0.90**	0.79	**0.99**	0.84	**1.05**	0.93	**1.16**	0.91	**1.14**
Heart	0.23	**0.24**	0.29	**0.31**	0.35	**0.36**	0.39	**0.41**	0.48	**0.50**	0.43	**0.45**
Stomach	0.15	**0.20**	0.20	**0.26**	0.23	**0.30**	0.26	**0.34**	0.32	**0.42**	0.29	**0.38**
Spleen	0.03	**0.03**	0.04	**0.04**	0.04	**0.05**	0.05	**0.06**	0.07	**0.08**	0.05	**0.07**
Liver	0.21	**0.28**	0.27	**0.36**	0.32	**0.42**	0.35	**0.47**	0.43	**0.57**	0.39	**0.53**
Gall bladder	0.08	**0.21**	0.11	**0.28**	0.13	**0.34**	0.15	**0.38**	0.19	**0.47**	0.17	**0.42**
Pancreas	0.07	**0.08**	0.10	**0.11**	0.12	**0.13**	0.13	**0.15**	0.17	**0.20**	0.14	**0.16**
Small intestine		**0.08**		**0.10**		**0.12**		**0.14**	0.01	**0.17**		**0.15**
Large intestine		**0.12**		**0.16**	0.01	**0.19**	0.01	**0.21**	0.02	**0.26**	0.01	**0.24**
Sigma/rectum		**0.02**		**0.03**		**0.03**		**0.04**		**0.05**		**0.04**
Lumbar spine	0.02	**0.04**	0.02	**0.06**	0.03	**0.07**	0.04	**0.08**	0.05	**0.11**	0.03	**0.08**
Pelvis		**0.02**		**0.03**		**0.03**		**0.04**		**0.05**		**0.04**
Adrenals	0.03	**0.03**	0.05	**0.05**	0.06	**0.06**	0.07	**0.07**	0.09	**0.09**	0.07	**0.07**
Kidneys	0.02	**0.03**	0.02	**0.04**	0.03	**0.04**	0.03	**0.05**	0.05	**0.07**	0.03	**0.05**
Urinary bladder												
Uterus				**0.01**		**0.02**		**0.02**		**0.02**		**0.02**
Ovaries				**0.01**		**0.02**		**0.02**		**0.02**		**0.02**
Prostate												
Testes												
Upper arm	0.27	**0.50**	0.34	**0.62**	0.39	**0.72**	0.44	**0.79**	0.52	**0.94**	0.48	**0.88**
Elbow	0.26	**0.54**	0.33	**0.67**	0.38	**0.77**	0.42	**0.85**	0.51	**1.01**	0.47	**0.95**
Forearm	0.02	**0.25**	0.03	**0.32**	0.04	**0.36**	0.04	**0.40**	0.05	**0.47**	0.05	**0.45**
Upper leg												
Knee												
Lower leg												
Lymph nodes	0.11	**0.17**	0.13	**0.21**	0.16	**0.25**	0.17	**0.28**	0.21	**0.33**	0.19	**0.30**
Skin	0.14	**0.22**	0.14	**0.24**	0.15	**0.25**	0.16	**0.25**	0.16	**0.27**	0.16	**0.26**
Muscle	0.07	**0.10**	0.08	**0.12**	0.09	**0.13**	0.10	**0.15**	0.12	**0.17**	0.11	**0.16**
Skeleton	0.14	**0.22**	0.18	**0.27**	0.20	**0.31**	0.22	**0.34**	0.26	**0.39**	0.24	**0.37**
Red bone marrow	0.07	**0.08**	0.08	**0.10**	0.09	**0.11**	0.10	**0.12**	0.12	**0.15**	0.11	**0.13**
Whole body dose	0.08	**0.12**	0.10	**0.15**	0.12	**0.17**	0.13	**0.19**	0.15	**0.22**	0.14	**0.21**
ED (ICRP 60)	0.12	**0.16**	0.15	**0.19**	0.18	**0.22**	0.19	**0.25**	0.23	**0.29**	0.21	**0.27**
ED (ICRP 103)	0.17	**0.22**	0.21	**0.26**	0.23	**0.30**	0.25	**0.33**	0.29	**0.38**	0.28	**0.36**
Total filtration (mm) **Al**	**1.5**		**2.5**		**3.5**		**4.5**		**3.5**		**3.5**	
Cu	**0**		**0**		**0**		**0**		**0.1**		**0.2**	

Table 10.50 Normalized organ doses—Bony thorax RAO— >17 years

Beam projection:			RAO			**Table cells:**						
Focus-to-detector distance (cm):			115			(+) = normal font = optimal field setting						
Tube voltage (kVp):			70			**(−) = bold font = suboptimal field setting**						
Total filtrations:			See last table row			Empty table cells correspond to values <0.01						
Organ or tissue	**Organ dose D_T: Entrance dose K_E**						**(mSv * mGy^{-1})**					
	(+)	(−)	(+)	(−)	(+)	(−)	(+)	(−)	(+)	(−)	(+)	(−)
Skull	0.03	**0.25**	0.03	**0.32**	0.04	**0.36**	0.04	**0.40**	0.05	**0.47**	0.06	**0.53**
Brain		**0.01**		**0.02**		**0.02**		**0.03**		**0.03**		**0.04**
Oral mucosa	0.02	**0.27**	0.02	**0.33**	0.02	**0.37**	0.03	**0.40**	0.03	**0.46**	0.04	**0.52**
Salivary glands	0.06	**0.30**	0.07	**0.35**	0.08	**0.39**	0.09	**0.42**	0.11	**0.48**	0.12	**0.53**
Cervical spine	0.34	**0.41**	0.44	**0.54**	0.52	**0.63**	0.59	**0.72**	0.72	**0.89**	0.85	**1.05**
Thyroid	0.58	**0.59**	0.68	**0.69**	0.75	**0.77**	0.80	**0.82**	0.90	**0.92**	0.98	**1.01**
Thymus	0.55	**0.57**	0.68	**0.70**	0.78	**0.80**	0.85	**0.88**	0.99	**1.02**	1.12	**1.15**
Thoracic spine	0.09	**0.09**	0.12	**0.13**	0.14	**0.15**	0.17	**0.18**	0.22	**0.23**	0.27	**0.29**
Claviculae	1.14	**1.19**	1.39	**1.45**	1.57	**1.64**	1.71	**1.79**	1.96	**2.06**	2.18	**2.29**
Scapulae	0.11	**0.12**	0.14	**0.16**	0.17	**0.19**	0.20	**0.22**	0.25	**0.28**	0.31	**0.34**
Ribs	1.01	**1.07**	1.19	**1.26**	1.31	**1.40**	1.41	**1.50**	1.57	**1.68**	1.71	**1.83**
Oesophagus	0.09	**0.10**	0.12	**0.13**	0.15	**0.16**	0.17	**0.18**	0.22	**0.23**	0.27	**0.28**
Trachea	0.21	**0.29**	0.26	**0.36**	0.29	**0.41**	0.31	**0.45**	0.35	**0.52**	0.39	**0.59**
Lung	0.23	**0.24**	0.29	**0.30**	0.33	**0.35**	0.37	**0.39**	0.45	**0.47**	0.52	**0.55**
Mamma	0.68	**0.85**	0.79	**0.98**	0.85	**1.06**	0.91	**1.13**	1.00	**1.24**	1.07	**1.33**
Heart	0.29	**0.30**	0.36	**0.38**	0.42	**0.44**	0.47	**0.49**	0.57	**0.59**	0.66	**0.69**
Stomach	0.19	**0.25**	0.24	**0.32**	0.28	**0.37**	0.32	**0.41**	0.38	**0.50**	0.44	**0.58**
Spleen	0.04	**0.05**	0.05	**0.06**	0.06	**0.08**	0.07	**0.09**	0.10	**0.12**	0.12	**0.14**
Liver	0.26	**0.35**	0.33	**0.44**	0.38	**0.51**	0.42	**0.56**	0.51	**0.67**	0.59	**0.78**
Gall bladder	0.11	**0.28**	0.14	**0.35**	0.17	**0.42**	0.19	**0.47**	0.24	**0.57**	0.28	**0.67**
Pancreas	0.10	**0.11**	0.13	**0.15**	0.16	**0.18**	0.18	**0.21**	0.23	**0.26**	0.28	**0.32**
Small intestine		**0.10**		**0.13**	0.01	**0.16**	0.01	**0.18**	0.02	**0.22**	0.02	**0.26**
Large intestine	0.01	**0.15**	0.01	**0.20**	0.02	**0.23**	0.02	**0.26**	0.03	**0.32**	0.03	**0.38**
Sigma/rectum		**0.03**		**0.04**		**0.04**		**0.05**		**0.06**		**0.07**
Lumbar spine	0.03	**0.07**	0.04	**0.09**	0.05	**0.11**	0.06	**0.13**	0.07	**0.17**	0.09	**0.21**
Pelvis		**0.03**		**0.04**		**0.05**		**0.05**		**0.07**		**0.09**
Adrenals	0.05	**0.05**	0.07	**0.07**	0.08	**0.09**	0.09	**0.10**	0.12	**0.13**	0.15	**0.16**
Kidneys	0.03	**0.04**	0.04	**0.05**	0.04	**0.07**	0.05	**0.08**	0.06	**0.10**	0.08	**0.12**
Urinary bladder												**0.01**
Uterus		**0.01**		**0.02**		**0.02**		**0.03**		**0.04**		**0.04**
Ovaries		**0.01**		**0.02**		**0.02**		**0.03**		**0.03**		**0.04**
Prostate												
Testes												
Upper arm	0.32	**0.59**	0.40	**0.72**	0.46	**0.83**	0.51	**0.91**	0.60	**1.06**	0.69	**1.20**
Elbow	0.31	**0.64**	0.39	**0.78**	0.45	**0.89**	0.49	**0.97**	0.58	**1.13**	0.67	**1.28**
Forearm	0.03	**0.30**	0.04	**0.37**	0.05	**0.42**	0.05	**0.46**	0.06	**0.53**	0.07	**0.60**
Upper leg												
Knee												
Lower leg												
Lymph nodes	0.13	**0.21**	0.16	**0.26**	0.19	**0.30**	0.21	**0.33**	0.24	**0.40**	0.28	**0.46**
Skin	0.14	**0.23**	0.15	**0.25**	0.16	**0.26**	0.16	**0.27**	0.17	**0.28**	0.18	**0.30**
Muscle	0.08	**0.11**	0.09	**0.14**	0.11	**0.16**	0.12	**0.17**	0.14	**0.20**	0.16	**0.23**
Skeleton	0.17	**0.25**	0.20	**0.31**	0.23	**0.35**	0.25	**0.38**	0.29	**0.44**	0.32	**0.50**
Red bone marrow	0.08	**0.10**	0.09	**0.12**	0.11	**0.13**	0.12	**0.15**	0.14	**0.17**	0.16	**0.20**
Whole body dose	0.10	**0.15**	0.12	**0.18**	0.14	**0.20**	0.15	**0.22**	0.17	**0.26**	0.20	**0.29**
ED (ICRP 60)	0.15	**0.19**	0.18	**0.23**	0.20	**0.26**	0.22	**0.29**	0.26	**0.34**	0.30	**0.38**
ED (ICRP 103)	0.20	**0.25**	0.24	**0.31**	0.27	**0.35**	0.29	**0.38**	0.33	**0.44**	0.37	**0.49**
Total filtration (mm) Al	**1.5**		**2.5**		**3.5**		**4.5**		**3.5**		**3.5**	
Cu	**0**		**0**		**0**		**0**		**0.1**		**0.2**	

Table 10.51 Normalized organ doses—Bony thorax LAO—1–2 years

Beam projection:			LAO			**Table cells:**						
Focus-to-detector distance (cm):			115			(+) = normal font = optimal field setting						
Tube voltage (kVp):			50			**(−) = bold font = suboptimal field setting**						
Total filtrations:			See last table row			Empty table cells correspond to values <0.01						
Organ or tissue	**Organ dose D_T: Entrance dose K_E**						**(mSv * mGy^{-1})**					
	(+)	(−)	(+)	(−)	(+)	(−)	(+)	(−)	(+)	(−)	(+)	(−)
Skull	0.01	**0.20**	0.02	**0.25**	0.02	**0.28**	0.02	**0.30**	0.03	**0.35**	0.03	**0.39**
Brain		**0.01**		**0.02**		**0.02**		**0.02**		**0.03**		**0.04**
Oral mucosa	0.02	**0.39**	0.02	**0.47**	0.03	**0.52**	0.03	**0.56**	0.04	**0.63**	0.05	**0.70**
Salivary glands	0.04	**0.36**	0.05	**0.42**	0.06	**0.47**	0.06	**0.50**	0.08	**0.56**	0.09	**0.62**
Cervical spine	0.23	**0.34**	0.30	**0.46**	0.36	**0.55**	0.41	**0.62**	0.49	**0.77**	0.58	**0.91**
Thyroid	0.75	**0.78**	0.86	**0.90**	0.92	**0.98**	0.97	**1.03**	1.05	**1.13**	1.11	**1.21**
Thymus	0.56	**0.58**	0.68	**0.70**	0.77	**0.79**	0.83	**0.85**	0.94	**0.97**	1.04	**1.07**
Thoracic spine	0.16	**0.17**	0.22	**0.23**	0.26	**0.28**	0.31	**0.33**	0.39	**0.42**	0.48	**0.52**
Claviculae	1.39	**1.46**	1.68	**1.78**	1.89	**2.00**	2.05	**2.17**	2.32	**2.47**	2.55	**2.72**
Scapulae	0.14	**0.17**	0.19	**0.23**	0.23	**0.28**	0.27	**0.33**	0.35	**0.42**	0.42	**0.51**
Ribs	1.12	**1.16**	1.32	**1.38**	1.46	**1.52**	1.56	**1.63**	1.74	**1.82**	1.89	**1.99**
Oesophagus	0.14	**0.15**	0.19	**0.21**	0.23	**0.25**	0.26	**0.29**	0.33	**0.35**	0.39	**0.42**
Trachea	0.07	**0.36**	0.08	**0.44**	0.09	**0.50**	0.10	**0.54**	0.11	**0.61**	0.12	**0.68**
Lung	0.25	**0.27**	0.32	**0.34**	0.37	**0.39**	0.41	**0.43**	0.49	**0.51**	0.55	**0.58**
Mamma	1.06	**1.07**	1.14	**1.15**	1.19	**1.20**	1.22	**1.22**	1.27	**1.27**	1.31	**1.31**
Heart	0.39	**0.40**	0.49	**0.50**	0.56	**0.58**	0.62	**0.64**	0.72	**0.74**	0.82	**0.84**
Stomach	0.26	**0.46**	0.33	**0.57**	0.37	**0.65**	0.41	**0.71**	0.47	**0.82**	0.53	**0.91**
Spleen	0.14	**0.18**	0.19	**0.24**	0.23	**0.29**	0.26	**0.32**	0.31	**0.40**	0.37	**0.47**
Liver	0.17	**0.25**	0.22	**0.32**	0.25	**0.37**	0.28	**0.41**	0.33	**0.48**	0.38	**0.55**
Gall bladder	0.07	**0.25**	0.10	**0.33**	0.12	**0.39**	0.13	**0.44**	0.16	**0.53**	0.19	**0.62**
Pancreas	0.16	**0.19**	0.21	**0.25**	0.25	**0.30**	0.29	**0.34**	0.35	**0.42**	0.42	**0.50**
Small intestine		**0.12**	0.01	**0.15**	0.01	**0.18**	0.02	**0.20**	0.02	**0.25**	0.03	**0.29**
Large intestine	0.01	**0.17**	0.02	**0.22**	0.02	**0.26**	0.02	**0.29**	0.03	**0.35**	0.04	**0.40**
Sigma/rectum		**0.04**		**0.05**		**0.06**		**0.07**		**0.09**		**0.10**
Lumbar spine	0.04	**0.12**	0.05	**0.16**	0.06	**0.20**	0.07	**0.24**	0.10	**0.31**	0.12	**0.38**
Pelvis		**0.04**		**0.06**		**0.07**		**0.08**	0.01	**0.11**	0.02	**0.13**
Adrenals	0.06	**0.07**	0.08	**0.09**	0.09	**0.11**	0.11	**0.13**	0.14	**0.16**	0.17	**0.20**
Kidneys	0.03	**0.06**	0.04	**0.08**	0.05	**0.10**	0.06	**0.11**	0.07	**0.14**	0.09	**0.18**
Urinary bladder						**0.01**		**0.01**		**0.02**		**0.02**
Uterus		**0.02**		**0.02**		**0.03**		**0.03**		**0.04**		**0.05**
Ovaries		**0.02**		**0.03**		**0.03**		**0.04**		**0.05**		**0.06**
Prostate												**0.01**
Testes												
Upper arm	0.54	**0.78**	0.67	**0.96**	0.77	**1.09**	0.84	**1.19**	0.98	**1.38**	1.11	**1.55**
Elbow	0.57	**0.85**	0.71	**1.04**	0.80	**1.18**	0.88	**1.29**	1.02	**1.49**	1.15	**1.66**
Forearm	0.03	**0.43**	0.05	**0.53**	0.05	**0.59**	0.06	**0.65**	0.07	**0.74**	0.09	**0.83**
Upper leg												
Knee												
Lower leg												
Lymph nodes	0.13	**0.26**	0.17	**0.33**	0.19	**0.37**	0.21	**0.41**	0.25	**0.48**	0.28	**0.54**
Skin	0.15	**0.27**	0.15	**0.29**	0.16	**0.30**	0.16	**0.30**	0.17	**0.32**	0.18	**0.33**
Muscle	0.09	**0.15**	0.10	**0.18**	0.12	**0.20**	0.13	**0.22**	0.15	**0.25**	0.16	**0.28**
Skeleton	0.24	**0.35**	0.30	**0.43**	0.34	**0.49**	0.37	**0.54**	0.42	**0.62**	0.48	**0.70**
Red bone marrow	0.05	**0.08**	0.07	**0.10**	0.07	**0.12**	0.08	**0.13**	0.10	**0.15**	0.11	**0.17**
Whole body dose	0.10	**0.17**	0.13	**0.21**	0.14	**0.23**	0.16	**0.25**	0.18	**0.29**	0.20	**0.33**
ED (ICRP 60)	0.18	**0.24**	0.22	**0.29**	0.24	**0.32**	0.26	**0.35**	0.29	**0.39**	0.32	**0.44**
ED (ICRP 103)	0.26	**0.32**	0.30	**0.38**	0.33	**0.42**	0.35	**0.45**	0.38	**0.50**	0.42	**0.55**
Total filtration (mm) Al	**1.5**		**2.5**		**3.5**		**4.5**		**3.5**		**3.5**	
Cu	**0**		**0**		**0**		**0**		**0.1**		**0.2**	

Table 10.52 Normalized organ doses—Bony thorax LAO—1–2 years

Beam projection:				LAO			**Table cells:**					
Focus-to-detector distance (cm):				115			(+) = normal font = optimal field setting					
Tube voltage (kVp):				60			**(−) = bold font = suboptimal field setting**					
Total filtrations:				See last table row			Empty table cells correspond to values <0.01					
Organ or tissue	**Organ dose D_T: Entrance dose K_E**							**(mSv * mGy^{-1})**				
	(+)	(−)	(+)	(−)	(+)	(−)	(+)	(−)	(+)	(−)	(+)	(−)
Skull	0.02	**0.24**	0.02	**0.29**	0.03	**0.33**	0.03	**0.35**	0.04	**0.40**	0.03	**0.39**
Brain		**0.02**		**0.02**		**0.03**		**0.03**		**0.04**		**0.04**
Oral mucosa	0.02	**0.45**	0.03	**0.54**	0.04	**0.59**	0.04	**0.64**	0.05	**0.71**	0.05	**0.70**
Salivary glands	0.05	**0.41**	0.06	**0.48**	0.07	**0.53**	0.08	**0.57**	0.09	**0.63**	0.09	**0.62**
Cervical spine	0.30	**0.47**	0.39	**0.61**	0.46	**0.72**	0.51	**0.81**	0.62	**0.98**	0.58	**0.91**
Thyroid	0.82	**0.86**	0.93	**0.99**	1.00	**1.07**	1.04	**1.12**	1.13	**1.22**	1.11	**1.21**
Thymus	0.66	**0.67**	0.78	**0.81**	0.87	**0.90**	0.94	**0.97**	1.06	**1.09**	1.04	**1.07**
Thoracic spine	0.23	**0.25**	0.31	**0.34**	0.38	**0.41**	0.43	**0.47**	0.55	**0.59**	0.48	**0.52**
Claviculae	1.60	**1.69**	1.91	**2.03**	2.12	**2.26**	2.27	**2.43**	2.54	**2.72**	2.55	**2.72**
Scapulae	0.21	**0.25**	0.27	**0.33**	0.33	**0.40**	0.38	**0.45**	0.47	**0.57**	0.42	**0.51**
Ribs	1.27	**1.32**	1.48	**1.55**	1.62	**1.70**	1.72	**1.81**	1.90	**2.01**	1.89	**1.99**
Oesophagus	0.20	**0.22**	0.26	**0.28**	0.31	**0.33**	0.35	**0.38**	0.43	**0.46**	0.39	**0.42**
Trachea	0.08	**0.43**	0.09	**0.51**	0.10	**0.57**	0.11	**0.61**	0.13	**0.69**	0.12	**0.68**
Lung	0.32	**0.33**	0.39	**0.41**	0.45	**0.47**	0.49	**0.52**	0.58	**0.61**	0.55	**0.58**
Mamma	1.11	**1.12**	1.19	**1.20**	1.23	**1.24**	1.26	**1.27**	1.31	**1.32**	1.31	**1.31**
Heart	0.48	**0.49**	0.59	**0.60**	0.67	**0.69**	0.73	**0.75**	0.84	**0.87**	0.82	**0.84**
Stomach	0.32	**0.55**	0.39	**0.67**	0.44	**0.76**	0.48	**0.82**	0.55	**0.94**	0.53	**0.91**
Spleen	0.19	**0.24**	0.25	**0.31**	0.29	**0.37**	0.32	**0.41**	0.39	**0.50**	0.37	**0.47**
Liver	0.21	**0.31**	0.27	**0.39**	0.31	**0.45**	0.34	**0.50**	0.40	**0.58**	0.38	**0.55**
Gall bladder	0.10	**0.33**	0.13	**0.42**	0.15	**0.49**	0.17	**0.55**	0.21	**0.65**	0.19	**0.62**
Pancreas	0.22	**0.26**	0.28	**0.34**	0.33	**0.40**	0.37	**0.45**	0.45	**0.54**	0.42	**0.50**
Small intestine	0.01	**0.15**	0.02	**0.20**	0.02	**0.23**	0.02	**0.26**	0.03	**0.31**	0.03	**0.29**
Large intestine	0.02	**0.22**	0.02	**0.28**	0.03	**0.32**	0.03	**0.36**	0.04	**0.43**	0.04	**0.40**
Sigma/rectum		**0.05**		**0.07**		**0.08**		**0.09**	0.01	**0.11**		**0.10**
Lumbar spine	0.06	**0.19**	0.08	**0.25**	0.10	**0.30**	0.12	**0.35**	0.15	**0.45**	0.12	**0.38**
Pelvis		**0.06**	0.01	**0.09**	0.01	**0.10**	0.02	**0.12**	0.02	**0.15**	0.02	**0.13**
Adrenals	0.08	**0.10**	0.11	**0.13**	0.13	**0.16**	0.15	**0.18**	0.19	**0.23**	0.17	**0.20**
Kidneys	0.05	**0.09**	0.06	**0.12**	0.07	**0.14**	0.08	**0.16**	0.10	**0.20**	0.09	**0.18**
Urinary bladder		**0.01**		**0.01**		**0.02**		**0.02**		**0.03**		**0.02**
Uterus		**0.03**		**0.04**		**0.04**		**0.05**	0.01	**0.06**		**0.05**
Ovaries		**0.03**		**0.04**		**0.05**		**0.06**		**0.07**		**0.06**
Prostate										**0.01**		**0.01**
Testes												
Upper arm	0.65	**0.93**	0.80	**1.13**	0.90	**1.27**	0.99	**1.38**	1.14	**1.58**	1.11	**1.55**
Elbow	0.69	**1.01**	0.84	**1.22**	0.94	**1.37**	1.03	**1.49**	1.18	**1.70**	1.15	**1.66**
Forearm	0.05	**0.51**	0.06	**0.61**	0.07	**0.69**	0.08	**0.74**	0.09	**0.84**	0.09	**0.83**
Upper leg												
Knee												
Lower leg												
Lymph nodes	0.17	**0.32**	0.20	**0.39**	0.23	**0.45**	0.25	**0.49**	0.29	**0.57**	0.28	**0.54**
Skin	0.15	**0.28**	0.16	**0.30**	0.17	**0.31**	0.18	**0.32**	0.19	**0.34**	0.18	**0.33**
Muscle	0.10	**0.17**	0.12	**0.21**	0.14	**0.23**	0.15	**0.25**	0.17	**0.29**	0.16	**0.28**
Skeleton	0.29	**0.43**	0.35	**0.52**	0.39	**0.58**	0.43	**0.63**	0.49	**0.73**	0.48	**0.70**
Red bone marrow	0.07	**0.10**	0.08	**0.13**	0.09	**0.15**	0.10	**0.16**	0.12	**0.19**	0.11	**0.17**
Whole body dose	0.12	**0.20**	0.15	**0.24**	0.17	**0.27**	0.18	**0.30**	0.21	**0.34**	0.20	**0.33**
ED (ICRP 60)	0.21	**0.28**	0.25	**0.33**	0.28	**0.37**	0.30	**0.40**	0.33	**0.46**	0.32	**0.44**
ED (ICRP 103)	0.29	**0.37**	0.33	**0.43**	0.36	**0.48**	0.39	**0.51**	0.43	**0.57**	0.42	**0.55**
Total filtration (mm) Al	**1.5**		**2.5**		**3.5**		**4.5**		**3.5**		**3.5**	
Cu	**0**		**0**		**0**		**0**		**0.1**		**0.2**	

Table 10.53 Normalized organ doses—Bony thorax LAO—1–2 years

Beam projection:				LAO				**Table cells:**				
Focus-to-detector distance (cm):				115				(+) = normal font = optimal field setting				
Tube voltage (kVp):				70				**(−) = bold font = suboptimal field setting**				
Total filtrations:				See last table row				Empty table cells correspond to values <0.01				
Organ or tissue	**Organ dose D_T: Entrance dose K_E**							**(mSv * mGy^{-1})**				
	(+)	(−)	(+)	(−)	(+)	(−)	(+)	(−)	(+)	(−)	(+)	(−)
Skull	0.02	**0.27**	0.03	**0.32**	0.03	**0.36**	0.04	**0.39**	0.05	**0.44**	0.05	**0.48**
Brain		**0.02**		**0.03**		**0.04**		**0.04**		**0.05**		**0.06**
Oral mucosa	0.03	**0.50**	0.04	**0.59**	0.04	**0.65**	0.05	**0.69**	0.06	**0.77**	0.07	**0.84**
Salivary glands	0.06	**0.45**	0.07	**0.53**	0.08	**0.58**	0.09	**0.62**	0.10	**0.68**	0.12	**0.74**
Cervical spine	0.37	**0.58**	0.46	**0.73**	0.53	**0.85**	0.59	**0.95**	0.70	**1.13**	0.79	**1.30**
Thyroid	0.88	**0.93**	0.98	**1.05**	1.05	**1.13**	1.10	**1.19**	1.18	**1.29**	1.25	**1.37**
Thymus	0.73	**0.75**	0.86	**0.89**	0.95	**0.98**	1.02	**1.05**	1.14	**1.18**	1.24	**1.28**
Thoracic spine	0.31	**0.33**	0.40	**0.44**	0.47	**0.52**	0.54	**0.58**	0.66	**0.72**	0.79	**0.86**
Claviculae	1.75	**1.86**	2.06	**2.19**	2.26	**2.41**	2.41	**2.58**	2.65	**2.85**	2.84	**3.05**
Scapulae	0.27	**0.32**	0.34	**0.42**	0.41	**0.49**	0.46	**0.55**	0.56	**0.68**	0.66	**0.80**
Ribs	1.38	**1.44**	1.58	**1.66**	1.72	**1.81**	1.82	**1.92**	1.99	**2.11**	2.12	**2.26**
Oesophagus	0.25	**0.27**	0.32	**0.35**	0.37	**0.41**	0.42	**0.45**	0.50	**0.55**	0.58	**0.64**
Trachea	0.09	**0.48**	0.10	**0.56**	0.11	**0.62**	0.12	**0.67**	0.14	**0.75**	0.15	**0.82**
Lung	0.37	**0.39**	0.45	**0.47**	0.51	**0.54**	0.55	**0.59**	0.64	**0.68**	0.71	**0.76**
Mamma	1.15	**1.16**	1.22	**1.23**	1.27	**1.28**	1.29	**1.30**	1.34	**1.35**	1.38	**1.39**
Heart	0.55	**0.57**	0.66	**0.69**	0.75	**0.77**	0.81	**0.84**	0.93	**0.96**	1.03	**1.07**
Stomach	0.36	**0.63**	0.44	**0.75**	0.49	**0.84**	0.53	**0.91**	0.60	**1.03**	0.66	**1.13**
Spleen	0.23	**0.30**	0.29	**0.38**	0.34	**0.44**	0.38	**0.49**	0.45	**0.58**	0.52	**0.66**
Liver	0.25	**0.37**	0.31	**0.45**	0.35	**0.52**	0.39	**0.56**	0.45	**0.65**	0.51	**0.74**
Gall bladder	0.12	**0.40**	0.16	**0.50**	0.18	**0.57**	0.20	**0.63**	0.24	**0.74**	0.28	**0.84**
Pancreas	0.27	**0.32**	0.34	**0.41**	0.39	**0.47**	0.43	**0.53**	0.52	**0.63**	0.59	**0.73**
Small intestine	0.02	**0.19**	0.02	**0.24**	0.03	**0.27**	0.03	**0.30**	0.04	**0.36**	0.05	**0.41**
Large intestine	0.02	**0.26**	0.03	**0.33**	0.04	**0.38**	0.04	**0.41**	0.05	**0.49**	0.06	**0.55**
Sigma/rectum		**0.07**		**0.08**		**0.09**	0.01	**0.10**	0.01	**0.12**	0.02	**0.14**
Lumbar spine	0.08	**0.25**	0.11	**0.33**	0.13	**0.39**	0.15	**0.45**	0.19	**0.56**	0.23	**0.67**
Pelvis	0.01	**0.09**	0.02	**0.11**	0.02	**0.13**	0.02	**0.15**	0.03	**0.19**	0.03	**0.22**
Adrenals	0.11	**0.13**	0.14	**0.17**	0.17	**0.20**	0.19	**0.23**	0.23	**0.28**	0.28	**0.33**
Kidneys	0.06	**0.11**	0.08	**0.15**	0.09	**0.18**	0.11	**0.20**	0.13	**0.25**	0.16	**0.29**
Urinary bladder		**0.01**		**0.02**		**0.02**		**0.03**		**0.03**		**0.04**
Uterus		**0.04**		**0.05**	0.01	**0.06**	0.01	**0.06**	0.02	**0.08**	0.02	**0.10**
Ovaries		**0.04**		**0.05**		**0.06**	0.01	**0.07**	0.01	**0.09**	0.02	**0.11**
Prostate						**0.01**		**0.01**		**0.02**		**0.02**
Testes												**0.01**
Upper arm	0.74	**1.05**	0.89	**1.26**	1.00	**1.40**	1.08	**1.51**	1.24	**1.71**	1.36	**1.87**
Elbow	0.78	**1.13**	0.93	**1.35**	1.04	**1.50**	1.13	**1.62**	1.28	**1.82**	1.41	**1.99**
Forearm	0.06	**0.57**	0.07	**0.67**	0.08	**0.75**	0.09	**0.80**	0.11	**0.90**	0.13	**0.98**
Upper leg										**0.01**		**0.01**
Knee												
Lower leg												
Lymph nodes	0.19	**0.37**	0.23	**0.45**	0.26	**0.51**	0.29	**0.55**	0.33	**0.63**	0.37	**0.71**
Skin	0.16	**0.30**	0.17	**0.31**	0.18	**0.33**	0.18	**0.34**	0.20	**0.36**	0.21	**0.38**
Muscle	0.12	**0.20**	0.14	**0.23**	0.16	**0.26**	0.17	**0.28**	0.19	**0.32**	0.21	**0.35**
Skeleton	0.33	**0.48**	0.39	**0.58**	0.44	**0.64**	0.47	**0.70**	0.54	**0.79**	0.59	**0.88**
Red bone marrow	0.08	**0.12**	0.09	**0.15**	0.11	**0.17**	0.12	**0.18**	0.13	**0.22**	0.15	**0.24**
Whole body dose	0.14	**0.23**	0.17	**0.27**	0.19	**0.30**	0.20	**0.33**	0.23	**0.38**	0.26	**0.42**
ED (ICRP 60)	0.24	**0.32**	0.28	**0.37**	0.30	**0.41**	0.33	**0.44**	0.36	**0.50**	0.40	**0.55**
ED (ICRP 103)	0.32	**0.41**	0.36	**0.48**	0.39	**0.52**	0.42	**0.56**	0.46	**0.62**	0.50	**0.68**
Total filtration (mm) **Al**	**1.5**		**2.5**		**3.5**		**4.5**		**3.5**		**3.5**	
Cu	**0**		**0**		**0**		**0**		**0.1**		**0.2**	

Table 10.54 Normalized organ doses—Bony thorax LAO—3–7 years

Beam projection:				LAO			**Table cells:**					
Focus-to-detector distance (cm):				115			(+) = normal font = optimal field setting					
Tube voltage (kVp):				60			**(−) = bold font = suboptimal field setting**					
Total filtrations:				See last table row			Empty table cells correspond to values <0.01					
Organ or tissue	**Organ dose D_T: Entrance dose K_E**							**(mSv * mGy^{-1})**				
	(+)	(−)	(+)	(−)	(+)	(−)	(+)	(−)	(+)	(−)	(+)	(−)
Skull	0.02	**0.25**	0.03	**0.31**	0.03	**0.35**	0.04	**0.38**	0.05	**0.44**	0.04	**0.43**
Brain		**0.01**		**0.02**		**0.02**		**0.02**		**0.03**		**0.02**
Oral mucosa	0.02	**0.35**	0.03	**0.42**	0.04	**0.47**	0.04	**0.51**	0.05	**0.59**	0.05	**0.56**
Salivary glands	0.05	**0.34**	0.07	**0.40**	0.08	**0.44**	0.09	**0.48**	0.10	**0.54**	0.10	**0.52**
Cervical spine	0.30	**0.39**	0.39	**0.51**	0.46	**0.60**	0.51	**0.68**	0.62	**0.84**	0.58	**0.77**
Thyroid	0.73	**0.76**	0.84	**0.88**	0.91	**0.96**	0.96	**1.02**	1.05	**1.12**	1.03	**1.10**
Thymus	0.61	**0.62**	0.74	**0.76**	0.83	**0.85**	0.90	**0.93**	1.03	**1.06**	1.00	**1.03**
Thoracic spine	0.16	**0.17**	0.22	**0.24**	0.26	**0.29**	0.30	**0.33**	0.39	**0.43**	0.33	**0.36**
Claviculae	1.43	**1.50**	1.73	**1.82**	1.94	**2.04**	2.09	**2.21**	2.37	**2.51**	2.35	**2.48**
Scapulae	0.14	**0.18**	0.19	**0.25**	0.23	**0.30**	0.27	**0.35**	0.34	**0.44**	0.29	**0.38**
Ribs	1.17	**1.23**	1.38	**1.45**	1.51	**1.60**	1.62	**1.72**	1.80	**1.91**	1.78	**1.89**
Oesophagus	0.11	**0.13**	0.15	**0.17**	0.18	**0.20**	0.21	**0.23**	0.26	**0.29**	0.23	**0.25**
Trachea	0.10	**0.29**	0.12	**0.36**	0.13	**0.41**	0.14	**0.45**	0.16	**0.53**	0.15	**0.51**
Lung	0.28	**0.29**	0.35	**0.37**	0.40	**0.43**	0.45	**0.47**	0.53	**0.56**	0.50	**0.53**
Mamma	1.07	**1.11**	1.15	**1.20**	1.20	**1.25**	1.23	**1.29**	1.29	**1.35**	1.28	**1.33**
Heart	0.41	**0.43**	0.52	**0.54**	0.60	**0.62**	0.66	**0.68**	0.77	**0.80**	0.74	**0.76**
Stomach	0.30	**0.54**	0.37	**0.66**	0.42	**0.75**	0.47	**0.82**	0.54	**0.95**	0.52	**0.91**
Spleen	0.15	**0.19**	0.19	**0.25**	0.23	**0.30**	0.26	**0.34**	0.31	**0.42**	0.29	**0.38**
Liver	0.18	**0.27**	0.23	**0.34**	0.27	**0.40**	0.30	**0.44**	0.36	**0.52**	0.34	**0.49**
Gall bladder	0.09	**0.29**	0.12	**0.38**	0.14	**0.44**	0.16	**0.50**	0.20	**0.60**	0.18	**0.56**
Pancreas	0.18	**0.21**	0.23	**0.28**	0.28	**0.34**	0.31	**0.38**	0.38	**0.47**	0.35	**0.42**
Small intestine		**0.16**	0.01	**0.21**	0.01	**0.24**	0.02	**0.27**	0.02	**0.33**	0.02	**0.31**
Large intestine	0.01	**0.21**	0.02	**0.27**	0.02	**0.31**	0.02	**0.35**	0.03	**0.42**	0.02	**0.39**
Sigma/rectum		**0.06**		**0.08**		**0.09**		**0.10**		**0.12**		**0.12**
Lumbar spine	0.04	**0.13**	0.05	**0.17**	0.07	**0.21**	0.08	**0.25**	0.10	**0.32**	0.08	**0.26**
Pelvis		**0.06**		**0.09**		**0.10**		**0.12**	0.01	**0.15**		**0.13**
Adrenals	0.06	**0.07**	0.09	**0.10**	0.10	**0.12**	0.12	**0.14**	0.15	**0.18**	0.13	**0.15**
Kidneys	0.03	**0.06**	0.04	**0.09**	0.05	**0.11**	0.06	**0.12**	0.08	**0.15**	0.07	**0.13**
Urinary bladder				**0.01**		**0.01**		**0.02**		**0.02**		**0.02**
Uterus		**0.03**		**0.04**		**0.04**		**0.05**		**0.06**		**0.05**
Ovaries		**0.03**		**0.04**		**0.05**		**0.05**		**0.07**		**0.06**
Prostate												
Testes												
Upper arm	0.30	**0.79**	0.38	**0.97**	0.44	**1.10**	0.49	**1.20**	0.58	**1.39**	0.55	**1.34**
Elbow	0.27	**0.85**	0.35	**1.05**	0.40	**1.18**	0.45	**1.29**	0.54	**1.49**	0.50	**1.45**
Forearm	0.02	**0.47**	0.03	**0.57**	0.04	**0.64**	0.04	**0.70**	0.05	**0.81**	0.05	**0.79**
Upper leg												
Knee												
Lower leg												
Lymph nodes	0.15	**0.27**	0.18	**0.34**	0.21	**0.39**	0.23	**0.43**	0.27	**0.51**	0.25	**0.48**
Skin	0.13	**0.25**	0.14	**0.27**	0.15	**0.28**	0.15	**0.28**	0.16	**0.30**	0.16	**0.29**
Muscle	0.09	**0.14**	0.11	**0.17**	0.12	**0.20**	0.13	**0.21**	0.15	**0.25**	0.14	**0.23**
Skeleton	0.18	**0.33**	0.22	**0.40**	0.25	**0.46**	0.28	**0.50**	0.32	**0.58**	0.30	**0.55**
Red bone marrow	0.06	**0.09**	0.07	**0.11**	0.08	**0.13**	0.08	**0.14**	0.10	**0.17**	0.09	**0.15**
Whole body dose	0.10	**0.17**	0.12	**0.21**	0.14	**0.24**	0.15	**0.26**	0.18	**0.30**	0.17	**0.28**
ED (ICRP 60)	0.19	**0.26**	0.23	**0.31**	0.25	**0.35**	0.27	**0.37**	0.31	**0.43**	0.29	**0.41**
ED (ICRP 103)	0.27	**0.34**	0.31	**0.40**	0.34	**0.45**	0.36	**0.48**	0.40	**0.54**	0.39	**0.52**
Total filtration (mm) Al	**1.5**		**2.5**		**3.5**		**4.5**		**3.5**		**3.5**	
Cu	**0**		**0**		**0**		**0**		**0.1**		**0.2**	

Table 10.55 Normalized organ doses—Bony thorax LAO—3–7 years

Beam projection:			LAO			**Table cells:**						
Focus-to-detector distance (cm):			115			(+) = normal font = optimal field setting						
Tube voltage (kVp):			70			**(−) = bold font = suboptimal field setting**						
Total filtrations:			See last table row			Empty table cells correspond to values <0.01						
Organ or tissue	**Organ dose D_T: Entrance dose K_E**							**(mSv * mGy^{-1})**				
	(+)	(−)	(+)	(−)	(+)	(−)	(+)	(−)	(+)	(−)	(+)	(−)
Skull	0.03	**0.29**	0.04	**0.35**	0.04	**0.39**	0.05	**0.43**	0.06	**0.49**	0.07	**0.54**
Brain		**0.02**		**0.02**		**0.03**		**0.03**		**0.04**		**0.05**
Oral mucosa	0.03	**0.40**	0.04	**0.48**	0.05	**0.53**	0.05	**0.57**	0.06	**0.65**	0.07	**0.72**
Salivary glands	0.06	**0.38**	0.08	**0.44**	0.09	**0.49**	0.10	**0.52**	0.12	**0.59**	0.13	**0.65**
Cervical spine	0.37	**0.49**	0.47	**0.63**	0.54	**0.73**	0.60	**0.82**	0.72	**0.99**	0.83	**1.16**
Thyroid	0.79	**0.83**	0.90	**0.95**	0.97	**1.03**	1.02	**1.09**	1.11	**1.20**	1.18	**1.29**
Thymus	0.69	**0.71**	0.83	**0.85**	0.92	**0.95**	0.99	**1.02**	1.12	**1.16**	1.24	**1.27**
Thoracic spine	0.22	**0.24**	0.29	**0.32**	0.35	**0.38**	0.39	**0.43**	0.50	**0.55**	0.60	**0.66**
Claviculae	1.59	**1.68**	1.89	**1.99**	2.09	**2.21**	2.25	**2.38**	2.51	**2.66**	2.71	**2.89**
Scapulae	0.19	**0.25**	0.25	**0.32**	0.29	**0.38**	0.33	**0.43**	0.41	**0.54**	0.49	**0.65**
Ribs	1.28	**1.36**	1.48	**1.58**	1.62	**1.72**	1.72	**1.83**	1.89	**2.03**	2.02	**2.18**
Oesophagus	0.15	**0.16**	0.19	**0.21**	0.23	**0.25**	0.26	**0.28**	0.32	**0.35**	0.37	**0.41**
Trachea	0.11	**0.34**	0.13	**0.41**	0.14	**0.47**	0.15	**0.51**	0.17	**0.59**	0.19	**0.66**
Lung	0.33	**0.35**	0.41	**0.43**	0.46	**0.49**	0.51	**0.54**	0.59	**0.63**	0.67	**0.71**
Mamma	1.11	**1.16**	1.19	**1.24**	1.24	**1.30**	1.27	**1.33**	1.32	**1.39**	1.37	**1.44**
Heart	0.49	**0.51**	0.60	**0.62**	0.68	**0.71**	0.75	**0.77**	0.87	**0.90**	0.97	**1.01**
Stomach	0.35	**0.62**	0.43	**0.75**	0.48	**0.85**	0.52	**0.92**	0.60	**1.06**	0.67	**1.17**
Spleen	0.18	**0.24**	0.24	**0.31**	0.28	**0.37**	0.31	**0.41**	0.37	**0.50**	0.43	**0.58**
Liver	0.22	**0.32**	0.28	**0.40**	0.32	**0.46**	0.35	**0.51**	0.41	**0.60**	0.47	**0.68**
Gall bladder	0.12	**0.36**	0.15	**0.45**	0.18	**0.53**	0.20	**0.59**	0.24	**0.70**	0.28	**0.80**
Pancreas	0.22	**0.27**	0.29	**0.35**	0.34	**0.41**	0.38	**0.46**	0.46	**0.56**	0.53	**0.66**
Small intestine	0.01	**0.20**	0.02	**0.25**	0.02	**0.29**	0.02	**0.33**	0.03	**0.39**	0.04	**0.45**
Large intestine	0.02	**0.25**	0.02	**0.32**	0.03	**0.37**	0.03	**0.41**	0.04	**0.49**	0.05	**0.56**
Sigma/rectum		**0.08**		**0.10**		**0.11**		**0.12**		**0.15**	0.01	**0.17**
Lumbar spine	0.06	**0.18**	0.08	**0.24**	0.09	**0.29**	0.11	**0.33**	0.14	**0.42**	0.17	**0.52**
Pelvis		**0.09**		**0.11**	0.01	**0.14**	0.01	**0.15**	0.02	**0.19**	0.02	**0.23**
Adrenals	0.09	**0.10**	0.11	**0.13**	0.14	**0.16**	0.16	**0.18**	0.20	**0.23**	0.23	**0.28**
Kidneys	0.05	**0.09**	0.06	**0.12**	0.07	**0.14**	0.08	**0.16**	0.10	**0.20**	0.13	**0.24**
Urinary bladder		**0.01**		**0.02**		**0.02**		**0.02**		**0.03**		**0.04**
Uterus		**0.04**		**0.05**		**0.06**		**0.07**		**0.08**	0.01	**0.10**
Ovaries		**0.04**		**0.05**		**0.06**		**0.07**		**0.09**	0.01	**0.11**
Prostate										**0.01**		**0.02**
Testes												
Upper arm	0.36	**0.90**	0.45	**1.09**	0.51	**1.22**	0.56	**1.33**	0.66	**1.52**	0.75	**1.69**
Elbow	0.33	**0.98**	0.41	**1.17**	0.47	**1.32**	0.52	**1.43**	0.61	**1.63**	0.70	**1.80**
Forearm	0.03	**0.53**	0.04	**0.64**	0.05	**0.71**	0.05	**0.77**	0.06	**0.88**	0.08	**0.97**
Upper leg												
Knee												
Lower leg												
Lymph nodes	0.17	**0.32**	0.21	**0.40**	0.24	**0.45**	0.26	**0.49**	0.30	**0.58**	0.34	**0.66**
Skin	0.14	**0.26**	0.15	**0.28**	0.15	**0.29**	0.16	**0.30**	0.17	**0.31**	0.18	**0.33**
Muscle	0.10	**0.17**	0.12	**0.20**	0.13	**0.22**	0.15	**0.24**	0.17	**0.28**	0.19	**0.31**
Skeleton	0.21	**0.38**	0.25	**0.46**	0.28	**0.51**	0.31	**0.56**	0.35	**0.64**	0.39	**0.71**
Red bone marrow	0.07	**0.11**	0.08	**0.13**	0.09	**0.15**	0.10	**0.17**	0.11	**0.19**	0.13	**0.22**
Whole body dose	0.12	**0.20**	0.14	**0.24**	0.16	**0.27**	0.17	**0.29**	0.20	**0.33**	0.22	**0.37**
ED (ICRP 60)	0.21	**0.29**	0.25	**0.35**	0.28	**0.39**	0.30	**0.42**	0.34	**0.48**	0.37	**0.53**
ED (ICRP 103)	0.29	**0.39**	0.34	**0.45**	0.37	**0.50**	0.39	**0.53**	0.43	**0.60**	0.47	**0.66**
Total filtration (mm) Al	**1.5**		**2.5**		**3.5**		**4.5**		**3.5**		**3.5**	
Cu	**0**		**0**		**0**		**0**		**0.1**		**0.2**	

Table 10.56 Normalized organ doses—Bony thorax LAO—8–12 years

Beam projection:			LAO			**Table cells:**						
Focus-to-detector distance (cm):			115			(+) = normal font = optimal field setting						
Tube voltage (kVp):			60			**(−) = bold font = suboptimal field setting**						
Total filtrations:			See last table row			Empty table cells correspond to values <0.01						
Organ or tissue	**Organ dose D_T: Entrance dose K_E**						**($mSv * mGy^{-1}$)**					
	(+)	(−)	(+)	(−)	(+)	(−)	(+)	(−)	(+)	(−)	(+)	(−)
Skull	0.03	**0.25**	0.04	**0.31**	0.05	**0.36**	0.05	**0.39**	0.06	**0.45**	0.06	**0.43**
Brain		**0.01**		**0.01**		**0.02**		**0.02**		**0.03**		**0.02**
Oral mucosa	0.02	**0.32**	0.03	**0.39**	0.03	**0.43**	0.04	**0.47**	0.05	**0.54**	0.04	**0.51**
Salivary glands	0.07	**0.32**	0.09	**0.38**	0.10	**0.42**	0.11	**0.46**	0.13	**0.52**	0.13	**0.50**
Cervical spine	0.30	**0.37**	0.39	**0.49**	0.47	**0.58**	0.53	**0.66**	0.65	**0.82**	0.59	**0.74**
Thyroid	0.67	**0.70**	0.78	**0.81**	0.85	**0.89**	0.90	**0.95**	1.00	**1.05**	0.98	**1.03**
Thymus	0.56	**0.58**	0.69	**0.71**	0.79	**0.81**	0.86	**0.89**	0.99	**1.02**	0.96	**0.99**
Thoracic spine	0.11	**0.12**	0.16	**0.17**	0.19	**0.21**	0.22	**0.24**	0.29	**0.31**	0.23	**0.25**
Claviculae	1.28	**1.35**	1.56	**1.65**	1.77	**1.87**	1.93	**2.04**	2.20	**2.34**	2.17	**2.30**
Scapulae	0.12	**0.14**	0.16	**0.19**	0.19	**0.23**	0.22	**0.27**	0.29	**0.35**	0.24	**0.29**
Ribs	1.08	**1.16**	1.27	**1.37**	1.41	**1.51**	1.50	**1.62**	1.68	**1.82**	1.66	**1.79**
Oesophagus	0.07	**0.08**	0.10	**0.11**	0.12	**0.13**	0.14	**0.15**	0.18	**0.19**	0.15	**0.16**
Trachea	0.22	**0.41**	0.26	**0.50**	0.30	**0.58**	0.32	**0.63**	0.36	**0.74**	0.35	**0.71**
Lung	0.24	**0.26**	0.31	**0.33**	0.36	**0.38**	0.40	**0.43**	0.48	**0.51**	0.45	**0.48**
Mamma	0.96	**1.07**	1.04	**1.16**	1.09	**1.22**	1.12	**1.25**	1.18	**1.31**	1.16	**1.30**
Heart	0.36	**0.38**	0.46	**0.48**	0.53	**0.55**	0.59	**0.61**	0.70	**0.73**	0.66	**0.68**
Stomach	0.30	**0.48**	0.37	**0.60**	0.42	**0.69**	0.47	**0.76**	0.55	**0.89**	0.52	**0.85**
Spleen	0.13	**0.15**	0.17	**0.20**	0.20	**0.24**	0.23	**0.28**	0.28	**0.35**	0.25	**0.31**
Liver	0.16	**0.21**	0.20	**0.27**	0.24	**0.32**	0.27	**0.36**	0.32	**0.43**	0.30	**0.40**
Gall bladder	0.09	**0.23**	0.12	**0.30**	0.14	**0.36**	0.16	**0.41**	0.20	**0.50**	0.18	**0.45**
Pancreas	0.15	**0.18**	0.20	**0.24**	0.24	**0.28**	0.27	**0.32**	0.34	**0.41**	0.30	**0.36**
Small intestine		**0.09**		**0.12**	0.01	**0.14**	0.01	**0.16**	0.02	**0.20**	0.01	**0.17**
Large intestine		**0.15**	0.01	**0.19**	0.02	**0.23**	0.02	**0.26**	0.02	**0.31**	0.02	**0.29**
Sigma/rectum		**0.03**		**0.03**		**0.04**		**0.04**		**0.06**		**0.05**
Lumbar spine	0.03	**0.08**	0.04	**0.12**	0.05	**0.14**	0.06	**0.17**	0.08	**0.22**	0.06	**0.17**
Pelvis		**0.03**		**0.03**		**0.04**		**0.05**		**0.06**		**0.05**
Adrenals	0.05	**0.06**	0.07	**0.08**	0.08	**0.10**	0.10	**0.11**	0.13	**0.15**	0.10	**0.12**
Kidneys	0.03	**0.05**	0.04	**0.06**	0.05	**0.08**	0.05	**0.09**	0.07	**0.12**	0.06	**0.10**
Urinary bladder												
Uterus		**0.01**		**0.02**		**0.02**		**0.02**		**0.03**		**0.02**
Ovaries		**0.01**		**0.02**		**0.02**		**0.02**		**0.03**		**0.02**
Prostate												
Testes												
Upper arm	0.29	**0.68**	0.36	**0.85**	0.42	**0.96**	0.46	**1.06**	0.55	**1.24**	0.52	**1.19**
Elbow	0.27	**0.74**	0.34	**0.92**	0.39	**1.04**	0.44	**1.15**	0.52	**1.33**	0.49	**1.28**
Forearm	0.02	**0.33**	0.03	**0.41**	0.03	**0.47**	0.04	**0.51**	0.05	**0.60**	0.04	**0.57**
Upper leg												
Knee												
Lower leg												
Lymph nodes	0.15	**0.24**	0.19	**0.31**	0.21	**0.35**	0.23	**0.39**	0.28	**0.46**	0.26	**0.43**
Skin	0.12	**0.22**	0.13	**0.23**	0.14	**0.24**	0.14	**0.25**	0.15	**0.26**	0.14	**0.26**
Muscle	0.08	**0.12**	0.10	**0.15**	0.11	**0.17**	0.12	**0.18**	0.14	**0.21**	0.13	**0.20**
Skeleton	0.17	**0.28**	0.20	**0.34**	0.23	**0.39**	0.25	**0.42**	0.29	**0.49**	0.28	**0.47**
Red bone marrow	0.06	**0.08**	0.07	**0.10**	0.08	**0.12**	0.09	**0.13**	0.11	**0.15**	0.10	**0.14**
Whole body dose	0.09	**0.15**	0.12	**0.18**	0.13	**0.20**	0.14	**0.22**	0.17	**0.26**	0.16	**0.24**
ED (ICRP 60)	0.17	**0.23**	0.21	**0.27**	0.23	**0.31**	0.25	**0.33**	0.29	**0.38**	0.27	**0.36**
ED (ICRP 103)	0.24	**0.31**	0.28	**0.37**	0.31	**0.41**	0.33	**0.44**	0.37	**0.49**	0.36	**0.47**
Total filtration (mm) **Al**	**1.5**		**2.5**		**3.5**		**4.5**		**3.5**		**3.5**	
Cu	**0**		**0**		**0**		**0**		**0.1**		**0.2**	

Table 10.57 Normalized organ doses—Bony thorax LAO—8–12 years

Beam projection:	LAO	**Table cells:**
Focus-to-detector distance (cm):	115	(+) = normal font = optimal field setting
Tube voltage (kVp):	70	**(−) = bold font = suboptimal field setting**
Total filtrations:	See last table row	Empty table cells correspond to values <0.01

Organ or tissue	**Organ dose D_T: Entrance dose K_E**							**(mSv * mGy^{-1})**				
	(+)	(−)	(+)	(−)	(+)	(−)	(+)	(−)	(+)	(−)	(+)	(−)
Skull	0.04	**0.29**	0.05	**0.36**	0.05	**0.40**	0.06	**0.44**	0.07	**0.50**	0.08	**0.56**
Brain		**0.02**		**0.02**		**0.03**		**0.03**		**0.04**		**0.05**
Oral mucosa	0.03	**0.37**	0.04	**0.44**	0.04	**0.49**	0.05	**0.53**	0.06	**0.60**	0.07	**0.67**
Salivary glands	0.08	**0.36**	0.10	**0.43**	0.12	**0.47**	0.13	**0.51**	0.15	**0.57**	0.17	**0.63**
Cervical spine	0.38	**0.47**	0.48	**0.61**	0.56	**0.72**	0.63	**0.80**	0.76	**0.98**	0.88	**1.15**
Thyroid	0.73	**0.77**	0.84	**0.89**	0.91	**0.97**	0.97	**1.03**	1.06	**1.13**	1.14	**1.22**
Thymus	0.65	**0.67**	0.79	**0.81**	0.88	**0.91**	0.96	**0.99**	1.10	**1.13**	1.22	**1.26**
Thoracic spine	0.16	**0.18**	0.22	**0.23**	0.26	**0.28**	0.30	**0.32**	0.38	**0.42**	0.47	**0.51**
Claviculae	1.44	**1.52**	1.73	**1.83**	1.94	**2.05**	2.09	**2.22**	2.36	**2.51**	2.58	**2.75**
Scapulae	0.16	**0.19**	0.21	**0.26**	0.25	**0.31**	0.29	**0.35**	0.36	**0.44**	0.44	**0.53**
Ribs	1.19	**1.28**	1.38	**1.49**	1.51	**1.64**	1.61	**1.75**	1.78	**1.94**	1.91	**2.10**
Oesophagus	0.10	**0.11**	0.13	**0.14**	0.16	**0.17**	0.18	**0.20**	0.22	**0.24**	0.27	**0.29**
Trachea	0.25	**0.48**	0.30	**0.58**	0.33	**0.66**	0.35	**0.72**	0.40	**0.84**	0.44	**0.94**
Lung	0.30	**0.31**	0.37	**0.39**	0.42	**0.45**	0.47	**0.50**	0.55	**0.59**	0.63	**0.67**
Mamma	1.00	**1.12**	1.08	**1.21**	1.13	**1.26**	1.16	**1.29**	1.22	**1.35**	1.26	**1.41**
Heart	0.44	**0.45**	0.54	**0.56**	0.62	**0.64**	0.68	**0.71**	0.80	**0.84**	0.91	**0.95**
Stomach	0.35	**0.57**	0.43	**0.70**	0.49	**0.79**	0.53	**0.86**	0.62	**1.00**	0.70	**1.12**
Spleen	0.16	**0.20**	0.21	**0.26**	0.25	**0.31**	0.28	**0.35**	0.34	**0.43**	0.40	**0.51**
Liver	0.19	**0.26**	0.24	**0.33**	0.28	**0.38**	0.31	**0.42**	0.38	**0.51**	0.43	**0.58**
Gall bladder	0.12	**0.29**	0.15	**0.37**	0.18	**0.44**	0.20	**0.49**	0.25	**0.59**	0.29	**0.69**
Pancreas	0.20	**0.23**	0.26	**0.30**	0.30	**0.36**	0.34	**0.40**	0.42	**0.50**	0.49	**0.59**
Small intestine		**0.11**	0.01	**0.15**	0.02	**0.17**	0.02	**0.19**	0.02	**0.24**	0.03	**0.28**
Large intestine	0.01	**0.19**	0.02	**0.24**	0.02	**0.28**	0.03	**0.31**	0.03	**0.37**	0.04	**0.43**
Sigma/rectum		**0.03**		**0.04**		**0.05**		**0.06**		**0.07**		**0.08**
Lumbar spine	0.05	**0.12**	0.06	**0.17**	0.08	**0.20**	0.09	**0.23**	0.12	**0.30**	0.15	**0.37**
Pelvis		**0.04**		**0.05**		**0.06**		**0.07**	0.01	**0.09**	0.01	**0.11**
Adrenals	0.07	**0.08**	0.10	**0.11**	0.11	**0.13**	0.13	**0.15**	0.17	**0.19**	0.20	**0.24**
Kidneys	0.04	**0.07**	0.05	**0.09**	0.06	**0.11**	0.07	**0.12**	0.09	**0.16**	0.11	**0.19**
Urinary bladder										**0.01**		**0.02**
Uterus		**0.02**		**0.02**		**0.03**		**0.03**		**0.04**		**0.05**
Ovaries		**0.02**		**0.03**		**0.03**		**0.04**		**0.05**		**0.06**
Prostate												
Testes												
Upper arm	0.34	**0.79**	0.43	**0.96**	0.49	**1.09**	0.53	**1.19**	0.63	**1.37**	0.71	**1.53**
Elbow	0.32	**0.86**	0.40	**1.04**	0.46	**1.17**	0.50	**1.28**	0.59	**1.47**	0.68	**1.64**
Forearm	0.03	**0.38**	0.04	**0.46**	0.04	**0.52**	0.05	**0.57**	0.06	**0.65**	0.07	**0.73**
Upper leg												
Knee												
Lower leg												
Lymph nodes	0.18	**0.29**	0.22	**0.36**	0.25	**0.41**	0.27	**0.45**	0.32	**0.53**	0.36	**0.61**
Skin	0.13	**0.23**	0.14	**0.24**	0.14	**0.25**	0.15	**0.26**	0.16	**0.27**	0.16	**0.29**
Muscle	0.09	**0.14**	0.11	**0.17**	0.13	**0.19**	0.14	**0.21**	0.16	**0.24**	0.18	**0.27**
Skeleton	0.19	**0.32**	0.23	**0.39**	0.26	**0.44**	0.28	**0.48**	0.32	**0.55**	0.36	**0.61**
Red bone marrow	0.07	**0.10**	0.09	**0.12**	0.10	**0.14**	0.11	**0.15**	0.12	**0.18**	0.14	**0.21**
Whole body dose	0.11	**0.17**	0.13	**0.21**	0.15	**0.23**	0.16	**0.25**	0.19	**0.29**	0.21	**0.33**
ED (ICRP 60)	0.20	**0.26**	0.23	**0.31**	0.26	**0.35**	0.28	**0.37**	0.32	**0.43**	0.35	**0.48**
ED (ICRP 103)	0.27	**0.35**	0.31	**0.41**	0.34	**0.45**	0.37	**0.49**	0.41	**0.55**	0.45	**0.60**
Total filtration (mm) Al	**1.5**		**2.5**		**3.5**		**4.5**		**3.5**		**3.5**	
Cu	**0**		**0**		**0**		**0**		**0.1**		**0.2**	

Table 10.58 Normalized organ doses—Bony thorax LAO—13–17 years

Beam projection:				LAO		**Table cells:**						
Focus-to-detector distance (cm):				115		(+) = normal font = optimal field setting						
Tube voltage (kVp):				60		**(−) = bold font = suboptimal field setting**						
Total filtrations:				See last table row		Empty table cells correspond to values <0.01						
Organ or tissue	**Organ dose D_T: Entrance dose K_E**						**(mSv * mGy^{-1})**					
	(+)	(−)	(+)	(−)	(+)	(−)	(+)	(−)	(+)	(−)	(+)	(−)
Skull	0.02	**0.26**	0.03	**0.32**	0.04	**0.37**	0.04	**0.41**	0.05	**0.49**	0.04	**0.46**
Brain		**0.02**		**0.02**		**0.03**		**0.03**		**0.04**		**0.03**
Oral mucosa	0.02	**0.25**	0.02	**0.31**	0.02	**0.35**	0.03	**0.38**	0.03	**0.44**	0.03	**0.41**
Salivary glands	0.06	**0.27**	0.07	**0.32**	0.08	**0.36**	0.09	**0.39**	0.11	**0.44**	0.10	**0.43**
Cervical spine	0.31	**0.37**	0.41	**0.48**	0.49	**0.58**	0.56	**0.66**	0.69	**0.83**	0.62	**0.73**
Thyroid	0.57	**0.57**	0.67	**0.68**	0.74	**0.75**	0.79	**0.80**	0.88	**0.90**	0.86	**0.88**
Thymus	0.48	**0.48**	0.60	**0.60**	0.68	**0.69**	0.75	**0.76**	0.88	**0.90**	0.84	**0.85**
Thoracic spine	0.06	**0.06**	0.08	**0.08**	0.10	**0.10**	0.11	**0.12**	0.15	**0.16**	0.11	**0.12**
Claviculae	1.01	**1.03**	1.25	**1.28**	1.42	**1.46**	1.56	**1.60**	1.81	**1.86**	1.76	**1.81**
Scapulae	0.07	**0.08**	0.09	**0.11**	0.11	**0.13**	0.13	**0.16**	0.17	**0.21**	0.14	**0.16**
Ribs	0.90	**0.93**	1.07	**1.11**	1.19	**1.24**	1.28	**1.34**	1.45	**1.51**	1.42	**1.48**
Oesophagus	0.07	**0.07**	0.10	**0.10**	0.12	**0.12**	0.13	**0.14**	0.17	**0.18**	0.14	**0.15**
Trachea	0.20	**0.28**	0.24	**0.34**	0.27	**0.39**	0.30	**0.43**	0.34	**0.50**	0.33	**0.48**
Lung	0.17	**0.17**	0.22	**0.22**	0.25	**0.26**	0.29	**0.30**	0.35	**0.36**	0.32	**0.33**
Mamma	5.61	**7.00**	6.51	**8.13**	7.11	**8.89**	7.56	**9.45**	8.34	**10.4**	8.18	**10.2**
Heart	0.26	**0.26**	0.33	**0.34**	0.39	**0.39**	0.43	**0.44**	0.53	**0.54**	0.48	**0.49**
Stomach	0.25	**0.37**	0.32	**0.47**	0.37	**0.55**	0.41	**0.61**	0.49	**0.73**	0.46	**0.68**
Spleen	0.09	**0.11**	0.12	**0.14**	0.14	**0.17**	0.17	**0.20**	0.21	**0.25**	0.18	**0.22**
Liver	0.12	**0.14**	0.16	**0.19**	0.19	**0.22**	0.21	**0.25**	0.26	**0.31**	0.23	**0.28**
Gall bladder	0.07	**0.15**	0.10	**0.20**	0.12	**0.24**	0.14	**0.28**	0.17	**0.35**	0.15	**0.31**
Pancreas	0.11	**0.12**	0.15	**0.17**	0.18	**0.20**	0.21	**0.23**	0.26	**0.29**	0.22	**0.25**
Small intestine		**0.08**		**0.10**		**0.12**		**0.14**	0.01	**0.17**		**0.15**
Large intestine		**0.12**	0.01	**0.15**	0.01	**0.18**	0.01	**0.21**	0.02	**0.26**	0.01	**0.23**
Sigma/rectum		**0.03**		**0.03**		**0.04**		**0.05**		**0.06**		**0.05**
Lumbar spine	0.02	**0.05**	0.03	**0.06**	0.03	**0.08**	0.04	**0.09**	0.05	**0.13**	0.04	**0.09**
Pelvis		**0.02**		**0.03**		**0.03**		**0.04**		**0.05**		**0.04**
Adrenals	0.03	**0.03**	0.04	**0.05**	0.05	**0.06**	0.06	**0.07**	0.08	**0.09**	0.06	**0.07**
Kidneys	0.02	**0.03**	0.02	**0.04**	0.03	**0.05**	0.04	**0.06**	0.05	**0.07**	0.04	**0.06**
Urinary bladder												
Uterus		**0.03**		**0.03**		**0.04**		**0.05**		**0.07**		**0.05**
Ovaries				**0.01**		**0.02**		**0.02**		**0.03**		**0.02**
Prostate												
Testes												
Upper arm	0.18	**0.52**	0.23	**0.65**	0.27	**0.75**	0.30	**0.82**	0.36	**0.97**	0.33	**0.92**
Elbow	0.15	**0.56**	0.20	**0.70**	0.23	**0.80**	0.26	**0.88**	0.32	**1.04**	0.29	**0.98**
Forearm	0.02	**0.27**	0.02	**0.33**	0.03	**0.38**	0.03	**0.42**	0.04	**0.49**	0.03	**0.47**
Upper leg												
Knee												
Lower leg												
Lymph nodes	0.12	**0.18**	0.15	**0.23**	0.17	**0.27**	0.19	**0.30**	0.23	**0.36**	0.21	**0.33**
Skin	0.13	**0.22**	0.14	**0.24**	0.15	**0.25**	0.15	**0.26**	0.16	**0.27**	0.16	**0.26**
Muscle	0.07	**0.10**	0.09	**0.13**	0.10	**0.14**	0.11	**0.16**	0.13	**0.18**	0.12	**0.17**
Skeleton	0.14	**0.23**	0.17	**0.29**	0.19	**0.33**	0.21	**0.36**	0.25	**0.42**	0.23	**0.40**
Red bone marrow	0.06	**0.07**	0.07	**0.09**	0.08	**0.11**	0.09	**0.12**	0.10	**0.14**	0.09	**0.13**
Whole body dose	0.08	**0.13**	0.10	**0.16**	0.12	**0.18**	0.13	**0.19**	0.15	**0.23**	0.14	**0.21**
ED (ICRP 60)	0.38	**0.48**	0.45	**0.57**	0.50	**0.64**	0.54	**0.69**	0.61	**0.78**	0.59	**0.75**
ED (ICRP 103)	0.78	**0.98**	0.91	**1.15**	1.00	**1.27**	1.07	**1.36**	1.20	**1.52**	1.16	**1.48**
Total filtration (mm) **Al**	**1.5**		**2.5**		**3.5**		**4.5**		**3.5**		**3.5**	
Cu	**0**		**0**		**0**		**0**		**0.1**		**0.2**	

Table 10.59 Normalized organ doses—Bony thorax LAO—13–17 years

Beam projection:				LAO		**Table cells:**						
Focus-to-detector distance (cm):				115		(+) = normal font = optimal field setting						
Tube voltage (kVp):				70		**(−) = bold font = suboptimal field setting**						
Total filtrations:				See last table row		Empty table cells correspond to values <0.01						
Organ or tissue	**Organ dose D_T: Entrance dose K_E**						**($mSv * mGy^{-1}$)**					
	(+)	(−)	(+)	(−)	(+)	(−)	(+)	(−)	(+)	(−)	(+)	(−)
Skull	0.03	**0.31**	0.04	**0.38**	0.04	**0.43**	0.05	**0.47**	0.06	**0.55**	0.07	**0.62**
Brain		**0.02**		**0.03**		**0.04**		**0.04**		**0.05**		**0.07**
Oral mucosa	0.02	**0.29**	0.03	**0.35**	0.03	**0.40**	0.03	**0.43**	0.04	**0.49**	0.05	**0.56**
Salivary glands	0.07	**0.31**	0.09	**0.36**	0.10	**0.40**	0.11	**0.44**	0.13	**0.49**	0.14	**0.55**
Cervical spine	0.40	**0.48**	0.51	**0.61**	0.60	**0.73**	0.68	**0.82**	0.83	**1.01**	0.97	**1.19**
Thyroid	0.63	**0.64**	0.74	**0.75**	0.81	**0.82**	0.86	**0.87**	0.95	**0.97**	1.03	**1.06**
Thymus	0.56	**0.57**	0.69	**0.70**	0.78	**0.79**	0.86	**0.87**	0.99	**1.01**	1.12	**1.13**
Thoracic spine	0.09	**0.09**	0.12	**0.12**	0.14	**0.15**	0.17	**0.18**	0.22	**0.23**	0.27	**0.29**
Claviculae	1.16	**1.19**	1.41	**1.44**	1.58	**1.63**	1.72	**1.77**	1.97	**2.03**	2.17	**2.25**
Scapulae	0.10	**0.12**	0.13	**0.15**	0.16	**0.19**	0.18	**0.21**	0.23	**0.27**	0.29	**0.34**
Ribs	1.00	**1.04**	1.18	**1.22**	1.30	**1.35**	1.39	**1.45**	1.55	**1.63**	1.69	**1.78**
Oesophagus	0.10	**0.10**	0.13	**0.14**	0.16	**0.16**	0.18	**0.19**	0.23	**0.24**	0.28	**0.29**
Trachea	0.23	**0.32**	0.27	**0.40**	0.30	**0.45**	0.33	**0.49**	0.37	**0.57**	0.40	**0.64**
Lung	0.21	**0.22**	0.27	**0.28**	0.31	**0.32**	0.35	**0.36**	0.42	**0.44**	0.49	**0.51**
Mamma	6.15	**7.67**	7.06	**8.81**	7.66	**9.57**	8.12	**10.1**	8.91	**11.1**	9.58	**12.0**
Heart	0.32	**0.32**	0.40	**0.41**	0.46	**0.48**	0.52	**0.53**	0.62	**0.64**	0.72	**0.74**
Stomach	0.30	**0.45**	0.37	**0.56**	0.43	**0.65**	0.48	**0.71**	0.56	**0.84**	0.64	**0.96**
Spleen	0.12	**0.14**	0.16	**0.19**	0.19	**0.23**	0.21	**0.26**	0.27	**0.32**	0.32	**0.39**
Liver	0.15	**0.18**	0.20	**0.23**	0.23	**0.27**	0.26	**0.31**	0.31	**0.37**	0.37	**0.44**
Gall bladder	0.10	**0.20**	0.13	**0.26**	0.15	**0.31**	0.17	**0.35**	0.21	**0.43**	0.25	**0.50**
Pancreas	0.15	**0.17**	0.19	**0.22**	0.23	**0.26**	0.26	**0.30**	0.33	**0.38**	0.39	**0.45**
Small intestine		**0.10**		**0.13**	0.01	**0.15**	0.01	**0.17**	0.02	**0.22**	0.02	**0.26**
Large intestine	0.01	**0.15**	0.01	**0.19**	0.02	**0.23**	0.02	**0.26**	0.03	**0.31**	0.03	**0.37**
Sigma/rectum		**0.03**		**0.04**		**0.05**		**0.06**		**0.07**		**0.08**
Lumbar spine	0.03	**0.07**	0.04	**0.10**	0.05	**0.12**	0.06	**0.14**	0.08	**0.18**	0.10	**0.23**
Pelvis		**0.03**		**0.04**		**0.05**		**0.06**		**0.07**		**0.09**
Adrenals	0.05	**0.05**	0.06	**0.07**	0.08	**0.08**	0.09	**0.10**	0.11	**0.13**	0.14	**0.16**
Kidneys	0.03	**0.04**	0.04	**0.06**	0.04	**0.07**	0.05	**0.08**	0.07	**0.10**	0.08	**0.13**
Urinary bladder												**0.01**
Uterus		**0.04**		**0.05**		**0.06**		**0.07**		**0.10**	0.01	**0.12**
Ovaries		**0.01**		**0.02**		**0.02**		**0.03**		**0.04**		**0.05**
Prostate												
Testes												
Upper arm	0.22	**0.61**	0.28	**0.75**	0.32	**0.85**	0.36	**0.93**	0.43	**1.09**	0.49	**1.22**
Elbow	0.19	**0.66**	0.24	**0.80**	0.28	**0.91**	0.31	**1.00**	0.38	**1.16**	0.44	**1.30**
Forearm	0.02	**0.31**	0.03	**0.38**	0.03	**0.43**	0.04	**0.47**	0.05	**0.55**	0.06	**0.62**
Upper leg												
Knee												
Lower leg												
Lymph nodes	0.15	**0.22**	0.18	**0.28**	0.21	**0.32**	0.23	**0.35**	0.27	**0.42**	0.31	**0.48**
Skin	0.14	**0.23**	0.15	**0.25**	0.16	**0.26**	0.16	**0.27**	0.17	**0.28**	0.18	**0.30**
Muscle	0.08	**0.12**	0.10	**0.15**	0.11	**0.17**	0.13	**0.18**	0.15	**0.21**	0.17	**0.24**
Skeleton	0.16	**0.27**	0.20	**0.33**	0.22	**0.37**	0.24	**0.41**	0.28	**0.47**	0.31	**0.53**
Red bone marrow	0.07	**0.09**	0.08	**0.11**	0.09	**0.13**	0.10	**0.14**	0.12	**0.17**	0.14	**0.19**
Whole body dose	0.10	**0.15**	0.12	**0.18**	0.14	**0.21**	0.15	**0.22**	0.17	**0.26**	0.20	**0.29**
ED (ICRP 60)	0.43	**0.54**	0.50	**0.64**	0.55	**0.71**	0.59	**0.76**	0.67	**0.86**	0.73	**0.94**
ED (ICRP 103)	0.86	**1.09**	1.00	**1.27**	1.10	**1.39**	1.17	**1.49**	1.30	**1.65**	1.41	**1.80**
Total filtration (mm) **Al**	**1.5**		**2.5**		**3.5**		**4.5**		**3.5**		**3.5**	
Cu	**0**		**0**		**0**		**0**		**0.1**		**0.2**	

Table 10.60 Normalized organ doses—Bony thorax LAO— >17 years

Beam projection:				LAO			**Table cells:**					
Focus-to-detector distance (cm):				115			(+) = normal font = optimal field setting					
Tube voltage (kVp):				60			**(−) = bold font = suboptimal field setting**					
Total filtrations:				See last table row			Empty table cells correspond to values <0.01					
Organ or tissue	**Organ dose D_T: Entrance dose K_E**							**(mSv * mGy^{-1})**				
	(+)	(−)	(+)	(−)	(+)	(−)	(+)	(−)	(+)	(−)	(+)	(−)
Skull	0.02	**0.21**	0.03	**0.27**	0.03	**0.31**	0.03	**0.34**	0.04	**0.41**	0.04	**0.38**
Brain				**0.01**		**0.02**		**0.02**		**0.02**		**0.02**
Oral mucosa	0.01	**0.24**	0.02	**0.29**	0.02	**0.32**	0.02	**0.35**	0.03	**0.41**	0.02	**0.38**
Salivary glands	0.05	**0.26**	0.06	**0.31**	0.07	**0.35**	0.08	**0.38**	0.09	**0.43**	0.09	**0.41**
Cervical spine	0.26	**0.32**	0.35	**0.42**	0.42	**0.50**	0.48	**0.58**	0.59	**0.72**	0.53	**0.63**
Thyroid	0.52	**0.53**	0.62	**0.63**	0.69	**0.70**	0.74	**0.75**	0.83	**0.85**	0.81	**0.83**
Thymus	0.47	**0.48**	0.58	**0.60**	0.67	**0.69**	0.74	**0.77**	0.87	**0.90**	0.83	**0.86**
Thoracic spine	0.05	**0.05**	0.07	**0.08**	0.09	**0.09**	0.10	**0.11**	0.14	**0.15**	0.10	**0.11**
Claviculae	0.99	**1.03**	1.23	**1.28**	1.40	**1.47**	1.54	**1.62**	1.79	**1.88**	1.74	**1.83**
Scapulae	0.08	**0.08**	0.10	**0.11**	0.13	**0.14**	0.15	**0.16**	0.19	**0.21**	0.16	**0.17**
Ribs	0.91	**0.96**	1.08	**1.14**	1.20	**1.27**	1.30	**1.38**	1.47	**1.56**	1.44	**1.52**
Oesophagus	0.07	**0.07**	0.10	**0.10**	0.12	**0.12**	0.14	**0.14**	0.18	**0.19**	0.14	**0.15**
Trachea	0.19	**0.25**	0.23	**0.31**	0.26	**0.35**	0.28	**0.39**	0.32	**0.46**	0.31	**0.43**
Lung	0.17	**0.18**	0.23	**0.24**	0.27	**0.28**	0.30	**0.31**	0.37	**0.38**	0.33	**0.35**
Mamma	0.62	**0.77**	0.73	**0.90**	0.79	**0.99**	0.84	**1.05**	0.93	**1.16**	0.91	**1.14**
Heart	0.25	**0.26**	0.32	**0.34**	0.38	**0.39**	0.42	**0.44**	0.51	**0.54**	0.47	**0.49**
Stomach	0.25	**0.37**	0.32	**0.47**	0.37	**0.55**	0.41	**0.61**	0.50	**0.73**	0.46	**0.68**
Spleen	0.09	**0.10**	0.12	**0.14**	0.14	**0.16**	0.16	**0.19**	0.21	**0.24**	0.18	**0.20**
Liver	0.12	**0.14**	0.16	**0.19**	0.18	**0.22**	0.21	**0.25**	0.26	**0.31**	0.23	**0.28**
Gall bladder	0.08	**0.15**	0.10	**0.20**	0.13	**0.24**	0.14	**0.27**	0.18	**0.34**	0.16	**0.30**
Pancreas	0.11	**0.12**	0.14	**0.16**	0.17	**0.19**	0.20	**0.22**	0.25	**0.28**	0.22	**0.24**
Small intestine		**0.08**		**0.10**		**0.12**		**0.14**	0.01	**0.17**		**0.15**
Large intestine		**0.12**		**0.15**	0.01	**0.18**	0.01	**0.21**	0.02	**0.26**	0.01	**0.23**
Sigma/rectum		**0.03**		**0.03**		**0.04**		**0.05**		**0.06**		**0.05**
Lumbar spine	0.02	**0.04**	0.02	**0.06**	0.03	**0.07**	0.04	**0.08**	0.05	**0.11**	0.04	**0.08**
Pelvis		**0.02**		**0.03**		**0.03**		**0.04**		**0.05**		**0.04**
Adrenals	0.03	**0.03**	0.04	**0.05**	0.05	**0.06**	0.06	**0.07**	0.08	**0.09**	0.06	**0.07**
Kidneys	0.02	**0.03**	0.02	**0.04**	0.03	**0.04**	0.03	**0.05**	0.05	**0.07**	0.03	**0.05**
Urinary bladder												
Uterus				**0.01**		**0.02**		**0.02**		**0.02**		**0.02**
Ovaries				**0.01**		**0.01**		**0.02**		**0.02**		**0.02**
Prostate												
Testes												
Upper arm	0.27	**0.50**	0.34	**0.62**	0.39	**0.72**	0.44	**0.79**	0.52	**0.94**	0.48	**0.89**
Elbow	0.26	**0.54**	0.33	**0.67**	0.38	**0.77**	0.43	**0.85**	0.51	**1.01**	0.47	**0.95**
Forearm	0.02	**0.25**	0.03	**0.31**	0.04	**0.36**	0.04	**0.40**	0.05	**0.47**	0.05	**0.45**
Upper leg												
Knee												
Lower leg												
Lymph nodes	0.12	**0.17**	0.15	**0.22**	0.17	**0.26**	0.19	**0.29**	0.23	**0.35**	0.21	**0.31**
Skin	0.14	**0.22**	0.14	**0.24**	0.15	**0.25**	0.16	**0.25**	0.16	**0.27**	0.16	**0.26**
Muscle	0.07	**0.10**	0.08	**0.12**	0.10	**0.14**	0.11	**0.15**	0.12	**0.18**	0.11	**0.17**
Skeleton	0.14	**0.22**	0.18	**0.27**	0.20	**0.30**	0.22	**0.33**	0.26	**0.39**	0.24	**0.37**
Red bone marrow	0.07	**0.08**	0.08	**0.10**	0.09	**0.11**	0.10	**0.12**	0.12	**0.15**	0.11	**0.13**
Whole body dose	0.08	**0.12**	0.10	**0.15**	0.12	**0.17**	0.13	**0.19**	0.15	**0.22**	0.14	**0.21**
ED (ICRP 60)	0.13	**0.17**	0.16	**0.21**	0.19	**0.24**	0.20	**0.27**	0.24	**0.31**	0.22	**0.29**
ED (ICRP 103)	0.18	**0.23**	0.22	**0.28**	0.24	**0.32**	0.27	**0.35**	0.31	**0.41**	0.29	**0.39**
Total filtration (mm) **Al**	**1.5**		**2.5**		**3.5**		**4.5**		**3.5**		**3.5**	
Cu	**0**		**0**		**0**		**0**		**0.1**		**0.2**	

Table 10.61 Normalized organ doses—Bony thorax LAO— >17 years

Beam projection:			LAO			**Table cells:**						
Focus-to-detector distance (cm):			115			(+) = normal font = optimal field setting						
Tube voltage (kVp):			70			**(−) = bold font = suboptimal field setting**						
Total filtrations:			See last table row			Empty table cells correspond to values <0.01						
Organ or tissue	**Organ dose D_T: Entrance dose K_E**							**(mSv * mGy^{-1})**				
	(+)	(−)	(+)	(−)	(+)	(−)	(+)	(−)	(+)	(−)	(+)	(−)
Skull	0.03	**0.25**	0.03	**0.31**	0.04	**0.36**	0.04	**0.40**	0.05	**0.47**	0.06	**0.53**
Brain		**0.01**		**0.02**		**0.02**		**0.03**		**0.03**		**0.04**
Oral mucosa	0.02	**0.28**	0.02	**0.33**	0.02	**0.37**	0.03	**0.40**	0.03	**0.46**	0.04	**0.52**
Salivary glands	0.06	**0.30**	0.07	**0.35**	0.08	**0.39**	0.09	**0.42**	0.11	**0.48**	0.12	**0.53**
Cervical spine	0.34	**0.42**	0.44	**0.54**	0.52	**0.64**	0.59	**0.72**	0.72	**0.89**	0.85	**1.06**
Thyroid	0.58	**0.59**	0.68	**0.70**	0.75	**0.77**	0.80	**0.82**	0.90	**0.92**	0.98	**1.01**
Thymus	0.55	**0.57**	0.68	**0.70**	0.77	**0.80**	0.85	**0.88**	0.99	**1.02**	1.11	**1.15**
Thoracic spine	0.08	**0.09**	0.11	**0.12**	0.13	**0.14**	0.15	**0.16**	0.20	**0.21**	0.25	**0.27**
Claviculae	1.14	**1.19**	1.39	**1.46**	1.57	**1.65**	1.71	**1.79**	1.96	**2.06**	2.17	**2.29**
Scapulae	0.11	**0.12**	0.15	**0.16**	0.18	**0.19**	0.20	**0.22**	0.26	**0.28**	0.32	**0.35**
Ribs	1.01	**1.07**	1.19	**1.26**	1.31	**1.39**	1.41	**1.50**	1.58	**1.68**	1.71	**1.83**
Oesophagus	0.10	**0.11**	0.13	**0.14**	0.16	**0.17**	0.18	**0.20**	0.23	**0.25**	0.29	**0.30**
Trachea	0.21	**0.29**	0.26	**0.36**	0.29	**0.41**	0.31	**0.45**	0.35	**0.52**	0.39	**0.59**
Lung	0.22	**0.23**	0.28	**0.29**	0.32	**0.34**	0.36	**0.38**	0.43	**0.46**	0.50	**0.53**
Mamma	0.68	**0.85**	0.79	**0.98**	0.86	**1.06**	0.91	**1.13**	1.00	**1.24**	1.07	**1.33**
Heart	0.31	**0.33**	0.39	**0.41**	0.46	**0.48**	0.51	**0.53**	0.61	**0.64**	0.71	**0.74**
Stomach	0.30	**0.45**	0.38	**0.56**	0.44	**0.65**	0.48	**0.71**	0.57	**0.84**	0.66	**0.97**
Spleen	0.12	**0.14**	0.15	**0.18**	0.18	**0.22**	0.21	**0.25**	0.26	**0.31**	0.32	**0.38**
Liver	0.15	**0.18**	0.19	**0.24**	0.23	**0.28**	0.25	**0.31**	0.31	**0.37**	0.36	**0.44**
Gall bladder	0.10	**0.20**	0.14	**0.26**	0.16	**0.30**	0.18	**0.34**	0.23	**0.42**	0.27	**0.50**
Pancreas	0.15	**0.16**	0.19	**0.21**	0.23	**0.26**	0.26	**0.29**	0.32	**0.37**	0.39	**0.44**
Small intestine		**0.10**		**0.13**	0.01	**0.16**	0.01	**0.18**	0.02	**0.22**	0.02	**0.26**
Large intestine	0.01	**0.15**	0.01	**0.19**	0.02	**0.23**	0.02	**0.26**	0.03	**0.32**	0.03	**0.37**
Sigma/rectum		**0.03**		**0.04**		**0.05**		**0.06**		**0.07**		**0.08**
Lumbar spine	0.03	**0.07**	0.04	**0.09**	0.05	**0.11**	0.06	**0.13**	0.07	**0.17**	0.09	**0.21**
Pelvis		**0.03**		**0.04**		**0.05**		**0.05**		**0.07**		**0.09**
Adrenals	0.05	**0.05**	0.06	**0.07**	0.07	**0.09**	0.09	**0.10**	0.11	**0.13**	0.14	**0.16**
Kidneys	0.03	**0.04**	0.04	**0.05**	0.04	**0.07**	0.05	**0.08**	0.06	**0.10**	0.08	**0.12**
Urinary bladder												
Uterus		**0.01**		**0.02**		**0.02**		**0.03**		**0.04**		**0.04**
Ovaries		**0.01**		**0.02**		**0.02**		**0.02**		**0.03**		**0.04**
Prostate												
Testes												
Upper arm	0.32	**0.59**	0.40	**0.73**	0.46	**0.83**	0.51	**0.91**	0.60	**1.06**	0.69	**1.20**
Elbow	0.32	**0.63**	0.39	**0.78**	0.45	**0.89**	0.50	**0.97**	0.59	**1.13**	0.67	**1.28**
Forearm	0.03	**0.30**	0.04	**0.36**	0.05	**0.42**	0.05	**0.46**	0.06	**0.53**	0.07	**0.60**
Upper leg												
Knee												
Lower leg												
Lymph nodes	0.14	**0.22**	0.18	**0.27**	0.20	**0.31**	0.22	**0.34**	0.26	**0.41**	0.30	**0.47**
Skin	0.14	**0.23**	0.15	**0.25**	0.16	**0.26**	0.16	**0.27**	0.17	**0.28**	0.18	**0.30**
Muscle	0.08	**0.12**	0.10	**0.14**	0.11	**0.16**	0.12	**0.18**	0.14	**0.21**	0.16	**0.24**
Skeleton	0.17	**0.25**	0.20	**0.31**	0.23	**0.35**	0.25	**0.38**	0.29	**0.44**	0.32	**0.50**
Red bone marrow	0.08	**0.10**	0.09	**0.12**	0.11	**0.13**	0.12	**0.15**	0.14	**0.17**	0.16	**0.20**
Whole body dose	0.10	**0.15**	0.12	**0.18**	0.14	**0.20**	0.15	**0.22**	0.17	**0.26**	0.20	**0.29**
ED (ICRP 60)	0.16	**0.20**	0.19	**0.25**	0.22	**0.28**	0.24	**0.31**	0.27	**0.36**	0.31	**0.41**
ED (ICRP 103)	0.21	**0.27**	0.25	**0.33**	0.28	**0.37**	0.30	**0.40**	0.35	**0.47**	0.39	**0.53**
Total filtration (mm) Al	**1.5**		**2.5**		**3.5**		**4.5**		**3.5**		**3.5**	
Cu	**0**		**0**		**0**		**0**		**0.1**		**0.2**	

Reference

1. Bontrager KL. Textbook of radiographic positioning and related anatomy. St. Louis, MO: Mosby Year Book; 1993.

11 Thoraco-abdominal Babygram

11.1 Introduction

The thoraco-abdominal babygram is performed in severely ill newborn babies in an intensive care unit. Table 11.1 shows indications and Table 11.2 contraindications for the thoraco-abdominal babygram. Image criteria as required for radiological findings in babygrams are listed in Table 11.3. In order to fulfil these image criteria, the field settings listed in Tables 11.4 and 11.5 should be achieved. These field settings go along with the patient preparation as listed in Table 11.6. Exposure parameter combinations for various age groups and patient diameters are listed in Tables 11.7, 11.8, 11.9, 11.10, and 11.11. The age groups in Tables 11.8 and 11.10 refer to age groups defined by the authors and to the standard age groups of the mathematical MIRD phantom family which has been used to calculate the normalized organ doses in this book. Tables 11.9 and 11.11 show radiation qualities (i.e. various combinations of tube voltage and total filtration) which may be applied in clinical practice. Please note that not all of these radiation qualities are recommended

Table 11.1 Indications for the thoraco-abdominal babygram

Control of the position of umbilical catheters
Hydrops fetalis
Control of interventional procedures, e.g. tubes and catheters in the chest and abdomen
Congenital diaphragmatic hernia[a]
Esophageal atresia[a]
Some lethal variants of skeletal dyplasias
Annotation:
[a]Initial X-ray

Table 11.2 Contraindications for the thoraco-abdominal babygram

None

Table 11.3 Image criteria for the thoraco-abdominal babygram

Thoraco-abdominal babygram ap
Symmetrical reproduction of the trunk without rotation and tilting
Visually sharp reproduction of the trachea and the main bronchi
Reproduction of the vascular pattern of the inner 2/3 of the lungs
Visually sharp reproduction of the heart border and the diaphragm with the costophrenic angles
Reproduction of the spine and the paraspinal structures, the retrocardiac lung, and the mediastinum
Reproduction of the chest and the abdomen from the mandible to the ossa ischia
Reproduction of the clavicles, scapulae, ribs, pelvic bones, thoracic and lumbar spine
Reproduction of the proximal femoral metaphyses

Table 11.4 Patient positions and beam projections for thoraco-abdominal babygrams

Patient position	Beam projections
Supine	ap

The original version of this chapter was revised. The correction to this chapter can be found at https://doi.org/10.1007/978-3-030-18504-6_31

M. Seidenbusch et al., *Imaging Practice and Radiation Protection in Pediatric Radiology*, https://doi.org/10.1007/978-3-030-18504-6_11

Table 11.5 Optimal field settings for thoraco-abdominal babygrams

Thoraco-abdominal babygram ap	
Upper field edge:	Tip of mandible
Field centring:	Midline between the xiphoid process and the navel
Lower field edge:	Perineum
Lateral field edges:	Close to the chest and abdominal wall including the proximal parts of the arms and thighs

Table 11.6 Patient preparation

General	Remove clothes, neck laces, and all metallic devices from the neck, chest, and abdomen of the patient
	Immobilize infants in the supine position. Put sandbags over the arms and legs. Alternatively, accompanying persons or staff members could hold the patient (in this case also take care for the radiation protection of the accompanying persons or staff members)
	Rotation of the trunk in the longitudinal and/or horizontal axis should be avoided
	Legs should be extended or placed in frog-leg position in very small premature babies
	Place eventually small X-ray positive markers at ileo-, colostomies, at palpable tumours or on pain points
Positioning	Supine, the arms should be slightly abducted from the chest and abdomen
Attention	On the Bucky table the patient should always be positioned on a foam mat of 10 to 20 cm thickness. So, the lower part of the body will not be cut if lateral projection X-rays (cross table views) in prone or in decubitus position are performed
Shielding	No gonad shielding should be used

Table 11.7 Radiographic procedure

X-ray source	The mAs product used in free-hand exposure is chosen on body weight
	Focal spot size between 0.6 and 1.0
	Check the use of additional filtration
	Check the focus-to-detector distance
Field setting	Check the field size
	Check the field centring
Image detector system	No antiscatter grid is used
Patient	The arms should be positioned alongside the chest
	The exposure should be performed in deep inspiration

Table 11.8 Radiographic techniques of bedside babygram

Age group (years)	Beam direction	Patient position	Beam projection	Typical patient diameter (cm)		Optimal field size (cm x cm)	Focus-to-detector distance (cm)	Voltage (kVp)						Antiscatter grid
				With arms	No arms			50	60	70	80	110	120	
24. gw[1]	Vertical	Supine	ap	5.2	-	7 x 16	100							
28. gw				6.0	-	8 x 18								
0				9.8	-	11 x 26								
1				13.0	-	16 x 36								

Annotations:

1 gw = gestational week.

Parameter combination not recommended, normalized organ doses not provided

Parameter combination not recommended, normalized organ doses provided

Parameter combination recommended, normalized organ doses provided

by the authors. The standard projection of the thoraco-abdominal babygram is depicted in Fig. 11.1. Also the optimal field (marked in green) and the suboptimal field (marked in red) are shown. Tables 11.13 and 11.14 outline the virtual field settings on the skin surface of mathematical MIRD phantoms of various age groups and the resulting virtual radiographies with and without bone structures, respectively. Normalized organ doses resulting from these virtual field settings have been calculated for all age groups and have been listed in Tables 11.15, 11.16, 11.17, 11.18, 11.19, 11.20, 11.21, 11.22, 11.23, 11.24, 11.25, 11.26, 11.27, 11.28, 11.29, 11.30, 11.31, and 11.32. Table cells with normal font refer to optimal field settings and table cells with bold font refer to suboptimal field settings. Please note that normalized organ doses have been provided also

Table 11.9 Radiation qualities of bedside babygram

Voltage (kVp)	Total filtration					
	Without additional filtration		With additional filtration			
	1.5 mm Al	2.5 Al	3.5 Al	4.5 Al	3.5 Al + 0.1 Cu	3.5 + 0.2 Cu
50						
60						
70						
80						
110						
120						

Annotations:
- Parameter combination not recommended, normalized organ doses not provided
- Parameter combination not recommended, normalized organ doses provided
- Parameter combination recommended, normalized organ doses provided

Table 11.10 Radiographic techniques of Bucky table babygram

Age group (years)	Beam direction	Patient position	Beam projection	Typical patient diameter (cm)		Optimal field size (cm x cm)	Focus-to-detector distance (cm)	Voltage (kVp)						Antiscatter grid
				With arms	No arms			50	60	70	80	110	120	
24. gw[1]	This radiographic technique is not applied in neonates													
28. gw														
0	Vertical	Supine	ap	9.8	-	11 x 26	115							
1				13.0	-	16 x 36								

Annotations:
- [1] gw = gestational week
- Parameter combination not recommended, normalized organ doses not provided
- Parameter combination not recommended, normalized organ doses provided
- Parameter combination recommended, normalized organ doses provided

Table 11.11 Radiation qualities of Bucky table babygram

Voltage (kVp)	Total filtration					
	Without additional filtration		With additional filtration			
	1.5 mm Al	2.5 Al	3.5 Al	4.5 Al	3.5 Al + 0.1 Cu	3.5 + 0.2 Cu
50						
60						
70						
80						
110						
120						

Annotations:
- Parameter combination not recommended, normalized organ doses not provided
- Parameter combination not recommended, normalized organ doses provided
- Parameter combination recommended, normalized organ doses provided

for exposure parameter combinations, which are not recommended by the authors.

11.2 Indications and Contraindications

Indications and contraindications for thoraco-abdominal babygrams are given in Tables 11.1 and 11.2.

11.3 Image Criteria

The image criteria required for the thoraco-abdominal babygram are listed in Table 11.3.

11.4 Field Settings

In Table 11.4, patient positions and beam projections used for thoraco-abdominal babygrams are listed, and in Table 11.5 the optimal field settings are listed which are required in order to fulfil the image criteria mentioned in Table 11.3. Antero-posterior beam projection is always used.

For more details, see e.g. [2].

11.5 Patient Preparation and Radiographic Procedure

See Tables 11.6, 11.7 and [1, 3].

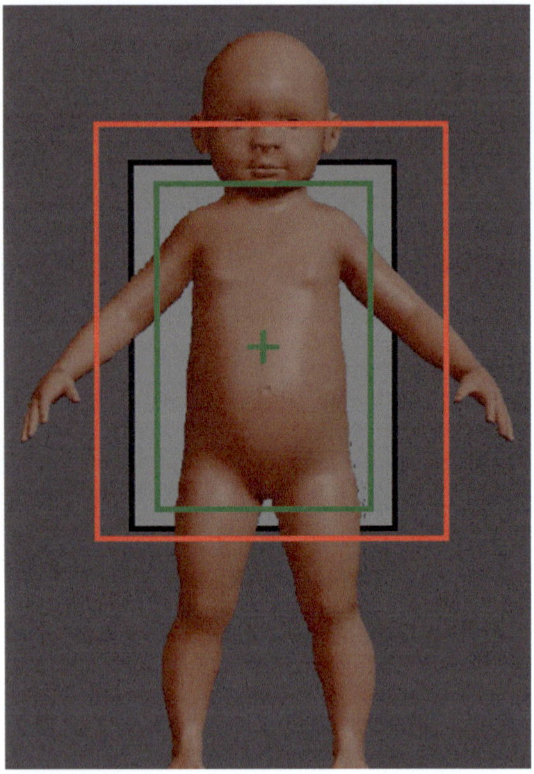

Fig. 11.1 **Field settings for the antero-posterior thoraco-abdominal babygram in a newborn baby** (© V. Rösenberger and M. Seidenbusch 2018). Green lines = optimal field settings; green cross = centring point; red lines = suboptimal field settings; grey shaded area = image detector; dark grey shaded area = Bucky table

11.6 Exposure Parameters

In Tables 11.8, 11.9, 11.10, and 11.11, exposure parameter combinations used for thoraco-abdominal babygrams in various age groups are listed. Dark grey shaded table cells refer to exposure parameter combinations which are not recommended for thoraco-abdominal babygrams and for which no normalized organ doses have been provided. Light grey and light blue shaded table cells refer to exposure parameter combinations for which normalized organ doses have been provided with light grey shaded table cells referring to exposure parameter combinations being not recommended and light blue shaded table cells referring to exposure parameter combinations being recommended for thoraco-abdominal babygrams. Note that due to the fact that the radiation dose is an exponential function of the patient diameter, it may make sense to use the patient diameters instead of the patient's age groups for determining the adequate exposure parameters from Tables 11.8 and 11.10. In this case, for determining the corresponding normalized organ doses from Tables 11.15, 11.16, 11.17, 11.18, 11.19, 11.20, 11.21, 11.22, 11.23, 11.24, 11.25, 11.26, 11.27, 11.28, 11.29, 11.30, 11.31, and 11.32, the age group related to the patient diameter should be used.

11.7 Normalized Organ Doses

For calculating normalized organ doses, the field settings described in Sect. 11.4 were transferred to the mathematical MIRD phantoms. Inconsistencies between the geometrical anatomy of the mathematical phantoms and the real human anatomy have been listed in Table 11.12 whenever they are of relevance for the validity of the normalized organ doses.

In Tables 11.13 and 11.14, optimal and suboptimal field settings are listed for the different age groups and the beam projections given in Tables 11.8, 11.9, 11.10, and 11.11. Green lines refer to optimal field settings, red lines refer to suboptimal field settings, respectively. In the tables, the optimal and suboptimal field settings which were used for performing the calculations of the normalized organ doses provided in Tables 11.15, 11.16, 11.17, 11.18, 11.19, 11.20, 11.21, 11.22, 11.23, 11.24, 11.25, 11.26, 11.27, 11.28, 11.29, 11.30, 11.31, and 11.32 were drawn in on the skin surface of the mathematical phantoms. The virtual mathematical radiographs resulting from these field settings were depicted in the tables with and without bone structures, respectively, in order to enable the reader to visually compare the virtual mathematical radiographs with the real radiographs made under clinical conditions. For comparing the virtual mathematical field settings with the real clinical field settings, the virtual mathematical field edges were also listed in the tables.

As the normalized organ doses provided in Tables 11.15, 11.16, 11.17, 11.18, 11.19, 11.20, 11.21, 11.22, 11.23, 11.24, 11.25, 11.26, 11.27,

11.28, 11.29, 11.30, 11.31, and 11.32 refer to entrance kerma values, optimal and suboptimal field sizes were provided in the tables in order to enable the reader to calculate entrance air kerma dose values from clinically measured dose area product values. Please note that these optimal and suboptimal field sizes refer to the mathematical phantoms of an age group only and not to real patients with corresponding ages.

Table 11.12 Inconsistencies between the mathematical MIRD phantoms and real human anatomy

Anatomical region	Inconsistencies
Chest	No variation of lung density with age
	Mass of glandular tissue is underestimated in newborn infants
Abdomen	No navel has been inserted
Gonads	No variation of gonad situs with age
Active bone marrow	Active bone marrow within the liver, spleen is not inserted in newborn infants
Skeleton	The shape of the spine is unrealistic (no lordosis, no kyphosis considered), the vertebral column is not clearly depicted
	Ribs are drawn as complete circles and thicker than in real humans, the course of the ribs is horizontal
	No xiphoid has been inserted

For details, see the text

Please note also that the anatomy of the mathematical phantoms has been idealized. Therefore, the virtual field settings for calculating the normalized organ doses have been defined with a view to ensuring that the resulting virtual organ doses would be consistent with the real organ doses which would be achieved during real X-ray examinations being performed with the real field settings recommended in Sect. 11.4. Thus, the field settings shown in the following tables may not be consistent with the clinical field settings as recommended in Sect. 11.4.

Normalized organ doses for thoraco-abdominal babygrams are also listed in e.g. [4, 5].

Table 11.13 Field settings—Thoraco-abdominal babygram ap

Age level (gestational weeks)		Anatomical landmarks: Optimal field setting		Anatomical landmarks: Suboptimal field setting	
24	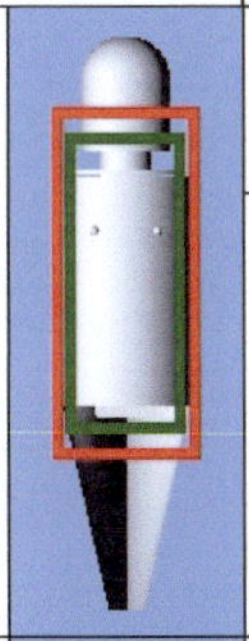	Upper field edge : Tip of the mandible Lower field edge : Just below the perineum including both proximal femoral metaphyses Lateral field edges: Proximal metaphysis of the humeri Field size = field width x field height: 7 cm x 16 cm = 112 cm²		Upper field edge : At the upper margins of the parotid glands Lower field edge : Upper third of tighs (in case of extended legs) Lateral field edges: Both arms included Field size = field width x field height: 8 cm x 18 cm = 144 cm²	
				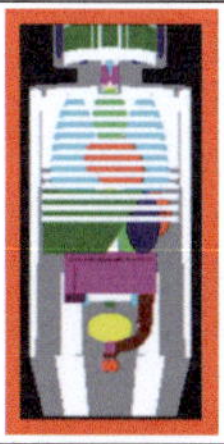	
28	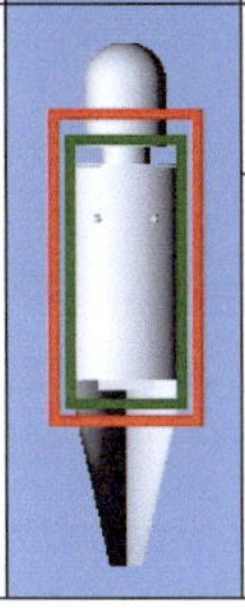	Upper field edge : Tip of the mandible Lower field edge : Just below the perineum including both proximal femoral metaphyses Lateral field edges: Proximal metaphysis of the humeri Field size = field width x field height: 8 cm x 18 cm = 144cm²		Upper field edge : At the upper margins of the parotid glands Lower field edge : Upper third of tighs (in case of extended legs) Lateral field edges: Both arms included Field size = field width x field height: 10 cm x 21 cm = 210 cm²	
		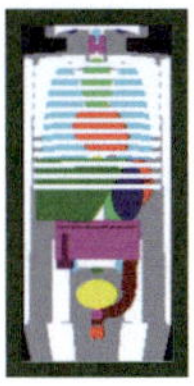		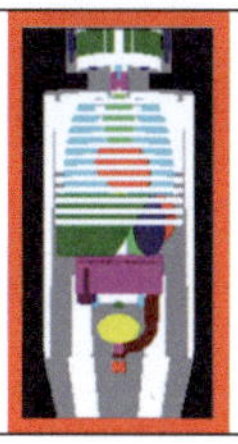	

Table 11.14 Field settings—Thoraco-abdominal babygram ap

Age level (gestational weeks)		Anatomical landmarks			
		Optimal field setting		Suboptimal field setting	
0	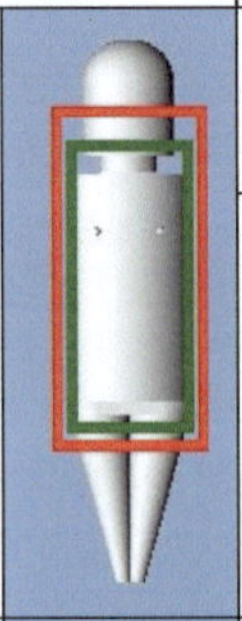	Upper field edge: Tip of the mandible Lower field edge: Just below the perineum including both proximal femoral metaphyses Lateral field edges: Proximal metaphysis of the humeri Field size = field width x field height: 11cm x 26 cm = 286 cm²		Upper field edge: At the upper margins of the parotid glands Lower field edge: Upper third of tighs (in case of extended legs) Lateral field edges: Both arms included Field size = field width x field height: 14 cm x 31 cm = 434 cm²	
					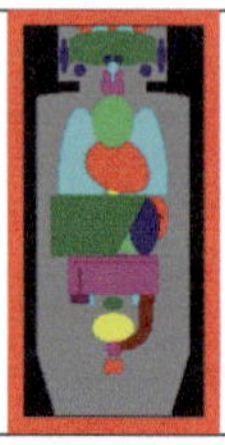
1	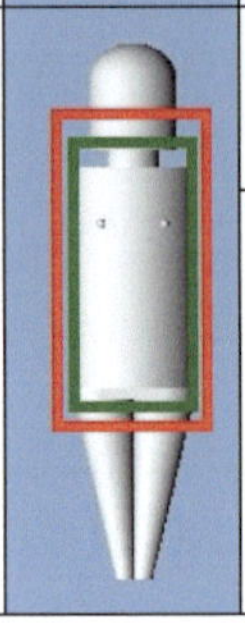	Upper field edge: Tip of the mandible Lower field edge: Just below the perineum including both proximal femoral metaphyses Lateral field edges: Proximal metaphysis of the humeri Field size = field width x field height: 16 cm x 36 cm = 576 cm²		Upper field edge: At the upper margins of the parotid glands Lower field edge: Upper third of tighs (in case of extended legs) Lateral field edges: Both arms included Field size = field width x field height: 20 cm x 42 cm = 840 cm²	

Table 11.15 Normalized organ doses—Thoraco-abdominal babygram ap—24th gestation week

Beam projection:			ap									**Table cells:**
Focus-to-detector distance (cm):			100						(+) = normal font = optimal field setting			
Tube voltage (kVp):			50						**(−) = bold font = suboptimal field setting**			
Total filtrations:			See last table row						Empty table cells correspond to values <0.01			
Organ or tissue	**Organ dose D_T : Entrance dose K_E**								**(mSv * mGy^{-1})**			
	(+)	(−)	(+)	(−)	(+)	(−)	(+)	(−)	(+)	(−)	(+)	(−)
Skull		**0.02**		**0.02**		**0.03**		**0.03**		**0.03**	0.01	**0.03**
Brain												
Oral mucosa		**0.05**		**0.05**		**0.06**		**0.06**		**0.06**		**0.07**
Salivary glands	0.02	**0.04**	0.02	**0.05**	0.02	**0.05**	0.02	**0.05**	0.03	**0.06**	0.03	**0.06**
Cervical spine	0.05	**0.07**	0.06	**0.08**	0.07	**0.10**	0.08	**0.11**	0.09	**0.12**	0.10	**0.14**
Thyroid	0.08	**0.08**	0.08	**0.09**	0.09	**0.09**	0.09	**0.09**	0.09	**0.10**	0.09	**0.10**
Thymus	0.07	**0.07**	0.08	**0.08**	0.08	**0.08**	0.08	**0.08**	0.09	**0.09**	0.09	**0.09**
Thoracic spine	0.04	**0.05**	0.06	**0.06**	0.07	**0.07**	0.08	**0.08**	0.09	**0.09**	0.10	**0.11**
Claviculae	0.19	**0.19**	0.21	**0.22**	0.23	**0.23**	0.24	**0.24**	0.25	**0.26**	0.27	**0.27**
Scapulae	0.06	**0.06**	0.07	**0.08**	0.09	**0.09**	0.09	**0.10**	0.11	**0.11**	0.12	**0.13**
Ribs	0.15	**0.15**	0.17	**0.17**	0.18	**0.18**	0.19	**0.19**	0.20	**0.21**	0.21	**0.22**
Oesophagus	0.03	**0.03**	0.04	**0.04**	0.04	**0.05**	0.05	**0.05**	0.06	**0.06**	0.06	**0.06**
Trachea		**0.03**	0.01	**0.03**	0.01	**0.03**	0.01	**0.04**	0.01	**0.04**	0.02	**0.04**
Lung	0.05	**0.05**	0.05	**0.05**	0.06	**0.06**	0.06	**0.06**	0.07	**0.07**	0.07	**0.07**
Mamma	0.09	**0.09**	0.09	**0.09**	0.10	**0.10**	0.10	**0.10**	0.10	**0.10**	0.10	**0.10**
Heart	0.05	**0.05**	0.06	**0.06**	0.07	**0.07**	0.07	**0.07**	0.08	**0.08**	0.08	**0.08**
Stomach	0.05	**0.05**	0.06	**0.06**	0.07	**0.07**	0.07	**0.07**	0.08	**0.08**	0.08	**0.08**
Spleen	0.03	**0.03**	0.03	**0.03**	0.04	**0.04**	0.04	**0.04**	0.05	**0.05**	0.05	**0.05**
Liver	0.05	**0.05**	0.06	**0.06**	0.06	**0.06**	0.06	**0.07**	0.07	**0.07**	0.08	**0.08**
Gall bladder	0.05	**0.05**	0.06	**0.06**	0.06	**0.06**	0.07	**0.07**	0.07	**0.07**	0.08	**0.08**
Pancreas	0.03	**0.03**	0.04	**0.04**	0.05	**0.05**	0.05	**0.05**	0.06	**0.06**	0.06	**0.07**
Small intestine	0.04	**0.04**	0.05	**0.05**	0.05	**0.05**	0.06	**0.06**	0.07	**0.07**	0.07	**0.07**
Large intestine	0.05	**0.05**	0.06	**0.06**	0.06	**0.06**	0.07	**0.07**	0.07	**0.07**	0.08	**0.08**
Sigma/Rectum	0.04	**0.04**	0.05	**0.05**	0.05	**0.05**	0.06	**0.06**	0.06	**0.06**	0.07	**0.07**
Lumbar spine	0.05	**0.05**	0.06	**0.06**	0.07	**0.08**	0.08	**0.09**	0.10	**0.10**	0.12	**0.12**
Pelvis	0.07	**0.07**	0.09	**0.09**	0.10	**0.11**	0.12	**0.12**	0.13	**0.14**	0.15	**0.15**
Adrenals	0.02	**0.02**	0.02	**0.02**	0.03	**0.03**	0.03	**0.03**	0.04	**0.04**	0.04	**0.04**
Kidneys	0.02	**0.02**	0.02	**0.02**	0.03	**0.03**	0.03	**0.03**	0.04	**0.04**	0.04	**0.04**
Urinary bladder	0.06	**0.06**	0.06	**0.06**	0.07	**0.07**	0.07	**0.07**	0.08	**0.08**	0.08	**0.09**
Uterus	0.04	**0.04**	0.05	**0.05**	0.06	**0.06**	0.06	**0.06**	0.07	**0.07**	0.07	**0.08**
Ovaries	0.04	**0.03**	0.04	**0.04**	0.05	**0.05**	0.05	**0.05**	0.06	**0.06**	0.07	**0.06**
Prostate	0.04	**0.04**	0.05	**0.05**	0.06	**0.06**	0.06	**0.06**	0.07	**0.07**	0.07	**0.07**
Testes	0.08	**0.08**	0.09	**0.09**	0.09	**0.09**	0.09	**0.10**	0.10	**0.10**	0.10	**0.10**
Upper arm	0.03	**0.11**	0.04	**0.13**	0.04	**0.15**	0.05	**0.16**	0.06	**0.18**	0.06	**0.19**
Elbow	0.03	**0.12**	0.03	**0.15**	0.04	**0.16**	0.04	**0.17**	0.05	**0.19**	0.05	**0.21**
Forearm	0.02	**0.13**	0.03	**0.16**	0.03	**0.17**	0.03	**0.18**	0.04	**0.20**	0.04	**0.22**
Upper leg	0.09	**0.10**	0.11	**0.12**	0.12	**0.14**	0.14	**0.15**	0.15	**0.17**	0.17	**0.19**
Knee		**0.05**		**0.06**		**0.06**		**0.07**	0.01	**0.08**	0.01	**0.09**
Lower leg												
Lymph nodes	0.04	**0.04**	0.04	**0.05**	0.05	**0.05**	0.05	**0.06**	0.06	**0.06**	0.06	**0.07**
Skin	0.03	**0.04**	0.03	**0.04**	0.03	**0.04**	0.03	**0.04**	0.03	**0.05**	0.03	**0.05**
Muscle	0.03	**0.03**	0.03	**0.04**	0.04	**0.04**	0.04	**0.04**	0.04	**0.05**	0.04	**0.05**
Skeleton	0.05	**0.07**	0.06	**0.09**	0.07	**0.10**	0.07	**0.11**	0.08	**0.12**	0.09	**0.13**
Red bone marrow	0.01	**0.02**	0.02	**0.02**	0.02	**0.02**	0.02	**0.03**	0.02	**0.03**	0.02	**0.03**
Whole body dose	0.03	**0.04**	0.03	**0.04**	0.04	**0.05**	0.04	**0.05**	0.04	**0.05**	0.05	**0.06**
ED (ICRP 60)	0.05	**0.05**	0.05	**0.06**	0.06	**0.06**	0.06	**0.06**	0.07	**0.07**	0.07	**0.07**
ED (ICRP 103)	0.05	**0.05**	0.05	**0.06**	0.06	**0.06**	0.06	**0.06**	0.07	**0.07**	0.07	**0.07**
Total filtration (mm) Al	**1.5**		**2.5**		**3.5**		**4.5**		**3.5**		**3.5**	
Cu	**0**		**0**		**0**		**0**		**0.1**		**0.2**	

Table 11.16 Normalized organ doses—Thoraco-abdominal babygram ap—24th gestation week

Beam projection:			ap									**Table cells:**
Focus-to-detector distance (cm):			100									(+) = normal font = optimal field setting
Tube voltage (kVp):			60									**(−) = bold font = suboptimal field setting**
Total filtrations:			See last table row									Empty table cells correspond to values <0.01
Organ or tissue	**Organ dose D_T : Entrance dose K_E**								**(mSv * mGy^{-1})**			
	(+)	(−)	(+)	(−)	(+)	(−)	(+)	(−)	(+)	(−)	(+)	(−)
Skull		**0.02**		**0.03**		**0.03**		**0.03**	0.01	**0.03**	0.01	**0.03**
Brain												
Oral mucosa		**0.05**		**0.06**		**0.06**		**0.06**		**0.07**		**0.07**
Salivary glands	0.02	**0.05**	0.02	**0.05**	0.02	**0.06**	0.03	**0.06**	0.03	**0.06**	0.03	**0.06**
Cervical spine	0.06	**0.08**	0.07	**0.10**	0.08	**0.11**	0.09	**0.12**	0.10	**0.14**	0.10	**0.14**
Thyroid	0.08	**0.08**	0.09	**0.09**	0.09	**0.09**	0.09	**0.10**	0.09	**0.10**	0.09	**0.10**
Thymus	0.07	**0.07**	0.08	**0.08**	0.08	**0.09**	0.09	**0.09**	0.09	**0.09**	0.09	**0.09**
Thoracic spine	0.06	**0.06**	0.07	**0.07**	0.08	**0.08**	0.09	**0.09**	0.11	**0.11**	0.10	**0.11**
Claviculae	0.20	**0.20**	0.22	**0.23**	0.24	**0.24**	0.25	**0.25**	0.26	**0.27**	0.27	**0.27**
Scapulae	0.07	**0.07**	0.09	**0.09**	0.10	**0.10**	0.11	**0.11**	0.12	**0.12**	0.12	**0.13**
Ribs	0.16	**0.16**	0.18	**0.18**	0.19	**0.19**	0.20	**0.20**	0.21	**0.21**	0.21	**0.22**
Oesophagus	0.04	**0.04**	0.05	**0.05**	0.05	**0.05**	0.06	**0.06**	0.06	**0.06**	0.06	**0.06**
Trachea	0.01	**0.03**	0.01	**0.03**	0.01	**0.04**	0.01	**0.04**	0.02	**0.04**	0.02	**0.04**
Lung	0.05	**0.05**	0.06	**0.06**	0.06	**0.06**	0.07	**0.07**	0.07	**0.07**	0.07	**0.07**
Mamma	0.09	**0.09**	0.10	**0.10**	0.10	**0.10**	0.10	**0.10**	0.10	**0.10**	0.10	**0.10**
Heart	0.06	**0.06**	0.07	**0.07**	0.07	**0.07**	0.08	**0.08**	0.08	**0.08**	0.08	**0.08**
Stomach	0.06	**0.06**	0.07	**0.07**	0.07	**0.07**	0.08	**0.08**	0.08	**0.08**	0.08	**0.08**
Spleen	0.03	**0.03**	0.04	**0.04**	0.04	**0.04**	0.05	**0.05**	0.05	**0.06**	0.05	**0.05**
Liver	0.05	**0.05**	0.06	**0.06**	0.07	**0.07**	0.07	**0.07**	0.08	**0.08**	0.08	**0.08**
Gall bladder	0.05	**0.05**	0.06	**0.06**	0.07	**0.07**	0.07	**0.07**	0.08	**0.08**	0.08	**0.08**
Pancreas	0.04	**0.04**	0.05	**0.05**	0.05	**0.05**	0.06	**0.06**	0.07	**0.07**	0.06	**0.07**
Small intestine	0.05	**0.05**	0.05	**0.06**	0.06	**0.06**	0.06	**0.07**	0.07	**0.07**	0.07	**0.07**
Large intestine	0.05	**0.05**	0.06	**0.06**	0.07	**0.07**	0.07	**0.07**	0.08	**0.08**	0.08	**0.08**
Sigma/Rectum	0.04	**0.04**	0.05	**0.05**	0.06	**0.06**	0.06	**0.06**	0.07	**0.07**	0.07	**0.07**
Lumbar spine	0.06	**0.06**	0.08	**0.08**	0.09	**0.09**	0.10	**0.10**	0.12	**0.12**	0.12	**0.12**
Pelvis	0.09	**0.09**	0.11	**0.11**	0.12	**0.12**	0.13	**0.13**	0.15	**0.15**	0.15	**0.15**
Adrenals	0.02	**0.02**	0.03	**0.03**	0.03	**0.03**	0.04	**0.04**	0.04	**0.04**	0.04	**0.04**
Kidneys	0.02	**0.02**	0.03	**0.03**	0.03	**0.03**	0.04	**0.04**	0.04	**0.04**	0.04	**0.04**
Urinary bladder	0.06	**0.06**	0.07	**0.07**	0.07	**0.08**	0.08	**0.08**	0.08	**0.09**	0.08	**0.09**
Uterus	0.05	**0.05**	0.06	**0.06**	0.06	**0.06**	0.07	**0.07**	0.07	**0.08**	0.07	**0.08**
Ovaries	0.04	**0.04**	0.05	**0.05**	0.06	**0.05**	0.06	**0.06**	0.07	**0.06**	0.07	**0.06**
Prostate	0.05	**0.05**	0.06	**0.06**	0.06	**0.06**	0.06	**0.07**	0.07	**0.07**	0.07	**0.07**
Testes	0.09	**0.09**	0.09	**0.09**	0.09	**0.10**	0.10	**0.10**	0.10	**0.10**	0.10	**0.10**
Upper arm	0.04	**0.13**	0.05	**0.15**	0.05	**0.16**	0.05	**0.17**	0.06	**0.19**	0.06	**0.19**
Elbow	0.03	**0.14**	0.04	**0.16**	0.04	**0.18**	0.05	**0.19**	0.05	**0.20**	0.05	**0.21**
Forearm	0.02	**0.15**	0.03	**0.17**	0.03	**0.19**	0.04	**0.20**	0.04	**0.21**	0.04	**0.22**
Upper leg	0.10	**0.11**	0.12	**0.14**	0.14	**0.15**	0.15	**0.17**	0.17	**0.19**	0.17	**0.19**
Knee		**0.05**		**0.06**		**0.07**	0.01	**0.08**	0.01	**0.09**	0.01	**0.09**
Lower leg												
Lymph nodes	0.04	**0.05**	0.05	**0.05**	0.05	**0.06**	0.06	**0.06**	0.06	**0.07**	0.06	**0.07**
Skin	0.03	**0.04**	0.03	**0.04**	0.03	**0.04**	0.03	**0.05**	0.03	**0.05**	0.03	**0.05**
Muscle	0.03	**0.04**	0.04	**0.04**	0.04	**0.05**	0.04	**0.05**	0.05	**0.05**	0.04	**0.05**
Skeleton	0.06	**0.08**	0.07	**0.10**	0.07	**0.11**	0.08	**0.12**	0.09	**0.13**	0.09	**0.13**
Red bone marrow	0.02	**0.02**	0.02	**0.02**	0.02	**0.03**	0.02	**0.03**	0.03	**0.03**	0.02	**0.03**
Whole body dose	0.03	**0.04**	0.04	**0.05**	0.04	**0.05**	0.04	**0.05**	0.05	**0.06**	0.05	**0.06**
ED (ICRP 60)	0.05	**0.05**	0.06	**0.06**	0.06	**0.07**	0.07	**0.07**	0.07	**0.07**	0.07	**0.07**
ED (ICRP 103)	0.05	**0.05**	0.06	**0.06**	0.06	**0.07**	0.07	**0.07**	0.07	**0.07**	0.07	**0.07**
Total filtration (mm) Al	**1.5**		**2.5**		**3.5**		**4.5**		**3.5**		**3.5**	
Cu	**0**		**0**		**0**		**0**		**0.1**		**0.2**	

Table 11.17 Normalized organ doses—Thoraco-abdominal babygram ap—24th gestation week

Beam projection:	ap											**Table cells:**
Focus-to-detector distance (cm):	100											(+) = normal font = optimal field setting
Tube voltage (kVp):	70											**(−) = bold font = suboptimal field setting**
Total filtrations:	See last table row											Empty table cells correspond to values <0.01
Organ or tissue	**Organ dose D_T : Entrance dose K_E**								**(mSv * mGy^{-1})**			
	(+)	(−)	(+)	(−)	(+)	(−)	(+)	(−)	(+)	(−)	(+)	(−)
Skull		**0.02**		**0.03**		**0.03**	0.01	**0.03**	0.01	**0.03**	0.01	**0.03**
Brain												
Oral mucosa		**0.05**		**0.06**		**0.06**		**0.07**		**0.07**		**0.07**
Salivary glands	0.02	**0.05**	0.02	**0.05**	0.03	**0.06**	0.03	**0.06**	0.03	**0.06**	0.03	**0.07**
Cervical spine	0.07	**0.09**	0.08	**0.11**	0.09	**0.12**	0.09	**0.13**	0.10	**0.14**	0.11	**0.16**
Thyroid	0.08	**0.09**	0.09	**0.09**	0.09	**0.10**	0.09	**0.10**	0.10	**0.10**	0.10	**0.10**
Thymus	0.08	**0.08**	0.08	**0.08**	0.09	**0.09**	0.09	**0.09**	0.10	**0.10**	0.10	**0.10**
Thoracic spine	0.06	**0.07**	0.08	**0.08**	0.09	**0.09**	0.10	**0.10**	0.11	**0.12**	0.13	**0.13**
Claviculae	0.21	**0.21**	0.23	**0.24**	0.24	**0.25**	0.25	**0.26**	0.26	**0.27**	0.26	**0.27**
Scapulae	0.08	**0.08**	0.09	**0.10**	0.11	**0.11**	0.11	**0.12**	0.13	**0.13**	0.14	**0.14**
Ribs	0.16	**0.17**	0.18	**0.19**	0.19	**0.20**	0.20	**0.20**	0.21	**0.21**	0.22	**0.22**
Oesophagus	0.04	**0.04**	0.05	**0.05**	0.06	**0.06**	0.06	**0.06**	0.07	**0.07**	0.07	**0.07**
Trachea	0.01	**0.03**	0.01	**0.04**	0.01	**0.04**	0.02	**0.04**	0.02	**0.04**	0.02	**0.05**
Lung	0.05	**0.06**	0.06	**0.06**	0.07	**0.07**	0.07	**0.07**	0.08	**0.08**	0.08	**0.08**
Mamma	0.09	**0.09**	0.10	**0.10**	0.10	**0.10**	0.10	**0.10**	0.10	**0.10**	0.10	**0.10**
Heart	0.06	**0.06**	0.07	**0.07**	0.08	**0.08**	0.08	**0.08**	0.09	**0.09**	0.09	**0.09**
Stomach	0.06	**0.06**	0.07	**0.07**	0.08	**0.08**	0.08	**0.08**	0.08	**0.09**	0.09	**0.09**
Spleen	0.04	**0.04**	0.04	**0.04**	0.05	**0.05**	0.05	**0.05**	0.06	**0.06**	0.06	**0.06**
Liver	0.06	**0.06**	0.06	**0.07**	0.07	**0.07**	0.07	**0.07**	0.08	**0.08**	0.08	**0.09**
Gall bladder	0.06	**0.06**	0.07	**0.07**	0.07	**0.07**	0.08	**0.08**	0.08	**0.09**	0.09	**0.09**
Pancreas	0.04	**0.05**	0.05	**0.05**	0.06	**0.06**	0.06	**0.06**	0.07	**0.07**	0.08	**0.08**
Small intestine	0.05	**0.05**	0.06	**0.06**	0.06	**0.07**	0.07	**0.07**	0.08	**0.08**	0.08	**0.08**
Large intestine	0.06	**0.06**	0.07	**0.07**	0.07	**0.07**	0.08	**0.08**	0.08	**0.08**	0.09	**0.09**
Sigma/Rectum	0.05	**0.05**	0.06	**0.06**	0.06	**0.06**	0.07	**0.07**	0.07	**0.07**	0.08	**0.08**
Lumbar spine	0.07	**0.07**	0.09	**0.09**	0.10	**0.10**	0.11	**0.11**	0.13	**0.13**	0.14	**0.14**
Pelvis	0.10	**0.10**	0.12	**0.12**	0.13	**0.13**	0.14	**0.14**	0.16	**0.16**	0.17	**0.17**
Adrenals	0.03	**0.03**	0.03	**0.03**	0.04	**0.04**	0.04	**0.04**	0.05	**0.05**	0.05	**0.05**
Kidneys	0.03	**0.03**	0.03	**0.03**	0.04	**0.04**	0.04	**0.04**	0.05	**0.05**	0.05	**0.05**
Urinary bladder	0.06	**0.07**	0.07	**0.07**	0.08	**0.08**	0.08	**0.08**	0.09	**0.09**	0.09	**0.09**
Uterus	0.05	**0.05**	0.06	**0.06**	0.07	**0.07**	0.07	**0.07**	0.08	**0.08**	0.08	**0.09**
Ovaries	0.05	**0.04**	0.05	**0.05**	0.06	**0.06**	0.06	**0.06**	0.07	**0.07**	0.08	**0.08**
Prostate	0.05	**0.05**	0.06	**0.06**	0.06	**0.07**	0.07	**0.07**	0.07	**0.08**	0.08	**0.08**
Testes	0.09	**0.09**	0.09	**0.10**	0.10	**0.10**	0.10	**0.10**	0.10	**0.10**	0.10	**0.10**
Upper arm	0.04	**0.14**	0.05	**0.16**	0.05	**0.17**	0.06	**0.18**	0.06	**0.20**	0.07	**0.21**
Elbow	0.03	**0.15**	0.04	**0.17**	0.05	**0.18**	0.05	**0.19**	0.06	**0.21**	0.06	**0.22**
Forearm	0.03	**0.16**	0.03	**0.18**	0.04	**0.19**	0.04	**0.20**	0.04	**0.22**	0.05	**0.22**
Upper leg	0.11	**0.13**	0.13	**0.15**	0.15	**0.16**	0.16	**0.17**	0.17	**0.19**	0.19	**0.21**
Knee		**0.06**		**0.07**	0.01	**0.08**	0.01	**0.08**	0.01	**0.09**	0.02	**0.10**
Lower leg												
Lymph nodes	0.05	**0.05**	0.05	**0.06**	0.06	**0.06**	0.06	**0.07**	0.06	**0.07**	0.07	**0.08**
Skin	0.03	**0.04**	0.03	**0.04**	0.03	**0.05**	0.03	**0.05**	0.03	**0.05**	0.04	**0.05**
Muscle	0.03	**0.04**	0.04	**0.04**	0.04	**0.05**	0.04	**0.05**	0.05	**0.05**	0.05	**0.06**
Skeleton	0.06	**0.09**	0.07	**0.11**	0.08	**0.12**	0.08	**0.12**	0.09	**0.13**	0.10	**0.14**
Red bone marrow	0.02	**0.02**	0.02	**0.03**	0.02	**0.03**	0.02	**0.03**	0.03	**0.04**	0.03	**0.04**
Whole body dose	0.04	**0.04**	0.04	**0.05**	0.04	**0.05**	0.05	**0.06**	0.05	**0.06**	0.05	**0.06**
ED (ICRP 60)	0.06	**0.06**	0.06	**0.06**	0.07	**0.07**	0.07	**0.07**	0.07	**0.08**	0.08	**0.08**
ED (ICRP 103)	0.05	**0.06**	0.06	**0.06**	0.07	**0.07**	0.07	**0.07**	0.07	**0.08**	0.08	**0.08**
Total filtration (mm) Al	**1.5**		**2.5**		**3.5**		**4.5**		**3.5**		**3.5**	
Cu	**0**		**0**		**0**		**0**		**0.1**		**0.2**	

Table 11.18 Normalized organ doses—Thoraco-abdominal babygram ap—28th gestation week

Beam projection:		ap										**Table cells:**
Focus-to-detector distance (cm):		100										(+) = normal font = optimal field setting
Tube voltage (kVp):		50										**(−) = bold font = suboptimal field setting**
Total filtrations:		See last table row										Empty table cells correspond to values <0.01
Organ or tissue	**Organ dose D_T : Entrance dose K_E**							**($mSv * mGy^{-1}$)**				
	(+)	(−)	(+)	(−)	(+)	(−)	(+)	(−)	(+)	(−)	(+)	(−)
Skull		**0.03**		**0.03**		**0.04**		**0.04**		**0.04**	0.01	**0.05**
Brain												
Oral mucosa		**0.06**		**0.07**		**0.08**		**0.08**		**0.09**		**0.09**
Salivary glands	0.01	**0.06**	0.01	**0.07**	0.02	**0.07**	0.02	**0.08**	0.02	**0.08**	0.02	**0.09**
Cervical spine	0.06	**0.09**	0.08	**0.11**	0.09	**0.13**	0.10	**0.14**	0.11	**0.17**	0.12	**0.19**
Thyroid	0.11	**0.11**	0.12	**0.12**	0.12	**0.13**	0.12	**0.13**	0.13	**0.14**	0.13	**0.14**
Thymus	0.09	**0.09**	0.11	**0.11**	0.11	**0.11**	0.12	**0.12**	0.13	**0.13**	0.13	**0.14**
Thoracic spine	0.06	**0.06**	0.07	**0.07**	0.09	**0.09**	0.10	**0.10**	0.12	**0.12**	0.14	**0.14**
Claviculae	0.26	**0.26**	0.29	**0.30**	0.32	**0.33**	0.33	**0.34**	0.36	**0.37**	0.38	**0.39**
Scapulae	0.08	**0.08**	0.10	**0.10**	0.11	**0.11**	0.12	**0.13**	0.14	**0.15**	0.16	**0.17**
Ribs	0.20	**0.20**	0.23	**0.23**	0.25	**0.25**	0.26	**0.26**	0.28	**0.29**	0.30	**0.30**
Oesophagus	0.04	**0.04**	0.05	**0.05**	0.06	**0.06**	0.07	**0.07**	0.08	**0.08**	0.09	**0.09**
Trachea	0.01	**0.04**	0.01	**0.05**	0.01	**0.05**	0.02	**0.05**	0.02	**0.06**	0.02	**0.06**
Lung	0.06	**0.06**	0.07	**0.07**	0.08	**0.08**	0.09	**0.09**	0.09	**0.10**	0.10	**0.10**
Mamma	0.13	**0.13**	0.13	**0.14**	0.14	**0.14**	0.14	**0.14**	0.14	**0.14**	0.14	**0.15**
Heart	0.07	**0.07**	0.08	**0.08**	0.09	**0.09**	0.10	**0.10**	0.11	**0.11**	0.12	**0.12**
Stomach	0.07	**0.07**	0.09	**0.09**	0.09	**0.09**	0.10	**0.10**	0.11	**0.11**	0.12	**0.12**
Spleen	0.04	**0.04**	0.04	**0.04**	0.05	**0.05**	0.06	**0.06**	0.07	**0.07**	0.07	**0.07**
Liver	0.07	**0.07**	0.08	**0.08**	0.08	**0.08**	0.09	**0.09**	0.10	**0.10**	0.11	**0.11**
Gall bladder	0.06	**0.07**	0.08	**0.08**	0.09	**0.09**	0.09	**0.09**	0.10	**0.11**	0.11	**0.11**
Pancreas	0.05	**0.05**	0.06	**0.06**	0.06	**0.06**	0.07	**0.07**	0.08	**0.08**	0.09	**0.09**
Small intestine	0.05	**0.05**	0.07	**0.07**	0.07	**0.07**	0.08	**0.08**	0.09	**0.09**	0.10	**0.10**
Large intestine	0.06	**0.06**	0.08	**0.08**	0.08	**0.09**	0.09	**0.09**	0.10	**0.10**	0.11	**0.11**
Sigma/Rectum	0.05	**0.05**	0.06	**0.06**	0.07	**0.07**	0.08	**0.08**	0.09	**0.09**	0.10	**0.10**
Lumbar spine	0.06	**0.06**	0.08	**0.08**	0.10	**0.10**	0.11	**0.11**	0.13	**0.13**	0.15	**0.15**
Pelvis	0.09	**0.10**	0.12	**0.12**	0.14	**0.14**	0.15	**0.15**	0.18	**0.18**	0.20	**0.20**
Adrenals	0.02	**0.02**	0.03	**0.03**	0.04	**0.04**	0.04	**0.04**	0.05	**0.05**	0.05	**0.05**
Kidneys	0.02	**0.02**	0.03	**0.03**	0.04	**0.04**	0.04	**0.04**	0.05	**0.05**	0.05	**0.05**
Urinary bladder	0.08	**0.08**	0.09	**0.09**	0.10	**0.10**	0.10	**0.10**	0.11	**0.11**	0.12	**0.12**
Uterus	0.06	**0.06**	0.07	**0.07**	0.08	**0.08**	0.08	**0.08**	0.09	**0.09**	0.10	**0.10**
Ovaries	0.05	**0.05**	0.06	**0.06**	0.07	**0.07**	0.07	**0.07**	0.08	**0.08**	0.09	**0.09**
Prostate	0.06	**0.06**	0.07	**0.07**	0.08	**0.08**	0.08	**0.09**	0.09	**0.10**	0.10	**0.11**
Testes	0.12	**0.12**	0.13	**0.13**	0.13	**0.13**	0.14	**0.14**	0.14	**0.14**	0.15	**0.14**
Upper arm	0.15	**0.15**	0.18	**0.18**	0.20	**0.20**	0.21	**0.22**	0.24	**0.24**	0.26	**0.27**
Elbow	0.16	**0.16**	0.19	**0.19**	0.22	**0.22**	0.23	**0.23**	0.26	**0.26**	0.28	**0.28**
Forearm	0.18	**0.18**	0.21	**0.21**	0.23	**0.23**	0.25	**0.25**	0.27	**0.28**	0.30	**0.30**
Upper leg	0.12	**0.13**	0.15	**0.16**	0.18	**0.18**	0.19	**0.20**	0.22	**0.23**	0.24	**0.26**
Knee	0.02	**0.06**	0.02	**0.08**	0.02	**0.09**	0.03	**0.10**	0.03	**0.11**	0.03	**0.12**
Lower leg												
Lymph nodes	0.05	**0.06**	0.06	**0.07**	0.06	**0.08**	0.07	**0.08**	0.08	**0.09**	0.08	**0.10**
Skin	0.05	**0.05**	0.05	**0.06**	0.05	**0.06**	0.05	**0.06**	0.05	**0.06**	0.06	**0.07**
Muscle	0.04	**0.05**	0.05	**0.05**	0.05	**0.06**	0.05	**0.06**	0.06	**0.07**	0.06	**0.07**
Skeleton	0.09	**0.10**	0.11	**0.12**	0.12	**0.13**	0.13	**0.15**	0.14	**0.16**	0.16	**0.18**
Red bone marrow	0.02	**0.02**	0.03	**0.03**	0.03	**0.03**	0.03	**0.04**	0.04	**0.04**	0.04	**0.04**
Whole body dose	0.04	**0.05**	0.05	**0.06**	0.06	**0.06**	0.06	**0.07**	0.07	**0.07**	0.07	**0.08**
ED (ICRP 60)	0.07	**0.07**	0.08	**0.08**	0.08	**0.08**	0.09	**0.09**	0.09	**0.10**	0.10	**0.10**
ED (ICRP 103)	0.07	**0.07**	0.08	**0.08**	0.08	**0.08**	0.09	**0.09**	0.09	**0.10**	0.10	**0.10**
Total filtration (mm) Al	**1.5**		**2.5**		**3.5**		**4.5**		**3.5**		**3.5**	
Cu	**0**		**0**		**0**		**0**		**0.1**		**0.2**	

Table 11.19 Normalized organ doses—Thoraco-abdominal babygram ap—28th gestation week

Beam projection:			ap									**Table cells:**
Focus-to-detector distance (cm):			100									(+) = normal font = optimal field setting
Tube voltage (kVp):			60									**(−) = bold font = suboptimal field setting**
Total filtrations:			See last table row									Empty table cells correspond to values <0.01
Organ or tissue	**Organ dose D_T : Entrance dose K_E**								**(mSv * mGy^{-1})**			
	(+)	(−)	(+)	(−)	(+)	(−)	(+)	(−)	(+)	(−)	(+)	(−)
Skull		**0.03**		**0.04**		**0.04**		**0.04**	0.01	**0.05**	0.01	**0.05**
Brain												
Oral mucosa		**0.07**		**0.08**		**0.08**		**0.09**		**0.10**		**0.09**
Salivary glands	0.01	**0.06**	0.02	**0.07**	0.02	**0.08**	0.02	**0.08**	0.02	**0.09**	0.02	**0.09**
Cervical spine	0.07	**0.11**	0.09	**0.13**	0.10	**0.15**	0.11	**0.16**	0.12	**0.19**	0.12	**0.19**
Thyroid	0.11	**0.12**	0.12	**0.13**	0.12	**0.13**	0.13	**0.14**	0.13	**0.14**	0.13	**0.14**
Thymus	0.10	**0.10**	0.11	**0.11**	0.12	**0.12**	0.12	**0.13**	0.13	**0.14**	0.13	**0.14**
Thoracic spine	0.07	**0.07**	0.09	**0.09**	0.11	**0.11**	0.12	**0.12**	0.14	**0.14**	0.14	**0.14**
Claviculae	0.28	**0.28**	0.31	**0.32**	0.33	**0.34**	0.35	**0.36**	0.37	**0.38**	0.38	**0.39**
Scapulae	0.09	**0.09**	0.11	**0.12**	0.13	**0.13**	0.14	**0.14**	0.16	**0.17**	0.16	**0.17**
Ribs	0.22	**0.22**	0.25	**0.25**	0.26	**0.27**	0.28	**0.28**	0.30	**0.30**	0.30	**0.30**
Oesophagus	0.05	**0.05**	0.06	**0.06**	0.07	**0.07**	0.08	**0.08**	0.09	**0.09**	0.09	**0.09**
Trachea	0.01	**0.05**	0.02	**0.05**	0.02	**0.06**	0.02	**0.06**	0.02	**0.06**	0.02	**0.06**
Lung	0.07	**0.07**	0.08	**0.08**	0.09	**0.09**	0.09	**0.09**	0.10	**0.10**	0.10	**0.10**
Mamma	0.13	**0.13**	0.14	**0.14**	0.14	**0.14**	0.14	**0.14**	0.14	**0.15**	0.14	**0.15**
Heart	0.08	**0.08**	0.09	**0.09**	0.10	**0.10**	0.11	**0.11**	0.12	**0.12**	0.12	**0.12**
Stomach	0.08	**0.08**	0.09	**0.09**	0.10	**0.10**	0.11	**0.11**	0.12	**0.12**	0.12	**0.12**
Spleen	0.04	**0.04**	0.05	**0.05**	0.06	**0.06**	0.07	**0.07**	0.07	**0.07**	0.07	**0.07**
Liver	0.07	**0.07**	0.09	**0.09**	0.09	**0.09**	0.10	**0.10**	0.11	**0.11**	0.11	**0.11**
Gall bladder	0.07	**0.07**	0.09	**0.09**	0.10	**0.10**	0.10	**0.10**	0.11	**0.12**	0.11	**0.11**
Pancreas	0.05	**0.05**	0.07	**0.07**	0.07	**0.07**	0.08	**0.08**	0.09	**0.09**	0.09	**0.09**
Small intestine	0.06	**0.06**	0.08	**0.08**	0.08	**0.08**	0.09	**0.09**	0.10	**0.10**	0.10	**0.10**
Large intestine	0.07	**0.07**	0.09	**0.09**	0.09	**0.10**	0.10	**0.10**	0.11	**0.11**	0.11	**0.11**
Sigma/Rectum	0.06	**0.06**	0.07	**0.07**	0.08	**0.08**	0.09	**0.09**	0.10	**0.10**	0.10	**0.10**
Lumbar spine	0.08	**0.08**	0.10	**0.10**	0.12	**0.12**	0.13	**0.13**	0.16	**0.16**	0.15	**0.15**
Pelvis	0.11	**0.12**	0.14	**0.14**	0.16	**0.16**	0.18	**0.18**	0.20	**0.20**	0.20	**0.20**
Adrenals	0.03	**0.03**	0.04	**0.04**	0.04	**0.04**	0.05	**0.05**	0.06	**0.06**	0.05	**0.05**
Kidneys	0.03	**0.03**	0.04	**0.04**	0.04	**0.04**	0.05	**0.05**	0.06	**0.06**	0.05	**0.05**
Urinary bladder	0.08	**0.09**	0.10	**0.10**	0.11	**0.11**	0.11	**0.11**	0.12	**0.12**	0.12	**0.12**
Uterus	0.06	**0.06**	0.08	**0.08**	0.08	**0.09**	0.09	**0.09**	0.10	**0.10**	0.10	**0.10**
Ovaries	0.06	**0.06**	0.07	**0.07**	0.08	**0.08**	0.08	**0.08**	0.09	**0.09**	0.09	**0.09**
Prostate	0.07	**0.07**	0.08	**0.08**	0.09	**0.09**	0.09	**0.10**	0.10	**0.11**	0.10	**0.11**
Testes	0.12	**0.12**	0.13	**0.13**	0.14	**0.14**	0.14	**0.14**	0.14	**0.15**	0.15	**0.14**
Upper arm	0.17	**0.17**	0.20	**0.20**	0.22	**0.22**	0.23	**0.24**	0.26	**0.26**	0.26	**0.27**
Elbow	0.18	**0.18**	0.21	**0.22**	0.24	**0.24**	0.25	**0.25**	0.28	**0.28**	0.28	**0.28**
Forearm	0.20	**0.20**	0.23	**0.23**	0.25	**0.25**	0.27	**0.27**	0.29	**0.29**	0.30	**0.30**
Upper leg	0.15	**0.15**	0.18	**0.18**	0.20	**0.21**	0.22	**0.22**	0.24	**0.25**	0.24	**0.26**
Knee	0.02	**0.08**	0.02	**0.09**	0.03	**0.10**	0.03	**0.11**	0.03	**0.12**	0.03	**0.12**
Lower leg												
Lymph nodes	0.06	**0.06**	0.07	**0.08**	0.07	**0.08**	0.08	**0.09**	0.09	**0.10**	0.08	**0.10**
Skin	0.05	**0.06**	0.05	**0.06**	0.05	**0.06**	0.05	**0.06**	0.06	**0.07**	0.06	**0.07**
Muscle	0.05	**0.05**	0.05	**0.06**	0.06	**0.06**	0.06	**0.07**	0.07	**0.07**	0.06	**0.07**
Skeleton	0.10	**0.11**	0.12	**0.14**	0.13	**0.15**	0.14	**0.16**	0.16	**0.18**	0.16	**0.18**
Red bone marrow	0.02	**0.03**	0.03	**0.03**	0.03	**0.04**	0.04	**0.04**	0.04	**0.05**	0.04	**0.04**
Whole body dose	0.05	**0.06**	0.06	**0.06**	0.06	**0.07**	0.07	**0.07**	0.07	**0.08**	0.07	**0.08**
ED (ICRP 60)	0.07	**0.07**	0.08	**0.08**	0.09	**0.09**	0.09	**0.10**	0.10	**0.10**	0.10	**0.10**
ED (ICRP 103)	0.07	**0.08**	0.08	**0.09**	0.09	**0.09**	0.09	**0.10**	0.10	**0.10**	0.10	**0.10**
Total filtration (mm) Al	**1.5**		**2.5**		**3.5**		**4.5**		**3.5**		**3.5**	
Cu	**0**		**0**		**0**		**0**		**0.1**		**0.2**	

Table 11.20 Normalized organ doses—Thoraco-abdominal babygram ap—28th gestation week

Beam projection:	ap											**Table cells:**
Focus-to-detector distance (cm):	100											(+) = normal font = optimal field setting
Tube voltage (kVp):	70											**(−) = bold font = suboptimal field setting**
Total filtrations:	See last table row											Empty table cells correspond to values <0.01
Organ or tissue	**Organ dose D_T : Entrance dose K_E (mSv * mGy^{-1})**											
	(+)	(−)	(+)	(−)	(+)	(−)	(+)	(−)	(+)	(−)	(+)	(−)
Skull		**0.03**		**0.04**		**0.04**	0.01	**0.04**	0.01	**0.05**	0.01	**0.05**
Brain												
Oral mucosa		**0.07**		**0.08**		**0.09**		**0.09**		**0.10**		**0.10**
Salivary glands	0.02	**0.07**	0.02	**0.08**	0.02	**0.08**	0.02	**0.09**	0.02	**0.09**	0.02	**0.10**
Cervical spine	0.08	**0.12**	0.10	**0.14**	0.11	**0.16**	0.12	**0.18**	0.13	**0.20**	0.14	**0.21**
Thyroid	0.11	**0.12**	0.12	**0.13**	0.13	**0.14**	0.13	**0.14**	0.14	**0.15**	0.14	**0.15**
Thymus	0.11	**0.11**	0.12	**0.12**	0.12	**0.13**	0.13	**0.13**	0.14	**0.14**	0.14	**0.15**
Thoracic spine	0.08	**0.09**	0.11	**0.11**	0.12	**0.12**	0.13	**0.13**	0.15	**0.16**	0.17	**0.17**
Claviculae	0.29	**0.30**	0.32	**0.33**	0.34	**0.35**	0.35	**0.36**	0.37	**0.38**	0.37	**0.39**
Scapulae	0.10	**0.11**	0.13	**0.13**	0.14	**0.14**	0.15	**0.16**	0.17	**0.18**	0.19	**0.19**
Ribs	0.23	**0.23**	0.26	**0.26**	0.27	**0.27**	0.28	**0.29**	0.30	**0.30**	0.31	**0.31**
Oesophagus	0.06	**0.06**	0.07	**0.07**	0.08	**0.08**	0.08	**0.08**	0.09	**0.10**	0.10	**0.11**
Trachea	0.01	**0.05**	0.02	**0.05**	0.02	**0.06**	0.02	**0.06**	0.02	**0.07**	0.02	**0.07**
Lung	0.07	**0.08**	0.09	**0.09**	0.09	**0.09**	0.10	**0.10**	0.11	**0.11**	0.11	**0.11**
Mamma	0.13	**0.14**	0.14	**0.14**	0.14	**0.14**	0.14	**0.14**	0.15	**0.15**	0.15	**0.15**
Heart	0.09	**0.09**	0.10	**0.10**	0.11	**0.11**	0.11	**0.11**	0.12	**0.12**	0.13	**0.13**
Stomach	0.09	**0.09**	0.10	**0.10**	0.11	**0.11**	0.11	**0.11**	0.12	**0.12**	0.13	**0.13**
Spleen	0.05	**0.05**	0.06	**0.06**	0.07	**0.07**	0.07	**0.07**	0.08	**0.08**	0.09	**0.09**
Liver	0.08	**0.08**	0.09	**0.09**	0.10	**0.10**	0.10	**0.11**	0.11	**0.12**	0.12	**0.12**
Gall bladder	0.08	**0.08**	0.09	**0.10**	0.10	**0.10**	0.11	**0.11**	0.12	**0.12**	0.13	**0.13**
Pancreas	0.06	**0.06**	0.07	**0.07**	0.08	**0.08**	0.09	**0.09**	0.10	**0.10**	0.11	**0.11**
Small intestine	0.07	**0.07**	0.08	**0.08**	0.09	**0.09**	0.10	**0.10**	0.11	**0.11**	0.12	**0.12**
Large intestine	0.08	**0.08**	0.09	**0.09**	0.10	**0.10**	0.11	**0.11**	0.12	**0.12**	0.13	**0.13**
Sigma/Rectum	0.07	**0.07**	0.08	**0.08**	0.09	**0.09**	0.09	**0.09**	0.10	**0.10**	0.11	**0.11**
Lumbar spine	0.09	**0.09**	0.12	**0.12**	0.13	**0.13**	0.15	**0.15**	0.17	**0.17**	0.19	**0.19**
Pelvis	0.13	**0.13**	0.16	**0.16**	0.18	**0.18**	0.19	**0.19**	0.22	**0.22**	0.23	**0.24**
Adrenals	0.04	**0.04**	0.04	**0.04**	0.05	**0.05**	0.05	**0.05**	0.06	**0.06**	0.07	**0.07**
Kidneys	0.03	**0.03**	0.04	**0.04**	0.05	**0.05**	0.05	**0.05**	0.06	**0.06**	0.07	**0.07**
Urinary bladder	0.09	**0.09**	0.10	**0.11**	0.11	**0.11**	0.12	**0.12**	0.13	**0.13**	0.13	**0.14**
Uterus	0.07	**0.07**	0.08	**0.09**	0.09	**0.09**	0.10	**0.10**	0.11	**0.11**	0.12	**0.12**
Ovaries	0.06	**0.06**	0.07	**0.08**	0.08	**0.08**	0.09	**0.09**	0.10	**0.10**	0.11	**0.11**
Prostate	0.07	**0.08**	0.08	**0.09**	0.09	**0.10**	0.10	**0.10**	0.11	**0.11**	0.11	**0.12**
Testes	0.13	**0.13**	0.14	**0.14**	0.14	**0.14**	0.14	**0.14**	0.15	**0.15**	0.15	**0.15**
Upper arm	0.18	**0.18**	0.21	**0.21**	0.23	**0.23**	0.25	**0.25**	0.27	**0.27**	0.28	**0.29**
Elbow	0.20	**0.20**	0.23	**0.23**	0.25	**0.25**	0.26	**0.27**	0.29	**0.29**	0.30	**0.30**
Forearm	0.21	**0.21**	0.24	**0.24**	0.26	**0.26**	0.28	**0.28**	0.30	**0.30**	0.31	**0.31**
Upper leg	0.16	**0.17**	0.19	**0.20**	0.21	**0.22**	0.23	**0.24**	0.25	**0.27**	0.27	**0.29**
Knee	0.02	**0.08**	0.03	**0.10**	0.03	**0.11**	0.03	**0.12**	0.04	**0.13**	0.04	**0.14**
Lower leg												
Lymph nodes	0.06	**0.07**	0.07	**0.08**	0.08	**0.09**	0.08	**0.09**	0.09	**0.10**	0.10	**0.11**
Skin	0.05	**0.06**	0.05	**0.06**	0.05	**0.06**	0.06	**0.07**	0.06	**0.07**	0.06	**0.07**
Muscle	0.05	**0.05**	0.06	**0.06**	0.06	**0.07**	0.06	**0.07**	0.07	**0.08**	0.07	**0.08**
Skeleton	0.11	**0.12**	0.13	**0.15**	0.14	**0.16**	0.15	**0.17**	0.16	**0.19**	0.17	**0.20**
Red bone marrow	0.03	**0.03**	0.03	**0.04**	0.04	**0.04**	0.04	**0.04**	0.04	**0.05**	0.05	**0.06**
Whole body dose	0.05	**0.06**	0.06	**0.07**	0.07	**0.07**	0.07	**0.08**	0.08	**0.09**	0.08	**0.09**
ED (ICRP 60)	0.08	**0.08**	0.09	**0.09**	0.09	**0.10**	0.10	**0.10**	0.11	**0.11**	0.11	**0.12**
ED (ICRP 103)	0.08	**0.08**	0.09	**0.09**	0.09	**0.10**	0.10	**0.10**	0.10	**0.11**	0.11	**0.12**
Total filtration (mm) **Al**	**1.5**		**2.5**		**3.5**		**4.5**		**3.5**		**3.5**	
Cu	**0**		**0**		**0**		**0**		**0.1**		**0.2**	

Table 11.21 Normalized organ doses—Thoraco-abdominal babygram ap—<1 year

Beam projection:			ap									**Table cells:**
Focus-to-detector distance (cm):			100									(+) = normal font = optimal field setting
Tube voltage (kVp):			50									**(−) = bold font = suboptimal field setting**
Total filtrations:			See last table row									Empty table cells correspond to values <0.01
Organ or tissue	**Organ dose D_T : Entrance dose K_E**								**(mSv * mGy^{-1})**			
	(+)	(−)	(+)	(−)	(+)	(−)	(+)	(−)	(+)	(−)	(+)	(−)
Skull	0.03	**0.22**	0.04	**0.27**	0.04	**0.30**	0.05	**0.33**	0.06	**0.38**	0.07	**0.42**
Brain		**0.02**		**0.03**		**0.03**		**0.04**		**0.05**	0.01	**0.06**
Oral mucosa	0.03	**0.51**	0.04	**0.60**	0.05	**0.66**	0.06	**0.71**	0.07	**0.79**	0.08	**0.85**
Salivary glands	0.14	**0.44**	0.18	**0.54**	0.20	**0.60**	0.21	**0.64**	0.24	**0.72**	0.27	**0.79**
Cervical spine	0.34	**0.48**	0.44	**0.63**	0.52	**0.74**	0.59	**0.84**	0.70	**1.01**	0.81	**1.18**
Thyroid	0.81	**0.86**	0.93	**0.98**	0.99	**1.06**	1.04	**1.12**	1.12	**1.21**	1.18	**1.29**
Thymus	0.72	**0.73**	0.84	**0.86**	0.92	**0.94**	0.98	**1.01**	1.08	**1.11**	1.16	**1.19**
Thoracic spine	0.24	**0.24**	0.32	**0.33**	0.39	**0.40**	0.45	**0.46**	0.56	**0.58**	0.68	**0.70**
Claviculae	1.79	**1.85**	2.10	**2.18**	2.31	**2.39**	2.46	**2.55**	2.70	**2.81**	2.89	**3.01**
Scapulae	0.42	**0.43**	0.55	**0.57**	0.66	**0.67**	0.74	**0.76**	0.90	**0.93**	1.05	**1.09**
Ribs	1.35	**1.38**	1.57	**1.60**	1.71	**1.75**	1.81	**1.85**	1.99	**2.03**	2.14	**2.19**
Oesophagus	0.19	**0.19**	0.25	**0.25**	0.30	**0.30**	0.34	**0.34**	0.41	**0.41**	0.48	**0.48**
Trachea	0.11	**0.47**	0.13	**0.55**	0.14	**0.60**	0.15	**0.64**	0.17	**0.71**	0.19	**0.77**
Lung	0.39	**0.39**	0.48	**0.49**	0.54	**0.55**	0.59	**0.60**	0.67	**0.69**	0.75	**0.76**
Mamma	1.13	**1.13**	1.20	**1.20**	1.23	**1.23**	1.26	**1.25**	1.30	**1.29**	1.34	**1.32**
Heart	0.50	**0.51**	0.61	**0.62**	0.69	**0.70**	0.75	**0.76**	0.86	**0.87**	0.95	**0.96**
Stomach	0.52	**0.53**	0.64	**0.64**	0.72	**0.73**	0.78	**0.79**	0.89	**0.90**	0.99	**1.00**
Spleen	0.18	**0.18**	0.24	**0.24**	0.28	**0.29**	0.32	**0.33**	0.39	**0.40**	0.45	**0.46**
Liver	0.46	**0.47**	0.56	**0.57**	0.63	**0.64**	0.69	**0.70**	0.79	**0.80**	0.87	**0.89**
Gall bladder	0.37	**0.38**	0.47	**0.48**	0.55	**0.55**	0.61	**0.61**	0.71	**0.72**	0.81	**0.82**
Pancreas	0.25	**0.26**	0.33	**0.34**	0.39	**0.40**	0.44	**0.45**	0.53	**0.54**	0.62	**0.63**
Small intestine	0.33	**0.33**	0.42	**0.42**	0.48	**0.49**	0.54	**0.54**	0.63	**0.64**	0.72	**0.73**
Large intestine	0.40	**0.41**	0.51	**0.51**	0.58	**0.59**	0.64	**0.65**	0.75	**0.76**	0.84	**0.86**
Sigma/Rectum	0.30	**0.30**	0.38	**0.39**	0.45	**0.45**	0.50	**0.50**	0.59	**0.60**	0.67	**0.68**
Lumbar spine	0.25	**0.25**	0.34	**0.35**	0.41	**0.42**	0.48	**0.48**	0.60	**0.61**	0.73	**0.74**
Pelvis	0.45	**0.46**	0.60	**0.61**	0.71	**0.72**	0.80	**0.82**	0.97	**0.99**	1.13	**1.16**
Adrenals	0.11	**0.11**	0.14	**0.15**	0.17	**0.18**	0.20	**0.20**	0.24	**0.25**	0.29	**0.30**
Kidneys	0.10	**0.11**	0.14	**0.14**	0.17	**0.17**	0.19	**0.20**	0.24	**0.24**	0.29	**0.29**
Urinary bladder	0.53	**0.54**	0.64	**0.65**	0.72	**0.74**	0.78	**0.80**	0.89	**0.91**	0.98	**1.00**
Uterus	0.32	**0.33**	0.42	**0.42**	0.48	**0.49**	0.54	**0.54**	0.64	**0.64**	0.73	**0.73**
Ovaries	0.27	**0.27**	0.35	**0.36**	0.41	**0.42**	0.46	**0.47**	0.56	**0.56**	0.64	**0.65**
Prostate	0.40	**0.40**	0.50	**0.50**	0.58	**0.57**	0.63	**0.63**	0.74	**0.74**	0.83	**0.83**
Testes	0.97	**0.97**	1.07	**1.08**	1.13	**1.14**	1.17	**1.18**	1.24	**1.25**	1.30	**1.30**
Upper arm	0.41	**0.95**	0.52	**1.17**	0.61	**1.33**	0.67	**1.46**	0.79	**1.67**	0.90	**1.86**
Elbow	0.39	**1.03**	0.50	**1.27**	0.58	**1.44**	0.64	**1.57**	0.76	**1.80**	0.87	**2.00**
Forearm	0.30	**1.11**	0.38	**1.37**	0.45	**1.54**	0.50	**1.68**	0.59	**1.91**	0.68	**2.11**
Upper leg	0.73	**0.76**	0.94	**0.98**	1.10	**1.14**	1.22	**1.27**	1.44	**1.50**	1.64	**1.72**
Knee	0.21	**0.55**	0.28	**0.69**	0.32	**0.80**	0.36	**0.89**	0.43	**1.04**	0.49	**1.18**
Lower leg				**0.01**		**0.02**		**0.02**	0.01	**0.02**	0.02	**0.03**
Lymph nodes	0.32	**0.40**	0.39	**0.49**	0.45	**0.55**	0.49	**0.60**	0.57	**0.69**	0.64	**0.77**
Skin	0.34	**0.48**	0.36	**0.51**	0.37	**0.53**	0.38	**0.54**	0.40	**0.57**	0.42	**0.60**
Muscle	0.31	**0.36**	0.37	**0.43**	0.41	**0.48**	0.44	**0.51**	0.50	**0.58**	0.55	**0.64**
Skeleton	0.46	**0.67**	0.57	**0.82**	0.65	**0.94**	0.72	**1.02**	0.83	**1.18**	0.94	**1.33**
Red bone marrow	0.11	**0.17**	0.14	**0.20**	0.16	**0.23**	0.17	**0.26**	0.20	**0.30**	0.23	**0.34**
Whole body dose	0.30	**0.36**	0.36	**0.43**	0.40	**0.48**	0.43	**0.52**	0.49	**0.59**	0.54	**0.65**
ED (ICRP 60)	0.46	**0.48**	0.55	**0.57**	0.61	**0.63**	0.65	**0.68**	0.73	**0.76**	0.80	**0.83**
ED (ICRP 103)	0.47	**0.50**	0.55	**0.59**	0.61	**0.64**	0.65	**0.69**	0.73	**0.77**	0.79	**0.84**
Total filtration (mm) **Al**	**1.5**		**2.5**		**3.5**		**4.5**		**3.5**		**3.5**	
Cu	**0**		**0**		**0**		**0**		**0.1**		**0.2**	

Table 11.22 Normalized organ doses—Thoraco-abdominal babygram ap—<1 year

Beam projection:	ap											**Table cells:**
Focus-to-detector distance (cm):	100											(+) = normal font = optimal field setting
Tube voltage (kVp):	60											**(−) = bold font = suboptimal field setting**
Total filtrations:	See last table row											Empty table cells correspond to values <0.01
Organ or tissue	**Organ dose D_T : Entrance dose K_E (mSv * mGy^{-1})**											
	(+)	(−)	(+)	(−)	(+)	(−)	(+)	(−)	(+)	(−)	(+)	(−)
Skull	0.04	**0.26**	0.05	**0.31**	0.05	**0.35**	0.06	**0.38**	0.07	**0.43**	0.07	**0.42**
Brain		**0.03**		**0.04**		**0.05**		**0.05**	0.01	**0.06**	0.01	**0.06**
Oral mucosa	0.04	**0.58**	0.05	**0.67**	0.06	**0.74**	0.07	**0.78**	0.08	**0.86**	0.08	**0.85**
Salivary glands	0.17	**0.51**	0.20	**0.60**	0.23	**0.67**	0.24	**0.72**	0.28	**0.80**	0.27	**0.79**
Cervical spine	0.44	**0.62**	0.55	**0.80**	0.64	**0.93**	0.71	**1.03**	0.84	**1.23**	0.81	**1.18**
Thyroid	0.88	**0.94**	1.00	**1.07**	1.06	**1.15**	1.11	**1.20**	1.19	**1.30**	1.18	**1.29**
Thymus	0.80	**0.82**	0.93	**0.95**	1.01	**1.03**	1.07	**1.10**	1.16	**1.20**	1.16	**1.19**
Thoracic spine	0.33	**0.34**	0.44	**0.45**	0.52	**0.54**	0.59	**0.61**	0.73	**0.76**	0.68	**0.70**
Claviculae	1.98	**2.05**	2.29	**2.37**	2.49	**2.58**	2.63	**2.73**	2.85	**2.97**	2.89	**3.01**
Scapulae	0.54	**0.56**	0.70	**0.72**	0.82	**0.85**	0.91	**0.94**	1.09	**1.13**	1.05	**1.09**
Ribs	1.49	**1.53**	1.71	**1.75**	1.85	**1.90**	1.96	**2.00**	2.13	**2.18**	2.14	**2.19**
Oesophagus	0.25	**0.25**	0.32	**0.33**	0.38	**0.38**	0.42	**0.43**	0.50	**0.51**	0.48	**0.48**
Trachea	0.12	**0.52**	0.15	**0.61**	0.16	**0.67**	0.17	**0.71**	0.20	**0.78**	0.19	**0.77**
Lung	0.46	**0.47**	0.55	**0.56**	0.62	**0.63**	0.67	**0.69**	0.76	**0.78**	0.75	**0.76**
Mamma	1.17	**1.17**	1.24	**1.23**	1.27	**1.26**	1.30	**1.28**	1.35	**1.32**	1.34	**1.32**
Heart	0.59	**0.60**	0.71	**0.72**	0.79	**0.80**	0.85	**0.87**	0.96	**0.98**	0.95	**0.96**
Stomach	0.61	**0.62**	0.73	**0.74**	0.82	**0.83**	0.89	**0.90**	1.00	**1.02**	0.99	**1.00**
Spleen	0.24	**0.24**	0.30	**0.31**	0.36	**0.37**	0.40	**0.41**	0.48	**0.49**	0.45	**0.46**
Liver	0.54	**0.55**	0.65	**0.66**	0.73	**0.74**	0.79	**0.80**	0.89	**0.91**	0.87	**0.89**
Gall bladder	0.46	**0.46**	0.57	**0.58**	0.66	**0.66**	0.72	**0.73**	0.84	**0.85**	0.81	**0.82**
Pancreas	0.33	**0.34**	0.42	**0.43**	0.49	**0.50**	0.54	**0.56**	0.65	**0.66**	0.62	**0.63**
Small intestine	0.41	**0.41**	0.51	**0.51**	0.58	**0.59**	0.64	**0.65**	0.75	**0.76**	0.72	**0.73**
Large intestine	0.49	**0.50**	0.60	**0.61**	0.69	**0.70**	0.75	**0.76**	0.87	**0.88**	0.84	**0.86**
Sigma/Rectum	0.37	**0.38**	0.47	**0.48**	0.54	**0.55**	0.60	**0.60**	0.70	**0.71**	0.67	**0.68**
Lumbar spine	0.36	**0.36**	0.47	**0.48**	0.57	**0.57**	0.64	**0.65**	0.80	**0.81**	0.73	**0.74**
Pelvis	0.59	**0.61**	0.76	**0.78**	0.89	**0.91**	0.99	**1.01**	1.19	**1.21**	1.13	**1.16**
Adrenals	0.15	**0.15**	0.19	**0.20**	0.23	**0.23**	0.26	**0.26**	0.31	**0.32**	0.29	**0.30**
Kidneys	0.14	**0.15**	0.19	**0.19**	0.22	**0.23**	0.25	**0.26**	0.31	**0.32**	0.29	**0.29**
Urinary bladder	0.62	**0.63**	0.74	**0.75**	0.82	**0.84**	0.89	**0.90**	1.00	**1.02**	0.98	**1.00**
Uterus	0.41	**0.41**	0.51	**0.51**	0.59	**0.59**	0.65	**0.65**	0.76	**0.76**	0.73	**0.73**
Ovaries	0.34	**0.35**	0.44	**0.44**	0.51	**0.52**	0.56	**0.57**	0.67	**0.68**	0.64	**0.65**
Prostate	0.48	**0.48**	0.59	**0.59**	0.67	**0.67**	0.74	**0.74**	0.84	**0.84**	0.83	**0.83**
Testes	1.03	**1.03**	1.13	**1.14**	1.19	**1.20**	1.23	**1.24**	1.30	**1.30**	1.30	**1.30**
Upper arm	0.51	**1.12**	0.63	**1.36**	0.72	**1.52**	0.79	**1.65**	0.92	**1.87**	0.90	**1.86**
Elbow	0.48	**1.21**	0.60	**1.46**	0.69	**1.64**	0.76	**1.77**	0.89	**2.01**	0.87	**2.00**
Forearm	0.37	**1.30**	0.47	**1.56**	0.54	**1.74**	0.59	**1.87**	0.69	**2.10**	0.68	**2.11**
Upper leg	0.91	**0.94**	1.14	**1.18**	1.31	**1.36**	1.44	**1.50**	1.68	**1.75**	1.64	**1.72**
Knee	0.27	**0.67**	0.34	**0.83**	0.39	**0.94**	0.43	**1.03**	0.50	**1.19**	0.49	**1.18**
Lower leg		**0.01**		**0.02**	0.01	**0.02**	0.01	**0.03**	0.02	**0.03**	0.02	**0.03**
Lymph nodes	0.38	**0.47**	0.47	**0.57**	0.53	**0.64**	0.57	**0.70**	0.66	**0.79**	0.64	**0.77**
Skin	0.36	**0.50**	0.38	**0.54**	0.39	**0.56**	0.41	**0.57**	0.43	**0.61**	0.42	**0.60**
Muscle	0.36	**0.42**	0.42	**0.49**	0.47	**0.54**	0.50	**0.58**	0.56	**0.65**	0.55	**0.64**
Skeleton	0.56	**0.79**	0.68	**0.97**	0.77	**1.09**	0.84	**1.18**	0.96	**1.35**	0.94	**1.33**
Red bone marrow	0.14	**0.20**	0.17	**0.25**	0.19	**0.28**	0.21	**0.31**	0.25	**0.36**	0.23	**0.34**
Whole body dose	0.35	**0.42**	0.41	**0.49**	0.46	**0.55**	0.49	**0.59**	0.55	**0.66**	0.54	**0.65**
ED (ICRP 60)	0.53	**0.55**	0.62	**0.64**	0.68	**0.71**	0.73	**0.76**	0.82	**0.85**	0.80	**0.83**
ED (ICRP 103)	0.54	**0.57**	0.62	**0.66**	0.68	**0.72**	0.73	**0.77**	0.81	**0.85**	0.79	**0.84**
Total filtration (mm) Al	**1.5**		**2.5**		**3.5**		**4.5**		**3.5**		**3.5**	
Cu	**0**		**0**		**0**		**0**		**0.1**		**0.2**	

Table 11.23 Normalized organ doses—Thoraco-abdominal babygram ap—<1 year

Beam projection:			ap									**Table cells:**
Focus-to-detector distance (cm):			100									(+) = normal font = optimal field setting
Tube voltage (kVp):			70									**(−) = bold font = suboptimal field setting**
Total filtrations:			See last table row									Empty table cells correspond to values <0.01
Organ or tissue	**Organ dose D_T : Entrance dose K_E**								**(mSv * mGy^{-1})**			
	(+)	(−)	(+)	(−)	(+)	(−)	(+)	(−)	(+)	(−)	(+)	(−)
Skull	0.05	**0.29**	0.06	**0.34**	0.06	**0.38**	0.07	**0.41**	0.08	**0.45**	0.09	**0.49**
Brain		**0.04**		**0.05**	0.01	**0.05**	0.01	**0.06**	0.02	**0.07**	0.02	**0.09**
Oral mucosa	0.05	**0.63**	0.06	**0.73**	0.07	**0.79**	0.08	**0.83**	0.09	**0.91**	0.11	**0.98**
Salivary glands	0.19	**0.56**	0.22	**0.66**	0.25	**0.72**	0.26	**0.77**	0.30	**0.85**	0.32	**0.91**
Cervical spine	0.51	**0.74**	0.63	**0.92**	0.72	**1.06**	0.79	**1.16**	0.92	**1.36**	1.03	**1.53**
Thyroid	0.94	**1.01**	1.05	**1.13**	1.11	**1.21**	1.16	**1.26**	1.24	**1.35**	1.30	**1.43**
Thymus	0.87	**0.89**	0.99	**1.02**	1.07	**1.10**	1.13	**1.16**	1.22	**1.26**	1.29	**1.34**
Thoracic spine	0.42	**0.43**	0.54	**0.55**	0.63	**0.65**	0.70	**0.73**	0.85	**0.88**	0.99	**1.02**
Claviculae	2.11	**2.19**	2.40	**2.50**	2.59	**2.69**	2.71	**2.83**	2.90	**3.03**	3.03	**3.17**
Scapulae	0.65	**0.67**	0.81	**0.84**	0.93	**0.97**	1.03	**1.07**	1.21	**1.26**	1.36	**1.42**
Ribs	1.60	**1.63**	1.81	**1.85**	1.94	**1.99**	2.04	**2.09**	2.19	**2.25**	2.30	**2.37**
Oesophagus	0.30	**0.31**	0.38	**0.39**	0.44	**0.45**	0.48	**0.49**	0.57	**0.58**	0.65	**0.66**
Trachea	0.14	**0.57**	0.16	**0.66**	0.18	**0.71**	0.19	**0.75**	0.21	**0.82**	0.23	**0.88**
Lung	0.51	**0.53**	0.61	**0.63**	0.68	**0.70**	0.73	**0.75**	0.82	**0.84**	0.90	**0.92**
Mamma	1.20	**1.20**	1.27	**1.25**	1.30	**1.28**	1.33	**1.30**	1.37	**1.33**	1.41	**1.36**
Heart	0.66	**0.67**	0.78	**0.79**	0.86	**0.88**	0.93	**0.94**	1.04	**1.05**	1.13	**1.15**
Stomach	0.68	**0.69**	0.81	**0.82**	0.90	**0.91**	0.96	**0.98**	1.08	**1.10**	1.18	**1.20**
Spleen	0.29	**0.29**	0.36	**0.37**	0.42	**0.43**	0.46	**0.47**	0.54	**0.56**	0.62	**0.64**
Liver	0.61	**0.62**	0.72	**0.73**	0.80	**0.81**	0.86	**0.87**	0.96	**0.98**	1.05	**1.07**
Gall bladder	0.53	**0.54**	0.65	**0.66**	0.74	**0.74**	0.80	**0.81**	0.92	**0.93**	1.03	**1.03**
Pancreas	0.39	**0.40**	0.49	**0.50**	0.56	**0.58**	0.62	**0.64**	0.73	**0.75**	0.82	**0.85**
Small intestine	0.47	**0.48**	0.58	**0.59**	0.66	**0.67**	0.72	**0.73**	0.83	**0.84**	0.93	**0.94**
Large intestine	0.56	**0.57**	0.68	**0.69**	0.77	**0.78**	0.83	**0.84**	0.95	**0.96**	1.05	**1.06**
Sigma/Rectum	0.44	**0.45**	0.54	**0.55**	0.61	**0.62**	0.67	**0.68**	0.77	**0.79**	0.86	**0.88**
Lumbar spine	0.45	**0.46**	0.58	**0.59**	0.68	**0.70**	0.77	**0.78**	0.93	**0.95**	1.09	**1.11**
Pelvis	0.71	**0.72**	0.89	**0.91**	1.02	**1.04**	1.13	**1.15**	1.32	**1.36**	1.50	**1.53**
Adrenals	0.18	**0.19**	0.23	**0.24**	0.27	**0.28**	0.30	**0.31**	0.37	**0.38**	0.43	**0.44**
Kidneys	0.18	**0.19**	0.23	**0.24**	0.27	**0.28**	0.30	**0.31**	0.37	**0.38**	0.43	**0.44**
Urinary bladder	0.69	**0.70**	0.81	**0.83**	0.90	**0.91**	0.96	**0.98**	1.07	**1.09**	1.17	**1.18**
Uterus	0.48	**0.48**	0.58	**0.59**	0.66	**0.67**	0.73	**0.73**	0.84	**0.85**	0.94	**0.94**
Ovaries	0.41	**0.42**	0.51	**0.52**	0.58	**0.59**	0.64	**0.65**	0.74	**0.76**	0.84	**0.86**
Prostate	0.55	**0.55**	0.66	**0.67**	0.74	**0.75**	0.81	**0.81**	0.91	**0.92**	1.00	**1.01**
Testes	1.08	**1.08**	1.18	**1.18**	1.23	**1.24**	1.27	**1.28**	1.34	**1.34**	1.39	**1.39**
Upper arm	0.58	**1.25**	0.71	**1.49**	0.80	**1.65**	0.87	**1.77**	0.99	**1.98**	1.09	**2.14**
Elbow	0.56	**1.34**	0.68	**1.60**	0.77	**1.77**	0.84	**1.90**	0.96	**2.12**	1.07	**2.29**
Forearm	0.43	**1.43**	0.53	**1.69**	0.60	**1.86**	0.66	**1.99**	0.76	**2.20**	0.84	**2.36**
Upper leg	1.05	**1.09**	1.28	**1.34**	1.45	**1.52**	1.58	**1.65**	1.82	**1.90**	2.00	**2.10**
Knee	0.31	**0.76**	0.38	**0.92**	0.44	**1.04**	0.48	**1.12**	0.55	**1.27**	0.61	**1.40**
Lower leg		**0.02**	0.01	**0.02**	0.02	**0.03**	0.02	**0.03**	0.02	**0.04**	0.03	**0.05**
Lymph nodes	0.44	**0.53**	0.52	**0.64**	0.59	**0.71**	0.63	**0.77**	0.72	**0.87**	0.79	**0.95**
Skin	0.37	**0.53**	0.39	**0.56**	0.41	**0.58**	0.42	**0.60**	0.45	**0.63**	0.47	**0.66**
Muscle	0.40	**0.46**	0.46	**0.54**	0.51	**0.59**	0.54	**0.63**	0.61	**0.70**	0.66	**0.76**
Skeleton	0.63	**0.89**	0.76	**1.07**	0.85	**1.19**	0.91	**1.28**	1.04	**1.45**	1.14	**1.58**
Red bone marrow	0.16	**0.23**	0.19	**0.28**	0.22	**0.32**	0.24	**0.35**	0.28	**0.40**	0.31	**0.45**
Whole body dose	0.39	**0.46**	0.45	**0.54**	0.50	**0.60**	0.53	**0.64**	0.60	**0.72**	0.65	**0.78**
ED (ICRP 60)	0.58	**0.61**	0.68	**0.70**	0.74	**0.77**	0.79	**0.82**	0.88	**0.91**	0.95	**0.99**
ED (ICRP 103)	0.59	**0.62**	0.68	**0.72**	0.74	**0.78**	0.78	**0.83**	0.86	**0.91**	0.93	**0.99**
Total filtration (mm) Al	**1.5**		**2.5**		**3.5**		**4.5**		**3.5**		**3.5**	
Cu	**0**		**0**		**0**		**0**		**0.1**		**0.2**	

Table 11.24 Normalized organ doses—Thoraco-abdominal babygram ap—1–2 years

Beam projection:	ap											**Table cells:**
Focus-to-detector distance (cm):	100											(+) = normal font = optimal field setting
Tube voltage (kVp):	60											**(−) = bold font = suboptimal field setting**
Total filtrations:	See last table row											Empty table cells correspond to values <0.01
Organ or tissue	**Organ dose D_T : Entrance dose K_E**						**(mSv * mGy^{-1})**					
	(+)	(−)	(+)	(−)	(+)	(−)	(+)	(−)	(+)	(−)	(+)	(−)
Skull	0.03	**0.24**	0.04	**0.29**	0.05	**0.33**	0.06	**0.36**	0.07	**0.41**	0.06	**0.40**
Brain		**0.02**		**0.03**		**0.03**		**0.04**		**0.04**		**0.04**
Oral mucosa	0.04	**0.46**	0.05	**0.55**	0.06	**0.60**	0.06	**0.65**	0.08	**0.73**	0.07	**0.71**
Salivary glands	0.12	**0.39**	0.15	**0.47**	0.17	**0.53**	0.18	**0.57**	0.21	**0.65**	0.20	**0.64**
Cervical spine	0.34	**0.44**	0.44	**0.58**	0.52	**0.68**	0.58	**0.77**	0.70	**0.94**	0.66	**0.87**
Thyroid	0.78	**0.83**	0.89	**0.95**	0.96	**1.02**	1.01	**1.08**	1.10	**1.18**	1.09	**1.16**
Thymus	0.73	**0.75**	0.87	**0.89**	0.96	**0.99**	1.03	**1.06**	1.15	**1.19**	1.13	**1.17**
Thoracic spine	0.21	**0.22**	0.28	**0.29**	0.34	**0.35**	0.39	**0.41**	0.50	**0.52**	0.43	**0.45**
Claviculae	1.77	**1.85**	2.12	**2.21**	2.35	**2.46**	2.52	**2.64**	2.82	**2.95**	2.83	**2.96**
Scapulae	0.30	**0.31**	0.39	**0.40**	0.46	**0.48**	0.52	**0.54**	0.64	**0.67**	0.60	**0.62**
Ribs	1.43	**1.45**	1.66	**1.70**	1.82	**1.86**	1.94	**1.98**	2.14	**2.19**	2.13	**2.17**
Oesophagus	0.22	**0.22**	0.28	**0.29**	0.34	**0.34**	0.38	**0.39**	0.47	**0.48**	0.43	**0.44**
Trachea	0.11	**0.41**	0.13	**0.49**	0.14	**0.54**	0.15	**0.59**	0.17	**0.66**	0.17	**0.65**
Lung	0.38	**0.39**	0.47	**0.48**	0.53	**0.54**	0.58	**0.60**	0.67	**0.69**	0.65	**0.67**
Mamma	1.16	**1.18**	1.24	**1.26**	1.29	**1.31**	1.32	**1.34**	1.37	**1.39**	1.36	**1.39**
Heart	0.51	**0.52**	0.63	**0.64**	0.71	**0.73**	0.78	**0.80**	0.90	**0.92**	0.87	**0.89**
Stomach	0.52	**0.53**	0.64	**0.65**	0.72	**0.73**	0.79	**0.80**	0.90	**0.92**	0.88	**0.89**
Spleen	0.17	**0.17**	0.22	**0.23**	0.27	**0.27**	0.30	**0.31**	0.37	**0.38**	0.34	**0.34**
Liver	0.44	**0.45**	0.54	**0.55**	0.62	**0.63**	0.68	**0.69**	0.79	**0.80**	0.75	**0.77**
Gall bladder	0.41	**0.42**	0.53	**0.53**	0.61	**0.62**	0.68	**0.69**	0.80	**0.82**	0.76	**0.77**
Pancreas	0.25	**0.26**	0.33	**0.34**	0.39	**0.40**	0.44	**0.45**	0.54	**0.54**	0.50	**0.50**
Small intestine	0.33	**0.33**	0.42	**0.42**	0.49	**0.49**	0.54	**0.55**	0.65	**0.65**	0.61	**0.61**
Large intestine	0.40	**0.40**	0.51	**0.51**	0.58	**0.59**	0.65	**0.65**	0.76	**0.77**	0.73	**0.73**
Sigma/Rectum	0.29	**0.29**	0.37	**0.38**	0.43	**0.44**	0.48	**0.49**	0.58	**0.59**	0.55	**0.55**
Lumbar spine	0.22	**0.22**	0.30	**0.30**	0.36	**0.37**	0.42	**0.42**	0.54	**0.54**	0.46	**0.46**
Pelvis	0.42	**0.43**	0.55	**0.56**	0.65	**0.66**	0.74	**0.75**	0.91	**0.92**	0.84	**0.85**
Adrenals	0.10	**0.10**	0.14	**0.14**	0.16	**0.16**	0.19	**0.19**	0.23	**0.24**	0.20	**0.21**
Kidneys	0.09	**0.10**	0.12	**0.13**	0.15	**0.15**	0.17	**0.18**	0.22	**0.22**	0.19	**0.19**
Urinary bladder	0.54	**0.54**	0.66	**0.66**	0.74	**0.75**	0.81	**0.82**	0.93	**0.94**	0.90	**0.91**
Uterus	0.33	**0.33**	0.42	**0.42**	0.49	**0.50**	0.55	**0.55**	0.66	**0.66**	0.62	**0.62**
Ovaries	0.26	**0.26**	0.34	**0.34**	0.40	**0.40**	0.45	**0.45**	0.54	**0.55**	0.50	**0.51**
Prostate	0.35	**0.36**	0.45	**0.46**	0.52	**0.53**	0.57	**0.59**	0.67	**0.70**	0.65	**0.67**
Testes	0.94	**0.94**	1.05	**1.06**	1.11	**1.13**	1.16	**1.18**	1.24	**1.26**	1.23	**1.24**
Upper arm	0.54	**0.81**	0.68	**1.01**	0.79	**1.15**	0.88	**1.27**	1.04	**1.48**	1.00	**1.44**
Elbow	0.57	**0.89**	0.72	**1.10**	0.84	**1.27**	0.93	**1.39**	1.10	**1.62**	1.06	**1.58**
Forearm	0.56	**0.97**	0.71	**1.20**	0.82	**1.37**	0.90	**1.50**	1.06	**1.74**	1.03	**1.71**
Upper leg	0.58	**0.61**	0.75	**0.79**	0.87	**0.92**	0.98	**1.03**	1.18	**1.25**	1.11	**1.17**
Knee	0.07	**0.23**	0.09	**0.29**	0.11	**0.34**	0.13	**0.38**	0.16	**0.46**	0.14	**0.43**
Lower leg												
Lymph nodes	0.32	**0.39**	0.39	**0.48**	0.45	**0.54**	0.50	**0.60**	0.58	**0.69**	0.55	**0.66**
Skin	0.33	**0.44**	0.35	**0.47**	0.37	**0.48**	0.38	**0.50**	0.40	**0.53**	0.39	**0.52**
Muscle	0.27	**0.31**	0.32	**0.37**	0.36	**0.41**	0.39	**0.44**	0.44	**0.50**	0.43	**0.48**
Skeleton	0.43	**0.56**	0.53	**0.69**	0.61	**0.78**	0.67	**0.86**	0.78	**1.00**	0.75	**0.96**
Red bone marrow	0.11	**0.14**	0.13	**0.17**	0.15	**0.20**	0.17	**0.22**	0.20	**0.26**	0.19	**0.24**
Whole body dose	0.27	**0.32**	0.33	**0.38**	0.37	**0.43**	0.40	**0.47**	0.46	**0.54**	0.44	**0.51**
ED (ICRP 60)	0.46	**0.47**	0.54	**0.56**	0.60	**0.62**	0.65	**0.67**	0.74	**0.76**	0.71	**0.74**
ED (ICRP 103)	0.47	**0.49**	0.55	**0.58**	0.61	**0.64**	0.66	**0.69**	0.74	**0.78**	0.72	**0.75**
Total filtration (mm) Al	**1.5**		**2.5**		**3.5**		**4.5**		**3.5**		**3.5**	
Cu	**0**		**0**		**0**		**0**		**0.1**		**0.2**	

Table 11.25 Normalized organ doses—Thoraco-abdominal babygram ap—1–2 years

Beam projection:				ap								**Table cells:**
Focus-to-detector distance (cm):				100								(+) = normal font = optimal field setting
Tube voltage (kVp):				70								**(−) = bold font = suboptimal field setting**
Total filtrations:				See last table row								Empty table cells correspond to values <0.01
Organ or tissue	**Organ dose D_T : Entrance dose K_E**								**(mSv * mGy^{-1})**			
	(+)	(−)	(+)	(−)	(+)	(−)	(+)	(−)	(+)	(−)	(+)	(−)
Skull	0.04	**0.27**	0.05	**0.33**	0.06	**0.36**	0.06	**0.39**	0.08	**0.44**	0.09	**0.49**
Brain		**0.03**		**0.03**		**0.04**		**0.05**	0.01	**0.06**	0.01	**0.07**
Oral mucosa	0.05	**0.51**	0.06	**0.60**	0.07	**0.66**	0.07	**0.71**	0.09	**0.79**	0.10	**0.86**
Salivary glands	0.14	**0.44**	0.17	**0.52**	0.19	**0.58**	0.20	**0.63**	0.23	**0.71**	0.26	**0.77**
Cervical spine	0.41	**0.55**	0.52	**0.70**	0.61	**0.81**	0.67	**0.91**	0.80	**1.09**	0.92	**1.26**
Thyroid	0.84	**0.90**	0.95	**1.02**	1.02	**1.09**	1.07	**1.15**	1.16	**1.25**	1.24	**1.34**
Thymus	0.81	**0.83**	0.95	**0.98**	1.04	**1.07**	1.11	**1.15**	1.24	**1.27**	1.34	**1.38**
Thoracic spine	0.28	**0.29**	0.37	**0.38**	0.44	**0.45**	0.50	**0.52**	0.62	**0.64**	0.74	**0.77**
Claviculae	1.95	**2.03**	2.28	**2.38**	2.50	**2.62**	2.66	**2.79**	2.93	**3.08**	3.12	**3.29**
Scapulae	0.37	**0.38**	0.47	**0.49**	0.55	**0.57**	0.62	**0.64**	0.75	**0.78**	0.87	**0.90**
Ribs	1.55	**1.58**	1.78	**1.82**	1.93	**1.98**	2.05	**2.09**	2.24	**2.29**	2.38	**2.44**
Oesophagus	0.27	**0.28**	0.35	**0.36**	0.41	**0.42**	0.46	**0.47**	0.55	**0.57**	0.64	**0.66**
Trachea	0.12	**0.46**	0.14	**0.54**	0.16	**0.59**	0.17	**0.64**	0.19	**0.71**	0.21	**0.77**
Lung	0.44	**0.45**	0.53	**0.54**	0.60	**0.61**	0.65	**0.67**	0.74	**0.77**	0.83	**0.85**
Mamma	1.20	**1.22**	1.28	**1.30**	1.32	**1.34**	1.35	**1.38**	1.40	**1.43**	1.44	**1.47**
Heart	0.59	**0.60**	0.71	**0.73**	0.80	**0.82**	0.87	**0.89**	0.99	**1.02**	1.10	**1.13**
Stomach	0.60	**0.61**	0.72	**0.73**	0.80	**0.82**	0.87	**0.89**	0.99	**1.01**	1.10	**1.12**
Spleen	0.22	**0.22**	0.28	**0.28**	0.33	**0.33**	0.36	**0.37**	0.44	**0.45**	0.51	**0.52**
Liver	0.51	**0.52**	0.62	**0.63**	0.70	**0.71**	0.76	**0.77**	0.87	**0.89**	0.97	**0.99**
Gall bladder	0.50	**0.50**	0.62	**0.62**	0.71	**0.72**	0.78	**0.79**	0.91	**0.93**	1.03	**1.05**
Pancreas	0.32	**0.32**	0.40	**0.41**	0.47	**0.48**	0.52	**0.53**	0.63	**0.64**	0.73	**0.74**
Small intestine	0.40	**0.40**	0.49	**0.50**	0.57	**0.57**	0.63	**0.63**	0.74	**0.75**	0.84	**0.85**
Large intestine	0.48	**0.48**	0.59	**0.59**	0.67	**0.68**	0.74	**0.75**	0.86	**0.87**	0.97	**0.98**
Sigma/Rectum	0.35	**0.36**	0.44	**0.45**	0.51	**0.51**	0.56	**0.57**	0.67	**0.67**	0.76	**0.77**
Lumbar spine	0.30	**0.30**	0.40	**0.40**	0.47	**0.48**	0.54	**0.54**	0.67	**0.68**	0.81	**0.82**
Pelvis	0.53	**0.53**	0.67	**0.68**	0.79	**0.80**	0.88	**0.89**	1.06	**1.08**	1.23	**1.26**
Adrenals	0.13	**0.14**	0.18	**0.18**	0.21	**0.21**	0.24	**0.24**	0.29	**0.29**	0.35	**0.35**
Kidneys	0.12	**0.13**	0.16	**0.16**	0.19	**0.20**	0.22	**0.22**	0.27	**0.27**	0.32	**0.33**
Urinary bladder	0.62	**0.62**	0.74	**0.75**	0.83	**0.84**	0.90	**0.91**	1.02	**1.03**	1.13	**1.14**
Uterus	0.40	**0.40**	0.50	**0.50**	0.58	**0.58**	0.64	**0.64**	0.76	**0.76**	0.87	**0.87**
Ovaries	0.32	**0.32**	0.41	**0.41**	0.47	**0.48**	0.53	**0.53**	0.64	**0.64**	0.73	**0.73**
Prostate	0.42	**0.43**	0.52	**0.54**	0.59	**0.61**	0.65	**0.68**	0.76	**0.79**	0.86	**0.89**
Testes	1.00	**1.01**	1.10	**1.11**	1.17	**1.18**	1.21	**1.23**	1.29	**1.31**	1.36	**1.38**
Upper arm	0.64	**0.94**	0.79	**1.14**	0.90	**1.29**	0.99	**1.41**	1.15	**1.63**	1.30	**1.81**
Elbow	0.68	**1.03**	0.84	**1.25**	0.96	**1.41**	1.05	**1.54**	1.23	**1.77**	1.38	**1.97**
Forearm	0.66	**1.11**	0.81	**1.35**	0.92	**1.52**	1.01	**1.65**	1.17	**1.88**	1.31	**2.07**
Upper leg	0.70	**0.74**	0.88	**0.93**	1.02	**1.08**	1.13	**1.20**	1.34	**1.42**	1.53	**1.62**
Knee	0.09	**0.28**	0.12	**0.35**	0.14	**0.40**	0.15	**0.44**	0.19	**0.53**	0.22	**0.60**
Lower leg												**0.01**
Lymph nodes	0.37	**0.45**	0.46	**0.55**	0.52	**0.62**	0.56	**0.67**	0.65	**0.78**	0.73	**0.87**
Skin	0.35	**0.46**	0.37	**0.49**	0.38	**0.51**	0.40	**0.52**	0.42	**0.55**	0.44	**0.58**
Muscle	0.31	**0.35**	0.36	**0.41**	0.40	**0.45**	0.43	**0.49**	0.49	**0.55**	0.54	**0.61**
Skeleton	0.50	**0.64**	0.61	**0.78**	0.69	**0.88**	0.75	**0.96**	0.87	**1.11**	0.97	**1.24**
Red bone marrow	0.13	**0.17**	0.16	**0.21**	0.18	**0.23**	0.20	**0.26**	0.24	**0.30**	0.27	**0.35**
Whole body dose	0.31	**0.36**	0.37	**0.43**	0.41	**0.48**	0.45	**0.52**	0.51	**0.59**	0.56	**0.65**
ED (ICRP 60)	0.51	**0.53**	0.60	**0.62**	0.67	**0.69**	0.72	**0.74**	0.81	**0.84**	0.89	**0.92**
ED (ICRP 103)	0.53	**0.55**	0.61	**0.64**	0.67	**0.71**	0.72	**0.76**	0.81	**0.85**	0.88	**0.93**
Total filtration (mm) **Al**	**1.5**		**2.5**		**3.5**		**4.5**		**3.5**		**3.5**	
Cu	**0**		**0**		**0**		**0**		**0.1**		**0.2**	

Table 11.26 Normalized organ doses—Thoraco-abdominal babygram ap—1–2 years

Beam projection:	ap											**Table cells:**
Focus-to-detector distance (cm):	100											(+) = normal font = optimal field setting
Tube voltage (kVp):	80											**(−) = bold font = suboptimal field setting**
Total filtrations:	See last table row											Empty table cells correspond to values <0.01
Organ or tissue	**Organ dose D_T : Entrance dose K_E**							**(mSv * mGy^{-1})**				
	(+)	(−)	(+)	(−)	(+)	(−)	(+)	(−)	(+)	(−)	(+)	(−)
Skull	0.05	**0.30**	0.06	**0.35**	0.07	**0.39**	0.07	**0.42**	0.08	**0.47**	0.10	**0.50**
Brain		**0.03**		**0.04**		**0.05**	0.01	**0.05**	0.01	**0.07**	0.02	**0.08**
Oral mucosa	0.05	**0.56**	0.07	**0.65**	0.08	**0.71**	0.08	**0.76**	0.10	**0.84**	0.11	**0.90**
Salivary glands	0.16	**0.49**	0.18	**0.57**	0.20	**0.63**	0.22	**0.67**	0.25	**0.75**	0.27	**0.82**
Cervical spine	0.48	**0.65**	0.60	**0.80**	0.68	**0.92**	0.75	**1.02**	0.87	**1.20**	0.98	**1.36**
Thyroid	0.90	**0.96**	1.00	**1.08**	1.07	**1.15**	1.12	**1.21**	1.21	**1.31**	1.28	**1.39**
Thymus	0.88	**0.91**	1.02	**1.05**	1.11	**1.15**	1.18	**1.22**	1.30	**1.34**	1.39	**1.44**
Thoracic spine	0.35	**0.36**	0.45	**0.46**	0.53	**0.54**	0.59	**0.61**	0.72	**0.75**	0.84	**0.87**
Claviculae	2.08	**2.18**	2.39	**2.51**	2.60	**2.73**	2.74	**2.88**	2.97	**3.13**	3.12	**3.30**
Scapulae	0.44	**0.46**	0.55	**0.57**	0.63	**0.66**	0.70	**0.72**	0.82	**0.86**	0.94	**0.98**
Ribs	1.65	**1.69**	1.87	**1.91**	2.01	**2.06**	2.11	**2.16**	2.28	**2.34**	2.40	**2.46**
Oesophagus	0.33	**0.34**	0.41	**0.42**	0.47	**0.49**	0.52	**0.54**	0.62	**0.64**	0.72	**0.74**
Trachea	0.13	**0.50**	0.16	**0.58**	0.17	**0.64**	0.18	**0.68**	0.20	**0.75**	0.22	**0.81**
Lung	0.49	**0.50**	0.59	**0.60**	0.65	**0.67**	0.71	**0.73**	0.80	**0.83**	0.88	**0.91**
Mamma	1.23	**1.25**	1.31	**1.33**	1.35	**1.37**	1.38	**1.40**	1.43	**1.45**	1.47	**1.49**
Heart	0.66	**0.68**	0.79	**0.80**	0.88	**0.89**	0.95	**0.96**	1.07	**1.09**	1.17	**1.20**
Stomach	0.67	**0.68**	0.79	**0.80**	0.88	**0.89**	0.95	**0.96**	1.07	**1.09**	1.17	**1.19**
Spleen	0.26	**0.27**	0.33	**0.34**	0.38	**0.39**	0.42	**0.43**	0.50	**0.51**	0.57	**0.59**
Liver	0.58	**0.59**	0.69	**0.70**	0.77	**0.78**	0.83	**0.85**	0.94	**0.96**	1.04	**1.06**
Gall bladder	0.57	**0.58**	0.70	**0.71**	0.79	**0.80**	0.86	**0.88**	1.00	**1.02**	1.12	**1.14**
Pancreas	0.38	**0.38**	0.47	**0.48**	0.54	**0.55**	0.60	**0.61**	0.70	**0.72**	0.80	**0.82**
Small intestine	0.46	**0.47**	0.57	**0.57**	0.64	**0.65**	0.70	**0.71**	0.82	**0.83**	0.92	**0.93**
Large intestine	0.55	**0.55**	0.67	**0.67**	0.75	**0.76**	0.82	**0.83**	0.94	**0.95**	1.05	**1.06**
Sigma/Rectum	0.41	**0.42**	0.51	**0.51**	0.58	**0.58**	0.63	**0.64**	0.74	**0.75**	0.83	**0.84**
Lumbar spine	0.38	**0.39**	0.49	**0.49**	0.57	**0.58**	0.64	**0.65**	0.79	**0.80**	0.93	**0.94**
Pelvis	0.63	**0.64**	0.78	**0.79**	0.90	**0.91**	0.99	**1.01**	1.18	**1.20**	1.34	**1.37**
Adrenals	0.17	**0.17**	0.22	**0.22**	0.25	**0.25**	0.28	**0.28**	0.34	**0.35**	0.40	**0.41**
Kidneys	0.16	**0.16**	0.20	**0.20**	0.23	**0.24**	0.26	**0.27**	0.32	**0.32**	0.37	**0.38**
Urinary bladder	0.69	**0.69**	0.82	**0.82**	0.90	**0.91**	0.97	**0.98**	1.09	**1.11**	1.19	**1.21**
Uterus	0.47	**0.47**	0.58	**0.58**	0.66	**0.66**	0.72	**0.73**	0.84	**0.85**	0.95	**0.96**
Ovaries	0.38	**0.38**	0.48	**0.48**	0.55	**0.55**	0.60	**0.60**	0.71	**0.71**	0.81	**0.80**
Prostate	0.48	**0.50**	0.59	**0.61**	0.66	**0.69**	0.72	**0.75**	0.83	**0.86**	0.92	**0.96**
Testes	1.05	**1.06**	1.15	**1.17**	1.21	**1.23**	1.26	**1.28**	1.33	**1.36**	1.39	**1.42**
Upper arm	0.72	**1.05**	0.88	**1.25**	0.99	**1.40**	1.08	**1.51**	1.23	**1.72**	1.36	**1.88**
Elbow	0.77	**1.15**	0.93	**1.37**	1.05	**1.53**	1.14	**1.65**	1.31	**1.87**	1.45	**2.04**
Forearm	0.74	**1.23**	0.89	**1.46**	1.01	**1.62**	1.09	**1.75**	1.24	**1.96**	1.37	**2.12**
Upper leg	0.81	**0.86**	1.00	**1.06**	1.14	**1.21**	1.25	**1.33**	1.46	**1.55**	1.63	**1.73**
Knee	0.11	**0.32**	0.14	**0.39**	0.16	**0.45**	0.18	**0.49**	0.21	**0.58**	0.25	**0.65**
Lower leg										**0.01**		**0.01**
Lymph nodes	0.43	**0.51**	0.51	**0.61**	0.57	**0.68**	0.62	**0.74**	0.71	**0.85**	0.79	**0.94**
Skin	0.36	**0.48**	0.38	**0.51**	0.40	**0.53**	0.41	**0.54**	0.44	**0.57**	0.46	**0.60**
Muscle	0.34	**0.39**	0.40	**0.45**	0.44	**0.50**	0.47	**0.53**	0.53	**0.59**	0.58	**0.65**
Skeleton	0.56	**0.72**	0.67	**0.86**	0.75	**0.96**	0.81	**1.04**	0.93	**1.18**	1.02	**1.30**
Red bone marrow	0.15	**0.20**	0.18	**0.24**	0.21	**0.27**	0.23	**0.29**	0.27	**0.34**	0.30	**0.39**
Whole body dose	0.35	**0.41**	0.41	**0.48**	0.45	**0.53**	0.49	**0.56**	0.55	**0.64**	0.60	**0.70**
ED (ICRP 60)	0.57	**0.59**	0.66	**0.68**	0.73	**0.75**	0.78	**0.80**	0.87	**0.90**	0.94	**0.98**
ED (ICRP 103)	0.58	**0.61**	0.67	**0.70**	0.73	**0.77**	0.77	**0.82**	0.86	**0.91**	0.93	**0.99**
Total filtration (mm) Al	**1.5**		**2.5**		**3.5**		**4.5**		**3.5**		**3.5**	
Cu	**0**		**0**		**0**		**0**		**0.1**		**0.2**	

Table 11.27 Normalized organ doses—Thoraco-abdominal babygram ap—<1 year

Beam projection:			ap									**Table cells:**
Focus-to-detector distance (cm):			115									(+) = normal font = optimal field setting
Tube voltage (kVp):			50									**(−) = bold font = suboptimal field setting**
Total filtrations:			See last table row									Empty table cells correspond to values <0.01
Organ or tissue	**Organ dose D_T : Entrance dose K_E**								**(mSv * mGy^{-1})**			
	(+)	(−)	(+)	(−)	(+)	(−)	(+)	(−)	(+)	(−)	(+)	(−)
Skull	0.03	**0.22**	0.04	**0.27**	0.04	**0.30**	0.05	**0.32**	0.06	**0.37**	0.06	**0.41**
Brain		**0.02**		**0.03**		**0.03**		**0.04**		**0.04**	0.01	**0.05**
Oral mucosa	0.03	**0.52**	0.04	**0.61**	0.05	**0.67**	0.05	**0.71**	0.07	**0.79**	0.08	**0.86**
Salivary glands	0.13	**0.45**	0.16	**0.54**	0.18	**0.60**	0.20	**0.65**	0.22	**0.73**	0.25	**0.80**
Cervical spine	0.34	**0.48**	0.44	**0.64**	0.52	**0.75**	0.58	**0.85**	0.69	**1.02**	0.80	**1.19**
Thyroid	0.83	**0.87**	0.94	**1.00**	1.01	**1.08**	1.06	**1.14**	1.14	**1.23**	1.20	**1.31**
Thymus	0.72	**0.74**	0.84	**0.87**	0.92	**0.95**	0.98	**1.01**	1.08	**1.11**	1.16	**1.20**
Thoracic spine	0.24	**0.25**	0.33	**0.34**	0.40	**0.41**	0.45	**0.47**	0.57	**0.59**	0.69	**0.71**
Claviculae	1.81	**1.87**	2.13	**2.20**	2.34	**2.42**	2.48	**2.57**	2.73	**2.83**	2.92	**3.04**
Scapulae	0.42	**0.43**	0.56	**0.57**	0.66	**0.68**	0.75	**0.77**	0.91	**0.94**	1.07	**1.10**
Ribs	1.36	**1.38**	1.58	**1.61**	1.72	**1.76**	1.83	**1.86**	2.00	**2.05**	2.15	**2.20**
Oesophagus	0.19	**0.20**	0.25	**0.26**	0.30	**0.31**	0.34	**0.35**	0.41	**0.42**	0.48	**0.49**
Trachea	0.10	**0.47**	0.12	**0.55**	0.14	**0.60**	0.15	**0.64**	0.17	**0.71**	0.18	**0.77**
Lung	0.39	**0.40**	0.48	**0.49**	0.55	**0.56**	0.59	**0.61**	0.68	**0.70**	0.76	**0.78**
Mamma	1.13	**1.12**	1.20	**1.18**	1.23	**1.22**	1.26	**1.24**	1.30	**1.28**	1.34	**1.32**
Heart	0.50	**0.51**	0.62	**0.62**	0.69	**0.70**	0.75	**0.77**	0.86	**0.87**	0.95	**0.97**
Stomach	0.52	**0.53**	0.64	**0.65**	0.72	**0.74**	0.79	**0.80**	0.90	**0.91**	0.99	**1.01**
Spleen	0.18	**0.19**	0.24	**0.25**	0.29	**0.29**	0.32	**0.33**	0.39	**0.40**	0.46	**0.47**
Liver	0.46	**0.47**	0.57	**0.57**	0.64	**0.65**	0.70	**0.70**	0.79	**0.80**	0.88	**0.89**
Gall bladder	0.37	**0.38**	0.48	**0.48**	0.55	**0.56**	0.61	**0.62**	0.72	**0.73**	0.82	**0.83**
Pancreas	0.26	**0.26**	0.34	**0.34**	0.40	**0.41**	0.45	**0.46**	0.54	**0.55**	0.62	**0.64**
Small intestine	0.33	**0.33**	0.42	**0.43**	0.49	**0.49**	0.54	**0.55**	0.64	**0.65**	0.73	**0.74**
Large intestine	0.41	**0.41**	0.51	**0.52**	0.59	**0.59**	0.65	**0.65**	0.75	**0.76**	0.85	**0.86**
Sigma/Rectum	0.30	**0.31**	0.39	**0.40**	0.45	**0.46**	0.50	**0.51**	0.60	**0.61**	0.68	**0.70**
Lumbar spine	0.25	**0.26**	0.35	**0.35**	0.42	**0.43**	0.48	**0.49**	0.61	**0.62**	0.74	**0.75**
Pelvis	0.46	**0.47**	0.61	**0.62**	0.72	**0.74**	0.81	**0.83**	0.99	**1.01**	1.15	**1.18**
Adrenals	0.11	**0.11**	0.14	**0.15**	0.17	**0.18**	0.20	**0.20**	0.25	**0.25**	0.29	**0.30**
Kidneys	0.11	**0.11**	0.14	**0.14**	0.17	**0.17**	0.20	**0.20**	0.24	**0.25**	0.29	**0.30**
Urinary bladder	0.53	**0.54**	0.65	**0.66**	0.73	**0.74**	0.79	**0.80**	0.90	**0.91**	0.99	**1.00**
Uterus	0.33	**0.33**	0.42	**0.42**	0.49	**0.49**	0.54	**0.55**	0.64	**0.65**	0.74	**0.74**
Ovaries	0.27	**0.27**	0.35	**0.36**	0.41	**0.42**	0.46	**0.47**	0.56	**0.56**	0.64	**0.65**
Prostate	0.40	**0.41**	0.50	**0.51**	0.58	**0.59**	0.64	**0.65**	0.74	**0.75**	0.83	**0.84**
Testes	0.99	**0.99**	1.09	**1.10**	1.15	**1.17**	1.20	**1.21**	1.26	**1.28**	1.32	**1.34**
Upper arm	0.37	**0.97**	0.48	**1.20**	0.55	**1.36**	0.61	**1.49**	0.73	**1.71**	0.83	**1.90**
Elbow	0.35	**1.05**	0.44	**1.30**	0.52	**1.47**	0.58	**1.60**	0.68	**1.84**	0.78	**2.04**
Forearm	0.26	**1.15**	0.33	**1.41**	0.39	**1.59**	0.43	**1.72**	0.52	**1.96**	0.60	**2.16**
Upper leg	0.75	**0.78**	0.97	**1.00**	1.12	**1.17**	1.25	**1.30**	1.48	**1.54**	1.68	**1.76**
Knee	0.20	**0.54**	0.26	**0.68**	0.30	**0.79**	0.33	**0.87**	0.40	**1.02**	0.46	**1.15**
Lower leg				**0.01**		**0.02**		**0.02**	0.01	**0.02**	0.01	**0.03**
Lymph nodes	0.32	**0.40**	0.40	**0.49**	0.45	**0.56**	0.49	**0.61**	0.57	**0.70**	0.64	**0.78**
Skin	0.34	**0.48**	0.36	**0.51**	0.37	**0.53**	0.38	**0.55**	0.40	**0.57**	0.42	**0.60**
Muscle	0.31	**0.36**	0.37	**0.43**	0.41	**0.48**	0.44	**0.52**	0.50	**0.58**	0.55	**0.64**
Skeleton	0.46	**0.68**	0.57	**0.83**	0.65	**0.95**	0.71	**1.04**	0.82	**1.20**	0.93	**1.34**
Red bone marrow	0.11	**0.17**	0.14	**0.21**	0.16	**0.24**	0.17	**0.26**	0.20	**0.30**	0.23	**0.34**
Whole body dose	0.30	**0.36**	0.36	**0.43**	0.40	**0.48**	0.43	**0.52**	0.49	**0.59**	0.54	**0.65**
ED (ICRP 60)	0.47	**0.48**	0.55	**0.57**	0.61	**0.64**	0.66	**0.68**	0.74	**0.77**	0.81	**0.84**
ED (ICRP 103)	0.48	**0.50**	0.56	**0.59**	0.61	**0.65**	0.66	**0.69**	0.73	**0.78**	0.80	**0.85**
Total filtration (mm) **Al**	**1.5**		**2.5**		**3.5**		**4.5**		**3.5**		**3.5**	
Cu	**0**		**0**		**0**		**0**		**0.1**		**0.2**	

Table 11.28 Normalized organ doses—Thoraco-abdominal babygram ap—<1 year

Beam projection:	ap											**Table cells:**
Focus-to-detector distance (cm):	115											(+) = normal font = optimal field setting
Tube voltage (kVp):	60											**(−) = bold font = suboptimal field setting**
Total filtrations:	See last table row											Empty table cells correspond to values <0.01
Organ or tissue	**Organ dose D_T : Entrance dose K_E**							**(mSv * mGy^{-1})**				
	(+)	(−)	(+)	(−)	(+)	(−)	(+)	(−)	(+)	(−)	(+)	(−)
Skull	0.04	**0.26**	0.05	**0.31**	0.05	**0.34**	0.06	**0.37**	0.07	**0.42**	0.06	**0.41**
Brain		**0.03**		**0.04**		**0.04**		**0.05**	0.01	**0.06**	0.01	**0.05**
Oral mucosa	0.04	**0.58**	0.05	**0.68**	0.06	**0.74**	0.07	**0.79**	0.08	**0.87**	0.08	**0.86**
Salivary glands	0.15	**0.52**	0.18	**0.61**	0.21	**0.68**	0.22	**0.73**	0.25	**0.81**	0.25	**0.80**
Cervical spine	0.43	**0.63**	0.55	**0.80**	0.63	**0.93**	0.70	**1.04**	0.83	**1.24**	0.80	**1.19**
Thyroid	0.90	**0.96**	1.01	**1.08**	1.08	**1.16**	1.13	**1.22**	1.21	**1.31**	1.20	**1.31**
Thymus	0.80	**0.82**	0.93	**0.95**	1.01	**1.04**	1.07	**1.10**	1.17	**1.20**	1.16	**1.20**
Thoracic spine	0.34	**0.35**	0.45	**0.46**	0.53	**0.55**	0.60	**0.62**	0.75	**0.77**	0.69	**0.71**
Claviculae	2.00	**2.07**	2.32	**2.40**	2.52	**2.61**	2.66	**2.76**	2.88	**3.00**	2.92	**3.04**
Scapulae	0.55	**0.57**	0.71	**0.73**	0.83	**0.86**	0.93	**0.96**	1.11	**1.15**	1.07	**1.10**
Ribs	1.50	**1.53**	1.73	**1.76**	1.87	**1.91**	1.97	**2.02**	2.14	**2.20**	2.15	**2.20**
Oesophagus	0.25	**0.26**	0.32	**0.33**	0.38	**0.39**	0.42	**0.43**	0.51	**0.52**	0.48	**0.49**
Trachea	0.12	**0.52**	0.14	**0.61**	0.15	**0.67**	0.17	**0.71**	0.19	**0.78**	0.18	**0.77**
Lung	0.46	**0.47**	0.56	**0.57**	0.63	**0.64**	0.68	**0.70**	0.77	**0.79**	0.76	**0.78**
Mamma	1.17	**1.15**	1.24	**1.22**	1.27	**1.25**	1.30	**1.28**	1.35	**1.31**	1.34	**1.32**
Heart	0.59	**0.60**	0.71	**0.72**	0.79	**0.81**	0.86	**0.87**	0.97	**0.99**	0.95	**0.97**
Stomach	0.61	**0.63**	0.74	**0.75**	0.83	**0.84**	0.89	**0.91**	1.01	**1.03**	0.99	**1.01**
Spleen	0.24	**0.25**	0.31	**0.32**	0.36	**0.37**	0.40	**0.42**	0.49	**0.50**	0.46	**0.47**
Liver	0.55	**0.55**	0.66	**0.67**	0.73	**0.74**	0.79	**0.81**	0.90	**0.91**	0.88	**0.89**
Gall bladder	0.46	**0.47**	0.58	**0.59**	0.66	**0.67**	0.73	**0.74**	0.85	**0.86**	0.82	**0.83**
Pancreas	0.33	**0.34**	0.43	**0.43**	0.49	**0.50**	0.55	**0.56**	0.65	**0.67**	0.62	**0.64**
Small intestine	0.41	**0.42**	0.51	**0.52**	0.59	**0.60**	0.65	**0.66**	0.76	**0.77**	0.73	**0.74**
Large intestine	0.49	**0.50**	0.61	**0.62**	0.69	**0.70**	0.76	**0.77**	0.87	**0.89**	0.85	**0.86**
Sigma/Rectum	0.38	**0.39**	0.48	**0.49**	0.55	**0.56**	0.60	**0.62**	0.71	**0.72**	0.68	**0.70**
Lumbar spine	0.36	**0.37**	0.48	**0.49**	0.57	**0.58**	0.65	**0.66**	0.81	**0.82**	0.74	**0.75**
Pelvis	0.60	**0.62**	0.77	**0.79**	0.90	**0.92**	1.01	**1.03**	1.20	**1.23**	1.15	**1.18**
Adrenals	0.15	**0.15**	0.19	**0.20**	0.23	**0.24**	0.26	**0.27**	0.32	**0.33**	0.29	**0.30**
Kidneys	0.15	**0.15**	0.19	**0.20**	0.23	**0.23**	0.26	**0.26**	0.32	**0.32**	0.29	**0.30**
Urinary bladder	0.62	**0.63**	0.74	**0.75**	0.83	**0.84**	0.89	**0.91**	1.01	**1.02**	0.99	**1.00**
Uterus	0.41	**0.41**	0.51	**0.52**	0.59	**0.60**	0.65	**0.66**	0.76	**0.77**	0.74	**0.74**
Ovaries	0.34	**0.35**	0.44	**0.44**	0.51	**0.52**	0.56	**0.57**	0.67	**0.68**	0.64	**0.65**
Prostate	0.48	**0.49**	0.60	**0.61**	0.68	**0.69**	0.74	**0.75**	0.85	**0.86**	0.83	**0.84**
Testes	1.05	**1.06**	1.15	**1.17**	1.21	**1.23**	1.26	**1.27**	1.32	**1.34**	1.32	**1.34**
Upper arm	0.46	**1.14**	0.58	**1.39**	0.66	**1.56**	0.73	**1.69**	0.84	**1.91**	0.83	**1.90**
Elbow	0.43	**1.23**	0.54	**1.49**	0.62	**1.67**	0.69	**1.81**	0.80	**2.05**	0.78	**2.04**
Forearm	0.32	**1.33**	0.41	**1.60**	0.47	**1.78**	0.52	**1.92**	0.62	**2.15**	0.60	**2.16**
Upper leg	0.93	**0.97**	1.17	**1.22**	1.34	**1.40**	1.47	**1.54**	1.72	**1.80**	1.68	**1.76**
Knee	0.25	**0.65**	0.32	**0.81**	0.36	**0.92**	0.40	**1.01**	0.47	**1.17**	0.46	**1.15**
Lower leg		**0.01**		**0.02**	0.01	**0.02**	0.01	**0.03**	0.02	**0.03**	0.01	**0.03**
Lymph nodes	0.38	**0.48**	0.47	**0.58**	0.53	**0.65**	0.58	**0.70**	0.66	**0.80**	0.64	**0.78**
Skin	0.35	**0.51**	0.38	**0.54**	0.39	**0.56**	0.40	**0.58**	0.43	**0.61**	0.42	**0.60**
Muscle	0.36	**0.42**	0.42	**0.49**	0.47	**0.55**	0.50	**0.59**	0.56	**0.66**	0.55	**0.64**
Skeleton	0.55	**0.80**	0.67	**0.98**	0.76	**1.10**	0.83	**1.19**	0.95	**1.36**	0.93	**1.34**
Red bone marrow	0.13	**0.20**	0.17	**0.25**	0.19	**0.28**	0.21	**0.31**	0.24	**0.36**	0.23	**0.34**
Whole body dose	0.35	**0.42**	0.41	**0.50**	0.46	**0.55**	0.49	**0.59**	0.55	**0.67**	0.54	**0.65**
ED (ICRP 60)	0.53	**0.55**	0.63	**0.65**	0.69	**0.72**	0.74	**0.77**	0.82	**0.86**	0.81	**0.84**
ED (ICRP 103)	0.54	**0.57**	0.63	**0.66**	0.69	**0.73**	0.73	**0.78**	0.81	**0.86**	0.80	**0.85**
Total filtration (mm) Al	**1.5**		**2.5**		**3.5**		**4.5**		**3.5**		**3.5**	
Cu	**0**		**0**		**0**		**0**		**0.1**		**0.2**	

Table 11.29 Normalized organ doses—Thoraco-abdominal babygram ap—<1 year

Beam projection:	ap	**Table cells:**
Focus-to-detector distance (cm):	115	(+) = normal font = optimal field setting
Tube voltage (kVp):	70	**(−) = bold font = suboptimal field setting**
Total filtrations:	See last table row	Empty table cells correspond to values <0.01

Organ or tissue	**Organ dose D_T : Entrance dose K_E**							**(mSv * mGy^{-1})**				
	(+)	(−)	(+)	(−)	(+)	(−)	(+)	(−)	(+)	(−)	(+)	(−)
Skull	0.04	**0.29**	0.05	**0.34**	0.06	**0.37**	0.07	**0.40**	0.08	**0.45**	0.09	**0.48**
Brain		**0.03**		**0.04**	0.01	**0.05**	0.01	**0.06**	0.01	**0.07**	0.02	**0.08**
Oral mucosa	0.05	**0.63**	0.06	**0.73**	0.07	**0.79**	0.08	**0.84**	0.09	**0.92**	0.10	**0.98**
Salivary glands	0.17	**0.57**	0.20	**0.67**	0.23	**0.73**	0.24	**0.78**	0.27	**0.86**	0.30	**0.93**
Cervical spine	0.51	**0.74**	0.63	**0.93**	0.72	**1.07**	0.79	**1.17**	0.91	**1.37**	1.02	**1.54**
Thyroid	0.96	**1.02**	1.06	**1.14**	1.13	**1.22**	1.18	**1.28**	1.25	**1.37**	1.31	**1.44**
Thymus	0.87	**0.89**	0.99	**1.02**	1.07	**1.10**	1.13	**1.16**	1.22	**1.26**	1.30	**1.34**
Thoracic spine	0.43	**0.44**	0.55	**0.56**	0.64	**0.66**	0.72	**0.74**	0.87	**0.90**	1.01	**1.04**
Claviculae	2.14	**2.21**	2.43	**2.52**	2.61	**2.72**	2.74	**2.85**	2.93	**3.06**	3.06	**3.20**
Scapulae	0.66	**0.68**	0.82	**0.85**	0.95	**0.98**	1.05	**1.09**	1.23	**1.28**	1.38	**1.44**
Ribs	1.61	**1.64**	1.82	**1.86**	1.96	**2.00**	2.05	**2.10**	2.21	**2.27**	2.32	**2.39**
Oesophagus	0.31	**0.31**	0.38	**0.39**	0.44	**0.45**	0.49	**0.50**	0.58	**0.59**	0.66	**0.67**
Trachea	0.13	**0.57**	0.15	**0.66**	0.17	**0.71**	0.18	**0.75**	0.20	**0.82**	0.22	**0.88**
Lung	0.52	**0.53**	0.62	**0.64**	0.69	**0.71**	0.74	**0.76**	0.83	**0.86**	0.91	**0.93**
Mamma	1.20	**1.18**	1.27	**1.24**	1.30	**1.27**	1.33	**1.29**	1.37	**1.33**	1.41	**1.35**
Heart	0.66	**0.67**	0.78	**0.80**	0.87	**0.88**	0.93	**0.95**	1.04	**1.06**	1.13	**1.16**
Stomach	0.69	**0.70**	0.82	**0.83**	0.90	**0.92**	0.97	**0.99**	1.09	**1.11**	1.18	**1.21**
Spleen	0.29	**0.30**	0.37	**0.38**	0.42	**0.44**	0.47	**0.48**	0.55	**0.57**	0.63	**0.65**
Liver	0.61	**0.62**	0.73	**0.74**	0.81	**0.82**	0.87	**0.88**	0.97	**0.99**	1.06	**1.08**
Gall bladder	0.54	**0.55**	0.66	**0.67**	0.74	**0.76**	0.81	**0.82**	0.93	**0.95**	1.04	**1.05**
Pancreas	0.40	**0.41**	0.50	**0.51**	0.57	**0.58**	0.63	**0.64**	0.74	**0.75**	0.83	**0.85**
Small intestine	0.48	**0.49**	0.59	**0.60**	0.67	**0.68**	0.73	**0.74**	0.84	**0.85**	0.94	**0.95**
Large intestine	0.57	**0.57**	0.69	**0.70**	0.77	**0.78**	0.84	**0.85**	0.96	**0.97**	1.06	**1.07**
Sigma/Rectum	0.44	**0.45**	0.55	**0.56**	0.62	**0.63**	0.68	**0.69**	0.78	**0.80**	0.87	**0.89**
Lumbar spine	0.46	**0.47**	0.59	**0.60**	0.69	**0.71**	0.78	**0.80**	0.95	**0.97**	1.11	**1.13**
Pelvis	0.72	**0.74**	0.90	**0.92**	1.03	**1.06**	1.14	**1.17**	1.34	**1.38**	1.52	**1.56**
Adrenals	0.19	**0.19**	0.24	**0.24**	0.28	**0.28**	0.31	**0.32**	0.37	**0.38**	0.43	**0.44**
Kidneys	0.18	**0.19**	0.24	**0.24**	0.28	**0.28**	0.31	**0.32**	0.37	**0.38**	0.43	**0.44**
Urinary bladder	0.69	**0.70**	0.82	**0.83**	0.90	**0.92**	0.97	**0.98**	1.08	**1.10**	1.18	**1.19**
Uterus	0.48	**0.48**	0.59	**0.59**	0.67	**0.68**	0.73	**0.74**	0.85	**0.86**	0.95	**0.96**
Ovaries	0.41	**0.42**	0.51	**0.52**	0.58	**0.59**	0.64	**0.65**	0.74	**0.76**	0.84	**0.86**
Prostate	0.55	**0.56**	0.67	**0.68**	0.75	**0.76**	0.81	**0.82**	0.92	**0.94**	1.01	**1.03**
Testes	1.10	**1.11**	1.20	**1.21**	1.26	**1.27**	1.30	**1.31**	1.36	**1.38**	1.41	**1.43**
Upper arm	0.53	**1.28**	0.65	**1.52**	0.73	**1.68**	0.80	**1.81**	0.91	**2.02**	1.01	**2.18**
Elbow	0.50	**1.37**	0.61	**1.63**	0.70	**1.81**	0.76	**1.94**	0.87	**2.16**	0.97	**2.33**
Forearm	0.38	**1.47**	0.47	**1.73**	0.53	**1.91**	0.58	**2.04**	0.68	**2.25**	0.75	**2.41**
Upper leg	1.07	**1.12**	1.31	**1.37**	1.49	**1.55**	1.62	**1.69**	1.85	**1.94**	2.05	**2.15**
Knee	0.29	**0.74**	0.36	**0.90**	0.41	**1.02**	0.45	**1.10**	0.52	**1.25**	0.58	**1.37**
Lower leg		**0.02**	0.01	**0.02**	0.01	**0.03**	0.02	**0.03**	0.02	**0.04**	0.03	**0.05**
Lymph nodes	0.44	**0.54**	0.53	**0.64**	0.59	**0.72**	0.64	**0.77**	0.72	**0.87**	0.80	**0.96**
Skin	0.37	**0.53**	0.39	**0.56**	0.41	**0.58**	0.42	**0.60**	0.45	**0.63**	0.47	**0.66**
Muscle	0.40	**0.47**	0.46	**0.54**	0.51	**0.60**	0.55	**0.64**	0.61	**0.71**	0.66	**0.77**
Skeleton	0.62	**0.90**	0.75	**1.08**	0.84	**1.20**	0.91	**1.30**	1.03	**1.46**	1.13	**1.60**
Red bone marrow	0.16	**0.24**	0.19	**0.29**	0.22	**0.32**	0.24	**0.35**	0.28	**0.40**	0.31	**0.45**
Whole body dose	0.39	**0.47**	0.45	**0.55**	0.50	**0.60**	0.54	**0.65**	0.60	**0.72**	0.65	**0.79**
ED (ICRP 60)	0.59	**0.61**	0.68	**0.71**	0.75	**0.78**	0.80	**0.83**	0.88	**0.92**	0.96	**1.00**
ED (ICRP 103)	0.59	**0.63**	0.68	**0.72**	0.74	**0.79**	0.79	**0.83**	0.87	**0.92**	0.94	**1.00**
Total filtration (mm) **Al**	**1.5**		**2.5**		**3.5**		**4.5**		**3.5**		**3.5**	
Cu	**0**		**0**		**0**		**0**		**0.1**		**0.2**	

Table 11.30 Normalized organ doses—Thoraco-abdominal babygram ap—1–2 years

Beam projection:	ap											**Table cells:**
Focus-to-detector distance (cm):	115											(+) = normal font = optimal field setting
Tube voltage (kVp):	60											**(−) = bold font = suboptimal field setting**
Total filtrations:	See last table row											Empty table cells correspond to values <0.01
Organ or tissue	**Organ dose D_T : Entrance dose K_E**								**(mSv * mGy^{-1})**			
	(+)	(−)	(+)	(−)	(+)	(−)	(+)	(−)	(+)	(−)	(+)	(−)
Skull	0.03	**0.24**	0.04	**0.29**	0.05	**0.33**	0.05	**0.35**	0.06	**0.40**	0.06	**0.39**
Brain		**0.02**		**0.02**		**0.03**		**0.03**		**0.04**		**0.04**
Oral mucosa	0.03	**0.47**	0.04	**0.56**	0.05	**0.62**	0.06	**0.66**	0.07	**0.74**	0.07	**0.72**
Salivary glands	0.11	**0.39**	0.13	**0.47**	0.15	**0.53**	0.16	**0.58**	0.19	**0.66**	0.18	**0.64**
Cervical spine	0.34	**0.45**	0.44	**0.59**	0.51	**0.69**	0.58	**0.78**	0.70	**0.96**	0.65	**0.89**
Thyroid	0.83	**0.87**	0.94	**0.99**	1.02	**1.07**	1.07	**1.13**	1.16	**1.23**	1.15	**1.22**
Thymus	0.73	**0.75**	0.87	**0.89**	0.97	**0.99**	1.04	**1.06**	1.16	**1.19**	1.14	**1.17**
Thoracic spine	0.21	**0.22**	0.29	**0.30**	0.35	**0.36**	0.40	**0.41**	0.51	**0.53**	0.44	**0.46**
Claviculae	1.80	**1.87**	2.14	**2.24**	2.38	**2.49**	2.55	**2.67**	2.85	**2.99**	2.86	**2.99**
Scapulae	0.30	**0.31**	0.40	**0.41**	0.47	**0.49**	0.53	**0.55**	0.66	**0.68**	0.61	**0.63**
Ribs	1.43	**1.46**	1.67	**1.71**	1.83	**1.87**	1.95	**1.99**	2.16	**2.20**	2.14	**2.19**
Oesophagus	0.22	**0.22**	0.29	**0.29**	0.34	**0.35**	0.38	**0.39**	0.47	**0.48**	0.43	**0.44**
Trachea	0.10	**0.41**	0.12	**0.49**	0.13	**0.54**	0.15	**0.59**	0.16	**0.66**	0.16	**0.65**
Lung	0.38	**0.39**	0.47	**0.48**	0.54	**0.55**	0.59	**0.60**	0.68	**0.70**	0.66	**0.68**
Mamma	1.16	**1.17**	1.24	**1.25**	1.29	**1.29**	1.32	**1.32**	1.38	**1.37**	1.37	**1.36**
Heart	0.51	**0.52**	0.63	**0.64**	0.72	**0.73**	0.79	**0.80**	0.91	**0.92**	0.88	**0.89**
Stomach	0.52	**0.53**	0.64	**0.65**	0.72	**0.73**	0.79	**0.80**	0.91	**0.92**	0.88	**0.89**
Spleen	0.17	**0.18**	0.23	**0.23**	0.27	**0.28**	0.31	**0.31**	0.38	**0.39**	0.35	**0.35**
Liver	0.44	**0.45**	0.55	**0.56**	0.62	**0.63**	0.68	**0.69**	0.79	**0.80**	0.76	**0.77**
Gall bladder	0.42	**0.42**	0.53	**0.54**	0.62	**0.62**	0.68	**0.69**	0.81	**0.82**	0.77	**0.78**
Pancreas	0.26	**0.26**	0.33	**0.34**	0.40	**0.40**	0.44	**0.45**	0.54	**0.55**	0.50	**0.51**
Small intestine	0.33	**0.33**	0.42	**0.43**	0.49	**0.50**	0.55	**0.55**	0.66	**0.66**	0.62	**0.62**
Large intestine	0.41	**0.41**	0.51	**0.52**	0.59	**0.60**	0.65	**0.66**	0.77	**0.78**	0.73	**0.74**
Sigma/Rectum	0.30	**0.30**	0.38	**0.38**	0.44	**0.45**	0.49	**0.50**	0.59	**0.60**	0.55	**0.56**
Lumbar spine	0.23	**0.23**	0.30	**0.31**	0.37	**0.38**	0.43	**0.43**	0.55	**0.56**	0.47	**0.47**
Pelvis	0.43	**0.44**	0.56	**0.57**	0.67	**0.68**	0.75	**0.77**	0.93	**0.95**	0.85	**0.87**
Adrenals	0.10	**0.10**	0.14	**0.14**	0.17	**0.17**	0.19	**0.19**	0.24	**0.24**	0.21	**0.21**
Kidneys	0.10	**0.10**	0.13	**0.13**	0.15	**0.16**	0.18	**0.18**	0.22	**0.23**	0.19	**0.20**
Urinary bladder	0.54	**0.54**	0.66	**0.67**	0.75	**0.75**	0.82	**0.82**	0.94	**0.95**	0.91	**0.91**
Uterus	0.34	**0.33**	0.43	**0.43**	0.51	**0.50**	0.57	**0.56**	0.68	**0.67**	0.64	**0.62**
Ovaries	0.26	**0.26**	0.33	**0.34**	0.39	**0.40**	0.44	**0.45**	0.54	**0.55**	0.50	**0.51**
Prostate	0.36	**0.37**	0.46	**0.47**	0.53	**0.54**	0.58	**0.60**	0.69	**0.71**	0.66	**0.68**
Testes	0.98	**0.98**	1.09	**1.09**	1.15	**1.16**	1.20	**1.20**	1.28	**1.28**	1.27	**1.27**
Upper arm	0.50	**0.84**	0.63	**1.04**	0.73	**1.19**	0.81	**1.31**	0.97	**1.53**	0.92	**1.48**
Elbow	0.52	**0.92**	0.66	**1.14**	0.76	**1.30**	0.85	**1.43**	1.01	**1.67**	0.96	**1.62**
Forearm	0.48	**1.00**	0.61	**1.24**	0.71	**1.42**	0.78	**1.55**	0.93	**1.79**	0.89	**1.76**
Upper leg	0.60	**0.63**	0.77	**0.82**	0.90	**0.96**	1.01	**1.07**	1.21	**1.30**	1.14	**1.22**
Knee	0.06	**0.21**	0.07	**0.27**	0.09	**0.32**	0.10	**0.36**	0.13	**0.43**	0.11	**0.41**
Lower leg												
Lymph nodes	0.32	**0.39**	0.40	**0.48**	0.46	**0.55**	0.50	**0.60**	0.59	**0.70**	0.56	**0.67**
Skin	0.33	**0.44**	0.35	**0.47**	0.36	**0.49**	0.37	**0.50**	0.40	**0.53**	0.39	**0.52**
Muscle	0.27	**0.31**	0.33	**0.37**	0.36	**0.41**	0.39	**0.45**	0.45	**0.51**	0.43	**0.49**
Skeleton	0.42	**0.57**	0.52	**0.70**	0.60	**0.80**	0.65	**0.87**	0.77	**1.02**	0.73	**0.98**
Red bone marrow	0.11	**0.14**	0.13	**0.18**	0.15	**0.20**	0.17	**0.22**	0.20	**0.26**	0.19	**0.25**
Whole body dose	0.27	**0.32**	0.33	**0.39**	0.37	**0.44**	0.40	**0.47**	0.46	**0.54**	0.44	**0.52**
ED (ICRP 60)	0.46	**0.48**	0.55	**0.57**	0.61	**0.63**	0.66	**0.68**	0.75	**0.77**	0.72	**0.74**
ED (ICRP 103)	0.48	**0.50**	0.56	**0.59**	0.62	**0.65**	0.66	**0.70**	0.75	**0.78**	0.72	**0.76**
Total filtration (mm) **Al**	**1.5**		**2.5**		**3.5**		**4.5**		**3.5**		**3.5**	
Cu	**0**		**0**		**0**		**0**		**0.1**		**0.2**	

Table 11.31 Normalized organ doses—Thoraco-abdominal babygram ap—1–2 years

Beam projection:			ap									**Table cells:**
Focus-to-detector distance (cm):			115									(+) = normal font = optimal field setting
Tube voltage (kVp):			70									**(−) = bold font = suboptimal field setting**
Total filtrations:			See last table row									Empty table cells correspond to values <0.01
Organ or tissue	**Organ dose D_T : Entrance dose K_E**								**(mSv * mGy^{-1})**			
	(+)	(−)	(+)	(−)	(+)	(−)	(+)	(−)	(+)	(−)	(+)	(−)
Skull	0.04	**0.27**	0.05	**0.32**	0.06	**0.36**	0.06	**0.39**	0.07	**0.44**	0.08	**0.48**
Brain		**0.02**		**0.03**		**0.04**		**0.04**	0.01	**0.05**	0.01	**0.06**
Oral mucosa	0.04	**0.52**	0.05	**0.61**	0.06	**0.67**	0.07	**0.72**	0.08	**0.80**	0.10	**0.87**
Salivary glands	0.12	**0.44**	0.15	**0.53**	0.17	**0.59**	0.18	**0.63**	0.21	**0.71**	0.23	**0.78**
Cervical spine	0.41	**0.56**	0.52	**0.71**	0.60	**0.83**	0.67	**0.92**	0.80	**1.11**	0.91	**1.28**
Thyroid	0.89	**0.94**	1.00	**1.06**	1.08	**1.14**	1.13	**1.20**	1.22	**1.30**	1.29	**1.39**
Thymus	0.81	**0.83**	0.95	**0.98**	1.05	**1.07**	1.12	**1.15**	1.24	**1.27**	1.34	**1.38**
Thoracic spine	0.29	**0.30**	0.38	**0.39**	0.45	**0.46**	0.51	**0.53**	0.63	**0.66**	0.76	**0.79**
Claviculae	1.97	**2.06**	2.30	**2.41**	2.53	**2.65**	2.69	**2.82**	2.96	**3.11**	3.16	**3.33**
Scapulae	0.38	**0.39**	0.48	**0.50**	0.57	**0.59**	0.63	**0.66**	0.76	**0.79**	0.89	**0.92**
Ribs	1.56	**1.59**	1.79	**1.83**	1.95	**1.99**	2.06	**2.11**	2.25	**2.31**	2.40	**2.46**
Oesophagus	0.28	**0.28**	0.35	**0.36**	0.41	**0.42**	0.46	**0.47**	0.56	**0.57**	0.65	**0.67**
Trachea	0.11	**0.46**	0.13	**0.54**	0.15	**0.60**	0.16	**0.64**	0.18	**0.71**	0.20	**0.78**
Lung	0.44	**0.45**	0.54	**0.55**	0.61	**0.62**	0.66	**0.68**	0.76	**0.78**	0.84	**0.86**
Mamma	1.20	**1.21**	1.28	**1.28**	1.33	**1.33**	1.36	**1.36**	1.42	**1.41**	1.46	**1.45**
Heart	0.59	**0.60**	0.72	**0.73**	0.81	**0.82**	0.88	**0.89**	1.00	**1.02**	1.11	**1.13**
Stomach	0.60	**0.61**	0.72	**0.73**	0.81	**0.82**	0.88	**0.89**	1.00	**1.02**	1.11	**1.13**
Spleen	0.22	**0.23**	0.28	**0.29**	0.33	**0.34**	0.37	**0.38**	0.45	**0.46**	0.52	**0.54**
Liver	0.52	**0.52**	0.63	**0.63**	0.70	**0.71**	0.77	**0.78**	0.88	**0.89**	0.98	**1.00**
Gall bladder	0.50	**0.51**	0.62	**0.63**	0.71	**0.72**	0.78	**0.80**	0.92	**0.93**	1.04	**1.06**
Pancreas	0.32	**0.32**	0.41	**0.41**	0.47	**0.48**	0.53	**0.54**	0.64	**0.64**	0.74	**0.75**
Small intestine	0.40	**0.40**	0.50	**0.51**	0.58	**0.58**	0.64	**0.64**	0.75	**0.76**	0.85	**0.86**
Large intestine	0.48	**0.49**	0.60	**0.60**	0.68	**0.69**	0.75	**0.76**	0.87	**0.88**	0.98	**1.00**
Sigma/Rectum	0.36	**0.36**	0.45	**0.45**	0.52	**0.52**	0.57	**0.58**	0.68	**0.69**	0.77	**0.78**
Lumbar spine	0.31	**0.31**	0.40	**0.41**	0.48	**0.49**	0.55	**0.56**	0.68	**0.70**	0.82	**0.84**
Pelvis	0.54	**0.55**	0.69	**0.70**	0.80	**0.82**	0.90	**0.91**	1.08	**1.11**	1.26	**1.28**
Adrenals	0.14	**0.14**	0.18	**0.18**	0.21	**0.21**	0.24	**0.24**	0.30	**0.30**	0.36	**0.36**
Kidneys	0.13	**0.13**	0.16	**0.17**	0.20	**0.20**	0.22	**0.23**	0.27	**0.28**	0.33	**0.33**
Urinary bladder	0.62	**0.62**	0.75	**0.75**	0.84	**0.84**	0.90	**0.91**	1.03	**1.04**	1.14	**1.15**
Uterus	0.41	**0.41**	0.52	**0.51**	0.59	**0.59**	0.66	**0.65**	0.78	**0.77**	0.89	**0.89**
Ovaries	0.32	**0.32**	0.40	**0.41**	0.47	**0.48**	0.52	**0.53**	0.63	**0.64**	0.72	**0.74**
Prostate	0.43	**0.44**	0.53	**0.54**	0.60	**0.62**	0.66	**0.68**	0.77	**0.79**	0.87	**0.89**
Testes	1.03	**1.04**	1.14	**1.14**	1.21	**1.21**	1.25	**1.26**	1.33	**1.33**	1.39	**1.40**
Upper arm	0.59	**0.97**	0.73	**1.18**	0.84	**1.33**	0.92	**1.45**	1.08	**1.67**	1.21	**1.86**
Elbow	0.61	**1.06**	0.76	**1.29**	0.87	**1.45**	0.96	**1.58**	1.12	**1.82**	1.27	**2.02**
Forearm	0.57	**1.15**	0.70	**1.39**	0.80	**1.56**	0.88	**1.70**	1.03	**1.93**	1.15	**2.13**
Upper leg	0.72	**0.77**	0.91	**0.97**	1.05	**1.12**	1.16	**1.24**	1.38	**1.47**	1.57	**1.68**
Knee	0.07	**0.26**	0.10	**0.32**	0.11	**0.38**	0.13	**0.42**	0.16	**0.50**	0.19	**0.57**
Lower leg												**0.01**
Lymph nodes	0.38	**0.46**	0.46	**0.55**	0.52	**0.63**	0.57	**0.68**	0.66	**0.79**	0.74	**0.88**
Skin	0.34	**0.46**	0.37	**0.49**	0.38	**0.51**	0.39	**0.53**	0.42	**0.56**	0.44	**0.58**
Muscle	0.31	**0.35**	0.36	**0.41**	0.40	**0.46**	0.44	**0.49**	0.49	**0.56**	0.54	**0.61**
Skeleton	0.49	**0.65**	0.60	**0.79**	0.68	**0.90**	0.74	**0.98**	0.85	**1.12**	0.96	**1.25**
Red bone marrow	0.13	**0.17**	0.16	**0.21**	0.18	**0.24**	0.20	**0.26**	0.24	**0.31**	0.27	**0.35**
Whole body dose	0.31	**0.37**	0.37	**0.44**	0.42	**0.49**	0.45	**0.53**	0.51	**0.60**	0.57	**0.66**
ED (ICRP 60)	0.52	**0.54**	0.61	**0.63**	0.68	**0.70**	0.73	**0.75**	0.82	**0.85**	0.90	**0.93**
ED (ICRP 103)	0.53	**0.56**	0.62	**0.65**	0.68	**0.71**	0.73	**0.76**	0.82	**0.86**	0.89	**0.94**
Total filtration (mm) **Al**	**1.5**		**2.5**		**3.5**		**4.5**		**3.5**		**3.5**	
Cu	**0**		**0**		**0**		**0**		**0.1**		**0.2**	

Table 11.32 Normalized organ doses—Thoraco-abdominal babygram ap—1–2 years

Beam projection:			ap									**Table cells:**
Focus-to-detector distance (cm):			115						(+) = normal font = optimal field setting			
Tube voltage (kVp):			80						**(−) = bold font = suboptimal field setting**			
Total filtrations:			See last table row						Empty table cells correspond to values <0.01			
Organ or tissue	**Organ dose D_T : Entrance dose K_E**								**(mSv * mGy^{-1})**			
	(+)	(−)	(+)	(−)	(+)	(−)	(+)	(−)	(+)	(−)	(+)	(−)
Skull	0.05	**0.30**	0.06	**0.35**	0.06	**0.39**	0.07	**0.41**	0.08	**0.46**	0.09	**0.50**
Brain		**0.03**		**0.04**		**0.05**	0.01	**0.05**	0.01	**0.06**	0.02	**0.07**
Oral mucosa	0.05	**0.57**	0.06	**0.66**	0.07	**0.72**	0.08	**0.77**	0.09	**0.85**	0.11	**0.92**
Salivary glands	0.14	**0.49**	0.16	**0.58**	0.18	**0.64**	0.20	**0.68**	0.22	**0.76**	0.24	**0.83**
Cervical spine	0.48	**0.66**	0.59	**0.82**	0.68	**0.94**	0.74	**1.03**	0.87	**1.22**	0.98	**1.38**
Thyroid	0.95	**1.00**	1.06	**1.12**	1.13	**1.20**	1.18	**1.26**	1.26	**1.36**	1.33	**1.44**
Thymus	0.89	**0.91**	1.03	**1.05**	1.12	**1.15**	1.19	**1.22**	1.30	**1.34**	1.40	**1.44**
Thoracic spine	0.36	**0.37**	0.46	**0.47**	0.54	**0.56**	0.60	**0.62**	0.73	**0.76**	0.86	**0.89**
Claviculae	2.11	**2.21**	2.42	**2.54**	2.63	**2.76**	2.77	**2.91**	3.00	**3.16**	3.16	**3.33**
Scapulae	0.45	**0.46**	0.56	**0.58**	0.64	**0.67**	0.71	**0.74**	0.84	**0.88**	0.96	**1.00**
Ribs	1.66	**1.70**	1.88	**1.92**	2.03	**2.07**	2.13	**2.18**	2.30	**2.36**	2.42	**2.48**
Oesophagus	0.33	**0.34**	0.42	**0.43**	0.48	**0.49**	0.53	**0.55**	0.63	**0.65**	0.72	**0.75**
Trachea	0.13	**0.50**	0.15	**0.58**	0.16	**0.64**	0.17	**0.68**	0.19	**0.76**	0.21	**0.82**
Lung	0.50	**0.51**	0.60	**0.61**	0.67	**0.68**	0.72	**0.74**	0.81	**0.84**	0.90	**0.92**
Mamma	1.24	**1.24**	1.32	**1.32**	1.36	**1.36**	1.40	**1.39**	1.45	**1.44**	1.50	**1.48**
Heart	0.67	**0.68**	0.79	**0.80**	0.88	**0.90**	0.95	**0.97**	1.08	**1.09**	1.18	**1.20**
Stomach	0.67	**0.68**	0.80	**0.81**	0.88	**0.90**	0.95	**0.97**	1.07	**1.09**	1.18	**1.20**
Spleen	0.27	**0.27**	0.34	**0.34**	0.39	**0.40**	0.43	**0.44**	0.51	**0.52**	0.59	**0.60**
Liver	0.58	**0.59**	0.70	**0.71**	0.78	**0.79**	0.84	**0.85**	0.95	**0.97**	1.05	**1.07**
Gall bladder	0.58	**0.59**	0.70	**0.71**	0.80	**0.81**	0.87	**0.88**	1.01	**1.02**	1.13	**1.14**
Pancreas	0.38	**0.39**	0.48	**0.48**	0.55	**0.56**	0.60	**0.61**	0.71	**0.73**	0.81	**0.83**
Small intestine	0.47	**0.47**	0.57	**0.58**	0.65	**0.66**	0.71	**0.72**	0.83	**0.84**	0.93	**0.94**
Large intestine	0.55	**0.56**	0.67	**0.68**	0.76	**0.77**	0.83	**0.84**	0.95	**0.97**	1.06	**1.08**
Sigma/Rectum	0.42	**0.43**	0.52	**0.52**	0.59	**0.59**	0.64	**0.65**	0.75	**0.76**	0.85	**0.86**
Lumbar spine	0.39	**0.39**	0.49	**0.50**	0.58	**0.59**	0.65	**0.67**	0.80	**0.82**	0.94	**0.96**
Pelvis	0.64	**0.65**	0.80	**0.81**	0.92	**0.94**	1.01	**1.04**	1.20	**1.23**	1.37	**1.40**
Adrenals	0.17	**0.17**	0.22	**0.22**	0.26	**0.26**	0.29	**0.29**	0.35	**0.35**	0.41	**0.42**
Kidneys	0.16	**0.16**	0.20	**0.21**	0.24	**0.24**	0.27	**0.27**	0.32	**0.33**	0.38	**0.39**
Urinary bladder	0.69	**0.70**	0.82	**0.83**	0.91	**0.92**	0.98	**0.99**	1.10	**1.12**	1.21	**1.23**
Uterus	0.48	**0.48**	0.59	**0.59**	0.67	**0.67**	0.74	**0.74**	0.86	**0.86**	0.97	**0.97**
Ovaries	0.38	**0.39**	0.47	**0.48**	0.54	**0.55**	0.60	**0.61**	0.70	**0.72**	0.80	**0.81**
Prostate	0.49	**0.50**	0.60	**0.61**	0.67	**0.69**	0.73	**0.75**	0.84	**0.86**	0.93	**0.96**
Testes	1.09	**1.09**	1.19	**1.19**	1.25	**1.26**	1.30	**1.30**	1.37	**1.38**	1.43	**1.44**
Upper arm	0.67	**1.08**	0.82	**1.29**	0.92	**1.44**	1.00	**1.56**	1.15	**1.76**	1.28	**1.93**
Elbow	0.70	**1.18**	0.85	**1.41**	0.96	**1.57**	1.05	**1.70**	1.20	**1.92**	1.33	**2.09**
Forearm	0.64	**1.27**	0.78	**1.51**	0.88	**1.67**	0.96	**1.80**	1.10	**2.02**	1.21	**2.18**
Upper leg	0.84	**0.89**	1.03	**1.10**	1.17	**1.25**	1.28	**1.37**	1.49	**1.60**	1.67	**1.79**
Knee	0.09	**0.30**	0.12	**0.37**	0.13	**0.42**	0.15	**0.46**	0.18	**0.54**	0.21	**0.61**
Lower leg										**0.01**		**0.01**
Lymph nodes	0.43	**0.52**	0.52	**0.62**	0.58	**0.69**	0.63	**0.75**	0.72	**0.85**	0.80	**0.95**
Skin	0.36	**0.48**	0.38	**0.51**	0.40	**0.53**	0.41	**0.55**	0.43	**0.58**	0.46	**0.61**
Muscle	0.34	**0.39**	0.40	**0.45**	0.44	**0.50**	0.47	**0.54**	0.53	**0.60**	0.58	**0.66**
Skeleton	0.55	**0.73**	0.66	**0.87**	0.74	**0.97**	0.80	**1.05**	0.91	**1.19**	1.01	**1.31**
Red bone marrow	0.15	**0.20**	0.18	**0.24**	0.21	**0.27**	0.23	**0.30**	0.27	**0.35**	0.30	**0.39**
Whole body dose	0.35	**0.41**	0.41	**0.48**	0.45	**0.53**	0.49	**0.57**	0.55	**0.64**	0.60	**0.70**
ED (ICRP 60)	0.58	**0.60**	0.67	**0.69**	0.74	**0.76**	0.79	**0.81**	0.88	**0.91**	0.96	**0.99**
ED (ICRP 103)	0.58	**0.61**	0.67	**0.71**	0.74	**0.77**	0.78	**0.82**	0.87	**0.92**	0.95	**0.99**
Total filtration (mm) **Al**	**1.5**		**2.5**		**3.5**		**4.5**		**3.5**		**3.5**	
Cu	**0**		**0**		**0**		**0**		**0.1**		**0.2**	

References

1. Bontrager KL. Textbook of radiographic positioning and related anatomy. St. Louis, MO: Mosby Year Book; 1993.
2. Datz H, Ben-Shlomo A, Bader D, Sadetzki S, Juster-Reicher A, Marks K, Smolkin T, Zangen S, Margaliot M. The additional dose to radiosensitive organs caused by using under-collimated x-ray beams in neonatal intensive care radiography. Radiat Prot Dosim. 2008;130:518–24.
3. Ebel K-D, Willich E. Die Röntgenuntersuchung im Kindesalter—Technik und Indikation. Berlin: Springer; 1968.
4. Elbakri IA. Estimation of dose-area product-to-effective dose conversion factors for neonatal radiography using PCXMC. Radiat Prot Dosim. 2014;158:43–50.
5. Seidenbusch MC, Regulla D, Schneider K. Radiation exposure of children in pediatric radiology. Part 7: conversion factors for reconstruction of organ dose during thoracoabdominal babygrams. Fortschr Röntgenstr. 2010;182:415–21.

Abdomen

12

12.1 Introduction

The type of abdomen radiography is determined by the clinical indication. Table 12.1 shows indications and Table 12.2 contraindications for abdomen radiography. Image criteria as required for radiological findings in abdomen radiographs are listed in Table 12.3. In order to fulfil these image criteria, the field settings listed in Tables 12.4 and 12.5 should be achieved. These field settings go along with the patient preparation as listed in Table 12.6. Exposure parameter combinations for various age groups and patient diameters are

Table 12.1 Indications for abdominal radiographs

Acute abdomen
Necrotizing enterocolitis
Blunt abdominal trauma
Ingestion of foreign bodies
Ileus/small/large bowel obstruction
Peritonitis
Bowel perforation
Severe enterocolitis
Severe meteorism (e.g. M. Hirschsprung, total agangliosis, severe obstipation)
Anorectal malformation, e.g. anal/rectal atresia
Localization of ventriculo-peritoneal shunt
Nephrolithiasis/urolithiasis
Position of catheters/tubes
Lost surgical instruments
Large abdominal tumours
Intoxication[a]
Annotations:
[a] Mainly in adolescents and adults

Table 12.2 Contraindications for abdominal radiographs

Pregnancy
Severe polytrauma[a]
Annotations:
[a] Classical indication for CT. Severe polytrauma has been defined as a polytrauma with two compartments (chest and abdomen/retroperitoneum) being involved

Table 12.3 Image criteria for abdomen radiographs

Abdomen ap/pa/lat
Symmetrical reproduction of the trunk without rotation and tilting
Reproduction of the abdomen from the diaphragm till to 1–5 cm below the tubera ossi ischii dependent on age
Reproduction of the lateral abdominal walls
Visually sharp reproduction of the diaphragm with the costophrenic angles
Visually sharp reproduction of the preperitoneal fat lines consistent with age
Reproduction of the lower thoracic spine with the adjacent ribs, the lumbar spine and the pelvic bones
Reproduction of the paraspinal and psoas outlines consistent with age and depending on bowel content

Table 12.4 Patient positions and beam projections for abdomen radiographs

Patient position	Beam projections
Supine	ap + lat
Prone	lat[a]
Decubitus left	ap + pa + lat
Annotations:	
[a]	Also known as "Columbia technique" (potbellied location). This technique is applied in special cases only, i.e. anorectal malformation, anal/rectal atresia.

M. Seidenbusch et al., *Imaging Practice and Radiation Protection in Pediatric Radiology*,
https://doi.org/10.1007/978-3-030-18504-6_12

listed in Tables 12.7, 12.8, 12.9, 12.10, and 12.11. The age groups in Tables 12.8 and 12.10 refer to the standard age groups of the mathematical MIRD phantom family which has been used to calculate the normalized organ doses in this book. Tables 12.9 and 12.11 show radiation qualities (i.e. various combinations of tube voltage and total filtration) which may be applied in clinical practice. Please note that not all of these radiation qualities are recommended by the authors. The standard projections of abdomen radiographies of the patient who is correctly positioned are depicted in Figs. 12.1, 12.2, and 12.3. Also the optimal field (marked in green), the suboptimal field (marked in red) and the correct shielding are shown. Tables 12.13, 12.14, 12.15, 12.16, 12.17, 12.18, 12.19, 12.20, 12.21, 12.22, and 12.23 outline the virtual field settings on the skin surface of mathematical MIRD phantoms of various age groups and the resulting virtual radiographies with and without bone structures, respectively. Normalized organ doses resulting from these virtual field settings have been calculated for all age groups and have been listed in Tables 12.24, 12.25, 12.26, 12.27, 12.28, 12.29, 12.30, 12.31, 12.32, 12.33, 12.34, 12.35, 12.36,

Table 12.5 Optimal field settings for abdomen radiographs

Abdomen ap	
Upper field edge:	Just below the mamillae
Field centring:	Just above the navel
Lower field edge:	Perineum
Lateral field edges:	Lateral to abdominal wall
Abdomen pa	
Upper field edge:	Just below the lower end of the scapulae
Field centring:	Just above the navel
Lower field edge:	Perineum
Lateral field edges:	Lateral to abdominal wall
Abdomen lat	
Upper field edge:	Just below the mamillae
Field centring:	Midaxillary line at the level of navel
Lower field edge:	Perineum
Anterior/posterior field edge:	Abdominal wall on both sides

Table 12.6 Patient preparation

General	Remove all clothes, external metallic devices, especially piercings, and jewellery from the chest and the abdomen of the patient.
	Immobilize infants and uncooperative patients in the supine position, e.g. with sandbags or restrain devices. Alternatively, accompanying persons or staff members could hold the patient (in this case also take care of the radiation protection of the accompanying persons or staff members).
	Place eventually small X-ray positive markers at ileo-, colostomies, at palpable tumours or on pain points.
Positioning	In supine position, the arms should be slightly abducted from the abdomen and positioned alongside the trunk. In decubitus, prone and left lateral position the arms should be positioned up.
	The legs should be extended and slightly flexed. Put sandbags under the knees.
	Rotation of the trunk in the longitudinal and/or horizontal axis should be avoided.
Attention	On the Bucky table the patient should always be positioned on a foam mat of 10 to 20 cm thickness. So, the lower part of the body will not be cut if lateral projection X-rays (cross table views) in prone or in decubitus position are performed
Shielding	In males, testicular capsules should be employed if possible. In females, the lower chest region should be shielded with lead masks immediate at the lower field edge.

Table 12.7 Radiographic procedure

X-ray source	Automatic exposure control (AEC) with the grid table is used in all age groups. However, bedside radiographs are performed until 10 years of age without grid and using free hand exposure
	Focal spot size between 0.6 and 1.3 dependent on the age of the patient
	Check the use of additional filtration
	Check the focus-to-detector distance
Field setting	Check the field size
	Check the field centring
Image detector system	Check the use of antiscatter grids
Patient	The exposure should be performed in expiration

Table 12.8 Radiographic techniques of bedside abdomen radiographs

Age group (years)	Beam direction	Patient position	Beam projection	Typical patient diameter (cm) with arms	Typical patient diameter (cm) no arms	Optimal field size (cm x cm)	Focus-to-detector distance (cm)	Voltage (kVp) 50	60	70	80	110	120	Antiscatter grid
24 gw	vertical	supine	ap	5.2		7 x 9	100							
28 gw	vertical	supine	ap	6.0		9 x 11								
	horizontal	l-decubitus	ap											
0	vertical	supine	ap	9.80		14 x 16								
		prone	pa											
	horizontal	l-decubitus	ap											
			pa											
		supine	r-lat	12.6	10.9	10 x 16								
			l-lat											
		prone	r-lat											
			l-lat											
1 – 2	vertical	supine	ap	13.00		17 x 22								
	horizontal	l-decubitus	pa											
		supine	r-lat	17.1	15.1	14 x 22								
			l-lat											
3 – 7	vertical	supine	ap	15.00		21 x 29								
	horizontal	l-decubitus	pa											
		supine	r-lat	22.6	19.6	16 x 19								
			l-lat											
8 – 12	vertical	supine	ap	16.80		26 x 35								
	horizontal	l-decubitus	pa											
13 – 17	vertical	supine	ap	19.60		30 x 44								
	horizontal	l-decubitus	pa											
18	vertical	supine	ap	20.00		36 x 52								
	horizontal	l-decubitus	pa											

Annotations:
- (black) Parameter combination not recommended, normalized organ doses not provided
- (blue) Parameter combination recommended, normalized organ doses provided

Table 12.9 Radiation qualities of bedside abdomen radiographs

Voltage (kVp)	Total filtration: Without additional filtration 1.5 mm Al	2.5 Al	With additional filtration 3.5 Al	4.5 Al	3.5 Al + 0.1 Cu[1]	3.5 Al + 0.2 Cu[1]
50						
60						
70						
80						
110						
120						

Annotations:
- 1 The total beam filtration is usually 3.5 mm Al + 0.1 mm Cu, in adolescents optionally 3.5 mm Al + 0.2 mm Cu
- (grey) Parameter combination not recommended, normalized organ doses provided
- (blue) Parameter combination recommended, normalized organ doses provided

12.37, 12.38, 12.39, 12.40, 12.41, 12.42, 12.43, 12.44, 12.45, 12.46, 12.47, 12.48, 12.49, 12.50, 12.51, 12.52, 12.53, 12.54, 12.55, 12.56, 12.57, 12.58, 12.59, 12.60, 12.61, 12.62, 12.63, 12.64, 12.65, 12.66, 12.67, 12.68, 12.69, 12.70, 12.71, 12.72, 12.73, 12.74, 12.75, 12.76, 12.77, 12.78, 12.79, 12.80, 12.81, 12.82, 12.83, 12.84, 12.85, 12.86, 12.87, 12.88, 12.89, 12.90, 12.91, 12.92, 12.93, 12.94, 12.95, 12.96, 12.97, 12.98, 12.99, 12.100, 12.101, 12.102, 12.103, 12.104, 12.105, 12.106, 12.107, 12.108, 12.109, 12.110, 12.111, 12.112, 12.113, 12.114, 12.115, 12.116, 12.117, 12.118, 12.119, 12.120, 12.121, 12.122, 12.123, 12.124, 12.125, 12.126, 12.127, 12.128, 12.129, 12.130, 12.131, 12.132, 12.133, 12.134, and 12.135. Table cells with normal font refer to optimal field settings and table cells with bold font refer to suboptimal field settings. Please note that normalized organ doses have been provided also for exposure parameter combinations, which are not recommended by the auth ors.

Table 12.10 Radiographic techniques of Bucky table abdomen radiographs

Age group (years)	Beam direction	Patient position	Beam projection	Typical patient diameter (cm) with arms	Typical patient diameter (cm) no arms	Optimal field size (cm x cm)	Focus-to-detector distance (cm)	Voltage (kVp) 50	60	70	80	110	120	Antiscatter grid
24 gw	vertical	supine	ap	5.2	7 x 9	5.2	115							
28 gw	vertical	supine	ap	6.0		9 x 11								
	horizontal	l-decubitus	ap											
0	vertical	supine	ap	9.80		14 x 16								
		prone	pa											
	horizontal	l-decubitus	ap											
			pa											
		supine	r-lat	12.6	10.9	12.6								
			l-lat											
		prone	r-lat											
			l-lat											
1 – 2	vertical	supine	ap	13.00		17 x 22								
	horizontal	l-decubitus	pa											
		supine	r-lat	17.1	15.1	17.1								
			l-lat											
3 – 7	vertical	supine	ap	15.00		21 x 29								
	horizontal	l-decubitus	pa											
		supine	r-lat	22.6	19.6	22.6								
			l-lat											
8 – 12	vertical	supine	ap	16.80		26 x 35								
	horizontal	l-decubitus	pa											
13 – 17	vertical	supine	ap	19.60		30 x 44								
	horizontal	l-decubitus	pa											
18	vertical	supine	ap	20.00		36 x 52								
	horizontal	l-decubitus	pa											

Annotations:
- (dark) Parameter combination not recommended, normalized organ doses not provided
- (light) Parameter combination recommended, normalized organ doses provided

Table 12.11 Radiation qualities of Bucky table abdomen radiographs

Voltage (kVp)	Total filtration: Without additional filtration 1.5 mm Al	2.5 Al	With additional filtration 3.5 Al	4.5 Al	3.5 Al + 0.1 Cu[1]	3.5 Al + 0.2 Cu[1]
50						
60						
70						
80						
110						
120						

Annotations:
- 1 The total beam filtration is usually 3.5 mm Al + 0.1 mm Cu, in adolescents optionally 3.5 mm Al + 0.2 mm Cu
- (grey) Parameter combination not recommended, normalized organ doses provided
- (light) Parameter combination recommended, normalized organ doses provided

Table 12.12 Inconsistencies between the mathematical MIRD phantoms and real human anatomy

Anatomical region	Inconsistencies
Abdomen	No navel has been inserted
Gonads	No variation of gonad situs with age
Active bone marrow	Active bone marrow within the liver, spleen is not inserted in newborn infants
Skeleton	The shape of the spine is unrealistic (no lordosis, no kyphosis considered), the vertebral column is not clearly depicted
	Ribs are drawn as complete circles and thicker than in real humans, the course of the ribs is horizontal
	No xiphoid has been inserted

For details, see the text

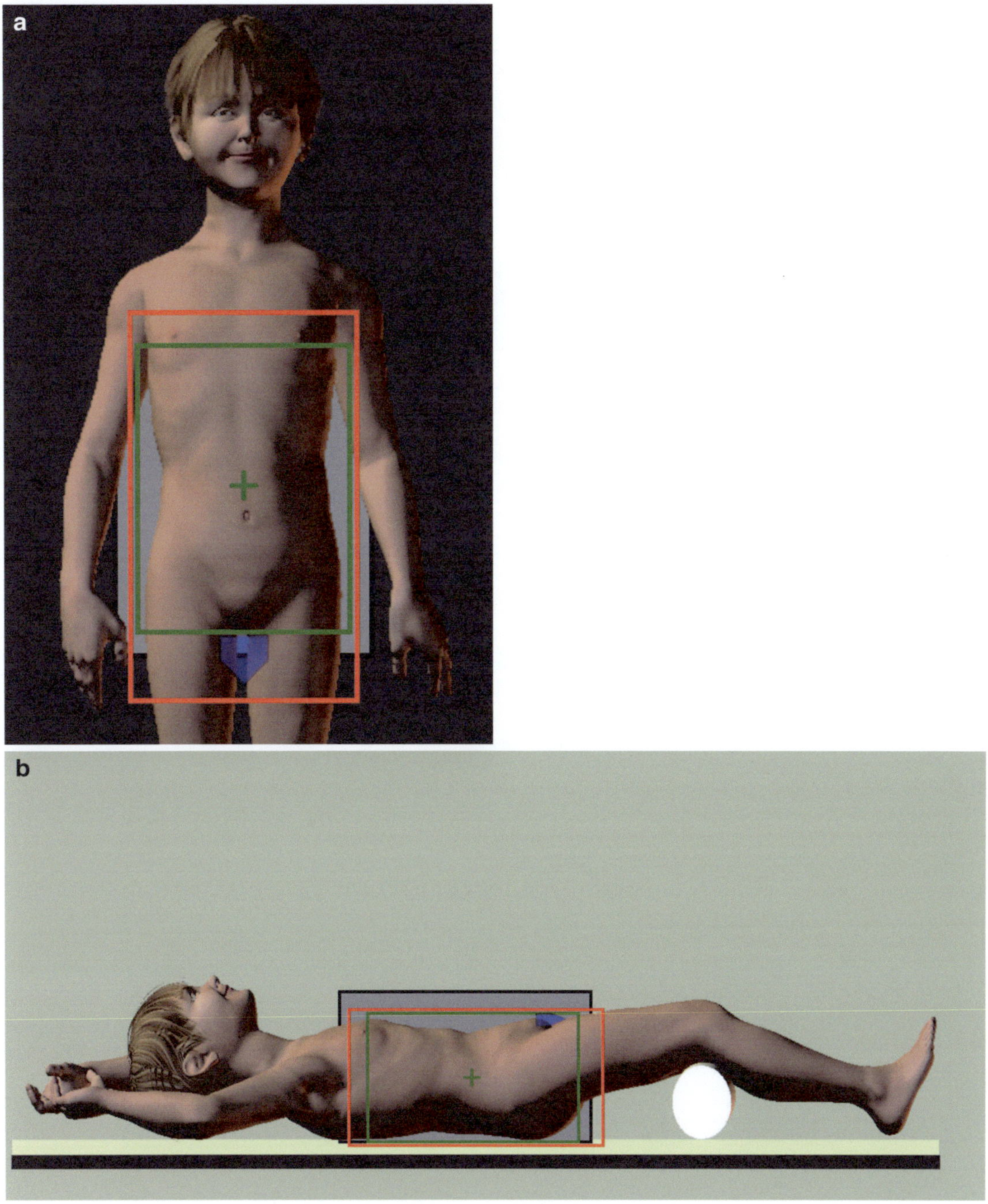

Fig. 12.1 Field settings for abdomen radiographs in supine position. (**a**) Antero-posterior projection. (**b**) Lateral projection (© V. Rösenberger & M. Seidenbusch 2018). Green lines = optimal field settings; green cross = centring point; red lines = suboptimal field settings; grey shaded area = image detector; blue shaded area = lead mask shielding of the patient; dark grey shaded area = Bucky table; white cylinder = foam roll; white foam mat on the top of the Bucky table

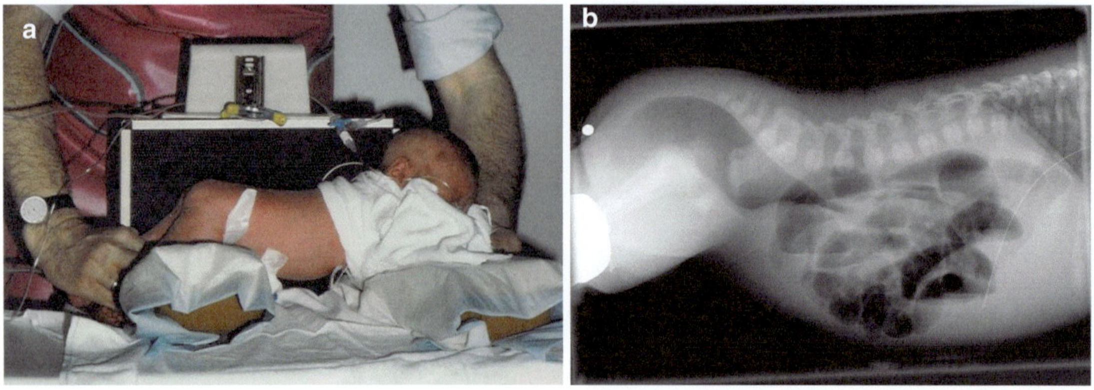

Fig. 12.2 Field settings for abdomen radiographs in lateral projection with the patient in prone position. (**a**) The assisting radiographer is standing behind the cassette in order to fix the patient properly and to secure the testicular capsule. Two assistants, one holding the legs and the other the arms of the patient would facilitate low occupational dose to the staff members (© V. Rösenberger). (**b**) The imperforate anus is marked by a beadlet (arrow). The rectosigmoid colon is distended with air and reaches almost the anus (© V. Rösenberger & K. Schneider)

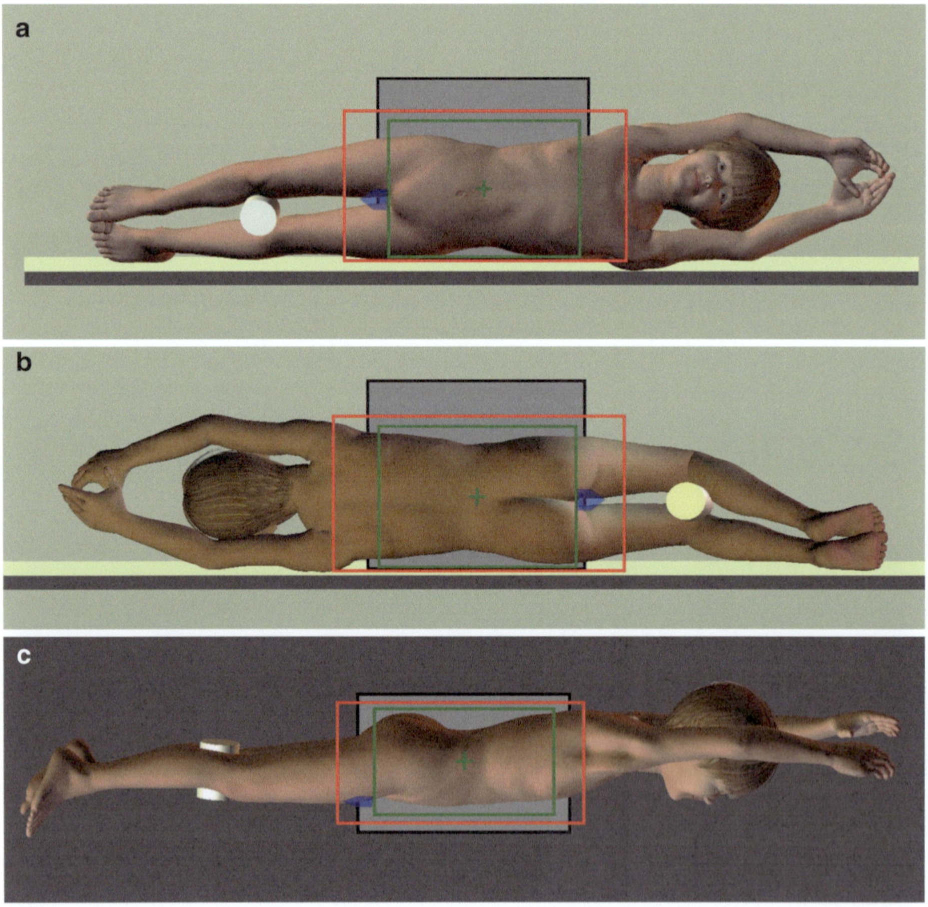

Fig. 12.3 Field settings for abdomen radiographs in left decubitus position. (**a**) Antero-posterior projection. (**b**) Postero-anterior projection. (**c**) Lateral projection (© V. Rösenberger & M. Seidenbusch 2018). Green lines = optimal field settings; green cross = centring point; red lines = suboptimal field settings; grey shaded area = image detector; blue shaded area = lead mask shielding of the patient; dark grey shaded area = Bucky table; white cylinder = foam roll; white foam mat on the top of the Bucky table

12.2 Indications and Contraindications

Indications and contraindications for abdomen radiographs are given in Tables 12.1 and 12.2.

12.3 Image Criteria

The image criteria required for abdomen radiographs are listed in Table 12.3.

See also [2, 3].

12.4 Field Settings

In Table 12.4, patient positions and beam projections used for abdomen radiographs are listed, and in Table 12.5, the optimal field settings are listed which are required in order to fulfil the image criteria mentioned in Table 12.3.

Abdominal X-rays in paediatrics are no longer performed in the upright position. For the detection of free air, left decubitus abdominal pa radiographs should be performed additionally to supine ap X-rays. The sensitivity to detect free air is considerably higher with the left decubitus radiographs than with supine ap radiographs alone. Furthermore, the site of bowel obstruction can often be detected by the combination of both views. The pa beam direction is preferred over the ap direction for the left decubitus radiographs because the mammarian and gonadal dose can be significantly reduced. However, the pa view requires great experience in clinical field setting. In very rare cases, abdominal X-rays are performed in severely ill patients lying supine using either right or left cross-table lateral views. A special technique in order to detect anorectal malformations in neonates is the lateral projection of the abdomen in prone position. The lateral projection (cross-table view) is the decisive radiograph. The pa projection of the abdomen in prone position is no longer used. This radiographic technique replaces the Wangensteen view/invertogram in neonates with anorectal malformation.

For reasons of radiation protection, the pa projection of the lateral decubitus abdominal radiographs should be preferred to the ap projection. Thereby, the organ doses of the breast and ovarian tissue can be reduced.

For more details, see also [4, 9–11, 13].

12.5 Patient Preparation and Radiographic Procedure

See Tables 12.6, 12.7 and [1, 6].

12.6 Exposure Parameters

In Tables 12.8, 12.9, 12.10, and 12.11, exposure parameter combinations used for abdomen radiographs in various age groups are listed. Dark grey shaded table cells refer to exposure parameter combinations which are not recommended for paediatric abdomen radiographs and for which no normalized organ doses have been provided. Light grey and light blue shaded table cells refer to exposure parameter combinations for which normalized organ doses have been provided with light grey shaded table cells referring to exposure parameter combinations being not recommended and light blue shaded table cells referring to exposure parameter combinations following the European Guidelines [2]. Note that due to the fact that the radiation dose is an exponential function of the patient diameter, it may make sense to use the patient diameters instead of the patient's age groups for determining the adequate exposure parameters from Tables 12.8 and 12.10. In this case, for determining the corresponding normalized organ doses from Tables 12.24, 12.25, 12.26, 12.27, 12.28, 12.29, 12.30, 12.31, 12.32, 12.33, 12.34, 12.35, 12.36, 12.37, 12.38, 12.39, 12.40, 12.41, 12.42, 12.43, 12.44, 12.45, 12.46, 12.47, 12.48, 12.49, 12.50, 12.51, 12.52, 12.53, 12.54, 12.55, 12.56, 12.57, 12.58, 12.59, 12.60, 12.61, 12.62, 12.63, 12.64, 12.65, 12.66, 12.67, 12.68, 12.69, 12.70, 12.71, 12.72, 12.73, 12.74, 12.75, 12.76, 12.77, 12.78, 12.79, 12.80, 12.81, 12.82, 12.83, 12.84, 12.85, 12.86, 12.87, 12.88, 12.89, 12.90, 12.91, 12.92, 12.93, 12.94, 12.95, 12.96, 12.97, 12.98, 12.99, 12.100, 12.101, 12.102, 12.103, 12.104, 12.105, 12.106, 12.107, 12.108,

12.109, 12.110, 12.111, 12.112, 12.113, 12.114, 12.115, 12.116, 12.117, 12.118, 12.119, 12.120, 12.121, 12.122, 12.123, 12.124, 12.125, 12.126, 12.127, 12.128, 12.129, 12.130, 12.131, 12.132, 12.133, 12.134, and 12.135, the age group related to the patient diameter should be used.

12.7 Normalized Organ Doses

For calculating normalized organ doses, the field settings described in Sect. 12.4 were transferred to the mathematical MIRD phantoms. Inconsistencies between the geometrical anatomy of the mathematical phantoms and the real human anatomy have been listed in Table 12.12 whenever they are of relevance for the validity of the normalized organ doses.

In Tables 12.13, 12.14, 12.15, 12.16, 12.17, 12.18, 12.19, 12.20, 12.21, 12.22, and 12.23, optimal and suboptimal field settings are listed for the different age groups and the beam projections given in Tables 12.8 and 12.10. Green lines refer to optimal field settings, red lines refer to suboptimal field settings, respectively. In the tables, the optimal and suboptimal field settings which were used for performing the calculations of the normalized organ doses provided in Tables 12.24, 12.25, 12.26, 12.27, 12.28, 12.29, 12.30, 12.31, 12.32, 12.33, 12.34, 12.35, 12.36, 12.37, 12.38, 12.39, 12.40, 12.41, 12.42, 12.43, 12.44, 12.45, 12.46, 12.47, 12.48, 12.49, 12.50, 12.51, 12.52, 12.53, 12.54, 12.55, 12.56, 12.57, 12.58, 12.59, 12.60, 12.61, 12.62, 12.63, 12.64, 12.65, 12.66, 12.67, 12.68, 12.69, 12.70, 12.71, 12.72, 12.73, 12.74, 12.75, 12.76, 12.77, 12.78, 12.79, 12.80, 12.81, 12.82, 12.83, 12.84, 12.85, 12.86, 12.87, 12.88, 12.89, 12.90, 12.91, 12.92, 12.93, 12.94, 12.95, 12.96, 12.97, 12.98, 12.99, 12.100, 12.101, 12.102, 12.103, 12.104, 12.105, 12.106, 12.107, 12.108, 12.109, 12.110, 12.111, 12.112, 12.113, 12.114, 12.115, 12.116, 12.117, 12.118, 12.119, 12.120, 12.121, 12.122, 12.123, 12.124, 12.125, 12.126, 12.127, 12.128, 12.129, 12.130, 12.131, 12.132, 12.133, 12.134, and 12.135 were drawn in on the skin surface of the mathematical phantoms. The virtual mathematical radiographs resulting from these field settings were depicted in the tables with and without bone structures, respectively, in order to enable the reader to visually compare the virtual mathematical radiographs with the real radiographs made under clinical conditions. For comparing the virtual mathematical field settings with the real clinical field settings, the virtual mathematical field edges were also listed in the tables.

As the normalized organ doses provided in Tables 12.24, 12.25, 12.26, 12.27, 12.28, 12.29, 12.30, 12.31, 12.32, 12.33, 12.34, 12.35, 12.36, 12.37, 12.38, 12.39, 12.40, 12.41, 12.42, 12.43, 12.44, 12.45, 12.46, 12.47, 12.48, 12.49, 12.50, 12.51, 12.52, 12.53, 12.54, 12.55, 12.56, 12.57, 12.58, 12.59, 12.60, 12.61, 12.62, 12.63, 12.64, 12.65, 12.66, 12.67, 12.68, 12.69, 12.70, 12.71, 12.72, 12.73, 12.74, 12.75, 12.76, 12.77, 12.78, 12.79, 12.80, 12.81, 12.82, 12.83, 12.84, 12.85, 12.86, 12.87, 12.88, 12.89, 12.90, 12.91, 12.92, 12.93, 12.94, 12.95, 12.96, 12.97, 12.98, 12.99, 12.100, 12.101, 12.102, 12.103, 12.104, 12.105, 12.106, 12.107, 12.108, 12.109, 12.110, 12.111, 12.112, 12.113, 12.114, 12.115, 12.116, 12.117, 12.118, 12.119, 12.120, 12.121, 12.122, 12.123, 12.124, 12.125, 12.126, 12.127, 12.128, 12.129, 12.130, 12.131, 12.132, 12.133, 12.134, and 12.135 refer to entrance kerma values, optimal and suboptimal field sizes were provided in the tables in order to enable the reader to calculate entrance air kerma dose values from clinically measured dose area product values. Please note that these optimal and suboptimal field sizes refer to the mathematical phantoms of an age group only and not to real patients with corresponding ages.

Please note also that the anatomy of the mathematical phantoms has been idealized. Therefore, the virtual field settings for calculating the normalized organ doses have been defined with a view to ensuring that the resulting virtual organ doses would be consistent with the real organ doses which would be achieved during real X-ray examinations being performed with the real field settings recommended in Sect. 12.4. Thus, the field settings shown in the following tables may not be consistent with the clinical field settings as recommended in Sect. 12.4.

Normalized organ doses for abdomen radiographs are also listed in e.g. [5, 7, 8, 12].

Table 12.13 Field settings—Abdomen ap

Age group (years)		Anatomical landmarks			
		Optimal field setting		Suboptimal field setting	
24	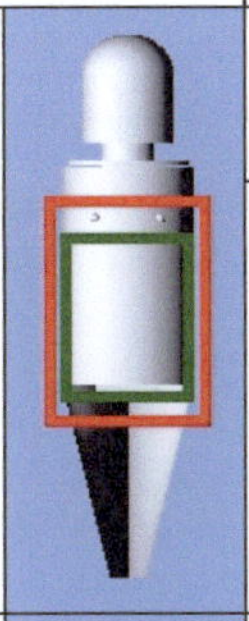	Upper field edge: Just below the mamillae Lower field edge: Just below the perineum Lateral field edges: Close to the abdominal wall Field size = field width × field height: 7 cm × 9 cm = 63 cm²		Upper field edge: Above mamillae Lower field edge: Upper third of thighs Lateral field edges: Lateral to the abdominal wall, both arms included Field size = field width × field height: 9 cm × 13 cm = 117 cm²	

28	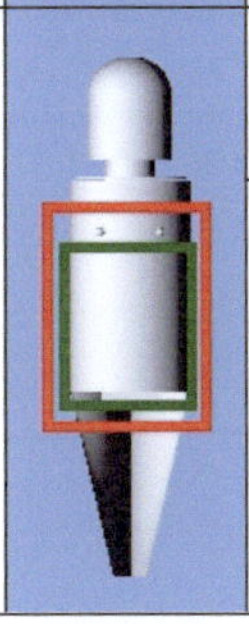	Upper field edge: Just below the mamillae Lower field edge: Just below the perineum Lateral field edges: Close to the abdominal wall Field size = field width × field height: 9 cm × 11cm = 99 cm²		Upper field edge: Above mamillae Lower field edge: Upper third of thighs Lateral field edges: Lateral to the abdominal wall, both arms included Field size = field width × field height: 11 cm × 15 cm = 165 cm²	
				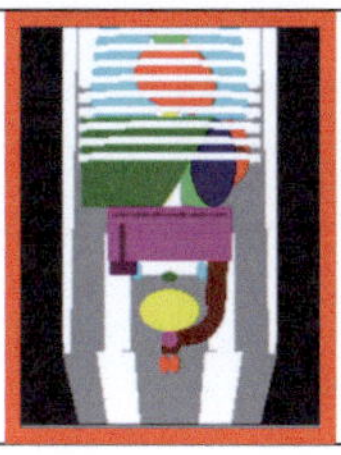	

Table 12.14 Field settings—Abdomen ap

Age group (years)		Anatomical landmarks			
		Optimal field setting		Suboptimal field setting	
0	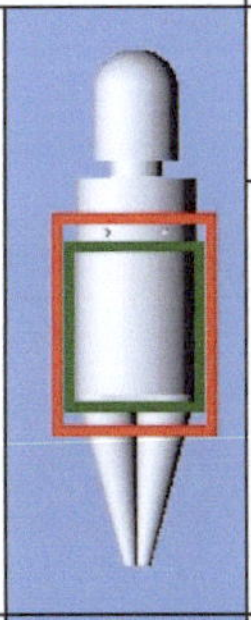	Upper field edge: Just below the mamillae Lower field edge: Just below the perineum Lateral field edges: Close to the abdominal wall Field size = field width × field height: 14 cm × 16 cm = 224 cm²		Upper field edge: Above mamillae Lower field edge: Upper third of thighs Lateral field edges: Lateral to the abdominal wall, both arms included Field size = field width × field height: 16 cm × 21 cm = 336 cm²	
				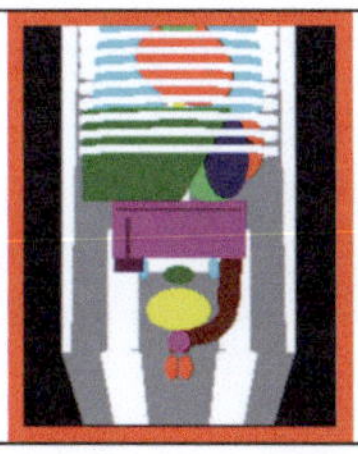	
1–2	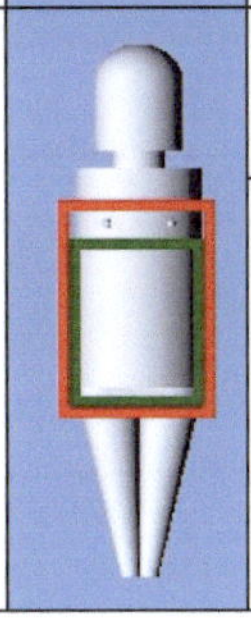	Upper field edge: Just below the mamillae Lower field edge: Just below the perineum Lateral field edges: Close to the abdominal wall Field size = field width × field height: 17 cm × 22 cm = 374 cm²		Upper field edge: Above mamillae Lower field edge: Upper third of thighs Lateral field edges: Lateral to the abdominal wall, both arms included Field size = field width × field height: 18 cm × 28 cm = 504 cm²	
		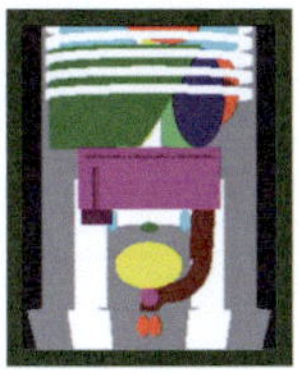		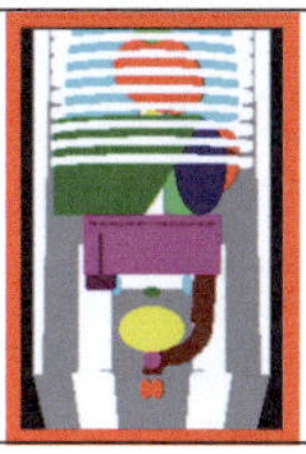	

Table 12.15 Field settings—Abdomen ap

Age group (years)		Anatomical landmarks			
		Optimal field setting		Suboptimal field setting	
3–7	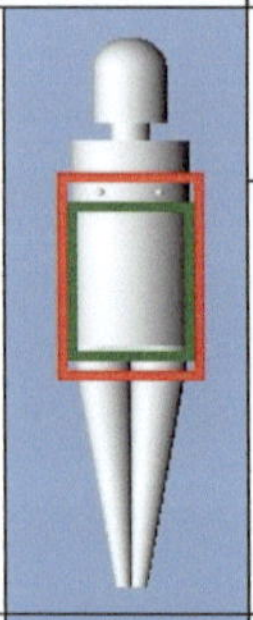	Upper field edge: Just below the mamillae Lower field edge: Just below the perineum Lateral field edges: Close to the abdominal wall Field size = field width × field height: 21 cm × 29 cm = 609 cm²		Upper field edge: Above mamillae Lower field edge: Upper third of thighs Lateral field edges: Lateral to the abdominal wall, both arms included Field size = field width × field height: 26 cm × 38 cm = 988 cm²	
		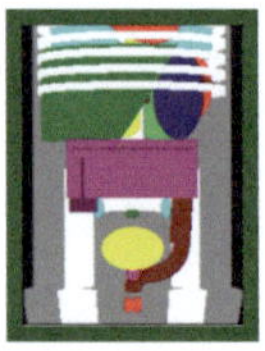			
8–12	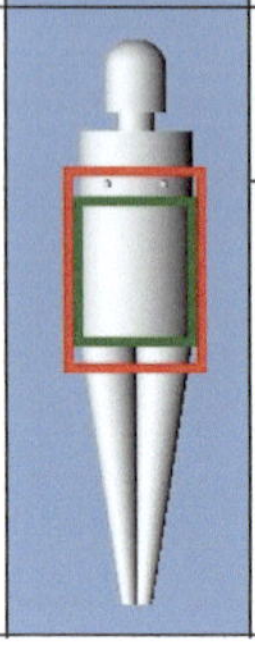	Upper field edge: Just below the mamillae Lower field edge: Just below the perineum Lateral field edges: Close to the abdominal wall Field size = field width × field height: 26 cm × 35 cm = 910 cm²		Upper field edge: Above mamillae Lower field edge : Upper third of thighs Lateral field edges: Lateral to the abdominal wall, both arms included Field size = field width × field height: 32 cm × 46 cm = 1472 cm²	
				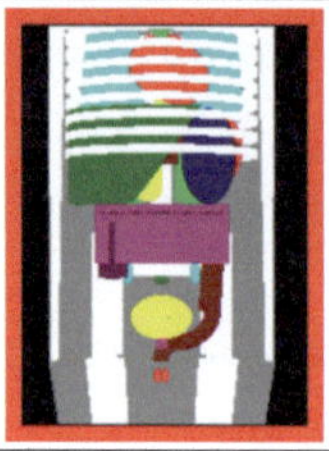	

Table 12.16 Field settings—Abdomen ap

Age group (years)		Anatomical landmarks			
		Optimal field setting		Suboptimal field setting	
13–17	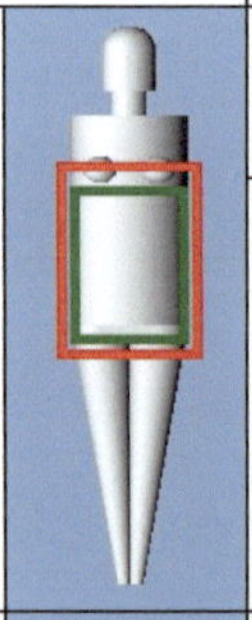	Upper field edge: Just below the mamillae Lower field edge: Just below the perineum Lateral field edges: Close to the abdominal wall Field size = field width × field height: 30 cm × 44 cm = 1320 cm²		Upper field edge: Above mamillae Lower field edge: Upper third of thighs Lateral field edges: Lateral to the abdominal wall, both arms included Field size = field width × field height: 40 cm × 56 cm = 2240 cm²	

18	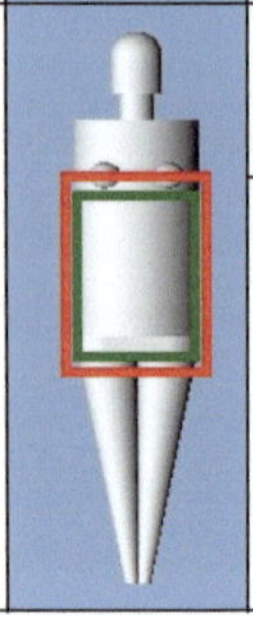	Upper field edge: Just below the mamillae Lower field edge: Just below the perineum Lateral field edges: Close to the abdominal wall Field size = field width × field height: 36 cm × 52 cm = 1872 cm²		Upper field edge: Above mamillae Lower field edge: Upper third of thighs Lateral field edges: Lateral to the abdominal wall, both arms included Field size = field width × field height: 44 cm × 62 cm = 2728 cm²	

Table 12.17 Field settings—Abdomen pa

Age group (years)		Anatomical landmarks			
		Optimal field setting		Suboptimal field setting	
0	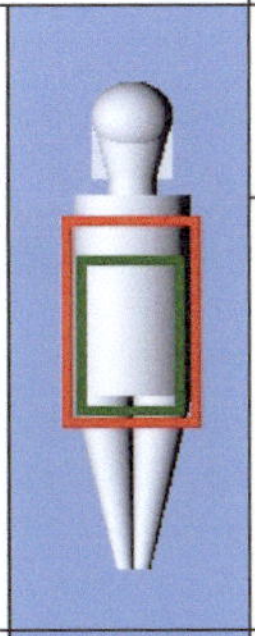	Upper field edge: Just below the scapulae Lower field edge: Just below the perineum Lateral field edges: Close to the lateral chest and abdominal wall Field size = field width × field height: 14 cm × 16 cm = 224 cm^2		Upper field edge: At the level of spinae scapulae Lower field edge: Upper third of thighs Lateral field edges: Distant to the lateral chest and abdominal wall, both arms included Field size = field width × field height: 16 cm × 21 cm = 336 cm^2	

1–2	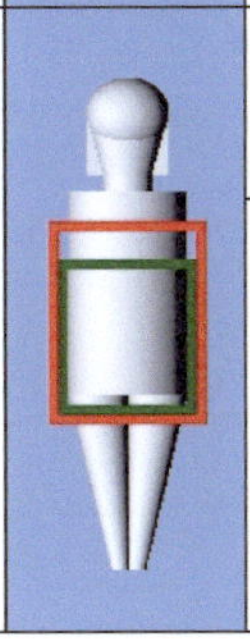	Upper field edge: Just below the scapulae Lower field edge: Just below the perineum Lateral field edges: Close to the lateral chest and abdominal wall Field size = field width × field height: 17 cm × 22 cm = 374 cm^2		Upper field edge: At the level of spinae scapulae Lower field edge: Upper third of thighs Lateral field edges: Distant to the lateral chest and abdominal wall, both arms included Field size = field width × field height: 18 cm × 28 cm = 504 cm^2	
		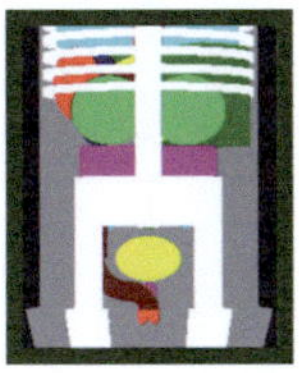	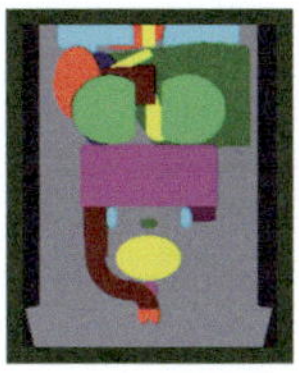	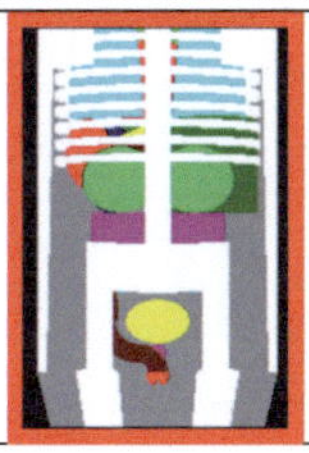	

Table 12.18 Field settings—Abdomen pa

Age group (years)		Anatomical landmarks			
		Optimal field setting		Suboptimal field setting	
3–7	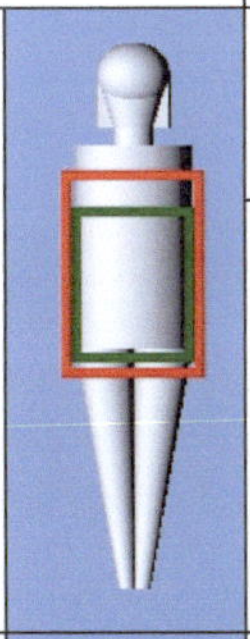	Upper field edge: Just below the scapulae Lower field edge: Just below the perineum Lateral field edges: Close to the lateral chest and abdominal wall Field size = field width × field height: 21 cm × 29 cm = 609 cm^2		Upper field edge: At the level of spinae scapulae Lower field edge: Upper third of thighs Lateral field edges: Distant to the lateral chest and abdominal wall, both arms included Field size = field width × field height: 26 cm × 38 cm = 988 cm^2	
		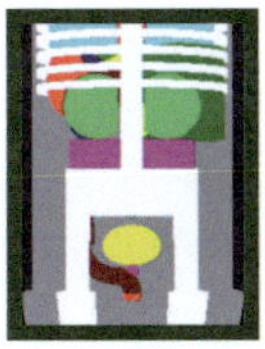	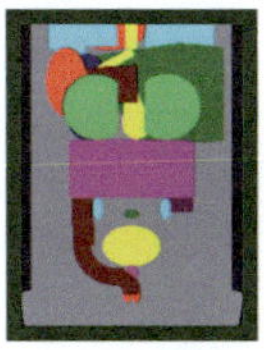	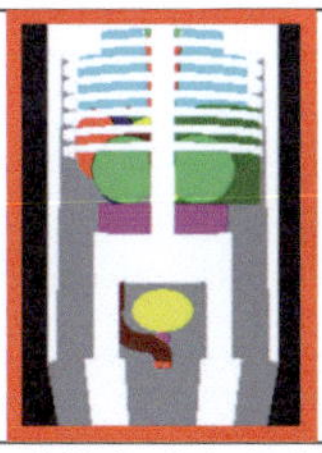	
8–12	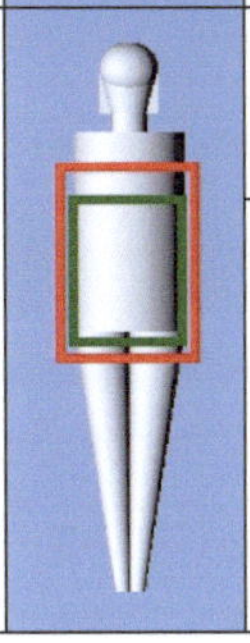	Upper field edge: Just below the scapulae Lower field edge: Just below the perineum Lateral field edges: Close to the lateral chest and abdominal wall Field size = field width × field height: 26 cm × 35 cm = 910 cm^2		Upper field edge: At the level of spinae scapulae Lower field edge: Upper third of thighs Lateral field edges: Distant to the lateral chest and abdominal wall, both arms included Field size = field width × field height: 32 cm × 46 cm = 1472 cm^2	
		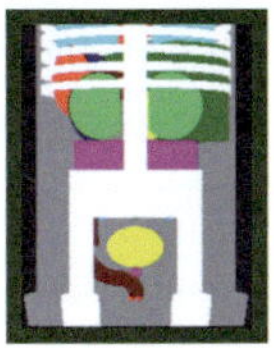	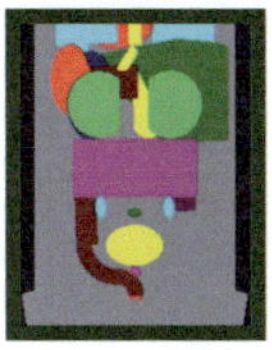	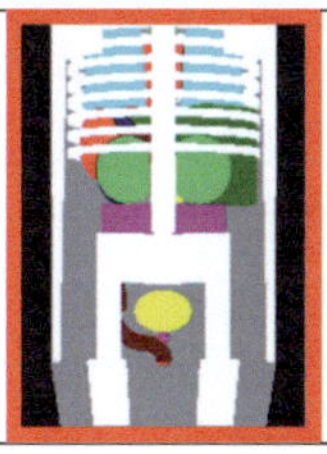	

Table 12.19 Field settings—Abdomen pa

Age group (years)		Anatomical landmarks			
		Optimal field setting		Suboptimal field setting	
13–17	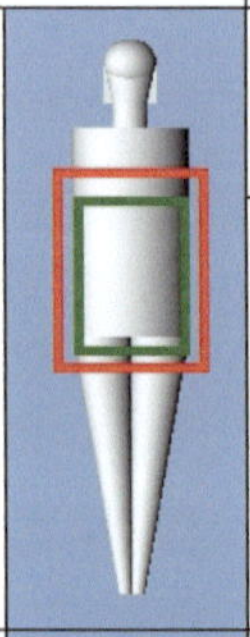	Upper field edge: Just below the scapulae Lower field edge: Just below the perineum Lateral field edges: Close to the lateral chest and abdominal wall Field size = field width × field height: 30 cm × 44 cm = 1320 cm^2		Upper field edge: At the level of spinae scapulae Lower field edge: Upper third of thighs Lateral field edges: Distant to the lateral chest and abdominal wall, both arms included Field size = field width × field height: 40 cm × 56 cm = 2240 cm^2	
		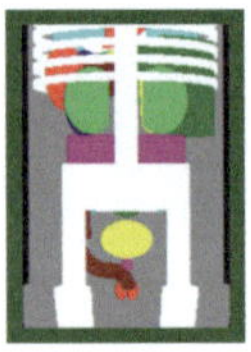		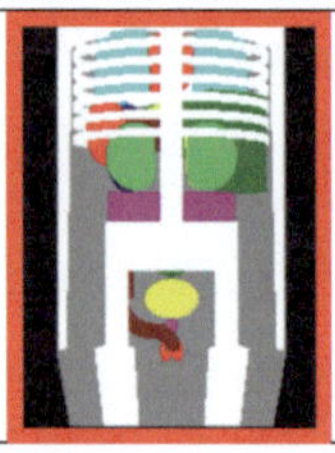	
18	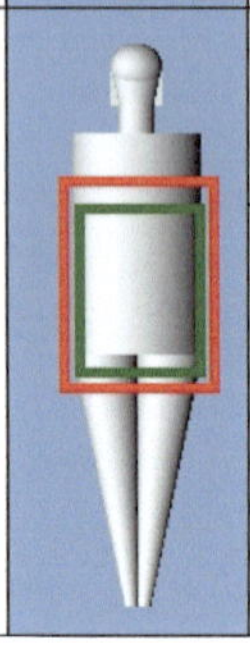	Upper field edge: Just below the scapulae Lower field edge: Just below the perineum Lateral field edges: Close to the lateral chest and abdominal wall Field size = field width × field height: 36 cm × 52 cm = 1872 cm^2		Upper field edge: At the level of spinae scapulae Lower field edge: Upper third of thighs Lateral field edges: Distant to the lateral chest and abdominal wall, both arms included Field size = field width × field height: 44 cm × 62 cm = 2728 cm^2	
				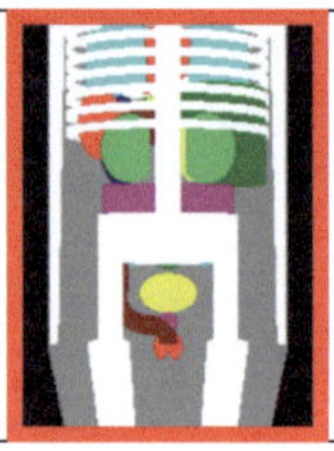	

Table 12.20 Field settings—Abdomen r-lat

Age group (years)		Anatomical landmarks			
		Optimal field setting		Suboptimal field setting	
0	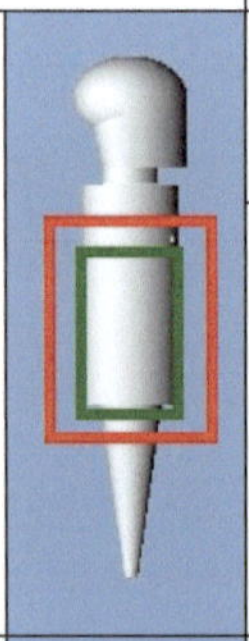	Upper field edge: Just below the mamillae Lower field edge: Just below the perineum Lateral field edges: Close the anterior abdominal wall and the back Field size = field width × field height: 10 cm × 16 cm= 160 cm^2		Upper field edge:Above mamillae Lower field edge: Upper third of thighs Lateral field edges: Distant to the anterior abdominal wall and the back, both arms included Field size = field width × field height: 16 cm × 21 cm= 336 cm^2	
				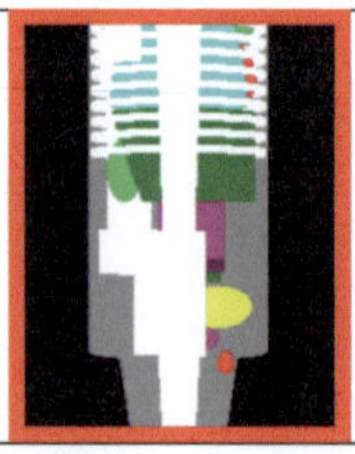	
1–2	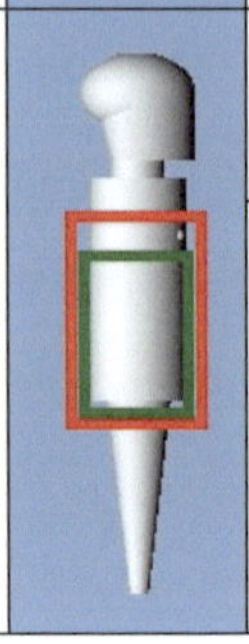	Upper field edge: Just below the mamillae Lower field edge: Just below the perineum Lateral field edges: Close the anterior abdominal wall and the back Field size = field width × field height: 14 cm × 22 cm = 308 cm^2		Upper field edge: Above mamillae Lower field edge: Upper third of thighs Lateral field edges: Distant to the anterior abdominal wall and the back, both arms included Field size = field width × field height: 18 cm × 28 cm = 504 cm^2	

Table 12.21 Field settings—Abdomen r-lat

Age group (years)		Anatomical landmarks			
		Optimal field setting		Suboptimal field setting	
3–7	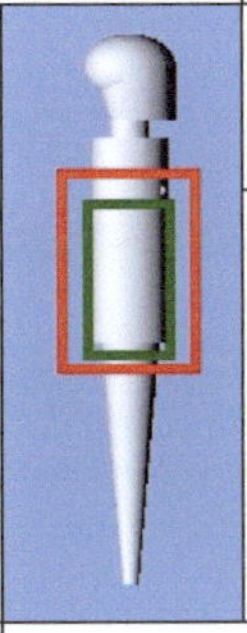	Upper field edge: Just below the mamillae Lower field edge: Just below the perineum Lateral field edges: Close the anterior abdominal wall and the back Field size = field width × field height: 16 cm × 29 cm = 464 cm^2		Upper field edge: Above mamillae Lower field edge: Upper third of thighs Lateral field edges: Distant to the anterior abdominal wall and the back, both arms included Field size = field width × field height: 26 cm × 38 cm = 988 cm^2	
			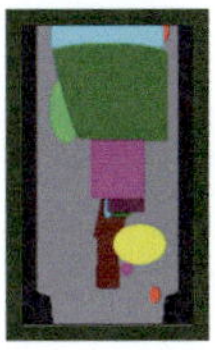	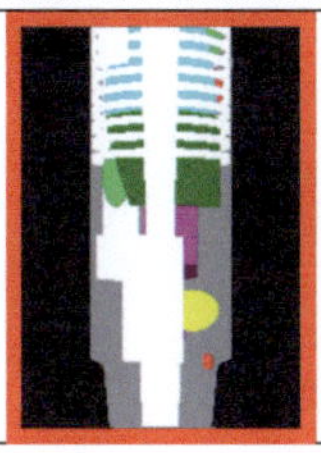	
8–12 13–17 18	This technique is usually not applied in these age groups.				

Table 12.22 Field settings—Abdomen l-lat

Age group (years)		Anatomical landmarks			
		Optimal field setting		Suboptimal field setting	
0	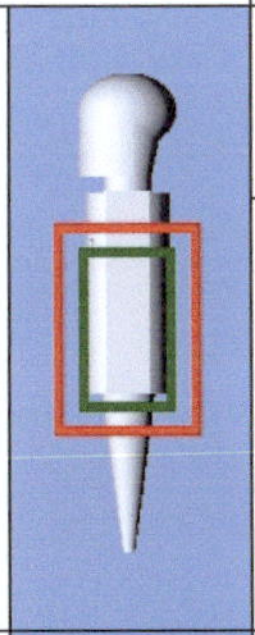	Upper field edge: Just below the mamillae Lower field edge: Just below the perineum Lateral field edges: Close the anterior abdominal wall and the back Field size = field width × field height: 10 cm × 16 cm= 160 cm^2		Upper field edge: Above mamillae Lower field edge: Upper third of thighs Lateral field edges: Distant to the anterior abdominal wall and the back, both arms included Field size = field width × field height: 16 cm × 21 cm= 336 cm^2	
				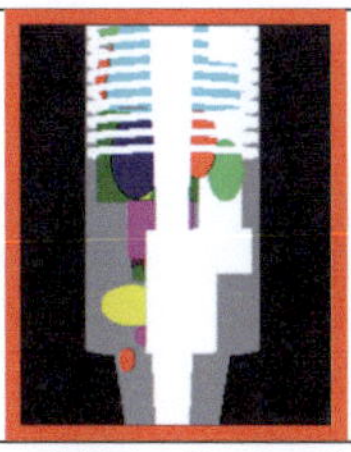	
1–2	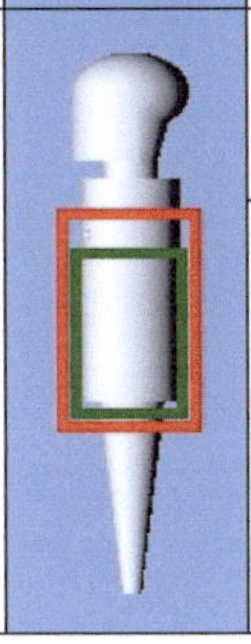	Upper field edge: Just below the mamillae Lower field edge: Just below the perineum Lateral field edges: Close the anterior abdominal wall and the back Field size = field width × field height: 14 cm × 22 cm = 308 cm^2		Upper field edge: Above mamillae Lower field edge: Upper third of thighs Lateral field edges: Distant to the anterior abdominal wall and the back, both arms included Field size = field width × field height: 18 cm × 28 cm = 504 cm^2	

Table 12.23 Field settings—Abdomen l-lat

Age group (years)		Anatomical landmarks			
		Optimal field setting		Suboptimal field setting	
3–7	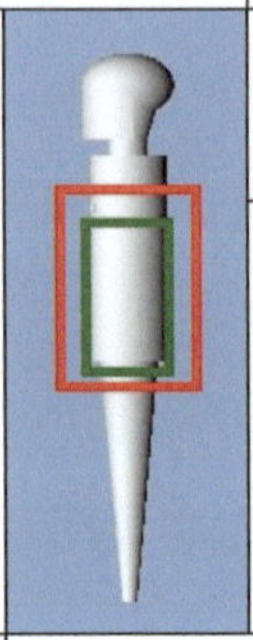	Upper field edge: Just below the mamillae Lower field edge: Just below the perineum Lateral field edges: Close the anterior abdominal wall and the back Field size = field width × field height: 16 cm × 29 cm = 464 cm^2		Upper field edge: Above mamillae Lower field edge: Upper third of thighs Lateral field edges: Distant to the anterior abdominal wall and the back, both arms included Field size = field width × field height: 26 cm × 38 cm = 988 cm^2	

8–12 13–17 18	This technique is usually not applied in these age groups.				

Table 12.24 Normalized organ doses—Abdomen ap—24th gestation week

Beam projection:			ap									**Table cells:**
Focus-to-detector distance (cm):			100									(+) = normal font = optimal field setting
Tube voltage (kVp):			50									**(−) = bold font = suboptimal field setting**
Total filtrations:			See last table row.									Empty table cells correspond to values <0.01
Organ or tissue	**Organ dose D_T : Entrance dose K_E**							**(mSv * mGy^{-1})**				
	(+)	(−)	(+)	(−)	(+)	(−)	(+)	(−)	(+)	(−)	(+)	(−)
Skull												
Brain												
Oral mucosa												
Salivary glands												
Cervical spine												
Thyroid												
Thymus												**0.01**
Thoracic spine	0.02	**0.03**	0.02	**0.04**	0.02	**0.05**	0.03	**0.05**	0.03	**0.06**	0.04	**0.07**
Claviculae												
Scapulae		**0.03**		**0.04**		**0.04**		**0.05**		**0.05**		**0.06**
Ribs	0.05	**0.10**	0.06	**0.11**	0.06	**0.12**	0.06	**0.13**	0.07	**0.14**	0.07	**0.15**
Oesophagus	0.01	**0.02**	0.02	**0.03**	0.02	**0.03**	0.02	**0.03**	0.02	**0.04**	0.02	**0.04**
Trachea												
Lung		**0.03**	0.01	**0.04**	0.01	**0.04**	0.01	**0.04**	0.01	**0.05**	0.01	**0.05**
Mamma		**0.09**		**0.09**		**0.09**		**0.09**		**0.10**		**0.10**
Heart		**0.05**	0.01	**0.06**	0.01	**0.06**	0.01	**0.07**	0.02	**0.07**	0.02	**0.08**
Stomach	0.05	**0.05**	0.06	**0.06**	0.06	**0.07**	0.07	**0.07**	0.07	**0.08**	0.08	**0.08**
Spleen	0.03	**0.03**	0.03	**0.03**	0.04	**0.04**	0.04	**0.04**	0.04	**0.05**	0.05	**0.05**
Liver	0.04	**0.05**	0.05	**0.06**	0.06	**0.06**	0.06	**0.06**	0.07	**0.07**	0.07	**0.08**
Gall bladder	0.05	**0.05**	0.06	**0.06**	0.06	**0.06**	0.07	**0.07**	0.07	**0.07**	0.08	**0.08**
Pancreas	0.03	**0.03**	0.04	**0.04**	0.04	**0.05**	0.05	**0.05**	0.05	**0.06**	0.06	**0.06**
Small intestine	0.04	**0.04**	0.05	**0.05**	0.05	**0.05**	0.06	**0.06**	0.06	**0.07**	0.07	**0.07**
Large intestine	0.05	**0.05**	0.05	**0.06**	0.06	**0.06**	0.06	**0.07**	0.07	**0.07**	0.08	**0.08**
Sigma/rectum	0.04	**0.04**	0.05	**0.05**	0.05	**0.05**	0.05	**0.06**	0.06	**0.06**	0.07	**0.07**
Lumbar spine	0.05	**0.05**	0.06	**0.06**	0.07	**0.07**	0.08	**0.08**	0.10	**0.10**	0.11	**0.11**
Pelvis	0.07	**0.07**	0.09	**0.09**	0.10	**0.11**	0.11	**0.12**	0.13	**0.14**	0.15	**0.15**
Adrenals	0.02	**0.02**	0.02	**0.02**	0.02	**0.03**	0.03	**0.03**	0.03	**0.04**	0.04	**0.04**
Kidneys	0.02	**0.02**	0.02	**0.02**	0.03	**0.03**	0.03	**0.03**	0.03	**0.04**	0.04	**0.04**
Urinary bladder	0.05	**0.06**	0.06	**0.06**	0.07	**0.07**	0.07	**0.07**	0.08	**0.08**	0.08	**0.09**
Uterus	0.04	**0.04**	0.05	**0.05**	0.05	**0.06**	0.06	**0.06**	0.07	**0.07**	0.07	**0.08**
Ovaries	0.04	**0.03**	0.04	**0.04**	0.05	**0.05**	0.05	**0.05**	0.06	**0.06**	0.06	**0.07**
Prostate	0.04	**0.04**	0.05	**0.05**	0.06	**0.06**	0.06	**0.06**	0.07	**0.07**	0.07	**0.07**
Testes	0.08	**0.09**	0.09	**0.09**	0.09	**0.10**	0.09	**0.10**	0.09	**0.10**	0.10	**0.10**
Upper arm		**0.03**		**0.04**		**0.05**		**0.05**		**0.05**		**0.06**
Elbow		**0.12**		**0.14**		**0.16**		**0.17**		**0.19**		**0.20**
Forearm		**0.13**		**0.16**		**0.17**		**0.18**		**0.20**		**0.22**
Upper leg	0.06	**0.10**	0.07	**0.12**	0.08	**0.14**	0.09	**0.15**	0.10	**0.17**	0.11	**0.19**
Knee		**0.04**		**0.04**		**0.05**		**0.05**		**0.06**		**0.07**
Lower leg												
Lymph nodes	0.02	**0.03**	0.03	**0.03**	0.03	**0.04**	0.03	**0.04**	0.04	**0.05**	0.04	**0.05**
Skin	0.02	**0.03**	0.02	**0.03**	0.02	**0.03**	0.02	**0.03**	0.02	**0.03**	0.02	**0.03**
Muscle	0.02	**0.03**	0.02	**0.03**	0.02	**0.03**	0.02	**0.03**	0.03	**0.04**	0.03	**0.04**
Skeleton	0.02	**0.05**	0.03	**0.06**	0.03	**0.07**	0.03	**0.07**	0.04	**0.08**	0.04	**0.09**
Red bone marrow		**0.01**		**0.01**		**0.02**		**0.02**	0.01	**0.02**	0.01	**0.02**
Whole body dose	0.02	**0.03**	0.02	**0.03**	0.02	**0.03**	0.02	**0.04**	0.03	**0.04**	0.03	**0.04**
ED (ICRP 60)	0.03	**0.04**	0.04	**0.05**	0.04	**0.05**	0.04	**0.05**	0.05	**0.06**	0.05	**0.06**
ED (ICRP 103)	0.03	**0.04**	0.03	**0.05**	0.03	**0.05**	0.03	**0.05**	0.04	**0.06**	0.04	**0.06**
Total filtration (mm) Al	**1.5**		**2.5**		**3.5**		**4.5**		**3.5**		**3.5**	
Cu	**0**		**0**		**0**		**0**		**0.1**		**0.2**	

Table 12.25 Normalized organ doses—Abdomen ap—24th gestation week

Beam projection:			ap									**Table cells:**
Focus-to-detector distance (cm):			100						(+) = normal font = optimal field setting			
Tube voltage (kVp):			60						**(−) = bold font = suboptimal field setting**			
Total filtrations:			See last table row.						Empty table cells correspond to values <0.01			
Organ or tissue	**Organ dose D_T : Entrance dose K_E**								**(mSv * mGy^{-1})**			
	(+)	(−)	(+)	(−)	(+)	(−)	(+)	(−)	(+)	(−)	(+)	(−)
Skull												
Brain												
Oral mucosa												
Salivary glands												
Cervical spine												
Thyroid												
Thymus										**0.01**		**0.01**
Thoracic spine	0.02	**0.04**	0.02	**0.05**	0.03	**0.05**	0.03	**0.06**	0.04	**0.07**	0.04	**0.07**
Claviculae												
Scapulae		**0.03**		**0.04**		**0.05**		**0.05**		**0.06**		**0.06**
Ribs	0.05	**0.11**	0.06	**0.12**	0.06	**0.13**	0.07	**0.14**	0.07	**0.14**	0.07	**0.15**
Oesophagus	0.01	**0.03**	0.02	**0.03**	0.02	**0.04**	0.02	**0.04**	0.02	**0.04**	0.02	**0.04**
Trachea												
Lung		**0.04**	0.01	**0.04**	0.01	**0.04**	0.01	**0.05**	0.01	**0.05**	0.01	**0.05**
Mamma		**0.09**		**0.09**		**0.09**		**0.10**		**0.10**		**0.10**
Heart	0.01	**0.05**	0.01	**0.06**	0.01	**0.07**	0.02	**0.07**	0.02	**0.08**	0.02	**0.08**
Stomach	0.06	**0.06**	0.06	**0.07**	0.07	**0.07**	0.07	**0.07**	0.08	**0.08**	0.08	**0.08**
Spleen	0.03	**0.03**	0.04	**0.04**	0.04	**0.04**	0.04	**0.05**	0.05	**0.05**	0.05	**0.05**
Liver	0.05	**0.05**	0.06	**0.06**	0.06	**0.07**	0.07	**0.07**	0.07	**0.08**	0.07	**0.08**
Gall bladder	0.05	**0.05**	0.06	**0.06**	0.07	**0.07**	0.07	**0.07**	0.08	**0.08**	0.08	**0.08**
Pancreas	0.04	**0.04**	0.04	**0.05**	0.05	**0.05**	0.05	**0.06**	0.06	**0.06**	0.06	**0.06**
Small intestine	0.05	**0.05**	0.05	**0.06**	0.06	**0.06**	0.06	**0.07**	0.07	**0.07**	0.07	**0.07**
Large intestine	0.05	**0.05**	0.06	**0.06**	0.07	**0.07**	0.07	**0.07**	0.08	**0.08**	0.08	**0.08**
Sigma/rectum	0.04	**0.04**	0.05	**0.05**	0.06	**0.06**	0.06	**0.06**	0.07	**0.07**	0.07	**0.07**
Lumbar spine	0.06	**0.06**	0.07	**0.08**	0.09	**0.09**	0.10	**0.10**	0.11	**0.12**	0.11	**0.11**
Pelvis	0.08	**0.09**	0.10	**0.11**	0.12	**0.12**	0.13	**0.13**	0.14	**0.15**	0.15	**0.15**
Adrenals	0.02	**0.02**	0.03	**0.03**	0.03	**0.03**	0.03	**0.04**	0.04	**0.04**	0.04	**0.04**
Kidneys	0.02	**0.02**	0.03	**0.03**	0.03	**0.03**	0.03	**0.04**	0.04	**0.04**	0.04	**0.04**
Urinary bladder	0.06	**0.06**	0.07	**0.07**	0.07	**0.08**	0.08	**0.08**	0.08	**0.09**	0.08	**0.09**
Uterus	0.05	**0.05**	0.06	**0.06**	0.06	**0.06**	0.07	**0.07**	0.07	**0.08**	0.07	**0.08**
Ovaries	0.04	**0.04**	0.05	**0.05**	0.05	**0.05**	0.06	**0.06**	0.06	**0.07**	0.06	**0.07**
Prostate	0.05	**0.05**	0.06	**0.06**	0.06	**0.06**	0.06	**0.07**	0.07	**0.07**	0.07	**0.07**
Testes	0.09	**0.09**	0.09	**0.09**	0.09	**0.10**	0.09	**0.10**	0.10	**0.10**	0.10	**0.10**
Upper arm		**0.04**		**0.05**		**0.05**		**0.05**		**0.06**		**0.06**
Elbow		**0.13**		**0.16**		**0.17**		**0.18**		**0.20**		**0.20**
Forearm		**0.15**		**0.17**		**0.18**		**0.20**		**0.21**		**0.22**
Upper leg	0.07	**0.11**	0.08	**0.14**	0.09	**0.15**	0.10	**0.17**	0.11	**0.19**	0.11	**0.19**
Knee		**0.04**		**0.05**		**0.06**		**0.06**		**0.07**		**0.07**
Lower leg												
Lymph nodes	0.03	**0.03**	0.03	**0.04**	0.04	**0.04**	0.04	**0.05**	0.04	**0.05**	0.04	**0.05**
Skin	0.02	**0.03**	0.02	**0.03**	0.02	**0.03**	0.02	**0.03**	0.02	**0.03**	0.02	**0.03**
Muscle	0.02	**0.03**	0.02	**0.03**	0.02	**0.03**	0.03	**0.04**	0.03	**0.04**	0.03	**0.04**
Skeleton	0.02	**0.06**	0.03	**0.07**	0.03	**0.08**	0.03	**0.08**	0.04	**0.09**	0.04	**0.09**
Red bone marrow		**0.01**		**0.02**	0.01	**0.02**	0.01	**0.02**	0.01	**0.02**	0.01	**0.02**
Whole body dose	0.02	**0.03**	0.02	**0.03**	0.02	**0.04**	0.03	**0.04**	0.03	**0.04**	0.03	**0.04**
ED (ICRP 60)	0.03	**0.05**	0.04	**0.05**	0.04	**0.06**	0.05	**0.06**	0.05	**0.06**	0.05	**0.06**
ED (ICRP 103)	0.03	**0.05**	0.03	**0.05**	0.04	**0.05**	0.04	**0.06**	0.04	**0.06**	0.04	**0.06**
Total filtration (mm) Al	**1.5**		**2.5**		**3.5**		**4.5**		**3.5**		**3.5**	
Cu	**0**		**0**		**0**		**0**		**0.1**		**0.2**	

Table 12.26 Normalized organ doses—Abdomen ap—24th gestation week

Beam projection:			ap									**Table cells:**
Focus-to-detector distance (cm):			100									(+) = normal font = optimal field setting
Tube voltage (kVp):			70									**(−) = bold font = suboptimal field setting**
Total filtrations:			See last table row.									Empty table cells correspond to values <0.01
Organ or tissue	**Organ dose D_T : Entrance dose K_E**								**(mSv * mGy^{-1})**			
	(+)	(−)	(+)	(−)	(+)	(−)	(+)	(−)	(+)	(−)	(+)	(−)
Skull												
Brain												
Oral mucosa												
Salivary glands												
Cervical spine												
Thyroid												
Thymus								**0.01**		**0.01**		**0.01**
Thoracic spine	0.02	**0.04**	0.03	**0.05**	0.03	**0.06**	0.04	**0.07**	0.04	**0.08**	0.05	**0.09**
Claviculae												**0.01**
Scapulae		**0.04**		**0.05**		**0.05**		**0.06**		**0.06**		**0.07**
Ribs	0.06	**0.11**	0.06	**0.13**	0.07	**0.13**	0.07	**0.14**	0.07	**0.15**	0.07	**0.15**
Oesophagus	0.02	**0.03**	0.02	**0.04**	0.02	**0.04**	0.02	**0.04**	0.03	**0.05**	0.03	**0.05**
Trachea												
Lung	0.01	**0.04**	0.01	**0.04**	0.01	**0.05**	0.01	**0.05**	0.02	**0.05**	0.02	**0.06**
Mamma		**0.09**		**0.09**		**0.10**		**0.10**		**0.10**		**0.10**
Heart	0.01	**0.06**	0.01	**0.07**	0.02	**0.07**	0.02	**0.07**	0.02	**0.08**	0.02	**0.08**
Stomach	0.06	**0.06**	0.07	**0.07**	0.07	**0.07**	0.08	**0.08**	0.08	**0.08**	0.08	**0.09**
Spleen	0.03	**0.04**	0.04	**0.04**	0.04	**0.05**	0.05	**0.05**	0.05	**0.06**	0.06	**0.06**
Liver	0.05	**0.06**	0.06	**0.06**	0.07	**0.07**	0.07	**0.07**	0.07	**0.08**	0.08	**0.08**
Gall bladder	0.06	**0.06**	0.07	**0.07**	0.07	**0.07**	0.08	**0.08**	0.08	**0.08**	0.09	**0.09**
Pancreas	0.04	**0.04**	0.05	**0.05**	0.05	**0.06**	0.06	**0.06**	0.06	**0.07**	0.07	**0.07**
Small intestine	0.05	**0.05**	0.06	**0.06**	0.06	**0.07**	0.07	**0.07**	0.07	**0.08**	0.08	**0.08**
Large intestine	0.06	**0.06**	0.06	**0.07**	0.07	**0.07**	0.07	**0.08**	0.08	**0.08**	0.09	**0.09**
Sigma/rectum	0.05	**0.05**	0.06	**0.06**	0.06	**0.06**	0.06	**0.07**	0.07	**0.07**	0.08	**0.08**
Lumbar spine	0.07	**0.07**	0.08	**0.09**	0.10	**0.10**	0.10	**0.11**	0.12	**0.13**	0.13	**0.14**
Pelvis	0.09	**0.10**	0.11	**0.12**	0.13	**0.13**	0.14	**0.14**	0.15	**0.16**	0.16	**0.17**
Adrenals	0.02	**0.03**	0.03	**0.03**	0.03	**0.04**	0.04	**0.04**	0.04	**0.04**	0.04	**0.05**
Kidneys	0.02	**0.03**	0.03	**0.03**	0.03	**0.04**	0.04	**0.04**	0.04	**0.04**	0.05	**0.05**
Urinary bladder	0.06	**0.07**	0.07	**0.07**	0.08	**0.08**	0.08	**0.08**	0.09	**0.09**	0.09	**0.09**
Uterus	0.05	**0.05**	0.06	**0.06**	0.07	**0.07**	0.07	**0.07**	0.08	**0.08**	0.08	**0.08**
Ovaries	0.05	**0.05**	0.05	**0.05**	0.06	**0.06**	0.06	**0.06**	0.07	**0.07**	0.07	**0.08**
Prostate	0.05	**0.05**	0.06	**0.06**	0.06	**0.07**	0.07	**0.07**	0.07	**0.08**	0.08	**0.08**
Testes	0.09	**0.09**	0.09	**0.10**	0.09	**0.10**	0.09	**0.10**	0.10	**0.10**	0.10	**0.10**
Upper arm		**0.04**		**0.05**		**0.05**		**0.06**		**0.06**		**0.06**
Elbow		**0.14**		**0.16**		**0.18**		**0.19**		**0.20**		**0.21**
Forearm		**0.16**		**0.18**		**0.19**		**0.20**		**0.21**		**0.22**
Upper leg	0.07	**0.13**	0.09	**0.15**	0.10	**0.16**	0.10	**0.17**	0.12	**0.19**	0.12	**0.21**
Knee		**0.05**		**0.05**		**0.06**		**0.06**		**0.07**		**0.07**
Lower leg												
Lymph nodes	0.03	**0.04**	0.03	**0.04**	0.04	**0.05**	0.04	**0.05**	0.04	**0.05**	0.05	**0.06**
Skin	0.02	**0.03**	0.02	**0.03**	0.02	**0.03**	0.02	**0.03**	0.02	**0.03**	0.02	**0.03**
Muscle	0.02	**0.03**	0.02	**0.03**	0.03	**0.04**	0.03	**0.04**	0.03	**0.04**	0.03	**0.04**
Skeleton	0.03	**0.06**	0.03	**0.07**	0.03	**0.08**	0.04	**0.09**	0.04	**0.09**	0.04	**0.10**
Red bone marrow		**0.02**	0.01	**0.02**	0.01	**0.02**	0.01	**0.02**	0.01	**0.03**	0.02	**0.03**
Whole body dose	0.02	**0.03**	0.02	**0.04**	0.03	**0.04**	0.03	**0.04**	0.03	**0.05**	0.03	**0.05**
ED (ICRP 60)	0.04	**0.05**	0.04	**0.05**	0.05	**0.06**	0.05	**0.06**	0.05	**0.07**	0.05	**0.07**
ED (ICRP 103)	0.03	**0.05**	0.03	**0.05**	0.04	**0.06**	0.04	**0.06**	0.04	**0.06**	0.05	**0.07**
Total filtration (mm) Al	**1.5**		**2.5**		**3.5**		**4.5**		**3.5**		**3.5**	
Cu	**0**		**0**		**0**		**0**		**0.1**		**0.2**	

Table 12.27 Normalized organ doses—Abdomen ap—28th gestation week

Beam projection:			ap									**Table cells:**
Focus-to-detector distance (cm):			100						(+) = normal font = optimal field setting			
Tube voltage (kVp):			50						**(−) = bold font = suboptimal field setting**			
Total filtrations:			See last table row.						Empty table cells correspond to values <0.01			
Organ or tissue	**Organ dose D_T : Entrance dose K_E**								**(mSv * mGy^{-1})**			
	(+)	(−)	(+)	(−)	(+)	(−)	(+)	(−)	(+)	(−)	(+)	(−)
Skull												
Brain												
Oral mucosa												
Salivary glands												
Cervical spine												
Thyroid												
Thymus		**0.03**		**0.03**		**0.03**		**0.03**		**0.04**		**0.04**
Thoracic spine	0.02	**0.04**	0.03	**0.05**	0.04	**0.06**	0.04	**0.07**	0.05	**0.09**	0.06	**0.10**
Claviculae								**0.01**		**0.01**		**0.02**
Scapulae		**0.05**		**0.06**		**0.07**		**0.08**		**0.09**	0.01	**0.11**
Ribs	0.08	**0.15**	0.09	**0.18**	0.10	**0.19**	0.10	**0.20**	0.11	**0.22**	0.12	**0.23**
Oesophagus	0.02	**0.03**	0.02	**0.04**	0.03	**0.05**	0.03	**0.05**	0.03	**0.06**	0.04	**0.06**
Trachea												
Lung	0.02	**0.05**	0.02	**0.06**	0.02	**0.07**	0.03	**0.07**	0.03	**0.08**	0.03	**0.08**
Mamma		**0.12**		**0.13**		**0.13**		**0.13**		**0.13**		**0.14**
Heart	0.03	**0.07**	0.03	**0.08**	0.03	**0.09**	0.04	**0.09**	0.04	**0.10**	0.04	**0.11**
Stomach	0.07	**0.07**	0.08	**0.08**	0.09	**0.09**	0.10	**0.10**	0.10	**0.11**	0.11	**0.12**
Spleen	0.03	**0.04**	0.04	**0.04**	0.05	**0.05**	0.05	**0.06**	0.06	**0.06**	0.07	**0.07**
Liver	0.06	**0.06**	0.07	**0.08**	0.08	**0.08**	0.09	**0.09**	0.09	**0.10**	0.10	**0.11**
Gall bladder	0.06	**0.06**	0.08	**0.08**	0.08	**0.09**	0.09	**0.09**	0.10	**0.10**	0.11	**0.11**
Pancreas	0.04	**0.05**	0.05	**0.06**	0.06	**0.06**	0.07	**0.07**	0.08	**0.08**	0.08	**0.09**
Small intestine	0.05	**0.05**	0.06	**0.07**	0.07	**0.07**	0.08	**0.08**	0.09	**0.09**	0.10	**0.10**
Large intestine	0.06	**0.06**	0.07	**0.08**	0.08	**0.08**	0.09	**0.09**	0.10	**0.10**	0.11	**0.11**
Sigma/rectum	0.05	**0.05**	0.06	**0.06**	0.07	**0.07**	0.07	**0.08**	0.08	**0.09**	0.09	**0.09**
Lumbar spine	0.06	**0.06**	0.08	**0.08**	0.09	**0.10**	0.10	**0.11**	0.13	**0.13**	0.14	**0.15**
Pelvis	0.09	**0.10**	0.12	**0.12**	0.13	**0.14**	0.15	**0.15**	0.17	**0.18**	0.20	**0.20**
Adrenals	0.02	**0.02**	0.03	**0.03**	0.03	**0.04**	0.04	**0.04**	0.04	**0.05**	0.05	**0.05**
Kidneys	0.02	**0.02**	0.03	**0.03**	0.03	**0.04**	0.04	**0.04**	0.04	**0.05**	0.05	**0.05**
Urinary bladder	0.08	**0.08**	0.09	**0.09**	0.10	**0.10**	0.10	**0.10**	0.11	**0.11**	0.12	**0.12**
Uterus	0.05	**0.06**	0.07	**0.07**	0.08	**0.08**	0.08	**0.08**	0.09	**0.09**	0.10	**0.10**
Ovaries	0.05	**0.05**	0.06	**0.06**	0.06	**0.06**	0.07	**0.07**	0.08	**0.08**	0.09	**0.09**
Prostate	0.06	**0.06**	0.07	**0.07**	0.08	**0.08**	0.08	**0.08**	0.09	**0.09**	0.10	**0.10**
Testes	0.12	**0.12**	0.13	**0.13**	0.13	**0.13**	0.14	**0.14**	0.14	**0.14**	0.14	**0.14**
Upper arm		**0.07**		**0.08**		**0.09**		**0.10**		**0.11**		**0.12**
Elbow		**0.16**		**0.19**		**0.21**		**0.23**		**0.26**		**0.28**
Forearm		**0.18**		**0.21**		**0.23**		**0.25**		**0.28**		**0.30**
Upper leg	0.09	**0.13**	0.11	**0.16**	0.12	**0.18**	0.13	**0.20**	0.15	**0.23**	0.17	**0.25**
Knee		**0.04**		**0.05**		**0.06**		**0.07**		**0.08**	0.01	**0.09**
Lower leg												
Lymph nodes	0.03	**0.04**	0.04	**0.05**	0.05	**0.05**	0.05	**0.06**	0.06	**0.07**	0.06	**0.07**
Skin	0.02	**0.04**	0.03	**0.04**	0.03	**0.04**	0.03	**0.04**	0.03	**0.04**	0.03	**0.05**
Muscle	0.03	**0.04**	0.03	**0.04**	0.03	**0.04**	0.03	**0.05**	0.04	**0.05**	0.04	**0.06**
Skeleton	0.03	**0.07**	0.04	**0.09**	0.04	**0.10**	0.05	**0.11**	0.05	**0.12**	0.06	**0.13**
Red bone marrow		**0.02**	0.01	**0.02**	0.01	**0.02**	0.01	**0.03**	0.02	**0.03**	0.02	**0.03**
Whole body dose	0.02	**0.04**	0.03	**0.04**	0.03	**0.05**	0.03	**0.05**	0.04	**0.06**	0.04	**0.06**
ED (ICRP 60)	0.04	**0.06**	0.05	**0.07**	0.06	**0.07**	0.06	**0.08**	0.07	**0.08**	0.07	**0.09**
ED (ICRP 103)	0.04	**0.06**	0.04	**0.07**	0.05	**0.07**	0.05	**0.08**	0.05	**0.08**	0.06	**0.09**
Total filtration (mm) **Al**	**1.5**		**2.5**		**3.5**		**4.5**		**3.5**		**3.5**	
Cu	**0**		**0**		**0**		**0**		**0.1**		**0.2**	

Table 12.28 Normalized organ doses—Abdomen ap—28th gestation week

Beam projection:			ap									**Table cells:**
Focus-to-detector distance (cm):			100									(+) = normal font = optimal field setting
Tube voltage (kVp):			60									**(−) = bold font = suboptimal field setting**
Total filtrations:			See last table row.									Empty table cells correspond to values <0.01
Organ or tissue	**Organ dose D_T : Entrance dose K_E**								**(mSv * mGy^{-1})**			
	(+)	(−)	(+)	(−)	(+)	(−)	(+)	(−)	(+)	(−)	(+)	(−)
Skull												
Brain												
Oral mucosa												
Salivary glands												
Cervical spine												
Thyroid												
Thymus		**0.03**		**0.03**		**0.04**		**0.04**		**0.04**		**0.04**
Thoracic spine	0.03	**0.05**	0.04	**0.07**	0.04	**0.08**	0.05	**0.09**	0.06	**0.11**	0.06	**0.10**
Claviculae				**0.01**		**0.01**		**0.01**		**0.02**		**0.02**
Scapulae		**0.06**		**0.07**		**0.08**		**0.09**	0.01	**0.11**	0.01	**0.11**
Ribs	0.09	**0.17**	0.10	**0.19**	0.11	**0.20**	0.11	**0.21**	0.12	**0.23**	0.12	**0.23**
Oesophagus	0.02	**0.04**	0.03	**0.05**	0.03	**0.05**	0.03	**0.06**	0.04	**0.07**	0.04	**0.06**
Trachea												
Lung	0.02	**0.06**	0.02	**0.07**	0.03	**0.07**	0.03	**0.08**	0.03	**0.08**	0.03	**0.08**
Mamma		**0.13**		**0.13**		**0.13**		**0.13**		**0.14**		**0.14**
Heart	0.03	**0.08**	0.03	**0.09**	0.04	**0.10**	0.04	**0.10**	0.04	**0.11**	0.04	**0.11**
Stomach	0.08	**0.08**	0.09	**0.09**	0.10	**0.10**	0.10	**0.11**	0.11	**0.12**	0.11	**0.12**
Spleen	0.04	**0.04**	0.05	**0.05**	0.05	**0.06**	0.06	**0.06**	0.07	**0.07**	0.07	**0.07**
Liver	0.07	**0.07**	0.08	**0.08**	0.09	**0.09**	0.09	**0.10**	0.10	**0.11**	0.10	**0.11**
Gall bladder	0.07	**0.07**	0.09	**0.09**	0.09	**0.10**	0.10	**0.10**	0.11	**0.11**	0.11	**0.11**
Pancreas	0.05	**0.05**	0.06	**0.07**	0.07	**0.07**	0.08	**0.08**	0.09	**0.09**	0.08	**0.09**
Small intestine	0.06	**0.06**	0.07	**0.08**	0.08	**0.08**	0.09	**0.09**	0.10	**0.10**	0.10	**0.10**
Large intestine	0.07	**0.07**	0.08	**0.09**	0.09	**0.09**	0.10	**0.10**	0.11	**0.11**	0.11	**0.11**
Sigma/rectum	0.06	**0.06**	0.07	**0.07**	0.08	**0.08**	0.08	**0.09**	0.09	**0.10**	0.09	**0.09**
Lumbar spine	0.08	**0.08**	0.10	**0.10**	0.11	**0.12**	0.13	**0.13**	0.15	**0.15**	0.14	**0.15**
Pelvis	0.11	**0.12**	0.14	**0.14**	0.16	**0.16**	0.17	**0.18**	0.20	**0.20**	0.20	**0.20**
Adrenals	0.03	**0.03**	0.03	**0.04**	0.04	**0.04**	0.04	**0.05**	0.05	**0.06**	0.05	**0.05**
Kidneys	0.03	**0.03**	0.04	**0.04**	0.04	**0.04**	0.04	**0.05**	0.05	**0.05**	0.05	**0.05**
Urinary bladder	0.08	**0.09**	0.10	**0.10**	0.10	**0.11**	0.11	**0.11**	0.12	**0.12**	0.12	**0.12**
Uterus	0.06	**0.06**	0.08	**0.08**	0.08	**0.09**	0.09	**0.09**	0.10	**0.10**	0.10	**0.10**
Ovaries	0.05	**0.05**	0.07	**0.07**	0.07	**0.07**	0.08	**0.08**	0.09	**0.09**	0.09	**0.09**
Prostate	0.07	**0.07**	0.08	**0.08**	0.08	**0.09**	0.09	**0.09**	0.10	**0.10**	0.10	**0.10**
Testes	0.12	**0.12**	0.13	**0.13**	0.14	**0.14**	0.14	**0.14**	0.14	**0.14**	0.14	**0.14**
Upper arm		**0.08**		**0.09**		**0.10**		**0.11**		**0.12**		**0.12**
Elbow		**0.18**		**0.21**		**0.23**		**0.25**		**0.28**		**0.28**
Forearm		**0.20**		**0.23**		**0.25**		**0.27**		**0.29**		**0.30**
Upper leg	0.10	**0.15**	0.12	**0.18**	0.14	**0.21**	0.15	**0.22**	0.17	**0.25**	0.17	**0.25**
Knee		**0.05**		**0.06**		**0.07**		**0.08**	0.01	**0.09**	0.01	**0.09**
Lower leg												
Lymph nodes	0.04	**0.05**	0.05	**0.05**	0.05	**0.06**	0.05	**0.06**	0.06	**0.07**	0.06	**0.07**
Skin	0.03	**0.04**	0.03	**0.04**	0.03	**0.04**	0.03	**0.04**	0.03	**0.05**	0.03	**0.05**
Muscle	0.03	**0.04**	0.03	**0.05**	0.04	**0.05**	0.04	**0.05**	0.04	**0.06**	0.04	**0.06**
Skeleton	0.03	**0.08**	0.04	**0.10**	0.05	**0.11**	0.05	**0.12**	0.06	**0.13**	0.06	**0.13**
Red bone marrow	0.01	**0.02**	0.01	**0.02**	0.02	**0.03**	0.02	**0.03**	0.02	**0.03**	0.02	**0.03**
Whole body dose	0.03	**0.04**	0.03	**0.05**	0.04	**0.05**	0.04	**0.06**	0.04	**0.06**	0.04	**0.06**
ED (ICRP 60)	0.05	**0.06**	0.06	**0.07**	0.06	**0.08**	0.07	**0.08**	0.07	**0.09**	0.07	**0.09**
ED (ICRP 103)	0.04	**0.06**	0.05	**0.07**	0.05	**0.08**	0.05	**0.08**	0.06	**0.09**	0.06	**0.09**
Total filtration (mm) Al	**1.5**		**2.5**		**3.5**		**4.5**		**3.5**		**3.5**	
Cu	**0**		**0**		**0**		**0**		**0.1**		**0.2**	

Table 12.29 Normalized organ doses—Abdomen ap—28th gestation week

Beam projection:			ap									**Table cells:**
Focus-to-detector distance (cm):			100									(+) = normal font = optimal field setting
Tube voltage (kVp):			70									**(−) = bold font = suboptimal field setting**
Total filtrations:			See last table row.									Empty table cells correspond to values <0.01
Organ or tissue	**Organ dose D_T : Entrance dose K_E**								**(mSv * mGy^-1)**			
	(+)	(−)	(+)	(−)	(+)	(−)	(+)	(−)	(+)	(−)	(+)	(−)
Skull												
Brain												
Oral mucosa												
Salivary glands												
Cervical spine												
Thyroid												
Thymus		**0.03**		**0.03**		**0.04**		**0.04**		**0.04**		**0.04**
Thoracic spine	0.03	**0.06**	0.04	**0.08**	0.05	**0.09**	0.05	**0.10**	0.06	**0.11**	0.07	**0.13**
Claviculae				**0.01**		**0.01**		**0.01**		**0.02**		**0.02**
Scapulae		**0.07**		**0.08**		**0.09**	0.01	**0.10**	0.01	**0.11**	0.01	**0.12**
Ribs	0.09	**0.18**	0.10	**0.20**	0.11	**0.21**	0.11	**0.22**	0.12	**0.23**	0.12	**0.24**
Oesophagus	0.03	**0.04**	0.03	**0.05**	0.03	**0.06**	0.04	**0.06**	0.04	**0.07**	0.05	**0.08**
Trachea												
Lung	0.02	**0.06**	0.03	**0.07**	0.03	**0.08**	0.03	**0.08**	0.03	**0.09**	0.03	**0.09**
Mamma		**0.13**		**0.13**		**0.13**		**0.14**		**0.14**		**0.14**
Heart	0.03	**0.08**	0.04	**0.10**	0.04	**0.10**	0.04	**0.11**	0.05	**0.12**	0.05	**0.12**
Stomach	0.08	**0.09**	0.10	**0.10**	0.10	**0.11**	0.11	**0.11**	0.12	**0.12**	0.12	**0.13**
Spleen	0.04	**0.05**	0.05	**0.06**	0.06	**0.07**	0.07	**0.07**	0.07	**0.08**	0.08	**0.09**
Liver	0.08	**0.08**	0.09	**0.09**	0.09	**0.10**	0.10	**0.10**	0.11	**0.11**	0.11	**0.12**
Gall bladder	0.08	**0.08**	0.09	**0.09**	0.10	**0.10**	0.11	**0.11**	0.12	**0.12**	0.12	**0.13**
Pancreas	0.06	**0.06**	0.07	**0.07**	0.08	**0.08**	0.08	**0.09**	0.09	**0.10**	0.10	**0.11**
Small intestine	0.07	**0.07**	0.08	**0.08**	0.09	**0.09**	0.09	**0.10**	0.10	**0.11**	0.11	**0.12**
Large intestine	0.08	**0.08**	0.09	**0.09**	0.10	**0.10**	0.10	**0.11**	0.11	**0.12**	0.12	**0.13**
Sigma/rectum	0.06	**0.07**	0.08	**0.08**	0.08	**0.09**	0.09	**0.09**	0.10	**0.10**	0.11	**0.11**
Lumbar spine	0.09	**0.09**	0.11	**0.12**	0.13	**0.13**	0.14	**0.15**	0.16	**0.17**	0.18	**0.19**
Pelvis	0.12	**0.13**	0.15	**0.16**	0.17	**0.18**	0.18	**0.19**	0.21	**0.22**	0.22	**0.24**
Adrenals	0.03	**0.03**	0.04	**0.04**	0.04	**0.05**	0.05	**0.05**	0.06	**0.06**	0.06	**0.07**
Kidneys	0.03	**0.03**	0.04	**0.04**	0.05	**0.05**	0.05	**0.05**	0.06	**0.06**	0.06	**0.07**
Urinary bladder	0.09	**0.09**	0.10	**0.10**	0.11	**0.11**	0.12	**0.12**	0.12	**0.13**	0.13	**0.13**
Uterus	0.07	**0.07**	0.08	**0.08**	0.09	**0.09**	0.10	**0.10**	0.11	**0.11**	0.12	**0.12**
Ovaries	0.06	**0.06**	0.07	**0.07**	0.08	**0.08**	0.09	**0.09**	0.10	**0.10**	0.10	**0.11**
Prostate	0.07	**0.07**	0.08	**0.08**	0.09	**0.09**	0.10	**0.10**	0.10	**0.11**	0.11	**0.11**
Testes	0.13	**0.13**	0.13	**0.14**	0.14	**0.14**	0.14	**0.14**	0.14	**0.15**	0.15	**0.15**
Upper arm		**0.09**		**0.10**		**0.11**		**0.12**		**0.13**		**0.13**
Elbow		**0.20**		**0.23**		**0.25**		**0.26**		**0.28**		**0.30**
Forearm		**0.21**		**0.24**		**0.26**		**0.28**		**0.30**		**0.31**
Upper leg	0.11	**0.17**	0.13	**0.20**	0.15	**0.22**	0.16	**0.24**	0.18	**0.26**	0.19	**0.28**
Knee		**0.06**		**0.07**		**0.08**	0.01	**0.08**	0.01	**0.09**	0.01	**0.10**
Lower leg												
Lymph nodes	0.04	**0.05**	0.05	**0.06**	0.05	**0.06**	0.06	**0.07**	0.06	**0.08**	0.07	**0.08**
Skin	0.03	**0.04**	0.03	**0.04**	0.03	**0.04**	0.03	**0.05**	0.03	**0.05**	0.03	**0.05**
Muscle	0.03	**0.04**	0.03	**0.05**	0.04	**0.05**	0.04	**0.05**	0.04	**0.06**	0.05	**0.06**
Skeleton	0.04	**0.09**	0.05	**0.11**	0.05	**0.12**	0.05	**0.12**	0.06	**0.14**	0.07	**0.15**
Red bone marrow	0.01	**0.02**	0.02	**0.03**	0.02	**0.03**	0.02	**0.03**	0.02	**0.04**	0.02	**0.04**
Whole body dose	0.03	**0.05**	0.04	**0.05**	0.04	**0.06**	0.04	**0.06**	0.04	**0.07**	0.05	**0.07**
ED (ICRP 60)	0.05	**0.07**	0.06	**0.08**	0.07	**0.08**	0.07	**0.09**	0.07	**0.09**	0.08	**0.10**
ED (ICRP 103)	0.04	**0.07**	0.05	**0.08**	0.05	**0.08**	0.06	**0.09**	0.06	**0.09**	0.07	**0.10**
Total filtration (mm) **Al**	**1.5**		**2.5**		**3.5**		**4.5**		**3.5**		**3.5**	
Cu	**0**		**0**		**0**		**0**		**0.1**		**0.2**	

Table 12.30 Normalized organ doses—Abdomen ap—<1 year

Beam projection:			ap									**Table cells:**
Focus-to-detector distance (cm):			100						(+) = normal font = optimal field setting			
Tube voltage (kVp):			60						**(−) = bold font = suboptimal field setting**			
Total filtrations:			See last table row.						Empty table cells correspond to values <0.01			
Organ or tissue	**Organ dose D_T : Entrance dose K_E**							**(mSv * mGy^{-1})**				
	(+)	(−)	(+)	(−)	(+)	(−)	(+)	(−)	(+)	(−)	(+)	(−)
Skull												
Brain												
Oral mucosa												
Salivary glands						**0.01**		**0.01**		**0.01**		**0.01**
Cervical spine		**0.02**		**0.03**	0.01	**0.03**	0.01	**0.04**	0.02	**0.05**	0.01	**0.04**
Thyroid		**0.01**		**0.02**		**0.02**		**0.02**	0.01	**0.03**		**0.02**
Thymus	0.01	**0.08**	0.02	**0.10**	0.02	**0.11**	0.02	**0.12**	0.03	**0.14**	0.02	**0.14**
Thoracic spine	0.14	**0.24**	0.19	**0.31**	0.22	**0.38**	0.25	**0.43**	0.31	**0.53**	0.29	**0.48**
Claviculae	0.01	**0.05**	0.02	**0.06**	0.02	**0.07**	0.03	**0.08**	0.03	**0.10**	0.03	**0.09**
Scapulae	0.05	**0.34**	0.06	**0.44**	0.07	**0.51**	0.08	**0.57**	0.10	**0.68**	0.09	**0.65**
Ribs	0.61	**1.05**	0.70	**1.20**	0.76	**1.31**	0.81	**1.38**	0.88	**1.51**	0.88	**1.51**
Oesophagus	0.12	**0.18**	0.15	**0.24**	0.18	**0.27**	0.20	**0.31**	0.23	**0.37**	0.22	**0.35**
Trachea												
Lung	0.15	**0.35**	0.19	**0.42**	0.21	**0.47**	0.23	**0.51**	0.26	**0.58**	0.25	**0.56**
Mamma	0.03	**1.15**	0.04	**1.20**	0.05	**1.23**	0.06	**1.24**	0.07	**1.27**	0.07	**1.27**
Heart	0.21	**0.54**	0.25	**0.65**	0.29	**0.73**	0.31	**0.78**	0.35	**0.88**	0.35	**0.87**
Stomach	0.59	**0.61**	0.71	**0.73**	0.79	**0.82**	0.85	**0.88**	0.96	**1.00**	0.94	**0.98**
Spleen	0.21	**0.23**	0.27	**0.30**	0.32	**0.35**	0.36	**0.39**	0.42	**0.47**	0.40	**0.45**
Liver	0.51	**0.54**	0.61	**0.65**	0.68	**0.72**	0.74	**0.78**	0.83	**0.88**	0.82	**0.87**
Gall bladder	0.44	**0.46**	0.55	**0.57**	0.63	**0.65**	0.69	**0.72**	0.80	**0.83**	0.78	**0.81**
Pancreas	0.30	**0.33**	0.39	**0.42**	0.45	**0.48**	0.50	**0.54**	0.59	**0.64**	0.57	**0.61**
Small intestine	0.40	**0.41**	0.50	**0.51**	0.57	**0.59**	0.63	**0.64**	0.73	**0.75**	0.71	**0.73**
Large intestine	0.48	**0.49**	0.59	**0.61**	0.67	**0.69**	0.73	**0.75**	0.84	**0.87**	0.82	**0.85**
Sigma/rectum	0.37	**0.38**	0.46	**0.48**	0.53	**0.55**	0.58	**0.60**	0.68	**0.71**	0.65	**0.68**
Lumbar spine	0.34	**0.36**	0.45	**0.47**	0.53	**0.56**	0.61	**0.64**	0.75	**0.79**	0.69	**0.73**
Pelvis	0.57	**0.60**	0.74	**0.78**	0.86	**0.91**	0.96	**1.01**	1.14	**1.21**	1.10	**1.16**
Adrenals	0.13	**0.14**	0.16	**0.19**	0.19	**0.22**	0.22	**0.25**	0.27	**0.30**	0.25	**0.28**
Kidneys	0.13	**0.14**	0.17	**0.19**	0.21	**0.22**	0.23	**0.25**	0.29	**0.31**	0.26	**0.28**
Urinary bladder	0.61	**0.62**	0.73	**0.75**	0.82	**0.83**	0.88	**0.90**	0.99	**1.01**	0.98	**0.99**
Uterus	0.40	**0.41**	0.50	**0.52**	0.57	**0.59**	0.63	**0.66**	0.74	**0.77**	0.71	**0.74**
Ovaries	0.33	**0.34**	0.42	**0.43**	0.49	**0.51**	0.54	**0.56**	0.64	**0.67**	0.62	**0.64**
Prostate	0.47	**0.49**	0.57	**0.60**	0.65	**0.68**	0.71	**0.74**	0.81	**0.85**	0.79	**0.84**
Testes	1.04	**1.07**	1.13	**1.17**	1.19	**1.23**	1.23	**1.27**	1.29	**1.34**	1.28	**1.34**
Upper arm		**0.43**		**0.53**		**0.60**		**0.65**		**0.74**		**0.73**
Elbow		**1.17**		**1.42**		**1.59**		**1.72**		**1.94**		**1.94**
Forearm		**1.29**		**1.55**		**1.73**		**1.87**		**2.10**		**2.10**
Upper leg	0.72	**0.96**	0.90	**1.20**	1.03	**1.38**	1.14	**1.52**	1.32	**1.77**	1.30	**1.74**
Knee	0.05	**0.45**	0.07	**0.56**	0.09	**0.64**	0.10	**0.71**	0.12	**0.82**	0.11	**0.80**
Lower leg				**0.01**		**0.02**		**0.02**	0.01	**0.02**		**0.02**
Lymph nodes	0.25	**0.30**	0.31	**0.37**	0.36	**0.42**	0.39	**0.46**	0.45	**0.53**	0.44	**0.51**
Skin	0.22	**0.33**	0.24	**0.35**	0.25	**0.36**	0.26	**0.37**	0.27	**0.39**	0.27	**0.39**
Muscle	0.23	**0.31**	0.27	**0.36**	0.29	**0.40**	0.32	**0.43**	0.36	**0.48**	0.35	**0.47**
Skeleton	0.23	**0.54**	0.29	**0.66**	0.33	**0.74**	0.36	**0.81**	0.42	**0.93**	0.40	**0.91**
Red bone marrow	0.06	**0.13**	0.07	**0.16**	0.09	**0.18**	0.09	**0.20**	0.11	**0.24**	0.10	**0.22**
Whole body dose	0.21	**0.31**	0.25	**0.36**	0.28	**0.40**	0.30	**0.43**	0.34	**0.49**	0.33	**0.48**
ED (ICRP 60)	0.36	**0.47**	0.43	**0.55**	0.48	**0.60**	0.51	**0.65**	0.57	**0.72**	0.56	**0.71**
ED (ICRP 103)	0.29	**0.47**	0.35	**0.55**	0.39	**0.60**	0.42	**0.64**	0.48	**0.71**	0.46	**0.70**
Total filtration (mm) **Al**	**1.5**		**2.5**		**3.5**		**4.5**		**3.5**		**3.5**	
Cu	**0**		**0**		**0**		**0**		**0.1**		**0.2**	

Table 12.31 Normalized organ doses—Abdomen ap—<1 year

Beam projection:			ap									**Table cells:**
Focus-to-detector distance (cm):			100									(+) = normal font = optimal field setting
Tube voltage (kVp):			70									**(−) = bold font = suboptimal field setting**
Total filtrations:			See last table row.									Empty table cells correspond to values <0.01
Organ or tissue	**Organ dose D_T : Entrance dose K_E**								**(mSv * mGy^{-1})**			
	(+)	(−)	(+)	(−)	(+)	(−)	(+)	(−)	(+)	(−)	(+)	(−)
Skull										**0.01**		**0.01**
Brain												
Oral mucosa										**0.01**		**0.01**
Salivary glands				**0.01**		**0.01**		**0.01**		**0.02**		**0.02**
Cervical spine		**0.03**	0.01	**0.03**	0.01	**0.04**	0.02	**0.05**	0.02	**0.06**	0.03	**0.07**
Thyroid		**0.02**		**0.02**		**0.03**	0.01	**0.03**	0.01	**0.04**	0.02	**0.04**
Thymus	0.01	**0.09**	0.02	**0.11**	0.02	**0.13**	0.03	**0.14**	0.03	**0.16**	0.04	**0.17**
Thoracic spine	0.18	**0.30**	0.23	**0.39**	0.27	**0.45**	0.30	**0.51**	0.37	**0.61**	0.43	**0.71**
Claviculae	0.02	**0.06**	0.02	**0.07**	0.03	**0.09**	0.03	**0.10**	0.04	**0.12**	0.05	**0.14**
Scapulae	0.06	**0.40**	0.07	**0.51**	0.09	**0.58**	0.10	**0.64**	0.12	**0.75**	0.14	**0.84**
Ribs	0.66	**1.12**	0.74	**1.27**	0.80	**1.37**	0.84	**1.44**	0.91	**1.56**	0.96	**1.64**
Oesophagus	0.14	**0.22**	0.18	**0.28**	0.20	**0.32**	0.23	**0.35**	0.27	**0.42**	0.30	**0.47**
Trachea										**0.01**		**0.01**
Lung	0.17	**0.39**	0.21	**0.47**	0.23	**0.52**	0.25	**0.55**	0.28	**0.62**	0.31	**0.68**
Mamma	0.04	**1.17**	0.05	**1.22**	0.06	**1.24**	0.07	**1.26**	0.08	**1.28**	0.09	**1.29**
Heart	0.24	**0.61**	0.28	**0.72**	0.31	**0.79**	0.34	**0.85**	0.38	**0.95**	0.42	**1.03**
Stomach	0.66	**0.68**	0.78	**0.81**	0.86	**0.89**	0.92	**0.96**	1.03	**1.07**	1.11	**1.17**
Spleen	0.26	**0.28**	0.32	**0.36**	0.37	**0.41**	0.41	**0.46**	0.48	**0.54**	0.55	**0.61**
Liver	0.57	**0.60**	0.68	**0.71**	0.75	**0.79**	0.80	**0.85**	0.90	**0.95**	0.97	**1.04**
Gall bladder	0.51	**0.53**	0.63	**0.65**	0.71	**0.73**	0.77	**0.80**	0.88	**0.92**	0.98	**1.02**
Pancreas	0.36	**0.39**	0.45	**0.48**	0.51	**0.56**	0.57	**0.61**	0.66	**0.72**	0.74	**0.81**
Small intestine	0.46	**0.48**	0.57	**0.58**	0.64	**0.66**	0.70	**0.72**	0.81	**0.84**	0.90	**0.93**
Large intestine	0.55	**0.56**	0.66	**0.68**	0.74	**0.77**	0.81	**0.84**	0.92	**0.95**	1.01	**1.05**
Sigma/rectum	0.43	**0.45**	0.52	**0.55**	0.59	**0.62**	0.65	**0.68**	0.75	**0.79**	0.83	**0.88**
Lumbar spine	0.43	**0.45**	0.55	**0.58**	0.64	**0.68**	0.72	**0.77**	0.87	**0.93**	1.02	**1.08**
Pelvis	0.68	**0.72**	0.85	**0.90**	0.98	**1.04**	1.08	**1.15**	1.27	**1.35**	1.44	**1.52**
Adrenals	0.16	**0.18**	0.20	**0.23**	0.23	**0.26**	0.26	**0.29**	0.31	**0.35**	0.36	**0.41**
Kidneys	0.17	**0.18**	0.21	**0.23**	0.25	**0.27**	0.28	**0.30**	0.34	**0.36**	0.39	**0.42**
Urinary bladder	0.68	**0.70**	0.81	**0.82**	0.89	**0.91**	0.95	**0.97**	1.06	**1.08**	1.15	**1.17**
Uterus	0.46	**0.48**	0.57	**0.59**	0.65	**0.67**	0.71	**0.74**	0.82	**0.85**	0.91	**0.95**
Ovaries	0.39	**0.41**	0.49	**0.51**	0.56	**0.58**	0.61	**0.64**	0.72	**0.75**	0.80	**0.84**
Prostate	0.53	**0.56**	0.64	**0.67**	0.72	**0.76**	0.78	**0.82**	0.88	**0.93**	0.97	**1.02**
Testes	1.08	**1.12**	1.17	**1.22**	1.22	**1.28**	1.26	**1.32**	1.32	**1.39**	1.36	**1.44**
Upper arm		**0.49**		**0.58**		**0.65**		**0.70**		**0.78**		**0.85**
Elbow		**1.30**		**1.55**		**1.71**		**1.84**		**2.04**		**2.20**
Forearm		**1.43**		**1.68**		**1.85**		**1.98**		**2.19**		**2.35**
Upper leg	0.83	**1.11**	1.01	**1.36**	1.15	**1.53**	1.25	**1.67**	1.43	**1.91**	1.59	**2.11**
Knee	0.07	**0.52**	0.09	**0.63**	0.10	**0.71**	0.12	**0.77**	0.14	**0.88**	0.17	**0.97**
Lower leg		**0.01**		**0.02**		**0.02**	0.01	**0.02**	0.01	**0.03**	0.02	**0.03**
Lymph nodes	0.29	**0.34**	0.35	**0.41**	0.40	**0.47**	0.44	**0.51**	0.50	**0.58**	0.55	**0.65**
Skin	0.23	**0.34**	0.25	**0.36**	0.26	**0.38**	0.27	**0.39**	0.29	**0.41**	0.30	**0.43**
Muscle	0.25	**0.34**	0.29	**0.40**	0.32	**0.44**	0.34	**0.47**	0.38	**0.52**	0.42	**0.57**
Skeleton	0.27	**0.61**	0.32	**0.73**	0.36	**0.81**	0.40	**0.88**	0.45	**0.99**	0.50	**1.09**
Red bone marrow	0.07	**0.15**	0.09	**0.19**	0.10	**0.21**	0.11	**0.23**	0.13	**0.27**	0.15	**0.30**
Whole body dose	0.24	**0.34**	0.28	**0.40**	0.31	**0.44**	0.33	**0.47**	0.37	**0.53**	0.40	**0.57**
ED (ICRP 60)	0.40	**0.52**	0.47	**0.60**	0.52	**0.66**	0.55	**0.70**	0.62	**0.78**	0.67	**0.84**
ED (ICRP 103)	0.32	**0.52**	0.38	**0.59**	0.43	**0.65**	0.46	**0.69**	0.52	**0.76**	0.56	**0.82**
Total filtration (mm) **Al**	**1.5**		**2.5**		**3.5**		**4.5**		**3.5**		**3.5**	
Cu	**0**		**0**		**0**		**0**		**0.1**		**0.2**	

Table 12.32 Normalized organ doses—Abdomen ap—1–2 years

Beam projection:			ap									**Table cells:**
Focus-to-detector distance (cm):			100				(+) = normal font = optimal field setting					
Tube voltage (kVp):			60				**(−) = bold font = suboptimal field setting**					
Total filtrations:			See last table row.				Empty table cells correspond to values <0.01					
Organ or tissue	**Organ dose D_T : Entrance dose K_E**							**(mSv * mGy^{-1})**				
	(+)	(−)	(+)	(−)	(+)	(−)	(+)	(−)	(+)	(−)	(+)	(−)
Skull												
Brain												
Oral mucosa												
Salivary glands												
Cervical spine		**0.01**		**0.02**		**0.02**		**0.02**		**0.03**		**0.02**
Thyroid				**0.01**		**0.01**		**0.02**		**0.02**		**0.02**
Thymus		**0.11**	0.01	**0.13**	0.01	**0.15**	0.02	**0.16**	0.02	**0.19**	0.02	**0.18**
Thoracic spine	0.08	**0.15**	0.11	**0.20**	0.13	**0.24**	0.15	**0.28**	0.19	**0.36**	0.16	**0.31**
Claviculae		**0.03**		**0.05**	0.01	**0.06**	0.01	**0.06**	0.02	**0.08**	0.02	**0.07**
Scapulae	0.02	**0.19**	0.03	**0.25**	0.04	**0.30**	0.04	**0.34**	0.05	**0.42**	0.04	**0.39**
Ribs	0.49	**0.98**	0.57	**1.15**	0.62	**1.26**	0.67	**1.34**	0.74	**1.49**	0.73	**1.48**
Oesophagus	0.09	**0.16**	0.12	**0.21**	0.14	**0.25**	0.16	**0.28**	0.20	**0.34**	0.18	**0.31**
Trachea												
Lung	0.10	**0.29**	0.12	**0.35**	0.14	**0.40**	0.15	**0.44**	0.17	**0.51**	0.17	**0.49**
Mamma	0.02	**1.11**	0.02	**1.18**	0.03	**1.22**	0.03	**1.25**	0.04	**1.30**	0.03	**1.29**
Heart	0.13	**0.47**	0.17	**0.58**	0.19	**0.66**	0.21	**0.72**	0.25	**0.83**	0.24	**0.80**
Stomach	0.49	**0.51**	0.60	**0.62**	0.67	**0.71**	0.73	**0.77**	0.84	**0.89**	0.82	**0.86**
Spleen	0.15	**0.17**	0.19	**0.22**	0.23	**0.26**	0.26	**0.29**	0.32	**0.36**	0.29	**0.33**
Liver	0.40	**0.43**	0.50	**0.53**	0.56	**0.61**	0.62	**0.67**	0.71	**0.77**	0.69	**0.74**
Gall bladder	0.40	**0.41**	0.50	**0.52**	0.58	**0.61**	0.65	**0.67**	0.77	**0.80**	0.73	**0.76**
Pancreas	0.23	**0.25**	0.29	**0.32**	0.35	**0.38**	0.39	**0.43**	0.47	**0.52**	0.44	**0.48**
Small intestine	0.32	**0.33**	0.41	**0.42**	0.47	**0.49**	0.53	**0.54**	0.63	**0.65**	0.59	**0.61**
Large intestine	0.39	**0.40**	0.49	**0.51**	0.57	**0.58**	0.63	**0.65**	0.74	**0.76**	0.71	**0.73**
Sigma/rectum	0.28	**0.29**	0.36	**0.37**	0.42	**0.44**	0.47	**0.49**	0.57	**0.58**	0.53	**0.55**
Lumbar spine	0.21	**0.22**	0.28	**0.30**	0.34	**0.36**	0.39	**0.41**	0.50	**0.53**	0.43	**0.45**
Pelvis	0.40	**0.43**	0.53	**0.56**	0.63	**0.66**	0.71	**0.75**	0.87	**0.92**	0.80	**0.85**
Adrenals	0.08	**0.10**	0.11	**0.13**	0.13	**0.16**	0.15	**0.18**	0.19	**0.22**	0.17	**0.20**
Kidneys	0.08	**0.09**	0.11	**0.12**	0.14	**0.15**	0.15	**0.17**	0.19	**0.21**	0.17	**0.19**
Urinary bladder	0.54	**0.54**	0.66	**0.66**	0.74	**0.75**	0.81	**0.82**	0.92	**0.94**	0.90	**0.91**
Uterus	0.32	**0.33**	0.42	**0.43**	0.49	**0.50**	0.54	**0.55**	0.65	**0.66**	0.61	**0.63**
Ovaries	0.25	**0.26**	0.32	**0.33**	0.38	**0.39**	0.43	**0.44**	0.52	**0.54**	0.48	**0.50**
Prostate	0.36	**0.37**	0.45	**0.46**	0.52	**0.54**	0.58	**0.59**	0.68	**0.70**	0.65	**0.67**
Testes	0.99	**0.99**	1.09	**1.10**	1.15	**1.16**	1.20	**1.21**	1.27	**1.28**	1.26	**1.27**
Upper arm		**0.31**		**0.39**		**0.45**		**0.50**		**0.58**		**0.56**
Elbow		**0.85**		**1.06**		**1.21**		**1.33**		**1.55**		**1.51**
Forearm		**0.96**		**1.19**		**1.36**		**1.49**		**1.72**		**1.69**
Upper leg	0.47	**0.61**	0.61	**0.79**	0.72	**0.93**	0.81	**1.04**	0.97	**1.25**	0.92	**1.18**
Knee	0.03	**0.09**	0.04	**0.12**	0.05	**0.14**	0.05	**0.16**	0.07	**0.20**	0.06	**0.18**
Lower leg												
Lymph nodes	0.20	**0.24**	0.25	**0.31**	0.29	**0.35**	0.33	**0.39**	0.39	**0.46**	0.37	**0.44**
Skin	0.20	**0.28**	0.21	**0.30**	0.22	**0.31**	0.23	**0.32**	0.24	**0.34**	0.24	**0.33**
Muscle	0.17	**0.23**	0.20	**0.27**	0.23	**0.30**	0.25	**0.32**	0.28	**0.37**	0.27	**0.36**
Skeleton	0.15	**0.35**	0.19	**0.44**	0.22	**0.50**	0.24	**0.56**	0.29	**0.65**	0.27	**0.62**
Red bone marrow	0.04	**0.09**	0.06	**0.11**	0.07	**0.13**	0.07	**0.14**	0.09	**0.17**	0.08	**0.16**
Whole body dose	0.16	**0.23**	0.19	**0.28**	0.22	**0.31**	0.24	**0.34**	0.27	**0.39**	0.26	**0.37**
ED (ICRP 60)	0.30	**0.40**	0.36	**0.47**	0.41	**0.53**	0.44	**0.57**	0.50	**0.65**	0.48	**0.62**
ED (ICRP 103)	0.23	**0.41**	0.29	**0.48**	0.32	**0.53**	0.35	**0.57**	0.41	**0.64**	0.39	**0.62**
Total filtration (mm) Al	**1.5**		**2.5**		**3.5**		**4.5**		**3.5**		**3.5**	
Cu	**0**		**0**		**0**		**0**		**0.1**		**0.2**	

Table 12.33 Normalized organ doses—Abdomen ap—1–2 years

Beam projection:			ap									**Table cells:**
Focus-to-detector distance (cm):			100				(+) = normal font = optimal field setting					
Tube voltage (kVp):			70				**(−) = bold font = suboptimal field setting**					
Total filtrations:			See last table row.				Empty table cells correspond to values <0.01					
Organ or tissue	**Organ dose D_T : Entrance dose K_E**								**(mSv * mGy⁻¹)**			
	(+)	(−)	(+)	(−)	(+)	(−)	(+)	(−)	(+)	(−)	(+)	(−)
Skull												
Brain												
Oral mucosa												
Salivary glands										**0.01**		**0.01**
Cervical spine		**0.02**		**0.02**		**0.03**		**0.03**		**0.04**	0.01	**0.05**
Thyroid		**0.01**		**0.02**		**0.02**		**0.02**		**0.03**		**0.03**
Thymus	0.01	**0.12**	0.01	**0.15**	0.02	**0.17**	0.02	**0.18**	0.02	**0.21**	0.03	**0.24**
Thoracic spine	0.11	**0.20**	0.14	**0.26**	0.17	**0.31**	0.19	**0.36**	0.24	**0.44**	0.29	**0.53**
Claviculae	0.01	**0.05**	0.01	**0.06**	0.02	**0.07**	0.02	**0.08**	0.02	**0.10**	0.03	**0.12**
Scapulae	0.03	**0.24**	0.04	**0.31**	0.05	**0.36**	0.05	**0.40**	0.06	**0.48**	0.08	**0.55**
Ribs	0.53	**1.07**	0.61	**1.23**	0.67	**1.34**	0.71	**1.42**	0.78	**1.56**	0.83	**1.67**
Oesophagus	0.12	**0.20**	0.15	**0.26**	0.17	**0.30**	0.19	**0.33**	0.23	**0.40**	0.27	**0.47**
Trachea												
Lung	0.11	**0.33**	0.14	**0.40**	0.15	**0.45**	0.17	**0.49**	0.20	**0.56**	0.22	**0.62**
Mamma	0.02	**1.15**	0.03	**1.21**	0.03	**1.25**	0.04	**1.28**	0.05	**1.33**	0.06	**1.36**
Heart	0.16	**0.54**	0.20	**0.66**	0.22	**0.74**	0.24	**0.80**	0.28	**0.91**	0.32	**1.00**
Stomach	0.56	**0.59**	0.67	**0.70**	0.75	**0.79**	0.81	**0.86**	0.92	**0.98**	1.01	**1.08**
Spleen	0.19	**0.21**	0.24	**0.27**	0.28	**0.32**	0.31	**0.35**	0.37	**0.43**	0.43	**0.50**
Liver	0.47	**0.50**	0.56	**0.61**	0.63	**0.69**	0.69	**0.74**	0.78	**0.86**	0.87	**0.95**
Gall bladder	0.47	**0.49**	0.59	**0.61**	0.67	**0.70**	0.74	**0.77**	0.86	**0.90**	0.97	**1.02**
Pancreas	0.28	**0.31**	0.35	**0.39**	0.41	**0.46**	0.46	**0.51**	0.55	**0.61**	0.63	**0.71**
Small intestine	0.39	**0.40**	0.48	**0.49**	0.55	**0.57**	0.61	**0.63**	0.72	**0.74**	0.82	**0.84**
Large intestine	0.46	**0.48**	0.57	**0.59**	0.65	**0.67**	0.71	**0.74**	0.83	**0.86**	0.94	**0.97**
Sigma/rectum	0.34	**0.35**	0.43	**0.44**	0.50	**0.51**	0.55	**0.57**	0.65	**0.67**	0.74	**0.77**
Lumbar spine	0.28	**0.30**	0.37	**0.39**	0.44	**0.47**	0.50	**0.53**	0.62	**0.67**	0.75	**0.80**
Pelvis	0.51	**0.53**	0.65	**0.68**	0.75	**0.80**	0.84	**0.89**	1.02	**1.08**	1.18	**1.25**
Adrenals	0.11	**0.13**	0.14	**0.17**	0.17	**0.20**	0.19	**0.22**	0.23	**0.28**	0.28	**0.33**
Kidneys	0.11	**0.12**	0.14	**0.16**	0.17	**0.19**	0.19	**0.21**	0.24	**0.26**	0.28	**0.31**
Urinary bladder	0.61	**0.62**	0.74	**0.75**	0.82	**0.84**	0.89	**0.91**	1.01	**1.03**	1.12	**1.14**
Uterus	0.39	**0.40**	0.49	**0.50**	0.57	**0.58**	0.63	**0.64**	0.75	**0.76**	0.86	**0.87**
Ovaries	0.31	**0.32**	0.39	**0.41**	0.45	**0.47**	0.51	**0.52**	0.61	**0.63**	0.70	**0.72**
Prostate	0.43	**0.44**	0.53	**0.54**	0.60	**0.62**	0.66	**0.68**	0.76	**0.79**	0.86	**0.89**
Testes	1.04	**1.04**	1.14	**1.15**	1.20	**1.21**	1.24	**1.26**	1.32	**1.34**	1.38	**1.40**
Upper arm		**0.37**		**0.45**		**0.51**		**0.56**		**0.64**		**0.72**
Elbow		**0.98**		**1.20**		**1.35**		**1.47**		**1.69**		**1.87**
Forearm		**1.10**		**1.33**		**1.50**		**1.63**		**1.86**		**2.04**
Upper leg	0.58	**0.74**	0.73	**0.93**	0.84	**1.08**	0.93	**1.20**	1.10	**1.42**	1.26	**1.61**
Knee	0.04	**0.12**	0.05	**0.15**	0.06	**0.17**	0.07	**0.19**	0.09	**0.23**	0.11	**0.27**
Lower leg												
Lymph nodes	0.24	**0.29**	0.30	**0.36**	0.34	**0.41**	0.37	**0.45**	0.44	**0.53**	0.50	**0.60**
Skin	0.21	**0.30**	0.22	**0.31**	0.23	**0.33**	0.24	**0.34**	0.25	**0.35**	0.27	**0.37**
Muscle	0.19	**0.26**	0.23	**0.30**	0.25	**0.33**	0.27	**0.36**	0.31	**0.41**	0.34	**0.45**
Skeleton	0.18	**0.41**	0.22	**0.51**	0.25	**0.57**	0.28	**0.63**	0.33	**0.72**	0.37	**0.81**
Red bone marrow	0.06	**0.11**	0.07	**0.13**	0.08	**0.15**	0.09	**0.17**	0.11	**0.20**	0.13	**0.23**
Whole body dose	0.18	**0.26**	0.22	**0.31**	0.24	**0.35**	0.26	**0.38**	0.30	**0.43**	0.33	**0.47**
ED (ICRP 60)	0.34	**0.45**	0.41	**0.53**	0.45	**0.58**	0.49	**0.63**	0.55	**0.71**	0.61	**0.78**
ED (ICRP 103)	0.27	**0.46**	0.32	**0.53**	0.36	**0.58**	0.40	**0.62**	0.45	**0.70**	0.50	**0.77**
Total filtration (mm) **Al**	**1.5**		**2.5**		**3.5**		**4.5**		**3.5**		**3.5**	
Cu	**0**		**0**		**0**		**0**		**0.1**		**0.2**	

Table 12.34 Normalized organ doses—Abdomen ap—3–7 years

Beam projection:			ap									**Table cells:**
Focus-to-detector distance (cm):			100									(+) = normal font = optimal field setting
Tube voltage (kVp):			60									**(−) = bold font = suboptimal field setting**
Total filtrations:			See last table row.									Empty table cells correspond to values <0.01
Organ or tissue	**Organ dose D_T : Entrance dose K_E**							**(mSv * mGy^{-1})**				
	(+)	(−)	(+)	(−)	(+)	(−)	(+)	(−)	(+)	(−)	(+)	(−)
Skull												
Brain												
Oral mucosa												
Salivary glands												
Cervical spine						**0.01**		**0.01**		**0.02**		**0.01**
Thyroid								**0.01**		**0.02**		**0.01**
Thymus		**0.14**		**0.17**	0.01	**0.20**	0.01	**0.21**	0.02	**0.25**	0.01	**0.24**
Thoracic spine	0.06	**0.11**	0.08	**0.15**	0.10	**0.18**	0.11	**0.21**	0.15	**0.27**	0.12	**0.23**
Claviculae		**0.02**		**0.03**		**0.04**		**0.04**	0.01	**0.05**		**0.05**
Scapulae	0.02	**0.15**	0.02	**0.20**	0.03	**0.24**	0.03	**0.28**	0.04	**0.34**	0.03	**0.31**
Ribs	0.46	**0.91**	0.55	**1.07**	0.60	**1.18**	0.64	**1.27**	0.72	**1.41**	0.71	**1.39**
Oesophagus	0.06	**0.09**	0.08	**0.12**	0.09	**0.15**	0.11	**0.17**	0.13	**0.21**	0.12	**0.19**
Trachea												
Lung	0.09	**0.26**	0.11	**0.32**	0.13	**0.37**	0.15	**0.40**	0.17	**0.47**	0.16	**0.45**
Mamma	0.02	**1.09**	0.02	**1.17**	0.03	**1.22**	0.03	**1.25**	0.04	**1.30**	0.04	**1.29**
Heart	0.14	**0.42**	0.17	**0.52**	0.20	**0.59**	0.22	**0.65**	0.26	**0.76**	0.25	**0.73**
Stomach	0.48	**0.50**	0.59	**0.62**	0.67	**0.70**	0.73	**0.77**	0.85	**0.90**	0.82	**0.86**
Spleen	0.12	**0.13**	0.15	**0.18**	0.18	**0.21**	0.21	**0.24**	0.26	**0.30**	0.23	**0.27**
Liver	0.36	**0.39**	0.45	**0.49**	0.52	**0.56**	0.57	**0.61**	0.67	**0.72**	0.63	**0.68**
Gall bladder	0.35	**0.36**	0.45	**0.47**	0.53	**0.55**	0.59	**0.61**	0.71	**0.74**	0.66	**0.69**
Pancreas	0.19	**0.21**	0.25	**0.28**	0.30	**0.33**	0.34	**0.37**	0.42	**0.46**	0.38	**0.42**
Small intestine	0.28	**0.29**	0.36	**0.37**	0.42	**0.44**	0.47	**0.49**	0.57	**0.59**	0.53	**0.55**
Large intestine	0.35	**0.36**	0.44	**0.46**	0.51	**0.53**	0.57	**0.59**	0.68	**0.71**	0.64	**0.67**
Sigma/rectum	0.24	**0.26**	0.32	**0.33**	0.37	**0.39**	0.42	**0.44**	0.51	**0.53**	0.47	**0.49**
Lumbar spine	0.15	**0.16**	0.21	**0.22**	0.25	**0.27**	0.29	**0.31**	0.38	**0.40**	0.31	**0.33**
Pelvis	0.32	**0.34**	0.42	**0.45**	0.51	**0.54**	0.58	**0.61**	0.72	**0.76**	0.64	**0.68**
Adrenals	0.06	**0.08**	0.09	**0.10**	0.11	**0.12**	0.12	**0.14**	0.16	**0.18**	0.13	**0.16**
Kidneys	0.06	**0.07**	0.09	**0.10**	0.11	**0.12**	0.12	**0.13**	0.16	**0.17**	0.13	**0.14**
Urinary bladder	0.49	**0.50**	0.61	**0.62**	0.70	**0.71**	0.76	**0.78**	0.89	**0.91**	0.85	**0.87**
Uterus	0.28	**0.29**	0.37	**0.38**	0.44	**0.45**	0.49	**0.50**	0.60	**0.61**	0.55	**0.56**
Ovaries	0.21	**0.22**	0.28	**0.29**	0.33	**0.35**	0.37	**0.39**	0.46	**0.48**	0.42	**0.44**
Prostate	0.30	**0.31**	0.39	**0.40**	0.45	**0.47**	0.50	**0.52**	0.60	**0.63**	0.57	**0.59**
Testes	0.97	**1.01**	1.08	**1.13**	1.15	**1.21**	1.20	**1.26**	1.28	**1.34**	1.27	**1.33**
Upper arm		**0.26**		**0.33**		**0.38**		**0.42**		**0.50**		**0.47**
Elbow		**0.71**		**0.89**		**1.03**		**1.13**		**1.34**		**1.28**
Forearm		**0.82**		**1.02**		**1.18**		**1.30**		**1.52**		**1.47**
Upper leg	0.24	**0.43**	0.32	**0.57**	0.38	**0.67**	0.43	**0.76**	0.53	**0.94**	0.48	**0.86**
Knee		**0.03**		**0.04**	0.01	**0.05**	0.01	**0.05**	0.02	**0.07**	0.01	**0.06**
Lower leg												
Lymph nodes	0.18	**0.21**	0.23	**0.27**	0.27	**0.32**	0.30	**0.35**	0.36	**0.42**	0.33	**0.39**
Skin	0.17	**0.27**	0.17	**0.29**	0.18	**0.30**	0.19	**0.31**	0.19	**0.32**	0.19	**0.31**
Muscle	0.15	**0.21**	0.18	**0.25**	0.20	**0.28**	0.22	**0.30**	0.25	**0.35**	0.24	**0.33**
Skeleton	0.10	**0.25**	0.13	**0.32**	0.15	**0.37**	0.17	**0.40**	0.20	**0.48**	0.18	**0.45**
Red bone marrow	0.05	**0.08**	0.06	**0.10**	0.07	**0.12**	0.08	**0.14**	0.10	**0.17**	0.09	**0.15**
Whole body dose	0.14	**0.21**	0.17	**0.25**	0.19	**0.29**	0.21	**0.31**	0.24	**0.36**	0.23	**0.34**
ED (ICRP 60)	0.28	**0.38**	0.34	**0.45**	0.39	**0.50**	0.42	**0.54**	0.48	**0.62**	0.46	**0.60**
ED (ICRP 103)	0.22	**0.39**	0.27	**0.46**	0.30	**0.51**	0.33	**0.54**	0.39	**0.62**	0.37	**0.59**
Total filtration (mm) **Al**	**1.5**		**2.5**		**3.5**		**4.5**		**3.5**		**3.5**	
Cu	**0**		**0**		**0**		**0**		**0.1**		**0.2**	

Table 12.35 Normalized organ doses—Abdomen ap—3–7 years

Beam projection:			ap									**Table cells:**
Focus-to-detector distance (cm):			100									(+) = normal font = optimal field setting
Tube voltage (kVp):			70									**(−) = bold font = suboptimal field setting**
Total filtrations:			See last table row.									Empty table cells correspond to values <0.01
Organ or tissue	**Organ dose D_T : Entrance dose K_E**								**(mSv * mGy^{-1})**			
	(+)	(−)	(+)	(−)	(+)	(−)	(+)	(−)	(+)	(−)	(+)	(−)
Skull												
Brain												
Oral mucosa												
Salivary glands												
Cervical spine		**0.01**		**0.01**		**0.02**		**0.02**		**0.03**		**0.03**
Thyroid				**0.01**		**0.01**		**0.02**		**0.02**		**0.03**
Thymus		**0.16**	0.01	**0.20**	0.02	**0.22**	0.02	**0.24**	0.02	**0.28**	0.03	**0.32**
Thoracic spine	0.08	**0.15**	0.11	**0.20**	0.13	**0.24**	0.15	**0.28**	0.19	**0.36**	0.24	**0.43**
Claviculae		**0.03**		**0.04**	0.01	**0.05**	0.01	**0.06**	0.02	**0.07**	0.02	**0.09**
Scapulae	0.02	**0.20**	0.03	**0.25**	0.04	**0.30**	0.04	**0.33**	0.05	**0.41**	0.06	**0.48**
Ribs	0.51	**1.00**	0.59	**1.16**	0.65	**1.27**	0.69	**1.35**	0.76	**1.50**	0.82	**1.61**
Oesophagus	0.08	**0.12**	0.10	**0.16**	0.12	**0.19**	0.13	**0.21**	0.16	**0.26**	0.19	**0.31**
Trachea												
Lung	0.11	**0.30**	0.13	**0.37**	0.15	**0.42**	0.17	**0.46**	0.19	**0.53**	0.22	**0.59**
Mamma	0.02	**1.13**	0.03	**1.21**	0.04	**1.25**	0.04	**1.28**	0.05	**1.33**	0.06	**1.37**
Heart	0.16	**0.49**	0.20	**0.60**	0.23	**0.68**	0.26	**0.74**	0.30	**0.85**	0.34	**0.96**
Stomach	0.55	**0.58**	0.67	**0.71**	0.76	**0.80**	0.82	**0.87**	0.95	**1.00**	1.05	**1.12**
Spleen	0.15	**0.17**	0.19	**0.22**	0.23	**0.27**	0.26	**0.30**	0.32	**0.37**	0.37	**0.44**
Liver	0.43	**0.46**	0.52	**0.56**	0.59	**0.64**	0.64	**0.70**	0.75	**0.81**	0.84	**0.92**
Gall bladder	0.43	**0.45**	0.54	**0.56**	0.62	**0.65**	0.69	**0.72**	0.81	**0.85**	0.93	**0.97**
Pancreas	0.24	**0.27**	0.31	**0.34**	0.37	**0.41**	0.41	**0.45**	0.50	**0.55**	0.58	**0.65**
Small intestine	0.34	**0.36**	0.43	**0.45**	0.50	**0.52**	0.56	**0.58**	0.67	**0.69**	0.77	**0.80**
Large intestine	0.42	**0.43**	0.52	**0.54**	0.60	**0.63**	0.67	**0.69**	0.79	**0.82**	0.90	**0.93**
Sigma/rectum	0.30	**0.32**	0.38	**0.40**	0.45	**0.47**	0.50	**0.52**	0.59	**0.63**	0.69	**0.72**
Lumbar spine	0.21	**0.23**	0.28	**0.30**	0.34	**0.36**	0.39	**0.41**	0.49	**0.53**	0.60	**0.64**
Pelvis	0.41	**0.44**	0.53	**0.57**	0.63	**0.67**	0.70	**0.75**	0.86	**0.92**	1.01	**1.08**
Adrenals	0.09	**0.10**	0.12	**0.14**	0.14	**0.16**	0.16	**0.19**	0.20	**0.24**	0.24	**0.29**
Kidneys	0.09	**0.10**	0.12	**0.13**	0.14	**0.15**	0.16	**0.17**	0.20	**0.22**	0.24	**0.27**
Urinary bladder	0.57	**0.59**	0.70	**0.72**	0.79	**0.81**	0.86	**0.88**	0.99	**1.01**	1.10	**1.13**
Uterus	0.35	**0.36**	0.45	**0.46**	0.52	**0.54**	0.58	**0.60**	0.70	**0.72**	0.81	**0.83**
Ovaries	0.27	**0.28**	0.34	**0.36**	0.40	**0.42**	0.45	**0.48**	0.55	**0.58**	0.65	**0.68**
Prostate	0.37	**0.38**	0.46	**0.48**	0.53	**0.55**	0.59	**0.61**	0.70	**0.73**	0.79	**0.83**
Testes	1.03	**1.08**	1.14	**1.19**	1.21	**1.26**	1.26	**1.31**	1.34	**1.40**	1.40	**1.47**
Upper arm		**0.31**		**0.38**		**0.44**		**0.48**		**0.56**		**0.64**
Elbow		**0.83**		**1.02**		**1.16**		**1.28**		**1.48**		**1.66**
Forearm		**0.95**		**1.17**		**1.32**		**1.45**		**1.67**		**1.86**
Upper leg	0.31	**0.54**	0.39	**0.70**	0.46	**0.81**	0.52	**0.91**	0.63	**1.10**	0.74	**1.28**
Knee		**0.04**	0.01	**0.05**	0.02	**0.06**	0.02	**0.07**	0.02	**0.09**	0.03	**0.11**
Lower leg												
Lymph nodes	0.22	**0.26**	0.27	**0.33**	0.31	**0.37**	0.35	**0.41**	0.41	**0.49**	0.47	**0.56**
Skin	0.17	**0.28**	0.18	**0.30**	0.19	**0.31**	0.19	**0.32**	0.20	**0.34**	0.21	**0.35**
Muscle	0.17	**0.24**	0.21	**0.28**	0.23	**0.31**	0.25	**0.34**	0.28	**0.39**	0.32	**0.43**
Skeleton	0.12	**0.30**	0.15	**0.37**	0.18	**0.42**	0.19	**0.46**	0.23	**0.54**	0.26	**0.62**
Red bone marrow	0.06	**0.10**	0.08	**0.13**	0.09	**0.15**	0.10	**0.17**	0.12	**0.20**	0.14	**0.23**
Whole body dose	0.16	**0.24**	0.20	**0.29**	0.22	**0.32**	0.24	**0.35**	0.28	**0.40**	0.31	**0.45**
ED (ICRP 60)	0.33	**0.43**	0.39	**0.51**	0.43	**0.56**	0.47	**0.61**	0.54	**0.69**	0.60	**0.76**
ED (ICRP 103)	0.25	**0.44**	0.31	**0.51**	0.35	**0.56**	0.38	**0.60**	0.44	**0.68**	0.49	**0.75**
Total filtration (mm) Al	**1.5**		**2.5**		**3.5**		**4.5**		**3.5**		**3.5**	
Cu	**0**		**0**		**0**		**0**		**0.1**		**0.2**	

Table 12.36 Normalized organ doses—Abdomen ap—3–7 years

Beam projection:			ap									**Table cells:**
Focus-to-detector distance (cm):			100					(+) = normal font = optimal field setting				
Tube voltage (kVp):			80					**(−) = bold font = suboptimal field setting**				
Total filtrations:			See last table row.					Empty table cells correspond to values <0.01				
Organ or tissue	**Organ dose D_T : Entrance dose K_E**								**(mSv * mGy^{-1})**			
	(+)	(−)	(+)	(−)	(+)	(−)	(+)	(−)	(+)	(−)	(+)	(−)
Skull												
Brain												
Oral mucosa												
Salivary glands												**0.01**
Cervical spine		**0.02**		**0.02**		**0.02**		**0.03**		**0.03**	0.01	**0.04**
Thyroid		**0.01**		**0.01**		**0.02**		**0.02**		**0.03**		**0.03**
Thymus	0.01	**0.18**	0.02	**0.22**	0.02	**0.25**	0.02	**0.27**	0.03	**0.31**	0.03	**0.34**
Thoracic spine	0.11	**0.20**	0.14	**0.26**	0.17	**0.30**	0.19	**0.34**	0.23	**0.43**	0.28	**0.51**
Claviculae		**0.04**	0.01	**0.05**	0.01	**0.06**	0.02	**0.07**	0.02	**0.09**	0.02	**0.10**
Scapulae	0.03	**0.24**	0.04	**0.30**	0.05	**0.34**	0.05	**0.38**	0.06	**0.46**	0.08	**0.52**
Ribs	0.55	**1.08**	0.63	**1.23**	0.68	**1.34**	0.72	**1.41**	0.79	**1.54**	0.84	**1.64**
Oesophagus	0.09	**0.15**	0.12	**0.19**	0.14	**0.22**	0.15	**0.25**	0.18	**0.30**	0.21	**0.35**
Trachea												
Lung	0.13	**0.34**	0.15	**0.41**	0.17	**0.46**	0.19	**0.50**	0.21	**0.58**	0.24	**0.64**
Mamma	0.03	**1.17**	0.04	**1.24**	0.04	**1.28**	0.05	**1.31**	0.06	**1.36**	0.06	**1.40**
Heart	0.19	**0.56**	0.23	**0.67**	0.26	**0.75**	0.29	**0.81**	0.33	**0.93**	0.38	**1.03**
Stomach	0.63	**0.66**	0.75	**0.79**	0.83	**0.88**	0.90	**0.95**	1.02	**1.09**	1.13	**1.20**
Spleen	0.19	**0.22**	0.24	**0.27**	0.27	**0.32**	0.30	**0.35**	0.37	**0.43**	0.43	**0.50**
Liver	0.49	**0.53**	0.58	**0.64**	0.66	**0.71**	0.71	**0.78**	0.81	**0.89**	0.90	**0.99**
Gall bladder	0.50	**0.53**	0.62	**0.65**	0.70	**0.74**	0.77	**0.81**	0.90	**0.95**	1.02	**1.07**
Pancreas	0.30	**0.33**	0.37	**0.41**	0.43	**0.48**	0.48	**0.53**	0.57	**0.64**	0.66	**0.73**
Small intestine	0.41	**0.43**	0.51	**0.53**	0.58	**0.60**	0.64	**0.66**	0.75	**0.78**	0.85	**0.89**
Large intestine	0.49	**0.51**	0.60	**0.62**	0.68	**0.71**	0.75	**0.78**	0.87	**0.91**	0.98	**1.02**
Sigma/rectum	0.36	**0.38**	0.45	**0.47**	0.51	**0.54**	0.57	**0.60**	0.67	**0.71**	0.76	**0.81**
Lumbar spine	0.28	**0.30**	0.36	**0.38**	0.42	**0.45**	0.48	**0.51**	0.60	**0.64**	0.71	**0.76**
Pelvis	0.50	**0.53**	0.63	**0.67**	0.73	**0.78**	0.81	**0.87**	0.98	**1.05**	1.13	**1.21**
Adrenals	0.12	**0.14**	0.15	**0.17**	0.17	**0.21**	0.20	**0.23**	0.24	**0.29**	0.28	**0.34**
Kidneys	0.12	**0.13**	0.15	**0.16**	0.17	**0.19**	0.20	**0.22**	0.24	**0.27**	0.29	**0.31**
Urinary bladder	0.65	**0.67**	0.78	**0.80**	0.87	**0.89**	0.94	**0.97**	1.07	**1.10**	1.18	**1.22**
Uterus	0.42	**0.43**	0.53	**0.54**	0.60	**0.62**	0.67	**0.68**	0.79	**0.81**	0.90	**0.92**
Ovaries	0.33	**0.34**	0.41	**0.43**	0.48	**0.50**	0.53	**0.55**	0.63	**0.66**	0.73	**0.76**
Prostate	0.43	**0.44**	0.53	**0.55**	0.60	**0.63**	0.66	**0.69**	0.77	**0.81**	0.87	**0.91**
Testes	1.08	**1.13**	1.19	**1.24**	1.25	**1.31**	1.30	**1.36**	1.38	**1.44**	1.44	**1.50**
Upper arm		**0.35**		**0.43**		**0.48**		**0.53**		**0.61**		**0.68**
Elbow		**0.94**		**1.13**		**1.27**		**1.38**		**1.58**		**1.74**
Forearm		**1.07**		**1.28**		**1.44**		**1.56**		**1.77**		**1.93**
Upper leg	0.37	**0.65**	0.46	**0.81**	0.54	**0.94**	0.60	**1.04**	0.71	**1.23**	0.82	**1.40**
Knee	0.01	**0.05**	0.02	**0.06**	0.02	**0.08**	0.02	**0.09**	0.03	**0.11**	0.04	**0.13**
Lower leg												
Lymph nodes	0.26	**0.31**	0.31	**0.38**	0.36	**0.43**	0.39	**0.47**	0.46	**0.55**	0.52	**0.62**
Skin	0.18	**0.29**	0.19	**0.31**	0.20	**0.32**	0.20	**0.33**	0.21	**0.35**	0.22	**0.37**
Muscle	0.20	**0.27**	0.23	**0.31**	0.25	**0.35**	0.27	**0.37**	0.31	**0.42**	0.34	**0.47**
Skeleton	0.14	**0.35**	0.18	**0.42**	0.20	**0.47**	0.22	**0.51**	0.26	**0.59**	0.29	**0.66**
Red bone marrow	0.07	**0.12**	0.09	**0.15**	0.10	**0.18**	0.12	**0.19**	0.14	**0.23**	0.16	**0.26**
Whole body dose	0.19	**0.27**	0.22	**0.32**	0.24	**0.36**	0.26	**0.39**	0.30	**0.44**	0.33	**0.49**
ED (ICRP 60)	0.37	**0.48**	0.43	**0.56**	0.48	**0.62**	0.52	**0.66**	0.59	**0.75**	0.65	**0.82**
ED (ICRP 103)	0.29	**0.48**	0.35	**0.56**	0.39	**0.61**	0.42	**0.66**	0.48	**0.74**	0.54	**0.80**
Total filtration (mm) **Al**	**1.5**		**2.5**		**3.5**		**4.5**		**3.5**		**3.5**	
Cu	**0**		**0**		**0**		**0**		**0.1**		**0.2**	

Table 12.37 Normalized organ doses—Abdomen ap—8–12 years

Beam projection:			ap									**Table cells:**
Focus-to-detector distance (cm):			100									(+) = normal font = optimal field setting
Tube voltage (kVp):			60									**(−) = bold font = suboptimal field setting**
Total filtrations:			See last table row.									Empty table cells correspond to values <0.01
Organ or tissue	**Organ dose D_T : Entrance dose K_E**								**(mSv * mGy^{-1})**			
	(+)	(−)	(+)	(−)	(+)	(−)	(+)	(−)	(+)	(−)	(+)	(−)
Skull												
Brain												
Oral mucosa												
Salivary glands												
Cervical spine										**0.01**		
Thyroid												
Thymus		**0.11**		**0.13**		**0.16**		**0.17**	0.01	**0.20**		**0.19**
Thoracic spine	0.04	**0.08**	0.06	**0.11**	0.07	**0.13**	0.08	**0.15**	0.11	**0.20**	0.08	**0.16**
Claviculae		**0.01**		**0.02**		**0.02**		**0.02**		**0.03**		**0.03**
Scapulae	0.01	**0.11**	0.01	**0.14**	0.02	**0.17**	0.02	**0.20**	0.03	**0.25**	0.02	**0.22**
Ribs	0.42	**0.82**	0.49	**0.97**	0.54	**1.07**	0.58	**1.15**	0.65	**1.29**	0.64	**1.27**
Oesophagus	0.03	**0.06**	0.04	**0.08**	0.05	**0.09**	0.06	**0.10**	0.07	**0.13**	0.06	**0.11**
Trachea												
Lung	0.07	**0.22**	0.09	**0.28**	0.10	**0.32**	0.11	**0.35**	0.13	**0.42**	0.13	**0.40**
Mamma	0.01	**1.07**	0.01	**1.16**	0.02	**1.20**	0.02	**1.23**	0.03	**1.28**	0.02	**1.27**
Heart	0.10	**0.36**	0.13	**0.46**	0.15	**0.53**	0.17	**0.59**	0.21	**0.69**	0.19	**0.66**
Stomach	0.44	**0.46**	0.54	**0.57**	0.62	**0.66**	0.69	**0.73**	0.80	**0.85**	0.77	**0.81**
Spleen	0.09	**0.11**	0.13	**0.15**	0.15	**0.18**	0.17	**0.20**	0.22	**0.26**	0.19	**0.22**
Liver	0.30	**0.33**	0.38	**0.42**	0.44	**0.49**	0.49	**0.54**	0.58	**0.64**	0.55	**0.60**
Gall bladder	0.31	**0.33**	0.40	**0.42**	0.48	**0.50**	0.53	**0.56**	0.65	**0.68**	0.60	**0.63**
Pancreas	0.16	**0.18**	0.22	**0.24**	0.26	**0.29**	0.30	**0.33**	0.37	**0.41**	0.33	**0.36**
Small intestine	0.25	**0.25**	0.32	**0.33**	0.38	**0.39**	0.43	**0.44**	0.52	**0.54**	0.47	**0.49**
Large intestine	0.31	**0.32**	0.40	**0.42**	0.47	**0.49**	0.52	**0.55**	0.63	**0.66**	0.59	**0.61**
Sigma/rectum	0.21	**0.22**	0.28	**0.29**	0.33	**0.34**	0.37	**0.39**	0.45	**0.48**	0.41	**0.43**
Lumbar spine	0.11	**0.12**	0.15	**0.16**	0.19	**0.20**	0.22	**0.23**	0.29	**0.31**	0.23	**0.24**
Pelvis	0.26	**0.27**	0.34	**0.37**	0.41	**0.44**	0.47	**0.51**	0.60	**0.64**	0.52	**0.56**
Adrenals	0.05	**0.06**	0.07	**0.08**	0.08	**0.10**	0.10	**0.12**	0.13	**0.16**	0.10	**0.13**
Kidneys	0.05	**0.06**	0.07	**0.08**	0.08	**0.09**	0.10	**0.11**	0.13	**0.14**	0.10	**0.11**
Urinary bladder	0.45	**0.47**	0.57	**0.58**	0.65	**0.67**	0.72	**0.74**	0.84	**0.87**	0.80	**0.83**
Uterus	0.25	**0.25**	0.33	**0.33**	0.39	**0.40**	0.44	**0.45**	0.54	**0.55**	0.49	**0.50**
Ovaries	0.18	**0.18**	0.24	**0.24**	0.29	**0.29**	0.32	**0.33**	0.41	**0.41**	0.36	**0.37**
Prostate	0.28	**0.29**	0.36	**0.38**	0.43	**0.45**	0.48	**0.51**	0.58	**0.61**	0.54	**0.57**
Testes	0.95	**0.97**	1.07	**1.09**	1.14	**1.17**	1.19	**1.22**	1.28	**1.31**	1.26	**1.30**
Upper arm		**0.19**		**0.24**		**0.28**		**0.32**		**0.38**		**0.36**
Elbow		**0.60**		**0.76**		**0.88**		**0.98**		**1.17**		**1.11**
Forearm		**0.70**		**0.89**		**1.03**		**1.14**		**1.34**		**1.29**
Upper leg	0.15	**0.29**	0.19	**0.38**	0.23	**0.45**	0.27	**0.52**	0.34	**0.65**	0.29	**0.57**
Knee				**0.01**		**0.02**		**0.02**		**0.02**		**0.02**
Lower leg												
Lymph nodes	0.16	**0.19**	0.20	**0.24**	0.23	**0.28**	0.26	**0.32**	0.32	**0.38**	0.29	**0.35**
Skin	0.16	**0.26**	0.17	**0.27**	0.17	**0.28**	0.18	**0.29**	0.19	**0.30**	0.18	**0.29**
Muscle	0.14	**0.19**	0.16	**0.23**	0.18	**0.25**	0.20	**0.28**	0.23	**0.32**	0.22	**0.30**
Skeleton	0.08	**0.21**	0.10	**0.26**	0.12	**0.30**	0.13	**0.34**	0.16	**0.40**	0.15	**0.37**
Red bone marrow	0.05	**0.08**	0.06	**0.10**	0.07	**0.11**	0.08	**0.13**	0.10	**0.16**	0.09	**0.14**
Whole body dose	0.13	**0.19**	0.16	**0.23**	0.18	**0.26**	0.19	**0.28**	0.22	**0.33**	0.21	**0.31**
ED (ICRP 60)	0.26	**0.35**	0.31	**0.41**	0.35	**0.47**	0.39	**0.50**	0.45	**0.58**	0.42	**0.55**
ED (ICRP 103)	0.19	**0.36**	0.24	**0.42**	0.28	**0.47**	0.30	**0.51**	0.36	**0.58**	0.34	**0.55**
Total filtration (mm) Al	**1.5**		**2.5**		**3.5**		**4.5**		**3.5**		**3.5**	
Cu	**0**		**0**		**0**		**0**		**0.1**		**0.2**	

Table 12.38 Normalized organ doses—Abdomen ap—8–12 years

Beam projection:		ap										**Table cells:**
Focus-to-detector distance (cm):		100						(+) = normal font = optimal field setting				
Tube voltage (kVp):		70						**(−) = bold font = suboptimal field setting**				
Total filtrations:		See last table row.						Empty table cells correspond to values <0.01				
Organ or tissue	**Organ dose D_T : Entrance dose K_E**							**(mSv * mGy⁻¹)**				
	(+)	(−)	(+)	(−)	(+)	(−)	(+)	(−)	(+)	(−)	(+)	(−)
Skull												
Brain												
Oral mucosa												
Salivary glands												
Cervical spine						**0.01**		**0.01**		**0.02**		**0.02**
Thyroid										**0.01**		**0.01**
Thymus		**0.13**		**0.16**	0.01	**0.18**	0.01	**0.20**	0.02	**0.23**	0.02	**0.27**
Thoracic spine	0.06	**0.11**	0.08	**0.15**	0.10	**0.18**	0.11	**0.21**	0.14	**0.27**	0.18	**0.34**
Claviculae		**0.02**		**0.02**		**0.03**		**0.03**		**0.04**	0.01	**0.06**
Scapulae	0.01	**0.14**	0.02	**0.18**	0.02	**0.22**	0.03	**0.24**	0.04	**0.30**	0.04	**0.36**
Ribs	0.46	**0.91**	0.53	**1.06**	0.58	**1.16**	0.62	**1.24**	0.69	**1.37**	0.75	**1.49**
Oesophagus	0.04	**0.08**	0.05	**0.10**	0.06	**0.12**	0.07	**0.14**	0.09	**0.17**	0.11	**0.20**
Trachea												
Lung	0.08	**0.26**	0.10	**0.32**	0.12	**0.37**	0.13	**0.41**	0.16	**0.48**	0.18	**0.54**
Mamma	0.01	**1.11**	0.02	**1.19**	0.02	**1.24**	0.03	**1.27**	0.03	**1.32**	0.04	**1.36**
Heart	0.13	**0.43**	0.16	**0.54**	0.18	**0.61**	0.21	**0.67**	0.25	**0.79**	0.28	**0.89**
Stomach	0.51	**0.54**	0.63	**0.67**	0.71	**0.76**	0.78	**0.83**	0.90	**0.96**	1.02	**1.09**
Spleen	0.13	**0.15**	0.16	**0.19**	0.19	**0.23**	0.22	**0.26**	0.27	**0.32**	0.33	**0.39**
Liver	0.36	**0.40**	0.45	**0.50**	0.52	**0.57**	0.57	**0.63**	0.67	**0.74**	0.76	**0.84**
Gall bladder	0.39	**0.41**	0.49	**0.51**	0.57	**0.60**	0.63	**0.67**	0.76	**0.80**	0.87	**0.92**
Pancreas	0.21	**0.24**	0.28	**0.31**	0.33	**0.36**	0.37	**0.41**	0.45	**0.51**	0.54	**0.60**
Small intestine	0.31	**0.32**	0.39	**0.41**	0.46	**0.48**	0.51	**0.53**	0.62	**0.64**	0.72	**0.75**
Large intestine	0.38	**0.40**	0.48	**0.50**	0.56	**0.58**	0.62	**0.65**	0.74	**0.77**	0.85	**0.89**
Sigma/rectum	0.27	**0.28**	0.34	**0.36**	0.40	**0.42**	0.45	**0.47**	0.54	**0.57**	0.63	**0.67**
Lumbar spine	0.16	**0.17**	0.22	**0.23**	0.26	**0.28**	0.30	**0.33**	0.39	**0.42**	0.48	**0.52**
Pelvis	0.34	**0.36**	0.44	**0.47**	0.52	**0.56**	0.59	**0.64**	0.73	**0.79**	0.87	**0.94**
Adrenals	0.07	**0.09**	0.09	**0.12**	0.11	**0.14**	0.13	**0.16**	0.17	**0.20**	0.20	**0.25**
Kidneys	0.07	**0.08**	0.10	**0.11**	0.11	**0.13**	0.13	**0.15**	0.17	**0.19**	0.20	**0.23**
Urinary bladder	0.54	**0.55**	0.66	**0.68**	0.75	**0.77**	0.82	**0.84**	0.95	**0.98**	1.07	**1.10**
Uterus	0.32	**0.33**	0.40	**0.42**	0.47	**0.49**	0.53	**0.55**	0.64	**0.67**	0.75	**0.78**
Ovaries	0.23	**0.24**	0.30	**0.31**	0.36	**0.37**	0.40	**0.41**	0.49	**0.51**	0.58	**0.60**
Prostate	0.35	**0.36**	0.44	**0.46**	0.51	**0.54**	0.56	**0.60**	0.67	**0.71**	0.78	**0.82**
Testes	1.01	**1.04**	1.13	**1.16**	1.20	**1.23**	1.25	**1.29**	1.33	**1.38**	1.41	**1.45**
Upper arm		**0.23**		**0.29**		**0.33**		**0.37**		**0.44**		**0.50**
Elbow		**0.71**		**0.89**		**1.01**		**1.12**		**1.31**		**1.48**
Forearm		**0.83**		**1.02**		**1.17**		**1.28**		**1.49**		**1.68**
Upper leg	0.19	**0.37**	0.25	**0.48**	0.30	**0.57**	0.34	**0.64**	0.42	**0.79**	0.49	**0.93**
Knee		**0.01**		**0.02**		**0.02**		**0.03**		**0.03**	0.01	**0.04**
Lower leg												
Lymph nodes	0.19	**0.23**	0.24	**0.29**	0.28	**0.34**	0.31	**0.38**	0.38	**0.45**	0.43	**0.52**
Skin	0.17	**0.27**	0.17	**0.28**	0.18	**0.29**	0.19	**0.30**	0.20	**0.32**	0.20	**0.33**
Muscle	0.16	**0.22**	0.19	**0.26**	0.21	**0.29**	0.23	**0.31**	0.26	**0.36**	0.29	**0.40**
Skeleton	0.10	**0.25**	0.12	**0.31**	0.14	**0.36**	0.16	**0.39**	0.19	**0.46**	0.22	**0.53**
Red bone marrow	0.06	**0.10**	0.08	**0.12**	0.09	**0.14**	0.10	**0.16**	0.12	**0.19**	0.15	**0.23**
Whole body dose	0.15	**0.22**	0.18	**0.27**	0.20	**0.30**	0.22	**0.32**	0.25	**0.38**	0.29	**0.42**
ED (ICRP 60)	0.30	**0.40**	0.36	**0.47**	0.40	**0.53**	0.44	**0.57**	0.51	**0.65**	0.57	**0.72**
ED (ICRP 103)	0.23	**0.41**	0.28	**0.48**	0.32	**0.53**	0.35	**0.57**	0.41	**0.64**	0.46	**0.72**
Total filtration (mm) Al	**1.5**		**2.5**		**3.5**		**4.5**		**3.5**		**3.5**	
Cu	**0**		**0**		**0**		**0**		**0.1**		**0.2**	

Table 12.39 Normalized organ doses—Abdomen ap—8–12 years

Beam projection:		ap										**Table cells:**
Focus-to-detector distance (cm):		100										(+) = normal font = optimal field setting
Tube voltage (kVp):		80										**(−) = bold font = suboptimal field setting**
Total filtrations:		See last table row.										Empty table cells correspond to values <0.01
Organ or tissue	**Organ dose D_T : Entrance dose K_E (mSv * mGy^{-1})**											
	(+)	(−)	(+)	(−)	(+)	(−)	(+)	(−)	(+)	(−)	(+)	(−)
Skull												
Brain												
Oral mucosa												
Salivary glands												
Cervical spine				**0.01**		**0.01**		**0.02**		**0.02**		**0.03**
Thyroid						**0.01**		**0.01**		**0.02**		**0.02**
Thymus		**0.15**	0.01	**0.18**	0.01	**0.21**	0.02	**0.22**	0.02	**0.26**	0.03	**0.29**
Thoracic spine	0.08	**0.15**	0.11	**0.20**	0.13	**0.24**	0.14	**0.27**	0.18	**0.34**	0.22	**0.41**
Claviculae		**0.02**		**0.03**		**0.04**		**0.04**	0.01	**0.06**	0.01	**0.07**
Scapulae	0.02	**0.17**	0.03	**0.22**	0.03	**0.26**	0.03	**0.29**	0.04	**0.35**	0.05	**0.40**
Ribs	0.50	**0.98**	0.57	**1.13**	0.62	**1.23**	0.65	**1.30**	0.72	**1.43**	0.77	**1.52**
Oesophagus	0.05	**0.10**	0.07	**0.12**	0.08	**0.15**	0.09	**0.16**	0.11	**0.20**	0.13	**0.24**
Trachea												
Lung	0.10	**0.30**	0.12	**0.37**	0.14	**0.42**	0.15	**0.45**	0.17	**0.52**	0.20	**0.59**
Mamma	0.02	**1.15**	0.02	**1.23**	0.03	**1.27**	0.03	**1.30**	0.04	**1.35**	0.05	**1.39**
Heart	0.15	**0.50**	0.19	**0.61**	0.21	**0.69**	0.24	**0.75**	0.28	**0.87**	0.32	**0.97**
Stomach	0.59	**0.62**	0.71	**0.75**	0.79	**0.85**	0.86	**0.92**	0.99	**1.06**	1.10	**1.18**
Spleen	0.16	**0.19**	0.20	**0.24**	0.24	**0.28**	0.27	**0.31**	0.32	**0.38**	0.38	**0.45**
Liver	0.42	**0.47**	0.52	**0.57**	0.58	**0.65**	0.64	**0.71**	0.74	**0.82**	0.83	**0.92**
Gall bladder	0.46	**0.49**	0.57	**0.60**	0.66	**0.69**	0.72	**0.76**	0.85	**0.90**	0.97	**1.03**
Pancreas	0.27	**0.30**	0.34	**0.37**	0.39	**0.44**	0.44	**0.49**	0.53	**0.59**	0.61	**0.69**
Small intestine	0.37	**0.39**	0.46	**0.48**	0.53	**0.56**	0.59	**0.62**	0.70	**0.73**	0.81	**0.84**
Large intestine	0.45	**0.47**	0.56	**0.58**	0.64	**0.67**	0.70	**0.74**	0.83	**0.87**	0.94	**0.99**
Sigma/rectum	0.32	**0.34**	0.41	**0.43**	0.47	**0.49**	0.52	**0.55**	0.62	**0.65**	0.71	**0.75**
Lumbar spine	0.22	**0.24**	0.29	**0.31**	0.34	**0.37**	0.39	**0.42**	0.49	**0.52**	0.59	**0.63**
Pelvis	0.42	**0.45**	0.54	**0.58**	0.62	**0.67**	0.70	**0.75**	0.85	**0.91**	0.99	**1.07**
Adrenals	0.10	**0.12**	0.12	**0.15**	0.15	**0.18**	0.16	**0.20**	0.21	**0.25**	0.25	**0.30**
Kidneys	0.10	**0.11**	0.12	**0.14**	0.15	**0.16**	0.17	**0.18**	0.21	**0.23**	0.25	**0.27**
Urinary bladder	0.61	**0.63**	0.74	**0.76**	0.83	**0.86**	0.91	**0.93**	1.04	**1.07**	1.16	**1.20**
Uterus	0.38	**0.40**	0.48	**0.50**	0.56	**0.58**	0.62	**0.64**	0.74	**0.77**	0.85	**0.89**
Ovaries	0.29	**0.30**	0.37	**0.38**	0.43	**0.44**	0.48	**0.49**	0.58	**0.59**	0.67	**0.69**
Prostate	0.41	**0.43**	0.51	**0.54**	0.58	**0.62**	0.64	**0.68**	0.75	**0.80**	0.86	**0.91**
Testes	1.07	**1.10**	1.18	**1.21**	1.25	**1.29**	1.30	**1.34**	1.38	**1.43**	1.45	**1.50**
Upper arm		**0.27**		**0.33**		**0.37**		**0.41**		**0.48**		**0.54**
Elbow		**0.82**		**0.99**		**1.12**		**1.23**		**1.41**		**1.57**
Forearm		**0.94**		**1.14**		**1.28**		**1.40**		**1.60**		**1.77**
Upper leg	0.24	**0.46**	0.30	**0.58**	0.35	**0.67**	0.40	**0.75**	0.48	**0.90**	0.56	**1.05**
Knee		**0.02**		**0.03**		**0.03**		**0.03**	0.01	**0.04**	0.02	**0.05**
Lower leg												
Lymph nodes	0.23	**0.28**	0.29	**0.34**	0.33	**0.39**	0.36	**0.43**	0.43	**0.51**	0.49	**0.58**
Skin	0.17	**0.28**	0.18	**0.29**	0.19	**0.30**	0.19	**0.31**	0.20	**0.33**	0.21	**0.35**
Muscle	0.18	**0.25**	0.21	**0.29**	0.23	**0.32**	0.25	**0.35**	0.29	**0.39**	0.32	**0.44**
Skeleton	0.12	**0.29**	0.14	**0.35**	0.16	**0.40**	0.18	**0.44**	0.21	**0.51**	0.24	**0.57**
Red bone marrow	0.08	**0.12**	0.10	**0.15**	0.11	**0.17**	0.12	**0.19**	0.15	**0.23**	0.17	**0.26**
Whole body dose	0.17	**0.25**	0.20	**0.30**	0.23	**0.33**	0.24	**0.36**	0.28	**0.41**	0.31	**0.46**
ED (ICRP 60)	0.34	**0.45**	0.40	**0.53**	0.45	**0.58**	0.49	**0.63**	0.56	**0.71**	0.62	**0.79**
ED (ICRP 103)	0.27	**0.45**	0.32	**0.53**	0.36	**0.58**	0.39	**0.62**	0.45	**0.70**	0.51	**0.77**
Total filtration (mm) Al	**1.5**		**2.5**		**3.5**		**4.5**		**3.5**		**3.5**	
Cu	**0**		**0**		**0**		**0**		**0.1**		**0.2**	

Table 12.40 Normalized organ doses—Abdomen ap—13–17 years

Beam projection:		ap										**Table cells:**
Focus-to-detector distance (cm):		100										(+) = normal font = optimal field setting
Tube voltage (kVp):		70										**(−) = bold font = suboptimal field setting**
Total filtrations:		See last table row.										Empty table cells correspond to values <0.01
Organ or tissue	**Organ dose D_T : Entrance dose K_E (mSv * mGy^{-1})**											
	(+)	(−)	(+)	(−)	(+)	(−)	(+)	(−)	(+)	(−)	(+)	(−)
Skull												
Brain												
Oral mucosa												
Salivary glands												
Cervical spine												
Thyroid												
Thymus		**0.05**		**0.06**		**0.07**	0.01	**0.08**	0.01	**0.11**	0.02	**0.13**
Thoracic spine	0.04	**0.06**	0.05	**0.09**	0.06	**0.10**	0.07	**0.12**	0.09	**0.16**	0.12	**0.20**
Claviculae								**0.01**		**0.02**		**0.02**
Scapulae		**0.05**	0.01	**0.06**	0.01	**0.08**	0.02	**0.09**	0.02	**0.11**	0.03	**0.14**
Ribs	0.40	**0.66**	0.46	**0.78**	0.51	**0.87**	0.55	**0.93**	0.61	**1.04**	0.66	**1.14**
Oesophagus	0.05	**0.08**	0.07	**0.10**	0.08	**0.13**	0.09	**0.14**	0.12	**0.18**	0.14	**0.22**
Trachea												
Lung	0.06	**0.16**	0.08	**0.21**	0.09	**0.24**	0.10	**0.27**	0.13	**0.32**	0.15	**0.37**
Mamma	0.15	**4.63**	0.19	**5.30**	0.22	**5.75**	0.25	**6.09**	0.31	**6.68**	0.37	**7.17**
Heart	0.10	**0.36**	0.12	**0.45**	0.15	**0.52**	0.16	**0.58**	0.20	**0.68**	0.23	**0.78**
Stomach	0.43	**0.46**	0.53	**0.57**	0.61	**0.66**	0.67	**0.73**	0.80	**0.86**	0.91	**0.98**
Spleen	0.09	**0.11**	0.12	**0.14**	0.15	**0.17**	0.17	**0.19**	0.21	**0.25**	0.26	**0.30**
Liver	0.29	**0.32**	0.37	**0.41**	0.42	**0.47**	0.47	**0.53**	0.56	**0.63**	0.64	**0.73**
Gall bladder	0.33	**0.35**	0.42	**0.45**	0.50	**0.53**	0.56	**0.59**	0.67	**0.72**	0.79	**0.84**
Pancreas	0.16	**0.18**	0.21	**0.24**	0.25	**0.28**	0.29	**0.32**	0.36	**0.40**	0.43	**0.49**
Small intestine	0.26	**0.27**	0.33	**0.35**	0.39	**0.41**	0.44	**0.46**	0.53	**0.56**	0.63	**0.66**
Large intestine	0.32	**0.34**	0.41	**0.43**	0.48	**0.51**	0.54	**0.57**	0.65	**0.69**	0.76	**0.80**
Sigma/rectum	0.22	**0.23**	0.28	**0.30**	0.34	**0.36**	0.38	**0.40**	0.46	**0.49**	0.55	**0.58**
Lumbar spine	0.11	**0.11**	0.14	**0.15**	0.18	**0.19**	0.20	**0.22**	0.27	**0.28**	0.33	**0.36**
Pelvis	0.25	**0.27**	0.33	**0.36**	0.40	**0.43**	0.45	**0.49**	0.57	**0.61**	0.68	**0.74**
Adrenals	0.05	**0.06**	0.07	**0.08**	0.08	**0.10**	0.09	**0.11**	0.12	**0.15**	0.15	**0.18**
Kidneys	0.05	**0.06**	0.07	**0.07**	0.08	**0.09**	0.09	**0.10**	0.12	**0.13**	0.15	**0.17**
Urinary bladder	0.48	**0.50**	0.59	**0.62**	0.68	**0.71**	0.75	**0.78**	0.88	**0.92**	1.01	**1.05**
Uterus	0.76	**0.79**	0.98	**1.02**	1.16	**1.20**	1.31	**1.36**	1.60	**1.67**	1.89	**1.97**
Ovaries	0.19	**0.20**	0.25	**0.26**	0.29	**0.31**	0.33	**0.35**	0.41	**0.44**	0.50	**0.52**
Prostate	0.32	**0.34**	0.41	**0.43**	0.48	**0.51**	0.53	**0.57**	0.65	**0.69**	0.76	**0.81**
Testes	0.89	**0.93**	1.01	**1.07**	1.10	**1.16**	1.16	**1.23**	1.27	**1.35**	1.36	**1.45**
Upper arm		**0.10**		**0.13**		**0.15**		**0.17**		**0.20**		**0.24**
Elbow		**0.54**		**0.68**		**0.78**		**0.87**		**1.03**		**1.18**
Forearm		**0.64**		**0.81**		**0.93**		**1.03**		**1.21**		**1.38**
Upper leg	0.17	**0.29**	0.23	**0.38**	0.27	**0.45**	0.31	**0.51**	0.39	**0.64**	0.47	**0.77**
Knee		**0.03**		**0.04**		**0.05**		**0.05**		**0.07**	0.01	**0.08**
Lower leg												
Lymph nodes	0.16	**0.19**	0.21	**0.24**	0.24	**0.29**	0.27	**0.32**	0.33	**0.39**	0.38	**0.45**
Skin	0.18	**0.28**	0.20	**0.30**	0.20	**0.31**	0.21	**0.32**	0.22	**0.34**	0.24	**0.35**
Muscle	0.15	**0.20**	0.18	**0.24**	0.20	**0.27**	0.22	**0.29**	0.25	**0.34**	0.29	**0.38**
Skeleton	0.09	**0.20**	0.11	**0.25**	0.13	**0.29**	0.14	**0.32**	0.17	**0.39**	0.20	**0.45**
Red bone marrow	0.06	**0.08**	0.07	**0.10**	0.08	**0.12**	0.10	**0.13**	0.12	**0.16**	0.14	**0.19**
Whole body dose	0.14	**0.20**	0.17	**0.24**	0.19	**0.28**	0.21	**0.30**	0.25	**0.35**	0.28	**0.40**
ED (ICRP 60)	0.26	**0.52**	0.32	**0.62**	0.37	**0.69**	0.40	**0.75**	0.47	**0.86**	0.54	**0.95**
ED (ICRP 103)	0.22	**0.79**	0.27	**0.92**	0.31	**1.02**	0.34	**1.09**	0.41	**1.23**	0.47	**1.36**
Total filtration (mm) **Al**	**1.5**		**2.5**		**3.5**		**4.5**		**3.5**		**3.5**	
Cu	**0**		**0**		**0**		**0**		**0.1**		**0.2**	

Table 12.41 Normalized organ doses—Abdomen ap—13–17 years

Beam projection:			ap									**Table cells:**
Focus-to-detector distance (cm):			100									(+) = normal font = optimal field setting
Tube voltage (kVp):			80									**(−) = bold font = suboptimal field setting**
Total filtrations:			See last table row.									Empty table cells correspond to values <0.01
Organ or tissue	**Organ dose D_T : Entrance dose K_E**								**(mSv * mGy^{-1})**			
	(+)	(−)	(+)	(−)	(+)	(−)	(+)	(−)	(+)	(−)	(+)	(−)
Skull												
Brain												
Oral mucosa												
Salivary glands												
Cervical spine												
Thyroid												
Thymus		**0.06**	0.01	**0.08**	0.01	**0.09**	0.01	**0.10**	0.02	**0.13**	0.02	**0.15**
Thoracic spine	0.05	**0.09**	0.07	**0.12**	0.08	**0.14**	0.09	**0.16**	0.12	**0.20**	0.15	**0.25**
Claviculae				**0.01**		**0.01**		**0.02**		**0.02**		**0.03**
Scapulae	0.01	**0.06**	0.02	**0.08**	0.02	**0.10**	0.02	**0.11**	0.03	**0.14**	0.03	**0.17**
Ribs	0.43	**0.73**	0.50	**0.84**	0.54	**0.93**	0.58	**0.99**	0.64	**1.09**	0.68	**1.18**
Oesophagus	0.07	**0.10**	0.09	**0.13**	0.10	**0.16**	0.12	**0.18**	0.14	**0.22**	0.17	**0.27**
Trachea												
Lung	0.08	**0.20**	0.10	**0.24**	0.11	**0.28**	0.12	**0.31**	0.15	**0.36**	0.17	**0.42**
Mamma	0.19	**5.00**	0.24	**5.67**	0.27	**6.11**	0.31	**6.44**	0.37	**7.02**	0.44	**7.50**
Heart	0.12	**0.43**	0.15	**0.52**	0.17	**0.60**	0.19	**0.65**	0.23	**0.77**	0.27	**0.87**
Stomach	0.50	**0.54**	0.61	**0.66**	0.69	**0.75**	0.76	**0.82**	0.88	**0.96**	1.00	**1.08**
Spleen	0.12	**0.14**	0.16	**0.18**	0.18	**0.21**	0.21	**0.24**	0.26	**0.30**	0.30	**0.36**
Liver	0.35	**0.39**	0.43	**0.48**	0.49	**0.55**	0.54	**0.60**	0.63	**0.71**	0.72	**0.81**
Gall bladder	0.40	**0.43**	0.50	**0.54**	0.58	**0.62**	0.65	**0.69**	0.77	**0.82**	0.89	**0.95**
Pancreas	0.21	**0.23**	0.27	**0.30**	0.31	**0.35**	0.35	**0.39**	0.43	**0.48**	0.51	**0.57**
Small intestine	0.32	**0.33**	0.40	**0.42**	0.46	**0.49**	0.51	**0.54**	0.62	**0.65**	0.72	**0.76**
Large intestine	0.39	**0.41**	0.49	**0.52**	0.56	**0.59**	0.62	**0.66**	0.74	**0.79**	0.86	**0.90**
Sigma/rectum	0.27	**0.29**	0.35	**0.37**	0.40	**0.43**	0.45	**0.48**	0.54	**0.58**	0.63	**0.67**
Lumbar spine	0.15	**0.16**	0.20	**0.21**	0.24	**0.25**	0.27	**0.29**	0.34	**0.37**	0.42	**0.45**
Pelvis	0.32	**0.35**	0.41	**0.45**	0.49	**0.53**	0.55	**0.59**	0.67	**0.73**	0.79	**0.86**
Adrenals	0.07	**0.08**	0.09	**0.11**	0.11	**0.13**	0.12	**0.15**	0.15	**0.19**	0.19	**0.23**
Kidneys	0.07	**0.08**	0.09	**0.10**	0.11	**0.12**	0.12	**0.14**	0.15	**0.17**	0.19	**0.21**
Urinary bladder	0.56	**0.58**	0.68	**0.71**	0.77	**0.81**	0.84	**0.88**	0.98	**1.03**	1.10	**1.16**
Uterus	0.95	**0.99**	1.19	**1.24**	1.39	**1.45**	1.55	**1.61**	1.87	**1.95**	2.18	**2.27**
Ovaries	0.24	**0.25**	0.31	**0.32**	0.36	**0.38**	0.40	**0.42**	0.50	**0.52**	0.58	**0.61**
Prostate	0.39	**0.41**	0.48	**0.52**	0.56	**0.60**	0.62	**0.67**	0.74	**0.80**	0.86	**0.92**
Testes	0.96	**1.01**	1.08	**1.15**	1.16	**1.24**	1.22	**1.30**	1.33	**1.42**	1.41	**1.52**
Upper arm		**0.12**		**0.15**		**0.18**		**0.19**		**0.23**		**0.26**
Elbow		**0.63**		**0.77**		**0.88**		**0.97**		**1.13**		**1.27**
Forearm		**0.74**		**0.91**		**1.04**		**1.13**		**1.32**		**1.47**
Upper leg	0.22	**0.37**	0.29	**0.47**	0.34	**0.55**	0.38	**0.62**	0.47	**0.76**	0.55	**0.89**
Knee		**0.04**		**0.05**		**0.06**	0.01	**0.07**	0.01	**0.08**	0.02	**0.10**
Lower leg												
Lymph nodes	0.20	**0.23**	0.25	**0.29**	0.28	**0.34**	0.31	**0.37**	0.38	**0.45**	0.43	**0.51**
Skin	0.19	**0.29**	0.20	**0.31**	0.21	**0.32**	0.22	**0.33**	0.23	**0.35**	0.25	**0.37**
Muscle	0.17	**0.23**	0.20	**0.27**	0.23	**0.30**	0.25	**0.33**	0.28	**0.38**	0.32	**0.42**
Skeleton	0.11	**0.24**	0.13	**0.29**	0.15	**0.34**	0.16	**0.37**	0.20	**0.43**	0.22	**0.50**
Red bone marrow	0.07	**0.10**	0.09	**0.12**	0.10	**0.14**	0.12	**0.16**	0.14	**0.19**	0.17	**0.23**
Whole body dose	0.16	**0.23**	0.20	**0.28**	0.22	**0.31**	0.24	**0.34**	0.27	**0.39**	0.31	**0.44**
ED (ICRP 60)	0.31	**0.59**	0.37	**0.69**	0.42	**0.77**	0.46	**0.83**	0.53	**0.94**	0.60	**1.03**
ED (ICRP 103)	0.26	**0.87**	0.31	**1.01**	0.36	**1.11**	0.39	**1.19**	0.46	**1.33**	0.53	**1.45**
Total filtration (mm) Al	**1.5**		**2.5**		**3.5**		**4.5**		**3.5**		**3.5**	
Cu	**0**		**0**		**0**		**0**		**0.1**		**0.2**	

Table 12.42 Normalized organ doses—Abdomen ap—13–17 years

Beam projection:			ap									**Table cells:**
Focus-to-detector distance (cm):			100				(+) = normal font = optimal field setting					
Tube voltage (kVp):			110				**(−) = bold font = suboptimal field setting**					
Total filtrations:			See last table row.				Empty table cells correspond to values <0.01					
Organ or tissue	**Organ dose D_T : Entrance dose K_E**								**(mSv * mGy^{-1})**			
	(+)	(−)	(+)	(−)	(+)	(−)	(+)	(−)	(+)	(−)	(+)	(−)
Skull												
Brain												
Oral mucosa												
Salivary glands												
Cervical spine						**0.01**		**0.01**		**0.01**		**0.02**
Thyroid												**0.01**
Thymus	0.02	**0.10**	0.02	**0.11**	0.02	**0.13**	0.02	**0.14**	0.03	**0.16**	0.03	**0.18**
Thoracic spine	0.10	**0.16**	0.12	**0.20**	0.13	**0.23**	0.15	**0.25**	0.18	**0.30**	0.20	**0.34**
Claviculae		**0.02**		**0.02**		**0.02**		**0.03**		**0.03**		**0.04**
Scapulae	0.02	**0.11**	0.03	**0.13**	0.03	**0.14**	0.03	**0.16**	0.04	**0.18**	0.04	**0.21**
Ribs	0.51	**0.86**	0.56	**0.96**	0.59	**1.02**	0.62	**1.07**	0.66	**1.15**	0.69	**1.20**
Oesophagus	0.11	**0.18**	0.14	**0.21**	0.15	**0.24**	0.17	**0.26**	0.20	**0.31**	0.22	**0.35**
Trachea												
Lung	0.11	**0.28**	0.13	**0.33**	0.15	**0.37**	0.16	**0.39**	0.18	**0.45**	0.20	**0.49**
Mamma	0.29	**5.89**	0.35	**6.48**	0.39	**6.86**	0.42	**7.14**	0.49	**7.61**	0.55	**7.98**
Heart	0.18	**0.60**	0.21	**0.69**	0.23	**0.76**	0.25	**0.82**	0.29	**0.92**	0.32	**1.00**
Stomach	0.69	**0.74**	0.80	**0.86**	0.87	**0.95**	0.93	**1.02**	1.05	**1.14**	1.14	**1.24**
Spleen	0.20	**0.24**	0.24	**0.28**	0.27	**0.32**	0.30	**0.35**	0.35	**0.41**	0.39	**0.47**
Liver	0.49	**0.55**	0.57	**0.65**	0.63	**0.71**	0.67	**0.77**	0.76	**0.87**	0.83	**0.95**
Gall bladder	0.59	**0.63**	0.70	**0.75**	0.77	**0.83**	0.84	**0.89**	0.95	**1.02**	1.05	**1.13**
Pancreas	0.34	**0.38**	0.40	**0.45**	0.45	**0.51**	0.49	**0.55**	0.57	**0.64**	0.64	**0.73**
Small intestine	0.48	**0.51**	0.57	**0.60**	0.63	**0.67**	0.68	**0.72**	0.78	**0.83**	0.87	**0.92**
Large intestine	0.57	**0.61**	0.68	**0.71**	0.75	**0.79**	0.81	**0.85**	0.92	**0.97**	1.01	**1.08**
Sigma/rectum	0.42	**0.45**	0.50	**0.53**	0.56	**0.59**	0.60	**0.64**	0.69	**0.74**	0.77	**0.83**
Lumbar spine	0.27	**0.30**	0.33	**0.36**	0.38	**0.41**	0.42	**0.45**	0.50	**0.54**	0.57	**0.62**
Pelvis	0.51	**0.55**	0.60	**0.66**	0.68	**0.74**	0.74	**0.81**	0.85	**0.93**	0.96	**1.05**
Adrenals	0.12	**0.15**	0.15	**0.18**	0.17	**0.21**	0.19	**0.23**	0.23	**0.27**	0.26	**0.31**
Kidneys	0.13	**0.14**	0.15	**0.17**	0.17	**0.19**	0.19	**0.21**	0.23	**0.25**	0.26	**0.29**
Urinary bladder	0.76	**0.80**	0.88	**0.93**	0.97	**1.02**	1.04	**1.09**	1.16	**1.22**	1.26	**1.32**
Uterus	1.45	**1.51**	1.71	**1.79**	1.91	**2.00**	2.07	**2.17**	2.38	**2.49**	2.65	**2.78**
Ovaries	0.39	**0.40**	0.46	**0.48**	0.52	**0.54**	0.56	**0.59**	0.65	**0.69**	0.74	**0.77**
Prostate	0.57	**0.61**	0.67	**0.72**	0.75	**0.80**	0.81	**0.86**	0.92	**0.98**	1.02	**1.09**
Testes	1.12	**1.19**	1.23	**1.31**	1.30	**1.38**	1.35	**1.44**	1.43	**1.53**	1.50	**1.61**
Upper arm		**0.17**		**0.20**		**0.22**		**0.24**		**0.27**		**0.29**
Elbow		**0.84**		**0.97**		**1.07**		**1.14**		**1.27**		**1.37**
Forearm		**0.98**		**1.13**		**1.23**		**1.31**		**1.45**		**1.56**
Upper leg	0.35	**0.57**	0.42	**0.68**	0.47	**0.76**	0.51	**0.83**	0.60	**0.96**	0.67	**1.07**
Knee	0.01	**0.06**	0.01	**0.08**	0.02	**0.09**	0.02	**0.10**	0.02	**0.11**	0.03	**0.13**
Lower leg												
Lymph nodes	0.29	**0.35**	0.34	**0.41**	0.38	**0.45**	0.41	**0.49**	0.47	**0.56**	0.52	**0.62**
Skin	0.21	**0.32**	0.23	**0.34**	0.23	**0.35**	0.24	**0.36**	0.25	**0.38**	0.26	**0.39**
Muscle	0.23	**0.30**	0.26	**0.35**	0.28	**0.38**	0.30	**0.40**	0.34	**0.45**	0.36	**0.49**
Skeleton	0.15	**0.33**	0.18	**0.38**	0.19	**0.42**	0.21	**0.45**	0.24	**0.51**	0.26	**0.56**
Red bone marrow	0.12	**0.16**	0.14	**0.19**	0.15	**0.21**	0.17	**0.23**	0.20	**0.26**	0.22	**0.30**
Whole body dose	0.22	**0.31**	0.25	**0.36**	0.28	**0.39**	0.29	**0.42**	0.33	**0.46**	0.36	**0.51**
ED (ICRP 60)	0.42	**0.76**	0.49	**0.86**	0.53	**0.93**	0.57	**0.98**	0.64	**1.08**	0.69	**1.17**
ED (ICRP 103)	0.36	**1.09**	0.42	**1.22**	0.47	**1.32**	0.50	**1.38**	0.56	**1.51**	0.62	**1.61**
Total filtration (mm) **Al**	**1.5**		**2.5**		**3.5**		**4.5**		**3.5**		**3.5**	
Cu	**0**		**0**		**0**		**0**		**0.1**		**0.2**	

Table 12.43 Normalized organ doses—Abdomen ap—13–17 years

Beam projection:		ap									**Table cells:**	
Focus-to-detector distance (cm):		100					(+) = normal font = optimal field setting					
Tube voltage (kVp):		120					**(−) = bold font = suboptimal field setting**					
Total filtrations:		See last table row.					Empty table cells correspond to values <0.01					
Organ or tissue	**Organ dose D_T : Entrance dose K_E**							**(mSv * mGy^{-1})**				
	(+)	(−)	(+)	(−)	(+)	(−)	(+)	(−)	(+)	(−)	(+)	(−)
Skull												
Brain												
Oral mucosa												
Salivary glands												
Cervical spine				**0.01**		**0.01**		**0.01**		**0.02**		**0.02**
Thyroid										**0.01**		**0.01**
Thymus	0.02	**0.11**	0.02	**0.12**	0.02	**0.14**	0.03	**0.15**	0.03	**0.17**	0.03	**0.19**
Thoracic spine	0.11	**0.18**	0.13	**0.22**	0.15	**0.25**	0.16	**0.27**	0.19	**0.32**	0.21	**0.36**
Claviculae		**0.02**		**0.02**		**0.03**		**0.03**		**0.03**	0.01	**0.04**
Scapulae	0.02	**0.12**	0.03	**0.14**	0.03	**0.15**	0.04	**0.17**	0.04	**0.19**	0.05	**0.21**
Ribs	0.52	**0.89**	0.57	**0.98**	0.60	**1.04**	0.62	**1.08**	0.66	**1.14**	0.68	**1.19**
Oesophagus	0.13	**0.20**	0.15	**0.23**	0.17	**0.26**	0.18	**0.28**	0.21	**0.32**	0.23	**0.36**
Trachea												
Lung	0.12	**0.31**	0.14	**0.35**	0.16	**0.39**	0.17	**0.41**	0.19	**0.46**	0.21	**0.51**
Mamma	0.32	**6.12**	0.38	**6.68**	0.42	**7.03**	0.45	**7.29**	0.51	**7.72**	0.57	**8.05**
Heart	0.19	**0.64**	0.23	**0.73**	0.25	**0.80**	0.27	**0.85**	0.30	**0.95**	0.33	**1.02**
Stomach	0.74	**0.80**	0.84	**0.91**	0.92	**1.00**	0.97	**1.06**	1.08	**1.17**	1.16	**1.27**
Spleen	0.22	**0.26**	0.26	**0.31**	0.30	**0.35**	0.32	**0.38**	0.37	**0.44**	0.41	**0.49**
Liver	0.53	**0.60**	0.61	**0.69**	0.66	**0.75**	0.71	**0.80**	0.79	**0.90**	0.85	**0.98**
Gall bladder	0.64	**0.69**	0.75	**0.80**	0.82	**0.88**	0.88	**0.94**	0.99	**1.06**	1.08	**1.16**
Pancreas	0.37	**0.42**	0.44	**0.49**	0.48	**0.55**	0.52	**0.59**	0.60	**0.68**	0.67	**0.75**
Small intestine	0.53	**0.55**	0.61	**0.65**	0.67	**0.71**	0.72	**0.76**	0.82	**0.87**	0.90	**0.95**
Large intestine	0.62	**0.66**	0.72	**0.76**	0.79	**0.84**	0.85	**0.90**	0.95	**1.01**	1.04	**1.11**
Sigma/rectum	0.46	**0.49**	0.54	**0.57**	0.59	**0.63**	0.64	**0.68**	0.72	**0.78**	0.80	**0.86**
Lumbar spine	0.31	**0.33**	0.37	**0.40**	0.41	**0.45**	0.45	**0.49**	0.53	**0.57**	0.60	**0.65**
Pelvis	0.55	**0.60**	0.65	**0.71**	0.72	**0.79**	0.78	**0.85**	0.89	**0.97**	0.98	**1.07**
Adrenals	0.14	**0.17**	0.17	**0.20**	0.19	**0.23**	0.21	**0.25**	0.24	**0.29**	0.28	**0.33**
Kidneys	0.14	**0.16**	0.17	**0.19**	0.19	**0.21**	0.21	**0.23**	0.24	**0.27**	0.28	**0.31**
Urinary bladder	0.82	**0.86**	0.93	**0.98**	1.02	**1.07**	1.08	**1.13**	1.19	**1.25**	1.29	**1.35**
Uterus	1.58	**1.66**	1.85	**1.93**	2.04	**2.14**	2.19	**2.30**	2.48	**2.61**	2.74	**2.87**
Ovaries	0.43	**0.45**	0.50	**0.52**	0.56	**0.58**	0.60	**0.63**	0.69	**0.72**	0.76	**0.80**
Prostate	0.62	**0.66**	0.72	**0.77**	0.79	**0.85**	0.85	**0.91**	0.96	**1.02**	1.04	**1.12**
Testes	1.16	**1.23**	1.26	**1.35**	1.33	**1.42**	1.37	**1.47**	1.45	**1.55**	1.51	**1.62**
Upper arm		**0.18**		**0.21**		**0.23**		**0.25**		**0.27**		**0.30**
Elbow		**0.89**		**1.02**		**1.10**		**1.17**		**1.28**		**1.37**
Forearm		**1.03**		**1.17**		**1.27**		**1.34**		**1.46**		**1.55**
Upper leg	0.39	**0.62**	0.45	**0.73**	0.50	**0.81**	0.54	**0.87**	0.62	**0.99**	0.69	**1.10**
Knee	0.01	**0.07**	0.02	**0.08**	0.02	**0.09**	0.02	**0.10**	0.02	**0.12**	0.03	**0.13**
Lower leg												
Lymph nodes	0.32	**0.38**	0.37	**0.44**	0.41	**0.48**	0.44	**0.52**	0.49	**0.58**	0.54	**0.64**
Skin	0.22	**0.33**	0.23	**0.35**	0.24	**0.36**	0.24	**0.37**	0.26	**0.39**	0.27	**0.40**
Muscle	0.24	**0.33**	0.28	**0.37**	0.30	**0.40**	0.31	**0.42**	0.35	**0.46**	0.37	**0.50**
Skeleton	0.16	**0.35**	0.19	**0.40**	0.20	**0.44**	0.22	**0.47**	0.24	**0.52**	0.26	**0.56**
Red bone marrow	0.13	**0.17**	0.15	**0.20**	0.17	**0.23**	0.18	**0.24**	0.21	**0.28**	0.23	**0.31**
Whole body dose	0.24	**0.33**	0.27	**0.38**	0.29	**0.41**	0.31	**0.43**	0.34	**0.48**	0.37	**0.52**
ED (ICRP 60)	0.45	**0.80**	0.51	**0.90**	0.56	**0.97**	0.59	**1.02**	0.66	**1.11**	0.71	**1.19**
ED (ICRP 103)	0.39	**1.15**	0.45	**1.28**	0.49	**1.36**	0.52	**1.43**	0.58	**1.54**	0.64	**1.64**
Total filtration (mm) Al	**1.5**		**2.5**		**3.5**		**4.5**		**3.5**		**3.5**	
Cu	**0**		**0**		**0**		**0**		**0.1**		**0.2**	

Table 12.44 Normalized organ doses—Abdomen ap—>17 years

Beam projection:			ap									Table cells:
Focus-to-detector distance (cm):			100									(+) = normal font = optimal field setting
Tube voltage (kVp):			70									(−) = bold font = suboptimal field setting
Total filtrations:			See last table row.									Empty table cells correspond to values <0.01
Organ or tissue	Organ dose D_T : Entrance dose K_E								(mSv * mGy^{-1})			
	(+)	(−)	(+)	(−)	(+)	(−)	(+)	(−)	(+)	(−)	(+)	(−)
Skull												
Brain												
Oral mucosa												
Salivary glands												
Cervical spine												
Thyroid												
Thymus		0.04		0.05	0.01	0.06	0.01	0.07	0.02	0.08	0.02	0.10
Thoracic spine	0.04	0.06	0.05	0.08	0.06	0.09	0.07	0.11	0.09	0.14	0.11	0.18
Claviculae										0.01		0.02
Scapulae	0.01	0.05	0.01	0.07	0.02	0.08	0.02	0.09	0.03	0.12	0.03	0.14
Ribs	0.42	0.64	0.49	0.76	0.54	0.84	0.58	0.90	0.65	1.01	0.70	1.10
Oesophagus	0.06	0.08	0.08	0.11	0.09	0.13	0.10	0.15	0.13	0.18	0.16	0.22
Trachea												
Lung	0.07	0.16	0.09	0.20	0.11	0.23	0.12	0.26	0.14	0.31	0.17	0.35
Mamma	0.02	0.37	0.02	0.42	0.03	0.46	0.03	0.49	0.04	0.54	0.05	0.58
Heart	0.12	0.32	0.15	0.41	0.17	0.47	0.19	0.52	0.23	0.62	0.28	0.71
Stomach	0.43	0.45	0.53	0.57	0.61	0.66	0.68	0.73	0.80	0.86	0.92	0.99
Spleen	0.09	0.11	0.12	0.14	0.15	0.17	0.17	0.19	0.21	0.24	0.26	0.30
Liver	0.29	0.32	0.37	0.41	0.43	0.47	0.48	0.52	0.57	0.63	0.66	0.73
Gall bladder	0.33	0.35	0.42	0.44	0.49	0.52	0.55	0.58	0.67	0.71	0.79	0.84
Pancreas	0.16	0.18	0.21	0.23	0.25	0.28	0.29	0.31	0.36	0.40	0.43	0.48
Small intestine	0.26	0.27	0.33	0.35	0.39	0.41	0.44	0.46	0.54	0.56	0.64	0.66
Large intestine	0.33	0.34	0.42	0.43	0.49	0.51	0.54	0.57	0.66	0.69	0.77	0.81
Sigma/rectum	0.22	0.23	0.28	0.30	0.34	0.35	0.38	0.40	0.47	0.49	0.55	0.58
Lumbar spine	0.10	0.11	0.14	0.14	0.17	0.18	0.19	0.21	0.25	0.27	0.32	0.34
Pelvis	0.25	0.26	0.32	0.35	0.39	0.41	0.44	0.47	0.56	0.59	0.67	0.72
Adrenals	0.05	0.06	0.07	0.08	0.08	0.09	0.09	0.11	0.12	0.14	0.15	0.17
Kidneys	0.05	0.05	0.07	0.07	0.08	0.09	0.09	0.10	0.12	0.13	0.15	0.16
Urinary bladder	0.48	0.50	0.60	0.62	0.69	0.72	0.76	0.79	0.90	0.94	1.03	1.07
Uterus	0.30	0.31	0.38	0.40	0.45	0.47	0.51	0.53	0.63	0.65	0.75	0.77
Ovaries	0.19	0.19	0.25	0.25	0.29	0.30	0.33	0.34	0.42	0.43	0.50	0.52
Prostate	0.30	0.31	0.38	0.39	0.45	0.46	0.50	0.52	0.61	0.63	0.72	0.74
Testes	0.87	0.90	1.01	1.04	1.10	1.14	1.17	1.21	1.30	1.34	1.40	1.45
Upper arm		0.08		0.10		0.12		0.13		0.16		0.19
Elbow		0.51		0.64		0.75		0.83		0.99		1.13
Forearm		0.64		0.80		0.93		1.03		1.22		1.39
Upper leg	0.20	0.24	0.26	0.31	0.31	0.37	0.36	0.43	0.45	0.54	0.54	0.66
Knee		0.04	0.01	0.05	0.01	0.06	0.02	0.07	0.02	0.09	0.03	0.12
Lower leg												
Lymph nodes	0.16	0.19	0.21	0.24	0.24	0.28	0.27	0.31	0.33	0.38	0.39	0.45
Skin	0.20	0.29	0.21	0.30	0.22	0.32	0.23	0.33	0.24	0.35	0.25	0.36
Muscle	0.16	0.20	0.19	0.24	0.22	0.27	0.24	0.30	0.27	0.34	0.31	0.39
Skeleton	0.09	0.19	0.12	0.24	0.14	0.27	0.15	0.31	0.18	0.37	0.21	0.42
Red bone marrow	0.06	0.08	0.08	0.10	0.09	0.12	0.10	0.13	0.13	0.16	0.15	0.19
Whole body dose	0.15	0.20	0.18	0.24	0.21	0.28	0.23	0.30	0.27	0.35	0.30	0.40
ED (ICRP 60)	0.26	0.30	0.32	0.37	0.36	0.42	0.40	0.46	0.47	0.54	0.53	0.62
ED (ICRP 103)	0.20	0.27	0.25	0.33	0.28	0.37	0.31	0.41	0.37	0.48	0.43	0.55
Total filtration (mm) Al	1.5		2.5		3.5		4.5		3.5		3.5	
Cu	0		0		0		0		0.1		0.2	

Table 12.45 Normalized organ doses—Abdomen ap—>17 years

Beam projection:			ap									**Table cells:**
Focus-to-detector distance (cm):			100									(+) = normal font = optimal field setting
Tube voltage (kVp):			80									**(−) = bold font = suboptimal field setting**
Total filtrations:			See last table row.									Empty table cells correspond to values <0.01
Organ or tissue	**Organ dose D_T : Entrance dose K_E**								**(mSv * mGy^{-1})**			
	(+)	(−)	(+)	(−)	(+)	(−)	(+)	(−)	(+)	(−)	(+)	(−)
Skull												
Brain												
Oral mucosa												
Salivary glands												
Cervical spine												
Thyroid												
Thymus		**0.05**	0.01	**0.06**	0.01	**0.07**	0.02	**0.08**	0.02	**0.10**	0.03	**0.12**
Thoracic spine	0.05	**0.08**	0.07	**0.10**	0.08	**0.13**	0.09	**0.14**	0.12	**0.18**	0.15	**0.23**
Claviculae						**0.01**		**0.01**		**0.02**		**0.02**
Scapulae	0.01	**0.07**	0.02	**0.08**	0.02	**0.10**	0.03	**0.11**	0.03	**0.14**	0.04	**0.17**
Ribs	0.46	**0.70**	0.53	**0.82**	0.58	**0.90**	0.61	**0.96**	0.68	**1.06**	0.73	**1.15**
Oesophagus	0.08	**0.11**	0.10	**0.14**	0.11	**0.16**	0.13	**0.18**	0.16	**0.23**	0.19	**0.27**
Trachea												
Lung	0.09	**0.19**	0.11	**0.23**	0.13	**0.27**	0.14	**0.29**	0.16	**0.35**	0.19	**0.39**
Mamma	0.02	**0.40**	0.03	**0.45**	0.03	**0.49**	0.04	**0.52**	0.05	**0.57**	0.05	**0.61**
Heart	0.14	**0.38**	0.18	**0.47**	0.20	**0.54**	0.23	**0.59**	0.27	**0.70**	0.31	**0.79**
Stomach	0.50	**0.54**	0.62	**0.66**	0.70	**0.75**	0.77	**0.82**	0.90	**0.96**	1.01	**1.09**
Spleen	0.12	**0.14**	0.16	**0.18**	0.19	**0.21**	0.21	**0.24**	0.26	**0.30**	0.31	**0.36**
Liver	0.35	**0.39**	0.43	**0.48**	0.50	**0.55**	0.55	**0.60**	0.64	**0.71**	0.73	**0.81**
Gall bladder	0.40	**0.43**	0.50	**0.53**	0.58	**0.62**	0.65	**0.68**	0.78	**0.82**	0.90	**0.95**
Pancreas	0.21	**0.23**	0.27	**0.29**	0.31	**0.34**	0.35	**0.39**	0.43	**0.48**	0.51	**0.57**
Small intestine	0.32	**0.33**	0.40	**0.42**	0.47	**0.49**	0.52	**0.54**	0.63	**0.66**	0.73	**0.77**
Large intestine	0.40	**0.42**	0.50	**0.52**	0.57	**0.60**	0.63	**0.66**	0.76	**0.79**	0.87	**0.92**
Sigma/rectum	0.28	**0.29**	0.35	**0.37**	0.41	**0.43**	0.45	**0.48**	0.55	**0.58**	0.64	**0.67**
Lumbar spine	0.14	**0.15**	0.19	**0.20**	0.23	**0.24**	0.26	**0.28**	0.33	**0.35**	0.41	**0.43**
Pelvis	0.32	**0.34**	0.41	**0.44**	0.48	**0.51**	0.54	**0.58**	0.67	**0.71**	0.79	**0.85**
Adrenals	0.07	**0.08**	0.09	**0.10**	0.11	**0.12**	0.12	**0.14**	0.16	**0.18**	0.19	**0.22**
Kidneys	0.07	**0.08**	0.09	**0.10**	0.11	**0.12**	0.12	**0.13**	0.16	**0.17**	0.19	**0.21**
Urinary bladder	0.57	**0.59**	0.69	**0.72**	0.79	**0.82**	0.86	**0.89**	1.01	**1.04**	1.14	**1.18**
Uterus	0.37	**0.38**	0.47	**0.48**	0.55	**0.56**	0.61	**0.63**	0.74	**0.77**	0.86	**0.89**
Ovaries	0.24	**0.25**	0.31	**0.32**	0.36	**0.38**	0.41	**0.42**	0.50	**0.52**	0.59	**0.62**
Prostate	0.37	**0.38**	0.46	**0.47**	0.53	**0.55**	0.59	**0.61**	0.71	**0.73**	0.82	**0.84**
Testes	0.95	**0.98**	1.08	**1.12**	1.18	**1.22**	1.25	**1.29**	1.37	**1.42**	1.47	**1.53**
Upper arm		**0.10**		**0.12**		**0.14**		**0.15**		**0.18**		**0.21**
Elbow		**0.60**		**0.74**		**0.84**		**0.93**		**1.09**		**1.23**
Forearm		**0.75**		**0.91**		**1.04**		**1.14**		**1.33**		**1.49**
Upper leg	0.26	**0.31**	0.33	**0.40**	0.39	**0.47**	0.44	**0.53**	0.54	**0.65**	0.64	**0.78**
Knee	0.01	**0.05**	0.02	**0.07**	0.02	**0.08**	0.02	**0.09**	0.03	**0.12**	0.03	**0.14**
Lower leg												
Lymph nodes	0.20	**0.23**	0.25	**0.29**	0.29	**0.33**	0.32	**0.37**	0.38	**0.44**	0.44	**0.51**
Skin	0.21	**0.30**	0.22	**0.32**	0.23	**0.33**	0.24	**0.34**	0.25	**0.36**	0.26	**0.38**
Muscle	0.18	**0.23**	0.22	**0.27**	0.24	**0.31**	0.27	**0.33**	0.31	**0.38**	0.34	**0.43**
Skeleton	0.11	**0.23**	0.14	**0.28**	0.16	**0.32**	0.18	**0.35**	0.21	**0.41**	0.24	**0.47**
Red bone marrow	0.08	**0.10**	0.10	**0.12**	0.11	**0.14**	0.12	**0.16**	0.15	**0.19**	0.18	**0.23**
Whole body dose	0.18	**0.23**	0.21	**0.28**	0.24	**0.31**	0.26	**0.34**	0.30	**0.39**	0.33	**0.44**
ED (ICRP 60)	0.30	**0.35**	0.36	**0.42**	0.41	**0.48**	0.45	**0.52**	0.52	**0.61**	0.59	**0.68**
ED (ICRP 103)	0.24	**0.31**	0.29	**0.38**	0.33	**0.43**	0.36	**0.47**	0.42	**0.54**	0.48	**0.62**
Total filtration (mm) Al	**1.5**		**2.5**		**3.5**		**4.5**		**3.5**		**3.5**	
Cu	**0**		**0**		**0**		**0**		**0.1**		**0.2**	

Table 12.46 Normalized organ doses—Abdomen ap—>17 years

Beam projection:			ap									**Table cells:**
Focus-to-detector distance (cm):			100									(+) = normal font = optimal field setting
Tube voltage (kVp):			110									**(−) = bold font = suboptimal field setting**
Total filtrations:			See last table row.									Empty table cells correspond to values <0.01
Organ or tissue	**Organ dose D_T : Entrance dose K_E**								**(mSv * mGy^{-1})**			
	(+)	(−)	(+)	(−)	(+)	(−)	(+)	(−)	(+)	(−)	(+)	(−)
Skull												
Brain												
Oral mucosa												
Salivary glands												
Cervical spine												**0.01**
Thyroid												
Thymus	0.02	**0.08**	0.02	**0.10**	0.02	**0.11**	0.03	**0.12**	0.03	**0.14**	0.04	**0.16**
Thoracic spine	0.10	**0.15**	0.12	**0.18**	0.13	**0.21**	0.15	**0.23**	0.18	**0.27**	0.21	**0.31**
Claviculae		**0.01**		**0.02**		**0.02**		**0.02**		**0.03**		**0.03**
Scapulae	0.03	**0.11**	0.03	**0.13**	0.04	**0.14**	0.04	**0.16**	0.05	**0.18**	0.05	**0.21**
Ribs	0.54	**0.84**	0.59	**0.93**	0.63	**0.99**	0.66	**1.04**	0.70	**1.11**	0.73	**1.16**
Oesophagus	0.13	**0.18**	0.15	**0.22**	0.17	**0.24**	0.19	**0.27**	0.22	**0.31**	0.25	**0.36**
Trachea												
Lung	0.13	**0.27**	0.15	**0.31**	0.17	**0.35**	0.18	**0.37**	0.20	**0.42**	0.22	**0.46**
Mamma	0.04	**0.47**	0.04	**0.52**	0.05	**0.55**	0.05	**0.58**	0.06	**0.62**	0.07	**0.65**
Heart	0.21	**0.54**	0.25	**0.63**	0.28	**0.70**	0.30	**0.75**	0.34	**0.84**	0.38	**0.93**
Stomach	0.70	**0.75**	0.81	**0.87**	0.89	**0.96**	0.95	**1.02**	1.07	**1.15**	1.17	**1.26**
Spleen	0.20	**0.24**	0.25	**0.28**	0.28	**0.32**	0.30	**0.35**	0.36	**0.41**	0.40	**0.47**
Liver	0.50	**0.55**	0.58	**0.65**	0.64	**0.72**	0.69	**0.77**	0.78	**0.87**	0.86	**0.96**
Gall bladder	0.60	**0.63**	0.71	**0.75**	0.78	**0.83**	0.85	**0.90**	0.97	**1.02**	1.07	**1.14**
Pancreas	0.34	**0.37**	0.40	**0.45**	0.46	**0.50**	0.50	**0.55**	0.58	**0.64**	0.65	**0.72**
Small intestine	0.49	**0.51**	0.58	**0.61**	0.65	**0.68**	0.70	**0.73**	0.80	**0.84**	0.90	**0.94**
Large intestine	0.59	**0.61**	0.69	**0.72**	0.77	**0.80**	0.83	**0.87**	0.94	**0.99**	1.04	**1.10**
Sigma/rectum	0.43	**0.45**	0.51	**0.53**	0.57	**0.60**	0.62	**0.65**	0.71	**0.75**	0.79	**0.84**
Lumbar spine	0.27	**0.29**	0.33	**0.35**	0.37	**0.40**	0.41	**0.44**	0.49	**0.52**	0.56	**0.60**
Pelvis	0.51	**0.54**	0.61	**0.65**	0.68	**0.73**	0.74	**0.80**	0.86	**0.93**	0.97	**1.04**
Adrenals	0.13	**0.15**	0.16	**0.18**	0.18	**0.20**	0.19	**0.22**	0.23	**0.27**	0.27	**0.31**
Kidneys	0.13	**0.14**	0.15	**0.17**	0.18	**0.19**	0.19	**0.21**	0.23	**0.25**	0.27	**0.29**
Urinary bladder	0.79	**0.81**	0.91	**0.94**	1.00	**1.04**	1.07	**1.11**	1.20	**1.24**	1.31	**1.35**
Uterus	0.57	**0.59**	0.68	**0.71**	0.76	**0.79**	0.83	**0.86**	0.95	**0.99**	1.06	**1.10**
Ovaries	0.39	**0.41**	0.47	**0.49**	0.53	**0.55**	0.57	**0.60**	0.67	**0.69**	0.75	**0.78**
Prostate	0.54	**0.56**	0.64	**0.67**	0.71	**0.74**	0.77	**0.80**	0.88	**0.91**	0.98	**1.01**
Testes	1.13	**1.18**	1.25	**1.30**	1.33	**1.39**	1.39	**1.45**	1.49	**1.55**	1.56	**1.63**
Upper arm		**0.14**		**0.16**		**0.18**		**0.19**		**0.21**		**0.24**
Elbow		**0.81**		**0.94**		**1.03**		**1.10**		**1.23**		**1.33**
Forearm		**0.98**		**1.14**		**1.25**		**1.33**		**1.47**		**1.58**
Upper leg	0.41	**0.50**	0.50	**0.60**	0.56	**0.67**	0.61	**0.73**	0.71	**0.85**	0.80	**0.96**
Knee	0.02	**0.09**	0.03	**0.11**	0.03	**0.12**	0.04	**0.14**	0.04	**0.16**	0.05	**0.18**
Lower leg												
Lymph nodes	0.30	**0.35**	0.35	**0.41**	0.39	**0.45**	0.42	**0.49**	0.49	**0.56**	0.54	**0.62**
Skin	0.23	**0.33**	0.24	**0.35**	0.25	**0.36**	0.26	**0.37**	0.27	**0.39**	0.28	**0.40**
Muscle	0.25	**0.31**	0.28	**0.35**	0.31	**0.39**	0.33	**0.41**	0.36	**0.46**	0.40	**0.50**
Skeleton	0.16	**0.32**	0.19	**0.37**	0.21	**0.41**	0.23	**0.44**	0.26	**0.49**	0.29	**0.54**
Red bone marrow	0.12	**0.16**	0.15	**0.19**	0.17	**0.21**	0.18	**0.23**	0.21	**0.26**	0.24	**0.29**
Whole body dose	0.24	**0.32**	0.27	**0.36**	0.30	**0.40**	0.32	**0.42**	0.36	**0.47**	0.39	**0.51**
ED (ICRP 60)	0.41	**0.48**	0.48	**0.56**	0.52	**0.61**	0.56	**0.65**	0.63	**0.73**	0.69	**0.80**
ED (ICRP 103)	0.33	**0.43**	0.39	**0.50**	0.43	**0.55**	0.46	**0.59**	0.52	**0.66**	0.57	**0.72**
Total filtration (mm) Al	**1.5**		**2.5**		**3.5**		**4.5**		**3.5**		**3.5**	
Cu	**0**		**0**		**0**		**0**		**0.1**		**0.2**	

Table 12.47 Normalized organ doses—Abdomen ap—>17 years

Beam projection:			ap									**Table cells:**
Focus-to-detector distance (cm):			100									(+) = normal font = optimal field setting
Tube voltage (kVp):			120									**(−) = bold font = suboptimal field setting**
Total filtrations:			See last table row.									Empty table cells correspond to values <0.01
Organ or tissue	**Organ dose D_T : Entrance dose K_E**							**(mSv * mGy^{-1})**				
	(+)	(−)	(+)	(−)	(+)	(−)	(+)	(−)	(+)	(−)	(+)	(−)
Skull												
Brain												
Oral mucosa												
Salivary glands												
Cervical spine										**0.01**		**0.01**
Thyroid												
Thymus	0.02	**0.09**	0.02	**0.10**	0.03	**0.12**	0.03	**0.13**	0.03	**0.14**	0.04	**0.16**
Thoracic spine	0.11	**0.17**	0.13	**0.20**	0.15	**0.23**	0.16	**0.25**	0.19	**0.29**	0.22	**0.33**
Claviculae		**0.02**		**0.02**		**0.02**		**0.02**		**0.03**	0.01	**0.03**
Scapulae	0.03	**0.12**	0.03	**0.14**	0.04	**0.15**	0.04	**0.17**	0.05	**0.19**	0.05	**0.21**
Ribs	0.55	**0.86**	0.61	**0.95**	0.64	**1.01**	0.66	**1.05**	0.70	**1.11**	0.72	**1.15**
Oesophagus	0.14	**0.20**	0.17	**0.24**	0.19	**0.27**	0.20	**0.29**	0.23	**0.33**	0.26	**0.37**
Trachea												
Lung	0.14	**0.29**	0.16	**0.34**	0.18	**0.37**	0.19	**0.39**	0.21	**0.44**	0.23	**0.48**
Mamma	0.04	**0.49**	0.05	**0.54**	0.05	**0.57**	0.05	**0.59**	0.06	**0.63**	0.07	**0.65**
Heart	0.23	**0.59**	0.27	**0.67**	0.29	**0.74**	0.32	**0.78**	0.36	**0.87**	0.39	**0.95**
Stomach	0.75	**0.80**	0.86	**0.92**	0.93	**1.01**	0.99	**1.07**	1.10	**1.19**	1.19	**1.28**
Spleen	0.23	**0.26**	0.27	**0.31**	0.30	**0.35**	0.33	**0.38**	0.38	**0.44**	0.42	**0.49**
Liver	0.54	**0.60**	0.62	**0.69**	0.68	**0.76**	0.73	**0.81**	0.81	**0.90**	0.88	**0.98**
Gall bladder	0.65	**0.69**	0.76	**0.80**	0.83	**0.88**	0.89	**0.95**	1.01	**1.07**	1.10	**1.17**
Pancreas	0.38	**0.41**	0.44	**0.49**	0.49	**0.54**	0.53	**0.59**	0.61	**0.67**	0.68	**0.75**
Small intestine	0.54	**0.56**	0.62	**0.65**	0.69	**0.72**	0.74	**0.78**	0.84	**0.88**	0.92	**0.97**
Large intestine	0.64	**0.67**	0.74	**0.78**	0.81	**0.85**	0.87	**0.91**	0.98	**1.03**	1.07	**1.13**
Sigma/rectum	0.47	**0.49**	0.55	**0.58**	0.61	**0.64**	0.65	**0.69**	0.74	**0.78**	0.82	**0.87**
Lumbar spine	0.30	**0.32**	0.36	**0.38**	0.41	**0.43**	0.44	**0.47**	0.52	**0.56**	0.59	**0.63**
Pelvis	0.56	**0.60**	0.65	**0.70**	0.73	**0.78**	0.78	**0.84**	0.90	**0.96**	0.99	**1.07**
Adrenals	0.15	**0.17**	0.17	**0.20**	0.19	**0.22**	0.21	**0.25**	0.25	**0.29**	0.28	**0.33**
Kidneys	0.14	**0.16**	0.17	**0.19**	0.19	**0.21**	0.21	**0.23**	0.25	**0.27**	0.28	**0.31**
Urinary bladder	0.84	**0.87**	0.96	**1.00**	1.05	**1.09**	1.12	**1.16**	1.24	**1.28**	1.33	**1.38**
Uterus	0.63	**0.65**	0.74	**0.76**	0.81	**0.84**	0.88	**0.91**	1.00	**1.03**	1.10	**1.14**
Ovaries	0.43	**0.45**	0.51	**0.53**	0.57	**0.59**	0.61	**0.64**	0.70	**0.73**	0.78	**0.81**
Prostate	0.59	**0.62**	0.69	**0.71**	0.76	**0.79**	0.81	**0.84**	0.92	**0.95**	1.01	**1.04**
Testes	1.18	**1.23**	1.29	**1.35**	1.37	**1.42**	1.42	**1.48**	1.51	**1.57**	1.58	**1.65**
Upper arm		**0.15**		**0.17**		**0.18**		**0.20**		**0.22**		**0.24**
Elbow		**0.86**		**0.98**		**1.07**		**1.13**		**1.25**		**1.33**
Forearm		**1.04**		**1.19**		**1.29**		**1.36**		**1.49**		**1.58**
Upper leg	0.46	**0.55**	0.54	**0.65**	0.60	**0.72**	0.65	**0.78**	0.74	**0.89**	0.82	**0.99**
Knee	0.03	**0.10**	0.03	**0.12**	0.04	**0.13**	0.04	**0.15**	0.05	**0.17**	0.05	**0.19**
Lower leg												
Lymph nodes	0.33	**0.38**	0.38	**0.44**	0.42	**0.48**	0.45	**0.52**	0.51	**0.58**	0.56	**0.64**
Skin	0.24	**0.34**	0.25	**0.36**	0.26	**0.37**	0.26	**0.38**	0.27	**0.39**	0.28	**0.41**
Muscle	0.26	**0.33**	0.30	**0.37**	0.32	**0.40**	0.34	**0.43**	0.38	**0.47**	0.41	**0.51**
Skeleton	0.18	**0.34**	0.20	**0.39**	0.22	**0.43**	0.24	**0.45**	0.27	**0.51**	0.29	**0.55**
Red bone marrow	0.14	**0.17**	0.16	**0.20**	0.18	**0.23**	0.19	**0.24**	0.22	**0.28**	0.25	**0.31**
Whole body dose	0.26	**0.34**	0.29	**0.38**	0.31	**0.42**	0.33	**0.44**	0.37	**0.49**	0.40	**0.53**
ED (ICRP 60)	0.45	**0.52**	0.51	**0.59**	0.55	**0.64**	0.59	**0.68**	0.65	**0.75**	0.70	**0.82**
ED (ICRP 103)	0.36	**0.46**	0.41	**0.53**	0.45	**0.58**	0.48	**0.61**	0.54	**0.68**	0.58	**0.74**
Total filtration (mm) Al	**1.5**		**2.5**		**3.5**		**4.5**		**3.5**		**3.5**	
Cu	**0**		**0**		**0**		**0**		**0.1**		**0.2**	

Table 12.48 Normalized organ doses—Abdomen pa—<1 year

Beam projection:			pa									**Table cells:**
Focus-to-detector distance (cm):			100				(+) = normal font = optimal field setting					
Tube voltage (kVp):			60				**(−) = bold font = suboptimal field setting**					
Total filtrations:			See last table row.				Empty table cells correspond to values <0.01					
Organ or tissue	**Organ dose D_T : Entrance dose K_E**								**(mSv * mGy⁻¹)**			
	(+)	(−)	(+)	(−)	(+)	(−)	(+)	(−)	(+)	(−)	(+)	(−)
Skull												
Brain												
Oral mucosa												
Salivary glands								**0.01**		**0.01**		**0.01**
Cervical spine		**0.02**		**0.03**	0.01	**0.03**	0.01	**0.04**	0.02	**0.05**	0.01	**0.04**
Thyroid		**0.01**		**0.02**		**0.02**		**0.03**	0.01	**0.03**		**0.03**
Thymus		**0.04**	0.01	**0.05**	0.02	**0.06**	0.02	**0.07**	0.02	**0.09**	0.02	**0.08**
Thoracic spine	0.49	**0.84**	0.61	**1.05**	0.69	**1.19**	0.76	**1.30**	0.88	**1.51**	0.86	**1.48**
Claviculae	0.01	**0.04**	0.02	**0.06**	0.02	**0.07**	0.03	**0.08**	0.03	**0.09**	0.03	**0.08**
Scapulae	0.05	**1.29**	0.06	**1.46**	0.07	**1.56**	0.08	**1.63**	0.10	**1.74**	0.09	**1.76**
Ribs	0.61	**1.01**	0.70	**1.17**	0.76	**1.27**	0.80	**1.35**	0.87	**1.47**	0.87	**1.47**
Oesophagus	0.11	**0.18**	0.14	**0.24**	0.16	**0.28**	0.18	**0.31**	0.22	**0.38**	0.21	**0.35**
Trachea												
Lung	0.18	**0.41**	0.21	**0.49**	0.24	**0.55**	0.25	**0.59**	0.29	**0.66**	0.28	**0.65**
Mamma	0.02	**0.19**	0.03	**0.24**	0.03	**0.27**	0.04	**0.29**	0.05	**0.34**	0.05	**0.33**
Heart	0.09	**0.19**	0.12	**0.24**	0.14	**0.29**	0.15	**0.32**	0.19	**0.38**	0.17	**0.36**
Stomach	0.18	**0.20**	0.24	**0.26**	0.28	**0.30**	0.31	**0.34**	0.38	**0.41**	0.35	**0.38**
Spleen	0.50	**0.53**	0.61	**0.65**	0.69	**0.73**	0.74	**0.79**	0.84	**0.90**	0.83	**0.88**
Liver	0.25	**0.27**	0.31	**0.34**	0.36	**0.39**	0.40	**0.43**	0.47	**0.51**	0.45	**0.49**
Gall bladder	0.17	**0.18**	0.22	**0.24**	0.26	**0.28**	0.30	**0.32**	0.37	**0.39**	0.33	**0.36**
Pancreas	0.27	**0.29**	0.34	**0.37**	0.40	**0.43**	0.44	**0.48**	0.53	**0.58**	0.51	**0.54**
Small intestine	0.26	**0.27**	0.33	**0.34**	0.39	**0.40**	0.43	**0.45**	0.52	**0.54**	0.49	**0.51**
Large intestine	0.22	**0.23**	0.28	**0.30**	0.33	**0.35**	0.37	**0.39**	0.45	**0.48**	0.42	**0.44**
Sigma/rectum	0.27	**0.28**	0.34	**0.36**	0.40	**0.42**	0.44	**0.47**	0.53	**0.56**	0.50	**0.53**
Lumbar spine	1.37	**1.39**	1.69	**1.72**	1.91	**1.95**	2.09	**2.14**	2.41	**2.46**	2.37	**2.42**
Pelvis	1.42	**1.46**	1.72	**1.77**	1.93	**1.99**	2.09	**2.15**	2.37	**2.44**	2.36	**2.42**
Adrenals	0.51	**0.54**	0.62	**0.65**	0.69	**0.72**	0.74	**0.78**	0.84	**0.89**	0.82	**0.87**
Kidneys	0.75	**0.76**	0.88	**0.89**	0.96	**0.98**	1.02	**1.04**	1.13	**1.15**	1.11	**1.14**
Urinary bladder	0.17	**0.18**	0.22	**0.23**	0.26	**0.28**	0.29	**0.31**	0.35	**0.38**	0.33	**0.35**
Uterus	0.25	**0.25**	0.32	**0.33**	0.37	**0.39**	0.42	**0.43**	0.51	**0.52**	0.48	**0.49**
Ovaries	0.27	**0.29**	0.35	**0.37**	0.41	**0.44**	0.46	**0.49**	0.56	**0.59**	0.52	**0.56**
Prostate	0.19	**0.20**	0.24	**0.26**	0.29	**0.31**	0.32	**0.35**	0.39	**0.42**	0.36	**0.39**
Testes	0.21	**0.23**	0.27	**0.29**	0.31	**0.34**	0.34	**0.37**	0.40	**0.44**	0.38	**0.42**
Upper arm		**0.43**		**0.52**		**0.59**		**0.64**		**0.73**		**0.72**
Elbow		**1.17**		**1.41**		**1.58**		**1.71**		**1.94**		**1.93**
Forearm		**1.29**		**1.54**		**1.72**		**1.86**		**2.08**		**2.09**
Upper leg	0.72	**0.96**	0.90	**1.20**	1.03	**1.38**	1.14	**1.52**	1.33	**1.77**	1.30	**1.74**
Knee	0.05	**0.45**	0.07	**0.56**	0.09	**0.65**	0.10	**0.71**	0.12	**0.82**	0.11	**0.81**
Lower leg				**0.01**		**0.02**		**0.02**	0.01	**0.02**		**0.02**
Lymph nodes	0.16	**0.20**	0.20	**0.25**	0.24	**0.29**	0.27	**0.32**	0.32	**0.38**	0.30	**0.36**
Skin	0.22	**0.33**	0.24	**0.35**	0.25	**0.36**	0.26	**0.37**	0.27	**0.39**	0.27	**0.39**
Muscle	0.22	**0.30**	0.26	**0.35**	0.29	**0.39**	0.31	**0.42**	0.35	**0.47**	0.34	**0.46**
Skeleton	0.40	**0.76**	0.49	**0.92**	0.55	**1.03**	0.60	**1.11**	0.68	**1.26**	0.67	**1.25**
Red bone marrow	0.11	**0.20**	0.14	**0.24**	0.16	**0.28**	0.17	**0.30**	0.20	**0.34**	0.19	**0.33**
Whole body dose	0.21	**0.31**	0.26	**0.37**	0.28	**0.41**	0.31	**0.44**	0.35	**0.49**	0.34	**0.48**
ED (ICRP 60)	0.18	**0.24**	0.23	**0.31**	0.26	**0.35**	0.29	**0.39**	0.34	**0.46**	0.33	**0.44**
ED (ICRP 103)	0.16	**0.24**	0.20	**0.30**	0.24	**0.34**	0.26	**0.37**	0.31	**0.44**	0.29	**0.42**
Total filtration (mm) **Al**	**1.5**		**2.5**		**3.5**		**4.5**		**3.5**		**3.5**	
Cu	**0**		**0**		**0**		**0**		**0.1**		**0.2**	

Table 12.49 Normalized organ doses—Abdomen pa—<1 year

Beam projection:			pa									**Table cells:**
Focus-to-detector distance (cm):			100						(+) = normal font = optimal field setting			
Tube voltage (kVp):			70						**(−) = bold font = suboptimal field setting**			
Total filtrations:			See last table row.						Empty table cells correspond to values <0.01			
Organ or tissue	**Organ dose D_T : Entrance dose K_E**								**(mSv * mGy^{-1})**			
	(+)	(−)	(+)	(−)	(+)	(−)	(+)	(−)	(+)	(−)	(+)	(−)
Skull										**0.01**		**0.01**
Brain												
Oral mucosa										**0.01**		**0.01**
Salivary glands				**0.01**		**0.01**		**0.01**		**0.02**		**0.02**
Cervical spine		**0.03**	0.01	**0.03**	0.01	**0.04**	0.02	**0.05**	0.02	**0.06**	0.03	**0.07**
Thyroid		**0.02**		**0.02**		**0.03**	0.01	**0.03**	0.01	**0.04**	0.02	**0.04**
Thymus	0.01	**0.05**	0.02	**0.06**	0.02	**0.08**	0.02	**0.08**	0.03	**0.10**	0.03	**0.12**
Thoracic spine	0.56	**0.96**	0.68	**1.17**	0.77	**1.31**	0.83	**1.43**	0.95	**1.62**	1.04	**1.78**
Claviculae	0.02	**0.05**	0.02	**0.07**	0.03	**0.08**	0.03	**0.09**	0.04	**0.11**	0.05	**0.13**
Scapulae	0.06	**1.36**	0.08	**1.52**	0.09	**1.61**	0.10	**1.67**	0.12	**1.76**	0.14	**1.82**
Ribs	0.65	**1.09**	0.74	**1.24**	0.79	**1.34**	0.83	**1.41**	0.90	**1.52**	0.95	**1.61**
Oesophagus	0.13	**0.22**	0.17	**0.28**	0.19	**0.33**	0.22	**0.36**	0.26	**0.44**	0.30	**0.50**
Trachea										**0.01**		**0.01**
Lung	0.20	**0.46**	0.23	**0.54**	0.26	**0.59**	0.28	**0.64**	0.31	**0.71**	0.34	**0.77**
Mamma	0.03	**0.22**	0.04	**0.27**	0.04	**0.30**	0.05	**0.33**	0.06	**0.37**	0.08	**0.41**
Heart	0.11	**0.23**	0.14	**0.29**	0.16	**0.33**	0.18	**0.37**	0.21	**0.44**	0.25	**0.50**
Stomach	0.22	**0.24**	0.28	**0.31**	0.33	**0.36**	0.36	**0.40**	0.43	**0.47**	0.50	**0.54**
Spleen	0.57	**0.60**	0.68	**0.72**	0.75	**0.80**	0.81	**0.86**	0.91	**0.97**	0.99	**1.06**
Liver	0.29	**0.32**	0.36	**0.40**	0.41	**0.45**	0.45	**0.49**	0.52	**0.58**	0.59	**0.65**
Gall bladder	0.21	**0.23**	0.27	**0.29**	0.32	**0.34**	0.36	**0.38**	0.43	**0.47**	0.51	**0.54**
Pancreas	0.32	**0.35**	0.40	**0.44**	0.46	**0.50**	0.51	**0.56**	0.61	**0.66**	0.69	**0.75**
Small intestine	0.31	**0.32**	0.39	**0.41**	0.46	**0.47**	0.50	**0.53**	0.60	**0.63**	0.68	**0.72**
Large intestine	0.27	**0.28**	0.34	**0.36**	0.40	**0.41**	0.44	**0.46**	0.52	**0.55**	0.60	**0.63**
Sigma/rectum	0.32	**0.34**	0.40	**0.43**	0.46	**0.49**	0.51	**0.54**	0.60	**0.64**	0.68	**0.73**
Lumbar spine	1.55	**1.58**	1.87	**1.92**	2.10	**2.15**	2.28	**2.33**	2.58	**2.64**	2.82	**2.90**
Pelvis	1.59	**1.63**	1.88	**1.94**	2.09	**2.15**	2.24	**2.31**	2.51	**2.59**	2.71	**2.81**
Adrenals	0.57	**0.60**	0.68	**0.72**	0.75	**0.79**	0.80	**0.85**	0.90	**0.95**	0.98	**1.04**
Kidneys	0.82	**0.84**	0.94	**0.96**	1.03	**1.05**	1.09	**1.11**	1.19	**1.22**	1.27	**1.31**
Urinary bladder	0.21	**0.22**	0.27	**0.28**	0.31	**0.33**	0.34	**0.37**	0.41	**0.44**	0.47	**0.51**
Uterus	0.30	**0.31**	0.38	**0.39**	0.44	**0.46**	0.49	**0.51**	0.58	**0.60**	0.67	**0.69**
Ovaries	0.33	**0.35**	0.42	**0.45**	0.49	**0.52**	0.54	**0.57**	0.64	**0.68**	0.74	**0.78**
Prostate	0.23	**0.25**	0.29	**0.31**	0.34	**0.36**	0.37	**0.40**	0.44	**0.48**	0.51	**0.55**
Testes	0.25	**0.27**	0.31	**0.34**	0.35	**0.39**	0.39	**0.42**	0.45	**0.49**	0.51	**0.56**
Upper arm		**0.48**		**0.57**		**0.64**		**0.69**		**0.77**		**0.84**
Elbow		**1.30**		**1.54**		**1.71**		**1.83**		**2.04**		**2.19**
Forearm		**1.42**		**1.67**		**1.84**		**1.97**		**2.18**		**2.33**
Upper leg	0.83	**1.11**	1.01	**1.35**	1.15	**1.53**	1.25	**1.67**	1.44	**1.91**	1.59	**2.11**
Knee	0.07	**0.52**	0.09	**0.63**	0.11	**0.71**	0.12	**0.78**	0.14	**0.89**	0.17	**0.98**
Lower leg		**0.01**		**0.02**		**0.02**	0.01	**0.02**	0.01	**0.03**	0.02	**0.03**
Lymph nodes	0.19	**0.24**	0.24	**0.29**	0.28	**0.34**	0.31	**0.37**	0.36	**0.44**	0.41	**0.50**
Skin	0.23	**0.34**	0.25	**0.36**	0.26	**0.38**	0.27	**0.39**	0.28	**0.41**	0.30	**0.43**
Muscle	0.24	**0.33**	0.29	**0.39**	0.31	**0.43**	0.34	**0.45**	0.38	**0.51**	0.41	**0.55**
Skeleton	0.45	**0.85**	0.54	**1.00**	0.60	**1.11**	0.64	**1.19**	0.72	**1.33**	0.78	**1.44**
Red bone marrow	0.13	**0.23**	0.16	**0.28**	0.18	**0.31**	0.19	**0.33**	0.22	**0.38**	0.24	**0.42**
Whole body dose	0.24	**0.34**	0.28	**0.40**	0.31	**0.44**	0.33	**0.48**	0.38	**0.53**	0.41	**0.58**
ED (ICRP 60)	0.21	**0.29**	0.27	**0.35**	0.30	**0.40**	0.33	**0.44**	0.39	**0.51**	0.44	**0.57**
ED (ICRP 103)	0.19	**0.28**	0.24	**0.34**	0.27	**0.39**	0.30	**0.42**	0.35	**0.49**	0.39	**0.55**
Total filtration (mm) **Al**	**1.5**		**2.5**		**3.5**		**4.5**		**3.5**		**3.5**	
Cu	**0**		**0**		**0**		**0**		**0.1**		**0.2**	

Table 12.50 Normalized organ doses—Abdomen pa—1–2 years

Beam projection:			pa									**Table cells:**
Focus-to-detector distance (cm):			100									(+) = normal font = optimal field setting
Tube voltage (kVp):			60									**(−) = bold font = suboptimal field setting**
Total filtrations:			See last table row.									Empty table cells correspond to values <0.01
Organ or tissue	**Organ dose D_T : Entrance dose K_E**								**(mSv * mGy^{-1})**			
	(+)	(−)	(+)	(−)	(+)	(−)	(+)	(−)	(+)	(−)	(+)	(−)
Skull												
Brain												
Oral mucosa												
Salivary glands												
Cervical spine		**0.01**		**0.01**		**0.02**		**0.02**		**0.03**		**0.02**
Thyroid				**0.01**		**0.01**		**0.02**		**0.02**		**0.02**
Thymus		**0.03**		**0.04**		**0.05**	0.01	**0.06**	0.01	**0.07**	0.01	**0.06**
Thoracic spine	0.33	**0.64**	0.42	**0.81**	0.49	**0.95**	0.54	**1.05**	0.65	**1.26**	0.61	**1.20**
Claviculae		**0.03**	0.01	**0.04**	0.01	**0.05**	0.02	**0.06**	0.02	**0.08**	0.02	**0.06**
Scapulae	0.02	**0.92**	0.03	**1.06**	0.03	**1.16**	0.04	**1.23**	0.05	**1.34**	0.04	**1.34**
Ribs	0.48	**0.94**	0.56	**1.10**	0.62	**1.21**	0.66	**1.29**	0.73	**1.43**	0.72	**1.42**
Oesophagus	0.08	**0.14**	0.10	**0.19**	0.12	**0.22**	0.14	**0.25**	0.17	**0.31**	0.15	**0.28**
Trachea												
Lung	0.11	**0.32**	0.13	**0.40**	0.15	**0.45**	0.16	**0.49**	0.19	**0.56**	0.18	**0.55**
Mamma	0.01	**0.13**	0.02	**0.17**	0.02	**0.19**	0.03	**0.22**	0.03	**0.26**	0.03	**0.24**
Heart	0.05	**0.12**	0.07	**0.16**	0.08	**0.19**	0.09	**0.22**	0.11	**0.27**	0.10	**0.24**
Stomach	0.11	**0.12**	0.15	**0.17**	0.18	**0.20**	0.20	**0.22**	0.25	**0.28**	0.22	**0.25**
Spleen	0.40	**0.43**	0.50	**0.54**	0.57	**0.61**	0.62	**0.68**	0.72	**0.79**	0.70	**0.76**
Liver	0.18	**0.20**	0.23	**0.26**	0.27	**0.30**	0.30	**0.34**	0.36	**0.41**	0.34	**0.38**
Gall bladder	0.11	**0.12**	0.15	**0.16**	0.18	**0.20**	0.21	**0.22**	0.26	**0.28**	0.23	**0.25**
Pancreas	0.18	**0.20**	0.24	**0.27**	0.29	**0.32**	0.32	**0.36**	0.40	**0.44**	0.36	**0.40**
Small intestine	0.18	**0.18**	0.23	**0.24**	0.28	**0.29**	0.31	**0.33**	0.39	**0.41**	0.35	**0.37**
Large intestine	0.15	**0.15**	0.19	**0.20**	0.23	**0.24**	0.26	**0.28**	0.33	**0.35**	0.29	**0.31**
Sigma/rectum	0.18	**0.19**	0.24	**0.25**	0.29	**0.30**	0.32	**0.34**	0.40	**0.42**	0.36	**0.38**
Lumbar spine	1.08	**1.10**	1.37	**1.39**	1.58	**1.62**	1.76	**1.79**	2.08	**2.13**	1.99	**2.03**
Pelvis	1.15	**1.17**	1.43	**1.46**	1.63	**1.67**	1.79	**1.84**	2.09	**2.14**	2.03	**2.08**
Adrenals	0.41	**0.44**	0.51	**0.55**	0.59	**0.63**	0.65	**0.69**	0.75	**0.80**	0.72	**0.77**
Kidneys	0.66	**0.68**	0.79	**0.81**	0.88	**0.90**	0.94	**0.97**	1.06	**1.09**	1.03	**1.06**
Urinary bladder	0.11	**0.12**	0.15	**0.16**	0.18	**0.19**	0.21	**0.22**	0.26	**0.27**	0.23	**0.24**
Uterus	0.15	**0.16**	0.21	**0.21**	0.25	**0.25**	0.28	**0.28**	0.35	**0.36**	0.31	**0.32**
Ovaries	0.19	**0.18**	0.25	**0.25**	0.30	**0.30**	0.34	**0.34**	0.42	**0.42**	0.38	**0.38**
Prostate	0.13	**0.14**	0.17	**0.18**	0.20	**0.21**	0.23	**0.24**	0.28	**0.30**	0.26	**0.27**
Testes	0.10	**0.10**	0.13	**0.13**	0.16	**0.16**	0.18	**0.18**	0.22	**0.22**	0.20	**0.20**
Upper arm		**0.30**		**0.38**		**0.44**		**0.48**		**0.57**		**0.54**
Elbow		**0.85**		**1.05**		**1.20**		**1.32**		**1.54**		**1.50**
Forearm		**0.96**		**1.18**		**1.35**		**1.48**		**1.71**		**1.68**
Upper leg	0.47	**0.61**	0.61	**0.79**	0.71	**0.92**	0.80	**1.03**	0.97	**1.24**	0.91	**1.17**
Knee	0.03	**0.09**	0.04	**0.12**	0.05	**0.14**	0.05	**0.16**	0.07	**0.20**	0.06	**0.18**
Lower leg												
Lymph nodes	0.11	**0.14**	0.14	**0.18**	0.17	**0.21**	0.19	**0.24**	0.23	**0.29**	0.21	**0.27**
Skin	0.20	**0.28**	0.21	**0.30**	0.22	**0.31**	0.22	**0.32**	0.24	**0.33**	0.23	**0.32**
Muscle	0.16	**0.21**	0.19	**0.25**	0.22	**0.28**	0.23	**0.31**	0.27	**0.35**	0.25	**0.33**
Skeleton	0.28	**0.52**	0.34	**0.65**	0.39	**0.74**	0.43	**0.81**	0.50	**0.94**	0.49	**0.91**
Red bone marrow	0.10	**0.15**	0.12	**0.19**	0.14	**0.22**	0.16	**0.24**	0.19	**0.29**	0.17	**0.27**
Whole body dose	0.16	**0.23**	0.20	**0.28**	0.22	**0.32**	0.24	**0.34**	0.28	**0.39**	0.26	**0.38**
ED (ICRP 60)	0.12	**0.17**	0.16	**0.22**	0.18	**0.25**	0.20	**0.28**	0.25	**0.34**	0.23	**0.31**
ED (ICRP 103)	0.11	**0.17**	0.14	**0.22**	0.17	**0.25**	0.19	**0.28**	0.23	**0.33**	0.21	**0.31**
Total filtration (mm) **Al**	**1.5**		**2.5**		**3.5**		**4.5**		**3.5**		**3.5**	
Cu	**0**		**0**		**0**		**0**		**0.1**		**0.2**	

Table 12.51 Normalized organ doses—Abdomen pa—1–2 years

Beam projection:			pa									**Table cells:**
Focus-to-detector distance (cm):			100									(+) = normal font = optimal field setting
Tube voltage (kVp):			70									**(−) = bold font = suboptimal field setting**
Total filtrations:			See last table row.									Empty table cells correspond to values <0.01
Organ or tissue	**Organ dose D_T : Entrance dose K_E**								**(mSv * mGy^{-1})**			
	(+)	(−)	(+)	(−)	(+)	(−)	(+)	(−)	(+)	(−)	(+)	(−)
Skull												
Brain												
Oral mucosa												
Salivary glands										**0.01**		**0.01**
Cervical spine		**0.01**		**0.02**		**0.02**		**0.03**	0.01	**0.04**	0.01	**0.04**
Thyroid		**0.01**		**0.02**		**0.02**		**0.02**		**0.03**	0.01	**0.03**
Thymus		**0.04**	0.01	**0.05**	0.01	**0.06**	0.02	**0.07**	0.02	**0.09**	0.02	**0.11**
Thoracic spine	0.39	**0.76**	0.49	**0.95**	0.56	**1.09**	0.62	**1.20**	0.73	**1.42**	0.83	**1.60**
Claviculae	0.01	**0.04**	0.01	**0.06**	0.02	**0.07**	0.02	**0.07**	0.03	**0.09**	0.03	**0.11**
Scapulae	0.03	**0.99**	0.04	**1.13**	0.04	**1.22**	0.05	**1.28**	0.06	**1.38**	0.08	**1.46**
Ribs	0.52	**1.02**	0.60	**1.18**	0.66	**1.29**	0.70	**1.37**	0.77	**1.51**	0.82	**1.62**
Oesophagus	0.10	**0.18**	0.13	**0.24**	0.15	**0.28**	0.17	**0.31**	0.21	**0.38**	0.25	**0.45**
Trachea												
Lung	0.12	**0.37**	0.15	**0.45**	0.17	**0.50**	0.18	**0.54**	0.21	**0.62**	0.23	**0.68**
Mamma	0.02	**0.16**	0.02	**0.20**	0.03	**0.23**	0.03	**0.25**	0.04	**0.30**	0.04	**0.34**
Heart	0.07	**0.16**	0.09	**0.20**	0.10	**0.23**	0.11	**0.26**	0.14	**0.32**	0.17	**0.38**
Stomach	0.14	**0.16**	0.19	**0.21**	0.22	**0.25**	0.24	**0.28**	0.30	**0.34**	0.35	**0.40**
Spleen	0.46	**0.50**	0.57	**0.61**	0.64	**0.70**	0.70	**0.76**	0.80	**0.88**	0.89	**0.98**
Liver	0.22	**0.25**	0.28	**0.31**	0.32	**0.36**	0.35	**0.40**	0.42	**0.47**	0.48	**0.55**
Gall bladder	0.15	**0.16**	0.19	**0.21**	0.23	**0.25**	0.26	**0.29**	0.33	**0.35**	0.39	**0.42**
Pancreas	0.23	**0.26**	0.30	**0.33**	0.35	**0.39**	0.39	**0.43**	0.47	**0.53**	0.55	**0.62**
Small intestine	0.23	**0.24**	0.29	**0.30**	0.34	**0.36**	0.39	**0.40**	0.47	**0.49**	0.55	**0.58**
Large intestine	0.19	**0.20**	0.25	**0.26**	0.29	**0.31**	0.33	**0.34**	0.40	**0.42**	0.47	**0.50**
Sigma/rectum	0.23	**0.24**	0.30	**0.31**	0.35	**0.37**	0.39	**0.41**	0.48	**0.50**	0.55	**0.58**
Lumbar spine	1.28	**1.31**	1.58	**1.62**	1.81	**1.85**	1.99	**2.04**	2.32	**2.38**	2.62	**2.69**
Pelvis	1.33	**1.36**	1.62	**1.66**	1.83	**1.87**	1.99	**2.04**	2.28	**2.35**	2.53	**2.61**
Adrenals	0.48	**0.51**	0.59	**0.63**	0.66	**0.71**	0.72	**0.78**	0.83	**0.90**	0.92	**1.00**
Kidneys	0.74	**0.76**	0.87	**0.89**	0.95	**0.98**	1.02	**1.05**	1.14	**1.17**	1.24	**1.28**
Urinary bladder	0.15	**0.16**	0.19	**0.20**	0.23	**0.24**	0.26	**0.27**	0.32	**0.33**	0.37	**0.39**
Uterus	0.20	**0.20**	0.26	**0.27**	0.31	**0.31**	0.35	**0.35**	0.43	**0.44**	0.51	**0.52**
Ovaries	0.24	**0.24**	0.31	**0.31**	0.37	**0.37**	0.41	**0.42**	0.51	**0.52**	0.60	**0.61**
Prostate	0.17	**0.17**	0.21	**0.22**	0.25	**0.26**	0.28	**0.30**	0.34	**0.36**	0.40	**0.43**
Testes	0.13	**0.13**	0.17	**0.16**	0.19	**0.19**	0.22	**0.21**	0.26	**0.26**	0.30	**0.30**
Upper arm		**0.35**		**0.44**		**0.49**		**0.54**		**0.63**		**0.70**
Elbow		**0.98**		**1.19**		**1.34**		**1.46**		**1.68**		**1.86**
Forearm		**1.09**		**1.32**		**1.49**		**1.62**		**1.84**		**2.02**
Upper leg	0.57	**0.74**	0.72	**0.93**	0.84	**1.07**	0.93	**1.19**	1.10	**1.41**	1.25	**1.60**
Knee	0.04	**0.12**	0.05	**0.15**	0.06	**0.17**	0.07	**0.19**	0.09	**0.23**	0.11	**0.27**
Lower leg												
Lymph nodes	0.14	**0.17**	0.17	**0.22**	0.20	**0.26**	0.23	**0.29**	0.28	**0.35**	0.32	**0.40**
Skin	0.20	**0.29**	0.22	**0.31**	0.23	**0.32**	0.24	**0.33**	0.25	**0.35**	0.26	**0.37**
Muscle	0.18	**0.24**	0.22	**0.29**	0.24	**0.32**	0.26	**0.34**	0.29	**0.39**	0.33	**0.43**
Skeleton	0.32	**0.60**	0.39	**0.73**	0.44	**0.82**	0.48	**0.89**	0.56	**1.02**	0.62	**1.13**
Red bone marrow	0.12	**0.18**	0.15	**0.22**	0.17	**0.25**	0.18	**0.28**	0.21	**0.33**	0.24	**0.37**
Whole body dose	0.19	**0.27**	0.22	**0.32**	0.25	**0.35**	0.27	**0.38**	0.31	**0.44**	0.34	**0.48**
ED (ICRP 60)	0.15	**0.21**	0.19	**0.26**	0.22	**0.30**	0.24	**0.33**	0.29	**0.40**	0.34	**0.46**
ED (ICRP 103)	0.14	**0.21**	0.17	**0.26**	0.20	**0.30**	0.22	**0.33**	0.27	**0.39**	0.31	**0.45**
Total filtration (mm) Al	**1.5**		**2.5**		**3.5**		**4.5**		**3.5**		**3.5**	
Cu	**0**		**0**		**0**		**0**		**0.1**		**0.2**	

Table 12.52 Normalized organ doses—Abdomen pa—3–7 years

Beam projection:			pa									**Table cells:**
Focus-to-detector distance (cm):			100						(+) = normal font = optimal field setting			
Tube voltage (kVp):			60						**(−) = bold font = suboptimal field setting**			
Total filtrations:			See last table row.						Empty table cells correspond to values <0.01			
Organ or tissue	**Organ dose D_T : Entrance dose K_E**								**(mSv * mGy^{-1})**			
	(+)	(−)	(+)	(−)	(+)	(−)	(+)	(−)	(+)	(−)	(+)	(−)
Skull												
Brain												
Oral mucosa												
Salivary glands												
Cervical spine						**0.01**		**0.01**		**0.02**		**0.01**
Thyroid										**0.01**		
Thymus		**0.03**		**0.04**		**0.05**	0.01	**0.05**	0.01	**0.07**	0.01	**0.06**
Thoracic spine	0.29	**0.54**	0.37	**0.70**	0.44	**0.82**	0.49	**0.92**	0.59	**1.12**	0.55	**1.04**
Claviculae		**0.02**		**0.03**		**0.04**		**0.04**	0.01	**0.05**	0.01	**0.04**
Scapulae	0.02	**0.83**	0.02	**0.97**	0.03	**1.06**	0.03	**1.13**	0.04	**1.25**	0.03	**1.24**
Ribs	0.46	**0.87**	0.54	**1.03**	0.59	**1.14**	0.64	**1.22**	0.71	**1.36**	0.70	**1.34**
Oesophagus	0.05	**0.09**	0.07	**0.12**	0.08	**0.14**	0.09	**0.16**	0.11	**0.20**	0.10	**0.18**
Trachea												
Lung	0.10	**0.29**	0.13	**0.36**	0.14	**0.41**	0.16	**0.45**	0.19	**0.52**	0.18	**0.50**
Mamma	0.01	**0.10**	0.02	**0.13**	0.02	**0.15**	0.02	**0.17**	0.03	**0.21**	0.02	**0.19**
Heart	0.05	**0.10**	0.06	**0.14**	0.08	**0.16**	0.09	**0.18**	0.11	**0.23**	0.10	**0.20**
Stomach	0.09	**0.11**	0.12	**0.14**	0.15	**0.17**	0.17	**0.19**	0.22	**0.24**	0.19	**0.21**
Spleen	0.37	**0.40**	0.46	**0.50**	0.54	**0.58**	0.59	**0.64**	0.70	**0.75**	0.67	**0.72**
Liver	0.16	**0.18**	0.21	**0.23**	0.25	**0.27**	0.28	**0.31**	0.33	**0.38**	0.31	**0.34**
Gall bladder	0.09	**0.09**	0.12	**0.13**	0.14	**0.15**	0.17	**0.18**	0.21	**0.23**	0.18	**0.19**
Pancreas	0.15	**0.17**	0.20	**0.22**	0.24	**0.27**	0.28	**0.31**	0.35	**0.38**	0.31	**0.34**
Small intestine	0.15	**0.15**	0.19	**0.20**	0.23	**0.24**	0.27	**0.28**	0.33	**0.35**	0.29	**0.30**
Large intestine	0.12	**0.12**	0.16	**0.17**	0.19	**0.20**	0.22	**0.23**	0.28	**0.29**	0.24	**0.25**
Sigma/rectum	0.15	**0.16**	0.20	**0.21**	0.24	**0.25**	0.27	**0.29**	0.34	**0.36**	0.30	**0.32**
Lumbar spine	0.94	**0.96**	1.20	**1.23**	1.39	**1.43**	1.56	**1.60**	1.87	**1.92**	1.75	**1.80**
Pelvis	1.01	**1.04**	1.27	**1.31**	1.46	**1.51**	1.61	**1.67**	1.90	**1.97**	1.82	**1.88**
Adrenals	0.38	**0.41**	0.48	**0.51**	0.55	**0.59**	0.61	**0.65**	0.72	**0.77**	0.68	**0.73**
Kidneys	0.62	**0.64**	0.75	**0.77**	0.84	**0.86**	0.91	**0.93**	1.03	**1.06**	1.00	**1.03**
Urinary bladder	0.09	**0.09**	0.12	**0.13**	0.15	**0.15**	0.17	**0.18**	0.21	**0.23**	0.18	**0.19**
Uterus	0.12	**0.12**	0.16	**0.17**	0.20	**0.20**	0.23	**0.23**	0.29	**0.30**	0.24	**0.25**
Ovaries	0.15	**0.15**	0.20	**0.20**	0.24	**0.25**	0.28	**0.28**	0.36	**0.36**	0.31	**0.31**
Prostate	0.11	**0.12**	0.15	**0.15**	0.18	**0.19**	0.21	**0.21**	0.26	**0.27**	0.23	**0.23**
Testes	0.05	**0.06**	0.07	**0.08**	0.08	**0.10**	0.09	**0.11**	0.12	**0.14**	0.10	**0.12**
Upper arm		**0.25**		**0.31**		**0.36**		**0.40**		**0.48**		**0.46**
Elbow		**0.70**		**0.88**		**1.02**		**1.13**		**1.33**		**1.28**
Forearm		**0.81**		**1.02**		**1.17**		**1.29**		**1.52**		**1.47**
Upper leg	0.24	**0.43**	0.31	**0.57**	0.38	**0.67**	0.43	**0.76**	0.53	**0.94**	0.48	**0.85**
Knee		**0.03**		**0.04**	0.01	**0.05**	0.01	**0.05**	0.02	**0.07**	0.01	**0.06**
Lower leg												
Lymph nodes	0.09	**0.12**	0.12	**0.15**	0.14	**0.18**	0.16	**0.20**	0.20	**0.25**	0.18	**0.22**
Skin	0.17	**0.27**	0.17	**0.28**	0.18	**0.29**	0.18	**0.30**	0.19	**0.32**	0.19	**0.31**
Muscle	0.14	**0.20**	0.17	**0.24**	0.19	**0.26**	0.21	**0.29**	0.24	**0.33**	0.23	**0.31**
Skeleton	0.19	**0.38**	0.24	**0.47**	0.28	**0.54**	0.31	**0.60**	0.37	**0.70**	0.35	**0.67**
Red bone marrow	0.12	**0.16**	0.15	**0.20**	0.17	**0.24**	0.19	**0.26**	0.23	**0.31**	0.21	**0.29**
Whole body dose	0.14	**0.21**	0.17	**0.26**	0.20	**0.29**	0.21	**0.32**	0.25	**0.37**	0.24	**0.35**
ED (ICRP 60)	0.10	**0.14**	0.13	**0.18**	0.15	**0.22**	0.17	**0.24**	0.22	**0.30**	0.19	**0.27**
ED (ICRP 103)	0.10	**0.14**	0.12	**0.19**	0.15	**0.22**	0.17	**0.24**	0.20	**0.30**	0.18	**0.27**
Total filtration (mm) **Al**	**1.5**		**2.5**		**3.5**		**4.5**		**3.5**		**3.5**	
Cu	**0**		**0**		**0**		**0**		**0.1**		**0.2**	

Table 12.53 Normalized organ doses—Abdomen pa—3–7 years

Beam projection:	pa	**Table cells:**
Focus-to-detector distance (cm):	100	(+) = normal font = optimal field setting
Tube voltage (kVp):	70	**(−) = bold font = suboptimal field setting**
Total filtrations:	See last table row.	Empty table cells correspond to values <0.01

Organ or tissue	**Organ dose D_T : Entrance dose K_E**								**(mSv * mGy^{-1})**			
	(+)	(−)	(+)	(−)	(+)	(−)	(+)	(−)	(+)	(−)	(+)	(−)
Skull												
Brain												
Oral mucosa												
Salivary glands												
Cervical spine				**0.01**		**0.02**		**0.02**		**0.02**		**0.03**
Thyroid						**0.01**		**0.01**		**0.02**		**0.02**
Thymus		**0.04**	0.01	**0.05**	0.01	**0.06**	0.01	**0.07**	0.02	**0.09**	0.02	**0.11**
Thoracic spine	0.35	**0.67**	0.45	**0.84**	0.52	**0.97**	0.57	**1.08**	0.68	**1.29**	0.79	**1.48**
Claviculae		**0.03**		**0.04**	0.01	**0.05**	0.01	**0.06**	0.02	**0.07**	0.02	**0.09**
Scapulae	0.02	**0.90**	0.03	**1.04**	0.03	**1.13**	0.04	**1.19**	0.05	**1.30**	0.06	**1.38**
Ribs	0.50	**0.96**	0.58	**1.12**	0.64	**1.22**	0.68	**1.30**	0.75	**1.44**	0.81	**1.56**
Oesophagus	0.07	**0.12**	0.09	**0.15**	0.10	**0.18**	0.11	**0.21**	0.14	**0.26**	0.17	**0.31**
Trachea												
Lung	0.12	**0.33**	0.15	**0.41**	0.17	**0.46**	0.18	**0.50**	0.21	**0.58**	0.24	**0.65**
Mamma	0.02	**0.13**	0.02	**0.16**	0.02	**0.19**	0.03	**0.21**	0.03	**0.25**	0.04	**0.30**
Heart	0.06	**0.13**	0.08	**0.17**	0.10	**0.21**	0.11	**0.23**	0.14	**0.29**	0.16	**0.34**
Stomach	0.12	**0.14**	0.16	**0.18**	0.19	**0.22**	0.22	**0.25**	0.27	**0.31**	0.32	**0.36**
Spleen	0.44	**0.47**	0.54	**0.58**	0.61	**0.66**	0.67	**0.73**	0.78	**0.85**	0.88	**0.96**
Liver	0.20	**0.23**	0.25	**0.29**	0.30	**0.33**	0.33	**0.37**	0.39	**0.45**	0.46	**0.52**
Gall bladder	0.12	**0.13**	0.16	**0.17**	0.19	**0.21**	0.22	**0.24**	0.28	**0.30**	0.34	**0.36**
Pancreas	0.20	**0.22**	0.26	**0.29**	0.31	**0.34**	0.34	**0.38**	0.42	**0.47**	0.50	**0.56**
Small intestine	0.19	**0.20**	0.25	**0.26**	0.30	**0.31**	0.34	**0.35**	0.42	**0.44**	0.50	**0.52**
Large intestine	0.16	**0.17**	0.21	**0.22**	0.25	**0.26**	0.28	**0.30**	0.35	**0.37**	0.41	**0.44**
Sigma/rectum	0.20	**0.21**	0.26	**0.27**	0.30	**0.32**	0.34	**0.36**	0.42	**0.45**	0.50	**0.53**
Lumbar spine	1.13	**1.16**	1.41	**1.45**	1.63	**1.67**	1.80	**1.85**	2.13	**2.19**	2.43	**2.51**
Pelvis	1.18	**1.22**	1.46	**1.51**	1.66	**1.72**	1.82	**1.88**	2.11	**2.19**	2.37	**2.47**
Adrenals	0.45	**0.48**	0.56	**0.60**	0.63	**0.68**	0.69	**0.75**	0.81	**0.88**	0.91	**0.99**
Kidneys	0.70	**0.72**	0.83	**0.86**	0.93	**0.96**	1.00	**1.03**	1.12	**1.16**	1.23	**1.28**
Urinary bladder	0.12	**0.13**	0.16	**0.17**	0.19	**0.20**	0.21	**0.23**	0.27	**0.29**	0.32	**0.34**
Uterus	0.16	**0.17**	0.22	**0.22**	0.26	**0.27**	0.29	**0.31**	0.37	**0.38**	0.45	**0.46**
Ovaries	0.20	**0.21**	0.26	**0.27**	0.32	**0.32**	0.36	**0.36**	0.45	**0.46**	0.54	**0.55**
Prostate	0.15	**0.15**	0.19	**0.20**	0.23	**0.24**	0.26	**0.27**	0.32	**0.33**	0.38	**0.40**
Testes	0.07	**0.08**	0.09	**0.11**	0.10	**0.13**	0.12	**0.14**	0.15	**0.18**	0.17	**0.21**
Upper arm		**0.30**		**0.37**		**0.42**		**0.46**		**0.54**		**0.62**
Elbow		**0.83**		**1.02**		**1.16**		**1.27**		**1.47**		**1.65**
Forearm		**0.95**		**1.16**		**1.32**		**1.44**		**1.66**		**1.85**
Upper leg	0.30	**0.54**	0.39	**0.69**	0.46	**0.81**	0.52	**0.91**	0.63	**1.10**	0.73	**1.28**
Knee		**0.04**	0.01	**0.05**	0.02	**0.06**	0.02	**0.07**	0.02	**0.09**	0.03	**0.11**
Lower leg												
Lymph nodes	0.12	**0.15**	0.15	**0.19**	0.18	**0.22**	0.20	**0.25**	0.25	**0.31**	0.29	**0.36**
Skin	0.17	**0.28**	0.18	**0.29**	0.19	**0.31**	0.19	**0.31**	0.20	**0.33**	0.21	**0.35**
Muscle	0.17	**0.22**	0.20	**0.27**	0.22	**0.30**	0.24	**0.32**	0.27	**0.37**	0.30	**0.41**
Skeleton	0.23	**0.44**	0.28	**0.54**	0.32	**0.61**	0.35	**0.67**	0.41	**0.78**	0.46	**0.88**
Red bone marrow	0.14	**0.20**	0.17	**0.24**	0.20	**0.28**	0.22	**0.31**	0.26	**0.36**	0.30	**0.42**
Whole body dose	0.17	**0.24**	0.20	**0.29**	0.22	**0.33**	0.24	**0.36**	0.28	**0.41**	0.31	**0.46**
ED (ICRP 60)	0.13	**0.18**	0.16	**0.23**	0.19	**0.26**	0.21	**0.29**	0.26	**0.35**	0.31	**0.41**
ED (ICRP 103)	0.12	**0.18**	0.15	**0.23**	0.18	**0.26**	0.20	**0.29**	0.24	**0.35**	0.28	**0.41**
Total filtration (mm) Al	**1.5**		**2.5**		**3.5**		**4.5**		**3.5**		**3.5**	
Cu	**0**		**0**		**0**		**0**		**0.1**		**0.2**	

Table 12.54 Normalized organ doses—Abdomen pa—3–7 years

Beam projection:			pa									**Table cells:**
Focus-to-detector distance (cm):			100									(+) = normal font = optimal field setting
Tube voltage (kVp):			80									**(−) = bold font = suboptimal field setting**
Total filtrations:			See last table row.									Empty table cells correspond to values <0.01
Organ or tissue	**Organ dose D_T : Entrance dose K_E**							**(mSv * mGy^{-1})**				
	(+)	(−)	(+)	(−)	(+)	(−)	(+)	(−)	(+)	(−)	(+)	(−)
Skull												
Brain												
Oral mucosa												
Salivary glands												**0.01**
Cervical spine		**0.01**		**0.02**		**0.02**		**0.02**		**0.03**	0.01	**0.04**
Thyroid				**0.01**		**0.02**		**0.02**		**0.02**		**0.03**
Thymus	0.01	**0.05**	0.01	**0.07**	0.02	**0.08**	0.02	**0.09**	0.02	**0.11**	0.03	**0.13**
Thoracic spine	0.41	**0.78**	0.51	**0.96**	0.58	**1.10**	0.64	**1.20**	0.75	**1.41**	0.85	**1.59**
Claviculae		**0.04**	0.01	**0.05**	0.02	**0.06**	0.02	**0.07**	0.02	**0.09**	0.03	**0.10**
Scapulae	0.03	**0.96**	0.04	**1.09**	0.04	**1.17**	0.05	**1.23**	0.06	**1.33**	0.07	**1.39**
Ribs	0.54	**1.04**	0.62	**1.19**	0.67	**1.29**	0.71	**1.36**	0.77	**1.49**	0.82	**1.59**
Oesophagus	0.08	**0.15**	0.11	**0.19**	0.12	**0.22**	0.14	**0.25**	0.17	**0.30**	0.20	**0.36**
Trachea												
Lung	0.14	**0.38**	0.16	**0.45**	0.18	**0.51**	0.20	**0.55**	0.23	**0.63**	0.26	**0.69**
Mamma	0.02	**0.15**	0.03	**0.19**	0.03	**0.22**	0.03	**0.24**	0.04	**0.29**	0.05	**0.33**
Heart	0.08	**0.17**	0.10	**0.21**	0.12	**0.25**	0.13	**0.28**	0.16	**0.34**	0.19	**0.39**
Stomach	0.16	**0.18**	0.20	**0.23**	0.23	**0.26**	0.26	**0.30**	0.32	**0.36**	0.37	**0.42**
Spleen	0.50	**0.54**	0.61	**0.66**	0.68	**0.74**	0.74	**0.81**	0.85	**0.93**	0.95	**1.04**
Liver	0.24	**0.27**	0.30	**0.34**	0.34	**0.39**	0.38	**0.43**	0.45	**0.51**	0.51	**0.58**
Gall bladder	0.16	**0.17**	0.20	**0.22**	0.24	**0.26**	0.27	**0.29**	0.34	**0.37**	0.40	**0.44**
Pancreas	0.25	**0.28**	0.31	**0.35**	0.37	**0.41**	0.41	**0.46**	0.49	**0.55**	0.57	**0.64**
Small intestine	0.24	**0.26**	0.31	**0.32**	0.36	**0.38**	0.40	**0.42**	0.49	**0.52**	0.58	**0.61**
Large intestine	0.20	**0.21**	0.26	**0.27**	0.30	**0.32**	0.34	**0.36**	0.41	**0.44**	0.48	**0.52**
Sigma/rectum	0.25	**0.26**	0.31	**0.33**	0.36	**0.39**	0.41	**0.43**	0.49	**0.52**	0.57	**0.61**
Lumbar spine	1.31	**1.35**	1.61	**1.65**	1.83	**1.88**	2.00	**2.06**	2.33	**2.40**	2.61	**2.70**
Pelvis	1.34	**1.39**	1.62	**1.68**	1.82	**1.89**	1.98	**2.05**	2.26	**2.35**	2.50	**2.61**
Adrenals	0.52	**0.56**	0.63	**0.68**	0.71	**0.77**	0.77	**0.84**	0.89	**0.97**	0.99	**1.08**
Kidneys	0.78	**0.80**	0.91	**0.94**	1.00	**1.04**	1.07	**1.11**	1.20	**1.25**	1.30	**1.36**
Urinary bladder	0.16	**0.17**	0.20	**0.21**	0.23	**0.25**	0.26	**0.28**	0.32	**0.34**	0.38	**0.40**
Uterus	0.21	**0.22**	0.27	**0.28**	0.32	**0.33**	0.36	**0.37**	0.44	**0.46**	0.52	**0.55**
Ovaries	0.26	**0.26**	0.33	**0.34**	0.38	**0.39**	0.43	**0.44**	0.53	**0.54**	0.62	**0.64**
Prostate	0.19	**0.19**	0.24	**0.25**	0.28	**0.29**	0.31	**0.32**	0.38	**0.39**	0.44	**0.46**
Testes	0.08	**0.10**	0.11	**0.13**	0.13	**0.15**	0.14	**0.17**	0.18	**0.21**	0.21	**0.25**
Upper arm		**0.34**		**0.41**		**0.47**		**0.51**		**0.59**		**0.66**
Elbow		**0.93**		**1.13**		**1.27**		**1.38**		**1.57**		**1.73**
Forearm		**1.06**		**1.28**		**1.43**		**1.55**		**1.76**		**1.93**
Upper leg	0.37	**0.65**	0.46	**0.81**	0.53	**0.93**	0.59	**1.03**	0.71	**1.23**	0.81	**1.40**
Knee	0.01	**0.05**	0.02	**0.07**	0.02	**0.08**	0.02	**0.09**	0.03	**0.11**	0.04	**0.13**
Lower leg												
Lymph nodes	0.15	**0.18**	0.18	**0.23**	0.21	**0.27**	0.24	**0.30**	0.29	**0.36**	0.33	**0.41**
Skin	0.18	**0.29**	0.19	**0.31**	0.19	**0.32**	0.20	**0.33**	0.21	**0.35**	0.22	**0.36**
Muscle	0.19	**0.25**	0.22	**0.30**	0.24	**0.33**	0.26	**0.36**	0.30	**0.40**	0.33	**0.45**
Skeleton	0.26	**0.50**	0.31	**0.60**	0.35	**0.67**	0.38	**0.73**	0.44	**0.84**	0.49	**0.92**
Red bone marrow	0.17	**0.23**	0.20	**0.28**	0.23	**0.32**	0.25	**0.35**	0.30	**0.41**	0.34	**0.46**
Whole body dose	0.19	**0.28**	0.22	**0.33**	0.25	**0.36**	0.27	**0.39**	0.31	**0.45**	0.34	**0.50**
ED (ICRP 60)	0.16	**0.22**	0.20	**0.27**	0.23	**0.31**	0.25	**0.34**	0.30	**0.41**	0.35	**0.47**
ED (ICRP 103)	0.15	**0.22**	0.18	**0.27**	0.21	**0.31**	0.23	**0.34**	0.28	**0.40**	0.32	**0.46**
Total filtration (mm) **Al**	**1.5**		**2.5**		**3.5**		**4.5**		**3.5**		**3.5**	
Cu	**0**		**0**		**0**		**0**		**0.1**		**0.2**	

Table 12.55 Normalized organ doses—Abdomen pa—8–12 years

Beam projection:			pa									**Table cells:**
Focus-to-detector distance (cm):			100									(+) = normal font = optimal field setting
Tube voltage (kVp):			60									**(−) = bold font = suboptimal field setting**
Total filtrations:			See last table row.									Empty table cells correspond to values <0.01
Organ or tissue	**Organ dose D_T : Entrance dose K_E**								**(mSv * mGy^{-1})**			
	(+)	(−)	(+)	(−)	(+)	(−)	(+)	(−)	(+)	(−)	(+)	(−)
Skull												
Brain												
Oral mucosa												
Salivary glands												
Cervical spine												
Thyroid												
Thymus		**0.02**		**0.03**		**0.03**		**0.04**		**0.05**		**0.04**
Thoracic spine	0.23	**0.45**	0.30	**0.59**	0.36	**0.70**	0.40	**0.79**	0.49	**0.96**	0.45	**0.88**
Claviculae		**0.01**		**0.02**		**0.02**		**0.03**		**0.03**		**0.03**
Scapulae		**0.67**	0.01	**0.79**	0.02	**0.87**	0.02	**0.93**	0.02	**1.04**	0.02	**1.03**
Ribs	0.41	**0.79**	0.49	**0.93**	0.54	**1.03**	0.57	**1.11**	0.64	**1.24**	0.63	**1.22**
Oesophagus	0.03	**0.06**	0.04	**0.08**	0.05	**0.10**	0.05	**0.11**	0.07	**0.14**	0.06	**0.12**
Trachea												
Lung	0.08	**0.24**	0.10	**0.31**	0.11	**0.35**	0.12	**0.39**	0.14	**0.46**	0.14	**0.44**
Mamma		**0.07**	0.01	**0.10**	0.02	**0.11**	0.02	**0.13**	0.02	**0.16**	0.02	**0.14**
Heart	0.03	**0.08**	0.05	**0.11**	0.06	**0.13**	0.06	**0.15**	0.08	**0.19**	0.07	**0.16**
Stomach	0.08	**0.08**	0.10	**0.11**	0.12	**0.14**	0.14	**0.16**	0.18	**0.20**	0.15	**0.17**
Spleen	0.33	**0.36**	0.42	**0.45**	0.49	**0.53**	0.54	**0.59**	0.64	**0.70**	0.61	**0.66**
Liver	0.15	**0.17**	0.19	**0.21**	0.22	**0.25**	0.25	**0.29**	0.31	**0.35**	0.28	**0.32**
Gall bladder	0.08	**0.08**	0.10	**0.11**	0.13	**0.14**	0.15	**0.16**	0.19	**0.21**	0.16	**0.17**
Pancreas	0.13	**0.15**	0.17	**0.19**	0.21	**0.23**	0.24	**0.27**	0.30	**0.34**	0.26	**0.29**
Small intestine	0.12	**0.13**	0.16	**0.17**	0.20	**0.21**	0.23	**0.24**	0.29	**0.30**	0.24	**0.26**
Large intestine	0.10	**0.10**	0.13	**0.14**	0.16	**0.17**	0.18	**0.19**	0.23	**0.25**	0.20	**0.21**
Sigma/rectum	0.13	**0.13**	0.17	**0.18**	0.20	**0.22**	0.23	**0.25**	0.29	**0.31**	0.25	**0.27**
Lumbar spine	0.83	**0.85**	1.07	**1.10**	1.25	**1.29**	1.40	**1.45**	1.70	**1.75**	1.57	**1.62**
Pelvis	0.89	**0.93**	1.13	**1.17**	1.31	**1.36**	1.45	**1.51**	1.73	**1.80**	1.64	**1.70**
Adrenals	0.35	**0.38**	0.44	**0.48**	0.51	**0.56**	0.57	**0.62**	0.68	**0.74**	0.64	**0.70**
Kidneys	0.58	**0.60**	0.71	**0.74**	0.80	**0.83**	0.87	**0.91**	1.00	**1.04**	0.97	**1.01**
Urinary bladder	0.07	**0.08**	0.10	**0.10**	0.12	**0.12**	0.14	**0.14**	0.18	**0.18**	0.15	**0.15**
Uterus	0.10	**0.10**	0.14	**0.14**	0.17	**0.17**	0.19	**0.20**	0.25	**0.26**	0.21	**0.22**
Ovaries	0.13	**0.13**	0.17	**0.17**	0.21	**0.21**	0.24	**0.24**	0.31	**0.31**	0.26	**0.26**
Prostate	0.08	**0.09**	0.11	**0.12**	0.13	**0.14**	0.15	**0.16**	0.19	**0.21**	0.16	**0.18**
Testes	0.03	**0.03**	0.04	**0.05**	0.05	**0.06**	0.06	**0.07**	0.08	**0.09**	0.06	**0.07**
Upper arm		**0.18**		**0.23**		**0.27**		**0.30**		**0.37**		**0.34**
Elbow		**0.60**		**0.76**		**0.88**		**0.97**		**1.16**		**1.10**
Forearm		**0.70**		**0.89**		**1.02**		**1.13**		**1.34**		**1.29**
Upper leg	0.14	**0.28**	0.19	**0.38**	0.23	**0.45**	0.27	**0.52**	0.33	**0.64**	0.29	**0.57**
Knee				**0.01**		**0.02**		**0.02**		**0.02**		**0.02**
Lower leg												
Lymph nodes	0.07	**0.10**	0.10	**0.13**	0.12	**0.15**	0.14	**0.17**	0.17	**0.22**	0.15	**0.19**
Skin	0.16	**0.25**	0.17	**0.27**	0.17	**0.28**	0.18	**0.28**	0.18	**0.30**	0.18	**0.29**
Muscle	0.13	**0.18**	0.15	**0.21**	0.17	**0.24**	0.19	**0.26**	0.22	**0.30**	0.20	**0.28**
Skeleton	0.16	**0.32**	0.21	**0.40**	0.24	**0.46**	0.26	**0.51**	0.31	**0.60**	0.29	**0.57**
Red bone marrow	0.13	**0.17**	0.16	**0.21**	0.19	**0.25**	0.21	**0.28**	0.25	**0.33**	0.23	**0.30**
Whole body dose	0.13	**0.19**	0.16	**0.23**	0.18	**0.26**	0.20	**0.29**	0.23	**0.34**	0.21	**0.32**
ED (ICRP 60)	0.09	**0.12**	0.11	**0.16**	0.13	**0.19**	0.15	**0.21**	0.19	**0.26**	0.16	**0.23**
ED (ICRP 103)	0.08	**0.12**	0.11	**0.16**	0.13	**0.19**	0.14	**0.21**	0.18	**0.26**	0.16	**0.23**
Total filtration (mm) **Al**	**1.5**		**2.5**		**3.5**		**4.5**		**3.5**		**3.5**	
Cu	**0**		**0**		**0**		**0**		**0.1**		**0.2**	

Table 12.56 Normalized organ doses—Abdomen pa—8–12 years

Beam projection:			pa									**Table cells:**
Focus-to-detector distance (cm):			100				(+) = normal font = optimal field setting					
Tube voltage (kVp):			70				**(−) = bold font = suboptimal field setting**					
Total filtrations:			See last table row.				Empty table cells correspond to values <0.01					
Organ or tissue	**Organ dose D_T : Entrance dose K_E**								**(mSv * mGy^{-1})**			
	(+)	(−)	(+)	(−)	(+)	(−)	(+)	(−)	(+)	(−)	(+)	(−)
Skull												
Brain												
Oral mucosa												
Salivary glands												
Cervical spine								**0.01**		**0.01**		**0.02**
Thyroid										**0.01**		**0.01**
Thymus		**0.03**		**0.04**		**0.05**	0.01	**0.06**	0.01	**0.07**	0.02	**0.09**
Thoracic spine	0.29	**0.57**	0.37	**0.72**	0.43	**0.84**	0.48	**0.93**	0.58	**1.12**	0.68	**1.30**
Claviculae		**0.02**		**0.03**		**0.03**		**0.04**		**0.05**	0.01	**0.06**
Scapulae	0.01	**0.73**	0.02	**0.85**	0.02	**0.93**	0.03	**0.99**	0.03	**1.09**	0.04	**1.17**
Ribs	0.45	**0.87**	0.53	**1.02**	0.58	**1.12**	0.62	**1.19**	0.68	**1.33**	0.74	**1.44**
Oesophagus	0.04	**0.08**	0.05	**0.11**	0.06	**0.13**	0.07	**0.15**	0.09	**0.19**	0.11	**0.23**
Trachea												
Lung	0.09	**0.29**	0.11	**0.36**	0.13	**0.41**	0.14	**0.44**	0.17	**0.52**	0.19	**0.58**
Mamma	0.01	**0.09**	0.02	**0.12**	0.02	**0.14**	0.02	**0.16**	0.03	**0.20**	0.03	**0.24**
Heart	0.05	**0.11**	0.06	**0.14**	0.08	**0.17**	0.09	**0.19**	0.11	**0.24**	0.13	**0.29**
Stomach	0.10	**0.12**	0.14	**0.15**	0.16	**0.18**	0.18	**0.21**	0.23	**0.26**	0.28	**0.32**
Spleen	0.40	**0.43**	0.49	**0.54**	0.57	**0.62**	0.62	**0.68**	0.73	**0.81**	0.83	**0.92**
Liver	0.18	**0.21**	0.23	**0.27**	0.27	**0.31**	0.30	**0.35**	0.37	**0.42**	0.43	**0.49**
Gall bladder	0.11	**0.12**	0.15	**0.16**	0.17	**0.19**	0.20	**0.22**	0.25	**0.28**	0.31	**0.34**
Pancreas	0.17	**0.20**	0.23	**0.26**	0.27	**0.30**	0.31	**0.34**	0.38	**0.43**	0.45	**0.51**
Small intestine	0.16	**0.17**	0.22	**0.23**	0.26	**0.27**	0.29	**0.31**	0.37	**0.39**	0.44	**0.46**
Large intestine	0.13	**0.14**	0.18	**0.19**	0.21	**0.22**	0.24	**0.25**	0.30	**0.32**	0.36	**0.39**
Sigma/rectum	0.17	**0.18**	0.22	**0.23**	0.26	**0.28**	0.30	**0.32**	0.37	**0.39**	0.44	**0.47**
Lumbar spine	1.02	**1.05**	1.28	**1.32**	1.48	**1.53**	1.65	**1.70**	1.97	**2.04**	2.27	**2.35**
Pelvis	1.07	**1.11**	1.32	**1.37**	1.51	**1.57**	1.66	**1.73**	1.95	**2.04**	2.21	**2.31**
Adrenals	0.42	**0.46**	0.52	**0.57**	0.60	**0.65**	0.66	**0.72**	0.78	**0.85**	0.89	**0.97**
Kidneys	0.67	**0.69**	0.80	**0.83**	0.89	**0.93**	0.97	**1.01**	1.10	**1.15**	1.22	**1.27**
Urinary bladder	0.10	**0.11**	0.13	**0.14**	0.16	**0.17**	0.18	**0.19**	0.23	**0.24**	0.28	**0.29**
Uterus	0.14	**0.15**	0.19	**0.19**	0.22	**0.23**	0.26	**0.27**	0.32	**0.34**	0.39	**0.41**
Ovaries	0.18	**0.18**	0.23	**0.24**	0.28	**0.28**	0.31	**0.32**	0.40	**0.41**	0.48	**0.49**
Prostate	0.11	**0.12**	0.14	**0.15**	0.17	**0.18**	0.20	**0.21**	0.25	**0.26**	0.30	**0.31**
Testes	0.04	**0.05**	0.06	**0.07**	0.07	**0.08**	0.08	**0.09**	0.10	**0.12**	0.12	**0.14**
Upper arm		**0.22**		**0.28**		**0.32**		**0.35**		**0.42**		**0.48**
Elbow		**0.71**		**0.88**		**1.01**		**1.11**		**1.30**		**1.48**
Forearm		**0.83**		**1.02**		**1.17**		**1.28**		**1.49**		**1.68**
Upper leg	0.19	**0.37**	0.25	**0.48**	0.30	**0.57**	0.33	**0.64**	0.41	**0.79**	0.49	**0.93**
Knee		**0.01**		**0.02**		**0.02**		**0.03**		**0.03**	0.01	**0.04**
Lower leg												
Lymph nodes	0.10	**0.13**	0.13	**0.17**	0.15	**0.19**	0.17	**0.22**	0.21	**0.27**	0.26	**0.32**
Skin	0.16	**0.26**	0.17	**0.28**	0.18	**0.29**	0.18	**0.30**	0.19	**0.31**	0.20	**0.33**
Muscle	0.15	**0.20**	0.18	**0.24**	0.20	**0.27**	0.21	**0.29**	0.25	**0.34**	0.28	**0.38**
Skeleton	0.19	**0.38**	0.24	**0.47**	0.27	**0.53**	0.30	**0.58**	0.35	**0.68**	0.40	**0.77**
Red bone marrow	0.16	**0.21**	0.19	**0.26**	0.22	**0.30**	0.25	**0.33**	0.30	**0.39**	0.34	**0.45**
Whole body dose	0.15	**0.22**	0.18	**0.27**	0.21	**0.30**	0.22	**0.33**	0.26	**0.38**	0.29	**0.43**
ED (ICRP 60)	0.11	**0.16**	0.14	**0.20**	0.17	**0.23**	0.19	**0.26**	0.23	**0.32**	0.27	**0.37**
ED (ICRP 103)	0.11	**0.16**	0.14	**0.20**	0.16	**0.23**	0.18	**0.26**	0.22	**0.31**	0.26	**0.37**
Total filtration (mm) **Al**	**1.5**		**2.5**		**3.5**		**4.5**		**3.5**		**3.5**	
Cu	**0**		**0**		**0**		**0**		**0.1**		**0.2**	

Table 12.57 Normalized organ doses—Abdomen pa—8–12 years

Beam projection:		pa										**Table cells:**
Focus-to-detector distance (cm):		100				(+) = normal font = optimal field setting						
Tube voltage (kVp):		80				**(−) = bold font = suboptimal field setting**						
Total filtrations:		See last table row.				Empty table cells correspond to values <0.01						
Organ or tissue	**Organ dose D_T : Entrance dose K_E**							**(mSv * mGy^{-1})**				
	(+)	(−)	(+)	(−)	(+)	(−)	(+)	(−)	(+)	(−)	(+)	(−)
Skull												
Brain												
Oral mucosa												
Salivary glands												
Cervical spine				**0.01**		**0.01**		**0.02**		**0.02**		**0.02**
Thyroid						**0.01**		**0.01**		**0.02**		**0.02**
Thymus		**0.04**	0.01	**0.05**	0.01	**0.06**	0.01	**0.07**	0.02	**0.09**	0.02	**0.11**
Thoracic spine	0.35	**0.67**	0.43	**0.84**	0.50	**0.96**	0.55	**1.06**	0.65	**1.25**	0.74	**1.43**
Claviculae		**0.03**		**0.03**		**0.04**	0.01	**0.05**	0.01	**0.06**	0.02	**0.07**
Scapulae	0.02	**0.79**	0.02	**0.90**	0.03	**0.97**	0.03	**1.03**	0.04	**1.12**	0.05	**1.19**
Ribs	0.49	**0.95**	0.56	**1.09**	0.61	**1.18**	0.65	**1.26**	0.71	**1.38**	0.75	**1.48**
Oesophagus	0.05	**0.11**	0.07	**0.14**	0.08	**0.16**	0.09	**0.18**	0.11	**0.23**	0.13	**0.27**
Trachea												
Lung	0.11	**0.33**	0.13	**0.40**	0.15	**0.45**	0.16	**0.49**	0.19	**0.57**	0.21	**0.63**
Mamma	0.02	**0.12**	0.02	**0.15**	0.02	**0.17**	0.03	**0.19**	0.03	**0.23**	0.04	**0.27**
Heart	0.06	**0.14**	0.08	**0.18**	0.09	**0.21**	0.11	**0.24**	0.13	**0.29**	0.16	**0.34**
Stomach	0.13	**0.15**	0.17	**0.19**	0.20	**0.23**	0.23	**0.26**	0.28	**0.32**	0.33	**0.38**
Spleen	0.46	**0.50**	0.56	**0.62**	0.64	**0.70**	0.70	**0.77**	0.81	**0.89**	0.91	**1.01**
Liver	0.22	**0.26**	0.28	**0.32**	0.32	**0.37**	0.35	**0.41**	0.42	**0.49**	0.48	**0.56**
Gall bladder	0.15	**0.16**	0.19	**0.20**	0.22	**0.24**	0.25	**0.27**	0.31	**0.34**	0.37	**0.41**
Pancreas	0.22	**0.25**	0.28	**0.32**	0.33	**0.37**	0.37	**0.42**	0.45	**0.51**	0.53	**0.60**
Small intestine	0.21	**0.22**	0.27	**0.29**	0.32	**0.33**	0.36	**0.38**	0.44	**0.46**	0.52	**0.55**
Large intestine	0.17	**0.18**	0.22	**0.24**	0.26	**0.28**	0.30	**0.31**	0.36	**0.39**	0.43	**0.46**
Sigma/rectum	0.21	**0.23**	0.27	**0.29**	0.32	**0.34**	0.36	**0.38**	0.44	**0.47**	0.51	**0.55**
Lumbar spine	1.20	**1.24**	1.48	**1.53**	1.69	**1.74**	1.85	**1.92**	2.18	**2.26**	2.47	**2.56**
Pelvis	1.22	**1.28**	1.49	**1.55**	1.68	**1.75**	1.83	**1.91**	2.11	**2.21**	2.36	**2.47**
Adrenals	0.49	**0.53**	0.60	**0.65**	0.68	**0.74**	0.74	**0.81**	0.86	**0.95**	0.97	**1.07**
Kidneys	0.75	**0.78**	0.88	**0.92**	0.98	**1.02**	1.05	**1.10**	1.19	**1.24**	1.30	**1.36**
Urinary bladder	0.13	**0.14**	0.17	**0.18**	0.20	**0.21**	0.23	**0.24**	0.28	**0.29**	0.33	**0.35**
Uterus	0.19	**0.19**	0.24	**0.25**	0.28	**0.29**	0.32	**0.33**	0.40	**0.41**	0.47	**0.49**
Ovaries	0.23	**0.23**	0.29	**0.30**	0.34	**0.35**	0.39	**0.40**	0.48	**0.50**	0.57	**0.59**
Prostate	0.14	**0.15**	0.18	**0.19**	0.22	**0.23**	0.24	**0.26**	0.30	**0.32**	0.36	**0.37**
Testes	0.06	**0.07**	0.07	**0.09**	0.09	**0.10**	0.10	**0.12**	0.12	**0.14**	0.15	**0.17**
Upper arm		**0.26**		**0.32**		**0.36**		**0.40**		**0.46**		**0.52**
Elbow		**0.81**		**0.99**		**1.12**		**1.22**		**1.41**		**1.57**
Forearm		**0.94**		**1.14**		**1.28**		**1.39**		**1.59**		**1.76**
Upper leg	0.24	**0.46**	0.30	**0.58**	0.35	**0.67**	0.40	**0.75**	0.48	**0.90**	0.56	**1.05**
Knee		**0.02**		**0.03**		**0.03**		**0.03**	0.01	**0.04**	0.02	**0.05**
Lower leg												
Lymph nodes	0.13	**0.16**	0.16	**0.20**	0.19	**0.24**	0.21	**0.26**	0.26	**0.32**	0.30	**0.38**
Skin	0.17	**0.27**	0.18	**0.29**	0.18	**0.30**	0.19	**0.31**	0.20	**0.33**	0.21	**0.34**
Muscle	0.17	**0.23**	0.20	**0.27**	0.22	**0.30**	0.24	**0.33**	0.27	**0.37**	0.30	**0.42**
Skeleton	0.22	**0.43**	0.27	**0.52**	0.31	**0.59**	0.33	**0.64**	0.39	**0.74**	0.43	**0.82**
Red bone marrow	0.19	**0.25**	0.23	**0.30**	0.26	**0.34**	0.29	**0.38**	0.34	**0.45**	0.39	**0.51**
Whole body dose	0.17	**0.25**	0.21	**0.30**	0.23	**0.34**	0.25	**0.37**	0.29	**0.42**	0.32	**0.47**
ED (ICRP 60)	0.14	**0.19**	0.17	**0.24**	0.20	**0.28**	0.22	**0.31**	0.27	**0.37**	0.32	**0.43**
ED (ICRP 103)	0.13	**0.19**	0.16	**0.24**	0.19	**0.28**	0.21	**0.31**	0.25	**0.36**	0.29	**0.42**
Total filtration (mm) Al	**1.5**		**2.5**		**3.5**		**4.5**		**3.5**		**3.5**	
Cu	**0**		**0**		**0**		**0**		**0.1**		**0.2**	

Table 12.58 Normalized organ doses—Abdomen pa—13–17 years

Beam projection:			pa									**Table cells:**
Focus-to-detector distance (cm):			100						(+) = normal font = optimal field setting			
Tube voltage (kVp):			70						**(−) = bold font = suboptimal field setting**			
Total filtrations:			See last table row.						Empty table cells correspond to values <0.01			
Organ or tissue	**Organ dose D_T : Entrance dose K_E**								**(mSv * mGy^{-1})**			
	(+)	(−)	(+)	(−)	(+)	(−)	(+)	(−)	(+)	(−)	(+)	(−)
Skull												
Brain												
Oral mucosa												
Salivary glands												
Cervical spine												
Thyroid												
Thymus		**0.02**		**0.02**		**0.03**		**0.03**		**0.04**	0.01	**0.06**
Thoracic spine	0.23	**0.41**	0.29	**0.52**	0.34	**0.61**	0.38	**0.68**	0.47	**0.83**	0.55	**0.98**
Claviculae						**0.01**		**0.01**		**0.02**		**0.02**
Scapulae		**0.35**	0.01	**0.42**	0.01	**0.46**	0.01	**0.49**	0.02	**0.55**	0.02	**0.60**
Ribs	0.39	**0.70**	0.46	**0.82**	0.51	**0.90**	0.54	**0.96**	0.60	**1.08**	0.65	**1.17**
Oesophagus	0.04	**0.08**	0.06	**0.10**	0.07	**0.12**	0.08	**0.14**	0.10	**0.18**	0.13	**0.22**
Trachea												
Lung	0.07	**0.21**	0.09	**0.26**	0.10	**0.30**	0.11	**0.33**	0.13	**0.39**	0.16	**0.45**
Mamma	0.14	**0.41**	0.18	**0.54**	0.22	**0.65**	0.25	**0.74**	0.32	**0.93**	0.39	**1.13**
Heart	0.03	**0.07**	0.04	**0.09**	0.05	**0.11**	0.06	**0.13**	0.08	**0.16**	0.09	**0.20**
Stomach	0.08	**0.09**	0.10	**0.12**	0.12	**0.14**	0.14	**0.16**	0.18	**0.20**	0.22	**0.25**
Spleen	0.34	**0.37**	0.42	**0.47**	0.49	**0.54**	0.55	**0.60**	0.65	**0.72**	0.75	**0.83**
Liver	0.15	**0.17**	0.19	**0.22**	0.22	**0.26**	0.25	**0.29**	0.30	**0.36**	0.36	**0.42**
Gall bladder	0.09	**0.10**	0.12	**0.13**	0.15	**0.16**	0.17	**0.19**	0.22	**0.24**	0.27	**0.29**
Pancreas	0.13	**0.15**	0.17	**0.19**	0.21	**0.23**	0.24	**0.27**	0.30	**0.34**	0.36	**0.41**
Small intestine	0.13	**0.14**	0.17	**0.18**	0.20	**0.21**	0.23	**0.25**	0.29	**0.31**	0.36	**0.38**
Large intestine	0.10	**0.11**	0.13	**0.14**	0.16	**0.17**	0.18	**0.20**	0.23	**0.25**	0.29	**0.31**
Sigma/rectum	0.13	**0.14**	0.17	**0.19**	0.21	**0.22**	0.24	**0.25**	0.30	**0.32**	0.36	**0.39**
Lumbar spine	0.85	**0.89**	1.08	**1.14**	1.26	**1.32**	1.41	**1.48**	1.70	**1.79**	1.98	**2.08**
Pelvis	0.92	**0.97**	1.15	**1.21**	1.32	**1.39**	1.47	**1.54**	1.74	**1.83**	1.99	**2.10**
Adrenals	0.37	**0.39**	0.46	**0.50**	0.53	**0.58**	0.59	**0.64**	0.70	**0.76**	0.81	**0.88**
Kidneys	0.61	**0.64**	0.74	**0.78**	0.83	**0.88**	0.91	**0.96**	1.05	**1.10**	1.17	**1.23**
Urinary bladder	0.07	**0.08**	0.10	**0.10**	0.12	**0.12**	0.13	**0.14**	0.17	**0.18**	0.21	**0.23**
Uterus	0.37	**0.38**	0.48	**0.50**	0.58	**0.61**	0.67	**0.70**	0.85	**0.89**	1.04	**1.09**
Ovaries	0.14	**0.14**	0.18	**0.19**	0.22	**0.23**	0.25	**0.26**	0.32	**0.33**	0.39	**0.41**
Prostate	0.09	**0.10**	0.12	**0.13**	0.15	**0.16**	0.17	**0.18**	0.22	**0.23**	0.26	**0.29**
Testes	0.03	**0.04**	0.05	**0.05**	0.06	**0.06**	0.06	**0.07**	0.08	**0.09**	0.10	**0.12**
Upper arm		**0.09**		**0.12**		**0.13**		**0.15**		**0.18**		**0.21**
Elbow		**0.46**		**0.58**		**0.66**		**0.74**		**0.88**		**1.00**
Forearm		**0.52**		**0.65**		**0.75**		**0.82**		**0.97**		**1.11**
Upper leg	0.18	**0.29**	0.23	**0.38**	0.28	**0.45**	0.31	**0.52**	0.39	**0.64**	0.48	**0.77**
Knee		**0.03**		**0.04**		**0.05**		**0.05**		**0.07**	0.01	**0.08**
Lower leg												
Lymph nodes	0.08	**0.10**	0.10	**0.13**	0.12	**0.15**	0.14	**0.17**	0.18	**0.22**	0.21	**0.26**
Skin	0.18	**0.25**	0.20	**0.27**	0.21	**0.28**	0.21	**0.29**	0.22	**0.30**	0.24	**0.32**
Muscle	0.14	**0.19**	0.17	**0.23**	0.19	**0.26**	0.21	**0.28**	0.24	**0.32**	0.27	**0.37**
Skeleton	0.18	**0.31**	0.22	**0.39**	0.26	**0.44**	0.28	**0.49**	0.34	**0.58**	0.38	**0.66**
Red bone marrow	0.16	**0.20**	0.20	**0.25**	0.23	**0.29**	0.26	**0.32**	0.31	**0.38**	0.36	**0.45**
Whole body dose	0.14	**0.20**	0.17	**0.25**	0.20	**0.28**	0.21	**0.30**	0.25	**0.35**	0.28	**0.40**
ED (ICRP 60)	0.10	**0.14**	0.13	**0.18**	0.15	**0.22**	0.17	**0.25**	0.21	**0.30**	0.26	**0.36**
ED (ICRP 103)	0.11	**0.17**	0.14	**0.22**	0.16	**0.26**	0.18	**0.29**	0.23	**0.36**	0.27	**0.43**
Total filtration (mm) **Al**	**1.5**		**2.5**		**3.5**		**4.5**		**3.5**		**3.5**	
Cu	**0**		**0**		**0**		**0**		**0.1**		**0.2**	

Table 12.59 Normalized organ doses—Abdomen pa—13–17 years

Beam projection:		pa										**Table cells:**
Focus-to-detector distance (cm):		100										(+) = normal font = optimal field setting
Tube voltage (kVp):		80										**(−) = bold font = suboptimal field setting**
Total filtrations:		See last table row.										Empty table cells correspond to values <0.01
Organ or tissue	**Organ dose D_T : Entrance dose K_E**							**(mSv * mGy^{-1})**				
	(+)	(−)	(+)	(−)	(+)	(−)	(+)	(−)	(+)	(−)	(+)	(−)
Skull												
Brain												
Oral mucosa												
Salivary glands												
Cervical spine												
Thyroid												
Thymus		**0.03**		**0.03**		**0.04**	0.01	**0.05**	0.01	**0.06**	0.02	**0.07**
Thoracic spine	0.27	**0.49**	0.34	**0.62**	0.40	**0.71**	0.44	**0.79**	0.53	**0.95**	0.61	**1.10**
Claviculae		**0.01**		**0.01**		**0.02**		**0.02**		**0.02**		**0.03**
Scapulae	0.01	**0.38**	0.01	**0.45**	0.02	**0.49**	0.02	**0.52**	0.02	**0.58**	0.03	**0.63**
Ribs	0.43	**0.76**	0.49	**0.88**	0.54	**0.96**	0.57	**1.02**	0.63	**1.13**	0.67	**1.21**
Oesophagus	0.06	**0.11**	0.08	**0.14**	0.09	**0.16**	0.10	**0.18**	0.13	**0.23**	0.16	**0.28**
Trachea												
Lung	0.08	**0.24**	0.10	**0.30**	0.12	**0.34**	0.13	**0.38**	0.15	**0.44**	0.18	**0.50**
Mamma	0.18	**0.54**	0.24	**0.69**	0.28	**0.81**	0.32	**0.92**	0.39	**1.14**	0.47	**1.36**
Heart	0.04	**0.09**	0.06	**0.12**	0.07	**0.14**	0.08	**0.16**	0.10	**0.20**	0.12	**0.24**
Stomach	0.10	**0.12**	0.13	**0.15**	0.16	**0.18**	0.18	**0.20**	0.22	**0.25**	0.26	**0.30**
Spleen	0.40	**0.44**	0.49	**0.55**	0.56	**0.62**	0.62	**0.69**	0.73	**0.81**	0.83	**0.93**
Liver	0.18	**0.21**	0.23	**0.27**	0.26	**0.31**	0.29	**0.35**	0.36	**0.42**	0.41	**0.49**
Gall bladder	0.12	**0.14**	0.16	**0.18**	0.19	**0.21**	0.22	**0.24**	0.27	**0.30**	0.33	**0.36**
Pancreas	0.17	**0.19**	0.22	**0.25**	0.26	**0.29**	0.29	**0.33**	0.36	**0.41**	0.43	**0.49**
Small intestine	0.17	**0.18**	0.22	**0.23**	0.26	**0.27**	0.29	**0.31**	0.36	**0.39**	0.43	**0.46**
Large intestine	0.13	**0.14**	0.17	**0.19**	0.20	**0.22**	0.23	**0.25**	0.29	**0.31**	0.35	**0.38**
Sigma/rectum	0.17	**0.19**	0.22	**0.24**	0.26	**0.28**	0.30	**0.32**	0.37	**0.39**	0.44	**0.47**
Lumbar spine	1.02	**1.07**	1.27	**1.33**	1.46	**1.53**	1.62	**1.70**	1.92	**2.01**	2.20	**2.31**
Pelvis	1.07	**1.13**	1.31	**1.38**	1.49	**1.58**	1.64	**1.73**	1.91	**2.02**	2.15	**2.28**
Adrenals	0.43	**0.47**	0.53	**0.58**	0.61	**0.67**	0.67	**0.73**	0.79	**0.86**	0.90	**0.98**
Kidneys	0.69	**0.73**	0.83	**0.87**	0.92	**0.97**	1.00	**1.05**	1.14	**1.20**	1.26	**1.33**
Urinary bladder	0.10	**0.11**	0.13	**0.14**	0.15	**0.16**	0.17	**0.18**	0.22	**0.23**	0.26	**0.28**
Uterus	0.49	**0.51**	0.63	**0.66**	0.75	**0.78**	0.84	**0.88**	1.06	**1.11**	1.27	**1.33**
Ovaries	0.18	**0.19**	0.24	**0.25**	0.28	**0.29**	0.32	**0.33**	0.40	**0.42**	0.48	**0.50**
Prostate	0.12	**0.13**	0.16	**0.17**	0.19	**0.20**	0.22	**0.23**	0.27	**0.29**	0.32	**0.35**
Testes	0.05	**0.05**	0.06	**0.07**	0.07	**0.08**	0.08	**0.09**	0.10	**0.12**	0.12	**0.14**
Upper arm		**0.11**		**0.14**		**0.16**		**0.17**		**0.21**		**0.24**
Elbow		**0.54**		**0.66**		**0.75**		**0.82**		**0.96**		**1.09**
Forearm		**0.60**		**0.73**		**0.83**		**0.91**		**1.06**		**1.19**
Upper leg	0.23	**0.37**	0.29	**0.47**	0.34	**0.55**	0.38	**0.62**	0.47	**0.76**	0.56	**0.90**
Knee		**0.04**		**0.05**		**0.06**	0.01	**0.07**	0.01	**0.08**	0.02	**0.10**
Lower leg												
Lymph nodes	0.10	**0.13**	0.13	**0.16**	0.15	**0.19**	0.17	**0.22**	0.21	**0.27**	0.25	**0.31**
Skin	0.19	**0.26**	0.21	**0.28**	0.21	**0.29**	0.22	**0.30**	0.23	**0.31**	0.25	**0.33**
Muscle	0.16	**0.22**	0.19	**0.26**	0.21	**0.29**	0.23	**0.31**	0.26	**0.36**	0.30	**0.40**
Skeleton	0.21	**0.36**	0.26	**0.44**	0.29	**0.50**	0.32	**0.55**	0.37	**0.64**	0.42	**0.72**
Red bone marrow	0.19	**0.24**	0.24	**0.30**	0.27	**0.34**	0.30	**0.37**	0.36	**0.44**	0.42	**0.51**
Whole body dose	0.17	**0.23**	0.20	**0.28**	0.22	**0.31**	0.24	**0.34**	0.28	**0.39**	0.31	**0.44**
ED (ICRP 60)	0.13	**0.18**	0.16	**0.23**	0.19	**0.27**	0.21	**0.30**	0.26	**0.36**	0.30	**0.43**
ED (ICRP 103)	0.14	**0.21**	0.17	**0.27**	0.20	**0.32**	0.22	**0.35**	0.27	**0.43**	0.32	**0.51**
Total filtration (mm) Al	**1.5**		**2.5**		**3.5**		**4.5**		**3.5**		**3.5**	
Cu	**0**		**0**		**0**		**0**		**0.1**		**0.2**	

Table 12.60 Normalized organ doses—Abdomen pa—13–17 years

Beam projection:			pa									**Table cells:**
Focus-to-detector distance (cm):			100									(+) = normal font = optimal field setting
Tube voltage (kVp):			110									**(−) = bold font = suboptimal field setting**
Total filtrations:			See last table row.									Empty table cells correspond to values <0.01
Organ or tissue	**Organ dose D_T : Entrance dose K_E**							**(mSv * mGy⁻¹)**				
	(+)	(−)	(+)	(−)	(+)	(−)	(+)	(−)	(+)	(−)	(+)	(−)
Skull												
Brain												
Oral mucosa												
Salivary glands												
Cervical spine						**0.01**		**0.01**		**0.01**		**0.02**
Thyroid												**0.01**
Thymus	0.01	**0.05**	0.01	**0.06**	0.02	**0.07**	0.02	**0.08**	0.02	**0.09**	0.03	**0.10**
Thoracic spine	0.40	**0.71**	0.47	**0.83**	0.52	**0.93**	0.56	**1.00**	0.64	**1.13**	0.70	**1.25**
Claviculae		**0.02**		**0.02**		**0.03**		**0.03**		**0.04**	0.01	**0.04**
Scapulae	0.02	**0.46**	0.02	**0.51**	0.03	**0.54**	0.03	**0.57**	0.04	**0.60**	0.04	**0.63**
Ribs	0.50	**0.89**	0.55	**0.99**	0.59	**1.06**	0.61	**1.10**	0.65	**1.18**	0.68	**1.23**
Oesophagus	0.10	**0.18**	0.13	**0.22**	0.14	**0.25**	0.16	**0.28**	0.19	**0.32**	0.21	**0.37**
Trachea												
Lung	0.12	**0.34**	0.14	**0.40**	0.16	**0.44**	0.17	**0.47**	0.19	**0.53**	0.21	**0.58**
Mamma	0.32	**0.91**	0.39	**1.10**	0.44	**1.24**	0.48	**1.36**	0.56	**1.59**	0.64	**1.81**
Heart	0.08	**0.16**	0.09	**0.20**	0.11	**0.22**	0.12	**0.24**	0.14	**0.29**	0.16	**0.33**
Stomach	0.18	**0.20**	0.21	**0.24**	0.24	**0.28**	0.26	**0.30**	0.31	**0.36**	0.35	**0.41**
Spleen	0.56	**0.63**	0.66	**0.73**	0.73	**0.81**	0.78	**0.87**	0.88	**0.98**	0.96	**1.08**
Liver	0.28	**0.33**	0.33	**0.39**	0.37	**0.43**	0.40	**0.47**	0.46	**0.54**	0.51	**0.60**
Gall bladder	0.22	**0.24**	0.26	**0.29**	0.30	**0.33**	0.33	**0.36**	0.38	**0.42**	0.44	**0.48**
Pancreas	0.29	**0.33**	0.34	**0.39**	0.39	**0.44**	0.43	**0.48**	0.50	**0.57**	0.57	**0.64**
Small intestine	0.29	**0.31**	0.35	**0.37**	0.39	**0.42**	0.43	**0.46**	0.51	**0.54**	0.58	**0.61**
Large intestine	0.23	**0.25**	0.28	**0.30**	0.32	**0.34**	0.35	**0.38**	0.41	**0.45**	0.47	**0.51**
Sigma/rectum	0.29	**0.31**	0.35	**0.37**	0.39	**0.42**	0.43	**0.46**	0.50	**0.54**	0.57	**0.61**
Lumbar spine	1.43	**1.51**	1.68	**1.76**	1.85	**1.95**	1.99	**2.09**	2.25	**2.36**	2.45	**2.59**
Pelvis	1.43	**1.51**	1.66	**1.75**	1.82	**1.92**	1.94	**2.05**	2.15	**2.29**	2.33	**2.47**
Adrenals	0.61	**0.67**	0.72	**0.79**	0.79	**0.87**	0.85	**0.94**	0.95	**1.06**	1.04	**1.16**
Kidneys	0.90	**0.95**	1.03	**1.09**	1.12	**1.18**	1.19	**1.25**	1.31	**1.38**	1.40	**1.49**
Urinary bladder	0.18	**0.19**	0.21	**0.23**	0.24	**0.26**	0.27	**0.28**	0.31	**0.34**	0.36	**0.39**
Uterus	0.84	**0.88**	1.01	**1.06**	1.15	**1.20**	1.26	**1.32**	1.48	**1.56**	1.69	**1.78**
Ovaries	0.32	**0.33**	0.38	**0.40**	0.43	**0.45**	0.47	**0.50**	0.56	**0.58**	0.63	**0.67**
Prostate	0.22	**0.23**	0.26	**0.28**	0.29	**0.32**	0.32	**0.35**	0.38	**0.41**	0.43	**0.46**
Testes	0.08	**0.10**	0.10	**0.12**	0.12	**0.13**	0.13	**0.15**	0.15	**0.17**	0.17	**0.20**
Upper arm		**0.15**		**0.18**		**0.20**		**0.21**		**0.24**		**0.27**
Elbow		**0.72**		**0.83**		**0.91**		**0.97**		**1.08**		**1.17**
Forearm		**0.78**		**0.91**		**0.99**		**1.06**		**1.17**		**1.25**
Upper leg	0.36	**0.57**	0.42	**0.68**	0.48	**0.77**	0.52	**0.83**	0.60	**0.96**	0.68	**1.08**
Knee	0.01	**0.07**	0.01	**0.08**	0.02	**0.09**	0.02	**0.10**	0.02	**0.11**	0.03	**0.13**
Lower leg												
Lymph nodes	0.17	**0.21**	0.20	**0.25**	0.23	**0.28**	0.25	**0.31**	0.29	**0.36**	0.33	**0.41**
Skin	0.21	**0.29**	0.23	**0.31**	0.23	**0.32**	0.24	**0.32**	0.25	**0.34**	0.26	**0.35**
Muscle	0.21	**0.29**	0.25	**0.34**	0.27	**0.37**	0.28	**0.39**	0.32	**0.43**	0.34	**0.47**
Skeleton	0.28	**0.48**	0.32	**0.56**	0.35	**0.61**	0.38	**0.65**	0.42	**0.72**	0.45	**0.78**
Red bone marrow	0.29	**0.35**	0.34	**0.41**	0.37	**0.46**	0.40	**0.49**	0.46	**0.56**	0.51	**0.62**
Whole body dose	0.22	**0.32**	0.26	**0.36**	0.28	**0.40**	0.30	**0.42**	0.33	**0.47**	0.36	**0.51**
ED (ICRP 60)	0.21	**0.29**	0.24	**0.34**	0.27	**0.39**	0.30	**0.42**	0.35	**0.49**	0.39	**0.55**
ED (ICRP 103)	0.22	**0.34**	0.26	**0.41**	0.29	**0.46**	0.32	**0.50**	0.37	**0.58**	0.41	**0.65**
Total filtration (mm) Al	**1.5**		**2.5**		**3.5**		**4.5**		**3.5**		**3.5**	
Cu	**0**		**0**		**0**		**0**		**0.1**		**0.2**	

Table 12.61 Normalized organ doses—Abdomen pa—13–17 years

Beam projection:			pa									**Table cells:**
Focus-to-detector distance (cm):			100									(+) = normal font = optimal field setting
Tube voltage (kVp):			120									**(−) = bold font = suboptimal field setting**
Total filtrations:			See last table row.									Empty table cells correspond to values <0.01
Organ or tissue	**Organ dose D_T : Entrance dose K_E**								**(mSv * mGy^{-1})**			
	(+)	(−)	(+)	(−)	(+)	(−)	(+)	(−)	(+)	(−)	(+)	(−)
Skull												
Brain												
Oral mucosa												
Salivary glands												
Cervical spine				**0.01**		**0.01**		**0.01**		**0.02**		**0.02**
Thyroid										**0.01**		**0.01**
Thymus	0.01	**0.06**	0.02	**0.07**	0.02	**0.08**	0.02	**0.08**	0.02	**0.10**	0.03	**0.11**
Thoracic spine	0.43	**0.76**	0.50	**0.88**	0.54	**0.97**	0.58	**1.04**	0.65	**1.16**	0.71	**1.26**
Claviculae		**0.02**		**0.03**		**0.03**		**0.03**	0.01	**0.04**	0.01	**0.04**
Scapulae	0.02	**0.47**	0.03	**0.52**	0.03	**0.55**	0.03	**0.57**	0.04	**0.60**	0.04	**0.62**
Ribs	0.52	**0.92**	0.56	**1.01**	0.59	**1.07**	0.62	**1.11**	0.65	**1.17**	0.67	**1.21**
Oesophagus	0.12	**0.21**	0.14	**0.24**	0.16	**0.27**	0.17	**0.30**	0.20	**0.35**	0.22	**0.39**
Trachea												
Lung	0.13	**0.37**	0.15	**0.42**	0.16	**0.46**	0.18	**0.49**	0.20	**0.55**	0.22	**0.59**
Mamma	0.36	**1.03**	0.43	**1.21**	0.48	**1.35**	0.52	**1.47**	0.60	**1.70**	0.68	**1.91**
Heart	0.09	**0.18**	0.11	**0.22**	0.12	**0.24**	0.13	**0.26**	0.15	**0.31**	0.17	**0.35**
Stomach	0.20	**0.23**	0.23	**0.27**	0.26	**0.30**	0.28	**0.33**	0.33	**0.38**	0.37	**0.43**
Spleen	0.61	**0.68**	0.70	**0.78**	0.77	**0.85**	0.82	**0.91**	0.91	**1.02**	0.98	**1.10**
Liver	0.31	**0.36**	0.36	**0.42**	0.39	**0.46**	0.42	**0.50**	0.48	**0.56**	0.53	**0.62**
Gall bladder	0.24	**0.27**	0.29	**0.32**	0.32	**0.36**	0.35	**0.39**	0.41	**0.45**	0.46	**0.51**
Pancreas	0.32	**0.36**	0.38	**0.43**	0.42	**0.48**	0.46	**0.52**	0.53	**0.60**	0.59	**0.67**
Small intestine	0.32	**0.35**	0.38	**0.41**	0.43	**0.46**	0.46	**0.49**	0.54	**0.57**	0.60	**0.64**
Large intestine	0.26	**0.28**	0.31	**0.34**	0.35	**0.38**	0.38	**0.41**	0.44	**0.48**	0.50	**0.54**
Sigma/rectum	0.32	**0.35**	0.38	**0.41**	0.42	**0.46**	0.46	**0.50**	0.53	**0.57**	0.59	**0.64**
Lumbar spine	1.53	**1.61**	1.77	**1.86**	1.93	**2.03**	2.06	**2.17**	2.29	**2.41**	2.47	**2.61**
Pelvis	1.52	**1.60**	1.73	**1.83**	1.88	**1.99**	1.99	**2.11**	2.18	**2.32**	2.33	**2.48**
Adrenals	0.66	**0.73**	0.76	**0.84**	0.83	**0.92**	0.89	**0.98**	0.99	**1.10**	1.07	**1.19**
Kidneys	0.96	**1.01**	1.08	**1.14**	1.16	**1.23**	1.23	**1.30**	1.34	**1.42**	1.43	**1.51**
Urinary bladder	0.20	**0.21**	0.24	**0.25**	0.27	**0.28**	0.29	**0.31**	0.34	**0.36**	0.38	**0.41**
Uterus	0.94	**0.99**	1.12	**1.17**	1.25	**1.31**	1.36	**1.43**	1.57	**1.65**	1.77	**1.86**
Ovaries	0.35	**0.37**	0.42	**0.44**	0.47	**0.49**	0.51	**0.54**	0.59	**0.62**	0.66	**0.70**
Prostate	0.24	**0.26**	0.29	**0.31**	0.32	**0.34**	0.35	**0.37**	0.40	**0.43**	0.45	**0.49**
Testes	0.10	**0.11**	0.11	**0.13**	0.13	**0.15**	0.14	**0.16**	0.16	**0.19**	0.18	**0.21**
Upper arm		**0.17**		**0.19**		**0.21**		**0.22**		**0.25**		**0.27**
Elbow		**0.76**		**0.87**		**0.94**		**1.00**		**1.09**		**1.17**
Forearm		**0.83**		**0.94**		**1.02**		**1.08**		**1.18**		**1.25**
Upper leg	0.39	**0.63**	0.46	**0.73**	0.51	**0.81**	0.55	**0.88**	0.63	**1.00**	0.69	**1.10**
Knee	0.01	**0.07**	0.02	**0.09**	0.02	**0.10**	0.02	**0.10**	0.02	**0.12**	0.03	**0.13**
Lower leg												
Lymph nodes	0.19	**0.23**	0.22	**0.28**	0.25	**0.31**	0.27	**0.33**	0.31	**0.38**	0.35	**0.43**
Skin	0.22	**0.30**	0.23	**0.31**	0.24	**0.32**	0.25	**0.33**	0.26	**0.34**	0.27	**0.36**
Muscle	0.23	**0.31**	0.26	**0.35**	0.28	**0.38**	0.30	**0.41**	0.33	**0.45**	0.35	**0.48**
Skeleton	0.30	**0.51**	0.34	**0.58**	0.37	**0.63**	0.39	**0.67**	0.43	**0.73**	0.45	**0.78**
Red bone marrow	0.31	**0.39**	0.36	**0.44**	0.40	**0.49**	0.42	**0.52**	0.48	**0.59**	0.52	**0.65**
Whole body dose	0.24	**0.34**	0.27	**0.38**	0.29	**0.41**	0.31	**0.44**	0.34	**0.48**	0.37	**0.52**
ED (ICRP 60)	0.23	**0.32**	0.27	**0.37**	0.30	**0.42**	0.32	**0.45**	0.37	**0.51**	0.41	**0.57**
ED (ICRP 103)	0.24	**0.38**	0.28	**0.44**	0.31	**0.49**	0.34	**0.53**	0.39	**0.61**	0.43	**0.68**
Total filtration (mm) Al	**1.5**		**2.5**		**3.5**		**4.5**		**3.5**		**3.5**	
Cu	**0**		**0**		**0**		**0**		**0.1**		**0.2**	

Table 12.62 Normalized organ doses—Abdomen pa—>17 years

Beam projection:			pa									**Table cells:**
Focus-to-detector distance (cm):			100					(+) = normal font = optimal field setting				
Tube voltage (kVp):			70					**(−) = bold font = suboptimal field setting**				
Total filtrations:			See last table row.					Empty table cells correspond to values <0.01				
Organ or tissue	**Organ dose D_T : Entrance dose K_E**							**(mSv * mGy^{-1})**				
	(+)	(−)	(+)	(−)	(+)	(−)	(+)	(−)	(+)	(−)	(+)	(−)
Skull												
Brain												
Oral mucosa												
Salivary glands												
Cervical spine												
Thyroid												
Thymus		**0.02**		**0.02**		**0.03**		**0.03**	0.01	**0.04**	0.01	**0.05**
Thoracic spine	0.23	**0.36**	0.30	**0.47**	0.35	**0.55**	0.39	**0.62**	0.48	**0.76**	0.56	**0.89**
Claviculae								**0.01**		**0.01**		**0.02**
Scapulae		**0.24**	0.01	**0.29**	0.01	**0.32**	0.02	**0.35**	0.02	**0.39**	0.03	**0.43**
Ribs	0.41	**0.65**	0.49	**0.77**	0.53	**0.85**	0.57	**0.91**	0.64	**1.02**	0.69	**1.11**
Oesophagus	0.05	**0.08**	0.07	**0.10**	0.08	**0.12**	0.09	**0.14**	0.12	**0.18**	0.15	**0.23**
Trachea												
Lung	0.08	**0.19**	0.10	**0.24**	0.12	**0.28**	0.13	**0.30**	0.15	**0.36**	0.18	**0.41**
Mamma	0.02	**0.05**	0.03	**0.07**	0.04	**0.08**	0.04	**0.09**	0.05	**0.11**	0.07	**0.14**
Heart	0.04	**0.07**	0.05	**0.09**	0.06	**0.11**	0.07	**0.12**	0.09	**0.16**	0.11	**0.19**
Stomach	0.08	**0.09**	0.10	**0.11**	0.12	**0.14**	0.14	**0.16**	0.18	**0.20**	0.22	**0.25**
Spleen	0.33	**0.36**	0.42	**0.46**	0.49	**0.54**	0.55	**0.60**	0.66	**0.72**	0.76	**0.83**
Liver	0.15	**0.17**	0.20	**0.22**	0.23	**0.26**	0.26	**0.29**	0.32	**0.36**	0.37	**0.42**
Gall bladder	0.09	**0.10**	0.12	**0.13**	0.14	**0.15**	0.16	**0.18**	0.21	**0.23**	0.26	**0.28**
Pancreas	0.13	**0.14**	0.17	**0.19**	0.20	**0.22**	0.23	**0.26**	0.29	**0.32**	0.36	**0.40**
Small intestine	0.13	**0.13**	0.17	**0.17**	0.20	**0.21**	0.23	**0.24**	0.29	**0.31**	0.36	**0.38**
Large intestine	0.10	**0.10**	0.13	**0.14**	0.16	**0.17**	0.18	**0.19**	0.23	**0.25**	0.28	**0.30**
Sigma/rectum	0.13	**0.14**	0.17	**0.18**	0.21	**0.22**	0.24	**0.25**	0.30	**0.32**	0.37	**0.39**
Lumbar spine	0.83	**0.87**	1.06	**1.11**	1.24	**1.30**	1.38	**1.45**	1.68	**1.75**	1.95	**2.04**
Pelvis	0.91	**0.95**	1.14	**1.19**	1.32	**1.38**	1.46	**1.53**	1.74	**1.81**	1.99	**2.08**
Adrenals	0.37	**0.40**	0.47	**0.50**	0.54	**0.58**	0.60	**0.65**	0.72	**0.77**	0.83	**0.89**
Kidneys	0.61	**0.64**	0.74	**0.78**	0.84	**0.88**	0.92	**0.96**	1.06	**1.11**	1.19	**1.25**
Urinary bladder	0.07	**0.08**	0.10	**0.10**	0.12	**0.12**	0.14	**0.14**	0.18	**0.18**	0.22	**0.23**
Uterus	0.14	**0.14**	0.18	**0.19**	0.22	**0.23**	0.25	**0.26**	0.32	**0.34**	0.40	**0.41**
Ovaries	0.13	**0.14**	0.18	**0.19**	0.21	**0.22**	0.25	**0.26**	0.32	**0.33**	0.39	**0.41**
Prostate	0.09	**0.09**	0.12	**0.12**	0.14	**0.15**	0.16	**0.17**	0.21	**0.21**	0.26	**0.26**
Testes	0.03	**0.03**	0.04	**0.05**	0.05	**0.05**	0.06	**0.06**	0.08	**0.08**	0.10	**0.10**
Upper arm		**0.08**		**0.10**		**0.12**		**0.13**		**0.16**		**0.19**
Elbow		**0.51**		**0.64**		**0.74**		**0.83**		**0.98**		**1.13**
Forearm		**0.64**		**0.80**		**0.93**		**1.03**		**1.22**		**1.39**
Upper leg	0.20	**0.24**	0.26	**0.32**	0.32	**0.38**	0.36	**0.43**	0.46	**0.55**	0.55	**0.66**
Knee		**0.04**	0.01	**0.05**	0.01	**0.07**	0.02	**0.08**	0.02	**0.10**	0.03	**0.12**
Lower leg												
Lymph nodes	0.08	**0.10**	0.10	**0.13**	0.12	**0.15**	0.14	**0.17**	0.18	**0.21**	0.21	**0.26**
Skin	0.20	**0.28**	0.21	**0.30**	0.22	**0.31**	0.23	**0.32**	0.24	**0.33**	0.25	**0.35**
Muscle	0.15	**0.19**	0.18	**0.23**	0.20	**0.26**	0.22	**0.28**	0.26	**0.33**	0.29	**0.37**
Skeleton	0.18	**0.30**	0.23	**0.37**	0.26	**0.43**	0.29	**0.48**	0.35	**0.57**	0.40	**0.65**
Red bone marrow	0.17	**0.20**	0.21	**0.25**	0.25	**0.29**	0.27	**0.32**	0.33	**0.39**	0.39	**0.45**
Whole body dose	0.15	**0.20**	0.19	**0.25**	0.21	**0.28**	0.23	**0.31**	0.27	**0.36**	0.31	**0.41**
ED (ICRP 60)	0.10	**0.12**	0.12	**0.16**	0.15	**0.18**	0.16	**0.21**	0.20	**0.26**	0.24	**0.30**
ED (ICRP 103)	0.09	**0.12**	0.12	**0.16**	0.14	**0.18**	0.16	**0.21**	0.20	**0.25**	0.24	**0.30**
Total filtration (mm) **Al**	**1.5**		**2.5**		**3.5**		**4.5**		**3.5**		**3.5**	
Cu	**0**		**0**		**0**		**0**		**0.1**		**0.2**	

Table 12.63 Normalized organ doses—Abdomen pa—>17 years

Beam projection:			pa									**Table cells:**
Focus-to-detector distance (cm):			100									(+) = normal font = optimal field setting
Tube voltage (kVp):			80									**(−) = bold font = suboptimal field setting**
Total filtrations:			See last table row.									Empty table cells correspond to values <0.01
Organ or tissue	**Organ dose D_T : Entrance dose K_E**								**(mSv * mGy^{-1})**			
	(+)	(−)	(+)	(−)	(+)	(−)	(+)	(−)	(+)	(−)	(+)	(−)
Skull												
Brain												
Oral mucosa												
Salivary glands												
Cervical spine												
Thyroid												
Thymus		**0.02**		**0.03**	0.01	**0.04**	0.01	**0.04**	0.02	**0.05**	0.02	**0.07**
Thoracic spine	0.28	**0.45**	0.35	**0.56**	0.41	**0.65**	0.46	**0.72**	0.55	**0.87**	0.63	**1.00**
Claviculae				**0.01**		**0.01**		**0.02**		**0.02**		**0.02**
Scapulae	0.01	**0.27**	0.02	**0.32**	0.02	**0.35**	0.02	**0.37**	0.03	**0.42**	0.03	**0.45**
Ribs	0.45	**0.71**	0.52	**0.83**	0.57	**0.91**	0.60	**0.96**	0.67	**1.07**	0.72	**1.15**
Oesophagus	0.07	**0.11**	0.09	**0.14**	0.10	**0.16**	0.12	**0.19**	0.15	**0.23**	0.18	**0.28**
Trachea												
Lung	0.10	**0.23**	0.12	**0.28**	0.13	**0.31**	0.15	**0.34**	0.17	**0.40**	0.20	**0.45**
Mamma	0.03	**0.07**	0.04	**0.08**	0.05	**0.10**	0.05	**0.11**	0.07	**0.14**	0.08	**0.16**
Heart	0.05	**0.09**	0.06	**0.12**	0.08	**0.14**	0.09	**0.16**	0.11	**0.20**	0.13	**0.24**
Stomach	0.10	**0.11**	0.13	**0.15**	0.16	**0.18**	0.18	**0.20**	0.22	**0.25**	0.27	**0.30**
Spleen	0.40	**0.44**	0.49	**0.54**	0.57	**0.62**	0.63	**0.69**	0.74	**0.81**	0.84	**0.92**
Liver	0.19	**0.21**	0.24	**0.27**	0.27	**0.31**	0.31	**0.35**	0.37	**0.42**	0.43	**0.49**
Gall bladder	0.12	**0.13**	0.16	**0.17**	0.19	**0.20**	0.21	**0.23**	0.27	**0.29**	0.32	**0.35**
Pancreas	0.17	**0.19**	0.22	**0.24**	0.26	**0.28**	0.29	**0.32**	0.36	**0.40**	0.43	**0.48**
Small intestine	0.17	**0.18**	0.22	**0.23**	0.26	**0.27**	0.29	**0.31**	0.36	**0.38**	0.44	**0.46**
Large intestine	0.13	**0.14**	0.17	**0.18**	0.20	**0.22**	0.23	**0.25**	0.29	**0.31**	0.35	**0.38**
Sigma/rectum	0.17	**0.18**	0.22	**0.24**	0.26	**0.28**	0.30	**0.32**	0.37	**0.39**	0.44	**0.47**
Lumbar spine	1.00	**1.05**	1.25	**1.31**	1.44	**1.50**	1.59	**1.67**	1.89	**1.98**	2.17	**2.27**
Pelvis	1.07	**1.12**	1.31	**1.37**	1.50	**1.56**	1.64	**1.71**	1.92	**2.01**	2.17	**2.27**
Adrenals	0.44	**0.47**	0.55	**0.59**	0.62	**0.67**	0.69	**0.74**	0.81	**0.88**	0.92	**1.00**
Kidneys	0.70	**0.73**	0.84	**0.87**	0.94	**0.98**	1.02	**1.06**	1.16	**1.22**	1.29	**1.35**
Urinary bladder	0.10	**0.10**	0.13	**0.14**	0.16	**0.16**	0.18	**0.18**	0.22	**0.23**	0.27	**0.28**
Uterus	0.19	**0.19**	0.24	**0.25**	0.29	**0.30**	0.32	**0.34**	0.41	**0.42**	0.49	**0.51**
Ovaries	0.18	**0.19**	0.23	**0.24**	0.28	**0.29**	0.31	**0.33**	0.39	**0.41**	0.48	**0.50**
Prostate	0.12	**0.12**	0.15	**0.16**	0.18	**0.19**	0.21	**0.21**	0.26	**0.27**	0.31	**0.32**
Testes	0.04	**0.05**	0.06	**0.06**	0.07	**0.07**	0.08	**0.08**	0.10	**0.11**	0.12	**0.13**
Upper arm		**0.10**		**0.12**		**0.14**		**0.15**		**0.18**		**0.21**
Elbow		**0.60**		**0.74**		**0.84**		**0.93**		**1.09**		**1.23**
Forearm		**0.74**		**0.91**		**1.04**		**1.14**		**1.33**		**1.49**
Upper leg	0.26	**0.31**	0.33	**0.40**	0.39	**0.47**	0.44	**0.53**	0.55	**0.66**	0.65	**0.79**
Knee	0.01	**0.05**	0.02	**0.07**	0.02	**0.08**	0.02	**0.09**	0.03	**0.12**	0.04	**0.14**
Lower leg												
Lymph nodes	0.10	**0.12**	0.13	**0.16**	0.16	**0.19**	0.18	**0.21**	0.22	**0.26**	0.26	**0.31**
Skin	0.21	**0.29**	0.22	**0.31**	0.23	**0.32**	0.24	**0.33**	0.25	**0.35**	0.26	**0.37**
Muscle	0.17	**0.22**	0.21	**0.26**	0.23	**0.29**	0.25	**0.32**	0.29	**0.36**	0.32	**0.41**
Skeleton	0.22	**0.35**	0.27	**0.43**	0.30	**0.49**	0.33	**0.54**	0.39	**0.63**	0.44	**0.71**
Red bone marrow	0.20	**0.24**	0.25	**0.30**	0.29	**0.34**	0.32	**0.38**	0.38	**0.45**	0.44	**0.52**
Whole body dose	0.18	**0.24**	0.21	**0.28**	0.24	**0.32**	0.26	**0.34**	0.30	**0.40**	0.34	**0.45**
ED (ICRP 60)	0.12	**0.15**	0.15	**0.19**	0.18	**0.22**	0.20	**0.25**	0.25	**0.31**	0.29	**0.36**
ED (ICRP 103)	0.12	**0.15**	0.15	**0.19**	0.17	**0.22**	0.19	**0.25**	0.24	**0.30**	0.28	**0.35**
Total filtration (mm) **Al**	**1.5**		**2.5**		**3.5**		**4.5**		**3.5**		**3.5**	
Cu	**0**		**0**		**0**		**0**		**0.1**		**0.2**	

Table 12.64 Normalized organ doses—Abdomen pa—>17 years

Beam projection:			pa									**Table cells:**
Focus-to-detector distance (cm):			100									(+) = normal font = optimal field setting
Tube voltage (kVp):			110									**(−) = bold font = suboptimal field setting**
Total filtrations:			See last table row.									Empty table cells correspond to values <0.01
Organ or tissue	**Organ dose D_T : Entrance dose K_E**							**(mSv * mGy^{-1})**				
	(+)	(−)	(+)	(−)	(+)	(−)	(+)	(−)	(+)	(−)	(+)	(−)
Skull												
Brain												
Oral mucosa												
Salivary glands												
Cervical spine										**0.01**		**0.01**
Thyroid												
Thymus	0.01	**0.05**	0.02	**0.06**	0.02	**0.06**	0.02	**0.07**	0.03	**0.08**	0.03	**0.10**
Thoracic spine	0.41	**0.65**	0.48	**0.76**	0.54	**0.85**	0.58	**0.92**	0.66	**1.04**	0.73	**1.15**
Claviculae		**0.02**		**0.02**		**0.02**		**0.02**	0.01	**0.03**	0.01	**0.03**
Scapulae	0.02	**0.33**	0.03	**0.36**	0.03	**0.39**	0.03	**0.41**	0.04	**0.44**	0.05	**0.46**
Ribs	0.53	**0.84**	0.59	**0.94**	0.62	**1.00**	0.65	**1.04**	0.69	**1.12**	0.72	**1.17**
Oesophagus	0.12	**0.19**	0.15	**0.23**	0.17	**0.26**	0.18	**0.28**	0.22	**0.33**	0.25	**0.38**
Trachea												
Lung	0.14	**0.31**	0.16	**0.37**	0.18	**0.40**	0.19	**0.43**	0.21	**0.48**	0.24	**0.53**
Mamma	0.05	**0.11**	0.06	**0.13**	0.07	**0.15**	0.08	**0.16**	0.09	**0.19**	0.11	**0.22**
Heart	0.09	**0.16**	0.11	**0.19**	0.12	**0.22**	0.14	**0.24**	0.16	**0.28**	0.18	**0.32**
Stomach	0.18	**0.20**	0.22	**0.24**	0.25	**0.28**	0.27	**0.30**	0.32	**0.36**	0.36	**0.41**
Spleen	0.57	**0.63**	0.67	**0.73**	0.74	**0.81**	0.79	**0.87**	0.89	**0.99**	0.98	**1.08**
Liver	0.29	**0.33**	0.34	**0.39**	0.38	**0.44**	0.41	**0.47**	0.48	**0.55**	0.53	**0.61**
Gall bladder	0.22	**0.23**	0.26	**0.28**	0.30	**0.32**	0.33	**0.35**	0.38	**0.41**	0.44	**0.47**
Pancreas	0.29	**0.32**	0.34	**0.38**	0.39	**0.43**	0.43	**0.47**	0.50	**0.56**	0.57	**0.63**
Small intestine	0.29	**0.31**	0.35	**0.37**	0.40	**0.42**	0.44	**0.46**	0.51	**0.54**	0.58	**0.62**
Large intestine	0.23	**0.25**	0.28	**0.30**	0.32	**0.34**	0.35	**0.38**	0.42	**0.45**	0.48	**0.51**
Sigma/rectum	0.30	**0.31**	0.35	**0.38**	0.40	**0.42**	0.44	**0.46**	0.51	**0.55**	0.59	**0.62**
Lumbar spine	1.42	**1.49**	1.66	**1.74**	1.84	**1.93**	1.98	**2.07**	2.24	**2.34**	2.45	**2.57**
Pelvis	1.44	**1.51**	1.67	**1.75**	1.83	**1.92**	1.96	**2.05**	2.18	**2.29**	2.36	**2.48**
Adrenals	0.63	**0.68**	0.73	**0.80**	0.81	**0.88**	0.87	**0.95**	0.98	**1.07**	1.08	**1.18**
Kidneys	0.92	**0.96**	1.05	**1.10**	1.14	**1.20**	1.21	**1.27**	1.34	**1.41**	1.44	**1.51**
Urinary bladder	0.18	**0.19**	0.22	**0.23**	0.25	**0.26**	0.28	**0.29**	0.33	**0.34**	0.37	**0.39**
Uterus	0.33	**0.34**	0.39	**0.41**	0.45	**0.46**	0.49	**0.51**	0.58	**0.60**	0.66	**0.69**
Ovaries	0.32	**0.33**	0.38	**0.40**	0.43	**0.45**	0.47	**0.50**	0.56	**0.59**	0.64	**0.67**
Prostate	0.21	**0.22**	0.25	**0.26**	0.29	**0.30**	0.31	**0.33**	0.37	**0.38**	0.42	**0.44**
Testes	0.08	**0.09**	0.10	**0.11**	0.11	**0.12**	0.12	**0.13**	0.15	**0.16**	0.17	**0.18**
Upper arm		**0.14**		**0.16**		**0.18**		**0.19**		**0.22**		**0.24**
Elbow		**0.81**		**0.94**		**1.03**		**1.10**		**1.23**		**1.33**
Forearm		**0.98**		**1.14**		**1.24**		**1.33**		**1.47**		**1.58**
Upper leg	0.42	**0.51**	0.50	**0.61**	0.57	**0.68**	0.62	**0.74**	0.72	**0.87**	0.81	**0.98**
Knee	0.02	**0.09**	0.03	**0.11**	0.03	**0.13**	0.04	**0.14**	0.05	**0.16**	0.05	**0.18**
Lower leg												
Lymph nodes	0.17	**0.21**	0.21	**0.25**	0.24	**0.28**	0.26	**0.30**	0.30	**0.36**	0.34	**0.40**
Skin	0.23	**0.32**	0.24	**0.34**	0.25	**0.35**	0.25	**0.36**	0.27	**0.38**	0.28	**0.39**
Muscle	0.23	**0.30**	0.27	**0.34**	0.29	**0.37**	0.31	**0.39**	0.35	**0.44**	0.38	**0.48**
Skeleton	0.29	**0.48**	0.34	**0.55**	0.38	**0.60**	0.40	**0.65**	0.45	**0.72**	0.49	**0.78**
Red bone marrow	0.31	**0.36**	0.36	**0.42**	0.40	**0.47**	0.43	**0.50**	0.49	**0.58**	0.54	**0.64**
Whole body dose	0.24	**0.32**	0.28	**0.37**	0.30	**0.40**	0.32	**0.43**	0.36	**0.48**	0.39	**0.52**
ED (ICRP 60)	0.20	**0.24**	0.23	**0.29**	0.26	**0.32**	0.29	**0.35**	0.33	**0.41**	0.38	**0.46**
ED (ICRP 103)	0.19	**0.24**	0.22	**0.28**	0.25	**0.32**	0.27	**0.34**	0.32	**0.40**	0.36	**0.45**
Total filtration (mm) **Al**	**1.5**		**2.5**		**3.5**		**4.5**		**3.5**		**3.5**	
Cu	**0**		**0**		**0**		**0**		**0.1**		**0.2**	

Table 12.65 Normalized organ doses—Abdomen pa—>17 years

Beam projection:			pa									**Table cells:**
Focus-to-detector distance (cm):			100						(+) = normal font = optimal field setting			
Tube voltage (kVp):			120						**(−) = bold font = suboptimal field setting**			
Total filtrations:			See last table row.						Empty table cells correspond to values <0.01			
Organ or tissue	**Organ dose D_T : Entrance dose K_E**								**(mSv * mGy^{-1})**			
	(+)	(−)	(+)	(−)	(+)	(−)	(+)	(−)	(+)	(−)	(+)	(−)
Skull												
Brain												
Oral mucosa												
Salivary glands												
Cervical spine										**0.01**		**0.01**
Thyroid												
Thymus	0.02	**0.05**	0.02	**0.06**	0.02	**0.07**	0.02	**0.08**	0.03	**0.09**	0.03	**0.10**
Thoracic spine	0.44	**0.70**	0.51	**0.81**	0.57	**0.89**	0.61	**0.95**	0.68	**1.07**	0.74	**1.17**
Claviculae		**0.02**		**0.02**		**0.02**		**0.03**	0.01	**0.03**	0.01	**0.04**
Scapulae	0.02	**0.34**	0.03	**0.37**	0.03	**0.40**	0.04	**0.42**	0.04	**0.44**	0.05	**0.46**
Ribs	0.55	**0.87**	0.60	**0.96**	0.63	**1.01**	0.65	**1.05**	0.69	**1.11**	0.71	**1.15**
Oesophagus	0.14	**0.21**	0.16	**0.25**	0.18	**0.28**	0.20	**0.31**	0.23	**0.36**	0.26	**0.40**
Trachea												
Lung	0.15	**0.34**	0.17	**0.39**	0.19	**0.42**	0.20	**0.45**	0.22	**0.50**	0.24	**0.54**
Mamma	0.06	**0.12**	0.07	**0.15**	0.08	**0.16**	0.09	**0.18**	0.10	**0.20**	0.11	**0.23**
Heart	0.10	**0.18**	0.12	**0.21**	0.14	**0.24**	0.15	**0.26**	0.17	**0.30**	0.20	**0.34**
Stomach	0.20	**0.23**	0.24	**0.27**	0.27	**0.30**	0.29	**0.33**	0.34	**0.38**	0.38	**0.43**
Spleen	0.62	**0.68**	0.71	**0.78**	0.78	**0.86**	0.83	**0.91**	0.92	**1.02**	1.00	**1.11**
Liver	0.32	**0.36**	0.37	**0.42**	0.41	**0.47**	0.44	**0.50**	0.50	**0.57**	0.55	**0.63**
Gall bladder	0.24	**0.26**	0.29	**0.31**	0.32	**0.35**	0.35	**0.38**	0.41	**0.44**	0.47	**0.50**
Pancreas	0.32	**0.36**	0.38	**0.42**	0.42	**0.47**	0.46	**0.51**	0.53	**0.59**	0.60	**0.66**
Small intestine	0.33	**0.35**	0.39	**0.41**	0.43	**0.46**	0.47	**0.50**	0.55	**0.58**	0.61	**0.65**
Large intestine	0.26	**0.28**	0.31	**0.34**	0.35	**0.38**	0.38	**0.41**	0.45	**0.48**	0.50	**0.54**
Sigma/rectum	0.33	**0.35**	0.39	**0.41**	0.44	**0.46**	0.47	**0.50**	0.55	**0.58**	0.61	**0.65**
Lumbar spine	1.52	**1.59**	1.76	**1.84**	1.92	**2.01**	2.05	**2.15**	2.29	**2.39**	2.48	**2.59**
Pelvis	1.53	**1.60**	1.75	**1.83**	1.90	**1.99**	2.01	**2.11**	2.21	**2.33**	2.37	**2.49**
Adrenals	0.68	**0.74**	0.78	**0.85**	0.85	**0.93**	0.91	**0.99**	1.02	**1.11**	1.10	**1.21**
Kidneys	0.97	**1.02**	1.10	**1.16**	1.19	**1.25**	1.25	**1.32**	1.37	**1.44**	1.46	**1.54**
Urinary bladder	0.21	**0.22**	0.24	**0.26**	0.27	**0.29**	0.30	**0.31**	0.35	**0.36**	0.40	**0.41**
Uterus	0.37	**0.38**	0.44	**0.45**	0.49	**0.51**	0.53	**0.55**	0.62	**0.64**	0.70	**0.72**
Ovaries	0.35	**0.37**	0.42	**0.44**	0.47	**0.50**	0.51	**0.54**	0.60	**0.63**	0.67	**0.71**
Prostate	0.24	**0.24**	0.28	**0.29**	0.31	**0.32**	0.34	**0.35**	0.40	**0.41**	0.45	**0.46**
Testes	0.09	**0.10**	0.11	**0.12**	0.12	**0.13**	0.14	**0.15**	0.16	**0.17**	0.18	**0.20**
Upper arm		**0.15**		**0.17**		**0.19**		**0.20**		**0.22**		**0.24**
Elbow		**0.86**		**0.98**		**1.07**		**1.13**		**1.24**		**1.33**
Forearm		**1.04**		**1.18**		**1.28**		**1.36**		**1.48**		**1.58**
Upper leg	0.46	**0.56**	0.55	**0.65**	0.61	**0.73**	0.66	**0.79**	0.75	**0.90**	0.84	**1.00**
Knee	0.03	**0.10**	0.03	**0.12**	0.04	**0.14**	0.04	**0.15**	0.05	**0.17**	0.06	**0.19**
Lower leg												
Lymph nodes	0.19	**0.23**	0.23	**0.27**	0.26	**0.30**	0.28	**0.33**	0.32	**0.38**	0.36	**0.42**
Skin	0.23	**0.33**	0.24	**0.35**	0.25	**0.36**	0.26	**0.37**	0.27	**0.38**	0.28	**0.40**
Muscle	0.25	**0.32**	0.28	**0.36**	0.31	**0.39**	0.33	**0.41**	0.36	**0.45**	0.39	**0.49**
Skeleton	0.31	**0.51**	0.36	**0.58**	0.39	**0.63**	0.41	**0.67**	0.46	**0.73**	0.49	**0.79**
Red bone marrow	0.33	**0.39**	0.39	**0.45**	0.42	**0.50**	0.45	**0.53**	0.51	**0.60**	0.56	**0.66**
Whole body dose	0.26	**0.34**	0.29	**0.39**	0.32	**0.42**	0.34	**0.45**	0.37	**0.49**	0.40	**0.53**
ED (ICRP 60)	0.22	**0.27**	0.26	**0.32**	0.28	**0.35**	0.31	**0.38**	0.35	**0.43**	0.39	**0.48**
ED (ICRP 103)	0.21	**0.26**	0.24	**0.31**	0.27	**0.34**	0.29	**0.37**	0.33	**0.42**	0.37	**0.46**
Total filtration (mm) Al	**1.5**		**2.5**		**3.5**		**4.5**		**3.5**		**3.5**	
Cu	**0**		**0**		**0**		**0**		**0.1**		**0.2**	

Table 12.66 Normalized organ doses—Abdomen r-lat—<1 year

Beam projection:			r-lat									**Table cells:**
Focus-to-detector distance (cm):			100					(+) = normal font = optimal field setting				
Tube voltage (kVp):			60					**(−) = bold font = suboptimal field setting**				
Total filtrations:			See last table row.					Empty table cells correspond to values <0.01				
Organ or tissue	**Organ dose D_T : Entrance dose K_E**							**(mSv * mGy^{-1})**				
	(+)	(−)	(+)	(−)	(+)	(−)	(+)	(−)	(+)	(−)	(+)	(−)
Skull												
Brain												
Oral mucosa												
Salivary glands										**0.01**		**0.01**
Cervical spine		**0.02**		**0.02**	0.01	**0.03**	0.01	**0.03**	0.02	**0.04**	0.01	**0.04**
Thyroid		**0.01**		**0.02**		**0.02**		**0.02**		**0.03**		**0.02**
Thymus	0.01	**0.07**	0.01	**0.09**	0.02	**0.10**	0.02	**0.11**	0.02	**0.13**	0.02	**0.13**
Thoracic spine	0.27	**0.44**	0.35	**0.57**	0.41	**0.66**	0.46	**0.74**	0.54	**0.89**	0.52	**0.85**
Claviculae	0.01	**0.04**	0.02	**0.05**	0.02	**0.06**	0.02	**0.07**	0.03	**0.08**	0.03	**0.08**
Scapulae	0.04	**0.70**	0.05	**0.81**	0.06	**0.87**	0.07	**0.93**	0.09	**1.01**	0.08	**1.01**
Ribs	0.55	**0.81**	0.64	**0.95**	0.69	**1.05**	0.74	**1.13**	0.81	**1.25**	0.81	**1.25**
Oesophagus	0.12	**0.17**	0.15	**0.22**	0.17	**0.26**	0.19	**0.29**	0.23	**0.34**	0.22	**0.33**
Trachea												
Lung	0.15	**0.26**	0.18	**0.32**	0.20	**0.37**	0.21	**0.40**	0.24	**0.46**	0.24	**0.45**
Mamma	0.02	**0.55**	0.03	**0.59**	0.04	**0.61**	0.04	**0.63**	0.05	**0.67**	0.05	**0.66**
Heart	0.13	**0.23**	0.16	**0.29**	0.19	**0.34**	0.21	**0.38**	0.24	**0.44**	0.23	**0.42**
Stomach	0.12	**0.12**	0.16	**0.15**	0.19	**0.18**	0.22	**0.21**	0.27	**0.25**	0.25	**0.23**
Spleen	0.08	**0.08**	0.11	**0.10**	0.13	**0.12**	0.15	**0.14**	0.19	**0.18**	0.17	**0.16**
Liver	0.55	**0.45**	0.65	**0.55**	0.73	**0.62**	0.78	**0.68**	0.88	**0.78**	0.86	**0.76**
Gall bladder	0.26	**0.22**	0.33	**0.29**	0.39	**0.34**	0.43	**0.38**	0.52	**0.46**	0.49	**0.43**
Pancreas	0.18	**0.15**	0.24	**0.19**	0.28	**0.23**	0.31	**0.26**	0.38	**0.32**	0.35	**0.29**
Small intestine	0.28	**0.22**	0.35	**0.29**	0.41	**0.34**	0.45	**0.38**	0.54	**0.45**	0.51	**0.43**
Large intestine	0.41	**0.33**	0.50	**0.41**	0.57	**0.48**	0.63	**0.53**	0.73	**0.62**	0.71	**0.59**
Sigma/rectum	0.13	**0.11**	0.17	**0.14**	0.20	**0.17**	0.23	**0.20**	0.28	**0.24**	0.26	**0.22**
Lumbar spine	0.60	**0.57**	0.78	**0.74**	0.91	**0.86**	1.02	**0.97**	1.23	**1.16**	1.17	**1.11**
Pelvis	0.88	**0.74**	1.09	**0.93**	1.23	**1.07**	1.35	**1.18**	1.56	**1.38**	1.53	**1.34**
Adrenals	0.21	**0.21**	0.27	**0.27**	0.31	**0.31**	0.35	**0.35**	0.41	**0.41**	0.39	**0.39**
Kidneys	0.30	**0.29**	0.37	**0.36**	0.42	**0.41**	0.46	**0.45**	0.53	**0.52**	0.51	**0.50**
Urinary bladder	0.26	**0.24**	0.33	**0.30**	0.38	**0.35**	0.43	**0.39**	0.50	**0.47**	0.48	**0.44**
Uterus	0.25	**0.21**	0.32	**0.27**	0.38	**0.32**	0.42	**0.36**	0.51	**0.43**	0.48	**0.40**
Ovaries	0.27	**0.21**	0.35	**0.27**	0.40	**0.32**	0.45	**0.36**	0.54	**0.44**	0.51	**0.41**
Prostate	0.22	**0.20**	0.28	**0.25**	0.33	**0.30**	0.37	**0.34**	0.44	**0.40**	0.41	**0.38**
Testes	0.20	**0.20**	0.25	**0.26**	0.29	**0.30**	0.32	**0.33**	0.38	**0.40**	0.37	**0.38**
Upper arm		**0.42**		**0.48**		**0.53**		**0.56**		**0.61**		**0.60**
Elbow		**1.41**		**1.59**		**1.70**		**1.78**		**1.91**		**1.92**
Forearm		**1.50**		**1.67**		**1.78**		**1.85**		**1.96**		**1.98**
Upper leg	0.45	**0.58**	0.56	**0.73**	0.65	**0.85**	0.72	**0.94**	0.85	**1.11**	0.82	**1.07**
Knee	0.04	**0.25**	0.05	**0.32**	0.06	**0.37**	0.07	**0.41**	0.09	**0.48**	0.08	**0.46**
Lower leg						**0.01**		**0.01**		**0.02**		**0.01**
Lymph nodes	0.16	**0.15**	0.20	**0.20**	0.23	**0.23**	0.26	**0.26**	0.31	**0.31**	0.29	**0.29**
Skin	0.22	**0.29**	0.24	**0.30**	0.25	**0.31**	0.25	**0.32**	0.27	**0.34**	0.26	**0.34**
Muscle	0.18	**0.22**	0.21	**0.26**	0.24	**0.29**	0.26	**0.31**	0.29	**0.36**	0.28	**0.34**
Skeleton	0.26	**0.54**	0.32	**0.64**	0.36	**0.72**	0.40	**0.78**	0.46	**0.88**	0.45	**0.86**
Red bone marrow	0.07	**0.14**	0.09	**0.17**	0.11	**0.19**	0.12	**0.21**	0.13	**0.24**	0.13	**0.23**
Whole body dose	0.18	**0.23**	0.22	**0.28**	0.24	**0.31**	0.26	**0.34**	0.30	**0.38**	0.29	**0.37**
ED (ICRP 60)	0.18	**0.22**	0.23	**0.27**	0.26	**0.31**	0.29	**0.34**	0.34	**0.39**	0.32	**0.37**
ED (ICRP 103)	0.16	**0.23**	0.20	**0.28**	0.23	**0.32**	0.25	**0.34**	0.29	**0.40**	0.28	**0.38**
Total filtration (mm) **Al**	**1.5**		**2.5**		**3.5**		**4.5**		**3.5**		**3.5**	
Cu	**0**		**0**		**0**		**0**		**0.1**		**0.2**	

Table 12.67 Normalized organ doses—Abdomen r-lat—<1 year

Beam projection:			r-lat									**Table cells:**
Focus-to-detector distance (cm):			100									(+) = normal font = optimal field setting
Tube voltage (kVp):			70									**(−) = bold font = suboptimal field setting**
Total filtrations:			See last table row.									Empty table cells correspond to values <0.01
Organ or tissue	**Organ dose D_T : Entrance dose K_E**								**(mSv * mGy^{-1})**			
	(+)	(−)	(+)	(−)	(+)	(−)	(+)	(−)	(+)	(−)	(+)	(−)
Skull												**0.01**
Brain												
Oral mucosa										**0.01**		**0.01**
Salivary glands						**0.01**		**0.01**		**0.02**		**0.02**
Cervical spine		**0.02**	0.01	**0.03**	0.01	**0.04**	0.02	**0.04**	0.02	**0.05**	0.02	**0.06**
Thyroid		**0.02**		**0.02**		**0.03**		**0.03**	0.01	**0.04**	0.01	**0.04**
Thymus	0.01	**0.08**	0.02	**0.10**	0.02	**0.12**	0.02	**0.13**	0.03	**0.15**	0.03	**0.17**
Thoracic spine	0.32	**0.53**	0.41	**0.66**	0.47	**0.76**	0.52	**0.84**	0.61	**0.98**	0.68	**1.11**
Claviculae	0.02	**0.05**	0.02	**0.06**	0.03	**0.07**	0.03	**0.08**	0.04	**0.10**	0.04	**0.12**
Scapulae	0.05	**0.76**	0.07	**0.86**	0.08	**0.92**	0.09	**0.97**	0.11	**1.05**	0.13	**1.12**
Ribs	0.59	**0.88**	0.68	**1.02**	0.73	**1.12**	0.77	**1.19**	0.84	**1.31**	0.89	**1.40**
Oesophagus	0.14	**0.21**	0.17	**0.26**	0.20	**0.30**	0.22	**0.33**	0.26	**0.39**	0.29	**0.44**
Trachea										**0.01**		**0.01**
Lung	0.16	**0.30**	0.20	**0.37**	0.22	**0.41**	0.23	**0.44**	0.26	**0.51**	0.29	**0.56**
Mamma	0.03	**0.57**	0.04	**0.61**	0.04	**0.64**	0.05	**0.66**	0.05	**0.70**	0.06	**0.73**
Heart	0.15	**0.27**	0.19	**0.34**	0.21	**0.39**	0.23	**0.43**	0.27	**0.50**	0.31	**0.57**
Stomach	0.16	**0.15**	0.20	**0.19**	0.23	**0.22**	0.26	**0.25**	0.31	**0.30**	0.37	**0.35**
Spleen	0.11	**0.10**	0.14	**0.13**	0.17	**0.16**	0.19	**0.18**	0.23	**0.22**	0.27	**0.26**
Liver	0.61	**0.51**	0.72	**0.62**	0.79	**0.69**	0.84	**0.75**	0.94	**0.85**	1.01	**0.94**
Gall bladder	0.31	**0.27**	0.39	**0.35**	0.45	**0.40**	0.50	**0.44**	0.58	**0.53**	0.66	**0.60**
Pancreas	0.22	**0.19**	0.28	**0.24**	0.33	**0.28**	0.37	**0.31**	0.44	**0.38**	0.50	**0.44**
Small intestine	0.33	**0.27**	0.41	**0.34**	0.47	**0.40**	0.52	**0.44**	0.61	**0.52**	0.69	**0.60**
Large intestine	0.47	**0.39**	0.57	**0.48**	0.64	**0.54**	0.70	**0.60**	0.80	**0.69**	0.89	**0.78**
Sigma/rectum	0.16	**0.14**	0.21	**0.18**	0.25	**0.21**	0.28	**0.24**	0.34	**0.29**	0.39	**0.34**
Lumbar spine	0.73	**0.69**	0.91	**0.87**	1.05	**1.00**	1.16	**1.11**	1.37	**1.31**	1.55	**1.49**
Pelvis	1.00	**0.86**	1.21	**1.06**	1.36	**1.20**	1.47	**1.31**	1.68	**1.51**	1.84	**1.68**
Adrenals	0.25	**0.25**	0.32	**0.31**	0.36	**0.36**	0.40	**0.40**	0.46	**0.46**	0.52	**0.53**
Kidneys	0.34	**0.33**	0.42	**0.41**	0.47	**0.46**	0.51	**0.50**	0.59	**0.58**	0.66	**0.65**
Urinary bladder	0.31	**0.28**	0.38	**0.35**	0.44	**0.41**	0.49	**0.45**	0.57	**0.53**	0.64	**0.60**
Uterus	0.31	**0.25**	0.38	**0.32**	0.44	**0.38**	0.49	**0.42**	0.58	**0.50**	0.67	**0.58**
Ovaries	0.33	**0.26**	0.41	**0.33**	0.47	**0.38**	0.52	**0.43**	0.60	**0.51**	0.68	**0.59**
Prostate	0.27	**0.24**	0.33	**0.30**	0.38	**0.35**	0.42	**0.39**	0.50	**0.46**	0.57	**0.53**
Testes	0.23	**0.24**	0.29	**0.30**	0.33	**0.35**	0.37	**0.38**	0.43	**0.45**	0.49	**0.50**
Upper arm		**0.45**		**0.52**		**0.56**		**0.59**		**0.64**		**0.68**
Elbow		**1.49**		**1.66**		**1.76**		**1.83**		**1.95**		**2.02**
Forearm		**1.57**		**1.73**		**1.83**		**1.89**		**1.99**		**2.06**
Upper leg	0.52	**0.68**	0.65	**0.84**	0.74	**0.96**	0.81	**1.05**	0.94	**1.22**	1.06	**1.36**
Knee	0.05	**0.30**	0.06	**0.37**	0.07	**0.42**	0.08	**0.46**	0.10	**0.53**	0.12	**0.60**
Lower leg				**0.01**		**0.01**		**0.02**	0.01	**0.02**	0.01	**0.02**
Lymph nodes	0.19	**0.19**	0.23	**0.23**	0.27	**0.27**	0.30	**0.30**	0.35	**0.35**	0.39	**0.40**
Skin	0.23	**0.30**	0.25	**0.31**	0.26	**0.33**	0.26	**0.34**	0.28	**0.36**	0.29	**0.37**
Muscle	0.20	**0.25**	0.24	**0.29**	0.26	**0.32**	0.28	**0.35**	0.32	**0.39**	0.35	**0.43**
Skeleton	0.30	**0.60**	0.36	**0.71**	0.40	**0.78**	0.44	**0.84**	0.50	**0.93**	0.55	**1.01**
Red bone marrow	0.09	**0.16**	0.11	**0.19**	0.12	**0.21**	0.13	**0.23**	0.15	**0.26**	0.17	**0.29**
Whole body dose	0.20	**0.26**	0.24	**0.31**	0.27	**0.34**	0.29	**0.37**	0.32	**0.42**	0.36	**0.46**
ED (ICRP 60)	0.21	**0.25**	0.26	**0.31**	0.30	**0.35**	0.33	**0.38**	0.38	**0.44**	0.42	**0.49**
ED (ICRP 103)	0.19	**0.27**	0.23	**0.32**	0.26	**0.36**	0.28	**0.39**	0.33	**0.44**	0.37	**0.49**
Total filtration (mm) **Al**	**1.5**		**2.5**		**3.5**		**4.5**		**3.5**		**3.5**	
Cu	**0**		**0**		**0**		**0**		**0.1**		**0.2**	

Table 12.68 Normalized organ doses—Abdomen r-lat—1–2 years

Beam projection:			r-lat									**Table cells:**
Focus-to-detector distance (cm):			100				(+) = normal font = optimal field setting					
Tube voltage (kVp):			60				**(−) = bold font = suboptimal field setting**					
Total filtrations:			See last table row.				Empty table cells correspond to values <0.01					
Organ or tissue	**Organ dose D_T : Entrance dose K_E**							**(mSv * mGy^{-1})**				
	(+)	(−)	(+)	(−)	(+)	(−)	(+)	(−)	(+)	(−)	(+)	(−)
Skull												
Brain												
Oral mucosa												
Salivary glands												
Cervical spine				**0.01**		**0.02**		**0.02**		**0.02**		**0.02**
Thyroid				**0.01**		**0.01**		**0.02**		**0.02**		**0.02**
Thymus		**0.07**		**0.09**	0.01	**0.10**	0.01	**0.11**	0.02	**0.14**	0.01	**0.13**
Thoracic spine	0.15	**0.27**	0.20	**0.36**	0.24	**0.43**	0.27	**0.49**	0.33	**0.60**	0.30	**0.55**
Claviculae		**0.03**		**0.03**	0.01	**0.04**	0.01	**0.05**	0.02	**0.06**	0.01	**0.05**
Scapulae	0.02	**0.47**	0.02	**0.55**	0.03	**0.61**	0.03	**0.65**	0.04	**0.72**	0.04	**0.71**
Ribs	0.41	**0.65**	0.49	**0.79**	0.54	**0.88**	0.58	**0.96**	0.65	**1.09**	0.64	**1.07**
Oesophagus	0.08	**0.12**	0.10	**0.16**	0.12	**0.19**	0.14	**0.21**	0.17	**0.26**	0.15	**0.24**
Trachea												
Lung	0.08	**0.19**	0.11	**0.23**	0.12	**0.27**	0.13	**0.30**	0.15	**0.35**	0.15	**0.34**
Mamma	0.01	**0.50**	0.02	**0.54**	0.02	**0.57**	0.02	**0.58**	0.03	**0.61**	0.02	**0.61**
Heart	0.08	**0.15**	0.10	**0.20**	0.12	**0.23**	0.13	**0.26**	0.16	**0.32**	0.15	**0.29**
Stomach	0.07	**0.06**	0.09	**0.08**	0.11	**0.10**	0.12	**0.11**	0.15	**0.14**	0.13	**0.12**
Spleen	0.04	**0.04**	0.05	**0.05**	0.07	**0.06**	0.08	**0.07**	0.10	**0.09**	0.08	**0.07**
Liver	0.45	**0.34**	0.55	**0.42**	0.62	**0.49**	0.67	**0.54**	0.77	**0.64**	0.74	**0.61**
Gall bladder	0.20	**0.16**	0.27	**0.22**	0.31	**0.26**	0.35	**0.29**	0.43	**0.36**	0.40	**0.33**
Pancreas	0.11	**0.08**	0.15	**0.11**	0.18	**0.14**	0.20	**0.16**	0.25	**0.20**	0.23	**0.17**
Small intestine	0.19	**0.14**	0.25	**0.19**	0.29	**0.22**	0.33	**0.25**	0.40	**0.31**	0.37	**0.28**
Large intestine	0.30	**0.23**	0.39	**0.29**	0.45	**0.34**	0.50	**0.39**	0.59	**0.47**	0.56	**0.43**
Sigma/rectum	0.07	**0.06**	0.10	**0.07**	0.12	**0.09**	0.13	**0.11**	0.17	**0.14**	0.15	**0.11**
Lumbar spine	0.37	**0.34**	0.50	**0.45**	0.59	**0.54**	0.67	**0.61**	0.84	**0.76**	0.76	**0.69**
Pelvis	0.66	**0.51**	0.83	**0.65**	0.96	**0.77**	1.06	**0.86**	1.25	**1.03**	1.20	**0.97**
Adrenals	0.13	**0.13**	0.17	**0.17**	0.20	**0.20**	0.23	**0.22**	0.28	**0.28**	0.26	**0.25**
Kidneys	0.21	**0.20**	0.27	**0.25**	0.31	**0.29**	0.34	**0.33**	0.40	**0.39**	0.38	**0.36**
Urinary bladder	0.17	**0.15**	0.23	**0.20**	0.27	**0.23**	0.30	**0.26**	0.37	**0.32**	0.34	**0.29**
Uterus	0.16	**0.12**	0.21	**0.17**	0.26	**0.20**	0.29	**0.23**	0.36	**0.28**	0.33	**0.25**
Ovaries	0.18	**0.12**	0.24	**0.17**	0.28	**0.20**	0.32	**0.23**	0.39	**0.29**	0.35	**0.25**
Prostate	0.13	**0.12**	0.17	**0.15**	0.21	**0.18**	0.24	**0.21**	0.29	**0.26**	0.26	**0.23**
Testes	0.12	**0.11**	0.16	**0.15**	0.19	**0.18**	0.21	**0.20**	0.25	**0.24**	0.24	**0.22**
Upper arm		**0.33**		**0.39**		**0.43**		**0.46**		**0.52**		**0.51**
Elbow		**1.23**		**1.42**		**1.55**		**1.64**		**1.80**		**1.80**
Forearm		**1.36**		**1.55**		**1.68**		**1.77**		**1.91**		**1.92**
Upper leg	0.28	**0.35**	0.36	**0.45**	0.43	**0.53**	0.48	**0.59**	0.59	**0.72**	0.54	**0.67**
Knee	0.02	**0.04**	0.02	**0.06**	0.03	**0.07**	0.03	**0.08**	0.04	**0.10**	0.03	**0.08**
Lower leg												
Lymph nodes	0.11	**0.10**	0.14	**0.13**	0.16	**0.15**	0.18	**0.17**	0.22	**0.22**	0.21	**0.19**
Skin	0.19	**0.24**	0.20	**0.25**	0.21	**0.26**	0.21	**0.27**	0.23	**0.28**	0.22	**0.28**
Muscle	0.13	**0.14**	0.15	**0.17**	0.17	**0.20**	0.18	**0.21**	0.21	**0.24**	0.20	**0.23**
Skeleton	0.16	**0.37**	0.21	**0.44**	0.24	**0.50**	0.26	**0.54**	0.31	**0.63**	0.29	**0.61**
Red bone marrow	0.06	**0.09**	0.07	**0.12**	0.08	**0.13**	0.09	**0.15**	0.11	**0.17**	0.10	**0.16**
Whole body dose	0.13	**0.16**	0.16	**0.20**	0.18	**0.22**	0.19	**0.24**	0.22	**0.28**	0.21	**0.27**
ED (ICRP 60)	0.12	**0.15**	0.16	**0.19**	0.18	**0.21**	0.20	**0.24**	0.24	**0.28**	0.23	**0.26**
ED (ICRP 103)	0.11	**0.17**	0.14	**0.21**	0.16	**0.23**	0.18	**0.26**	0.21	**0.30**	0.20	**0.28**
Total filtration (mm) **Al**	**1.5**		**2.5**		**3.5**		**4.5**		**3.5**		**3.5**	
Cu	**0**		**0**		**0**		**0**		**0.1**		**0.2**	

Table 12.69 Normalized organ doses—Abdomen r-lat—1–2 years

Beam projection:			r-lat									**Table cells:**
Focus-to-detector distance (cm):			100						(+) = normal font = optimal field setting			
Tube voltage (kVp):			70						**(−) = bold font = suboptimal field setting**			
Total filtrations:			See last table row.						Empty table cells correspond to values <0.01			
Organ or tissue	**Organ dose D_T : Entrance dose K_E**								**(mSv * mGy^{-1})**			
	(+)	(−)	(+)	(−)	(+)	(−)	(+)	(−)	(+)	(−)	(+)	(−)
Skull												
Brain												
Oral mucosa												
Salivary glands												**0.01**
Cervical spine		**0.01**		**0.02**		**0.02**		**0.02**		**0.03**	0.01	**0.04**
Thyroid		**0.01**		**0.02**		**0.02**		**0.02**		**0.03**		**0.03**
Thymus		**0.08**	0.01	**0.10**	0.01	**0.12**	0.02	**0.14**	0.02	**0.16**	0.03	**0.19**
Thoracic spine	0.19	**0.34**	0.25	**0.44**	0.29	**0.52**	0.32	**0.58**	0.39	**0.71**	0.46	**0.83**
Claviculae		**0.03**	0.01	**0.05**	0.02	**0.05**	0.02	**0.06**	0.02	**0.08**	0.03	**0.09**
Scapulae	0.02	**0.51**	0.03	**0.60**	0.04	**0.65**	0.04	**0.69**	0.05	**0.76**	0.07	**0.82**
Ribs	0.45	**0.73**	0.53	**0.87**	0.58	**0.97**	0.62	**1.04**	0.68	**1.17**	0.73	**1.27**
Oesophagus	0.10	**0.15**	0.13	**0.20**	0.15	**0.23**	0.16	**0.26**	0.20	**0.32**	0.23	**0.37**
Trachea												
Lung	0.10	**0.22**	0.12	**0.27**	0.14	**0.31**	0.15	**0.34**	0.17	**0.40**	0.19	**0.45**
Mamma	0.02	**0.52**	0.02	**0.56**	0.02	**0.59**	0.03	**0.60**	0.03	**0.64**	0.04	**0.66**
Heart	0.10	**0.19**	0.12	**0.24**	0.14	**0.28**	0.16	**0.31**	0.19	**0.37**	0.22	**0.43**
Stomach	0.09	**0.08**	0.11	**0.10**	0.14	**0.12**	0.15	**0.14**	0.19	**0.18**	0.23	**0.21**
Spleen	0.06	**0.05**	0.07	**0.07**	0.09	**0.08**	0.10	**0.09**	0.13	**0.12**	0.16	**0.15**
Liver	0.51	**0.40**	0.61	**0.49**	0.68	**0.56**	0.74	**0.62**	0.84	**0.72**	0.92	**0.81**
Gall bladder	0.25	**0.21**	0.32	**0.27**	0.38	**0.32**	0.42	**0.36**	0.51	**0.43**	0.59	**0.51**
Pancreas	0.15	**0.11**	0.19	**0.15**	0.22	**0.18**	0.25	**0.20**	0.31	**0.25**	0.36	**0.31**
Small intestine	0.24	**0.18**	0.30	**0.23**	0.35	**0.28**	0.39	**0.31**	0.47	**0.38**	0.55	**0.44**
Large intestine	0.36	**0.28**	0.45	**0.35**	0.52	**0.41**	0.57	**0.46**	0.67	**0.54**	0.76	**0.63**
Sigma/rectum	0.10	**0.08**	0.13	**0.10**	0.15	**0.12**	0.17	**0.14**	0.21	**0.17**	0.26	**0.21**
Lumbar spine	0.48	**0.44**	0.61	**0.56**	0.72	**0.66**	0.81	**0.75**	0.99	**0.91**	1.16	**1.07**
Pelvis	0.77	**0.61**	0.96	**0.77**	1.09	**0.89**	1.20	**0.99**	1.39	**1.17**	1.56	**1.33**
Adrenals	0.17	**0.16**	0.21	**0.21**	0.25	**0.24**	0.28	**0.27**	0.33	**0.33**	0.39	**0.39**
Kidneys	0.25	**0.24**	0.31	**0.30**	0.36	**0.34**	0.39	**0.38**	0.46	**0.44**	0.52	**0.50**
Urinary bladder	0.22	**0.19**	0.28	**0.24**	0.32	**0.28**	0.36	**0.31**	0.43	**0.38**	0.50	**0.44**
Uterus	0.21	**0.16**	0.27	**0.21**	0.32	**0.25**	0.36	**0.28**	0.44	**0.35**	0.51	**0.41**
Ovaries	0.23	**0.16**	0.29	**0.21**	0.34	**0.25**	0.38	**0.28**	0.47	**0.35**	0.54	**0.42**
Prostate	0.17	**0.15**	0.22	**0.19**	0.25	**0.23**	0.29	**0.25**	0.35	**0.31**	0.41	**0.36**
Testes	0.15	**0.14**	0.19	**0.18**	0.22	**0.21**	0.25	**0.24**	0.30	**0.29**	0.34	**0.33**
Upper arm		**0.37**		**0.43**		**0.47**		**0.50**		**0.55**		**0.59**
Elbow		**1.33**		**1.51**		**1.63**		**1.72**		**1.86**		**1.97**
Forearm		**1.45**		**1.63**		**1.75**		**1.83**		**1.96**		**2.06**
Upper leg	0.35	**0.43**	0.44	**0.54**	0.51	**0.62**	0.57	**0.69**	0.68	**0.83**	0.78	**0.95**
Knee	0.02	**0.06**	0.03	**0.07**	0.04	**0.09**	0.04	**0.10**	0.06	**0.12**	0.07	**0.14**
Lower leg												
Lymph nodes	0.13	**0.13**	0.17	**0.16**	0.20	**0.19**	0.22	**0.21**	0.26	**0.26**	0.31	**0.30**
Skin	0.20	**0.25**	0.21	**0.26**	0.22	**0.27**	0.22	**0.28**	0.24	**0.30**	0.25	**0.31**
Muscle	0.14	**0.17**	0.17	**0.20**	0.19	**0.22**	0.20	**0.24**	0.23	**0.27**	0.26	**0.31**
Skeleton	0.19	**0.42**	0.24	**0.50**	0.27	**0.56**	0.30	**0.60**	0.35	**0.68**	0.39	**0.76**
Red bone marrow	0.07	**0.11**	0.09	**0.14**	0.10	**0.16**	0.11	**0.17**	0.13	**0.20**	0.15	**0.23**
Whole body dose	0.15	**0.19**	0.18	**0.23**	0.20	**0.25**	0.21	**0.27**	0.25	**0.31**	0.27	**0.35**
ED (ICRP 60)	0.15	**0.18**	0.19	**0.22**	0.22	**0.25**	0.24	**0.28**	0.28	**0.33**	0.33	**0.37**
ED (ICRP 103)	0.13	**0.20**	0.16	**0.24**	0.19	**0.27**	0.21	**0.29**	0.25	**0.34**	0.28	**0.38**
Total filtration (mm) Al	**1.5**		**2.5**		**3.5**		**4.5**		**3.5**		**3.5**	
Cu	**0**		**0**		**0**		**0**		**0.1**		**0.2**	

Table 12.70 Normalized organ doses—Abdomen r-lat—3–7 years

Beam projection:			r-lat									**Table cells:**
Focus-to-detector distance (cm):			100						(+) = normal font = optimal field setting			
Tube voltage (kVp):			60						**(−) = bold font = suboptimal field setting**			
Total filtrations:			See last table row.						Empty table cells correspond to values <0.01			
Organ or tissue	**Organ dose D_T : Entrance dose K_E**								**(mSv * mGy^{-1})**			
	(+)	(−)	(+)	(−)	(+)	(−)	(+)	(−)	(+)	(−)	(+)	(−)
Skull												
Brain												
Oral mucosa												
Salivary glands												
Cervical spine								**0.01**		**0.01**		**0.01**
Thyroid										**0.01**		
Thymus		**0.06**		**0.08**		**0.10**	0.01	**0.11**	0.01	**0.14**	0.01	**0.12**
Thoracic spine	0.10	**0.16**	0.13	**0.21**	0.16	**0.26**	0.18	**0.30**	0.23	**0.38**	0.20	**0.33**
Claviculae		**0.02**		**0.02**		**0.03**		**0.03**		**0.04**		**0.03**
Scapulae	0.01	**0.39**	0.02	**0.46**	0.02	**0.51**	0.02	**0.55**	0.03	**0.62**	0.03	**0.61**
Ribs	0.34	**0.49**	0.41	**0.61**	0.46	**0.69**	0.49	**0.75**	0.55	**0.87**	0.54	**0.85**
Oesophagus	0.04	**0.06**	0.05	**0.07**	0.06	**0.09**	0.07	**0.10**	0.09	**0.13**	0.08	**0.11**
Trachea												
Lung	0.07	**0.14**	0.09	**0.17**	0.11	**0.20**	0.12	**0.23**	0.14	**0.27**	0.13	**0.25**
Mamma		**0.45**	0.01	**0.49**	0.02	**0.52**	0.02	**0.53**	0.02	**0.56**	0.02	**0.55**
Heart	0.06	**0.09**	0.08	**0.12**	0.09	**0.14**	0.11	**0.16**	0.13	**0.20**	0.12	**0.18**
Stomach	0.03	**0.03**	0.05	**0.04**	0.06	**0.05**	0.07	**0.06**	0.09	**0.08**	0.07	**0.06**
Spleen	0.02	**0.02**	0.03	**0.02**	0.03	**0.03**	0.04	**0.03**	0.05	**0.04**	0.04	**0.03**
Liver	0.38	**0.26**	0.47	**0.33**	0.53	**0.39**	0.58	**0.43**	0.67	**0.52**	0.65	**0.49**
Gall bladder	0.12	**0.09**	0.16	**0.12**	0.19	**0.15**	0.22	**0.17**	0.27	**0.22**	0.24	**0.19**
Pancreas	0.07	**0.05**	0.09	**0.06**	0.11	**0.08**	0.12	**0.09**	0.16	**0.12**	0.13	**0.09**
Small intestine	0.13	**0.09**	0.17	**0.12**	0.20	**0.15**	0.23	**0.17**	0.29	**0.21**	0.25	**0.18**
Large intestine	0.22	**0.16**	0.29	**0.21**	0.34	**0.25**	0.38	**0.28**	0.46	**0.34**	0.43	**0.31**
Sigma/rectum	0.04	**0.03**	0.05	**0.04**	0.06	**0.05**	0.07	**0.06**	0.09	**0.07**	0.08	**0.06**
Lumbar spine	0.21	**0.19**	0.29	**0.25**	0.35	**0.31**	0.40	**0.35**	0.51	**0.45**	0.44	**0.38**
Pelvis	0.47	**0.34**	0.60	**0.44**	0.70	**0.53**	0.78	**0.59**	0.94	**0.73**	0.89	**0.67**
Adrenals	0.08	**0.08**	0.11	**0.11**	0.13	**0.13**	0.15	**0.14**	0.19	**0.18**	0.17	**0.16**
Kidneys	0.15	**0.14**	0.19	**0.18**	0.22	**0.21**	0.25	**0.23**	0.30	**0.28**	0.28	**0.26**
Urinary bladder	0.11	**0.09**	0.14	**0.12**	0.17	**0.15**	0.19	**0.17**	0.24	**0.21**	0.22	**0.18**
Uterus	0.10	**0.07**	0.13	**0.10**	0.16	**0.12**	0.18	**0.13**	0.23	**0.17**	0.20	**0.14**
Ovaries	0.13	**0.07**	0.17	**0.10**	0.20	**0.12**	0.23	**0.14**	0.29	**0.19**	0.25	**0.15**
Prostate	0.08	**0.06**	0.11	**0.09**	0.13	**0.11**	0.15	**0.12**	0.19	**0.16**	0.16	**0.13**
Testes	0.07	**0.07**	0.09	**0.09**	0.11	**0.11**	0.13	**0.13**	0.15	**0.16**	0.14	**0.14**
Upper arm		**0.27**		**0.32**		**0.35**		**0.38**		**0.43**		**0.42**
Elbow		**1.09**		**1.28**		**1.40**		**1.50**		**1.66**		**1.65**
Forearm		**1.26**		**1.46**		**1.59**		**1.69**		**1.85**		**1.85**
Upper leg	0.14	**0.22**	0.18	**0.29**	0.22	**0.35**	0.25	**0.39**	0.31	**0.49**	0.27	**0.44**
Knee		**0.01**		**0.02**		**0.02**		**0.03**		**0.03**		**0.03**
Lower leg												
Lymph nodes	0.07	**0.06**	0.09	**0.08**	0.11	**0.10**	0.12	**0.11**	0.15	**0.14**	0.14	**0.12**
Skin	0.13	**0.21**	0.14	**0.22**	0.14	**0.23**	0.15	**0.24**	0.15	**0.25**	0.15	**0.24**
Muscle	0.10	**0.11**	0.12	**0.14**	0.13	**0.16**	0.14	**0.17**	0.17	**0.20**	0.16	**0.18**
Skeleton	0.10	**0.24**	0.12	**0.29**	0.14	**0.33**	0.16	**0.36**	0.19	**0.42**	0.18	**0.40**
Red bone marrow	0.05	**0.08**	0.07	**0.10**	0.08	**0.11**	0.09	**0.12**	0.11	**0.15**	0.10	**0.14**
Whole body dose	0.10	**0.13**	0.12	**0.16**	0.14	**0.18**	0.15	**0.19**	0.17	**0.23**	0.16	**0.21**
ED (ICRP 60)	0.09	**0.11**	0.11	**0.13**	0.13	**0.15**	0.15	**0.17**	0.18	**0.21**	0.17	**0.19**
ED (ICRP 103)	0.08	**0.13**	0.10	**0.16**	0.12	**0.18**	0.13	**0.19**	0.16	**0.23**	0.14	**0.21**
Total filtration (mm) **Al**	**1.5**		**2.5**		**3.5**		**4.5**		**3.5**		**3.5**	
Cu	**0**		**0**		**0**		**0**		**0.1**		**0.2**	

Table 12.71 Normalized organ doses—Abdomen r-lat—3–7 years

Beam projection:	r-lat											**Table cells:**
Focus-to-detector distance (cm):	100											(+) = normal font = optimal field setting
Tube voltage (kVp):	70											**(−) = bold font = suboptimal field setting**
Total filtrations:	See last table row.											Empty table cells correspond to values <0.01
Organ or tissue	**Organ dose D_T : Entrance dose K_E**							**(mSv * mGy^{-1})**				
	(+)	(−)	(+)	(−)	(+)	(−)	(+)	(−)	(+)	(−)	(+)	(−)
Skull												
Brain												
Oral mucosa												
Salivary glands												
Cervical spine				**0.01**		**0.01**		**0.02**		**0.02**		**0.03**
Thyroid						**0.01**		**0.01**		**0.02**		**0.02**
Thymus		**0.08**	0.01	**0.10**	0.01	**0.12**	0.01	**0.13**	0.02	**0.16**	0.02	**0.19**
Thoracic spine	0.13	**0.21**	0.17	**0.28**	0.20	**0.33**	0.22	**0.38**	0.28	**0.47**	0.33	**0.56**
Claviculae		**0.02**		**0.03**		**0.04**	0.01	**0.04**	0.01	**0.05**	0.02	**0.06**
Scapulae	0.02	**0.43**	0.02	**0.50**	0.03	**0.55**	0.03	**0.59**	0.04	**0.66**	0.05	**0.71**
Ribs	0.38	**0.56**	0.45	**0.68**	0.49	**0.77**	0.53	**0.83**	0.59	**0.95**	0.64	**1.05**
Oesophagus	0.05	**0.08**	0.07	**0.10**	0.08	**0.12**	0.09	**0.13**	0.11	**0.17**	0.13	**0.20**
Trachea												
Lung	0.09	**0.17**	0.11	**0.21**	0.12	**0.24**	0.13	**0.27**	0.16	**0.32**	0.18	**0.36**
Mamma	0.01	**0.47**	0.02	**0.52**	0.02	**0.54**	0.02	**0.56**	0.03	**0.59**	0.04	**0.62**
Heart	0.08	**0.12**	0.10	**0.15**	0.12	**0.18**	0.13	**0.20**	0.16	**0.25**	0.19	**0.30**
Stomach	0.05	**0.04**	0.07	**0.06**	0.08	**0.07**	0.09	**0.08**	0.12	**0.10**	0.14	**0.13**
Spleen	0.03	**0.02**	0.04	**0.03**	0.04	**0.04**	0.05	**0.05**	0.07	**0.06**	0.08	**0.08**
Liver	0.44	**0.32**	0.53	**0.40**	0.60	**0.46**	0.65	**0.51**	0.75	**0.60**	0.84	**0.69**
Gall bladder	0.16	**0.13**	0.20	**0.16**	0.24	**0.19**	0.27	**0.22**	0.34	**0.28**	0.40	**0.33**
Pancreas	0.09	**0.07**	0.12	**0.09**	0.14	**0.11**	0.16	**0.12**	0.20	**0.16**	0.24	**0.19**
Small intestine	0.17	**0.12**	0.21	**0.16**	0.25	**0.19**	0.28	**0.21**	0.35	**0.26**	0.41	**0.32**
Large intestine	0.28	**0.20**	0.35	**0.26**	0.41	**0.30**	0.45	**0.34**	0.54	**0.42**	0.62	**0.49**
Sigma/rectum	0.05	**0.04**	0.07	**0.06**	0.09	**0.07**	0.10	**0.08**	0.13	**0.10**	0.15	**0.12**
Lumbar spine	0.29	**0.25**	0.37	**0.33**	0.45	**0.40**	0.51	**0.45**	0.63	**0.56**	0.76	**0.68**
Pelvis	0.57	**0.43**	0.71	**0.54**	0.82	**0.63**	0.91	**0.71**	1.07	**0.86**	1.22	**0.99**
Adrenals	0.11	**0.11**	0.14	**0.14**	0.17	**0.16**	0.19	**0.18**	0.24	**0.23**	0.28	**0.27**
Kidneys	0.18	**0.17**	0.23	**0.21**	0.27	**0.25**	0.30	**0.28**	0.35	**0.33**	0.41	**0.38**
Urinary bladder	0.14	**0.12**	0.18	**0.16**	0.22	**0.19**	0.24	**0.21**	0.30	**0.26**	0.35	**0.31**
Uterus	0.13	**0.10**	0.17	**0.13**	0.21	**0.15**	0.24	**0.18**	0.30	**0.22**	0.36	**0.27**
Ovaries	0.16	**0.11**	0.21	**0.14**	0.25	**0.17**	0.28	**0.19**	0.35	**0.24**	0.42	**0.30**
Prostate	0.11	**0.09**	0.14	**0.12**	0.17	**0.14**	0.19	**0.16**	0.24	**0.20**	0.29	**0.24**
Testes	0.09	**0.09**	0.12	**0.12**	0.14	**0.14**	0.15	**0.16**	0.19	**0.19**	0.22	**0.23**
Upper arm		**0.30**		**0.35**		**0.39**		**0.41**		**0.46**		**0.50**
Elbow		**1.19**		**1.37**		**1.49**		**1.58**		**1.73**		**1.84**
Forearm		**1.36**		**1.55**		**1.67**		**1.76**		**1.91**		**2.02**
Upper leg	0.18	**0.28**	0.23	**0.36**	0.27	**0.43**	0.30	**0.48**	0.37	**0.58**	0.43	**0.68**
Knee		**0.02**		**0.03**		**0.03**		**0.04**	0.01	**0.05**	0.02	**0.06**
Lower leg												
Lymph nodes	0.09	**0.08**	0.12	**0.11**	0.14	**0.13**	0.15	**0.14**	0.19	**0.18**	0.22	**0.21**
Skin	0.13	**0.22**	0.14	**0.23**	0.15	**0.24**	0.15	**0.25**	0.16	**0.26**	0.17	**0.27**
Muscle	0.11	**0.13**	0.13	**0.16**	0.15	**0.18**	0.16	**0.19**	0.19	**0.22**	0.21	**0.25**
Skeleton	0.12	**0.28**	0.14	**0.33**	0.17	**0.37**	0.18	**0.41**	0.22	**0.47**	0.25	**0.52**
Red bone marrow	0.07	**0.09**	0.09	**0.12**	0.10	**0.13**	0.11	**0.15**	0.13	**0.18**	0.15	**0.20**
Whole body dose	0.12	**0.15**	0.14	**0.18**	0.16	**0.20**	0.17	**0.22**	0.20	**0.25**	0.22	**0.29**
ED (ICRP 60)	0.11	**0.13**	0.14	**0.16**	0.16	**0.18**	0.18	**0.20**	0.22	**0.24**	0.25	**0.28**
ED (ICRP 103)	0.10	**0.15**	0.12	**0.18**	0.14	**0.21**	0.16	**0.23**	0.19	**0.26**	0.22	**0.30**
Total filtration (mm) **Al**	**1.5**		**2.5**		**3.5**		**4.5**		**3.5**		**3.5**	
Cu	**0**		**0**		**0**		**0**		**0.1**		**0.2**	

Table 12.72 Normalized organ doses—Abdomen r-lat—3–7 years

Beam projection:			r-lat									**Table cells:**
Focus-to-detector distance (cm):			100						(+) = normal font = optimal field setting			
Tube voltage (kVp):			80						**(−) = bold font = suboptimal field setting**			
Total filtrations:			See last table row.						Empty table cells correspond to values <0.01			
Organ or tissue	**Organ dose D_T : Entrance dose K_E**								**(mSv * mGy^{-1})**			
	(+)	(−)	(+)	(−)	(+)	(−)	(+)	(−)	(+)	(−)	(+)	(−)
Skull												
Brain												
Oral mucosa												
Salivary glands												
Cervical spine		**0.01**		**0.02**		**0.02**		**0.02**		**0.03**		**0.03**
Thyroid				**0.01**		**0.01**		**0.02**		**0.02**		**0.03**
Thymus	0.01	**0.10**	0.01	**0.12**	0.02	**0.14**	0.02	**0.16**	0.02	**0.19**	0.03	**0.22**
Thoracic spine	0.16	**0.27**	0.20	**0.34**	0.24	**0.40**	0.27	**0.45**	0.32	**0.54**	0.38	**0.64**
Claviculae		**0.03**		**0.04**	0.01	**0.04**	0.01	**0.05**	0.02	**0.06**	0.02	**0.08**
Scapulae	0.02	**0.47**	0.03	**0.54**	0.04	**0.59**	0.04	**0.62**	0.05	**0.69**	0.06	**0.73**
Ribs	0.41	**0.63**	0.48	**0.74**	0.52	**0.82**	0.56	**0.89**	0.61	**1.00**	0.66	**1.09**
Oesophagus	0.07	**0.10**	0.08	**0.12**	0.10	**0.14**	0.11	**0.16**	0.13	**0.20**	0.15	**0.23**
Trachea												
Lung	0.10	**0.20**	0.12	**0.24**	0.14	**0.28**	0.15	**0.30**	0.17	**0.35**	0.19	**0.40**
Mamma	0.02	**0.50**	0.02	**0.54**	0.03	**0.56**	0.03	**0.58**	0.04	**0.61**	0.04	**0.63**
Heart	0.09	**0.15**	0.12	**0.19**	0.14	**0.22**	0.15	**0.24**	0.18	**0.29**	0.21	**0.34**
Stomach	0.07	**0.06**	0.08	**0.08**	0.10	**0.09**	0.11	**0.10**	0.14	**0.13**	0.17	**0.16**
Spleen	0.04	**0.03**	0.05	**0.05**	0.06	**0.05**	0.07	**0.06**	0.09	**0.08**	0.10	**0.10**
Liver	0.50	**0.37**	0.59	**0.46**	0.66	**0.52**	0.72	**0.58**	0.82	**0.67**	0.90	**0.76**
Gall bladder	0.20	**0.16**	0.25	**0.20**	0.29	**0.24**	0.32	**0.27**	0.39	**0.33**	0.46	**0.39**
Pancreas	0.12	**0.09**	0.15	**0.12**	0.17	**0.14**	0.19	**0.15**	0.24	**0.19**	0.28	**0.23**
Small intestine	0.21	**0.15**	0.26	**0.20**	0.30	**0.23**	0.33	**0.26**	0.40	**0.31**	0.47	**0.37**
Large intestine	0.33	**0.25**	0.41	**0.31**	0.47	**0.36**	0.52	**0.40**	0.61	**0.48**	0.69	**0.55**
Sigma/rectum	0.07	**0.06**	0.09	**0.07**	0.11	**0.09**	0.12	**0.10**	0.16	**0.13**	0.19	**0.15**
Lumbar spine	0.36	**0.32**	0.46	**0.41**	0.54	**0.48**	0.60	**0.54**	0.74	**0.66**	0.86	**0.78**
Pelvis	0.66	**0.51**	0.81	**0.63**	0.92	**0.73**	1.01	**0.81**	1.17	**0.96**	1.31	**1.09**
Adrenals	0.14	**0.13**	0.18	**0.17**	0.21	**0.20**	0.23	**0.22**	0.28	**0.27**	0.33	**0.32**
Kidneys	0.22	**0.20**	0.27	**0.25**	0.31	**0.29**	0.34	**0.32**	0.40	**0.37**	0.45	**0.43**
Urinary bladder	0.18	**0.15**	0.22	**0.19**	0.26	**0.23**	0.29	**0.25**	0.35	**0.31**	0.41	**0.36**
Uterus	0.17	**0.13**	0.22	**0.16**	0.26	**0.19**	0.29	**0.22**	0.35	**0.27**	0.42	**0.32**
Ovaries	0.21	**0.14**	0.26	**0.18**	0.30	**0.21**	0.34	**0.24**	0.41	**0.30**	0.48	**0.36**
Prostate	0.14	**0.11**	0.18	**0.15**	0.21	**0.17**	0.23	**0.19**	0.29	**0.24**	0.34	**0.28**
Testes	0.11	**0.11**	0.14	**0.14**	0.16	**0.17**	0.18	**0.19**	0.22	**0.23**	0.25	**0.26**
Upper arm		**0.33**		**0.38**		**0.41**		**0.44**		**0.48**		**0.52**
Elbow		**1.27**		**1.44**		**1.55**		**1.63**		**1.77**		**1.86**
Forearm		**1.43**		**1.61**		**1.73**		**1.81**		**1.93**		**2.02**
Upper leg	0.22	**0.34**	0.27	**0.43**	0.32	**0.50**	0.35	**0.55**	0.42	**0.66**	0.49	**0.76**
Knee		**0.03**		**0.03**	0.01	**0.04**	0.01	**0.05**	0.02	**0.06**	0.02	**0.07**
Lower leg												
Lymph nodes	0.11	**0.10**	0.14	**0.13**	0.16	**0.15**	0.18	**0.17**	0.22	**0.21**	0.25	**0.24**
Skin	0.14	**0.23**	0.15	**0.24**	0.16	**0.25**	0.16	**0.26**	0.17	**0.27**	0.18	**0.28**
Muscle	0.13	**0.15**	0.15	**0.18**	0.17	**0.20**	0.18	**0.22**	0.21	**0.25**	0.23	**0.27**
Skeleton	0.13	**0.31**	0.16	**0.37**	0.19	**0.41**	0.20	**0.44**	0.24	**0.50**	0.27	**0.55**
Red bone marrow	0.08	**0.11**	0.10	**0.14**	0.12	**0.16**	0.13	**0.17**	0.15	**0.20**	0.17	**0.23**
Whole body dose	0.13	**0.17**	0.16	**0.20**	0.17	**0.23**	0.19	**0.24**	0.22	**0.28**	0.24	**0.31**
ED (ICRP 60)	0.13	**0.15**	0.16	**0.19**	0.19	**0.22**	0.21	**0.24**	0.25	**0.28**	0.28	**0.32**
ED (ICRP 103)	0.12	**0.18**	0.14	**0.21**	0.16	**0.23**	0.18	**0.25**	0.22	**0.30**	0.25	**0.33**
Total filtration (mm) **Al**	**1.5**		**2.5**		**3.5**		**4.5**		**3.5**		**3.5**	
Cu	**0**		**0**		**0**		**0**		**0.1**		**0.2**	

Table 12.73 Normalized organ doses—Abdomen l-lat—<1 year

Beam projection:			l-lat									**Table cells:**
Focus-to-detector distance (cm):			100									(+) = normal font = optimal field setting
Tube voltage (kVp):			60									**(−) = bold font = suboptimal field setting**
Total filtrations:			See last table row.									Empty table cells correspond to values <0.01
Organ or tissue	**Organ dose D_T : Entrance dose K_E**							**(mSv * mGy^{-1})**				
	(+)	(−)	(+)	(−)	(+)	(−)	(+)	(−)	(+)	(−)	(+)	(−)
Skull												
Brain												
Oral mucosa												
Salivary glands										**0.01**		
Cervical spine		**0.02**		**0.02**		**0.03**	0.01	**0.03**	0.01	**0.04**	0.01	**0.03**
Thyroid		**0.01**		**0.01**		**0.02**		**0.02**		**0.03**		**0.02**
Thymus		**0.07**	0.01	**0.08**	0.02	**0.10**	0.02	**0.11**	0.02	**0.13**	0.02	**0.12**
Thoracic spine	0.27	**0.44**	0.35	**0.57**	0.41	**0.66**	0.46	**0.74**	0.54	**0.88**	0.52	**0.85**
Claviculae	0.01	**0.04**	0.02	**0.05**	0.02	**0.06**	0.02	**0.07**	0.03	**0.08**	0.02	**0.07**
Scapulae	0.04	**0.71**	0.05	**0.81**	0.07	**0.88**	0.07	**0.93**	0.09	**1.02**	0.08	**1.01**
Ribs	0.55	**0.80**	0.64	**0.95**	0.69	**1.05**	0.74	**1.12**	0.81	**1.25**	0.81	**1.25**
Oesophagus	0.15	**0.20**	0.19	**0.25**	0.21	**0.29**	0.24	**0.33**	0.28	**0.39**	0.27	**0.37**
Trachea												
Lung	0.13	**0.24**	0.15	**0.29**	0.17	**0.33**	0.19	**0.36**	0.22	**0.42**	0.21	**0.40**
Mamma	0.02	**0.58**	0.03	**0.62**	0.04	**0.65**	0.04	**0.67**	0.05	**0.70**	0.05	**0.69**
Heart	0.16	**0.31**	0.20	**0.39**	0.23	**0.44**	0.25	**0.49**	0.29	**0.57**	0.28	**0.55**
Stomach	0.52	**0.46**	0.64	**0.57**	0.71	**0.65**	0.77	**0.71**	0.88	**0.81**	0.87	**0.79**
Spleen	0.68	**0.55**	0.80	**0.67**	0.88	**0.76**	0.95	**0.82**	1.05	**0.93**	1.04	**0.92**
Liver	0.14	**0.13**	0.18	**0.17**	0.22	**0.20**	0.24	**0.23**	0.29	**0.28**	0.27	**0.26**
Gall bladder	0.24	**0.21**	0.31	**0.27**	0.36	**0.32**	0.40	**0.36**	0.49	**0.44**	0.46	**0.41**
Pancreas	0.37	**0.27**	0.46	**0.35**	0.53	**0.41**	0.58	**0.45**	0.67	**0.54**	0.65	**0.51**
Small intestine	0.30	**0.24**	0.37	**0.30**	0.43	**0.36**	0.48	**0.40**	0.57	**0.48**	0.54	**0.45**
Large intestine	0.22	**0.18**	0.29	**0.24**	0.33	**0.28**	0.37	**0.31**	0.44	**0.38**	0.42	**0.35**
Sigma/rectum	0.45	**0.34**	0.55	**0.44**	0.63	**0.50**	0.69	**0.56**	0.80	**0.66**	0.78	**0.63**
Lumbar spine	0.60	**0.57**	0.78	**0.73**	0.91	**0.86**	1.02	**0.96**	1.22	**1.16**	1.16	**1.10**
Pelvis	0.88	**0.74**	1.09	**0.93**	1.23	**1.07**	1.35	**1.18**	1.55	**1.38**	1.53	**1.35**
Adrenals	0.21	**0.21**	0.27	**0.27**	0.31	**0.31**	0.34	**0.34**	0.41	**0.41**	0.39	**0.39**
Kidneys	0.30	**0.29**	0.36	**0.35**	0.42	**0.40**	0.46	**0.44**	0.53	**0.52**	0.51	**0.50**
Urinary bladder	0.26	**0.24**	0.33	**0.30**	0.38	**0.35**	0.42	**0.39**	0.50	**0.47**	0.48	**0.44**
Uterus	0.25	**0.21**	0.32	**0.27**	0.38	**0.32**	0.42	**0.36**	0.51	**0.43**	0.48	**0.40**
Ovaries	0.28	**0.21**	0.35	**0.27**	0.41	**0.32**	0.45	**0.36**	0.54	**0.44**	0.51	**0.41**
Prostate	0.22	**0.20**	0.28	**0.26**	0.33	**0.30**	0.37	**0.34**	0.44	**0.41**	0.42	**0.38**
Testes	0.20	**0.21**	0.26	**0.26**	0.30	**0.31**	0.33	**0.34**	0.39	**0.40**	0.37	**0.39**
Upper arm		**0.42**		**0.48**		**0.52**		**0.56**		**0.61**		**0.60**
Elbow		**1.41**		**1.59**		**1.70**		**1.78**		**1.91**		**1.92**
Forearm		**1.50**		**1.67**		**1.78**		**1.85**		**1.97**		**1.98**
Upper leg	0.45	**0.58**	0.56	**0.73**	0.65	**0.85**	0.72	**0.94**	0.85	**1.10**	0.82	**1.06**
Knee	0.04	**0.25**	0.05	**0.32**	0.06	**0.37**	0.07	**0.41**	0.09	**0.48**	0.08	**0.46**
Lower leg						**0.01**		**0.01**		**0.02**		**0.01**
Lymph nodes	0.20	**0.19**	0.25	**0.24**	0.29	**0.27**	0.32	**0.30**	0.38	**0.36**	0.36	**0.34**
Skin	0.22	**0.29**	0.24	**0.30**	0.24	**0.31**	0.25	**0.32**	0.27	**0.34**	0.26	**0.34**
Muscle	0.20	**0.23**	0.23	**0.28**	0.26	**0.31**	0.28	**0.33**	0.31	**0.38**	0.31	**0.37**
Skeleton	0.26	**0.54**	0.32	**0.64**	0.36	**0.72**	0.40	**0.78**	0.46	**0.88**	0.45	**0.86**
Red bone marrow	0.07	**0.14**	0.09	**0.17**	0.11	**0.19**	0.12	**0.21**	0.13	**0.24**	0.13	**0.23**
Whole body dose	0.18	**0.23**	0.22	**0.28**	0.24	**0.31**	0.26	**0.34**	0.30	**0.38**	0.29	**0.37**
ED (ICRP 60)	0.22	**0.25**	0.27	**0.31**	0.31	**0.35**	0.34	**0.38**	0.39	**0.44**	0.38	**0.42**
ED (ICRP 103)	0.20	**0.27**	0.25	**0.33**	0.28	**0.37**	0.31	**0.40**	0.36	**0.46**	0.35	**0.44**
Total filtration (mm) Al	**1.5**		**2.5**		**3.5**		**4.5**		**3.5**		**3.5**	
Cu	**0**		**0**		**0**		**0**		**0.1**		**0.2**	

Table 12.74 Normalized organ doses—Abdomen l-lat—<1 year

Beam projection:			l-lat									**Table cells:**
Focus-to-detector distance (cm):			100				(+) = normal font = optimal field setting					
Tube voltage (kVp):			70				**(−) = bold font = suboptimal field setting**					
Total filtrations:			See last table row.				Empty table cells correspond to values <0.01					
Organ or tissue	**Organ dose D_T : Entrance dose K_E**								**(mSv * mGy^{-1})**			
	(+)	(−)	(+)	(−)	(+)	(−)	(+)	(−)	(+)	(−)	(+)	(−)
Skull												**0.01**
Brain												
Oral mucosa										**0.01**		**0.01**
Salivary glands						**0.01**		**0.01**		**0.01**		**0.02**
Cervical spine		**0.02**	0.01	**0.03**	0.01	**0.04**	0.02	**0.04**	0.02	**0.05**	0.02	**0.06**
Thyroid		**0.01**		**0.02**		**0.02**		**0.03**	0.01	**0.03**	0.01	**0.04**
Thymus	0.01	**0.08**	0.02	**0.10**	0.02	**0.11**	0.02	**0.13**	0.03	**0.15**	0.03	**0.17**
Thoracic spine	0.33	**0.53**	0.41	**0.66**	0.47	**0.76**	0.52	**0.84**	0.61	**0.98**	0.68	**1.11**
Claviculae	0.02	**0.05**	0.02	**0.06**	0.02	**0.07**	0.03	**0.08**	0.04	**0.10**	0.04	**0.11**
Scapulae	0.05	**0.76**	0.07	**0.87**	0.08	**0.93**	0.09	**0.98**	0.11	**1.06**	0.13	**1.13**
Ribs	0.59	**0.88**	0.68	**1.02**	0.73	**1.12**	0.77	**1.19**	0.84	**1.31**	0.89	**1.39**
Oesophagus	0.17	**0.24**	0.21	**0.30**	0.24	**0.34**	0.27	**0.37**	0.31	**0.44**	0.35	**0.50**
Trachea										**0.01**		**0.01**
Lung	0.15	**0.27**	0.17	**0.33**	0.19	**0.37**	0.21	**0.40**	0.24	**0.46**	0.26	**0.51**
Mamma	0.03	**0.60**	0.04	**0.65**	0.04	**0.67**	0.05	**0.69**	0.05	**0.72**	0.06	**0.75**
Heart	0.18	**0.36**	0.23	**0.44**	0.25	**0.50**	0.28	**0.55**	0.32	**0.63**	0.36	**0.70**
Stomach	0.59	**0.53**	0.70	**0.64**	0.78	**0.72**	0.84	**0.78**	0.95	**0.89**	1.03	**0.98**
Spleen	0.75	**0.62**	0.87	**0.74**	0.95	**0.83**	1.01	**0.89**	1.11	**1.00**	1.19	**1.10**
Liver	0.17	**0.17**	0.22	**0.21**	0.26	**0.24**	0.28	**0.27**	0.34	**0.33**	0.39	**0.38**
Gall bladder	0.29	**0.26**	0.37	**0.33**	0.42	**0.38**	0.47	**0.42**	0.55	**0.50**	0.63	**0.58**
Pancreas	0.43	**0.33**	0.52	**0.41**	0.59	**0.47**	0.65	**0.52**	0.74	**0.62**	0.83	**0.71**
Small intestine	0.35	**0.29**	0.43	**0.36**	0.50	**0.42**	0.54	**0.46**	0.64	**0.55**	0.72	**0.62**
Large intestine	0.27	**0.23**	0.34	**0.29**	0.39	**0.33**	0.43	**0.37**	0.50	**0.44**	0.57	**0.50**
Sigma/rectum	0.52	**0.41**	0.63	**0.50**	0.71	**0.58**	0.77	**0.63**	0.88	**0.74**	0.97	**0.83**
Lumbar spine	0.72	**0.68**	0.91	**0.86**	1.05	**1.00**	1.16	**1.11**	1.37	**1.31**	1.55	**1.48**
Pelvis	1.00	**0.86**	1.21	**1.06**	1.36	**1.20**	1.47	**1.31**	1.67	**1.51**	1.84	**1.68**
Adrenals	0.25	**0.25**	0.31	**0.31**	0.36	**0.36**	0.39	**0.40**	0.46	**0.47**	0.52	**0.53**
Kidneys	0.34	**0.33**	0.42	**0.41**	0.47	**0.46**	0.51	**0.50**	0.59	**0.58**	0.65	**0.64**
Urinary bladder	0.31	**0.28**	0.38	**0.36**	0.44	**0.41**	0.48	**0.45**	0.57	**0.53**	0.64	**0.60**
Uterus	0.30	**0.25**	0.38	**0.32**	0.44	**0.38**	0.49	**0.42**	0.58	**0.50**	0.66	**0.58**
Ovaries	0.33	**0.26**	0.41	**0.33**	0.47	**0.38**	0.52	**0.43**	0.61	**0.51**	0.70	**0.59**
Prostate	0.27	**0.24**	0.33	**0.31**	0.38	**0.36**	0.42	**0.39**	0.50	**0.47**	0.57	**0.54**
Testes	0.24	**0.25**	0.30	**0.31**	0.34	**0.35**	0.37	**0.39**	0.44	**0.45**	0.49	**0.51**
Upper arm		**0.45**		**0.52**		**0.56**		**0.59**		**0.64**		**0.68**
Elbow		**1.49**		**1.66**		**1.76**		**1.84**		**1.95**		**2.03**
Forearm		**1.58**		**1.74**		**1.83**		**1.90**		**2.00**		**2.07**
Upper leg	0.52	**0.68**	0.65	**0.84**	0.74	**0.96**	0.81	**1.05**	0.94	**1.22**	1.06	**1.36**
Knee	0.05	**0.30**	0.06	**0.37**	0.07	**0.42**	0.08	**0.46**	0.10	**0.53**	0.12	**0.60**
Lower leg				**0.01**		**0.01**		**0.02**	0.01	**0.02**	0.01	**0.02**
Lymph nodes	0.24	**0.22**	0.29	**0.28**	0.33	**0.32**	0.36	**0.35**	0.42	**0.41**	0.47	**0.47**
Skin	0.23	**0.30**	0.25	**0.31**	0.25	**0.33**	0.26	**0.34**	0.28	**0.36**	0.29	**0.37**
Muscle	0.22	**0.26**	0.26	**0.31**	0.29	**0.34**	0.31	**0.37**	0.34	**0.41**	0.37	**0.45**
Skeleton	0.30	**0.60**	0.36	**0.71**	0.40	**0.78**	0.44	**0.84**	0.50	**0.93**	0.55	**1.01**
Red bone marrow	0.09	**0.16**	0.11	**0.19**	0.12	**0.21**	0.13	**0.23**	0.15	**0.26**	0.17	**0.29**
Whole body dose	0.20	**0.26**	0.24	**0.31**	0.27	**0.34**	0.29	**0.37**	0.32	**0.42**	0.36	**0.46**
ED (ICRP 60)	0.25	**0.29**	0.31	**0.35**	0.35	**0.39**	0.38	**0.43**	0.44	**0.49**	0.49	**0.55**
ED (ICRP 103)	0.23	**0.31**	0.28	**0.37**	0.32	**0.41**	0.35	**0.44**	0.40	**0.51**	0.44	**0.56**
Total filtration (mm) **Al**	**1.5**		**2.5**		**3.5**		**4.5**		**3.5**		**3.5**	
Cu	**0**		**0**		**0**		**0**		**0.1**		**0.2**	

Table 12.75 Normalized organ doses—Abdomen l-lat—1–2 years

Beam projection:		l-lat										**Table cells:**
Focus-to-detector distance (cm):		100										(+) = normal font = optimal field setting
Tube voltage (kVp):		60										**(−) = bold font = suboptimal field setting**
Total filtrations:		See last table row.										Empty table cells correspond to values <0.01
Organ or tissue	**Organ dose D_T : Entrance dose K_E**								**(mSv * mGy⁻¹)**			
	(+)	(−)	(+)	(−)	(+)	(−)	(+)	(−)	(+)	(−)	(+)	(−)
Skull												
Brain												
Oral mucosa												
Salivary glands												
Cervical spine				**0.01**		**0.01**		**0.02**		**0.02**		**0.02**
Thyroid				**0.01**		**0.01**		**0.02**		**0.02**		**0.02**
Thymus		**0.06**		**0.08**	0.01	**0.10**	0.01	**0.11**	0.02	**0.13**	0.01	**0.12**
Thoracic spine	0.15	**0.27**	0.20	**0.36**	0.24	**0.43**	0.27	**0.48**	0.33	**0.60**	0.30	**0.54**
Claviculae		**0.02**		**0.03**	0.01	**0.04**	0.01	**0.05**	0.02	**0.06**	0.01	**0.05**
Scapulae	0.02	**0.47**	0.02	**0.55**	0.03	**0.61**	0.03	**0.65**	0.04	**0.73**	0.04	**0.72**
Ribs	0.41	**0.65**	0.49	**0.79**	0.54	**0.88**	0.58	**0.96**	0.65	**1.09**	0.64	**1.07**
Oesophagus	0.10	**0.14**	0.14	**0.18**	0.16	**0.21**	0.18	**0.24**	0.22	**0.30**	0.20	**0.27**
Trachea												
Lung	0.07	**0.17**	0.09	**0.21**	0.10	**0.24**	0.12	**0.27**	0.14	**0.32**	0.13	**0.30**
Mamma	0.01	**0.50**	0.02	**0.54**	0.02	**0.57**	0.02	**0.59**	0.03	**0.62**	0.03	**0.61**
Heart	0.10	**0.21**	0.13	**0.27**	0.15	**0.32**	0.16	**0.36**	0.20	**0.43**	0.18	**0.40**
Stomach	0.40	**0.34**	0.50	**0.43**	0.57	**0.49**	0.62	**0.55**	0.72	**0.64**	0.70	**0.61**
Spleen	0.57	**0.43**	0.69	**0.54**	0.78	**0.62**	0.85	**0.68**	0.96	**0.79**	0.94	**0.76**
Liver	0.08	**0.07**	0.10	**0.09**	0.12	**0.11**	0.14	**0.13**	0.17	**0.16**	0.15	**0.14**
Gall bladder	0.15	**0.12**	0.20	**0.16**	0.24	**0.20**	0.27	**0.22**	0.34	**0.28**	0.31	**0.25**
Pancreas	0.27	**0.18**	0.35	**0.23**	0.40	**0.28**	0.45	**0.31**	0.53	**0.39**	0.50	**0.35**
Small intestine	0.21	**0.15**	0.27	**0.20**	0.31	**0.24**	0.35	**0.27**	0.43	**0.34**	0.40	**0.30**
Large intestine	0.15	**0.12**	0.20	**0.15**	0.23	**0.18**	0.26	**0.21**	0.32	**0.26**	0.29	**0.23**
Sigma/rectum	0.33	**0.23**	0.42	**0.30**	0.49	**0.35**	0.55	**0.40**	0.65	**0.49**	0.61	**0.45**
Lumbar spine	0.38	**0.34**	0.50	**0.45**	0.59	**0.54**	0.67	**0.61**	0.84	**0.76**	0.76	**0.69**
Pelvis	0.66	**0.51**	0.83	**0.65**	0.96	**0.77**	1.06	**0.86**	1.25	**1.03**	1.20	**0.97**
Adrenals	0.13	**0.13**	0.17	**0.17**	0.20	**0.20**	0.23	**0.22**	0.28	**0.28**	0.25	**0.25**
Kidneys	0.21	**0.20**	0.27	**0.25**	0.31	**0.29**	0.34	**0.32**	0.40	**0.39**	0.38	**0.36**
Urinary bladder	0.17	**0.15**	0.23	**0.20**	0.27	**0.23**	0.30	**0.26**	0.37	**0.32**	0.34	**0.29**
Uterus	0.16	**0.12**	0.21	**0.16**	0.25	**0.20**	0.29	**0.22**	0.36	**0.28**	0.33	**0.25**
Ovaries	0.18	**0.12**	0.24	**0.17**	0.29	**0.20**	0.32	**0.23**	0.39	**0.29**	0.36	**0.25**
Prostate	0.13	**0.11**	0.18	**0.15**	0.21	**0.18**	0.24	**0.20**	0.30	**0.25**	0.27	**0.22**
Testes	0.12	**0.12**	0.16	**0.15**	0.19	**0.18**	0.21	**0.20**	0.25	**0.24**	0.23	**0.23**
Upper arm		**0.33**		**0.39**		**0.43**		**0.46**		**0.52**		**0.51**
Elbow		**1.24**		**1.43**		**1.55**		**1.65**		**1.80**		**1.80**
Forearm		**1.35**		**1.55**		**1.68**		**1.77**		**1.91**		**1.92**
Upper leg	0.28	**0.35**	0.36	**0.45**	0.43	**0.53**	0.48	**0.59**	0.59	**0.72**	0.54	**0.67**
Knee	0.02	**0.04**	0.02	**0.06**	0.03	**0.07**	0.03	**0.08**	0.04	**0.10**	0.03	**0.08**
Lower leg												
Lymph nodes	0.14	**0.12**	0.18	**0.16**	0.21	**0.19**	0.24	**0.21**	0.28	**0.26**	0.26	**0.24**
Skin	0.19	**0.24**	0.20	**0.25**	0.21	**0.26**	0.21	**0.27**	0.23	**0.28**	0.22	**0.28**
Muscle	0.14	**0.15**	0.17	**0.19**	0.19	**0.21**	0.20	**0.23**	0.23	**0.26**	0.22	**0.25**
Skeleton	0.16	**0.37**	0.21	**0.44**	0.24	**0.50**	0.26	**0.54**	0.31	**0.63**	0.29	**0.61**
Red bone marrow	0.06	**0.09**	0.07	**0.12**	0.08	**0.13**	0.09	**0.15**	0.11	**0.17**	0.10	**0.16**
Whole body dose	0.13	**0.16**	0.16	**0.20**	0.18	**0.22**	0.19	**0.24**	0.22	**0.28**	0.21	**0.27**
ED (ICRP 60)	0.15	**0.17**	0.19	**0.21**	0.22	**0.25**	0.25	**0.27**	0.30	**0.32**	0.28	**0.30**
ED (ICRP 103)	0.14	**0.20**	0.18	**0.24**	0.21	**0.27**	0.23	**0.30**	0.27	**0.35**	0.26	**0.33**
Total filtration (mm) Al	**1.5**		**2.5**		**3.5**		**4.5**		**3.5**		**3.5**	
Cu	**0**		**0**		**0**		**0**		**0.1**		**0.2**	

Table 12.76 Normalized organ doses—Abdomen l-lat—1–2 years

Beam projection:			l-lat									**Table cells:**
Focus-to-detector distance (cm):			100									(+) = normal font = optimal field setting
Tube voltage (kVp):			70									**(−) = bold font = suboptimal field setting**
Total filtrations:			See last table row.									Empty table cells correspond to values <0.01
Organ or tissue	**Organ dose D_T : Entrance dose K_E**								**(mSv * mGy^{-1})**			
	(+)	(−)	(+)	(−)	(+)	(−)	(+)	(−)	(+)	(−)	(+)	(−)
Skull												
Brain												
Oral mucosa												
Salivary glands												**0.01**
Cervical spine		**0.01**		**0.02**		**0.02**		**0.02**		**0.03**	0.01	**0.04**
Thyroid		**0.01**		**0.02**		**0.02**		**0.02**		**0.03**		**0.03**
Thymus		**0.08**	0.01	**0.10**	0.01	**0.12**	0.02	**0.13**	0.02	**0.16**	0.02	**0.18**
Thoracic spine	0.19	**0.34**	0.25	**0.44**	0.29	**0.52**	0.32	**0.58**	0.39	**0.71**	0.46	**0.82**
Claviculae		**0.03**	0.01	**0.04**	0.01	**0.05**	0.02	**0.06**	0.02	**0.08**	0.03	**0.09**
Scapulae	0.02	**0.52**	0.03	**0.60**	0.04	**0.65**	0.04	**0.70**	0.05	**0.77**	0.07	**0.82**
Ribs	0.46	**0.73**	0.53	**0.87**	0.58	**0.96**	0.62	**1.04**	0.68	**1.17**	0.73	**1.27**
Oesophagus	0.13	**0.18**	0.16	**0.23**	0.19	**0.26**	0.21	**0.30**	0.25	**0.36**	0.29	**0.42**
Trachea												
Lung	0.09	**0.20**	0.11	**0.25**	0.12	**0.28**	0.13	**0.31**	0.15	**0.37**	0.17	**0.41**
Mamma	0.02	**0.52**	0.02	**0.57**	0.03	**0.59**	0.03	**0.61**	0.04	**0.65**	0.05	**0.68**
Heart	0.12	**0.26**	0.15	**0.33**	0.17	**0.38**	0.19	**0.42**	0.23	**0.49**	0.26	**0.56**
Stomach	0.46	**0.40**	0.57	**0.50**	0.64	**0.57**	0.70	**0.62**	0.80	**0.72**	0.89	**0.81**
Spleen	0.65	**0.50**	0.77	**0.62**	0.86	**0.70**	0.93	**0.77**	1.04	**0.88**	1.14	**0.99**
Liver	0.10	**0.09**	0.13	**0.12**	0.15	**0.14**	0.17	**0.16**	0.21	**0.20**	0.25	**0.24**
Gall bladder	0.20	**0.16**	0.25	**0.21**	0.30	**0.25**	0.33	**0.28**	0.40	**0.34**	0.47	**0.40**
Pancreas	0.33	**0.22**	0.41	**0.29**	0.47	**0.34**	0.52	**0.38**	0.61	**0.47**	0.69	**0.55**
Small intestine	0.26	**0.20**	0.32	**0.25**	0.38	**0.30**	0.42	**0.33**	0.50	**0.40**	0.58	**0.47**
Large intestine	0.19	**0.15**	0.24	**0.19**	0.28	**0.23**	0.31	**0.25**	0.38	**0.31**	0.44	**0.36**
Sigma/rectum	0.40	**0.29**	0.50	**0.36**	0.57	**0.43**	0.63	**0.47**	0.74	**0.57**	0.84	**0.66**
Lumbar spine	0.48	**0.44**	0.62	**0.56**	0.72	**0.66**	0.81	**0.75**	0.99	**0.91**	1.16	**1.07**
Pelvis	0.77	**0.61**	0.96	**0.77**	1.09	**0.89**	1.20	**0.99**	1.39	**1.17**	1.56	**1.33**
Adrenals	0.17	**0.16**	0.21	**0.21**	0.25	**0.24**	0.28	**0.27**	0.33	**0.33**	0.39	**0.39**
Kidneys	0.25	**0.24**	0.31	**0.30**	0.36	**0.34**	0.39	**0.38**	0.46	**0.44**	0.52	**0.50**
Urinary bladder	0.22	**0.19**	0.27	**0.24**	0.32	**0.28**	0.36	**0.31**	0.43	**0.38**	0.50	**0.44**
Uterus	0.21	**0.16**	0.27	**0.21**	0.31	**0.25**	0.35	**0.28**	0.43	**0.35**	0.51	**0.41**
Ovaries	0.23	**0.16**	0.30	**0.21**	0.35	**0.25**	0.39	**0.28**	0.47	**0.35**	0.55	**0.42**
Prostate	0.17	**0.14**	0.22	**0.18**	0.26	**0.22**	0.29	**0.25**	0.36	**0.30**	0.42	**0.36**
Testes	0.15	**0.14**	0.19	**0.18**	0.22	**0.21**	0.25	**0.24**	0.29	**0.29**	0.34	**0.33**
Upper arm		**0.37**		**0.43**		**0.47**		**0.50**		**0.55**		**0.59**
Elbow		**1.33**		**1.52**		**1.64**		**1.72**		**1.87**		**1.97**
Forearm		**1.45**		**1.63**		**1.75**		**1.83**		**1.97**		**2.06**
Upper leg	0.35	**0.42**	0.44	**0.54**	0.51	**0.62**	0.57	**0.69**	0.68	**0.83**	0.78	**0.95**
Knee	0.02	**0.06**	0.03	**0.07**	0.04	**0.09**	0.04	**0.10**	0.06	**0.12**	0.07	**0.14**
Lower leg												
Lymph nodes	0.17	**0.15**	0.22	**0.20**	0.25	**0.23**	0.28	**0.25**	0.33	**0.31**	0.37	**0.36**
Skin	0.20	**0.25**	0.21	**0.26**	0.22	**0.27**	0.22	**0.28**	0.24	**0.30**	0.25	**0.31**
Muscle	0.16	**0.18**	0.19	**0.21**	0.21	**0.24**	0.22	**0.26**	0.25	**0.29**	0.28	**0.33**
Skeleton	0.19	**0.42**	0.24	**0.50**	0.27	**0.56**	0.30	**0.60**	0.35	**0.68**	0.39	**0.76**
Red bone marrow	0.07	**0.11**	0.09	**0.14**	0.10	**0.16**	0.11	**0.17**	0.13	**0.20**	0.15	**0.23**
Whole body dose	0.15	**0.19**	0.18	**0.23**	0.20	**0.25**	0.21	**0.27**	0.25	**0.31**	0.27	**0.35**
ED (ICRP 60)	0.18	**0.21**	0.23	**0.25**	0.26	**0.29**	0.29	**0.32**	0.34	**0.37**	0.39	**0.43**
ED (ICRP 103)	0.17	**0.23**	0.21	**0.28**	0.24	**0.31**	0.27	**0.34**	0.31	**0.40**	0.36	**0.45**
Total filtration (mm) **Al**	**1.5**		**2.5**		**3.5**		**4.5**		**3.5**		**3.5**	
Cu	**0**		**0**		**0**		**0**		**0.1**		**0.2**	

Table 12.77 Normalized organ doses—Abdomen l-lat—3–7 years

Beam projection:			l-lat									**Table cells:**
Focus-to-detector distance (cm):			100									(+) = normal font = optimal field setting
Tube voltage (kVp):			60									**(−) = bold font = suboptimal field setting**
Total filtrations:			See last table row.									Empty table cells correspond to values <0.01
Organ or tissue	**Organ dose D_T : Entrance dose K_E**								**(mSv * mGy^{-1})**			
	(+)	(−)	(+)	(−)	(+)	(−)	(+)	(−)	(+)	(−)	(+)	(−)
Skull												
Brain												
Oral mucosa												
Salivary glands												
Cervical spine								**0.01**		**0.01**		**0.01**
Thyroid										**0.01**		
Thymus		**0.06**		**0.08**		**0.09**		**0.10**	0.01	**0.13**		**0.12**
Thoracic spine	0.10	**0.16**	0.13	**0.21**	0.16	**0.26**	0.18	**0.30**	0.23	**0.38**	0.20	**0.33**
Claviculae		**0.01**		**0.02**		**0.03**		**0.03**		**0.04**		**0.03**
Scapulae	0.01	**0.39**	0.02	**0.46**	0.02	**0.51**	0.02	**0.55**	0.03	**0.62**	0.02	**0.61**
Ribs	0.34	**0.49**	0.41	**0.60**	0.45	**0.69**	0.49	**0.75**	0.55	**0.87**	0.54	**0.85**
Oesophagus	0.06	**0.07**	0.07	**0.09**	0.09	**0.11**	0.10	**0.12**	0.12	**0.16**	0.11	**0.14**
Trachea												
Lung	0.06	**0.12**	0.08	**0.15**	0.09	**0.18**	0.10	**0.20**	0.12	**0.24**	0.11	**0.22**
Mamma		**0.45**	0.01	**0.49**	0.02	**0.51**	0.02	**0.53**	0.02	**0.56**	0.02	**0.55**
Heart	0.09	**0.14**	0.11	**0.19**	0.13	**0.22**	0.15	**0.25**	0.18	**0.31**	0.16	**0.28**
Stomach	0.33	**0.27**	0.41	**0.34**	0.48	**0.40**	0.53	**0.44**	0.63	**0.53**	0.60	**0.50**
Spleen	0.47	**0.34**	0.59	**0.43**	0.67	**0.50**	0.73	**0.56**	0.85	**0.66**	0.82	**0.63**
Liver	0.04	**0.04**	0.05	**0.05**	0.07	**0.06**	0.08	**0.07**	0.10	**0.09**	0.08	**0.07**
Gall bladder	0.10	**0.08**	0.13	**0.10**	0.16	**0.12**	0.18	**0.14**	0.23	**0.18**	0.20	**0.15**
Pancreas	0.20	**0.12**	0.26	**0.16**	0.30	**0.19**	0.34	**0.22**	0.41	**0.28**	0.38	**0.24**
Small intestine	0.14	**0.10**	0.19	**0.13**	0.22	**0.16**	0.25	**0.18**	0.31	**0.23**	0.28	**0.20**
Large intestine	0.10	**0.07**	0.13	**0.10**	0.16	**0.11**	0.18	**0.13**	0.22	**0.17**	0.19	**0.14**
Sigma/rectum	0.25	**0.16**	0.32	**0.22**	0.38	**0.26**	0.42	**0.29**	0.52	**0.37**	0.48	**0.32**
Lumbar spine	0.21	**0.19**	0.29	**0.25**	0.35	**0.31**	0.40	**0.35**	0.51	**0.45**	0.44	**0.38**
Pelvis	0.47	**0.34**	0.60	**0.44**	0.70	**0.53**	0.78	**0.60**	0.94	**0.73**	0.89	**0.67**
Adrenals	0.09	**0.08**	0.11	**0.10**	0.14	**0.12**	0.16	**0.14**	0.19	**0.18**	0.17	**0.16**
Kidneys	0.15	**0.14**	0.19	**0.18**	0.22	**0.21**	0.25	**0.23**	0.30	**0.28**	0.28	**0.26**
Urinary bladder	0.11	**0.09**	0.14	**0.12**	0.17	**0.14**	0.19	**0.16**	0.24	**0.21**	0.22	**0.18**
Uterus	0.10	**0.07**	0.13	**0.10**	0.16	**0.12**	0.18	**0.13**	0.23	**0.17**	0.20	**0.14**
Ovaries	0.12	**0.08**	0.16	**0.10**	0.20	**0.13**	0.23	**0.14**	0.28	**0.18**	0.25	**0.16**
Prostate	0.08	**0.06**	0.11	**0.09**	0.13	**0.11**	0.15	**0.12**	0.19	**0.16**	0.17	**0.13**
Testes	0.07	**0.07**	0.10	**0.10**	0.12	**0.12**	0.13	**0.13**	0.16	**0.16**	0.15	**0.15**
Upper arm		**0.27**		**0.32**		**0.36**		**0.38**		**0.43**		**0.42**
Elbow		**1.09**		**1.28**		**1.41**		**1.50**		**1.66**		**1.65**
Forearm		**1.26**		**1.46**		**1.59**		**1.69**		**1.85**		**1.85**
Upper leg	0.14	**0.22**	0.18	**0.29**	0.22	**0.35**	0.25	**0.39**	0.31	**0.49**	0.27	**0.44**
Knee		**0.01**		**0.02**		**0.02**		**0.03**		**0.03**		**0.03**
Lower leg												
Lymph nodes	0.10	**0.08**	0.13	**0.11**	0.15	**0.13**	0.17	**0.15**	0.21	**0.18**	0.19	**0.16**
Skin	0.13	**0.21**	0.14	**0.22**	0.14	**0.23**	0.15	**0.24**	0.15	**0.25**	0.15	**0.24**
Muscle	0.11	**0.12**	0.13	**0.15**	0.15	**0.17**	0.16	**0.18**	0.18	**0.21**	0.18	**0.20**
Skeleton	0.10	**0.24**	0.12	**0.29**	0.14	**0.33**	0.16	**0.36**	0.19	**0.42**	0.18	**0.40**
Red bone marrow	0.05	**0.08**	0.07	**0.10**	0.08	**0.11**	0.09	**0.12**	0.11	**0.15**	0.10	**0.14**
Whole body dose	0.10	**0.13**	0.12	**0.16**	0.14	**0.18**	0.15	**0.19**	0.17	**0.23**	0.16	**0.21**
ED (ICRP 60)	0.11	**0.12**	0.14	**0.16**	0.17	**0.18**	0.19	**0.20**	0.23	**0.24**	0.21	**0.22**
ED (ICRP 103)	0.11	**0.15**	0.14	**0.19**	0.16	**0.21**	0.18	**0.23**	0.21	**0.27**	0.20	**0.25**
Total filtration (mm) Al	**1.5**		**2.5**		**3.5**		**4.5**		**3.5**		**3.5**	
Cu	**0**		**0**		**0**		**0**		**0.1**		**0.2**	

Table 12.78 Normalized organ doses—Abdomen l-lat—3–7 years

Beam projection:			l-lat									**Table cells:**
Focus-to-detector distance (cm):			100					(+) = normal font = optimal field setting				
Tube voltage (kVp):			70					**(−) = bold font = suboptimal field setting**				
Total filtrations:			See last table row.					Empty table cells correspond to values <0.01				
Organ or tissue	**Organ dose D_T : Entrance dose K_E**							**(mSv * mGy^{-1})**				
	(+)	(−)	(+)	(−)	(+)	(−)	(+)	(−)	(+)	(−)	(+)	(−)
Skull												
Brain												
Oral mucosa												
Salivary glands												
Cervical spine				**0.01**		**0.01**		**0.02**		**0.02**		**0.03**
Thyroid								**0.01**		**0.01**		**0.02**
Thymus		**0.08**		**0.10**	0.01	**0.11**	0.01	**0.13**	0.02	**0.16**	0.02	**0.19**
Thoracic spine	0.13	**0.21**	0.17	**0.28**	0.20	**0.33**	0.22	**0.37**	0.28	**0.46**	0.33	**0.55**
Claviculae		**0.02**		**0.03**		**0.04**		**0.04**	0.01	**0.05**	0.02	**0.06**
Scapulae	0.02	**0.43**	0.02	**0.50**	0.03	**0.55**	0.03	**0.59**	0.04	**0.66**	0.05	**0.71**
Ribs	0.38	**0.56**	0.45	**0.68**	0.49	**0.76**	0.53	**0.83**	0.59	**0.94**	0.64	**1.04**
Oesophagus	0.07	**0.09**	0.09	**0.12**	0.11	**0.14**	0.12	**0.16**	0.15	**0.19**	0.17	**0.23**
Trachea												
Lung	0.07	**0.15**	0.09	**0.19**	0.10	**0.21**	0.12	**0.24**	0.14	**0.28**	0.15	**0.33**
Mamma	0.01	**0.47**	0.02	**0.51**	0.02	**0.53**	0.02	**0.55**	0.03	**0.58**	0.04	**0.61**
Heart	0.11	**0.18**	0.14	**0.23**	0.16	**0.27**	0.18	**0.30**	0.21	**0.37**	0.25	**0.43**
Stomach	0.39	**0.33**	0.48	**0.41**	0.55	**0.47**	0.61	**0.52**	0.71	**0.62**	0.81	**0.71**
Spleen	0.55	**0.41**	0.67	**0.51**	0.75	**0.58**	0.82	**0.64**	0.94	**0.76**	1.04	**0.86**
Liver	0.06	**0.05**	0.07	**0.07**	0.09	**0.08**	0.10	**0.09**	0.13	**0.12**	0.15	**0.14**
Gall bladder	0.13	**0.10**	0.17	**0.14**	0.20	**0.16**	0.23	**0.19**	0.29	**0.23**	0.34	**0.28**
Pancreas	0.25	**0.16**	0.31	**0.21**	0.36	**0.25**	0.40	**0.28**	0.49	**0.35**	0.56	**0.42**
Small intestine	0.18	**0.13**	0.23	**0.17**	0.27	**0.20**	0.31	**0.23**	0.37	**0.29**	0.44	**0.34**
Large intestine	0.13	**0.10**	0.17	**0.13**	0.19	**0.15**	0.22	**0.17**	0.27	**0.21**	0.32	**0.25**
Sigma/rectum	0.31	**0.21**	0.39	**0.27**	0.45	**0.32**	0.51	**0.36**	0.61	**0.45**	0.70	**0.53**
Lumbar spine	0.29	**0.25**	0.37	**0.33**	0.44	**0.40**	0.51	**0.45**	0.63	**0.56**	0.76	**0.68**
Pelvis	0.57	**0.43**	0.71	**0.54**	0.82	**0.63**	0.91	**0.71**	1.07	**0.86**	1.22	**0.99**
Adrenals	0.11	**0.10**	0.15	**0.14**	0.17	**0.16**	0.19	**0.18**	0.24	**0.23**	0.28	**0.27**
Kidneys	0.18	**0.17**	0.23	**0.22**	0.27	**0.25**	0.30	**0.28**	0.35	**0.33**	0.41	**0.38**
Urinary bladder	0.14	**0.12**	0.18	**0.16**	0.21	**0.18**	0.24	**0.21**	0.30	**0.26**	0.35	**0.31**
Uterus	0.13	**0.10**	0.17	**0.13**	0.21	**0.16**	0.24	**0.18**	0.29	**0.23**	0.35	**0.27**
Ovaries	0.16	**0.11**	0.21	**0.14**	0.25	**0.17**	0.28	**0.19**	0.35	**0.24**	0.42	**0.29**
Prostate	0.11	**0.09**	0.14	**0.12**	0.17	**0.14**	0.19	**0.16**	0.24	**0.20**	0.29	**0.25**
Testes	0.10	**0.09**	0.12	**0.12**	0.14	**0.14**	0.16	**0.16**	0.20	**0.20**	0.23	**0.23**
Upper arm		**0.30**		**0.35**		**0.39**		**0.41**		**0.46**		**0.50**
Elbow		**1.19**		**1.37**		**1.50**		**1.59**		**1.73**		**1.85**
Forearm		**1.36**		**1.55**		**1.67**		**1.76**		**1.91**		**2.02**
Upper leg	0.18	**0.28**	0.23	**0.36**	0.27	**0.43**	0.30	**0.48**	0.37	**0.58**	0.43	**0.68**
Knee		**0.02**		**0.03**		**0.03**		**0.04**	0.01	**0.05**	0.02	**0.06**
Lower leg												
Lymph nodes	0.13	**0.11**	0.16	**0.14**	0.19	**0.16**	0.21	**0.18**	0.25	**0.22**	0.29	**0.27**
Skin	0.13	**0.22**	0.14	**0.23**	0.15	**0.24**	0.15	**0.25**	0.16	**0.26**	0.17	**0.27**
Muscle	0.13	**0.14**	0.15	**0.17**	0.17	**0.19**	0.18	**0.21**	0.21	**0.24**	0.23	**0.27**
Skeleton	0.12	**0.28**	0.14	**0.33**	0.17	**0.37**	0.18	**0.41**	0.22	**0.47**	0.25	**0.52**
Red bone marrow	0.07	**0.09**	0.08	**0.12**	0.10	**0.13**	0.11	**0.15**	0.13	**0.18**	0.15	**0.20**
Whole body dose	0.12	**0.15**	0.14	**0.18**	0.16	**0.20**	0.17	**0.22**	0.20	**0.26**	0.22	**0.29**
ED (ICRP 60)	0.14	**0.15**	0.17	**0.19**	0.20	**0.22**	0.22	**0.24**	0.27	**0.29**	0.31	**0.33**
ED (ICRP 103)	0.13	**0.18**	0.16	**0.22**	0.19	**0.25**	0.21	**0.27**	0.25	**0.31**	0.29	**0.36**
Total filtration (mm) Al	**1.5**		**2.5**		**3.5**		**4.5**		**3.5**		**3.5**	
Cu	**0**		**0**		**0**		**0**		**0.1**		**0.2**	

Table 12.79 Normalized organ doses—Abdomen l-lat—3–7 years

Beam projection:			l-lat									**Table cells:**
Focus-to-detector distance (cm):			100						(+) = normal font = optimal field setting			
Tube voltage (kVp):			80						**(−) = bold font = suboptimal field setting**			
Total filtrations:			See last table row.						Empty table cells correspond to values <0.01			
Organ or tissue	**Organ dose D_T : Entrance dose K_E**								**(mSv * mGy⁻¹)**			
	(+)	(−)	(+)	(−)	(+)	(−)	(+)	(−)	(+)	(−)	(+)	(−)
Skull												
Brain												
Oral mucosa												
Salivary glands												
Cervical spine		**0.01**		**0.02**		**0.02**		**0.02**		**0.03**		**0.03**
Thyroid				**0.01**		**0.01**		**0.01**		**0.02**		**0.02**
Thymus		**0.09**	0.01	**0.12**	0.01	**0.14**	0.02	**0.15**	0.02	**0.18**	0.02	**0.21**
Thoracic spine	0.16	**0.27**	0.20	**0.34**	0.24	**0.39**	0.26	**0.44**	0.32	**0.54**	0.38	**0.63**
Claviculae		**0.03**		**0.04**	0.01	**0.04**	0.01	**0.05**	0.02	**0.06**	0.02	**0.08**
Scapulae	0.02	**0.47**	0.03	**0.54**	0.03	**0.59**	0.04	**0.62**	0.05	**0.68**	0.06	**0.73**
Ribs	0.41	**0.62**	0.48	**0.74**	0.52	**0.82**	0.56	**0.88**	0.61	**0.99**	0.66	**1.08**
Oesophagus	0.09	**0.11**	0.11	**0.14**	0.13	**0.17**	0.14	**0.19**	0.17	**0.23**	0.20	**0.27**
Trachea												
Lung	0.09	**0.18**	0.11	**0.22**	0.12	**0.25**	0.13	**0.27**	0.15	**0.32**	0.17	**0.36**
Mamma	0.02	**0.49**	0.02	**0.53**	0.03	**0.55**	0.03	**0.57**	0.04	**0.60**	0.04	**0.63**
Heart	0.13	**0.22**	0.16	**0.28**	0.18	**0.32**	0.20	**0.35**	0.24	**0.42**	0.27	**0.48**
Stomach	0.45	**0.38**	0.55	**0.47**	0.62	**0.54**	0.68	**0.59**	0.78	**0.69**	0.87	**0.78**
Spleen	0.62	**0.48**	0.74	**0.58**	0.83	**0.66**	0.89	**0.72**	1.01	**0.84**	1.11	**0.94**
Liver	0.07	**0.07**	0.09	**0.09**	0.11	**0.10**	0.12	**0.11**	0.15	**0.14**	0.18	**0.17**
Gall bladder	0.17	**0.13**	0.21	**0.17**	0.25	**0.20**	0.28	**0.23**	0.34	**0.28**	0.40	**0.33**
Pancreas	0.30	**0.20**	0.37	**0.26**	0.42	**0.30**	0.46	**0.34**	0.55	**0.41**	0.63	**0.48**
Small intestine	0.22	**0.17**	0.28	**0.21**	0.32	**0.25**	0.36	**0.28**	0.43	**0.34**	0.50	**0.40**
Large intestine	0.16	**0.12**	0.20	**0.16**	0.23	**0.18**	0.26	**0.20**	0.31	**0.25**	0.36	**0.29**
Sigma/rectum	0.37	**0.26**	0.46	**0.33**	0.52	**0.38**	0.58	**0.43**	0.68	**0.52**	0.78	**0.60**
Lumbar spine	0.36	**0.32**	0.46	**0.41**	0.54	**0.48**	0.60	**0.54**	0.74	**0.66**	0.86	**0.78**
Pelvis	0.66	**0.51**	0.81	**0.63**	0.92	**0.73**	1.01	**0.81**	1.17	**0.95**	1.31	**1.09**
Adrenals	0.14	**0.13**	0.18	**0.17**	0.21	**0.20**	0.23	**0.22**	0.28	**0.27**	0.33	**0.32**
Kidneys	0.22	**0.20**	0.27	**0.25**	0.31	**0.29**	0.34	**0.32**	0.40	**0.37**	0.45	**0.43**
Urinary bladder	0.18	**0.15**	0.22	**0.19**	0.26	**0.22**	0.29	**0.25**	0.35	**0.30**	0.40	**0.35**
Uterus	0.17	**0.13**	0.22	**0.17**	0.25	**0.20**	0.29	**0.22**	0.35	**0.28**	0.41	**0.33**
Ovaries	0.20	**0.14**	0.26	**0.18**	0.30	**0.21**	0.34	**0.23**	0.41	**0.29**	0.48	**0.34**
Prostate	0.14	**0.12**	0.18	**0.15**	0.21	**0.18**	0.24	**0.20**	0.29	**0.25**	0.34	**0.29**
Testes	0.12	**0.12**	0.15	**0.15**	0.17	**0.17**	0.19	**0.19**	0.23	**0.23**	0.26	**0.26**
Upper arm		**0.33**		**0.38**		**0.41**		**0.44**		**0.48**		**0.52**
Elbow		**1.27**		**1.44**		**1.56**		**1.64**		**1.77**		**1.86**
Forearm		**1.43**		**1.62**		**1.73**		**1.81**		**1.94**		**2.02**
Upper leg	0.22	**0.34**	0.27	**0.43**	0.32	**0.50**	0.35	**0.55**	0.42	**0.66**	0.49	**0.76**
Knee		**0.03**		**0.03**	0.01	**0.04**	0.01	**0.05**	0.02	**0.06**	0.02	**0.07**
Lower leg												
Lymph nodes	0.15	**0.13**	0.19	**0.17**	0.22	**0.19**	0.24	**0.22**	0.29	**0.26**	0.33	**0.30**
Skin	0.14	**0.23**	0.15	**0.24**	0.15	**0.25**	0.16	**0.26**	0.17	**0.27**	0.18	**0.28**
Muscle	0.14	**0.16**	0.17	**0.19**	0.19	**0.21**	0.20	**0.23**	0.23	**0.26**	0.25	**0.29**
Skeleton	0.13	**0.31**	0.16	**0.37**	0.19	**0.41**	0.20	**0.44**	0.24	**0.50**	0.27	**0.55**
Red bone marrow	0.08	**0.11**	0.10	**0.14**	0.12	**0.16**	0.13	**0.17**	0.15	**0.20**	0.17	**0.23**
Whole body dose	0.13	**0.17**	0.16	**0.20**	0.17	**0.23**	0.19	**0.25**	0.22	**0.28**	0.24	**0.31**
ED (ICRP 60)	0.17	**0.18**	0.20	**0.22**	0.23	**0.25**	0.26	**0.28**	0.30	**0.33**	0.35	**0.37**
ED (ICRP 103)	0.16	**0.21**	0.19	**0.25**	0.22	**0.28**	0.24	**0.30**	0.28	**0.35**	0.32	**0.40**
Total filtration (mm) Al	**1.5**		**2.5**		**3.5**		**4.5**		**3.5**		**3.5**	
Cu	**0**		**0**		**0**		**0**		**0.1**		**0.2**	

Table 12.80 Normalized organ doses—Abdomen ap—24th gestation week

Beam projection:			ap									**Table cells:**
Focus-to-detector distance (cm):			115					(+) = normal font = optimal field setting				
Tube voltage (kVp):			50					**(−) = bold font = suboptimal field setting**				
Total filtrations:			See last table row.					Empty table cells correspond to values <0.01				
Organ or tissue	**Organ dose D_T : Entrance dose K_E**							**(mSv * mGy^{-1})**				
	(+)	(−)	(+)	(−)	(+)	(−)	(+)	(−)	(+)	(−)	(+)	(−)
Skull												
Brain												
Oral mucosa												
Salivary glands												
Cervical spine												
Thyroid												
Thymus												**0.01**
Thoracic spine	0.02	**0.03**	0.02	**0.04**	0.02	**0.05**	0.03	**0.05**	0.03	**0.06**	0.04	**0.07**
Claviculae												
Scapulae		**0.03**		**0.04**		**0.04**		**0.05**		**0.05**		**0.06**
Ribs	0.05	**0.10**	0.06	**0.11**	0.06	**0.12**	0.06	**0.13**	0.07	**0.14**	0.07	**0.15**
Oesophagus	0.01	**0.02**	0.02	**0.03**	0.02	**0.03**	0.02	**0.03**	0.02	**0.04**	0.02	**0.04**
Trachea												
Lung		**0.03**		**0.04**	0.01	**0.04**	0.01	**0.04**	0.01	**0.05**	0.01	**0.05**
Mamma		**0.09**		**0.09**		**0.09**		**0.09**		**0.10**		**0.10**
Heart		**0.05**	0.01	**0.06**	0.01	**0.06**	0.01	**0.07**	0.01	**0.07**	0.02	**0.08**
Stomach	0.05	**0.05**	0.06	**0.06**	0.06	**0.07**	0.07	**0.07**	0.07	**0.08**	0.08	**0.08**
Spleen	0.03	**0.03**	0.03	**0.03**	0.04	**0.04**	0.04	**0.04**	0.04	**0.05**	0.05	**0.05**
Liver	0.05	**0.05**	0.05	**0.06**	0.06	**0.06**	0.06	**0.06**	0.07	**0.07**	0.07	**0.08**
Gall bladder	0.05	**0.05**	0.06	**0.06**	0.06	**0.06**	0.07	**0.07**	0.07	**0.07**	0.08	**0.08**
Pancreas	0.03	**0.03**	0.04	**0.04**	0.04	**0.05**	0.05	**0.05**	0.05	**0.06**	0.06	**0.06**
Small intestine	0.04	**0.04**	0.05	**0.05**	0.05	**0.05**	0.06	**0.06**	0.06	**0.07**	0.07	**0.07**
Large intestine	0.05	**0.05**	0.05	**0.06**	0.06	**0.06**	0.06	**0.07**	0.07	**0.07**	0.08	**0.08**
Sigma/rectum	0.04	**0.04**	0.05	**0.05**	0.05	**0.05**	0.05	**0.06**	0.06	**0.06**	0.07	**0.07**
Lumbar spine	0.05	**0.05**	0.06	**0.06**	0.07	**0.08**	0.08	**0.08**	0.10	**0.10**	0.11	**0.12**
Pelvis	0.07	**0.07**	0.09	**0.09**	0.10	**0.11**	0.11	**0.12**	0.13	**0.14**	0.15	**0.15**
Adrenals	0.02	**0.02**	0.02	**0.02**	0.03	**0.03**	0.03	**0.03**	0.03	**0.04**	0.04	**0.04**
Kidneys	0.02	**0.02**	0.02	**0.02**	0.03	**0.03**	0.03	**0.03**	0.03	**0.04**	0.04	**0.04**
Urinary bladder	0.06	**0.06**	0.06	**0.06**	0.07	**0.07**	0.07	**0.07**	0.08	**0.08**	0.08	**0.09**
Uterus	0.04	**0.04**	0.05	**0.05**	0.06	**0.06**	0.06	**0.06**	0.07	**0.07**	0.07	**0.08**
Ovaries	0.04	**0.04**	0.04	**0.04**	0.05	**0.05**	0.05	**0.05**	0.06	**0.06**	0.06	**0.07**
Prostate	0.04	**0.04**	0.05	**0.05**	0.06	**0.06**	0.06	**0.06**	0.07	**0.07**	0.07	**0.08**
Testes	0.08	**0.09**	0.09	**0.09**	0.09	**0.09**	0.09	**0.10**	0.09	**0.10**	0.10	**0.10**
Upper arm		**0.03**		**0.04**		**0.04**		**0.05**		**0.05**		**0.06**
Elbow		**0.11**		**0.14**		**0.15**		**0.16**		**0.18**		**0.19**
Forearm		**0.12**		**0.15**		**0.16**		**0.17**		**0.19**		**0.20**
Upper leg	0.06	**0.10**	0.07	**0.12**	0.08	**0.14**	0.09	**0.15**	0.10	**0.17**	0.11	**0.19**
Knee		**0.04**		**0.04**		**0.05**		**0.05**		**0.06**		**0.07**
Lower leg												
Lymph nodes	0.02	**0.03**	0.03	**0.03**	0.03	**0.04**	0.03	**0.04**	0.04	**0.05**	0.04	**0.05**
Skin	0.02	**0.02**	0.02	**0.03**	0.02	**0.03**	0.02	**0.03**	0.02	**0.03**	0.02	**0.03**
Muscle	0.02	**0.03**	0.02	**0.03**	0.02	**0.03**	0.02	**0.03**	0.03	**0.04**	0.03	**0.04**
Skeleton	0.02	**0.05**	0.03	**0.06**	0.03	**0.07**	0.03	**0.07**	0.04	**0.08**	0.04	**0.09**
Red bone marrow		**0.01**		**0.01**		**0.02**	0.01	**0.02**	0.01	**0.02**	0.01	**0.02**
Whole body dose	0.02	**0.03**	0.02	**0.03**	0.02	**0.03**	0.02	**0.04**	0.03	**0.04**	0.03	**0.04**
ED (ICRP 60)	0.03	**0.04**	0.04	**0.05**	0.04	**0.05**	0.04	**0.05**	0.05	**0.06**	0.05	**0.06**
ED (ICRP 103)	0.03	**0.04**	0.03	**0.05**	0.03	**0.05**	0.03	**0.05**	0.04	**0.06**	0.04	**0.06**
Total filtration (mm) **Al**	**1.5**		**2.5**		**3.5**		**4.5**		**3.5**		**3.5**	
Cu	**0**		**0**		**0**		**0**		**0.1**		**0.2**	

Table 12.81 Normalized organ doses—Abdomen ap—24th gestation week

Beam projection:	ap											**Table cells:**
Focus-to-detector distance (cm):	115											(+) = normal font = optimal field setting
Tube voltage (kVp):	60											**(−) = bold font = suboptimal field setting**
Total filtrations:	See last table row.											Empty table cells correspond to values <0.01
Organ or tissue	**Organ dose D_T : Entrance dose K_E**								**(mSv * mGy^{-1})**			
	(+)	(−)	(+)	(−)	(+)	(−)	(+)	(−)	(+)	(−)	(+)	(−)
Skull												
Brain												
Oral mucosa												
Salivary glands												
Cervical spine												
Thyroid												
Thymus										**0.01**		**0.01**
Thoracic spine	0.02	**0.04**	0.02	**0.05**	0.03	**0.05**	0.03	**0.06**	0.04	**0.07**	0.04	**0.07**
Claviculae												
Scapulae		**0.03**		**0.04**		**0.05**		**0.05**		**0.06**		**0.06**
Ribs	0.05	**0.11**	0.06	**0.12**	0.06	**0.13**	0.07	**0.14**	0.07	**0.14**	0.07	**0.15**
Oesophagus	0.01	**0.03**	0.02	**0.03**	0.02	**0.04**	0.02	**0.04**	0.02	**0.04**	0.02	**0.04**
Trachea												
Lung		**0.04**	0.01	**0.04**	0.01	**0.04**	0.01	**0.05**	0.01	**0.05**	0.01	**0.05**
Mamma		**0.09**		**0.09**		**0.09**		**0.10**		**0.10**		**0.10**
Heart	0.01	**0.05**	0.01	**0.06**	0.01	**0.07**	0.01	**0.07**	0.02	**0.08**	0.02	**0.08**
Stomach	0.06	**0.06**	0.06	**0.07**	0.07	**0.07**	0.07	**0.08**	0.08	**0.08**	0.08	**0.08**
Spleen	0.03	**0.03**	0.04	**0.04**	0.04	**0.04**	0.04	**0.05**	0.05	**0.05**	0.05	**0.05**
Liver	0.05	**0.05**	0.06	**0.06**	0.06	**0.07**	0.07	**0.07**	0.07	**0.08**	0.07	**0.08**
Gall bladder	0.05	**0.05**	0.06	**0.06**	0.07	**0.07**	0.07	**0.07**	0.08	**0.08**	0.08	**0.08**
Pancreas	0.04	**0.04**	0.05	**0.05**	0.05	**0.05**	0.05	**0.06**	0.06	**0.07**	0.06	**0.06**
Small intestine	0.05	**0.05**	0.05	**0.06**	0.06	**0.06**	0.06	**0.07**	0.07	**0.07**	0.07	**0.07**
Large intestine	0.05	**0.05**	0.06	**0.06**	0.07	**0.07**	0.07	**0.07**	0.08	**0.08**	0.08	**0.08**
Sigma/rectum	0.04	**0.04**	0.05	**0.05**	0.06	**0.06**	0.06	**0.06**	0.07	**0.07**	0.07	**0.07**
Lumbar spine	0.06	**0.06**	0.08	**0.08**	0.09	**0.09**	0.10	**0.10**	0.11	**0.12**	0.11	**0.12**
Pelvis	0.09	**0.09**	0.10	**0.11**	0.12	**0.12**	0.13	**0.13**	0.15	**0.15**	0.15	**0.15**
Adrenals	0.02	**0.02**	0.03	**0.03**	0.03	**0.03**	0.03	**0.04**	0.04	**0.04**	0.04	**0.04**
Kidneys	0.02	**0.02**	0.03	**0.03**	0.03	**0.03**	0.03	**0.04**	0.04	**0.04**	0.04	**0.04**
Urinary bladder	0.06	**0.06**	0.07	**0.07**	0.07	**0.08**	0.08	**0.08**	0.08	**0.09**	0.08	**0.09**
Uterus	0.05	**0.05**	0.06	**0.06**	0.06	**0.06**	0.07	**0.07**	0.07	**0.08**	0.07	**0.08**
Ovaries	0.04	**0.04**	0.05	**0.05**	0.05	**0.06**	0.06	**0.06**	0.06	**0.07**	0.06	**0.07**
Prostate	0.05	**0.05**	0.06	**0.06**	0.06	**0.06**	0.06	**0.07**	0.07	**0.07**	0.07	**0.08**
Testes	0.09	**0.09**	0.09	**0.09**	0.09	**0.10**	0.09	**0.10**	0.10	**0.10**	0.10	**0.10**
Upper arm		**0.04**		**0.04**		**0.05**		**0.05**		**0.06**		**0.06**
Elbow		**0.13**		**0.15**		**0.16**		**0.17**		**0.19**		**0.19**
Forearm		**0.14**		**0.16**		**0.17**		**0.18**		**0.20**		**0.20**
Upper leg	0.07	**0.12**	0.08	**0.14**	0.09	**0.16**	0.10	**0.17**	0.11	**0.19**	0.11	**0.19**
Knee		**0.04**		**0.05**		**0.05**		**0.06**		**0.07**		**0.07**
Lower leg												
Lymph nodes	0.03	**0.03**	0.03	**0.04**	0.04	**0.04**	0.04	**0.05**	0.04	**0.05**	0.04	**0.05**
Skin	0.02	**0.03**	0.02	**0.03**	0.02	**0.03**	0.02	**0.03**	0.02	**0.03**	0.02	**0.03**
Muscle	0.02	**0.03**	0.02	**0.03**	0.02	**0.03**	0.03	**0.04**	0.03	**0.04**	0.03	**0.04**
Skeleton	0.02	**0.06**	0.03	**0.07**	0.03	**0.08**	0.04	**0.08**	0.04	**0.09**	0.04	**0.09**
Red bone marrow		**0.01**		**0.02**	0.01	**0.02**	0.01	**0.02**	0.01	**0.02**	0.01	**0.02**
Whole body dose	0.02	**0.03**	0.02	**0.03**	0.02	**0.04**	0.03	**0.04**	0.03	**0.04**	0.03	**0.04**
ED (ICRP 60)	0.04	**0.05**	0.04	**0.05**	0.04	**0.06**	0.05	**0.06**	0.05	**0.06**	0.05	**0.06**
ED (ICRP 103)	0.03	**0.05**	0.03	**0.05**	0.04	**0.05**	0.04	**0.06**	0.04	**0.06**	0.04	**0.06**
Total filtration (mm) **Al**	**1.5**		**2.5**		**3.5**		**4.5**		**3.5**		**3.5**	
Cu	**0**		**0**		**0**		**0**		**0.1**		**0.2**	

Table 12.82 Normalized organ doses—Abdomen ap—24th gestation week

Beam projection:			ap									**Table cells:**
Focus-to-detector distance (cm):			115						(+) = normal font = optimal field setting			
Tube voltage (kVp):			70						**(−) = bold font = suboptimal field setting**			
Total filtrations:			See last table row.						Empty table cells correspond to values <0.01			
Organ or tissue	**Organ dose D_T : Entrance dose K_E**								**(mSv * mGy⁻¹)**			
	(+)	(−)	(+)	(−)	(+)	(−)	(+)	(−)	(+)	(−)	(+)	(−)
Skull												
Brain												
Oral mucosa												
Salivary glands												
Cervical spine												
Thyroid												
Thymus										**0.01**		**0.01**
Thoracic spine	0.02	**0.04**	0.03	**0.05**	0.03	**0.06**	0.04	**0.07**	0.04	**0.08**	0.05	**0.09**
Claviculae												**0.01**
Scapulae		**0.04**		**0.05**		**0.05**		**0.05**		**0.06**		**0.07**
Ribs	0.06	**0.11**	0.06	**0.13**	0.07	**0.13**	0.07	**0.14**	0.07	**0.15**	0.07	**0.15**
Oesophagus	0.02	**0.03**	0.02	**0.03**	0.02	**0.04**	0.02	**0.04**	0.03	**0.05**	0.03	**0.05**
Trachea												
Lung	0.01	**0.04**	0.01	**0.04**	0.01	**0.05**	0.01	**0.05**	0.02	**0.05**	0.02	**0.06**
Mamma		**0.09**		**0.09**		**0.09**		**0.10**		**0.10**		**0.10**
Heart	0.01	**0.06**	0.01	**0.07**	0.01	**0.07**	0.02	**0.07**	0.02	**0.08**	0.02	**0.08**
Stomach	0.06	**0.06**	0.07	**0.07**	0.07	**0.08**	0.08	**0.08**	0.08	**0.08**	0.08	**0.09**
Spleen	0.03	**0.04**	0.04	**0.04**	0.04	**0.05**	0.05	**0.05**	0.05	**0.06**	0.06	**0.06**
Liver	0.05	**0.06**	0.06	**0.06**	0.07	**0.07**	0.07	**0.07**	0.07	**0.08**	0.08	**0.08**
Gall bladder	0.06	**0.06**	0.07	**0.07**	0.07	**0.07**	0.08	**0.08**	0.08	**0.08**	0.09	**0.09**
Pancreas	0.04	**0.04**	0.05	**0.05**	0.05	**0.06**	0.06	**0.06**	0.06	**0.07**	0.07	**0.08**
Small intestine	0.05	**0.05**	0.06	**0.06**	0.06	**0.07**	0.07	**0.07**	0.07	**0.08**	0.08	**0.08**
Large intestine	0.06	**0.06**	0.06	**0.07**	0.07	**0.07**	0.07	**0.08**	0.08	**0.08**	0.09	**0.09**
Sigma/rectum	0.05	**0.05**	0.06	**0.06**	0.06	**0.06**	0.06	**0.07**	0.07	**0.07**	0.08	**0.08**
Lumbar spine	0.07	**0.07**	0.09	**0.09**	0.10	**0.10**	0.11	**0.11**	0.12	**0.13**	0.13	**0.14**
Pelvis	0.10	**0.10**	0.11	**0.12**	0.13	**0.13**	0.14	**0.14**	0.15	**0.16**	0.16	**0.17**
Adrenals	0.02	**0.03**	0.03	**0.03**	0.03	**0.04**	0.04	**0.04**	0.04	**0.04**	0.05	**0.05**
Kidneys	0.03	**0.03**	0.03	**0.03**	0.03	**0.04**	0.04	**0.04**	0.04	**0.05**	0.05	**0.05**
Urinary bladder	0.06	**0.07**	0.07	**0.07**	0.08	**0.08**	0.08	**0.08**	0.09	**0.09**	0.09	**0.09**
Uterus	0.05	**0.05**	0.06	**0.06**	0.07	**0.07**	0.07	**0.07**	0.08	**0.08**	0.08	**0.08**
Ovaries	0.05	**0.05**	0.05	**0.05**	0.06	**0.06**	0.06	**0.06**	0.07	**0.07**	0.07	**0.08**
Prostate	0.05	**0.05**	0.06	**0.06**	0.06	**0.07**	0.07	**0.07**	0.07	**0.08**	0.08	**0.08**
Testes	0.09	**0.09**	0.09	**0.10**	0.09	**0.10**	0.10	**0.10**	0.10	**0.10**	0.10	**0.11**
Upper arm		**0.04**		**0.05**		**0.05**		**0.05**		**0.06**		**0.06**
Elbow		**0.14**		**0.16**		**0.17**		**0.18**		**0.19**		**0.20**
Forearm		**0.15**		**0.17**		**0.18**		**0.19**		**0.20**		**0.21**
Upper leg	0.07	**0.13**	0.09	**0.15**	0.10	**0.17**	0.10	**0.18**	0.12	**0.19**	0.12	**0.21**
Knee		**0.04**		**0.05**		**0.06**		**0.06**		**0.07**		**0.07**
Lower leg												
Lymph nodes	0.03	**0.04**	0.04	**0.04**	0.04	**0.05**	0.04	**0.05**	0.04	**0.05**	0.05	**0.06**
Skin	0.02	**0.03**	0.02	**0.03**	0.02	**0.03**	0.02	**0.03**	0.02	**0.03**	0.02	**0.03**
Muscle	0.02	**0.03**	0.02	**0.03**	0.03	**0.04**	0.03	**0.04**	0.03	**0.04**	0.03	**0.04**
Skeleton	0.03	**0.06**	0.03	**0.07**	0.03	**0.08**	0.04	**0.08**	0.04	**0.09**	0.04	**0.10**
Red bone marrow		**0.02**	0.01	**0.02**	0.01	**0.02**	0.01	**0.02**	0.01	**0.03**	0.02	**0.03**
Whole body dose	0.02	**0.03**	0.02	**0.04**	0.03	**0.04**	0.03	**0.04**	0.03	**0.05**	0.03	**0.05**
ED (ICRP 60)	0.04	**0.05**	0.04	**0.05**	0.05	**0.06**	0.05	**0.06**	0.05	**0.07**	0.05	**0.07**
ED (ICRP 103)	0.03	**0.05**	0.03	**0.05**	0.04	**0.06**	0.04	**0.06**	0.04	**0.06**	0.05	**0.07**
Total filtration (mm) Al	**1.5**		**2.5**		**3.5**		**4.5**		**3.5**		**3.5**	
Cu	**0**		**0**		**0**		**0**		**0.1**		**0.2**	

Table 12.83 Normalized organ doses—Abdomen ap—28th gestation week

Beam projection:			ap									**Table cells:**
Focus-to-detector distance (cm):			115									(+) = normal font = optimal field setting
Tube voltage (kVp):			50									**(−) = bold font = suboptimal field setting**
Total filtrations:			See last table row.									Empty table cells correspond to values <0.01
Organ or tissue	**Organ dose D_T : Entrance dose K_E**								**(mSv * mGy^{-1})**			
	(+)	(−)	(+)	(−)	(+)	(−)	(+)	(−)	(+)	(−)	(+)	(−)
Skull												
Brain												
Oral mucosa												
Salivary glands												
Cervical spine												
Thyroid												
Thymus		**0.03**		**0.03**		**0.03**		**0.03**		**0.04**		**0.04**
Thoracic spine	0.02	**0.04**	0.03	**0.06**	0.03	**0.07**	0.04	**0.07**	0.05	**0.09**	0.06	**0.10**
Claviculae								**0.01**		**0.01**		**0.02**
Scapulae		**0.05**		**0.06**		**0.07**		**0.08**		**0.09**	0.01	**0.11**
Ribs	0.08	**0.15**	0.09	**0.18**	0.10	**0.19**	0.10	**0.20**	0.11	**0.22**	0.12	**0.23**
Oesophagus	0.02	**0.03**	0.02	**0.04**	0.03	**0.05**	0.03	**0.05**	0.03	**0.06**	0.04	**0.06**
Trachea												
Lung	0.02	**0.05**	0.02	**0.06**	0.02	**0.07**	0.02	**0.07**	0.03	**0.08**	0.03	**0.08**
Mamma		**0.13**		**0.13**		**0.14**		**0.14**		**0.14**		**0.14**
Heart	0.03	**0.07**	0.03	**0.08**	0.03	**0.09**	0.04	**0.10**	0.04	**0.10**	0.04	**0.11**
Stomach	0.07	**0.07**	0.08	**0.09**	0.09	**0.09**	0.10	**0.10**	0.10	**0.11**	0.11	**0.12**
Spleen	0.03	**0.04**	0.04	**0.04**	0.05	**0.05**	0.05	**0.06**	0.06	**0.07**	0.07	**0.07**
Liver	0.06	**0.06**	0.07	**0.08**	0.08	**0.08**	0.09	**0.09**	0.09	**0.10**	0.10	**0.11**
Gall bladder	0.06	**0.06**	0.08	**0.08**	0.08	**0.09**	0.09	**0.09**	0.10	**0.10**	0.11	**0.11**
Pancreas	0.04	**0.05**	0.05	**0.06**	0.06	**0.06**	0.07	**0.07**	0.08	**0.08**	0.08	**0.09**
Small intestine	0.05	**0.06**	0.07	**0.07**	0.07	**0.08**	0.08	**0.08**	0.09	**0.09**	0.10	**0.10**
Large intestine	0.06	**0.06**	0.08	**0.08**	0.08	**0.08**	0.09	**0.09**	0.10	**0.10**	0.11	**0.11**
Sigma/rectum	0.05	**0.05**	0.06	**0.06**	0.07	**0.07**	0.07	**0.08**	0.08	**0.09**	0.09	**0.10**
Lumbar spine	0.06	**0.06**	0.08	**0.08**	0.09	**0.10**	0.10	**0.11**	0.13	**0.13**	0.15	**0.15**
Pelvis	0.09	**0.10**	0.12	**0.12**	0.14	**0.14**	0.15	**0.16**	0.18	**0.18**	0.20	**0.21**
Adrenals	0.02	**0.02**	0.03	**0.03**	0.03	**0.04**	0.04	**0.04**	0.04	**0.05**	0.05	**0.05**
Kidneys	0.02	**0.02**	0.03	**0.03**	0.03	**0.04**	0.04	**0.04**	0.04	**0.05**	0.05	**0.05**
Urinary bladder	0.08	**0.08**	0.09	**0.09**	0.10	**0.10**	0.10	**0.10**	0.11	**0.11**	0.12	**0.12**
Uterus	0.06	**0.06**	0.07	**0.07**	0.08	**0.08**	0.08	**0.08**	0.09	**0.09**	0.10	**0.10**
Ovaries	0.05	**0.05**	0.06	**0.06**	0.07	**0.07**	0.07	**0.08**	0.08	**0.09**	0.09	**0.10**
Prostate	0.06	**0.06**	0.07	**0.07**	0.08	**0.08**	0.08	**0.08**	0.09	**0.09**	0.10	**0.10**
Testes	0.12	**0.12**	0.13	**0.13**	0.13	**0.14**	0.13	**0.14**	0.14	**0.14**	0.14	**0.15**
Upper arm		**0.07**		**0.08**		**0.09**		**0.10**		**0.11**		**0.12**
Elbow		**0.16**		**0.19**		**0.22**		**0.23**		**0.26**		**0.28**
Forearm		**0.18**		**0.21**		**0.23**		**0.25**		**0.28**		**0.30**
Upper leg	0.09	**0.13**	0.11	**0.16**	0.12	**0.18**	0.13	**0.20**	0.15	**0.23**	0.17	**0.26**
Knee		**0.04**		**0.05**		**0.06**		**0.07**		**0.08**	0.01	**0.08**
Lower leg												
Lymph nodes	0.03	**0.04**	0.04	**0.05**	0.05	**0.05**	0.05	**0.06**	0.06	**0.07**	0.06	**0.07**
Skin	0.03	**0.04**	0.03	**0.04**	0.03	**0.04**	0.03	**0.04**	0.03	**0.04**	0.03	**0.05**
Muscle	0.03	**0.04**	0.03	**0.04**	0.03	**0.04**	0.03	**0.05**	0.04	**0.05**	0.04	**0.06**
Skeleton	0.03	**0.07**	0.04	**0.09**	0.04	**0.10**	0.05	**0.11**	0.05	**0.12**	0.06	**0.13**
Red bone marrow		**0.02**	0.01	**0.02**	0.01	**0.02**	0.01	**0.03**	0.02	**0.03**	0.02	**0.03**
Whole body dose	0.03	**0.04**	0.03	**0.04**	0.03	**0.05**	0.03	**0.05**	0.04	**0.06**	0.04	**0.06**
ED (ICRP 60)	0.04	**0.06**	0.05	**0.07**	0.06	**0.07**	0.06	**0.08**	0.07	**0.09**	0.07	**0.09**
ED (ICRP 103)	0.04	**0.06**	0.04	**0.07**	0.05	**0.07**	0.05	**0.08**	0.05	**0.08**	0.06	**0.09**
Total filtration (mm) Al	**1.5**		**2.5**		**3.5**		**4.5**		**3.5**		**3.5**	
Cu	**0**		**0**		**0**		**0**		**0.1**		**0.2**	

Table 12.84 Normalized organ doses—Abdomen ap—28th gestation week

Beam projection:			ap									**Table cells:**
Focus-to-detector distance (cm):			115						(+) = normal font = optimal field setting			
Tube voltage (kVp):			60						**(−) = bold font = suboptimal field setting**			
Total filtrations:			See last table row.						Empty table cells correspond to values <0.01			
Organ or tissue	**Organ dose D_T : Entrance dose K_E**								**(mSv * mGy^{-1})**			
	(+)	(−)	(+)	(−)	(+)	(−)	(+)	(−)	(+)	(−)	(+)	(−)
Skull												
Brain												
Oral mucosa												
Salivary glands												
Cervical spine												
Thyroid												
Thymus		**0.03**		**0.03**		**0.03**		**0.04**		**0.04**		**0.04**
Thoracic spine	0.03	**0.05**	0.04	**0.07**	0.04	**0.08**	0.05	**0.09**	0.06	**0.11**	0.06	**0.10**
Claviculae				**0.01**		**0.01**		**0.01**		**0.02**		**0.02**
Scapulae		**0.06**		**0.07**		**0.08**		**0.09**	0.01	**0.11**	0.01	**0.11**
Ribs	0.09	**0.17**	0.10	**0.19**	0.11	**0.20**	0.11	**0.21**	0.12	**0.23**	0.12	**0.23**
Oesophagus	0.02	**0.04**	0.03	**0.05**	0.03	**0.05**	0.03	**0.06**	0.04	**0.07**	0.04	**0.06**
Trachea												
Lung	0.02	**0.06**	0.02	**0.07**	0.03	**0.07**	0.03	**0.08**	0.03	**0.08**	0.03	**0.08**
Mamma		**0.13**		**0.13**		**0.14**		**0.14**		**0.14**		**0.14**
Heart	0.03	**0.08**	0.03	**0.09**	0.04	**0.10**	0.04	**0.10**	0.04	**0.11**	0.04	**0.11**
Stomach	0.08	**0.08**	0.09	**0.09**	0.10	**0.10**	0.10	**0.11**	0.11	**0.12**	0.11	**0.12**
Spleen	0.04	**0.04**	0.05	**0.05**	0.06	**0.06**	0.06	**0.06**	0.07	**0.07**	0.07	**0.07**
Liver	0.07	**0.07**	0.08	**0.08**	0.09	**0.09**	0.09	**0.10**	0.10	**0.11**	0.10	**0.11**
Gall bladder	0.07	**0.07**	0.08	**0.09**	0.09	**0.10**	0.10	**0.10**	0.11	**0.11**	0.11	**0.11**
Pancreas	0.05	**0.05**	0.06	**0.07**	0.07	**0.07**	0.08	**0.08**	0.09	**0.09**	0.08	**0.09**
Small intestine	0.06	**0.06**	0.07	**0.08**	0.08	**0.08**	0.09	**0.09**	0.10	**0.10**	0.10	**0.10**
Large intestine	0.07	**0.07**	0.08	**0.09**	0.09	**0.09**	0.10	**0.10**	0.11	**0.11**	0.11	**0.11**
Sigma/rectum	0.06	**0.06**	0.07	**0.07**	0.08	**0.08**	0.08	**0.09**	0.09	**0.10**	0.09	**0.10**
Lumbar spine	0.08	**0.08**	0.10	**0.10**	0.11	**0.12**	0.13	**0.13**	0.15	**0.16**	0.15	**0.15**
Pelvis	0.11	**0.12**	0.14	**0.14**	0.16	**0.16**	0.17	**0.18**	0.20	**0.21**	0.20	**0.21**
Adrenals	0.03	**0.03**	0.03	**0.04**	0.04	**0.04**	0.04	**0.05**	0.05	**0.06**	0.05	**0.05**
Kidneys	0.03	**0.03**	0.04	**0.04**	0.04	**0.04**	0.04	**0.05**	0.05	**0.06**	0.05	**0.05**
Urinary bladder	0.08	**0.09**	0.10	**0.10**	0.10	**0.11**	0.11	**0.11**	0.12	**0.12**	0.12	**0.12**
Uterus	0.06	**0.06**	0.08	**0.08**	0.09	**0.09**	0.09	**0.09**	0.10	**0.11**	0.10	**0.10**
Ovaries	0.06	**0.06**	0.07	**0.07**	0.08	**0.08**	0.08	**0.09**	0.09	**0.10**	0.09	**0.10**
Prostate	0.06	**0.07**	0.08	**0.08**	0.08	**0.09**	0.09	**0.09**	0.10	**0.10**	0.10	**0.10**
Testes	0.12	**0.13**	0.13	**0.14**	0.13	**0.14**	0.14	**0.14**	0.14	**0.15**	0.14	**0.15**
Upper arm		**0.08**		**0.09**		**0.10**		**0.11**		**0.12**		**0.12**
Elbow		**0.18**		**0.22**		**0.24**		**0.25**		**0.28**		**0.28**
Forearm		**0.20**		**0.23**		**0.26**		**0.27**		**0.30**		**0.30**
Upper leg	0.10	**0.15**	0.12	**0.19**	0.14	**0.21**	0.15	**0.23**	0.17	**0.26**	0.17	**0.26**
Knee		**0.05**		**0.06**		**0.07**		**0.07**	0.01	**0.08**	0.01	**0.08**
Lower leg												
Lymph nodes	0.04	**0.05**	0.05	**0.05**	0.05	**0.06**	0.05	**0.06**	0.06	**0.07**	0.06	**0.07**
Skin	0.03	**0.04**	0.03	**0.04**	0.03	**0.04**	0.03	**0.04**	0.03	**0.05**	0.03	**0.05**
Muscle	0.03	**0.04**	0.03	**0.05**	0.04	**0.05**	0.04	**0.05**	0.04	**0.06**	0.04	**0.06**
Skeleton	0.03	**0.08**	0.04	**0.10**	0.05	**0.11**	0.05	**0.12**	0.06	**0.13**	0.06	**0.13**
Red bone marrow	0.01	**0.02**	0.01	**0.02**	0.02	**0.03**	0.02	**0.03**	0.02	**0.03**	0.02	**0.03**
Whole body dose	0.03	**0.04**	0.03	**0.05**	0.04	**0.05**	0.04	**0.06**	0.04	**0.06**	0.04	**0.06**
ED (ICRP 60)	0.05	**0.06**	0.06	**0.07**	0.06	**0.08**	0.07	**0.08**	0.07	**0.09**	0.07	**0.09**
ED (ICRP 103)	0.04	**0.06**	0.05	**0.07**	0.05	**0.08**	0.05	**0.08**	0.06	**0.09**	0.06	**0.09**
Total filtration (mm) Al	**1.5**		**2.5**		**3.5**		**4.5**		**3.5**		**3.5**	
Cu	**0**		**0**		**0**		**0**		**0.1**		**0.2**	

Table 12.85 Normalized organ doses—Abdomen ap—28th gestation week

Beam projection:		ap										**Table cells:**
Focus-to-detector distance (cm):		115						(+) = normal font = optimal field setting				
Tube voltage (kVp):		70						**(−) = bold font = suboptimal field setting**				
Total filtrations:		See last table row.						Empty table cells correspond to values <0.01				
Organ or tissue	**Organ dose D_T : Entrance dose K_E**							**(mSv * mGy⁻¹)**				
	(+)	(−)	(+)	(−)	(+)	(−)	(+)	(−)	(+)	(−)	(+)	(−)
Skull												
Brain												
Oral mucosa												
Salivary glands												
Cervical spine												
Thyroid												
Thymus		**0.03**		**0.03**		**0.04**		**0.04**		**0.04**		**0.04**
Thoracic spine	0.03	**0.06**	0.04	**0.08**	0.05	**0.09**	0.05	**0.10**	0.06	**0.12**	0.07	**0.13**
Claviculae				**0.01**		**0.01**		**0.01**		**0.02**		**0.02**
Scapulae		**0.07**		**0.08**		**0.09**		**0.10**	0.01	**0.11**	0.01	**0.12**
Ribs	0.09	**0.18**	0.10	**0.20**	0.11	**0.21**	0.11	**0.22**	0.12	**0.23**	0.12	**0.24**
Oesophagus	0.03	**0.04**	0.03	**0.05**	0.03	**0.06**	0.04	**0.06**	0.04	**0.07**	0.05	**0.08**
Trachea												
Lung	0.02	**0.06**	0.03	**0.07**	0.03	**0.08**	0.03	**0.08**	0.03	**0.09**	0.03	**0.09**
Mamma		**0.13**		**0.14**		**0.14**		**0.14**		**0.14**		**0.14**
Heart	0.03	**0.08**	0.04	**0.10**	0.04	**0.10**	0.04	**0.11**	0.04	**0.12**	0.05	**0.12**
Stomach	0.08	**0.09**	0.10	**0.10**	0.10	**0.11**	0.11	**0.11**	0.12	**0.12**	0.12	**0.13**
Spleen	0.05	**0.05**	0.05	**0.06**	0.06	**0.07**	0.07	**0.07**	0.07	**0.08**	0.08	**0.09**
Liver	0.08	**0.08**	0.09	**0.09**	0.09	**0.10**	0.10	**0.10**	0.11	**0.11**	0.11	**0.12**
Gall bladder	0.08	**0.08**	0.09	**0.09**	0.10	**0.10**	0.11	**0.11**	0.12	**0.12**	0.12	**0.13**
Pancreas	0.06	**0.06**	0.07	**0.07**	0.08	**0.08**	0.08	**0.09**	0.09	**0.10**	0.10	**0.11**
Small intestine	0.07	**0.07**	0.08	**0.08**	0.09	**0.09**	0.10	**0.10**	0.11	**0.11**	0.11	**0.12**
Large intestine	0.08	**0.08**	0.09	**0.09**	0.10	**0.10**	0.11	**0.11**	0.12	**0.12**	0.12	**0.13**
Sigma/rectum	0.07	**0.07**	0.08	**0.08**	0.08	**0.09**	0.09	**0.09**	0.10	**0.10**	0.11	**0.11**
Lumbar spine	0.09	**0.09**	0.11	**0.12**	0.13	**0.13**	0.14	**0.15**	0.16	**0.17**	0.18	**0.19**
Pelvis	0.13	**0.13**	0.15	**0.16**	0.17	**0.18**	0.19	**0.19**	0.21	**0.22**	0.23	**0.24**
Adrenals	0.03	**0.04**	0.04	**0.04**	0.04	**0.05**	0.05	**0.05**	0.06	**0.06**	0.06	**0.07**
Kidneys	0.03	**0.03**	0.04	**0.04**	0.05	**0.05**	0.05	**0.05**	0.06	**0.06**	0.06	**0.07**
Urinary bladder	0.09	**0.09**	0.10	**0.10**	0.11	**0.11**	0.12	**0.12**	0.12	**0.13**	0.13	**0.14**
Uterus	0.07	**0.07**	0.08	**0.09**	0.09	**0.09**	0.10	**0.10**	0.11	**0.11**	0.12	**0.12**
Ovaries	0.06	**0.07**	0.07	**0.08**	0.08	**0.09**	0.09	**0.09**	0.10	**0.10**	0.11	**0.11**
Prostate	0.07	**0.07**	0.08	**0.09**	0.09	**0.09**	0.09	**0.10**	0.10	**0.11**	0.11	**0.12**
Testes	0.13	**0.13**	0.13	**0.14**	0.14	**0.14**	0.14	**0.15**	0.14	**0.15**	0.14	**0.16**
Upper arm		**0.09**		**0.10**		**0.11**		**0.11**		**0.13**		**0.13**
Elbow		**0.20**		**0.23**		**0.25**		**0.26**		**0.29**		**0.30**
Forearm		**0.21**		**0.25**		**0.27**		**0.28**		**0.30**		**0.32**
Upper leg	0.11	**0.17**	0.13	**0.20**	0.15	**0.22**	0.16	**0.24**	0.18	**0.27**	0.19	**0.29**
Knee		**0.06**		**0.07**		**0.07**	0.01	**0.08**	0.01	**0.09**	0.01	**0.10**
Lower leg												
Lymph nodes	0.04	**0.05**	0.05	**0.06**	0.05	**0.07**	0.06	**0.07**	0.06	**0.08**	0.07	**0.08**
Skin	0.03	**0.04**	0.03	**0.04**	0.03	**0.04**	0.03	**0.05**	0.03	**0.05**	0.03	**0.05**
Muscle	0.03	**0.04**	0.04	**0.05**	0.04	**0.05**	0.04	**0.06**	0.04	**0.06**	0.05	**0.06**
Skeleton	0.04	**0.09**	0.05	**0.11**	0.05	**0.12**	0.05	**0.12**	0.06	**0.14**	0.07	**0.15**
Red bone marrow	0.01	**0.02**	0.02	**0.03**	0.02	**0.03**	0.02	**0.03**	0.02	**0.04**	0.02	**0.04**
Whole body dose	0.03	**0.05**	0.04	**0.05**	0.04	**0.06**	0.04	**0.06**	0.04	**0.07**	0.05	**0.07**
ED (ICRP 60)	0.05	**0.07**	0.06	**0.08**	0.07	**0.08**	0.07	**0.09**	0.07	**0.10**	0.08	**0.10**
ED (ICRP 103)	0.04	**0.07**	0.05	**0.08**	0.05	**0.08**	0.06	**0.09**	0.06	**0.09**	0.07	**0.10**
Total filtration (mm) Al	**1.5**		**2.5**		**3.5**		**4.5**		**3.5**		**3.5**	
Cu	**0**		**0**		**0**		**0**		**0.1**		**0.2**	

Table 12.86 Normalized organ doses—Abdomen ap—<1 year

Beam projection:			ap									**Table cells:**
Focus-to-detector distance (cm):			115									(+) = normal font = optimal field setting
Tube voltage (kVp):			60									**(−) = bold font = suboptimal field setting**
Total filtrations:			See last table row.									Empty table cells correspond to values <0.01
Organ or tissue	**Organ dose D_T : Entrance dose K_E**								**(mSv * mGy^{-1})**			
	(+)	(−)	(+)	(−)	(+)	(−)	(+)	(−)	(+)	(−)	(+)	(−)
Skull												
Brain												
Oral mucosa												
Salivary glands								**0.01**		**0.01**		**0.01**
Cervical spine		**0.02**		**0.03**	0.01	**0.03**	0.01	**0.04**	0.02	**0.05**	0.01	**0.04**
Thyroid		**0.01**		**0.02**		**0.02**		**0.03**	0.01	**0.03**		**0.03**
Thymus	0.01	**0.08**	0.02	**0.10**	0.02	**0.11**	0.02	**0.12**	0.03	**0.14**	0.02	**0.13**
Thoracic spine	0.14	**0.24**	0.19	**0.32**	0.22	**0.38**	0.25	**0.43**	0.32	**0.53**	0.29	**0.49**
Claviculae	0.01	**0.04**	0.02	**0.06**	0.02	**0.07**	0.03	**0.08**	0.03	**0.10**	0.03	**0.09**
Scapulae	0.04	**0.33**	0.06	**0.43**	0.07	**0.50**	0.08	**0.56**	0.10	**0.66**	0.09	**0.64**
Ribs	0.61	**1.05**	0.70	**1.21**	0.76	**1.31**	0.81	**1.39**	0.88	**1.52**	0.88	**1.52**
Oesophagus	0.12	**0.18**	0.15	**0.23**	0.18	**0.27**	0.20	**0.31**	0.24	**0.37**	0.22	**0.35**
Trachea												
Lung	0.15	**0.35**	0.18	**0.42**	0.21	**0.47**	0.22	**0.51**	0.26	**0.58**	0.25	**0.57**
Mamma	0.03	**1.16**	0.04	**1.21**	0.05	**1.24**	0.06	**1.26**	0.07	**1.30**	0.07	**1.29**
Heart	0.21	**0.55**	0.25	**0.65**	0.28	**0.73**	0.31	**0.79**	0.35	**0.89**	0.34	**0.87**
Stomach	0.59	**0.61**	0.71	**0.74**	0.79	**0.82**	0.86	**0.89**	0.96	**1.01**	0.95	**0.99**
Spleen	0.22	**0.24**	0.28	**0.31**	0.33	**0.36**	0.36	**0.40**	0.43	**0.48**	0.41	**0.45**
Liver	0.52	**0.54**	0.62	**0.65**	0.69	**0.73**	0.74	**0.79**	0.84	**0.89**	0.83	**0.87**
Gall bladder	0.45	**0.46**	0.56	**0.58**	0.64	**0.66**	0.70	**0.72**	0.81	**0.84**	0.79	**0.82**
Pancreas	0.31	**0.33**	0.39	**0.42**	0.46	**0.49**	0.51	**0.54**	0.60	**0.64**	0.57	**0.61**
Small intestine	0.40	**0.41**	0.50	**0.52**	0.58	**0.59**	0.63	**0.65**	0.74	**0.76**	0.71	**0.73**
Large intestine	0.48	**0.50**	0.60	**0.61**	0.68	**0.70**	0.74	**0.76**	0.85	**0.88**	0.83	**0.85**
Sigma/rectum	0.37	**0.39**	0.46	**0.48**	0.53	**0.56**	0.59	**0.61**	0.68	**0.72**	0.66	**0.69**
Lumbar spine	0.34	**0.36**	0.45	**0.48**	0.54	**0.57**	0.62	**0.65**	0.76	**0.81**	0.70	**0.74**
Pelvis	0.58	**0.62**	0.75	**0.79**	0.87	**0.92**	0.97	**1.03**	1.16	**1.23**	1.12	**1.18**
Adrenals	0.13	**0.14**	0.17	**0.19**	0.20	**0.22**	0.22	**0.25**	0.27	**0.31**	0.25	**0.29**
Kidneys	0.14	**0.15**	0.18	**0.19**	0.21	**0.23**	0.24	**0.26**	0.29	**0.31**	0.27	**0.29**
Urinary bladder	0.62	**0.63**	0.74	**0.75**	0.83	**0.84**	0.89	**0.90**	1.00	**1.01**	0.99	**1.00**
Uterus	0.40	**0.42**	0.50	**0.52**	0.58	**0.60**	0.64	**0.66**	0.75	**0.78**	0.72	**0.75**
Ovaries	0.33	**0.35**	0.42	**0.45**	0.49	**0.52**	0.55	**0.58**	0.65	**0.69**	0.62	**0.65**
Prostate	0.47	**0.50**	0.58	**0.61**	0.66	**0.70**	0.72	**0.76**	0.82	**0.87**	0.80	**0.86**
Testes	1.05	**1.08**	1.14	**1.18**	1.20	**1.24**	1.23	**1.28**	1.29	**1.35**	1.29	**1.34**
Upper arm		**0.43**		**0.52**		**0.59**		**0.64**		**0.73**		**0.72**
Elbow		**1.20**		**1.45**		**1.63**		**1.76**		**1.99**		**1.98**
Forearm		**1.33**		**1.59**		**1.78**		**1.91**		**2.14**		**2.15**
Upper leg	0.71	**0.98**	0.89	**1.23**	1.02	**1.41**	1.13	**1.55**	1.31	**1.81**	1.29	**1.78**
Knee	0.05	**0.44**	0.07	**0.55**	0.08	**0.63**	0.10	**0.69**	0.12	**0.80**	0.11	**0.79**
Lower leg				**0.01**		**0.02**		**0.02**	0.01	**0.02**		**0.02**
Lymph nodes	0.25	**0.30**	0.32	**0.37**	0.36	**0.42**	0.39	**0.46**	0.46	**0.53**	0.44	**0.52**
Skin	0.23	**0.33**	0.24	**0.35**	0.25	**0.36**	0.26	**0.38**	0.28	**0.40**	0.27	**0.39**
Muscle	0.23	**0.31**	0.27	**0.37**	0.30	**0.40**	0.32	**0.43**	0.36	**0.49**	0.35	**0.47**
Skeleton	0.23	**0.54**	0.29	**0.66**	0.33	**0.75**	0.36	**0.82**	0.42	**0.93**	0.40	**0.92**
Red bone marrow	0.06	**0.13**	0.07	**0.16**	0.09	**0.19**	0.10	**0.20**	0.11	**0.24**	0.11	**0.23**
Whole body dose	0.21	**0.31**	0.25	**0.36**	0.28	**0.41**	0.30	**0.44**	0.34	**0.49**	0.33	**0.48**
ED (ICRP 60)	0.36	**0.47**	0.43	**0.55**	0.48	**0.61**	0.51	**0.65**	0.58	**0.73**	0.57	**0.71**
ED (ICRP 103)	0.29	**0.48**	0.35	**0.55**	0.39	**0.61**	0.42	**0.65**	0.48	**0.72**	0.47	**0.70**
Total filtration (mm) Al	**1.5**		**2.5**		**3.5**		**4.5**		**3.5**		**3.5**	
Cu	**0**		**0**		**0**		**0**		**0.1**		**0.2**	

Table 12.87 Normalized organ doses—Abdomen ap—<1 year

Beam projection:			ap									**Table cells:**
Focus-to-detector distance (cm):			115									(+) = normal font = optimal field setting
Tube voltage (kVp):			70									**(−) = bold font = suboptimal field setting**
Total filtrations:			See last table row.									Empty table cells correspond to values <0.01
Organ or tissue	**Organ dose D_T : Entrance dose K_E**								**(mSv * mGy^{-1})**			
	(+)	(−)	(+)	(−)	(+)	(−)	(+)	(−)	(+)	(−)	(+)	(−)
Skull										**0.01**		**0.01**
Brain												
Oral mucosa										**0.01**		**0.01**
Salivary glands				**0.01**		**0.01**		**0.01**		**0.02**		**0.02**
Cervical spine		**0.03**	0.01	**0.03**	0.01	**0.04**	0.02	**0.05**	0.02	**0.06**	0.03	**0.07**
Thyroid		**0.02**		**0.02**		**0.03**	0.01	**0.03**	0.01	**0.04**	0.02	**0.04**
Thymus	0.01	**0.09**	0.02	**0.11**	0.02	**0.12**	0.03	**0.13**	0.03	**0.15**	0.04	**0.17**
Thoracic spine	0.18	**0.30**	0.23	**0.39**	0.27	**0.46**	0.30	**0.51**	0.37	**0.62**	0.43	**0.72**
Claviculae	0.02	**0.06**	0.02	**0.07**	0.03	**0.09**	0.03	**0.10**	0.04	**0.12**	0.05	**0.14**
Scapulae	0.06	**0.40**	0.07	**0.50**	0.09	**0.57**	0.10	**0.63**	0.12	**0.73**	0.14	**0.82**
Ribs	0.66	**1.13**	0.75	**1.28**	0.80	**1.38**	0.84	**1.45**	0.91	**1.57**	0.96	**1.65**
Oesophagus	0.14	**0.22**	0.18	**0.28**	0.21	**0.32**	0.23	**0.35**	0.27	**0.42**	0.30	**0.47**
Trachea										**0.01**		**0.01**
Lung	0.17	**0.39**	0.21	**0.47**	0.23	**0.52**	0.25	**0.56**	0.28	**0.62**	0.31	**0.68**
Mamma	0.04	**1.18**	0.05	**1.24**	0.06	**1.26**	0.07	**1.28**	0.08	**1.31**	0.09	**1.34**
Heart	0.23	**0.61**	0.28	**0.72**	0.31	**0.80**	0.34	**0.85**	0.38	**0.95**	0.42	**1.03**
Stomach	0.66	**0.69**	0.78	**0.81**	0.86	**0.90**	0.93	**0.97**	1.03	**1.08**	1.12	**1.18**
Spleen	0.26	**0.29**	0.33	**0.36**	0.38	**0.42**	0.42	**0.46**	0.49	**0.54**	0.56	**0.62**
Liver	0.58	**0.61**	0.68	**0.72**	0.75	**0.80**	0.81	**0.86**	0.90	**0.96**	0.98	**1.05**
Gall bladder	0.52	**0.54**	0.63	**0.65**	0.71	**0.74**	0.78	**0.81**	0.89	**0.93**	0.99	**1.03**
Pancreas	0.37	**0.39**	0.46	**0.49**	0.52	**0.56**	0.57	**0.62**	0.67	**0.73**	0.76	**0.82**
Small intestine	0.47	**0.48**	0.57	**0.59**	0.65	**0.67**	0.71	**0.73**	0.82	**0.85**	0.91	**0.94**
Large intestine	0.55	**0.57**	0.67	**0.69**	0.75	**0.78**	0.82	**0.84**	0.93	**0.96**	1.03	**1.06**
Sigma/rectum	0.43	**0.45**	0.53	**0.56**	0.60	**0.63**	0.66	**0.69**	0.76	**0.80**	0.84	**0.89**
Lumbar spine	0.43	**0.46**	0.56	**0.59**	0.65	**0.69**	0.73	**0.78**	0.89	**0.95**	1.03	**1.10**
Pelvis	0.69	**0.73**	0.87	**0.92**	1.00	**1.06**	1.10	**1.17**	1.29	**1.37**	1.46	**1.55**
Adrenals	0.16	**0.18**	0.20	**0.23**	0.24	**0.27**	0.26	**0.30**	0.32	**0.36**	0.37	**0.42**
Kidneys	0.17	**0.18**	0.22	**0.23**	0.26	**0.27**	0.29	**0.31**	0.34	**0.37**	0.40	**0.43**
Urinary bladder	0.69	**0.70**	0.81	**0.83**	0.90	**0.91**	0.96	**0.98**	1.07	**1.09**	1.16	**1.18**
Uterus	0.47	**0.49**	0.58	**0.60**	0.66	**0.68**	0.72	**0.74**	0.83	**0.86**	0.92	**0.96**
Ovaries	0.39	**0.42**	0.49	**0.52**	0.56	**0.60**	0.62	**0.66**	0.72	**0.77**	0.82	**0.87**
Prostate	0.54	**0.57**	0.65	**0.69**	0.73	**0.77**	0.79	**0.84**	0.89	**0.95**	0.98	**1.04**
Testes	1.09	**1.13**	1.18	**1.23**	1.23	**1.29**	1.27	**1.33**	1.32	**1.39**	1.37	**1.44**
Upper arm		**0.48**		**0.57**		**0.64**		**0.69**		**0.77**		**0.84**
Elbow		**1.34**		**1.58**		**1.75**		**1.88**		**2.09**		**2.25**
Forearm		**1.46**		**1.72**		**1.90**		**2.03**		**2.24**		**2.40**
Upper leg	0.82	**1.13**	1.00	**1.39**	1.14	**1.57**	1.24	**1.71**	1.42	**1.95**	1.57	**2.16**
Knee	0.07	**0.51**	0.09	**0.62**	0.10	**0.70**	0.12	**0.76**	0.14	**0.87**	0.17	**0.96**
Lower leg		**0.01**		**0.02**	0.01	**0.02**	0.01	**0.02**	0.01	**0.03**	0.02	**0.03**
Lymph nodes	0.29	**0.35**	0.36	**0.42**	0.40	**0.47**	0.44	**0.51**	0.50	**0.59**	0.56	**0.65**
Skin	0.24	**0.35**	0.25	**0.37**	0.26	**0.38**	0.27	**0.39**	0.29	**0.41**	0.30	**0.43**
Muscle	0.25	**0.34**	0.30	**0.40**	0.32	**0.44**	0.35	**0.47**	0.39	**0.53**	0.42	**0.57**
Skeleton	0.27	**0.61**	0.32	**0.74**	0.37	**0.82**	0.40	**0.89**	0.46	**1.00**	0.51	**1.10**
Red bone marrow	0.07	**0.16**	0.09	**0.19**	0.10	**0.21**	0.11	**0.23**	0.13	**0.27**	0.15	**0.30**
Whole body dose	0.24	**0.34**	0.28	**0.40**	0.31	**0.44**	0.33	**0.47**	0.37	**0.53**	0.41	**0.58**
ED (ICRP 60)	0.41	**0.52**	0.47	**0.61**	0.52	**0.66**	0.56	**0.71**	0.62	**0.78**	0.68	**0.85**
ED (ICRP 103)	0.33	**0.52**	0.39	**0.60**	0.43	**0.65**	0.46	**0.70**	0.52	**0.77**	0.57	**0.83**
Total filtration (mm) Al	**1.5**		**2.5**		**3.5**		**4.5**		**3.5**		**3.5**	
Cu	**0**		**0**		**0**		**0**		**0.1**		**0.2**	

Table 12.88 Normalized organ doses—Abdomen ap—1–2 years

Beam projection:			ap									**Table cells:**
Focus-to-detector distance (cm):			115									(+) = normal font = optimal field setting
Tube voltage (kVp):			60									**(−) = bold font = suboptimal field setting**
Total filtrations:			See last table row.									Empty table cells correspond to values <0.01
Organ or tissue	**Organ dose D_T : Entrance dose K_E**								**(mSv * mGy^{-1})**			
	(+)	(−)	(+)	(−)	(+)	(−)	(+)	(−)	(+)	(−)	(+)	(−)
Skull												
Brain												
Oral mucosa												
Salivary glands												
Cervical spine		**0.01**		**0.01**		**0.02**		**0.02**		**0.03**		**0.02**
Thyroid		**0.01**		**0.01**		**0.02**		**0.02**		**0.03**		**0.02**
Thymus		**0.10**	0.01	**0.13**	0.01	**0.14**	0.02	**0.16**	0.02	**0.19**	0.02	**0.18**
Thoracic spine	0.08	**0.15**	0.11	**0.20**	0.13	**0.25**	0.15	**0.28**	0.19	**0.36**	0.16	**0.31**
Claviculae		**0.03**		**0.04**	0.01	**0.05**	0.01	**0.06**	0.02	**0.08**	0.01	**0.07**
Scapulae	0.02	**0.19**	0.03	**0.25**	0.03	**0.30**	0.04	**0.33**	0.05	**0.41**	0.04	**0.38**
Ribs	0.49	**0.99**	0.57	**1.15**	0.63	**1.27**	0.67	**1.35**	0.74	**1.50**	0.73	**1.49**
Oesophagus	0.09	**0.16**	0.12	**0.21**	0.15	**0.25**	0.16	**0.28**	0.20	**0.34**	0.18	**0.31**
Trachea												
Lung	0.09	**0.29**	0.12	**0.36**	0.13	**0.41**	0.15	**0.44**	0.17	**0.51**	0.16	**0.50**
Mamma	0.02	**1.12**	0.02	**1.20**	0.03	**1.24**	0.03	**1.27**	0.04	**1.31**	0.03	**1.31**
Heart	0.13	**0.48**	0.16	**0.59**	0.19	**0.67**	0.21	**0.73**	0.25	**0.84**	0.23	**0.81**
Stomach	0.49	**0.52**	0.60	**0.63**	0.68	**0.72**	0.74	**0.78**	0.85	**0.90**	0.83	**0.87**
Spleen	0.15	**0.17**	0.20	**0.23**	0.23	**0.27**	0.26	**0.30**	0.32	**0.37**	0.30	**0.34**
Liver	0.41	**0.44**	0.50	**0.54**	0.57	**0.62**	0.62	**0.67**	0.72	**0.78**	0.70	**0.75**
Gall bladder	0.41	**0.41**	0.51	**0.53**	0.59	**0.61**	0.66	**0.68**	0.78	**0.80**	0.74	**0.76**
Pancreas	0.23	**0.25**	0.30	**0.33**	0.35	**0.39**	0.40	**0.44**	0.48	**0.53**	0.45	**0.49**
Small intestine	0.32	**0.33**	0.41	**0.42**	0.48	**0.49**	0.54	**0.55**	0.64	**0.66**	0.60	**0.62**
Large intestine	0.40	**0.41**	0.50	**0.51**	0.58	**0.59**	0.64	**0.65**	0.75	**0.77**	0.72	**0.74**
Sigma/rectum	0.29	**0.30**	0.37	**0.38**	0.43	**0.44**	0.48	**0.50**	0.58	**0.59**	0.54	**0.56**
Lumbar spine	0.21	**0.22**	0.29	**0.30**	0.35	**0.37**	0.40	**0.42**	0.51	**0.54**	0.44	**0.46**
Pelvis	0.41	**0.44**	0.54	**0.57**	0.64	**0.68**	0.73	**0.77**	0.89	**0.94**	0.82	**0.87**
Adrenals	0.09	**0.10**	0.11	**0.13**	0.14	**0.16**	0.16	**0.18**	0.20	**0.23**	0.17	**0.20**
Kidneys	0.09	**0.09**	0.12	**0.13**	0.14	**0.15**	0.16	**0.17**	0.20	**0.22**	0.17	**0.19**
Urinary bladder	0.54	**0.55**	0.66	**0.67**	0.75	**0.76**	0.82	**0.82**	0.94	**0.95**	0.91	**0.92**
Uterus	0.33	**0.34**	0.42	**0.43**	0.49	**0.51**	0.55	**0.57**	0.66	**0.68**	0.62	**0.64**
Ovaries	0.26	**0.26**	0.34	**0.34**	0.40	**0.41**	0.45	**0.46**	0.54	**0.56**	0.50	**0.52**
Prostate	0.37	**0.38**	0.47	**0.48**	0.54	**0.55**	0.59	**0.61**	0.70	**0.72**	0.67	**0.69**
Testes	1.01	**1.02**	1.12	**1.13**	1.18	**1.20**	1.22	**1.24**	1.30	**1.32**	1.29	**1.31**
Upper arm		**0.31**		**0.39**		**0.45**		**0.49**		**0.58**		**0.56**
Elbow		**0.89**		**1.10**		**1.26**		**1.38**		**1.61**		**1.57**
Forearm		**1.00**		**1.24**		**1.41**		**1.54**		**1.78**		**1.75**
Upper leg	0.47	**0.63**	0.61	**0.81**	0.71	**0.95**	0.80	**1.06**	0.97	**1.28**	0.91	**1.21**
Knee	0.03	**0.08**	0.04	**0.10**	0.05	**0.12**	0.05	**0.14**	0.07	**0.17**	0.06	**0.16**
Lower leg												
Lymph nodes	0.21	**0.25**	0.26	**0.31**	0.30	**0.36**	0.33	**0.40**	0.39	**0.47**	0.37	**0.44**
Skin	0.20	**0.29**	0.21	**0.30**	0.22	**0.31**	0.23	**0.32**	0.24	**0.34**	0.24	**0.33**
Muscle	0.17	**0.23**	0.21	**0.27**	0.23	**0.30**	0.25	**0.33**	0.28	**0.37**	0.27	**0.36**
Skeleton	0.15	**0.36**	0.19	**0.45**	0.22	**0.51**	0.24	**0.56**	0.29	**0.66**	0.27	**0.63**
Red bone marrow	0.04	**0.09**	0.06	**0.11**	0.07	**0.13**	0.07	**0.14**	0.09	**0.17**	0.08	**0.16**
Whole body dose	0.16	**0.23**	0.20	**0.28**	0.22	**0.31**	0.24	**0.34**	0.27	**0.39**	0.26	**0.38**
ED (ICRP 60)	0.31	**0.41**	0.37	**0.48**	0.41	**0.54**	0.45	**0.58**	0.51	**0.66**	0.49	**0.63**
ED (ICRP 103)	0.24	**0.42**	0.29	**0.49**	0.33	**0.54**	0.36	**0.58**	0.41	**0.65**	0.40	**0.63**
Total filtration (mm) **Al**	**1.5**		**2.5**		**3.5**		**4.5**		**3.5**		**3.5**	
Cu	**0**		**0**		**0**		**0**		**0.1**		**0.2**	

Table 12.89 Normalized organ doses—Abdomen ap—1–2 years

Beam projection:			ap									**Table cells:**
Focus-to-detector distance (cm):			115									(+) = normal font = optimal field setting
Tube voltage (kVp):			70									**(−) = bold font = suboptimal field setting**
Total filtrations:			See last table row.									Empty table cells correspond to values <0.01
Organ or tissue	**Organ dose D_T : Entrance dose K_E**							**(mSv * mGy^{-1})**				
	(+)	(−)	(+)	(−)	(+)	(−)	(+)	(−)	(+)	(−)	(+)	(−)
Skull												
Brain												
Oral mucosa												
Salivary glands										**0.01**		**0.01**
Cervical spine		**0.02**		**0.02**		**0.03**		**0.03**		**0.04**	0.01	**0.05**
Thyroid		**0.01**		**0.02**		**0.02**		**0.03**		**0.03**		**0.04**
Thymus	0.01	**0.12**	0.01	**0.15**	0.02	**0.16**	0.02	**0.18**	0.02	**0.21**	0.03	**0.23**
Thoracic spine	0.11	**0.20**	0.14	**0.27**	0.17	**0.32**	0.19	**0.36**	0.24	**0.45**	0.29	**0.54**
Claviculae	0.01	**0.04**	0.01	**0.06**	0.02	**0.07**	0.02	**0.08**	0.02	**0.10**	0.03	**0.12**
Scapulae	0.03	**0.24**	0.04	**0.30**	0.04	**0.35**	0.05	**0.39**	0.06	**0.47**	0.08	**0.54**
Ribs	0.53	**1.07**	0.61	**1.24**	0.67	**1.35**	0.71	**1.43**	0.78	**1.57**	0.84	**1.68**
Oesophagus	0.12	**0.20**	0.15	**0.26**	0.18	**0.30**	0.20	**0.33**	0.24	**0.40**	0.28	**0.47**
Trachea												
Lung	0.11	**0.34**	0.13	**0.41**	0.15	**0.46**	0.17	**0.49**	0.19	**0.57**	0.22	**0.63**
Mamma	0.02	**1.16**	0.03	**1.23**	0.03	**1.27**	0.04	**1.30**	0.05	**1.34**	0.06	**1.37**
Heart	0.16	**0.55**	0.19	**0.66**	0.22	**0.74**	0.24	**0.81**	0.28	**0.92**	0.31	**1.01**
Stomach	0.56	**0.59**	0.68	**0.71**	0.76	**0.80**	0.82	**0.87**	0.93	**0.99**	1.03	**1.09**
Spleen	0.19	**0.22**	0.24	**0.28**	0.28	**0.33**	0.32	**0.36**	0.38	**0.44**	0.44	**0.51**
Liver	0.47	**0.51**	0.57	**0.62**	0.64	**0.69**	0.69	**0.76**	0.79	**0.87**	0.88	**0.96**
Gall bladder	0.48	**0.49**	0.60	**0.62**	0.68	**0.70**	0.75	**0.78**	0.88	**0.91**	0.99	**1.03**
Pancreas	0.29	**0.32**	0.36	**0.40**	0.42	**0.47**	0.47	**0.52**	0.56	**0.62**	0.65	**0.72**
Small intestine	0.39	**0.40**	0.49	**0.50**	0.56	**0.58**	0.62	**0.64**	0.73	**0.75**	0.83	**0.85**
Large intestine	0.47	**0.48**	0.58	**0.60**	0.66	**0.68**	0.72	**0.75**	0.84	**0.87**	0.95	**0.98**
Sigma/rectum	0.35	**0.36**	0.44	**0.45**	0.50	**0.52**	0.56	**0.58**	0.66	**0.68**	0.75	**0.78**
Lumbar spine	0.29	**0.30**	0.38	**0.40**	0.45	**0.48**	0.51	**0.54**	0.64	**0.68**	0.77	**0.82**
Pelvis	0.52	**0.54**	0.66	**0.70**	0.77	**0.81**	0.86	**0.91**	1.04	**1.10**	1.20	**1.28**
Adrenals	0.11	**0.13**	0.15	**0.17**	0.17	**0.20**	0.19	**0.23**	0.24	**0.29**	0.28	**0.34**
Kidneys	0.11	**0.12**	0.15	**0.16**	0.18	**0.19**	0.20	**0.22**	0.25	**0.27**	0.29	**0.32**
Urinary bladder	0.62	**0.63**	0.75	**0.75**	0.83	**0.85**	0.90	**0.92**	1.03	**1.04**	1.13	**1.15**
Uterus	0.40	**0.41**	0.50	**0.52**	0.58	**0.60**	0.64	**0.66**	0.76	**0.78**	0.86	**0.89**
Ovaries	0.32	**0.33**	0.41	**0.42**	0.47	**0.48**	0.53	**0.54**	0.63	**0.65**	0.73	**0.74**
Prostate	0.44	**0.45**	0.54	**0.55**	0.61	**0.63**	0.67	**0.69**	0.78	**0.81**	0.88	**0.90**
Testes	1.07	**1.08**	1.17	**1.19**	1.23	**1.25**	1.27	**1.30**	1.34	**1.38**	1.40	**1.44**
Upper arm		**0.36**		**0.44**		**0.50**		**0.55**		**0.64**		**0.71**
Elbow		**1.02**		**1.24**		**1.40**		**1.52**		**1.75**		**1.93**
Forearm		**1.14**		**1.38**		**1.55**		**1.68**		**1.91**		**2.10**
Upper leg	0.57	**0.76**	0.72	**0.96**	0.83	**1.11**	0.92	**1.23**	1.10	**1.45**	1.25	**1.65**
Knee	0.04	**0.10**	0.05	**0.13**	0.06	**0.15**	0.07	**0.17**	0.09	**0.21**	0.10	**0.24**
Lower leg												
Lymph nodes	0.24	**0.29**	0.30	**0.36**	0.35	**0.41**	0.38	**0.46**	0.45	**0.53**	0.50	**0.60**
Skin	0.21	**0.30**	0.22	**0.31**	0.23	**0.33**	0.24	**0.34**	0.25	**0.36**	0.27	**0.37**
Muscle	0.20	**0.26**	0.23	**0.31**	0.26	**0.34**	0.27	**0.36**	0.31	**0.41**	0.34	**0.45**
Skeleton	0.18	**0.42**	0.22	**0.51**	0.25	**0.58**	0.28	**0.63**	0.33	**0.73**	0.37	**0.82**
Red bone marrow	0.06	**0.11**	0.07	**0.13**	0.08	**0.15**	0.09	**0.17**	0.11	**0.20**	0.13	**0.23**
Whole body dose	0.19	**0.27**	0.22	**0.32**	0.25	**0.35**	0.27	**0.38**	0.30	**0.43**	0.34	**0.48**
ED (ICRP 60)	0.35	**0.46**	0.41	**0.54**	0.46	**0.59**	0.50	**0.64**	0.56	**0.72**	0.62	**0.79**
ED (ICRP 103)	0.27	**0.46**	0.33	**0.54**	0.37	**0.59**	0.40	**0.63**	0.46	**0.71**	0.51	**0.78**
Total filtration (mm) Al	**1.5**		**2.5**		**3.5**		**4.5**		**3.5**		**3.5**	
Cu	**0**		**0**		**0**		**0**		**0.1**		**0.2**	

Table 12.90 Normalized organ doses—Abdomen ap—3–7 years

Beam projection:			ap									**Table cells:**
Focus-to-detector distance (cm):			115									(+) = normal font = optimal field setting
Tube voltage (kVp):			60									**(−) = bold font = suboptimal field setting**
Total filtrations:			See last table row.									Empty table cells correspond to values <0.01
Organ or tissue	**Organ dose D_T : Entrance dose K_E**								**(mSv * mGy⁻¹)**			
	(+)	(−)	(+)	(−)	(+)	(−)	(+)	(−)	(+)	(−)	(+)	(−)
Skull												
Brain												
Oral mucosa												
Salivary glands												
Cervical spine						**0.01**		**0.01**		**0.02**		**0.01**
Thyroid										**0.01**		
Thymus		**0.13**		**0.16**	0.01	**0.19**	0.01	**0.21**	0.02	**0.24**	0.01	**0.23**
Thoracic spine	0.06	**0.11**	0.08	**0.15**	0.10	**0.19**	0.11	**0.22**	0.15	**0.28**	0.12	**0.23**
Claviculae		**0.02**		**0.03**		**0.04**		**0.04**	0.01	**0.05**		**0.04**
Scapulae	0.02	**0.15**	0.02	**0.20**	0.03	**0.24**	0.03	**0.27**	0.04	**0.34**	0.03	**0.31**
Ribs	0.46	**0.92**	0.55	**1.08**	0.60	**1.19**	0.64	**1.28**	0.72	**1.42**	0.71	**1.40**
Oesophagus	0.06	**0.10**	0.08	**0.13**	0.09	**0.15**	0.11	**0.17**	0.13	**0.22**	0.12	**0.19**
Trachea												
Lung	0.09	**0.26**	0.11	**0.32**	0.13	**0.37**	0.14	**0.41**	0.17	**0.48**	0.16	**0.46**
Mamma	0.02	**1.11**	0.02	**1.19**	0.02	**1.24**	0.03	**1.27**	0.04	**1.32**	0.03	**1.31**
Heart	0.13	**0.42**	0.17	**0.53**	0.20	**0.60**	0.22	**0.66**	0.26	**0.78**	0.24	**0.74**
Stomach	0.48	**0.50**	0.60	**0.62**	0.68	**0.71**	0.74	**0.78**	0.86	**0.91**	0.83	**0.87**
Spleen	0.12	**0.14**	0.16	**0.18**	0.19	**0.22**	0.22	**0.25**	0.27	**0.31**	0.24	**0.28**
Liver	0.37	**0.39**	0.46	**0.49**	0.53	**0.56**	0.58	**0.62**	0.68	**0.73**	0.65	**0.69**
Gall bladder	0.35	**0.36**	0.45	**0.47**	0.53	**0.55**	0.59	**0.61**	0.71	**0.74**	0.67	**0.69**
Pancreas	0.19	**0.21**	0.26	**0.28**	0.30	**0.33**	0.34	**0.38**	0.43	**0.47**	0.39	**0.42**
Small intestine	0.28	**0.29**	0.37	**0.38**	0.43	**0.44**	0.48	**0.50**	0.58	**0.60**	0.54	**0.55**
Large intestine	0.35	**0.36**	0.45	**0.46**	0.52	**0.54**	0.58	**0.60**	0.70	**0.72**	0.65	**0.67**
Sigma/rectum	0.25	**0.26**	0.32	**0.34**	0.38	**0.40**	0.43	**0.45**	0.52	**0.54**	0.48	**0.50**
Lumbar spine	0.15	**0.16**	0.21	**0.22**	0.26	**0.27**	0.30	**0.32**	0.39	**0.41**	0.32	**0.34**
Pelvis	0.33	**0.35**	0.43	**0.46**	0.52	**0.55**	0.59	**0.63**	0.74	**0.78**	0.66	**0.70**
Adrenals	0.07	**0.08**	0.09	**0.11**	0.11	**0.13**	0.12	**0.15**	0.16	**0.19**	0.14	**0.16**
Kidneys	0.07	**0.07**	0.09	**0.10**	0.11	**0.12**	0.13	**0.14**	0.16	**0.18**	0.14	**0.15**
Urinary bladder	0.50	**0.51**	0.62	**0.63**	0.70	**0.72**	0.77	**0.79**	0.90	**0.92**	0.86	**0.88**
Uterus	0.29	**0.29**	0.37	**0.38**	0.44	**0.45**	0.49	**0.51**	0.60	**0.62**	0.55	**0.57**
Ovaries	0.22	**0.22**	0.29	**0.29**	0.34	**0.35**	0.39	**0.40**	0.48	**0.49**	0.43	**0.45**
Prostate	0.31	**0.31**	0.39	**0.40**	0.46	**0.47**	0.52	**0.53**	0.62	**0.64**	0.58	**0.60**
Testes	0.98	**1.02**	1.09	**1.14**	1.15	**1.21**	1.20	**1.27**	1.28	**1.36**	1.26	**1.34**
Upper arm		**0.26**		**0.33**		**0.38**		**0.42**		**0.50**		**0.47**
Elbow		**0.74**		**0.93**		**1.07**		**1.19**		**1.40**		**1.34**
Forearm		**0.86**		**1.07**		**1.23**		**1.36**		**1.59**		**1.54**
Upper leg	0.24	**0.44**	0.31	**0.58**	0.37	**0.69**	0.42	**0.78**	0.52	**0.96**	0.47	**0.88**
Knee		**0.02**		**0.03**	0.01	**0.04**	0.01	**0.05**	0.02	**0.06**	0.01	**0.05**
Lower leg												
Lymph nodes	0.18	**0.22**	0.23	**0.28**	0.27	**0.32**	0.30	**0.36**	0.36	**0.43**	0.33	**0.40**
Skin	0.17	**0.27**	0.17	**0.29**	0.18	**0.30**	0.19	**0.31**	0.19	**0.32**	0.19	**0.32**
Muscle	0.15	**0.21**	0.18	**0.25**	0.21	**0.28**	0.22	**0.31**	0.26	**0.35**	0.24	**0.33**
Skeleton	0.10	**0.26**	0.13	**0.32**	0.15	**0.37**	0.17	**0.41**	0.20	**0.49**	0.18	**0.46**
Red bone marrow	0.05	**0.08**	0.06	**0.11**	0.07	**0.12**	0.08	**0.14**	0.10	**0.17**	0.09	**0.15**
Whole body dose	0.14	**0.21**	0.17	**0.26**	0.20	**0.29**	0.21	**0.32**	0.25	**0.36**	0.23	**0.35**
ED (ICRP 60)	0.29	**0.38**	0.35	**0.46**	0.39	**0.51**	0.42	**0.55**	0.49	**0.63**	0.46	**0.60**
ED (ICRP 103)	0.22	**0.39**	0.27	**0.46**	0.31	**0.51**	0.34	**0.55**	0.39	**0.63**	0.37	**0.60**
Total filtration (mm) **Al**	**1.5**		**2.5**		**3.5**		**4.5**		**3.5**		**3.5**	
Cu	**0**		**0**		**0**		**0**		**0.1**		**0.2**	

Table 12.91 Normalized organ doses—Abdomen ap—3–7 years

Beam projection:		ap										**Table cells:**
Focus-to-detector distance (cm):		115										(+) = normal font = optimal field setting
Tube voltage (kVp):		70										**(−) = bold font = suboptimal field setting**
Total filtrations:		See last table row.										Empty table cells correspond to values <0.01
Organ or tissue	**Organ dose D_T : Entrance dose K_E**								**(mSv * mGy^{-1})**			
	(+)	(−)	(+)	(−)	(+)	(−)	(+)	(−)	(+)	(−)	(+)	(−)
Skull												
Brain												
Oral mucosa												
Salivary glands												
Cervical spine		**0.01**		**0.01**		**0.02**		**0.02**		**0.03**		**0.03**
Thyroid						**0.01**		**0.01**		**0.02**		**0.02**
Thymus		**0.16**	0.01	**0.19**	0.02	**0.22**	0.02	**0.24**	0.02	**0.27**	0.03	**0.31**
Thoracic spine	0.08	**0.16**	0.11	**0.21**	0.13	**0.25**	0.15	**0.28**	0.19	**0.36**	0.24	**0.44**
Claviculae		**0.03**		**0.04**	0.01	**0.05**	0.01	**0.05**	0.02	**0.07**	0.02	**0.08**
Scapulae	0.02	**0.19**	0.03	**0.25**	0.04	**0.29**	0.04	**0.33**	0.05	**0.40**	0.06	**0.47**
Ribs	0.51	**1.01**	0.59	**1.17**	0.65	**1.28**	0.69	**1.36**	0.76	**1.51**	0.82	**1.62**
Oesophagus	0.08	**0.13**	0.10	**0.16**	0.12	**0.19**	0.13	**0.22**	0.16	**0.26**	0.19	**0.31**
Trachea												
Lung	0.11	**0.31**	0.13	**0.37**	0.15	**0.42**	0.16	**0.46**	0.19	**0.53**	0.22	**0.60**
Mamma	0.02	**1.15**	0.03	**1.23**	0.03	**1.27**	0.03	**1.30**	0.04	**1.35**	0.05	**1.39**
Heart	0.16	**0.49**	0.20	**0.61**	0.23	**0.69**	0.25	**0.75**	0.30	**0.87**	0.34	**0.97**
Stomach	0.56	**0.59**	0.68	**0.71**	0.77	**0.81**	0.83	**0.88**	0.96	**1.01**	1.07	**1.13**
Spleen	0.16	**0.18**	0.20	**0.23**	0.24	**0.27**	0.27	**0.31**	0.33	**0.38**	0.39	**0.45**
Liver	0.43	**0.46**	0.53	**0.57**	0.60	**0.65**	0.66	**0.71**	0.76	**0.82**	0.85	**0.93**
Gall bladder	0.43	**0.45**	0.54	**0.56**	0.63	**0.65**	0.69	**0.72**	0.82	**0.86**	0.94	**0.98**
Pancreas	0.25	**0.27**	0.32	**0.35**	0.37	**0.41**	0.42	**0.46**	0.51	**0.56**	0.59	**0.66**
Small intestine	0.35	**0.36**	0.44	**0.46**	0.51	**0.53**	0.57	**0.59**	0.68	**0.70**	0.78	**0.81**
Large intestine	0.43	**0.44**	0.53	**0.55**	0.61	**0.63**	0.68	**0.70**	0.80	**0.83**	0.91	**0.94**
Sigma/rectum	0.31	**0.32**	0.39	**0.41**	0.46	**0.48**	0.51	**0.53**	0.61	**0.64**	0.70	**0.74**
Lumbar spine	0.22	**0.23**	0.29	**0.31**	0.35	**0.37**	0.40	**0.42**	0.51	**0.54**	0.62	**0.66**
Pelvis	0.42	**0.45**	0.55	**0.58**	0.64	**0.68**	0.72	**0.77**	0.88	**0.94**	1.04	**1.11**
Adrenals	0.09	**0.11**	0.12	**0.14**	0.14	**0.17**	0.16	**0.19**	0.20	**0.24**	0.25	**0.29**
Kidneys	0.09	**0.10**	0.12	**0.13**	0.14	**0.16**	0.16	**0.18**	0.21	**0.23**	0.25	**0.27**
Urinary bladder	0.58	**0.59**	0.71	**0.72**	0.80	**0.81**	0.87	**0.89**	1.00	**1.02**	1.11	**1.14**
Uterus	0.36	**0.37**	0.45	**0.47**	0.53	**0.54**	0.59	**0.61**	0.71	**0.73**	0.82	**0.84**
Ovaries	0.28	**0.28**	0.36	**0.37**	0.42	**0.43**	0.47	**0.48**	0.57	**0.59**	0.66	**0.68**
Prostate	0.37	**0.38**	0.47	**0.48**	0.54	**0.56**	0.60	**0.62**	0.71	**0.74**	0.82	**0.85**
Testes	1.03	**1.08**	1.14	**1.20**	1.21	**1.27**	1.25	**1.33**	1.33	**1.42**	1.40	**1.50**
Upper arm		**0.31**		**0.38**		**0.43**		**0.48**		**0.56**		**0.63**
Elbow		**0.87**		**1.07**		**1.22**		**1.33**		**1.55**		**1.73**
Forearm		**0.99**		**1.22**		**1.38**		**1.51**		**1.74**		**1.93**
Upper leg	0.30	**0.56**	0.39	**0.71**	0.46	**0.83**	0.51	**0.93**	0.62	**1.13**	0.73	**1.31**
Knee		**0.03**	0.01	**0.04**	0.02	**0.05**	0.02	**0.06**	0.02	**0.08**	0.03	**0.09**
Lower leg												
Lymph nodes	0.22	**0.26**	0.28	**0.33**	0.32	**0.38**	0.35	**0.42**	0.42	**0.50**	0.48	**0.57**
Skin	0.17	**0.28**	0.18	**0.30**	0.19	**0.31**	0.19	**0.32**	0.20	**0.34**	0.21	**0.36**
Muscle	0.18	**0.24**	0.21	**0.29**	0.23	**0.32**	0.25	**0.34**	0.29	**0.39**	0.32	**0.44**
Skeleton	0.12	**0.31**	0.15	**0.38**	0.18	**0.43**	0.20	**0.47**	0.23	**0.55**	0.27	**0.63**
Red bone marrow	0.06	**0.10**	0.08	**0.13**	0.09	**0.15**	0.10	**0.17**	0.12	**0.20**	0.14	**0.24**
Whole body dose	0.17	**0.24**	0.20	**0.29**	0.22	**0.33**	0.24	**0.36**	0.28	**0.41**	0.31	**0.46**
ED (ICRP 60)	0.33	**0.43**	0.39	**0.51**	0.44	**0.57**	0.48	**0.61**	0.54	**0.70**	0.61	**0.77**
ED (ICRP 103)	0.26	**0.44**	0.31	**0.52**	0.35	**0.57**	0.38	**0.61**	0.44	**0.69**	0.50	**0.76**
Total filtration (mm) Al	**1.5**		**2.5**		**3.5**		**4.5**		**3.5**		**3.5**	
Cu	**0**		**0**		**0**		**0**		**0.1**		**0.2**	

Table 12.92 Normalized organ doses—Abdomen ap—3–7 years

Beam projection:			ap									**Table cells:**
Focus-to-detector distance (cm):			115					(+) = normal font = optimal field setting				
Tube voltage (kVp):			80					**(−) = bold font = suboptimal field setting**				
Total filtrations:			See last table row.					Empty table cells correspond to values <0.01				
Organ or tissue	**Organ dose D_T : Entrance dose K_E**							**(mSv * mGy^{-1})**				
	(+)	(−)	(+)	(−)	(+)	(−)	(+)	(−)	(+)	(−)	(+)	(−)
Skull												
Brain												
Oral mucosa												
Salivary glands												**0.01**
Cervical spine		**0.01**		**0.02**		**0.02**		**0.03**		**0.03**		**0.04**
Thyroid				**0.01**		**0.02**		**0.02**		**0.02**		**0.03**
Thymus	0.01	**0.18**	0.02	**0.21**	0.02	**0.24**	0.02	**0.26**	0.03	**0.30**	0.03	**0.33**
Thoracic spine	0.11	**0.20**	0.14	**0.26**	0.17	**0.31**	0.19	**0.35**	0.24	**0.43**	0.28	**0.52**
Claviculae		**0.04**	0.01	**0.05**	0.01	**0.06**	0.02	**0.07**	0.02	**0.08**	0.02	**0.10**
Scapulae	0.03	**0.23**	0.04	**0.29**	0.04	**0.34**	0.05	**0.38**	0.06	**0.45**	0.07	**0.51**
Ribs	0.55	**1.09**	0.63	**1.24**	0.68	**1.35**	0.72	**1.42**	0.79	**1.55**	0.84	**1.65**
Oesophagus	0.10	**0.16**	0.12	**0.20**	0.14	**0.23**	0.16	**0.25**	0.19	**0.31**	0.22	**0.36**
Trachea												
Lung	0.12	**0.35**	0.15	**0.42**	0.17	**0.47**	0.18	**0.51**	0.21	**0.58**	0.24	**0.65**
Mamma	0.02	**1.19**	0.03	**1.26**	0.04	**1.30**	0.04	**1.33**	0.05	**1.38**	0.06	**1.42**
Heart	0.19	**0.56**	0.23	**0.68**	0.26	**0.76**	0.28	**0.82**	0.33	**0.94**	0.37	**1.04**
Stomach	0.63	**0.67**	0.76	**0.80**	0.84	**0.89**	0.91	**0.97**	1.04	**1.10**	1.14	**1.22**
Spleen	0.19	**0.22**	0.24	**0.28**	0.28	**0.33**	0.32	**0.36**	0.38	**0.44**	0.44	**0.51**
Liver	0.49	**0.53**	0.59	**0.64**	0.67	**0.72**	0.72	**0.79**	0.83	**0.90**	0.92	**1.01**
Gall bladder	0.51	**0.53**	0.62	**0.65**	0.71	**0.74**	0.78	**0.81**	0.91	**0.95**	1.03	**1.08**
Pancreas	0.30	**0.33**	0.38	**0.42**	0.44	**0.49**	0.49	**0.54**	0.58	**0.65**	0.67	**0.75**
Small intestine	0.42	**0.43**	0.51	**0.53**	0.59	**0.61**	0.65	**0.67**	0.76	**0.79**	0.87	**0.90**
Large intestine	0.50	**0.51**	0.61	**0.63**	0.69	**0.72**	0.76	**0.79**	0.88	**0.92**	0.99	**1.03**
Sigma/rectum	0.37	**0.39**	0.46	**0.48**	0.52	**0.55**	0.58	**0.61**	0.68	**0.72**	0.78	**0.82**
Lumbar spine	0.29	**0.30**	0.37	**0.39**	0.43	**0.46**	0.49	**0.52**	0.61	**0.65**	0.73	**0.78**
Pelvis	0.51	**0.55**	0.65	**0.69**	0.75	**0.80**	0.83	**0.89**	1.00	**1.07**	1.16	**1.24**
Adrenals	0.12	**0.14**	0.15	**0.18**	0.18	**0.21**	0.20	**0.24**	0.24	**0.29**	0.29	**0.35**
Kidneys	0.12	**0.13**	0.15	**0.17**	0.18	**0.20**	0.20	**0.22**	0.25	**0.27**	0.29	**0.32**
Urinary bladder	0.66	**0.67**	0.79	**0.80**	0.88	**0.90**	0.95	**0.97**	1.08	**1.11**	1.19	**1.23**
Uterus	0.43	**0.44**	0.53	**0.55**	0.61	**0.63**	0.68	**0.69**	0.80	**0.82**	0.91	**0.94**
Ovaries	0.34	**0.35**	0.42	**0.43**	0.49	**0.50**	0.54	**0.56**	0.65	**0.67**	0.75	**0.77**
Prostate	0.44	**0.45**	0.54	**0.56**	0.62	**0.64**	0.68	**0.70**	0.79	**0.82**	0.90	**0.93**
Testes	1.09	**1.14**	1.19	**1.25**	1.25	**1.33**	1.30	**1.38**	1.38	**1.47**	1.44	**1.54**
Upper arm		**0.35**		**0.42**		**0.48**		**0.52**		**0.60**		**0.67**
Elbow		**0.98**		**1.19**		**1.33**		**1.44**		**1.65**		**1.82**
Forearm		**1.12**		**1.34**		**1.50**		**1.62**		**1.83**		**2.01**
Upper leg	0.37	**0.67**	0.46	**0.83**	0.53	**0.96**	0.59	**1.06**	0.70	**1.26**	0.81	**1.43**
Knee	0.01	**0.04**	0.02	**0.06**	0.02	**0.07**	0.02	**0.07**	0.03	**0.09**	0.04	**0.11**
Lower leg												
Lymph nodes	0.26	**0.31**	0.32	**0.38**	0.36	**0.43**	0.40	**0.47**	0.47	**0.55**	0.53	**0.63**
Skin	0.18	**0.30**	0.19	**0.31**	0.20	**0.33**	0.20	**0.34**	0.21	**0.35**	0.22	**0.37**
Muscle	0.20	**0.27**	0.23	**0.32**	0.26	**0.35**	0.28	**0.38**	0.31	**0.43**	0.34	**0.47**
Skeleton	0.14	**0.35**	0.18	**0.42**	0.20	**0.48**	0.22	**0.52**	0.26	**0.60**	0.29	**0.67**
Red bone marrow	0.07	**0.13**	0.09	**0.16**	0.11	**0.18**	0.12	**0.20**	0.14	**0.23**	0.17	**0.27**
Whole body dose	0.19	**0.28**	0.22	**0.33**	0.25	**0.36**	0.27	**0.39**	0.30	**0.45**	0.34	**0.49**
ED (ICRP 60)	0.37	**0.48**	0.44	**0.57**	0.48	**0.63**	0.52	**0.67**	0.59	**0.76**	0.65	**0.83**
ED (ICRP 103)	0.29	**0.49**	0.35	**0.57**	0.39	**0.62**	0.43	**0.66**	0.49	**0.74**	0.54	**0.81**
Total filtration (mm) Al	**1.5**		**2.5**		**3.5**		**4.5**		**3.5**		**3.5**	
Cu	**0**		**0**		**0**		**0**		**0.1**		**0.2**	

Table 12.93 Normalized organ doses—Abdomen ap—8–12 years

Beam projection:			ap									**Table cells:**
Focus-to-detector distance (cm):			115									(+) = normal font = optimal field setting
Tube voltage (kVp):			60									**(−) = bold font = suboptimal field setting**
Total filtrations:			See last table row.									Empty table cells correspond to values <0.01
Organ or tissue	**Organ dose D_T : Entrance dose K_E**								**(mSv * mGy⁻¹)**			
	(+)	(−)	(+)	(−)	(+)	(−)	(+)	(−)	(+)	(−)	(+)	(−)
Skull												
Brain												
Oral mucosa												
Salivary glands												
Cervical spine										**0.01**		
Thyroid												
Thymus		**0.10**		**0.13**		**0.15**		**0.16**	0.01	**0.19**		**0.18**
Thoracic spine	0.04	**0.08**	0.06	**0.11**	0.07	**0.13**	0.08	**0.16**	0.11	**0.21**	0.08	**0.16**
Claviculae		**0.01**		**0.02**		**0.02**		**0.02**		**0.03**		**0.02**
Scapulae		**0.10**	0.01	**0.14**	0.02	**0.17**	0.02	**0.19**	0.03	**0.24**	0.02	**0.21**
Ribs	0.42	**0.83**	0.49	**0.98**	0.54	**1.08**	0.58	**1.16**	0.65	**1.30**	0.64	**1.28**
Oesophagus	0.03	**0.06**	0.04	**0.08**	0.05	**0.09**	0.06	**0.11**	0.07	**0.14**	0.06	**0.12**
Trachea												
Lung	0.07	**0.22**	0.08	**0.28**	0.10	**0.33**	0.11	**0.36**	0.13	**0.43**	0.12	**0.40**
Mamma	0.01	**1.07**	0.02	**1.16**	0.02	**1.21**	0.02	**1.25**	0.03	**1.30**	0.02	**1.30**
Heart	0.10	**0.37**	0.13	**0.47**	0.15	**0.54**	0.17	**0.60**	0.20	**0.71**	0.18	**0.67**
Stomach	0.44	**0.46**	0.55	**0.58**	0.63	**0.66**	0.70	**0.73**	0.81	**0.86**	0.78	**0.82**
Spleen	0.10	**0.11**	0.13	**0.15**	0.16	**0.18**	0.18	**0.21**	0.23	**0.26**	0.20	**0.23**
Liver	0.31	**0.34**	0.39	**0.43**	0.45	**0.49**	0.50	**0.55**	0.60	**0.65**	0.56	**0.61**
Gall bladder	0.32	**0.33**	0.41	**0.42**	0.48	**0.50**	0.54	**0.56**	0.66	**0.68**	0.61	**0.63**
Pancreas	0.17	**0.18**	0.22	**0.24**	0.27	**0.29**	0.30	**0.33**	0.38	**0.42**	0.34	**0.37**
Small intestine	0.25	**0.26**	0.32	**0.33**	0.38	**0.39**	0.43	**0.44**	0.53	**0.54**	0.48	**0.49**
Large intestine	0.31	**0.32**	0.40	**0.42**	0.47	**0.49**	0.53	**0.55**	0.64	**0.67**	0.59	**0.62**
Sigma/rectum	0.22	**0.23**	0.28	**0.29**	0.34	**0.35**	0.38	**0.40**	0.47	**0.49**	0.42	**0.44**
Lumbar spine	0.12	**0.12**	0.16	**0.17**	0.19	**0.21**	0.23	**0.24**	0.30	**0.32**	0.24	**0.25**
Pelvis	0.27	**0.28**	0.35	**0.38**	0.43	**0.45**	0.49	**0.52**	0.61	**0.65**	0.54	**0.57**
Adrenals	0.05	**0.06**	0.07	**0.08**	0.09	**0.10**	0.10	**0.12**	0.13	**0.15**	0.10	**0.13**
Kidneys	0.05	**0.06**	0.07	**0.08**	0.09	**0.10**	0.10	**0.11**	0.13	**0.14**	0.11	**0.12**
Urinary bladder	0.46	**0.47**	0.57	**0.59**	0.66	**0.67**	0.72	**0.74**	0.85	**0.87**	0.81	**0.83**
Uterus	0.25	**0.26**	0.33	**0.34**	0.39	**0.40**	0.44	**0.46**	0.55	**0.56**	0.49	**0.51**
Ovaries	0.19	**0.19**	0.24	**0.25**	0.29	**0.30**	0.33	**0.34**	0.41	**0.42**	0.37	**0.37**
Prostate	0.29	**0.29**	0.38	**0.38**	0.44	**0.45**	0.50	**0.51**	0.60	**0.61**	0.56	**0.57**
Testes	0.98	**1.02**	1.09	**1.14**	1.16	**1.22**	1.21	**1.28**	1.29	**1.38**	1.28	**1.36**
Upper arm		**0.19**		**0.24**		**0.28**		**0.32**		**0.38**		**0.36**
Elbow		**0.63**		**0.80**		**0.93**		**1.03**		**1.23**		**1.17**
Forearm		**0.74**		**0.94**		**1.08**		**1.20**		**1.42**		**1.36**
Upper leg	0.14	**0.29**	0.19	**0.38**	0.23	**0.46**	0.26	**0.52**	0.33	**0.65**	0.29	**0.58**
Knee				**0.01**		**0.01**		**0.02**		**0.02**		**0.02**
Lower leg												
Lymph nodes	0.16	**0.19**	0.20	**0.24**	0.24	**0.29**	0.27	**0.32**	0.32	**0.39**	0.30	**0.36**
Skin	0.16	**0.26**	0.17	**0.27**	0.17	**0.28**	0.18	**0.29**	0.19	**0.30**	0.18	**0.30**
Muscle	0.14	**0.19**	0.17	**0.23**	0.19	**0.26**	0.20	**0.28**	0.23	**0.32**	0.22	**0.31**
Skeleton	0.08	**0.21**	0.10	**0.27**	0.12	**0.31**	0.13	**0.35**	0.16	**0.41**	0.15	**0.38**
Red bone marrow	0.05	**0.08**	0.06	**0.10**	0.07	**0.12**	0.08	**0.13**	0.10	**0.16**	0.09	**0.14**
Whole body dose	0.13	**0.19**	0.16	**0.23**	0.18	**0.26**	0.19	**0.29**	0.23	**0.34**	0.21	**0.32**
ED (ICRP 60)	0.26	**0.35**	0.32	**0.42**	0.36	**0.47**	0.39	**0.52**	0.46	**0.59**	0.43	**0.56**
ED (ICRP 103)	0.20	**0.36**	0.24	**0.43**	0.28	**0.48**	0.31	**0.52**	0.36	**0.59**	0.34	**0.56**
Total filtration (mm) **Al**	**1.5**		**2.5**		**3.5**		**4.5**		**3.5**		**3.5**	
Cu	**0**		**0**		**0**		**0**		**0.1**		**0.2**	

Table 12.94 Normalized organ doses—Abdomen ap—8–12 years

Beam projection:		ap										**Table cells:**
Focus-to-detector distance (cm):		115										(+) = normal font = optimal field setting
Tube voltage (kVp):		70										**(−) = bold font = suboptimal field setting**
Total filtrations:		See last table row.										Empty table cells correspond to values <0.01
Organ or tissue	**Organ dose D_T : Entrance dose K_E (mSv * mGy^{-1})**											
	(+)	(−)	(+)	(−)	(+)	(−)	(+)	(−)	(+)	(−)	(+)	(−)
Skull												
Brain												
Oral mucosa												
Salivary glands												
Cervical spine						**0.01**		**0.01**		**0.02**		**0.02**
Thyroid										**0.01**		**0.01**
Thymus		**0.12**		**0.15**	0.01	**0.17**	0.01	**0.19**	0.02	**0.22**	0.02	**0.25**
Thoracic spine	0.06	**0.12**	0.08	**0.16**	0.10	**0.19**	0.11	**0.22**	0.14	**0.28**	0.18	**0.34**
Claviculae		**0.02**		**0.02**		**0.03**		**0.03**		**0.04**		**0.05**
Scapulae	0.01	**0.14**	0.02	**0.18**	0.02	**0.21**	0.03	**0.24**	0.03	**0.30**	0.04	**0.35**
Ribs	0.46	**0.91**	0.54	**1.07**	0.59	**1.17**	0.63	**1.25**	0.69	**1.39**	0.75	**1.50**
Oesophagus	0.04	**0.08**	0.05	**0.10**	0.06	**0.12**	0.07	**0.14**	0.09	**0.17**	0.11	**0.21**
Trachea												
Lung	0.08	**0.27**	0.10	**0.33**	0.12	**0.38**	0.13	**0.41**	0.15	**0.48**	0.17	**0.55**
Mamma	0.02	**1.11**	0.02	**1.20**	0.02	**1.25**	0.03	**1.28**	0.03	**1.34**	0.04	**1.38**
Heart	0.12	**0.44**	0.15	**0.55**	0.18	**0.62**	0.20	**0.69**	0.24	**0.80**	0.27	**0.91**
Stomach	0.52	**0.55**	0.64	**0.67**	0.72	**0.76**	0.79	**0.84**	0.92	**0.97**	1.03	**1.10**
Spleen	0.13	**0.15**	0.17	**0.20**	0.20	**0.24**	0.23	**0.27**	0.28	**0.33**	0.34	**0.40**
Liver	0.37	**0.41**	0.46	**0.50**	0.53	**0.58**	0.58	**0.64**	0.68	**0.75**	0.77	**0.85**
Gall bladder	0.39	**0.41**	0.50	**0.52**	0.58	**0.60**	0.64	**0.67**	0.77	**0.81**	0.89	**0.93**
Pancreas	0.22	**0.24**	0.28	**0.31**	0.34	**0.37**	0.38	**0.42**	0.47	**0.51**	0.55	**0.61**
Small intestine	0.31	**0.32**	0.40	**0.41**	0.46	**0.48**	0.52	**0.54**	0.63	**0.65**	0.73	**0.75**
Large intestine	0.39	**0.40**	0.49	**0.51**	0.56	**0.59**	0.63	**0.65**	0.75	**0.78**	0.86	**0.90**
Sigma/rectum	0.27	**0.29**	0.35	**0.37**	0.41	**0.43**	0.46	**0.48**	0.55	**0.58**	0.65	**0.68**
Lumbar spine	0.17	**0.18**	0.23	**0.24**	0.27	**0.29**	0.31	**0.33**	0.40	**0.43**	0.50	**0.53**
Pelvis	0.35	**0.37**	0.45	**0.49**	0.54	**0.58**	0.61	**0.65**	0.75	**0.81**	0.90	**0.96**
Adrenals	0.07	**0.09**	0.10	**0.12**	0.12	**0.14**	0.13	**0.16**	0.17	**0.21**	0.21	**0.25**
Kidneys	0.07	**0.08**	0.10	**0.11**	0.12	**0.13**	0.14	**0.15**	0.17	**0.19**	0.21	**0.23**
Urinary bladder	0.54	**0.55**	0.66	**0.68**	0.75	**0.77**	0.83	**0.85**	0.96	**0.99**	1.08	**1.11**
Uterus	0.32	**0.33**	0.41	**0.42**	0.48	**0.49**	0.54	**0.55**	0.65	**0.67**	0.76	**0.78**
Ovaries	0.24	**0.25**	0.31	**0.32**	0.37	**0.38**	0.41	**0.42**	0.51	**0.52**	0.60	**0.62**
Prostate	0.36	**0.37**	0.46	**0.46**	0.53	**0.54**	0.59	**0.60**	0.70	**0.72**	0.81	**0.82**
Testes	1.04	**1.09**	1.15	**1.21**	1.22	**1.29**	1.27	**1.35**	1.35	**1.44**	1.42	**1.52**
Upper arm		**0.23**		**0.29**		**0.33**		**0.37**		**0.43**		**0.49**
Elbow		**0.75**		**0.93**		**1.07**		**1.18**		**1.38**		**1.56**
Forearm		**0.87**		**1.08**		**1.23**		**1.35**		**1.57**		**1.76**
Upper leg	0.19	**0.38**	0.25	**0.49**	0.29	**0.57**	0.33	**0.65**	0.41	**0.80**	0.49	**0.94**
Knee		**0.01**		**0.02**		**0.02**		**0.02**		**0.03**	0.01	**0.04**
Lower leg												
Lymph nodes	0.20	**0.24**	0.25	**0.30**	0.29	**0.34**	0.32	**0.38**	0.38	**0.46**	0.44	**0.53**
Skin	0.17	**0.27**	0.17	**0.28**	0.18	**0.29**	0.19	**0.30**	0.20	**0.32**	0.20	**0.33**
Muscle	0.16	**0.22**	0.19	**0.26**	0.21	**0.29**	0.23	**0.32**	0.26	**0.36**	0.30	**0.41**
Skeleton	0.10	**0.26**	0.13	**0.32**	0.14	**0.36**	0.16	**0.40**	0.19	**0.47**	0.22	**0.54**
Red bone marrow	0.06	**0.10**	0.08	**0.12**	0.09	**0.14**	0.10	**0.16**	0.13	**0.20**	0.15	**0.23**
Whole body dose	0.15	**0.22**	0.18	**0.27**	0.20	**0.30**	0.22	**0.33**	0.26	**0.38**	0.29	**0.43**
ED (ICRP 60)	0.31	**0.41**	0.37	**0.48**	0.41	**0.54**	0.45	**0.58**	0.51	**0.66**	0.57	**0.74**
ED (ICRP 103)	0.23	**0.41**	0.29	**0.48**	0.32	**0.54**	0.36	**0.58**	0.41	**0.65**	0.47	**0.73**
Total filtration (mm) Al	**1.5**		**2.5**		**3.5**		**4.5**		**3.5**		**3.5**	
Cu	**0**		**0**		**0**		**0**		**0.1**		**0.2**	

Table 12.95 Normalized organ doses—Abdomen ap—8–12 years

Beam projection:			ap									**Table cells:**
Focus-to-detector distance (cm):			115									(+) = normal font = optimal field setting
Tube voltage (kVp):			80									**(−) = bold font = suboptimal field setting**
Total filtrations:			See last table row.									Empty table cells correspond to values <0.01
Organ or tissue	**Organ dose D_T : Entrance dose K_E**								**(mSv * mGy^{-1})**			
	(+)	(−)	(+)	(−)	(+)	(−)	(+)	(−)	(+)	(−)	(+)	(−)
Skull												
Brain												
Oral mucosa												
Salivary glands												
Cervical spine				**0.01**		**0.01**		**0.02**		**0.02**		**0.03**
Thyroid						**0.01**		**0.01**		**0.02**		**0.02**
Thymus		**0.14**	0.01	**0.17**	0.01	**0.19**	0.02	**0.21**	0.02	**0.25**	0.03	**0.28**
Thoracic spine	0.08	**0.16**	0.11	**0.20**	0.13	**0.24**	0.14	**0.27**	0.18	**0.34**	0.22	**0.41**
Claviculae		**0.02**		**0.03**		**0.04**		**0.04**	0.01	**0.05**	0.01	**0.07**
Scapulae	0.02	**0.17**	0.02	**0.21**	0.03	**0.25**	0.03	**0.28**	0.04	**0.34**	0.05	**0.39**
Ribs	0.50	**0.99**	0.57	**1.14**	0.62	**1.24**	0.66	**1.31**	0.72	**1.44**	0.77	**1.54**
Oesophagus	0.05	**0.10**	0.07	**0.13**	0.08	**0.15**	0.09	**0.17**	0.11	**0.20**	0.13	**0.24**
Trachea												
Lung	0.10	**0.31**	0.12	**0.37**	0.13	**0.42**	0.15	**0.46**	0.17	**0.53**	0.19	**0.59**
Mamma	0.02	**1.15**	0.03	**1.23**	0.03	**1.28**	0.03	**1.31**	0.04	**1.37**	0.05	**1.41**
Heart	0.15	**0.51**	0.18	**0.62**	0.21	**0.70**	0.23	**0.77**	0.27	**0.89**	0.31	**0.99**
Stomach	0.60	**0.63**	0.72	**0.76**	0.80	**0.85**	0.87	**0.93**	1.00	**1.07**	1.11	**1.19**
Spleen	0.17	**0.19**	0.21	**0.25**	0.25	**0.29**	0.28	**0.32**	0.34	**0.39**	0.39	**0.46**
Liver	0.43	**0.47**	0.53	**0.58**	0.59	**0.65**	0.65	**0.72**	0.75	**0.83**	0.84	**0.94**
Gall bladder	0.47	**0.49**	0.58	**0.61**	0.67	**0.70**	0.73	**0.77**	0.87	**0.91**	0.99	**1.04**
Pancreas	0.27	**0.30**	0.34	**0.38**	0.40	**0.44**	0.45	**0.50**	0.54	**0.60**	0.63	**0.70**
Small intestine	0.38	**0.39**	0.47	**0.49**	0.54	**0.56**	0.60	**0.62**	0.71	**0.74**	0.82	**0.85**
Large intestine	0.46	**0.48**	0.57	**0.59**	0.65	**0.67**	0.71	**0.74**	0.84	**0.87**	0.95	**0.99**
Sigma/rectum	0.33	**0.35**	0.42	**0.44**	0.48	**0.50**	0.53	**0.56**	0.63	**0.67**	0.73	**0.77**
Lumbar spine	0.23	**0.24**	0.30	**0.31**	0.35	**0.37**	0.40	**0.42**	0.50	**0.53**	0.61	**0.64**
Pelvis	0.43	**0.47**	0.55	**0.59**	0.64	**0.69**	0.72	**0.77**	0.87	**0.94**	1.01	**1.09**
Adrenals	0.10	**0.12**	0.13	**0.15**	0.15	**0.18**	0.17	**0.20**	0.21	**0.25**	0.25	**0.31**
Kidneys	0.10	**0.11**	0.13	**0.14**	0.15	**0.17**	0.17	**0.19**	0.21	**0.23**	0.25	**0.28**
Urinary bladder	0.62	**0.64**	0.75	**0.77**	0.84	**0.87**	0.91	**0.94**	1.05	**1.08**	1.17	**1.21**
Uterus	0.39	**0.40**	0.49	**0.50**	0.56	**0.58**	0.63	**0.64**	0.75	**0.77**	0.86	**0.89**
Ovaries	0.30	**0.31**	0.38	**0.39**	0.44	**0.45**	0.49	**0.50**	0.59	**0.61**	0.68	**0.71**
Prostate	0.43	**0.44**	0.53	**0.54**	0.61	**0.62**	0.67	**0.68**	0.79	**0.81**	0.90	**0.92**
Testes	1.10	**1.15**	1.20	**1.27**	1.27	**1.35**	1.32	**1.40**	1.40	**1.49**	1.46	**1.57**
Upper arm		**0.27**		**0.33**		**0.37**		**0.41**		**0.47**		**0.53**
Elbow		**0.86**		**1.05**		**1.18**		**1.29**		**1.48**		**1.65**
Forearm		**0.99**		**1.20**		**1.35**		**1.46**		**1.67**		**1.85**
Upper leg	0.24	**0.46**	0.30	**0.58**	0.35	**0.68**	0.39	**0.76**	0.48	**0.91**	0.56	**1.06**
Knee		**0.02**		**0.02**		**0.03**		**0.03**	0.01	**0.04**	0.02	**0.05**
Lower leg												
Lymph nodes	0.23	**0.28**	0.29	**0.35**	0.33	**0.40**	0.37	**0.44**	0.43	**0.52**	0.49	**0.59**
Skin	0.17	**0.28**	0.18	**0.29**	0.19	**0.31**	0.19	**0.31**	0.20	**0.33**	0.21	**0.35**
Muscle	0.18	**0.25**	0.21	**0.29**	0.24	**0.32**	0.25	**0.35**	0.29	**0.40**	0.32	**0.44**
Skeleton	0.12	**0.30**	0.15	**0.36**	0.17	**0.41**	0.18	**0.45**	0.21	**0.52**	0.24	**0.59**
Red bone marrow	0.08	**0.12**	0.10	**0.15**	0.11	**0.17**	0.13	**0.19**	0.15	**0.23**	0.18	**0.27**
Whole body dose	0.17	**0.25**	0.20	**0.30**	0.23	**0.34**	0.25	**0.37**	0.28	**0.42**	0.32	**0.47**
ED (ICRP 60)	0.35	**0.46**	0.41	**0.54**	0.46	**0.59**	0.49	**0.64**	0.56	**0.72**	0.63	**0.80**
ED (ICRP 103)	0.27	**0.46**	0.32	**0.53**	0.37	**0.59**	0.40	**0.63**	0.46	**0.71**	0.51	**0.78**
Total filtration (mm) **Al**	**1.5**		**2.5**		**3.5**		**4.5**		**3.5**		**3.5**	
Cu	**0**		**0**		**0**		**0**		**0.1**		**0.2**	

Table 12.96 Normalized organ doses—Abdomen ap—13–17 years

Beam projection:		ap										**Table cells:**
Focus-to-detector distance (cm):		115										(+) = normal font = optimal field setting
Tube voltage (kVp):		70										**(−) = bold font = suboptimal field setting**
Total filtrations:		See last table row.										Empty table cells correspond to values <0.01
Organ or tissue	**Organ dose D_T : Entrance dose K_E**								**(mSv * mGy⁻¹)**			
	(+)	(−)	(+)	(−)	(+)	(−)	(+)	(−)	(+)	(−)	(+)	(−)
Skull												
Brain												
Oral mucosa												
Salivary glands												
Cervical spine												
Thyroid												
Thymus		**0.05**		**0.06**		**0.07**	0.01	**0.08**	0.01	**0.10**	0.02	**0.12**
Thoracic spine	0.04	**0.06**	0.05	**0.09**	0.06	**0.11**	0.07	**0.12**	0.09	**0.16**	0.12	**0.20**
Claviculae								**0.01**		**0.01**		**0.02**
Scapulae		**0.05**	0.01	**0.06**	0.01	**0.07**	0.02	**0.08**	0.02	**0.11**	0.02	**0.13**
Ribs	0.40	**0.67**	0.47	**0.79**	0.52	**0.87**	0.55	**0.94**	0.62	**1.05**	0.67	**1.15**
Oesophagus	0.05	**0.08**	0.07	**0.11**	0.08	**0.13**	0.10	**0.15**	0.12	**0.19**	0.15	**0.23**
Trachea												
Lung	0.06	**0.17**	0.08	**0.21**	0.09	**0.24**	0.10	**0.27**	0.12	**0.32**	0.14	**0.38**
Mamma	0.15	**4.70**	0.19	**5.39**	0.22	**5.85**	0.25	**6.19**	0.31	**6.79**	0.37	**7.29**
Heart	0.09	**0.37**	0.12	**0.46**	0.14	**0.53**	0.16	**0.59**	0.19	**0.69**	0.22	**0.79**
Stomach	0.44	**0.46**	0.54	**0.58**	0.62	**0.66**	0.69	**0.73**	0.81	**0.87**	0.93	**0.99**
Spleen	0.10	**0.11**	0.13	**0.15**	0.15	**0.18**	0.18	**0.20**	0.22	**0.26**	0.27	**0.31**
Liver	0.30	**0.33**	0.38	**0.42**	0.44	**0.48**	0.48	**0.54**	0.58	**0.64**	0.66	**0.74**
Gall bladder	0.34	**0.35**	0.43	**0.45**	0.51	**0.53**	0.57	**0.59**	0.69	**0.72**	0.81	**0.84**
Pancreas	0.17	**0.19**	0.22	**0.24**	0.26	**0.29**	0.30	**0.33**	0.37	**0.41**	0.45	**0.50**
Small intestine	0.26	**0.27**	0.34	**0.35**	0.39	**0.41**	0.44	**0.46**	0.54	**0.56**	0.64	**0.66**
Large intestine	0.33	**0.34**	0.42	**0.44**	0.49	**0.51**	0.55	**0.57**	0.66	**0.69**	0.77	**0.81**
Sigma/rectum	0.23	**0.24**	0.29	**0.31**	0.34	**0.36**	0.39	**0.41**	0.48	**0.50**	0.56	**0.59**
Lumbar spine	0.11	**0.12**	0.15	**0.16**	0.18	**0.19**	0.21	**0.22**	0.27	**0.29**	0.34	**0.36**
Pelvis	0.26	**0.28**	0.34	**0.37**	0.41	**0.44**	0.47	**0.50**	0.58	**0.63**	0.70	**0.76**
Adrenals	0.05	**0.06**	0.07	**0.08**	0.08	**0.10**	0.10	**0.12**	0.13	**0.15**	0.16	**0.19**
Kidneys	0.05	**0.06**	0.07	**0.08**	0.08	**0.09**	0.10	**0.11**	0.13	**0.14**	0.16	**0.17**
Urinary bladder	0.48	**0.50**	0.60	**0.62**	0.69	**0.71**	0.76	**0.78**	0.89	**0.92**	1.02	**1.05**
Uterus	0.77	**0.80**	1.00	**1.03**	1.18	**1.21**	1.32	**1.37**	1.63	**1.68**	1.92	**1.99**
Ovaries	0.19	**0.20**	0.25	**0.26**	0.30	**0.31**	0.34	**0.35**	0.42	**0.44**	0.50	**0.53**
Prostate	0.32	**0.34**	0.41	**0.44**	0.48	**0.51**	0.54	**0.57**	0.66	**0.70**	0.77	**0.81**
Testes	0.91	**0.95**	1.04	**1.09**	1.12	**1.18**	1.18	**1.25**	1.29	**1.37**	1.38	**1.47**
Upper arm		**0.10**		**0.13**		**0.15**		**0.16**		**0.20**		**0.23**
Elbow		**0.58**		**0.73**		**0.84**		**0.93**		**1.10**		**1.26**
Forearm		**0.69**		**0.86**		**0.99**		**1.09**		**1.28**		**1.46**
Upper leg	0.17	**0.30**	0.23	**0.40**	0.27	**0.47**	0.31	**0.54**	0.39	**0.67**	0.47	**0.80**
Knee		**0.02**		**0.03**		**0.04**		**0.04**		**0.06**	0.01	**0.07**
Lower leg												
Lymph nodes	0.16	**0.19**	0.21	**0.25**	0.24	**0.29**	0.27	**0.32**	0.33	**0.39**	0.39	**0.46**
Skin	0.18	**0.28**	0.20	**0.30**	0.20	**0.31**	0.21	**0.32**	0.22	**0.34**	0.24	**0.36**
Muscle	0.15	**0.20**	0.18	**0.24**	0.21	**0.27**	0.22	**0.30**	0.26	**0.34**	0.29	**0.39**
Skeleton	0.09	**0.21**	0.11	**0.26**	0.13	**0.30**	0.14	**0.33**	0.17	**0.40**	0.20	**0.46**
Red bone marrow	0.06	**0.08**	0.07	**0.10**	0.09	**0.12**	0.10	**0.13**	0.12	**0.16**	0.14	**0.19**
Whole body dose	0.14	**0.20**	0.17	**0.25**	0.20	**0.28**	0.21	**0.31**	0.25	**0.36**	0.28	**0.40**
ED (ICRP 60)	0.27	**0.53**	0.33	**0.63**	0.37	**0.70**	0.41	**0.76**	0.48	**0.87**	0.55	**0.97**
ED (ICRP 103)	0.22	**0.80**	0.27	**0.94**	0.31	**1.03**	0.35	**1.11**	0.41	**1.25**	0.48	**1.38**
Total filtration (mm) Al	**1.5**		**2.5**		**3.5**		**4.5**		**3.5**		**3.5**	
Cu	**0**		**0**		**0**		**0**		**0.1**		**0.2**	

Table 12.97 Normalized organ doses—Abdomen ap—13–17 years

Beam projection:			ap									**Table cells:**
Focus-to-detector distance (cm):			115									(+) = normal font = optimal field setting
Tube voltage (kVp):			80									**(−) = bold font = suboptimal field setting**
Total filtrations:			See last table row.									Empty table cells correspond to values <0.01
Organ or tissue	**Organ dose D_T : Entrance dose K_E**							**(mSv * mGy^{-1})**				
	(+)	(−)	(+)	(−)	(+)	(−)	(+)	(−)	(+)	(−)	(+)	(−)
Skull												
Brain												
Oral mucosa												
Salivary glands												
Cervical spine												
Thyroid												
Thymus		**0.06**	0.01	**0.07**	0.01	**0.09**	0.01	**0.10**	0.02	**0.12**	0.02	**0.14**
Thoracic spine	0.05	**0.09**	0.07	**0.12**	0.08	**0.14**	0.10	**0.16**	0.12	**0.21**	0.15	**0.26**
Claviculae				**0.01**		**0.01**		**0.02**		**0.02**		**0.02**
Scapulae	0.01	**0.06**	0.01	**0.08**	0.02	**0.09**	0.02	**0.10**	0.03	**0.13**	0.03	**0.16**
Ribs	0.44	**0.73**	0.50	**0.85**	0.55	**0.94**	0.58	**1.00**	0.64	**1.11**	0.69	**1.20**
Oesophagus	0.07	**0.11**	0.09	**0.14**	0.11	**0.16**	0.12	**0.18**	0.15	**0.23**	0.17	**0.27**
Trachea												
Lung	0.07	**0.20**	0.09	**0.25**	0.11	**0.28**	0.12	**0.31**	0.14	**0.37**	0.16	**0.42**
Mamma	0.18	**5.08**	0.23	**5.76**	0.27	**6.21**	0.30	**6.55**	0.37	**7.14**	0.43	**7.62**
Heart	0.11	**0.43**	0.14	**0.53**	0.17	**0.60**	0.18	**0.66**	0.22	**0.78**	0.26	**0.88**
Stomach	0.51	**0.54**	0.62	**0.66**	0.71	**0.76**	0.78	**0.83**	0.90	**0.97**	1.02	**1.09**
Spleen	0.13	**0.15**	0.16	**0.19**	0.19	**0.22**	0.22	**0.25**	0.27	**0.31**	0.32	**0.37**
Liver	0.36	**0.40**	0.44	**0.49**	0.50	**0.56**	0.55	**0.62**	0.65	**0.73**	0.73	**0.83**
Gall bladder	0.41	**0.43**	0.52	**0.54**	0.60	**0.62**	0.66	**0.69**	0.79	**0.82**	0.91	**0.95**
Pancreas	0.22	**0.24**	0.28	**0.30**	0.32	**0.36**	0.36	**0.40**	0.45	**0.50**	0.53	**0.59**
Small intestine	0.32	**0.34**	0.41	**0.42**	0.47	**0.49**	0.52	**0.55**	0.63	**0.66**	0.73	**0.76**
Large intestine	0.40	**0.42**	0.50	**0.52**	0.57	**0.60**	0.64	**0.66**	0.76	**0.79**	0.87	**0.91**
Sigma/rectum	0.28	**0.30**	0.35	**0.37**	0.41	**0.43**	0.46	**0.48**	0.55	**0.59**	0.65	**0.68**
Lumbar spine	0.16	**0.16**	0.20	**0.22**	0.24	**0.26**	0.28	**0.30**	0.35	**0.38**	0.43	**0.46**
Pelvis	0.33	**0.36**	0.43	**0.46**	0.50	**0.54**	0.56	**0.61**	0.69	**0.75**	0.82	**0.88**
Adrenals	0.07	**0.09**	0.09	**0.11**	0.11	**0.13**	0.13	**0.15**	0.16	**0.19**	0.19	**0.23**
Kidneys	0.07	**0.08**	0.09	**0.10**	0.11	**0.12**	0.13	**0.14**	0.16	**0.18**	0.19	**0.21**
Urinary bladder	0.56	**0.58**	0.69	**0.71**	0.78	**0.80**	0.85	**0.88**	0.99	**1.02**	1.12	**1.15**
Uterus	0.96	**0.99**	1.21	**1.25**	1.41	**1.46**	1.57	**1.62**	1.90	**1.96**	2.21	**2.29**
Ovaries	0.24	**0.26**	0.31	**0.33**	0.36	**0.38**	0.41	**0.43**	0.50	**0.53**	0.59	**0.62**
Prostate	0.39	**0.42**	0.49	**0.52**	0.57	**0.60**	0.63	**0.67**	0.75	**0.80**	0.87	**0.92**
Testes	0.97	**1.03**	1.10	**1.16**	1.18	**1.26**	1.25	**1.32**	1.35	**1.44**	1.44	**1.54**
Upper arm		**0.12**		**0.15**		**0.17**		**0.19**		**0.22**		**0.25**
Elbow		**0.68**		**0.83**		**0.95**		**1.04**		**1.21**		**1.36**
Forearm		**0.79**		**0.97**		**1.10**		**1.20**		**1.39**		**1.55**
Upper leg	0.22	**0.38**	0.29	**0.49**	0.34	**0.57**	0.38	**0.65**	0.47	**0.79**	0.55	**0.93**
Knee		**0.03**		**0.04**		**0.05**		**0.06**	0.01	**0.07**	0.02	**0.09**
Lower leg												
Lymph nodes	0.20	**0.24**	0.25	**0.29**	0.29	**0.34**	0.32	**0.38**	0.38	**0.45**	0.44	**0.52**
Skin	0.19	**0.29**	0.20	**0.31**	0.21	**0.32**	0.22	**0.33**	0.23	**0.35**	0.25	**0.37**
Muscle	0.17	**0.23**	0.21	**0.27**	0.23	**0.31**	0.25	**0.33**	0.29	**0.38**	0.32	**0.43**
Skeleton	0.11	**0.25**	0.13	**0.30**	0.15	**0.34**	0.17	**0.38**	0.20	**0.45**	0.23	**0.51**
Red bone marrow	0.07	**0.10**	0.09	**0.13**	0.11	**0.15**	0.12	**0.16**	0.15	**0.20**	0.17	**0.23**
Whole body dose	0.17	**0.24**	0.20	**0.28**	0.22	**0.32**	0.24	**0.34**	0.28	**0.40**	0.31	**0.44**
ED (ICRP 60)	0.31	**0.60**	0.38	**0.70**	0.43	**0.78**	0.46	**0.84**	0.54	**0.95**	0.61	**1.05**
ED (ICRP 103)	0.26	**0.89**	0.32	**1.03**	0.36	**1.13**	0.40	**1.21**	0.47	**1.35**	0.53	**1.48**
Total filtration (mm) Al	**1.5**		**2.5**		**3.5**		**4.5**		**3.5**		**3.5**	
Cu	**0**		**0**		**0**		**0**		**0.1**		**0.2**	

Table 12.98 Normalized organ doses—Abdomen ap—13–17 years

Beam projection:			ap									**Table cells:**
Focus-to-detector distance (cm):			115									(+) = normal font = optimal field setting
Tube voltage (kVp):			110									**(−) = bold font = suboptimal field setting**
Total filtrations:			See last table row.									Empty table cells correspond to values <0.01
Organ or tissue	**Organ dose D_T : Entrance dose K_E**								**(mSv * mGy⁻¹)**			
	(+)	(−)	(+)	(−)	(+)	(−)	(+)	(−)	(+)	(−)	(+)	(−)
Skull												
Brain												
Oral mucosa												
Salivary glands												
Cervical spine								**0.01**		**0.01**		**0.02**
Thyroid												**0.01**
Thymus	0.01	**0.09**	0.02	**0.11**	0.02	**0.12**	0.02	**0.13**	0.03	**0.15**	0.03	**0.17**
Thoracic spine	0.10	**0.17**	0.12	**0.20**	0.14	**0.23**	0.15	**0.25**	0.18	**0.30**	0.20	**0.35**
Claviculae		**0.02**		**0.02**		**0.02**		**0.03**		**0.03**		**0.04**
Scapulae	0.02	**0.10**	0.02	**0.12**	0.03	**0.14**	0.03	**0.15**	0.04	**0.17**	0.04	**0.20**
Ribs	0.51	**0.87**	0.57	**0.97**	0.60	**1.04**	0.63	**1.08**	0.67	**1.16**	0.69	**1.21**
Oesophagus	0.12	**0.18**	0.14	**0.22**	0.16	**0.24**	0.17	**0.27**	0.20	**0.31**	0.22	**0.36**
Trachea												
Lung	0.11	**0.29**	0.13	**0.33**	0.14	**0.37**	0.16	**0.40**	0.18	**0.45**	0.20	**0.50**
Mamma	0.29	**5.99**	0.34	**6.58**	0.38	**6.97**	0.42	**7.25**	0.48	**7.73**	0.54	**8.10**
Heart	0.17	**0.60**	0.20	**0.70**	0.23	**0.77**	0.24	**0.83**	0.28	**0.93**	0.31	**1.01**
Stomach	0.70	**0.75**	0.81	**0.87**	0.89	**0.96**	0.95	**1.03**	1.07	**1.15**	1.16	**1.25**
Spleen	0.21	**0.25**	0.25	**0.29**	0.28	**0.33**	0.31	**0.36**	0.36	**0.43**	0.41	**0.48**
Liver	0.50	**0.56**	0.59	**0.66**	0.65	**0.73**	0.69	**0.78**	0.78	**0.88**	0.86	**0.97**
Gall bladder	0.61	**0.63**	0.71	**0.75**	0.79	**0.83**	0.86	**0.89**	0.97	**1.02**	1.08	**1.13**
Pancreas	0.35	**0.39**	0.41	**0.46**	0.46	**0.52**	0.51	**0.57**	0.59	**0.66**	0.66	**0.74**
Small intestine	0.49	**0.51**	0.58	**0.60**	0.64	**0.67**	0.70	**0.73**	0.80	**0.83**	0.89	**0.93**
Large intestine	0.58	**0.61**	0.69	**0.72**	0.76	**0.80**	0.82	**0.86**	0.93	**0.98**	1.03	**1.08**
Sigma/rectum	0.43	**0.45**	0.51	**0.54**	0.57	**0.60**	0.62	**0.65**	0.71	**0.75**	0.79	**0.84**
Lumbar spine	0.28	**0.30**	0.34	**0.37**	0.39	**0.42**	0.43	**0.46**	0.51	**0.55**	0.59	**0.63**
Pelvis	0.52	**0.57**	0.62	**0.67**	0.70	**0.76**	0.76	**0.83**	0.88	**0.96**	0.99	**1.07**
Adrenals	0.13	**0.16**	0.16	**0.19**	0.18	**0.21**	0.20	**0.24**	0.23	**0.28**	0.27	**0.32**
Kidneys	0.13	**0.14**	0.16	**0.17**	0.18	**0.20**	0.20	**0.22**	0.23	**0.26**	0.27	**0.30**
Urinary bladder	0.77	**0.80**	0.90	**0.92**	0.98	**1.02**	1.05	**1.09**	1.17	**1.22**	1.28	**1.32**
Uterus	1.47	**1.52**	1.74	**1.80**	1.94	**2.01**	2.10	**2.18**	2.41	**2.51**	2.69	**2.80**
Ovaries	0.39	**0.41**	0.47	**0.49**	0.52	**0.55**	0.57	**0.60**	0.66	**0.70**	0.74	**0.79**
Prostate	0.58	**0.62**	0.68	**0.73**	0.76	**0.81**	0.82	**0.87**	0.93	**0.99**	1.03	**1.10**
Testes	1.14	**1.21**	1.25	**1.33**	1.32	**1.40**	1.37	**1.46**	1.45	**1.56**	1.52	**1.63**
Upper arm		**0.17**		**0.19**		**0.22**		**0.23**		**0.26**		**0.28**
Elbow		**0.90**		**1.04**		**1.14**		**1.22**		**1.35**		**1.46**
Forearm		**1.03**		**1.19**		**1.30**		**1.38**		**1.53**		**1.64**
Upper leg	0.35	**0.59**	0.42	**0.71**	0.47	**0.79**	0.52	**0.86**	0.60	**1.00**	0.67	**1.12**
Knee	0.01	**0.06**	0.01	**0.07**	0.02	**0.08**	0.02	**0.08**	0.02	**0.10**	0.02	**0.11**
Lower leg												
Lymph nodes	0.30	**0.35**	0.35	**0.41**	0.39	**0.46**	0.42	**0.50**	0.48	**0.57**	0.53	**0.63**
Skin	0.21	**0.32**	0.23	**0.34**	0.23	**0.35**	0.24	**0.36**	0.25	**0.38**	0.26	**0.40**
Muscle	0.23	**0.31**	0.26	**0.35**	0.29	**0.38**	0.31	**0.41**	0.34	**0.45**	0.37	**0.49**
Skeleton	0.15	**0.34**	0.18	**0.39**	0.20	**0.43**	0.21	**0.47**	0.24	**0.52**	0.26	**0.57**
Red bone marrow	0.12	**0.16**	0.14	**0.19**	0.16	**0.21**	0.17	**0.23**	0.20	**0.27**	0.23	**0.30**
Whole body dose	0.22	**0.32**	0.26	**0.36**	0.28	**0.40**	0.30	**0.42**	0.33	**0.47**	0.36	**0.51**
ED (ICRP 60)	0.43	**0.77**	0.49	**0.87**	0.54	**0.94**	0.58	**1.00**	0.65	**1.10**	0.70	**1.18**
ED (ICRP 103)	0.37	**1.11**	0.43	**1.24**	0.47	**1.33**	0.51	**1.40**	0.57	**1.53**	0.63	**1.63**
Total filtration (mm) Al	**1.5**		**2.5**		**3.5**		**4.5**		**3.5**		**3.5**	
Cu	**0**		**0**		**0**		**0**		**0.1**		**0.2**	

Table 12.99 Normalized organ doses—Abdomen ap—13–17 years

Beam projection:			ap									**Table cells:**
Focus-to-detector distance (cm):			115									(+) = normal font = optimal field setting
Tube voltage (kVp):			120									**(−) = bold font = suboptimal field setting**
Total filtrations:			See last table row.									Empty table cells correspond to values <0.01
Organ or tissue	**Organ dose D_T : Entrance dose K_E**								**(mSv * mGy^{-1})**			
	(+)	(−)	(+)	(−)	(+)	(−)	(+)	(−)	(+)	(−)	(+)	(−)
Skull												
Brain												
Oral mucosa												
Salivary glands												
Cervical spine						**0.01**		**0.01**		**0.01**		**0.02**
Thyroid										**0.01**		**0.01**
Thymus	0.02	**0.10**	0.02	**0.12**	0.02	**0.13**	0.02	**0.14**	0.03	**0.16**	0.03	**0.18**
Thoracic spine	0.11	**0.19**	0.13	**0.22**	0.15	**0.25**	0.16	**0.28**	0.19	**0.32**	0.21	**0.37**
Claviculae		**0.02**		**0.02**		**0.03**		**0.03**		**0.03**	0.01	**0.04**
Scapulae	0.02	**0.11**	0.03	**0.13**	0.03	**0.15**	0.03	**0.16**	0.04	**0.18**	0.04	**0.20**
Ribs	0.53	**0.90**	0.58	**0.99**	0.61	**1.05**	0.63	**1.09**	0.66	**1.16**	0.68	**1.20**
Oesophagus	0.13	**0.20**	0.15	**0.24**	0.17	**0.27**	0.18	**0.29**	0.21	**0.33**	0.23	**0.37**
Trachea												
Lung	0.12	**0.31**	0.14	**0.36**	0.15	**0.39**	0.16	**0.42**	0.18	**0.47**	0.20	**0.51**
Mamma	0.32	**6.22**	0.37	**6.78**	0.41	**7.14**	0.44	**7.40**	0.51	**7.83**	0.56	**8.16**
Heart	0.19	**0.65**	0.22	**0.74**	0.24	**0.81**	0.26	**0.86**	0.29	**0.96**	0.32	**1.04**
Stomach	0.75	**0.81**	0.86	**0.92**	0.94	**1.01**	0.99	**1.07**	1.10	**1.18**	1.18	**1.28**
Spleen	0.23	**0.27**	0.28	**0.32**	0.31	**0.36**	0.33	**0.39**	0.38	**0.45**	0.43	**0.50**
Liver	0.54	**0.61**	0.62	**0.70**	0.68	**0.77**	0.73	**0.82**	0.81	**0.91**	0.88	**0.99**
Gall bladder	0.66	**0.69**	0.76	**0.80**	0.84	**0.88**	0.90	**0.94**	1.01	**1.06**	1.10	**1.16**
Pancreas	0.38	**0.43**	0.45	**0.50**	0.50	**0.56**	0.54	**0.60**	0.62	**0.69**	0.69	**0.77**
Small intestine	0.53	**0.56**	0.62	**0.65**	0.68	**0.72**	0.74	**0.77**	0.83	**0.87**	0.92	**0.96**
Large intestine	0.63	**0.66**	0.73	**0.77**	0.81	**0.85**	0.86	**0.91**	0.97	**1.02**	1.06	**1.11**
Sigma/rectum	0.47	**0.50**	0.55	**0.58**	0.61	**0.64**	0.65	**0.69**	0.74	**0.79**	0.82	**0.87**
Lumbar spine	0.32	**0.34**	0.38	**0.41**	0.43	**0.46**	0.47	**0.50**	0.54	**0.58**	0.62	**0.66**
Pelvis	0.57	**0.62**	0.67	**0.73**	0.74	**0.81**	0.80	**0.87**	0.91	**0.99**	1.01	**1.10**
Adrenals	0.15	**0.18**	0.18	**0.21**	0.20	**0.24**	0.21	**0.26**	0.25	**0.30**	0.28	**0.34**
Kidneys	0.15	**0.16**	0.18	**0.19**	0.20	**0.22**	0.21	**0.24**	0.25	**0.28**	0.28	**0.32**
Urinary bladder	0.83	**0.86**	0.95	**0.98**	1.03	**1.07**	1.09	**1.13**	1.21	**1.25**	1.30	**1.35**
Uterus	1.61	**1.67**	1.87	**1.95**	2.07	**2.15**	2.22	**2.31**	2.52	**2.62**	2.78	**2.89**
Ovaries	0.43	**0.45**	0.51	**0.53**	0.56	**0.59**	0.61	**0.64**	0.69	**0.73**	0.77	**0.82**
Prostate	0.63	**0.67**	0.73	**0.78**	0.80	**0.86**	0.86	**0.92**	0.97	**1.03**	1.06	**1.13**
Testes	1.18	**1.25**	1.28	**1.37**	1.35	**1.44**	1.39	**1.49**	1.47	**1.58**	1.53	**1.64**
Upper arm		**0.18**		**0.21**		**0.22**		**0.24**		**0.26**		**0.29**
Elbow		**0.95**		**1.09**		**1.18**		**1.25**		**1.37**		**1.46**
Forearm		**1.09**		**1.24**		**1.34**		**1.41**		**1.54**		**1.63**
Upper leg	0.39	**0.65**	0.45	**0.76**	0.50	**0.84**	0.54	**0.91**	0.62	**1.03**	0.69	**1.14**
Knee	0.01	**0.06**	0.02	**0.07**	0.02	**0.08**	0.02	**0.09**	0.02	**0.10**	0.03	**0.12**
Lower leg												
Lymph nodes	0.32	**0.38**	0.38	**0.44**	0.41	**0.49**	0.44	**0.52**	0.50	**0.59**	0.55	**0.65**
Skin	0.22	**0.33**	0.23	**0.35**	0.24	**0.36**	0.25	**0.37**	0.26	**0.39**	0.27	**0.40**
Muscle	0.25	**0.33**	0.28	**0.37**	0.30	**0.40**	0.32	**0.42**	0.35	**0.47**	0.38	**0.50**
Skeleton	0.16	**0.36**	0.19	**0.42**	0.21	**0.45**	0.22	**0.48**	0.24	**0.53**	0.27	**0.58**
Red bone marrow	0.13	**0.18**	0.15	**0.21**	0.17	**0.23**	0.19	**0.25**	0.21	**0.28**	0.24	**0.32**
Whole body dose	0.24	**0.34**	0.27	**0.38**	0.29	**0.42**	0.31	**0.44**	0.34	**0.49**	0.37	**0.52**
ED (ICRP 60)	0.46	**0.82**	0.52	**0.91**	0.57	**0.98**	0.60	**1.03**	0.67	**1.13**	0.72	**1.20**
ED (ICRP 103)	0.40	**1.17**	0.46	**1.29**	0.50	**1.38**	0.53	**1.45**	0.59	**1.56**	0.64	**1.66**
Total filtration (mm) Al	**1.5**		**2.5**		**3.5**		**4.5**		**3.5**		**3.5**	
Cu	**0**		**0**		**0**		**0**		**0.1**		**0.2**	

Table 12.100 Normalized organ doses—Abdomen ap—>17 years

Beam projection:			ap									**Table cells:**
Focus-to-detector distance (cm):			115						(+) = normal font = optimal field setting			
Tube voltage (kVp):			70						**(−) = bold font = suboptimal field setting**			
Total filtrations:			See last table row.						Empty table cells correspond to values <0.01			
Organ or tissue	**Organ dose D_T : Entrance dose K_E**								**(mSv * mGy^{-1})**			
	(+)	(−)	(+)	(−)	(+)	(−)	(+)	(−)	(+)	(−)	(+)	(−)
Skull												
Brain												
Oral mucosa												
Salivary glands												
Cervical spine												
Thyroid												
Thymus		**0.04**		**0.05**		**0.06**	0.01	**0.07**	0.01	**0.08**	0.02	**0.10**
Thoracic spine	0.04	**0.06**	0.05	**0.08**	0.06	**0.09**	0.07	**0.11**	0.09	**0.14**	0.12	**0.18**
Claviculae										**0.01**		**0.01**
Scapulae		**0.05**	0.01	**0.06**	0.01	**0.07**	0.02	**0.09**	0.02	**0.11**	0.03	**0.13**
Ribs	0.42	**0.65**	0.50	**0.77**	0.55	**0.85**	0.59	**0.91**	0.65	**1.02**	0.71	**1.12**
Oesophagus	0.06	**0.08**	0.08	**0.11**	0.09	**0.13**	0.11	**0.15**	0.13	**0.19**	0.16	**0.23**
Trachea												
Lung	0.07	**0.16**	0.09	**0.20**	0.11	**0.23**	0.12	**0.26**	0.14	**0.31**	0.16	**0.36**
Mamma	0.02	**0.37**	0.02	**0.43**	0.03	**0.47**	0.03	**0.49**	0.04	**0.54**	0.04	**0.59**
Heart	0.11	**0.33**	0.14	**0.42**	0.17	**0.48**	0.19	**0.53**	0.23	**0.63**	0.27	**0.73**
Stomach	0.44	**0.46**	0.55	**0.57**	0.63	**0.66**	0.69	**0.73**	0.82	**0.87**	0.94	**0.99**
Spleen	0.10	**0.11**	0.13	**0.14**	0.15	**0.17**	0.18	**0.20**	0.22	**0.25**	0.27	**0.31**
Liver	0.30	**0.33**	0.38	**0.42**	0.44	**0.48**	0.49	**0.54**	0.59	**0.64**	0.67	**0.74**
Gall bladder	0.33	**0.35**	0.43	**0.45**	0.50	**0.52**	0.56	**0.59**	0.69	**0.71**	0.80	**0.84**
Pancreas	0.17	**0.18**	0.22	**0.24**	0.26	**0.28**	0.30	**0.32**	0.37	**0.40**	0.45	**0.49**
Small intestine	0.26	**0.27**	0.34	**0.35**	0.39	**0.41**	0.44	**0.46**	0.54	**0.56**	0.64	**0.67**
Large intestine	0.33	**0.34**	0.42	**0.44**	0.49	**0.51**	0.55	**0.57**	0.67	**0.69**	0.78	**0.81**
Sigma/rectum	0.22	**0.23**	0.29	**0.30**	0.34	**0.36**	0.39	**0.40**	0.48	**0.50**	0.57	**0.59**
Lumbar spine	0.10	**0.11**	0.14	**0.15**	0.17	**0.18**	0.20	**0.21**	0.26	**0.27**	0.33	**0.34**
Pelvis	0.26	**0.27**	0.33	**0.35**	0.40	**0.42**	0.46	**0.48**	0.57	**0.61**	0.69	**0.74**
Adrenals	0.05	**0.06**	0.07	**0.08**	0.09	**0.10**	0.10	**0.11**	0.13	**0.15**	0.16	**0.18**
Kidneys	0.05	**0.06**	0.07	**0.07**	0.08	**0.09**	0.10	**0.10**	0.13	**0.14**	0.16	**0.17**
Urinary bladder	0.49	**0.50**	0.60	**0.62**	0.69	**0.71**	0.77	**0.78**	0.91	**0.93**	1.03	**1.06**
Uterus	0.30	**0.31**	0.39	**0.40**	0.46	**0.47**	0.52	**0.53**	0.64	**0.65**	0.75	**0.77**
Ovaries	0.19	**0.19**	0.25	**0.25**	0.30	**0.30**	0.34	**0.35**	0.42	**0.43**	0.51	**0.52**
Prostate	0.30	**0.31**	0.39	**0.40**	0.46	**0.47**	0.51	**0.53**	0.62	**0.64**	0.73	**0.75**
Testes	0.88	**0.91**	1.02	**1.06**	1.12	**1.15**	1.19	**1.23**	1.31	**1.36**	1.42	**1.47**
Upper arm		**0.07**		**0.09**		**0.11**		**0.12**		**0.15**		**0.17**
Elbow		**0.56**		**0.70**		**0.81**		**0.90**		**1.07**		**1.23**
Forearm		**0.69**		**0.86**		**0.99**		**1.10**		**1.30**		**1.48**
Upper leg	0.20	**0.25**	0.26	**0.33**	0.32	**0.39**	0.36	**0.45**	0.46	**0.56**	0.55	**0.68**
Knee		**0.04**	0.01	**0.05**	0.01	**0.06**	0.01	**0.07**	0.02	**0.09**	0.02	**0.11**
Lower leg												
Lymph nodes	0.17	**0.19**	0.21	**0.24**	0.25	**0.28**	0.28	**0.32**	0.34	**0.38**	0.39	**0.45**
Skin	0.20	**0.29**	0.21	**0.30**	0.22	**0.32**	0.23	**0.33**	0.24	**0.35**	0.25	**0.36**
Muscle	0.16	**0.20**	0.20	**0.24**	0.22	**0.27**	0.24	**0.30**	0.28	**0.35**	0.31	**0.39**
Skeleton	0.10	**0.20**	0.12	**0.25**	0.14	**0.28**	0.15	**0.31**	0.19	**0.38**	0.22	**0.44**
Red bone marrow	0.06	**0.08**	0.08	**0.10**	0.09	**0.12**	0.10	**0.13**	0.13	**0.16**	0.15	**0.19**
Whole body dose	0.15	**0.20**	0.19	**0.25**	0.21	**0.28**	0.23	**0.31**	0.27	**0.36**	0.31	**0.41**
ED (ICRP 60)	0.26	**0.30**	0.32	**0.37**	0.37	**0.42**	0.40	**0.47**	0.47	**0.55**	0.54	**0.62**
ED (ICRP 103)	0.20	**0.27**	0.25	**0.33**	0.29	**0.38**	0.32	**0.42**	0.38	**0.49**	0.44	**0.56**
Total filtration (mm) **Al**	**1.5**		**2.5**		**3.5**		**4.5**		**3.5**		**3.5**	
Cu	**0**		**0**		**0**		**0**		**0.1**		**0.2**	

Table 12.101 Normalized organ doses—Abdomen ap—>17 years

Beam projection:			ap									**Table cells:**
Focus-to-detector distance (cm):			115									(+) = normal font = optimal field setting
Tube voltage (kVp):			80									**(−) = bold font = suboptimal field setting**
Total filtrations:			See last table row.									Empty table cells correspond to values <0.01
Organ or tissue	**Organ dose D_T : Entrance dose K_E**								**(mSv * mGy^{-1})**			
	(+)	(−)	(+)	(−)	(+)	(−)	(+)	(−)	(+)	(−)	(+)	(−)
Skull												
Brain												
Oral mucosa												
Salivary glands												
Cervical spine												
Thyroid												
Thymus		**0.05**	0.01	**0.06**	0.01	**0.07**	0.02	**0.08**	0.02	**0.10**	0.03	**0.12**
Thoracic spine	0.05	**0.08**	0.07	**0.11**	0.08	**0.13**	0.09	**0.15**	0.12	**0.19**	0.15	**0.23**
Claviculae								**0.01**		**0.01**		**0.02**
Scapulae	0.01	**0.06**	0.02	**0.08**	0.02	**0.09**	0.02	**0.11**	0.03	**0.13**	0.04	**0.16**
Ribs	0.46	**0.71**	0.53	**0.83**	0.58	**0.91**	0.62	**0.97**	0.68	**1.07**	0.73	**1.16**
Oesophagus	0.08	**0.11**	0.10	**0.14**	0.12	**0.17**	0.13	**0.19**	0.16	**0.23**	0.20	**0.28**
Trachea												
Lung	0.09	**0.19**	0.11	**0.24**	0.12	**0.27**	0.14	**0.30**	0.16	**0.35**	0.18	**0.40**
Mamma	0.02	**0.40**	0.03	**0.46**	0.03	**0.50**	0.04	**0.52**	0.04	**0.57**	0.05	**0.62**
Heart	0.14	**0.39**	0.17	**0.48**	0.20	**0.55**	0.22	**0.61**	0.26	**0.71**	0.31	**0.81**
Stomach	0.51	**0.54**	0.63	**0.66**	0.72	**0.76**	0.78	**0.83**	0.92	**0.97**	1.04	**1.10**
Spleen	0.13	**0.14**	0.16	**0.19**	0.19	**0.22**	0.22	**0.25**	0.27	**0.31**	0.32	**0.37**
Liver	0.36	**0.39**	0.45	**0.49**	0.51	**0.56**	0.56	**0.61**	0.66	**0.73**	0.75	**0.83**
Gall bladder	0.41	**0.43**	0.51	**0.53**	0.59	**0.62**	0.66	**0.68**	0.79	**0.82**	0.91	**0.95**
Pancreas	0.22	**0.23**	0.28	**0.30**	0.32	**0.35**	0.36	**0.40**	0.45	**0.49**	0.53	**0.58**
Small intestine	0.32	**0.33**	0.41	**0.42**	0.47	**0.49**	0.53	**0.55**	0.64	**0.66**	0.74	**0.77**
Large intestine	0.40	**0.42**	0.50	**0.52**	0.58	**0.60**	0.64	**0.67**	0.76	**0.80**	0.88	**0.92**
Sigma/rectum	0.28	**0.29**	0.36	**0.37**	0.41	**0.43**	0.46	**0.48**	0.56	**0.58**	0.66	**0.68**
Lumbar spine	0.15	**0.16**	0.19	**0.20**	0.23	**0.24**	0.27	**0.28**	0.34	**0.36**	0.42	**0.44**
Pelvis	0.33	**0.35**	0.42	**0.45**	0.49	**0.52**	0.56	**0.59**	0.69	**0.73**	0.81	**0.86**
Adrenals	0.07	**0.09**	0.09	**0.11**	0.11	**0.13**	0.13	**0.15**	0.16	**0.19**	0.20	**0.23**
Kidneys	0.07	**0.08**	0.09	**0.10**	0.11	**0.12**	0.13	**0.14**	0.16	**0.17**	0.20	**0.21**
Urinary bladder	0.57	**0.58**	0.70	**0.71**	0.79	**0.81**	0.87	**0.89**	1.01	**1.04**	1.14	**1.17**
Uterus	0.37	**0.38**	0.47	**0.48**	0.55	**0.56**	0.62	**0.63**	0.75	**0.76**	0.87	**0.89**
Ovaries	0.25	**0.25**	0.31	**0.32**	0.37	**0.38**	0.41	**0.42**	0.51	**0.52**	0.60	**0.62**
Prostate	0.37	**0.38**	0.46	**0.48**	0.54	**0.55**	0.60	**0.62**	0.72	**0.74**	0.83	**0.86**
Testes	0.96	**0.99**	1.10	**1.14**	1.19	**1.23**	1.26	**1.31**	1.38	**1.44**	1.48	**1.54**
Upper arm		**0.09**		**0.11**		**0.13**		**0.14**		**0.17**		**0.19**
Elbow		**0.66**		**0.80**		**0.92**		**1.01**		**1.18**		**1.32**
Forearm		**0.80**		**0.98**		**1.11**		**1.21**		**1.41**		**1.58**
Upper leg	0.26	**0.32**	0.33	**0.41**	0.39	**0.49**	0.44	**0.55**	0.55	**0.68**	0.65	**0.81**
Knee	0.01	**0.05**	0.01	**0.07**	0.02	**0.08**	0.02	**0.09**	0.03	**0.11**	0.03	**0.13**
Lower leg												
Lymph nodes	0.20	**0.23**	0.25	**0.29**	0.29	**0.34**	0.33	**0.37**	0.39	**0.45**	0.45	**0.52**
Skin	0.21	**0.30**	0.22	**0.32**	0.23	**0.33**	0.24	**0.34**	0.25	**0.36**	0.26	**0.38**
Muscle	0.19	**0.23**	0.22	**0.28**	0.25	**0.31**	0.27	**0.34**	0.31	**0.39**	0.35	**0.43**
Skeleton	0.12	**0.23**	0.14	**0.29**	0.16	**0.33**	0.18	**0.36**	0.21	**0.43**	0.25	**0.49**
Red bone marrow	0.08	**0.10**	0.10	**0.13**	0.11	**0.15**	0.13	**0.16**	0.16	**0.20**	0.18	**0.23**
Whole body dose	0.18	**0.24**	0.21	**0.28**	0.24	**0.32**	0.26	**0.35**	0.30	**0.40**	0.34	**0.45**
ED (ICRP 60)	0.31	**0.35**	0.37	**0.43**	0.42	**0.48**	0.46	**0.53**	0.53	**0.61**	0.60	**0.69**
ED (ICRP 103)	0.24	**0.32**	0.29	**0.38**	0.33	**0.43**	0.37	**0.47**	0.43	**0.55**	0.49	**0.62**
Total filtration (mm) Al	**1.5**		**2.5**		**3.5**		**4.5**		**3.5**		**3.5**	
Cu	**0**		**0**		**0**		**0**		**0.1**		**0.2**	

Table 12.102 Normalized organ doses—Abdomen ap—>17 years

Beam projection:			ap									**Table cells:**
Focus-to-detector distance (cm):			115						(+) = normal font = optimal field setting			
Tube voltage (kVp):			110						**(−) = bold font = suboptimal field setting**			
Total filtrations:			See last table row.						Empty table cells correspond to values <0.01			
Organ or tissue	**Organ dose D_T : Entrance dose K_E**								**(mSv * mGy^{-1})**			
	(+)	(−)	(+)	(−)	(+)	(−)	(+)	(−)	(+)	(−)	(+)	(−)
Skull												
Brain												
Oral mucosa												
Salivary glands												
Cervical spine												**0.01**
Thyroid												
Thymus	0.02	**0.08**	0.02	**0.09**	0.02	**0.11**	0.03	**0.12**	0.03	**0.13**	0.04	**0.15**
Thoracic spine	0.10	**0.15**	0.12	**0.18**	0.14	**0.21**	0.15	**0.23**	0.18	**0.28**	0.21	**0.32**
Claviculae		**0.01**		**0.02**		**0.02**		**0.02**		**0.02**	0.01	**0.03**
Scapulae	0.02	**0.10**	0.03	**0.12**	0.03	**0.14**	0.04	**0.15**	0.04	**0.17**	0.05	**0.20**
Ribs	0.54	**0.85**	0.60	**0.94**	0.64	**1.00**	0.67	**1.05**	0.71	**1.12**	0.74	**1.18**
Oesophagus	0.13	**0.18**	0.16	**0.22**	0.18	**0.25**	0.19	**0.27**	0.23	**0.32**	0.26	**0.37**
Trachea												
Lung	0.13	**0.27**	0.15	**0.32**	0.16	**0.35**	0.18	**0.38**	0.20	**0.43**	0.22	**0.47**
Mamma	0.03	**0.48**	0.04	**0.53**	0.05	**0.56**	0.05	**0.58**	0.06	**0.63**	0.06	**0.66**
Heart	0.20	**0.55**	0.24	**0.64**	0.27	**0.71**	0.29	**0.76**	0.33	**0.86**	0.37	**0.94**
Stomach	0.71	**0.75**	0.83	**0.88**	0.91	**0.97**	0.97	**1.03**	1.09	**1.16**	1.19	**1.27**
Spleen	0.21	**0.24**	0.26	**0.29**	0.29	**0.33**	0.32	**0.36**	0.37	**0.43**	0.42	**0.48**
Liver	0.51	**0.56**	0.60	**0.66**	0.66	**0.73**	0.71	**0.78**	0.80	**0.89**	0.88	**0.98**
Gall bladder	0.61	**0.63**	0.72	**0.75**	0.80	**0.83**	0.86	**0.90**	0.98	**1.03**	1.09	**1.14**
Pancreas	0.35	**0.38**	0.42	**0.46**	0.47	**0.51**	0.51	**0.56**	0.60	**0.65**	0.67	**0.74**
Small intestine	0.49	**0.51**	0.58	**0.61**	0.65	**0.68**	0.71	**0.73**	0.81	**0.84**	0.90	**0.94**
Large intestine	0.59	**0.62**	0.70	**0.73**	0.77	**0.81**	0.83	**0.87**	0.95	**0.99**	1.05	**1.10**
Sigma/rectum	0.44	**0.46**	0.52	**0.54**	0.58	**0.60**	0.63	**0.66**	0.72	**0.76**	0.81	**0.85**
Lumbar spine	0.28	**0.29**	0.33	**0.35**	0.38	**0.40**	0.42	**0.44**	0.50	**0.53**	0.58	**0.61**
Pelvis	0.52	**0.55**	0.62	**0.66**	0.70	**0.75**	0.76	**0.81**	0.89	**0.94**	1.00	**1.06**
Adrenals	0.13	**0.16**	0.16	**0.19**	0.18	**0.21**	0.20	**0.24**	0.24	**0.28**	0.27	**0.32**
Kidneys	0.13	**0.14**	0.16	**0.17**	0.18	**0.20**	0.20	**0.22**	0.24	**0.26**	0.27	**0.30**
Urinary bladder	0.79	**0.81**	0.91	**0.94**	1.00	**1.03**	1.07	**1.10**	1.20	**1.24**	1.31	**1.35**
Uterus	0.58	**0.59**	0.69	**0.70**	0.77	**0.79**	0.83	**0.85**	0.96	**0.98**	1.07	**1.10**
Ovaries	0.40	**0.41**	0.48	**0.49**	0.53	**0.55**	0.58	**0.60**	0.68	**0.70**	0.76	**0.79**
Prostate	0.55	**0.57**	0.65	**0.67**	0.73	**0.75**	0.78	**0.81**	0.89	**0.93**	0.99	**1.03**
Testes	1.15	**1.19**	1.27	**1.31**	1.35	**1.40**	1.40	**1.46**	1.50	**1.56**	1.58	**1.64**
Upper arm		**0.13**		**0.15**		**0.17**		**0.18**		**0.20**		**0.22**
Elbow		**0.87**		**1.01**		**1.11**		**1.19**		**1.32**		**1.43**
Forearm		**1.05**		**1.21**		**1.32**		**1.41**		**1.56**		**1.67**
Upper leg	0.42	**0.52**	0.50	**0.62**	0.57	**0.70**	0.62	**0.76**	0.72	**0.89**	0.81	**1.00**
Knee	0.02	**0.09**	0.03	**0.10**	0.03	**0.12**	0.03	**0.13**	0.04	**0.15**	0.05	**0.17**
Lower leg												
Lymph nodes	0.30	**0.35**	0.36	**0.41**	0.40	**0.46**	0.43	**0.49**	0.49	**0.56**	0.55	**0.63**
Skin	0.23	**0.33**	0.24	**0.35**	0.25	**0.36**	0.26	**0.37**	0.27	**0.39**	0.28	**0.41**
Muscle	0.25	**0.31**	0.29	**0.36**	0.31	**0.39**	0.33	**0.41**	0.37	**0.46**	0.40	**0.50**
Skeleton	0.17	**0.33**	0.19	**0.38**	0.22	**0.42**	0.23	**0.45**	0.26	**0.51**	0.29	**0.56**
Red bone marrow	0.13	**0.16**	0.15	**0.19**	0.17	**0.21**	0.18	**0.23**	0.21	**0.27**	0.24	**0.30**
Whole body dose	0.24	**0.32**	0.28	**0.37**	0.30	**0.40**	0.32	**0.43**	0.36	**0.48**	0.39	**0.52**
ED (ICRP 60)	0.42	**0.49**	0.48	**0.56**	0.53	**0.62**	0.57	**0.66**	0.64	**0.74**	0.70	**0.80**
ED (ICRP 103)	0.34	**0.44**	0.39	**0.50**	0.43	**0.55**	0.46	**0.59**	0.52	**0.66**	0.58	**0.72**
Total filtration (mm) **Al**	**1.5**		**2.5**		**3.5**		**4.5**		**3.5**		**3.5**	
Cu	**0**		**0**		**0**		**0**		**0.1**		**0.2**	

Table 12.103 Normalized organ doses—Abdomen ap—>17 years

Beam projection:			ap									**Table cells:**
Focus-to-detector distance (cm):			115									(+) = normal font = optimal field setting
Tube voltage (kVp):			120									**(−) = bold font = suboptimal field setting**
Total filtrations:			See last table row.									Empty table cells correspond to values <0.01
Organ or tissue	**Organ dose D_T : Entrance dose K_E**								**(mSv * mGy⁻¹)**			
	(+)	(−)	(+)	(−)	(+)	(−)	(+)	(−)	(+)	(−)	(+)	(−)
Skull												
Brain												
Oral mucosa												
Salivary glands												
Cervical spine										**0.01**		**0.01**
Thyroid												
Thymus	0.02	**0.09**	0.02	**0.10**	0.03	**0.11**	0.03	**0.12**	0.03	**0.14**	0.04	**0.16**
Thoracic spine	0.11	**0.17**	0.13	**0.21**	0.15	**0.23**	0.16	**0.25**	0.19	**0.30**	0.22	**0.34**
Claviculae		**0.01**		**0.02**		**0.02**		**0.02**		**0.03**	0.01	**0.03**
Scapulae	0.03	**0.11**	0.03	**0.13**	0.03	**0.15**	0.04	**0.16**	0.04	**0.18**	0.05	**0.20**
Ribs	0.56	**0.87**	0.61	**0.96**	0.65	**1.02**	0.67	**1.06**	0.71	**1.12**	0.73	**1.16**
Oesophagus	0.15	**0.21**	0.17	**0.24**	0.19	**0.27**	0.21	**0.29**	0.24	**0.34**	0.27	**0.38**
Trachea												
Lung	0.14	**0.29**	0.16	**0.34**	0.17	**0.37**	0.19	**0.40**	0.21	**0.44**	0.23	**0.48**
Mamma	0.04	**0.50**	0.04	**0.55**	0.05	**0.58**	0.05	**0.60**	0.06	**0.63**	0.07	**0.66**
Heart	0.22	**0.60**	0.26	**0.69**	0.29	**0.75**	0.31	**0.80**	0.35	**0.89**	0.38	**0.96**
Stomach	0.76	**0.81**	0.88	**0.93**	0.95	**1.01**	1.01	**1.08**	1.12	**1.20**	1.21	**1.30**
Spleen	0.24	**0.27**	0.28	**0.32**	0.31	**0.36**	0.34	**0.39**	0.39	**0.45**	0.44	**0.51**
Liver	0.56	**0.61**	0.64	**0.70**	0.70	**0.77**	0.75	**0.82**	0.83	**0.92**	0.90	**1.00**
Gall bladder	0.66	**0.69**	0.77	**0.80**	0.85	**0.88**	0.91	**0.95**	1.02	**1.07**	1.12	**1.17**
Pancreas	0.39	**0.42**	0.46	**0.50**	0.51	**0.55**	0.55	**0.60**	0.63	**0.69**	0.70	**0.77**
Small intestine	0.54	**0.56**	0.63	**0.65**	0.70	**0.72**	0.75	**0.78**	0.85	**0.88**	0.93	**0.97**
Large intestine	0.64	**0.67**	0.75	**0.78**	0.82	**0.86**	0.88	**0.92**	0.99	**1.03**	1.08	**1.13**
Sigma/rectum	0.48	**0.50**	0.56	**0.59**	0.62	**0.65**	0.67	**0.70**	0.76	**0.79**	0.84	**0.88**
Lumbar spine	0.31	**0.33**	0.37	**0.39**	0.42	**0.44**	0.46	**0.48**	0.53	**0.56**	0.61	**0.64**
Pelvis	0.57	**0.61**	0.67	**0.71**	0.75	**0.79**	0.81	**0.86**	0.92	**0.98**	1.02	**1.09**
Adrenals	0.15	**0.18**	0.18	**0.21**	0.20	**0.24**	0.22	**0.26**	0.26	**0.30**	0.29	**0.34**
Kidneys	0.15	**0.16**	0.18	**0.19**	0.20	**0.22**	0.22	**0.24**	0.26	**0.28**	0.29	**0.32**
Urinary bladder	0.85	**0.87**	0.97	**0.99**	1.05	**1.08**	1.12	**1.15**	1.24	**1.27**	1.34	**1.38**
Uterus	0.64	**0.65**	0.74	**0.76**	0.82	**0.84**	0.89	**0.91**	1.01	**1.03**	1.11	**1.14**
Ovaries	0.44	**0.45**	0.52	**0.53**	0.58	**0.59**	0.62	**0.64**	0.71	**0.74**	0.79	**0.82**
Prostate	0.60	**0.62**	0.70	**0.72**	0.77	**0.80**	0.83	**0.85**	0.93	**0.96**	1.02	**1.06**
Testes	1.19	**1.24**	1.31	**1.36**	1.38	**1.43**	1.43	**1.49**	1.52	**1.59**	1.59	**1.66**
Upper arm		**0.14**		**0.16**		**0.17**		**0.18**		**0.21**		**0.22**
Elbow		**0.93**		**1.06**		**1.15**		**1.22**		**1.34**		**1.43**
Forearm		**1.10**		**1.26**		**1.36**		**1.44**		**1.57**		**1.67**
Upper leg	0.46	**0.57**	0.54	**0.67**	0.61	**0.75**	0.66	**0.81**	0.75	**0.93**	0.84	**1.03**
Knee	0.03	**0.10**	0.03	**0.11**	0.03	**0.13**	0.04	**0.14**	0.04	**0.16**	0.05	**0.18**
Lower leg												
Lymph nodes	0.33	**0.38**	0.39	**0.44**	0.42	**0.49**	0.46	**0.52**	0.51	**0.59**	0.57	**0.65**
Skin	0.24	**0.34**	0.25	**0.36**	0.26	**0.37**	0.26	**0.38**	0.27	**0.40**	0.28	**0.41**
Muscle	0.27	**0.33**	0.30	**0.38**	0.33	**0.41**	0.35	**0.43**	0.38	**0.48**	0.41	**0.51**
Skeleton	0.18	**0.35**	0.21	**0.40**	0.23	**0.44**	0.24	**0.47**	0.27	**0.52**	0.30	**0.56**
Red bone marrow	0.14	**0.18**	0.17	**0.21**	0.18	**0.23**	0.20	**0.25**	0.23	**0.28**	0.25	**0.31**
Whole body dose	0.26	**0.34**	0.29	**0.39**	0.32	**0.42**	0.34	**0.45**	0.37	**0.49**	0.40	**0.53**
ED (ICRP 60)	0.45	**0.52**	0.51	**0.60**	0.56	**0.65**	0.59	**0.69**	0.66	**0.76**	0.71	**0.82**
ED (ICRP 103)	0.36	**0.47**	0.42	**0.53**	0.46	**0.58**	0.49	**0.62**	0.54	**0.69**	0.59	**0.74**
Total filtration (mm) Al	**1.5**		**2.5**		**3.5**		**4.5**		**3.5**		**3.5**	
Cu	**0**		**0**		**0**		**0**		**0.1**		**0.2**	

Table 12.104 Normalized organ doses—Abdomen pa—<1 year

Beam projection:			pa									**Table cells:**
Focus-to-detector distance (cm):			115									(+) = normal font = optimal field setting
Tube voltage (kVp):			60									**(−) = bold font = suboptimal field setting**
Total filtrations:			See last table row.									Empty table cells correspond to values <0.01
Organ or tissue	**Organ dose D_T : Entrance dose K_E**								**(mSv * mGy^{-1})**			
	(+)	(−)	(+)	(−)	(+)	(−)	(+)	(−)	(+)	(−)	(+)	(−)
Skull												
Brain												
Oral mucosa												
Salivary glands								**0.01**		**0.01**		**0.01**
Cervical spine		**0.02**		**0.02**	0.01	**0.03**	0.01	**0.03**	0.02	**0.04**	0.01	**0.04**
Thyroid		**0.01**		**0.02**		**0.02**		**0.03**	0.01	**0.03**		**0.03**
Thymus		**0.04**	0.01	**0.05**	0.01	**0.06**	0.02	**0.07**	0.02	**0.08**	0.02	**0.07**
Thoracic spine	0.49	**0.85**	0.61	**1.05**	0.69	**1.19**	0.76	**1.31**	0.88	**1.51**	0.86	**1.48**
Claviculae	0.01	**0.04**	0.02	**0.05**	0.02	**0.06**	0.02	**0.07**	0.03	**0.09**	0.03	**0.08**
Scapulae	0.05	**1.29**	0.06	**1.45**	0.07	**1.55**	0.08	**1.62**	0.10	**1.73**	0.09	**1.75**
Ribs	0.61	**1.02**	0.70	**1.17**	0.76	**1.28**	0.80	**1.35**	0.87	**1.48**	0.87	**1.48**
Oesophagus	0.11	**0.18**	0.14	**0.23**	0.17	**0.28**	0.19	**0.31**	0.23	**0.38**	0.21	**0.35**
Trachea												
Lung	0.18	**0.41**	0.21	**0.49**	0.23	**0.55**	0.25	**0.59**	0.29	**0.66**	0.28	**0.65**
Mamma	0.02	**0.17**	0.03	**0.21**	0.03	**0.25**	0.04	**0.27**	0.05	**0.32**	0.05	**0.30**
Heart	0.09	**0.19**	0.11	**0.25**	0.13	**0.29**	0.15	**0.32**	0.18	**0.39**	0.17	**0.37**
Stomach	0.19	**0.20**	0.24	**0.26**	0.28	**0.31**	0.32	**0.34**	0.38	**0.42**	0.36	**0.39**
Spleen	0.51	**0.54**	0.62	**0.65**	0.69	**0.73**	0.75	**0.80**	0.85	**0.91**	0.84	**0.89**
Liver	0.25	**0.28**	0.32	**0.35**	0.37	**0.40**	0.40	**0.44**	0.48	**0.52**	0.45	**0.50**
Gall bladder	0.17	**0.18**	0.22	**0.24**	0.27	**0.28**	0.30	**0.32**	0.37	**0.40**	0.34	**0.36**
Pancreas	0.27	**0.29**	0.35	**0.37**	0.41	**0.43**	0.45	**0.48**	0.54	**0.58**	0.51	**0.55**
Small intestine	0.26	**0.27**	0.34	**0.35**	0.39	**0.41**	0.44	**0.46**	0.53	**0.55**	0.50	**0.52**
Large intestine	0.22	**0.23**	0.29	**0.30**	0.34	**0.36**	0.38	**0.40**	0.46	**0.48**	0.43	**0.45**
Sigma/rectum	0.27	**0.29**	0.35	**0.37**	0.40	**0.43**	0.45	**0.48**	0.54	**0.57**	0.51	**0.54**
Lumbar spine	1.37	**1.40**	1.69	**1.73**	1.92	**1.97**	2.10	**2.15**	2.42	**2.48**	2.38	**2.44**
Pelvis	1.43	**1.47**	1.74	**1.78**	1.95	**2.00**	2.11	**2.17**	2.39	**2.47**	2.38	**2.45**
Adrenals	0.52	**0.54**	0.62	**0.65**	0.69	**0.73**	0.75	**0.79**	0.84	**0.89**	0.83	**0.88**
Kidneys	0.76	**0.77**	0.88	**0.90**	0.97	**0.99**	1.03	**1.05**	1.13	**1.16**	1.12	**1.15**
Urinary bladder	0.17	**0.18**	0.22	**0.24**	0.27	**0.28**	0.30	**0.32**	0.36	**0.38**	0.34	**0.36**
Uterus	0.25	**0.26**	0.32	**0.33**	0.38	**0.39**	0.43	**0.44**	0.51	**0.53**	0.48	**0.50**
Ovaries	0.28	**0.29**	0.36	**0.38**	0.42	**0.44**	0.47	**0.50**	0.57	**0.60**	0.53	**0.56**
Prostate	0.19	**0.21**	0.25	**0.27**	0.29	**0.32**	0.33	**0.36**	0.39	**0.43**	0.37	**0.41**
Testes	0.22	**0.25**	0.28	**0.31**	0.33	**0.36**	0.36	**0.39**	0.42	**0.46**	0.41	**0.44**
Upper arm		**0.42**		**0.52**		**0.58**		**0.63**		**0.72**		**0.71**
Elbow		**1.20**		**1.45**		**1.62**		**1.75**		**1.98**		**1.98**
Forearm		**1.32**		**1.58**		**1.77**		**1.90**		**2.13**		**2.14**
Upper leg	0.71	**0.98**	0.89	**1.23**	1.03	**1.41**	1.13	**1.56**	1.32	**1.81**	1.29	**1.78**
Knee	0.05	**0.44**	0.07	**0.55**	0.08	**0.63**	0.10	**0.69**	0.12	**0.80**	0.11	**0.79**
Lower leg				**0.01**		**0.02**		**0.02**	0.01	**0.02**		**0.02**
Lymph nodes	0.16	**0.20**	0.21	**0.25**	0.24	**0.29**	0.27	**0.32**	0.32	**0.39**	0.30	**0.37**
Skin	0.22	**0.33**	0.24	**0.35**	0.25	**0.36**	0.26	**0.37**	0.27	**0.39**	0.27	**0.39**
Muscle	0.22	**0.30**	0.26	**0.35**	0.29	**0.39**	0.31	**0.42**	0.35	**0.47**	0.34	**0.46**
Skeleton	0.40	**0.77**	0.49	**0.93**	0.55	**1.04**	0.60	**1.12**	0.68	**1.27**	0.67	**1.26**
Red bone marrow	0.11	**0.20**	0.14	**0.25**	0.16	**0.28**	0.17	**0.30**	0.20	**0.35**	0.19	**0.34**
Whole body dose	0.22	**0.31**	0.26	**0.37**	0.29	**0.41**	0.31	**0.44**	0.35	**0.50**	0.34	**0.49**
ED (ICRP 60)	0.18	**0.25**	0.23	**0.31**	0.27	**0.36**	0.30	**0.39**	0.35	**0.46**	0.33	**0.44**
ED (ICRP 103)	0.16	**0.24**	0.21	**0.30**	0.24	**0.34**	0.26	**0.37**	0.31	**0.44**	0.30	**0.42**
Total filtration (mm) Al	**1.5**		**2.5**		**3.5**		**4.5**		**3.5**		**3.5**	
Cu	**0**		**0**		**0**		**0**		**0.1**		**0.2**	

Table 12.105 Normalized organ doses—Abdomen pa—<1 year

Beam projection:			pa									**Table cells:**
Focus-to-detector distance (cm):			115									(+) = normal font = optimal field setting
Tube voltage (kVp):			70									**(−) = bold font = suboptimal field setting**
Total filtrations:			See last table row.									Empty table cells correspond to values <0.01
Organ or tissue	**Organ dose D_T : Entrance dose K_E**								**(mSv * mGy^{-1})**			
	(+)	(−)	(+)	(−)	(+)	(−)	(+)	(−)	(+)	(−)	(+)	(−)
Skull										**0.01**		**0.01**
Brain												
Oral mucosa										**0.01**		**0.01**
Salivary glands						**0.01**		**0.01**		**0.02**		**0.02**
Cervical spine		**0.03**	0.01	**0.03**	0.01	**0.04**	0.02	**0.04**	0.02	**0.06**	0.03	**0.07**
Thyroid		**0.02**		**0.03**		**0.03**	0.01	**0.03**	0.01	**0.04**	0.02	**0.05**
Thymus	0.01	**0.05**	0.02	**0.06**	0.02	**0.07**	0.02	**0.08**	0.03	**0.10**	0.03	**0.12**
Thoracic spine	0.56	**0.96**	0.68	**1.17**	0.77	**1.32**	0.83	**1.43**	0.95	**1.62**	1.04	**1.78**
Claviculae	0.02	**0.05**	0.02	**0.07**	0.03	**0.08**	0.03	**0.09**	0.04	**0.11**	0.05	**0.13**
Scapulae	0.06	**1.36**	0.08	**1.51**	0.09	**1.60**	0.10	**1.66**	0.12	**1.75**	0.14	**1.81**
Ribs	0.65	**1.09**	0.74	**1.25**	0.80	**1.34**	0.84	**1.41**	0.90	**1.53**	0.95	**1.62**
Oesophagus	0.13	**0.22**	0.17	**0.28**	0.20	**0.33**	0.22	**0.36**	0.26	**0.44**	0.30	**0.50**
Trachea										**0.01**		**0.01**
Lung	0.20	**0.46**	0.23	**0.54**	0.26	**0.60**	0.28	**0.64**	0.31	**0.71**	0.34	**0.77**
Mamma	0.03	**0.21**	0.04	**0.25**	0.04	**0.29**	0.05	**0.32**	0.06	**0.37**	0.08	**0.42**
Heart	0.11	**0.23**	0.14	**0.29**	0.16	**0.34**	0.18	**0.38**	0.21	**0.44**	0.24	**0.51**
Stomach	0.23	**0.25**	0.29	**0.31**	0.33	**0.36**	0.37	**0.40**	0.44	**0.48**	0.51	**0.55**
Spleen	0.57	**0.61**	0.68	**0.73**	0.76	**0.81**	0.82	**0.87**	0.92	**0.98**	1.00	**1.07**
Liver	0.30	**0.33**	0.37	**0.40**	0.42	**0.46**	0.46	**0.50**	0.53	**0.58**	0.60	**0.66**
Gall bladder	0.22	**0.23**	0.28	**0.30**	0.32	**0.35**	0.36	**0.39**	0.44	**0.47**	0.51	**0.55**
Pancreas	0.33	**0.35**	0.41	**0.44**	0.47	**0.51**	0.52	**0.56**	0.62	**0.66**	0.70	**0.75**
Small intestine	0.32	**0.33**	0.40	**0.42**	0.46	**0.48**	0.51	**0.53**	0.61	**0.63**	0.69	**0.72**
Large intestine	0.27	**0.29**	0.35	**0.36**	0.40	**0.42**	0.45	**0.47**	0.53	**0.56**	0.61	**0.65**
Sigma/rectum	0.33	**0.34**	0.41	**0.43**	0.47	**0.50**	0.52	**0.55**	0.61	**0.65**	0.69	**0.74**
Lumbar spine	1.56	**1.59**	1.88	**1.93**	2.11	**2.16**	2.29	**2.34**	2.59	**2.66**	2.84	**2.92**
Pelvis	1.60	**1.64**	1.90	**1.96**	2.11	**2.17**	2.26	**2.33**	2.53	**2.61**	2.74	**2.83**
Adrenals	0.58	**0.61**	0.68	**0.72**	0.76	**0.80**	0.81	**0.86**	0.91	**0.96**	0.98	**1.05**
Kidneys	0.83	**0.84**	0.95	**0.97**	1.03	**1.06**	1.09	**1.12**	1.20	**1.23**	1.28	**1.32**
Urinary bladder	0.21	**0.23**	0.27	**0.29**	0.32	**0.34**	0.35	**0.37**	0.42	**0.45**	0.48	**0.52**
Uterus	0.31	**0.32**	0.39	**0.40**	0.45	**0.46**	0.50	**0.51**	0.59	**0.61**	0.68	**0.70**
Ovaries	0.34	**0.36**	0.42	**0.45**	0.49	**0.53**	0.55	**0.58**	0.65	**0.70**	0.74	**0.80**
Prostate	0.24	**0.26**	0.30	**0.33**	0.34	**0.38**	0.38	**0.42**	0.45	**0.50**	0.52	**0.57**
Testes	0.26	**0.29**	0.33	**0.36**	0.37	**0.41**	0.41	**0.44**	0.47	**0.52**	0.53	**0.58**
Upper arm		**0.48**		**0.57**		**0.63**		**0.68**		**0.77**		**0.83**
Elbow		**1.33**		**1.58**		**1.75**		**1.87**		**2.08**		**2.24**
Forearm		**1.46**		**1.71**		**1.89**		**2.02**		**2.23**		**2.39**
Upper leg	0.82	**1.13**	1.01	**1.39**	1.14	**1.57**	1.25	**1.71**	1.43	**1.96**	1.58	**2.16**
Knee	0.07	**0.51**	0.09	**0.62**	0.10	**0.70**	0.12	**0.76**	0.14	**0.86**	0.17	**0.95**
Lower leg		**0.01**		**0.02**		**0.02**	0.01	**0.02**	0.01	**0.03**	0.02	**0.03**
Lymph nodes	0.20	**0.24**	0.24	**0.30**	0.28	**0.34**	0.31	**0.37**	0.37	**0.44**	0.42	**0.50**
Skin	0.23	**0.34**	0.25	**0.36**	0.26	**0.38**	0.27	**0.39**	0.29	**0.41**	0.30	**0.43**
Muscle	0.25	**0.33**	0.29	**0.39**	0.32	**0.43**	0.34	**0.46**	0.38	**0.51**	0.41	**0.56**
Skeleton	0.45	**0.85**	0.54	**1.01**	0.60	**1.12**	0.64	**1.20**	0.72	**1.34**	0.78	**1.45**
Red bone marrow	0.13	**0.23**	0.16	**0.28**	0.18	**0.31**	0.19	**0.34**	0.22	**0.38**	0.24	**0.42**
Whole body dose	0.24	**0.35**	0.28	**0.41**	0.31	**0.45**	0.34	**0.48**	0.38	**0.54**	0.41	**0.59**
ED (ICRP 60)	0.22	**0.29**	0.27	**0.36**	0.31	**0.41**	0.34	**0.45**	0.40	**0.52**	0.45	**0.58**
ED (ICRP 103)	0.20	**0.28**	0.24	**0.34**	0.27	**0.39**	0.30	**0.43**	0.35	**0.49**	0.40	**0.55**
Total filtration (mm) Al	**1.5**		**2.5**		**3.5**		**4.5**		**3.5**		**3.5**	
Cu	**0**		**0**		**0**		**0**		**0.1**		**0.2**	

Table 12.106 Normalized organ doses—Abdomen pa—1–2 years

Beam projection:			pa									**Table cells:**
Focus-to-detector distance (cm):			115				(+) = normal font = optimal field setting					
Tube voltage (kVp):			60				**(−) = bold font = suboptimal field setting**					
Total filtrations:			See last table row.				Empty table cells correspond to values <0.01					
Organ or tissue	**Organ dose D_T : Entrance dose K_E**								**(mSv * mGy⁻¹)**			
	(+)	(−)	(+)	(−)	(+)	(−)	(+)	(−)	(+)	(−)	(+)	(−)
Skull												
Brain												
Oral mucosa												
Salivary glands												
Cervical spine				**0.01**		**0.02**		**0.02**		**0.03**		**0.02**
Thyroid				**0.01**		**0.01**		**0.02**		**0.02**		**0.02**
Thymus		**0.03**		**0.04**		**0.05**	0.01	**0.06**	0.01	**0.07**	0.01	**0.06**
Thoracic spine	0.33	**0.64**	0.42	**0.82**	0.49	**0.95**	0.54	**1.06**	0.65	**1.26**	0.62	**1.20**
Claviculae		**0.03**	0.01	**0.04**	0.01	**0.05**	0.02	**0.06**	0.02	**0.07**	0.02	**0.06**
Scapulae	0.02	**0.92**	0.03	**1.06**	0.03	**1.15**	0.04	**1.22**	0.05	**1.33**	0.04	**1.34**
Ribs	0.48	**0.94**	0.56	**1.10**	0.62	**1.21**	0.66	**1.30**	0.73	**1.44**	0.72	**1.43**
Oesophagus	0.08	**0.14**	0.10	**0.19**	0.12	**0.23**	0.14	**0.26**	0.17	**0.32**	0.15	**0.28**
Trachea												
Lung	0.10	**0.33**	0.13	**0.40**	0.14	**0.45**	0.16	**0.49**	0.18	**0.57**	0.18	**0.55**
Mamma	0.01	**0.13**	0.02	**0.17**	0.02	**0.20**	0.03	**0.22**	0.03	**0.27**	0.03	**0.25**
Heart	0.05	**0.12**	0.07	**0.16**	0.08	**0.20**	0.09	**0.22**	0.11	**0.27**	0.10	**0.25**
Stomach	0.11	**0.13**	0.15	**0.17**	0.18	**0.20**	0.21	**0.23**	0.25	**0.29**	0.23	**0.26**
Spleen	0.41	**0.44**	0.51	**0.55**	0.58	**0.63**	0.63	**0.69**	0.74	**0.80**	0.71	**0.77**
Liver	0.19	**0.21**	0.24	**0.27**	0.28	**0.31**	0.31	**0.35**	0.37	**0.42**	0.34	**0.39**
Gall bladder	0.11	**0.12**	0.15	**0.17**	0.18	**0.20**	0.21	**0.23**	0.27	**0.29**	0.23	**0.25**
Pancreas	0.19	**0.21**	0.24	**0.27**	0.29	**0.32**	0.33	**0.37**	0.40	**0.45**	0.37	**0.41**
Small intestine	0.18	**0.19**	0.24	**0.25**	0.28	**0.29**	0.32	**0.33**	0.40	**0.41**	0.36	**0.37**
Large intestine	0.15	**0.16**	0.20	**0.21**	0.24	**0.25**	0.27	**0.29**	0.34	**0.36**	0.30	**0.32**
Sigma/rectum	0.19	**0.20**	0.25	**0.26**	0.29	**0.31**	0.33	**0.35**	0.41	**0.43**	0.37	**0.39**
Lumbar spine	1.09	**1.11**	1.38	**1.40**	1.60	**1.62**	1.77	**1.80**	2.10	**2.14**	2.00	**2.04**
Pelvis	1.16	**1.19**	1.44	**1.48**	1.65	**1.69**	1.81	**1.86**	2.11	**2.17**	2.05	**2.11**
Adrenals	0.42	**0.44**	0.52	**0.55**	0.60	**0.63**	0.66	**0.70**	0.76	**0.81**	0.74	**0.78**
Kidneys	0.67	**0.68**	0.80	**0.81**	0.88	**0.90**	0.95	**0.97**	1.06	**1.09**	1.04	**1.07**
Urinary bladder	0.12	**0.12**	0.16	**0.16**	0.19	**0.19**	0.21	**0.22**	0.27	**0.28**	0.24	**0.24**
Uterus	0.16	**0.16**	0.21	**0.22**	0.25	**0.26**	0.29	**0.30**	0.36	**0.37**	0.32	**0.33**
Ovaries	0.19	**0.20**	0.25	**0.26**	0.30	**0.31**	0.34	**0.36**	0.43	**0.45**	0.38	**0.40**
Prostate	0.14	**0.14**	0.18	**0.18**	0.21	**0.22**	0.24	**0.25**	0.30	**0.31**	0.27	**0.27**
Testes	0.12	**0.11**	0.15	**0.15**	0.18	**0.17**	0.20	**0.19**	0.24	**0.24**	0.23	**0.22**
Upper arm		**0.30**		**0.38**		**0.44**		**0.48**		**0.57**		**0.54**
Elbow		**0.88**		**1.09**		**1.25**		**1.37**		**1.60**		**1.55**
Forearm		**0.99**		**1.23**		**1.39**		**1.53**		**1.76**		**1.73**
Upper leg	0.47	**0.63**	0.60	**0.81**	0.71	**0.95**	0.79	**1.06**	0.96	**1.28**	0.90	**1.21**
Knee	0.03	**0.08**	0.04	**0.10**	0.05	**0.12**	0.05	**0.14**	0.07	**0.17**	0.06	**0.16**
Lower leg												
Lymph nodes	0.11	**0.14**	0.14	**0.18**	0.17	**0.22**	0.19	**0.24**	0.24	**0.30**	0.21	**0.27**
Skin	0.20	**0.28**	0.21	**0.30**	0.22	**0.31**	0.23	**0.32**	0.24	**0.33**	0.23	**0.33**
Muscle	0.16	**0.22**	0.20	**0.26**	0.22	**0.29**	0.24	**0.31**	0.27	**0.35**	0.26	**0.34**
Skeleton	0.28	**0.53**	0.35	**0.65**	0.40	**0.74**	0.43	**0.81**	0.51	**0.95**	0.49	**0.92**
Red bone marrow	0.10	**0.16**	0.12	**0.19**	0.14	**0.22**	0.16	**0.25**	0.19	**0.29**	0.18	**0.27**
Whole body dose	0.16	**0.24**	0.20	**0.28**	0.22	**0.32**	0.24	**0.35**	0.28	**0.40**	0.27	**0.38**
ED (ICRP 60)	0.12	**0.17**	0.16	**0.22**	0.19	**0.26**	0.21	**0.29**	0.25	**0.35**	0.23	**0.32**
ED (ICRP 103)	0.11	**0.17**	0.15	**0.22**	0.17	**0.26**	0.19	**0.29**	0.23	**0.34**	0.21	**0.32**
Total filtration (mm) **Al**	**1.5**		**2.5**		**3.5**		**4.5**		**3.5**		**3.5**	
Cu	**0**		**0**		**0**		**0**		**0.1**		**0.2**	

Table 12.107 Normalized organ doses—Abdomen pa—1–2 years

Beam projection:			pa									**Table cells:**
Focus-to-detector distance (cm):			115				(+) = normal font = optimal field setting					
Tube voltage (kVp):			70				**(−) = bold font = suboptimal field setting**					
Total filtrations:			See last table row.				Empty table cells correspond to values <0.01					
Organ or tissue	**Organ dose D_T : Entrance dose K_E (mSv * mGy^{-1})**											
	(+)	(−)	(+)	(−)	(+)	(−)	(+)	(−)	(+)	(−)	(+)	(−)
Skull												
Brain												
Oral mucosa												
Salivary glands										**0.01**		**0.01**
Cervical spine		**0.01**		**0.02**		**0.02**		**0.03**		**0.04**	0.01	**0.04**
Thyroid		**0.01**		**0.02**		**0.02**		**0.02**		**0.03**		**0.03**
Thymus		**0.04**	0.01	**0.05**	0.01	**0.06**	0.02	**0.07**	0.02	**0.09**	0.02	**0.11**
Thoracic spine	0.39	**0.77**	0.49	**0.95**	0.56	**1.10**	0.62	**1.21**	0.73	**1.42**	0.83	**1.61**
Claviculae	0.01	**0.04**	0.01	**0.05**	0.02	**0.06**	0.02	**0.07**	0.03	**0.09**	0.03	**0.11**
Scapulae	0.03	**0.99**	0.04	**1.12**	0.04	**1.21**	0.05	**1.27**	0.06	**1.37**	0.07	**1.45**
Ribs	0.53	**1.03**	0.61	**1.19**	0.66	**1.30**	0.70	**1.38**	0.77	**1.52**	0.82	**1.63**
Oesophagus	0.10	**0.18**	0.13	**0.24**	0.15	**0.28**	0.17	**0.32**	0.21	**0.39**	0.25	**0.46**
Trachea												
Lung	0.12	**0.37**	0.15	**0.45**	0.16	**0.51**	0.18	**0.55**	0.21	**0.62**	0.23	**0.69**
Mamma	0.02	**0.16**	0.02	**0.20**	0.03	**0.24**	0.03	**0.26**	0.04	**0.31**	0.05	**0.36**
Heart	0.07	**0.16**	0.08	**0.20**	0.10	**0.24**	0.11	**0.27**	0.14	**0.33**	0.16	**0.38**
Stomach	0.15	**0.17**	0.19	**0.21**	0.22	**0.25**	0.25	**0.28**	0.31	**0.35**	0.36	**0.41**
Spleen	0.47	**0.51**	0.58	**0.63**	0.65	**0.71**	0.71	**0.77**	0.82	**0.89**	0.91	**0.99**
Liver	0.23	**0.25**	0.28	**0.32**	0.33	**0.37**	0.36	**0.41**	0.43	**0.49**	0.49	**0.56**
Gall bladder	0.15	**0.17**	0.20	**0.22**	0.24	**0.26**	0.27	**0.29**	0.33	**0.36**	0.40	**0.44**
Pancreas	0.24	**0.26**	0.30	**0.34**	0.35	**0.40**	0.40	**0.45**	0.48	**0.54**	0.56	**0.63**
Small intestine	0.23	**0.24**	0.30	**0.31**	0.35	**0.36**	0.39	**0.41**	0.48	**0.50**	0.56	**0.59**
Large intestine	0.19	**0.21**	0.25	**0.27**	0.30	**0.31**	0.33	**0.35**	0.41	**0.43**	0.48	**0.51**
Sigma/rectum	0.24	**0.25**	0.31	**0.32**	0.36	**0.38**	0.40	**0.42**	0.49	**0.51**	0.57	**0.59**
Lumbar spine	1.29	**1.31**	1.60	**1.63**	1.82	**1.86**	2.01	**2.05**	2.34	**2.40**	2.64	**2.71**
Pelvis	1.34	**1.38**	1.63	**1.68**	1.85	**1.90**	2.01	**2.07**	2.31	**2.38**	2.56	**2.65**
Adrenals	0.49	**0.52**	0.60	**0.63**	0.67	**0.72**	0.73	**0.78**	0.84	**0.90**	0.94	**1.01**
Kidneys	0.74	**0.76**	0.87	**0.90**	0.96	**0.99**	1.03	**1.06**	1.15	**1.18**	1.25	**1.29**
Urinary bladder	0.15	**0.16**	0.20	**0.21**	0.23	**0.24**	0.26	**0.27**	0.32	**0.34**	0.38	**0.40**
Uterus	0.21	**0.21**	0.27	**0.28**	0.32	**0.33**	0.36	**0.37**	0.44	**0.46**	0.52	**0.54**
Ovaries	0.25	**0.25**	0.32	**0.33**	0.37	**0.39**	0.42	**0.44**	0.51	**0.54**	0.60	**0.63**
Prostate	0.17	**0.18**	0.22	**0.23**	0.26	**0.27**	0.29	**0.31**	0.36	**0.37**	0.42	**0.44**
Testes	0.15	**0.14**	0.18	**0.18**	0.22	**0.21**	0.24	**0.23**	0.29	**0.28**	0.33	**0.33**
Upper arm		**0.35**		**0.43**		**0.49**		**0.54**		**0.63**		**0.70**
Elbow		**1.01**		**1.23**		**1.39**		**1.51**		**1.74**		**1.92**
Forearm		**1.13**		**1.37**		**1.54**		**1.67**		**1.90**		**2.09**
Upper leg	0.57	**0.76**	0.72	**0.96**	0.83	**1.10**	0.92	**1.22**	1.09	**1.44**	1.25	**1.65**
Knee	0.04	**0.10**	0.05	**0.13**	0.06	**0.15**	0.07	**0.17**	0.09	**0.21**	0.10	**0.24**
Lower leg												
Lymph nodes	0.14	**0.18**	0.18	**0.23**	0.21	**0.26**	0.23	**0.29**	0.28	**0.35**	0.33	**0.41**
Skin	0.21	**0.29**	0.22	**0.31**	0.23	**0.32**	0.24	**0.33**	0.25	**0.35**	0.26	**0.37**
Muscle	0.19	**0.24**	0.22	**0.29**	0.24	**0.32**	0.26	**0.35**	0.30	**0.39**	0.33	**0.43**
Skeleton	0.32	**0.61**	0.39	**0.74**	0.44	**0.83**	0.48	**0.90**	0.56	**1.03**	0.62	**1.14**
Red bone marrow	0.12	**0.18**	0.15	**0.23**	0.17	**0.26**	0.18	**0.28**	0.21	**0.33**	0.24	**0.37**
Whole body dose	0.19	**0.27**	0.22	**0.32**	0.25	**0.36**	0.27	**0.39**	0.31	**0.44**	0.34	**0.49**
ED (ICRP 60)	0.15	**0.21**	0.19	**0.27**	0.22	**0.31**	0.25	**0.34**	0.30	**0.41**	0.35	**0.47**
ED (ICRP 103)	0.14	**0.21**	0.18	**0.26**	0.20	**0.30**	0.23	**0.34**	0.27	**0.40**	0.31	**0.46**
Total filtration (mm) Al	**1.5**		**2.5**		**3.5**		**4.5**		**3.5**		**3.5**	
Cu	**0**		**0**		**0**		**0**		**0.1**		**0.2**	

Table 12.108 Normalized organ doses—Abdomen pa—3–7 years

Beam projection:			pa									**Table cells:**
Focus-to-detector distance (cm):			115				(+) = normal font = optimal field setting					
Tube voltage (kVp):			60				**(−) = bold font = suboptimal field setting**					
Total filtrations:			See last table row.				Empty table cells correspond to values <0.01					
Organ or tissue	**Organ dose D_T : Entrance dose K_E**						**(mSv * mGy^{-1})**					
	(+)	(−)	(+)	(−)	(+)	(−)	(+)	(−)	(+)	(−)	(+)	(−)
Skull												
Brain												
Oral mucosa												
Salivary glands												
Cervical spine						**0.01**		**0.01**		**0.02**		**0.01**
Thyroid										**0.01**		
Thymus		**0.03**		**0.04**		**0.04**		**0.05**	0.01	**0.07**	0.01	**0.06**
Thoracic spine	0.29	**0.55**	0.37	**0.70**	0.44	**0.83**	0.49	**0.93**	0.60	**1.12**	0.55	**1.04**
Claviculae		**0.02**		**0.03**		**0.04**		**0.04**	0.01	**0.05**		**0.04**
Scapulae	0.02	**0.83**	0.02	**0.97**	0.03	**1.06**	0.03	**1.13**	0.04	**1.24**	0.03	**1.24**
Ribs	0.46	**0.88**	0.54	**1.04**	0.59	**1.14**	0.64	**1.23**	0.71	**1.37**	0.70	**1.35**
Oesophagus	0.05	**0.09**	0.07	**0.12**	0.08	**0.14**	0.09	**0.16**	0.12	**0.21**	0.10	**0.18**
Trachea												
Lung	0.10	**0.29**	0.12	**0.36**	0.14	**0.41**	0.16	**0.45**	0.18	**0.52**	0.17	**0.50**
Mamma	0.01	**0.11**	0.02	**0.14**	0.02	**0.17**	0.02	**0.19**	0.03	**0.23**	0.03	**0.21**
Heart	0.05	**0.11**	0.06	**0.14**	0.07	**0.17**	0.09	**0.19**	0.11	**0.24**	0.09	**0.21**
Stomach	0.10	**0.11**	0.13	**0.14**	0.15	**0.17**	0.18	**0.20**	0.22	**0.25**	0.19	**0.21**
Spleen	0.38	**0.40**	0.48	**0.51**	0.55	**0.59**	0.61	**0.65**	0.71	**0.77**	0.68	**0.73**
Liver	0.17	**0.19**	0.21	**0.24**	0.25	**0.28**	0.28	**0.32**	0.34	**0.39**	0.31	**0.35**
Gall bladder	0.09	**0.09**	0.12	**0.13**	0.15	**0.16**	0.17	**0.18**	0.22	**0.23**	0.18	**0.19**
Pancreas	0.16	**0.17**	0.21	**0.23**	0.25	**0.28**	0.29	**0.31**	0.36	**0.39**	0.32	**0.35**
Small intestine	0.15	**0.16**	0.20	**0.21**	0.24	**0.25**	0.27	**0.28**	0.34	**0.36**	0.30	**0.31**
Large intestine	0.12	**0.13**	0.16	**0.17**	0.20	**0.21**	0.22	**0.24**	0.28	**0.30**	0.24	**0.26**
Sigma/rectum	0.16	**0.16**	0.21	**0.22**	0.25	**0.26**	0.28	**0.30**	0.35	**0.37**	0.31	**0.33**
Lumbar spine	0.94	**0.96**	1.20	**1.23**	1.41	**1.44**	1.57	**1.60**	1.88	**1.93**	1.77	**1.80**
Pelvis	1.02	**1.05**	1.28	**1.32**	1.48	**1.52**	1.63	**1.68**	1.92	**1.99**	1.84	**1.89**
Adrenals	0.38	**0.41**	0.48	**0.51**	0.55	**0.59**	0.61	**0.66**	0.72	**0.78**	0.69	**0.73**
Kidneys	0.62	**0.64**	0.75	**0.77**	0.84	**0.87**	0.91	**0.94**	1.04	**1.07**	1.01	**1.04**
Urinary bladder	0.09	**0.10**	0.12	**0.13**	0.15	**0.16**	0.17	**0.18**	0.22	**0.23**	0.19	**0.19**
Uterus	0.13	**0.13**	0.17	**0.17**	0.21	**0.21**	0.24	**0.24**	0.31	**0.30**	0.26	**0.26**
Ovaries	0.15	**0.16**	0.21	**0.21**	0.25	**0.26**	0.29	**0.29**	0.37	**0.37**	0.32	**0.32**
Prostate	0.11	**0.12**	0.15	**0.16**	0.18	**0.19**	0.21	**0.21**	0.26	**0.27**	0.23	**0.23**
Testes	0.06	**0.06**	0.07	**0.08**	0.09	**0.10**	0.10	**0.11**	0.13	**0.14**	0.11	**0.12**
Upper arm		**0.25**		**0.31**		**0.36**		**0.40**		**0.48**		**0.45**
Elbow		**0.74**		**0.93**		**1.07**		**1.18**		**1.39**		**1.34**
Forearm		**0.85**		**1.07**		**1.22**		**1.35**		**1.58**		**1.53**
Upper leg	0.24	**0.44**	0.31	**0.58**	0.37	**0.69**	0.42	**0.78**	0.52	**0.95**	0.47	**0.87**
Knee		**0.02**		**0.03**	0.01	**0.04**	0.01	**0.05**	0.02	**0.06**	0.01	**0.05**
Lower leg												
Lymph nodes	0.09	**0.12**	0.12	**0.15**	0.14	**0.18**	0.16	**0.21**	0.20	**0.26**	0.18	**0.23**
Skin	0.17	**0.27**	0.17	**0.28**	0.18	**0.29**	0.18	**0.30**	0.19	**0.32**	0.19	**0.31**
Muscle	0.15	**0.20**	0.17	**0.24**	0.20	**0.27**	0.21	**0.29**	0.24	**0.33**	0.23	**0.32**
Skeleton	0.20	**0.38**	0.24	**0.48**	0.28	**0.55**	0.31	**0.60**	0.37	**0.71**	0.35	**0.68**
Red bone marrow	0.12	**0.16**	0.15	**0.21**	0.17	**0.24**	0.19	**0.26**	0.23	**0.32**	0.21	**0.29**
Whole body dose	0.15	**0.21**	0.18	**0.26**	0.20	**0.29**	0.22	**0.32**	0.25	**0.37**	0.24	**0.35**
ED (ICRP 60)	0.10	**0.15**	0.13	**0.19**	0.16	**0.22**	0.18	**0.25**	0.22	**0.30**	0.20	**0.27**
ED (ICRP 103)	0.10	**0.15**	0.13	**0.19**	0.15	**0.22**	0.17	**0.25**	0.21	**0.30**	0.19	**0.27**
Total filtration (mm) Al	**1.5**		**2.5**		**3.5**		**4.5**		**3.5**		**3.5**	
Cu	**0**		**0**		**0**		**0**		**0.1**		**0.2**	

Table 12.109 Normalized organ doses—Abdomen pa—3–7 years

Beam projection:	pa	**Table cells:**
Focus-to-detector distance (cm):	115	(+) = normal font = optimal field setting
Tube voltage (kVp):	70	**(−) = bold font = suboptimal field setting**
Total filtrations:	See last table row.	Empty table cells correspond to values <0.01

Organ or tissue	**Organ dose D_T : Entrance dose K_E (mSv * mGy^{-1})**											
	(+)	(−)	(+)	(−)	(+)	(−)	(+)	(−)	(+)	(−)	(+)	(−)
Skull												
Brain												
Oral mucosa												
Salivary glands												
Cervical spine				**0.01**		**0.02**		**0.02**		**0.02**		**0.03**
Thyroid						**0.01**		**0.01**		**0.02**		**0.02**
Thymus		**0.04**		**0.05**	0.01	**0.06**	0.01	**0.07**	0.02	**0.09**	0.02	**0.11**
Thoracic spine	0.35	**0.67**	0.45	**0.84**	0.52	**0.97**	0.58	**1.08**	0.69	**1.29**	0.79	**1.48**
Claviculae		**0.03**		**0.04**	0.01	**0.05**	0.01	**0.05**	0.02	**0.07**	0.02	**0.08**
Scapulae	0.02	**0.90**	0.03	**1.03**	0.03	**1.12**	0.04	**1.19**	0.05	**1.29**	0.06	**1.37**
Ribs	0.50	**0.97**	0.58	**1.13**	0.64	**1.23**	0.68	**1.31**	0.75	**1.45**	0.81	**1.57**
Oesophagus	0.07	**0.12**	0.09	**0.16**	0.10	**0.18**	0.12	**0.21**	0.15	**0.26**	0.17	**0.31**
Trachea												
Lung	0.12	**0.34**	0.14	**0.41**	0.16	**0.46**	0.18	**0.51**	0.21	**0.58**	0.23	**0.65**
Mamma	0.02	**0.13**	0.02	**0.17**	0.03	**0.20**	0.03	**0.22**	0.04	**0.27**	0.04	**0.31**
Heart	0.06	**0.14**	0.08	**0.18**	0.10	**0.21**	0.11	**0.24**	0.13	**0.30**	0.16	**0.35**
Stomach	0.13	**0.14**	0.17	**0.19**	0.20	**0.22**	0.22	**0.25**	0.28	**0.31**	0.33	**0.37**
Spleen	0.45	**0.48**	0.55	**0.59**	0.63	**0.68**	0.69	**0.74**	0.80	**0.86**	0.90	**0.98**
Liver	0.21	**0.23**	0.26	**0.29**	0.30	**0.34**	0.34	**0.38**	0.40	**0.46**	0.47	**0.53**
Gall bladder	0.12	**0.13**	0.16	**0.18**	0.20	**0.21**	0.22	**0.24**	0.28	**0.31**	0.35	**0.37**
Pancreas	0.21	**0.23**	0.27	**0.29**	0.31	**0.35**	0.35	**0.39**	0.44	**0.48**	0.51	**0.57**
Small intestine	0.20	**0.21**	0.26	**0.27**	0.30	**0.32**	0.34	**0.36**	0.43	**0.45**	0.51	**0.53**
Large intestine	0.16	**0.17**	0.21	**0.22**	0.25	**0.27**	0.29	**0.30**	0.36	**0.38**	0.43	**0.45**
Sigma/rectum	0.20	**0.22**	0.26	**0.28**	0.31	**0.33**	0.35	**0.37**	0.43	**0.46**	0.51	**0.54**
Lumbar spine	1.14	**1.17**	1.42	**1.46**	1.64	**1.68**	1.82	**1.86**	2.15	**2.20**	2.45	**2.52**
Pelvis	1.20	**1.24**	1.48	**1.52**	1.68	**1.74**	1.84	**1.90**	2.14	**2.22**	2.40	**2.49**
Adrenals	0.45	**0.49**	0.56	**0.60**	0.64	**0.69**	0.70	**0.75**	0.82	**0.88**	0.92	**1.00**
Kidneys	0.71	**0.72**	0.84	**0.86**	0.93	**0.96**	1.00	**1.03**	1.13	**1.17**	1.24	**1.28**
Urinary bladder	0.13	**0.13**	0.16	**0.17**	0.20	**0.20**	0.22	**0.23**	0.28	**0.29**	0.33	**0.35**
Uterus	0.18	**0.17**	0.23	**0.23**	0.27	**0.27**	0.31	**0.31**	0.39	**0.39**	0.47	**0.47**
Ovaries	0.21	**0.21**	0.27	**0.28**	0.32	**0.33**	0.37	**0.38**	0.46	**0.47**	0.55	**0.56**
Prostate	0.15	**0.16**	0.20	**0.20**	0.23	**0.24**	0.26	**0.27**	0.33	**0.34**	0.39	**0.41**
Testes	0.07	**0.08**	0.10	**0.11**	0.12	**0.13**	0.13	**0.15**	0.16	**0.18**	0.19	**0.22**
Upper arm		**0.30**		**0.37**		**0.42**		**0.46**		**0.54**		**0.61**
Elbow		**0.86**		**1.06**		**1.21**		**1.32**		**1.54**		**1.72**
Forearm		**0.99**		**1.21**		**1.37**		**1.50**		**1.73**		**1.92**
Upper leg	0.30	**0.56**	0.39	**0.71**	0.45	**0.83**	0.51	**0.93**	0.62	**1.12**	0.73	**1.31**
Knee		**0.03**	0.01	**0.04**	0.02	**0.05**	0.02	**0.06**	0.02	**0.08**	0.03	**0.09**
Lower leg												
Lymph nodes	0.12	**0.15**	0.15	**0.19**	0.18	**0.23**	0.20	**0.26**	0.25	**0.31**	0.30	**0.37**
Skin	0.17	**0.28**	0.18	**0.30**	0.19	**0.31**	0.19	**0.32**	0.20	**0.33**	0.21	**0.35**
Muscle	0.17	**0.23**	0.20	**0.27**	0.22	**0.30**	0.24	**0.33**	0.27	**0.37**	0.30	**0.41**
Skeleton	0.23	**0.45**	0.28	**0.55**	0.32	**0.62**	0.35	**0.68**	0.41	**0.79**	0.46	**0.88**
Red bone marrow	0.14	**0.20**	0.18	**0.25**	0.20	**0.28**	0.22	**0.31**	0.27	**0.37**	0.30	**0.42**
Whole body dose	0.17	**0.25**	0.20	**0.30**	0.23	**0.33**	0.25	**0.36**	0.28	**0.41**	0.32	**0.46**
ED (ICRP 60)	0.13	**0.18**	0.17	**0.23**	0.20	**0.27**	0.22	**0.30**	0.27	**0.36**	0.31	**0.42**
ED (ICRP 103)	0.12	**0.18**	0.16	**0.23**	0.18	**0.27**	0.20	**0.30**	0.25	**0.36**	0.29	**0.42**
Total filtration (mm) Al	**1.5**		**2.5**		**3.5**		**4.5**		**3.5**		**3.5**	
Cu	**0**		**0**		**0**		**0**		**0.1**		**0.2**	

Table 12.110 Normalized organ doses—Abdomen pa—3–7 years

Beam projection:			pa									**Table cells:**
Focus-to-detector distance (cm):			115									(+) = normal font = optimal field setting
Tube voltage (kVp):			80									**(−) = bold font = suboptimal field setting**
Total filtrations:			See last table row.									Empty table cells correspond to values <0.01
Organ or tissue	**Organ dose D_T : Entrance dose K_E**								**(mSv * mGy^{-1})**			
	(+)	(−)	(+)	(−)	(+)	(−)	(+)	(−)	(+)	(−)	(+)	(−)
Skull												
Brain												
Oral mucosa												
Salivary glands												**0.01**
Cervical spine		**0.01**		**0.02**		**0.02**		**0.02**		**0.03**	0.01	**0.04**
Thyroid				**0.01**		**0.01**		**0.02**		**0.02**		**0.03**
Thymus	0.01	**0.05**	0.01	**0.06**	0.02	**0.08**	0.02	**0.09**	0.02	**0.11**	0.03	**0.13**
Thoracic spine	0.42	**0.78**	0.51	**0.96**	0.59	**1.10**	0.64	**1.21**	0.76	**1.42**	0.85	**1.60**
Claviculae		**0.04**	0.01	**0.05**	0.01	**0.06**	0.02	**0.07**	0.02	**0.08**	0.03	**0.10**
Scapulae	0.03	**0.95**	0.04	**1.08**	0.04	**1.16**	0.05	**1.22**	0.06	**1.32**	0.07	**1.38**
Ribs	0.54	**1.04**	0.62	**1.20**	0.67	**1.30**	0.71	**1.37**	0.77	**1.50**	0.82	**1.60**
Oesophagus	0.08	**0.15**	0.11	**0.19**	0.13	**0.22**	0.14	**0.25**	0.17	**0.31**	0.20	**0.36**
Trachea												
Lung	0.13	**0.38**	0.16	**0.46**	0.18	**0.51**	0.20	**0.56**	0.23	**0.63**	0.25	**0.70**
Mamma	0.02	**0.16**	0.03	**0.20**	0.03	**0.23**	0.04	**0.26**	0.04	**0.30**	0.05	**0.35**
Heart	0.08	**0.17**	0.10	**0.22**	0.12	**0.26**	0.13	**0.29**	0.16	**0.35**	0.19	**0.40**
Stomach	0.16	**0.18**	0.21	**0.23**	0.24	**0.27**	0.27	**0.30**	0.33	**0.37**	0.38	**0.43**
Spleen	0.51	**0.55**	0.62	**0.67**	0.70	**0.76**	0.76	**0.82**	0.87	**0.95**	0.97	**1.06**
Liver	0.25	**0.28**	0.31	**0.35**	0.35	**0.40**	0.39	**0.44**	0.46	**0.52**	0.52	**0.60**
Gall bladder	0.16	**0.17**	0.21	**0.22**	0.25	**0.27**	0.28	**0.30**	0.35	**0.37**	0.41	**0.45**
Pancreas	0.26	**0.28**	0.32	**0.36**	0.38	**0.42**	0.42	**0.47**	0.51	**0.56**	0.59	**0.66**
Small intestine	0.25	**0.26**	0.32	**0.33**	0.37	**0.39**	0.41	**0.43**	0.50	**0.53**	0.59	**0.62**
Large intestine	0.21	**0.22**	0.26	**0.28**	0.31	**0.33**	0.34	**0.37**	0.42	**0.45**	0.50	**0.53**
Sigma/rectum	0.25	**0.27**	0.32	**0.34**	0.37	**0.39**	0.41	**0.44**	0.50	**0.53**	0.58	**0.62**
Lumbar spine	1.32	**1.35**	1.62	**1.66**	1.84	**1.89**	2.01	**2.07**	2.34	**2.41**	2.63	**2.70**
Pelvis	1.36	**1.41**	1.64	**1.70**	1.84	**1.91**	2.00	**2.08**	2.29	**2.38**	2.53	**2.63**
Adrenals	0.52	**0.56**	0.63	**0.68**	0.71	**0.77**	0.78	**0.84**	0.90	**0.97**	1.00	**1.09**
Kidneys	0.78	**0.81**	0.92	**0.95**	1.01	**1.04**	1.08	**1.12**	1.21	**1.25**	1.31	**1.36**
Urinary bladder	0.16	**0.17**	0.20	**0.21**	0.24	**0.25**	0.27	**0.28**	0.33	**0.35**	0.39	**0.41**
Uterus	0.22	**0.23**	0.29	**0.29**	0.34	**0.34**	0.38	**0.38**	0.46	**0.47**	0.55	**0.56**
Ovaries	0.26	**0.27**	0.34	**0.34**	0.39	**0.40**	0.44	**0.45**	0.54	**0.56**	0.64	**0.65**
Prostate	0.19	**0.20**	0.24	**0.25**	0.28	**0.30**	0.32	**0.33**	0.38	**0.41**	0.45	**0.48**
Testes	0.10	**0.11**	0.12	**0.13**	0.14	**0.16**	0.16	**0.18**	0.19	**0.22**	0.23	**0.26**
Upper arm		**0.34**		**0.41**		**0.46**		**0.51**		**0.58**		**0.65**
Elbow		**0.98**		**1.18**		**1.32**		**1.43**		**1.64**		**1.80**
Forearm		**1.11**		**1.33**		**1.49**		**1.61**		**1.82**		**2.00**
Upper leg	0.36	**0.66**	0.46	**0.83**	0.53	**0.95**	0.59	**1.06**	0.70	**1.25**	0.81	**1.43**
Knee	0.01	**0.04**	0.02	**0.06**	0.02	**0.07**	0.02	**0.07**	0.03	**0.09**	0.04	**0.11**
Lower leg												
Lymph nodes	0.15	**0.19**	0.19	**0.23**	0.22	**0.27**	0.24	**0.30**	0.29	**0.36**	0.34	**0.42**
Skin	0.18	**0.29**	0.19	**0.31**	0.19	**0.32**	0.20	**0.33**	0.21	**0.35**	0.22	**0.37**
Muscle	0.19	**0.26**	0.22	**0.30**	0.25	**0.33**	0.26	**0.36**	0.30	**0.41**	0.33	**0.45**
Skeleton	0.26	**0.51**	0.32	**0.61**	0.36	**0.68**	0.39	**0.74**	0.44	**0.84**	0.49	**0.93**
Red bone marrow	0.17	**0.23**	0.20	**0.28**	0.23	**0.32**	0.25	**0.35**	0.30	**0.41**	0.34	**0.47**
Whole body dose	0.19	**0.28**	0.23	**0.33**	0.25	**0.37**	0.27	**0.40**	0.31	**0.45**	0.34	**0.50**
ED (ICRP 60)	0.16	**0.22**	0.20	**0.27**	0.23	**0.31**	0.26	**0.35**	0.31	**0.41**	0.36	**0.48**
ED (ICRP 103)	0.15	**0.22**	0.19	**0.27**	0.22	**0.31**	0.24	**0.35**	0.28	**0.41**	0.33	**0.47**
Total filtration (mm) **Al**	**1.5**		**2.5**		**3.5**		**4.5**		**3.5**		**3.5**	
Cu	**0**		**0**		**0**		**0**		**0.1**		**0.2**	

Table 12.111 Normalized organ doses—Abdomen pa—8–12 years

Beam projection:			pa									**Table cells:**
Focus-to-detector distance (cm):			115									(+) = normal font = optimal field setting
Tube voltage (kVp):			60									**(−) = bold font = suboptimal field setting**
Total filtrations:			See last table row.									Empty table cells correspond to values <0.01
Organ or tissue	**Organ dose D_T : Entrance dose K_E**								**(mSv * mGy^{-1})**			
	(+)	(−)	(+)	(−)	(+)	(−)	(+)	(−)	(+)	(−)	(+)	(−)
Skull												
Brain												
Oral mucosa												
Salivary glands												
Cervical spine												
Thyroid												
Thymus		**0.02**		**0.03**		**0.03**		**0.04**		**0.05**		**0.04**
Thoracic spine	0.23	**0.46**	0.30	**0.59**	0.36	**0.70**	0.41	**0.79**	0.50	**0.96**	0.45	**0.88**
Claviculae		**0.01**		**0.02**		**0.02**		**0.02**		**0.03**		**0.03**
Scapulae		**0.66**	0.01	**0.78**	0.02	**0.86**	0.02	**0.92**	0.02	**1.02**	0.02	**1.02**
Ribs	0.41	**0.79**	0.49	**0.94**	0.54	**1.04**	0.58	**1.11**	0.64	**1.25**	0.63	**1.23**
Oesophagus	0.03	**0.06**	0.04	**0.08**	0.05	**0.10**	0.05	**0.11**	0.07	**0.15**	0.06	**0.12**
Trachea												
Lung	0.07	**0.25**	0.09	**0.31**	0.11	**0.36**	0.12	**0.40**	0.14	**0.46**	0.13	**0.44**
Mamma		**0.09**	0.01	**0.12**	0.02	**0.14**	0.02	**0.15**	0.02	**0.19**	0.02	**0.17**
Heart	0.03	**0.08**	0.05	**0.11**	0.06	**0.14**	0.06	**0.16**	0.08	**0.20**	0.07	**0.17**
Stomach	0.08	**0.09**	0.10	**0.12**	0.13	**0.14**	0.15	**0.16**	0.19	**0.21**	0.16	**0.18**
Spleen	0.34	**0.36**	0.43	**0.46**	0.50	**0.54**	0.56	**0.60**	0.66	**0.72**	0.62	**0.68**
Liver	0.15	**0.17**	0.20	**0.22**	0.23	**0.26**	0.26	**0.29**	0.32	**0.36**	0.29	**0.33**
Gall bladder	0.08	**0.08**	0.11	**0.12**	0.13	**0.14**	0.15	**0.16**	0.20	**0.21**	0.16	**0.17**
Pancreas	0.13	**0.15**	0.18	**0.20**	0.22	**0.24**	0.25	**0.28**	0.31	**0.35**	0.27	**0.30**
Small intestine	0.12	**0.13**	0.17	**0.17**	0.20	**0.21**	0.23	**0.24**	0.29	**0.31**	0.25	**0.26**
Large intestine	0.10	**0.10**	0.13	**0.14**	0.16	**0.17**	0.19	**0.20**	0.24	**0.25**	0.20	**0.21**
Sigma/rectum	0.13	**0.14**	0.17	**0.18**	0.21	**0.22**	0.24	**0.25**	0.30	**0.32**	0.26	**0.28**
Lumbar spine	0.83	**0.85**	1.07	**1.10**	1.26	**1.29**	1.41	**1.44**	1.71	**1.75**	1.58	**1.62**
Pelvis	0.91	**0.93**	1.15	**1.18**	1.33	**1.37**	1.47	**1.52**	1.75	**1.82**	1.66	**1.71**
Adrenals	0.36	**0.38**	0.45	**0.48**	0.52	**0.56**	0.58	**0.62**	0.69	**0.74**	0.65	**0.70**
Kidneys	0.59	**0.60**	0.72	**0.73**	0.81	**0.83**	0.88	**0.90**	1.01	**1.04**	0.98	**1.00**
Urinary bladder	0.07	**0.08**	0.10	**0.10**	0.12	**0.13**	0.14	**0.15**	0.18	**0.19**	0.15	**0.16**
Uterus	0.10	**0.11**	0.14	**0.14**	0.17	**0.17**	0.20	**0.20**	0.25	**0.26**	0.21	**0.22**
Ovaries	0.13	**0.13**	0.17	**0.18**	0.21	**0.22**	0.24	**0.25**	0.31	**0.32**	0.26	**0.27**
Prostate	0.08	**0.09**	0.11	**0.12**	0.14	**0.15**	0.16	**0.17**	0.20	**0.22**	0.17	**0.18**
Testes	0.03	**0.04**	0.05	**0.05**	0.06	**0.06**	0.07	**0.07**	0.08	**0.09**	0.07	**0.07**
Upper arm		**0.18**		**0.23**		**0.27**		**0.30**		**0.36**		**0.34**
Elbow		**0.63**		**0.80**		**0.93**		**1.03**		**1.22**		**1.16**
Forearm		**0.74**		**0.93**		**1.08**		**1.19**		**1.41**		**1.35**
Upper leg	0.14	**0.29**	0.19	**0.38**	0.23	**0.46**	0.26	**0.52**	0.33	**0.65**	0.29	**0.58**
Knee				**0.01**		**0.01**		**0.02**		**0.02**		**0.02**
Lower leg												
Lymph nodes	0.08	**0.10**	0.10	**0.13**	0.12	**0.16**	0.14	**0.18**	0.17	**0.22**	0.15	**0.19**
Skin	0.16	**0.26**	0.17	**0.27**	0.17	**0.28**	0.18	**0.29**	0.18	**0.30**	0.18	**0.29**
Muscle	0.13	**0.18**	0.16	**0.21**	0.17	**0.24**	0.19	**0.26**	0.22	**0.30**	0.21	**0.29**
Skeleton	0.16	**0.32**	0.21	**0.41**	0.24	**0.47**	0.26	**0.52**	0.31	**0.61**	0.30	**0.58**
Red bone marrow	0.13	**0.17**	0.16	**0.22**	0.19	**0.25**	0.21	**0.28**	0.25	**0.33**	0.23	**0.31**
Whole body dose	0.13	**0.19**	0.16	**0.24**	0.18	**0.27**	0.20	**0.29**	0.23	**0.34**	0.22	**0.32**
ED (ICRP 60)	0.09	**0.13**	0.11	**0.16**	0.14	**0.19**	0.15	**0.22**	0.19	**0.27**	0.17	**0.24**
ED (ICRP 103)	0.08	**0.13**	0.11	**0.17**	0.13	**0.19**	0.15	**0.22**	0.18	**0.27**	0.16	**0.24**
Total filtration (mm) Al	**1.5**		**2.5**		**3.5**		**4.5**		**3.5**		**3.5**	
Cu	**0**		**0**		**0**		**0**		**0.1**		**0.2**	

Table 12.112 Normalized organ doses—Abdomen pa—8–12 years

Beam projection:			pa									**Table cells:**
Focus-to-detector distance (cm):			115									(+) = normal font = optimal field setting
Tube voltage (kVp):			70									**(−) = bold font = suboptimal field setting**
Total filtrations:			See last table row.									Empty table cells correspond to values <0.01
Organ or tissue	**Organ dose D_T : Entrance dose K_E**								**(mSv * mGy^{-1})**			
	(+)	(−)	(+)	(−)	(+)	(−)	(+)	(−)	(+)	(−)	(+)	(−)
Skull												
Brain												
Oral mucosa												
Salivary glands												
Cervical spine								**0.01**		**0.01**		**0.02**
Thyroid										**0.01**		**0.01**
Thymus		**0.03**		**0.04**		**0.05**	0.01	**0.05**	0.01	**0.07**	0.02	**0.09**
Thoracic spine	0.29	**0.57**	0.37	**0.72**	0.43	**0.84**	0.48	**0.94**	0.58	**1.13**	0.68	**1.31**
Claviculae		**0.02**		**0.02**		**0.03**		**0.03**		**0.04**	0.01	**0.06**
Scapulae	0.01	**0.73**	0.02	**0.84**	0.02	**0.92**	0.03	**0.98**	0.03	**1.08**	0.04	**1.16**
Ribs	0.46	**0.88**	0.53	**1.02**	0.58	**1.13**	0.62	**1.20**	0.68	**1.34**	0.74	**1.45**
Oesophagus	0.04	**0.08**	0.05	**0.11**	0.06	**0.13**	0.07	**0.15**	0.09	**0.19**	0.11	**0.23**
Trachea												
Lung	0.09	**0.29**	0.11	**0.36**	0.13	**0.41**	0.14	**0.45**	0.16	**0.52**	0.19	**0.59**
Mamma	0.01	**0.11**	0.02	**0.14**	0.02	**0.17**	0.02	**0.19**	0.03	**0.22**	0.03	**0.26**
Heart	0.05	**0.11**	0.06	**0.15**	0.07	**0.18**	0.08	**0.20**	0.11	**0.25**	0.13	**0.30**
Stomach	0.11	**0.12**	0.14	**0.16**	0.17	**0.19**	0.19	**0.21**	0.24	**0.27**	0.29	**0.33**
Spleen	0.41	**0.44**	0.51	**0.55**	0.58	**0.63**	0.64	**0.70**	0.75	**0.82**	0.86	**0.94**
Liver	0.19	**0.22**	0.24	**0.27**	0.28	**0.32**	0.31	**0.36**	0.38	**0.43**	0.44	**0.51**
Gall bladder	0.11	**0.12**	0.15	**0.16**	0.18	**0.19**	0.20	**0.22**	0.26	**0.28**	0.32	**0.34**
Pancreas	0.18	**0.20**	0.23	**0.26**	0.28	**0.31**	0.32	**0.35**	0.39	**0.44**	0.47	**0.52**
Small intestine	0.17	**0.18**	0.22	**0.23**	0.26	**0.27**	0.30	**0.31**	0.37	**0.39**	0.45	**0.47**
Large intestine	0.14	**0.14**	0.18	**0.19**	0.22	**0.23**	0.25	**0.26**	0.31	**0.33**	0.37	**0.39**
Sigma/rectum	0.17	**0.18**	0.23	**0.24**	0.27	**0.29**	0.30	**0.32**	0.38	**0.40**	0.45	**0.48**
Lumbar spine	1.02	**1.05**	1.29	**1.32**	1.49	**1.53**	1.66	**1.70**	1.98	**2.04**	2.28	**2.35**
Pelvis	1.08	**1.12**	1.34	**1.39**	1.53	**1.59**	1.69	**1.75**	1.98	**2.06**	2.24	**2.33**
Adrenals	0.43	**0.46**	0.53	**0.57**	0.61	**0.66**	0.67	**0.73**	0.79	**0.86**	0.90	**0.98**
Kidneys	0.67	**0.69**	0.81	**0.83**	0.90	**0.93**	0.98	**1.01**	1.11	**1.15**	1.23	**1.27**
Urinary bladder	0.10	**0.11**	0.14	**0.14**	0.16	**0.17**	0.19	**0.19**	0.23	**0.25**	0.29	**0.30**
Uterus	0.14	**0.15**	0.19	**0.20**	0.23	**0.24**	0.26	**0.27**	0.33	**0.34**	0.40	**0.42**
Ovaries	0.18	**0.18**	0.23	**0.24**	0.28	**0.29**	0.32	**0.33**	0.40	**0.42**	0.48	**0.51**
Prostate	0.11	**0.12**	0.15	**0.16**	0.18	**0.19**	0.21	**0.22**	0.26	**0.28**	0.31	**0.34**
Testes	0.05	**0.05**	0.06	**0.07**	0.08	**0.08**	0.09	**0.10**	0.11	**0.12**	0.13	**0.15**
Upper arm		**0.22**		**0.28**		**0.32**		**0.35**		**0.42**		**0.48**
Elbow		**0.75**		**0.93**		**1.06**		**1.17**		**1.37**		**1.55**
Forearm		**0.87**		**1.07**		**1.22**		**1.34**		**1.56**		**1.75**
Upper leg	0.19	**0.37**	0.25	**0.48**	0.29	**0.57**	0.33	**0.64**	0.41	**0.79**	0.48	**0.94**
Knee		**0.01**		**0.02**		**0.02**		**0.02**		**0.03**	0.01	**0.04**
Lower leg												
Lymph nodes	0.10	**0.13**	0.13	**0.17**	0.16	**0.20**	0.18	**0.22**	0.22	**0.28**	0.26	**0.33**
Skin	0.16	**0.27**	0.17	**0.28**	0.18	**0.29**	0.18	**0.30**	0.19	**0.31**	0.20	**0.33**
Muscle	0.15	**0.21**	0.18	**0.25**	0.20	**0.27**	0.22	**0.30**	0.25	**0.34**	0.28	**0.38**
Skeleton	0.20	**0.38**	0.24	**0.47**	0.28	**0.54**	0.30	**0.59**	0.36	**0.69**	0.40	**0.78**
Red bone marrow	0.16	**0.21**	0.20	**0.26**	0.23	**0.30**	0.25	**0.33**	0.30	**0.39**	0.35	**0.45**
Whole body dose	0.15	**0.23**	0.18	**0.27**	0.21	**0.31**	0.23	**0.33**	0.26	**0.39**	0.29	**0.43**
ED (ICRP 60)	0.11	**0.16**	0.15	**0.20**	0.17	**0.24**	0.19	**0.27**	0.24	**0.32**	0.28	**0.38**
ED (ICRP 103)	0.11	**0.16**	0.14	**0.21**	0.16	**0.24**	0.18	**0.27**	0.22	**0.32**	0.26	**0.38**
Total filtration (mm) **Al**	**1.5**		**2.5**		**3.5**		**4.5**		**3.5**		**3.5**	
Cu	**0**		**0**		**0**		**0**		**0.1**		**0.2**	

Table 12.113 Normalized organ doses—Abdomen pa—8–12 years

Beam projection:			pa									**Table cells:**
Focus-to-detector distance (cm):			115									(+) = normal font = optimal field setting
Tube voltage (kVp):			80									**(−) = bold font = suboptimal field setting**
Total filtrations:			See last table row.									Empty table cells correspond to values <0.01
Organ or tissue	**Organ dose D_T : Entrance dose K_E**								**(mSv * mGy^{-1})**			
	(+)	(−)	(+)	(−)	(+)	(−)	(+)	(−)	(+)	(−)	(+)	(−)
Skull												
Brain												
Oral mucosa												
Salivary glands												
Cervical spine				**0.01**		**0.01**		**0.01**		**0.02**		**0.02**
Thyroid								**0.01**		**0.01**		**0.02**
Thymus		**0.04**		**0.05**	0.01	**0.06**	0.01	**0.07**	0.02	**0.09**	0.02	**0.11**
Thoracic spine	0.35	**0.68**	0.43	**0.84**	0.50	**0.96**	0.55	**1.06**	0.65	**1.26**	0.74	**1.43**
Claviculae		**0.03**		**0.03**		**0.04**		**0.04**	0.01	**0.06**	0.01	**0.07**
Scapulae	0.02	**0.78**	0.02	**0.89**	0.03	**0.97**	0.03	**1.02**	0.04	**1.11**	0.05	**1.17**
Ribs	0.49	**0.95**	0.56	**1.09**	0.61	**1.19**	0.65	**1.26**	0.71	**1.39**	0.76	**1.49**
Oesophagus	0.05	**0.11**	0.07	**0.14**	0.08	**0.16**	0.09	**0.19**	0.11	**0.23**	0.13	**0.27**
Trachea												
Lung	0.10	**0.34**	0.13	**0.41**	0.14	**0.46**	0.16	**0.50**	0.18	**0.58**	0.20	**0.64**
Mamma	0.02	**0.14**	0.02	**0.17**	0.03	**0.19**	0.03	**0.22**	0.03	**0.26**	0.04	**0.30**
Heart	0.06	**0.15**	0.08	**0.19**	0.09	**0.22**	0.10	**0.24**	0.13	**0.30**	0.15	**0.35**
Stomach	0.14	**0.16**	0.18	**0.20**	0.21	**0.23**	0.23	**0.26**	0.29	**0.33**	0.34	**0.39**
Spleen	0.47	**0.52**	0.58	**0.63**	0.66	**0.72**	0.72	**0.78**	0.83	**0.91**	0.93	**1.03**
Liver	0.23	**0.26**	0.29	**0.33**	0.33	**0.38**	0.36	**0.42**	0.43	**0.50**	0.50	**0.57**
Gall bladder	0.15	**0.16**	0.19	**0.21**	0.23	**0.25**	0.26	**0.28**	0.32	**0.35**	0.38	**0.42**
Pancreas	0.23	**0.26**	0.29	**0.32**	0.34	**0.38**	0.38	**0.43**	0.46	**0.52**	0.55	**0.61**
Small intestine	0.22	**0.23**	0.28	**0.29**	0.33	**0.34**	0.37	**0.38**	0.45	**0.47**	0.53	**0.56**
Large intestine	0.18	**0.19**	0.23	**0.24**	0.27	**0.28**	0.30	**0.32**	0.37	**0.40**	0.44	**0.47**
Sigma/rectum	0.22	**0.23**	0.28	**0.30**	0.33	**0.35**	0.37	**0.39**	0.45	**0.48**	0.53	**0.56**
Lumbar spine	1.20	**1.24**	1.48	**1.53**	1.69	**1.74**	1.86	**1.92**	2.19	**2.26**	2.48	**2.56**
Pelvis	1.24	**1.29**	1.51	**1.56**	1.70	**1.77**	1.86	**1.93**	2.14	**2.23**	2.39	**2.49**
Adrenals	0.50	**0.54**	0.61	**0.66**	0.69	**0.75**	0.76	**0.82**	0.88	**0.96**	0.99	**1.08**
Kidneys	0.75	**0.78**	0.89	**0.92**	0.99	**1.02**	1.06	**1.10**	1.20	**1.24**	1.31	**1.36**
Urinary bladder	0.13	**0.14**	0.17	**0.18**	0.20	**0.22**	0.23	**0.24**	0.29	**0.30**	0.34	**0.36**
Uterus	0.19	**0.20**	0.24	**0.25**	0.29	**0.30**	0.33	**0.34**	0.41	**0.42**	0.48	**0.50**
Ovaries	0.23	**0.24**	0.30	**0.31**	0.35	**0.36**	0.39	**0.41**	0.48	**0.51**	0.57	**0.60**
Prostate	0.15	**0.16**	0.19	**0.21**	0.22	**0.24**	0.25	**0.27**	0.31	**0.34**	0.37	**0.40**
Testes	0.06	**0.07**	0.08	**0.09**	0.10	**0.11**	0.11	**0.12**	0.13	**0.15**	0.16	**0.19**
Upper arm		**0.26**		**0.32**		**0.36**		**0.39**		**0.46**		**0.51**
Elbow		**0.86**		**1.04**		**1.18**		**1.28**		**1.48**		**1.64**
Forearm		**0.99**		**1.19**		**1.34**		**1.46**		**1.66**		**1.84**
Upper leg	0.24	**0.46**	0.30	**0.58**	0.35	**0.68**	0.39	**0.75**	0.47	**0.91**	0.55	**1.06**
Knee		**0.02**		**0.02**		**0.03**		**0.03**	0.01	**0.04**	0.02	**0.05**
Lower leg												
Lymph nodes	0.13	**0.16**	0.16	**0.21**	0.19	**0.24**	0.21	**0.27**	0.26	**0.33**	0.31	**0.38**
Skin	0.17	**0.28**	0.18	**0.29**	0.18	**0.30**	0.19	**0.31**	0.20	**0.33**	0.21	**0.34**
Muscle	0.17	**0.23**	0.20	**0.28**	0.22	**0.31**	0.24	**0.33**	0.27	**0.38**	0.31	**0.42**
Skeleton	0.22	**0.44**	0.27	**0.53**	0.31	**0.60**	0.34	**0.65**	0.39	**0.75**	0.43	**0.83**
Red bone marrow	0.19	**0.25**	0.23	**0.30**	0.26	**0.35**	0.29	**0.38**	0.34	**0.45**	0.39	**0.51**
Whole body dose	0.17	**0.26**	0.21	**0.31**	0.23	**0.34**	0.25	**0.37**	0.29	**0.42**	0.32	**0.47**
ED (ICRP 60)	0.14	**0.20**	0.18	**0.24**	0.21	**0.28**	0.23	**0.31**	0.28	**0.38**	0.32	**0.44**
ED (ICRP 103)	0.13	**0.20**	0.17	**0.25**	0.19	**0.28**	0.21	**0.31**	0.26	**0.37**	0.30	**0.43**
Total filtration (mm) Al	**1.5**		**2.5**		**3.5**		**4.5**		**3.5**		**3.5**	
Cu	**0**		**0**		**0**		**0**		**0.1**		**0.2**	

Table 12.114 Normalized organ doses—Abdomen pa—13–17 years

Beam projection:		pa										**Table cells:**
Focus-to-detector distance (cm):		115										(+) = normal font = optimal field setting
Tube voltage (kVp):		70										**(−) = bold font = suboptimal field setting**
Total filtrations:		See last table row.										Empty table cells correspond to values <0.01
Organ or tissue	**Organ dose D_T : Entrance dose K_E (mSv * mGy^{-1})**											
	(+)	(−)	(+)	(−)	(+)	(−)	(+)	(−)	(+)	(−)	(+)	(−)
Skull												
Brain												
Oral mucosa												
Salivary glands												
Cervical spine												
Thyroid												
Thymus		**0.02**		**0.02**		**0.03**		**0.03**		**0.04**	0.01	**0.06**
Thoracic spine	0.23	**0.41**	0.29	**0.52**	0.34	**0.61**	0.38	**0.69**	0.47	**0.84**	0.55	**0.99**
Claviculae						**0.01**		**0.01**		**0.02**		**0.02**
Scapulae		**0.34**		**0.40**	0.01	**0.44**	0.01	**0.48**	0.02	**0.53**	0.02	**0.58**
Ribs	0.40	**0.70**	0.46	**0.82**	0.51	**0.91**	0.55	**0.97**	0.61	**1.09**	0.66	**1.19**
Oesophagus	0.05	**0.08**	0.06	**0.11**	0.07	**0.13**	0.08	**0.15**	0.11	**0.19**	0.13	**0.23**
Trachea												
Lung	0.07	**0.21**	0.08	**0.27**	0.10	**0.31**	0.11	**0.34**	0.13	**0.40**	0.15	**0.46**
Mamma	0.11	**0.43**	0.14	**0.57**	0.17	**0.68**	0.20	**0.78**	0.25	**0.99**	0.31	**1.20**
Heart	0.03	**0.07**	0.04	**0.10**	0.05	**0.12**	0.06	**0.13**	0.07	**0.17**	0.09	**0.21**
Stomach	0.08	**0.09**	0.11	**0.12**	0.13	**0.14**	0.15	**0.16**	0.18	**0.21**	0.23	**0.26**
Spleen	0.35	**0.38**	0.44	**0.48**	0.51	**0.55**	0.56	**0.61**	0.67	**0.73**	0.77	**0.85**
Liver	0.15	**0.18**	0.20	**0.23**	0.23	**0.27**	0.26	**0.30**	0.32	**0.37**	0.37	**0.43**
Gall bladder	0.10	**0.10**	0.13	**0.14**	0.15	**0.16**	0.18	**0.19**	0.22	**0.24**	0.27	**0.30**
Pancreas	0.14	**0.15**	0.18	**0.20**	0.21	**0.24**	0.24	**0.27**	0.31	**0.34**	0.37	**0.42**
Small intestine	0.13	**0.14**	0.17	**0.18**	0.21	**0.22**	0.24	**0.25**	0.30	**0.32**	0.37	**0.39**
Large intestine	0.10	**0.11**	0.14	**0.15**	0.16	**0.17**	0.19	**0.20**	0.24	**0.26**	0.29	**0.32**
Sigma/rectum	0.14	**0.14**	0.18	**0.19**	0.22	**0.23**	0.25	**0.26**	0.31	**0.33**	0.38	**0.40**
Lumbar spine	0.86	**0.89**	1.09	**1.13**	1.27	**1.31**	1.42	**1.47**	1.71	**1.77**	1.99	**2.06**
Pelvis	0.93	**0.97**	1.16	**1.21**	1.34	**1.40**	1.48	**1.55**	1.76	**1.84**	2.01	**2.10**
Adrenals	0.37	**0.41**	0.47	**0.51**	0.54	**0.59**	0.60	**0.66**	0.71	**0.78**	0.82	**0.90**
Kidneys	0.61	**0.64**	0.75	**0.78**	0.84	**0.88**	0.92	**0.96**	1.06	**1.11**	1.18	**1.24**
Urinary bladder	0.08	**0.08**	0.10	**0.11**	0.12	**0.13**	0.14	**0.15**	0.18	**0.19**	0.22	**0.23**
Uterus	0.38	**0.39**	0.50	**0.51**	0.60	**0.62**	0.69	**0.71**	0.87	**0.90**	1.07	**1.11**
Ovaries	0.14	**0.15**	0.18	**0.19**	0.22	**0.23**	0.25	**0.27**	0.32	**0.34**	0.40	**0.42**
Prostate	0.10	**0.10**	0.13	**0.14**	0.15	**0.16**	0.17	**0.19**	0.22	**0.24**	0.27	**0.29**
Testes	0.04	**0.04**	0.05	**0.06**	0.06	**0.07**	0.07	**0.08**	0.09	**0.10**	0.11	**0.13**
Upper arm		**0.10**		**0.13**		**0.15**		**0.17**		**0.20**		**0.23**
Elbow		**0.58**		**0.73**		**0.84**		**0.93**		**1.11**		**1.27**
Forearm		**0.69**		**0.86**		**0.99**		**1.09**		**1.28**		**1.45**
Upper leg	0.18	**0.31**	0.23	**0.40**	0.28	**0.48**	0.31	**0.54**	0.39	**0.67**	0.48	**0.81**
Knee		**0.02**		**0.03**		**0.04**		**0.04**		**0.06**	0.01	**0.07**
Lower leg												
Lymph nodes	0.08	**0.10**	0.10	**0.13**	0.13	**0.16**	0.14	**0.18**	0.18	**0.22**	0.22	**0.27**
Skin	0.18	**0.27**	0.19	**0.29**	0.20	**0.30**	0.21	**0.31**	0.22	**0.32**	0.23	**0.34**
Muscle	0.14	**0.19**	0.17	**0.23**	0.19	**0.26**	0.21	**0.28**	0.24	**0.33**	0.27	**0.37**
Skeleton	0.18	**0.33**	0.22	**0.41**	0.26	**0.47**	0.29	**0.51**	0.34	**0.61**	0.39	**0.70**
Red bone marrow	0.16	**0.20**	0.20	**0.25**	0.23	**0.29**	0.26	**0.32**	0.32	**0.39**	0.37	**0.45**
Whole body dose	0.15	**0.21**	0.18	**0.25**	0.20	**0.28**	0.22	**0.31**	0.25	**0.36**	0.29	**0.41**
ED (ICRP 60)	0.10	**0.15**	0.13	**0.19**	0.15	**0.22**	0.17	**0.25**	0.21	**0.31**	0.26	**0.37**
ED (ICRP 103)	0.10	**0.18**	0.13	**0.23**	0.16	**0.27**	0.18	**0.30**	0.22	**0.37**	0.27	**0.44**
Total filtration (mm) Al	**1.5**		**2.5**		**3.5**		**4.5**		**3.5**		**3.5**	
Cu	**0**		**0**		**0**		**0**		**0.1**		**0.2**	

Table 12.115 Normalized organ doses—Abdomen pa—13–17 years

Beam projection:		pa										**Table cells:**
Focus-to-detector distance (cm):		115										(+) = normal font = optimal field setting
Tube voltage (kVp):		80										**(−) = bold font = suboptimal field setting**
Total filtrations:		See last table row.										Empty table cells correspond to values <0.01
Organ or tissue	**Organ dose D_T : Entrance dose K_E**								**(mSv * mGy^{-1})**			
	(+)	(−)	(+)	(−)	(+)	(−)	(+)	(−)	(+)	(−)	(+)	(−)
Skull												
Brain												
Oral mucosa												
Salivary glands												
Cervical spine												
Thyroid												
Thymus		**0.03**		**0.03**		**0.04**		**0.05**	0.01	**0.06**	0.02	**0.07**
Thoracic spine	0.28	**0.50**	0.35	**0.62**	0.40	**0.72**	0.45	**0.80**	0.53	**0.96**	0.62	**1.10**
Claviculae				**0.01**		**0.02**		**0.02**		**0.02**		**0.03**
Scapulae	0.01	**0.37**	0.01	**0.43**	0.02	**0.47**	0.02	**0.50**	0.02	**0.56**	0.03	**0.60**
Ribs	0.43	**0.77**	0.50	**0.89**	0.54	**0.97**	0.57	**1.03**	0.63	**1.14**	0.68	**1.23**
Oesophagus	0.06	**0.11**	0.08	**0.14**	0.09	**0.17**	0.11	**0.19**	0.13	**0.24**	0.16	**0.28**
Trachea												
Lung	0.08	**0.25**	0.10	**0.31**	0.11	**0.35**	0.13	**0.38**	0.15	**0.45**	0.17	**0.51**
Mamma	0.14	**0.57**	0.19	**0.73**	0.22	**0.86**	0.25	**0.97**	0.31	**1.21**	0.37	**1.44**
Heart	0.04	**0.10**	0.06	**0.13**	0.07	**0.15**	0.08	**0.17**	0.10	**0.21**	0.12	**0.26**
Stomach	0.11	**0.12**	0.14	**0.16**	0.16	**0.18**	0.18	**0.21**	0.23	**0.26**	0.27	**0.31**
Spleen	0.41	**0.45**	0.51	**0.56**	0.58	**0.64**	0.64	**0.70**	0.75	**0.83**	0.85	**0.94**
Liver	0.19	**0.22**	0.24	**0.28**	0.28	**0.32**	0.31	**0.36**	0.37	**0.43**	0.43	**0.50**
Gall bladder	0.13	**0.14**	0.17	**0.18**	0.20	**0.21**	0.22	**0.24**	0.28	**0.30**	0.34	**0.36**
Pancreas	0.18	**0.20**	0.23	**0.25**	0.27	**0.30**	0.30	**0.34**	0.38	**0.42**	0.45	**0.50**
Small intestine	0.17	**0.18**	0.22	**0.23**	0.26	**0.28**	0.30	**0.31**	0.37	**0.39**	0.44	**0.47**
Large intestine	0.14	**0.15**	0.18	**0.19**	0.21	**0.23**	0.24	**0.26**	0.30	**0.32**	0.36	**0.39**
Sigma/rectum	0.18	**0.19**	0.23	**0.24**	0.27	**0.29**	0.31	**0.33**	0.38	**0.40**	0.45	**0.48**
Lumbar spine	1.03	**1.06**	1.28	**1.32**	1.47	**1.52**	1.63	**1.68**	1.93	**2.00**	2.21	**2.29**
Pelvis	1.09	**1.13**	1.33	**1.39**	1.51	**1.58**	1.66	**1.73**	1.93	**2.03**	2.18	**2.29**
Adrenals	0.44	**0.48**	0.54	**0.60**	0.62	**0.68**	0.68	**0.75**	0.80	**0.88**	0.91	**1.00**
Kidneys	0.70	**0.73**	0.84	**0.87**	0.93	**0.97**	1.01	**1.06**	1.15	**1.21**	1.28	**1.34**
Urinary bladder	0.10	**0.11**	0.13	**0.14**	0.16	**0.17**	0.18	**0.19**	0.23	**0.24**	0.27	**0.29**
Uterus	0.50	**0.52**	0.65	**0.67**	0.76	**0.79**	0.87	**0.90**	1.08	**1.12**	1.30	**1.35**
Ovaries	0.18	**0.20**	0.24	**0.25**	0.28	**0.30**	0.32	**0.34**	0.40	**0.43**	0.48	**0.51**
Prostate	0.13	**0.14**	0.16	**0.18**	0.19	**0.21**	0.22	**0.24**	0.28	**0.30**	0.33	**0.36**
Testes	0.05	**0.06**	0.07	**0.08**	0.08	**0.09**	0.09	**0.10**	0.11	**0.13**	0.13	**0.16**
Upper arm		**0.12**		**0.15**		**0.17**		**0.19**		**0.23**		**0.26**
Elbow		**0.68**		**0.83**		**0.95**		**1.04**		**1.21**		**1.36**
Forearm		**0.79**		**0.97**		**1.10**		**1.20**		**1.39**		**1.55**
Upper leg	0.23	**0.39**	0.29	**0.49**	0.34	**0.58**	0.38	**0.65**	0.47	**0.79**	0.56	**0.94**
Knee		**0.03**		**0.04**		**0.05**		**0.06**	0.01	**0.07**	0.02	**0.09**
Lower leg												
Lymph nodes	0.10	**0.13**	0.13	**0.17**	0.16	**0.20**	0.18	**0.22**	0.22	**0.27**	0.26	**0.32**
Skin	0.19	**0.28**	0.20	**0.30**	0.21	**0.31**	0.22	**0.32**	0.23	**0.34**	0.24	**0.36**
Muscle	0.16	**0.22**	0.19	**0.26**	0.22	**0.29**	0.23	**0.32**	0.27	**0.36**	0.30	**0.41**
Skeleton	0.21	**0.38**	0.26	**0.46**	0.29	**0.53**	0.32	**0.58**	0.37	**0.67**	0.42	**0.76**
Red bone marrow	0.20	**0.24**	0.24	**0.30**	0.28	**0.34**	0.31	**0.38**	0.36	**0.45**	0.42	**0.51**
Whole body dose	0.17	**0.24**	0.20	**0.29**	0.23	**0.32**	0.24	**0.35**	0.28	**0.40**	0.32	**0.45**
ED (ICRP 60)	0.13	**0.19**	0.16	**0.24**	0.19	**0.28**	0.21	**0.31**	0.26	**0.37**	0.31	**0.44**
ED (ICRP 103)	0.13	**0.22**	0.17	**0.28**	0.20	**0.33**	0.22	**0.37**	0.27	**0.45**	0.31	**0.52**
Total filtration (mm) **Al**	**1.5**		**2.5**		**3.5**		**4.5**		**3.5**		**3.5**	
Cu	**0**		**0**		**0**		**0**		**0.1**		**0.2**	

Table 12.116 Normalized organ doses—Abdomen pa—13–17 years

Beam projection:			pa									**Table cells:**
Focus-to-detector distance (cm):			115						(+) = normal font = optimal field setting			
Tube voltage (kVp):			110						**(−) = bold font = suboptimal field setting**			
Total filtrations:			See last table row.						Empty table cells correspond to values <0.01			
Organ or tissue	**Organ dose D_T : Entrance dose K_E**								**(mSv * mGy^{-1})**			
	(+)	(−)	(+)	(−)	(+)	(−)	(+)	(−)	(+)	(−)	(+)	(−)
Skull												
Brain												
Oral mucosa												
Salivary glands												
Cervical spine						**0.01**		**0.01**		**0.01**		**0.02**
Thyroid												**0.01**
Thymus	0.01	**0.05**	0.01	**0.06**	0.02	**0.07**	0.02	**0.07**	0.02	**0.09**	0.02	**0.10**
Thoracic spine	0.40	**0.71**	0.47	**0.84**	0.52	**0.93**	0.56	**1.00**	0.64	**1.14**	0.71	**1.25**
Claviculae		**0.02**		**0.02**		**0.03**		**0.03**		**0.04**	0.01	**0.04**
Scapulae	0.02	**0.44**	0.02	**0.49**	0.03	**0.52**	0.03	**0.54**	0.04	**0.58**	0.04	**0.61**
Ribs	0.50	**0.90**	0.56	**1.00**	0.59	**1.07**	0.62	**1.11**	0.66	**1.19**	0.68	**1.24**
Oesophagus	0.11	**0.19**	0.13	**0.23**	0.14	**0.26**	0.16	**0.28**	0.19	**0.33**	0.21	**0.38**
Trachea												
Lung	0.12	**0.35**	0.14	**0.41**	0.15	**0.45**	0.16	**0.48**	0.18	**0.54**	0.20	**0.59**
Mamma	0.25	**0.97**	0.31	**1.16**	0.35	**1.31**	0.38	**1.43**	0.45	**1.68**	0.52	**1.91**
Heart	0.08	**0.17**	0.09	**0.21**	0.11	**0.23**	0.12	**0.25**	0.14	**0.30**	0.16	**0.34**
Stomach	0.18	**0.21**	0.22	**0.25**	0.25	**0.29**	0.27	**0.31**	0.32	**0.37**	0.36	**0.42**
Spleen	0.58	**0.64**	0.68	**0.75**	0.75	**0.83**	0.80	**0.89**	0.90	**1.00**	0.99	**1.10**
Liver	0.29	**0.34**	0.34	**0.40**	0.38	**0.45**	0.41	**0.48**	0.47	**0.55**	0.53	**0.62**
Gall bladder	0.22	**0.24**	0.27	**0.29**	0.30	**0.33**	0.33	**0.36**	0.39	**0.43**	0.45	**0.49**
Pancreas	0.30	**0.33**	0.36	**0.40**	0.40	**0.45**	0.44	**0.49**	0.51	**0.58**	0.58	**0.66**
Small intestine	0.30	**0.31**	0.36	**0.37**	0.40	**0.42**	0.44	**0.46**	0.52	**0.55**	0.59	**0.62**
Large intestine	0.24	**0.26**	0.29	**0.31**	0.33	**0.35**	0.36	**0.39**	0.42	**0.46**	0.48	**0.52**
Sigma/rectum	0.30	**0.32**	0.36	**0.38**	0.40	**0.43**	0.44	**0.47**	0.52	**0.55**	0.59	**0.63**
Lumbar spine	1.44	**1.49**	1.69	**1.75**	1.86	**1.94**	2.00	**2.08**	2.26	**2.35**	2.47	**2.57**
Pelvis	1.45	**1.52**	1.68	**1.76**	1.84	**1.93**	1.96	**2.06**	2.18	**2.30**	2.36	**2.48**
Adrenals	0.62	**0.69**	0.73	**0.80**	0.80	**0.89**	0.86	**0.95**	0.97	**1.07**	1.06	**1.18**
Kidneys	0.91	**0.95**	1.04	**1.09**	1.13	**1.18**	1.20	**1.26**	1.32	**1.38**	1.42	**1.49**
Urinary bladder	0.18	**0.19**	0.22	**0.23**	0.25	**0.27**	0.27	**0.29**	0.32	**0.35**	0.37	**0.40**
Uterus	0.86	**0.89**	1.04	**1.08**	1.17	**1.22**	1.28	**1.34**	1.51	**1.58**	1.72	**1.80**
Ovaries	0.32	**0.34**	0.39	**0.41**	0.44	**0.46**	0.48	**0.51**	0.56	**0.59**	0.64	**0.68**
Prostate	0.22	**0.24**	0.26	**0.29**	0.30	**0.32**	0.33	**0.35**	0.39	**0.42**	0.44	**0.48**
Testes	0.09	**0.11**	0.11	**0.13**	0.12	**0.14**	0.14	**0.16**	0.16	**0.19**	0.18	**0.22**
Upper arm		**0.17**		**0.20**		**0.22**		**0.24**		**0.27**		**0.29**
Elbow		**0.90**		**1.05**		**1.15**		**1.22**		**1.36**		**1.46**
Forearm		**1.03**		**1.19**		**1.30**		**1.38**		**1.52**		**1.63**
Upper leg	0.36	**0.60**	0.43	**0.71**	0.48	**0.80**	0.52	**0.87**	0.60	**1.00**	0.68	**1.12**
Knee	0.01	**0.06**	0.01	**0.07**	0.02	**0.08**	0.02	**0.08**	0.02	**0.10**	0.02	**0.11**
Lower leg												
Lymph nodes	0.17	**0.21**	0.21	**0.26**	0.23	**0.29**	0.26	**0.31**	0.30	**0.37**	0.34	**0.42**
Skin	0.21	**0.31**	0.22	**0.33**	0.23	**0.34**	0.24	**0.35**	0.25	**0.37**	0.26	**0.38**
Muscle	0.22	**0.29**	0.25	**0.34**	0.27	**0.37**	0.29	**0.39**	0.32	**0.44**	0.35	**0.47**
Skeleton	0.28	**0.51**	0.33	**0.58**	0.36	**0.64**	0.38	**0.68**	0.42	**0.76**	0.46	**0.82**
Red bone marrow	0.29	**0.36**	0.34	**0.42**	0.38	**0.46**	0.40	**0.50**	0.46	**0.57**	0.51	**0.63**
Whole body dose	0.23	**0.32**	0.26	**0.37**	0.28	**0.40**	0.30	**0.43**	0.34	**0.48**	0.37	**0.52**
ED (ICRP 60)	0.21	**0.30**	0.25	**0.35**	0.28	**0.40**	0.30	**0.43**	0.35	**0.50**	0.39	**0.56**
ED (ICRP 103)	0.21	**0.35**	0.25	**0.42**	0.28	**0.47**	0.31	**0.51**	0.36	**0.60**	0.41	**0.67**
Total filtration (mm) **Al**	**1.5**		**2.5**		**3.5**		**4.5**		**3.5**		**3.5**	
Cu	**0**		**0**		**0**		**0**		**0.1**		**0.2**	

Table 12.117 Normalized organ doses—Abdomen pa—13–17 years

Beam projection:			pa									**Table cells:**
Focus-to-detector distance (cm):			115									(+) = normal font = optimal field setting
Tube voltage (kVp):			120									**(−) = bold font = suboptimal field setting**
Total filtrations:			See last table row.									Empty table cells correspond to values <0.01
Organ or tissue	**Organ dose D_T : Entrance dose K_E**								**(mSv * mGy^{-1})**			
	(+)	(−)	(+)	(−)	(+)	(−)	(+)	(−)	(+)	(−)	(+)	(−)
Skull												
Brain												
Oral mucosa												
Salivary glands												
Cervical spine				**0.01**		**0.01**		**0.01**		**0.01**		**0.02**
Thyroid										**0.01**		**0.01**
Thymus	0.01	**0.06**	0.02	**0.07**	0.02	**0.08**	0.02	**0.08**	0.02	**0.10**	0.03	**0.11**
Thoracic spine	0.43	**0.77**	0.50	**0.89**	0.55	**0.97**	0.59	**1.04**	0.66	**1.17**	0.71	**1.27**
Claviculae		**0.02**		**0.03**		**0.03**		**0.03**	0.01	**0.04**	0.01	**0.04**
Scapulae	0.02	**0.45**	0.03	**0.50**	0.03	**0.53**	0.03	**0.55**	0.04	**0.58**	0.04	**0.60**
Ribs	0.52	**0.93**	0.57	**1.02**	0.60	**1.08**	0.62	**1.12**	0.65	**1.19**	0.67	**1.23**
Oesophagus	0.12	**0.21**	0.14	**0.25**	0.16	**0.28**	0.17	**0.30**	0.20	**0.35**	0.22	**0.40**
Trachea												
Lung	0.13	**0.38**	0.15	**0.43**	0.16	**0.47**	0.17	**0.50**	0.19	**0.56**	0.21	**0.60**
Mamma	0.29	**1.09**	0.34	**1.28**	0.38	**1.43**	0.42	**1.55**	0.48	**1.79**	0.55	**2.02**
Heart	0.09	**0.19**	0.10	**0.23**	0.12	**0.25**	0.13	**0.28**	0.15	**0.32**	0.17	**0.36**
Stomach	0.20	**0.23**	0.24	**0.28**	0.27	**0.31**	0.29	**0.34**	0.34	**0.39**	0.38	**0.44**
Spleen	0.63	**0.69**	0.72	**0.79**	0.79	**0.87**	0.84	**0.93**	0.93	**1.03**	1.01	**1.12**
Liver	0.32	**0.37**	0.37	**0.43**	0.41	**0.48**	0.44	**0.51**	0.50	**0.58**	0.55	**0.64**
Gall bladder	0.25	**0.27**	0.30	**0.32**	0.33	**0.36**	0.36	**0.39**	0.42	**0.46**	0.47	**0.52**
Pancreas	0.33	**0.37**	0.39	**0.44**	0.43	**0.49**	0.47	**0.53**	0.54	**0.61**	0.61	**0.69**
Small intestine	0.33	**0.35**	0.39	**0.41**	0.44	**0.46**	0.47	**0.50**	0.55	**0.58**	0.62	**0.65**
Large intestine	0.27	**0.29**	0.32	**0.34**	0.36	**0.38**	0.39	**0.42**	0.45	**0.49**	0.51	**0.55**
Sigma/rectum	0.33	**0.36**	0.39	**0.42**	0.44	**0.47**	0.48	**0.51**	0.55	**0.59**	0.61	**0.66**
Lumbar spine	1.54	**1.60**	1.78	**1.85**	1.94	**2.02**	2.07	**2.15**	2.30	**2.40**	2.49	**2.59**
Pelvis	1.54	**1.61**	1.75	**1.84**	1.90	**2.00**	2.01	**2.12**	2.21	**2.33**	2.36	**2.49**
Adrenals	0.67	**0.74**	0.77	**0.86**	0.84	**0.94**	0.90	**1.00**	1.00	**1.11**	1.09	**1.21**
Kidneys	0.97	**1.01**	1.09	**1.14**	1.18	**1.23**	1.24	**1.30**	1.35	**1.42**	1.44	**1.51**
Urinary bladder	0.21	**0.22**	0.24	**0.26**	0.27	**0.29**	0.30	**0.32**	0.35	**0.37**	0.39	**0.42**
Uterus	0.97	**1.01**	1.14	**1.19**	1.28	**1.33**	1.39	**1.45**	1.61	**1.68**	1.81	**1.89**
Ovaries	0.36	**0.38**	0.42	**0.45**	0.48	**0.50**	0.52	**0.54**	0.60	**0.63**	0.68	**0.71**
Prostate	0.25	**0.27**	0.29	**0.32**	0.33	**0.35**	0.36	**0.38**	0.41	**0.44**	0.46	**0.50**
Testes	0.10	**0.12**	0.12	**0.14**	0.14	**0.16**	0.15	**0.17**	0.17	**0.20**	0.19	**0.23**
Upper arm		**0.18**		**0.21**		**0.23**		**0.25**		**0.27**		**0.30**
Elbow		**0.96**		**1.09**		**1.18**		**1.25**		**1.37**		**1.46**
Forearm		**1.09**		**1.23**		**1.34**		**1.41**		**1.54**		**1.63**
Upper leg	0.39	**0.65**	0.46	**0.76**	0.51	**0.85**	0.55	**0.91**	0.63	**1.04**	0.69	**1.14**
Knee	0.01	**0.06**	0.02	**0.07**	0.02	**0.08**	0.02	**0.09**	0.02	**0.10**	0.03	**0.12**
Lower leg												
Lymph nodes	0.19	**0.24**	0.23	**0.28**	0.25	**0.31**	0.28	**0.34**	0.32	**0.39**	0.36	**0.43**
Skin	0.22	**0.32**	0.23	**0.34**	0.24	**0.35**	0.24	**0.36**	0.25	**0.37**	0.26	**0.39**
Muscle	0.23	**0.32**	0.26	**0.36**	0.29	**0.39**	0.30	**0.41**	0.33	**0.45**	0.36	**0.49**
Skeleton	0.30	**0.54**	0.34	**0.61**	0.37	**0.66**	0.39	**0.70**	0.43	**0.77**	0.46	**0.82**
Red bone marrow	0.32	**0.39**	0.36	**0.45**	0.40	**0.49**	0.43	**0.53**	0.48	**0.59**	0.53	**0.65**
Whole body dose	0.24	**0.34**	0.27	**0.39**	0.30	**0.42**	0.32	**0.45**	0.35	**0.49**	0.38	**0.53**
ED (ICRP 60)	0.23	**0.33**	0.27	**0.39**	0.30	**0.43**	0.32	**0.46**	0.37	**0.53**	0.41	**0.59**
ED (ICRP 103)	0.24	**0.39**	0.28	**0.46**	0.31	**0.51**	0.33	**0.55**	0.38	**0.63**	0.42	**0.70**
Total filtration (mm) Al	**1.5**		**2.5**		**3.5**		**4.5**		**3.5**		**3.5**	
Cu	**0**		**0**		**0**		**0**		**0.1**		**0.2**	

Table 12.118 Normalized organ doses—Abdomen pa—>17 years

Beam projection:			pa									**Table cells:**
Focus-to-detector distance (cm):			115					(+) = normal font = optimal field setting				
Tube voltage (kVp):			70					**(−) = bold font = suboptimal field setting**				
Total filtrations:			See last table row.					Empty table cells correspond to values <0.01				
Organ or tissue	**Organ dose D_T : Entrance dose K_E**							**(mSv * mGy^{-1})**				
	(+)	(−)	(+)	(−)	(+)	(−)	(+)	(−)	(+)	(−)	(+)	(−)
Skull												
Brain												
Oral mucosa												
Salivary glands												
Cervical spine												
Thyroid												
Thymus		**0.02**		**0.02**		**0.03**		**0.03**	0.01	**0.04**	0.01	**0.05**
Thoracic spine	0.23	**0.37**	0.30	**0.47**	0.35	**0.56**	0.39	**0.63**	0.48	**0.76**	0.57	**0.90**
Claviculae								**0.01**		**0.01**		**0.02**
Scapulae		**0.23**	0.01	**0.27**	0.01	**0.30**	0.02	**0.33**	0.02	**0.37**	0.03	**0.41**
Ribs	0.42	**0.66**	0.49	**0.78**	0.54	**0.86**	0.58	**0.92**	0.65	**1.03**	0.70	**1.13**
Oesophagus	0.05	**0.08**	0.07	**0.11**	0.08	**0.13**	0.10	**0.15**	0.12	**0.19**	0.15	**0.23**
Trachea												
Lung	0.08	**0.20**	0.10	**0.24**	0.11	**0.28**	0.13	**0.31**	0.15	**0.37**	0.17	**0.42**
Mamma	0.02	**0.05**	0.03	**0.07**	0.03	**0.08**	0.03	**0.10**	0.04	**0.12**	0.05	**0.15**
Heart	0.04	**0.07**	0.05	**0.09**	0.06	**0.11**	0.07	**0.13**	0.09	**0.17**	0.11	**0.20**
Stomach	0.08	**0.09**	0.11	**0.12**	0.13	**0.14**	0.15	**0.16**	0.19	**0.21**	0.23	**0.25**
Spleen	0.35	**0.37**	0.44	**0.47**	0.51	**0.55**	0.57	**0.61**	0.68	**0.73**	0.78	**0.85**
Liver	0.16	**0.18**	0.20	**0.23**	0.24	**0.27**	0.27	**0.30**	0.33	**0.37**	0.39	**0.44**
Gall bladder	0.09	**0.10**	0.12	**0.13**	0.15	**0.16**	0.17	**0.18**	0.22	**0.23**	0.27	**0.28**
Pancreas	0.13	**0.15**	0.18	**0.19**	0.21	**0.23**	0.24	**0.26**	0.30	**0.33**	0.37	**0.41**
Small intestine	0.13	**0.13**	0.17	**0.18**	0.20	**0.21**	0.23	**0.24**	0.30	**0.31**	0.36	**0.38**
Large intestine	0.10	**0.11**	0.13	**0.14**	0.16	**0.17**	0.19	**0.20**	0.24	**0.25**	0.29	**0.31**
Sigma/rectum	0.14	**0.14**	0.18	**0.19**	0.21	**0.22**	0.25	**0.26**	0.31	**0.32**	0.38	**0.40**
Lumbar spine	0.83	**0.86**	1.06	**1.09**	1.24	**1.28**	1.38	**1.43**	1.68	**1.73**	1.95	**2.02**
Pelvis	0.92	**0.95**	1.15	**1.19**	1.33	**1.38**	1.47	**1.52**	1.75	**1.81**	2.01	**2.08**
Adrenals	0.38	**0.40**	0.48	**0.51**	0.55	**0.59**	0.61	**0.66**	0.73	**0.79**	0.84	**0.91**
Kidneys	0.61	**0.63**	0.75	**0.78**	0.85	**0.88**	0.93	**0.96**	1.07	**1.11**	1.20	**1.25**
Urinary bladder	0.08	**0.08**	0.10	**0.10**	0.12	**0.13**	0.14	**0.15**	0.18	**0.19**	0.22	**0.23**
Uterus	0.14	**0.14**	0.19	**0.19**	0.22	**0.23**	0.26	**0.26**	0.33	**0.34**	0.41	**0.42**
Ovaries	0.14	**0.14**	0.18	**0.19**	0.22	**0.23**	0.25	**0.26**	0.33	**0.34**	0.40	**0.42**
Prostate	0.09	**0.10**	0.12	**0.13**	0.15	**0.15**	0.17	**0.17**	0.22	**0.22**	0.27	**0.27**
Testes	0.04	**0.04**	0.05	**0.05**	0.06	**0.06**	0.07	**0.07**	0.09	**0.09**	0.11	**0.11**
Upper arm		**0.07**		**0.09**		**0.11**		**0.12**		**0.15**		**0.17**
Elbow		**0.56**		**0.70**		**0.81**		**0.90**		**1.07**		**1.23**
Forearm		**0.69**		**0.86**		**0.99**		**1.09**		**1.29**		**1.47**
Upper leg	0.20	**0.25**	0.27	**0.33**	0.32	**0.40**	0.37	**0.45**	0.46	**0.57**	0.56	**0.69**
Knee		**0.04**	0.01	**0.05**	0.01	**0.06**	0.01	**0.07**	0.02	**0.09**	0.02	**0.11**
Lower leg												
Lymph nodes	0.08	**0.10**	0.11	**0.13**	0.13	**0.15**	0.14	**0.17**	0.18	**0.22**	0.22	**0.26**
Skin	0.20	**0.28**	0.21	**0.30**	0.22	**0.31**	0.23	**0.32**	0.24	**0.34**	0.25	**0.35**
Muscle	0.15	**0.19**	0.18	**0.23**	0.21	**0.26**	0.23	**0.28**	0.26	**0.33**	0.30	**0.37**
Skeleton	0.19	**0.31**	0.23	**0.38**	0.27	**0.44**	0.30	**0.48**	0.35	**0.58**	0.41	**0.66**
Red bone marrow	0.17	**0.20**	0.21	**0.25**	0.25	**0.29**	0.28	**0.32**	0.33	**0.39**	0.39	**0.45**
Whole body dose	0.16	**0.21**	0.19	**0.25**	0.21	**0.28**	0.23	**0.31**	0.27	**0.36**	0.31	**0.41**
ED (ICRP 60)	0.10	**0.12**	0.13	**0.16**	0.15	**0.19**	0.17	**0.21**	0.21	**0.26**	0.25	**0.31**
ED (ICRP 103)	0.10	**0.12**	0.12	**0.16**	0.14	**0.19**	0.16	**0.21**	0.20	**0.26**	0.24	**0.31**
Total filtration (mm) **Al**	**1.5**		**2.5**		**3.5**		**4.5**		**3.5**		**3.5**	
Cu	**0**		**0**		**0**		**0**		**0.1**		**0.2**	

Table 12.119 Normalized organ doses—Abdomen pa—>17 years

Beam projection:			pa									**Table cells:**
Focus-to-detector distance (cm):			115									(+) = normal font = optimal field setting
Tube voltage (kVp):			80									**(−) = bold font = suboptimal field setting**
Total filtrations:			See last table row.									Empty table cells correspond to values <0.01
Organ or tissue	**Organ dose D_T : Entrance dose K_E**								**(mSv * mGy^{-1})**			
	(+)	(−)	(+)	(−)	(+)	(−)	(+)	(−)	(+)	(−)	(+)	(−)
Skull												
Brain												
Oral mucosa												
Salivary glands												
Cervical spine												
Thyroid												
Thymus		**0.02**		**0.03**	0.01	**0.04**	0.01	**0.04**	0.02	**0.05**	0.02	**0.06**
Thoracic spine	0.28	**0.45**	0.36	**0.57**	0.41	**0.66**	0.46	**0.73**	0.55	**0.87**	0.64	**1.01**
Claviculae				**0.01**		**0.01**		**0.02**		**0.02**		**0.02**
Scapulae	0.01	**0.25**	0.02	**0.30**	0.02	**0.33**	0.02	**0.35**	0.03	**0.39**	0.03	**0.43**
Ribs	0.46	**0.72**	0.53	**0.84**	0.58	**0.92**	0.61	**0.98**	0.67	**1.08**	0.72	**1.17**
Oesophagus	0.07	**0.11**	0.09	**0.14**	0.11	**0.17**	0.12	**0.19**	0.15	**0.24**	0.19	**0.29**
Trachea												
Lung	0.09	**0.23**	0.12	**0.28**	0.13	**0.32**	0.15	**0.35**	0.17	**0.41**	0.20	**0.46**
Mamma	0.03	**0.07**	0.03	**0.09**	0.04	**0.11**	0.04	**0.12**	0.05	**0.15**	0.06	**0.17**
Heart	0.05	**0.10**	0.06	**0.12**	0.08	**0.15**	0.09	**0.17**	0.11	**0.21**	0.13	**0.25**
Stomach	0.11	**0.12**	0.14	**0.15**	0.16	**0.18**	0.19	**0.21**	0.23	**0.26**	0.28	**0.31**
Spleen	0.41	**0.45**	0.51	**0.55**	0.59	**0.64**	0.65	**0.70**	0.76	**0.83**	0.87	**0.95**
Liver	0.20	**0.22**	0.25	**0.28**	0.29	**0.32**	0.32	**0.36**	0.38	**0.43**	0.45	**0.51**
Gall bladder	0.13	**0.13**	0.16	**0.17**	0.19	**0.20**	0.22	**0.23**	0.27	**0.29**	0.33	**0.35**
Pancreas	0.18	**0.19**	0.23	**0.25**	0.27	**0.29**	0.30	**0.33**	0.37	**0.41**	0.45	**0.49**
Small intestine	0.17	**0.18**	0.22	**0.23**	0.26	**0.27**	0.30	**0.31**	0.37	**0.39**	0.44	**0.46**
Large intestine	0.14	**0.14**	0.18	**0.19**	0.21	**0.22**	0.24	**0.25**	0.30	**0.32**	0.36	**0.38**
Sigma/rectum	0.18	**0.19**	0.23	**0.24**	0.27	**0.28**	0.31	**0.32**	0.38	**0.40**	0.46	**0.48**
Lumbar spine	1.00	**1.04**	1.25	**1.29**	1.44	**1.49**	1.59	**1.64**	1.89	**1.96**	2.17	**2.25**
Pelvis	1.08	**1.12**	1.32	**1.37**	1.51	**1.56**	1.65	**1.71**	1.93	**2.01**	2.18	**2.27**
Adrenals	0.45	**0.48**	0.56	**0.60**	0.63	**0.68**	0.70	**0.75**	0.82	**0.89**	0.93	**1.01**
Kidneys	0.70	**0.73**	0.84	**0.87**	0.94	**0.98**	1.02	**1.06**	1.17	**1.22**	1.30	**1.35**
Urinary bladder	0.10	**0.11**	0.13	**0.14**	0.16	**0.17**	0.18	**0.19**	0.23	**0.24**	0.28	**0.29**
Uterus	0.19	**0.19**	0.24	**0.25**	0.29	**0.30**	0.33	**0.34**	0.41	**0.42**	0.50	**0.51**
Ovaries	0.19	**0.19**	0.24	**0.25**	0.29	**0.30**	0.32	**0.34**	0.41	**0.42**	0.49	**0.51**
Prostate	0.12	**0.13**	0.16	**0.17**	0.19	**0.20**	0.22	**0.22**	0.27	**0.28**	0.33	**0.34**
Testes	0.05	**0.05**	0.06	**0.07**	0.08	**0.08**	0.09	**0.09**	0.11	**0.12**	0.13	**0.14**
Upper arm		**0.09**		**0.11**		**0.13**		**0.14**		**0.17**		**0.20**
Elbow		**0.66**		**0.80**		**0.92**		**1.01**		**1.18**		**1.33**
Forearm		**0.80**		**0.97**		**1.11**		**1.21**		**1.41**		**1.57**
Upper leg	0.27	**0.33**	0.34	**0.42**	0.40	**0.50**	0.45	**0.56**	0.56	**0.69**	0.67	**0.82**
Knee	0.01	**0.05**	0.01	**0.07**	0.02	**0.08**	0.02	**0.09**	0.03	**0.11**	0.03	**0.13**
Lower leg												
Lymph nodes	0.11	**0.13**	0.14	**0.16**	0.16	**0.19**	0.18	**0.21**	0.22	**0.26**	0.26	**0.31**
Skin	0.21	**0.29**	0.22	**0.31**	0.23	**0.32**	0.24	**0.33**	0.25	**0.35**	0.26	**0.37**
Muscle	0.18	**0.22**	0.21	**0.26**	0.23	**0.30**	0.25	**0.32**	0.29	**0.37**	0.33	**0.41**
Skeleton	0.22	**0.36**	0.27	**0.44**	0.30	**0.50**	0.33	**0.55**	0.39	**0.64**	0.45	**0.72**
Red bone marrow	0.21	**0.24**	0.26	**0.30**	0.29	**0.34**	0.32	**0.38**	0.39	**0.45**	0.45	**0.52**
Whole body dose	0.18	**0.24**	0.22	**0.29**	0.24	**0.32**	0.26	**0.35**	0.31	**0.40**	0.34	**0.45**
ED (ICRP 60)	0.12	**0.16**	0.16	**0.20**	0.18	**0.23**	0.21	**0.26**	0.25	**0.31**	0.30	**0.37**
ED (ICRP 103)	0.12	**0.16**	0.15	**0.20**	0.18	**0.23**	0.20	**0.25**	0.24	**0.31**	0.28	**0.36**
Total filtration (mm) Al	**1.5**		**2.5**		**3.5**		**4.5**		**3.5**		**3.5**	
Cu	**0**		**0**		**0**		**0**		**0.1**		**0.2**	

Table 12.120 Normalized organ doses—Abdomen pa—>17 years

Beam projection:			pa									**Table cells:**
Focus-to-detector distance (cm):			115									(+) = normal font = optimal field setting
Tube voltage (kVp):			110									**(−) = bold font = suboptimal field setting**
Total filtrations:			See last table row.									Empty table cells correspond to values <0.01
Organ or tissue	**Organ dose D_T : Entrance dose K_E**								**(mSv * mGy⁻¹)**			
	(+)	(−)	(+)	(−)	(+)	(−)	(+)	(−)	(+)	(−)	(+)	(−)
Skull												
Brain												
Oral mucosa												
Salivary glands												
Cervical spine												**0.01**
Thyroid												
Thymus	0.01	**0.04**	0.02	**0.05**	0.02	**0.06**	0.02	**0.07**	0.03	**0.08**	0.03	**0.09**
Thoracic spine	0.41	**0.65**	0.49	**0.77**	0.54	**0.86**	0.59	**0.92**	0.67	**1.05**	0.74	**1.16**
Claviculae		**0.02**		**0.02**		**0.02**		**0.02**		**0.03**	0.01	**0.03**
Scapulae	0.02	**0.31**	0.03	**0.34**	0.03	**0.37**	0.03	**0.39**	0.04	**0.42**	0.04	**0.44**
Ribs	0.54	**0.86**	0.59	**0.95**	0.63	**1.01**	0.66	**1.06**	0.70	**1.13**	0.73	**1.18**
Oesophagus	0.12	**0.19**	0.15	**0.23**	0.17	**0.26**	0.19	**0.29**	0.22	**0.34**	0.25	**0.39**
Trachea												
Lung	0.13	**0.32**	0.16	**0.37**	0.17	**0.41**	0.19	**0.44**	0.21	**0.49**	0.23	**0.54**
Mamma	0.04	**0.12**	0.05	**0.14**	0.06	**0.16**	0.06	**0.17**	0.08	**0.20**	0.09	**0.23**
Heart	0.09	**0.17**	0.11	**0.20**	0.12	**0.23**	0.13	**0.25**	0.16	**0.30**	0.18	**0.34**
Stomach	0.19	**0.21**	0.23	**0.25**	0.26	**0.28**	0.28	**0.31**	0.33	**0.37**	0.38	**0.42**
Spleen	0.59	**0.64**	0.69	**0.75**	0.76	**0.83**	0.82	**0.89**	0.92	**1.01**	1.01	**1.11**
Liver	0.30	**0.34**	0.36	**0.40**	0.40	**0.45**	0.43	**0.49**	0.49	**0.56**	0.55	**0.63**
Gall bladder	0.22	**0.24**	0.27	**0.29**	0.30	**0.32**	0.33	**0.36**	0.40	**0.42**	0.45	**0.48**
Pancreas	0.30	**0.33**	0.36	**0.39**	0.40	**0.44**	0.44	**0.49**	0.52	**0.57**	0.59	**0.65**
Small intestine	0.30	**0.31**	0.36	**0.37**	0.40	**0.42**	0.44	**0.46**	0.52	**0.55**	0.60	**0.62**
Large intestine	0.24	**0.26**	0.29	**0.31**	0.33	**0.35**	0.36	**0.38**	0.43	**0.45**	0.49	**0.52**
Sigma/rectum	0.30	**0.32**	0.36	**0.38**	0.41	**0.43**	0.45	**0.47**	0.53	**0.56**	0.60	**0.63**
Lumbar spine	1.42	**1.47**	1.66	**1.72**	1.84	**1.91**	1.98	**2.05**	2.24	**2.32**	2.45	**2.54**
Pelvis	1.45	**1.51**	1.68	**1.75**	1.85	**1.92**	1.97	**2.05**	2.20	**2.29**	2.38	**2.48**
Adrenals	0.64	**0.69**	0.75	**0.81**	0.82	**0.89**	0.88	**0.96**	1.00	**1.09**	1.09	**1.19**
Kidneys	0.92	**0.96**	1.06	**1.10**	1.15	**1.20**	1.22	**1.27**	1.35	**1.40**	1.45	**1.51**
Urinary bladder	0.19	**0.19**	0.23	**0.23**	0.26	**0.27**	0.28	**0.29**	0.34	**0.35**	0.39	**0.40**
Uterus	0.33	**0.34**	0.40	**0.41**	0.45	**0.47**	0.50	**0.51**	0.59	**0.61**	0.67	**0.69**
Ovaries	0.32	**0.34**	0.39	**0.41**	0.44	**0.46**	0.49	**0.51**	0.57	**0.60**	0.65	**0.68**
Prostate	0.22	**0.22**	0.26	**0.27**	0.30	**0.31**	0.33	**0.34**	0.38	**0.40**	0.44	**0.45**
Testes	0.09	**0.10**	0.11	**0.12**	0.12	**0.13**	0.13	**0.15**	0.16	**0.17**	0.18	**0.20**
Upper arm		**0.13**		**0.15**		**0.17**		**0.18**		**0.20**		**0.22**
Elbow		**0.88**		**1.01**		**1.11**		**1.19**		**1.32**		**1.43**
Forearm		**1.04**		**1.20**		**1.32**		**1.40**		**1.55**		**1.66**
Upper leg	0.43	**0.53**	0.51	**0.63**	0.58	**0.71**	0.63	**0.78**	0.73	**0.90**	0.83	**1.02**
Knee	0.02	**0.09**	0.03	**0.10**	0.03	**0.12**	0.04	**0.13**	0.04	**0.15**	0.05	**0.17**
Lower leg												
Lymph nodes	0.18	**0.21**	0.21	**0.25**	0.24	**0.28**	0.26	**0.31**	0.31	**0.36**	0.35	**0.41**
Skin	0.23	**0.32**	0.24	**0.34**	0.25	**0.35**	0.25	**0.36**	0.27	**0.38**	0.28	**0.39**
Muscle	0.24	**0.30**	0.27	**0.34**	0.30	**0.37**	0.32	**0.40**	0.35	**0.44**	0.38	**0.48**
Skeleton	0.30	**0.48**	0.34	**0.56**	0.38	**0.61**	0.40	**0.66**	0.45	**0.73**	0.49	**0.79**
Red bone marrow	0.31	**0.36**	0.36	**0.42**	0.40	**0.47**	0.43	**0.51**	0.49	**0.58**	0.55	**0.64**
Whole body dose	0.25	**0.32**	0.28	**0.37**	0.31	**0.41**	0.33	**0.43**	0.37	**0.48**	0.40	**0.53**
ED (ICRP 60)	0.20	**0.25**	0.24	**0.30**	0.27	**0.33**	0.29	**0.36**	0.34	**0.42**	0.38	**0.47**
ED (ICRP 103)	0.19	**0.24**	0.23	**0.29**	0.25	**0.32**	0.28	**0.35**	0.32	**0.40**	0.36	**0.45**
Total filtration (mm) **Al**	**1.5**		**2.5**		**3.5**		**4.5**		**3.5**		**3.5**	
Cu	**0**		**0**		**0**		**0**		**0.1**		**0.2**	

Table 12.121 Normalized organ doses—Abdomen pa—>17 years

Beam projection:			pa									**Table cells:**
Focus-to-detector distance (cm):			115				(+) = normal font = optimal field setting					
Tube voltage (kVp):			120				**(−) = bold font = suboptimal field setting**					
Total filtrations:			See last table row.				Empty table cells correspond to values <0.01					
Organ or tissue	**Organ dose D_T : Entrance dose K_E**								**(mSv * mGy^{-1})**			
	(+)	(−)	(+)	(−)	(+)	(−)	(+)	(−)	(+)	(−)	(+)	(−)
Skull												
Brain												
Oral mucosa												
Salivary glands												
Cervical spine										**0.01**		**0.01**
Thyroid												
Thymus	0.02	**0.05**	0.02	**0.06**	0.02	**0.07**	0.02	**0.07**	0.03	**0.09**	0.03	**0.10**
Thoracic spine	0.45	**0.71**	0.52	**0.82**	0.57	**0.90**	0.61	**0.96**	0.68	**1.08**	0.75	**1.18**
Claviculae		**0.02**		**0.02**		**0.02**		**0.03**	0.01	**0.03**	0.01	**0.04**
Scapulae	0.02	**0.32**	0.03	**0.35**	0.03	**0.38**	0.03	**0.39**	0.04	**0.42**	0.05	**0.44**
Ribs	0.55	**0.88**	0.61	**0.97**	0.64	**1.03**	0.66	**1.07**	0.70	**1.13**	0.72	**1.17**
Oesophagus	0.14	**0.22**	0.17	**0.26**	0.19	**0.29**	0.20	**0.31**	0.24	**0.37**	0.27	**0.41**
Trachea												
Lung	0.15	**0.34**	0.17	**0.39**	0.18	**0.43**	0.20	**0.46**	0.22	**0.51**	0.24	**0.55**
Mamma	0.05	**0.13**	0.06	**0.15**	0.06	**0.17**	0.07	**0.19**	0.08	**0.22**	0.09	**0.24**
Heart	0.10	**0.19**	0.12	**0.22**	0.13	**0.25**	0.15	**0.27**	0.17	**0.32**	0.19	**0.36**
Stomach	0.21	**0.23**	0.25	**0.28**	0.28	**0.31**	0.30	**0.34**	0.35	**0.39**	0.40	**0.44**
Spleen	0.64	**0.69**	0.73	**0.80**	0.80	**0.87**	0.86	**0.93**	0.96	**1.04**	1.04	**1.13**
Liver	0.33	**0.37**	0.38	**0.44**	0.43	**0.48**	0.46	**0.52**	0.52	**0.59**	0.57	**0.65**
Gall bladder	0.25	**0.27**	0.30	**0.32**	0.33	**0.36**	0.36	**0.39**	0.42	**0.45**	0.48	**0.51**
Pancreas	0.33	**0.37**	0.39	**0.43**	0.44	**0.48**	0.48	**0.52**	0.55	**0.61**	0.62	**0.68**
Small intestine	0.33	**0.35**	0.39	**0.41**	0.44	**0.46**	0.48	**0.50**	0.56	**0.58**	0.63	**0.65**
Large intestine	0.27	**0.29**	0.32	**0.34**	0.36	**0.38**	0.39	**0.42**	0.46	**0.49**	0.52	**0.55**
Sigma/rectum	0.34	**0.36**	0.40	**0.42**	0.45	**0.47**	0.49	**0.51**	0.56	**0.59**	0.63	**0.66**
Lumbar spine	1.52	**1.58**	1.76	**1.82**	1.92	**1.99**	2.05	**2.13**	2.29	**2.37**	2.48	**2.57**
Pelvis	1.54	**1.60**	1.76	**1.83**	1.91	**1.99**	2.03	**2.11**	2.23	**2.33**	2.39	**2.49**
Adrenals	0.69	**0.75**	0.79	**0.86**	0.87	**0.95**	0.93	**1.01**	1.03	**1.13**	1.12	**1.23**
Kidneys	0.98	**1.02**	1.11	**1.15**	1.20	**1.25**	1.26	**1.31**	1.38	**1.44**	1.47	**1.54**
Urinary bladder	0.21	**0.22**	0.25	**0.26**	0.28	**0.29**	0.31	**0.32**	0.36	**0.37**	0.41	**0.42**
Uterus	0.37	**0.38**	0.44	**0.46**	0.50	**0.51**	0.54	**0.56**	0.63	**0.65**	0.71	**0.73**
Ovaries	0.36	**0.38**	0.43	**0.45**	0.48	**0.50**	0.53	**0.55**	0.61	**0.64**	0.69	**0.72**
Prostate	0.24	**0.25**	0.29	**0.30**	0.32	**0.34**	0.35	**0.36**	0.41	**0.42**	0.46	**0.48**
Testes	0.10	**0.11**	0.12	**0.13**	0.14	**0.15**	0.15	**0.16**	0.17	**0.19**	0.20	**0.21**
Upper arm		**0.14**		**0.16**		**0.17**		**0.19**		**0.21**		**0.23**
Elbow		**0.93**		**1.06**		**1.15**		**1.22**		**1.34**		**1.43**
Forearm		**1.10**		**1.25**		**1.36**		**1.43**		**1.57**		**1.66**
Upper leg	0.47	**0.58**	0.55	**0.68**	0.62	**0.76**	0.67	**0.82**	0.76	**0.94**	0.85	**1.05**
Knee	0.03	**0.10**	0.03	**0.11**	0.04	**0.13**	0.04	**0.14**	0.05	**0.16**	0.05	**0.18**
Lower leg												
Lymph nodes	0.20	**0.24**	0.23	**0.28**	0.26	**0.31**	0.28	**0.33**	0.33	**0.38**	0.37	**0.43**
Skin	0.23	**0.33**	0.25	**0.35**	0.25	**0.36**	0.26	**0.37**	0.27	**0.39**	0.28	**0.40**
Muscle	0.25	**0.32**	0.29	**0.36**	0.31	**0.39**	0.33	**0.42**	0.36	**0.46**	0.39	**0.50**
Skeleton	0.32	**0.51**	0.36	**0.59**	0.39	**0.64**	0.42	**0.68**	0.46	**0.74**	0.50	**0.80**
Red bone marrow	0.34	**0.39**	0.39	**0.45**	0.43	**0.50**	0.46	**0.54**	0.52	**0.60**	0.57	**0.66**
Whole body dose	0.26	**0.35**	0.30	**0.39**	0.32	**0.43**	0.34	**0.45**	0.38	**0.50**	0.41	**0.54**
ED (ICRP 60)	0.22	**0.28**	0.26	**0.32**	0.29	**0.36**	0.31	**0.38**	0.36	**0.44**	0.40	**0.49**
ED (ICRP 103)	0.21	**0.27**	0.25	**0.31**	0.27	**0.35**	0.30	**0.37**	0.34	**0.43**	0.38	**0.47**
Total filtration (mm) Al	**1.5**		**2.5**		**3.5**		**4.5**		**3.5**		**3.5**	
Cu	**0**		**0**		**0**		**0**		**0.1**		**0.2**	

Table 12.122 Normalized organ doses—Abdomen r-lat—<1 year

Beam projection:			r-lat									**Table cells:**
Focus-to-detector distance (cm):			115									(+) = normal font = optimal field setting
Tube voltage (kVp):			60									**(−) = bold font = suboptimal field setting**
Total filtrations:			See last table row.									Empty table cells correspond to values <0.01
Organ or tissue	**Organ dose D_T : Entrance dose K_E**								**(mSv * mGy^{-1})**			
	(+)	(−)	(+)	(−)	(+)	(−)	(+)	(−)	(+)	(−)	(+)	(−)
Skull												
Brain												
Oral mucosa												
Salivary glands										**0.01**		
Cervical spine		**0.02**		**0.02**	0.01	**0.03**	0.01	**0.03**	0.02	**0.04**	0.01	**0.04**
Thyroid		**0.01**		**0.01**		**0.02**		**0.02**		**0.03**		**0.02**
Thymus	0.01	**0.06**	0.01	**0.08**	0.02	**0.09**	0.02	**0.10**	0.02	**0.12**	0.02	**0.12**
Thoracic spine	0.27	**0.45**	0.35	**0.58**	0.41	**0.67**	0.46	**0.75**	0.55	**0.89**	0.52	**0.86**
Claviculae	0.01	**0.04**	0.02	**0.05**	0.02	**0.06**	0.02	**0.07**	0.03	**0.08**	0.03	**0.07**
Scapulae	0.04	**0.70**	0.05	**0.81**	0.06	**0.87**	0.07	**0.92**	0.09	**1.01**	0.08	**1.01**
Ribs	0.55	**0.81**	0.64	**0.96**	0.70	**1.06**	0.74	**1.13**	0.81	**1.26**	0.81	**1.26**
Oesophagus	0.12	**0.18**	0.15	**0.23**	0.17	**0.26**	0.19	**0.29**	0.23	**0.35**	0.22	**0.33**
Trachea												
Lung	0.14	**0.27**	0.17	**0.33**	0.20	**0.37**	0.21	**0.40**	0.24	**0.46**	0.23	**0.45**
Mamma	0.02	**0.57**	0.03	**0.61**	0.04	**0.64**	0.04	**0.66**	0.05	**0.69**	0.05	**0.69**
Heart	0.12	**0.23**	0.16	**0.30**	0.18	**0.35**	0.20	**0.38**	0.24	**0.46**	0.23	**0.44**
Stomach	0.13	**0.12**	0.17	**0.16**	0.20	**0.19**	0.22	**0.21**	0.27	**0.26**	0.25	**0.24**
Spleen	0.09	**0.08**	0.11	**0.11**	0.14	**0.13**	0.15	**0.15**	0.19	**0.18**	0.17	**0.16**
Liver	0.55	**0.46**	0.66	**0.56**	0.73	**0.63**	0.79	**0.69**	0.88	**0.79**	0.87	**0.77**
Gall bladder	0.26	**0.23**	0.34	**0.29**	0.39	**0.35**	0.44	**0.39**	0.52	**0.47**	0.50	**0.44**
Pancreas	0.19	**0.15**	0.24	**0.20**	0.28	**0.23**	0.32	**0.27**	0.39	**0.33**	0.36	**0.30**
Small intestine	0.28	**0.23**	0.36	**0.29**	0.41	**0.34**	0.46	**0.38**	0.55	**0.46**	0.52	**0.43**
Large intestine	0.41	**0.33**	0.51	**0.42**	0.58	**0.48**	0.63	**0.53**	0.73	**0.63**	0.71	**0.60**
Sigma/rectum	0.13	**0.11**	0.17	**0.15**	0.21	**0.18**	0.24	**0.20**	0.29	**0.25**	0.26	**0.22**
Lumbar spine	0.62	**0.58**	0.79	**0.75**	0.93	**0.88**	1.04	**0.99**	1.25	**1.19**	1.19	**1.13**
Pelvis	0.89	**0.76**	1.10	**0.95**	1.25	**1.09**	1.37	**1.20**	1.57	**1.41**	1.55	**1.37**
Adrenals	0.22	**0.21**	0.27	**0.27**	0.32	**0.32**	0.35	**0.35**	0.42	**0.42**	0.40	**0.40**
Kidneys	0.30	**0.29**	0.37	**0.36**	0.42	**0.41**	0.46	**0.45**	0.54	**0.53**	0.52	**0.51**
Urinary bladder	0.26	**0.24**	0.34	**0.31**	0.39	**0.36**	0.43	**0.40**	0.51	**0.47**	0.49	**0.45**
Uterus	0.25	**0.21**	0.33	**0.27**	0.38	**0.32**	0.43	**0.36**	0.52	**0.44**	0.49	**0.41**
Ovaries	0.28	**0.21**	0.36	**0.27**	0.42	**0.32**	0.46	**0.36**	0.55	**0.44**	0.53	**0.41**
Prostate	0.22	**0.20**	0.29	**0.26**	0.34	**0.31**	0.38	**0.34**	0.45	**0.42**	0.43	**0.39**
Testes	0.21	**0.21**	0.26	**0.26**	0.30	**0.31**	0.34	**0.34**	0.40	**0.40**	0.38	**0.38**
Upper arm		**0.42**		**0.48**		**0.52**		**0.55**		**0.60**		**0.60**
Elbow		**1.42**		**1.60**		**1.71**		**1.79**		**1.91**		**1.92**
Forearm		**1.50**		**1.67**		**1.78**		**1.85**		**1.97**		**1.98**
Upper leg	0.44	**0.59**	0.56	**0.75**	0.65	**0.86**	0.72	**0.96**	0.85	**1.13**	0.81	**1.09**
Knee	0.04	**0.25**	0.05	**0.31**	0.06	**0.36**	0.07	**0.40**	0.09	**0.47**	0.07	**0.45**
Lower leg						**0.01**		**0.01**		**0.02**		**0.01**
Lymph nodes	0.16	**0.16**	0.20	**0.20**	0.24	**0.23**	0.26	**0.26**	0.31	**0.31**	0.30	**0.29**
Skin	0.23	**0.29**	0.24	**0.31**	0.25	**0.32**	0.25	**0.33**	0.27	**0.34**	0.26	**0.34**
Muscle	0.18	**0.22**	0.22	**0.26**	0.24	**0.29**	0.26	**0.32**	0.29	**0.36**	0.28	**0.35**
Skeleton	0.26	**0.54**	0.32	**0.65**	0.37	**0.72**	0.40	**0.78**	0.46	**0.88**	0.45	**0.87**
Red bone marrow	0.08	**0.14**	0.09	**0.17**	0.11	**0.19**	0.12	**0.21**	0.14	**0.24**	0.13	**0.23**
Whole body dose	0.18	**0.23**	0.22	**0.28**	0.24	**0.31**	0.26	**0.34**	0.30	**0.39**	0.29	**0.37**
ED (ICRP 60)	0.18	**0.22**	0.23	**0.27**	0.27	**0.31**	0.29	**0.34**	0.34	**0.40**	0.33	**0.38**
ED (ICRP 103)	0.16	**0.24**	0.20	**0.29**	0.23	**0.32**	0.25	**0.35**	0.30	**0.40**	0.28	**0.39**
Total filtration (mm) **Al**	**1.5**		**2.5**		**3.5**		**4.5**		**3.5**		**3.5**	
Cu	**0**		**0**		**0**		**0**		**0.1**		**0.2**	

Table 12.123 Normalized organ doses—Abdomen r-lat—<1 year

Beam projection:		r-lat										**Table cells:**
Focus-to-detector distance (cm):		115										(+) = normal font = optimal field setting
Tube voltage (kVp):		70										**(−) = bold font = suboptimal field setting**
Total filtrations:		See last table row.										Empty table cells correspond to values <0.01
Organ or tissue	**Organ dose D_T : Entrance dose K_E**								**(mSv * mGy⁻¹)**			
	(+)	(−)	(+)	(−)	(+)	(−)	(+)	(−)	(+)	(−)	(+)	(−)
Skull												**0.01**
Brain												
Oral mucosa										**0.01**		**0.01**
Salivary glands						**0.01**		**0.01**		**0.01**		**0.02**
Cervical spine		**0.02**	0.01	**0.03**	0.01	**0.04**	0.02	**0.04**	0.02	**0.05**	0.02	**0.06**
Thyroid		**0.01**		**0.02**		**0.02**		**0.03**	0.01	**0.03**	0.01	**0.04**
Thymus	0.01	**0.08**	0.02	**0.09**	0.02	**0.11**	0.02	**0.12**	0.03	**0.14**	0.03	**0.16**
Thoracic spine	0.33	**0.53**	0.41	**0.67**	0.47	**0.77**	0.52	**0.85**	0.61	**0.99**	0.68	**1.12**
Claviculae	0.02	**0.05**	0.02	**0.06**	0.03	**0.07**	0.03	**0.08**	0.04	**0.10**	0.04	**0.11**
Scapulae	0.05	**0.75**	0.07	**0.86**	0.08	**0.92**	0.09	**0.97**	0.11	**1.05**	0.13	**1.11**
Ribs	0.60	**0.89**	0.68	**1.03**	0.74	**1.13**	0.78	**1.20**	0.84	**1.32**	0.89	**1.41**
Oesophagus	0.14	**0.21**	0.18	**0.27**	0.20	**0.31**	0.22	**0.34**	0.26	**0.40**	0.29	**0.45**
Trachea												**0.01**
Lung	0.16	**0.30**	0.19	**0.37**	0.22	**0.41**	0.23	**0.45**	0.26	**0.51**	0.29	**0.56**
Mamma	0.03	**0.59**	0.04	**0.64**	0.04	**0.67**	0.05	**0.69**	0.05	**0.72**	0.06	**0.75**
Heart	0.15	**0.28**	0.18	**0.35**	0.21	**0.40**	0.23	**0.44**	0.27	**0.51**	0.30	**0.58**
Stomach	0.16	**0.15**	0.20	**0.19**	0.24	**0.23**	0.27	**0.25**	0.32	**0.31**	0.37	**0.36**
Spleen	0.11	**0.11**	0.14	**0.14**	0.17	**0.16**	0.19	**0.18**	0.23	**0.23**	0.28	**0.27**
Liver	0.62	**0.52**	0.72	**0.63**	0.80	**0.70**	0.85	**0.76**	0.95	**0.86**	1.02	**0.95**
Gall bladder	0.32	**0.28**	0.40	**0.35**	0.46	**0.41**	0.50	**0.45**	0.59	**0.54**	0.67	**0.61**
Pancreas	0.23	**0.19**	0.29	**0.24**	0.34	**0.29**	0.37	**0.32**	0.45	**0.39**	0.51	**0.45**
Small intestine	0.34	**0.28**	0.42	**0.35**	0.48	**0.40**	0.53	**0.45**	0.62	**0.53**	0.70	**0.61**
Large intestine	0.47	**0.39**	0.57	**0.48**	0.65	**0.55**	0.70	**0.60**	0.81	**0.70**	0.90	**0.79**
Sigma/rectum	0.17	**0.14**	0.22	**0.19**	0.25	**0.22**	0.28	**0.25**	0.34	**0.30**	0.40	**0.35**
Lumbar spine	0.74	**0.70**	0.93	**0.88**	1.07	**1.02**	1.18	**1.13**	1.39	**1.34**	1.57	**1.52**
Pelvis	1.02	**0.88**	1.23	**1.08**	1.38	**1.22**	1.49	**1.33**	1.69	**1.54**	1.86	**1.71**
Adrenals	0.26	**0.26**	0.32	**0.32**	0.37	**0.37**	0.40	**0.40**	0.47	**0.48**	0.53	**0.54**
Kidneys	0.35	**0.34**	0.42	**0.42**	0.48	**0.47**	0.52	**0.51**	0.60	**0.59**	0.67	**0.66**
Urinary bladder	0.31	**0.29**	0.39	**0.36**	0.45	**0.42**	0.49	**0.46**	0.58	**0.54**	0.65	**0.61**
Uterus	0.31	**0.26**	0.39	**0.33**	0.45	**0.38**	0.50	**0.43**	0.59	**0.51**	0.67	**0.59**
Ovaries	0.34	**0.26**	0.42	**0.33**	0.48	**0.38**	0.53	**0.43**	0.62	**0.51**	0.71	**0.59**
Prostate	0.27	**0.25**	0.34	**0.31**	0.39	**0.36**	0.44	**0.40**	0.51	**0.48**	0.59	**0.55**
Testes	0.25	**0.25**	0.30	**0.31**	0.35	**0.35**	0.38	**0.39**	0.45	**0.45**	0.50	**0.51**
Upper arm		**0.45**		**0.51**		**0.55**		**0.58**		**0.63**		**0.67**
Elbow		**1.50**		**1.66**		**1.77**		**1.84**		**1.95**		**2.03**
Forearm		**1.57**		**1.74**		**1.83**		**1.90**		**2.00**		**2.07**
Upper leg	0.52	**0.69**	0.64	**0.86**	0.73	**0.98**	0.80	**1.07**	0.93	**1.24**	1.05	**1.39**
Knee	0.05	**0.29**	0.06	**0.36**	0.07	**0.41**	0.08	**0.45**	0.10	**0.52**	0.12	**0.59**
Lower leg				**0.01**		**0.01**		**0.02**	0.01	**0.02**	0.01	**0.02**
Lymph nodes	0.19	**0.19**	0.24	**0.24**	0.27	**0.27**	0.30	**0.30**	0.35	**0.36**	0.40	**0.41**
Skin	0.23	**0.30**	0.25	**0.32**	0.26	**0.33**	0.26	**0.34**	0.28	**0.36**	0.29	**0.38**
Muscle	0.20	**0.25**	0.24	**0.29**	0.27	**0.33**	0.28	**0.35**	0.32	**0.39**	0.35	**0.43**
Skeleton	0.30	**0.60**	0.36	**0.71**	0.41	**0.79**	0.44	**0.84**	0.50	**0.94**	0.55	**1.02**
Red bone marrow	0.09	**0.16**	0.11	**0.19**	0.12	**0.22**	0.13	**0.23**	0.15	**0.27**	0.17	**0.30**
Whole body dose	0.21	**0.26**	0.24	**0.31**	0.27	**0.35**	0.29	**0.37**	0.33	**0.42**	0.36	**0.46**
ED (ICRP 60)	0.22	**0.26**	0.27	**0.31**	0.30	**0.35**	0.33	**0.38**	0.38	**0.45**	0.43	**0.50**
ED (ICRP 103)	0.19	**0.27**	0.23	**0.32**	0.26	**0.36**	0.29	**0.39**	0.33	**0.45**	0.37	**0.50**
Total filtration (mm) Al	**1.5**		**2.5**		**3.5**		**4.5**		**3.5**		**3.5**	
Cu	**0**		**0**		**0**		**0**		**0.1**		**0.2**	

Table 12.124 Normalized organ doses—Abdomen r-lat—1–2 years

Beam projection:		r-lat										**Table cells:**
Focus-to-detector distance (cm):		115										(+) = normal font = optimal field setting
Tube voltage (kVp):		60										**(−) = bold font = suboptimal field setting**
Total filtrations:		See last table row.										Empty table cells correspond to values <0.01
Organ or tissue	**Organ dose D_T : Entrance dose K_E**								**(mSv * mGy^{-1})**			
	(+)	(−)	(+)	(−)	(+)	(−)	(+)	(−)	(+)	(−)	(+)	(−)
Skull												
Brain												
Oral mucosa												
Salivary glands												
Cervical spine				**0.01**		**0.01**		**0.02**		**0.02**		**0.02**
Thyroid				**0.01**		**0.01**		**0.02**		**0.02**		**0.02**
Thymus		**0.06**		**0.08**	0.01	**0.09**	0.01	**0.11**	0.02	**0.13**	0.01	**0.12**
Thoracic spine	0.15	**0.28**	0.20	**0.37**	0.24	**0.44**	0.27	**0.49**	0.33	**0.61**	0.30	**0.56**
Claviculae		**0.02**		**0.03**	0.01	**0.04**	0.01	**0.05**	0.02	**0.06**	0.01	**0.05**
Scapulae	0.02	**0.47**	0.02	**0.55**	0.03	**0.60**	0.03	**0.65**	0.04	**0.72**	0.04	**0.71**
Ribs	0.42	**0.66**	0.49	**0.80**	0.54	**0.90**	0.58	**0.97**	0.65	**1.11**	0.64	**1.09**
Oesophagus	0.08	**0.12**	0.10	**0.16**	0.12	**0.19**	0.13	**0.22**	0.16	**0.27**	0.15	**0.24**
Trachea												
Lung	0.08	**0.19**	0.10	**0.24**	0.12	**0.27**	0.13	**0.30**	0.15	**0.36**	0.14	**0.34**
Mamma	0.01	**0.50**	0.02	**0.54**	0.02	**0.57**	0.02	**0.58**	0.03	**0.62**	0.03	**0.60**
Heart	0.07	**0.16**	0.10	**0.20**	0.11	**0.24**	0.13	**0.27**	0.16	**0.33**	0.14	**0.30**
Stomach	0.07	**0.06**	0.09	**0.08**	0.11	**0.10**	0.12	**0.11**	0.16	**0.14**	0.14	**0.12**
Spleen	0.04	**0.04**	0.06	**0.05**	0.07	**0.06**	0.08	**0.07**	0.10	**0.10**	0.09	**0.08**
Liver	0.46	**0.34**	0.55	**0.43**	0.62	**0.50**	0.68	**0.55**	0.78	**0.65**	0.75	**0.62**
Gall bladder	0.21	**0.17**	0.27	**0.22**	0.32	**0.26**	0.36	**0.30**	0.44	**0.37**	0.41	**0.34**
Pancreas	0.11	**0.09**	0.15	**0.12**	0.18	**0.14**	0.21	**0.16**	0.26	**0.21**	0.23	**0.18**
Small intestine	0.20	**0.15**	0.25	**0.19**	0.30	**0.23**	0.34	**0.26**	0.41	**0.32**	0.38	**0.29**
Large intestine	0.31	**0.23**	0.39	**0.30**	0.45	**0.35**	0.50	**0.39**	0.60	**0.48**	0.57	**0.44**
Sigma/rectum	0.07	**0.06**	0.10	**0.08**	0.12	**0.09**	0.14	**0.11**	0.17	**0.14**	0.15	**0.12**
Lumbar spine	0.39	**0.35**	0.51	**0.47**	0.61	**0.56**	0.69	**0.63**	0.86	**0.79**	0.78	**0.71**
Pelvis	0.67	**0.52**	0.84	**0.67**	0.97	**0.78**	1.08	**0.88**	1.27	**1.06**	1.22	**0.99**
Adrenals	0.13	**0.13**	0.18	**0.17**	0.21	**0.20**	0.23	**0.23**	0.29	**0.29**	0.26	**0.26**
Kidneys	0.22	**0.21**	0.27	**0.26**	0.32	**0.30**	0.35	**0.33**	0.41	**0.40**	0.39	**0.37**
Urinary bladder	0.18	**0.16**	0.23	**0.20**	0.27	**0.24**	0.31	**0.27**	0.38	**0.33**	0.35	**0.30**
Uterus	0.17	**0.12**	0.22	**0.17**	0.26	**0.20**	0.30	**0.23**	0.37	**0.29**	0.33	**0.25**
Ovaries	0.19	**0.13**	0.25	**0.17**	0.30	**0.20**	0.34	**0.23**	0.42	**0.29**	0.39	**0.26**
Prostate	0.14	**0.12**	0.18	**0.16**	0.22	**0.19**	0.25	**0.21**	0.31	**0.26**	0.28	**0.24**
Testes	0.12	**0.12**	0.16	**0.16**	0.19	**0.19**	0.22	**0.21**	0.26	**0.26**	0.24	**0.24**
Upper arm		**0.33**		**0.39**		**0.43**		**0.46**		**0.51**		**0.50**
Elbow		**1.24**		**1.43**		**1.56**		**1.65**		**1.81**		**1.81**
Forearm		**1.35**		**1.55**		**1.67**		**1.76**		**1.91**		**1.92**
Upper leg	0.28	**0.36**	0.36	**0.46**	0.43	**0.54**	0.48	**0.61**	0.58	**0.74**	0.54	**0.69**
Knee	0.02	**0.04**	0.02	**0.05**	0.03	**0.06**	0.03	**0.07**	0.04	**0.09**	0.03	**0.07**
Lower leg												
Lymph nodes	0.11	**0.10**	0.14	**0.13**	0.17	**0.16**	0.19	**0.18**	0.23	**0.22**	0.21	**0.20**
Skin	0.19	**0.24**	0.20	**0.26**	0.21	**0.27**	0.22	**0.27**	0.23	**0.29**	0.22	**0.28**
Muscle	0.13	**0.15**	0.15	**0.18**	0.17	**0.20**	0.18	**0.22**	0.21	**0.25**	0.20	**0.24**
Skeleton	0.17	**0.37**	0.21	**0.45**	0.24	**0.50**	0.26	**0.55**	0.31	**0.63**	0.30	**0.61**
Red bone marrow	0.06	**0.10**	0.07	**0.12**	0.09	**0.14**	0.10	**0.15**	0.11	**0.18**	0.10	**0.16**
Whole body dose	0.13	**0.17**	0.16	**0.20**	0.18	**0.23**	0.19	**0.25**	0.22	**0.29**	0.21	**0.27**
ED (ICRP 60)	0.13	**0.15**	0.16	**0.19**	0.19	**0.22**	0.21	**0.24**	0.25	**0.29**	0.23	**0.27**
ED (ICRP 103)	0.11	**0.17**	0.14	**0.21**	0.16	**0.24**	0.18	**0.26**	0.22	**0.30**	0.20	**0.28**
Total filtration (mm) **Al**	**1.5**		**2.5**		**3.5**		**4.5**		**3.5**		**3.5**	
Cu	**0**		**0**		**0**		**0**		**0.1**		**0.2**	

Table 12.125 Normalized organ doses—Abdomen r-lat—1–2 years

Beam projection:			r-lat									**Table cells:**
Focus-to-detector distance (cm):			115						(+) = normal font = optimal field setting			
Tube voltage (kVp):			70						**(−) = bold font = suboptimal field setting**			
Total filtrations:			See last table row.						Empty table cells correspond to values <0.01			
Organ or tissue	**Organ dose D_T : Entrance dose K_E**								**(mSv * mGy^{-1})**			
	(+)	(−)	(+)	(−)	(+)	(−)	(+)	(−)	(+)	(−)	(+)	(−)
Skull												
Brain												
Oral mucosa												
Salivary glands												**0.01**
Cervical spine		**0.01**		**0.02**		**0.02**		**0.02**		**0.03**	0.01	**0.04**
Thyroid		**0.01**		**0.02**		**0.02**		**0.02**		**0.03**		**0.03**
Thymus		**0.08**	0.01	**0.10**	0.01	**0.11**	0.02	**0.13**	0.02	**0.15**	0.03	**0.18**
Thoracic spine	0.19	**0.35**	0.25	**0.45**	0.29	**0.53**	0.33	**0.59**	0.40	**0.72**	0.46	**0.84**
Claviculae		**0.03**	0.01	**0.04**	0.02	**0.05**	0.02	**0.06**	0.02	**0.08**	0.03	**0.09**
Scapulae	0.02	**0.51**	0.03	**0.59**	0.04	**0.65**	0.04	**0.69**	0.05	**0.76**	0.07	**0.81**
Ribs	0.46	**0.74**	0.53	**0.88**	0.58	**0.98**	0.62	**1.05**	0.69	**1.18**	0.74	**1.29**
Oesophagus	0.10	**0.16**	0.12	**0.20**	0.14	**0.23**	0.16	**0.26**	0.20	**0.32**	0.23	**0.38**
Trachea												
Lung	0.10	**0.23**	0.12	**0.28**	0.13	**0.32**	0.15	**0.35**	0.17	**0.41**	0.19	**0.46**
Mamma	0.02	**0.52**	0.02	**0.56**	0.03	**0.59**	0.03	**0.61**	0.04	**0.64**	0.05	**0.68**
Heart	0.09	**0.19**	0.12	**0.25**	0.14	**0.29**	0.15	**0.32**	0.18	**0.39**	0.21	**0.45**
Stomach	0.09	**0.08**	0.12	**0.11**	0.14	**0.13**	0.16	**0.15**	0.20	**0.18**	0.23	**0.22**
Spleen	0.06	**0.05**	0.08	**0.07**	0.09	**0.09**	0.10	**0.10**	0.13	**0.13**	0.16	**0.15**
Liver	0.52	**0.41**	0.62	**0.50**	0.69	**0.57**	0.75	**0.63**	0.85	**0.73**	0.93	**0.82**
Gall bladder	0.26	**0.22**	0.33	**0.28**	0.38	**0.32**	0.43	**0.36**	0.52	**0.44**	0.60	**0.52**
Pancreas	0.15	**0.12**	0.19	**0.15**	0.23	**0.18**	0.26	**0.21**	0.31	**0.26**	0.37	**0.31**
Small intestine	0.24	**0.19**	0.31	**0.24**	0.36	**0.28**	0.40	**0.32**	0.48	**0.39**	0.56	**0.46**
Large intestine	0.37	**0.29**	0.46	**0.36**	0.53	**0.42**	0.58	**0.47**	0.68	**0.56**	0.77	**0.64**
Sigma/rectum	0.10	**0.08**	0.13	**0.10**	0.16	**0.13**	0.18	**0.14**	0.22	**0.18**	0.26	**0.22**
Lumbar spine	0.49	**0.45**	0.63	**0.58**	0.74	**0.69**	0.83	**0.77**	1.01	**0.94**	1.19	**1.10**
Pelvis	0.79	**0.63**	0.97	**0.79**	1.11	**0.91**	1.22	**1.01**	1.42	**1.20**	1.59	**1.37**
Adrenals	0.17	**0.17**	0.22	**0.21**	0.25	**0.25**	0.28	**0.28**	0.34	**0.34**	0.40	**0.40**
Kidneys	0.26	**0.25**	0.32	**0.31**	0.37	**0.35**	0.40	**0.39**	0.47	**0.46**	0.54	**0.52**
Urinary bladder	0.22	**0.20**	0.28	**0.25**	0.33	**0.29**	0.37	**0.33**	0.44	**0.39**	0.51	**0.46**
Uterus	0.21	**0.16**	0.27	**0.21**	0.32	**0.25**	0.36	**0.29**	0.44	**0.35**	0.52	**0.42**
Ovaries	0.24	**0.17**	0.31	**0.22**	0.37	**0.26**	0.41	**0.29**	0.49	**0.37**	0.57	**0.44**
Prostate	0.18	**0.15**	0.23	**0.20**	0.27	**0.23**	0.30	**0.26**	0.36	**0.32**	0.42	**0.37**
Testes	0.16	**0.15**	0.20	**0.20**	0.23	**0.23**	0.26	**0.25**	0.31	**0.30**	0.36	**0.35**
Upper arm		**0.37**		**0.42**		**0.46**		**0.49**		**0.55**		**0.59**
Elbow		**1.33**		**1.52**		**1.64**		**1.73**		**1.87**		**1.98**
Forearm		**1.44**		**1.63**		**1.75**		**1.83**		**1.96**		**2.06**
Upper leg	0.35	**0.44**	0.44	**0.55**	0.51	**0.64**	0.57	**0.71**	0.68	**0.85**	0.78	**0.97**
Knee	0.02	**0.05**	0.03	**0.06**	0.04	**0.08**	0.04	**0.09**	0.05	**0.11**	0.07	**0.13**
Lower leg												
Lymph nodes	0.14	**0.13**	0.17	**0.17**	0.20	**0.19**	0.22	**0.22**	0.27	**0.26**	0.31	**0.31**
Skin	0.20	**0.25**	0.21	**0.27**	0.22	**0.28**	0.23	**0.29**	0.24	**0.30**	0.25	**0.32**
Muscle	0.15	**0.17**	0.17	**0.20**	0.19	**0.23**	0.21	**0.24**	0.24	**0.28**	0.26	**0.31**
Skeleton	0.20	**0.42**	0.24	**0.50**	0.27	**0.56**	0.30	**0.61**	0.35	**0.69**	0.39	**0.76**
Red bone marrow	0.07	**0.11**	0.09	**0.14**	0.10	**0.16**	0.11	**0.17**	0.13	**0.20**	0.15	**0.23**
Whole body dose	0.15	**0.19**	0.18	**0.23**	0.20	**0.26**	0.22	**0.28**	0.25	**0.32**	0.28	**0.36**
ED (ICRP 60)	0.15	**0.18**	0.19	**0.22**	0.22	**0.26**	0.25	**0.28**	0.29	**0.34**	0.33	**0.38**
ED (ICRP 103)	0.13	**0.20**	0.17	**0.24**	0.19	**0.27**	0.21	**0.30**	0.25	**0.35**	0.29	**0.39**
Total filtration (mm) Al	**1.5**		**2.5**		**3.5**		**4.5**		**3.5**		**3.5**	
Cu	**0**		**0**		**0**		**0**		**0.1**		**0.2**	

Table 12.126 Normalized organ doses—Abdomen r-lat—3–7 years

Beam projection:			r-lat									**Table cells:**
Focus-to-detector distance (cm):			115									(+) = normal font = optimal field setting
Tube voltage (kVp):			60									**(−) = bold font = suboptimal field setting**
Total filtrations:			See last table row.									Empty table cells correspond to values <0.01
Organ or tissue	**Organ dose D_T : Entrance dose K_E**								**(mSv * mGy^{-1})**			
	(+)	(−)	(+)	(−)	(+)	(−)	(+)	(−)	(+)	(−)	(+)	(−)
Skull												
Brain												
Oral mucosa												
Salivary glands												
Cervical spine								**0.01**		**0.01**		
Thyroid										**0.01**		
Thymus		**0.06**		**0.08**		**0.09**	0.01	**0.10**	0.01	**0.13**	0.01	**0.11**
Thoracic spine	0.10	**0.17**	0.13	**0.22**	0.16	**0.27**	0.18	**0.31**	0.23	**0.39**	0.20	**0.34**
Claviculae		**0.01**		**0.02**		**0.02**		**0.03**		**0.04**		**0.03**
Scapulae	0.01	**0.38**	0.02	**0.46**	0.02	**0.51**	0.02	**0.55**	0.03	**0.61**	0.02	**0.60**
Ribs	0.35	**0.50**	0.41	**0.62**	0.46	**0.71**	0.50	**0.77**	0.56	**0.89**	0.55	**0.87**
Oesophagus	0.04	**0.06**	0.05	**0.08**	0.06	**0.09**	0.07	**0.11**	0.09	**0.14**	0.08	**0.12**
Trachea												
Lung	0.07	**0.14**	0.09	**0.18**	0.10	**0.21**	0.11	**0.23**	0.14	**0.28**	0.13	**0.26**
Mamma		**0.47**	0.01	**0.51**	0.02	**0.53**	0.02	**0.55**	0.02	**0.58**	0.02	**0.57**
Heart	0.06	**0.10**	0.08	**0.13**	0.09	**0.15**	0.10	**0.17**	0.13	**0.22**	0.12	**0.19**
Stomach	0.04	**0.03**	0.05	**0.04**	0.06	**0.05**	0.07	**0.06**	0.09	**0.08**	0.07	**0.06**
Spleen	0.02	**0.02**	0.03	**0.02**	0.03	**0.03**	0.04	**0.03**	0.05	**0.04**	0.04	**0.03**
Liver	0.38	**0.27**	0.47	**0.34**	0.54	**0.40**	0.59	**0.44**	0.68	**0.53**	0.66	**0.50**
Gall bladder	0.12	**0.10**	0.16	**0.13**	0.20	**0.15**	0.23	**0.18**	0.28	**0.23**	0.25	**0.19**
Pancreas	0.07	**0.05**	0.09	**0.07**	0.11	**0.08**	0.13	**0.09**	0.16	**0.12**	0.14	**0.10**
Small intestine	0.13	**0.09**	0.17	**0.12**	0.21	**0.15**	0.24	**0.17**	0.29	**0.22**	0.26	**0.19**
Large intestine	0.23	**0.16**	0.29	**0.21**	0.34	**0.25**	0.39	**0.29**	0.47	**0.35**	0.43	**0.32**
Sigma/rectum	0.04	**0.03**	0.05	**0.04**	0.06	**0.05**	0.08	**0.06**	0.10	**0.08**	0.08	**0.06**
Lumbar spine	0.22	**0.20**	0.30	**0.26**	0.36	**0.32**	0.41	**0.37**	0.53	**0.47**	0.45	**0.40**
Pelvis	0.48	**0.35**	0.61	**0.46**	0.72	**0.54**	0.80	**0.61**	0.96	**0.75**	0.91	**0.69**
Adrenals	0.09	**0.08**	0.12	**0.11**	0.14	**0.13**	0.16	**0.15**	0.20	**0.19**	0.17	**0.17**
Kidneys	0.15	**0.14**	0.20	**0.18**	0.23	**0.21**	0.26	**0.24**	0.31	**0.29**	0.29	**0.27**
Urinary bladder	0.11	**0.10**	0.15	**0.13**	0.18	**0.15**	0.20	**0.18**	0.25	**0.22**	0.23	**0.19**
Uterus	0.10	**0.07**	0.14	**0.10**	0.16	**0.12**	0.19	**0.14**	0.24	**0.18**	0.21	**0.15**
Ovaries	0.13	**0.08**	0.17	**0.11**	0.20	**0.14**	0.23	**0.16**	0.29	**0.20**	0.25	**0.17**
Prostate	0.08	**0.06**	0.11	**0.09**	0.14	**0.11**	0.16	**0.12**	0.20	**0.16**	0.17	**0.13**
Testes	0.07	**0.07**	0.10	**0.09**	0.12	**0.11**	0.13	**0.13**	0.17	**0.16**	0.15	**0.14**
Upper arm		**0.27**		**0.32**		**0.35**		**0.38**		**0.43**		**0.42**
Elbow		**1.09**		**1.28**		**1.41**		**1.50**		**1.66**		**1.66**
Forearm		**1.25**		**1.45**		**1.58**		**1.67**		**1.83**		**1.84**
Upper leg	0.14	**0.23**	0.18	**0.30**	0.22	**0.36**	0.24	**0.40**	0.30	**0.50**	0.27	**0.45**
Knee		**0.01**		**0.02**		**0.02**		**0.02**		**0.03**		**0.02**
Lower leg												
Lymph nodes	0.07	**0.06**	0.09	**0.09**	0.11	**0.10**	0.13	**0.12**	0.16	**0.15**	0.14	**0.13**
Skin	0.13	**0.22**	0.14	**0.23**	0.15	**0.24**	0.15	**0.24**	0.16	**0.25**	0.15	**0.25**
Muscle	0.10	**0.12**	0.12	**0.14**	0.13	**0.16**	0.15	**0.17**	0.17	**0.20**	0.16	**0.19**
Skeleton	0.10	**0.24**	0.12	**0.30**	0.14	**0.34**	0.16	**0.37**	0.19	**0.43**	0.18	**0.41**
Red bone marrow	0.06	**0.08**	0.07	**0.10**	0.08	**0.11**	0.09	**0.13**	0.11	**0.15**	0.10	**0.14**
Whole body dose	0.10	**0.13**	0.12	**0.16**	0.14	**0.18**	0.15	**0.20**	0.18	**0.23**	0.17	**0.22**
ED (ICRP 60)	0.09	**0.11**	0.12	**0.14**	0.14	**0.16**	0.15	**0.18**	0.19	**0.21**	0.17	**0.19**
ED (ICRP 103)	0.08	**0.13**	0.10	**0.16**	0.12	**0.18**	0.13	**0.20**	0.16	**0.23**	0.15	**0.21**
Total filtration (mm) **Al**	**1.5**		**2.5**		**3.5**		**4.5**		**3.5**		**3.5**	
Cu	**0**		**0**		**0**		**0**		**0.1**		**0.2**	

Table 12.127 Normalized organ doses—Abdomen r-lat—3–7 years

Beam projection:			r-lat									**Table cells:**
Focus-to-detector distance (cm):			115						(+) = normal font = optimal field setting			
Tube voltage (kVp):			70						**(−) = bold font = suboptimal field setting**			
Total filtrations:			See last table row.						Empty table cells correspond to values <0.01			
Organ or tissue	**Organ dose D_T : Entrance dose K_E**								**(mSv * mGy^{-1})**			
	(+)	(−)	(+)	(−)	(+)	(−)	(+)	(−)	(+)	(−)	(+)	(−)
Skull												
Brain												
Oral mucosa												
Salivary glands												
Cervical spine				**0.01**		**0.01**		**0.02**		**0.02**		**0.03**
Thyroid								**0.01**		**0.01**		**0.02**
Thymus		**0.07**	0.01	**0.10**	0.01	**0.11**	0.01	**0.13**	0.02	**0.16**	0.02	**0.18**
Thoracic spine	0.13	**0.22**	0.17	**0.29**	0.20	**0.34**	0.23	**0.39**	0.28	**0.48**	0.34	**0.58**
Claviculae		**0.02**		**0.03**		**0.03**	0.01	**0.04**	0.01	**0.05**	0.02	**0.06**
Scapulae	0.02	**0.43**	0.02	**0.50**	0.03	**0.55**	0.03	**0.59**	0.04	**0.65**	0.05	**0.71**
Ribs	0.39	**0.58**	0.45	**0.70**	0.50	**0.78**	0.54	**0.85**	0.60	**0.97**	0.65	**1.07**
Oesophagus	0.05	**0.08**	0.07	**0.10**	0.08	**0.12**	0.09	**0.14**	0.11	**0.17**	0.13	**0.20**
Trachea												
Lung	0.09	**0.17**	0.11	**0.21**	0.12	**0.25**	0.13	**0.27**	0.15	**0.32**	0.17	**0.37**
Mamma	0.01	**0.49**	0.02	**0.53**	0.02	**0.56**	0.02	**0.57**	0.03	**0.60**	0.04	**0.63**
Heart	0.08	**0.12**	0.10	**0.16**	0.11	**0.19**	0.13	**0.21**	0.16	**0.26**	0.18	**0.31**
Stomach	0.05	**0.05**	0.07	**0.06**	0.08	**0.07**	0.09	**0.08**	0.12	**0.11**	0.15	**0.13**
Spleen	0.03	**0.03**	0.04	**0.03**	0.05	**0.04**	0.05	**0.05**	0.07	**0.06**	0.09	**0.08**
Liver	0.45	**0.32**	0.54	**0.41**	0.61	**0.47**	0.66	**0.52**	0.76	**0.62**	0.85	**0.71**
Gall bladder	0.16	**0.13**	0.21	**0.17**	0.25	**0.20**	0.28	**0.23**	0.35	**0.29**	0.41	**0.34**
Pancreas	0.09	**0.07**	0.12	**0.09**	0.14	**0.11**	0.16	**0.13**	0.21	**0.16**	0.25	**0.20**
Small intestine	0.17	**0.12**	0.22	**0.16**	0.26	**0.19**	0.29	**0.22**	0.36	**0.27**	0.42	**0.33**
Large intestine	0.28	**0.21**	0.36	**0.27**	0.41	**0.31**	0.46	**0.35**	0.55	**0.43**	0.63	**0.50**
Sigma/rectum	0.06	**0.04**	0.07	**0.06**	0.09	**0.07**	0.10	**0.08**	0.13	**0.10**	0.16	**0.13**
Lumbar spine	0.30	**0.26**	0.39	**0.35**	0.46	**0.41**	0.53	**0.47**	0.66	**0.59**	0.79	**0.71**
Pelvis	0.58	**0.44**	0.73	**0.56**	0.84	**0.65**	0.93	**0.73**	1.10	**0.88**	1.25	**1.03**
Adrenals	0.12	**0.11**	0.15	**0.14**	0.18	**0.17**	0.20	**0.19**	0.25	**0.24**	0.29	**0.29**
Kidneys	0.19	**0.18**	0.24	**0.22**	0.27	**0.26**	0.30	**0.29**	0.36	**0.34**	0.42	**0.39**
Urinary bladder	0.15	**0.13**	0.19	**0.17**	0.22	**0.20**	0.25	**0.22**	0.31	**0.27**	0.37	**0.32**
Uterus	0.14	**0.10**	0.18	**0.14**	0.21	**0.17**	0.24	**0.19**	0.30	**0.24**	0.36	**0.29**
Ovaries	0.16	**0.11**	0.21	**0.15**	0.25	**0.18**	0.29	**0.20**	0.35	**0.26**	0.42	**0.31**
Prostate	0.11	**0.09**	0.15	**0.12**	0.18	**0.14**	0.20	**0.16**	0.25	**0.20**	0.30	**0.25**
Testes	0.10	**0.09**	0.12	**0.12**	0.15	**0.14**	0.16	**0.16**	0.20	**0.20**	0.24	**0.24**
Upper arm		**0.30**		**0.35**		**0.38**		**0.41**		**0.46**		**0.50**
Elbow		**1.19**		**1.38**		**1.50**		**1.59**		**1.74**		**1.85**
Forearm		**1.35**		**1.54**		**1.66**		**1.75**		**1.90**		**2.01**
Upper leg	0.18	**0.29**	0.23	**0.37**	0.27	**0.44**	0.30	**0.49**	0.37	**0.59**	0.43	**0.70**
Knee		**0.02**		**0.02**		**0.03**		**0.03**	0.01	**0.04**	0.02	**0.05**
Lower leg												
Lymph nodes	0.09	**0.09**	0.12	**0.11**	0.14	**0.13**	0.16	**0.15**	0.19	**0.18**	0.23	**0.22**
Skin	0.14	**0.22**	0.15	**0.24**	0.15	**0.25**	0.16	**0.25**	0.17	**0.27**	0.17	**0.28**
Muscle	0.12	**0.14**	0.14	**0.16**	0.15	**0.18**	0.17	**0.20**	0.19	**0.23**	0.21	**0.26**
Skeleton	0.12	**0.28**	0.15	**0.34**	0.17	**0.38**	0.19	**0.41**	0.22	**0.47**	0.25	**0.53**
Red bone marrow	0.07	**0.10**	0.09	**0.12**	0.10	**0.14**	0.11	**0.15**	0.13	**0.18**	0.16	**0.21**
Whole body dose	0.12	**0.15**	0.14	**0.18**	0.16	**0.21**	0.17	**0.22**	0.20	**0.26**	0.22	**0.29**
ED (ICRP 60)	0.11	**0.13**	0.14	**0.17**	0.16	**0.19**	0.18	**0.21**	0.22	**0.25**	0.26	**0.29**
ED (ICRP 103)	0.10	**0.16**	0.12	**0.19**	0.14	**0.21**	0.16	**0.23**	0.19	**0.27**	0.22	**0.31**
Total filtration (mm) Al	**1.5**		**2.5**		**3.5**		**4.5**		**3.5**		**3.5**	
Cu	**0**		**0**		**0**		**0**		**0.1**		**0.2**	

Table 12.128 Normalized organ doses—Abdomen r-lat—3–7 years

Beam projection:			r-lat									**Table cells:**
Focus-to-detector distance (cm):			115					(+) = normal font = optimal field setting				
Tube voltage (kVp):			80					**(−) = bold font = suboptimal field setting**				
Total filtrations:			See last table row.					Empty table cells correspond to values <0.01				
Organ or tissue	**Organ dose D_T : Entrance dose K_E**							**(mSv * mGy^{-1})**				
	(+)	(−)	(+)	(−)	(+)	(−)	(+)	(−)	(+)	(−)	(+)	(−)
Skull												
Brain												
Oral mucosa												
Salivary glands												
Cervical spine		**0.01**		**0.01**		**0.02**		**0.02**		**0.03**		**0.03**
Thyroid				**0.01**		**0.01**		**0.01**		**0.02**		**0.02**
Thymus	0.01	**0.09**	0.01	**0.12**	0.02	**0.13**	0.02	**0.15**	0.02	**0.18**	0.03	**0.21**
Thoracic spine	0.16	**0.28**	0.21	**0.35**	0.24	**0.41**	0.27	**0.46**	0.33	**0.56**	0.38	**0.66**
Claviculae		**0.03**		**0.04**	0.01	**0.04**	0.01	**0.05**	0.02	**0.06**	0.02	**0.07**
Scapulae	0.02	**0.46**	0.03	**0.53**	0.03	**0.58**	0.04	**0.62**	0.05	**0.68**	0.06	**0.73**
Ribs	0.42	**0.64**	0.49	**0.76**	0.53	**0.84**	0.56	**0.91**	0.62	**1.02**	0.67	**1.11**
Oesophagus	0.07	**0.10**	0.08	**0.13**	0.10	**0.15**	0.11	**0.17**	0.13	**0.20**	0.15	**0.24**
Trachea												
Lung	0.10	**0.20**	0.12	**0.25**	0.14	**0.28**	0.15	**0.31**	0.17	**0.36**	0.19	**0.41**
Mamma	0.02	**0.51**	0.02	**0.55**	0.03	**0.58**	0.03	**0.59**	0.04	**0.62**	0.04	**0.65**
Heart	0.09	**0.16**	0.12	**0.20**	0.14	**0.23**	0.15	**0.26**	0.18	**0.31**	0.21	**0.36**
Stomach	0.07	**0.06**	0.09	**0.08**	0.10	**0.09**	0.12	**0.11**	0.15	**0.13**	0.18	**0.16**
Spleen	0.04	**0.04**	0.05	**0.05**	0.06	**0.06**	0.07	**0.07**	0.09	**0.08**	0.11	**0.10**
Liver	0.50	**0.38**	0.60	**0.47**	0.67	**0.54**	0.73	**0.59**	0.83	**0.69**	0.92	**0.78**
Gall bladder	0.20	**0.17**	0.26	**0.21**	0.30	**0.25**	0.34	**0.28**	0.41	**0.34**	0.48	**0.40**
Pancreas	0.12	**0.09**	0.15	**0.12**	0.18	**0.14**	0.20	**0.16**	0.25	**0.20**	0.29	**0.24**
Small intestine	0.21	**0.16**	0.27	**0.20**	0.31	**0.24**	0.34	**0.26**	0.41	**0.32**	0.48	**0.38**
Large intestine	0.34	**0.25**	0.41	**0.32**	0.48	**0.37**	0.52	**0.41**	0.62	**0.49**	0.70	**0.57**
Sigma/rectum	0.07	**0.06**	0.10	**0.08**	0.11	**0.09**	0.13	**0.10**	0.16	**0.13**	0.19	**0.16**
Lumbar spine	0.37	**0.33**	0.48	**0.43**	0.56	**0.50**	0.63	**0.56**	0.76	**0.69**	0.90	**0.81**
Pelvis	0.67	**0.52**	0.83	**0.65**	0.94	**0.75**	1.03	**0.83**	1.20	**0.98**	1.34	**1.12**
Adrenals	0.14	**0.14**	0.18	**0.18**	0.21	**0.21**	0.24	**0.23**	0.29	**0.28**	0.34	**0.33**
Kidneys	0.23	**0.21**	0.28	**0.26**	0.32	**0.30**	0.35	**0.33**	0.41	**0.39**	0.47	**0.44**
Urinary bladder	0.18	**0.16**	0.23	**0.20**	0.27	**0.24**	0.30	**0.26**	0.36	**0.32**	0.42	**0.37**
Uterus	0.17	**0.14**	0.22	**0.18**	0.26	**0.21**	0.29	**0.23**	0.36	**0.29**	0.42	**0.35**
Ovaries	0.21	**0.15**	0.26	**0.19**	0.30	**0.22**	0.34	**0.25**	0.41	**0.31**	0.48	**0.37**
Prostate	0.15	**0.12**	0.19	**0.15**	0.22	**0.18**	0.24	**0.20**	0.30	**0.25**	0.35	**0.29**
Testes	0.12	**0.12**	0.15	**0.15**	0.17	**0.17**	0.19	**0.19**	0.23	**0.23**	0.27	**0.27**
Upper arm		**0.32**		**0.37**		**0.41**		**0.43**		**0.48**		**0.52**
Elbow		**1.27**		**1.45**		**1.56**		**1.64**		**1.77**		**1.87**
Forearm		**1.42**		**1.60**		**1.72**		**1.80**		**1.92**		**2.01**
Upper leg	0.21	**0.35**	0.27	**0.44**	0.31	**0.51**	0.35	**0.56**	0.42	**0.67**	0.48	**0.77**
Knee		**0.02**		**0.03**	0.01	**0.04**	0.01	**0.04**	0.02	**0.05**	0.02	**0.06**
Lower leg												
Lymph nodes	0.11	**0.11**	0.14	**0.14**	0.17	**0.16**	0.19	**0.18**	0.22	**0.21**	0.26	**0.25**
Skin	0.14	**0.23**	0.15	**0.25**	0.16	**0.25**	0.16	**0.26**	0.17	**0.28**	0.18	**0.29**
Muscle	0.13	**0.15**	0.15	**0.18**	0.17	**0.20**	0.18	**0.22**	0.21	**0.25**	0.23	**0.28**
Skeleton	0.14	**0.31**	0.17	**0.37**	0.19	**0.41**	0.21	**0.45**	0.24	**0.51**	0.27	**0.56**
Red bone marrow	0.08	**0.11**	0.10	**0.14**	0.12	**0.16**	0.13	**0.18**	0.15	**0.21**	0.18	**0.24**
Whole body dose	0.13	**0.17**	0.16	**0.21**	0.18	**0.23**	0.19	**0.25**	0.22	**0.29**	0.24	**0.32**
ED (ICRP 60)	0.14	**0.16**	0.17	**0.19**	0.19	**0.22**	0.21	**0.24**	0.25	**0.29**	0.29	**0.33**
ED (ICRP 103)	0.12	**0.18**	0.15	**0.22**	0.17	**0.24**	0.19	**0.26**	0.22	**0.30**	0.25	**0.34**
Total filtration (mm) **Al**	**1.5**		**2.5**		**3.5**		**4.5**		**3.5**		**3.5**	
Cu	**0**		**0**		**0**		**0**		**0.1**		**0.2**	

Table 12.129 Normalized organ doses—Abdomen l-lat—<1 year

Beam projection:			l-lat									**Table cells:**
Focus-to-detector distance (cm):			115									(+) = normal font = optimal field setting
Tube voltage (kVp):			60									**(−) = bold font = suboptimal field setting**
Total filtrations:			See last table row.									Empty table cells correspond to values <0.01
Organ or tissue	**Organ dose D_T : Entrance dose K_E**								**(mSv * mGy^{-1})**			
	(+)	(−)	(+)	(−)	(+)	(−)	(+)	(−)	(+)	(−)	(+)	(−)
Skull												
Brain												
Oral mucosa												
Salivary glands										**0.01**		
Cervical spine		**0.02**		**0.02**		**0.03**	0.01	**0.03**	0.01	**0.04**	0.01	**0.03**
Thyroid		**0.01**		**0.01**		**0.02**		**0.02**		**0.03**		**0.02**
Thymus		**0.06**	0.01	**0.08**	0.02	**0.09**	0.02	**0.10**	0.02	**0.12**	0.02	**0.11**
Thoracic spine	0.27	**0.45**	0.35	**0.57**	0.41	**0.67**	0.46	**0.75**	0.55	**0.89**	0.52	**0.86**
Claviculae	0.01	**0.04**	0.02	**0.05**	0.02	**0.06**	0.02	**0.07**	0.03	**0.08**	0.02	**0.07**
Scapulae	0.04	**0.70**	0.05	**0.81**	0.06	**0.88**	0.07	**0.93**	0.09	**1.01**	0.08	**1.01**
Ribs	0.56	**0.81**	0.64	**0.96**	0.70	**1.06**	0.74	**1.13**	0.81	**1.26**	0.81	**1.26**
Oesophagus	0.15	**0.20**	0.18	**0.25**	0.21	**0.29**	0.23	**0.33**	0.27	**0.39**	0.26	**0.37**
Trachea												
Lung	0.13	**0.24**	0.15	**0.29**	0.17	**0.33**	0.19	**0.36**	0.22	**0.42**	0.21	**0.41**
Mamma	0.02	**0.57**	0.03	**0.61**	0.04	**0.64**	0.04	**0.66**	0.05	**0.69**	0.05	**0.69**
Heart	0.16	**0.32**	0.19	**0.39**	0.22	**0.45**	0.24	**0.50**	0.28	**0.58**	0.28	**0.56**
Stomach	0.53	**0.47**	0.64	**0.58**	0.72	**0.66**	0.78	**0.72**	0.89	**0.83**	0.87	**0.81**
Spleen	0.68	**0.56**	0.80	**0.68**	0.89	**0.76**	0.95	**0.83**	1.06	**0.94**	1.05	**0.92**
Liver	0.14	**0.14**	0.19	**0.18**	0.22	**0.21**	0.25	**0.24**	0.30	**0.29**	0.28	**0.26**
Gall bladder	0.25	**0.21**	0.32	**0.27**	0.37	**0.32**	0.41	**0.36**	0.49	**0.44**	0.47	**0.41**
Pancreas	0.37	**0.27**	0.46	**0.35**	0.53	**0.41**	0.59	**0.46**	0.68	**0.55**	0.66	**0.52**
Small intestine	0.30	**0.24**	0.38	**0.31**	0.44	**0.36**	0.48	**0.40**	0.57	**0.48**	0.55	**0.46**
Large intestine	0.23	**0.19**	0.29	**0.24**	0.34	**0.28**	0.37	**0.32**	0.45	**0.38**	0.42	**0.36**
Sigma/rectum	0.45	**0.35**	0.56	**0.44**	0.64	**0.51**	0.70	**0.56**	0.81	**0.67**	0.79	**0.64**
Lumbar spine	0.62	**0.58**	0.79	**0.75**	0.93	**0.88**	1.04	**0.99**	1.24	**1.19**	1.19	**1.13**
Pelvis	0.89	**0.76**	1.10	**0.95**	1.25	**1.09**	1.37	**1.20**	1.58	**1.41**	1.55	**1.37**
Adrenals	0.22	**0.21**	0.27	**0.27**	0.32	**0.32**	0.35	**0.35**	0.41	**0.42**	0.39	**0.39**
Kidneys	0.30	**0.29**	0.37	**0.36**	0.42	**0.41**	0.46	**0.45**	0.54	**0.53**	0.52	**0.51**
Urinary bladder	0.26	**0.24**	0.34	**0.31**	0.39	**0.36**	0.43	**0.40**	0.51	**0.48**	0.49	**0.45**
Uterus	0.25	**0.21**	0.33	**0.27**	0.38	**0.32**	0.43	**0.36**	0.52	**0.44**	0.49	**0.41**
Ovaries	0.28	**0.22**	0.35	**0.28**	0.41	**0.33**	0.45	**0.37**	0.54	**0.45**	0.51	**0.42**
Prostate	0.22	**0.20**	0.29	**0.26**	0.34	**0.31**	0.37	**0.35**	0.45	**0.42**	0.43	**0.39**
Testes	0.21	**0.22**	0.26	**0.27**	0.31	**0.32**	0.34	**0.35**	0.40	**0.41**	0.38	**0.39**
Upper arm		**0.42**		**0.48**		**0.52**		**0.55**		**0.60**		**0.60**
Elbow		**1.42**		**1.60**		**1.71**		**1.79**		**1.92**		**1.93**
Forearm		**1.50**		**1.68**		**1.78**		**1.85**		**1.97**		**1.98**
Upper leg	0.44	**0.59**	0.56	**0.74**	0.65	**0.86**	0.72	**0.95**	0.85	**1.12**	0.81	**1.08**
Knee	0.04	**0.25**	0.05	**0.31**	0.06	**0.36**	0.07	**0.40**	0.08	**0.47**	0.07	**0.45**
Lower leg						**0.01**		**0.01**		**0.02**		**0.01**
Lymph nodes	0.20	**0.19**	0.26	**0.24**	0.29	**0.28**	0.32	**0.31**	0.38	**0.37**	0.36	**0.35**
Skin	0.22	**0.29**	0.24	**0.31**	0.25	**0.32**	0.25	**0.33**	0.27	**0.34**	0.26	**0.34**
Muscle	0.20	**0.23**	0.24	**0.28**	0.26	**0.31**	0.28	**0.34**	0.32	**0.38**	0.31	**0.37**
Skeleton	0.26	**0.54**	0.32	**0.65**	0.37	**0.72**	0.40	**0.78**	0.46	**0.88**	0.45	**0.87**
Red bone marrow	0.08	**0.14**	0.09	**0.17**	0.11	**0.19**	0.12	**0.21**	0.14	**0.24**	0.13	**0.23**
Whole body dose	0.18	**0.23**	0.22	**0.28**	0.24	**0.31**	0.26	**0.34**	0.30	**0.39**	0.29	**0.37**
ED (ICRP 60)	0.22	**0.25**	0.27	**0.31**	0.31	**0.35**	0.34	**0.39**	0.40	**0.45**	0.38	**0.43**
ED (ICRP 103)	0.20	**0.27**	0.25	**0.33**	0.29	**0.37**	0.31	**0.40**	0.36	**0.46**	0.35	**0.45**
Total filtration (mm) Al	**1.5**		**2.5**		**3.5**		**4.5**		**3.5**		**3.5**	
Cu	**0**		**0**		**0**		**0**		**0.1**		**0.2**	

Table 12.130 Normalized organ doses—Abdomen l-lat—<1 year

Beam projection:			l-lat									**Table cells:**
Focus-to-detector distance (cm):			115					(+) = normal font = optimal field setting				
Tube voltage (kVp):			70					**(−) = bold font = suboptimal field setting**				
Total filtrations:			See last table row.					Empty table cells correspond to values <0.01				
Organ or tissue	**Organ dose D_T : Entrance dose K_E**							**(mSv * mGy^{-1})**				
	(+)	(−)	(+)	(−)	(+)	(−)	(+)	(−)	(+)	(−)	(+)	(−)
Skull												**0.01**
Brain												
Oral mucosa										**0.01**		**0.01**
Salivary glands						**0.01**		**0.01**		**0.01**		**0.02**
Cervical spine		**0.02**	0.01	**0.03**	0.01	**0.04**	0.01	**0.04**	0.02	**0.05**	0.02	**0.06**
Thyroid		**0.01**		**0.02**		**0.02**		**0.03**	0.01	**0.03**	0.01	**0.04**
Thymus	0.01	**0.07**	0.02	**0.09**	0.02	**0.11**	0.02	**0.12**	0.03	**0.14**	0.03	**0.16**
Thoracic spine	0.33	**0.53**	0.41	**0.67**	0.47	**0.77**	0.52	**0.85**	0.61	**0.99**	0.69	**1.12**
Claviculae	0.02	**0.05**	0.02	**0.06**	0.02	**0.07**	0.03	**0.08**	0.04	**0.10**	0.04	**0.11**
Scapulae	0.05	**0.76**	0.07	**0.86**	0.08	**0.92**	0.09	**0.97**	0.11	**1.05**	0.13	**1.11**
Ribs	0.60	**0.89**	0.68	**1.03**	0.74	**1.13**	0.78	**1.20**	0.84	**1.32**	0.89	**1.41**
Oesophagus	0.17	**0.24**	0.21	**0.30**	0.24	**0.34**	0.26	**0.38**	0.31	**0.44**	0.34	**0.50**
Trachea										**0.01**		**0.01**
Lung	0.14	**0.27**	0.17	**0.33**	0.19	**0.37**	0.21	**0.40**	0.24	**0.46**	0.26	**0.51**
Mamma	0.03	**0.59**	0.04	**0.64**	0.04	**0.67**	0.05	**0.69**	0.05	**0.72**	0.06	**0.75**
Heart	0.18	**0.37**	0.22	**0.45**	0.25	**0.51**	0.27	**0.56**	0.32	**0.64**	0.35	**0.71**
Stomach	0.60	**0.54**	0.71	**0.65**	0.79	**0.73**	0.85	**0.79**	0.96	**0.90**	1.04	**0.99**
Spleen	0.75	**0.63**	0.87	**0.75**	0.96	**0.84**	1.02	**0.90**	1.12	**1.02**	1.20	**1.11**
Liver	0.18	**0.17**	0.22	**0.22**	0.26	**0.25**	0.29	**0.28**	0.35	**0.34**	0.40	**0.39**
Gall bladder	0.30	**0.26**	0.37	**0.33**	0.43	**0.38**	0.48	**0.42**	0.56	**0.51**	0.64	**0.58**
Pancreas	0.43	**0.33**	0.53	**0.42**	0.60	**0.48**	0.66	**0.53**	0.76	**0.63**	0.84	**0.72**
Small intestine	0.36	**0.29**	0.44	**0.37**	0.50	**0.42**	0.55	**0.47**	0.64	**0.56**	0.73	**0.63**
Large intestine	0.27	**0.23**	0.34	**0.29**	0.39	**0.34**	0.43	**0.37**	0.51	**0.45**	0.58	**0.51**
Sigma/rectum	0.52	**0.41**	0.63	**0.51**	0.71	**0.58**	0.78	**0.64**	0.89	**0.75**	0.98	**0.84**
Lumbar spine	0.74	**0.70**	0.93	**0.88**	1.07	**1.02**	1.18	**1.13**	1.39	**1.34**	1.58	**1.52**
Pelvis	1.02	**0.88**	1.23	**1.08**	1.38	**1.22**	1.49	**1.33**	1.70	**1.54**	1.86	**1.71**
Adrenals	0.26	**0.26**	0.32	**0.32**	0.36	**0.37**	0.40	**0.40**	0.47	**0.47**	0.53	**0.54**
Kidneys	0.35	**0.34**	0.42	**0.42**	0.48	**0.47**	0.52	**0.51**	0.60	**0.59**	0.66	**0.66**
Urinary bladder	0.31	**0.29**	0.39	**0.36**	0.45	**0.42**	0.49	**0.46**	0.58	**0.54**	0.65	**0.62**
Uterus	0.31	**0.26**	0.39	**0.33**	0.45	**0.38**	0.50	**0.43**	0.59	**0.51**	0.67	**0.59**
Ovaries	0.33	**0.27**	0.41	**0.34**	0.47	**0.39**	0.52	**0.44**	0.61	**0.52**	0.70	**0.60**
Prostate	0.27	**0.25**	0.34	**0.31**	0.39	**0.36**	0.43	**0.40**	0.51	**0.48**	0.58	**0.55**
Testes	0.25	**0.26**	0.31	**0.32**	0.35	**0.36**	0.38	**0.40**	0.45	**0.46**	0.50	**0.52**
Upper arm		**0.45**		**0.51**		**0.55**		**0.58**		**0.63**		**0.67**
Elbow		**1.50**		**1.67**		**1.77**		**1.84**		**1.96**		**2.04**
Forearm		**1.58**		**1.74**		**1.83**		**1.90**		**2.00**		**2.07**
Upper leg	0.52	**0.69**	0.64	**0.85**	0.73	**0.97**	0.80	**1.07**	0.93	**1.24**	1.05	**1.38**
Knee	0.05	**0.29**	0.06	**0.36**	0.07	**0.41**	0.08	**0.45**	0.10	**0.52**	0.12	**0.59**
Lower leg				**0.01**		**0.01**		**0.02**	0.01	**0.02**	0.01	**0.02**
Lymph nodes	0.24	**0.23**	0.29	**0.28**	0.33	**0.32**	0.37	**0.35**	0.42	**0.42**	0.47	**0.47**
Skin	0.23	**0.30**	0.25	**0.32**	0.26	**0.33**	0.26	**0.34**	0.28	**0.36**	0.29	**0.38**
Muscle	0.22	**0.26**	0.26	**0.31**	0.29	**0.34**	0.31	**0.37**	0.35	**0.42**	0.38	**0.46**
Skeleton	0.30	**0.60**	0.36	**0.71**	0.41	**0.79**	0.44	**0.84**	0.50	**0.94**	0.55	**1.02**
Red bone marrow	0.09	**0.16**	0.11	**0.19**	0.12	**0.22**	0.13	**0.23**	0.15	**0.27**	0.17	**0.30**
Whole body dose	0.21	**0.26**	0.24	**0.31**	0.27	**0.35**	0.29	**0.37**	0.33	**0.42**	0.36	**0.46**
ED (ICRP 60)	0.26	**0.29**	0.31	**0.35**	0.35	**0.40**	0.38	**0.43**	0.44	**0.50**	0.49	**0.56**
ED (ICRP 103)	0.23	**0.31**	0.29	**0.37**	0.32	**0.42**	0.35	**0.45**	0.40	**0.51**	0.45	**0.56**
Total filtration (mm) **Al**	**1.5**		**2.5**		**3.5**		**4.5**		**3.5**		**3.5**	
Cu	**0**		**0**		**0**		**0**		**0.1**		**0.2**	

Table 12.131 Normalized organ doses—Abdomen l-lat—1–2 years

Beam projection:			l-lat									**Table cells:**
Focus-to-detector distance (cm):			115									(+) = normal font = optimal field setting
Tube voltage (kVp):			60									**(−) = bold font = suboptimal field setting**
Total filtrations:			See last table row.									Empty table cells correspond to values <0.01
Organ or tissue	**Organ dose D_T : Entrance dose K_E**								**(mSv * mGy^{-1})**			
	(+)	(−)	(+)	(−)	(+)	(−)	(+)	(−)	(+)	(−)	(+)	(−)
Skull												
Brain												
Oral mucosa												
Salivary glands												
Cervical spine				**0.01**		**0.01**		**0.02**		**0.02**		**0.02**
Thyroid				**0.01**		**0.01**		**0.02**		**0.02**		**0.02**
Thymus		**0.06**		**0.08**	0.01	**0.09**	0.01	**0.10**	0.01	**0.13**	0.01	**0.12**
Thoracic spine	0.15	**0.28**	0.20	**0.36**	0.24	**0.43**	0.27	**0.49**	0.33	**0.61**	0.30	**0.56**
Claviculae		**0.02**		**0.03**	0.01	**0.04**	0.01	**0.05**	0.02	**0.06**	0.01	**0.05**
Scapulae	0.02	**0.47**	0.02	**0.55**	0.03	**0.60**	0.03	**0.65**	0.04	**0.72**	0.04	**0.71**
Ribs	0.42	**0.66**	0.49	**0.80**	0.54	**0.90**	0.58	**0.97**	0.65	**1.11**	0.64	**1.09**
Oesophagus	0.11	**0.14**	0.14	**0.18**	0.16	**0.22**	0.18	**0.25**	0.22	**0.30**	0.20	**0.27**
Trachea												
Lung	0.07	**0.17**	0.09	**0.21**	0.10	**0.25**	0.11	**0.27**	0.13	**0.32**	0.13	**0.31**
Mamma	0.01	**0.51**	0.02	**0.55**	0.02	**0.58**	0.02	**0.60**	0.03	**0.63**	0.03	**0.63**
Heart	0.09	**0.22**	0.12	**0.28**	0.14	**0.33**	0.16	**0.37**	0.19	**0.44**	0.18	**0.42**
Stomach	0.40	**0.35**	0.50	**0.44**	0.57	**0.50**	0.63	**0.55**	0.73	**0.65**	0.71	**0.62**
Spleen	0.58	**0.44**	0.70	**0.54**	0.79	**0.62**	0.85	**0.69**	0.97	**0.80**	0.95	**0.77**
Liver	0.08	**0.07**	0.10	**0.10**	0.13	**0.12**	0.14	**0.13**	0.18	**0.17**	0.16	**0.15**
Gall bladder	0.16	**0.13**	0.21	**0.17**	0.24	**0.20**	0.28	**0.23**	0.34	**0.29**	0.31	**0.26**
Pancreas	0.28	**0.18**	0.35	**0.23**	0.41	**0.28**	0.45	**0.32**	0.54	**0.39**	0.51	**0.36**
Small intestine	0.21	**0.16**	0.27	**0.21**	0.32	**0.25**	0.36	**0.28**	0.44	**0.34**	0.40	**0.31**
Large intestine	0.16	**0.12**	0.20	**0.16**	0.24	**0.19**	0.27	**0.21**	0.33	**0.26**	0.30	**0.23**
Sigma/rectum	0.33	**0.23**	0.43	**0.30**	0.50	**0.36**	0.55	**0.40**	0.66	**0.49**	0.62	**0.45**
Lumbar spine	0.39	**0.35**	0.51	**0.47**	0.61	**0.56**	0.69	**0.63**	0.86	**0.79**	0.78	**0.71**
Pelvis	0.67	**0.52**	0.84	**0.67**	0.97	**0.78**	1.08	**0.88**	1.27	**1.06**	1.22	**0.99**
Adrenals	0.13	**0.13**	0.17	**0.17**	0.21	**0.20**	0.23	**0.23**	0.29	**0.28**	0.26	**0.26**
Kidneys	0.22	**0.21**	0.27	**0.26**	0.32	**0.30**	0.35	**0.33**	0.41	**0.39**	0.39	**0.37**
Urinary bladder	0.18	**0.16**	0.23	**0.20**	0.27	**0.24**	0.31	**0.27**	0.38	**0.33**	0.35	**0.30**
Uterus	0.17	**0.12**	0.22	**0.17**	0.26	**0.20**	0.30	**0.23**	0.37	**0.29**	0.33	**0.26**
Ovaries	0.19	**0.13**	0.25	**0.17**	0.30	**0.21**	0.33	**0.24**	0.41	**0.30**	0.37	**0.26**
Prostate	0.14	**0.12**	0.18	**0.16**	0.22	**0.19**	0.25	**0.21**	0.30	**0.26**	0.28	**0.24**
Testes	0.13	**0.12**	0.16	**0.15**	0.19	**0.18**	0.22	**0.20**	0.27	**0.25**	0.25	**0.23**
Upper arm		**0.33**		**0.39**		**0.43**		**0.46**		**0.51**		**0.50**
Elbow		**1.24**		**1.43**		**1.56**		**1.65**		**1.81**		**1.81**
Forearm		**1.35**		**1.54**		**1.67**		**1.76**		**1.91**		**1.92**
Upper leg	0.28	**0.35**	0.36	**0.46**	0.43	**0.54**	0.48	**0.61**	0.58	**0.73**	0.54	**0.68**
Knee	0.02	**0.04**	0.02	**0.05**	0.03	**0.06**	0.03	**0.07**	0.04	**0.09**	0.03	**0.07**
Lower leg												
Lymph nodes	0.14	**0.13**	0.18	**0.16**	0.21	**0.19**	0.24	**0.22**	0.29	**0.26**	0.27	**0.24**
Skin	0.19	**0.24**	0.20	**0.26**	0.21	**0.27**	0.22	**0.27**	0.23	**0.29**	0.22	**0.28**
Muscle	0.14	**0.16**	0.17	**0.19**	0.19	**0.21**	0.20	**0.23**	0.23	**0.27**	0.22	**0.25**
Skeleton	0.17	**0.37**	0.21	**0.45**	0.24	**0.50**	0.26	**0.55**	0.31	**0.63**	0.30	**0.61**
Red bone marrow	0.06	**0.10**	0.07	**0.12**	0.09	**0.14**	0.10	**0.15**	0.11	**0.18**	0.11	**0.16**
Whole body dose	0.13	**0.17**	0.16	**0.20**	0.18	**0.23**	0.19	**0.25**	0.22	**0.29**	0.21	**0.27**
ED (ICRP 60)	0.15	**0.17**	0.20	**0.22**	0.23	**0.25**	0.25	**0.28**	0.30	**0.33**	0.28	**0.31**
ED (ICRP 103)	0.14	**0.20**	0.18	**0.25**	0.21	**0.28**	0.23	**0.31**	0.28	**0.36**	0.26	**0.34**
Total filtration (mm) **Al**	**1.5**		**2.5**		**3.5**		**4.5**		**3.5**		**3.5**	
Cu	**0**		**0**		**0**		**0**		**0.1**		**0.2**	

Table 12.132 Normalized organ doses—Abdomen l-lat—1–2 years

Beam projection:			l-lat									**Table cells:**
Focus-to-detector distance (cm):			115						(+) = normal font = optimal field setting			
Tube voltage (kVp):			70						**(−) = bold font = suboptimal field setting**			
Total filtrations:			See last table row.						Empty table cells correspond to values <0.01			
Organ or tissue	**Organ dose D_T : Entrance dose K_E**								**(mSv * mGy^{-1})**			
	(+)	(−)	(+)	(−)	(+)	(−)	(+)	(−)	(+)	(−)	(+)	(−)
Skull												
Brain												
Oral mucosa												
Salivary glands												
Cervical spine		**0.01**		**0.02**		**0.02**		**0.02**		**0.03**	0.01	**0.04**
Thyroid		**0.01**		**0.02**		**0.02**		**0.02**		**0.03**		**0.03**
Thymus		**0.08**	0.01	**0.10**	0.01	**0.11**	0.02	**0.12**	0.02	**0.15**	0.02	**0.17**
Thoracic spine	0.19	**0.35**	0.25	**0.45**	0.29	**0.53**	0.32	**0.59**	0.39	**0.72**	0.46	**0.84**
Claviculae		**0.03**	0.01	**0.04**	0.01	**0.05**	0.02	**0.06**	0.02	**0.07**	0.03	**0.09**
Scapulae	0.02	**0.51**	0.03	**0.59**	0.04	**0.65**	0.04	**0.69**	0.05	**0.76**	0.07	**0.81**
Ribs	0.46	**0.74**	0.53	**0.88**	0.58	**0.98**	0.62	**1.05**	0.69	**1.18**	0.74	**1.29**
Oesophagus	0.13	**0.18**	0.17	**0.23**	0.19	**0.27**	0.21	**0.30**	0.26	**0.36**	0.29	**0.43**
Trachea												
Lung	0.09	**0.20**	0.10	**0.25**	0.12	**0.29**	0.13	**0.32**	0.15	**0.37**	0.17	**0.42**
Mamma	0.02	**0.53**	0.02	**0.58**	0.03	**0.60**	0.03	**0.62**	0.04	**0.66**	0.05	**0.69**
Heart	0.12	**0.27**	0.14	**0.34**	0.17	**0.39**	0.18	**0.43**	0.22	**0.51**	0.25	**0.58**
Stomach	0.47	**0.41**	0.57	**0.51**	0.65	**0.58**	0.71	**0.63**	0.81	**0.73**	0.90	**0.82**
Spleen	0.65	**0.51**	0.78	**0.62**	0.87	**0.71**	0.93	**0.77**	1.05	**0.89**	1.15	**1.00**
Liver	0.10	**0.10**	0.13	**0.13**	0.16	**0.15**	0.18	**0.17**	0.22	**0.21**	0.26	**0.25**
Gall bladder	0.20	**0.17**	0.26	**0.22**	0.30	**0.25**	0.34	**0.29**	0.41	**0.35**	0.48	**0.42**
Pancreas	0.33	**0.23**	0.41	**0.29**	0.48	**0.35**	0.53	**0.39**	0.62	**0.47**	0.70	**0.56**
Small intestine	0.26	**0.20**	0.33	**0.26**	0.38	**0.30**	0.43	**0.34**	0.51	**0.41**	0.59	**0.48**
Large intestine	0.20	**0.15**	0.25	**0.20**	0.29	**0.23**	0.32	**0.26**	0.39	**0.32**	0.45	**0.37**
Sigma/rectum	0.40	**0.29**	0.50	**0.37**	0.58	**0.43**	0.64	**0.48**	0.75	**0.58**	0.85	**0.67**
Lumbar spine	0.49	**0.45**	0.63	**0.58**	0.74	**0.69**	0.84	**0.77**	1.02	**0.94**	1.19	**1.11**
Pelvis	0.79	**0.63**	0.97	**0.79**	1.11	**0.91**	1.22	**1.01**	1.42	**1.20**	1.59	**1.37**
Adrenals	0.17	**0.17**	0.22	**0.21**	0.25	**0.25**	0.28	**0.28**	0.34	**0.34**	0.40	**0.40**
Kidneys	0.26	**0.25**	0.32	**0.31**	0.37	**0.35**	0.40	**0.39**	0.47	**0.45**	0.54	**0.52**
Urinary bladder	0.22	**0.20**	0.28	**0.25**	0.33	**0.29**	0.37	**0.33**	0.44	**0.39**	0.51	**0.46**
Uterus	0.21	**0.16**	0.27	**0.22**	0.32	**0.26**	0.36	**0.29**	0.44	**0.36**	0.52	**0.43**
Ovaries	0.24	**0.17**	0.31	**0.22**	0.36	**0.26**	0.40	**0.30**	0.49	**0.37**	0.57	**0.43**
Prostate	0.18	**0.15**	0.23	**0.20**	0.27	**0.23**	0.30	**0.26**	0.36	**0.32**	0.42	**0.37**
Testes	0.16	**0.15**	0.20	**0.19**	0.23	**0.22**	0.26	**0.24**	0.31	**0.29**	0.36	**0.34**
Upper arm		**0.37**		**0.42**		**0.46**		**0.50**		**0.55**		**0.59**
Elbow		**1.33**		**1.52**		**1.64**		**1.73**		**1.87**		**1.98**
Forearm		**1.44**		**1.63**		**1.74**		**1.83**		**1.96**		**2.06**
Upper leg	0.34	**0.44**	0.44	**0.55**	0.51	**0.64**	0.56	**0.71**	0.67	**0.85**	0.77	**0.97**
Knee	0.02	**0.05**	0.03	**0.06**	0.04	**0.08**	0.04	**0.09**	0.05	**0.11**	0.07	**0.13**
Lower leg												
Lymph nodes	0.17	**0.16**	0.22	**0.20**	0.25	**0.23**	0.28	**0.26**	0.33	**0.31**	0.38	**0.36**
Skin	0.20	**0.25**	0.21	**0.27**	0.22	**0.28**	0.23	**0.29**	0.24	**0.30**	0.25	**0.31**
Muscle	0.16	**0.18**	0.19	**0.21**	0.21	**0.24**	0.23	**0.26**	0.26	**0.30**	0.29	**0.33**
Skeleton	0.20	**0.42**	0.24	**0.50**	0.27	**0.56**	0.30	**0.61**	0.35	**0.69**	0.39	**0.76**
Red bone marrow	0.07	**0.11**	0.09	**0.14**	0.10	**0.16**	0.11	**0.17**	0.13	**0.20**	0.15	**0.23**
Whole body dose	0.15	**0.19**	0.18	**0.23**	0.20	**0.26**	0.22	**0.28**	0.25	**0.32**	0.28	**0.36**
ED (ICRP 60)	0.19	**0.21**	0.23	**0.26**	0.27	**0.29**	0.29	**0.32**	0.35	**0.38**	0.40	**0.44**
ED (ICRP 103)	0.17	**0.24**	0.21	**0.28**	0.25	**0.32**	0.27	**0.35**	0.32	**0.40**	0.36	**0.45**
Total filtration (mm) **Al**	**1.5**		**2.5**		**3.5**		**4.5**		**3.5**		**3.5**	
Cu	**0**		**0**		**0**		**0**		**0.1**		**0.2**	

Table 12.133 Normalized organ doses—Abdomen l-lat—3–7 years

Beam projection:	l-lat											Table cells:
Focus-to-detector distance (cm):	115											(+) = normal font = optimal field setting
Tube voltage (kVp):	60											(−) = bold font = suboptimal field setting
Total filtrations:	See last table row.											Empty table cells correspond to values <0.01
Organ or tissue	**Organ dose D_T : Entrance dose K_E**								**(mSv * mGy^{-1})**			
	(+)	(−)	(+)	(−)	(+)	(−)	(+)	(−)	(+)	(−)	(+)	(−)
Skull												
Brain												
Oral mucosa												
Salivary glands												
Cervical spine										**0.01**		
Thyroid										**0.01**		
Thymus		**0.06**		**0.07**		**0.09**		**0.10**	0.01	**0.12**		**0.11**
Thoracic spine	0.10	**0.16**	0.13	**0.22**	0.16	**0.27**	0.18	**0.31**	0.23	**0.39**	0.20	**0.34**
Claviculae		**0.01**		**0.02**		**0.02**		**0.03**		**0.04**		**0.03**
Scapulae	0.01	**0.38**	0.02	**0.46**	0.02	**0.51**	0.02	**0.54**	0.03	**0.61**	0.02	**0.60**
Ribs	0.35	**0.50**	0.42	**0.62**	0.46	**0.71**	0.50	**0.77**	0.56	**0.89**	0.55	**0.87**
Oesophagus	0.06	**0.07**	0.07	**0.09**	0.09	**0.11**	0.10	**0.12**	0.12	**0.16**	0.11	**0.14**
Trachea												
Lung	0.06	**0.12**	0.08	**0.16**	0.09	**0.18**	0.10	**0.21**	0.12	**0.25**	0.11	**0.23**
Mamma		**0.46**	0.01	**0.50**	0.02	**0.53**	0.02	**0.54**	0.02	**0.57**	0.02	**0.56**
Heart	0.08	**0.15**	0.11	**0.20**	0.13	**0.23**	0.14	**0.26**	0.17	**0.32**	0.16	**0.29**
Stomach	0.33	**0.27**	0.42	**0.35**	0.49	**0.41**	0.54	**0.46**	0.64	**0.55**	0.61	**0.51**
Spleen	0.48	**0.34**	0.59	**0.44**	0.68	**0.51**	0.74	**0.57**	0.86	**0.68**	0.83	**0.64**
Liver	0.04	**0.04**	0.06	**0.05**	0.07	**0.06**	0.08	**0.07**	0.10	**0.09**	0.09	**0.08**
Gall bladder	0.10	**0.08**	0.14	**0.11**	0.16	**0.13**	0.19	**0.15**	0.24	**0.19**	0.21	**0.16**
Pancreas	0.20	**0.12**	0.26	**0.16**	0.31	**0.20**	0.35	**0.23**	0.42	**0.29**	0.39	**0.25**
Small intestine	0.14	**0.10**	0.19	**0.14**	0.23	**0.16**	0.26	**0.19**	0.32	**0.24**	0.28	**0.20**
Large intestine	0.10	**0.07**	0.13	**0.10**	0.16	**0.12**	0.18	**0.14**	0.22	**0.17**	0.20	**0.15**
Sigma/rectum	0.25	**0.17**	0.33	**0.22**	0.38	**0.26**	0.43	**0.30**	0.53	**0.37**	0.48	**0.33**
Lumbar spine	0.22	**0.20**	0.30	**0.26**	0.36	**0.32**	0.41	**0.37**	0.53	**0.47**	0.45	**0.40**
Pelvis	0.48	**0.35**	0.61	**0.46**	0.72	**0.54**	0.80	**0.61**	0.96	**0.75**	0.90	**0.69**
Adrenals	0.09	**0.08**	0.12	**0.11**	0.14	**0.13**	0.16	**0.15**	0.20	**0.19**	0.18	**0.17**
Kidneys	0.15	**0.14**	0.20	**0.18**	0.23	**0.21**	0.26	**0.24**	0.31	**0.29**	0.29	**0.27**
Urinary bladder	0.11	**0.10**	0.15	**0.13**	0.18	**0.15**	0.20	**0.17**	0.25	**0.22**	0.22	**0.19**
Uterus	0.10	**0.08**	0.14	**0.10**	0.17	**0.13**	0.19	**0.15**	0.25	**0.19**	0.21	**0.16**
Ovaries	0.13	**0.08**	0.17	**0.11**	0.20	**0.13**	0.23	**0.15**	0.29	**0.19**	0.25	**0.17**
Prostate	0.08	**0.07**	0.11	**0.09**	0.13	**0.11**	0.15	**0.13**	0.20	**0.17**	0.17	**0.14**
Testes	0.08	**0.08**	0.10	**0.10**	0.12	**0.12**	0.14	**0.14**	0.17	**0.17**	0.15	**0.15**
Upper arm		**0.27**		**0.32**		**0.36**		**0.38**		**0.43**		**0.42**
Elbow		**1.09**		**1.28**		**1.41**		**1.50**		**1.66**		**1.65**
Forearm		**1.25**		**1.45**		**1.58**		**1.67**		**1.83**		**1.83**
Upper leg	0.14	**0.23**	0.18	**0.30**	0.21	**0.36**	0.24	**0.40**	0.30	**0.50**	0.27	**0.45**
Knee		**0.01**		**0.02**		**0.02**		**0.02**		**0.03**		**0.02**
Lower leg												
Lymph nodes	0.10	**0.09**	0.13	**0.11**	0.16	**0.13**	0.18	**0.15**	0.22	**0.19**	0.20	**0.17**
Skin	0.13	**0.22**	0.14	**0.23**	0.15	**0.24**	0.15	**0.24**	0.16	**0.25**	0.16	**0.25**
Muscle	0.11	**0.12**	0.13	**0.15**	0.15	**0.17**	0.16	**0.19**	0.19	**0.21**	0.18	**0.20**
Skeleton	0.10	**0.24**	0.12	**0.30**	0.14	**0.33**	0.16	**0.37**	0.19	**0.43**	0.18	**0.41**
Red bone marrow	0.06	**0.08**	0.07	**0.10**	0.08	**0.11**	0.09	**0.13**	0.11	**0.15**	0.10	**0.14**
Whole body dose	0.10	**0.13**	0.12	**0.16**	0.14	**0.18**	0.15	**0.20**	0.18	**0.23**	0.17	**0.22**
ED (ICRP 60)	0.11	**0.13**	0.15	**0.16**	0.17	**0.19**	0.19	**0.21**	0.23	**0.25**	0.21	**0.23**
ED (ICRP 103)	0.11	**0.16**	0.14	**0.19**	0.16	**0.22**	0.18	**0.24**	0.22	**0.28**	0.20	**0.26**
Total filtration (mm) Al	**1.5**		**2.5**		**3.5**		**4.5**		**3.5**		**3.5**	
Cu	**0**		**0**		**0**		**0**		**0.1**		**0.2**	

Table 12.134 Normalized organ doses—Abdomen l-lat—3–7 years

Beam projection:			l-lat									**Table cells:**
Focus-to-detector distance (cm):			115						(+) = normal font = optimal field setting			
Tube voltage (kVp):			70						**(−) = bold font = suboptimal field setting**			
Total filtrations:			See last table row.						Empty table cells correspond to values <0.01			
Organ or tissue	**Organ dose D_T : Entrance dose K_E**								**(mSv * mGy^{-1})**			
	(+)	(−)	(+)	(−)	(+)	(−)	(+)	(−)	(+)	(−)	(+)	(−)
Skull												
Brain												
Oral mucosa												
Salivary glands												
Cervical spine				**0.01**		**0.01**		**0.01**		**0.02**		**0.02**
Thyroid						**0.01**		**0.01**		**0.02**		**0.02**
Thymus		**0.07**		**0.09**	0.01	**0.11**	0.01	**0.12**	0.02	**0.15**	0.02	**0.18**
Thoracic spine	0.13	**0.22**	0.17	**0.29**	0.20	**0.34**	0.23	**0.39**	0.28	**0.48**	0.34	**0.57**
Claviculae		**0.02**		**0.03**		**0.03**		**0.04**	0.01	**0.05**	0.02	**0.06**
Scapulae	0.02	**0.43**	0.02	**0.50**	0.03	**0.55**	0.03	**0.59**	0.04	**0.65**	0.05	**0.71**
Ribs	0.39	**0.58**	0.45	**0.70**	0.50	**0.78**	0.54	**0.85**	0.60	**0.97**	0.65	**1.07**
Oesophagus	0.07	**0.09**	0.09	**0.12**	0.11	**0.14**	0.12	**0.16**	0.15	**0.20**	0.18	**0.23**
Trachea												
Lung	0.07	**0.15**	0.09	**0.19**	0.10	**0.22**	0.11	**0.24**	0.13	**0.29**	0.15	**0.33**
Mamma	0.01	**0.49**	0.02	**0.53**	0.02	**0.55**	0.02	**0.57**	0.03	**0.59**	0.03	**0.62**
Heart	0.10	**0.19**	0.13	**0.24**	0.15	**0.28**	0.17	**0.32**	0.21	**0.38**	0.24	**0.45**
Stomach	0.40	**0.33**	0.49	**0.42**	0.56	**0.48**	0.62	**0.53**	0.73	**0.63**	0.82	**0.73**
Spleen	0.56	**0.41**	0.68	**0.52**	0.76	**0.59**	0.83	**0.65**	0.95	**0.77**	1.06	**0.88**
Liver	0.06	**0.05**	0.08	**0.07**	0.09	**0.08**	0.10	**0.10**	0.13	**0.12**	0.16	**0.15**
Gall bladder	0.14	**0.11**	0.18	**0.14**	0.21	**0.17**	0.24	**0.19**	0.30	**0.24**	0.35	**0.29**
Pancreas	0.25	**0.16**	0.32	**0.21**	0.37	**0.25**	0.41	**0.29**	0.50	**0.36**	0.57	**0.43**
Small intestine	0.19	**0.14**	0.24	**0.18**	0.28	**0.21**	0.31	**0.24**	0.38	**0.29**	0.45	**0.35**
Large intestine	0.13	**0.10**	0.17	**0.13**	0.20	**0.15**	0.22	**0.17**	0.28	**0.22**	0.32	**0.26**
Sigma/rectum	0.31	**0.22**	0.40	**0.28**	0.46	**0.33**	0.51	**0.37**	0.62	**0.46**	0.71	**0.54**
Lumbar spine	0.30	**0.26**	0.39	**0.35**	0.46	**0.41**	0.52	**0.47**	0.65	**0.59**	0.78	**0.71**
Pelvis	0.58	**0.44**	0.73	**0.56**	0.84	**0.65**	0.93	**0.73**	1.10	**0.88**	1.25	**1.02**
Adrenals	0.12	**0.11**	0.15	**0.14**	0.18	**0.17**	0.20	**0.19**	0.25	**0.24**	0.29	**0.28**
Kidneys	0.19	**0.18**	0.24	**0.22**	0.28	**0.26**	0.31	**0.29**	0.36	**0.34**	0.42	**0.39**
Urinary bladder	0.15	**0.13**	0.19	**0.16**	0.22	**0.19**	0.25	**0.22**	0.31	**0.27**	0.37	**0.32**
Uterus	0.14	**0.11**	0.18	**0.14**	0.22	**0.17**	0.25	**0.19**	0.31	**0.24**	0.37	**0.29**
Ovaries	0.16	**0.11**	0.21	**0.15**	0.25	**0.17**	0.29	**0.20**	0.35	**0.25**	0.42	**0.30**
Prostate	0.11	**0.09**	0.15	**0.12**	0.18	**0.15**	0.20	**0.17**	0.25	**0.21**	0.30	**0.26**
Testes	0.10	**0.10**	0.13	**0.13**	0.15	**0.15**	0.17	**0.17**	0.20	**0.21**	0.24	**0.25**
Upper arm		**0.30**		**0.35**		**0.39**		**0.41**		**0.46**		**0.50**
Elbow		**1.19**		**1.37**		**1.50**		**1.59**		**1.74**		**1.85**
Forearm		**1.34**		**1.54**		**1.66**		**1.75**		**1.90**		**2.01**
Upper leg	0.18	**0.29**	0.23	**0.37**	0.27	**0.44**	0.30	**0.49**	0.36	**0.59**	0.43	**0.70**
Knee		**0.02**		**0.02**		**0.03**		**0.03**	0.01	**0.04**	0.02	**0.05**
Lower leg												
Lymph nodes	0.13	**0.11**	0.16	**0.14**	0.19	**0.17**	0.21	**0.19**	0.26	**0.23**	0.30	**0.27**
Skin	0.14	**0.22**	0.15	**0.24**	0.15	**0.25**	0.16	**0.25**	0.17	**0.27**	0.17	**0.28**
Muscle	0.13	**0.14**	0.15	**0.17**	0.17	**0.19**	0.18	**0.21**	0.21	**0.24**	0.24	**0.27**
Skeleton	0.12	**0.28**	0.15	**0.34**	0.17	**0.38**	0.19	**0.41**	0.22	**0.47**	0.25	**0.53**
Red bone marrow	0.07	**0.10**	0.09	**0.12**	0.10	**0.14**	0.11	**0.15**	0.13	**0.18**	0.16	**0.21**
Whole body dose	0.12	**0.15**	0.14	**0.18**	0.16	**0.21**	0.17	**0.22**	0.20	**0.26**	0.22	**0.29**
ED (ICRP 60)	0.14	**0.16**	0.18	**0.20**	0.21	**0.22**	0.23	**0.25**	0.27	**0.30**	0.32	**0.34**
ED (ICRP 103)	0.13	**0.19**	0.17	**0.22**	0.19	**0.25**	0.21	**0.28**	0.26	**0.32**	0.30	**0.37**
Total filtration (mm) **Al**	**1.5**		**2.5**		**3.5**		**4.5**		**3.5**		**3.5**	
Cu	**0**		**0**		**0**		**0**		**0.1**		**0.2**	

Table 12.135 Normalized organ doses—Abdomen l-lat—3–7 years

Beam projection:			l-lat									**Table cells:**
Focus-to-detector distance (cm):			115									(+) = normal font = optimal field setting
Tube voltage (kVp):			80									**(−) = bold font = suboptimal field setting**
Total filtrations:			See last table row.									Empty table cells correspond to values <0.01
Organ or tissue	**Organ dose D_T : Entrance dose K_E**								**(mSv * mGy^{-1})**			
	(+)	(−)	(+)	(−)	(+)	(−)	(+)	(−)	(+)	(−)	(+)	(−)
Skull												
Brain												
Oral mucosa												
Salivary glands												
Cervical spine		**0.01**		**0.01**		**0.02**		**0.02**		**0.03**		**0.03**
Thyroid				**0.01**		**0.01**		**0.02**		**0.02**		**0.03**
Thymus		**0.09**	0.01	**0.11**	0.01	**0.13**	0.02	**0.15**	0.02	**0.18**	0.02	**0.21**
Thoracic spine	0.16	**0.27**	0.20	**0.35**	0.24	**0.41**	0.27	**0.46**	0.33	**0.56**	0.38	**0.65**
Claviculae		**0.03**		**0.04**	0.01	**0.04**	0.01	**0.05**	0.02	**0.06**	0.02	**0.07**
Scapulae	0.02	**0.46**	0.03	**0.53**	0.03	**0.58**	0.04	**0.62**	0.05	**0.68**	0.06	**0.72**
Ribs	0.42	**0.64**	0.49	**0.76**	0.53	**0.84**	0.57	**0.91**	0.62	**1.02**	0.67	**1.11**
Oesophagus	0.09	**0.11**	0.11	**0.15**	0.13	**0.17**	0.14	**0.19**	0.17	**0.23**	0.20	**0.27**
Trachea												
Lung	0.09	**0.18**	0.10	**0.22**	0.12	**0.25**	0.13	**0.28**	0.15	**0.32**	0.17	**0.37**
Mamma	0.02	**0.51**	0.02	**0.55**	0.03	**0.57**	0.03	**0.59**	0.03	**0.62**	0.04	**0.64**
Heart	0.13	**0.23**	0.16	**0.29**	0.18	**0.33**	0.20	**0.37**	0.24	**0.44**	0.27	**0.50**
Stomach	0.46	**0.39**	0.56	**0.48**	0.63	**0.55**	0.69	**0.60**	0.80	**0.71**	0.89	**0.80**
Spleen	0.63	**0.48**	0.75	**0.59**	0.84	**0.67**	0.90	**0.73**	1.03	**0.85**	1.13	**0.96**
Liver	0.08	**0.07**	0.10	**0.09**	0.11	**0.11**	0.13	**0.12**	0.16	**0.15**	0.19	**0.18**
Gall bladder	0.17	**0.14**	0.22	**0.18**	0.26	**0.21**	0.29	**0.24**	0.35	**0.29**	0.41	**0.35**
Pancreas	0.30	**0.21**	0.38	**0.26**	0.43	**0.31**	0.47	**0.34**	0.56	**0.42**	0.64	**0.49**
Small intestine	0.23	**0.17**	0.29	**0.22**	0.33	**0.25**	0.37	**0.29**	0.44	**0.35**	0.51	**0.41**
Large intestine	0.16	**0.13**	0.21	**0.16**	0.24	**0.19**	0.27	**0.21**	0.32	**0.26**	0.37	**0.30**
Sigma/rectum	0.37	**0.27**	0.46	**0.34**	0.53	**0.39**	0.59	**0.44**	0.70	**0.53**	0.79	**0.61**
Lumbar spine	0.37	**0.34**	0.48	**0.43**	0.56	**0.50**	0.62	**0.56**	0.76	**0.69**	0.89	**0.81**
Pelvis	0.67	**0.52**	0.83	**0.65**	0.94	**0.75**	1.03	**0.83**	1.20	**0.98**	1.34	**1.12**
Adrenals	0.15	**0.14**	0.18	**0.18**	0.21	**0.21**	0.24	**0.23**	0.29	**0.28**	0.34	**0.33**
Kidneys	0.23	**0.21**	0.28	**0.26**	0.32	**0.30**	0.35	**0.33**	0.41	**0.39**	0.47	**0.44**
Urinary bladder	0.18	**0.16**	0.23	**0.20**	0.27	**0.23**	0.30	**0.26**	0.36	**0.32**	0.42	**0.37**
Uterus	0.18	**0.14**	0.23	**0.18**	0.27	**0.21**	0.30	**0.24**	0.37	**0.29**	0.43	**0.35**
Ovaries	0.21	**0.14**	0.26	**0.19**	0.30	**0.22**	0.34	**0.25**	0.41	**0.30**	0.48	**0.36**
Prostate	0.14	**0.12**	0.18	**0.15**	0.22	**0.18**	0.24	**0.21**	0.30	**0.25**	0.35	**0.30**
Testes	0.12	**0.12**	0.15	**0.16**	0.18	**0.18**	0.20	**0.20**	0.24	**0.24**	0.27	**0.28**
Upper arm		**0.33**		**0.38**		**0.41**		**0.44**		**0.48**		**0.52**
Elbow		**1.27**		**1.44**		**1.56**		**1.64**		**1.77**		**1.87**
Forearm		**1.42**		**1.60**		**1.72**		**1.80**		**1.92**		**2.01**
Upper leg	0.21	**0.35**	0.27	**0.44**	0.31	**0.51**	0.35	**0.56**	0.42	**0.67**	0.48	**0.77**
Knee		**0.02**		**0.03**	0.01	**0.04**	0.01	**0.04**	0.02	**0.05**	0.02	**0.06**
Lower leg												
Lymph nodes	0.16	**0.14**	0.19	**0.17**	0.22	**0.20**	0.25	**0.22**	0.29	**0.27**	0.34	**0.31**
Skin	0.14	**0.23**	0.15	**0.25**	0.16	**0.25**	0.16	**0.26**	0.17	**0.28**	0.18	**0.29**
Muscle	0.14	**0.16**	0.17	**0.19**	0.19	**0.22**	0.20	**0.24**	0.23	**0.27**	0.26	**0.30**
Skeleton	0.14	**0.31**	0.17	**0.37**	0.19	**0.41**	0.21	**0.45**	0.24	**0.51**	0.27	**0.56**
Red bone marrow	0.08	**0.11**	0.10	**0.14**	0.12	**0.16**	0.13	**0.17**	0.15	**0.21**	0.18	**0.24**
Whole body dose	0.13	**0.17**	0.16	**0.21**	0.18	**0.23**	0.19	**0.25**	0.22	**0.29**	0.24	**0.32**
ED (ICRP 60)	0.17	**0.19**	0.21	**0.23**	0.24	**0.26**	0.26	**0.29**	0.31	**0.34**	0.35	**0.38**
ED (ICRP 103)	0.16	**0.21**	0.20	**0.26**	0.22	**0.29**	0.25	**0.31**	0.29	**0.36**	0.33	**0.41**
Total filtration (mm) Al	**1.5**		**2.5**		**3.5**		**4.5**		**3.5**		**3.5**	
Cu	**0**		**0**		**0**		**0**		**0.1**		**0.2**	

References

1. Bontrager KL. Textbook of radiographic positioning and related anatomy. St. Louis: Mosby Year Book; 1993.
2. Commission of the European Communities. European guidelines on quality criteria for diagnostic radiographic images in paediatrics. EUR 16261; SBN 92-827-7843-6; 1996.
3. Cook JV, Shah K, Pablot S, Kyriou J, Pettett A, Fitzgerald M. Guidelines on best practice in the x-ray imaging of children. A manual for all x-ray departments. London: Queen Mary's Hospital for Children, Surrey, and The Radiological Protection Centre; 1998.
4. Datz H, Ben-Shlomo A, Bader D, Sadetzki S, Juster-Reicher A, Marks K, Smolkin T, Zangen S, Margaliot M. The additional dose to radiosensitive organs caused by using under-collimated x-ray beams in neonatal intensive care radiography. Radiat Prot Dosim. 2008;130:518–24.
5. Drexler G, Panzer W, Stieve FE. Die Bestimmung von Organdosen in der Röntgendiagnostik. Berlin: Hoffmann Verlag; 1993.
6. Ebel K-D, Willich E. Die Röntgenuntersuchung im Kindesalter – Technik und Indikation. Berlin: Springer; 1968.
7. Hart D, Jones DJ, Wall BF. Estimation of effective dose in diagnostic radiology from entrance surface dose and dose-area-product measurements. National Radiological Protection Board NRPB-R262; 1994.
8. Hart D, Jones DJ, Wall BF. Coefficients for estimating effective doses from pediatric x-ray examinations. National Radiological Protection Board NRPB-R279; 1996.
9. Hösle M. Untersuchung über die Patientendosis und die Bildqualität bei häufigen Röntgenaufnahmen im Säuglingsalter in Kliniken der alten Bundesländer mit Leitung durch einen Pädiater oder einen Radiologen. Dissertation, Universität München; 1993.
10. Pehe J. Untersuchung über die Patientendosis und die Bildqualität bei häufigen Röntgenaufnahmen in der Kinderradiologie. Dissertation, Universität München; 1993.
11. Schneider K, Fendel H, Bakowski C, Stein E, Kohn M, Kellner M, Schweighofer K, Cartagena G, Padovani R, Panzer W, Scheurer C, Wall B. Results of a dosimetry study in the European Community on frequent x ray examinations in infants. Radiat Prot Dosim. 1992;43:31–6.
12. Seidenbusch MC, Regulla D, Schneider K. Radiation exposure of children in pediatric radiology. Part 6: conversion coefficients for reconstruction of organ dose in abdominal radiography. Fortschr Röntgenstr. 2009;181:945–61.
13. Winter W. Untersuchung über die Patientendosis und die Bildqualität bei häufigen Röntgenaufnahmen im Säuglingsalter bei niedergelassenen Ärzten (Pädiater, Orthopäden, Chirurgen). Dissertation, Universität München; 1994.

Full Spine

13

13.1 Introduction

The type of full spine radiography is determined by the clinical indication. Table 13.1 shows indications and Table 13.2 contraindications for full spine radiography. Image criteria as required for radiological findings in full spine radiographs are listed in Table 13.3. In order to fulfil these image criteria, the field settings listed in Tables 13.4 and 13.5 should be achieved. These field settings go along with the patient preparation as listed in Table 13.6. Exposure parameter combinations for various age groups and patient diameters are listed in Tables 13.7, 13.8, 13.9, 13.10, and 13.11. The age groups in Tables 13.8 and 13.10 refer to the

Table 13.1 Indications for full spine radiographs

Skeletal dysplasia
Idiopathic scoliosis
Battered child
Multiple vertebral anomalies[a]
Caudal regression syndrome[a]
Metabolic diseases, e.g. mucopolysaccharidosis

Annotations:
[a]MRI is the preferred imaging modality in patients with these spinal and spinal cord anomalies

Table 13.2 Contraindications for full spine radiographs

Pregnancy

Table 13.3 Image criteria for full spine radiographs

Full spine ap/pa/lat
Symmetrical reproduction of the whole vertical column
Reproduction of the vertebral bodies, the pedicles, and the medial parts of the ribs
Visualization of the posterior articular processes
Visualization of the spinous and transverse processes
Reproduction of the symphysis pubis
Reproduction of the apophyses of the pelvic bones

Table 13.4 Patient positions and beam projections for full spine radiographs

Patient position	Beam projections
Standing	ap + pa + lat
Supine	ap
Prone	pa[a]
Decubitus left	lat

Annotation:
[a]In special cases only, e.g. myelomeningocele postoperative

Table 13.5 Optimal field settings for full spine radiographs

Full spine ap	
Upper field edge	Skull base
Field centring	Midline at the level of the xiphoid
Lower field edge	Perineum
Lateral field edges	Medioclavicular lines (ap), midpoint of the scapulae (pa)
Full spine lat	
Upper field edge	Skull base
Field centring	Midaxillary line at the level of the xiphoid
Lower field edge	Perineum
Anterior field edge	Anterior axillary lines
Posterior field edge	Close to the skin surface on the back

M. Seidenbusch et al., *Imaging Practice and Radiation Protection in Pediatric Radiology*,
https://doi.org/10.1007/978-3-030-18504-6_13

Table 13.6 Patient preparation

General	Remove all clothes, external metallic devices, especially piercings, and jewellery
	Immobilize infants and uncooperative patients in the supine position, e.g. with sandbags or restrain devices. Alternatively, accompanying persons or staff members could hold the patient (in this case also take care of the radiation protection of the accompanying persons or staff members)
	Cooperative patients are positioned upright at the vertical stand for pa projection of the spine
	Place eventually small X-ray positive markers at palpable tumours or on pain points
Positioning	The arms should be kept up or positioned alongside the trunk
	In supine position, the legs should be extended using a sandbag
	Rotation of the trunk in the longitudinal and/or horizontal axis should be avoided
Attention	On the Bucky table the patient should always be positioned on a foam mat of 10 to 20 cm thickness. So, the lower part of the body will not be cut if lateral projection X-rays (cross table views) in prone or in decubitus position are performed
Shielding	In males, testicular capsules should be used if possible

Table 13.7 Radiographic procedure

X-ray source	Automatic exposure control (AEC) is not used
	Focal spot size between 0.6 and 1.3 dependent on the age of the patient
	Check the use of additional filtration
	Check the focus-to-detector distance
Field setting	Check the field size
	Check the field centring
Image detector system	Check the use of antiscatter grids
Patient	The exposure should be performed during breath-holding

Table 13.8 Radiographic techniques Bucky table full spine radiographs

Age group (years)	Beam direction	Patient position	Beam projection	Typical patient diameter (cm)		Optimal field size (cm × cm)	Focus-to-detector distance (cm)	Voltage (kVp)						Antiscatter grid
				With arms	No arms			50	60	70	80	90	100	
0	Vertical	Supine	ap	9.8		4 × 27	100[2]							
		Decubitus	lat[1]	12.6	10.9	–								
1–2		Supine	ap	13.0		6 × 38	115							
		Decubitus	lat[1]	17.1	15.1	–								
3–7		Supine	ap	15.0		7 × 50	115–200							
		Decubitus	lat[1]	22.6	19.6	–								
8–12	This technique is not applied in these age groups													
13–17														
18														

Annotations:	
1	Full spine radiographs in lateral projection could not be considered in this book due to the unrealistic construction of the full spine in the mathematical STUK phantoms.
2	The patient is lying on the table top directly on the detector
	Parameter combination not recommended, normalized organ doses not provided
	Parameter combination not recommended, but normalized organ doses provided
	Parameter combination recommended, normalized organ doses provided

Table 13.9 Radiation qualities of Bucky table full spine radiographs

Voltage (kVp)	Total filtration					
	Without additional filtration		With additional filtration			
	1.5 mm Al	2.5 Al	3.5 Al	4.5 Al	3.5 Al + 0.1 Cu	3.5 Al + 0.2 Cu
50	This technique is not suitable for full spine radiographs					
60						
70						
80						
90						
100						

Annotations:	
	Parameter combination not recommended, normalized organ doses not provided
	Parameter combination not recommended, but normalized organ doses provided
	Parameter combination recommended, normalized organ doses provided

Table 13.10 Radiographic techniques of full spine vertical stand radiographs

Age group (years)	Beam direction	Patient position	Beam projection	Typical patient diameter (cm) with arms	Typical patient diameter (cm) no arms	Optimal field size (cm x cm)	Focus-to-detector distance (cm)	Voltage (kVp) 50	60	70	80	90	100	Antiscatter grid
0	This technique is not applied in these age groups.													
1–2														
3-7														
8–12	horizontal	standing	ap	16.80		8 x 64	200 (2)							
			pa											
			lat (1)	27.3	23.8	-								
13–17			ap	19.60		9 x 80								
			pa											
			lat (1)	34.4	29.6	-								
18			ap	20.00		10 x 90								
			pa											
			lat (1)	38.9	34.4	-								

Annotations:
(1) Full spine radiographs in lateral projection could not be considered due to the unrealistic construction of the full spine in the mathematical STUK phantoms.
(2) The focus-to-detector distance should be kept as large as possible.
Parameter combination not recommended, normalized organ doses not provided
Parameter combination recommended, normalized organ doses provided

Table 13.11 Radiation qualities of full spine vertical stand radiographs

Voltage (kVp)	Total filtration: Without additional filtration 1.5 mm Al	2.5 Al	With additional filtration 3.5 Al	4.5 Al	3.5 Al + 0.1 Cu[1]	3.5 Al + 0.2 Cu[1]
50						
60						
70	This technique is not suitable for full spine radiographs					
80						
90						
100						

Annotations:
1 The total beam filtration is usually 3.5 mm Al + 0.1 mm Cu, in adolescents optionally 3.5 mm Al + 0.2 mm Cu
Parameter combination not recommended, normalized organ doses not provided
Parameter combination not recommended, but normalized organ doses provided
Parameter combination recommended, normalized organ doses provided

standard age groups of the mathematical MIRD phantom family which has been used to calculate the normalized organ doses in this book. Tables 13.9 and 13.11 show radiation qualities (i.e. various combinations of tube voltage and total filtration) which may be applied in clinical practice. Please note that not all of these radiation qualities are recommended by the authors. The standard projections of full spine radiographies of the patient who is correctly positioned are depicted in Figs. 13.1, 13.2, and 13.3. Also the optimal field (marked in green), the suboptimal field (marked in red), and the correct shielding are shown. Tables 13.13, 13.14, 13.15, 13.16, and 13.17 outline the virtual field settings on the skin surface of mathematical MIRD phantoms of various age groups and the resulting virtual radiographies with and without bone structures, respectively. In patients with severe scoliosis, the suboptimal field size must be used. Otherwise, the pathology of the spine cannot be visualized. Normalized organ doses resulting from these virtual field settings have been calculated for all age groups and have been listed in Tables 13.19, 13.20, 13.21, 13.22, 13.23, 13.24, 13.25, 13.26, 13.27, 13.28, 13.29, 13.30, 13.31, 13.32, 13.33, 13.34, 13.35, 13.36, 13.37, 13.38 13.39, 13.40, 13.41, 13.42, 13.43, 13.44, 13.45, 13.46, 13.47, and 13.48. Table cells with normal font refer to optimal field settings and table cells with bold font refer to suboptimal field settings. Please note that normalized organ doses have been provided also for exposure parameter combinations, which are not recommended by the authors.

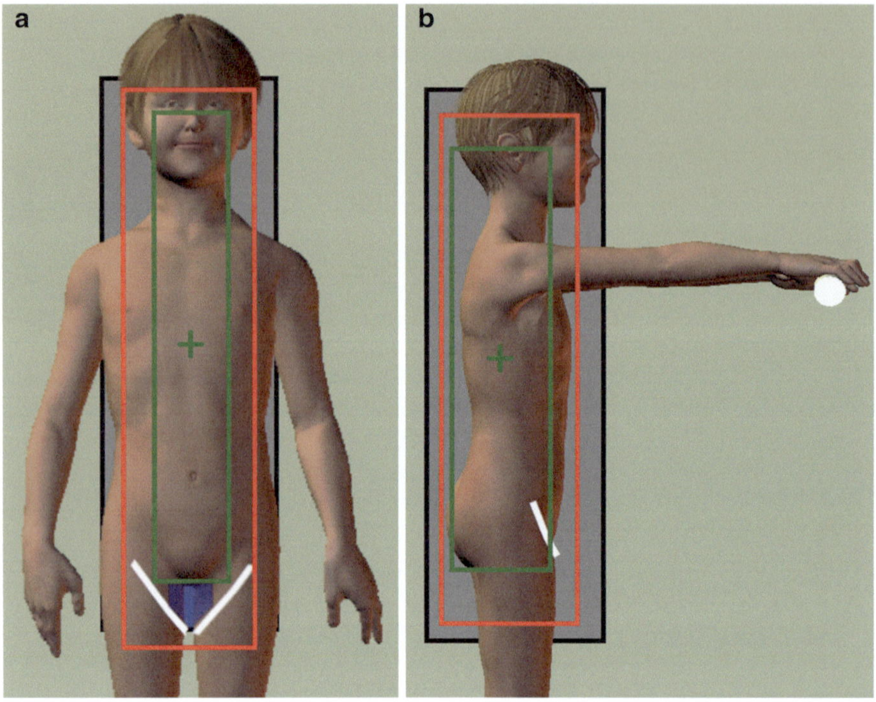

Fig. 13.1 Field settings for full spine radiographs in standing position. (**a**) Antero-posterior projection. (**b**) Lateral projection; the patient holds on a handrail. (© V. Rösenberger & M. Seidenbusch 2018). Green lines = optimal field settings; green cross = centring point; red lines = suboptimal field settings; grey shaded area = image detector; blue shaded area = testicular capsule; white lines = sticking plaster for fixing the gonad shielding

Fig. 13.2 Field settings for full spine radiographs in supine position. Antero-posterior projection. (© V. Rösenberger & M. Seidenbusch 2018). Green lines = optimal field settings; green cross = centring point; red lines = suboptimal field settings; grey shaded area = image detector; blue shaded area = testicular capsule; dark grey shaded area = Bucky table

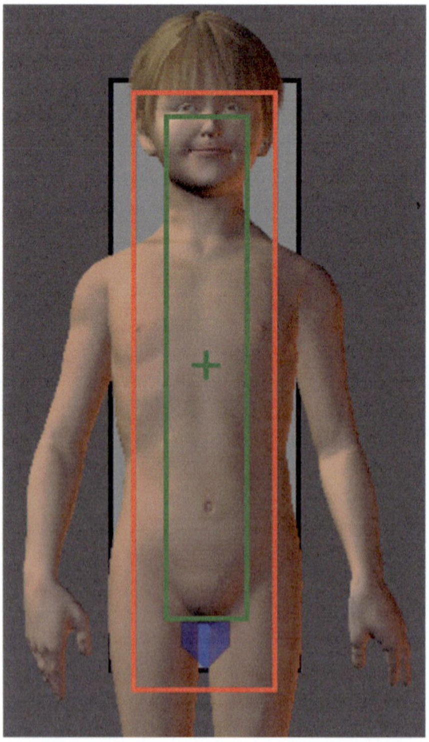

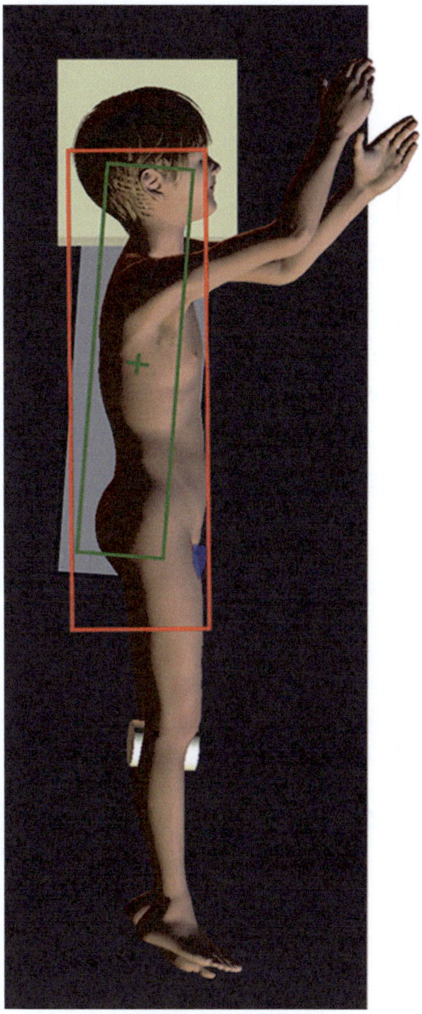

Fig. 13.3 Field settings for full spine radiographs in left decubitus position. Lateral projection. (© V. Rösenberger & M. Seidenbusch 2018). Green lines = optimal field settings; green cross = centring point; red lines = suboptimal field settings; grey shaded area = image detector; blue shaded area = testicular capsule; dark grey shaded area = Bucky table; white cylinder = foam roll; white foam mat for the head on the top of the Bucky table

Table 13.12 Inconsistencies between the mathematical MIRD phantoms and real human anatomy

Anatomical region	Inconsistencies
Head	No eyes, nose, mouth, and ears have been inserted
	No eye lens has been inserted
Skeleton	The shape of the spine is unrealistic (no lordosis, no kyphosis considered), the vertebral column is not clearly depicted. The transverse processes are not outlined

For details, see the text

13.2 Indications and Contraindications

Indications and contraindications for full spine radiographs are given in Tables 13.1 and 13.2.

13.3 Image Criteria

The image criteria required for full spine radiographs are listed in Table 13.3.

See also [2, 3].

13.4 Field Settings

In Table 13.4, patient positions and beam projections used for full spine radiographs are listed, and in Table 13.5, the optimal field settings are listed which are required in order to fulfil the image criteria mentioned in Table 13.3.

For more details, see also [8–11].

13.5 Patient Preparation and Radiographic Procedure

See Tables 13.6, 13.7 and [1, 5].

13.6 Exposure Parameters

In Tables 13.8 and 13.10, exposure parameter combinations used for full spine radiographs in various age groups are listed. Dark grey shaded table cells refer to exposure parameter combinations which are not recommended for paediatric full spine radiographs and for which no normalized organ doses have been provided. Light grey and light blue shaded table cells refer to exposure parameter combinations for which normalized organ doses have been provided with light grey shaded table cells referring to exposure parameter combinations being not recommended and light blue shaded table cells referring to exposure parameter combinations following the European Guidelines [2]. Note that due to the fact that the radiation dose is an exponential function of the patient diameter, it may make sense to use the patient diameters instead of the patient's age groups for determining the adequate exposure parameters from Tables 13.8 and 13.10. In this case, for determining the corresponding normalized organ doses from Tables 13.19, 13.20, 13.21, 13.22, 13.23, 13.24, 13.25, 13.26, 13.27, 13.28, 13.29, 13.30, 13.31, 13.32, 13.33, 13.34, 13.35, 13.36, 13.37, 13.38, 13.39, 13.40, 13.41, 13.42, 13.43, 13.44, 13.45, 13.46, 13.47, and 13.48, the age group related to the patient diameter should be used.

13.7 Normalized Organ Doses

For calculating normalized organ doses, the field settings described in Sect. 13.4 were transferred to the mathematical MIRD phantoms. Inconsistencies between the geometrical anatomy of the mathematical phantoms and the real human anatomy have been listed in Table 13.12 whenever they are of relevance for the validity of the normalized organ doses.

In Tables 13.13, 13.14, 13.15, 13.16, 13.17, and 13.18, optimal and suboptimal field settings are listed for the different age groups and the beam projections given in Tables 13.8, 13.9, 13.10, and 13.11. Green lines refer to optimal field settings, red lines refer to suboptimal field settings, respectively. In the tables, the optimal and suboptimal field settings which were used for performing the calculations of the normalized organ doses provided in Tables 13.19, 13.20, 13.21, 13.22, 13.23, 13.24, 13.25, 13.26, 13.27, 13.28, 13.29, 13.30, 13.31, 13.32, 13.33, 13.34, 13.35, 13.36, 13.37, 13.38, 13.39, 13.40, 13.41, 13.42, 13.43, 13.44, 13.45, 13.46, 13.47, and 13.48 were drawn in on the skin surface of the mathematical phantoms. The virtual mathematical radiographs resulting from these field settings were depicted in the tables with and without bone structures, respectively, in order to enable the reader to visually compare the virtual mathematical radiographs with the real radiographs made under clinical conditions. For better visualization of the spine in the skeletal view, thymus, heart, lungs, oesophagus, kidneys, liver, gall bladder, small intestine, and colon have been removed. For comparing the virtual mathematical field settings with the real clinical field settings, the virtual mathematical field edges were also listed in the tables.

As the normalized organ doses provided in Tables 13.19, 13.20, 13.21, 13.22, 13.23, 13.24, 13.25, 13.26, 13.27, 13.28, 13.29, 13.30, 13.31, 13.32, 13.33, 13.34, 13.35, 13.36, 13.37, 13.38, 13.39, 13.40, 13.41, 13.42, 13.43, 13.44, 13.45, 13.46, 13.47, and 13.48 refer to entrance kerma values, optimal and suboptimal field sizes were provided in the tables in order to enable the reader to calculate entrance air kerma dose values from clinically measured dose area product values. Please note that these optimal and suboptimal field sizes refer to the mathematical phantoms of an age group only and not to real patients with corresponding ages.

Please note also that the anatomy of the mathematical phantoms has been idealized. Therefore, the virtual field settings for calculating the normalized organ doses have been defined with a view to ensuring that the resulting virtual organ doses would be consistent with the real organ doses which would be achieved during real X-ray examinations being performed with the real field settings recommended in Sect. 13.4. Thus, the field settings shown in the following tables may not be consistent with the clinical field settings as recommended in Sect. 13.4.

Normalized organ doses for full spine radiographs are also listed in e.g. [4, 6, 7].

Table 13.13 Field settings—Full spine ap

Age group (years)		Anatomical landmarks			
		Optimal field setting		Suboptimal field setting	
0	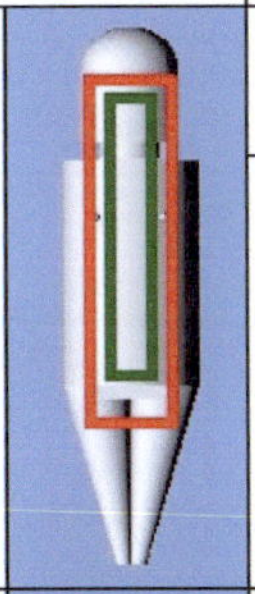	Upper field edge: Just below the skull base Lower field edge: Just below the perineum Lateral field edges: Just medial to the medioclavicular line Field size = field width × field height: 4 cm × 27 cm = 108 cm^2		Upper field edge: At the midpoint of the skull Lower field edge: Upper third of thighs Lateral field edges: At the mamillary line Field size = field width × field height: 8 cm × 32 cm = 256 cm^2	

1–2	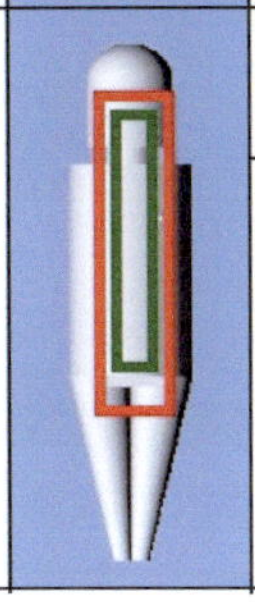	Upper field edge: Just below the skull base Lower field edge: Just below the perineum Lateral field edges: Just medial to the medioclavicular line Field size = field width × field height: 6 cm × 38 cm = 228 cm^2		Upper field edge: At the midpoint of the skull Lower field edge: Upper third of thighs Lateral field edges: At the mamillary line Field size = field width × field height: 10 cm × 44 cm = 440 cm^2	
		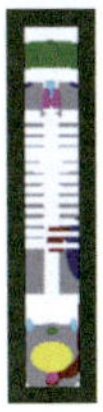			

Table 13.14 Field settings—Full spine ap

Age group (years)		Anatomical landmarks			
		Optimal field setting		Suboptimal field setting	
3–7	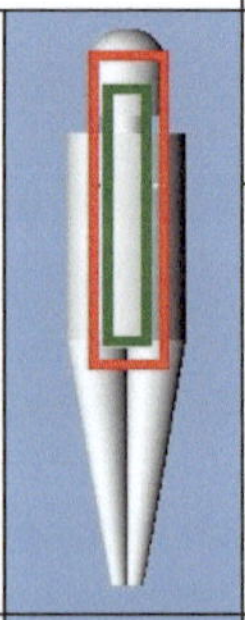	Upper field edge: Just below the skull base Lower field edge: Just below the perineum Lateral field edges: Just medial to the medioclavicular line Field size = field width × field height: 7 cm × 50 cm = 350 cm^2		Upper field edge: At the midpoint of the skull Lower field edge: Upper third of thighs Lateral field edges: At the mamillary line Field size = field width × field height: 13 cm × 60 cm = 780 cm^2	

8–12	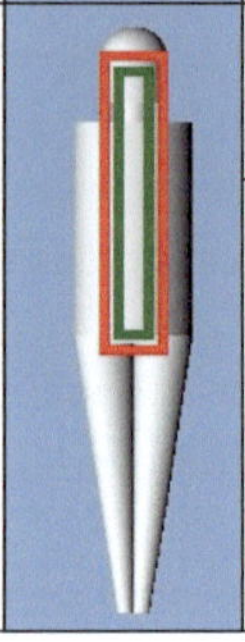	Upper field edge: Just below the skull base Lower field edge: Just below the perineum Lateral field edges: Just medial to the medioclavicular line Field size = field width × field height: 8 cm × 64 cm = 512 cm^2		Upper field edge: At the midpoint of the skull Lower field edge: Upper third of thighs Lateral field edges: At the mamillary line Field size = field width × field height: 14 cm × 70 cm = 980 cm^2	
		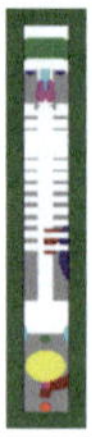		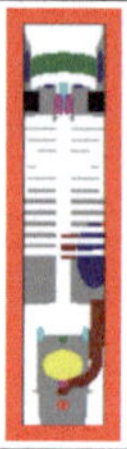	

Table 13.15 Field settings—Full spine pa

Age group (years)		Anatomical landmarks			
		Optimal field setting		Suboptimal field setting	
13–17	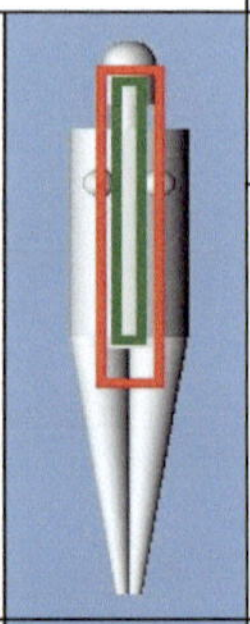	Upper field edge: Just below the skull base Lower field edge: Just below the perineum Lateral field edges: Just medial to the medioclavicular line Field size = field width × field height: 9 cm × 80 cm = 720 cm^2		Upper field edge: At the midpoint of the skull Lower field edge: Upper third of thighs Lateral field edges: At the mamillary line Field size = field width × field height: 17 cm × 94 cm = 1,598 cm^2	

18	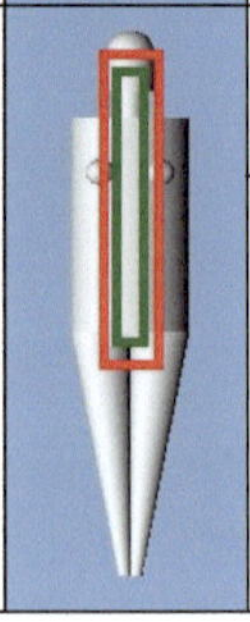	Upper field edge: Just below the skull base Lower field edge: Just below the perineum Lateral field edges: Just medial to the medioclavicular line Field size = field width × field height: 10 cm × 90 cm = 900 cm^2		Upper field edge: At the midpoint of the skull Lower field edge: Upper third of thighs Lateral field edges: At the mamillary line Field size = field width × field height: 18 cm × 100 cm = 1800 cm^2	

Table 13.16 Field settings—Full spine pa

Age group (years)		Anatomical landmarks			
		Optimal field setting		Suboptimal field setting	
8–12	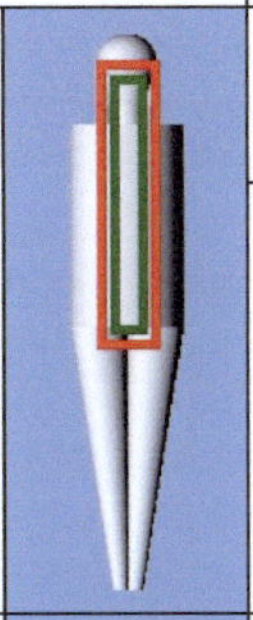	Upper field edge: Just below the skull base Lower field edge: Just below the perineum Lateral field edges: Just medial to the medioclavicular line Field size = field width × field height: 8 cm × 64 cm = 512 cm^2		Upper field edge: At the midpoint of the skull Lower field edge: Upper third of thighs Lateral field edges: At the mamillary line Field size = field width × field height: 14 cm × 70 cm = 980 cm^2	

13–17	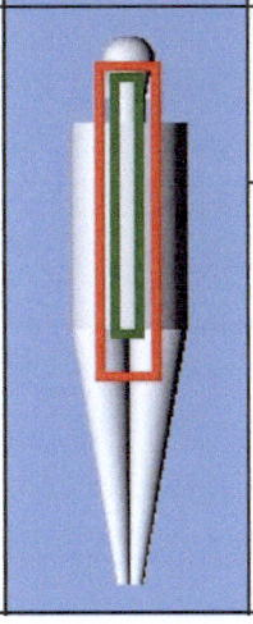	Upper field edge: Just below the skull base Lower field edge: Just below the perineum Lateral field edges: Just medial to the medioclavicular line Field size = field width × field height: 9 cm × 80 cm = 720 cm^2		Upper field edge: At the midpoint of the skull Lower field edge: Upper third of thighs Lateral field edges: At the mamillary line Field size = field width × field height: 17 cm × 94 cm = 1598 cm^2	

Table 13.17 Field settings—Full spine pa

Age group (years)		Anatomical landmarks			
		Optimal field setting		Suboptimal field setting	
18	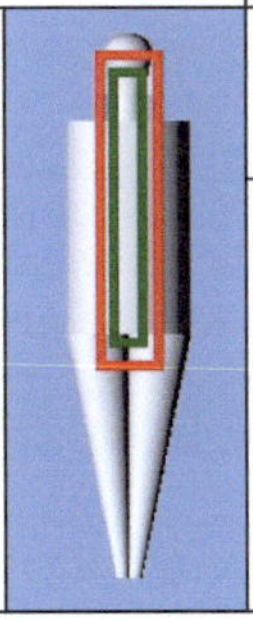	Upper field edge: Just below the skull base Lower field edge: Just below the perineum Lateral field edges: Just medial to the medioclavicular line Field size = field width × field height: 10 cm × 90 cm = 900 cm^2		Upper field edge: At the midpoint of the skull Lower field edge: Upper third of thighs Lateral field edges: At the mamillary line Field size = field width × field height: 18 cm × 100 cm = 1800 cm^2	

Table 13.18 Field settings—Full spine lat

Age group (years)	Anatomical landmarks	
	Optimal field setting	Suboptimal field setting
0	As the full spine has been approximated geometrically by bars representing the spinal segments and as the transitions between the spinal segments were not properly designed in the STUK MIRD phantoms, normalized organ doses for full spine radiographs in lateral projection were not considered in this book	
1–2		
3–7		
8–12		
13–17		
18		

Table 13.19 Normalized organ doses—Full spine ap—<1 year

Beam projection:	ap											**Table cells:**
Focus-to-detector distance (cm):	100											(+) = normal font = optimal field setting
Tube voltage (kVp):	50											**(−) = bold font = suboptimal field setting**
Total filtrations:	See last table row											Empty table cells correspond to values <0.01
Organ or tissue	**Organ dose D_T: Entrance dose K_E**						**(mSv * mGy^{-1})**					
	(+)	(−)	(+)	(−)	(+)	(−)	(+)	(−)	(+)	(−)	(+)	(−)
Skull	0.15	**0.47**	0.18	**0.56**	0.20	**0.63**	0.22	**0.68**	0.25	**0.77**	0.28	**0.85**
Brain	0.02	**0.08**	0.03	**0.11**	0.03	**0.13**	0.04	**0.14**	0.05	**0.17**	0.06	**0.20**
Oral mucosa	0.30	**0.53**	0.36	**0.62**	0.39	**0.68**	0.41	**0.73**	0.46	**0.81**	0.49	**0.88**
Salivary glands	0.11	**0.40**	0.13	**0.49**	0.14	**0.55**	0.16	**0.59**	0.18	**0.67**	0.20	**0.73**
Cervical spine	0.42	**0.48**	0.55	**0.63**	0.64	**0.75**	0.72	**0.84**	0.87	**1.02**	1.01	**1.19**
Thyroid	0.83	**0.86**	0.94	**0.99**	1.01	**1.06**	1.06	**1.12**	1.13	**1.20**	1.20	**1.28**
Thymus	0.69	**0.73**	0.80	**0.85**	0.87	**0.93**	0.93	**0.99**	1.01	**1.09**	1.08	**1.17**
Thoracic spine	0.17	**0.22**	0.23	**0.30**	0.28	**0.36**	0.33	**0.42**	0.41	**0.53**	0.49	**0.63**
Claviculae	0.60	**1.56**	0.70	**1.83**	0.76	**2.00**	0.81	**2.13**	0.89	**2.34**	0.95	**2.50**
Scapulae	0.05	**0.25**	0.07	**0.33**	0.08	**0.40**	0.09	**0.45**	0.12	**0.54**	0.14	**0.63**
Ribs	0.47	**0.99**	0.53	**1.13**	0.56	**1.22**	0.59	**1.29**	0.64	**1.40**	0.68	**1.49**
Oesophagus	0.15	**0.18**	0.19	**0.23**	0.23	**0.28**	0.26	**0.31**	0.31	**0.38**	0.37	**0.44**
Trachea	0.48	**0.53**	0.56	**0.62**	0.62	**0.68**	0.65	**0.72**	0.72	**0.80**	0.77	**0.86**
Lung	0.07	**0.35**	0.10	**0.43**	0.11	**0.49**	0.13	**0.53**	0.15	**0.61**	0.17	**0.67**
Mamma	0.04	**1.13**	0.05	**1.19**	0.05	**1.22**	0.06	**1.24**	0.07	**1.27**	0.09	**1.30**
Heart	0.37	**0.49**	0.46	**0.60**	0.51	**0.68**	0.56	**0.73**	0.63	**0.84**	0.70	**0.92**
Stomach	0.16	**0.50**	0.20	**0.60**	0.23	**0.68**	0.26	**0.74**	0.30	**0.84**	0.34	**0.93**
Spleen	0.03	**0.15**	0.04	**0.19**	0.04	**0.23**	0.05	**0.26**	0.06	**0.32**	0.08	**0.37**
Liver	0.22	**0.40**	0.27	**0.49**	0.30	**0.55**	0.33	**0.60**	0.37	**0.68**	0.41	**0.76**
Gall bladder	0.31	**0.36**	0.40	**0.45**	0.46	**0.53**	0.50	**0.58**	0.59	**0.69**	0.66	**0.78**
Pancreas	0.16	**0.24**	0.21	**0.31**	0.25	**0.37**	0.28	**0.41**	0.34	**0.49**	0.39	**0.57**
Small intestine	0.20	**0.31**	0.26	**0.39**	0.30	**0.46**	0.33	**0.51**	0.39	**0.60**	0.44	**0.68**
Large intestine	0.16	**0.38**	0.21	**0.48**	0.24	**0.55**	0.26	**0.60**	0.31	**0.70**	0.35	**0.79**
Sigma/rectum	0.09	**0.28**	0.11	**0.36**	0.13	**0.42**	0.15	**0.46**	0.18	**0.55**	0.21	**0.62**
Lumbar spine	0.18	**0.23**	0.25	**0.32**	0.30	**0.39**	0.35	**0.44**	0.43	**0.56**	0.52	**0.68**
Pelvis	0.09	**0.40**	0.13	**0.52**	0.15	**0.62**	0.17	**0.70**	0.22	**0.85**	0.26	**1.00**
Adrenals	0.07	**0.10**	0.09	**0.13**	0.11	**0.16**	0.12	**0.18**	0.15	**0.22**	0.18	**0.26**
Kidneys	0.05	**0.09**	0.07	**0.13**	0.08	**0.15**	0.09	**0.17**	0.12	**0.22**	0.14	**0.26**
Urinary bladder	0.45	**0.52**	0.54	**0.63**	0.61	**0.71**	0.65	**0.77**	0.74	**0.87**	0.80	**0.96**
Uterus	0.27	**0.31**	0.34	**0.40**	0.39	**0.47**	0.43	**0.52**	0.51	**0.62**	0.58	**0.70**
Ovaries	0.18	**0.25**	0.23	**0.32**	0.27	**0.38**	0.30	**0.42**	0.36	**0.51**	0.41	**0.59**
Prostate	0.30	**0.38**	0.38	**0.48**	0.43	**0.55**	0.47	**0.61**	0.55	**0.71**	0.61	**0.79**
Testes	0.05	**0.94**	0.06	**1.05**	0.07	**1.11**	0.08	**1.15**	0.10	**1.22**	0.11	**1.27**
Upper arm	0.03	**0.07**	0.04	**0.10**	0.05	**0.12**	0.06	**0.14**	0.07	**0.17**	0.09	**0.21**
Elbow	0.03	**0.08**	0.04	**0.10**	0.05	**0.12**	0.06	**0.14**	0.07	**0.18**	0.09	**0.21**
Forearm	0.02	**0.07**	0.03	**0.10**	0.04	**0.12**	0.04	**0.14**	0.05	**0.17**	0.07	**0.21**
Upper leg	0.03	**0.67**	0.04	**0.87**	0.05	**1.01**	0.06	**1.12**	0.08	**1.33**	0.09	**1.52**
Knee		**0.31**		**0.40**	0.01	**0.46**	0.01	**0.51**	0.02	**0.60**	0.02	**0.69**
Lower leg								**0.01**		**0.01**		**0.02**
Lymph nodes	0.26	**0.38**	0.31	**0.47**	0.35	**0.53**	0.38	**0.58**	0.44	**0.66**	0.49	**0.74**
Skin	0.11	**0.30**	0.12	**0.31**	0.12	**0.33**	0.13	**0.34**	0.14	**0.35**	0.14	**0.37**
Muscle	0.13	**0.27**	0.15	**0.33**	0.17	**0.36**	0.18	**0.39**	0.21	**0.44**	0.23	**0.49**
Skeleton	0.17	**0.45**	0.21	**0.55**	0.23	**0.63**	0.26	**0.69**	0.30	**0.80**	0.34	**0.90**
Red bone marrow	0.05	**0.13**	0.06	**0.16**	0.07	**0.18**	0.08	**0.20**	0.09	**0.24**	0.10	**0.27**
Whole body dose	0.13	**0.28**	0.15	**0.33**	0.17	**0.37**	0.18	**0.41**	0.21	**0.46**	0.24	**0.51**
ED (ICRP 60)	0.17	**0.45**	0.21	**0.53**	0.23	**0.58**	0.25	**0.63**	0.29	**0.70**	0.32	**0.77**
ED (ICRP 103)	0.16	**0.47**	0.20	**0.55**	0.23	**0.60**	0.25	**0.64**	0.28	**0.72**	0.32	**0.78**
Total filtration (mm) Al	**1.5**		**2.5**		**3.5**		**4.5**		**3.5**		**3.5**	
Cu	**0**		**0**		**0**		**0**		**0.1**		**0.2**	

Table 13.20 Normalized organ doses—Full spine ap—<1 year

Beam projection:		ap										**Table cells:**
Focus-to-detector distance (cm):		100						(+) = normal font = optimal field setting				
Tube voltage (kVp):		60						**(−) = bold font = suboptimal field setting**				
Total filtrations:		See last table row						Empty table cells correspond to values <0.01				
Organ or tissue	**Organ dose D_T: Entrance dose K_E**							**(mSv * mGy^{-1})**				
	(+)	(−)	(+)	(−)	(+)	(−)	(+)	(−)	(+)	(−)	(+)	(−)
Skull	0.18	**0.54**	0.21	**0.64**	0.23	**0.71**	0.25	**0.76**	0.28	**0.85**	0.28	**0.85**
Brain	0.03	**0.11**	0.04	**0.14**	0.04	**0.16**	0.05	**0.18**	0.06	**0.21**	0.06	**0.20**
Oral mucosa	0.34	**0.59**	0.39	**0.69**	0.43	**0.76**	0.46	**0.81**	0.50	**0.89**	0.49	**0.88**
Salivary glands	0.12	**0.46**	0.15	**0.55**	0.17	**0.62**	0.18	**0.66**	0.20	**0.74**	0.20	**0.73**
Cervical spine	0.54	**0.62**	0.68	**0.80**	0.79	**0.93**	0.88	**1.04**	1.05	**1.23**	1.01	**1.19**
Thyroid	0.90	**0.94**	1.01	**1.07**	1.08	**1.14**	1.12	**1.20**	1.20	**1.29**	1.20	**1.28**
Thymus	0.76	**0.81**	0.87	**0.94**	0.95	**1.02**	1.00	**1.08**	1.08	**1.18**	1.08	**1.17**
Thoracic spine	0.24	**0.31**	0.32	**0.41**	0.38	**0.49**	0.43	**0.56**	0.54	**0.69**	0.49	**0.63**
Claviculae	0.66	**1.72**	0.76	**1.99**	0.83	**2.16**	0.87	**2.28**	0.95	**2.47**	0.95	**2.50**
Scapulae	0.07	**0.33**	0.09	**0.42**	0.11	**0.50**	0.13	**0.55**	0.15	**0.66**	0.14	**0.63**
Ribs	0.51	**1.08**	0.57	**1.22**	0.61	**1.31**	0.63	**1.38**	0.68	**1.49**	0.68	**1.49**
Oesophagus	0.19	**0.23**	0.25	**0.30**	0.29	**0.35**	0.32	**0.39**	0.38	**0.47**	0.37	**0.44**
Trachea	0.54	**0.59**	0.62	**0.68**	0.67	**0.75**	0.71	**0.79**	0.77	**0.87**	0.77	**0.86**
Lung	0.10	**0.41**	0.12	**0.50**	0.14	**0.56**	0.15	**0.61**	0.18	**0.69**	0.17	**0.67**
Mamma	0.05	**1.16**	0.06	**1.22**	0.07	**1.24**	0.08	**1.26**	0.09	**1.29**	0.09	**1.30**
Heart	0.44	**0.58**	0.52	**0.69**	0.58	**0.77**	0.63	**0.83**	0.71	**0.94**	0.70	**0.92**
Stomach	0.20	**0.58**	0.24	**0.70**	0.27	**0.78**	0.30	**0.84**	0.34	**0.94**	0.34	**0.93**
Spleen	0.04	**0.19**	0.05	**0.25**	0.06	**0.29**	0.07	**0.33**	0.08	**0.39**	0.08	**0.37**
Liver	0.26	**0.47**	0.31	**0.57**	0.35	**0.63**	0.37	**0.68**	0.42	**0.77**	0.41	**0.76**
Gall bladder	0.38	**0.44**	0.47	**0.55**	0.54	**0.63**	0.59	**0.69**	0.68	**0.80**	0.66	**0.78**
Pancreas	0.21	**0.31**	0.27	**0.39**	0.31	**0.46**	0.34	**0.51**	0.41	**0.60**	0.39	**0.57**
Small intestine	0.25	**0.38**	0.31	**0.48**	0.35	**0.55**	0.39	**0.60**	0.45	**0.70**	0.44	**0.68**
Large intestine	0.20	**0.46**	0.25	**0.57**	0.29	**0.64**	0.31	**0.70**	0.36	**0.81**	0.35	**0.79**
Sigma/rectum	0.11	**0.35**	0.14	**0.44**	0.16	**0.50**	0.18	**0.55**	0.22	**0.64**	0.21	**0.62**
Lumbar spine	0.26	**0.33**	0.34	**0.44**	0.41	**0.53**	0.46	**0.60**	0.57	**0.74**	0.52	**0.68**
Pelvis	0.13	**0.52**	0.17	**0.67**	0.20	**0.78**	0.23	**0.87**	0.28	**1.04**	0.26	**1.00**
Adrenals	0.09	**0.13**	0.12	**0.18**	0.14	**0.21**	0.16	**0.23**	0.20	**0.29**	0.18	**0.26**
Kidneys	0.07	**0.13**	0.09	**0.17**	0.11	**0.20**	0.13	**0.23**	0.16	**0.28**	0.14	**0.26**
Urinary bladder	0.52	**0.60**	0.62	**0.72**	0.68	**0.80**	0.73	**0.86**	0.81	**0.97**	0.80	**0.96**
Uterus	0.33	**0.39**	0.41	**0.49**	0.47	**0.56**	0.51	**0.62**	0.59	**0.73**	0.58	**0.70**
Ovaries	0.22	**0.31**	0.28	**0.40**	0.33	**0.46**	0.36	**0.51**	0.43	**0.61**	0.41	**0.59**
Prostate	0.36	**0.46**	0.44	**0.57**	0.50	**0.65**	0.54	**0.70**	0.62	**0.81**	0.61	**0.79**
Testes	0.06	**1.01**	0.08	**1.11**	0.09	**1.17**	0.10	**1.21**	0.11	**1.28**	0.11	**1.27**
Upper arm	0.04	**0.10**	0.06	**0.13**	0.07	**0.16**	0.08	**0.18**	0.10	**0.22**	0.09	**0.21**
Elbow	0.04	**0.11**	0.06	**0.14**	0.07	**0.17**	0.08	**0.19**	0.10	**0.23**	0.09	**0.21**
Forearm	0.03	**0.10**	0.04	**0.14**	0.05	**0.16**	0.06	**0.18**	0.07	**0.22**	0.07	**0.21**
Upper leg	0.05	**0.84**	0.06	**1.05**	0.07	**1.20**	0.08	**1.32**	0.10	**1.54**	0.09	**1.52**
Knee	0.01	**0.39**	0.01	**0.48**	0.02	**0.55**	0.02	**0.60**	0.03	**0.70**	0.02	**0.69**
Lower leg				**0.01**		**0.01**		**0.01**		**0.02**		**0.02**
Lymph nodes	0.30	**0.45**	0.36	**0.55**	0.41	**0.61**	0.44	**0.66**	0.50	**0.76**	0.49	**0.74**
Skin	0.12	**0.31**	0.13	**0.33**	0.13	**0.35**	0.14	**0.36**	0.15	**0.38**	0.14	**0.37**
Muscle	0.15	**0.32**	0.18	**0.37**	0.20	**0.41**	0.21	**0.45**	0.24	**0.50**	0.23	**0.49**
Skeleton	0.20	**0.54**	0.25	**0.65**	0.28	**0.74**	0.31	**0.80**	0.36	**0.92**	0.34	**0.90**
Red bone marrow	0.06	**0.16**	0.08	**0.20**	0.09	**0.22**	0.10	**0.24**	0.11	**0.28**	0.10	**0.27**
Whole body dose	0.15	**0.32**	0.18	**0.39**	0.20	**0.43**	0.21	**0.46**	0.24	**0.52**	0.24	**0.51**
ED (ICRP 60)	0.20	**0.51**	0.24	**0.60**	0.27	**0.66**	0.29	**0.70**	0.33	**0.78**	0.32	**0.77**
ED (ICRP 103)	0.19	**0.53**	0.23	**0.62**	0.26	**0.67**	0.28	**0.72**	0.32	**0.79**	0.32	**0.78**
Total filtration (mm) **Al**	**1.5**		**2.5**		**3.5**		**4.5**		**3.5**		**3.5**	
Cu	**0**		**0**		**0**		**0**		**0.1**		**0.2**	

Table 13.21 Normalized organ doses—Full spine ap—<1 year

Beam projection:	ap											**Table cells:**
Focus-to-detector distance (cm):	100											(+) = normal font = optimal field setting
Tube voltage (kVp):	70											**(−) = bold font = suboptimal field setting**
Total filtrations:	See last table row											Empty table cells correspond to values <0.01
Organ or tissue	**Organ dose D_T: Entrance dose K_E**							**(mSv ∗ mGy⁻¹)**				
	(+)	(−)	(+)	(−)	(+)	(−)	(+)	(−)	(+)	(−)	(+)	(−)
Skull	0.20	**0.59**	0.23	**0.69**	0.25	**0.76**	0.27	**0.81**	0.30	**0.90**	0.33	**0.96**
Brain	0.04	**0.13**	0.04	**0.16**	0.05	**0.18**	0.06	**0.20**	0.07	**0.24**	0.08	**0.27**
Oral mucosa	0.37	**0.65**	0.42	**0.75**	0.46	**0.81**	0.49	**0.86**	0.53	**0.95**	0.57	**1.01**
Salivary glands	0.14	**0.52**	0.16	**0.61**	0.18	**0.67**	0.19	**0.71**	0.22	**0.79**	0.24	**0.85**
Cervical spine	0.63	**0.74**	0.79	**0.93**	0.90	**1.06**	0.99	**1.17**	1.15	**1.37**	1.29	**1.55**
Thyroid	0.95	**1.01**	1.06	**1.13**	1.12	**1.20**	1.17	**1.25**	1.24	**1.34**	1.30	**1.41**
Thymus	0.82	**0.88**	0.93	**1.00**	1.00	**1.08**	1.05	**1.14**	1.13	**1.23**	1.19	**1.31**
Thoracic spine	0.31	**0.39**	0.39	**0.50**	0.46	**0.59**	0.51	**0.66**	0.62	**0.80**	0.72	**0.93**
Claviculae	0.71	**1.84**	0.80	**2.09**	0.86	**2.24**	0.90	**2.35**	0.97	**2.51**	1.01	**2.62**
Scapulae	0.09	**0.39**	0.11	**0.49**	0.13	**0.57**	0.15	**0.63**	0.18	**0.73**	0.21	**0.83**
Ribs	0.54	**1.15**	0.60	**1.28**	0.63	**1.37**	0.66	**1.43**	0.70	**1.53**	0.74	**1.60**
Oesophagus	0.23	**0.28**	0.29	**0.36**	0.34	**0.41**	0.37	**0.45**	0.44	**0.54**	0.49	**0.61**
Trachea	0.58	**0.64**	0.66	**0.74**	0.71	**0.80**	0.75	**0.84**	0.81	**0.92**	0.86	**0.98**
Lung	0.11	**0.47**	0.14	**0.55**	0.16	**0.62**	0.18	**0.66**	0.20	**0.74**	0.23	**0.81**
Mamma	0.06	**1.19**	0.07	**1.24**	0.08	**1.26**	0.09	**1.28**	0.11	**1.31**	0.12	**1.33**
Heart	0.49	**0.64**	0.58	**0.76**	0.64	**0.84**	0.68	**0.90**	0.76	**1.01**	0.82	**1.09**
Stomach	0.23	**0.65**	0.27	**0.77**	0.30	**0.85**	0.33	**0.91**	0.38	**1.01**	0.41	**1.10**
Spleen	0.05	**0.24**	0.06	**0.30**	0.07	**0.34**	0.08	**0.38**	0.10	**0.45**	0.12	**0.51**
Liver	0.29	**0.53**	0.34	**0.63**	0.38	**0.69**	0.40	**0.74**	0.45	**0.83**	0.49	**0.91**
Gall bladder	0.44	**0.51**	0.53	**0.62**	0.60	**0.71**	0.65	**0.77**	0.74	**0.88**	0.82	**0.98**
Pancreas	0.25	**0.37**	0.31	**0.46**	0.36	**0.53**	0.39	**0.58**	0.46	**0.68**	0.52	**0.77**
Small intestine	0.29	**0.45**	0.35	**0.55**	0.40	**0.62**	0.44	**0.68**	0.50	**0.78**	0.56	**0.87**
Large intestine	0.23	**0.53**	0.28	**0.64**	0.32	**0.72**	0.35	**0.78**	0.40	**0.88**	0.45	**0.97**
Sigma/Rectum	0.13	**0.41**	0.16	**0.50**	0.19	**0.57**	0.21	**0.62**	0.24	**0.71**	0.27	**0.80**
Lumbar spine	0.32	**0.42**	0.42	**0.54**	0.49	**0.63**	0.54	**0.71**	0.66	**0.86**	0.76	**1.01**
Pelvis	0.16	**0.62**	0.21	**0.78**	0.24	**0.90**	0.27	**0.99**	0.33	**1.16**	0.38	**1.31**
Adrenals	0.12	**0.17**	0.15	**0.21**	0.17	**0.25**	0.19	**0.28**	0.23	**0.34**	0.27	**0.39**
Kidneys	0.09	**0.16**	0.12	**0.21**	0.14	**0.25**	0.15	**0.28**	0.18	**0.33**	0.21	**0.39**
Urinary bladder	0.57	**0.67**	0.67	**0.79**	0.73	**0.87**	0.78	**0.93**	0.86	**1.04**	0.93	**1.13**
Uterus	0.38	**0.46**	0.46	**0.56**	0.52	**0.64**	0.57	**0.70**	0.65	**0.80**	0.72	**0.90**
Ovaries	0.26	**0.37**	0.33	**0.46**	0.37	**0.53**	0.41	**0.58**	0.48	**0.68**	0.53	**0.76**
Prostate	0.41	**0.53**	0.49	**0.64**	0.55	**0.71**	0.59	**0.77**	0.67	**0.88**	0.73	**0.97**
Testes	0.07	**1.06**	0.09	**1.15**	0.10	**1.21**	0.11	**1.25**	0.12	**1.31**	0.14	**1.36**
Upper arm	0.06	**0.13**	0.07	**0.16**	0.08	**0.19**	0.10	**0.21**	0.12	**0.26**	0.14	**0.30**
Elbow	0.06	**0.13**	0.07	**0.17**	0.09	**0.20**	0.10	**0.22**	0.12	**0.27**	0.14	**0.32**
Forearm	0.04	**0.13**	0.05	**0.17**	0.06	**0.19**	0.07	**0.22**	0.09	**0.26**	0.11	**0.30**
Upper leg	0.06	**0.96**	0.07	**1.18**	0.09	**1.34**	0.10	**1.46**	0.12	**1.67**	0.14	**1.84**
Knee	0.01	**0.44**	0.02	**0.54**	0.02	**0.61**	0.03	**0.66**	0.03	**0.76**	0.04	**0.83**
Lower leg		**0.01**		**0.01**		**0.02**		**0.02**		**0.02**		**0.03**
Lymph nodes	0.34	**0.51**	0.40	**0.61**	0.45	**0.68**	0.48	**0.73**	0.54	**0.82**	0.59	**0.90**
Skin	0.12	**0.33**	0.13	**0.35**	0.14	**0.36**	0.14	**0.37**	0.15	**0.40**	0.16	**0.42**
Muscle	0.16	**0.35**	0.19	**0.41**	0.21	**0.45**	0.23	**0.49**	0.26	**0.54**	0.28	**0.59**
Skeleton	0.23	**0.61**	0.28	**0.73**	0.31	**0.81**	0.34	**0.88**	0.39	**0.99**	0.43	**1.09**
Red bone marrow	0.07	**0.19**	0.09	**0.22**	0.10	**0.25**	0.11	**0.27**	0.13	**0.32**	0.14	**0.35**
Whole body dose	0.17	**0.36**	0.20	**0.43**	0.22	**0.47**	0.23	**0.51**	0.26	**0.57**	0.29	**0.62**
ED (ICRP 60)	0.22	**0.56**	0.27	**0.65**	0.30	**0.71**	0.32	**0.76**	0.36	**0.84**	0.40	**0.90**
ED (ICRP 103)	0.22	**0.58**	0.26	**0.67**	0.29	**0.72**	0.31	**0.77**	0.35	**0.85**	0.39	**0.91**
Total filtration (mm) Al	**1.5**		**2.5**		**3.5**		**4.5**		**3.5**		**3.5**	
Cu	**0**		**0**		**0**		**0**		**0.1**		**0.2**	

Table 13.22 Normalized organ doses—Full spine ap—1–2 years

Beam projection:		ap										**Table cells:**
Focus-to-detector distance (cm):		115										(+) = normal font = optimal field setting
Tube voltage (kVp):		50										**(−) = bold font = suboptimal field setting**
Total filtrations:		See last table row										Empty table cells correspond to values <0.01
Organ or tissue	**Organ dose D_T: Entrance dose K_E**							**(mSv * mGy^{-1})**				
	(+)	(−)	(+)	(−)	(+)	(−)	(+)	(−)	(+)	(−)	(+)	(−)
Skull	0.16	**0.33**	0.20	**0.40**	0.22	**0.45**	0.24	**0.50**	0.27	**0.57**	0.30	**0.63**
Brain	0.01	**0.03**	0.02	**0.05**	0.02	**0.05**	0.03	**0.06**	0.04	**0.08**	0.04	**0.09**
Oral mucosa	0.27	**0.41**	0.32	**0.49**	0.35	**0.54**	0.38	**0.58**	0.43	**0.66**	0.47	**0.73**
Salivary glands	0.12	**0.19**	0.15	**0.24**	0.17	**0.27**	0.19	**0.30**	0.22	**0.35**	0.25	**0.40**
Cervical spine	0.30	**0.33**	0.39	**0.43**	0.47	**0.52**	0.53	**0.59**	0.66	**0.73**	0.78	**0.87**
Thyroid	0.79	**0.79**	0.90	**0.91**	0.98	**0.99**	1.03	**1.04**	1.11	**1.14**	1.18	**1.22**
Thymus	0.62	**0.64**	0.74	**0.77**	0.82	**0.86**	0.88	**0.93**	0.99	**1.05**	1.08	**1.15**
Thoracic spine	0.10	**0.12**	0.14	**0.17**	0.17	**0.21**	0.19	**0.24**	0.25	**0.31**	0.31	**0.39**
Claviculae	0.64	**1.25**	0.76	**1.51**	0.85	**1.68**	0.92	**1.82**	1.03	**2.04**	1.13	**2.24**
Scapulae	0.03	**0.11**	0.04	**0.15**	0.05	**0.18**	0.06	**0.20**	0.08	**0.25**	0.10	**0.30**
Ribs	0.51	**0.87**	0.59	**1.01**	0.64	**1.10**	0.67	**1.17**	0.73	**1.28**	0.79	**1.38**
Oesophagus	0.12	**0.14**	0.16	**0.19**	0.20	**0.23**	0.22	**0.26**	0.28	**0.33**	0.33	**0.39**
Trachea	0.41	**0.44**	0.49	**0.53**	0.54	**0.59**	0.59	**0.64**	0.66	**0.72**	0.72	**0.79**
Lung	0.08	**0.25**	0.10	**0.31**	0.12	**0.36**	0.13	**0.40**	0.16	**0.46**	0.19	**0.52**
Mamma	0.03	**1.09**	0.05	**1.17**	0.05	**1.21**	0.06	**1.23**	0.07	**1.27**	0.09	**1.31**
Heart	0.35	**0.41**	0.44	**0.51**	0.50	**0.59**	0.55	**0.64**	0.64	**0.75**	0.72	**0.84**
Stomach	0.16	**0.39**	0.20	**0.48**	0.23	**0.55**	0.26	**0.60**	0.31	**0.70**	0.35	**0.78**
Spleen	0.02	**0.08**	0.03	**0.11**	0.04	**0.13**	0.04	**0.15**	0.05	**0.19**	0.07	**0.22**
Liver	0.19	**0.29**	0.23	**0.36**	0.26	**0.42**	0.29	**0.46**	0.33	**0.53**	0.38	**0.60**
Gall bladder	0.28	**0.31**	0.36	**0.40**	0.42	**0.47**	0.47	**0.53**	0.56	**0.63**	0.64	**0.73**
Pancreas	0.12	**0.16**	0.16	**0.22**	0.19	**0.26**	0.22	**0.30**	0.27	**0.37**	0.32	**0.44**
Small intestine	0.16	**0.23**	0.21	**0.31**	0.25	**0.36**	0.28	**0.40**	0.34	**0.48**	0.40	**0.56**
Large intestine	0.15	**0.29**	0.20	**0.37**	0.23	**0.43**	0.26	**0.48**	0.31	**0.58**	0.36	**0.66**
Sigma/rectum	0.08	**0.20**	0.10	**0.26**	0.12	**0.31**	0.14	**0.35**	0.17	**0.42**	0.20	**0.49**
Lumbar spine	0.10	**0.13**	0.14	**0.18**	0.18	**0.22**	0.21	**0.26**	0.27	**0.34**	0.34	**0.42**
Pelvis	0.07	**0.24**	0.09	**0.32**	0.11	**0.39**	0.13	**0.45**	0.17	**0.56**	0.21	**0.67**
Adrenals	0.05	**0.06**	0.07	**0.08**	0.08	**0.10**	0.09	**0.12**	0.12	**0.15**	0.15	**0.19**
Kidneys	0.03	**0.05**	0.05	**0.07**	0.06	**0.09**	0.06	**0.10**	0.08	**0.13**	0.10	**0.16**
Urinary bladder	0.39	**0.44**	0.48	**0.54**	0.54	**0.62**	0.59	**0.68**	0.68	**0.78**	0.75	**0.88**
Uterus	0.21	**0.24**	0.27	**0.31**	0.32	**0.37**	0.36	**0.41**	0.44	**0.50**	0.51	**0.59**
Ovaries	0.13	**0.17**	0.18	**0.22**	0.21	**0.27**	0.24	**0.30**	0.29	**0.37**	0.34	**0.44**
Prostate	0.22	**0.28**	0.28	**0.36**	0.33	**0.42**	0.36	**0.47**	0.43	**0.56**	0.49	**0.64**
Testes	0.05	**0.88**	0.06	**0.99**	0.07	**1.06**	0.08	**1.11**	0.09	**1.19**	0.11	**1.25**
Upper arm	0.02	**0.04**	0.03	**0.06**	0.04	**0.07**	0.04	**0.08**	0.06	**0.11**	0.07	**0.13**
Elbow	0.02	**0.04**	0.03	**0.06**	0.04	**0.07**	0.04	**0.09**	0.06	**0.11**	0.07	**0.14**
Forearm	0.01	**0.04**	0.02	**0.05**	0.03	**0.07**	0.03	**0.08**	0.04	**0.10**	0.05	**0.13**
Upper leg	0.02	**0.39**	0.03	**0.51**	0.04	**0.61**	0.04	**0.69**	0.05	**0.84**	0.07	**0.99**
Knee		**0.13**		**0.18**		**0.21**		**0.24**		**0.29**		**0.34**
Lower leg												
Lymph nodes	0.22	**0.30**	0.28	**0.37**	0.32	**0.42**	0.35	**0.46**	0.41	**0.54**	0.46	**0.61**
Skin	0.11	**0.24**	0.12	**0.25**	0.12	**0.26**	0.13	**0.27**	0.14	**0.28**	0.14	**0.30**
Muscle	0.10	**0.18**	0.12	**0.22**	0.14	**0.25**	0.15	**0.27**	0.17	**0.31**	0.19	**0.35**
Skeleton	0.14	**0.29**	0.17	**0.36**	0.20	**0.42**	0.21	**0.46**	0.25	**0.54**	0.29	**0.62**
Red bone marrow	0.04	**0.08**	0.05	**0.10**	0.06	**0.12**	0.06	**0.13**	0.07	**0.15**	0.09	**0.18**
Whole body dose	0.10	**0.19**	0.13	**0.23**	0.14	**0.26**	0.16	**0.29**	0.18	**0.33**	0.20	**0.37**
ED (ICRP 60)	0.15	**0.37**	0.19	**0.44**	0.21	**0.49**	0.23	**0.53**	0.27	**0.60**	0.31	**0.66**
ED (ICRP 103)	0.15	**0.40**	0.18	**0.46**	0.21	**0.51**	0.23	**0.55**	0.27	**0.62**	0.30	**0.68**
Total filtration (mm) **Al**	**1.5**		**2.5**		**3.5**		**4.5**		**3.5**		**3.5**	
Cu	**0**		**0**		**0**		**0**		**0.1**		**0.2**	

Table 13.23 Normalized organ doses—Full spine ap—1–2 years

Beam projection:	ap											**Table cells:**
Focus-to-detector distance (cm):	115											(+) = normal font = optimal field setting
Tube voltage (kVp):	60											**(−) = bold font = suboptimal field setting**
Total filtrations:	See last table row											Empty table cells correspond to values <0.01
Organ or tissue	**Organ dose D_T: Entrance dose K_E**							**(mSv ∗ mGy^{-1})**				
	(+)	(−)	(+)	(−)	(+)	(−)	(+)	(−)	(+)	(−)	(+)	(−)
Skull	0.19	**0.39**	0.23	**0.47**	0.25	**0.53**	0.27	**0.57**	0.31	**0.65**	0.30	**0.63**
Brain	0.02	**0.05**	0.03	**0.06**	0.03	**0.08**	0.04	**0.09**	0.05	**0.11**	0.04	**0.09**
Oral mucosa	0.31	**0.47**	0.36	**0.56**	0.40	**0.62**	0.43	**0.66**	0.48	**0.75**	0.47	**0.73**
Salivary glands	0.15	**0.23**	0.18	**0.29**	0.20	**0.32**	0.22	**0.36**	0.26	**0.41**	0.25	**0.40**
Cervical spine	0.40	**0.45**	0.52	**0.58**	0.61	**0.69**	0.69	**0.77**	0.84	**0.94**	0.78	**0.87**
Thyroid	0.86	**0.88**	0.98	**1.00**	1.05	**1.08**	1.11	**1.13**	1.20	**1.23**	1.18	**1.22**
Thymus	0.70	**0.74**	0.83	**0.88**	0.92	**0.97**	0.98	**1.04**	1.09	**1.16**	1.08	**1.15**
Thoracic spine	0.15	**0.19**	0.20	**0.25**	0.25	**0.31**	0.28	**0.35**	0.36	**0.45**	0.31	**0.39**
Claviculae	0.73	**1.43**	0.86	**1.70**	0.95	**1.87**	1.01	**2.01**	1.13	**2.23**	1.13	**2.24**
Scapulae	0.05	**0.15**	0.06	**0.20**	0.08	**0.24**	0.09	**0.27**	0.11	**0.33**	0.10	**0.30**
Ribs	0.56	**0.97**	0.64	**1.11**	0.69	**1.20**	0.73	**1.27**	0.79	**1.39**	0.79	**1.38**
Oesophagus	0.17	**0.20**	0.22	**0.26**	0.26	**0.31**	0.29	**0.35**	0.36	**0.43**	0.33	**0.39**
Trachea	0.47	**0.51**	0.55	**0.60**	0.61	**0.67**	0.66	**0.72**	0.73	**0.81**	0.72	**0.79**
Lung	0.10	**0.30**	0.13	**0.38**	0.15	**0.43**	0.17	**0.47**	0.20	**0.54**	0.19	**0.52**
Mamma	0.05	**1.13**	0.06	**1.20**	0.07	**1.24**	0.08	**1.27**	0.09	**1.31**	0.09	**1.31**
Heart	0.43	**0.50**	0.52	**0.61**	0.59	**0.69**	0.64	**0.75**	0.74	**0.86**	0.72	**0.84**
Stomach	0.20	**0.47**	0.25	**0.57**	0.28	**0.64**	0.31	**0.70**	0.36	**0.80**	0.35	**0.78**
Spleen	0.03	**0.11**	0.04	**0.15**	0.05	**0.18**	0.06	**0.20**	0.08	**0.25**	0.07	**0.22**
Liver	0.22	**0.36**	0.28	**0.44**	0.31	**0.50**	0.34	**0.54**	0.39	**0.63**	0.38	**0.60**
Gall bladder	0.35	**0.40**	0.44	**0.50**	0.51	**0.58**	0.57	**0.65**	0.67	**0.76**	0.64	**0.73**
Pancreas	0.16	**0.22**	0.21	**0.29**	0.25	**0.35**	0.29	**0.39**	0.35	**0.47**	0.32	**0.44**
Small intestine	0.21	**0.30**	0.27	**0.39**	0.32	**0.45**	0.35	**0.50**	0.42	**0.59**	0.40	**0.56**
Large intestine	0.19	**0.37**	0.25	**0.46**	0.29	**0.53**	0.32	**0.59**	0.38	**0.69**	0.36	**0.66**
Sigma/rectum	0.10	**0.26**	0.13	**0.34**	0.16	**0.39**	0.17	**0.44**	0.21	**0.52**	0.20	**0.49**
Lumbar spine	0.16	**0.20**	0.22	**0.27**	0.26	**0.33**	0.30	**0.38**	0.39	**0.49**	0.34	**0.42**
Pelvis	0.10	**0.34**	0.14	**0.44**	0.16	**0.52**	0.19	**0.59**	0.24	**0.73**	0.21	**0.67**
Adrenals	0.07	**0.09**	0.10	**0.12**	0.12	**0.15**	0.13	**0.17**	0.17	**0.21**	0.15	**0.19**
Kidneys	0.05	**0.08**	0.07	**0.11**	0.08	**0.13**	0.09	**0.15**	0.12	**0.19**	0.10	**0.16**
Urinary bladder	0.46	**0.53**	0.56	**0.64**	0.62	**0.72**	0.68	**0.79**	0.77	**0.90**	0.75	**0.88**
Uterus	0.27	**0.31**	0.35	**0.40**	0.41	**0.47**	0.45	**0.52**	0.54	**0.63**	0.51	**0.59**
Ovaries	0.18	**0.23**	0.23	**0.29**	0.27	**0.35**	0.30	**0.39**	0.36	**0.48**	0.34	**0.44**
Prostate	0.27	**0.35**	0.34	**0.44**	0.39	**0.51**	0.43	**0.57**	0.50	**0.67**	0.49	**0.64**
Testes	0.06	**0.95**	0.07	**1.07**	0.09	**1.13**	0.10	**1.18**	0.11	**1.26**	0.11	**1.25**
Upper arm	0.03	**0.06**	0.05	**0.09**	0.06	**0.10**	0.07	**0.12**	0.08	**0.15**	0.07	**0.13**
Elbow	0.03	**0.07**	0.05	**0.09**	0.06	**0.11**	0.06	**0.13**	0.08	**0.16**	0.07	**0.14**
Forearm	0.02	**0.06**	0.03	**0.08**	0.04	**0.10**	0.05	**0.11**	0.06	**0.15**	0.05	**0.13**
Upper leg	0.03	**0.52**	0.04	**0.66**	0.05	**0.78**	0.06	**0.87**	0.08	**1.05**	0.07	**0.99**
Knee		**0.18**		**0.23**		**0.27**		**0.30**	0.01	**0.36**		**0.34**
Lower leg												
Lymph nodes	0.27	**0.36**	0.33	**0.44**	0.38	**0.51**	0.41	**0.55**	0.48	**0.64**	0.46	**0.61**
Skin	0.12	**0.25**	0.13	**0.27**	0.13	**0.28**	0.14	**0.29**	0.15	**0.31**	0.14	**0.30**
Muscle	0.12	**0.22**	0.15	**0.26**	0.16	**0.29**	0.18	**0.32**	0.20	**0.36**	0.19	**0.35**
Skeleton	0.17	**0.36**	0.21	**0.44**	0.24	**0.51**	0.26	**0.55**	0.30	**0.65**	0.29	**0.62**
Red bone marrow	0.05	**0.10**	0.06	**0.13**	0.07	**0.15**	0.08	**0.16**	0.09	**0.19**	0.09	**0.18**
Whole body dose	0.12	**0.23**	0.15	**0.28**	0.17	**0.31**	0.19	**0.34**	0.21	**0.39**	0.20	**0.37**
ED (ICRP 60)	0.18	**0.43**	0.22	**0.51**	0.25	**0.57**	0.28	**0.61**	0.32	**0.69**	0.31	**0.66**
ED (ICRP 103)	0.18	**0.45**	0.22	**0.53**	0.25	**0.58**	0.27	**0.62**	0.32	**0.70**	0.30	**0.68**
Total filtration (mm) **Al**	**1.5**		**2.5**		**3.5**		**4.5**		**3.5**		**3.5**	
Cu	**0**		**0**		**0**		**0**		**0.1**		**0.2**	

Table 13.24 Normalized organ doses—Full spine ap—1–2 years

Beam projection:	ap											Table cells:
Focus-to-detector distance (cm):	115											(+) = normal = optimal field setting
Tube voltage (kVp):	70											**(−) = bold font = suboptimal field setting**
Total filtrations:	See last table row											Empty table cells correspond to values <0.01
Organ or tissue	**Organ dose D_T: Entrance dose K_E**								**(mSv ∗ mGy^{-1})**			
	(+)	(−)	(+)	(−)	(+)	(−)	(+)	(−)	(+)	(−)	(+)	(−)
Skull	0.21	**0.44**	0.25	**0.52**	0.28	**0.58**	0.30	**0.62**	0.34	**0.70**	0.37	**0.76**
Brain	0.03	**0.06**	0.04	**0.08**	0.04	**0.09**	0.05	**0.10**	0.06	**0.13**	0.07	**0.15**
Oral mucosa	0.34	**0.53**	0.40	**0.62**	0.44	**0.68**	0.47	**0.72**	0.52	**0.81**	0.56	**0.88**
Salivary glands	0.17	**0.27**	0.20	**0.32**	0.23	**0.37**	0.25	**0.40**	0.29	**0.46**	0.32	**0.51**
Cervical spine	0.49	**0.55**	0.62	**0.70**	0.73	**0.81**	0.81	**0.91**	0.96	**1.09**	1.11	**1.26**
Thyroid	0.93	**0.94**	1.04	**1.07**	1.11	**1.14**	1.16	**1.20**	1.25	**1.30**	1.32	**1.38**
Thymus	0.78	**0.82**	0.90	**0.96**	0.99	**1.05**	1.05	**1.12**	1.16	**1.24**	1.25	**1.34**
Thoracic spine	0.20	**0.25**	0.26	**0.33**	0.32	**0.39**	0.36	**0.45**	0.45	**0.56**	0.53	**0.67**
Claviculae	0.79	**1.56**	0.92	**1.82**	1.01	**1.99**	1.07	**2.12**	1.18	**2.32**	1.26	**2.47**
Scapulae	0.06	**0.19**	0.08	**0.24**	0.10	**0.29**	0.11	**0.32**	0.14	**0.39**	0.17	**0.45**
Ribs	0.61	**1.04**	0.68	**1.18**	0.73	**1.27**	0.77	**1.33**	0.83	**1.44**	0.88	**1.53**
Oesophagus	0.21	**0.25**	0.27	**0.32**	0.31	**0.38**	0.35	**0.42**	0.42	**0.51**	0.49	**0.59**
Trachea	0.52	**0.56**	0.61	**0.66**	0.67	**0.73**	0.71	**0.78**	0.79	**0.87**	0.85	**0.95**
Lung	0.12	**0.35**	0.15	**0.43**	0.18	**0.48**	0.20	**0.52**	0.23	**0.60**	0.26	**0.67**
Mamma	0.06	**1.17**	0.07	**1.24**	0.08	**1.28**	0.09	**1.30**	0.11	**1.34**	0.12	**1.38**
Heart	0.49	**0.57**	0.58	**0.69**	0.65	**0.77**	0.71	**0.83**	0.81	**0.95**	0.89	**1.05**
Stomach	0.23	**0.53**	0.28	**0.64**	0.32	**0.72**	0.35	**0.78**	0.40	**0.88**	0.45	**0.97**
Spleen	0.04	**0.14**	0.06	**0.19**	0.07	**0.22**	0.08	**0.24**	0.10	**0.30**	0.12	**0.35**
Liver	0.26	**0.41**	0.31	**0.50**	0.35	**0.56**	0.38	**0.61**	0.43	**0.69**	0.48	**0.77**
Gall bladder	0.42	**0.47**	0.51	**0.59**	0.59	**0.67**	0.64	**0.74**	0.75	**0.86**	0.84	**0.97**
Pancreas	0.21	**0.28**	0.26	**0.36**	0.30	**0.41**	0.34	**0.46**	0.41	**0.55**	0.47	**0.64**
Small intestine	0.26	**0.37**	0.32	**0.45**	0.37	**0.52**	0.40	**0.58**	0.47	**0.68**	0.54	**0.77**
Large intestine	0.23	**0.43**	0.29	**0.53**	0.33	**0.61**	0.36	**0.67**	0.43	**0.78**	0.48	**0.87**
Sigma/rectum	0.13	**0.32**	0.16	**0.40**	0.18	**0.46**	0.21	**0.50**	0.25	**0.60**	0.28	**0.68**
Lumbar spine	0.22	**0.27**	0.28	**0.36**	0.34	**0.43**	0.38	**0.49**	0.48	**0.61**	0.57	**0.73**
Pelvis	0.13	**0.42**	0.17	**0.54**	0.21	**0.63**	0.23	**0.71**	0.29	**0.86**	0.35	**0.99**
Adrenals	0.09	**0.12**	0.12	**0.16**	0.15	**0.19**	0.17	**0.21**	0.20	**0.26**	0.24	**0.31**
Kidneys	0.07	**0.11**	0.09	**0.14**	0.10	**0.17**	0.12	**0.19**	0.15	**0.23**	0.17	**0.28**
Urinary bladder	0.52	**0.60**	0.62	**0.72**	0.69	**0.81**	0.74	**0.87**	0.83	**0.99**	0.91	**1.09**
Uterus	0.33	**0.38**	0.41	**0.47**	0.47	**0.55**	0.52	**0.61**	0.61	**0.72**	0.69	**0.82**
Ovaries	0.22	**0.28**	0.28	**0.36**	0.32	**0.41**	0.35	**0.46**	0.42	**0.55**	0.48	**0.64**
Prostate	0.32	**0.42**	0.39	**0.52**	0.44	**0.59**	0.48	**0.65**	0.56	**0.75**	0.62	**0.85**
Testes	0.07	**1.01**	0.09	**1.12**	0.10	**1.19**	0.11	**1.23**	0.13	**1.31**	0.14	**1.37**
Upper arm	0.05	**0.08**	0.06	**0.11**	0.07	**0.13**	0.08	**0.15**	0.10	**0.18**	0.13	**0.22**
Elbow	0.05	**0.09**	0.06	**0.12**	0.07	**0.14**	0.08	**0.16**	0.10	**0.20**	0.13	**0.24**
Forearm	0.03	**0.08**	0.04	**0.11**	0.05	**0.13**	0.06	**0.15**	0.08	**0.18**	0.09	**0.22**
Upper leg	0.04	**0.62**	0.06	**0.79**	0.07	**0.91**	0.08	**1.01**	0.10	**1.19**	0.12	**1.36**
Knee		**0.22**		**0.27**	0.01	**0.31**	0.01	**0.35**	0.02	**0.41**	0.02	**0.47**
Lower leg												
Lymph nodes	0.31	**0.42**	0.38	**0.51**	0.43	**0.57**	0.46	**0.62**	0.53	**0.72**	0.59	**0.80**
Skin	0.13	**0.27**	0.13	**0.28**	0.14	**0.29**	0.15	**0.30**	0.16	**0.32**	0.16	**0.34**
Muscle	0.14	**0.25**	0.16	**0.29**	0.18	**0.33**	0.20	**0.35**	0.23	**0.40**	0.25	**0.44**
Skeleton	0.20	**0.42**	0.24	**0.51**	0.27	**0.57**	0.29	**0.62**	0.34	**0.72**	0.38	**0.81**
Red bone marrow	0.06	**0.13**	0.08	**0.15**	0.09	**0.17**	0.09	**0.19**	0.11	**0.23**	0.13	**0.26**
Whole body dose	0.14	**0.26**	0.17	**0.31**	0.19	**0.35**	0.21	**0.38**	0.24	**0.43**	0.27	**0.48**
ED (ICRP 60)	0.21	**0.48**	0.26	**0.56**	0.29	**0.62**	0.31	**0.67**	0.36	**0.75**	0.40	**0.82**
ED (ICRP 103)	0.21	**0.50**	0.25	**0.58**	0.28	**0.64**	0.31	**0.68**	0.35	**0.76**	0.39	**0.83**
Total filtration (mm) Al	**1.5**		**2.5**		**3.5**		**4.5**		**3.5**		**3.5**	
Cu	**0**		**0**		**0**		**0**		**0.1**		**0.2**	

Table 13.25 Normalized organ doses—Full spine ap—3–7 years

Beam projection:			ap									**Table cells:**
Focus-to-detector distance (cm):			115									(+) = normal font =optimal field setting
Tube voltage (kVp):			60									**(−) = bold font = suboptimal field setting**
Total filtrations:			See last table row									Empty table cells correspond to values <0.01
Organ or tissue	**Organ dose D_T : Entrance dose K_E**								**(mSv ∗ mGy^{-1})**			
	(+)	(−)	(+)	(−)	(+)	(−)	(+)	(−)	(+)	(−)	(+)	(−)
Skull	0.17	**0.55**	0.21	**0.67**	0.23	**0.76**	0.25	**0.83**	0.29	**0.95**	0.28	**0.92**
Brain	0.01	**0.08**	0.01	**0.11**	0.02	**0.13**	0.02	**0.15**	0.03	**0.18**	0.02	**0.16**
Oral mucosa	0.24	**0.38**	0.29	**0.46**	0.32	**0.51**	0.35	**0.56**	0.40	**0.64**	0.38	**0.61**
Salivary glands	0.13	**0.31**	0.16	**0.39**	0.19	**0.44**	0.20	**0.48**	0.24	**0.56**	0.23	**0.53**
Cervical spine	0.35	**0.39**	0.45	**0.51**	0.54	**0.60**	0.60	**0.68**	0.74	**0.84**	0.68	**0.76**
Thyroid	0.74	**0.77**	0.85	**0.90**	0.92	**0.98**	0.97	**1.03**	1.06	**1.14**	1.05	**1.12**
Thymus	0.66	**0.71**	0.79	**0.85**	0.88	**0.95**	0.95	**1.03**	1.08	**1.17**	1.06	**1.14**
Thoracic spine	0.11	**0.14**	0.14	**0.19**	0.18	**0.24**	0.20	**0.27**	0.26	**0.35**	0.22	**0.29**
Claviculae	0.57	**1.33**	0.68	**1.60**	0.76	**1.78**	0.82	**1.92**	0.93	**2.16**	0.92	**2.15**
Scapulae	0.03	**0.13**	0.04	**0.17**	0.05	**0.20**	0.06	**0.23**	0.08	**0.29**	0.07	**0.26**
Ribs	0.50	**0.96**	0.57	**1.11**	0.62	**1.20**	0.65	**1.28**	0.71	**1.40**	0.70	**1.39**
Oesophagus	0.10	**0.12**	0.13	**0.16**	0.15	**0.19**	0.17	**0.22**	0.21	**0.27**	0.19	**0.24**
Trachea	0.30	**0.34**	0.37	**0.43**	0.43	**0.49**	0.47	**0.53**	0.54	**0.62**	0.52	**0.60**
Lung	0.07	**0.28**	0.10	**0.35**	0.11	**0.40**	0.13	**0.44**	0.15	**0.52**	0.14	**0.49**
Mamma	0.03	**1.14**	0.04	**1.22**	0.05	**1.27**	0.05	**1.30**	0.07	**1.34**	0.06	**1.34**
Heart	0.35	**0.44**	0.44	**0.55**	0.50	**0.63**	0.55	**0.70**	0.64	**0.81**	0.61	**0.78**
Stomach	0.16	**0.46**	0.20	**0.56**	0.23	**0.64**	0.26	**0.70**	0.30	**0.82**	0.29	**0.78**
Spleen	0.02	**0.09**	0.03	**0.12**	0.04	**0.15**	0.04	**0.17**	0.06	**0.21**	0.05	**0.19**
Liver	0.18	**0.32**	0.23	**0.40**	0.26	**0.46**	0.28	**0.51**	0.33	**0.60**	0.31	**0.57**
Gall bladder	0.31	**0.36**	0.39	**0.46**	0.46	**0.54**	0.51	**0.60**	0.61	**0.72**	0.58	**0.67**
Pancreas	0.13	**0.19**	0.17	**0.25**	0.20	**0.30**	0.23	**0.34**	0.29	**0.42**	0.26	**0.38**
Small intestine	0.17	**0.27**	0.22	**0.34**	0.26	**0.40**	0.29	**0.45**	0.35	**0.55**	0.33	**0.51**
Large intestine	0.14	**0.33**	0.19	**0.42**	0.22	**0.48**	0.24	**0.54**	0.29	**0.64**	0.27	**0.60**
Sigma/rectum	0.08	**0.22**	0.10	**0.29**	0.12	**0.34**	0.13	**0.38**	0.17	**0.46**	0.15	**0.43**
Lumbar spine	0.11	**0.15**	0.16	**0.20**	0.19	**0.25**	0.22	**0.29**	0.28	**0.37**	0.24	**0.31**
Pelvis	0.07	**0.26**	0.09	**0.35**	0.11	**0.42**	0.13	**0.47**	0.17	**0.59**	0.14	**0.53**
Adrenals	0.05	**0.07**	0.07	**0.10**	0.09	**0.12**	0.10	**0.14**	0.13	**0.18**	0.11	**0.15**
Kidneys	0.03	**0.06**	0.05	**0.08**	0.06	**0.10**	0.07	**0.12**	0.08	**0.15**	0.07	**0.13**
Urinary bladder	0.42	**0.48**	0.51	**0.59**	0.58	**0.68**	0.63	**0.74**	0.73	**0.86**	0.71	**0.83**
Uterus	0.23	**0.27**	0.30	**0.36**	0.35	**0.42**	0.39	**0.47**	0.47	**0.58**	0.44	**0.53**
Ovaries	0.13	**0.20**	0.17	**0.26**	0.21	**0.31**	0.24	**0.35**	0.29	**0.43**	0.26	**0.39**
Prostate	0.24	**0.28**	0.30	**0.37**	0.35	**0.43**	0.39	**0.48**	0.47	**0.58**	0.44	**0.54**
Testes	0.31	**0.91**	0.36	**1.03**	0.39	**1.10**	0.42	**1.15**	0.46	**1.24**	0.46	**1.23**
Upper arm	0.02	**0.05**	0.03	**0.06**	0.03	**0.08**	0.04	**0.09**	0.05	**0.12**	0.04	**0.10**
Elbow	0.02	**0.05**	0.03	**0.07**	0.03	**0.08**	0.04	**0.09**	0.05	**0.12**	0.04	**0.10**
Forearm	0.01	**0.04**	0.02	**0.06**	0.02	**0.07**	0.03	**0.08**	0.04	**0.11**	0.03	**0.09**
Upper leg	0.02	**0.35**	0.03	**0.45**	0.04	**0.54**	0.05	**0.61**	0.06	**0.75**	0.05	**0.69**
Knee		**0.03**		**0.04**		**0.05**		**0.06**		**0.07**		**0.06**
Lower leg												
Lymph nodes	0.22	**0.32**	0.27	**0.39**	0.31	**0.45**	0.34	**0.50**	0.40	**0.59**	0.38	**0.56**
Skin	0.10	**0.25**	0.11	**0.27**	0.11	**0.28**	0.12	**0.28**	0.12	**0.30**	0.12	**0.29**
Muscle	0.09	**0.19**	0.11	**0.22**	0.13	**0.25**	0.14	**0.27**	0.16	**0.32**	0.15	**0.30**
Skeleton	0.13	**0.34**	0.16	**0.41**	0.18	**0.47**	0.20	**0.52**	0.23	**0.61**	0.21	**0.57**
Red bone marrow	0.04	**0.10**	0.05	**0.13**	0.06	**0.15**	0.06	**0.16**	0.08	**0.19**	0.07	**0.17**
Whole body dose	0.10	**0.21**	0.12	**0.26**	0.14	**0.29**	0.15	**0.32**	0.17	**0.37**	0.16	**0.35**
ED (ICRP 60)	0.17	**0.40**	0.21	**0.48**	0.24	**0.53**	0.26	**0.58**	0.30	**0.65**	0.29	**0.63**
ED (ICRP 103)	0.15	**0.43**	0.19	**0.50**	0.22	**0.55**	0.24	**0.60**	0.28	**0.67**	0.26	**0.65**
Total filtration (mm) Al	**1.5**		**2.5**		**3.5**		**4.5**		**3.5**		**3.5**	
Cu	**0**		**0**		**0**		**0**		**0.1**		**0.2**	

Table 13.26 Normalized organ doses—Full spine ap—3–7 years

Beam projection:	ap						**Table cells:**					
Focus-to-detector distance (cm):	115						(+) = normal font = optimal field setting					
Tube voltage (kVp):	70						**(−) = bold font = suboptimal field setting**					
Total filtrations:	See last table row						Empty table cells correspond to values <0.01					
Organ or tissue	**Organ dose D_T: Entrance dose K_E**						**(mSv * mGy^{-1})**					
	(+)	(−)	(+)	(−)	(+)	(−)	(+)	(−)	(+)	(−)	(+)	(−)
Skull	0.20	**0.63**	0.23	**0.75**	0.26	**0.84**	0.28	**0.91**	0.32	**1.04**	0.35	**1.15**
Brain	0.01	**0.11**	0.02	**0.14**	0.02	**0.16**	0.03	**0.18**	0.03	**0.22**	0.04	**0.27**
Oral mucosa	0.27	**0.44**	0.32	**0.52**	0.36	**0.58**	0.39	**0.62**	0.44	**0.71**	0.49	**0.79**
Salivary glands	0.15	**0.36**	0.19	**0.44**	0.21	**0.50**	0.23	**0.54**	0.27	**0.62**	0.31	**0.69**
Cervical spine	0.43	**0.49**	0.55	**0.62**	0.64	**0.73**	0.72	**0.82**	0.87	**1.00**	1.01	**1.16**
Thyroid	0.80	**0.85**	0.91	**0.97**	0.98	**1.05**	1.03	**1.11**	1.12	**1.21**	1.20	**1.30**
Thymus	0.74	**0.79**	0.87	**0.94**	0.97	**1.05**	1.03	**1.13**	1.16	**1.26**	1.26	**1.38**
Thoracic spine	0.15	**0.20**	0.19	**0.26**	0.23	**0.32**	0.27	**0.36**	0.34	**0.46**	0.41	**0.56**
Claviculae	0.64	**1.48**	0.75	**1.74**	0.82	**1.92**	0.88	**2.05**	0.98	**2.28**	1.06	**2.46**
Scapulae	0.05	**0.16**	0.06	**0.21**	0.07	**0.25**	0.08	**0.28**	0.10	**0.34**	0.13	**0.41**
Ribs	0.54	**1.04**	0.61	**1.18**	0.66	**1.28**	0.69	**1.35**	0.75	**1.47**	0.79	**1.56**
Oesophagus	0.12	**0.16**	0.16	**0.20**	0.19	**0.24**	0.21	**0.27**	0.26	**0.33**	0.31	**0.39**
Trachea	0.35	**0.40**	0.43	**0.49**	0.48	**0.55**	0.52	**0.60**	0.60	**0.70**	0.66	**0.78**
Lung	0.09	**0.33**	0.12	**0.40**	0.14	**0.46**	0.15	**0.50**	0.18	**0.58**	0.21	**0.65**
Mamma	0.04	**1.18**	0.05	**1.26**	0.06	**1.30**	0.07	**1.33**	0.08	**1.38**	0.09	**1.42**
Heart	0.41	**0.52**	0.50	**0.64**	0.57	**0.72**	0.62	**0.79**	0.71	**0.91**	0.79	**1.02**
Stomach	0.19	**0.53**	0.23	**0.64**	0.27	**0.73**	0.29	**0.79**	0.35	**0.91**	0.39	**1.01**
Spleen	0.03	**0.12**	0.04	**0.16**	0.05	**0.19**	0.06	**0.21**	0.07	**0.26**	0.09	**0.31**
Liver	0.21	**0.38**	0.26	**0.47**	0.29	**0.53**	0.32	**0.58**	0.37	**0.67**	0.42	**0.75**
Gall bladder	0.37	**0.44**	0.46	**0.55**	0.53	**0.63**	0.59	**0.70**	0.69	**0.83**	0.79	**0.94**
Pancreas	0.17	**0.25**	0.21	**0.32**	0.25	**0.37**	0.28	**0.42**	0.34	**0.51**	0.40	**0.60**
Small intestine	0.21	**0.33**	0.27	**0.41**	0.31	**0.48**	0.34	**0.53**	0.41	**0.64**	0.47	**0.73**
Large intestine	0.18	**0.39**	0.22	**0.49**	0.26	**0.56**	0.29	**0.62**	0.34	**0.74**	0.39	**0.84**
Sigma/rectum	0.10	**0.28**	0.12	**0.35**	0.15	**0.41**	0.16	**0.45**	0.20	**0.54**	0.23	**0.62**
Lumbar spine	0.16	**0.21**	0.21	**0.28**	0.25	**0.33**	0.29	**0.38**	0.37	**0.48**	0.45	**0.59**
Pelvis	0.09	**0.34**	0.12	**0.44**	0.15	**0.52**	0.17	**0.58**	0.22	**0.71**	0.26	**0.84**
Adrenals	0.07	**0.10**	0.10	**0.13**	0.11	**0.16**	0.13	**0.18**	0.16	**0.23**	0.20	**0.27**
Kidneys	0.05	**0.09**	0.06	**0.11**	0.08	**0.14**	0.09	**0.15**	0.11	**0.19**	0.13	**0.24**
Urinary bladder	0.48	**0.56**	0.58	**0.68**	0.65	**0.76**	0.70	**0.83**	0.80	**0.96**	0.89	**1.07**
Uterus	0.28	**0.34**	0.36	**0.43**	0.41	**0.50**	0.46	**0.56**	0.55	**0.67**	0.63	**0.78**
Ovaries	0.17	**0.25**	0.22	**0.32**	0.25	**0.38**	0.28	**0.42**	0.34	**0.51**	0.40	**0.60**
Prostate	0.28	**0.35**	0.36	**0.44**	0.41	**0.51**	0.45	**0.56**	0.53	**0.67**	0.60	**0.77**
Testes	0.33	**0.97**	0.39	**1.09**	0.42	**1.16**	0.45	**1.21**	0.49	**1.30**	0.53	**1.37**
Upper arm	0.03	**0.06**	0.04	**0.09**	0.05	**0.10**	0.05	**0.12**	0.07	**0.15**	0.08	**0.18**
Elbow	0.03	**0.07**	0.04	**0.09**	0.05	**0.11**	0.05	**0.12**	0.07	**0.16**	0.08	**0.19**
Forearm	0.02	**0.06**	0.03	**0.08**	0.03	**0.10**	0.04	**0.11**	0.05	**0.14**	0.06	**0.17**
Upper leg	0.03	**0.44**	0.04	**0.56**	0.05	**0.65**	0.06	**0.73**	0.08	**0.88**	0.09	**1.03**
Knee		**0.04**		**0.05**		**0.06**		**0.07**		**0.09**		**0.11**
Lower leg												
Lymph nodes	0.26	**0.37**	0.31	**0.46**	0.36	**0.52**	0.39	**0.57**	0.46	**0.67**	0.51	**0.76**
Skin	0.11	**0.26**	0.12	**0.28**	0.12	**0.29**	0.12	**0.30**	0.13	**0.32**	0.14	**0.33**
Muscle	0.11	**0.21**	0.13	**0.26**	0.15	**0.29**	0.16	**0.31**	0.18	**0.35**	0.21	**0.40**
Skeleton	0.15	**0.39**	0.18	**0.48**	0.21	**0.54**	0.22	**0.59**	0.26	**0.68**	0.29	**0.76**
Red bone marrow	0.05	**0.12**	0.06	**0.15**	0.07	**0.17**	0.08	**0.19**	0.09	**0.23**	0.11	**0.26**
Whole body dose	0.12	**0.24**	0.14	**0.29**	0.16	**0.33**	0.17	**0.36**	0.20	**0.41**	0.22	**0.46**
ED (ICRP 60)	0.20	**0.45**	0.24	**0.54**	0.27	**0.59**	0.30	**0.64**	0.34	**0.72**	0.38	**0.80**
ED (ICRP 103)	0.18	**0.48**	0.22	**0.56**	0.25	**0.61**	0.27	**0.66**	0.31	**0.74**	0.35	**0.81**
Total filtration (mm) Al	**1.5**		**2.5**		**3.5**		**4.5**		**3.5**		**3.5**	
Cu	**0**		**0**		**0**		**0**		**0.1**		**0.2**	

Table 13.27 Normalized organ doses—Full spine ap—3–7 years

Beam projection:			ap									**Table cells:**
Focus-to-detector distance (cm):			115						(+) = normal font = optimal field setting			
Tube voltage (kVp):			80						**(−) = bold font = suboptimal field setting**			
Total filtrations:			See last table row						Empty table cells correspond to values <0.01			
Organ or tissue	**Organ dose D_T: Entrance dose K_E**								**(mSv * mGy^{-1})**			
	(+)	(−)	(+)	(−)	(+)	(−)	(+)	(−)	(+)	(−)	(+)	(−)
Skull	0.22	**0.70**	0.25	**0.82**	0.28	**0.91**	0.30	**0.98**	0.34	**1.10**	0.37	**1.20**
Brain	0.02	**0.13**	0.03	**0.17**	0.03	**0.19**	0.03	**0.22**	0.04	**0.26**	0.05	**0.31**
Oral mucosa	0.31	**0.49**	0.36	**0.57**	0.39	**0.63**	0.42	**0.68**	0.48	**0.77**	0.52	**0.85**
Salivary glands	0.18	**0.41**	0.21	**0.49**	0.24	**0.55**	0.26	**0.59**	0.30	**0.67**	0.33	**0.74**
Cervical spine	0.51	**0.59**	0.64	**0.73**	0.74	**0.85**	0.81	**0.94**	0.96	**1.12**	1.10	**1.28**
Thyroid	0.86	**0.91**	0.96	**1.03**	1.03	**1.11**	1.08	**1.17**	1.17	**1.28**	1.24	**1.36**
Thymus	0.81	**0.88**	0.94	**1.03**	1.04	**1.13**	1.10	**1.20**	1.22	**1.34**	1.31	**1.45**
Thoracic spine	0.19	**0.26**	0.25	**0.33**	0.29	**0.39**	0.33	**0.44**	0.40	**0.55**	0.48	**0.65**
Claviculae	0.69	**1.59**	0.80	**1.85**	0.87	**2.02**	0.92	**2.14**	1.01	**2.34**	1.08	**2.48**
Scapulae	0.06	**0.20**	0.07	**0.25**	0.09	**0.29**	0.10	**0.32**	0.12	**0.39**	0.15	**0.45**
Ribs	0.57	**1.10**	0.64	**1.24**	0.68	**1.33**	0.71	**1.39**	0.77	**1.50**	0.81	**1.58**
Oesophagus	0.15	**0.20**	0.19	**0.25**	0.22	**0.29**	0.25	**0.32**	0.30	**0.39**	0.35	**0.45**
Trachea	0.40	**0.46**	0.47	**0.55**	0.53	**0.61**	0.57	**0.66**	0.65	**0.76**	0.71	**0.84**
Lung	0.11	**0.38**	0.14	**0.45**	0.16	**0.51**	0.17	**0.55**	0.21	**0.63**	0.23	**0.70**
Mamma	0.05	**1.22**	0.06	**1.29**	0.07	**1.33**	0.08	**1.36**	0.09	**1.41**	0.11	**1.45**
Heart	0.47	**0.59**	0.56	**0.71**	0.62	**0.80**	0.68	**0.87**	0.77	**0.99**	0.85	**1.10**
Stomach	0.22	**0.60**	0.27	**0.72**	0.30	**0.80**	0.33	**0.86**	0.38	**0.98**	0.43	**1.08**
Spleen	0.04	**0.15**	0.05	**0.19**	0.06	**0.23**	0.07	**0.25**	0.09	**0.31**	0.11	**0.36**
Liver	0.24	**0.44**	0.29	**0.53**	0.33	**0.59**	0.36	**0.64**	0.41	**0.73**	0.45	**0.82**
Gall bladder	0.43	**0.51**	0.53	**0.63**	0.60	**0.72**	0.66	**0.79**	0.76	**0.92**	0.85	**1.04**
Pancreas	0.20	**0.30**	0.25	**0.38**	0.29	**0.44**	0.33	**0.49**	0.39	**0.58**	0.45	**0.67**
Small intestine	0.25	**0.39**	0.31	**0.48**	0.35	**0.55**	0.39	**0.61**	0.46	**0.71**	0.52	**0.81**
Large intestine	0.21	**0.46**	0.26	**0.56**	0.30	**0.64**	0.33	**0.70**	0.38	**0.82**	0.43	**0.92**
Sigma/rectum	0.12	**0.33**	0.15	**0.41**	0.17	**0.47**	0.19	**0.52**	0.23	**0.61**	0.26	**0.69**
Lumbar spine	0.21	**0.27**	0.27	**0.35**	0.31	**0.42**	0.35	**0.47**	0.44	**0.58**	0.52	**0.70**
Pelvis	0.12	**0.41**	0.16	**0.52**	0.18	**0.61**	0.21	**0.68**	0.26	**0.81**	0.30	**0.94**
Adrenals	0.09	**0.13**	0.12	**0.17**	0.14	**0.20**	0.16	**0.22**	0.20	**0.27**	0.23	**0.32**
Kidneys	0.06	**0.11**	0.08	**0.14**	0.09	**0.17**	0.11	**0.19**	0.13	**0.24**	0.16	**0.28**
Urinary bladder	0.54	**0.63**	0.64	**0.75**	0.71	**0.84**	0.76	**0.91**	0.86	**1.04**	0.94	**1.14**
Uterus	0.33	**0.41**	0.41	**0.51**	0.47	**0.58**	0.52	**0.64**	0.61	**0.76**	0.69	**0.86**
Ovaries	0.20	**0.31**	0.26	**0.38**	0.30	**0.44**	0.33	**0.49**	0.39	**0.59**	0.45	**0.68**
Prostate	0.33	**0.41**	0.41	**0.51**	0.46	**0.58**	0.50	**0.64**	0.59	**0.75**	0.66	**0.84**
Testes	0.36	**1.03**	0.41	**1.14**	0.45	**1.21**	0.47	**1.26**	0.52	**1.34**	0.55	**1.41**
Upper arm	0.04	**0.08**	0.05	**0.11**	0.06	**0.13**	0.07	**0.14**	0.08	**0.18**	0.10	**0.21**
Elbow	0.04	**0.09**	0.05	**0.11**	0.06	**0.13**	0.07	**0.15**	0.08	**0.19**	0.10	**0.22**
Forearm	0.03	**0.08**	0.04	**0.10**	0.04	**0.12**	0.05	**0.14**	0.06	**0.17**	0.07	**0.20**
Upper leg	0.04	**0.52**	0.06	**0.65**	0.07	**0.75**	0.07	**0.83**	0.09	**0.98**	0.11	**1.13**
Knee		**0.05**		**0.07**		**0.08**		**0.09**		**0.11**	0.01	**0.13**
Lower leg												
Lymph nodes	0.29	**0.43**	0.35	**0.52**	0.40	**0.59**	0.43	**0.64**	0.50	**0.74**	0.56	**0.83**
Skin	0.11	**0.27**	0.12	**0.29**	0.13	**0.30**	0.13	**0.31**	0.14	**0.33**	0.15	**0.35**
Muscle	0.12	**0.24**	0.15	**0.29**	0.16	**0.32**	0.18	**0.34**	0.20	**0.39**	0.22	**0.43**
Skeleton	0.17	**0.44**	0.20	**0.53**	0.23	**0.59**	0.25	**0.64**	0.28	**0.73**	0.32	**0.81**
Red bone marrow	0.06	**0.15**	0.07	**0.18**	0.08	**0.20**	0.09	**0.22**	0.11	**0.26**	0.12	**0.30**
Whole body dose	0.13	**0.28**	0.16	**0.33**	0.17	**0.37**	0.19	**0.39**	0.22	**0.45**	0.24	**0.50**
ED (ICRP 60)	0.23	**0.51**	0.27	**0.59**	0.30	**0.65**	0.33	**0.69**	0.37	**0.78**	0.41	**0.85**
ED (ICRP 103)	0.21	**0.53**	0.25	**0.61**	0.28	**0.67**	0.30	**0.71**	0.35	**0.80**	0.38	**0.87**
Total filtration (mm) **Al**	**1.5**		**2.5**		**3.5**		**4.5**		**3.5**		**3.5**	
Cu	**0**		**0**		**0**		**0**		**0.1**		**0.2**	

Table 13.28 Normalized organ doses—Full spine ap—3–7 years

Beam projection:		ap										**Table cells:**
Focus-to-detector distance (cm):		200										(+) = normal font = optimal field setting
Tube voltage (kVp):		60										**(−) = bold font = suboptimal field setting**
Total filtrations:		See last table row										Empty table cells correspond to values <0.01
Organ or tissue	**Organ dose D_T: Entrance dose K_E**							**(mSv * mGy^{-1})**				
	(+)	(−)	(+)	(−)	(+)	(−)	(+)	(−)	(+)	(−)	(+)	(−)
Skull	0.17	**0.56**	0.21	**0.69**	0.23	**0.78**	0.25	**0.85**	0.29	**0.97**	0.28	**0.94**
Brain		**0.08**	0.01	**0.11**	0.01	**0.13**	0.02	**0.15**	0.02	**0.18**	0.02	**0.16**
Oral mucosa	0.24	**0.39**	0.29	**0.47**	0.33	**0.53**	0.35	**0.57**	0.40	**0.66**	0.38	**0.63**
Salivary glands	0.11	**0.31**	0.14	**0.38**	0.16	**0.43**	0.18	**0.47**	0.21	**0.55**	0.20	**0.52**
Cervical spine	0.36	**0.40**	0.47	**0.52**	0.56	**0.62**	0.63	**0.70**	0.77	**0.87**	0.71	**0.78**
Thyroid	0.84	**0.87**	0.96	**1.00**	1.04	**1.08**	1.09	**1.15**	1.19	**1.25**	1.17	**1.23**
Thymus	0.66	**0.70**	0.80	**0.85**	0.89	**0.95**	0.96	**1.02**	1.08	**1.16**	1.06	**1.13**
Thoracic spine	0.11	**0.15**	0.15	**0.20**	0.19	**0.25**	0.22	**0.29**	0.28	**0.37**	0.23	**0.31**
Claviculae	0.57	**1.32**	0.68	**1.58**	0.76	**1.77**	0.82	**1.90**	0.92	**2.14**	0.91	**2.13**
Scapulae	0.03	**0.11**	0.04	**0.15**	0.05	**0.18**	0.06	**0.21**	0.08	**0.26**	0.07	**0.23**
Ribs	0.49	**0.93**	0.56	**1.08**	0.61	**1.17**	0.64	**1.24**	0.70	**1.37**	0.69	**1.35**
Oesophagus	0.10	**0.13**	0.13	**0.17**	0.16	**0.20**	0.18	**0.23**	0.22	**0.28**	0.20	**0.25**
Trachea	0.31	**0.34**	0.38	**0.43**	0.43	**0.49**	0.47	**0.54**	0.55	**0.63**	0.53	**0.60**
Lung	0.07	**0.28**	0.09	**0.35**	0.11	**0.40**	0.12	**0.44**	0.15	**0.52**	0.14	**0.49**
Mamma	0.03	**1.12**	0.04	**1.20**	0.05	**1.25**	0.05	**1.27**	0.07	**1.32**	0.06	**1.31**
Heart	0.35	**0.44**	0.44	**0.55**	0.50	**0.63**	0.55	**0.70**	0.64	**0.82**	0.61	**0.78**
Stomach	0.16	**0.45**	0.20	**0.55**	0.23	**0.63**	0.25	**0.69**	0.30	**0.80**	0.28	**0.77**
Spleen	0.02	**0.09**	0.03	**0.12**	0.04	**0.14**	0.04	**0.16**	0.06	**0.21**	0.05	**0.18**
Liver	0.18	**0.32**	0.22	**0.40**	0.26	**0.46**	0.28	**0.50**	0.33	**0.59**	0.31	**0.56**
Gall bladder	0.32	**0.36**	0.40	**0.47**	0.47	**0.54**	0.52	**0.61**	0.62	**0.73**	0.59	**0.68**
Pancreas	0.13	**0.20**	0.17	**0.26**	0.21	**0.31**	0.23	**0.35**	0.29	**0.43**	0.26	**0.39**
Small intestine	0.18	**0.28**	0.23	**0.35**	0.27	**0.42**	0.30	**0.47**	0.36	**0.56**	0.34	**0.52**
Large intestine	0.14	**0.34**	0.18	**0.43**	0.22	**0.50**	0.24	**0.55**	0.29	**0.66**	0.27	**0.62**
Sigma/rectum	0.08	**0.24**	0.10	**0.31**	0.12	**0.36**	0.14	**0.41**	0.17	**0.49**	0.15	**0.46**
Lumbar spine	0.12	**0.16**	0.16	**0.21**	0.20	**0.26**	0.23	**0.30**	0.30	**0.39**	0.25	**0.32**
Pelvis	0.07	**0.26**	0.09	**0.35**	0.12	**0.42**	0.13	**0.48**	0.17	**0.60**	0.14	**0.54**
Adrenals	0.06	**0.08**	0.08	**0.10**	0.09	**0.12**	0.11	**0.14**	0.14	**0.18**	0.12	**0.15**
Kidneys	0.03	**0.07**	0.05	**0.09**	0.06	**0.11**	0.07	**0.12**	0.08	**0.16**	0.07	**0.13**
Urinary bladder	0.44	**0.50**	0.54	**0.62**	0.61	**0.70**	0.66	**0.77**	0.76	**0.89**	0.74	**0.86**
Uterus	0.25	**0.29**	0.32	**0.37**	0.38	**0.44**	0.42	**0.49**	0.51	**0.59**	0.48	**0.55**
Ovaries	0.13	**0.21**	0.17	**0.28**	0.21	**0.33**	0.23	**0.37**	0.29	**0.46**	0.26	**0.41**
Prostate	0.25	**0.31**	0.32	**0.39**	0.37	**0.46**	0.42	**0.52**	0.50	**0.62**	0.47	**0.58**
Testes	0.17	**1.03**	0.21	**1.15**	0.23	**1.22**	0.25	**1.28**	0.28	**1.37**	0.27	**1.35**
Upper arm	0.02	**0.05**	0.03	**0.06**	0.03	**0.08**	0.04	**0.09**	0.05	**0.11**	0.04	**0.09**
Elbow	0.02	**0.05**	0.03	**0.06**	0.03	**0.08**	0.04	**0.09**	0.05	**0.12**	0.04	**0.10**
Forearm	0.01	**0.04**	0.02	**0.06**	0.02	**0.07**	0.03	**0.08**	0.04	**0.11**	0.03	**0.09**
Upper leg	0.02	**0.35**	0.03	**0.46**	0.03	**0.55**	0.04	**0.62**	0.05	**0.76**	0.04	**0.70**
Knee		**0.02**		**0.02**		**0.03**		**0.03**		**0.04**		**0.03**
Lower leg												
Lymph nodes	0.22	**0.32**	0.27	**0.40**	0.32	**0.46**	0.35	**0.51**	0.41	**0.60**	0.39	**0.57**
Skin	0.10	**0.24**	0.11	**0.26**	0.11	**0.27**	0.12	**0.28**	0.12	**0.29**	0.12	**0.29**
Muscle	0.09	**0.19**	0.11	**0.23**	0.13	**0.25**	0.14	**0.28**	0.16	**0.32**	0.15	**0.30**
Skeleton	0.13	**0.34**	0.16	**0.41**	0.18	**0.47**	0.20	**0.52**	0.23	**0.61**	0.22	**0.57**
Red bone marrow	0.04	**0.10**	0.05	**0.13**	0.06	**0.15**	0.06	**0.16**	0.08	**0.19**	0.07	**0.18**
Whole body dose	0.10	**0.21**	0.12	**0.26**	0.14	**0.29**	0.15	**0.32**	0.17	**0.37**	0.16	**0.35**
ED (ICRP 60)	0.16	**0.42**	0.20	**0.50**	0.23	**0.55**	0.25	**0.60**	0.29	**0.68**	0.28	**0.65**
ED (ICRP 103)	0.15	**0.43**	0.19	**0.51**	0.22	**0.57**	0.24	**0.61**	0.28	**0.68**	0.26	**0.66**
Total filtration (mm) Al	**1.5**		**2.5**		**3.5**		**4.5**		**3.5**		**3.5**	
Cu	**0**		**0**		**0**		**0**		**0.1**		**0.2**	

Table 13.29 Normalized organ doses—Full spine ap—3–7 years

Beam projection:	ap											**Table cells:**
Focus-to-detector distance (cm):	200											(+) = normal font = optimal field setting
Tube voltage (kVp):	70											**(−) = bold font = suboptimal field setting**
Total filtrations:	See last table row											Empty table cells correspond to values <0.01
Organ or tissue	**Organ dose D_T: Entrance dose K_E**							**(mSv * mGy^{-1})**				
	(+)	(−)	(+)	(−)	(+)	(−)	(+)	(−)	(+)	(−)	(+)	(−)
Skull	0.20	**0.64**	0.23	**0.77**	0.26	**0.86**	0.28	**0.93**	0.32	**1.06**	0.35	**1.17**
Brain	0.01	**0.11**	0.02	**0.14**	0.02	**0.16**	0.02	**0.18**	0.03	**0.23**	0.03	**0.27**
Oral mucosa	0.28	**0.45**	0.33	**0.53**	0.36	**0.60**	0.39	**0.64**	0.45	**0.73**	0.49	**0.81**
Salivary glands	0.13	**0.36**	0.16	**0.43**	0.19	**0.49**	0.21	**0.53**	0.24	**0.61**	0.28	**0.68**
Cervical spine	0.45	**0.50**	0.57	**0.65**	0.67	**0.76**	0.75	**0.85**	0.90	**1.03**	1.04	**1.20**
Thyroid	0.91	**0.94**	1.02	**1.07**	1.10	**1.16**	1.15	**1.22**	1.25	**1.33**	1.32	**1.42**
Thymus	0.74	**0.79**	0.88	**0.94**	0.97	**1.04**	1.04	**1.12**	1.16	**1.25**	1.26	**1.37**
Thoracic spine	0.16	**0.21**	0.21	**0.27**	0.25	**0.33**	0.28	**0.38**	0.36	**0.48**	0.44	**0.58**
Claviculae	0.63	**1.46**	0.74	**1.72**	0.82	**1.90**	0.88	**2.03**	0.97	**2.26**	1.05	**2.44**
Scapulae	0.05	**0.15**	0.06	**0.19**	0.07	**0.23**	0.08	**0.26**	0.10	**0.31**	0.13	**0.37**
Ribs	0.53	**1.01**	0.60	**1.15**	0.65	**1.25**	0.68	**1.31**	0.74	**1.43**	0.78	**1.53**
Oesophagus	0.13	**0.16**	0.17	**0.21**	0.20	**0.25**	0.22	**0.28**	0.27	**0.35**	0.32	**0.41**
Trachea	0.35	**0.40**	0.43	**0.49**	0.48	**0.56**	0.53	**0.61**	0.60	**0.71**	0.67	**0.79**
Lung	0.09	**0.33**	0.11	**0.40**	0.13	**0.46**	0.15	**0.50**	0.18	**0.58**	0.21	**0.65**
Mamma	0.04	**1.16**	0.05	**1.24**	0.06	**1.28**	0.07	**1.31**	0.08	**1.35**	0.09	**1.39**
Heart	0.41	**0.52**	0.50	**0.64**	0.57	**0.72**	0.62	**0.79**	0.71	**0.91**	0.80	**1.02**
Stomach	0.19	**0.52**	0.23	**0.63**	0.26	**0.71**	0.29	**0.78**	0.34	**0.89**	0.39	**1.00**
Spleen	0.03	**0.12**	0.04	**0.15**	0.05	**0.18**	0.06	**0.21**	0.07	**0.26**	0.09	**0.30**
Liver	0.21	**0.38**	0.26	**0.46**	0.29	**0.52**	0.32	**0.57**	0.37	**0.66**	0.42	**0.75**
Gall bladder	0.38	**0.44**	0.47	**0.55**	0.55	**0.64**	0.60	**0.71**	0.71	**0.84**	0.80	**0.96**
Pancreas	0.17	**0.25**	0.22	**0.32**	0.25	**0.38**	0.29	**0.43**	0.35	**0.52**	0.41	**0.61**
Small intestine	0.22	**0.34**	0.27	**0.43**	0.32	**0.49**	0.35	**0.55**	0.42	**0.65**	0.48	**0.75**
Large intestine	0.18	**0.41**	0.22	**0.51**	0.26	**0.58**	0.29	**0.64**	0.34	**0.76**	0.39	**0.86**
Sigma/rectum	0.10	**0.29**	0.13	**0.37**	0.15	**0.43**	0.17	**0.48**	0.20	**0.58**	0.24	**0.66**
Lumbar spine	0.17	**0.22**	0.22	**0.29**	0.27	**0.35**	0.31	**0.40**	0.39	**0.51**	0.47	**0.62**
Pelvis	0.10	**0.34**	0.13	**0.45**	0.15	**0.53**	0.17	**0.59**	0.22	**0.73**	0.27	**0.86**
Adrenals	0.08	**0.10**	0.10	**0.14**	0.12	**0.16**	0.14	**0.19**	0.17	**0.23**	0.21	**0.28**
Kidneys	0.05	**0.09**	0.06	**0.12**	0.08	**0.14**	0.09	**0.16**	0.11	**0.20**	0.13	**0.25**
Urinary bladder	0.50	**0.58**	0.60	**0.70**	0.68	**0.79**	0.73	**0.86**	0.84	**0.99**	0.92	**1.11**
Uterus	0.30	**0.35**	0.38	**0.45**	0.44	**0.52**	0.49	**0.58**	0.58	**0.70**	0.67	**0.80**
Ovaries	0.17	**0.27**	0.21	**0.34**	0.25	**0.40**	0.28	**0.45**	0.34	**0.54**	0.39	**0.64**
Prostate	0.30	**0.37**	0.38	**0.47**	0.44	**0.54**	0.48	**0.60**	0.57	**0.72**	0.64	**0.82**
Testes	0.19	**1.09**	0.23	**1.21**	0.25	**1.28**	0.27	**1.34**	0.30	**1.43**	0.33	**1.50**
Upper arm	0.03	**0.06**	0.04	**0.08**	0.05	**0.10**	0.05	**0.12**	0.07	**0.15**	0.09	**0.18**
Elbow	0.03	**0.07**	0.04	**0.09**	0.05	**0.11**	0.05	**0.12**	0.07	**0.15**	0.08	**0.19**
Forearm	0.02	**0.06**	0.03	**0.08**	0.03	**0.10**	0.04	**0.11**	0.05	**0.14**	0.06	**0.17**
Upper leg	0.03	**0.44**	0.04	**0.56**	0.05	**0.66**	0.05	**0.74**	0.07	**0.89**	0.08	**1.04**
Knee		**0.02**		**0.03**		**0.03**		**0.04**		**0.05**		**0.06**
Lower leg												
Lymph nodes	0.26	**0.38**	0.32	**0.47**	0.36	**0.54**	0.40	**0.59**	0.46	**0.69**	0.52	**0.77**
Skin	0.11	**0.26**	0.11	**0.27**	0.12	**0.28**	0.12	**0.29**	0.13	**0.31**	0.14	**0.33**
Muscle	0.11	**0.21**	0.13	**0.26**	0.15	**0.29**	0.16	**0.31**	0.18	**0.36**	0.21	**0.40**
Skeleton	0.15	**0.39**	0.18	**0.48**	0.21	**0.54**	0.22	**0.59**	0.26	**0.68**	0.29	**0.76**
Red bone marrow	0.05	**0.12**	0.06	**0.15**	0.07	**0.18**	0.08	**0.19**	0.09	**0.23**	0.11	**0.27**
Whole body dose	0.12	**0.24**	0.14	**0.29**	0.16	**0.33**	0.17	**0.36**	0.20	**0.41**	0.22	**0.46**
ED (ICRP 60)	0.19	**0.47**	0.23	**0.56**	0.26	**0.62**	0.29	**0.66**	0.33	**0.75**	0.37	**0.82**
ED (ICRP 103)	0.18	**0.49**	0.22	**0.57**	0.25	**0.62**	0.27	**0.67**	0.31	**0.75**	0.35	**0.83**
Total filtration (mm) Al	**1.5**		**2.5**		**3.5**		**4.5**		**3.5**		**3.5**	
Cu	**0**		**0**		**0**		**0**		**0.1**		**0.2**	

Table 13.30 Normalized organ doses—Full spine ap—3–7 years

Beam projection:					ap							**Table cells:**
Focus-to-detector distance (cm):					200							(+) = normal font = optimal field setting
Tube voltage (kVp):					80							**(−) = bold font = suboptimal field setting**
Total filtrations:					See last table row							Empty table cells correspond to values <0.01
Organ or tissue	**Organ dose D_T: Entrance dose K_E**							**(mSv ∗ mGy⁻¹)**				
	(+)	(−)	(+)	(−)	(+)	(−)	(+)	(−)	(+)	(−)	(+)	(−)
Skull	0.22	**0.71**	0.25	**0.84**	0.28	**0.93**	0.30	**1.00**	0.33	**1.12**	0.36	**1.22**
Brain	0.02	**0.13**	0.02	**0.17**	0.02	**0.20**	0.03	**0.22**	0.03	**0.26**	0.04	**0.31**
Oral mucosa	0.31	**0.51**	0.36	**0.59**	0.40	**0.66**	0.43	**0.70**	0.48	**0.80**	0.53	**0.88**
Salivary glands	0.15	**0.41**	0.19	**0.48**	0.21	**0.54**	0.23	**0.59**	0.27	**0.67**	0.30	**0.74**
Cervical spine	0.53	**0.61**	0.66	**0.76**	0.76	**0.87**	0.84	**0.97**	1.00	**1.15**	1.14	**1.32**
Thyroid	0.97	**1.01**	1.08	**1.14**	1.15	**1.22**	1.21	**1.28**	1.30	**1.39**	1.37	**1.47**
Thymus	0.82	**0.87**	0.95	**1.02**	1.04	**1.12**	1.11	**1.20**	1.22	**1.33**	1.32	**1.44**
Thoracic spine	0.20	**0.27**	0.26	**0.35**	0.31	**0.41**	0.35	**0.46**	0.43	**0.57**	0.51	**0.68**
Claviculae	0.68	**1.58**	0.79	**1.83**	0.86	**2.00**	0.92	**2.12**	1.00	**2.32**	1.07	**2.46**
Scapulae	0.06	**0.18**	0.08	**0.23**	0.09	**0.27**	0.10	**0.30**	0.12	**0.36**	0.15	**0.41**
Ribs	0.56	**1.08**	0.63	**1.21**	0.67	**1.30**	0.70	**1.36**	0.76	**1.47**	0.80	**1.55**
Oesophagus	0.16	**0.20**	0.20	**0.26**	0.23	**0.30**	0.26	**0.33**	0.31	**0.40**	0.36	**0.46**
Trachea	0.40	**0.46**	0.48	**0.55**	0.53	**0.62**	0.57	**0.67**	0.65	**0.77**	0.72	**0.85**
Lung	0.11	**0.38**	0.14	**0.45**	0.16	**0.51**	0.17	**0.55**	0.20	**0.63**	0.23	**0.70**
Mamma	0.05	**1.20**	0.06	**1.27**	0.07	**1.31**	0.08	**1.34**	0.09	**1.38**	0.11	**1.42**
Heart	0.47	**0.59**	0.56	**0.71**	0.63	**0.80**	0.68	**0.87**	0.77	**0.99**	0.86	**1.10**
Stomach	0.22	**0.59**	0.26	**0.71**	0.30	**0.79**	0.33	**0.85**	0.38	**0.97**	0.42	**1.07**
Spleen	0.04	**0.15**	0.05	**0.19**	0.06	**0.22**	0.07	**0.25**	0.09	**0.30**	0.11	**0.35**
Liver	0.24	**0.43**	0.29	**0.52**	0.33	**0.58**	0.35	**0.63**	0.40	**0.73**	0.45	**0.81**
Gall bladder	0.44	**0.52**	0.54	**0.64**	0.61	**0.73**	0.67	**0.80**	0.78	**0.93**	0.87	**1.05**
Pancreas	0.21	**0.31**	0.26	**0.39**	0.30	**0.45**	0.33	**0.50**	0.40	**0.59**	0.46	**0.69**
Small intestine	0.26	**0.40**	0.32	**0.50**	0.36	**0.57**	0.40	**0.62**	0.47	**0.73**	0.53	**0.83**
Large intestine	0.21	**0.47**	0.26	**0.58**	0.29	**0.66**	0.32	**0.72**	0.38	**0.84**	0.43	**0.94**
Sigma/rectum	0.12	**0.35**	0.15	**0.43**	0.17	**0.50**	0.19	**0.55**	0.23	**0.65**	0.27	**0.73**
Lumbar spine	0.22	**0.29**	0.28	**0.37**	0.33	**0.44**	0.37	**0.49**	0.46	**0.61**	0.55	**0.73**
Pelvis	0.12	**0.42**	0.16	**0.53**	0.19	**0.62**	0.21	**0.69**	0.26	**0.83**	0.31	**0.97**
Adrenals	0.10	**0.13**	0.13	**0.17**	0.15	**0.20**	0.17	**0.23**	0.21	**0.28**	0.24	**0.33**
Kidneys	0.06	**0.12**	0.08	**0.15**	0.10	**0.18**	0.11	**0.20**	0.13	**0.25**	0.16	**0.29**
Urinary bladder	0.56	**0.66**	0.67	**0.78**	0.74	**0.88**	0.80	**0.95**	0.90	**1.07**	0.98	**1.18**
Uterus	0.36	**0.42**	0.44	**0.53**	0.50	**0.60**	0.55	**0.66**	0.65	**0.78**	0.73	**0.89**
Ovaries	0.20	**0.33**	0.25	**0.41**	0.29	**0.47**	0.32	**0.52**	0.38	**0.62**	0.44	**0.71**
Prostate	0.35	**0.44**	0.43	**0.54**	0.49	**0.62**	0.54	**0.68**	0.62	**0.80**	0.70	**0.90**
Testes	0.21	**1.15**	0.25	**1.26**	0.27	**1.33**	0.29	**1.38**	0.32	**1.47**	0.35	**1.54**
Upper arm	0.04	**0.08**	0.05	**0.11**	0.06	**0.12**	0.07	**0.14**	0.08	**0.17**	0.10	**0.21**
Elbow	0.04	**0.09**	0.05	**0.11**	0.06	**0.13**	0.07	**0.15**	0.08	**0.18**	0.10	**0.22**
Forearm	0.03	**0.08**	0.04	**0.10**	0.04	**0.12**	0.05	**0.13**	0.06	**0.17**	0.07	**0.20**
Upper leg	0.04	**0.53**	0.05	**0.66**	0.06	**0.76**	0.06	**0.84**	0.08	**0.99**	0.10	**1.13**
Knee		**0.03**		**0.04**		**0.04**		**0.05**		**0.06**		**0.07**
Lower leg												
Lymph nodes	0.30	**0.44**	0.36	**0.53**	0.41	**0.60**	0.44	**0.65**	0.51	**0.76**	0.57	**0.84**
Skin	0.11	**0.27**	0.12	**0.28**	0.12	**0.30**	0.13	**0.30**	0.14	**0.32**	0.14	**0.34**
Muscle	0.12	**0.24**	0.15	**0.29**	0.16	**0.32**	0.18	**0.34**	0.20	**0.39**	0.23	**0.43**
Skeleton	0.17	**0.45**	0.21	**0.53**	0.23	**0.59**	0.25	**0.64**	0.28	**0.74**	0.32	**0.82**
Red bone marrow	0.06	**0.15**	0.07	**0.18**	0.08	**0.20**	0.09	**0.22**	0.11	**0.26**	0.12	**0.30**
Whole body dose	0.13	**0.28**	0.16	**0.33**	0.17	**0.37**	0.19	**0.40**	0.22	**0.45**	0.24	**0.50**
ED (ICRP 60)	0.22	**0.53**	0.26	**0.61**	0.29	**0.67**	0.32	**0.72**	0.36	**0.80**	0.40	**0.88**
ED (ICRP 103)	0.21	**0.54**	0.25	**0.62**	0.28	**0.68**	0.30	**0.72**	0.34	**0.81**	0.38	**0.88**
Total filtration (mm) Al	**1.5**		**2.5**		**3.5**		**4.5**		**3.5**		**3.5**	
Cu	**0**		**0**		**0**		**0**		**0.1**		**0.2**	

Table 13.31 Normalized organ doses—Full spine ap—8–12 years

Beam projection:	ap	**Table cells:**
Focus-to-detector distance (cm):	200	(+) = normal font = optimal field setting
Tube voltage (kVp):	70	**(−) = bold font = suboptimal field setting**
Total filtrations:	See last table row	Empty table cells correspond to values <0.01

Organ or tissue	**Organ dose D_T: Entrance dose K_E (mSv * mGy^{-1})**											
	(+)	(−)	(+)	(−)	(+)	(−)	(+)	(−)	(+)	(−)	(+)	(−)
Skull	0.25	**0.54**	0.30	**0.65**	0.34	**0.73**	0.37	**0.80**	0.42	**0.91**	0.46	**1.02**
Brain	0.02	**0.07**	0.02	**0.09**	0.03	**0.11**	0.03	**0.13**	0.04	**0.16**	0.05	**0.19**
Oral mucosa	0.28	**0.41**	0.34	**0.49**	0.37	**0.55**	0.40	**0.59**	0.46	**0.68**	0.51	**0.76**
Salivary glands	0.13	**0.36**	0.16	**0.43**	0.19	**0.49**	0.21	**0.53**	0.25	**0.61**	0.28	**0.68**
Cervical spine	0.46	**0.50**	0.59	**0.64**	0.69	**0.76**	0.78	**0.85**	0.94	**1.04**	1.10	**1.22**
Thyroid	0.84	**0.87**	0.96	**1.00**	1.04	**1.09**	1.10	**1.15**	1.20	**1.26**	1.28	**1.35**
Thymus	0.71	**0.75**	0.85	**0.90**	0.95	**1.01**	1.02	**1.09**	1.15	**1.24**	1.26	**1.36**
Thoracic spine	0.12	**0.16**	0.16	**0.21**	0.19	**0.25**	0.22	**0.29**	0.28	**0.37**	0.35	**0.45**
Claviculae	0.55	**1.19**	0.65	**1.42**	0.72	**1.57**	0.77	**1.69**	0.87	**1.89**	0.94	**2.05**
Scapulae	0.03	**0.09**	0.05	**0.11**	0.05	**0.13**	0.06	**0.15**	0.08	**0.19**	0.10	**0.23**
Ribs	0.49	**0.87**	0.56	**0.99**	0.60	**1.08**	0.63	**1.14**	0.69	**1.24**	0.73	**1.32**
Oesophagus	0.09	**0.11**	0.12	**0.15**	0.14	**0.17**	0.16	**0.20**	0.20	**0.24**	0.23	**0.29**
Trachea	0.51	**0.55**	0.62	**0.67**	0.70	**0.76**	0.76	**0.83**	0.88	**0.96**	0.98	**1.08**
Lung	0.07	**0.26**	0.09	**0.32**	0.11	**0.37**	0.12	**0.40**	0.15	**0.47**	0.18	**0.54**
Mamma	0.03	**0.60**	0.04	**0.65**	0.05	**0.68**	0.05	**0.70**	0.07	**0.74**	0.08	**0.77**
Heart	0.37	**0.46**	0.45	**0.57**	0.51	**0.65**	0.56	**0.72**	0.66	**0.84**	0.74	**0.95**
Stomach	0.15	**0.41**	0.19	**0.50**	0.22	**0.57**	0.25	**0.62**	0.29	**0.73**	0.34	**0.82**
Spleen	0.03	**0.08**	0.03	**0.10**	0.04	**0.12**	0.05	**0.14**	0.06	**0.18**	0.07	**0.21**
Liver	0.16	**0.29**	0.20	**0.36**	0.23	**0.41**	0.25	**0.45**	0.29	**0.53**	0.33	**0.60**
Gall bladder	0.34	**0.39**	0.42	**0.50**	0.49	**0.58**	0.54	**0.64**	0.64	**0.77**	0.74	**0.89**
Pancreas	0.15	**0.21**	0.19	**0.27**	0.22	**0.32**	0.25	**0.37**	0.31	**0.45**	0.37	**0.54**
Small intestine	0.18	**0.28**	0.24	**0.36**	0.27	**0.42**	0.31	**0.47**	0.37	**0.56**	0.43	**0.65**
Large intestine	0.15	**0.34**	0.19	**0.42**	0.22	**0.49**	0.24	**0.54**	0.30	**0.65**	0.34	**0.74**
Sigma/rectum	0.08	**0.23**	0.11	**0.30**	0.13	**0.35**	0.14	**0.39**	0.18	**0.47**	0.21	**0.55**
Lumbar spine	0.13	**0.16**	0.17	**0.22**	0.21	**0.27**	0.24	**0.31**	0.30	**0.39**	0.37	**0.48**
Pelvis	0.07	**0.20**	0.10	**0.26**	0.12	**0.31**	0.13	**0.35**	0.17	**0.44**	0.21	**0.54**
Adrenals	0.06	**0.09**	0.08	**0.11**	0.10	**0.14**	0.11	**0.16**	0.14	**0.20**	0.17	**0.24**
Kidneys	0.04	**0.07**	0.05	**0.09**	0.06	**0.11**	0.07	**0.12**	0.09	**0.16**	0.11	**0.19**
Urinary bladder	0.47	**0.53**	0.57	**0.65**	0.64	**0.74**	0.70	**0.81**	0.81	**0.94**	0.90	**1.05**
Uterus	0.27	**0.31**	0.34	**0.40**	0.40	**0.47**	0.45	**0.53**	0.54	**0.64**	0.62	**0.75**
Ovaries	0.12	**0.22**	0.16	**0.29**	0.18	**0.34**	0.21	**0.38**	0.25	**0.47**	0.30	**0.55**
Prostate	0.30	**0.36**	0.38	**0.45**	0.44	**0.53**	0.48	**0.58**	0.57	**0.70**	0.65	**0.80**
Testes	0.91	**1.06**	1.01	**1.17**	1.07	**1.24**	1.11	**1.29**	1.17	**1.38**	1.23	**1.45**
Upper arm	0.02	**0.04**	0.03	**0.05**	0.03	**0.06**	0.04	**0.07**	0.05	**0.09**	0.06	**0.12**
Elbow	0.02	**0.04**	0.03	**0.05**	0.03	**0.07**	0.04	**0.08**	0.05	**0.10**	0.06	**0.12**
Forearm	0.01	**0.03**	0.02	**0.04**	0.02	**0.05**	0.02	**0.06**	0.03	**0.08**	0.04	**0.10**
Upper leg	0.02	**0.18**	0.03	**0.23**	0.04	**0.28**	0.04	**0.31**	0.06	**0.39**	0.07	**0.46**
Knee										**0.01**		**0.01**
Lower leg												
Lymph nodes	0.26	**0.36**	0.32	**0.44**	0.37	**0.50**	0.40	**0.55**	0.47	**0.65**	0.53	**0.74**
Skin	0.10	**0.21**	0.11	**0.22**	0.11	**0.23**	0.12	**0.24**	0.12	**0.25**	0.13	**0.26**
Muscle	0.10	**0.17**	0.12	**0.21**	0.13	**0.23**	0.15	**0.25**	0.17	**0.29**	0.19	**0.33**
Skeleton	0.14	**0.27**	0.17	**0.33**	0.19	**0.37**	0.20	**0.41**	0.24	**0.48**	0.27	**0.54**
Red bone marrow	0.05	**0.10**	0.06	**0.12**	0.07	**0.14**	0.08	**0.16**	0.10	**0.19**	0.11	**0.22**
Whole body dose	0.10	**0.19**	0.13	**0.23**	0.14	**0.26**	0.16	**0.28**	0.18	**0.33**	0.20	**0.37**
ED (ICRP 60)	0.24	**0.39**	0.28	**0.46**	0.31	**0.51**	0.34	**0.55**	0.38	**0.63**	0.42	**0.70**
ED (ICRP 103)	0.19	**0.37**	0.23	**0.44**	0.26	**0.49**	0.28	**0.53**	0.32	**0.60**	0.36	**0.67**
Total filtration (mm) Al	**1.5**		**2.5**		**3.5**		**4.5**		**3.5**		**3.5**	
Cu	**0**		**0**		**0**		**0**		**0.1**		**0.2**	

Table 13.32 Normalized organ doses—Full spine ap—8–12 years

Beam projection:		ap										**Table cells:**
Focus-to-detector distance (cm):		200										(+) = normal font = optimal field setting
Tube voltage (kVp):		80										**(−) = bold font = suboptimal field setting**
Total filtrations:		See last table row										Empty table cells correspond to values <0.01
Organ or tissue	**Organ dose D_T: Entrance dose K_E**							**(mSv ∗ mGy^{-1})**				
	(+)	(−)	(+)	(−)	(+)	(−)	(+)	(−)	(+)	(−)	(+)	(−)
Skull	0.28	**0.60**	0.33	**0.72**	0.37	**0.80**	0.39	**0.86**	0.44	**0.97**	0.49	**1.07**
Brain	0.02	**0.09**	0.03	**0.12**	0.04	**0.14**	0.04	**0.15**	0.05	**0.19**	0.06	**0.22**
Oral mucosa	0.32	**0.47**	0.37	**0.55**	0.41	**0.61**	0.44	**0.66**	0.50	**0.74**	0.56	**0.82**
Salivary glands	0.15	**0.40**	0.19	**0.48**	0.22	**0.54**	0.24	**0.58**	0.28	**0.66**	0.31	**0.73**
Cervical spine	0.56	**0.61**	0.69	**0.76**	0.80	**0.88**	0.89	**0.98**	1.06	**1.18**	1.22	**1.36**
Thyroid	0.90	**0.94**	1.02	**1.07**	1.10	**1.15**	1.16	**1.21**	1.25	**1.32**	1.33	**1.41**
Thymus	0.79	**0.84**	0.93	**0.99**	1.03	**1.10**	1.10	**1.18**	1.23	**1.32**	1.33	**1.44**
Thoracic spine	0.16	**0.21**	0.21	**0.27**	0.24	**0.32**	0.28	**0.36**	0.35	**0.45**	0.42	**0.54**
Claviculae	0.60	**1.30**	0.70	**1.51**	0.76	**1.66**	0.81	**1.77**	0.90	**1.95**	0.97	**2.09**
Scapulae	0.05	**0.11**	0.06	**0.14**	0.07	**0.16**	0.08	**0.18**	0.10	**0.23**	0.12	**0.27**
Ribs	0.52	**0.93**	0.59	**1.05**	0.63	**1.12**	0.66	**1.18**	0.71	**1.28**	0.75	**1.35**
Oesophagus	0.11	**0.14**	0.14	**0.18**	0.17	**0.21**	0.19	**0.24**	0.23	**0.29**	0.27	**0.34**
Trachea	0.58	**0.63**	0.69	**0.76**	0.78	**0.85**	0.84	**0.92**	0.96	**1.05**	1.06	**1.17**
Lung	0.09	**0.30**	0.11	**0.37**	0.13	**0.41**	0.14	**0.45**	0.17	**0.52**	0.20	**0.59**
Mamma	0.04	**0.63**	0.05	**0.67**	0.06	**0.70**	0.06	**0.72**	0.08	**0.76**	0.09	**0.79**
Heart	0.42	**0.54**	0.51	**0.65**	0.58	**0.73**	0.63	**0.80**	0.72	**0.92**	0.81	**1.03**
Stomach	0.18	**0.47**	0.22	**0.57**	0.25	**0.64**	0.28	**0.69**	0.33	**0.80**	0.37	**0.89**
Spleen	0.03	**0.10**	0.04	**0.13**	0.05	**0.15**	0.06	**0.17**	0.07	**0.21**	0.09	**0.25**
Liver	0.19	**0.33**	0.23	**0.41**	0.26	**0.46**	0.28	**0.50**	0.32	**0.59**	0.37	**0.66**
Gall bladder	0.40	**0.47**	0.49	**0.58**	0.56	**0.67**	0.61	**0.73**	0.72	**0.86**	0.81	**0.98**
Pancreas	0.18	**0.26**	0.23	**0.33**	0.27	**0.39**	0.30	**0.44**	0.36	**0.53**	0.42	**0.62**
Small intestine	0.22	**0.34**	0.28	**0.42**	0.32	**0.49**	0.35	**0.54**	0.42	**0.64**	0.48	**0.73**
Large intestine	0.18	**0.40**	0.22	**0.49**	0.26	**0.56**	0.28	**0.62**	0.34	**0.72**	0.38	**0.82**
Sigma/rectum	0.10	**0.28**	0.13	**0.35**	0.15	**0.41**	0.17	**0.45**	0.21	**0.54**	0.24	**0.61**
Lumbar spine	0.17	**0.22**	0.22	**0.29**	0.26	**0.34**	0.30	**0.39**	0.37	**0.49**	0.45	**0.58**
Pelvis	0.10	**0.25**	0.12	**0.32**	0.15	**0.38**	0.17	**0.43**	0.21	**0.53**	0.25	**0.62**
Adrenals	0.08	**0.11**	0.11	**0.15**	0.13	**0.17**	0.14	**0.19**	0.18	**0.24**	0.21	**0.29**
Kidneys	0.05	**0.09**	0.06	**0.12**	0.08	**0.14**	0.09	**0.16**	0.11	**0.20**	0.13	**0.23**
Urinary bladder	0.53	**0.61**	0.64	**0.73**	0.71	**0.82**	0.77	**0.89**	0.87	**1.02**	0.96	**1.14**
Uterus	0.32	**0.38**	0.40	**0.48**	0.46	**0.55**	0.51	**0.61**	0.61	**0.73**	0.69	**0.84**
Ovaries	0.15	**0.28**	0.19	**0.35**	0.22	**0.40**	0.25	**0.45**	0.30	**0.54**	0.34	**0.63**
Prostate	0.35	**0.43**	0.43	**0.53**	0.50	**0.60**	0.54	**0.66**	0.64	**0.78**	0.72	**0.88**
Testes	0.96	**1.11**	1.05	**1.22**	1.10	**1.29**	1.14	**1.34**	1.20	**1.42**	1.25	**1.49**
Upper arm	0.03	**0.05**	0.04	**0.07**	0.04	**0.08**	0.05	**0.09**	0.06	**0.12**	0.07	**0.14**
Elbow	0.03	**0.05**	0.03	**0.07**	0.04	**0.08**	0.05	**0.10**	0.06	**0.12**	0.07	**0.15**
Forearm	0.02	**0.05**	0.02	**0.06**	0.03	**0.07**	0.03	**0.08**	0.04	**0.10**	0.05	**0.12**
Upper leg	0.03	**0.22**	0.04	**0.28**	0.05	**0.33**	0.06	**0.37**	0.07	**0.45**	0.09	**0.52**
Knee						**0.01**		**0.01**		**0.02**		**0.02**
Lower leg												
Lymph nodes	0.30	**0.42**	0.37	**0.51**	0.41	**0.57**	0.45	**0.62**	0.52	**0.72**	0.59	**0.81**
Skin	0.11	**0.22**	0.11	**0.23**	0.12	**0.24**	0.12	**0.25**	0.13	**0.26**	0.14	**0.28**
Muscle	0.11	**0.20**	0.13	**0.23**	0.15	**0.26**	0.16	**0.28**	0.19	**0.32**	0.21	**0.36**
Skeleton	0.16	**0.31**	0.19	**0.37**	0.21	**0.42**	0.23	**0.45**	0.26	**0.52**	0.29	**0.58**
Red bone marrow	0.06	**0.12**	0.07	**0.15**	0.09	**0.17**	0.09	**0.18**	0.11	**0.22**	0.13	**0.25**
Whole body dose	0.12	**0.22**	0.14	**0.26**	0.16	**0.29**	0.17	**0.31**	0.20	**0.36**	0.22	**0.40**
ED (ICRP 60)	0.27	**0.44**	0.31	**0.51**	0.34	**0.56**	0.37	**0.61**	0.41	**0.69**	0.45	**0.76**
ED (ICRP 103)	0.22	**0.41**	0.26	**0.49**	0.28	**0.54**	0.31	**0.58**	0.35	**0.65**	0.39	**0.72**
Total filtration (mm) **Al**	**1.5**		**2.5**		**3.5**		**4.5**		**3.5**		**3.5**	
Cu	**0**		**0**		**0**		**0**		**0.1**		**0.2**	

Table 13.33 Normalized organ doses—Full spine ap—8–12 years

Beam projection:	ap											**Table cells:**
Focus-to-detector distance (cm):	200											(+) = normal font = optimal field setting
Tube voltage (kVp):	90											**(−) = bold font = suboptimal field setting**
Total filtrations:	See last table row											Empty table cells correspond to values <0.01
Organ or tissue	**Organ dose D_T: Entrance dose K_E**								**(mSv * mGy⁻¹)**			
	(+)	(−)	(+)	(−)	(+)	(−)	(+)	(−)	(+)	(−)	(+)	(−)
Skull	0.30	**0.66**	0.35	**0.77**	0.39	**0.85**	0.41	**0.91**	0.46	**1.01**	0.50	**1.09**
Brain	0.03	**0.11**	0.04	**0.14**	0.04	**0.16**	0.05	**0.18**	0.06	**0.21**	0.07	**0.25**
Oral mucosa	0.35	**0.52**	0.41	**0.60**	0.45	**0.66**	0.48	**0.71**	0.54	**0.79**	0.59	**0.87**
Salivary glands	0.18	**0.45**	0.21	**0.53**	0.24	**0.59**	0.26	**0.63**	0.30	**0.71**	0.34	**0.78**
Cervical spine	0.64	**0.71**	0.78	**0.87**	0.89	**0.99**	0.98	**1.09**	1.14	**1.28**	1.29	**1.45**
Thyroid	0.96	**1.00**	1.07	**1.13**	1.15	**1.21**	1.20	**1.27**	1.29	**1.37**	1.37	**1.45**
Thymus	0.86	**0.92**	1.00	**1.07**	1.09	**1.17**	1.16	**1.25**	1.28	**1.38**	1.38	**1.49**
Thoracic spine	0.20	**0.26**	0.25	**0.32**	0.29	**0.38**	0.32	**0.42**	0.40	**0.52**	0.46	**0.61**
Claviculae	0.64	**1.39**	0.73	**1.59**	0.80	**1.72**	0.84	**1.82**	0.92	**1.98**	0.98	**2.09**
Scapulae	0.06	**0.13**	0.07	**0.16**	0.08	**0.19**	0.09	**0.21**	0.11	**0.25**	0.13	**0.29**
Ribs	0.55	**0.98**	0.61	**1.09**	0.65	**1.16**	0.67	**1.21**	0.72	**1.29**	0.75	**1.35**
Oesophagus	0.14	**0.17**	0.17	**0.21**	0.20	**0.25**	0.22	**0.27**	0.26	**0.33**	0.30	**0.38**
Trachea	0.65	**0.70**	0.76	**0.83**	0.84	**0.92**	0.90	**0.99**	1.02	**1.12**	1.11	**1.24**
Lung	0.10	**0.34**	0.13	**0.41**	0.15	**0.45**	0.16	**0.49**	0.19	**0.56**	0.22	**0.63**
Mamma	0.05	**0.65**	0.06	**0.69**	0.07	**0.72**	0.07	**0.74**	0.09	**0.78**	0.10	**0.81**
Heart	0.47	**0.60**	0.56	**0.72**	0.63	**0.80**	0.68	**0.87**	0.77	**0.99**	0.85	**1.09**
Stomach	0.21	**0.53**	0.25	**0.62**	0.28	**0.70**	0.31	**0.75**	0.35	**0.85**	0.40	**0.94**
Spleen	0.04	**0.12**	0.05	**0.15**	0.06	**0.18**	0.07	**0.20**	0.09	**0.24**	0.10	**0.28**
Liver	0.21	**0.38**	0.25	**0.45**	0.28	**0.51**	0.31	**0.55**	0.35	**0.63**	0.39	**0.70**
Gall bladder	0.45	**0.54**	0.55	**0.66**	0.62	**0.74**	0.67	**0.81**	0.78	**0.94**	0.87	**1.05**
Pancreas	0.22	**0.32**	0.27	**0.39**	0.31	**0.45**	0.34	**0.49**	0.40	**0.59**	0.46	**0.68**
Small intestine	0.26	**0.39**	0.32	**0.48**	0.36	**0.55**	0.39	**0.60**	0.46	**0.70**	0.52	**0.79**
Large intestine	0.21	**0.46**	0.25	**0.55**	0.29	**0.62**	0.31	**0.68**	0.37	**0.79**	0.41	**0.88**
Sigma/rectum	0.12	**0.33**	0.15	**0.40**	0.18	**0.46**	0.19	**0.50**	0.23	**0.59**	0.26	**0.67**
Lumbar spine	0.21	**0.28**	0.27	**0.35**	0.31	**0.41**	0.35	**0.46**	0.43	**0.56**	0.50	**0.66**
Pelvis	0.12	**0.30**	0.15	**0.38**	0.17	**0.44**	0.20	**0.49**	0.24	**0.59**	0.28	**0.68**
Adrenals	0.10	**0.14**	0.13	**0.18**	0.15	**0.20**	0.17	**0.23**	0.20	**0.28**	0.24	**0.33**
Kidneys	0.06	**0.11**	0.08	**0.14**	0.09	**0.17**	0.10	**0.19**	0.13	**0.23**	0.15	**0.27**
Urinary bladder	0.59	**0.68**	0.69	**0.81**	0.77	**0.89**	0.82	**0.96**	0.92	**1.09**	1.01	**1.20**
Uterus	0.38	**0.45**	0.46	**0.54**	0.52	**0.62**	0.57	**0.68**	0.66	**0.80**	0.74	**0.90**
Ovaries	0.18	**0.33**	0.22	**0.40**	0.25	**0.46**	0.28	**0.51**	0.33	**0.60**	0.38	**0.68**
Prostate	0.40	**0.49**	0.49	**0.59**	0.55	**0.67**	0.59	**0.73**	0.68	**0.84**	0.76	**0.94**
Testes	1.00	**1.16**	1.08	**1.27**	1.13	**1.33**	1.17	**1.38**	1.23	**1.46**	1.27	**1.52**
Upper arm	0.03	**0.07**	0.04	**0.08**	0.05	**0.10**	0.06	**0.11**	0.07	**0.13**	0.08	**0.16**
Elbow	0.03	**0.07**	0.04	**0.09**	0.05	**0.10**	0.06	**0.11**	0.07	**0.14**	0.08	**0.16**
Forearm	0.02	**0.06**	0.03	**0.07**	0.03	**0.08**	0.04	**0.10**	0.05	**0.12**	0.06	**0.14**
Upper leg	0.04	**0.27**	0.05	**0.33**	0.06	**0.38**	0.07	**0.41**	0.08	**0.49**	0.10	**0.56**
Knee				**0.01**		**0.01**		**0.01**		**0.02**		**0.02**
Lower leg												
Lymph nodes	0.34	**0.47**	0.41	**0.56**	0.46	**0.63**	0.49	**0.68**	0.56	**0.78**	0.62	**0.87**
Skin	0.11	**0.23**	0.12	**0.24**	0.12	**0.25**	0.13	**0.26**	0.14	**0.27**	0.14	**0.29**
Muscle	0.13	**0.22**	0.15	**0.26**	0.16	**0.28**	0.18	**0.30**	0.20	**0.34**	0.22	**0.38**
Skeleton	0.18	**0.34**	0.21	**0.41**	0.23	**0.45**	0.24	**0.48**	0.28	**0.55**	0.31	**0.60**
Red bone marrow	0.07	**0.14**	0.09	**0.17**	0.10	**0.19**	0.11	**0.21**	0.13	**0.24**	0.14	**0.28**
Whole body dose	0.13	**0.24**	0.16	**0.29**	0.18	**0.32**	0.19	**0.34**	0.22	**0.39**	0.24	**0.43**
ED (ICRP 60)	0.29	**0.48**	0.34	**0.56**	0.37	**0.61**	0.39	**0.65**	0.44	**0.73**	0.48	**0.80**
ED (ICRP 103)	0.24	**0.46**	0.28	**0.53**	0.31	**0.58**	0.33	**0.62**	0.38	**0.70**	0.41	**0.76**
Total filtration (mm) Al	**1.5**		**2.5**		**3.5**		**4.5**		**3.5**		**3.5**	
Cu	**0**		**0**		**0**		**0**		**0.1**		**0.2**	

Table 13.34 Normalized organ doses—Full spine ap—13–17 years

Beam projection:		ap										**Table cells:**
Focus-to-detector distance (cm):		200						(+) = normal font = optimal field setting				
Tube voltage (kVp):		70						**(−) = bold font = suboptimal field setting**				
Total filtrations:		See last table row						Empty table cells correspond to values <0.01				
Organ or tissue	**Organ dose D_T: Entrance dose K_E**							**(mSv * mGy^{-1})**				
	(+)	(−)	(+)	(−)	(+)	(−)	(+)	(−)	(+)	(−)	(+)	(−)
Skull	0.25	**0.46**	0.30	**0.56**	0.34	**0.64**	0.38	**0.70**	0.44	**0.82**	0.49	**0.92**
Brain	0.01	**0.06**	0.02	**0.07**	0.02	**0.09**	0.02	**0.10**	0.03	**0.12**	0.04	**0.15**
Oral mucosa	0.27	**0.35**	0.32	**0.43**	0.35	**0.48**	0.38	**0.52**	0.44	**0.60**	0.49	**0.67**
Salivary glands	0.12	**0.33**	0.15	**0.40**	0.17	**0.45**	0.19	**0.49**	0.23	**0.57**	0.27	**0.64**
Cervical spine	0.51	**0.55**	0.66	**0.71**	0.78	**0.83**	0.88	**0.94**	1.07	**1.16**	1.26	**1.37**
Thyroid	0.77	**0.77**	0.89	**0.90**	0.97	**0.98**	1.03	**1.05**	1.13	**1.16**	1.22	**1.25**
Thymus	0.66	**0.70**	0.80	**0.85**	0.90	**0.96**	0.98	**1.05**	1.12	**1.21**	1.24	**1.34**
Thoracic spine	0.08	**0.10**	0.10	**0.14**	0.12	**0.17**	0.14	**0.19**	0.19	**0.25**	0.23	**0.31**
Claviculae	0.34	**0.97**	0.41	**1.16**	0.46	**1.30**	0.50	**1.40**	0.57	**1.59**	0.63	**1.75**
Scapulae	0.02	**0.05**	0.03	**0.07**	0.03	**0.08**	0.04	**0.09**	0.05	**0.12**	0.06	**0.14**
Ribs	0.43	**0.78**	0.49	**0.90**	0.53	**0.98**	0.56	**1.04**	0.61	**1.14**	0.65	**1.23**
Oesophagus	0.09	**0.12**	0.12	**0.16**	0.15	**0.19**	0.17	**0.22**	0.21	**0.27**	0.25	**0.33**
Trachea	0.38	**0.40**	0.46	**0.48**	0.51	**0.55**	0.56	**0.60**	0.64	**0.69**	0.72	**0.78**
Lung	0.05	**0.19**	0.06	**0.24**	0.08	**0.28**	0.09	**0.31**	0.11	**0.37**	0.13	**0.43**
Mamma	0.31	**3.83**	0.39	**4.38**	0.46	**4.76**	0.51	**5.04**	0.62	**5.54**	0.73	**5.97**
Heart	0.32	**0.42**	0.40	**0.52**	0.46	**0.60**	0.51	**0.67**	0.60	**0.79**	0.69	**0.90**
Stomach	0.10	**0.35**	0.12	**0.44**	0.15	**0.51**	0.16	**0.56**	0.20	**0.66**	0.23	**0.76**
Spleen	0.02	**0.06**	0.02	**0.08**	0.03	**0.09**	0.03	**0.10**	0.04	**0.13**	0.05	**0.16**
Liver	0.10	**0.22**	0.13	**0.28**	0.15	**0.32**	0.17	**0.36**	0.20	**0.43**	0.23	**0.49**
Gall bladder	0.20	**0.34**	0.26	**0.43**	0.30	**0.51**	0.34	**0.57**	0.41	**0.68**	0.48	**0.79**
Pancreas	0.11	**0.17**	0.14	**0.22**	0.17	**0.26**	0.19	**0.30**	0.24	**0.37**	0.29	**0.45**
Small intestine	0.14	**0.24**	0.18	**0.30**	0.21	**0.36**	0.24	**0.40**	0.29	**0.49**	0.34	**0.57**
Large intestine	0.11	**0.28**	0.14	**0.36**	0.17	**0.42**	0.19	**0.47**	0.23	**0.57**	0.28	**0.66**
Sigma/rectum	0.06	**0.19**	0.08	**0.24**	0.10	**0.29**	0.11	**0.32**	0.14	**0.40**	0.16	**0.47**
Lumbar spine	0.08	**0.11**	0.11	**0.15**	0.13	**0.18**	0.15	**0.21**	0.20	**0.27**	0.25	**0.34**
Pelvis	0.04	**0.14**	0.06	**0.18**	0.07	**0.22**	0.08	**0.25**	0.11	**0.33**	0.14	**0.40**
Adrenals	0.04	**0.06**	0.05	**0.08**	0.07	**0.10**	0.08	**0.11**	0.10	**0.14**	0.12	**0.18**
Kidneys	0.02	**0.05**	0.03	**0.07**	0.04	**0.08**	0.04	**0.09**	0.06	**0.12**	0.07	**0.15**
Urinary bladder	0.41	**0.48**	0.50	**0.60**	0.57	**0.68**	0.63	**0.75**	0.73	**0.89**	0.82	**1.01**
Uterus	0.63	**0.77**	0.81	**0.99**	0.95	**1.16**	1.07	**1.31**	1.30	**1.60**	1.52	**1.88**
Ovaries	0.07	**0.18**	0.09	**0.23**	0.11	**0.27**	0.13	**0.31**	0.16	**0.39**	0.19	**0.46**
Prostate	0.28	**0.34**	0.35	**0.43**	0.41	**0.50**	0.46	**0.56**	0.55	**0.68**	0.64	**0.80**
Testes	0.74	**0.96**	0.84	**1.10**	0.90	**1.19**	0.95	**1.26**	1.03	**1.38**	1.10	**1.48**
Upper arm	0.01	**0.02**	0.01	**0.03**	0.02	**0.04**	0.02	**0.04**	0.03	**0.06**	0.03	**0.07**
Elbow		**0.02**	0.01	**0.03**	0.02	**0.04**	0.02	**0.05**	0.02	**0.06**	0.03	**0.08**
Forearm		**0.02**		**0.03**	0.01	**0.03**	0.01	**0.04**	0.02	**0.05**	0.02	**0.06**
Upper leg	0.02	**0.22**	0.02	**0.28**	0.03	**0.34**	0.03	**0.38**	0.04	**0.48**	0.05	**0.58**
Knee		**0.07**		**0.09**		**0.11**		**0.12**		**0.15**		**0.18**
Lower leg												
Lymph nodes	0.20	**0.30**	0.25	**0.37**	0.29	**0.43**	0.32	**0.47**	0.37	**0.56**	0.43	**0.64**
Skin	0.10	**0.24**	0.10	**0.25**	0.11	**0.27**	0.11	**0.28**	0.12	**0.29**	0.13	**0.31**
Muscle	0.08	**0.17**	0.10	**0.21**	0.12	**0.24**	0.13	**0.26**	0.15	**0.30**	0.17	**0.34**
Skeleton	0.11	**0.24**	0.14	**0.29**	0.16	**0.33**	0.17	**0.37**	0.20	**0.43**	0.23	**0.50**
Red bone marrow	0.05	**0.09**	0.06	**0.11**	0.07	**0.13**	0.07	**0.14**	0.09	**0.17**	0.10	**0.20**
Whole body dose	0.09	**0.19**	0.11	**0.22**	0.12	**0.25**	0.14	**0.28**	0.16	**0.32**	0.18	**0.37**
ED (ICRP 60)	0.21	**0.50**	0.25	**0.60**	0.28	**0.67**	0.31	**0.72**	0.35	**0.82**	0.40	**0.92**
ED (ICRP 103)	0.19	**0.72**	0.23	**0.84**	0.27	**0.93**	0.29	**1.00**	0.34	**1.13**	0.39	**1.25**
Total filtration (mm) Al	**1.5**		**2.5**		**3.5**		**4.5**		**3.5**		**3.5**	
Cu	**0**		**0**		**0**		**0**		**0.1**		**0.2**	

Table 13.35 Normalized organ doses—Full spine ap—13–17 years

Beam projection:	ap											**Table cells:**
Focus-to-detector distance (cm):	200											(+) = normal font = optimal field setting
Tube voltage (kVp):	80											**(−) = bold font = suboptimal field setting**
Total filtrations:	See last table row											Empty table cells correspond to values <0.01
Organ or tissue	**Organ dose D_T: Entrance dose K_E**							**(mSv * mGy⁻¹)**				
	(+)	(−)	(+)	(−)	(+)	(−)	(+)	(−)	(+)	(−)	(+)	(−)
Skull	0.28	**0.52**	0.34	**0.63**	0.38	**0.71**	0.41	**0.77**	0.47	**0.88**	0.52	**0.98**
Brain	0.02	**0.07**	0.02	**0.09**	0.03	**0.11**	0.03	**0.12**	0.04	**0.15**	0.05	**0.18**
Oral mucosa	0.30	**0.40**	0.36	**0.48**	0.39	**0.53**	0.43	**0.58**	0.49	**0.66**	0.54	**0.73**
Salivary glands	0.14	**0.37**	0.17	**0.45**	0.20	**0.50**	0.22	**0.55**	0.26	**0.63**	0.30	**0.70**
Cervical spine	0.63	**0.68**	0.79	**0.85**	0.92	**0.99**	1.02	**1.10**	1.22	**1.33**	1.41	**1.54**
Thyroid	0.84	**0.84**	0.95	**0.97**	1.03	**1.05**	1.09	**1.12**	1.19	**1.23**	1.27	**1.32**
Thymus	0.74	**0.79**	0.88	**0.95**	0.99	**1.06**	1.06	**1.15**	1.20	**1.31**	1.32	**1.44**
Thoracic spine	0.11	**0.14**	0.14	**0.18**	0.16	**0.22**	0.19	**0.25**	0.24	**0.32**	0.29	**0.39**
Claviculae	0.38	**1.07**	0.45	**1.26**	0.50	**1.39**	0.53	**1.49**	0.60	**1.66**	0.66	**1.80**
Scapulae	0.03	**0.07**	0.03	**0.09**	0.04	**0.10**	0.05	**0.12**	0.06	**0.15**	0.07	**0.18**
Ribs	0.46	**0.84**	0.51	**0.95**	0.55	**1.03**	0.58	**1.09**	0.63	**1.18**	0.66	**1.26**
Oesophagus	0.12	**0.16**	0.15	**0.20**	0.18	**0.24**	0.20	**0.27**	0.25	**0.33**	0.30	**0.40**
Trachea	0.43	**0.46**	0.51	**0.55**	0.57	**0.61**	0.62	**0.66**	0.70	**0.76**	0.78	**0.85**
Lung	0.06	**0.23**	0.08	**0.28**	0.09	**0.32**	0.11	**0.36**	0.13	**0.43**	0.15	**0.49**
Mamma	0.38	**4.14**	0.47	**4.70**	0.54	**5.07**	0.60	**5.35**	0.72	**5.85**	0.84	**6.27**
Heart	0.38	**0.49**	0.46	**0.60**	0.52	**0.69**	0.57	**0.75**	0.67	**0.88**	0.75	**1.00**
Stomach	0.12	**0.41**	0.15	**0.51**	0.17	**0.58**	0.19	**0.63**	0.23	**0.74**	0.27	**0.83**
Spleen	0.02	**0.08**	0.03	**0.10**	0.04	**0.12**	0.04	**0.13**	0.05	**0.17**	0.06	**0.20**
Liver	0.12	**0.26**	0.15	**0.33**	0.18	**0.37**	0.19	**0.41**	0.23	**0.48**	0.26	**0.55**
Gall bladder	0.25	**0.41**	0.31	**0.51**	0.36	**0.59**	0.39	**0.65**	0.47	**0.78**	0.54	**0.89**
Pancreas	0.14	**0.21**	0.18	**0.27**	0.21	**0.32**	0.23	**0.36**	0.29	**0.44**	0.34	**0.52**
Small intestine	0.17	**0.29**	0.22	**0.36**	0.25	**0.42**	0.28	**0.47**	0.34	**0.57**	0.39	**0.66**
Large intestine	0.14	**0.34**	0.17	**0.43**	0.20	**0.49**	0.23	**0.54**	0.27	**0.65**	0.32	**0.74**
Sigma/rectum	0.08	**0.24**	0.10	**0.30**	0.12	**0.35**	0.13	**0.39**	0.16	**0.47**	0.19	**0.54**
Lumbar spine	0.11	**0.15**	0.15	**0.20**	0.18	**0.24**	0.20	**0.28**	0.25	**0.35**	0.31	**0.43**
Pelvis	0.06	**0.18**	0.08	**0.24**	0.10	**0.28**	0.11	**0.32**	0.14	**0.40**	0.17	**0.48**
Adrenals	0.06	**0.08**	0.07	**0.11**	0.09	**0.13**	0.10	**0.14**	0.12	**0.18**	0.15	**0.22**
Kidneys	0.03	**0.07**	0.04	**0.09**	0.05	**0.10**	0.06	**0.12**	0.07	**0.15**	0.09	**0.18**
Urinary bladder	0.47	**0.56**	0.57	**0.68**	0.64	**0.77**	0.70	**0.85**	0.80	**0.98**	0.89	**1.10**
Uterus	0.77	**0.95**	0.97	**1.19**	1.12	**1.38**	1.24	**1.54**	1.49	**1.86**	1.72	**2.16**
Ovaries	0.09	**0.22**	0.12	**0.28**	0.14	**0.33**	0.16	**0.37**	0.19	**0.46**	0.23	**0.54**
Prostate	0.33	**0.41**	0.41	**0.51**	0.47	**0.59**	0.52	**0.65**	0.62	**0.78**	0.71	**0.90**
Testes	0.79	**1.03**	0.89	**1.17**	0.95	**1.26**	1.00	**1.33**	1.07	**1.45**	1.14	**1.54**
Upper arm	0.02	**0.03**	0.02	**0.04**	0.02	**0.05**	0.03	**0.06**	0.04	**0.07**	0.04	**0.09**
Elbow	0.01	**0.03**	0.02	**0.04**	0.02	**0.05**	0.03	**0.06**	0.03	**0.08**	0.04	**0.10**
Forearm		**0.03**	0.01	**0.04**	0.01	**0.05**	0.02	**0.05**	0.02	**0.07**	0.03	**0.08**
Upper leg	0.02	**0.28**	0.03	**0.35**	0.04	**0.41**	0.04	**0.46**	0.06	**0.57**	0.07	**0.67**
Knee		**0.09**		**0.11**		**0.13**		**0.15**		**0.18**		**0.21**
Lower leg												
Lymph nodes	0.24	**0.35**	0.29	**0.43**	0.33	**0.49**	0.36	**0.54**	0.42	**0.63**	0.47	**0.71**
Skin	0.10	**0.25**	0.11	**0.27**	0.11	**0.28**	0.12	**0.29**	0.13	**0.31**	0.13	**0.32**
Muscle	0.10	**0.20**	0.12	**0.24**	0.13	**0.27**	0.14	**0.29**	0.17	**0.33**	0.19	**0.37**
Skeleton	0.13	**0.28**	0.16	**0.33**	0.18	**0.38**	0.19	**0.41**	0.22	**0.48**	0.25	**0.55**
Red bone marrow	0.06	**0.11**	0.07	**0.14**	0.08	**0.16**	0.09	**0.17**	0.10	**0.21**	0.12	**0.24**
Whole body dose	0.10	**0.21**	0.12	**0.26**	0.14	**0.29**	0.15	**0.31**	0.18	**0.36**	0.20	**0.41**
ED (ICRP 60)	0.24	**0.57**	0.28	**0.67**	0.32	**0.74**	0.34	**0.79**	0.39	**0.90**	0.44	**0.99**
ED (ICRP 103)	0.22	**0.80**	0.27	**0.93**	0.30	**1.02**	0.33	**1.09**	0.39	**1.23**	0.44	**1.34**
Total filtration (mm) Al	**1.5**		**2.5**		**3.5**		**4.5**		**3.5**		**3.5**	
Cu	**0**		**0**		**0**		**0**		**0.1**		**0.2**	

Table 13.36 Normalized organ doses—Full spine ap—13–17 years

Beam projection:	ap	**Table cells:**
Focus-to-detector distance (cm):	200	(+) = normal font = optimal field setting
Tube voltage (kVp):	90	**(−) = bold font = suboptimal field setting**
Total filtrations:	See last table row	Empty table cells correspond to values <0.01

Organ or tissue	**Organ dose D_T: Entrance dose K_E**							**(mSv * mGy^{-1})**				
	(+)	(−)	(+)	(−)	(+)	(−)	(+)	(−)	(+)	(−)	(+)	(−)
Skull	0.31	**0.58**	0.37	**0.68**	0.41	**0.76**	0.44	**0.82**	0.49	**0.92**	0.54	**1.01**
Brain	0.02	**0.09**	0.03	**0.11**	0.03	**0.13**	0.04	**0.14**	0.05	**0.17**	0.06	**0.20**
Oral mucosa	0.34	**0.45**	0.39	**0.53**	0.43	**0.58**	0.46	**0.63**	0.52	**0.71**	0.58	**0.78**
Salivary glands	0.16	**0.42**	0.20	**0.50**	0.22	**0.55**	0.24	**0.59**	0.28	**0.67**	0.32	**0.74**
Cervical spine	0.73	**0.79**	0.90	**0.97**	1.03	**1.12**	1.13	**1.23**	1.33	**1.45**	1.51	**1.66**
Thyroid	0.89	**0.91**	1.01	**1.03**	1.08	**1.11**	1.14	**1.17**	1.23	**1.28**	1.31	**1.36**
Thymus	0.82	**0.88**	0.96	**1.04**	1.06	**1.15**	1.13	**1.23**	1.26	**1.38**	1.37	**1.51**
Thoracic spine	0.13	**0.18**	0.17	**0.23**	0.20	**0.27**	0.23	**0.30**	0.28	**0.38**	0.33	**0.45**
Claviculae	0.41	**1.15**	0.48	**1.33**	0.53	**1.46**	0.56	**1.55**	0.62	**1.70**	0.67	**1.81**
Scapulae	0.03	**0.08**	0.04	**0.10**	0.05	**0.12**	0.06	**0.14**	0.07	**0.17**	0.09	**0.20**
Ribs	0.48	**0.89**	0.54	**1.00**	0.57	**1.07**	0.60	**1.12**	0.64	**1.21**	0.67	**1.27**
Oesophagus	0.15	**0.19**	0.18	**0.24**	0.21	**0.28**	0.24	**0.31**	0.29	**0.38**	0.34	**0.44**
Trachea	0.48	**0.51**	0.56	**0.60**	0.62	**0.67**	0.66	**0.72**	0.75	**0.82**	0.82	**0.90**
Lung	0.08	**0.26**	0.10	**0.32**	0.11	**0.36**	0.12	**0.40**	0.15	**0.47**	0.17	**0.53**
Mamma	0.45	**4.43**	0.54	**4.98**	0.62	**5.34**	0.68	**5.61**	0.80	**6.09**	0.92	**6.48**
Heart	0.43	**0.56**	0.52	**0.68**	0.58	**0.76**	0.63	**0.83**	0.72	**0.95**	0.80	**1.06**
Stomach	0.14	**0.47**	0.17	**0.57**	0.20	**0.64**	0.22	**0.69**	0.26	**0.80**	0.29	**0.89**
Spleen	0.03	**0.10**	0.04	**0.12**	0.04	**0.14**	0.05	**0.16**	0.06	**0.19**	0.07	**0.23**
Liver	0.14	**0.31**	0.18	**0.37**	0.20	**0.42**	0.22	**0.46**	0.25	**0.53**	0.29	**0.60**
Gall bladder	0.29	**0.48**	0.35	**0.59**	0.40	**0.67**	0.44	**0.73**	0.52	**0.85**	0.58	**0.96**
Pancreas	0.17	**0.26**	0.21	**0.32**	0.24	**0.38**	0.27	**0.42**	0.32	**0.50**	0.38	**0.58**
Small intestine	0.21	**0.34**	0.25	**0.42**	0.29	**0.48**	0.32	**0.53**	0.38	**0.63**	0.43	**0.72**
Large intestine	0.16	**0.40**	0.20	**0.49**	0.23	**0.55**	0.26	**0.61**	0.30	**0.71**	0.34	**0.80**
Sigma/rectum	0.10	**0.28**	0.12	**0.35**	0.14	**0.40**	0.15	**0.44**	0.18	**0.52**	0.21	**0.60**
Lumbar spine	0.14	**0.20**	0.18	**0.25**	0.21	**0.30**	0.24	**0.33**	0.30	**0.41**	0.36	**0.49**
Pelvis	0.08	**0.23**	0.10	**0.29**	0.12	**0.34**	0.13	**0.37**	0.16	**0.46**	0.20	**0.54**
Adrenals	0.07	**0.11**	0.09	**0.13**	0.11	**0.16**	0.12	**0.18**	0.15	**0.22**	0.18	**0.26**
Kidneys	0.04	**0.09**	0.05	**0.11**	0.06	**0.13**	0.07	**0.14**	0.09	**0.18**	0.10	**0.21**
Urinary bladder	0.53	**0.64**	0.63	**0.76**	0.70	**0.85**	0.75	**0.92**	0.86	**1.06**	0.94	**1.17**
Uterus	0.91	**1.12**	1.11	**1.38**	1.27	**1.58**	1.39	**1.74**	1.64	**2.06**	1.86	**2.36**
Ovaries	0.11	**0.27**	0.14	**0.34**	0.16	**0.39**	0.18	**0.43**	0.22	**0.51**	0.26	**0.59**
Prostate	0.38	**0.48**	0.47	**0.58**	0.53	**0.66**	0.58	**0.73**	0.67	**0.86**	0.76	**0.97**
Testes	0.83	**1.10**	0.93	**1.23**	0.99	**1.32**	1.03	**1.38**	1.10	**1.49**	1.16	**1.58**
Upper arm	0.02	**0.04**	0.03	**0.05**	0.03	**0.06**	0.03	**0.07**	0.04	**0.09**	0.05	**0.11**
Elbow	0.02	**0.04**	0.02	**0.06**	0.03	**0.07**	0.03	**0.08**	0.04	**0.09**	0.05	**0.11**
Forearm	0.01	**0.04**	0.02	**0.05**	0.02	**0.06**	0.02	**0.06**	0.03	**0.08**	0.03	**0.10**
Upper leg	0.03	**0.33**	0.04	**0.41**	0.05	**0.48**	0.05	**0.53**	0.07	**0.64**	0.08	**0.74**
Knee		**0.11**		**0.13**		**0.15**		**0.17**		**0.20**		**0.23**
Lower leg												
Lymph nodes	0.27	**0.40**	0.33	**0.49**	0.36	**0.55**	0.40	**0.59**	0.46	**0.69**	0.51	**0.77**
Skin	0.11	**0.26**	0.11	**0.28**	0.12	**0.29**	0.12	**0.30**	0.13	**0.32**	0.14	**0.33**
Muscle	0.11	**0.22**	0.13	**0.26**	0.15	**0.29**	0.16	**0.32**	0.18	**0.36**	0.20	**0.40**
Skeleton	0.15	**0.31**	0.17	**0.37**	0.19	**0.42**	0.21	**0.45**	0.24	**0.52**	0.27	**0.58**
Red bone marrow	0.07	**0.13**	0.08	**0.16**	0.09	**0.18**	0.10	**0.20**	0.12	**0.23**	0.13	**0.27**
Whole body dose	0.12	**0.24**	0.14	**0.29**	0.16	**0.32**	0.17	**0.34**	0.19	**0.39**	0.21	**0.44**
ED (ICRP 60)	0.27	**0.63**	0.31	**0.73**	0.34	**0.80**	0.37	**0.85**	0.42	**0.96**	0.46	**1.05**
ED (ICRP 103)	0.25	**0.88**	0.30	**1.01**	0.34	**1.10**	0.36	**1.17**	0.42	**1.30**	0.47	**1.41**
Total filtration (mm) Al	**1.5**		**2.5**		**3.5**		**4.5**		**3.5**		**3.5**	
Cu	**0**		**0**		**0**		**0**		**0.1**		**0.2**	

Table 13.37 Normalized organ doses—Full spine ap—>17 years

Beam projection:	ap	**Table cells:**
Focus-to-detector distance (cm):	200	(+) = normal font = optimal field setting
Tube voltage (kVp):	70	**(−) = bold font = suboptimal field setting**
Total filtrations:	See last table row	Empty table cells correspond to values <0.01

Organ or tissue	**Organ dose D_T: Entrance dose K_E**							**(mSv ∗ mGy^{-1})**				
	(+)	(−)	(+)	(−)	(+)	(−)	(+)	(−)	(+)	(−)	(+)	(−)
Skull	0.25	**0.41**	0.31	**0.51**	0.35	**0.58**	0.39	**0.64**	0.46	**0.75**	0.52	**0.85**
Brain	0.01	**0.05**	0.01	**0.06**	0.02	**0.07**	0.02	**0.08**	0.03	**0.11**	0.03	**0.13**
Oral mucosa	0.27	**0.34**	0.32	**0.40**	0.35	**0.45**	0.38	**0.49**	0.44	**0.57**	0.50	**0.64**
Salivary glands	0.12	**0.31**	0.15	**0.38**	0.17	**0.43**	0.19	**0.47**	0.23	**0.54**	0.27	**0.61**
Cervical spine	0.46	**0.48**	0.59	**0.63**	0.70	**0.74**	0.79	**0.84**	0.97	**1.03**	1.15	**1.23**
Thyroid	0.73	**0.74**	0.85	**0.86**	0.93	**0.95**	0.99	**1.01**	1.10	**1.13**	1.19	**1.22**
Thymus	0.66	**0.70**	0.80	**0.85**	0.91	**0.96**	0.99	**1.05**	1.13	**1.21**	1.26	**1.35**
Thoracic spine	0.07	**0.09**	0.10	**0.13**	0.12	**0.15**	0.14	**0.18**	0.18	**0.23**	0.22	**0.29**
Claviculae	0.34	**0.89**	0.41	**1.07**	0.46	**1.20**	0.50	**1.29**	0.57	**1.47**	0.63	**1.61**
Scapulae	0.02	**0.04**	0.02	**0.05**	0.03	**0.06**	0.03	**0.07**	0.04	**0.10**	0.05	**0.12**
Ribs	0.42	**0.74**	0.48	**0.86**	0.52	**0.93**	0.55	**0.99**	0.60	**1.08**	0.64	**1.16**
Oesophagus	0.10	**0.12**	0.13	**0.16**	0.15	**0.20**	0.17	**0.22**	0.22	**0.28**	0.26	**0.34**
Trachea	0.34	**0.36**	0.42	**0.44**	0.47	**0.50**	0.52	**0.55**	0.60	**0.64**	0.67	**0.72**
Lung	0.04	**0.17**	0.06	**0.21**	0.07	**0.25**	0.08	**0.28**	0.10	**0.33**	0.12	**0.39**
Mamma	0.02	**0.30**	0.03	**0.35**	0.04	**0.38**	0.04	**0.41**	0.05	**0.45**	0.06	**0.49**
Heart	0.31	**0.40**	0.38	**0.50**	0.44	**0.58**	0.49	**0.64**	0.58	**0.76**	0.66	**0.87**
Stomach	0.08	**0.30**	0.11	**0.38**	0.13	**0.44**	0.14	**0.48**	0.18	**0.58**	0.21	**0.66**
Spleen	0.01	**0.04**	0.02	**0.06**	0.02	**0.07**	0.03	**0.08**	0.03	**0.10**	0.04	**0.13**
Liver	0.10	**0.20**	0.13	**0.26**	0.15	**0.30**	0.17	**0.33**	0.20	**0.40**	0.23	**0.46**
Gall bladder	0.23	**0.33**	0.29	**0.42**	0.34	**0.50**	0.38	**0.55**	0.46	**0.67**	0.54	**0.78**
Pancreas	0.11	**0.16**	0.14	**0.20**	0.17	**0.24**	0.19	**0.28**	0.24	**0.35**	0.28	**0.42**
Small intestine	0.14	**0.22**	0.18	**0.28**	0.21	**0.34**	0.23	**0.38**	0.28	**0.46**	0.34	**0.55**
Large intestine	0.11	**0.25**	0.14	**0.32**	0.16	**0.37**	0.18	**0.42**	0.22	**0.50**	0.26	**0.59**
Sigma/rectum	0.06	**0.16**	0.08	**0.21**	0.09	**0.24**	0.10	**0.27**	0.13	**0.34**	0.16	**0.40**
Lumbar spine	0.08	**0.10**	0.11	**0.14**	0.13	**0.17**	0.15	**0.20**	0.20	**0.26**	0.24	**0.32**
Pelvis	0.04	**0.10**	0.05	**0.14**	0.06	**0.17**	0.07	**0.19**	0.09	**0.25**	0.12	**0.31**
Adrenals	0.04	**0.06**	0.06	**0.08**	0.07	**0.09**	0.08	**0.11**	0.10	**0.14**	0.12	**0.18**
Kidneys	0.02	**0.05**	0.03	**0.06**	0.04	**0.07**	0.04	**0.08**	0.05	**0.11**	0.07	**0.14**
Urinary bladder	0.41	**0.47**	0.50	**0.59**	0.57	**0.67**	0.63	**0.74**	0.74	**0.87**	0.83	**0.99**
Uterus	0.25	**0.29**	0.32	**0.38**	0.37	**0.44**	0.42	**0.50**	0.51	**0.61**	0.60	**0.72**
Ovaries	0.06	**0.17**	0.08	**0.22**	0.10	**0.26**	0.12	**0.30**	0.15	**0.37**	0.18	**0.44**
Prostate	0.26	**0.30**	0.33	**0.38**	0.38	**0.45**	0.43	**0.50**	0.52	**0.61**	0.60	**0.72**
Testes	0.81	**0.89**	0.93	**1.03**	1.01	**1.12**	1.07	**1.20**	1.17	**1.32**	1.25	**1.43**
Upper arm		**0.02**	0.01	**0.02**	0.01	**0.03**	0.01	**0.03**	0.02	**0.04**	0.02	**0.05**
Elbow		**0.02**		**0.02**	0.01	**0.03**	0.01	**0.03**	0.02	**0.04**	0.02	**0.05**
Forearm		**0.01**		**0.02**		**0.02**		**0.02**	0.01	**0.03**	0.01	**0.04**
Upper leg	0.02	**0.14**	0.02	**0.18**	0.03	**0.21**	0.03	**0.24**	0.05	**0.31**	0.06	**0.37**
Knee		**0.03**		**0.03**		**0.04**		**0.05**		**0.06**		**0.07**
Lower leg												
Lymph nodes	0.20	**0.28**	0.24	**0.35**	0.28	**0.40**	0.31	**0.44**	0.37	**0.53**	0.42	**0.61**
Skin	0.10	**0.21**	0.11	**0.23**	0.11	**0.24**	0.12	**0.25**	0.13	**0.26**	0.13	**0.28**
Muscle	0.08	**0.15**	0.10	**0.18**	0.11	**0.21**	0.12	**0.23**	0.15	**0.27**	0.17	**0.30**
Skeleton	0.11	**0.20**	0.13	**0.24**	0.15	**0.27**	0.16	**0.30**	0.19	**0.35**	0.22	**0.41**
Red bone marrow	0.05	**0.09**	0.06	**0.11**	0.07	**0.12**	0.08	**0.13**	0.09	**0.16**	0.11	**0.19**
Whole body dose	0.09	**0.16**	0.11	**0.20**	0.12	**0.22**	0.13	**0.25**	0.15	**0.29**	0.18	**0.33**
ED (ICRP 60)	0.19	**0.30**	0.23	**0.37**	0.26	**0.41**	0.28	**0.45**	0.33	**0.53**	0.37	**0.60**
ED (ICRP 103)	0.15	**0.27**	0.18	**0.33**	0.21	**0.38**	0.23	**0.41**	0.27	**0.48**	0.30	**0.55**
Total filtration (mm) Al	**1.5**		**2.5**		**3.5**		**4.5**		**3.5**		**3.5**	
Cu	**0**		**0**		**0**		**0**		**0.1**		**0.2**	

Table 13.38 Normalized organ doses—Full spine ap—>17 years

Beam projection:			ap									**Table cells:**
Focus-to-detector distance (cm):			200					(+) = normal font = optimal field setting				
Tube voltage (kVp):			80					**(−) = bold font = suboptimal field setting**				
Total filtrations:			See last table row					Empty table cells correspond to values <0.01				
Organ or tissue	**Organ dose D_T: Entrance dose K_E**							**(mSv * mGy^{-1})**				
	(+)	(−)	(+)	(−)	(+)	(−)	(+)	(−)	(+)	(−)	(+)	(−)
Skull	0.29	**0.47**	0.35	**0.57**	0.39	**0.65**	0.43	**0.71**	0.50	**0.82**	0.55	**0.91**
Brain	0.02	**0.06**	0.02	**0.08**	0.03	**0.09**	0.03	**0.11**	0.04	**0.13**	0.05	**0.16**
Oral mucosa	0.30	**0.38**	0.36	**0.46**	0.40	**0.51**	0.43	**0.55**	0.49	**0.63**	0.55	**0.70**
Salivary glands	0.14	**0.35**	0.17	**0.43**	0.20	**0.48**	0.22	**0.52**	0.26	**0.60**	0.30	**0.67**
Cervical spine	0.57	**0.60**	0.71	**0.76**	0.83	**0.88**	0.93	**0.99**	1.12	**1.20**	1.30	**1.39**
Thyroid	0.79	**0.81**	0.91	**0.93**	0.99	**1.02**	1.05	**1.08**	1.16	**1.19**	1.24	**1.29**
Thymus	0.75	**0.79**	0.89	**0.95**	1.00	**1.07**	1.07	**1.15**	1.22	**1.31**	1.34	**1.45**
Thoracic spine	0.10	**0.13**	0.13	**0.17**	0.16	**0.21**	0.18	**0.24**	0.23	**0.30**	0.28	**0.37**
Claviculae	0.38	**0.98**	0.45	**1.16**	0.50	**1.28**	0.53	**1.37**	0.60	**1.54**	0.66	**1.66**
Scapulae	0.02	**0.05**	0.03	**0.07**	0.04	**0.08**	0.04	**0.09**	0.05	**0.12**	0.07	**0.14**
Ribs	0.45	**0.80**	0.51	**0.91**	0.55	**0.98**	0.57	**1.03**	0.62	**1.12**	0.65	**1.19**
Oesophagus	0.12	**0.16**	0.16	**0.21**	0.19	**0.25**	0.21	**0.28**	0.27	**0.35**	0.32	**0.41**
Trachea	0.39	**0.42**	0.47	**0.50**	0.53	**0.56**	0.57	**0.61**	0.66	**0.71**	0.73	**0.79**
Lung	0.06	**0.20**	0.07	**0.25**	0.09	**0.29**	0.10	**0.32**	0.12	**0.38**	0.14	**0.44**
Mamma	0.03	**0.33**	0.04	**0.38**	0.05	**0.41**	0.05	**0.43**	0.06	**0.48**	0.07	**0.51**
Heart	0.36	**0.47**	0.44	**0.58**	0.50	**0.66**	0.55	**0.73**	0.65	**0.85**	0.73	**0.97**
Stomach	0.10	**0.36**	0.13	**0.44**	0.15	**0.50**	0.17	**0.55**	0.20	**0.65**	0.24	**0.74**
Spleen	0.02	**0.06**	0.03	**0.08**	0.03	**0.09**	0.03	**0.10**	0.04	**0.13**	0.05	**0.16**
Liver	0.12	**0.24**	0.15	**0.30**	0.18	**0.35**	0.19	**0.38**	0.23	**0.45**	0.26	**0.51**
Gall bladder	0.28	**0.40**	0.35	**0.50**	0.40	**0.58**	0.44	**0.64**	0.53	**0.77**	0.60	**0.88**
Pancreas	0.14	**0.20**	0.17	**0.26**	0.20	**0.30**	0.23	**0.34**	0.28	**0.42**	0.33	**0.50**
Small intestine	0.17	**0.27**	0.21	**0.35**	0.25	**0.40**	0.28	**0.45**	0.33	**0.54**	0.39	**0.63**
Large intestine	0.13	**0.30**	0.17	**0.38**	0.19	**0.44**	0.21	**0.48**	0.26	**0.58**	0.30	**0.67**
Sigma/rectum	0.07	**0.20**	0.10	**0.25**	0.11	**0.29**	0.13	**0.33**	0.16	**0.40**	0.18	**0.47**
Lumbar spine	0.11	**0.15**	0.14	**0.19**	0.17	**0.23**	0.20	**0.26**	0.25	**0.33**	0.30	**0.41**
Pelvis	0.05	**0.14**	0.07	**0.18**	0.08	**0.22**	0.10	**0.25**	0.12	**0.31**	0.15	**0.38**
Adrenals	0.06	**0.08**	0.07	**0.11**	0.09	**0.13**	0.10	**0.14**	0.13	**0.18**	0.15	**0.22**
Kidneys	0.03	**0.06**	0.04	**0.08**	0.05	**0.10**	0.05	**0.11**	0.07	**0.14**	0.08	**0.17**
Urinary bladder	0.47	**0.55**	0.57	**0.67**	0.64	**0.76**	0.70	**0.83**	0.81	**0.97**	0.91	**1.09**
Uterus	0.30	**0.36**	0.38	**0.46**	0.44	**0.53**	0.49	**0.59**	0.59	**0.71**	0.68	**0.83**
Ovaries	0.08	**0.22**	0.11	**0.28**	0.13	**0.32**	0.14	**0.36**	0.18	**0.44**	0.21	**0.52**
Prostate	0.31	**0.36**	0.39	**0.46**	0.45	**0.53**	0.49	**0.59**	0.59	**0.70**	0.67	**0.81**
Testes	0.87	**0.96**	0.99	**1.10**	1.07	**1.20**	1.12	**1.27**	1.22	**1.39**	1.30	**1.49**
Upper arm	0.01	**0.02**	0.01	**0.03**	0.02	**0.04**	0.02	**0.04**	0.03	**0.05**	0.03	**0.07**
Elbow		**0.02**	0.01	**0.03**	0.02	**0.04**	0.02	**0.04**	0.02	**0.05**	0.03	**0.07**
Forearm		**0.02**		**0.02**		**0.03**	0.01	**0.03**	0.01	**0.04**	0.02	**0.05**
Upper leg	0.03	**0.18**	0.03	**0.23**	0.04	**0.27**	0.05	**0.30**	0.06	**0.37**	0.07	**0.44**
Knee		**0.04**		**0.05**		**0.05**		**0.06**		**0.08**		**0.09**
Lower leg												
Lymph nodes	0.23	**0.33**	0.28	**0.41**	0.32	**0.46**	0.35	**0.51**	0.41	**0.60**	0.47	**0.68**
Skin	0.11	**0.22**	0.11	**0.24**	0.12	**0.25**	0.12	**0.26**	0.13	**0.28**	0.14	**0.29**
Muscle	0.10	**0.18**	0.11	**0.21**	0.13	**0.24**	0.14	**0.26**	0.16	**0.30**	0.18	**0.33**
Skeleton	0.12	**0.23**	0.15	**0.28**	0.17	**0.31**	0.18	**0.34**	0.21	**0.40**	0.24	**0.45**
Red bone marrow	0.06	**0.10**	0.07	**0.13**	0.08	**0.14**	0.09	**0.16**	0.11	**0.19**	0.12	**0.22**
Whole body dose	0.10	**0.19**	0.12	**0.23**	0.14	**0.25**	0.15	**0.28**	0.17	**0.32**	0.20	**0.36**
ED (ICRP 60)	0.22	**0.35**	0.26	**0.42**	0.29	**0.47**	0.32	**0.51**	0.36	**0.59**	0.40	**0.66**
ED (ICRP 103)	0.17	**0.31**	0.21	**0.38**	0.24	**0.43**	0.26	**0.46**	0.30	**0.54**	0.33	**0.60**
Total filtration (mm) **Al**	**1.5**		**2.5**		**3.5**		**4.5**		**3.5**		**3.5**	
Cu	**0**		**0**		**0**		**0**		**0.1**		**0.2**	

Table 13.39 Normalized organ doses—Full spine ap—>17 years

Beam projection:			ap									**Table cells:**
Focus-to-detector distance (cm):			200				(+) = normal font = optimal field setting					
Tube voltage (kVp):			90				**(−) = bold font = suboptimal field setting**					
Total filtrations:			See last table row				Empty table cells correspond to values <0.01					
Organ or tissue	**Organ dose D_T: Entrance dose K_E**						**(mSv * mGy^{-1})**					
	(+)	(−)	(+)	(−)	(+)	(−)	(+)	(−)	(+)	(−)	(+)	(−)
Skull	0.32	**0.53**	0.38	**0.63**	0.43	**0.70**	0.46	**0.76**	0.52	**0.86**	0.58	**0.95**
Brain	0.02	**0.08**	0.03	**0.10**	0.03	**0.11**	0.04	**0.13**	0.04	**0.16**	0.05	**0.18**
Oral mucosa	0.34	**0.43**	0.40	**0.50**	0.44	**0.56**	0.47	**0.60**	0.53	**0.68**	0.58	**0.75**
Salivary glands	0.16	**0.40**	0.20	**0.47**	0.23	**0.52**	0.25	**0.57**	0.29	**0.64**	0.33	**0.71**
Cervical spine	0.67	**0.71**	0.82	**0.88**	0.94	**1.00**	1.04	**1.11**	1.23	**1.32**	1.40	**1.51**
Thyroid	0.85	**0.87**	0.97	**0.99**	1.05	**1.08**	1.11	**1.14**	1.20	**1.24**	1.28	**1.33**
Thymus	0.83	**0.88**	0.97	**1.04**	1.07	**1.15**	1.15	**1.24**	1.29	**1.39**	1.40	**1.52**
Thoracic spine	0.13	**0.17**	0.17	**0.22**	0.20	**0.26**	0.22	**0.29**	0.27	**0.36**	0.33	**0.43**
Claviculae	0.41	**1.06**	0.48	**1.23**	0.53	**1.34**	0.56	**1.43**	0.62	**1.57**	0.67	**1.68**
Scapulae	0.03	**0.07**	0.04	**0.09**	0.05	**0.10**	0.05	**0.11**	0.06	**0.14**	0.08	**0.16**
Ribs	0.47	**0.85**	0.53	**0.95**	0.56	**1.01**	0.59	**1.06**	0.63	**1.14**	0.66	**1.20**
Oesophagus	0.15	**0.20**	0.19	**0.25**	0.23	**0.29**	0.25	**0.33**	0.31	**0.40**	0.36	**0.47**
Trachea	0.44	**0.47**	0.52	**0.55**	0.58	**0.62**	0.62	**0.67**	0.70	**0.76**	0.77	**0.84**
Lung	0.07	**0.24**	0.09	**0.29**	0.10	**0.33**	0.11	**0.36**	0.14	**0.42**	0.16	**0.47**
Mamma	0.04	**0.36**	0.05	**0.40**	0.05	**0.43**	0.06	**0.46**	0.07	**0.50**	0.08	**0.53**
Heart	0.41	**0.54**	0.50	**0.65**	0.56	**0.74**	0.61	**0.80**	0.70	**0.93**	0.78	**1.04**
Stomach	0.12	**0.41**	0.15	**0.50**	0.17	**0.56**	0.19	**0.61**	0.23	**0.70**	0.26	**0.79**
Spleen	0.03	**0.07**	0.03	**0.09**	0.04	**0.11**	0.04	**0.12**	0.05	**0.15**	0.06	**0.18**
Liver	0.15	**0.28**	0.18	**0.34**	0.20	**0.39**	0.22	**0.42**	0.25	**0.49**	0.29	**0.56**
Gall bladder	0.32	**0.47**	0.39	**0.58**	0.45	**0.66**	0.49	**0.72**	0.58	**0.85**	0.65	**0.96**
Pancreas	0.17	**0.25**	0.21	**0.31**	0.24	**0.36**	0.27	**0.40**	0.32	**0.48**	0.37	**0.56**
Small intestine	0.20	**0.33**	0.25	**0.40**	0.28	**0.46**	0.31	**0.51**	0.37	**0.60**	0.42	**0.69**
Large intestine	0.16	**0.36**	0.19	**0.43**	0.22	**0.49**	0.24	**0.54**	0.29	**0.64**	0.33	**0.73**
Sigma/rectum	0.09	**0.24**	0.11	**0.30**	0.13	**0.34**	0.15	**0.38**	0.18	**0.45**	0.21	**0.52**
Lumbar spine	0.14	**0.19**	0.18	**0.24**	0.21	**0.28**	0.24	**0.32**	0.29	**0.39**	0.35	**0.47**
Pelvis	0.07	**0.18**	0.09	**0.23**	0.10	**0.26**	0.12	**0.30**	0.15	**0.36**	0.17	**0.43**
Adrenals	0.07	**0.10**	0.09	**0.13**	0.11	**0.15**	0.12	**0.17**	0.15	**0.21**	0.18	**0.26**
Kidneys	0.04	**0.08**	0.05	**0.10**	0.06	**0.12**	0.07	**0.14**	0.08	**0.17**	0.10	**0.20**
Urinary bladder	0.53	**0.63**	0.63	**0.75**	0.71	**0.84**	0.76	**0.91**	0.87	**1.04**	0.96	**1.16**
Uterus	0.36	**0.43**	0.44	**0.53**	0.50	**0.61**	0.55	**0.67**	0.65	**0.80**	0.74	**0.91**
Ovaries	0.10	**0.26**	0.13	**0.33**	0.15	**0.38**	0.17	**0.42**	0.20	**0.50**	0.24	**0.58**
Prostate	0.36	**0.43**	0.44	**0.52**	0.50	**0.60**	0.55	**0.66**	0.64	**0.77**	0.72	**0.88**
Testes	0.93	**1.03**	1.04	**1.17**	1.11	**1.26**	1.17	**1.33**	1.26	**1.44**	1.33	**1.54**
Upper arm	0.01	**0.03**	0.02	**0.04**	0.02	**0.05**	0.03	**0.05**	0.03	**0.06**	0.04	**0.08**
Elbow	0.01	**0.03**	0.02	**0.04**	0.02	**0.05**	0.02	**0.05**	0.03	**0.07**	0.04	**0.08**
Forearm		**0.02**	0.01	**0.03**	0.01	**0.04**	0.01	**0.04**	0.02	**0.05**	0.02	**0.06**
Upper leg	0.03	**0.22**	0.04	**0.27**	0.05	**0.31**	0.06	**0.35**	0.07	**0.43**	0.08	**0.50**
Knee		**0.04**		**0.05**		**0.06**		**0.07**		**0.09**		**0.10**
Lower leg												
Lymph nodes	0.27	**0.38**	0.32	**0.46**	0.36	**0.52**	0.39	**0.56**	0.45	**0.65**	0.51	**0.74**
Skin	0.11	**0.23**	0.12	**0.25**	0.12	**0.26**	0.13	**0.27**	0.14	**0.29**	0.14	**0.30**
Muscle	0.11	**0.20**	0.13	**0.23**	0.14	**0.26**	0.15	**0.28**	0.18	**0.32**	0.20	**0.36**
Skeleton	0.14	**0.26**	0.17	**0.31**	0.18	**0.34**	0.20	**0.37**	0.23	**0.43**	0.25	**0.48**
Red bone marrow	0.07	**0.12**	0.08	**0.15**	0.09	**0.17**	0.10	**0.18**	0.12	**0.21**	0.14	**0.25**
Whole body dose	0.12	**0.21**	0.14	**0.25**	0.15	**0.28**	0.16	**0.30**	0.19	**0.35**	0.21	**0.39**
ED (ICRP 60)	0.25	**0.39**	0.29	**0.46**	0.32	**0.51**	0.34	**0.55**	0.39	**0.63**	0.43	**0.70**
ED (ICRP 103)	0.20	**0.36**	0.23	**0.42**	0.26	**0.47**	0.28	**0.51**	0.32	**0.58**	0.36	**0.65**
Total filtration (mm) Al	**1.5**		**2.5**		**3.5**		**4.5**		**3.5**		**3.5**	
Cu	**0**		**0**		**0**		**0**		**0.1**		**0.2**	

Table 13.40 Normalized organ doses—Full spine pa—8–12 years

Beam projection:	pa											**Table cells:**
Focus-to-detector distance (cm):	200											(+) = normal font = optimal field setting
Tube voltage (kVp):	70											**(−) = bold font = suboptimal field setting**
Total filtrations:	See last table row											Empty table cells correspond to values <0.01
Organ or tissue	**Organ dose D_T: Entrance dose K_E (mSv ∗ mGy^{-1})**											
	(+)	(−)	(+)	(−)	(+)	(−)	(+)	(−)	(+)	(−)	(+)	(−)
Skull	0.14	**0.43**	0.17	**0.52**	0.20	**0.59**	0.22	**0.64**	0.26	**0.74**	0.30	**0.83**
Brain	0.03	**0.11**	0.04	**0.14**	0.05	**0.16**	0.06	**0.18**	0.07	**0.22**	0.09	**0.26**
Oral mucosa	0.08	**0.17**	0.10	**0.21**	0.12	**0.24**	0.14	**0.27**	0.17	**0.32**	0.20	**0.37**
Salivary glands	0.09	**0.44**	0.12	**0.51**	0.14	**0.56**	0.15	**0.60**	0.18	**0.67**	0.21	**0.73**
Cervical spine	0.91	**0.97**	1.14	**1.22**	1.32	**1.40**	1.45	**1.56**	1.72	**1.85**	1.96	**2.11**
Thyroid	0.10	**0.11**	0.13	**0.15**	0.15	**0.18**	0.17	**0.20**	0.21	**0.25**	0.25	**0.31**
Thymus	0.03	**0.05**	0.04	**0.07**	0.05	**0.08**	0.06	**0.09**	0.08	**0.12**	0.10	**0.15**
Thoracic spine	0.85	**0.93**	1.08	**1.17**	1.25	**1.36**	1.38	**1.51**	1.65	**1.81**	1.89	**2.09**
Claviculae	0.06	**0.12**	0.08	**0.16**	0.09	**0.19**	0.10	**0.22**	0.13	**0.27**	0.16	**0.33**
Scapulae	0.04	**0.14**	0.05	**0.17**	0.07	**0.20**	0.08	**0.23**	0.10	**0.27**	0.12	**0.32**
Ribs	0.48	**0.86**	0.54	**0.98**	0.58	**1.06**	0.61	**1.12**	0.66	**1.22**	0.71	**1.30**
Oesophagus	0.10	**0.13**	0.13	**0.17**	0.16	**0.20**	0.18	**0.23**	0.23	**0.29**	0.28	**0.35**
Trachea	0.10	**0.13**	0.14	**0.17**	0.16	**0.20**	0.18	**0.22**	0.23	**0.28**	0.28	**0.34**
Lung	0.08	**0.29**	0.10	**0.35**	0.12	**0.40**	0.13	**0.44**	0.16	**0.52**	0.19	**0.58**
Mamma	0.01	**0.11**	0.01	**0.14**	0.02	**0.16**	0.02	**0.18**	0.03	**0.22**	0.03	**0.26**
Heart	0.08	**0.12**	0.10	**0.15**	0.12	**0.18**	0.14	**0.21**	0.17	**0.26**	0.21	**0.31**
Stomach	0.04	**0.09**	0.05	**0.12**	0.06	**0.14**	0.07	**0.16**	0.09	**0.20**	0.11	**0.24**
Spleen	0.04	**0.19**	0.05	**0.24**	0.06	**0.27**	0.07	**0.30**	0.08	**0.37**	0.10	**0.42**
Liver	0.04	**0.11**	0.06	**0.15**	0.07	**0.17**	0.08	**0.19**	0.10	**0.24**	0.12	**0.28**
Gall bladder	0.08	**0.11**	0.11	**0.15**	0.13	**0.17**	0.15	**0.20**	0.19	**0.25**	0.23	**0.31**
Pancreas	0.11	**0.17**	0.14	**0.22**	0.17	**0.27**	0.19	**0.30**	0.24	**0.38**	0.28	**0.45**
Small intestine	0.09	**0.15**	0.12	**0.20**	0.14	**0.24**	0.16	**0.27**	0.21	**0.34**	0.25	**0.41**
Large intestine	0.05	**0.12**	0.07	**0.15**	0.09	**0.18**	0.10	**0.21**	0.12	**0.26**	0.15	**0.32**
Sigma/rectum	0.07	**0.15**	0.09	**0.20**	0.11	**0.24**	0.12	**0.27**	0.15	**0.33**	0.18	**0.39**
Lumbar spine	0.93	**1.02**	1.17	**1.28**	1.35	**1.49**	1.49	**1.65**	1.77	**1.97**	2.02	**2.26**
Pelvis	0.48	**0.90**	0.59	**1.11**	0.66	**1.27**	0.72	**1.39**	0.83	**1.62**	0.93	**1.82**
Adrenals	0.39	**0.46**	0.48	**0.57**	0.55	**0.65**	0.61	**0.72**	0.71	**0.85**	0.81	**0.96**
Kidneys	0.28	**0.63**	0.35	**0.76**	0.39	**0.85**	0.42	**0.92**	0.49	**1.04**	0.54	**1.15**
Urinary bladder	0.08	**0.10**	0.10	**0.13**	0.12	**0.16**	0.14	**0.18**	0.17	**0.23**	0.21	**0.27**
Uterus	0.11	**0.14**	0.14	**0.18**	0.17	**0.22**	0.20	**0.25**	0.25	**0.32**	0.30	**0.39**
Ovaries	0.09	**0.16**	0.12	**0.21**	0.14	**0.25**	0.16	**0.29**	0.20	**0.36**	0.24	**0.44**
Prostate	0.09	**0.12**	0.12	**0.16**	0.14	**0.19**	0.16	**0.21**	0.20	**0.27**	0.24	**0.32**
Testes	0.03	**0.05**	0.04	**0.07**	0.05	**0.08**	0.06	**0.09**	0.07	**0.11**	0.09	**0.14**
Upper arm	0.02	**0.04**	0.02	**0.05**	0.03	**0.06**	0.03	**0.07**	0.04	**0.09**	0.06	**0.11**
Elbow	0.02	**0.04**	0.02	**0.05**	0.03	**0.06**	0.03	**0.07**	0.04	**0.09**	0.05	**0.11**
Forearm	0.01	**0.03**	0.01	**0.04**	0.02	**0.05**	0.02	**0.06**	0.03	**0.07**	0.03	**0.09**
Upper leg	0.03	**0.18**	0.03	**0.23**	0.04	**0.28**	0.05	**0.31**	0.06	**0.39**	0.07	**0.46**
Knee										**0.01**		**0.01**
Lower leg												
Lymph nodes	0.09	**0.17**	0.12	**0.22**	0.14	**0.26**	0.16	**0.29**	0.20	**0.35**	0.24	**0.41**
Skin	0.10	**0.20**	0.10	**0.21**	0.11	**0.22**	0.11	**0.23**	0.12	**0.24**	0.12	**0.25**
Muscle	0.09	**0.16**	0.11	**0.19**	0.12	**0.21**	0.13	**0.23**	0.15	**0.27**	0.17	**0.30**
Skeleton	0.23	**0.39**	0.28	**0.48**	0.32	**0.54**	0.35	**0.59**	0.41	**0.69**	0.46	**0.78**
Red bone marrow	0.13	**0.21**	0.16	**0.26**	0.19	**0.30**	0.21	**0.33**	0.24	**0.39**	0.28	**0.45**
Whole body dose	0.11	**0.19**	0.13	**0.23**	0.15	**0.26**	0.16	**0.29**	0.19	**0.33**	0.21	**0.38**
ED (ICRP 60)	0.07	**0.15**	0.09	**0.19**	0.11	**0.22**	0.12	**0.25**	0.15	**0.30**	0.18	**0.35**
ED (ICRP 103)	0.07	**0.16**	0.09	**0.20**	0.11	**0.23**	0.12	**0.26**	0.15	**0.31**	0.18	**0.37**
Total filtration (mm) **Al**	**1.5**		**2.5**		**3.5**		**4.5**		**3.5**		**3.5**	
Cu	**0**		**0**		**0**		**0**		**0.1**		**0.2**	

Table 13.41 Normalized organ doses—Full spine pa—8–12 years

Beam projection:			pa									**Table cells:**
Focus-to-detector distance (cm):			200									(+) = normal font = optimal field setting
Tube voltage (kVp):			80									**(−) = bold font = suboptimal field setting**
Total filtrations:			See last table row									Empty table cells correspond to values <0.01
Organ or tissue	**Organ dose D_T: Entrance dose K_E**							**(mSv ∗ mGy^{-1})**				
	(+)	(−)	(+)	(−)	(+)	(−)	(+)	(−)	(+)	(−)	(+)	(−)
Skull	0.16	**0.48**	0.20	**0.58**	0.22	**0.65**	0.24	**0.70**	0.29	**0.80**	0.32	**0.88**
Brain	0.04	**0.13**	0.05	**0.17**	0.06	**0.19**	0.07	**0.21**	0.09	**0.26**	0.10	**0.30**
Oral mucosa	0.10	**0.20**	0.13	**0.25**	0.15	**0.28**	0.16	**0.31**	0.20	**0.37**	0.23	**0.42**
Salivary glands	0.11	**0.48**	0.14	**0.56**	0.16	**0.61**	0.18	**0.65**	0.21	**0.72**	0.24	**0.78**
Cervical spine	1.06	**1.13**	1.29	**1.39**	1.47	**1.58**	1.61	**1.73**	1.87	**2.02**	2.09	**2.27**
Thyroid	0.12	**0.15**	0.16	**0.19**	0.18	**0.22**	0.21	**0.25**	0.26	**0.31**	0.30	**0.36**
Thymus	0.05	**0.07**	0.06	**0.09**	0.07	**0.10**	0.08	**0.12**	0.10	**0.15**	0.12	**0.18**
Thoracic spine	1.00	**1.09**	1.23	**1.35**	1.40	**1.54**	1.54	**1.70**	1.81	**2.00**	2.04	**2.26**
Claviculae	0.07	**0.15**	0.10	**0.20**	0.11	**0.23**	0.13	**0.26**	0.16	**0.32**	0.19	**0.38**
Scapulae	0.05	**0.17**	0.07	**0.21**	0.08	**0.24**	0.09	**0.26**	0.12	**0.31**	0.14	**0.36**
Ribs	0.51	**0.92**	0.57	**1.03**	0.61	**1.11**	0.64	**1.16**	0.68	**1.25**	0.72	**1.32**
Oesophagus	0.13	**0.17**	0.17	**0.21**	0.20	**0.25**	0.23	**0.28**	0.28	**0.35**	0.33	**0.41**
Trachea	0.13	**0.16**	0.17	**0.21**	0.20	**0.25**	0.23	**0.28**	0.28	**0.34**	0.34	**0.41**
Lung	0.10	**0.33**	0.12	**0.40**	0.14	**0.45**	0.15	**0.49**	0.18	**0.57**	0.21	**0.63**
Mamma	0.01	**0.13**	0.02	**0.17**	0.02	**0.19**	0.03	**0.21**	0.03	**0.25**	0.04	**0.29**
Heart	0.10	**0.15**	0.13	**0.19**	0.15	**0.23**	0.17	**0.25**	0.21	**0.31**	0.25	**0.37**
Stomach	0.05	**0.12**	0.07	**0.15**	0.08	**0.18**	0.09	**0.20**	0.11	**0.24**	0.13	**0.29**
Spleen	0.05	**0.22**	0.06	**0.28**	0.07	**0.32**	0.08	**0.35**	0.10	**0.41**	0.12	**0.47**
Liver	0.06	**0.14**	0.07	**0.18**	0.08	**0.21**	0.10	**0.23**	0.12	**0.28**	0.14	**0.33**
Gall bladder	0.11	**0.15**	0.14	**0.19**	0.16	**0.22**	0.18	**0.25**	0.23	**0.31**	0.27	**0.37**
Pancreas	0.14	**0.22**	0.18	**0.28**	0.21	**0.33**	0.23	**0.37**	0.28	**0.45**	0.34	**0.53**
Small intestine	0.12	**0.20**	0.15	**0.25**	0.18	**0.29**	0.20	**0.33**	0.25	**0.40**	0.29	**0.48**
Large intestine	0.07	**0.15**	0.09	**0.19**	0.11	**0.23**	0.12	**0.26**	0.15	**0.32**	0.18	**0.38**
Sigma/rectum	0.09	**0.19**	0.11	**0.24**	0.13	**0.29**	0.15	**0.32**	0.18	**0.39**	0.21	**0.46**
Lumbar spine	1.09	**1.20**	1.33	**1.47**	1.51	**1.68**	1.66	**1.85**	1.93	**2.17**	2.17	**2.45**
Pelvis	0.55	**1.03**	0.65	**1.24**	0.73	**1.40**	0.79	**1.52**	0.89	**1.74**	0.98	**1.93**
Adrenals	0.45	**0.53**	0.55	**0.65**	0.62	**0.74**	0.68	**0.81**	0.79	**0.94**	0.88	**1.05**
Kidneys	0.32	**0.71**	0.39	**0.84**	0.43	**0.93**	0.47	**1.00**	0.53	**1.12**	0.59	**1.23**
Urinary bladder	0.10	**0.13**	0.13	**0.17**	0.15	**0.20**	0.17	**0.22**	0.21	**0.27**	0.24	**0.33**
Uterus	0.14	**0.18**	0.18	**0.23**	0.21	**0.28**	0.24	**0.31**	0.30	**0.39**	0.36	**0.46**
Ovaries	0.12	**0.21**	0.15	**0.27**	0.17	**0.31**	0.20	**0.35**	0.24	**0.44**	0.29	**0.52**
Prostate	0.12	**0.15**	0.15	**0.20**	0.17	**0.23**	0.19	**0.26**	0.24	**0.32**	0.28	**0.38**
Testes	0.04	**0.07**	0.05	**0.08**	0.06	**0.10**	0.07	**0.11**	0.09	**0.14**	0.11	**0.16**
Upper arm	0.03	**0.05**	0.03	**0.07**	0.04	**0.08**	0.04	**0.09**	0.06	**0.11**	0.07	**0.14**
Elbow	0.02	**0.05**	0.03	**0.07**	0.04	**0.08**	0.04	**0.09**	0.05	**0.12**	0.07	**0.14**
Forearm	0.02	**0.04**	0.02	**0.05**	0.02	**0.06**	0.03	**0.07**	0.04	**0.09**	0.04	**0.11**
Upper leg	0.03	**0.22**	0.04	**0.28**	0.05	**0.33**	0.06	**0.37**	0.07	**0.45**	0.09	**0.52**
Knee						**0.01**		**0.01**		**0.02**		**0.02**
Lower leg												
Lymph nodes	0.12	**0.21**	0.15	**0.27**	0.18	**0.31**	0.20	**0.34**	0.24	**0.41**	0.28	**0.48**
Skin	0.10	**0.21**	0.11	**0.22**	0.11	**0.23**	0.12	**0.24**	0.12	**0.25**	0.13	**0.27**
Muscle	0.10	**0.18**	0.12	**0.22**	0.14	**0.24**	0.15	**0.26**	0.17	**0.30**	0.19	**0.33**
Skeleton	0.26	**0.44**	0.32	**0.53**	0.36	**0.60**	0.39	**0.65**	0.44	**0.74**	0.49	**0.83**
Red bone marrow	0.16	**0.25**	0.19	**0.30**	0.21	**0.34**	0.24	**0.37**	0.28	**0.44**	0.31	**0.50**
Whole body dose	0.12	**0.22**	0.15	**0.26**	0.16	**0.30**	0.18	**0.32**	0.21	**0.37**	0.23	**0.41**
ED (ICRP 60)	0.09	**0.18**	0.12	**0.23**	0.13	**0.26**	0.15	**0.29**	0.18	**0.35**	0.21	**0.41**
ED (ICRP 103)	0.09	**0.19**	0.11	**0.24**	0.13	**0.27**	0.15	**0.30**	0.18	**0.36**	0.21	**0.42**
Total filtration (mm) Al	**1.5**		**2.5**		**3.5**		**4.5**		**3.5**		**3.5**	
Cu	**0**		**0**		**0**		**0**		**0.1**		**0.2**	

Table 13.42 Normalized organ doses—Full spine pa—8–12 years

Beam projection:		pa										**Table cells:**
Focus-to-detector distance (cm):		200										(+) = normal font = optimal field setting
Tube voltage (kVp):		90										**(−) = bold font = suboptimal field setting**
Total filtrations:		See last table row										Empty table cells correspond to values <0.01
Organ or tissue	**Organ dose D_T: Entrance dose K_E**						**(mSv * mGy^{-1})**					
	(+)	(−)	(+)	(−)	(+)	(−)	(+)	(−)	(+)	(−)	(+)	(−)
Skull	0.18	**0.53**	0.22	**0.63**	0.24	**0.69**	0.26	**0.74**	0.30	**0.83**	0.34	**0.91**
Brain	0.05	**0.16**	0.06	**0.19**	0.07	**0.22**	0.08	**0.24**	0.10	**0.29**	0.11	**0.33**
Oral mucosa	0.12	**0.23**	0.15	**0.28**	0.17	**0.32**	0.19	**0.35**	0.22	**0.40**	0.25	**0.46**
Salivary glands	0.13	**0.52**	0.16	**0.60**	0.18	**0.65**	0.20	**0.69**	0.23	**0.76**	0.26	**0.82**
Cervical spine	1.19	**1.27**	1.42	**1.53**	1.59	**1.72**	1.72	**1.86**	1.97	**2.14**	2.17	**2.36**
Thyroid	0.15	**0.18**	0.19	**0.22**	0.22	**0.26**	0.24	**0.29**	0.29	**0.35**	0.34	**0.40**
Thymus	0.06	**0.09**	0.07	**0.11**	0.09	**0.13**	0.10	**0.14**	0.12	**0.18**	0.15	**0.21**
Thoracic spine	1.13	**1.24**	1.36	**1.50**	1.53	**1.69**	1.66	**1.84**	1.91	**2.12**	2.12	**2.37**
Claviculae	0.09	**0.19**	0.11	**0.23**	0.13	**0.27**	0.15	**0.30**	0.18	**0.36**	0.21	**0.41**
Scapulae	0.07	**0.19**	0.08	**0.23**	0.10	**0.27**	0.11	**0.29**	0.13	**0.34**	0.16	**0.39**
Ribs	0.53	**0.96**	0.59	**1.07**	0.63	**1.14**	0.65	**1.19**	0.69	**1.26**	0.72	**1.32**
Oesophagus	0.16	**0.20**	0.20	**0.26**	0.24	**0.30**	0.27	**0.33**	0.32	**0.40**	0.38	**0.47**
Trachea	0.17	**0.20**	0.21	**0.25**	0.24	**0.29**	0.27	**0.33**	0.32	**0.39**	0.38	**0.46**
Lung	0.11	**0.37**	0.14	**0.44**	0.16	**0.49**	0.17	**0.53**	0.20	**0.61**	0.23	**0.67**
Mamma	0.02	**0.16**	0.02	**0.19**	0.03	**0.22**	0.03	**0.24**	0.04	**0.28**	0.05	**0.32**
Heart	0.12	**0.18**	0.15	**0.23**	0.18	**0.26**	0.20	**0.29**	0.24	**0.35**	0.28	**0.41**
Stomach	0.06	**0.14**	0.08	**0.18**	0.09	**0.21**	0.10	**0.23**	0.13	**0.28**	0.15	**0.33**
Spleen	0.06	**0.26**	0.08	**0.31**	0.09	**0.35**	0.10	**0.39**	0.12	**0.45**	0.14	**0.51**
Liver	0.07	**0.17**	0.09	**0.21**	0.10	**0.24**	0.11	**0.26**	0.14	**0.31**	0.16	**0.36**
Gall bladder	0.13	**0.18**	0.17	**0.23**	0.19	**0.27**	0.22	**0.30**	0.27	**0.36**	0.31	**0.42**
Pancreas	0.17	**0.27**	0.21	**0.33**	0.24	**0.38**	0.27	**0.42**	0.33	**0.51**	0.38	**0.59**
Small intestine	0.15	**0.24**	0.18	**0.30**	0.21	**0.34**	0.23	**0.38**	0.28	**0.46**	0.33	**0.53**
Large intestine	0.09	**0.19**	0.11	**0.23**	0.13	**0.27**	0.15	**0.30**	0.18	**0.37**	0.21	**0.43**
Sigma/rectum	0.11	**0.23**	0.13	**0.29**	0.15	**0.33**	0.17	**0.37**	0,20	**0.44**	0.23	**0.51**
Lumbar spine	1.22	**1.36**	1.47	**1.63**	1.64	**1.84**	1.78	**2.00**	2.04	**2.31**	2.26	**2.56**
Pelvis	0.60	**1.14**	0.70	**1.35**	0.77	**1.50**	0.83	**1.61**	0.93	**1.82**	1.01	**1.99**
Adrenals	0.51	**0.61**	0.61	**0.73**	0.69	**0.81**	0.74	**0.88**	0.85	**1.01**	0.94	**1.12**
Kidneys	0.36	**0.78**	0.42	**0.91**	0.47	**0.99**	0.50	**1.06**	0.56	**1.18**	0.62	**1.28**
Urinary bladder	0.12	**0.16**	0.15	**0.20**	0.18	**0.23**	0.20	**0.26**	0.24	**0.31**	0.27	**0.37**
Uterus	0.17	**0.22**	0.22	**0.28**	0.25	**0.33**	0.28	**0.37**	0.34	**0.44**	0.40	**0.52**
Ovaries	0.14	**0.26**	0.18	**0.32**	0.21	**0.37**	0.23	**0.41**	0.28	**0.50**	0.32	**0.58**
Prostate	0.14	**0.19**	0.17	**0.23**	0.20	**0.27**	0.22	**0.30**	0.27	**0.36**	0.31	**0.42**
Testes	0.05	**0.08**	0.07	**0.10**	0.08	**0.12**	0.09	**0.13**	0.11	**0.16**	0.12	**0.18**
Upper arm	0.03	**0.06**	0.04	**0.08**	0.05	**0.09**	0.05	**0.11**	0.07	**0.13**	0.08	**0.15**
Elbow	0.03	**0.07**	0.04	**0.08**	0.05	**0.10**	0.05	**0.11**	0.06	**0.13**	0.08	**0.16**
Forearm	0.02	**0.05**	0.03	**0.07**	0.03	**0.08**	0.03	**0.09**	0.04	**0.11**	0.05	**0.13**
Upper leg	0.04	**0.26**	0.05	**0.33**	0.06	**0.37**	0.07	**0.41**	0.08	**0.49**	0.10	**0.56**
Knee				**0.01**		**0.01**		**0.01**		**0.02**		**0.02**
Lower leg												
Lymph nodes	0.14	**0.25**	0.18	**0.31**	0.21	**0.35**	0.23	**0.39**	0.27	**0.46**	0.32	**0.52**
Skin	0.11	**0.22**	0.11	**0.23**	0.12	**0.24**	0.12	**0.25**	0.13	**0.26**	0.14	**0.28**
Muscle	0.11	**0.20**	0.13	**0.24**	0.15	**0.26**	0.16	**0.28**	0.18	**0.32**	0.20	**0.35**
Skeleton	0.29	**0.49**	0.35	**0.58**	0.38	**0.64**	0.41	**0.69**	0.47	**0.78**	0.51	**0.85**
Red bone marrow	0.18	**0.28**	0.21	**0.34**	0.24	**0.38**	0.26	**0.42**	0.30	**0.48**	0.34	**0.54**
Whole body dose	0.14	**0.25**	0.16	**0.29**	0.18	**0.32**	0.19	**0.35**	0.22	**0.40**	0.25	**0.44**
ED (ICRP 60)	0.11	**0.22**	0.14	**0.26**	0.16	**0.30**	0.17	**0.33**	0.20	**0.39**	0.24	**0.44**
ED (ICRP 103)	0.11	**0.23**	0.13	**0.27**	0.15	**0.31**	0.17	**0.34**	0.20	**0.40**	0.23	**0.46**
Total filtration (mm) Al	**1.5**		**2.5**		**3.5**		**4.5**		**3.5**		**3.5**	
Cu	**0**		**0**		**0**		**0**		**0.1**		**0.2**	

Table 13.43 Normalized organ doses—Full spine pa—13–17 years

Beam projection:	pa											**Table cells:**
Focus-to-detector distance (cm):	200											(+) = normal font = optimal field setting
Tube voltage (kVp):	70											**(−) = bold font = suboptimal field setting**
Total filtrations:	See last table row											Empty table cells correspond to values <0.01
Organ or tissue	**Organ dose D_T: Entrance dose K_E**						**(mSv * mGy^{-1})**					
	(+)	(−)	(+)	(−)	(+)	(−)	(+)	(−)	(+)	(−)	(+)	(−)
Skull	0.12	**0.34**	0.15	**0.42**	0.18	**0.48**	0.20	**0.53**	0.24	**0.62**	0.28	**0.70**
Brain	0.03	**0.09**	0.04	**0.12**	0.04	**0.14**	0.05	**0.16**	0.06	**0.19**	0.07	**0.23**
Oral mucosa	0.07	**0.13**	0.09	**0.16**	0.11	**0.19**	0.13	**0.21**	0.15	**0.26**	0.18	**0.30**
Salivary glands	0.08	**0.39**	0.11	**0.46**	0.12	**0.51**	0.14	**0.55**	0.17	**0.61**	0.20	**0.68**
Cervical spine	0.95	**0.99**	1.20	**1.25**	1.39	**1.45**	1.54	**1.61**	1.84	**1.93**	2.12	**2.23**
Thyroid	0.06	**0.07**	0.08	**0.09**	0.10	**0.11**	0.12	**0.13**	0.15	**0.17**	0.18	**0.20**
Thymus	0.02	**0.03**	0.03	**0.05**	0.03	**0.06**	0.04	**0.07**	0.05	**0.09**	0.06	**0.11**
Thoracic spine	0.72	**0.79**	0.91	**1.00**	1.06	**1.17**	1.18	**1.31**	1.43	**1.59**	1.66	**1.86**
Claviculae	0.03	**0.07**	0.04	**0.09**	0.05	**0.11**	0.06	**0.13**	0.08	**0.17**	0.09	**0.20**
Scapulae	0.02	**0.09**	0.03	**0.12**	0.04	**0.14**	0.05	**0.16**	0.06	**0.20**	0.07	**0.24**
Ribs	0.42	**0.81**	0.47	**0.92**	0.51	**1.00**	0.54	**1.06**	0.58	**1.16**	0.62	**1.24**
Oesophagus	0.10	**0.13**	0.13	**0.17**	0.16	**0.21**	0.18	**0.24**	0.23	**0.30**	0.28	**0.37**
Trachea	0.06	**0.07**	0.08	**0.10**	0.10	**0.12**	0.11	**0.14**	0.15	**0.17**	0.18	**0.21**
Lung	0.05	**0.24**	0.06	**0.30**	0.08	**0.35**	0.09	**0.39**	0.11	**0.46**	0.13	**0.52**
Mamma	0.05	**0.37**	0.07	**0.48**	0.09	**0.58**	0.10	**0.66**	0.13	**0.83**	0.17	**1.01**
Heart	0.05	**0.08**	0.06	**0.11**	0.08	**0.13**	0.09	**0.15**	0.11	**0.19**	0.14	**0.23**
Stomach	0.02	**0.07**	0.03	**0.09**	0.04	**0.11**	0.04	**0.13**	0.06	**0.16**	0.07	**0.20**
Spleen	0.03	**0.15**	0.03	**0.19**	0.04	**0.23**	0.05	**0.25**	0.06	**0.31**	0.07	**0.36**
Liver	0.03	**0.09**	0.03	**0.11**	0.04	**0.13**	0.05	**0.15**	0.06	**0.19**	0.08	**0.23**
Gall bladder	0.05	**0.09**	0.07	**0.12**	0.08	**0.15**	0.10	**0.17**	0.12	**0.22**	0.15	**0.26**
Pancreas	0.07	**0.13**	0.10	**0.17**	0.12	**0.21**	0.13	**0.24**	0.17	**0.30**	0.21	**0.37**
Small intestine	0.06	**0.12**	0.09	**0.16**	0.10	**0.19**	0.12	**0.22**	0.15	**0.27**	0.18	**0.33**
Large intestine	0.04	**0.09**	0.05	**0.12**	0.06	**0.14**	0.07	**0.16**	0.09	**0.20**	0.11	**0.25**
Sigma/rectum	0.05	**0.12**	0.07	**0.16**	0.08	**0.19**	0.09	**0.22**	0.12	**0.27**	0.14	**0.33**
Lumbar spine	0.78	**0.87**	0.99	**1.11**	1.15	**1.29**	1.28	**1.44**	1.53	**1.74**	1.77	**2.02**
Pelvis	0.39	**0.78**	0.48	**0.97**	0.55	**1.12**	0.60	**1.23**	0.70	**1.45**	0.78	**1.64**
Adrenals	0.33	**0.40**	0.42	**0.51**	0.48	**0.59**	0.53	**0.65**	0.63	**0.78**	0.72	**0.90**
Kidneys	0.21	**0.59**	0.26	**0.72**	0.30	**0.81**	0.33	**0.89**	0.39	**1.02**	0.44	**1.14**
Urinary bladder	0.05	**0.07**	0.07	**0.10**	0.09	**0.12**	0.10	**0.14**	0.13	**0.18**	0.15	**0.22**
Uterus	0.27	**0.36**	0.36	**0.47**	0.43	**0.57**	0.50	**0.65**	0.63	**0.83**	0.77	**1.02**
Ovaries	0.05	**0.13**	0.07	**0.17**	0.09	**0.20**	0.10	**0.23**	0.13	**0.29**	0.16	**0.36**
Prostate	0.07	**0.10**	0.10	**0.13**	0.12	**0.16**	0.13	**0.18**	0.17	**0.23**	0.20	**0.29**
Testes	0.03	**0.05**	0.03	**0.06**	0.04	**0.08**	0.05	**0.09**	0.06	**0.11**	0.07	**0.14**
Upper arm		**0.02**	0.01	**0.03**	0.01	**0.04**	0.02	**0.04**	0.02	**0.05**	0.03	**0.07**
Elbow		**0.02**	0.01	**0.03**	0.01	**0.04**	0.02	**0.04**	0.02	**0.05**	0.03	**0.07**
Forearm		**0.02**		**0.02**		**0.03**		**0.03**	0.01	**0.04**	0.01	**0.05**
Upper leg	0.02	**0.22**	0.03	**0.29**	0.03	**0.34**	0.04	**0.39**	0.05	**0.49**	0.06	**0.58**
Knee		**0.07**		**0.09**		**0.11**		**0.12**		**0.15**		**0.18**
Lower leg												
Lymph nodes	0.07	**0.14**	0.09	**0.18**	0.10	**0.21**	0.12	**0.24**	0.15	**0.29**	0.18	**0.34**
Skin	0.10	**0.23**	0.10	**0.24**	0.10	**0.25**	0.11	**0.26**	0.11	**0.27**	0.12	**0.29**
Muscle	0.08	**0.16**	0.09	**0.19**	0.10	**0.22**	0.11	**0.24**	0.13	**0.28**	0.15	**0.31**
Skeleton	0.20	**0.36**	0.25	**0.44**	0.29	**0.50**	0.32	**0.56**	0.37	**0.66**	0.42	**0.75**
Red bone marrow	0.13	**0.21**	0.16	**0.27**	0.19	**0.31**	0.21	**0.34**	0.25	**0.41**	0.29	**0.47**
Whole body dose	0.09	**0.19**	0.11	**0.23**	0.13	**0.26**	0.14	**0.28**	0.16	**0.33**	0.18	**0.37**
ED (ICRP 60)	0.06	**0.14**	0.08	**0.18**	0.09	**0.22**	0.10	**0.24**	0.13	**0.30**	0.16	**0.36**
ED (ICRP 103)	0.06	**0.17**	0.08	**0.22**	0.10	**0.26**	0.11	**0.29**	0.14	**0.36**	0.17	**0.43**
Total filtration (mm) Al	**1.5**		**2.5**		**3.5**		**4.5**		**3.5**		**3.5**	
Cu	**0**		**0**		**0**		**0**		**0.1**		**0.2**	

Table 13.44 Normalized organ doses—Full spine pa—13–17 years

Beam projection:			pa									**Table cells:**
Focus-to-detector distance (cm):			200									(+) = normal font = optimal field setting
Tube voltage (kVp):			80									**(−) = bold font = suboptimal field setting**
Total filtrations:			See last table row									Empty table cells correspond to values <0.01
Organ or tissue	**Organ dose D_T: Entrance dose K_E**						**(mSv * mGy^{-1})**					
	(+)	(−)	(+)	(−)	(+)	(−)	(+)	(−)	(+)	(−)	(+)	(−)
Skull	0.15	**0.39**	0.18	**0.48**	0.21	**0.54**	0.23	**0.59**	0.27	**0.68**	0.31	**0.76**
Brain	0.04	**0.11**	0.05	**0.14**	0.05	**0.17**	0.06	**0.19**	0.07	**0.23**	0.09	**0.26**
Oral mucosa	0.09	**0.16**	0.12	**0.20**	0.14	**0.22**	0.15	**0.25**	0.18	**0.30**	0.21	**0.34**
Salivary glands	0.10	**0.43**	0.13	**0.50**	0.15	**0.55**	0.16	**0.59**	0.19	**0.66**	0.22	**0.73**
Cervical spine	1.12	**1.17**	1.38	**1.44**	1.57	**1.65**	1.73	**1.82**	2.03	**2.15**	2.30	**2.44**
Thyroid	0.08	**0.10**	0.11	**0.12**	0.13	**0.15**	0.15	**0.16**	0.18	**0.21**	0.22	**0.25**
Thymus	0.03	**0.05**	0.04	**0.06**	0.05	**0.08**	0.05	**0.09**	0.07	**0.11**	0.08	**0.14**
Thoracic spine	0.85	**0.95**	1.06	**1.18**	1.22	**1.36**	1.34	**1.51**	1.59	**1.79**	1.82	**2.05**
Claviculae	0.04	**0.09**	0.06	**0.12**	0.07	**0.14**	0.08	**0.16**	0.09	**0.20**	0.11	**0.24**
Scapulae	0.03	**0.12**	0.04	**0.15**	0.05	**0.17**	0.06	**0.19**	0.07	**0.24**	0.09	**0.28**
Ribs	0.44	**0.86**	0.50	**0.98**	0.53	**1.05**	0.56	**1.10**	0.60	**1.20**	0.63	**1.27**
Oesophagus	0.13	**0.17**	0.17	**0.23**	0.20	**0.27**	0.23	**0.30**	0.29	**0.38**	0.34	**0.46**
Trachea	0.08	**0.10**	0.11	**0.13**	0.13	**0.15**	0.15	**0.17**	0.18	**0.21**	0.22	**0.26**
Lung	0.06	**0.29**	0.08	**0.35**	0.09	**0.40**	0.10	**0.44**	0.13	**0.51**	0.15	**0.58**
Mamma	0.08	**0.48**	0.10	**0.62**	0.12	**0.73**	0.14	**0.82**	0.18	**1.02**	0.22	**1.21**
Heart	0.07	**0.11**	0.08	**0.14**	0.10	**0.17**	0.11	**0.19**	0.14	**0.24**	0.17	**0.29**
Stomach	0.03	**0.09**	0.04	**0.12**	0.05	**0.14**	0.06	**0.16**	0.07	**0.20**	0.09	**0.24**
Spleen	0.03	**0.18**	0.04	**0.23**	0.05	**0.27**	0.06	**0.30**	0.07	**0.35**	0.09	**0.41**
Liver	0.04	**0.11**	0.05	**0.14**	0.06	**0.17**	0.06	**0.19**	0.08	**0.23**	0.10	**0.27**
Gall bladder	0.07	**0.12**	0.09	**0.16**	0.11	**0.19**	0.12	**0.22**	0.16	**0.27**	0.19	**0.32**
Pancreas	0.10	**0.17**	0.13	**0.22**	0.15	**0.26**	0.17	**0.30**	0.21	**0.37**	0.26	**0.44**
Small intestine	0.09	**0.16**	0.11	**0.20**	0.13	**0.24**	0.15	**0.27**	0.19	**0.34**	0.22	**0.40**
Large intestine	0.05	**0.12**	0.06	**0.15**	0.08	**0.18**	0.09	**0.20**	0.11	**0.25**	0.13	**0.31**
Sigma/rectum	0.07	**0.16**	0.09	**0.20**	0.10	**0.24**	0.11	**0.27**	0.14	**0.33**	0.17	**0.39**
Lumbar spine	0.93	**1.04**	1.15	**1.30**	1.31	**1.49**	1.45	**1.65**	1.70	**1.95**	1.93	**2.23**
Pelvis	0.45	**0.91**	0.54	**1.10**	0.61	**1.25**	0.66	**1.36**	0.75	**1.58**	0.84	**1.77**
Adrenals	0.39	**0.48**	0.48	**0.59**	0.55	**0.68**	0.60	**0.75**	0.70	**0.88**	0.79	**1.00**
Kidneys	0.25	**0.67**	0.30	**0.81**	0.34	**0.90**	0.37	**0.97**	0.43	**1.11**	0.48	**1.22**
Urinary bladder	0.07	**0.10**	0.09	**0.13**	0.11	**0.16**	0.12	**0.18**	0.16	**0.22**	0.19	**0.27**
Uterus	0.36	**0.48**	0.46	**0.62**	0.55	**0.73**	0.62	**0.83**	0.77	**1.03**	0.93	**1.24**
Ovaries	0.07	**0.17**	0.09	**0.22**	0.11	**0.26**	0.13	**0.29**	0.16	**0.36**	0.19	**0.44**
Prostate	0.10	**0.13**	0.12	**0.17**	0.15	**0.20**	0.16	**0.23**	0.20	**0.29**	0.24	**0.35**
Testes	0.04	**0.06**	0.05	**0.08**	0.05	**0.10**	0.06	**0.11**	0.08	**0.14**	0.09	**0.17**
Upper arm	0.01	**0.03**	0.02	**0.04**	0.02	**0.05**	0.02	**0.05**	0.03	**0.07**	0.04	**0.08**
Elbow	0.01	**0.03**	0.02	**0.04**	0.02	**0.05**	0.02	**0.06**	0.03	**0.07**	0.03	**0.09**
Forearm		**0.02**		**0.03**	0.01	**0.04**	0.01	**0.04**	0.02	**0.05**	0.02	**0.06**
Upper leg	0.03	**0.28**	0.04	**0.36**	0.04	**0.42**	0.05	**0.47**	0.06	**0.57**	0.07	**0.68**
Knee		**0.09**		**0.11**		**0.13**		**0.15**		**0.18**		**0.21**
Lower leg												
Lymph nodes	0.09	**0.18**	0.11	**0.22**	0.13	**0.25**	0.15	**0.28**	0.18	**0.34**	0.21	**0.40**
Skin	0.10	**0.24**	0.11	**0.25**	0.11	**0.26**	0.11	**0.27**	0.12	**0.29**	0.13	**0.30**
Muscle	0.09	**0.18**	0.10	**0.22**	0.12	**0.25**	0.13	**0.27**	0.15	**0.31**	0.17	**0.35**
Skeleton	0.24	**0.41**	0.28	**0.50**	0.32	**0.57**	0.35	**0.62**	0.41	**0.72**	0.46	**0.81**
Red bone marrow	0.16	**0.26**	0.19	**0.31**	0.22	**0.36**	0.24	**0.40**	0.29	**0.47**	0.33	**0.54**
Whole body dose	0.11	**0.22**	0.13	**0.26**	0.14	**0.29**	0.16	**0.32**	0.18	**0.37**	0.20	**0.41**
ED (ICRP 60)	0.08	**0.18**	0.10	**0.23**	0.11	**0.26**	0.13	**0.29**	0.16	**0.36**	0.19	**0.42**
ED (ICRP 103)	0.08	**0.22**	0.10	**0.27**	0.12	**0.31**	0.14	**0.35**	0.17	**0.43**	0.20	**0.50**
Total filtration (mm) **Al**	**1.5**		**2.5**		**3.5**		**4.5**		**3.5**		**3.5**	
Cu	**0**		**0**		**0**		**0**		**0.1**		**0.2**	

Table 13.45 Normalized organ doses—Full spine pa—13–17 years

Beam projection:	pa	**Table cells:**
Focus-to-detector distance (cm):	200	(+) = normal font = optimal field setting
Tube voltage (kVp):	90	**(−) = bold font = suboptimal field setting**
Total filtrations:	See last table row	Empty table cells correspond to values <0.01

Organ or tissue	**Organ dose D_T: Entrance dose K_E**							**(mSv * mGy^{-1})**				
	(+)	(−)	(+)	(−)	(+)	(−)	(+)	(−)	(+)	(−)	(+)	(−)
Skull	0.17	**0.44**	0.20	**0.52**	0.23	**0.58**	0.25	**0.63**	0.29	**0.72**	0.33	**0.79**
Brain	0.04	**0.14**	0.05	**0.17**	0.06	**0.19**	0.07	**0.21**	0.08	**0.25**	0.10	**0.29**
Oral mucosa	0.11	**0.19**	0.14	**0.23**	0.16	**0.26**	0.17	**0.28**	0.21	**0.33**	0.24	**0.38**
Salivary glands	0.12	**0.47**	0.15	**0.55**	0.17	**0.60**	0.18	**0.63**	0.22	**0.70**	0.24	**0.76**
Cervical spine	1.27	**1.33**	1.53	**1.61**	1.72	**1.82**	1.87	**1.98**	2.16	**2.30**	2.40	**2.57**
Thyroid	0.11	**0.12**	0.13	**0.15**	0.15	**0.18**	0.17	**0.20**	0.21	**0.24**	0.25	**0.28**
Thymus	0.04	**0.06**	0.05	**0.08**	0.06	**0.10**	0.07	**0.11**	0.08	**0.14**	0.10	**0.16**
Thoracic spine	0.98	**1.09**	1.19	**1.33**	1.34	**1.51**	1.47	**1.66**	1.71	**1.93**	1.91	**2.18**
Claviculae	0.05	**0.12**	0.07	**0.15**	0.08	**0.17**	0.09	**0.19**	0.11	**0.23**	0.13	**0.27**
Scapulae	0.04	**0.14**	0.05	**0.17**	0.06	**0.20**	0.07	**0.22**	0.09	**0.27**	0.10	**0.31**
Ribs	0.47	**0.91**	0.52	**1.02**	0.55	**1.08**	0.57	**1.13**	0.61	**1.21**	0.64	**1.27**
Oesophagus	0.17	**0.22**	0.21	**0.28**	0.24	**0.32**	0.27	**0.36**	0.34	**0.44**	0.40	**0.52**
Trachea	0.11	**0.13**	0.13	**0.16**	0.15	**0.18**	0.17	**0.21**	0.21	**0.25**	0.25	**0.30**
Lung	0.08	**0.33**	0.09	**0.39**	0.11	**0.44**	0.12	**0.48**	0.14	**0.56**	0.17	**0.62**
Mamma	0.10	**0.60**	0.13	**0.75**	0.16	**0.87**	0.18	**0.97**	0.22	**1.17**	0.27	**1.38**
Heart	0.08	**0.14**	0.11	**0.17**	0.12	**0.20**	0.14	**0.23**	0.17	**0.28**	0.20	**0.33**
Stomach	0.04	**0.12**	0.05	**0.15**	0.06	**0.17**	0.07	**0.19**	0.09	**0.23**	0.10	**0.27**
Spleen	0.04	**0.22**	0.05	**0.27**	0.06	**0.30**	0.07	**0.33**	0.09	**0.39**	0.10	**0.45**
Liver	0.05	**0.14**	0.06	**0.17**	0.07	**0.19**	0.08	**0.22**	0.09	**0.26**	0.11	**0.30**
Gall bladder	0.09	**0.16**	0.11	**0.20**	0.13	**0.23**	0.15	**0.26**	0.18	**0.31**	0.22	**0.37**
Pancreas	0.12	**0.22**	0.16	**0.27**	0.18	**0.31**	0.20	**0.35**	0.25	**0.43**	0.29	**0.50**
Small intestine	0.11	**0.20**	0.14	**0.25**	0.16	**0.29**	0.18	**0.32**	0.22	**0.39**	0.26	**0.46**
Large intestine	0.06	**0.15**	0.08	**0.19**	0.09	**0.22**	0.11	**0.24**	0.13	**0.30**	0.15	**0.35**
Sigma/rectum	0.08	**0.19**	0.10	**0.24**	0.12	**0.28**	0.13	**0.31**	0.16	**0.38**	0.19	**0.44**
Lumbar spine	1.06	**1.20**	1.28	**1.46**	1.44	**1.65**	1.57	**1.81**	1.82	**2.10**	2.03	**2.36**
Pelvis	0.50	**1.02**	0.59	**1.21**	0.65	**1.35**	0.70	**1.46**	0.79	**1.67**	0.87	**1.84**
Adrenals	0.45	**0.55**	0.54	**0.67**	0.61	**0.76**	0.66	**0.83**	0.76	**0.96**	0.85	**1.08**
Kidneys	0.28	**0.75**	0.33	**0.88**	0.37	**0.97**	0.40	**1.05**	0.46	**1.18**	0.51	**1.28**
Urinary bladder	0.09	**0.13**	0.11	**0.16**	0.13	**0.19**	0.15	**0.21**	0.18	**0.26**	0.21	**0.31**
Uterus	0.45	**0.60**	0.56	**0.75**	0.66	**0.88**	0.73	**0.98**	0.89	**1.20**	1.05	**1.41**
Ovaries	0.09	**0.21**	0.11	**0.27**	0.13	**0.31**	0.15	**0.35**	0.18	**0.42**	0.22	**0.50**
Prostate	0.12	**0.17**	0.15	**0.21**	0.17	**0.24**	0.19	**0.27**	0.23	**0.33**	0.27	**0.39**
Testes	0.04	**0.08**	0.06	**0.10**	0.06	**0.12**	0.07	**0.13**	0.09	**0.16**	0.10	**0.19**
Upper arm	0.02	**0.04**	0.02	**0.05**	0.03	**0.06**	0.03	**0.07**	0.04	**0.08**	0.04	**0.10**
Elbow	0.02	**0.04**	0.02	**0.05**	0.02	**0.06**	0.03	**0.07**	0.03	**0.08**	0.04	**0.10**
Forearm		**0.03**	0.01	**0.04**	0.01	**0.04**	0.02	**0.05**	0.02	**0.06**	0.02	**0.08**
Upper leg	0.03	**0.34**	0.04	**0.42**	0.05	**0.48**	0.06	**0.54**	0.07	**0.64**	0.08	**0.74**
Knee		**0.11**		**0.13**		**0.15**		**0.17**		**0.20**		**0.23**
Lower leg												
Lymph nodes	0.11	**0.21**	0.13	**0.26**	0.15	**0.30**	0.17	**0.33**	0.21	**0.39**	0.24	**0.45**
Skin	0.10	**0.24**	0.11	**0.26**	0.11	**0.27**	0.12	**0.28**	0.12	**0.30**	0.13	**0.31**
Muscle	0.10	**0.21**	0.12	**0.25**	0.13	**0.27**	0.14	**0.29**	0.16	**0.34**	0.18	**0.37**
Skeleton	0.26	**0.47**	0.31	**0.55**	0.35	**0.62**	0.38	**0.67**	0.43	**0.77**	0.48	**0.85**
Red bone marrow	0.18	**0.30**	0.22	**0.36**	0.25	**0.41**	0.27	**0.44**	0.32	**0.52**	0.36	**0.59**
Whole body dose	0.12	**0.25**	0.14	**0.29**	0.16	**0.32**	0.17	**0.35**	0.20	**0.40**	0.22	**0.44**
ED (ICRP 60)	0.09	**0.22**	0.12	**0.27**	0.13	**0.31**	0.15	**0.34**	0.18	**0.41**	0.21	**0.47**
ED (ICRP 103)	0.10	**0.26**	0.13	**0.32**	0.15	**0.37**	0.16	**0.41**	0.19	**0.48**	0.23	**0.56**
Total filtration (mm) Al	**1.5**		**2.5**		**3.5**		**4.5**		**3.5**		**3.5**	
Cu	**0**		**0**		**0**		**0**		**0.1**		**0.2**	

Table 13.46 Normalized organ doses—Full spine pa—>17 years

Beam projection:			pa									**Table cells:**
Focus-to-detector distance (cm):			200									(+) = normal font = optimal field setting
Tube voltage (kVp):			70									**(−) = bold font = suboptimal field setting**
Total filtrations:			See last table row									Empty table cells correspond to values <0.01
Organ or tissue	**Organ dose D_T: Entrance dose K_E**							**(mSv * mGy^{-1})**				
	(+)	(−)	(+)	(−)	(+)	(−)	(+)	(−)	(+)	(−)	(+)	(−)
Skull	0.13	**0.30**	0.16	**0.37**	0.19	**0.43**	0.21	**0.47**	0.26	**0.56**	0.30	**0.64**
Brain	0.03	**0.08**	0.03	**0.10**	0.04	**0.12**	0.05	**0.14**	0.06	**0.17**	0.07	**0.21**
Oral mucosa	0.07	**0.12**	0.09	**0.15**	0.10	**0.17**	0.12	**0.19**	0.15	**0.23**	0.17	**0.28**
Salivary glands	0.08	**0.36**	0.10	**0.43**	0.11	**0.48**	0.13	**0.51**	0.16	**0.58**	0.18	**0.64**
Cervical spine	1.00	**1.03**	1.26	**1.30**	1.45	**1.51**	1.61	**1.68**	1.92	**2.01**	2.21	**2.31**
Thyroid	0.07	**0.07**	0.09	**0.10**	0.10	**0.12**	0.12	**0.13**	0.15	**0.17**	0.18	**0.21**
Thymus	0.02	**0.03**	0.03	**0.04**	0.03	**0.05**	0.04	**0.06**	0.05	**0.08**	0.06	**0.10**
Thoracic spine	0.70	**0.76**	0.89	**0.97**	1.04	**1.13**	1.16	**1.27**	1.40	**1.54**	1.63	**1.80**
Claviculae	0.03	**0.06**	0.04	**0.09**	0.05	**0.10**	0.05	**0.12**	0.07	**0.15**	0.09	**0.18**
Scapulae	0.02	**0.06**	0.03	**0.08**	0.03	**0.09**	0.04	**0.11**	0.05	**0.14**	0.06	**0.17**
Ribs	0.41	**0.75**	0.47	**0.86**	0.50	**0.94**	0.53	**0.99**	0.58	**1.08**	0.61	**1.16**
Oesophagus	0.10	**0.14**	0.14	**0.18**	0.17	**0.22**	0.19	**0.25**	0.25	**0.32**	0.31	**0.40**
Trachea	0.07	**0.08**	0.09	**0.10**	0.10	**0.12**	0.12	**0.14**	0.15	**0.17**	0.19	**0.21**
Lung	0.04	**0.21**	0.06	**0.26**	0.07	**0.30**	0.08	**0.33**	0.10	**0.39**	0.12	**0.45**
Mamma		**0.04**		**0.05**		**0.06**	0.01	**0.06**	0.01	**0.08**	0.02	**0.10**
Heart	0.05	**0.08**	0.07	**0.11**	0.08	**0.13**	0.09	**0.15**	0.12	**0.19**	0.14	**0.24**
Stomach	0.02	**0.06**	0.03	**0.08**	0.03	**0.10**	0.04	**0.11**	0.05	**0.14**	0.06	**0.17**
Spleen	0.02	**0.10**	0.03	**0.13**	0.03	**0.16**	0.04	**0.18**	0.05	**0.22**	0.06	**0.26**
Liver	0.02	**0.08**	0.03	**0.10**	0.04	**0.12**	0.05	**0.13**	0.06	**0.17**	0.07	**0.20**
Gall bladder	0.06	**0.09**	0.07	**0.12**	0.09	**0.14**	0.10	**0.16**	0.13	**0.21**	0.16	**0.26**
Pancreas	0.07	**0.12**	0.09	**0.16**	0.11	**0.19**	0.13	**0.22**	0.16	**0.28**	0.20	**0.34**
Small intestine	0.06	**0.11**	0.08	**0.15**	0.10	**0.17**	0.11	**0.20**	0.14	**0.25**	0.18	**0.31**
Large intestine	0.03	**0.08**	0.04	**0.10**	0.05	**0.12**	0.06	**0.14**	0.08	**0.18**	0.10	**0.22**
Sigma/rectum	0.05	**0.11**	0.07	**0.14**	0.08	**0.17**	0.09	**0.19**	0.11	**0.24**	0.14	**0.29**
Lumbar spine	0.77	**0.84**	0.97	**1.07**	1.13	**1.25**	1.26	**1.40**	1.52	**1.69**	1.75	**1.96**
Pelvis	0.37	**0.69**	0.45	**0.86**	0.52	**0.98**	0.56	**1.08**	0.66	**1.27**	0.74	**1.44**
Adrenals	0.34	**0.41**	0.42	**0.51**	0.49	**0.59**	0.54	**0.66**	0.64	**0.78**	0.73	**0.90**
Kidneys	0.19	**0.56**	0.24	**0.68**	0.27	**0.77**	0.30	**0.84**	0.35	**0.97**	0.40	**1.08**
Urinary bladder	0.05	**0.07**	0.07	**0.10**	0.09	**0.12**	0.10	**0.13**	0.13	**0.17**	0.15	**0.21**
Uterus	0.10	**0.13**	0.14	**0.17**	0.16	**0.21**	0.19	**0.24**	0.24	**0.30**	0.30	**0.37**
Ovaries	0.05	**0.12**	0.06	**0.16**	0.08	**0.19**	0.09	**0.22**	0.12	**0.28**	0.14	**0.34**
Prostate	0.07	**0.09**	0.09	**0.12**	0.11	**0.15**	0.13	**0.17**	0.16	**0.22**	0.20	**0.26**
Testes	0.03	**0.04**	0.03	**0.05**	0.04	**0.06**	0.05	**0.07**	0.06	**0.09**	0.08	**0.11**
Upper arm		**0.02**		**0.02**	0.01	**0.03**	0.01	**0.03**	0.02	**0.04**	0.02	**0.05**
Elbow		**0.02**		**0.02**		**0.03**	0.01	**0.03**	0.02	**0.04**	0.02	**0.05**
Forearm		**0.01**		**0.01**		**0.02**		**0.02**		**0.03**	0.01	**0.03**
Upper leg	0.02	**0.14**	0.03	**0.18**	0.04	**0.22**	0.04	**0.25**	0.05	**0.32**	0.06	**0.39**
Knee		**0.03**		**0.04**		**0.04**		**0.05**		**0.06**		**0.08**
Lower leg												
Lymph nodes	0.06	**0.13**	0.08	**0.17**	0.10	**0.20**	0.11	**0.22**	0.15	**0.27**	0.18	**0.32**
Skin	0.10	**0.20**	0.11	**0.22**	0.11	**0.23**	0.12	**0.23**	0.12	**0.25**	0.13	**0.26**
Muscle	0.07	**0.14**	0.09	**0.17**	0.10	**0.19**	0.11	**0.21**	0.13	**0.24**	0.14	**0.28**
Skeleton	0.20	**0.31**	0.24	**0.38**	0.28	**0.44**	0.31	**0.48**	0.36	**0.57**	0.41	**0.65**
Red bone marrow	0.14	**0.21**	0.18	**0.26**	0.20	**0.30**	0.23	**0.34**	0.27	**0.40**	0.31	**0.47**
Whole body dose	0.09	**0.16**	0.11	**0.20**	0.12	**0.23**	0.13	**0.25**	0.16	**0.29**	0.18	**0.33**
ED (ICRP 60)	0.06	**0.11**	0.07	**0.15**	0.09	**0.17**	0.10	**0.19**	0.12	**0.24**	0.14	**0.29**
ED (ICRP 103)	0.06	**0.12**	0.07	**0.15**	0.09	**0.18**	0.10	**0.20**	0.12	**0.24**	0.14	**0.29**
Total filtration (mm) **Al**	**1.5**		**2.5**		**3.5**		**4.5**		**3.5**		**3.5**	
Cu	**0**		**0**		**0**		**0**		**0.1**		**0.2**	

Table 13.47 Normalized organ doses—Full spine pa—>17 years

Beam projection:			pa									**Table cells:**
Focus-to-detector distance (cm):			200									(+) = normal font = optimal field setting
Tube voltage (kVp):			80									**(−) = bold font = suboptimal field setting**
Total filtrations:			See last table row									Empty table cells correspond to values <0.01
Organ or tissue	**Organ dose D_T: Entrance dose K_E**								**(mSv * mGy^{-1})**			
	(+)	(−)	(+)	(−)	(+)	(−)	(+)	(−)	(+)	(−)	(+)	(−)
Skull	0.15	**0.35**	0.19	**0.43**	0.22	**0.48**	0.24	**0.53**	0.29	**0.61**	0.33	**0.69**
Brain	0.03	**0.10**	0.04	**0.13**	0.05	**0.15**	0.06	**0.17**	0.07	**0.20**	0.09	**0.24**
Oral mucosa	0.09	**0.14**	0.11	**0.18**	0.13	**0.21**	0.14	**0.23**	0.18	**0.28**	0.21	**0.32**
Salivary glands	0.09	**0.41**	0.12	**0.47**	0.14	**0.52**	0.15	**0.56**	0.18	**0.63**	0.21	**0.69**
Cervical spine	1.17	**1.22**	1.44	**1.50**	1.65	**1.72**	1.81	**1.89**	2.12	**2.22**	2.40	**2.52**
Thyroid	0.09	**0.10**	0.11	**0.13**	0.13	**0.15**	0.15	**0.17**	0.19	**0.21**	0.22	**0.25**
Thymus	0.03	**0.05**	0.04	**0.06**	0.04	**0.07**	0.05	**0.08**	0.07	**0.11**	0.08	**0.13**
Thoracic spine	0.83	**0.91**	1.04	**1.14**	1.20	**1.32**	1.32	**1.46**	1.57	**1.74**	1.80	**2.00**
Claviculae	0.04	**0.09**	0.05	**0.11**	0.06	**0.13**	0.07	**0.15**	0.09	**0.19**	0.11	**0.22**
Scapulae	0.03	**0.08**	0.03	**0.10**	0.04	**0.12**	0.05	**0.13**	0.06	**0.17**	0.07	**0.20**
Ribs	0.44	**0.81**	0.49	**0.91**	0.53	**0.98**	0.55	**1.03**	0.59	**1.12**	0.63	**1.18**
Oesophagus	0.14	**0.19**	0.19	**0.24**	0.22	**0.29**	0.25	**0.32**	0.32	**0.41**	0.38	**0.49**
Trachea	0.09	**0.10**	0.11	**0.13**	0.13	**0.15**	0.15	**0.17**	0.19	**0.22**	0.23	**0.26**
Lung	0.06	**0.25**	0.07	**0.30**	0.09	**0.34**	0.10	**0.38**	0.12	**0.44**	0.14	**0.50**
Mamma		**0.05**	0.01	**0.06**	0.01	**0.07**	0.01	**0.08**	0.02	**0.10**	0.02	**0.12**
Heart	0.07	**0.11**	0.09	**0.14**	0.10	**0.17**	0.12	**0.19**	0.15	**0.24**	0.18	**0.29**
Stomach	0.03	**0.08**	0.04	**0.10**	0.05	**0.12**	0.05	**0.14**	0.07	**0.18**	0.08	**0.21**
Spleen	0.03	**0.13**	0.04	**0.16**	0.04	**0.19**	0.05	**0.21**	0.06	**0.26**	0.07	**0.30**
Liver	0.03	**0.10**	0.04	**0.12**	0.05	**0.15**	0.06	**0.16**	0.08	**0.20**	0.09	**0.24**
Gall bladder	0.08	**0.12**	0.10	**0.16**	0.12	**0.18**	0.13	**0.21**	0.16	**0.26**	0.20	**0.31**
Pancreas	0.09	**0.16**	0.12	**0.21**	0.15	**0.24**	0.16	**0.28**	0.21	**0.34**	0.25	**0.41**
Small intestine	0.08	**0.15**	0.11	**0.19**	0.13	**0.22**	0.14	**0.25**	0.18	**0.32**	0.22	**0.38**
Large intestine	0.05	**0.10**	0.06	**0.13**	0.07	**0.16**	0.08	**0.18**	0.10	**0.23**	0.13	**0.27**
Sigma/rectum	0.07	**0.14**	0.08	**0.18**	0.10	**0.21**	0.11	**0.24**	0.14	**0.30**	0.16	**0.35**
Lumbar spine	0.92	**1.01**	1.13	**1.26**	1.30	**1.45**	1.43	**1.60**	1.69	**1.90**	1.93	**2.17**
Pelvis	0.42	**0.80**	0.51	**0.97**	0.57	**1.10**	0.62	**1.20**	0.71	**1.38**	0.79	**1.55**
Adrenals	0.40	**0.48**	0.49	**0.60**	0.56	**0.68**	0.61	**0.75**	0.72	**0.88**	0.81	**1.01**
Kidneys	0.23	**0.64**	0.27	**0.76**	0.31	**0.85**	0.34	**0.92**	0.39	**1.05**	0.44	**1.17**
Urinary bladder	0.07	**0.10**	0.09	**0.13**	0.11	**0.15**	0.13	**0.17**	0.16	**0.22**	0.19	**0.26**
Uterus	0.14	**0.17**	0.18	**0.23**	0.21	**0.27**	0.24	**0.30**	0.30	**0.38**	0.36	**0.46**
Ovaries	0.07	**0.16**	0.09	**0.21**	0.10	**0.24**	0.12	**0.28**	0.14	**0.35**	0.17	**0.42**
Prostate	0.09	**0.12**	0.12	**0.16**	0.14	**0.19**	0.16	**0.21**	0.20	**0.27**	0.24	**0.32**
Testes	0.04	**0.05**	0.05	**0.07**	0.05	**0.08**	0.06	**0.09**	0.08	**0.12**	0.09	**0.14**
Upper arm	0.01	**0.02**	0.01	**0.03**	0.02	**0.04**	0.02	**0.04**	0.02	**0.05**	0.03	**0.06**
Elbow		**0.02**	0.01	**0.03**	0.01	**0.04**	0.02	**0.04**	0.02	**0.05**	0.03	**0.07**
Forearm		**0.02**		**0.02**		**0.02**		**0.03**	0.01	**0.04**	0.02	**0.05**
Upper leg	0.03	**0.18**	0.04	**0.23**	0.05	**0.27**	0.05	**0.31**	0.07	**0.38**	0.08	**0.46**
Knee		**0.04**		**0.05**		**0.05**		**0.06**		**0.08**		**0.09**
Lower leg												
Lymph nodes	0.08	**0.16**	0.11	**0.21**	0.13	**0.24**	0.14	**0.27**	0.18	**0.33**	0.21	**0.38**
Skin	0.11	**0.21**	0.11	**0.23**	0.12	**0.24**	0.12	**0.25**	0.13	**0.26**	0.13	**0.27**
Muscle	0.08	**0.16**	0.10	**0.19**	0.11	**0.22**	0.12	**0.24**	0.14	**0.27**	0.16	**0.31**
Skeleton	0.23	**0.36**	0.28	**0.44**	0.31	**0.49**	0.34	**0.54**	0.40	**0.63**	0.44	**0.71**
Red bone marrow	0.17	**0.25**	0.21	**0.31**	0.24	**0.35**	0.26	**0.39**	0.31	**0.46**	0.36	**0.53**
Whole body dose	0.10	**0.19**	0.12	**0.23**	0.14	**0.26**	0.15	**0.28**	0.18	**0.33**	0.20	**0.37**
ED (ICRP 60)	0.07	**0.14**	0.09	**0.18**	0.11	**0.21**	0.12	**0.24**	0.15	**0.29**	0.17	**0.34**
ED (ICRP 103)	0.07	**0.15**	0.09	**0.19**	0.11	**0.22**	0.12	**0.24**	0.15	**0.29**	0.17	**0.34**
Total filtration (mm) **Al**	**1.5**		**2.5**		**3.5**		**4.5**		**3.5**		**3.5**	
Cu	**0**		**0**		**0**		**0**		**0.1**		**0.2**	

Table 13.48 Normalized organ doses—Full spine pa—>17 years

Beam projection:			pa									**Table cells:**
Focus-to-detector distance (cm):			200									(+) = normal font = optimal field setting
Tube voltage (kVp):			90									**(−) = bold font = suboptimal field setting**
Total filtrations:			See last table row									Empty table cells correspond to values <0.01
Organ or tissue	**Organ dose D_T: Entrance dose K_E**								**(mSv * mGy^{-1})**			
	(+)	(−)	(+)	(−)	(+)	(−)	(+)	(−)	(+)	(−)	(+)	(−)
Skull	0.18	**0.39**	0.22	**0.47**	0.25	**0.53**	0.27	**0.57**	0.32	**0.66**	0.36	**0.73**
Brain	0.04	**0.12**	0.05	**0.15**	0.06	**0.17**	0.07	**0.19**	0.08	**0.23**	0.10	**0.27**
Oral mucosa	0.11	**0.17**	0.13	**0.21**	0.15	**0.24**	0.17	**0.26**	0.20	**0.31**	0.23	**0.36**
Salivary glands	0.11	**0.45**	0.14	**0.52**	0.16	**0.56**	0.17	**0.60**	0.21	**0.67**	0.24	**0.73**
Cervical spine	1.33	**1.39**	1.60	**1.67**	1.80	**1.89**	1.96	**2.06**	2.26	**2.37**	2.51	**2.65**
Thyroid	0.11	**0.12**	0.14	**0.15**	0.16	**0.18**	0.18	**0.20**	0.22	**0.24**	0.26	**0.29**
Thymus	0.04	**0.06**	0.05	**0.08**	0.06	**0.09**	0.07	**0.10**	0.08	**0.13**	0.10	**0.16**
Thoracic spine	0.96	**1.06**	1.17	**1.29**	1.33	**1.47**	1.45	**1.61**	1.69	**1.89**	1.90	**2.13**
Claviculae	0.05	**0.11**	0.06	**0.13**	0.08	**0.16**	0.08	**0.17**	0.10	**0.21**	0.12	**0.25**
Scapulae	0.03	**0.10**	0.04	**0.12**	0.05	**0.14**	0.06	**0.15**	0.07	**0.19**	0.09	**0.22**
Ribs	0.46	**0.85**	0.51	**0.95**	0.54	**1.01**	0.57	**1.06**	0.60	**1.14**	0.63	**1.19**
Oesophagus	0.18	**0.24**	0.23	**0.30**	0.27	**0.35**	0.30	**0.39**	0.37	**0.48**	0.44	**0.56**
Trachea	0.11	**0.13**	0.14	**0.16**	0.16	**0.18**	0.18	**0.21**	0.22	**0.25**	0.26	**0.30**
Lung	0.07	**0.28**	0.09	**0.34**	0.10	**0.38**	0.11	**0.42**	0.13	**0.48**	0.15	**0.54**
Mamma	0.01	**0.06**	0.01	**0.07**	0.02	**0.08**	0.02	**0.09**	0.02	**0.11**	0.03	**0.13**
Heart	0.09	**0.14**	0.11	**0.18**	0.13	**0.21**	0.14	**0.23**	0.18	**0.28**	0.21	**0.33**
Stomach	0.04	**0.10**	0.05	**0.13**	0.06	**0.15**	0.06	**0.17**	0.08	**0.21**	0.10	**0.24**
Spleen	0.04	**0.15**	0.05	**0.19**	0.05	**0.22**	0.06	**0.24**	0.07	**0.29**	0.09	**0.33**
Liver	0.04	**0.12**	0.06	**0.15**	0.06	**0.17**	0.07	**0.19**	0.09	**0.23**	0.11	**0.27**
Gall bladder	0.10	**0.15**	0.12	**0.19**	0.14	**0.22**	0.16	**0.25**	0.19	**0.31**	0.23	**0.36**
Pancreas	0.12	**0.20**	0.15	**0.25**	0.18	**0.29**	0.20	**0.33**	0.24	**0.40**	0.29	**0.47**
Small intestine	0.10	**0.18**	0.13	**0.23**	0.15	**0.27**	0.17	**0.30**	0.21	**0.37**	0.25	**0.43**
Large intestine	0.06	**0.13**	0.08	**0.17**	0.09	**0.19**	0.10	**0.22**	0.12	**0.27**	0.15	**0.32**
Sigma/rectum	0.08	**0.17**	0.10	**0.22**	0.12	**0.25**	0.13	**0.28**	0.16	**0.34**	0.18	**0.40**
Lumbar spine	1.05	**1.17**	1.27	**1.42**	1.43	**1.61**	1.57	**1.76**	1.82	**2.05**	2.03	**2.31**
Pelvis	0.47	**0.89**	0.55	**1.06**	0.62	**1.19**	0.66	**1.29**	0.75	**1.47**	0.82	**1.62**
Adrenals	0.46	**0.56**	0.55	**0.68**	0.62	**0.76**	0.67	**0.83**	0.78	**0.96**	0.87	**1.08**
Kidneys	0.26	**0.71**	0.31	**0.84**	0.34	**0.93**	0.37	**1.00**	0.42	**1.12**	0.47	**1.23**
Urinary bladder	0.09	**0.12**	0.11	**0.16**	0.13	**0.18**	0.15	**0.21**	0.18	**0.25**	0.22	**0.30**
Uterus	0.17	**0.22**	0.22	**0.28**	0.25	**0.32**	0.28	**0.36**	0.35	**0.44**	0.41	**0.53**
Ovaries	0.08	**0.20**	0.11	**0.25**	0.12	**0.29**	0.14	**0.33**	0.17	**0.40**	0.20	**0.48**
Prostate	0.12	**0.15**	0.15	**0.19**	0.17	**0.23**	0.19	**0.25**	0.23	**0.31**	0.27	**0.36**
Testes	0.04	**0.07**	0.06	**0.08**	0.07	**0.10**	0.07	**0.11**	0.09	**0.14**	0.11	**0.16**
Upper arm	0.01	**0.03**	0.02	**0.04**	0.02	**0.04**	0.02	**0.05**	0.03	**0.06**	0.04	**0.08**
Elbow	0.01	**0.03**	0.02	**0.04**	0.02	**0.05**	0.02	**0.05**	0.03	**0.06**	0.03	**0.08**
Forearm		**0.02**		**0.03**	0.01	**0.03**	0.01	**0.04**	0.02	**0.05**	0.02	**0.06**
Upper leg	0.04	**0.22**	0.05	**0.28**	0.06	**0.32**	0.06	**0.36**	0.08	**0.44**	0.09	**0.51**
Knee		**0.04**		**0.06**		**0.06**		**0.07**		**0.09**		**0.10**
Lower leg												
Lymph nodes	0.11	**0.20**	0.13	**0.24**	0.15	**0.28**	0.17	**0.31**	0.21	**0.37**	0.25	**0.43**
Skin	0.11	**0.22**	0.12	**0.24**	0.12	**0.25**	0.13	**0.25**	0.13	**0.27**	0.14	**0.28**
Muscle	0.10	**0.18**	0.11	**0.22**	0.13	**0.24**	0.14	**0.26**	0.16	**0.30**	0.18	**0.33**
Skeleton	0.26	**0.41**	0.30	**0.48**	0.34	**0.54**	0.37	**0.59**	0.42	**0.67**	0.47	**0.74**
Red bone marrow	0.20	**0.29**	0.24	**0.36**	0.27	**0.40**	0.30	**0.44**	0.35	**0.51**	0.39	**0.58**
Whole body dose	0.12	**0.22**	0.14	**0.26**	0.16	**0.29**	0.17	**0.31**	0.19	**0.35**	0.22	**0.40**
ED (ICRP 60)	0.09	**0.17**	0.11	**0.21**	0.13	**0.25**	0.14	**0.27**	0.17	**0.33**	0.20	**0.38**
ED (ICRP 103)	0.09	**0.18**	0.11	**0.22**	0.13	**0.25**	0.14	**0.28**	0.17	**0.33**	0.19	**0.38**
Total filtration (mm) **Al**	**1.5**		**2.5**		**3.5**		**4.5**		**3.5**		**3.5**	
Cu	**0**		**0**		**0**		**0**		**0.1**		**0.2**	

References

1. Bontrager KL. Textbook of radiographic positioning and related anatomy. St. Louis: Mosby Year Book; 1993.
2. Commission of the European Communities. European guidelines on quality criteria for diagnostic radiographic images in paediatrics. EUR 16261, ISBN 92-827-7843-6; 1996.
3. Cook JV, Shah K, Pablot S, Kyriou J, Pettett A, Fitzgerald M. Guidelines on best practice in the x-ray imaging of children. A manual for all x-ray departments. London: Queen Mary's Hospital for Children, Surrey, and The Radiological Protection Centre; 1998.
4. Drexler G, Panzer W, Stieve FE. Die Bestimmung von Organdosen in der Röntgendiagnostik. Berlin: Hoffmann Verlag; 1993.
5. Ebel K-D, Willich E. Die Röntgenuntersuchung im Kindesalter – Technik und Indikation. Berlin: Springer; 1968.
6. Hart D, Jones DJ, Wall BF. Estimation of effective dose in diagnostic radiology from entrance surface dose and dose-area-product measurements. National Radiological Protection Board NRPB-R262; 1994.
7. Hart D, Jones DJ, Wall BF. Coefficients for estimating effective doses from pediatric x-ray examinations. National Radiological Protection Board NRPB-R279; 1996.
8. Hösle M. Untersuchung über die Patientendosis und die Bildqualität bei häufigen Röntgenaufnahmen im Säuglingsalter in Kliniken der alten Bundesländer mit Leitung durch einen Pädiater oder einen Radiologen. Dissertation, Universität München; 1993.
9. Pehe J. Untersuchung über die Patientendosis und die Bildqualität bei häufigen Röntgenaufnahmen in der Kinderradiologie. Dissertation, Universität München; 1993.
10. Schneider K, Fendel H, Bakowski C, Stein E, Kohn M, Kellner M, Schweighofer K, Cartagena G, Padovani R, Panzer W, Scheurer C, Wall B. Results of a dosimetry study in the European Community on frequent x ray examinations in infants. Radiat Prot Dosimetry. 1992;43:31–6.
11. Winter W. Untersuchung über die Patientendosis und die Bildqualität bei häufigen Röntgenaufnahmen im Säuglingsalter bei niedergelassenen Ärzten (Pädiater, Orthopäden, Chirurgen). Dissertation, Universität München; 1994.

14 Cervical Spine

14.1 Introduction

The type of cervical spine radiography is determined by the clinical indication. Table 14.1 shows indications and Table 14.2 contraindications for cervical spine radiography. Image criteria as required for radiological findings in cervical spine radiographs are listed in Table 14.3. In order to fulfil these image criteria, the field settings listed in Table 14.4 and Table 14.5 should be achieved. These field settings go along with the patient preparation as listed in Table 14.6. Exposure parameter combinations for various age groups and patient diameters are listed in Tables 14.7, 14.8, 14.9, 14.10, and 14.11. The age groups in Tables 14.8 and 14.10 refer to the standard age groups of the mathematical MIRD phantom family which has been used to calculate the normalized organ doses in this book. Tables 14.9 and 14.11 show radiation qualities (i.e. various combinations of tube voltage and total filtration) which may be applied in clinical practice. Please note that not all of these radiation qualities are recommended by the authors. The standard projections of cervical spine radiographies of the patient who is correctly positioned are depicted in Figs. 14.1 to 14.2. Also the optimal field (marked in green), the suboptimal field (marked in red), and the correct shielding are shown. Tables 14.13, 14.14, 14.15, 14.16, 14.17, and 14.18 outline the virtual field settings on the skin surface of mathematical MIRD phantoms of various age groups and the resulting virtual radiographies with and without bone structures, respectively. Normalized organ doses resulting from these virtual field settings have been calculated for all age groups and have been listed in Tables 14.19, 14.20, 14.21, 14.22, 14.23, 14.24, 14.25, 14.26, 14.27, 14.28, 14.29, 14.30, 14.31, 14.32, 14.33, 14.34, 14.35, 14.36, 14.37, 14.38, 14.39, 14.40, 14.41, 14.42, 14.43, 14.44, 14.45, 14.46, 14.47, 14.48, 14.49, 14.50, 14.51, 14.52, 14.53, 14.54, 14.55, 14.56, 14.57, and 14.58. Table cells with normal font refer to optimal field settings and table cells with bold font refer to suboptimal field settings. Please note that normalized organ doses have been provided also for exposure parameter combinations, which are not recommended by the authors.

Table 14.1 Indications for cervical spine radiographs

Trauma
Torticollis
Juvenile rheumatoid arthritis
Unclear neck pain
Skeletal dysplasia
Metabolic diseases, e.g. mucopolysaccharidosis
Chromosomal aberrations, e.g. Down syndrome
Bone tumours
Soft tissue tumours

Table 14.2 Contraindications for cervical spine radiographs

None

M. Seidenbusch et al., *Imaging Practice and Radiation Protection in Pediatric Radiology*,
https://doi.org/10.1007/978-3-030-18504-6_14

Table 14.3 Image criteria for cervical spine radiographs

Cervical spine ap
Symmetrical reproduction of the cervical spine
Visually sharp reproduction of the vertebral bodies and the articular processes consistent with age
Visualization of the spinous and transverse processes consistent with age
Visualization of the intervertebral spaces in the centre of the beam area consistent with age
Visually sharp reproduction of the cortex and trabecular structures of the vertebral bodies and the transverse processes consistent with age
Reproduction of the adjacent soft tissues
Cervical spine lat
Visualization of the entire cervical spine from the foramen magnum till T2 vertebra
Full superimposition of the margins of the vertebral bodies
Reproduction of the pedicles and the intervertebral bodies
Visualization of the posterior articular processes
Visualization of the spinous processes consistent with age
Visually sharp reproduction of the cortex and trabecular structures of the vertebral bodies consistent with age
Reproduction of the adjacent soft tissues

Table 14.4 Patient positions and beam projections for cervical spine radiographs

Patient position	Beam projections
Standing, sitting	ap + lat
Supine	ap + lat[a]
Decubitus left[b]	lat
Annotations:	
[a]Lateral projection in emergency cases only	
[b]Rather rare technique which has been neglected in this book	

Table 14.5 Optimal field settings for cervical spine radiographs

Cervical spine ap	
Upper field edge:	Tip of the nose
Field centring:	Just below the tip of the mandible
Lower field edge:	Jugular fossa
Lateral field edges:	At the skin surfaces of the neck
Cervical spine lat	
Upper field edge:	At the ear lobe/acoustic meatus
Field centring:	At the level of the tip of the mandible and the longitudinal line through the acoustic meatus
Lower field edge:	Spinous process of T1
Anterior field edge:	Midpoint of the mandible
Posterior field edge:	Close to the skin surface of the lower neck
Annotations:	
Adjust the angulation of the tube dependent on the position of the head and neck of patient	

Table 14.6 Patient preparation

General	Remove clothes, neck laces, and all metallic devices, especially piercings, from the head, neck, and upper chest of the patient.
	Immobilize infants and uncooperative patients in the supine position, e.g. with sandbags or restrain devices. Immobilize infants and uncooperative patients either in the right or left lateral decubitus position for the lateral view[a]. Alternatively, staff members could hold the patient (in this case also take care of the radiation protection of the staff members)
	Cooperative patients >8 years of age can be positioned upright at the vertical stand
	Place eventually small X-ray positive markers at ileo-, colostomies, at palpable tumours or on pain points.
Positioning	The head should be slightly extended.
	Rotation of neck in the decubitus position can be avoided, if the patient's head is positioned on a rubber foam pad of appropriate size for the patient's head.
	The arms can be kept in front of the chest or behind the back of the patient and should be extended by staff members.
Attention	On the Bucky table the patient should always be positioned on a foam mat of 10 to 20 cm thickness. So, the lower part of the body will not be cut if lateral projection X-rays (cross table views) in prone or in decubitus position are performed
Shielding	Put lead aprons and lead masks immediately at the adjacent field edges.
Annotation:	
[a]True lateral radiographs in patients with severe head or neck trauma lying supine supplied with immobilization device/stiff neck should be obtained with horizontal beam direction	

Table 14.7 Radiographic procedure

X-ray source	Use automatic exposure control (AEC) in patients >5 years of age
	Focal spot size between 0.6 and 1.3 dependent on the age of the patient
	Check the use of additional filtration
	Check the focus-to-detector distance
Field setting	Check the field size
	Check the field centring
Image detector system	Check the use of antiscatter grids
Patient	The exposure should be performed during breath-holding

Table 14.8 Radiographic techniques of Bucky table cervical spine radiographs

Age group (years)	Beam direction	Patient position	Beam projection	Typical patient diameter (cm)		Optimal field size (cm x cm)	Focus-to-detector distance (cm)	Voltage (kVp)						Antiscatter grid
				With arms	No arms			50	60	70	80	110	120	
0	Vertical	Supine	ap	9.8		5 × 7	100[1]							
		Decubitus	lat	12.6	10.9	5 × 7								
	Horizontal	Supine												
1–2	Vertical	Supine	ap	13.0		5 × 9	115							
		Decubitus	lat	17.1	15.1	6 × 10								
	Horizontal	Supine												
3–7	Vertical	Supine	ap	15.0		6 × 11								
		Decubitus	lat	22.6	19.6	7 × 11								
	Horizontal	Supine												
8–12	Vertical	Supine	ap	16.8		7 × 15								
	Horizontal	Sitting												
	Vertical	Decubitus	lat	27.3	23.8	8 × 15								
	Horizontal	Supine												
13–17	Vertical	Supine	ap	19.6		8 × 20								
	Horizontal	Sitting												
	Vertical	Decubitus	lat	34.4	29.6	9 × 20								
	Horizontal	Supine												
18	Vertical	Supine	ap	20.0		9 × 22								
	Horizontal	Sitting												
	Vertical	Decubitus	lat	38.9	34.4	10 × 22								
	Horizontal	Supine												

Annotations:

1 The patient is lying on the table top directly on the detector

Parameter combination not recommended, normalized organ doses not provided

Parameter combination not recommended, normalized organ doses provided

Parameter combination recommended,normalized organ doses provided

Table 14.9 Radiation qualities of Bucky table cervical spine radiographs

Voltage (kVp)	Total filtration					
	Without additional filtration		With additional filtration			
	1.5 mm Al	2.5 Al	3.5 Al	4.5 Al	3.5 Al + 0.1 Cu	3.5 Al + 0.2 Cu
50	This technique is not suitable for cervical spine radiographs.					
60						
70						
80						
110						
120						

Annotations:

Parameter combination not recommended, normalized organ doses not provided

Parameter combination not recommended, normalized organ doses provided

Parameter combination recommended, normalized organ doses provided

Table 14.10 Radiographic techniques of vertical stand cervical spine radiographs

Age group (years)	Beam direction	Patient position	Beam projection	Typical patient diameter (cm)		Optimal field size (cm × cm)	Focus-to-detector distance (cm)	Voltage (kVp)						Antiscatter grid
				With arms	No arms			50	60	70	80	110	120	
0	This technique is not applied in these age groups.													
1–2														
3–7														
8–12	Horizontal	Standing sitting	ap	16.8		7 × 15	115–150							
			lat	27.3	23.8	8 × 15								
13–17			ap	19.6		8 × 20								
			lat	34.4	29.6	9 × 20								
18			ap	20.0		9 × 22								
			lat	38.9	34.4	10 × 22								

Annotations:

Parameter combination not recommended, normalized organ doses not provided

Parameter combination recommended, normalized organ doses provided

Table 14.11 Radiation qualities of vertical stand cervical spine radiographs

Voltage (kVp)	Total filtration					
	Without additional filtration		With additional filtration			
	1.5 mm Al	2.5 Al	3.5 Al	4.5 Al	3.5 Al + 0.1 Cu	3.5 Al + 0.2 Cu
50						
60	This technique is not suitable for cervical spine radiographs.					
70						
80						
110						
120						
Annotations:						
	Parameter combination not recommended, normalized organ doses not provided					
	Parameter combination not recommended, normalized organ doses provided					
	Parameter combination recommended, normalized organ doses provided					

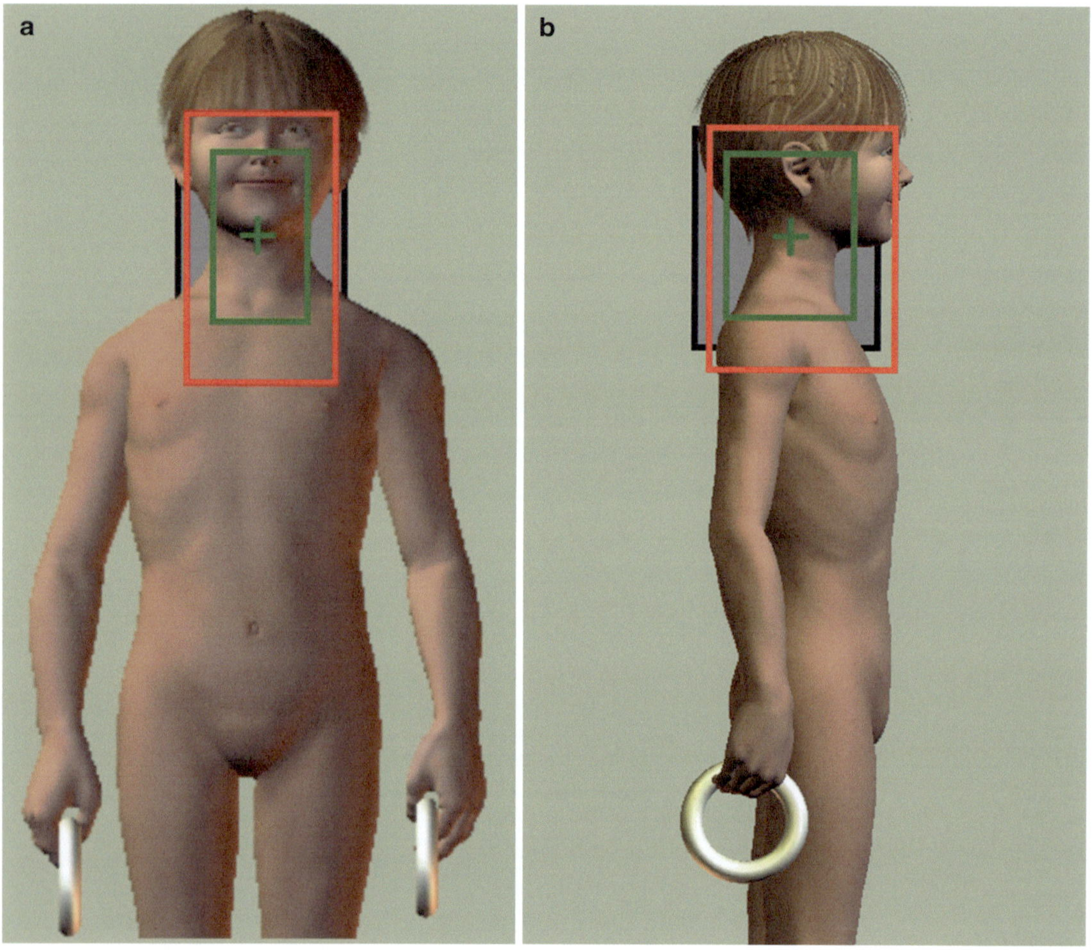

Fig. 14.1 Field settings for cervical spine radiographs in standing position. (**a**) Antero-posterior projection. (**b**) Lateral projection; for lowering both shoulders and thus to make the lower cervical spine visible, the patient holds a weight in each hand. Note that a gonad shielding is just optional due to the large distance of the gonads to the primary radiation field (© V. Rösenberger & M. Seidenbusch 2018). Green lines = optimal field settings; green cross = centring point; red lines = suboptimal field settings; grey shaded area = image detector

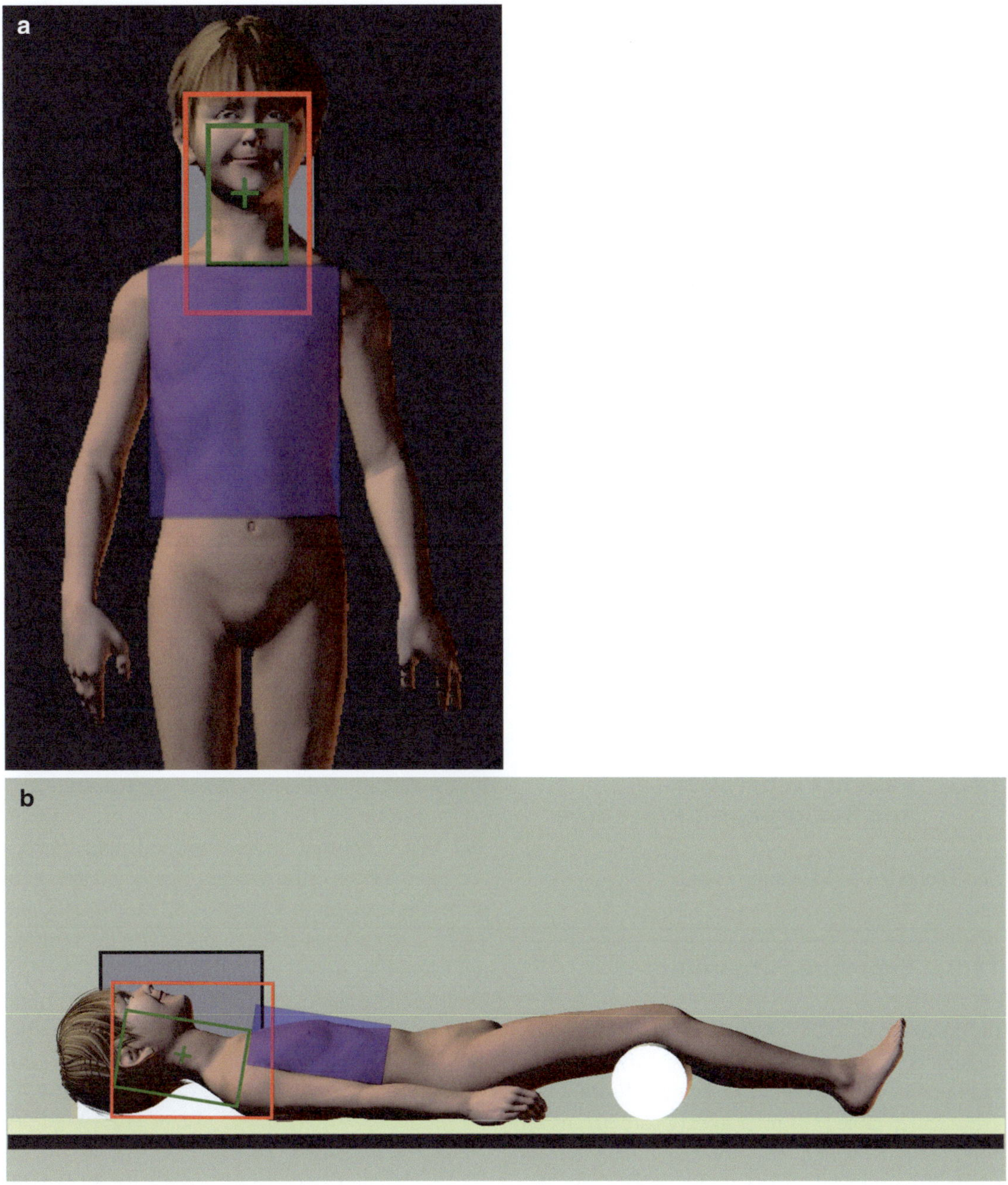

Fig. 14.2 Field settings for cervical spine radiographs in supine position. (**a**) Antero-posterior projection. (**b**) Lateral projection. For lowering both shoulders and thus to make the lower cervical spine visible, the patient's hands should be pulled down by an assisting person. Note that a gonad shielding is just optional due to the large distance of the gonads to the primary radiation field (© V. Rösenberger & M. Seidenbusch 2018). Green lines = optimal field settings; green cross = centring point; red lines = suboptimal field settings; grey shaded area = image detector; blue shaded area = lead mask shielding; dark grey shaded area = Bucky table; white cylinder = foam roll; white device behind the head and neck = foam pillow; white foam mat on the top of the Bucky table

14.2 Indications and contraindications

In recent years, cervical spine radiographs have been widely replaced by MR examinations. Indications and contraindications for cervical spine radiographs are given in Tables 14.1 and 14.2.

14.3 Image Criteria

The image criteria required for cervical spine radiographs are listed in Table 14.3.

See also [2].

14.4 Field Settings

In Table 14.4, patient positions and beam projections used for cervical spine radiographs are listed, and in Table 14.5, the optimal field settings are listed which are required in order to fulfil the image criteria mentioned in Table 14.3.

14.5 Patient Preparation and Radiographic Procedure

See Tables 14.6, 14.7 and [1, 4].

14.6 Exposure Parameters

In Tables 14.8, 14.9, 14.10, and 14.11, exposure parameter combinations used for cervical spine radiographs in various age groups are listed. Dark grey shaded table cells refer to exposure parameter combinations which are not recommended for paediatric cervical spine radiographs and for which no normalized organ doses have been provided. Light grey and light blue shaded table cells refer to exposure parameter combinations for which normalized organ doses have been provided with light grey shaded table cells referring to exposure parameter combinations being not recommended and light blue shaded table cells referring to exposure parameter combinations following the European Guidelines [2]. Note that due to the fact that the radiation dose is an exponential function of the patient diameter, it may make sense to use the patient diameters instead of the patient's age groups for determining the adequate exposure parameters from Tables 14.8 and 14.10. In this case, for determining the corresponding normalized organ doses from Tables 14.19, 14.20, 14.21, 14.22, 14.23, 14.24, 14.25, 14.26, 14.27, 14.28, 14.29, 14.30, 14.31, 14.32, 14.33, 14.34, 14.35, 14.36, 14.37, 14.38, 14.39, 14.40, 14.41, 14.42, 14.43, 14.44, 14.45, 14.46, 14.47, 14.48, 14.49, 14.50, 14.51, 14.52, 14.53, 14.54, 14.55, 14.56, 14.57, and 14.58, the age group related to the patient diameter should be used.

14.7 Normalized Organ Doses

For calculating normalized organ doses, the field settings described in Sect. 14.4 were transferred to the mathematical MIRD phantoms. Inconsistencies between the geometrical anatomy of the mathematical phantoms and the real human anatomy have been listed in Table 14.12 whenever they are of relevance for the validity of the normalized organ doses.

In Tables 14.13, 14.14, 14.15, 14.16, 14.17, and 14.18, optimal and suboptimal field settings are listed for the different age groups and the beam projections given in Tables 14.8, 14.9, 14.10, and 14.11. Green lines refer to optimal field settings, red lines refer to suboptimal field settings, respectively. In the tables, the optimal and suboptimal field settings which were used for performing the calculations of the normalized organ doses provided in Tables 14.19, 14.20, 14.21, 14.22, 14.23, 14.24, 14.25, 14.26, 14.27, 14.28, 14.29, 14.30, 14.31, 14.32, 14.33, 14.34, 14.35, 14.36, 14.37, 14.38, 14.39, 14.40, 14.41, 14.42, 14.43, 14.44, 14.45, 14.46, 14.47, 14.48, 14.49, 14.50, 14.51, 14.52, 14.53, 14.54, 14.55, 14.56, 14.57, and 14.58 were drawn in on the skin surface of the mathematical phantoms. The virtual mathematical radiographs resulting from these field settings were depicted in the tables with and without bone structures, respectively, in order to enable the reader to visually compare the virtual mathemati-

Table 14.12 Inconsistencies between the mathematical MIRD phantoms and real human anatomy

Anatomical region	Inconsistencies
Head	No eyes, nose, mouth, and ears have been inserted
	No eye lens has been inserted
Skeleton	The shape of the spine is unrealistic (no lordosis, no kyphosis considered), the vertebral column is not clearly depicted. The transverse processes are not outlined

For details see the text

cal radiographs with the real radiographs made under clinical conditions. For better visualization of the spine in the skeletal view, oral mucosa, pharynx, paranasal sinus, thymus, and oesophagus have been removed. For comparing the virtual mathematical field settings with the real clinical field settings, the virtual mathematical field edges were also listed in the tables.

As the normalized organ doses provided in Tables 14.19, 14.20, 14.21, 14.22, 14.23, 14.24, 14.25, 14.26, 14.27, 14.28, 14.29, 14.30, 14.31, 14.32, 14.33, 14.34, 14.35, 14.36, 14.37, 14.38, 14.39, 14.40, 14.41, 14.42, 14.43, 14.44, 14.45, 14.46, 14.47, 14.48, 14.49, 14.50, 14.51, 14.52, 14.53, 14.54, 14.55, 14.56, 14.57, and 14.58 refer to entrance kerma values, optimal and suboptimal field sizes were provided in the tables in order to enable the reader to calculate entrance air kerma dose values from clinically measured dose area product values. Please note that these optimal and suboptimal field sizes refer to the mathematical phantoms of an age group only and not to real patients with corresponding ages.

Please note also that the anatomy of the mathematical phantoms has been idealized. Therefore, the virtual field settings for calculating the normalized organ doses have been defined with a view to ensuring that the resulting virtual organ doses would be consistent with the real organ doses which would be achieved during real X-ray examinations being performed with the real field settings recommended in Sect. 14.4. Thus, the field settings shown in the following tables may not be consistent with the clinical field settings as recommended in Sect. 14.4.

Normalized organ doses for full spine radiographs are also listed in e.g. [3, 5–7].

Table 14.13 Field settings—Cervical spine ap

Age group (years)		Anatomical landmarks			
		Optimal field setting		**Suboptimal field setting**	
0		Upper field edge: At the skull base[a] Lower field edge: Upper edge of the manubrium sterni Lateral field edges: Close tothe sternoclavicular joints Field size = field width × field height: 5 cm × 7 cm = 35 cm^2		Upper field edge: At the midpoint of the skull [(1)] Lower field edge: Below the sternoclavicular joints Lateral field edges: Close tothe skin surfaces of the cheeks Field size = field width × field height: 9 cm × 10 cm= 90 cm^2	

1 - 2	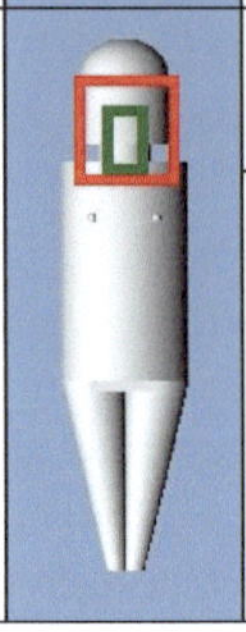	Upper field edge: At the skull base[a] Lower field edge: Upper edge of the manubrium sterni Lateral field edges: Close tothe sternoclavicular joints Field size = field width × field height: 5 cm × 9 cm = 45 cm^2		Upper field edge: At the midpoint of the skull[a] Lower field edge: Below the sternoclavicular jointsdge Lateral field edges: Close to the skin surfaces of the cheeks Field size = field width × field height: 13 cm × 14 cm = 182 cm^2	

Annotation: [a] The skull base cannot be seen in the ap view, but in the lateral view of the MIRD phantom					

Table 14.14 Field settings—Cervical spine ap

Age group (years)		Anatomical landmarks			
		Optimal field setting		**Suboptimal field setting**	
3–7	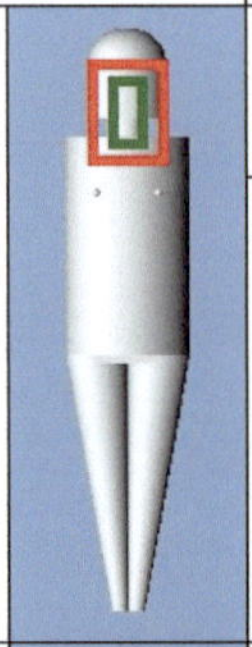	Upper field edge: At the skull base[a] Lower field edge: Upper edge of the manubrium sterni Lateral field edges: Close to the sternoclavicular joints Field size = field width × field height: 6 cm × 11 cm = 66 cm^2		Upper field edge: At the midpoint of the skull[a] Lower field edge: Below the sternoclavicular joints Lateral field edges: Close to the skin surfaces of the cheeks Field size = field width × field height: 14 cm × 16 cm = 224 cm^2	
				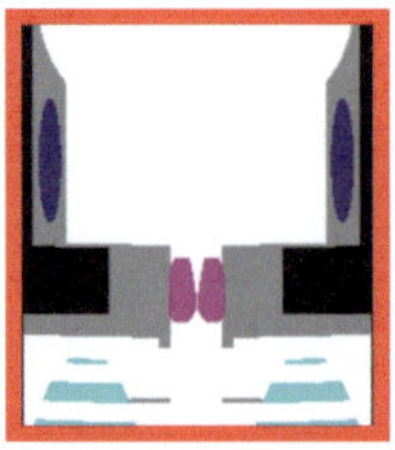	
8–12	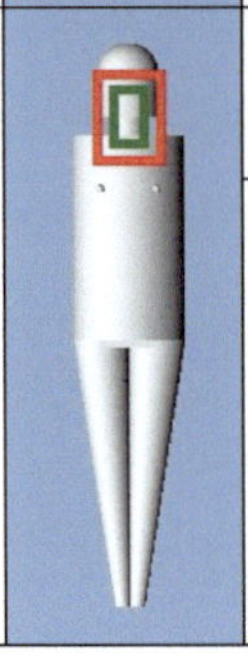	Upper field edge: At the skull base[a] Lower field edge: Upper edge of the manubrium sterni Lateral field edges: Close to the sternoclavicular joints Field size = field width × field height: 7 cm × 15 cm = 105 cm^2		Upper field edge: At the midpoint of the skull[a] Lower field edge: Below the sternoclavicular joints Lateral field edges: Close to the skin surfaces of the cheeks Field size = field width × field height: 15 cm × 20 cm = 300 cm^2	

Annotation: [a] The skull base cannot be seen in the ap view, but in the lateral view of the MIRD phantom					

Table 14.15 Field settings—Cervical spine ap

Age group (years)		Anatomical landmarks			
		Optimal field setting		Suboptimal field setting	
13–17	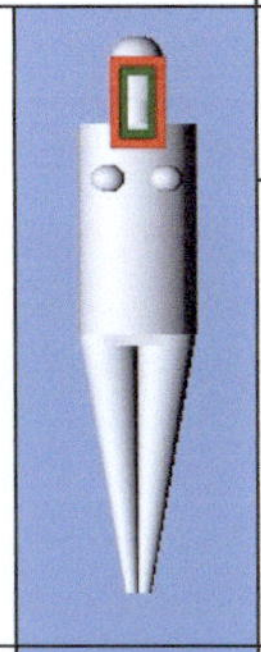	Upper field edge: At the skull base[a] Lower field edge: Upper edge of the manubrium sterni Lateral field edges: Close to the sternoclavicular joints Field size = field width × field height: 8 cm × 20 cm = 160 cm^2		Upper field edge: At the midpoint of the skull[a] Lower field edge: Below the sternoclavicular joints Lateral field edges: Close to the skin surfaces of the cheeks Field size = field width × field height: 15 cm × 22 cm = 330 cm^2	
		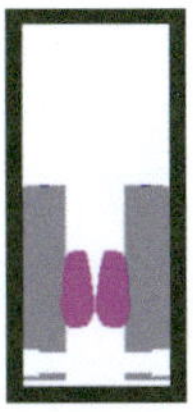		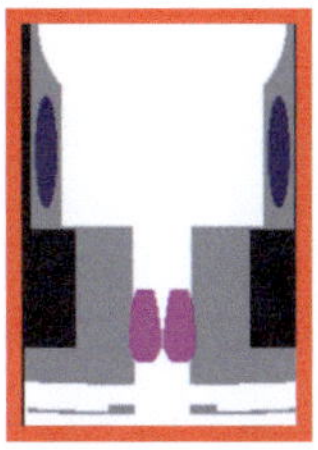	
18	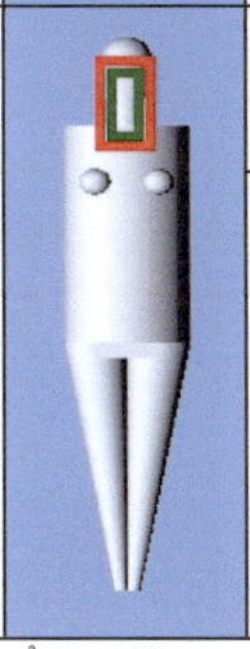	Upper field edge: At the skull base[a] Lower field edge: Upper edge of the manubrium sterni Lateral field edges: Close to the sternoclavicular joints Field size = field width × field height: 9 cm × 22 cm = 198 cm^2		Upper field edge: At the midpoint of the skull[a] Lower field edge: Below the sternoclavicular joints Lateral field edges: Close to the skin surfaces of the cheeks Field size = field width × field height: 16 cm × 24 cm = 384 cm^2	
				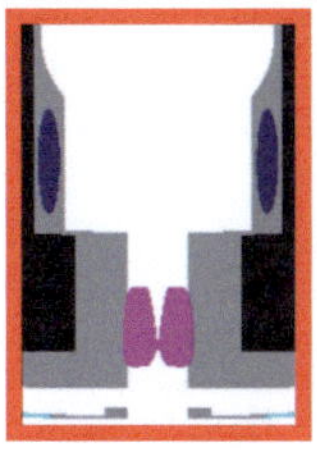	

Annotation: [a] The skull base cannot be seen in the ap view, but in the lateral view of the MIRD phantom

Table 14.16 Field settings—Cervical spine lat

Age group (years)		Anatomical landmarks			
		Optimal field setting		Suboptimal field setting	
0	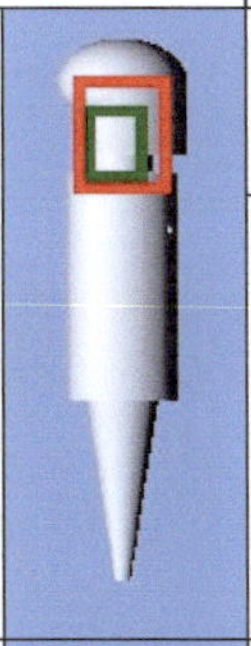	Upper field edge: Close to the lower border of the skull base[a] Lower field edge: Sternoclavicular joint Posterior field edge: At the skin surface of the occiput Anterior field edge: Midpoint of the mandible Field size = field width × field height: 5 cm × 7 cm = 35 cm^2		Upper field edge: At the midpoint of the skull[a] Lower field edge: Below the sternoclavicular joint Posterior field edge: At the skin surface of the occiput Anterior field edge: Close to the front of the face Field size = field width × field height: 8 cm × 10 cm = 80 cm^2	

1–2	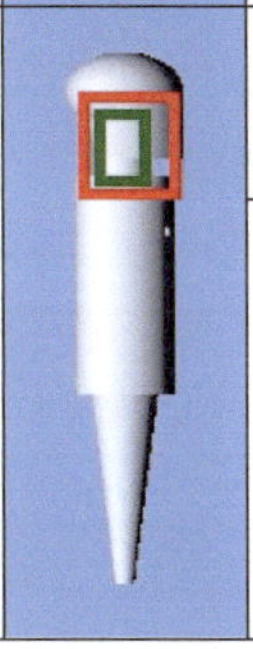	Upper field edge: Close to the lower border of the skull base[a] Lower field edge: Sternoclavicular joint Posterior field edge: At the skin surface of the occiput Anterior field edge: Midpoint of the mandible Field size = field width × field height: 6 cm × 10 cm = 60 cm^2		Upper field edge: At the midpoint of the skull[a] Lower field edge: Below the sternoclavicular joint Posterior field edge: At the skin surface of the occiput Anterior field edge: Close to the front of the face Field size = field width × field height:12 cm × 14 cm = 168 cm^2	

Annotation: [a] The skull base cannot be seen in the ap view, but in the lateral view of the MIRD phantom

Table 14.17 Field settings—Cervical spine lat

Age group (years)		Anatomical landmarks			
		Optimal field setting		Suboptimal field setting	
3–7	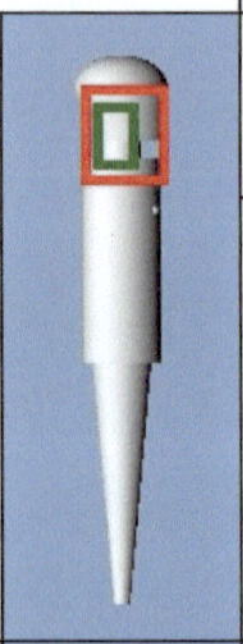	Upper field edge: Close to the lower border of the skull base[a] Lower field edge: Sternoclavicular joint Posterior field edge: At the skin surface of the occiput Anterior field edge: Midpoint of the mandible Field size = field width × field height: 7 cm × 11 cm = 77 cm^2		Upper field edge: At the midpoint of the skull[a] Lower field edge: Below the sternoclavicular joint Posterior field edge: At the skin surface of the occiput Anterior field edge: Close to the front of the face Field size = field width × field height: 14 cm × 17 cm = 238 cm^2	

8–12	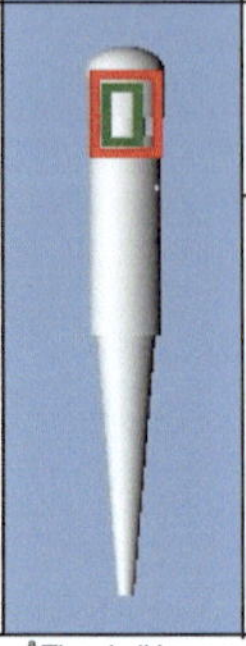	Upper field edge: Close to the lower border of the skull base[a] Lower field edge: Sternoclavicular joint Posterior field edge: At the skin surface of the occiput Anterior field edge: Midpoint of the mandible Field size = field width × field height: 8 cm × 15 cm = 120 cm^2		Upper field edge: At the midpoint of the skull[a] Lower field edge: Below the sternoclavicular joint Posterior field edge: At the skin surface of the occiput Anterior field edge: Close to the front of the face Field size = field width × field height: 15 cm × 19 cm = 285 cm^2	

Annotation: [a] The skull base cannot be seen in the ap view, but in the lateral view of the MIRD phantom					

Table 14.18 Field settings—Cervical spine lat

Age group (years)		Anatomical landmarks			
		Optimal field setting		Suboptimal field setting	
13–17	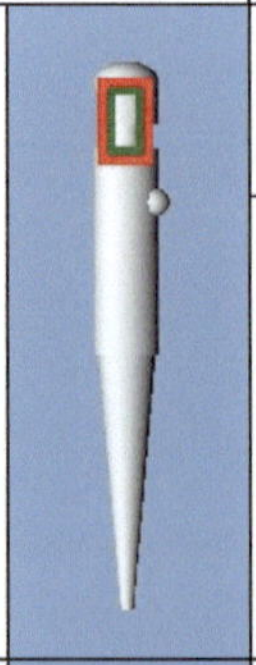	Upper field edge: Close to the lower border of the skull base[a] Lower field edge: Sternoclavicular joint Posterior field edge: At the skin surface of the occiput Anterior field edge: Midpoint of the mandible Field size = field width × field height: 9 cm × 20 cm = 180 cm^2		Upper field edge: At the midpoint of the skull[a] Lower field edge: Below the sternoclavicular joint Posterior field edge: At the skin surface of the occiput Anterior field edge: Close to the front of the face Field size = field width × field height: 16 cm × 24 cm = 384 cm^2	
			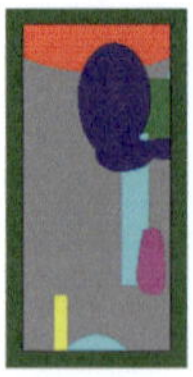		
18	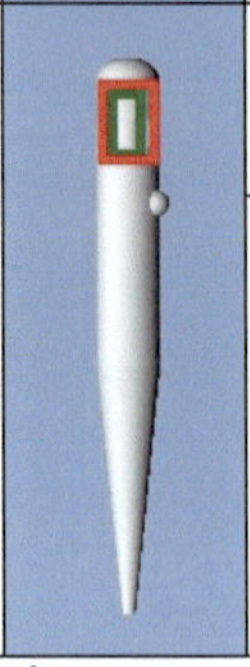	Upper field edge: Close to the lower border of the skull base[a] Lower field edge: Sternoclavicular joint Posterior field edge: At the skin surface of the occiput Anterior field edge: Midpoint of the mandible Field size = field width × field height: 10 cm × 22 cm = 220 cm^2		Upper field edge: At the midpoint of the skull[a] Lower field edge: Below the sternoclavicular joint Posterior field edge: At the skin surface of the occiput Anterior field edge: Close to the front of the face Field size = field width × field height: 18 cm × 26 cm = 468 cm^2	

Annotation: [a]The skull base cannot be seen in the ap view, but in the lateral view of the MIRD phantom					

Table 14.19 Normalized organ doses—Cervical spine ap—<1 year

Beam projection:			ap									**Table cells:**
Focus-to-detector distance (cm):			100					(+) = normal font = optimal field setting				
Tube voltage (kVp):			50					**(−) = bold font = suboptimal field setting**				
Total filtrations:			See last table row.					Empty table cells correspond to values <0.01				
Organ or tissue	**Organ dose D_T : Entrance dose K_E**							**(mSv * mGy⁻¹)**				
	(+)	(−)	(+)	(−)	(+)	(−)	(+)	(−)	(+)	(−)	(+)	(−)
Skull	0.16	**0.51**	0.19	**0.61**	0.21	**0.68**	0.23	**0.73**	0.26	**0.82**	0.29	**0.90**
Brain	0.01	**0.09**	0.02	**0.12**	0.02	**0.14**	0.03	**0.16**	0.03	**0.19**	0.04	**0.21**
Oral mucosa	0.37	**0.52**	0.43	**0.61**	0.47	**0.68**	0.50	**0.72**	0.55	**0.80**	0.59	**0.87**
Salivary glands	0.13	**0.42**	0.16	**0.51**	0.18	**0.57**	0.20	**0.62**	0.22	**0.70**	0.25	**0.76**
Cervical spine	0.40	**0.44**	0.52	**0.58**	0.61	**0.69**	0.68	**0.78**	0.82	**0.94**	0.95	**1.09**
Thyroid	0.78	**0.81**	0.88	**0.92**	0.95	**0.99**	0.99	**1.04**	1.06	**1.13**	1.11	**1.19**
Thymus	0.08	**0.33**	0.09	**0.39**	0.10	**0.43**	0.11	**0.46**	0.12	**0.50**	0.14	**0.54**
Thoracic spine	0.02	**0.05**	0.03	**0.07**	0.04	**0.09**	0.05	**0.10**	0.06	**0.13**	0.07	**0.16**
Claviculae	0.74	**1.69**	0.86	**1.99**	0.93	**2.17**	0.97	**2.30**	1.05	**2.52**	1.11	**2.69**
Scapulae	0.01	**0.09**	0.02	**0.12**	0.02	**0.14**	0.03	**0.16**	0.03	**0.20**	0.04	**0.24**
Ribs	0.02	**0.18**	0.02	**0.21**	0.03	**0.23**	0.03	**0.24**	0.04	**0.27**	0.04	**0.29**
Oesophagus	0.02	**0.04**	0.02	**0.05**	0.03	**0.06**	0.03	**0.07**	0.04	**0.08**	0.04	**0.10**
Trachea	0.46	**0.50**	0.54	**0.59**	0.59	**0.65**	0.63	**0.69**	0.69	**0.77**	0.74	**0.83**
Lung		**0.04**	0.01	**0.05**	0.01	**0.06**	0.01	**0.06**	0.02	**0.08**	0.02	**0.09**
Mamma		**0.01**		**0.01**		**0.02**		**0.02**		**0.02**		**0.03**
Heart		**0.01**		**0.02**		**0.02**		**0.02**		**0.03**	0.01	**0.03**
Stomach												
Spleen												
Liver												
Gall bladder												
Pancreas												
Small intestine												
Large intestine												
Sigma/rectum												
Lumbar spine												
Pelvis												
Adrenals												
Kidneys												
Urinary bladder												
Uterus												
Ovaries												
Prostate												
Testes												
Upper arm	0.01	**0.04**	0.02	**0.05**	0.02	**0.06**	0.02	**0.07**	0.03	**0.09**	0.03	**0.11**
Elbow						**0.01**		**0.01**		**0.02**		**0.02**
Forearm												
Upper leg												
Knee												
Lower leg												
Lymph nodes	0.11	**0.15**	0.13	**0.18**	0.14	**0.19**	0.15	**0.21**	0.17	**0.23**	0.18	**0.25**
Skin	0.05	**0.14**	0.05	**0.15**	0.05	**0.16**	0.06	**0.16**	0.06	**0.17**	0.06	**0.18**
Muscle	0.04	**0.08**	0.04	**0.09**	0.05	**0.10**	0.05	**0.11**	0.06	**0.12**	0.07	**0.13**
Skeleton	0.07	**0.18**	0.08	**0.22**	0.09	**0.25**	0.10	**0.27**	0.12	**0.31**	0.14	**0.34**
Red bone marrow	0.02	**0.07**	0.03	**0.09**	0.03	**0.10**	0.04	**0.11**	0.04	**0.12**	0.05	**0.14**
Whole body dose	0.04	**0.09**	0.04	**0.10**	0.05	**0.11**	0.05	**0.12**	0.06	**0.14**	0.07	**0.16**
ED (ICRP 60)	0.05	**0.06**	0.05	**0.07**	0.06	**0.08**	0.06	**0.08**	0.07	**0.09**	0.07	**0.10**
ED (ICRP 103)	0.05	**0.07**	0.06	**0.08**	0.06	**0.09**	0.06	**0.10**	0.07	**0.11**	0.08	**0.12**
Total filtration (mm) **Al**	**1.5**		**2.5**		**3.5**		**4.5**		**3.5**		**3.5**	
Cu	**0**		**0**		**0**		**0**		**0.1**		**0.2**	

Table 14.20 Normalized organ doses—Cervical spine ap—<1 year

Beam projection:			ap									**Table cells:**
Focus-to-detector distance (cm):			100						(+) = normal font = optimal field setting			
Tube voltage (kVp):			60						**(−) = bold font = suboptimal field setting**			
Total filtrations:			See last table row.				Empty table cells correspond to values <0.01					
Organ or tissue	**Organ dose D_T : Entrance dose K_E**								**(mSv * mGy^{-1})**			
	(+)	(−)	(+)	(−)	(+)	(−)	(+)	(−)	(+)	(−)	(+)	(−)
Skull	0.18	**0.58**	0.22	**0.69**	0.24	**0.76**	0.26	**0.81**	0.29	**0.90**	0.29	**0.90**
Brain	0.02	**0.12**	0.02	**0.15**	0.03	**0.17**	0.03	**0.19**	0.04	**0.22**	0.04	**0.21**
Oral mucosa	0.41	**0.59**	0.47	**0.68**	0.51	**0.75**	0.55	**0.80**	0.60	**0.88**	0.59	**0.87**
Salivary glands	0.16	**0.49**	0.19	**0.58**	0.21	**0.64**	0.22	**0.69**	0.25	**0.77**	0.25	**0.76**
Cervical spine	0.51	**0.58**	0.64	**0.74**	0.75	**0.86**	0.83	**0.96**	0.98	**1.14**	0.95	**1.09**
Thyroid	0.84	**0.88**	0.94	**1.00**	1.00	**1.07**	1.05	**1.12**	1.11	**1.20**	1.11	**1.19**
Thymus	0.09	**0.37**	0.10	**0.43**	0.12	**0.47**	0.12	**0.50**	0.14	**0.54**	0.14	**0.54**
Thoracic spine	0.03	**0.08**	0.05	**0.10**	0.05	**0.12**	0.06	**0.14**	0.08	**0.17**	0.07	**0.16**
Claviculae	0.81	**1.86**	0.91	**2.15**	0.98	**2.32**	1.02	**2.45**	1.09	**2.64**	1.11	**2.69**
Scapulae	0.02	**0.12**	0.03	**0.16**	0.03	**0.19**	0.04	**0.21**	0.04	**0.25**	0.04	**0.24**
Ribs	0.02	**0.20**	0.03	**0.23**	0.03	**0.25**	0.04	**0.27**	0.04	**0.29**	0.04	**0.29**
Oesophagus	0.02	**0.05**	0.03	**0.06**	0.03	**0.08**	0.04	**0.09**	0.05	**0.10**	0.04	**0.10**
Trachea	0.52	**0.56**	0.60	**0.66**	0.65	**0.72**	0.68	**0.76**	0.74	**0.84**	0.74	**0.83**
Lung	0.01	**0.05**	0.01	**0.06**	0.02	**0.07**	0.02	**0.08**	0.02	**0.09**	0.02	**0.09**
Mamma		**0.01**		**0.02**		**0.02**		**0.02**		**0.03**		**0.03**
Heart		**0.02**		**0.02**		**0.03**	0.01	**0.03**	0.01	**0.04**	0.01	**0.03**
Stomach												
Spleen												
Liver												
Gall bladder												
Pancreas												
Small intestine												
Large intestine												
Sigma/rectum												
Lumbar spine												
Pelvis												
Adrenals												
Kidneys												
Urinary bladder												
Uterus												
Ovaries												
Prostate												
Testes												
Upper arm	0.02	**0.05**	0.02	**0.07**	0.03	**0.08**	0.03	**0.09**	0.04	**0.12**	0.03	**0.11**
Elbow				**0.01**		**0.02**		**0.02**		**0.02**		**0.02**
Forearm												
Upper leg												
Knee												
Lower leg												
Lymph nodes	0.13	**0.17**	0.14	**0.20**	0.16	**0.22**	0.17	**0.23**	0.18	**0.25**	0.18	**0.25**
Skin	0.05	**0.15**	0.06	**0.16**	0.06	**0.17**	0.06	**0.17**	0.06	**0.18**	0.06	**0.18**
Muscle	0.04	**0.09**	0.05	**0.10**	0.06	**0.11**	0.06	**0.12**	0.07	**0.13**	0.07	**0.13**
Skeleton	0.08	**0.21**	0.10	**0.26**	0.11	**0.29**	0.12	**0.31**	0.14	**0.35**	0.14	**0.34**
Red bone marrow	0.03	**0.09**	0.04	**0.10**	0.04	**0.11**	0.04	**0.12**	0.05	**0.14**	0.05	**0.14**
Whole body dose	0.04	**0.10**	0.05	**0.12**	0.06	**0.13**	0.06	**0.14**	0.07	**0.16**	0.07	**0.16**
ED (ICRP 60)	0.05	**0.07**	0.06	**0.08**	0.06	**0.09**	0.07	**0.09**	0.07	**0.10**	0.07	**0.10**
ED (ICRP 103)	0.05	**0.08**	0.06	**0.09**	0.07	**0.10**	0.07	**0.11**	0.08	**0.12**	0.08	**0.12**
Total filtration (mm) Al	**1.5**		**2.5**		**3.5**		**4.5**		**3.5**		**3.5**	
Cu	**0**		**0**		**0**		**0**		**0.1**		**0.2**	

Table 14.21 Normalized organ doses—Cervical spine ap—<1 year

Beam projection:			ap									**Table cells:**
Focus-to-detector distance (cm):			100					(+) = normal font = optimal field setting				
Tube voltage (kVp):			70					**(−) = bold font = suboptimal field setting**				
Total filtrations:			See last table row.					Empty table cells correspond to values <0.01				
Organ or tissue	**Organ dose D_T : Entrance dose K_E**							**(mSv * mGy⁻¹)**				
	(+)	(−)	(+)	(−)	(+)	(−)	(+)	(−)	(+)	(−)	(+)	(−)
Skull	0.20	**0.64**	0.24	**0.74**	0.26	**0.81**	0.28	**0.86**	0.31	**0.94**	0.33	**1.01**
Brain	0.02	**0.14**	0.03	**0.17**	0.03	**0.20**	0.04	**0.22**	0.05	**0.25**	0.05	**0.28**
Oral mucosa	0.44	**0.64**	0.51	**0.74**	0.55	**0.80**	0.58	**0.85**	0.63	**0.93**	0.67	**0.99**
Salivary glands	0.17	**0.54**	0.20	**0.63**	0.23	**0.69**	0.24	**0.74**	0.27	**0.82**	0.29	**0.88**
Cervical spine	0.59	**0.68**	0.74	**0.85**	0.85	**0.98**	0.93	**1.08**	1.08	**1.26**	1.21	**1.42**
Thyroid	0.89	**0.94**	0.99	**1.05**	1.05	**1.12**	1.09	**1.17**	1.15	**1.25**	1.20	**1.31**
Thymus	0.10	**0.40**	0.11	**0.46**	0.12	**0.50**	0.13	**0.53**	0.15	**0.57**	0.16	**0.61**
Thoracic spine	0.04	**0.10**	0.06	**0.12**	0.07	**0.15**	0.07	**0.16**	0.09	**0.20**	0.10	**0.23**
Claviculae	0.85	**1.98**	0.95	**2.24**	1.00	**2.40**	1.04	**2.51**	1.10	**2.68**	1.13	**2.78**
Scapulae	0.03	**0.15**	0.03	**0.19**	0.04	**0.22**	0.04	**0.24**	0.05	**0.29**	0.06	**0.33**
Ribs	0.03	**0.22**	0.03	**0.25**	0.04	**0.27**	0.04	**0.28**	0.05	**0.30**	0.05	**0.32**
Oesophagus	0.03	**0.06**	0.04	**0.08**	0.04	**0.09**	0.05	**0.10**	0.05	**0.12**	0.06	**0.14**
Trachea	0.56	**0.61**	0.64	**0.71**	0.68	**0.76**	0.72	**0.81**	0.78	**0.88**	0.83	**0.94**
Lung	0.01	**0.06**	0.02	**0.07**	0.02	**0.08**	0.02	**0.09**	0.03	**0.10**	0.03	**0.11**
Mamma		**0.02**		**0.02**		**0.02**		**0.03**	0.01	**0.03**	0.01	**0.03**
Heart		**0.02**		**0.03**	0.01	**0.03**	0.01	**0.04**	0.02	**0.04**	0.02	**0.05**
Stomach												
Spleen										**0.01**		**0.01**
Liver												**0.01**
Gall bladder												
Pancreas										**0.01**		**0.01**
Small intestine												
Large intestine												
Sigma/rectum												
Lumbar spine												**0.01**
Pelvis												
Adrenals										**0.01**		**0.01**
Kidneys												
Urinary bladder												
Uterus												
Ovaries												
Prostate												
Testes												
Upper arm	0.02	**0.07**	0.03	**0.09**	0.03	**0.10**	0.03	**0.11**	0.04	**0.13**	0.05	**0.15**
Elbow		**0.01**		**0.02**		**0.02**		**0.02**	0.01	**0.03**	0.01	**0.03**
Forearm												
Upper leg												
Knee												
Lower leg												
Lymph nodes	0.14	**0.18**	0.15	**0.21**	0.17	**0.23**	0.18	**0.24**	0.19	**0.27**	0.20	**0.29**
Skin	0.05	**0.16**	0.06	**0.17**	0.06	**0.17**	0.06	**0.18**	0.07	**0.19**	0.07	**0.20**
Muscle	0.05	**0.10**	0.06	**0.11**	0.06	**0.12**	0.07	**0.13**	0.07	**0.15**	0.08	**0.16**
Skeleton	0.09	**0.24**	0.11	**0.28**	0.12	**0.31**	0.13	**0.33**	0.15	**0.37**	0.16	**0.40**
Red bone marrow	0.03	**0.10**	0.04	**0.11**	0.05	**0.13**	0.05	**0.14**	0.06	**0.16**	0.06	**0.17**
Whole body dose	0.05	**0.11**	0.05	**0.13**	0.06	**0.14**	0.06	**0.15**	0.07	**0.17**	0.08	**0.19**
ED (ICRP 60)	0.05	**0.08**	0.06	**0.09**	0.07	**0.10**	0.07	**0.10**	0.07	**0.11**	0.08	**0.12**
ED (ICRP 103)	0.06	**0.09**	0.07	**0.10**	0.07	**0.11**	0.07	**0.12**	0.08	**0.13**	0.09	**0.14**
Total filtration (mm) **Al**	**1.5**		**2.5**		**3.5**		**4.5**		**3.5**		**3.5**	
Cu	**0**		**0**		**0**		**0**		**0.1**		**0.2**	

Table 14.22 Normalized organ doses—Cervical spine r-lat—<1 year

Beam projection:			r-lat									**Table cells:**
Focus-to-detector distance (cm):			100									(+) = normal font = optimal field setting
Tube voltage (kVp):			50									**(−) = bold font = suboptimal field setting**
Total filtrations:			See last table row.									Empty table cells correspond to values <0.01
Organ or tissue	**Organ dose D_T : Entrance dose K_E**								**(mSv ∗ mGy^{-1})**			
	(+)	(−)	(+)	(−)	(+)	(−)	(+)	(−)	(+)	(−)	(+)	(−)
Skull	0.09	**0.47**	0.12	**0.56**	0.13	**0.62**	0.15	**0.66**	0.17	**0.74**	0.19	**0.82**
Brain	0.01	**0.13**	0.02	**0.17**	0.02	**0.19**	0.02	**0.21**	0.03	**0.25**	0.03	**0.28**
Oral mucosa	0.14	**0.28**	0.17	**0.34**	0.19	**0.39**	0.20	**0.42**	0.23	**0.47**	0.26	**0.52**
Salivary glands	0.37	**0.46**	0.42	**0.52**	0.44	**0.56**	0.46	**0.59**	0.49	**0.64**	0.52	**0.69**
Cervical spine	0.70	**0.75**	0.90	**0.96**	1.04	**1.12**	1.15	**1.24**	1.35	**1.47**	1.53	**1.67**
Thyroid	0.33	**0.49**	0.39	**0.58**	0.44	**0.65**	0.47	**0.70**	0.53	**0.79**	0.58	**0.86**
Thymus	0.01	**0.08**	0.01	**0.11**	0.02	**0.12**	0.02	**0.14**	0.02	**0.17**	0.03	**0.19**
Thoracic spine	0.04	**0.08**	0.05	**0.11**	0.07	**0.13**	0.07	**0.15**	0.09	**0.18**	0.11	**0.21**
Claviculae	0.24	**0.69**	0.28	**0.84**	0.31	**0.95**	0.33	**1.03**	0.37	**1.17**	0.41	**1.30**
Scapulae	0.06	**0.33**	0.08	**0.38**	0.09	**0.42**	0.10	**0.44**	0.11	**0.49**	0.12	**0.52**
Ribs	0.03	**0.09**	0.04	**0.11**	0.04	**0.12**	0.05	**0.14**	0.05	**0.16**	0.06	**0.18**
Oesophagus	0.02	**0.03**	0.03	**0.05**	0.04	**0.05**	0.04	**0.06**	0.05	**0.07**	0.06	**0.09**
Trachea	0.10	**0.16**	0.13	**0.20**	0.15	**0.24**	0.16	**0.26**	0.19	**0.31**	0.22	**0.35**
Lung	0.01	**0.02**	0.01	**0.03**	0.02	**0.03**	0.02	**0.04**	0.02	**0.05**	0.03	**0.06**
Mamma										**0.01**		**0.02**
Heart				**0.01**		**0.01**		**0.01**		**0.02**	0.01	**0.02**
Stomach												
Spleen												
Liver												
Gall bladder												
Pancreas												
Small intestine												
Large intestine												
Sigma/rectum												
Lumbar spine												
Pelvis												
Adrenals												
Kidneys												
Urinary bladder												
Uterus												
Ovaries												
Prostate												
Testes												
Upper arm		**0.40**		**0.46**		**0.49**		**0.52**		**0.57**		**0.61**
Elbow										**0.01**		**0.02**
Forearm												
Upper leg												
Knee												
Lower leg												
Lymph nodes	0.06	**0.09**	0.07	**0.11**	0.08	**0.12**	0.09	**0.13**	0.10	**0.15**	0.11	**0.16**
Skin	0.05	**0.12**	0.06	**0.12**	0.06	**0.13**	0.06	**0.13**	0.06	**0.14**	0.07	**0.15**
Muscle	0.04	**0.06**	0.04	**0.07**	0.05	**0.08**	0.05	**0.09**	0.06	**0.10**	0.06	**0.11**
Skeleton	0.07	**0.20**	0.08	**0.24**	0.10	**0.27**	0.11	**0.30**	0.13	**0.34**	0.14	**0.37**
Red bone marrow	0.02	**0.07**	0.02	**0.09**	0.03	**0.10**	0.03	**0.10**	0.04	**0.12**	0.04	**0.13**
Whole body dose	0.03	**0.08**	0.04	**0.10**	0.05	**0.11**	0.05	**0.12**	0.06	**0.13**	0.06	**0.15**
ED (ICRP 60)	0.02	**0.04**	0.03	**0.05**	0.03	**0.06**	0.03	**0.06**	0.04	**0.07**	0.04	**0.08**
ED (ICRP 103)	0.03	**0.05**	0.03	**0.06**	0.04	**0.06**	0.04	**0.07**	0.04	**0.08**	0.05	**0.09**
Total filtration (mm) Al	**1.5**		**2.5**		**3.5**		**4.5**		**3.5**		**3.5**	
Cu	**0**		**0**		**0**		**0**		**0.1**		**0.2**	

Table 14.23 Normalized organ doses—Cervical spine r-lat—<1 year

Beam projection:			r-lat									**Table cells:**
Focus-to-detector distance (cm):			100				(+) = normal font = optimal field setting					
Tube voltage (kVp):			60				**(−) = bold font = suboptimal field setting**					
Total filtrations:			See last table row.				Empty table cells correspond to values <0.01					
Organ or tissue	**Organ dose D_T : Entrance dose K_E**								**(mSv ∗ mGy⁻¹)**			
	(+)	(−)	(+)	(−)	(+)	(−)	(+)	(−)	(+)	(−)	(+)	(−)
Skull	0.11	**0.53**	0.14	**0.63**	0.16	**0.69**	0.17	**0.74**	0.20	**0.82**	0.19	**0.82**
Brain	0.02	**0.16**	0.02	**0.20**	0.02	**0.23**	0.03	**0.25**	0.03	**0.29**	0.03	**0.28**
Oral mucosa	0.16	**0.33**	0.19	**0.40**	0.22	**0.44**	0.23	**0.48**	0.26	**0.54**	0.26	**0.52**
Salivary glands	0.40	**0.51**	0.44	**0.57**	0.47	**0.61**	0.49	**0.64**	0.52	**0.70**	0.52	**0.69**
Cervical spine	0.86	**0.93**	1.07	**1.16**	1.22	**1.33**	1.34	**1.46**	1.54	**1.70**	1.53	**1.67**
Thyroid	0.38	**0.56**	0.45	**0.66**	0.49	**0.73**	0.53	**0.78**	0.59	**0.87**	0.58	**0.86**
Thymus	0.02	**0.10**	0.02	**0.13**	0.02	**0.15**	0.03	**0.17**	0.03	**0.20**	0.03	**0.19**
Thoracic spine	0.05	**0.11**	0.07	**0.14**	0.08	**0.17**	0.09	**0.19**	0.11	**0.22**	0.11	**0.21**
Claviculae	0.27	**0.80**	0.31	**0.96**	0.34	**1.07**	0.37	**1.15**	0.41	**1.29**	0.41	**1.30**
Scapulae	0.08	**0.36**	0.09	**0.42**	0.10	**0.45**	0.11	**0.48**	0.13	**0.53**	0.12	**0.52**
Ribs	0.04	**0.11**	0.04	**0.13**	0.05	**0.14**	0.05	**0.16**	0.06	**0.18**	0.06	**0.18**
Oesophagus	0.03	**0.05**	0.04	**0.06**	0.05	**0.07**	0.05	**0.08**	0.06	**0.09**	0.06	**0.09**
Trachea	0.12	**0.20**	0.15	**0.25**	0.18	**0.28**	0.19	**0.31**	0.22	**0.36**	0.22	**0.35**
Lung	0.01	**0.03**	0.02	**0.04**	0.02	**0.04**	0.02	**0.05**	0.03	**0.06**	0.03	**0.06**
Mamma						**0.01**		**0.01**		**0.02**		**0.02**
Heart		**0.01**		**0.02**		**0.02**		**0.02**	0.01	**0.03**	0.01	**0.02**
Stomach												
Spleen												
Liver												
Gall bladder												
Pancreas												
Small intestine												
Large intestine												
Sigma/rectum												
Lumbar spine												
Pelvis												
Adrenals												
Kidneys												
Urinary bladder												
Uterus												
Ovaries												
Prostate												
Testes												
Upper arm		**0.44**		**0.50**		**0.53**		**0.56**		**0.61**		**0.61**
Elbow				**0.01**		**0.01**		**0.01**		**0.02**		**0.02**
Forearm												
Upper leg												
Knee												
Lower leg												
Lymph nodes	0.07	**0.10**	0.08	**0.12**	0.09	**0.14**	0.10	**0.15**	0.11	**0.16**	0.11	**0.16**
Skin	0.05	**0.12**	0.06	**0.13**	0.06	**0.14**	0.06	**0.14**	0.07	**0.15**	0.07	**0.15**
Muscle	0.04	**0.07**	0.05	**0.09**	0.05	**0.09**	0.06	**0.10**	0.06	**0.11**	0.06	**0.11**
Skeleton	0.08	**0.23**	0.10	**0.28**	0.11	**0.31**	0.13	**0.33**	0.14	**0.38**	0.14	**0.37**
Red bone marrow	0.02	**0.08**	0.03	**0.10**	0.03	**0.11**	0.04	**0.12**	0.04	**0.14**	0.04	**0.13**
Whole body dose	0.04	**0.09**	0.05	**0.11**	0.05	**0.12**	0.06	**0.13**	0.06	**0.15**	0.06	**0.15**
ED (ICRP 60)	0.03	**0.05**	0.03	**0.06**	0.04	**0.07**	0.04	**0.07**	0.04	**0.08**	0.04	**0.08**
ED (ICRP 103)	0.03	**0.06**	0.04	**0.07**	0.04	**0.07**	0.04	**0.08**	0.05	**0.09**	0.05	**0.09**
Total filtration (mm) Al	**1.5**		**2.5**		**3.5**		**4.5**		**3.5**		**3.5**	
Cu	**0**		**0**		**0**		**0**		**0.1**		**0.2**	

Table 14.24 Normalized organ doses—Cervical spine r-lat—<1 year

Beam projection:			r-lat									**Table cells:**
Focus-to-detector distance (cm):			100									(+) = normal font = optimal field setting
Tube voltage (kVp):			70									**(−) = bold font = suboptimal field setting**
Total filtrations:			See last table row.									Empty table cells correspond to values <0.01
Organ or tissue	**Organ dose D_T : Entrance dose K_E**								**(mSv * mGy^{-1})**			
	(+)	(−)	(+)	(−)	(+)	(−)	(+)	(−)	(+)	(−)	(+)	(−)
Skull	0.13	**0.58**	0.15	**0.67**	0.17	**0.74**	0.19	**0.78**	0.21	**0.86**	0.23	**0.92**
Brain	0.02	**0.19**	0.03	**0.23**	0.03	**0.25**	0.03	**0.28**	0.04	**0.32**	0.04	**0.35**
Oral mucosa	0.18	**0.37**	0.21	**0.44**	0.24	**0.48**	0.25	**0.52**	0.28	**0.58**	0.31	**0.63**
Salivary glands	0.42	**0.54**	0.47	**0.61**	0.49	**0.65**	0.51	**0.68**	0.54	**0.73**	0.57	**0.78**
Cervical spine	0.98	**1.06**	1.19	**1.30**	1.34	**1.47**	1.45	**1.60**	1.65	**1.83**	1.80	**2.01**
Thyroid	0.42	**0.61**	0.49	**0.72**	0.53	**0.79**	0.57	**0.84**	0.63	**0.92**	0.68	**1.00**
Thymus	0.02	**0.12**	0.02	**0.15**	0.03	**0.18**	0.03	**0.19**	0.04	**0.23**	0.04	**0.25**
Thoracic spine	0.07	**0.13**	0.08	**0.17**	0.10	**0.19**	0.11	**0.21**	0.12	**0.25**	0.14	**0.29**
Claviculae	0.29	**0.88**	0.34	**1.04**	0.37	**1.15**	0.39	**1.22**	0.42	**1.36**	0.45	**1.45**
Scapulae	0.09	**0.39**	0.10	**0.45**	0.11	**0.48**	0.12	**0.50**	0.14	**0.55**	0.15	**0.58**
Ribs	0.04	**0.12**	0.05	**0.14**	0.05	**0.16**	0.06	**0.17**	0.07	**0.19**	0.07	**0.21**
Oesophagus	0.04	**0.06**	0.05	**0.07**	0.05	**0.08**	0.06	**0.09**	0.07	**0.11**	0.08	**0.12**
Trachea	0.14	**0.23**	0.17	**0.28**	0.20	**0.32**	0.21	**0.35**	0.25	**0.40**	0.27	**0.44**
Lung	0.02	**0.04**	0.02	**0.05**	0.02	**0.05**	0.03	**0.06**	0.03	**0.07**	0.04	**0.08**
Mamma				**0.01**		**0.02**		**0.02**		**0.02**	0.01	**0.03**
Heart		**0.01**		**0.02**	0.01	**0.02**	0.01	**0.03**	0.01	**0.03**	0.02	**0.04**
Stomach												
Spleen												
Liver												**0.01**
Gall bladder												
Pancreas												**0.01**
Small intestine												
Large intestine												
Sigma/rectum												
Lumbar spine												
Pelvis												
Adrenals												**0.01**
Kidneys												
Urinary bladder												
Uterus												
Ovaries												
Prostate												
Testes												
Upper arm		**0.46**		**0.52**		**0.56**		**0.58**		**0.63**		**0.66**
Elbow		**0.01**		**0.01**		**0.02**		**0.02**		**0.02**		**0.03**
Forearm												
Upper leg												
Knee												
Lower leg												
Lymph nodes	0.08	**0.12**	0.09	**0.14**	0.10	**0.15**	0.11	**0.16**	0.12	**0.18**	0.13	**0.19**
Skin	0.06	**0.13**	0.06	**0.14**	0.06	**0.14**	0.07	**0.15**	0.07	**0.16**	0.07	**0.17**
Muscle	0.05	**0.08**	0.05	**0.09**	0.06	**0.10**	0.06	**0.11**	0.07	**0.12**	0.07	**0.13**
Skeleton	0.09	**0.26**	0.11	**0.30**	0.13	**0.33**	0.14	**0.36**	0.16	**0.40**	0.17	**0.43**
Red bone marrow	0.03	**0.09**	0.04	**0.11**	0.04	**0.12**	0.04	**0.13**	0.05	**0.15**	0.06	**0.17**
Whole body dose	0.04	**0.10**	0.05	**0.12**	0.06	**0.13**	0.06	**0.14**	0.07	**0.16**	0.07	**0.18**
ED (ICRP 60)	0.03	**0.06**	0.04	**0.07**	0.04	**0.07**	0.04	**0.08**	0.05	**0.09**	0.05	**0.10**
ED (ICRP 103)	0.03	**0.06**	0.04	**0.07**	0.04	**0.08**	0.05	**0.09**	0.05	**0.10**	0.06	**0.11**
Total filtration (mm) **Al**	**1.5**		**2.5**		**3.5**		**4.5**		**3.5**		**3.5**	
Cu	**0**		**0**		**0**		**0**		**0.1**		**0.2**	

Table 14.25 Normalized organ doses—Cervical spine ap—1–2 years

Beam projection:			ap									**Table cells:**
Focus-to-detector distance (cm):			115									(+) = normal font = optimal field setting
Tube voltage (kVp):			50									**(−) = bold font = suboptimal field setting**
Total filtrations:			See last table row.									Empty table cells correspond to values <0.01
Organ or tissue	**Organ dose D_T : Entrance dose K_E**								**(mSv * mGy⁻¹)**			
	(+)	(−)	(+)	(−)	(+)	(−)	(+)	(−)	(+)	(−)	(+)	(−)
Skull	0.11	**0.47**	0.14	**0.57**	0.15	**0.64**	0.16	**0.69**	0.18	**0.79**	0.20	**0.88**
Brain		**0.06**		**0.09**	0.01	**0.10**	0.01	**0.11**	0.02	**0.14**	0.02	**0.17**
Oral mucosa	0.22	**0.41**	0.26	**0.50**	0.29	**0.55**	0.31	**0.60**	0.35	**0.67**	0.38	**0.74**
Salivary glands	0.08	**0.31**	0.10	**0.39**	0.12	**0.44**	0.13	**0.48**	0.15	**0.55**	0.17	**0.61**
Cervical spine	0.25	**0.30**	0.34	**0.40**	0.40	**0.48**	0.46	**0.55**	0.56	**0.68**	0.66	**0.81**
Thyroid	0.79	**0.83**	0.89	**0.95**	0.95	**1.02**	1.00	**1.07**	1.06	**1.15**	1.12	**1.23**
Thymus	0.04	**0.22**	0.05	**0.27**	0.05	**0.30**	0.06	**0.32**	0.07	**0.37**	0.08	**0.41**
Thoracic spine	0.01	**0.03**	0.02	**0.04**	0.02	**0.05**	0.02	**0.06**	0.03	**0.08**	0.04	**0.10**
Claviculae	0.41	**1.51**	0.49	**1.83**	0.54	**2.04**	0.58	**2.21**	0.64	**2.48**	0.69	**2.71**
Scapulae		**0.05**		**0.07**		**0.08**	0.01	**0.09**	0.01	**0.12**	0.02	**0.15**
Ribs	0.01	**0.18**	0.01	**0.21**	0.02	**0.23**	0.02	**0.24**	0.02	**0.27**	0.02	**0.30**
Oesophagus	0.01	**0.03**	0.02	**0.04**	0.02	**0.05**	0.02	**0.06**	0.03	**0.07**	0.03	**0.08**
Trachea	0.37	**0.42**	0.44	**0.51**	0.49	**0.57**	0.53	**0.62**	0.59	**0.70**	0.65	**0.77**
Lung		**0.03**		**0.04**		**0.04**		**0.05**	0.01	**0.06**	0.01	**0.07**
Mamma										**0.01**		**0.01**
Heart				**0.01**		**0.01**		**0.01**		**0.02**		**0.02**
Stomach												
Spleen												
Liver												
Gall bladder												
Pancreas												
Small intestine												
Large intestine												
Sigma/rectum												
Lumbar spine												
Pelvis												
Adrenals												
Kidneys												
Urinary bladder												
Uterus												
Ovaries												
Prostate												
Testes												
Upper arm		**0.03**		**0.04**		**0.05**		**0.06**	0.01	**0.07**	0.02	**0.09**
Elbow												**0.01**
Forearm												
Upper leg												
Knee												
Lower leg												
Lymph nodes	0.10	**0.13**	0.11	**0.15**	0.12	**0.17**	0.13	**0.18**	0.15	**0.21**	0.16	**0.23**
Skin	0.03	**0.13**	0.03	**0.14**	0.03	**0.14**	0.03	**0.15**	0.04	**0.15**	0.04	**0.16**
Muscle	0.02	**0.05**	0.02	**0.06**	0.03	**0.07**	0.03	**0.08**	0.03	**0.09**	0.04	**0.10**
Skeleton	0.04	**0.16**	0.05	**0.19**	0.06	**0.22**	0.07	**0.24**	0.08	**0.27**	0.09	**0.31**
Red bone marrow	0.01	**0.05**	0.02	**0.06**	0.02	**0.07**	0.02	**0.08**	0.02	**0.09**	0.03	**0.11**
Whole body dose	0.02	**0.07**	0.02	**0.08**	0.03	**0.09**	0.03	**0.10**	0.03	**0.11**	0.04	**0.13**
ED (ICRP 60)	0.04	**0.06**	0.05	**0.07**	0.05	**0.07**	0.06	**0.08**	0.06	**0.09**	0.06	**0.09**
ED (ICRP 103)	0.04	**0.06**	0.05	**0.07**	0.05	**0.08**	0.06	**0.09**	0.06	**0.10**	0.07	**0.11**
Total filtration (mm) Al	**1.5**		**2.5**		**3.5**		**4.5**		**3.5**		**3.5**	
Cu	**0**		**0**		**0**		**0**		**0.1**		**0.2**	

Table 14.26 Normalized organ doses—Cervical spine ap—1–2 years

Beam projection:			ap									**Table cells:**
Focus-to-detector distance (cm):			115									(+) = normal font = optimal field setting
Tube voltage (kVp):			60									**(−) = bold font = suboptimal field setting**
Total filtrations:			See last table row.									Empty table cells correspond to values <0.01
Organ or tissue	**Organ dose D_T : Entrance dose K_E**							**(mSv * mGy^{-1})**				
	(+)	(−)	(+)	(−)	(+)	(−)	(+)	(−)	(+)	(−)	(+)	(−)
Skull	0.13	**0.55**	0.16	**0.66**	0.17	**0.73**	0.19	**0.79**	0.21	**0.89**	0.20	**0.88**
Brain		**0.09**	0.01	**0.11**	0.02	**0.13**	0.02	**0.15**	0.02	**0.18**	0.02	**0.17**
Oral mucosa	0.25	**0.48**	0.30	**0.57**	0.33	**0.63**	0.35	**0.68**	0.39	**0.76**	0.38	**0.74**
Salivary glands	0.10	**0.37**	0.12	**0.45**	0.14	**0.51**	0.15	**0.55**	0.18	**0.63**	0.17	**0.61**
Cervical spine	0.34	**0.41**	0.44	**0.54**	0.52	**0.63**	0.58	**0.72**	0.71	**0.87**	0.66	**0.81**
Thyroid	0.85	**0.90**	0.95	**1.02**	1.01	**1.10**	1.06	**1.15**	1.12	**1.24**	1.12	**1.23**
Thymus	0.05	**0.26**	0.06	**0.31**	0.06	**0.34**	0.07	**0.37**	0.08	**0.41**	0.08	**0.41**
Thoracic spine	0.02	**0.05**	0.03	**0.06**	0.03	**0.08**	0.04	**0.09**	0.04	**0.11**	0.04	**0.10**
Claviculae	0.46	**1.72**	0.54	**2.05**	0.59	**2.27**	0.63	**2.43**	0.69	**2.69**	0.69	**2.71**
Scapulae		**0.07**	0.01	**0.09**	0.01	**0.11**	0.02	**0.13**	0.02	**0.16**	0.02	**0.15**
Ribs	0.01	**0.20**	0.02	**0.23**	0.02	**0.25**	0.02	**0.27**	0.03	**0.30**	0.02	**0.30**
Oesophagus	0.02	**0.04**	0.02	**0.06**	0.03	**0.07**	0.03	**0.07**	0.04	**0.09**	0.03	**0.08**
Trachea	0.42	**0.49**	0.50	**0.58**	0.55	**0.65**	0.59	**0.70**	0.66	**0.78**	0.65	**0.77**
Lung		**0.04**		**0.05**		**0.06**	0.01	**0.06**	0.01	**0.08**	0.01	**0.07**
Mamma						**0.01**		**0.01**		**0.02**		**0.01**
Heart		**0.01**		**0.02**		**0.02**		**0.02**		**0.03**		**0.02**
Stomach												
Spleen												
Liver												
Gall bladder												
Pancreas												
Small intestine												
Large intestine												
Sigma/rectum												
Lumbar spine												
Pelvis												
Adrenals												
Kidneys												
Urinary bladder												
Uterus												
Ovaries												
Prostate												
Testes												
Upper arm		**0.04**	0.01	**0.06**	0.01	**0.07**	0.01	**0.08**	0.02	**0.10**	0.02	**0.09**
Elbow								**0.01**		**0.01**		**0.01**
Forearm												
Upper leg												
Knee												
Lower leg												
Lymph nodes	0.11	**0.15**	0.13	**0.17**	0.14	**0.19**	0.15	**0.21**	0.16	**0.23**	0.16	**0.23**
Skin	0.03	**0.13**	0.03	**0.14**	0.04	**0.15**	0.04	**0.16**	0.04	**0.16**	0.04	**0.16**
Muscle	0.02	**0.06**	0.03	**0.07**	0.03	**0.08**	0.03	**0.09**	0.04	**0.10**	0.04	**0.10**
Skeleton	0.05	**0.19**	0.06	**0.22**	0.07	**0.25**	0.08	**0.27**	0.09	**0.31**	0.09	**0.31**
Red bone marrow	0.02	**0.06**	0.02	**0.08**	0.02	**0.09**	0.02	**0.10**	0.03	**0.11**	0.03	**0.11**
Whole body dose	0.02	**0.08**	0.03	**0.09**	0.03	**0.11**	0.04	**0.11**	0.04	**0.13**	0.04	**0.13**
ED (ICRP 60)	0.05	**0.06**	0.05	**0.08**	0.06	**0.08**	0.06	**0.09**	0.07	**0.10**	0.06	**0.09**
ED (ICRP 103)	0.05	**0.07**	0.05	**0.08**	0.06	**0.09**	0.06	**0.10**	0.07	**0.11**	0.07	**0.11**
Total filtration (mm) Al	**1.5**		**2.5**		**3.5**		**4.5**		**3.5**		**3.5**	
Cu	**0**		**0**		**0**		**0**		**0.1**		**0.2**	

Table 14.27 Normalized organ doses—Cervical spine ap—1–2 years

Beam projection:		ap										**Table cells:**
Focus-to-detector distance (cm):		115										(+) = normal font = optimal field setting
Tube voltage (kVp):		70										**(−) = bold font = suboptimal field setting**
Total filtrations:		See last table row.										Empty table cells correspond to values <0.01
Organ or tissue	**Organ dose D_T : Entrance dose K_E**							**(mSv ∗ mGy⁻¹)**				
	(+)	(−)	(+)	(−)	(+)	(−)	(+)	(−)	(+)	(−)	(+)	(−)
Skull	0.15	**0.61**	0.17	**0.72**	0.19	**0.79**	0.20	**0.85**	0.22	**0.95**	0.24	**1.03**
Brain	0.01	**0.11**	0.02	**0.14**	0.02	**0.16**	0.02	**0.18**	0.03	**0.21**	0.03	**0.24**
Oral mucosa	0.28	**0.53**	0.32	**0.63**	0.35	**0.69**	0.38	**0.74**	0.42	**0.82**	0.45	**0.90**
Salivary glands	0.11	**0.42**	0.14	**0.51**	0.16	**0.56**	0.17	**0.61**	0.19	**0.69**	0.21	**0.75**
Cervical spine	0.42	**0.51**	0.53	**0.65**	0.61	**0.76**	0.68	**0.84**	0.80	**1.01**	0.92	**1.17**
Thyroid	0.90	**0.97**	1.00	**1.08**	1.06	**1.16**	1.10	**1.21**	1.16	**1.30**	1.21	**1.37**
Thymus	0.05	**0.29**	0.06	**0.34**	0.07	**0.37**	0.08	**0.40**	0.09	**0.45**	0.10	**0.49**
Thoracic spine	0.03	**0.06**	0.03	**0.08**	0.04	**0.10**	0.04	**0.11**	0.06	**0.14**	0.07	**0.17**
Claviculae	0.50	**1.88**	0.57	**2.19**	0.62	**2.40**	0.65	**2.54**	0.71	**2.78**	0.75	**2.96**
Scapulae	0.01	**0.09**	0.01	**0.12**	0.02	**0.14**	0.02	**0.16**	0.02	**0.19**	0.03	**0.23**
Ribs	0.02	**0.22**	0.02	**0.25**	0.02	**0.27**	0.03	**0.29**	0.03	**0.32**	0.04	**0.34**
Oesophagus	0.02	**0.05**	0.03	**0.07**	0.03	**0.08**	0.04	**0.09**	0.05	**0.11**	0.05	**0.13**
Trachea	0.47	**0.54**	0.54	**0.64**	0.60	**0.71**	0.63	**0.76**	0.70	**0.85**	0.75	**0.92**
Lung		**0.05**	0.01	**0.06**	0.01	**0.07**	0.01	**0.07**	0.02	**0.09**	0.02	**0.10**
Mamma				**0.01**		**0.01**		**0.02**		**0.02**		**0.02**
Heart		**0.02**		**0.02**		**0.02**		**0.03**		**0.03**	0.01	**0.04**
Stomach												
Spleen												
Liver												
Gall bladder												
Pancreas												
Small intestine												
Large intestine												
Sigma/rectum												
Lumbar spine												
Pelvis												
Adrenals												
Kidneys												
Urinary bladder												
Uterus												
Ovaries												
Prostate												
Testes												
Upper arm	0.01	**0.06**	0.01	**0.07**	0.02	**0.09**	0.02	**0.10**	0.02	**0.12**	0.03	**0.14**
Elbow				**0.01**		**0.01**		**0.01**		**0.02**		**0.02**
Forearm												
Upper leg												
Knee												
Lower leg												
Lymph nodes	0.12	**0.16**	0.14	**0.19**	0.15	**0.21**	0.16	**0.23**	0.17	**0.25**	0.18	**0.27**
Skin	0.03	**0.14**	0.04	**0.15**	0.04	**0.16**	0.04	**0.16**	0.04	**0.17**	0.04	**0.18**
Muscle	0.03	**0.07**	0.03	**0.08**	0.03	**0.09**	0.04	**0.10**	0.04	**0.11**	0.05	**0.12**
Skeleton	0.06	**0.21**	0.07	**0.25**	0.08	**0.28**	0.09	**0.30**	0.10	**0.34**	0.11	**0.37**
Red bone marrow	0.02	**0.07**	0.02	**0.09**	0.03	**0.10**	0.03	**0.11**	0.03	**0.12**	0.04	**0.14**
Whole body dose	0.03	**0.09**	0.03	**0.11**	0.04	**0.12**	0.04	**0.13**	0.05	**0.14**	0.05	**0.16**
ED (ICRP 60)	0.05	**0.07**	0.06	**0.08**	0.06	**0.09**	0.06	**0.10**	0.07	**0.11**	0.07	**0.11**
ED (ICRP 103)	0.05	**0.08**	0.06	**0.09**	0.06	**0.10**	0.07	**0.11**	0.07	**0.12**	0.08	**0.13**
Total filtration (mm) Al	**1.5**		**2.5**		**3.5**		**4.5**		**3.5**		**3.5**	
Cu	**0**		**0**		**0**		**0**		**0.1**		**0.2**	

Table 14.28 Normalized organ doses—Cervical spine ap—3–7 years

Beam projection:			ap									**Table cells:**
Focus-to-detector distance (cm):			115									(+) = normal font = optimal field setting
Tube voltage (kVp):			60									**(−) = bold font = suboptimal field setting**
Total filtrations:			See last table row.									Empty table cells correspond to values <0.01
Organ or tissue	**Organ dose D_T : Entrance dose K_E**								**(mSv ∗ mGy⁻¹)**			
	(+)	(−)	(+)	(−)	(+)	(−)	(+)	(−)	(+)	(−)	(+)	(−)
Skull	0.16	**0.44**	0.19	**0.53**	0.21	**0.60**	0.23	**0.66**	0.26	**0.76**	0.25	**0.74**
Brain		**0.04**		**0.06**	0.01	**0.07**	0.01	**0.08**	0.02	**0.10**	0.01	**0.09**
Oral mucosa	0.21	**0.37**	0.25	**0.44**	0.28	**0.50**	0.30	**0.54**	0.34	**0.62**	0.32	**0.59**
Salivary glands	0.08	**0.27**	0.10	**0.34**	0.12	**0.39**	0.13	**0.42**	0.16	**0.49**	0.15	**0.47**
Cervical spine	0.30	**0.34**	0.39	**0.45**	0.46	**0.53**	0.52	**0.60**	0.63	**0.74**	0.58	**0.67**
Thyroid	0.80	**0.84**	0.91	**0.96**	0.97	**1.03**	1.02	**1.09**	1.09	**1.18**	1.09	**1.16**
Thymus	0.01	**0.16**	0.02	**0.19**	0.02	**0.22**	0.02	**0.24**	0.03	**0.27**	0.02	**0.27**
Thoracic spine	0.01	**0.03**	0.02	**0.05**	0.02	**0.06**	0.02	**0.07**	0.03	**0.08**	0.03	**0.07**
Claviculae	0.36	**1.35**	0.42	**1.62**	0.46	**1.80**	0.50	**1.94**	0.55	**2.17**	0.55	**2.17**
Scapulae		**0.04**		**0.05**		**0.06**		**0.07**	0.01	**0.08**		**0.07**
Ribs		**0.16**		**0.19**	0.01	**0.20**	0.01	**0.22**	0.02	**0.24**	0.01	**0.24**
Oesophagus		**0.02**	0.01	**0.03**	0.01	**0.04**	0.02	**0.04**	0.02	**0.05**	0.02	**0.05**
Trachea	0.28	**0.31**	0.34	**0.39**	0.39	**0.44**	0.43	**0.49**	0.49	**0.57**	0.48	**0.55**
Lung		**0.03**		**0.04**		**0.05**		**0.05**		**0.06**		**0.06**
Mamma										**0.01**		
Heart						**0.01**		**0.01**		**0.02**		**0.01**
Stomach												
Spleen												
Liver												
Gall bladder												
Pancreas												
Small intestine												
Large intestine												
Sigma/rectum												
Lumbar spine												
Pelvis												
Adrenals												
Kidneys												
Urinary bladder												
Uterus												
Ovaries												
Prostate												
Testes												
Upper arm		**0.02**		**0.03**		**0.03**		**0.04**	0.01	**0.05**		**0.04**
Elbow												
Forearm												
Upper leg												
Knee												
Lower leg												
Lymph nodes	0.08	**0.11**	0.10	**0.13**	0.11	**0.15**	0.12	**0.16**	0.13	**0.19**	0.13	**0.18**
Skin	0.03	**0.10**	0.03	**0.10**	0.03	**0.11**	0.03	**0.11**	0.03	**0.12**	0.03	**0.11**
Muscle	0.01	**0.04**	0.02	**0.05**	0.02	**0.05**	0.02	**0.06**	0.02	**0.06**	0.02	**0.06**
Skeleton	0.05	**0.15**	0.07	**0.19**	0.08	**0.21**	0.08	**0.23**	0.10	**0.27**	0.09	**0.26**
Red bone marrow	0.01	**0.04**	0.01	**0.05**	0.02	**0.05**	0.02	**0.06**	0.02	**0.07**	0.02	**0.06**
Whole body dose	0.02	**0.05**	0.02	**0.07**	0.03	**0.07**	0.03	**0.08**	0.03	**0.09**	0.03	**0.09**
ED (ICRP 60)	0.04	**0.06**	0.05	**0.06**	0.05	**0.07**	0.06	**0.08**	0.06	**0.08**	0.06	**0.08**
ED (ICRP 103)	0.04	**0.06**	0.05	**0.07**	0.05	**0.08**	0.05	**0.08**	0.06	**0.09**	0.06	**0.09**
Total filtration (mm) Al	**1.5**		**2.5**		**3.5**		**4.5**		**3.5**		**3.5**	
Cu	**0**		**0**		**0**		**0**		**0.1**		**0.2**	

Table 14.29 Normalized organ doses—Cervical spine ap—3–7 years

Beam projection:			ap									**Table cells:**
Focus-to-detector distance (cm):			115									(+) = normal font = optimal field setting
Tube voltage (kVp):			70									**(−) = bold font = suboptimal field setting**
Total filtrations:			See last table row.									Empty table cells correspond to values <0.01
Organ or tissue	**Organ dose D_T : Entrance dose K_E**								**(mSv ∗ mGy⁻¹)**			
	(+)	(−)	(+)	(−)	(+)	(−)	(+)	(−)	(+)	(−)	(+)	(−)
Skull	0.18	**0.50**	0.21	**0.60**	0.23	**0.67**	0.25	**0.73**	0.28	**0.83**	0.31	**0.91**
Brain	0.01	**0.06**	0.01	**0.08**	0.02	**0.09**	0.02	**0.10**	0.02	**0.12**	0.03	**0.15**
Oral mucosa	0.24	**0.42**	0.28	**0.50**	0.31	**0.56**	0.33	**0.60**	0.37	**0.68**	0.41	**0.76**
Salivary glands	0.09	**0.32**	0.12	**0.39**	0.14	**0.44**	0.15	**0.48**	0.18	**0.55**	0.20	**0.62**
Cervical spine	0.37	**0.43**	0.47	**0.55**	0.55	**0.65**	0.61	**0.72**	0.73	**0.88**	0.85	**1.02**
Thyroid	0.85	**0.91**	0.96	**1.02**	1.02	**1.10**	1.06	**1.15**	1.14	**1.24**	1.20	**1.32**
Thymus	0.02	**0.18**	0.02	**0.22**	0.02	**0.24**	0.03	**0.26**	0.03	**0.30**	0.04	**0.33**
Thoracic spine	0.02	**0.05**	0.02	**0.06**	0.03	**0.08**	0.03	**0.09**	0.04	**0.11**	0.05	**0.13**
Claviculae	0.39	**1.49**	0.45	**1.75**	0.49	**1.93**	0.52	**2.05**	0.57	**2.27**	0.61	**2.43**
Scapulae		**0.05**		**0.06**	0.01	**0.07**	0.01	**0.08**	0.02	**0.11**	0.02	**0.13**
Ribs		**0.18**	0.01	**0.20**	0.01	**0.22**	0.02	**0.23**	0.02	**0.25**	0.02	**0.27**
Oesophagus	0.01	**0.03**	0.02	**0.04**	0.02	**0.05**	0.02	**0.05**	0.03	**0.07**	0.03	**0.08**
Trachea	0.32	**0.36**	0.39	**0.45**	0.44	**0.51**	0.48	**0.55**	0.54	**0.64**	0.60	**0.72**
Lung		**0.04**		**0.05**		**0.06**		**0.06**	0.01	**0.08**	0.01	**0.09**
Mamma						**0.01**		**0.01**		**0.02**		**0.02**
Heart				**0.01**		**0.02**		**0.02**		**0.02**		**0.03**
Stomach												
Spleen												
Liver												
Gall bladder												
Pancreas												
Small intestine												
Large intestine												
Sigma/rectum												
Lumbar spine												
Pelvis												
Adrenals												
Kidneys												
Urinary bladder												
Uterus												
Ovaries												
Prostate												
Testes												
Upper arm		**0.03**		**0.04**		**0.05**	0.01	**0.05**	0.01	**0.07**	0.02	**0.08**
Elbow										**0.01**		**0.01**
Forearm												
Upper leg												
Knee												
Lower leg												
Lymph nodes	0.09	**0.13**	0.11	**0.15**	0.12	**0.17**	0.13	**0.18**	0.15	**0.20**	0.16	**0.23**
Skin	0.03	**0.10**	0.03	**0.11**	0.03	**0.11**	0.03	**0.11**	0.03	**0.12**	0.04	**0.13**
Muscle	0.02	**0.04**	0.02	**0.05**	0.02	**0.06**	0.02	**0.06**	0.03	**0.07**	0.03	**0.08**
Skeleton	0.06	**0.18**	0.08	**0.21**	0.08	**0.24**	0.09	**0.26**	0.11	**0.30**	0.12	**0.33**
Red bone marrow	0.01	**0.04**	0.02	**0.05**	0.02	**0.06**	0.02	**0.07**	0.03	**0.08**	0.03	**0.09**
Whole body dose	0.02	**0.06**	0.03	**0.07**	0.03	**0.08**	0.03	**0.09**	0.04	**0.10**	0.04	**0.12**
ED (ICRP 60)	0.05	**0.06**	0.05	**0.07**	0.06	**0.08**	0.06	**0.08**	0.06	**0.09**	0.07	**0.10**
ED (ICRP 103)	0.04	**0.06**	0.05	**0.08**	0.06	**0.08**	0.06	**0.09**	0.06	**0.10**	0.07	**0.11**
Total filtration (mm) **Al**	**1.5**		**2.5**		**3.5**		**4.5**		**3.5**		**3.5**	
Cu	**0**		**0**		**0**		**0**		**0.1**		**0.2**	

Table 14.30 Normalized organ doses—Cervical spine ap—8–12 years

Beam projection:			ap									**Table cells:**
Focus-to-detector distance (cm):			115									(+) = normal font = optimal field setting
Tube voltage (kVp):			60									**(−) = bold font = suboptimal field setting**
Total filtrations:			See last table row.									Empty table cells correspond to values <0.01
Organ or tissue	**Organ dose D_T : Entrance dose K_E**								**(mSv * mGy^{-1})**			
	(+)	(−)	(+)	(−)	(+)	(−)	(+)	(−)	(+)	(−)	(+)	(−)
Skull	0.17	**0.44**	0.21	**0.53**	0.23	**0.61**	0.25	**0.66**	0.29	**0.77**	0.28	**0.74**
Brain		**0.05**		**0.06**		**0.07**	0.01	**0.09**	0.01	**0.11**	0.01	**0.09**
Oral mucosa	0.23	**0.35**	0.27	**0.42**	0.30	**0.47**	0.33	**0.50**	0.37	**0.58**	0.35	**0.55**
Salivary glands	0.08	**0.25**	0.11	**0.31**	0.12	**0.36**	0.14	**0.39**	0.17	**0.46**	0.16	**0.44**
Cervical spine	0.34	**0.35**	0.45	**0.45**	0.53	**0.54**	0.60	**0.61**	0.74	**0.76**	0.67	**0.68**
Thyroid	0.84	**0.79**	0.96	**0.91**	1.04	**0.99**	1.10	**1.05**	1.19	**1.15**	1.18	**1.13**
Thymus	0.01	**0.07**	0.02	**0.09**	0.02	**0.10**	0.03	**0.11**	0.03	**0.14**	0.03	**0.13**
Thoracic spine	0.01	**0.02**	0.02	**0.03**	0.02	**0.04**	0.03	**0.05**	0.03	**0.06**	0.03	**0.05**
Claviculae	0.35	**1.10**	0.42	**1.33**	0.47	**1.48**	0.50	**1.60**	0.57	**1.81**	0.56	**1.80**
Scapulae		**0.02**		**0.03**		**0.03**		**0.04**	0.01	**0.05**		**0.04**
Ribs	0.03	**0.13**	0.04	**0.16**	0.04	**0.17**	0.05	**0.18**	0.05	**0.20**	0.05	**0.19**
Oesophagus	0.01	**0.02**	0.01	**0.02**	0.02	**0.03**	0.02	**0.03**	0.03	**0.04**	0.02	**0.04**
Trachea	0.42	**0.43**	0.52	**0.53**	0.59	**0.61**	0.65	**0.67**	0.75	**0.78**	0.72	**0.75**
Lung		**0.02**		**0.03**		**0.03**		**0.04**		**0.05**		**0.04**
Mamma												
Heart												
Stomach												
Spleen												
Liver												
Gall bladder												
Pancreas												
Small intestine												
Large intestine												
Sigma/rectum												
Lumbar spine												
Pelvis												
Adrenals												
Kidneys												
Urinary bladder												
Uterus												
Ovaries												
Prostate												
Testes												
Upper arm		**0.01**		**0.02**		**0.02**		**0.02**		**0.03**		**0.02**
Elbow												
Forearm												
Upper leg												
Knee												
Lower leg												
Lymph nodes	0.10	**0.12**	0.13	**0.15**	0.14	**0.16**	0.15	**0.18**	0.17	**0.21**	0.17	**0.20**
Skin	0.03	**0.08**	0.03	**0.08**	0.03	**0.09**	0.03	**0.09**	0.03	**0.09**	0.03	**0.09**
Muscle	0.01	**0.03**	0.02	**0.03**	0.02	**0.04**	0.02	**0.04**	0.02	**0.05**	0.02	**0.05**
Skeleton	0.05	**0.12**	0.06	**0.15**	0.07	**0.17**	0.08	**0.18**	0.09	**0.21**	0.09	**0.21**
Red bone marrow	0.01	**0.03**	0.01	**0.04**	0.02	**0.04**	0.02	**0.05**	0.02	**0.05**	0.02	**0.05**
Whole body dose	0.02	**0.04**	0.02	**0.05**	0.02	**0.06**	0.03	**0.06**	0.03	**0.07**	0.03	**0.07**
ED (ICRP 60)	0.05	**0.05**	0.05	**0.06**	0.06	**0.06**	0.06	**0.07**	0.07	**0.08**	0.07	**0.07**
ED (ICRP 103)	0.04	**0.05**	0.05	**0.06**	0.06	**0.07**	0.06	**0.07**	0.07	**0.08**	0.07	**0.08**
Total filtration (mm) Al	**1.5**		**2.5**		**3.5**		**4.5**		**3.5**		**3.5**	
Cu	**0**		**0**		**0**		**0**		**0.1**		**0.2**	

Table 14.31 Normalized organ doses—Cervical spine ap—8–12 years

Beam projection:			ap									**Table cells:**
Focus-to-detector distance (cm):			115									(+) = normal font = optimal field setting
Tube voltage (kVp):			70									**(−) = bold font = suboptimal field setting**
Total filtrations:			See last table row.									Empty table cells correspond to values <0.01
Organ or tissue	**Organ dose D_T : Entrance dose K_E**								**(mSv * mGy⁻¹)**			
	(+)	(−)	(+)	(−)	(+)	(−)	(+)	(−)	(+)	(−)	(+)	(−)
Skull	0.19	**0.50**	0.23	**0.61**	0.26	**0.68**	0.28	**0.74**	0.31	**0.85**	0.35	**0.94**
Brain		**0.06**	0.01	**0.08**	0.01	**0.10**	0.02	**0.11**	0.02	**0.14**	0.03	**0.16**
Oral mucosa	0.26	**0.40**	0.30	**0.47**	0.34	**0.52**	0.36	**0.57**	0.41	**0.65**	0.46	**0.72**
Salivary glands	0.10	**0.30**	0.13	**0.36**	0.15	**0.41**	0.17	**0.45**	0.20	**0.52**	0.23	**0.58**
Cervical spine	0.43	**0.44**	0.55	**0.57**	0.64	**0.67**	0.72	**0.75**	0.87	**0.91**	1.01	**1.07**
Thyroid	0.90	**0.86**	1.02	**0.98**	1.10	**1.06**	1.16	**1.12**	1.25	**1.22**	1.33	**1.30**
Thymus	0.02	**0.08**	0.02	**0.10**	0.03	**0.12**	0.03	**0.13**	0.04	**0.16**	0.05	**0.18**
Thoracic spine	0.02	**0.03**	0.03	**0.05**	0.03	**0.06**	0.04	**0.06**	0.05	**0.08**	0.06	**0.10**
Claviculae	0.39	**1.22**	0.46	**1.44**	0.50	**1.60**	0.54	**1.71**	0.60	**1.91**	0.65	**2.06**
Scapulae		**0.03**		**0.04**	0.01	**0.04**	0.01	**0.05**	0.02	**0.06**	0.02	**0.08**
Ribs	0.04	**0.15**	0.04	**0.17**	0.05	**0.18**	0.05	**0.19**	0.06	**0.21**	0.06	**0.22**
Oesophagus	0.01	**0.03**	0.02	**0.03**	0.02	**0.04**	0.03	**0.04**	0.03	**0.06**	0.04	**0.07**
Trachea	0.49	**0.50**	0.59	**0.61**	0.67	**0.70**	0.73	**0.76**	0.84	**0.88**	0.93	**0.99**
Lung		**0.03**		**0.04**		**0.04**		**0.05**	0.01	**0.06**	0.02	**0.07**
Mamma												
Heart								**0.01**		**0.01**		**0.02**
Stomach												
Spleen												
Liver												
Gall bladder												
Pancreas												
Small intestine												
Large intestine												
Sigma/rectum												
Lumbar spine												
Pelvis												
Adrenals												
Kidneys												
Urinary bladder												
Uterus												
Ovaries												
Prostate												
Testes												
Upper arm		**0.02**		**0.02**		**0.03**		**0.03**	0.01	**0.04**	0.02	**0.05**
Elbow												
Forearm												
Upper leg												
Knee												
Lower leg												
Lymph nodes	0.12	**0.14**	0.14	**0.17**	0.16	**0.18**	0.17	**0.20**	0.19	**0.23**	0.21	**0.25**
Skin	0.03	**0.08**	0.03	**0.09**	0.03	**0.09**	0.03	**0.09**	0.03	**0.10**	0.04	**0.10**
Muscle	0.02	**0.03**	0.02	**0.04**	0.02	**0.04**	0.02	**0.05**	0.03	**0.05**	0.03	**0.06**
Skeleton	0.06	**0.14**	0.07	**0.17**	0.08	**0.19**	0.09	**0.21**	0.10	**0.24**	0.11	**0.27**
Red bone marrow	0.01	**0.04**	0.02	**0.04**	0.02	**0.05**	0.02	**0.05**	0.03	**0.06**	0.03	**0.07**
Whole body dose	0.02	**0.05**	0.03	**0.06**	0.03	**0.07**	0.03	**0.07**	0.04	**0.08**	0.04	**0.09**
ED (ICRP 60)	0.05	**0.05**	0.06	**0.06**	0.06	**0.07**	0.06	**0.07**	0.07	**0.08**	0.08	**0.09**
ED (ICRP 103)	0.05	**0.06**	0.06	**0.07**	0.06	**0.08**	0.07	**0.08**	0.07	**0.09**	0.08	**0.10**
Total filtration (mm) **Al**	**1.5**		**2.5**		**3.5**		**4.5**		**3.5**		**3.5**	
Cu	**0**		**0**		**0**		**0**		**0.1**		**0.2**	

Table 14.32 Normalized organ doses—Cervical spine ap—13–17 years

Beam projection:			ap									**Table cells:**
Focus-to-detector distance (cm):			115									(+) = normal font = optimal field setting
Tube voltage (kVp):			60									**(−) = bold font = suboptimal field setting**
Total filtrations:			See last table row.									Empty table cells correspond to values <0.01
Organ or tissue	**Organ dose D_T : Entrance dose K_E**								**(mSv * mGy⁻¹)**			
	(+)	(−)	(+)	(−)	(+)	(−)	(+)	(−)	(+)	(−)	(+)	(−)
Skull	0.23	**0.39**	0.28	**0.48**	0.33	**0.56**	0.36	**0.61**	0.42	**0.72**	0.40	**0.69**
Brain	0.01	**0.04**	0.01	**0.06**	0.02	**0.07**	0.02	**0.08**	0.03	**0.10**	0.02	**0.09**
Oral mucosa	0.22	**0.30**	0.27	**0.36**	0.30	**0.41**	0.32	**0.45**	0.37	**0.51**	0.35	**0.48**
Salivary glands	0.07	**0.22**	0.09	**0.28**	0.10	**0.32**	0.12	**0.35**	0.15	**0.41**	0.13	**0.39**
Cervical spine	0.39	**0.39**	0.52	**0.51**	0.61	**0.61**	0.70	**0.69**	0.87	**0.86**	0.78	**0.76**
Thyroid	0.80	**0.75**	0.93	**0.88**	1.01	**0.96**	1.07	**1.02**	1.18	**1.12**	1.16	**1.11**
Thymus						**0.01**		**0.01**		**0.02**		**0.01**
Thoracic spine		**0.01**	0.01	**0.01**	0.01	**0.02**	0.02	**0.02**	0.02	**0.03**	0.02	**0.02**
Claviculae	0.18	**0.66**	0.22	**0.80**	0.25	**0.90**	0.27	**0.98**	0.31	**1.11**	0.31	**1.10**
Scapulae										**0.01**		
Ribs		**0.03**	0.01	**0.04**	0.01	**0.04**	0.01	**0.05**	0.02	**0.05**	0.02	**0.05**
Oesophagus			0.01	**0.01**	0.01	**0.02**	0.02	**0.02**	0.02	**0.02**	0.02	**0.02**
Trachea	0.32	**0.31**	0.40	**0.39**	0.45	**0.44**	0.50	**0.48**	0.58	**0.56**	0.55	**0.53**
Lung												
Mamma		**0.01**		**0.01**		**0.02**	0.01	**0.02**	0.01	**0.02**	0.01	**0.02**
Heart												
Stomach												
Spleen												
Liver												
Gall bladder												
Pancreas												
Small intestine												
Large intestine												
Sigma/rectum												
Lumbar spine												
Pelvis												
Adrenals												
Kidneys												
Urinary bladder												
Uterus												
Ovaries												
Prostate												
Testes												
Upper arm										**0.01**		
Elbow												
Forearm												
Upper leg												
Knee												
Lower leg												
Lymph nodes	0.09	**0.10**	0.11	**0.12**	0.12	**0.13**	0.13	**0.15**	0.15	**0.17**	0.14	**0.16**
Skin	0.03	**0.06**	0.03	**0.06**	0.03	**0.07**	0.03	**0.07**	0.03	**0.07**	0.03	**0.07**
Muscle	0.01	**0.02**	0.01	**0.02**	0.02	**0.02**	0.02	**0.03**	0.02	**0.03**	0.02	**0.03**
Skeleton	0.05	**0.08**	0.06	**0.10**	0.07	**0.11**	0.08	**0.13**	0.09	**0.15**	0.09	**0.14**
Red bone marrow	0.01	**0.02**	0.01	**0.02**	0.02	**0.03**	0.02	**0.03**	0.02	**0.04**	0.02	**0.03**
Whole body dose	0.02	**0.03**	0.02	**0.03**	0.02	**0.04**	0.02	**0.04**	0.03	**0.05**	0.03	**0.05**
ED (ICRP 60)	0.04	**0.04**	0.05	**0.05**	0.06	**0.06**	0.06	**0.06**	0.06	**0.07**	0.06	**0.06**
ED (ICRP 103)	0.04	**0.04**	0.05	**0.05**	0.05	**0.06**	0.06	**0.06**	0.07	**0.07**	0.06	**0.07**
Total filtration (mm) Al	**1.5**		**2.5**		**3.5**		**4.5**		**3.5**		**3.5**	
Cu	**0**		**0**		**0**		**0**		**0.1**		**0.2**	

Table 14.33 Normalized organ doses—Cervical spine ap—13–17 years

Beam projection:	ap											**Table cells:**
Focus-to-detector distance (cm):	115											(+) = normal font = optimal field setting
Tube voltage (kVp):	70											**(−) = bold font = suboptimal field setting**
Total filtrations:	See last table row.											Empty table cells correspond to values <0.01
Organ or tissue	**Organ dose D_T : Entrance dose K_E**							**(mSv ∗ mGy⁻¹)**				
	(+)	(−)	(+)	(−)	(+)	(−)	(+)	(−)	(+)	(−)	(+)	(−)
Skull	0.27	**0.46**	0.33	**0.56**	0.37	**0.63**	0.40	**0.69**	0.46	**0.81**	0.52	**0.91**
Brain	0.02	**0.06**	0.02	**0.08**	0.03	**0.09**	0.03	**0.10**	0.04	**0.13**	0.05	**0.15**
Oral mucosa	0.26	**0.35**	0.30	**0.42**	0.34	**0.47**	0.37	**0.51**	0.42	**0.58**	0.47	**0.65**
Salivary glands	0.09	**0.26**	0.11	**0.32**	0.13	**0.37**	0.15	**0.40**	0.18	**0.47**	0.21	**0.53**
Cervical spine	0.50	**0.50**	0.64	**0.64**	0.76	**0.76**	0.85	**0.85**	1.04	**1.04**	1.22	**1.23**
Thyroid	0.87	**0.82**	1.00	**0.94**	1.08	**1.02**	1.14	**1.08**	1.24	**1.18**	1.32	**1.27**
Thymus				**0.01**		**0.01**	0.01	**0.02**	0.01	**0.02**	0.02	**0.03**
Thoracic spine	0.01	**0.01**	0.02	**0.02**	0.02	**0.02**	0.02	**0.03**	0.03	**0.04**	0.03	**0.05**
Claviculae	0.21	**0.74**	0.25	**0.88**	0.28	**0.98**	0.30	**1.05**	0.34	**1.18**	0.37	**1.28**
Scapulae						**0.01**		**0.01**		**0.02**	0.01	**0.02**
Ribs	0.01	**0.04**	0.01	**0.04**	0.02	**0.05**	0.02	**0.05**	0.02	**0.06**	0.02	**0.06**
Oesophagus	0.01	**0.01**	0.01	**0.02**	0.02	**0.02**	0.02	**0.02**	0.03	**0.03**	0.03	**0.04**
Trachea	0.38	**0.37**	0.46	**0.45**	0.51	**0.50**	0.56	**0.55**	0.64	**0.64**	0.72	**0.72**
Lung										**0.01**		**0.02**
Mamma		**0.01**	0.01	**0.02**	0.01	**0.02**	0.01	**0.03**	0.02	**0.03**	0.02	**0.04**
Heart												
Stomach												
Spleen												
Liver												
Gall bladder												
Pancreas												
Small intestine												
Large intestine												
Sigma/rectum												
Lumbar spine												
Pelvis												
Adrenals												
Kidneys												
Urinary bladder												
Uterus												
Ovaries												
Prostate												
Testes												
Upper arm								**0.01**		**0.01**		**0.02**
Elbow												
Forearm												
Upper leg												
Knee												
Lower leg												
Lymph nodes	0.10	**0.11**	0.12	**0.13**	0.13	**0.15**	0.14	**0.16**	0.16	**0.18**	0.18	**0.20**
Skin	0.03	**0.06**	0.03	**0.07**	0.03	**0.07**	0.03	**0.07**	0.03	**0.08**	0.04	**0.08**
Muscle	0.01	**0.02**	0.02	**0.02**	0.02	**0.03**	0.02	**0.03**	0.02	**0.03**	0.02	**0.04**
Skeleton	0.06	**0.09**	0.07	**0.12**	0.08	**0.13**	0.09	**0.14**	0.11	**0.17**	0.12	**0.19**
Red bone marrow	0.01	**0.02**	0.02	**0.03**	0.02	**0.03**	0.02	**0.04**	0.03	**0.05**	0.03	**0.05**
Whole body dose	0.02	**0.03**	0.02	**0.04**	0.03	**0.04**	0.03	**0.05**	0.03	**0.05**	0.04	**0.06**
ED (ICRP 60)	0.05	**0.05**	0.05	**0.06**	0.06	**0.06**	0.06	**0.07**	0.07	**0.07**	0.08	**0.08**
ED (ICRP 103)	0.05	**0.05**	0.05	**0.06**	0.06	**0.07**	0.06	**0.07**	0.07	**0.08**	0.08	**0.09**
Total filtration (mm) Al	**1.5**		**2.5**		**3.5**		**4.5**		**3.5**		**3.5**	
Cu	**0**		**0**		**0**		**0**		**0.1**		**0.2**	

Table 14.34 Normalized organ doses—Cervical spine ap—>17 years

Beam projection:			ap									**Table cells:**
Focus-to-detector distance (cm):			115					(+) = normal font = optimal field setting				
Tube voltage (kVp):			60					**(−) = bold font = suboptimal field setting**				
Total filtrations:			See last table row.					Empty table cells correspond to values <0.01				
Organ or tissue	**Organ dose D_T : Entrance dose K_E**								**(mSv * mGy^{-1})**			
	(+)	(−)	(+)	(−)	(+)	(−)	(+)	(−)	(+)	(−)	(+)	(−)
Skull	0.23	**0.37**	0.28	**0.47**	0.33	**0.54**	0.36	**0.60**	0.43	**0.71**	0.40	**0.67**
Brain		**0.05**	0.01	**0.06**	0.01	**0.08**	0.02	**0.09**	0.02	**0.11**	0.02	**0.10**
Oral mucosa	0.23	**0.29**	0.27	**0.34**	0.30	**0.39**	0.33	**0.42**	0.38	**0.49**	0.36	**0.46**
Salivary glands	0.06	**0.21**	0.08	**0.26**	0.10	**0.30**	0.11	**0.33**	0.14	**0.39**	0.12	**0.36**
Cervical spine	0.35	**0.34**	0.46	**0.45**	0.56	**0.53**	0.63	**0.61**	0.79	**0.76**	0.70	**0.67**
Thyroid	0.76	**0.70**	0.89	**0.83**	0.97	**0.91**	1.04	**0.97**	1.14	**1.07**	1.13	**1.06**
Thymus												
Thoracic spine			0.01	**0.01**	0.01	**0.01**	0.02	**0.02**	0.02	**0.02**	0.02	**0.02**
Claviculae	0.20	**0.61**	0.25	**0.74**	0.28	**0.83**	0.30	**0.89**	0.35	**1.02**	0.34	**1.01**
Scapulae												
Ribs	0.02	**0.02**	0.02	**0.02**	0.02	**0.03**	0.03	**0.03**	0.03	**0.03**	0.03	**0.03**
Oesophagus			0.01	**0.01**	0.01	**0.02**	0.02	**0.02**	0.02	**0.02**	0.02	**0.02**
Trachea	0.30	**0.28**	0.37	**0.35**	0.42	**0.40**	0.46	**0.44**	0.54	**0.52**	0.51	**0.49**
Lung												
Mamma												
Heart												
Stomach												
Spleen												
Liver												
Gall bladder												
Pancreas												
Small intestine												
Large intestine												
Sigma/rectum												
Lumbar spine												
Pelvis												
Adrenals												
Kidneys												
Urinary bladder												
Uterus												
Ovaries												
Prostate												
Testes												
Upper arm												
Elbow												
Forearm												
Upper leg												
Knee												
Lower leg												
Lymph nodes	0.08	**0.09**	0.10	**0.11**	0.11	**0.12**	0.12	**0.14**	0.14	**0.16**	0.14	**0.15**
Skin	0.03	**0.06**	0.03	**0.06**	0.03	**0.06**	0.03	**0.07**	0.03	**0.07**	0.03	**0.07**
Muscle	0.01	**0.01**	0.01	**0.02**	0.01	**0.02**	0.02	**0.02**	0.02	**0.02**	0.02	**0.02**
Skeleton	0.04	**0.07**	0.06	**0.09**	0.07	**0.10**	0.07	**0.11**	0.09	**0.13**	0.08	**0.12**
Red bone marrow	0.01	**0.02**	0.01	**0.02**	0.02	**0.02**	0.02	**0.03**	0.02	**0.03**	0.02	**0.03**
Whole body dose	0.02	**0.02**	0.02	**0.03**	0.02	**0.03**	0.02	**0.04**	0.03	**0.04**	0.03	**0.04**
ED (ICRP 60)	0.04	**0.04**	0.05	**0.05**	0.05	**0.05**	0.06	**0.06**	0.06	**0.06**	0.06	**0.06**
ED (ICRP 103)	0.04	**0.04**	0.05	**0.05**	0.05	**0.05**	0.06	**0.06**	0.06	**0.07**	0.06	**0.06**
Total filtration (mm) Al	**1.5**		**2.5**		**3.5**		**4.5**		**3.5**		**3.5**	
Cu	**0**		**0**		**0**		**0**		**0.1**		**0.2**	

Table 14.35 Normalized organ doses—Cervical spine ap—>17 years

Beam projection:			ap									**Table cells:**
Focus-to-detector distance (cm):			115									(+) = normal font = optimal field setting
Tube voltage (kVp):			70									**(−) = bold font = suboptimal field setting**
Total filtrations:			See last table row.									Empty table cells correspond to values <0.01
Organ or tissue	**Organ dose D_T : Entrance dose K_E**								**(mSv ∗ mGy⁻¹)**			
	(+)	(−)	(+)	(−)	(+)	(−)	(+)	(−)	(+)	(−)	(+)	(−)
Skull	0.27	**0.44**	0.33	**0.55**	0.37	**0.62**	0.41	**0.69**	0.48	**0.80**	0.54	**0.91**
Brain	0.01	**0.06**	0.02	**0.08**	0.02	**0.10**	0.03	**0.11**	0.03	**0.14**	0.04	**0.17**
Oral mucosa	0.26	**0.33**	0.31	**0.40**	0.35	**0.44**	0.38	**0.48**	0.43	**0.55**	0.49	**0.62**
Salivary glands	0.08	**0.25**	0.11	**0.30**	0.13	**0.35**	0.14	**0.38**	0.17	**0.45**	0.21	**0.51**
Cervical spine	0.46	**0.44**	0.59	**0.57**	0.69	**0.67**	0.78	**0.76**	0.96	**0.93**	1.13	**1.11**
Thyroid	0.83	**0.77**	0.96	**0.90**	1.04	**0.98**	1.11	**1.04**	1.22	**1.14**	1.30	**1.23**
Thymus									0.01	**0.01**	0.01	**0.01**
Thoracic spine	0.01	**0.01**	0.02	**0.02**	0.02	**0.02**	0.02	**0.02**	0.03	**0.03**	0.04	**0.04**
Claviculae	0.23	**0.68**	0.28	**0.81**	0.31	**0.90**	0.33	**0.96**	0.38	**1.08**	0.42	**1.17**
Scapulae										**0.01**		**0.02**
Ribs	0.02	**0.02**	0.03	**0.03**	0.03	**0.03**	0.03	**0.03**	0.03	**0.04**	0.04	**0.04**
Oesophagus	0.01	**0.01**	0.02	**0.02**	0.02	**0.02**	0.02	**0.02**	0.03	**0.03**	0.03	**0.04**
Trachea	0.35	**0.33**	0.43	**0.41**	0.48	**0.46**	0.53	**0.50**	0.61	**0.59**	0.68	**0.66**
Lung										**0.01**		**0.01**
Mamma												
Heart												
Stomach												
Spleen												
Liver												
Gall bladder												
Pancreas												
Small intestine												
Large intestine												
Sigma/rectum												
Lumbar spine												
Pelvis												
Adrenals												
Kidneys												
Urinary bladder												
Uterus												
Ovaries												
Prostate												
Testes												
Upper arm												**0.01**
Elbow												
Forearm												
Upper leg												
Knee												
Lower leg												
Lymph nodes	0.10	**0.10**	0.11	**0.12**	0.13	**0.14**	0.14	**0.15**	0.16	**0.17**	0.17	**0.19**
Skin	0.03	**0.06**	0.03	**0.06**	0.03	**0.07**	0.03	**0.07**	0.04	**0.07**	0.04	**0.08**
Muscle	0.01	**0.02**	0.01	**0.02**	0.02	**0.02**	0.02	**0.02**	0.02	**0.03**	0.02	**0.03**
Skeleton	0.05	**0.08**	0.07	**0.10**	0.08	**0.12**	0.09	**0.13**	0.10	**0.15**	0.12	**0.17**
Red bone marrow	0.01	**0.02**	0.02	**0.02**	0.02	**0.03**	0.02	**0.03**	0.03	**0.04**	0.03	**0.04**
Whole body dose	0.02	**0.03**	0.02	**0.03**	0.02	**0.04**	0.03	**0.04**	0.03	**0.05**	0.04	**0.05**
ED (ICRP 60)	0.04	**0.04**	0.05	**0.05**	0.06	**0.06**	0.06	**0.06**	0.07	**0.07**	0.07	**0.07**
ED (ICRP 103)	0.04	**0.05**	0.05	**0.05**	0.06	**0.06**	0.06	**0.06**	0.07	**0.07**	0.07	**0.08**
Total filtration (mm) Al	**1.5**		**2.5**		**3.5**		**4.5**		**3.5**		**3.5**	
Cu	**0**		**0**		**0**		**0**		**0.1**		**0.2**	

Table 14.36 Normalized organ doses—Cervical spine r-lat—1–2 years

Beam projection:			r-lat									**Table cells:**
Focus-to-detector distance (cm):			115						(+) = normal font = optimal field setting			
Tube voltage (kVp):			50						**(−) = bold font = suboptimal field setting**			
Total filtrations:			See last table row.						Empty table cells correspond to values <0.01			
Organ or tissue	**Organ dose D_T : Entrance dose K_E**								**(mSv * mGy^{-1})**			
	(+)	(−)	(+)	(−)	(+)	(−)	(+)	(−)	(+)	(−)	(+)	(−)
Skull	0.11	**0.45**	0.13	**0.55**	0.15	**0.61**	0.17	**0.67**	0.19	**0.76**	0.22	**0.84**
Brain	0.01	**0.10**	0.02	**0.13**	0.02	**0.15**	0.03	**0.17**	0.03	**0.20**	0.04	**0.23**
Oral mucosa	0.11	**0.25**	0.14	**0.31**	0.16	**0.35**	0.17	**0.38**	0.20	**0.43**	0.22	**0.49**
Salivary glands	0.33	**0.38**	0.37	**0.43**	0.40	**0.47**	0.41	**0.50**	0.45	**0.55**	0.47	**0.59**
Cervical spine	0.43	**0.47**	0.57	**0.61**	0.67	**0.73**	0.76	**0.82**	0.92	**1.00**	1.07	**1.17**
Thyroid	0.39	**0.43**	0.47	**0.52**	0.52	**0.58**	0.56	**0.63**	0.63	**0.72**	0.69	**0.79**
Thymus		**0.05**	0.01	**0.07**	0.01	**0.08**	0.02	**0.09**	0.02	**0.12**	0.03	**0.14**
Thoracic spine	0.02	**0.05**	0.03	**0.07**	0.03	**0.08**	0.04	**0.10**	0.05	**0.12**	0.06	**0.15**
Claviculae	0.25	**0.58**	0.30	**0.73**	0.34	**0.84**	0.37	**0.93**	0.43	**1.09**	0.47	**1.23**
Scapulae	0.02	**0.23**	0.03	**0.27**	0.04	**0.30**	0.04	**0.33**	0.05	**0.36**	0.05	**0.40**
Ribs	0.03	**0.09**	0.04	**0.12**	0.04	**0.13**	0.05	**0.15**	0.05	**0.17**	0.06	**0.20**
Oesophagus	0.02	**0.03**	0.03	**0.03**	0.03	**0.04**	0.04	**0.05**	0.05	**0.06**	0.05	**0.07**
Trachea	0.07	**0.21**	0.09	**0.26**	0.10	**0.31**	0.11	**0.34**	0.13	**0.40**	0.16	**0.46**
Lung		**0.02**	0.01	**0.02**	0.01	**0.03**	0.02	**0.03**	0.02	**0.04**	0.02	**0.04**
Mamma												
Heart										**0.01**		**0.02**
Stomach												
Spleen												
Liver												
Gall bladder												
Pancreas												
Small intestine												
Large intestine												
Sigma/rectum												
Lumbar spine												
Pelvis												
Adrenals												
Kidneys												
Urinary bladder												
Uterus												
Ovaries												
Prostate												
Testes												
Upper arm		**0.33**		**0.39**		**0.44**		**0.47**		**0.52**		**0.57**
Elbow												
Forearm												
Upper leg												
Knee												
Lower leg												
Lymph nodes	0.06	**0.08**	0.07	**0.10**	0.08	**0.12**	0.08	**0.13**	0.09	**0.14**	0.10	**0.16**
Skin	0.04	**0.11**	0.04	**0.12**	0.05	**0.13**	0.05	**0.13**	0.05	**0.14**	0.05	**0.15**
Muscle	0.02	**0.05**	0.03	**0.05**	0.03	**0.06**	0.04	**0.07**	0.04	**0.08**	0.04	**0.08**
Skeleton	0.05	**0.17**	0.07	**0.21**	0.08	**0.23**	0.08	**0.26**	0.10	**0.30**	0.11	**0.33**
Red bone marrow	0.01	**0.05**	0.02	**0.06**	0.02	**0.07**	0.02	**0.08**	0.03	**0.09**	0.03	**0.10**
Whole body dose	0.03	**0.06**	0.03	**0.08**	0.04	**0.09**	0.04	**0.10**	0.05	**0.11**	0.05	**0.12**
ED (ICRP 60)	0.02	**0.03**	0.03	**0.04**	0.03	**0.05**	0.04	**0.05**	0.04	**0.06**	0.05	**0.07**
ED (ICRP 103)	0.03	**0.04**	0.03	**0.05**	0.04	**0.05**	0.04	**0.06**	0.04	**0.07**	0.05	**0.08**
Total filtration (mm) Al	**1.5**		**2.5**		**3.5**		**4.5**		**3.5**		**3.5**	
Cu	**0**		**0**		**0**		**0**		**0.1**		**0.2**	

Table 14.37 Normalized organ doses—Cervical spine r-lat—1–2 years

Beam projection:			r-lat									**Table cells:**
Focus-to-detector distance (cm):			115				(+) = normal font = optimal field setting					
Tube voltage (kVp):			60				**(−) = bold font = suboptimal field setting**					
Total filtrations:			See last table row.				Empty table cells correspond to values <0.01					
Organ or tissue	**Organ dose D_T : Entrance dose K_E**								**(mSv ∗ mGy⁻¹)**			
	(+)	(−)	(+)	(−)	(+)	(−)	(+)	(−)	(+)	(−)	(+)	(−)
Skull	0.13	**0.53**	0.16	**0.63**	0.18	**0.70**	0.20	**0.76**	0.23	**0.85**	0.22	**0.84**
Brain	0.02	**0.13**	0.03	**0.16**	0.03	**0.19**	0.03	**0.21**	0.04	**0.25**	0.04	**0.23**
Oral mucosa	0.14	**0.30**	0.17	**0.36**	0.19	**0.41**	0.20	**0.44**	0.23	**0.50**	0.22	**0.49**
Salivary glands	0.36	**0.42**	0.40	**0.48**	0.43	**0.52**	0.45	**0.55**	0.48	**0.61**	0.47	**0.59**
Cervical spine	0.56	**0.61**	0.72	**0.79**	0.84	**0.92**	0.94	**1.03**	1.12	**1.24**	1.07	**1.17**
Thyroid	0.45	**0.50**	0.53	**0.60**	0.59	**0.66**	0.63	**0.71**	0.70	**0.80**	0.69	**0.79**
Thymus	0.01	**0.07**	0.02	**0.09**	0.02	**0.11**	0.02	**0.12**	0.03	**0.15**	0.03	**0.14**
Thoracic spine	0.03	**0.07**	0.04	**0.10**	0.05	**0.12**	0.05	**0.13**	0.06	**0.16**	0.06	**0.15**
Claviculae	0.29	**0.71**	0.35	**0.87**	0.39	**0.99**	0.42	**1.09**	0.48	**1.25**	0.47	**1.23**
Scapulae	0.03	**0.26**	0.04	**0.31**	0.04	**0.34**	0.05	**0.36**	0.06	**0.40**	0.05	**0.40**
Ribs	0.04	**0.11**	0.05	**0.14**	0.05	**0.16**	0.05	**0.18**	0.06	**0.20**	0.06	**0.20**
Oesophagus	0.03	**0.04**	0.04	**0.05**	0.04	**0.06**	0.05	**0.07**	0.06	**0.08**	0.05	**0.07**
Trachea	0.08	**0.26**	0.11	**0.32**	0.12	**0.37**	0.14	**0.41**	0.16	**0.48**	0.16	**0.46**
Lung	0.01	**0.02**	0.02	**0.03**	0.02	**0.03**	0.02	**0.04**	0.02	**0.05**	0.02	**0.04**
Mamma												
Heart						**0.01**		**0.01**		**0.02**		**0.02**
Stomach												
Spleen												
Liver												
Gall bladder												
Pancreas												
Small intestine												
Large intestine												
Sigma/rectum												
Lumbar spine												
Pelvis												
Adrenals												
Kidneys												
Urinary bladder												
Uterus												
Ovaries												
Prostate												
Testes												
Upper arm		**0.38**		**0.44**		**0.48**		**0.51**		**0.57**		**0.57**
Elbow										**0.01**		
Forearm												
Upper leg												
Knee												
Lower leg												
Lymph nodes	0.07	**0.10**	0.08	**0.12**	0.09	**0.13**	0.09	**0.15**	0.10	**0.17**	0.10	**0.16**
Skin	0.04	**0.12**	0.05	**0.13**	0.05	**0.14**	0.05	**0.14**	0.06	**0.15**	0.05	**0.15**
Muscle	0.03	**0.05**	0.03	**0.06**	0.04	**0.07**	0.04	**0.08**	0.05	**0.09**	0.04	**0.08**
Skeleton	0.06	**0.20**	0.08	**0.24**	0.09	**0.27**	0.10	**0.30**	0.12	**0.34**	0.11	**0.33**
Red bone marrow	0.02	**0.06**	0.02	**0.08**	0.03	**0.09**	0.03	**0.10**	0.03	**0.11**	0.03	**0.10**
Whole body dose	0.03	**0.08**	0.04	**0.09**	0.04	**0.10**	0.05	**0.11**	0.05	**0.13**	0.05	**0.12**
ED (ICRP 60)	0.03	**0.04**	0.03	**0.05**	0.04	**0.06**	0.04	**0.06**	0.05	**0.07**	0.05	**0.07**
ED (ICRP 103)	0.03	**0.05**	0.04	**0.06**	0.04	**0.06**	0.04	**0.07**	0.05	**0.08**	0.05	**0.08**
Total filtration (mm) **Al**	**1.5**		**2.5**		**3.5**		**4.5**		**3.5**		**3.5**	
Cu	**0**		**0**		**0**		**0**		**0.1**		**0.2**	

Table 14.38 Normalized organ doses—Cervical spine r-lat—1–2 years

Beam projection:			r-lat									**Table cells:**
Focus-to-detector distance (cm):			115				(+) = normal font = optimal field setting					
Tube voltage (kVp):			70				**(−) = bold font = suboptimal field setting**					
Total filtrations:			See last table row.				Empty table cells correspond to values <0.01					
Organ or tissue	**Organ dose D_T : Entrance dose K_E**							**(mSv * mGy⁻¹)**				
	(+)	(−)	(+)	(−)	(+)	(−)	(+)	(−)	(+)	(−)	(+)	(−)
Skull	0.15	**0.59**	0.18	**0.69**	0.20	**0.76**	0.22	**0.82**	0.25	**0.91**	0.27	**0.99**
Brain	0.02	**0.15**	0.03	**0.19**	0.04	**0.22**	0.04	**0.24**	0.05	**0.28**	0.06	**0.32**
Oral mucosa	0.16	**0.34**	0.19	**0.41**	0.21	**0.45**	0.23	**0.49**	0.26	**0.56**	0.28	**0.61**
Salivary glands	0.38	**0.46**	0.42	**0.52**	0.45	**0.56**	0.47	**0.59**	0.51	**0.65**	0.54	**0.70**
Cervical spine	0.67	**0.74**	0.84	**0.93**	0.96	**1.07**	1.06	**1.18**	1.24	**1.39**	1.40	**1.58**
Thyroid	0.50	**0.56**	0.58	**0.66**	0.64	**0.73**	0.68	**0.78**	0.75	**0.87**	0.81	**0.94**
Thymus	0.02	**0.09**	0.02	**0.11**	0.03	**0.13**	0.03	**0.15**	0.04	**0.17**	0.04	**0.20**
Thoracic spine	0.04	**0.09**	0.05	**0.12**	0.06	**0.14**	0.06	**0.16**	0.08	**0.20**	0.09	**0.23**
Claviculae	0.32	**0.80**	0.38	**0.97**	0.42	**1.09**	0.46	**1.19**	0.51	**1.35**	0.55	**1.48**
Scapulae	0.04	**0.29**	0.05	**0.33**	0.05	**0.36**	0.06	**0.39**	0.07	**0.43**	0.08	**0.46**
Ribs	0.04	**0.13**	0.05	**0.16**	0.06	**0.18**	0.06	**0.19**	0.07	**0.22**	0.07	**0.25**
Oesophagus	0.03	**0.05**	0.04	**0.06**	0.05	**0.07**	0.06	**0.08**	0.07	**0.10**	0.08	**0.11**
Trachea	0.10	**0.30**	0.13	**0.37**	0.14	**0.42**	0.16	**0.46**	0.19	**0.54**	0.21	**0.60**
Lung	0.01	**0.03**	0.02	**0.04**	0.02	**0.04**	0.02	**0.05**	0.03	**0.06**	0.03	**0.07**
Mamma										**0.01**		**0.02**
Heart		**0.01**		**0.01**		**0.02**		**0.02**	0.01	**0.02**	0.01	**0.03**
Stomach												
Spleen												
Liver												
Gall bladder												
Pancreas												
Small intestine												
Large intestine												
Sigma/rectum												
Lumbar spine												
Pelvis												
Adrenals												
Kidneys												
Urinary bladder												
Uterus												
Ovaries												
Prostate												
Testes												
Upper arm		**0.41**		**0.47**		**0.51**		**0.54**		**0.60**		**0.64**
Elbow						**0.01**		**0.01**		**0.02**		**0.02**
Forearm												
Upper leg												
Knee												
Lower leg												
Lymph nodes	0.07	**0.11**	0.09	**0.13**	0.09	**0.15**	0.10	**0.16**	0.11	**0.18**	0.12	**0.20**
Skin	0.05	**0.13**	0.05	**0.14**	0.05	**0.14**	0.05	**0.15**	0.06	**0.16**	0.06	**0.16**
Muscle	0.03	**0.06**	0.04	**0.07**	0.04	**0.08**	0.05	**0.09**	0.05	**0.10**	0.06	**0.11**
Skeleton	0.07	**0.23**	0.09	**0.27**	0.10	**0.30**	0.11	**0.33**	0.13	**0.37**	0.15	**0.40**
Red bone marrow	0.02	**0.07**	0.03	**0.09**	0.03	**0.10**	0.03	**0.11**	0.04	**0.12**	0.05	**0.14**
Whole body dose	0.04	**0.09**	0.04	**0.10**	0.05	**0.12**	0.05	**0.13**	0.06	**0.14**	0.06	**0.16**
ED (ICRP 60)	0.03	**0.05**	0.04	**0.06**	0.04	**0.06**	0.05	**0.07**	0.05	**0.08**	0.06	**0.09**
ED (ICRP 103)	0.03	**0.05**	0.04	**0.06**	0.05	**0.07**	0.05	**0.08**	0.05	**0.09**	0.06	**0.10**
Total filtration (mm) Al	**1.5**		**2.5**		**3.5**		**4.5**		**3.5**		**3.5**	
Cu	**0**		**0**		**0**		**0**		**0.1**		**0.2**	

Table 14.39 Normalized organ doses—Cervical spine r-lat—3–7 years

Beam projection:			r-lat									**Table cells:**
Focus-to-detector distance (cm):			115						(+) = normal font = optimal field setting			
Tube voltage (kVp):			60						**(−) = bold font = suboptimal field setting**			
Total filtrations:			See last table row.						Empty table cells correspond to values <0.01			
Organ or tissue	**Organ dose D_T : Entrance dose K_E**								**(mSv * mGy^{-1})**			
	(+)	(−)	(+)	(−)	(+)	(−)	(+)	(−)	(+)	(−)	(+)	(−)
Skull	0.08	**0.45**	0.10	**0.55**	0.11	**0.62**	0.13	**0.67**	0.15	**0.77**	0.14	**0.75**
Brain		**0.10**	0.01	**0.13**	0.01	**0.15**	0.02	**0.17**	0.02	**0.21**	0.02	**0.19**
Oral mucosa	0.08	**0.23**	0.10	**0.28**	0.11	**0.32**	0.12	**0.35**	0.14	**0.41**	0.13	**0.38**
Salivary glands	0.29	**0.36**	0.33	**0.42**	0.35	**0.45**	0.37	**0.48**	0.40	**0.53**	0.39	**0.51**
Cervical spine	0.56	**0.60**	0.72	**0.77**	0.84	**0.90**	0.93	**1.00**	1.12	**1.21**	1.06	**1.14**
Thyroid	0.28	**0.46**	0.34	**0.55**	0.38	**0.62**	0.42	**0.66**	0.47	**0.75**	0.46	**0.73**
Thymus		**0.03**		**0.04**		**0.04**		**0.05**	0.01	**0.06**	0.01	**0.05**
Thoracic spine	0.02	**0.05**	0.03	**0.06**	0.04	**0.07**	0.04	**0.08**	0.05	**0.10**	0.05	**0.09**
Claviculae	0.13	**0.52**	0.16	**0.65**	0.18	**0.75**	0.19	**0.83**	0.22	**0.98**	0.22	**0.95**
Scapulae	0.02	**0.19**	0.03	**0.22**	0.03	**0.25**	0.04	**0.27**	0.04	**0.30**	0.04	**0.30**
Ribs	0.01	**0.07**	0.01	**0.09**	0.02	**0.11**	0.02	**0.12**	0.02	**0.14**	0.02	**0.13**
Oesophagus	0.01	**0.02**	0.02	**0.02**	0.02	**0.03**	0.02	**0.03**	0.03	**0.04**	0.02	**0.03**
Trachea	0.07	**0.17**	0.09	**0.22**	0.10	**0.26**	0.11	**0.29**	0.14	**0.35**	0.13	**0.32**
Lung		**0.01**		**0.02**		**0.02**	0.01	**0.02**	0.01	**0.03**	0.01	**0.03**
Mamma												
Heart												
Stomach												
Spleen												
Liver												
Gall bladder												
Pancreas												
Small intestine												
Large intestine												
Sigma/rectum												
Lumbar spine												
Pelvis												
Adrenals												
Kidneys												
Urinary bladder												
Uterus												
Ovaries												
Prostate												
Testes												
Upper arm		**0.29**		**0.35**		**0.38**		**0.41**		**0.46**		**0.45**
Elbow												
Forearm												
Upper leg												
Knee												
Lower leg												
Lymph nodes	0.05	**0.08**	0.06	**0.10**	0.06	**0.11**	0.07	**0.12**	0.08	**0.13**	0.07	**0.13**
Skin	0.03	**0.10**	0.04	**0.10**	0.04	**0.11**	0.04	**0.11**	0.04	**0.12**	0.04	**0.11**
Muscle	0.02	**0.03**	0.02	**0.04**	0.02	**0.04**	0.02	**0.05**	0.03	**0.05**	0.03	**0.05**
Skeleton	0.04	**0.17**	0.05	**0.21**	0.06	**0.23**	0.07	**0.26**	0.08	**0.30**	0.08	**0.29**
Red bone marrow		**0.04**	0.01	**0.05**	0.01	**0.05**	0.01	**0.06**	0.02	**0.07**	0.02	**0.06**
Whole body dose	0.02	**0.06**	0.02	**0.07**	0.03	**0.08**	0.03	**0.08**	0.03	**0.10**	0.03	**0.09**
ED (ICRP 60)	0.02	**0.03**	0.02	**0.04**	0.02	**0.05**	0.03	**0.05**	0.03	**0.06**	0.03	**0.05**
ED (ICRP 103)	0.02	**0.04**	0.02	**0.05**	0.03	**0.05**	0.03	**0.06**	0.03	**0.06**	0.03	**0.06**
Total filtration (mm) **Al**	**1.5**		**2.5**		**3.5**		**4.5**		**3.5**		**3.5**	
Cu	**0**		**0**		**0**		**0**		**0.1**		**0.2**	

Table 14.40 Normalized organ doses—Cervical spine r-lat—3–7 years

Beam projection:			r-lat									**Table cells:**
Focus-to-detector distance (cm):			115					(+) = normal font = optimal field setting				
Tube voltage (kVp):			70					**(−) = bold font = suboptimal field setting**				
Total filtrations:			See last table row.					Empty table cells correspond to values <0.01				
Organ or tissue	**Organ dose D_T : Entrance dose K_E**							**(mSv * mGy^{-1})**				
	(+)	(−)	(+)	(−)	(+)	(−)	(+)	(−)	(+)	(−)	(+)	(−)
Skull	0.09	**0.51**	0.11	**0.61**	0.13	**0.69**	0.14	**0.74**	0.17	**0.85**	0.19	**0.93**
Brain	0.01	**0.13**	0.01	**0.16**	0.02	**0.19**	0.02	**0.21**	0.03	**0.25**	0.03	**0.29**
Oral mucosa	0.09	**0.27**	0.11	**0.32**	0.13	**0.36**	0.14	**0.40**	0.16	**0.46**	0.18	**0.52**
Salivary glands	0.32	**0.40**	0.35	**0.45**	0.37	**0.49**	0.39	**0.52**	0.42	**0.58**	0.45	**0.62**
Cervical spine	0.67	**0.72**	0.84	**0.91**	0.97	**1.05**	1.07	**1.16**	1.26	**1.38**	1.42	**1.57**
Thyroid	0.32	**0.52**	0.38	**0.61**	0.42	**0.68**	0.46	**0.73**	0.51	**0.82**	0.56	**0.89**
Thymus		**0.04**		**0.05**	0.01	**0.05**	0.01	**0.06**	0.02	**0.08**	0.02	**0.09**
Thoracic spine	0.03	**0.06**	0.04	**0.08**	0.05	**0.09**	0.05	**0.10**	0.06	**0.13**	0.07	**0.15**
Claviculae	0.15	**0.61**	0.18	**0.75**	0.20	**0.85**	0.21	**0.93**	0.24	**1.07**	0.27	**1.20**
Scapulae	0.03	**0.21**	0.03	**0.24**	0.04	**0.27**	0.04	**0.29**	0.05	**0.32**	0.06	**0.35**
Ribs	0.01	**0.09**	0.02	**0.11**	0.02	**0.12**	0.02	**0.13**	0.03	**0.16**	0.03	**0.18**
Oesophagus	0.02	**0.02**	0.02	**0.03**	0.02	**0.04**	0.03	**0.04**	0.03	**0.05**	0.04	**0.06**
Trachea	0.08	**0.21**	0.10	**0.26**	0.12	**0.30**	0.13	**0.34**	0.16	**0.40**	0.18	**0.46**
Lung		**0.02**	0.01	**0.02**	0.01	**0.03**	0.01	**0.03**	0.02	**0.04**	0.02	**0.05**
Mamma												**0.01**
Heart										**0.01**		**0.02**
Stomach												
Spleen												
Liver												
Gall bladder												
Pancreas												
Small intestine												
Large intestine												
Sigma/rectum												
Lumbar spine												
Pelvis												
Adrenals												
Kidneys												
Urinary bladder												
Uterus												
Ovaries												
Prostate												
Testes												
Upper arm		**0.32**		**0.37**		**0.41**		**0.44**		**0.48**		**0.52**
Elbow												
Forearm												
Upper leg												
Knee												
Lower leg												
Lymph nodes	0.05	**0.09**	0.06	**0.11**	0.07	**0.12**	0.08	**0.13**	0.08	**0.15**	0.09	**0.17**
Skin	0.04	**0.10**	0.04	**0.11**	0.04	**0.11**	0.04	**0.12**	0.05	**0.12**	0.05	**0.13**
Muscle	0.02	**0.04**	0.02	**0.04**	0.02	**0.05**	0.03	**0.05**	0.03	**0.06**	0.03	**0.07**
Skeleton	0.05	**0.19**	0.06	**0.23**	0.07	**0.26**	0.08	**0.28**	0.09	**0.33**	0.10	**0.36**
Red bone marrow	0.01	**0.04**	0.01	**0.05**	0.02	**0.06**	0.02	**0.07**	0.02	**0.08**	0.02	**0.09**
Whole body dose	0.02	**0.06**	0.03	**0.08**	0.03	**0.09**	0.03	**0.09**	0.04	**0.11**	0.04	**0.12**
ED (ICRP 60)	0.02	**0.04**	0.02	**0.05**	0.03	**0.05**	0.03	**0.06**	0.03	**0.06**	0.04	**0.07**
ED (ICRP 103)	0.02	**0.04**	0.03	**0.05**	0.03	**0.06**	0.03	**0.06**	0.04	**0.07**	0.04	**0.08**
Total filtration (mm) Al	**1.5**		**2.5**		**3.5**		**4.5**		**3.5**		**3.5**	
Cu	**0**		**0**		**0**		**0**		**0.1**		**0.2**	

Table 14.41 Normalized organ doses—Cervical spine r-lat—8–12 years

Beam projection:			r-lat									**Table cells:**
Focus-to-detector distance (cm):			115									(+) = normal font = optimal field setting
Tube voltage (kVp):			60									**(−) = bold font = suboptimal field setting**
Total filtrations:			See last table row.									Empty table cells correspond to values <0.01
Organ or tissue	**Organ dose D_T : Entrance dose K_E**							**(mSv * mGy^{-1})**				
	(+)	(−)	(+)	(−)	(+)	(−)	(+)	(−)	(+)	(−)	(+)	(−)
Skull	0.12	**0.41**	0.15	**0.51**	0.17	**0.58**	0.19	**0.63**	0.23	**0.73**	0.22	**0.71**
Brain	0.01	**0.08**	0.02	**0.11**	0.02	**0.13**	0.03	**0.15**	0.03	**0.18**	0.03	**0.17**
Oral mucosa	0.11	**0.21**	0.14	**0.26**	0.16	**0.30**	0.17	**0.33**	0.20	**0.38**	0.19	**0.36**
Salivary glands	0.33	**0.36**	0.37	**0.41**	0.40	**0.44**	0.42	**0.47**	0.46	**0.52**	0.45	**0.50**
Cervical spine	0.56	**0.59**	0.73	**0.76**	0.85	**0.89**	0.96	**1.00**	1.16	**1.22**	1.08	**1.13**
Thyroid	0.39	**0.41**	0.48	**0.50**	0.54	**0.57**	0.58	**0.61**	0.66	**0.70**	0.64	**0.68**
Thymus				**0.01**		**0.02**		**0.02**	0.01	**0.02**		**0.02**
Thoracic spine	0.02	**0.03**	0.02	**0.04**	0.02	**0.05**	0.03	**0.05**	0.04	**0.07**	0.03	**0.06**
Claviculae	0.21	**0.40**	0.26	**0.51**	0.30	**0.59**	0.33	**0.66**	0.38	**0.78**	0.37	**0.75**
Scapulae	0.02	**0.10**	0.02	**0.12**	0.03	**0.13**	0.03	**0.14**	0.04	**0.16**	0.03	**0.16**
Ribs	0.02	**0.05**	0.03	**0.06**	0.03	**0.07**	0.04	**0.08**	0.04	**0.09**	0.04	**0.08**
Oesophagus	0.01	**0.01**	0.01	**0.02**	0.02	**0.02**	0.02	**0.02**	0.02	**0.03**	0.02	**0.02**
Trachea	0.16	**0.25**	0.20	**0.32**	0.24	**0.38**	0.26	**0.42**	0.31	**0.50**	0.29	**0.47**
Lung						**0.01**	0.01	**0.01**	0.01	**0.02**	0.01	**0.01**
Mamma												
Heart												
Stomach												
Spleen												
Liver												
Gall bladder												
Pancreas												
Small intestine												
Large intestine												
Sigma/rectum												
Lumbar spine												
Pelvis												
Adrenals												
Kidneys												
Urinary bladder												
Uterus												
Ovaries												
Prostate												
Testes												
Upper arm		**0.19**		**0.22**		**0.25**		**0.27**		**0.30**		**0.30**
Elbow												
Forearm												
Upper leg												
Knee												
Lower leg												
Lymph nodes	0.07	**0.08**	0.08	**0.10**	0.09	**0.12**	0.10	**0.13**	0.11	**0.15**	0.11	**0.14**
Skin	0.03	**0.08**	0.04	**0.08**	0.04	**0.09**	0.04	**0.09**	0.04	**0.09**	0.04	**0.09**
Muscle	0.02	**0.02**	0.02	**0.03**	0.02	**0.03**	0.02	**0.03**	0.03	**0.04**	0.02	**0.04**
Skeleton	0.05	**0.12**	0.06	**0.15**	0.07	**0.17**	0.07	**0.19**	0.09	**0.22**	0.08	**0.21**
Red bone marrow	0.01	**0.03**	0.01	**0.03**	0.02	**0.04**	0.02	**0.04**	0.02	**0.05**	0.02	**0.05**
Whole body dose	0.02	**0.04**	0.02	**0.05**	0.03	**0.05**	0.03	**0.06**	0.03	**0.07**	0.03	**0.07**
ED (ICRP 60)	0.02	**0.03**	0.03	**0.03**	0.03	**0.04**	0.03	**0.04**	0.04	**0.05**	0.04	**0.05**
ED (ICRP 103)	0.03	**0.03**	0.03	**0.04**	0.03	**0.04**	0.04	**0.05**	0.04	**0.06**	0.04	**0.05**
Total filtration (mm) Al	**1.5**		**2.5**		**3.5**		**4.5**		**3.5**		**3.5**	
Cu	**0**		**0**		**0**		**0**		**0.1**		**0.2**	

Table 14.42 Normalized organ doses—Cervical spine r-lat—8–12 years

Beam projection:			r-lat									**Table cells:**
Focus-to-detector distance (cm):			115						(+) = normal font = optimal field setting			
Tube voltage (kVp):			70						**(−) = bold font = suboptimal field setting**			
Total filtrations:			See last table row.						Empty table cells correspond to values <0.01			
Organ or tissue	**Organ dose D_T : Entrance dose K_E**								**(mSv * mGy^{-1})**			
	(+)	(−)	(+)	(−)	(+)	(−)	(+)	(−)	(+)	(−)	(+)	(−)
Skull	0.14	**0.48**	0.18	**0.58**	0.20	**0.65**	0.22	**0.71**	0.25	**0.81**	0.29	**0.90**
Brain	0.02	**0.11**	0.02	**0.14**	0.03	**0.16**	0.03	**0.18**	0.04	**0.22**	0.05	**0.26**
Oral mucosa	0.13	**0.25**	0.16	**0.31**	0.18	**0.35**	0.20	**0.38**	0.23	**0.44**	0.26	**0.50**
Salivary glands	0.36	**0.39**	0.40	**0.45**	0.43	**0.48**	0.45	**0.51**	0.49	**0.56**	0.53	**0.61**
Cervical spine	0.68	**0.72**	0.86	**0.91**	1.00	**1.06**	1.11	**1.18**	1.32	**1.41**	1.52	**1.62**
Thyroid	0.45	**0.47**	0.53	**0.56**	0.59	**0.63**	0.64	**0.68**	0.72	**0.77**	0.78	**0.84**
Thymus		**0.01**		**0.02**	0.01	**0.02**	0.01	**0.02**	0.02	**0.03**	0.02	**0.04**
Thoracic spine	0.02	**0.04**	0.03	**0.05**	0.03	**0.06**	0.04	**0.07**	0.05	**0.08**	0.05	**0.10**
Claviculae	0.24	**0.47**	0.29	**0.59**	0.33	**0.68**	0.36	**0.75**	0.42	**0.87**	0.46	**0.98**
Scapulae	0.02	**0.11**	0.03	**0.13**	0.03	**0.14**	0.04	**0.16**	0.04	**0.18**	0.05	**0.19**
Ribs	0.03	**0.06**	0.03	**0.07**	0.04	**0.08**	0.04	**0.09**	0.05	**0.10**	0.05	**0.12**
Oesophagus	0.01	**0.02**	0.02	**0.02**	0.02	**0.02**	0.02	**0.03**	0.03	**0.03**	0.04	**0.04**
Trachea	0.19	**0.31**	0.24	**0.39**	0.28	**0.45**	0.30	**0.49**	0.36	**0.59**	0.41	**0.68**
Lung		**0.01**	0.01	**0.01**	0.01	**0.02**	0.01	**0.02**	0.02	**0.02**	0.02	**0.03**
Mamma												
Heart												
Stomach												
Spleen												
Liver												
Gall bladder												
Pancreas												
Small intestine												
Large intestine												
Sigma/rectum												
Lumbar spine												
Pelvis												
Adrenals												
Kidneys												
Urinary bladder												
Uterus												
Ovaries												
Prostate												
Testes												
Upper arm		**0.21**		**0.25**		**0.27**		**0.29**		**0.32**		**0.35**
Elbow												
Forearm												
Upper leg												
Knee												
Lower leg												
Lymph nodes	0.08	**0.10**	0.09	**0.12**	0.10	**0.13**	0.11	**0.15**	0.13	**0.17**	0.14	**0.19**
Skin	0.03	**0.08**	0.04	**0.09**	0.04	**0.09**	0.04	**0.09**	0.04	**0.10**	0.05	**0.10**
Muscle	0.02	**0.03**	0.02	**0.03**	0.02	**0.04**	0.03	**0.04**	0.03	**0.04**	0.03	**0.05**
Skeleton	0.05	**0.14**	0.07	**0.17**	0.08	**0.19**	0.08	**0.21**	0.10	**0.24**	0.11	**0.27**
Red bone marrow	0.01	**0.03**	0.02	**0.04**	0.02	**0.05**	0.02	**0.05**	0.02	**0.06**	0.03	**0.07**
Whole body dose	0.02	**0.05**	0.03	**0.06**	0.03	**0.06**	0.03	**0.07**	0.04	**0.08**	0.04	**0.09**
ED (ICRP 60)	0.03	**0.03**	0.03	**0.04**	0.04	**0.04**	0.04	**0.05**	0.04	**0.05**	0.05	**0.06**
ED (ICRP 103)	0.03	**0.04**	0.03	**0.05**	0.04	**0.05**	0.04	**0.06**	0.05	**0.06**	0.05	**0.07**
Total filtration (mm) Al	**1.5**		**2.5**		**3.5**		**4.5**		**3.5**		**3.5**	
Cu	**0**		**0**		**0**		**0**		**0.1**		**0.2**	

Table 14.43 Normalized organ doses—Cervical spine r-lat—13–17 years

Beam projection:			r-lat									**Table cells:**
Focus-to-detector distance (cm):			115									(+) = normal font = optimal field setting
Tube voltage (kVp):			60									**(−) = bold font = suboptimal field setting**
Total filtrations:			See last table row.									Empty table cells correspond to values <0.01
Organ or tissue	**Organ dose D_T : Entrance dose K_E**							**(mSv * mGy^{-1})**				
	(+)	(−)	(+)	(−)	(+)	(−)	(+)	(−)	(+)	(−)	(+)	(−)
Skull	0.15	**0.36**	0.19	**0.45**	0.22	**0.52**	0.24	**0.57**	0.29	**0.67**	0.27	**0.64**
Brain	0.02	**0.07**	0.03	**0.09**	0.04	**0.10**	0.04	**0.12**	0.05	**0.15**	0.04	**0.13**
Oral mucosa	0.11	**0.20**	0.14	**0.25**	0.16	**0.28**	0.17	**0.31**	0.21	**0.37**	0.19	**0.34**
Salivary glands	0.34	**0.36**	0.39	**0.41**	0.42	**0.45**	0.45	**0.48**	0.49	**0.53**	0.47	**0.51**
Cervical spine	0.63	**0.66**	0.82	**0.86**	0.97	**1.02**	1.09	**1.15**	1.34	**1.41**	1.23	**1.29**
Thyroid	0.36	**0.39**	0.44	**0.47**	0.50	**0.54**	0.55	**0.59**	0.63	**0.67**	0.61	**0.65**
Thymus										**0.01**		
Thoracic spine		**0.02**	0.01	**0.03**	0.02	**0.03**	0.02	**0.04**	0.02	**0.05**	0.02	**0.04**
Claviculae	0.16	**0.30**	0.20	**0.38**	0.23	**0.45**	0.25	**0.51**	0.29	**0.61**	0.28	**0.58**
Scapulae		**0.09**		**0.11**	0.01	**0.13**	0.01	**0.14**	0.02	**0.16**	0.01	**0.15**
Ribs		**0.03**		**0.04**		**0.05**	0.01	**0.06**	0.01	**0.07**	0.01	**0.06**
Oesophagus			0.01	**0.01**	0.01	**0.02**	0.02	**0.02**	0.02	**0.02**	0.02	**0.02**
Trachea	0.15	**0.19**	0.19	**0.25**	0.22	**0.29**	0.25	**0.32**	0.30	**0.39**	0.28	**0.36**
Lung										**0.01**		
Mamma				**0.01**		**0.01**		**0.02**		**0.02**		**0.02**
Heart												
Stomach												
Spleen												
Liver												
Gall bladder												
Pancreas												
Small intestine												
Large intestine												
Sigma/rectum												
Lumbar spine												
Pelvis												
Adrenals												
Kidneys												
Urinary bladder												
Uterus												
Ovaries												
Prostate												
Testes												
Upper arm		**0.21**		**0.25**		**0.28**		**0.31**		**0.35**		**0.34**
Elbow												
Forearm												
Upper leg												
Knee												
Lower leg												
Lymph nodes	0.07	**0.08**	0.08	**0.09**	0.09	**0.10**	0.10	**0.11**	0.11	**0.13**	0.11	**0.12**
Skin	0.03	**0.07**	0.04	**0.08**	0.04	**0.08**	0.04	**0.08**	0.04	**0.09**	0.04	**0.08**
Muscle	0.01	**0.02**	0.02	**0.02**	0.02	**0.03**	0.02	**0.03**	0.02	**0.03**	0.02	**0.03**
Skeleton	0.04	**0.09**	0.05	**0.12**	0.06	**0.14**	0.07	**0.15**	0.08	**0.18**	0.08	**0.17**
Red bone marrow		**0.02**	0.01	**0.03**	0.01	**0.03**	0.02	**0.04**	0.02	**0.04**	0.02	**0.04**
Whole body dose	0.02	**0.03**	0.02	**0.04**	0.02	**0.04**	0.03	**0.05**	0.03	**0.06**	0.03	**0.05**
ED (ICRP 60)	0.02	**0.03**	0.03	**0.03**	0.03	**0.04**	0.03	**0.04**	0.04	**0.05**	0.04	**0.04**
ED (ICRP 103)	0.02	**0.03**	0.03	**0.04**	0.03	**0.04**	0.04	**0.05**	0.04	**0.05**	0.04	**0.05**
Total filtration (mm) Al	**1.5**		**2.5**		**3.5**		**4.5**		**3.5**		**3.5**	
Cu	**0**		**0**		**0**		**0**		**0.1**		**0.2**	

Table 14.44 Normalized organ doses—Cervical spine r-lat—13–17 years

Beam projection:	r-lat											**Table cells:**
Focus-to-detector distance (cm):	115											(+) = normal font = optimal field setting
Tube voltage (kVp):	70											**(−) = bold font = suboptimal field setting**
Total filtrations:	See last table row.											Empty table cells correspond to values <0.01
Organ or tissue	**Organ dose D_T : Entrance dose K_E (mSv * mGy^{-1})**											
	(+)	(−)	(+)	(−)	(+)	(−)	(+)	(−)	(+)	(−)	(+)	(−)
Skull	0.18	**0.42**	0.22	**0.52**	0.25	**0.59**	0.28	**0.65**	0.32	**0.76**	0.37	**0.85**
Brain	0.03	**0.09**	0.04	**0.11**	0.05	**0.13**	0.05	**0.15**	0.06	**0.18**	0.08	**0.22**
Oral mucosa	0.13	**0.24**	0.16	**0.29**	0.18	**0.33**	0.20	**0.36**	0.24	**0.43**	0.27	**0.49**
Salivary glands	0.37	**0.39**	0.42	**0.45**	0.46	**0.49**	0.48	**0.52**	0.53	**0.58**	0.57	**0.63**
Cervical spine	0.78	**0.82**	1.00	**1.05**	1.16	**1.23**	1.30	**1.37**	1.56	**1.65**	1.81	**1.92**
Thyroid	0.41	**0.44**	0.50	**0.54**	0.56	**0.60**	0.61	**0.65**	0.69	**0.74**	0.76	**0.82**
Thymus								**0.01**		**0.01**		**0.02**
Thoracic spine	0.01	**0.03**	0.02	**0.03**	0.02	**0.04**	0.02	**0.05**	0.03	**0.06**	0.04	**0.07**
Claviculae	0.18	**0.36**	0.22	**0.46**	0.26	**0.53**	0.28	**0.59**	0.33	**0.70**	0.36	**0.80**
Scapulae		**0.10**	0.01	**0.13**	0.01	**0.14**	0.02	**0.15**	0.02	**0.18**	0.02	**0.20**
Ribs		**0.04**		**0.05**	0.01	**0.06**	0.01	**0.07**	0.02	**0.08**	0.02	**0.09**
Oesophagus	0.01	**0.01**	0.01	**0.02**	0.02	**0.02**	0.02	**0.02**	0.02	**0.03**	0.03	**0.04**
Trachea	0.18	**0.24**	0.23	**0.30**	0.26	**0.35**	0.29	**0.38**	0.34	**0.46**	0.39	**0.53**
Lung						**0.01**		**0.01**		**0.01**	0.01	**0.02**
Mamma		**0.01**		**0.02**		**0.02**		**0.02**	0.01	**0.03**	0.01	**0.04**
Heart												
Stomach												
Spleen												
Liver												
Gall bladder												
Pancreas												
Small intestine												
Large intestine												
Sigma/rectum												
Lumbar spine												
Pelvis												
Adrenals												
Kidneys												
Urinary bladder												
Uterus												
Ovaries												
Prostate												
Testes												
Upper arm		**0.24**		**0.28**		**0.31**		**0.33**		**0.37**		**0.41**
Elbow												
Forearm												
Upper leg												
Knee												
Lower leg												
Lymph nodes	0.08	**0.09**	0.09	**0.11**	0.10	**0.12**	0.11	**0.13**	0.12	**0.15**	0.14	**0.17**
Skin	0.03	**0.07**	0.04	**0.08**	0.04	**0.08**	0.04	**0.09**	0.04	**0.09**	0.05	**0.10**
Muscle	0.02	**0.02**	0.02	**0.03**	0.02	**0.03**	0.02	**0.03**	0.02	**0.04**	0.03	**0.04**
Skeleton	0.05	**0.11**	0.06	**0.14**	0.07	**0.16**	0.08	**0.17**	0.10	**0.20**	0.11	**0.23**
Red bone marrow	0.01	**0.03**	0.02	**0.03**	0.02	**0.04**	0.02	**0.04**	0.02	**0.05**	0.03	**0.06**
Whole body dose	0.02	**0.04**	0.02	**0.04**	0.03	**0.05**	0.03	**0.06**	0.04	**0.06**	0.04	**0.07**
ED (ICRP 60)	0.02	**0.03**	0.03	**0.04**	0.03	**0.04**	0.04	**0.04**	0.04	**0.05**	0.05	**0.06**
ED (ICRP 103)	0.03	**0.04**	0.03	**0.04**	0.04	**0.05**	0.04	**0.05**	0.05	**0.06**	0.05	**0.07**
Total filtration (mm) Al	**1.5**		**2.5**		**3.5**		**4.5**		**3.5**		**3.5**	
Cu	**0**		**0**		**0**		**0**		**0.1**		**0.2**	

Table 14.45 Normalized organ doses—Cervical spine r-lat—>17 years

Beam projection:			r-lat									**Table cells:**
Focus-to-detector distance (cm):			115						(+) = normal font = optimal field setting			
Tube voltage (kVp):			60						**(−) = bold font = suboptimal field setting**			
Total filtrations:			See last table row.						Empty table cells correspond to values <0.01			
Organ or tissue	**Organ dose D_T : Entrance dose K_E**								**(mSv * mGy^{-1})**			
	(+)	(−)	(+)	(−)	(+)	(−)	(+)	(−)	(+)	(−)	(+)	(−)
Skull	0.12	**0.37**	0.16	**0.46**	0.18	**0.53**	0.20	**0.59**	0.25	**0.70**	0.23	**0.66**
Brain	0.01	**0.07**	0.01	**0.10**	0.02	**0.12**	0.02	**0.13**	0.03	**0.17**	0.02	**0.15**
Oral mucosa	0.11	**0.20**	0.14	**0.25**	0.16	**0.28**	0.17	**0.31**	0.21	**0.37**	0.19	**0.34**
Salivary glands	0.33	**0.34**	0.37	**0.40**	0.41	**0.43**	0.43	**0.46**	0.48	**0.51**	0.46	**0.49**
Cervical spine	0.61	**0.63**	0.80	**0.83**	0.95	**0.98**	1.07	**1.11**	1.31	**1.36**	1.20	**1.24**
Thyroid	0.36	**0.37**	0.44	**0.46**	0.50	**0.52**	0.55	**0.57**	0.63	**0.66**	0.61	**0.64**
Thymus												
Thoracic spine		**0.02**	0.01	**0.02**	0.02	**0.03**	0.02	**0.03**	0.02	**0.04**	0.02	**0.03**
Claviculae	0.16	**0.25**	0.21	**0.33**	0.24	**0.39**	0.27	**0.43**	0.32	**0.53**	0.30	**0.49**
Scapulae	0.02	**0.06**	0.02	**0.07**	0.03	**0.08**	0.03	**0.09**	0.03	**0.11**	0.03	**0.10**
Ribs	0.03	**0.03**	0.03	**0.03**	0.04	**0.04**	0.04	**0.05**	0.04	**0.05**	0.04	**0.05**
Oesophagus			0.01	**0.01**	0.01	**0.01**	0.01	**0.02**	0.02	**0.02**	0.02	**0.02**
Trachea	0.16	**0.19**	0.21	**0.24**	0.24	**0.28**	0.27	**0.31**	0.32	**0.38**	0.30	**0.35**
Lung												
Mamma												
Heart												
Stomach												
Spleen												
Liver												
Gall bladder												
Pancreas												
Small intestine												
Large intestine												
Sigma/rectum												
Lumbar spine												
Pelvis												
Adrenals												
Kidneys												
Urinary bladder												
Uterus												
Ovaries												
Prostate												
Testes												
Upper arm		**0.15**		**0.18**		**0.20**		**0.22**		**0.25**		**0.24**
Elbow												
Forearm												
Upper leg												
Knee												
Lower leg												
Lymph nodes	0.07	**0.07**	0.08	**0.09**	0.09	**0.10**	0.10	**0.11**	0.11	**0.13**	0.11	**0.12**
Skin	0.03	**0.07**	0.04	**0.08**	0.04	**0.08**	0.04	**0.09**	0.04	**0.09**	0.04	**0.09**
Muscle	0.01	**0.02**	0.02	**0.02**	0.02	**0.02**	0.02	**0.02**	0.02	**0.03**	0.02	**0.03**
Skeleton	0.04	**0.08**	0.05	**0.11**	0.06	**0.12**	0.06	**0.14**	0.08	**0.16**	0.07	**0.15**
Red bone marrow		**0.02**	0.01	**0.02**	0.01	**0.03**	0.02	**0.03**	0.02	**0.04**	0.02	**0.03**
Whole body dose	0.02	**0.03**	0.02	**0.04**	0.02	**0.04**	0.02	**0.04**	0.03	**0.05**	0.03	**0.05**
ED (ICRP 60)	0.02	**0.02**	0.03	**0.03**	0.03	**0.03**	0.03	**0.04**	0.04	**0.04**	0.04	**0.04**
ED (ICRP 103)	0.02	**0.03**	0.03	**0.03**	0.03	**0.04**	0.04	**0.04**	0.04	**0.05**	0.04	**0.05**
Total filtration (mm) **Al**	**1.5**		**2.5**		**3.5**		**4.5**		**3.5**		**3.5**	
Cu	**0**		**0**		**0**		**0**		**0.1**		**0.2**	

Table 14.46 Normalized organ doses—Cervical spine r-lat—>17 years

Beam projection:			r-lat									**Table cells:**
Focus-to-detector distance (cm):			115									(+) = normal font = optimal field setting
Tube voltage (kVp):			70									**(−) = bold font = suboptimal field setting**
Total filtrations:			See last table row.									Empty table cells correspond to values <0.01
Organ or tissue	**Organ dose D_T : Entrance dose K_E**							**(mSv * mGy^{-1})**				
	(+)	(−)	(+)	(−)	(+)	(−)	(+)	(−)	(+)	(−)	(+)	(−)
Skull	0.15	**0.44**	0.19	**0.54**	0.22	**0.61**	0.24	**0.67**	0.28	**0.79**	0.32	**0.89**
Brain	0.02	**0.10**	0.02	**0.13**	0.02	**0.15**	0.03	**0.17**	0.04	**0.21**	0.04	**0.25**
Oral mucosa	0.13	**0.24**	0.16	**0.29**	0.19	**0.33**	0.21	**0.36**	0.24	**0.43**	0.28	**0.48**
Salivary glands	0.36	**0.38**	0.41	**0.44**	0.44	**0.47**	0.47	**0.50**	0.52	**0.56**	0.56	**0.61**
Cervical spine	0.77	**0.80**	0.98	**1.02**	1.14	**1.19**	1.27	**1.33**	1.54	**1.61**	1.78	**1.87**
Thyroid	0.41	**0.43**	0.50	**0.52**	0.56	**0.59**	0.61	**0.64**	0.69	**0.73**	0.77	**0.81**
Thymus												**0.01**
Thoracic spine	0.01	**0.02**	0.02	**0.03**	0.02	**0.03**	0.02	**0.04**	0.03	**0.05**	0.04	**0.06**
Claviculae	0.19	**0.31**	0.24	**0.39**	0.28	**0.46**	0.30	**0.51**	0.36	**0.61**	0.40	**0.71**
Scapulae	0.02	**0.07**	0.03	**0.09**	0.03	**0.10**	0.03	**0.11**	0.04	**0.12**	0.05	**0.14**
Ribs	0.03	**0.03**	0.04	**0.04**	0.04	**0.05**	0.04	**0.05**	0.05	**0.06**	0.05	**0.07**
Oesophagus	0.01	**0.01**	0.01	**0.02**	0.02	**0.02**	0.02	**0.02**	0.02	**0.03**	0.03	**0.03**
Trachea	0.20	**0.23**	0.25	**0.29**	0.28	**0.33**	0.31	**0.37**	0.37	**0.44**	0.42	**0.51**
Lung							0.01	**0.01**	0.01	**0.01**	0.02	**0.02**
Mamma												
Heart												
Stomach												
Spleen												
Liver												
Gall bladder												
Pancreas												
Small intestine												
Large intestine												
Sigma/rectum												
Lumbar spine												
Pelvis												
Adrenals												
Kidneys												
Urinary bladder												
Uterus												
Ovaries												
Prostate												
Testes												
Upper arm		**0.16**		**0.20**		**0.22**		**0.24**		**0.27**		**0.29**
Elbow												
Forearm												
Upper leg												
Knee												
Lower leg												
Lymph nodes	0.08	**0.08**	0.09	**0.10**	0.10	**0.11**	0.11	**0.12**	0.13	**0.14**	0.14	**0.16**
Skin	0.04	**0.08**	0.04	**0.08**	0.04	**0.09**	0.04	**0.09**	0.04	**0.09**	0.05	**0.10**
Muscle	0.01	**0.02**	0.02	**0.02**	0.02	**0.03**	0.02	**0.03**	0.02	**0.03**	0.03	**0.04**
Skeleton	0.05	**0.10**	0.06	**0.12**	0.07	**0.14**	0.08	**0.16**	0.09	**0.19**	0.10	**0.21**
Red bone marrow	0.01	**0.02**	0.02	**0.03**	0.02	**0.03**	0.02	**0.04**	0.03	**0.05**	0.03	**0.05**
Whole body dose	0.02	**0.03**	0.02	**0.04**	0.03	**0.05**	0.03	**0.05**	0.03	**0.06**	0.04	**0.07**
ED (ICRP 60)	0.02	**0.03**	0.03	**0.03**	0.03	**0.04**	0.04	**0.04**	0.04	**0.05**	0.05	**0.05**
ED (ICRP 103)	0.03	**0.03**	0.03	**0.04**	0.04	**0.04**	0.04	**0.05**	0.05	**0.06**	0.05	**0.06**
Total filtration (mm) Al	**1.5**		**2.5**		**3.5**		**4.5**		**3.5**		**3.5**	
Cu	**0**		**0**		**0**		**0**		**0.1**		**0.2**	

Table 14.47 Normalized organ doses—Cervical spine ap—8–12 years

Beam projection:			ap									**Table cells:**
Focus-to-detector distance (cm):			150				(+) = normal font = optimal field setting					
Tube voltage (kVp):			60				**(−) = bold font = suboptimal field setting**					
Total filtrations:			See last table row.				Empty table cells correspond to values <0.01					
Organ or tissue	**Organ dose D_T : Entrance dose K_E**								**(mSv ∗ mGy^{-1})**			
	(+)	(−)	(+)	(−)	(+)	(−)	(+)	(−)	(+)	(−)	(+)	(−)
Skull	0.17	**0.44**	0.21	**0.54**	0.23	**0.62**	0.25	**0.68**	0.29	**0.78**	0.28	**0.75**
Brain		**0.05**		**0.06**		**0.07**	0.01	**0.08**	0.01	**0.11**	0.01	**0.09**
Oral mucosa	0.23	**0.35**	0.27	**0.42**	0.30	**0.48**	0.32	**0.52**	0.37	**0.59**	0.35	**0.56**
Salivary glands	0.08	**0.27**	0.11	**0.33**	0.12	**0.38**	0.14	**0.42**	0.17	**0.49**	0.16	**0.46**
Cervical spine	0.35	**0.36**	0.45	**0.47**	0.54	**0.56**	0.61	**0.64**	0.75	**0.79**	0.68	**0.71**
Thyroid	0.84	**0.81**	0.96	**0.94**	1.04	**1.02**	1.10	**1.08**	1.20	**1.18**	1.19	**1.16**
Thymus	0.01	**0.07**	0.02	**0.08**	0.02	**0.10**	0.03	**0.11**	0.03	**0.13**	0.03	**0.12**
Thoracic spine	0.01	**0.02**	0.02	**0.03**	0.02	**0.04**	0.03	**0.05**	0.03	**0.06**	0.03	**0.05**
Claviculae	0.35	**1.10**	0.42	**1.33**	0.47	**1.49**	0.51	**1.61**	0.57	**1.81**	0.56	**1.80**
Scapulae		**0.02**		**0.02**		**0.03**		**0.03**	0.01	**0.04**		**0.04**
Ribs	0.03	**0.13**	0.04	**0.15**	0.04	**0.17**	0.05	**0.18**	0.05	**0.20**	0.05	**0.19**
Oesophagus	0.01	**0.02**	0.01	**0.02**	0.02	**0.03**	0.02	**0.03**	0.03	**0.04**	0.02	**0.04**
Trachea	0.42	**0.44**	0.52	**0.55**	0.59	**0.62**	0.65	**0.69**	0.75	**0.80**	0.72	**0.76**
Lung		**0.02**		**0.03**		**0.03**		**0.04**		**0.05**		**0.04**
Mamma												
Heart												
Stomach												
Spleen												
Liver												
Gall bladder												
Pancreas												
Small intestine												
Large intestine												
Sigma/rectum												
Lumbar spine												
Pelvis												
Adrenals												
Kidneys												
Urinary bladder												
Uterus												
Ovaries												
Prostate												
Testes												
Upper arm		**0.01**		**0.02**		**0.02**		**0.02**		**0.03**		**0.02**
Elbow												
Forearm												
Upper leg												
Knee												
Lower leg												
Lymph nodes	0.11	**0.12**	0.13	**0.15**	0.14	**0.17**	0.15	**0.18**	0.17	**0.21**	0.17	**0.20**
Skin	0.03	**0.08**	0.03	**0.08**	0.03	**0.09**	0.03	**0.09**	0.03	**0.10**	0.03	**0.09**
Muscle	0.01	**0.03**	0.02	**0.03**	0.02	**0.04**	0.02	**0.04**	0.02	**0.05**	0.02	**0.05**
Skeleton	0.05	**0.12**	0.06	**0.15**	0.07	**0.17**	0.08	**0.19**	0.09	**0.22**	0.09	**0.21**
Red bone marrow	0.01	**0.03**	0.01	**0.04**	0.02	**0.04**	0.02	**0.05**	0.02	**0.06**	0.02	**0.05**
Whole body dose	0.02	**0.04**	0.02	**0.05**	0.02	**0.06**	0.03	**0.06**	0.03	**0.07**	0.03	**0.07**
ED (ICRP 60)	0.05	**0.05**	0.05	**0.06**	0.06	**0.07**	0.06	**0.07**	0.07	**0.08**	0.07	**0.08**
ED (ICRP 103)	0.04	**0.05**	0.05	**0.06**	0.06	**0.07**	0.06	**0.08**	0.07	**0.09**	0.07	**0.08**
Total filtration (mm) Al	**1.5**		**2.5**		**3.5**		**4.5**		**3.5**		**3.5**	
Cu	**0**		**0**		**0**		**0**		**0.1**		**0.2**	

Table 14.48 Normalized organ doses—Cervical spine ap—8–12 years

Beam projection:			ap									**Table cells:**
Focus-to-detector distance (cm):			150									(+) = normal font = optimal field setting
Tube voltage (kVp):			70									**(−) = bold font = suboptimal field setting**
Total filtrations:			See last table row.									Empty table cells correspond to values <0.01
Organ or tissue	**Organ dose D_T : Entrance dose K_E**							**(mSv * mGy^{-1})**				
	(+)	(−)	(+)	(−)	(+)	(−)	(+)	(−)	(+)	(−)	(+)	(−)
Skull	0.19	**0.51**	0.23	**0.62**	0.26	**0.69**	0.28	**0.75**	0.31	**0.86**	0.35	**0.96**
Brain		**0.06**	0.01	**0.08**	0.01	**0.10**	0.02	**0.11**	0.02	**0.14**	0.02	**0.16**
Oral mucosa	0.26	**0.41**	0.30	**0.48**	0.34	**0.54**	0.36	**0.58**	0.41	**0.66**	0.46	**0.74**
Salivary glands	0.10	**0.32**	0.13	**0.38**	0.15	**0.43**	0.17	**0.47**	0.20	**0.55**	0.23	**0.61**
Cervical spine	0.44	**0.46**	0.56	**0.59**	0.65	**0.69**	0.73	**0.78**	0.88	**0.95**	1.03	**1.12**
Thyroid	0.91	**0.88**	1.03	**1.01**	1.11	**1.09**	1.16	**1.15**	1.26	**1.25**	1.33	**1.33**
Thymus	0.02	**0.08**	0.02	**0.10**	0.03	**0.12**	0.03	**0.13**	0.04	**0.15**	0.05	**0.17**
Thoracic spine	0.02	**0.03**	0.03	**0.05**	0.03	**0.05**	0.03	**0.06**	0.04	**0.08**	0.05	**0.10**
Claviculae	0.39	**1.23**	0.46	**1.45**	0.51	**1.60**	0.54	**1.72**	0.60	**1.91**	0.65	**2.06**
Scapulae		**0.03**		**0.03**	0.01	**0.04**	0.01	**0.05**	0.02	**0.06**	0.02	**0.07**
Ribs	0.04	**0.14**	0.04	**0.16**	0.05	**0.18**	0.05	**0.19**	0.05	**0.21**	0.06	**0.22**
Oesophagus	0.01	**0.02**	0.02	**0.03**	0.02	**0.04**	0.03	**0.04**	0.03	**0.05**	0.04	**0.06**
Trachea	0.49	**0.52**	0.60	**0.63**	0.67	**0.71**	0.73	**0.78**	0.84	**0.91**	0.93	**1.02**
Lung		**0.03**		**0.03**		**0.04**		**0.05**	0.01	**0.06**	0.01	**0.06**
Mamma												**0.01**
Heart								**0.01**		**0.01**		**0.02**
Stomach												
Spleen												
Liver												
Gall bladder												
Pancreas												
Small intestine												
Large intestine												
Sigma/rectum												
Lumbar spine												
Pelvis												
Adrenals												
Kidneys												
Urinary bladder												
Uterus												
Ovaries												
Prostate												
Testes												
Upper arm		**0.02**		**0.02**		**0.03**		**0.03**	0.01	**0.04**	0.02	**0.05**
Elbow												
Forearm												
Upper leg												
Knee												
Lower leg												
Lymph nodes	0.12	**0.14**	0.14	**0.17**	0.16	**0.19**	0.17	**0.21**	0.19	**0.24**	0.21	**0.26**
Skin	0.03	**0.08**	0.03	**0.09**	0.03	**0.09**	0.03	**0.09**	0.03	**0.10**	0.04	**0.11**
Muscle	0.02	**0.03**	0.02	**0.04**	0.02	**0.04**	0.02	**0.05**	0.03	**0.06**	0.03	**0.06**
Skeleton	0.06	**0.14**	0.07	**0.17**	0.08	**0.19**	0.09	**0.21**	0.10	**0.24**	0.11	**0.27**
Red bone marrow	0.01	**0.04**	0.02	**0.04**	0.02	**0.05**	0.02	**0.05**	0.03	**0.06**	0.03	**0.07**
Whole body dose	0.02	**0.05**	0.03	**0.06**	0.03	**0.07**	0.03	**0.07**	0.04	**0.08**	0.04	**0.09**
ED (ICRP 60)	0.05	**0.06**	0.06	**0.07**	0.06	**0.07**	0.06	**0.08**	0.07	**0.08**	0.08	**0.09**
ED (ICRP 103)	0.05	**0.06**	0.06	**0.07**	0.06	**0.08**	0.07	**0.08**	0.07	**0.09**	0.08	**0.10**
Total filtration (mm) Al	**1.5**		**2.5**		**3.5**		**4.5**		**3.5**		**3.5**	
Cu	**0**		**0**		**0**		**0**		**0.1**		**0.2**	

Table 14.49 Normalized organ doses—Cervical spine ap—13–17 years

Beam projection:			ap									**Table cells:**
Focus-to-detector distance (cm):			150									(+) = normal font = optimal field setting
Tube voltage (kVp):			60									**(−) = bold font = suboptimal field setting**
Total filtrations:			See last table row.									Empty table cells correspond to values <0.01
Organ or tissue	**Organ dose D_T : Entrance dose K_E**								**(mSv * mGy^{-1})**			
	(+)	(−)	(+)	(−)	(+)	(−)	(+)	(−)	(+)	(−)	(+)	(−)
Skull	0.23	**0.39**	0.28	**0.49**	0.33	**0.57**	0.36	**0.62**	0.42	**0.73**	0.40	**0.70**
Brain	0.01	**0.04**	0.01	**0.06**	0.02	**0.07**	0.02	**0.08**	0.03	**0.10**	0.02	**0.08**
Oral mucosa	0.22	**0.31**	0.26	**0.37**	0.30	**0.42**	0.32	**0.45**	0.37	**0.53**	0.35	**0.49**
Salivary glands	0.07	**0.24**	0.09	**0.30**	0.10	**0.34**	0.12	**0.37**	0.15	**0.44**	0.13	**0.41**
Cervical spine	0.40	**0.40**	0.52	**0.53**	0.62	**0.63**	0.71	**0.72**	0.88	**0.90**	0.79	**0.80**
Thyroid	0.81	**0.77**	0.93	**0.89**	1.02	**0.98**	1.08	**1.04**	1.18	**1.14**	1.17	**1.13**
Thymus								**0.01**		**0.01**		**0.01**
Thoracic spine			0.01	**0.01**	0.01	**0.02**	0.01	**0.02**	0.02	**0.03**	0.02	**0.02**
Claviculae	0.19	**0.67**	0.23	**0.81**	0.26	**0.91**	0.28	**0.98**	0.32	**1.11**	0.31	**1.10**
Scapulae										**0.01**		
Ribs	0.01	**0.03**	0.01	**0.04**	0.02	**0.04**	0.02	**0.05**	0.02	**0.05**	0.02	**0.05**
Oesophagus			0.01	**0.01**	0.01	**0.02**	0.01	**0.02**	0.02	**0.02**	0.02	**0.02**
Trachea	0.33	**0.32**	0.40	**0.40**	0.46	**0.45**	0.50	**0.50**	0.58	**0.58**	0.55	**0.55**
Lung												
Mamma				**0.01**		**0.02**	0.01	**0.02**	0.01	**0.02**	0.01	**0.02**
Heart												
Stomach												
Spleen												
Liver												
Gall bladder												
Pancreas												
Small intestine												
Large intestine												
Sigma/rectum												
Lumbar spine												
Pelvis												
Adrenals												
Kidneys												
Urinary bladder												
Uterus												
Ovaries												
Prostate												
Testes												
Upper arm										**0.01**		
Elbow												
Forearm												
Upper leg												
Knee												
Lower leg												
Lymph nodes	0.09	**0.10**	0.11	**0.12**	0.12	**0.14**	0.13	**0.15**	0.15	**0.17**	0.14	**0.16**
Skin	0.03	**0.06**	0.03	**0.07**	0.03	**0.07**	0.03	**0.07**	0.03	**0.07**	0.03	**0.07**
Muscle	0.01	**0.02**	0.01	**0.02**	0.02	**0.02**	0.02	**0.03**	0.02	**0.03**	0.02	**0.03**
Skeleton	0.05	**0.08**	0.06	**0.10**	0.07	**0.12**	0.08	**0.13**	0.09	**0.15**	0.09	**0.14**
Red bone marrow	0.01	**0.02**	0.01	**0.03**	0.02	**0.03**	0.02	**0.03**	0.02	**0.04**	0.02	**0.04**
Whole body dose	0.02	**0.03**	0.02	**0.03**	0.02	**0.04**	0.02	**0.04**	0.03	**0.05**	0.03	**0.05**
ED (ICRP 60)	0.04	**0.04**	0.05	**0.05**	0.06	**0.06**	0.06	**0.06**	0.07	**0.07**	0.06	**0.07**
ED (ICRP 103)	0.04	**0.05**	0.05	**0.05**	0.05	**0.06**	0.06	**0.07**	0.07	**0.07**	0.06	**0.07**
Total filtration (mm) **Al**	**1.5**		**2.5**		**3.5**		**4.5**		**3.5**		**3.5**	
Cu	**0**		**0**		**0**		**0**		**0.1**		**0.2**	

Table 14.50 Normalized organ doses—Cervical spine ap—13–17 years

Beam projection:			ap									**Table cells:**
Focus-to-detector distance (cm):			150									(+) = normal font = optimal field setting
Tube voltage (kVp):			70									**(−) = bold font = suboptimal field setting**
Total filtrations:			See last table row.									Empty table cells correspond to values <0.01
Organ or tissue	**Organ dose D_T : Entrance dose K_E**								**(mSv * mGy^{-1})**			
	(+)	(−)	(+)	(−)	(+)	(−)	(+)	(−)	(+)	(−)	(+)	(−)
Skull	0.27	**0.46**	0.33	**0.57**	0.37	**0.65**	0.40	**0.71**	0.46	**0.82**	0.52	**0.92**
Brain	0.02	**0.06**	0.02	**0.07**	0.03	**0.09**	0.03	**0.10**	0.04	**0.13**	0.05	**0.15**
Oral mucosa	0.25	**0.36**	0.30	**0.43**	0.34	**0.48**	0.37	**0.52**	0.42	**0.59**	0.47	**0.67**
Salivary glands	0.09	**0.28**	0.11	**0.35**	0.13	**0.39**	0.15	**0.43**	0.18	**0.50**	0.21	**0.56**
Cervical spine	0.51	**0.52**	0.65	**0.67**	0.77	**0.79**	0.86	**0.89**	1.06	**1.09**	1.24	**1.28**
Thyroid	0.88	**0.84**	1.00	**0.97**	1.09	**1.05**	1.15	**1.11**	1.25	**1.21**	1.33	**1.30**
Thymus				**0.01**		**0.01**	0.01	**0.02**	0.01	**0.02**	0.02	**0.02**
Thoracic spine	0.01	**0.01**	0.01	**0.02**	0.02	**0.02**	0.02	**0.03**	0.03	**0.03**	0.03	**0.04**
Claviculae	0.21	**0.74**	0.25	**0.88**	0.28	**0.98**	0.30	**1.06**	0.34	**1.18**	0.38	**1.28**
Scapulae						**0.01**		**0.01**		**0.02**	0.01	**0.02**
Ribs	0.01	**0.04**	0.02	**0.04**	0.02	**0.05**	0.02	**0.05**	0.02	**0.06**	0.03	**0.06**
Oesophagus	0.01	**0.01**	0.01	**0.02**	0.02	**0.02**	0.02	**0.02**	0.02	**0.03**	0.03	**0.04**
Trachea	0.38	**0.38**	0.46	**0.46**	0.52	**0.52**	0.56	**0.57**	0.65	**0.66**	0.73	**0.74**
Lung										**0.01**		**0.01**
Mamma		**0.01**	0.01	**0.02**	0.01	**0.02**	0.01	**0.02**	0.02	**0.03**	0.02	**0.04**
Heart												
Stomach												
Spleen												
Liver												
Gall bladder												
Pancreas												
Small intestine												
Large intestine												
Sigma/rectum												
Lumbar spine												
Pelvis												
Adrenals												
Kidneys												
Urinary bladder												
Uterus												
Ovaries												
Prostate												
Testes												
Upper arm								**0.01**		**0.01**		**0.02**
Elbow												
Forearm												
Upper leg												
Knee												
Lower leg												
Lymph nodes	0.10	**0.12**	0.12	**0.14**	0.13	**0.15**	0.14	**0.17**	0.16	**0.19**	0.18	**0.21**
Skin	0.03	**0.06**	0.03	**0.07**	0.03	**0.07**	0.03	**0.07**	0.03	**0.08**	0.04	**0.08**
Muscle	0.01	**0.02**	0.02	**0.02**	0.02	**0.03**	0.02	**0.03**	0.02	**0.03**	0.02	**0.04**
Skeleton	0.06	**0.10**	0.07	**0.12**	0.08	**0.13**	0.09	**0.15**	0.11	**0.17**	0.12	**0.19**
Red bone marrow	0.01	**0.02**	0.02	**0.03**	0.02	**0.03**	0.02	**0.04**	0.03	**0.05**	0.03	**0.05**
Whole body dose	0.02	**0.03**	0.02	**0.04**	0.03	**0.04**	0.03	**0.05**	0.03	**0.06**	0.04	**0.06**
ED (ICRP 60)	0.05	**0.05**	0.05	**0.06**	0.06	**0.06**	0.06	**0.07**	0.07	**0.07**	0.08	**0.08**
ED (ICRP 103)	0.05	**0.05**	0.05	**0.06**	0.06	**0.07**	0.06	**0.07**	0.07	**0.08**	0.08	**0.09**
Total filtration (mm) Al	**1.5**		**2.5**		**3.5**		**4.5**		**3.5**		**3.5**	
Cu	**0**		**0**		**0**		**0**		**0.1**		**0.2**	

Table 14.51 Normalized organ doses—Cervical spine ap—>17 years

Beam projection:			ap									**Table cells:**
Focus-to-detector distance (cm):			150				(+) = normal font = optimal field setting					
Tube voltage (kVp):			60				**(−) = bold font = suboptimal field setting**					
Total filtrations:			See last table row.				Empty table cells correspond to values <0.01					
Organ or tissue	**Organ dose D_T : Entrance dose K_E**								**(mSv ∗ mGy^{-1})**			
	(+)	(−)	(+)	(−)	(+)	(−)	(+)	(−)	(+)	(−)	(+)	(−)
Skull	0.23	**0.38**	0.28	**0.48**	0.33	**0.55**	0.36	**0.61**	0.43	**0.72**	0.40	**0.68**
Brain		**0.05**	0.01	**0.06**	0.01	**0.08**	0.02	**0.09**	0.02	**0.11**	0.02	**0.10**
Oral mucosa	0.22	**0.29**	0.27	**0.35**	0.30	**0.40**	0.33	**0.43**	0.38	**0.50**	0.35	**0.47**
Salivary glands	0.06	**0.22**	0.08	**0.28**	0.10	**0.32**	0.11	**0.35**	0.14	**0.41**	0.12	**0.39**
Cervical spine	0.36	**0.35**	0.47	**0.47**	0.57	**0.56**	0.64	**0.64**	0.80	**0.80**	0.71	**0.70**
Thyroid	0.76	**0.72**	0.89	**0.85**	0.98	**0.93**	1.04	**0.99**	1.15	**1.10**	1.14	**1.09**
Thymus												
Thoracic spine			0.01	**0.01**	0.01	**0.01**	0.01	**0.02**	0.02	**0.02**	0.02	**0.02**
Claviculae	0.21	**0.61**	0.25	**0.74**	0.28	**0.83**	0.31	**0.90**	0.35	**1.02**	0.34	**1.01**
Scapulae												
Ribs	0.02	**0.02**	0.02	**0.02**	0.03	**0.02**	0.03	**0.03**	0.03	**0.03**	0.03	**0.03**
Oesophagus			0.01	**0.01**	0.01	**0.01**	0.02	**0.02**	0.02	**0.02**	0.02	**0.02**
Trachea	0.30	**0.29**	0.37	**0.36**	0.42	**0.41**	0.47	**0.46**	0.54	**0.53**	0.52	**0.50**
Lung												
Mamma												
Heart												
Stomach												
Spleen												
Liver												
Gall bladder												
Pancreas												
Small intestine												
Large intestine												
Sigma/rectum												
Lumbar spine												
Pelvis												
Adrenals												
Kidneys												
Urinary bladder												
Uterus												
Ovaries												
Prostate												
Testes												
Upper arm												
Elbow												
Forearm												
Upper leg												
Knee												
Lower leg												
Lymph nodes	0.08	**0.09**	0.10	**0.11**	0.11	**0.13**	0.12	**0.14**	0.14	**0.16**	0.14	**0.15**
Skin	0.03	**0.06**	0.03	**0.06**	0.03	**0.06**	0.03	**0.07**	0.03	**0.07**	0.03	**0.07**
Muscle	0.01	**0.01**	0.01	**0.02**	0.01	**0.02**	0.02	**0.02**	0.02	**0.03**	0.02	**0.02**
Skeleton	0.05	**0.07**	0.06	**0.09**	0.07	**0.10**	0.07	**0.11**	0.09	**0.14**	0.08	**0.13**
Red bone marrow	0.01	**0.02**	0.01	**0.02**	0.02	**0.02**	0.02	**0.03**	0.02	**0.03**	0.02	**0.03**
Whole body dose	0.02	**0.02**	0.02	**0.03**	0.02	**0.03**	0.02	**0.04**	0.03	**0.04**	0.03	**0.04**
ED (ICRP 60)	0.04	**0.04**	0.05	**0.05**	0.05	**0.05**	0.06	**0.06**	0.06	**0.06**	0.06	**0.06**
ED (ICRP 103)	0.04	**0.04**	0.05	**0.05**	0.05	**0.06**	0.06	**0.06**	0.06	**0.07**	0.06	**0.07**
Total filtration (mm) Al	**1.5**		**2.5**		**3.5**		**4.5**		**3.5**		**3.5**	
Cu	**0**		**0**		**0**		**0**		**0.1**		**0.2**	

Table 14.52 Normalized organ doses—Cervical spine ap—>17 years

Beam projection:			ap									**Table cells:**
Focus-to-detector distance (cm):			150									(+) = normal font = optimal field setting
Tube voltage (kVp):			70									**(−) = bold font = suboptimal field setting**
Total filtrations:			See last table row.									Empty table cells correspond to values <0.01
Organ or tissue	**Organ dose D_T : Entrance dose K_E**								**(mSv * mGy^{-1})**			
	(+)	(−)	(+)	(−)	(+)	(−)	(+)	(−)	(+)	(−)	(+)	(−)
Skull	0.27	**0.45**	0.33	**0.56**	0.37	**0.64**	0.41	**0.70**	0.48	**0.82**	0.54	**0.92**
Brain	0.01	**0.06**	0.02	**0.08**	0.02	**0.10**	0.02	**0.11**	0.03	**0.14**	0.04	**0.17**
Oral mucosa	0.26	**0.34**	0.31	**0.41**	0.35	**0.46**	0.37	**0.49**	0.43	**0.57**	0.48	**0.64**
Salivary glands	0.08	**0.26**	0.11	**0.32**	0.13	**0.37**	0.14	**0.40**	0.17	**0.47**	0.21	**0.53**
Cervical spine	0.46	**0.46**	0.60	**0.59**	0.70	**0.70**	0.79	**0.79**	0.97	**0.98**	1.15	**1.16**
Thyroid	0.83	**0.79**	0.96	**0.92**	1.05	**1.00**	1.11	**1.07**	1.22	**1.17**	1.31	**1.26**
Thymus									0.01	**0.01**	0.01	**0.01**
Thoracic spine	0.01	**0.01**	0.02	**0.02**	0.02	**0.02**	0.02	**0.02**	0.03	**0.03**	0.03	**0.04**
Claviculae	0.23	**0.68**	0.28	**0.81**	0.31	**0.90**	0.34	**0.97**	0.38	**1.08**	0.42	**1.17**
Scapulae										**0.01**		**0.02**
Ribs	0.02	**0.02**	0.03	**0.02**	0.03	**0.03**	0.03	**0.03**	0.04	**0.03**	0.04	**0.04**
Oesophagus	0.01	**0.01**	0.02	**0.02**	0.02	**0.02**	0.02	**0.02**	0.03	**0.03**	0.03	**0.03**
Trachea	0.35	**0.35**	0.43	**0.42**	0.49	**0.48**	0.53	**0.52**	0.61	**0.61**	0.69	**0.69**
Lung										**0.01**		**0.01**
Mamma												
Heart												
Stomach												
Spleen												
Liver												
Gall bladder												
Pancreas												
Small intestine												
Large intestine												
Sigma/rectum												
Lumbar spine												
Pelvis												
Adrenals												
Kidneys												
Urinary bladder												
Uterus												
Ovaries												
Prostate												
Testes												
Upper arm												**0.01**
Elbow												
Forearm												
Upper leg												
Knee												
Lower leg												
Lymph nodes	0.10	**0.11**	0.11	**0.13**	0.13	**0.15**	0.14	**0.16**	0.16	**0.18**	0.18	**0.20**
Skin	0.03	**0.06**	0.03	**0.06**	0.03	**0.07**	0.03	**0.07**	0.03	**0.07**	0.04	**0.08**
Muscle	0.01	**0.02**	0.01	**0.02**	0.02	**0.02**	0.02	**0.02**	0.02	**0.03**	0.02	**0.03**
Skeleton	0.05	**0.08**	0.07	**0.10**	0.08	**0.12**	0.09	**0.13**	0.10	**0.16**	0.12	**0.18**
Red bone marrow	0.01	**0.02**	0.02	**0.03**	0.02	**0.03**	0.02	**0.03**	0.03	**0.04**	0.03	**0.05**
Whole body dose	0.02	**0.03**	0.02	**0.03**	0.02	**0.04**	0.03	**0.04**	0.03	**0.05**	0.04	**0.06**
ED (ICRP 60)	0.05	**0.05**	0.05	**0.05**	0.06	**0.06**	0.06	**0.06**	0.07	**0.07**	0.07	**0.08**
ED (ICRP 103)	0.04	**0.05**	0.05	**0.06**	0.06	**0.06**	0.06	**0.07**	0.07	**0.07**	0.07	**0.08**
Total filtration (mm) Al	**1.5**		**2.5**		**3.5**		**4.5**		**3.5**		**3.5**	
Cu	**0**		**0**		**0**		**0**		**0.1**		**0.2**	

Table 14.53 Normalized organ doses—Cervical spine r-lat—8–12 years

Beam projection:			r-lat									**Table cells:**
Focus-to-detector distance (cm):			150						(+) = normal font = optimal field setting			
Tube voltage (kVp):			60						**(−) = bold font = suboptimal field setting**			
Total filtrations:			See last table row.						Empty table cells correspond to values <0.01			
Organ or tissue	**Organ dose D_T : Entrance dose K_E**								**(mSv ∗ mGy^{-1})**			
	(+)	(−)	(+)	(−)	(+)	(−)	(+)	(−)	(+)	(−)	(+)	(−)
Skull	0.12	**0.41**	0.15	**0.51**	0.17	**0.58**	0.19	**0.63**	0.23	**0.73**	0.21	**0.70**
Brain	0.01	**0.08**	0.02	**0.11**	0.02	**0.13**	0.02	**0.15**	0.03	**0.18**	0.03	**0.16**
Oral mucosa	0.11	**0.22**	0.14	**0.27**	0.16	**0.30**	0.17	**0.33**	0.20	**0.39**	0.19	**0.36**
Salivary glands	0.33	**0.36**	0.38	**0.41**	0.41	**0.45**	0.43	**0.48**	0.47	**0.53**	0.45	**0.51**
Cervical spine	0.58	**0.60**	0.75	**0.78**	0.88	**0.92**	0.99	**1.04**	1.19	**1.26**	1.12	**1.17**
Thyroid	0.41	**0.43**	0.49	**0.52**	0.55	**0.59**	0.60	**0.63**	0.68	**0.72**	0.67	**0.70**
Thymus				**0.01**		**0.01**		**0.02**	0.01	**0.02**		**0.02**
Thoracic spine	0.01	**0.03**	0.02	**0.04**	0.02	**0.04**	0.03	**0.05**	0.03	**0.06**	0.03	**0.05**
Claviculae	0.21	**0.40**	0.26	**0.51**	0.30	**0.60**	0.32	**0.67**	0.38	**0.79**	0.37	**0.76**
Scapulae	0.02	**0.10**	0.02	**0.12**	0.03	**0.14**	0.03	**0.15**	0.04	**0.17**	0.04	**0.16**
Ribs	0.03	**0.05**	0.03	**0.06**	0.04	**0.07**	0.04	**0.08**	0.04	**0.09**	0.04	**0.09**
Oesophagus	0.01	**0.01**	0.01	**0.01**	0.02	**0.02**	0.02	**0.02**	0.02	**0.03**	0.02	**0.02**
Trachea	0.17	**0.26**	0.21	**0.33**	0.24	**0.39**	0.27	**0.43**	0.32	**0.52**	0.30	**0.48**
Lung						**0.01**	0.01	**0.01**	0.01	**0.02**	0.01	**0.01**
Mamma												
Heart												
Stomach												
Spleen												
Liver												
Gall bladder												
Pancreas												
Small intestine												
Large intestine												
Sigma/rectum												
Lumbar spine												
Pelvis												
Adrenals												
Kidneys												
Urinary bladder												
Uterus												
Ovaries												
Prostate												
Testes												
Upper arm		**0.19**		**0.23**		**0.26**		**0.28**		**0.31**		**0.31**
Elbow												
Forearm												
Upper leg												
Knee												
Lower leg												
Lymph nodes	0.07	**0.09**	0.08	**0.11**	0.09	**0.12**	0.10	**0.13**	0.12	**0.15**	0.11	**0.14**
Skin	0.03	**0.08**	0.03	**0.08**	0.04	**0.09**	0.04	**0.09**	0.04	**0.09**	0.04	**0.09**
Muscle	0.02	**0.02**	0.02	**0.03**	0.02	**0.03**	0.02	**0.03**	0.03	**0.04**	0.02	**0.04**
Skeleton	0.05	**0.12**	0.06	**0.15**	0.07	**0.17**	0.07	**0.19**	0.09	**0.22**	0.08	**0.21**
Red bone marrow	0.01	**0.03**	0.01	**0.03**	0.02	**0.04**	0.02	**0.04**	0.02	**0.05**	0.02	**0.05**
Whole body dose	0.02	**0.04**	0.02	**0.05**	0.03	**0.05**	0.03	**0.06**	0.03	**0.07**	0.03	**0.07**
ED (ICRP 60)	0.02	**0.03**	0.03	**0.03**	0.03	**0.04**	0.04	**0.04**	0.04	**0.05**	0.04	**0.05**
ED (ICRP 103)	0.03	**0.03**	0.03	**0.04**	0.04	**0.05**	0.04	**0.05**	0.04	**0.06**	0.04	**0.06**
Total filtration (mm) Al	**1.5**		**2.5**		**3.5**		**4.5**		**3.5**		**3.5**	
Cu	**0**		**0**		**0**		**0**		**0.1**		**0.2**	

Table 14.54 Normalized organ doses—Cervical spine r-lat—8–12 years

Beam projection:		r-lat										**Table cells:**
Focus-to-detector distance (cm):		150						(+) = normal font = optimal field setting				
Tube voltage (kVp):		70						**(−) = bold font = suboptimal field setting**				
Total filtrations:		See last table row.						Empty table cells correspond to values <0.01				
Organ or tissue	**Organ dose D_T : Entrance dose K_E**							**(mSv * mGy⁻¹)**				
	(+)	(−)	(+)	(−)	(+)	(−)	(+)	(−)	(+)	(−)	(+)	(−)
Skull	0.14	**0.48**	0.17	**0.58**	0.20	**0.65**	0.22	**0.70**	0.25	**0.81**	0.29	**0.90**
Brain	0.02	**0.11**	0.02	**0.14**	0.03	**0.16**	0.03	**0.18**	0.04	**0.22**	0.05	**0.26**
Oral mucosa	0.13	**0.25**	0.16	**0.31**	0.18	**0.35**	0.20	**0.38**	0.23	**0.44**	0.26	**0.50**
Salivary glands	0.36	**0.40**	0.41	**0.45**	0.44	**0.49**	0.46	**0.52**	0.50	**0.57**	0.54	**0.62**
Cervical spine	0.71	**0.74**	0.89	**0.94**	1.03	**1.09**	1.15	**1.21**	1.36	**1.45**	1.56	**1.67**
Thyroid	0.46	**0.49**	0.55	**0.58**	0.61	**0.65**	0.66	**0.70**	0.74	**0.79**	0.81	**0.87**
Thymus		**0.01**		**0.02**	0.01	**0.02**	0.01	**0.02**	0.02	**0.03**	0.02	**0.04**
Thoracic spine	0.02	**0.04**	0.03	**0.05**	0.03	**0.05**	0.03	**0.06**	0.04	**0.08**	0.05	**0.09**
Claviculae	0.24	**0.48**	0.29	**0.60**	0.33	**0.68**	0.36	**0.75**	0.41	**0.88**	0.46	**0.99**
Scapulae	0.02	**0.11**	0.03	**0.13**	0.03	**0.15**	0.04	**0.16**	0.05	**0.18**	0.05	**0.20**
Ribs	0.03	**0.06**	0.04	**0.07**	0.04	**0.08**	0.04	**0.09**	0.05	**0.10**	0.05	**0.12**
Oesophagus	0.01	**0.01**	0.02	**0.02**	0.02	**0.02**	0.02	**0.03**	0.03	**0.03**	0.03	**0.04**
Trachea	0.20	**0.32**	0.25	**0.40**	0.28	**0.46**	0.31	**0.51**	0.37	**0.61**	0.42	**0.70**
Lung				**0.01**	0.01	**0.02**	0.01	**0.02**	0.02	**0.02**	0.02	**0.03**
Mamma												
Heart												
Stomach												
Spleen												
Liver												
Gall bladder												
Pancreas												
Small intestine												
Large intestine												
Sigma/rectum												
Lumbar spine												
Pelvis												
Adrenals												
Kidneys												
Urinary bladder												
Uterus												
Ovaries												
Prostate												
Testes												
Upper arm		**0.22**		**0.25**		**0.28**		**0.30**		**0.33**		**0.36**
Elbow												
Forearm												
Upper leg												
Knee												
Lower leg												
Lymph nodes	0.08	**0.10**	0.10	**0.12**	0.11	**0.14**	0.11	**0.15**	0.13	**0.17**	0.14	**0.19**
Skin	0.03	**0.08**	0.04	**0.09**	0.04	**0.09**	0.04	**0.09**	0.04	**0.10**	0.04	**0.10**
Muscle	0.02	**0.03**	0.02	**0.03**	0.02	**0.04**	0.03	**0.04**	0.03	**0.04**	0.03	**0.05**
Skeleton	0.05	**0.14**	0.07	**0.17**	0.08	**0.19**	0.09	**0.21**	0.10	**0.25**	0.11	**0.28**
Red bone marrow	0.01	**0.03**	0.02	**0.04**	0.02	**0.05**	0.02	**0.05**	0.02	**0.06**	0.03	**0.07**
Whole body dose	0.02	**0.05**	0.03	**0.06**	0.03	**0.06**	0.03	**0.07**	0.04	**0.08**	0.04	**0.09**
ED (ICRP 60)	0.03	**0.03**	0.03	**0.04**	0.04	**0.04**	0.04	**0.05**	0.05	**0.06**	0.05	**0.06**
ED (ICRP 103)	0.03	**0.04**	0.04	**0.05**	0.04	**0.05**	0.04	**0.06**	0.05	**0.07**	0.05	**0.07**
Total filtration (mm) Al	**1.5**		**2.5**		**3.5**		**4.5**		**3.5**		**3.5**	
Cu	**0**		**0**		**0**		**0**		**0.1**		**0.2**	

Table 14.55 Normalized organ doses—Cervical spine r-lat—13–17 years

Beam projection:			r-lat									**Table cells:**
Focus-to-detector distance (cm):			150						(+) = normal font = optimal field setting			
Tube voltage (kVp):			60						**(−) = bold font = suboptimal field setting**			
Total filtrations:			See last table row.						Empty table cells correspond to values <0.01			
Organ or tissue	**Organ dose D_T : Entrance dose K_E**								**(mSv * mGy⁻¹)**			
	(+)	(−)	(+)	(−)	(+)	(−)	(+)	(−)	(+)	(−)	(+)	(−)
Skull	0.15	**0.36**	0.19	**0.45**	0.22	**0.52**	0.24	**0.57**	0.29	**0.68**	0.27	**0.64**
Brain	0.02	**0.06**	0.03	**0.09**	0.03	**0.10**	0.04	**0.12**	0.05	**0.15**	0.04	**0.13**
Oral mucosa	0.11	**0.20**	0.14	**0.25**	0.16	**0.28**	0.17	**0.31**	0.21	**0.37**	0.19	**0.34**
Salivary glands	0.34	**0.36**	0.39	**0.41**	0.42	**0.45**	0.44	**0.48**	0.49	**0.53**	0.47	**0.51**
Cervical spine	0.64	**0.67**	0.83	**0.87**	0.99	**1.03**	1.11	**1.17**	1.36	**1.43**	1.25	**1.31**
Thyroid	0.37	**0.39**	0.45	**0.48**	0.51	**0.55**	0.56	**0.60**	0.64	**0.69**	0.62	**0.67**
Thymus												
Thoracic spine		**0.02**	0.01	**0.02**	0.01	**0.03**	0.02	**0.03**	0.02	**0.04**	0.02	**0.03**
Claviculae	0.16	**0.30**	0.20	**0.38**	0.23	**0.45**	0.25	**0.50**	0.29	**0.61**	0.28	**0.57**
Scapulae		**0.10**		**0.13**	0.01	**0.14**	0.01	**0.16**	0.02	**0.18**	0.01	**0.17**
Ribs		**0.03**		**0.04**		**0.05**	0.01	**0.06**	0.01	**0.07**	0.01	**0.06**
Oesophagus			0.01	**0.01**	0.01	**0.01**	0.01	**0.02**	0.02	**0.02**	0.02	**0.02**
Trachea	0.15	**0.20**	0.20	**0.25**	0.23	**0.29**	0.25	**0.33**	0.30	**0.40**	0.28	**0.37**
Lung										**0.01**		
Mamma				**0.01**		**0.01**		**0.02**		**0.02**		**0.02**
Heart												
Stomach												
Spleen												
Liver												
Gall bladder												
Pancreas												
Small intestine												
Large intestine												
Sigma/rectum												
Lumbar spine												
Pelvis												
Adrenals												
Kidneys												
Urinary bladder												
Uterus												
Ovaries												
Prostate												
Testes												
Upper arm		**0.20**		**0.24**		**0.27**		**0.29**		**0.33**		**0.32**
Elbow												
Forearm												
Upper leg												
Knee												
Lower leg												
Lymph nodes	0.07	**0.08**	0.08	**0.09**	0.09	**0.10**	0.10	**0.11**	0.11	**0.13**	0.11	**0.13**
Skin	0.03	**0.07**	0.03	**0.07**	0.04	**0.07**	0.04	**0.08**	0.04	**0.08**	0.04	**0.08**
Muscle	0.01	**0.02**	0.02	**0.02**	0.02	**0.03**	0.02	**0.03**	0.02	**0.03**	0.02	**0.03**
Skeleton	0.04	**0.09**	0.05	**0.12**	0.06	**0.14**	0.07	**0.15**	0.09	**0.18**	0.08	**0.17**
Red bone marrow		**0.02**	0.01	**0.03**	0.01	**0.03**	0.02	**0.04**	0.02	**0.04**	0.02	**0.04**
Whole body dose	0.02	**0.03**	0.02	**0.04**	0.02	**0.04**	0.03	**0.05**	0.03	**0.06**	0.03	**0.05**
ED (ICRP 60)	0.02	**0.03**	0.03	**0.03**	0.03	**0.04**	0.03	**0.04**	0.04	**0.05**	0.04	**0.04**
ED (ICRP 103)	0.02	**0.03**	0.03	**0.04**	0.03	**0.04**	0.04	**0.05**	0.04	**0.05**	0.04	**0.05**
Total filtration (mm) Al	**1.5**		**2.5**		**3.5**		**4.5**		**3.5**		**3.5**	
Cu	**0**		**0**		**0**		**0**		**0.1**		**0.2**	

Table 14.56 Normalized organ doses—Cervical spine r-lat—13–17 years

Beam projection:			r-lat									**Table cells:**
Focus-to-detector distance (cm):			150									(+) = normal font = optimal field setting
Tube voltage (kVp):			70									**(−) = bold font = suboptimal field setting**
Total filtrations:			See last table row.									Empty table cells correspond to values <0.01
Organ or tissue	**Organ dose D_T : Entrance dose K_E**								**(mSv ∗ mGy^{-1})**			
	(+)	(−)	(+)	(−)	(+)	(−)	(+)	(−)	(+)	(−)	(+)	(−)
Skull	0.18	**0.43**	0.22	**0.52**	0.25	**0.59**	0.28	**0.65**	0.32	**0.76**	0.37	**0.85**
Brain	0.03	**0.09**	0.04	**0.11**	0.04	**0.13**	0.05	**0.15**	0.06	**0.18**	0.08	**0.21**
Oral mucosa	0.13	**0.24**	0.16	**0.29**	0.19	**0.33**	0.20	**0.36**	0.24	**0.43**	0.28	**0.48**
Salivary glands	0.37	**0.39**	0.42	**0.45**	0.45	**0.49**	0.48	**0.52**	0.53	**0.58**	0.57	**0.63**
Cervical spine	0.80	**0.84**	1.01	**1.07**	1.18	**1.24**	1.32	**1.39**	1.59	**1.68**	1.84	**1.95**
Thyroid	0.42	**0.45**	0.51	**0.55**	0.57	**0.61**	0.62	**0.66**	0.70	**0.76**	0.77	**0.84**
Thymus								**0.01**		**0.01**		**0.02**
Thoracic spine	0.01	**0.02**	0.02	**0.03**	0.02	**0.04**	0.02	**0.04**	0.03	**0.05**	0.03	**0.07**
Claviculae	0.18	**0.36**	0.23	**0.45**	0.26	**0.53**	0.28	**0.58**	0.33	**0.70**	0.36	**0.80**
Scapulae		**0.12**	0.01	**0.14**	0.02	**0.16**	0.02	**0.17**	0.02	**0.20**	0.03	**0.22**
Ribs		**0.04**	0.01	**0.05**	0.01	**0.06**	0.01	**0.07**	0.02	**0.08**	0.02	**0.09**
Oesophagus	0.01	**0.01**	0.01	**0.02**	0.02	**0.02**	0.02	**0.02**	0.02	**0.03**	0.03	**0.03**
Trachea	0.19	**0.24**	0.23	**0.30**	0.27	**0.35**	0.29	**0.39**	0.35	**0.46**	0.39	**0.53**
Lung						**0.01**		**0.01**		**0.02**		**0.02**
Mamma		**0.01**		**0.02**		**0.02**		**0.02**	0.01	**0.03**	0.01	**0.04**
Heart												
Stomach												
Spleen												
Liver												
Gall bladder												
Pancreas												
Small intestine												
Large intestine												
Sigma/rectum												
Lumbar spine												
Pelvis												
Adrenals												
Kidneys												
Urinary bladder												
Uterus												
Ovaries												
Prostate												
Testes												
Upper arm		**0.22**		**0.26**		**0.29**		**0.31**		**0.35**		**0.38**
Elbow												
Forearm												
Upper leg												
Knee												
Lower leg												
Lymph nodes	0.08	**0.09**	0.09	**0.11**	0.10	**0.12**	0.11	**0.13**	0.13	**0.15**	0.14	**0.17**
Skin	0.03	**0.07**	0.04	**0.07**	0.04	**0.08**	0.04	**0.08**	0.04	**0.09**	0.05	**0.09**
Muscle	0.02	**0.02**	0.02	**0.03**	0.02	**0.03**	0.02	**0.03**	0.02	**0.04**	0.03	**0.04**
Skeleton	0.05	**0.11**	0.06	**0.14**	0.07	**0.16**	0.08	**0.17**	0.10	**0.20**	0.11	**0.23**
Red bone marrow	0.01	**0.03**	0.02	**0.03**	0.02	**0.04**	0.02	**0.04**	0.02	**0.05**	0.03	**0.06**
Whole body dose	0.02	**0.04**	0.02	**0.04**	0.03	**0.05**	0.03	**0.06**	0.04	**0.06**	0.04	**0.07**
ED (ICRP 60)	0.02	**0.03**	0.03	**0.04**	0.03	**0.04**	0.04	**0.05**	0.04	**0.05**	0.05	**0.06**
ED (ICRP 103)	0.03	**0.04**	0.03	**0.04**	0.04	**0.05**	0.04	**0.05**	0.05	**0.06**	0.05	**0.07**
Total filtration (mm) **Al**	**1.5**		**2.5**		**3.5**		**4.5**		**3.5**		**3.5**	
Cu	**0**		**0**		**0**		**0**		**0.1**		**0.2**	

Table 14.57 Normalized organ doses—Cervical spine r-lat—>17 years

Beam projection:			r-lat									**Table cells:**
Focus-to-detector distance (cm):			150				(+) = normal font = optimal field setting					
Tube voltage (kVp):			60				**(−) = bold font = suboptimal field setting**					
Total filtrations:			See last table row.				Empty table cells correspond to values <0.01					
Organ or tissue	**Organ dose D_T : Entrance dose K_E**						**(mSv * mGy^{-1})**					
	(+)	(−)	(+)	(−)	(+)	(−)	(+)	(−)	(+)	(−)	(+)	(−)
Skull	0.12	**0.37**	0.16	**0.46**	0.18	**0.54**	0.20	**0.59**	0.24	**0.70**	0.23	**0.66**
Brain	0.01	**0.07**	0.01	**0.10**	0.02	**0.12**	0.02	**0.13**	0.03	**0.17**	0.02	**0.15**
Oral mucosa	0.11	**0.20**	0.14	**0.25**	0.16	**0.29**	0.18	**0.31**	0.21	**0.37**	0.19	**0.34**
Salivary glands	0.32	**0.34**	0.37	**0.39**	0.41	**0.43**	0.43	**0.46**	0.48	**0.51**	0.46	**0.49**
Cervical spine	0.62	**0.65**	0.81	**0.84**	0.96	**1.00**	1.09	**1.13**	1.33	**1.39**	1.22	**1.26**
Thyroid	0.36	**0.38**	0.45	**0.47**	0.51	**0.53**	0.56	**0.58**	0.64	**0.67**	0.62	**0.65**
Thymus												
Thoracic spine		**0.01**	0.01	**0.02**	0.01	**0.02**	0.02	**0.03**	0.02	**0.03**	0.02	**0.03**
Claviculae	0.16	**0.25**	0.20	**0.32**	0.24	**0.38**	0.26	**0.43**	0.31	**0.53**	0.30	**0.49**
Scapulae	0.02	**0.07**	0.03	**0.09**	0.03	**0.10**	0.03	**0.11**	0.04	**0.13**	0.04	**0.13**
Ribs	0.03	**0.03**	0.04	**0.04**	0.04	**0.04**	0.04	**0.05**	0.05	**0.06**	0.05	**0.06**
Oesophagus				**0.01**	0.01	**0.01**	0.01	**0.01**	0.02	**0.02**	0.01	**0.02**
Trachea	0.16	**0.19**	0.21	**0.24**	0.24	**0.28**	0.27	**0.32**	0.32	**0.38**	0.30	**0.35**
Lung												
Mamma												
Heart												
Stomach												
Spleen												
Liver												
Gall bladder												
Pancreas												
Small intestine												
Large intestine												
Sigma/rectum												
Lumbar spine												
Pelvis												
Adrenals												
Kidneys												
Urinary bladder												
Uterus												
Ovaries												
Prostate												
Testes												
Upper arm		**0.16**		**0.19**		**0.22**		**0.24**		**0.27**		**0.26**
Elbow												
Forearm												
Upper leg												
Knee												
Lower leg												
Lymph nodes	0.07	**0.07**	0.08	**0.09**	0.09	**0.10**	0.10	**0.11**	0.12	**0.13**	0.11	**0.12**
Skin	0.03	**0.07**	0.03	**0.08**	0.04	**0.08**	0.04	**0.08**	0.04	**0.09**	0.04	**0.09**
Muscle	0.01	**0.02**	0.02	**0.02**	0.02	**0.02**	0.02	**0.02**	0.02	**0.03**	0.02	**0.03**
Skeleton	0.04	**0.09**	0.05	**0.11**	0.06	**0.13**	0.06	**0.14**	0.08	**0.17**	0.07	**0.16**
Red bone marrow	0.01	**0.02**	0.01	**0.02**	0.02	**0.03**	0.02	**0.03**	0.02	**0.04**	0.02	**0.04**
Whole body dose	0.02	**0.03**	0.02	**0.04**	0.02	**0.04**	0.03	**0.04**	0.03	**0.05**	0.03	**0.05**
ED (ICRP 60)	0.02	**0.02**	0.03	**0.03**	0.03	**0.03**	0.03	**0.04**	0.04	**0.04**	0.04	**0.04**
ED (ICRP 103)	0.02	**0.03**	0.03	**0.03**	0.03	**0.04**	0.04	**0.04**	0.04	**0.05**	0.04	**0.05**
Total filtration (mm) Al	**1.5**		**2.5**		**3.5**		**4.5**		**3.5**		**3.5**	
Cu	**0**		**0**		**0**		**0**		**0.1**		**0.2**	

Table 14.58 Normalized organ doses—Cervical spine r-lat—>17 years

Beam projection:			r-lat									**Table cells:**
Focus-to-detector distance (cm):			150						(+) = normal font = optimal field setting			
Tube voltage (kVp):			70						**(−) = bold font = suboptimal field setting**			
Total filtrations:			See last table row.				Empty table cells correspond to values <0.01					
Organ or tissue	**Organ dose D_T : Entrance dose K_E**							**(mSv * mGy^{-1})**				
	(+)	(−)	(+)	(−)	(+)	(−)	(+)	(−)	(+)	(−)	(+)	(−)
Skull	0.15	**0.44**	0.19	**0.54**	0.21	**0.62**	0.24	**0.68**	0.28	**0.80**	0.32	**0.90**
Brain	0.01	**0.10**	0.02	**0.13**	0.02	**0.15**	0.03	**0.17**	0.03	**0.21**	0.04	**0.25**
Oral mucosa	0.13	**0.24**	0.17	**0.30**	0.19	**0.34**	0.21	**0.37**	0.25	**0.43**	0.28	**0.49**
Salivary glands	0.36	**0.38**	0.41	**0.43**	0.44	**0.47**	0.47	**0.50**	0.52	**0.56**	0.56	**0.61**
Cervical spine	0.78	**0.81**	0.99	**1.03**	1.16	**1.21**	1.30	**1.35**	1.56	**1.64**	1.81	**1.90**
Thyroid	0.42	**0.44**	0.51	**0.53**	0.57	**0.60**	0.62	**0.65**	0.71	**0.74**	0.78	**0.82**
Thymus												**0.01**
Thoracic spine	0.01	**0.02**	0.02	**0.03**	0.02	**0.03**	0.02	**0.03**	0.03	**0.04**	0.03	**0.05**
Claviculae	0.19	**0.31**	0.24	**0.39**	0.27	**0.45**	0.30	**0.51**	0.35	**0.61**	0.40	**0.70**
Scapulae	0.03	**0.08**	0.03	**0.10**	0.04	**0.11**	0.04	**0.13**	0.05	**0.14**	0.06	**0.16**
Ribs	0.03	**0.04**	0.04	**0.05**	0.04	**0.05**	0.05	**0.06**	0.05	**0.07**	0.06	**0.08**
Oesophagus	0.01	**0.01**	0.01	**0.01**	0.02	**0.02**	0.02	**0.02**	0.02	**0.02**	0.03	**0.03**
Trachea	0.20	**0.23**	0.25	**0.29**	0.29	**0.34**	0.32	**0.38**	0.38	**0.45**	0.43	**0.52**
Lung							0.01	**0.01**	0.01	**0.01**	0.02	**0.02**
Mamma												
Heart												
Stomach												
Spleen												
Liver												
Gall bladder												
Pancreas												
Small intestine												
Large intestine												
Sigma/rectum												
Lumbar spine												
Pelvis												
Adrenals												
Kidneys												
Urinary bladder												
Uterus												
Ovaries												
Prostate												
Testes												
Upper arm		**0.18**		**0.21**		**0.24**		**0.26**		**0.29**		**0.32**
Elbow												
Forearm												
Upper leg												
Knee												
Lower leg												
Lymph nodes	0.08	**0.08**	0.09	**0.10**	0.10	**0.12**	0.11	**0.13**	0.13	**0.15**	0.14	**0.16**
Skin	0.03	**0.07**	0.04	**0.08**	0.04	**0.08**	0.04	**0.09**	0.04	**0.09**	0.05	**0.10**
Muscle	0.01	**0.02**	0.02	**0.02**	0.02	**0.03**	0.02	**0.03**	0.02	**0.03**	0.03	**0.04**
Skeleton	0.05	**0.10**	0.06	**0.13**	0.07	**0.15**	0.08	**0.16**	0.09	**0.19**	0.10	**0.22**
Red bone marrow	0.01	**0.02**	0.02	**0.03**	0.02	**0.04**	0.02	**0.04**	0.03	**0.05**	0.03	**0.06**
Whole body dose	0.02	**0.03**	0.02	**0.04**	0.03	**0.05**	0.03	**0.05**	0.03	**0.06**	0.04	**0.07**
ED (ICRP 60)	0.02	**0.03**	0.03	**0.03**	0.03	**0.04**	0.04	**0.04**	0.04	**0.05**	0.05	**0.06**
ED (ICRP 103)	0.03	**0.03**	0.03	**0.04**	0.04	**0.05**	0.04	**0.05**	0.05	**0.06**	0.05	**0.06**
Total filtration (mm) Al	**1.5**		**2.5**		**3.5**		**4.5**		**3.5**		**3.5**	
Cu	**0**		**0**		**0**		**0**		**0.1**		**0.2**	

References

1. Bontrager KL. Textbook of radiographic positioning and related anatomy. St. Louis: Mosby Year Book; 1993.
2. Commission of the European Communities. European guidelines on quality criteria for diagnostic radiographic images in paediatrics. EUR 16261, SBN 92-827-7843-6; 1996
3. Drexler G, Panzer W, Stieve FE. Die Bestimmung von Organdosen in der Röntgendiagnostik. Berlin: Hoffmann Verlag; 1993.
4. Ebel K-D, Willich E. Die Röntgenuntersuchung im Kindesalter – Technik und Indikation. Berlin: Springer; 1968.
5. Hart D, Jones DJ, Wall BF. Estimation of effective dose in diagnostic radiology from entrance surface dose and dose-area-product measurements. National Radiological Protection Board NRPB-R262; 1994
6. Hart D, Jones DJ, Wall BF. Coefficients for estimating effective doses from pediatric x-ray examinations. National Radiological Protection Board NRPB-R279; 1996
7. Seidenbusch MC, Schneider K. Conversion coefficients for determining organ doses in paediatric spine radiography. Pediatr Radiol. 2014;44: 434–56.

15 Thoracic Spine

15.1 Introduction

The type of thoracic spine radiography is determined by the clinical indication. Table 15.1 shows indications and Table 15.2 contraindications for thoracic spine radiography. Image criteria as required for radiological findings in thoracic spine radiographs are listed in Table 15.3. In order to fulfil these image criteria, the field settings listed in Tables 15.4 and 15.5 should be achieved. These field settings go along with the patient preparation as listed in Table 15.6. Exposure parameter combinations for various age groups and patient diameters are listed in Tables 15.7, 15.8, 15.9, 15.10 and 15.11. The age groups in Tables 15.8 and 15.10 refer to the standard age groups of the mathematical MIRD phantom family which has been used to calculate the normalized organ doses in this book. Tables 15.9 and 15.11 show radiation qualities (i.e. various combinations of tube voltage and total filtration) which may be applied in clinical practice. Please note that not all of these radiation qualities are recommended by the authors. The standard projections of thoracic spine radiographies of the patient who is correctly positioned are depicted in Figs. 15.1, 15.2 and 15.3. Also the optimal field (marked in green), the suboptimal field (marked in red) and the correct shielding are shown. Tables 15.13, 15.14, 15.15, 15.16, 15.17 and 15.18 outline the virtual field settings on the skin surface of mathematical MIRD phantoms of

Table 15.1 Indications for thoracic spine radiographs

Trauma
Juvenile rheumatoid arthritis
Localized back pain
Discitis/spondylitis
Soft tissue tumours

Table 15.2 Contraindications for thoracic spine radiographs

Pregnancy

Table 15.3 Image criteria for thoracic spine radiographs

Thoracic spine ap
Symmetrical reproduction of the thoracic spine
Visually sharp reproduction of the vertebral bodies and the articular processes consistent with age
Visualization of the spinous and transverse processes consistent with age
Visualization of the intervertebral spaces in the centre of the beam area consistent with age
Visually sharp reproduction of the cortex and trabecular structures of the vertebral bodies consistent with age
Reproduction of the adjacent soft tissues
Thoracic spine lat
Full superimposition of the margins of the vertebral bodies
Reproduction of the pedicles and the vertebral bodies
Visualization of the intervertebral spaces in the centre of the beam area consistent with age
Visualization of the posterior articular processes
Visualization of the spinous processes consistent with age
Visually sharp reproduction of the cortex and trabecular structures of the vertebral bodies consistent with age
Reproduction of the adjacent soft tissues

M. Seidenbusch et al., *Imaging Practice and Radiation Protection in Pediatric Radiology*,
https://doi.org/10.1007/978-3-030-18504-6_15

Table 15.4 Patient positions and beam projections for thoracic spine radiographs

Patient position	Beam projections
Standing, sitting	ap + lat
Supine	ap + lat[a]
Decubitus left	lat
Annotation: [a]Lateral projection in emergency cases only	

Table 15.5 Optimal field settings for thoracic spine radiographs

Thoracic spine ap	
Upper field edge:	Jugular fossa
Field centring:	Midline at the level of the mamillae
Lower field edge:	Midpoint between the processus xiphoideus and the navel
Lateral field edges:	Midpoints between parasternal line and medioclavicular line
Thoracic spine lat	
Upper field edge:	6th cervical vertebra
Field centring:	Posterior axillary line at the level of the mamillae
Lower field edge:	At the 12th rib
Anterior field edge:	Anterior axillary line
Posterior field edge:	Close to the skin surface of the back

various age groups and the resulting virtual radiographies with and without bone structures, respectively. Normalized organ doses resulting from these virtual field settings have been calculated for all age groups and have been listed in Tables 15.19, 15.20, 15.21, 15.22, 15.23, 15.24, 15.25, 15.26, 15.27, 15.28, 15.29, 15.30, 15.31, 15.32, 15.33, 15.34, 15.35, 15.36, 15.37, 15.38, 15.39, 15.40, 15.41, 15.42, 15.43, 15.44, 15.45, 15.46, 15.47, 15.48, 15.49, 15.50, 15.51, 15.52, 15.53, 15.54, 15.55, 15.56, 15.57, 15.58, 15.59, 15.60, 15.61, 15.62, 15.63, 15.64, 15.65, 15.66, 15.67, 15.68, 15.69 and 15.70. Table cells with normal font refer to optimal field settings and table cells with bold font refer to suboptimal field settings. Please note that normalized organ doses have been provided also for exposure parameter combinations, which are not recommended by the authors.

Table 15.6 Patient preparation

General	Remove all clothes, external metallic devices, especially piercings, and jewellery from the neck and the chest of the patient
	Immobilize infants and uncooperative patients in the supine position, e.g. with sandbags or restrain devices. Immobilize infants and uncooperative patients either in the right or left lateral decubitus position for the lateral view. Alternatively, staff members could hold the patient (in this case also take care of the radiation protection of the staff members)
	Cooperative patients >8 years of age can be positioned upright at the vertical stand
	Place eventually small X-ray positive markers at palpable tumours or on pain points
Positioning	The arms should be extended upwards
	Rotation of the thoracic spine in the decubitus position can be avoided, if the patient's head is positioned on a large rubber foam pad between the rib cage and the upper pelvic rim to relieve discomfort and bowing of the thoraco-lumbar segment
Attention	On the Bucky table the patient should always be positioned on a foam mat of 10 to 20 cm thickness. So, the lower part of the body will not be cut if lateral projection X-rays (cross table views) in prone or in decubitus position are performed
Shielding	Lead masks should be placed at the upper and lower field edges if supine ap and lateral decubitus projections are used. Lead aprons should be used for upright views. In males, use testicular capsules

15.2 Indications and Contraindications

In recent years, thoracic spine radiographs have been widely replaced by MR examinations. Indications and contraindications for thoracic spine radiographs are given in Tables 15.1 and 15.2.

15.3 Image Criteria

The image criteria required for thoracic spine radiographs are listed in Table 15.3.

See also [2].

15.4 Field Settings

In Table 15.4, patient positions and beam projections used for thoracic spine radiographs are listed, and in Table 15.5, the optimal field settings are listed which are required in order to fulfil the image criteria mentioned in Table 15.3.

Table 15.7 Radiographic procedure

X-ray source	Use automatic exposure control (AEC) in patients >1 year of age
	Focal spot size between 0.6 and 1.3 dependent on the age of the patient
	Check the use of additional filtration
	Check the focus-to-detector distance
Field setting	Check the field size
	Check the field centring
Image detector system	Check the use of antiscatter grids
Patient	In sagittal view, the exposure should be performed during breath-holding; in lateral view, the exposure should be performed during slight breathing

15.5 Patient Preparation and Radiographic Procedure

See Tables 15.6, 15.7 and [1, 4].

15.6 Exposure Parameters

In Tables 15.8, 15.9, 15.10 and 15.11, exposure parameter combinations used for thoracic spine radiographs in various age groups are listed. Dark grey shaded table cells refer to exposure parameter combinations which are not recommended for paediatric thoracic spine radiographs and for which no normalized organ doses have been provided. Light grey and light blue shaded table cells refer to exposure parameter combinations for which normalized organ doses have been provided with light grey shaded table cells referring to exposure parameter combinations being not recommended and light blue shaded table cells referring to exposure parameter combinations following the European Guidelines [2]. Note that due to the fact that the radiation dose is an exponential function of the patient diameter, it may make sense to use the patient diameters instead of the patient's age groups for determining the adequate exposure parameters from Tables 15.8 and 15.10. In this case, for determining the corre-

Table 15.8 Radiographic techniques of Bucky table thoracic spine radiographs

Age group (years)	Beam direction	Patient position	Beam Projection	Typical patient diameter (cm)		Optimal field size (cm x cm)	Focus-to-detector distance (cm)	Voltage (kVp)						Antiscatter grid
				With arms	No arms			50	60	70	80	90	100	
0	Vertical	Supine	ap	9.80		3 × 10	100[1]							
		Decubitus	lat	12.6	10.9	5 × 11								
	Horizontal	Supine												
1–2	Vertical	Supine	ap	13.0		4 × 14	115							
		Decubitus	lat	17.1	15.1	7 × 16								
	Horizontal	Supine												
3–7	Vertical	Supine	ap	15.0		6 × 20								
		Decubitus	lat	22.6	19.6	9 × 22								
	Horizontal	Supine												
8–12	Vertical	Supine	ap	16.8		8 × 25								
	Horizontal	Sitting												
	Vertical	Decubitus	lat	27.3	23.8	12 × 26								
13–17	Vertical	Supine	ap	19.6		10 × 31								
	Horizontal	Sitting												
	Vertical	Decubitus	lat	34.4	29.6	14 × 33								
18	Vertical	Supine	ap	20.0		10 × 34								
	Horizontal	Sitting												
	Vertical	Decubitus	lat	38.9	34.4	15 × 33								

Annotations:

1	The patient is lying on the table top directly on the detector
(dark grey)	Parameter combination not recommended, normalized organ doses not provided
(light grey)	Parameter combination not recommended, but normalized organ doses provided
(light blue)	Parameter combination recommended, normalized organ doses provided

Table 15.9 Radiation qualities of Bucky table thoracic spine radiographs

Voltage (kVp)	Total filtration					
	Without additional filtration		With additional filtration			
	1.5 mm Al	2.5 Al	3.5 Al	4.5 Al	3.5 Al + 0.1 Cu	3.5 Al + 0.2 Cu
50	This technique is not suitable for thoracic spine radiographs					
60						
70						
80						
90						
100						

Annotations:
- Parameter combination not recommended, normalized organ doses not provided
- Parameter combination not recommended, but normalized organ doses provided
- Parameter combination recommended, normalized organ doses provided

Table 15.10 Radiographic techniques of vertical stand thoracic spine radiographs

Age group (years)	Beam direction	Patient position	Beam projection	Typical patient diameter (cm)		Optimal field size (cm x cm)	Focus-to-detector distance (cm)	Voltage (kVp)						Antiscatter grid
				With arms	No arms			50	60	70	80	90	100	
0	This technique is not applied in these age groups													
1-2														
3-7														
8-12	Horizontal	Standing Sitting	ap	16.8		8 × 25	150							
			lat	27.3	23.8	12 × 26								
13-17			ap	19.6		10 × 31								
			lat	34.4	29.6	14 × 33								
18			ap	20.0		10 × 34								
			lat	38.9	34.4	15 × 33								

Annotations:
- Parameter combination not recommended, normalized organ doses not provided
- Parameter combination not recommended, normalized organ doses provided
- Parameter combination recommended, normalized organ doses provided

Table 15.11 Radiation qualities of vertical stand thoracic spine radiographs

Voltage (kVp)	Total filtration					
	Without additional filtration		With additional filtration			
	1.5 mm Al	2.5 Al	3.5 Al	4.5 Al	3.5 Al + 0.1 Cu	3.5 Al + 0.2 Cu
50						
60	This technique is not suitable for thoracic spine radiographs					
70						
80						
90						
100						

Annotations:
- Parameter combination not recommended, normalized organ doses not provided
- Parameter combination not recommended, normalized organ doses provided
- Parameter combination recommended, normalized organ doses provided

sponding normalized organ doses from Tables 15.19, 15.20, 15.21, 15.22, 15.23, 15.24, 15.25, 15.26, 15.27, 15.28, 15.29, 15.30, 15.31, 15.32, 15.33, 15.34, 15.35, 15.36, 15.37, 15.38, 15.39, 15.40, 15.41, 15.42, 15.43, 15.44, 15.45, 15.46, 15.47, 15.48, 15.49, 15.50, 15.51, 15.52, 15.53, 15.54, 15.55, 15.56, 15.57, 15.58, 15.59, 15.60, 15.61, 15.62, 15.63, 15.64, 15.65, 15.66, 15.67, 15.68, 15.69 and 15.70, the age group related to the patient diameter should be used.

15.7 Normalized Organ Doses

For calculating normalized organ doses, the field settings described in Sect. 15.4 were transferred to the mathematical MIRD phantoms. Inconsistencies between the geometrical anatomy of the mathematical phantoms and the real human anatomy have been listed in Table 15.12 whenever they are of relevance for the validity of the normalized organ doses.

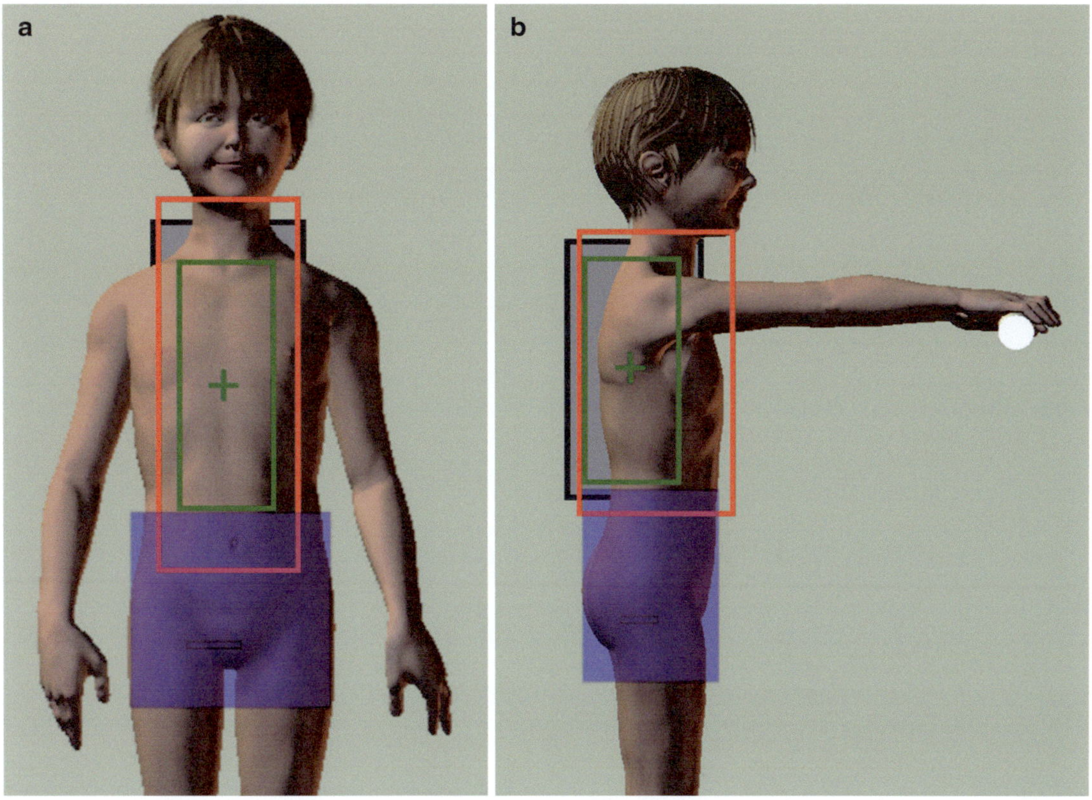

Fig. 15.1 **Field settings for thoracic spine radiographs in standing or sitting position.** (**a**) Antero-posterior projection. (**b**) Lateral projection; the patient holds on a handrail. (© V. Rösenberger and M. Seidenbusch 2018). Green lines = optimal field settings; green cross = centring point; red lines = suboptimal field settings; grey shaded area = image detector; blue shaded area = lead mask shielding

In Tables 15.13, 15.14, 15.15, 15.16, 15.17 and 15.18, optimal and suboptimal field settings are listed for the different age groups and the beam projections given in Tables 15.8 and 15.10. Green lines refer to optimal field settings, red lines refer to suboptimal field settings, respectively. In the tables, the optimal and suboptimal field settings which were used for performing the calculations of the normalized organ doses provided in Tables 15.19, 15.20, 15.21, 15.22, 15.23, 15.24, 15.25, 15.26, 15.27, 15.28, 15.29, 15.30, 15.31, 15.32, 15.33, 15.34, 15.35, 15.36, 15.37, 15.38, 15.39, 15.40, 15.41, 15.42, 15.43, 15.44, 15.45, 15.46, 15.47, 15.48, 15.49, 15.50, 15.51, 15.52, 15.53, 15.54, 15.55, 15.56, 15.57, 15.58, 15.59, 15.60, 15.61, 15.62, 15.63, 15.64, 15.65, 15.66, 15.67, 15.68, 15.69 and 15.70 were drawn in on the skin surface of the mathematical phantoms. The virtual mathematical radiographs resulting from these field settings were depicted in the tables with and without bone structures, respectively, in order to enable the reader to visually compare the virtual mathematical radiographs with the real radiographs made under clinical conditions. For better visualization of the spine in the sagittal skeletal view, thymus, heart, oesophagus, liver, small intestine, and colon have been removed, and for better visualization of the spine in the lateral skeletal view, lungs, kidneys, and liver have been removed. For comparing the virtual mathematical field settings with the real clinical field settings, the virtual mathematical field edges were also listed in the tables.

As the normalized organ doses provided in Tables 15.19, 15.20, 15.21, 15.22, 15.23, 15.24, 15.25, 15.26, 15.27, 15.28, 15.29, 15.30, 15.31,

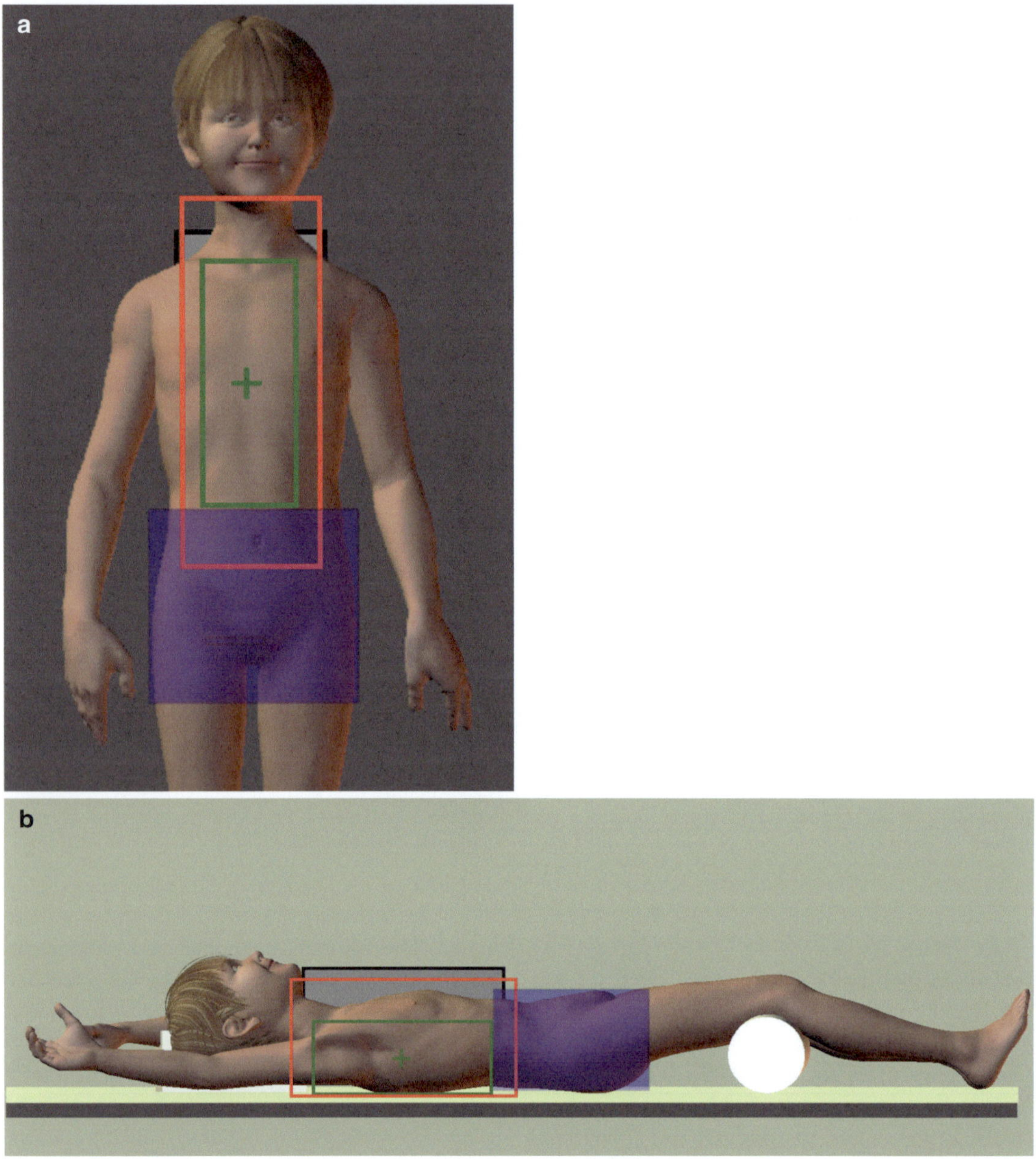

Fig. 15.2 **Field settings for thoracic spine radiographs in supine position**. (**a**) Antero-posterior projection. (**b**) Lateral projection. (© V. Rösenberger and M. Seidenbusch 2018). Green lines = optimal field settings; green cross = centring point; red lines = suboptimal field settings; grey shaded area = image detector; blue shaded area = lead mask shielding; dark grey shaded area = Bucky table; white cylinder = foam roll; white device behind the head and neck = foam pillow; white foam mat on the top of the Bucky table

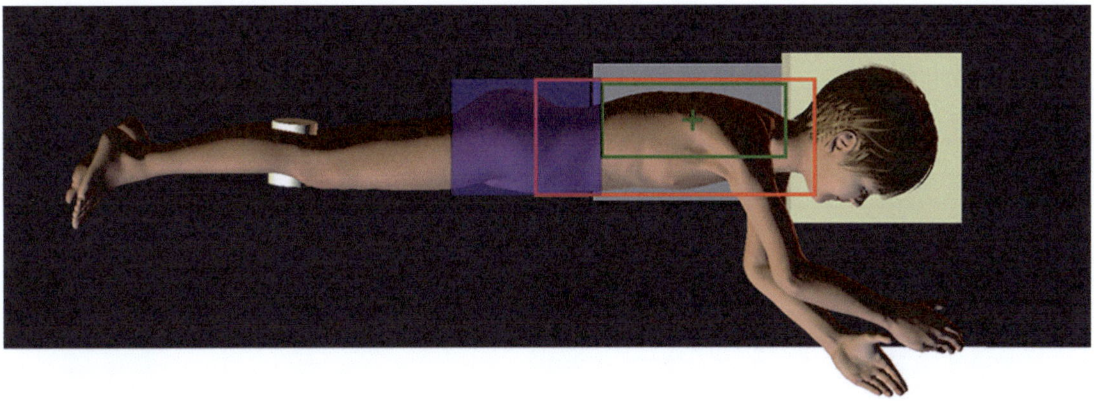

Fig. 15.3 **Field settings for thoracic spine radiographs in decubitus position.** Lateral projection. (© V. Rösenberger and M. Seidenbusch 2018). *Green lines* optimal field settings; Green lines = optimal field settings; green cross = centring point; red lines = suboptimal field settings; grey shaded area = image detector; blue shaded area = lead mask shielding; dark grey shaded area = Bucky table; white cylinder = foam roll; white device behind the head and neck = foam pillow

Table 15.12 Inconsistencies between the mathematical MIRD phantoms and real human anatomy

Anatomical region	Inconsistencies
Skeleton	The shape of the spine is unrealistic (no lordosis, no kyphosis considered), the vertebral column is not clearly depicted. The transverse processes are not outlined
	Ribs are drawn as complete circles and thicker than in real humans

For details, see the text

15.32, 15.33, 15.34, 15.35, 15.36, 15.37, 15.38, 15.39, 15.40, 15.41, 15.42, 15.43, 15.44, 15.45, 15.46, 15.47, 15.48, 15.49, 15.50, 15.51, 15.52, 15.53, 15.54, 15.55, 15.56, 15.57, 15.58, 15.59, 15.60, 15.61, 15.62, 15.63, 15.64, 15.65, 15.66, 15.67, 15.68, 15.69 and 15.70 refer to entrance kerma values, optimal and suboptimal field sizes were provided in the tables in order to enable the reader to calculate entrance air kerma dose values from clinically measured dose area product values. Please note that these optimal and suboptimal field sizes refer to the mathematical phantoms of an age grouonly and not to real patients with corresponding ages.

Please note also that the anatomy of the mathematical phantoms has been idealized. Therefore, the virtual field settings for calculating the normalized organ doses have been defined with a view to ensuring that the resulting virtual organ doses would be consistent with the real organ doses which would be achieved during real X-ray examinations being performed with the real field settings recommended in Sect. 15.4. Thus, the field settings shown in the following tables may not be consistent with the clinical field settings as recommended in Sect. 15.4.

Normalized organ doses for thoracic spine radiographs are also listed in e.g. [3, 5–7].

Table 15.13 Field settings—Thoracic spine ap

Age group (years)		Anatomical landmarks			
		Optimal field setting		Suboptimal field setting	
0	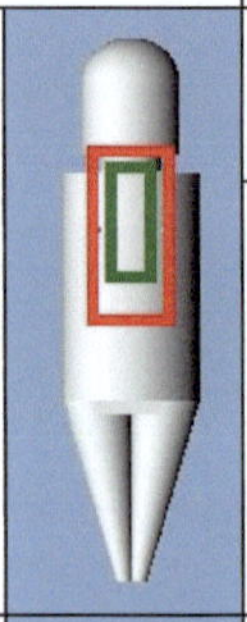	Upper field edge: Below the tip of the mandible Lower field edge: Just below 12th rib Lateral field edges: Close to thesternoclavicular joints Field size = field width x field height: 3 cm x 10 cm = 30 cm^2		Upper field edge: Tip of the mandible Lower field edge: Upper edge of the iliac crest Lateral field edges: At the mamillary line Field size = field width x field height: 7 cm x 16 cm = 112 cm^2	
		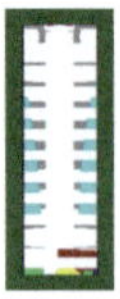	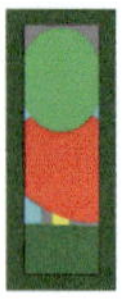	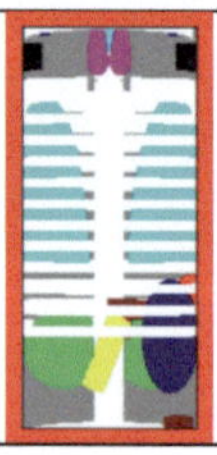	
1 - 2	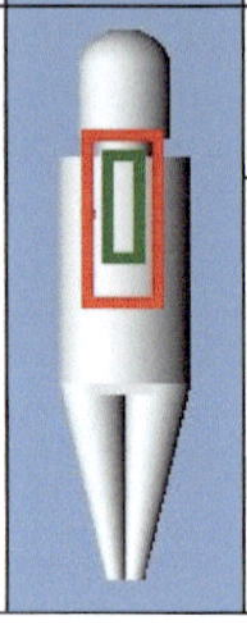	Upper field edge: Below the tip of the mandible Lower field edge: Just below 12th rib Lateral field edges: Close to thesternoclavicular joints Field size = field width x field height: 4 cm x 14 cm = 56 cm^2		Upper field edge: Tip of the mandible Lower field edge: Upper edge of the iliac crest Lateral field edges: At the mamillary line Field size = field width x field height: 9 cm x 22 cm = 198 cm^2	
		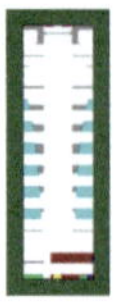		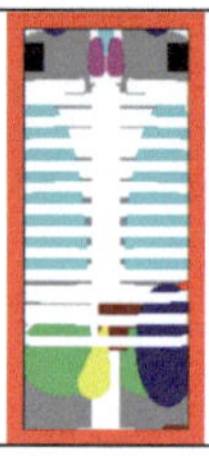	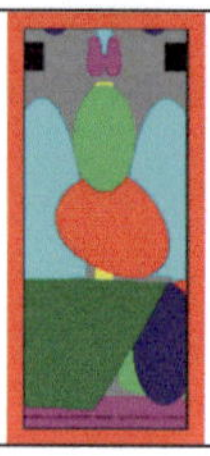

Table 15.14 Field settings—Thoracic spine ap

Age group (years)		Anatomical landmarks			
		Optimal field setting		Suboptimal field setting	
3 - 7	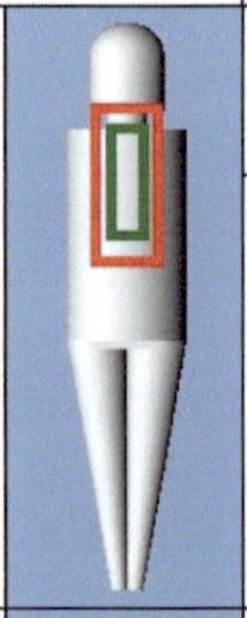	Upper field edge: Below the tip of the mandible Lower field edge: Just below 12th rib Lateral field edges: Close to thesternoclavicular joints Field size = field width x field height: 6 cm x 20 cm = 120 cm^2		Upper field edge: Tip of the mandible Lower field edge: Upper edge of the iliac crest Lateral field edges: At the mamillary line Field size = field width x field height: 11 cm x 29 cm = 319 cm^2	
		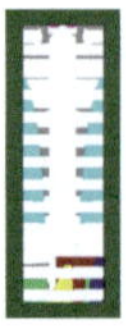		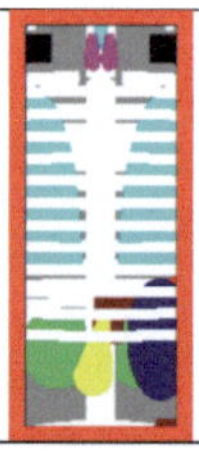	
8 - 12	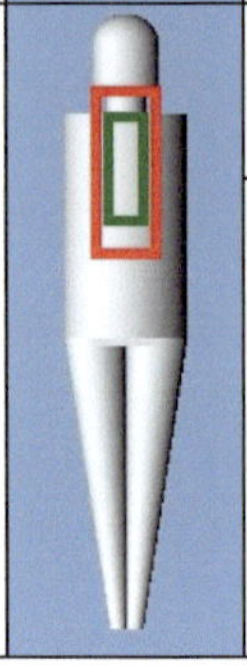	Upper field edge: Below the tip of the mandible Lower field edge: Just below 12th rib Lateral field edges: Close to thesternoclavicular joints Field size = field width x field height: 8 cm x 25 cm = 200 cm^2		Upper field edge: Tip of the mandible Lower field edge: Upper edge of the iliac crest Lateral field edges: At the mamillary line Field size = field width x field height: 13 cm x 36 cm = 468 cm^2	
		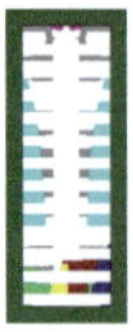		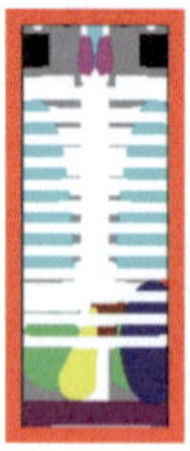	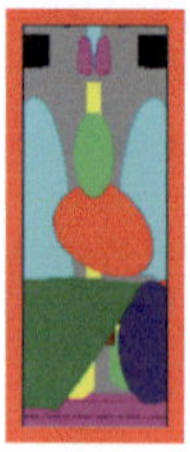

Table 15.15 Field settings—Thoracic spine ap

Age group (years)		Anatomical landmarks			
		Optimal field setting		Suboptimal field setting	
13 - 17	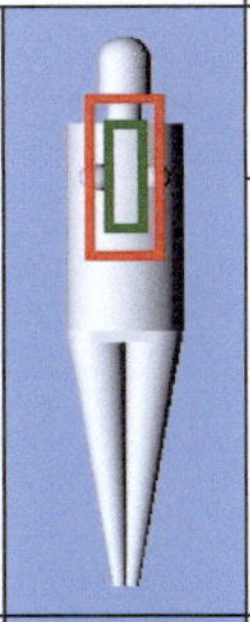	Upper field edge: Below the tip of the mandible Lower field edge: Just below 12th rib Lateral field edges: Close to thesternoclavicular joints Field size = field width x field height: 10 cm x 31 cm = 310 cm^2		Upper field edge: Tip of the mandible Lower field edge: Upper edge of the iliac crest Lateral field edges: At the mamillary line Field size = field width x field height: 19 cm x 46 cm = 874 cm^2	
		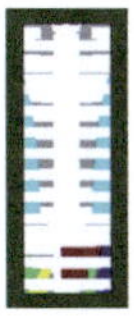		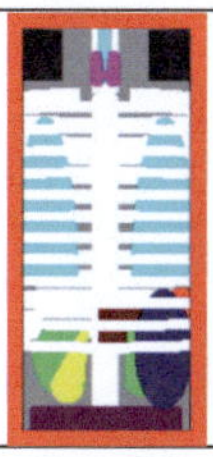	
18	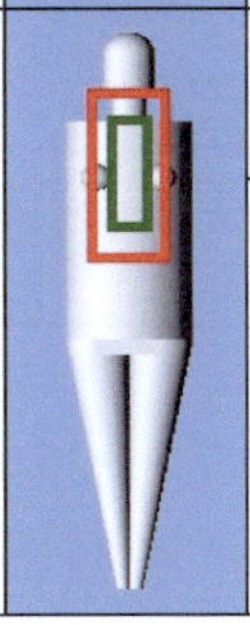	Upper field edge: Below the tip of the mandible Lower field edge: Just below 12th rib Lateral field edges: Close to thesternoclavicular joints Field size = field width x field height: 10 cm x 34 cm = 340 cm^2		Upper field edge: Tip of the mandible Lower field edge: Upper edge of the iliac crest Lateral field edges: At the mamillary line Field size = field width x field height: 22 cm x 51cm = 1,122 cm^2	
		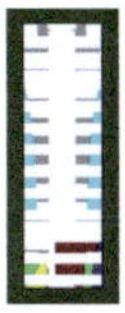			

Table 15.16 Field settings—Thoracic spine lat

Age group (years)		Anatomical landmarks			
		Optimal field setting		Suboptimal field setting	
0	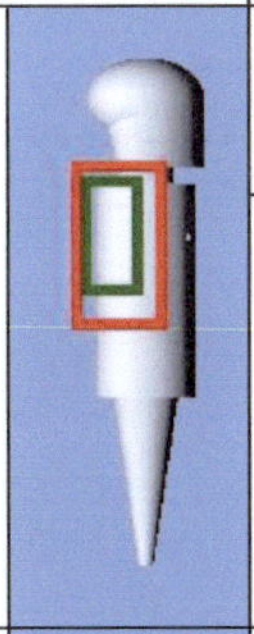	Upper field edge: Just abve the sternoc lavicular joints Lower field edge: Just below 12th rib Posterior field edge: Close to the skin of the back Anterior field edge: Anterior axillary line Field size = field width x field height: 5 cm x 11 cm = 55 cm^2		Upper field edge: Tip of the mandible Lower field edge: Upper edge of the iliac crest Posterior field edge: Close to the skin of the back Anterior field edge: Close the anterior thoracic wall Field size = field width x field height: 8 cm x 15 cm = 120 cm^2	
		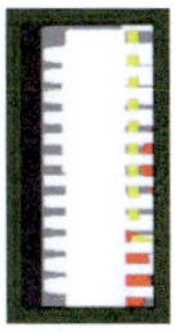	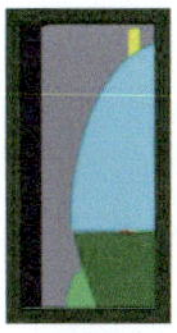	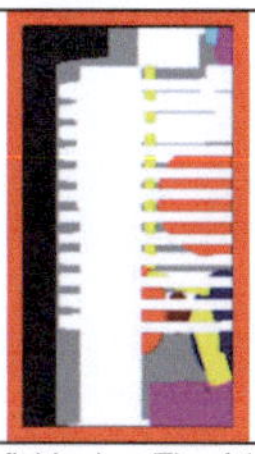	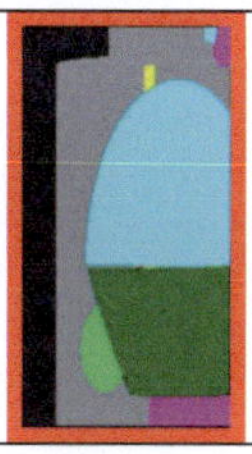
1–2	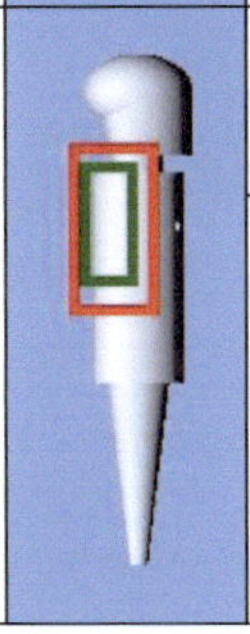	Upper field edge: Just abve the sternoc lavicular joints Lower field edge: Just below 12th rib Posteriorfield edge: Close to the skin of the back Anterior field edge: Anterior axillary line Field size = field width x field height: 7 cm x 16 cm = 112 cm^2		Upper field edge: Tip of the mandible Lower field edge: Upper edge of the iliac crest Posterior field edge: Close to the skin of the back Anterior field edge: Close the anterior thoracic wall Field size = field width x field height: 9 cm x 22 cm = 198 cm^2	
		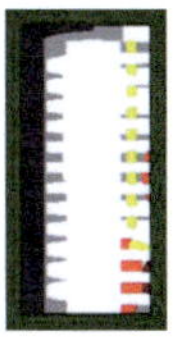	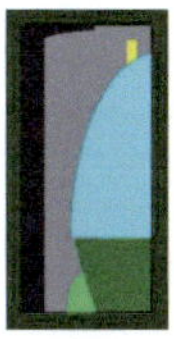	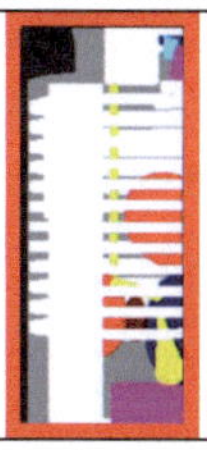	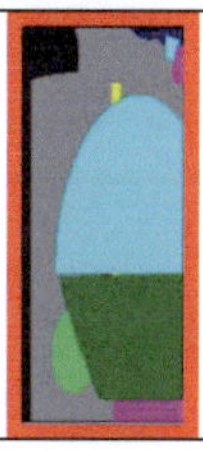

Table 15.17 Field settings—Thoracic spine lat

Age group (years)		Anatomical landmarks			
		Optimal field setting		Suboptimal field setting	
3–7	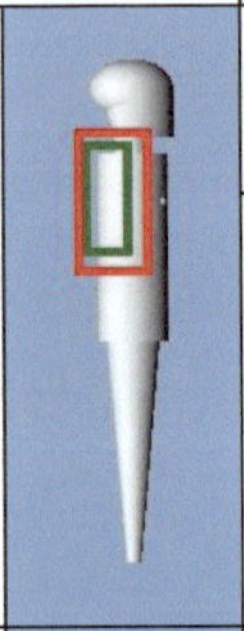	Upper field edge: Just abve the sternoc lavicular joints Lower field edge: Just below 12th rib Posterior field edge: Close to the skin of the back Anterior field edge: Anterior axillary line Field size = field width x field height: 9 cm x 22 cm = 198 cm^2		Upper field edge: Tip of the mandible Lower field edge: Upper edge of the iliac crest Posterior field edge: Close to the skin of the back Anterior field edge: Close the anterior thoracic wall Field size = field width x field height: 11 cm x 28 cm = 308 cm^2	
			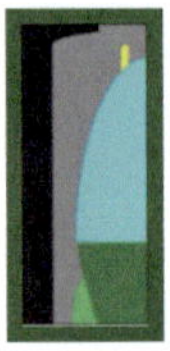	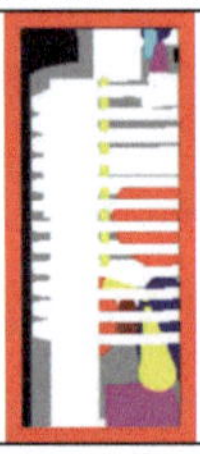	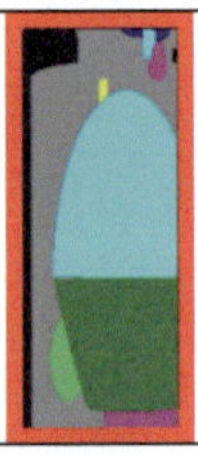
8–12	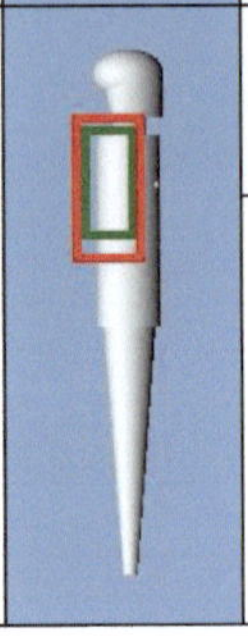	Upper field edge: Just abve the sternoc lavicular joints Lower field edge: Just below 12th rib Posterior field edge: Close to the skin of the back Anterior field edge: Anterior axillary line Field size = field width x field height: 12 cm x 26 cm = 312 cm^2		Upper field edge: Tip of the mandible Lower field edge: Upper edge of the iliac crest Posterior field edge: Close to the skin of the back Anterior field edge: Close the anteriorthoracic wall Field size = field width x field height: 14 cm x 36 cm = 504 cm^2	
		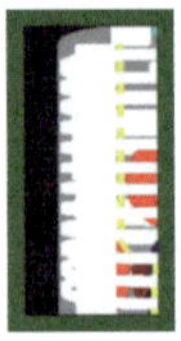	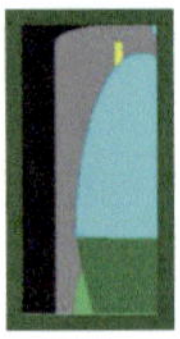	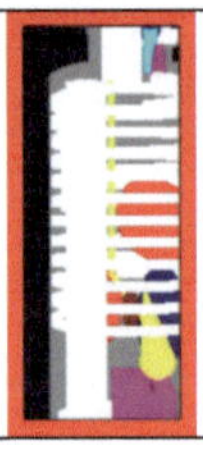	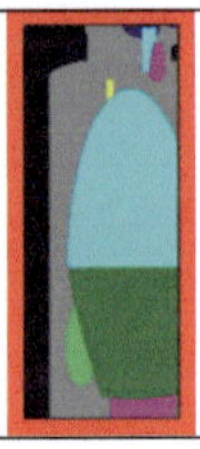

Table 15.18 Field settings—Thoracic spine lat

Age group (years)		Anatomical landmarks			
		Optimal field setting		Suboptimal field setting	
13–17	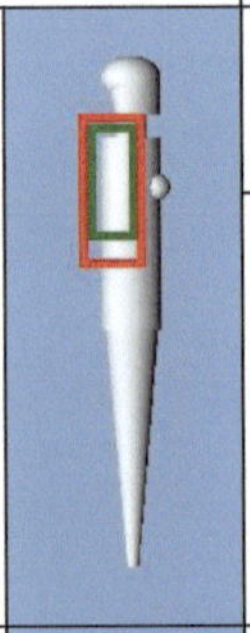	Upper field edge: Just abve the sternoclavicular joints Lower field edge: Just below 12th rib Posterior field edge: Close to the skin of the back Anterior field edge: Anterior axillary line Field size = field width x field height: 14 cm x 33cm = 462 cm^2		Upper field edge: Tip of the mandible Lower field edge: Upper edge of the iliac crest Posterior field edge: Close to the skin of the back Anterior field edge: Close the anterior thoracic wall Field size = field width x field height: 16cm x 43 cm = 688 cm^2	
		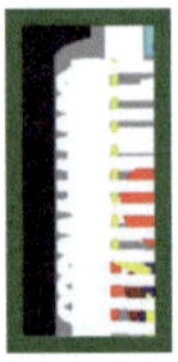	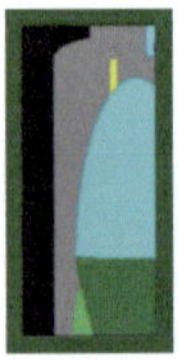		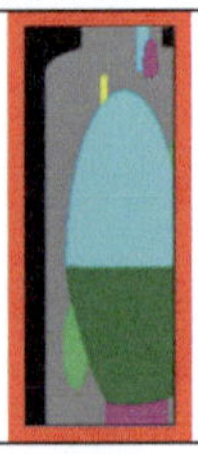
18	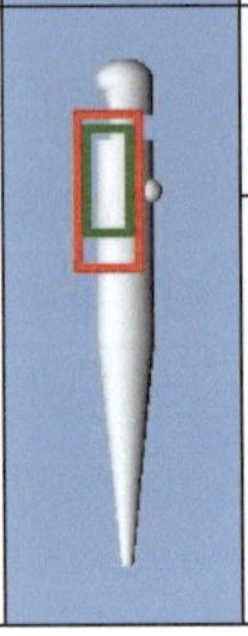	Upper field edge: Just abve the sternoc lavicular joints Lower field edge: Just below 12th rib Posterior field edge: Close to the skin of the back Anterior field edge: Anterior axillary line Field size = field width x field height: 15 cm x 33 cm = 495 cm^2		Upper field edge: Tip of the mandible Lower field edge: Upper edge of the iliac crest Posterior field edge: Close to the skin of the back Anterior field edge: Close the anterior thoracic wall Field size = field width x field height: 18 cm x 48 cm = 864 cm^2	
		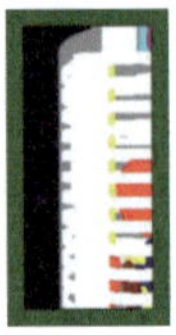			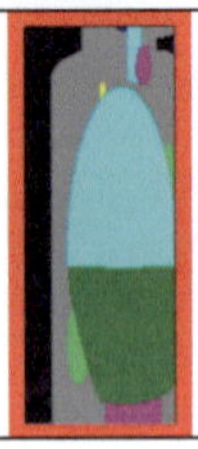

Table 15.19 Normalized organ doses—Thoracic spine ap—<1 year

Beam projection:			ap									**Table cells:**
Focus-to-detector distance (cm):			100									(+) = normal font = optimal field setting
Tube voltage (kVp):			50									**(−) = bold font = suboptimal field setting**
Total filtrations:			See last table row.									Empty table cells correspond to values <0.01
Organ or tissue	**Organ dose D_T : Entrance dose K_E**							**(mSv ∗ mGy^{-1})**				
	(+)	(−)	(+)	(−)	(+)	(−)	(+)	(−)	(+)	(−)	(+)	(−)
Skull		**0.02**		**0.02**		**0.03**		**0.03**		**0.04**		**0.04**
Brain												
Oral mucosa		**0.02**		**0.03**		**0.03**		**0.04**	0.01	**0.05**	0.01	**0.05**
Salivary glands		**0.07**		**0.08**	0.01	**0.09**	0.01	**0.10**	0.01	**0.12**	0.02	**0.13**
Cervical spine	0.04	**0.30**	0.05	**0.39**	0.06	**0.45**	0.07	**0.51**	0.09	**0.61**	0.10	**0.70**
Thyroid	0.04	**0.81**	0.05	**0.93**	0.06	**0.99**	0.07	**1.04**	0.08	**1.11**	0.09	**1.17**
Thymus	0.63	**0.71**	0.74	**0.83**	0.80	**0.90**	0.85	**0.96**	0.92	**1.05**	0.98	**1.13**
Thoracic spine	0.14	**0.20**	0.19	**0.28**	0.23	**0.33**	0.26	**0.38**	0.33	**0.48**	0.40	**0.58**
Claviculae	0.31	**1.27**	0.36	**1.48**	0.39	**1.61**	0.42	**1.70**	0.45	**1.85**	0.49	**1.97**
Scapulae	0.03	**0.16**	0.04	**0.21**	0.05	**0.25**	0.05	**0.28**	0.07	**0.34**	0.08	**0.40**
Ribs	0.31	**0.83**	0.34	**0.95**	0.37	**1.02**	0.38	**1.07**	0.41	**1.16**	0.43	**1.23**
Oesophagus	0.12	**0.17**	0.16	**0.22**	0.18	**0.26**	0.21	**0.29**	0.25	**0.36**	0.29	**0.42**
Trachea		**0.09**		**0.10**		**0.11**	0.01	**0.12**	0.01	**0.14**	0.02	**0.15**
Lung	0.04	**0.28**	0.05	**0.35**	0.06	**0.39**	0.06	**0.43**	0.08	**0.49**	0.09	**0.55**
Mamma	0.02	**1.10**	0.03	**1.16**	0.03	**1.19**	0.04	**1.21**	0.04	**1.24**	0.05	**1.27**
Heart	0.28	**0.48**	0.34	**0.58**	0.38	**0.65**	0.42	**0.71**	0.47	**0.80**	0.52	**0.88**
Stomach	0.02	**0.45**	0.03	**0.55**	0.04	**0.61**	0.04	**0.67**	0.05	**0.76**	0.06	**0.83**
Spleen		**0.10**	0.01	**0.14**	0.02	**0.16**	0.02	**0.18**	0.02	**0.22**	0.03	**0.26**
Liver	0.09	**0.35**	0.10	**0.42**	0.12	**0.48**	0.13	**0.52**	0.14	**0.58**	0.16	**0.64**
Gall bladder	0.04	**0.33**	0.05	**0.42**	0.05	**0.49**	0.06	**0.54**	0.07	**0.63**	0.09	**0.71**
Pancreas	0.08	**0.22**	0.11	**0.28**	0.13	**0.33**	0.14	**0.37**	0.17	**0.45**	0.20	**0.52**
Small intestine		**0.11**		**0.15**		**0.17**		**0.19**		**0.23**		**0.26**
Large intestine		**0.21**		**0.26**		**0.30**		**0.33**		**0.38**	0.01	**0.43**
Sigma/Rectum		**0.03**		**0.04**		**0.04**		**0.05**		**0.06**		**0.07**
Lumbar spine	0.01	**0.19**	0.02	**0.26**	0.02	**0.31**	0.02	**0.36**	0.03	**0.45**	0.04	**0.54**
Pelvis		**0.03**		**0.05**		**0.06**		**0.06**		**0.08**		**0.10**
Adrenals	0.04	**0.09**	0.05	**0.12**	0.07	**0.14**	0.07	**0.16**	0.09	**0.20**	0.11	**0.24**
Kidneys		**0.08**	0.01	**0.11**	0.01	**0.13**	0.02	**0.15**	0.02	**0.19**	0.02	**0.22**
Urinary bladder						**0.01**		**0.01**		**0.02**		**0.02**
Uterus		**0.02**		**0.02**		**0.03**		**0.03**		**0.04**		**0.05**
Ovaries		**0.02**		**0.02**		**0.03**		**0.03**		**0.04**		**0.05**
Prostate												
Testes												
Upper arm	0.02	**0.05**	0.02	**0.07**	0.03	**0.08**	0.03	**0.10**	0.04	**0.12**	0.05	**0.14**
Elbow	0.02	**0.06**	0.02	**0.08**	0.03	**0.09**	0.03	**0.11**	0.04	**0.13**	0.04	**0.16**
Forearm		**0.03**		**0.03**		**0.04**		**0.05**		**0.06**		**0.07**
Upper leg												
Knee												
Lower leg												
Lymph nodes	0.04	**0.22**	0.05	**0.27**	0.06	**0.30**	0.07	**0.33**	0.08	**0.38**	0.09	**0.43**
Skin	0.03	**0.13**	0.03	**0.14**	0.03	**0.14**	0.03	**0.15**	0.03	**0.15**	0.04	**0.16**
Muscle	0.03	**0.11**	0.04	**0.13**	0.04	**0.15**	0.05	**0.16**	0.05	**0.18**	0.06	**0.20**
Skeleton	0.07	**0.19**	0.08	**0.23**	0.09	**0.26**	0.10	**0.28**	0.11	**0.32**	0.12	**0.36**
Red bone marrow	0.02	**0.05**	0.02	**0.06**	0.02	**0.07**	0.03	**0.07**	0.03	**0.08**	0.03	**0.10**
Whole body dose	0.04	**0.12**	0.04	**0.15**	0.05	**0.17**	0.05	**0.18**	0.06	**0.20**	0.07	**0.23**
ED (ICRP 60)	0.03	**0.24**	0.04	**0.29**	0.04	**0.31**	0.05	**0.34**	0.05	**0.38**	0.06	**0.41**
ED (ICRP 103)	0.03	**0.32**	0.04	**0.37**	0.05	**0.40**	0.05	**0.43**	0.06	**0.47**	0.07	**0.50**
Total filtration (mm) Al	**1.5**		**2.5**		**3.5**		**4.5**		**3.5**		**3.5**	
Cu	**0**		**0**		**0**		**0**		**0.1**		**0.2**	

Table 15.20 Normalized organ doses—Thoracic spine ap—<1 year

Beam projection:			ap									**Table cells:**
Focus-to-detector distance (cm):			100									(+) = normal font = optimal field setting
Tube voltage (kVp):			60									**(−) = bold font = suboptimal field setting**
Total filtrations:			See last table row.									Empty table cells correspond to values <0.01
Organ or tissue	**Organ dose D_T : Entrance dose K_E**								**(mSv * mGy^{-1})**			
	(+)	(−)	(+)	(−)	(+)	(−)	(+)	(−)	(+)	(−)	(+)	(−)
Skull		**0.02**		**0.03**		**0.03**		**0.04**		**0.04**		**0.04**
Brain												
Oral mucosa		**0.03**		**0.04**	0.01	**0.04**	0.01	**0.05**	0.01	**0.06**	0.01	**0.05**
Salivary glands		**0.08**	0.01	**0.10**	0.01	**0.11**	0.01	**0.12**	0.02	**0.14**	0.02	**0.13**
Cervical spine	0.05	**0.38**	0.07	**0.48**	0.08	**0.55**	0.09	**0.61**	0.11	**0.72**	0.10	**0.70**
Thyroid	0.05	**0.88**	0.07	**0.99**	0.08	**1.05**	0.08	**1.10**	0.10	**1.17**	0.09	**1.17**
Thymus	0.70	**0.78**	0.80	**0.90**	0.86	**0.98**	0.91	**1.04**	0.98	**1.13**	0.98	**1.13**
Thoracic spine	0.20	**0.29**	0.26	**0.38**	0.31	**0.45**	0.35	**0.51**	0.43	**0.63**	0.40	**0.58**
Claviculae	0.34	**1.39**	0.39	**1.59**	0.42	**1.71**	0.45	**1.80**	0.48	**1.94**	0.49	**1.97**
Scapulae	0.04	**0.21**	0.05	**0.27**	0.06	**0.31**	0.07	**0.35**	0.09	**0.42**	0.08	**0.40**
Ribs	0.33	**0.91**	0.37	**1.02**	0.39	**1.09**	0.41	**1.14**	0.43	**1.23**	0.43	**1.23**
Oesophagus	0.15	**0.22**	0.20	**0.28**	0.23	**0.33**	0.26	**0.37**	0.31	**0.44**	0.29	**0.42**
Trachea		**0.10**	0.01	**0.12**	0.01	**0.13**	0.01	**0.14**	0.02	**0.15**	0.02	**0.15**
Lung	0.05	**0.33**	0.06	**0.40**	0.07	**0.45**	0.08	**0.49**	0.09	**0.56**	0.09	**0.55**
Mamma	0.03	**1.13**	0.03	**1.19**	0.04	**1.22**	0.04	**1.24**	0.05	**1.26**	0.05	**1.27**
Heart	0.33	**0.55**	0.39	**0.66**	0.44	**0.74**	0.47	**0.80**	0.53	**0.90**	0.52	**0.88**
Stomach	0.03	**0.52**	0.04	**0.63**	0.05	**0.70**	0.05	**0.75**	0.06	**0.84**	0.06	**0.83**
Spleen	0.01	**0.14**	0.02	**0.18**	0.02	**0.21**	0.02	**0.23**	0.03	**0.28**	0.03	**0.26**
Liver	0.10	**0.41**	0.12	**0.49**	0.13	**0.54**	0.14	**0.58**	0.16	**0.66**	0.16	**0.64**
Gall bladder	0.05	**0.41**	0.06	**0.50**	0.07	**0.57**	0.07	**0.63**	0.09	**0.73**	0.09	**0.71**
Pancreas	0.11	**0.28**	0.14	**0.36**	0.16	**0.41**	0.18	**0.46**	0.21	**0.54**	0.20	**0.52**
Small intestine		**0.14**		**0.18**		**0.21**		**0.23**		**0.27**		**0.26**
Large intestine		**0.25**		**0.31**		**0.35**		**0.38**	0.01	**0.44**	0.01	**0.43**
Sigma/Rectum		**0.04**		**0.05**		**0.05**		**0.06**		**0.07**		**0.07**
Lumbar spine	0.02	**0.26**	0.03	**0.35**	0.03	**0.42**	0.04	**0.47**	0.05	**0.58**	0.04	**0.54**
Pelvis		**0.05**		**0.06**		**0.08**		**0.09**		**0.11**		**0.10**
Adrenals	0.06	**0.12**	0.07	**0.16**	0.09	**0.19**	0.10	**0.21**	0.12	**0.26**	0.11	**0.24**
Kidneys	0.01	**0.11**	0.02	**0.15**	0.02	**0.17**	0.02	**0.20**	0.03	**0.24**	0.02	**0.22**
Urinary bladder		**0.01**		**0.01**		**0.02**		**0.02**		**0.02**		**0.02**
Uterus		**0.02**		**0.03**		**0.04**		**0.04**		**0.05**		**0.05**
Ovaries		**0.02**		**0.03**		**0.04**		**0.04**		**0.05**		**0.05**
Prostate										**0.01**		
Testes												
Upper arm	0.02	**0.07**	0.03	**0.09**	0.04	**0.11**	0.04	**0.13**	0.05	**0.16**	0.05	**0.14**
Elbow	0.02	**0.08**	0.03	**0.11**	0.03	**0.13**	0.04	**0.14**	0.05	**0.18**	0.04	**0.16**
Forearm		**0.04**		**0.05**		**0.06**		**0.07**	0.01	**0.08**		**0.07**
Upper leg										**0.01**		
Knee												
Lower leg												
Lymph nodes	0.05	**0.26**	0.06	**0.31**	0.07	**0.35**	0.08	**0.38**	0.10	**0.44**	0.09	**0.43**
Skin	0.03	**0.14**	0.03	**0.14**	0.03	**0.15**	0.04	**0.16**	0.04	**0.17**	0.04	**0.16**
Muscle	0.04	**0.13**	0.04	**0.15**	0.05	**0.17**	0.05	**0.19**	0.06	**0.21**	0.06	**0.20**
Skeleton	0.08	**0.23**	0.09	**0.27**	0.10	**0.30**	0.11	**0.33**	0.13	**0.37**	0.12	**0.36**
Red bone marrow	0.02	**0.06**	0.03	**0.07**	0.03	**0.08**	0.03	**0.09**	0.04	**0.10**	0.03	**0.10**
Whole body dose	0.04	**0.14**	0.05	**0.17**	0.06	**0.19**	0.06	**0.21**	0.07	**0.23**	0.07	**0.23**
ED (ICRP 60)	0.04	**0.28**	0.04	**0.32**	0.05	**0.35**	0.05	**0.38**	0.06	**0.42**	0.06	**0.41**
ED (ICRP 103)	0.04	**0.36**	0.05	**0.41**	0.06	**0.44**	0.06	**0.47**	0.07	**0.51**	0.07	**0.50**
Total filtration (mm) Al	**1.5**		**2.5**		**3.5**		**4.5**		**3.5**		**3.5**	
Cu	**0**		**0**		**0**		**0**		**0.1**		**0.2**	

Table 15.21 Normalized organ doses—Thoracic spine ap—<1 year

Beam projection:			ap									**Table cells:**
Focus-to-detector distance (cm):			100						(+) = normal font = optimal field setting			
Tube voltage (kVp):			70						**(−) = bold font = suboptimal field setting**			
Total filtrations:			See last table row.						Empty table cells correspond to values <0.01			
Organ or tissue	**Organ dose D_T : Entrance dose K_E**								**(mSv * mGy^{-1})**			
	(+)	(−)	(+)	(−)	(+)	(−)	(+)	(−)	(+)	(−)	(+)	(−)
Skull		**0.03**		**0.03**		**0.04**		**0.04**	0.01	**0.05**	0.01	**0.06**
Brain										**0.01**		**0.01**
Oral mucosa		**0.03**	0.01	**0.04**	0.01	**0.05**	0.01	**0.05**	0.02	**0.06**	0.02	**0.07**
Salivary glands	0.01	**0.09**	0.01	**0.11**	0.02	**0.12**	0.02	**0.13**	0.02	**0.15**	0.02	**0.16**
Cervical spine	0.06	**0.44**	0.08	**0.54**	0.10	**0.62**	0.11	**0.68**	0.13	**0.78**	0.15	**0.87**
Thyroid	0.06	**0.93**	0.07	**1.03**	0.08	**1.10**	0.09	**1.14**	0.11	**1.21**	0.12	**1.26**
Thymus	0.75	**0.85**	0.84	**0.96**	0.90	**1.04**	0.95	**1.09**	1.01	**1.18**	1.06	**1.24**
Thoracic spine	0.24	**0.36**	0.31	**0.46**	0.36	**0.54**	0.41	**0.60**	0.49	**0.72**	0.57	**0.84**
Claviculae	0.37	**1.47**	0.41	**1.66**	0.44	**1.77**	0.46	**1.85**	0.49	**1.96**	0.51	**2.03**
Scapulae	0.05	**0.25**	0.06	**0.31**	0.07	**0.36**	0.08	**0.40**	0.10	**0.47**	0.12	**0.53**
Ribs	0.35	**0.96**	0.38	**1.07**	0.40	**1.13**	0.42	**1.18**	0.44	**1.25**	0.46	**1.31**
Oesophagus	0.19	**0.26**	0.23	**0.33**	0.27	**0.38**	0.29	**0.42**	0.35	**0.50**	0.39	**0.56**
Trachea	0.01	**0.11**	0.01	**0.13**	0.01	**0.14**	0.02	**0.15**	0.02	**0.16**	0.02	**0.17**
Lung	0.06	**0.38**	0.07	**0.45**	0.08	**0.50**	0.09	**0.54**	0.11	**0.60**	0.12	**0.66**
Mamma	0.03	**1.16**	0.04	**1.21**	0.05	**1.23**	0.05	**1.25**	0.06	**1.28**	0.06	**1.29**
Heart	0.36	**0.62**	0.43	**0.73**	0.47	**0.80**	0.51	**0.86**	0.56	**0.96**	0.61	**1.04**
Stomach	0.04	**0.58**	0.05	**0.69**	0.05	**0.76**	0.06	**0.81**	0.07	**0.90**	0.08	**0.98**
Spleen	0.02	**0.17**	0.02	**0.21**	0.03	**0.24**	0.03	**0.27**	0.04	**0.32**	0.04	**0.36**
Liver	0.11	**0.45**	0.13	**0.53**	0.15	**0.59**	0.16	**0.63**	0.18	**0.70**	0.19	**0.76**
Gall bladder	0.05	**0.47**	0.07	**0.57**	0.08	**0.64**	0.08	**0.70**	0.10	**0.79**	0.11	**0.88**
Pancreas	0.13	**0.33**	0.16	**0.41**	0.18	**0.47**	0.20	**0.52**	0.23	**0.61**	0.26	**0.69**
Small intestine		**0.17**		**0.21**		**0.24**		**0.26**	0.01	**0.30**	0.01	**0.34**
Large intestine		**0.29**		**0.35**	0.01	**0.39**	0.01	**0.42**	0.01	**0.48**	0.02	**0.53**
Sigma/Rectum		**0.04**		**0.05**		**0.06**		**0.07**		**0.08**		**0.09**
Lumbar spine	0.03	**0.33**	0.03	**0.43**	0.04	**0.50**	0.04	**0.56**	0.06	**0.68**	0.07	**0.79**
Pelvis		**0.06**		**0.08**		**0.10**		**0.11**	0.01	**0.13**	0.01	**0.15**
Adrenals	0.07	**0.15**	0.09	**0.19**	0.10	**0.23**	0.11	**0.25**	0.14	**0.30**	0.16	**0.35**
Kidneys	0.02	**0.14**	0.02	**0.18**	0.02	**0.21**	0.03	**0.23**	0.03	**0.28**	0.04	**0.33**
Urinary bladder		**0.01**		**0.02**		**0.02**		**0.02**		**0.03**		**0.03**
Uterus		**0.03**		**0.04**		**0.05**		**0.05**		**0.06**		**0.08**
Ovaries		**0.03**		**0.04**		**0.05**		**0.05**		**0.06**		**0.08**
Prostate						**0.01**		**0.01**		**0.01**		**0.02**
Testes												
Upper arm	0.03	**0.09**	0.04	**0.11**	0.04	**0.13**	0.05	**0.15**	0.06	**0.18**	0.07	**0.21**
Elbow	0.03	**0.10**	0.04	**0.13**	0.04	**0.15**	0.05	**0.17**	0.06	**0.20**	0.07	**0.24**
Forearm		**0.05**		**0.06**		**0.07**	0.01	**0.08**	0.01	**0.10**	0.02	**0.11**
Upper leg						**0.01**		**0.01**		**0.01**		**0.02**
Knee												
Lower leg												
Lymph nodes	0.06	**0.29**	0.07	**0.35**	0.08	**0.39**	0.09	**0.42**	0.11	**0.48**	0.12	**0.53**
Skin	0.03	**0.14**	0.03	**0.15**	0.04	**0.16**	0.04	**0.16**	0.04	**0.17**	0.04	**0.18**
Muscle	0.04	**0.15**	0.05	**0.17**	0.06	**0.19**	0.06	**0.20**	0.07	**0.23**	0.07	**0.25**
Skeleton	0.09	**0.25**	0.10	**0.30**	0.12	**0.33**	0.12	**0.36**	0.14	**0.40**	0.15	**0.44**
Red bone marrow	0.02	**0.07**	0.03	**0.08**	0.03	**0.09**	0.04	**0.10**	0.04	**0.11**	0.05	**0.13**
Whole body dose	0.05	**0.16**	0.06	**0.19**	0.06	**0.21**	0.07	**0.23**	0.08	**0.25**	0.08	**0.28**
ED (ICRP 60)	0.04	**0.30**	0.05	**0.35**	0.06	**0.38**	0.06	**0.40**	0.07	**0.45**	0.08	**0.48**
ED (ICRP 103)	0.05	**0.39**	0.06	**0.44**	0.07	**0.47**	0.07	**0.50**	0.08	**0.54**	0.09	**0.58**
Total filtration (mm) Al	**1.5**		**2.5**		**3.5**		**4.5**		**3.5**		**3.5**	
Cu	**0**		**0**		**0**		**0**		**0.1**		**0.2**	

Table 15.22 Normalized organ doses—Thoracic spine r-lat—<1 year

Beam projection:		r-lat										**Table cells:**
Focus-to-detector distance (cm):		100										(+) = normal font = optimal field setting
Tube voltage (kVp):		50										**(−) = bold font = suboptimal field setting**
Total filtrations:		See last table row.										Empty table cells correspond to values <0.01
Organ or tissue	**Organ dose D_T : Entrance dose K_E**						**($mSv * mGy^{-1}$)**					
	(+)	(−)	(+)	(−)	(+)	(−)	(+)	(−)	(+)	(−)	(+)	(−)
Skull				**0.01**		**0.01**		**0.02**		**0.02**		**0.02**
Brain												
Oral mucosa		**0.01**		**0.01**		**0.02**		**0.02**		**0.02**		**0.03**
Salivary glands		**0.02**		**0.03**		**0.03**		**0.04**	0.01	**0.04**	0.01	**0.05**
Cervical spine	0.03	**0.29**	0.04	**0.37**	0.05	**0.43**	0.06	**0.47**	0.07	**0.56**	0.09	**0.63**
Thyroid		**0.33**	0.01	**0.40**	0.01	**0.44**	0.02	**0.48**	0.02	**0.53**	0.02	**0.58**
Thymus		**0.03**	0.01	**0.04**	0.01	**0.05**	0.01	**0.06**	0.02	**0.07**	0.02	**0.08**
Thoracic spine	0.44	**0.49**	0.58	**0.64**	0.68	**0.76**	0.77	**0.86**	0.93	**1.03**	1.07	**1.20**
Claviculae	0.02	**0.27**	0.02	**0.32**	0.03	**0.36**	0.03	**0.39**	0.04	**0.44**	0.05	**0.48**
Scapulae	1.52	**1.54**	1.74	**1.77**	1.87	**1.90**	1.96	**2.00**	2.11	**2.17**	2.23	**2.30**
Ribs	0.52	**0.80**	0.61	**0.93**	0.67	**1.01**	0.71	**1.07**	0.79	**1.18**	0.85	**1.27**
Oesophagus	0.19	**0.24**	0.24	**0.31**	0.28	**0.36**	0.31	**0.40**	0.36	**0.47**	0.41	**0.54**
Trachea		**0.04**		**0.05**		**0.06**		**0.06**		**0.07**		**0.08**
Lung	0.17	**0.33**	0.21	**0.40**	0.24	**0.45**	0.26	**0.49**	0.29	**0.55**	0.32	**0.61**
Mamma		**0.02**		**0.03**	0.01	**0.04**	0.01	**0.04**	0.01	**0.05**	0.02	**0.06**
Heart	0.02	**0.15**	0.03	**0.19**	0.04	**0.22**	0.04	**0.24**	0.05	**0.29**	0.06	**0.33**
Stomach		**0.06**	0.01	**0.08**	0.01	**0.09**	0.02	**0.11**	0.02	**0.13**	0.02	**0.16**
Spleen	0.04	**0.06**	0.05	**0.08**	0.06	**0.09**	0.07	**0.11**	0.09	**0.14**	0.11	**0.17**
Liver	0.09	**0.36**	0.11	**0.43**	0.13	**0.48**	0.14	**0.51**	0.15	**0.58**	0.17	**0.63**
Gall bladder	0.01	**0.16**	0.02	**0.21**	0.02	**0.25**	0.02	**0.28**	0.03	**0.34**	0.04	**0.40**
Pancreas	0.03	**0.13**	0.04	**0.17**	0.05	**0.20**	0.06	**0.23**	0.07	**0.28**	0.09	**0.33**
Small intestine		**0.09**		**0.11**		**0.13**		**0.15**	0.01	**0.18**	0.01	**0.20**
Large intestine		**0.13**		**0.17**		**0.19**		**0.22**	0.01	**0.25**	0.01	**0.29**
Sigma/Rectum		**0.01**		**0.02**		**0.02**		**0.02**		**0.03**		**0.03**
Lumbar spine	0.08	**0.42**	0.11	**0.55**	0.13	**0.65**	0.15	**0.73**	0.19	**0.89**	0.22	**1.04**
Pelvis		**0.04**		**0.06**		**0.07**		**0.08**	0.01	**0.10**	0.01	**0.12**
Adrenals	0.15	**0.18**	0.20	**0.23**	0.23	**0.27**	0.26	**0.30**	0.31	**0.36**	0.35	**0.41**
Kidneys	0.09	**0.24**	0.12	**0.30**	0.13	**0.34**	0.15	**0.37**	0.17	**0.43**	0.20	**0.49**
Urinary bladder										**0.01**		**0.01**
Uterus		**0.01**		**0.02**		**0.02**		**0.02**		**0.03**		**0.04**
Ovaries		**0.01**		**0.02**		**0.02**		**0.03**		**0.03**		**0.04**
Prostate												
Testes												
Upper arm												
Elbow												
Forearm												
Upper leg												
Knee												
Lower leg												
Lymph nodes	0.03	**0.13**	0.04	**0.16**	0.05	**0.19**	0.05	**0.21**	0.06	**0.24**	0.07	**0.28**
Skin	0.08	**0.14**	0.08	**0.15**	0.08	**0.16**	0.08	**0.16**	0.09	**0.17**	0.09	**0.18**
Muscle	0.04	**0.09**	0.05	**0.11**	0.05	**0.13**	0.06	**0.14**	0.07	**0.16**	0.07	**0.18**
Skeleton	0.17	**0.25**	0.21	**0.30**	0.23	**0.34**	0.25	**0.37**	0.29	**0.42**	0.32	**0.47**
Red bone marrow	0.05	**0.06**	0.06	**0.08**	0.06	**0.09**	0.07	**0.10**	0.08	**0.11**	0.09	**0.13**
Whole body dose	0.05	**0.11**	0.06	**0.14**	0.07	**0.15**	0.08	**0.17**	0.09	**0.19**	0.10	**0.21**
ED (ICRP 60)	0.05	**0.12**	0.06	**0.15**	0.07	**0.17**	0.07	**0.18**	0.08	**0.21**	0.09	**0.24**
ED (ICRP 103)	0.05	**0.12**	0.06	**0.15**	0.06	**0.17**	0.07	**0.18**	0.08	**0.21**	0.09	**0.24**
Total filtration (mm) **Al**	**1.5**		**2.5**		**3.5**		**4.5**		**3.5**		**3.5**	
Cu	**0**		**0**		**0**		**0**		**0.1**		**0.2**	

Table 15.23 Normalized organ doses—Thoracic spine r-lat—<1 year

Beam projection:			r-lat									**Table cells:**
Focus-to-detector distance (cm):			100				(+) = normal font = optimal field setting					
Tube voltage (kVp):			60				**(−) = bold font = suboptimal field setting**					
Total filtrations:			See last table row.				Empty table cells correspond to values <0.01					
Organ or tissue	**Organ dose D_T : Entrance dose K_E**							**(mSv * mGy^{-1})**				
	(+)	(−)	(+)	(−)	(+)	(−)	(+)	(−)	(+)	(−)	(+)	(−)
Skull		**0.01**		**0.02**		**0.02**		**0.02**	0.01	**0.03**		**0.02**
Brain												
Oral mucosa		**0.01**		**0.02**		**0.02**		**0.03**		**0.03**		**0.03**
Salivary glands		**0.03**		**0.04**		**0.04**	0.01	**0.05**	0.01	**0.06**	0.01	**0.05**
Cervical spine	0.04	**0.35**	0.06	**0.44**	0.07	**0.51**	0.08	**0.55**	0.10	**0.64**	0.09	**0.63**
Thyroid	0.01	**0.38**	0.02	**0.45**	0.02	**0.49**	0.02	**0.53**	0.03	**0.59**	0.02	**0.58**
Thymus	0.01	**0.04**	0.01	**0.06**	0.02	**0.07**	0.02	**0.07**	0.02	**0.09**	0.02	**0.08**
Thoracic spine	0.57	**0.63**	0.72	**0.81**	0.84	**0.94**	0.93	**1.05**	1.10	**1.24**	1.07	**1.20**
Claviculae	0.02	**0.31**	0.03	**0.37**	0.04	**0.41**	0.04	**0.44**	0.05	**0.49**	0.05	**0.48**
Scapulae	1.65	**1.68**	1.85	**1.90**	1.98	**2.03**	2.07	**2.13**	2.20	**2.28**	2.23	**2.30**
Ribs	0.58	**0.88**	0.67	**1.02**	0.73	**1.10**	0.77	**1.16**	0.85	**1.27**	0.85	**1.27**
Oesophagus	0.24	**0.30**	0.29	**0.38**	0.33	**0.43**	0.37	**0.48**	0.43	**0.56**	0.41	**0.54**
Trachea		**0.05**		**0.06**		**0.07**		**0.07**		**0.08**		**0.08**
Lung	0.20	**0.38**	0.24	**0.46**	0.27	**0.51**	0.29	**0.55**	0.33	**0.62**	0.32	**0.61**
Mamma		**0.03**	0.01	**0.04**	0.01	**0.05**	0.01	**0.05**	0.02	**0.07**	0.02	**0.06**
Heart	0.03	**0.18**	0.04	**0.23**	0.05	**0.26**	0.06	**0.29**	0.07	**0.34**	0.06	**0.33**
Stomach	0.01	**0.08**	0.02	**0.11**	0.02	**0.13**	0.02	**0.14**	0.03	**0.18**	0.02	**0.16**
Spleen	0.06	**0.08**	0.07	**0.11**	0.09	**0.13**	0.10	**0.15**	0.12	**0.19**	0.11	**0.17**
Liver	0.11	**0.41**	0.13	**0.49**	0.14	**0.54**	0.15	**0.57**	0.17	**0.64**	0.17	**0.63**
Gall bladder	0.02	**0.21**	0.02	**0.27**	0.03	**0.31**	0.03	**0.35**	0.04	**0.42**	0.04	**0.40**
Pancreas	0.05	**0.17**	0.06	**0.22**	0.07	**0.26**	0.08	**0.29**	0.10	**0.35**	0.09	**0.33**
Small intestine		**0.11**		**0.14**	0.01	**0.16**	0.01	**0.18**	0.01	**0.21**	0.01	**0.20**
Large intestine		**0.16**		**0.20**	0.01	**0.23**	0.01	**0.26**	0.02	**0.30**	0.01	**0.29**
Sigma/Rectum		**0.02**		**0.02**		**0.03**		**0.03**		**0.04**		**0.03**
Lumbar spine	0.11	**0.54**	0.15	**0.69**	0.17	**0.81**	0.19	**0.90**	0.23	**1.08**	0.22	**1.04**
Pelvis		**0.06**		**0.08**	0.01	**0.09**	0.01	**0.11**	0.02	**0.13**	0.01	**0.12**
Adrenals	0.19	**0.23**	0.24	**0.29**	0.28	**0.33**	0.31	**0.37**	0.36	**0.43**	0.35	**0.41**
Kidneys	0.11	**0.29**	0.14	**0.35**	0.16	**0.40**	0.18	**0.44**	0.21	**0.51**	0.20	**0.49**
Urinary bladder						**0.01**		**0.01**		**0.02**		**0.01**
Uterus		**0.02**		**0.03**		**0.03**		**0.03**		**0.04**		**0.04**
Ovaries		**0.02**		**0.03**		**0.03**		**0.04**		**0.05**		**0.04**
Prostate										**0.01**		
Testes												
Upper arm												
Elbow												
Forearm												
Upper leg												
Knee												
Lower leg												
Lymph nodes	0.04	**0.16**	0.05	**0.20**	0.06	**0.22**	0.06	**0.25**	0.07	**0.29**	0.07	**0.28**
Skin	0.08	**0.15**	0.08	**0.16**	0.09	**0.16**	0.09	**0.17**	0.09	**0.18**	0.09	**0.18**
Muscle	0.05	**0.11**	0.06	**0.13**	0.06	**0.15**	0.07	**0.16**	0.08	**0.18**	0.07	**0.18**
Skeleton	0.20	**0.29**	0.24	**0.35**	0.26	**0.39**	0.28	**0.42**	0.32	**0.48**	0.32	**0.47**
Red bone marrow	0.05	**0.08**	0.07	**0.09**	0.07	**0.11**	0.08	**0.12**	0.09	**0.13**	0.09	**0.13**
Whole body dose	0.06	**0.13**	0.07	**0.16**	0.08	**0.18**	0.09	**0.19**	0.10	**0.22**	0.10	**0.21**
ED (ICRP 60)	0.06	**0.14**	0.07	**0.18**	0.08	**0.20**	0.08	**0.22**	0.10	**0.25**	0.09	**0.24**
ED (ICRP 103)	0.06	**0.14**	0.07	**0.18**	0.08	**0.20**	0.08	**0.22**	0.10	**0.25**	0.09	**0.24**
Total filtration (mm) **Al**	**1.5**		**2.5**		**3.5**		**4.5**		**3.5**		**3.5**	
Cu	**0**		**0**		**0**		**0**		**0.1**		**0.2**	

Table 15.24 Normalized organ doses—Thoracic spine r-lat—<1 year

Beam projection:			r-lat									**Table cells:**
Focus-to-detector distance (cm):			100						(+) = normal font = optimal field setting			
Tube voltage (kVp):			70						**(−) = bold font = suboptimal field setting**			
Total filtrations:			See last table row.						Empty table cells correspond to values <0.01			
Organ or tissue		**Organ dose D_T : Entrance dose K_E**						**($mSv * mGy^{-1}$)**				
	(+)	(−)	(+)	(−)	(+)	(−)	(+)	(−)	(+)	(−)	(+)	(−)
Skull		**0.02**		**0.02**		**0.02**	0.01	**0.03**	0.01	**0.03**	0.02	**0.04**
Brain												**0.01**
Oral mucosa		**0.02**		**0.02**		**0.03**		**0.03**	0.01	**0.04**	0.01	**0.04**
Salivary glands		**0.03**	0.01	**0.04**	0.01	**0.05**	0.01	**0.05**	0.02	**0.06**	0.02	**0.07**
Cervical spine	0.06	**0.41**	0.07	**0.50**	0.08	**0.56**	0.09	**0.61**	0.11	**0.69**	0.13	**0.77**
Thyroid	0.02	**0.42**	0.02	**0.49**	0.02	**0.53**	0.03	**0.57**	0.03	**0.63**	0.04	**0.67**
Thymus	0.01	**0.05**	0.02	**0.07**	0.02	**0.08**	0.02	**0.09**	0.03	**0.10**	0.03	**0.12**
Thoracic spine	0.66	**0.75**	0.83	**0.93**	0.95	**1.07**	1.04	**1.18**	1.21	**1.37**	1.35	**1.54**
Claviculae	0.03	**0.34**	0.04	**0.40**	0.05	**0.44**	0.05	**0.47**	0.06	**0.52**	0.07	**0.56**
Scapulae	1.73	**1.77**	1.92	**1.98**	2.04	**2.10**	2.12	**2.19**	2.23	**2.33**	2.31	**2.42**
Ribs	0.62	**0.95**	0.71	**1.07**	0.77	**1.15**	0.81	**1.21**	0.87	**1.31**	0.92	**1.38**
Oesophagus	0.27	**0.35**	0.33	**0.43**	0.38	**0.49**	0.41	**0.54**	0.47	**0.62**	0.52	**0.69**
Trachea		**0.05**		**0.07**		**0.07**		**0.08**	0.01	**0.09**	0.01	**0.10**
Lung	0.23	**0.43**	0.27	**0.50**	0.30	**0.56**	0.32	**0.59**	0.36	**0.66**	0.39	**0.72**
Mamma	0.01	**0.04**	0.01	**0.05**	0.01	**0.06**	0.02	**0.06**	0.02	**0.08**	0.02	**0.09**
Heart	0.04	**0.21**	0.05	**0.26**	0.06	**0.30**	0.06	**0.33**	0.08	**0.38**	0.09	**0.43**
Stomach	0.02	**0.10**	0.02	**0.13**	0.02	**0.15**	0.03	**0.17**	0.03	**0.21**	0.04	**0.24**
Spleen	0.07	**0.11**	0.09	**0.14**	0.11	**0.16**	0.12	**0.18**	0.15	**0.22**	0.18	**0.26**
Liver	0.12	**0.45**	0.14	**0.53**	0.16	**0.58**	0.17	**0.62**	0.18	**0.68**	0.20	**0.74**
Gall bladder	0.02	**0.25**	0.03	**0.32**	0.03	**0.36**	0.04	**0.40**	0.05	**0.47**	0.06	**0.53**
Pancreas	0.06	**0.21**	0.07	**0.27**	0.08	**0.31**	0.09	**0.34**	0.11	**0.41**	0.13	**0.47**
Small intestine		**0.13**	0.01	**0.16**	0.01	**0.19**	0.01	**0.21**	0.02	**0.24**	0.02	**0.27**
Large intestine		**0.19**	0.01	**0.23**	0.01	**0.26**	0.02	**0.29**	0.02	**0.33**	0.02	**0.37**
Sigma/Rectum		**0.02**		**0.03**		**0.03**		**0.04**		**0.05**		**0.06**
Lumbar spine	0.14	**0.64**	0.17	**0.80**	0.20	**0.92**	0.22	**1.02**	0.26	**1.19**	0.30	**1.34**
Pelvis		**0.08**	0.01	**0.10**	0.01	**0.11**	0.02	**0.13**	0.02	**0.15**	0.02	**0.18**
Adrenals	0.23	**0.27**	0.28	**0.33**	0.32	**0.38**	0.35	**0.42**	0.41	**0.49**	0.45	**0.55**
Kidneys	0.13	**0.33**	0.16	**0.40**	0.18	**0.45**	0.20	**0.49**	0.23	**0.56**	0.26	**0.62**
Urinary bladder				**0.01**		**0.01**		**0.02**		**0.02**		**0.02**
Uterus		**0.02**		**0.03**		**0.04**		**0.04**		**0.05**		**0.06**
Ovaries		**0.03**		**0.04**		**0.04**		**0.05**		**0.06**	0.01	**0.07**
Prostate								**0.01**		**0.01**		**0.02**
Testes												
Upper arm												
Elbow												
Forearm												
Upper leg								**0.01**		**0.01**		**0.02**
Knee												
Lower leg												
Lymph nodes	0.05	**0.18**	0.06	**0.22**	0.07	**0.25**	0.07	**0.28**	0.08	**0.32**	0.09	**0.36**
Skin	0.08	**0.15**	0.09	**0.16**	0.09	**0.17**	0.09	**0.18**	0.10	**0.19**	0.10	**0.20**
Muscle	0.05	**0.13**	0.06	**0.15**	0.07	**0.16**	0.07	**0.18**	0.08	**0.20**	0.09	**0.22**
Skeleton	0.22	**0.32**	0.26	**0.38**	0.28	**0.42**	0.30	**0.45**	0.34	**0.51**	0.36	**0.55**
Red bone marrow	0.06	**0.09**	0.08	**0.11**	0.08	**0.12**	0.09	**0.13**	0.10	**0.15**	0.11	**0.17**
Whole body dose	0.07	**0.15**	0.08	**0.18**	0.09	**0.19**	0.09	**0.21**	0.11	**0.23**	0.12	**0.26**
ED (ICRP 60)	0.06	**0.17**	0.08	**0.20**	0.09	**0.22**	0.09	**0.24**	0.11	**0.27**	0.12	**0.30**
ED (ICRP 103)	0.06	**0.16**	0.08	**0.20**	0.09	**0.22**	0.09	**0.24**	0.11	**0.27**	0.12	**0.30**
Total filtration (mm) Al	**1.5**		**2.5**		**3.5**		**4.5**		**3.5**		**3.5**	
Cu	**0**		**0**		**0**		**0**		**0.1**		**0.2**	

Table 15.25 Normalized organ doses—Thoracic spine ap—1–2 years

Beam projection:		ap										**Table cells:**
Focus-to-detector distance (cm):		115										(+) = normal font = optimal field setting
Tube voltage (kVp):		50										**(−) = bold font = suboptimal field setting**
Total filtrations:		See last table row.										Empty table cells correspond to values <0.01
Organ or tissue	**Organ dose D_T : Entrance dose K_E**							**(mSv * mGy^{-1})**				
	(+)	(−)	(+)	(−)	(+)	(−)	(+)	(−)	(+)	(−)	(+)	(−)
Skull		**0.01**		**0.01**		**0.02**		**0.02**		**0.02**		**0.03**
Brain												
Oral mucosa		**0.02**		**0.02**		**0.02**		**0.03**		**0.03**		**0.04**
Salivary glands		**0.04**		**0.05**		**0.06**		**0.06**	0.01	**0.07**	0.01	**0.08**
Cervical spine	0.02	**0.22**	0.03	**0.29**	0.04	**0.35**	0.05	**0.39**	0.06	**0.48**	0.07	**0.56**
Thyroid	0.04	**0.77**	0.05	**0.89**	0.06	**0.95**	0.07	**1.00**	0.08	**1.08**	0.09	**1.14**
Thymus	0.57	**0.62**	0.68	**0.75**	0.75	**0.84**	0.81	**0.90**	0.90	**1.01**	0.98	**1.11**
Thoracic spine	0.07	**0.11**	0.10	**0.16**	0.12	**0.19**	0.14	**0.22**	0.18	**0.29**	0.23	**0.36**
Claviculae	0.27	**1.05**	0.32	**1.26**	0.35	**1.40**	0.38	**1.50**	0.43	**1.68**	0.47	**1.82**
Scapulae	0.02	**0.07**	0.02	**0.10**	0.03	**0.12**	0.03	**0.14**	0.04	**0.17**	0.05	**0.21**
Ribs	0.30	**0.76**	0.34	**0.88**	0.37	**0.96**	0.39	**1.01**	0.42	**1.11**	0.45	**1.19**
Oesophagus	0.09	**0.13**	0.12	**0.18**	0.14	**0.22**	0.16	**0.25**	0.20	**0.30**	0.24	**0.36**
Trachea		**0.07**		**0.08**		**0.09**		**0.10**		**0.11**	0.01	**0.12**
Lung	0.03	**0.20**	0.04	**0.25**	0.05	**0.29**	0.06	**0.32**	0.07	**0.38**	0.08	**0.43**
Mamma	0.02	**0.64**	0.02	**0.68**	0.03	**0.71**	0.03	**0.73**	0.04	**0.76**	0.04	**0.79**
Heart	0.24	**0.40**	0.30	**0.49**	0.35	**0.56**	0.38	**0.62**	0.44	**0.72**	0.49	**0.81**
Stomach	0.02	**0.31**	0.03	**0.39**	0.03	**0.44**	0.04	**0.48**	0.04	**0.56**	0.05	**0.63**
Spleen		**0.06**		**0.08**	0.01	**0.09**	0.01	**0.11**	0.02	**0.13**	0.02	**0.16**
Liver	0.06	**0.26**	0.08	**0.32**	0.09	**0.36**	0.10	**0.40**	0.12	**0.46**	0.13	**0.52**
Gall bladder	0.02	**0.28**	0.02	**0.36**	0.02	**0.42**	0.03	**0.47**	0.04	**0.56**	0.04	**0.64**
Pancreas	0.05	**0.15**	0.07	**0.20**	0.08	**0.24**	0.10	**0.27**	0.12	**0.33**	0.14	**0.39**
Small intestine		**0.06**		**0.08**		**0.09**		**0.10**		**0.13**		**0.15**
Large intestine		**0.14**		**0.18**		**0.20**		**0.23**		**0.27**		**0.31**
Sigma/Rectum		**0.01**		**0.01**		**0.02**		**0.02**		**0.02**		**0.03**
Lumbar spine		**0.10**		**0.13**	0.01	**0.17**	0.01	**0.19**	0.02	**0.25**	0.02	**0.31**
Pelvis		**0.01**		**0.02**		**0.02**		**0.03**		**0.04**		**0.05**
Adrenals	0.03	**0.06**	0.04	**0.08**	0.05	**0.09**	0.06	**0.11**	0.08	**0.14**	0.09	**0.17**
Kidneys		**0.05**		**0.06**		**0.08**		**0.09**	0.01	**0.11**	0.01	**0.14**
Urinary bladder												**0.01**
Uterus				**0.01**		**0.02**		**0.02**		**0.03**		**0.03**
Ovaries				**0.01**		**0.02**		**0.02**		**0.02**		**0.03**
Prostate												
Testes												
Upper arm	0.01	**0.03**	0.01	**0.04**	0.02	**0.05**	0.02	**0.06**	0.03	**0.08**	0.03	**0.10**
Elbow		**0.03**	0.01	**0.05**	0.02	**0.06**	0.02	**0.07**	0.03	**0.09**	0.03	**0.11**
Forearm		**0.01**		**0.02**		**0.02**		**0.03**		**0.04**		**0.04**
Upper leg												
Knee												
Lower leg												
Lymph nodes	0.03	**0.16**	0.04	**0.21**	0.05	**0.24**	0.05	**0.26**	0.06	**0.30**	0.07	**0.34**
Skin	0.02	**0.11**	0.03	**0.11**	0.03	**0.12**	0.03	**0.12**	0.03	**0.13**	0.03	**0.13**
Muscle	0.02	**0.07**	0.03	**0.09**	0.03	**0.10**	0.03	**0.11**	0.04	**0.13**	0.04	**0.15**
Skeleton	0.05	**0.13**	0.06	**0.16**	0.06	**0.18**	0.07	**0.20**	0.08	**0.23**	0.09	**0.26**
Red bone marrow	0.01	**0.03**	0.01	**0.04**	0.02	**0.05**	0.02	**0.05**	0.02	**0.06**	0.02	**0.07**
Whole body dose	0.03	**0.09**	0.03	**0.10**	0.04	**0.12**	0.04	**0.13**	0.05	**0.15**	0.05	**0.17**
ED (ICRP 60)	0.02	**0.17**	0.03	**0.21**	0.03	**0.23**	0.04	**0.25**	0.04	**0.29**	0.05	**0.32**
ED (ICRP 103)	0.03	**0.22**	0.03	**0.26**	0.04	**0.28**	0.04	**0.30**	0.05	**0.34**	0.06	**0.38**
Total filtration (mm) Al	**1.5**		**2.5**		**3.5**		**4.5**		**3.5**		**3.5**	
Cu	**0**		**0**		**0**		**0**		**0.1**		**0.2**	

Table 15.26 Normalized organ doses—Thoracic spine ap—1–2 years

Beam projection:			ap									**Table cells:**
Focus-to-detector distance (cm):			115									(+) = normal font = optimal field setting
Tube voltage (kVp):			60									**(−) = bold font = suboptimal field setting**
Total filtrations:			See last table row.									Empty table cells correspond to values <0.01
Organ or tissue	**Organ dose D_T : Entrance dose K_E**								**(mSv ∗ mGy^{-1})**			
	(+)	(−)	(+)	(−)	(+)	(−)	(+)	(−)	(+)	(−)	(+)	(−)
Skull		**0.01**		**0.02**		**0.02**		**0.03**		**0.03**		**0.03**
Brain												
Oral mucosa		**0.02**		**0.03**		**0.03**		**0.04**	0.01	**0.04**		**0.04**
Salivary glands		**0.05**		**0.06**	0.01	**0.07**	0.01	**0.07**	0.01	**0.09**	0.01	**0.08**
Cervical spine	0.03	**0.29**	0.05	**0.38**	0.06	**0.44**	0.06	**0.49**	0.08	**0.59**	0.07	**0.56**
Thyroid	0.05	**0.85**	0.07	**0.96**	0.08	**1.02**	0.08	**1.07**	0.10	**1.15**	0.09	**1.14**
Thymus	0.65	**0.72**	0.76	**0.85**	0.83	**0.94**	0.89	**1.00**	0.98	**1.12**	0.98	**1.11**
Thoracic spine	0.11	**0.17**	0.15	**0.23**	0.18	**0.28**	0.21	**0.32**	0.26	**0.41**	0.23	**0.36**
Claviculae	0.30	**1.19**	0.36	**1.40**	0.39	**1.54**	0.42	**1.64**	0.47	**1.81**	0.47	**1.82**
Scapulae	0.02	**0.10**	0.03	**0.13**	0.04	**0.16**	0.04	**0.18**	0.06	**0.23**	0.05	**0.21**
Ribs	0.33	**0.84**	0.37	**0.96**	0.40	**1.04**	0.42	**1.10**	0.45	**1.19**	0.45	**1.19**
Oesophagus	0.12	**0.18**	0.16	**0.24**	0.19	**0.29**	0.21	**0.32**	0.26	**0.40**	0.24	**0.36**
Trachea		**0.08**		**0.10**		**0.11**	0.01	**0.11**	0.01	**0.13**	0.01	**0.12**
Lung	0.04	**0.25**	0.05	**0.30**	0.06	**0.35**	0.07	**0.38**	0.09	**0.44**	0.08	**0.43**
Mamma	0.02	**0.67**	0.03	**0.71**	0.03	**0.74**	0.04	**0.76**	0.05	**0.79**	0.04	**0.79**
Heart	0.29	**0.48**	0.36	**0.59**	0.41	**0.66**	0.44	**0.72**	0.51	**0.83**	0.49	**0.81**
Stomach	0.03	**0.37**	0.03	**0.46**	0.04	**0.52**	0.05	**0.56**	0.05	**0.64**	0.05	**0.63**
Spleen		**0.08**	0.01	**0.11**	0.02	**0.13**	0.02	**0.14**	0.02	**0.18**	0.02	**0.16**
Liver	0.08	**0.31**	0.10	**0.38**	0.11	**0.43**	0.12	**0.47**	0.14	**0.54**	0.13	**0.52**
Gall bladder	0.02	**0.35**	0.03	**0.45**	0.03	**0.51**	0.04	**0.57**	0.05	**0.67**	0.04	**0.64**
Pancreas	0.07	**0.20**	0.09	**0.26**	0.11	**0.31**	0.12	**0.35**	0.15	**0.42**	0.14	**0.39**
Small intestine		**0.08**		**0.10**		**0.12**		**0.13**		**0.16**		**0.15**
Large intestine		**0.17**		**0.22**		**0.25**		**0.28**		**0.33**		**0.31**
Sigma/Rectum		**0.02**		**0.02**		**0.02**		**0.03**		**0.03**		**0.03**
Lumbar spine	0.01	**0.15**	0.02	**0.20**	0.02	**0.25**	0.02	**0.28**	0.03	**0.36**	0.02	**0.31**
Pelvis		**0.02**		**0.03**		**0.04**		**0.04**		**0.06**		**0.05**
Adrenals	0.05	**0.08**	0.06	**0.11**	0.07	**0.13**	0.08	**0.15**	0.11	**0.19**	0.09	**0.17**
Kidneys		**0.07**		**0.09**	0.01	**0.11**	0.01	**0.12**	0.02	**0.15**	0.01	**0.14**
Urinary bladder								**0.01**		**0.01**		**0.01**
Uterus		**0.02**		**0.02**		**0.03**		**0.03**		**0.04**		**0.03**
Ovaries		**0.01**		**0.02**		**0.02**		**0.03**		**0.03**		**0.03**
Prostate												
Testes												
Upper arm	0.02	**0.05**	0.02	**0.06**	0.03	**0.08**	0.03	**0.09**	0.04	**0.11**	0.03	**0.10**
Elbow	0.02	**0.05**	0.02	**0.07**	0.03	**0.09**	0.03	**0.10**	0.04	**0.13**	0.03	**0.11**
Forearm		**0.02**		**0.03**		**0.04**		**0.04**		**0.05**		**0.04**
Upper leg												
Knee												
Lower leg												
Lymph nodes	0.04	**0.20**	0.05	**0.25**	0.06	**0.28**	0.06	**0.31**	0.08	**0.36**	0.07	**0.34**
Skin	0.03	**0.11**	0.03	**0.12**	0.03	**0.12**	0.03	**0.13**	0.03	**0.14**	0.03	**0.13**
Muscle	0.03	**0.09**	0.03	**0.11**	0.04	**0.12**	0.04	**0.13**	0.05	**0.15**	0.04	**0.15**
Skeleton	0.06	**0.16**	0.07	**0.19**	0.08	**0.22**	0.08	**0.24**	0.09	**0.27**	0.09	**0.26**
Red bone marrow	0.01	**0.04**	0.02	**0.05**	0.02	**0.06**	0.02	**0.06**	0.03	**0.07**	0.02	**0.07**
Whole body dose	0.03	**0.10**	0.04	**0.12**	0.04	**0.14**	0.05	**0.15**	0.05	**0.18**	0.05	**0.17**
ED (ICRP 60)	0.03	**0.20**	0.04	**0.24**	0.04	**0.27**	0.05	**0.29**	0.06	**0.33**	0.05	**0.32**
ED (ICRP 103)	0.03	**0.25**	0.04	**0.29**	0.05	**0.32**	0.05	**0.34**	0.06	**0.39**	0.06	**0.38**
Total filtration (mm) Al	**1.5**		**2.5**		**3.5**		**4.5**		**3.5**		**3.5**	
Cu	**0**		**0**		**0**		**0**		**0.1**		**0.2**	

Table 15.27 Normalized organ doses—Thoracic spine ap—1–2 years

Beam projection:			ap									**Table cells:**
Focus-to-detector distance (cm):			115									(+) = normal font = optimal field setting
Tube voltage (kVp):			70									**(−) = bold font = suboptimal field setting**
Total filtrations:			See last table row.									Empty table cells correspond to values <0.01
Organ or tissue	**Organ dose D_T : Entrance dose K_E**								**(mSv ∗ mGy^{-1})**			
	(+)	(−)	(+)	(−)	(+)	(−)	(+)	(−)	(+)	(−)	(+)	(−)
Skull		**0.02**		**0.02**		**0.03**		**0.03**		**0.04**	0.01	**0.04**
Brain												
Oral mucosa		**0.03**		**0.03**		**0.04**	0.01	**0.04**	0.01	**0.05**	0.02	**0.06**
Salivary glands		**0.06**	0.01	**0.07**	0.01	**0.08**	0.01	**0.08**	0.02	**0.10**	0.02	**0.11**
Cervical spine	0.04	**0.35**	0.06	**0.44**	0.07	**0.51**	0.08	**0.57**	0.10	**0.67**	0.11	**0.76**
Thyroid	0.06	**0.90**	0.08	**1.01**	0.09	**1.08**	0.10	**1.12**	0.11	**1.20**	0.13	**1.27**
Thymus	0.71	**0.79**	0.82	**0.92**	0.89	**1.01**	0.95	**1.08**	1.04	**1.19**	1.11	**1.28**
Thoracic spine	0.15	**0.23**	0.19	**0.30**	0.23	**0.36**	0.26	**0.41**	0.32	**0.51**	0.38	**0.61**
Claviculae	0.33	**1.29**	0.38	**1.49**	0.42	**1.62**	0.45	**1.72**	0.49	**1.87**	0.52	**1.99**
Scapulae	0.03	**0.13**	0.04	**0.17**	0.05	**0.19**	0.06	**0.22**	0.07	**0.27**	0.08	**0.31**
Ribs	0.35	**0.90**	0.39	**1.02**	0.41	**1.09**	0.43	**1.15**	0.46	**1.24**	0.49	**1.31**
Oesophagus	0.15	**0.23**	0.20	**0.30**	0.23	**0.35**	0.25	**0.39**	0.31	**0.47**	0.35	**0.54**
Trachea		**0.09**		**0.10**	0.01	**0.12**	0.01	**0.12**	0.02	**0.14**	0.02	**0.15**
Lung	0.05	**0.29**	0.06	**0.35**	0.08	**0.39**	0.08	**0.43**	0.10	**0.49**	0.12	**0.55**
Mamma	0.03	**0.69**	0.03	**0.74**	0.04	**0.76**	0.04	**0.78**	0.05	**0.82**	0.06	**0.84**
Heart	0.34	**0.55**	0.40	**0.66**	0.45	**0.74**	0.49	**0.80**	0.55	**0.91**	0.61	**1.00**
Stomach	0.03	**0.43**	0.04	**0.51**	0.05	**0.57**	0.05	**0.62**	0.06	**0.71**	0.07	**0.78**
Spleen	0.01	**0.10**	0.02	**0.13**	0.02	**0.16**	0.02	**0.17**	0.03	**0.21**	0.04	**0.25**
Liver	0.09	**0.35**	0.11	**0.43**	0.12	**0.48**	0.13	**0.52**	0.15	**0.59**	0.17	**0.65**
Gall bladder	0.03	**0.42**	0.03	**0.52**	0.04	**0.59**	0.05	**0.64**	0.06	**0.75**	0.06	**0.84**
Pancreas	0.09	**0.25**	0.11	**0.32**	0.13	**0.37**	0.15	**0.41**	0.17	**0.49**	0.20	**0.56**
Small intestine		**0.10**		**0.12**		**0.14**		**0.16**		**0.19**	0.01	**0.22**
Large intestine		**0.20**		**0.25**		**0.29**		**0.31**	0.01	**0.37**	0.01	**0.41**
Sigma/Rectum		**0.02**		**0.03**		**0.03**		**0.03**		**0.04**		**0.05**
Lumbar spine	0.02	**0.20**	0.02	**0.26**	0.03	**0.31**	0.03	**0.36**	0.04	**0.45**	0.05	**0.53**
Pelvis		**0.03**		**0.04**		**0.05**		**0.06**		**0.07**		**0.09**
Adrenals	0.06	**0.11**	0.08	**0.14**	0.09	**0.17**	0.11	**0.19**	0.13	**0.23**	0.15	**0.28**
Kidneys	0.01	**0.09**	0.01	**0.11**	0.02	**0.14**	0.02	**0.15**	0.02	**0.19**	0.03	**0.23**
Urinary bladder				**0.01**		**0.01**		**0.01**		**0.02**		**0.02**
Uterus		**0.02**		**0.03**		**0.03**		**0.04**		**0.05**		**0.06**
Ovaries		**0.02**		**0.02**		**0.03**		**0.03**		**0.04**		**0.05**
Prostate												
Testes												
Upper arm	0.02	**0.06**	0.03	**0.08**	0.03	**0.10**	0.04	**0.11**	0.05	**0.14**	0.06	**0.17**
Elbow	0.02	**0.07**	0.03	**0.09**	0.03	**0.11**	0.04	**0.13**	0.05	**0.16**	0.06	**0.19**
Forearm		**0.03**		**0.04**		**0.05**		**0.05**		**0.07**		**0.08**
Upper leg												
Knee												
Lower leg												
Lymph nodes	0.05	**0.23**	0.06	**0.28**	0.07	**0.32**	0.08	**0.35**	0.09	**0.40**	0.10	**0.45**
Skin	0.03	**0.12**	0.03	**0.13**	0.03	**0.13**	0.03	**0.14**	0.03	**0.14**	0.04	**0.15**
Muscle	0.03	**0.10**	0.04	**0.12**	0.04	**0.14**	0.05	**0.15**	0.05	**0.17**	0.06	**0.19**
Skeleton	0.07	**0.18**	0.08	**0.22**	0.09	**0.24**	0.09	**0.26**	0.11	**0.30**	0.12	**0.33**
Red bone marrow	0.02	**0.05**	0.02	**0.06**	0.02	**0.07**	0.03	**0.07**	0.03	**0.08**	0.03	**0.10**
Whole body dose	0.04	**0.12**	0.04	**0.14**	0.05	**0.16**	0.05	**0.17**	0.06	**0.19**	0.07	**0.22**
ED (ICRP 60)	0.03	**0.23**	0.04	**0.27**	0.05	**0.30**	0.05	**0.32**	0.06	**0.36**	0.07	**0.39**
ED (ICRP 103)	0.04	**0.28**	0.05	**0.32**	0.06	**0.35**	0.06	**0.38**	0.07	**0.42**	0.08	**0.46**
Total filtration (mm) Al	**1.5**		**2.5**		**3.5**		**4.5**		**3.5**		**3.5**	
Cu	**0**		**0**		**0**		**0**		**0.1**		**0.2**	

Table 15.28 Normalized organ doses—Thoracic spine ap—3–7 years

Beam projection:			ap									**Table cells:**
Focus-to-detector distance (cm):			115									(+) = normal font = optimal field setting
Tube voltage (kVp):			60									**(−) = bold font = suboptimal field setting**
Total filtrations:			See last table row.									Empty table cells correspond to values <0.01
Organ or tissue	**Organ dose D_T : Entrance dose K_E**							**($mSv * mGy^{-1}$)**				
	(+)	(−)	(+)	(−)	(+)	(−)	(+)	(−)	(+)	(−)	(+)	(−)
Skull		**0.01**		**0.02**		**0.02**		**0.02**		**0.03**		**0.02**
Brain												
Oral mucosa		**0.02**		**0.02**		**0.03**		**0.03**	0.01	**0.04**		**0.03**
Salivary glands		**0.03**		**0.04**	0.01	**0.04**	0.01	**0.05**	0.01	**0.06**	0.01	**0.05**
Cervical spine	0.03	**0.26**	0.05	**0.34**	0.06	**0.40**	0.06	**0.45**	0.08	**0.55**	0.07	**0.51**
Thyroid	0.06	**0.75**	0.07	**0.86**	0.08	**0.93**	0.09	**0.98**	0.11	**1.06**	0.11	**1.05**
Thymus	0.63	**0.68**	0.76	**0.81**	0.84	**0.91**	0.91	**0.98**	1.02	**1.11**	1.00	**1.08**
Thoracic spine	0.09	**0.13**	0.12	**0.17**	0.15	**0.21**	0.17	**0.24**	0.22	**0.31**	0.18	**0.26**
Claviculae	0.39	**1.02**	0.46	**1.22**	0.51	**1.36**	0.55	**1.46**	0.61	**1.63**	0.61	**1.63**
Scapulae	0.02	**0.07**	0.03	**0.09**	0.04	**0.11**	0.04	**0.13**	0.06	**0.16**	0.05	**0.14**
Ribs	0.41	**0.78**	0.46	**0.89**	0.50	**0.97**	0.52	**1.02**	0.57	**1.12**	0.56	**1.11**
Oesophagus	0.08	**0.11**	0.11	**0.15**	0.13	**0.17**	0.15	**0.20**	0.18	**0.25**	0.16	**0.22**
Trachea		**0.09**		**0.10**	0.01	**0.12**	0.01	**0.12**	0.01	**0.14**	0.01	**0.14**
Lung	0.05	**0.20**	0.06	**0.25**	0.08	**0.29**	0.09	**0.32**	0.11	**0.38**	0.10	**0.36**
Mamma	0.02	**0.31**	0.03	**0.34**	0.04	**0.36**	0.04	**0.38**	0.05	**0.40**	0.04	**0.40**
Heart	0.30	**0.42**	0.37	**0.52**	0.43	**0.59**	0.47	**0.65**	0.54	**0.76**	0.52	**0.73**
Stomach	0.06	**0.33**	0.07	**0.41**	0.09	**0.46**	0.10	**0.51**	0.11	**0.59**	0.11	**0.57**
Spleen	0.01	**0.06**	0.02	**0.08**	0.02	**0.09**	0.02	**0.10**	0.03	**0.13**	0.02	**0.11**
Liver	0.10	**0.26**	0.13	**0.33**	0.14	**0.37**	0.16	**0.41**	0.19	**0.48**	0.18	**0.46**
Gall bladder	0.06	**0.31**	0.07	**0.40**	0.09	**0.47**	0.10	**0.52**	0.12	**0.62**	0.11	**0.59**
Pancreas	0.09	**0.16**	0.12	**0.22**	0.14	**0.26**	0.16	**0.29**	0.20	**0.36**	0.18	**0.33**
Small intestine		**0.07**		**0.09**		**0.11**		**0.12**		**0.15**		**0.14**
Large intestine		**0.15**		**0.19**		**0.22**		**0.24**	0.01	**0.29**		**0.27**
Sigma/Rectum		**0.01**		**0.02**		**0.02**		**0.02**		**0.03**		**0.03**
Lumbar spine	0.02	**0.11**	0.03	**0.15**	0.03	**0.18**	0.04	**0.21**	0.05	**0.27**	0.04	**0.23**
Pelvis		**0.02**		**0.02**		**0.03**		**0.03**		**0.04**		**0.03**
Adrenals	0.04	**0.06**	0.06	**0.09**	0.07	**0.10**	0.08	**0.12**	0.10	**0.15**	0.09	**0.13**
Kidneys	0.01	**0.05**	0.02	**0.07**	0.02	**0.08**	0.02	**0.09**	0.03	**0.12**	0.02	**0.10**
Urinary bladder												
Uterus		**0.01**		**0.01**		**0.02**		**0.02**		**0.03**		**0.02**
Ovaries		**0.01**		**0.01**		**0.02**		**0.02**		**0.03**		**0.02**
Prostate												
Testes												
Upper arm	0.01	**0.03**	0.02	**0.04**	0.02	**0.05**	0.03	**0.06**	0.03	**0.08**	0.03	**0.06**
Elbow	0.01	**0.03**	0.02	**0.05**	0.02	**0.06**	0.03	**0.07**	0.03	**0.09**	0.03	**0.07**
Forearm		**0.01**		**0.02**		**0.02**		**0.03**		**0.03**		**0.03**
Upper leg												
Knee												
Lower leg												
Lymph nodes	0.05	**0.17**	0.06	**0.22**	0.07	**0.25**	0.08	**0.27**	0.10	**0.32**	0.09	**0.30**
Skin	0.03	**0.09**	0.03	**0.10**	0.03	**0.10**	0.04	**0.10**	0.04	**0.11**	0.04	**0.11**
Muscle	0.03	**0.07**	0.04	**0.09**	0.04	**0.10**	0.04	**0.11**	0.05	**0.13**	0.05	**0.12**
Skeleton	0.05	**0.11**	0.06	**0.14**	0.07	**0.15**	0.08	**0.17**	0.09	**0.19**	0.08	**0.18**
Red bone marrow	0.02	**0.04**	0.02	**0.05**	0.02	**0.05**	0.03	**0.06**	0.03	**0.07**	0.03	**0.06**
Whole body dose	0.03	**0.09**	0.04	**0.10**	0.05	**0.12**	0.05	**0.13**	0.06	**0.15**	0.06	**0.14**
ED (ICRP 60)	0.03	**0.16**	0.04	**0.19**	0.05	**0.22**	0.05	**0.24**	0.07	**0.27**	0.06	**0.26**
ED (ICRP 103)	0.04	**0.18**	0.05	**0.22**	0.06	**0.25**	0.06	**0.27**	0.08	**0.30**	0.07	**0.29**
Total filtration (mm) Al	**1.5**		**2.5**		**3.5**		**4.5**		**3.5**		**3.5**	
Cu	**0**		**0**		**0**		**0**		**0.1**		**0.2**	

Table 15.29 Normalized organ doses—Thoracic spine ap—3–7 years

Beam projection:			ap									**Table cells:**
Focus-to-detector distance (cm):			115									(+) = normal font = optimal field setting
Tube voltage (kVp):			70									**(−) = bold font = suboptimal field setting**
Total filtrations:			See last table row.									Empty table cells correspond to values <0.01
Organ or tissue	**Organ dose D_T : Entrance dose K_E**							**($mSv * mGy^{-1}$)**				
	(+)	(−)	(+)	(−)	(+)	(−)	(+)	(−)	(+)	(−)	(+)	(−)
Skull		**0.02**		**0.02**		**0.03**		**0.03**	0.01	**0.03**	0.01	**0.04**
Brain												
Oral mucosa		**0.02**		**0.03**		**0.03**	0.01	**0.04**	0.01	**0.04**	0.02	**0.05**
Salivary glands		**0.04**	0.01	**0.05**	0.01	**0.05**	0.01	**0.06**	0.02	**0.07**	0.02	**0.08**
Cervical spine	0.05	**0.32**	0.06	**0.41**	0.07	**0.48**	0.08	**0.53**	0.10	**0.63**	0.12	**0.73**
Thyroid	0.07	**0.81**	0.09	**0.92**	0.10	**0.98**	0.11	**1.03**	0.13	**1.12**	0.15	**1.19**
Thymus	0.70	**0.76**	0.83	**0.90**	0.91	**0.99**	0.98	**1.07**	1.09	**1.19**	1.18	**1.30**
Thoracic spine	0.12	**0.17**	0.16	**0.23**	0.19	**0.28**	0.22	**0.32**	0.28	**0.40**	0.34	**0.49**
Claviculae	0.43	**1.13**	0.50	**1.32**	0.55	**1.45**	0.58	**1.54**	0.64	**1.70**	0.69	**1.82**
Scapulae	0.03	**0.09**	0.04	**0.12**	0.05	**0.14**	0.06	**0.16**	0.07	**0.20**	0.09	**0.23**
Ribs	0.44	**0.84**	0.49	**0.95**	0.52	**1.02**	0.55	**1.08**	0.59	**1.17**	0.62	**1.24**
Oesophagus	0.10	**0.14**	0.13	**0.18**	0.16	**0.22**	0.18	**0.24**	0.22	**0.30**	0.26	**0.35**
Trachea		**0.10**	0.01	**0.11**	0.01	**0.13**	0.01	**0.13**	0.02	**0.15**	0.02	**0.16**
Lung	0.06	**0.24**	0.08	**0.29**	0.09	**0.33**	0.10	**0.36**	0.13	**0.42**	0.15	**0.48**
Mamma	0.03	**0.33**	0.04	**0.36**	0.04	**0.38**	0.05	**0.40**	0.06	**0.43**	0.07	**0.45**
Heart	0.35	**0.49**	0.43	**0.59**	0.48	**0.67**	0.52	**0.73**	0.60	**0.85**	0.67	**0.95**
Stomach	0.07	**0.38**	0.09	**0.47**	0.10	**0.53**	0.11	**0.57**	0.13	**0.66**	0.15	**0.74**
Spleen	0.02	**0.08**	0.02	**0.10**	0.03	**0.12**	0.03	**0.13**	0.04	**0.16**	0.04	**0.19**
Liver	0.12	**0.31**	0.15	**0.37**	0.16	**0.42**	0.18	**0.46**	0.21	**0.54**	0.23	**0.60**
Gall bladder	0.07	**0.38**	0.09	**0.47**	0.10	**0.54**	0.12	**0.60**	0.14	**0.71**	0.16	**0.80**
Pancreas	0.12	**0.21**	0.15	**0.27**	0.17	**0.32**	0.20	**0.36**	0.24	**0.43**	0.28	**0.51**
Small intestine		**0.09**		**0.11**		**0.13**		**0.15**	0.01	**0.18**	0.01	**0.21**
Large intestine		**0.18**		**0.22**		**0.26**	0.01	**0.28**	0.01	**0.33**	0.02	**0.38**
Sigma/Rectum		**0.02**		**0.02**		**0.03**		**0.03**		**0.04**		**0.04**
Lumbar spine	0.03	**0.15**	0.04	**0.20**	0.05	**0.24**	0.05	**0.28**	0.07	**0.35**	0.08	**0.43**
Pelvis		**0.02**		**0.03**		**0.04**		**0.04**		**0.06**		**0.07**
Adrenals	0.06	**0.09**	0.08	**0.12**	0.09	**0.14**	0.10	**0.16**	0.13	**0.20**	0.15	**0.24**
Kidneys	0.02	**0.07**	0.02	**0.09**	0.03	**0.11**	0.03	**0.12**	0.04	**0.15**	0.04	**0.18**
Urinary bladder										**0.01**		**0.01**
Uterus		**0.01**		**0.02**		**0.02**		**0.03**		**0.03**		**0.04**
Ovaries		**0.01**		**0.02**		**0.02**		**0.03**		**0.03**		**0.04**
Prostate												
Testes												
Upper arm	0.02	**0.04**	0.03	**0.06**	0.03	**0.07**	0.04	**0.08**	0.05	**0.10**	0.06	**0.12**
Elbow	0.02	**0.05**	0.03	**0.06**	0.03	**0.08**	0.04	**0.09**	0.04	**0.11**	0.05	**0.14**
Forearm		**0.02**		**0.02**		**0.03**		**0.03**		**0.04**		**0.05**
Upper leg												
Knee												
Lower leg												
Lymph nodes	0.06	**0.21**	0.07	**0.25**	0.08	**0.29**	0.09	**0.31**	0.11	**0.36**	0.13	**0.41**
Skin	0.03	**0.10**	0.03	**0.10**	0.04	**0.11**	0.04	**0.11**	0.04	**0.12**	0.04	**0.12**
Muscle	0.03	**0.09**	0.04	**0.10**	0.05	**0.12**	0.05	**0.13**	0.06	**0.14**	0.07	**0.16**
Skeleton	0.06	**0.13**	0.07	**0.16**	0.08	**0.17**	0.09	**0.19**	0.10	**0.22**	0.11	**0.24**
Red bone marrow	0.02	**0.04**	0.03	**0.05**	0.03	**0.06**	0.03	**0.07**	0.04	**0.08**	0.04	**0.09**
Whole body dose	0.04	**0.10**	0.05	**0.12**	0.05	**0.13**	0.06	**0.14**	0.07	**0.17**	0.08	**0.19**
ED (ICRP 60)	0.04	**0.18**	0.05	**0.22**	0.06	**0.24**	0.06	**0.26**	0.08	**0.30**	0.09	**0.34**
ED (ICRP 103)	0.05	**0.21**	0.06	**0.25**	0.07	**0.27**	0.07	**0.30**	0.09	**0.34**	0.10	**0.38**
Total filtration (mm) Al	**1.5**		**2.5**		**3.5**		**4.5**		**3.5**		**3.5**	
Cu	**0**		**0**		**0**		**0**		**0.1**		**0.2**	

Table 15.30 Normalized organ doses–Thoracic spine ap—8–12 years

Beam projection:			ap									**Table cells:**
Focus-to-detector distance (cm):			115									(+) = normal font = optimal field setting
Tube voltage (kVp):			60									**(−) = bold font = suboptimal field setting**
Total filtrations:			See last table row.									Empty table cells correspond to values <0.01
Organ or tissue	**Organ dose D_T : Entrance dose K_E (mSv * mGy^{-1})**											
	(+)	(−)	(+)	(−)	(+)	(−)	(+)	(−)	(+)	(−)	(+)	(−)
Skull		**0.01**		**0.02**		**0.02**		**0.02**		**0.03**		**0.02**
Brain												
Oral mucosa		**0.01**		**0.02**		**0.02**		**0.02**		**0.03**		**0.03**
Salivary glands		**0.03**		**0.04**		**0.05**		**0.05**	0.01	**0.06**	0.01	**0.06**
Cervical spine	0.03	**0.26**	0.05	**0.34**	0.06	**0.40**	0.06	**0.46**	0.08	**0.56**	0.07	**0.51**
Thyroid	0.06	**0.68**	0.08	**0.79**	0.09	**0.86**	0.10	**0.92**	0.13	**1.01**	0.12	**0.99**
Thymus	0.61	**0.64**	0.74	**0.78**	0.83	**0.88**	0.90	**0.96**	1.03	**1.10**	1.01	**1.07**
Thoracic spine	0.07	**0.09**	0.10	**0.13**	0.12	**0.16**	0.14	**0.18**	0.18	**0.24**	0.14	**0.19**
Claviculae	0.44	**0.93**	0.52	**1.12**	0.58	**1.25**	0.63	**1.35**	0.71	**1.52**	0.70	**1.51**
Scapulae	0.02	**0.05**	0.03	**0.07**	0.03	**0.09**	0.04	**0.10**	0.05	**0.13**	0.04	**0.11**
Ribs	0.44	**0.74**	0.50	**0.85**	0.54	**0.92**	0.57	**0.98**	0.62	**1.07**	0.62	**1.06**
Oesophagus	0.06	**0.07**	0.08	**0.10**	0.09	**0.12**	0.11	**0.14**	0.13	**0.17**	0.12	**0.15**
Trachea	0.01	**0.20**	0.02	**0.24**	0.02	**0.26**	0.02	**0.28**	0.03	**0.32**	0.03	**0.31**
Lung	0.05	**0.18**	0.07	**0.23**	0.08	**0.27**	0.09	**0.29**	0.12	**0.35**	0.11	**0.33**
Mamma	0.02	**0.19**	0.03	**0.22**	0.04	**0.23**	0.04	**0.25**	0.05	**0.27**	0.04	**0.26**
Heart	0.30	**0.37**	0.37	**0.47**	0.43	**0.54**	0.47	**0.59**	0.56	**0.70**	0.53	**0.67**
Stomach	0.06	**0.29**	0.08	**0.36**	0.09	**0.41**	0.11	**0.46**	0.13	**0.54**	0.12	**0.51**
Spleen	0.01	**0.05**	0.01	**0.06**	0.02	**0.07**	0.02	**0.09**	0.03	**0.11**	0.02	**0.09**
Liver	0.09	**0.21**	0.12	**0.26**	0.14	**0.30**	0.15	**0.34**	0.18	**0.40**	0.17	**0.38**
Gall bladder	0.05	**0.28**	0.07	**0.36**	0.08	**0.42**	0.09	**0.47**	0.11	**0.56**	0.10	**0.53**
Pancreas	0.09	**0.14**	0.11	**0.19**	0.14	**0.23**	0.16	**0.26**	0.19	**0.32**	0.17	**0.29**
Small intestine		**0.05**		**0.07**		**0.08**		**0.09**		**0.11**		**0.10**
Large intestine		**0.12**		**0.15**		**0.18**		**0.20**		**0.24**		**0.22**
Sigma/Rectum				**0.01**		**0.01**		**0.02**		**0.02**		**0.02**
Lumbar spine	0.02	**0.08**	0.02	**0.11**	0.03	**0.13**	0.03	**0.16**	0.04	**0.20**	0.03	**0.16**
Pelvis		**0.01**		**0.01**		**0.02**		**0.02**		**0.03**		**0.02**
Adrenals	0.04	**0.05**	0.05	**0.07**	0.06	**0.09**	0.07	**0.10**	0.09	**0.13**	0.07	**0.11**
Kidneys	0.01	**0.04**	0.01	**0.05**	0.02	**0.06**	0.02	**0.07**	0.03	**0.09**	0.02	**0.08**
Urinary bladder												
Uterus						**0.01**		**0.01**		**0.02**		**0.01**
Ovaries						**0.01**		**0.01**		**0.02**		**0.01**
Prostate												
Testes												
Upper arm	0.01	**0.02**	0.01	**0.03**	0.02	**0.04**	0.02	**0.04**	0.03	**0.05**	0.02	**0.04**
Elbow	0.01	**0.02**	0.01	**0.03**	0.02	**0.04**	0.02	**0.05**	0.03	**0.06**	0.02	**0.05**
Forearm				**0.01**		**0.01**		**0.02**		**0.02**		**0.02**
Upper leg												
Knee												
Lower leg												
Lymph nodes	0.05	**0.17**	0.06	**0.21**	0.07	**0.24**	0.08	**0.26**	0.10	**0.31**	0.09	**0.29**
Skin	0.03	**0.09**	0.04	**0.09**	0.04	**0.10**	0.04	**0.10**	0.04	**0.10**	0.04	**0.10**
Muscle	0.03	**0.07**	0.04	**0.08**	0.04	**0.09**	0.05	**0.10**	0.05	**0.12**	0.05	**0.11**
Skeleton	0.05	**0.10**	0.06	**0.12**	0.07	**0.14**	0.08	**0.15**	0.09	**0.17**	0.08	**0.16**
Red bone marrow	0.02	**0.04**	0.03	**0.05**	0.03	**0.06**	0.03	**0.06**	0.04	**0.07**	0.03	**0.07**
Whole body dose	0.04	**0.08**	0.04	**0.09**	0.05	**0.11**	0.05	**0.12**	0.06	**0.14**	0.06	**0.13**
ED (ICRP 60)	0.03	**0.13**	0.04	**0.17**	0.05	**0.19**	0.06	**0.21**	0.07	**0.24**	0.06	**0.23**
ED (ICRP 103)	0.04	**0.15**	0.05	**0.18**	0.06	**0.21**	0.07	**0.23**	0.08	**0.27**	0.07	**0.25**
Total filtration (mm) Al	**1.5**		**2.5**		**3.5**		**4.5**		**3.5**		**3.5**	
Cu	**0**		**0**		**0**		**0**		**0.1**		**0.2**	

Table 15.31 Normalized organ doses—Thoracic spine ap—8–12 years

Beam projection:			ap									**Table cells:**
Focus-to-detector distance (cm):			115				(+) = normal font = optimal field setting					
Tube voltage (kVp):			70				**(−) = bold font = suboptimal field setting**					
Total filtrations:			See last table row.				Empty table cells correspond to values <0.01					
Organ or tissue	**Organ dose D_T : Entrance dose K_E**								**(mSv ∗ mGy^{-1})**			
	(+)	(−)	(+)	(−)	(+)	(−)	(+)	(−)	(+)	(−)	(+)	(−)
Skull		**0.02**		**0.02**		**0.02**		**0.03**	0.01	**0.03**	0.01	**0.04**
Brain												
Oral mucosa		**0.02**		**0.02**		**0.03**		**0.03**	0.01	**0.04**	0.01	**0.04**
Salivary glands		**0.04**		**0.05**	0.01	**0.06**	0.01	**0.06**	0.02	**0.07**	0.02	**0.08**
Cervical spine	0.05	**0.33**	0.06	**0.42**	0.07	**0.49**	0.08	**0.54**	0.10	**0.65**	0.13	**0.76**
Thyroid	0.08	**0.75**	0.09	**0.85**	0.11	**0.93**	0.12	**0.98**	0.15	**1.07**	0.17	**1.15**
Thymus	0.69	**0.73**	0.82	**0.88**	0.92	**0.98**	0.99	**1.06**	1.11	**1.19**	1.22	**1.31**
Thoracic spine	0.10	**0.14**	0.13	**0.18**	0.16	**0.22**	0.19	**0.25**	0.24	**0.32**	0.29	**0.39**
Claviculae	0.48	**1.03**	0.57	**1.22**	0.63	**1.35**	0.67	**1.45**	0.75	**1.61**	0.81	**1.74**
Scapulae	0.03	**0.07**	0.04	**0.09**	0.05	**0.11**	0.05	**0.13**	0.07	**0.16**	0.08	**0.19**
Ribs	0.47	**0.80**	0.53	**0.91**	0.57	**0.98**	0.60	**1.03**	0.65	**1.12**	0.69	**1.20**
Oesophagus	0.08	**0.10**	0.10	**0.13**	0.12	**0.15**	0.13	**0.17**	0.17	**0.22**	0.20	**0.26**
Trachea	0.02	**0.22**	0.02	**0.26**	0.03	**0.29**	0.03	**0.31**	0.04	**0.34**	0.04	**0.38**
Lung	0.07	**0.22**	0.09	**0.27**	0.10	**0.31**	0.12	**0.34**	0.14	**0.40**	0.16	**0.46**
Mamma	0.03	**0.21**	0.04	**0.23**	0.04	**0.25**	0.05	**0.27**	0.06	**0.29**	0.07	**0.32**
Heart	0.35	**0.44**	0.43	**0.54**	0.49	**0.62**	0.54	**0.68**	0.63	**0.80**	0.71	**0.90**
Stomach	0.08	**0.34**	0.10	**0.42**	0.11	**0.48**	0.13	**0.52**	0.15	**0.61**	0.17	**0.68**
Spleen	0.02	**0.06**	0.02	**0.08**	0.03	**0.10**	0.03	**0.11**	0.04	**0.14**	0.05	**0.17**
Liver	0.11	**0.25**	0.14	**0.31**	0.16	**0.35**	0.18	**0.39**	0.21	**0.46**	0.23	**0.52**
Gall bladder	0.07	**0.34**	0.08	**0.43**	0.10	**0.49**	0.11	**0.55**	0.13	**0.65**	0.16	**0.75**
Pancreas	0.11	**0.19**	0.14	**0.24**	0.17	**0.28**	0.19	**0.32**	0.24	**0.40**	0.28	**0.47**
Small intestine		**0.07**		**0.09**		**0.10**		**0.11**		**0.14**	0.01	**0.16**
Large intestine		**0.15**		**0.18**		**0.21**		**0.24**	0.01	**0.28**	0.01	**0.33**
Sigma/Rectum		**0.01**		**0.01**		**0.02**		**0.02**		**0.03**		**0.03**
Lumbar spine	0.02	**0.11**	0.03	**0.15**	0.04	**0.18**	0.05	**0.21**	0.06	**0.27**	0.07	**0.34**
Pelvis		**0.02**		**0.02**		**0.02**		**0.03**		**0.04**		**0.05**
Adrenals	0.05	**0.07**	0.07	**0.10**	0.08	**0.12**	0.09	**0.13**	0.11	**0.17**	0.14	**0.21**
Kidneys	0.02	**0.05**	0.02	**0.07**	0.02	**0.09**	0.03	**0.10**	0.04	**0.12**	0.04	**0.15**
Urinary bladder												
Uterus				**0.01**		**0.02**		**0.02**		**0.02**		**0.03**
Ovaries				**0.01**		**0.02**		**0.02**		**0.02**		**0.03**
Prostate												
Testes												
Upper arm	0.02	**0.03**	0.02	**0.04**	0.02	**0.05**	0.03	**0.06**	0.04	**0.07**	0.05	**0.09**
Elbow	0.02	**0.03**	0.02	**0.05**	0.02	**0.06**	0.03	**0.06**	0.04	**0.08**	0.05	**0.10**
Forearm		**0.01**		**0.02**		**0.02**		**0.02**		**0.03**		**0.04**
Upper leg												
Knee												
Lower leg												
Lymph nodes	0.06	**0.20**	0.07	**0.24**	0.08	**0.28**	0.09	**0.31**	0.11	**0.36**	0.13	**0.41**
Skin	0.04	**0.09**	0.04	**0.10**	0.04	**0.10**	0.04	**0.10**	0.04	**0.11**	0.05	**0.12**
Muscle	0.04	**0.08**	0.04	**0.09**	0.05	**0.11**	0.05	**0.12**	0.06	**0.13**	0.07	**0.15**
Skeleton	0.06	**0.12**	0.07	**0.14**	0.08	**0.15**	0.09	**0.17**	0.10	**0.19**	0.11	**0.21**
Red bone marrow	0.03	**0.05**	0.03	**0.06**	0.04	**0.07**	0.04	**0.07**	0.04	**0.09**	0.05	**0.10**
Whole body dose	0.04	**0.09**	0.05	**0.11**	0.06	**0.12**	0.06	**0.13**	0.07	**0.15**	0.08	**0.17**
ED (ICRP 60)	0.04	**0.16**	0.05	**0.19**	0.06	**0.21**	0.07	**0.23**	0.08	**0.27**	0.09	**0.30**
ED (ICRP 103)	0.05	**0.18**	0.06	**0.21**	0.07	**0.24**	0.08	**0.26**	0.09	**0.30**	0.11	**0.34**
Total filtration (mm) **Al**	**1.5**		**2.5**		**3.5**		**4.5**		**3.5**		**3.5**	
Cu	**0**		**0**		**0**		**0**		**0.1**		**0.2**	

Table 15.32 Normalized organ doses—Thoracic spine ap—8–12 years

Beam projection:	ap											**Table cells:**
Focus-to-detector distance (cm):	115								(+) = normal font = optimal field setting			
Tube voltage (kVp):	80								**(−) = bold font = suboptimal field setting**			
Total filtrations:	See last table row.								Empty table cells correspond to values <0.01			
Organ or tissue	**Organ dose D_T : Entrance dose K_E**								**(mSv ∗ mGy^{-1})**			
	(+)	(−)	(+)	(−)	(+)	(−)	(+)	(−)	(+)	(−)	(+)	(−)
Skull		**0.02**		**0.03**		**0.03**	0.01	**0.03**	0.01	**0.04**	0.02	**0.05**
Brain												
Oral mucosa		**0.02**		**0.03**		**0.03**	0.01	**0.04**	0.01	**0.04**	0.02	**0.05**
Salivary glands		**0.05**	0.01	**0.06**	0.01	**0.06**	0.02	**0.07**	0.02	**0.08**	0.02	**0.09**
Cervical spine	0.06	**0.39**	0.08	**0.48**	0.09	**0.56**	0.10	**0.62**	0.13	**0.73**	0.15	**0.83**
Thyroid	0.09	**0.80**	0.11	**0.91**	0.13	**0.98**	0.14	**1.03**	0.16	**1.12**	0.19	**1.19**
Thymus	0.77	**0.81**	0.90	**0.96**	0.99	**1.06**	1.06	**1.14**	1.18	**1.27**	1.28	**1.38**
Thoracic spine	0.13	**0.18**	0.17	**0.23**	0.20	**0.28**	0.23	**0.31**	0.29	**0.39**	0.35	**0.47**
Claviculae	0.53	**1.12**	0.61	**1.30**	0.66	**1.42**	0.70	**1.51**	0.77	**1.66**	0.82	**1.77**
Scapulae	0.04	**0.09**	0.05	**0.12**	0.06	**0.14**	0.07	**0.15**	0.08	**0.19**	0.10	**0.22**
Ribs	0.50	**0.85**	0.56	**0.95**	0.59	**1.02**	0.62	**1.07**	0.66	**1.15**	0.70	**1.22**
Oesophagus	0.10	**0.13**	0.12	**0.16**	0.14	**0.19**	0.16	**0.21**	0.20	**0.26**	0.23	**0.30**
Trachea	0.02	**0.25**	0.03	**0.28**	0.03	**0.31**	0.03	**0.33**	0.04	**0.37**	0.05	**0.40**
Lung	0.08	**0.25**	0.10	**0.31**	0.12	**0.35**	0.13	**0.38**	0.16	**0.45**	0.19	**0.50**
Mamma	0.04	**0.22**	0.05	**0.25**	0.05	**0.27**	0.06	**0.29**	0.07	**0.31**	0.08	**0.34**
Heart	0.40	**0.51**	0.49	**0.62**	0.55	**0.69**	0.60	**0.76**	0.69	**0.87**	0.77	**0.97**
Stomach	0.09	**0.39**	0.11	**0.47**	0.13	**0.53**	0.14	**0.58**	0.17	**0.67**	0.19	**0.74**
Spleen	0.02	**0.08**	0.03	**0.10**	0.03	**0.12**	0.04	**0.14**	0.05	**0.17**	0.06	**0.20**
Liver	0.13	**0.29**	0.16	**0.35**	0.18	**0.40**	0.20	**0.44**	0.23	**0.51**	0.26	**0.57**
Gall bladder	0.08	**0.40**	0.10	**0.49**	0.12	**0.57**	0.13	**0.62**	0.15	**0.73**	0.18	**0.83**
Pancreas	0.14	**0.23**	0.17	**0.29**	0.20	**0.34**	0.23	**0.38**	0.27	**0.46**	0.32	**0.54**
Small intestine		**0.08**		**0.10**		**0.12**		**0.13**	0.01	**0.16**	0.01	**0.19**
Large intestine		**0.17**		**0.21**	0.01	**0.25**	0.01	**0.27**	0.02	**0.32**	0.02	**0.36**
Sigma/Rectum		**0.01**		**0.02**		**0.02**		**0.02**		**0.03**		**0.04**
Lumbar spine	0.03	**0.15**	0.04	**0.20**	0.05	**0.24**	0.06	**0.27**	0.07	**0.33**	0.09	**0.40**
Pelvis		**0.02**		**0.03**		**0.03**		**0.04**		**0.05**		**0.06**
Adrenals	0.07	**0.10**	0.08	**0.12**	0.10	**0.15**	0.11	**0.17**	0.14	**0.21**	0.17	**0.25**
Kidneys	0.02	**0.07**	0.03	**0.09**	0.03	**0.11**	0.04	**0.12**	0.04	**0.15**	0.05	**0.18**
Urinary bladder												**0.01**
Uterus		**0.01**		**0.02**		**0.02**		**0.02**		**0.03**		**0.04**
Ovaries		**0.01**		**0.02**		**0.02**		**0.02**		**0.03**		**0.04**
Prostate												
Testes												
Upper arm	0.02	**0.04**	0.03	**0.05**	0.03	**0.06**	0.04	**0.07**	0.05	**0.09**	0.06	**0.11**
Elbow	0.02	**0.05**	0.03	**0.06**	0.03	**0.07**	0.04	**0.08**	0.05	**0.10**	0.06	**0.12**
Forearm		**0.02**		**0.02**		**0.03**		**0.03**		**0.04**		**0.05**
Upper leg												
Knee												
Lower leg												
Lymph nodes	0.07	**0.23**	0.09	**0.28**	0.10	**0.31**	0.11	**0.34**	0.13	**0.40**	0.15	**0.45**
Skin	0.04	**0.10**	0.04	**0.10**	0.04	**0.11**	0.04	**0.11**	0.05	**0.12**	0.05	**0.12**
Muscle	0.04	**0.09**	0.05	**0.11**	0.06	**0.12**	0.06	**0.13**	0.07	**0.15**	0.08	**0.17**
Skeleton	0.07	**0.13**	0.08	**0.15**	0.09	**0.17**	0.09	**0.18**	0.11	**0.21**	0.12	**0.23**
Red bone marrow	0.03	**0.06**	0.04	**0.07**	0.04	**0.08**	0.04	**0.08**	0.05	**0.10**	0.06	**0.11**
Whole body dose	0.05	**0.10**	0.06	**0.12**	0.06	**0.14**	0.07	**0.15**	0.08	**0.17**	0.09	**0.19**
ED (ICRP 60)	0.05	**0.18**	0.06	**0.21**	0.07	**0.24**	0.08	**0.26**	0.09	**0.30**	0.10	**0.33**
ED (ICRP 103)	0.06	**0.20**	0.07	**0.24**	0.08	**0.27**	0.09	**0.29**	0.10	**0.33**	0.12	**0.37**
Total filtration (mm) Al	**1.5**		**2.5**		**3.5**		**4.5**		**3.5**		**3.5**	
Cu	**0**		**0**		**0**		**0**		**0.1**		**0.2**	

Table 15.33 Normalized organ doses—Thoracic spine ap—13–17 years

Beam projection:			ap									**Table cells:**
Focus-to-detector distance (cm):			115				(+) = normal font = optimal field setting					
Tube voltage (kVp):			60				**(−) = bold font = suboptimal field setting**					
Total filtrations:			See last table row.				Empty table cells correspond to values <0.01					
Organ or tissue	**Organ dose D_T : Entrance dose K_E**								**(mSv ∗ mGy^{-1})**			
	(+)	(−)	(+)	(−)	(+)	(−)	(+)	(−)	(+)	(−)	(+)	(−)
Skull		**0.01**		**0.02**		**0.02**		**0.02**		**0.03**		**0.03**
Brain												
Oral mucosa		**0.01**		**0.01**		**0.02**		**0.02**		**0.03**		**0.02**
Salivary glands		**0.03**		**0.04**		**0.04**		**0.05**		**0.06**		**0.05**
Cervical spine	0.03	**0.30**	0.04	**0.39**	0.04	**0.46**	0.05	**0.53**	0.07	**0.65**	0.05	**0.58**
Thyroid	0.05	**0.62**	0.07	**0.73**	0.08	**0.80**	0.09	**0.86**	0.11	**0.96**	0.10	**0.94**
Thymus	0.56	**0.60**	0.69	**0.74**	0.79	**0.84**	0.86	**0.93**	0.99	**1.07**	0.96	**1.03**
Thoracic spine	0.04	**0.06**	0.06	**0.09**	0.08	**0.11**	0.09	**0.13**	0.12	**0.17**	0.09	**0.13**
Claviculae	0.33	**0.94**	0.40	**1.15**	0.45	**1.30**	0.49	**1.42**	0.56	**1.63**	0.54	**1.60**
Scapulae	0.01	**0.04**	0.02	**0.06**	0.02	**0.07**	0.02	**0.09**	0.03	**0.11**	0.02	**0.09**
Ribs	0.42	**0.77**	0.48	**0.90**	0.53	**0.99**	0.56	**1.06**	0.61	**1.18**	0.60	**1.16**
Oesophagus	0.06	**0.08**	0.08	**0.11**	0.10	**0.13**	0.12	**0.16**	0.15	**0.20**	0.12	**0.17**
Trachea		**0.20**	0.01	**0.24**	0.01	**0.26**	0.02	**0.29**	0.02	**0.32**	0.02	**0.32**
Lung	0.04	**0.17**	0.06	**0.22**	0.07	**0.26**	0.08	**0.29**	0.10	**0.35**	0.09	**0.32**
Mamma	0.39	**4.68**	0.48	**5.41**	0.55	**5.90**	0.60	**6.26**	0.71	**6.89**	0.65	**6.76**
Heart	0.28	**0.34**	0.35	**0.43**	0.40	**0.50**	0.45	**0.55**	0.54	**0.66**	0.50	**0.62**
Stomach	0.05	**0.32**	0.07	**0.40**	0.08	**0.47**	0.09	**0.52**	0.11	**0.62**	0.10	**0.58**
Spleen		**0.05**	0.01	**0.07**	0.01	**0.08**	0.02	**0.09**	0.02	**0.12**	0.02	**0.10**
Liver	0.08	**0.19**	0.10	**0.25**	0.11	**0.29**	0.13	**0.32**	0.15	**0.39**	0.14	**0.36**
Gall bladder	0.04	**0.23**	0.05	**0.30**	0.06	**0.36**	0.06	**0.41**	0.08	**0.50**	0.07	**0.45**
Pancreas	0.07	**0.12**	0.09	**0.16**	0.11	**0.19**	0.12	**0.22**	0.15	**0.28**	0.13	**0.24**
Small intestine		**0.05**		**0.06**		**0.08**		**0.09**		**0.11**		**0.10**
Large intestine		**0.11**		**0.14**		**0.17**		**0.19**		**0.23**		**0.21**
Sigma/Rectum				**0.01**		**0.01**		**0.02**		**0.02**		**0.02**
Lumbar spine	0.01	**0.05**	0.02	**0.07**	0.02	**0.09**	0.02	**0.10**	0.03	**0.14**	0.02	**0.10**
Pelvis				**0.01**		**0.01**		**0.02**		**0.02**		**0.02**
Adrenals	0.02	**0.04**	0.03	**0.05**	0.04	**0.06**	0.05	**0.08**	0.06	**0.10**	0.05	**0.08**
Kidneys		**0.03**	0.01	**0.04**	0.01	**0.05**	0.02	**0.05**	0.02	**0.07**	0.02	**0.06**
Urinary bladder												
Uterus		**0.01**		**0.02**		**0.02**		**0.03**		**0.04**		**0.03**
Ovaries										**0.01**		
Prostate												
Testes												
Upper arm		**0.02**		**0.02**		**0.03**	0.01	**0.03**	0.02	**0.04**	0.01	**0.03**
Elbow		**0.02**		**0.03**	0.01	**0.03**	0.01	**0.04**	0.02	**0.05**	0.01	**0.04**
Forearm				**0.01**		**0.01**		**0.02**		**0.02**		**0.02**
Upper leg												
Knee												
Lower leg												
Lymph nodes	0.04	**0.15**	0.05	**0.19**	0.06	**0.22**	0.07	**0.25**	0.08	**0.30**	0.07	**0.28**
Skin	0.04	**0.12**	0.04	**0.12**	0.04	**0.13**	0.04	**0.13**	0.05	**0.14**	0.04	**0.14**
Muscle	0.03	**0.07**	0.04	**0.09**	0.04	**0.10**	0.05	**0.11**	0.05	**0.13**	0.05	**0.12**
Skeleton	0.05	**0.11**	0.06	**0.13**	0.07	**0.15**	0.07	**0.16**	0.08	**0.19**	0.08	**0.17**
Red bone marrow	0.03	**0.05**	0.03	**0.06**	0.03	**0.07**	0.04	**0.07**	0.04	**0.09**	0.04	**0.08**
Whole body dose	0.03	**0.08**	0.04	**0.10**	0.05	**0.12**	0.05	**0.13**	0.06	**0.15**	0.06	**0.14**
ED (ICRP 60)	0.05	**0.36**	0.06	**0.43**	0.07	**0.47**	0.08	**0.51**	0.09	**0.58**	0.08	**0.56**
ED (ICRP 103)	0.08	**0.69**	0.10	**0.81**	0.11	**0.89**	0.13	**0.95**	0.15	**1.07**	0.14	**1.03**
Total filtration (mm) Al	**1.5**		**2.5**		**3.5**		**4.5**		**3.5**		**3.5**	
Cu	**0**		**0**		**0**		**0**		**0.1**		**0.2**	

Table 15.34 Normalized organ doses—Thoracic spine ap—13–17 years

Beam projection:			ap									**Table cells:**
Focus-to-detector distance (cm):			115									(+) = normal font = optimal field setting
Tube voltage (kVp):			70									**(−) = bold font = suboptimal field setting**
Total filtrations:			See last table row.									Empty table cells correspond to values <0.01
Organ or tissue	**Organ dose D_T : Entrance dose K_E**							**(mSv * mGy⁻¹)**				
	(+)	(−)	(+)	(−)	(+)	(−)	(+)	(−)	(+)	(−)	(+)	(−)
Skull		**0.02**		**0.02**		**0.03**		**0.03**		**0.04**		**0.04**
Brain												
Oral mucosa		**0.01**		**0.02**		**0.02**		**0.03**		**0.03**		**0.04**
Salivary glands		**0.04**		**0.04**		**0.05**		**0.06**		**0.07**	0.01	**0.08**
Cervical spine	0.04	**0.38**	0.05	**0.48**	0.06	**0.57**	0.07	**0.64**	0.09	**0.78**	0.11	**0.92**
Thyroid	0.06	**0.69**	0.08	**0.80**	0.10	**0.87**	0.11	**0.93**	0.13	**1.03**	0.15	**1.11**
Thymus	0.65	**0.70**	0.78	**0.85**	0.88	**0.95**	0.96	**1.04**	1.09	**1.19**	1.21	**1.33**
Thoracic spine	0.07	**0.10**	0.09	**0.13**	0.11	**0.16**	0.13	**0.18**	0.16	**0.24**	0.20	**0.29**
Claviculae	0.37	**1.06**	0.44	**1.28**	0.49	**1.43**	0.53	**1.55**	0.60	**1.76**	0.65	**1.93**
Scapulae	0.02	**0.06**	0.02	**0.08**	0.03	**0.10**	0.04	**0.12**	0.05	**0.15**	0.06	**0.18**
Ribs	0.46	**0.85**	0.52	**0.98**	0.56	**1.07**	0.59	**1.13**	0.64	**1.25**	0.68	**1.34**
Oesophagus	0.08	**0.11**	0.11	**0.15**	0.13	**0.18**	0.15	**0.20**	0.19	**0.26**	0.23	**0.31**
Trachea	0.01	**0.22**	0.02	**0.26**	0.02	**0.29**	0.02	**0.31**	0.03	**0.35**	0.03	**0.39**
Lung	0.06	**0.21**	0.08	**0.27**	0.09	**0.31**	0.10	**0.34**	0.12	**0.41**	0.15	**0.48**
Mamma	0.47	**5.11**	0.57	**5.85**	0.64	**6.35**	0.71	**6.71**	0.83	**7.36**	0.95	**7.91**
Heart	0.33	**0.41**	0.41	**0.51**	0.47	**0.59**	0.52	**0.65**	0.62	**0.77**	0.70	**0.88**
Stomach	0.07	**0.38**	0.08	**0.48**	0.10	**0.55**	0.11	**0.60**	0.13	**0.71**	0.16	**0.81**
Spleen	0.01	**0.07**	0.02	**0.09**	0.02	**0.11**	0.02	**0.13**	0.03	**0.16**	0.04	**0.20**
Liver	0.09	**0.24**	0.12	**0.30**	0.14	**0.34**	0.15	**0.38**	0.18	**0.45**	0.21	**0.52**
Gall bladder	0.05	**0.29**	0.06	**0.38**	0.07	**0.44**	0.08	**0.49**	0.10	**0.59**	0.12	**0.69**
Pancreas	0.09	**0.16**	0.12	**0.21**	0.14	**0.25**	0.16	**0.28**	0.19	**0.35**	0.23	**0.42**
Small intestine		**0.06**		**0.08**		**0.10**		**0.11**		**0.14**		**0.17**
Large intestine		**0.14**		**0.17**		**0.20**		**0.23**		**0.27**	0.01	**0.32**
Sigma/Rectum		**0.01**		**0.02**		**0.02**		**0.02**		**0.03**		**0.03**
Lumbar spine	0.02	**0.08**	0.02	**0.11**	0.03	**0.13**	0.03	**0.15**	0.04	**0.19**	0.05	**0.24**
Pelvis		**0.01**		**0.02**		**0.02**		**0.02**		**0.03**		**0.04**
Adrenals	0.04	**0.06**	0.05	**0.08**	0.06	**0.09**	0.07	**0.10**	0.09	**0.13**	0.11	**0.17**
Kidneys	0.01	**0.04**	0.02	**0.06**	0.02	**0.07**	0.02	**0.08**	0.03	**0.10**	0.04	**0.12**
Urinary bladder												
Uterus		**0.02**		**0.03**		**0.03**		**0.04**		**0.05**		**0.07**
Ovaries				**0.01**		**0.01**		**0.02**		**0.02**		**0.03**
Prostate												
Testes												
Upper arm		**0.03**	0.01	**0.03**	0.01	**0.04**	0.02	**0.05**	0.02	**0.06**	0.03	**0.08**
Elbow		**0.03**	0.01	**0.04**	0.02	**0.05**	0.02	**0.05**	0.02	**0.07**	0.03	**0.09**
Forearm		**0.01**		**0.02**		**0.02**		**0.02**		**0.03**		**0.04**
Upper leg												
Knee												
Lower leg												
Lymph nodes	0.05	**0.19**	0.06	**0.23**	0.07	**0.27**	0.08	**0.29**	0.10	**0.35**	0.12	**0.40**
Skin	0.04	**0.12**	0.04	**0.13**	0.04	**0.14**	0.05	**0.14**	0.05	**0.15**	0.05	**0.16**
Muscle	0.03	**0.09**	0.04	**0.10**	0.05	**0.12**	0.05	**0.13**	0.06	**0.15**	0.07	**0.17**
Skeleton	0.06	**0.13**	0.07	**0.15**	0.08	**0.17**	0.08	**0.18**	0.09	**0.21**	0.10	**0.23**
Red bone marrow	0.03	**0.06**	0.03	**0.07**	0.04	**0.08**	0.04	**0.09**	0.05	**0.10**	0.05	**0.12**
Whole body dose	0.04	**0.10**	0.05	**0.12**	0.05	**0.14**	0.06	**0.15**	0.07	**0.17**	0.08	**0.20**
ED (ICRP 60)	0.06	**0.40**	0.07	**0.48**	0.08	**0.53**	0.09	**0.57**	0.11	**0.64**	0.13	**0.70**
ED (ICRP 103)	0.10	**0.77**	0.12	**0.89**	0.14	**0.98**	0.15	**1.04**	0.18	**1.16**	0.21	**1.27**
Total filtration (mm) Al	**1.5**		**2.5**		**3.5**		**4.5**		**3.5**		**3.5**	
Cu	**0**		**0**		**0**		**0**		**0.1**		**0.2**	

Table 15.35 Normalized organ doses—Thoracic spine ap—13–17 years

Beam projection:			ap									**Table cells:**
Focus-to-detector distance (cm):			115				(+) = normal font = optimal field setting					
Tube voltage (kVp):			80				**(−) = bold font = suboptimal field setting**					
Total filtrations:			See last table row.				Empty table cells correspond to values <0.01					
Organ or tissue	**Organ dose D_T : Entrance dose K_E**							**(mSv * mGy^{-1})**				
	(+)	(−)	(+)	(−)	(+)	(−)	(+)	(−)	(+)	(−)	(+)	(−)
Skull		**0.02**		**0.03**		**0.03**		**0.04**		**0.04**	0.01	**0.05**
Brain												
Oral mucosa		**0.02**		**0.02**		**0.03**		**0.03**		**0.04**		**0.05**
Salivary glands		**0.04**		**0.05**		**0.06**		**0.07**	0.01	**0.08**	0.01	**0.09**
Cervical spine	0.05	**0.46**	0.07	**0.57**	0.08	**0.66**	0.09	**0.74**	0.11	**0.88**	0.13	**1.02**
Thyroid	0.08	**0.75**	0.10	**0.86**	0.11	**0.93**	0.12	**0.99**	0.15	**1.09**	0.17	**1.17**
Thymus	0.73	**0.79**	0.87	**0.94**	0.96	**1.05**	1.04	**1.14**	1.18	**1.29**	1.29	**1.43**
Thoracic spine	0.09	**0.13**	0.12	**0.17**	0.14	**0.21**	0.16	**0.24**	0.21	**0.30**	0.25	**0.37**
Claviculae	0.41	**1.17**	0.48	**1.38**	0.53	**1.53**	0.56	**1.64**	0.62	**1.83**	0.67	**1.98**
Scapulae	0.03	**0.08**	0.03	**0.11**	0.04	**0.13**	0.05	**0.15**	0.06	**0.18**	0.07	**0.22**
Ribs	0.48	**0.91**	0.54	**1.04**	0.58	**1.12**	0.61	**1.18**	0.66	**1.29**	0.69	**1.37**
Oesophagus	0.11	**0.15**	0.14	**0.19**	0.16	**0.23**	0.19	**0.25**	0.23	**0.31**	0.27	**0.38**
Trachea	0.02	**0.25**	0.02	**0.29**	0.02	**0.32**	0.03	**0.34**	0.03	**0.38**	0.04	**0.41**
Lung	0.07	**0.25**	0.09	**0.31**	0.11	**0.36**	0.12	**0.40**	0.15	**0.47**	0.17	**0.54**
Mamma	0.55	**5.52**	0.65	**6.25**	0.74	**6.74**	0.80	**7.11**	0.94	**7.75**	1.06	**8.27**
Heart	0.39	**0.48**	0.47	**0.59**	0.54	**0.67**	0.59	**0.74**	0.68	**0.86**	0.77	**0.98**
Stomach	0.08	**0.45**	0.10	**0.55**	0.12	**0.62**	0.13	**0.68**	0.15	**0.79**	0.18	**0.89**
Spleen	0.02	**0.09**	0.02	**0.12**	0.03	**0.14**	0.03	**0.16**	0.04	**0.20**	0.05	**0.24**
Liver	0.11	**0.28**	0.14	**0.35**	0.16	**0.40**	0.17	**0.44**	0.20	**0.51**	0.23	**0.58**
Gall bladder	0.06	**0.36**	0.07	**0.44**	0.09	**0.51**	0.10	**0.57**	0.12	**0.67**	0.14	**0.77**
Pancreas	0.11	**0.20**	0.14	**0.26**	0.17	**0.30**	0.19	**0.34**	0.23	**0.42**	0.27	**0.50**
Small intestine		**0.08**		**0.10**		**0.12**		**0.14**		**0.17**	0.01	**0.19**
Large intestine		**0.17**		**0.21**		**0.24**		**0.26**	0.01	**0.31**	0.01	**0.36**
Sigma/Rectum		**0.02**		**0.02**		**0.02**		**0.03**		**0.03**		**0.04**
Lumbar spine	0.02	**0.11**	0.03	**0.14**	0.04	**0.17**	0.04	**0.20**	0.06	**0.25**	0.07	**0.30**
Pelvis		**0.02**		**0.02**		**0.03**		**0.03**		**0.04**		**0.05**
Adrenals	0.05	**0.08**	0.06	**0.10**	0.08	**0.12**	0.09	**0.14**	0.11	**0.17**	0.13	**0.21**
Kidneys	0.02	**0.06**	0.02	**0.07**	0.03	**0.09**	0.03	**0.10**	0.04	**0.13**	0.04	**0.15**
Urinary bladder												
Uterus		**0.03**		**0.04**		**0.05**		**0.06**		**0.07**		**0.09**
Ovaries		**0.01**		**0.01**		**0.02**		**0.02**		**0.03**		**0.03**
Prostate												
Testes												
Upper arm	0.01	**0.04**	0.02	**0.05**	0.02	**0.06**	0.02	**0.06**	0.03	**0.08**	0.04	**0.10**
Elbow	0.01	**0.04**	0.02	**0.05**	0.02	**0.06**	0.02	**0.07**	0.03	**0.09**	0.04	**0.11**
Forearm		**0.02**		**0.02**		**0.03**		**0.03**		**0.04**		**0.05**
Upper leg												
Knee												
Lower leg												
Lymph nodes	0.06	**0.22**	0.07	**0.27**	0.09	**0.31**	0.10	**0.34**	0.12	**0.39**	0.13	**0.45**
Skin	0.04	**0.13**	0.04	**0.14**	0.05	**0.14**	0.05	**0.15**	0.05	**0.16**	0.05	**0.17**
Muscle	0.04	**0.10**	0.05	**0.12**	0.06	**0.13**	0.06	**0.15**	0.07	**0.17**	0.08	**0.19**
Skeleton	0.07	**0.14**	0.08	**0.17**	0.08	**0.19**	0.09	**0.20**	0.10	**0.23**	0.11	**0.25**
Red bone marrow	0.03	**0.07**	0.04	**0.08**	0.04	**0.09**	0.05	**0.10**	0.06	**0.12**	0.06	**0.14**
Whole body dose	0.05	**0.11**	0.06	**0.14**	0.06	**0.15**	0.07	**0.17**	0.08	**0.19**	0.09	**0.22**
ED (ICRP 60)	0.07	**0.45**	0.09	**0.52**	0.10	**0.58**	0.11	**0.62**	0.13	**0.69**	0.15	**0.76**
ED (ICRP 103)	0.11	**0.84**	0.14	**0.97**	0.16	**1.05**	0.17	**1.12**	0.20	**1.24**	0.23	**1.35**
Total filtration (mm) **Al**	**1.5**		**2.5**		**3.5**		**4.5**		**3.5**		**3.5**	
Cu	**0**		**0**		**0**		**0**		**0.1**		**0.2**	

Table 15.36 Normalized organ doses—Thoracic spine ap—>17 years

Beam projection:	ap											**Table cells:**
Focus-to-detector distance (cm):	115											(+) = normal font = optimal field setting
Tube voltage (kVp):	60											**(−) = bold font = suboptimal field setting**
Total filtrations:	See last table row.											Empty table cells correspond to values <0.01
Organ or tissue	**Organ dose D_T : Entrance dose K_E**						**(mSv ∗ mGy^{-1})**					
	(+)	(−)	(+)	(−)	(+)	(−)	(+)	(−)	(+)	(−)	(+)	(−)
Skull		**0.01**		**0.02**		**0.02**		**0.03**		**0.03**		**0.03**
Brain												
Oral mucosa				**0.01**		**0.02**		**0.02**		**0.02**		**0.02**
Salivary glands		**0.03**		**0.04**		**0.05**		**0.05**		**0.06**		**0.06**
Cervical spine	0.02	**0.26**	0.03	**0.35**	0.03	**0.42**	0.04	**0.47**	0.05	**0.59**	0.04	**0.52**
Thyroid	0.03	**0.58**	0.04	**0.69**	0.05	**0.77**	0.06	**0.82**	0.07	**0.92**	0.07	**0.90**
Thymus	0.56	**0.60**	0.69	**0.74**	0.78	**0.85**	0.86	**0.93**	0.99	**1.08**	0.96	**1.04**
Thoracic spine	0.04	**0.06**	0.05	**0.08**	0.07	**0.10**	0.08	**0.12**	0.11	**0.16**	0.08	**0.12**
Claviculae	0.26	**0.95**	0.31	**1.17**	0.35	**1.33**	0.38	**1.45**	0.44	**1.67**	0.42	**1.64**
Scapulae		**0.05**	0.01	**0.07**	0.02	**0.08**	0.02	**0.09**	0.03	**0.12**	0.02	**0.10**
Ribs	0.37	**0.80**	0.43	**0.94**	0.47	**1.03**	0.50	**1.10**	0.54	**1.22**	0.54	**1.20**
Oesophagus	0.06	**0.08**	0.08	**0.11**	0.10	**0.14**	0.11	**0.16**	0.15	**0.21**	0.12	**0.17**
Trachea		**0.19**	0.01	**0.23**	0.01	**0.26**	0.02	**0.29**	0.02	**0.33**	0.02	**0.32**
Lung	0.03	**0.18**	0.04	**0.24**	0.05	**0.28**	0.06	**0.31**	0.07	**0.38**	0.06	**0.35**
Mamma	0.02	**0.53**	0.02	**0.62**	0.03	**0.67**	0.03	**0.71**	0.04	**0.79**	0.03	**0.77**
Heart	0.24	**0.32**	0.31	**0.41**	0.36	**0.48**	0.40	**0.54**	0.47	**0.65**	0.44	**0.60**
Stomach	0.03	**0.32**	0.04	**0.40**	0.05	**0.47**	0.06	**0.52**	0.08	**0.62**	0.07	**0.58**
Spleen		**0.05**		**0.07**		**0.08**	0.01	**0.09**	0.01	**0.12**	0.01	**0.10**
Liver	0.07	**0.19**	0.08	**0.25**	0.10	**0.29**	0.11	**0.33**	0.13	**0.39**	0.12	**0.36**
Gall bladder	0.03	**0.23**	0.04	**0.30**	0.05	**0.36**	0.05	**0.40**	0.07	**0.49**	0.06	**0.45**
Pancreas	0.06	**0.11**	0.08	**0.15**	0.09	**0.19**	0.11	**0.21**	0.14	**0.27**	0.12	**0.23**
Small intestine		**0.05**		**0.06**		**0.08**		**0.09**		**0.11**		**0.09**
Large intestine		**0.11**		**0.14**		**0.16**		**0.19**		**0.23**		**0.21**
Sigma/Rectum				**0.01**		**0.01**		**0.02**		**0.02**		**0.02**
Lumbar spine		**0.05**	0.01	**0.07**	0.02	**0.08**	0.02	**0.10**	0.03	**0.13**	0.02	**0.10**
Pelvis				**0.01**		**0.01**		**0.01**		**0.02**		**0.01**
Adrenals	0.02	**0.04**	0.03	**0.05**	0.04	**0.06**	0.04	**0.07**	0.06	**0.10**	0.05	**0.08**
Kidneys		**0.03**		**0.04**	0.01	**0.05**	0.01	**0.05**	0.02	**0.07**	0.01	**0.06**
Urinary bladder												
Uterus										**0.01**		
Ovaries										**0.01**		
Prostate												
Testes												
Upper arm		**0.01**		**0.02**		**0.02**		**0.03**		**0.04**		**0.03**
Elbow		**0.01**		**0.02**		**0.03**		**0.03**		**0.04**		**0.03**
Forearm								**0.01**		**0.02**		**0.01**
Upper leg												
Knee												
Lower leg												
Lymph nodes	0.03	**0.15**	0.04	**0.19**	0.05	**0.22**	0.06	**0.25**	0.07	**0.30**	0.06	**0.27**
Skin	0.03	**0.12**	0.03	**0.13**	0.04	**0.13**	0.04	**0.14**	0.04	**0.14**	0.04	**0.14**
Muscle	0.02	**0.07**	0.03	**0.09**	0.03	**0.10**	0.04	**0.11**	0.05	**0.13**	0.04	**0.12**
Skeleton	0.04	**0.11**	0.05	**0.13**	0.06	**0.15**	0.06	**0.16**	0.07	**0.18**	0.07	**0.17**
Red bone marrow	0.03	**0.06**	0.03	**0.07**	0.03	**0.08**	0.04	**0.09**	0.04	**0.10**	0.04	**0.09**
Whole body dose	0.03	**0.08**	0.03	**0.10**	0.04	**0.12**	0.04	**0.13**	0.05	**0.15**	0.05	**0.14**
ED (ICRP 60)	0.02	**0.15**	0.03	**0.19**	0.04	**0.21**	0.04	**0.24**	0.05	**0.28**	0.04	**0.26**
ED (ICRP 103)	0.03	**0.19**	0.04	**0.23**	0.04	**0.27**	0.05	**0.29**	0.06	**0.34**	0.05	**0.32**
Total filtration (mm) Al	**1.5**		**2.5**		**3.5**		**4.5**		**3.5**		**3.5**	
Cu	**0**		**0**		**0**		**0**		**0.1**		**0.2**	

Table 15.37 Normalized organ doses—Thoracic spine ap—>17 years

Beam projection:			ap									**Table cells:**
Focus-to-detector distance (cm):			115									(+) = normal font = optimal field setting
Tube voltage (kVp):			70									**(−) = bold font = suboptimal field setting**
Total filtrations:			See last table row.									Empty table cells correspond to values <0.01
Organ or tissue	**Organ dose D_T : Entrance dose K_E**							**(mSv * mGy^{-1})**				
	(+)	(−)	(+)	(−)	(+)	(−)	(+)	(−)	(+)	(−)	(+)	(−)
Skull		**0.02**		**0.02**		**0.03**		**0.03**		**0.04**		**0.05**
Brain												
Oral mucosa		**0.01**		**0.02**		**0.02**		**0.02**		**0.03**		**0.04**
Salivary glands		**0.04**		**0.05**		**0.06**		**0.06**		**0.07**		**0.08**
Cervical spine	0.03	**0.34**	0.04	**0.44**	0.05	**0.52**	0.05	**0.58**	0.07	**0.71**	0.09	**0.84**
Thyroid	0.04	**0.65**	0.06	**0.76**	0.07	**0.84**	0.07	**0.89**	0.09	**0.99**	0.11	**1.08**
Thymus	0.64	**0.70**	0.78	**0.85**	0.88	**0.96**	0.95	**1.05**	1.09	**1.21**	1.21	**1.35**
Thoracic spine	0.06	**0.09**	0.08	**0.12**	0.10	**0.15**	0.12	**0.17**	0.15	**0.23**	0.19	**0.29**
Claviculae	0.29	**1.08**	0.35	**1.31**	0.39	**1.47**	0.42	**1.59**	0.47	**1.80**	0.52	**1.98**
Scapulae	0.01	**0.07**	0.02	**0.09**	0.02	**0.11**	0.03	**0.13**	0.04	**0.16**	0.05	**0.20**
Ribs	0.40	**0.88**	0.46	**1.01**	0.50	**1.11**	0.52	**1.17**	0.57	**1.29**	0.61	**1.39**
Oesophagus	0.08	**0.12**	0.11	**0.16**	0.13	**0.19**	0.15	**0.22**	0.19	**0.27**	0.23	**0.33**
Trachea	0.01	**0.22**	0.01	**0.26**	0.02	**0.30**	0.02	**0.32**	0.02	**0.36**	0.03	**0.40**
Lung	0.04	**0.23**	0.06	**0.29**	0.07	**0.33**	0.07	**0.37**	0.09	**0.44**	0.11	**0.51**
Mamma	0.02	**0.58**	0.03	**0.67**	0.03	**0.72**	0.04	**0.77**	0.05	**0.84**	0.06	**0.90**
Heart	0.29	**0.40**	0.36	**0.50**	0.42	**0.57**	0.46	**0.63**	0.55	**0.75**	0.63	**0.87**
Stomach	0.04	**0.38**	0.06	**0.48**	0.07	**0.55**	0.08	**0.60**	0.09	**0.71**	0.11	**0.81**
Spleen		**0.07**	0.01	**0.09**	0.01	**0.11**	0.02	**0.13**	0.02	**0.16**	0.03	**0.20**
Liver	0.08	**0.24**	0.10	**0.30**	0.12	**0.35**	0.13	**0.39**	0.16	**0.46**	0.18	**0.53**
Gall bladder	0.04	**0.29**	0.05	**0.37**	0.06	**0.44**	0.07	**0.49**	0.09	**0.59**	0.10	**0.69**
Pancreas	0.08	**0.16**	0.10	**0.20**	0.12	**0.24**	0.14	**0.28**	0.18	**0.35**	0.21	**0.42**
Small intestine		**0.06**		**0.08**		**0.10**		**0.11**		**0.14**		**0.16**
Large intestine		**0.13**		**0.17**		**0.20**		**0.23**		**0.27**		**0.32**
Sigma/Rectum		**0.01**		**0.02**		**0.02**		**0.02**		**0.03**		**0.03**
Lumbar spine	0.01	**0.07**	0.02	**0.10**	0.02	**0.12**	0.03	**0.14**	0.04	**0.19**	0.05	**0.23**
Pelvis		**0.01**		**0.02**		**0.02**		**0.02**		**0.03**		**0.04**
Adrenals	0.03	**0.06**	0.04	**0.08**	0.05	**0.09**	0.06	**0.11**	0.08	**0.14**	0.10	**0.17**
Kidneys		**0.04**	0.01	**0.05**	0.01	**0.07**	0.02	**0.08**	0.02	**0.10**	0.03	**0.12**
Urinary bladder												
Uterus						**0.01**		**0.01**		**0.02**		**0.02**
Ovaries								**0.01**		**0.02**		**0.02**
Prostate												
Testes												
Upper arm		**0.02**		**0.03**		**0.03**	0.01	**0.04**	0.01	**0.05**	0.02	**0.06**
Elbow		**0.02**		**0.03**		**0.04**	0.01	**0.04**	0.01	**0.06**	0.02	**0.07**
Forearm				**0.01**		**0.01**		**0.02**		**0.02**		**0.03**
Upper leg												
Knee												
Lower leg												
Lymph nodes	0.04	**0.19**	0.05	**0.23**	0.06	**0.27**	0.07	**0.29**	0.09	**0.35**	0.10	**0.40**
Skin	0.03	**0.12**	0.04	**0.13**	0.04	**0.14**	0.04	**0.14**	0.04	**0.15**	0.04	**0.16**
Muscle	0.03	**0.08**	0.04	**0.10**	0.04	**0.11**	0.05	**0.13**	0.05	**0.15**	0.06	**0.17**
Skeleton	0.05	**0.12**	0.06	**0.15**	0.07	**0.16**	0.07	**0.18**	0.08	**0.21**	0.09	**0.23**
Red bone marrow	0.03	**0.07**	0.04	**0.08**	0.04	**0.09**	0.04	**0.10**	0.05	**0.12**	0.06	**0.14**
Whole body dose	0.03	**0.10**	0.04	**0.12**	0.05	**0.14**	0.05	**0.15**	0.06	**0.17**	0.07	**0.20**
ED (ICRP 60)	0.03	**0.18**	0.04	**0.22**	0.04	**0.25**	0.05	**0.27**	0.06	**0.32**	0.07	**0.36**
ED (ICRP 103)	0.04	**0.22**	0.05	**0.27**	0.05	**0.30**	0.06	**0.33**	0.07	**0.38**	0.08	**0.43**
Total filtration (mm) Al	**1.5**		**2.5**		**3.5**		**4.5**		**3.5**		**3.5**	
Cu	**0**		**0**		**0**		**0**		**0.1**		**0.2**	

Table 15.38 Normalized organ doses—Thoracic spine ap—>17 years

Beam projection:	ap											**Table cells:**
Focus-to-detector distance (cm):	115											(+) = normal font = optimal field setting
Tube voltage (kVp):	80											**(−) = bold font = suboptimal field setting**
Total filtrations:	See last table row.											Empty table cells correspond to values <0.01
Organ or tissue	**Organ dose D_T : Entrance dose K_E**							**(mSv * mGy^{-1})**				
	(+)	(−)	(+)	(−)	(+)	(−)	(+)	(−)	(+)	(−)	(+)	(−)
Skull		**0.02**		**0.03**		**0.03**		**0.04**		**0.05**		**0.05**
Brain												
Oral mucosa		**0.02**		**0.02**		**0.03**		**0.03**		**0.04**		**0.04**
Salivary glands		**0.05**		**0.06**		**0.06**		**0.07**		**0.08**		**0.09**
Cervical spine	0.04	**0.42**	0.05	**0.53**	0.06	**0.61**	0.07	**0.68**	0.09	**0.82**	0.10	**0.95**
Thyroid	0.05	**0.71**	0.07	**0.82**	0.08	**0.90**	0.09	**0.95**	0.11	**1.05**	0.12	**1.13**
Thymus	0.72	**0.79**	0.86	**0.95**	0.96	**1.06**	1.04	**1.15**	1.18	**1.31**	1.29	**1.45**
Thoracic spine	0.08	**0.13**	0.11	**0.17**	0.13	**0.20**	0.15	**0.23**	0.19	**0.29**	0.23	**0.36**
Claviculae	0.32	**1.20**	0.37	**1.42**	0.41	**1.57**	0.44	**1.69**	0.49	**1.89**	0.53	**2.04**
Scapulae	0.02	**0.09**	0.03	**0.12**	0.03	**0.14**	0.04	**0.16**	0.05	**0.20**	0.06	**0.24**
Ribs	0.43	**0.94**	0.48	**1.08**	0.52	**1.16**	0.54	**1.23**	0.58	**1.34**	0.62	**1.43**
Oesophagus	0.11	**0.16**	0.14	**0.20**	0.16	**0.24**	0.19	**0.27**	0.23	**0.34**	0.27	**0.40**
Trachea	0.01	**0.25**	0.02	**0.29**	0.02	**0.32**	0.02	**0.35**	0.03	**0.39**	0.04	**0.43**
Lung	0.05	**0.27**	0.07	**0.33**	0.08	**0.38**	0.09	**0.42**	0.11	**0.50**	0.13	**0.57**
Mamma	0.03	**0.63**	0.04	**0.71**	0.04	**0.77**	0.05	**0.81**	0.06	**0.88**	0.07	**0.94**
Heart	0.34	**0.47**	0.42	**0.58**	0.48	**0.66**	0.52	**0.72**	0.61	**0.85**	0.69	**0.96**
Stomach	0.06	**0.45**	0.07	**0.55**	0.08	**0.62**	0.09	**0.68**	0.11	**0.80**	0.13	**0.90**
Spleen	0.01	**0.09**	0.02	**0.12**	0.02	**0.14**	0.02	**0.16**	0.03	**0.20**	0.03	**0.24**
Liver	0.10	**0.29**	0.12	**0.35**	0.14	**0.40**	0.15	**0.44**	0.18	**0.52**	0.20	**0.60**
Gall bladder	0.05	**0.35**	0.06	**0.44**	0.08	**0.51**	0.08	**0.57**	0.10	**0.68**	0.12	**0.78**
Pancreas	0.10	**0.20**	0.13	**0.26**	0.15	**0.30**	0.17	**0.34**	0.21	**0.42**	0.25	**0.50**
Small intestine		**0.08**		**0.10**		**0.12**		**0.13**		**0.16**		**0.19**
Large intestine		**0.16**		**0.21**		**0.24**		**0.26**		**0.31**	0.01	**0.36**
Sigma/Rectum		**0.02**		**0.02**		**0.02**		**0.03**		**0.03**		**0.04**
Lumbar spine	0.02	**0.11**	0.03	**0.14**	0.03	**0.16**	0.04	**0.19**	0.05	**0.24**	0.06	**0.29**
Pelvis		**0.02**		**0.02**		**0.03**		**0.03**		**0.04**		**0.05**
Adrenals	0.05	**0.08**	0.06	**0.10**	0.07	**0.12**	0.08	**0.14**	0.10	**0.17**	0.12	**0.21**
Kidneys	0.01	**0.06**	0.02	**0.07**	0.02	**0.09**	0.02	**0.10**	0.03	**0.13**	0.04	**0.16**
Urinary bladder												
Uterus		**0.01**		**0.01**		**0.02**		**0.02**		**0.02**		**0.03**
Ovaries				**0.01**		**0.01**		**0.02**		**0.02**		**0.03**
Prostate												
Testes												
Upper arm		**0.03**	0.01	**0.04**	0.01	**0.05**	0.02	**0.05**	0.02	**0.07**	0.02	**0.08**
Elbow		**0.03**	0.01	**0.04**	0.01	**0.05**	0.02	**0.06**	0.02	**0.08**	0.02	**0.09**
Forearm		**0.01**		**0.02**		**0.02**		**0.02**		**0.03**		**0.04**
Upper leg												
Knee												
Lower leg												
Lymph nodes	0.05	**0.22**	0.06	**0.27**	0.07	**0.31**	0.08	**0.34**	0.10	**0.40**	0.12	**0.45**
Skin	0.04	**0.13**	0.04	**0.14**	0.04	**0.15**	0.04	**0.15**	0.04	**0.16**	0.05	**0.17**
Muscle	0.03	**0.10**	0.04	**0.12**	0.05	**0.13**	0.05	**0.14**	0.06	**0.17**	0.07	**0.19**
Skeleton	0.06	**0.14**	0.07	**0.16**	0.07	**0.18**	0.08	**0.20**	0.09	**0.22**	0.09	**0.25**
Red bone marrow	0.03	**0.08**	0.04	**0.10**	0.05	**0.11**	0.05	**0.12**	0.06	**0.14**	0.06	**0.16**
Whole body dose	0.04	**0.11**	0.05	**0.14**	0.05	**0.15**	0.06	**0.17**	0.07	**0.19**	0.08	**0.22**
ED (ICRP 60)	0.04	**0.21**	0.05	**0.25**	0.05	**0.28**	0.06	**0.31**	0.07	**0.35**	0.08	**0.40**
ED (ICRP 103)	0.04	**0.26**	0.05	**0.31**	0.06	**0.34**	0.07	**0.37**	0.08	**0.42**	0.10	**0.47**
Total filtration (mm) Al	**1.5**		**2.5**		**3.5**		**4.5**		**3.5**		**3.5**	
Cu	**0**		**0**		**0**		**0**		**0.1**		**0.2**	

Table 15.39 Normalized organ doses—Thoracic spine r-lat—1–2 years

Beam projection:		r-lat									**Table cells:**
Focus-to-detector distance (cm):		115									(+) = normal font = optimal field setting
Tube voltage (kVp):		50									**(−) = bold font = suboptimal field setting**
Total filtrations:		See last table row.									Empty table cells correspond to values <0.01
Organ or tissue	**Organ dose D_T : Entrance dose K_E**							**(mSv * mGy^{-1})**				
	(+)	(−)	(+)	(−)	(+)	(−)	(+)	(−)	(+)	(−)	(+)	(−)
Skull				**0.01**		**0.01**		**0.02**		**0.02**		**0.02**
Brain												
Oral mucosa		**0.01**		**0.01**		**0.02**		**0.02**		**0.02**		**0.03**
Salivary glands		**0.03**		**0.04**		**0.05**		**0.05**		**0.06**		**0.07**
Cervical spine	0.04	**0.26**	0.05	**0.34**	0.06	**0.40**	0.06	**0.45**	0.08	**0.55**	0.10	**0.64**
Thyroid		**0.25**		**0.31**	0.01	**0.35**	0.01	**0.38**	0.02	**0.43**	0.02	**0.48**
Thymus		**0.03**	0.01	**0.04**	0.01	**0.04**	0.01	**0.05**	0.02	**0.06**	0.02	**0.08**
Thoracic spine	0.26	**0.29**	0.35	**0.39**	0.43	**0.47**	0.49	**0.54**	0.61	**0.68**	0.73	**0.82**
Claviculae	0.02	**0.21**	0.02	**0.26**	0.03	**0.29**	0.03	**0.32**	0.04	**0.37**	0.05	**0.42**
Scapulae	1.07	**1.07**	1.25	**1.26**	1.37	**1.38**	1.46	**1.47**	1.60	**1.63**	1.72	**1.76**
Ribs	0.45	**0.69**	0.54	**0.82**	0.60	**0.91**	0.65	**0.97**	0.73	**1.08**	0.80	**1.18**
Oesophagus	0.15	**0.19**	0.20	**0.25**	0.23	**0.30**	0.26	**0.33**	0.31	**0.40**	0.36	**0.47**
Trachea		**0.04**		**0.05**		**0.06**		**0.06**		**0.07**		**0.08**
Lung	0.13	**0.25**	0.17	**0.31**	0.19	**0.35**	0.21	**0.38**	0.24	**0.45**	0.27	**0.50**
Mamma		**0.01**		**0.02**		**0.02**		**0.02**	0.01	**0.03**	0.01	**0.04**
Heart	0.02	**0.09**	0.03	**0.12**	0.03	**0.15**	0.03	**0.17**	0.04	**0.20**	0.05	**0.24**
Stomach		**0.03**		**0.04**		**0.05**		**0.05**	0.01	**0.07**	0.02	**0.08**
Spleen	0.02	**0.03**	0.02	**0.04**	0.03	**0.05**	0.03	**0.05**	0.04	**0.07**	0.06	**0.09**
Liver	0.08	**0.30**	0.10	**0.36**	0.12	**0.41**	0.13	**0.44**	0.15	**0.51**	0.17	**0.57**
Gall bladder		**0.11**	0.01	**0.15**	0.01	**0.18**	0.02	**0.20**	0.02	**0.25**	0.03	**0.29**
Pancreas	0.03	**0.08**	0.04	**0.10**	0.05	**0.13**	0.05	**0.15**	0.07	**0.18**	0.08	**0.22**
Small intestine		**0.05**		**0.06**		**0.08**		**0.09**		**0.11**		**0.13**
Large intestine		**0.08**		**0.10**		**0.12**		**0.13**		**0.16**		**0.19**
Sigma/Rectum										**0.01**		**0.02**
Lumbar spine	0.05	**0.24**	0.07	**0.32**	0.08	**0.39**	0.10	**0.44**	0.12	**0.55**	0.15	**0.67**
Pelvis		**0.02**		**0.03**		**0.04**		**0.04**		**0.05**		**0.07**
Adrenals	0.09	**0.11**	0.12	**0.14**	0.14	**0.17**	0.16	**0.20**	0.20	**0.24**	0.24	**0.29**
Kidneys	0.06	**0.17**	0.08	**0.21**	0.09	**0.25**	0.11	**0.28**	0.13	**0.33**	0.15	**0.38**
Urinary bladder												
Uterus				**0.01**		**0.01**		**0.01**		**0.02**		**0.02**
Ovaries				**0.01**		**0.01**		**0.02**		**0.02**		**0.03**
Prostate												
Testes												
Upper arm												
Elbow												
Forearm												
Upper leg												
Knee												
Lower leg												
Lymph nodes	0.02	**0.09**	0.03	**0.11**	0.04	**0.13**	0.04	**0.15**	0.05	**0.18**	0.06	**0.21**
Skin	0.07	**0.13**	0.07	**0.13**	0.08	**0.14**	0.08	**0.14**	0.08	**0.15**	0.09	**0.16**
Muscle	0.03	**0.06**	0.03	**0.08**	0.04	**0.09**	0.04	**0.10**	0.05	**0.11**	0.06	**0.12**
Skeleton	0.12	**0.17**	0.15	**0.21**	0.17	**0.24**	0.18	**0.26**	0.21	**0.31**	0.23	**0.35**
Red bone marrow	0.03	**0.04**	0.04	**0.05**	0.04	**0.06**	0.05	**0.07**	0.06	**0.08**	0.06	**0.09**
Whole body dose	0.04	**0.08**	0.05	**0.10**	0.06	**0.11**	0.06	**0.12**	0.07	**0.14**	0.08	**0.16**
ED (ICRP 60)	0.04	**0.09**	0.04	**0.11**	0.05	**0.13**	0.06	**0.14**	0.07	**0.16**	0.08	**0.19**
ED (ICRP 103)	0.03	**0.09**	0.04	**0.11**	0.05	**0.12**	0.06	**0.14**	0.07	**0.16**	0.08	**0.18**
Total filtration (mm) Al	**1.5**		**2.5**		**3.5**		**4.5**		**3.5**		**3.5**	
Cu	**0**		**0**		**0**		**0**		**0.1**		**0.2**	

Table 15.40 Normalized organ doses—Thoracic spine r-lat—1–2 years

Beam projection:			r-lat									**Table cells:**
Focus-to-detector distance (cm):			115									(+) = normal font = optimal field setting
Tube voltage (kVp):			60									**(−) = bold font = suboptimal field setting**
Total filtrations:			See last table row.									Empty table cells correspond to values <0.01
Organ or tissue	**Organ dose D_T : Entrance dose K_E**							**(mSv ∗ mGy^{-1})**				
	(+)	(−)	(+)	(−)	(+)	(−)	(+)	(−)	(+)	(−)	(+)	(−)
Skull		**0.01**		**0.02**		**0.02**		**0.02**		**0.03**		**0.02**
Brain												
Oral mucosa		**0.01**		**0.02**		**0.02**		**0.03**		**0.03**		**0.03**
Salivary glands		**0.04**		**0.05**		**0.06**		**0.06**	0.01	**0.07**		**0.07**
Cervical spine	0.05	**0.34**	0.06	**0.43**	0.08	**0.50**	0.09	**0.56**	0.11	**0.67**	0.10	**0.64**
Thyroid	0.01	**0.30**	0.01	**0.36**	0.02	**0.40**	0.02	**0.43**	0.02	**0.49**	0.02	**0.48**
Thymus	0.01	**0.04**	0.01	**0.05**	0.02	**0.06**	0.02	**0.07**	0.03	**0.08**	0.02	**0.08**
Thoracic spine	0.36	**0.41**	0.48	**0.54**	0.57	**0.64**	0.64	**0.72**	0.79	**0.89**	0.73	**0.82**
Claviculae	0.02	**0.25**	0.03	**0.31**	0.04	**0.35**	0.04	**0.38**	0.05	**0.43**	0.05	**0.42**
Scapulae	1.19	**1.20**	1.37	**1.39**	1.49	**1.51**	1.57	**1.60**	1.71	**1.76**	1.72	**1.76**
Ribs	0.52	**0.78**	0.61	**0.92**	0.68	**1.01**	0.72	**1.08**	0.81	**1.19**	0.80	**1.18**
Oesophagus	0.19	**0.25**	0.25	**0.32**	0.29	**0.38**	0.32	**0.42**	0.38	**0.50**	0.36	**0.47**
Trachea		**0.05**		**0.06**		**0.07**		**0.08**		**0.09**		**0.08**
Lung	0.16	**0.30**	0.20	**0.37**	0.23	**0.41**	0.25	**0.45**	0.28	**0.52**	0.27	**0.50**
Mamma		**0.02**		**0.02**	0.01	**0.03**	0.01	**0.03**	0.02	**0.04**	0.01	**0.04**
Heart	0.03	**0.13**	0.03	**0.16**	0.04	**0.19**	0.05	**0.21**	0.06	**0.25**	0.05	**0.24**
Stomach		**0.04**	0.01	**0.06**	0.01	**0.07**	0.02	**0.08**	0.02	**0.10**	0.02	**0.08**
Spleen	0.03	**0.04**	0.04	**0.06**	0.04	**0.07**	0.05	**0.08**	0.07	**0.10**	0.06	**0.09**
Liver	0.10	**0.35**	0.12	**0.42**	0.14	**0.47**	0.15	**0.51**	0.17	**0.58**	0.17	**0.57**
Gall bladder	0.01	**0.15**	0.02	**0.20**	0.02	**0.23**	0.02	**0.26**	0.03	**0.32**	0.03	**0.29**
Pancreas	0.04	**0.11**	0.06	**0.15**	0.07	**0.18**	0.08	**0.20**	0.09	**0.25**	0.08	**0.22**
Small intestine		**0.07**		**0.09**		**0.10**		**0.11**	0.01	**0.14**		**0.13**
Large intestine		**0.10**		**0.13**		**0.15**		**0.17**	0.01	**0.20**		**0.19**
Sigma/Rectum				**0.01**		**0.01**		**0.02**		**0.02**		**0.02**
Lumbar spine	0.07	**0.33**	0.10	**0.44**	0.12	**0.52**	0.13	**0.59**	0.17	**0.73**	0.15	**0.67**
Pelvis		**0.03**		**0.04**		**0.05**		**0.06**		**0.08**		**0.07**
Adrenals	0.12	**0.15**	0.16	**0.19**	0.19	**0.23**	0.21	**0.26**	0.26	**0.32**	0.24	**0.29**
Kidneys	0.08	**0.21**	0.10	**0.27**	0.12	**0.31**	0.13	**0.34**	0.16	**0.40**	0.15	**0.38**
Urinary bladder												
Uterus		**0.01**		**0.02**		**0.02**		**0.02**		**0.03**		**0.02**
Ovaries		**0.01**		**0.02**		**0.02**		**0.02**		**0.03**		**0.03**
Prostate												
Testes												
Upper arm												
Elbow												
Forearm												
Upper leg												
Knee												
Lower leg												
Lymph nodes	0.03	**0.11**	0.04	**0.15**	0.05	**0.17**	0.05	**0.19**	0.06	**0.22**	0.06	**0.21**
Skin	0.07	**0.13**	0.08	**0.14**	0.08	**0.15**	0.08	**0.15**	0.09	**0.16**	0.09	**0.16**
Muscle	0.03	**0.08**	0.04	**0.09**	0.05	**0.10**	0.05	**0.11**	0.06	**0.13**	0.06	**0.12**
Skeleton	0.14	**0.21**	0.17	**0.25**	0.19	**0.28**	0.21	**0.31**	0.24	**0.36**	0.23	**0.35**
Red bone marrow	0.04	**0.06**	0.05	**0.07**	0.05	**0.08**	0.06	**0.09**	0.07	**0.10**	0.06	**0.09**
Whole body dose	0.05	**0.10**	0.06	**0.12**	0.07	**0.13**	0.07	**0.15**	0.08	**0.17**	0.08	**0.16**
ED (ICRP 60)	0.04	**0.11**	0.06	**0.13**	0.06	**0.15**	0.07	**0.17**	0.08	**0.20**	0.08	**0.19**
ED (ICRP 103)	0.04	**0.11**	0.05	**0.13**	0.06	**0.15**	0.07	**0.17**	0.08	**0.20**	0.08	**0.18**
Total filtration (mm) **Al**	**1.5**		**2.5**		**3.5**		**4.5**		**3.5**		**3.5**	
Cu	**0**		**0**		**0**		**0**		**0.1**		**0.2**	

Table 15.41 Normalized organ doses—Thoracic spine r-lat—1–2 years

Beam projection:			r-lat									**Table cells:**
Focus-to-detector distance (cm):			115						(+) = normal font = optimal field setting			
Tube voltage (kVp):			70						**(−) = bold font = suboptimal field setting**			
Total filtrations:			See last table row.						Empty table cells correspond to values <0.01			
Organ or tissue	**Organ dose D_T : Entrance dose K_E**								**(mSv * mGy^{-1})**			
	(+)	(−)	(+)	(−)	(+)	(−)	(+)	(−)	(+)	(−)	(+)	(−)
Skull		**0.02**		**0.02**		**0.02**		**0.03**	0.01	**0.03**	0.01	**0.04**
Brain												
Oral mucosa		**0.02**		**0.02**		**0.03**		**0.03**		**0.04**	0.01	**0.05**
Salivary glands		**0.05**		**0.06**		**0.07**	0.01	**0.07**	0.01	**0.09**	0.02	**0.10**
Cervical spine	0.06	**0.40**	0.08	**0.50**	0.09	**0.58**	0.10	**0.64**	0.12	**0.74**	0.14	**0.84**
Thyroid	0.01	**0.34**	0.02	**0.40**	0.02	**0.44**	0.02	**0.48**	0.03	**0.53**	0.04	**0.59**
Thymus	0.01	**0.05**	0.02	**0.06**	0.02	**0.07**	0.03	**0.08**	0.03	**0.10**	0.04	**0.12**
Thoracic spine	0.45	**0.51**	0.58	**0.66**	0.68	**0.77**	0.76	**0.86**	0.91	**1.04**	1.05	**1.21**
Claviculae	0.03	**0.29**	0.04	**0.34**	0.05	**0.38**	0.05	**0.41**	0.07	**0.47**	0.08	**0.52**
Scapulae	1.27	**1.29**	1.45	**1.47**	1.55	**1.59**	1.63	**1.68**	1.76	**1.82**	1.84	**1.92**
Ribs	0.57	**0.86**	0.66	**0.99**	0.72	**1.07**	0.77	**1.14**	0.85	**1.25**	0.91	**1.34**
Oesophagus	0.23	**0.31**	0.29	**0.38**	0.33	**0.44**	0.37	**0.49**	0.43	**0.58**	0.49	**0.66**
Trachea		**0.06**		**0.07**		**0.08**		**0.09**		**0.10**		**0.11**
Lung	0.19	**0.34**	0.22	**0.41**	0.25	**0.46**	0.27	**0.50**	0.31	**0.57**	0.35	**0.63**
Mamma		**0.02**	0.01	**0.03**	0.01	**0.04**	0.02	**0.04**	0.02	**0.05**	0.02	**0.06**
Heart	0.03	**0.15**	0.04	**0.19**	0.05	**0.22**	0.06	**0.25**	0.07	**0.29**	0.08	**0.34**
Stomach	0.01	**0.06**	0.01	**0.07**	0.02	**0.09**	0.02	**0.10**	0.03	**0.12**	0.03	**0.15**
Spleen	0.04	**0.06**	0.05	**0.08**	0.06	**0.09**	0.07	**0.11**	0.09	**0.13**	0.10	**0.16**
Liver	0.12	**0.40**	0.14	**0.47**	0.16	**0.52**	0.17	**0.56**	0.19	**0.63**	0.21	**0.70**
Gall bladder	0.02	**0.19**	0.02	**0.24**	0.03	**0.28**	0.03	**0.31**	0.04	**0.37**	0.05	**0.43**
Pancreas	0.05	**0.14**	0.07	**0.18**	0.08	**0.22**	0.09	**0.24**	0.11	**0.30**	0.14	**0.35**
Small intestine		**0.08**		**0.10**		**0.12**	0.01	**0.13**	0.01	**0.16**	0.02	**0.19**
Large intestine		**0.12**		**0.15**	0.01	**0.18**	0.01	**0.20**	0.02	**0.23**	0.02	**0.27**
Sigma/Rectum		**0.01**		**0.02**		**0.02**		**0.02**		**0.03**		**0.03**
Lumbar spine	0.09	**0.42**	0.12	**0.54**	0.14	**0.63**	0.16	**0.70**	0.20	**0.85**	0.24	**0.99**
Pelvis		**0.04**		**0.06**		**0.07**		**0.08**	0.01	**0.10**	0.02	**0.12**
Adrenals	0.15	**0.19**	0.20	**0.24**	0.23	**0.28**	0.25	**0.31**	0.30	**0.37**	0.35	**0.44**
Kidneys	0.10	**0.25**	0.12	**0.31**	0.14	**0.35**	0.15	**0.39**	0.18	**0.45**	0.21	**0.51**
Urinary bladder								**0.01**		**0.01**		**0.02**
Uterus		**0.02**		**0.02**		**0.02**		**0.03**		**0.04**		**0.04**
Ovaries		**0.02**		**0.02**		**0.03**		**0.03**		**0.04**		**0.05**
Prostate												
Testes												
Upper arm												
Elbow												
Forearm												
Upper leg												
Knee												
Lower leg												
Lymph nodes	0.04	**0.14**	0.05	**0.17**	0.06	**0.20**	0.06	**0.22**	0.08	**0.26**	0.09	**0.29**
Skin	0.08	**0.14**	0.08	**0.15**	0.08	**0.15**	0.09	**0.16**	0.09	**0.17**	0.10	**0.18**
Muscle	0.04	**0.09**	0.05	**0.11**	0.05	**0.12**	0.06	**0.13**	0.07	**0.15**	0.07	**0.16**
Skeleton	0.16	**0.24**	0.19	**0.28**	0.21	**0.32**	0.23	**0.34**	0.26	**0.39**	0.29	**0.43**
Red bone marrow	0.05	**0.07**	0.05	**0.08**	0.06	**0.09**	0.07	**0.10**	0.08	**0.12**	0.09	**0.13**
Whole body dose	0.05	**0.11**	0.07	**0.13**	0.07	**0.15**	0.08	**0.16**	0.09	**0.19**	0.10	**0.21**
ED (ICRP 60)	0.05	**0.13**	0.06	**0.15**	0.07	**0.18**	0.08	**0.19**	0.09	**0.22**	0.10	**0.25**
ED (ICRP 103)	0.05	**0.13**	0.06	**0.15**	0.07	**0.17**	0.08	**0.19**	0.09	**0.22**	0.10	**0.25**
Total filtration (mm) **Al**	**1.5**		**2.5**		**3.5**		**4.5**		**3.5**		**3.5**	
Cu	**0**		**0**		**0**		**0**		**0.1**		**0.2**	

Table 15.42 Normalized organ doses—Thoracic spine r-lat—3–7 years

Beam projection:			r-lat									**Table cells:**
Focus-to-detector distance (cm):			115						(+) = normal font = optimal field setting			
Tube voltage (kVp):			60						**(−) = bold font = suboptimal field setting**			
Total filtrations:			See last table row.						Empty table cells correspond to values <0.01			
Organ or tissue	**Organ dose D_T : Entrance dose K_E**								**(mSv * mGy^{-1})**			
	(+)	(−)	(+)	(−)	(+)	(−)	(+)	(−)	(+)	(−)	(+)	(−)
Skull		**0.01**		**0.01**		**0.02**		**0.02**		**0.02**		**0.02**
Brain												
Oral mucosa		**0.01**		**0.02**		**0.02**		**0.02**		**0.03**		**0.02**
Salivary glands		**0.03**		**0.04**		**0.04**		**0.05**		**0.06**		**0.05**
Cervical spine	0.07	**0.30**	0.10	**0.38**	0.11	**0.44**	0.13	**0.50**	0.16	**0.59**	0.14	**0.56**
Thyroid		**0.27**	0.01	**0.33**	0.02	**0.37**	0.02	**0.41**	0.02	**0.46**	0.02	**0.45**
Thymus	0.01	**0.04**	0.01	**0.05**	0.02	**0.06**	0.02	**0.07**	0.03	**0.09**	0.02	**0.08**
Thoracic spine	0.23	**0.25**	0.30	**0.34**	0.37	**0.40**	0.42	**0.46**	0.53	**0.58**	0.47	**0.51**
Claviculae	0.02	**0.27**	0.03	**0.33**	0.03	**0.37**	0.04	**0.41**	0.05	**0.48**	0.04	**0.46**
Scapulae	1.05	**1.03**	1.23	**1.21**	1.35	**1.34**	1.44	**1.43**	1.58	**1.58**	1.59	**1.58**
Ribs	0.44	**0.69**	0.52	**0.81**	0.58	**0.90**	0.63	**0.96**	0.71	**1.08**	0.70	**1.06**
Oesophagus	0.10	**0.12**	0.13	**0.16**	0.15	**0.19**	0.17	**0.22**	0.21	**0.27**	0.19	**0.24**
Trachea		**0.06**		**0.07**		**0.08**		**0.09**		**0.10**		**0.10**
Lung	0.13	**0.25**	0.16	**0.31**	0.19	**0.35**	0.21	**0.38**	0.24	**0.45**	0.23	**0.43**
Mamma		**0.02**		**0.03**		**0.03**	0.01	**0.04**	0.01	**0.04**	0.01	**0.04**
Heart	0.03	**0.11**	0.03	**0.14**	0.04	**0.17**	0.05	**0.19**	0.06	**0.23**	0.05	**0.21**
Stomach		**0.03**		**0.04**		**0.04**	0.01	**0.05**	0.02	**0.07**	0.01	**0.05**
Spleen	0.01	**0.02**	0.02	**0.03**	0.02	**0.03**	0.03	**0.04**	0.03	**0.05**	0.03	**0.04**
Liver	0.10	**0.33**	0.12	**0.40**	0.14	**0.45**	0.15	**0.50**	0.18	**0.57**	0.17	**0.55**
Gall bladder	0.01	**0.09**	0.02	**0.12**	0.02	**0.15**	0.02	**0.17**	0.03	**0.21**	0.02	**0.19**
Pancreas	0.03	**0.07**	0.04	**0.09**	0.05	**0.11**	0.06	**0.12**	0.07	**0.16**	0.06	**0.13**
Small intestine		**0.04**		**0.06**		**0.07**		**0.08**		**0.10**		**0.08**
Large intestine		**0.07**		**0.09**		**0.11**		**0.12**		**0.15**		**0.14**
Sigma/Rectum										**0.01**		
Lumbar spine	0.06	**0.18**	0.08	**0.25**	0.09	**0.30**	0.11	**0.34**	0.14	**0.43**	0.12	**0.38**
Pelvis		**0.02**		**0.03**		**0.03**		**0.04**		**0.05**		**0.04**
Adrenals	0.09	**0.10**	0.11	**0.13**	0.13	**0.16**	0.15	**0.18**	0.19	**0.22**	0.17	**0.20**
Kidneys	0.07	**0.15**	0.09	**0.19**	0.11	**0.22**	0.12	**0.25**	0.15	**0.30**	0.14	**0.28**
Urinary bladder												
Uterus						**0.01**		**0.01**		**0.02**		**0.01**
Ovaries				**0.01**		**0.01**		**0.02**		**0.02**		**0.02**
Prostate												
Testes												
Upper arm												
Elbow												
Forearm												
Upper leg												
Knee												
Lower leg												
Lymph nodes	0.02	**0.08**	0.03	**0.11**	0.04	**0.13**	0.04	**0.14**	0.05	**0.17**	0.04	**0.16**
Skin	0.06	**0.08**	0.06	**0.09**	0.06	**0.09**	0.06	**0.10**	0.07	**0.10**	0.07	**0.10**
Muscle	0.03	**0.06**	0.03	**0.07**	0.04	**0.08**	0.04	**0.09**	0.05	**0.10**	0.05	**0.10**
Skeleton	0.09	**0.14**	0.12	**0.17**	0.13	**0.19**	0.14	**0.20**	0.16	**0.24**	0.16	**0.23**
Red bone marrow	0.03	**0.05**	0.04	**0.06**	0.04	**0.06**	0.05	**0.07**	0.06	**0.08**	0.05	**0.08**
Whole body dose	0.04	**0.08**	0.05	**0.09**	0.05	**0.11**	0.06	**0.12**	0.07	**0.14**	0.06	**0.13**
ED (ICRP 60)	0.03	**0.09**	0.04	**0.11**	0.05	**0.12**	0.05	**0.14**	0.06	**0.16**	0.06	**0.15**
ED (ICRP 103)	0.03	**0.08**	0.04	**0.11**	0.05	**0.12**	0.05	**0.13**	0.06	**0.16**	0.06	**0.15**
Total filtration (mm) Al	**1.5**		**2.5**		**3.5**		**4.5**		**3.5**		**3.5**	
Cu	**0**		**0**		**0**		**0**		**0.1**		**0.2**	

Table 15.43 Normalized organ doses—Thoracic spine r-lat—3–7 years

Beam projection:			r-lat									**Table cells:**
Focus-to-detector distance (cm):			115						(+) = normal font = optimal field setting			
Tube voltage (kVp):			70						**(−) = bold font = suboptimal field setting**			
Total filtrations:			See last table row.						Empty table cells correspond to values <0.01			
Organ or tissue	**Organ dose D_T : Entrance dose K_E**								**(mSv * mGy^{-1})**			
	(+)	(−)	(+)	(−)	(+)	(−)	(+)	(−)	(+)	(−)	(+)	(−)
Skull		**0.01**		**0.02**		**0.02**		**0.02**		**0.03**	0.01	**0.04**
Brain												
Oral mucosa		**0.02**		**0.02**		**0.02**		**0.03**		**0.03**		**0.04**
Salivary glands		**0.04**		**0.05**		**0.05**		**0.06**	0.01	**0.07**	0.01	**0.08**
Cervical spine	0.09	**0.36**	0.12	**0.45**	0.14	**0.52**	0.15	**0.57**	0.18	**0.67**	0.21	**0.77**
Thyroid	0.01	**0.31**	0.02	**0.37**	0.02	**0.42**	0.02	**0.45**	0.03	**0.51**	0.04	**0.57**
Thymus	0.01	**0.05**	0.02	**0.06**	0.02	**0.08**	0.03	**0.09**	0.03	**0.11**	0.04	**0.13**
Thoracic spine	0.30	**0.33**	0.39	**0.43**	0.46	**0.51**	0.52	**0.58**	0.64	**0.71**	0.75	**0.85**
Claviculae	0.03	**0.31**	0.03	**0.37**	0.04	**0.42**	0.05	**0.46**	0.06	**0.52**	0.07	**0.58**
Scapulae	1.14	**1.12**	1.31	**1.30**	1.43	**1.42**	1.51	**1.51**	1.64	**1.65**	1.74	**1.76**
Ribs	0.49	**0.76**	0.57	**0.88**	0.63	**0.97**	0.68	**1.03**	0.75	**1.14**	0.82	**1.23**
Oesophagus	0.12	**0.16**	0.16	**0.20**	0.18	**0.24**	0.20	**0.26**	0.24	**0.32**	0.28	**0.37**
Trachea		**0.07**		**0.08**		**0.09**		**0.10**		**0.11**		**0.12**
Lung	0.15	**0.29**	0.19	**0.35**	0.21	**0.40**	0.23	**0.43**	0.27	**0.50**	0.30	**0.56**
Mamma		**0.03**	0.01	**0.03**	0.01	**0.04**	0.02	**0.04**	0.02	**0.06**	0.02	**0.07**
Heart	0.03	**0.14**	0.04	**0.18**	0.05	**0.20**	0.06	**0.23**	0.07	**0.28**	0.08	**0.32**
Stomach		**0.04**	0.01	**0.05**	0.01	**0.06**	0.02	**0.07**	0.02	**0.09**	0.03	**0.11**
Spleen	0.02	**0.03**	0.03	**0.04**	0.03	**0.05**	0.04	**0.06**	0.05	**0.07**	0.06	**0.09**
Liver	0.11	**0.38**	0.14	**0.46**	0.16	**0.51**	0.17	**0.56**	0.20	**0.64**	0.22	**0.71**
Gall bladder	0.02	**0.12**	0.02	**0.16**	0.03	**0.19**	0.03	**0.21**	0.04	**0.26**	0.04	**0.31**
Pancreas	0.04	**0.09**	0.05	**0.12**	0.07	**0.14**	0.07	**0.16**	0.09	**0.20**	0.11	**0.24**
Small intestine		**0.06**		**0.07**		**0.08**		**0.10**	0.01	**0.12**	0.01	**0.14**
Large intestine		**0.09**		**0.11**		**0.13**		**0.15**	0.01	**0.18**	0.02	**0.21**
Sigma/Rectum								**0.01**		**0.01**		**0.02**
Lumbar spine	0.08	**0.24**	0.10	**0.32**	0.12	**0.38**	0.14	**0.43**	0.17	**0.53**	0.20	**0.63**
Pelvis		**0.03**		**0.03**		**0.04**		**0.05**		**0.06**		**0.07**
Adrenals	0.11	**0.13**	0.14	**0.17**	0.17	**0.20**	0.19	**0.22**	0.23	**0.27**	0.27	**0.32**
Kidneys	0.09	**0.18**	0.11	**0.23**	0.13	**0.26**	0.15	**0.29**	0.17	**0.35**	0.20	**0.40**
Urinary bladder												
Uterus				**0.01**		**0.01**		**0.02**		**0.02**		**0.03**
Ovaries		**0.01**		**0.02**		**0.02**		**0.02**		**0.03**		**0.03**
Prostate												
Testes												
Upper arm												
Elbow												
Forearm												
Upper leg												
Knee												
Lower leg												
Lymph nodes	0.03	**0.10**	0.04	**0.13**	0.04	**0.15**	0.05	**0.17**	0.06	**0.20**	0.07	**0.23**
Skin	0.06	**0.09**	0.06	**0.09**	0.07	**0.10**	0.07	**0.10**	0.07	**0.11**	0.07	**0.12**
Muscle	0.03	**0.07**	0.04	**0.08**	0.04	**0.09**	0.05	**0.10**	0.06	**0.12**	0.06	**0.13**
Skeleton	0.11	**0.16**	0.13	**0.19**	0.15	**0.21**	0.16	**0.23**	0.18	**0.26**	0.20	**0.29**
Red bone marrow	0.04	**0.05**	0.05	**0.07**	0.05	**0.08**	0.06	**0.08**	0.07	**0.10**	0.08	**0.11**
Whole body dose	0.05	**0.09**	0.05	**0.11**	0.06	**0.12**	0.07	**0.13**	0.08	**0.15**	0.09	**0.17**
ED (ICRP 60)	0.04	**0.10**	0.05	**0.13**	0.06	**0.14**	0.06	**0.16**	0.07	**0.18**	0.08	**0.21**
ED (ICRP 103)	0.04	**0.10**	0.05	**0.12**	0.06	**0.14**	0.06	**0.16**	0.07	**0.18**	0.09	**0.21**
Total filtration (mm) Al	**1.5**		**2.5**		**3.5**		**4.5**		**3.5**		**3.5**	
Cu	**0**		**0**		**0**		**0**		**0.1**		**0.2**	

Table 15.44 Normalized organ doses—Thoracic spine r-lat—8–12 years

Beam projection:			r-lat									**Table cells:**
Focus-to-detector distance (cm):			115				(+) = normal font = optimal field setting					
Tube voltage (kVp):			60				**(−) = bold font = suboptimal field setting**					
Total filtrations:			See last table row.				Empty table cells correspond to values <0.01					
Organ or tissue	**Organ dose D_T : Entrance dose K_E**								**(mSv * mGy^{-1})**			
	(+)	(−)	(+)	(−)	(+)	(−)	(+)	(−)	(+)	(−)	(+)	(−)
Skull		**0.01**		**0.01**		**0.02**		**0.02**		**0.02**		**0.02**
Brain												
Oral mucosa		**0.01**		**0.01**		**0.02**		**0.02**		**0.02**		**0.02**
Salivary glands		**0.04**		**0.05**		**0.05**		**0.06**		**0.07**		**0.07**
Cervical spine	0.05	**0.28**	0.07	**0.36**	0.08	**0.42**	0.09	**0.47**	0.12	**0.57**	0.10	**0.53**
Thyroid	0.01	**0.20**	0.02	**0.24**	0.02	**0.28**	0.02	**0.31**	0.03	**0.35**	0.03	**0.34**
Thymus	0.01	**0.03**	0.02	**0.04**	0.02	**0.05**	0.03	**0.06**	0.03	**0.07**	0.03	**0.06**
Thoracic spine	0.15	**0.16**	0.20	**0.22**	0.24	**0.26**	0.28	**0.30**	0.36	**0.39**	0.31	**0.33**
Claviculae	0.03	**0.22**	0.04	**0.28**	0.04	**0.32**	0.05	**0.36**	0.06	**0.42**	0.05	**0.40**
Scapulae	0.94	**0.93**	1.11	**1.10**	1.23	**1.22**	1.32	**1.31**	1.47	**1.47**	1.47	**1.46**
Ribs	0.45	**0.61**	0.54	**0.72**	0.60	**0.80**	0.65	**0.86**	0.73	**0.97**	0.72	**0.95**
Oesophagus	0.07	**0.08**	0.09	**0.11**	0.11	**0.13**	0.13	**0.14**	0.16	**0.18**	0.14	**0.16**
Trachea		**0.10**		**0.13**	0.01	**0.15**	0.01	**0.16**	0.02	**0.19**	0.01	**0.18**
Lung	0.14	**0.19**	0.17	**0.24**	0.20	**0.28**	0.22	**0.31**	0.26	**0.37**	0.25	**0.35**
Mamma		**0.02**	0.01	**0.02**	0.01	**0.02**	0.01	**0.03**	0.02	**0.04**	0.02	**0.03**
Heart	0.04	**0.08**	0.05	**0.11**	0.06	**0.13**	0.07	**0.14**	0.09	**0.18**	0.08	**0.16**
Stomach		**0.02**		**0.02**	0.01	**0.03**	0.01	**0.03**	0.02	**0.04**	0.01	**0.03**
Spleen		**0.01**	0.01	**0.01**	0.01	**0.02**	0.02	**0.02**	0.02	**0.03**	0.02	**0.02**
Liver	0.13	**0.30**	0.16	**0.38**	0.18	**0.43**	0.20	**0.47**	0.24	**0.55**	0.23	**0.53**
Gall bladder	0.01	**0.08**	0.02	**0.10**	0.02	**0.12**	0.03	**0.14**	0.04	**0.18**	0.03	**0.15**
Pancreas	0.03	**0.04**	0.04	**0.05**	0.05	**0.07**	0.05	**0.08**	0.07	**0.10**	0.06	**0.08**
Small intestine		**0.04**		**0.05**		**0.06**		**0.07**		**0.08**		**0.07**
Large intestine		**0.06**		**0.08**		**0.09**		**0.10**		**0.13**		**0.11**
Sigma/Rectum												
Lumbar spine	0.03	**0.12**	0.04	**0.16**	0.05	**0.20**	0.06	**0.23**	0.08	**0.29**	0.06	**0.25**
Pelvis		**0.02**		**0.02**		**0.03**		**0.03**		**0.04**		**0.03**
Adrenals	0.06	**0.07**	0.08	**0.09**	0.10	**0.11**	0.12	**0.13**	0.15	**0.17**	0.13	**0.14**
Kidneys	0.05	**0.11**	0.06	**0.15**	0.07	**0.17**	0.08	**0.19**	0.10	**0.24**	0.09	**0.22**
Urinary bladder												
Uterus										**0.01**		
Ovaries										**0.01**		
Prostate												
Testes												
Upper arm												
Elbow												
Forearm												
Upper leg												
Knee												
Lower leg												
Lymph nodes	0.02	**0.07**	0.03	**0.09**	0.04	**0.11**	0.04	**0.12**	0.05	**0.15**	0.05	**0.14**
Skin	0.05	**0.08**	0.05	**0.08**	0.06	**0.09**	0.06	**0.09**	0.06	**0.10**	0.06	**0.09**
Muscle	0.03	**0.05**	0.03	**0.06**	0.04	**0.07**	0.04	**0.08**	0.05	**0.09**	0.05	**0.09**
Skeleton	0.08	**0.11**	0.10	**0.14**	0.12	**0.16**	0.13	**0.17**	0.15	**0.20**	0.14	**0.19**
Red bone marrow	0.03	**0.05**	0.04	**0.06**	0.05	**0.07**	0.05	**0.07**	0.06	**0.09**	0.06	**0.08**
Whole body dose	0.04	**0.07**	0.05	**0.08**	0.05	**0.09**	0.06	**0.10**	0.07	**0.12**	0.06	**0.11**
ED (ICRP 60)	0.03	**0.07**	0.04	**0.09**	0.05	**0.10**	0.06	**0.11**	0.07	**0.13**	0.06	**0.12**
ED (ICRP 103)	0.03	**0.07**	0.04	**0.09**	0.05	**0.10**	0.06	**0.11**	0.07	**0.13**	0.06	**0.12**
Total filtration (mm) **Al**	**1.5**		**2.5**		**3.5**		**4.5**		**3.5**		**3.5**	
Cu	**0**		**0**		**0**		**0**		**0.1**		**0.2**	

Table 15.45 Normalized organ doses—Thoracic spine r-lat—8–12 years

Beam projection:			r-lat									**Table cells:**
Focus-to-detector distance (cm):			115				(+) = normal font = optimal field setting					
Tube voltage (kVp):			70				**(−) = bold font = suboptimal field setting**					
Total filtrations:			See last table row.				Empty table cells correspond to values <0.01					
Organ or tissue	**Organ dose D_T : Entrance dose K_E**								**(mSv ∗ mGy⁻¹)**			
	(+)	(−)	(+)	(−)	(+)	(−)	(+)	(−)	(+)	(−)	(+)	(−)
Skull		**0.01**		**0.02**		**0.02**		**0.02**		**0.03**	0.01	**0.04**
Brain												
Oral mucosa		**0.01**		**0.02**		**0.02**		**0.02**		**0.03**		**0.04**
Salivary glands		**0.04**		**0.05**		**0.06**	0.01	**0.07**	0.01	**0.08**	0.01	**0.09**
Cervical spine	0.07	**0.34**	0.09	**0.43**	0.10	**0.50**	0.12	**0.55**	0.14	**0.66**	0.17	**0.76**
Thyroid	0.02	**0.23**	0.02	**0.28**	0.03	**0.32**	0.03	**0.34**	0.04	**0.40**	0.05	**0.44**
Thymus	0.02	**0.04**	0.03	**0.06**	0.03	**0.07**	0.03	**0.08**	0.04	**0.10**	0.05	**0.12**
Thoracic spine	0.20	**0.22**	0.27	**0.29**	0.32	**0.35**	0.36	**0.39**	0.46	**0.50**	0.55	**0.60**
Claviculae	0.04	**0.26**	0.05	**0.32**	0.06	**0.37**	0.06	**0.40**	0.08	**0.47**	0.10	**0.53**
Scapulae	1.03	**1.02**	1.20	**1.19**	1.32	**1.31**	1.40	**1.40**	1.54	**1.54**	1.65	**1.66**
Ribs	0.50	**0.67**	0.59	**0.79**	0.65	**0.87**	0.70	**0.93**	0.78	**1.04**	0.85	**1.12**
Oesophagus	0.09	**0.10**	0.12	**0.14**	0.14	**0.16**	0.16	**0.18**	0.19	**0.22**	0.23	**0.26**
Trachea	0.01	**0.12**	0.01	**0.15**	0.02	**0.17**	0.02	**0.18**	0.02	**0.21**	0.03	**0.24**
Lung	0.16	**0.23**	0.20	**0.29**	0.23	**0.33**	0.26	**0.36**	0.30	**0.42**	0.34	**0.48**
Mamma	0.01	**0.02**	0.01	**0.03**	0.02	**0.03**	0.02	**0.04**	0.02	**0.05**	0.03	**0.06**
Heart	0.05	**0.11**	0.07	**0.14**	0.08	**0.16**	0.09	**0.18**	0.11	**0.22**	0.14	**0.26**
Stomach		**0.02**	0.01	**0.03**	0.01	**0.04**	0.02	**0.04**	0.02	**0.06**	0.03	**0.07**
Spleen	0.01	**0.02**	0.02	**0.02**	0.02	**0.03**	0.02	**0.03**	0.03	**0.04**	0.04	**0.05**
Liver	0.15	**0.36**	0.19	**0.43**	0.21	**0.49**	0.23	**0.54**	0.27	**0.62**	0.31	**0.70**
Gall bladder	0.02	**0.10**	0.03	**0.13**	0.03	**0.16**	0.04	**0.18**	0.05	**0.23**	0.06	**0.27**
Pancreas	0.04	**0.06**	0.05	**0.08**	0.06	**0.09**	0.07	**0.11**	0.09	**0.14**	0.11	**0.17**
Small intestine		**0.05**		**0.06**		**0.07**		**0.08**		**0.10**		**0.12**
Large intestine		**0.08**		**0.10**		**0.11**		**0.13**	0.01	**0.16**	0.01	**0.18**
Sigma/Rectum												**0.01**
Lumbar spine	0.04	**0.17**	0.06	**0.22**	0.07	**0.26**	0.08	**0.30**	0.10	**0.38**	0.13	**0.45**
Pelvis		**0.02**		**0.03**		**0.04**		**0.04**		**0.05**		**0.07**
Adrenals	0.08	**0.10**	0.11	**0.12**	0.13	**0.15**	0.15	**0.17**	0.18	**0.21**	0.22	**0.25**
Kidneys	0.06	**0.14**	0.08	**0.18**	0.09	**0.21**	0.10	**0.23**	0.12	**0.28**	0.14	**0.32**
Urinary bladder												
Uterus						**0.01**		**0.01**		**0.02**		**0.02**
Ovaries				**0.01**		**0.01**		**0.01**		**0.02**		**0.02**
Prostate												
Testes												
Upper arm												
Elbow												
Forearm												
Upper leg												
Knee												
Lower leg												
Lymph nodes	0.03	**0.09**	0.04	**0.11**	0.05	**0.13**	0.05	**0.15**	0.06	**0.18**	0.08	**0.21**
Skin	0.05	**0.08**	0.06	**0.09**	0.06	**0.09**	0.06	**0.10**	0.07	**0.10**	0.07	**0.11**
Muscle	0.03	**0.06**	0.04	**0.07**	0.04	**0.08**	0.05	**0.09**	0.06	**0.10**	0.06	**0.12**
Skeleton	0.10	**0.13**	0.12	**0.16**	0.13	**0.18**	0.14	**0.19**	0.16	**0.22**	0.18	**0.25**
Red bone marrow	0.04	**0.06**	0.05	**0.07**	0.06	**0.08**	0.06	**0.09**	0.07	**0.10**	0.08	**0.12**
Whole body dose	0.05	**0.08**	0.05	**0.10**	0.06	**0.11**	0.07	**0.12**	0.08	**0.14**	0.09	**0.15**
ED (ICRP 60)	0.04	**0.08**	0.05	**0.10**	0.06	**0.12**	0.07	**0.13**	0.08	**0.15**	0.09	**0.18**
ED (ICRP 103)	0.04	**0.08**	0.05	**0.10**	0.06	**0.12**	0.07	**0.13**	0.08	**0.16**	0.09	**0.18**
Total filtration (mm) **Al**	**1.5**		**2.5**		**3.5**		**4.5**		**3.5**		**3.5**	
Cu	**0**		**0**		**0**		**0**		**0.1**		**0.2**	

Table 15.46 Normalized organ doses—Thoracic spine r-lat—8–12 years

Beam projection:		r-lat										**Table cells:**
Focus-to-detector distance (cm):		115						(+) = normal font = optimal field setting				
Tube voltage (kVp):		80						**(−) = bold font = suboptimal field setting**				
Total filtrations:		See last table row.						Empty table cells correspond to values <0.01				
Organ or tissue	**Organ dose D_T : Entrance dose K_E**							**($mSv * mGy^{-1}$)**				
	(+)	(−)	(+)	(−)	(+)	(−)	(+)	(−)	(+)	(−)	(+)	(−)
Skull		**0.02**		**0.02**		**0.03**		**0.03**	0.01	**0.04**	0.01	**0.04**
Brain												
Oral mucosa		**0.02**		**0.02**		**0.03**		**0.03**		**0.04**	0.01	**0.04**
Salivary glands		**0.05**		**0.06**	0.01	**0.07**	0.01	**0.08**	0.01	**0.09**	0.02	**0.10**
Cervical spine	0.08	**0.40**	0.11	**0.49**	0.12	**0.56**	0.14	**0.62**	0.17	**0.72**	0.19	**0.82**
Thyroid	0.02	**0.26**	0.03	**0.31**	0.03	**0.35**	0.04	**0.38**	0.05	**0.43**	0.06	**0.48**
Thymus	0.03	**0.05**	0.03	**0.07**	0.04	**0.08**	0.04	**0.09**	0.05	**0.11**	0.06	**0.14**
Thoracic spine	0.26	**0.28**	0.33	**0.36**	0.39	**0.42**	0.44	**0.48**	0.54	**0.59**	0.63	**0.70**
Claviculae	0.05	**0.30**	0.06	**0.36**	0.07	**0.40**	0.08	**0.44**	0.09	**0.50**	0.11	**0.56**
Scapulae	1.10	**1.10**	1.27	**1.26**	1.37	**1.37**	1.45	**1.45**	1.57	**1.58**	1.66	**1.68**
Ribs	0.55	**0.73**	0.63	**0.84**	0.69	**0.92**	0.74	**0.98**	0.81	**1.08**	0.87	**1.15**
Oesophagus	0.11	**0.13**	0.14	**0.17**	0.17	**0.19**	0.19	**0.22**	0.23	**0.26**	0.26	**0.31**
Trachea	0.01	**0.14**	0.02	**0.17**	0.02	**0.19**	0.02	**0.20**	0.03	**0.23**	0.03	**0.26**
Lung	0.19	**0.27**	0.23	**0.33**	0.26	**0.37**	0.29	**0.40**	0.33	**0.47**	0.37	**0.52**
Mamma	0.01	**0.03**	0.02	**0.03**	0.02	**0.04**	0.02	**0.05**	0.03	**0.06**	0.04	**0.07**
Heart	0.07	**0.13**	0.08	**0.17**	0.10	**0.19**	0.11	**0.22**	0.13	**0.26**	0.16	**0.30**
Stomach	0.01	**0.03**	0.02	**0.04**	0.02	**0.05**	0.02	**0.06**	0.03	**0.07**	0.04	**0.09**
Spleen	0.02	**0.02**	0.02	**0.03**	0.03	**0.04**	0.03	**0.04**	0.04	**0.06**	0.05	**0.07**
Liver	0.17	**0.41**	0.21	**0.49**	0.24	**0.55**	0.26	**0.59**	0.30	**0.68**	0.33	**0.76**
Gall bladder	0.03	**0.13**	0.04	**0.17**	0.04	**0.20**	0.05	**0.22**	0.06	**0.27**	0.07	**0.32**
Pancreas	0.05	**0.08**	0.07	**0.10**	0.08	**0.12**	0.09	**0.13**	0.11	**0.17**	0.14	**0.20**
Small intestine		**0.06**		**0.08**		**0.09**		**0.10**	0.01	**0.12**	0.01	**0.14**
Large intestine		**0.09**		**0.12**		**0.14**	0.01	**0.15**	0.01	**0.18**	0.02	**0.21**
Sigma/Rectum										**0.01**		**0.02**
Lumbar spine	0.06	**0.21**	0.08	**0.27**	0.09	**0.32**	0.10	**0.36**	0.13	**0.45**	0.15	**0.53**
Pelvis		**0.03**		**0.04**		**0.05**		**0.05**		**0.07**		**0.08**
Adrenals	0.11	**0.12**	0.14	**0.16**	0.16	**0.18**	0.18	**0.21**	0.22	**0.25**	0.26	**0.30**
Kidneys	0.07	**0.17**	0.09	**0.21**	0.11	**0.24**	0.12	**0.27**	0.14	**0.32**	0.16	**0.37**
Urinary bladder												
Uterus				**0.01**		**0.01**		**0.02**		**0.02**		**0.03**
Ovaries		**0.01**		**0.01**		**0.02**		**0.02**		**0.03**		**0.03**
Prostate												
Testes												
Upper arm												
Elbow												
Forearm												
Upper leg												
Knee												
Lower leg												
Lymph nodes	0.04	**0.11**	0.05	**0.14**	0.06	**0.16**	0.06	**0.17**	0.08	**0.21**	0.09	**0.24**
Skin	0.06	**0.09**	0.06	**0.09**	0.06	**0.10**	0.06	**0.10**	0.07	**0.11**	0.07	**0.11**
Muscle	0.04	**0.07**	0.04	**0.08**	0.05	**0.09**	0.05	**0.10**	0.06	**0.12**	0.07	**0.13**
Skeleton	0.11	**0.15**	0.13	**0.17**	0.14	**0.19**	0.15	**0.21**	0.17	**0.24**	0.19	**0.26**
Red bone marrow	0.05	**0.07**	0.06	**0.08**	0.07	**0.09**	0.07	**0.10**	0.08	**0.12**	0.09	**0.13**
Whole body dose	0.05	**0.09**	0.06	**0.11**	0.07	**0.12**	0.07	**0.13**	0.09	**0.15**	0.10	**0.17**
ED (ICRP 60)	0.05	**0.10**	0.06	**0.12**	0.07	**0.14**	0.08	**0.15**	0.09	**0.17**	0.10	**0.20**
ED (ICRP 103)	0.05	**0.10**	0.06	**0.12**	0.07	**0.14**	0.08	**0.15**	0.09	**0.18**	0.10	**0.20**
Total filtration (mm) **Al**	**1.5**		**2.5**		**3.5**		**4.5**		**3.5**		**3.5**	
Cu	**0**		**0**		**0**		**0**		**0.1**		**0.2**	

Table 15.47 Normalized organ doses—Thoracic spine r-lat—13–17 years

Beam projection:			r-lat									**Table cells:**
Focus-to-detector distance (cm):			115					(+) = normal font = optimal field setting				
Tube voltage (kVp):			60					**(−) = bold font = suboptimal field setting**				
Total filtrations:			See last table row.					Empty table cells correspond to values <0.01				
Organ or tissue	**Organ dose D_T : Entrance dose K_E**							**(mSv ∗ mGy^{-1})**				
	(+)	(−)	(+)	(−)	(+)	(−)	(+)	(−)	(+)	(−)	(+)	(−)
Skull										**0.01**		
Brain												
Oral mucosa												
Salivary glands				**0.01**		**0.01**		**0.01**		**0.02**		**0.02**
Cervical spine	0.10	**0.16**	0.13	**0.21**	0.16	**0.25**	0.18	**0.29**	0.22	**0.36**	0.20	**0.32**
Thyroid	0.02	**0.09**	0.03	**0.11**	0.03	**0.13**	0.03	**0.15**	0.04	**0.18**	0.04	**0.16**
Thymus		**0.02**	0.01	**0.03**	0.02	**0.04**	0.02	**0.04**	0.03	**0.06**	0.02	**0.04**
Thoracic spine	0.08	**0.08**	0.10	**0.11**	0.13	**0.14**	0.15	**0.16**	0.20	**0.21**	0.16	**0.16**
Claviculae	0.02	**0.18**	0.03	**0.23**	0.04	**0.27**	0.05	**0.30**	0.06	**0.35**	0.05	**0.34**
Scapulae	0.81	**0.78**	0.97	**0.95**	1.08	**1.06**	1.17	**1.15**	1.32	**1.30**	1.31	**1.28**
Ribs	0.39	**0.51**	0.47	**0.62**	0.52	**0.69**	0.56	**0.75**	0.64	**0.85**	0.63	**0.83**
Oesophagus	0.04	**0.05**	0.06	**0.07**	0.07	**0.08**	0.08	**0.10**	0.11	**0.12**	0.09	**0.10**
Trachea	0.02	**0.05**	0.02	**0.06**	0.02	**0.07**	0.03	**0.07**	0.03	**0.09**	0.03	**0.08**
Lung	0.10	**0.14**	0.13	**0.18**	0.16	**0.21**	0.18	**0.24**	0.21	**0.29**	0.20	**0.27**
Mamma	0.04	**0.11**	0.06	**0.15**	0.07	**0.18**	0.08	**0.21**	0.11	**0.27**	0.09	**0.22**
Heart	0.03	**0.05**	0.04	**0.07**	0.04	**0.08**	0.05	**0.09**	0.07	**0.12**	0.06	**0.10**
Stomach						**0.01**		**0.01**		**0.02**		**0.01**
Spleen										**0.01**		
Liver	0.12	**0.27**	0.15	**0.34**	0.18	**0.40**	0.20	**0.44**	0.24	**0.52**	0.22	**0.49**
Gall bladder	0.01	**0.08**	0.02	**0.10**	0.02	**0.13**	0.03	**0.14**	0.03	**0.18**	0.03	**0.16**
Pancreas	0.01	**0.02**	0.02	**0.03**	0.02	**0.04**	0.03	**0.04**	0.04	**0.06**	0.03	**0.04**
Small intestine		**0.02**		**0.03**		**0.04**		**0.04**		**0.06**		**0.05**
Large intestine		**0.04**		**0.05**		**0.06**		**0.07**		**0.09**		**0.08**
Sigma/Rectum												
Lumbar spine	0.02	**0.06**	0.02	**0.08**	0.03	**0.10**	0.03	**0.12**	0.04	**0.16**	0.03	**0.12**
Pelvis		**0.01**		**0.01**		**0.02**		**0.02**		**0.03**		**0.02**
Adrenals	0.04	**0.04**	0.05	**0.06**	0.06	**0.07**	0.07	**0.08**	0.09	**0.11**	0.08	**0.09**
Kidneys	0.03	**0.07**	0.04	**0.10**	0.05	**0.12**	0.06	**0.13**	0.07	**0.17**	0.06	**0.15**
Urinary bladder												
Uterus						**0.01**		**0.01**		**0.02**		**0.01**
Ovaries												
Prostate												
Testes												
Upper arm												
Elbow												
Forearm												
Upper leg												
Knee												
Lower leg												
Lymph nodes	0.02	**0.04**	0.02	**0.06**	0.03	**0.07**	0.03	**0.08**	0.04	**0.10**	0.04	**0.09**
Skin	0.06	**0.09**	0.07	**0.10**	0.07	**0.10**	0.07	**0.11**	0.07	**0.11**	0.07	**0.11**
Muscle	0.02	**0.04**	0.03	**0.05**	0.03	**0.06**	0.04	**0.07**	0.04	**0.08**	0.04	**0.07**
Skeleton	0.07	**0.09**	0.09	**0.11**	0.10	**0.13**	0.11	**0.14**	0.13	**0.17**	0.13	**0.16**
Red bone marrow	0.03	**0.04**	0.04	**0.05**	0.05	**0.06**	0.05	**0.07**	0.06	**0.08**	0.06	**0.07**
Whole body dose	0.03	**0.06**	0.04	**0.07**	0.05	**0.08**	0.05	**0.09**	0.06	**0.10**	0.06	**0.10**
ED (ICRP 60)	0.03	**0.05**	0.04	**0.07**	0.05	**0.08**	0.05	**0.09**	0.06	**0.11**	0.06	**0.10**
ED (ICRP 103)	0.03	**0.06**	0.04	**0.08**	0.05	**0.09**	0.06	**0.11**	0.07	**0.13**	0.06	**0.11**
Total filtration (mm) Al	**1.5**		**2.5**		**3.5**		**4.5**		**3.5**		**3.5**	
Cu	**0**		**0**		**0**		**0**		**0.1**		**0.2**	

Table 15.48 Normalized organ doses—Thoracic spine r-lat—13–17 years

Beam projection:			r-lat									**Table cells:**
Focus-to-detector distance (cm):			115						(+) = normal font = optimal field setting			
Tube voltage (kVp):			70						**(−) = bold font = suboptimal field setting**			
Total filtrations:			See last table row.						Empty table cells correspond to values <0.01			
Organ or tissue	**Organ dose D_T : Entrance dose K_E**								**(mSv * mGy^{-1})**			
	(+)	(−)	(+)	(−)	(+)	(−)	(+)	(−)	(+)	(−)	(+)	(−)
Skull						**0.01**		**0.01**		**0.02**		**0.02**
Brain												
Oral mucosa										**0.01**		**0.01**
Salivary glands		**0.01**		**0.01**		**0.02**		**0.02**		**0.02**	0.01	**0.03**
Cervical spine	0.13	**0.21**	0.16	**0.27**	0.19	**0.31**	0.21	**0.35**	0.26	**0.43**	0.31	**0.50**
Thyroid	0.03	**0.11**	0.03	**0.14**	0.04	**0.16**	0.04	**0.17**	0.05	**0.21**	0.06	**0.24**
Thymus	0.01	**0.03**	0.02	**0.04**	0.02	**0.05**	0.03	**0.06**	0.04	**0.07**	0.04	**0.09**
Thoracic spine	0.11	**0.12**	0.15	**0.16**	0.18	**0.19**	0.21	**0.22**	0.26	**0.28**	0.32	**0.35**
Claviculae	0.03	**0.22**	0.04	**0.27**	0.05	**0.31**	0.06	**0.34**	0.07	**0.40**	0.09	**0.46**
Scapulae	0.90	**0.88**	1.06	**1.04**	1.17	**1.15**	1.26	**1.24**	1.40	**1.38**	1.52	**1.50**
Ribs	0.43	**0.58**	0.52	**0.68**	0.57	**0.76**	0.61	**0.81**	0.69	**0.92**	0.76	**1.00**
Oesophagus	0.06	**0.07**	0.08	**0.09**	0.10	**0.11**	0.11	**0.13**	0.14	**0.16**	0.17	**0.20**
Trachea	0.02	**0.05**	0.02	**0.07**	0.03	**0.08**	0.03	**0.09**	0.04	**0.10**	0.04	**0.12**
Lung	0.13	**0.17**	0.16	**0.22**	0.19	**0.25**	0.21	**0.28**	0.25	**0.34**	0.29	**0.39**
Mamma	0.06	**0.16**	0.09	**0.21**	0.10	**0.25**	0.12	**0.28**	0.15	**0.36**	0.19	**0.43**
Heart	0.04	**0.07**	0.05	**0.09**	0.06	**0.11**	0.07	**0.12**	0.09	**0.15**	0.10	**0.18**
Stomach		**0.01**		**0.02**		**0.02**	0.01	**0.02**	0.01	**0.03**	0.02	**0.04**
Spleen				**0.01**		**0.01**	0.01	**0.01**	0.01	**0.02**	0.02	**0.03**
Liver	0.15	**0.33**	0.18	**0.40**	0.21	**0.46**	0.23	**0.51**	0.27	**0.59**	0.31	**0.68**
Gall bladder	0.02	**0.11**	0.03	**0.14**	0.03	**0.16**	0.04	**0.19**	0.05	**0.23**	0.06	**0.28**
Pancreas	0.02	**0.03**	0.03	**0.04**	0.04	**0.05**	0.04	**0.06**	0.05	**0.08**	0.07	**0.10**
Small intestine		**0.03**		**0.04**		**0.05**		**0.06**		**0.07**		**0.09**
Large intestine		**0.05**		**0.07**		**0.08**		**0.09**		**0.12**		**0.14**
Sigma/Rectum												
Lumbar spine	0.03	**0.09**	0.03	**0.12**	0.04	**0.14**	0.05	**0.16**	0.06	**0.21**	0.08	**0.26**
Pelvis		**0.02**		**0.02**		**0.03**		**0.03**		**0.04**		**0.05**
Adrenals	0.05	**0.06**	0.07	**0.08**	0.09	**0.10**	0.10	**0.11**	0.12	**0.14**	0.15	**0.17**
Kidneys	0.04	**0.10**	0.05	**0.12**	0.06	**0.15**	0.07	**0.16**	0.09	**0.20**	0.10	**0.24**
Urinary bladder												
Uterus		**0.01**		**0.02**		**0.02**		**0.02**		**0.03**		**0.04**
Ovaries										**0.01**		**0.02**
Prostate												
Testes												
Upper arm												
Elbow												
Forearm												
Upper leg												
Knee												
Lower leg												
Lymph nodes	0.02	**0.06**	0.03	**0.07**	0.04	**0.09**	0.04	**0.10**	0.05	**0.12**	0.06	**0.14**
Skin	0.06	**0.10**	0.07	**0.11**	0.07	**0.11**	0.07	**0.11**	0.08	**0.12**	0.08	**0.13**
Muscle	0.03	**0.05**	0.04	**0.06**	0.04	**0.07**	0.04	**0.08**	0.05	**0.09**	0.06	**0.11**
Skeleton	0.09	**0.11**	0.10	**0.13**	0.12	**0.15**	0.13	**0.16**	0.15	**0.19**	0.16	**0.21**
Red bone marrow	0.04	**0.05**	0.05	**0.06**	0.06	**0.07**	0.06	**0.08**	0.07	**0.10**	0.08	**0.11**
Whole body dose	0.04	**0.07**	0.05	**0.08**	0.06	**0.09**	0.06	**0.10**	0.07	**0.12**	0.08	**0.13**
ED (ICRP 60)	0.04	**0.07**	0.05	**0.09**	0.06	**0.10**	0.06	**0.11**	0.08	**0.14**	0.09	**0.16**
ED (ICRP 103)	0.04	**0.08**	0.05	**0.10**	0.06	**0.12**	0.07	**0.13**	0.08	**0.16**	0.10	**0.19**
Total filtration (mm) Al	**1.5**		**2.5**		**3.5**		**4.5**		**3.5**		**3.5**	
Cu	**0**		**0**		**0**		**0**		**0.1**		**0.2**	

Table 15.49 Normalized organ doses—Thoracic spine r-lat—13–17 years

Beam projection:			r-lat									**Table cells:**
Focus-to-detector distance (cm):			115				(+) = normal font = optimal field setting					
Tube voltage (kVp):			80				**(−) = bold font = suboptimal field setting**					
Total filtrations:			See last table row.				Empty table cells correspond to values <0.01					
Organ or tissue	**Organ dose D_T : Entrance dose K_E**								**(mSv ∗ mGy^{-1})**			
	(+)	(−)	(+)	(−)	(+)	(−)	(+)	(−)	(+)	(−)	(+)	(−)
Skull				**0.01**		**0.01**		**0.01**		**0.02**	0.01	**0.02**
Brain												
Oral mucosa						**0.01**		**0.01**		**0.01**		**0.02**
Salivary glands		**0.01**		**0.02**		**0.02**		**0.02**	0.01	**0.03**	0.01	**0.03**
Cervical spine	0.15	**0.25**	0.19	**0.31**	0.22	**0.36**	0.25	**0.40**	0.30	**0.48**	0.34	**0.55**
Thyroid	0.03	**0.13**	0.04	**0.16**	0.05	**0.18**	0.05	**0.20**	0.06	**0.23**	0.07	**0.26**
Thymus	0.02	**0.04**	0.03	**0.06**	0.03	**0.07**	0.04	**0.07**	0.05	**0.09**	0.06	**0.11**
Thoracic spine	0.15	**0.16**	0.19	**0.21**	0.23	**0.25**	0.26	**0.28**	0.32	**0.35**	0.39	**0.42**
Claviculae	0.04	**0.25**	0.05	**0.30**	0.06	**0.34**	0.07	**0.38**	0.09	**0.44**	0.10	**0.49**
Scapulae	0.97	**0.95**	1.13	**1.11**	1.24	**1.22**	1.32	**1.30**	1.45	**1.43**	1.54	**1.54**
Ribs	0.48	**0.63**	0.56	**0.74**	0.61	**0.81**	0.65	**0.86**	0.72	**0.96**	0.78	**1.04**
Oesophagus	0.08	**0.09**	0.10	**0.12**	0.12	**0.14**	0.14	**0.16**	0.17	**0.20**	0.20	**0.24**
Trachea	0.02	**0.06**	0.03	**0.08**	0.03	**0.09**	0.04	**0.10**	0.04	**0.11**	0.05	**0.13**
Lung	0.15	**0.21**	0.19	**0.25**	0.22	**0.29**	0.24	**0.32**	0.28	**0.38**	0.32	**0.43**
Mamma	0.09	**0.21**	0.11	**0.26**	0.14	**0.31**	0.16	**0.35**	0.20	**0.44**	0.24	**0.53**
Heart	0.05	**0.09**	0.06	**0.11**	0.08	**0.13**	0.08	**0.15**	0.11	**0.18**	0.13	**0.22**
Stomach		**0.02**	0.01	**0.02**	0.01	**0.03**	0.01	**0.03**	0.02	**0.04**	0.02	**0.05**
Spleen		**0.01**	0.01	**0.02**	0.01	**0.02**	0.02	**0.02**	0.02	**0.03**	0.03	**0.03**
Liver	0.17	**0.38**	0.21	**0.46**	0.24	**0.52**	0.26	**0.57**	0.30	**0.66**	0.34	**0.74**
Gall bladder	0.03	**0.14**	0.03	**0.17**	0.04	**0.20**	0.05	**0.23**	0.06	**0.28**	0.07	**0.33**
Pancreas	0.03	**0.05**	0.04	**0.06**	0.05	**0.07**	0.05	**0.08**	0.07	**0.10**	0.08	**0.12**
Small intestine		**0.04**		**0.05**		**0.06**		**0.07**		**0.09**		**0.10**
Large intestine		**0.07**		**0.09**		**0.10**		**0.11**		**0.14**	0.01	**0.16**
Sigma/Rectum												
Lumbar spine	0.03	**0.12**	0.05	**0.16**	0.05	**0.18**	0.06	**0.21**	0.08	**0.26**	0.09	**0.32**
Pelvis		**0.02**		**0.03**		**0.03**		**0.04**		**0.05**		**0.06**
Adrenals	0.07	**0.08**	0.09	**0.10**	0.11	**0.12**	0.12	**0.14**	0.15	**0.17**	0.18	**0.21**
Kidneys	0.05	**0.12**	0.06	**0.15**	0.08	**0.18**	0.08	**0.20**	0.10	**0.24**	0.12	**0.27**
Urinary bladder												
Uterus		**0.02**		**0.02**		**0.03**		**0.03**		**0.04**		**0.05**
Ovaries						**0.01**		**0.01**		**0.02**		**0.02**
Prostate												
Testes												
Upper arm												
Elbow												
Forearm												
Upper leg												
Knee												
Lower leg												
Lymph nodes	0.03	**0.07**	0.04	**0.09**	0.05	**0.11**	0.05	**0.12**	0.06	**0.14**	0.07	**0.17**
Skin	0.07	**0.10**	0.07	**0.11**	0.07	**0.11**	0.08	**0.12**	0.08	**0.13**	0.09	**0.13**
Muscle	0.03	**0.06**	0.04	**0.07**	0.05	**0.08**	0.05	**0.09**	0.06	**0.10**	0.07	**0.12**
Skeleton	0.10	**0.12**	0.12	**0.15**	0.13	**0.16**	0.14	**0.18**	0.16	**0.20**	0.17	**0.22**
Red bone marrow	0.05	**0.06**	0.06	**0.08**	0.06	**0.09**	0.07	**0.09**	0.08	**0.11**	0.09	**0.13**
Whole body dose	0.05	**0.08**	0.06	**0.09**	0.06	**0.10**	0.07	**0.11**	0.08	**0.13**	0.09	**0.15**
ED (ICRP 60)	0.05	**0.08**	0.06	**0.10**	0.07	**0.12**	0.07	**0.13**	0.09	**0.16**	0.10	**0.18**
ED (ICRP 103)	0.05	**0.10**	0.06	**0.12**	0.07	**0.14**	0.08	**0.16**	0.10	**0.19**	0.12	**0.22**
Total filtration (mm) **Al**	**1.5**		**2.5**		**3.5**		**4.5**		**3.5**		**3.5**	
Cu	**0**		**0**		**0**		**0**		**0.1**		**0.2**	

Table 15.50 Normalized organ doses—Thoracic spine r-lat—>17 years

Beam projection:			r-lat									**Table cells:**
Focus-to-detector distance (cm):			115									(+) = normal font = optimal field setting
Tube voltage (kVp):			60									**(−) = bold font = suboptimal field setting**
Total filtrations:			See last table row.									Empty table cells correspond to values <0.01
Organ or tissue	**Organ dose D_T : Entrance dose K_E**							**(mSv ∗ mGy^{-1})**				
	(+)	(−)	(+)	(−)	(+)	(−)	(+)	(−)	(+)	(−)	(+)	(−)
Skull				**0.01**		**0.01**		**0.02**		**0.02**		**0.02**
Brain												
Oral mucosa										**0.01**		
Salivary glands		**0.01**		**0.02**		**0.02**		**0.02**		**0.03**		**0.03**
Cervical spine	0.05	**0.16**	0.06	**0.22**	0.07	**0.26**	0.08	**0.29**	0.10	**0.36**	0.09	**0.32**
Thyroid	0.02	**0.06**	0.02	**0.08**	0.03	**0.10**	0.03	**0.11**	0.04	**0.13**	0.03	**0.12**
Thymus		**0.02**	0.01	**0.03**	0.01	**0.04**	0.02	**0.05**	0.02	**0.06**	0.02	**0.05**
Thoracic spine	0.05	**0.05**	0.07	**0.07**	0.09	**0.09**	0.10	**0.11**	0.13	**0.14**	0.10	**0.11**
Claviculae	0.02	**0.17**	0.03	**0.22**	0.03	**0.25**	0.04	**0.29**	0.05	**0.35**	0.04	**0.32**
Scapulae	0.73	**0.69**	0.89	**0.85**	1.00	**0.95**	1.08	**1.04**	1.23	**1.19**	1.21	**1.17**
Ribs	0.31	**0.47**	0.38	**0.57**	0.42	**0.64**	0.46	**0.69**	0.53	**0.79**	0.51	**0.77**
Oesophagus	0.03	**0.04**	0.05	**0.05**	0.06	**0.06**	0.06	**0.07**	0.08	**0.10**	0.07	**0.08**
Trachea		**0.05**	0.01	**0.06**	0.01	**0.07**	0.02	**0.08**	0.02	**0.10**	0.02	**0.09**
Lung	0.08	**0.12**	0.11	**0.15**	0.13	**0.18**	0.14	**0.20**	0.18	**0.25**	0.16	**0.23**
Mamma		**0.02**		**0.02**		**0.03**		**0.03**	0.01	**0.04**		**0.03**
Heart	0.02	**0.04**	0.03	**0.05**	0.04	**0.06**	0.04	**0.07**	0.05	**0.09**	0.04	**0.08**
Stomach										**0.01**		
Spleen												
Liver	0.07	**0.23**	0.09	**0.29**	0.11	**0.34**	0.12	**0.38**	0.15	**0.45**	0.13	**0.42**
Gall bladder		**0.05**		**0.06**		**0.08**		**0.09**	0.01	**0.12**		**0.10**
Pancreas		**0.01**	0.01	**0.02**	0.01	**0.02**	0.02	**0.03**	0.02	**0.04**	0.02	**0.03**
Small intestine		**0.02**		**0.02**		**0.03**		**0.03**		**0.04**		**0.03**
Large intestine		**0.03**		**0.04**		**0.05**		**0.06**		**0.07**		**0.06**
Sigma/Rectum												
Lumbar spine		**0.04**		**0.05**		**0.06**		**0.07**	0.01	**0.09**		**0.07**
Pelvis				**0.01**		**0.01**		**0.02**		**0.02**		**0.02**
Adrenals	0.03	**0.03**	0.03	**0.04**	0.04	**0.05**	0.05	**0.06**	0.06	**0.08**	0.05	**0.06**
Kidneys	0.01	**0.05**	0.01	**0.07**	0.02	**0.09**	0.02	**0.10**	0.03	**0.12**	0.02	**0.11**
Urinary bladder												
Uterus												
Ovaries												
Prostate												
Testes												
Upper arm												
Elbow												
Forearm												
Upper leg												
Knee												
Lower leg												
Lymph nodes	0.01	**0.04**	0.02	**0.05**	0.02	**0.06**	0.02	**0.06**	0.03	**0.08**	0.03	**0.07**
Skin	0.04	**0.08**	0.05	**0.08**	0.05	**0.09**	0.05	**0.09**	0.05	**0.10**	0.05	**0.09**
Muscle	0.02	**0.04**	0.02	**0.05**	0.03	**0.06**	0.03	**0.06**	0.04	**0.07**	0.03	**0.07**
Skeleton	0.06	**0.08**	0.07	**0.10**	0.08	**0.11**	0.09	**0.13**	0.11	**0.15**	0.10	**0.14**
Red bone marrow	0.03	**0.04**	0.04	**0.05**	0.04	**0.06**	0.05	**0.07**	0.05	**0.08**	0.05	**0.07**
Whole body dose	0.03	**0.05**	0.03	**0.06**	0.04	**0.07**	0.04	**0.08**	0.05	**0.09**	0.05	**0.09**
ED (ICRP 60)	0.02	**0.04**	0.03	**0.05**	0.03	**0.06**	0.04	**0.07**	0.04	**0.09**	0.04	**0.08**
ED (ICRP 103)	0.02	**0.04**	0.03	**0.05**	0.03	**0.06**	0.04	**0.07**	0.04	**0.09**	0.04	**0.08**
Total filtration (mm) Al	**1.5**		**2.5**		**3.5**		**4.5**		**3.5**		**3.5**	
Cu	**0**		**0**		**0**		**0**		**0.1**		**0.2**	

Table 15.51 Normalized organ doses—Thoracic spine r-lat—>17 years

Beam projection:			r-lat									**Table cells:**
Focus-to-detector distance (cm):			115									(+) = normal font = optimal field setting
Tube voltage (kVp):			70									**(−) = bold font = suboptimal field setting**
Total filtrations:			See last table row.									Empty table cells correspond to values <0.01
Organ or tissue	**Organ dose D_T : Entrance dose K_E**								**(mSv * mGy^{-1})**			
	(+)	(−)	(+)	(−)	(+)	(−)	(+)	(−)	(+)	(−)	(+)	(−)
Skull		**0.01**		**0.02**		**0.02**		**0.02**		**0.02**		**0.03**
Brain												
Oral mucosa						**0.01**		**0.01**		**0.01**		**0.02**
Salivary glands		**0.02**		**0.02**		**0.03**		**0.03**		**0.03**		**0.04**
Cervical spine	0.06	**0.21**	0.08	**0.27**	0.09	**0.32**	0.11	**0.35**	0.13	**0.43**	0.16	**0.50**
Thyroid	0.02	**0.08**	0.03	**0.10**	0.03	**0.12**	0.04	**0.13**	0.05	**0.15**	0.06	**0.18**
Thymus	0.01	**0.03**	0.02	**0.05**	0.02	**0.06**	0.02	**0.06**	0.03	**0.08**	0.04	**0.10**
Thoracic spine	0.08	**0.08**	0.10	**0.11**	0.12	**0.13**	0.14	**0.15**	0.18	**0.20**	0.23	**0.25**
Claviculae	0.03	**0.21**	0.04	**0.26**	0.05	**0.30**	0.05	**0.33**	0.07	**0.40**	0.08	**0.46**
Scapulae	0.82	**0.79**	0.98	**0.94**	1.09	**1.05**	1.17	**1.13**	1.31	**1.28**	1.43	**1.40**
Ribs	0.35	**0.53**	0.42	**0.63**	0.47	**0.71**	0.51	**0.76**	0.57	**0.86**	0.63	**0.95**
Oesophagus	0.05	**0.06**	0.06	**0.07**	0.08	**0.09**	0.09	**0.10**	0.11	**0.13**	0.14	**0.16**
Trachea	0.01	**0.06**	0.02	**0.08**	0.02	**0.09**	0.02	**0.10**	0.02	**0.11**	0.03	**0.13**
Lung	0.10	**0.15**	0.13	**0.19**	0.15	**0.22**	0.17	**0.24**	0.21	**0.29**	0.24	**0.34**
Mamma		**0.02**		**0.03**	0.01	**0.03**	0.01	**0.04**	0.02	**0.05**	0.02	**0.06**
Heart	0.03	**0.05**	0.04	**0.07**	0.05	**0.08**	0.06	**0.10**	0.07	**0.12**	0.09	**0.15**
Stomach						**0.01**		**0.01**		**0.02**		**0.02**
Spleen										**0.01**		**0.01**
Liver	0.09	**0.28**	0.11	**0.35**	0.13	**0.40**	0.14	**0.45**	0.17	**0.53**	0.20	**0.60**
Gall bladder		**0.07**	0.01	**0.09**	0.01	**0.11**	0.01	**0.12**	0.02	**0.16**	0.02	**0.19**
Pancreas	0.01	**0.02**	0.02	**0.03**	0.02	**0.03**	0.02	**0.04**	0.03	**0.05**	0.04	**0.07**
Small intestine		**0.02**		**0.03**		**0.04**		**0.04**		**0.06**		**0.07**
Large intestine		**0.04**		**0.06**		**0.07**		**0.07**		**0.09**		**0.11**
Sigma/Rectum												
Lumbar spine		**0.05**	0.01	**0.07**	0.01	**0.09**	0.02	**0.10**	0.02	**0.13**	0.03	**0.17**
Pelvis		**0.01**		**0.02**		**0.02**		**0.02**		**0.03**		**0.04**
Adrenals	0.04	**0.04**	0.05	**0.06**	0.06	**0.07**	0.07	**0.08**	0.08	**0.10**	0.10	**0.13**
Kidneys	0.01	**0.07**	0.02	**0.09**	0.02	**0.11**	0.03	**0.12**	0.03	**0.15**	0.04	**0.18**
Urinary bladder												
Uterus												**0.01**
Ovaries												**0.01**
Prostate												
Testes												
Upper arm												
Elbow												
Forearm												
Upper leg												
Knee												
Lower leg												
Lymph nodes	0.02	**0.05**	0.02	**0.06**	0.03	**0.07**	0.03	**0.08**	0.04	**0.10**	0.05	**0.12**
Skin	0.05	**0.08**	0.05	**0.09**	0.05	**0.09**	0.05	**0.10**	0.06	**0.10**	0.06	**0.11**
Muscle	0.02	**0.05**	0.03	**0.06**	0.03	**0.07**	0.04	**0.07**	0.04	**0.08**	0.05	**0.10**
Skeleton	0.07	**0.10**	0.08	**0.12**	0.09	**0.13**	0.10	**0.14**	0.12	**0.16**	0.13	**0.19**
Red bone marrow	0.03	**0.05**	0.04	**0.06**	0.05	**0.07**	0.05	**0.08**	0.06	**0.09**	0.07	**0.11**
Whole body dose	0.03	**0.06**	0.04	**0.08**	0.04	**0.09**	0.05	**0.09**	0.06	**0.11**	0.06	**0.12**
ED (ICRP 60)	0.03	**0.05**	0.03	**0.07**	0.04	**0.08**	0.04	**0.09**	0.05	**0.10**	0.06	**0.12**
ED (ICRP 103)	0.03	**0.05**	0.03	**0.07**	0.04	**0.08**	0.04	**0.09**	0.05	**0.11**	0.06	**0.12**
Total filtration (mm) **Al**	**1.5**		**2.5**		**3.5**		**4.5**		**3.5**		**3.5**	
Cu	**0**		**0**		**0**		**0**		**0.1**		**0.2**	

Table 15.52 Normalized organ doses—Thoracic spine r-lat—>17 years

Beam projection:			r-lat									**Table cells:**
Focus-to-detector distance (cm):			115									(+) = normal font = optimal field setting
Tube voltage (kVp):			80									**(−) = bold font = suboptimal field setting**
Total filtrations:			See last table row.									Empty table cells correspond to values <0.01
Organ or tissue	**Organ dose D_T : Entrance dose K_E**								**(mSv * mGy^{-1})**			
	(+)	(−)	(+)	(−)	(+)	(−)	(+)	(−)	(+)	(−)	(+)	(−)
Skull		**0.01**		**0.02**		**0.02**		**0.02**		**0.03**		**0.03**
Brain												
Oral mucosa				**0.01**		**0.01**		**0.01**		**0.02**		**0.02**
Salivary glands		**0.02**		**0.03**		**0.03**		**0.03**		**0.04**	0.01	**0.05**
Cervical spine	0.08	**0.25**	0.10	**0.32**	0.11	**0.37**	0.13	**0.41**	0.15	**0.49**	0.18	**0.56**
Thyroid	0.03	**0.10**	0.04	**0.12**	0.04	**0.14**	0.05	**0.15**	0.06	**0.18**	0.07	**0.20**
Thymus	0.02	**0.05**	0.02	**0.06**	0.03	**0.07**	0.03	**0.08**	0.04	**0.10**	0.05	**0.13**
Thoracic spine	0.10	**0.11**	0.13	**0.15**	0.16	**0.18**	0.18	**0.20**	0.23	**0.25**	0.28	**0.31**
Claviculae	0.04	**0.24**	0.05	**0.30**	0.06	**0.34**	0.06	**0.38**	0.08	**0.44**	0.09	**0.50**
Scapulae	0.90	**0.86**	1.05	**1.01**	1.15	**1.12**	1.23	**1.20**	1.36	**1.33**	1.46	**1.44**
Ribs	0.39	**0.59**	0.46	**0.69**	0.51	**0.76**	0.54	**0.81**	0.60	**0.91**	0.66	**0.98**
Oesophagus	0.06	**0.08**	0.08	**0.10**	0.10	**0.12**	0.11	**0.13**	0.14	**0.17**	0.17	**0.20**
Trachea	0.01	**0.07**	0.02	**0.09**	0.02	**0.10**	0.02	**0.11**	0.03	**0.13**	0.03	**0.14**
Lung	0.13	**0.18**	0.16	**0.22**	0.18	**0.25**	0.20	**0.28**	0.24	**0.33**	0.27	**0.38**
Mamma		**0.03**	0.01	**0.04**	0.01	**0.04**	0.02	**0.05**	0.02	**0.06**	0.03	**0.07**
Heart	0.04	**0.07**	0.05	**0.09**	0.06	**0.11**	0.07	**0.12**	0.09	**0.15**	0.10	**0.18**
Stomach		**0.01**		**0.01**		**0.02**		**0.02**	0.01	**0.02**	0.01	**0.03**
Spleen						**0.01**		**0.01**	0.01	**0.02**	0.01	**0.02**
Liver	0.11	**0.33**	0.13	**0.41**	0.15	**0.46**	0.17	**0.51**	0.20	**0.59**	0.22	**0.67**
Gall bladder	0.01	**0.09**	0.01	**0.12**	0.02	**0.14**	0.02	**0.16**	0.02	**0.19**	0.03	**0.23**
Pancreas	0.02	**0.03**	0.02	**0.04**	0.03	**0.05**	0.03	**0.05**	0.04	**0.07**	0.05	**0.09**
Small intestine		**0.03**		**0.04**		**0.05**		**0.06**		**0.07**		**0.08**
Large intestine		**0.05**		**0.07**		**0.08**		**0.09**		**0.11**		**0.13**
Sigma/Rectum												
Lumbar spine	0.01	**0.08**	0.02	**0.10**	0.02	**0.12**	0.02	**0.13**	0.03	**0.17**	0.03	**0.21**
Pelvis		**0.02**		**0.02**		**0.03**		**0.03**		**0.04**		**0.05**
Adrenals	0.05	**0.06**	0.06	**0.08**	0.07	**0.09**	0.08	**0.10**	0.11	**0.13**	0.13	**0.16**
Kidneys	0.02	**0.09**	0.02	**0.11**	0.03	**0.13**	0.03	**0.15**	0.04	**0.18**	0.05	**0.21**
Urinary bladder												
Uterus										**0.01**		**0.01**
Ovaries								**0.01**		**0.01**		**0.02**
Prostate												
Testes												
Upper arm												
Elbow												
Forearm												
Upper leg												
Knee												
Lower leg												
Lymph nodes	0.02	**0.06**	0.03	**0.08**	0.03	**0.09**	0.04	**0.10**	0.05	**0.12**	0.05	**0.14**
Skin	0.05	**0.09**	0.05	**0.09**	0.05	**0.10**	0.06	**0.10**	0.06	**0.11**	0.06	**0.11**
Muscle	0.03	**0.06**	0.03	**0.07**	0.04	**0.08**	0.04	**0.08**	0.05	**0.09**	0.05	**0.11**
Skeleton	0.08	**0.11**	0.09	**0.13**	0.10	**0.14**	0.11	**0.16**	0.13	**0.18**	0.14	**0.20**
Red bone marrow	0.04	**0.06**	0.05	**0.07**	0.06	**0.08**	0.06	**0.09**	0.07	**0.11**	0.08	**0.12**
Whole body dose	0.04	**0.07**	0.04	**0.09**	0.05	**0.10**	0.05	**0.10**	0.06	**0.12**	0.07	**0.14**
ED (ICRP 60)	0.03	**0.06**	0.04	**0.08**	0.05	**0.09**	0.05	**0.10**	0.06	**0.12**	0.07	**0.14**
ED (ICRP 103)	0.03	**0.06**	0.04	**0.08**	0.05	**0.09**	0.05	**0.10**	0.06	**0.12**	0.07	**0.14**
Total filtration (mm) Al	**1.5**		**2.5**		**3.5**		**4.5**		**3.5**		**3.5**	
Cu	**0**		**0**		**0**		**0**		**0.1**		**0.2**	

Table 15.53 Normalized organ doses—Thoracic spine ap—8–12 years

Beam projection:		ap										**Table cells:**
Focus-to-detector distance (cm):		150										(+) = normal font = optimal field setting
Tube voltage (kVp):		60										**(−) = bold font = suboptimal field setting**
Total filtrations:		See last table row.										Empty table cells correspond to values <0.01
Organ or tissue	**Organ dose D_T : Entrance dose K_E**								**(mSv * mGy^{-1})**			
	(+)	(−)	(+)	(−)	(+)	(−)	(+)	(−)	(+)	(−)	(+)	(−)
Skull		**0.01**		**0.01**		**0.02**		**0.02**		**0.02**		**0.02**
Brain												
Oral mucosa		**0.01**		**0.02**		**0.02**		**0.02**		**0.03**		**0.02**
Salivary glands		**0.03**		**0.03**		**0.04**		**0.04**	0.01	**0.05**	0.01	**0.05**
Cervical spine	0.03	**0.26**	0.04	**0.34**	0.05	**0.40**	0.06	**0.45**	0.07	**0.56**	0.06	**0.51**
Thyroid	0.05	**0.74**	0.07	**0.85**	0.08	**0.92**	0.09	**0.98**	0.11	**1.07**	0.10	**1.06**
Thymus	0.62	**0.65**	0.75	**0.79**	0.84	**0.89**	0.91	**0.96**	1.04	**1.10**	1.01	**1.07**
Thoracic spine	0.07	**0.10**	0.10	**0.13**	0.12	**0.16**	0.14	**0.19**	0.19	**0.25**	0.15	**0.20**
Claviculae	0.44	**0.93**	0.52	**1.12**	0.58	**1.25**	0.63	**1.35**	0.71	**1.53**	0.70	**1.52**
Scapulae	0.02	**0.05**	0.03	**0.07**	0.03	**0.08**	0.04	**0.09**	0.05	**0.12**	0.04	**0.10**
Ribs	0.44	**0.73**	0.50	**0.85**	0.54	**0.92**	0.57	**0.98**	0.62	**1.07**	0.62	**1.06**
Oesophagus	0.06	**0.08**	0.08	**0.10**	0.10	**0.12**	0.11	**0.14**	0.14	**0.18**	0.12	**0.16**
Trachea	0.01	**0.19**	0.02	**0.23**	0.02	**0.26**	0.02	**0.27**	0.03	**0.31**	0.02	**0.30**
Lung	0.05	**0.18**	0.07	**0.23**	0.08	**0.26**	0.09	**0.29**	0.12	**0.35**	0.10	**0.33**
Mamma	0.02	**0.18**	0.03	**0.21**	0.04	**0.22**	0.04	**0.24**	0.05	**0.26**	0.04	**0.25**
Heart	0.30	**0.38**	0.38	**0.47**	0.43	**0.55**	0.48	**0.60**	0.56	**0.71**	0.54	**0.67**
Stomach	0.06	**0.29**	0.08	**0.36**	0.09	**0.42**	0.10	**0.46**	0.13	**0.54**	0.12	**0.51**
Spleen	0.01	**0.04**	0.01	**0.06**	0.02	**0.07**	0.02	**0.08**	0.03	**0.11**	0.02	**0.09**
Liver	0.09	**0.21**	0.12	**0.27**	0.14	**0.31**	0.15	**0.34**	0.18	**0.40**	0.17	**0.38**
Gall bladder	0.05	**0.28**	0.06	**0.37**	0.08	**0.43**	0.09	**0.48**	0.11	**0.58**	0.10	**0.54**
Pancreas	0.09	**0.15**	0.12	**0.19**	0.14	**0.23**	0.16	**0.26**	0.20	**0.33**	0.18	**0.29**
Small intestine		**0.05**		**0.06**		**0.07**		**0.08**		**0.10**		**0.09**
Large intestine		**0.11**		**0.14**		**0.17**		**0.19**		**0.23**		**0.21**
Sigma/Rectum						**0.01**		**0.01**		**0.02**		**0.01**
Lumbar spine	0.01	**0.08**	0.02	**0.11**	0.02	**0.14**	0.03	**0.16**	0.04	**0.21**	0.03	**0.17**
Pelvis				**0.01**		**0.02**		**0.02**		**0.02**		**0.02**
Adrenals	0.04	**0.05**	0.05	**0.07**	0.06	**0.09**	0.07	**0.10**	0.09	**0.14**	0.08	**0.11**
Kidneys		**0.04**	0.01	**0.05**	0.02	**0.07**	0.02	**0.08**	0.02	**0.10**	0.02	**0.08**
Urinary bladder												
Uterus						**0.01**		**0.01**		**0.02**		**0.01**
Ovaries						**0.01**		**0.01**		**0.02**		**0.01**
Prostate												
Testes												
Upper arm	0.01	**0.02**	0.01	**0.03**	0.02	**0.04**	0.02	**0.04**	0.03	**0.05**	0.02	**0.04**
Elbow	0.01	**0.02**	0.01	**0.03**	0.02	**0.04**	0.02	**0.05**	0.03	**0.06**	0.02	**0.05**
Forearm				**0.01**		**0.01**		**0.02**		**0.02**		**0.02**
Upper leg												
Knee												
Lower leg												
Lymph nodes	0.04	**0.17**	0.06	**0.21**	0.07	**0.24**	0.08	**0.27**	0.09	**0.31**	0.09	**0.29**
Skin	0.03	**0.09**	0.04	**0.09**	0.04	**0.09**	0.04	**0.10**	0.04	**0.10**	0.04	**0.10**
Muscle	0.03	**0.07**	0.04	**0.08**	0.04	**0.09**	0.05	**0.10**	0.05	**0.12**	0.05	**0.11**
Skeleton	0.05	**0.10**	0.06	**0.12**	0.07	**0.14**	0.08	**0.15**	0.09	**0.17**	0.08	**0.16**
Red bone marrow	0.02	**0.04**	0.03	**0.05**	0.03	**0.06**	0.03	**0.06**	0.04	**0.07**	0.03	**0.07**
Whole body dose	0.04	**0.08**	0.04	**0.09**	0.05	**0.11**	0.05	**0.12**	0.06	**0.14**	0.06	**0.13**
ED (ICRP 60)	0.03	**0.14**	0.04	**0.17**	0.05	**0.19**	0.05	**0.21**	0.07	**0.24**	0.06	**0.23**
ED (ICRP 103)	0.04	**0.15**	0.05	**0.19**	0.06	**0.21**	0.06	**0.23**	0.08	**0.27**	0.07	**0.25**
Total filtration (mm) **Al**	**1.5**		**2.5**		**3.5**		**4.5**		**3.5**		**3.5**	
Cu	**0**		**0**		**0**		**0**		**0.1**		**0.2**	

Table 15.54 Normalized organ doses—Thoracic spine ap—8–12 years

Beam projection:			ap									**Table cells:**
Focus-to-detector distance (cm):			150									(+) = normal font = optimal field setting
Tube voltage (kVp):			70									**(−) = bold font = suboptimal field setting**
Total filtrations:			See last table row.									Empty table cells correspond to values <0.01
Organ or tissue	**Organ dose D_T : Entrance dose K_E**							**($mSv * mGy^{-1}$)**				
	(+)	(−)	(+)	(−)	(+)	(−)	(+)	(−)	(+)	(−)	(+)	(−)
Skull		**0.01**		**0.02**		**0.02**		**0.02**	0.01	**0.03**	0.01	**0.03**
Brain												
Oral mucosa		**0.02**		**0.02**		**0.02**		**0.03**	0.01	**0.03**	0.01	**0.04**
Salivary glands		**0.03**		**0.04**	0.01	**0.05**	0.01	**0.05**	0.02	**0.06**	0.02	**0.07**
Cervical spine	0.04	**0.33**	0.05	**0.41**	0.07	**0.48**	0.07	**0.54**	0.09	**0.65**	0.11	**0.76**
Thyroid	0.06	**0.80**	0.08	**0.91**	0.09	**0.99**	0.10	**1.04**	0.13	**1.13**	0.15	**1.21**
Thymus	0.70	**0.74**	0.83	**0.88**	0.92	**0.98**	1.00	**1.06**	1.12	**1.20**	1.23	**1.32**
Thoracic spine	0.10	**0.14**	0.14	**0.19**	0.17	**0.23**	0.19	**0.26**	0.25	**0.33**	0.30	**0.41**
Claviculae	0.49	**1.04**	0.57	**1.23**	0.63	**1.36**	0.68	**1.45**	0.75	**1.62**	0.81	**1.75**
Scapulae	0.03	**0.07**	0.04	**0.09**	0.05	**0.11**	0.05	**0.12**	0.07	**0.15**	0.08	**0.18**
Ribs	0.47	**0.80**	0.53	**0.91**	0.57	**0.98**	0.60	**1.03**	0.65	**1.12**	0.69	**1.20**
Oesophagus	0.08	**0.10**	0.10	**0.13**	0.12	**0.16**	0.14	**0.18**	0.17	**0.23**	0.21	**0.27**
Trachea	0.02	**0.22**	0.02	**0.25**	0.02	**0.28**	0.03	**0.30**	0.03	**0.33**	0.04	**0.36**
Lung	0.07	**0.22**	0.09	**0.27**	0.10	**0.31**	0.11	**0.34**	0.14	**0.40**	0.16	**0.46**
Mamma	0.03	**0.20**	0.04	**0.22**	0.04	**0.24**	0.05	**0.26**	0.06	**0.28**	0.07	**0.31**
Heart	0.36	**0.45**	0.44	**0.55**	0.50	**0.63**	0.55	**0.69**	0.64	**0.81**	0.72	**0.91**
Stomach	0.08	**0.34**	0.10	**0.42**	0.11	**0.48**	0.12	**0.52**	0.15	**0.61**	0.17	**0.69**
Spleen	0.02	**0.06**	0.02	**0.08**	0.02	**0.10**	0.03	**0.11**	0.04	**0.14**	0.04	**0.16**
Liver	0.11	**0.25**	0.14	**0.31**	0.16	**0.36**	0.18	**0.39**	0.21	**0.46**	0.23	**0.52**
Gall bladder	0.06	**0.35**	0.08	**0.44**	0.09	**0.51**	0.11	**0.56**	0.13	**0.67**	0.15	**0.77**
Pancreas	0.11	**0.19**	0.15	**0.25**	0.17	**0.29**	0.19	**0.33**	0.24	**0.41**	0.28	**0.48**
Small intestine		**0.06**		**0.08**		**0.09**		**0.10**		**0.13**	0.01	**0.15**
Large intestine		**0.14**		**0.18**		**0.20**		**0.23**	0.01	**0.27**	0.01	**0.31**
Sigma/Rectum				**0.01**		**0.02**		**0.02**		**0.02**		**0.03**
Lumbar spine	0.02	**0.12**	0.03	**0.16**	0.04	**0.19**	0.04	**0.22**	0.05	**0.28**	0.07	**0.34**
Pelvis		**0.01**		**0.02**		**0.02**		**0.03**		**0.03**		**0.04**
Adrenals	0.05	**0.08**	0.07	**0.10**	0.08	**0.12**	0.09	**0.14**	0.12	**0.18**	0.15	**0.21**
Kidneys	0.01	**0.06**	0.02	**0.07**	0.02	**0.09**	0.03	**0.10**	0.03	**0.13**	0.04	**0.16**
Urinary bladder												
Uterus				**0.01**		**0.01**		**0.02**		**0.02**		**0.03**
Ovaries				**0.01**		**0.02**		**0.02**		**0.02**		**0.03**
Prostate												
Testes												
Upper arm	0.02	**0.03**	0.02	**0.04**	0.02	**0.05**	0.03	**0.06**	0.04	**0.07**	0.05	**0.09**
Elbow	0.02	**0.03**	0.02	**0.05**	0.02	**0.06**	0.03	**0.06**	0.04	**0.08**	0.05	**0.10**
Forearm		**0.01**		**0.02**		**0.02**		**0.02**		**0.03**		**0.04**
Upper leg												
Knee												
Lower leg												
Lymph nodes	0.06	**0.20**	0.07	**0.25**	0.08	**0.28**	0.09	**0.31**	0.11	**0.36**	0.13	**0.41**
Skin	0.04	**0.09**	0.04	**0.09**	0.04	**0.10**	0.04	**0.10**	0.04	**0.11**	0.05	**0.11**
Muscle	0.04	**0.08**	0.04	**0.09**	0.05	**0.11**	0.05	**0.12**	0.06	**0.13**	0.07	**0.15**
Skeleton	0.06	**0.12**	0.07	**0.14**	0.08	**0.15**	0.09	**0.17**	0.10	**0.19**	0.11	**0.21**
Red bone marrow	0.03	**0.05**	0.03	**0.06**	0.04	**0.07**	0.04	**0.07**	0.04	**0.09**	0.05	**0.10**
Whole body dose	0.04	**0.09**	0.05	**0.11**	0.06	**0.12**	0.06	**0.13**	0.07	**0.15**	0.08	**0.17**
ED (ICRP 60)	0.04	**0.16**	0.05	**0.19**	0.06	**0.22**	0.07	**0.24**	0.08	**0.27**	0.09	**0.31**
ED (ICRP 103)	0.05	**0.18**	0.06	**0.21**	0.07	**0.24**	0.08	**0.26**	0.09	**0.30**	0.11	**0.34**
Total filtration (mm) Al	**1.5**		**2.5**		**3.5**		**4.5**		**3.5**		**3.5**	
Cu	**0**		**0**		**0**		**0**		**0.1**		**0.2**	

Table 15.55 Normalized organ doses—Thoracic spine ap—8–12 years

Beam projection:			ap									**Table cells:**
Focus-to-detector distance (cm):			150				(+) = normal font = optimal field setting					
Tube voltage (kVp):			80				**(−) = bold font = suboptimal field setting**					
Total filtrations:			See last table row.				Empty table cells correspond to values <0.01					
Organ or tissue	**Organ dose D_T : Entrance dose K_E**								**(mSv * mGy^{-1})**			
	(+)	(−)	(+)	(−)	(+)	(−)	(+)	(−)	(+)	(−)	(+)	(−)
Skull		**0.02**		**0.02**		**0.03**	0.01	**0.03**	0.01	**0.03**	0.01	**0.04**
Brain												
Oral mucosa		**0.02**		**0.03**		**0.03**	0.01	**0.03**	0.01	**0.04**	0.02	**0.05**
Salivary glands		**0.04**	0.01	**0.05**	0.01	**0.05**	0.01	**0.06**	0.02	**0.07**	0.02	**0.08**
Cervical spine	0.05	**0.39**	0.07	**0.48**	0.08	**0.55**	0.09	**0.61**	0.11	**0.73**	0.13	**0.83**
Thyroid	0.08	**0.86**	0.09	**0.97**	0.11	**1.04**	0.12	**1.09**	0.14	**1.18**	0.16	**1.26**
Thymus	0.77	**0.82**	0.91	**0.96**	1.00	**1.07**	1.07	**1.14**	1.19	**1.28**	1.29	**1.39**
Thoracic spine	0.14	**0.19**	0.18	**0.24**	0.21	**0.29**	0.24	**0.33**	0.30	**0.41**	0.36	**0.49**
Claviculae	0.53	**1.12**	0.61	**1.31**	0.66	**1.43**	0.71	**1.52**	0.77	**1.67**	0.82	**1.78**
Scapulae	0.04	**0.09**	0.05	**0.11**	0.06	**0.13**	0.07	**0.15**	0.08	**0.18**	0.10	**0.21**
Ribs	0.50	**0.85**	0.56	**0.95**	0.59	**1.02**	0.62	**1.07**	0.66	**1.15**	0.70	**1.21**
Oesophagus	0.10	**0.13**	0.13	**0.17**	0.15	**0.19**	0.17	**0.22**	0.20	**0.27**	0.24	**0.31**
Trachea	0.02	**0.24**	0.03	**0.28**	0.03	**0.30**	0.03	**0.32**	0.04	**0.36**	0.05	**0.39**
Lung	0.08	**0.25**	0.10	**0.31**	0.12	**0.35**	0.13	**0.38**	0.16	**0.45**	0.18	**0.50**
Mamma	0.04	**0.21**	0.05	**0.24**	0.05	**0.26**	0.06	**0.27**	0.07	**0.30**	0.08	**0.32**
Heart	0.41	**0.52**	0.49	**0.62**	0.56	**0.70**	0.61	**0.77**	0.70	**0.89**	0.78	**0.99**
Stomach	0.09	**0.39**	0.11	**0.48**	0.13	**0.54**	0.14	**0.58**	0.17	**0.67**	0.19	**0.75**
Spleen	0.02	**0.08**	0.03	**0.10**	0.03	**0.12**	0.04	**0.13**	0.05	**0.16**	0.05	**0.20**
Liver	0.13	**0.29**	0.16	**0.36**	0.18	**0.40**	0.20	**0.44**	0.23	**0.51**	0.26	**0.57**
Gall bladder	0.08	**0.41**	0.10	**0.51**	0.11	**0.58**	0.12	**0.64**	0.15	**0.75**	0.17	**0.85**
Pancreas	0.14	**0.24**	0.18	**0.30**	0.21	**0.35**	0.23	**0.39**	0.28	**0.47**	0.32	**0.55**
Small intestine		**0.07**		**0.09**		**0.11**		**0.12**	0.01	**0.15**	0.01	**0.17**
Large intestine		**0.17**		**0.20**	0.01	**0.23**	0.01	**0.26**	0.01	**0.30**	0.02	**0.34**
Sigma/Rectum		**0.01**		**0.02**		**0.02**		**0.02**		**0.03**		**0.03**
Lumbar spine	0.03	**0.16**	0.04	**0.20**	0.05	**0.24**	0.05	**0.27**	0.07	**0.34**	0.08	**0.41**
Pelvis		**0.02**		**0.03**		**0.03**		**0.03**		**0.04**		**0.05**
Adrenals	0.07	**0.10**	0.09	**0.13**	0.10	**0.15**	0.12	**0.17**	0.15	**0.22**	0.17	**0.26**
Kidneys	0.02	**0.07**	0.03	**0.09**	0.03	**0.11**	0.03	**0.13**	0.04	**0.16**	0.05	**0.19**
Urinary bladder												**0.01**
Uterus		**0.01**		**0.02**		**0.02**		**0.02**		**0.03**		**0.03**
Ovaries		**0.01**		**0.02**		**0.02**		**0.02**		**0.03**		**0.04**
Prostate												
Testes												
Upper arm	0.02	**0.04**	0.03	**0.05**	0.03	**0.06**	0.04	**0.07**	0.05	**0.09**	0.06	**0.11**
Elbow	0.02	**0.05**	0.03	**0.06**	0.03	**0.07**	0.04	**0.08**	0.05	**0.10**	0.06	**0.12**
Forearm		**0.02**		**0.02**		**0.03**		**0.03**		**0.04**		**0.04**
Upper leg												
Knee												
Lower leg												
Lymph nodes	0.07	**0.23**	0.08	**0.28**	0.10	**0.32**	0.11	**0.35**	0.13	**0.40**	0.15	**0.45**
Skin	0.04	**0.09**	0.04	**0.10**	0.04	**0.10**	0.04	**0.11**	0.05	**0.11**	0.05	**0.12**
Muscle	0.04	**0.09**	0.05	**0.11**	0.06	**0.12**	0.06	**0.13**	0.07	**0.15**	0.08	**0.17**
Skeleton	0.07	**0.13**	0.08	**0.15**	0.09	**0.17**	0.09	**0.18**	0.11	**0.21**	0.12	**0.23**
Red bone marrow	0.03	**0.06**	0.04	**0.07**	0.04	**0.08**	0.04	**0.08**	0.05	**0.10**	0.06	**0.11**
Whole body dose	0.05	**0.10**	0.06	**0.12**	0.06	**0.14**	0.07	**0.15**	0.08	**0.17**	0.09	**0.19**
ED (ICRP 60)	0.05	**0.18**	0.06	**0.22**	0.07	**0.24**	0.08	**0.26**	0.09	**0.30**	0.10	**0.33**
ED (ICRP 103)	0.06	**0.20**	0.07	**0.24**	0.08	**0.27**	0.09	**0.29**	0.10	**0.33**	0.12	**0.37**
Total filtration (mm) Al	**1.5**		**2.5**		**3.5**		**4.5**		**3.5**		**3.5**	
Cu	**0**		**0**		**0**		**0**		**0.1**		**0.2**	

Table 15.56 Normalized organ doses—Thoracic spine ap—13–17 years

Beam projection:	ap											**Table cells:**
Focus-to-detector distance (cm):	150											(+) = normal font = optimal field setting
Tube voltage (kVp):	60											**(−) = bold font = suboptimal field setting**
Total filtrations:	See last table row.											Empty table cells correspond to values <0.01
Organ or tissue	**Organ dose D_T : Entrance dose K_E**							**(mSv ∗ mGy^{-1})**				
	(+)	(−)	(+)	(−)	(+)	(−)	(+)	(−)	(+)	(−)	(+)	(−)
Skull		**0.01**		**0.01**		**0.02**		**0.02**		**0.02**		**0.02**
Brain												
Oral mucosa				**0.01**		**0.02**		**0.02**		**0.02**		**0.02**
Salivary glands		**0.02**		**0.03**		**0.03**		**0.04**		**0.05**		**0.04**
Cervical spine	0.02	**0.30**	0.03	**0.39**	0.04	**0.47**	0.05	**0.53**	0.06	**0.66**	0.05	**0.59**
Thyroid	0.04	**0.69**	0.05	**0.81**	0.06	**0.88**	0.07	**0.94**	0.09	**1.05**	0.08	**1.03**
Thymus	0.57	**0.60**	0.70	**0.74**	0.79	**0.85**	0.87	**0.93**	1.00	**1.08**	0.97	**1.04**
Thoracic spine	0.05	**0.06**	0.06	**0.09**	0.08	**0.11**	0.09	**0.13**	0.12	**0.17**	0.09	**0.13**
Claviculae	0.33	**0.95**	0.40	**1.16**	0.45	**1.31**	0.49	**1.43**	0.55	**1.64**	0.54	**1.61**
Scapulae	0.01	**0.04**	0.02	**0.06**	0.02	**0.07**	0.02	**0.08**	0.03	**0.11**	0.02	**0.09**
Ribs	0.42	**0.77**	0.48	**0.90**	0.53	**0.98**	0.56	**1.05**	0.61	**1.17**	0.60	**1.15**
Oesophagus	0.06	**0.08**	0.09	**0.11**	0.10	**0.14**	0.12	**0.16**	0.15	**0.21**	0.13	**0.17**
Trachea		**0.20**	0.01	**0.24**	0.01	**0.27**	0.02	**0.29**	0.02	**0.32**	0.02	**0.32**
Lung	0.04	**0.17**	0.06	**0.22**	0.07	**0.26**	0.08	**0.29**	0.10	**0.36**	0.09	**0.33**
Mamma	0.39	**4.62**	0.48	**5.34**	0.55	**5.82**	0.60	**6.18**	0.71	**6.80**	0.65	**6.67**
Heart	0.28	**0.34**	0.35	**0.43**	0.41	**0.50**	0.46	**0.56**	0.54	**0.67**	0.51	**0.62**
Stomach	0.05	**0.32**	0.07	**0.41**	0.08	**0.48**	0.09	**0.53**	0.11	**0.63**	0.10	**0.59**
Spleen		**0.05**	0.01	**0.07**	0.01	**0.08**	0.02	**0.10**	0.02	**0.12**	0.02	**0.10**
Liver	0.08	**0.19**	0.10	**0.25**	0.11	**0.29**	0.13	**0.32**	0.15	**0.39**	0.14	**0.36**
Gall bladder	0.03	**0.24**	0.04	**0.32**	0.05	**0.38**	0.06	**0.42**	0.07	**0.52**	0.07	**0.47**
Pancreas	0.07	**0.12**	0.09	**0.16**	0.11	**0.19**	0.12	**0.22**	0.16	**0.28**	0.14	**0.24**
Small intestine		**0.04**		**0.06**		**0.07**		**0.08**		**0.10**		**0.09**
Large intestine		**0.11**		**0.14**		**0.17**		**0.19**		**0.23**		**0.21**
Sigma/Rectum						**0.01**		**0.01**		**0.02**		**0.01**
Lumbar spine		**0.05**	0.01	**0.07**	0.02	**0.09**	0.02	**0.11**	0.03	**0.14**	0.02	**0.11**
Pelvis						**0.01**		**0.01**		**0.02**		**0.01**
Adrenals	0.03	**0.04**	0.04	**0.05**	0.04	**0.07**	0.05	**0.08**	0.07	**0.10**	0.05	**0.08**
Kidneys		**0.03**		**0.04**	0.01	**0.05**	0.01	**0.06**	0.02	**0.08**	0.01	**0.06**
Urinary bladder												
Uterus		**0.01**		**0.02**		**0.02**		**0.03**		**0.03**		**0.02**
Ovaries										**0.01**		
Prostate												
Testes												
Upper arm		**0.02**		**0.02**		**0.03**	0.01	**0.03**	0.02	**0.04**	0.01	**0.03**
Elbow		**0.02**		**0.03**	0.01	**0.03**	0.01	**0.04**	0.02	**0.05**	0.01	**0.04**
Forearm				**0.01**		**0.01**		**0.01**		**0.02**		**0.01**
Upper leg												
Knee												
Lower leg												
Lymph nodes	0.04	**0.16**	0.05	**0.20**	0.06	**0.23**	0.07	**0.26**	0.08	**0.30**	0.07	**0.28**
Skin	0.04	**0.11**	0.04	**0.12**	0.04	**0.13**	0.04	**0.13**	0.05	**0.14**	0.04	**0.14**
Muscle	0.03	**0.07**	0.04	**0.09**	0.04	**0.10**	0.05	**0.11**	0.05	**0.13**	0.05	**0.12**
Skeleton	0.05	**0.11**	0.06	**0.13**	0.07	**0.15**	0.07	**0.16**	0.08	**0.18**	0.08	**0.17**
Red bone marrow	0.03	**0.05**	0.03	**0.06**	0.03	**0.07**	0.04	**0.07**	0.04	**0.09**	0.04	**0.08**
Whole body dose	0.03	**0.08**	0.04	**0.10**	0.05	**0.12**	0.05	**0.13**	0.06	**0.15**	0.06	**0.14**
ED (ICRP 60)	0.05	**0.36**	0.06	**0.43**	0.07	**0.48**	0.08	**0.51**	0.09	**0.58**	0.08	**0.56**
ED (ICRP 103)	0.08	**0.68**	0.10	**0.80**	0.11	**0.89**	0.13	**0.95**	0.15	**1.06**	0.14	**1.03**
Total filtration (mm) Al	**1.5**		**2.5**		**3.5**		**4.5**		**3.5**		**3.5**	
Cu	**0**		**0**		**0**		**0**		**0.1**		**0.2**	

Table 15.57 Normalized organ doses—Thoracic spine ap—13–17 years

Beam projection:		ap										**Table cells:**
Focus-to-detector distance (cm):		150										(+) = normal font = optimal field setting
Tube voltage (kVp):		70										**(−) = bold font = suboptimal field setting**
Total filtrations:		See last table row.										Empty table cells correspond to values <0.01
Organ or tissue	**Organ dose D_T : Entrance dose K_E**								**(mSv ∗ mGy^{-1})**			
	(+)	(−)	(+)	(−)	(+)	(−)	(+)	(−)	(+)	(−)	(+)	(−)
Skull		**0.01**		**0.02**		**0.02**		**0.02**		**0.03**		**0.04**
Brain												
Oral mucosa		**0.01**		**0.02**		**0.02**		**0.02**		**0.03**		**0.03**
Salivary glands		**0.03**		**0.04**		**0.04**		**0.05**		**0.06**		**0.06**
Cervical spine	0.03	**0.38**	0.05	**0.49**	0.05	**0.58**	0.06	**0.65**	0.08	**0.79**	0.10	**0.92**
Thyroid	0.05	**0.76**	0.07	**0.88**	0.08	**0.96**	0.09	**1.01**	0.11	**1.12**	0.13	**1.20**
Thymus	0.65	**0.70**	0.79	**0.85**	0.89	**0.96**	0.97	**1.04**	1.11	**1.20**	1.23	**1.34**
Thoracic spine	0.07	**0.10**	0.09	**0.13**	0.11	**0.16**	0.13	**0.19**	0.17	**0.25**	0.21	**0.31**
Claviculae	0.37	**1.07**	0.44	**1.29**	0.49	**1.44**	0.53	**1.56**	0.60	**1.77**	0.65	**1.93**
Scapulae	0.02	**0.06**	0.03	**0.08**	0.03	**0.10**	0.04	**0.11**	0.05	**0.15**	0.06	**0.18**
Ribs	0.45	**0.84**	0.52	**0.97**	0.56	**1.06**	0.59	**1.12**	0.64	**1.24**	0.68	**1.33**
Oesophagus	0.09	**0.12**	0.12	**0.16**	0.14	**0.19**	0.16	**0.21**	0.20	**0.27**	0.24	**0.33**
Trachea	0.01	**0.22**	0.01	**0.26**	0.02	**0.29**	0.02	**0.31**	0.02	**0.35**	0.03	**0.39**
Lung	0.06	**0.21**	0.07	**0.27**	0.09	**0.31**	0.10	**0.35**	0.12	**0.42**	0.15	**0.49**
Mamma	0.47	**5.05**	0.57	**5.78**	0.64	**6.27**	0.71	**6.63**	0.83	**7.27**	0.95	**7.81**
Heart	0.34	**0.41**	0.42	**0.51**	0.48	**0.59**	0.53	**0.65**	0.62	**0.78**	0.71	**0.89**
Stomach	0.06	**0.39**	0.08	**0.49**	0.10	**0.56**	0.11	**0.61**	0.13	**0.72**	0.15	**0.82**
Spleen	0.01	**0.07**	0.02	**0.09**	0.02	**0.11**	0.02	**0.13**	0.03	**0.16**	0.04	**0.20**
Liver	0.09	**0.24**	0.12	**0.30**	0.14	**0.35**	0.15	**0.38**	0.18	**0.46**	0.21	**0.53**
Gall bladder	0.04	**0.31**	0.06	**0.39**	0.07	**0.46**	0.08	**0.51**	0.09	**0.62**	0.11	**0.72**
Pancreas	0.09	**0.16**	0.12	**0.21**	0.14	**0.25**	0.16	**0.29**	0.20	**0.36**	0.24	**0.44**
Small intestine		**0.06**		**0.08**		**0.09**		**0.10**		**0.13**		**0.15**
Large intestine		**0.14**		**0.17**		**0.20**		**0.23**		**0.27**	0.01	**0.32**
Sigma/Rectum		**0.01**		**0.01**		**0.02**		**0.02**		**0.02**		**0.03**
Lumbar spine	0.02	**0.08**	0.02	**0.11**	0.03	**0.13**	0.03	**0.15**	0.04	**0.20**	0.05	**0.25**
Pelvis		**0.01**		**0.01**		**0.02**		**0.02**		**0.03**		**0.04**
Adrenals	0.04	**0.06**	0.05	**0.08**	0.06	**0.09**	0.07	**0.11**	0.09	**0.14**	0.11	**0.17**
Kidneys	0.01	**0.04**	0.01	**0.06**	0.02	**0.07**	0.02	**0.08**	0.03	**0.11**	0.03	**0.13**
Urinary bladder												
Uterus		**0.02**		**0.03**		**0.03**		**0.04**		**0.05**		**0.06**
Ovaries						**0.01**		**0.01**		**0.02**		**0.02**
Prostate												
Testes												
Upper arm		**0.02**	0.01	**0.03**	0.01	**0.04**	0.02	**0.05**	0.02	**0.06**	0.03	**0.08**
Elbow		**0.03**	0.01	**0.04**	0.02	**0.05**	0.02	**0.05**	0.02	**0.07**	0.03	**0.09**
Forearm		**0.01**		**0.01**		**0.02**		**0.02**		**0.03**		**0.03**
Upper leg												
Knee												
Lower leg												
Lymph nodes	0.05	**0.19**	0.06	**0.24**	0.07	**0.27**	0.08	**0.30**	0.10	**0.36**	0.12	**0.41**
Skin	0.04	**0.12**	0.04	**0.13**	0.04	**0.13**	0.04	**0.14**	0.05	**0.15**	0.05	**0.16**
Muscle	0.03	**0.09**	0.04	**0.10**	0.05	**0.12**	0.05	**0.13**	0.06	**0.15**	0.07	**0.17**
Skeleton	0.06	**0.12**	0.07	**0.15**	0.08	**0.17**	0.08	**0.18**	0.09	**0.21**	0.10	**0.23**
Red bone marrow	0.03	**0.06**	0.03	**0.07**	0.04	**0.08**	0.04	**0.09**	0.05	**0.10**	0.06	**0.12**
Whole body dose	0.04	**0.10**	0.05	**0.12**	0.05	**0.14**	0.06	**0.15**	0.07	**0.17**	0.08	**0.20**
ED (ICRP 60)	0.06	**0.41**	0.07	**0.48**	0.08	**0.53**	0.09	**0.57**	0.11	**0.64**	0.13	**0.71**
ED (ICRP 103)	0.10	**0.76**	0.12	**0.89**	0.14	**0.97**	0.15	**1.04**	0.18	**1.16**	0.21	**1.26**
Total filtration (mm) Al	**1.5**		**2.5**		**3.5**		**4.5**		**3.5**		**3.5**	
Cu	**0**		**0**		**0**		**0**		**0.1**		**0.2**	

Table 15.58 Normalized organ doses—Thoracic spine ap—13–17 years

Beam projection:		ap										**Table cells:**
Focus-to-detector distance (cm):		150										(+) = normal font = optimal field setting
Tube voltage (kVp):		80										**(−) = bold font = suboptimal field setting**
Total filtrations:		See last table row.										Empty table cells correspond to values <0.01
Organ or tissue	**Organ dose D_T : Entrance dose K_E**								**(mSv * mGy^{-1})**			
	(+)	(−)	(+)	(−)	(+)	(−)	(+)	(−)	(+)	(−)	(+)	(−)
Skull		**0.02**		**0.02**		**0.03**		**0.03**		**0.04**	0.01	**0.04**
Brain												
Oral mucosa		**0.02**		**0.02**		**0.03**		**0.03**		**0.03**		**0.04**
Salivary glands		**0.03**		**0.04**		**0.05**		**0.05**	0.01	**0.06**	0.01	**0.07**
Cervical spine	0.05	**0.46**	0.06	**0.58**	0.07	**0.67**	0.08	**0.75**	0.10	**0.89**	0.12	**1.03**
Thyroid	0.07	**0.82**	0.08	**0.94**	0.10	**1.02**	0.11	**1.08**	0.13	**1.18**	0.15	**1.26**
Thymus	0.74	**0.79**	0.88	**0.94**	0.98	**1.06**	1.05	**1.14**	1.19	**1.30**	1.30	**1.43**
Thoracic spine	0.10	**0.14**	0.13	**0.18**	0.15	**0.22**	0.17	**0.25**	0.22	**0.31**	0.26	**0.38**
Claviculae	0.41	**1.18**	0.48	**1.39**	0.52	**1.54**	0.56	**1.65**	0.62	**1.84**	0.67	**1.99**
Scapulae	0.03	**0.08**	0.03	**0.11**	0.04	**0.13**	0.05	**0.14**	0.06	**0.18**	0.07	**0.21**
Ribs	0.48	**0.91**	0.54	**1.03**	0.58	**1.12**	0.61	**1.18**	0.66	**1.28**	0.69	**1.37**
Oesophagus	0.11	**0.15**	0.15	**0.20**	0.17	**0.23**	0.19	**0.26**	0.24	**0.33**	0.28	**0.39**
Trachea	0.01	**0.25**	0.02	**0.29**	0.02	**0.32**	0.02	**0.34**	0.03	**0.38**	0.03	**0.41**
Lung	0.07	**0.26**	0.09	**0.32**	0.11	**0.36**	0.12	**0.40**	0.15	**0.48**	0.17	**0.55**
Mamma	0.55	**5.46**	0.66	**6.18**	0.74	**6.66**	0.81	**7.02**	0.94	**7.65**	1.07	**8.17**
Heart	0.39	**0.48**	0.48	**0.59**	0.54	**0.68**	0.59	**0.74**	0.69	**0.87**	0.78	**0.98**
Stomach	0.08	**0.46**	0.10	**0.56**	0.11	**0.63**	0.13	**0.69**	0.15	**0.80**	0.17	**0.90**
Spleen	0.02	**0.09**	0.02	**0.12**	0.03	**0.14**	0.03	**0.16**	0.04	**0.20**	0.05	**0.24**
Liver	0.11	**0.28**	0.14	**0.35**	0.16	**0.40**	0.17	**0.44**	0.20	**0.52**	0.23	**0.59**
Gall bladder	0.05	**0.37**	0.07	**0.46**	0.08	**0.53**	0.09	**0.59**	0.11	**0.70**	0.13	**0.80**
Pancreas	0.11	**0.21**	0.15	**0.27**	0.17	**0.31**	0.19	**0.35**	0.23	**0.43**	0.28	**0.51**
Small intestine		**0.07**		**0.09**		**0.11**		**0.12**		**0.15**		**0.18**
Large intestine		**0.16**		**0.20**		**0.24**		**0.26**	0.01	**0.31**	0.01	**0.36**
Sigma/Rectum		**0.01**		**0.02**		**0.02**		**0.02**		**0.03**		**0.03**
Lumbar spine	0.02	**0.11**	0.03	**0.15**	0.04	**0.18**	0.04	**0.20**	0.05	**0.26**	0.06	**0.31**
Pelvis		**0.02**		**0.02**		**0.03**		**0.03**		**0.04**		**0.05**
Adrenals	0.05	**0.08**	0.07	**0.10**	0.08	**0.12**	0.09	**0.14**	0.11	**0.18**	0.13	**0.22**
Kidneys	0.02	**0.06**	0.02	**0.08**	0.02	**0.09**	0.03	**0.11**	0.03	**0.13**	0.04	**0.16**
Urinary bladder												
Uterus		**0.03**		**0.04**		**0.04**		**0.05**		**0.07**		**0.08**
Ovaries				**0.01**		**0.02**		**0.02**		**0.02**		**0.03**
Prostate												
Testes												
Upper arm	0.01	**0.03**	0.02	**0.05**	0.02	**0.05**	0.02	**0.06**	0.03	**0.08**	0.04	**0.10**
Elbow	0.01	**0.04**	0.02	**0.05**	0.02	**0.06**	0.02	**0.07**	0.03	**0.09**	0.04	**0.11**
Forearm		**0.02**		**0.02**		**0.02**		**0.03**		**0.04**		**0.04**
Upper leg												
Knee												
Lower leg												
Lymph nodes	0.06	**0.23**	0.07	**0.28**	0.09	**0.31**	0.10	**0.34**	0.12	**0.40**	0.13	**0.45**
Skin	0.04	**0.13**	0.04	**0.13**	0.05	**0.14**	0.05	**0.15**	0.05	**0.15**	0.05	**0.16**
Muscle	0.04	**0.10**	0.05	**0.12**	0.06	**0.14**	0.06	**0.15**	0.07	**0.17**	0.08	**0.19**
Skeleton	0.07	**0.14**	0.08	**0.17**	0.08	**0.19**	0.09	**0.20**	0.10	**0.23**	0.11	**0.25**
Red bone marrow	0.03	**0.07**	0.04	**0.08**	0.04	**0.09**	0.05	**0.10**	0.06	**0.12**	0.06	**0.14**
Whole body dose	0.05	**0.12**	0.06	**0.14**	0.06	**0.15**	0.07	**0.17**	0.08	**0.19**	0.09	**0.22**
ED (ICRP 60)	0.07	**0.45**	0.09	**0.53**	0.10	**0.58**	0.11	**0.62**	0.13	**0.69**	0.15	**0.76**
ED (ICRP 103)	0.11	**0.84**	0.14	**0.96**	0.16	**1.05**	0.17	**1.12**	0.20	**1.24**	0.23	**1.34**
Total filtration (mm) **Al**	**1.5**		**2.5**		**3.5**		**4.5**		**3.5**		**3.5**	
Cu	**0**		**0**		**0**		**0**		**0.1**		**0.2**	

Table 15.59 Normalized organ doses—Thoracic spine ap—>17 years

Beam projection:			ap									**Table cells:**
Focus-to-detector distance (cm):			150									(+) = normal font = optimal field setting
Tube voltage (kVp):			60									**(−) = bold font = suboptimal field setting**
Total filtrations:			See last table row.									Empty table cells correspond to values <0.01
Organ or tissue	**Organ dose D_T : Entrance dose K_E**								**(mSv * mGy^{-1})**			
	(+)	(−)	(+)	(−)	(+)	(−)	(+)	(−)	(+)	(−)	(+)	(−)
Skull		**0.01**		**0.01**		**0.02**		**0.02**		**0.02**		**0.02**
Brain												
Oral mucosa				**0.01**		**0.01**		**0.02**		**0.02**		**0.02**
Salivary glands		**0.02**		**0.03**		**0.03**		**0.04**		**0.05**		**0.04**
Cervical spine	0.02	**0.27**	0.02	**0.36**	0.03	**0.42**	0.03	**0.48**	0.04	**0.60**	0.03	**0.54**
Thyroid	0.03	**0.64**	0.04	**0.76**	0.05	**0.84**	0.05	**0.90**	0.07	**1.01**	0.06	**0.99**
Thymus	0.56	**0.60**	0.69	**0.74**	0.79	**0.84**	0.86	**0.93**	1.00	**1.08**	0.97	**1.03**
Thoracic spine	0.04	**0.06**	0.06	**0.09**	0.07	**0.11**	0.08	**0.12**	0.11	**0.17**	0.08	**0.12**
Claviculae	0.26	**0.96**	0.31	**1.18**	0.35	**1.34**	0.38	**1.46**	0.44	**1.69**	0.43	**1.65**
Scapulae		**0.05**	0.01	**0.06**	0.02	**0.08**	0.02	**0.09**	0.03	**0.12**	0.02	**0.10**
Ribs	0.37	**0.79**	0.43	**0.93**	0.47	**1.02**	0.50	**1.09**	0.54	**1.21**	0.54	**1.19**
Oesophagus	0.06	**0.09**	0.08	**0.12**	0.10	**0.15**	0.12	**0.17**	0.15	**0.22**	0.13	**0.18**
Trachea		**0.19**	0.01	**0.24**	0.01	**0.27**	0.01	**0.29**	0.02	**0.33**	0.02	**0.32**
Lung	0.03	**0.18**	0.04	**0.24**	0.05	**0.28**	0.06	**0.31**	0.07	**0.38**	0.06	**0.35**
Mamma	0.02	**0.52**	0.02	**0.61**	0.03	**0.66**	0.03	**0.70**	0.04	**0.77**	0.03	**0.76**
Heart	0.24	**0.32**	0.31	**0.42**	0.36	**0.48**	0.40	**0.54**	0.48	**0.65**	0.45	**0.60**
Stomach	0.03	**0.32**	0.04	**0.41**	0.05	**0.48**	0.06	**0.53**	0.07	**0.63**	0.06	**0.59**
Spleen		**0.05**		**0.07**		**0.08**	0.01	**0.09**	0.01	**0.12**	0.01	**0.10**
Liver	0.07	**0.20**	0.09	**0.25**	0.10	**0.29**	0.11	**0.33**	0.13	**0.40**	0.12	**0.37**
Gall bladder	0.03	**0.24**	0.04	**0.31**	0.04	**0.37**	0.05	**0.42**	0.06	**0.51**	0.05	**0.47**
Pancreas	0.06	**0.12**	0.08	**0.16**	0.10	**0.19**	0.11	**0.22**	0.14	**0.28**	0.12	**0.24**
Small intestine		**0.04**		**0.06**		**0.07**		**0.08**		**0.10**		**0.08**
Large intestine		**0.11**		**0.14**		**0.16**		**0.19**		**0.23**		**0.21**
Sigma/Rectum						**0.01**		**0.01**		**0.02**		**0.01**
Lumbar spine		**0.05**	0.01	**0.07**	0.01	**0.09**	0.02	**0.10**	0.02	**0.13**	0.02	**0.10**
Pelvis						**0.01**		**0.01**		**0.02**		**0.01**
Adrenals	0.02	**0.04**	0.03	**0.05**	0.04	**0.07**	0.05	**0.08**	0.06	**0.10**	0.05	**0.08**
Kidneys		**0.03**		**0.04**		**0.05**	0.01	**0.06**	0.01	**0.08**	0.01	**0.06**
Urinary bladder												
Uterus										**0.01**		
Ovaries										**0.01**		
Prostate												
Testes												
Upper arm		**0.01**		**0.02**		**0.02**		**0.03**		**0.03**		**0.03**
Elbow		**0.01**		**0.02**		**0.02**		**0.03**		**0.04**		**0.03**
Forearm								**0.01**		**0.01**		**0.01**
Upper leg												
Knee												
Lower leg												
Lymph nodes	0.03	**0.15**	0.04	**0.20**	0.05	**0.23**	0.06	**0.25**	0.07	**0.30**	0.06	**0.28**
Skin	0.03	**0.12**	0.03	**0.13**	0.04	**0.13**	0.04	**0.13**	0.04	**0.14**	0.04	**0.14**
Muscle	0.02	**0.07**	0.03	**0.09**	0.03	**0.10**	0.04	**0.11**	0.05	**0.13**	0.04	**0.12**
Skeleton	0.04	**0.11**	0.05	**0.13**	0.06	**0.14**	0.06	**0.16**	0.07	**0.18**	0.07	**0.17**
Red bone marrow	0.03	**0.06**	0.03	**0.07**	0.03	**0.08**	0.04	**0.09**	0.04	**0.10**	0.04	**0.09**
Whole body dose	0.03	**0.08**	0.04	**0.10**	0.04	**0.12**	0.04	**0.13**	0.05	**0.15**	0.05	**0.14**
ED (ICRP 60)	0.02	**0.15**	0.03	**0.19**	0.04	**0.22**	0.04	**0.24**	0.05	**0.28**	0.04	**0.26**
ED (ICRP 103)	0.03	**0.19**	0.04	**0.24**	0.04	**0.27**	0.05	**0.29**	0.06	**0.34**	0.05	**0.32**
Total filtration (mm) **Al**	**1.5**		**2.5**		**3.5**		**4.5**		**3.5**		**3.5**	
Cu	**0**		**0**		**0**		**0**		**0.1**		**0.2**	

Table 15.60 Normalized organ doses—Thoracic spine ap—>17 years

Beam projection:			ap									**Table cells:**
Focus-to-detector distance (cm):			150									(+) = normal font = optimal field setting
Tube voltage (kVp):			70									**(−) = bold font = suboptimal field setting**
Total filtrations:			See last table row.									Empty table cells correspond to values <0.01
Organ or tissue	**Organ dose D_T : Entrance dose K_E**							**(mSv *mGy^{-1})**				
	(+)	(−)	(+)	(−)	(+)	(−)	(+)	(−)	(+)	(−)	(+)	(−)
Skull		**0.01**		**0.02**		**0.02**		**0.02**		**0.03**		**0.04**
Brain												
Oral mucosa		**0.01**		**0.02**		**0.02**		**0.02**		**0.03**		**0.03**
Salivary glands		**0.03**		**0.04**		**0.04**		**0.05**		**0.06**		**0.07**
Cervical spine	0.03	**0.35**	0.03	**0.45**	0.04	**0.53**	0.05	**0.59**	0.06	**0.73**	0.08	**0.86**
Thyroid	0.04	**0.72**	0.05	**0.83**	0.06	**0.91**	0.07	**0.98**	0.08	**1.08**	0.10	**1.17**
Thymus	0.65	**0.70**	0.79	**0.85**	0.89	**0.96**	0.96	**1.05**	1.10	**1.21**	1.22	**1.35**
Thoracic spine	0.06	**0.10**	0.09	**0.13**	0.10	**0.16**	0.12	**0.18**	0.16	**0.24**	0.20	**0.30**
Claviculae	0.29	**1.10**	0.35	**1.32**	0.39	**1.48**	0.42	**1.60**	0.47	**1.82**	0.52	**2.00**
Scapulae	0.01	**0.07**	0.02	**0.09**	0.02	**0.11**	0.03	**0.12**	0.04	**0.16**	0.05	**0.19**
Ribs	0.40	**0.87**	0.46	**1.00**	0.50	**1.10**	0.52	**1.16**	0.57	**1.28**	0.61	**1.38**
Oesophagus	0.09	**0.12**	0.11	**0.16**	0.14	**0.20**	0.15	**0.22**	0.20	**0.28**	0.24	**0.35**
Trachea	0.01	**0.22**	0.01	**0.27**	0.02	**0.30**	0.02	**0.32**	0.02	**0.36**	0.03	**0.40**
Lung	0.04	**0.23**	0.05	**0.29**	0.07	**0.34**	0.07	**0.37**	0.09	**0.45**	0.11	**0.52**
Mamma	0.02	**0.57**	0.03	**0.66**	0.03	**0.71**	0.04	**0.75**	0.05	**0.83**	0.06	**0.89**
Heart	0.30	**0.40**	0.37	**0.50**	0.42	**0.58**	0.47	**0.64**	0.55	**0.76**	0.63	**0.87**
Stomach	0.04	**0.39**	0.06	**0.49**	0.07	**0.56**	0.07	**0.62**	0.09	**0.73**	0.11	**0.83**
Spleen		**0.07**	0.01	**0.09**	0.01	**0.11**	0.02	**0.13**	0.02	**0.16**	0.03	**0.20**
Liver	0.08	**0.24**	0.10	**0.30**	0.12	**0.35**	0.13	**0.39**	0.16	**0.47**	0.18	**0.54**
Gall bladder	0.04	**0.30**	0.05	**0.39**	0.06	**0.45**	0.06	**0.51**	0.08	**0.61**	0.10	**0.72**
Pancreas	0.08	**0.16**	0.11	**0.21**	0.13	**0.25**	0.14	**0.29**	0.18	**0.36**	0.21	**0.44**
Small intestine		**0.06**		**0.07**		**0.09**		**0.10**		**0.13**		**0.15**
Large intestine		**0.13**		**0.17**		**0.20**		**0.23**		**0.27**		**0.32**
Sigma/Rectum				**0.01**		**0.01**		**0.02**		**0.02**		**0.03**
Lumbar spine	0.01	**0.08**	0.02	**0.10**	0.02	**0.13**	0.03	**0.15**	0.03	**0.19**	0.04	**0.24**
Pelvis				**0.01**		**0.02**		**0.02**		**0.02**		**0.03**
Adrenals	0.03	**0.06**	0.05	**0.08**	0.06	**0.10**	0.06	**0.11**	0.08	**0.14**	0.10	**0.18**
Kidneys		**0.04**	0.01	**0.06**	0.01	**0.07**	0.02	**0.08**	0.02	**0.11**	0.03	**0.13**
Urinary bladder												
Uterus						**0.01**		**0.01**		**0.02**		**0.02**
Ovaries								**0.01**		**0.02**		**0.02**
Prostate												
Testes												
Upper arm		**0.02**		**0.03**		**0.03**	0.01	**0.04**	0.01	**0.05**	0.02	**0.06**
Elbow		**0.02**		**0.03**		**0.04**	0.01	**0.04**	0.01	**0.06**	0.02	**0.07**
Forearm				**0.01**		**0.01**		**0.02**		**0.02**		**0.03**
Upper leg												
Knee												
Lower leg												
Lymph nodes	0.04	**0.19**	0.05	**0.24**	0.06	**0.27**	0.07	**0.30**	0.09	**0.36**	0.10	**0.41**
Skin	0.03	**0.12**	0.04	**0.13**	0.04	**0.14**	0.04	**0.14**	0.04	**0.15**	0.04	**0.16**
Muscle	0.03	**0.08**	0.04	**0.10**	0.04	**0.12**	0.05	**0.13**	0.05	**0.15**	0.06	**0.17**
Skeleton	0.05	**0.12**	0.06	**0.15**	0.07	**0.16**	0.07	**0.18**	0.08	**0.20**	0.09	**0.23**
Red bone marrow	0.03	**0.07**	0.04	**0.08**	0.04	**0.09**	0.04	**0.10**	0.05	**0.12**	0.06	**0.14**
Whole body dose	0.03	**0.10**	0.04	**0.12**	0.05	**0.14**	0.05	**0.15**	0.06	**0.17**	0.07	**0.20**
ED (ICRP 60)	0.03	**0.18**	0.04	**0.22**	0.04	**0.25**	0.05	**0.28**	0.06	**0.32**	0.07	**0.37**
ED (ICRP 103)	0.04	**0.23**	0.05	**0.27**	0.05	**0.31**	0.06	**0.34**	0.07	**0.39**	0.08	**0.44**
Total filtration (mm) Al	**1.5**		**2.5**		**3.5**		**4.5**		**3.5**		**3.5**	
Cu	**0**		**0**		**0**		**0**		**0.1**		**0.2**	

Table 15.61 Normalized organ doses—Thoracic spine ap—>17 years

Beam projection:			ap									**Table cells:**
Focus-to-detector distance (cm):			150					(+) = normal font = optimal field setting				
Tube voltage (kVp):			80					**(−) = bold font = suboptimal field setting**				
Total filtrations:			See last table row.					Empty table cells correspond to values <0.01				
Organ or tissue	**Organ dose D_T : Entrance dose K_E**							**(mSv * mGy^{-1})**				
	(+)	(−)	(+)	(−)	(+)	(−)	(+)	(−)	(+)	(−)	(+)	(−)
Skull		**0.02**		**0.02**		**0.03**		**0.03**		**0.04**		**0.04**
Brain												
Oral mucosa		**0.02**		**0.02**		**0.02**		**0.03**		**0.03**		**0.04**
Salivary glands		**0.03**		**0.04**		**0.05**		**0.05**		**0.07**		**0.08**
Cervical spine	0.03	**0.43**	0.04	**0.53**	0.05	**0.62**	0.06	**0.69**	0.08	**0.83**	0.09	**0.96**
Thyroid	0.05	**0.78**	0.06	**0.90**	0.07	**0.98**	0.08	**1.04**	0.10	**1.14**	0.11	**1.23**
Thymus	0.73	**0.79**	0.87	**0.95**	0.97	**1.06**	1.05	**1.15**	1.19	**1.31**	1.30	**1.45**
Thoracic spine	0.09	**0.13**	0.12	**0.18**	0.14	**0.21**	0.16	**0.24**	0.20	**0.31**	0.24	**0.38**
Claviculae	0.32	**1.21**	0.37	**1.43**	0.41	**1.59**	0.44	**1.70**	0.49	**1.90**	0.53	**2.06**
Scapulae	0.02	**0.09**	0.03	**0.11**	0.03	**0.13**	0.04	**0.15**	0.05	**0.19**	0.06	**0.23**
Ribs	0.43	**0.94**	0.48	**1.07**	0.52	**1.15**	0.54	**1.22**	0.58	**1.33**	0.62	**1.41**
Oesophagus	0.11	**0.16**	0.14	**0.21**	0.17	**0.25**	0.19	**0.28**	0.24	**0.35**	0.28	**0.42**
Trachea	0.01	**0.25**	0.02	**0.29**	0.02	**0.32**	0.02	**0.35**	0.03	**0.39**	0.03	**0.43**
Lung	0.05	**0.27**	0.07	**0.34**	0.08	**0.39**	0.09	**0.43**	0.11	**0.50**	0.13	**0.58**
Mamma	0.03	**0.62**	0.04	**0.70**	0.04	**0.76**	0.05	**0.80**	0.06	**0.87**	0.07	**0.93**
Heart	0.35	**0.47**	0.42	**0.58**	0.48	**0.66**	0.53	**0.73**	0.62	**0.85**	0.70	**0.97**
Stomach	0.05	**0.46**	0.07	**0.56**	0.08	**0.64**	0.09	**0.70**	0.11	**0.81**	0.12	**0.91**
Spleen	0.01	**0.09**	0.02	**0.12**	0.02	**0.14**	0.02	**0.16**	0.03	**0.20**	0.03	**0.24**
Liver	0.10	**0.29**	0.12	**0.36**	0.14	**0.41**	0.15	**0.45**	0.18	**0.53**	0.20	**0.60**
Gall bladder	0.05	**0.37**	0.06	**0.46**	0.07	**0.53**	0.08	**0.59**	0.10	**0.70**	0.11	**0.81**
Pancreas	0.10	**0.21**	0.13	**0.27**	0.15	**0.31**	0.17	**0.35**	0.21	**0.43**	0.25	**0.51**
Small intestine		**0.07**		**0.09**		**0.11**		**0.12**		**0.15**		**0.18**
Large intestine		**0.16**		**0.21**		**0.24**		**0.26**		**0.31**	0.01	**0.36**
Sigma/Rectum		**0.01**		**0.02**		**0.02**		**0.02**		**0.03**		**0.03**
Lumbar spine	0.02	**0.11**	0.03	**0.14**	0.03	**0.17**	0.04	**0.20**	0.05	**0.25**	0.06	**0.31**
Pelvis		**0.01**		**0.02**		**0.02**		**0.03**		**0.03**		**0.04**
Adrenals	0.05	**0.08**	0.06	**0.11**	0.07	**0.13**	0.08	**0.14**	0.11	**0.18**	0.13	**0.22**
Kidneys	0.01	**0.06**	0.02	**0.08**	0.02	**0.09**	0.02	**0.11**	0.03	**0.13**	0.03	**0.16**
Urinary bladder												
Uterus				**0.01**		**0.01**		**0.02**		**0.02**		**0.03**
Ovaries				**0.01**		**0.01**		**0.02**		**0.02**		**0.03**
Prostate												
Testes												
Upper arm		**0.03**	0.01	**0.04**	0.01	**0.04**	0.02	**0.05**	0.02	**0.07**	0.02	**0.08**
Elbow		**0.03**	0.01	**0.04**	0.01	**0.05**	0.02	**0.06**	0.02	**0.07**	0.02	**0.09**
Forearm		**0.01**		**0.02**		**0.02**		**0.02**		**0.03**		**0.03**
Upper leg												
Knee												
Lower leg												
Lymph nodes	0.05	**0.22**	0.06	**0.27**	0.07	**0.31**	0.08	**0.34**	0.10	**0.40**	0.12	**0.46**
Skin	0.04	**0.13**	0.04	**0.14**	0.04	**0.14**	0.04	**0.15**	0.04	**0.16**	0.05	**0.17**
Muscle	0.03	**0.10**	0.04	**0.12**	0.05	**0.13**	0.05	**0.14**	0.06	**0.17**	0.07	**0.19**
Skeleton	0.06	**0.14**	0.07	**0.16**	0.07	**0.18**	0.08	**0.20**	0.09	**0.22**	0.09	**0.25**
Red bone marrow	0.03	**0.08**	0.04	**0.10**	0.05	**0.11**	0.05	**0.12**	0.06	**0.14**	0.06	**0.16**
Whole body dose	0.04	**0.11**	0.05	**0.14**	0.05	**0.15**	0.06	**0.17**	0.07	**0.19**	0.08	**0.22**
ED (ICRP 60)	0.04	**0.21**	0.05	**0.25**	0.05	**0.29**	0.06	**0.31**	0.07	**0.36**	0.08	**0.40**
ED (ICRP 103)	0.04	**0.26**	0.05	**0.31**	0.06	**0.35**	0.07	**0.37**	0.08	**0.43**	0.10	**0.48**
Total filtration (mm) **Al**	**1.5**		**2.5**		**3.5**		**4.5**		**3.5**		**3.5**	
Cu	**0**		**0**		**0**		**0**		**0.1**		**0.2**	

Table 15.62 Normalized organ doses—Thoracic spine r-lat—8–12 years

Beam projection:			r-lat									**Table cells:**
Focus-to-detector distance (cm):			150									(+) = normal font = optimal field setting
Tube voltage (kVp):			60									**(−) = bold font = suboptimal field setting**
Total filtrations:			See last table row.									Empty table cells correspond to values <0.01
Organ or tissue	**Organ dose D_T : Entrance dose K_E**								**(mSv ∗ mGy^{-1})**			
	(+)	(−)	(+)	(−)	(+)	(−)	(+)	(−)	(+)	(−)	(+)	(−)
Skull				**0.01**		**0.01**		**0.02**		**0.02**		**0.02**
Brain												
Oral mucosa				**0.01**		**0.01**		**0.02**		**0.02**		**0.02**
Salivary glands		**0.02**		**0.03**		**0.03**		**0.04**		**0.04**		**0.04**
Cervical spine	0.05	**0.28**	0.06	**0.36**	0.07	**0.42**	0.08	**0.47**	0.11	**0.57**	0.09	**0.53**
Thyroid	0.01	**0.26**	0.02	**0.32**	0.02	**0.36**	0.02	**0.40**	0.03	**0.46**	0.02	**0.44**
Thymus	0.01	**0.03**	0.02	**0.04**	0.02	**0.05**	0.03	**0.06**	0.03	**0.07**	0.03	**0.06**
Thoracic spine	0.16	**0.17**	0.21	**0.23**	0.26	**0.28**	0.30	**0.32**	0.38	**0.41**	0.32	**0.35**
Claviculae	0.03	**0.22**	0.04	**0.28**	0.04	**0.32**	0.05	**0.36**	0.06	**0.42**	0.05	**0.40**
Scapulae	0.95	**0.94**	1.12	**1.12**	1.24	**1.24**	1.33	**1.33**	1.48	**1.48**	1.48	**1.48**
Ribs	0.46	**0.62**	0.55	**0.73**	0.62	**0.82**	0.66	**0.88**	0.75	**0.99**	0.74	**0.97**
Oesophagus	0.07	**0.08**	0.10	**0.11**	0.11	**0.13**	0.13	**0.15**	0.16	**0.19**	0.14	**0.17**
Trachea		**0.10**		**0.13**		**0.14**	0.01	**0.16**	0.01	**0.18**	0.01	**0.18**
Lung	0.14	**0.20**	0.18	**0.25**	0.20	**0.29**	0.23	**0.32**	0.27	**0.38**	0.25	**0.36**
Mamma		**0.02**	0.01	**0.02**	0.01	**0.02**	0.01	**0.03**	0.02	**0.04**	0.02	**0.03**
Heart	0.04	**0.08**	0.05	**0.11**	0.06	**0.13**	0.07	**0.15**	0.09	**0.18**	0.08	**0.16**
Stomach		**0.02**		**0.02**		**0.03**	0.01	**0.03**	0.02	**0.04**	0.01	**0.03**
Spleen		**0.01**	0.01	**0.02**	0.01	**0.02**	0.02	**0.02**	0.02	**0.03**	0.02	**0.02**
Liver	0.13	**0.31**	0.16	**0.38**	0.18	**0.44**	0.20	**0.48**	0.24	**0.56**	0.22	**0.53**
Gall bladder	0.01	**0.08**	0.02	**0.11**	0.02	**0.13**	0.03	**0.15**	0.03	**0.19**	0.03	**0.16**
Pancreas	0.03	**0.04**	0.04	**0.06**	0.05	**0.07**	0.05	**0.08**	0.07	**0.11**	0.06	**0.09**
Small intestine		**0.03**		**0.04**		**0.05**		**0.06**		**0.08**		**0.07**
Large intestine		**0.06**		**0.07**		**0.09**		**0.10**		**0.12**		**0.11**
Sigma/Rectum												
Lumbar spine	0.03	**0.13**	0.04	**0.17**	0.05	**0.21**	0.06	**0.24**	0.07	**0.31**	0.06	**0.26**
Pelvis		**0.01**		**0.02**		**0.02**		**0.03**		**0.04**		**0.03**
Adrenals	0.07	**0.07**	0.09	**0.10**	0.11	**0.12**	0.12	**0.14**	0.15	**0.18**	0.13	**0.15**
Kidneys	0.04	**0.12**	0.06	**0.16**	0.07	**0.18**	0.08	**0.21**	0.10	**0.25**	0.09	**0.23**
Urinary bladder												
Uterus												
Ovaries										**0.01**		
Prostate												
Testes												
Upper arm												
Elbow												
Forearm												
Upper leg												
Knee												
Lower leg												
Lymph nodes	0.02	**0.07**	0.03	**0.10**	0.04	**0.11**	0.04	**0.13**	0.05	**0.15**	0.05	**0.14**
Skin	0.05	**0.08**	0.06	**0.09**	0.06	**0.09**	0.06	**0.09**	0.06	**0.10**	0.06	**0.10**
Muscle	0.03	**0.05**	0.03	**0.06**	0.04	**0.07**	0.04	**0.08**	0.05	**0.09**	0.05	**0.09**
Skeleton	0.09	**0.11**	0.10	**0.14**	0.12	**0.16**	0.13	**0.17**	0.15	**0.20**	0.14	**0.19**
Red bone marrow	0.03	**0.05**	0.04	**0.06**	0.05	**0.07**	0.05	**0.07**	0.06	**0.09**	0.06	**0.08**
Whole body dose	0.04	**0.07**	0.05	**0.08**	0.05	**0.09**	0.06	**0.10**	0.07	**0.12**	0.06	**0.11**
ED (ICRP 60)	0.03	**0.07**	0.04	**0.09**	0.05	**0.11**	0.06	**0.12**	0.07	**0.14**	0.06	**0.13**
ED (ICRP 103)	0.03	**0.07**	0.04	**0.09**	0.05	**0.10**	0.06	**0.12**	0.07	**0.14**	0.06	**0.13**
Total filtration (mm) Al	**1.5**		**2.5**		**3.5**		**4.5**		**3.5**		**3.5**	
Cu	**0**		**0**		**0**		**0**		**0.1**		**0.2**	

Table 15.63 Normalized organ doses—Thoracic spine r-lat—8–12 years

Beam projection:			r-lat									**Table cells:**
Focus-to-detector distance (cm):			150									(+) = normal font = optimal field setting
Tube voltage (kVp):			70									**(–) = bold font = suboptimal field setting**
Total filtrations:			See last table row.									Empty table cells correspond to values <0.01
Organ or tissue	**Organ dose D_T : Entrance dose K_E**								**(mSv * mGy^{-1})**			
	(+)	(–)	(+)	(–)	(+)	(–)	(+)	(–)	(+)	(–)	(+)	(–)
Skull		**0.01**		**0.02**		**0.02**		**0.02**		**0.03**	0.01	**0.03**
Brain												
Oral mucosa		**0.01**		**0.02**		**0.02**		**0.02**		**0.03**		**0.03**
Salivary glands		**0.03**		**0.03**		**0.04**		**0.04**	0.01	**0.05**	0.01	**0.06**
Cervical spine	0.06	**0.34**	0.08	**0.43**	0.09	**0.50**	0.11	**0.55**	0.13	**0.66**	0.16	**0.75**
Thyroid	0.02	**0.30**	0.02	**0.36**	0.03	**0.41**	0.03	**0.44**	0.04	**0.50**	0.04	**0.56**
Thymus	0.02	**0.04**	0.03	**0.05**	0.03	**0.07**	0.03	**0.08**	0.04	**0.09**	0.05	**0.11**
Thoracic spine	0.21	**0.23**	0.28	**0.31**	0.33	**0.37**	0.38	**0.42**	0.48	**0.53**	0.58	**0.64**
Claviculae	0.04	**0.26**	0.05	**0.32**	0.06	**0.37**	0.06	**0.40**	0.08	**0.47**	0.10	**0.53**
Scapulae	1.04	**1.03**	1.21	**1.21**	1.33	**1.33**	1.41	**1.41**	1.55	**1.56**	1.66	**1.68**
Ribs	0.51	**0.69**	0.61	**0.80**	0.67	**0.89**	0.72	**0.95**	0.80	**1.06**	0.87	**1.15**
Oesophagus	0.09	**0.11**	0.12	**0.14**	0.14	**0.17**	0.16	**0.19**	0.20	**0.23**	0.24	**0.28**
Trachea		**0.12**	0.01	**0.15**	0.01	**0.16**	0.01	**0.18**	0.02	**0.21**	0.02	**0.23**
Lung	0.17	**0.23**	0.21	**0.29**	0.24	**0.33**	0.26	**0.37**	0.31	**0.43**	0.35	**0.49**
Mamma	0.01	**0.02**	0.01	**0.03**	0.02	**0.03**	0.02	**0.04**	0.02	**0.05**	0.03	**0.06**
Heart	0.05	**0.11**	0.07	**0.14**	0.08	**0.16**	0.09	**0.18**	0.11	**0.23**	0.13	**0.27**
Stomach		**0.02**	0.01	**0.03**	0.01	**0.04**	0.02	**0.04**	0.02	**0.06**	0.03	**0.07**
Spleen	0.01	**0.02**	0.02	**0.02**	0.02	**0.03**	0.02	**0.04**	0.03	**0.05**	0.04	**0.06**
Liver	0.15	**0.36**	0.18	**0.44**	0.21	**0.50**	0.23	**0.54**	0.27	**0.63**	0.31	**0.71**
Gall bladder	0.02	**0.11**	0.03	**0.14**	0.03	**0.17**	0.04	**0.19**	0.04	**0.23**	0.05	**0.28**
Pancreas	0.04	**0.06**	0.05	**0.08**	0.06	**0.10**	0.07	**0.11**	0.09	**0.14**	0.12	**0.17**
Small intestine		**0.04**		**0.06**		**0.07**		**0.08**		**0.10**		**0.11**
Large intestine		**0.07**		**0.09**		**0.11**		**0.12**		**0.15**	0.01	**0.18**
Sigma/Rectum												**0.01**
Lumbar spine	0.04	**0.17**	0.05	**0.23**	0.07	**0.28**	0.08	**0.31**	0.10	**0.40**	0.12	**0.48**
Pelvis		**0.02**		**0.03**		**0.03**		**0.04**		**0.05**		**0.06**
Adrenals	0.09	**0.10**	0.12	**0.13**	0.14	**0.16**	0.16	**0.18**	0.19	**0.22**	0.23	**0.27**
Kidneys	0.06	**0.15**	0.07	**0.19**	0.09	**0.22**	0.10	**0.25**	0.12	**0.30**	0.14	**0.34**
Urinary bladder												
Uterus								**0.01**		**0.01**		**0.02**
Ovaries						**0.01**		**0.01**		**0.02**		**0.02**
Prostate												
Testes												
Upper arm												
Elbow												
Forearm												
Upper leg												
Knee												
Lower leg												
Lymph nodes	0.03	**0.09**	0.04	**0.12**	0.05	**0.14**	0.05	**0.15**	0.06	**0.18**	0.08	**0.21**
Skin	0.05	**0.08**	0.06	**0.09**	0.06	**0.10**	0.06	**0.10**	0.07	**0.10**	0.07	**0.11**
Muscle	0.03	**0.06**	0.04	**0.07**	0.04	**0.08**	0.05	**0.09**	0.06	**0.11**	0.06	**0.12**
Skeleton	0.10	**0.13**	0.12	**0.16**	0.13	**0.18**	0.14	**0.20**	0.16	**0.23**	0.18	**0.25**
Red bone marrow	0.04	**0.06**	0.05	**0.07**	0.06	**0.08**	0.06	**0.09**	0.07	**0.10**	0.09	**0.12**
Whole body dose	0.05	**0.08**	0.05	**0.10**	0.06	**0.11**	0.07	**0.12**	0.08	**0.14**	0.09	**0.15**
ED (ICRP 60)	0.04	**0.09**	0.05	**0.11**	0.06	**0.12**	0.07	**0.14**	0.08	**0.16**	0.09	**0.19**
ED (ICRP 103)	0.04	**0.09**	0.05	**0.11**	0.06	**0.12**	0.07	**0.14**	0.08	**0.16**	0.09	**0.19**
Total filtration (mm) Al	**1.5**		**2.5**		**3.5**		**4.5**		**3.5**		**3.5**	
Cu	**0**		**0**		**0**		**0**		**0.1**		**0.2**	

Table 15.64 Normalized organ doses—Thoracic spine r-lat—8–12 years

Beam projection:			r-lat									**Table cells:**
Focus-to-detector distance (cm):			150						(+) = normal font = optimal field setting			
Tube voltage (kVp):			80						**(−) = bold font = suboptimal field setting**			
Total filtrations:			See last table row.						Empty table cells correspond to values <0.01			
Organ or tissue	**Organ dose D_T : Entrance dose K_E**								**(mSv * mGy^{-1})**			
	(+)	(−)	(+)	(−)	(+)	(−)	(+)	(−)	(+)	(−)	(+)	(−)
Skull		**0.01**		**0.02**		**0.02**		**0.02**	0.01	**0.03**	0.01	**0.04**
Brain												
Oral mucosa		**0.01**		**0.02**		**0.02**		**0.02**		**0.03**	0.01	**0.04**
Salivary glands		**0.03**		**0.04**	0.01	**0.05**	0.01	**0.05**	0.01	**0.06**	0.02	**0.07**
Cervical spine	0.08	**0.40**	0.10	**0.49**	0.11	**0.56**	0.13	**0.61**	0.15	**0.72**	0.18	**0.81**
Thyroid	0.02	**0.34**	0.03	**0.40**	0.03	**0.45**	0.04	**0.48**	0.04	**0.54**	0.05	**0.60**
Thymus	0.03	**0.05**	0.03	**0.07**	0.04	**0.08**	0.04	**0.09**	0.05	**0.11**	0.06	**0.14**
Thoracic spine	0.27	**0.30**	0.35	**0.38**	0.41	**0.45**	0.46	**0.51**	0.56	**0.62**	0.66	**0.74**
Claviculae	0.05	**0.30**	0.06	**0.36**	0.07	**0.41**	0.08	**0.44**	0.09	**0.50**	0.11	**0.56**
Scapulae	1.11	**1.11**	1.28	**1.28**	1.38	**1.39**	1.46	**1.47**	1.59	**1.60**	1.67	**1.70**
Ribs	0.56	**0.74**	0.65	**0.86**	0.71	**0.94**	0.75	**1.00**	0.83	**1.10**	0.89	**1.18**
Oesophagus	0.12	**0.14**	0.15	**0.17**	0.17	**0.20**	0.19	**0.23**	0.23	**0.27**	0.27	**0.32**
Trachea	0.01	**0.14**	0.01	**0.16**	0.02	**0.18**	0.02	**0.20**	0.02	**0.23**	0.03	**0.25**
Lung	0.19	**0.27**	0.24	**0.33**	0.27	**0.38**	0.29	**0.41**	0.34	**0.48**	0.38	**0.53**
Mamma	0.01	**0.03**	0.02	**0.03**	0.02	**0.04**	0.02	**0.05**	0.03	**0.06**	0.04	**0.07**
Heart	0.07	**0.13**	0.08	**0.17**	0.10	**0.20**	0.11	**0.22**	0.13	**0.27**	0.16	**0.31**
Stomach	0.01	**0.03**	0.02	**0.04**	0.02	**0.05**	0.02	**0.06**	0.03	**0.07**	0.03	**0.09**
Spleen	0.02	**0.03**	0.02	**0.03**	0.03	**0.04**	0.03	**0.05**	0.04	**0.06**	0.05	**0.08**
Liver	0.17	**0.41**	0.21	**0.49**	0.24	**0.55**	0.26	**0.60**	0.30	**0.69**	0.33	**0.77**
Gall bladder	0.03	**0.14**	0.03	**0.17**	0.04	**0.20**	0.04	**0.23**	0.06	**0.28**	0.07	**0.33**
Pancreas	0.05	**0.08**	0.07	**0.10**	0.08	**0.12**	0.09	**0.14**	0.12	**0.18**	0.14	**0.21**
Small intestine		**0.06**		**0.07**		**0.08**		**0.09**		**0.11**	0.01	**0.13**
Large intestine		**0.09**		**0.11**		**0.13**		**0.15**	0.01	**0.18**	0.02	**0.20**
Sigma/Rectum										**0.01**		**0.01**
Lumbar spine	0.05	**0.22**	0.07	**0.29**	0.08	**0.34**	0.09	**0.38**	0.12	**0.47**	0.14	**0.56**
Pelvis		**0.03**		**0.03**		**0.04**		**0.05**		**0.06**		**0.07**
Adrenals	0.11	**0.13**	0.14	**0.17**	0.17	**0.19**	0.19	**0.22**	0.23	**0.27**	0.27	**0.32**
Kidneys	0.07	**0.18**	0.09	**0.22**	0.10	**0.26**	0.11	**0.28**	0.13	**0.34**	0.16	**0.38**
Urinary bladder												
Uterus				**0.01**		**0.01**		**0.02**		**0.02**		**0.02**
Ovaries				**0.01**		**0.02**		**0.02**		**0.02**		**0.03**
Prostate												
Testes												
Upper arm												
Elbow												
Forearm												
Upper leg												
Knee												
Lower leg												
Lymph nodes	0.04	**0.11**	0.05	**0.14**	0.06	**0.16**	0.06	**0.18**	0.08	**0.21**	0.09	**0.24**
Skin	0.06	**0.09**	0.06	**0.10**	0.06	**0.10**	0.07	**0.10**	0.07	**0.11**	0.07	**0.12**
Muscle	0.04	**0.07**	0.05	**0.08**	0.05	**0.09**	0.05	**0.10**	0.06	**0.12**	0.07	**0.13**
Skeleton	0.11	**0.15**	0.13	**0.18**	0.14	**0.20**	0.16	**0.21**	0.18	**0.24**	0.19	**0.27**
Red bone marrow	0.05	**0.07**	0.06	**0.08**	0.07	**0.09**	0.07	**0.10**	0.08	**0.12**	0.10	**0.14**
Whole body dose	0.05	**0.09**	0.06	**0.11**	0.07	**0.12**	0.08	**0.13**	0.09	**0.15**	0.10	**0.17**
ED (ICRP 60)	0.05	**0.10**	0.06	**0.13**	0.07	**0.14**	0.08	**0.16**	0.09	**0.18**	0.10	**0.21**
ED (ICRP 103)	0.05	**0.10**	0.06	**0.13**	0.07	**0.14**	0.08	**0.16**	0.09	**0.18**	0.10	**0.21**
Total filtration (mm) **Al**	**1.5**		**2.5**		**3.5**		**4.5**		**3.5**		**3.5**	
Cu	**0**		**0**		**0**		**0**		**0.1**		**0.2**	

Table 15.65 Normalized organ doses—Thoracic spine r-lat—13–17 years

Beam projection:			r-lat									**Table cells:**
Focus-to-detector distance (cm):			150				(+) = normal font = optimal field setting					
Tube voltage (kVp):			60				**(−) = bold font = suboptimal field setting**					
Total filtrations:			See last table row.				Empty table cells correspond to values <0.01					
Organ or tissue	**Organ dose D_T : Entrance dose K_E**								**(mSv ∗ mGy^{-1})**			
	(+)	(−)	(+)	(−)	(+)	(−)	(+)	(−)	(+)	(−)	(+)	(−)
Skull										**0.01**		
Brain												
Oral mucosa												
Salivary glands						**0.01**		**0.01**		**0.01**		**0.01**
Cervical spine	0.10	**0.16**	0.13	**0.22**	0.15	**0.26**	0.17	**0.29**	0.21	**0.36**	0.19	**0.32**
Thyroid	0.02	**0.15**	0.02	**0.19**	0.03	**0.22**	0.03	**0.24**	0.04	**0.28**	0.03	**0.27**
Thymus		**0.02**	0.01	**0.03**	0.02	**0.03**	0.02	**0.04**	0.03	**0.05**	0.02	**0.04**
Thoracic spine	0.08	**0.09**	0.11	**0.12**	0.14	**0.15**	0.16	**0.17**	0.21	**0.23**	0.17	**0.18**
Claviculae	0.02	**0.18**	0.03	**0.23**	0.04	**0.27**	0.05	**0.30**	0.06	**0.36**	0.05	**0.34**
Scapulae	0.81	**0.80**	0.98	**0.96**	1.09	**1.08**	1.18	**1.17**	1.34	**1.32**	1.32	**1.30**
Ribs	0.40	**0.52**	0.48	**0.63**	0.53	**0.71**	0.58	**0.76**	0.66	**0.87**	0.64	**0.85**
Oesophagus	0.05	**0.05**	0.06	**0.07**	0.08	**0.09**	0.09	**0.10**	0.11	**0.13**	0.10	**0.11**
Trachea	0.01	**0.04**	0.02	**0.06**	0.02	**0.07**	0.02	**0.07**	0.02	**0.09**	0.02	**0.08**
Lung	0.10	**0.14**	0.13	**0.18**	0.16	**0.22**	0.18	**0.24**	0.22	**0.29**	0.20	**0.27**
Mamma	0.04	**0.11**	0.06	**0.15**	0.07	**0.18**	0.09	**0.21**	0.11	**0.27**	0.09	**0.22**
Heart	0.03	**0.05**	0.04	**0.07**	0.04	**0.08**	0.05	**0.09**	0.07	**0.12**	0.05	**0.10**
Stomach				**0.01**		**0.01**		**0.02**		**0.02**		**0.01**
Spleen										**0.01**		
Liver	0.12	**0.28**	0.15	**0.35**	0.18	**0.40**	0.20	**0.45**	0.24	**0.53**	0.22	**0.50**
Gall bladder	0.01	**0.08**	0.02	**0.11**	0.02	**0.13**	0.02	**0.15**	0.03	**0.19**	0.02	**0.17**
Pancreas	0.01	**0.02**	0.02	**0.03**	0.03	**0.04**	0.03	**0.04**	0.04	**0.06**	0.03	**0.05**
Small intestine		**0.02**		**0.03**		**0.04**		**0.04**		**0.05**		**0.04**
Large intestine		**0.04**		**0.05**		**0.06**		**0.07**		**0.09**		**0.08**
Sigma/Rectum												
Lumbar spine	0.02	**0.07**	0.02	**0.09**	0.03	**0.11**	0.03	**0.13**	0.04	**0.17**	0.03	**0.14**
Pelvis				**0.01**		**0.02**		**0.02**		**0.02**		**0.02**
Adrenals	0.04	**0.05**	0.06	**0.06**	0.07	**0.08**	0.08	**0.09**	0.10	**0.11**	0.08	**0.09**
Kidneys	0.03	**0.08**	0.04	**0.11**	0.05	**0.13**	0.05	**0.14**	0.07	**0.18**	0.06	**0.16**
Urinary bladder												
Uterus						**0.01**		**0.01**		**0.02**		**0.01**
Ovaries												
Prostate												
Testes												
Upper arm												
Elbow												
Forearm												
Upper leg												
Knee												
Lower leg												
Lymph nodes	0.02	**0.05**	0.02	**0.06**	0.03	**0.07**	0.03	**0.08**	0.04	**0.10**	0.04	**0.09**
Skin	0.06	**0.10**	0.07	**0.10**	0.07	**0.11**	0.07	**0.11**	0.08	**0.12**	0.07	**0.12**
Muscle	0.02	**0.05**	0.03	**0.06**	0.04	**0.06**	0.04	**0.07**	0.05	**0.08**	0.04	**0.08**
Skeleton	0.08	**0.10**	0.09	**0.12**	0.11	**0.13**	0.12	**0.15**	0.13	**0.17**	0.13	**0.16**
Red bone marrow	0.03	**0.04**	0.04	**0.05**	0.05	**0.06**	0.05	**0.07**	0.06	**0.08**	0.06	**0.07**
Whole body dose	0.04	**0.06**	0.04	**0.07**	0.05	**0.08**	0.05	**0.09**	0.06	**0.11**	0.06	**0.10**
ED (ICRP 60)	0.03	**0.06**	0.04	**0.08**	0.05	**0.09**	0.05	**0.10**	0.06	**0.12**	0.06	**0.11**
ED (ICRP 103)	0.03	**0.06**	0.04	**0.08**	0.05	**0.10**	0.06	**0.11**	0.07	**0.14**	0.06	**0.12**
Total filtration (mm) Al	**1.5**		**2.5**		**3.5**		**4.5**		**3.5**		**3.5**	
Cu	**0**		**0**		**0**		**0**		**0.1**		**0.2**	

Table 15.66 Normalized organ doses—Thoracic spine r-lat—13–17 years

Beam projection:		r-lat										**Table cells:**
Focus-to-detector distance (cm):		150						(+) = normal font = optimal field setting				
Tube voltage (kVp):		70						**(−) = bold font = suboptimal field setting**				
Total filtrations:		See last table row.						Empty table cells correspond to values <0.01				
Organ or tissue	**Organ dose D_T : Entrance dose K_E**							**(mSv * mGy^{-1})**				
	(+)	(−)	(+)	(−)	(+)	(−)	(+)	(−)	(+)	(−)	(+)	(−)
Skull								**0.01**		**0.01**		**0.02**
Brain												
Oral mucosa												**0.01**
Salivary glands				**0.01**		**0.01**		**0.01**		**0.02**	0.01	**0.02**
Cervical spine	0.12	**0.21**	0.16	**0.27**	0.19	**0.31**	0.21	**0.35**	0.25	**0.42**	0.30	**0.49**
Thyroid	0.02	**0.18**	0.03	**0.22**	0.04	**0.25**	0.04	**0.28**	0.05	**0.32**	0.06	**0.36**
Thymus	0.01	**0.03**	0.02	**0.04**	0.02	**0.05**	0.03	**0.06**	0.04	**0.07**	0.05	**0.09**
Thoracic spine	0.12	**0.13**	0.16	**0.17**	0.19	**0.21**	0.22	**0.24**	0.28	**0.31**	0.35	**0.38**
Claviculae	0.03	**0.22**	0.04	**0.27**	0.05	**0.31**	0.06	**0.34**	0.07	**0.41**	0.09	**0.46**
Scapulae	0.91	**0.89**	1.07	**1.06**	1.18	**1.17**	1.27	**1.26**	1.42	**1.41**	1.53	**1.53**
Ribs	0.45	**0.59**	0.53	**0.70**	0.59	**0.77**	0.63	**0.83**	0.71	**0.94**	0.78	**1.02**
Oesophagus	0.07	**0.08**	0.09	**0.10**	0.10	**0.12**	0.12	**0.14**	0.15	**0.17**	0.18	**0.21**
Trachea	0.01	**0.05**	0.02	**0.07**	0.02	**0.08**	0.02	**0.08**	0.03	**0.10**	0.04	**0.11**
Lung	0.13	**0.18**	0.16	**0.22**	0.19	**0.26**	0.21	**0.29**	0.25	**0.35**	0.29	**0.40**
Mamma	0.06	**0.15**	0.09	**0.20**	0.10	**0.24**	0.12	**0.28**	0.16	**0.35**	0.19	**0.43**
Heart	0.04	**0.07**	0.05	**0.09**	0.06	**0.11**	0.07	**0.12**	0.09	**0.16**	0.10	**0.19**
Stomach		**0.01**		**0.02**		**0.02**		**0.02**	0.01	**0.03**	0.02	**0.04**
Spleen				**0.01**		**0.01**	0.01	**0.02**	0.01	**0.02**	0.02	**0.03**
Liver	0.15	**0.33**	0.18	**0.41**	0.21	**0.47**	0.23	**0.52**	0.27	**0.61**	0.31	**0.69**
Gall bladder	0.02	**0.11**	0.02	**0.15**	0.03	**0.17**	0.03	**0.20**	0.04	**0.25**	0.05	**0.30**
Pancreas	0.02	**0.03**	0.03	**0.05**	0.04	**0.06**	0.04	**0.06**	0.06	**0.08**	0.07	**0.10**
Small intestine		**0.03**		**0.04**		**0.05**		**0.05**		**0.07**		**0.08**
Large intestine		**0.05**		**0.07**		**0.08**		**0.09**		**0.11**		**0.13**
Sigma/Rectum												
Lumbar spine	0.02	**0.10**	0.03	**0.13**	0.04	**0.16**	0.04	**0.18**	0.06	**0.23**	0.07	**0.29**
Pelvis		**0.01**		**0.02**		**0.02**		**0.02**		**0.03**		**0.04**
Adrenals	0.06	**0.06**	0.08	**0.09**	0.09	**0.10**	0.10	**0.12**	0.13	**0.15**	0.16	**0.18**
Kidneys	0.04	**0.11**	0.05	**0.14**	0.06	**0.16**	0.07	**0.18**	0.08	**0.22**	0.10	**0.26**
Urinary bladder												
Uterus		**0.01**		**0.01**		**0.02**		**0.02**		**0.03**		**0.03**
Ovaries										**0.01**		**0.01**
Prostate												
Testes												
Upper arm												
Elbow												
Forearm												
Upper leg												
Knee												
Lower leg												
Lymph nodes	0.02	**0.06**	0.03	**0.08**	0.04	**0.09**	0.04	**0.11**	0.05	**0.13**	0.06	**0.15**
Skin	0.07	**0.10**	0.07	**0.11**	0.07	**0.11**	0.08	**0.12**	0.08	**0.12**	0.08	**0.13**
Muscle	0.03	**0.05**	0.04	**0.07**	0.04	**0.07**	0.05	**0.08**	0.05	**0.09**	0.06	**0.11**
Skeleton	0.09	**0.11**	0.11	**0.13**	0.12	**0.15**	0.13	**0.16**	0.15	**0.19**	0.17	**0.21**
Red bone marrow	0.04	**0.05**	0.05	**0.07**	0.06	**0.07**	0.06	**0.08**	0.07	**0.10**	0.08	**0.11**
Whole body dose	0.04	**0.07**	0.05	**0.08**	0.06	**0.09**	0.06	**0.10**	0.07	**0.12**	0.08	**0.14**
ED (ICRP 60)	0.04	**0.07**	0.05	**0.09**	0.06	**0.11**	0.06	**0.12**	0.08	**0.14**	0.09	**0.17**
ED (ICRP 103)	0.04	**0.08**	0.05	**0.10**	0.06	**0.12**	0.07	**0.14**	0.09	**0.17**	0.10	**0.19**
Total filtration (mm) **Al**	**1.5**		**2.5**		**3.5**		**4.5**		**3.5**		**3.5**	
Cu	**0**		**0**		**0**		**0**		**0.1**		**0.2**	

Table 15.67 Normalized organ doses—Thoracic spine r-lat—13–17 years

Beam projection:			r-lat									**Table cells:**
Focus-to-detector distance (cm):			150									(+) = normal font = optimal field setting
Tube voltage (kVp):			80									**(−) = bold font = suboptimal field setting**
Total filtrations:			See last table row.									Empty table cells correspond to values <0.01
Organ or tissue	**Organ dose D_T : Entrance dose K_E**								**(mSv ∗ mGy^{-1})**			
	(+)	(−)	(+)	(−)	(+)	(−)	(+)	(−)	(+)	(−)	(+)	(−)
Skull				**0.01**		**0.01**		**0.01**		**0.02**	0.01	**0.02**
Brain												
Oral mucosa										**0.01**		**0.02**
Salivary glands		**0.01**		**0.01**		**0.02**		**0.02**	0.01	**0.02**	0.01	**0.03**
Cervical spine	0.15	**0.25**	0.19	**0.31**	0.22	**0.36**	0.24	**0.40**	0.29	**0.48**	0.33	**0.55**
Thyroid	0.03	**0.21**	0.04	**0.25**	0.04	**0.28**	0.05	**0.31**	0.06	**0.35**	0.07	**0.39**
Thymus	0.02	**0.04**	0.03	**0.05**	0.03	**0.06**	0.04	**0.07**	0.05	**0.09**	0.06	**0.11**
Thoracic spine	0.16	**0.17**	0.21	**0.22**	0.24	**0.27**	0.28	**0.30**	0.35	**0.38**	0.42	**0.46**
Claviculae	0.04	**0.25**	0.05	**0.31**	0.06	**0.35**	0.07	**0.38**	0.09	**0.44**	0.10	**0.49**
Scapulae	0.98	**0.97**	1.14	**1.13**	1.25	**1.24**	1.33	**1.32**	1.46	**1.46**	1.56	**1.56**
Ribs	0.49	**0.65**	0.57	**0.75**	0.63	**0.83**	0.67	**0.88**	0.75	**0.98**	0.80	**1.06**
Oesophagus	0.09	**0.10**	0.11	**0.13**	0.13	**0.15**	0.15	**0.17**	0.18	**0.21**	0.22	**0.25**
Trachea	0.02	**0.06**	0.02	**0.08**	0.03	**0.09**	0.03	**0.09**	0.03	**0.11**	0.04	**0.12**
Lung	0.15	**0.21**	0.19	**0.26**	0.22	**0.30**	0.24	**0.33**	0.29	**0.39**	0.33	**0.44**
Mamma	0.09	**0.20**	0.12	**0.26**	0.14	**0.31**	0.16	**0.35**	0.20	**0.43**	0.24	**0.52**
Heart	0.05	**0.09**	0.06	**0.11**	0.08	**0.14**	0.08	**0.15**	0.10	**0.19**	0.13	**0.22**
Stomach		**0.02**		**0.02**	0.01	**0.03**	0.01	**0.03**	0.02	**0.04**	0.02	**0.05**
Spleen		**0.01**	0.01	**0.02**	0.01	**0.02**	0.02	**0.02**	0.02	**0.03**	0.03	**0.04**
Liver	0.17	**0.39**	0.21	**0.47**	0.24	**0.53**	0.26	**0.58**	0.30	**0.67**	0.34	**0.75**
Gall bladder	0.02	**0.14**	0.03	**0.18**	0.04	**0.21**	0.04	**0.24**	0.05	**0.29**	0.06	**0.35**
Pancreas	0.03	**0.05**	0.04	**0.06**	0.05	**0.07**	0.06	**0.09**	0.07	**0.11**	0.09	**0.13**
Small intestine		**0.04**		**0.05**		**0.06**		**0.07**		**0.08**		**0.10**
Large intestine		**0.07**		**0.08**		**0.10**		**0.11**		**0.13**		**0.16**
Sigma/Rectum												
Lumbar spine	0.03	**0.13**	0.04	**0.17**	0.05	**0.20**	0.06	**0.23**	0.07	**0.29**	0.09	**0.35**
Pelvis		**0.02**		**0.02**		**0.03**		**0.03**		**0.04**		**0.05**
Adrenals	0.08	**0.09**	0.10	**0.11**	0.11	**0.13**	0.13	**0.15**	0.16	**0.18**	0.19	**0.22**
Kidneys	0.05	**0.13**	0.06	**0.16**	0.07	**0.19**	0.08	**0.21**	0.10	**0.25**	0.11	**0.29**
Urinary bladder												
Uterus		**0.02**		**0.02**		**0.02**		**0.03**		**0.04**		**0.05**
Ovaries								**0.01**		**0.01**		**0.02**
Prostate												
Testes												
Upper arm												
Elbow												
Forearm												
Upper leg												
Knee												
Lower leg												
Lymph nodes	0.03	**0.08**	0.04	**0.10**	0.05	**0.11**	0.05	**0.13**	0.06	**0.15**	0.07	**0.18**
Skin	0.07	**0.11**	0.07	**0.11**	0.08	**0.12**	0.08	**0.12**	0.08	**0.13**	0.09	**0.14**
Muscle	0.03	**0.06**	0.04	**0.08**	0.05	**0.08**	0.05	**0.09**	0.06	**0.11**	0.07	**0.12**
Skeleton	0.10	**0.13**	0.12	**0.15**	0.13	**0.17**	0.14	**0.18**	0.16	**0.21**	0.18	**0.23**
Red bone marrow	0.05	**0.06**	0.06	**0.08**	0.07	**0.09**	0.07	**0.10**	0.08	**0.11**	0.10	**0.13**
Whole body dose	0.05	**0.08**	0.06	**0.10**	0.06	**0.11**	0.07	**0.12**	0.08	**0.13**	0.09	**0.15**
ED (ICRP 60)	0.05	**0.09**	0.06	**0.11**	0.07	**0.13**	0.07	**0.14**	0.09	**0.17**	0.10	**0.19**
ED (ICRP 103)	0.05	**0.10**	0.06	**0.13**	0.08	**0.15**	0.08	**0.16**	0.10	**0.19**	0.12	**0.22**
Total filtration (mm) **Al**	**1.5**		**2.5**		**3.5**		**4.5**		**3.5**		**3.5**	
Cu	**0**		**0**		**0**		**0**		**0.1**		**0.2**	

Table 15.68 Normalized organ doses—Thoracic spine r-lat—>17 years

Beam projection:			r-lat									**Table cells:**
Focus-to-detector distance (cm):			150									(+) = normal font = optimal field setting
Tube voltage (kVp):			60									**(−) = bold font = suboptimal field setting**
Total filtrations:			See last table row.									Empty table cells correspond to values <0.01
Organ or tissue	**Organ dose D_T : Entrance dose K_E**								**(mSv * mGy^{-1})**			
	(+)	(−)	(+)	(−)	(+)	(−)	(+)	(−)	(+)	(−)	(+)	(−)
Skull										**0.01**		
Brain												
Oral mucosa												
Salivary glands						**0.01**		**0.01**		**0.02**		**0.01**
Cervical spine	0.04	**0.17**	0.06	**0.22**	0.07	**0.26**	0.08	**0.30**	0.10	**0.37**	0.08	**0.33**
Thyroid	0.01	**0.12**	0.02	**0.16**	0.02	**0.18**	0.02	**0.20**	0.03	**0.24**	0.02	**0.22**
Thymus		**0.02**	0.01	**0.03**	0.01	**0.04**	0.02	**0.04**	0.02	**0.06**	0.02	**0.05**
Thoracic spine	0.05	**0.06**	0.08	**0.08**	0.09	**0.10**	0.11	**0.12**	0.14	**0.16**	0.11	**0.12**
Claviculae	0.02	**0.17**	0.03	**0.23**	0.03	**0.26**	0.04	**0.30**	0.05	**0.36**	0.04	**0.33**
Scapulae	0.74	**0.71**	0.89	**0.87**	1.00	**0.98**	1.09	**1.06**	1.24	**1.21**	1.22	**1.19**
Ribs	0.32	**0.48**	0.39	**0.58**	0.43	**0.65**	0.47	**0.71**	0.54	**0.81**	0.52	**0.79**
Oesophagus	0.04	**0.04**	0.05	**0.06**	0.06	**0.07**	0.07	**0.08**	0.09	**0.10**	0.07	**0.08**
Trachea		**0.05**		**0.07**	0.01	**0.08**	0.01	**0.08**	0.02	**0.10**	0.01	**0.09**
Lung	0.09	**0.12**	0.11	**0.16**	0.13	**0.19**	0.15	**0.21**	0.18	**0.26**	0.16	**0.23**
Mamma		**0.02**		**0.02**		**0.02**		**0.03**	0.01	**0.04**		**0.03**
Heart	0.02	**0.04**	0.03	**0.05**	0.04	**0.06**	0.04	**0.07**	0.05	**0.10**	0.04	**0.08**
Stomach										**0.01**		
Spleen												
Liver	0.07	**0.24**	0.09	**0.30**	0.11	**0.35**	0.12	**0.39**	0.14	**0.47**	0.13	**0.43**
Gall bladder		**0.05**		**0.07**		**0.08**		**0.10**	0.01	**0.13**		**0.10**
Pancreas		**0.01**	0.01	**0.02**	0.01	**0.02**	0.02	**0.03**	0.02	**0.04**	0.02	**0.03**
Small intestine		**0.02**		**0.02**		**0.03**		**0.03**		**0.04**		**0.03**
Large intestine		**0.03**		**0.04**		**0.05**		**0.06**		**0.07**		**0.06**
Sigma/Rectum												
Lumbar spine		**0.04**		**0.06**		**0.07**		**0.08**	0.01	**0.11**		**0.08**
Pelvis						**0.01**		**0.01**		**0.02**		**0.01**
Adrenals	0.03	**0.03**	0.04	**0.04**	0.04	**0.05**	0.05	**0.06**	0.07	**0.08**	0.05	**0.07**
Kidneys		**0.06**	0.01	**0.08**	0.02	**0.10**	0.02	**0.11**	0.02	**0.14**	0.02	**0.12**
Urinary bladder												
Uterus												
Ovaries												
Prostate												
Testes												
Upper arm												
Elbow												
Forearm												
Upper leg												
Knee												
Lower leg												
Lymph nodes	0.01	**0.04**	0.02	**0.05**	0.02	**0.06**	0.02	**0.07**	0.03	**0.09**	0.03	**0.07**
Skin	0.04	**0.08**	0.05	**0.09**	0.05	**0.09**	0.05	**0.09**	0.06	**0.10**	0.05	**0.10**
Muscle	0.02	**0.04**	0.02	**0.05**	0.03	**0.06**	0.03	**0.06**	0.04	**0.07**	0.03	**0.07**
Skeleton	0.06	**0.08**	0.07	**0.10**	0.08	**0.12**	0.09	**0.13**	0.11	**0.15**	0.10	**0.14**
Red bone marrow	0.03	**0.04**	0.04	**0.05**	0.04	**0.06**	0.05	**0.07**	0.06	**0.08**	0.05	**0.07**
Whole body dose	0.03	**0.05**	0.03	**0.07**	0.04	**0.07**	0.04	**0.08**	0.05	**0.10**	0.05	**0.09**
ED (ICRP 60)	0.02	**0.05**	0.03	**0.06**	0.03	**0.07**	0.04	**0.08**	0.04	**0.09**	0.04	**0.08**
ED (ICRP 103)	0.02	**0.04**	0.03	**0.06**	0.03	**0.07**	0.04	**0.08**	0.04	**0.09**	0.04	**0.08**
Total filtration (mm) Al	**1.5**		**2.5**		**3.5**		**4.5**		**3.5**		**3.5**	
Cu	**0**		**0**		**0**		**0**		**0.1**		**0.2**	

Table 15.69 Normalized organ doses—Thoracic spine r-lat—>17 years

Beam projection:			r-lat									**Table cells:**
Focus-to-detector distance (cm):			150					(+) = normal font = optimal field setting				
Tube voltage (kVp):			70					**(−) = bold font = suboptimal field setting**				
Total filtrations:			See last table row.					Empty table cells correspond to values <0.01				
Organ or tissue	**Organ dose D_T : Entrance dose K_E**								**(mSv ∗ mGy⁻¹)**			
	(+)	(−)	(+)	(−)	(+)	(−)	(+)	(−)	(+)	(−)	(+)	(−)
Skull								**0.01**		**0.01**		**0.02**
Brain												
Oral mucosa										**0.01**		**0.01**
Salivary glands				**0.01**		**0.02**		**0.02**		**0.02**		**0.03**
Cervical spine	0.06	**0.21**	0.07	**0.27**	0.09	**0.32**	0.10	**0.36**	0.12	**0.44**	0.14	**0.51**
Thyroid	0.02	**0.15**	0.02	**0.18**	0.02	**0.21**	0.03	**0.23**	0.03	**0.27**	0.04	**0.30**
Thymus	0.01	**0.03**	0.02	**0.04**	0.02	**0.05**	0.02	**0.06**	0.03	**0.08**	0.04	**0.10**
Thoracic spine	0.08	**0.09**	0.11	**0.12**	0.13	**0.15**	0.15	**0.17**	0.20	**0.22**	0.24	**0.27**
Claviculae	0.03	**0.21**	0.04	**0.27**	0.05	**0.31**	0.05	**0.35**	0.07	**0.41**	0.08	**0.47**
Scapulae	0.83	**0.80**	0.98	**0.96**	1.09	**1.07**	1.18	**1.16**	1.32	**1.30**	1.44	**1.42**
Ribs	0.36	**0.54**	0.43	**0.65**	0.48	**0.72**	0.52	**0.78**	0.59	**0.88**	0.65	**0.97**
Oesophagus	0.05	**0.06**	0.07	**0.08**	0.08	**0.10**	0.09	**0.11**	0.12	**0.14**	0.15	**0.17**
Trachea		**0.06**	0.01	**0.08**	0.01	**0.09**	0.02	**0.10**	0.02	**0.12**	0.02	**0.13**
Lung	0.11	**0.15**	0.14	**0.19**	0.16	**0.23**	0.18	**0.25**	0.21	**0.30**	0.25	**0.35**
Mamma		**0.02**		**0.03**	0.01	**0.03**	0.01	**0.04**	0.02	**0.05**	0.02	**0.06**
Heart	0.03	**0.05**	0.04	**0.07**	0.05	**0.09**	0.06	**0.10**	0.07	**0.13**	0.09	**0.15**
Stomach						**0.01**		**0.01**		**0.02**		**0.02**
Spleen										**0.01**		**0.02**
Liver	0.09	**0.29**	0.11	**0.36**	0.13	**0.41**	0.14	**0.46**	0.17	**0.54**	0.20	**0.62**
Gall bladder		**0.07**		**0.10**	0.01	**0.12**	0.01	**0.13**	0.02	**0.17**	0.02	**0.21**
Pancreas	0.01	**0.02**	0.02	**0.03**	0.02	**0.04**	0.02	**0.04**	0.03	**0.06**	0.04	**0.07**
Small intestine		**0.02**		**0.03**		**0.04**		**0.04**		**0.05**		**0.07**
Large intestine		**0.04**		**0.05**		**0.06**		**0.07**		**0.09**		**0.11**
Sigma/Rectum												
Lumbar spine		**0.06**		**0.08**	0.01	**0.10**	0.01	**0.12**	0.02	**0.15**	0.02	**0.19**
Pelvis		**0.01**		**0.01**		**0.02**		**0.02**		**0.03**		**0.03**
Adrenals	0.04	**0.05**	0.05	**0.06**	0.06	**0.08**	0.07	**0.09**	0.09	**0.11**	0.11	**0.14**
Kidneys	0.01	**0.08**	0.02	**0.10**	0.02	**0.12**	0.02	**0.14**	0.03	**0.17**	0.04	**0.20**
Urinary bladder												
Uterus												
Ovaries												**0.01**
Prostate												
Testes												
Upper arm												
Elbow												
Forearm												
Upper leg												
Knee												
Lower leg												
Lymph nodes	0.02	**0.05**	0.02	**0.07**	0.03	**0.08**	0.03	**0.09**	0.04	**0.11**	0.04	**0.13**
Skin	0.05	**0.08**	0.05	**0.09**	0.05	**0.09**	0.06	**0.10**	0.06	**0.10**	0.06	**0.11**
Muscle	0.02	**0.05**	0.03	**0.06**	0.03	**0.07**	0.04	**0.07**	0.04	**0.09**	0.05	**0.10**
Skeleton	0.07	**0.10**	0.08	**0.12**	0.10	**0.13**	0.10	**0.15**	0.12	**0.17**	0.13	**0.19**
Red bone marrow	0.04	**0.05**	0.04	**0.06**	0.05	**0.07**	0.06	**0.08**	0.07	**0.10**	0.08	**0.11**
Whole body dose	0.03	**0.06**	0.04	**0.08**	0.04	**0.09**	0.05	**0.10**	0.06	**0.11**	0.06	**0.13**
ED (ICRP 60)	0.03	**0.06**	0.03	**0.07**	0.04	**0.08**	0.04	**0.09**	0.05	**0.11**	0.06	**0.13**
ED (ICRP 103)	0.03	**0.06**	0.03	**0.07**	0.04	**0.08**	0.04	**0.09**	0.05	**0.11**	0.06	**0.13**
Total filtration (mm) Al	**1.5**		**2.5**		**3.5**		**4.5**		**3.5**		**3.5**	
Cu	**0**		**0**		**0**		**0**		**0.1**		**0.2**	

Table 15.70 Normalized organ doses—Thoracic spine r-lat—>17 years

Beam projection:			r-lat									**Table cells:**
Focus-to-detector distance (cm):			150									(+) = normal font = optimal field setting
Tube voltage (kVp):			80									**(−) = bold font = suboptimal field setting**
Total filtrations:			See last table row.									Empty table cells correspond to values <0.01
Organ or tissue	**Organ dose D_T : Entrance dose K_E**							**(mSv ∗ mGy^{-1})**				
	(+)	(−)	(+)	(−)	(+)	(−)	(+)	(−)	(+)	(−)	(+)	(−)
Skull				**0.01**		**0.01**		**0.01**		**0.02**		**0.02**
Brain												
Oral mucosa								**0.01**		**0.01**		**0.02**
Salivary glands		**0.01**		**0.02**		**0.02**		**0.02**		**0.03**	0.01	**0.03**
Cervical spine	0.07	**0.26**	0.09	**0.32**	0.10	**0.37**	0.12	**0.41**	0.14	**0.49**	0.17	**0.57**
Thyroid	0.02	**0.17**	0.03	**0.21**	0.03	**0.24**	0.03	**0.26**	0.04	**0.30**	0.05	**0.33**
Thymus	0.02	**0.05**	0.02	**0.06**	0.03	**0.07**	0.03	**0.08**	0.04	**0.10**	0.05	**0.12**
Thoracic spine	0.11	**0.12**	0.14	**0.16**	0.17	**0.19**	0.20	**0.22**	0.25	**0.28**	0.30	**0.34**
Claviculae	0.04	**0.25**	0.05	**0.31**	0.06	**0.35**	0.06	**0.39**	0.08	**0.45**	0.09	**0.51**
Scapulae	0.90	**0.88**	1.06	**1.04**	1.16	**1.14**	1.24	**1.22**	1.37	**1.36**	1.47	**1.46**
Ribs	0.40	**0.60**	0.47	**0.70**	0.52	**0.77**	0.55	**0.83**	0.62	**0.93**	0.67	**1.01**
Oesophagus	0.07	**0.08**	0.09	**0.11**	0.10	**0.12**	0.12	**0.14**	0.15	**0.18**	0.18	**0.21**
Trachea	0.01	**0.07**	0.01	**0.09**	0.02	**0.10**	0.02	**0.11**	0.02	**0.13**	0.03	**0.14**
Lung	0.13	**0.18**	0.16	**0.23**	0.18	**0.26**	0.20	**0.29**	0.24	**0.34**	0.28	**0.39**
Mamma		**0.03**	0.01	**0.03**	0.01	**0.04**	0.02	**0.05**	0.02	**0.06**	0.03	**0.07**
Heart	0.04	**0.07**	0.05	**0.09**	0.06	**0.11**	0.07	**0.13**	0.09	**0.16**	0.10	**0.19**
Stomach		**0.01**		**0.01**		**0.02**		**0.02**		**0.03**	0.01	**0.03**
Spleen						**0.01**		**0.01**	0.01	**0.02**	0.01	**0.02**
Liver	0.11	**0.34**	0.13	**0.42**	0.15	**0.47**	0.16	**0.52**	0.19	**0.61**	0.22	**0.69**
Gall bladder		**0.10**	0.01	**0.12**	0.02	**0.15**	0.02	**0.17**	0.02	**0.21**	0.03	**0.25**
Pancreas	0.02	**0.03**	0.02	**0.04**	0.03	**0.05**	0.03	**0.06**	0.04	**0.07**	0.05	**0.09**
Small intestine		**0.03**		**0.04**		**0.05**		**0.05**		**0.07**		**0.08**
Large intestine		**0.05**		**0.07**		**0.08**		**0.09**		**0.11**		**0.13**
Sigma/Rectum												
Lumbar spine		**0.08**	0.01	**0.11**	0.02	**0.13**	0.02	**0.15**	0.02	**0.19**	0.03	**0.23**
Pelvis		**0.01**		**0.02**		**0.02**		**0.03**		**0.03**		**0.04**
Adrenals	0.05	**0.06**	0.07	**0.08**	0.08	**0.10**	0.09	**0.11**	0.11	**0.14**	0.13	**0.17**
Kidneys	0.02	**0.10**	0.02	**0.13**	0.03	**0.15**	0.03	**0.16**	0.04	**0.20**	0.04	**0.23**
Urinary bladder												
Uterus										**0.01**		**0.01**
Ovaries										**0.01**		**0.01**
Prostate												
Testes												
Upper arm												
Elbow												
Forearm												
Upper leg												
Knee												
Lower leg												
Lymph nodes	0.02	**0.06**	0.03	**0.08**	0.03	**0.09**	0.04	**0.10**	0.04	**0.13**	0.05	**0.15**
Skin	0.05	**0.09**	0.05	**0.09**	0.06	**0.10**	0.06	**0.10**	0.06	**0.11**	0.06	**0.12**
Muscle	0.03	**0.06**	0.03	**0.07**	0.04	**0.08**	0.04	**0.08**	0.05	**0.10**	0.05	**0.11**
Skeleton	0.08	**0.11**	0.09	**0.13**	0.11	**0.15**	0.11	**0.16**	0.13	**0.18**	0.14	**0.20**
Red bone marrow	0.04	**0.06**	0.05	**0.08**	0.06	**0.09**	0.06	**0.09**	0.08	**0.11**	0.09	**0.13**
Whole body dose	0.04	**0.07**	0.05	**0.09**	0.05	**0.10**	0.06	**0.11**	0.06	**0.12**	0.07	**0.14**
ED (ICRP 60)	0.03	**0.07**	0.04	**0.09**	0.05	**0.10**	0.05	**0.11**	0.06	**0.13**	0.07	**0.15**
ED (ICRP 103)	0.03	**0.07**	0.04	**0.09**	0.05	**0.10**	0.05	**0.11**	0.06	**0.13**	0.07	**0.15**
Total filtration (mm) Al	**1.5**		**2.5**		**3.5**		**4.5**		**3.5**		**3.5**	
Cu	**0**		**0**		**0**		**0**		**0.1**		**0.2**	

References

1. Bontrager KL. Textbook of radiographic positioning and related anatomy. St. Louis, MO: Mosby Year Book; 1993.
2. Commission of the European Communities. European guidelines on quality criteria for diagnostic radiographic images in paediatrics. EUR 16261, SBN 92-827-7843-6; 1996.
3. Drexler G, Panzer W, Stieve FE. Die Bestimmung von Organdosen in der Röntgendiagnostik. Berlin: Hoffmann Verlag; 1993.
4. Ebel K-D, Willich E. Die Röntgenuntersuchung im Kindesalter—Technik und Indikation. Berlin: Springer Verlag; 1968.
5. Hart D, Jones DJ, Wall BF. Estimation of effective dose in diagnostic radiology from entrance surface dose and dose-area-product measurements. Oxon: National Radiological Protection Board NRPB-R262; 1994.
6. Hart D, Jones DJ, Wall BF. Coefficients for estimating effective doses from pediatric x-ray examinations. Oxon: National Radiological Protection Board NRPB-R279; 1996.
7. Seidenbusch MC, Schneider K. Conversion coefficients for determining organ doses in paediatric spine radiography. Pediatr Radiol. 2014;44:434–56.

16 Lumbar Spine

16.1 Introduction

The type of lumbar spine radiography is determined by the clinical indication. Table 16.1 shows indications and Table 16.2 contraindications for lumbar spine radiography. Image criteria as required for radiological findings in lumbar spine radiographs are listed in Table 16.3. In order to fulfil these image criteria, the field settings listed in Tables 16.4 and 16.5 should be achieved. These field settings go along with the patient preparation as listed in Table 16.6. Exposure parameter combinations for various age groups and patient diameters are listed in Tables 16.7, 16.8, 16.9, 16.10, and 16.11. The age groups in Tables 16.8 and 16.10 refer to the standard age groups of the mathematical MIRD phantom family which has been used to calculate the normalized organ doses in this book. Tables 16.9 and 16.11 show radiation qualities (i.e. various combinations of tube voltage and total filtration) which may be applied in clinical practice. Please note that not all of these radiation qualities are recommended by the authors. The standard projections of lumbar spine radiographies of the patient who is correctly positioned are depicted in Figs. 16.1, 16.2, and 16.3. Also the optimal field (marked in green), the suboptimal field (marked in red), and the correct shielding are shown. Table 16.12, 16.13, 16.14, 16.15, 16.16, and 16.17 outline the virtual field settings on the skin surface of mathematical MIRD phantoms of various age groups and the resulting virtual radiographies with and without bone structures, respectively. Normalized organ doses resulting from these virtual field settings have been calculated for all age groups and have been listed in Tables 16.18, 16.19, 16.20, 16.21, 16.22, 16.23, 16.24, 16.25, 16.26, 16.27, 16.28, 16.29, 16.30, 16.31, 16.32, 16.33, 16.34, 16.35, 16.36, 16.37, 16.38, 16.39, 16.40, 16.41, 16.42, 16.43,

Table 16.1 Indications for lumbar spine radiographs

Trauma
Lumbo-sacral spinal dysrhaphia
Suspected vertebral anomalies
Congenital scoliosis
Urogenital/anorectal malformation
Spondylitis/discitis
Langerhans cell histiocytosis
Unclear low back pain
Juvenile rheumatoid arthritis
Spondylolisthesis
Bone tumours
Soft tissue tumours

Table 16.2 Contraindications for lumbar spine radiographs

Pregnancy

M. Seidenbusch et al., *Imaging Practice and Radiation Protection in Pediatric Radiology*,
https://doi.org/10.1007/978-3-030-18504-6_16

Table 16.3 Image criteria for lumbar spine radiographs

Lumbar spine ap
Symmetrical reproduction of the entire lumbar spine
Visually sharp reproduction of the vertebral bodies and the articular processes consistent with age
Visualization of the spinous and transverse processes consistent with age
Visualization of the intervertebral spaces in the centre of the beam area consistent with age
Visually sharp reproduction of the cortex and trabecular structures of the vertebral bodies consistent with age
Reproduction of the adjacent soft tissues
Lumbar spine lat
Full superimposition of the margins of the vertebral bodies
Reproduction of the pedicles and the vertebral bodies
Visualization of the posterior articular processes
Visualization of the spinous processes consistent with age
Visually sharp reproduction of the cortex and trabecular structures of the vertebral bodies consistent with age
Reproduction of the symphysis pubis
Reproduction of the adjacent soft tissues

Table 16.4 Patient positions and beam projections for lumbar spine radiographs

Patient position	Beam projections
Standing, sitting	ap + lat
Supine	ap + lat[a]
Decubitus left	lat

Annotation:
[a]Lateral projection in emergency cases only

Table 16.5 Optimal field settings for lumbar spine radiographs

Lumbar spine ap	
Upper field edge	Processus xiphoideus
Field centring	Centring point at the lower third between the xiphoid and the navel
Lower field edge	Midpoint between the navel and the symphysis pubis
Lateral field edges	Midpoints of parasternal and medioclavicular lines
Lumbar spine lat	
Upper field edge	11th thoracic vertebra (T11)
Field centring	Posterior axillary line at the level of the twelfth rib
Lower field edge	Second sacral vertebra (S2)
Anterior field edge	Anterior axillary line
Posterior field edge	Skin surface

Table 16.6 Patient preparation

General	Remove all clothes, external metallic devices, especially piercings, and jewellery from the abdomen and pelvis of the patient
	Immobilize infants and uncooperative patients in the supine position, e.g. with sandbags under the knees. Immobilize infants and uncooperative patients either in the right or left lateral decubitus position for the lateral view
	Cooperative patients >8 years of age can be positioned upright at the vertical stand
	Place eventually small X-ray positive markers at palpable tumours or on pain points
Positioning	The arms should be held upwards
	Rotation of lumbar spine in the decubitus position can be avoided, if the patient is positioned on a rubber foam pad of appropriate between the rib cage and the iliac crests
Attention	On the Bucky table the patient should always be positioned on a foam mat of 10 to 20 cm thickness. So, the lower part of the body will not be cut if lateral projection X-rays (cross table views) in prone or in decubitus position are performed
Shielding	In males, use testicular capsules if possible. In females, lead masks or lead aprons should be placed at the upper and lower field edges

Table 16.7 Radiographic procedure

X-ray source	Use automatic exposure control (AEC) in patients >1 year of age
	Focal spot size between 0.6 and 1.3 dependent on the age of the patient
	Check the focus-to-detector distance
	Check the field size
Field setting	Check the field centring
	Check the use of antiscatter grids
Image detector system	Check the use of antiscatter grids
Patient	The exposure should be performed during breath-holding

16.44, 16.45, 16.46, 16.47, 16.48, 16.49, 16.50, 16.51, 16.52, and 16.53. Table cells with normal font refer to optimal field settings and table cells with bold font refer to suboptimal field settings. Please note that normalized organ doses have been provided also for exposure parameter combinations, which are not recommended by the authors.

Table 16.8 Radiographic techniques Bucky table lumbar spine radiographs

Age group (years)	Beam direction	Patient position	Beam projection	Typical patient diameter (cm) With arms	Typical patient diameter (cm) No arms	Optimal field size (cm x cm)	Focus-to-detector distance (cm)	Voltage (kVp) 50	60	70	80	90	100	Antiscatter grid
0	Vertical	Supine	ap	9.80		3 × 9	100[1]							
	Vertical	Decubitus	lat	12.6	10.9	4 × 9								
1–2	Vertical	Supine	ap	13.00		4 × 13	115							
	Vertical	Decubitus	lat	17.1	15.1	6 × 13								
	Horizontal	Supine												
3–7	Vertical	Supine	ap	15.0		5 × 17								
	Vertical	Decubitus	lat	22.6	19.6	7 × 17								
	Horizontal	Supine												
8–12	Vertical	Supine	ap	16.8		6 × 20								
	Horizontal	Sitting												
	Vertical	Decubitus	lat	27.3	23.8	8 × 20								
	Horizontal	Supine												
13–17	Vertical	Supine	ap	19.6		8 × 26								
	Horizontal	Sitting												
	Vertical	Decubitus	lat	34.4	27.3	9 × 26								
	Horizontal	Supine												
18	Vertical	Supine	ap	20.0		9 × 29								
	Horizontal	Sitting												
	Vertical	Decubitus	lat	38.9	34.4	9 × 29								
	Horizontal	Supine												

Annotations:

1 The patient is lying on the table top directly on the detector

(black) Parameter combination not recommended, normalized organ doses not provided

(light blue) Parameter combination recommended, normalized organ doses provided

Table 16.9 Radiographic techniques of vertical stand lumbar spine radiographs

Voltage (kVp)	Total filtration: Without additional filtration 1.5 mm Al	2.5 Al	With additional filtration 3.5 Al	4.5 Al	3.5 Al + 0.1 Cu	3.5Al + 0.2 Cu
50						
60	This technique is not suitable for lumbar spine radiographs					
70						
80						
90						
100						

Annotations:

(black) Parameter combination not recommended, normalized organ doses not provided

(grey) Parameter combination not recommended, normalized organ doses provided

(light blue) Parameter combination recommended, normalized organ doses provided

Table 16.10 Radiation qualities of Bucky table lumbar spine radiographs

Age group (years)	Beam direction	Patient position	Beam projection	Typical patient diameter (cm) With arms	Typical patient diameter (cm) No arms	Optimal field size (cm x cm)	Focus-to-detector distance (cm)	Voltage (kVp) 50	60	70	80	90	100	Antiscatter grid
0	This technique is not applied in these age groups													
1–2														
3–7														
8–12	Horizontal	Standing	ap	16.8		6 × 20	150							
			lat	27.3	23.8	8 × 20								
13–17			ap	19.6		8 × 26								
			lat	34.4	27.3	9 × 26								
18			ap	20.0		9 × 29								
			lat	38.9	34.4	9 × 29								

Annotations:

(black) Parameter combination not recommended, normalized organ doses not provided

(light blue) Parameter combination recommended, normalized organ doses provided

Table 16.11 Radiation qualities of vertical stand table lumbar spine radiographs

Voltage (kVp)	Total filtration					
	Without additional filtration		With additional filtration			
	1.5 mm Al	2.5 Al	3.5 Al	4.5 Al	3.5 Al + 0.1 Cu[1]	3.5 Al + 0.2 Cu[1]
50						
60						
70	This technique is not suitable for lumbar spine radiographs					
80						
90						
100						

Annotations:	
1	The total beam filtration is usually 3.5 mm Al + 0.1 mm Cu, in adolescents optionally 3.5 mm Al + 0.2 mm Cu
	Parameter combination not recommended, normalized organ doses not provided
	Parameter combination not recommended, normalized organ doses provided
	Parameter combination recommended, normalized organ doses provided

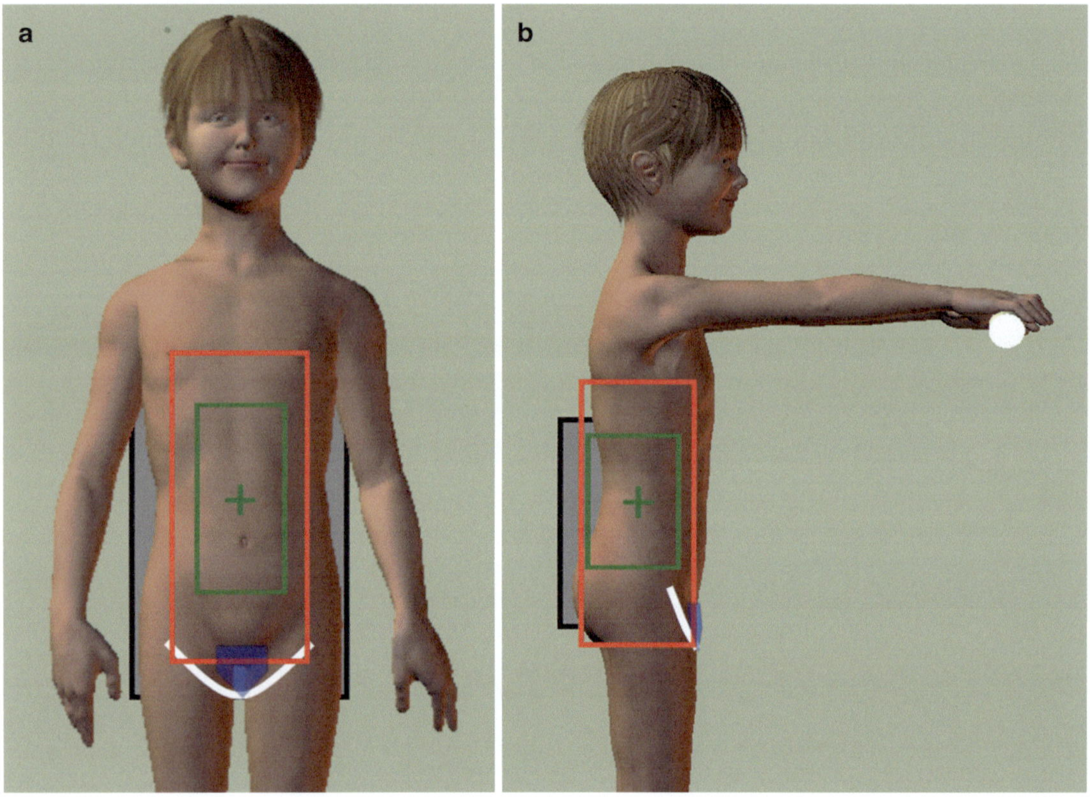

Fig. 16.1 Field settings for lumbar spine radiographs in standing position. (**a**) Antero-posterior projection. (**b**) Lateral projection; the patient holds on a handrail. (© V. Rösenberger & M. Seidenbusch 2018). Green lines = optimal field settings; green cross = centring point; red lines = suboptimal field settings; grey shaded area = image detector; blue shaded area = testicular capsule; white lines = sticking plaster for fixing the gonad shielding

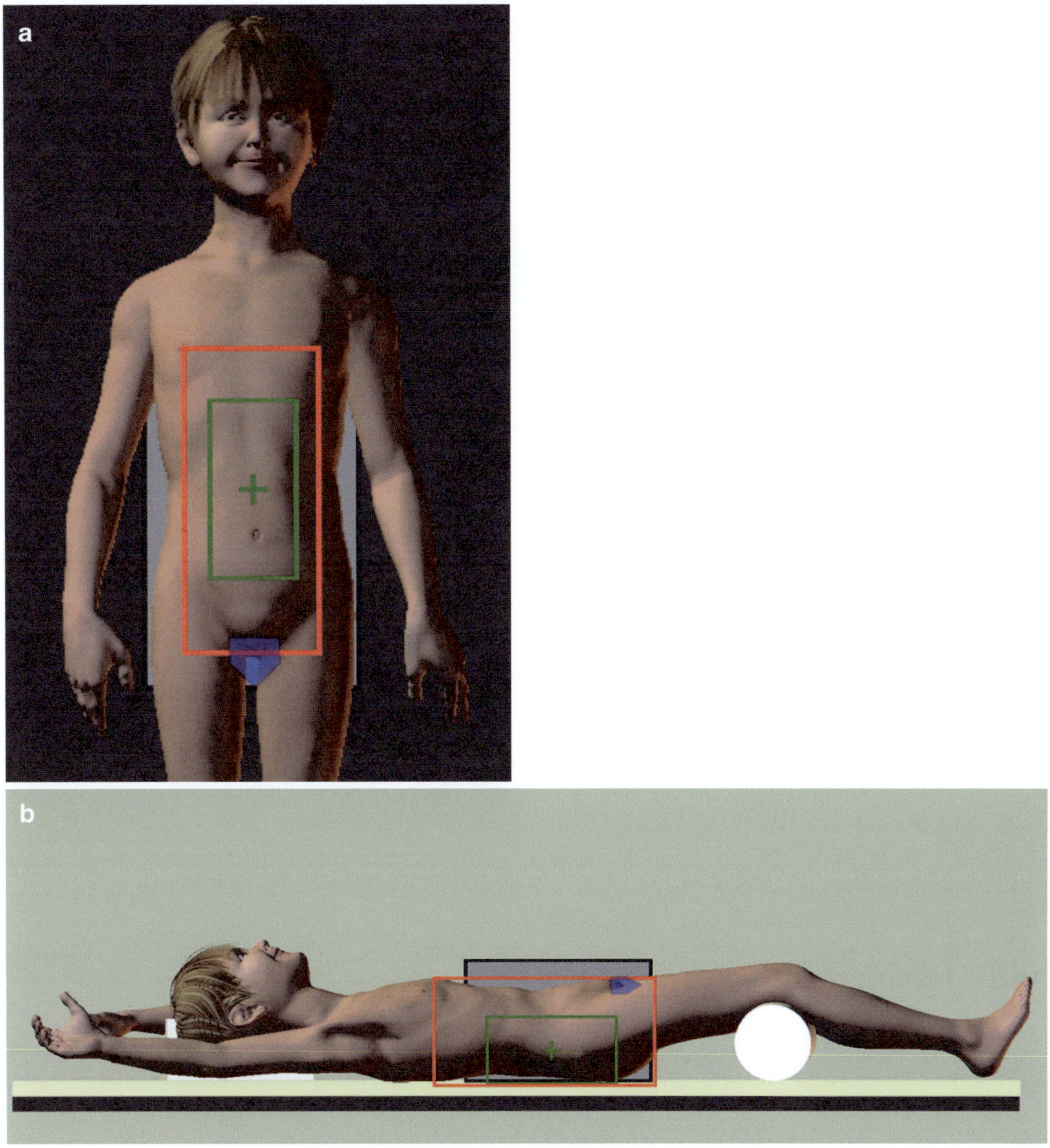

Fig. 16.2 Field settings for lumbar spine radiographs in supine position. (**a**) Antero-posterior projection. (**b**) Lateral projection. (© V. Rösenberger & M. Seidenbusch 2018). Green lines = optimal field settings; green cross = centring point; red lines = suboptimal field settings; grey shaded area = image detector; blue shaded area = testicular capsule; dark grey shaded area = Bucky table; white cylinder = foam roll; white device behind the head and neck = foam pillow; white foam mat on the top of the Bucky table

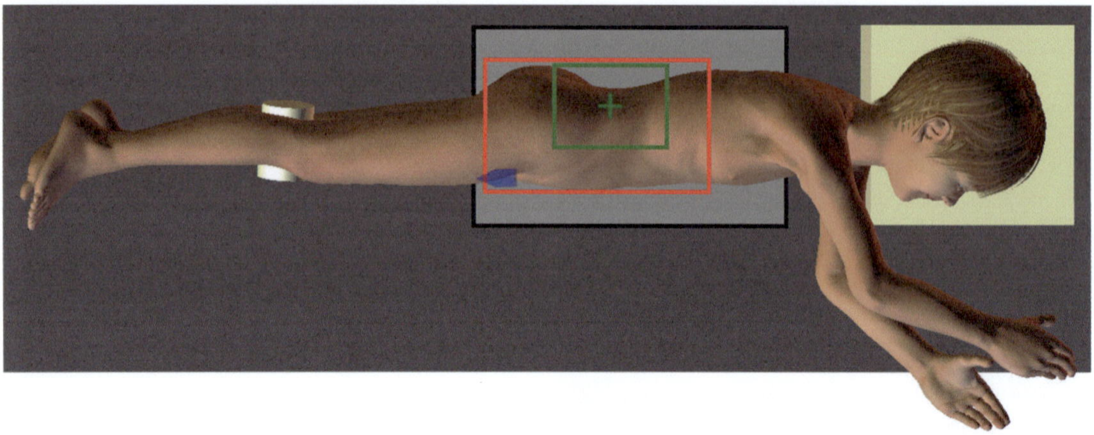

Fig. 16.3 Field settings for lumbar spine radiographs in decubitus position. Lateral projection. (© V. Rösenberger & M. Seidenbusch 2018). Green lines = optimal field settings; green cross = centring point; red lines = suboptimal field settings; grey shaded area = image detector; blue shaded area = testicular capsule; dark grey shaded area = Bucky table; white cylinder = foam roll; white device behind the head and neck = foam pillow

16.2 Indications and Contraindications

In recent years, lumbar spine radiographs have been widely replaced by MR examinations. Indications and contraindications for lumbar spine radiographs are given in Tables 16.1 and 16.2.

16.3 Image Criteria

The image criteria required for lumbar spine radiographs are listed in Table 16.3.

See also [2, 3].

16.4 Field Settings

In Table 16.4, patient positions and beam projections used for lumbar spine radiographs are listed, and in Table 16.5, the optimal field settings are listed which are required in order to fulfil the image criteria mentioned in Table 16.3.

16.5 Patient Preparation and Radiographic Procedure

See Tables 16.6, 16.7 and [1, 5].

16.6 Exposure Parameters

In Tables 16.8, 16.9, 16.10, and 16.11, exposure parameter combinations used for lumbar spine radiographs in various age groups are listed. Dark grey shaded table cells refer to exposure parameter combinations which are not recommended for paediatric lumbar spine radiographs and for which no normalized organ doses have been provided. Light grey and light blue shaded table cells refer to exposure parameter combinations for which normalized organ doses have been provided with light grey shaded table cells referring to exposure parameter combinations being not recommended and light blue shaded table cells referring to exposure parameter combinations following the European Guidelines [2]. Note that due to the fact that the radiation dose is an exponential function of the patient diameter, it may make sense to use the patient diameters instead of the patient's age groups for determining the adequate exposure parameters from Tables 16.8 and 16.10. In this case, for determining the corresponding normalized organ doses from Tables 16.18, 16.19, 16.20, 16.21, 16.22, 16.23, 16.24, 16.25, 16.26, 16.27, 16.28, 16.29, 16.30, 16.31, 16.32, 16.33, 16.34, 16.35, 16.36, 16.37, 16.38, 16.39, 16.40, 16.41, 16.42, 16.43, 16.44, 16.45, 16.46, 16.47, 16.48, 16.49, 16.50, 16.51, 16.52, and 16.53, the age group related to the patient diameter should be used.

16.7 Normalized Organ Doses

For calculating normalized organ doses, the field settings described in Sect. 16.4 were transferred to the mathematical MIRD phantoms. Inconsistencies between the geometrical anatomy of the mathematical phantoms and the real human anatomy have been listed in Table 16.54 whenever they are of relevance for the validity of the normalized organ doses.

In Tables 16.12, 16.13, 16.14, 16.15, 16.16, and 16.17, optimal and suboptimal field settings are listed for the different age groups and the beam projections given in Tables 16.8 and 16.10. Green lines refer to optimal field settings, red lines refer to suboptimal field settings, respectively. In the tables, the optimal and suboptimal field settings which were used for performing the calculations of the normalized organ doses provided in Tables 16.18, 16.19, 16.20, 16.21, 16.22, 16.23, 16.24, 16.25, 16.26, 16.27, 16.28, 16.29, 16.30, 16.31, 16.32, 16.33, 16.34, 16.35, 16.36, 16.37, 16.38, 16.39, 16.40, 16.41, 16.42, 16.43, 16.44, 16.45, 16.46, 16.47, 16.48, 16.49, 16.50, 16.51, 16.52, and 16.53 were drawn in on the skin surface of the mathematical phantoms. The virtual mathematical radiographs resulting from these field settings were depicted in the tables with and without bone structures, respectively, in order to enable the reader to visually compare the virtual mathematical radiographs with the real radiographs made under clinical conditions. For better visualization of the spine in the sagittal skeletal view, liver, gall bladder, pancreas, small intestine, and colon have been removed, and for better visualization of the spine in the lateral skeletal view, kidneys and liver have been removed. For comparing the virtual mathematical field settings with the real clinical field settings, the virtual mathematical field edges were also listed in the tables.

As the normalized organ doses provided in Tables 16.18, 16.19, 16.20, 16.21, 16.22, 16.23, 16.24, 16.25, 16.26, 16.27, 16.28, 16.29, 16.30, 16.31, 16.32, 16.33, 16.34, 16.35, 16.36, 16.37, 16.38, 16.39, 16.40, 16.41, 16.42, 16.43, 16.44, 16.45, 16.46, 16.47, 16.48, 16.49, 16.50, 16.51, 16.52, and 16.53 refer to entrance kerma values, optimal and suboptimal field sizes were provided in the tables in order to enable the reader to calculate entrance air kerma dose values from clinically measured dose area product values. Please note that these optimal and suboptimal field sizes refer to the mathematical phantoms of an age group only and not to real patients with corresponding ages.

Please note also that the anatomy of the mathematical phantoms has been idealized. Therefore, the virtual field settings for calculating the normalized organ doses have been defined with a view to ensuring that the resulting virtual organ doses would be consistent with the real organ doses which would be achieved during real X-ray examinations being performed with the real field settings recommended in Sect. 16.4. Thus, the field settings shown in the following tables may not be consistent with the clinical field settings as recommended in Sect. 16.4.

Normalized organ doses for lumbar spine radiographs are also listed in e.g. [4, 6–8].

Table 16.12 Field settings—Lumbar spine ap

Age group (years)		Anatomical landmarks			
		Optimal field setting		Suboptimal field setting	
0	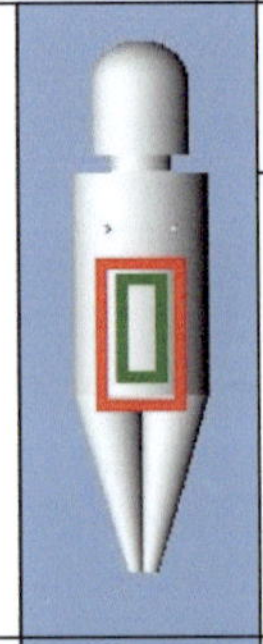	Upper field edge: 12th rib Lower field edge: midpoint between the navel and the symphysis Lateral field edges: Close to the spinal column Field size = field width × field height: 3 cm × 9 cm = 27 cm^2		Upper field edge: 10th thoracic vertebra Lower field edge: Perineum Lateral field edges: At the mammillary lines Field size = field width × field height: 7 cm × 14 cm = 98 cm^2	

1–2	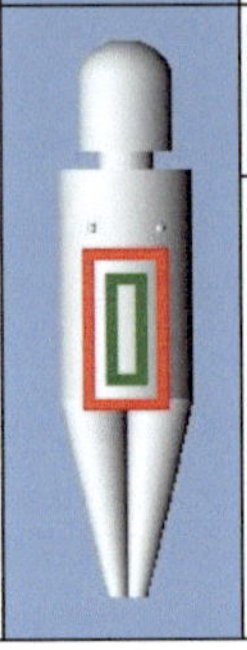	Upper field edge: 12th rib Lower field edge: midpoint between the navel and the symphysis Lateral field edges: Close to the spinal column Field size = field width × field height: 4 cm × 13 cm = 52 cm^2		Upper field edge: 10th thoracic vertebra Lower field edge: Perineum Lateral field edges: At the mammillary lines Field size = field width × field height: 9 cm × 20 cm = 180 cm^2	

Table 16.13 Field settings—Lumbar spine ap

Age group (years)		Anatomical landmarks			
		Optimal field setting		Suboptimal field setting	
3–7	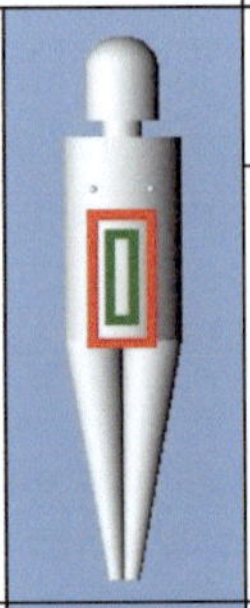	Upper field edge: 12th rib Lower field edge: midpoint between the navel and the symphysis Lateral field edges: Close to the spinal column Field size = field width × field height: 5 cm × 17 cm = 85 cm^2		Upper field edge: 10th thoracic vertebra Lower field edge: Perineum Lateral field edges: At the mammillary lines Field size = field width × field height: 12 cm × 26 cm = 312 cm^2	
			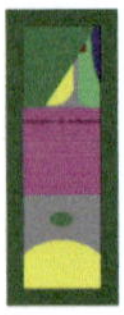		
8–12	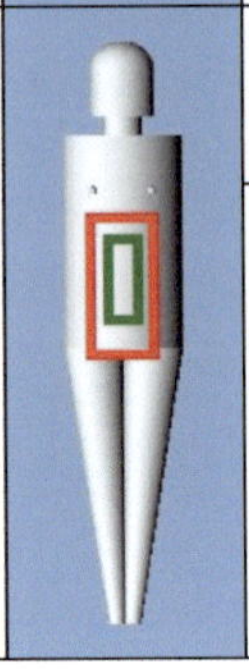	Upper field edge: 12th rib Lower field edge: midpoint between the navel and the symphysis Lateral field edges: Close to the spinal column Field size = field width × field height: 6 cm × 20 cm = 120 cm^2		Upper field edge: 10th thoracic vertebra Lower field edge: Perineum Lateral field edges: At the mammillary lines Field size = field width × field height: 14 cm × 32 cm = 448 cm^2	

Table 16.14 Field settings—Lumbar spine ap

Age group (years)		Anatomical landmarks			
		Optimal field setting		Suboptimal field setting	
13–17	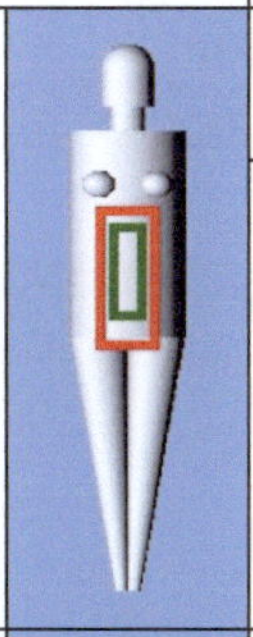	Upper field edge: 12th rib Lower field edge: midpoint between the navel and the symphysis Lateral field edges: Close to the spinal column Field size = field width × field height: 8 cm × 26 cm = 208 cm^2		Upper field edge: 10th thoracic vertebra Lower field edge: Perineum Lateral field edges: At the mammillary lines Field size = field width × field height: 16 cm × 41 cm = 656 cm^2	
			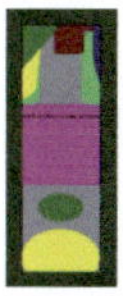	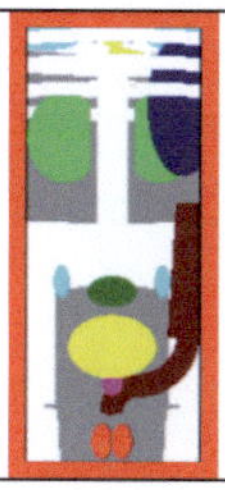	
18	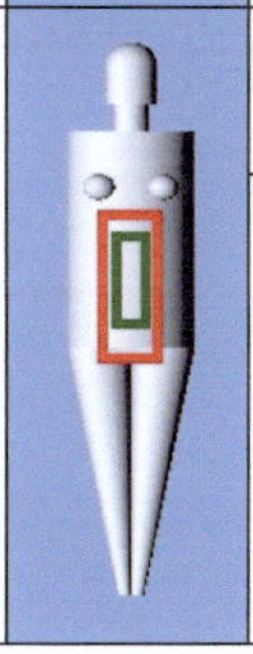	Upper field edge: 12th rib Lower field edge: midpoint between the navel and the symphysis Lateral field edges: Close to the spinal column Field size = field width × field height: 9 cm × 29 cm = 261 cm^2		Upper field edge: 10th thoracic vertebra Lower field edge: Perineum Lateral field edges: At the mammillary lines Field size = field width × field height: 18 cm × 46 cm = 828 cm^2	
					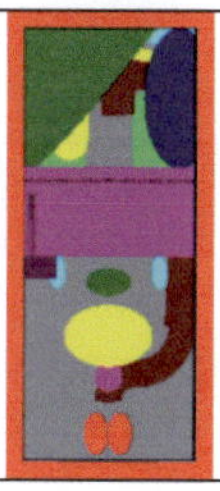

Table 16.15 Field settings—Lumbar spine lat

Age group (years)		Anatomical landmarks			
		Optimal field setting		Suboptimal field setting	
0	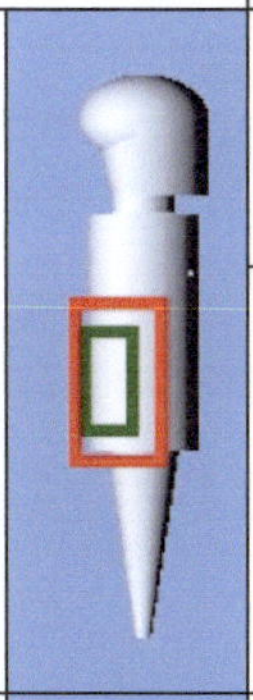	Upper field edge: 12th thoracic vertebra Lower field edge: Midpoint between the xyphoid and pelvic symphysis Posterior field edge: Skin surface of the back Anterior field edge: Posterior axillary line Field size = field width × field height: 4 cm × 9 cm = 36 cm^2		Upper field edge: 10th thoracic vertebra Lower field edge: Perineum Left lateral field edge: Skin surface of the back Right lateral field edge: close the anterior thoracic wall Field size = field width × field height: 7 cm × 14 cm = 98 cm^2	
		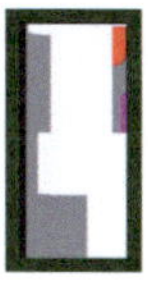	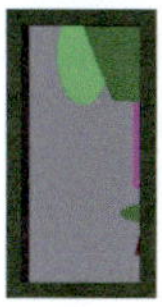		
1–2	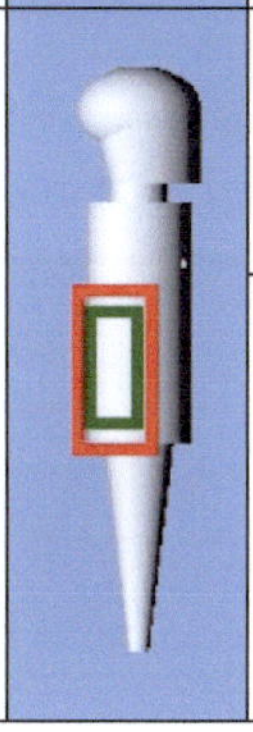	Upper field edge: 12th thoracic vertebra Lower field edge: Midpoint between the xyphoid and pelvic symphysis Posterior field edge: Skin surface of the back Anterior field edge: Posterior axillary line Field size = field width × field height: 6 cm × 13 cm = 78 cm^2		Upper field edge: 10th thoracic vertebra Lower field edge: Perineum Left lateral field edge: Skin surface of the back Right lateral field edge: close the anterior thoracic wall Field size = field width × field height: 8 cm × 20 cm = 160 cm^2	
			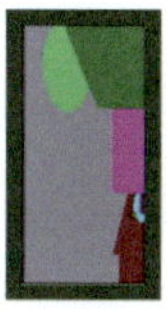		

Table 16.16 Field settings—Lumbar spine lat

Age group (years)		Anatomical landmarks			
		Optimal field setting		Suboptimal field setting	
3–7	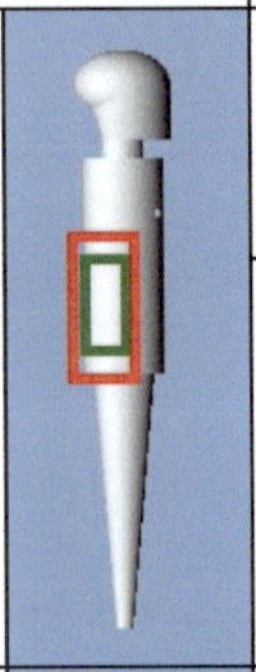	Upper field edge: 12th thoracic vertebra Lower field edge: Midpoint between the xyphoid and pelvic symphysis Posterior field edge: Skin surface of the back Anterior field edge: Posterior axillary line Field size = field width × field height: 7 cm × 17 cm = 119 cm^2		Upper field edge: 10th thoracic vertebra Lower field edge: Perineum Left lateral field edge: Skin surface of the back Right lateral field edge: close the anterior thoracic wall Field size = field width × field height: 9 m × 26 cm = 234 cm^2	
			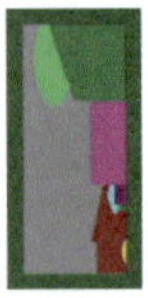		
8–12	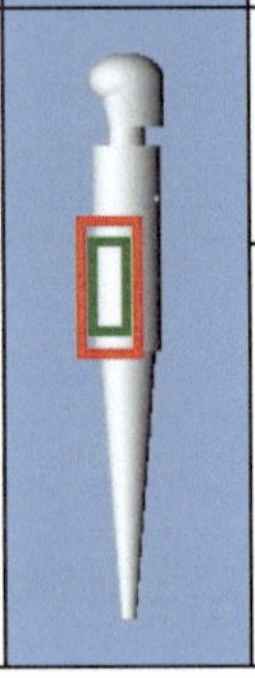	Upper field edge: 12th thoracic vertebra Lower field edge: Midpoint between the xyphoid and pelvic symphysis Posterior field edge: Skin surface of the back Anterior field edge: Posterior axillary line Field size = field width × field height: 8 cm × 20 cm = 160 cm^2		Upper field edge: 10th thoracic vertebra Lower field edge: Perineum Left lateral field edge: Skin surface of the back Right lateral field edge: close the anterior thoracic wall Field size = field width × field height: 10 cm × 32 cm = 320 cm^2	

Table 16.17 Field settings—Lumbar spine lat

Age group (years)		Anatomical landmarks			
		Optimal field setting		Suboptimal field setting	
13–17	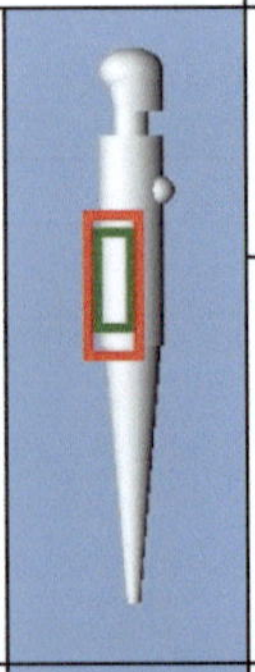	Upper field edge: 12th thoracic vertebra Lower field edge: Midpoint between the xyphoid and pelvic symphysis Posterior field edge: Skin surface of the back Anterior field edge: Posterior axillary line Field size = field width × field height: 9 cm × 26 cm = 234 cm^2		Upper field edge: 10th thoracic vertebra Lower field edge: Perineum Left lateral field edge: Skin surface of the back Right lateral field edge: close the anterior thoracic wall Field size = field width × field height: 12 cm × 41 cm = 492 cm^2	
				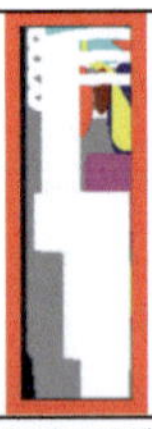	
18	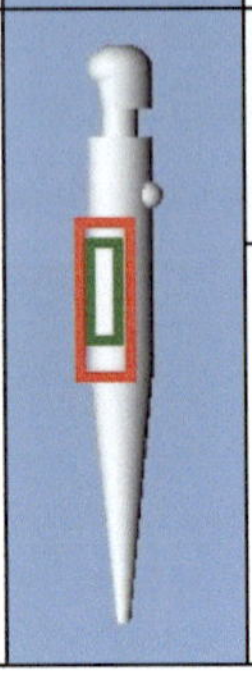	Upper field edge: 12th thoracic vertebra Lower field edge: Midpoint between the xyphoid and pelvic symphysis Posterior field edge: Skin surface of the back Anterior field edge: Posterior axillary line Field size = field width × field height: 9 cm × 29 cm = 261 cm^2		Upper field edge: 10th thoracic vertebra Lower field edge: Perineum Left lateral field edge: Skin surface of the back Right lateral field edge: close the anterior thoracic wall Field size = field width × field height: 13 cm × 46 cm = 598 cm^2	

Table 16.18 Normalized organ doses—Lumbar spine ap—<1 year

Beam projection:	ap											**Table cells:**
Focus-to-detector distance (cm):	100											(+) = normal font = optimal field setting
Tube voltage (kVp):	60											**(−) = bold font = suboptimal field setting**
Total filtrations:	See last table row											Empty table cells correspond to values <0.01
Organ or tissue	**Organ dose D_T: Entrance dose K_E**								**(mSv ∗ mGy^{-1})**			
	(+)	(−)	(+)	(−)	(+)	(−)	(+)	(−)	(+)	(−)	(+)	(−)
Skull												
Brain												
Oral mucosa												
Salivary glands												
Cervical spine												
Thyroid												
Thymus										**0.01**		
Thoracic spine	0.02	**0.08**	0.02	**0.10**	0.03	**0.12**	0.03	**0.14**	0.04	**0.18**	0.04	**0.16**
Claviculae								**0.01**		**0.01**		**0.01**
Scapulae		**0.02**		**0.02**		**0.03**		**0.03**		**0.04**		**0.04**
Ribs	0.02	**0.21**	0.02	**0.24**	0.03	**0.26**	0.03	**0.27**	0.04	**0.30**	0.03	**0.29**
Oesophagus		**0.07**	0.01	**0.09**	0.01	**0.11**	0.01	**0.12**	0.02	**0.14**	0.02	**0.13**
Trachea												
Lung		**0.02**		**0.03**		**0.03**		**0.04**		**0.05**		**0.04**
Mamma				**0.01**		**0.01**		**0.02**		**0.02**		**0.02**
Heart		**0.03**		**0.04**		**0.05**	0.01	**0.05**	0.01	**0.06**	0.01	**0.06**
Stomach	0.05	**0.50**	0.07	**0.60**	0.07	**0.67**	0.08	**0.72**	0.10	**0.81**	0.09	**0.80**
Spleen	0.01	**0.12**	0.02	**0.16**	0.02	**0.19**	0.02	**0.21**	0.03	**0.25**	0.02	**0.24**
Liver	0.09	**0.36**	0.10	**0.43**	0.11	**0.48**	0.12	**0.52**	0.14	**0.58**	0.13	**0.57**
Gall bladder	0.30	**0.41**	0.37	**0.51**	0.42	**0.58**	0.45	**0.64**	0.52	**0.73**	0.51	**0.72**
Pancreas	0.06	**0.26**	0.08	**0.32**	0.09	**0.38**	0.11	**0.42**	0.12	**0.49**	0.12	**0.47**
Small intestine	0.18	**0.36**	0.23	**0.45**	0.26	**0.52**	0.28	**0.57**	0.33	**0.66**	0.32	**0.64**
Large intestine	0.13	**0.43**	0.16	**0.53**	0.18	**0.60**	0.20	**0.66**	0.23	**0.75**	0.23	**0.74**
Sigma/rectum	0.04	**0.33**	0.05	**0.41**	0.05	**0.47**	0.06	**0.51**	0.07	**0.60**	0.07	**0.58**
Lumbar spine	0.21	**0.30**	0.28	**0.40**	0.33	**0.47**	0.37	**0.54**	0.46	**0.66**	0.43	**0.61**
Pelvis	0.08	**0.43**	0.10	**0.55**	0.12	**0.65**	0.14	**0.73**	0.17	**0.87**	0.16	**0.83**
Adrenals	0.01	**0.10**	0.02	**0.13**	0.02	**0.15**	0.02	**0.17**	0.03	**0.21**	0.02	**0.20**
Kidneys	0.04	**0.11**	0.05	**0.15**	0.06	**0.17**	0.07	**0.20**	0.08	**0.24**	0.07	**0.22**
Urinary bladder	0.35	**0.59**	0.41	**0.70**	0.45	**0.78**	0.48	**0.84**	0.54	**0.94**	0.53	**0.93**
Uterus	0.29	**0.38**	0.37	**0.47**	0.42	**0.54**	0.46	**0.59**	0.53	**0.69**	0.51	**0.67**
Ovaries	0.06	**0.30**	0.08	**0.38**	0.09	**0.44**	0.11	**0.48**	0.13	**0.57**	0.12	**0.55**
Prostate	0.04	**0.45**	0.05	**0.55**	0.05	**0.62**	0.06	**0.68**	0.07	**0.78**	0.07	**0.76**
Testes	0.01	**1.01**	0.02	**1.11**	0.02	**1.16**	0.02	**1.19**	0.03	**1.25**	0.03	**1.25**
Upper arm				**0.01**		**0.02**		**0.02**		**0.02**		**0.02**
Elbow		**0.04**		**0.05**		**0.06**		**0.07**		**0.09**		**0.08**
Forearm		**0.07**		**0.09**		**0.11**		**0.13**		**0.16**		**0.14**
Upper leg	0.01	**0.51**	0.02	**0.63**	0.02	**0.73**	0.03	**0.80**	0.03	**0.94**	0.03	**0.92**
Knee		**0.04**		**0.05**		**0.06**		**0.07**		**0.08**		**0.07**
Lower leg												
Lymph nodes	0.09	**0.21**	0.11	**0.26**	0.13	**0.29**	0.14	**0.32**	0.16	**0.37**	0.16	**0.36**
Skin	0.03	**0.10**	0.03	**0.10**	0.03	**0.11**	0.03	**0.11**	0.03	**0.12**	0.03	**0.12**
Muscle	0.04	**0.14**	0.05	**0.17**	0.05	**0.18**	0.06	**0.20**	0.07	**0.23**	0.06	**0.22**
Skeleton	0.02	**0.14**	0.03	**0.18**	0.04	**0.21**	0.04	**0.23**	0.05	**0.27**	0.05	**0.26**
Red bone marrow		**0.04**		**0.05**	0.01	**0.05**	0.01	**0.06**	0.01	**0.07**	0.01	**0.07**
Whole body dose	0.04	**0.13**	0.05	**0.16**	0.05	**0.18**	0.06	**0.19**	0.06	**0.22**	0.06	**0.21**
ED (ICRP 60)	0.05	**0.31**	0.06	**0.36**	0.07	**0.40**	0.08	**0.43**	0.09	**0.48**	0.09	**0.47**
ED (ICRP 103)	0.05	**0.23**	0.06	**0.28**	0.07	**0.31**	0.07	**0.34**	0.09	**0.38**	0.08	**0.37**
Total filtration (mm) **Al**	**1.5**		**2.5**		**3.5**		**4.5**		**3.5**		**3.5**	
Cu	**0**		**0**		**0**		**0**		**0.1**		**0.2**	

Table 16.19 Normalized organ doses—Lumbar spine ap—<1 year

Beam projection:	ap	**Table cells:**
Focus-to-detector distance (cm):	100	(+) = normal font = optimal field setting
Tube voltage (kVp):	70	**(−) = bold font = suboptimal field setting**
Total filtrations:	See last table row	Empty table cells correspond to values <0.01

Organ or tissue	**Organ dose D_T: Entrance dose K_E (mSv ∗ mGy^{-1})**											
	(+)	(−)	(+)	(−)	(+)	(−)	(+)	(−)	(+)	(−)	(+)	(−)
Skull												
Brain												
Oral mucosa												
Salivary glands												
Cervical spine												**0.01**
Thyroid												
Thymus								**0.01**		**0.01**		**0.02**
Thoracic spine	0.02	**0.10**	0.03	**0.13**	0.04	**0.15**	0.04	**0.17**	0.05	**0.21**	0.06	**0.24**
Claviculae						**0.01**		**0.01**		**0.02**		**0.02**
Scapulae		**0.02**		**0.03**		**0.04**		**0.04**	0.01	**0.05**	0.01	**0.06**
Ribs	0.02	**0.22**	0.03	**0.25**	0.03	**0.27**	0.03	**0.29**	0.04	**0.31**	0.04	**0.33**
Oesophagus		**0.09**	0.01	**0.11**	0.01	**0.12**	0.02	**0.14**	0.02	**0.16**	0.02	**0.18**
Trachea												
Lung		**0.03**		**0.03**		**0.04**		**0.04**	0.01	**0.05**	0.01	**0.06**
Mamma		**0.01**		**0.02**		**0.02**		**0.02**		**0.03**		**0.03**
Heart		**0.04**		**0.05**	0.01	**0.06**	0.01	**0.06**	0.01	**0.07**	0.02	**0.09**
Stomach	0.06	**0.56**	0.07	**0.66**	0.08	**0.73**	0.09	**0.78**	0.11	**0.86**	0.12	**0.94**
Spleen	0.02	**0.15**	0.02	**0.19**	0.02	**0.22**	0.03	**0.24**	0.03	**0.29**	0.04	**0.33**
Liver	0.10	**0.40**	0.11	**0.48**	0.12	**0.52**	0.13	**0.56**	0.15	**0.63**	0.16	**0.68**
Gall bladder	0.34	**0.47**	0.41	**0.57**	0.46	**0.65**	0.50	**0.70**	0.56	**0.80**	0.61	**0.89**
Pancreas	0.08	**0.30**	0.09	**0.38**	0.11	**0.43**	0.12	**0.47**	0.14	**0.55**	0.16	**0.62**
Small intestine	0.21	**0.42**	0.26	**0.51**	0.29	**0.58**	0.32	**0.63**	0.36	**0.73**	0.40	**0.81**
Large intestine	0.15	**0.49**	0.18	**0.60**	0.21	**0.67**	0.22	**0.72**	0.26	**0.82**	0.29	**0.90**
Sigma/rectum	0.04	**0.38**	0.05	**0.46**	0.06	**0.52**	0.07	**0.57**	0.08	**0.66**	0.09	**0.73**
Lumbar spine	0.26	**0.38**	0.34	**0.48**	0.39	**0.57**	0.44	**0.64**	0.53	**0.77**	0.61	**0.89**
Pelvis	0.10	**0.52**	0.13	**0.65**	0.15	**0.75**	0.17	**0.83**	0.20	**0.97**	0.23	**1.10**
Adrenals	0.02	**0.12**	0.02	**0.16**	0.02	**0.19**	0.03	**0.21**	0.03	**0.25**	0.04	**0.29**
Kidneys	0.05	**0.14**	0.06	**0.18**	0.07	**0.21**	0.08	**0.24**	0.09	**0.28**	0.11	**0.33**
Urinary bladder	0.38	**0.66**	0.45	**0.77**	0.49	**0.85**	0.52	**0.91**	0.57	**1.01**	0.61	**1.09**
Uterus	0.34	**0.44**	0.41	**0.53**	0.46	**0.61**	0.50	**0.66**	0.57	**0.76**	0.63	**0.85**
Ovaries	0.08	**0.35**	0.09	**0.44**	0.11	**0.50**	0.12	**0.55**	0.14	**0.64**	0.16	**0.72**
Prostate	0.04	**0.51**	0.05	**0.62**	0.06	**0.69**	0.07	**0.74**	0.08	**0.84**	0.09	**0.92**
Testes	0.02	**1.06**	0.02	**1.14**	0.02	**1.19**	0.03	**1.23**	0.03	**1.28**	0.04	**1.32**
Upper arm		**0.01**		**0.02**		**0.02**		**0.02**		**0.03**		**0.03**
Elbow		**0.05**		**0.06**		**0.08**		**0.09**		**0.10**		**0.12**
Forearm		**0.09**		**0.11**		**0.13**		**0.15**		**0.18**		**0.21**
Upper leg	0.02	**0.58**	0.03	**0.72**	0.03	**0.81**	0.03	**0.89**	0.04	**1.02**	0.05	**1.13**
Knee		**0.05**		**0.06**		**0.07**		**0.08**	0.01	**0.10**	0.01	**0.12**
Lower leg										**0.01**		**0.01**
Lymph nodes	0.10	**0.24**	0.13	**0.29**	0.14	**0.33**	0.15	**0.36**	0.18	**0.41**	0.20	**0.45**
Skin	0.03	**0.10**	0.03	**0.11**	0.03	**0.11**	0.03	**0.12**	0.03	**0.13**	0.03	**0.13**
Muscle	0.05	**0.16**	0.05	**0.18**	0.06	**0.20**	0.06	**0.22**	0.07	**0.25**	0.08	**0.27**
Skeleton	0.03	**0.17**	0.04	**0.21**	0.04	**0.23**	0.05	**0.26**	0.06	**0.30**	0.07	**0.34**
Red bone marrow		**0.05**	0.01	**0.06**	0.01	**0.07**	0.01	**0.07**	0.02	**0.09**	0.02	**0.10**
Whole body dose	0.04	**0.15**	0.05	**0.18**	0.06	**0.20**	0.06	**0.21**	0.07	**0.24**	0.08	**0.26**
ED (ICRP 60)	0.06	**0.34**	0.07	**0.39**	0.08	**0.43**	0.09	**0.46**	0.10	**0.51**	0.11	**0.56**
ED (ICRP 103)	0.06	**0.26**	0.07	**0.31**	0.08	**0.34**	0.08	**0.37**	0.09	**0.41**	0.10	**0.45**
Total filtration (mm) Al	**1.5**		**2.5**		**3.5**		**4.5**		**3.5**		**3.5**	
Cu	**0**		**0**		**0**		**0**		**0.1**		**0.2**	

Table 16.20 Normalized organ doses—Lumbar spine r-lat—<1 year

Beam projection:			r-lat									**Table cells:**
Focus-to-detector distance (cm):			100					(+) = normal font = optimal field setting				
Tube voltage (kVp):			60					**(−) = bold font = suboptimal field setting**				
Total filtrations:			See last table row					Empty table cells correspond to values <0.01				
Organ or tissue	**Organ dose D_T: Entrance dose K_E**							**(mSv * mGy⁻¹)**				
	(+)	(−)	(+)	(−)	(+)	(−)	(+)	(−)	(+)	(−)	(+)	(−)
Skull												
Brain												
Oral mucosa												
Salivary glands												
Cervical spine												
Thyroid												
Thymus												
Thoracic spine	0.03	**0.14**	0.05	**0.18**	0.05	**0.21**	0.06	**0.23**	0.07	**0.28**	0.07	**0.26**
Claviculae										**0.01**		
Scapulae		**0.02**		**0.02**		**0.02**		**0.03**		**0.04**		**0.03**
Ribs	0.03	**0.20**	0.04	**0.23**	0.05	**0.26**	0.05	**0.27**	0.06	**0.30**	0.06	**0.30**
Oesophagus		**0.05**		**0.06**		**0.07**		**0.08**	0.01	**0.10**	0.01	**0.09**
Trachea												
Lung		**0.02**		**0.03**		**0.03**		**0.03**		**0.04**		**0.04**
Mamma												
Heart		**0.01**		**0.02**		**0.02**		**0.03**		**0.03**		**0.03**
Stomach		**0.06**	0.01	**0.07**	0.01	**0.09**	0.02	**0.10**	0.02	**0.12**	0.02	**0.11**
Spleen	0.03	**0.07**	0.04	**0.09**	0.05	**0.11**	0.05	**0.12**	0.06	**0.15**	0.06	**0.14**
Liver	0.05	**0.32**	0.06	**0.38**	0.06	**0.42**	0.07	**0.45**	0.08	**0.50**	0.07	**0.49**
Gall bladder	0.02	**0.18**	0.02	**0.23**	0.03	**0.26**	0.03	**0.29**	0.04	**0.35**	0.03	**0.33**
Pancreas	0.02	**0.14**	0.02	**0.19**	0.03	**0.22**	0.03	**0.25**	0.04	**0.30**	0.04	**0.28**
Small intestine	0.04	**0.20**	0.06	**0.26**	0.07	**0.30**	0.08	**0.33**	0.09	**0.40**	0.09	**0.38**
Large intestine	0.03	**0.26**	0.04	**0.32**	0.04	**0.37**	0.05	**0.41**	0.06	**0.47**	0.05	**0.46**
Sigma/rectum	0.02	**0.11**	0.02	**0.14**	0.03	**0.17**	0.03	**0.19**	0.04	**0.23**	0.03	**0.21**
Lumbar spine	0.51	**0.58**	0.65	**0.75**	0.76	**0.88**	0.84	**0.98**	1.00	**1.17**	0.97	**1.12**
Pelvis	0.43	**0.84**	0.53	**1.04**	0.61	**1.17**	0.67	**1.28**	0.78	**1.47**	0.76	**1.45**
Adrenals	0.02	**0.18**	0.03	**0.23**	0.03	**0.27**	0.04	**0.30**	0.05	**0.35**	0.04	**0.33**
Kidneys	0.24	**0.29**	0.29	**0.35**	0.33	**0.40**	0.36	**0.44**	0.41	**0.50**	0.40	**0.49**
Urinary bladder	0.01	**0.10**	0.02	**0.13**	0.02	**0.15**	0.02	**0.17**	0.03	**0.20**	0.02	**0.19**
Uterus	0.04	**0.18**	0.05	**0.24**	0.06	**0.28**	0.06	**0.31**	0.08	**0.37**	0.07	**0.35**
Ovaries	0.04	**0.24**	0.05	**0.31**	0.06	**0.36**	0.07	**0.40**	0.08	**0.47**	0.08	**0.45**
Prostate		**0.13**	0.01	**0.17**	0.01	**0.20**	0.02	**0.22**	0.02	**0.26**	0.02	**0.25**
Testes		**0.03**		**0.04**		**0.05**		**0.05**		**0.07**		**0.06**
Upper arm												
Elbow												
Forearm												
Upper leg	0.01	**0.37**	0.02	**0.47**	0.02	**0.54**	0.03	**0.60**	0.03	**0.70**	0.03	**0.68**
Knee		**0.02**		**0.03**		**0.04**		**0.05**		**0.06**		**0.05**
Lower leg												
Lymph nodes	0.02	**0.10**	0.02	**0.13**	0.03	**0.15**	0.03	**0.17**	0.04	**0.20**	0.04	**0.19**
Skin	0.06	**0.12**	0.06	**0.13**	0.06	**0.13**	0.07	**0.13**	0.07	**0.14**	0.07	**0.14**
Muscle	0.05	**0.11**	0.05	**0.13**	0.06	**0.15**	0.07	**0.16**	0.07	**0.18**	0.07	**0.17**
Skeleton	0.08	**0.19**	0.10	**0.23**	0.11	**0.26**	0.12	**0.29**	0.15	**0.33**	0.14	**0.33**
Red bone marrow	0.02	**0.05**	0.03	**0.07**	0.04	**0.08**	0.04	**0.09**	0.05	**0.10**	0.04	**0.10**
Whole body dose	0.04	**0.11**	0.05	**0.13**	0.06	**0.15**	0.06	**0.16**	0.07	**0.19**	0.07	**0.18**
ED (ICRP 60)	0.02	**0.10**	0.02	**0.12**	0.03	**0.14**	0.03	**0.16**	0.04	**0.18**	0.03	**0.18**
ED (ICRP 103)	0.02	**0.09**	0.02	**0.11**	0.02	**0.12**	0.03	**0.14**	0.03	**0.16**	0.03	**0.15**
Total filtration (mm) Al	**1.5**		**2.5**		**3.5**		**4.5**		**3.5**		**3.5**	
Cu	**0**		**0**		**0**		**0**		**0.1**		**0.2**	

Table 16.21 Normalized organ doses—Lumbar spine r-lat—<1 year

Beam projection:				r-lat								**Table cells:**
Focus-to-detector distance (cm):				100								(+) = normal font = optimal field setting
Tube voltage (kVp):				70								**(−) = bold font = suboptimal field setting**
Total filtrations:				See last table row								Empty table cells correspond to values <0.01
Organ or tissue	**Organ dose D_T: Entrance dose K_E**							**(mSv ∗ mGy⁻¹)**				
	(+)	(−)	(+)	(−)	(+)	(−)	(+)	(−)	(+)	(−)	(+)	(−)
Skull												
Brain												
Oral mucosa												
Salivary glands												
Cervical spine												**0.01**
Thyroid												
Thymus												**0.01**
Thoracic spine	0.04	**0.16**	0.05	**0.21**	0.06	**0.24**	0.07	**0.27**	0.08	**0.31**	0.10	**0.36**
Claviculae								**0.01**		**0.01**		**0.02**
Scapulae		**0.02**		**0.03**		**0.03**		**0.04**	0.01	**0.04**	0.01	**0.05**
Ribs	0.04	**0.22**	0.05	**0.25**	0.05	**0.27**	0.06	**0.29**	0.06	**0.31**	0.07	**0.33**
Oesophagus		**0.06**		**0.07**	0.01	**0.09**	0.01	**0.09**	0.02	**0.11**	0.02	**0.13**
Trachea												
Lung		**0.03**		**0.03**		**0.04**		**0.04**	0.01	**0.05**	0.01	**0.06**
Mamma												
Heart		**0.02**		**0.02**		**0.03**		**0.03**		**0.04**	0.01	**0.05**
Stomach	0.01	**0.07**	0.01	**0.09**	0.02	**0.11**	0.02	**0.12**	0.02	**0.15**	0.03	**0.17**
Spleen	0.04	**0.09**	0.05	**0.11**	0.06	**0.13**	0.06	**0.15**	0.08	**0.19**	0.09	**0.22**
Liver	0.05	**0.35**	0.06	**0.41**	0.07	**0.45**	0.07	**0.48**	0.08	**0.53**	0.09	**0.57**
Gall bladder	0.02	**0.21**	0.03	**0.26**	0.03	**0.30**	0.04	**0.34**	0.05	**0.40**	0.05	**0.45**
Pancreas	0.02	**0.18**	0.03	**0.22**	0.03	**0.26**	0.04	**0.29**	0.05	**0.34**	0.06	**0.39**
Small intestine	0.05	**0.24**	0.07	**0.30**	0.08	**0.35**	0.09	**0.38**	0.11	**0.45**	0.12	**0.51**
Large intestine	0.03	**0.30**	0.04	**0.37**	0.05	**0.42**	0.06	**0.45**	0.07	**0.52**	0.08	**0.58**
Sigma/rectum	0.02	**0.14**	0.03	**0.17**	0.03	**0.20**	0.04	**0.23**	0.04	**0.28**	0.05	**0.32**
Lumbar spine	0.60	**0.69**	0.75	**0.87**	0.86	**1.00**	0.94	**1.10**	1.09	**1.29**	1.22	**1.46**
Pelvis	0.49	**0.95**	0.60	**1.15**	0.67	**1.29**	0.73	**1.39**	0.84	**1.57**	0.92	**1.72**
Adrenals	0.03	**0.22**	0.04	**0.27**	0.04	**0.31**	0.05	**0.34**	0.05	**0.39**	0.06	**0.44**
Kidneys	0.27	**0.33**	0.33	**0.40**	0.37	**0.45**	0.40	**0.49**	0.45	**0.56**	0.50	**0.62**
Urinary bladder	0.02	**0.12**	0.02	**0.15**	0.02	**0.18**	0.03	**0.19**	0.03	**0.23**	0.04	**0.27**
Uterus	0.05	**0.22**	0.06	**0.28**	0.07	**0.33**	0.08	**0.36**	0.09	**0.43**	0.11	**0.49**
Ovaries	0.05	**0.29**	0.06	**0.36**	0.07	**0.41**	0.08	**0.45**	0.09	**0.53**	0.11	**0.60**
Prostate	0.01	**0.16**	0.02	**0.20**	0.02	**0.23**	0.02	**0.25**	0.03	**0.30**	0.03	**0.35**
Testes		**0.04**		**0.05**		**0.06**		**0.06**	0.01	**0.08**	0.01	**0.09**
Upper arm												
Elbow												
Forearm												
Upper leg	0.02	**0.43**	0.02	**0.54**	0.03	**0.61**	0.03	**0.67**	0.04	**0.77**	0.05	**0.87**
Knee		**0.03**		**0.04**		**0.05**		**0.06**	0.01	**0.07**	0.01	**0.08**
Lower leg												
Lymph nodes	0.02	**0.12**	0.03	**0.15**	0.03	**0.18**	0.04	**0.19**	0.04	**0.23**	0.05	**0.26**
Skin	0.06	**0.12**	0.06	**0.13**	0.07	**0.14**	0.07	**0.14**	0.07	**0.15**	0.07	**0.15**
Muscle	0.05	**0.12**	0.06	**0.15**	0.07	**0.16**	0.07	**0.17**	0.08	**0.20**	0.09	**0.22**
Skeleton	0.09	**0.21**	0.11	**0.26**	0.13	**0.29**	0.14	**0.32**	0.16	**0.36**	0.18	**0.40**
Red bone marrow	0.03	**0.06**	0.04	**0.08**	0.04	**0.09**	0.05	**0.10**	0.05	**0.11**	0.06	**0.13**
Whole body dose	0.05	**0.13**	0.06	**0.15**	0.06	**0.17**	0.07	**0.18**	0.08	**0.20**	0.09	**0.22**
ED (ICRP 60)	0.02	**0.12**	0.03	**0.14**	0.03	**0.16**	0.03	**0.18**	0.04	**0.21**	0.05	**0.24**
ED (ICRP 103)	0.02	**0.10**	0.03	**0.12**	0.03	**0.14**	0.03	**0.16**	0.04	**0.18**	0.04	**0.20**
Total filtration (mm) Al	**1.5**		**2.5**		**3.5**		**4.5**		**3.5**		**3.5**	
Cu	**0**		**0**		**0**		**0**		**0.1**		**0.2**	

Table 16.22 Normalized organ doses—Lumbar spine ap—1–2 years

Beam projection:		ap										**Table cells:**
Focus-to-detector distance (cm):		115										(+) = normal font = optimal field setting
Tube voltage (kVp):		60										**(−) = bold font = suboptimal field setting**
Total filtrations:		See last table row										Empty table cells correspond to values <0.01
Organ or tissue	**Organ dose D_T: Entrance dose K_E**								**(mSv ∗ mGy^{-1})**			
	(+)	(−)	(+)	(−)	(+)	(−)	(+)	(−)	(+)	(−)	(+)	(−)
Skull												
Brain												
Oral mucosa												
Salivary glands												
Cervical spine												
Thyroid												
Thymus												
Thoracic spine	0.01	**0.05**	0.02	**0.07**	0.02	**0.09**	0.03	**0.10**	0.03	**0.13**	0.03	**0.11**
Claviculae												
Scapulae				**0.01**		**0.02**		**0.02**		**0.02**		**0.02**
Ribs	0.03	**0.22**	0.04	**0.25**	0.04	**0.27**	0.04	**0.29**	0.05	**0.32**	0.05	**0.31**
Oesophagus		**0.07**	0.01	**0.09**	0.01	**0.10**	0.01	**0.12**	0.02	**0.14**	0.02	**0.13**
Trachea												
Lung		**0.02**		**0.03**		**0.03**		**0.04**		**0.05**		**0.04**
Mamma				**0.01**		**0.01**		**0.01**		**0.02**		**0.02**
Heart		**0.04**		**0.05**		**0.06**		**0.06**	0.01	**0.08**	0.01	**0.07**
Stomach	0.05	**0.37**	0.07	**0.45**	0.08	**0.51**	0.09	**0.55**	0.10	**0.63**	0.10	**0.62**
Spleen		**0.07**	0.01	**0.10**	0.02	**0.12**	0.02	**0.13**	0.02	**0.16**	0.02	**0.15**
Liver	0.07	**0.29**	0.09	**0.36**	0.10	**0.40**	0.10	**0.44**	0.12	**0.50**	0.12	**0.49**
Gall bladder	0.29	**0.37**	0.36	**0.47**	0.41	**0.55**	0.46	**0.60**	0.53	**0.71**	0.51	**0.68**
Pancreas	0.06	**0.19**	0.08	**0.24**	0.10	**0.29**	0.11	**0.32**	0.13	**0.39**	0.12	**0.36**
Small intestine	0.14	**0.28**	0.18	**0.36**	0.21	**0.41**	0.23	**0.46**	0.28	**0.55**	0.26	**0.52**
Large intestine	0.11	**0.34**	0.14	**0.43**	0.16	**0.49**	0.17	**0.54**	0.21	**0.64**	0.20	**0.61**
Sigma/rectum	0.03	**0.24**	0.04	**0.31**	0.05	**0.36**	0.05	**0.40**	0.06	**0.48**	0.06	**0.45**
Lumbar spine	0.12	**0.18**	0.17	**0.25**	0.20	**0.30**	0.23	**0.35**	0.29	**0.44**	0.26	**0.38**
Pelvis	0.05	**0.24**	0.07	**0.32**	0.09	**0.38**	0.10	**0.43**	0.12	**0.54**	0.11	**0.49**
Adrenals	0.01	**0.07**	0.02	**0.09**	0.02	**0.11**	0.02	**0.13**	0.03	**0.16**	0.02	**0.14**
Kidneys	0.02	**0.07**	0.03	**0.09**	0.04	**0.11**	0.04	**0.13**	0.05	**0.16**	0.05	**0.14**
Urinary bladder	0.28	**0.52**	0.33	**0.63**	0.37	**0.71**	0.40	**0.77**	0.45	**0.88**	0.45	**0.86**
Uterus	0.23	**0.31**	0.30	**0.39**	0.34	**0.46**	0.38	**0.51**	0.45	**0.61**	0.43	**0.58**
Ovaries	0.05	**0.23**	0.07	**0.29**	0.08	**0.34**	0.09	**0.39**	0.11	**0.47**	0.11	**0.44**
Prostate	0.03	**0.34**	0.04	**0.43**	0.05	**0.49**	0.05	**0.54**	0.07	**0.64**	0.06	**0.61**
Testes	0.01	**0.98**	0.01	**1.08**	0.02	**1.14**	0.02	**1.18**	0.02	**1.25**	0.02	**1.24**
Upper arm								**0.01**		**0.02**		**0.01**
Elbow		**0.03**		**0.04**		**0.04**		**0.05**		**0.06**		**0.05**
Forearm		**0.05**		**0.06**		**0.07**		**0.09**		**0.11**		**0.09**
Upper leg		**0.23**	0.01	**0.30**	0.01	**0.36**	0.02	**0.40**	0.02	**0.49**	0.02	**0.46**
Knee		**0.01**		**0.02**		**0.02**		**0.03**		**0.04**		**0.03**
Lower leg												
Lymph nodes	0.08	**0.17**	0.10	**0.21**	0.11	**0.24**	0.13	**0.27**	0.15	**0.32**	0.14	**0.30**
Skin	0.02	**0.09**	0.02	**0.09**	0.03	**0.09**	0.03	**0.10**	0.03	**0.10**	0.03	**0.10**
Muscle	0.03	**0.10**	0.04	**0.12**	0.04	**0.14**	0.05	**0.15**	0.05	**0.17**	0.05	**0.17**
Skeleton	0.02	**0.08**	0.02	**0.11**	0.02	**0.12**	0.03	**0.14**	0.03	**0.17**	0.03	**0.15**
Red bone marrow		**0.03**		**0.03**		**0.04**	0.01	**0.04**	0.01	**0.05**	0.01	**0.05**
Whole body dose	0.03	**0.10**	0.04	**0.12**	0.04	**0.13**	0.04	**0.15**	0.05	**0.17**	0.05	**0.16**
ED (ICRP 60)	0.04	**0.26**	0.05	**0.31**	0.06	**0.34**	0.07	**0.37**	0.08	**0.42**	0.08	**0.40**
ED (ICRP 103)	0.04	**0.19**	0.05	**0.23**	0.06	**0.26**	0.07	**0.28**	0.08	**0.32**	0.07	**0.31**
Total filtration (mm) **Al**	**1.5**		**2.5**		**3.5**		**4.5**		**3.5**		**3.5**	
Cu	**0**		**0**		**0**		**0**		**0.1**		**0.2**	

Table 16.23 Normalized organ doses—Lumbar spine ap—1–2 years

Beam projection:			ap									**Table cells:**
Focus-to-detector distance (cm):			115									(+) = normal font = optimal field setting
Tube voltage (kVp):			70									**(−) = bold font = suboptimal field setting**
Total filtrations:			See last table row									Empty table cells correspond to values <0.01
Organ or tissue	**Organ dose D_T: Entrance dose K_E**							**(mSv * mGy^{-1})**				
	(+)	(−)	(+)	(−)	(+)	(−)	(+)	(−)	(+)	(−)	(+)	(−)
Skull												
Brain												
Oral mucosa												
Salivary glands												
Cervical spine												
Thyroid												
Thymus										**0.01**		**0.01**
Thoracic spine	0.02	**0.07**	0.02	**0.09**	0.03	**0.11**	0.03	**0.13**	0.04	**0.16**	0.05	**0.19**
Claviculae										**0.01**		**0.02**
Scapulae		**0.01**		**0.02**		**0.02**		**0.03**		**0.03**		**0.04**
Ribs	0.04	**0.24**	0.04	**0.27**	0.05	**0.29**	0.05	**0.31**	0.05	**0.34**	0.06	**0.36**
Oesophagus	0.01	**0.08**	0.01	**0.11**	0.02	**0.12**	0.02	**0.14**	0.02	**0.17**	0.03	**0.19**
Trachea												
Lung		**0.03**		**0.04**		**0.04**		**0.05**		**0.06**	0.01	**0.07**
Mamma		**0.01**		**0.01**		**0.02**		**0.02**		**0.02**		**0.03**
Heart		**0.05**		**0.06**	0.01	**0.07**	0.01	**0.08**	0.02	**0.09**	0.02	**0.11**
Stomach	0.06	**0.42**	0.08	**0.50**	0.09	**0.56**	0.10	**0.61**	0.11	**0.69**	0.13	**0.76**
Spleen	0.01	**0.09**	0.02	**0.12**	0.02	**0.14**	0.02	**0.16**	0.03	**0.19**	0.04	**0.23**
Liver	0.08	**0.33**	0.10	**0.40**	0.11	**0.45**	0.12	**0.49**	0.13	**0.56**	0.15	**0.61**
Gall bladder	0.34	**0.44**	0.41	**0.55**	0.47	**0.62**	0.51	**0.68**	0.59	**0.80**	0.65	**0.89**
Pancreas	0.08	**0.23**	0.10	**0.30**	0.12	**0.34**	0.13	**0.38**	0.15	**0.46**	0.18	**0.52**
Small intestine	0.17	**0.34**	0.21	**0.42**	0.24	**0.48**	0.27	**0.53**	0.31	**0.62**	0.35	**0.71**
Large intestine	0.13	**0.40**	0.16	**0.49**	0.18	**0.56**	0.20	**0.62**	0.23	**0.71**	0.27	**0.80**
Sigma/rectum	0.04	**0.29**	0.05	**0.37**	0.06	**0.42**	0.06	**0.46**	0.08	**0.55**	0.09	**0.62**
Lumbar spine	0.17	**0.25**	0.22	**0.32**	0.26	**0.39**	0.29	**0.44**	0.36	**0.55**	0.43	**0.66**
Pelvis	0.07	**0.31**	0.09	**0.39**	0.11	**0.46**	0.12	**0.52**	0.15	**0.64**	0.18	**0.74**
Adrenals	0.02	**0.09**	0.02	**0.12**	0.02	**0.14**	0.03	**0.16**	0.03	**0.19**	0.04	**0.23**
Kidneys	0.03	**0.09**	0.04	**0.12**	0.05	**0.14**	0.05	**0.16**	0.07	**0.20**	0.08	**0.23**
Urinary bladder	0.31	**0.59**	0.37	**0.70**	0.41	**0.78**	0.44	**0.85**	0.49	**0.96**	0.54	**1.05**
Uterus	0.28	**0.37**	0.34	**0.46**	0.39	**0.53**	0.43	**0.59**	0.50	**0.70**	0.57	**0.79**
Ovaries	0.07	**0.28**	0.09	**0.35**	0.10	**0.41**	0.11	**0.45**	0.14	**0.54**	0.16	**0.62**
Prostate	0.04	**0.40**	0.05	**0.49**	0.06	**0.56**	0.06	**0.61**	0.08	**0.71**	0.09	**0.79**
Testes	0.01	**1.03**	0.02	**1.13**	0.02	**1.18**	0.02	**1.22**	0.03	**1.29**	0.03	**1.34**
Upper arm				**0.01**		**0.01**		**0.02**		**0.02**		**0.02**
Elbow		**0.04**		**0.05**		**0.06**		**0.06**		**0.08**		**0.10**
Forearm		**0.06**		**0.08**		**0.10**		**0.11**		**0.14**		**0.16**
Upper leg	0.01	**0.29**	0.02	**0.36**	0.02	**0.42**	0.02	**0.47**	0.03	**0.56**	0.03	**0.65**
Knee		**0.02**		**0.03**		**0.03**		**0.04**		**0.05**		**0.06**
Lower leg												
Lymph nodes	0.09	**0.20**	0.11	**0.25**	0.13	**0.28**	0.14	**0.31**	0.17	**0.36**	0.19	**0.41**
Skin	0.02	**0.09**	0.03	**0.10**	0.03	**0.10**	0.03	**0.10**	0.03	**0.11**	0.03	**0.12**
Muscle	0.04	**0.12**	0.04	**0.14**	0.05	**0.16**	0.05	**0.17**	0.06	**0.19**	0.07	**0.21**
Skeleton	0.02	**0.10**	0.03	**0.13**	0.03	**0.15**	0.03	**0.16**	0.04	**0.20**	0.05	**0.23**
Red bone marrow		**0.03**		**0.04**	0.01	**0.05**	0.01	**0.05**	0.02	**0.07**	0.02	**0.08**
Whole body dose	0.03	**0.11**	0.04	**0.14**	0.05	**0.15**	0.05	**0.17**	0.06	**0.19**	0.06	**0.21**
ED (ICRP 60)	0.05	**0.29**	0.06	**0.34**	0.07	**0.38**	0.08	**0.41**	0.09	**0.46**	0.10	**0.50**
ED (ICRP 103)	0.05	**0.22**	0.06	**0.26**	0.07	**0.29**	0.08	**0.32**	0.09	**0.36**	0.10	**0.40**
Total filtration (mm) Al	**1.5**		**2.5**		**3.5**		**4.5**		**3.5**		**3.5**	
Cu	**0**		**0**		**0**		**0**		**0.1**		**0.2**	

Table 16.24 Normalized organ doses—Lumbar spine ap—3–7 years

Beam projection:			ap									**Table cells:**
Focus-to-detector distance (cm):			115									(+) = normal font = optimal field setting
Tube voltage (kVp):			60									**(−) = bold font = suboptimal field setting**
Total filtrations:			See last table row									Empty table cells correspond to values <0.01
Organ or tissue	**Organ dose D_T: Entrance dose K_E**								**(mSv * mGy^{-1})**			
	(+)	(−)	(+)	(−)	(+)	(−)	(+)	(−)	(+)	(−)	(+)	(−)
Skull												
Brain												
Oral mucosa												
Salivary glands												
Cervical spine												
Thyroid												
Thymus												
Thoracic spine		**0.04**	0.01	**0.05**	0.01	**0.06**	0.02	**0.07**	0.02	**0.09**	0.02	**0.08**
Claviculae												
Scapulae						**0.01**		**0.01**		**0.02**		**0.01**
Ribs	0.02	**0.21**	0.03	**0.25**	0.03	**0.27**	0.03	**0.28**	0.03	**0.31**	0.03	**0.31**
Oesophagus		**0.04**		**0.06**		**0.07**		**0.07**	0.01	**0.09**		**0.08**
Trachea												
Lung		**0.02**		**0.02**		**0.03**		**0.03**		**0.04**		**0.04**
Mamma								**0.01**		**0.01**		**0.01**
Heart		**0.03**		**0.04**		**0.05**		**0.05**		**0.07**		**0.06**
Stomach	0.05	**0.37**	0.06	**0.46**	0.07	**0.52**	0.08	**0.57**	0.09	**0.66**	0.09	**0.63**
Spleen		**0.06**		**0.08**	0.01	**0.10**	0.01	**0.12**	0.02	**0.15**	0.01	**0.13**
Liver	0.05	**0.26**	0.07	**0.33**	0.08	**0.37**	0.08	**0.41**	0.10	**0.48**	0.09	**0.46**
Gall bladder	0.25	**0.33**	0.32	**0.43**	0.37	**0.50**	0.40	**0.55**	0.48	**0.66**	0.46	**0.63**
Pancreas	0.05	**0.16**	0.06	**0.21**	0.07	**0.25**	0.08	**0.28**	0.10	**0.35**	0.09	**0.32**
Small intestine	0.12	**0.25**	0.16	**0.32**	0.18	**0.38**	0.21	**0.43**	0.25	**0.51**	0.23	**0.48**
Large intestine	0.09	**0.31**	0.12	**0.39**	0.14	**0.46**	0.15	**0.51**	0.19	**0.60**	0.17	**0.57**
Sigma/rectum	0.02	**0.21**	0.03	**0.27**	0.04	**0.32**	0.04	**0.36**	0.06	**0.44**	0.05	**0.41**
Lumbar spine	0.09	**0.14**	0.12	**0.19**	0.15	**0.23**	0.17	**0.26**	0.22	**0.34**	0.18	**0.28**
Pelvis	0.04	**0.20**	0.05	**0.27**	0.06	**0.33**	0.07	**0.37**	0.09	**0.47**	0.08	**0.42**
Adrenals		**0.05**	0.01	**0.07**	0.01	**0.09**	0.02	**0.10**	0.02	**0.13**	0.02	**0.11**
Kidneys	0.02	**0.05**	0.02	**0.07**	0.03	**0.09**	0.03	**0.10**	0.04	**0.13**	0.03	**0.11**
Urinary bladder	0.25	**0.48**	0.31	**0.59**	0.35	**0.67**	0.38	**0.74**	0.43	**0.85**	0.42	**0.82**
Uterus	0.20	**0.27**	0.26	**0.35**	0.31	**0.42**	0.34	**0.47**	0.41	**0.57**	0.39	**0.53**
Ovaries	0.05	**0.19**	0.06	**0.25**	0.07	**0.30**	0.08	**0.34**	0.10	**0.42**	0.09	**0.38**
Prostate	0.03	**0.29**	0.04	**0.37**	0.05	**0.43**	0.06	**0.48**	0.07	**0.57**	0.06	**0.54**
Testes		**0.96**	0.01	**1.07**	0.01	**1.13**	0.02	**1.18**	0.02	**1.25**	0.02	**1.24**
Upper arm												
Elbow		**0.02**		**0.03**		**0.03**		**0.04**		**0.05**		**0.04**
Forearm		**0.03**		**0.05**		**0.06**		**0.06**		**0.08**		**0.07**
Upper leg		**0.12**		**0.15**		**0.18**		**0.21**	0.01	**0.26**		**0.23**
Knee												
Lower leg												
Lymph nodes	0.07	**0.15**	0.08	**0.19**	0.10	**0.22**	0.11	**0.25**	0.13	**0.30**	0.12	**0.28**
Skin	0.02	**0.08**	0.02	**0.09**	0.02	**0.09**	0.02	**0.09**	0.02	**0.10**	0.02	**0.10**
Muscle	0.03	**0.09**	0.03	**0.11**	0.04	**0.13**	0.04	**0.14**	0.05	**0.16**	0.04	**0.15**
Skeleton		**0.06**	0.01	**0.07**	0.01	**0.08**	0.02	**0.09**	0.02	**0.12**	0.02	**0.10**
Red bone marrow		**0.03**		**0.04**		**0.04**	0.01	**0.05**	0.01	**0.06**	0.01	**0.05**
Whole body dose	0.02	**0.09**	0.03	**0.11**	0.03	**0.12**	0.04	**0.13**	0.04	**0.16**	0.04	**0.15**
ED (ICRP 60)	0.04	**0.24**	0.05	**0.29**	0.06	**0.33**	0.06	**0.35**	0.07	**0.41**	0.07	**0.39**
ED (ICRP 103)	0.04	**0.18**	0.05	**0.22**	0.05	**0.25**	0.06	**0.27**	0.07	**0.31**	0.07	**0.30**
Total filtration (mm) **Al**	**1.5**		**2.5**		**3.5**		**4.5**		**3.5**		**3.5**	
Cu	**0**		**0**		**0**		**0**		**0.1**		**0.2**	

Table 16.25 Normalized organ doses—Lumbar spine ap—3–7 years

Beam projection:			ap									**Table cells:**
Focus-to-detector distance (cm):			115									(+) = normal font = optimal field setting
Tube voltage (kVp):			70									**(−) = bold font = suboptimal field setting**
Total filtrations:			See last table row									Empty table cells correspond to values <0.01
Organ or tissue	**Organ dose D_T: Entrance dose K_E**						**(mSv ∗ mGy^{-1})**					
	(+)	(−)	(+)	(−)	(+)	(−)	(+)	(−)	(+)	(−)	(+)	(−)
Skull												
Brain												
Oral mucosa												
Salivary glands												
Cervical spine												
Thyroid												
Thymus												**0.01**
Thoracic spine	0.01	**0.05**	0.02	**0.07**	0.02	**0.08**	0.02	**0.10**	0.03	**0.12**	0.04	**0.15**
Claviculae												
Scapulae				**0.01**		**0.02**		**0.02**		**0.02**		**0.03**
Ribs	0.02	**0.23**	0.03	**0.26**	0.03	**0.29**	0.03	**0.30**	0.04	**0.33**	0.04	**0.36**
Oesophagus		**0.05**		**0.07**		**0.08**	0.01	**0.09**	0.01	**0.11**	0.02	**0.13**
Trachea												
Lung		**0.02**		**0.03**		**0.04**		**0.04**		**0.05**		**0.06**
Mamma				**0.01**		**0.01**		**0.01**		**0.02**		**0.02**
Heart		**0.04**		**0.05**		**0.06**		**0.07**	0.01	**0.08**	0.01	**0.10**
Stomach	0.06	**0.43**	0.07	**0.52**	0.08	**0.59**	0.09	**0.64**	0.11	**0.73**	0.12	**0.81**
Spleen		**0.08**	0.01	**0.11**	0.02	**0.13**	0.02	**0.15**	0.02	**0.18**	0.03	**0.21**
Liver	0.06	**0.31**	0.08	**0.38**	0.09	**0.43**	0.10	**0.46**	0.11	**0.54**	0.12	**0.60**
Gall bladder	0.30	**0.40**	0.37	**0.51**	0.42	**0.58**	0.46	**0.64**	0.54	**0.76**	0.61	**0.86**
Pancreas	0.06	**0.20**	0.08	**0.26**	0.09	**0.30**	0.10	**0.34**	0.12	**0.41**	0.14	**0.48**
Small intestine	0.15	**0.31**	0.19	**0.39**	0.22	**0.45**	0.24	**0.50**	0.29	**0.60**	0.33	**0.69**
Large intestine	0.11	**0.37**	0.14	**0.46**	0.16	**0.53**	0.18	**0.59**	0.22	**0.69**	0.25	**0.78**
Sigma/rectum	0.03	**0.26**	0.04	**0.33**	0.05	**0.38**	0.06	**0.43**	0.07	**0.51**	0.08	**0.59**
Lumbar spine	0.12	**0.19**	0.16	**0.25**	0.19	**0.30**	0.22	**0.35**	0.28	**0.44**	0.34	**0.54**
Pelvis	0.05	**0.27**	0.07	**0.35**	0.08	**0.41**	0.10	**0.47**	0.12	**0.58**	0.15	**0.68**
Adrenals	0.01	**0.07**	0.02	**0.10**	0.02	**0.11**	0.02	**0.13**	0.03	**0.16**	0.03	**0.20**
Kidneys	0.02	**0.07**	0.03	**0.10**	0.04	**0.12**	0.04	**0.13**	0.05	**0.17**	0.06	**0.20**
Urinary bladder	0.29	**0.55**	0.35	**0.67**	0.39	**0.76**	0.42	**0.82**	0.47	**0.94**	0.52	**1.05**
Uterus	0.25	**0.34**	0.31	**0.43**	0.36	**0.50**	0.40	**0.55**	0.47	**0.66**	0.54	**0.76**
Ovaries	0.06	**0.24**	0.08	**0.31**	0.09	**0.37**	0.10	**0.41**	0.13	**0.49**	0.15	**0.58**
Prostate	0.04	**0.35**	0.05	**0.43**	0.06	**0.50**	0.07	**0.55**	0.08	**0.65**	0.10	**0.74**
Testes	0.01	**1.02**	0.01	**1.12**	0.02	**1.18**	0.02	**1.23**	0.02	**1.30**	0.03	**1.36**
Upper arm										**0.01**		**0.02**
Elbow		**0.03**		**0.03**		**0.04**		**0.05**		**0.06**		**0.08**
Forearm		**0.05**		**0.06**		**0.08**		**0.09**		**0.11**		**0.13**
Upper leg		**0.15**		**0.19**		**0.23**	0.01	**0.26**	0.01	**0.32**	0.02	**0.37**
Knee										**0.01**		**0.02**
Lower leg												
Lymph nodes	0.08	**0.18**	0.10	**0.23**	0.12	**0.26**	0.13	**0.29**	0.15	**0.34**	0.17	**0.39**
Skin	0.02	**0.09**	0.02	**0.09**	0.02	**0.10**	0.02	**0.10**	0.03	**0.10**	0.03	**0.11**
Muscle	0.03	**0.11**	0.04	**0.13**	0.04	**0.14**	0.05	**0.16**	0.05	**0.18**	0.06	**0.20**
Skeleton	0.01	**0.07**	0.01	**0.09**	0.02	**0.10**	0.02	**0.11**	0.02	**0.14**	0.03	**0.16**
Red bone marrow		**0.04**		**0.05**	0.01	**0.05**	0.01	**0.06**	0.02	**0.08**	0.02	**0.09**
Whole body dose	0.03	**0.10**	0.03	**0.12**	0.04	**0.14**	0.04	**0.15**	0.05	**0.18**	0.06	**0.20**
ED (ICRP 60)	0.05	**0.28**	0.06	**0.33**	0.06	**0.37**	0.07	**0.40**	0.08	**0.45**	0.10	**0.50**
ED (ICRP 103)	0.04	**0.21**	0.05	**0.25**	0.06	**0.28**	0.07	**0.30**	0.08	**0.35**	0.09	**0.39**
Total filtration (mm) **Al**	**1.5**		**2.5**		**3.5**		**4.5**		**3.5**		**3.5**	
Cu	**0**		**0**		**0**		**0**		**0.1**		**0.2**	

Table 16.26 Normalized organ doses—Lumbar spine ap—8–12 years

Beam projection:			ap									**Table cells:**
Focus-to-detector distance (cm):			115									(+) = normal font = optimal field setting
Tube voltage (kVp):			70									**(−) = bold font = suboptimal field setting**
Total filtrations:			See last table row									Empty table cells correspond to values <0.01
Organ or tissue	**Organ dose D_T: Entrance dose K_E**								**(mSv ∗ mGy^{-1})**			
	(+)	(−)	(+)	(−)	(+)	(−)	(+)	(−)	(+)	(−)	(+)	(−)
Skull												
Brain												
Oral mucosa												
Salivary glands												
Cervical spine												
Thyroid												
Thymus												
Thoracic spine		**0.04**	0.01	**0.05**	0.01	**0.06**	0.02	**0.07**	0.02	**0.09**	0.03	**0.12**
Claviculae												
Scapulae								**0.01**		**0.01**		**0.02**
Ribs	0.02	**0.21**	0.03	**0.24**	0.03	**0.26**	0.03	**0.28**	0.03	**0.31**	0.04	**0.33**
Oesophagus		**0.03**		**0.04**		**0.04**	0.01	**0.05**	0.01	**0.06**	0.02	**0.07**
Trachea												
Lung		**0.02**		**0.02**		**0.03**		**0.03**		**0.04**		**0.05**
Mamma						**0.01**		**0.01**		**0.01**		**0.02**
Heart		**0.03**		**0.04**		**0.05**		**0.06**		**0.07**		**0.09**
Stomach	0.05	**0.38**	0.06	**0.46**	0.07	**0.52**	0.08	**0.57**	0.10	**0.66**	0.12	**0.74**
Spleen		**0.07**	0.01	**0.09**	0.01	**0.10**	0.01	**0.12**	0.02	**0.15**	0.02	**0.18**
Liver	0.04	**0.25**	0.05	**0.30**	0.06	**0.35**	0.06	**0.38**	0.07	**0.45**	0.08	**0.51**
Gall bladder	0.26	**0.37**	0.32	**0.46**	0.37	**0.54**	0.41	**0.60**	0.48	**0.71**	0.55	**0.82**
Pancreas	0.05	**0.18**	0.07	**0.23**	0.08	**0.27**	0.09	**0.31**	0.11	**0.38**	0.13	**0.44**
Small intestine	0.13	**0.27**	0.17	**0.34**	0.20	**0.40**	0.22	**0.45**	0.26	**0.54**	0.31	**0.62**
Large intestine	0.10	**0.33**	0.13	**0.41**	0.15	**0.48**	0.17	**0.53**	0.20	**0.63**	0.23	**0.72**
Sigma/rectum	0.02	**0.22**	0.03	**0.29**	0.04	**0.33**	0.04	**0.37**	0.06	**0.45**	0.07	**0.52**
Lumbar spine	0.09	**0.15**	0.13	**0.20**	0.15	**0.24**	0.17	**0.27**	0.22	**0.35**	0.27	**0.43**
Pelvis	0.04	**0.20**	0.05	**0.26**	0.06	**0.31**	0.07	**0.35**	0.09	**0.44**	0.11	**0.53**
Adrenals		**0.06**	0.01	**0.08**	0.02	**0.10**	0.02	**0.11**	0.02	**0.14**	0.03	**0.17**
Kidneys	0.02	**0.06**	0.03	**0.08**	0.03	**0.09**	0.03	**0.11**	0.04	**0.14**	0.05	**0.17**
Urinary bladder	0.21	**0.52**	0.26	**0.63**	0.29	**0.72**	0.32	**0.78**	0.36	**0.91**	0.40	**1.02**
Uterus	0.22	**0.30**	0.28	**0.39**	0.33	**0.45**	0.36	**0.50**	0.43	**0.61**	0.50	**0.71**
Ovaries	0.05	**0.21**	0.07	**0.27**	0.08	**0.32**	0.09	**0.36**	0.12	**0.44**	0.14	**0.52**
Prostate	0.02	**0.33**	0.03	**0.41**	0.04	**0.48**	0.04	**0.53**	0.05	**0.63**	0.06	**0.73**
Testes		**1.01**		**1.12**		**1.19**	0.01	**1.23**	0.01	**1.31**	0.02	**1.37**
Upper arm												
Elbow		**0.02**		**0.02**		**0.03**		**0.03**		**0.04**		**0.05**
Forearm		**0.03**		**0.04**		**0.05**		**0.06**		**0.08**		**0.10**
Upper leg		**0.09**		**0.12**		**0.15**		**0.17**		**0.21**		**0.25**
Knee												
Lower leg												
Lymph nodes	0.07	**0.16**	0.09	**0.20**	0.10	**0.23**	0.11	**0.26**	0.14	**0.31**	0.16	**0.36**
Skin	0.02	**0.08**	0.02	**0.09**	0.02	**0.09**	0.02	**0.09**	0.02	**0.10**	0.03	**0.10**
Muscle	0.03	**0.10**	0.03	**0.11**	0.04	**0.13**	0.04	**0.14**	0.05	**0.16**	0.05	**0.18**
Skeleton		**0.06**	0.01	**0.07**	0.01	**0.08**	0.01	**0.09**	0.02	**0.11**	0.02	**0.13**
Red bone marrow		**0.03**		**0.04**	0.01	**0.05**	0.01	**0.06**	0.02	**0.07**	0.02	**0.09**
Whole body dose	0.02	**0.09**	0.03	**0.11**	0.03	**0.13**	0.04	**0.14**	0.04	**0.16**	0.05	**0.18**
ED (ICRP 60)	0.04	**0.25**	0.05	**0.30**	0.05	**0.34**	0.06	**0.37**	0.07	**0.42**	0.08	**0.47**
ED (ICRP 103)	0.04	**0.18**	0.04	**0.22**	0.05	**0.25**	0.06	**0.28**	0.07	**0.32**	0.08	**0.36**
Total filtration (mm) Al	**1.5**		**2.5**		**3.5**		**4.5**		**3.5**		**3.5**	
Cu	**0**		**0**		**0**		**0**		**0.1**		**0.2**	

Table 16.27 Normalized organ doses—Lumbar spine ap—8–12 years

Beam projection:			ap									**Table cells:**
Focus-to-detector distance (cm):			115									(+) = normal font = optimal field setting
Tube voltage (kVp):			80									**(−) = bold font = suboptimal field setting**
Total filtrations:			See last table row									Empty table cells correspond to values <0.01
Organ or tissue	**Organ dose D_T: Entrance dose K_E**							**(mSv ∗ mGy^{-1})**				
	(+)	(−)	(+)	(−)	(+)	(−)	(+)	(−)	(+)	(−)	(+)	(−)
Skull												
Brain												
Oral mucosa												
Salivary glands												
Cervical spine												
Thyroid												
Thymus												**0.01**
Thoracic spine	0.01	**0.05**	0.02	**0.07**	0.02	**0.08**	0.02	**0.09**	0.03	**0.12**	0.03	**0.14**
Claviculae												
Scapulae				**0.01**		**0.01**		**0.01**		**0.02**		**0.02**
Ribs	0.02	**0.23**	0.03	**0.26**	0.03	**0.28**	0.03	**0.29**	0.04	**0.32**	0.04	**0.34**
Oesophagus		**0.03**	0.01	**0.04**	0.01	**0.05**	0.01	**0.06**	0.02	**0.07**	0.02	**0.08**
Trachea												
Lung		**0.02**		**0.03**		**0.04**		**0.04**		**0.05**		**0.06**
Mamma				**0.01**		**0.01**		**0.01**		**0.02**		**0.02**
Heart		**0.04**		**0.05**		**0.06**		**0.07**		**0.09**	0.01	**0.10**
Stomach	0.06	**0.43**	0.07	**0.52**	0.09	**0.58**	0.09	**0.63**	0.11	**0.72**	0.13	**0.81**
Spleen	0.01	**0.09**	0.01	**0.11**	0.02	**0.13**	0.02	**0.14**	0.02	**0.18**	0.03	**0.21**
Liver	0.05	**0.29**	0.06	**0.35**	0.06	**0.39**	0.07	**0.43**	0.08	**0.49**	0.09	**0.55**
Gall bladder	0.30	**0.43**	0.37	**0.54**	0.42	**0.61**	0.46	**0.68**	0.53	**0.79**	0.60	**0.90**
Pancreas	0.06	**0.22**	0.08	**0.28**	0.09	**0.32**	0.10	**0.36**	0.13	**0.44**	0.15	**0.51**
Small intestine	0.16	**0.33**	0.20	**0.40**	0.23	**0.47**	0.25	**0.51**	0.30	**0.61**	0.34	**0.70**
Large intestine	0.12	**0.39**	0.15	**0.48**	0.17	**0.55**	0.19	**0.60**	0.23	**0.70**	0.26	**0.80**
Sigma/rectum	0.03	**0.27**	0.04	**0.34**	0.05	**0.39**	0.05	**0.43**	0.07	**0.51**	0.08	**0.59**
Lumbar spine	0.12	**0.20**	0.16	**0.26**	0.19	**0.30**	0.22	**0.35**	0.27	**0.43**	0.32	**0.52**
Pelvis	0.05	**0.25**	0.07	**0.32**	0.08	**0.38**	0.09	**0.43**	0.11	**0.52**	0.14	**0.62**
Adrenals	0.01	**0.08**	0.02	**0.10**	0.02	**0.12**	0.02	**0.14**	0.03	**0.17**	0.04	**0.20**
Kidneys	0.03	**0.08**	0.03	**0.10**	0.04	**0.12**	0.04	**0.14**	0.05	**0.17**	0.07	**0.20**
Urinary bladder	0.24	**0.59**	0.29	**0.71**	0.32	**0.80**	0.35	**0.86**	0.39	**0.99**	0.43	**1.10**
Uterus	0.26	**0.36**	0.33	**0.46**	0.37	**0.53**	0.41	**0.58**	0.49	**0.69**	0.55	**0.80**
Ovaries	0.07	**0.26**	0.09	**0.33**	0.10	**0.38**	0.11	**0.43**	0.14	**0.51**	0.16	**0.59**
Prostate	0.03	**0.39**	0.04	**0.48**	0.05	**0.55**	0.05	**0.60**	0.06	**0.71**	0.07	**0.80**
Testes		**1.07**	0.01	**1.17**	0.01	**1.23**	0.01	**1.28**	0.02	**1.35**	0.02	**1.41**
Upper arm												**0.01**
Elbow		**0.02**		**0.03**		**0.04**		**0.04**		**0.05**		**0.06**
Forearm		**0.04**		**0.06**		**0.07**		**0.08**		**0.10**		**0.12**
Upper leg		**0.12**		**0.15**		**0.18**		**0.20**		**0.24**	0.01	**0.29**
Knee												
Lower leg												
Lymph nodes	0.08	**0.19**	0.10	**0.24**	0.12	**0.27**	0.13	**0.30**	0.15	**0.35**	0.18	**0.40**
Skin	0.02	**0.09**	0.02	**0.09**	0.02	**0.09**	0.02	**0.10**	0.03	**0.10**	0.03	**0.11**
Muscle	0.03	**0.11**	0.04	**0.13**	0.04	**0.14**	0.04	**0.16**	0.05	**0.18**	0.06	**0.20**
Skeleton	0.01	**0.07**	0.01	**0.08**	0.02	**0.09**	0.02	**0.10**	0.02	**0.12**	0.03	**0.14**
Red bone marrow		**0.04**	0.01	**0.06**	0.01	**0.06**	0.02	**0.07**	0.02	**0.09**	0.03	**0.11**
Whole body dose	0.03	**0.11**	0.03	**0.13**	0.04	**0.14**	0.04	**0.15**	0.05	**0.18**	0.05	**0.20**
ED (ICRP 60)	0.04	**0.29**	0.05	**0.34**	0.06	**0.37**	0.07	**0.40**	0.08	**0.46**	0.09	**0.50**
ED (ICRP 103)	0.04	**0.21**	0.05	**0.25**	0.06	**0.28**	0.07	**0.31**	0.08	**0.36**	0.09	**0.40**
Total filtration (mm) **Al**	**1.5**		**2.5**		**3.5**		**4.5**		**3.5**		**3.5**	
Cu	**0**		**0**		**0**		**0**		**0.1**		**0.2**	

Table 16.28 Normalized organ doses—Lumbar spine ap—13–17 years

Beam projection:			ap									**Table cells:**
Focus-to-detector distance (cm):			115						(+) = normal font = optimal field setting			
Tube voltage (kVp):			70						**(−) = bold font = suboptimal field setting**			
Total filtrations:			See last table row						Empty table cells correspond to values <0.01			
Organ or tissue	**Organ dose D_T: Entrance dose K_E**							**(mSv * mGy^{-1})**				
	(+)	(−)	(+)	(−)	(+)	(−)	(+)	(−)	(+)	(−)	(+)	(−)
Skull												
Brain												
Oral mucosa												
Salivary glands												
Cervical spine												
Thyroid												
Thymus												
Thoracic spine		**0.03**		**0.03**	0.01	**0.04**	0.01	**0.05**	0.02	**0.06**	0.02	**0.08**
Claviculae												
Scapulae												**0.01**
Ribs	0.02	**0.18**	0.03	**0.21**	0.03	**0.23**	0.03	**0.24**	0.04	**0.27**	0.04	**0.29**
Oesophagus		**0.04**		**0.05**		**0.06**		**0.07**	0.01	**0.09**	0.01	**0.11**
Trachea												
Lung		**0.02**		**0.02**		**0.02**		**0.03**		**0.04**		**0.04**
Mamma		**0.05**	0.01	**0.07**	0.01	**0.08**	0.02	**0.10**	0.02	**0.12**	0.02	**0.15**
Heart		**0.03**		**0.04**		**0.05**		**0.05**		**0.07**		**0.08**
Stomach	0.05	**0.29**	0.06	**0.36**	0.07	**0.41**	0.08	**0.46**	0.10	**0.54**	0.11	**0.62**
Spleen		**0.04**		**0.05**	0.01	**0.07**	0.01	**0.08**	0.02	**0.10**	0.02	**0.12**
Liver	0.02	**0.17**	0.03	**0.21**	0.04	**0.25**	0.04	**0.27**	0.05	**0.33**	0.06	**0.37**
Gall bladder	0.15	**0.30**	0.19	**0.39**	0.22	**0.45**	0.25	**0.50**	0.30	**0.61**	0.35	**0.70**
Pancreas	0.05	**0.13**	0.06	**0.17**	0.07	**0.20**	0.08	**0.23**	0.10	**0.29**	0.12	**0.35**
Small intestine	0.12	**0.21**	0.15	**0.28**	0.18	**0.32**	0.20	**0.36**	0.25	**0.44**	0.29	**0.52**
Large intestine	0.09	**0.25**	0.12	**0.32**	0.14	**0.37**	0.16	**0.42**	0.19	**0.50**	0.23	**0.59**
Sigma/rectum	0.02	**0.17**	0.03	**0.21**	0.04	**0.25**	0.04	**0.28**	0.06	**0.35**	0.07	**0.41**
Lumbar spine	0.07	**0.10**	0.09	**0.13**	0.11	**0.16**	0.12	**0.18**	0.16	**0.23**	0.20	**0.29**
Pelvis	0.03	**0.12**	0.04	**0.16**	0.05	**0.19**	0.06	**0.22**	0.07	**0.28**	0.09	**0.34**
Adrenals		**0.04**	0.01	**0.05**	0.02	**0.06**	0.02	**0.07**	0.02	**0.10**	0.03	**0.12**
Kidneys	0.02	**0.04**	0.02	**0.05**	0.02	**0.06**	0.03	**0.07**	0.04	**0.09**	0.05	**0.12**
Urinary bladder	0.26	**0.46**	0.32	**0.57**	0.36	**0.65**	0.40	**0.71**	0.46	**0.84**	0.52	**0.95**
Uterus	0.56	**0.72**	0.71	**0.93**	0.83	**1.09**	0.93	**1.23**	1.13	**1.50**	1.32	**1.76**
Ovaries	0.05	**0.16**	0.07	**0.21**	0.08	**0.25**	0.09	**0.29**	0.12	**0.36**	0.14	**0.42**
Prostate	0.04	**0.31**	0.05	**0.39**	0.06	**0.46**	0.06	**0.52**	0.08	**0.62**	0.09	**0.73**
Testes		**0.88**		**1.00**		**1.08**	0.01	**1.14**	0.01	**1.24**	0.02	**1.32**
Upper arm												
Elbow				**0.01**		**0.02**		**0.02**		**0.02**		**0.03**
Forearm		**0.02**		**0.02**		**0.03**		**0.03**		**0.04**		**0.05**
Upper leg		**0.08**		**0.11**		**0.13**		**0.15**		**0.19**		**0.22**
Knee												
Lower leg												
Lymph nodes	0.06	**0.13**	0.07	**0.16**	0.08	**0.19**	0.09	**0.21**	0.11	**0.26**	0.13	**0.30**
Skin	0.03	**0.08**	0.03	**0.09**	0.03	**0.09**	0.03	**0.10**	0.03	**0.10**	0.03	**0.11**
Muscle	0.03	**0.09**	0.04	**0.11**	0.04	**0.12**	0.04	**0.13**	0.05	**0.15**	0.06	**0.17**
Skeleton		**0.04**		**0.06**	0.01	**0.06**	0.01	**0.07**	0.02	**0.09**	0.02	**0.10**
Red bone marrow		**0.03**		**0.04**	0.01	**0.04**	0.01	**0.05**	0.02	**0.06**	0.02	**0.07**
Whole body dose	0.03	**0.08**	0.03	**0.10**	0.04	**0.11**	0.04	**0.12**	0.05	**0.15**	0.05	**0.17**
ED (ICRP 60)	0.04	**0.21**	0.05	**0.26**	0.06	**0.29**	0.06	**0.32**	0.08	**0.37**	0.09	**0.42**
ED (ICRP 103)	0.04	**0.16**	0.05	**0.20**	0.06	**0.23**	0.06	**0.25**	0.08	**0.29**	0.09	**0.34**
Total filtration (mm) **Al**	**1.5**		**2.5**		**3.5**		**4.5**		**3.5**		**3.5**	
Cu	**0**		**0**		**0**		**0**		**0.1**		**0.2**	

Table 16.29 Normalized organ doses—Lumbar spine ap—13–17 years

Beam projection:	ap	**Table cells:**
Focus-to-detector distance (cm):	115	(+) = normal font = optimal field setting
Tube voltage (kVp):	80	**(−) = bold font = suboptimal field setting**
Total filtrations:	See last table row	Empty table cells correspond to values <0.01

Organ or tissue	**Organ dose D_T: Entrance dose K_E**						**(mSv * mGy^{-1})**					
	(+)	(−)	(+)	(−)	(+)	(−)	(+)	(−)	(+)	(−)	(+)	(−)
Skull												
Brain												
Oral mucosa												
Salivary glands												
Cervical spine												
Thyroid												
Thymus												
Thoracic spine		**0.04**	0.01	**0.05**	0.02	**0.06**	0.02	**0.06**	0.02	**0.08**	0.03	**0.10**
Claviculae												
Scapulae										**0.01**		**0.01**
Ribs	0.03	**0.20**	0.03	**0.22**	0.03	**0.24**	0.04	**0.26**	0.04	**0.28**	0.04	**0.30**
Oesophagus		**0.05**		**0.06**		**0.08**	0.01	**0.09**	0.01	**0.11**	0.02	**0.12**
Trachea												
Lung		**0.02**		**0.03**		**0.03**		**0.04**		**0.04**		**0.05**
Mamma	0.01	**0.07**	0.01	**0.09**	0.02	**0.11**	0.02	**0.12**	0.02	**0.15**	0.03	**0.18**
Heart		**0.04**		**0.05**		**0.06**		**0.06**		**0.08**	0.01	**0.09**
Stomach	0.06	**0.34**	0.07	**0.41**	0.08	**0.47**	0.09	**0.52**	0.11	**0.60**	0.13	**0.68**
Spleen		**0.05**	0.01	**0.07**	0.01	**0.08**	0.02	**0.10**	0.02	**0.12**	0.03	**0.14**
Liver	0.03	**0.20**	0.04	**0.25**	0.04	**0.28**	0.05	**0.31**	0.06	**0.37**	0.07	**0.42**
Gall bladder	0.18	**0.37**	0.23	**0.46**	0.26	**0.52**	0.29	**0.58**	0.34	**0.69**	0.39	**0.79**
Pancreas	0.06	**0.17**	0.08	**0.21**	0.09	**0.25**	0.10	**0.28**	0.12	**0.34**	0.15	**0.41**
Small intestine	0.15	**0.26**	0.19	**0.33**	0.21	**0.38**	0.24	**0.43**	0.29	**0.51**	0.33	**0.60**
Large intestine	0.11	**0.30**	0.14	**0.38**	0.17	**0.44**	0.19	**0.48**	0.22	**0.57**	0.26	**0.66**
Sigma/rectum	0.03	**0.21**	0.04	**0.26**	0.05	**0.30**	0.06	**0.34**	0.07	**0.41**	0.08	**0.47**
Lumbar spine	0.09	**0.13**	0.12	**0.17**	0.14	**0.21**	0.16	**0.24**	0.20	**0.30**	0.24	**0.37**
Pelvis	0.04	**0.16**	0.06	**0.21**	0.07	**0.24**	0.08	**0.28**	0.10	**0.35**	0.12	**0.41**
Adrenals	0.01	**0.05**	0.02	**0.07**	0.02	**0.08**	0.02	**0.10**	0.03	**0.12**	0.04	**0.15**
Kidneys	0.02	**0.05**	0.03	**0.07**	0.03	**0.08**	0.04	**0.10**	0.05	**0.12**	0.06	**0.15**
Urinary bladder	0.30	**0.53**	0.36	**0.65**	0.41	**0.73**	0.44	**0.80**	0.51	**0.93**	0.56	**1.04**
Uterus	0.68	**0.89**	0.85	**1.12**	0.98	**1.30**	1.08	**1.44**	1.29	**1.74**	1.48	**2.01**
Ovaries	0.07	**0.21**	0.09	**0.26**	0.10	**0.31**	0.11	**0.35**	0.14	**0.42**	0.17	**0.49**
Prostate	0.05	**0.37**	0.06	**0.47**	0.07	**0.54**	0.08	**0.59**	0.09	**0.71**	0.11	**0.81**
Testes		**0.94**		**1.06**	0.01	**1.14**	0.01	**1.20**	0.02	**1.29**	0.02	**1.37**
Upper arm												
Elbow		**0.01**		**0.02**		**0.02**		**0.02**		**0.03**		**0.04**
Forearm		**0.02**		**0.03**		**0.04**		**0.04**		**0.06**		**0.07**
Upper leg		**0.11**		**0.14**		**0.16**		**0.18**		**0.22**		**0.27**
Knee												
Lower leg												
Lymph nodes	0.07	**0.16**	0.09	**0.20**	0.10	**0.23**	0.11	**0.25**	0.13	**0.30**	0.15	**0.34**
Skin	0.03	**0.09**	0.03	**0.09**	0.03	**0.10**	0.03	**0.10**	0.03	**0.11**	0.03	**0.11**
Muscle	0.03	**0.10**	0.04	**0.12**	0.04	**0.13**	0.05	**0.15**	0.06	**0.17**	0.06	**0.19**
Skeleton		**0.05**	0.01	**0.07**	0.01	**0.08**	0.02	**0.09**	0.02	**0.10**	0.02	**0.12**
Red bone marrow		**0.04**	0.01	**0.05**	0.01	**0.05**	0.02	**0.06**	0.02	**0.07**	0.03	**0.09**
Whole body dose	0.03	**0.10**	0.04	**0.11**	0.04	**0.13**	0.05	**0.14**	0.05	**0.16**	0.06	**0.18**
ED (ICRP 60)	0.05	**0.25**	0.06	**0.29**	0.07	**0.33**	0.08	**0.36**	0.09	**0.41**	0.10	**0.46**
ED (ICRP 103)	0.05	**0.19**	0.06	**0.23**	0.07	**0.26**	0.07	**0.28**	0.09	**0.33**	0.10	**0.38**
Total filtration (mm) Al	**1.5**		**2.5**		**3.5**		**4.5**		**3.5**		**3.5**	
Cu	**0**		**0**		**0**		**0**		**0.1**		**0.2**	

Table 16.30 Normalized organ doses—Lumbar spine ap—>17 years

Beam projection:		ap										**Table cells:**
Focus-to-detector distance (cm):		115										(+) = normal font = optimal field setting
Tube voltage (kVp):		70										**(−) = bold font = suboptimal field setting**
Total filtrations:		See last table row										Empty table cells correspond to values <0.01
Organ or tissue	**Organ dose D_T: Entrance dose K_E**							**(mSv ∗ mGy^{-1})**				
	(+)	(−)	(+)	(−)	(+)	(−)	(+)	(−)	(+)	(−)	(+)	(−)
Skull												
Brain												
Oral mucosa												
Salivary glands												
Cervical spine												
Thyroid												
Thymus												
Thoracic spine		**0.02**		**0.03**	0.01	**0.04**	0.01	**0.04**	0.02	**0.05**	0.02	**0.07**
Claviculae												
Scapulae												
Ribs	0.02	**0.17**	0.03	**0.19**	0.03	**0.21**	0.03	**0.22**	0.04	**0.24**	0.04	**0.26**
Oesophagus		**0.04**		**0.05**		**0.06**		**0.07**	0.01	**0.09**	0.01	**0.11**
Trachea												
Lung		**0.01**		**0.01**		**0.02**		**0.02**		**0.02**		**0.03**
Mamma										**0.01**		**0.01**
Heart		**0.02**		**0.03**		**0.03**		**0.04**		**0.05**		**0.06**
Stomach	0.04	**0.27**	0.05	**0.34**	0.06	**0.39**	0.07	**0.43**	0.09	**0.51**	0.10	**0.58**
Spleen		**0.04**		**0.05**		**0.06**	0.01	**0.07**	0.01	**0.08**	0.02	**0.10**
Liver	0.03	**0.16**	0.03	**0.20**	0.04	**0.24**	0.04	**0.26**	0.05	**0.31**	0.06	**0.36**
Gall bladder	0.18	**0.30**	0.22	**0.38**	0.26	**0.45**	0.29	**0.50**	0.35	**0.61**	0.41	**0.71**
Pancreas	0.05	**0.12**	0.06	**0.16**	0.07	**0.19**	0.08	**0.22**	0.10	**0.28**	0.12	**0.33**
Small intestine	0.12	**0.21**	0.15	**0.27**	0.18	**0.32**	0.20	**0.36**	0.24	**0.44**	0.29	**0.52**
Large intestine	0.09	**0.24**	0.11	**0.31**	0.14	**0.36**	0.15	**0.40**	0.19	**0.49**	0.22	**0.57**
Sigma/rectum	0.02	**0.16**	0.03	**0.20**	0.04	**0.24**	0.04	**0.27**	0.05	**0.33**	0.06	**0.39**
Lumbar spine	0.06	**0.09**	0.08	**0.12**	0.10	**0.15**	0.12	**0.17**	0.16	**0.22**	0.19	**0.28**
Pelvis	0.03	**0.11**	0.04	**0.14**	0.04	**0.17**	0.05	**0.20**	0.07	**0.25**	0.08	**0.31**
Adrenals		**0.04**	0.01	**0.05**	0.01	**0.06**	0.02	**0.07**	0.02	**0.09**	0.03	**0.12**
Kidneys	0.01	**0.04**	0.02	**0.05**	0.02	**0.06**	0.03	**0.07**	0.04	**0.09**	0.04	**0.11**
Urinary bladder	0.27	**0.46**	0.33	**0.57**	0.38	**0.66**	0.42	**0.72**	0.48	**0.85**	0.55	**0.97**
Uterus	0.22	**0.28**	0.28	**0.36**	0.33	**0.42**	0.37	**0.48**	0.45	**0.59**	0.53	**0.69**
Ovaries	0.05	**0.16**	0.06	**0.21**	0.08	**0.25**	0.09	**0.28**	0.11	**0.35**	0.14	**0.42**
Prostate	0.03	**0.28**	0.04	**0.36**	0.05	**0.42**	0.05	**0.48**	0.07	**0.58**	0.08	**0.68**
Testes		**0.84**		**0.97**		**1.06**		**1.13**	0.01	**1.24**	0.01	**1.33**
Upper arm												
Elbow								**0.01**		**0.02**		**0.02**
Forearm		**0.01**		**0.02**		**0.02**		**0.02**		**0.03**		**0.04**
Upper leg		**0.08**		**0.10**		**0.13**		**0.14**		**0.18**		**0.22**
Knee												
Lower leg												
Lymph nodes	0.06	**0.13**	0.07	**0.16**	0.09	**0.19**	0.10	**0.21**	0.12	**0.25**	0.14	**0.30**
Skin	0.03	**0.09**	0.03	**0.09**	0.03	**0.10**	0.03	**0.10**	0.03	**0.11**	0.03	**0.11**
Muscle	0.03	**0.09**	0.03	**0.10**	0.04	**0.12**	0.04	**0.13**	0.05	**0.15**	0.06	**0.17**
Skeleton		**0.04**		**0.05**	0.01	**0.06**	0.01	**0.07**	0.02	**0.08**	0.02	**0.10**
Red bone marrow		**0.03**		**0.03**	0.01	**0.04**	0.01	**0.05**	0.02	**0.06**	0.02	**0.07**
Whole body dose	0.03	**0.08**	0.03	**0.10**	0.04	**0.11**	0.04	**0.12**	0.05	**0.14**	0.05	**0.16**
ED (ICRP 60)	0.04	**0.20**	0.05	**0.24**	0.05	**0.28**	0.06	**0.30**	0.07	**0.35**	0.08	**0.40**
ED (ICRP 103)	0.03	**0.14**	0.04	**0.18**	0.05	**0.20**	0.06	**0.22**	0.07	**0.27**	0.08	**0.30**
Total filtration (mm) Al	**1.5**		**2.5**		**3.5**		**4.5**		**3.5**		**3.5**	
Cu	**0**		**0**		**0**		**0**		**0.1**		**0.2**	

Table 16.31 Normalized organ doses—Lumbar spine ap—>17 years

Beam projection:			ap									**Table cells:**
Focus-to-detector distance (cm):			115									(+) = normal font = optimal field setting
Tube voltage (kVp):			80									**(−) = bold font = suboptimal field setting**
Total filtrations:			See last table row									Empty table cells correspond to values <0.01
Organ or tissue	**Organ dose D_T: Entrance dose K_E**							**(mSv ∗ mGy⁻¹)**				
	(+)	(−)	(+)	(−)	(+)	(−)	(+)	(−)	(+)	(−)	(+)	(−)
Skull												
Brain												
Oral mucosa												
Salivary glands												
Cervical spine												
Thyroid												
Thymus												
Thoracic spine		**0.03**	0.01	**0.04**	0.01	**0.05**	0.02	**0.06**	0.02	**0.07**	0.03	**0.09**
Claviculae												
Scapulae												
Ribs	0.03	**0.18**	0.03	**0.20**	0.03	**0.22**	0.03	**0.23**	0.04	**0.25**	0.04	**0.27**
Oesophagus		**0.05**		**0.06**		**0.08**	0.01	**0.09**	0.01	**0.11**	0.02	**0.13**
Trachea												
Lung		**0.01**		**0.02**		**0.02**		**0.02**		**0.03**		**0.04**
Mamma								**0.01**		**0.01**		**0.02**
Heart		**0.03**		**0.04**		**0.04**		**0.05**		**0.06**		**0.07**
Stomach	0.05	**0.32**	0.06	**0.39**	0.07	**0.44**	0.08	**0.49**	0.10	**0.57**	0.12	**0.65**
Spleen		**0.05**	0.01	**0.06**	0.01	**0.07**	0.01	**0.08**	0.02	**0.11**	0.02	**0.13**
Liver	0.03	**0.19**	0.04	**0.24**	0.04	**0.27**	0.05	**0.30**	0.06	**0.35**	0.07	**0.40**
Gall bladder	0.21	**0.36**	0.26	**0.45**	0.30	**0.52**	0.34	**0.58**	0.40	**0.69**	0.46	**0.80**
Pancreas	0.06	**0.16**	0.07	**0.20**	0.09	**0.24**	0.10	**0.27**	0.12	**0.33**	0.14	**0.39**
Small intestine	0.15	**0.26**	0.18	**0.33**	0.21	**0.38**	0.24	**0.43**	0.28	**0.51**	0.33	**0.60**
Large intestine	0.11	**0.29**	0.14	**0.37**	0.16	**0.42**	0.18	**0.47**	0.22	**0.56**	0.25	**0.65**
Sigma/rectum	0.03	**0.20**	0.04	**0.25**	0.05	**0.29**	0.05	**0.32**	0.06	**0.39**	0.08	**0.46**
Lumbar spine	0.09	**0.13**	0.11	**0.17**	0.14	**0.20**	0.16	**0.23**	0.20	**0.29**	0.24	**0.35**
Pelvis	0.04	**0.14**	0.05	**0.18**	0.06	**0.22**	0.07	**0.25**	0.09	**0.31**	0.10	**0.38**
Adrenals	0.01	**0.05**	0.02	**0.07**	0.02	**0.08**	0.02	**0.09**	0.03	**0.12**	0.04	**0.15**
Kidneys	0.02	**0.05**	0.03	**0.07**	0.03	**0.08**	0.04	**0.09**	0.05	**0.12**	0.06	**0.14**
Urinary bladder	0.31	**0.54**	0.38	**0.65**	0.42	**0.74**	0.46	**0.81**	0.53	**0.94**	0.59	**1.06**
Uterus	0.27	**0.35**	0.34	**0.44**	0.39	**0.51**	0.43	**0.57**	0.52	**0.68**	0.60	**0.80**
Ovaries	0.06	**0.20**	0.08	**0.26**	0.10	**0.31**	0.11	**0.34**	0.14	**0.42**	0.16	**0.49**
Prostate	0.04	**0.34**	0.05	**0.43**	0.06	**0.50**	0.06	**0.55**	0.08	**0.66**	0.09	**0.76**
Testes		**0.91**		**1.04**		**1.13**	0.01	**1.19**	0.01	**1.30**	0.02	**1.39**
Upper arm												
Elbow				**0.01**		**0.01**		**0.02**		**0.02**		**0.03**
Forearm		**0.02**		**0.02**		**0.03**		**0.03**		**0.04**		**0.05**
Upper leg		**0.10**		**0.13**		**0.16**		**0.18**		**0.22**		**0.27**
Knee												**0.01**
Lower leg												
Lymph nodes	0.07	**0.15**	0.09	**0.19**	0.10	**0.22**	0.11	**0.25**	0.14	**0.29**	0.16	**0.34**
Skin	0.03	**0.09**	0.03	**0.10**	0.03	**0.10**	0.03	**0.11**	0.03	**0.11**	0.03	**0.12**
Muscle	0.03	**0.10**	0.04	**0.12**	0.04	**0.13**	0.05	**0.14**	0.06	**0.17**	0.06	**0.19**
Skeleton		**0.05**	0.01	**0.06**	0.01	**0.07**	0.02	**0.08**	0.02	**0.10**	0.02	**0.11**
Red bone marrow		**0.04**	0.01	**0.04**	0.01	**0.05**	0.02	**0.06**	0.02	**0.07**	0.03	**0.09**
Whole body dose	0.03	**0.09**	0.04	**0.11**	0.04	**0.13**	0.04	**0.14**	0.05	**0.16**	0.06	**0.18**
ED (ICRP 60)	0.04	**0.23**	0.05	**0.28**	0.06	**0.31**	0.07	**0.34**	0.08	**0.39**	0.10	**0.44**
ED (ICRP 103)	0.04	**0.17**	0.05	**0.21**	0.06	**0.23**	0.07	**0.26**	0.08	**0.30**	0.09	**0.34**
Total filtration (mm) Al	**1.5**		**2.5**		**3.5**		**4.5**		**3.5**		**3.5**	
Cu	**0**		**0**		**0**		**0**		**0.1**		**0.2**	

Table 16.32 Normalized organ doses—Lumbar spine r-lat—1–2 years

Beam projection:	r-lat	**Table cells:**
Focus-to-detector distance (cm):	115	(+) = normal font = optimal field setting
Tube voltage (kVp):	60	**(−) = bold font = suboptimal field setting**
Total filtrations:	See last table row	Empty table cells correspond to values <0.01

Organ or tissue	**Organ dose D_T: Entrance dose K_E (mSv ∗ mGy^{-1})**											
	(+)	(−)	(+)	(−)	(+)	(−)	(+)	(−)	(+)	(−)	(+)	(−)
Skull												
Brain												
Oral mucosa												
Salivary glands												
Cervical spine												
Thyroid												
Thymus												
Thoracic spine	0.03	**0.10**	0.04	**0.13**	0.05	**0.16**	0.06	**0.18**	0.07	**0.22**	0.06	**0.20**
Claviculae												
Scapulae				**0.01**		**0.01**		**0.02**		**0.02**		**0.02**
Ribs	0.05	**0.20**	0.06	**0.24**	0.07	**0.26**	0.08	**0.28**	0.09	**0.31**	0.08	**0.30**
Oesophagus		**0.04**		**0.05**	0.01	**0.06**	0.01	**0.07**	0.02	**0.09**	0.01	**0.08**
Trachea												
Lung		**0.02**		**0.03**		**0.03**		**0.04**		**0.04**		**0.04**
Mamma												
Heart		**0.01**		**0.02**		**0.02**		**0.02**		**0.03**		**0.03**
Stomach		**0.03**	0.01	**0.04**	0.01	**0.05**	0.02	**0.05**	0.02	**0.07**	0.02	**0.06**
Spleen	0.02	**0.03**	0.03	**0.04**	0.03	**0.05**	0.04	**0.06**	0.05	**0.08**	0.04	**0.07**
Liver	0.09	**0.28**	0.11	**0.34**	0.13	**0.38**	0.14	**0.41**	0.15	**0.46**	0.15	**0.45**
Gall bladder	0.03	**0.10**	0.04	**0.14**	0.05	**0.16**	0.05	**0.18**	0.06	**0.22**	0.06	**0.21**
Pancreas	0.04	**0.09**	0.05	**0.12**	0.06	**0.14**	0.07	**0.16**	0.09	**0.20**	0.08	**0.18**
Small intestine	0.08	**0.13**	0.11	**0.17**	0.13	**0.20**	0.14	**0.23**	0.17	**0.28**	0.16	**0.26**
Large intestine	0.05	**0.17**	0.07	**0.22**	0.08	**0.26**	0.09	**0.29**	0.11	**0.34**	0.10	**0.32**
Sigma/rectum	0.02	**0.06**	0.03	**0.07**	0.03	**0.09**	0.04	**0.10**	0.05	**0.13**	0.04	**0.11**
Lumbar spine	0.34	**0.37**	0.44	**0.49**	0.53	**0.59**	0.60	**0.67**	0.73	**0.82**	0.68	**0.75**
Pelvis	0.38	**0.62**	0.49	**0.78**	0.57	**0.90**	0.63	**1.00**	0.74	**1.17**	0.71	**1.13**
Adrenals	0.03	**0.12**	0.03	**0.15**	0.04	**0.18**	0.05	**0.20**	0.06	**0.25**	0.05	**0.23**
Kidneys	0.19	**0.21**	0.24	**0.27**	0.27	**0.31**	0.30	**0.34**	0.35	**0.40**	0.33	**0.38**
Urinary bladder	0.01	**0.06**	0.02	**0.08**	0.02	**0.09**	0.03	**0.10**	0.04	**0.13**	0.03	**0.11**
Uterus	0.04	**0.11**	0.05	**0.15**	0.07	**0.18**	0.07	**0.21**	0.09	**0.26**	0.08	**0.23**
Ovaries	0.11	**0.16**	0.14	**0.21**	0.17	**0.25**	0.19	**0.28**	0.24	**0.34**	0.22	**0.32**
Prostate		**0.07**	0.01	**0.09**	0.02	**0.11**	0.02	**0.12**	0.02	**0.15**	0.02	**0.13**
Testes		**0.02**		**0.02**		**0.02**		**0.03**		**0.04**		**0.03**
Upper arm												
Elbow												
Forearm												
Upper leg		**0.19**	0.01	**0.24**	0.02	**0.29**	0.02	**0.32**	0.02	**0.39**	0.02	**0.36**
Knee				**0.01**		**0.01**		**0.02**		**0.02**		**0.02**
Lower leg												
Lymph nodes	0.03	**0.07**	0.04	**0.09**	0.05	**0.10**	0.06	**0.11**	0.07	**0.14**	0.06	**0.13**
Skin	0.05	**0.10**	0.05	**0.11**	0.05	**0.11**	0.05	**0.12**	0.06	**0.12**	0.06	**0.12**
Muscle	0.04	**0.08**	0.05	**0.09**	0.05	**0.10**	0.06	**0.11**	0.07	**0.13**	0.06	**0.12**
Skeleton	0.06	**0.12**	0.08	**0.15**	0.09	**0.18**	0.10	**0.20**	0.12	**0.23**	0.11	**0.22**
Red bone marrow	0.03	**0.05**	0.03	**0.06**	0.04	**0.07**	0.04	**0.08**	0.05	**0.09**	0.05	**0.09**
Whole body dose	0.04	**0.08**	0.05	**0.10**	0.06	**0.11**	0.06	**0.12**	0.07	**0.14**	0.07	**0.13**
ED (ICRP 60)	0.03	**0.07**	0.04	**0.09**	0.04	**0.10**	0.05	**0.11**	0.06	**0.14**	0.06	**0.13**
ED (ICRP 103)	0.02	**0.06**	0.03	**0.08**	0.04	**0.09**	0.04	**0.10**	0.05	**0.12**	0.04	**0.11**
Total filtration (mm) Al	**1.5**		**2.5**		**3.5**		**4.5**		**3.5**		**3.5**	
Cu	**0**		**0**		**0**		**0**		**0.1**		**0.2**	

Table 16.33 Normalized organ doses—Lumbar spine r-lat—1–2 years

Beam projection:		r-lat										**Table cells:**
Focus-to-detector distance (cm):		115										(+) = normal font = optimal field setting
Tube voltage (kVp):		70										**(−) = bold font = suboptimal field setting**
Total filtrations:		See last table row										Empty table cells correspond to values <0.01
Organ or tissue	**Organ dose D_T: Entrance dose K_E**								**(mSv * mGy^{-1})**			
	(+)	(−)	(+)	(−)	(+)	(−)	(+)	(−)	(+)	(−)	(+)	(−)
Skull												
Brain												
Oral mucosa												
Salivary glands												
Cervical spine												
Thyroid												
Thymus												
Thoracic spine	0.04	**0.13**	0.05	**0.16**	0.06	**0.19**	0.07	**0.21**	0.09	**0.26**	0.10	**0.31**
Claviculae										**0.01**		**0.01**
Scapulae		**0.01**		**0.02**		**0.02**		**0.02**		**0.03**	0.01	**0.03**
Ribs	0.06	**0.22**	0.07	**0.25**	0.08	**0.28**	0.08	**0.30**	0.09	**0.33**	0.10	**0.35**
Oesophagus		**0.05**	0.01	**0.07**	0.01	**0.08**	0.02	**0.09**	0.02	**0.11**	0.03	**0.12**
Trachea												
Lung		**0.03**		**0.03**		**0.04**		**0.04**	0.01	**0.05**	0.01	**0.06**
Mamma										**0.01**		**0.01**
Heart		**0.02**		**0.02**		**0.03**		**0.03**	0.01	**0.04**	0.01	**0.04**
Stomach	0.01	**0.04**	0.02	**0.05**	0.02	**0.06**	0.02	**0.07**	0.03	**0.08**	0.03	**0.10**
Spleen	0.03	**0.05**	0.04	**0.06**	0.04	**0.07**	0.05	**0.08**	0.06	**0.11**	0.08	**0.13**
Liver	0.11	**0.32**	0.13	**0.38**	0.14	**0.42**	0.15	**0.45**	0.17	**0.50**	0.19	**0.55**
Gall bladder	0.04	**0.13**	0.05	**0.17**	0.06	**0.20**	0.06	**0.22**	0.08	**0.27**	0.09	**0.31**
Pancreas	0.05	**0.12**	0.07	**0.15**	0.08	**0.18**	0.09	**0.20**	0.11	**0.24**	0.13	**0.29**
Small intestine	0.10	**0.17**	0.13	**0.21**	0.15	**0.25**	0.17	**0.27**	0.20	**0.33**	0.24	**0.38**
Large intestine	0.07	**0.21**	0.09	**0.26**	0.10	**0.30**	0.11	**0.33**	0.14	**0.39**	0.16	**0.45**
Sigma/rectum	0.03	**0.07**	0.04	**0.10**	0.04	**0.12**	0.05	**0.13**	0.06	**0.17**	0.07	**0.20**
Lumbar spine	0.42	**0.47**	0.54	**0.61**	0.63	**0.71**	0.71	**0.80**	0.85	**0.96**	0.99	**1.12**
Pelvis	0.45	**0.73**	0.56	**0.90**	0.65	**1.02**	0.71	**1.12**	0.83	**1.30**	0.93	**1.45**
Adrenals	0.03	**0.15**	0.04	**0.19**	0.05	**0.22**	0.06	**0.25**	0.07	**0.30**	0.08	**0.35**
Kidneys	0.22	**0.25**	0.27	**0.31**	0.31	**0.35**	0.34	**0.39**	0.39	**0.45**	0.44	**0.51**
Urinary bladder	0.02	**0.07**	0.03	**0.10**	0.03	**0.11**	0.04	**0.13**	0.04	**0.16**	0.05	**0.19**
Uterus	0.05	**0.15**	0.07	**0.19**	0.08	**0.23**	0.09	**0.25**	0.12	**0.31**	0.14	**0.37**
Ovaries	0.14	**0.20**	0.18	**0.26**	0.21	**0.30**	0.23	**0.33**	0.28	**0.40**	0.32	**0.47**
Prostate	0.01	**0.09**	0.02	**0.11**	0.02	**0.13**	0.02	**0.15**	0.03	**0.18**	0.04	**0.22**
Testes		**0.02**		**0.03**		**0.03**		**0.04**	0.01	**0.05**	0.02	**0.06**
Upper arm												
Elbow												
Forearm												
Upper leg	0.01	**0.23**	0.02	**0.29**	0.02	**0.34**	0.03	**0.38**	0.03	**0.45**	0.04	**0.52**
Knee		**0.01**		**0.02**		**0.02**		**0.02**		**0.03**		**0.04**
Lower leg												
Lymph nodes	0.04	**0.08**	0.05	**0.11**	0.06	**0.12**	0.07	**0.14**	0.08	**0.17**	0.10	**0.19**
Skin	0.05	**0.11**	0.05	**0.12**	0.06	**0.12**	0.06	**0.12**	0.06	**0.13**	0.06	**0.14**
Muscle	0.05	**0.09**	0.06	**0.10**	0.06	**0.12**	0.07	**0.13**	0.08	**0.14**	0.08	**0.16**
Skeleton	0.07	**0.14**	0.09	**0.18**	0.10	**0.20**	0.11	**0.22**	0.13	**0.26**	0.15	**0.29**
Red bone marrow	0.03	**0.06**	0.04	**0.07**	0.05	**0.08**	0.05	**0.09**	0.06	**0.11**	0.07	**0.12**
Whole body dose	0.05	**0.09**	0.06	**0.11**	0.06	**0.13**	0.07	**0.14**	0.08	**0.16**	0.09	**0.18**
ED (ICRP 60)	0.04	**0.08**	0.05	**0.11**	0.05	**0.12**	0.06	**0.14**	0.07	**0.16**	0.08	**0.19**
ED (ICRP 103)	0.03	**0.07**	0.04	**0.09**	0.04	**0.11**	0.05	**0.12**	0.06	**0.14**	0.07	**0.16**
Total filtration (mm) **Al**	**1.5**		**2.5**		**3.5**		**4.5**		**3.5**		**3.5**	
Cu	**0**		**0**		**0**		**0**		**0.1**		**0.2**	

Table 16.34 Normalized organ doses—Lumbar spine r-lat—3–7 years

Beam projection:	r-lat	**Table cells:**
Focus-to-detector distance (cm):	115	(+) = normal font = optimal field setting
Tube voltage (kVp):	60	**(−) = bold font = suboptimal field setting**
Total filtrations:	See last table row	Empty table cells correspond to values <0.01

Organ or tissue	**Organ dose D_T: Entrance dose K_E (mSv * mGy⁻¹)**											
	(+)	(−)	(+)	(−)	(+)	(−)	(+)	(−)	(+)	(−)	(+)	(−)
Skull												
Brain												
Oral mucosa												
Salivary glands												
Cervical spine												
Thyroid												
Thymus												
Thoracic spine	0.02	**0.06**	0.02	**0.08**	0.03	**0.09**	0.03	**0.11**	0.04	**0.13**	0.03	**0.12**
Claviculae												
Scapulae										**0.01**		
Ribs	0.03	**0.17**	0.04	**0.20**	0.05	**0.22**	0.05	**0.23**	0.06	**0.26**	0.06	**0.26**
Oesophagus		**0.02**		**0.02**		**0.03**		**0.03**		**0.04**		**0.04**
Trachea												
Lung		**0.02**		**0.02**		**0.02**		**0.03**		**0.03**		**0.03**
Mamma												
Heart				**0.01**		**0.02**		**0.02**		**0.02**		**0.02**
Stomach		**0.02**		**0.02**	0.01	**0.03**	0.01	**0.03**	0.02	**0.04**	0.01	**0.03**
Spleen		**0.01**	0.01	**0.02**	0.01	**0.02**	0.02	**0.03**	0.02	**0.04**	0.02	**0.03**
Liver	0.10	**0.26**	0.12	**0.31**	0.14	**0.36**	0.15	**0.39**	0.17	**0.45**	0.16	**0.43**
Gall bladder	0.02	**0.07**	0.03	**0.10**	0.04	**0.12**	0.05	**0.13**	0.06	**0.17**	0.05	**0.15**
Pancreas	0.03	**0.05**	0.03	**0.07**	0.04	**0.08**	0.05	**0.10**	0.06	**0.12**	0.05	**0.10**
Small intestine	0.06	**0.09**	0.08	**0.12**	0.10	**0.15**	0.11	**0.17**	0.14	**0.21**	0.13	**0.18**
Large intestine	0.06	**0.14**	0.08	**0.18**	0.10	**0.21**	0.11	**0.24**	0.14	**0.29**	0.13	**0.26**
Sigma/rectum	0.01	**0.03**	0.02	**0.04**	0.02	**0.05**	0.02	**0.06**	0.03	**0.07**	0.02	**0.06**
Lumbar spine	0.19	**0.21**	0.25	**0.28**	0.30	**0.34**	0.35	**0.39**	0.44	**0.50**	0.38	**0.43**
Pelvis	0.28	**0.45**	0.37	**0.57**	0.43	**0.67**	0.48	**0.75**	0.58	**0.89**	0.54	**0.84**
Adrenals	0.02	**0.08**	0.02	**0.10**	0.03	**0.12**	0.03	**0.14**	0.04	**0.17**	0.03	**0.15**
Kidneys	0.13	**0.15**	0.17	**0.19**	0.19	**0.22**	0.22	**0.25**	0.26	**0.30**	0.24	**0.28**
Urinary bladder	0.01	**0.04**	0.02	**0.05**	0.02	**0.07**	0.03	**0.08**	0.03	**0.10**	0.03	**0.08**
Uterus	0.04	**0.07**	0.05	**0.10**	0.06	**0.12**	0.07	**0.14**	0.09	**0.17**	0.07	**0.15**
Ovaries	0.08	**0.10**	0.11	**0.14**	0.13	**0.16**	0.15	**0.19**	0.18	**0.24**	0.16	**0.21**
Prostate		**0.05**	0.01	**0.07**	0.02	**0.09**	0.02	**0.10**	0.02	**0.13**	0.02	**0.11**
Testes				**0.01**		**0.02**		**0.02**		**0.03**		**0.02**
Upper arm												
Elbow												
Forearm												
Upper leg		**0.09**		**0.12**		**0.14**		**0.16**	0.01	**0.20**		**0.18**
Knee												
Lower leg												
Lymph nodes	0.02	**0.04**	0.03	**0.06**	0.04	**0.07**	0.04	**0.08**	0.05	**0.10**	0.05	**0.09**
Skin	0.03	**0.05**	0.03	**0.05**	0.03	**0.06**	0.03	**0.06**	0.03	**0.06**	0.03	**0.06**
Muscle	0.03	**0.06**	0.04	**0.07**	0.04	**0.08**	0.05	**0.09**	0.06	**0.10**	0.05	**0.10**
Skeleton	0.03	**0.07**	0.04	**0.09**	0.05	**0.10**	0.06	**0.12**	0.07	**0.14**	0.06	**0.13**
Red bone marrow	0.03	**0.04**	0.03	**0.06**	0.04	**0.07**	0.05	**0.08**	0.06	**0.09**	0.05	**0.08**
Whole body dose	0.03	**0.06**	0.04	**0.08**	0.05	**0.09**	0.05	**0.10**	0.06	**0.11**	0.05	**0.10**
ED (ICRP 60)	0.03	**0.05**	0.03	**0.07**	0.04	**0.08**	0.04	**0.09**	0.05	**0.11**	0.05	**0.10**
ED (ICRP 103)	0.02	**0.04**	0.03	**0.06**	0.03	**0.07**	0.04	**0.08**	0.04	**0.09**	0.04	**0.08**
Total filtration (mm) Al	**1.5**		**2.5**		**3.5**		**4.5**		**3.5**		**3.5**	
Cu	**0**		**0**		**0**		**0**		**0.1**		**0.2**	

Table 16.35 Normalized organ doses—Lumbar spine r-lat—3–7 years

Beam projection:	r-lat											**Table cells:**
Focus-to-detector distance (cm):	115											(+) = normal font = optimal field setting
Tube voltage (kVp):	70											**(−) = bold font = suboptimal field setting**
Total filtrations:	See last table row											Empty table cells correspond to values <0.01
Organ or tissue	**Organ dose D_T: Entrance dose K_E (mSv * mGy⁻¹)**											
	(+)	(−)	(+)	(−)	(+)	(−)	(+)	(−)	(+)	(−)	(+)	(−)
Skull												
Brain												
Oral mucosa												
Salivary glands												
Cervical spine												
Thyroid												
Thymus												
Thoracic spine	0.02	**0.08**	0.03	**0.10**	0.04	**0.12**	0.04	**0.13**	0.05	**0.17**	0.06	**0.20**
Claviculae												
Scapulae						**0.01**		**0.01**		**0.02**		**0.02**
Ribs	0.04	**0.18**	0.05	**0.21**	0.05	**0.23**	0.06	**0.25**	0.06	**0.28**	0.07	**0.30**
Oesophagus		**0.02**		**0.03**		**0.04**		**0.04**	0.01	**0.05**	0.01	**0.06**
Trachea												
Lung		**0.02**		**0.03**		**0.03**		**0.03**		**0.04**	0.01	**0.05**
Mamma												**0.01**
Heart		**0.01**		**0.02**		**0.02**		**0.02**		**0.03**		**0.04**
Stomach		**0.02**	0.01	**0.03**	0.02	**0.04**	0.02	**0.04**	0.02	**0.06**	0.03	**0.07**
Spleen	0.01	**0.02**	0.02	**0.03**	0.02	**0.04**	0.02	**0.04**	0.03	**0.05**	0.04	**0.07**
Liver	0.11	**0.30**	0.14	**0.36**	0.15	**0.40**	0.17	**0.43**	0.19	**0.49**	0.21	**0.55**
Gall bladder	0.03	**0.10**	0.04	**0.12**	0.05	**0.15**	0.06	**0.17**	0.07	**0.21**	0.09	**0.24**
Pancreas	0.04	**0.07**	0.05	**0.09**	0.05	**0.11**	0.06	**0.12**	0.08	**0.15**	0.09	**0.19**
Small intestine	0.08	**0.12**	0.11	**0.15**	0.13	**0.18**	0.14	**0.20**	0.17	**0.25**	0.21	**0.30**
Large intestine	0.08	**0.17**	0.11	**0.22**	0.12	**0.25**	0.14	**0.28**	0.17	**0.34**	0.20	**0.40**
Sigma/rectum	0.02	**0.04**	0.02	**0.06**	0.03	**0.07**	0.03	**0.08**	0.04	**0.10**	0.05	**0.12**
Lumbar spine	0.25	**0.28**	0.32	**0.37**	0.38	**0.44**	0.43	**0.49**	0.53	**0.61**	0.63	**0.73**
Pelvis	0.34	**0.54**	0.43	**0.67**	0.50	**0.78**	0.55	**0.86**	0.66	**1.01**	0.75	**1.15**
Adrenals	0.02	**0.10**	0.03	**0.13**	0.03	**0.15**	0.04	**0.17**	0.05	**0.21**	0.06	**0.25**
Kidneys	0.16	**0.18**	0.20	**0.23**	0.23	**0.27**	0.25	**0.29**	0.29	**0.35**	0.34	**0.40**
Urinary bladder	0.02	**0.06**	0.02	**0.07**	0.03	**0.09**	0.03	**0.10**	0.04	**0.12**	0.05	**0.15**
Uterus	0.05	**0.10**	0.06	**0.13**	0.08	**0.15**	0.09	**0.18**	0.11	**0.22**	0.14	**0.26**
Ovaries	0.11	**0.14**	0.14	**0.18**	0.16	**0.21**	0.18	**0.24**	0.22	**0.29**	0.26	**0.34**
Prostate	0.01	**0.07**	0.02	**0.10**	0.02	**0.12**	0.03	**0.13**	0.03	**0.16**	0.04	**0.20**
Testes		**0.01**		**0.02**		**0.02**		**0.03**		**0.03**	0.01	**0.04**
Upper arm												
Elbow												
Forearm												
Upper leg		**0.11**		**0.15**	0.01	**0.17**	0.01	**0.19**	0.02	**0.24**	0.02	**0.28**
Knee												
Lower leg												
Lymph nodes	0.03	**0.06**	0.04	**0.07**	0.05	**0.09**	0.06	**0.10**	0.07	**0.12**	0.08	**0.14**
Skin	0.03	**0.05**	0.03	**0.06**	0.03	**0.06**	0.03	**0.06**	0.03	**0.07**	0.04	**0.07**
Muscle	0.04	**0.07**	0.05	**0.08**	0.05	**0.09**	0.05	**0.10**	0.06	**0.12**	0.07	**0.13**
Skeleton	0.04	**0.08**	0.05	**0.11**	0.06	**0.12**	0.07	**0.13**	0.08	**0.16**	0.09	**0.18**
Red bone marrow	0.03	**0.06**	0.04	**0.07**	0.05	**0.08**	0.06	**0.09**	0.07	**0.11**	0.08	**0.13**
Whole body dose	0.04	**0.07**	0.05	**0.09**	0.05	**0.10**	0.06	**0.11**	0.07	**0.13**	0.07	**0.14**
ED (ICRP 60)	0.03	**0.06**	0.04	**0.08**	0.05	**0.09**	0.05	**0.11**	0.07	**0.13**	0.08	**0.15**
ED (ICRP 103)	0.03	**0.06**	0.03	**0.07**	0.04	**0.08**	0.04	**0.09**	0.05	**0.11**	0.06	**0.13**
Total filtration (mm) Al	**1.5**		**2.5**		**3.5**		**4.5**		**3.5**		**3.5**	
Cu	**0**		**0**		**0**		**0**		**0.1**		**0.2**	

Table 16.36 Normalized organ doses—Lumbar spine r-lat—8–12 years

Beam projection:	r-lat											**Table cells:**
Focus-to-detector distance (cm):	115											(+) = normal font = optimal field setting
Tube voltage (kVp):	70											**(−) = bold font = suboptimal field setting**
Total filtrations:	See last table row											Empty table cells correspond to values <0.01
Organ or tissue	**Organ dose D_T: Entrance dose K_E**						**(mSv ∗ mGy^{-1})**					
	(+)	(−)	(+)	(−)	(+)	(−)	(+)	(−)	(+)	(−)	(+)	(−)
Skull												
Brain												
Oral mucosa												
Salivary glands												
Cervical spine												
Thyroid												
Thymus												
Thoracic spine	0.02	**0.05**	0.02	**0.06**	0.02	**0.08**	0.03	**0.09**	0.04	**0.11**	0.04	**0.13**
Claviculae												
Scapulae												**0.01**
Ribs	0.04	**0.16**	0.04	**0.18**	0.05	**0.20**	0.05	**0.21**	0.06	**0.24**	0.07	**0.26**
Oesophagus		**0.01**		**0.02**		**0.02**		**0.02**		**0.03**		**0.04**
Trachea												
Lung		**0.02**		**0.02**		**0.02**		**0.03**		**0.03**		**0.04**
Mamma												
Heart		**0.01**		**0.01**		**0.02**		**0.02**		**0.02**		**0.03**
Stomach		**0.01**		**0.02**		**0.02**		**0.03**	0.01	**0.04**	0.02	**0.05**
Spleen		**0.01**		**0.02**	0.01	**0.02**	0.01	**0.02**	0.02	**0.03**	0.02	**0.04**
Liver	0.09	**0.28**	0.11	**0.35**	0.12	**0.39**	0.13	**0.42**	0.15	**0.49**	0.17	**0.55**
Gall bladder	0.03	**0.09**	0.03	**0.11**	0.04	**0.13**	0.05	**0.15**	0.06	**0.19**	0.07	**0.23**
Pancreas	0.02	**0.04**	0.03	**0.06**	0.03	**0.07**	0.04	**0.08**	0.05	**0.10**	0.06	**0.12**
Small intestine	0.05	**0.09**	0.07	**0.12**	0.08	**0.14**	0.10	**0.16**	0.12	**0.20**	0.14	**0.23**
Large intestine	0.05	**0.14**	0.06	**0.18**	0.08	**0.22**	0.09	**0.24**	0.11	**0.30**	0.13	**0.35**
Sigma/rectum		**0.02**	0.01	**0.03**	0.01	**0.04**	0.02	**0.05**	0.02	**0.06**	0.03	**0.07**
Lumbar spine	0.17	**0.18**	0.22	**0.23**	0.26	**0.28**	0.30	**0.32**	0.37	**0.40**	0.45	**0.49**
Pelvis	0.24	**0.41**	0.30	**0.52**	0.35	**0.61**	0.40	**0.68**	0.48	**0.81**	0.55	**0.93**
Adrenals	0.02	**0.07**	0.02	**0.09**	0.03	**0.11**	0.03	**0.12**	0.04	**0.16**	0.05	**0.19**
Kidneys	0.12	**0.14**	0.16	**0.18**	0.18	**0.20**	0.20	**0.23**	0.24	**0.27**	0.28	**0.32**
Urinary bladder	0.01	**0.04**	0.01	**0.06**	0.02	**0.07**	0.02	**0.08**	0.03	**0.10**	0.03	**0.12**
Uterus	0.03	**0.07**	0.04	**0.09**	0.04	**0.11**	0.05	**0.12**	0.07	**0.15**	0.08	**0.19**
Ovaries	0.07	**0.10**	0.09	**0.13**	0.11	**0.15**	0.12	**0.18**	0.15	**0.22**	0.18	**0.26**
Prostate		**0.05**		**0.07**	0.01	**0.08**	0.01	**0.09**	0.02	**0.12**	0.02	**0.14**
Testes		**0.01**		**0.01**		**0.02**		**0.02**		**0.02**		**0.03**
Upper arm												
Elbow												
Forearm												
Upper leg		**0.08**		**0.10**		**0.12**		**0.13**		**0.16**		**0.19**
Knee												
Lower leg												
Lymph nodes	0.02	**0.04**	0.03	**0.06**	0.03	**0.07**	0.04	**0.08**	0.05	**0.09**	0.06	**0.11**
Skin	0.04	**0.05**	0.04	**0.05**	0.04	**0.05**	0.04	**0.05**	0.05	**0.06**	0.05	**0.06**
Muscle	0.03	**0.06**	0.04	**0.07**	0.04	**0.08**	0.04	**0.09**	0.05	**0.10**	0.06	**0.11**
Skeleton	0.03	**0.06**	0.03	**0.08**	0.04	**0.09**	0.05	**0.10**	0.05	**0.12**	0.06	**0.14**
Red bone marrow	0.03	**0.05**	0.04	**0.07**	0.04	**0.08**	0.05	**0.09**	0.06	**0.11**	0.07	**0.13**
Whole body dose	0.03	**0.06**	0.04	**0.08**	0.04	**0.09**	0.04	**0.09**	0.05	**0.11**	0.06	**0.12**
ED (ICRP 60)	0.02	**0.05**	0.03	**0.07**	0.03	**0.08**	0.04	**0.09**	0.05	**0.11**	0.05	**0.12**
ED (ICRP 103)	0.02	**0.05**	0.02	**0.06**	0.03	**0.07**	0.03	**0.08**	0.04	**0.09**	0.04	**0.11**
Total filtration (mm) **Al**	**1.5**		**2.5**		**3.5**		**4.5**		**3.5**		**3.5**	
Cu	**0**		**0**		**0**		**0**		**0.1**		**0.2**	

Table 16.37 Normalized organ doses—Lumbar spine r-lat—8–12 years

Beam projection:				r-lat								**Table cells:**
Focus-to-detector distance (cm):				115								(+) = normal font = optimal field setting
Tube voltage (kVp):				80								**(−) = bold font = suboptimal field setting**
Total filtrations:				See last table row								Empty table cells correspond to values <0.01
Organ or tissue	**Organ dose D_T: Entrance dose K_E**								**(mSv * mGy^{-1})**			
	(+)	(−)	(+)	(−)	(+)	(−)	(+)	(−)	(+)	(−)	(+)	(−)
Skull												
Brain												
Oral mucosa												
Salivary glands												
Cervical spine												
Thyroid												
Thymus												
Thoracic spine	0.02	**0.06**	0.03	**0.08**	0.03	**0.09**	0.04	**0.11**	0.04	**0.13**	0.05	**0.16**
Claviculae												
Scapulae										**0.01**		**0.02**
Ribs	0.04	**0.17**	0.05	**0.20**	0.05	**0.21**	0.06	**0.23**	0.06	**0.25**	0.07	**0.27**
Oesophagus		**0.02**		**0.02**		**0.03**		**0.03**		**0.04**		**0.04**
Trachea												
Lung		**0.02**		**0.02**		**0.03**		**0.03**		**0.04**		**0.05**
Mamma												
Heart		**0.01**		**0.02**		**0.02**		**0.03**		**0.03**		**0.04**
Stomach		**0.02**		**0.03**	0.01	**0.03**	0.01	**0.04**	0.02	**0.05**	0.02	**0.06**
Spleen		**0.02**	0.01	**0.02**	0.02	**0.03**	0.02	**0.03**	0.02	**0.04**	0.03	**0.05**
Liver	0.10	**0.32**	0.12	**0.39**	0.14	**0.43**	0.15	**0.47**	0.17	**0.53**	0.19	**0.59**
Gall bladder	0.03	**0.11**	0.04	**0.14**	0.05	**0.16**	0.06	**0.18**	0.07	**0.23**	0.09	**0.27**
Pancreas	0.03	**0.06**	0.03	**0.08**	0.04	**0.09**	0.05	**0.10**	0.06	**0.13**	0.07	**0.15**
Small intestine	0.07	**0.11**	0.09	**0.15**	0.10	**0.17**	0.12	**0.19**	0.14	**0.23**	0.17	**0.27**
Large intestine	0.06	**0.18**	0.08	**0.22**	0.09	**0.26**	0.10	**0.28**	0.13	**0.34**	0.15	**0.39**
Sigma/rectum	0.01	**0.03**	0.02	**0.04**	0.02	**0.05**	0.02	**0.06**	0.03	**0.08**	0.04	**0.09**
Lumbar spine	0.21	**0.23**	0.27	**0.29**	0.32	**0.34**	0.36	**0.39**	0.44	**0.48**	0.52	**0.57**
Pelvis	0.28	**0.49**	0.35	**0.60**	0.40	**0.69**	0.45	**0.76**	0.53	**0.89**	0.60	**1.00**
Adrenals	0.02	**0.09**	0.03	**0.11**	0.03	**0.13**	0.04	**0.15**	0.05	**0.19**	0.06	**0.22**
Kidneys	0.15	**0.17**	0.18	**0.21**	0.21	**0.24**	0.23	**0.26**	0.27	**0.31**	0.31	**0.35**
Urinary bladder	0.01	**0.06**	0.02	**0.07**	0.02	**0.08**	0.03	**0.10**	0.03	**0.12**	0.04	**0.14**
Uterus	0.04	**0.09**	0.05	**0.11**	0.06	**0.13**	0.07	**0.15**	0.08	**0.19**	0.10	**0.23**
Ovaries	0.09	**0.13**	0.11	**0.16**	0.13	**0.19**	0.15	**0.21**	0.18	**0.26**	0.21	**0.31**
Prostate		**0.07**	0.01	**0.08**	0.02	**0.10**	0.02	**0.11**	0.02	**0.14**	0.03	**0.17**
Testes		**0.01**		**0.02**		**0.02**		**0.03**		**0.03**		**0.04**
Upper arm												
Elbow												
Forearm												
Upper leg		**0.10**		**0.12**		**0.14**		**0.16**		**0.19**	0.01	**0.22**
Knee												
Lower leg												
Lymph nodes	0.03	**0.06**	0.03	**0.07**	0.04	**0.08**	0.05	**0.09**	0.06	**0.11**	0.07	**0.13**
Skin	0.04	**0.05**	0.04	**0.05**	0.04	**0.05**	0.04	**0.06**	0.05	**0.06**	0.05	**0.06**
Muscle	0.03	**0.07**	0.04	**0.08**	0.04	**0.09**	0.05	**0.10**	0.05	**0.11**	0.06	**0.12**
Skeleton	0.03	**0.07**	0.04	**0.09**	0.05	**0.10**	0.05	**0.11**	0.06	**0.13**	0.07	**0.15**
Red bone marrow	0.04	**0.07**	0.05	**0.08**	0.05	**0.09**	0.06	**0.10**	0.07	**0.12**	0.08	**0.14**
Whole body dose	0.03	**0.07**	0.04	**0.09**	0.05	**0.10**	0.05	**0.10**	0.06	**0.12**	0.06	**0.14**
ED (ICRP 60)	0.03	**0.06**	0.03	**0.08**	0.04	**0.09**	0.04	**0.10**	0.05	**0.12**	0.06	**0.14**
ED (ICRP 103)	0.02	**0.06**	0.03	**0.07**	0.03	**0.08**	0.04	**0.09**	0.04	**0.11**	0.05	**0.13**
Total filtration (mm) Al	**1.5**		**2.5**		**3.5**		**4.5**		**3.5**		**3.5**	
Cu	**0**		**0**		**0**		**0**		**0.1**		**0.2**	

Table 16.38 Normalized organ doses—Lumbar spine r-lat—13–17 years

Beam projection:			r-lat									**Table cells:**
Focus-to-detector distance (cm):			115									(+) = normal font = optimal field setting
Tube voltage (kVp):			70									**(−) = bold font = suboptimal field setting**
Total filtrations:			See last table row									Empty table cells correspond to values <0.01
Organ or tissue	**Organ dose D_T: Entrance dose K_E**							**(mSv ∗ mGy^{-1})**				
	(+)	(−)	(+)	(−)	(+)	(−)	(+)	(−)	(+)	(−)	(+)	(−)
Skull												
Brain												
Oral mucosa												
Salivary glands												
Cervical spine												
Thyroid												
Thymus												
Thoracic spine		**0.03**	0.01	**0.04**	0.02	**0.05**	0.02	**0.05**	0.02	**0.07**	0.03	**0.09**
Claviculae												
Scapulae												
Ribs	0.03	**0.13**	0.04	**0.16**	0.04	**0.17**	0.04	**0.19**	0.05	**0.21**	0.06	**0.23**
Oesophagus		**0.01**		**0.01**		**0.02**		**0.02**		**0.03**		**0.03**
Trachea												
Lung		**0.01**		**0.02**		**0.02**		**0.02**		**0.03**		**0.04**
Mamma		**0.01**		**0.02**		**0.02**		**0.03**		**0.03**		**0.04**
Heart				**0.01**		**0.01**		**0.01**		**0.02**		**0.02**
Stomach						**0.01**		**0.01**		**0.02**		**0.02**
Spleen										**0.01**	0.01	**0.02**
Liver	0.07	**0.25**	0.08	**0.30**	0.09	**0.34**	0.10	**0.37**	0.12	**0.44**	0.13	**0.49**
Gall bladder	0.02	**0.08**	0.03	**0.10**	0.04	**0.13**	0.04	**0.14**	0.05	**0.18**	0.07	**0.22**
Pancreas	0.01	**0.02**	0.01	**0.03**	0.02	**0.04**	0.02	**0.04**	0.03	**0.06**	0.03	**0.07**
Small intestine	0.03	**0.06**	0.04	**0.07**	0.05	**0.09**	0.06	**0.10**	0.07	**0.13**	0.09	**0.16**
Large intestine	0.03	**0.10**	0.04	**0.12**	0.05	**0.15**	0.05	**0.17**	0.07	**0.21**	0.08	**0.25**
Sigma/rectum		**0.01**		**0.02**		**0.02**		**0.02**	0.01	**0.03**	0.01	**0.04**
Lumbar spine	0.09	**0.10**	0.12	**0.13**	0.15	**0.16**	0.17	**0.19**	0.22	**0.24**	0.27	**0.30**
Pelvis	0.17	**0.30**	0.22	**0.39**	0.25	**0.45**	0.29	**0.51**	0.35	**0.61**	0.41	**0.71**
Adrenals	0.01	**0.04**	0.02	**0.06**	0.02	**0.07**	0.03	**0.08**	0.03	**0.10**	0.04	**0.12**
Kidneys	0.09	**0.10**	0.11	**0.13**	0.13	**0.15**	0.15	**0.17**	0.18	**0.20**	0.21	**0.24**
Urinary bladder		**0.02**		**0.03**	0.01	**0.04**	0.01	**0.04**	0.02	**0.06**	0.02	**0.07**
Uterus	0.05	**0.11**	0.07	**0.14**	0.09	**0.18**	0.10	**0.20**	0.13	**0.26**	0.17	**0.33**
Ovaries	0.04	**0.06**	0.05	**0.08**	0.06	**0.10**	0.06	**0.11**	0.08	**0.15**	0.10	**0.18**
Prostate		**0.03**		**0.04**		**0.05**		**0.05**	0.01	**0.07**	0.02	**0.09**
Testes								**0.01**		**0.01**		**0.02**
Upper arm												
Elbow												
Forearm												
Upper leg		**0.07**		**0.09**		**0.11**		**0.13**		**0.16**		**0.19**
Knee												
Lower leg												
Lymph nodes	0.01	**0.03**	0.02	**0.04**	0.02	**0.05**	0.02	**0.05**	0.03	**0.07**	0.04	**0.08**
Skin	0.04	**0.08**	0.05	**0.09**	0.05	**0.09**	0.05	**0.09**	0.05	**0.10**	0.06	**0.10**
Muscle	0.03	**0.05**	0.03	**0.06**	0.03	**0.07**	0.04	**0.08**	0.04	**0.09**	0.05	**0.10**
Skeleton	0.02	**0.05**	0.03	**0.07**	0.03	**0.08**	0.03	**0.08**	0.04	**0.10**	0.05	**0.12**
Red bone marrow	0.02	**0.05**	0.03	**0.06**	0.04	**0.07**	0.04	**0.08**	0.05	**0.10**	0.06	**0.11**
Whole body dose	0.03	**0.05**	0.03	**0.07**	0.03	**0.08**	0.04	**0.08**	0.04	**0.10**	0.05	**0.11**
ED (ICRP 60)	0.01	**0.04**	0.02	**0.05**	0.02	**0.06**	0.03	**0.07**	0.03	**0.08**	0.04	**0.10**
ED (ICRP 103)	0.01	**0.04**	0.02	**0.05**	0.02	**0.05**	0.02	**0.06**	0.03	**0.08**	0.03	**0.09**
Total filtration (mm) Al	**1.5**		**2.5**		**3.5**		**4.5**		**3.5**		**3.5**	
Cu	**0**		**0**		**0**		**0**		**0.1**		**0.2**	

Table 16.39 Normalized organ doses—Lumbar spine r-lat—13–17 years

Beam projection:	r-lat											**Table cells:**
Focus-to-detector distance (cm):	115											(+) = normal font = optimal field setting
Tube voltage (kVp):	80											**(−) = bold font = suboptimal field setting**
otal filtrations:	See last table row											Empty table cells correspond to values <0.01
Organ or tissue	**Organ dose D_T: Entrance dose K_E**								**(mSv * mGy^{-1})**			
	(+)	(−)	(+)	(−)	(+)	(−)	(+)	(−)	(+)	(−)	(+)	(−)
Skull												
Brain												
Oral mucosa												
Salivary glands												
Cervical spine												
Thyroid												
Thymus												
Thoracic spine	0.01	**0.04**	0.02	**0.05**	0.02	**0.06**	0.02	**0.07**	0.03	**0.09**	0.04	**0.11**
Claviculae												
Scapulae												
Ribs	0.03	**0.15**	0.04	**0.17**	0.04	**0.19**	0.05	**0.20**	0.06	**0.22**	0.06	**0.24**
Oesophagus		**0.01**		**0.02**		**0.02**		**0.03**		**0.03**		**0.04**
Trachea												
Lung		**0.02**		**0.02**		**0.03**		**0.03**		**0.04**		**0.04**
Mamma		**0.02**		**0.03**		**0.03**		**0.04**		**0.05**		**0.06**
Heart		**0.01**		**0.01**		**0.02**		**0.02**		**0.02**		**0.03**
Stomach		**0.01**		**0.01**		**0.02**		**0.02**		**0.02**	0.01	**0.03**
Spleen				**0.01**		**0.01**		**0.01**	0.01	**0.02**	0.01	**0.02**
Liver	0.08	**0.28**	0.09	**0.34**	0.10	**0.38**	0.11	**0.42**	0.13	**0.48**	0.15	**0.53**
Gall bladder	0.03	**0.10**	0.04	**0.13**	0.05	**0.16**	0.05	**0.18**	0.07	**0.22**	0.08	**0.26**
Pancreas	0.02	**0.03**	0.02	**0.04**	0.02	**0.05**	0.03	**0.06**	0.03	**0.07**	0.04	**0.09**
Small intestine	0.04	**0.07**	0.06	**0.10**	0.07	**0.11**	0.07	**0.13**	0.09	**0.16**	0.11	**0.19**
Large intestine	0.04	**0.12**	0.05	**0.15**	0.06	**0.18**	0.07	**0.20**	0.08	**0.24**	0.10	**0.29**
Sigma/rectum		**0.02**		**0.02**	0.01	**0.03**	0.01	**0.03**	0.02	**0.04**	0.02	**0.05**
Lumbar spine	0.12	**0.14**	0.16	**0.18**	0.19	**0.21**	0.21	**0.24**	0.27	**0.30**	0.32	**0.36**
Pelvis	0.20	**0.36**	0.26	**0.45**	0.30	**0.52**	0.33	**0.58**	0.40	**0.69**	0.46	**0.79**
Adrenals	0.02	**0.06**	0.02	**0.08**	0.03	**0.09**	0.03	**0.10**	0.04	**0.13**	0.05	**0.15**
Kidneys	0.11	**0.12**	0.14	**0.15**	0.16	**0.18**	0.17	**0.20**	0.21	**0.24**	0.24	**0.27**
Urinary bladder		**0.03**	0.01	**0.04**	0.02	**0.05**	0.02	**0.06**	0.02	**0.07**	0.03	**0.09**
Uterus	0.08	**0.15**	0.10	**0.20**	0.12	**0.24**	0.14	**0.27**	0.18	**0.34**	0.22	**0.42**
Ovaries	0.05	**0.08**	0.06	**0.11**	0.07	**0.13**	0.08	**0.14**	0.10	**0.18**	0.12	**0.22**
Prostate		**0.04**		**0.05**	0.01	**0.06**	0.01	**0.07**	0.02	**0.09**	0.02	**0.11**
Testes				**0.01**		**0.01**		**0.01**		**0.02**		**0.02**
Upper arm												
Elbow												
Forearm												
Upper leg		**0.09**		**0.12**		**0.14**		**0.15**		**0.19**		**0.22**
Knee												
Lower leg												
Lymph nodes	0.02	**0.04**	0.02	**0.05**	0.03	**0.06**	0.03	**0.07**	0.04	**0.08**	0.05	**0.10**
Skin	0.05	**0.08**	0.05	**0.09**	0.05	**0.09**	0.05	**0.10**	0.06	**0.10**	0.06	**0.11**
Muscle	0.03	**0.06**	0.03	**0.07**	0.04	**0.08**	0.04	**0.09**	0.05	**0.10**	0.05	**0.11**
Skeleton	0.02	**0.06**	0.03	**0.08**	0.04	**0.09**	0.04	**0.10**	0.05	**0.12**	0.06	**0.13**
Red bone marrow	0.03	**0.06**	0.04	**0.07**	0.04	**0.08**	0.05	**0.09**	0.06	**0.11**	0.07	**0.13**
Whole body dose	0.03	**0.06**	0.04	**0.08**	0.04	**0.09**	0.04	**0.09**	0.05	**0.11**	0.06	**0.12**
ED (ICRP 60)	0.02	**0.05**	0.02	**0.06**	0.03	**0.07**	0.03	**0.08**	0.04	**0.10**	0.05	**0.11**
ED (ICRP 103)	0.02	**0.05**	0.02	**0.06**	0.02	**0.07**	0.03	**0.07**	0.03	**0.09**	0.04	**0.11**
Total filtration (mm) **Al**	**1.5**		**2.5**		**3.5**		**4.5**		**3.5**		**3.5**	
Cu	**0**		**0**		**0**		**0**		**0.1**		**0.2**	

Table 16.40 Normalized organ doses—Lumbar spine r-lat—>17 years

Beam projection:		r-lat										**Table cells:**
Focus-to-detector distance (cm):		115										(+) = normal font = optimal field setting
Tube voltage (kVp):		70										**(−) = bold font = suboptimal field setting**
Total filtrations:		See last table row										Empty table cells correspond to values <0.01
Organ or tissue	**Organ dose D_T: Entrance dose K_E (mSv * mGy^{-1})**											
	(+)	(−)	(+)	(−)	(+)	(−)	(+)	(−)	(+)	(−)	(+)	(−)
Skull												
Brain												
Oral mucosa												
Salivary glands												
Cervical spine												
Thyroid												
Thymus												
Thoracic spine		**0.02**		**0.02**	0.01	**0.03**	0.01	**0.03**	0.02	**0.04**	0.02	**0.05**
Claviculae												
Scapulae												
Ribs	0.03	**0.11**	0.03	**0.13**	0.04	**0.14**	0.04	**0.15**	0.04	**0.17**	0.05	**0.19**
Oesophagus						**0.01**		**0.01**		**0.02**		**0.02**
Trachea												
Lung				**0.01**		**0.01**		**0.02**		**0.02**		**0.02**
Mamma												
Heart										**0.01**		**0.02**
Stomach												**0.01**
Spleen												
Liver	0.06	**0.20**	0.08	**0.25**	0.09	**0.28**	0.10	**0.31**	0.12	**0.37**	0.13	**0.42**
Gall bladder	0.02	**0.05**	0.02	**0.07**	0.03	**0.08**	0.03	**0.09**	0.04	**0.12**	0.05	**0.15**
Pancreas		**0.01**		**0.02**	0.01	**0.02**	0.01	**0.03**	0.02	**0.03**	0.02	**0.04**
Small intestine	0.02	**0.04**	0.03	**0.05**	0.04	**0.06**	0.04	**0.07**	0.06	**0.09**	0.07	**0.11**
Large intestine	0.02	**0.07**	0.03	**0.10**	0.04	**0.11**	0.04	**0.13**	0.06	**0.16**	0.07	**0.20**
Sigma/rectum						**0.01**		**0.01**		**0.02**		**0.02**
Lumbar spine	0.06	**0.06**	0.08	**0.08**	0.09	**0.10**	0.11	**0.12**	0.14	**0.15**	0.17	**0.19**
Pelvis	0.13	**0.23**	0.17	**0.30**	0.20	**0.35**	0.22	**0.40**	0.27	**0.49**	0.32	**0.57**
Adrenals	0.01	**0.03**	0.01	**0.04**	0.02	**0.05**	0.02	**0.05**	0.03	**0.07**	0.03	**0.08**
Kidneys	0.07	**0.07**	0.09	**0.09**	0.10	**0.11**	0.11	**0.13**	0.14	**0.16**	0.16	**0.18**
Urinary bladder		**0.02**		**0.02**		**0.03**		**0.03**	0.01	**0.04**	0.02	**0.05**
Uterus	0.01	**0.03**	0.02	**0.04**	0.02	**0.04**	0.03	**0.05**	0.03	**0.07**	0.04	**0.08**
Ovaries	0.03	**0.04**	0.03	**0.06**	0.04	**0.07**	0.05	**0.08**	0.06	**0.10**	0.08	**0.13**
Prostate		**0.02**		**0.02**		**0.03**		**0.03**		**0.04**	0.01	**0.05**
Testes												**0.01**
Upper arm												
Elbow												
Forearm												
Upper leg		**0.07**		**0.09**		**0.11**		**0.12**		**0.15**		**0.18**
Knee												
Lower leg												
Lymph nodes		**0.02**	0.01	**0.03**	0.02	**0.03**	0.02	**0.04**	0.02	**0.05**	0.03	**0.06**
Skin	0.04	**0.06**	0.04	**0.06**	0.04	**0.07**	0.04	**0.07**	0.05	**0.07**	0.05	**0.08**
Muscle	0.02	**0.05**	0.03	**0.06**	0.03	**0.06**	0.03	**0.07**	0.04	**0.08**	0.04	**0.09**
Skeleton	0.02	**0.04**	0.02	**0.05**	0.02	**0.06**	0.03	**0.07**	0.03	**0.08**	0.04	**0.10**
Red bone marrow	0.02	**0.04**	0.02	**0.05**	0.03	**0.06**	0.03	**0.06**	0.04	**0.08**	0.05	**0.09**
Whole body dose	0.02	**0.05**	0.03	**0.06**	0.03	**0.07**	0.03	**0.07**	0.04	**0.08**	0.04	**0.10**
ED (ICRP 60)	0.01	**0.03**	0.01	**0.04**	0.02	**0.04**	0.02	**0.05**	0.02	**0.06**	0.03	**0.07**
ED (ICRP 103)		**0.03**	0.01	**0.03**	0.02	**0.04**	0.02	**0.04**	0.02	**0.05**	0.03	**0.06**
Total filtration (mm) **Al**	**1.5**		**2.5**		**3.5**		**4.5**		**3.5**		**3.5**	
Cu	**0**		**0**		**0**		**0**		**0.1**		**0.2**	

Table 16.41 Normalized organ doses—Lumbar spine r-lat—>17 years

Beam projection:	r-lat											**Table cells:**
Focus-to-detector distance (cm):	115											(+) = normal font = optimal field setting
Tube voltage (kVp):	80											**(−) = bold font = suboptimal field setting**
Total filtrations:	See last table row											Empty table cells correspond to values <0.01
Organ or tissue	**Organ dose D_T: Entrance dose K_E**						**(mSv * mGy^{-1})**					
	(+)	(−)	(+)	(−)	(+)	(−)	(+)	(−)	(+)	(−)	(+)	(−)
Skull												
Brain												
Oral mucosa												
Salivary glands												
Cervical spine												
Thyroid												
Thymus												
Thoracic spine		**0.02**	0.01	**0.03**	0.01	**0.04**	0.02	**0.04**	0.02	**0.05**	0.03	**0.07**
Claviculae												
Scapulae												
Ribs	0.03	**0.12**	0.04	**0.14**	0.04	**0.15**	0.04	**0.16**	0.05	**0.18**	0.05	**0.20**
Oesophagus				**0.01**		**0.02**		**0.02**		**0.02**		**0.03**
Trachea												
Lung		**0.01**		**0.01**		**0.02**		**0.02**		**0.02**		**0.03**
Mamma												
Heart						**0.01**		**0.01**		**0.02**		**0.02**
Stomach								**0.01**		**0.01**		**0.02**
Spleen												**0.01**
Liver	0.07	**0.24**	0.09	**0.29**	0.10	**0.32**	0.11	**0.35**	0.13	**0.41**	0.15	**0.46**
Gall bladder	0.02	**0.07**	0.03	**0.09**	0.04	**0.10**	0.04	**0.12**	0.05	**0.15**	0.06	**0.18**
Pancreas		**0.02**	0.01	**0.03**	0.02	**0.03**	0.02	**0.03**	0.02	**0.04**	0.03	**0.06**
Small intestine	0.03	**0.05**	0.04	**0.07**	0.05	**0.08**	0.06	**0.09**	0.07	**0.12**	0.08	**0.14**
Large intestine	0.03	**0.09**	0.04	**0.12**	0.05	**0.14**	0.06	**0.16**	0.07	**0.20**	0.09	**0.23**
Sigma/rectum		**0.01**		**0.01**		**0.02**		**0.02**	0.01	**0.02**	0.01	**0.03**
Lumbar spine	0.08	**0.09**	0.10	**0.11**	0.12	**0.13**	0.14	**0.15**	0.17	**0.19**	0.21	**0.23**
Pelvis	0.16	**0.28**	0.20	**0.36**	0.23	**0.41**	0.26	**0.46**	0.31	**0.55**	0.36	**0.64**
Adrenals	0.01	**0.04**	0.02	**0.05**	0.02	**0.06**	0.03	**0.07**	0.03	**0.09**	0.04	**0.10**
Kidneys	0.08	**0.09**	0.10	**0.12**	0.12	**0.13**	0.13	**0.15**	0.16	**0.18**	0.18	**0.21**
Urinary bladder		**0.02**		**0.03**	0.01	**0.04**	0.01	**0.04**	0.02	**0.05**	0.02	**0.07**
Uterus	0.02	**0.04**	0.03	**0.05**	0.03	**0.06**	0.04	**0.07**	0.05	**0.09**	0.06	**0.11**
Ovaries	0.04	**0.06**	0.05	**0.08**	0.06	**0.09**	0.06	**0.10**	0.08	**0.13**	0.10	**0.16**
Prostate		**0.02**		**0.03**		**0.04**		**0.04**	0.01	**0.05**	0.01	**0.07**
Testes										**0.01**		**0.02**
Upper arm												
Elbow												
Forearm												
Upper leg		**0.09**		**0.11**		**0.13**		**0.15**		**0.18**		**0.22**
Knee												
Lower leg												
Lymph nodes	0.01	**0.03**	0.02	**0.04**	0.02	**0.04**	0.02	**0.05**	0.03	**0.06**	0.03	**0.07**
Skin	0.04	**0.06**	0.04	**0.07**	0.04	**0.07**	0.04	**0.07**	0.05	**0.08**	0.05	**0.08**
Muscle	0.03	**0.05**	0.03	**0.06**	0.03	**0.07**	0.04	**0.08**	0.04	**0.09**	0.05	**0.10**
Skeleton	0.02	**0.05**	0.02	**0.06**	0.03	**0.07**	0.03	**0.08**	0.04	**0.10**	0.04	**0.11**
Red bone marrow	0.02	**0.05**	0.03	**0.06**	0.04	**0.07**	0.04	**0.08**	0.05	**0.09**	0.06	**0.11**
Whole body dose	0.03	**0.06**	0.03	**0.07**	0.03	**0.08**	0.04	**0.08**	0.04	**0.09**	0.05	**0.11**
ED (ICRP 60)	0.01	**0.04**	0.02	**0.05**	0.02	**0.05**	0.02	**0.06**	0.03	**0.07**	0.04	**0.09**
ED (ICRP 103)	0.01	**0.03**	0.02	**0.04**	0.02	**0.05**	0.02	**0.05**	0.03	**0.06**	0.03	**0.08**
Total filtration (mm) Al	**1.5**		**2.5**		**3.5**		**4.5**		**3.5**		**3.5**	
Cu	**0**		**0**		**0**		**0**		**0.1**		**0.2**	

Table 16.42 Normalized organ doses—Lumbar spine ap—8–12 years

Beam projection:	ap	**Table cells:**
Focus-to-detector distance (cm):	150	(+) = normal font = optimal field setting
Tube voltage (kVp):	70	**(−) = bold font = suboptimal field setting**
Total filtrations:	See last table row	Empty table cells correspond to values <0.01

Organ or tissue	**Organ dose D_T: Entrance dose K_E**						**(mSv ∗ mGy^{-1})**					
	(+)	(−)	(+)	(−)	(+)	(−)	(+)	(−)	(+)	(−)	(+)	(−)
Skull												
Brain												
Oral mucosa												
Salivary glands												
Cervical spine												
Thyroid												
Thymus												
Thoracic spine		**0.04**	0.01	**0.05**	0.01	**0.06**	0.02	**0.07**	0.02	**0.09**	0.03	**0.12**
Claviculae												
Scapulae								**0.01**		**0.01**		**0.02**
Ribs	0.02	**0.21**	0.03	**0.24**	0.03	**0.26**	0.03	**0.28**	0.03	**0.31**	0.04	**0.33**
Oesophagus		**0.03**		**0.04**		**0.04**		**0.05**	0.01	**0.06**	0.01	**0.07**
Trachea												
Lung		**0.02**		**0.02**		**0.03**		**0.03**		**0.04**		**0.04**
Mamma										**0.01**		**0.01**
Heart		**0.03**		**0.04**		**0.05**		**0.06**		**0.07**		**0.08**
Stomach	0.05	**0.38**	0.06	**0.46**	0.07	**0.53**	0.08	**0.57**	0.10	**0.67**	0.11	**0.75**
Spleen		**0.07**	0.01	**0.09**	0.01	**0.10**	0.01	**0.12**	0.02	**0.15**	0.02	**0.18**
Liver	0.04	**0.25**	0.05	**0.30**	0.05	**0.35**	0.06	**0.38**	0.07	**0.45**	0.08	**0.51**
Gall bladder	0.26	**0.37**	0.33	**0.47**	0.38	**0.55**	0.42	**0.61**	0.49	**0.72**	0.56	**0.83**
Pancreas	0.05	**0.18**	0.06	**0.24**	0.07	**0.28**	0.08	**0.31**	0.10	**0.38**	0.12	**0.45**
Small intestine	0.13	**0.27**	0.17	**0.35**	0.20	**0.41**	0.22	**0.45**	0.27	**0.55**	0.31	**0.64**
Large intestine	0.10	**0.33**	0.13	**0.42**	0.15	**0.48**	0.17	**0.53**	0.20	**0.64**	0.24	**0.73**
Sigma/rectum	0.02	**0.23**	0.03	**0.29**	0.04	**0.34**	0.04	**0.38**	0.06	**0.46**	0.07	**0.53**
Lumbar spine	0.10	**0.15**	0.13	**0.20**	0.16	**0.25**	0.18	**0.28**	0.23	**0.37**	0.28	**0.45**
Pelvis	0.04	**0.20**	0.05	**0.26**	0.06	**0.31**	0.07	**0.35**	0.09	**0.44**	0.12	**0.53**
Adrenals		**0.06**	0.01	**0.08**	0.02	**0.10**	0.02	**0.11**	0.02	**0.14**	0.03	**0.17**
Kidneys	0.02	**0.06**	0.02	**0.08**	0.03	**0.10**	0.03	**0.11**	0.04	**0.14**	0.05	**0.17**
Urinary bladder	0.21	**0.53**	0.25	**0.64**	0.28	**0.73**	0.31	**0.80**	0.36	**0.92**	0.40	**1.04**
Uterus	0.23	**0.31**	0.29	**0.39**	0.33	**0.46**	0.37	**0.51**	0.44	**0.62**	0.51	**0.73**
Ovaries	0.05	**0.22**	0.07	**0.28**	0.08	**0.33**	0.09	**0.37**	0.12	**0.46**	0.14	**0.54**
Prostate	0.02	**0.34**	0.03	**0.43**	0.04	**0.50**	0.04	**0.55**	0.05	**0.66**	0.06	**0.76**
Testes		**1.03**		**1.14**		**1.21**	0.01	**1.25**	0.01	**1.33**	0.02	**1.39**
Upper arm												
Elbow		**0.02**		**0.02**		**0.03**		**0.03**		**0.04**		**0.05**
Forearm		**0.03**		**0.04**		**0.05**		**0.06**		**0.08**		**0.09**
Upper leg		**0.09**		**0.11**		**0.14**		**0.15**		**0.19**		**0.23**
Knee												
Lower leg												
Lymph nodes	0.07	**0.16**	0.09	**0.21**	0.10	**0.24**	0.11	**0.26**	0.14	**0.32**	0.16	**0.37**
Skin	0.02	**0.08**	0.02	**0.09**	0.02	**0.09**	0.02	**0.09**	0.02	**0.10**	0.03	**0.10**
Muscle	0.03	**0.10**	0.03	**0.12**	0.04	**0.13**	0.04	**0.14**	0.05	**0.16**	0.05	**0.18**
Skeleton		**0.05**	0.01	**0.07**	0.01	**0.08**	0.01	**0.09**	0.02	**0.11**	0.02	**0.12**
Red bone marrow		**0.03**		**0.04**	0.01	**0.05**	0.01	**0.06**	0.02	**0.07**	0.02	**0.09**
Whole body dose	0.02	**0.09**	0.03	**0.11**	0.03	**0.13**	0.04	**0.14**	0.04	**0.16**	0.05	**0.18**
ED (ICRP 60)	0.04	**0.26**	0.05	**0.31**	0.05	**0.34**	0.06	**0.37**	0.07	**0.42**	0.08	**0.47**
ED (ICRP 103)	0.04	**0.19**	0.04	**0.23**	0.05	**0.26**	0.06	**0.28**	0.07	**0.33**	0.08	**0.37**
Total filtration (mm) Al	**1.5**		**2.5**		**3.5**		**4.5**		**3.5**		**3.5**	
Cu	**0**		**0**		**0**		**0**		**0.1**		**0.2**	

Table 16.43 Normalized organ doses—Lumbar spine ap—8–12 years

Beam projection:		ap											**Table cells:**
Focus-to-detector distance (cm):		150											(+) = normal font = optimal field setting
Tube voltage (kVp):		80											**(−) = bold font = suboptimal field setting**
Total filtrations:		See last table row											Empty table cells correspond to values <0.01
Organ or tissue		**Organ dose D_T: Entrance dose K_E**								**(mSv ∗ mGy^{-1})**			
		(+)	(−)	(+)	(−)	(+)	(−)	(+)	(−)	(+)	(−)	(+)	(−)
Skull													
Brain													
Oral mucosa													
Salivary glands													
Cervical spine													
Thyroid													
Thymus													**0.01**
Thoracic spine		0.01	**0.05**	0.02	**0.07**	0.02	**0.08**	0.02	**0.09**	0.03	**0.12**	0.03	**0.14**
Claviculae													
Scapulae					**0.01**		**0.01**		**0.01**		**0.02**		**0.02**
Ribs		0.02	**0.23**	0.03	**0.26**	0.03	**0.28**	0.03	**0.29**	0.04	**0.32**	0.04	**0.34**
Oesophagus			**0.03**		**0.04**	0.01	**0.05**	0.01	**0.06**	0.01	**0.07**	0.02	**0.08**
Trachea													
Lung			**0.02**		**0.03**		**0.03**		**0.04**		**0.04**		**0.05**
Mamma									**0.01**		**0.01**		**0.02**
Heart			**0.04**		**0.05**		**0.06**		**0.07**		**0.08**	0.01	**0.10**
Stomach		0.06	**0.43**	0.07	**0.52**	0.08	**0.59**	0.09	**0.64**	0.11	**0.73**	0.13	**0.81**
Spleen		0.01	**0.08**	0.01	**0.11**	0.02	**0.13**	0.02	**0.14**	0.02	**0.17**	0.03	**0.21**
Liver		0.04	**0.29**	0.05	**0.35**	0.06	**0.39**	0.07	**0.43**	0.08	**0.49**	0.09	**0.55**
Gall bladder		0.30	**0.44**	0.37	**0.55**	0.42	**0.63**	0.47	**0.69**	0.54	**0.81**	0.61	**0.92**
Pancreas		0.06	**0.23**	0.08	**0.29**	0.09	**0.33**	0.10	**0.37**	0.12	**0.45**	0.14	**0.52**
Small intestine		0.16	**0.33**	0.20	**0.41**	0.23	**0.47**	0.25	**0.52**	0.30	**0.62**	0.35	**0.71**
Large intestine		0.12	**0.39**	0.15	**0.48**	0.17	**0.55**	0.19	**0.61**	0.23	**0.71**	0.26	**0.81**
Sigma/rectum		0.03	**0.28**	0.04	**0.34**	0.05	**0.40**	0.05	**0.44**	0.07	**0.52**	0.08	**0.60**
Lumbar spine		0.13	**0.21**	0.17	**0.27**	0.20	**0.32**	0.22	**0.36**	0.28	**0.45**	0.33	**0.54**
Pelvis		0.05	**0.25**	0.07	**0.32**	0.08	**0.38**	0.09	**0.42**	0.11	**0.52**	0.14	**0.61**
Adrenals		0.01	**0.08**	0.02	**0.11**	0.02	**0.12**	0.02	**0.14**	0.03	**0.17**	0.03	**0.21**
Kidneys		0.03	**0.08**	0.03	**0.10**	0.04	**0.12**	0.04	**0.14**	0.05	**0.17**	0.07	**0.21**
Urinary bladder		0.24	**0.60**	0.28	**0.72**	0.31	**0.81**	0.34	**0.88**	0.39	**1.01**	0.43	**1.12**
Uterus		0.27	**0.37**	0.33	**0.47**	0.38	**0.54**	0.42	**0.60**	0.50	**0.71**	0.57	**0.81**
Ovaries		0.07	**0.27**	0.09	**0.34**	0.10	**0.39**	0.11	**0.44**	0.14	**0.53**	0.16	**0.61**
Prostate		0.03	**0.40**	0.04	**0.50**	0.05	**0.57**	0.05	**0.63**	0.06	**0.74**	0.07	**0.84**
Testes			**1.08**		**1.19**	0.01	**1.25**	0.01	**1.29**	0.02	**1.37**	0.02	**1.43**
Upper arm													**0.01**
Elbow			**0.02**		**0.03**		**0.04**		**0.04**		**0.05**		**0.06**
Forearm			**0.04**		**0.06**		**0.07**		**0.08**		**0.10**		**0.12**
Upper leg			**0.11**		**0.14**		**0.16**		**0.18**		**0.23**	0.01	**0.27**
Knee													
Lower leg													
Lymph nodes		0.08	**0.19**	0.10	**0.24**	0.12	**0.27**	0.13	**0.30**	0.15	**0.36**	0.18	**0.41**
Skin		0.02	**0.09**	0.02	**0.09**	0.02	**0.09**	0.02	**0.10**	0.03	**0.10**	0.03	**0.11**
Muscle		0.03	**0.11**	0.04	**0.13**	0.04	**0.15**	0.04	**0.16**	0.05	**0.18**	0.06	**0.20**
Skeleton		0.01	**0.07**	0.01	**0.08**	0.02	**0.09**	0.02	**0.10**	0.02	**0.12**	0.03	**0.14**
Red bone marrow			**0.04**	0.01	**0.06**	0.02	**0.06**	0.02	**0.07**	0.02	**0.09**	0.03	**0.11**
Whole body dose		0.03	**0.11**	0.03	**0.13**	0.04	**0.14**	0.04	**0.15**	0.05	**0.18**	0.05	**0.20**
ED (ICRP 60)		0.04	**0.29**	0.05	**0.34**	0.06	**0.38**	0.07	**0.41**	0.08	**0.46**	0.09	**0.51**
ED (ICRP 103)		0.04	**0.21**	0.05	**0.26**	0.06	**0.29**	0.07	**0.31**	0.08	**0.36**	0.09	**0.40**
Total filtration (mm)	**Al**	**1.5**		**2.5**		**3.5**		**4.5**		**3.5**		**3.5**	
	Cu	**0**		**0**		**0**		**0**		**0.1**		**0.2**	

Table 16.44 Normalized organ doses—Lumbar spine ap—13–17 years

Beam projection:			ap									**Table cells:**
Focus-to-detector distance (cm):			150									(+) = normal font = optimal field setting
Tube voltage (kVp):			70									**(−) = bold font = suboptimal field setting**
Total filtrations:			See last table row									Empty table cells correspond to values <0.01
Organ or tissue	**Organ dose D_T: Entrance dose K_E**							**(mSv ∗ mGy⁻¹)**				
	(+)	(−)	(+)	(−)	(+)	(−)	(+)	(−)	(+)	(−)	(+)	(−)
Skull												
Brain												
Oral mucosa												
Salivary glands												
Cervical spine												
Thyroid												
Thymus												
Thoracic spine		**0.03**		**0.03**	0.01	**0.04**	0.01	**0.05**	0.02	**0.06**	0.02	**0.08**
Claviculae												
Scapulae												
Ribs	0.02	**0.18**	0.03	**0.21**	0.03	**0.23**	0.03	**0.24**	0.04	**0.27**	0.04	**0.29**
Oesophagus		**0.04**		**0.05**		**0.06**		**0.07**	0.01	**0.09**	0.01	**0.11**
Trachea												
Lung		**0.01**		**0.02**		**0.02**		**0.02**		**0.03**		**0.04**
Mamma		**0.05**	0.01	**0.07**	0.01	**0.08**	0.01	**0.09**	0.02	**0.12**	0.02	**0.14**
Heart		**0.03**		**0.04**		**0.04**		**0.05**		**0.06**		**0.08**
Stomach	0.05	**0.29**	0.06	**0.36**	0.07	**0.42**	0.08	**0.46**	0.09	**0.54**	0.11	**0.62**
Spleen		**0.04**		**0.05**	0.01	**0.06**	0.01	**0.07**	0.02	**0.09**	0.02	**0.11**
Liver	0.02	**0.17**	0.03	**0.21**	0.03	**0.25**	0.04	**0.27**	0.05	**0.33**	0.06	**0.38**
Gall bladder	0.15	**0.31**	0.19	**0.40**	0.22	**0.46**	0.25	**0.52**	0.30	**0.62**	0.35	**0.72**
Pancreas	0.04	**0.14**	0.06	**0.18**	0.07	**0.21**	0.08	**0.24**	0.10	**0.30**	0.12	**0.36**
Small intestine	0.12	**0.22**	0.16	**0.28**	0.18	**0.33**	0.21	**0.37**	0.25	**0.45**	0.30	**0.53**
Large intestine	0.09	**0.25**	0.12	**0.32**	0.14	**0.37**	0.16	**0.42**	0.20	**0.51**	0.23	**0.59**
Sigma/rectum	0.02	**0.17**	0.03	**0.21**	0.04	**0.25**	0.04	**0.28**	0.06	**0.35**	0.07	**0.41**
Lumbar spine	0.07	**0.10**	0.09	**0.13**	0.11	**0.16**	0.13	**0.19**	0.17	**0.24**	0.21	**0.31**
Pelvis	0.03	**0.12**	0.04	**0.16**	0.05	**0.19**	0.06	**0.22**	0.08	**0.28**	0.09	**0.34**
Adrenals		**0.04**	0.01	**0.06**	0.01	**0.07**	0.02	**0.08**	0.02	**0.10**	0.02	**0.13**
Kidneys	0.02	**0.04**	0.02	**0.06**	0.02	**0.07**	0.03	**0.08**	0.04	**0.10**	0.05	**0.12**
Urinary bladder	0.26	**0.47**	0.32	**0.58**	0.36	**0.66**	0.40	**0.73**	0.46	**0.86**	0.52	**0.97**
Uterus	0.57	**0.74**	0.73	**0.95**	0.86	**1.12**	0.96	**1.26**	1.16	**1.54**	1.36	**1.81**
Ovaries	0.05	**0.17**	0.07	**0.22**	0.08	**0.26**	0.09	**0.30**	0.12	**0.37**	0.14	**0.44**
Prostate	0.03	**0.32**	0.04	**0.41**	0.05	**0.48**	0.06	**0.53**	0.07	**0.65**	0.09	**0.75**
Testes		**0.91**		**1.03**		**1.11**		**1.17**	0.01	**1.27**	0.02	**1.36**
Upper arm												
Elbow				**0.01**		**0.01**		**0.02**		**0.02**		**0.03**
Forearm		**0.02**		**0.02**		**0.03**		**0.03**		**0.04**		**0.05**
Upper leg		**0.08**		**0.10**		**0.12**		**0.14**		**0.17**		**0.21**
Knee												
Lower leg												
Lymph nodes	0.06	**0.13**	0.07	**0.17**	0.08	**0.20**	0.09	**0.22**	0.11	**0.26**	0.13	**0.31**
Skin	0.03	**0.08**	0.03	**0.09**	0.03	**0.09**	0.03	**0.09**	0.03	**0.10**	0.03	**0.11**
Muscle	0.03	**0.09**	0.04	**0.11**	0.04	**0.12**	0.04	**0.13**	0.05	**0.15**	0.06	**0.17**
Skeleton		**0.04**		**0.06**	0.01	**0.06**	0.01	**0.07**	0.02	**0.09**	0.02	**0.10**
Red bone marrow		**0.03**		**0.04**	0.01	**0.04**	0.01	**0.05**	0.02	**0.06**	0.02	**0.07**
Whole body dose	0.03	**0.08**	0.03	**0.10**	0.04	**0.11**	0.04	**0.12**	0.05	**0.15**	0.05	**0.17**
ED (ICRP 60)	0.04	**0.22**	0.05	**0.26**	0.06	**0.30**	0.06	**0.32**	0.08	**0.38**	0.09	**0.43**
ED (ICRP 103)	0.04	**0.16**	0.05	**0.20**	0.06	**0.23**	0.06	**0.25**	0.08	**0.30**	0.09	**0.34**
Total filtration (mm) **Al**	**1.5**		**2.5**		**3.5**		**4.5**		**3.5**		**3.5**	
Cu	**0**		**0**		**0**		**0**		**0.1**		**0.2**	

Table 16.45 Normalized organ doses—Lumbar spine ap—13–17 years

Beam projection:			ap									**Table cells:**
Focus-to-detector distance (cm):			150									(+) = normal font = optimal field setting
Tube voltage (kVp):			80									**(−) = bold font = suboptimal field setting**
Total filtrations:			See last table row									Empty table cells correspond to values <0.01
Organ or tissue	**Organ dose D_T: Entrance dose K_E**						**(mSv ∗ mGy⁻¹)**					
	(+)	(−)	(+)	(−)	(+)	(−)	(+)	(−)	(+)	(−)	(+)	(−)
Skull												
Brain												
Oral mucosa												
Salivary glands												
Cervical spine												
Thyroid												
Thymus												
Thoracic spine		**0.04**	0.01	**0.05**	0.01	**0.06**	0.02	**0.06**	0.02	**0.08**	0.03	**0.10**
Claviculae												
Scapulae										**0.01**		**0.01**
Ribs	0.03	**0.20**	0.03	**0.22**	0.03	**0.24**	0.04	**0.26**	0.04	**0.28**	0.04	**0.30**
Oesophagus		**0.05**		**0.07**		**0.08**	0.01	**0.09**	0.01	**0.11**	0.02	**0.13**
Trachea												
Lung		**0.02**		**0.02**		**0.03**		**0.03**		**0.04**		**0.05**
Mamma	0.01	**0.07**	0.01	**0.09**	0.02	**0.10**	0.02	**0.12**	0.02	**0.14**	0.03	**0.17**
Heart		**0.04**		**0.05**		**0.05**		**0.06**		**0.08**	0.01	**0.09**
Stomach	0.06	**0.34**	0.07	**0.42**	0.08	**0.47**	0.09	**0.52**	0.11	**0.60**	0.13	**0.68**
Spleen		**0.05**	0.01	**0.07**	0.01	**0.08**	0.02	**0.09**	0.02	**0.12**	0.03	**0.14**
Liver	0.03	**0.20**	0.04	**0.25**	0.04	**0.28**	0.05	**0.31**	0.05	**0.37**	0.06	**0.42**
Gall bladder	0.18	**0.38**	0.23	**0.47**	0.26	**0.54**	0.29	**0.59**	0.34	**0.71**	0.39	**0.81**
Pancreas	0.06	**0.17**	0.07	**0.22**	0.08	**0.26**	0.09	**0.29**	0.12	**0.36**	0.14	**0.42**
Small intestine	0.15	**0.27**	0.19	**0.34**	0.22	**0.39**	0.24	**0.44**	0.29	**0.52**	0.34	**0.61**
Large intestine	0.12	**0.31**	0.15	**0.38**	0.17	**0.44**	0.19	**0.49**	0.23	**0.58**	0.26	**0.66**
Sigma/rectum	0.03	**0.21**	0.04	**0.26**	0.05	**0.30**	0.05	**0.34**	0.07	**0.41**	0.08	**0.47**
Lumbar spine	0.09	**0.14**	0.12	**0.18**	0.14	**0.22**	0.16	**0.25**	0.21	**0.31**	0.25	**0.38**
Pelvis	0.04	**0.16**	0.06	**0.20**	0.07	**0.24**	0.08	**0.27**	0.10	**0.34**	0.12	**0.41**
Adrenals	0.01	**0.06**	0.01	**0.08**	0.02	**0.09**	0.02	**0.10**	0.03	**0.13**	0.03	**0.16**
Kidneys	0.02	**0.06**	0.03	**0.07**	0.03	**0.09**	0.04	**0.10**	0.05	**0.13**	0.06	**0.15**
Urinary bladder	0.30	**0.54**	0.36	**0.66**	0.40	**0.75**	0.44	**0.82**	0.50	**0.95**	0.56	**1.06**
Uterus	0.70	**0.91**	0.87	**1.14**	1.00	**1.33**	1.11	**1.48**	1.33	**1.78**	1.53	**2.06**
Ovaries	0.07	**0.21**	0.09	**0.27**	0.10	**0.32**	0.11	**0.36**	0.14	**0.43**	0.17	**0.51**
Prostate	0.04	**0.39**	0.05	**0.48**	0.06	**0.56**	0.07	**0.61**	0.09	**0.73**	0.10	**0.84**
Testes		**0.97**		**1.09**	0.01	**1.17**	0.01	**1.23**	0.02	**1.33**	0.02	**1.41**
Upper arm												
Elbow		**0.01**		**0.02**		**0.02**		**0.02**		**0.03**		**0.04**
Forearm		**0.02**		**0.03**		**0.04**		**0.04**		**0.06**		**0.07**
Upper leg		**0.10**		**0.13**		**0.15**		**0.17**		**0.21**		**0.25**
Knee												
Lower leg												
Lymph nodes	0.07	**0.16**	0.09	**0.20**	0.10	**0.23**	0.11	**0.26**	0.13	**0.30**	0.15	**0.35**
Skin	0.03	**0.09**	0.03	**0.09**	0.03	**0.10**	0.03	**0.10**	0.03	**0.11**	0.03	**0.11**
Muscle	0.03	**0.10**	0.04	**0.12**	0.04	**0.13**	0.05	**0.15**	0.06	**0.17**	0.06	**0.19**
Skeleton		**0.05**	0.01	**0.07**	0.01	**0.08**	0.02	**0.08**	0.02	**0.10**	0.02	**0.12**
Red bone marrow		**0.04**	0.01	**0.05**	0.01	**0.05**	0.02	**0.06**	0.02	**0.07**	0.03	**0.09**
Whole body dose	0.03	**0.10**	0.04	**0.12**	0.04	**0.13**	0.05	**0.14**	0.05	**0.16**	0.06	**0.19**
ED (ICRP 60)	0.05	**0.25**	0.06	**0.30**	0.07	**0.33**	0.08	**0.36**	0.09	**0.42**	0.10	**0.47**
ED (ICRP 103)	0.05	**0.19**	0.06	**0.23**	0.07	**0.26**	0.07	**0.29**	0.09	**0.34**	0.10	**0.38**
Total filtration (mm) Al	**1.5**		**2.5**		**3.5**		**4.5**		**3.5**		**3.5**	
Cu	**0**		**0**		**0**		**0**		**0.1**		**0.2**	

Table 16.46 Normalized organ doses—Lumbar spine ap—>17 years

Beam projection:	ap	**Table cells:**
Focus-to-detector distance (cm):	150	(+) = normal font = optimal field setting
Tube voltage (kVp):	70	**(−) = bold font = suboptimal field setting**
Total filtrations:	See last table row	Empty table cells correspond to values <0.01

Organ or tissue	**Organ dose D_T: Entrance dose K_E**						**(mSv ∗ mGy^{-1})**					
	(+)	(−)	(+)	(−)	(+)	(−)	(+)	(−)	(+)	(−)	(+)	(−)
Skull												
Brain												
Oral mucosa												
Salivary glands												
Cervical spine												
Thyroid												
Thymus												
Thoracic spine		**0.02**		**0.03**		**0.04**	0.01	**0.04**	0.01	**0.05**	0.02	**0.07**
Claviculae												
Scapulae												
Ribs	0.02	**0.16**	0.03	**0.19**	0.03	**0.21**	0.03	**0.22**	0.03	**0.24**	0.04	**0.26**
Oesophagus		**0.04**		**0.05**		**0.06**		**0.07**	0.01	**0.09**	0.01	**0.11**
Trachea												
Lung				**0.01**		**0.02**		**0.02**		**0.02**		**0.03**
Mamma												**0.01**
Heart		**0.02**		**0.03**		**0.03**		**0.04**		**0.05**		**0.06**
Stomach	0.04	**0.27**	0.05	**0.34**	0.06	**0.39**	0.07	**0.43**	0.08	**0.51**	0.10	**0.59**
Spleen		**0.03**		**0.05**		**0.06**	0.01	**0.06**	0.01	**0.08**	0.02	**0.10**
Liver	0.02	**0.16**	0.03	**0.20**	0.04	**0.24**	0.04	**0.26**	0.05	**0.31**	0.06	**0.36**
Gall bladder	0.18	**0.31**	0.22	**0.39**	0.26	**0.46**	0.29	**0.52**	0.35	**0.63**	0.41	**0.73**
Pancreas	0.04	**0.13**	0.06	**0.17**	0.07	**0.20**	0.08	**0.23**	0.09	**0.29**	0.11	**0.35**
Small intestine	0.12	**0.21**	0.15	**0.28**	0.18	**0.32**	0.20	**0.37**	0.25	**0.45**	0.29	**0.53**
Large intestine	0.09	**0.24**	0.12	**0.31**	0.14	**0.36**	0.15	**0.40**	0.19	**0.49**	0.22	**0.57**
Sigma/rectum	0.02	**0.16**	0.03	**0.20**	0.03	**0.24**	0.04	**0.27**	0.05	**0.33**	0.06	**0.40**
Lumbar spine	0.07	**0.09**	0.09	**0.13**	0.11	**0.15**	0.13	**0.18**	0.16	**0.23**	0.20	**0.29**
Pelvis	0.03	**0.10**	0.04	**0.14**	0.04	**0.17**	0.05	**0.19**	0.07	**0.25**	0.08	**0.31**
Adrenals		**0.04**	0.01	**0.05**	0.01	**0.07**	0.01	**0.08**	0.02	**0.10**	0.02	**0.12**
Kidneys	0.01	**0.04**	0.02	**0.05**	0.02	**0.06**	0.03	**0.07**	0.04	**0.10**	0.04	**0.12**
Urinary bladder	0.27	**0.47**	0.33	**0.58**	0.38	**0.67**	0.41	**0.73**	0.48	**0.86**	0.54	**0.98**
Uterus	0.23	**0.29**	0.29	**0.37**	0.34	**0.43**	0.38	**0.49**	0.46	**0.60**	0.54	**0.71**
Ovaries	0.05	**0.17**	0.06	**0.22**	0.08	**0.26**	0.09	**0.29**	0.11	**0.37**	0.14	**0.44**
Prostate	0.03	**0.29**	0.04	**0.38**	0.04	**0.44**	0.05	**0.49**	0.06	**0.60**	0.08	**0.70**
Testes		**0.86**		**1.00**		**1.09**		**1.15**		**1.27**	0.01	**1.36**
Upper arm												
Elbow								**0.01**		**0.01**		**0.02**
Forearm		**0.01**		**0.02**		**0.02**		**0.02**		**0.03**		**0.04**
Upper leg		**0.07**		**0.10**		**0.12**		**0.14**		**0.17**		**0.21**
Knee												
Lower leg												
Lymph nodes	0.06	**0.13**	0.07	**0.16**	0.09	**0.19**	0.10	**0.21**	0.12	**0.26**	0.14	**0.30**
Skin	0.03	**0.09**	0.03	**0.09**	0.03	**0.10**	0.03	**0.10**	0.03	**0.11**	0.03	**0.11**
Muscle	0.03	**0.09**	0.03	**0.10**	0.04	**0.12**	0.04	**0.13**	0.05	**0.15**	0.06	**0.17**
Skeleton		**0.04**		**0.05**	0.01	**0.06**	0.01	**0.06**	0.02	**0.08**	0.02	**0.09**
Red bone marrow		**0.03**		**0.03**	0.01	**0.04**	0.01	**0.05**	0.02	**0.06**	0.02	**0.07**
Whole body dose	0.03	**0.08**	0.03	**0.10**	0.04	**0.11**	0.04	**0.12**	0.05	**0.14**	0.05	**0.16**
ED (ICRP 60)	0.04	**0.20**	0.05	**0.25**	0.05	**0.28**	0.06	**0.31**	0.07	**0.36**	0.08	**0.41**
ED (ICRP 103)	0.03	**0.14**	0.04	**0.18**	0.05	**0.21**	0.06	**0.23**	0.07	**0.27**	0.08	**0.31**
Total filtration (mm) **Al**	**1.5**		**2.5**		**3.5**		**4.5**		**3.5**		**3.5**	
Cu	**0**		**0**		**0**		**0**		**0.1**		**0.2**	

Table 16.47 Normalized organ doses—Lumbar spine ap—>17 years

Beam projection:	ap		**Table cells:**
Focus-to-detector distance (cm):	150		(+) = normal font = optimal field setting
Tube voltage (kVp):	80		**(−) = bold font = suboptimal field setting**
Total filtrations:	See last table row		Empty table cells correspond to values <0.01

Organ or tissue	**Organ dose D_T: Entrance dose K_E**						**(mSv ∗ mGy^{-1})**					
	(+)	(−)	(+)	(−)	(+)	(−)	(+)	(−)	(+)	(−)	(+)	(−)
Skull												
Brain												
Oral mucosa												
Salivary glands												
Cervical spine												
Thyroid												
Thymus												
Thoracic spine		**0.03**	0.01	**0.04**	0.01	**0.05**	0.02	**0.06**	0.02	**0.07**	0.02	**0.09**
Claviculae												
Scapulae												
Ribs	0.03	**0.18**	0.03	**0.20**	0.03	**0.22**	0.03	**0.23**	0.04	**0.25**	0.04	**0.27**
Oesophagus		**0.05**		**0.07**		**0.08**	0.01	**0.09**	0.01	**0.11**	0.02	**0.13**
Trachea												
Lung		**0.01**		**0.02**		**0.02**		**0.02**		**0.03**		**0.03**
Mamma										**0.01**		**0.01**
Heart		**0.03**		**0.03**		**0.04**		**0.05**		**0.06**		**0.07**
Stomach	0.05	**0.32**	0.06	**0.39**	0.07	**0.45**	0.08	**0.49**	0.10	**0.57**	0.11	**0.65**
Spleen		**0.05**		**0.06**	0.01	**0.07**	0.01	**0.08**	0.02	**0.10**	0.02	**0.12**
Liver	0.03	**0.19**	0.04	**0.24**	0.04	**0.27**	0.05	**0.30**	0.06	**0.35**	0.07	**0.40**
Gall bladder	0.21	**0.38**	0.27	**0.47**	0.31	**0.54**	0.34	**0.60**	0.40	**0.71**	0.46	**0.82**
Pancreas	0.05	**0.17**	0.07	**0.21**	0.08	**0.25**	0.09	**0.28**	0.11	**0.34**	0.13	**0.41**
Small intestine	0.15	**0.27**	0.19	**0.33**	0.21	**0.39**	0.24	**0.43**	0.29	**0.52**	0.33	**0.61**
Large intestine	0.11	**0.29**	0.14	**0.37**	0.16	**0.42**	0.18	**0.47**	0.22	**0.56**	0.25	**0.65**
Sigma/rectum	0.03	**0.20**	0.04	**0.25**	0.04	**0.29**	0.05	**0.32**	0.06	**0.39**	0.08	**0.46**
Lumbar spine	0.09	**0.13**	0.12	**0.17**	0.14	**0.21**	0.16	**0.24**	0.21	**0.30**	0.25	**0.37**
Pelvis	0.04	**0.14**	0.05	**0.18**	0.06	**0.22**	0.07	**0.25**	0.09	**0.31**	0.11	**0.38**
Adrenals	0.01	**0.06**	0.01	**0.07**	0.02	**0.09**	0.02	**0.10**	0.03	**0.13**	0.03	**0.15**
Kidneys	0.02	**0.06**	0.03	**0.07**	0.03	**0.09**	0.04	**0.10**	0.05	**0.12**	0.06	**0.15**
Urinary bladder	0.31	**0.55**	0.37	**0.67**	0.42	**0.75**	0.46	**0.82**	0.52	**0.96**	0.58	**1.08**
Uterus	0.28	**0.35**	0.35	**0.45**	0.40	**0.52**	0.44	**0.58**	0.53	**0.70**	0.61	**0.81**
Ovaries	0.06	**0.21**	0.08	**0.27**	0.10	**0.32**	0.11	**0.36**	0.14	**0.43**	0.16	**0.51**
Prostate	0.04	**0.36**	0.05	**0.45**	0.05	**0.52**	0.06	**0.57**	0.07	**0.68**	0.09	**0.79**
Testes		**0.94**		**1.07**		**1.15**		**1.22**	0.01	**1.33**	0.02	**1.42**
Upper arm												
Elbow				**0.01**		**0.01**		**0.02**		**0.02**		**0.02**
Forearm		**0.02**		**0.02**		**0.03**		**0.03**		**0.04**		**0.05**
Upper leg		**0.10**		**0.13**		**0.15**		**0.17**		**0.21**		**0.25**
Knee												
Lower leg												
Lymph nodes	0.07	**0.16**	0.09	**0.20**	0.10	**0.23**	0.11	**0.25**	0.14	**0.30**	0.16	**0.35**
Skin	0.03	**0.09**	0.03	**0.10**	0.03	**0.10**	0.03	**0.11**	0.03	**0.11**	0.03	**0.12**
Muscle	0.03	**0.10**	0.04	**0.12**	0.04	**0.13**	0.05	**0.15**	0.06	**0.17**	0.06	**0.19**
Skeleton		**0.05**	0.01	**0.06**	0.01	**0.07**	0.02	**0.08**	0.02	**0.09**	0.02	**0.11**
Red bone marrow		**0.04**	0.01	**0.04**	0.01	**0.05**	0.02	**0.06**	0.02	**0.07**	0.03	**0.09**
Whole body dose	0.03	**0.09**	0.04	**0.11**	0.04	**0.13**	0.04	**0.14**	0.05	**0.16**	0.06	**0.18**
ED (ICRP 60)	0.04	**0.24**	0.05	**0.28**	0.06	**0.32**	0.07	**0.35**	0.08	**0.40**	0.10	**0.45**
ED (ICRP 103)	0.04	**0.17**	0.05	**0.21**	0.06	**0.24**	0.06	**0.26**	0.08	**0.30**	0.09	**0.34**
Total filtration (mm) Al	**1.5**		**2.5**		**3.5**		**4.5**		**3.5**		**3.5**	
Cu	**0**		**0**		**0**		**0**		**0.1**		**0.2**	

Table 16.48 Normalized organ doses—Lumbar spine r-lat—8–12 years

Beam projection:			r-lat									**Table cells:**
Focus-to-detector distance (cm):			150									(+) = normal font = optimal field setting
Tube voltage (kVp):			70									**(−) = bold font = suboptimal field setting**
Total filtrations:			See last table row									Empty table cells correspond to values <0.01
Organ or tissue	**Organ dose D_T: Entrance dose K_E**						**(mSv ∗ mGy⁻¹)**					
	(+)	(−)	(+)	(−)	(+)	(−)	(+)	(−)	(+)	(−)	(+)	(−)
Skull												
Brain												
Oral mucosa												
Salivary glands												
Cervical spine												
Thyroid												
Thymus												
Thoracic spine	0.01	**0.05**	0.02	**0.06**	0.02	**0.07**	0.03	**0.09**	0.03	**0.11**	0.04	**0.13**
Claviculae												
Scapulae												**0.01**
Ribs	0.03	**0.15**	0.04	**0.18**	0.05	**0.20**	0.05	**0.21**	0.06	**0.23**	0.06	**0.25**
Oesophagus		**0.01**		**0.02**		**0.02**		**0.02**		**0.03**		**0.03**
Trachea												
Lung		**0.01**		**0.02**		**0.02**		**0.02**		**0.03**		**0.04**
Mamma												
Heart				**0.01**		**0.01**		**0.02**		**0.02**		**0.03**
Stomach		**0.02**		**0.02**		**0.02**		**0.03**	0.01	**0.04**	0.02	**0.05**
Spleen		**0.01**		**0.02**	0.01	**0.02**	0.01	**0.02**	0.02	**0.03**	0.02	**0.04**
Liver	0.09	**0.29**	0.11	**0.35**	0.12	**0.39**	0.13	**0.43**	0.15	**0.49**	0.17	**0.55**
Gall bladder	0.03	**0.09**	0.03	**0.11**	0.04	**0.13**	0.05	**0.15**	0.06	**0.19**	0.07	**0.23**
Pancreas	0.02	**0.05**	0.02	**0.06**	0.03	**0.07**	0.03	**0.08**	0.04	**0.11**	0.05	**0.13**
Small intestine	0.06	**0.09**	0.07	**0.12**	0.09	**0.14**	0.10	**0.16**	0.12	**0.20**	0.14	**0.24**
Large intestine	0.05	**0.14**	0.06	**0.19**	0.07	**0.22**	0.08	**0.24**	0.10	**0.30**	0.12	**0.35**
Sigma/rectum		**0.03**	0.01	**0.03**	0.01	**0.04**	0.02	**0.05**	0.02	**0.06**	0.03	**0.08**
Lumbar spine	0.18	**0.19**	0.23	**0.24**	0.28	**0.29**	0.32	**0.33**	0.40	**0.42**	0.48	**0.51**
Pelvis	0.25	**0.43**	0.31	**0.54**	0.36	**0.63**	0.41	**0.70**	0.49	**0.83**	0.57	**0.95**
Adrenals	0.02	**0.07**	0.02	**0.09**	0.03	**0.11**	0.03	**0.13**	0.04	**0.16**	0.04	**0.19**
Kidneys	0.13	**0.14**	0.16	**0.18**	0.19	**0.21**	0.21	**0.24**	0.25	**0.28**	0.29	**0.33**
Urinary bladder	0.01	**0.04**	0.01	**0.06**	0.02	**0.07**	0.02	**0.08**	0.02	**0.10**	0.03	**0.12**
Uterus	0.03	**0.07**	0.04	**0.09**	0.04	**0.11**	0.05	**0.13**	0.06	**0.16**	0.08	**0.20**
Ovaries	0.07	**0.10**	0.09	**0.13**	0.11	**0.16**	0.12	**0.18**	0.15	**0.23**	0.18	**0.27**
Prostate		**0.05**		**0.07**		**0.08**	0.01	**0.09**	0.02	**0.12**	0.02	**0.14**
Testes				**0.01**		**0.01**		**0.02**		**0.02**		**0.03**
Upper arm												
Elbow												
Forearm												
Upper leg		**0.07**		**0.09**		**0.11**		**0.12**		**0.15**		**0.18**
Knee												
Lower leg												
Lymph nodes	0.02	**0.04**	0.03	**0.06**	0.03	**0.07**	0.04	**0.08**	0.05	**0.10**	0.06	**0.11**
Skin	0.04	**0.05**	0.04	**0.05**	0.04	**0.05**	0.04	**0.05**	0.05	**0.06**	0.05	**0.06**
Muscle	0.03	**0.06**	0.04	**0.07**	0.04	**0.08**	0.04	**0.09**	0.05	**0.10**	0.06	**0.11**
Skeleton	0.03	**0.06**	0.04	**0.08**	0.04	**0.09**	0.05	**0.10**	0.06	**0.12**	0.07	**0.14**
Red bone marrow	0.03	**0.05**	0.04	**0.07**	0.04	**0.08**	0.05	**0.09**	0.06	**0.11**	0.07	**0.13**
Whole body dose	0.03	**0.06**	0.04	**0.08**	0.04	**0.09**	0.05	**0.09**	0.05	**0.11**	0.06	**0.12**
ED (ICRP 60)	0.02	**0.05**	0.03	**0.07**	0.03	**0.08**	0.04	**0.09**	0.05	**0.11**	0.05	**0.13**
ED (ICRP 103)	0.02	**0.05**	0.02	**0.06**	0.03	**0.07**	0.03	**0.08**	0.04	**0.09**	0.04	**0.11**
Total filtration (mm) Al	**1.5**		**2.5**		**3.5**		**4.5**		**3.5**		**3.5**	
Cu	**0**		**0**		**0**		**0**		**0.1**		**0.2**	

Table 16.49 Normalized organ doses—Lumbar spine r-lat—8–12 years

Beam projection:	r-lat											**Table cells:**
Focus-to-detector distance (cm):	150											(+) = normal font = optimal field setting
Tube voltage (kVp):	80											**(−) = bold font = suboptimal field setting**
Total filtrations:	See last table row											Empty table cells correspond to values <0.01
Organ or tissue	**Organ dose D_T: Entrance dose K_E**							**(mSv ∗ mGy^{-1})**				
	(+)	(−)	(+)	(−)	(+)	(−)	(+)	(−)	(+)	(−)	(+)	(−)
Skull												
Brain												
Oral mucosa												
Salivary glands												
Cervical spine												
Thyroid												
Thymus												
Thoracic spine	0.02	**0.06**	0.02	**0.08**	0.03	**0.09**	0.03	**0.11**	0.04	**0.13**	0.05	**0.16**
Claviculae												
Scapulae										**0.01**		**0.01**
Ribs	0.04	**0.17**	0.05	**0.19**	0.05	**0.21**	0.06	**0.22**	0.06	**0.24**	0.07	**0.26**
Oesophagus		**0.02**		**0.02**		**0.02**		**0.03**		**0.03**		**0.04**
Trachea												
Lung		**0.02**		**0.02**		**0.03**		**0.03**		**0.04**		**0.04**
Mamma												
Heart		**0.01**		**0.02**		**0.02**		**0.02**		**0.03**		**0.03**
Stomach		**0.02**		**0.03**	0.01	**0.03**	0.01	**0.04**	0.02	**0.05**	0.02	**0.06**
Spleen	0.01	**0.02**	0.01	**0.02**	0.02	**0.03**	0.02	**0.03**	0.02	**0.04**	0.03	**0.05**
Liver	0.10	**0.33**	0.12	**0.39**	0.13	**0.44**	0.14	**0.47**	0.16	**0.54**	0.18	**0.59**
Gall bladder	0.03	**0.11**	0.04	**0.14**	0.05	**0.17**	0.06	**0.19**	0.07	**0.23**	0.09	**0.27**
Pancreas	0.02	**0.06**	0.03	**0.08**	0.04	**0.09**	0.04	**0.10**	0.05	**0.13**	0.06	**0.16**
Small intestine	0.07	**0.12**	0.09	**0.15**	0.10	**0.17**	0.12	**0.19**	0.14	**0.24**	0.17	**0.28**
Large intestine	0.06	**0.18**	0.08	**0.22**	0.09	**0.26**	0.10	**0.29**	0.12	**0.34**	0.14	**0.40**
Sigma/rectum	0.01	**0.04**	0.02	**0.05**	0.02	**0.06**	0.02	**0.06**	0.03	**0.08**	0.03	**0.10**
Lumbar spine	0.23	**0.24**	0.29	**0.31**	0.34	**0.36**	0.38	**0.41**	0.47	**0.51**	0.55	**0.60**
Pelvis	0.29	**0.50**	0.36	**0.62**	0.42	**0.71**	0.46	**0.78**	0.54	**0.91**	0.62	**1.03**
Adrenals	0.02	**0.09**	0.03	**0.12**	0.03	**0.14**	0.04	**0.16**	0.04	**0.19**	0.05	**0.23**
Kidneys	0.15	**0.17**	0.19	**0.22**	0.22	**0.25**	0.24	**0.27**	0.28	**0.32**	0.32	**0.37**
Urinary bladder	0.01	**0.06**	0.02	**0.07**	0.02	**0.08**	0.02	**0.10**	0.03	**0.12**	0.04	**0.14**
Uterus	0.04	**0.09**	0.05	**0.12**	0.06	**0.14**	0.06	**0.16**	0.08	**0.20**	0.10	**0.24**
Ovaries	0.09	**0.13**	0.11	**0.17**	0.13	**0.20**	0.15	**0.22**	0.18	**0.27**	0.21	**0.32**
Prostate		**0.07**	0.01	**0.09**	0.01	**0.10**	0.02	**0.12**	0.02	**0.14**	0.03	**0.17**
Testes		**0.01**		**0.02**		**0.02**		**0.02**		**0.03**		**0.04**
Upper arm												
Elbow												
Forearm												
Upper leg		**0.09**		**0.11**		**0.13**		**0.15**		**0.18**	0.01	**0.21**
Knee												
Lower leg												
Lymph nodes	0.03	**0.06**	0.03	**0.07**	0.04	**0.08**	0.05	**0.09**	0.06	**0.11**	0.07	**0.14**
Skin	0.04	**0.05**	0.04	**0.05**	0.04	**0.05**	0.05	**0.05**	0.05	**0.06**	0.05	**0.06**
Muscle	0.03	**0.07**	0.04	**0.08**	0.05	**0.09**	0.05	**0.10**	0.06	**0.11**	0.06	**0.12**
Skeleton	0.03	**0.07**	0.04	**0.09**	0.05	**0.10**	0.05	**0.11**	0.06	**0.13**	0.07	**0.15**
Red bone marrow	0.04	**0.07**	0.05	**0.08**	0.05	**0.10**	0.06	**0.11**	0.07	**0.13**	0.09	**0.15**
Whole body dose	0.04	**0.07**	0.04	**0.09**	0.05	**0.10**	0.05	**0.11**	0.06	**0.12**	0.07	**0.14**
ED (ICRP 60)	0.03	**0.06**	0.03	**0.08**	0.04	**0.09**	0.04	**0.10**	0.05	**0.12**	0.06	**0.15**
ED (ICRP 103)	0.02	**0.06**	0.03	**0.07**	0.03	**0.08**	0.04	**0.09**	0.04	**0.11**	0.05	**0.13**
Total filtration (mm) **Al**	**1.5**		**2.5**		**3.5**		**4.5**		**3.5**		**3.5**	
Cu	**0**		**0**		**0**		**0**		**0.1**		**0.2**	

Table 16.50 Normalized organ doses—Lumbar spine r-lat—13–17 years

Beam projection:		r-lat										**Table cells:**
Focus-to-detector distance (cm):		150										(+) = normal font = optimal field setting
Tube voltage (kVp):		70										**(−) = bold font = suboptimal field setting**
Total filtrations:		See last table row										Empty table cells correspond to values <0.01
Organ or tissue	**Organ dose D_T: Entrance dose K_E (mSv * mGy^{-1})**											
	(+)	(−)	(+)	(−)	(+)	(−)	(+)	(−)	(+)	(−)	(+)	(−)
Skull												
Brain												
Oral mucosa												
Salivary glands												
Cervical spine												
Thyroid												
Thymus												
Thoracic spine		**0.03**	0.01	**0.04**	0.01	**0.05**	0.02	**0.05**	0.02	**0.07**	0.03	**0.09**
Claviculae												
Scapulae												
Ribs	0.03	**0.14**	0.04	**0.16**	0.04	**0.18**	0.04	**0.19**	0.05	**0.21**	0.06	**0.23**
Oesophagus				**0.01**		**0.02**		**0.02**		**0.02**		**0.03**
Trachea												
Lung		**0.01**		**0.02**		**0.02**		**0.02**		**0.03**		**0.03**
Mamma		**0.01**		**0.02**		**0.02**		**0.02**		**0.03**		**0.04**
Heart								**0.01**		**0.02**		**0.02**
Stomach						**0.01**		**0.01**		**0.02**		**0.02**
Spleen								**0.01**		**0.01**	0.01	**0.02**
Liver	0.06	**0.25**	0.08	**0.30**	0.09	**0.35**	0.10	**0.38**	0.12	**0.44**	0.13	**0.50**
Gall bladder	0.02	**0.08**	0.03	**0.11**	0.04	**0.13**	0.04	**0.14**	0.05	**0.18**	0.07	**0.22**
Pancreas		**0.02**	0.01	**0.03**	0.02	**0.04**	0.02	**0.05**	0.02	**0.06**	0.03	**0.07**
Small intestine	0.03	**0.06**	0.04	**0.08**	0.05	**0.09**	0.06	**0.10**	0.08	**0.13**	0.09	**0.16**
Large intestine	0.03	**0.10**	0.04	**0.12**	0.05	**0.15**	0.05	**0.17**	0.07	**0.21**	0.08	**0.25**
Sigma/rectum		**0.01**		**0.02**		**0.02**		**0.02**	0.01	**0.03**	0.01	**0.04**
Lumbar spine	0.10	**0.11**	0.13	**0.15**	0.16	**0.18**	0.18	**0.21**	0.24	**0.26**	0.29	**0.33**
Pelvis	0.17	**0.31**	0.22	**0.40**	0.26	**0.47**	0.30	**0.53**	0.36	**0.64**	0.42	**0.74**
Adrenals	0.01	**0.05**	0.02	**0.06**	0.02	**0.08**	0.02	**0.09**	0.03	**0.11**	0.04	**0.13**
Kidneys	0.09	**0.11**	0.12	**0.14**	0.14	**0.16**	0.16	**0.18**	0.19	**0.22**	0.22	**0.26**
Urinary bladder		**0.02**		**0.03**	0.01	**0.04**	0.01	**0.04**	0.02	**0.06**	0.02	**0.07**
Uterus	0.06	**0.11**	0.08	**0.15**	0.09	**0.18**	0.11	**0.21**	0.14	**0.27**	0.17	**0.34**
Ovaries	0.03	**0.06**	0.04	**0.09**	0.05	**0.10**	0.06	**0.12**	0.08	**0.15**	0.10	**0.18**
Prostate		**0.03**		**0.04**		**0.05**		**0.05**	0.01	**0.07**	0.02	**0.09**
Testes								**0.01**		**0.01**		**0.02**
Upper arm												
Elbow												
Forearm												
Upper leg		**0.07**		**0.09**		**0.11**		**0.12**		**0.15**		**0.18**
Knee												
Lower leg												
Lymph nodes	0.01	**0.03**	0.02	**0.04**	0.02	**0.05**	0.02	**0.05**	0.03	**0.07**	0.04	**0.08**
Skin	0.05	**0.08**	0.05	**0.09**	0.05	**0.09**	0.05	**0.09**	0.06	**0.10**	0.06	**0.10**
Muscle	0.03	**0.05**	0.03	**0.06**	0.04	**0.07**	0.04	**0.08**	0.04	**0.09**	0.05	**0.10**
Skeleton	0.02	**0.05**	0.03	**0.07**	0.03	**0.08**	0.04	**0.09**	0.04	**0.10**	0.05	**0.12**
Red bone marrow	0.02	**0.05**	0.03	**0.06**	0.04	**0.07**	0.04	**0.08**	0.05	**0.10**	0.06	**0.12**
Whole body dose	0.03	**0.06**	0.03	**0.07**	0.04	**0.08**	0.04	**0.08**	0.05	**0.10**	0.05	**0.11**
ED (ICRP 60)	0.01	**0.04**	0.02	**0.05**	0.02	**0.06**	0.02	**0.07**	0.03	**0.08**	0.04	**0.10**
ED (ICRP 103)	0.01	**0.04**	0.02	**0.05**	0.02	**0.05**	0.02	**0.06**	0.03	**0.08**	0.03	**0.09**
Total filtration (mm) **Al**	**1.5**		**2.5**		**3.5**		**4.5**		**3.5**		**3.5**	
Cu	**0**		**0**		**0**		**0**		**0.1**		**0.2**	

Table 16.51 Normalized organ doses—Lumbar spine r-lat—13–17 years

Beam projection:	r-lat											**Table cells:**
Focus-to-detector distance (cm):	150											(+) = normal font = optimal field setting
Tube voltage (kVp):	80											**(−) = bold font = suboptimal field setting**
Total filtrations:	See last table row											Empty table cells correspond to values <0.01
Organ or tissue	**Organ dose D_T: Entrance dose K_E (mSv * mGy^{-1})**											
	(+)	(−)	(+)	(−)	(+)	(−)	(+)	(−)	(+)	(−)	(+)	(−)
Skull												
Brain												
Oral mucosa												
Salivary glands												
Cervical spine												
Thyroid												
Thymus												
Thoracic spine	0.01	**0.04**	0.02	**0.05**	0.02	**0.06**	0.02	**0.07**	0.03	**0.09**	0.03	**0.11**
Claviculae												
Scapulae												
Ribs	0.03	**0.15**	0.04	**0.17**	0.04	**0.19**	0.05	**0.20**	0.05	**0.23**	0.06	**0.24**
Oesophagus		**0.01**		**0.02**		**0.02**		**0.02**		**0.03**		**0.04**
Trachea												
Lung		**0.01**		**0.02**		**0.02**		**0.02**		**0.03**		**0.04**
Mamma		**0.02**		**0.02**		**0.03**		**0.03**		**0.04**		**0.05**
Heart				**0.01**		**0.01**		**0.02**		**0.02**		**0.02**
Stomach		**0.01**		**0.01**		**0.02**		**0.02**		**0.03**	0.01	**0.03**
Spleen				**0.01**		**0.01**		**0.02**	0.01	**0.02**	0.01	**0.03**
Liver	0.08	**0.29**	0.09	**0.34**	0.10	**0.39**	0.11	**0.42**	0.13	**0.48**	0.14	**0.54**
Gall bladder	0.03	**0.10**	0.04	**0.13**	0.05	**0.16**	0.05	**0.18**	0.07	**0.22**	0.08	**0.26**
Pancreas	0.01	**0.03**	0.02	**0.04**	0.02	**0.05**	0.02	**0.06**	0.03	**0.08**	0.04	**0.09**
Small intestine	0.04	**0.08**	0.06	**0.10**	0.07	**0.12**	0.07	**0.13**	0.09	**0.16**	0.11	**0.19**
Large intestine	0.04	**0.12**	0.05	**0.15**	0.06	**0.18**	0.07	**0.20**	0.08	**0.25**	0.10	**0.29**
Sigma/rectum		**0.02**		**0.02**	0.01	**0.03**	0.01	**0.03**	0.02	**0.04**	0.02	**0.05**
Lumbar spine	0.13	**0.15**	0.17	**0.19**	0.20	**0.23**	0.23	**0.26**	0.29	**0.33**	0.35	**0.40**
Pelvis	0.21	**0.38**	0.27	**0.47**	0.31	**0.54**	0.34	**0.60**	0.41	**0.72**	0.47	**0.82**
Adrenals	0.02	**0.06**	0.02	**0.08**	0.03	**0.10**	0.03	**0.11**	0.04	**0.14**	0.04	**0.16**
Kidneys	0.11	**0.13**	0.14	**0.16**	0.16	**0.19**	0.18	**0.21**	0.22	**0.25**	0.25	**0.29**
Urinary bladder		**0.03**	0.01	**0.04**	0.01	**0.05**	0.02	**0.06**	0.02	**0.07**	0.03	**0.09**
Uterus	0.08	**0.16**	0.10	**0.20**	0.12	**0.24**	0.14	**0.28**	0.18	**0.35**	0.22	**0.43**
Ovaries	0.04	**0.09**	0.06	**0.11**	0.07	**0.13**	0.08	**0.15**	0.10	**0.19**	0.12	**0.22**
Prostate		**0.04**		**0.05**	0.01	**0.06**	0.01	**0.07**	0.02	**0.09**	0.02	**0.11**
Testes				**0.01**		**0.01**		**0.01**		**0.02**		**0.02**
Upper arm												
Elbow												
Forearm												
Upper leg		**0.09**		**0.11**		**0.13**		**0.15**		**0.18**		**0.21**
Knee												
Lower leg												
Lymph nodes	0.02	**0.04**	0.02	**0.05**	0.03	**0.06**	0.03	**0.07**	0.04	**0.08**	0.05	**0.10**
Skin	0.05	**0.09**	0.05	**0.09**	0.05	**0.10**	0.05	**0.10**	0.06	**0.10**	0.06	**0.11**
Muscle	0.03	**0.06**	0.04	**0.07**	0.04	**0.08**	0.04	**0.09**	0.05	**0.10**	0.06	**0.11**
Skeleton	0.03	**0.06**	0.03	**0.08**	0.04	**0.09**	0.04	**0.10**	0.05	**0.12**	0.06	**0.13**
Red bone marrow	0.03	**0.06**	0.04	**0.08**	0.05	**0.09**	0.05	**0.10**	0.06	**0.12**	0.08	**0.14**
Whole body dose	0.03	**0.06**	0.04	**0.08**	0.04	**0.09**	0.04	**0.09**	0.05	**0.11**	0.06	**0.12**
ED (ICRP 60)	0.02	**0.05**	0.02	**0.06**	0.03	**0.07**	0.03	**0.08**	0.04	**0.10**	0.04	**0.12**
ED (ICRP 103)	0.02	**0.05**	0.02	**0.06**	0.02	**0.07**	0.03	**0.07**	0.03	**0.09**	0.04	**0.11**
Total filtration (mm) Al	**1.5**		**2.5**		**3.5**		**4.5**		**3.5**		**3.5**	
Cu	**0**		**0**		**0**		**0**		**0.1**		**0.2**	

Table 16.52 Normalized organ doses—Lumbar spine r-lat—>17 years

Beam projection:			r-lat									**Table cells:**
Focus-to-detector distance (cm):			150									(+) = normal font = optimal field setting
Tube voltage (kVp):			70									**(−) = bold font = suboptimal field setting**
Total filtrations:			See last table row									Empty table cells correspond to values <0.01
Organ or tissue	**Organ dose D_T: Entrance dose K_E**							**(mSv ∗ mGy^{-1})**				
	(+)	(−)	(+)	(−)	(+)	(−)	(+)	(−)	(+)	(−)	(+)	(−)
Skull												
Brain												
Oral mucosa												
Salivary glands												
Cervical spine												
Thyroid												
Thymus												
Thoracic spine		**0.02**		**0.02**		**0.03**	0.01	**0.03**	0.01	**0.04**	0.02	**0.05**
Claviculae												
Scapulae												
Ribs	0.03	**0.11**	0.03	**0.13**	0.03	**0.14**	0.04	**0.16**	0.04	**0.18**	0.05	**0.19**
Oesophagus						**0.01**		**0.01**		**0.02**		**0.02**
Trachea												
Lung						**0.01**		**0.01**		**0.02**		**0.02**
Mamma												
Heart												**0.01**
Stomach										**0.01**		**0.01**
Spleen												
Liver	0.06	**0.20**	0.08	**0.25**	0.09	**0.29**	0.10	**0.32**	0.11	**0.37**	0.13	**0.42**
Gall bladder	0.02	**0.05**	0.02	**0.07**	0.03	**0.08**	0.03	**0.09**	0.04	**0.12**	0.05	**0.15**
Pancreas		**0.01**		**0.02**	0.01	**0.02**	0.01	**0.03**	0.02	**0.04**	0.02	**0.05**
Small intestine	0.02	**0.04**	0.03	**0.05**	0.04	**0.07**	0.04	**0.08**	0.06	**0.10**	0.07	**0.12**
Large intestine	0.02	**0.07**	0.03	**0.10**	0.04	**0.11**	0.04	**0.13**	0.06	**0.16**	0.07	**0.20**
Sigma/rectum						**0.01**		**0.01**		**0.02**		**0.02**
Lumbar spine	0.06	**0.07**	0.08	**0.09**	0.10	**0.11**	0.12	**0.13**	0.15	**0.17**	0.19	**0.21**
Pelvis	0.13	**0.24**	0.17	**0.31**	0.20	**0.37**	0.23	**0.42**	0.28	**0.51**	0.34	**0.60**
Adrenals		**0.03**	0.01	**0.04**	0.01	**0.05**	0.02	**0.06**	0.02	**0.08**	0.03	**0.09**
Kidneys	0.07	**0.08**	0.09	**0.10**	0.11	**0.12**	0.12	**0.14**	0.15	**0.17**	0.17	**0.20**
Urinary bladder		**0.02**		**0.02**		**0.03**		**0.03**	0.01	**0.04**	0.02	**0.05**
Uterus	0.01	**0.03**	0.02	**0.04**	0.02	**0.05**	0.03	**0.05**	0.04	**0.07**	0.05	**0.09**
Ovaries	0.02	**0.05**	0.03	**0.06**	0.04	**0.07**	0.05	**0.08**	0.06	**0.11**	0.07	**0.13**
Prostate		**0.02**		**0.02**		**0.03**		**0.03**		**0.04**	0.01	**0.05**
Testes												**0.01**
Upper arm												
Elbow												
Forearm												
Upper leg		**0.07**		**0.09**		**0.10**		**0.12**		**0.15**		**0.18**
Knee												
Lower leg												
Lymph nodes		**0.02**	0.01	**0.03**	0.02	**0.03**	0.02	**0.04**	0.02	**0.05**	0.03	**0.06**
Skin	0.04	**0.06**	0.04	**0.07**	0.04	**0.07**	0.04	**0.07**	0.05	**0.08**	0.05	**0.08**
Muscle	0.02	**0.05**	0.03	**0.06**	0.03	**0.06**	0.03	**0.07**	0.04	**0.08**	0.04	**0.09**
Skeleton	0.02	**0.04**	0.02	**0.05**	0.02	**0.06**	0.03	**0.07**	0.03	**0.09**	0.04	**0.10**
Red bone marrow	0.02	**0.04**	0.02	**0.05**	0.03	**0.06**	0.03	**0.06**	0.04	**0.08**	0.05	**0.10**
Whole body dose	0.02	**0.05**	0.03	**0.06**	0.03	**0.07**	0.03	**0.07**	0.04	**0.09**	0.04	**0.10**
ED (ICRP 60)	0.01	**0.03**	0.01	**0.04**	0.02	**0.04**	0.02	**0.05**	0.02	**0.06**	0.03	**0.07**
ED (ICRP 103)	0.01	**0.03**	0.01	**0.03**	0.02	**0.04**	0.02	**0.04**	0.02	**0.05**	0.03	**0.06**
Total filtration (mm) **Al**	**1.5**		**2.5**		**3.5**		**4.5**		**3.5**		**3.5**	
Cu	**0**		**0**		**0**		**0**		**0.1**		**0.2**	

Table 16.53 Normalized organ doses—Lumbar spine r-lat—>17 years

Beam projection:	r-lat											**Table cells:**
Focus-to-detector distance (cm):	150											(+) = normal font = optimal field setting
Tube voltage (kVp):	80											**(−) = bold font = suboptimal field setting**
Total filtrations:	See last table row											Empty table cells correspond to values <0.01
Organ or tissue	**Organ dose D_T: Entrance dose K_E (mSv * mGy⁻¹)**											
	(+)	(−)	(+)	(−)	(+)	(−)	(+)	(−)	(+)	(−)	(+)	(−)
Skull												
Brain												
Oral mucosa												
Salivary glands												
Cervical spine												
Thyroid												
Thymus												
Thoracic spine		**0.02**	0.01	**0.03**	0.01	**0.04**	0.02	**0.04**	0.02	**0.06**	0.02	**0.07**
Claviculae												
Scapulae												
Ribs	0.03	**0.12**	0.03	**0.14**	0.04	**0.16**	0.04	**0.17**	0.05	**0.19**	0.05	**0.20**
Oesophagus				**0.01**		**0.01**		**0.02**		**0.02**		**0.03**
Trachea												
Lung				**0.01**		**0.01**		**0.02**		**0.02**		**0.02**
Mamma												
Heart										**0.01**		**0.02**
Stomach								**0.01**		**0.01**		**0.02**
Spleen												**0.01**
Liver	0.07	**0.24**	0.09	**0.29**	0.10	**0.33**	0.11	**0.36**	0.13	**0.41**	0.14	**0.47**
Gall bladder	0.02	**0.07**	0.03	**0.09**	0.04	**0.11**	0.04	**0.12**	0.05	**0.15**	0.06	**0.18**
Pancreas		**0.02**	0.01	**0.03**	0.01	**0.03**	0.02	**0.04**	0.02	**0.05**	0.03	**0.06**
Small intestine	0.03	**0.06**	0.04	**0.07**	0.05	**0.08**	0.06	**0.10**	0.07	**0.12**	0.09	**0.14**
Large intestine	0.03	**0.09**	0.04	**0.12**	0.05	**0.14**	0.06	**0.16**	0.07	**0.20**	0.08	**0.23**
Sigma/rectum		**0.01**		**0.01**		**0.02**		**0.02**	0.01	**0.03**	0.01	**0.03**
Lumbar spine	0.09	**0.10**	0.11	**0.12**	0.13	**0.15**	0.15	**0.17**	0.19	**0.22**	0.23	**0.26**
Pelvis	0.16	**0.30**	0.21	**0.37**	0.24	**0.43**	0.27	**0.48**	0.33	**0.58**	0.38	**0.67**
Adrenals	0.01	**0.04**	0.02	**0.06**	0.02	**0.07**	0.02	**0.08**	0.03	**0.09**	0.03	**0.11**
Kidneys	0.09	**0.10**	0.11	**0.13**	0.13	**0.15**	0.14	**0.16**	0.17	**0.20**	0.20	**0.23**
Urinary bladder		**0.02**		**0.03**	0.01	**0.04**	0.01	**0.04**	0.02	**0.05**	0.02	**0.07**
Uterus	0.02	**0.04**	0.03	**0.05**	0.03	**0.06**	0.04	**0.07**	0.05	**0.09**	0.06	**0.11**
Ovaries	0.03	**0.06**	0.04	**0.08**	0.05	**0.10**	0.06	**0.11**	0.08	**0.14**	0.09	**0.17**
Prostate		**0.02**		**0.03**		**0.04**		**0.04**	0.01	**0.06**	0.01	**0.07**
Testes										**0.01**		**0.02**
Upper arm												
Elbow												
Forearm												
Upper leg		**0.09**		**0.11**		**0.13**		**0.15**		**0.18**		**0.21**
Knee												
Lower leg												
Lymph nodes	0.01	**0.03**	0.02	**0.04**	0.02	**0.04**	0.02	**0.05**	0.03	**0.06**	0.04	**0.07**
Skin	0.04	**0.07**	0.04	**0.07**	0.05	**0.07**	0.05	**0.08**	0.05	**0.08**	0.05	**0.09**
Muscle	0.03	**0.05**	0.03	**0.07**	0.03	**0.07**	0.04	**0.08**	0.04	**0.09**	0.05	**0.10**
Skeleton	0.02	**0.05**	0.02	**0.06**	0.03	**0.07**	0.03	**0.08**	0.04	**0.10**	0.04	**0.11**
Red bone marrow	0.02	**0.05**	0.03	**0.06**	0.04	**0.07**	0.04	**0.08**	0.05	**0.10**	0.06	**0.11**
Whole body dose	0.03	**0.06**	0.03	**0.07**	0.04	**0.08**	0.04	**0.08**	0.04	**0.10**	0.05	**0.11**
ED (ICRP 60)	0.01	**0.04**	0.02	**0.05**	0.02	**0.05**	0.02	**0.06**	0.03	**0.07**	0.04	**0.09**
ED (ICRP 103)	0.01	**0.03**	0.02	**0.04**	0.02	**0.05**	0.02	**0.05**	0.03	**0.07**	0.03	**0.08**
Total filtration (mm) **Al**	**1.5**		**2.5**		**3.5**		**4.5**		**3.5**		**3.5**	
Cu	**0**		**0**		**0**		**0**		**0.1**		**0.2**	

Table 16.54 Inconsistencies between the mathematical MIRD phantoms and real human anatomy

Anatomical region	Inconsistencies
Abdomen	The navel has not been inserted
Skeleton	The shape of the spine is unrealistic (no lordosis, no kyphosis considered), the vertebral column is not clearly depicted. The transverse processes are not outlined

For details, see the text

References

1. Bontrager KL. Textbook of radiographic positioning and related anatomy. St. Louis: Mosby Year Book; 1993.
2. Commission of the European Communities. European guidelines on quality criteria for diagnostic radiographic images in paediatrics. EUR 16261, SBN 92-827-7843-6; 1996.
3. Cook JV, Shah K, Pablot S, Kyriou J, Pettett A, Fitzgerald M. Guidelines on best practice in the x-ray imaging of children. A manual for all x-ray departments. London: Queen Mary's Hospital for Children, Surrey, and The Radiological Protection Centre; 1998.
4. Drexler G, Panzer W, Stieve FE. Die Bestimmung von Organdosen in der Röntgendiagnostik. Berlin: Hoffmann Verlag; 1993.
5. Ebel K-D, Willich E. Die Röntgenuntersuchung im Kindesalter – Technik und Indikation. Berlin: Springer; 1968.
6. Hart D, Jones DJ, Wall BF. Estimation of effective dose in diagnostic radiology from entrance surface dose and dose-area-product measurements. National Radiological Protection Board NRPB-R262; 1994.
7. Hart D, Jones DJ, Wall BF. Coefficients for estimating effective doses from pediatric x-ray examinations. National Radiological Protection Board NRPB-R279; 1996.
8. Seidenbusch MC, Schneider K. Conversion coefficients for determining organ doses in paediatric spine radiography. Pediatr Radiol. 2014;44:434–56.

Pelvis

17

17.1 Introduction

The type of pelvis radiography is determined by the clinical indication. Table 17.1 shows indications and Table 17.2 contraindications for pelvis radiography. Image criteria as required for radiological findings in pelvis radiographs are listed in Table 17.3. In order to fulfil these image criteria, the field settings listed in Tables 17.4 and 17.5 should be achieved. These field settings go along with the patient preparation as listed in Table 17.6. Exposure parameter combinations for various age groups and patient diameters are listed in Tables 17.7, 17.8 and 17.9. The age groups in Table 17.8 refer to the standard age groups of the mathematical MIRD phantom family which has been used to calculate the

Table 17.1 Indications for pelvis radiographs

Hip dysplasia[a]
Perthes disease
Coxitis
Epiphysiolysis capitis femoris
Sacroiliitis
Juvenile rheumatoid arthritis
Osteomyelitis
Spastic diplegia
Unclear pain in the pelvic region
Trauma
Suspected pelvic anomaly (e.g. bladder atrophy, cleido-cranial dysplasia)
Bone tumours
Soft tissue tumours
Annotation:
[a] In the first six months of life, hip sonography is the preferable diagnostic tool

Table 17.2 Contraindications for pelvis radiographs

Pregnancy

Table 17.3 Image criteria for pelvis radiographs

Pelvis ap
Symmetrical reproduction of the pelvis
Visualization of the sacrum and its intervertebral foramina depending on bowel content
Reproduction of the lower part of the sacroiliac joints
Reproduction of the Y-cartilage, the ischopubic synchondrosis, the symphysis and the trochanters, and consistent with age
Reproduction of the necks of the femora which should not be distorted by foreshortening or external rotation
Visually sharp reproduction of the cortex, the spongiosa and trabecular structures of the acetabula, the femoral heads and femoral neck
Visualization of the periarticular soft tissue planes

Table 17.4 Patient positions and beam projections for pelvis radiographs

Patient position	Beam projections
Supine	ap

Table 17.5 Optimal field settings for pelvis radiographs

Pelvis ap	
Cranial field edge:	Navel
Field centring:	Just above the symphysis pubis
Lower field edge:	Perineum
Lateral field edges:	Skin surface

M. Seidenbusch et al., *Imaging Practice and Radiation Protection in Pediatric Radiology*,
https://doi.org/10.1007/978-3-030-18504-6_17

Table 17.6 Patient preparation

General	Remove all clothes, external metallic devices, especially piercings, and jewellery from the abdomen and pelvis of the patient. Clearance of the rectum is reasonable in cases of "unclear pain in the pelvic region"
	Immobilize infants and uncooperative patients in the supine position, e.g. with sandbags over the arms and legs. Put sandbags under the knees to relieve exaggerated lumbar lordosis
	Place eventually small X-ray positive markers on palpable tumours or pain points
Positioning	The arms should be positioned alongside the trunk
	The legs should be extended and intoed rotated
Shielding	In males and females, lead aprons and lead masks should be employed at the upper field edge
	In males, use testicular capsules if possible. In females, ovarian shields or masks of appropriate size should only be used to look for hip dysplasia / hip luxation

normalized organ doses in this book. Table 17.9 shows radiation qualities (i.e. various combinations of tube voltage and total filtration) which may be applied in clinical practice. Please note that not all of these radiation qualities are recommended by the authors. The standard projections of pelvis radiographies of the patient who is correctly positioned are depicted in Fig. 17.1. Also the optimal field (marked in green), the

Table 17.7 Radiographic procedure

X-ray source	Use automatic exposure control (AEC) in patients >1 year of age
	Focal spot size between 0.6 and 1.3 dependent on the age of the patient
	Check the use of additional filtration
	Check the focus-to-detector distance
Field setting	Check the field size
	Check the field centring
Image detector system	Check the use of antiscatter grids
Patient	–

Table 17.8 Radiographic techniques of pelvic radiographs

Age group (years)	Beam direction	Patient position	Beam projection	Typical patient diameter (cm)	Optimal field size (cm x cm)	Focus-to-detector distance (cm)	Voltage (kVp) 50	60	70	80	90	100	Antiscatter grid
0	vertical	supine	ap	9.80	10 x 9	100⁽¹⁾							
1–2				13.00	12 x 12								
3–7				15.00	17 x 15	115							
8–12				16.80	21 x 19								
13–17				19.60	28 x 26								
18				20.00	32 x 29								

Annotations:
(1) The patient is lying on the table top directly on the detector.
Parameter combination not recommended, normalized organ doses not provided
Parameter combination not recommended, normalized organ doses provided
Parameter combination recommended, normalized organ doses provided

Table 17.9 Radiation qualities of pelvic radiographs

Voltage (kVp)	Total filtration: Without additional filtration 1.5 mm Al	2.5 Al	With additional filtration 3.5 Al	4.5 Al	3.5 Al + 0.1 Cu[1]	3.5 Al + 0.2 Cu[1]
50						
60	This technique is not suitable for pelvis radiographs					
70						
80						
90						
100						

Annotations:
1 The total beam filtration is usually 3.5 mm Al + 0.1 mm Cu, in adolescents optionally3.5 mm Al + 0.2 mm Cu
Parameter combination not recommended, normalized organ doses not provided
Parameter combination not recommended, normalized organ doses provided
Parameter combination recommended, normalized organ doses provided

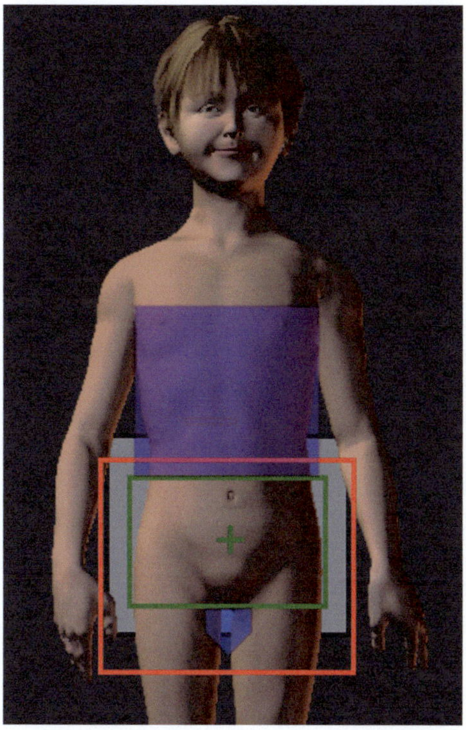

Fig. 17.1 **Field settings for pelvis radiograph in supine position**. Antero-posterior projection. (© V. Rösenberger and M. Seidenbusch 2018). Green lines = optimal field settings; green cross = centring point; red lines = suboptimal field settings; grey shaded area = image detector; blue shaded area = lead mask shielding and testicular capsule; dark grey shaded area = Bucky table

Table 17.10 Inconsistencies between the mathematical MIRD phantoms and real human anatomy

Anatomical region	Inconsistencies
Abdomen	The navel has not been inserted
Gonads	No variation of gonad situs with age
Skeleton	Hip joints have not been inserted

For details, see the text

suboptimal field (marked in red), and the correct shielding are shown. Tables 17.11, 17.12 and 17.13 outline the virtual field settings on the skin surface of mathematical MIRD phantoms of various age groups and the resulting virtual radiographies with and without bone structures, respectively. Normalized organ doses resulting from these virtual field settings have been calculated for all age groups and have been listed in Tables 17.14, 17.15, 17.16, 17.17, 17.18, 17.19, 17.20, 17.21, 17.22, 17.23, 17.24, 17.25, 17.26, 17.27 and 17.28. Table cells with normal font refer to optimal field settings and table cells with bold font refer to suboptimal field settings. Please note that normalized organ doses have been provided also for exposure parameter combinations, which are not recommended by the authors.

17.2 Indications and Contraindications

Indications and contraindications for pelvis radiographs are given in Tables 17.1 and 17.2.

17.3 Image Criteria

The image criteria required for pelvis radiographs are listed in Table 17.3.

See also [2, 3].

17.4 Field Settings

In Table 17.4, patient positions and beam projections used for pelvis radiographs are listed, and in Table 17.5, the optimal field settings are listed which are required in order to fulfil the image criteria mentioned in Table 17.3.

For more details, see e.g. [8–10].

17.5 Patient Preparation and Radiographic Procedure

See Tables 17.6, 17.7 and [1, 5].

17.6 Exposure Parameters

In Tables 17.8 and 17.9, exposure parameter combinations used for pelvic radiographs in various age groups are listed. Dark grey shaded table cells refer to exposure parameter combinations which are not recommended for paediatric pelvic radiographs and for which no normalized organ doses have been provided. Light grey and light blue shaded table cells refer to exposure parameter combinations for which normalized organ doses have been provided with light grey shaded table cells referring to exposure parameter combinations being not recommended and light blue shaded table cells referring to exposure parameter combinations following the European Guidelines [2]. Note that due to the fact that the radiation dose is an exponential function of the patient diameter, it may make sense to use the patient diameters instead of the patient's age groups for determining the adequate exposure parameters from Table 17.8. In this case, for determining the corresponding normalized organ doses from Tables 17.14, 17.15, 17.16, 17.17, 17.18, 17.19, 17.20, 17.21, 17.22, 17.23, 17.24, 17.25, 17.26, 17.27 and 17.28 the age group related to the patient diameter should be used.

17.7 Normalized Organ Doses

For calculating normalized organ doses, the field settings described in Sect. 17.4 were transferred to the mathematical MIRD phantoms. Inconsistencies between the geometrical anatomy of the mathematical phantoms and the real human anatomy have been listed in Table 17.10 whenever they are of relevance for the validity of the normalized organ doses.

In Tables 17.11, 17.12 and 17.13, optimal and suboptimal field settings are listed for the different age groups and the beam projections given in Table 17.8. Green lines refer to optimal field settings, red lines refer to suboptimal field settings, respectively. In the tables, the optimal and suboptimal field settings which were used for performing the calculations of the normalized organ doses provided in Tables 17.14, 17.15, 17.16, 17.17, 17.18, 17.19, 17.20, 17.21, 17.22, 17.23, 17.24, 17.25, 17.26, 17.27 and 17.28 were drawn in on the skin surface of the mathematical phantoms. The virtual mathematical radiographs resulting from these field settings were depicted in the tables with and without bone structures, respectively, in order to enable the reader to visually compare the virtual mathematical radiographs with the real radiographs made under clinical conditions. For comparing the virtual mathematical field settings with the real clinical field settings, the virtual mathematical field edges were also listed in the tables.

As the normalized organ doses provided in Tables 17.14, 17.15, 17.16, 17.17, 17.18, 17.19, 17.20, 17.21, 17.22, 17.23, 17.24, 17.25, 17.26, 17.27 and 17.28 refer to entrance kerma values, optimal and suboptimal field sizes were provided in the tables in order to enable the reader to calculate entrance air kerma dose values from clinically measured dose area product values. Please note that these optimal and suboptimal field sizes refer to the mathematical phantoms of an age group only and not to real patients with corresponding ages.

Please note also that the anatomy of the mathematical phantoms has been idealized. Therefore, the virtual field settings for calculating the normalized organ doses have been defined with a view to ensuring that the resulting virtual organ doses would be consistent with the real organ doses which would be achieved during real X-ray examinations being performed with the real field settings recommended in Sect. 17.4. Thus, the field settings shown in the following tables may not be consistent with the clinical field settings as recommended in Sect. 17.4.

Normalized organ doses for pelvic radiographs are also listed in e.g. [4, 6, 7, 11].

Table 17.11 Field settings—Pelvis ap

Age group (years)		Anatomical landmarks			
		Optimal field setting		**Suboptimal field setting**	
0	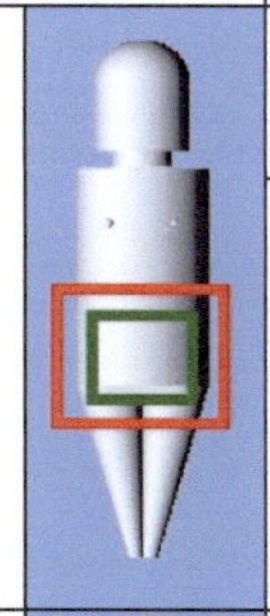	Upper field edge: Just below the umbilicus Lower field edge: Just below the perineum Lateral field edges: Close to the skin surface of the pelvis Field size = field width × field height: 9 cm × 9 cm = 81 cm²		Upper field edge: Above the umbilicus Lower field edge: Upper third of thighs Lateral field edges: Both forearms and hands included Field size = field width × field height: 16 cm × 13 cm = 208 cm²	
		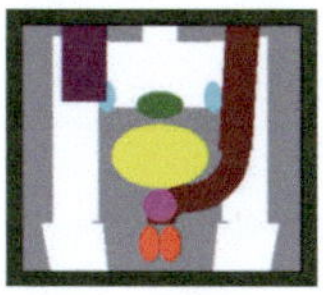	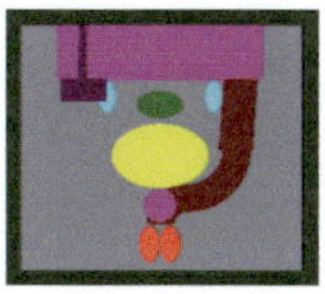		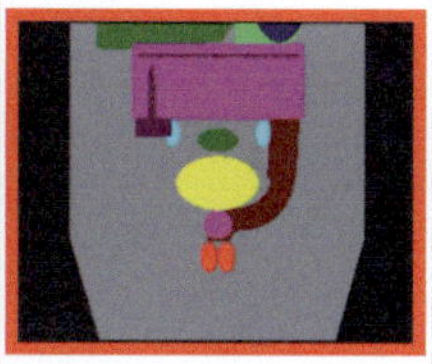
1 - 2	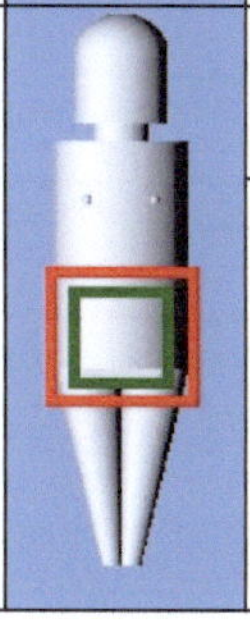	Upper field edge: Just below the umbilicus Lower field edge: Just below the perineum Lateral field edges: Close to the skin surface of the pelvis Field size = field width × field height: 12 cm × 12 cm = 144 cm²		Upper field edge: Above the umbilicus Lower field edge: Upper third of thighs Lateral field edges: Both forearms and hands included Field size = field width × field height: 18 cm × 17 cm = 306 cm²	

Table 17.12 Field settings—Pelvis ap

Age group (years)		Anatomical landmarks			
		Optimal field setting		**Suboptimal field setting**	
3 - 7	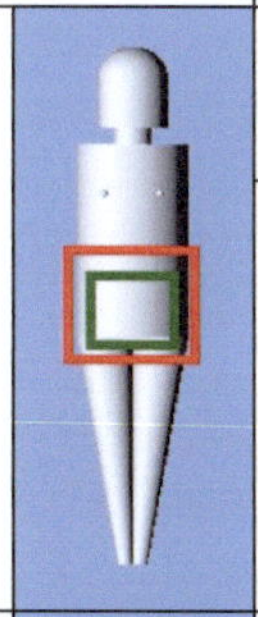	Upper field edge: Just below the umbilicus Lower field edge: Just below the perineum Lateral field edges: Close to the skin surface of the pelvis Field size = field width × field height: 17 cm × 15 cm = 255 cm²		Upper field edge: Above the umbilicus Lower field edge: Upper third of thighs Lateral field edges: Both forearms and hands included Field size = field width × field height: 26 cm × 22 cm = 572 cm²	
		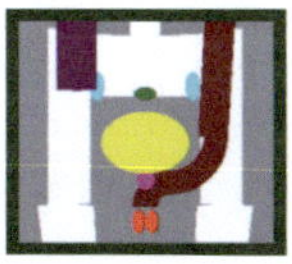			
8 - 12	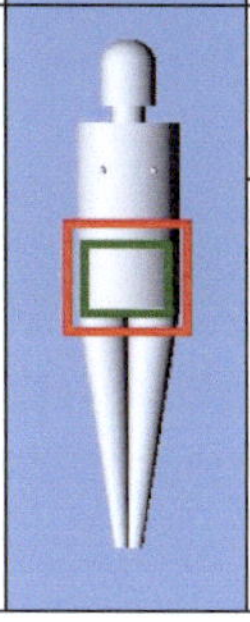	Upper field edge: Just below the umbilicus Lower field edge: Just below the perineum Lateral field edges: Close to the skin surface of the pelvis Field size = field width × field height: 21 cm × 19 cm = 399 cm²		Upper field edge: Above the umbilicus Lower field edge: Upper third of thighs Lateral field edges: Both forearms and hands included Field size = field width × field height: 32 cm × 29 cm = 928 cm²	

Table 17.13 Field settings—Pelvis ap

Age group (years)		Anatomical landmarks			
		Optimal field setting		Suboptimal field setting	
13 - 17	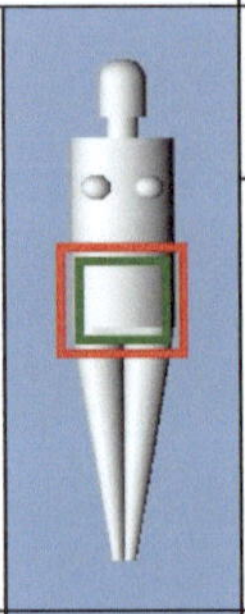	Upper field edge: Just below the umbilicus Lower field edge: Just below the perineum Lateral field edges: Close to the skin surface of the pelvis Field size = field width × field height: 28 cm × 26 cm = 728 cm^2		Upper field edge: Above the umbilicus Lower field edge: Upper third of thighs Lateral field edges: Both forearms and hands included Field size = field width × field height: 40 cm × 35 cm = 1,400 cm^2	

18	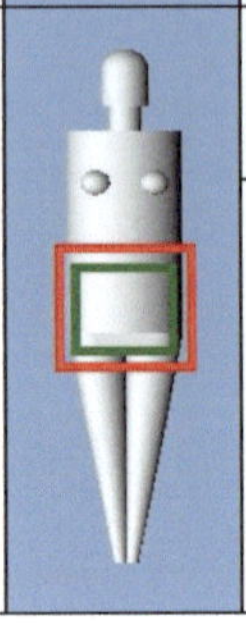	Upper field edge: Just below the umbilicus Lower field edge: Just below the perineum Lateral field edges: Close to the skin surface of the pelvis Field size = field width × field height: 32 cm × 29 cm = 928 cm^2		Upper field edge: Above the umbilicus Lower field edge: Upper third of thighs Lateral field edges: Both forearms and hands included Field size = field width × field height: 44 cm × 40 cm = 1,760 cm^2	
			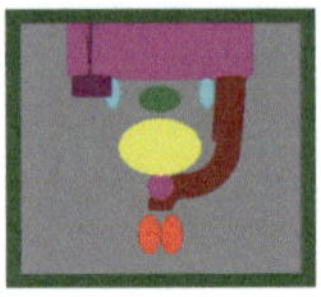		

Table 17.14 Normalized organ doses—Pelvis ap—<1 year

Beam projection:			ap									**Table cells:**
Focus-to-detector distance (cm):			100				(+) = normal font = optimal field setting					
Tube voltage (kVp):			60				**(−) = bold font = suboptimal field setting**					
Total filtrations:			See last table row.				Empty table cells correspond to values <0.01					
Organ or tissue	**Organ dose D_T : Entrance dose K_E**							**(mSv * mGy^{-1})**				
	(+)	(−)	(+)	(−)	(+)	(−)	(+)	(−)	(+)	(−)	(+)	(−)
Skull												
Brain												
Oral mucosa												
Salivary glands												
Cervical spine												
Thyroid												
Thymus												
Thoracic spine		**0.02**		**0.02**		**0.03**	0.01	**0.03**	0.01	**0.04**	0.01	**0.03**
Claviculae												
Scapulae								**0.01**		**0.02**		**0.01**
Ribs		**0.02**		**0.02**	0.01	**0.03**	0.01	**0.03**	0.01	**0.04**	0.01	**0.04**
Oesophagus				**0.01**		**0.01**		**0.02**		**0.02**		**0.02**
Trachea												
Lung								**0.01**		**0.01**		**0.01**
Mamma												
Heart						**0.01**		**0.01**		**0.02**		**0.01**
Stomach	0.02	**0.11**	0.02	**0.14**	0.03	**0.16**	0.03	**0.17**	0.04	**0.20**	0.04	**0.19**
Spleen		**0.03**	0.01	**0.04**	0.02	**0.05**	0.02	**0.05**	0.02	**0.07**	0.02	**0.06**
Liver	0.02	**0.10**	0.02	**0.12**	0.02	**0.14**	0.03	**0.15**	0.03	**0.17**	0.03	**0.17**
Gall bladder	0.03	**0.15**	0.04	**0.19**	0.04	**0.22**	0.05	**0.24**	0.06	**0.28**	0.06	**0.27**
Pancreas	0.01	**0.04**	0.02	**0.05**	0.02	**0.06**	0.02	**0.06**	0.03	**0.08**	0.02	**0.07**
Small intestine	0.27	**0.38**	0.33	**0.47**	0.38	**0.54**	0.41	**0.59**	0.48	**0.69**	0.47	**0.66**
Large intestine	0.22	**0.45**	0.27	**0.55**	0.31	**0.62**	0.34	**0.68**	0.39	**0.78**	0.38	**0.76**
Sigma/Rectum	0.33	**0.37**	0.41	**0.47**	0.47	**0.53**	0.52	**0.59**	0.60	**0.69**	0.58	**0.66**
Lumbar spine	0.08	**0.23**	0.11	**0.30**	0.14	**0.36**	0.15	**0.41**	0.19	**0.50**	0.17	**0.46**
Pelvis	0.51	**0.58**	0.65	**0.75**	0.76	**0.87**	0.85	**0.97**	1.01	**1.16**	0.97	**1.12**
Adrenals		**0.01**		**0.02**		**0.02**	0.01	**0.03**	0.01	**0.03**	0.01	**0.03**
Kidneys	0.01	**0.06**	0.02	**0.07**	0.02	**0.09**	0.03	**0.10**	0.03	**0.12**	0.03	**0.11**
Urinary bladder	0.60	**0.62**	0.71	**0.74**	0.79	**0.83**	0.85	**0.89**	0.96	**1.00**	0.94	**0.99**
Uterus	0.37	**0.40**	0.46	**0.50**	0.53	**0.57**	0.58	**0.63**	0.68	**0.73**	0.66	**0.71**
Ovaries	0.30	**0.32**	0.38	**0.41**	0.44	**0.48**	0.48	**0.53**	0.57	**0.63**	0.55	**0.60**
Prostate	0.46	**0.49**	0.57	**0.60**	0.64	**0.68**	0.70	**0.75**	0.80	**0.85**	0.78	**0.84**
Testes	1.04	**1.08**	1.14	**1.19**	1.19	**1.25**	1.23	**1.29**	1.28	**1.36**	1.28	**1.36**
Upper arm												
Elbow		**0.02**		**0.03**		**0.04**		**0.04**		**0.05**		**0.05**
Forearm		**1.08**		**1.29**		**1.44**		**1.55**		**1.74**		**1.74**
Upper leg	0.68	**0.99**	0.85	**1.24**	0.98	**1.43**	1.08	**1.57**	1.25	**1.82**	1.23	**1.79**
Knee	0.05	**0.52**	0.06	**0.65**	0.08	**0.74**	0.09	**0.81**	0.11	**0.93**	0.10	**0.92**
Lower leg		**0.01**		**0.01**		**0.02**		**0.02**		**0.02**		**0.02**
Lymph nodes	0.09	**0.14**	0.11	**0.17**	0.12	**0.20**	0.13	**0.21**	0.15	**0.25**	0.15	**0.24**
Skin	0.09	**0.21**	0.09	**0.22**	0.10	**0.23**	0.10	**0.23**	0.11	**0.25**	0.10	**0.24**
Muscle	0.13	**0.21**	0.15	**0.25**	0.16	**0.27**	0.18	**0.29**	0.20	**0.33**	0.19	**0.32**
Skeleton	0.11	**0.26**	0.14	**0.33**	0.16	**0.37**	0.18	**0.41**	0.22	**0.48**	0.21	**0.47**
Red bone marrow	0.03	**0.07**	0.04	**0.08**	0.04	**0.10**	0.05	**0.11**	0.06	**0.13**	0.05	**0.12**
Whole body dose	0.10	**0.18**	0.12	**0.22**	0.14	**0.24**	0.15	**0.26**	0.17	**0.29**	0.16	**0.28**
ED (ICRP 60)	0.21	**0.26**	0.24	**0.30**	0.27	**0.34**	0.28	**0.36**	0.31	**0.40**	0.31	**0.40**
ED (ICRP 103)	0.13	**0.18**	0.15	**0.21**	0.17	**0.24**	0.18	**0.26**	0.21	**0.29**	0.20	**0.28**
Total filtration (mm) **Al**	**1.5**		**2.5**		**3.5**		**4.5**		**3.5**		**3.5**	
Cu	**0**		**0**		**0**		**0**		**0.1**		**0.2**	

Table 17.15 Normalized organ doses—Pelvis ap—<1 year

Beam projection:	ap										**Table cells:**
Focus-to-detector distance (cm):	100										(+) = normal font = optimal field setting
Tube voltage (kVp):	70										**(−) = bold font = suboptimal field setting**
Total filtrations:	See last table row.										Empty table cells correspond to values <0.01

Organ or tissue	**Organ dose D_T : Entrance dose K_E (mSv * mGy^{-1})**											
	(+)	(−)	(+)	(−)	(+)	(−)	(+)	(−)	(+)	(−)	(+)	(−)
Skull												
Brain												
Oral mucosa												
Salivary glands												
Cervical spine												
Thyroid												
Thymus												
Thoracic spine		**0.02**	0.01	**0.03**	0.01	**0.03**	0.01	**0.04**	0.02	**0.05**	0.02	**0.06**
Claviculae												
Scapulae				**0.01**		**0.01**		**0.02**		**0.02**		**0.02**
Ribs		**0.02**	0.01	**0.03**	0.01	**0.04**	0.01	**0.04**	0.02	**0.05**	0.02	**0.06**
Oesophagus		**0.01**		**0.02**		**0.02**		**0.02**		**0.03**	0.01	**0.03**
Trachea												
Lung				**0.01**		**0.01**		**0.01**		**0.02**		**0.02**
Mamma												
Heart				**0.01**		**0.01**		**0.02**		**0.02**		**0.02**
Stomach	0.02	**0.13**	0.03	**0.16**	0.04	**0.18**	0.04	**0.19**	0.05	**0.22**	0.06	**0.24**
Spleen	0.01	**0.04**	0.02	**0.05**	0.02	**0.06**	0.02	**0.07**	0.03	**0.08**	0.03	**0.09**
Liver	0.02	**0.11**	0.02	**0.14**	0.03	**0.15**	0.03	**0.17**	0.04	**0.19**	0.05	**0.21**
Gall bladder	0.04	**0.18**	0.05	**0.22**	0.05	**0.24**	0.06	**0.27**	0.07	**0.31**	0.08	**0.34**
Pancreas	0.02	**0.05**	0.02	**0.06**	0.02	**0.07**	0.03	**0.08**	0.03	**0.09**	0.04	**0.11**
Small intestine	0.31	**0.44**	0.37	**0.53**	0.42	**0.60**	0.46	**0.66**	0.53	**0.76**	0.59	**0.84**
Large intestine	0.25	**0.51**	0.30	**0.62**	0.34	**0.69**	0.37	**0.75**	0.42	**0.85**	0.47	**0.93**
Sigma/Rectum	0.38	**0.43**	0.47	**0.53**	0.53	**0.60**	0.58	**0.66**	0.66	**0.76**	0.73	**0.85**
Lumbar spine	0.11	**0.29**	0.14	**0.37**	0.17	**0.43**	0.19	**0.48**	0.23	**0.59**	0.27	**0.68**
Pelvis	0.60	**0.70**	0.75	**0.87**	0.86	**1.00**	0.95	**1.10**	1.11	**1.29**	1.25	**1.45**
Adrenals		**0.02**		**0.03**	0.01	**0.03**	0.01	**0.03**	0.02	**0.04**	0.02	**0.05**
Kidneys	0.02	**0.07**	0.02	**0.09**	0.03	**0.11**	0.03	**0.12**	0.04	**0.15**	0.05	**0.17**
Urinary bladder	0.66	**0.69**	0.78	**0.82**	0.86	**0.90**	0.92	**0.96**	1.02	**1.07**	1.10	**1.16**
Uterus	0.43	**0.46**	0.53	**0.57**	0.60	**0.65**	0.65	**0.70**	0.75	**0.81**	0.83	**0.90**
Ovaries	0.35	**0.38**	0.43	**0.48**	0.50	**0.55**	0.54	**0.60**	0.63	**0.70**	0.71	**0.79**
Prostate	0.52	**0.56**	0.63	**0.68**	0.71	**0.76**	0.76	**0.82**	0.86	**0.93**	0.94	**1.02**
Testes	1.09	**1.13**	1.18	**1.23**	1.23	**1.29**	1.26	**1.33**	1.31	**1.40**	1.36	**1.44**
Upper arm										**0.01**		**0.01**
Elbow		**0.03**		**0.04**		**0.04**		**0.05**		**0.06**		**0.07**
Forearm		**1.19**		**1.40**		**1.54**		**1.64**		**1.81**		**1.94**
Upper leg	0.78	**1.14**	0.96	**1.40**	1.08	**1.58**	1.18	**1.72**	1.35	**1.97**	1.49	**2.17**
Knee	0.06	**0.59**	0.08	**0.72**	0.09	**0.81**	0.10	**0.88**	0.13	**1.00**	0.15	**1.10**
Lower leg		**0.01**		**0.02**		**0.02**		**0.02**	0.01	**0.03**	0.02	**0.04**
Lymph nodes	0.10	**0.16**	0.12	**0.19**	0.14	**0.22**	0.15	**0.24**	0.17	**0.27**	0.19	**0.30**
Skin	0.09	**0.22**	0.10	**0.23**	0.10	**0.24**	0.10	**0.24**	0.11	**0.26**	0.12	**0.27**
Muscle	0.14	**0.23**	0.16	**0.27**	0.18	**0.30**	0.19	**0.32**	0.21	**0.35**	0.23	**0.38**
Skeleton	0.13	**0.30**	0.16	**0.37**	0.19	**0.42**	0.20	**0.45**	0.24	**0.52**	0.27	**0.58**
Red bone marrow	0.03	**0.08**	0.04	**0.10**	0.05	**0.11**	0.06	**0.12**	0.07	**0.14**	0.08	**0.16**
Whole body dose	0.12	**0.20**	0.14	**0.24**	0.15	**0.26**	0.16	**0.28**	0.18	**0.31**	0.20	**0.34**
ED (ICRP 60)	0.23	**0.29**	0.26	**0.33**	0.29	**0.37**	0.31	**0.39**	0.34	**0.43**	0.36	**0.47**
ED (ICRP 103)	0.14	**0.20**	0.17	**0.24**	0.19	**0.26**	0.20	**0.28**	0.22	**0.32**	0.24	**0.35**
Total filtration (mm) Al	**1.5**		**2.5**		**3.5**		**4.5**		**3.5**		**3.5**	
Cu	**0**		**0**		**0**		**0**		**0.1**		**0.2**	

Table 17.16 Normalized organ doses—Pelvis ap—1–2 years

Beam projection:			ap									**Table cells:**
Focus-to-detector distance (cm):			100						(+) = normal font = optimal field setting			
Tube voltage (kVp):			60						**(−) = bold font = suboptimal field setting**			
Total filtrations:			See last table row.				Empty table cells correspond to values <0.01					
Organ or tissue	**Organ dose D_T : Entrance dose K_E**								**(mSv * mGy^{-1})**			
	(+)	(−)	(+)	(−)	(+)	(−)	(+)	(−)	(+)	(−)	(+)	(−)
Skull												
Brain												
Oral mucosa												
Salivary glands												
Cervical spine												
Thyroid												
Thymus												
Thoracic spine				**0.01**		**0.01**		**0.02**		**0.02**		**0.02**
Claviculae												
Scapulae												
Ribs		**0.01**		**0.02**		**0.02**		**0.02**	0.01	**0.03**		**0.03**
Oesophagus						**0.01**		**0.01**		**0.02**		**0.01**
Trachea												
Lung												
Mamma												
Heart										**0.01**		
Stomach	0.02	**0.09**	0.02	**0.11**	0.03	**0.13**	0.03	**0.14**	0.04	**0.17**	0.03	**0.16**
Spleen		**0.02**		**0.03**	0.01	**0.04**	0.01	**0.04**	0.02	**0.05**	0.01	**0.04**
Liver	0.01	**0.07**	0.02	**0.09**	0.02	**0.10**	0.02	**0.11**	0.03	**0.13**	0.03	**0.13**
Gall bladder	0.03	**0.15**	0.05	**0.20**	0.05	**0.23**	0.06	**0.25**	0.08	**0.30**	0.07	**0.29**
Pancreas		**0.03**	0.01	**0.04**	0.02	**0.04**	0.02	**0.05**	0.02	**0.06**	0.02	**0.05**
Small intestine	0.24	**0.30**	0.30	**0.38**	0.35	**0.44**	0.39	**0.49**	0.46	**0.59**	0.44	**0.55**
Large intestine	0.23	**0.36**	0.29	**0.46**	0.33	**0.52**	0.36	**0.58**	0.43	**0.68**	0.41	**0.65**
Sigma/Rectum	0.25	**0.29**	0.32	**0.37**	0.38	**0.43**	0.42	**0.48**	0.50	**0.57**	0.47	**0.54**
Lumbar spine	0.07	**0.14**	0.09	**0.19**	0.11	**0.23**	0.13	**0.26**	0.16	**0.33**	0.14	**0.28**
Pelvis	0.35	**0.41**	0.46	**0.54**	0.54	**0.64**	0.61	**0.73**	0.75	**0.89**	0.69	**0.82**
Adrenals		**0.01**		**0.01**		**0.02**		**0.02**	0.01	**0.03**		**0.02**
Kidneys	0.01	**0.04**	0.01	**0.05**	0.02	**0.06**	0.02	**0.07**	0.03	**0.09**	0.02	**0.07**
Urinary bladder	0.52	**0.54**	0.63	**0.66**	0.71	**0.75**	0.78	**0.82**	0.89	**0.94**	0.87	**0.91**
Uterus	0.30	**0.32**	0.39	**0.42**	0.45	**0.49**	0.50	**0.54**	0.60	**0.65**	0.57	**0.61**
Ovaries	0.22	**0.25**	0.29	**0.32**	0.34	**0.38**	0.38	**0.42**	0.46	**0.51**	0.43	**0.48**
Prostate	0.35	**0.37**	0.44	**0.47**	0.51	**0.54**	0.56	**0.60**	0.65	**0.71**	0.63	**0.68**
Testes	1.00	**1.05**	1.10	**1.16**	1.16	**1.22**	1.20	**1.27**	1.27	**1.35**	1.26	**1.34**
Upper arm												
Elbow		**0.01**		**0.02**		**0.02**		**0.02**		**0.03**		**0.03**
Forearm		**0.80**		**0.99**		**1.12**		**1.23**		**1.42**		**1.40**
Upper leg	0.33	**0.66**	0.43	**0.85**	0.51	**1.00**	0.57	**1.11**	0.69	**1.34**	0.64	**1.27**
Knee	0.02	**0.16**	0.02	**0.21**	0.03	**0.24**	0.03	**0.27**	0.04	**0.33**	0.04	**0.31**
Lower leg												
Lymph nodes	0.08	**0.12**	0.10	**0.14**	0.11	**0.17**	0.12	**0.18**	0.15	**0.22**	0.14	**0.21**
Skin	0.07	**0.18**	0.08	**0.19**	0.08	**0.19**	0.08	**0.20**	0.09	**0.21**	0.09	**0.20**
Muscle	0.09	**0.15**	0.11	**0.18**	0.12	**0.20**	0.13	**0.22**	0.15	**0.25**	0.14	**0.24**
Skeleton	0.06	**0.16**	0.08	**0.20**	0.10	**0.24**	0.11	**0.26**	0.14	**0.32**	0.12	**0.30**
Red bone marrow	0.02	**0.04**	0.03	**0.05**	0.03	**0.06**	0.04	**0.07**	0.05	**0.09**	0.04	**0.08**
Whole body dose	0.08	**0.14**	0.09	**0.16**	0.10	**0.18**	0.11	**0.20**	0.13	**0.23**	0.12	**0.22**
ED (ICRP 60)	0.19	**0.22**	0.22	**0.27**	0.24	**0.30**	0.26	**0.32**	0.29	**0.36**	0.28	**0.35**
ED (ICRP 103)	0.11	**0.15**	0.14	**0.18**	0.15	**0.20**	0.17	**0.22**	0.19	**0.26**	0.18	**0.24**
Total filtration (mm) Al	**1.5**		**2.5**		**3.5**		**4.5**		**3.5**		**3.5**	
Cu	**0**		**0**		**0**		**0**		**0.1**		**0.2**	

Table 17.17 Normalized organ doses—Pelvis ap—1–2 years

Beam projection:			ap									**Table cells:**
Focus-to-detector distance (cm):			100						(+) = normal font = optimal field setting			
Tube voltage (kVp):			70						**(−) = bold font = suboptimal field setting**			
Total filtrations:			See last table row.						Empty table cells correspond to values <0.01			
Organ or tissue	**Organ dose D_T : Entrance dose K_E**								**(mSv * mGy^{-1})**			
	(+)	(−)	(+)	(−)	(+)	(−)	(+)	(−)	(+)	(−)	(+)	(−)
Skull												
Brain												
Oral mucosa												
Salivary glands												
Cervical spine												
Thyroid												
Thymus												
Thoracic spine		**0.01**		**0.02**		**0.02**		**0.02**	0.01	**0.03**	0.02	**0.04**
Claviculae												
Scapulae												**0.01**
Ribs		**0.02**		**0.02**		**0.03**	0.01	**0.03**	0.01	**0.04**	0.02	**0.05**
Oesophagus				**0.01**		**0.02**		**0.02**		**0.02**	0.01	**0.03**
Trachea												
Lung										**0.01**		**0.01**
Mamma												
Heart						**0.01**		**0.01**		**0.01**		**0.02**
Stomach	0.02	**0.11**	0.03	**0.13**	0.03	**0.15**	0.04	**0.16**	0.05	**0.19**	0.06	**0.22**
Spleen		**0.03**	0.01	**0.04**	0.02	**0.05**	0.02	**0.05**	0.02	**0.06**	0.03	**0.08**
Liver	0.02	**0.09**	0.02	**0.11**	0.03	**0.12**	0.03	**0.13**	0.04	**0.15**	0.04	**0.17**
Gall bladder	0.04	**0.19**	0.06	**0.23**	0.07	**0.27**	0.08	**0.29**	0.09	**0.35**	0.11	**0.39**
Pancreas	0.01	**0.04**	0.02	**0.05**	0.02	**0.06**	0.02	**0.06**	0.03	**0.08**	0.04	**0.09**
Small intestine	0.28	**0.36**	0.35	**0.45**	0.40	**0.51**	0.44	**0.57**	0.52	**0.67**	0.59	**0.75**
Large intestine	0.27	**0.43**	0.33	**0.53**	0.38	**0.60**	0.41	**0.66**	0.48	**0.76**	0.54	**0.85**
Sigma/Rectum	0.31	**0.35**	0.38	**0.44**	0.44	**0.50**	0.48	**0.55**	0.57	**0.65**	0.65	**0.74**
Lumbar spine	0.09	**0.19**	0.12	**0.24**	0.14	**0.29**	0.17	**0.33**	0.21	**0.41**	0.25	**0.50**
Pelvis	0.43	**0.52**	0.55	**0.66**	0.64	**0.77**	0.72	**0.86**	0.86	**1.04**	0.99	**1.20**
Adrenals		**0.01**		**0.02**		**0.02**	0.01	**0.03**	0.01	**0.03**	0.02	**0.04**
Kidneys	0.02	**0.05**	0.02	**0.06**	0.02	**0.08**	0.03	**0.09**	0.03	**0.11**	0.04	**0.13**
Urinary bladder	0.59	**0.62**	0.71	**0.75**	0.79	**0.83**	0.86	**0.90**	0.97	**1.03**	1.06	**1.13**
Uterus	0.36	**0.39**	0.46	**0.49**	0.52	**0.57**	0.58	**0.63**	0.68	**0.74**	0.77	**0.85**
Ovaries	0.27	**0.31**	0.35	**0.39**	0.40	**0.45**	0.45	**0.50**	0.53	**0.60**	0.61	**0.69**
Prostate	0.41	**0.44**	0.50	**0.54**	0.57	**0.62**	0.63	**0.68**	0.73	**0.79**	0.81	**0.89**
Testes	1.05	**1.10**	1.15	**1.21**	1.21	**1.28**	1.25	**1.32**	1.31	**1.40**	1.37	**1.46**
Upper arm												
Elbow		**0.02**		**0.02**		**0.03**		**0.03**		**0.04**		**0.05**
Forearm		**0.91**		**1.10**		**1.24**		**1.34**		**1.53**		**1.67**
Upper leg	0.41	**0.80**	0.51	**1.00**	0.59	**1.16**	0.66	**1.28**	0.78	**1.51**	0.90	**1.72**
Knee	0.02	**0.20**	0.03	**0.25**	0.04	**0.29**	0.04	**0.32**	0.06	**0.38**	0.07	**0.44**
Lower leg												
Lymph nodes	0.09	**0.14**	0.11	**0.17**	0.13	**0.19**	0.14	**0.21**	0.17	**0.25**	0.19	**0.28**
Skin	0.08	**0.18**	0.08	**0.19**	0.08	**0.20**	0.09	**0.21**	0.09	**0.22**	0.10	**0.23**
Muscle	0.10	**0.17**	0.12	**0.21**	0.13	**0.23**	0.15	**0.24**	0.16	**0.28**	0.18	**0.30**
Skeleton	0.08	**0.19**	0.10	**0.24**	0.12	**0.28**	0.13	**0.30**	0.16	**0.36**	0.18	**0.41**
Red bone marrow	0.03	**0.05**	0.04	**0.07**	0.04	**0.08**	0.05	**0.09**	0.06	**0.11**	0.07	**0.12**
Whole body dose	0.09	**0.16**	0.10	**0.18**	0.12	**0.21**	0.13	**0.22**	0.14	**0.25**	0.16	**0.28**
ED (ICRP 60)	0.21	**0.25**	0.24	**0.30**	0.27	**0.33**	0.29	**0.35**	0.32	**0.40**	0.35	**0.44**
ED (ICRP 103)	0.13	**0.17**	0.16	**0.20**	0.17	**0.23**	0.19	**0.25**	0.21	**0.28**	0.24	**0.32**
Total filtration (mm) Al	**1.5**		**2.5**		**3.5**		**4.5**		**3.5**		**3.5**	
Cu	**0**		**0**		**0**		**0**		**0.1**		**0.2**	

Table 17.18 Normalized organ doses—Pelvis ap—3–7 years

Beam projection:			ap									**Table cells:**
Focus-to-detector distance (cm):			115						(+) = normal font = optimal field setting			
Tube voltage (kVp):			60						**(−) = bold font = suboptimal field setting**			
Total filtrations:			See last table row.						Empty table cells correspond to values <0.01			
Organ or tissue	**Organ dose D_T : Entrance dose K_E**								**(mSv * mGy^{-1})**			
	(+)	(−)	(+)	(−)	(+)	(−)	(+)	(−)	(+)	(−)	(+)	(−)
Skull												
Brain												
Oral mucosa												
Salivary glands												
Cervical spine												
Thyroid												
Thymus												
Thoracic spine								**0.01**		**0.02**		**0.01**
Claviculae												
Scapulae												
Ribs				**0.01**		**0.01**		**0.02**		**0.02**		**0.02**
Oesophagus												
Trachea												
Lung												
Mamma												
Heart												
Stomach		**0.11**	0.01	**0.14**	0.02	**0.16**	0.02	**0.18**	0.02	**0.21**	0.02	**0.20**
Spleen		**0.02**		**0.03**		**0.03**		**0.04**		**0.05**		**0.04**
Liver		**0.08**		**0.10**	0.01	**0.11**	0.01	**0.13**	0.02	**0.15**	0.01	**0.14**
Gall bladder	0.02	**0.17**	0.03	**0.22**	0.03	**0.26**	0.04	**0.29**	0.05	**0.35**	0.04	**0.33**
Pancreas		**0.02**		**0.03**		**0.04**		**0.04**	0.01	**0.06**		**0.05**
Small intestine	0.18	**0.27**	0.23	**0.35**	0.27	**0.41**	0.30	**0.45**	0.36	**0.55**	0.34	**0.51**
Large intestine	0.16	**0.33**	0.20	**0.42**	0.23	**0.49**	0.26	**0.54**	0.31	**0.65**	0.29	**0.61**
Sigma/Rectum	0.22	**0.25**	0.28	**0.32**	0.33	**0.38**	0.37	**0.43**	0.45	**0.52**	0.42	**0.48**
Lumbar spine	0.04	**0.11**	0.05	**0.15**	0.06	**0.18**	0.07	**0.21**	0.09	**0.27**	0.07	**0.22**
Pelvis	0.29	**0.33**	0.38	**0.43**	0.46	**0.52**	0.52	**0.59**	0.65	**0.73**	0.59	**0.66**
Adrenals				**0.01**		**0.01**		**0.01**		**0.02**		**0.02**
Kidneys		**0.03**		**0.04**	0.01	**0.05**	0.01	**0.06**	0.02	**0.08**	0.01	**0.07**
Urinary bladder	0.49	**0.50**	0.61	**0.62**	0.69	**0.71**	0.75	**0.78**	0.88	**0.91**	0.84	**0.87**
Uterus	0.27	**0.28**	0.35	**0.37**	0.41	**0.44**	0.46	**0.49**	0.56	**0.59**	0.52	**0.55**
Ovaries	0.20	**0.21**	0.26	**0.28**	0.30	**0.33**	0.34	**0.38**	0.42	**0.46**	0.38	**0.42**
Prostate	0.29	**0.31**	0.38	**0.40**	0.44	**0.47**	0.49	**0.52**	0.59	**0.63**	0.55	**0.59**
Testes	1.01	**1.04**	1.12	**1.16**	1.19	**1.23**	1.23	**1.28**	1.31	**1.37**	1.30	**1.35**
Upper arm												
Elbow												
Forearm												
Upper leg	0.19	**0.38**	0.25	**0.50**	0.30	**0.59**	0.34	**0.67**	0.42	**0.82**	0.38	**0.75**
Knee		**0.01**		**0.02**		**0.02**		**0.03**	0.01	**0.03**		**0.03**
Lower leg												
Lymph nodes	0.06	**0.11**	0.08	**0.14**	0.09	**0.16**	0.10	**0.18**	0.12	**0.21**	0.11	**0.20**
Skin	0.07	**0.14**	0.08	**0.14**	0.08	**0.15**	0.08	**0.15**	0.08	**0.16**	0.08	**0.16**
Muscle	0.08	**0.13**	0.10	**0.16**	0.11	**0.17**	0.12	**0.19**	0.14	**0.22**	0.13	**0.21**
Skeleton	0.04	**0.07**	0.06	**0.09**	0.07	**0.10**	0.07	**0.12**	0.09	**0.15**	0.08	**0.13**
Red bone marrow	0.02	**0.03**	0.03	**0.04**	0.04	**0.05**	0.05	**0.06**	0.06	**0.07**	0.05	**0.06**
Whole body dose	0.07	**0.11**	0.08	**0.13**	0.09	**0.15**	0.10	**0.16**	0.12	**0.19**	0.11	**0.18**
ED (ICRP 60)	0.17	**0.21**	0.21	**0.25**	0.23	**0.28**	0.25	**0.31**	0.28	**0.35**	0.27	**0.34**
ED (ICRP 103)	0.10	**0.14**	0.12	**0.17**	0.14	**0.19**	0.15	**0.21**	0.18	**0.25**	0.17	**0.23**
Total filtration (mm) Al	**1.5**		**2.5**		**3.5**		**4.5**		**3.5**		**3.5**	
Cu	**0**		**0**		**0**		**0**		**0.1**		**0.2**	

Table 17.19 Normalized organ doses—Pelvis ap—3–7 years

Beam projection:		ap										**Table cells:**
Focus-to-detector distance (cm):		115							(+) = normal font = optimal field setting			
Tube voltage (kVp):		70							**(−) = bold font = suboptimal field setting**			
Total filtrations:		See last table row.							Empty table cells correspond to values <0.01			
Organ or tissue	**Organ dose D_T : Entrance dose K_E**								**(mSv * mGy^{-1})**			
	(+)	(−)	(+)	(−)	(+)	(−)	(+)	(−)	(+)	(−)	(+)	(−)
Skull												
Brain												
Oral mucosa												
Salivary glands												
Cervical spine												
Thyroid												
Thymus												
Thoracic spine				**0.01**		**0.01**		**0.02**		**0.02**		**0.03**
Claviculae												
Scapulae												
Ribs		**0.01**		**0.02**		**0.02**		**0.02**		**0.03**		**0.04**
Oesophagus								**0.01**		**0.01**		**0.02**
Trachea												
Lung												
Mamma												
Heart										**0.01**		**0.01**
Stomach	0.01	**0.13**	0.02	**0.16**	0.02	**0.19**	0.02	**0.20**	0.03	**0.24**	0.04	**0.27**
Spleen		**0.03**		**0.03**		**0.04**	0.01	**0.05**	0.01	**0.06**	0.02	**0.07**
Liver		**0.09**	0.01	**0.12**	0.02	**0.13**	0.02	**0.15**	0.02	**0.17**	0.03	**0.19**
Gall bladder	0.03	**0.21**	0.03	**0.27**	0.04	**0.31**	0.05	**0.34**	0.06	**0.40**	0.07	**0.46**
Pancreas		**0.03**		**0.04**	0.01	**0.05**	0.01	**0.06**	0.02	**0.07**	0.02	**0.09**
Small intestine	0.22	**0.33**	0.28	**0.42**	0.32	**0.48**	0.35	**0.53**	0.42	**0.64**	0.48	**0.73**
Large intestine	0.19	**0.40**	0.24	**0.49**	0.27	**0.57**	0.30	**0.63**	0.35	**0.74**	0.41	**0.83**
Sigma/Rectum	0.27	**0.31**	0.34	**0.39**	0.40	**0.46**	0.44	**0.51**	0.53	**0.61**	0.61	**0.70**
Lumbar spine	0.05	**0.15**	0.07	**0.20**	0.08	**0.24**	0.10	**0.28**	0.12	**0.35**	0.15	**0.43**
Pelvis	0.37	**0.42**	0.48	**0.54**	0.56	**0.64**	0.63	**0.72**	0.77	**0.88**	0.90	**1.03**
Adrenals		**0.01**		**0.01**		**0.02**		**0.02**		**0.03**		**0.03**
Kidneys		**0.04**	0.01	**0.06**	0.01	**0.07**	0.02	**0.08**	0.02	**0.10**	0.03	**0.12**
Urinary bladder	0.57	**0.59**	0.69	**0.71**	0.78	**0.81**	0.84	**0.88**	0.97	**1.01**	1.08	**1.13**
Uterus	0.33	**0.35**	0.42	**0.45**	0.49	**0.52**	0.54	**0.58**	0.65	**0.70**	0.75	**0.80**
Ovaries	0.25	**0.27**	0.32	**0.35**	0.37	**0.41**	0.41	**0.46**	0.50	**0.55**	0.58	**0.65**
Prostate	0.36	**0.38**	0.45	**0.48**	0.52	**0.55**	0.57	**0.61**	0.68	**0.72**	0.77	**0.82**
Testes	1.06	**1.10**	1.17	**1.22**	1.24	**1.29**	1.28	**1.34**	1.36	**1.43**	1.42	**1.50**
Upper arm												
Elbow												
Forearm												
Upper leg	0.24	**0.48**	0.31	**0.61**	0.37	**0.72**	0.41	**0.80**	0.50	**0.97**	0.59	**1.13**
Knee		**0.02**		**0.03**	0.01	**0.03**	0.01	**0.04**	0.02	**0.05**	0.02	**0.06**
Lower leg												
Lymph nodes	0.07	**0.13**	0.09	**0.16**	0.10	**0.19**	0.11	**0.21**	0.14	**0.24**	0.15	**0.28**
Skin	0.07	**0.14**	0.08	**0.15**	0.08	**0.15**	0.08	**0.16**	0.09	**0.17**	0.09	**0.17**
Muscle	0.10	**0.15**	0.11	**0.18**	0.13	**0.20**	0.14	**0.21**	0.16	**0.24**	0.18	**0.27**
Skeleton	0.05	**0.08**	0.07	**0.11**	0.08	**0.13**	0.09	**0.14**	0.11	**0.17**	0.13	**0.20**
Red bone marrow	0.03	**0.04**	0.04	**0.05**	0.05	**0.06**	0.06	**0.07**	0.07	**0.09**	0.08	**0.11**
Whole body dose	0.08	**0.13**	0.10	**0.15**	0.11	**0.17**	0.12	**0.18**	0.14	**0.21**	0.15	**0.24**
ED (ICRP 60)	0.20	**0.24**	0.23	**0.29**	0.25	**0.32**	0.27	**0.34**	0.31	**0.39**	0.34	**0.43**
ED (ICRP 103)	0.12	**0.16**	0.14	**0.20**	0.16	**0.22**	0.17	**0.24**	0.20	**0.28**	0.22	**0.31**
Total filtration (mm) **Al**	**1.5**		**2.5**		**3.5**		**4.5**		**3.5**		**3.5**	
Cu	**0**		**0**		**0**		**0**		**0.1**		**0.2**	

Table 17.20 Normalized organ doses—Pelvis ap—8–12 years

Beam projection:			ap									**Table cells:**
Focus-to-detector distance (cm):			115									(+) = normal font = optimal field setting
Tube voltage (kVp):			60									**(−) = bold font = suboptimal field setting**
Total filtrations:			See last table row.									Empty table cells correspond to values <0.01
Organ or tissue	**Organ dose D_T : Entrance dose K_E**								**(mSv * mGy^{-1})**			
	(+)	(−)	(+)	(−)	(+)	(−)	(+)	(−)	(+)	(−)	(+)	(−)
Skull												
Brain												
Oral mucosa												
Salivary glands												
Cervical spine												
Thyroid												
Thymus												
Thoracic spine										**0.01**		
Claviculae												
Scapulae												
Ribs						**0.01**		**0.01**		**0.02**		**0.01**
Oesophagus												
Trachea												
Lung												
Mamma												
Heart												
Stomach		**0.11**	0.01	**0.14**	0.01	**0.16**	0.02	**0.17**	0.02	**0.21**	0.02	**0.19**
Spleen		**0.02**		**0.03**		**0.03**		**0.04**		**0.05**		**0.04**
Liver		**0.07**		**0.08**		**0.10**	0.01	**0.11**	0.01	**0.13**	0.01	**0.12**
Gall bladder	0.02	**0.17**	0.02	**0.22**	0.03	**0.25**	0.03	**0.29**	0.04	**0.35**	0.04	**0.32**
Pancreas		**0.02**		**0.03**		**0.03**		**0.04**		**0.05**		**0.04**
Small intestine	0.18	**0.24**	0.23	**0.31**	0.28	**0.37**	0.31	**0.41**	0.38	**0.51**	0.35	**0.46**
Large intestine	0.16	**0.30**	0.21	**0.38**	0.24	**0.45**	0.27	**0.50**	0.33	**0.61**	0.31	**0.57**
Sigma/Rectum	0.20	**0.22**	0.26	**0.29**	0.31	**0.34**	0.35	**0.39**	0.43	**0.48**	0.39	**0.43**
Lumbar spine	0.03	**0.08**	0.04	**0.11**	0.06	**0.14**	0.07	**0.16**	0.09	**0.22**	0.07	**0.17**
Pelvis	0.25	**0.28**	0.33	**0.37**	0.39	**0.45**	0.45	**0.51**	0.56	**0.64**	0.50	**0.56**
Adrenals						**0.01**		**0.01**		**0.02**		**0.01**
Kidneys		**0.03**		**0.04**		**0.05**	0.01	**0.05**	0.01	**0.07**	0.01	**0.06**
Urinary bladder	0.45	**0.47**	0.57	**0.59**	0.65	**0.67**	0.72	**0.74**	0.84	**0.87**	0.80	**0.83**
Uterus	0.24	**0.25**	0.31	**0.33**	0.37	**0.39**	0.42	**0.44**	0.51	**0.55**	0.47	**0.50**
Ovaries	0.17	**0.18**	0.23	**0.24**	0.27	**0.29**	0.31	**0.33**	0.39	**0.41**	0.34	**0.36**
Prostate	0.28	**0.31**	0.37	**0.40**	0.43	**0.47**	0.49	**0.52**	0.59	**0.63**	0.55	**0.59**
Testes	1.01	**1.05**	1.12	**1.18**	1.19	**1.26**	1.24	**1.31**	1.32	**1.41**	1.31	**1.39**
Upper arm												
Elbow						**0.01**		**0.01**		**0.02**		**0.01**
Forearm		**0.65**		**0.82**		**0.95**		**1.05**		**1.24**		**1.19**
Upper leg	0.12	**0.32**	0.16	**0.43**	0.19	**0.51**	0.22	**0.58**	0.28	**0.73**	0.24	**0.65**
Knee		**0.01**		**0.02**		**0.02**		**0.02**		**0.03**		**0.02**
Lower leg												
Lymph nodes	0.06	**0.10**	0.08	**0.13**	0.09	**0.15**	0.10	**0.17**	0.12	**0.20**	0.11	**0.18**
Skin	0.08	**0.17**	0.08	**0.18**	0.08	**0.18**	0.09	**0.19**	0.09	**0.20**	0.09	**0.19**
Muscle	0.08	**0.13**	0.10	**0.16**	0.11	**0.18**	0.12	**0.20**	0.14	**0.23**	0.13	**0.22**
Skeleton	0.03	**0.09**	0.04	**0.12**	0.05	**0.14**	0.06	**0.16**	0.08	**0.20**	0.07	**0.18**
Red bone marrow	0.03	**0.04**	0.03	**0.05**	0.04	**0.06**	0.05	**0.07**	0.06	**0.08**	0.05	**0.07**
Whole body dose	0.07	**0.12**	0.08	**0.14**	0.09	**0.16**	0.10	**0.18**	0.12	**0.21**	0.11	**0.20**
ED (ICRP 60)	0.17	**0.21**	0.20	**0.25**	0.22	**0.28**	0.24	**0.30**	0.27	**0.34**	0.26	**0.33**
ED (ICRP 103)	0.10	**0.13**	0.12	**0.16**	0.14	**0.19**	0.15	**0.21**	0.17	**0.24**	0.16	**0.23**
Total filtration (mm) Al	**1.5**		**2.5**		**3.5**		**4.5**		**3.5**		**3.5**	
Cu	**0**		**0**		**0**		**0**		**0.1**		**0.2**	

Table 17.21 Normalized organ doses—Pelvis ap—8–12 years

Beam projection:			ap									**Table cells:**
Focus-to-detector distance (cm):			115									(+) = normal font = optimal field setting
Tube voltage (kVp):			70									**(−) = bold font = suboptimal field setting**
Total filtrations:			See last table row.									Empty table cells correspond to values <0.01
Organ or tissue	**Organ dose D_T : Entrance dose K_E**							**(mSv * mGy^{-1})**				
	(+)	(−)	(+)	(−)	(+)	(−)	(+)	(−)	(+)	(−)	(+)	(−)
Skull												
Brain												
Oral mucosa												
Salivary glands												
Cervical spine												
Thyroid												
Thymus												
Thoracic spine						**0.01**		**0.01**		**0.02**		**0.02**
Claviculae												
Scapulae												
Ribs		**0.01**		**0.01**		**0.02**		**0.02**		**0.02**		**0.03**
Oesophagus												**0.01**
Trachea												
Lung												
Mamma												
Heart												
Stomach	0.01	**0.13**	0.02	**0.16**	0.02	**0.18**	0.02	**0.20**	0.03	**0.24**	0.04	**0.27**
Spleen		**0.03**		**0.03**		**0.04**		**0.05**	0.01	**0.06**	0.01	**0.07**
Liver		**0.08**	0.01	**0.10**	0.01	**0.12**	0.02	**0.13**	0.02	**0.15**	0.02	**0.18**
Gall bladder	0.02	**0.21**	0.03	**0.26**	0.04	**0.30**	0.04	**0.34**	0.06	**0.41**	0.07	**0.47**
Pancreas		**0.03**		**0.04**		**0.04**	0.01	**0.05**	0.01	**0.06**	0.02	**0.08**
Small intestine	0.22	**0.30**	0.29	**0.38**	0.33	**0.45**	0.37	**0.50**	0.44	**0.60**	0.51	**0.70**
Large intestine	0.20	**0.37**	0.25	**0.46**	0.29	**0.53**	0.32	**0.59**	0.39	**0.71**	0.44	**0.81**
Sigma/Rectum	0.25	**0.28**	0.32	**0.36**	0.38	**0.42**	0.42	**0.47**	0.51	**0.57**	0.59	**0.67**
Lumbar spine	0.05	**0.12**	0.07	**0.16**	0.08	**0.20**	0.09	**0.23**	0.12	**0.29**	0.15	**0.36**
Pelvis	0.32	**0.37**	0.42	**0.48**	0.49	**0.56**	0.56	**0.64**	0.69	**0.79**	0.81	**0.94**
Adrenals				**0.01**		**0.02**		**0.02**		**0.02**		**0.03**
Kidneys		**0.04**	0.01	**0.05**	0.01	**0.06**	0.02	**0.07**	0.02	**0.09**	0.03	**0.11**
Urinary bladder	0.54	**0.55**	0.66	**0.68**	0.74	**0.77**	0.81	**0.85**	0.95	**0.99**	1.06	**1.11**
Uterus	0.30	**0.32**	0.39	**0.41**	0.45	**0.48**	0.51	**0.54**	0.61	**0.66**	0.71	**0.77**
Ovaries	0.22	**0.24**	0.29	**0.31**	0.34	**0.37**	0.38	**0.41**	0.47	**0.51**	0.55	**0.60**
Prostate	0.35	**0.38**	0.44	**0.48**	0.52	**0.56**	0.57	**0.62**	0.68	**0.74**	0.79	**0.85**
Testes	1.07	**1.12**	1.18	**1.24**	1.25	**1.32**	1.29	**1.38**	1.37	**1.47**	1.44	**1.55**
Upper arm												
Elbow		**0.01**		**0.01**		**0.02**		**0.02**		**0.02**		**0.03**
Forearm		**0.76**		**0.94**		**1.08**		**1.18**		**1.37**		**1.53**
Upper leg	0.16	**0.42**	0.20	**0.54**	0.24	**0.64**	0.27	**0.72**	0.34	**0.88**	0.41	**1.04**
Knee		**0.02**		**0.02**		**0.03**		**0.03**		**0.04**		**0.05**
Lower leg												
Lymph nodes	0.07	**0.12**	0.09	**0.15**	0.11	**0.18**	0.12	**0.20**	0.14	**0.23**	0.16	**0.27**
Skin	0.08	**0.17**	0.08	**0.18**	0.09	**0.19**	0.09	**0.20**	0.09	**0.21**	0.10	**0.22**
Muscle	0.09	**0.15**	0.11	**0.18**	0.12	**0.21**	0.14	**0.22**	0.16	**0.26**	0.17	**0.28**
Skeleton	0.04	**0.12**	0.06	**0.15**	0.07	**0.17**	0.07	**0.19**	0.09	**0.24**	0.11	**0.28**
Red bone marrow	0.03	**0.05**	0.05	**0.06**	0.05	**0.07**	0.06	**0.09**	0.08	**0.11**	0.09	**0.13**
Whole body dose	0.08	**0.14**	0.10	**0.17**	0.11	**0.19**	0.12	**0.20**	0.14	**0.24**	0.15	**0.27**
ED (ICRP 60)	0.19	**0.24**	0.23	**0.28**	0.25	**0.31**	0.27	**0.34**	0.31	**0.39**	0.34	**0.43**
ED (ICRP 103)	0.12	**0.16**	0.14	**0.19**	0.16	**0.22**	0.17	**0.24**	0.20	**0.28**	0.22	**0.31**
Total filtration (mm) **Al**	**1.5**		**2.5**		**3.5**		**4.5**		**3.5**		**3.5**	
Cu	**0**		**0**		**0**		**0**		**0.1**		**0.2**	

Table 17.22 Normalized organ doses—Pelvis ap—8–12 years

Beam projection:			ap									**Table cells:**
Focus-to-detector distance (cm):			115									(+) = normal font = optimal field setting
Tube voltage (kVp):			80									**(−) = bold font = suboptimal field setting**
Total filtrations:			See last table row.									Empty table cells correspond to values <0.01
Organ or tissue	**Organ dose D_T : Entrance dose K_E**								**($mSv * mGy^{-1}$)**			
	(+)	(−)	(+)	(−)	(+)	(−)	(+)	(−)	(+)	(−)	(+)	(−)
Skull												
Brain												
Oral mucosa												
Salivary glands												
Cervical spine												
Thyroid												
Thymus												
Thoracic spine				**0.01**		**0.01**		**0.02**		**0.02**		**0.03**
Claviculae												
Scapulae												
Ribs		**0.01**		**0.02**		**0.02**		**0.02**		**0.03**		**0.04**
Oesophagus										**0.01**		**0.01**
Trachea												
Lung												
Mamma												
Heart										**0.01**		**0.01**
Stomach	0.02	**0.15**	0.02	**0.19**	0.03	**0.21**	0.03	**0.23**	0.04	**0.27**	0.04	**0.30**
Spleen		**0.03**		**0.04**	0.01	**0.05**	0.01	**0.06**	0.01	**0.07**	0.02	**0.09**
Liver	0.01	**0.10**	0.01	**0.12**	0.02	**0.14**	0.02	**0.15**	0.02	**0.17**	0.03	**0.20**
Gall bladder	0.03	**0.25**	0.04	**0.31**	0.05	**0.35**	0.06	**0.39**	0.07	**0.46**	0.08	**0.52**
Pancreas		**0.04**	0.01	**0.05**	0.01	**0.06**	0.01	**0.06**	0.02	**0.08**	0.02	**0.10**
Small intestine	0.27	**0.36**	0.33	**0.45**	0.38	**0.52**	0.42	**0.57**	0.50	**0.68**	0.57	**0.78**
Large intestine	0.24	**0.43**	0.29	**0.54**	0.33	**0.61**	0.37	**0.67**	0.43	**0.79**	0.49	**0.89**
Sigma/Rectum	0.30	**0.34**	0.38	**0.43**	0.44	**0.49**	0.48	**0.55**	0.58	**0.65**	0.66	**0.75**
Lumbar spine	0.07	**0.16**	0.09	**0.21**	0.10	**0.25**	0.12	**0.29**	0.15	**0.36**	0.18	**0.44**
Pelvis	0.40	**0.46**	0.50	**0.58**	0.58	**0.67**	0.65	**0.75**	0.79	**0.91**	0.92	**1.07**
Adrenals		**0.01**		**0.02**		**0.02**		**0.02**		**0.03**		**0.04**
Kidneys	0.01	**0.05**	0.02	**0.07**	0.02	**0.08**	0.02	**0.09**	0.03	**0.11**	0.03	**0.14**
Urinary bladder	0.61	**0.64**	0.74	**0.77**	0.83	**0.86**	0.90	**0.94**	1.03	**1.08**	1.15	**1.21**
Uterus	0.37	**0.39**	0.46	**0.49**	0.53	**0.56**	0.59	**0.63**	0.70	**0.75**	0.80	**0.86**
Ovaries	0.28	**0.30**	0.35	**0.38**	0.40	**0.44**	0.45	**0.49**	0.54	**0.59**	0.63	**0.69**
Prostate	0.42	**0.45**	0.52	**0.56**	0.59	**0.64**	0.65	**0.71**	0.76	**0.83**	0.87	**0.94**
Testes	1.12	**1.18**	1.23	**1.30**	1.29	**1.38**	1.34	**1.43**	1.42	**1.53**	1.48	**1.60**
Upper arm												
Elbow		**0.01**		**0.02**		**0.02**		**0.02**		**0.03**		**0.04**
Forearm		**0.87**		**1.05**		**1.18**		**1.28**		**1.46**		**1.61**
Upper leg	0.20	**0.51**	0.25	**0.65**	0.29	**0.75**	0.33	**0.84**	0.40	**1.01**	0.46	**1.17**
Knee		**0.02**		**0.03**		**0.03**		**0.04**		**0.05**	0.01	**0.06**
Lower leg												
Lymph nodes	0.09	**0.15**	0.11	**0.18**	0.12	**0.20**	0.13	**0.23**	0.16	**0.27**	0.18	**0.30**
Skin	0.08	**0.18**	0.09	**0.19**	0.09	**0.20**	0.09	**0.21**	0.10	**0.22**	0.10	**0.23**
Muscle	0.11	**0.17**	0.12	**0.21**	0.14	**0.23**	0.15	**0.25**	0.17	**0.28**	0.19	**0.31**
Skeleton	0.05	**0.14**	0.07	**0.17**	0.08	**0.20**	0.09	**0.22**	0.11	**0.27**	0.13	**0.31**
Red bone marrow	0.04	**0.06**	0.06	**0.08**	0.07	**0.09**	0.08	**0.10**	0.09	**0.13**	0.11	**0.15**
Whole body dose	0.09	**0.16**	0.11	**0.19**	0.12	**0.21**	0.13	**0.23**	0.15	**0.26**	0.17	**0.29**
ED (ICRP 60)	0.22	**0.27**	0.25	**0.31**	0.28	**0.35**	0.29	**0.37**	0.33	**0.43**	0.37	**0.47**
ED (ICRP 103)	0.13	**0.18**	0.16	**0.22**	0.18	**0.24**	0.19	**0.27**	0.22	**0.31**	0.24	**0.34**
Total filtration (mm) **Al**	**1.5**		**2.5**		**3.5**		**4.5**		**3.5**		**3.5**	
Cu	**0**		**0**		**0**		**0**		**0.1**		**0.2**	

Table 17.23 Normalized organ doses—Pelvis ap—13–17 years

Beam projection:			ap									**Table cells:**
Focus-to-detector distance (cm):			115									(+) = normal font = optimal field setting
Tube voltage (kVp):			60									**(−) = bold font = suboptimal field setting**
Total filtrations:			See last table row.									Empty table cells correspond to values <0.01
Organ or tissue	**Organ dose D_T : Entrance dose K_E**							**(mSv * mGy^{-1})**				
	(+)	(−)	(+)	(−)	(+)	(−)	(+)	(−)	(+)	(−)	(+)	(−)
Skull												
Brain												
Oral mucosa												
Salivary glands												
Cervical spine												
Thyroid												
Thymus												
Thoracic spine												
Claviculae												
Scapulae												
Ribs												
Oesophagus												
Trachea												
Lung												
Mamma				**0.01**		**0.02**		**0.02**	0.01	**0.02**		**0.02**
Heart												
Stomach		**0.06**		**0.08**	0.01	**0.09**	0.01	**0.10**	0.02	**0.12**	0.01	**0.11**
Spleen				**0.01**		**0.02**		**0.02**		**0.02**		**0.02**
Liver		**0.03**		**0.04**		**0.05**		**0.05**		**0.06**		**0.06**
Gall bladder	0.01	**0.10**	0.02	**0.14**	0.02	**0.16**	0.03	**0.18**	0.03	**0.23**	0.03	**0.20**
Pancreas				**0.01**		**0.02**		**0.02**		**0.03**		**0.02**
Small intestine	0.15	**0.19**	0.19	**0.25**	0.23	**0.30**	0.26	**0.34**	0.32	**0.42**	0.29	**0.37**
Large intestine	0.14	**0.24**	0.18	**0.31**	0.21	**0.37**	0.24	**0.42**	0.30	**0.51**	0.27	**0.47**
Sigma/Rectum	0.16	**0.18**	0.21	**0.23**	0.26	**0.28**	0.29	**0.32**	0.36	**0.40**	0.32	**0.35**
Lumbar spine	0.02	**0.05**	0.03	**0.07**	0.04	**0.08**	0.04	**0.10**	0.06	**0.13**	0.04	**0.09**
Pelvis	0.18	**0.20**	0.24	**0.27**	0.30	**0.33**	0.34	**0.38**	0.43	**0.48**	0.37	**0.41**
Adrenals												
Kidneys		**0.01**		**0.02**		**0.03**		**0.03**		**0.04**		**0.03**
Urinary bladder	0.40	**0.41**	0.51	**0.52**	0.59	**0.60**	0.65	**0.67**	0.77	**0.79**	0.72	**0.74**
Uterus	0.57	**0.59**	0.75	**0.78**	0.89	**0.93**	1.01	**1.06**	1.26	**1.32**	1.12	**1.17**
Ovaries	0.13	**0.14**	0.18	**0.19**	0.22	**0.23**	0.25	**0.27**	0.31	**0.34**	0.27	**0.29**
Prostate	0.26	**0.27**	0.34	**0.35**	0.41	**0.42**	0.46	**0.48**	0.57	**0.58**	0.52	**0.53**
Testes	0.87	**0.91**	1.01	**1.05**	1.09	**1.14**	1.16	**1.21**	1.27	**1.33**	1.25	**1.31**
Upper arm												
Elbow												
Forearm		**0.47**		**0.60**		**0.70**		**0.78**		**0.94**		**0.89**
Upper leg	0.14	**0.23**	0.19	**0.31**	0.23	**0.38**	0.27	**0.43**	0.34	**0.55**	0.29	**0.47**
Knee		**0.01**		**0.02**		**0.02**		**0.02**		**0.03**		**0.02**
Lower leg												
Lymph nodes	0.05	**0.08**	0.07	**0.10**	0.08	**0.12**	0.09	**0.13**	0.10	**0.16**	0.09	**0.14**
Skin	0.10	**0.17**	0.11	**0.18**	0.12	**0.18**	0.12	**0.19**	0.13	**0.20**	0.12	**0.20**
Muscle	0.08	**0.12**	0.10	**0.15**	0.11	**0.16**	0.12	**0.18**	0.14	**0.21**	0.14	**0.20**
Skeleton	0.03	**0.07**	0.04	**0.10**	0.05	**0.11**	0.06	**0.13**	0.08	**0.16**	0.07	**0.14**
Red bone marrow	0.02	**0.03**	0.03	**0.04**	0.04	**0.04**	0.04	**0.05**	0.05	**0.07**	0.05	**0.05**
Whole body dose	0.07	**0.11**	0.09	**0.13**	0.10	**0.15**	0.11	**0.16**	0.13	**0.19**	0.12	**0.18**
ED (ICRP 60)	0.15	**0.17**	0.18	**0.21**	0.20	**0.24**	0.22	**0.26**	0.26	**0.30**	0.24	**0.28**
ED (ICRP 103)	0.09	**0.11**	0.11	**0.14**	0.13	**0.16**	0.14	**0.18**	0.17	**0.21**	0.15	**0.19**
Total filtration (mm) **Al**	**1.5**		**2.5**		**3.5**		**4.5**		**3.5**		**3.5**	
Cu	**0**		**0**		**0**		**0**		**0.1**		**0.2**	

Table 17.24 Normalized organ doses—Pelvis ap—13–17 years

Beam projection:			ap									**Table cells:**
Focus-to-detector distance (cm):			115									(+) = normal font = optimal field setting
Tube voltage (kVp):			70									**(−) = bold font = suboptimal field setting**
Total filtrations:			See last table row.									Empty table cells correspond to values <0.01
Organ or tissue	**Organ dose D_T : Entrance dose K_E**								**(mSv * mGy^{-1})**			
	(+)	(−)	(+)	(−)	(+)	(−)	(+)	(−)	(+)	(−)	(+)	(−)
Skull												
Brain												
Oral mucosa												
Salivary glands												
Cervical spine												
Thyroid												
Thymus												
Thoracic spine												
Claviculae												
Scapulae												
Ribs										**0.01**		**0.02**
Oesophagus												
Trachea												
Lung												
Mamma		**0.01**		**0.02**		**0.02**	0.01	**0.02**	0.01	**0.03**	0.01	**0.03**
Heart												
Stomach		**0.07**	0.01	**0.09**	0.02	**0.11**	0.02	**0.12**	0.02	**0.15**	0.03	**0.17**
Spleen		**0.01**		**0.02**		**0.02**		**0.03**		**0.03**		**0.04**
Liver		**0.04**		**0.05**		**0.06**		**0.06**	0.01	**0.08**	0.01	**0.09**
Gall bladder	0.02	**0.13**	0.02	**0.17**	0.03	**0.20**	0.03	**0.22**	0.04	**0.27**	0.05	**0.32**
Pancreas		**0.02**		**0.02**		**0.02**		**0.03**		**0.04**	0.01	**0.05**
Small intestine	0.19	**0.25**	0.24	**0.32**	0.28	**0.37**	0.32	**0.42**	0.39	**0.51**	0.45	**0.60**
Large intestine	0.18	**0.30**	0.22	**0.39**	0.26	**0.45**	0.29	**0.50**	0.35	**0.61**	0.41	**0.70**
Sigma/Rectum	0.21	**0.23**	0.27	**0.30**	0.32	**0.35**	0.36	**0.40**	0.44	**0.49**	0.52	**0.57**
Lumbar spine	0.03	**0.07**	0.04	**0.10**	0.05	**0.12**	0.06	**0.14**	0.08	**0.18**	0.10	**0.23**
Pelvis	0.25	**0.27**	0.32	**0.36**	0.38	**0.43**	0.44	**0.49**	0.55	**0.61**	0.66	**0.73**
Adrenals										**0.01**		**0.02**
Kidneys		**0.02**		**0.03**		**0.04**	0.01	**0.04**	0.01	**0.05**	0.02	**0.07**
Urinary bladder	0.48	**0.49**	0.60	**0.61**	0.68	**0.70**	0.75	**0.78**	0.89	**0.92**	1.01	**1.04**
Uterus	0.74	**0.77**	0.95	**1.00**	1.12	**1.17**	1.26	**1.32**	1.54	**1.62**	1.82	**1.92**
Ovaries	0.18	**0.19**	0.23	**0.25**	0.28	**0.30**	0.31	**0.34**	0.39	**0.43**	0.47	**0.51**
Prostate	0.33	**0.34**	0.42	**0.44**	0.50	**0.52**	0.56	**0.58**	0.68	**0.70**	0.79	**0.82**
Testes	0.95	**0.99**	1.08	**1.13**	1.17	**1.23**	1.24	**1.30**	1.35	**1.42**	1.44	**1.52**
Upper arm												
Elbow										**0.01**		**0.01**
Forearm		**0.57**		**0.71**		**0.81**		**0.90**		**1.06**		**1.20**
Upper leg	0.19	**0.31**	0.25	**0.41**	0.30	**0.49**	0.34	**0.55**	0.43	**0.69**	0.52	**0.83**
Knee		**0.02**		**0.02**		**0.03**		**0.03**	0.01	**0.04**	0.01	**0.05**
Lower leg												
Lymph nodes	0.06	**0.10**	0.08	**0.12**	0.09	**0.14**	0.10	**0.16**	0.12	**0.19**	0.14	**0.22**
Skin	0.11	**0.17**	0.12	**0.19**	0.12	**0.19**	0.12	**0.20**	0.13	**0.21**	0.14	**0.22**
Muscle	0.10	**0.14**	0.12	**0.17**	0.13	**0.19**	0.14	**0.20**	0.16	**0.24**	0.18	**0.27**
Skeleton	0.04	**0.09**	0.06	**0.12**	0.07	**0.14**	0.08	**0.16**	0.10	**0.20**	0.12	**0.24**
Red bone marrow	0.03	**0.04**	0.04	**0.05**	0.05	**0.06**	0.06	**0.07**	0.07	**0.09**	0.09	**0.11**
Whole body dose	0.08	**0.13**	0.10	**0.15**	0.11	**0.17**	0.13	**0.19**	0.15	**0.22**	0.17	**0.25**
ED (ICRP 60)	0.17	**0.20**	0.21	**0.24**	0.23	**0.27**	0.25	**0.30**	0.29	**0.35**	0.33	**0.39**
ED (ICRP 103)	0.11	**0.13**	0.13	**0.17**	0.15	**0.19**	0.16	**0.21**	0.19	**0.25**	0.22	**0.28**
Total filtration (mm) Al	**1.5**		**2.5**		**3.5**		**4.5**		**3.5**		**3.5**	
Cu	**0**		**0**		**0**		**0**		**0.1**		**0.2**	

Table 17.25 Normalized organ doses—Pelvis ap—13–17 years

Beam projection:			ap									**Table cells:**
Focus-to-detector distance (cm):			115									(+) = normal font = optimal field setting
Tube voltage (kVp):			80									**(−) = bold font = suboptimal field setting**
Total filtrations:			See last table row.									Empty table cells correspond to values <0.01
Organ or tissue	**Organ dose D_T : Entrance dose K_E (mSv * mGy^{-1})**											
	(+)	(−)	(+)	(−)	(+)	(−)	(+)	(−)	(+)	(−)	(+)	(−)
Skull												
Brain												
Oral mucosa												
Salivary glands												
Cervical spine												
Thyroid												
Thymus												
Thoracic spine										**0.01**		**0.01**
Claviculae												
Scapulae												
Ribs						**0.01**		**0.01**		**0.02**		**0.02**
Oesophagus												
Trachea												
Lung												
Mamma		**0.02**		**0.02**	0.01	**0.02**	0.01	**0.03**	0.01	**0.03**	0.02	**0.04**
Heart												
Stomach	0.01	**0.09**	0.02	**0.11**	0.02	**0.13**	0.02	**0.14**	0.03	**0.17**	0.03	**0.20**
Spleen		**0.02**		**0.02**		**0.03**		**0.03**	0.01	**0.04**	0.01	**0.05**
Liver		**0.05**		**0.06**	0.01	**0.07**	0.01	**0.08**	0.01	**0.09**	0.02	**0.11**
Gall bladder	0.03	**0.16**	0.03	**0.20**	0.04	**0.24**	0.04	**0.26**	0.06	**0.32**	0.07	**0.37**
Pancreas		**0.02**		**0.03**		**0.03**		**0.04**	0.01	**0.05**	0.02	**0.06**
Small intestine	0.23	**0.30**	0.29	**0.38**	0.34	**0.44**	0.37	**0.49**	0.45	**0.59**	0.52	**0.68**
Large intestine	0.21	**0.37**	0.27	**0.46**	0.31	**0.52**	0.34	**0.58**	0.41	**0.69**	0.47	**0.79**
Sigma/Rectum	0.26	**0.29**	0.33	**0.36**	0.38	**0.42**	0.43	**0.47**	0.51	**0.57**	0.60	**0.66**
Lumbar spine	0.05	**0.10**	0.06	**0.13**	0.07	**0.16**	0.08	**0.18**	0.11	**0.23**	0.13	**0.28**
Pelvis	0.31	**0.35**	0.40	**0.45**	0.47	**0.52**	0.53	**0.59**	0.64	**0.72**	0.76	**0.85**
Adrenals						**0.01**		**0.01**		**0.02**		**0.02**
Kidneys		**0.03**	0.01	**0.04**	0.01	**0.05**	0.02	**0.06**	0.02	**0.07**	0.02	**0.09**
Urinary bladder	0.56	**0.58**	0.68	**0.70**	0.78	**0.80**	0.85	**0.87**	0.99	**1.02**	1.11	**1.15**
Uterus	0.91	**0.96**	1.15	**1.21**	1.33	**1.41**	1.49	**1.57**	1.79	**1.90**	2.08	**2.21**
Ovaries	0.23	**0.25**	0.29	**0.31**	0.34	**0.37**	0.38	**0.41**	0.46	**0.51**	0.54	**0.60**
Prostate	0.40	**0.42**	0.50	**0.53**	0.58	**0.61**	0.65	**0.67**	0.77	**0.81**	0.89	**0.93**
Testes	1.02	**1.07**	1.15	**1.21**	1.24	**1.30**	1.30	**1.37**	1.41	**1.49**	1.50	**1.59**
Upper arm												
Elbow						**0.01**		**0.01**		**0.02**		**0.02**
Forearm		**0.65**		**0.80**		**0.90**		**0.99**		**1.14**		**1.28**
Upper leg	0.25	**0.40**	0.31	**0.51**	0.37	**0.59**	0.41	**0.66**	0.51	**0.81**	0.60	**0.95**
Knee		**0.02**		**0.03**		**0.04**	0.01	**0.04**	0.01	**0.05**	0.02	**0.06**
Lower leg												
Lymph nodes	0.08	**0.12**	0.09	**0.14**	0.11	**0.17**	0.12	**0.18**	0.14	**0.22**	0.16	**0.25**
Skin	0.11	**0.18**	0.12	**0.19**	0.13	**0.20**	0.13	**0.21**	0.14	**0.22**	0.15	**0.23**
Muscle	0.11	**0.16**	0.13	**0.19**	0.15	**0.21**	0.16	**0.23**	0.18	**0.26**	0.20	**0.29**
Skeleton	0.06	**0.12**	0.07	**0.15**	0.08	**0.17**	0.09	**0.19**	0.12	**0.23**	0.14	**0.27**
Red bone marrow	0.04	**0.05**	0.05	**0.07**	0.06	**0.08**	0.07	**0.09**	0.09	**0.11**	0.11	**0.13**
Whole body dose	0.10	**0.15**	0.12	**0.17**	0.13	**0.19**	0.14	**0.21**	0.16	**0.24**	0.18	**0.27**
ED (ICRP 60)	0.20	**0.23**	0.23	**0.28**	0.26	**0.31**	0.28	**0.33**	0.32	**0.39**	0.36	**0.43**
ED (ICRP 103)	0.12	**0.16**	0.15	**0.19**	0.17	**0.22**	0.19	**0.24**	0.22	**0.28**	0.24	**0.32**
Total filtration (mm) **Al**	**1.5**		**2.5**		**3.5**		**4.5**		**3.5**		**3.5**	
Cu	**0**		**0**		**0**		**0**		**0.1**		**0.2**	

Table 17.26 Normalized organ doses—Pelvis ap—>17 years

Beam projection:			ap									**Table cells:**
Focus-to-detector distance (cm):			115				(+) = normal font = optimal field setting					
Tube voltage (kVp):			60				**(−) = bold font = suboptimal field setting**					
Total filtrations:			See last table row.				Empty table cells correspond to values <0.01					
Organ or tissue	**Organ dose D_T : Entrance dose K_E**								**(mSv * mGy^{-1})**			
	(+)	(−)	(+)	(−)	(+)	(−)	(+)	(−)	(+)	(−)	(+)	(−)
Skull												
Brain												
Oral mucosa												
Salivary glands												
Cervical spine												
Thyroid												
Thymus												
Thoracic spine												
Claviculae												
Scapulae												
Ribs												
Oesophagus												
Trachea												
Lung												
Mamma												
Heart												
Stomach		**0.05**		**0.07**		**0.08**		**0.09**	0.01	**0.12**		**0.10**
Spleen				**0.01**		**0.01**		**0.02**		**0.02**		**0.02**
Liver		**0.03**		**0.04**		**0.05**		**0.05**		**0.07**		**0.06**
Gall bladder		**0.10**	0.01	**0.13**	0.02	**0.15**	0.02	**0.17**	0.02	**0.21**	0.02	**0.19**
Pancreas				**0.01**		**0.01**		**0.02**		**0.02**		**0.02**
Small intestine	0.14	**0.19**	0.18	**0.25**	0.21	**0.29**	0.24	**0.33**	0.30	**0.41**	0.26	**0.37**
Large intestine	0.12	**0.24**	0.16	**0.31**	0.19	**0.37**	0.21	**0.41**	0.26	**0.51**	0.23	**0.46**
Sigma/Rectum	0.16	**0.17**	0.21	**0.23**	0.25	**0.27**	0.28	**0.31**	0.35	**0.39**	0.31	**0.34**
Lumbar spine	0.02	**0.04**	0.02	**0.06**	0.03	**0.07**	0.03	**0.09**	0.05	**0.12**	0.03	**0.09**
Pelvis	0.17	**0.19**	0.23	**0.26**	0.28	**0.31**	0.32	**0.36**	0.41	**0.46**	0.35	**0.38**
Adrenals												
Kidneys		**0.01**		**0.02**		**0.02**		**0.03**		**0.04**		**0.03**
Urinary bladder	0.40	**0.41**	0.50	**0.52**	0.58	**0.60**	0.65	**0.67**	0.77	**0.80**	0.72	**0.74**
Uterus	0.22	**0.22**	0.29	**0.30**	0.34	**0.36**	0.39	**0.41**	0.48	**0.51**	0.43	**0.45**
Ovaries	0.13	**0.14**	0.17	**0.19**	0.21	**0.22**	0.24	**0.26**	0.31	**0.33**	0.26	**0.28**
Prostate	0.24	**0.25**	0.31	**0.32**	0.37	**0.38**	0.42	**0.44**	0.52	**0.54**	0.47	**0.48**
Testes	0.82	**0.85**	0.96	**1.00**	1.05	**1.10**	1.12	**1.17**	1.24	**1.30**	1.22	**1.27**
Upper arm												
Elbow												
Forearm		**0.47**		**0.60**		**0.70**		**0.78**		**0.94**		**0.89**
Upper leg	0.13	**0.19**	0.17	**0.26**	0.21	**0.31**	0.25	**0.36**	0.31	**0.46**	0.26	**0.38**
Knee		**0.03**		**0.04**		**0.05**		**0.05**		**0.07**		**0.05**
Lower leg												
Lymph nodes	0.05	**0.07**	0.06	**0.09**	0.07	**0.11**	0.08	**0.13**	0.10	**0.15**	0.09	**0.14**
Skin	0.10	**0.18**	0.11	**0.19**	0.11	**0.20**	0.12	**0.20**	0.12	**0.21**	0.12	**0.21**
Muscle	0.08	**0.12**	0.10	**0.15**	0.12	**0.17**	0.13	**0.19**	0.15	**0.22**	0.14	**0.20**
Skeleton	0.03	**0.07**	0.04	**0.09**	0.05	**0.11**	0.06	**0.12**	0.07	**0.16**	0.06	**0.14**
Red bone marrow	0.02	**0.02**	0.03	**0.03**	0.03	**0.04**	0.04	**0.05**	0.05	**0.06**	0.04	**0.05**
Whole body dose	0.07	**0.11**	0.09	**0.13**	0.10	**0.15**	0.11	**0.17**	0.13	**0.20**	0.12	**0.18**
ED (ICRP 60)	0.14	**0.16**	0.17	**0.20**	0.19	**0.23**	0.21	**0.25**	0.24	**0.29**	0.23	**0.27**
ED (ICRP 103)	0.08	**0.10**	0.10	**0.13**	0.12	**0.15**	0.13	**0.17**	0.15	**0.20**	0.14	**0.18**
Total filtration (mm) **Al**	**1.5**		**2.5**		**3.5**		**4.5**		**3.5**		**3.5**	
Cu	**0**		**0**		**0**		**0**		**0.1**		**0.2**	

Table 17.27 Normalized organ doses—Pelvis ap-—>17 years

Beam projection:			ap									**Table cells:**
Focus-to-detector distance (cm):			115				(+) = normal font = optimal field setting					
Tube voltage (kVp):			70				**(−) = bold font = suboptimal field setting**					
Total filtrations:			See last table row.				Empty table cells correspond to values <0.01					
Organ or tissue	**Organ dose D_T : Entrance dose K_E (mSv * mGy^{-1})**											
	(+)	(−)	(+)	(−)	(+)	(−)	(+)	(−)	(+)	(−)	(+)	(−)
Skull												
Brain												
Oral mucosa												
Salivary glands												
Cervical spine												
Thyroid												
Thymus												
Thoracic spine												
Claviculae												
Scapulae												
Ribs												**0.01**
Oesophagus												
Trachea												
Lung												
Mamma												
Heart												
Stomach		**0.07**		**0.09**	0.01	**0.10**	0.01	**0.12**	0.02	**0.14**	0.02	**0.16**
Spleen		**0.01**		**0.02**		**0.02**		**0.02**		**0.03**		**0.04**
Liver		**0.04**		**0.05**		**0.06**		**0.07**		**0.08**	0.01	**0.09**
Gall bladder	0.01	**0.12**	0.02	**0.16**	0.02	**0.19**	0.03	**0.21**	0.03	**0.26**	0.04	**0.30**
Pancreas		**0.01**		**0.02**		**0.02**		**0.02**		**0.03**		**0.04**
Small intestine	0.17	**0.24**	0.22	**0.31**	0.26	**0.37**	0.30	**0.41**	0.36	**0.51**	0.43	**0.60**
Large intestine	0.15	**0.30**	0.20	**0.38**	0.23	**0.45**	0.26	**0.50**	0.31	**0.61**	0.37	**0.71**
Sigma/Rectum	0.20	**0.23**	0.26	**0.29**	0.31	**0.35**	0.35	**0.39**	0.43	**0.48**	0.51	**0.57**
Lumbar spine	0.03	**0.07**	0.04	**0.09**	0.04	**0.11**	0.05	**0.13**	0.07	**0.17**	0.08	**0.21**
Pelvis	0.24	**0.26**	0.31	**0.34**	0.37	**0.41**	0.42	**0.47**	0.53	**0.59**	0.63	**0.71**
Adrenals												**0.01**
Kidneys		**0.02**		**0.03**		**0.03**		**0.04**	0.01	**0.05**	0.01	**0.07**
Urinary bladder	0.48	**0.50**	0.60	**0.62**	0.68	**0.71**	0.75	**0.78**	0.89	**0.92**	1.02	**1.05**
Uterus	0.28	**0.30**	0.36	**0.38**	0.43	**0.45**	0.48	**0.51**	0.59	**0.63**	0.70	**0.74**
Ovaries	0.18	**0.19**	0.23	**0.25**	0.27	**0.29**	0.31	**0.33**	0.39	**0.42**	0.47	**0.50**
Prostate	0.30	**0.32**	0.39	**0.41**	0.46	**0.48**	0.51	**0.53**	0.62	**0.65**	0.73	**0.76**
Testes	0.90	**0.94**	1.04	**1.09**	1.14	**1.19**	1.21	**1.26**	1.33	**1.40**	1.43	**1.51**
Upper arm												
Elbow												**0.01**
Forearm		**0.57**		**0.71**		**0.81**		**0.90**		**1.06**		**1.21**
Upper leg	0.18	**0.26**	0.24	**0.34**	0.28	**0.41**	0.32	**0.47**	0.41	**0.59**	0.50	**0.72**
Knee		**0.04**		**0.05**		**0.06**	0.01	**0.07**	0.01	**0.09**	0.02	**0.11**
Lower leg												
Lymph nodes	0.06	**0.09**	0.07	**0.12**	0.09	**0.14**	0.10	**0.15**	0.12	**0.19**	0.13	**0.22**
Skin	0.11	**0.19**	0.11	**0.20**	0.12	**0.21**	0.12	**0.21**	0.13	**0.23**	0.14	**0.24**
Muscle	0.10	**0.14**	0.12	**0.17**	0.13	**0.20**	0.15	**0.21**	0.17	**0.25**	0.19	**0.28**
Skeleton	0.04	**0.09**	0.06	**0.12**	0.07	**0.14**	0.08	**0.16**	0.10	**0.20**	0.12	**0.23**
Red bone marrow	0.03	**0.03**	0.04	**0.05**	0.05	**0.06**	0.05	**0.06**	0.07	**0.08**	0.08	**0.10**
Whole body dose	0.08	**0.13**	0.10	**0.16**	0.12	**0.18**	0.13	**0.20**	0.15	**0.23**	0.17	**0.26**
ED (ICRP 60)	0.16	**0.19**	0.19	**0.23**	0.22	**0.26**	0.24	**0.29**	0.28	**0.34**	0.31	**0.38**
ED (ICRP 103)	0.10	**0.12**	0.12	**0.15**	0.14	**0.18**	0.15	**0.20**	0.18	**0.23**	0.20	**0.27**
Total filtration (mm) Al	**1.5**		**2.5**		**3.5**		**4.5**		**3.5**		**3.5**	
Cu	**0**		**0**		**0**		**0**		**0.1**		**0.2**	

Table 17.28 Normalized organ doses––Pelvis ap—>17 years

Beam projection:			ap									**Table cells:**
Focus-to-detector distance (cm):			115									(+) = normal font = optimal field setting
Tube voltage (kVp):			80									**(−) = bold font = suboptimal field setting**
Total filtrations:			See last table row.									Empty table cells correspond to values <0.01
Organ or tissue	**Organ dose D_T : Entrance dose K_E**								**(mSv * mGy⁻¹)**			
	(+)	(−)	(+)	(−)	(+)	(−)	(+)	(−)	(+)	(−)	(+)	(−)
Skull												
Brain												
Oral mucosa												
Salivary glands												
Cervical spine												
Thyroid												
Thymus												
Thoracic spine												**0.01**
Claviculae												
Scapulae												
Ribs								**0.01**		**0.01**		**0.02**
Oesophagus												
Trachea												
Lung												
Mamma												
Heart												
Stomach		**0.08**	0.01	**0.11**	0.01	**0.12**	0.02	**0.14**	0.02	**0.16**	0.03	**0.19**
Spleen		**0.02**		**0.02**		**0.03**		**0.03**		**0.04**		**0.05**
Liver		**0.05**		**0.06**		**0.07**		**0.08**	0.01	**0.09**	0.01	**0.11**
Gall bladder	0.02	**0.15**	0.02	**0.19**	0.03	**0.22**	0.03	**0.25**	0.04	**0.30**	0.05	**0.35**
Pancreas		**0.02**		**0.02**		**0.03**		**0.03**		**0.04**	0.01	**0.05**
Small intestine	0.22	**0.30**	0.27	**0.38**	0.31	**0.44**	0.35	**0.49**	0.42	**0.59**	0.49	**0.68**
Large intestine	0.19	**0.36**	0.24	**0.45**	0.27	**0.52**	0.30	**0.58**	0.36	**0.69**	0.42	**0.79**
Sigma/Rectum	0.25	**0.28**	0.32	**0.36**	0.37	**0.42**	0.42	**0.47**	0.50	**0.57**	0.59	**0.66**
Lumbar spine	0.04	**0.09**	0.05	**0.12**	0.06	**0.15**	0.07	**0.17**	0.09	**0.22**	0.11	**0.27**
Pelvis	0.30	**0.34**	0.38	**0.43**	0.45	**0.51**	0.51	**0.57**	0.63	**0.70**	0.74	**0.83**
Adrenals								**0.01**		**0.01**		**0.02**
Kidneys		**0.03**		**0.04**	0.01	**0.05**	0.01	**0.05**	0.02	**0.07**	0.02	**0.08**
Urinary bladder	0.56	**0.58**	0.69	**0.71**	0.78	**0.81**	0.85	**0.88**	0.99	**1.03**	1.12	**1.17**
Uterus	0.35	**0.37**	0.44	**0.47**	0.51	**0.54**	0.57	**0.61**	0.69	**0.74**	0.81	**0.86**
Ovaries	0.23	**0.24**	0.29	**0.31**	0.34	**0.36**	0.38	**0.41**	0.46	**0.50**	0.55	**0.59**
Prostate	0.37	**0.39**	0.47	**0.49**	0.54	**0.56**	0.60	**0.63**	0.72	**0.75**	0.83	**0.87**
Testes	0.98	**1.02**	1.12	**1.17**	1.21	**1.27**	1.28	**1.34**	1.40	**1.48**	1.50	**1.59**
Upper arm												
Elbow										**0.01**		**0.01**
Forearm		**0.65**		**0.80**		**0.91**		**0.99**		**1.15**		**1.29**
Upper leg	0.23	**0.34**	0.30	**0.44**	0.35	**0.51**	0.40	**0.58**	0.49	**0.71**	0.59	**0.85**
Knee		**0.05**	0.01	**0.07**	0.01	**0.08**	0.01	**0.09**	0.02	**0.11**	0.02	**0.13**
Lower leg												
Lymph nodes	0.07	**0.11**	0.09	**0.14**	0.10	**0.16**	0.11	**0.18**	0.13	**0.22**	0.15	**0.25**
Skin	0.11	**0.19**	0.12	**0.21**	0.12	**0.22**	0.13	**0.22**	0.13	**0.24**	0.14	**0.25**
Muscle	0.11	**0.17**	0.13	**0.20**	0.15	**0.22**	0.16	**0.24**	0.19	**0.27**	0.21	**0.31**
Skeleton	0.06	**0.11**	0.07	**0.14**	0.08	**0.17**	0.09	**0.19**	0.12	**0.23**	0.14	**0.27**
Red bone marrow	0.04	**0.05**	0.05	**0.06**	0.06	**0.07**	0.07	**0.08**	0.08	**0.10**	0.10	**0.12**
Whole body dose	0.10	**0.15**	0.12	**0.18**	0.13	**0.20**	0.14	**0.22**	0.17	**0.25**	0.19	**0.29**
ED (ICRP 60)	0.19	**0.22**	0.22	**0.26**	0.25	**0.30**	0.27	**0.32**	0.31	**0.38**	0.34	**0.42**
ED (ICRP 103)	0.11	**0.15**	0.14	**0.18**	0.16	**0.20**	0.17	**0.22**	0.20	**0.26**	0.23	**0.30**
Total filtration (mm) **Al**	**1.5**		**2.5**		**3.5**		**4.5**		**3.5**		**3.5**	
Cu	**0**		**0**		**0**		**0**		**0.1**		**0.2**	

References

1. Bontrager KL. Textbook of radiographic positioning and related anatomy. St. Louis, MO: Mosby Year Book; 1993.
2. Commission of the European Communities. European guidelines on quality criteria for diagnostic radiographic images in paediatrics. EUR 16261, SBN 92-827-7843-6; 1996.
3. Cook JV, Shah K, Pablot S, Kyriou J, Pettett A, Fitzgerald M. Guidelines on best practice in the x-ray imaging of children. A manual for all x-ray departments. London: Queen Mary's Hospital for Children, Surrey, and The Radiological Protection Centre; 1998.
4. Drexler G, Panzer W, Stieve FE. Die Bestimmung von Organdosen in der Röntgendiagnostik. Berlin: Hoffmann Verlag; 1993.
5. Ebel K-D, Willich E. Die Röntgenuntersuchung im Kindesalter—Technik und Indikation. Berlin: Springer Verlag; 1968.
6. Hart D, Jones DJ, Wall BF. Estimation of effective dose in diagnostic radiology from entrance surface dose and dose-area-product measurements. Oxon: National Radiological Protection Board NRPB-R262; 1994.
7. Hart D, Jones DJ, Wall BF. Coefficients for estimating effective doses from pediatric x-ray examinations. Oxon: National Radiological Protection Board NRPB-R279; 1996.
8. Hösle M. Untersuchung über die Patientendosis und die Bildqualität bei häufigen Röntgenaufnahmen im Säuglingsalter in Kliniken der alten Bundesländer mit Leitung durch einen Pädiater oder einen Radiologen. Dissertation, Universität München; 1993.
9. Pehe J. Untersuchung über die Patientendosis und die Bildqualität bei häufigen Röntgenaufnahmen in der Kinderradiologie. Dissertation, Universität München; 1993.
10. Schneider K, Fendel H, Bakowski C, Stein E, Kohn M, Kellner M, Schweighofer K, Cartagena G, Padovani R, Panzer W, Scheurer C, Wall B. Results of a dosimetry study in the European Community on frequent x ray examinations in infants. Radiat Prot Dosim. 1992;43:31–6.
11. Seidenbusch MC, Schneider K. Conversion coefficients for determining organ doses in paediatric pelvis and hip joint radiography. Pediatr Radiol. 2014;44:1110–23.

Hip Joint

18

18.1 Introduction

The type of hip joint radiography is determined by the clinical indication. Table 18.1 shows indications and Table 18.2 contraindications for hip joint radiography. Image criteria as required for radiological findings in hip joint radiographs are listed in Table 18.3. In order to fulfil these image criteria, the field settings listed in Tables 18.4 and 18.5 should be achieved. These field settings go along with the patient preparation as listed in Table 18.6. Exposure parameter combinations for various age groups and patient diameters are listed in Tables 18.7, 18.8, and 18.9. The age groups in Table 18.8 refer to the standard age groups of the mathematical MIRD phantom family which has been used to calculate the normalized organ doses in this book. Table 18.9 shows radiation qualities (i.e. various combinations of tube voltage and total filtration) which may be applied in clinical practice. Please note

Table 18.1 Indications for hip joint radiographs

Coxitis simplex (transitory synovitis)
Purulent coxitis
Simple bone cyst
Perthes disease
Epiphysiolysis capitis femoris
Trauma
Stress fracture
Bone tumours (osteoid osteoma)
Soft tissue tumours

Table 18.2 Contraindications for hip joint radiographs

Pregnancy

Table 18.3 Image criteria for hip joint radiographs

Hip joint ap and frog-leg lateral (Lauenstein)
Reproduction of the hip joint
Visually sharp reproduction of the cortex and trabecular structures of the acetabulum, the proximal femoral epiphysis, and the neck of the femur consistent with age
Visualization of the trochanters consistent with age
Visualization of the periarticular soft tissue planes

Table 18.4 Patient positions and beam projections for hip joint radiographs

Patient position	Beam projections
Supine	ap + frog-leg lateral (Lauenstein)

Table 18.5 Optimal field settings for hip joint radiographs

Hip joint ap	
Upper field edge	Spina iliaca anterior superior
Field centring	Midpoint of the inguinal groove
Lower field edge	Perineum
Medial field edge	Symphysis pubis
Lateral field edge	Skin surface of the upper thigh
Adjust the angulations of the tube for the ap and lateral frog-leg (Lauenstein) appropriately	

M. Seidenbusch et al., *Imaging Practice and Radiation Protection in Pediatric Radiology*,
https://doi.org/10.1007/978-3-030-18504-6_18

that not all of these radiation qualities are recommended by the authors. The standard projections of hip joint radiographies of the patient who is correctly positioned are depicted in Figs. 18.1 and 18.2. Also the optimal field (marked in green), the suboptimal field (marked in red), and the correct shielding are shown. Tables 18.10, 18.11, and 18.12 outline the virtual field settings on the skin surface of mathematical MIRD phantoms of various age groups and the resulting virtual radiographies with and without bone structures, respectively. Normalized organ doses resulting from these virtual field settings have been calculated for all age groups and have been listed in Tables 18.13, 18.14, 18.15, 18.16, 18.17, 18.18, 18.19, 18.20, 18.21, 18.22, 18.23, 18.24, 18.25, 18.26, and 18.27. Table cells with normal font refer to optimal field settings and table cells

Table 18.6 Patient preparation

General	Remove all clothes, external metallic devices, and jewellery from the pelvic region and the upper thigh
	Immobilize infants and uncooperative patients in the supine position for the ap and Lauenstein projection, e.g. with sandbags over the arms and legs
	Place eventually small X-ray positive markers at palpable tumours or on pain points
Positioning	The legs should be extended and rotated intoed
	In order to get the Lauenstein projection, the thigh is flexed and abducted in the hip joint by 45°. The knee is also flexed and the foot stands on the table
Shielding	In males, use testicular capsules if possible. In females, lead masks should be placed at the medial and upper field edges

Table 18.7 Radiographic procedure

X-ray source	Use automatic exposure control (AEC) in patients >1 year of age
	Focal spot size between 0.6 and 1.3 dependent on the age of the patient
	Check the use of additional filtration
	Check the focus-to-detector distance
Field setting	Check the field size
	Check the field centring
Image detector system	Check the use of antiscatter grids
Patient	–

Table 18.8 Radiographic techniques of hip joint radiographs

Age group (years)	Beam direction	Patient position	Beam projection	Typical patient diameter (cm)	Optimal field size (cm x cm)	Focus-to-detector distance (cm)	Voltage (kVp) 50	60	70	80	90	100	Antiscatter grid
0	vertical	supine	ap	9.8	4 x 6	100⁽¹⁾							
1 – 2				13.0	6 x 8								
3 – 7				15.0	8 x 11	115							
8 – 12				16.8	10 x 10								
13 – 17				19.6	12 x 12								
18				20.0	14 x 14								

Annotations:

(1)	The patient is lying on the table top directly on the detector.
	Parameter combination not recommended, normalized organ doses not provided
	Parameter combination not recommended, normalized organ doses provided
	Parameter combination recommended, normalized organ doses provided

Table 18.9 Radiation qualities of hip joint radiographs

Voltage (kVp)	Total filtration: Without additional		With additional filtration			
	1.5 mm Al	2.5 Al	3.5 Al	4.5 Al	3.5 Al + 0.1 Cu[1]	3.5 Al + 0.2 Cu[1]
50						
60						
70						
80						
90						
100						

Annotations:

1	The total beam filtration is usually 3.5 mm Al + 0.1 mm Cu, in adolescents optionally 3.5 mm Al + 0.2 mm Cu
	Parameter combination not recommended, normalized organ doses not provided
	Parameter combination not recommended, normalized organ doses provided
	Parameter combination recommended, normalized organ doses provided

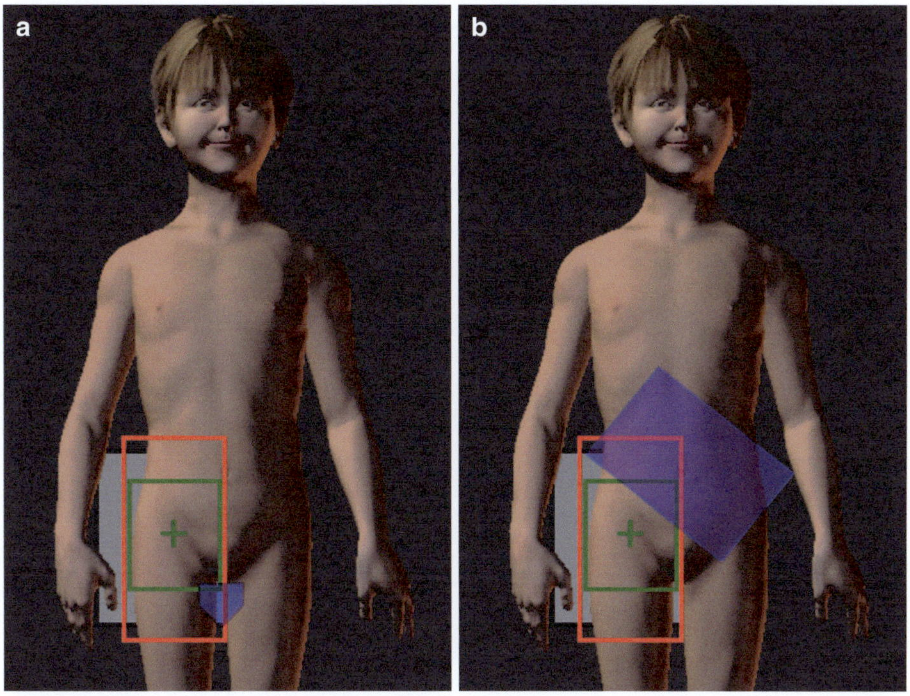

Fig. 18.1 Field settings for hip joint radiographs in supine position. Antero-posterior projection. (**a**) Male. (**b**) Female. (© V. Rösenberger & M. Seidenbusch 2018). Green lines = optimal field settings; green cross = centring point; red lines = suboptimal field settings; grey shaded area = image detector; blue shaded area = lead mask shielding and testicular capsule; dark grey shaded area = Bucky table

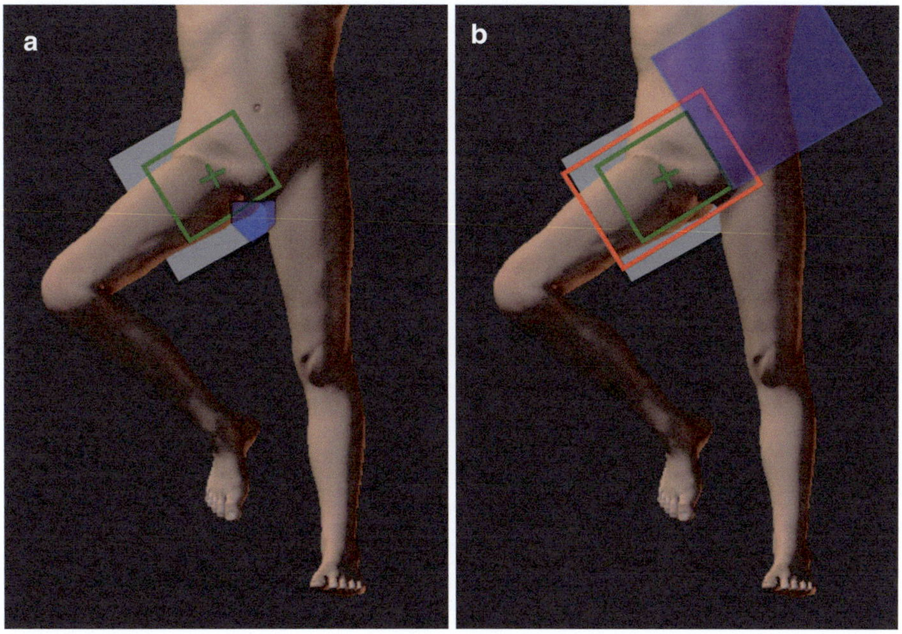

Fig. 18.2 Field settings for hip joint radiographs in supine position. Frog-leg lateral/Lauenstein. (**a**) Male. (**b**) Female. (© V. Rösenberger & M. Seidenbusch 2018). Green lines = optimal field settings; green cross = centring point; red lines = suboptimal field settings; grey shaded area = image detector; blue shaded area = lead mask shielding and testicular capsule; dark grey shaded area = Bucky table

with bold font refer to suboptimal field settings. Please note that normalized organ doses have been provided also for exposure parameter combinations, which are not recommended by the authors.

18.2 Indications and Contraindications

Indications and contraindications for hip joints radiographs are given in Tables 18.1 and 18.2.

18.3 Image Criteria

The image criteria required for hip joint radiographs are listed in Table 18.3.

See also [2].

18.4 Field Settings

In Table 18.4, patient positions and beam projections used for lumbar spine radiographs are listed, and in Table 18.5, the optimal field settings are listed which are required in order to fulfil the image criteria mentioned in Table 18.3.

For more details, see e.g. [8].

18.5 Patient Preparation and Radiographic Procedure

See Tables 18.6, 18.7 and [1, 4].

18.6 Exposure Parameters

In Table 18.8, exposure parameter combinations used for hip joint radiographs in various age groups are listed. Dark grey shaded table cells refer to exposure parameter combinations which are not recommended for paediatric hip joint radiographs and for which no normalized organ doses have been provided. Light grey and light blue shaded table cells refer to exposure parameter combinations for which normalized organ doses have been provided with light grey shaded table cells referring to exposure parameter combinations being not recommended and light blue shaded table cells referring to exposure parameter combinations being recommended for paediatric hip joint radiographs. Note that due to the fact that the radiation dose is an exponential function of the patient diameter, it may make sense to use the patient diameters instead of the patient's age groups for determining the adequate exposure parameters from Table 18.8. In this case, for determining the corresponding normalized organ doses from Tables 18.13, 18.14, 18.15, 18.16, 18.17, 18.18, 18.19, 18.20, 18.21, 18.22, 18.23, 18.24, 18.25, 18.26, and 18.27, the age group related to the patient diameter should be used.

18.7 Normalized Organ Doses

For calculating normalized organ doses, the field settings described in Sect. 18.4 were transferred to the mathematical MIRD phantoms. Inconsistencies between the geometrical anatomy of the mathematical phantoms and the real human anatomy have been listed in Table 18.28 whenever they are of relevance for the validity of the normalized organ doses.

In Tables 18.10, 18.11 and 18.12, optimal and suboptimal field settings are listed for the different age groups and the beam projections given in Table 18.8. Green lines refer to optimal field settings, red lines refer to suboptimal field settings, respectively. In the tables, the optimal and suboptimal field settings which were used for performing the calculations of the normalized organ doses provided in Tables 18.13, 18.14, 18.15, 18.16, 18.17, 18.18, 18.19, 18.20, 18.21, 18.22, 18.23, 18.24, 18.25, 18.26, and 18.27 were drawn in on the skin surface of the mathematical phantoms. The virtual mathematical radiographs resulting from these field settings were depicted in the tables with and without bone structures, respectively, in order to enable the reader to visually compare the virtual mathematical radiographs with the real radiographs made under clinical conditions. For better visualization of the hip joint in the skeletal view, the colon has been removed. For comparing the virtual mathematical field settings with the real clinical field settings, the virtual mathematical field edges were also listed in the tables.

As the normalized organ doses provided in Table 18.13, 18.14, 18.15, 18.16, 18.17, 18.18, 18.19, 18.20, 18.21, 18.22, 18.23, 18.24, 18.25, 18.26, and 18.27 refer to entrance kerma values, optimal and suboptimal field sizes were provided in the tables in order to enable the reader to calculate entrance air kerma dose values from clinically measured dose area product values. Please note that these optimal and suboptimal field sizes refer to the mathematical phantoms of an age group only and not to real patients with corresponding ages.

Please note also that the anatomy of the mathematical phantoms has been idealized. Therefore, the virtual field settings for calculating the normalized organ doses have been defined with a view to ensuring that the resulting virtual organ doses would be consistent with the real organ doses which would be achieved during real X-ray examinations being performed with the real field settings recommended in Sect. 18.4. Thus, the field settings shown in the following tables may not be consistent with the clinical field settings as recommended in Sect. 18.4.

Please note also that for the Lauenstein technique no normalized organ doses are provided. As an approximation, normalized organ doses for the ap projection may be used.

Normalized organ doses for hip joint radiographs are also listed in e.g. [3, 5–7].

Table 18.10 Field settings—hip joint ap + frog-leg (Lauenstein) projection

Age group (years)		Anatomical landmarks			
		Optimal field setting		Suboptimal field setting	
0	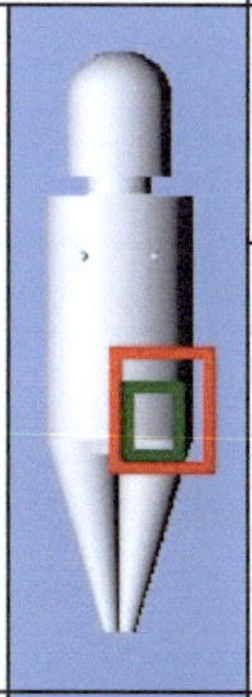	Upper field edge: Close to the spina iliaca anterior superior Lower field edge: Just above the perineum Medial field edge: 1 cm lateral to the symphysis pubis Lateral field edges: Close to the skin surface of the pelvis Field size = field width × field height: 4 cm × 6 cm = 24 cm^2		Upper field edge: Just below the iliac crest Lower field edge: Upper third of the thigh Medial field edge: Linea alba Lateral field edge: Aside the pelvis including forearm and hand Field size = field width × field height: 8 cm × 10 cm = 80 cm^2	

1 - 2	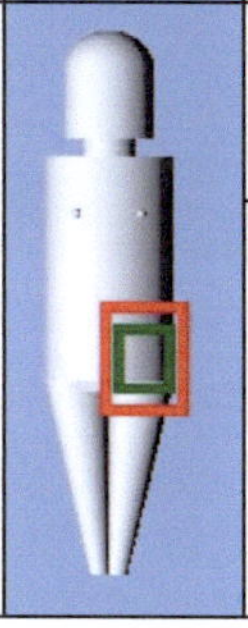	Upper field edge: Close to the spina iliaca anterior superior Lower field edge: Just above the perineum Medial field edge: 1 cm lateral to the symphysis pubis Lateral field edges: Close to the skin surface of the pelvis Field size = field width × field height: 6 cm × 8 cm = 48 cm^2		Upper field edge: Just below the iliac crest Lower field edge: Upper third of the thigh Medial field edge: Linea alba Lateral field edge: Aside the pelvis including forearm and hand Field size = field width × field height: 10 cm × 14 cm = 140 cm^2	
				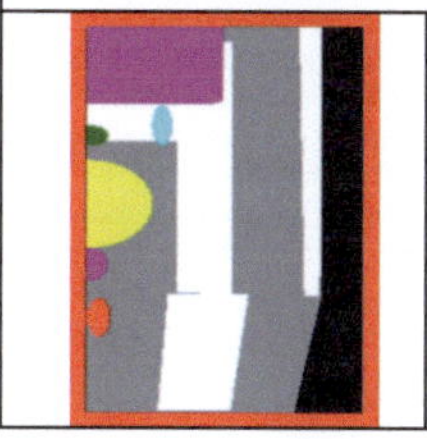	

Table 18.11 Field settings—hip joint ap + frog-leg (Lauenstein) projection

Age group (years)		Anatomical landmarks			
		Optimal field setting		Suboptimal field setting	
3 - 7	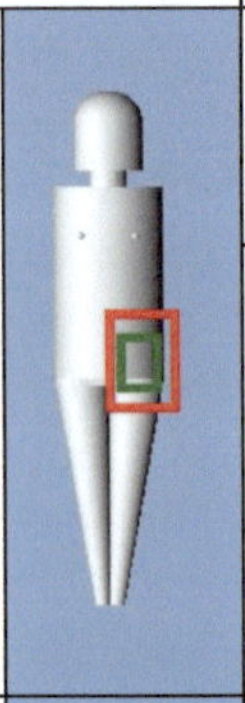	Upper field edge: Close to the spina iliaca anterior superior Lower field edge: Just above the perineum Medial field edge: 1 cm lateral to the symphysis pubis Lateral field edges: Close to the skin surface of the pelvis Field size = field width x field height: 8 cm x 11 cm = 88 cm^2		Upper field edge: Just below the iliac crest Lower field edge: Upper third of the thigh Medial field edge: Linea alba Lateral field edge: Aside the pelvis including forearm and hand Field size = field width x field height: 14 cm x 20 cm = 280 cm^2	
			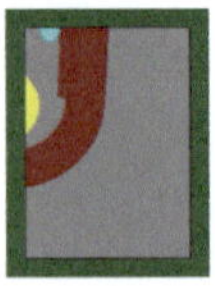		
8 - 12	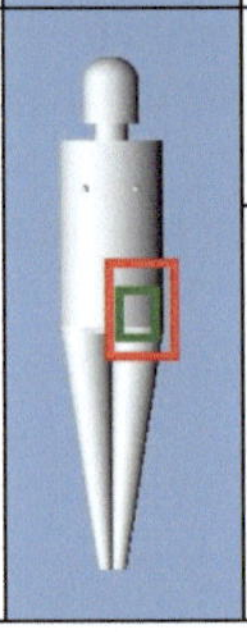	Upper field edge: Close to the spina iliaca anterior superior Lower field edge: Just above the perineum Medial field edge: 1 cm lateral to the symphysis pubis Lateral field edges: Close to the skin surface of the pelvis Field size = field width x field height: 10 cm x 10 cm = 100 cm^2		Upper field edge: Just below the iliac crest Lower field edge: Upper third of the thigh Medial field edge: Linea alba Lateral field edge: Aside the pelvis including forearm and hand Field size = field width x field height: 16 cm x 24 cm = 384 cm^2	

Table 18.12 Field settings—hip joint ap + frog-leg (Lauenstein) projection

Age group (years)		Anatomical landmarks			
		Optimal field setting		Suboptimal field setting	
13 - 17	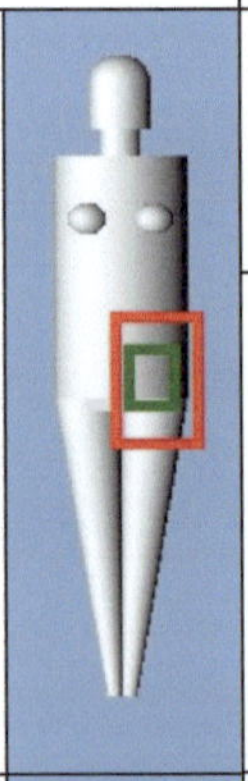	Upper field edge: Close to the spina iliaca anterior superior Lower field edge: Just above the perineum Medial field edge: 1 cm lateral to the symphysis pubis Lateral field edges: Close to the skin surface of the pelvis Field size = field width × field height: 12 cm × 12 cm = 144 cm^2		Upper field edge: Just below the iliac crest Lower field edge: Upper third of the thigh Medial field edge: Linea alba Lateral field edge: Aside the pelvis including forearm and hand Field size = field width × field height: 20 cm × 32 cm = 640 cm^2	

18	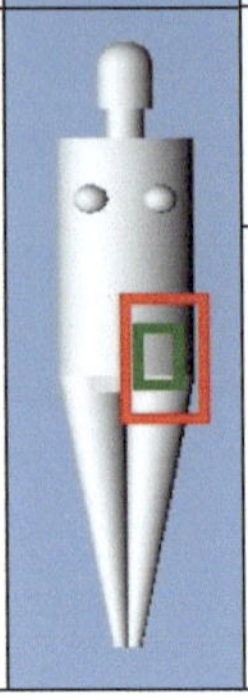	Upper field edge: Close to the spina iliaca anterior superior Lower field edge: Just above the perineum Medial field edge: 1 cm lateral to the symphysis pubis Lateral field edges: Close to the skin surface of the pelvis Field size = field width × field height: 14 cm × 14 cm = 196 cm^2		Upper field edge: Just below the iliac crest Lower field edge: Upper third of the thigh Medial field edge: Linea alba Lateral field edge: Aside the pelvis including forearm and hand Field size = field width × field height: 22 cm × 35 cm = 770 cm^2	
			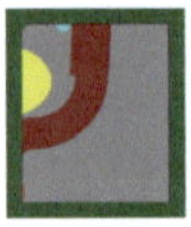		

Table 18.13 Normalized organ doses—Hip joint ap—<1 year

Beam projection:			ap									**Table cells:**
Focus-to-detector distance (cm):			100						(+) = normal font = optimal field setting			
Tube voltage (kVp):			60						**(−) = bold font = suboptimal field setting**			
Total filtrations:			See last table row						Empty table cells correspond to values <0.01			
Organ or tissue	**Organ dose D_T: Entrance dose K_E**								**(mSv * mGy^{-1})**			
	(+)	(−)	(+)	(−)	(+)	(−)	(+)	(−)	(+)	(−)	(+)	(−)
Skull												
Brain												
Oral mucosa												
Salivary glands												
Cervical spine												
Thyroid												
Thymus												
Thoracic spine												
Claviculae												
Scapulae												
Ribs												
Oesophagus												
Trachea												
Lung												
Mamma												
Heart												
Stomach		**0.02**		**0.02**		**0.02**		**0.03**		**0.03**		**0.03**
Spleen				**0.01**		**0.01**		**0.02**		**0.02**		**0.02**
Liver										**0.01**		**0.01**
Gall bladder		**0.01**		**0.02**		**0.02**		**0.03**		**0.03**		**0.03**
Pancreas				**0.01**		**0.01**		**0.02**		**0.02**		**0.02**
Small intestine	0.01	**0.15**	0.02	**0.19**	0.02	**0.22**	0.02	**0.24**	0.03	**0.27**	0.03	**0.27**
Large intestine		**0.04**		**0.05**	0.01	**0.05**	0.01	**0.06**	0.02	**0.07**	0.01	**0.07**
Sigma/rectum	0.16	**0.32**	0.20	**0.40**	0.23	**0.46**	0.25	**0.50**	0.29	**0.58**	0.28	**0.56**
Lumbar spine		**0.05**	0.01	**0.07**	0.01	**0.08**	0.02	**0.10**	0.02	**0.12**	0.02	**0.11**
Pelvis	0.16	**0.29**	0.20	**0.37**	0.23	**0.43**	0.26	**0.48**	0.30	**0.58**	0.30	**0.55**
Adrenals												
Kidneys				**0.01**		**0.01**		**0.02**		**0.02**		**0.02**
Urinary bladder	0.13	**0.32**	0.16	**0.38**	0.18	**0.43**	0.19	**0.46**	0.22	**0.52**	0.22	**0.51**
Uterus	0.04	**0.22**	0.05	**0.27**	0.05	**0.31**	0.06	**0.34**	0.07	**0.40**	0.07	**0.39**
Ovaries	0.07	**0.16**	0.09	**0.20**	0.11	**0.24**	0.12	**0.26**	0.14	**0.31**	0.13	**0.30**
Prostate	0.06	**0.28**	0.07	**0.35**	0.08	**0.39**	0.09	**0.43**	0.10	**0.49**	0.10	**0.48**
Testes	0.05	**0.57**	0.06	**0.62**	0.06	**0.65**	0.07	**0.67**	0.08	**0.71**	0.08	**0.70**
Upper arm												
Elbow										**0.01**		
Forearm		**0.42**		**0.50**		**0.55**		**0.59**		**0.66**		**0.66**
Upper leg	0.32	**0.50**	0.40	**0.63**	0.46	**0.72**	0.51	**0.79**	0.59	**0.92**	0.58	**0.90**
Knee	0.02	**0.14**	0.02	**0.17**	0.03	**0.20**	0.03	**0.22**	0.04	**0.26**	0.04	**0.25**
Lower leg												
Lymph nodes		**0.05**		**0.06**		**0.07**	0.01	**0.08**	0.01	**0.09**	0.01	**0.09**
Skin	0.03	**0.08**	0.03	**0.09**	0.03	**0.09**	0.03	**0.09**	0.03	**0.10**	0.03	**0.10**
Muscle	0.04	**0.09**	0.04	**0.10**	0.05	**0.11**	0.05	**0.12**	0.06	**0.13**	0.06	**0.13**
Skeleton	0.04	**0.11**	0.05	**0.13**	0.06	**0.15**	0.07	**0.17**	0.08	**0.20**	0.08	**0.19**
Red bone marrow		**0.03**	0.01	**0.03**	0.01	**0.04**	0.02	**0.04**	0.02	**0.05**	0.02	**0.05**
Whole body dose	0.03	**0.07**	0.04	**0.09**	0.04	**0.10**	0.04	**0.10**	0.05	**0.12**	0.05	**0.11**
ED (ICRP 60)	0.03	**0.12**	0.04	**0.14**	0.04	**0.15**	0.05	**0.16**	0.05	**0.18**	0.05	**0.18**
ED (ICRP 103)	0.02	**0.08**	0.03	**0.09**	0.03	**0.10**	0.03	**0.11**	0.04	**0.12**	0.04	**0.12**
Total filtration (mm) **Al**	**1.5**		**2.5**		**3.5**		**4.5**		**3.5**		**3.5**	
Cu	**0**		**0**		**0**		**0**		**0.1**		**0.2**	

Table 18.14 Normalized organ doses—Hip joint ap—<1 year

Beam projection:			ap									**Table cells:**
Focus-to-detector distance (cm):			100									(+) = normal font = optimal field setting
Tube voltage (kVp):			70									**(−) = bold font = suboptimal field setting**
Total filtrations:			See last table row									Empty table cells correspond to values <0.01
Organ or tissue	**Organ dose D_T: Entrance dose K_E**							**(mSv * mGy^{-1})**				
	(+)	(−)	(+)	(−)	(+)	(−)	(+)	(−)	(+)	(−)	(+)	(−)
Skull												
Brain												
Oral mucosa												
Salivary glands												
Cervical spine												
Thyroid												
Thymus												
Thoracic spine										**0.01**		**0.01**
Claviculae												
Scapulae												
Ribs										**0.01**		**0.01**
Oesophagus												
Trachea												
Lung												
Mamma												
Heart												
Stomach		**0.02**		**0.02**		**0.03**		**0.03**		**0.04**		**0.04**
Spleen		**0.01**		**0.01**		**0.02**		**0.02**		**0.02**		**0.03**
Liver						**0.01**		**0.01**		**0.02**		**0.02**
Gall bladder		**0.02**		**0.02**		**0.03**		**0.03**		**0.04**	0.01	**0.04**
Pancreas		**0.01**		**0.01**		**0.02**		**0.02**		**0.02**		**0.03**
Small intestine	0.02	**0.18**	0.02	**0.21**	0.03	**0.24**	0.03	**0.26**	0.04	**0.30**	0.04	**0.34**
Large intestine		**0.04**	0.01	**0.05**	0.01	**0.06**	0.01	**0.07**	0.02	**0.08**	0.02	**0.09**
Sigma/rectum	0.18	**0.37**	0.23	**0.45**	0.26	**0.51**	0.28	**0.56**	0.32	**0.64**	0.36	**0.71**
Lumbar spine	0.01	**0.07**	0.01	**0.09**	0.02	**0.10**	0.02	**0.12**	0.03	**0.14**	0.03	**0.17**
Pelvis	0.18	**0.34**	0.23	**0.43**	0.26	**0.49**	0.29	**0.54**	0.33	**0.64**	0.37	**0.72**
Adrenals										**0.01**		**0.01**
Kidneys		**0.01**		**0.01**		**0.02**		**0.02**		**0.02**		**0.03**
Urinary bladder	0.15	**0.36**	0.18	**0.42**	0.20	**0.46**	0.21	**0.50**	0.24	**0.55**	0.26	**0.60**
Uterus	0.04	**0.25**	0.05	**0.31**	0.06	**0.35**	0.07	**0.39**	0.08	**0.44**	0.09	**0.49**
Ovaries	0.09	**0.19**	0.11	**0.24**	0.12	**0.27**	0.13	**0.30**	0.16	**0.34**	0.17	**0.39**
Prostate	0.07	**0.32**	0.08	**0.39**	0.09	**0.44**	0.10	**0.47**	0.11	**0.53**	0.13	**0.59**
Testes	0.05	**0.59**	0.06	**0.64**	0.07	**0.67**	0.08	**0.69**	0.09	**0.72**	0.10	**0.75**
Upper arm												
Elbow								**0.01**		**0.01**		**0.02**
Forearm		**0.46**		**0.54**		**0.59**		**0.63**		**0.69**		**0.73**
Upper leg	0.37	**0.58**	0.45	**0.70**	0.51	**0.79**	0.55	**0.86**	0.63	**0.99**	0.69	**1.08**
Knee	0.02	**0.16**	0.03	**0.20**	0.03	**0.22**	0.04	**0.24**	0.05	**0.28**	0.06	**0.31**
Lower leg										**0.01**		**0.01**
Lymph nodes		**0.06**		**0.07**	0.01	**0.08**	0.01	**0.09**	0.01	**0.10**	0.02	**0.11**
Skin	0.03	**0.09**	0.03	**0.09**	0.03	**0.10**	0.03	**0.10**	0.03	**0.10**	0.04	**0.11**
Muscle	0.04	**0.10**	0.05	**0.11**	0.05	**0.12**	0.06	**0.13**	0.06	**0.15**	0.07	**0.16**
Skeleton	0.05	**0.12**	0.06	**0.15**	0.07	**0.17**	0.07	**0.19**	0.08	**0.21**	0.09	**0.24**
Red bone marrow	0.01	**0.03**	0.01	**0.04**	0.02	**0.05**	0.02	**0.05**	0.02	**0.06**	0.02	**0.07**
Whole body dose	0.03	**0.08**	0.04	**0.10**	0.04	**0.11**	0.05	**0.11**	0.05	**0.13**	0.06	**0.14**
ED (ICRP 60)	0.03	**0.13**	0.04	**0.15**	0.05	**0.16**	0.05	**0.17**	0.06	**0.19**	0.07	**0.21**
ED (ICRP 103)	0.03	**0.08**	0.03	**0.10**	0.04	**0.11**	0.04	**0.12**	0.04	**0.13**	0.05	**0.15**
Total filtration (mm) Al	**1.5**		**2.5**		**3.5**		**4.5**		**3.5**		**3.5**	
Cu	**0**		**0**		**0**		**0**		**0.1**		**0.2**	

Table 18.15 Normalized organ doses—Hip joint ap—1–2 years

Beam projection:			ap									**Table cells:**
Focus-to-detector distance (cm):			100						(+) = normal font = optimal field setting			
Tube voltage (kVp):			60						**(−) = bold font = suboptimal field setting**			
Total filtrations:			See last table row						Empty table cells correspond to values <0.01			
Organ or tissue	**Organ dose D_T: Entrance dose K_E**								**(mSv * mGy^{-1})**			
	(+)	(−)	(+)	(−)	(+)	(−)	(+)	(−)	(+)	(−)	(+)	(−)
Skull												
Brain												
Oral mucosa												
Salivary glands												
Cervical spine												
Thyroid												
Thymus												
Thoracic spine												
Claviculae												
Scapulae												
Ribs												
Oesophagus												
Trachea												
Lung												
Mamma												
Heart												
Stomach		**0.01**		**0.02**		**0.02**		**0.02**		**0.03**		**0.02**
Spleen								**0.01**		**0.01**		**0.01**
Liver												
Gall bladder		**0.01**		**0.02**		**0.02**		**0.03**		**0.03**		**0.03**
Pancreas								**0.01**		**0.01**		**0.01**
Small intestine	0.01	**0.13**	0.02	**0.16**	0.02	**0.19**	0.02	**0.21**	0.03	**0.25**	0.03	**0.23**
Large intestine		**0.03**		**0.04**		**0.05**	0.01	**0.06**	0.01	**0.07**	0.01	**0.06**
Sigma/rectum	0.12	**0.25**	0.16	**0.32**	0.19	**0.37**	0.21	**0.41**	0.25	**0.49**	0.23	**0.47**
Lumbar spine		**0.04**		**0.05**		**0.06**	0.01	**0.07**	0.01	**0.09**	0.01	**0.08**
Pelvis	0.12	**0.21**	0.15	**0.27**	0.18	**0.32**	0.20	**0.37**	0.25	**0.45**	0.23	**0.42**
Adrenals												
Kidneys								**0.01**		**0.01**		**0.01**
Urinary bladder	0.11	**0.29**	0.14	**0.35**	0.16	**0.39**	0.17	**0.43**	0.20	**0.49**	0.19	**0.48**
Uterus	0.03	**0.20**	0.04	**0.25**	0.05	**0.29**	0.06	**0.33**	0.07	**0.39**	0.06	**0.37**
Ovaries	0.06	**0.12**	0.08	**0.16**	0.10	**0.19**	0.11	**0.21**	0.13	**0.26**	0.12	**0.24**
Prostate	0.05	**0.23**	0.06	**0.30**	0.07	**0.34**	0.08	**0.38**	0.10	**0.44**	0.09	**0.42**
Testes	0.04	**0.56**	0.05	**0.62**	0.06	**0.66**	0.06	**0.68**	0.07	**0.72**	0.07	**0.71**
Upper arm												
Elbow												
Forearm		**0.33**		**0.40**		**0.46**		**0.50**		**0.57**		**0.56**
Upper leg	0.16	**0.34**	0.20	**0.44**	0.24	**0.51**	0.27	**0.57**	0.32	**0.69**	0.30	**0.65**
Knee		**0.06**		**0.08**	0.01	**0.09**	0.01	**0.10**	0.02	**0.12**	0.01	**0.11**
Lower leg												
Lymph nodes		**0.04**		**0.05**		**0.06**		**0.07**	0.01	**0.08**		**0.08**
Skin	0.03	**0.08**	0.03	**0.08**	0.03	**0.08**	0.03	**0.09**	0.04	**0.09**	0.04	**0.09**
Muscle	0.03	**0.07**	0.03	**0.08**	0.04	**0.09**	0.04	**0.10**	0.05	**0.11**	0.05	**0.11**
Skeleton	0.02	**0.07**	0.03	**0.09**	0.04	**0.11**	0.04	**0.12**	0.05	**0.14**	0.05	**0.13**
Red bone marrow		**0.02**		**0.02**	0.01	**0.03**	0.01	**0.03**	0.02	**0.04**	0.01	**0.04**
Whole body dose	0.02	**0.06**	0.03	**0.07**	0.03	**0.08**	0.03	**0.09**	0.04	**0.10**	0.04	**0.09**
ED (ICRP 60)	0.02	**0.11**	0.03	**0.12**	0.04	**0.14**	0.04	**0.15**	0.05	**0.17**	0.05	**0.16**
ED (ICRP 103)	0.02	**0.07**	0.02	**0.08**	0.03	**0.09**	0.03	**0.10**	0.04	**0.11**	0.03	**0.11**
Total filtration (mm) **Al**	**1.5**		**2.5**		**3.5**		**4.5**		**3.5**		**3.5**	
Cu	**0**		**0**		**0**		**0**		**0.1**		**0.2**	

Table 18.16 Normalized organ doses—Hip joint ap—1–2 years

Beam projection:			ap									**Table cells:**
Focus-to-detector distance (cm):			100									(+) = normal font = optimal field setting
Tube voltage (kVp):			70									**(−) = bold font = suboptimal field setting**
Total filtrations:			See last table row									Empty table cells correspond to values <0.01
Organ or tissue	**Organ dose D_T: Entrance dose K_E**						**(mSv ∗ mGy^{-1})**					
	(+)	(−)	(+)	(−)	(+)	(−)	(+)	(−)	(+)	(−)	(+)	(−)
Skull												
Brain												
Oral mucosa												
Salivary glands												
Cervical spine												
Thyroid												
Thymus												
Thoracic spine												
Claviculae												
Scapulae												
Ribs												**0.01**
Oesophagus												
Trachea												
Lung												
Mamma												
Heart												
Stomach		**0.02**		**0.02**		**0.02**		**0.03**		**0.03**		**0.04**
Spleen				**0.01**		**0.01**		**0.02**		**0.02**		**0.02**
Liver										**0.01**		**0.01**
Gall bladder		**0.02**		**0.02**		**0.03**		**0.03**		**0.04**		**0.05**
Pancreas				**0.01**		**0.01**		**0.01**		**0.02**		**0.02**
Small intestine	0.02	**0.15**	0.02	**0.19**	0.02	**0.22**	0.03	**0.24**	0.03	**0.28**	0.04	**0.32**
Large intestine		**0.04**		**0.05**	0.01	**0.06**	0.01	**0.07**	0.02	**0.08**	0.02	**0.10**
Sigma/rectum	0.15	**0.30**	0.19	**0.38**	0.22	**0.43**	0.24	**0.48**	0.28	**0.56**	0.32	**0.63**
Lumbar spine		**0.05**	0.01	**0.07**	0.01	**0.08**	0.01	**0.09**	0.02	**0.11**	0.02	**0.14**
Pelvis	0.15	**0.26**	0.18	**0.33**	0.21	**0.39**	0.24	**0.43**	0.29	**0.52**	0.33	**0.60**
Adrenals												
Kidneys				**0.01**		**0.01**		**0.01**		**0.02**		**0.02**
Urinary bladder	0.13	**0.33**	0.16	**0.39**	0.17	**0.44**	0.19	**0.47**	0.22	**0.54**	0.24	**0.59**
Uterus	0.04	**0.24**	0.05	**0.30**	0.06	**0.34**	0.07	**0.37**	0.08	**0.44**	0.10	**0.50**
Ovaries	0.08	**0.15**	0.10	**0.19**	0.11	**0.22**	0.13	**0.25**	0.15	**0.30**	0.17	**0.34**
Prostate	0.06	**0.28**	0.07	**0.34**	0.08	**0.39**	0.09	**0.42**	0.11	**0.49**	0.12	**0.55**
Testes	0.05	**0.59**	0.06	**0.65**	0.06	**0.68**	0.07	**0.71**	0.08	**0.75**	0.09	**0.78**
Upper arm												
Elbow												**0.01**
Forearm		**0.37**		**0.45**		**0.50**		**0.54**		**0.61**		**0.67**
Upper leg	0.19	**0.41**	0.24	**0.51**	0.28	**0.59**	0.31	**0.66**	0.37	**0.78**	0.42	**0.88**
Knee		**0.07**	0.01	**0.09**	0.02	**0.11**	0.02	**0.12**	0.02	**0.14**	0.03	**0.17**
Lower leg												
Lymph nodes		**0.05**		**0.06**		**0.07**	0.01	**0.08**	0.01	**0.09**	0.02	**0.10**
Skin	0.03	**0.08**	0.03	**0.08**	0.04	**0.09**	0.04	**0.09**	0.04	**0.10**	0.04	**0.10**
Muscle	0.03	**0.08**	0.04	**0.09**	0.04	**0.10**	0.05	**0.11**	0.05	**0.12**	0.06	**0.13**
Skeleton	0.03	**0.09**	0.04	**0.11**	0.04	**0.12**	0.05	**0.14**	0.06	**0.16**	0.07	**0.18**
Red bone marrow		**0.02**	0.01	**0.03**	0.01	**0.04**	0.02	**0.04**	0.02	**0.05**	0.02	**0.06**
Whole body dose	0.03	**0.07**	0.03	**0.08**	0.04	**0.09**	0.04	**0.10**	0.04	**0.11**	0.05	**0.12**
ED (ICRP 60)	0.03	**0.12**	0.04	**0.14**	0.04	**0.15**	0.05	**0.16**	0.05	**0.18**	0.06	**0.20**
ED (ICRP 103)	0.02	**0.08**	0.03	**0.09**	0.03	**0.10**	0.03	**0.11**	0.04	**0.12**	0.05	**0.14**
Total filtration (mm) Al	**1.5**		**2.5**		**3.5**		**4.5**		**3.5**		**3.5**	
Cu	**0**		**0**		**0**		**0**		**0.1**		**0.2**	

Table 18.17 Normalized organ doses—Hip joint ap—3–7 years

Beam projection:			ap									**Table cells:**
Focus-to-detector distance (cm):			115									(+) = normal font = optimal field setting
Tube voltage (kVp):			60									**(−) = bold font = suboptimal field setting**
Total filtrations:			See last table row									Empty table cells correspond to values <0.01
Organ or tissue	**Organ dose D_T: Entrance dose K_E**								**(mSv ∗ mGy^{-1})**			
	(+)	(−)	(+)	(−)	(+)	(−)	(+)	(−)	(+)	(−)	(+)	(−)
Skull												
Brain												
Oral mucosa												
Salivary glands												
Cervical spine												
Thyroid												
Thymus												
Thoracic spine												
Claviculae												
Scapulae												
Ribs												
Oesophagus												
Trachea												
Lung												
Mamma												
Heart												
Stomach				**0.01**		**0.01**		**0.02**		**0.02**		**0.02**
Spleen												
Liver												
Gall bladder				**0.01**		**0.02**		**0.02**		**0.02**		**0.02**
Pancreas												
Small intestine		**0.11**		**0.14**	0.01	**0.16**	0.01	**0.18**	0.02	**0.22**	0.01	**0.20**
Large intestine		**0.02**		**0.03**		**0.04**		**0.04**		**0.06**		**0.05**
Sigma/rectum	0.09	**0.22**	0.12	**0.29**	0.14	**0.34**	0.16	**0.38**	0.19	**0.45**	0.18	**0.42**
Lumbar spine		**0.02**		**0.03**		**0.04**		**0.05**		**0.06**		**0.05**
Pelvis	0.09	**0.17**	0.11	**0.23**	0.14	**0.27**	0.15	**0.31**	0.19	**0.38**	0.17	**0.35**
Adrenals												
Kidneys												
Urinary bladder	0.07	**0.27**	0.08	**0.33**	0.10	**0.38**	0.11	**0.41**	0.13	**0.48**	0.12	**0.46**
Uterus	0.02	**0.17**	0.03	**0.22**	0.03	**0.26**	0.04	**0.30**	0.05	**0.36**	0.04	**0.33**
Ovaries	0.03	**0.11**	0.04	**0.14**	0.05	**0.17**	0.06	**0.19**	0.07	**0.24**	0.06	**0.22**
Prostate	0.03	**0.20**	0.04	**0.26**	0.05	**0.30**	0.06	**0.34**	0.07	**0.41**	0.07	**0.38**
Testes	0.02	**0.56**	0.03	**0.63**	0.04	**0.67**	0.04	**0.70**	0.05	**0.74**	0.05	**0.74**
Upper arm												
Elbow												
Forearm		**0.29**		**0.36**		**0.42**		**0.46**		**0.53**		**0.52**
Upper leg	0.08	**0.26**	0.11	**0.33**	0.13	**0.40**	0.15	**0.45**	0.18	**0.55**	0.17	**0.50**
Knee		**0.03**		**0.04**		**0.05**		**0.05**		**0.07**		**0.06**
Lower leg												
Lymph nodes		**0.04**		**0.05**		**0.05**		**0.06**		**0.07**		**0.07**
Skin	0.02	**0.08**	0.02	**0.08**	0.02	**0.09**	0.02	**0.09**	0.02	**0.10**	0.02	**0.09**
Muscle	0.02	**0.07**	0.03	**0.08**	0.03	**0.09**	0.03	**0.10**	0.04	**0.11**	0.04	**0.10**
Skeleton	0.01	**0.05**	0.02	**0.07**	0.02	**0.08**	0.03	**0.09**	0.03	**0.11**	0.03	**0.11**
Red bone marrow		**0.02**		**0.03**	0.01	**0.03**	0.01	**0.03**	0.02	**0.04**	0.01	**0.04**
Whole body dose	0.02	**0.06**	0.02	**0.07**	0.03	**0.08**	0.03	**0.08**	0.03	**0.10**	0.03	**0.09**
ED (ICRP 60)	0.02	**0.10**	0.02	**0.12**	0.02	**0.13**	0.03	**0.14**	0.03	**0.16**	0.03	**0.16**
ED (ICRP 103)	0.01	**0.06**	0.02	**0.07**	0.02	**0.08**	0.02	**0.09**	0.02	**0.11**	0.02	**0.10**
Total filtration (mm) Al	**1.5**		**2.5**		**3.5**		**4.5**		**3.5**		**3.5**	
Cu	**0**		**0**		**0**		**0**		**0.1**		**0.2**	

Table 18.18 Normalized organ doses—Hip joint ap—3–7 years

Beam projection:	ap											**Table cells:**
Focus-to-detector distance (cm):	115											(+) = normal font = optimal field setting
Tube voltage (kVp):	70											**(−) = bold font = suboptimal field setting**
Total filtrations:	See last table row											Empty table cells correspond to values <0.01
Organ or tissue	**Organ dose D_T: Entrance dose K_E**							**(mSv * mGy^{-1})**				
	(+)	(−)	(+)	(−)	(+)	(−)	(+)	(−)	(+)	(−)	(+)	(−)
Skull												
Brain												
Oral mucosa												
Salivary glands												
Cervical spine												
Thyroid												
Thymus												
Thoracic spine												
Claviculae												
Scapulae												
Ribs												
Oesophagus												
Trachea												
Lung												
Mamma												
Heart												
Stomach		**0.01**		**0.02**		**0.02**		**0.02**		**0.03**		**0.03**
Spleen										**0.01**		**0.02**
Liver												
Gall bladder		**0.01**		**0.02**		**0.02**		**0.02**		**0.03**		**0.04**
Pancreas										**0.01**		**0.01**
Small intestine		**0.13**	0.01	**0.16**	0.01	**0.19**	0.02	**0.21**	0.02	**0.25**	0.03	**0.29**
Large intestine		**0.03**		**0.04**		**0.05**		**0.06**		**0.07**	0.01	**0.08**
Sigma/rectum	0.12	**0.27**	0.15	**0.34**	0.17	**0.40**	0.19	**0.44**	0.23	**0.53**	0.26	**0.60**
Lumbar spine		**0.03**		**0.05**		**0.05**		**0.06**		**0.08**	0.01	**0.10**
Pelvis	0.11	**0.22**	0.14	**0.28**	0.17	**0.33**	0.18	**0.38**	0.22	**0.46**	0.26	**0.54**
Adrenals												
Kidneys								**0.01**		**0.01**		**0.02**
Urinary bladder	0.08	**0.31**	0.10	**0.38**	0.11	**0.43**	0.12	**0.47**	0.14	**0.54**	0.16	**0.60**
Uterus	0.03	**0.21**	0.03	**0.27**	0.04	**0.31**	0.05	**0.35**	0.06	**0.42**	0.07	**0.48**
Ovaries	0.04	**0.14**	0.05	**0.18**	0.06	**0.21**	0.07	**0.23**	0.08	**0.28**	0.10	**0.33**
Prostate	0.04	**0.25**	0.05	**0.31**	0.06	**0.36**	0.07	**0.39**	0.09	**0.47**	0.10	**0.53**
Testes	0.03	**0.60**	0.04	**0.66**	0.04	**0.70**	0.05	**0.73**	0.06	**0.77**	0.07	**0.81**
Upper arm												
Elbow												
Forearm		**0.34**		**0.41**		**0.46**		**0.50**		**0.57**		**0.63**
Upper leg	0.11	**0.32**	0.14	**0.41**	0.16	**0.48**	0.18	**0.53**	0.22	**0.64**	0.25	**0.75**
Knee		**0.04**		**0.05**		**0.06**		**0.07**		**0.08**		**0.10**
Lower leg												
Lymph nodes		**0.04**		**0.05**		**0.06**		**0.07**		**0.08**		**0.09**
Skin	0.02	**0.08**	0.02	**0.09**	0.02	**0.09**	0.02	**0.09**	0.03	**0.10**	0.03	**0.10**
Muscle	0.03	**0.08**	0.03	**0.09**	0.03	**0.10**	0.04	**0.11**	0.04	**0.12**	0.05	**0.13**
Skeleton	0.02	**0.07**	0.02	**0.09**	0.03	**0.10**	0.03	**0.11**	0.04	**0.13**	0.04	**0.15**
Red bone marrow		**0.03**	0.01	**0.03**	0.02	**0.04**	0.02	**0.04**	0.02	**0.05**	0.02	**0.06**
Whole body dose	0.02	**0.07**	0.03	**0.08**	0.03	**0.09**	0.03	**0.10**	0.04	**0.11**	0.04	**0.12**
ED (ICRP 60)	0.02	**0.11**	0.02	**0.13**	0.03	**0.15**	0.03	**0.16**	0.04	**0.18**	0.04	**0.20**
ED (ICRP 103)	0.01	**0.07**	0.02	**0.08**	0.02	**0.10**	0.02	**0.10**	0.03	**0.12**	0.03	**0.13**
Total filtration (mm) **Al**	**1.5**		**2.5**		**3.5**		**4.5**		**3.5**		**3.5**	
Cu	**0**		**0**		**0**		**0**		**0.1**		**0.2**	

Table 18.19 Normalized organ doses—Hip joint ap—8–12 years

Beam projection:			ap									**Table cells:**
Focus-to-detector distance (cm):			115									(+) = normal font = optimal field setting
Tube voltage (kVp):			60									**(−) = bold font = suboptimal field setting**
Total filtrations:			See last table row									Empty table cells correspond to values <0.01
Organ or tissue	**Organ dose D_T: Entrance dose K_E**								**(mSv ∗ mGy⁻¹)**			
	(+)	(−)	(+)	(−)	(+)	(−)	(+)	(−)	(+)	(−)	(+)	(−)
Skull												
Brain												
Oral mucosa												
Salivary glands												
Cervical spine												
Thyroid												
Thymus												
Thoracic spine												
Claviculae												
Scapulae												
Ribs												
Oesophagus												
Trachea												
Lung												
Mamma												
Heart												
Stomach								**0.01**		**0.01**		**0.01**
Spleen												
Liver												
Gall bladder								**0.01**		**0.01**		**0.01**
Pancreas												
Small intestine		**0.09**		**0.12**		**0.14**	0.01	**0.16**	0.01	**0.19**	0.01	**0.18**
Large intestine		**0.02**		**0.03**		**0.03**		**0.04**		**0.05**		**0.04**
Sigma/rectum	0.09	**0.19**	0.11	**0.25**	0.13	**0.30**	0.15	**0.34**	0.19	**0.41**	0.17	**0.38**
Lumbar spine		**0.02**		**0.02**		**0.03**		**0.03**		**0.04**		**0.03**
Pelvis	0.07	**0.14**	0.10	**0.19**	0.12	**0.22**	0.13	**0.26**	0.16	**0.32**	0.15	**0.28**
Adrenals												
Kidneys												
Urinary bladder	0.11	**0.25**	0.14	**0.31**	0.16	**0.36**	0.17	**0.39**	0.21	**0.46**	0.20	**0.44**
Uterus	0.02	**0.15**	0.03	**0.20**	0.04	**0.24**	0.04	**0.27**	0.05	**0.33**	0.05	**0.30**
Ovaries	0.02	**0.09**	0.03	**0.12**	0.04	**0.15**	0.04	**0.17**	0.05	**0.21**	0.05	**0.18**
Prostate	0.05	**0.20**	0.06	**0.25**	0.07	**0.30**	0.08	**0.33**	0.10	**0.40**	0.09	**0.37**
Testes	0.04	**0.55**	0.05	**0.62**	0.06	**0.66**	0.06	**0.69**	0.08	**0.74**	0.07	**0.73**
Upper arm												
Elbow												
Forearm		**0.26**		**0.32**		**0.37**		**0.41**		**0.47**		**0.46**
Upper leg	0.06	**0.18**	0.08	**0.24**	0.10	**0.28**	0.11	**0.32**	0.14	**0.40**	0.12	**0.36**
Knee						**0.01**		**0.01**		**0.02**		**0.01**
Lower leg												
Lymph nodes		**0.03**		**0.04**		**0.05**		**0.05**		**0.06**		**0.06**
Skin	0.02	**0.08**	0.02	**0.08**	0.03	**0.08**	0.03	**0.09**	0.03	**0.09**	0.03	**0.09**
Muscle	0.02	**0.06**	0.03	**0.07**	0.03	**0.08**	0.03	**0.09**	0.04	**0.10**	0.04	**0.09**
Skeleton	0.01	**0.04**	0.02	**0.06**	0.02	**0.07**	0.02	**0.08**	0.03	**0.09**	0.02	**0.08**
Red bone marrow		**0.02**	0.01	**0.02**	0.01	**0.03**	0.01	**0.03**	0.02	**0.04**	0.02	**0.03**
Whole body dose	0.02	**0.05**	0.02	**0.06**	0.03	**0.07**	0.03	**0.08**	0.03	**0.09**	0.03	**0.08**
ED (ICRP 60)	0.02	**0.09**	0.02	**0.11**	0.03	**0.12**	0.03	**0.13**	0.04	**0.15**	0.03	**0.15**
ED (ICRP 103)	0.01	**0.06**	0.02	**0.07**	0.02	**0.08**	0.02	**0.08**	0.03	**0.10**	0.03	**0.09**
Total filtration (mm) **Al**	**1.5**		**2.5**		**3.5**		**4.5**		**3.5**		**3.5**	
Cu	**0**		**0**		**0**		**0**		**0.1**		**0.2**	

Table 18.20 Normalized organ doses—Hip joint ap—8–12 years

Beam projection:			ap									**Table cells:**
Focus-to-detector distance (cm):			115									(+) = normal font = optimal field setting
Tube voltage (kVp):			70									**(−) = bold font = suboptimal field setting**
Total filtrations:			See last table row									Empty table cells correspond to values <0.01
Organ or tissue	**Organ dose D_T: Entrance dose K_E**							**(mSv * mGy^{-1})**				
	(+)	(−)	(+)	(−)	(+)	(−)	(+)	(−)	(+)	(−)	(+)	(−)
Skull												
Brain												
Oral mucosa												
Salivary glands												
Cervical spine												
Thyroid												
Thymus												
Thoracic spine												
Claviculae												
Scapulae												
Ribs												
Oesophagus												
Trachea												
Lung												
Mamma												
Heart												
Stomach				**0.01**		**0.01**		**0.01**		**0.02**		**0.02**
Spleen												**0.01**
Liver												
Gall bladder				**0.01**		**0.01**		**0.02**		**0.02**		**0.02**
Pancreas												
Small intestine		**0.11**	0.01	**0.14**	0.01	**0.17**	0.01	**0.19**	0.02	**0.23**	0.02	**0.26**
Large intestine		**0.03**		**0.03**		**0.04**		**0.05**		**0.06**		**0.07**
Sigma/rectum	0.11	**0.24**	0.14	**0.31**	0.16	**0.36**	0.18	**0.40**	0.22	**0.48**	0.26	**0.56**
Lumbar spine		**0.02**		**0.03**		**0.04**		**0.05**		**0.06**		**0.07**
Pelvis	0.09	**0.18**	0.12	**0.24**	0.14	**0.28**	0.16	**0.32**	0.20	**0.40**	0.23	**0.47**
Adrenals												
Kidneys												**0.01**
Urinary bladder	0.13	**0.29**	0.16	**0.36**	0.18	**0.41**	0.20	**0.45**	0.24	**0.52**	0.27	**0.59**
Uterus	0.03	**0.19**	0.04	**0.25**	0.05	**0.29**	0.05	**0.32**	0.07	**0.39**	0.08	**0.45**
Ovaries	0.03	**0.12**	0.04	**0.15**	0.05	**0.18**	0.05	**0.21**	0.07	**0.25**	0.08	**0.30**
Prostate	0.06	**0.24**	0.08	**0.30**	0.09	**0.35**	0.10	**0.39**	0.12	**0.47**	0.14	**0.53**
Testes	0.05	**0.59**	0.06	**0.65**	0.07	**0.69**	0.07	**0.72**	0.09	**0.77**	0.10	**0.82**
Upper arm												
Elbow												
Forearm		**0.30**		**0.36**		**0.41**		**0.45**		**0.52**		**0.58**
Upper leg	0.08	**0.23**	0.10	**0.30**	0.12	**0.35**	0.14	**0.40**	0.17	**0.49**	0.20	**0.57**
Knee		**0.01**		**0.01**		**0.02**		**0.02**		**0.02**		**0.03**
Lower leg												
Lymph nodes		**0.04**		**0.05**		**0.06**		**0.06**		**0.07**		**0.08**
Skin	0.02	**0.08**	0.03	**0.08**	0.03	**0.09**	0.03	**0.09**	0.03	**0.09**	0.03	**0.10**
Muscle	0.03	**0.07**	0.03	**0.08**	0.04	**0.09**	0.04	**0.10**	0.04	**0.11**	0.05	**0.12**
Skeleton	0.02	**0.05**	0.02	**0.07**	0.02	**0.08**	0.03	**0.09**	0.03	**0.11**	0.04	**0.13**
Red bone marrow	0.01	**0.02**	0.01	**0.03**	0.02	**0.04**	0.02	**0.04**	0.02	**0.05**	0.03	**0.06**
Whole body dose	0.02	**0.06**	0.03	**0.07**	0.03	**0.08**	0.03	**0.09**	0.04	**0.10**	0.04	**0.11**
ED (ICRP 60)	0.02	**0.11**	0.03	**0.13**	0.03	**0.14**	0.04	**0.15**	0.04	**0.17**	0.05	**0.19**
ED (ICRP 103)	0.02	**0.07**	0.02	**0.08**	0.02	**0.09**	0.03	**0.10**	0.03	**0.11**	0.04	**0.13**
Total filtration (mm) **Al**	**1.5**		**2.5**		**3.5**		**4.5**		**3.5**		**3.5**	
Cu	**0**		**0**		**0**		**0**		**0.1**		**0.2**	

Table 18.21 Normalized organ doses—Hip joint ap—8–12 years

Beam projection:	ap		**Table cells:**
Focus-to-detector distance (cm):	115		(+) = normal font = optimal field setting
Tube voltage (kVp):	80		**(−) = bold font = suboptimal field setting**
Total filtrations:	See last table row		Empty table cells correspond to values <0.01

Organ or tissue	**Organ dose D_T: Entrance dose K_E**							**(mSv * mGy⁻¹)**				
	(+)	(−)	(+)	(−)	(+)	(−)	(+)	(−)	(+)	(−)	(+)	(−)
Skull												
Brain												
Oral mucosa												
Salivary glands												
Cervical spine												
Thyroid												
Thymus												
Thoracic spine												
Claviculae												
Scapulae												
Ribs												
Oesophagus												
Trachea												
Lung												
Mamma												
Heart												
Stomach		**0.01**		**0.01**		**0.02**		**0.02**		**0.02**		**0.03**
Spleen										**0.01**		**0.01**
Liver												
Gall bladder		**0.01**		**0.01**		**0.02**		**0.02**		**0.02**		**0.03**
Pancreas										**0.01**		**0.01**
Small intestine	0.01	**0.14**	0.01	**0.17**	0.02	**0.20**	0.02	**0.22**	0.02	**0.26**	0.03	**0.29**
Large intestine		**0.03**		**0.04**		**0.05**		**0.06**		**0.07**	0.01	**0.08**
Sigma/rectum	0.13	**0.29**	0.17	**0.36**	0.19	**0.42**	0.21	**0.46**	0.25	**0.55**	0.29	**0.63**
Lumbar spine		**0.03**		**0.04**		**0.05**		**0.06**		**0.08**	0.01	**0.09**
Pelvis	0.12	**0.23**	0.14	**0.29**	0.17	**0.34**	0.19	**0.38**	0.23	**0.45**	0.26	**0.53**
Adrenals												
Kidneys										**0.01**		**0.02**
Urinary bladder	0.15	**0.34**	0.18	**0.41**	0.21	**0.46**	0.22	**0.50**	0.26	**0.57**	0.29	**0.64**
Uterus	0.04	**0.24**	0.05	**0.29**	0.06	**0.34**	0.06	**0.37**	0.08	**0.44**	0.09	**0.51**
Ovaries	0.04	**0.15**	0.05	**0.19**	0.06	**0.22**	0.06	**0.24**	0.08	**0.29**	0.09	**0.34**
Prostate	0.07	**0.29**	0.09	**0.35**	0.10	**0.40**	0.12	**0.44**	0.14	**0.52**	0.16	**0.59**
Testes	0.05	**0.62**	0.07	**0.68**	0.08	**0.72**	0.08	**0.75**	0.10	**0.80**	0.11	**0.85**
Upper arm												
Elbow												
Forearm		**0.33**		**0.40**		**0.45**		**0.49**		**0.55**		**0.60**
Upper leg	0.10	**0.28**	0.12	**0.36**	0.15	**0.41**	0.16	**0.46**	0.20	**0.55**	0.23	**0.64**
Knee		**0.01**		**0.02**		**0.02**		**0.02**		**0.03**		**0.04**
Lower leg												
Lymph nodes		**0.05**		**0.06**		**0.06**		**0.07**		**0.08**	0.01	**0.09**
Skin	0.03	**0.08**	0.03	**0.09**	0.03	**0.09**	0.03	**0.09**	0.03	**0.10**	0.03	**0.10**
Muscle	0.03	**0.08**	0.04	**0.09**	0.04	**0.10**	0.04	**0.11**	0.05	**0.12**	0.05	**0.14**
Skeleton	0.02	**0.07**	0.02	**0.08**	0.03	**0.10**	0.03	**0.11**	0.04	**0.13**	0.04	**0.14**
Red bone marrow	0.01	**0.03**	0.02	**0.04**	0.02	**0.04**	0.02	**0.05**	0.03	**0.06**	0.03	**0.07**
Whole body dose	0.03	**0.07**	0.03	**0.08**	0.03	**0.09**	0.04	**0.10**	0.04	**0.11**	0.05	**0.13**
ED (ICRP 60)	0.03	**0.12**	0.03	**0.14**	0.04	**0.15**	0.04	**0.17**	0.05	**0.19**	0.06	**0.21**
ED (ICRP 103)	0.02	**0.07**	0.03	**0.09**	0.03	**0.10**	0.03	**0.11**	0.04	**0.12**	0.04	**0.14**
Total filtration (mm) Al	**1.5**		**2.5**		**3.5**		**4.5**		**3.5**		**3.5**	
Cu	**0**		**0**		**0**		**0**		**0.1**		**0.2**	

Table 18.22 Normalized organ doses—Hip joint ap—13–17 years

Beam projection:			ap									**Table cells:**
Focus-to-detector distance (cm):			115									(+) = normal font = optimal field setting
Tube voltage (kVp):			60									**(−) = bold font = suboptimal field setting**
Total filtrations:			See last table row									Empty table cells correspond to values <0.01
Organ or tissue	**Organ dose D_T: Entrance dose K_E**							**(mSv * mGy^{-1})**				
	(+)	(−)	(+)	(−)	(+)	(−)	(+)	(−)	(+)	(−)	(+)	(−)
Skull												
Brain												
Oral mucosa												
Salivary glands												
Cervical spine												
Thyroid												
Thymus												
Thoracic spine												
Claviculae												
Scapulae												
Ribs												
Oesophagus												
Trachea												
Lung												
Mamma												
Heart												
Stomach										**0.01**		
Spleen												
Liver												
Gall bladder												
Pancreas												
Small intestine		**0.08**		**0.10**		**0.12**		**0.14**	0.01	**0.17**		**0.15**
Large intestine		**0.02**		**0.02**		**0.03**		**0.03**		**0.04**		**0.03**
Sigma/rectum	0.07	**0.16**	0.09	**0.21**	0.11	**0.25**	0.13	**0.28**	0.16	**0.35**	0.14	**0.31**
Lumbar spine		**0.01**		**0.02**		**0.02**		**0.02**		**0.03**		**0.02**
Pelvis	0.05	**0.10**	0.07	**0.14**	0.09	**0.17**	0.10	**0.19**	0.13	**0.25**	0.11	**0.21**
Adrenals												
Kidneys												
Urinary bladder	0.11	**0.22**	0.14	**0.28**	0.16	**0.32**	0.18	**0.36**	0.22	**0.42**	0.20	**0.40**
Uterus	0.07	**0.34**	0.09	**0.45**	0.11	**0.53**	0.12	**0.60**	0.16	**0.75**	0.13	**0.67**
Ovaries	0.02	**0.07**	0.02	**0.10**	0.03	**0.12**	0.03	**0.13**	0.04	**0.17**	0.04	**0.15**
Prostate	0.05	**0.18**	0.06	**0.23**	0.08	**0.28**	0.09	**0.31**	0.11	**0.38**	0.10	**0.35**
Testes	0.05	**0.47**	0.06	**0.55**	0.07	**0.60**	0.08	**0.63**	0.10	**0.70**	0.09	**0.68**
Upper arm												
Elbow												
Forearm		**0.21**		**0.26**		**0.30**		**0.34**		**0.40**		**0.38**
Upper leg	0.04	**0.13**	0.05	**0.18**	0.07	**0.22**	0.08	**0.25**	0.10	**0.32**	0.08	**0.27**
Knee		**0.02**		**0.03**		**0.03**		**0.04**		**0.05**		**0.04**
Lower leg												
Lymph nodes		**0.03**		**0.03**		**0.04**		**0.05**		**0.06**		**0.05**
Skin	0.02	**0.08**	0.02	**0.09**	0.03	**0.09**	0.03	**0.10**	0.03	**0.10**	0.03	**0.10**
Muscle	0.02	**0.06**	0.03	**0.07**	0.03	**0.08**	0.03	**0.09**	0.04	**0.10**	0.03	**0.09**
Skeleton		**0.04**	0.01	**0.05**	0.01	**0.06**	0.02	**0.07**	0.02	**0.09**	0.02	**0.08**
Red bone marrow		**0.01**		**0.02**	0.01	**0.02**	0.01	**0.03**	0.02	**0.03**	0.01	**0.03**
Whole body dose	0.02	**0.05**	0.02	**0.06**	0.02	**0.07**	0.03	**0.08**	0.03	**0.09**	0.03	**0.08**
ED (ICRP 60)	0.02	**0.08**	0.02	**0.10**	0.03	**0.11**	0.03	**0.12**	0.04	**0.14**	0.03	**0.13**
ED (ICRP 103)	0.01	**0.05**	0.02	**0.06**	0.02	**0.07**	0.02	**0.08**	0.03	**0.09**	0.03	**0.09**
Total filtration (mm) Al	**1.5**		**2.5**		**3.5**		**4.5**		**3.5**		**3.5**	
Cu	**0**		**0**		**0**		**0**		**0.1**		**0.2**	

Table 18.23 Normalized organ doses—Hip joint ap—13–17 years

Beam projection:	ap											Table cells:
Focus-to-detector distance (cm):	115											(+) = normal font = optimal field setting
Tube voltage (kVp):	70											**(−) = bold font = suboptimal field setting**
Total filtrations:	See last table row											Empty table cells correspond to values <0.01
Organ or tissue	**Organ dose D_T: Entrance dose K_E**							**(mSv * mGy⁻¹)**				
	(+)	(−)	(+)	(−)	(+)	(−)	(+)	(−)	(+)	(−)	(+)	(−)
Skull												
Brain												
Oral mucosa												
Salivary glands												
Cervical spine												
Thyroid												
Thymus												
Thoracic spine												
Claviculae												
Scapulae												
Ribs												
Oesophagus												
Trachea												
Lung												
Mamma												
Heart												
Stomach						**0.01**		**0.01**		**0.02**		**0.02**
Spleen												
Liver												
Gall bladder										**0.01**		**0.01**
Pancreas												
Small intestine		**0.10**		**0.13**		**0.15**	0.01	**0.17**	0.01	**0.20**	0.02	**0.24**
Large intestine		**0.02**		**0.03**		**0.04**		**0.04**		**0.05**		**0.06**
Sigma/rectum	0.09	**0.20**	0.12	**0.26**	0.14	**0.31**	0.16	**0.35**	0.20	**0.43**	0.23	**0.50**
Lumbar spine		**0.02**		**0.02**		**0.03**		**0.03**		**0.04**		**0.05**
Pelvis	0.07	**0.14**	0.09	**0.18**	0.11	**0.22**	0.13	**0.25**	0.16	**0.31**	0.19	**0.38**
Adrenals												
Kidneys												
Urinary bladder	0.13	**0.26**	0.17	**0.33**	0.19	**0.38**	0.21	**0.41**	0.25	**0.49**	0.29	**0.56**
Uterus	0.09	**0.44**	0.12	**0.56**	0.14	**0.66**	0.16	**0.74**	0.20	**0.91**	0.24	**1.07**
Ovaries	0.03	**0.10**	0.03	**0.13**	0.04	**0.15**	0.05	**0.17**	0.06	**0.21**	0.07	**0.25**
Prostate	0.06	**0.22**	0.08	**0.29**	0.10	**0.34**	0.11	**0.38**	0.14	**0.45**	0.16	**0.53**
Testes	0.06	**0.52**	0.07	**0.59**	0.08	**0.64**	0.09	**0.68**	0.11	**0.74**	0.13	**0.80**
Upper arm												
Elbow												
Forearm		**0.25**		**0.30**		**0.35**		**0.38**		**0.45**		**0.50**
Upper leg	0.06	**0.18**	0.07	**0.24**	0.09	**0.28**	0.10	**0.32**	0.12	**0.40**	0.15	**0.48**
Knee		**0.03**		**0.04**		**0.04**		**0.05**		**0.06**		**0.08**
Lower leg												
Lymph nodes		**0.03**		**0.04**		**0.05**		**0.05**		**0.07**		**0.08**
Skin	0.02	**0.09**	0.03	**0.09**	0.03	**0.10**	0.03	**0.10**	0.03	**0.11**	0.03	**0.11**
Muscle	0.02	**0.07**	0.03	**0.08**	0.03	**0.09**	0.04	**0.10**	0.04	**0.11**	0.05	**0.13**
Skeleton	0.01	**0.05**	0.02	**0.07**	0.02	**0.08**	0.02	**0.09**	0.03	**0.11**	0.03	**0.13**
Red bone marrow		**0.02**	0.01	**0.03**	0.01	**0.03**	0.02	**0.03**	0.02	**0.04**	0.02	**0.05**
Whole body dose	0.02	**0.06**	0.02	**0.07**	0.03	**0.08**	0.03	**0.09**	0.04	**0.10**	0.04	**0.12**
ED (ICRP 60)	0.02	**0.09**	0.03	**0.11**	0.03	**0.13**	0.04	**0.14**	0.04	**0.16**	0.05	**0.18**
ED (ICRP 103)	0.02	**0.06**	0.02	**0.07**	0.02	**0.08**	0.03	**0.09**	0.03	**0.11**	0.04	**0.12**
Total filtration (mm) **Al**	**1.5**		**2.5**		**3.5**		**4.5**		**3.5**		**3.5**	
Cu	**0**		**0**		**0**		**0**		**0.1**		**0.2**	

Table 18.24 Normalized organ doses—Hip joint ap—13–17 years

Beam projection:		ap										**Table cells:**
Focus-to-detector distance (cm):		115										(+) = normal font = optimal field setting
Tube voltage (kVp):		80										**(−) = bold font = suboptimal field setting**
Total filtrations:		See last table row										Empty table cells correspond to values <0.01
Organ or tissue	**Organ dose D_T: Entrance dose K_E**								**(mSv ∗ mGy^{-1})**			
	(+)	(−)	(+)	(−)	(+)	(−)	(+)	(−)	(+)	(−)	(+)	(−)
Skull												
Brain												
Oral mucosa												
Salivary glands												
Cervical spine												
Thyroid												
Thymus												
Thoracic spine												
Claviculae												
Scapulae												
Ribs												
Oesophagus												
Trachea												
Lung												
Mamma												
Heart												
Stomach				**0.01**		**0.01**		**0.02**		**0.02**		**0.02**
Spleen												
Liver												
Gall bladder								**0.01**		**0.01**		**0.02**
Pancreas												
Small intestine		**0.12**	0.01	**0.15**	0.01	**0.18**	0.01	**0.20**	0.02	**0.24**	0.02	**0.27**
Large intestine		**0.03**		**0.04**		**0.04**		**0.05**		**0.06**		**0.08**
Sigma/rectum	0.12	**0.25**	0.15	**0.32**	0.17	**0.37**	0.19	**0.41**	0.23	**0.49**	0.27	**0.57**
Lumbar spine		**0.02**		**0.03**		**0.04**		**0.04**		**0.06**		**0.07**
Pelvis	0.09	**0.18**	0.12	**0.23**	0.14	**0.27**	0.15	**0.30**	0.18	**0.37**	0.22	**0.44**
Adrenals												
Kidneys												**0.01**
Urinary bladder	0.16	**0.31**	0.19	**0.38**	0.22	**0.43**	0.24	**0.47**	0.28	**0.54**	0.32	**0.61**
Uterus	0.12	**0.54**	0.15	**0.68**	0.17	**0.79**	0.19	**0.88**	0.24	**1.06**	0.28	**1.23**
Ovaries	0.03	**0.12**	0.04	**0.16**	0.05	**0.18**	0.06	**0.21**	0.07	**0.25**	0.08	**0.30**
Prostate	0.08	**0.27**	0.10	**0.34**	0.12	**0.39**	0.13	**0.43**	0.16	**0.52**	0.19	**0.59**
Testes	0.07	**0.56**	0.08	**0.63**	0.10	**0.68**	0.11	**0.72**	0.12	**0.78**	0.14	**0.83**
Upper arm												
Elbow												
Forearm		**0.28**		**0.34**		**0.38**		**0.42**		**0.48**		**0.53**
Upper leg	0.07	**0.23**	0.09	**0.29**	0.11	**0.34**	0.12	**0.38**	0.15	**0.47**	0.17	**0.55**
Knee		**0.04**		**0.05**		**0.05**		**0.06**		**0.08**		**0.09**
Lower leg												
Lymph nodes		**0.04**		**0.05**		**0.06**		**0.06**		**0.08**		**0.09**
Skin	0.02	**0.09**	0.03	**0.10**	0.03	**0.10**	0.03	**0.10**	0.03	**0.11**	0.03	**0.12**
Muscle	0.03	**0.08**	0.03	**0.09**	0.04	**0.10**	0.04	**0.11**	0.05	**0.12**	0.05	**0.14**
Skeleton	0.02	**0.06**	0.02	**0.08**	0.02	**0.09**	0.03	**0.10**	0.03	**0.12**	0.04	**0.14**
Red bone marrow	0.01	**0.03**	0.02	**0.03**	0.02	**0.04**	0.02	**0.04**	0.02	**0.05**	0.03	**0.07**
Whole body dose	0.02	**0.07**	0.03	**0.08**	0.03	**0.09**	0.03	**0.10**	0.04	**0.12**	0.04	**0.13**
ED (ICRP 60)	0.03	**0.11**	0.03	**0.13**	0.04	**0.14**	0.04	**0.16**	0.05	**0.18**	0.06	**0.20**
ED (ICRP 103)	0.02	**0.07**	0.03	**0.08**	0.03	**0.10**	0.03	**0.10**	0.04	**0.12**	0.05	**0.14**
Total filtration (mm) **Al**	**1.5**		**2.5**		**3.5**		**4.5**		**3.5**		**3.5**	
Cu	**0**		**0**		**0**		**0**		**0.1**		**0.2**	

Table 18.25 Normalized organ doses—Hip joint ap—>17 years

Beam projection:	ap											**Table cells:**
Focus-to-detector distance (cm):	115											(+) = normal font = optimal field setting
Tube voltage (kVp):	60											**(−) = bold font = suboptimal field setting**
Total filtrations:	See last table row											Empty table cells correspond to values <0.01
Organ or tissue	**Organ dose D_T: Entrance dose K_E (mSv ∗ mGy⁻¹)**											
	(+)	(−)	(+)	(−)	(+)	(−)	(+)	(−)	(+)	(−)	(+)	(−)
Skull												
Brain												
Oral mucosa												
Salivary glands												
Cervical spine												
Thyroid												
Thymus												
Thoracic spine												
Claviculae												
Scapulae												
Ribs												
Oesophagus												
Trachea												
Lung												
Mamma												
Heart												
Stomach												
Spleen												
Liver												
Gall bladder												
Pancreas												
Small intestine		**0.07**		**0.09**		**0.11**		**0.12**		**0.15**		**0.13**
Large intestine		**0.01**		**0.02**		**0.02**		**0.02**		**0.03**		**0.02**
Sigma/rectum	0.07	**0.15**	0.09	**0.20**	0.11	**0.24**	0.12	**0.27**	0.16	**0.34**	0.14	**0.30**
Lumbar spine				**0.01**		**0.01**		**0.02**		**0.02**		**0.02**
Pelvis	0.05	**0.10**	0.07	**0.13**	0.08	**0.16**	0.09	**0.18**	0.12	**0.23**	0.10	**0.20**
Adrenals												
Kidneys												
Urinary bladder	0.10	**0.22**	0.13	**0.27**	0.15	**0.32**	0.17	**0.35**	0.20	**0.42**	0.19	**0.39**
Uterus	0.02	**0.13**	0.03	**0.17**	0.03	**0.20**	0.04	**0.23**	0.05	**0.29**	0.04	**0.25**
Ovaries	0.02	**0.07**	0.02	**0.09**	0.03	**0.11**	0.03	**0.13**	0.04	**0.16**	0.04	**0.14**
Prostate	0.04	**0.15**	0.05	**0.20**	0.06	**0.24**	0.07	**0.27**	0.09	**0.33**	0.08	**0.30**
Testes	0.04	**0.44**	0.05	**0.51**	0.06	**0.56**	0.07	**0.60**	0.08	**0.67**	0.08	**0.65**
Upper arm												
Elbow												
Forearm		**0.20**		**0.26**		**0.30**		**0.33**		**0.39**		**0.37**
Upper leg	0.03	**0.10**	0.05	**0.14**	0.05	**0.17**	0.06	**0.20**	0.08	**0.25**	0.07	**0.21**
Knee		**0.02**		**0.03**		**0.04**		**0.04**		**0.06**		**0.05**
Lower leg												
Lymph nodes		**0.02**		**0.03**		**0.04**		**0.04**		**0.05**		**0.05**
Skin	0.02	**0.08**	0.02	**0.09**	0.02	**0.09**	0.02	**0.10**	0.03	**0.10**	0.03	**0.10**
Muscle	0.02	**0.06**	0.02	**0.07**	0.03	**0.08**	0.03	**0.09**	0.03	**0.10**	0.03	**0.09**
Skeleton		**0.03**	0.01	**0.05**	0.01	**0.06**	0.02	**0.06**	0.02	**0.08**	0.02	**0.07**
Red bone marrow		**0.01**		**0.02**		**0.02**	0.01	**0.02**	0.01	**0.03**	0.01	**0.02**
Whole body dose	0.02	**0.05**	0.02	**0.06**	0.02	**0.07**	0.03	**0.08**	0.03	**0.09**	0.03	**0.08**
ED (ICRP 60)	0.02	**0.07**	0.02	**0.09**	0.02	**0.10**	0.03	**0.11**	0.03	**0.13**	0.03	**0.12**
ED (ICRP 103)	0.01	**0.04**	0.02	**0.06**	0.02	**0.06**	0.02	**0.07**	0.02	**0.08**	0.02	**0.08**
Total filtration (mm) **Al**	**1.5**		**2.5**		**3.5**		**4.5**		**3.5**		**3.5**	
Cu	**0**		**0**		**0**		**0**		**0.1**		**0.2**	

Table 18.26 Normalized organ doses—Hip joint ap—>17 years

Beam projection:			ap									**Table cells:**
Focus-to-detector distance (cm):			115									(+) = normal font = optimal field setting
Tube voltage (kVp):			70									**(−) = bold font = suboptimal field setting**
Total filtrations:			See last table row									Empty table cells correspond to values <0.01
Organ or tissue	**Organ dose D_T: Entrance dose K_E**								**(mSv * mGy⁻¹)**			
	(+)	(−)	(+)	(−)	(+)	(−)	(+)	(−)	(+)	(−)	(+)	(−)
Skull												
Brain												
Oral mucosa												
Salivary glands												
Cervical spine												
Thyroid												
Thymus												
Thoracic spine												
Claviculae												
Scapulae												
Ribs												
Oesophagus												
Trachea												
Lung												
Mamma												
Heart												
Stomach										**0.01**		**0.01**
Spleen												
Liver												
Gall bladder												**0.01**
Pancreas												
Small intestine		**0.09**		**0.11**		**0.13**		**0.15**	0.01	**0.18**	0.02	**0.22**
Large intestine		**0.02**		**0.02**		**0.03**		**0.03**		**0.04**		**0.05**
Sigma/rectum	0.09	**0.20**	0.12	**0.25**	0.14	**0.30**	0.16	**0.34**	0.19	**0.41**	0.23	**0.49**
Lumbar spine		**0.01**		**0.02**		**0.02**		**0.03**		**0.03**		**0.04**
Pelvis	0.07	**0.13**	0.09	**0.17**	0.11	**0.21**	0.12	**0.24**	0.15	**0.30**	0.18	**0.36**
Adrenals												
Kidneys												
Urinary bladder	0.12	**0.26**	0.15	**0.33**	0.18	**0.37**	0.20	**0.41**	0.23	**0.49**	0.27	**0.56**
Uterus	0.03	**0.17**	0.04	**0.22**	0.05	**0.25**	0.05	**0.29**	0.07	**0.35**	0.08	**0.42**
Ovaries	0.02	**0.09**	0.03	**0.12**	0.04	**0.15**	0.04	**0.17**	0.06	**0.21**	0.07	**0.25**
Prostate	0.05	**0.20**	0.07	**0.25**	0.08	**0.29**	0.09	**0.33**	0.11	**0.40**	0.14	**0.47**
Testes	0.05	**0.48**	0.06	**0.56**	0.07	**0.61**	0.08	**0.65**	0.10	**0.72**	0.11	**0.78**
Upper arm												
Elbow												
Forearm		**0.24**		**0.30**		**0.34**		**0.37**		**0.44**		**0.49**
Upper leg	0.05	**0.14**	0.06	**0.19**	0.07	**0.23**	0.08	**0.26**	0.11	**0.32**	0.13	**0.39**
Knee		**0.03**		**0.04**		**0.05**		**0.06**		**0.07**		**0.09**
Lower leg												
Lymph nodes		**0.03**		**0.04**		**0.04**		**0.05**		**0.06**		**0.07**
Skin	0.02	**0.09**	0.02	**0.09**	0.03	**0.10**	0.03	**0.10**	0.03	**0.11**	0.03	**0.11**
Muscle	0.02	**0.07**	0.03	**0.08**	0.03	**0.09**	0.03	**0.10**	0.04	**0.11**	0.05	**0.13**
Skeleton	0.01	**0.05**	0.01	**0.06**	0.02	**0.07**	0.02	**0.08**	0.03	**0.10**	0.03	**0.12**
Red bone marrow		**0.02**	0.01	**0.02**	0.01	**0.03**	0.01	**0.03**	0.02	**0.04**	0.02	**0.05**
Whole body dose	0.02	**0.06**	0.02	**0.07**	0.03	**0.08**	0.03	**0.09**	0.03	**0.10**	0.04	**0.12**
ED (ICRP 60)	0.02	**0.09**	0.03	**0.11**	0.03	**0.12**	0.03	**0.13**	0.04	**0.15**	0.05	**0.17**
ED (ICRP 103)	0.01	**0.05**	0.02	**0.07**	0.02	**0.08**	0.02	**0.08**	0.03	**0.10**	0.04	**0.11**
Total filtration (mm) Al	**1.5**		**2.5**		**3.5**		**4.5**		**3.5**		**3.5**	
Cu	**0**		**0**		**0**		**0**		**0.1**		**0.2**	

Table 18.27 Normalized organ doses—Hip joint ap—>17 years

Beam projection:	ap											**Table cells:**
Focus-to-detector distance (cm):	115											(+) = normal font = optimal field setting
Tube voltage (kVp):	80											**(−) = bold font = suboptimal field setting**
Total filtrations:	See last table row											Empty table cells correspond to values <0.01
Organ or tissue	**Organ dose D_T: Entrance dose K_E (mSv * mGy⁻¹)**											
	(+)	(−)	(+)	(−)	(+)	(−)	(+)	(−)	(+)	(−)	(+)	(−)
Skull												
Brain												
Oral mucosa												
Salivary glands												
Cervical spine												
Thyroid												
Thymus												
Thoracic spine												
Claviculae												
Scapulae												
Ribs												
Oesophagus												
Trachea												
Lung												
Mamma												
Heart												
Stomach								**0.01**		**0.01**		**0.02**
Spleen												
Liver												
Gall bladder										**0.01**		**0.01**
Pancreas												
Small intestine		**0.11**		**0.14**	0.01	**0.16**	0.01	**0.18**	0.02	**0.21**	0.02	**0.25**
Large intestine		**0.02**		**0.03**		**0.04**		**0.04**		**0.05**		**0.06**
Sigma/rectum	0.11	**0.24**	0.14	**0.31**	0.17	**0.36**	0.19	**0.40**	0.22	**0.48**	0.26	**0.56**
Lumbar spine		**0.02**		**0.03**		**0.03**		**0.03**		**0.04**		**0.06**
Pelvis	0.09	**0.17**	0.11	**0.22**	0.13	**0.26**	0.14	**0.29**	0.18	**0.35**	0.21	**0.42**
Adrenals												
Kidneys												
Urinary bladder	0.14	**0.31**	0.18	**0.38**	0.20	**0.43**	0.22	**0.47**	0.26	**0.54**	0.30	**0.61**
Uterus	0.04	**0.21**	0.05	**0.26**	0.06	**0.30**	0.07	**0.34**	0.08	**0.41**	0.10	**0.48**
Ovaries	0.03	**0.12**	0.04	**0.15**	0.05	**0.18**	0.06	**0.20**	0.07	**0.25**	0.08	**0.29**
Prostate	0.07	**0.24**	0.08	**0.30**	0.10	**0.35**	0.11	**0.38**	0.13	**0.46**	0.16	**0.53**
Testes	0.06	**0.52**	0.07	**0.60**	0.08	**0.65**	0.09	**0.69**	0.11	**0.76**	0.12	**0.82**
Upper arm												
Elbow												
Forearm		**0.27**		**0.33**		**0.38**		**0.41**		**0.47**		**0.52**
Upper leg	0.06	**0.19**	0.08	**0.24**	0.09	**0.28**	0.10	**0.32**	0.13	**0.39**	0.15	**0.46**
Knee		**0.04**		**0.05**		**0.06**		**0.07**		**0.09**		**0.10**
Lower leg												
Lymph nodes		**0.04**		**0.05**		**0.05**		**0.06**		**0.07**		**0.08**
Skin	0.02	**0.09**	0.03	**0.10**	0.03	**0.10**	0.03	**0.11**	0.03	**0.11**	0.03	**0.12**
Muscle	0.03	**0.08**	0.03	**0.09**	0.04	**0.10**	0.04	**0.11**	0.04	**0.13**	0.05	**0.14**
Skeleton	0.01	**0.06**	0.02	**0.07**	0.02	**0.09**	0.02	**0.10**	0.03	**0.12**	0.04	**0.14**
Red bone marrow	0.01	**0.02**	0.01	**0.03**	0.02	**0.03**	0.02	**0.04**	0.02	**0.05**	0.03	**0.06**
Whole body dose	0.02	**0.07**	0.03	**0.08**	0.03	**0.09**	0.03	**0.10**	0.04	**0.12**	0.04	**0.13**
ED (ICRP 60)	0.02	**0.10**	0.03	**0.12**	0.04	**0.14**	0.04	**0.15**	0.05	**0.17**	0.05	**0.19**
ED (ICRP 103)	0.02	**0.06**	0.02	**0.08**	0.03	**0.09**	0.03	**0.09**	0.04	**0.11**	0.04	**0.13**
Total filtration (mm) Al	**1.5**		**2.5**		**3.5**		**4.5**		**3.5**		**3.5**	
Cu	**0**		**0**		**0**		**0**		**0.1**		**0.2**	

Table 18.28 Inconsistencies between the mathematical MIRD phantoms and real human anatomy

Anatomical region	Inconsistencies
Skeleton	No acetabulum, no symphysis pubis have been inserted

For details, see the text

References

1. Bontrager KL. Textbook of radiographic positioning and related anatomy. St. Louis: Mosby Year Book; 1993.
2. Cook JV, Shah K, Pablot S, Kyriou J, Pettett A, Fitzgerald M. Guidelines on best practice in the x-ray imaging of children. A manual for all x-rax departments. London: Queen Mary's Hospital for Children, Surrey, and The Radiological Protection Centre; 1998.
3. Drexler G, Panzer W, Stieve FE. Die Bestimmung von Organdosen in der Röntgendiagnostik. Berlin: Hoffmann Verlag; 1993.
4. Ebel K-D, Willich E. Die Röntgenuntersuchung im Kindesalter – Technik und Indikation: Berlin, Springer; 1968.
5. Hart D, Jones DJ, Wall BF. Estimation of effective dose in diagnostic radiology from entrance surface dose and dose-area-product measurements. National Radiological Protection Board NRPB-R262; 1994.
6. Hart D, Jones DJ, Wall BF. Coefficients for estimating effective doses from pediatric x-ray examinations. National Radiological Protection Board NRPB-R279; 1996.
7. Seidenbusch MC, Schneider K. Conversion coefficients for determining organ doses in paediatric pelvis and hip joint radiography. Pediatr Radiol. 2014;44:1110–23.
8. Winter W. Untersuchung über die Patientendosis und die Bildqualität bei häufigen Röntgenaufnahmen im Säuglingsalter bei niedergelassenen Ärzten (Pädiater, Orthopäden, Chirurgen). Dissertation, Universität München; 1994.

Thigh

19

19.1 Introduction

The type of thigh radiography is determined by the clinical indication. Table 19.1 shows indications and Table 19.2 contraindications for thigh radiography. Image criteria as required for radiological findings in thigh radiographs are listed in Table 19.3. In order to fulfil these image criteria, the field settings listed in Tables 19.4 and 19.5 should be achieved. These field settings go along with the patient preparation as listed in Table 19.6. Exposure parameter combinations for various age groups and thigh diameters are listed in Tables 19.7, 19.8, and 19.9. The age groups in Table 19.8 refer to the standard age groups of the mathematical MIRD phantom family which has been used to calculate the normalized organ doses in this book. Table 19.9 shows radiation qualities (i.e. various combinations of tube voltage and total filtration)

Table 19.1 Indications for thigh radiographs

Femur fracture
Osteomyelitis
Myositis
Soft tissue tumours
Bone tumours
Stress fracture
Metabolic disease
Femoral hypoplasia/aplasia

Table 19.2 Contraindications for thigh radiographs

None

Table 19.3 Image criteria for thigh radiographs

Thigh ap/lat
Visualization of the hip joint, the femur with adjacent soft tissue, and the knee joint
Visually sharp reproduction of the cortex and trabecular structures of the acetabulum, the femur with its femoral epiphysis, and the proximal tibial epiphysis consistent with age
Visually sharp reproduction the patella consistent with age
Visualization of the periarticular and the soft tissue planes of the thigh, the hip, and knee joints

Table 19.4 Patient positions and beam projections for thigh radiographs

Patient position	Beam projections
Supine	ap + lat

Table 19.5 Optimal field settings for thigh radiographs

Thigh ap	
Upper field edge	Spina iliaca anterior superior
Field centring	Midpoint of the thigh
Lower field edge	Just below the knee joint
Medial field edge	Somewhat medially to symphysis pubis
Lateral field edge	Skin surface of the upper thigh
Thigh lat	
Upper field edge	Inguinal groove
Field centring	Midpoint of the thigh
Lower field edge	Just at the patella
Anterior field edge	Skin surface of the upper thigh
Posterior field edge	Skin surface of the upper thigh
Adjust the angulation of the tube dependent on the position of the thigh	

M. Seidenbusch et al., *Imaging Practice and Radiation Protection in Pediatric Radiology*,
https://doi.org/10.1007/978-3-030-18504-6_19

Table 19.6 Patient preparation

General	Remove all clothes, external metallic devices, and jewellery from the pelvic region and the upper thigh
	Immobilize infants and uncooperative patients in the supine position for the ap projection with extended legs
	Place eventually small X-ray positive markers at palpable tumours or on pain points
Positioning	The legs should be extended
	For the lateral projection of the thigh, the patient is turned by about 30°–45° to ipsilateral side. The patient's back should be stabilized with foam pads or sandbags
Shielding	In males, use testicular capsules if possible. In females, lead masks should be placed at the medial and upper field edges

Table 19.7 Radiographic procedure

X-ray source	Use automatic exposure control (AEC) in patients >5 years of age
	Focal spot size between 0.6 and 1.3 dependent on the age of the patient
	Check the use of additional filtration
	Check the focus-to-detector distance
Field setting	Check the field size
	Check the field centring
Image detector system	Check the use of antiscatter grids
Patient	–

Table 19.8 Radiographic techniques of thigh radiographs

Age group (years)	Beam direction	Patient position	Beam projection	Typical patient diameter (cm)	Optimal field size (cm x cm)	Focus-to-detector distance (cm)	Voltage (kVp) 50	60	70	80	90	100	Antiscatter grid
0	Vertical	Supine	ap	–	5 × 10	100[1]							
	Horizontal		lat										
1–2	Vertical		ap	–	7 × 14								
	Horizontal		lat										
3–7	Vertical		ap	–	9 × 24								
	Horizontal		lat										
8–12	Vertical		ap	–	11 × 34	115							
	Horizontal		lat										
13–17	Vertical		ap	–	13 × 40								
	Horizontal		lat										
18	Vertical		ap	–	14 × 44								
	Horizontal		lat										

Annotations:
1 The patient is lying on the table top directly on the detector
(dark) Parameter combination not recommended, normalized organ doses not provided
(light blue) Parameter combination recommended, normalized organ doses provided

Table 19.9 Radiation qualities of thigh radiographs

Voltage (kVp)	Total filtration: Without additional filtration 1.5 mm Al	2.5 Al	With additional filtration 3.5 Al	4.5 Al	3.5 Al + 0.1 Cu[1]	3.5 Al + 0.2 Cu[1]
50						
60						
70						
80						
90						
100						

Annotations:
1 The total beam filtration is usually 3.5 mm Al + 0.1 mm Cu, in adolescents optionally 3.5 mm Al + 0.2 mm Cu.
(dark) Parameter combination not recommended, normalized organ doses not provided
(grey) Parameter combination not recommended, normalized organ doses provided
(light blue) Parameter combination recommended, normalized organ doses provided

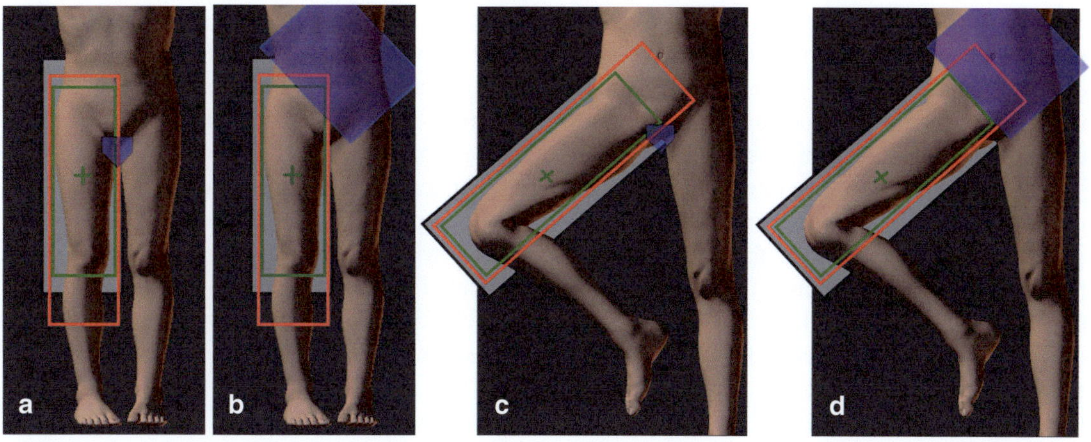

Fig. 19.1 Field settings for thigh radiographs in supine position. (**a**) Antero-posterior projection, male. (**b**) Antero-posterior projection, female. (**c**) Lateral projection, male. (**d**) Lateral projection, female. (© V. Rösenberger & M. Seidenbusch 2018). Green lines = optimal field settings; green cross = centring point; red lines = suboptimal field settings; grey shaded area = image detector; blue shaded area = lead mask shielding and testicular capsule; dark grey shaded area = Bucky table

which may be applied in clinical practice. Please note that not all of these radiation qualities are recommended by the authors. The standard projections of thigh radiographies of the patient who is correctly positioned are depicted in Fig. 19.1. Also the optimal field (marked in green), the suboptimal field (marked in red), and the correct shielding are shown. Tables 19.10, 19.11, and 19.12 outline the virtual field settings on the skin surface of mathematical MIRD phantoms of various age groups and the resulting virtual radiographies with and without bone structures, respectively. Normalized organ doses resulting from these virtual field settings have been calculated for all age groups and have been listed in Tables 19.14, 19.15, 19.16, 19.17, 19.18, 19.19, 19.20, 19.21, 19.22, 19.23, 19.24, 19.25, 19.26, 19.27, and 19.28. Table cells with normal font refer to optimal field settings and table cells with bold font refer to suboptimal field settings. Please note that normalized organ doses have been provided also for exposure parameter combinations, which are not recommended by the authors.

19.2 Indications and Contraindications

Indications and contraindications for thigh radiographs are given in Tables 19.1 and 19.2.

19.3 Image Criteria

The image criteria required for hip joint radiographs are listed in Table 19.3.

19.4 Field Settings

In Table 19.4, patient positions and beam projections used for thigh radiographs are listed, and in Table 19.5, the optimal field settings are listed which are required in order to fulfil the image criteria mentioned in Table 19.3.

19.5 Patient Preparation and Radiographic Procedure

See Tables 19.6, 19.7.

19.6 Exposure Parameters

In Tables 19.8 and 19.9, exposure parameter combinations used for thigh radiographs in various age groups are listed. Dark grey shaded table cells refer to exposure parameter combinations which are not recommended for paediatric thigh radiographs and for which no normalized organ doses have been provided. Light grey and light blue shaded table cells refer to exposure parameter combinations for which normalized organ doses have been provided with light grey shaded table cells referring to exposure parameter combinations being not recommended and light blue shaded table cells referring to exposure parameter combinations being recommended for paediatric thigh radiographs. Note that due to the fact that the radiation dose is an exponential function of the patient diameter, it may make sense to use the patient diameters instead of the patient's age groups for determining the adequate exposure parameters from Table 19.8. In this case, for determining the corresponding normalized organ doses from Tables 19.14, 19.15, 19.16, 19.17, 19.18, 19.19, 19.20, 19.21, 19.22, 19.23, 19.24, 19.25, 19.26, 19.27, and 19.28, the age group related to the patient diameter should be used.

19.7 Normalized Organ Doses

For calculating normalized organ doses, the field settings described in Sect. 19.4 were transferred to the mathematical MIRD phantoms. Inconsistencies between the geometrical anatomy of the mathematical phantoms and the real human anatomy have been listed in Table 19.13 whenever they are of relevance for the validity of the normalized organ doses.

In Table 19.10, 19.11, and 19.12, optimal and suboptimal field settings are listed for the different age groups and the beam projections given in Table 19.8. Green lines refer to optimal field settings, red lines refer to suboptimal field settings, respectively. In the tables, the optimal and suboptimal field settings which were used for performing the calculations of the normalized organ doses provided in Tables 19.14, 19.15, 19.16, 19.17, 19.18, 19.19, 19.20, 19.21, 19.22, 19.23, 19.24, 19.25, 19.26, 19.27, and 19.28 were drawn in on the skin surface of the mathematical phantoms. The virtual mathematical radiographs resulting from these field settings were depicted in the tables with and without bone structures, respectively, in order to enable the reader to visually compare the virtual mathematical radiographs with the real radiographs made under clinical conditions. For comparing the virtual mathematical field settings with the real clinical field settings, the virtual mathematical field edges were also listed in the tables.

As the normalized organ doses provided in Tables 19.14, 19.15, 19.16, 19.17, 19.18, 19.19, 19.20, 19.21, 19.22, 19.23, 19.24, 19.25, 19.26, 19.27, and 19.28 refer to entrance kerma values, optimal and suboptimal field sizes were provided in the tables in order to enable the reader to calculate entrance air kerma dose values from clinically measured dose area product values. Please note that these optimal and suboptimal field sizes refer to the mathematical phantoms of an age group only and not to real patients with corresponding ages.

Please note also that the anatomy of the mathematical phantoms has been idealized. Therefore, the virtual field settings for calculating the normalized organ doses have been defined with a view to ensuring that the resulting virtual organ doses would be consistent with the real organ doses which would be achieved during real X-ray examinations being performed with the real field settings recommended in Sect. 19.4. Thus, the field settings shown in the following tables may not be consistent with the clinical field settings as recommended in Sect. 19.4.

Please note also that normalized organ doses are provided for the ap projection only. However, normalized organ doses for the lateral projection equal to those of ap projection.

Table 19.10 Field settings—thigh ap + lat

Age group (years)		Anatomical landmarks: Optimal field setting	Anatomical landmarks: Suboptimal field setting
0	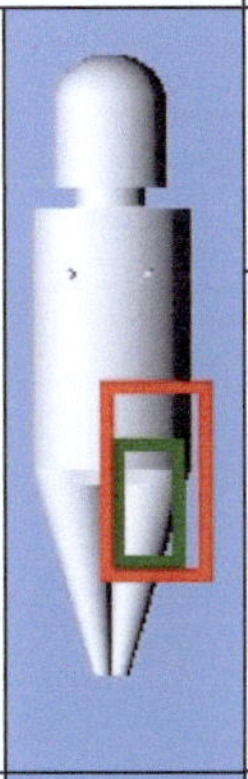	Upper field edge: About 2 cm below the spina iliaca ant. sup. Lower field edge: Just below the inferior edge of the patella Medial field edge: About 2 cm lateral to the symphysis pubis Lateral field edge: Close to the skin surface of the pelvis Field size = field width × field height: 5 cm × 10 cm = 50 cm² 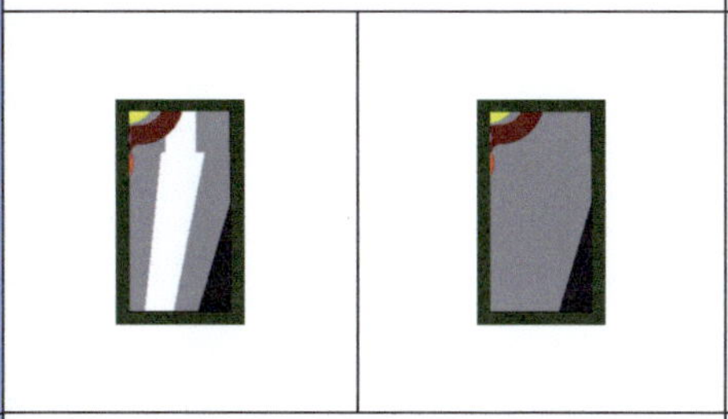	Upper field edge: Just below the iliac crest Lower field edge: Upper third of the shank Medial field edge: Just at the symphysis Lateral field edge: Aside the pelvis including forearm and hand Field size = field width × field height: 8 cm × 16 cm = 128 cm² 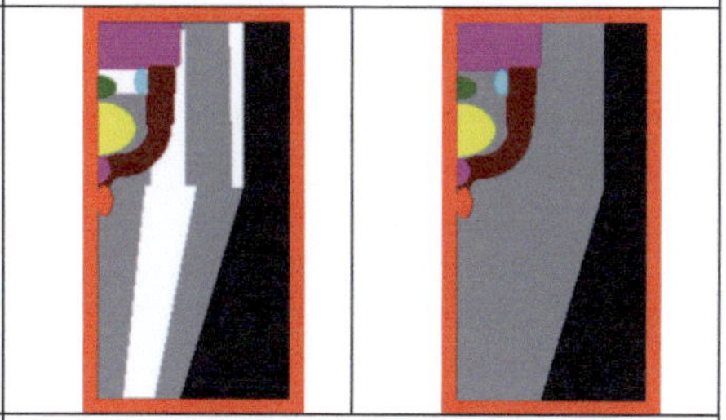
1 - 2	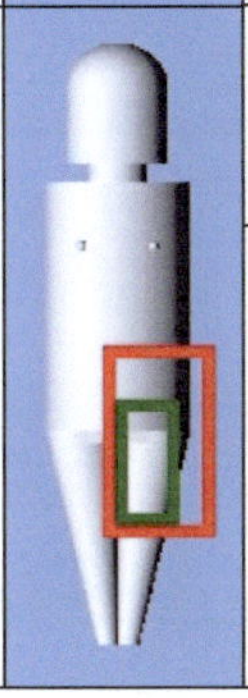	Upper field edge: About 2 cm below the spina iliaca ant. sup. Lower field edge: Just below the inferior edge of the patella Medial field edge: About 2 cm lateral to the symphysis pubis Lateral field edge: Close to the skin surface of the pelvis Field size = field width × field height: 6 cm × 14 cm = 84 cm² 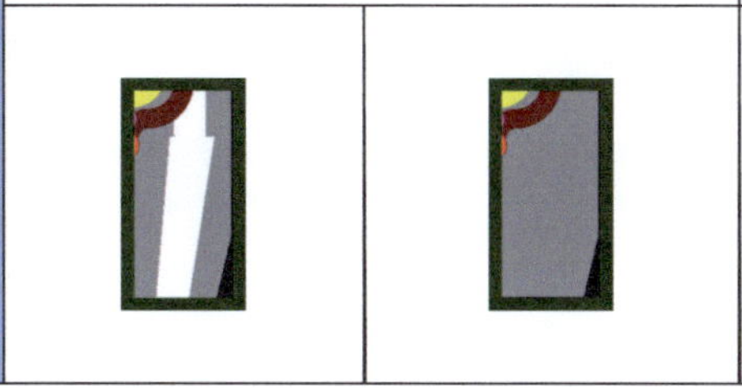	Upper field edge: Just below the iliac crest Lower field edge: Upper third of the shank Medial field edge: Just at the symphysis Lateral field edge: Aside the pelvis including forearm and hand Field size = field width × field height: 11 cm × 22 cm = 242 cm² 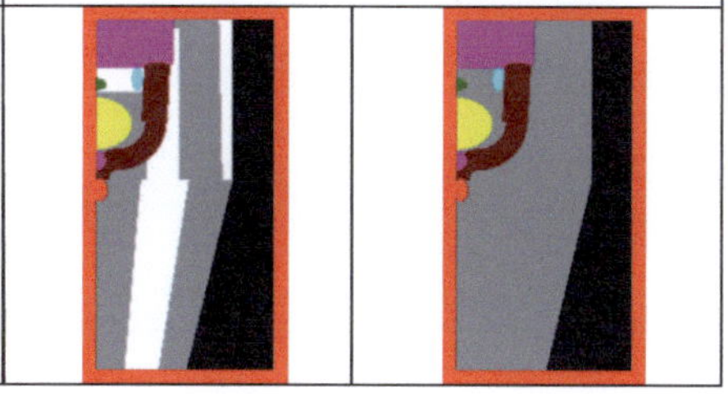

Table 19.11 Field settings—thigh ap + lat

Age group (years)		Anatomical landmarks: Optimal field setting	Anatomical landmarks: Suboptimal field setting
3 - 7	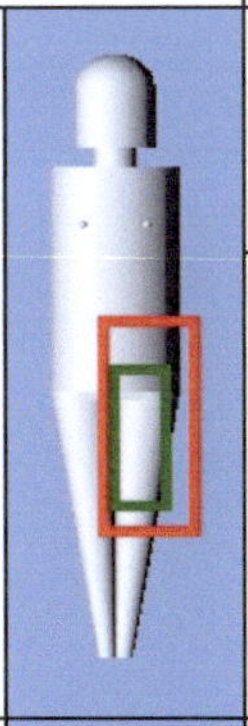	Upper field edge: About 2 cm below the spina iliaca ant. sup. Lower field edge: Just below the inferior edge of the patella Medial field edge: About 3 cm lateral to the symphysis pubis Lateral field edge: Close to the skin surface of the pelvis Field size = field width × field height: 8 cm × 24 cm = 192 cm² 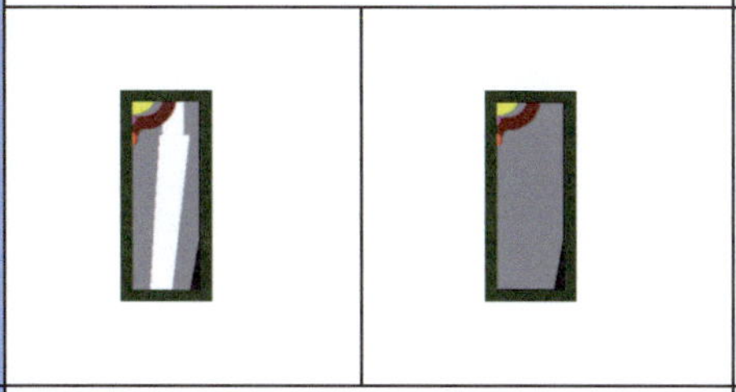	Upper field edge: Just below the iliac crest Lower field edge: Upper third of the shank Medial field edge: Just at the symphysis Lateral field edge: Aside the pelvis including forearm and hand Field size = field width × field height: 14 cm × 38 cm = 532 cm² 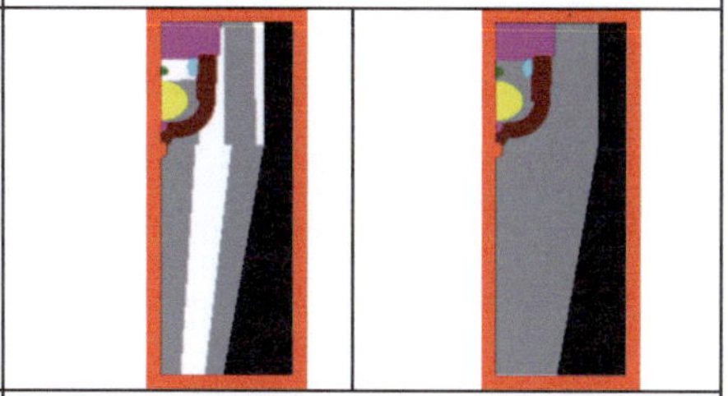
8 - 12	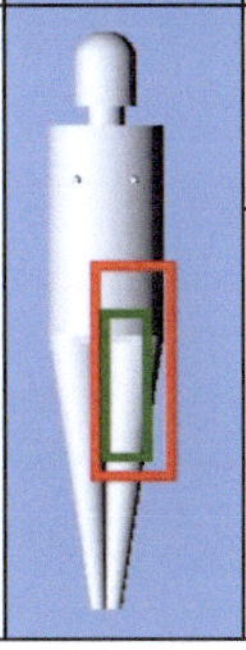	Upper field edge: About 4 cm below the spina iliaca ant. sup. Lower field edge: Just below the inferior edge of the patella Medial field edge: About 4 cm lateral to the symphysis pubis Lateral field edge: Close to the skin surface of the pelvis Field size = field width × field height: 10 cm × 34 cm = 340 cm² 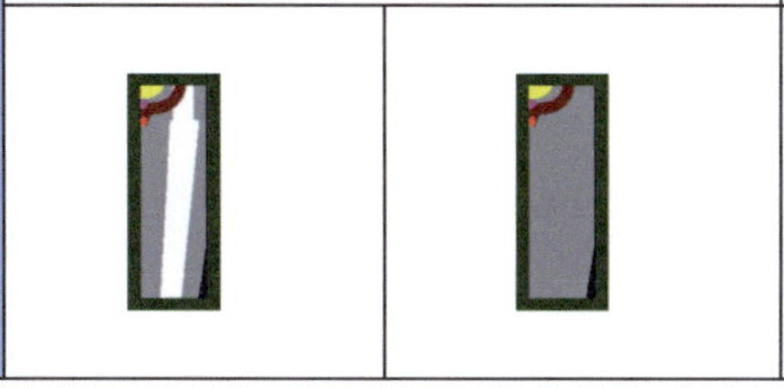	Upper field edge: Just below the iliac crest Lower field edge: Upper third of the shank Medial field edge: Just at the symphysis Lateral field edge: Aside the pelvis including forearm and hand Field size = field width × field height: 16 cm × 50 cm = 800 cm² 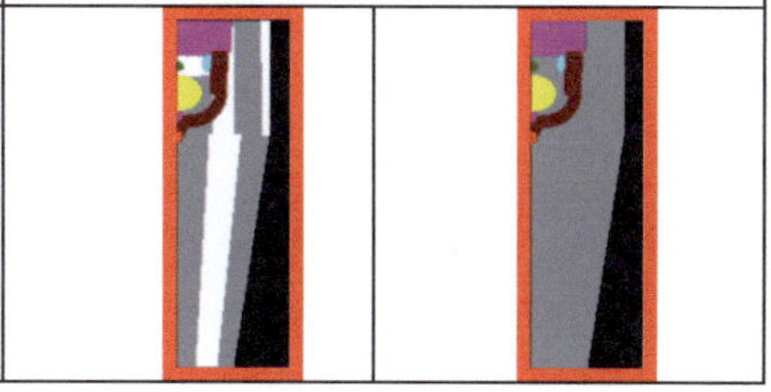

Table 19.12 Field settings—thigh ap + lat

<table>
<tr><th rowspan="2" colspan="2">Age group (years)</th><th colspan="4">Anatomical landmarks</th></tr>
<tr><th colspan="2">Optimal field setting</th><th colspan="2">Suboptimal field setting</th></tr>
<tr><td rowspan="2">13 - 17</td><td rowspan="2">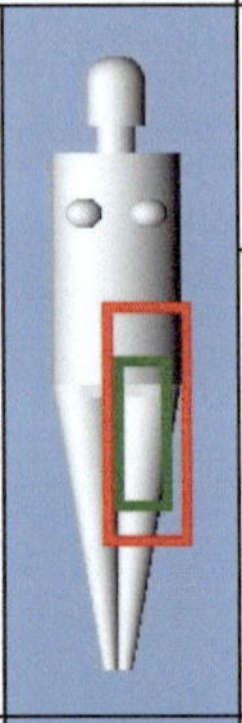</td><td colspan="2">Upper field edge: About 4 cm below the spina iliaca ant. sup.
Lower field edge: Just below the inferior edge of the patella
Medial field edge: About 4 cm lateral to the symphysis pubis
Lateral field edge: Close to the skin surface of the pelvis
Field size = field width x field height: 12 cm x 40 cm = 480 cm²</td><td colspan="2">Upper field edge: Just below the iliac crest
Lower field edge: Upper third of the shank
Medial field edge: Just at the symphysis
Lateral field edge: Aside the pelvis including forearm and hand
Field size = field width x field height: 20 cm x 62 cm = 1,240 cm²</td></tr>
<tr><td>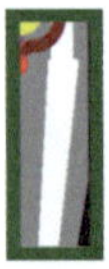</td><td></td><td></td><td>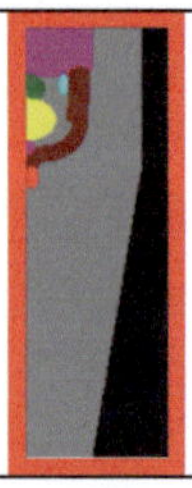</td></tr>
<tr><td rowspan="2">18</td><td rowspan="2">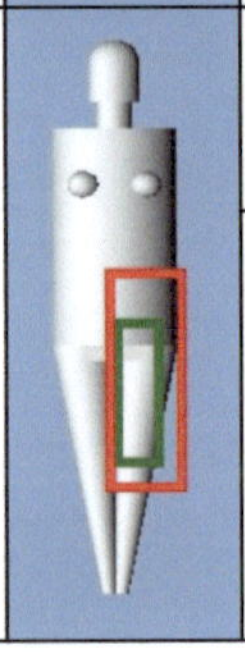</td><td colspan="2">Upper field edge: About 5 cm below the spina iliaca ant. sup.
Lower field edge: Just below the inferior edge of the patella
Medial field edge: About 5 cm lateral to the symphysis pubis
Lateral field edge: Close to the skin surface of the pelvis
Field size = field width x field height: 13 cm x 44 cm = 572 cm²</td><td colspan="2">Upper field edge: Just below the iliac crest
Lower field edge: Upper third of the shank
Medial field edge: Just at the symphysis
Lateral field edge: Aside the pelvis including forearm and hand
Field size = field width x field height: 22 cm x 66 cm = 1,452 cm²</td></tr>
<tr><td></td><td></td><td></td><td></td></tr>
</table>

Table 19.13 Inconsistencies between the mathematical MIRD phantoms and real human anatomy

Anatomical region	Inconsistencies
Extremities	No hip joint and no knee have been inserted

For details, see the text

Table 19.14 Normalized organ doses—Thigh ap—<1 year

Beam projection:			ap									**Table cells:**
Focus-to-detector distance (cm):			100									(+) = normal font = optimal field setting
Tube voltage (kVp):			60									**(−) = bold font = suboptimal field setting**
Total filtrations:			See last table row									Empty table cells correspond to values <0.01
Organ or tissue	**Organ dose D_T: Entrance dose K_E**								**(mSv ∗ mGy⁻¹)**			
	(+)	(−)	(+)	(−)	(+)	(−)	(+)	(−)	(+)	(−)	(+)	(−)
Skull												
Brain												
Oral mucosa												
Salivary glands												
Cervical spine												
Thyroid												
Thymus												
Thoracic spine												
Claviculae												
Scapulae												
Ribs												
Oesophagus												
Trachea												
Lung												
Mamma												
Heart												
Stomach		**0.02**		**0.02**		**0.02**		**0.03**		**0.03**		**0.03**
Spleen				**0.01**		**0.01**		**0.02**		**0.02**		**0.02**
Liver								**0.01**		**0.01**		**0.01**
Gall bladder		**0.01**		**0.02**		**0.02**		**0.03**		**0.03**		**0.03**
Pancreas				**0.01**		**0.01**		**0.02**		**0.02**		**0.02**
Small intestine		**0.15**		**0.19**		**0.22**		**0.24**	0.01	**0.28**	0.01	**0.27**
Large intestine		**0.04**		**0.05**		**0.05**		**0.06**		**0.07**		**0.07**
Sigma/rectum	0.09	**0.33**	0.12	**0.41**	0.13	**0.47**	0.15	**0.51**	0.17	**0.60**	0.17	**0.58**
Lumbar spine		**0.06**		**0.07**		**0.09**		**0.10**		**0.13**		**0.12**
Pelvis	0.09	**0.30**	0.11	**0.38**	0.13	**0.45**	0.14	**0.50**	0.17	**0.59**	0.16	**0.57**
Adrenals												
Kidneys				**0.01**		**0.01**		**0.02**		**0.02**		**0.02**
Urinary bladder	0.08	**0.32**	0.09	**0.39**	0.11	**0.43**	0.12	**0.46**	0.13	**0.52**	0.13	**0.51**
Uterus	0.01	**0.22**	0.02	**0.27**	0.02	**0.31**	0.03	**0.34**	0.03	**0.40**	0.03	**0.38**
Ovaries	0.01	**0.17**	0.01	**0.21**	0.02	**0.25**	0.02	**0.27**	0.02	**0.32**	0.02	**0.31**
Prostate	0.08	**0.28**	0.10	**0.34**	0.12	**0.39**	0.13	**0.42**	0.15	**0.49**	0.14	**0.48**
Testes	0.19	**0.57**	0.21	**0.63**	0.23	**0.66**	0.24	**0.69**	0.26	**0.72**	0.26	**0.72**
Upper arm												
Elbow										**0.01**		
Forearm		**0.43**		**0.51**		**0.57**		**0.61**		**0.68**		**0.68**
Upper leg	0.51	**0.54**	0.64	**0.68**	0.73	**0.78**	0.80	**0.86**	0.92	**1.00**	0.91	**0.98**
Knee	0.59	**0.61**	0.73	**0.75**	0.82	**0.85**	0.90	**0.93**	1.02	**1.06**	1.02	**1.06**
Lower leg	0.13	**0.25**	0.15	**0.30**	0.17	**0.34**	0.19	**0.36**	0.21	**0.41**	0.21	**0.41**
Lymph nodes		**0.05**		**0.06**		**0.07**		**0.08**	0.01	**0.09**		**0.09**
Skin	0.06	**0.13**	0.07	**0.14**	0.07	**0.15**	0.07	**0.15**	0.08	**0.16**	0.08	**0.16**
Muscle	0.06	**0.11**	0.06	**0.13**	0.07	**0.14**	0.08	**0.15**	0.08	**0.17**	0.08	**0.17**
Skeleton	0.11	**0.18**	0.14	**0.22**	0.16	**0.25**	0.17	**0.27**	0.20	**0.31**	0.19	**0.31**
Red bone marrow	0.02	**0.05**	0.03	**0.06**	0.03	**0.07**	0.04	**0.08**	0.04	**0.09**	0.04	**0.09**
Whole body dose	0.05	**0.10**	0.06	**0.12**	0.06	**0.13**	0.07	**0.14**	0.08	**0.15**	0.07	**0.15**
ED (ICRP 60)	0.03	**0.12**	0.04	**0.14**	0.04	**0.16**	0.05	**0.17**	0.05	**0.19**	0.05	**0.19**
ED (ICRP 103)	0.02	**0.08**	0.03	**0.10**	0.03	**0.11**	0.03	**0.12**	0.04	**0.13**	0.04	**0.13**
Total filtration (mm) **Al**	**1.5**		**2.5**		**3.5**		**4.5**		**3.5**		**3.5**	
Cu	**0**		**0**		**0**		**0**		**0.1**		**0.2**	

Table 19.15 Normalized organ doses—Thigh ap—<1 year

Beam projection:	ap											**Table cells:**
Focus-to-detector distance (cm):	100											(+) = normal font = optimal field setting
Tube voltage (kVp):	70											**(−) = bold font = suboptimal field setting**
Total filtrations:	See last table row											Empty table cells correspond to values <0.01
Organ or tissue	**Organ dose D_T: Entrance dose K_E**							**(mSv * mGy⁻¹)**				
	(+)	(−)	(+)	(−)	(+)	(−)	(+)	(−)	(+)	(−)	(+)	(−)
Skull												
Brain												
Oral mucosa												
Salivary glands												
Cervical spine												
Thyroid												
Thymus												
Thoracic spine										**0.01**		**0.01**
Claviculae												
Scapulae												
Ribs										**0.01**		**0.01**
Oesophagus												
Trachea												
Lung												
Mamma												
Heart												
Stomach		**0.02**		**0.02**		**0.03**		**0.03**		**0.04**		**0.04**
Spleen		**0.01**		**0.01**		**0.02**		**0.02**		**0.02**		**0.03**
Liver						**0.01**		**0.01**		**0.02**		**0.02**
Gall bladder		**0.02**		**0.02**		**0.03**		**0.03**		**0.04**		**0.05**
Pancreas		**0.01**		**0.01**		**0.02**		**0.02**		**0.02**		**0.03**
Small intestine		**0.18**		**0.22**	0.01	**0.25**	0.01	**0.27**	0.01	**0.31**	0.02	**0.34**
Large intestine		**0.04**		**0.05**		**0.06**		**0.07**		**0.08**	0.01	**0.10**
Sigma/rectum	0.11	**0.38**	0.13	**0.46**	0.15	**0.52**	0.17	**0.57**	0.19	**0.66**	0.21	**0.73**
Lumbar spine		**0.07**		**0.09**		**0.11**	0.01	**0.12**	0.01	**0.15**	0.02	**0.18**
Pelvis	0.10	**0.36**	0.13	**0.44**	0.15	**0.51**	0.16	**0.56**	0.19	**0.66**	0.21	**0.74**
Adrenals										**0.01**		**0.01**
Kidneys		**0.01**		**0.02**		**0.02**		**0.02**		**0.03**		**0.03**
Urinary bladder	0.09	**0.36**	0.10	**0.42**	0.12	**0.47**	0.13	**0.50**	0.14	**0.56**	0.16	**0.60**
Uterus	0.02	**0.25**	0.02	**0.31**	0.03	**0.35**	0.03	**0.38**	0.04	**0.44**	0.04	**0.49**
Ovaries	0.01	**0.20**	0.02	**0.25**	0.02	**0.28**	0.02	**0.31**	0.03	**0.36**	0.03	**0.41**
Prostate	0.09	**0.32**	0.12	**0.38**	0.13	**0.43**	0.14	**0.46**	0.16	**0.53**	0.18	**0.58**
Testes	0.20	**0.60**	0.23	**0.65**	0.24	**0.69**	0.25	**0.71**	0.27	**0.74**	0.29	**0.77**
Upper arm												
Elbow								**0.01**		**0.01**		**0.02**
Forearm		**0.47**		**0.55**		**0.61**		**0.64**		**0.71**		**0.76**
Upper leg	0.58	**0.62**	0.71	**0.76**	0.80	**0.86**	0.87	**0.94**	0.99	**1.08**	1.08	**1.19**
Knee	0.67	**0.69**	0.80	**0.83**	0.89	**0.93**	0.96	**1.00**	1.08	**1.13**	1.17	**1.22**
Lower leg	0.14	**0.28**	0.17	**0.33**	0.19	**0.36**	0.20	**0.39**	0.22	**0.43**	0.24	**0.46**
Lymph nodes		**0.06**		**0.07**		**0.08**		**0.09**	0.01	**0.10**	0.01	**0.11**
Skin	0.07	**0.14**	0.07	**0.14**	0.07	**0.15**	0.08	**0.15**	0.08	**0.16**	0.08	**0.17**
Muscle	0.06	**0.12**	0.07	**0.14**	0.08	**0.16**	0.08	**0.17**	0.09	**0.18**	0.10	**0.20**
Skeleton	0.13	**0.20**	0.15	**0.24**	0.17	**0.27**	0.18	**0.30**	0.21	**0.34**	0.23	**0.37**
Red bone marrow	0.03	**0.06**	0.03	**0.07**	0.04	**0.08**	0.04	**0.09**	0.05	**0.10**	0.05	**0.11**
Whole body dose	0.05	**0.11**	0.06	**0.13**	0.07	**0.14**	0.07	**0.15**	0.08	**0.17**	0.09	**0.18**
ED (ICRP 60)	0.04	**0.14**	0.04	**0.16**	0.05	**0.17**	0.05	**0.19**	0.06	**0.21**	0.06	**0.22**
ED (ICRP 103)	0.02	**0.09**	0.03	**0.11**	0.03	**0.12**	0.04	**0.13**	0.04	**0.14**	0.04	**0.16**
Total filtration (mm) **Al**	**1.5**		**2.5**		**3.5**		**4.5**		**3.5**		**3.5**	
Cu	**0**		**0**		**0**		**0**		**0.1**		**0.2**	

Table 19.16 Normalized organ doses—Thigh ap—1–2 years

Beam projection:	ap											**Table cells:**
Focus-to-detector distance (cm):	100											(+) = normal font = optimal field setting
Tube voltage (kVp):	60											**(−) = bold font = suboptimal field setting**
Total filtrations:	See last table row											Empty table cells correspond to values <0.01
Organ or tissue	**Organ dose D_T: Entrance dose K_E (mSv ∗ mGy^{-1})**											
	(+)	(−)	(+)	(−)	(+)	(−)	(+)	(−)	(+)	(−)	(+)	(−)
Skull												
Brain												
Oral mucosa												
Salivary glands												
Cervical spine												
Thyroid												
Thymus												
Thoracic spine												
Claviculae												
Scapulae												
Ribs												
Oesophagus												
Trachea												
Lung												
Mamma												
Heart												
Stomach		**0.01**		**0.02**		**0.02**		**0.02**		**0.03**		**0.03**
Spleen						**0.01**		**0.01**		**0.02**		**0.01**
Liver												
Gall bladder		**0.01**		**0.02**		**0.02**		**0.03**		**0.03**		**0.03**
Pancreas								**0.01**		**0.01**		**0.01**
Small intestine		**0.13**		**0.16**		**0.19**		**0.21**		**0.25**		**0.24**
Large intestine		**0.03**		**0.04**		**0.05**		**0.06**		**0.07**		**0.06**
Sigma/rectum	0.08	**0.26**	0.10	**0.33**	0.12	**0.39**	0.13	**0.43**	0.16	**0.51**	0.15	**0.48**
Lumbar spine		**0.04**		**0.05**		**0.07**		**0.08**		**0.10**		**0.08**
Pelvis	0.07	**0.22**	0.09	**0.29**	0.10	**0.34**	0.12	**0.38**	0.14	**0.47**	0.13	**0.43**
Adrenals												
Kidneys						**0.01**		**0.01**		**0.02**		**0.01**
Urinary bladder	0.09	**0.29**	0.11	**0.35**	0.13	**0.40**	0.14	**0.43**	0.16	**0.50**	0.15	**0.48**
Uterus	0.01	**0.19**	0.02	**0.25**	0.02	**0.29**	0.02	**0.32**	0.03	**0.38**	0.03	**0.36**
Ovaries		**0.13**	0.01	**0.16**	0.02	**0.19**	0.02	**0.22**	0.02	**0.26**	0.02	**0.25**
Prostate	0.10	**0.23**	0.12	**0.29**	0.14	**0.34**	0.16	**0.37**	0.18	**0.44**	0.18	**0.42**
Testes	0.29	**0.57**	0.33	**0.64**	0.35	**0.67**	0.37	**0.70**	0.40	**0.74**	0.39	**0.74**
Upper arm												
Elbow												
Forearm		**0.34**		**0.42**		**0.48**		**0.52**		**0.60**		**0.59**
Upper leg	0.36	**0.38**	0.46	**0.49**	0.53	**0.57**	0.60	**0.64**	0.71	**0.77**	0.68	**0.73**
Knee	0.41	**0.43**	0.52	**0.55**	0.61	**0.63**	0.67	**0.70**	0.80	**0.83**	0.77	**0.80**
Lower leg	0.01	**0.07**	0.02	**0.08**	0.02	**0.10**	0.03	**0.11**	0.03	**0.13**	0.03	**0.12**
Lymph nodes		**0.04**		**0.06**		**0.06**		**0.07**	0.01	**0.08**		**0.08**
Skin	0.05	**0.12**	0.06	**0.13**	0.06	**0.14**	0.06	**0.14**	0.07	**0.15**	0.07	**0.15**
Muscle	0.05	**0.09**	0.05	**0.11**	0.06	**0.12**	0.06	**0.13**	0.07	**0.14**	0.07	**0.14**
Skeleton	0.08	**0.12**	0.10	**0.16**	0.12	**0.18**	0.13	**0.20**	0.16	**0.24**	0.15	**0.23**
Red bone marrow	0.01	**0.03**	0.01	**0.04**	0.02	**0.04**	0.02	**0.05**	0.02	**0.06**	0.02	**0.05**
Whole body dose	0.04	**0.08**	0.05	**0.10**	0.06	**0.11**	0.06	**0.12**	0.07	**0.13**	0.07	**0.13**
ED (ICRP 60)	0.04	**0.11**	0.05	**0.13**	0.05	**0.14**	0.06	**0.15**	0.06	**0.18**	0.06	**0.17**
ED (ICRP 103)	0.02	**0.07**	0.03	**0.08**	0.03	**0.09**	0.03	**0.10**	0.04	**0.12**	0.04	**0.11**
Total filtration (mm) Al	**1.5**		**2.5**		**3.5**		**4.5**		**3.5**		**3.5**	
Cu	**0**		**0**		**0**		**0**		**0.1**		**0.2**	

Table 19.17 Normalized organ doses—Thigh ap—1–2 years

Beam projection:	ap											**Table cells:**
Focus-to-detector distance (cm):	100											(+) = normal font = optimal field setting
Tube voltage (kVp):	70											**(−) = bold font = suboptimal field setting**
Total filtrations:	See last table row											Empty table cells correspond to values <0.01
Organ or tissue	**Organ dose D_T: Entrance dose K_E**							**(mSv * mGy⁻¹)**				
	(+)	(−)	(+)	(−)	(+)	(−)	(+)	(−)	(+)	(−)	(+)	(−)
Skull												
Brain												
Oral mucosa												
Salivary glands												
Cervical spine												
Thyroid												
Thymus												
Thoracic spine												
Claviculae												
Scapulae												
Ribs												**0.01**
Oesophagus												
Trachea												
Lung												
Mamma												
Heart												
Stomach		**0.02**		**0.02**		**0.02**		**0.03**		**0.03**		**0.04**
Spleen				**0.01**		**0.01**		**0.02**		**0.02**		**0.02**
Liver										**0.01**		**0.01**
Gall bladder		**0.02**		**0.02**		**0.03**		**0.03**		**0.04**		**0.05**
Pancreas				**0.01**		**0.01**		**0.01**		**0.02**		**0.02**
Small intestine		**0.15**		**0.19**		**0.22**	0.01	**0.24**	0.01	**0.28**	0.02	**0.32**
Large intestine		**0.04**		**0.05**		**0.06**		**0.07**		**0.08**		**0.10**
Sigma/rectum	0.10	**0.31**	0.12	**0.39**	0.14	**0.45**	0.15	**0.49**	0.18	**0.58**	0.21	**0.65**
Lumbar spine		**0.05**		**0.07**		**0.09**		**0.10**		**0.12**	0.01	**0.15**
Pelvis	0.08	**0.27**	0.10	**0.35**	0.12	**0.40**	0.14	**0.45**	0.16	**0.54**	0.19	**0.63**
Adrenals												
Kidneys				**0.01**		**0.01**		**0.02**		**0.02**		**0.02**
Urinary bladder	0.10	**0.33**	0.13	**0.40**	0.14	**0.44**	0.15	**0.48**	0.18	**0.54**	0.20	**0.60**
Uterus	0.02	**0.23**	0.02	**0.29**	0.03	**0.33**	0.03	**0.37**	0.04	**0.43**	0.05	**0.49**
Ovaries	0.01	**0.15**	0.02	**0.20**	0.02	**0.23**	0.02	**0.25**	0.03	**0.30**	0.03	**0.35**
Prostate	0.11	**0.27**	0.14	**0.34**	0.16	**0.38**	0.18	**0.42**	0.21	**0.49**	0.23	**0.55**
Testes	0.31	**0.61**	0.35	**0.67**	0.37	**0.70**	0.39	**0.73**	0.41	**0.77**	0.44	**0.81**
Upper arm												
Elbow												**0.01**
Forearm		**0.39**		**0.47**		**0.52**		**0.56**		**0.64**		**0.70**
Upper leg	0.43	**0.46**	0.54	**0.58**	0.62	**0.67**	0.68	**0.74**	0.80	**0.88**	0.91	**1.00**
Knee	0.49	**0.51**	0.60	**0.63**	0.69	**0.72**	0.75	**0.79**	0.88	**0.92**	0.98	**1.04**
Lower leg	0.02	**0.08**	0.02	**0.10**	0.03	**0.11**	0.03	**0.12**	0.04	**0.14**	0.05	**0.16**
Lymph nodes		**0.05**		**0.07**		**0.07**	0.01	**0.08**	0.01	**0.09**	0.01	**0.11**
Skin	0.06	**0.13**	0.06	**0.14**	0.06	**0.14**	0.07	**0.15**	0.07	**0.15**	0.07	**0.16**
Muscle	0.05	**0.10**	0.06	**0.12**	0.07	**0.13**	0.07	**0.14**	0.08	**0.16**	0.09	**0.17**
Skeleton	0.09	**0.15**	0.12	**0.18**	0.13	**0.21**	0.15	**0.23**	0.17	**0.27**	0.20	**0.31**
Red bone marrow	0.01	**0.03**	0.02	**0.04**	0.02	**0.05**	0.02	**0.06**	0.03	**0.07**	0.03	**0.08**
Whole body dose	0.05	**0.09**	0.06	**0.11**	0.06	**0.12**	0.07	**0.13**	0.08	**0.15**	0.08	**0.16**
ED (ICRP 60)	0.05	**0.12**	0.05	**0.14**	0.06	**0.16**	0.06	**0.17**	0.07	**0.19**	0.08	**0.21**
ED (ICRP 103)	0.03	**0.08**	0.03	**0.09**	0.04	**0.11**	0.04	**0.11**	0.04	**0.13**	0.05	**0.15**
Total filtration (mm) Al	**1.5**		**2.5**		**3.5**		**4.5**		**3.5**		**3.5**	
Cu	**0**		**0**		**0**		**0**		**0.1**		**0.2**	

Table 19.18 Normalized organ doses—Thigh ap—3–7 years

Beam projection:			ap									**Table cells:**
Focus-to-detector distance (cm):			100									(+) = normal font = optimal field setting
Tube voltage (kVp):			60									**(−) = bold font = suboptimal field setting**
Total filtrations:			See last table row									Empty table cells correspond to values <0.01
Organ or tissue	**Organ dose D_T: Entrance dose K_E**							**(mSv ∗ mGy^{-1})**				
	(+)	(−)	(+)	(−)	(+)	(−)	(+)	(−)	(+)	(−)	(+)	(−)
Skull												
Brain												
Oral mucosa												
Salivary glands												
Cervical spine												
Thyroid												
Thymus												
Thoracic spine												
Claviculae												
Scapulae												
Ribs												
Oesophagus												
Trachea												
Lung												
Mamma												
Heart												
Stomach				**0.01**		**0.01**		**0.02**		**0.02**		**0.02**
Spleen										**0.01**		
Liver												
Gall bladder				**0.01**		**0.02**		**0.02**		**0.02**		**0.02**
Pancreas												
Small intestine		**0.11**		**0.14**		**0.16**		**0.18**		**0.22**		**0.21**
Large intestine		**0.03**		**0.03**		**0.04**		**0.05**		**0.06**		**0.05**
Sigma/rectum	0.07	**0.23**	0.09	**0.29**	0.11	**0.34**	0.13	**0.38**	0.15	**0.46**	0.14	**0.43**
Lumbar spine		**0.03**		**0.04**		**0.05**		**0.06**		**0.07**		**0.06**
Pelvis	0.05	**0.17**	0.07	**0.23**	0.09	**0.27**	0.10	**0.31**	0.12	**0.39**	0.11	**0.35**
Adrenals												
Kidneys										**0.01**		
Urinary bladder	0.09	**0.27**	0.12	**0.33**	0.14	**0.38**	0.15	**0.42**	0.18	**0.48**	0.17	**0.47**
Uterus	0.01	**0.17**	0.02	**0.22**	0.02	**0.26**	0.02	**0.29**	0.03	**0.35**	0.02	**0.32**
Ovaries		**0.11**	0.01	**0.14**	0.01	**0.17**	0.02	**0.19**	0.02	**0.23**	0.02	**0.21**
Prostate	0.13	**0.21**	0.16	**0.27**	0.19	**0.31**	0.21	**0.35**	0.25	**0.42**	0.24	**0.40**
Testes	0.46	**0.58**	0.52	**0.65**	0.55	**0.69**	0.57	**0.72**	0.61	**0.77**	0.60	**0.76**
Upper arm												
Elbow												
Forearm		**0.30**		**0.37**		**0.43**		**0.47**		**0.54**		**0.53**
Upper leg	0.26	**0.28**	0.34	**0.37**	0.40	**0.44**	0.45	**0.49**	0.55	**0.61**	0.51	**0.55**
Knee	0.32	**0.33**	0.41	**0.43**	0.48	**0.51**	0.54	**0.57**	0.65	**0.69**	0.61	**0.64**
Lower leg	0.01	**0.12**	0.01	**0.15**	0.02	**0.18**	0.02	**0.20**	0.02	**0.24**	0.02	**0.22**
Lymph nodes		**0.04**		**0.05**	0.01	**0.06**	0.01	**0.06**	0.01	**0.08**	0.01	**0.07**
Skin	0.07	**0.15**	0.08	**0.16**	0.08	**0.17**	0.09	**0.17**	0.09	**0.18**	0.09	**0.18**
Muscle	0.06	**0.10**	0.07	**0.12**	0.07	**0.13**	0.08	**0.15**	0.09	**0.16**	0.09	**0.16**
Skeleton	0.07	**0.11**	0.09	**0.15**	0.11	**0.17**	0.12	**0.19**	0.15	**0.23**	0.14	**0.22**
Red bone marrow	0.01	**0.03**	0.02	**0.04**	0.02	**0.05**	0.02	**0.05**	0.03	**0.07**	0.03	**0.06**
Whole body dose	0.05	**0.09**	0.06	**0.11**	0.07	**0.12**	0.07	**0.13**	0.08	**0.15**	0.08	**0.15**
ED (ICRP 60)	0.06	**0.10**	0.07	**0.12**	0.07	**0.14**	0.08	**0.15**	0.09	**0.17**	0.08	**0.16**
ED (ICRP 103)	0.03	**0.06**	0.04	**0.08**	0.04	**0.09**	0.04	**0.10**	0.05	**0.11**	0.05	**0.11**
Total filtration (mm) **Al**	**1.5**		**2.5**		**3.5**		**4.5**		**3.5**		**3.5**	
Cu	**0**		**0**		**0**		**0**		**0.1**		**0.2**	

Table 19.19 Normalized organ doses—Thigh ap—3–7 years

Beam projection:	ap	**Table cells:**
Focus-to-detector distance (cm):	100	(+) = normal font = optimal field setting
Tube voltage (kVp):	70	**(−) = bold font = suboptimal field setting**
Total filtrations:	See last table row	Empty table cells correspond to values <0.01

Organ or tissue	**Organ dose D_T: Entrance dose K_E**								**(mSv * mGy^{-1})**			
	(+)	(−)	(+)	(−)	(+)	(−)	(+)	(−)	(+)	(−)	(+)	(−)
Skull												
Brain												
Oral mucosa												
Salivary glands												
Cervical spine												
Thyroid												
Thymus												
Thoracic spine												
Claviculae												
Scapulae												
Ribs												
Oesophagus												
Trachea												
Lung												
Mamma												
Heart												
Stomach		**0.01**		**0.02**		**0.02**		**0.02**		**0.03**		**0.03**
Spleen								**0.01**		**0.01**		**0.02**
Liver												
Gall bladder		**0.01**		**0.02**		**0.02**		**0.02**		**0.03**		**0.04**
Pancreas										**0.01**		**0.02**
Small intestine		**0.13**		**0.17**		**0.20**		**0.22**		**0.26**	0.01	**0.30**
Large intestine		**0.03**		**0.04**		**0.05**		**0.06**		**0.07**		**0.09**
Sigma/rectum	0.09	**0.28**	0.12	**0.35**	0.13	**0.41**	0.15	**0.45**	0.18	**0.54**	0.21	**0.62**
Lumbar spine		**0.04**		**0.06**		**0.07**		**0.08**		**0.10**		**0.12**
Pelvis	0.07	**0.22**	0.09	**0.29**	0.11	**0.34**	0.12	**0.38**	0.15	**0.46**	0.17	**0.55**
Adrenals												
Kidneys						**0.01**		**0.01**		**0.02**		**0.02**
Urinary bladder	0.11	**0.31**	0.14	**0.38**	0.16	**0.43**	0.17	**0.47**	0.20	**0.54**	0.22	**0.60**
Uterus	0.02	**0.21**	0.02	**0.26**	0.03	**0.31**	0.03	**0.34**	0.04	**0.41**	0.04	**0.47**
Ovaries	0.01	**0.13**	0.02	**0.17**	0.02	**0.20**	0.02	**0.23**	0.03	**0.28**	0.03	**0.32**
Prostate	0.15	**0.25**	0.19	**0.32**	0.22	**0.37**	0.24	**0.41**	0.29	**0.48**	0.33	**0.55**
Testes	0.49	**0.61**	0.54	**0.68**	0.58	**0.72**	0.60	**0.75**	0.64	**0.81**	0.67	**0.85**
Upper arm												
Elbow												
Forearm		**0.35**		**0.42**		**0.47**		**0.52**		**0.59**		**0.65**
Upper leg	0.32	**0.35**	0.41	**0.45**	0.48	**0.53**	0.54	**0.59**	0.65	**0.72**	0.75	**0.83**
Knee	0.38	**0.41**	0.49	**0.51**	0.56	**0.60**	0.62	**0.66**	0.74	**0.79**	0.85	**0.91**
Lower leg	0.01	**0.14**	0.02	**0.18**	0.02	**0.20**	0.02	**0.23**	0.03	**0.26**	0.03	**0.30**
Lymph nodes		**0.05**	0.01	**0.06**	0.01	**0.07**	0.01	**0.07**	0.01	**0.09**	0.02	**0.10**
Skin	0.08	**0.16**	0.08	**0.17**	0.09	**0.18**	0.09	**0.18**	0.10	**0.19**	0.10	**0.20**
Muscle	0.06	**0.11**	0.07	**0.14**	0.08	**0.15**	0.09	**0.16**	0.10	**0.18**	0.11	**0.20**
Skeleton	0.09	**0.14**	0.11	**0.18**	0.13	**0.20**	0.14	**0.23**	0.17	**0.27**	0.20	**0.31**
Red bone marrow	0.02	**0.04**	0.02	**0.05**	0.03	**0.06**	0.03	**0.07**	0.04	**0.08**	0.04	**0.10**
Whole body dose	0.06	**0.10**	0.07	**0.12**	0.08	**0.14**	0.08	**0.15**	0.09	**0.17**	0.10	**0.19**
ED (ICRP 60)	0.06	**0.12**	0.07	**0.14**	0.08	**0.16**	0.09	**0.17**	0.09	**0.19**	0.10	**0.21**
ED (ICRP 103)	0.04	**0.07**	0.04	**0.09**	0.05	**0.10**	0.05	**0.11**	0.06	**0.13**	0.06	**0.14**
Total filtration (mm) **Al**	**1.5**		**2.5**		**3.5**		**4.5**		**3.5**		**3.5**	
Cu	**0**		**0**		**0**		**0**		**0.1**		**0.2**	

Table 19.20 Normalized organ doses—Thigh ap—8–12 years

Beam projection:			ap									**Table cells:**
Focus-to-detector distance (cm):			115									(+) = normal font = optimal field setting
Tube voltage (kVp):			60									**(−) = bold font = suboptimal field setting**
Total filtrations:			See last table row									Empty table cells correspond to values <0.01
Organ or tissue	**Organ dose D_T: Entrance dose K_E (mSv ∗ mGy^{-1})**											
	(+)	(−)	(+)	(−)	(+)	(−)	(+)	(−)	(+)	(−)	(+)	(−)
Skull												
Brain												
Oral mucosa												
Salivary glands												
Cervical spine												
Thyroid												
Thymus												
Thoracic spine												
Claviculae												
Scapulae												
Ribs												
Oesophagus												
Trachea												
Lung												
Mamma												
Heart												
Stomach						**0.01**		**0.01**		**0.02**		**0.01**
Spleen												
Liver												
Gall bladder								**0.01**		**0.02**		**0.01**
Pancreas												
Small intestine		**0.09**		**0.12**		**0.14**		**0.16**		**0.20**		**0.18**
Large intestine		**0.02**		**0.03**		**0.03**		**0.04**		**0.05**		**0.04**
Sigma/rectum	0.07	**0.20**	0.09	**0.26**	0.10	**0.30**	0.12	**0.34**	0.15	**0.42**	0.13	**0.38**
Lumbar spine		**0.02**		**0.03**		**0.03**		**0.04**		**0.05**		**0.04**
Pelvis	0.04	**0.14**	0.06	**0.19**	0.07	**0.23**	0.08	**0.26**	0.10	**0.33**	0.09	**0.29**
Adrenals												
Kidneys												
Urinary bladder	0.10	**0.25**	0.13	**0.31**	0.15	**0.35**	0.16	**0.39**	0.20	**0.46**	0.18	**0.43**
Uterus	0.01	**0.15**	0.01	**0.20**	0.02	**0.23**	0.02	**0.26**	0.02	**0.32**	0.02	**0.29**
Ovaries		**0.09**		**0.12**	0.01	**0.14**	0.01	**0.16**	0.02	**0.20**	0.01	**0.18**
Prostate	0.19	**0.20**	0.24	**0.26**	0.29	**0.30**	0.32	**0.34**	0.38	**0.41**	0.36	**0.38**
Testes	0.64	**0.57**	0.72	**0.64**	0.76	**0.68**	0.79	**0.71**	0.85	**0.76**	0.84	**0.75**
Upper arm												
Elbow												
Forearm		**0.27**		**0.33**		**0.38**		**0.42**		**0.49**		**0.48**
Upper leg	0.20	**0.21**	0.26	**0.28**	0.31	**0.34**	0.35	**0.39**	0.44	**0.48**	0.39	**0.43**
Knee	0.25	**0.26**	0.33	**0.34**	0.39	**0.40**	0.44	**0.46**	0.55	**0.56**	0.50	**0.51**
Lower leg	0.02	**0.09**	0.03	**0.12**	0.04	**0.14**	0.04	**0.15**	0.05	**0.19**	0.05	**0.17**
Lymph nodes		**0.03**	0.01	**0.04**	0.01	**0.05**	0.01	**0.06**	0.01	**0.07**	0.01	**0.06**
Skin	0.10	**0.16**	0.10	**0.17**	0.11	**0.18**	0.11	**0.18**	0.12	**0.19**	0.11	**0.19**
Muscle	0.06	**0.10**	0.07	**0.12**	0.08	**0.13**	0.09	**0.14**	0.10	**0.16**	0.09	**0.16**
Skeleton	0.07	**0.10**	0.09	**0.13**	0.11	**0.16**	0.12	**0.18**	0.15	**0.22**	0.14	**0.20**
Red bone marrow	0.01	**0.02**	0.02	**0.03**	0.02	**0.04**	0.02	**0.04**	0.03	**0.06**	0.02	**0.05**
Whole body dose	0.05	**0.09**	0.07	**0.11**	0.07	**0.12**	0.08	**0.13**	0.09	**0.16**	0.09	**0.15**
ED (ICRP 60)	0.08	**0.10**	0.09	**0.12**	0.10	**0.13**	0.10	**0.14**	0.11	**0.16**	0.11	**0.15**
ED (ICRP 103)	0.04	**0.06**	0.05	**0.07**	0.05	**0.08**	0.05	**0.09**	0.06	**0.10**	0.06	**0.10**
Total filtration (mm) Al	**1.5**		**2.5**		**3.5**		**4.5**		**3.5**		**3.5**	
Cu	**0**		**0**		**0**		**0**		**0.1**		**0.2**	

Table 19.21 Normalized organ doses—Thigh ap—8–12 years

Beam projection:	ap											**Table cells:**
Focus-to-detector distance (cm):	115											(+) = normal font = optimal field setting
Tube voltage (kVp):	70											**(−) = bold font = suboptimal field setting**
Total filtrations:	See last table row											Empty table cells correspond to values <0.01
Organ or tissue	**Organ dose D_T: Entrance dose K_E**							**(mSv * mGy^{-1})**				
	(+)	(−)	(+)	(−)	(+)	(−)	(+)	(−)	(+)	(−)	(+)	(−)
Skull												
Brain												
Oral mucosa												
Salivary glands												
Cervical spine												
Thyroid												
Thymus												
Thoracic spine												
Claviculae												
Scapulae												
Ribs												
Oesophagus												
Trachea												
Lung												
Mamma												
Heart												
Stomach				**0.01**		**0.01**		**0.02**		**0.02**		**0.03**
Spleen												**0.01**
Liver												
Gall bladder				**0.01**		**0.01**		**0.02**		**0.02**		**0.03**
Pancreas												**0.01**
Small intestine		**0.12**		**0.15**		**0.17**		**0.19**		**0.23**		**0.27**
Large intestine		**0.03**		**0.04**		**0.04**		**0.05**		**0.06**		**0.07**
Sigma/rectum	0.09	**0.25**	0.11	**0.32**	0.13	**0.37**	0.14	**0.41**	0.18	**0.50**	0.21	**0.58**
Lumbar spine		**0.03**		**0.04**		**0.05**		**0.06**		**0.07**		**0.09**
Pelvis	0.06	**0.19**	0.08	**0.24**	0.09	**0.29**	0.10	**0.32**	0.13	**0.40**	0.15	**0.47**
Adrenals												
Kidneys										**0.01**		**0.01**
Urinary bladder	0.12	**0.29**	0.15	**0.36**	0.17	**0.41**	0.19	**0.45**	0.22	**0.52**	0.25	**0.59**
Uterus	0.01	**0.19**	0.02	**0.24**	0.02	**0.28**	0.03	**0.32**	0.03	**0.38**	0.04	**0.44**
Ovaries		**0.12**	0.01	**0.15**	0.02	**0.18**	0.02	**0.20**	0.02	**0.25**	0.03	**0.30**
Prostate	0.23	**0.24**	0.29	**0.31**	0.34	**0.36**	0.37	**0.39**	0.44	**0.47**	0.50	**0.54**
Testes	0.68	**0.61**	0.75	**0.67**	0.80	**0.72**	0.83	**0.75**	0.88	**0.80**	0.93	**0.84**
Upper arm												
Elbow												
Forearm		**0.31**		**0.38**		**0.43**		**0.47**		**0.54**		**0.61**
Upper leg	0.25	**0.28**	0.33	**0.36**	0.38	**0.42**	0.43	**0.48**	0.53	**0.58**	0.63	**0.69**
Knee	0.32	**0.33**	0.41	**0.42**	0.48	**0.49**	0.53	**0.55**	0.64	**0.66**	0.75	**0.77**
Lower leg	0.03	**0.11**	0.04	**0.14**	0.04	**0.16**	0.05	**0.18**	0.06	**0.21**	0.07	**0.25**
Lymph nodes	0.01	**0.04**	0.01	**0.05**	0.01	**0.06**	0.01	**0.07**	0.02	**0.08**	0.02	**0.09**
Skin	0.10	**0.17**	0.11	**0.18**	0.11	**0.19**	0.12	**0.19**	0.12	**0.20**	0.13	**0.21**
Muscle	0.07	**0.11**	0.08	**0.13**	0.09	**0.15**	0.10	**0.16**	0.11	**0.18**	0.12	**0.20**
Skeleton	0.09	**0.13**	0.11	**0.17**	0.13	**0.19**	0.15	**0.22**	0.18	**0.26**	0.21	**0.30**
Red bone marrow	0.02	**0.03**	0.02	**0.04**	0.03	**0.05**	0.03	**0.06**	0.04	**0.07**	0.04	**0.09**
Whole body dose	0.06	**0.10**	0.07	**0.13**	0.08	**0.14**	0.09	**0.15**	0.10	**0.18**	0.12	**0.20**
ED (ICRP 60)	0.08	**0.11**	0.10	**0.13**	0.10	**0.15**	0.11	**0.16**	0.12	**0.18**	0.13	**0.20**
ED (ICRP 103)	0.04	**0.07**	0.05	**0.08**	0.06	**0.09**	0.06	**0.10**	0.07	**0.12**	0.07	**0.14**
Total filtration (mm) **Al**	**1.5**		**2.5**		**3.5**		**4.5**		**3.5**		**3.5**	
Cu	**0**		**0**		**0**		**0**		**0.1**		**0.2**	

Table 19.22 Normalized organ doses—Thigh ap—8–12 years

Beam projection:			ap									**Table cells:**
Focus-to-detector distance (cm):			115									(+) = normal font = optimal field setting
Tube voltage (kVp):			80									**(−) = bold font = suboptimal field setting**
Total filtrations:			See last table row									Empty table cells correspond to values <0.01
Organ or tissue	**Organ dose D_T: Entrance dose K_E**								**(mSv ∗ mGy^{-1})**			
	(+)	(−)	(+)	(−)	(+)	(−)	(+)	(−)	(+)	(−)	(+)	(−)
Skull												
Brain												
Oral mucosa												
Salivary glands												
Cervical spine												
Thyroid												
Thymus												
Thoracic spine												
Claviculae												
Scapulae												
Ribs												
Oesophagus												
Trachea												
Lung												
Mamma												
Heart												
Stomach		**0.01**		**0.02**		**0.02**		**0.02**		**0.03**		**0.03**
Spleen								**0.01**		**0.01**		**0.02**
Liver												
Gall bladder		**0.01**		**0.02**		**0.02**		**0.02**		**0.03**		**0.03**
Pancreas										**0.01**		**0.01**
Small intestine		**0.14**		**0.18**		**0.20**		**0.22**		**0.27**	0.01	**0.30**
Large intestine		**0.04**		**0.05**		**0.05**		**0.06**		**0.07**		**0.09**
Sigma/rectum	0.10	**0.30**	0.13	**0.37**	0.15	**0.43**	0.17	**0.47**	0.20	**0.56**	0.23	**0.64**
Lumbar spine		**0.04**		**0.05**		**0.06**		**0.07**		**0.09**		**0.11**
Pelvis	0.07	**0.23**	0.09	**0.29**	0.11	**0.34**	0.12	**0.38**	0.15	**0.46**	0.17	**0.54**
Adrenals												
Kidneys								**0.01**		**0.01**		**0.02**
Urinary bladder	0.14	**0.34**	0.17	**0.40**	0.19	**0.46**	0.21	**0.50**	0.25	**0.57**	0.28	**0.64**
Uterus	0.02	**0.23**	0.02	**0.29**	0.03	**0.33**	0.03	**0.36**	0.04	**0.43**	0.05	**0.50**
Ovaries	0.01	**0.15**	0.02	**0.19**	0.02	**0.22**	0.02	**0.24**	0.03	**0.29**	0.03	**0.34**
Prostate	0.27	**0.29**	0.33	**0.36**	0.38	**0.41**	0.42	**0.45**	0.49	**0.52**	0.55	**0.59**
Testes	0.72	**0.64**	0.78	**0.71**	0.83	**0.75**	0.86	**0.78**	0.91	**0.83**	0.95	**0.87**
Upper arm												
Elbow												
Forearm		**0.35**		**0.42**		**0.47**		**0.51**		**0.58**		**0.63**
Upper leg	0.31	**0.34**	0.39	**0.43**	0.45	**0.50**	0.50	**0.56**	0.61	**0.67**	0.70	**0.78**
Knee	0.38	**0.39**	0.48	**0.49**	0.55	**0.57**	0.61	**0.63**	0.72	**0.74**	0.82	**0.85**
Lower leg	0.04	**0.13**	0.04	**0.16**	0.05	**0.18**	0.06	**0.20**	0.07	**0.23**	0.08	**0.27**
Lymph nodes	0.01	**0.05**	0.01	**0.06**	0.01	**0.07**	0.02	**0.08**	0.02	**0.09**	0.02	**0.10**
Skin	0.10	**0.17**	0.11	**0.18**	0.12	**0.19**	0.12	**0.20**	0.13	**0.21**	0.13	**0.22**
Muscle	0.08	**0.13**	0.09	**0.15**	0.10	**0.16**	0.10	**0.18**	0.12	**0.20**	0.13	**0.22**
Skeleton	0.11	**0.16**	0.13	**0.19**	0.15	**0.22**	0.17	**0.25**	0.20	**0.29**	0.23	**0.34**
Red bone marrow	0.02	**0.04**	0.03	**0.05**	0.03	**0.06**	0.04	**0.07**	0.04	**0.09**	0.05	**0.10**
Whole body dose	0.07	**0.12**	0.08	**0.14**	0.09	**0.16**	0.10	**0.17**	0.11	**0.19**	0.13	**0.21**
ED (ICRP 60)	0.09	**0.13**	0.10	**0.15**	0.11	**0.16**	0.12	**0.17**	0.13	**0.20**	0.13	**0.22**
ED (ICRP 103)	0.05	**0.08**	0.06	**0.10**	0.06	**0.11**	0.06	**0.12**	0.07	**0.13**	0.08	**0.15**
Total filtration (mm) **Al**	**1.5**		**2.5**		**3.5**		**4.5**		**3.5**		**3.5**	
Cu	**0**		**0**		**0**		**0**		**0.1**		**0.2**	

Table 19.23 Normalized organ doses—Thigh ap—13–17 years

Beam projection:	ap											**Table cells:**
Focus-to-detector distance (cm):	115											(+) = normal font = optimal field setting
Tube voltage (kVp):	60											**(−) = bold font = suboptimal field setting**
Total filtrations:	See last table row											Empty table cells correspond to values <0.01
Organ or tissue	**Organ dose D_T: Entrance dose K_E (mSv * mGy^{-1})**											
	(+)	(−)	(+)	(−)	(+)	(−)	(+)	(−)	(+)	(−)	(+)	(−)
Skull												
Brain												
Oral mucosa												
Salivary glands												
Cervical spine												
Thyroid												
Thymus												
Thoracic spine												
Claviculae												
Scapulae												
Ribs												
Oesophagus												
Trachea												
Lung												
Mamma												
Heart												
Stomach										**0.01**		
Spleen												
Liver												
Gall bladder												
Pancreas												
Small intestine		**0.08**		**0.10**		**0.12**		**0.14**		**0.17**		**0.15**
Large intestine		**0.02**		**0.02**		**0.03**		**0.03**		**0.04**		**0.04**
Sigma/rectum	0.05	**0.16**	0.06	**0.21**	0.08	**0.25**	0.09	**0.29**	0.11	**0.35**	0.10	**0.32**
Lumbar spine		**0.01**		**0.02**		**0.02**		**0.03**		**0.04**		**0.03**
Pelvis	0.03	**0.10**	0.04	**0.14**	0.05	**0.17**	0.06	**0.19**	0.07	**0.24**	0.06	**0.21**
Adrenals												
Kidneys												
Urinary bladder	0.07	**0.21**	0.09	**0.27**	0.11	**0.32**	0.12	**0.35**	0.15	**0.42**	0.14	**0.39**
Uterus	0.02	**0.32**	0.03	**0.42**	0.03	**0.51**	0.04	**0.58**	0.05	**0.72**	0.04	**0.64**
Ovaries		**0.07**		**0.09**		**0.11**		**0.13**	0.01	**0.17**		**0.14**
Prostate	0.12	**0.18**	0.16	**0.23**	0.19	**0.28**	0.21	**0.31**	0.26	**0.38**	0.24	**0.35**
Testes	0.45	**0.50**	0.52	**0.58**	0.56	**0.63**	0.60	**0.67**	0.66	**0.74**	0.65	**0.72**
Upper arm												
Elbow												
Forearm		**0.22**		**0.27**		**0.32**		**0.35**		**0.41**		**0.40**
Upper leg	0.13	**0.15**	0.18	**0.20**	0.22	**0.24**	0.25	**0.28**	0.31	**0.35**	0.27	**0.30**
Knee	0.18	**0.19**	0.24	**0.26**	0.29	**0.31**	0.33	**0.35**	0.41	**0.44**	0.37	**0.39**
Lower leg	0.02	**0.09**	0.03	**0.12**	0.03	**0.14**	0.03	**0.16**	0.04	**0.20**	0.04	**0.18**
Lymph nodes		**0.03**		**0.04**		**0.04**	0.01	**0.05**	0.01	**0.06**	0.01	**0.05**
Skin	0.09	**0.16**	0.09	**0.17**	0.10	**0.18**	0.10	**0.19**	0.11	**0.20**	0.10	**0.19**
Muscle	0.05	**0.09**	0.06	**0.11**	0.07	**0.12**	0.07	**0.13**	0.09	**0.15**	0.08	**0.15**
Skeleton	0.05	**0.08**	0.07	**0.11**	0.08	**0.13**	0.10	**0.15**	0.12	**0.19**	0.11	**0.17**
Red bone marrow		**0.01**		**0.02**	0.01	**0.02**	0.01	**0.03**	0.02	**0.04**	0.01	**0.03**
Whole body dose	0.05	**0.08**	0.06	**0.10**	0.06	**0.12**	0.07	**0.13**	0.08	**0.15**	0.08	**0.14**
ED (ICRP 60)	0.05	**0.08**	0.06	**0.10**	0.07	**0.12**	0.08	**0.13**	0.08	**0.15**	0.08	**0.14**
ED (ICRP 103)	0.03	**0.05**	0.03	**0.06**	0.04	**0.07**	0.04	**0.08**	0.05	**0.10**	0.04	**0.09**
Total filtration (mm) **Al**	**1.5**		**2.5**		**3.5**		**4.5**		**3.5**		**3.5**	
Cu	**0**		**0**		**0**		**0**		**0.1**		**0.2**	

Table 19.24 Normalized organ doses—Thigh ap—13–17 years

Beam projection:	ap											**Table cells:**
Focus-to-detector distance (cm):	115											(+) = normal font = optimal field setting
Tube voltage (kVp):	70											**(−) = bold font = suboptimal field setting**
Total filtrations:	See last table row											Empty table cells correspond to values <0.01
Organ or tissue	**Organ dose D_T: Entrance dose K_E**						**(mSv ∗ mGy^{-1})**					
	(+)	(−)	(+)	(−)	(+)	(−)	(+)	(−)	(+)	(−)	(+)	(−)
Skull												
Brain												
Oral mucosa												
Salivary glands												
Cervical spine												
Thyroid												
Thymus												
Thoracic spine												
Claviculae												
Scapulae												
Ribs												
Oesophagus												
Trachea												
Lung												
Mamma												
Heart												
Stomach						**0.01**		**0.01**		**0.02**		**0.02**
Spleen												
Liver												
Gall bladder										**0.01**		**0.01**
Pancreas												
Small intestine		**0.10**		**0.13**		**0.15**		**0.17**		**0.21**		**0.24**
Large intestine		**0.03**		**0.03**		**0.04**		**0.05**		**0.06**		**0.07**
Sigma/rectum	0.06	**0.21**	0.08	**0.27**	0.10	**0.31**	0.11	**0.35**	0.14	**0.43**	0.16	**0.51**
Lumbar spine		**0.02**		**0.03**		**0.03**		**0.04**		**0.05**		**0.07**
Pelvis	0.04	**0.14**	0.05	**0.18**	0.07	**0.22**	0.07	**0.25**	0.09	**0.31**	0.11	**0.37**
Adrenals												
Kidneys												**0.01**
Urinary bladder	0.09	**0.26**	0.11	**0.32**	0.13	**0.37**	0.15	**0.41**	0.18	**0.48**	0.20	**0.55**
Uterus	0.03	**0.42**	0.04	**0.54**	0.04	**0.63**	0.05	**0.71**	0.07	**0.87**	0.08	**1.03**
Ovaries		**0.09**		**0.12**	0.01	**0.15**	0.01	**0.17**	0.02	**0.21**	0.02	**0.25**
Prostate	0.15	**0.22**	0.20	**0.29**	0.23	**0.34**	0.26	**0.37**	0.31	**0.45**	0.36	**0.53**
Testes	0.49	**0.55**	0.56	**0.63**	0.60	**0.68**	0.64	**0.72**	0.70	**0.79**	0.75	**0.85**
Upper arm												
Elbow												
Forearm		**0.25**		**0.32**		**0.36**		**0.40**		**0.46**		**0.52**
Upper leg	0.18	**0.20**	0.23	**0.26**	0.28	**0.31**	0.32	**0.36**	0.39	**0.45**	0.47	**0.53**
Knee	0.24	**0.25**	0.31	**0.32**	0.36	**0.38**	0.41	**0.43**	0.50	**0.53**	0.59	**0.63**
Lower leg	0.02	**0.12**	0.03	**0.15**	0.04	**0.17**	0.04	**0.19**	0.05	**0.24**	0.06	**0.27**
Lymph nodes		**0.04**		**0.05**	0.01	**0.05**	0.01	**0.06**	0.01	**0.07**	0.01	**0.08**
Skin	0.09	**0.17**	0.10	**0.18**	0.10	**0.19**	0.11	**0.20**	0.11	**0.21**	0.12	**0.22**
Muscle	0.06	**0.10**	0.07	**0.13**	0.08	**0.14**	0.08	**0.15**	0.10	**0.17**	0.11	**0.19**
Skeleton	0.07	**0.11**	0.09	**0.14**	0.11	**0.17**	0.12	**0.19**	0.15	**0.23**	0.18	**0.27**
Red bone marrow		**0.02**	0.01	**0.03**	0.01	**0.03**	0.02	**0.04**	0.02	**0.05**	0.03	**0.06**
Whole body dose	0.05	**0.10**	0.07	**0.12**	0.07	**0.13**	0.08	**0.15**	0.09	**0.17**	0.11	**0.19**
ED (ICRP 60)	0.06	**0.10**	0.07	**0.12**	0.08	**0.13**	0.08	**0.15**	0.09	**0.17**	0.10	**0.19**
ED (ICRP 103)	0.03	**0.06**	0.04	**0.08**	0.04	**0.09**	0.05	**0.10**	0.05	**0.11**	0.06	**0.13**
Total filtration (mm) **Al**	**1.5**		**2.5**		**3.5**		**4.5**		**3.5**		**3.5**	
Cu	**0**		**0**		**0**		**0**		**0.1**		**0.2**	

Table 19.25 Normalized organ doses—Thigh ap—13–17 years

Beam projection:	ap											**Table cells:**
Focus-to-detector distance (cm):	115											(+) = normal font = optimal field setting
Tube voltage (kVp):	80											**(−) = bold font = suboptimal field setting**
Total filtrations:	See last table row											Empty table cells correspond to values <0.01
Organ or tissue	**Organ dose D_T: Entrance dose K_E**							**(mSv * mGy^{-1})**				
	(+)	(−)	(+)	(−)	(+)	(−)	(+)	(−)	(+)	(−)	(+)	(−)
Skull												
Brain												
Oral mucosa												
Salivary glands												
Cervical spine												
Thyroid												
Thymus												
Thoracic spine												
Claviculae												
Scapulae												
Ribs												
Oesophagus												
Trachea												
Lung												
Mamma												
Heart												
Stomach				**0.01**		**0.01**		**0.02**		**0.02**		**0.03**
Spleen												**0.01**
Liver												
Gall bladder						**0.01**		**0.01**		**0.01**		**0.02**
Pancreas												
Small intestine		**0.12**		**0.15**		**0.18**		**0.20**		**0.24**		**0.28**
Large intestine		**0.03**		**0.04**		**0.05**		**0.06**		**0.07**		**0.08**
Sigma/rectum	0.08	**0.26**	0.10	**0.32**	0.12	**0.37**	0.13	**0.41**	0.16	**0.50**	0.19	**0.58**
Lumbar spine		**0.03**		**0.04**		**0.05**		**0.05**		**0.07**		**0.08**
Pelvis	0.05	**0.18**	0.07	**0.23**	0.08	**0.27**	0.09	**0.30**	0.11	**0.37**	0.13	**0.43**
Adrenals												
Kidneys										**0.01**		**0.01**
Urinary bladder	0.11	**0.30**	0.13	**0.37**	0.15	**0.42**	0.17	**0.46**	0.20	**0.54**	0.22	**0.61**
Uterus	0.04	**0.52**	0.05	**0.65**	0.06	**0.76**	0.07	**0.84**	0.09	**1.02**	0.10	**1.18**
Ovaries		**0.12**	0.01	**0.15**	0.01	**0.18**	0.02	**0.20**	0.02	**0.25**	0.03	**0.29**
Prostate	0.19	**0.27**	0.23	**0.34**	0.27	**0.39**	0.30	**0.43**	0.35	**0.52**	0.41	**0.59**
Testes	0.53	**0.59**	0.59	**0.67**	0.64	**0.72**	0.67	**0.76**	0.73	**0.83**	0.78	**0.89**
Upper arm												
Elbow												
Forearm		**0.29**		**0.35**		**0.40**		**0.43**		**0.50**		**0.55**
Upper leg	0.23	**0.26**	0.29	**0.33**	0.34	**0.38**	0.38	**0.43**	0.46	**0.53**	0.54	**0.62**
Knee	0.29	**0.31**	0.37	**0.39**	0.43	**0.45**	0.48	**0.51**	0.58	**0.61**	0.67	**0.71**
Lower leg	0.03	**0.14**	0.04	**0.17**	0.05	**0.20**	0.05	**0.22**	0.06	**0.26**	0.07	**0.30**
Lymph nodes		**0.04**	0.01	**0.05**	0.01	**0.06**	0.01	**0.07**	0.01	**0.08**	0.02	**0.09**
Skin	0.09	**0.18**	0.10	**0.19**	0.11	**0.20**	0.11	**0.20**	0.12	**0.22**	0.12	**0.23**
Muscle	0.07	**0.12**	0.08	**0.14**	0.09	**0.16**	0.09	**0.17**	0.11	**0.19**	0.12	**0.21**
Skeleton	0.09	**0.13**	0.11	**0.17**	0.13	**0.20**	0.14	**0.22**	0.17	**0.26**	0.20	**0.31**
Red bone marrow	0.01	**0.03**	0.02	**0.04**	0.02	**0.04**	0.02	**0.05**	0.03	**0.06**	0.03	**0.07**
Whole body dose	0.06	**0.11**	0.07	**0.13**	0.08	**0.15**	0.09	**0.16**	0.10	**0.19**	0.12	**0.21**
ED (ICRP 60)	0.07	**0.11**	0.08	**0.13**	0.08	**0.15**	0.09	**0.16**	0.10	**0.19**	0.11	**0.21**
ED (ICRP 103)	0.04	**0.07**	0.04	**0.09**	0.05	**0.10**	0.05	**0.11**	0.06	**0.13**	0.06	**0.15**
Total filtration (mm) Al	**1.5**		**2.5**		**3.5**		**4.5**		**3.5**		**3.5**	
Cu	**0**		**0**		**0**		**0**		**0.1**		**0.2**	

Table 19.26 Normalized organ doses—Thigh ap—>17 years

Beam projection:			ap									**Table cells:**
Focus-to-detector distance (cm):			115									(+) = normal font = optimal field setting
Tube voltage (kVp):			60									**(−) = bold font = suboptimal field setting**
Total filtrations:			See last table row									Empty table cells correspond to values <0.01
Organ or tissue	**Organ dose D_T: Entrance dose K_E**								**(mSv ∗ mGy^{-1})**			
	(+)	(−)	(+)	(−)	(+)	(−)	(+)	(−)	(+)	(−)	(+)	(−)
Skull												
Brain												
Oral mucosa												
Salivary glands												
Cervical spine												
Thyroid												
Thymus												
Thoracic spine												
Claviculae												
Scapulae												
Ribs												
Oesophagus												
Trachea												
Lung												
Mamma												
Heart												
Stomach										**0.01**		
Spleen												
Liver												
Gall bladder												
Pancreas												
Small intestine		**0.08**		**0.10**		**0.12**		**0.13**		**0.17**		**0.15**
Large intestine		**0.02**		**0.02**		**0.03**		**0.03**		**0.04**		**0.04**
Sigma/rectum	0.05	**0.15**	0.06	**0.20**	0.07	**0.24**	0.08	**0.27**	0.11	**0.34**	0.09	**0.30**
Lumbar spine		**0.01**		**0.02**		**0.02**		**0.02**		**0.03**		**0.02**
Pelvis	0.03	**0.09**	0.04	**0.13**	0.05	**0.15**	0.05	**0.18**	0.07	**0.23**	0.06	**0.19**
Adrenals												
Kidneys												
Urinary bladder	0.08	**0.21**	0.10	**0.27**	0.11	**0.31**	0.13	**0.35**	0.15	**0.42**	0.14	**0.39**
Uterus		**0.12**		**0.16**	0.01	**0.20**	0.01	**0.22**	0.02	**0.28**	0.01	**0.24**
Ovaries		**0.07**		**0.09**		**0.11**		**0.12**	0.01	**0.16**		**0.13**
Prostate	0.09	**0.16**	0.12	**0.20**	0.14	**0.24**	0.16	**0.27**	0.19	**0.34**	0.17	**0.31**
Testes	0.35	**0.46**	0.41	**0.54**	0.45	**0.59**	0.49	**0.63**	0.54	**0.71**	0.53	**0.69**
Upper arm												
Elbow												
Forearm		**0.21**		**0.27**		**0.31**		**0.35**		**0.41**		**0.39**
Upper leg	0.10	**0.11**	0.13	**0.15**	0.16	**0.18**	0.19	**0.21**	0.24	**0.27**	0.20	**0.23**
Knee	0.14	**0.15**	0.19	**0.20**	0.23	**0.24**	0.26	**0.28**	0.33	**0.35**	0.29	**0.30**
Lower leg	0.03	**0.09**	0.04	**0.11**	0.05	**0.13**	0.06	**0.15**	0.07	**0.19**	0.06	**0.17**
Lymph nodes		**0.03**		**0.04**		**0.04**		**0.05**		**0.06**		**0.05**
Skin	0.08	**0.16**	0.08	**0.17**	0.09	**0.18**	0.09	**0.18**	0.10	**0.20**	0.09	**0.19**
Muscle	0.05	**0.09**	0.06	**0.11**	0.07	**0.12**	0.07	**0.13**	0.08	**0.15**	0.08	**0.14**
Skeleton	0.04	**0.07**	0.06	**0.09**	0.07	**0.11**	0.08	**0.13**	0.10	**0.16**	0.09	**0.14**
Red bone marrow		**0.01**		**0.02**		**0.02**		**0.02**	0.01	**0.03**		**0.02**
Whole body dose	0.04	**0.08**	0.05	**0.10**	0.06	**0.11**	0.07	**0.12**	0.08	**0.14**	0.07	**0.13**
ED (ICRP 60)	0.04	**0.08**	0.05	**0.10**	0.06	**0.11**	0.06	**0.12**	0.07	**0.14**	0.07	**0.13**
ED (ICRP 103)	0.02	**0.05**	0.03	**0.06**	0.03	**0.07**	0.03	**0.07**	0.04	**0.09**	0.04	**0.08**
Total filtration (mm) **Al**	**1.5**		**2.5**		**3.5**		**4.5**		**3.5**		**3.5**	
Cu	**0**		**0**		**0**		**0**		**0.1**		**0.2**	

Table 19.27 Normalized organ doses—Thigh ap—>17 years

Beam projection:		ap										**Table cells:**
Focus-to-detector distance (cm):		115										(+) = normal font = optimal field setting
Tube voltage (kVp):		70										**(−) = bold font = suboptimal field setting**
Total filtrations:		See last table row										Empty table cells correspond to values <0.01
Organ or tissue	**Organ dose D_T: Entrance dose K_E**							**(mSv * mGy^{-1})**				
	(+)	(−)	(+)	(−)	(+)	(−)	(+)	(−)	(+)	(−)	(+)	(−)
Skull												
Brain												
Oral mucosa												
Salivary glands												
Cervical spine												
Thyroid												
Thymus												
Thoracic spine												
Claviculae												
Scapulae												
Ribs												
Oesophagus												
Trachea												
Lung												
Mamma												
Heart												
Stomach						**0.01**		**0.01**		**0.02**		**0.02**
Spleen												
Liver												
Gall bladder										**0.01**		**0.01**
Pancreas												
Small intestine		**0.10**		**0.13**		**0.15**		**0.17**		**0.20**		**0.24**
Large intestine		**0.03**		**0.03**		**0.04**		**0.05**		**0.06**		**0.07**
Sigma/rectum	0.06	**0.20**	0.08	**0.26**	0.09	**0.30**	0.11	**0.34**	0.13	**0.42**	0.16	**0.49**
Lumbar spine		**0.02**		**0.03**		**0.03**		**0.04**		**0.05**		**0.06**
Pelvis	0.04	**0.13**	0.05	**0.17**	0.06	**0.20**	0.07	**0.23**	0.09	**0.29**	0.11	**0.35**
Adrenals												
Kidneys												
Urinary bladder	0.09	**0.26**	0.12	**0.32**	0.14	**0.37**	0.15	**0.41**	0.18	**0.48**	0.21	**0.55**
Uterus	0.01	**0.16**	0.01	**0.21**	0.02	**0.25**	0.02	**0.28**	0.02	**0.34**	0.03	**0.40**
Ovaries		**0.09**		**0.12**	0.01	**0.14**	0.01	**0.16**	0.02	**0.20**	0.02	**0.24**
Prostate	0.11	**0.20**	0.15	**0.26**	0.17	**0.30**	0.19	**0.34**	0.23	**0.41**	0.28	**0.48**
Testes	0.39	**0.51**	0.45	**0.59**	0.49	**0.65**	0.53	**0.69**	0.58	**0.76**	0.63	**0.82**
Upper arm												
Elbow												
Forearm		**0.25**		**0.31**		**0.36**		**0.40**		**0.46**		**0.52**
Upper leg	0.14	**0.16**	0.18	**0.21**	0.21	**0.25**	0.24	**0.28**	0.31	**0.35**	0.37	**0.43**
Knee	0.19	**0.20**	0.25	**0.26**	0.29	**0.31**	0.33	**0.35**	0.41	**0.44**	0.49	**0.52**
Lower leg	0.04	**0.11**	0.05	**0.14**	0.06	**0.17**	0.07	**0.19**	0.09	**0.23**	0.11	**0.27**
Lymph nodes		**0.04**		**0.04**		**0.05**		**0.06**	0.01	**0.07**	0.01	**0.08**
Skin	0.08	**0.17**	0.09	**0.18**	0.09	**0.19**	0.10	**0.19**	0.10	**0.21**	0.11	**0.22**
Muscle	0.06	**0.10**	0.07	**0.12**	0.08	**0.14**	0.08	**0.15**	0.10	**0.17**	0.11	**0.19**
Skeleton	0.06	**0.09**	0.08	**0.12**	0.09	**0.14**	0.10	**0.16**	0.13	**0.20**	0.16	**0.24**
Red bone marrow		**0.02**		**0.02**		**0.03**	0.01	**0.03**	0.01	**0.04**	0.02	**0.05**
Whole body dose	0.05	**0.10**	0.06	**0.11**	0.07	**0.13**	0.08	**0.14**	0.09	**0.16**	0.10	**0.19**
ED (ICRP 60)	0.05	**0.09**	0.06	**0.11**	0.07	**0.13**	0.07	**0.14**	0.08	**0.16**	0.09	**0.18**
ED (ICRP 103)	0.03	**0.06**	0.03	**0.07**	0.04	**0.08**	0.04	**0.09**	0.05	**0.10**	0.05	**0.12**
Total filtration (mm) **Al**	**1.5**		**2.5**		**3.5**		**4.5**		**3.5**		**3.5**	
Cu	**0**		**0**		**0**		**0**		**0.1**		**0.2**	

Table 19.28 Normalized organ doses—Thigh ap—>17 years

Beam projection:			ap									**Table cells:**
Focus-to-detector distance (cm):			115									(+) = normal font = optimal field setting
Tube voltage (kVp):			80									**(−) = bold font = suboptimal field setting**
Total filtrations:			See last table row									Empty table cells correspond to values <0.01
Organ or tissue	**Organ dose D_T: Entrance dose K_E (mSv * mGy^{-1})**											
	(+)	(−)	(+)	(−)	(+)	(−)	(+)	(−)	(+)	(−)	(+)	(−)
Skull												
Brain												
Oral mucosa												
Salivary glands												
Cervical spine												
Thyroid												
Thymus												
Thoracic spine												
Claviculae												
Scapulae												
Ribs												
Oesophagus												
Trachea												
Lung												
Mamma												
Heart												
Stomach				**0.01**		**0.01**		**0.02**		**0.02**		**0.02**
Spleen												**0.01**
Liver												
Gall bladder								**0.01**		**0.01**		**0.02**
Pancreas												
Small intestine		**0.12**		**0.15**		**0.18**		**0.20**		**0.24**		**0.27**
Large intestine		**0.03**		**0.04**		**0.05**		**0.06**		**0.07**		**0.08**
Sigma/rectum	0.08	**0.25**	0.10	**0.31**	0.11	**0.36**	0.13	**0.40**	0.16	**0.49**	0.18	**0.57**
Lumbar spine		**0.03**		**0.04**		**0.04**		**0.05**		**0.06**		**0.08**
Pelvis	0.05	**0.17**	0.06	**0.21**	0.08	**0.25**	0.09	**0.28**	0.11	**0.35**	0.12	**0.41**
Adrenals												
Kidneys										**0.01**		**0.01**
Urinary bladder	0.11	**0.30**	0.14	**0.37**	0.16	**0.42**	0.17	**0.46**	0.20	**0.54**	0.23	**0.61**
Uterus	0.01	**0.20**	0.02	**0.25**	0.02	**0.30**	0.02	**0.33**	0.03	**0.40**	0.04	**0.47**
Ovaries		**0.12**	0.01	**0.15**	0.01	**0.17**	0.02	**0.19**	0.02	**0.24**	0.02	**0.28**
Prostate	0.14	**0.24**	0.17	**0.31**	0.20	**0.35**	0.23	**0.39**	0.27	**0.47**	0.31	**0.54**
Testes	0.42	**0.55**	0.49	**0.64**	0.53	**0.69**	0.56	**0.73**	0.62	**0.80**	0.66	**0.86**
Upper arm												
Elbow												
Forearm		**0.29**		**0.35**		**0.40**		**0.43**		**0.50**		**0.56**
Upper leg	0.18	**0.20**	0.22	**0.26**	0.26	**0.31**	0.30	**0.35**	0.37	**0.43**	0.44	**0.51**
Knee	0.24	**0.25**	0.30	**0.32**	0.35	**0.38**	0.40	**0.42**	0.48	**0.51**	0.57	**0.60**
Lower leg	0.05	**0.13**	0.07	**0.17**	0.08	**0.20**	0.09	**0.22**	0.10	**0.26**	0.12	**0.30**
Lymph nodes		**0.04**		**0.05**	0.01	**0.06**	0.01	**0.07**	0.01	**0.08**	0.01	**0.09**
Skin	0.09	**0.17**	0.09	**0.19**	0.10	**0.20**	0.10	**0.20**	0.11	**0.21**	0.11	**0.23**
Muscle	0.06	**0.12**	0.08	**0.14**	0.08	**0.15**	0.09	**0.17**	0.10	**0.19**	0.12	**0.21**
Skeleton	0.07	**0.12**	0.09	**0.15**	0.11	**0.17**	0.12	**0.19**	0.15	**0.23**	0.18	**0.27**
Red bone marrow		**0.02**	0.01	**0.03**	0.01	**0.03**	0.01	**0.04**	0.02	**0.05**	0.02	**0.06**
Whole body dose	0.06	**0.11**	0.07	**0.13**	0.08	**0.15**	0.09	**0.16**	0.10	**0.18**	0.11	**0.20**
ED (ICRP 60)	0.06	**0.11**	0.07	**0.13**	0.07	**0.14**	0.08	**0.15**	0.09	**0.18**	0.10	**0.20**
ED (ICRP 103)	0.03	**0.07**	0.04	**0.08**	0.04	**0.09**	0.04	**0.10**	0.05	**0.12**	0.06	**0.13**
Total filtration (mm) Al	**1.5**		**2.5**		**3.5**		**4.5**		**3.5**		**3.5**	
Cu	**0**		**0**		**0**		**0**		**0.1**		**0.2**	

Shoulder

20

20.1 Introduction

The type of shoulder radiography is determined by the clinical indication. Table 20.1 shows indications and Table 20.2 contraindications for shoulder radiography. Image criteria as required for radiological findings in shoulder radiographs are listed in Table 20.3. In order to fulfil these image criteria, the field settings listed in Tables 20.4 and 20.5 should be achieved. These field settings go along with the patient preparation as listed in Table 20.6. Exposure parameter combinations for various age groups and patient diameters are listed in Tables 20.7, 20.8 and 20.9. The age groups in Table 20.8 refer to the standard age groups of the mathematical MIRD phantom family which has been used to calculate the normalized organ doses in this book. Table 20.9 shows

Table 20.1 Indications for shoulder radiographs

Birth trauma
Fracture
Luxation
Simple bone cyst
Osteomyelitis
Battered child syndrome
Soft tissue tumours
Bone tumour

Table 20.2 Contraindications for shoulder radiographs

None

Table 20.3 Image criteria for shoulder radiographs

Shoulder ap and Y projection
Reproduction of the acromion, the lateral part of the clavicle, the coracoid, the head of the humerus with the proximal humeral epi- and metaphysis consistent with age
Visually sharp reproduction of the fossa glenoidalis and the acromioclavicular joint
Visually sharp reproduction of the glenohumeral joint space
Visually sharp reproduction of the cortex and spongiosa of the humeral head

Table 20.4 Patient positions and beam projections for shoulder radiographs

Patient position	Beam projections
Supine	ap + Y projection[a]
Annotation:	
[a] The Y projection is not considered in this book	

Table 20.5 Optimal field settings for thigh radiographs

Shoulder ap and Y projection	
Upper field edge:	Above the acromioclavicular joint (including the skin)
Field centring:	Centre of the axilla
Lower field edge:	Upper third of the proximal humerus
Medial field edge:	Medial skin surface of the upper arm
Lateral field edge:	Lateral skin surface of the upper arm
Adjust angulation of the tube between 12° and 20° dependent on age of the patient	

M. Seidenbusch et al., *Imaging Practice and Radiation Protection in Pediatric Radiology*,
https://doi.org/10.1007/978-3-030-18504-6_20

radiation qualities (i.e. various combinations of tube voltage and total filtration) which may be applied in clinical practice. Please note that not all of these radiation qualities are recommended by the authors. The standard projections of shoulder radiographies of the patient who is correctly positioned are depicted in Fig. 20.1. Also the optimal field (marked in green), the suboptimal field (marked in red), and the correct shielding are shown. Tables 20.11, 20.12 and 20.13 outline the virtual field settings on the skin surface of mathematical MIRD phantoms of various age groups and the resulting virtual radiographies with and without bone structures, respectively. Normalized organ doses resulting from these virtual field settings have been calculated for all age groups and have been listed in Tables 20.14, 20.15, 20.16, 20.17, 20.18, 20.19, 20.20, 20.21,

Table 20.6 Patient preparation

General	Remove all clothes, external metallic devices, and jewellery from the neck and chest of the patient
	Immobilize infants and uncooperative patients
	Cooperative patients can be positioned upright at the vertical stand
	Place eventually small X-ray positive markers at palpable tumours or on pain points
Positioning	For the ap projection, the arm should be slightly positioned alongside the chest in the supine position for the ap view
	The Y projection of the shoulder should be done in supine position or at vertical stands in upright position with 30° oblique ap/pa projection
Shielding	Lead mask shielding should be used at the entrance side at the medial field edge in ap supine views

Table 20.7 Radiographic procedure

X-ray source	Use automatic exposure control (AEC) in patients >10 years of age
	Focal spot size between 0.6 and 1.3 dependent on the age of the patient
	Check the use of additional filtration
	Check the focus-to-detector distance
Field setting	Check the field size
	Check the field centring
Image detector system	Check the use of antiscatter grids
Patient	–

Table 20.8 Radiographic techniques of shoulder radiographs

Age group (years)	Beam direction	Patient position	Beam projection	Typical patient diameter (cm)	Optimal field size (cm x cm)	Focus-to-detector distance (cm)	Voltage (kVp)						Antiscatter grid
							50	60	70	80	90	100	
0	Vertical	Supine	ap	9.8	3 x 4	100[1]							
1 – 2				13.0	4 x 6								
3 – 7				15.0	5 x 8								
8 – 12				16.8	6 x 10								
13 – 17				19.6	7 x 12	115							
18				20.0	7 x 14								

Annotations:
1 The patient is lying on the table top directly on the detector
(dark) Parameter combination not recommended, normalized organ doses not provided
(light) Parameter combination recommended, normalized organ doses provided

Table 20.9 Radiation qualities of shoulder radiographs

Voltage (kVp)	Total filtration					
	Without additional filtration		With additional filtration			
	1.5 mm Al	2.5 Al	3.5 Al	4.5 Al	3.5 Al + 0.1 Cu[1]	3.5 Al + 0.2 Cu[1]
50						
60						
70						
80						
90						
100						

Annotations:
1 The total beam filtration is usually 3.5 mm Al + 0.1 mm Cu, in adolescents optionally 3.5 mm Al + 0.2 mm Cu.
(dark) Parameter combination not recommended, normalized organ doses not provided
(grey) Parameter combination not recommended, normalized organ doses provided
(light) Parameter combination recommended, normalized organ doses provided

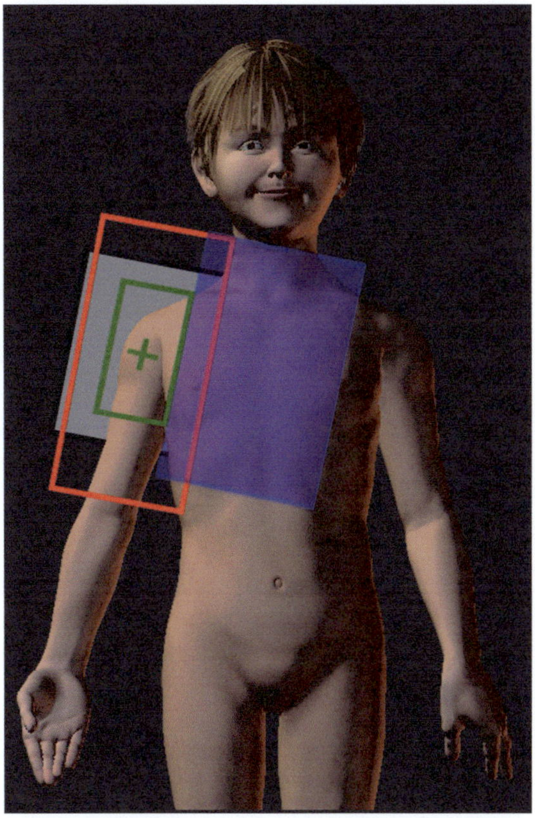

Fig. 20.1 **Field settings for shoulder radiographs in supine position**. Antero-posterior projection. (© V. Rösenberger and M. Seidenbusch 2018). Green lines = optimal field settings; green cross = centring point; red lines = suboptimal field settings; grey shaded area = image detector; blue shaded area = lead mask shielding; dark grey shaded area = Bucky table

Table 20.10 Inconsistencies between the mathematical MIRD phantoms and real human anatomy

Anatomical region	Inconsistencies
Extremities	No shoulder joint has been inserted

For details, see the text

20.22, 20.23, 20.24, 20.25, 20.26, 20.27 and 20.28. Table cells with normal font refer to optimal field settings and table cells with bold font refer to suboptimal field settings. Please note that normalized organ doses have been provided also for exposure parameter combinations, which are not recommended by the authors.

20.2 Indications and Contraindications

Indications and contraindications for shoulder radiographs are given in Tables 20.1 and 20.2.

20.3 Image Criteria

The image criteria required for shoulder radiographs are listed in Table 20.3.

See also [2].

20.4 Field Settings

In Table 20.4, patient positions and beam projections used for shoulder radiographs are listed, and in Table 20.5, the optimal field settings are listed which are required in order to fulfil the image criteria mentioned in Table 20.3.

20.5 Patient Preparation and Radiographic Procedure

See Tables 20.6, 20.7 and [1].

20.6 Exposure Parameters

In Tables 20.8 and 20.9, exposure parameter combinations used for shoulder radiographs in various age groups are listed. Dark grey shaded table cells refer to exposure parameter combinations which are not recommended for paediatric shoulder radiographs and for which no normalized organ doses have been provided. Light grey and light blue shaded table cells refer to exposure parameter combinations for which normalized organ doses have been provided with light grey shaded table cells referring to exposure parameter combinations being not recommended and light blue shaded table cells referring to exposure parameter combinations being recommended for paediatric shoulder radiographs. Note that due to the fact that the radiation dose is an exponential function of the patient diameter, it may make sense to use the patient diameters instead of the patient's age groups for determining the adequate exposure parameters from Table 20.8. In this case, for determining the corresponding normalized organ doses from Tables 20.14, 20.15, 20.16, 20.17, 20.18, 20.19, 20.20, 20.21, 20.22, 20.23, 20.24, 20.25, 20.26, 20.27 and 20.28, the age group related to the patient diameter should be used.

20.7 Normalized Organ Doses

For calculating normalized organ doses, the field settings described in Sect. 20.4 were transferred to the mathematical MIRD phantoms. Inconsistencies between the geometrical anatomy of the mathematical phantoms and the real human anatomy have been listed in Table 20.10 whenever they are of relevance for the validity of the normalized organ doses.

In Tables 20.11, 20.12 and 20.13, optimal and suboptimal field settings are listed for the different age groups and the beam projections given in Tables 20.8 and 20.9. Green lines refer to optimal field settings, red lines refer to suboptimal field settings, respectively. In the tables, the optimal and suboptimal field settings which were used for performing the calculations of the normalized organ doses provided in Tables 20.14, 20.15, 20.16, 20.17, 20.18, 20.19, 20.20, 20.21, 20.22, 20.23, 20.24, 20.25, 20.26, 20.27 and 20.28 were drawn in on the skin surface of the mathematical phantoms. The virtual mathematical radiographs resulting from these field settings were depicted in the tables with and without bone structures, respectively, in order to enable the reader to visually compare the virtual mathematical radiographs with the real radiographs made under clinical conditions. For comparing the virtual mathematical field settings with the real clinical field settings, the virtual mathematical field edges were also listed in the tables.

As the normalized organ doses provided in Tables 20.14, 20.15, 20.16, 20.17, 20.18, 20.19, 20.20, 20.21, 20.22, 20.23, 20.24, 20.25, 20.26, 20.27 and 20.28 refer to entrance kerma values, optimal and suboptimal field sizes were provided in the tables in order to enable the reader to calculate entrance air kerma dose values from clinically measured dose area product values. Please note that these optimal and suboptimal field sizes refer to the mathematical phantoms of an age group only and not to real patients with corresponding ages.

Please note also that the anatomy of the mathematical phantoms has been idealized. Therefore, the virtual field settings for calculating the normalized organ doses have been defined with a view to ensuring that the resulting virtual organ doses would be consistent with the real organ doses which would be achieved during real X-ray examinations being performed with the real field settings recommended in Sect. 20.4. Thus, the field settings shown in the following tables may not be consistent with the clinical field settings as recommended in Sect. 20.4.

Table 20.11 Field settings—shoulder ap

Age group (years)		Anatomical landmarks			
		Optimal field setting		Suboptimal field setting	
0	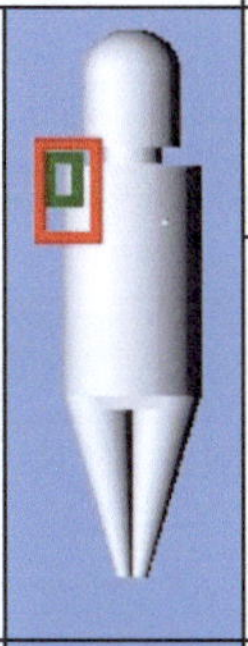	Upper field edge: Just above the skin surface of the shoulder Lower field edge: Upper third of the upper arm Medial field edge: Between the anterior axillary line and the mammillary line Lateral field edge: Just beneath the skin surface of the upper arm Field size = field width × field height: 3 cm × 4 cm = 12 cm²		Upper field edge: Just below the tip of the mandible Lower field edge: Elbow Medial field edge: Close to the mammillary line Lateral field edge: Above the skin surface of the upper arm Field size = field width × field height: 5 cm × 9 cm = 45 cm²	
			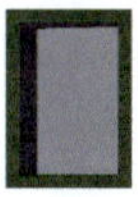		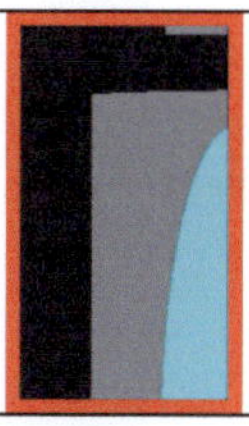
1 - 2	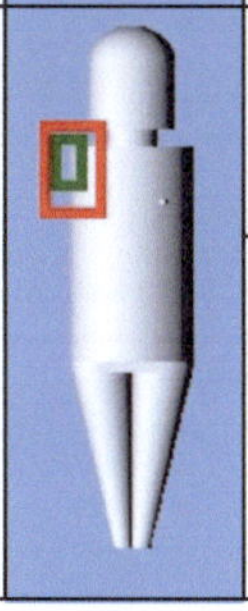	Upper field edge: Just above the skin surface of the shoulder Lower field edge: Upper third of the upper arm Medial field edge: Between the anterior axillary line and the mammillary line Lateral field edge: Just beneath the skin surface of the upper arm Field size = field width × field height: 3 cm × 6 cm = 18 cm²		Upper field edge: Just below the tip of the mandible Lower field edge: Elbow Medial field edge: Close to the mammillary line Lateral field edge: Above the skin surface of the upper arm Field size = field width × field height: 7 cm × 12 cm = 84 cm²	
					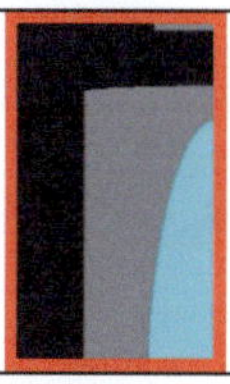

Table 20.12 Field settings—shoulder ap

Age group (years)		Anatomical landmarks			
		Optimal field setting		Suboptimal field setting	
3 - 7	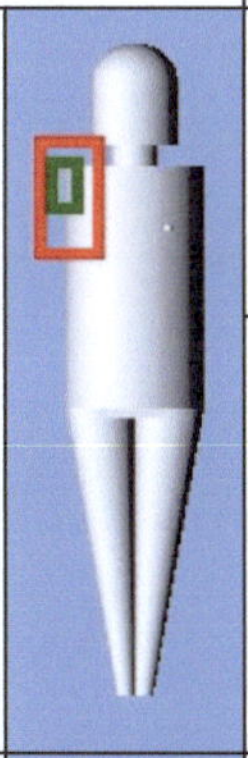	Upper field edge: Just above the skin surface of the shoulder Lower field edge: Upper third of the upper arm Medial field edge: Between the anterior axillary line and the mammillary line Lateral field edge: Just beneath the skin surface of the upper arm Field size = field width × field height: 5 cm × 8 cm = 40 cm²		Upper field edge: Just below the tip of the mandible Lower field edge: Elbow Medial field edge: Close to the mammillary line Lateral field edge: Above the skin surface of the upper arm Field size = field width × field height: 9 cm × 18 cm = 162 cm²	
				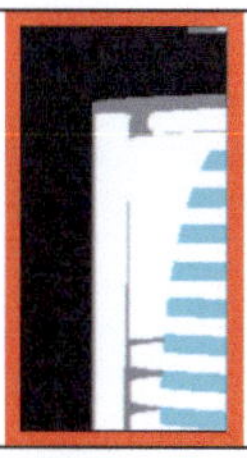	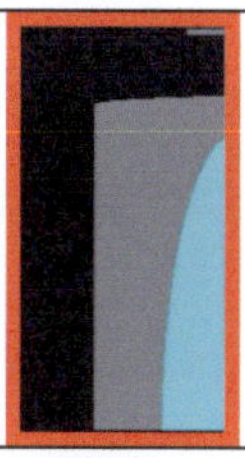
8 - 12	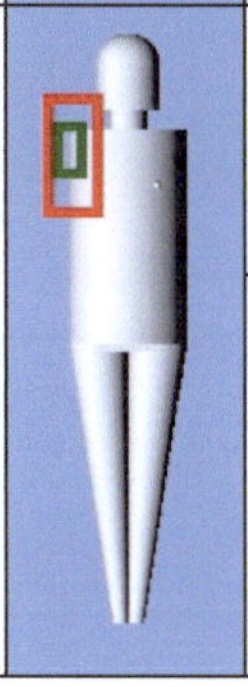	Upper field edge: Just above the skin surface of the shoulder Lower field edge: Upper third of the upper arm Medial field edge: Between the anterior axillary line and the mammillary line Lateral field edge: Just beneath the skin surface of the upper arm Field size = field width × field height: 6 cm × 10 cm = 60 cm²		Upper field edge: Just below the tip of the mandible Lower field edge: Elbow Medial field edge: Close to the mammillary line Lateral field edge: Above the skin surface of the upper arm Field size = field width x field height: 10 cm x 26 cm = 260 cm²	
					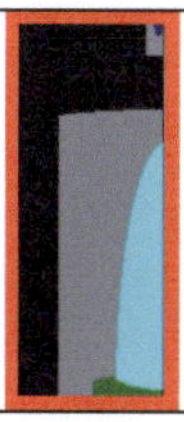

Table 20.13 Field settings—shoulder ap

Age group (years)		Anatomical landmarks			
		Optimal field setting		Suboptimal field setting	
13 - 17	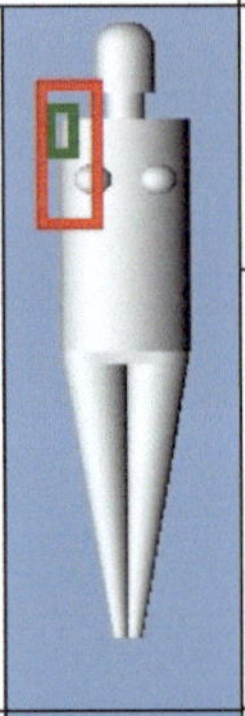	Upper field edge: Just above the skin surface of the shoulder Lower field edge: Upper third of the upper arm Medial field edge: Between the anterior axillary line and the mammillary line Lateral field edge: Just beneath the skin surface of the upper arm Field size = field width × field height: 7 cm × 12 cm = 84 cm^2		Upper field edge: Just below the tip of the mandible Lower field edge: Elbow Medial field edge: Close to the mammillary line Lateral field edge: Above the skin surface of the upper arm Field size = field width × field height: 11 cm × 36 cm = 396 cm^2	

18	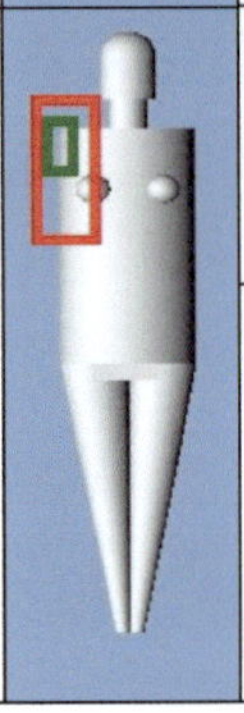	Upper field edge: Just above the skin surface of the shoulder Lower field edge: Upper third of the upper arm Medial field edge: Between the anterior axillary line and the mammillary line Lateral field edge: Just beneath the skin surface of the upper arm Field size = field width × field height: 7 cm × 14 cm = 98 cm^2		Upper field edge: Just below the tip of the mandible Lower field edge: Elbow Medial field edge: Close to the mammillary line Lateral field edge: Above the skin surface of the upper arm Field size = field width × field height: 12 cm × 40 cm = 480 cm^2	
				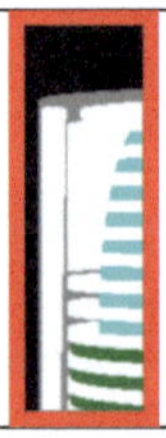	

Table 20.14 Normalized organ doses—Shoulder ap—<1 year

Beam projection:			ap									**Table cells:**
Focus-to-detector distance (cm):			100									(+) = normal font = optimal field setting
Tube voltage (kVp):			60									**(−) = bold font = suboptimal field setting**
Total filtrations:			See last table row.									Empty table cells correspond to values <0.01
Organ or tissue	**Organ dose D_T : Entrance dose K_E**								**(mSv * mGy^{-1})**			
	(+)	(−)	(+)	(−)	(+)	(−)	(+)	(−)	(+)	(−)	(+)	(−)
Skull												
Brain												
Oral mucosa												
Salivary glands						**0.01**		**0.01**		**0.01**		**0.01**
Cervical spine		**0.01**		**0.02**		**0.02**		**0.02**		**0.03**		**0.03**
Thyroid				**0.01**		**0.01**		**0.02**		**0.02**		**0.02**
Thymus		**0.01**		**0.02**		**0.02**		**0.02**		**0.02**		**0.02**
Thoracic spine		**0.02**		**0.03**		**0.04**		**0.04**		**0.05**		**0.05**
Claviculae	0.05	**0.49**	0.06	**0.57**	0.06	**0.63**	0.07	**0.67**	0.07	**0.73**	0.08	**0.74**
Scapulae	0.06	**0.23**	0.07	**0.29**	0.09	**0.34**	0.10	**0.38**	0.11	**0.45**	0.11	**0.44**
Ribs	0.05	**0.27**	0.06	**0.32**	0.07	**0.35**	0.07	**0.37**	0.08	**0.41**	0.08	**0.42**
Oesophagus		**0.01**		**0.02**		**0.02**		**0.02**		**0.03**		**0.02**
Trachea												
Lung		**0.12**		**0.14**		**0.15**		**0.16**		**0.18**		**0.18**
Mamma		**0.55**		**0.57**		**0.58**		**0.59**		**0.59**		**0.59**
Heart		**0.01**		**0.02**		**0.02**		**0.02**		**0.02**		**0.02**
Stomach												
Spleen												
Liver				**0.01**		**0.01**		**0.01**		**0.02**		**0.02**
Gall bladder												
Pancreas												
Small intestine												
Large intestine												
Sigma/Rectum												
Lumbar spine												
Pelvis												
Adrenals								**0.01**		**0.01**		**0.01**
Kidneys												
Urinary bladder												
Uterus												
Ovaries												
Prostate												
Testes												
Upper arm	0.40	**0.60**	0.48	**0.72**	0.53	**0.81**	0.57	**0.87**	0.64	**0.97**	0.64	**0.98**
Elbow		**0.23**		**0.28**		**0.31**		**0.33**		**0.37**		**0.38**
Forearm												
Upper leg												
Knee												
Lower leg												
Lymph nodes		**0.01**		**0.02**		**0.02**		**0.02**		**0.02**		**0.02**
Skin	0.01	**0.05**	0.02	**0.05**	0.02	**0.05**	0.02	**0.05**	0.02	**0.06**	0.02	**0.06**
Muscle		**0.02**		**0.03**		**0.03**		**0.03**		**0.04**		**0.03**
Skeleton	0.04	**0.11**	0.04	**0.13**	0.05	**0.14**	0.05	**0.15**	0.06	**0.17**	0.06	**0.17**
Red bone marrow		**0.02**		**0.03**		**0.03**		**0.03**	0.01	**0.04**	0.01	**0.03**
Whole body dose		**0.03**		**0.03**		**0.04**	0.01	**0.04**	0.01	**0.05**	0.01	**0.04**
ED (ICRP 60)		**0.05**		**0.05**		**0.06**		**0.06**		**0.06**		**0.06**
ED (ICRP 103)		**0.09**		**0.09**		**0.10**		**0.10**		**0.10**		**0.10**
Total filtration (mm) Al	**1.5**		**2.5**		**3.5**		**4.5**		**3.5**		**3.5**	
Cu	**0**		**0**		**0**		**0**		**0.1**		**0.2**	

Table 20.15 Normalized organ doses—Shoulder ap—<1 year

Beam projection:			ap									**Table cells:**
Focus-to-detector distance (cm):			100						(+) = normal font = optimal field setting			
Tube voltage (kVp):			70						**(−) = bold font = suboptimal field setting**			
Total filtrations:			See last table row.						Empty table cells correspond to values <0.01			
Organ or tissue	**Organ dose D_T : Entrance dose K_E**							**(mSv * mGy^{-1})**				
	(+)	(−)	(+)	(−)	(+)	(−)	(+)	(−)	(+)	(−)	(+)	(−)
Skull												**0.01**
Brain												
Oral mucosa												**0.01**
Salivary glands				**0.01**		**0.01**		**0.01**		**0.02**		**0.02**
Cervical spine		**0.02**		**0.02**		**0.03**		**0.03**		**0.04**		**0.04**
Thyroid		**0.01**		**0.01**		**0.02**		**0.02**		**0.02**		**0.03**
Thymus		**0.01**		**0.02**		**0.02**		**0.02**		**0.03**		**0.03**
Thoracic spine		**0.03**		**0.04**		**0.04**		**0.05**		**0.06**	0.01	**0.07**
Claviculae	0.05	**0.53**	0.06	**0.61**	0.07	**0.65**	0.07	**0.69**	0.08	**0.74**	0.08	**0.77**
Scapulae	0.07	**0.27**	0.09	**0.34**	0.10	**0.39**	0.11	**0.42**	0.13	**0.49**	0.14	**0.55**
Ribs	0.06	**0.30**	0.07	**0.34**	0.07	**0.37**	0.08	**0.39**	0.09	**0.43**	0.09	**0.45**
Oesophagus		**0.01**		**0.02**		**0.02**		**0.02**		**0.03**		**0.03**
Trachea												
Lung		**0.13**		**0.15**		**0.17**		**0.18**		**0.19**		**0.21**
Mamma		**0.56**		**0.57**		**0.58**		**0.59**		**0.60**		**0.60**
Heart		**0.01**		**0.02**		**0.02**		**0.02**		**0.03**		**0.03**
Stomach												
Spleen												
Liver		**0.01**		**0.01**		**0.02**		**0.02**		**0.02**		**0.02**
Gall bladder												**0.01**
Pancreas										**0.01**		**0.01**
Small intestine												
Large intestine												
Sigma/Rectum												
Lumbar spine										**0.01**		**0.01**
Pelvis												
Adrenals						**0.01**		**0.01**		**0.02**		**0.02**
Kidneys												
Urinary bladder												
Uterus												
Ovaries												
Prostate												
Testes												
Upper arm	0.44	**0.66**	0.51	**0.78**	0.56	**0.86**	0.60	**0.92**	0.66	**1.01**	0.70	**1.08**
Elbow		**0.26**		**0.30**		**0.33**		**0.35**		**0.39**		**0.42**
Forearm												
Upper leg												
Knee												
Lower leg												
Lymph nodes		**0.02**		**0.02**		**0.02**		**0.02**		**0.03**		**0.03**
Skin	0.02	**0.05**	0.02	**0.05**	0.02	**0.05**	0.02	**0.06**	0.02	**0.06**	0.02	**0.06**
Muscle		**0.03**		**0.03**		**0.03**		**0.03**		**0.04**		**0.04**
Skeleton	0.04	**0.12**	0.05	**0.14**	0.05	**0.15**	0.06	**0.16**	0.06	**0.18**	0.07	**0.19**
Red bone marrow		**0.02**		**0.03**		**0.03**	0.01	**0.03**	0.01	**0.04**	0.01	**0.04**
Whole body dose		**0.03**		**0.04**	0.01	**0.04**	0.01	**0.04**	0.01	**0.05**	0.01	**0.05**
ED (ICRP 60)		**0.05**		**0.06**		**0.06**		**0.06**		**0.06**		**0.07**
ED (ICRP 103)		**0.09**		**0.10**		**0.10**		**0.10**		**0.11**		**0.11**
Total filtration (mm) Al	**1.5**		**2.5**		**3.5**		**4.5**		**3.5**		**3.5**	
Cu	**0**		**0**		**0**		**0**		**0.1**		**0.2**	

Table 20.16 Normalized organ doses—Shoulder ap—1–2 years

Beam projection:			ap									**Table cells:**
Focus-to-detector distance (cm):			100									(+) = normal font = optimal field setting
Tube voltage (kVp):			60									**(−) = bold font = suboptimal field setting**
Total filtrations:			See last table row.									Empty table cells correspond to values <0.01
Organ or tissue	**Organ dose D_T : Entrance dose K_E**								**(mSv * mGy^{-1})**			
	(+)	(−)	(+)	(−)	(+)	(−)	(+)	(−)	(+)	(−)	(+)	(−)
Skull												
Brain												
Oral mucosa												
Salivary glands										**0.01**		
Cervical spine				**0.01**		**0.02**		**0.02**		**0.02**		**0.02**
Thyroid				**0.01**		**0.01**		**0.01**		**0.02**		**0.02**
Thymus		**0.01**		**0.02**		**0.02**		**0.02**		**0.03**		**0.02**
Thoracic spine		**0.02**		**0.02**		**0.03**		**0.03**		**0.04**		**0.04**
Claviculae	0.08	**0.44**	0.10	**0.53**	0.11	**0.59**	0.12	**0.64**	0.13	**0.72**	0.13	**0.72**
Scapulae	0.04	**0.11**	0.05	**0.15**	0.06	**0.18**	0.06	**0.20**	0.08	**0.25**	0.07	**0.23**
Ribs	0.05	**0.23**	0.06	**0.27**	0.07	**0.30**	0.08	**0.32**	0.09	**0.36**	0.09	**0.36**
Oesophagus		**0.01**		**0.02**		**0.02**		**0.02**		**0.03**		**0.02**
Trachea												
Lung		**0.09**		**0.11**		**0.12**		**0.13**		**0.15**		**0.15**
Mamma		**0.53**		**0.56**		**0.57**		**0.58**		**0.59**		**0.59**
Heart		**0.01**		**0.01**		**0.02**		**0.02**		**0.02**		**0.02**
Stomach												
Spleen												
Liver						**0.01**		**0.01**		**0.01**		**0.01**
Gall bladder												
Pancreas												
Small intestine												
Large intestine												
Sigma/Rectum												
Lumbar spine												
Pelvis												
Adrenals										**0.01**		
Kidneys												
Urinary bladder												
Uterus												
Ovaries												
Prostate												
Testes												
Upper arm	0.32	**0.45**	0.39	**0.55**	0.44	**0.63**	0.48	**0.69**	0.55	**0.79**	0.55	**0.78**
Elbow		**0.14**		**0.17**		**0.19**		**0.21**		**0.24**		**0.23**
Forearm												
Upper leg												
Knee												
Lower leg												
Lymph nodes				**0.01**		**0.01**		**0.02**		**0.02**		**0.02**
Skin	0.01	**0.04**	0.02	**0.04**	0.02	**0.04**	0.02	**0.04**	0.02	**0.05**	0.02	**0.05**
Muscle		**0.02**		**0.02**		**0.02**		**0.02**		**0.03**		**0.03**
Skeleton	0.03	**0.07**	0.04	**0.09**	0.04	**0.10**	0.04	**0.11**	0.05	**0.13**	0.05	**0.12**
Red bone marrow		**0.01**		**0.02**		**0.02**		**0.02**		**0.03**		**0.02**
Whole body dose		**0.02**		**0.03**		**0.03**	0.01	**0.03**	0.01	**0.04**	0.01	**0.04**
ED (ICRP 60)		**0.04**		**0.05**		**0.05**		**0.05**		**0.06**		**0.05**
ED (ICRP 103)		**0.08**		**0.09**		**0.09**		**0.09**		**0.10**		**0.10**
Total filtration (mm) **Al**	**1.5**		**2.5**		**3.5**		**4.5**		**3.5**		**3.5**	
Cu	**0**		**0**		**0**		**0**		**0.1**		**0.2**	

Table 20.17 Normalized organ doses—Shoulder ap—1–2 years

Beam projection:			ap									**Table cells:**
Focus-to-detector distance (cm):			100									(+) = normal font = optimal field setting
Tube voltage (kVp):			70									**(−) = bold font = suboptimal field setting**
Total filtrations:			See last table row.									Empty table cells correspond to values <0.01
Organ or tissue	**Organ dose D_T : Entrance dose K_E (mSv * mGy^{-1})**											
	(+)	(−)	(+)	(−)	(+)	(−)	(+)	(−)	(+)	(−)	(+)	(−)
Skull												
Brain												
Oral mucosa												
Salivary glands										**0.01**		**0.01**
Cervical spine		**0.01**		**0.02**		**0.02**		**0.02**		**0.03**		**0.03**
Thyroid		**0.01**		**0.01**		**0.02**		**0.02**		**0.02**		**0.02**
Thymus		**0.01**		**0.02**		**0.02**		**0.02**		**0.03**		**0.03**
Thoracic spine		**0.02**		**0.03**		**0.04**		**0.04**		**0.05**	0.01	**0.06**
Claviculae	0.09	**0.49**	0.10	**0.57**	0.12	**0.63**	0.13	**0.67**	0.14	**0.74**	0.15	**0.79**
Scapulae	0.05	**0.14**	0.06	**0.18**	0.07	**0.21**	0.08	**0.24**	0.09	**0.28**	0.11	**0.33**
Ribs	0.06	**0.25**	0.07	**0.29**	0.08	**0.32**	0.08	**0.35**	0.09	**0.38**	0.10	**0.41**
Oesophagus		**0.02**		**0.02**		**0.02**		**0.03**		**0.03**		**0.04**
Trachea												
Lung		**0.10**		**0.12**		**0.13**		**0.14**	0.01	**0.16**	0.01	**0.18**
Mamma		**0.54**		**0.57**		**0.58**		**0.59**		**0.60**		**0.61**
Heart		**0.01**		**0.02**		**0.02**		**0.02**		**0.03**		**0.03**
Stomach												
Spleen												
Liver				**0.01**		**0.01**		**0.01**		**0.02**		**0.02**
Gall bladder												
Pancreas												**0.01**
Small intestine												
Large intestine												
Sigma/Rectum												
Lumbar spine												
Pelvis												
Adrenals								**0.01**		**0.01**		**0.02**
Kidneys												
Urinary bladder												
Uterus												
Ovaries												
Prostate												
Testes												
Upper arm	0.36	**0.51**	0.43	**0.62**	0.48	**0.69**	0.52	**0.75**	0.59	**0.85**	0.65	**0.93**
Elbow		**0.15**		**0.19**		**0.21**		**0.23**		**0.26**		**0.29**
Forearm												
Upper leg												
Knee												
Lower leg												
Lymph nodes		**0.01**		**0.01**		**0.02**		**0.02**		**0.02**		**0.02**
Skin	0.01	**0.04**	0.02	**0.04**	0.02	**0.04**	0.02	**0.05**	0.02	**0.05**	0.02	**0.05**
Muscle		**0.02**		**0.02**		**0.02**		**0.03**		**0.03**		**0.03**
Skeleton	0.03	**0.08**	0.04	**0.10**	0.04	**0.11**	0.05	**0.12**	0.06	**0.14**	0.06	**0.15**
Red bone marrow		**0.02**		**0.02**		**0.02**		**0.03**	0.01	**0.03**	0.01	**0.03**
Whole body dose		**0.03**		**0.03**	0.01	**0.03**	0.01	**0.04**	0.01	**0.04**	0.01	**0.04**
ED (ICRP 60)		**0.04**		**0.05**		**0.05**		**0.05**		**0.06**		**0.06**
ED (ICRP 103)		**0.08**		**0.09**		**0.09**		**0.10**		**0.10**		**0.11**
Total filtration (mm) **Al**	**1.5**		**2.5**		**3.5**		**4.5**		**3.5**		**3.5**	
Cu	**0**		**0**		**0**		**0**		**0.1**		**0.2**	

Table 20.18 Normalized organ doses—Shoulder ap—3–7 years

Beam projection:			ap									**Table cells:**
Focus-to-detector distance (cm):			100									(+) = normal font = optimal field setting
Tube voltage (kVp):			60									**(−) = bold font = suboptimal field setting**
Total filtrations:			See last table row.									Empty table cells correspond to values <0.01
Organ or tissue	**Organ dose D_T : Entrance dose K_E**							**(mSv * mGy^{-1})**				
	(+)	(−)	(+)	(−)	(+)	(−)	(+)	(−)	(+)	(−)	(+)	(−)
Skull												
Brain												
Oral mucosa												
Salivary glands												
Cervical spine				**0.01**		**0.01**		**0.01**		**0.02**		**0.02**
Thyroid								**0.01**		**0.01**		**0.01**
Thymus				**0.01**		**0.02**		**0.02**		**0.02**		**0.02**
Thoracic spine		**0.02**		**0.02**		**0.03**		**0.03**		**0.04**		**0.03**
Claviculae	0.03	**0.35**	0.04	**0.43**	0.05	**0.49**	0.05	**0.53**	0.06	**0.61**	0.06	**0.60**
Scapulae	0.02	**0.09**	0.03	**0.12**	0.04	**0.14**	0.05	**0.16**	0.06	**0.20**	0.05	**0.18**
Ribs	0.04	**0.23**	0.05	**0.28**	0.05	**0.31**	0.06	**0.34**	0.07	**0.38**	0.07	**0.38**
Oesophagus				**0.01**		**0.01**		**0.02**		**0.02**		**0.02**
Trachea												
Lung		**0.09**		**0.11**		**0.13**		**0.14**		**0.16**		**0.15**
Mamma		**0.39**		**0.41**		**0.43**		**0.44**		**0.45**		**0.45**
Heart				**0.01**		**0.02**		**0.02**		**0.02**		**0.02**
Stomach												
Spleen												
Liver				**0.01**		**0.01**		**0.02**		**0.02**		**0.02**
Gall bladder												
Pancreas												
Small intestine												
Large intestine												
Sigma/Rectum												
Lumbar spine												
Pelvis												
Adrenals										**0.01**		
Kidneys												
Urinary bladder												
Uterus												
Ovaries												
Prostate												
Testes												
Upper arm	0.30	**0.40**	0.37	**0.50**	0.42	**0.57**	0.46	**0.63**	0.53	**0.73**	0.52	**0.71**
Elbow		**0.20**		**0.25**		**0.29**		**0.32**		**0.37**		**0.36**
Forearm												
Upper leg												
Knee												
Lower leg												
Lymph nodes				**0.01**		**0.01**		**0.02**		**0.02**		**0.02**
Skin	0.01	**0.04**	0.01	**0.04**	0.01	**0.04**	0.01	**0.04**	0.01	**0.04**	0.01	**0.04**
Muscle		**0.02**		**0.02**		**0.02**		**0.02**		**0.03**		**0.03**
Skeleton	0.02	**0.06**	0.03	**0.07**	0.03	**0.08**	0.03	**0.09**	0.04	**0.10**	0.04	**0.10**
Red bone marrow		**0.01**		**0.02**		**0.02**		**0.02**		**0.03**		**0.03**
Whole body dose		**0.02**		**0.03**		**0.03**		**0.03**		**0.04**		**0.04**
ED (ICRP 60)		**0.03**		**0.04**		**0.04**		**0.04**		**0.05**		**0.05**
ED (ICRP 103)		**0.06**		**0.07**		**0.07**		**0.08**		**0.08**		**0.08**
Total filtration (mm) Al	**1.5**		**2.5**		**3.5**		**4.5**		**3.5**		**3.5**	
Cu	**0**		**0**		**0**		**0**		**0.1**		**0.2**	

Table 20.19 Normalized organ doses—Shoulder ap—3–7 years

Beam projection:			ap									**Table cells:**
Focus-to-detector distance (cm):			100						(+) = normal font = optimal field setting			
Tube voltage (kVp):			70						**(−) = bold font = suboptimal field setting**			
Total filtrations:			See last table row.						Empty table cells correspond to values <0.01			
Organ or tissue	**Organ dose D_T : Entrance dose K_E**								**(mSv * mGy^{-1})**			
	(+)	(−)	(+)	(−)	(+)	(−)	(+)	(−)	(+)	(−)	(+)	(−)
Skull												
Brain												
Oral mucosa												
Salivary glands										**0.01**		**0.01**
Cervical spine		**0.01**		**0.01**		**0.02**		**0.02**		**0.02**		**0.03**
Thyroid				**0.01**		**0.01**		**0.01**		**0.02**		**0.02**
Thymus		**0.01**		**0.02**		**0.02**		**0.02**		**0.03**		**0.03**
Thoracic spine		**0.02**		**0.03**		**0.03**		**0.04**		**0.05**		**0.06**
Claviculae	0.04	**0.40**	0.05	**0.47**	0.06	**0.53**	0.06	**0.57**	0.07	**0.64**	0.08	**0.69**
Scapulae	0.03	**0.12**	0.04	**0.15**	0.05	**0.18**	0.06	**0.20**	0.07	**0.24**	0.08	**0.28**
Ribs	0.04	**0.26**	0.05	**0.30**	0.06	**0.34**	0.06	**0.36**	0.07	**0.41**	0.08	**0.44**
Oesophagus		**0.01**		**0.01**		**0.02**		**0.02**		**0.02**		**0.03**
Trachea												
Lung		**0.10**		**0.13**		**0.14**		**0.15**		**0.18**	0.01	**0.19**
Mamma		**0.40**		**0.42**		**0.44**		**0.45**		**0.46**		**0.47**
Heart		**0.01**		**0.02**		**0.02**		**0.02**		**0.03**		**0.03**
Stomach												
Spleen												
Liver		**0.01**		**0.01**		**0.02**		**0.02**		**0.02**		**0.03**
Gall bladder												
Pancreas												**0.01**
Small intestine												
Large intestine												
Sigma/Rectum												
Lumbar spine												
Pelvis												
Adrenals						**0.01**		**0.01**		**0.01**		**0.02**
Kidneys												
Urinary bladder												
Uterus												
Ovaries												
Prostate												
Testes												
Upper arm	0.34	**0.46**	0.41	**0.57**	0.46	**0.64**	0.50	**0.70**	0.58	**0.80**	0.64	**0.89**
Elbow		**0.24**		**0.29**		**0.32**		**0.35**		**0.41**		**0.45**
Forearm												
Upper leg												
Knee												
Lower leg												
Lymph nodes		**0.01**		**0.01**		**0.02**		**0.02**		**0.02**		**0.02**
Skin	0.01	**0.04**	0.01	**0.04**	0.01	**0.04**	0.01	**0.04**	0.01	**0.05**	0.02	**0.05**
Muscle		**0.02**		**0.02**		**0.02**		**0.03**		**0.03**		**0.03**
Skeleton	0.02	**0.07**	0.03	**0.08**	0.03	**0.09**	0.04	**0.10**	0.04	**0.11**	0.05	**0.13**
Red bone marrow		**0.02**		**0.02**		**0.02**		**0.03**		**0.03**		**0.03**
Whole body dose		**0.02**		**0.03**		**0.03**		**0.04**	0.01	**0.04**	0.01	**0.04**
ED (ICRP 60)		**0.04**		**0.04**		**0.05**		**0.05**		**0.05**		**0.06**
ED (ICRP 103)		**0.07**		**0.07**		**0.08**		**0.08**		**0.09**		**0.09**
Total filtration (mm) **Al**	**1.5**		**2.5**		**3.5**		**4.5**		**3.5**		**3.5**	
Cu	**0**		**0**		**0**		**0**		**0.1**		**0.2**	

Table 20.20 Normalized organ doses—Shoulder ap—8–12 years

Beam projection:			ap									**Table cells:**
Focus-to-detector distance (cm):			100						(+) = normal font = optimal field setting			
Tube voltage (kVp):			60						**(−) = bold font = suboptimal field setting**			
Total filtrations:			See last table row.						Empty table cells correspond to values <0.01			
Organ or tissue	**Organ dose D_T : Entrance dose K_E**							**($mSv * mGy^{-1}$)**				
	(+)	(−)	(+)	(−)	(+)	(−)	(+)	(−)	(+)	(−)	(+)	(−)
Skull												
Brain												
Oral mucosa												
Salivary glands		**0.01**		**0.01**		**0.02**		**0.02**		**0.02**		**0.02**
Cervical spine						**0.01**		**0.01**		**0.02**		**0.01**
Thyroid										**0.01**		
Thymus				**0.01**		**0.01**		**0.02**		**0.02**		**0.02**
Thoracic spine		**0.01**		**0.02**		**0.02**		**0.03**		**0.04**		**0.03**
Claviculae	0.04	**0.33**	0.05	**0.41**	0.05	**0.47**	0.06	**0.51**	0.07	**0.59**	0.07	**0.58**
Scapulae	0.02	**0.07**	0.03	**0.09**	0.04	**0.11**	0.04	**0.13**	0.05	**0.16**	0.05	**0.14**
Ribs	0.04	**0.24**	0.04	**0.29**	0.05	**0.32**	0.06	**0.35**	0.07	**0.40**	0.06	**0.39**
Oesophagus						**0.01**		**0.01**		**0.02**		**0.01**
Trachea												
Lung		**0.09**		**0.11**		**0.13**		**0.14**		**0.17**		**0.16**
Mamma		**0.46**		**0.50**		**0.52**		**0.53**		**0.55**		**0.55**
Heart				**0.01**		**0.02**		**0.02**		**0.02**		**0.02**
Stomach												
Spleen												
Liver		**0.02**		**0.02**		**0.03**		**0.03**		**0.04**		**0.03**
Gall bladder												
Pancreas												
Small intestine												
Large intestine												
Sigma/Rectum												
Lumbar spine												
Pelvis												
Adrenals								**0.01**		**0.01**		**0.01**
Kidneys												
Urinary bladder												
Uterus												
Ovaries												
Prostate												
Testes												
Upper arm	0.27	**0.36**	0.34	**0.45**	0.39	**0.51**	0.43	**0.57**	0.50	**0.67**	0.48	**0.64**
Elbow		**0.23**		**0.29**		**0.33**		**0.36**		**0.43**		**0.41**
Forearm												
Upper leg												
Knee												
Lower leg												
Lymph nodes		**0.01**		**0.01**		**0.01**		**0.02**		**0.02**		**0.02**
Skin	0.01	**0.04**	0.01	**0.04**	0.01	**0.04**	0.01	**0.04**	0.01	**0.04**	0.01	**0.04**
Muscle		**0.02**		**0.02**		**0.02**		**0.02**		**0.03**		**0.03**
Skeleton	0.02	**0.06**	0.02	**0.07**	0.03	**0.08**	0.03	**0.09**	0.03	**0.10**	0.03	**0.10**
Red bone marrow		**0.02**		**0.02**		**0.02**		**0.03**		**0.03**		**0.03**
Whole body dose		**0.02**		**0.03**		**0.03**		**0.03**		**0.04**		**0.04**
ED (ICRP 60)		**0.04**		**0.04**		**0.05**		**0.05**		**0.06**		**0.05**
ED (ICRP 103)		**0.07**		**0.08**		**0.08**		**0.09**		**0.09**		**0.09**
Total filtration (mm) Al	**1.5**		**2.5**		**3.5**		**4.5**		**3.5**		**3.5**	
Cu	**0**		**0**		**0**		**0**		**0.1**		**0.2**	

Table 20.21 Normalized organ doses—Shoulder ap—8–12 years

Beam projection:	ap											**Table cells:**
Focus-to-detector distance (cm):	100											(+) = normal font = optimal field setting
Tube voltage (kVp):	70											**(−) = bold font = suboptimal field setting**
Total filtrations:	See last table row.											Empty table cells correspond to values <0.01
Organ or tissue	**Organ dose D_T : Entrance dose K_E**							**(mSv * mGy^{-1})**				
	(+)	(−)	(+)	(−)	(+)	(−)	(+)	(−)	(+)	(−)	(+)	(−)
Skull												
Brain												
Oral mucosa												
Salivary glands		**0.01**		**0.02**		**0.02**		**0.02**		**0.02**		**0.02**
Cervical spine				**0.01**		**0.02**		**0.02**		**0.02**		**0.03**
Thyroid						**0.01**		**0.01**		**0.02**		**0.02**
Thymus		**0.01**		**0.02**		**0.02**		**0.02**		**0.03**		**0.03**
Thoracic spine		**0.02**		**0.03**		**0.03**		**0.04**		**0.05**		**0.06**
Claviculae	0.04	**0.38**	0.05	**0.46**	0.06	**0.51**	0.07	**0.56**	0.08	**0.63**	0.09	**0.69**
Scapulae	0.03	**0.09**	0.04	**0.12**	0.05	**0.14**	0.05	**0.16**	0.06	**0.20**	0.08	**0.24**
Ribs	0.04	**0.26**	0.05	**0.32**	0.06	**0.35**	0.06	**0.38**	0.07	**0.43**	0.08	**0.47**
Oesophagus		**0.01**		**0.01**		**0.02**		**0.02**		**0.02**		**0.03**
Trachea												**0.01**
Lung		**0.11**		**0.13**		**0.15**		**0.16**		**0.19**	0.01	**0.21**
Mamma		**0.48**		**0.51**		**0.53**		**0.54**		**0.56**		**0.57**
Heart		**0.01**		**0.02**		**0.02**		**0.02**		**0.03**		**0.03**
Stomach												
Spleen												
Liver		**0.02**		**0.03**		**0.03**		**0.04**		**0.05**		**0.05**
Gall bladder												
Pancreas												**0.01**
Small intestine												
Large intestine												
Sigma/Rectum												
Lumbar spine												
Pelvis												
Adrenals						**0.01**		**0.01**		**0.02**		**0.02**
Kidneys												
Urinary bladder												
Uterus												
Ovaries												
Prostate												
Testes												
Upper arm	0.31	**0.42**	0.38	**0.51**	0.43	**0.58**	0.47	**0.64**	0.55	**0.74**	0.61	**0.83**
Elbow		**0.27**		**0.33**		**0.37**		**0.41**		**0.47**		**0.53**
Forearm												
Upper leg												
Knee												
Lower leg												
Lymph nodes		**0.01**		**0.02**		**0.02**		**0.02**		**0.02**		**0.03**
Skin	0.01	**0.04**	0.01	**0.04**	0.01	**0.04**	0.01	**0.04**	0.01	**0.05**	0.01	**0.05**
Muscle		**0.02**		**0.02**		**0.03**		**0.03**		**0.03**		**0.03**
Skeleton	0.02	**0.07**	0.03	**0.08**	0.03	**0.09**	0.03	**0.10**	0.04	**0.11**	0.04	**0.13**
Red bone marrow		**0.02**		**0.02**		**0.03**		**0.03**		**0.03**	0.01	**0.04**
Whole body dose		**0.03**		**0.03**		**0.03**		**0.04**	0.01	**0.04**	0.01	**0.05**
ED (ICRP 60)		**0.04**		**0.05**		**0.05**		**0.05**		**0.06**		**0.07**
ED (ICRP 103)		**0.08**		**0.08**		**0.09**		**0.09**		**0.10**		**0.11**
Total filtration (mm) Al	**1.5**		**2.5**		**3.5**		**4.5**		**3.5**		**3.5**	
Cu	**0**		**0**		**0**		**0**		**0.1**		**0.2**	

Table 20.22 Normalized organ doses—Shoulder ap—8–12 years

Beam projection:			ap									**Table cells:**
Focus-to-detector distance (cm):			100									(+) = normal font = optimal field setting
Tube voltage (kVp):			80									**(−) = bold font = suboptimal field setting**
Total filtrations:			See last table row.									Empty table cells correspond to values <0.01
Organ or tissue	**Organ dose D_T : Entrance dose K_E**								**(mSv * mGy^{-1})**			
	(+)	(−)	(+)	(−)	(+)	(−)	(+)	(−)	(+)	(−)	(+)	(−)
Skull												**0.01**
Brain												
Oral mucosa												
Salivary glands		**0.01**		**0.02**		**0.02**		**0.02**		**0.02**		**0.03**
Cervical spine		**0.01**		**0.02**		**0.02**		**0.02**		**0.03**		**0.03**
Thyroid				**0.01**		**0.01**		**0.02**		**0.02**		**0.02**
Thymus		**0.02**		**0.02**		**0.02**		**0.03**		**0.03**		**0.04**
Thoracic spine		**0.03**		**0.03**		**0.04**		**0.05**		**0.06**		**0.07**
Claviculae	0.05	**0.42**	0.06	**0.49**	0.07	**0.55**	0.07	**0.59**	0.08	**0.66**	0.09	**0.71**
Scapulae	0.04	**0.11**	0.05	**0.14**	0.05	**0.17**	0.06	**0.19**	0.07	**0.23**	0.09	**0.27**
Ribs	0.05	**0.29**	0.06	**0.34**	0.06	**0.37**	0.07	**0.40**	0.08	**0.44**	0.09	**0.48**
Oesophagus		**0.01**		**0.02**		**0.02**		**0.02**		**0.03**		**0.03**
Trachea										**0.01**		**0.01**
Lung		**0.12**		**0.14**		**0.16**		**0.18**	0.01	**0.20**	0.01	**0.22**
Mamma		**0.49**		**0.52**		**0.54**		**0.55**		**0.57**		**0.58**
Heart		**0.02**		**0.02**		**0.02**		**0.03**		**0.03**		**0.04**
Stomach												
Spleen												
Liver		**0.03**		**0.03**		**0.04**		**0.04**		**0.05**		**0.06**
Gall bladder												**0.01**
Pancreas										**0.01**		**0.01**
Small intestine												
Large intestine												
Sigma/Rectum												
Lumbar spine												
Pelvis												
Adrenals				**0.01**		**0.01**		**0.02**		**0.02**		**0.02**
Kidneys												
Urinary bladder												
Uterus												
Ovaries												
Prostate												
Testes												
Upper arm	0.35	**0.47**	0.42	**0.57**	0.47	**0.63**	0.51	**0.69**	0.58	**0.78**	0.63	**0.87**
Elbow		**0.30**		**0.36**		**0.41**		**0.44**		**0.50**		**0.55**
Forearm												
Upper leg												
Knee												
Lower leg												
Lymph nodes		**0.01**		**0.02**		**0.02**		**0.02**		**0.03**		**0.03**
Skin	0.01	**0.04**	0.01	**0.04**	0.01	**0.04**	0.01	**0.05**	0.01	**0.05**	0.02	**0.05**
Muscle		**0.02**		**0.03**		**0.03**		**0.03**		**0.03**		**0.04**
Skeleton	0.02	**0.07**	0.03	**0.09**	0.03	**0.10**	0.04	**0.11**	0.04	**0.12**	0.05	**0.13**
Red bone marrow		**0.02**		**0.03**		**0.03**		**0.03**	0.01	**0.04**	0.01	**0.04**
Whole body dose		**0.03**		**0.03**		**0.04**		**0.04**	0.01	**0.05**	0.01	**0.05**
ED (ICRP 60)		**0.05**		**0.05**		**0.06**		**0.06**		**0.06**		**0.07**
ED (ICRP 103)		**0.08**		**0.09**		**0.09**		**0.10**		**0.10**		**0.11**
Total filtration (mm) **Al**	**1.5**		**2.5**		**3.5**		**4.5**		**3.5**		**3.5**	
Cu	**0**		**0**		**0**		**0**		**0.1**		**0.2**	

Table 20.23 Normalized organ doses—Shoulder ap—13–17 years

Beam projection:			ap									**Table cells:**
Focus-to-detector distance (cm):			115									(+) = normal font = optimal field setting
Tube voltage (kVp):			60									**(−) = bold font = suboptimal field setting**
Total filtrations:			See last table row.									Empty table cells correspond to values <0.01
Organ or tissue	**Organ dose D_T : Entrance dose K_E (mSv * mGy^{-1})**											
	(+)	(−)	(+)	(−)	(+)	(−)	(+)	(−)	(+)	(−)	(+)	(−)
Skull												
Brain												
Oral mucosa												
Salivary glands												
Cervical spine										**0.01**		
Thyroid												
Thymus				**0.01**		**0.01**		**0.02**		**0.02**		**0.02**
Thoracic spine				**0.01**		**0.02**		**0.02**		**0.02**		**0.02**
Claviculae	0.04	**0.32**	0.05	**0.40**	0.06	**0.46**	0.06	**0.50**	0.07	**0.59**	0.07	**0.57**
Scapulae	0.01	**0.04**	0.02	**0.05**	0.02	**0.07**	0.02	**0.08**	0.03	**0.10**	0.03	**0.08**
Ribs	0.03	**0.23**	0.03	**0.28**	0.04	**0.32**	0.04	**0.34**	0.05	**0.40**	0.05	**0.39**
Oesophagus						**0.01**		**0.01**		**0.02**		**0.01**
Trachea												
Lung		**0.06**		**0.08**		**0.10**		**0.11**		**0.13**		**0.12**
Mamma		**2.94**		**3.41**	0.01	**3.71**	0.01	**3.94**	0.02	**4.32**	0.01	**4.27**
Heart				**0.01**		**0.01**		**0.02**		**0.02**		**0.02**
Stomach												
Spleen												
Liver		**0.05**		**0.07**		**0.08**		**0.09**		**0.11**		**0.10**
Gall bladder												
Pancreas												
Small intestine												
Large intestine												
Sigma/Rectum												
Lumbar spine												
Pelvis												
Adrenals										**0.01**		
Kidneys												
Urinary bladder												
Uterus												
Ovaries												
Prostate												
Testes												
Upper arm	0.19	**0.30**	0.24	**0.37**	0.28	**0.43**	0.31	**0.48**	0.37	**0.57**	0.35	**0.54**
Elbow		**0.26**		**0.33**		**0.38**		**0.42**		**0.50**		**0.48**
Forearm												
Upper leg												
Knee												
Lower leg												
Lymph nodes				**0.01**		**0.01**		**0.01**		**0.02**		**0.01**
Skin		**0.05**	0.01	**0.05**	0.01	**0.05**	0.01	**0.06**	0.01	**0.06**	0.01	**0.06**
Muscle		**0.02**		**0.02**		**0.02**		**0.03**		**0.03**		**0.03**
Skeleton	0.01	**0.06**	0.02	**0.07**	0.02	**0.08**	0.02	**0.09**	0.03	**0.11**	0.03	**0.10**
Red bone marrow		**0.02**		**0.02**		**0.02**		**0.03**		**0.03**		**0.03**
Whole body dose		**0.03**		**0.03**		**0.04**		**0.04**		**0.05**		**0.04**
ED (ICRP 60)		**0.16**		**0.19**		**0.21**		**0.22**		**0.24**		**0.24**
ED (ICRP 103)		**0.37**		**0.43**		**0.47**		**0.50**		**0.55**		**0.54**
Total filtration (mm) Al	**1.5**		**2.5**		**3.5**		**4.5**		**3.5**		**3.5**	
Cu	**0**		**0**		**0**		**0**		**0.1**		**0.2**	

Table 20.24 Normalized organ doses—Shoulder ap—13–17 years

Beam projection:			ap									**Table cells:**
Focus-to-detector distance (cm):			115					(+) = normal font = optimal field setting				
Tube voltage (kVp):			70					**(−) = bold font = suboptimal field setting**				
Total filtrations:			See last table row.					Empty table cells correspond to values <0.01				
Organ or tissue	**Organ dose D_T : Entrance dose K_E**							**(mSv * mGy^{-1})**				
	(+)	(−)	(+)	(−)	(+)	(−)	(+)	(−)	(+)	(−)	(+)	(−)
Skull												
Brain												
Oral mucosa												
Salivary glands												
Cervical spine				**0.01**		**0.01**		**0.01**		**0.02**		**0.02**
Thyroid										**0.01**		**0.02**
Thymus		**0.01**		**0.02**		**0.02**		**0.02**		**0.03**		**0.03**
Thoracic spine		**0.01**		**0.02**		**0.02**		**0.03**		**0.04**		**0.04**
Claviculae	0.05	**0.37**	0.06	**0.45**	0.06	**0.51**	0.07	**0.56**	0.08	**0.64**	0.09	**0.71**
Scapulae	0.02	**0.06**	0.02	**0.07**	0.03	**0.09**	0.03	**0.10**	0.04	**0.13**	0.05	**0.16**
Ribs	0.03	**0.26**	0.04	**0.31**	0.04	**0.35**	0.05	**0.38**	0.05	**0.44**	0.06	**0.48**
Oesophagus				**0.01**		**0.02**		**0.02**		**0.02**		**0.03**
Trachea												
Lung		**0.08**		**0.10**		**0.12**		**0.13**		**0.15**		**0.18**
Mamma		**3.20**	0.01	**3.66**	0.01	**3.97**	0.02	**4.19**	0.02	**4.57**	0.02	**4.88**
Heart		**0.01**		**0.02**		**0.02**		**0.02**		**0.03**		**0.03**
Stomach												
Spleen												
Liver		**0.07**		**0.08**		**0.10**		**0.11**		**0.13**		**0.15**
Gall bladder								**0.01**		**0.01**		**0.02**
Pancreas												**0.01**
Small intestine												
Large intestine												
Sigma/Rectum												
Lumbar spine												
Pelvis												
Adrenals						**0.01**		**0.01**		**0.02**		**0.02**
Kidneys												
Urinary bladder												
Uterus												
Ovaries												
Prostate												
Testes												
Upper arm	0.23	**0.35**	0.28	**0.44**	0.32	**0.50**	0.35	**0.55**	0.41	**0.64**	0.46	**0.73**
Elbow		**0.31**		**0.38**		**0.44**		**0.48**		**0.56**		**0.64**
Forearm												
Upper leg												
Knee												
Lower leg												
Lymph nodes				**0.01**		**0.01**		**0.02**		**0.02**		**0.02**
Skin	0.01	**0.05**	0.01	**0.05**	0.01	**0.06**	0.01	**0.06**	0.01	**0.06**	0.01	**0.07**
Muscle		**0.02**		**0.02**		**0.03**		**0.03**		**0.03**		**0.04**
Skeleton	0.02	**0.07**	0.02	**0.08**	0.02	**0.09**	0.03	**0.10**	0.03	**0.12**	0.04	**0.13**
Red bone marrow		**0.02**		**0.03**		**0.03**		**0.03**		**0.04**		**0.04**
Whole body dose		**0.03**		**0.04**		**0.04**		**0.04**		**0.05**		**0.06**
ED (ICRP 60)		**0.18**		**0.20**		**0.22**		**0.24**		**0.26**		**0.28**
ED (ICRP 103)		**0.40**		**0.46**		**0.50**		**0.53**		**0.58**		**0.62**
Total filtration (mm) Al	**1.5**		**2.5**		**3.5**		**4.5**		**3.5**		**3.5**	
Cu	**0**		**0**		**0**		**0**		**0.1**		**0.2**	

Table 20.25 Normalized organ doses—Shoulder ap—13–17 years

Beam projection:			ap									**Table cells:**
Focus-to-detector distance (cm):			115						(+) = normal font = optimal field setting			
Tube voltage (kVp):			80						**(−) = bold font = suboptimal field setting**			
Total filtrations:			See last table row.						Empty table cells correspond to values <0.01			
Organ or tissue	**Organ dose D_T : Entrance dose K_E**						**(mSv * mGy^{-1})**					
	(+)	(−)	(+)	(−)	(+)	(−)	(+)	(−)	(+)	(−)	(+)	(−)
Skull												
Brain												
Oral mucosa												
Salivary glands												
Cervical spine		**0.01**		**0.01**		**0.02**		**0.02**		**0.02**		**0.03**
Thyroid						**0.01**		**0.01**		**0.02**		**0.02**
Thymus		**0.02**		**0.02**		**0.02**		**0.03**		**0.03**		**0.04**
Thoracic spine		**0.02**		**0.03**		**0.03**		**0.04**		**0.05**		**0.06**
Claviculae	0.05	**0.41**	0.06	**0.49**	0.07	**0.55**	0.08	**0.60**	0.09	**0.67**	0.10	**0.73**
Scapulae	0.02	**0.07**	0.03	**0.09**	0.03	**0.11**	0.04	**0.13**	0.05	**0.16**	0.06	**0.19**
Ribs	0.03	**0.29**	0.04	**0.34**	0.05	**0.38**	0.05	**0.41**	0.06	**0.46**	0.06	**0.51**
Oesophagus		**0.01**		**0.02**		**0.02**		**0.02**		**0.03**		**0.04**
Trachea												
Lung		**0.09**		**0.12**		**0.13**		**0.15**		**0.17**		**0.20**
Mamma	0.01	**3.44**	0.01	**3.89**	0.02	**4.19**	0.02	**4.40**	0.02	**4.77**	0.03	**5.06**
Heart		**0.02**		**0.02**		**0.02**		**0.03**		**0.03**		**0.04**
Stomach												
Spleen												
Liver		**0.08**		**0.10**		**0.11**		**0.13**		**0.15**		**0.17**
Gall bladder				**0.01**		**0.01**		**0.01**		**0.02**		**0.02**
Pancreas										**0.01**		**0.01**
Small intestine												
Large intestine												
Sigma/Rectum												
Lumbar spine												
Pelvis												
Adrenals				**0.01**		**0.02**		**0.02**		**0.02**		**0.03**
Kidneys												**0.01**
Urinary bladder												
Uterus												
Ovaries												
Prostate												
Testes												
Upper arm	0.26	**0.40**	0.31	**0.49**	0.35	**0.55**	0.38	**0.60**	0.44	**0.70**	0.48	**0.78**
Elbow		**0.35**		**0.43**		**0.48**		**0.53**		**0.61**		**0.68**
Forearm												
Upper leg												
Knee												
Lower leg												
Lymph nodes		**0.01**		**0.02**		**0.02**		**0.02**		**0.02**		**0.03**
Skin	0.01	**0.05**	0.01	**0.06**	0.01	**0.06**	0.01	**0.06**	0.01	**0.07**	0.01	**0.07**
Muscle		**0.02**		**0.03**		**0.03**		**0.03**		**0.04**		**0.04**
Skeleton	0.02	**0.08**	0.02	**0.09**	0.03	**0.10**	0.03	**0.11**	0.03	**0.13**	0.04	**0.14**
Red bone marrow		**0.02**		**0.03**		**0.03**		**0.04**		**0.04**	0.01	**0.05**
Whole body dose		**0.03**		**0.04**		**0.05**		**0.05**		**0.06**	0.01	**0.06**
ED (ICRP 60)		**0.19**		**0.22**		**0.24**		**0.25**		**0.28**		**0.30**
ED (ICRP 103)		**0.43**		**0.49**		**0.53**		**0.56**		**0.61**		**0.65**
Total filtration (mm) **Al**	**1.5**		**2.5**		**3.5**		**4.5**		**3.5**		**3.5**	
Cu	**0**		**0**		**0**		**0**		**0.1**		**0.2**	

Table 20.26 Normalized organ doses—Shoulder ap—>17 years

Beam projection:			ap									**Table cells:**
Focus-to-detector distance (cm):			115									(+) = normal font = optimal field setting
Tube voltage (kVp):			60									**(−) = bold font = suboptimal field setting**
Total filtrations:			See last table row.									Empty table cells correspond to values <0.01
Organ or tissue	**Organ dose D_T : Entrance dose K_E**								**(mSv * mGy^{-1})**			
	(+)	(−)	(+)	(−)	(+)	(−)	(+)	(−)	(+)	(−)	(+)	(−)
Skull												
Brain												
Oral mucosa												
Salivary glands												
Cervical spine										**0.01**		
Thyroid												
Thymus						**0.01**		**0.01**		**0.02**		**0.01**
Thoracic spine				**0.01**		**0.01**		**0.02**		**0.02**		**0.02**
Claviculae	0.07	**0.29**	0.09	**0.36**	0.11	**0.42**	0.12	**0.46**	0.14	**0.54**	0.13	**0.52**
Scapulae	0.02	**0.04**	0.02	**0.06**	0.03	**0.07**	0.03	**0.08**	0.05	**0.11**	0.04	**0.09**
Ribs	0.04	**0.23**	0.05	**0.29**	0.06	**0.33**	0.06	**0.36**	0.08	**0.42**	0.07	**0.40**
Oesophagus						**0.01**		**0.01**		**0.02**		**0.01**
Trachea												
Lung		**0.07**		**0.08**		**0.10**		**0.11**		**0.13**		**0.12**
Mamma		**0.31**		**0.36**		**0.39**		**0.42**		**0.46**		**0.45**
Heart						**0.01**		**0.01**		**0.02**		**0.01**
Stomach												
Spleen												
Liver		**0.06**		**0.08**		**0.09**		**0.10**		**0.13**		**0.12**
Gall bladder												
Pancreas												
Small intestine												
Large intestine												
Sigma/Rectum												
Lumbar spine												
Pelvis												
Adrenals										**0.01**		
Kidneys												
Urinary bladder												
Uterus												
Ovaries												
Prostate												
Testes												
Upper arm	0.22	**0.29**	0.27	**0.37**	0.32	**0.43**	0.35	**0.48**	0.42	**0.57**	0.40	**0.54**
Elbow		**0.29**		**0.36**		**0.42**		**0.47**		**0.56**		**0.53**
Forearm												
Upper leg												
Knee												
Lower leg												
Lymph nodes						**0.01**		**0.01**		**0.02**		**0.01**
Skin	0.01	**0.05**	0.01	**0.05**	0.01	**0.05**	0.01	**0.06**	0.01	**0.06**	0.01	**0.06**
Muscle		**0.02**		**0.02**		**0.02**		**0.03**		**0.03**		**0.03**
Skeleton	0.02	**0.06**	0.02	**0.07**	0.03	**0.08**	0.03	**0.09**	0.03	**0.11**	0.03	**0.10**
Red bone marrow		**0.02**		**0.02**		**0.03**		**0.03**		**0.03**		**0.03**
Whole body dose		**0.03**		**0.03**		**0.04**		**0.04**	0.01	**0.05**		**0.04**
ED (ICRP 60)		**0.03**		**0.04**		**0.04**		**0.05**		**0.05**		**0.05**
ED (ICRP 103)		**0.05**		**0.06**		**0.07**		**0.07**		**0.08**		**0.08**
Total filtration (mm) Al	**1.5**		**2.5**		**3.5**		**4.5**		**3.5**		**3.5**	
Cu	**0**		**0**		**0**		**0**		**0.1**		**0.2**	

Table 20.27 Normalized organ doses—Shoulder ap—>17 years

Beam projection:	ap											**Table cells:**
Focus-to-detector distance (cm):	115											(+) = normal font = optimal field setting
Tube voltage (kVp):	70											**(−) = bold font = suboptimal field setting**
Total filtrations:	See last table row.											Empty table cells correspond to values <0.01
Organ or tissue	**Organ dose D_T : Entrance dose K_E (mSv * mGy^{-1})**											
	(+)	(−)	(+)	(−)	(+)	(−)	(+)	(−)	(+)	(−)	(+)	(−)
Skull												
Brain												
Oral mucosa												
Salivary glands												
Cervical spine								**0.01**		**0.01**		**0.02**
Thyroid												**0.01**
Thymus				**0.01**		**0.01**		**0.02**		**0.02**		**0.03**
Thoracic spine		**0.01**		**0.02**		**0.02**		**0.02**		**0.03**		**0.04**
Claviculae	0.08	**0.33**	0.11	**0.41**	0.12	**0.47**	0.13	**0.51**	0.16	**0.59**	0.18	**0.66**
Scapulae	0.03	**0.06**	0.03	**0.08**	0.04	**0.09**	0.05	**0.11**	0.06	**0.14**	0.07	**0.17**
Ribs	0.05	**0.27**	0.06	**0.33**	0.07	**0.37**	0.07	**0.40**	0.08	**0.46**	0.09	**0.51**
Oesophagus				**0.01**		**0.01**		**0.02**		**0.02**		**0.03**
Trachea												
Lung		**0.08**		**0.10**		**0.12**		**0.13**	0.01	**0.16**	0.01	**0.18**
Mamma		**0.34**		**0.39**		**0.42**		**0.44**		**0.49**		**0.52**
Heart				**0.01**		**0.02**		**0.02**		**0.02**		**0.03**
Stomach												
Spleen												
Liver		**0.08**		**0.10**		**0.11**		**0.13**		**0.15**		**0.18**
Gall bladder								**0.01**		**0.01**		**0.02**
Pancreas												
Small intestine												
Large intestine												
Sigma/Rectum												
Lumbar spine												
Pelvis												
Adrenals						**0.01**		**0.01**		**0.02**		**0.02**
Kidneys												**0.01**
Urinary bladder												
Uterus												
Ovaries												
Prostate												
Testes												
Upper arm	0.26	**0.35**	0.32	**0.43**	0.36	**0.50**	0.40	**0.55**	0.47	**0.64**	0.53	**0.73**
Elbow		**0.34**		**0.42**		**0.49**		**0.54**		**0.63**		**0.71**
Forearm												
Upper leg												
Knee												
Lower leg												
Lymph nodes				**0.01**		**0.01**		**0.02**		**0.02**		**0.02**
Skin	0.01	**0.05**	0.01	**0.05**	0.01	**0.06**	0.01	**0.06**	0.02	**0.06**	0.02	**0.07**
Muscle		**0.02**		**0.02**		**0.03**		**0.03**		**0.03**		**0.04**
Skeleton	0.02	**0.07**	0.03	**0.08**	0.03	**0.09**	0.03	**0.10**	0.04	**0.12**	0.04	**0.14**
Red bone marrow		**0.02**		**0.03**		**0.03**		**0.03**	0.01	**0.04**	0.01	**0.05**
Whole body dose		**0.03**		**0.04**		**0.04**	0.01	**0.05**	0.01	**0.05**	0.01	**0.06**
ED (ICRP 60)		**0.04**		**0.04**		**0.05**		**0.05**		**0.06**		**0.07**
ED (ICRP 103)		**0.06**		**0.07**		**0.08**		**0.08**		**0.09**		**0.10**
Total filtration (mm) Al	**1.5**		**2.5**		**3.5**		**4.5**		**3.5**		**3.5**	
Cu	**0**		**0**		**0**		**0**		**0.1**		**0.2**	

Table 20.28 Normalized organ doses—Shoulder ap—>17 years

Beam projection:			ap									**Table cells:**
Focus-to-detector distance (cm):			115						(+) = normal font = optimal field setting			
Tube voltage (kVp):			80						**(−) = bold font = suboptimal field setting**			
Total filtrations:			See last table row.						Empty table cells correspond to values <0.01			
Organ or tissue	**Organ dose D_T : Entrance dose K_E**								**(mSv * mGy^{-1})**			
	(+)	(−)	(+)	(−)	(+)	(−)	(+)	(−)	(+)	(−)	(+)	(−)
Skull												
Brain												
Oral mucosa												
Salivary glands												
Cervical spine				**0.01**		**0.01**		**0.02**		**0.02**		**0.02**
Thyroid										**0.01**		**0.02**
Thymus		**0.01**		**0.02**		**0.02**		**0.02**		**0.03**		**0.03**
Thoracic spine		**0.02**		**0.02**		**0.03**		**0.03**		**0.04**		**0.05**
Claviculae	0.10	**0.38**	0.12	**0.45**	0.13	**0.51**	0.15	**0.55**	0.17	**0.62**	0.19	**0.68**
Scapulae	0.03	**0.08**	0.04	**0.10**	0.05	**0.12**	0.06	**0.13**	0.07	**0.16**	0.08	**0.20**
Ribs	0.05	**0.30**	0.06	**0.36**	0.07	**0.40**	0.08	**0.43**	0.09	**0.48**	0.10	**0.53**
Oesophagus		**0.01**		**0.02**		**0.02**		**0.02**		**0.03**		**0.03**
Trachea												
Lung		**0.10**		**0.12**		**0.14**		**0.15**	0.01	**0.18**	0.01	**0.20**
Mamma		**0.36**		**0.41**		**0.44**		**0.47**		**0.51**		**0.54**
Heart		**0.01**		**0.02**		**0.02**		**0.02**		**0.03**		**0.03**
Stomach												
Spleen												
Liver		**0.09**		**0.12**		**0.13**		**0.15**		**0.18**		**0.20**
Gall bladder				**0.01**		**0.01**		**0.01**		**0.02**		**0.02**
Pancreas										**0.01**		**0.01**
Small intestine												
Large intestine												
Sigma/Rectum												
Lumbar spine												
Pelvis												
Adrenals				**0.01**		**0.01**		**0.02**		**0.02**		**0.03**
Kidneys										**0.01**		**0.01**
Urinary bladder												
Uterus												
Ovaries												
Prostate												
Testes												
Upper arm	0.30	**0.40**	0.36	**0.49**	0.40	**0.55**	0.44	**0.61**	0.51	**0.70**	0.56	**0.78**
Elbow		**0.39**		**0.48**		**0.54**		**0.59**		**0.68**		**0.76**
Forearm												**0.01**
Upper leg												
Knee												
Lower leg												
Lymph nodes		**0.01**		**0.02**		**0.02**		**0.02**		**0.02**		**0.03**
Skin	0.01	**0.05**	0.01	**0.06**	0.01	**0.06**	0.02	**0.06**	0.02	**0.07**	0.02	**0.07**
Muscle		**0.02**		**0.03**		**0.03**		**0.03**		**0.04**		**0.04**
Skeleton	0.02	**0.08**	0.03	**0.09**	0.03	**0.10**	0.04	**0.11**	0.04	**0.13**	0.05	**0.15**
Red bone marrow		**0.03**		**0.03**		**0.04**	0.01	**0.04**	0.01	**0.05**	0.01	**0.05**
Whole body dose		**0.04**		**0.04**	0.01	**0.05**	0.01	**0.05**	0.01	**0.06**	0.01	**0.06**
ED (ICRP 60)		**0.04**		**0.05**		**0.05**		**0.06**		**0.07**		**0.07**
ED (ICRP 103)		**0.06**		**0.08**		**0.08**		**0.09**		**0.10**		**0.11**
Total filtration (mm) **Al**	**1.5**		**2.5**		**3.5**		**4.5**		**3.5**		**3.5**	
Cu	**0**		**0**		**0**		**0**		**0.1**		**0.2**	

References

1. Bontrager KL. Textbook of radiographic positioning and related anatomy. St. Louis, MO: Mosby Year Book; 1993.
2. Cook JV, Shah K, Pablot S, Kyriou J, Pettett A, Fitzgerald M. Guidelines on best practice in the x-ray imaging of children. A manual for all x-ray departments. London: Queen Mary's Hospital for Children, Surrey, and The Radiological Protection Centre; 1998.

21 Clavicle

21.1 Introduction

The type of clavicle radiography is determined by the clinical indication. Table 21.1 shows indications and Table 21.2 contraindications for clavicle radiography. Image criteria as required for radiological findings in clavicle radiographs are listed in Table 21.3. In order to fulfil these image criteria, the field settings listed in Tables 21.4 and 21.5 should be achieved. These field settings go along with the patient preparation as listed in Table 21.6. Exposure parameter combinations for various age groups and patient diameters are listed in Tables 21.7, 21.8, and 21.9. The age groups in Table 21.8 refer to the standard age groups of the mathematical MIRD phantom family which has been used to calculate the normalized organ doses in this book. Table 21.9 shows radiation qualities (i.e. various combinations of tube voltage and total filtration) which may be applied in clinical practice. Please note that not all of these radiation qualities are recommended by the authors. The standard pro-

Table 21.1 Indications for clavicle radiographs

Fracture
Aplasia/hypoplasia
Congenital pseudarthrosis
Osteomyelitis
Non bacterial osteitis
Bone tumours
Soft tissue tumours

Table 21.2 Contraindications for clavicle radiographs

None

Table 21.3 Image criteria for clavicle radiographs

Clavicle ap
Visualization of the entire clavicle
Visually sharp reproduction of the sternoclavicular and the acromioclavicular joints
Visually sharp reproduction of the cortex and spongiosa of the clavicle
Visualization of the companion shadow (linear soft tissue density) along the upper cortex of the clavicle

Table 21.4 Patient positions and beam projections for clavicle radiographs

Patient position	Beam projections
Supine	ap

Table 21.5 Optimal field settings for clavicle radiographs

Clavicle ap	
Upper field edge:	1–2.5 cm above the sternoclavicular joint[a]
Field centring:	
Lower field edge:	1–2.5 cm below of the sternoclavicular joint[a]
Medial field edge:	Midpoint of the jugular fossa
Lateral field edge:	Including the acromion
Annotations:	
Adjust angulation of the tube dependent on the course of the clavicle.	
[a] Dependent on the patient's age	

M. Seidenbusch et al., *Imaging Practice and Radiation Protection in Pediatric Radiology*,
https://doi.org/10.1007/978-3-030-18504-6_21

Table 21.6 Patient preparation

General	Remove all clothes, external metallic devices, and jewellery from the neck and chest of the patient.
	Immobilize infants and uncooperative patients using sandbags.
	Cooperative patients can be positioned upright at the vertical stand.
Positioning	For the ap projection, the arm should be positioned alongside the chest in the supine position.
Shielding	Lead mask shielding should be used close to the medial and lower field edges.

Table 21.7 Radiographic procedure

X-ray source	Use automatic exposure control (AEC) in patients >10 years of age.
	Focal spot size between 0.6 and 1.3 dependent on the age of the patient
	Check the use of additional filtration
	Check the focus-to-detector distance
Field setting	Check the field size
	Check the field centring
Image detector system	Check the use of antiscatter grids
Patient	-

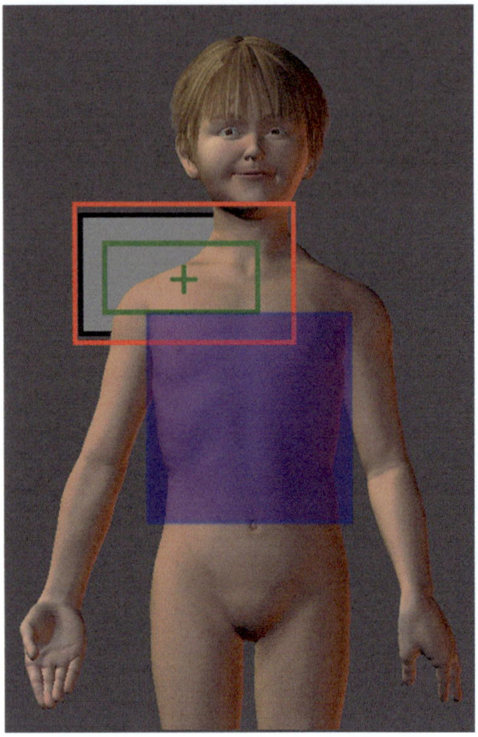

Fig. 21.1 Field settings for clavicle radiographs in supine position. Antero-posterior projection (© V. Rösenberger & M. Seidenbusch 2018). Green lines = optimal field settings; green cross = centring point; red lines = suboptimal field settings; grey shaded area = image detector; blue shaded area = lead mask shielding; dark grey shaded area = Bucky table

Table 21.8 Radiographic techniques of clavicle radiographs

Age group (years)	Beam direction	Patient position	Beam projection	Typical patient diameter (cm)	Optimal field size (cm x cm)	Focus-to-detector distance (cm)	Voltage (kVp) 50	60	70	80	90	100	Antiscatter grid
0	Vertical	Supine	ap	9.8	7 × 3	100[1]							
1–2				13.0	9 × 5								
3–7				15.0	12 × 6	115							
8–12				16.8	14 × 8								
13–17				19.6	17 × 9								
18				20.0	20 × 11								

Annotations:
1 The patient is lying on the table top directly on the detector
(dark grey) Parameter combination not recommended, normalized organ doses not provided
(light blue) Parameter combination recommended, normalized organ doses provided

Table 21.9 Radiation qualities of clavicle radiographs

Voltage (kVp)	Total filtration: Without additional filtration 1.5 mm Al	2.5 Al	With additional filtration 3.5 Al	4.5 Al	3.5 Al + 0.1 Cu	3.5 Al + 0.2 Cu
50						
60						
70						
80						
90						
100						

Annotations:
(dark grey) Parameter combination not recommended, normalized organ doses not provided
(grey) Parameter combination not recommended, but normalized organ doses provided
(light blue) Parameter combination recommended, normalized organ doses provided

Table 21.10 Inconsistencies between the mathematical MIRD phantoms and real human anatomy

Anatomical region	Inconsistencies
Extremities	No shoulder joint has been inserted

jections of clavicle radiographies of the patient who is correctly positioned are depicted in Fig. 21.1. Also the optimal field (marked in green), the suboptimal field (marked in red) and the correct shielding are shown. Tables 21.11, 21.12, and 21.13 outline the virtual field settings on the skin surface of mathematical MIRD phantoms of various age groups and the resulting virtual radiographies with and without bone structures, respectively. Normalized organ doses resulting from these virtual field settings have been calculated for all age groups and have been listed in Tables 21.14, 21.15, 21.16, 21.17, 21.18, 21.19, 21.20, 21.21, 21.22, 21.23, 21.24, 21.25, 21.26, 21.27, and 21.28. Table cells with normal font refer to optimal field settings and table cells with bold font refer to suboptimal field settings. Please note that normalized organ doses have been provided also for exposure parameter combinations, which are not recommended by the authors.

21.2 Indications and Contraindications

Indications and contraindications for clavicle radiographs are given in Tables 21.1 and 21.2.

21.3 Image Criteria

The image criteria required for clavicle radiographs are listed in Table 21.3.

21.4 Field Settings

In Table 21.4, patient positions and beam projections used for clavicle radiographs are listed, and in Table 21.5, the optimal field settings are listed which are required in order to fulfil the image criteria mentioned in Table 21.3.

21.5 Patient Preparation and Radiographic Procedure

See Tables 21.6, 21.7 and [1].

21.6 Exposure Parameters

In Tables 21.8 and 21.9, exposure parameter combinations used for clavicle radiographs in various age groups are listed. Dark grey shaded table cells refer to exposure parameter combinations which are not recommended for paediatric clavicle radiographs and for which no normalized organ doses have been provided. Light grey and light blue shaded table cells refer to exposure parameter combinations for which normalized organ doses have been provided with light grey shaded table cells referring to exposure parameter combinations being not recommended and light blue shaded table cells referring to exposure parameter combinations being recommended for paediatric clavicle radiographs. Note that due to the fact that the radiation dose is an exponential function of the patient diameter, it may make sense to use the patient diameters instead of the patient's age groups for determining the adequate exposure parameters from Table 21.8. In this case, for determining the corresponding normalized organ doses from Tables 21.14, 21.15, 21.16, 21.17, 21.18, 21.19, 21.20, 21.21, 21.22, 21.23, 21.24, 21.25, 21.26, 21.27, and 21.28, the age group related to the patient diameter should be used.

21.7 Normalized Organ Doses

For calculating normalized organ doses, the field settings described in Sect. 21.4 were transferred to the mathematical MIRD phantoms. Inconsistencies between the geometrical anatomy of the mathematical phantoms and the real human anatomy have been listed in Table 21.10 whenever they are of relevance for the validity of the normalized organ doses.

In Tables 21.11, 21.12, and 21.13, optimal and suboptimal field settings are listed for the different age groups and the beam projections given in Tables 21.8 and 21.9. Green lines refer to optimal field settings, red lines refer to suboptimal field settings, respectively. In the tables, the optimal and suboptimal field settings which were used for performing the calculations of the normalized organ doses provided in Tables 21.14, 21.15, 21.16, 21.17, 21.18, 21.19, 21.20, 21.21, 21.22, 21.23, 21.24, 21.25, 21.26, 21.27, and 21.28 were drawn in on the skin surface of the mathematical phantoms. The virtual mathematical radiographs resulting from these field settings were depicted in the tables with and without bone structures, respectively, in order to enable the reader to visually compare the virtual mathematical radiographs with the real radiographs made under clinical conditions. For comparing the virtual mathematical field settings with the real clinical field settings, the virtual mathematical field edges were also listed in the tables.

As the normalized organ doses provided in Tables 21.14, 21.15, 21.16, 21.17, 21.18, 21.19, 21.20, 21.21, 21.22, 21.23, 21.24, 21.25, 21.26, 21.27, and 21.28 refer to entrance kerma values, optimal and suboptimal field sizes were provided in the tables in order to enable the reader to calculate entrance air kerma dose values from clinically measured dose area product values. Please note that these optimal and suboptimal field sizes refer to the mathematical phantoms of an age group only and not to real patients with corresponding ages.

Please note also that the anatomy of the mathematical phantoms has been idealized. Therefore, the virtual field settings for calculating the normalized organ doses have been defined with a view to ensuring that the resulting virtual organ doses would be consistent with the real organ doses which would be achieved during real X-ray examinations being performed with the real field settings recommended in Sect. 21.4. Thus, the field settings shown in the following tables may not be consistent with the clinical field settings as recommended in Sect. 21.4.

Table 21.11 Field settings—clavicle ap

Age group (years)		Anatomical landmarks: Optimal field setting		Anatomical landmarks: Suboptimal field setting	
0	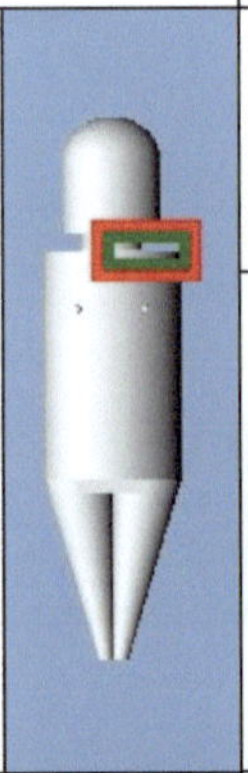	Upper field edge: Just below the tip of the mandible Lower field edge: About 2 cm below the sternoclavicular joint Medial field edge: Median line Lateral field edge: Close to the skin surface of the shoulder Field size = field width × field height: 7 cm × 3 cm = 21 cm^2		Upper field edge: Lower mandible Lower field edge: At level between shoulder height and mamilla Medial field edge: Close to the opposite parasternal line Lateral field edge: Close to the skin surface of the shoulder Field size = field width × field height: 8 cm × 5 cm = 40 cm^2	
		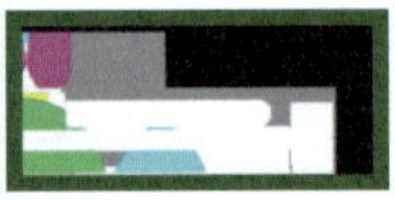	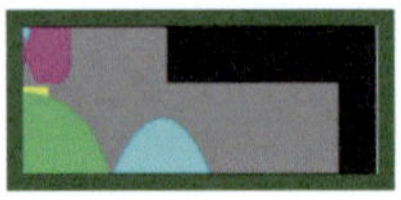	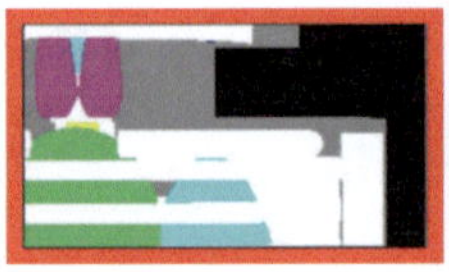	
1–2	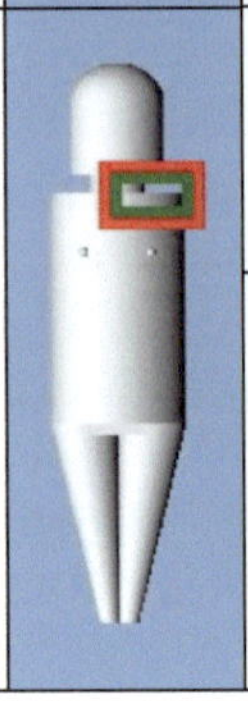	Upper field edge: Just below the tip of the mandible Lower field edge: About 2 cm below the sternoclavicular joint Medial field edge: Median line Lateral field edge: Close to the skin surface of the shoulder Field size = field width × field height: 9 cm × 4 cm = 36 cm^2		Upper field edge: Lower mandible Lower field edge: At level between shoulder height and mamilla Medial field edge: Close to the opposite parasternal line Lateral field edge: Close to the skin surface of the shoulder Field size = field width × field height: 11 cm × 7 cm = 77 cm^2	
		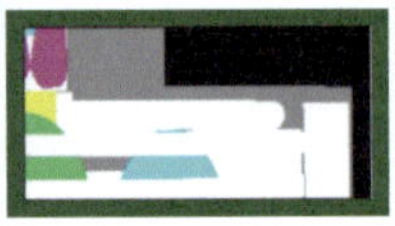	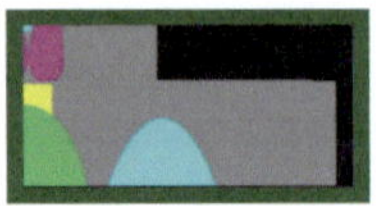		

Table 21.12 Field settings—clavicle ap

Age group (years)		Anatomical landmarks: Optimal field setting	Anatomical landmarks: Suboptimal field setting
3–7	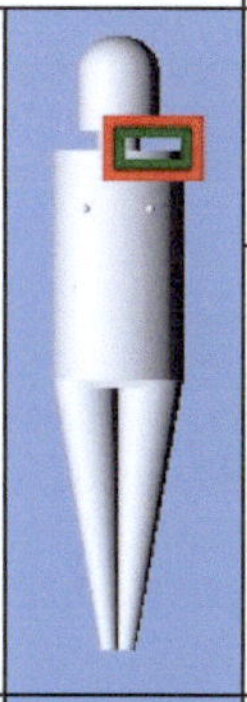	Upper field edge: Just below the tip of the mandible Lower field edge: About 3 cm below the sternoclavicular joint Medial field edge: Median line Lateral field edge: Close to the skin surface of the shoulder Field size = field width × field height: 12 cm × 6 cm = 72 cm^2	Upper field edge: Lower mandible Lower field edge: At level between shoulder height and mamilla Medial field edge: Close to the opposite parasternal line Lateral field edge: Close to the skin surface of the shoulder Field size = field width × field height: 14 cm × 9 cm = 126 cm^2
		 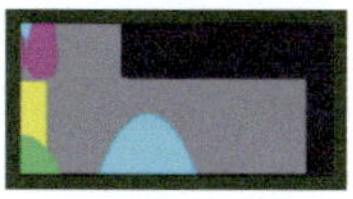	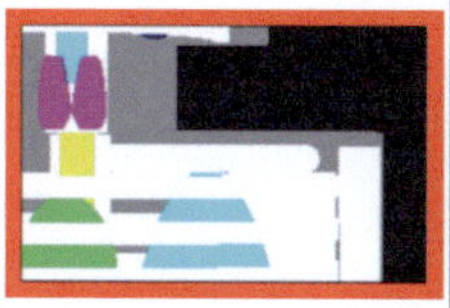 
8–12	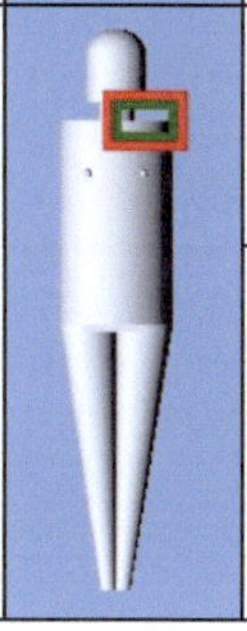	Upper field edge: Just below the tip of the mandible Lower field edge: About 4 cm below the sternoclavicular joint Medial field edge: Median line Lateral field edge: Close to the skin surface of the shoulder Field size = field width × field height: 14 cm × 8 cm = 112 cm^2	Upper field edge: Lower mandible Lower field edge: At level between shoulder height and mamilla Medial field edge: Close to the opposite parasternal line Lateral field edge: Close to the skin surface of the shoulder Field size = field width × field height: 17 cm × 11 cm = 187 cm^2
		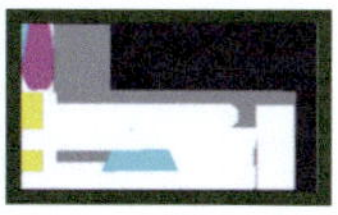 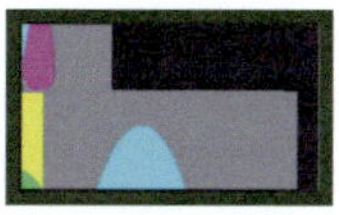	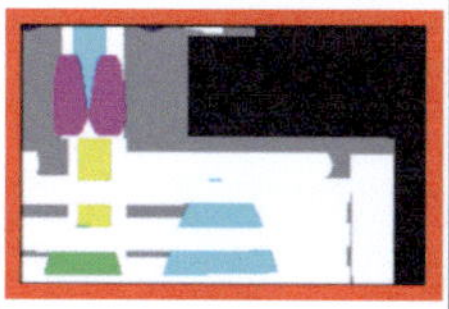 

Table 21.13 Field settings—clavicle ap

Age group (years)		Anatomical landmarks: Optimal field setting	Anatomical landmarks: Suboptimal field setting
13–17	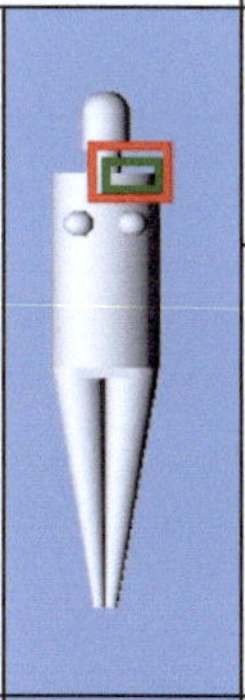	Upper field edge: Just below the tip of the mandible Lower field edge: About 4 cm below the sternoclavicular joint Medial field edge: Median line Lateral field edge: Close to the skin surface of the shoulder Field size = field width × field height: 17 cm × 9 cm = 153 cm^2	Upper field edge: Lower mandible Lower field edge: At level between shoulder height and mamilla Medial field edge: Close to the opposite parasternal line Lateral field edge: Close to the skin surface of the shoulder Field size = field width × field height: 22 cm × 16 cm = 352 cm^2
		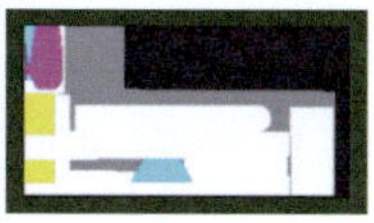 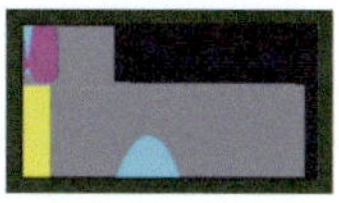	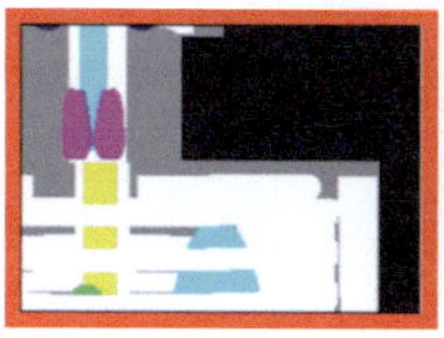 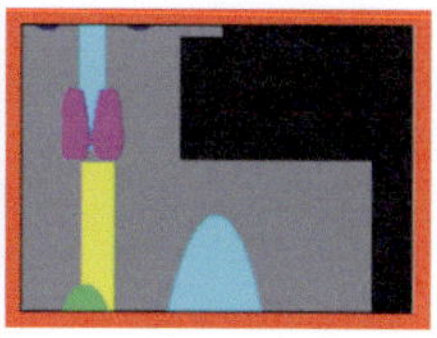
18	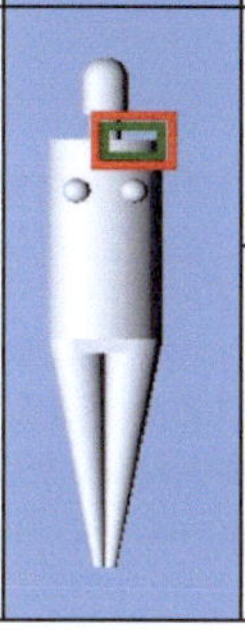	Upper field edge: Just below the tip of the mandible Lower field edge: About 4 cm below the sternoclavicular joint Medial field edge: Median line Lateral field edge: Close to the skin surface of the shoulder Field size = field width × field height: 20 cm × 11 cm = 220 cm^2	Upper field edge: Lower mandible Lower field edge: At level between shoulder height and mamilla Medial field edge: Close to the opposite parasternal line Lateral field edge: Close to the skin surface of the shoulder Field size = field width × field height: 24 cm × 17 cm = 408 cm^2
		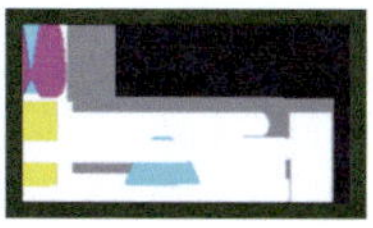 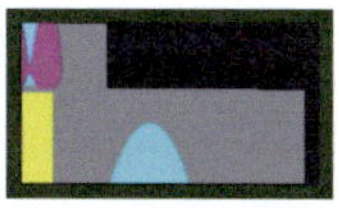	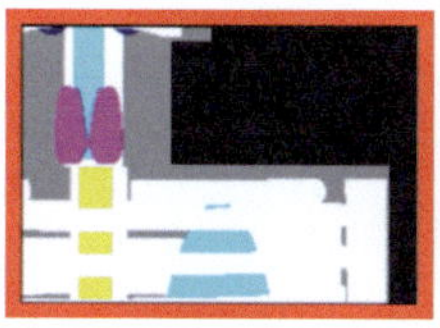 

Table 21.14 Normalized organ doses—Clavicle ap—<1 year

Beam projection:	ap											**Table cells:**
Focus-to-detector distance (cm):	100											(+) = normal font = optimal field setting
Tube voltage (kVp):	60											**(−) = bold font = suboptimal field setting**
Total filtrations:	See last table row.											Empty table cells correspond to values <0.01
Organ or tissue	**Organ dose D_T : Entrance dose K_E (mSv * mGy^{-1})**											
	(+)	(−)	(+)	(−)	(+)	(−)	(+)	(−)	(+)	(−)	(+)	(−)
Skull		**0.02**		**0.02**		**0.02**		**0.03**	0.01	**0.03**	0.01	**0.03**
Brain												
Oral mucosa		**0.02**	0.01	**0.02**	0.01	**0.03**	0.01	**0.03**	0.02	**0.04**	0.02	**0.04**
Salivary glands	0.01	**0.04**	0.02	**0.05**	0.02	**0.06**	0.02	**0.07**	0.03	**0.08**	0.02	**0.08**
Cervical spine	0.15	**0.32**	0.19	**0.41**	0.22	**0.47**	0.24	**0.52**	0.29	**0.60**	0.28	**0.59**
Thyroid	0.36	**0.86**	0.40	**0.96**	0.42	**1.02**	0.44	**1.06**	0.46	**1.12**	0.46	**1.12**
Thymus	0.14	**0.41**	0.16	**0.47**	0.18	**0.51**	0.19	**0.54**	0.21	**0.58**	0.21	**0.58**
Thoracic spine	0.03	**0.07**	0.05	**0.09**	0.05	**0.10**	0.06	**0.12**	0.08	**0.14**	0.07	**0.13**
Claviculae	0.98	**1.06**	1.12	**1.22**	1.21	**1.32**	1.27	**1.39**	1.37	**1.50**	1.39	**1.53**
Scapulae	0.05	**0.09**	0.07	**0.12**	0.08	**0.14**	0.09	**0.16**	0.11	**0.19**	0.10	**0.18**
Ribs	0.09	**0.16**	0.10	**0.18**	0.11	**0.20**	0.12	**0.21**	0.13	**0.23**	0.13	**0.23**
Oesophagus	0.02	**0.04**	0.03	**0.06**	0.04	**0.07**	0.04	**0.08**	0.05	**0.09**	0.05	**0.09**
Trachea	0.02	**0.08**	0.03	**0.09**	0.03	**0.10**	0.03	**0.11**	0.04	**0.12**	0.04	**0.12**
Lung	0.02	**0.03**	0.02	**0.04**	0.02	**0.05**	0.02	**0.05**	0.03	**0.06**	0.03	**0.06**
Mamma										**0.01**		**0.01**
Heart		**0.01**		**0.02**	0.01	**0.02**	0.01	**0.02**	0.01	**0.03**	0.01	**0.03**
Stomach												
Spleen												
Liver												
Gall bladder												
Pancreas												
Small intestine												
Large intestine												
Sigma/rectum												
Lumbar spine												
Pelvis												
Adrenals												
Kidneys												
Urinary bladder												
Uterus												
Ovaries												
Prostate												
Testes												
Upper arm	0.18	**0.27**	0.22	**0.33**	0.25	**0.37**	0.27	**0.40**	0.30	**0.45**	0.30	**0.45**
Elbow						**0.01**		**0.01**		**0.02**		**0.01**
Forearm												
Upper leg												
Knee												
Lower leg												
Lymph nodes	0.03	**0.06**	0.03	**0.07**	0.03	**0.08**	0.03	**0.08**	0.04	**0.09**	0.04	**0.09**
Skin	0.03	**0.05**	0.03	**0.06**	0.03	**0.06**	0.03	**0.06**	0.04	**0.07**	0.04	**0.07**
Muscle	0.02	**0.03**	0.02	**0.04**	0.02	**0.04**	0.03	**0.05**	0.03	**0.05**	0.03	**0.05**
Skeleton	0.05	**0.08**	0.06	**0.10**	0.07	**0.11**	0.07	**0.12**	0.08	**0.14**	0.08	**0.13**
Red bone marrow	0.01	**0.02**	0.01	**0.02**	0.02	**0.03**	0.02	**0.03**	0.02	**0.03**	0.02	**0.03**
Whole body dose	0.02	**0.03**	0.02	**0.04**	0.02	**0.04**	0.03	**0.05**	0.03	**0.05**	0.03	**0.05**
ED (ICRP 60)	0.02	**0.06**	0.03	**0.06**	0.03	**0.07**	0.03	**0.07**	0.03	**0.08**	0.03	**0.08**
ED (ICRP 103)	0.02	**0.05**	0.03	**0.06**	0.03	**0.06**	0.03	**0.07**	0.03	**0.07**	0.03	**0.07**
Total filtration (mm) Al	**1.5**		**2.5**		**3.5**		**4.5**		**3.5**		**3.5**	
Cu	**0**		**0**		**0**		**0**		**0.1**		**0.2**	

Table 21.15 Normalized organ doses—Clavicle ap—<1 year

Beam projection:			ap									**Table cells:**
Focus-to-detector distance (cm):			100									(+) = normal font = optimal field setting
Tube voltage (kVp):			70									**(−) = bold font = suboptimal field setting**
Total filtrations:			See last table row.									Empty table cells correspond to values <0.01
Organ or tissue	**Organ dose D_T: Entrance dose K_E**								**(mSv * mGy^{-1})**			
	(+)	(−)	(+)	(−)	(+)	(−)	(+)	(−)	(+)	(−)	(+)	(−)
Skull		**0.02**		**0.02**		**0.03**	0.01	**0.03**	0.01	**0.03**	0.01	**0.04**
Brain												
Oral mucosa	0.01	**0.02**	0.01	**0.03**	0.01	**0.03**	0.02	**0.04**	0.02	**0.04**	0.02	**0.05**
Salivary glands	0.02	**0.05**	0.02	**0.06**	0.02	**0.07**	0.02	**0.07**	0.03	**0.08**	0.03	**0.09**
Cervical spine	0.18	**0.37**	0.22	**0.46**	0.25	**0.52**	0.27	**0.57**	0.31	**0.65**	0.34	**0.72**
Thyroid	0.37	**0.91**	0.41	**1.00**	0.43	**1.06**	0.45	**1.09**	0.47	**1.15**	0.49	**1.19**
Thymus	0.15	**0.44**	0.18	**0.50**	0.19	**0.54**	0.20	**0.56**	0.22	**0.61**	0.23	**0.64**
Thoracic spine	0.04	**0.08**	0.05	**0.11**	0.06	**0.12**	0.07	**0.14**	0.09	**0.17**	0.10	**0.19**
Claviculae	1.04	**1.13**	1.17	**1.27**	1.25	**1.36**	1.30	**1.43**	1.38	**1.52**	1.42	**1.57**
Scapulae	0.06	**0.11**	0.08	**0.14**	0.09	**0.16**	0.10	**0.18**	0.12	**0.21**	0.14	**0.24**
Ribs	0.09	**0.17**	0.11	**0.20**	0.12	**0.21**	0.12	**0.22**	0.13	**0.24**	0.14	**0.25**
Oesophagus	0.03	**0.05**	0.04	**0.07**	0.04	**0.08**	0.05	**0.09**	0.06	**0.10**	0.06	**0.12**
Trachea	0.03	**0.09**	0.03	**0.10**	0.03	**0.11**	0.04	**0.12**	0.04	**0.13**	0.04	**0.14**
Lung	0.02	**0.04**	0.02	**0.05**	0.03	**0.06**	0.03	**0.06**	0.03	**0.07**	0.04	**0.08**
Mamma								**0.01**		**0.01**		**0.01**
Heart		**0.02**	0.01	**0.02**	0.01	**0.03**	0.01	**0.03**	0.02	**0.03**	0.02	**0.04**
Stomach												
Spleen												
Liver												
Gall bladder												
Pancreas												
Small intestine												
Large intestine												
c												
Lumbar spine												
Pelvis												
Adrenals												
Kidneys												
Urinary bladder												
Uterus												
Ovaries												
Prostate												
Testes												
Upper arm	0.20	**0.30**	0.24	**0.36**	0.27	**0.40**	0.29	**0.43**	0.32	**0.48**	0.34	**0.51**
Elbow				**0.01**		**0.01**		**0.01**		**0.02**	0.01	**0.02**
Forearm												
Upper leg												
Knee												
Lower leg												
Lymph nodes	0.03	**0.07**	0.03	**0.08**	0.03	**0.08**	0.04	**0.09**	0.04	**0.10**	0.04	**0.10**
Skin	0.03	**0.06**	0.03	**0.06**	0.03	**0.06**	0.04	**0.07**	0.04	**0.07**	0.04	**0.07**
Muscle	0.02	**0.04**	0.02	**0.04**	0.03	**0.05**	0.03	**0.05**	0.03	**0.06**	0.03	**0.06**
Skeleton	0.05	**0.09**	0.06	**0.11**	0.07	**0.12**	0.08	**0.13**	0.09	**0.15**	0.09	**0.16**
Red bone marrow	0.01	**0.02**	0.02	**0.03**	0.02	**0.03**	0.02	**0.03**	0.02	**0.04**	0.02	**0.04**
Whole body dose	0.02	**0.04**	0.02	**0.04**	0.03	**0.05**	0.03	**0.05**	0.03	**0.06**	0.03	**0.06**
ED (ICRP 60)	0.03	**0.06**	0.03	**0.07**	0.03	**0.07**	0.03	**0.08**	0.04	**0.08**	0.04	**0.09**
ED (ICRP 103)	0.02	**0.06**	0.03	**0.06**	0.03	**0.07**	0.03	**0.07**	0.03	**0.08**	0.04	**0.08**
Total filtration (mm) Al	**1.5**		**2.5**		**3.5**		**4.5**		**3.5**		**3.5**	
Cu	**0**		**0**		**0**		**0**		**0.1**		**0.2**	

Table 21.16 Normalized organ doses—Clavicle ap—1–2 years

Beam projection:			ap									**Table cells:**
Focus-to-detector distance (cm):			100									(+) = normal font = optimal field setting
Tube voltage (kVp):			60									**(−) = bold font = suboptimal field setting**
Total filtrations:			See last table row.									Empty table cells correspond to values <0.01
Organ or tissue	**Organ dose D_T : Entrance dose K_E**								**(mSv * mGy⁻¹)**			
	(+)	(−)	(+)	(−)	(+)	(−)	(+)	(−)	(+)	(−)	(+)	(−)
Skull		**0.01**		**0.01**		**0.02**		**0.02**	0.01	**0.02**		**0.02**
Brain												
Oral mucosa		**0.01**		**0.02**	0.01	**0.02**	0.01	**0.03**	0.01	**0.03**	0.01	**0.03**
Salivary glands	0.01	**0.03**	0.01	**0.03**	0.02	**0.04**	0.02	**0.04**	0.02	**0.05**	0.02	**0.05**
Cervical spine	0.12	**0.26**	0.15	**0.33**	0.18	**0.38**	0.20	**0.42**	0.24	**0.51**	0.22	**0.48**
Thyroid	0.40	**0.86**	0.44	**0.97**	0.47	**1.03**	0.49	**1.07**	0.53	**1.14**	0.52	**1.13**
Thymus	0.12	**0.38**	0.15	**0.45**	0.16	**0.49**	0.17	**0.52**	0.20	**0.58**	0.19	**0.58**
Thoracic spine	0.02	**0.04**	0.03	**0.06**	0.04	**0.07**	0.04	**0.08**	0.06	**0.10**	0.05	**0.09**
Claviculae	0.87	**0.97**	1.03	**1.15**	1.14	**1.27**	1.22	**1.36**	1.34	**1.51**	1.36	**1.52**
Scapulae	0.03	**0.05**	0.04	**0.07**	0.05	**0.08**	0.06	**0.09**	0.07	**0.12**	0.06	**0.11**
Ribs	0.11	**0.18**	0.12	**0.21**	0.14	**0.23**	0.14	**0.24**	0.16	**0.27**	0.16	**0.26**
Oesophagus	0.02	**0.04**	0.03	**0.05**	0.04	**0.06**	0.04	**0.07**	0.05	**0.09**	0.04	**0.08**
Trachea	0.02	**0.07**	0.02	**0.08**	0.02	**0.09**	0.02	**0.09**	0.03	**0.10**	0.03	**0.10**
Lung	0.02	**0.03**	0.02	**0.04**	0.02	**0.05**	0.03	**0.05**	0.03	**0.06**	0.03	**0.06**
Mamma								**0.01**		**0.01**		**0.01**
Heart		**0.01**		**0.01**		**0.02**		**0.02**	0.01	**0.03**	0.01	**0.02**
Stomach												
Spleen												
Liver												
Gall bladder												
Pancreas												
Small intestine												
Large intestine												
Sigma/rectum												
Lumbar spine												
Pelvis												
Adrenals												
Kidneys												
Urinary bladder												
Uterus												
Ovaries												
Prostate												
Testes												
Upper arm	0.15	**0.22**	0.19	**0.27**	0.22	**0.31**	0.24	**0.34**	0.28	**0.39**	0.27	**0.38**
Elbow										**0.01**		
Forearm												
Upper leg												
Knee												
Lower leg												
Lymph nodes	0.03	**0.06**	0.03	**0.07**	0.03	**0.07**	0.03	**0.08**	0.04	**0.09**	0.04	**0.08**
Skin	0.03	**0.05**	0.03	**0.05**	0.03	**0.05**	0.03	**0.06**	0.03	**0.06**	0.03	**0.06**
Muscle	0.01	**0.02**	0.02	**0.03**	0.02	**0.03**	0.02	**0.04**	0.02	**0.04**	0.02	**0.04**
Skeleton	0.04	**0.06**	0.05	**0.08**	0.06	**0.09**	0.06	**0.10**	0.07	**0.11**	0.07	**0.11**
Red bone marrow		**0.01**	0.01	**0.02**	0.01	**0.02**	0.01	**0.02**	0.02	**0.03**	0.01	**0.02**
Whole body dose	0.02	**0.03**	0.02	**0.03**	0.02	**0.04**	0.02	**0.04**	0.03	**0.05**	0.03	**0.04**
ED (ICRP 60)	0.03	**0.05**	0.03	**0.06**	0.03	**0.07**	0.03	**0.07**	0.04	**0.08**	0.04	**0.08**
ED (ICRP 103)	0.02	**0.05**	0.03	**0.06**	0.03	**0.06**	0.03	**0.06**	0.03	**0.07**	0.03	**0.07**
Total filtration (mm) Al	**1.5**		**2.5**		**3.5**		**4.5**		**3.5**		**3.5**	
Cu	**0**		**0**		**0**		**0**		**0.1**		**0.2**	

Table 21.17 Normalized organ doses—Clavicle ap—1–2 years

Beam projection:			ap									**Table cells:**
Focus-to-detector distance (cm):			100									(+) = normal font = optimal field setting
Tube voltage (kVp):			70									**(−) = bold font = suboptimal field setting**
Total filtrations:			See last table row.									Empty table cells correspond to values <0.01
Organ or tissue	**Organ dose D_T : Entrance dose K_E**							**(mSv * mGy^{-1})**				
	(+)	(−)	(+)	(−)	(+)	(−)	(+)	(−)	(+)	(−)	(+)	(−)
Skull		**0.01**		**0.02**		**0.02**	0.01	**0.02**	0.01	**0.03**	0.01	**0.03**
Brain												
Oral mucosa		**0.02**	0.01	**0.02**	0.01	**0.03**	0.01	**0.03**	0.02	**0.04**	0.02	**0.04**
Salivary glands	0.01	**0.03**	0.02	**0.04**	0.02	**0.04**	0.02	**0.05**	0.03	**0.06**	0.03	**0.06**
Cervical spine	0.14	**0.31**	0.18	**0.38**	0.20	**0.44**	0.22	**0.48**	0.26	**0.57**	0.30	**0.64**
Thyroid	0.42	**0.91**	0.47	**1.01**	0.49	**1.07**	0.51	**1.11**	0.55	**1.17**	0.57	**1.22**
Thymus	0.14	**0.42**	0.16	**0.48**	0.18	**0.53**	0.19	**0.56**	0.21	**0.62**	0.23	**0.66**
Thoracic spine	0.03	**0.06**	0.04	**0.08**	0.05	**0.09**	0.06	**0.10**	0.07	**0.13**	0.08	**0.15**
Claviculae	0.95	**1.06**	1.10	**1.23**	1.20	**1.35**	1.27	**1.43**	1.38	**1.56**	1.46	**1.65**
Scapulae	0.04	**0.07**	0.05	**0.09**	0.06	**0.10**	0.07	**0.11**	0.08	**0.14**	0.10	**0.16**
Ribs	0.12	**0.19**	0.13	**0.22**	0.14	**0.24**	0.15	**0.26**	0.17	**0.28**	0.18	**0.30**
Oesophagus	0.03	**0.05**	0.04	**0.07**	0.04	**0.08**	0.05	**0.09**	0.06	**0.10**	0.07	**0.12**
Trachea	0.02	**0.07**	0.02	**0.09**	0.02	**0.09**	0.03	**0.10**	0.03	**0.11**	0.03	**0.12**
Lung	0.02	**0.04**	0.02	**0.05**	0.03	**0.06**	0.03	**0.06**	0.04	**0.07**	0.04	**0.08**
Mamma						**0.01**		**0.01**		**0.02**		**0.02**
Heart		**0.01**		**0.02**	0.01	**0.02**	0.01	**0.02**	0.01	**0.03**	0.02	**0.04**
Stomach												
Spleen												
Liver												
Gall bladder												
Pancreas												
Small intestine												
Large intestine												
Sigma/rectum												
Lumbar spine												
Pelvis												
Adrenals												
Kidneys												
Urinary bladder												
Uterus												
Ovaries												
Prostate												
Testes												
Upper arm	0.18	**0.25**	0.21	**0.30**	0.24	**0.34**	0.26	**0.37**	0.30	**0.43**	0.33	**0.47**
Elbow						**0.01**		**0.01**		**0.01**		**0.02**
Forearm												
Upper leg												
Knee												
Lower leg												
Lymph nodes	0.03	**0.07**	0.03	**0.07**	0.04	**0.08**	0.04	**0.08**	0.04	**0.09**	0.04	**0.10**
Skin	0.03	**0.05**	0.03	**0.05**	0.03	**0.06**	0.03	**0.06**	0.03	**0.06**	0.04	**0.07**
Muscle	0.02	**0.03**	0.02	**0.03**	0.02	**0.04**	0.02	**0.04**	0.03	**0.05**	0.03	**0.05**
Skeleton	0.05	**0.07**	0.05	**0.09**	0.06	**0.10**	0.07	**0.11**	0.08	**0.12**	0.08	**0.14**
Red bone marrow	0.01	**0.02**	0.01	**0.02**	0.01	**0.02**	0.02	**0.03**	0.02	**0.03**	0.02	**0.03**
Whole body dose	0.02	**0.03**	0.02	**0.04**	0.02	**0.04**	0.03	**0.04**	0.03	**0.05**	0.03	**0.06**
ED (ICRP 60)	0.03	**0.06**	0.03	**0.07**	0.03	**0.07**	0.04	**0.08**	0.04	**0.08**	0.04	**0.09**
ED (ICRP 103)	0.02	**0.05**	0.03	**0.06**	0.03	**0.07**	0.03	**0.07**	0.04	**0.08**	0.04	**0.08**
Total filtration (mm) **Al**	**1.5**		**2.5**		**3.5**		**4.5**		**3.5**		**3.5**	
Cu	**0**		**0**		**0**		**0**		**0.1**		**0.2**	

Table 21.18 Normalized organ doses—Clavicle ap—3–7 years

Beam projection:			ap									**Table cells:**
Focus-to-detector distance (cm):			115					(+) = normal font = optimal field setting				
Tube voltage (kVp):			60					**(−) = bold font = suboptimal field setting**				
Total filtrations:			See last table row.					Empty table cells correspond to values <0.01				
Organ or tissue	**Organ dose D_T : Entrance dose K_E**							**(mSv * mGy^{-1})**				
	(+)	(−)	(+)	(−)	(+)	(−)	(+)	(−)	(+)	(−)	(+)	(−)
Skull		**0.02**		**0.02**		**0.02**		**0.03**	0.01	**0.03**	0.01	**0.03**
Brain												
Oral mucosa		**0.02**		**0.02**	0.01	**0.02**	0.01	**0.03**	0.01	**0.03**	0.01	**0.03**
Salivary glands	0.01	**0.03**	0.01	**0.03**	0.02	**0.04**	0.02	**0.04**	0.02	**0.05**	0.02	**0.05**
Cervical spine	0.12	**0.26**	0.16	**0.34**	0.19	**0.40**	0.21	**0.45**	0.25	**0.54**	0.24	**0.51**
Thyroid	0.42	**0.82**	0.47	**0.93**	0.51	**0.99**	0.53	**1.04**	0.57	**1.12**	0.57	**1.11**
Thymus	0.06	**0.26**	0.07	**0.31**	0.08	**0.35**	0.09	**0.37**	0.10	**0.42**	0.10	**0.41**
Thoracic spine	0.02	**0.03**	0.02	**0.04**	0.03	**0.05**	0.03	**0.06**	0.04	**0.08**	0.04	**0.07**
Claviculae	0.81	**0.86**	0.97	**1.03**	1.08	**1.15**	1.16	**1.24**	1.30	**1.40**	1.30	**1.39**
Scapulae	0.02	**0.04**	0.03	**0.05**	0.04	**0.07**	0.04	**0.08**	0.05	**0.10**	0.05	**0.08**
Ribs	0.09	**0.17**	0.11	**0.20**	0.12	**0.21**	0.13	**0.23**	0.14	**0.25**	0.14	**0.25**
Oesophagus	0.01	**0.02**	0.02	**0.03**	0.02	**0.04**	0.02	**0.04**	0.03	**0.05**	0.03	**0.05**
Trachea	0.03	**0.09**	0.03	**0.10**	0.03	**0.11**	0.04	**0.12**	0.04	**0.14**	0.04	**0.13**
Lung	0.01	**0.03**	0.02	**0.04**	0.02	**0.04**	0.02	**0.05**	0.03	**0.06**	0.02	**0.05**
Mamma												
Heart				**0.01**		**0.01**		**0.01**		**0.02**		**0.02**
Stomach												
Spleen												
Liver												
Gall bladder												
Pancreas												
Small intestine												
Large intestine												
Sigma/rectum												
Lumbar spine												
Pelvis												
Adrenals												
Kidneys												
Urinary bladder												
Uterus												
Ovaries												
Prostate												
Testes												
Upper arm	0.13	**0.19**	0.17	**0.24**	0.19	**0.28**	0.21	**0.31**	0.25	**0.36**	0.24	**0.35**
Elbow												
Forearm												
Upper leg												
Knee												
Lower leg												
Lymph nodes	0.03	**0.06**	0.03	**0.07**	0.03	**0.07**	0.04	**0.08**	0.04	**0.09**	0.04	**0.09**
Skin	0.02	**0.04**	0.02	**0.04**	0.03	**0.05**	0.03	**0.05**	0.03	**0.05**	0.03	**0.05**
Muscle	0.01	**0.02**	0.01	**0.03**	0.02	**0.03**	0.02	**0.03**	0.02	**0.04**	0.02	**0.03**
Skeleton	0.03	**0.05**	0.04	**0.06**	0.04	**0.07**	0.04	**0.08**	0.05	**0.09**	0.05	**0.08**
Red bone marrow		**0.01**		**0.02**	0.01	**0.02**	0.01	**0.02**	0.01	**0.02**	0.01	**0.02**
Whole body dose	0.01	**0.02**	0.02	**0.03**	0.02	**0.03**	0.02	**0.03**	0.02	**0.04**	0.02	**0.04**
ED (ICRP 60)	0.02	**0.05**	0.03	**0.06**	0.03	**0.06**	0.03	**0.07**	0.04	**0.07**	0.04	**0.07**
ED (ICRP 103)	0.02	**0.04**	0.03	**0.05**	0.03	**0.06**	0.03	**0.06**	0.03	**0.07**	0.03	**0.06**
Total filtration (mm) Al	**1.5**		**2.5**		**3.5**		**4.5**		**3.5**		**3.5**	
Cu	**0**		**0**		**0**		**0**		**0.1**		**0.2**	

Table 21.19 Normalized organ doses—Clavicle ap—3–7 years

Beam projection:			ap									**Table cells:**
Focus-to-detector distance (cm):			115									(+) = normal font = optimal field setting
Tube voltage (kVp):			70									**(−) = bold font = suboptimal field setting**
Total filtrations:			See last table row.									Empty table cells correspond to values <0.01
Organ or tissue	**Organ dose D_T : Entrance dose K_E**							**(mSv * mGy^{-1})**				
	(+)	(−)	(+)	(−)	(+)	(−)	(+)	(−)	(+)	(−)	(+)	(−)
Skull		**0.02**		**0.02**	0.01	**0.03**	0.01	**0.03**	0.01	**0.04**	0.02	**0.04**
Brain												
Oral mucosa		**0.02**	0.01	**0.02**	0.01	**0.03**	0.01	**0.03**	0.02	**0.04**	0.02	**0.05**
Salivary glands	0.01	**0.03**	0.02	**0.04**	0.02	**0.05**	0.02	**0.05**	0.03	**0.06**	0.03	**0.07**
Cervical spine	0.15	**0.32**	0.19	**0.40**	0.22	**0.47**	0.24	**0.52**	0.29	**0.62**	0.33	**0.70**
Thyroid	0.45	**0.87**	0.50	**0.98**	0.53	**1.04**	0.56	**1.09**	0.60	**1.17**	0.63	**1.23**
Thymus	0.07	**0.29**	0.08	**0.34**	0.09	**0.38**	0.10	**0.41**	0.11	**0.45**	0.12	**0.49**
Thoracic spine	0.02	**0.04**	0.03	**0.06**	0.04	**0.07**	0.04	**0.08**	0.05	**0.10**	0.07	**0.12**
Claviculae	0.89	**0.95**	1.05	**1.12**	1.15	**1.23**	1.23	**1.32**	1.35	**1.46**	1.44	**1.57**
Scapulae	0.03	**0.05**	0.04	**0.07**	0.05	**0.08**	0.05	**0.09**	0.07	**0.12**	0.08	**0.14**
Ribs	0.10	**0.18**	0.12	**0.21**	0.13	**0.23**	0.14	**0.24**	0.15	**0.27**	0.16	**0.29**
Oesophagus	0.02	**0.03**	0.02	**0.04**	0.03	**0.05**	0.03	**0.05**	0.04	**0.06**	0.04	**0.07**
Trachea	0.03	**0.10**	0.03	**0.11**	0.04	**0.12**	0.04	**0.13**	0.05	**0.15**	0.05	**0.16**
Lung	0.02	**0.03**	0.02	**0.04**	0.02	**0.05**	0.03	**0.05**	0.03	**0.07**	0.04	**0.08**
Mamma										**0.01**		**0.01**
Heart		**0.01**		**0.01**		**0.02**		**0.02**	0.01	**0.02**	0.01	**0.03**
Stomach												
Spleen												
Liver												
Gall bladder												
Pancreas												
Small intestine												
Large intestine												
Sigma/rectum												
Lumbar spine												
Pelvis												
Adrenals												
Kidneys												
Urinary bladder												
Uterus												
Ovaries												
Prostate												
Testes												
Upper arm	0.15	**0.23**	0.19	**0.28**	0.21	**0.31**	0.23	**0.34**	0.27	**0.40**	0.30	**0.44**
Elbow												**0.01**
Forearm												
Upper leg												
Knee												
Lower leg												
Lymph nodes	0.03	**0.06**	0.03	**0.07**	0.04	**0.08**	0.04	**0.09**	0.04	**0.09**	0.05	**0.10**
Skin	0.02	**0.04**	0.03	**0.05**	0.03	**0.05**	0.03	**0.05**	0.03	**0.05**	0.03	**0.06**
Muscle	0.01	**0.02**	0.02	**0.03**	0.02	**0.03**	0.02	**0.03**	0.02	**0.04**	0.03	**0.04**
Skeleton	0.03	**0.06**	0.04	**0.07**	0.05	**0.08**	0.05	**0.08**	0.06	**0.10**	0.06	**0.11**
Red bone marrow		**0.02**	0.01	**0.02**	0.01	**0.02**	0.01	**0.02**	0.02	**0.03**	0.02	**0.03**
Whole body dose	0.02	**0.03**	0.02	**0.03**	0.02	**0.04**	0.02	**0.04**	0.03	**0.04**	0.03	**0.05**
ED (ICRP 60)	0.03	**0.05**	0.03	**0.06**	0.03	**0.07**	0.04	**0.07**	0.04	**0.08**	0.04	**0.08**
ED (ICRP 103)	0.02	**0.05**	0.03	**0.06**	0.03	**0.06**	0.03	**0.06**	0.04	**0.07**	0.04	**0.08**
Total filtration (mm) **Al**	**1.5**		**2.5**		**3.5**		**4.5**		**3.5**		**3.5**	
Cu	**0**		**0**		**0**		**0**		**0.1**		**0.2**	

Table 21.20 Normalized organ doses—Clavicle ap—8–12 years

Beam projection:			ap									**Table cells:**
Focus-to-detector distance (cm):			115									(+) = normal font = optimal field setting
Tube voltage (kVp):			60									(−) = **bold font = suboptimal field setting**
Total filtrations:			See last table row.									Empty table cells correspond to values < 0.01
Organ or tissue	**Organ dose D_T : Entrance dose K_E**								**(mSv * mGy^{-1})**			
	(+)	(−)	(+)	(−)	(+)	(−)	(+)	(−)	(+)	(−)	(+)	(−)
Skull		**0.01**		**0.01**		**0.02**		**0.02**	0.01	**0.02**		**0.02**
Brain												
Oral mucosa		**0.01**		**0.02**		**0.02**		**0.02**	0.01	**0.03**	0.01	**0.02**
Salivary glands		**0.03**	0.01	**0.04**	0.01	**0.04**	0.02	**0.05**	0.02	**0.06**	0.02	**0.05**
Cervical spine	0.12	**0.28**	0.15	**0.36**	0.18	**0.43**	0.20	**0.48**	0.25	**0.58**	0.23	**0.54**
Thyroid	0.38	**0.77**	0.43	**0.89**	0.47	**0.96**	0.49	**1.01**	0.54	**1.10**	0.53	**1.09**
Thymus	0.02	**0.17**	0.03	**0.20**	0.04	**0.23**	0.04	**0.25**	0.05	**0.29**	0.05	**0.28**
Thoracic spine	0.01	**0.02**	0.02	**0.03**	0.02	**0.04**	0.02	**0.05**	0.03	**0.06**	0.03	**0.05**
Claviculae	0.75	**0.84**	0.91	**1.02**	1.02	**1.15**	1.10	**1.24**	1.25	**1.41**	1.24	**1.40**
Scapulae	0.02	**0.03**	0.02	**0.04**	0.03	**0.05**	0.03	**0.06**	0.04	**0.07**	0.04	**0.06**
Ribs	0.09	**0.15**	0.11	**0.18**	0.12	**0.20**	0.13	**0.21**	0.14	**0.24**	0.14	**0.23**
Oesophagus		**0.02**	0.01	**0.03**	0.02	**0.03**	0.02	**0.03**	0.02	**0.04**	0.02	**0.04**
Trachea	0.07	**0.21**	0.09	**0.25**	0.10	**0.27**	0.10	**0.29**	0.12	**0.33**	0.11	**0.32**
Lung	0.01	**0.02**	0.01	**0.03**	0.02	**0.04**	0.02	**0.04**	0.02	**0.05**	0.02	**0.04**
Mamma												
Heart										**0.01**		**0.01**
Stomach												
Spleen												
Liver												
Gall bladder												
Pancreas												
Small intestine												
Large intestine												
Sigma/rectum												
Lumbar spine												
Pelvis												
Adrenals												
Kidneys												
Urinary bladder												
Uterus												
Ovaries												
Prostate												
Testes												
Upper arm	0.12	**0.16**	0.15	**0.20**	0.17	**0.23**	0.19	**0.26**	0.23	**0.31**	0.22	**0.29**
Elbow												
Forearm												
Upper leg												
Knee												
Lower leg												
Lymph nodes	0.03	**0.07**	0.04	**0.08**	0.04	**0.09**	0.04	**0.10**	0.05	**0.11**	0.05	**0.11**
Skin	0.02	**0.04**	0.02	**0.04**	0.02	**0.04**	0.02	**0.04**	0.03	**0.05**	0.03	**0.04**
Muscle	0.01	**0.02**	0.01	**0.02**	0.01	**0.03**	0.02	**0.03**	0.02	**0.03**	0.02	**0.03**
Skeleton	0.03	**0.04**	0.03	**0.05**	0.04	**0.06**	0.04	**0.07**	0.05	**0.08**	0.05	**0.07**
Red bone marrow		**0.01**	0.01	**0.02**	0.01	**0.02**	0.01	**0.02**	0.02	**0.02**	0.01	**0.02**
Whole body dose	0.01	**0.02**	0.01	**0.03**	0.02	**0.03**	0.02	**0.03**	0.02	**0.04**	0.02	**0.04**
ED (ICRP 60)	0.02	**0.05**	0.03	**0.05**	0.03	**0.06**	0.03	**0.06**	0.03	**0.07**	0.03	**0.07**
ED (ICRP 103)	0.02	**0.04**	0.02	**0.05**	0.03	**0.05**	0.03	**0.06**	0.03	**0.06**	0.03	**0.06**
Total filtration (mm) Al	**1.5**		**2.5**		**3.5**		**4.5**		**3.5**		**3.5**	
Cu	**0**		**0**		**0**		**0**		**0.1**		**0.2**	

Table 21.21 Normalized organ doses—Clavicle ap—8–12 years

Beam projection:			ap									**Table cells:**
Focus-to-detector distance (cm):			115									(+) = normal font = optimal field setting
Tube voltage (kVp):			70									**(−) = bold font = suboptimal field setting**
Total filtrations:			See last table row.									Empty table cells correspond to values <0.01
Organ or tissue	**Organ dose D_T : Entrance dose K_E**								**(mSv * mGy⁻¹)**			
	(+)	(−)	(+)	(−)	(+)	(−)	(+)	(−)	(+)	(−)	(+)	(−)
Skull		**0.01**		**0.02**		**0.02**	0.01	**0.02**	0.01	**0.03**	0.02	**0.03**
Brain												
Oral mucosa		**0.02**		**0.02**	0.01	**0.02**	0.01	**0.03**	0.01	**0.03**	0.02	**0.04**
Salivary glands	0.01	**0.04**	0.02	**0.04**	0.02	**0.05**	0.02	**0.06**	0.02	**0.07**	0.03	**0.07**
Cervical spine	0.15	**0.34**	0.18	**0.44**	0.22	**0.51**	0.24	**0.56**	0.29	**0.68**	0.33	**0.78**
Thyroid	0.41	**0.83**	0.46	**0.95**	0.50	**1.02**	0.52	**1.07**	0.56	**1.16**	0.60	**1.22**
Thymus	0.03	**0.19**	0.04	**0.23**	0.04	**0.26**	0.05	**0.28**	0.06	**0.32**	0.07	**0.35**
Thoracic spine	0.02	**0.03**	0.02	**0.05**	0.03	**0.06**	0.03	**0.06**	0.04	**0.08**	0.05	**0.10**
Claviculae	0.83	**0.94**	0.99	**1.12**	1.10	**1.24**	1.18	**1.33**	1.31	**1.49**	1.42	**1.62**
Scapulae	0.02	**0.04**	0.03	**0.05**	0.04	**0.06**	0.04	**0.07**	0.06	**0.09**	0.07	**0.11**
Ribs	0.10	**0.17**	0.12	**0.20**	0.13	**0.21**	0.14	**0.23**	0.15	**0.25**	0.16	**0.27**
Oesophagus	0.01	**0.03**	0.02	**0.03**	0.02	**0.04**	0.02	**0.04**	0.03	**0.06**	0.03	**0.07**
Trachea	0.08	**0.23**	0.10	**0.27**	0.11	**0.30**	0.11	**0.32**	0.13	**0.35**	0.14	**0.38**
Lung	0.01	**0.03**	0.02	**0.04**	0.02	**0.04**	0.02	**0.05**	0.03	**0.06**	0.03	**0.07**
Mamma												**0.01**
Heart						**0.01**		**0.01**		**0.02**		**0.02**
Stomach												
Spleen												
Liver												
Gall bladder												
Pancreas												
Small intestine												
Large intestine												
Sigma/rectum												
Lumbar spine												
Pelvis												
Adrenals												
Kidneys												
Urinary bladder												
Uterus												
Ovaries												
Prostate												
Testes												
Upper arm	0.14	**0.19**	0.17	**0.23**	0.20	**0.27**	0.22	**0.29**	0.25	**0.34**	0.28	**0.39**
Elbow												
Forearm												
Upper leg												
Knee												
Lower leg												
Lymph nodes	0.03	**0.08**	0.04	**0.09**	0.04	**0.10**	0.05	**0.11**	0.05	**0.12**	0.06	**0.13**
Skin	0.02	**0.04**	0.02	**0.04**	0.03	**0.04**	0.03	**0.05**	0.03	**0.05**	0.03	**0.05**
Muscle	0.01	**0.02**	0.01	**0.03**	0.02	**0.03**	0.02	**0.03**	0.02	**0.04**	0.02	**0.04**
Skeleton	0.03	**0.05**	0.04	**0.06**	0.04	**0.07**	0.05	**0.07**	0.05	**0.09**	0.06	**0.10**
Red bone marrow		**0.02**	0.01	**0.02**	0.01	**0.02**	0.01	**0.02**	0.02	**0.03**	0.02	**0.03**
Whole body dose	0.01	**0.02**	0.02	**0.03**	0.02	**0.03**	0.02	**0.04**	0.02	**0.04**	0.03	**0.05**
ED (ICRP 60)	0.02	**0.05**	0.03	**0.06**	0.03	**0.06**	0.03	**0.07**	0.04	**0.07**	0.04	**0.08**
ED (ICRP 103)	0.02	**0.05**	0.03	**0.05**	0.03	**0.06**	0.03	**0.06**	0.03	**0.07**	0.04	**0.08**
Total filtration (mm) **Al**	**1.5**		**2.5**		**3.5**		**4.5**		**3.5**		**3.5**	
Cu	**0**		**0**		**0**		**0**		**0.1**		**0.2**	

Table 21.22 Normalized organ doses—Clavicle ap—8–12 years

Beam projection:	ap											**Table cells:**
Focus-to-detector distance (cm):	115											(+) = normal font = optimal field setting
Tube voltage (kVp):	80											**(−) = bold font = suboptimal field setting**
Total filtrations:	See last table row.											Empty table cells correspond to values <0.01
Organ or tissue	**Organ dose D_T : Entrance dose K_E**								**(mSv * mGy^{-1})**			
	(+)	(−)	(+)	(−)	(+)	(−)	(+)	(−)	(+)	(−)	(+)	(−)
Skull		**0.02**		**0.02**	0.01	**0.02**	0.01	**0.03**	0.02	**0.03**	0.02	**0.04**
Brain												
Oral mucosa		**0.02**	0.01	**0.03**	0.01	**0.03**	0.01	**0.03**	0.02	**0.04**	0.02	**0.05**
Salivary glands	0.01	**0.04**	0.02	**0.05**	0.02	**0.06**	0.02	**0.06**	0.03	**0.07**	0.03	**0.08**
Cervical spine	0.17	**0.40**	0.21	**0.50**	0.25	**0.57**	0.27	**0.63**	0.32	**0.74**	0.36	**0.84**
Thyroid	0.43	**0.89**	0.49	**1.00**	0.52	**1.07**	0.54	**1.12**	0.58	**1.20**	0.62	**1.26**
Thymus	0.04	**0.21**	0.04	**0.25**	0.05	**0.28**	0.06	**0.30**	0.07	**0.34**	0.08	**0.37**
Thoracic spine	0.02	**0.05**	0.03	**0.06**	0.04	**0.07**	0.04	**0.08**	0.05	**0.10**	0.06	**0.12**
Claviculae	0.90	**1.02**	1.05	**1.19**	1.15	**1.31**	1.22	**1.39**	1.34	**1.54**	1.43	**1.64**
Scapulae	0.03	**0.05**	0.04	**0.07**	0.05	**0.08**	0.05	**0.09**	0.06	**0.11**	0.08	**0.13**
Ribs	0.11	**0.18**	0.13	**0.21**	0.14	**0.23**	0.14	**0.24**	0.16	**0.26**	0.17	**0.28**
Oesophagus	0.02	**0.03**	0.02	**0.04**	0.02	**0.05**	0.03	**0.05**	0.03	**0.07**	0.04	**0.08**
Trachea	0.09	**0.25**	0.10	**0.29**	0.11	**0.32**	0.12	**0.34**	0.14	**0.37**	0.15	**0.40**
Lung	0.02	**0.03**	0.02	**0.04**	0.02	**0.05**	0.03	**0.06**	0.03	**0.07**	0.04	**0.08**
Mamma										**0.01**		**0.01**
Heart				**0.01**		**0.01**		**0.02**		**0.02**	0.01	**0.03**
Stomach												
Spleen												
Liver												
Gall bladder												
Pancreas												
Small intestine												
Large intestine												
Sigma/rectum												
Lumbar spine												
Pelvis												
Adrenals												
Kidneys												
Urinary bladder												
Uterus												
Ovaries												
Prostate												
Testes												
Upper arm	0.16	**0.22**	0.19	**0.26**	0.22	**0.30**	0.24	**0.32**	0.27	**0.37**	0.30	**0.41**
Elbow												
Forearm												
Upper leg												
Knee												
Lower leg												
Lymph nodes	0.04	**0.09**	0.04	**0.10**	0.05	**0.11**	0.05	**0.11**	0.05	**0.13**	0.06	**0.13**
Skin	0.02	**0.04**	0.03	**0.04**	0.03	**0.05**	0.03	**0.05**	0.03	**0.05**	0.03	**0.05**
Muscle	0.01	**0.03**	0.02	**0.03**	0.02	**0.03**	0.02	**0.04**	0.02	**0.04**	0.03	**0.05**
Skeleton	0.03	**0.06**	0.04	**0.07**	0.05	**0.08**	0.05	**0.08**	0.06	**0.09**	0.06	**0.10**
Red bone marrow	0.01	**0.02**	0.01	**0.02**	0.02	**0.02**	0.02	**0.03**	0.02	**0.03**	0.02	**0.04**
Whole body dose	0.02	**0.03**	0.02	**0.03**	0.02	**0.04**	0.02	**0.04**	0.03	**0.05**	0.03	**0.05**
ED (ICRP 60)	0.03	**0.05**	0.03	**0.06**	0.03	**0.07**	0.04	**0.07**	0.04	**0.08**	0.04	**0.09**
ED (ICRP 103)	0.02	**0.05**	0.03	**0.06**	0.03	**0.06**	0.03	**0.07**	0.04	**0.07**	0.04	**0.08**
Total filtration (mm) Al	**1.5**		**2.5**		**3.5**		**4.5**		**3.5**		**3.5**	
Cu	**0**		**0**		**0**		**0**		**0.1**		**0.2**	

Table 21.23 Normalized organ doses—Clavicle ap—13–17 years

Beam projection:			ap									**Table cells:**
Focus-to-detector distance (cm):			115									(+) = normal font = optimal field setting
Tube voltage (kVp):			60									**(−) = bold font = suboptimal field setting**
Total filtrations:			See last table row.									Empty table cells correspond to values <0.01
Organ or tissue	**Organ dose D_T : Entrance dose K_E**								**(mSv * mGy⁻¹)**			
	(+)	(−)	(+)	(−)	(+)	(−)	(+)	(−)	(+)	(−)	(+)	(−)
Skull		**0.01**		**0.02**		**0.02**		**0.02**		**0.03**		**0.03**
Brain												
Oral mucosa		**0.01**		**0.02**		**0.02**		**0.02**		**0.03**		**0.02**
Salivary glands		**0.03**		**0.04**		**0.05**		**0.06**	0.01	**0.07**		**0.06**
Cervical spine	0.11	**0.34**	0.14	**0.45**	0.17	**0.53**	0.19	**0.60**	0.23	**0.74**	0.21	**0.67**
Thyroid	0.36	**0.75**	0.42	**0.87**	0.45	**0.95**	0.48	**1.01**	0.53	**1.11**	0.52	**1.09**
Thymus		**0.07**	0.01	**0.10**	0.01	**0.11**	0.02	**0.12**	0.02	**0.15**	0.02	**0.14**
Thoracic spine		**0.02**	0.01	**0.02**	0.01	**0.03**	0.02	**0.03**	0.02	**0.04**	0.02	**0.03**
Claviculae	0.63	**0.71**	0.77	**0.88**	0.87	**0.99**	0.95	**1.09**	1.09	**1.25**	1.08	**1.23**
Scapulae	0.01	**0.02**	0.01	**0.03**	0.02	**0.04**	0.02	**0.04**	0.03	**0.05**	0.02	**0.04**
Ribs	0.07	**0.17**	0.09	**0.20**	0.10	**0.22**	0.10	**0.23**	0.12	**0.26**	0.11	**0.26**
Oesophagus		**0.02**	0.01	**0.02**	0.01	**0.03**	0.01	**0.03**	0.02	**0.04**	0.02	**0.04**
Trachea	0.06	**0.23**	0.07	**0.27**	0.08	**0.30**	0.09	**0.32**	0.10	**0.37**	0.10	**0.36**
Lung		**0.02**		**0.02**		**0.03**	0.01	**0.03**	0.01	**0.04**	0.01	**0.04**
Mamma	0.01	**0.04**	0.02	**0.05**	0.02	**0.06**	0.02	**0.07**	0.03	**0.09**	0.03	**0.07**
Heart												
Stomach												
Spleen												
Liver												
Gall bladder												
Pancreas												
Small intestine												
Large intestine												
Sigma/rectum												
Lumbar spine												
Pelvis												
Adrenals												
Kidneys												
Urinary bladder												
Uterus												
Ovaries												
Prostate												
Testes												
Upper arm	0.09	**0.14**	0.12	**0.18**	0.13	**0.21**	0.15	**0.24**	0.18	**0.28**	0.17	**0.26**
Elbow												
Forearm												
Upper leg												
Knee												
Lower leg												
Lymph nodes	0.03	**0.07**	0.03	**0.09**	0.04	**0.10**	0.04	**0.10**	0.04	**0.12**	0.04	**0.11**
Skin	0.02	**0.04**	0.02	**0.04**	0.02	**0.05**	0.02	**0.05**	0.03	**0.05**	0.02	**0.05**
Muscle		**0.02**	0.01	**0.03**	0.01	**0.03**	0.01	**0.03**	0.02	**0.04**	0.02	**0.03**
Skeleton	0.02	**0.04**	0.03	**0.06**	0.03	**0.06**	0.03	**0.07**	0.04	**0.08**	0.04	**0.08**
Red bone marrow		**0.02**		**0.02**	0.01	**0.02**	0.01	**0.02**	0.01	**0.03**	0.01	**0.03**
Whole body dose	0.01	**0.02**	0.01	**0.03**	0.01	**0.03**	0.02	**0.03**	0.02	**0.04**	0.02	**0.04**
ED (ICRP 60)	0.02	**0.05**	0.02	**0.05**	0.03	**0.06**	0.03	**0.06**	0.03	**0.07**	0.03	**0.07**
ED (ICRP 103)	0.02	**0.04**	0.02	**0.05**	0.03	**0.06**	0.03	**0.06**	0.03	**0.07**	0.03	**0.07**
Total filtration (mm) Al	**1.5**		**2.5**		**3.5**		**4.5**		**3.5**		**3.5**	
Cu	**0**		**0**		**0**		**0**		**0.1**		**0.2**	

Table 21.24 Normalized organ doses—Clavicle ap—13–17 years

Beam projection:		ap										**Table cells:**
Focus-to-detector distance (cm):		115										(+) = normal font = optimal field setting
Tube voltage (kVp):		70										**(−) = bold font = suboptimal field setting**
Total filtrations:		See last table row.										Empty table cells correspond to values <0.01
Organ or tissue	**Organ dose D_T : Entrance dose K_E**								**(mSv * mGy^{-1})**			
	(+)	(−)	(+)	(−)	(+)	(−)	(+)	(−)	(+)	(−)	(+)	(−)
Skull		**0.02**		**0.02**		**0.03**		**0.03**	0.01	**0.04**	0.01	**0.04**
Brain												
Oral mucosa		**0.02**		**0.02**		**0.02**		**0.03**		**0.03**		**0.04**
Salivary glands		**0.04**		**0.05**		**0.06**	0.01	**0.07**	0.01	**0.08**	0.02	**0.09**
Cervical spine	0.13	**0.43**	0.17	**0.55**	0.20	**0.64**	0.23	**0.72**	0.28	**0.87**	0.32	**1.02**
Thyroid	0.39	**0.81**	0.45	**0.94**	0.49	**1.01**	0.51	**1.07**	0.56	**1.17**	0.60	**1.25**
Thymus	0.01	**0.09**	0.01	**0.11**	0.02	**0.13**	0.02	**0.14**	0.02	**0.17**	0.03	**0.20**
Thoracic spine	0.01	**0.03**	0.02	**0.03**	0.02	**0.04**	0.02	**0.05**	0.03	**0.06**	0.03	**0.08**
Claviculae	0.71	**0.81**	0.85	**0.98**	0.95	**1.10**	1.03	**1.19**	1.16	**1.35**	1.27	**1.48**
Scapulae	0.01	**0.03**	0.02	**0.04**	0.02	**0.05**	0.03	**0.06**	0.04	**0.07**	0.04	**0.09**
Ribs	0.08	**0.18**	0.09	**0.21**	0.10	**0.23**	0.11	**0.25**	0.12	**0.28**	0.13	**0.30**
Oesophagus	0.01	**0.02**	0.01	**0.03**	0.02	**0.04**	0.02	**0.04**	0.03	**0.06**	0.03	**0.07**
Trachea	0.07	**0.25**	0.08	**0.30**	0.09	**0.33**	0.10	**0.35**	0.11	**0.39**	0.12	**0.43**
Lung		**0.02**	0.01	**0.03**	0.01	**0.04**	0.01	**0.04**	0.02	**0.05**	0.02	**0.06**
Mamma	0.02	**0.05**	0.02	**0.07**	0.03	**0.08**	0.03	**0.09**	0.04	**0.11**	0.05	**0.14**
Heart								**0.01**		**0.01**		**0.02**
Stomach												
Spleen												
Liver												
Gall bladder												
Pancreas												
Small intestine												
Large intestine												
Sigma/rectum												
Lumbar spine												
Pelvis												
Adrenals												
Kidneys												
Urinary bladder												
Uterus												
Ovaries												
Prostate												
Testes												
Upper arm	0.11	**0.17**	0.14	**0.22**	0.16	**0.25**	0.17	**0.27**	0.20	**0.32**	0.23	**0.37**
Elbow												
Forearm												
Upper leg												
Knee												
Lower leg												
Lymph nodes	0.03	**0.08**	0.04	**0.10**	0.04	**0.10**	0.04	**0.11**	0.05	**0.13**	0.05	**0.14**
Skin	0.02	**0.04**	0.02	**0.05**	0.02	**0.05**	0.02	**0.05**	0.03	**0.05**	0.03	**0.06**
Muscle	0.01	**0.02**	0.01	**0.03**	0.01	**0.03**	0.02	**0.04**	0.02	**0.04**	0.02	**0.05**
Skeleton	0.03	**0.05**	0.03	**0.06**	0.04	**0.07**	0.04	**0.08**	0.05	**0.09**	0.05	**0.11**
Red bone marrow		**0.02**	0.01	**0.02**	0.01	**0.03**	0.01	**0.03**	0.02	**0.03**	0.02	**0.04**
Whole body dose	0.01	**0.03**	0.01	**0.03**	0.02	**0.04**	0.02	**0.04**	0.02	**0.05**	0.02	**0.05**
ED (ICRP 60)	0.02	**0.05**	0.03	**0.06**	0.03	**0.07**	0.03	**0.07**	0.04	**0.08**	0.04	**0.09**
ED (ICRP 103)	0.02	**0.05**	0.03	**0.06**	0.03	**0.07**	0.03	**0.07**	0.03	**0.08**	0.04	**0.09**
Total filtration (mm) **Al**	**1.5**		**2.5**		**3.5**		**4.5**		**3.5**		**3.5**	
Cu	**0**		**0**		**0**		**0**		**0.1**		**0.2**	

Table 21.25 Normalized organ doses—Clavicle ap—13–17 years

Beam projection:	ap											**Table cells:**
Focus-to-detector distance (cm):	115											(+) = normal font = optimal field setting
Tube voltage (kVp):	80											**(−) = bold font = suboptimal field setting**
Total filtrations:	See last table row.											Empty table cells correspond to values <0.01
Organ or tissue	**Organ dose D_T : Entrance dose K_E**						**(mSv * mGy⁻¹)**					
	(+)	(−)	(+)	(−)	(+)	(−)	(+)	(−)	(+)	(−)	(+)	(−)
Skull		**0.02**		**0.03**		**0.03**		**0.03**	0.01	**0.04**	0.01	**0.05**
Brain												
Oral mucosa		**0.02**		**0.02**		**0.03**		**0.03**	0.01	**0.04**	0.01	**0.05**
Salivary glands		**0.05**		**0.06**	0.01	**0.07**	0.01	**0.07**	0.02	**0.09**	0.02	**0.10**
Cervical spine	0.16	**0.52**	0.20	**0.64**	0.23	**0.74**	0.26	**0.82**	0.31	**0.98**	0.36	**1.12**
Thyroid	0.42	**0.88**	0.48	**1.00**	0.51	**1.07**	0.54	**1.13**	0.58	**1.23**	0.62	**1.30**
Thymus	0.01	**0.11**	0.02	**0.13**	0.02	**0.15**	0.02	**0.16**	0.03	**0.19**	0.04	**0.22**
Thoracic spine	0.02	**0.04**	0.02	**0.05**	0.02	**0.05**	0.03	**0.06**	0.04	**0.08**	0.04	**0.10**
Claviculae	0.78	**0.89**	0.92	**1.06**	1.01	**1.17**	1.08	**1.26**	1.21	**1.41**	1.30	**1.53**
Scapulae	0.02	**0.04**	0.03	**0.05**	0.03	**0.06**	0.03	**0.07**	0.04	**0.09**	0.05	**0.11**
Ribs	0.09	**0.20**	0.10	**0.23**	0.11	**0.25**	0.12	**0.26**	0.13	**0.29**	0.14	**0.31**
Oesophagus	0.01	**0.03**	0.02	**0.04**	0.02	**0.05**	0.02	**0.05**	0.03	**0.07**	0.04	**0.08**
Trachea	0.08	**0.28**	0.09	**0.32**	0.10	**0.35**	0.10	**0.38**	0.12	**0.42**	0.13	**0.45**
Lung	0.01	**0.03**	0.01	**0.04**	0.02	**0.04**	0.02	**0.05**	0.02	**0.06**	0.02	**0.07**
Mamma	0.02	**0.07**	0.03	**0.09**	0.04	**0.10**	0.04	**0.11**	0.05	**0.14**	0.06	**0.16**
Heart				**0.01**		**0.01**		**0.01**		**0.02**		**0.02**
Stomach												
Spleen												
Liver												
Gall bladder												
Pancreas												
Small intestine												
Large intestine												
Sigma/rectum												
Lumbar spine												
Pelvis												
Adrenals												
Kidneys												
Urinary bladder												
Uterus												
Ovaries												
Prostate												
Testes												
Upper arm	0.13	**0.20**	0.15	**0.24**	0.17	**0.28**	0.19	**0.30**	0.22	**0.35**	0.25	**0.40**
Elbow												
Forearm												
Upper leg												
Knee												
Lower leg												
Lymph nodes	0.03	**0.09**	0.04	**0.10**	0.04	**0.11**	0.04	**0.12**	0.05	**0.13**	0.05	**0.14**
Skin	0.02	**0.05**	0.02	**0.05**	0.03	**0.05**	0.03	**0.05**	0.03	**0.06**	0.03	**0.06**
Muscle	0.01	**0.03**	0.01	**0.03**	0.02	**0.04**	0.02	**0.04**	0.02	**0.05**	0.02	**0.05**
Skeleton	0.03	**0.06**	0.04	**0.07**	0.04	**0.08**	0.04	**0.09**	0.05	**0.10**	0.06	**0.11**
Red bone marrow	0.01	**0.02**	0.01	**0.03**	0.01	**0.03**	0.02	**0.03**	0.02	**0.04**	0.02	**0.04**
Whole body dose	0.01	**0.03**	0.02	**0.04**	0.02	**0.04**	0.02	**0.04**	0.02	**0.05**	0.03	**0.06**
ED (ICRP 60)	0.03	**0.06**	0.03	**0.07**	0.03	**0.07**	0.04	**0.08**	0.04	**0.09**	0.04	**0.09**
ED (ICRP 103)	0.02	**0.06**	0.03	**0.07**	0.03	**0.07**	0.03	**0.08**	0.04	**0.09**	0.04	**0.10**
Total filtration (mm) Al	**1.5**		**2.5**		**3.5**		**4.5**		**3.5**		**3.5**	
Cu	**0**		**0**		**0**		**0**		**0.1**		**0.2**	

Table 21.26 Normalized organ doses—Clavicle ap—>17 years

Beam projection:			ap										**Table cells:**
Focus-to-detector distance (cm):			115										(+) = normal font = optimal field setting
Tube voltage (kVp):			60										**(−) = bold font = suboptimal field setting**
Total filtrations:			See last table row.										Empty table cells correspond to values <0.01
Organ or tissue		**Organ dose D_T : Entrance dose K_E**							**(mSv * mGy^{-1})**				
		(+)	(−)	(+)	(−)	(+)	(−)	(+)	(−)	(+)	(−)	(+)	(−)
Skull			**0.02**		**0.02**		**0.02**		**0.03**		**0.03**		**0.03**
Brain													
Oral mucosa			**0.01**		**0.01**		**0.02**		**0.02**		**0.02**		**0.02**
Salivary glands			**0.04**		**0.05**		**0.05**		**0.06**	0.01	**0.07**	0.01	**0.07**
Cervical spine		0.13	**0.30**	0.17	**0.40**	0.20	**0.48**	0.22	**0.54**	0.28	**0.67**	0.25	**0.60**
Thyroid		0.39	**0.70**	0.45	**0.82**	0.50	**0.90**	0.53	**0.96**	0.58	**1.07**	0.58	**1.05**
Thymus			**0.02**		**0.03**	0.01	**0.04**	0.01	**0.04**	0.02	**0.05**	0.01	**0.05**
Thoracic spine			**0.01**	0.01	**0.02**	0.01	**0.02**	0.02	**0.03**	0.02	**0.04**	0.02	**0.03**
Claviculae		0.63	**0.69**	0.78	**0.85**	0.88	**0.97**	0.97	**1.06**	1.11	**1.23**	1.09	**1.20**
Scapulae		0.01	**0.02**	0.02	**0.03**	0.02	**0.03**	0.02	**0.04**	0.03	**0.05**	0.02	**0.04**
Ribs		0.08	**0.14**	0.10	**0.16**	0.11	**0.18**	0.12	**0.19**	0.13	**0.21**	0.13	**0.21**
Oesophagus			**0.02**	0.01	**0.02**	0.01	**0.03**	0.02	**0.03**	0.02	**0.04**	0.02	**0.03**
Trachea		0.09	**0.23**	0.11	**0.27**	0.12	**0.31**	0.13	**0.33**	0.15	**0.38**	0.14	**0.37**
Lung			**0.02**	0.01	**0.02**	0.01	**0.03**	0.01	**0.03**	0.02	**0.04**	0.02	**0.03**
Mamma													
Heart													
Stomach													
Spleen													
Liver													
Gall bladder													
Pancreas													
Small intestine													
Large intestine													
Sigma/rectum													
Lumbar spine													
Pelvis													
Adrenals													
Kidneys													
Urinary bladder													
Uterus													
Ovaries													
Prostate													
Testes													
Upper arm		0.09	**0.13**	0.12	**0.16**	0.14	**0.19**	0.16	**0.21**	0.19	**0.25**	0.17	**0.24**
Elbow													
Forearm													
Upper leg													
Knee													
Lower leg													
Lymph nodes		0.03	**0.07**	0.04	**0.08**	0.04	**0.09**	0.05	**0.10**	0.05	**0.11**	0.05	**0.11**
Skin		0.02	**0.04**	0.02	**0.04**	0.02	**0.04**	0.03	**0.05**	0.03	**0.05**	0.03	**0.05**
Muscle			**0.02**	0.01	**0.02**	0.01	**0.03**	0.01	**0.03**	0.02	**0.03**	0.02	**0.03**
Skeleton		0.02	**0.04**	0.03	**0.05**	0.03	**0.05**	0.04	**0.06**	0.04	**0.07**	0.04	**0.07**
Red bone marrow			**0.01**	0.01	**0.02**	0.01	**0.02**	0.01	**0.02**	0.02	**0.03**	0.01	**0.02**
Whole body dose		0.01	**0.02**	0.01	**0.03**	0.02	**0.03**	0.02	**0.03**	0.02	**0.04**	0.02	**0.03**
ED (ICRP 60)		0.02	**0.04**	0.03	**0.05**	0.03	**0.05**	0.03	**0.06**	0.04	**0.06**	0.03	**0.06**
ED (ICRP 103)		0.02	**0.04**	0.02	**0.04**	0.03	**0.05**	0.03	**0.05**	0.03	**0.06**	0.03	**0.06**
Total filtration (mm)	**Al**	**1.5**		**2.5**		**3.5**		**4.5**		**3.5**		**3.5**	
	Cu	**0**		**0**		**0**		**0**		**0.1**		**0.2**	

Table 21.27 Normalized organ doses—Clavicle ap—>17 years

Beam projection:		ap										**Table cells:**
Focus-to-detector distance (cm):		115						(+) = normal font = optimal field setting				
Tube voltage (kVp):		70						**(−) = bold font = suboptimal field setting**				
Total filtrations:		See last table row.						Empty table cells correspond to values <0.01				
Organ or tissue	**Organ dose D_T : Entrance dose K_E**							**(mSv * mGy⁻¹)**				
	(+)	(−)	(+)	(−)	(+)	(−)	(+)	(−)	(+)	(−)	(+)	(−)
Skull		**0.02**		**0.03**		**0.03**		**0.03**	0.01	**0.04**	0.01	**0.05**
Brain												
Oral mucosa		**0.01**		**0.02**		**0.02**		**0.02**		**0.03**	0.01	**0.04**
Salivary glands		**0.04**		**0.05**	0.01	**0.06**	0.01	**0.07**	0.02	**0.08**	0.02	**0.09**
Cervical spine	0.16	**0.39**	0.21	**0.50**	0.24	**0.59**	0.27	**0.66**	0.33	**0.80**	0.39	**0.94**
Thyroid	0.43	**0.77**	0.49	**0.89**	0.53	**0.97**	0.57	**1.03**	0.62	**1.14**	0.67	**1.22**
Thymus		**0.03**	0.01	**0.04**	0.01	**0.05**	0.02	**0.06**	0.02	**0.07**	0.03	**0.08**
Thoracic spine	0.01	**0.02**	0.02	**0.03**	0.02	**0.04**	0.02	**0.04**	0.03	**0.05**	0.04	**0.07**
Claviculae	0.72	**0.79**	0.87	**0.95**	0.97	**1.07**	1.05	**1.16**	1.20	**1.33**	1.31	**1.46**
Scapulae	0.02	**0.03**	0.02	**0.04**	0.03	**0.04**	0.03	**0.05**	0.04	**0.06**	0.05	**0.08**
Ribs	0.09	**0.15**	0.11	**0.18**	0.12	**0.19**	0.13	**0.21**	0.14	**0.23**	0.15	**0.25**
Oesophagus	0.01	**0.02**	0.02	**0.03**	0.02	**0.04**	0.02	**0.04**	0.03	**0.05**	0.04	**0.06**
Trachea	0.10	**0.26**	0.12	**0.30**	0.13	**0.34**	0.14	**0.36**	0.16	**0.41**	0.17	**0.45**
Lung	0.01	**0.02**	0.01	**0.03**	0.02	**0.03**	0.02	**0.04**	0.02	**0.05**	0.03	**0.06**
Mamma												**0.01**
Heart												**0.01**
Stomach												
Spleen												
Liver												
Gall bladder												
Pancreas												
Small intestine												
Large intestine												
Sigma/rectum												
Lumbar spine												
Pelvis												
Adrenals												
Kidneys												
Urinary bladder												
Uterus												
Ovaries												
Prostate												
Testes												
Upper arm	0.11	**0.15**	0.14	**0.19**	0.16	**0.22**	0.18	**0.24**	0.21	**0.29**	0.24	**0.33**
Elbow												
Forearm												
Upper leg												
Knee												
Lower leg												
Lymph nodes	0.04	**0.08**	0.04	**0.09**	0.05	**0.10**	0.05	**0.11**	0.06	**0.12**	0.06	**0.14**
Skin	0.02	**0.04**	0.02	**0.04**	0.03	**0.05**	0.03	**0.05**	0.03	**0.05**	0.03	**0.05**
Muscle	0.01	**0.02**	0.01	**0.03**	0.02	**0.03**	0.02	**0.03**	0.02	**0.04**	0.02	**0.04**
Skeleton	0.03	**0.05**	0.03	**0.06**	0.04	**0.06**	0.04	**0.07**	0.05	**0.08**	0.06	**0.09**
Red bone marrow	0.01	**0.02**	0.01	**0.02**	0.01	**0.02**	0.02	**0.03**	0.02	**0.03**	0.02	**0.04**
Whole body dose	0.01	**0.02**	0.02	**0.03**	0.02	**0.03**	0.02	**0.04**	0.02	**0.04**	0.03	**0.05**
ED (ICRP 60)	0.03	**0.05**	0.03	**0.05**	0.03	**0.06**	0.03	**0.06**	0.04	**0.07**	0.04	**0.08**
ED (ICRP 103)	0.02	**0.04**	0.03	**0.05**	0.03	**0.05**	0.03	**0.06**	0.03	**0.07**	0.04	**0.07**
Total filtration (mm) **Al**	**1.5**		**2.5**		**3.5**		**4.5**		**3.5**		**3.5**	
Cu	**0**		**0**		**0**		**0**		**0.1**		**0.2**	

Table 21.28 Normalized organ doses—Clavicle ap—>17 years

Beam projection:			ap									**Table cells:**
Focus-to-detector distance (cm):			115					(+) = normal font = optimal field setting				
Tube voltage (kVp):			80					**(−) = bold font = suboptimal field setting**				
Total filtrations:			See last table row.					Empty table cells correspond to values <0.01				
Organ or tissue	**Organ dose D_T : Entrance dose K_E**							**(mSv * mGy^{-1})**				
	(+)	(−)	(+)	(−)	(+)	(−)	(+)	(−)	(+)	(−)	(+)	(−)
Skull		**0.02**		**0.03**		**0.03**	0.01	**0.04**	0.01	**0.05**	0.02	**0.05**
Brain												
Oral mucosa		**0.02**		**0.02**		**0.03**		**0.03**	0.01	**0.04**	0.01	**0.04**
Salivary glands		**0.05**	0.01	**0.06**	0.01	**0.07**	0.01	**0.08**	0.02	**0.09**	0.02	**0.10**
Cervical spine	0.20	**0.47**	0.25	**0.59**	0.28	**0.68**	0.31	**0.76**	0.38	**0.91**	0.43	**1.05**
Thyroid	0.46	**0.83**	0.52	**0.95**	0.57	**1.03**	0.60	**1.09**	0.65	**1.19**	0.69	**1.27**
Thymus	0.01	**0.04**	0.02	**0.05**	0.02	**0.06**	0.02	**0.07**	0.03	**0.08**	0.03	**0.09**
Thoracic spine	0.02	**0.03**	0.02	**0.04**	0.03	**0.05**	0.03	**0.05**	0.04	**0.07**	0.05	**0.08**
Claviculae	0.79	**0.87**	0.94	**1.04**	1.04	**1.15**	1.12	**1.24**	1.25	**1.39**	1.34	**1.51**
Scapulae	0.02	**0.04**	0.03	**0.05**	0.03	**0.06**	0.04	**0.06**	0.05	**0.08**	0.06	**0.09**
Ribs	0.10	**0.16**	0.12	**0.19**	0.13	**0.21**	0.13	**0.22**	0.15	**0.24**	0.16	**0.26**
Oesophagus	0.02	**0.03**	0.02	**0.04**	0.03	**0.05**	0.03	**0.05**	0.04	**0.06**	0.04	**0.08**
Trachea	0.11	**0.28**	0.13	**0.33**	0.14	**0.37**	0.15	**0.39**	0.17	**0.44**	0.19	**0.47**
Lung	0.01	**0.03**	0.02	**0.04**	0.02	**0.04**	0.02	**0.05**	0.03	**0.06**	0.03	**0.06**
Mamma										**0.01**		**0.01**
Heart								**0.01**		**0.01**		**0.02**
Stomach												
Spleen												
Liver												
Gall bladder												
Pancreas												
Small intestine												
Large intestine												
Sigma/rectum												
Lumbar spine												
Pelvis												
Adrenals												
Kidneys												
Urinary bladder												
Uterus												
Ovaries												
Prostate												
Testes												
Upper arm	0.13	**0.18**	0.16	**0.22**	0.18	**0.25**	0.20	**0.27**	0.23	**0.32**	0.26	**0.36**
Elbow												
Forearm												
Upper leg												
Knee												
Lower leg												
Lymph nodes	0.04	**0.09**	0.05	**0.10**	0.05	**0.11**	0.06	**0.12**	0.06	**0.13**	0.07	**0.14**
Skin	0.02	**0.04**	0.03	**0.05**	0.03	**0.05**	0.03	**0.05**	0.03	**0.05**	0.03	**0.06**
Muscle	0.01	**0.02**	0.02	**0.03**	0.02	**0.03**	0.02	**0.04**	0.02	**0.04**	0.02	**0.05**
Skeleton	0.03	**0.05**	0.04	**0.06**	0.04	**0.07**	0.05	**0.08**	0.05	**0.09**	0.06	**0.10**
Red bone marrow	0.01	**0.02**	0.01	**0.02**	0.02	**0.03**	0.02	**0.03**	0.02	**0.04**	0.02	**0.04**
Whole body dose	0.02	**0.03**	0.02	**0.03**	0.02	**0.04**	0.02	**0.04**	0.03	**0.05**	0.03	**0.05**
ED (ICRP 60)	0.03	**0.05**	0.03	**0.06**	0.04	**0.06**	0.04	**0.07**	0.04	**0.08**	0.05	**0.08**
ED (ICRP 103)	0.02	**0.05**	0.03	**0.05**	0.03	**0.06**	0.03	**0.06**	0.04	**0.07**	0.04	**0.08**
Total filtration (mm) **Al**	**1.5**		**2.5**		**3.5**		**4.5**		**3.5**		**3.5**	
Cu	**0**		**0**		**0**		**0**		**0.1**		**0.2**	

Reference

1. Bontrager KL. Textbook of radiographic positioning and related anatomy. St. Louis: Mosby Year Book; 1993.

22 Upper Arm

22.1 Introduction

The type of upper arm radiography is determined by the clinical indication. Table 22.1 shows indications and Table 22.2 contraindications for upper arm radiography. Image criteria as required for radiological findings in upper arm radiographs are listed in Table 22.3. In order to fulfil these image criteria, the field settings listed in Tables 22.4 and 22.5 should be achieved. These field settings go along with the patient preparation as listed in Table 22.6. Exposure parameter combinations for various age groups and upper arm diameters are listed in Tables 22.7, 22.8, and 22.9. The age groups in Table 22.8 refer to the standard age groups of the mathematical MIRD phantom family which has been used to calculate the normalized organ doses in this book. Table 22.9 shows radiation qualities (i.e. various combinations of tube voltage and total filtration) which may be applied in clinical practice. Please note that not all of these radiation qualities are recommended by the authors. The standard projections of upper arm radiographies of the patient who is correctly positioned are depicted in Fig. 22.1. Also the optimal field (marked in green), the suboptimal field (marked in red), and the correct shielding are shown. Tables 22.11, 22.12, and 22.13 outline the virtual field settings on the skin surface of mathematical MIRD phantoms of various age groups and the resulting virtual radiographies with and without bone structures, respectively. Normalized organ doses

Table 22.1 Indications for upper arm radiographs

Birth trauma
Battered child syndrome
Congenital malformation of upper extremity
Soft tissue tumours
Bone tumours
Congenital vascular malformation
Skeletal dysplasia
Vanishing bone disease

Table 22.2 Contraindications for upper arm radiographs

None

Table 22.3 Image criteria for upper arm radiographs

Upper arm ap and lat
Visualization of the entire humerus
Reproduction of the glenohumeral joint and the elbow joint
Visualization of the radial head and neck, as well as proximal ulna with the coronoid process and the olecranon consistent with age
Visually sharp reproduction of the cortex and spongiosa of the humerus, the proximal radius, and ulna

Table 22.4 Patient positions and beam projections for upper arm radiographs

Patient position	Beam projections
Supine	ap + lat

M. Seidenbusch et al., *Imaging Practice and Radiation Protection in Pediatric Radiology*, https://doi.org/10.1007/978-3-030-18504-6_22

resulting from these virtual field settings have been calculated for all age groups and have been listed in Tables 22.14, 22.15, 22.16, 22.17, 22.18, 22.19, 22.20, 22.21, 22.22, 22.23, 22.24, 22.25, 22.26, 22.27, 22.28, and 22.29. Table cells with normal font refer to optimal field settings and table cells with bold font refer to suboptimal field

Table 22.5 Optimal field settings for upper arm radiographs

Upper arm ap	
Upper field edge:	Tip of acromion (including the adjacent skin)
Field centring:	Midpoint of the upper extremity
Lower field edge:	Elbow joint
Medial lateral field edge:	Skin surfaces of the upper arm
Outer lateral field edges:	Close to the skin of the upper arm
Upper arm lat	
Upper field edge:	Midpoint between the tip of mandible and the jugular fossa
Field centring:	Midpoint of the upper extremity
Lower field edge:	Proximal forearm
Medial lateral field edge:	Close to the skin of the upper arm, includes a part of the chest wall medial to the mammilla
Outer lateral field edges:	Close to the skin of the upper arm
Annotations:	
Adjust angulation of the tube dependent on the position of the patient.	

Table 22.6 Patient preparation

General	Remove all clothes, external metallic devices, and jewellery from the neck and chest of the patient.
	Immobilize infants and uncooperative patients.
	Cooperative patients can be positioned upright at the vertical stand.
	Place eventually small X-ray positive markers at palpable tumours or on pain points.
Positioning	For the ap view, the arm should be positioned alongside the chest in the supine position.
	For the lateral view, the upper arm is angled by 90° in the shoulder and the elbow joint.
Shielding	Lead mask shielding should be used at the entrance side beneath the medial field edge in ap/lateral supine and upright views.

Table 22.7 Radiographic procedure

X-ray source	Use automatic exposure control (AEC) in patients >10 years of age
	Focal spot size between 0.6 and 1.3 dependent on the age of the patient
	Check the use of additional filtration
	Check the focus-to-detector distance
Field setting	Check the field size
	Check the field centring
Image detector system	Check the use of antiscatter grids
Patient	–

Table 22.8 Radiographic techniques of upper arm radiographs

Age group (years)	Beam direction	Patient position	Beam projection	Typical patient diameter (cm)	Optimal field size (cm x cm)	Focus-to-detector distance (cm)	Voltage (kVp) 50	60	70	80	90	100	Antiscatter grid
0	Vertical	Supine	ap	–	3 × 9	100[1]	□	□	□	■	■	■	■
			lat				□	□	□	■	■	■	
1–2			ap	–	3 × 12		■	□	□	■	■	■	■
			lat				■	□	□	■	■	■	
3–7			ap	–	4 × 16		■	□	□	■	■	■	■
			lat				■	□	□	■	■	■	
8–12			ap	–	5 × 20		■	□	□	□	■	■	■
			lat				■	□	□	□	■	■	
13–17			ap	–	6 × 30	115	■	□	□	□	■	■	□
			lat				■	□	□	□	■	■	
18			ap	–	6 × 35		■	□	□	□	■	■	□
			lat				■	□	□	□	■	■	

Annotations:

1	The patient is lying on the table top directly on the detector
■	Parameter combination not recommended, normalized organ doses not provided
□	Parameter combination recommended, normalized organ doses provided

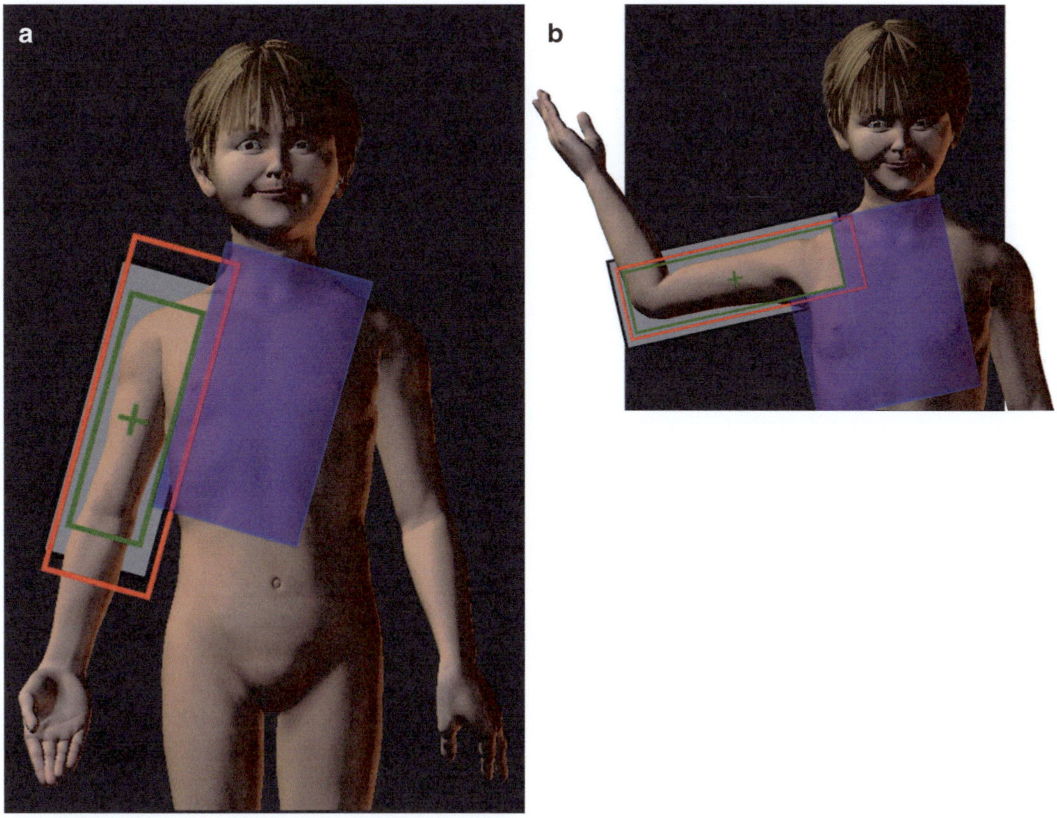

Fig. 22.1 Field settings for upper arm radiographs in supine position. (**a**) Antero-posterior projection. (**b**) Lateral projection. Note that an exact hand position is a prerequisite for correctly depicting the elbow joint (© V. Rösenberger & M. Seidenbusch 2018). Green lines = optimal field settings; green cross = centring point; red lines = suboptimal field settings; grey shaded area = image detector; blue shaded area = lead mask shielding; dark grey shaded area = Bucky table

Table 22.9 Radiation qualities of upper arm radiographs

Voltage (kVp)	Total filtration					
	Without additional filtration		With additional filtration			
	1.5 mm Al	2.5 Al	3.5 Al	4.5 Al	3.5 Al + 0.1 Cu	3.5 Al + 0.2 Cu
50						
60						
70						
80						
90						
100						

Annotations:

(black)	Parameter combination not recommended, normalized organ doses not provided
(grey)	Parameter combination not recommended, normalized organ doses provided
(light blue)	Parameter combination recommended, normalized organ doses provided

Table 22.10 Inconsistencies between the mathematical MIRD phantoms and real human anatomy

Anatomical region	Inconsistencies
Extremities	Position of the extremities cannot be changed
	Arms constructed too long in the ages 0, 1, and 5 years

For details, see the text

settings. Please note that normalized organ doses have been provided also for exposure parameter combinations, which are not recommended by the authors.

22.2 Indications and Contraindications

Indications and contraindications for upper arm radiographs are given in Tables 22.1 and 22.2.

22.3 Image Criteria

The image criteria required for clavicle radiographs are listed in Table 22.3.

22.4 Field Settings

In Table 22.4, patient positions and beam projections used for upper arm radiographs are listed, and in Table 22.5, the optimal field settings are listed which are required in order to fulfil the image criteria mentioned in Table 22.3.

22.5 Patient Preparation and Radiographic Procedure

See Tables 22.6, 22.7 and [1].

22.6 Exposure Parameters

In Tables 22.8 and 22.9, exposure parameter combinations used for upper arm radiographs in various age groups are listed. Dark grey shaded table cells refer to exposure parameter combinations which are not recommended for paediatric upper arm radiographs and for which no normalized organ doses have been provided. Light grey and light blue shaded table cells refer to exposure parameter combinations for which normalized organ doses have been provided with light grey shaded table cells referring to exposure parameter combinations being not recommended and light blue shaded table cells referring to exposure parameter combinations being recommended for paediatric upper arm radiographs. Note that due to the fact that the radiation dose is an exponential function of the patient diameter, it may make sense to use the patient diameters instead of the patient's age groups for determining the adequate exposure parameters from Table 22.8. In this case, for determining the corresponding normalized organ doses from Tables 22.14, 22.15, 22.16, 22.17, 22.18, 22.19, 22.20, 22.21, 22.22, 22.23, 22.24, 22.25, 22.26, 22.27, 22.28, and 22.29, the age group related to the patient diameter should be used.

22.7 Normalized Organ Doses

For calculating normalized organ doses, the field settings described in Sect. 22.4 were transferred to the mathematical MIRD phantoms. Inconsistencies between the geometrical anatomy of the mathematical phantoms and the real human anatomy have been listed in Table 22.10 whenever they are of relevance for the validity of the normalized organ doses.

In Tables 22.11, 22.12, and 22.13, optimal and suboptimal field settings are listed for the different age groups and the beam projections given in Tables 22.8 and 22.9. Green lines refer to optimal field settings, red lines refer to suboptimal field settings, respectively. In the tables, the optimal and suboptimal field settings which were used for performing the calculations of the normalized organ doses provided in Tables 22.14, 22.15, 22.16, 22.17, 22.18, 22.19, 22.20, 22.21, 22.22, 22.23, 22.24, 22.25, 22.26, 22.27, 22.28, and 22.29 were drawn in on the skin surface of the mathematical phantoms. The virtual mathematical radiographs resulting from these field settings were depicted in the tables with and without bone structures, respectively, in order to enable the reader to visually compare the virtual mathematical radiographs with the real radiographs made under clinical conditions.

For comparing the virtual mathematical field settings with the real clinical field settings, the virtual mathematical field edges were also listed in the tables.

As the normalized organ doses provided in Tables 22.14, 22.15, 22.16, 22.17, 22.18, 22.19, 22.20, 22.21, 22.22, 22.23, 22.24, 22.25, 22.26, 22.27, 22.28, and 22.29 refer to entrance kerma values, optimal and suboptimal field sizes were provided in the tables in order to enable the reader to calculate entrance air kerma dose values from clinically measured dose area product values. Please note that these optimal and suboptimal field sizes refer to the mathematical phantoms of an age group only and not to real patients with corresponding ages.

Please note also that the anatomy of the mathematical phantoms has been idealized. Therefore, the virtual field settings for calculating the normalized organ doses have been defined with a view to ensuring that the resulting virtual organ doses would be consistent with the real organ doses which would be achieved during real X-ray examinations being performed with the real field settings recommended in Sect. 22.4. Thus, the field settings shown in the following tables may not be consistent with the clinical field settings as recommended in Sect. 22.4.

Please note also that normalized organ doses are provided for the ap projection only. However, normalized organ doses for the lateral projection equal to those of ap projection.

Table 22.11 Field settings—upper arm ap + lat

Age group (years)		Anatomical landmarks			
		Optimal field setting		Suboptimal field setting	
0	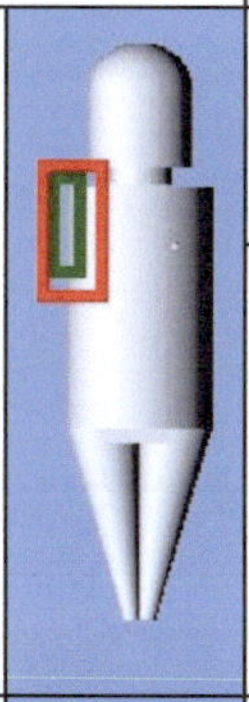	Upper field edge: Just above the skin surface of the shoulder Lower field edge: Just below the elbow Medial field edge: Just beneath the chest wall Lateral field edge: Just beneath the skin surface of the upper arm Field size = field width × field height: 3 cm × 9 cm = 27 cm^2		Upper field edge: Just below the tip of the mandible Lower field edge: Lower third of the forearm Medial field edge: Just beneath the mammillary line Lateral field edge: Above the skin surface of the upper arm Field size = field width × field height: 5 cm × 12 cm = 60 cm^2	
				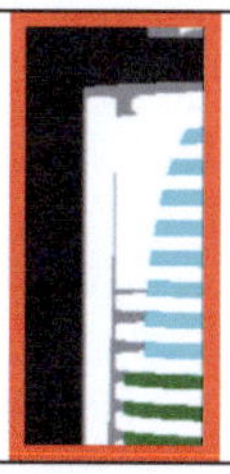	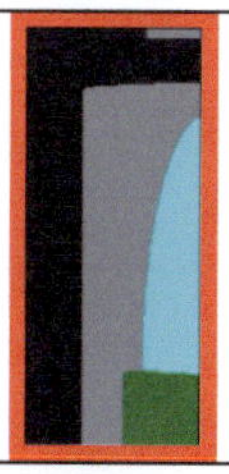
1–2	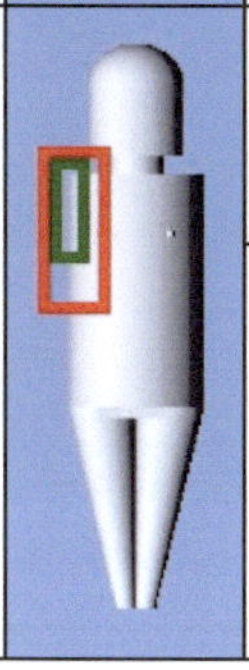	Upper field edge: Just above the skin surface of the shoulder Lower field edge: Just below the elbow Medial field edge: Just beneath the chest wall Lateral field edge: Just beneath the skin surface of the upper arm Field size = field width × field height: 3 cm × 12 cm = 36 cm^2		Upper field edge: Just below the tip of the mandible Lower field edge: Lower third of the forearm Medial field edge: Just beneath the mammillary line Lateral field edge: Above the skin surface of the upper arm Field size = field width × field height: 7 cm × 20 cm = 140 cm^2	
				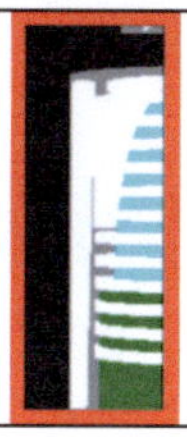	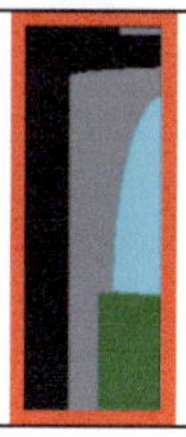

Table 22.12 Field settings—upper arm ap + lat

Age group (years)		Anatomical landmarks			
		Optimal field setting		Suboptimal field setting	
3–7	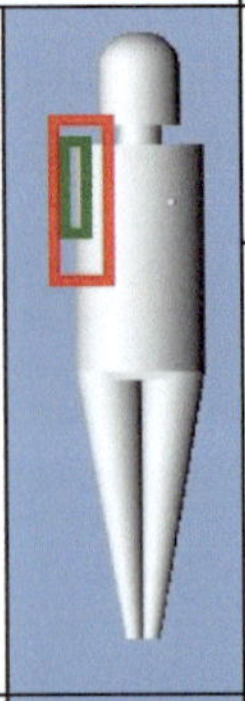	Upper field edge: Just above the skin surface of the shoulder Lower field edge: Just below the elbow Medial field edge: Just beneath the chest wall Lateral field edge: Just beneath the skin surface of the upper arm Field size = field width × field height: 4 cm × 16 cm = 64 cm^2		Upper field edge: Just below the tip of the mandible Lower field edge: Lower third of the forearm Medial field edge: Just beneath the mammillary line Lateral field edge: Above the skin surface of the upper arm Field size = field width × field height: 8 cm × 28 cm = 224 cm^2	

8–12	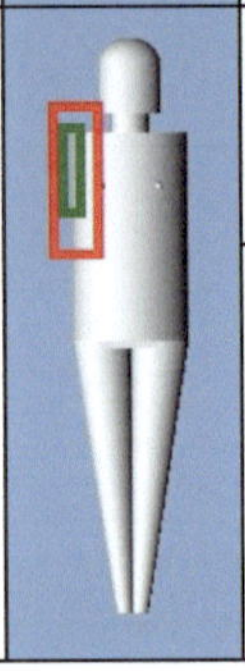	Upper field edge: Just above the skin surface of the shoulder Lower field edge: Just below the elbow Medial field edge: Just beneath the chest wall Lateral field edge: Just beneath the skin surface of the upper arm Field size = field width × field height: 4 cm × 20 cm = 80 cm^2		Upper field edge: Just below the tip of the mandible Lower field edge: Lower third of the forearm Medial field edge: Just beneath the mammillary line Lateral field edge: Above the skin surface of the upper arm Field size = field width × field height: 9 cm × 35 cm = 315 cm^2	

Table 22.13 Field settings—upper arm ap + lat

Age group (years)		Anatomical landmarks			
		Optimal field setting		Suboptimal field setting	
13–17	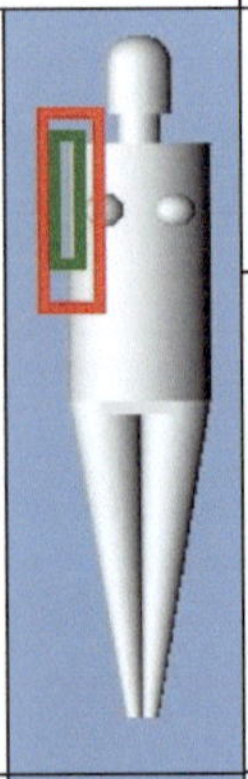	Upper field edge: Just above the skin surface of the shoulder Lower field edge: Just below the elbow Medial field edge: Just beneath the chest wall Lateral field edge: Just beneath the skin surface of the upper arm Field size = field width × field height: 6 cm × 30 cm = 180 cm^2		Upper field edge: Just below the tip of the mandible Lower field edge: Lower third of the forearm Medial field edge: Just beneath the mammillary line Lateral field edge: Above the skin surface of the upper arm Field size = field width × field height: 10 cm × 45 cm = 450 cm^2	

18	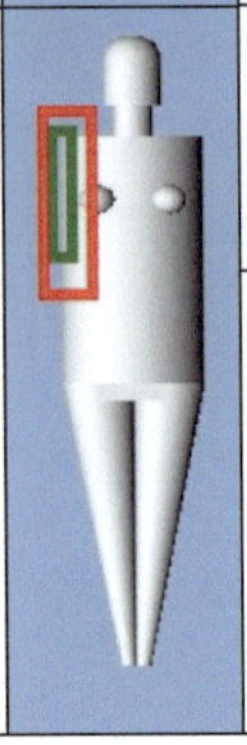	Upper field edge: Just above the skin surface of the shoulder Lower field edge: Just below the elbow Medial field edge: Just beneath the chest wall Lateral field edge: Just beneath the skin surface of the upper arm Field size = field width × field height: 6 cm × 35 cm = 210 cm^2		Upper field edge: Just below the tip of the mandible Lower field edge: Lower third of the forearm Medial field edge: Just beneath the mammillary line Lateral field edge: Above the skin surface of the upper arm Field size = field width × field height: 11 cm × 50 cm = 550 cm^2	

Table 22.14 Normalized organ doses—Upper arm ap—<1 year

Beam projection:		**ap**										**Table cells:**
Focus-to-detector distance (cm):		100										(+) = normal font = optimal field setting
Tube voltage (kVp):		50										**(−) = bold font = suboptimal field setting**
Total filtrations:		See last table row.										Empty table cells correspond to values <0.01
Organ or tissue	**Organ dose D_T : Entrance dose K_E**								**(mSv * mGy^{-1})**			
	(+)	(−)	(+)	(−)	(+)	(−)	(+)	(−)	(+)	(−)	(+)	(−)
Skull												
Brain												
Oral mucosa												
Salivary glands								**0.01**		**0.01**		**0.02**
Cervical spine		**0.01**		**0.01**		**0.02**		**0.02**		**0.02**		**0.03**
Thyroid						**0.01**		**0.01**		**0.02**		**0.02**
Thymus		**0.01**		**0.01**		**0.02**		**0.02**		**0.02**		**0.03**
Thoracic spine		**0.02**		**0.03**		**0.04**		**0.04**	0.01	**0.05**	0.01	**0.06**
Claviculae	0.01	**0.44**	0.02	**0.53**	0.02	**0.58**	0.02	**0.62**	0.03	**0.69**	0.03	**0.75**
Scapulae	0.04	**0.18**	0.06	**0.24**	0.07	**0.28**	0.08	**0.32**	0.09	**0.39**	0.11	**0.45**
Ribs	0.07	**0.34**	0.08	**0.41**	0.09	**0.45**	0.10	**0.49**	0.12	**0.54**	0.13	**0.59**
Oesophagus		**0.01**		**0.02**		**0.02**		**0.02**		**0.03**		**0.03**
Trachea												
Lung		**0.13**		**0.15**	0.01	**0.17**	0.01	**0.19**	0.01	**0.21**	0.02	**0.23**
Mamma		**0.53**		**0.56**		**0.57**		**0.57**		**0.58**		**0.59**
Heart		**0.01**		**0.02**		**0.02**		**0.02**		**0.03**		**0.03**
Stomach												
Spleen												
Liver		**0.07**		**0.09**		**0.10**		**0.11**		**0.13**		**0.14**
Gall bladder								**0.01**		**0.01**		**0.02**
Pancreas								**0.01**		**0.01**		**0.02**
Small intestine												
Large intestine												
Sigma/rectum												
Lumbar spine						**0.01**		**0.01**		**0.02**		**0.02**
Pelvis												
Adrenals				**0.01**		**0.02**		**0.02**		**0.02**		**0.03**
Kidneys										**0.01**		**0.02**
Urinary bladder												
Uterus												
Ovaries												
Prostate												
Testes												
Upper arm	0.51	**0.52**	0.62	**0.64**	0.70	**0.72**	0.76	**0.78**	0.86	**0.89**	0.94	**0.98**
Elbow	0.27	**0.50**	0.32	**0.61**	0.36	**0.69**	0.39	**0.75**	0.44	**0.85**	0.48	**0.93**
Forearm										**0.01**		**0.01**
Upper leg												
Knee												
Lower leg												
Lymph nodes		**0.01**		**0.02**		**0.02**		**0.02**		**0.03**		**0.03**
Skin	0.02	**0.06**	0.03	**0.06**	0.03	**0.07**	0.03	**0.07**	0.03	**0.07**	0.03	**0.07**
Muscle		**0.03**		**0.03**		**0.04**		**0.04**	0.01	**0.04**	0.01	**0.05**
Skeleton	0.06	**0.12**	0.07	**0.15**	0.08	**0.17**	0.09	**0.18**	0.10	**0.21**	0.11	**0.23**
Red bone marrow		**0.02**	0.01	**0.03**	0.01	**0.03**	0.01	**0.04**	0.02	**0.04**	0.02	**0.05**
Whole body dose	0.01	**0.04**	0.01	**0.04**	0.01	**0.05**	0.02	**0.05**	0.02	**0.06**	0.02	**0.06**
ED (ICRP 60)		**0.05**		**0.06**		**0.06**		**0.07**		**0.07**		**0.08**
ED (ICRP 103)		**0.09**		**0.10**		**0.10**		**0.11**		**0.11**		**0.12**
Total filtration (mm) Al	**1.5**		**2.5**		**3.5**		**4.5**		**3.5**		**3.5**	
Cu	**0**		**0**		**0**		**0**		**0.1**		**0.2**	

Table 22.15 Normalized organ doses—Upper arm ap—<1 year

Beam projection:		**ap**										**Table cells:**
Focus-to-detector distance (cm):		100					(+) = normal font = optimal field setting					
Tube voltage (kVp):		60					**(−) = bold font = suboptimal field setting**					
Total filtrations:		See last table row.				Empty table cells correspond to values <0.01						
Organ or tissue	**Organ dose D_T : Entrance dose K_E**							**(mSv * mGy^{-1})**				
	(+)	(−)	(+)	(−)	(+)	(−)	(+)	(−)	(+)	(−)	(+)	(−)
Skull												
Brain												
Oral mucosa												
Salivary glands				**0.01**		**0.01**		**0.01**		**0.02**		**0.02**
Cervical spine		**0.01**		**0.02**		**0.02**		**0.03**		**0.03**		**0.03**
Thyroid		**0.01**		**0.01**		**0.02**		**0.02**		**0.02**		**0.02**
Thymus		**0.01**		**0.02**		**0.02**		**0.02**		**0.03**		**0.03**
Thoracic spine		**0.03**		**0.04**		**0.05**	0.01	**0.06**	0.01	**0.07**	0.01	**0.06**
Claviculae	0.02	**0.49**	0.02	**0.58**	0.02	**0.63**	0.03	**0.67**	0.03	**0.73**	0.03	**0.75**
Scapulae	0.06	**0.23**	0.07	**0.30**	0.09	**0.35**	0.10	**0.39**	0.11	**0.46**	0.11	**0.45**
Ribs	0.08	**0.39**	0.10	**0.45**	0.11	**0.50**	0.12	**0.53**	0.13	**0.58**	0.13	**0.59**
Oesophagus		**0.02**		**0.02**		**0.03**		**0.03**		**0.03**		**0.03**
Trachea												
Lung		**0.15**	0.01	**0.17**	0.01	**0.19**	0.01	**0.21**	0.02	**0.23**	0.02	**0.23**
Mamma		**0.55**		**0.57**		**0.58**		**0.59**		**0.60**		**0.59**
Heart		**0.02**		**0.02**		**0.03**		**0.03**		**0.03**		**0.03**
Stomach												
Spleen												
Liver		**0.08**		**0.10**		**0.12**		**0.13**		**0.15**		**0.14**
Gall bladder				**0.01**		**0.01**		**0.02**		**0.02**		**0.02**
Pancreas				**0.01**		**0.01**		**0.01**		**0.02**		**0.02**
Small intestine												
Large intestine												
Sigma/rectum												
Lumbar spine		**0.01**		**0.01**		**0.02**		**0.02**		**0.02**		**0.02**
Pelvis												
Adrenals		**0.01**		**0.02**		**0.02**		**0.02**		**0.03**		**0.03**
Kidneys						**0.01**		**0.01**		**0.02**		**0.02**
Urinary bladder												
Uterus												
Ovaries												
Prostate												
Testes												
Upper arm	0.59	**0.61**	0.70	**0.73**	0.78	**0.81**	0.84	**0.87**	0.93	**0.98**	0.94	**0.98**
Elbow	0.30	**0.58**	0.36	**0.69**	0.40	**0.77**	0.43	**0.83**	0.48	**0.93**	0.48	**0.93**
Forearm						**0.01**		**0.01**		**0.02**		**0.01**
Upper leg												
Knee												
Lower leg												
Lymph nodes		**0.02**		**0.02**		**0.03**		**0.03**		**0.03**		**0.03**
Skin	0.03	**0.06**	0.03	**0.07**	0.03	**0.07**	0.03	**0.07**	0.03	**0.08**	0.03	**0.07**
Muscle		**0.03**		**0.04**		**0.04**	0.01	**0.04**	0.01	**0.05**	0.01	**0.05**
Skeleton	0.07	**0.14**	0.08	**0.17**	0.09	**0.19**	0.10	**0.20**	0.11	**0.23**	0.11	**0.23**
Red bone marrow	0.01	**0.03**	0.01	**0.03**	0.02	**0.04**	0.02	**0.04**	0.02	**0.05**	0.02	**0.05**
Whole body dose	0.01	**0.04**	0.01	**0.05**	0.02	**0.05**	0.02	**0.06**	0.02	**0.06**	0.02	**0.06**
ED (ICRP 60)		**0.06**		**0.06**		**0.07**		**0.07**		**0.08**		**0.08**
ED (ICRP 103)		**0.09**		**0.10**		**0.11**		**0.11**		**0.12**		**0.12**
Total filtration (mm) **Al**	**1.5**		**2.5**		**3.5**		**4.5**		**3.5**		**3.5**	
Cu	**0**		**0**		**0**		**0**		**0.1**		**0.2**	

Table 22.16 Normalized organ doses—Upper arm ap—<1 year

Beam projection:			ap									**Table cells:**
Focus-to-detector distance (cm):			100					(+) = normal font = optimal field setting				
Tube voltage (kVp):			70					**(−) = bold font = suboptimal field setting**				
Total filtrations:			See last table row.					Empty table cells correspond to values <0.01				
Organ or tissue	**Organ dose D_T : Entrance dose K_E**							**(mSv * mGy^{-1})**				
	(+)	(−)	(+)	(−)	(+)	(−)	(+)	(−)	(+)	(−)	(+)	(−)
Skull												**0.01**
Brain												
Oral mucosa										**0.01**		**0.01**
Salivary glands				**0.01**		**0.01**		**0.02**		**0.02**		**0.02**
Cervical spine		**0.02**		**0.02**		**0.03**		**0.03**		**0.04**	0.01	**0.05**
Thyroid		**0.01**		**0.02**		**0.02**		**0.02**		**0.02**		**0.03**
Thymus		**0.02**		**0.02**		**0.02**		**0.03**		**0.03**		**0.04**
Thoracic spine		**0.04**	0.01	**0.05**	0.01	**0.06**	0.01	**0.07**	0.02	**0.08**	0.02	**0.09**
Claviculae	0.02	**0.53**	0.02	**0.61**	0.03	**0.66**	0.03	**0.69**	0.03	**0.75**	0.04	**0.78**
Scapulae	0.07	**0.28**	0.09	**0.35**	0.10	**0.40**	0.11	**0.44**	0.13	**0.51**	0.14	**0.57**
Ribs	0.09	**0.42**	0.10	**0.48**	0.12	**0.52**	0.12	**0.55**	0.14	**0.60**	0.15	**0.63**
Oesophagus		**0.02**		**0.03**		**0.03**		**0.03**		**0.04**		**0.04**
Trachea												**0.01**
Lung		**0.16**	0.01	**0.19**	0.01	**0.21**	0.02	**0.22**	0.02	**0.25**	0.02	**0.26**
Mamma		**0.56**		**0.58**		**0.59**		**0.59**		**0.60**		**0.61**
Heart		**0.02**		**0.03**		**0.03**		**0.03**		**0.04**		**0.04**
Stomach										**0.01**		**0.01**
Spleen												
Liver		**0.10**		**0.12**		**0.13**		**0.14**		**0.16**		**0.17**
Gall bladder		**0.01**		**0.01**		**0.02**		**0.02**		**0.02**		**0.03**
Pancreas		**0.01**		**0.01**		**0.02**		**0.02**		**0.02**		**0.03**
Small intestine												
Large intestine										**0.01**		**0.01**
Sigma/rectum												
Lumbar spine		**0.01**		**0.02**		**0.02**		**0.02**		**0.03**		**0.03**
Pelvis												
Adrenals		**0.02**		**0.02**		**0.02**		**0.03**		**0.03**		**0.04**
Kidneys				**0.01**		**0.01**		**0.02**		**0.02**		**0.02**
Urinary bladder												
Uterus												
Ovaries												
Prostate												
Testes												
Upper arm	0.64	**0.67**	0.75	**0.79**	0.83	**0.87**	0.88	**0.92**	0.96	**1.02**	1.03	**1.09**
Elbow	0.33	**0.64**	0.39	**0.75**	0.43	**0.82**	0.45	**0.88**	0.50	**0.96**	0.53	**1.03**
Forearm				**0.01**		**0.01**		**0.02**		**0.02**		**0.02**
Upper leg												
Knee												
Lower leg												
Lymph nodes		**0.02**		**0.03**		**0.03**		**0.03**		**0.04**		**0.04**
Skin	0.03	**0.06**	0.03	**0.07**	0.03	**0.07**	0.03	**0.07**	0.03	**0.08**	0.03	**0.08**
Muscle		**0.03**		**0.04**	0.01	**0.04**	0.01	**0.05**	0.01	**0.05**	0.01	**0.05**
Skeleton	0.08	**0.16**	0.09	**0.18**	0.10	**0.20**	0.11	**0.22**	0.12	**0.24**	0.13	**0.26**
Red bone marrow	0.01	**0.03**	0.02	**0.04**	0.02	**0.04**	0.02	**0.05**	0.02	**0.05**	0.02	**0.06**
Whole body dose	0.01	**0.05**	0.02	**0.05**	0.02	**0.06**	0.02	**0.06**	0.02	**0.07**	0.02	**0.07**
ED (ICRP 60)		**0.06**		**0.07**		**0.07**		**0.08**		**0.08**		**0.09**
ED (ICRP 103)		**0.10**		**0.11**		**0.11**		**0.12**		**0.12**		**0.13**
Total filtration (mm) **Al**	**1.5**		**2.5**		**3.5**		**4.5**		**3.5**		**3.5**	
Cu	**0**		**0**		**0**		**0**		**0.1**		**0.2**	

Table 22.17 Normalized organ doses—Upper arm ap—1–2 years

Beam projection:		ap										**Table cells:**
Focus-to-detector distance (cm):		100										(+) = normal font = optimal field setting
Tube voltage (kVp):		60										**(−) = bold font = suboptimal field setting**
Total filtrations:		See last table row.										Empty table cells correspond to values <0.01
Organ or tissue	**Organ dose D_T : Entrance dose K_E (mSv * mGy^{-1})**											
	(+)	(−)	(+)	(−)	(+)	(−)	(+)	(−)	(+)	(−)	(+)	(−)
Skull												
Brain												
Oral mucosa												
Salivary glands										**0.01**		**0.01**
Cervical spine		**0.01**		**0.01**		**0.02**		**0.02**		**0.03**		**0.02**
Thyroid				**0.01**		**0.01**		**0.02**		**0.02**		**0.02**
Thymus		**0.01**		**0.02**		**0.02**		**0.02**		**0.03**		**0.03**
Thoracic spine		**0.03**		**0.04**		**0.05**		**0.05**	0.01	**0.07**		**0.06**
Claviculae	0.02	**0.44**	0.02	**0.53**	0.02	**0.59**	0.02	**0.64**	0.03	**0.72**	0.03	**0.72**
Scapulae	0.03	**0.12**	0.03	**0.16**	0.04	**0.19**	0.05	**0.22**	0.06	**0.27**	0.05	**0.25**
Ribs	0.06	**0.38**	0.07	**0.46**	0.08	**0.51**	0.09	**0.55**	0.10	**0.61**	0.10	**0.61**
Oesophagus		**0.02**		**0.03**		**0.03**		**0.03**		**0.04**		**0.04**
Trachea												
Lung		**0.12**		**0.15**	0.01	**0.17**	0.01	**0.19**	0.01	**0.21**	0.01	**0.21**
Mamma		**0.54**		**0.57**		**0.59**		**0.60**		**0.61**		**0.61**
Heart		**0.02**		**0.02**		**0.03**		**0.03**		**0.04**		**0.03**
Stomach												
Spleen												
Liver		**0.15**		**0.18**		**0.21**		**0.23**		**0.26**		**0.25**
Gall bladder		**0.02**		**0.02**		**0.03**		**0.03**		**0.04**		**0.03**
Pancreas		**0.01**		**0.01**		**0.02**		**0.02**		**0.02**		**0.02**
Small intestine								**0.01**		**0.01**		**0.01**
Large intestine				**0.01**		**0.01**		**0.02**		**0.02**		**0.02**
Sigma/rectum												
Lumbar spine		**0.01**		**0.02**		**0.02**		**0.03**		**0.03**		**0.03**
Pelvis												
Adrenals		**0.01**		**0.02**		**0.02**		**0.02**		**0.03**		**0.03**
Kidneys		**0.01**		**0.02**		**0.02**		**0.03**		**0.03**		**0.03**
Urinary bladder												
Uterus												
Ovaries												
Prostate												
Testes												
Upper arm	0.44	**0.45**	0.54	**0.56**	0.61	**0.63**	0.66	**0.69**	0.76	**0.80**	0.75	**0.78**
Elbow	0.22	**0.49**	0.27	**0.61**	0.30	**0.69**	0.33	**0.75**	0.38	**0.86**	0.37	**0.85**
Forearm		**0.12**		**0.14**		**0.16**		**0.18**		**0.21**		**0.20**
Upper leg												
Knee												
Lower leg												
Lymph nodes		**0.02**		**0.02**		**0.03**		**0.03**		**0.04**		**0.03**
Skin	0.02	**0.07**	0.02	**0.07**	0.02	**0.07**	0.02	**0.07**	0.02	**0.08**	0.02	**0.08**
Muscle		**0.03**		**0.03**		**0.04**		**0.04**		**0.05**		**0.04**
Skeleton	0.05	**0.12**	0.06	**0.14**	0.06	**0.16**	0.07	**0.18**	0.08	**0.20**	0.08	**0.20**
Red bone marrow		**0.02**		**0.03**	0.01	**0.03**	0.01	**0.04**	0.01	**0.04**	0.01	**0.04**
Whole body dose		**0.04**	0.01	**0.05**	0.01	**0.06**	0.01	**0.06**	0.02	**0.07**	0.02	**0.07**
ED (ICRP 60)		**0.06**		**0.07**		**0.07**		**0.08**		**0.08**		**0.08**
ED (ICRP 103)		**0.09**		**0.10**		**0.11**		**0.12**		**0.13**		**0.12**
Total filtration (mm) Al	**1.5**		**2.5**		**3.5**		**4.5**		**3.5**		**3.5**	
Cu	**0**		**0**		**0**		**0**		**0.1**		**0.2**	

Table 22.18 Normalized organ doses—Upper arm ap—1–2 years

Beam projection:	ap											**Table cells:**
Focus-to-detector distance (cm):	100											(+) = normal font = optimal field setting
Tube voltage (kVp):	70											**(−) = bold font = suboptimal field setting**
Total filtrations:	See last table row.											Empty table cells correspond to values <0.01
Organ or tissue	**Organ dose D_T : Entrance dose K_E**							**(mSv * mGy^{-1})**				
	(+)	(−)	(+)	(−)	(+)	(−)	(+)	(−)	(+)	(−)	(+)	(−)
Skull												
Brain												
Oral mucosa												**0.01**
Salivary glands								**0.01**		**0.01**		**0.02**
Cervical spine		**0.01**		**0.02**		**0.02**		**0.03**		**0.03**		**0.04**
Thyroid		**0.01**		**0.01**		**0.02**		**0.02**		**0.02**		**0.03**
Thymus		**0.02**		**0.02**		**0.03**		**0.03**		**0.04**		**0.04**
Thoracic spine		**0.04**		**0.05**		**0.06**	0.01	**0.07**	0.01	**0.08**	0.02	**0.10**
Claviculae	0.02	**0.49**	0.02	**0.57**	0.03	**0.63**	0.03	**0.68**	0.03	**0.75**	0.04	**0.80**
Scapulae	0.03	**0.15**	0.04	**0.20**	0.05	**0.23**	0.05	**0.25**	0.07	**0.31**	0.08	**0.35**
Ribs	0.06	**0.42**	0.08	**0.50**	0.09	**0.54**	0.10	**0.58**	0.11	**0.64**	0.12	**0.69**
Oesophagus		**0.02**		**0.03**		**0.04**		**0.04**		**0.05**		**0.06**
Trachea												
Lung		**0.14**	0.01	**0.17**	0.01	**0.19**	0.01	**0.21**	0.02	**0.23**	0.02	**0.25**
Mamma		**0.55**		**0.58**		**0.60**		**0.61**		**0.62**		**0.63**
Heart		**0.02**		**0.03**		**0.03**		**0.04**		**0.04**		**0.05**
Stomach								**0.01**		**0.01**		**0.02**
Spleen												**0.01**
Liver		**0.17**		**0.21**		**0.23**		**0.25**		**0.29**		**0.32**
Gall bladder		**0.02**		**0.03**		**0.03**		**0.04**		**0.04**		**0.05**
Pancreas		**0.01**		**0.02**		**0.02**		**0.02**		**0.03**		**0.04**
Small intestine						**0.01**		**0.01**		**0.02**		**0.02**
Large intestine		**0.01**		**0.02**		**0.02**		**0.02**		**0.03**		**0.03**
Sigma/rectum												
Lumbar spine		**0.02**		**0.03**		**0.03**		**0.03**		**0.04**		**0.05**
Pelvis								**0.01**		**0.01**		**0.02**
Adrenals		**0.02**		**0.02**		**0.03**		**0.03**		**0.04**		**0.04**
Kidneys		**0.02**		**0.03**		**0.03**		**0.03**		**0.04**		**0.05**
Urinary bladder												
Uterus												
Ovaries												
Prostate												
Testes												
Upper arm	0.49	**0.52**	0.59	**0.62**	0.66	**0.70**	0.71	**0.76**	0.81	**0.86**	0.88	**0.95**
Elbow	0.25	**0.56**	0.30	**0.67**	0.33	**0.75**	0.36	**0.82**	0.40	**0.93**	0.44	**1.01**
Forearm		**0.13**		**0.16**		**0.18**		**0.20**		**0.22**		**0.24**
Upper leg												
Knee												
Lower leg												
Lymph nodes		**0.02**		**0.03**		**0.03**		**0.03**		**0.04**		**0.05**
Skin	0.02	**0.07**	0.02	**0.07**	0.02	**0.08**	0.02	**0.08**	0.03	**0.08**	0.03	**0.09**
Muscle		**0.03**		**0.04**		**0.04**		**0.04**		**0.05**		**0.05**
Skeleton	0.05	**0.13**	0.06	**0.16**	0.07	**0.18**	0.08	**0.19**	0.09	**0.22**	0.10	**0.24**
Red bone marrow		**0.03**	0.01	**0.03**	0.01	**0.04**	0.01	**0.04**	0.02	**0.05**	0.02	**0.05**
Whole body dose	0.01	**0.05**	0.01	**0.06**	0.01	**0.06**	0.02	**0.07**	0.02	**0.07**	0.02	**0.08**
ED (ICRP 60)		**0.06**		**0.07**		**0.08**		**0.08**		**0.09**		**0.10**
ED (ICRP 103)		**0.10**		**0.11**		**0.12**		**0.12**		**0.13**		**0.14**
Total filtration (mm) **Al**	**1.5**		**2.5**		**3.5**		**4.5**		**3.5**		**3.5**	
Cu	**0**		**0**		**0**		**0**		**0.1**		**0.2**	

Table 22.19 Normalized organ doses—Upper arm ap—3–7 years

Beam projection:			ap									**Table cells:**
Focus-to-detector distance (cm):			100									(+) = normal font = optimal field setting
Tube voltage (kVp):			60									**(−) = bold font = suboptimal field setting**
Total filtrations:			See last table row.									Empty table cells correspond to values <0.01
Organ or tissue	**Organ dose D_T : Entrance dose K_E (mSv * mGy^{-1})**											
	(+)	(−)	(+)	(−)	(+)	(−)	(+)	(−)	(+)	(−)	(+)	(−)
Skull												
Brain												
Oral mucosa												
Salivary glands										**0.01**		
Cervical spine				**0.01**		**0.02**		**0.02**		**0.02**		**0.02**
Thyroid				**0.01**		**0.01**		**0.01**		**0.02**		**0.02**
Thymus		**0.01**		**0.02**		**0.02**		**0.02**		**0.03**		**0.03**
Thoracic spine		**0.02**		**0.03**		**0.04**		**0.05**		**0.06**		**0.05**
Claviculae		**0.41**	0.01	**0.50**	0.01	**0.56**	0.01	**0.61**	0.02	**0.70**	0.02	**0.69**
Scapulae	0.02	**0.10**	0.02	**0.13**	0.02	**0.15**	0.03	**0.17**	0.04	**0.22**	0.03	**0.20**
Ribs	0.04	**0.36**	0.05	**0.44**	0.06	**0.49**	0.07	**0.52**	0.08	**0.59**	0.08	**0.59**
Oesophagus		**0.01**		**0.02**		**0.02**		**0.02**		**0.03**		**0.02**
Trachea												
Lung		**0.12**		**0.15**		**0.17**	0.01	**0.19**	0.01	**0.22**	0.01	**0.21**
Mamma		**0.57**		**0.61**		**0.63**		**0.64**		**0.66**		**0.66**
Heart		**0.02**		**0.02**		**0.02**		**0.03**		**0.03**		**0.03**
Stomach												
Spleen												
Liver		**0.16**		**0.20**		**0.23**		**0.25**		**0.30**		**0.28**
Gall bladder		**0.02**		**0.02**		**0.02**		**0.03**		**0.04**		**0.03**
Pancreas				**0.01**		**0.01**		**0.02**		**0.02**		**0.02**
Small intestine						**0.01**		**0.01**		**0.02**		**0.01**
Large intestine		**0.01**		**0.02**		**0.02**		**0.02**		**0.03**		**0.03**
Sigma/rectum												
Lumbar spine		**0.01**		**0.02**		**0.02**		**0.02**		**0.03**		**0.02**
Pelvis												
Adrenals		**0.01**		**0.02**		**0.02**		**0.02**		**0.03**		**0.02**
Kidneys		**0.01**		**0.02**		**0.02**		**0.03**		**0.03**		**0.03**
Urinary bladder												
Uterus												
Ovaries												
Prostate												
Testes												
Upper arm	0.39	**0.39**	0.48	**0.49**	0.54	**0.56**	0.59	**0.61**	0.68	**0.72**	0.67	**0.69**
Elbow	0.28	**0.43**	0.34	**0.53**	0.39	**0.61**	0.42	**0.67**	0.49	**0.78**	0.48	**0.75**
Forearm		**0.14**		**0.18**		**0.20**		**0.22**		**0.26**		**0.25**
Upper leg												
Knee												
Lower leg												
Lymph nodes		**0.02**		**0.02**		**0.03**		**0.03**		**0.03**		**0.03**
Skin	0.02	**0.06**	0.02	**0.07**	0.02	**0.07**	0.02	**0.07**	0.02	**0.07**	0.02	**0.07**
Muscle		**0.03**		**0.03**		**0.04**		**0.04**		**0.05**		**0.04**
Skeleton	0.04	**0.09**	0.04	**0.11**	0.05	**0.12**	0.06	**0.13**	0.06	**0.16**	0.06	**0.15**
Red bone marrow		**0.02**		**0.03**		**0.03**	0.01	**0.04**	0.01	**0.04**	0.01	**0.04**
Whole body dose		**0.04**	0.01	**0.05**	0.01	**0.05**	0.01	**0.06**	0.01	**0.07**	0.01	**0.06**
ED (ICRP 60)		**0.06**		**0.07**		**0.07**		**0.08**		**0.09**		**0.08**
ED (ICRP 103)		**0.10**		**0.11**		**0.12**		**0.12**		**0.13**		**0.13**
Total filtration (mm) Al	**1.5**		**2.5**		**3.5**		**4.5**		**3.5**		**3.5**	
Cu	**0**		**0**		**0**		**0**		**0.1**		**0.2**	

Table 22.20 Normalized organ doses—Upper arm ap—3–7 years

Beam projection:			**ap**									**Table cells:**
Focus-to-detector distance (cm):			100									(+) = normal font = optimal field setting
Tube voltage (kVp):			70									**(−) = bold font = suboptimal field setting**
Total filtrations:			See last table row.									Empty table cells correspond to values <0.01
Organ or tissue	**Organ dose D_T : Entrance dose K_E (mSv * mGy⁻¹)**											
	(+)	(−)	(+)	(−)	(+)	(−)	(+)	(−)	(+)	(−)	(+)	(−)
Skull										**0.01**		**0.01**
Brain												
Oral mucosa												
Salivary glands								**0.01**		**0.01**		**0.01**
Cervical spine		**0.01**		**0.02**		**0.02**		**0.02**		**0.03**		**0.04**
Thyroid		**0.01**		**0.01**		**0.02**		**0.02**		**0.02**		**0.03**
Thymus		**0.02**		**0.02**		**0.03**		**0.03**		**0.04**		**0.04**
Thoracic spine		**0.03**		**0.04**		**0.05**		**0.06**	0.01	**0.08**	0.01	**0.09**
Claviculae	0.01	**0.46**	0.01	**0.55**	0.01	**0.61**	0.02	**0.66**	0.02	**0.74**	0.02	**0.80**
Scapulae	0.02	**0.12**	0.03	**0.16**	0.03	**0.19**	0.04	**0.21**	0.04	**0.26**	0.05	**0.30**
Ribs	0.05	**0.40**	0.06	**0.48**	0.07	**0.53**	0.08	**0.56**	0.09	**0.63**	0.10	**0.68**
Oesophagus		**0.02**		**0.02**		**0.03**		**0.03**		**0.04**		**0.04**
Trachea												
Lung		**0.14**		**0.17**	0.01	**0.19**	0.01	**0.21**	0.01	**0.24**	0.02	**0.26**
Mamma		**0.58**		**0.62**		**0.64**		**0.65**		**0.67**		**0.69**
Heart		**0.02**		**0.03**		**0.03**		**0.03**		**0.04**		**0.05**
Stomach										**0.01**		**0.01**
Spleen												
Liver		**0.19**		**0.23**		**0.26**		**0.29**		**0.34**		**0.38**
Gall bladder		**0.02**		**0.03**		**0.03**		**0.04**		**0.04**		**0.05**
Pancreas		**0.01**		**0.02**		**0.02**		**0.02**		**0.03**		**0.03**
Small intestine				**0.01**		**0.01**		**0.02**		**0.02**		**0.03**
Large intestine		**0.02**		**0.02**		**0.03**		**0.03**		**0.04**		**0.04**
Sigma/rectum												
Lumbar spine		**0.02**		**0.02**		**0.03**		**0.03**		**0.04**		**0.05**
Pelvis								**0.01**		**0.01**		**0.02**
Adrenals		**0.02**		**0.02**		**0.03**		**0.03**		**0.04**		**0.04**
Kidneys		**0.02**		**0.02**		**0.03**		**0.03**		**0.04**		**0.05**
Urinary bladder												
Uterus												
Ovaries												
Prostate												
Testes												
Upper arm	0.44	**0.45**	0.53	**0.55**	0.60	**0.63**	0.65	**0.68**	0.74	**0.78**	0.81	**0.87**
Elbow	0.31	**0.49**	0.38	**0.60**	0.43	**0.68**	0.46	**0.74**	0.52	**0.85**	0.58	**0.94**
Forearm		**0.16**		**0.20**		**0.22**		**0.24**		**0.28**		**0.31**
Upper leg												
Knee												
Lower leg												
Lymph nodes		**0.02**		**0.03**		**0.03**		**0.03**		**0.04**		**0.05**
Skin	0.02	**0.07**	0.02	**0.07**	0.02	**0.07**	0.02	**0.07**	0.02	**0.08**	0.02	**0.08**
Muscle		**0.03**		**0.04**		**0.04**		**0.04**		**0.05**		**0.06**
Skeleton	0.04	**0.10**	0.05	**0.12**	0.06	**0.14**	0.06	**0.15**	0.07	**0.17**	0.08	**0.19**
Red bone marrow		**0.03**		**0.03**	0.01	**0.04**	0.01	**0.04**	0.01	**0.05**	0.02	**0.05**
Whole body dose		**0.05**	0.01	**0.05**	0.01	**0.06**	0.01	**0.07**	0.02	**0.07**	0.02	**0.08**
ED (ICRP 60)		**0.06**		**0.07**		**0.08**		**0.09**		**0.10**		**0.10**
ED (ICRP 103)		**0.10**		**0.12**		**0.12**		**0.13**		**0.14**		**0.15**
Total filtration (mm) **Al**	**1.5**		**2.5**		**3.5**		**4.5**		**3.5**		**3.5**	
Cu	**0**		**0**		**0**		**0**		**0.1**		**0.2**	

Table 22.21 Normalized organ doses—Upper arm ap—8–12 years

Beam projection:			ap									**Table cells:**
Focus-to-detector distance (cm):			100						(+) = normal font = optimal field setting			
Tube voltage (kVp):			60						**(−) = bold font = suboptimal field setting**			
Total filtrations:			See last table row.						Empty table cells correspond to values <0.01			
Organ or tissue	**Organ dose D_T : Entrance dose K_E**								**(mSv * mGy⁻¹)**			
	(+)	(−)	(+)	(−)	(+)	(−)	(+)	(−)	(+)	(−)	(+)	(−)
Skull												
Brain												
Oral mucosa												
Salivary glands												
Cervical spine						**0.01**		**0.01**		**0.02**		**0.01**
Thyroid								**0.01**		**0.01**		**0.01**
Thymus				**0.01**		**0.02**		**0.02**		**0.02**		**0.02**
Thoracic spine		**0.02**		**0.02**		**0.03**		**0.03**		**0.04**		**0.03**
Claviculae		**0.32**	0.01	**0.40**	0.01	**0.46**	0.02	**0.50**	0.02	**0.58**	0.02	**0.57**
Scapulae	0.01	**0.07**	0.02	**0.10**	0.02	**0.12**	0.03	**0.13**	0.03	**0.17**	0.03	**0.15**
Ribs	0.04	**0.31**	0.05	**0.38**	0.06	**0.42**	0.06	**0.46**	0.07	**0.52**	0.07	**0.51**
Oesophagus				**0.01**		**0.01**		**0.02**		**0.02**		**0.02**
Trachea												
Lung		**0.10**		**0.12**		**0.14**		**0.15**	0.01	**0.18**		**0.17**
Mamma		**0.48**		**0.52**		**0.54**		**0.55**		**0.57**		**0.57**
Heart		**0.01**		**0.02**		**0.02**		**0.02**		**0.03**		**0.02**
Stomach												
Spleen												
Liver		**0.14**		**0.17**		**0.20**		**0.22**		**0.27**		**0.25**
Gall bladder		**0.01**		**0.02**		**0.02**		**0.03**		**0.03**		**0.03**
Pancreas								**0.01**		**0.01**		**0.01**
Small intestine												
Large intestine				**0.01**		**0.01**		**0.01**		**0.02**		**0.02**
Sigma/rectum												
Lumbar spine						**0.01**		**0.01**		**0.02**		**0.01**
Pelvis												
Adrenals				**0.01**		**0.01**		**0.02**		**0.02**		**0.02**
Kidneys				**0.01**		**0.02**		**0.02**		**0.02**		**0.02**
Urinary bladder												
Uterus												
Ovaries												
Prostate												
Testes												
Upper arm	0.36	**0.35**	0.44	**0.44**	0.50	**0.51**	0.55	**0.56**	0.65	**0.66**	0.63	**0.63**
Elbow	0.22	**0.38**	0.27	**0.48**	0.31	**0.55**	0.34	**0.61**	0.39	**0.72**	0.38	**0.69**
Forearm		**0.12**		**0.14**		**0.17**		**0.18**		**0.22**		**0.21**
Upper leg												
Knee												
Lower leg												
Lymph nodes		**0.01**		**0.02**		**0.02**		**0.02**		**0.03**		**0.02**
Skin	0.02	**0.05**	0.02	**0.06**	0.02	**0.06**	0.02	**0.06**	0.02	**0.06**	0.02	**0.06**
Muscle		**0.02**		**0.03**		**0.03**		**0.04**		**0.04**		**0.04**
Skeleton	0.03	**0.07**	0.04	**0.09**	0.04	**0.11**	0.05	**0.12**	0.06	**0.14**	0.06	**0.13**
Red bone marrow		**0.02**		**0.03**		**0.03**		**0.03**	0.01	**0.04**	0.01	**0.04**
Whole body dose		**0.03**		**0.04**	0.01	**0.05**	0.01	**0.05**	0.01	**0.06**	0.01	**0.06**
ED (ICRP 60)		**0.05**		**0.06**		**0.06**		**0.07**		**0.07**		**0.07**
ED (ICRP 103)		**0.08**		**0.09**		**0.10**		**0.10**		**0.11**		**0.11**
Total filtration (mm) **Al**	**1.5**		**2.5**		**3.5**		**4.5**		**3.5**		**3.5**	
Cu	**0**		**0**		**0**		**0**		**0.1**		**0.2**	

Table 22.22 Normalized organ doses—Upper arm ap—8–12 years

Beam projection:			ap									**Table cells:**
Focus-to-detector distance (cm):			100									(+) = normal font = optimal field setting
Tube voltage (kVp):			70									**(−) = bold font = suboptimal field setting**
Total filtrations:			See last table row.									Empty table cells correspond to values <0.01
Organ or tissue	**Organ dose D_T : Entrance dose K_E**								**(mSv * mGy⁻¹)**			
	(+)	(−)	(+)	(−)	(+)	(−)	(+)	(−)	(+)	(−)	(+)	(−)
Skull												
Brain												
Oral mucosa												
Salivary glands												**0.01**
Cervical spine				**0.01**		**0.02**		**0.02**		**0.02**		**0.03**
Thyroid						**0.01**		**0.01**		**0.02**		**0.02**
Thymus		**0.01**		**0.02**		**0.02**		**0.02**		**0.03**		**0.03**
Thoracic spine		**0.02**		**0.03**		**0.04**		**0.04**		**0.06**		**0.07**
Claviculae	0.01	**0.37**	0.02	**0.45**	0.02	**0.50**	0.02	**0.55**	0.02	**0.62**	0.03	**0.68**
Scapulae	0.02	**0.09**	0.03	**0.12**	0.03	**0.15**	0.03	**0.16**	0.04	**0.20**	0.05	**0.24**
Ribs	0.05	**0.35**	0.06	**0.42**	0.06	**0.46**	0.07	**0.50**	0.08	**0.56**	0.09	**0.61**
Oesophagus		**0.01**		**0.02**		**0.02**		**0.02**		**0.03**		**0.03**
Trachea												**0.01**
Lung		**0.11**		**0.14**		**0.16**	0.01	**0.17**	0.01	**0.20**	0.02	**0.22**
Mamma		**0.50**		**0.53**		**0.55**		**0.56**		**0.58**		**0.60**
Heart		**0.02**		**0.02**		**0.02**		**0.03**		**0.04**		**0.04**
Stomach												
Spleen												
Liver		**0.16**		**0.20**		**0.23**		**0.26**		**0.30**		**0.35**
Gall bladder		**0.02**		**0.02**		**0.03**		**0.03**		**0.04**		**0.05**
Pancreas				**0.01**		**0.01**		**0.02**		**0.02**		**0.02**
Small intestine								**0.01**		**0.01**		**0.02**
Large intestine		**0.01**		**0.01**		**0.02**		**0.02**		**0.02**		**0.03**
Sigma/rectum												
Lumbar spine		**0.01**		**0.01**		**0.02**		**0.02**		**0.03**		**0.03**
Pelvis												
Adrenals		**0.01**		**0.02**		**0.02**		**0.02**		**0.03**		**0.04**
Kidneys		**0.01**		**0.02**		**0.02**		**0.02**		**0.03**		**0.04**
Urinary bladder												
Uterus												
Ovaries												
Prostate												
Testes												
Upper arm	0.41	**0.41**	0.50	**0.51**	0.56	**0.57**	0.61	**0.63**	0.71	**0.73**	0.78	**0.82**
Elbow	0.25	**0.45**	0.30	**0.55**	0.34	**0.62**	0.37	**0.68**	0.43	**0.79**	0.48	**0.88**
Forearm		**0.13**		**0.17**		**0.19**		**0.21**		**0.24**		**0.27**
Upper leg												
Knee												
Lower leg												
Lymph nodes		**0.02**		**0.02**		**0.02**		**0.03**		**0.03**		**0.04**
Skin	0.02	**0.06**	0.02	**0.06**	0.02	**0.06**	0.02	**0.06**	0.02	**0.07**	0.02	**0.07**
Muscle		**0.03**		**0.03**		**0.04**		**0.04**		**0.05**		**0.05**
Skeleton	0.04	**0.09**	0.04	**0.11**	0.05	**0.12**	0.05	**0.13**	0.06	**0.15**	0.07	**0.17**
Red bone marrow		**0.02**		**0.03**	0.01	**0.03**	0.01	**0.04**	0.01	**0.04**	0.02	**0.05**
Whole body dose		**0.04**	0.01	**0.05**	0.01	**0.05**	0.01	**0.06**	0.01	**0.07**	0.02	**0.07**
ED (ICRP 60)		**0.05**		**0.06**		**0.07**		**0.07**		**0.08**		**0.09**
ED (ICRP 103)		**0.09**		**0.10**		**0.10**		**0.11**		**0.12**		**0.13**
Total filtration (mm) **Al**	**1.5**		**2.5**		**3.5**		**4.5**		**3.5**		**3.5**	
Cu	**0**		**0**		**0**		**0**		**0.1**		**0.2**	

Table 22.23 Normalized organ doses—Upper arm ap—8–12 years

Beam projection:			ap									**Table cells:**
Focus-to-detector distance (cm):			100						(+) = normal font = optimal field setting			
Tube voltage (kVp):			80						**(−) = bold font = suboptimal field setting**			
Total filtrations:			See last table row.						Empty table cells correspond to values <0.01			
Organ or tissue	**Organ dose D_T : Entrance dose K_E**								**(mSv * mGy⁻¹)**			
	(+)	(−)	(+)	(−)	(+)	(−)	(+)	(−)	(+)	(−)	(+)	(−)
Skull												
Brain												
Oral mucosa												
Salivary glands										**0.01**		**0.01**
Cervical spine		**0.01**		**0.02**		**0.02**		**0.02**		**0.03**		**0.03**
Thyroid				**0.01**		**0.01**		**0.02**		**0.02**		**0.02**
Thymus		**0.02**		**0.02**		**0.02**		**0.03**		**0.03**		**0.04**
Thoracic spine		**0.03**		**0.04**		**0.05**		**0.06**		**0.07**	0.01	**0.08**
Claviculae	0.01	**0.41**	0.02	**0.48**	0.02	**0.54**	0.02	**0.58**	0.03	**0.64**	0.03	**0.70**
Scapulae	0.02	**0.12**	0.03	**0.15**	0.04	**0.17**	0.04	**0.19**	0.05	**0.23**	0.06	**0.27**
Ribs	0.05	**0.38**	0.06	**0.45**	0.07	**0.49**	0.08	**0.53**	0.09	**0.59**	0.10	**0.63**
Oesophagus		**0.02**		**0.02**		**0.02**		**0.03**		**0.03**		**0.04**
Trachea										**0.01**		**0.01**
Lung		**0.13**		**0.15**	0.01	**0.17**	0.01	**0.19**	0.02	**0.22**	0.02	**0.24**
Mamma		**0.51**		**0.55**		**0.56**		**0.57**		**0.59**		**0.61**
Heart		**0.02**		**0.03**		**0.03**		**0.03**		**0.04**		**0.05**
Stomach												**0.01**
Spleen												
Liver		**0.19**		**0.23**		**0.26**		**0.29**		**0.34**		**0.38**
Gall bladder		**0.02**		**0.03**		**0.04**		**0.04**		**0.05**		**0.06**
Pancreas		**0.01**		**0.01**		**0.02**		**0.02**		**0.03**		**0.03**
Small intestine						**0.01**		**0.01**		**0.02**		**0.02**
Large intestine		**0.01**		**0.02**		**0.02**		**0.02**		**0.03**		**0.03**
Sigma/rectum												
Lumbar spine		**0.02**		**0.02**		**0.02**		**0.03**		**0.03**		**0.04**
Pelvis												**0.01**
Adrenals		**0.02**		**0.02**		**0.03**		**0.03**		**0.04**		**0.04**
Kidneys		**0.02**		**0.02**		**0.03**		**0.03**		**0.04**		**0.04**
Urinary bladder												
Uterus												
Ovaries												
Prostate												
Testes												
Upper arm	0.46	**0.47**	0.55	**0.56**	0.61	**0.63**	0.66	**0.68**	0.74	**0.78**	0.81	**0.86**
Elbow	0.28	**0.50**	0.33	**0.61**	0.37	**0.68**	0.40	**0.74**	0.45	**0.84**	0.49	**0.92**
Forearm		**0.15**		**0.18**		**0.20**		**0.22**		**0.25**		**0.28**
Upper leg												
Knee												
Lower leg												
Lymph nodes		**0.02**		**0.03**		**0.03**		**0.03**		**0.04**		**0.04**
Skin	0.02	**0.06**	0.02	**0.06**	0.02	**0.06**	0.02	**0.07**	0.02	**0.07**	0.02	**0.07**
Muscle		**0.03**		**0.04**		**0.04**		**0.04**		**0.05**		**0.05**
Skeleton	0.04	**0.10**	0.05	**0.12**	0.05	**0.13**	0.06	**0.14**	0.07	**0.16**	0.07	**0.18**
Red bone marrow		**0.03**	0.01	**0.03**	0.01	**0.04**	0.01	**0.04**	0.01	**0.05**	0.02	**0.06**
Whole body dose	0.01	**0.04**	0.01	**0.05**	0.01	**0.06**	0.01	**0.06**	0.02	**0.07**	0.02	**0.08**
ED (ICRP 60)		**0.06**		**0.07**		**0.07**		**0.08**		**0.09**		**0.10**
ED (ICRP 103)		**0.09**		**0.10**		**0.11**		**0.12**		**0.13**		**0.14**
Total filtration (mm) Al	**1.5**		**2.5**		**3.5**		**4.5**		**3.5**		**3.5**	
Cu	**0**		**0**		**0**		**0**		**0.1**		**0.2**	

Table 22.24 Normalized organ doses—Upper arm ap—13–17 years

Beam projection:			ap									**Table cells:**
Focus-to-detector distance (cm):			115									(+) = normal font = optimal field setting
Tube voltage (kVp):			60									**(−) = bold font = suboptimal field setting**
Total filtrations:			See last table row.									Empty table cells correspond to values <0.01
Organ or tissue	**Organ dose D_T : Entrance dose K_E**								**(mSv * mGy^{-1})**			
	(+)	(−)	(+)	(−)	(+)	(−)	(+)	(−)	(+)	(−)	(+)	(−)
Skull												
Brain												
Oral mucosa												
Salivary glands												
Cervical spine												
Thyroid												
Thymus								**0.01**		**0.01**		**0.01**
Thoracic spine				**0.01**		**0.01**		**0.02**		**0.02**		**0.01**
Claviculae		**0.22**	0.01	**0.28**	0.01	**0.32**	0.01	**0.36**	0.02	**0.42**	0.02	**0.41**
Scapulae	0.01	**0.04**	0.01	**0.05**	0.02	**0.06**	0.02	**0.07**	0.03	**0.09**	0.02	**0.08**
Ribs	0.04	**0.21**	0.05	**0.26**	0.06	**0.30**	0.06	**0.33**	0.07	**0.39**	0.07	**0.37**
Oesophagus								**0.01**		**0.01**		**0.01**
Trachea												
Lung		**0.04**		**0.05**		**0.06**		**0.07**		**0.09**		**0.08**
Mamma	0.02	**2.09**	0.03	**2.42**	0.03	**2.64**	0.04	**2.80**	0.05	**3.08**	0.04	**3.04**
Heart						**0.01**		**0.01**		**0.02**		**0.01**
Stomach												
Spleen												
Liver		**0.11**		**0.14**		**0.17**		**0.19**	0.01	**0.23**		**0.21**
Gall bladder		**0.02**		**0.02**		**0.03**		**0.03**		**0.04**		**0.03**
Pancreas												
Small intestine												
Large intestine						**0.01**		**0.01**		**0.02**		**0.01**
Sigma/rectum												
Lumbar spine												
Pelvis												
Adrenals										**0.01**		
Kidneys										**0.01**		
Urinary bladder												
Uterus												
Ovaries												
Prostate												
Testes												
Upper arm	0.31	**0.30**	0.39	**0.38**	0.45	**0.44**	0.50	**0.49**	0.59	**0.58**	0.56	**0.55**
Elbow	0.28	**0.33**	0.36	**0.42**	0.41	**0.48**	0.45	**0.54**	0.53	**0.64**	0.51	**0.60**
Forearm		**0.11**		**0.14**		**0.16**		**0.18**		**0.22**		**0.21**
Upper leg												
Knee												
Lower leg												
Lymph nodes				**0.01**		**0.01**		**0.01**		**0.02**		**0.02**
Skin	0.02	**0.06**	0.02	**0.06**	0.02	**0.06**	0.02	**0.06**	0.02	**0.07**	0.02	**0.07**
Muscle		**0.02**		**0.02**		**0.03**		**0.03**		**0.04**		**0.03**
Skeleton	0.03	**0.06**	0.04	**0.08**	0.05	**0.09**	0.06	**0.10**	0.07	**0.12**	0.06	**0.11**
Red bone marrow		**0.02**		**0.02**		**0.02**		**0.03**	0.01	**0.03**	0.01	**0.03**
Whole body dose		**0.03**	0.01	**0.04**	0.01	**0.04**	0.01	**0.05**	0.02	**0.05**	0.01	**0.05**
ED (ICRP 60)		**0.12**		**0.14**		**0.15**		**0.17**		**0.18**		**0.18**
ED (ICRP 103)		**0.26**		**0.31**		**0.34**		**0.36**		**0.40**		**0.39**
Total filtration (mm) Al	**1.5**		**2.5**		**3.5**		**4.5**		**3.5**		**3.5**	
Cu	**0**		**0**		**0**		**0**		**0.1**		**0.2**	

Table 22.25 Normalized organ doses—Upper arm ap—13–17 years

Beam projection:			ap									**Table cells:**
Focus-to-detector distance (cm):			115				(+) = normal font = optimal field setting					
Tube voltage (kVp):			70				**(−) = bold font = suboptimal field setting**					
Total filtrations:			See last table row.				Empty table cells correspond to values <0.01					
Organ or tissue	**Organ dose D_T : Entrance dose K_E**						**(mSv * mGy^{-1})**					
	(+)	(−)	(+)	(−)	(+)	(−)	(+)	(−)	(+)	(−)	(+)	(−)
Skull												
Brain												
Oral mucosa												
Salivary glands												
Cervical spine								**0.01**		**0.01**		**0.02**
Thyroid												**0.01**
Thymus				**0.01**		**0.01**		**0.01**		**0.02**		**0.02**
Thoracic spine		**0.01**		**0.02**		**0.02**		**0.02**		**0.03**		**0.04**
Claviculae	0.01	**0.26**	0.01	**0.32**	0.02	**0.36**	0.02	**0.40**	0.02	**0.46**	0.02	**0.51**
Scapulae	0.01	**0.05**	0.02	**0.07**	0.02	**0.08**	0.03	**0.09**	0.03	**0.12**	0.04	**0.15**
Ribs	0.04	**0.25**	0.06	**0.30**	0.07	**0.34**	0.07	**0.37**	0.09	**0.42**	0.10	**0.47**
Oesophagus				**0.01**		**0.01**		**0.01**		**0.02**		**0.02**
Trachea												
Lung		**0.05**		**0.07**		**0.08**		**0.09**	0.01	**0.10**	0.01	**0.12**
Mamma	0.03	**2.28**	0.04	**2.61**	0.04	**2.83**	0.05	**2.99**	0.06	**3.27**	0.07	**3.50**
Heart				**0.01**		**0.02**		**0.02**		**0.02**		**0.03**
Stomach												
Spleen												
Liver		**0.14**		**0.17**	0.01	**0.20**	0.01	**0.22**	0.01	**0.27**	0.02	**0.31**
Gall bladder		**0.02**		**0.03**		**0.04**		**0.04**		**0.05**		**0.06**
Pancreas										**0.01**		**0.01**
Small intestine										**0.01**		**0.01**
Large intestine				**0.01**		**0.01**		**0.02**		**0.02**		**0.02**
Sigma/rectum												
Lumbar spine								**0.01**		**0.01**		**0.02**
Pelvis												
Adrenals						**0.01**		**0.01**		**0.02**		**0.02**
Kidneys						**0.01**		**0.01**		**0.02**		**0.02**
Urinary bladder												
Uterus												
Ovaries												
Prostate												
Testes												
Upper arm	0.37	**0.36**	0.45	**0.45**	0.51	**0.51**	0.56	**0.56**	0.65	**0.66**	0.73	**0.74**
Elbow	0.33	**0.39**	0.41	**0.49**	0.46	**0.55**	0.51	**0.61**	0.59	**0.71**	0.66	**0.80**
Forearm		**0.13**		**0.16**		**0.19**		**0.21**		**0.24**		**0.27**
Upper leg												
Knee												
Lower leg												
Lymph nodes		**0.01**		**0.01**		**0.02**		**0.02**		**0.02**		**0.03**
Skin	0.02	**0.06**	0.02	**0.06**	0.02	**0.07**	0.02	**0.07**	0.03	**0.07**	0.03	**0.07**
Muscle		**0.02**		**0.03**		**0.03**		**0.03**		**0.04**		**0.04**
Skeleton	0.04	**0.08**	0.05	**0.09**	0.06	**0.11**	0.06	**0.12**	0.07	**0.14**	0.08	**0.15**
Red bone marrow		**0.02**		**0.03**	0.01	**0.03**	0.01	**0.03**	0.01	**0.04**	0.02	**0.04**
Whole body dose	0.01	**0.03**	0.01	**0.04**	0.01	**0.05**	0.01	**0.05**	0.02	**0.06**	0.02	**0.07**
ED (ICRP 60)		**0.13**		**0.15**		**0.17**		**0.18**		**0.20**		**0.22**
ED (ICRP 103)		**0.29**		**0.34**		**0.37**	0.01	**0.39**	0.01	**0.43**	0.01	**0.46**
Total filtration (mm) Al	**1.5**		**2.5**		**3.5**		**4.5**		**3.5**		**3.5**	
Cu	**0**		**0**		**0**		**0**		**0.1**		**0.2**	

Table 22.26 Normalized organ doses—Upper arm ap—13–17 years

Beam projection:			ap									**Table cells:**
Focus-to-detector distance (cm):			115						(+) = normal font = optimal field setting			
Tube voltage (kVp):			80						**(−) = bold font = suboptimal field setting**			
Total filtrations:			See last table row.						Empty table cells correspond to values <0.01			
Organ or tissue	**Organ dose D_T : Entrance dose K_E**								**(mSv * mGy^{-1})**			
	(+)	(−)	(+)	(−)	(+)	(−)	(+)	(−)	(+)	(−)	(+)	(−)
Skull												
Brain												
Oral mucosa												
Salivary glands												
Cervical spine				**0.01**		**0.01**		**0.01**		**0.02**		**0.02**
Thyroid										**0.01**		**0.01**
Thymus		**0.01**		**0.01**		**0.02**		**0.02**		**0.02**		**0.03**
Thoracic spine		**0.02**		**0.02**		**0.03**		**0.03**		**0.04**		**0.05**
Claviculae	0.01	**0.29**	0.02	**0.35**	0.02	**0.39**	0.02	**0.43**	0.02	**0.48**	0.03	**0.53**
Scapulae	0.02	**0.07**	0.02	**0.09**	0.03	**0.10**	0.03	**0.12**	0.04	**0.14**	0.05	**0.17**
Ribs	0.05	**0.28**	0.06	**0.33**	0.07	**0.37**	0.08	**0.40**	0.09	**0.45**	0.11	**0.49**
Oesophagus		**0.01**		**0.01**		**0.02**		**0.02**		**0.02**		**0.03**
Trachea												
Lung		**0.06**		**0.08**	0.01	**0.09**	0.01	**0.10**	0.01	**0.12**	0.02	**0.14**
Mamma	0.04	**2.45**	0.04	**2.78**	0.05	**2.99**	0.06	**3.15**	0.07	**3.42**	0.08	**3.64**
Heart		**0.01**		**0.02**		**0.02**		**0.02**		**0.03**		**0.03**
Stomach												
Spleen												
Liver		**0.16**	0.01	**0.20**	0.01	**0.23**	0.01	**0.26**	0.02	**0.30**	0.02	**0.34**
Gall bladder		**0.03**		**0.04**		**0.04**		**0.05**		**0.06**		**0.07**
Pancreas						**0.01**		**0.01**		**0.02**		**0.02**
Small intestine								**0.01**		**0.01**		**0.02**
Large intestine		**0.01**		**0.01**		**0.02**		**0.02**		**0.02**		**0.03**
Sigma/rectum												
Lumbar spine				**0.01**		**0.01**		**0.02**		**0.02**		**0.02**
Pelvis												
Adrenals				**0.01**		**0.02**		**0.02**		**0.02**		**0.03**
Kidneys				**0.01**		**0.02**		**0.02**		**0.02**		**0.03**
Urinary bladder												
Uterus												
Ovaries												
Prostate												
Testes												
Upper arm	0.41	**0.41**	0.50	**0.50**	0.56	**0.57**	0.61	**0.62**	0.69	**0.71**	0.76	**0.79**
Elbow	0.37	**0.45**	0.45	**0.54**	0.51	**0.61**	0.55	**0.67**	0.62	**0.77**	0.69	**0.85**
Forearm		**0.15**		**0.18**		**0.21**		**0.23**		**0.26**		**0.29**
Upper leg												
Knee												
Lower leg												
Lymph nodes		**0.01**		**0.02**		**0.02**		**0.02**		**0.03**		**0.03**
Skin	0.02	**0.06**	0.02	**0.07**	0.02	**0.07**	0.03	**0.07**	0.03	**0.07**	0.03	**0.08**
Muscle		**0.03**		**0.03**		**0.04**		**0.04**		**0.04**	0.01	**0.05**
Skeleton	0.05	**0.09**	0.06	**0.10**	0.06	**0.12**	0.07	**0.13**	0.08	**0.15**	0.09	**0.16**
Red bone marrow		**0.02**	0.01	**0.03**	0.01	**0.03**	0.01	**0.04**	0.02	**0.04**	0.02	**0.05**
Whole body dose	0.01	**0.04**	0.01	**0.05**	0.02	**0.05**	0.02	**0.06**	0.02	**0.07**	0.02	**0.07**
ED (ICRP 60)		**0.14**		**0.17**		**0.18**		**0.19**		**0.21**	0.01	**0.23**
ED (ICRP 103)		**0.32**		**0.36**	0.01	**0.39**	0.01	**0.41**	0.01	**0.45**	0.02	**0.48**
Total filtration (mm) **Al**	**1.5**		**2.5**		**3.5**		**4.5**		**3.5**		**3.5**	
Cu	**0**		**0**		**0**		**0**		**0.1**		**0.2**	

Table 22.27 Normalized organ doses—Upper arm ap—>17 years

Beam projection:			ap									**Table cells:**
Focus-to-detector distance (cm):			115									(+) = normal font = optimal field setting
Tube voltage (kVp):			60									**(−) = bold font = suboptimal field setting**
Total filtrations:			See last table row.									Empty table cells correspond to values <0.01
Organ or tissue	**Organ dose D_T : Entrance dose K_E**								**(mSv * mGy^{-1})**			
	(+)	(−)	(+)	(−)	(+)	(−)	(+)	(−)	(+)	(−)	(+)	(−)
Skull												
Brain												
Oral mucosa												
Salivary glands												
Cervical spine												
Thyroid												
Thymus										**0.01**		
Thoracic spine						**0.01**		**0.01**		**0.02**		**0.01**
Claviculae		**0.20**		**0.26**		**0.30**		**0.33**	0.01	**0.39**	0.01	**0.38**
Scapulae		**0.04**	0.01	**0.05**	0.01	**0.07**	0.02	**0.08**	0.02	**0.10**	0.02	**0.08**
Ribs	0.03	**0.21**	0.04	**0.27**	0.04	**0.30**	0.05	**0.33**	0.06	**0.39**	0.06	**0.37**
Oesophagus										**0.01**		
Trachea												
Lung		**0.04**		**0.06**		**0.07**		**0.08**		**0.09**		**0.08**
Mamma		**0.22**		**0.25**		**0.27**		**0.29**		**0.32**		**0.32**
Heart								**0.01**		**0.01**		**0.01**
Stomach												
Spleen												
Liver		**0.10**		**0.13**		**0.16**		**0.18**	0.01	**0.22**		**0.20**
Gall bladder		**0.01**		**0.01**		**0.02**		**0.02**		**0.03**		**0.02**
Pancreas												
Small intestine												
Large intestine								**0.01**		**0.01**		**0.01**
Sigma/rectum												
Lumbar spine												
Pelvis												
Adrenals										**0.01**		
Kidneys										**0.01**		
Urinary bladder												
Uterus												
Ovaries												
Prostate												
Testes												
Upper arm	0.30	**0.30**	0.38	**0.38**	0.44	**0.44**	0.49	**0.49**	0.57	**0.58**	0.55	**0.55**
Elbow	0.32	**0.33**	0.40	**0.41**	0.46	**0.48**	0.51	**0.53**	0.60	**0.63**	0.57	**0.60**
Forearm		**0.11**		**0.14**		**0.16**		**0.18**		**0.22**		**0.21**
Upper leg												
Knee												
Lower leg												
Lymph nodes						**0.01**		**0.01**		**0.02**		**0.01**
Skin	0.02	**0.05**	0.02	**0.06**	0.02	**0.06**	0.02	**0.06**	0.02	**0.07**	0.02	**0.06**
Muscle		**0.02**		**0.02**		**0.03**		**0.03**		**0.03**		**0.03**
Skeleton	0.03	**0.06**	0.04	**0.08**	0.05	**0.09**	0.05	**0.10**	0.06	**0.12**	0.06	**0.11**
Red bone marrow		**0.02**		**0.02**		**0.02**		**0.03**		**0.03**		**0.03**
Whole body dose		**0.03**	0.01	**0.04**	0.01	**0.04**	0.01	**0.04**	0.02	**0.05**	0.01	**0.05**
ED (ICRP 60)		**0.03**		**0.03**		**0.04**		**0.04**		**0.05**		**0.04**
ED (ICRP 103)		**0.04**		**0.05**		**0.05**		**0.06**		**0.07**		**0.06**
Total filtration (mm) **Al**	**1.5**		**2.5**		**3.5**		**4.5**		**3.5**		**3.5**	
Cu	**0**		**0**		**0**		**0**		**0.1**		**0.2**	

Table 22.28 Normalized organ doses—Upper arm ap—>17 years

Beam projection:			ap									**Table cells:**
Focus-to-detector distance (cm):			115									(+) = normal font = optimal field setting
Tube voltage (kVp):			70									**(−) = bold font = suboptimal field setting**
Total filtrations:			See last table row.									Empty table cells correspond to values <0.01
Organ or tissue	**Organ dose D_T : Entrance dose K_E**							**(mSv * mGy⁻¹)**				
	(+)	(−)	(+)	(−)	(+)	(−)	(+)	(−)	(+)	(−)	(+)	(−)
Skull												
Brain												
Oral mucosa												
Salivary glands												
Cervical spine										**0.01**		**0.01**
Thyroid												
Thymus						**0.01**		**0.01**		**0.02**		**0.02**
Thoracic spine		**0.01**		**0.01**		**0.02**		**0.02**		**0.03**		**0.03**
Claviculae		**0.24**		**0.30**	0.01	**0.34**	0.01	**0.37**	0.02	**0.43**	0.02	**0.48**
Scapulae	0.01	**0.06**	0.01	**0.07**	0.02	**0.09**	0.02	**0.10**	0.03	**0.13**	0.03	**0.16**
Ribs	0.04	**0.25**	0.05	**0.30**	0.05	**0.34**	0.06	**0.37**	0.07	**0.42**	0.08	**0.47**
Oesophagus						**0.01**		**0.01**		**0.02**		**0.02**
Trachea												
Lung		**0.06**		**0.07**		**0.08**		**0.09**		**0.11**	0.01	**0.13**
Mamma		**0.24**		**0.27**		**0.29**		**0.31**		**0.34**		**0.36**
Heart				**0.01**		**0.01**		**0.01**		**0.02**		**0.02**
Stomach												
Spleen												
Liver		**0.13**		**0.16**		**0.19**	0.01	**0.21**	0.01	**0.25**	0.02	**0.29**
Gall bladder		**0.02**		**0.02**		**0.02**		**0.03**		**0.04**		**0.04**
Pancreas												**0.01**
Small intestine												**0.01**
Large intestine				**0.01**		**0.01**		**0.01**		**0.02**		**0.02**
Sigma/rectum												
Lumbar spine										**0.01**		**0.01**
Pelvis												
Adrenals						**0.01**		**0.01**		**0.02**		**0.02**
Kidneys						**0.01**		**0.01**		**0.02**		**0.02**
Urinary bladder												
Uterus												
Ovaries												
Prostate												
Testes												
Upper arm	0.36	**0.36**	0.44	**0.44**	0.50	**0.51**	0.55	**0.56**	0.64	**0.66**	0.72	**0.74**
Elbow	0.37	**0.39**	0.46	**0.48**	0.52	**0.55**	0.57	**0.60**	0.66	**0.71**	0.74	**0.80**
Forearm		**0.13**		**0.17**		**0.19**		**0.21**		**0.24**		**0.28**
Upper leg												
Knee												
Lower leg												
Lymph nodes				**0.01**		**0.01**		**0.02**		**0.02**		**0.02**
Skin	0.02	**0.06**	0.02	**0.06**	0.02	**0.06**	0.02	**0.07**	0.03	**0.07**	0.03	**0.07**
Muscle		**0.02**		**0.03**		**0.03**		**0.03**		**0.04**		**0.04**
Skeleton	0.04	**0.07**	0.05	**0.09**	0.06	**0.10**	0.06	**0.11**	0.07	**0.13**	0.08	**0.15**
Red bone marrow		**0.02**		**0.03**		**0.03**		**0.03**	0.01	**0.04**	0.01	**0.04**
Whole body dose	0.01	**0.03**	0.01	**0.04**	0.01	**0.05**	0.01	**0.05**	0.02	**0.06**	0.02	**0.07**
ED (ICRP 60)		**0.03**		**0.04**		**0.04**		**0.05**		**0.05**		**0.06**
ED (ICRP 103)		**0.05**		**0.05**		**0.06**		**0.07**		**0.07**		**0.08**
Total filtration (mm) Al	**1.5**		**2.5**		**3.5**		**4.5**		**3.5**		**3.5**	
Cu	**0**		**0**		**0**		**0**		**0.1**		**0.2**	

Table 22.29 Normalized organ doses—Upper arm ap—>17 years

Beam projection:			ap									**Table cells:**
Focus-to-detector distance (cm):			115					(+) = normal font = optimal field setting				
Tube voltage (kVp):			80					**(−) = bold font = suboptimal field setting**				
Total filtrations:			See last table row.					Empty table cells correspond to values <0.01				
Organ or tissue	**Organ dose D_T : Entrance dose K_E**							**(mSv * mGy^{-1})**				
	(+)	(−)	(+)	(−)	(+)	(−)	(+)	(−)	(+)	(−)	(+)	(−)
Skull												
Brain												
Oral mucosa												
Salivary glands												
Cervical spine						**0.01**		**0.01**		**0.02**		**0.02**
Thyroid												**0.01**
Thymus				**0.01**		**0.01**		**0.02**		**0.02**		**0.02**
Thoracic spine		**0.01**		**0.02**		**0.02**		**0.03**		**0.03**		**0.04**
Claviculae		**0.27**	0.01	**0.33**	0.01	**0.37**	0.01	**0.40**	0.02	**0.46**	0.02	**0.51**
Scapulae	0.02	**0.07**	0.02	**0.09**	0.02	**0.11**	0.03	**0.12**	0.03	**0.15**	0.04	**0.18**
Ribs	0.04	**0.28**	0.05	**0.33**	0.06	**0.37**	0.07	**0.40**	0.08	**0.45**	0.09	**0.50**
Oesophagus		**0.01**		**0.01**		**0.02**		**0.02**		**0.02**		**0.03**
Trachea												
Lung		**0.07**		**0.08**		**0.10**		**0.11**	0.01	**0.13**	0.01	**0.14**
Mamma		**0.25**		**0.29**		**0.31**		**0.33**		**0.36**		**0.38**
Heart		**0.01**		**0.01**		**0.02**		**0.02**		**0.02**		**0.03**
Stomach												
Spleen												
Liver		**0.16**	0.01	**0.19**	0.01	**0.22**	0.01	**0.24**	0.02	**0.29**	0.02	**0.33**
Gall bladder		**0.02**		**0.03**		**0.03**		**0.04**		**0.05**		**0.05**
Pancreas										**0.01**		**0.01**
Small intestine										**0.01**		**0.01**
Large intestine		**0.01**		**0.01**		**0.02**		**0.02**		**0.02**		**0.03**
Sigma/rectum												
Lumbar spine						**0.01**		**0.01**		**0.02**		**0.02**
Pelvis												
Adrenals				**0.01**		**0.01**		**0.02**		**0.02**		**0.02**
Kidneys				**0.01**		**0.01**		**0.02**		**0.02**		**0.03**
Urinary bladder												
Uterus												
Ovaries												
Prostate												
Testes												
Upper arm	0.41	**0.41**	0.49	**0.50**	0.55	**0.56**	0.60	**0.62**	0.68	**0.71**	0.76	**0.79**
Elbow	0.42	**0.44**	0.51	**0.54**	0.57	**0.61**	0.62	**0.66**	0.70	**0.77**	0.78	**0.86**
Forearm		**0.15**		**0.19**		**0.21**		**0.23**		**0.26**		**0.29**
Upper leg												
Knee												
Lower leg												
Lymph nodes		**0.01**		**0.02**		**0.02**		**0.02**		**0.02**		**0.03**
Skin	0.02	**0.06**	0.02	**0.06**	0.02	**0.07**	0.02	**0.07**	0.03	**0.07**	0.03	**0.08**
Muscle		**0.03**		**0.03**		**0.04**		**0.04**		**0.04**		**0.05**
Skeleton	0.05	**0.08**	0.05	**0.10**	0.06	**0.11**	0.07	**0.12**	0.08	**0.14**	0.09	**0.16**
Red bone marrow		**0.02**		**0.03**		**0.03**	0.01	**0.04**	0.01	**0.04**	0.01	**0.05**
Whole body dose	0.01	**0.04**	0.01	**0.05**	0.01	**0.05**	0.02	**0.06**	0.02	**0.07**	0.02	**0.07**
ED (ICRP 60)		**0.03**		**0.04**		**0.05**		**0.05**		**0.06**		**0.07**
ED (ICRP 103)		**0.05**		**0.06**		**0.07**		**0.07**		**0.08**		**0.09**
Total filtration (mm) Al	**1.5**		**2.5**		**3.5**		**4.5**		**3.5**		**3.5**	
Cu	**0**		**0**		**0**		**0**		**0.1**		**0.2**	

Reference

1. Bontrager KL. Textbook of radiographic positioning and related anatomy. St. Louis: Mosby Year Book; 1993.

Upper Extremity 23

23.1 Introduction

The type of upper extremity radiography is determined by the clinical indication. Table 23.1 shows indications and Table 23.2 contraindications for upper extremity radiography. Image criteria as required for radiological findings in upper extremity radiographs are listed in Table 23.3. In order to fulfil these image criteria, the field settings listed in Tables 23.4 and 23.5 should be achieved. These field settings go along with the patient preparation as listed in Table 23.6. Exposure parameter combinations for various age groups and upper arm diameters are listed in Tables 23.7, 23.8, and 23.9. The age groups in Table 23.8 refer to the standard age groups of the mathematical MIRD phantom family which has been used to calculate the normalized organ doses in this book. Table 23.9 shows radiation qualities (i.e.

Table 23.1 Indications for upper extremity radiographs

Battered child syndrome
Congenital malformation of upper extremity
Skeletal dysplasia
Congenital vascular malformation
Soft tissue tumours
Bone tumours
Vanishing bone disease

Table 23.2 Contraindications for upper extremity radiographs

None

Table 23.3 Image criteria for upper extremity radiographs

Upper extremity ap
Visualization of the entire upper extremity
Reproduction of the glenohumeral joint, the elbow joint, the forearm, and the hand
Visualization of all bones of the upper extremity
Visually sharp reproduction of the cortex and spongiosa of the humerus, the radius and ulna, the carpal bones, and the small bones of the hand

Table 23.4 Patient positions and beam projections for upper extremity radiographs

Patient position	Beam projections
Supine	ap

Table 23.5 Optimal field settings for upper extremity radiographs

Upper extremity ap	
Upper field edge:	Tip of acromion (including the adjacent skin)
Field centring:	Midpoint of the upper extremity
Lower field edge:	Tips of the thumb and all fingers
Lateral field edges:	Skin surfaces of the arm
Annotation:	
Adjust angulation of the tube dependent on the position of the patient.	

M. Seidenbusch et al., *Imaging Practice and Radiation Protection in Pediatric Radiology*,
https://doi.org/10.1007/978-3-030-18504-6_23

various combinations of tube voltage and total filtration) which may be applied in clinical practice. Please note that not all of these radiation qualities are recommended by the authors. The standard projections of upper extremity radiographies of the patient who is correctly positioned are depicted in Fig. 23.1. Also the optimal field (marked in green), the suboptimal field (marked in red), and the correct shielding are shown. Tables 23.11, 23.12, and 23.13 outline the virtual field settings on the skin surface of mathematical MIRD phantoms of various age groups and the resulting virtual radiographies with and without bone structures, respectively. Normalized organ doses resulting from these virtual field settings have been calculated for all age groups and have been listed in Tables 23.14, 23.15, 23.16, 23.17, 23.18, 23.19, 23.20, 23.21, 23.22, 23.23, 23.24, 23.25, 23.26, 23.27, and 23.28. Table cells with normal font refer to optimal field settings and table cells with bold font refer to suboptimal

Table 23.6 Patient preparation

General	Remove all clothes, external metallic devices, and jewellery from the upper extremities of the patient.
	Immobilize infants and uncooperative patients using sandbags.
	Place eventually small X-ray positive markers at palpable tumours or on pain points.
Positioning	The upper extremity should be positioned slightly alongside the chest in the supine position. The arm should be extended in the elbow and wrist joint and all fingers and the thaws should be fully extended.
Shielding	Lead mask shielding should be used at the entrance side beneath the medial field edges.

Table 23.7 Radiographic procedure

X-ray source	Focal spot size between 0.6 and 1.3 dependent on the age of the patient
	Check the use of additional filtration
	Check the focus-to-detector distance
Field setting	Check the field size
	Check the field centring
Image detector system	Check the use of antiscatter grids
Patient	–

Table 23.8 Radiographic techniques of upper extremity radiographs

Age group (years)	Beam direction	Patient position	Beam projection	Typical patient diameter (cm)	Optimal field size (cm x cm)	Focus-to-detector distance (cm)	Voltage (kVp) 50	60	70	80	90	100	Antiscatter grid
0	vertical	supine	ap	-	2 x 23	100[(1)]							
1–2					3 x 33								
3–7					4 x 42	115							
8–12					4 x 52								
13–17					6 x 66								
18					6 x72								

Annotations:
(1) The patient is lying on the table top directly on the detector.
[dark grey] Parameter combination not recommended, normalized organ doses not provided
[light blue] Parameter combination recommended, normalized organ doses provided

Table 23.9 Radiation qualities of upper extremity radiographs

Voltage (kVp)	Total filtration: Without additional filtration 1.5 mm Al	2.5 Al	With additional filtration 3.5 Al	4.5 Al	3.5 Al + 0.1 Cu	3.5 Al + 0.2 Cu
50						
60						
70						
80						
90						
100						

Annotations:
[dark grey] Parameter combination not recommended, normalized organ doses not provided
[light grey] Parameter combination not recommended, normalized organ doses provided
[light blue] Parameter combination recommended, normalized organ doses provided

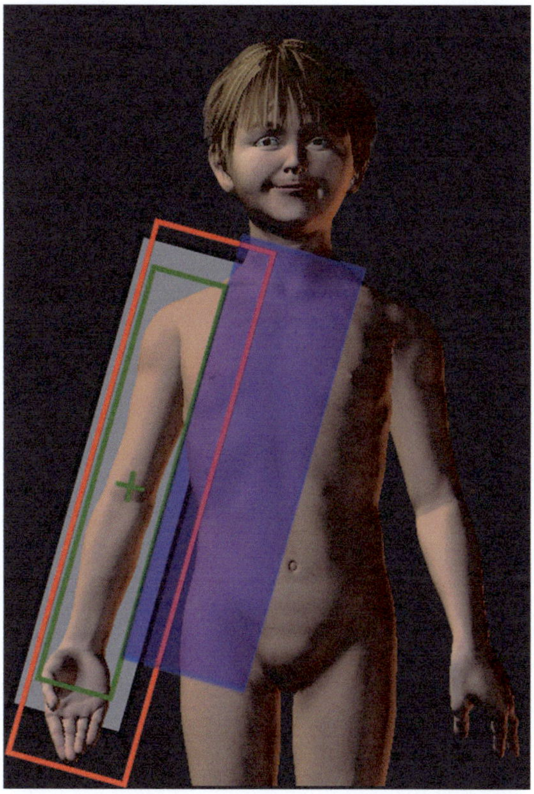

Fig. 23.1 Field settings for upper extremity radiographs in supine position. Antero-posterior projection (© V. Rösenberger & M. Seidenbusch 2018). Green lines = optimal field settings; green cross = centring point; red lines = suboptimal field settings; grey shaded area = image detector; blue shaded area = lead mask shielding; dark grey shaded area = Bucky table

Table 23.10 Inconsistencies between the mathematical MIRD phantoms and real human anatomy

Anatomical region	Inconsistencies
Extremities	Arms constructed too long in the ages 0, 1, and 5 years
	Position of the extremities cannot be changed

For details, see the text

field settings. Please note that normalized organ doses have been provided also for exposure parameter combinations, which are not recommended by the authors.

23.2 Indications and contraindications

Indications and contraindications for paediatric upper extremity radiographs are given in Tables 23.1 and 23.2.

23.3 Image Criteria

The image criteria required for upper extremity radiographs are listed in Table 23.3.

23.4 Field Settings

In Table 23.4, patient positions and beam projections used for upper extremity radiographs are listed, and in Table 23.5, the optimal field settings are listed which are required in order to fulfil the image criteria mentioned in Table 23.3.

23.5 Patient Preparation and Radiographic Procedure

See Tables 23.6, 23.7 and [1].

23.6 Exposure Parameters

In Tables 23.8 and 23.9, exposure parameter combinations used for upper extremity radiographs in various age groups are listed. Dark grey shaded table cells refer to exposure parameter combinations which are not recommended for paediatric upper extremity radiographs and for which no normalized organ doses have been provided. Light grey and light blue shaded table cells refer to exposure parameter combinations for which normalized organ doses have been provided with light grey shaded table cells referring to exposure parameter combinations being not recommended and light blue shaded table cells referring to exposure parameter combinations being recommended for paediatric upper extremity radiographs. Note that due to the fact that the radiation dose is an exponential function of the patient diameter, it may make sense to use the patient diameters instead of the patient's age groups for determining the adequate exposure parameters from Table 23.8. In this case, for determining the corresponding normalized organ doses from Tables 23.14, 23.15, 23.16, 23.17, 23.18, 23.19, 23.20, 23.21, 23.22, 23.23, 23.24, 23.25, 23.26, 23.27, and 23.28, the age group related to the patient diameter should be used.

23.7 Normalized Organ Doses

For calculating normalized organ doses, the field settings described in Sect. 23.4 were transferred to the mathematical MIRD phantoms. Inconsistencies between the geometrical anatomy of the mathematical phantoms and the real human anatomy have been listed in Table 23.10 whenever they are of relevance for the validity of the normalized organ doses.

In Tables 23.11, 23.12, and 23.13, optimal and suboptimal field settings are listed for the different age groups and the beam projections given in Tables 23.8. Green lines refer to optimal field settings, red lines refer to suboptimal field settings, respectively. In the tables, the optimal and suboptimal field settings which were used for performing the calculations of the normalized organ doses provided in Tables 23.14, 23.15, 23.16, 23.17, 23.18, 23.19, 23.20, 23.21, 23.22, 23.23, 23.24, 23.25, 23.26, 23.27, and 23.28 were drawn in on the skin surface of the mathematical phantoms. The virtual mathematical radiographs resulting from these field settings were depicted in the tables with and without bone structures, respectively, in order to enable the reader to visually compare the virtual mathematical radiographs with the real radiographs made under clinical conditions. For comparing the virtual mathematical field settings with the real clinical field settings, the virtual mathematical field edges were also listed in the tables.

As the normalized organ doses provided in Tables 23.14, 23.15, 23.16, 23.17, 23.18, 23.19, 23.20, 23.21, 23.22, 23.23, 23.24, 23.25, 23.26, 23.27, and 23.28 refer to entrance kerma values, optimal and suboptimal field sizes were provided in the tables in order to enable the reader to calculate entrance air kerma dose values from clinically measured dose area product values. Please note that these optimal and suboptimal field sizes refer to the mathematical phantoms of an age group only and not to real patients with corresponding ages.

Please note also that the anatomy of the mathematical phantoms has been idealized. Therefore, the virtual field settings for calculating the normalized organ doses have been defined with a view to ensuring that the resulting virtual organ doses would be consistent with the real organ doses which would be achieved during real X-ray examinations being performed with the real field settings recommended in Sect. 23.4. Thus, the field settings shown in the following tables may not be consistent with the clinical field settings as recommended in Sect. 23.4.

Table 23.11 Field settings—upper extremity ap

Age group (years)		Anatomical landmarks: Optimal field setting	Anatomical landmarks: Suboptimal field setting
0	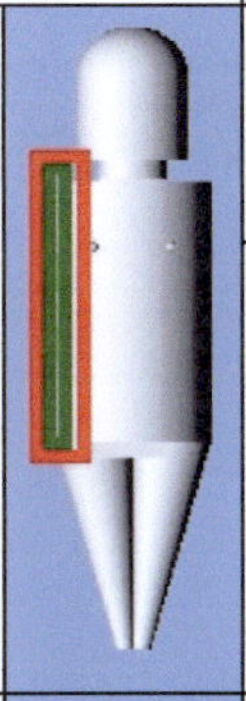	Upper field edge: Just above the skin surface of the shoulder Lower field edge: Just below the finger tips Medial field edge: Just close to the chest wall Lateral field edge: Just beneath the skin surface of the upper extremity Field size = field width × field height: 2 cm × 23 cm = 46 cm^2	Upper field edge: Just below the tip of the mandible Lower field edge: Just below the finger tips Medial field edge: Just beneath the mammillary line Lateral field edge: Above the skin surface of the upper arm Field size = field width × field height: 3 cm × 23 cm = 69 cm^2

1–2	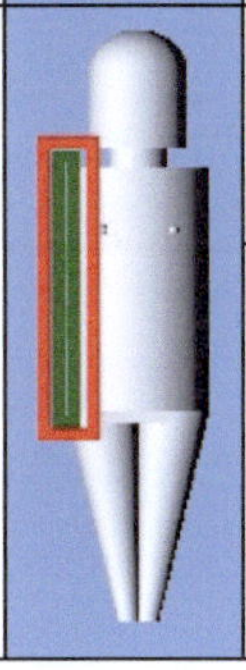	Upper field edge: Just above the skin surface of the shoulder Lower field edge: Just below the finger tips Medial field edge: Just close to the chest wall Lateral field edge: Just beneath the skin surface of the upper extremity Field size = field width × field height: 3 cm × 33 cm = 99 cm^2	Upper field edge: Just below the tip of the mandible Lower field edge: Just below the finger tips Medial field edge: Just beneath the mammillary line Lateral field edge: Above the skin surface of the upper arm Field size = field width × field height: 6 cm × 32 cm = 192 cm^2

Table 23.12 Field settings—upper extremity ap

Age group (years)		Anatomical landmarks: Optimal field setting	Anatomical landmarks: Suboptimal field setting
3–7	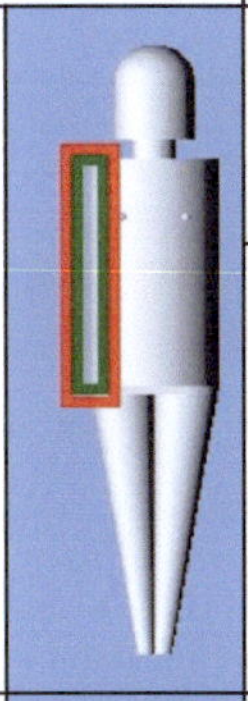	Upper field edge: Just above the skin surface of the shoulder Lower field edge: Just below the finger tips Medial field edge: Just close to the chest wall Lateral field edge: Just beneath the skin surface of the upper extremity Field size = field width × field height: 4 cm × 42 cm = 168 cm^2	Upper field edge: Just below the tip of the mandible Lower field edge: Just below the finger tips Medial field edge: Just beneath the mammillary line Lateral field edge: Above the skin surface of the upper arm Field size = field width × field height: 7 cm × 42 cm = 294 cm^2

8–12	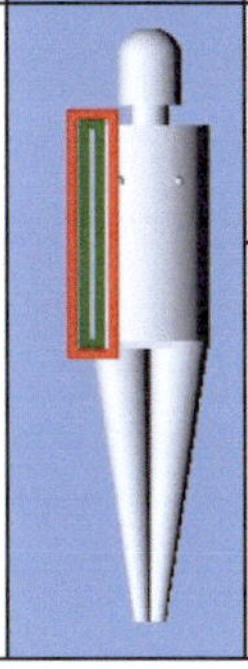	Upper field edge: Just above the skin surface of the shoulder Lower field edge: Just below the finger tips Medial field edge: Just close to the chest wall Lateral field edge: Just beneath the skin surface of the upper extremity Field size = field width × field height: 4 cm × 52 cm = 208 cm^2	Upper field edge: Just below the tip of the mandible Lower field edge: Just below the finger tips Medial field edge: Just beneath the mammillary line Lateral field edge: Above the skin surface of the upper arm Field size = field width × field height: 8 cm × 52 cm = 416 cm^2

Table 23.13 Field settings—upper extremity ap

Age group (years)		Anatomical landmarks			
		Optimal field setting		**Suboptimal field setting**	
13–17	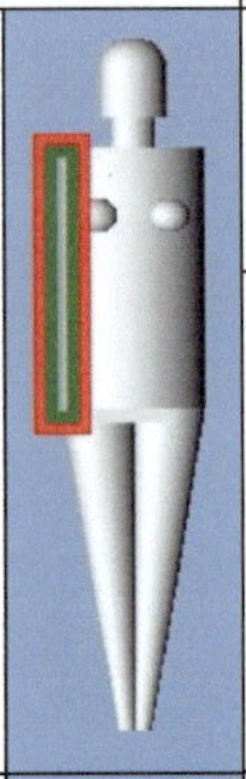	Upper field edge: Just above the skin surface of the shoulder Lower field edge: Just below the finger tips Medial field edge: Just close to the chest wall Lateral field edge: Just beneath the skin surface of the upper extremity Field size = field width × field height: 6 cm × 66 cm = 396 cm^2		Upper field edge: Just below the tip of the mandible Lower field edge: Just below the finger tips Medial field edge: Just beneath the mammillary line Lateral field edge: Above the skin surface of the upper arm Field size = field width × field height: 10 cm × 65 cm = 650 cm^2	

18	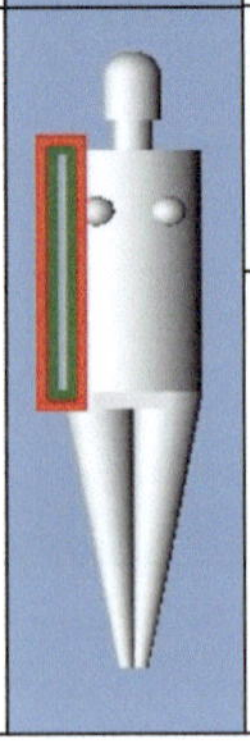	Upper field edge: Just above the skin surface of the shoulder Lower field edge: Just below the finger tips Medial field edge: Just close to the chest wall Lateral field edge: Just beneath the skin surface of the upper extremity Field size = field width × field height: 6 cm × 72 cm = 432 cm^2		Upper field edge: Just below the tip of the mandible Lower field edge: Just below the finger tips Medial field edge: Just beneath the mammillary line Lateral field edge: Above the skin surface of the upper arm Field size = field width × field height: 12 cm × 72 cm = 864 cm^2	

Table 23.14 Normalized organ doses—Upper extremities ap—<1 year

Beam projection:			ap									**Table cells:**
Focus-to-detector distance (cm):			100									(+) = normal font = optimal field setting
Tube voltage (kVp):			50									**(−) = bold font = suboptimal field setting**
Total filtrations:			See last table row.									Empty table cells correspond to values <0.01
Organ or tissue	**Organ dose D_T : Entrance dose K_E**							**(mSv * mGy⁻¹)**				
	(+)	(−)	(+)	(−)	(+)	(−)	(+)	(−)	(+)	(−)	(+)	(−)
Skull												
Brain												
Oral mucosa												
Salivary glands												
Cervical spine						**0.01**		**0.01**		**0.01**		**0.02**
Thyroid												**0.01**
Thymus										**0.01**		**0.01**
Thoracic spine		**0.02**		**0.02**		**0.03**		**0.03**	0.01	**0.04**	0.01	**0.05**
Claviculae		**0.21**	0.01	**0.25**	0.01	**0.28**	0.02	**0.30**	0.02	**0.34**	0.02	**0.37**
Scapulae	0.02	**0.13**	0.03	**0.18**	0.04	**0.21**	0.04	**0.24**	0.05	**0.29**	0.06	**0.34**
Ribs	0.06	**0.24**	0.07	**0.30**	0.08	**0.33**	0.09	**0.36**	0.10	**0.40**	0.12	**0.44**
Oesophagus				**0.01**		**0.01**		**0.02**		**0.02**		**0.02**
Trachea												
Lung		**0.04**		**0.05**		**0.06**	0.01	**0.06**	0.01	**0.07**	0.01	**0.08**
Mamma		**0.01**		**0.01**		**0.02**		**0.02**		**0.02**		**0.02**
Heart				**0.01**		**0.01**		**0.01**		**0.02**		**0.02**
Stomach												
Spleen												
Liver	0.01	**0.09**	0.01	**0.11**	0.02	**0.12**	0.02	**0.13**	0.02	**0.16**	0.03	**0.17**
Gall bladder				**0.01**		**0.01**		**0.02**		**0.02**		**0.03**
Pancreas								**0.01**		**0.01**		**0.02**
Small intestine		**0.01**		**0.02**		**0.02**		**0.02**		**0.03**		**0.03**
Large intestine		**0.02**		**0.03**		**0.04**	0.01	**0.04**	0.01	**0.05**	0.02	**0.06**
Sigma/rectum												
Lumbar spine		**0.01**		**0.02**		**0.02**		**0.03**	0.01	**0.03**	0.01	**0.04**
Pelvis	0.01	**0.04**	0.02	**0.06**	0.02	**0.07**	0.02	**0.07**	0.03	**0.09**	0.03	**0.11**
Adrenals				**0.01**		**0.01**		**0.01**		**0.02**		**0.02**
Kidneys				**0.01**		**0.01**		**0.01**		**0.02**		**0.02**
Urinary bladder						**0.01**		**0.01**		**0.02**		**0.02**
Uterus				**0.01**		**0.01**		**0.01**		**0.02**		**0.02**
Ovaries		**0.01**		**0.01**		**0.02**		**0.02**		**0.03**		**0.03**
Prostate								**0.01**		**0.01**		**0.02**
Testes												
Upper arm	0.49	**0.51**	0.60	**0.63**	0.67	**0.70**	0.73	**0.76**	0.82	**0.87**	0.90	**0.96**
Elbow	0.54	**0.56**	0.65	**0.68**	0.73	**0.76**	0.79	**0.83**	0.89	**0.94**	0.97	**1.03**
Forearm	0.59	**0.62**	0.71	**0.75**	0.79	**0.83**	0.85	**0.90**	0.95	**1.01**	1.03	**1.10**
Upper leg		**0.03**		**0.04**		**0.04**	0.01	**0.05**	0.01	**0.06**	0.02	**0.07**
Knee												
Lower leg												
Lymph nodes		**0.01**		**0.01**		**0.02**		**0.02**		**0.02**		**0.03**
Skin	0.06	**0.09**	0.06	**0.10**	0.07	**0.10**	0.07	**0.10**	0.07	**0.11**	0.07	**0.11**
Muscle	0.02	**0.05**	0.02	**0.06**	0.02	**0.06**	0.02	**0.07**	0.03	**0.07**	0.03	**0.08**
Skeleton	0.11	**0.15**	0.13	**0.18**	0.15	**0.21**	0.16	**0.22**	0.18	**0.26**	0.20	**0.28**
Red bone marrow	0.02	**0.03**	0.03	**0.04**	0.03	**0.05**	0.04	**0.05**	0.04	**0.06**	0.05	**0.07**
Whole body dose	0.02	**0.05**	0.03	**0.06**	0.03	**0.07**	0.03	**0.07**	0.04	**0.08**	0.04	**0.09**
ED (ICRP 60)		**0.02**		**0.03**	0.01	**0.03**	0.01	**0.03**	0.01	**0.04**	0.01	**0.04**
ED (ICRP 103)		**0.02**		**0.03**		**0.03**	0.01	**0.03**	0.01	**0.04**	0.01	**0.04**
Total filtration (mm) Al	**1.5**		**2.5**		**3.5**		**4.5**		**3.5**		**3.5**	
Cu	**0**		**0**		**0**		**0**		**0.1**		**0.2**	

Table 23.15 Normalized organ doses—Upper extremities ap—<1 year

Beam projection:			ap									**Table cells:**
Focus-to-detector distance (cm):			100				(+) = normal font = optimal field setting					
Tube voltage (kVp):			60				**(−) = bold font = suboptimal field setting**					
Total filtrations:			See last table row.				Empty table cells correspond to values <0.01					
Organ or tissue	**Organ dose D_T : Entrance dose K_E**								**(mSv * mGy^{-1})**			
	(+)	(−)	(+)	(−)	(+)	(−)	(+)	(−)	(+)	(−)	(+)	(−)
Skull												
Brain												
Oral mucosa												
Salivary glands												
Cervical spine				**0.01**		**0.01**		**0.02**		**0.02**		**0.02**
Thyroid										**0.01**		**0.01**
Thymus						**0.01**		**0.01**		**0.02**		**0.01**
Thoracic spine		**0.02**		**0.03**	0.01	**0.04**	0.01	**0.04**	0.02	**0.05**	0.01	**0.05**
Claviculae	0.01	**0.23**	0.02	**0.28**	0.02	**0.31**	0.02	**0.33**	0.02	**0.36**	0.02	**0.37**
Scapulae	0.03	**0.17**	0.04	**0.22**	0.05	**0.26**	0.05	**0.29**	0.06	**0.35**	0.06	**0.34**
Ribs	0.07	**0.28**	0.08	**0.33**	0.09	**0.37**	0.10	**0.39**	0.12	**0.44**	0.12	**0.44**
Oesophagus		**0.01**		**0.01**		**0.02**		**0.02**		**0.02**		**0.02**
Trachea												
Lung		**0.05**	0.01	**0.06**	0.01	**0.07**	0.01	**0.07**	0.02	**0.08**	0.01	**0.08**
Mamma		**0.01**		**0.02**		**0.02**		**0.02**		**0.02**		**0.02**
Heart		**0.01**		**0.01**		**0.02**		**0.02**		**0.02**		**0.02**
Stomach										**0.01**		
Spleen												
Liver	0.01	**0.10**	0.02	**0.13**	0.02	**0.14**	0.02	**0.16**	0.03	**0.18**	0.03	**0.17**
Gall bladder		**0.01**		**0.02**		**0.02**		**0.02**		**0.03**		**0.03**
Pancreas				**0.01**		**0.01**		**0.01**		**0.02**		**0.02**
Small intestine		**0.02**		**0.02**		**0.02**		**0.03**		**0.03**		**0.03**
Large intestine		**0.03**	0.01	**0.04**	0.01	**0.05**	0.01	**0.05**	0.02	**0.06**	0.02	**0.06**
Sigma/rectum												
Lumbar spine		**0.02**		**0.03**	0.01	**0.03**	0.01	**0.04**	0.01	**0.05**	0.01	**0.04**
Pelvis	0.02	**0.06**	0.02	**0.07**	0.02	**0.08**	0.03	**0.09**	0.03	**0.11**	0.03	**0.11**
Adrenals		**0.01**		**0.01**		**0.02**		**0.02**		**0.02**		**0.02**
Kidneys		**0.01**		**0.01**		**0.02**		**0.02**		**0.02**		**0.02**
Urinary bladder				**0.01**		**0.01**		**0.02**		**0.02**		**0.02**
Uterus		**0.01**		**0.01**		**0.02**		**0.02**		**0.02**		**0.02**
Ovaries		**0.02**		**0.02**		**0.02**		**0.03**		**0.03**		**0.03**
Prostate				**0.01**		**0.01**		**0.01**		**0.02**		**0.02**
Testes										**0.01**		
Upper arm	0.57	**0.59**	0.68	**0.71**	0.75	**0.79**	0.80	**0.85**	0.90	**0.95**	0.90	**0.96**
Elbow	0.61	**0.64**	0.73	**0.77**	0.81	**0.85**	0.86	**0.92**	0.96	**1.02**	0.97	**1.03**
Forearm	0.67	**0.70**	0.79	**0.83**	0.87	**0.92**	0.93	**0.98**	1.02	**1.09**	1.03	**1.10**
Upper leg		**0.04**	0.01	**0.05**	0.01	**0.05**	0.01	**0.06**	0.02	**0.07**	0.02	**0.07**
Knee												
Lower leg												
Lymph nodes		**0.01**		**0.02**		**0.02**		**0.02**		**0.03**		**0.03**
Skin	0.06	**0.10**	0.07	**0.10**	0.07	**0.11**	0.07	**0.11**	0.07	**0.12**	0.07	**0.11**
Muscle	0.02	**0.05**	0.02	**0.06**	0.02	**0.07**	0.03	**0.07**	0.03	**0.08**	0.03	**0.08**
Skeleton	0.12	**0.17**	0.15	**0.21**	0.16	**0.23**	0.18	**0.25**	0.20	**0.28**	0.20	**0.28**
Red bone marrow	0.03	**0.04**	0.03	**0.05**	0.04	**0.06**	0.04	**0.06**	0.05	**0.07**	0.05	**0.07**
Whole body dose	0.03	**0.06**	0.03	**0.07**	0.03	**0.08**	0.04	**0.08**	0.04	**0.09**	0.04	**0.09**
ED (ICRP 60)		**0.03**	0.01	**0.03**	0.01	**0.04**	0.01	**0.04**	0.02	**0.05**	0.01	**0.04**
ED (ICRP 103)		**0.03**	0.01	**0.03**	0.01	**0.04**	0.01	**0.04**	0.02	**0.04**	0.01	**0.04**
Total filtration (mm) Al	**1.5**		**2.5**		**3.5**		**4.5**		**3.5**		**3.5**	
Cu	**0**		**0**		**0**		**0**		**0.1**		**0.2**	

Table 23.16 Normalized organ doses—Upper extremities ap—<1 year

Beam projection:			ap									**Table cells:**
Focus-to-detector distance (cm):			100						(+) = normal font = optimal field setting			
Tube voltage (kVp):			70						**(−) = bold font = suboptimal field setting**			
Total filtrations:			See last table row.						Empty table cells correspond to values <0.01			
Organ or tissue	**Organ dose D_T : Entrance dose K_E**								**(mSv * mGy^{-1})**			
	(+)	(−)	(+)	(−)	(+)	(−)	(+)	(−)	(+)	(−)	(+)	(−)
Skull												
Brain												
Oral mucosa												
Salivary glands												**0.01**
Cervical spine		**0.01**		**0.01**		**0.02**		**0.02**		**0.02**		**0.03**
Thyroid								**0.01**		**0.01**		**0.02**
Thymus				**0.01**		**0.01**		**0.02**		**0.02**		**0.02**
Thoracic spine		**0.03**	0.01	**0.04**	0.01	**0.04**	0.02	**0.05**	0.02	**0.06**	0.02	**0.07**
Claviculae	0.01	**0.25**	0.02	**0.30**	0.02	**0.32**	0.02	**0.34**	0.02	**0.37**	0.03	**0.39**
Scapulae	0.04	**0.21**	0.05	**0.26**	0.05	**0.30**	0.06	**0.33**	0.07	**0.39**	0.08	**0.44**
Ribs	0.08	**0.31**	0.09	**0.36**	0.10	**0.39**	0.11	**0.41**	0.13	**0.45**	0.14	**0.48**
Oesophagus		**0.01**		**0.02**		**0.02**		**0.02**		**0.03**		**0.03**
Trachea												
Lung		**0.06**	0.01	**0.07**	0.01	**0.07**	0.01	**0.08**	0.02	**0.09**	0.02	**0.10**
Mamma		**0.02**		**0.02**		**0.02**		**0.02**		**0.02**	0.01	**0.03**
Heart		**0.01**		**0.02**		**0.02**		**0.02**		**0.03**		**0.03**
Stomach								**0.01**		**0.01**		**0.01**
Spleen												**0.01**
Liver	0.02	**0.12**	0.02	**0.14**	0.02	**0.16**	0.03	**0.17**	0.03	**0.19**	0.03	**0.21**
Gall bladder		**0.02**		**0.02**		**0.02**		**0.03**		**0.03**	0.01	**0.04**
Pancreas		**0.01**		**0.01**		**0.02**		**0.02**		**0.02**		**0.03**
Small intestine		**0.02**		**0.03**		**0.03**		**0.03**	0.01	**0.04**	0.01	**0.04**
Large intestine		**0.04**	0.01	**0.05**	0.01	**0.05**	0.02	**0.06**	0.02	**0.07**	0.02	**0.08**
Sigma/rectum										**0.01**		**0.01**
Lumbar spine		**0.03**	0.01	**0.03**	0.01	**0.04**	0.01	**0.05**	0.02	**0.06**	0.02	**0.07**
Pelvis	0.02	**0.07**	0.02	**0.08**	0.03	**0.09**	0.03	**0.10**	0.03	**0.12**	0.04	**0.14**
Adrenals		**0.01**		**0.02**		**0.02**		**0.02**		**0.03**		**0.03**
Kidneys		**0.01**		**0.02**		**0.02**		**0.02**		**0.03**	0.01	**0.03**
Urinary bladder		**0.01**		**0.01**		**0.02**		**0.02**		**0.02**		**0.03**
Uterus		**0.01**		**0.02**		**0.02**		**0.02**		**0.03**		**0.03**
Ovaries		**0.02**		**0.02**		**0.03**		**0.03**	0.01	**0.04**	0.01	**0.05**
Prostate				**0.01**		**0.01**		**0.02**		**0.02**		**0.02**
Testes								**0.01**		**0.01**		**0.01**
Upper arm	0.62	**0.65**	0.73	**0.77**	0.80	**0.84**	0.85	**0.90**	0.93	**0.99**	0.99	**1.06**
Elbow	0.67	**0.70**	0.78	**0.83**	0.85	**0.91**	0.90	**0.97**	0.99	**1.06**	1.05	**1.13**
Forearm	0.72	**0.76**	0.84	**0.89**	0.91	**0.97**	0.96	**1.03**	1.04	**1.12**	1.10	**1.19**
Upper leg		**0.04**	0.01	**0.05**	0.01	**0.06**	0.02	**0.07**	0.02	**0.08**	0.02	**0.09**
Knee										**0.01**		**0.01**
Lower leg												
Lymph nodes		**0.02**		**0.02**		**0.03**		**0.03**		**0.03**	0.01	**0.04**
Skin	0.06	**0.10**	0.07	**0.11**	0.07	**0.11**	0.07	**0.11**	0.08	**0.12**	0.08	**0.12**
Muscle	0.02	**0.06**	0.02	**0.07**	0.03	**0.07**	0.03	**0.08**	0.03	**0.09**	0.03	**0.09**
Skeleton	0.14	**0.19**	0.16	**0.23**	0.17	**0.25**	0.19	**0.27**	0.20	**0.30**	0.22	**0.32**
Red bone marrow	0.03	**0.05**	0.04	**0.06**	0.04	**0.06**	0.05	**0.07**	0.05	**0.08**	0.06	**0.08**
Whole body dose	0.03	**0.07**	0.03	**0.08**	0.04	**0.08**	0.04	**0.09**	0.04	**0.10**	0.05	**0.11**
ED (ICRP 60)		**0.03**	0.01	**0.04**	0.01	**0.04**	0.01	**0.04**	0.02	**0.05**	0.02	**0.06**
ED (ICRP 103)		**0.03**	0.01	**0.03**	0.01	**0.04**	0.01	**0.04**	0.02	**0.05**	0.02	**0.05**
Total filtration (mm) Al	**1.5**		**2.5**		**3.5**		**4.5**		**3.5**		**3.5**	
Cu	**0**		**0**		**0**		**0**		**0.1**		**0.2**	

Table 23.17 Normalized organ doses—Upper extremities ap—1–2 years

Beam projection:			ap									**Table cells:**
Focus-to-detector distance (cm):			100						(+) = normal font = optimal field setting			
Tube voltage (kVp):			60						**(−) = bold font = suboptimal field setting**			
Total filtrations:			See last table row.						Empty table cells correspond to values <0.01			
Organ or tissue	**Organ dose D_T : Entrance dose K_E**								**(mSv * mGy^{-1})**			
	(+)	(−)	(+)	(−)	(+)	(−)	(+)	(−)	(+)	(−)	(+)	(−)
Skull												
Brain												
Oral mucosa												
Salivary glands												
Cervical spine								**0.01**		**0.01**		**0.01**
Thyroid												
Thymus						**0.01**		**0.01**		**0.01**		**0.01**
Thoracic spine		**0.02**		**0.02**		**0.03**		**0.03**	0.01	**0.04**	0.01	**0.04**
Claviculae	0.01	**0.19**	0.01	**0.23**	0.02	**0.26**	0.02	**0.28**	0.02	**0.32**	0.02	**0.32**
Scapulae	0.01	**0.09**	0.02	**0.12**	0.02	**0.14**	0.03	**0.16**	0.03	**0.20**	0.03	**0.18**
Ribs	0.05	**0.24**	0.06	**0.30**	0.07	**0.33**	0.08	**0.36**	0.09	**0.41**	0.09	**0.41**
Oesophagus		**0.01**		**0.01**		**0.02**		**0.02**		**0.02**		**0.02**
Trachea												
Lung		**0.04**		**0.05**	0.01	**0.06**	0.01	**0.07**	0.01	**0.08**	0.01	**0.07**
Mamma		**0.01**		**0.02**		**0.02**		**0.02**		**0.03**		**0.02**
Heart				**0.01**		**0.01**		**0.02**		**0.02**		**0.02**
Stomach												
Spleen												
Liver	0.01	**0.09**	0.02	**0.11**	0.02	**0.13**	0.02	**0.14**	0.03	**0.17**	0.02	**0.16**
Gall bladder		**0.01**		**0.02**		**0.02**		**0.02**		**0.03**		**0.03**
Pancreas						**0.01**		**0.01**		**0.02**		**0.01**
Small intestine		**0.01**		**0.02**		**0.02**		**0.02**		**0.03**		**0.03**
Large intestine		**0.03**		**0.04**		**0.04**	0.01	**0.05**	0.01	**0.06**	0.01	**0.05**
Sigma/rectum												
Lumbar spine		**0.01**		**0.02**		**0.02**		**0.03**	0.01	**0.04**		**0.03**
Pelvis	0.01	**0.05**	0.02	**0.06**	0.02	**0.07**	0.02	**0.08**	0.03	**0.10**	0.03	**0.10**
Adrenals				**0.01**		**0.01**		**0.01**		**0.02**		**0.02**
Kidneys				**0.01**		**0.01**		**0.02**		**0.02**		**0.02**
Urinary bladder				**0.01**		**0.01**		**0.01**		**0.02**		**0.02**
Uterus				**0.01**		**0.01**		**0.02**		**0.02**		**0.02**
Ovaries		**0.01**		**0.02**		**0.02**		**0.02**		**0.03**		**0.03**
Prostate								**0.01**		**0.01**		**0.01**
Testes												
Upper arm	0.43	**0.46**	0.52	**0.57**	0.59	**0.64**	0.64	**0.70**	0.74	**0.81**	0.73	**0.80**
Elbow	0.48	**0.51**	0.58	**0.63**	0.65	**0.71**	0.71	**0.78**	0.81	**0.89**	0.80	**0.88**
Forearm	0.53	**0.58**	0.64	**0.70**	0.72	**0.79**	0.78	**0.86**	0.88	**0.98**	0.88	**0.97**
Upper leg		**0.03**		**0.03**	0.01	**0.04**	0.01	**0.05**	0.01	**0.06**	0.01	**0.05**
Knee												
Lower leg												
Lymph nodes		**0.01**		**0.02**		**0.02**		**0.02**		**0.03**		**0.02**
Skin	0.05	**0.09**	0.06	**0.09**	0.06	**0.09**	0.06	**0.10**	0.06	**0.10**	0.06	**0.10**
Muscle	0.01	**0.04**	0.02	**0.05**	0.02	**0.05**	0.02	**0.06**	0.02	**0.07**	0.02	**0.06**
Skeleton	0.09	**0.13**	0.11	**0.16**	0.12	**0.18**	0.13	**0.20**	0.15	**0.23**	0.15	**0.22**
Red bone marrow	0.02	**0.03**	0.02	**0.03**	0.02	**0.04**	0.03	**0.04**	0.03	**0.05**	0.03	**0.05**
Whole body dose	0.02	**0.05**	0.03	**0.06**	0.03	**0.07**	0.03	**0.07**	0.04	**0.08**	0.03	**0.08**
ED (ICRP 60)		**0.02**		**0.03**		**0.03**	0.01	**0.03**	0.01	**0.04**	0.01	**0.04**
ED (ICRP 103)		**0.02**		**0.03**		**0.03**		**0.03**	0.01	**0.04**	0.01	**0.04**
Total filtration (mm) Al	**1.5**		**2.5**		**3.5**		**4.5**		**3.5**		**3.5**	
Cu	**0**		**0**		**0**		**0**		**0.1**		**0.2**	

Table 23.18 Normalized organ doses—Upper extremities ap—1–2 years

Beam projection:			ap									**Table cells:**
Focus-to-detector distance (cm):			100									(+) = normal font = optimal field setting
Tube voltage (kVp):			70									**(−) = bold font = suboptimal field setting**
Total filtrations:			See last table row.									Empty table cells correspond to values <0.01
Organ or tissue	**Organ dose D_T : Entrance dose K_E**								**(mSv * mGy^{-1})**			
	(+)	(−)	(+)	(−)	(+)	(−)	(+)	(−)	(+)	(−)	(+)	(−)
Skull												
Brain												
Oral mucosa												
Salivary glands												
Cervical spine				**0.01**		**0.01**		**0.01**		**0.02**		**0.02**
Thyroid										**0.01**		**0.01**
Thymus				**0.01**		**0.01**		**0.01**		**0.02**		**0.02**
Thoracic spine		**0.02**		**0.03**	0.01	**0.04**	0.01	**0.04**	0.02	**0.05**	0.02	**0.06**
Claviculae	0.01	**0.21**	0.02	**0.25**	0.02	**0.28**	0.02	**0.30**	0.02	**0.34**	0.03	**0.37**
Scapulae	0.02	**0.11**	0.02	**0.14**	0.03	**0.17**	0.03	**0.19**	0.04	**0.23**	0.05	**0.27**
Ribs	0.06	**0.27**	0.07	**0.33**	0.08	**0.36**	0.09	**0.39**	0.10	**0.43**	0.11	**0.47**
Oesophagus		**0.01**		**0.02**		**0.02**		**0.02**		**0.03**	0.01	**0.04**
Trachea												
Lung		**0.05**	0.01	**0.06**	0.01	**0.07**	0.01	**0.07**	0.02	**0.09**	0.02	**0.10**
Mamma		**0.02**		**0.02**		**0.02**		**0.02**		**0.03**		**0.03**
Heart		**0.01**		**0.02**		**0.02**		**0.02**		**0.03**		**0.03**
Stomach												**0.01**
Spleen												
Liver	0.02	**0.11**	0.02	**0.13**	0.02	**0.15**	0.03	**0.16**	0.03	**0.19**	0.04	**0.21**
Gall bladder		**0.02**		**0.02**		**0.03**		**0.03**		**0.04**	0.01	**0.04**
Pancreas				**0.01**		**0.01**		**0.02**		**0.02**		**0.03**
Small intestine		**0.02**		**0.02**		**0.03**		**0.03**		**0.04**	0.01	**0.04**
Large intestine		**0.04**	0.01	**0.04**	0.01	**0.05**	0.01	**0.06**	0.02	**0.07**	0.02	**0.08**
Sigma/rectum												**0.01**
Lumbar spine		**0.02**		**0.03**		**0.03**	0.01	**0.04**	0.01	**0.05**	0.02	**0.06**
Pelvis	0.02	**0.06**	0.02	**0.08**	0.02	**0.09**	0.03	**0.10**	0.03	**0.12**	0.04	**0.14**
Adrenals		**0.01**		**0.01**		**0.02**		**0.02**		**0.02**		**0.03**
Kidneys		**0.01**		**0.02**		**0.02**		**0.02**		**0.03**		**0.03**
Urinary bladder		**0.01**		**0.01**		**0.02**		**0.02**		**0.02**		**0.03**
Uterus		**0.01**		**0.01**		**0.02**		**0.02**		**0.02**		**0.03**
Ovaries		**0.02**		**0.02**		**0.03**		**0.03**	0.01	**0.03**	0.01	**0.04**
Prostate						**0.01**		**0.01**		**0.02**		**0.02**
Testes												**0.01**
Upper arm	0.48	**0.52**	0.58	**0.63**	0.64	**0.71**	0.69	**0.77**	0.78	**0.87**	0.85	**0.95**
Elbow	0.53	**0.58**	0.64	**0.70**	0.71	**0.78**	0.76	**0.84**	0.86	**0.95**	0.93	**1.04**
Forearm	0.59	**0.64**	0.70	**0.77**	0.78	**0.86**	0.83	**0.92**	0.93	**1.03**	1.00	**1.12**
Upper leg		**0.03**	0.01	**0.04**	0.01	**0.05**	0.01	**0.06**	0.02	**0.07**	0.02	**0.08**
Knee												
Lower leg												
Lymph nodes		**0.02**		**0.02**		**0.02**		**0.03**		**0.03**		**0.04**
Skin	0.06	**0.09**	0.06	**0.09**	0.06	**0.10**	0.06	**0.10**	0.07	**0.11**	0.07	**0.11**
Muscle	0.02	**0.05**	0.02	**0.05**	0.02	**0.06**	0.02	**0.06**	0.02	**0.07**	0.03	**0.08**
Skeleton	0.10	**0.15**	0.12	**0.18**	0.13	**0.20**	0.14	**0.22**	0.16	**0.24**	0.18	**0.27**
Red bone marrow	0.02	**0.03**	0.02	**0.04**	0.03	**0.05**	0.03	**0.05**	0.03	**0.06**	0.04	**0.07**
Whole body dose	0.02	**0.06**	0.03	**0.07**	0.03	**0.07**	0.03	**0.08**	0.04	**0.09**	0.04	**0.10**
ED (ICRP 60)		**0.03**		**0.03**	0.01	**0.04**	0.01	**0.04**	0.01	**0.05**	0.02	**0.05**
ED (ICRP 103)		**0.02**		**0.03**	0.01	**0.03**	0.01	**0.04**	0.01	**0.04**	0.02	**0.05**
Total filtration (mm) **Al**	**1.5**		**2.5**		**3.5**		**4.5**		**3.5**		**3.5**	
Cu	**0**		**0**		**0**		**0**		**0.1**		**0.2**	

Table 23.19 Normalized organ doses—Upper extremities ap—3–7 years

Beam projection:			ap									**Table cells:**
Focus-to-detector distance (cm):			115									(+) = normal font = optimal field setting
Tube voltage (kVp):			60									**(−) = bold font = suboptimal field setting**
Total filtrations:			See last table row.									Empty table cells correspond to values <0.01
Organ or tissue	**Organ dose D_T : Entrance dose K_E**								**(mSv * mGy^{-1})**			
	(+)	(−)	(+)	(−)	(+)	(−)	(+)	(−)	(+)	(−)	(+)	(−)
Skull												
Brain												
Oral mucosa												
Salivary glands												
Cervical spine										**0.01**		
Thyroid												
Thymus										**0.01**		
Thoracic spine		**0.01**		**0.02**		**0.02**		**0.02**	0.01	**0.03**		**0.02**
Claviculae	0.01	**0.12**	0.02	**0.15**	0.02	**0.17**	0.02	**0.19**	0.03	**0.22**	0.03	**0.22**
Scapulae	0.02	**0.06**	0.03	**0.08**	0.04	**0.10**	0.04	**0.11**	0.05	**0.14**	0.05	**0.12**
Ribs	0.08	**0.19**	0.10	**0.24**	0.12	**0.27**	0.13	**0.29**	0.15	**0.34**	0.14	**0.33**
Oesophagus								**0.01**		**0.01**		**0.01**
Trachea												
Lung		**0.03**	0.01	**0.04**	0.01	**0.04**	0.02	**0.05**	0.02	**0.06**	0.02	**0.05**
Mamma				**0.01**		**0.01**		**0.02**		**0.02**		**0.02**
Heart						**0.01**		**0.01**		**0.02**		**0.01**
Stomach												
Spleen												
Liver	0.02	**0.07**	0.03	**0.09**	0.03	**0.10**	0.03	**0.11**	0.04	**0.14**	0.04	**0.13**
Gall bladder						**0.01**		**0.01**		**0.02**		**0.01**
Pancreas										**0.01**		
Small intestine				**0.01**		**0.01**		**0.02**		**0.02**		**0.02**
Large intestine		**0.02**		**0.03**	0.01	**0.03**	0.01	**0.04**	0.02	**0.05**	0.01	**0.04**
Sigma/rectum												
Lumbar spine				**0.01**		**0.01**		**0.02**		**0.02**		**0.02**
Pelvis	0.01	**0.04**	0.02	**0.05**	0.02	**0.06**	0.03	**0.06**	0.03	**0.08**	0.03	**0.07**
Adrenals										**0.01**		**0.01**
Kidneys								**0.01**		**0.01**		**0.01**
Urinary bladder										**0.01**		
Uterus										**0.01**		
Ovaries				**0.01**		**0.01**		**0.02**		**0.02**		**0.02**
Prostate												
Testes												
Upper arm	0.39	**0.41**	0.48	**0.51**	0.55	**0.58**	0.60	**0.63**	0.70	**0.74**	0.68	**0.72**
Elbow	0.43	**0.46**	0.54	**0.57**	0.61	**0.64**	0.67	**0.71**	0.77	**0.82**	0.76	**0.80**
Forearm	0.49	**0.52**	0.61	**0.64**	0.69	**0.73**	0.75	**0.79**	0.86	**0.91**	0.85	**0.90**
Upper leg		**0.01**		**0.02**		**0.02**		**0.02**	0.01	**0.03**		**0.02**
Knee												
Lower leg												
Lymph nodes				**0.01**		**0.01**		**0.01**		**0.02**		**0.02**
Skin	0.05	**0.07**	0.06	**0.08**	0.06	**0.08**	0.06	**0.08**	0.06	**0.08**	0.06	**0.08**
Muscle	0.02	**0.03**	0.02	**0.04**	0.02	**0.04**	0.02	**0.05**	0.03	**0.05**	0.03	**0.05**
Skeleton	0.07	**0.09**	0.09	**0.11**	0.10	**0.13**	0.11	**0.14**	0.13	**0.16**	0.12	**0.16**
Red bone marrow	0.01	**0.02**	0.02	**0.03**	0.02	**0.03**	0.02	**0.03**	0.03	**0.04**	0.03	**0.04**
Whole body dose	0.02	**0.04**	0.03	**0.05**	0.03	**0.05**	0.03	**0.06**	0.04	**0.07**	0.04	**0.06**
ED (ICRP 60)		**0.02**		**0.02**		**0.02**	0.01	**0.03**	0.01	**0.03**	0.01	**0.03**
ED (ICRP 103)		**0.01**		**0.02**		**0.02**	0.01	**0.02**	0.01	**0.03**	0.01	**0.03**
Total filtration (mm) **Al**	**1.5**		**2.5**		**3.5**		**4.5**		**3.5**		**3.5**	
Cu	**0**		**0**		**0**		**0**		**0.1**		**0.2**	

Table 23.20 Normalized organ doses—Upper extremities ap—3–7 years

Beam projection:			ap									**Table cells:**
Focus-to-detector distance (cm):			115									(+) = normal font = optimal field setting
Tube voltage (kVp):			70									**(−) = bold font = suboptimal field setting**
Total filtrations:			See last table row.									Empty table cells correspond to values <0.01
Organ or tissue	**Organ dose D_T : Entrance dose K_E**								**(mSv * mGy⁻¹)**			
	(+)	(−)	(+)	(−)	(+)	(−)	(+)	(−)	(+)	(−)	(+)	(−)
Skull												
Brain												
Oral mucosa												
Salivary glands												
Cervical spine								**0.01**		**0.01**		**0.02**
Thyroid												**0.01**
Thymus						**0.01**		**0.01**		**0.01**		**0.02**
Thoracic spine		**0.02**		**0.02**	0.01	**0.03**	0.01	**0.03**	0.02	**0.04**	0.02	**0.05**
Claviculae	0.02	**0.14**	0.02	**0.17**	0.03	**0.19**	0.03	**0.21**	0.03	**0.24**	0.04	**0.26**
Scapulae	0.03	**0.08**	0.04	**0.10**	0.05	**0.12**	0.05	**0.14**	0.07	**0.17**	0.08	**0.20**
Ribs	0.09	**0.22**	0.11	**0.26**	0.13	**0.29**	0.14	**0.32**	0.16	**0.36**	0.18	**0.40**
Oesophagus						**0.01**		**0.01**		**0.02**		**0.02**
Trachea												
Lung	0.01	**0.03**	0.01	**0.04**	0.02	**0.05**	0.02	**0.06**	0.02	**0.07**	0.03	**0.07**
Mamma		**0.01**		**0.01**		**0.02**		**0.02**		**0.02**		**0.03**
Heart				**0.01**		**0.01**		**0.02**		**0.02**		**0.02**
Stomach												
Spleen												
Liver	0.03	**0.08**	0.03	**0.10**	0.04	**0.12**	0.04	**0.13**	0.05	**0.15**	0.06	**0.18**
Gall bladder				**0.01**		**0.02**		**0.02**		**0.02**	0.01	**0.03**
Pancreas								**0.01**		**0.01**		**0.02**
Small intestine		**0.01**		**0.02**		**0.02**		**0.02**	0.01	**0.03**	0.01	**0.03**
Large intestine		**0.03**	0.01	**0.03**	0.02	**0.04**	0.02	**0.04**	0.02	**0.05**	0.03	**0.06**
Sigma/rectum												
Lumbar spine		**0.01**		**0.02**		**0.02**		**0.02**	0.01	**0.03**	0.02	**0.04**
Pelvis	0.02	**0.05**	0.02	**0.06**	0.03	**0.07**	0.03	**0.08**	0.04	**0.09**	0.05	**0.11**
Adrenals						**0.01**		**0.01**		**0.02**		**0.02**
Kidneys				**0.01**		**0.01**		**0.02**		**0.02**	0.01	**0.02**
Urinary bladder						**0.01**		**0.01**		**0.01**		**0.02**
Uterus						**0.01**		**0.01**		**0.02**		**0.02**
Ovaries		**0.01**		**0.02**		**0.02**		**0.02**	0.01	**0.03**	0.01	**0.03**
Prostate										**0.01**		**0.01**
Testes												
Upper arm	0.45	**0.47**	0.54	**0.57**	0.61	**0.64**	0.66	**0.70**	0.75	**0.80**	0.83	**0.89**
Elbow	0.50	**0.52**	0.60	**0.63**	0.67	**0.71**	0.73	**0.78**	0.83	**0.89**	0.91	**0.98**
Forearm	0.56	**0.59**	0.67	**0.71**	0.75	**0.80**	0.81	**0.86**	0.91	**0.98**	1.00	**1.07**
Upper leg		**0.02**		**0.02**		**0.02**	0.01	**0.03**	0.01	**0.03**	0.02	**0.04**
Knee												
Lower leg												
Lymph nodes		**0.01**		**0.01**		**0.02**		**0.02**		**0.02**	0.01	**0.03**
Skin	0.05	**0.08**	0.06	**0.08**	0.06	**0.08**	0.06	**0.08**	0.06	**0.09**	0.07	**0.09**
Muscle	0.02	**0.04**	0.02	**0.05**	0.03	**0.05**	0.03	**0.05**	0.03	**0.06**	0.03	**0.07**
Skeleton	0.08	**0.10**	0.10	**0.13**	0.11	**0.14**	0.12	**0.15**	0.14	**0.18**	0.15	**0.20**
Red bone marrow	0.02	**0.03**	0.02	**0.03**	0.03	**0.04**	0.03	**0.04**	0.03	**0.05**	0.04	**0.05**
Whole body dose	0.03	**0.05**	0.03	**0.05**	0.04	**0.06**	0.04	**0.07**	0.04	**0.07**	0.05	**0.08**
ED (ICRP 60)		**0.02**	0.01	**0.02**	0.01	**0.03**	0.01	**0.03**	0.02	**0.04**	0.02	**0.04**
ED (ICRP 103)		**0.02**		**0.02**	0.01	**0.03**	0.01	**0.03**	0.02	**0.03**	0.02	**0.04**
Total filtration (mm) **Al**	**1.5**		**2.5**		**3.5**		**4.5**		**3.5**		**3.5**	
Cu	**0**		**0**		**0**		**0**		**0.1**		**0.2**	

Table 23.21 Normalized organ doses—Upper extremities ap—8–12 years

Beam projection:			ap									**Table cells:**
Focus-to-detector distance (cm):			115									(+) = normal font = optimal field setting
Tube voltage (kVp):			60									**(−) = bold font = suboptimal field setting**
Total filtrations:			See last table row.									Empty table cells correspond to values <0.01
Organ or tissue	**Organ dose D_T : Entrance dose K_E**							**(mSv * mGy^{-1})**				
	(+)	(−)	(+)	(−)	(+)	(−)	(+)	(−)	(+)	(−)	(+)	(−)
Skull												
Brain												
Oral mucosa												
Salivary glands												
Cervical spine												
Thyroid												
Thymus												
Thoracic spine				**0.01**		**0.01**		**0.02**		**0.02**		**0.02**
Claviculae		**0.11**		**0.14**	0.01	**0.16**	0.01	**0.18**	0.02	**0.21**	0.01	**0.20**
Scapulae		**0.05**	0.01	**0.06**	0.02	**0.08**	0.02	**0.09**	0.02	**0.11**	0.02	**0.10**
Ribs	0.04	**0.18**	0.05	**0.22**	0.05	**0.25**	0.06	**0.27**	0.07	**0.32**	0.07	**0.31**
Oesophagus										**0.01**		
Trachea												
Lung		**0.02**		**0.03**		**0.04**		**0.04**	0.01	**0.05**		**0.05**
Mamma				**0.01**		**0.01**		**0.02**		**0.02**		**0.02**
Heart								**0.01**		**0.01**		**0.01**
Stomach												
Spleen												
Liver	0.01	**0.07**	0.02	**0.09**	0.02	**0.11**	0.02	**0.12**	0.03	**0.14**	0.02	**0.13**
Gall bladder						**0.01**		**0.01**		**0.02**		**0.01**
Pancreas												
Small intestine				**0.01**		**0.01**		**0.01**		**0.02**		**0.02**
Large intestine		**0.02**		**0.02**		**0.03**		**0.03**		**0.04**		**0.04**
Sigma/rectum												
Lumbar spine								**0.01**		**0.01**		**0.01**
Pelvis		**0.03**	0.01	**0.04**	0.01	**0.05**	0.02	**0.06**	0.02	**0.07**	0.02	**0.06**
Adrenals										**0.01**		
Kidneys										**0.01**		**0.01**
Urinary bladder												
Uterus												
Ovaries						**0.01**		**0.01**		**0.02**		**0.02**
Prostate												
Testes												
Upper arm	0.34	**0.36**	0.43	**0.45**	0.49	**0.52**	0.54	**0.57**	0.62	**0.67**	0.61	**0.65**
Elbow	0.39	**0.41**	0.49	**0.52**	0.56	**0.59**	0.61	**0.65**	0.71	**0.76**	0.69	**0.74**
Forearm	0.45	**0.47**	0.55	**0.59**	0.63	**0.67**	0.69	**0.74**	0.79	**0.85**	0.78	**0.84**
Upper leg				**0.01**		**0.01**		**0.02**		**0.02**		**0.02**
Knee												
Lower leg												
Lymph nodes						**0.01**		**0.01**		**0.02**		**0.01**
Skin	0.04	**0.07**	0.05	**0.07**	0.05	**0.08**	0.05	**0.08**	0.05	**0.08**	0.05	**0.08**
Muscle	0.01	**0.03**	0.01	**0.04**	0.02	**0.04**	0.02	**0.05**	0.02	**0.05**	0.02	**0.05**
Skeleton	0.06	**0.08**	0.07	**0.10**	0.08	**0.12**	0.09	**0.13**	0.11	**0.15**	0.10	**0.14**
Red bone marrow		**0.02**	0.01	**0.02**	0.01	**0.03**	0.01	**0.03**	0.02	**0.04**	0.02	**0.03**
Whole body dose	0.02	**0.04**	0.02	**0.05**	0.02	**0.05**	0.03	**0.06**	0.03	**0.07**	0.03	**0.06**
ED (ICRP 60)		**0.01**		**0.02**		**0.02**		**0.02**		**0.03**		**0.03**
ED (ICRP 103)		**0.01**		**0.02**		**0.02**		**0.02**		**0.03**		**0.02**
Total filtration (mm) **Al**	**1.5**		**2.5**		**3.5**		**4.5**		**3.5**		**3.5**	
Cu	**0**		**0**		**0**		**0**		**0.1**		**0.2**	

Table 23.22 Normalized organ doses—Upper extremities ap—8–12 years

Beam projection:			ap									**Table cells:**
Focus-to-detector distance (cm):			115				(+) = normal font = optimal field setting					
Tube voltage (kVp):			70				**(−) = bold font = suboptimal field setting**					
Total filtrations:			See last table row.				Empty table cells correspond to values <0.01					
Organ or tissue	**Organ dose D_T : Entrance dose K_E**								**(mSv * mGy^{-1})**			
	(+)	(−)	(+)	(−)	(+)	(−)	(+)	(−)	(+)	(−)	(+)	(−)
Skull												
Brain												
Oral mucosa												
Salivary glands												
Cervical spine										**0.01**		**0.01**
Thyroid												
Thymus								**0.01**		**0.01**		**0.02**
Thoracic spine		**0.01**		**0.02**		**0.02**		**0.02**		**0.03**	0.01	**0.04**
Claviculae		**0.13**	0.01	**0.16**	0.01	**0.18**	0.02	**0.20**	0.02	**0.23**	0.02	**0.26**
Scapulae	0.01	**0.06**	0.02	**0.08**	0.02	**0.10**	0.02	**0.11**	0.03	**0.14**	0.04	**0.17**
Ribs	0.04	**0.20**	0.05	**0.25**	0.06	**0.28**	0.07	**0.30**	0.08	**0.35**	0.09	**0.38**
Oesophagus								**0.01**		**0.01**		**0.02**
Trachea												
Lung		**0.03**		**0.04**		**0.05**	0.01	**0.05**	0.01	**0.06**	0.02	**0.07**
Mamma		**0.01**		**0.02**		**0.02**		**0.02**		**0.02**		**0.03**
Heart						**0.01**		**0.01**		**0.02**		**0.02**
Stomach												
Spleen												
Liver	0.01	**0.09**	0.02	**0.11**	0.02	**0.13**	0.02	**0.14**	0.03	**0.17**	0.04	**0.19**
Gall bladder		**0.01**		**0.01**		**0.02**		**0.02**		**0.02**		**0.03**
Pancreas										**0.01**		**0.01**
Small intestine		**0.01**		**0.01**		**0.02**		**0.02**		**0.02**		**0.03**
Large intestine		**0.02**		**0.03**		**0.04**		**0.04**	0.01	**0.05**	0.01	**0.06**
Sigma/rectum												
Lumbar spine				**0.01**		**0.01**		**0.02**		**0.02**		**0.03**
Pelvis	0.01	**0.04**	0.01	**0.05**	0.02	**0.06**	0.02	**0.07**	0.02	**0.09**	0.03	**0.11**
Adrenals						**0.01**		**0.01**		**0.02**		**0.02**
Kidneys						**0.01**		**0.01**		**0.02**		**0.02**
Urinary bladder										**0.01**		**0.02**
Uterus								**0.01**		**0.01**		**0.02**
Ovaries		**0.01**		**0.01**		**0.02**		**0.02**		**0.02**		**0.03**
Prostate												**0.01**
Testes												
Upper arm	0.40	**0.42**	0.48	**0.52**	0.54	**0.59**	0.59	**0.64**	0.68	**0.74**	0.76	**0.83**
Elbow	0.45	**0.48**	0.55	**0.59**	0.62	**0.67**	0.67	**0.73**	0.77	**0.84**	0.85	**0.93**
Forearm	0.51	**0.54**	0.61	**0.66**	0.69	**0.74**	0.75	**0.81**	0.85	**0.93**	0.93	**1.02**
Upper leg		**0.01**		**0.02**		**0.02**		**0.02**		**0.03**		**0.03**
Knee												
Lower leg												
Lymph nodes				**0.01**		**0.01**		**0.02**		**0.02**		**0.02**
Skin	0.04	**0.07**	0.05	**0.08**	0.05	**0.08**	0.05	**0.08**	0.05	**0.08**	0.05	**0.09**
Muscle	0.01	**0.04**	0.02	**0.04**	0.02	**0.05**	0.02	**0.05**	0.02	**0.06**	0.02	**0.07**
Skeleton	0.07	**0.09**	0.08	**0.12**	0.09	**0.13**	0.10	**0.14**	0.12	**0.17**	0.13	**0.19**
Red bone marrow	0.01	**0.02**	0.01	**0.03**	0.02	**0.03**	0.02	**0.04**	0.02	**0.04**	0.02	**0.05**
Whole body dose	0.02	**0.04**	0.02	**0.05**	0.03	**0.06**	0.03	**0.06**	0.03	**0.07**	0.04	**0.08**
ED (ICRP 60)		**0.02**		**0.02**		**0.03**		**0.03**		**0.03**	0.01	**0.04**
ED (ICRP 103)		**0.02**		**0.02**		**0.02**		**0.03**		**0.03**	0.01	**0.04**
Total filtration (mm) Al	**1.5**		**2.5**		**3.5**		**4.5**		**3.5**		**3.5**	
Cu	**0**		**0**		**0**		**0**		**0.1**		**0.2**	

Table 23.23 Normalized organ doses—Upper extremities ap—13–17 years

Beam projection:			ap									**Table cells:**
Focus-to-detector distance (cm):			115						(+) = normal font = optimal field setting			
Tube voltage (kVp):			60						**(−) = bold font = suboptimal field setting**			
Total filtrations:			See last table row.						Empty table cells correspond to values <0.01			
Organ or tissue	**Organ dose D_T : Entrance dose K_E**								**(mSv * mGy⁻¹)**			
	(+)	(−)	(+)	(−)	(+)	(−)	(+)	(−)	(+)	(−)	(+)	(−)
Skull												
Brain												
Oral mucosa												
Salivary glands												
Cervical spine												
Thyroid												
Thymus												
Thoracic spine												
Claviculae		**0.05**		**0.07**	0.01	**0.08**	0.01	**0.09**	0.02	**0.11**	0.01	**0.10**
Scapulae		**0.03**	0.01	**0.04**	0.02	**0.04**	0.02	**0.05**	0.02	**0.07**	0.02	**0.05**
Ribs	0.04	**0.12**	0.05	**0.15**	0.06	**0.17**	0.07	**0.19**	0.09	**0.22**	0.08	**0.21**
Oesophagus												
Trachea												
Lung		**0.01**		**0.02**		**0.02**		**0.02**	0.01	**0.03**		**0.02**
Mamma	0.02	**0.32**	0.03	**0.38**	0.04	**0.42**	0.04	**0.45**	0.05	**0.50**	0.05	**0.48**
Heart												
Stomach												
Spleen												
Liver	0.01	**0.05**	0.02	**0.07**	0.02	**0.08**	0.03	**0.09**	0.03	**0.11**	0.03	**0.10**
Gall bladder						**0.01**		**0.01**		**0.02**		**0.01**
Pancreas												
Small intestine										**0.01**		
Large intestine		**0.01**		**0.01**		**0.02**		**0.02**		**0.03**		**0.02**
Sigma/rectum												
Lumbar spine												
Pelvis		**0.02**	0.01	**0.02**	0.01	**0.03**	0.01	**0.03**	0.02	**0.04**	0.02	**0.04**
Adrenals												
Kidneys												
Urinary bladder												
Uterus								**0.01**		**0.01**		
Ovaries												
Prostate												
Testes												
Upper arm	0.30	**0.32**	0.38	**0.40**	0.43	**0.46**	0.48	**0.51**	0.56	**0.61**	0.54	**0.58**
Elbow	0.35	**0.37**	0.44	**0.47**	0.51	**0.54**	0.56	**0.60**	0.66	**0.70**	0.63	**0.67**
Forearm	0.39	**0.41**	0.49	**0.51**	0.56	**0.59**	0.62	**0.65**	0.72	**0.76**	0.70	**0.74**
Upper leg												
Knee												
Lower leg												
Lymph nodes										**0.01**		
Skin	0.05	**0.06**	0.05	**0.07**	0.05	**0.07**	0.05	**0.07**	0.05	**0.08**	0.05	**0.08**
Muscle	0.01	**0.02**	0.02	**0.03**	0.02	**0.03**	0.02	**0.04**	0.02	**0.04**	0.02	**0.04**
Skeleton	0.06	**0.07**	0.07	**0.09**	0.08	**0.10**	0.09	**0.12**	0.11	**0.14**	0.11	**0.13**
Red bone marrow		**0.01**		**0.02**	0.01	**0.02**	0.01	**0.02**	0.01	**0.03**	0.01	**0.02**
Whole body dose	0.02	**0.03**	0.02	**0.04**	0.03	**0.04**	0.03	**0.05**	0.03	**0.05**	0.03	**0.05**
ED (ICRP 60)		**0.02**		**0.03**		**0.03**		**0.04**	0.01	**0.04**		**0.04**
ED (ICRP 103)		**0.05**		**0.06**		**0.06**	0.01	**0.07**	0.01	**0.08**	0.01	**0.07**
Total filtration (mm) Al	**1.5**		**2.5**		**3.5**		**4.5**		**3.5**		**3.5**	
Cu	**0**		**0**		**0**		**0**		**0.1**		**0.2**	

Table 23.24 Normalized organ doses—Upper extremities ap—13–17 years

Beam projection:			ap									**Table cells:**
Focus-to-detector distance (cm):			115									(+) = normal font = optimal field setting
Tube voltage (kVp):			70									**(−) = bold font = suboptimal field setting**
Total filtrations:			See last table row.									Empty table cells correspond to values <0.01
Organ or tissue	**Organ dose D_T : Entrance dose K_E**								**(mSv * mGy^{-1})**			
	(+)	(−)	(+)	(−)	(+)	(−)	(+)	(−)	(+)	(−)	(+)	(−)
Skull												
Brain												
Oral mucosa												
Salivary glands												
Cervical spine												
Thyroid												
Thymus												
Thoracic spine								**0.01**		**0.01**		**0.02**
Claviculae		**0.06**	0.01	**0.08**	0.01	**0.09**	0.02	**0.10**	0.02	**0.12**	0.02	**0.14**
Scapulae	0.01	**0.04**	0.02	**0.05**	0.02	**0.06**	0.03	**0.07**	0.03	**0.09**	0.04	**0.10**
Ribs	0.05	**0.14**	0.06	**0.17**	0.07	**0.20**	0.08	**0.21**	0.10	**0.25**	0.11	**0.28**
Oesophagus												**0.01**
Trachea												
Lung		**0.02**		**0.02**		**0.02**	0.01	**0.03**	0.01	**0.03**	0.02	**0.04**
Mamma	0.03	**0.36**	0.04	**0.42**	0.05	**0.46**	0.05	**0.49**	0.06	**0.54**	0.07	**0.59**
Heart												**0.01**
Stomach												
Spleen												
Liver	0.02	**0.06**	0.02	**0.08**	0.03	**0.10**	0.03	**0.11**	0.04	**0.13**	0.05	**0.15**
Gall bladder		**0.01**		**0.01**		**0.02**		**0.02**		**0.02**	0.01	**0.03**
Pancreas												
Small intestine								**0.01**		**0.01**		**0.02**
Large intestine		**0.01**		**0.02**		**0.02**		**0.03**	0.01	**0.03**	0.01	**0.04**
Sigma/rectum												
Lumbar spine										**0.01**		**0.01**
Pelvis	0.01	**0.02**	0.01	**0.03**	0.02	**0.04**	0.02	**0.04**	0.02	**0.06**	0.03	**0.07**
Adrenals												
Kidneys												**0.01**
Urinary bladder												
Uterus				**0.01**		**0.01**		**0.02**		**0.02**	0.01	**0.03**
Ovaries										**0.01**		**0.02**
Prostate												
Testes												
Upper arm	0.35	**0.38**	0.43	**0.47**	0.49	**0.53**	0.54	**0.58**	0.63	**0.68**	0.70	**0.77**
Elbow	0.41	**0.44**	0.51	**0.54**	0.57	**0.61**	0.63	**0.67**	0.73	**0.78**	0.81	**0.88**
Forearm	0.45	**0.48**	0.56	**0.59**	0.63	**0.66**	0.69	**0.73**	0.79	**0.84**	0.88	**0.94**
Upper leg										**0.01**	0.01	**0.01**
Knee												
Lower leg												
Lymph nodes								**0.01**		**0.01**		**0.02**
Skin	0.05	**0.07**	0.05	**0.07**	0.05	**0.07**	0.05	**0.08**	0.06	**0.08**	0.06	**0.08**
Muscle	0.01	**0.03**	0.02	**0.03**	0.02	**0.04**	0.02	**0.04**	0.02	**0.05**	0.03	**0.05**
Skeleton	0.07	**0.08**	0.08	**0.10**	0.10	**0.12**	0.11	**0.13**	0.12	**0.15**	0.14	**0.17**
Red bone marrow		**0.02**	0.01	**0.02**	0.01	**0.02**	0.01	**0.03**	0.02	**0.03**	0.02	**0.04**
Whole body dose	0.02	**0.04**	0.03	**0.04**	0.03	**0.05**	0.03	**0.05**	0.04	**0.06**	0.04	**0.07**
ED (ICRP 60)		**0.03**		**0.03**		**0.04**	0.01	**0.04**	0.01	**0.05**	0.01	**0.06**
ED (ICRP 103)		**0.05**	0.01	**0.06**	0.01	**0.07**	0.01	**0.08**	0.02	**0.09**	0.02	**0.10**
Total filtration (mm) Al	**1.5**		**2.5**		**3.5**		**4.5**		**3.5**		**3.5**	
Cu	**0**		**0**		**0**		**0**		**0.1**		**0.2**	

Table 23.25 Normalized organ doses—Upper extremities ap—13–17 years

Beam projection:			ap									**Table cells:**
Focus-to-detector distance (cm):			115									(+) = normal font = optimal field setting
Tube voltage (kVp):			80									**(−) = bold font = suboptimal field setting**
Total filtrations:			See last table row.									Empty table cells correspond to values <0.01
Organ or tissue	**Organ dose D_T : Entrance dose K_E**								**(mSv * mGy^{-1})**			
	(+)	(−)	(+)	(−)	(+)	(−)	(+)	(−)	(+)	(−)	(+)	(−)
Skull												
Brain												
Oral mucosa												
Salivary glands												
Cervical spine												**0.01**
Thyroid												
Thymus												**0.01**
Thoracic spine				**0.01**		**0.01**		**0.01**		**0.02**		**0.02**
Claviculae	0.01	**0.07**	0.01	**0.09**	0.02	**0.10**	0.02	**0.11**	0.02	**0.13**	0.02	**0.15**
Scapulae	0.02	**0.05**	0.02	**0.06**	0.03	**0.07**	0.03	**0.08**	0.04	**0.10**	0.05	**0.12**
Ribs	0.06	**0.16**	0.07	**0.19**	0.08	**0.22**	0.09	**0.23**	0.11	**0.27**	0.12	**0.30**
Oesophagus										**0.01**		**0.01**
Trachea												
Lung		**0.02**		**0.03**	0.01	**0.03**	0.01	**0.03**	0.02	**0.04**	0.02	**0.05**
Mamma	0.04	**0.39**	0.05	**0.45**	0.06	**0.49**	0.06	**0.52**	0.07	**0.58**	0.09	**0.63**
Heart										**0.01**		**0.02**
Stomach												
Spleen												
Liver	0.02	**0.08**	0.03	**0.10**	0.03	**0.11**	0.04	**0.12**	0.05	**0.15**	0.05	**0.17**
Gall bladder		**0.01**		**0.02**		**0.02**		**0.02**	0.01	**0.03**	0.01	**0.04**
Pancreas												
Small intestine				**0.01**		**0.01**		**0.01**		**0.02**		**0.02**
Large intestine		**0.02**		**0.02**	0.01	**0.03**	0.01	**0.03**	0.02	**0.04**	0.02	**0.05**
Sigma/rectum												
Lumbar spine								**0.01**		**0.01**		**0.02**
Pelvis	0.01	**0.03**	0.02	**0.04**	0.02	**0.05**	0.02	**0.05**	0.03	**0.07**	0.03	**0.08**
Adrenals										**0.01**		**0.01**
Kidneys								**0.01**		**0.01**		**0.02**
Urinary bladder												**0.01**
Uterus		**0.01**		**0.02**		**0.02**		**0.02**	0.01	**0.03**	0.01	**0.03**
Ovaries						**0.01**		**0.01**		**0.02**		**0.02**
Prostate												
Testes												
Upper arm	0.40	**0.43**	0.48	**0.52**	0.54	**0.58**	0.59	**0.64**	0.67	**0.73**	0.74	**0.81**
Elbow	0.46	**0.50**	0.56	**0.60**	0.63	**0.67**	0.68	**0.73**	0.77	**0.84**	0.85	**0.93**
Forearm	0.51	**0.54**	0.61	**0.64**	0.68	**0.72**	0.73	**0.78**	0.83	**0.89**	0.91	**0.97**
Upper leg						**0.01**		**0.01**	0.01	**0.01**	0.01	**0.02**
Knee												
Lower leg												
Lymph nodes				**0.01**		**0.01**		**0.01**		**0.02**		**0.02**
Skin	0.05	**0.07**	0.05	**0.07**	0.06	**0.08**	0.06	**0.08**	0.06	**0.08**	0.06	**0.09**
Muscle	0.02	**0.03**	0.02	**0.04**	0.02	**0.04**	0.02	**0.04**	0.03	**0.05**	0.03	**0.06**
Skeleton	0.08	**0.10**	0.09	**0.12**	0.11	**0.13**	0.11	**0.14**	0.13	**0.16**	0.14	**0.18**
Red bone marrow	0.01	**0.02**	0.01	**0.02**	0.02	**0.03**	0.02	**0.03**	0.02	**0.04**	0.02	**0.04**
Whole body dose	0.02	**0.04**	0.03	**0.05**	0.03	**0.05**	0.03	**0.06**	0.04	**0.07**	0.04	**0.07**
ED (ICRP 60)		**0.03**		**0.04**	0.01	**0.04**	0.01	**0.05**	0.01	**0.05**	0.02	**0.06**
ED (ICRP 103)	0.01	**0.06**	0.01	**0.07**	0.01	**0.08**	0.02	**0.08**	0.02	**0.09**	0.02	**0.10**
Total filtration (mm) **Al**	**1.5**		**2.5**		**3.5**		**4.5**		**3.5**		**3.5**	
Cu	**0**		**0**		**0**		**0**		**0.1**		**0.2**	

Table 23.26 Normalized organ doses—Upper extremities ap—>17 years

Beam projection:			ap									**Table cells:**
Focus-to-detector distance (cm):			115									(+) = normal font = optimal field setting
Tube voltage (kVp):			60									**(−) = bold font = suboptimal field setting**
Total filtrations:			See last table row.									Empty table cells correspond to values <0.01
Organ or tissue	**Organ dose D_T : Entrance dose K_E**								**(mSv * mGy^{-1})**			
	(+)	(−)	(+)	(−)	(+)	(−)	(+)	(−)	(+)	(−)	(+)	(−)
Skull												
Brain												
Oral mucosa												
Salivary glands												
Cervical spine												
Thyroid												
Thymus												
Thoracic spine												
Claviculae		**0.06**		**0.08**		**0.09**		**0.11**	0.01	**0.13**		**0.12**
Scapulae		**0.03**		**0.04**	0.01	**0.05**	0.01	**0.06**	0.02	**0.07**	0.01	**0.06**
Ribs	0.03	**0.13**	0.04	**0.16**	0.05	**0.19**	0.05	**0.21**	0.06	**0.24**	0.06	**0.23**
Oesophagus												
Trachea												
Lung		**0.01**		**0.02**		**0.02**		**0.03**		**0.03**		**0.03**
Mamma		**0.03**		**0.03**		**0.04**		**0.04**		**0.05**		**0.04**
Heart												
Stomach												
Spleen												
Liver	0.01	**0.06**	0.01	**0.07**	0.02	**0.09**	0.02	**0.10**	0.02	**0.12**	0.02	**0.11**
Gall bladder								**0.01**		**0.01**		**0.01**
Pancreas												
Small intestine												
Large intestine		**0.01**		**0.01**		**0.02**		**0.02**		**0.03**		**0.02**
Sigma/rectum												
Lumbar spine												
Pelvis		**0.02**		**0.02**		**0.03**	0.01	**0.04**	0.01	**0.05**	0.01	**0.04**
Adrenals												
Kidneys												
Urinary bladder												
Uterus												
Ovaries												
Prostate												
Testes												
Upper arm	0.29	**0.31**	0.37	**0.40**	0.42	**0.46**	0.47	**0.51**	0.55	**0.60**	0.53	**0.57**
Elbow	0.35	**0.37**	0.43	**0.47**	0.50	**0.54**	0.55	**0.60**	0.65	**0.71**	0.62	**0.67**
Forearm	0.39	**0.42**	0.49	**0.52**	0.56	**0.60**	0.62	**0.67**	0.73	**0.79**	0.71	**0.76**
Upper leg										**0.01**		
Knee												
Lower leg												
Lymph nodes										**0.01**		
Skin	0.04	**0.07**	0.04	**0.07**	0.05	**0.07**	0.05	**0.07**	0.05	**0.08**	0.05	**0.08**
Muscle	0.01	**0.03**	0.01	**0.03**	0.01	**0.04**	0.02	**0.04**	0.02	**0.05**	0.02	**0.04**
Skeleton	0.05	**0.07**	0.07	**0.09**	0.08	**0.10**	0.09	**0.12**	0.10	**0.14**	0.10	**0.13**
Red bone marrow		**0.01**		**0.02**		**0.02**		**0.02**	0.01	**0.03**		**0.02**
Whole body dose	0.02	**0.03**	0.02	**0.04**	0.02	**0.05**	0.03	**0.05**	0.03	**0.06**	0.03	**0.06**
ED (ICRP 60)		**0.01**		**0.01**		**0.02**		**0.02**		**0.02**		**0.02**
ED (ICRP 103)		**0.01**		**0.01**		**0.02**		**0.02**		**0.02**		**0.02**
Total filtration (mm) **Al**	**1.5**		**2.5**		**3.5**		**4.5**		**3.5**		**3.5**	
Cu	**0**		**0**		**0**		**0**		**0.1**		**0.2**	

Table 23.27 Normalized organ doses—Upper extremities ap—>17 years

Beam projection:			ap									**Table cells:**
Focus-to-detector distance (cm):			115					(+) = normal font = optimal field setting				
Tube voltage (kVp):			70					**(−) = bold font = suboptimal field setting**				
Total filtrations:			See last table row.					Empty table cells correspond to values <0.01				
Organ or tissue	**Organ dose D_T : Entrance dose K_E**								**(mSv * mGy^{-1})**			
	(+)	(−)	(+)	(−)	(+)	(−)	(+)	(−)	(+)	(−)	(+)	(−)
Skull												
Brain												
Oral mucosa												
Salivary glands												
Cervical spine												
Thyroid												
Thymus												
Thoracic spine								**0.01**		**0.01**		**0.02**
Claviculae		**0.08**		**0.10**		**0.11**	0.01	**0.12**	0.01	**0.14**	0.02	**0.16**
Scapulae	0.01	**0.04**	0.01	**0.05**	0.02	**0.06**	0.02	**0.07**	0.03	**0.09**	0.03	**0.11**
Ribs	0.04	**0.15**	0.05	**0.19**	0.06	**0.21**	0.06	**0.23**	0.07	**0.27**	0.09	**0.31**
Oesophagus												**0.01**
Trachea												
Lung		**0.02**		**0.02**		**0.03**		**0.03**		**0.04**	0.01	**0.05**
Mamma		**0.03**		**0.04**		**0.04**		**0.04**		**0.05**		**0.06**
Heart												**0.01**
Stomach												
Spleen												
Liver	0.01	**0.07**	0.02	**0.09**	0.02	**0.10**	0.02	**0.12**	0.03	**0.14**	0.04	**0.16**
Gall bladder				**0.01**		**0.01**		**0.01**		**0.02**		**0.02**
Pancreas												
Small intestine								**0.01**		**0.01**		**0.02**
Large intestine		**0.01**		**0.02**		**0.02**		**0.03**		**0.03**	0.01	**0.04**
Sigma/rectum												
Lumbar spine												**0.01**
Pelvis		**0.03**		**0.03**	0.01	**0.04**	0.01	**0.05**	0.02	**0.06**	0.02	**0.07**
Adrenals												
Kidneys												**0.01**
Urinary bladder												
Uterus												
Ovaries										**0.01**		**0.02**
Prostate												
Testes												
Upper arm	0.34	**0.37**	0.42	**0.46**	0.48	**0.53**	0.53	**0.58**	0.61	**0.68**	0.69	**0.77**
Elbow	0.41	**0.44**	0.50	**0.54**	0.57	**0.62**	0.62	**0.68**	0.72	**0.79**	0.81	**0.89**
Forearm	0.46	**0.49**	0.56	**0.60**	0.63	**0.68**	0.69	**0.75**	0.80	**0.87**	0.89	**0.97**
Upper leg								**0.01**		**0.01**		**0.02**
Knee												
Lower leg												
Lymph nodes								**0.01**		**0.01**		**0.02**
Skin	0.04	**0.07**	0.05	**0.07**	0.05	**0.08**	0.05	**0.08**	0.05	**0.08**	0.05	**0.09**
Muscle	0.01	**0.03**	0.02	**0.04**	0.02	**0.04**	0.02	**0.05**	0.02	**0.05**	0.02	**0.06**
Skeleton	0.06	**0.08**	0.08	**0.10**	0.09	**0.12**	0.10	**0.13**	0.12	**0.15**	0.13	**0.18**
Red bone marrow		**0.02**		**0.02**		**0.02**	0.01	**0.03**	0.01	**0.03**	0.01	**0.04**
Whole body dose	0.02	**0.04**	0.02	**0.05**	0.03	**0.05**	0.03	**0.06**	0.03	**0.07**	0.04	**0.07**
ED (ICRP 60)		**0.01**		**0.02**		**0.02**		**0.02**		**0.03**		**0.03**
ED (ICRP 103)		**0.01**		**0.02**		**0.02**		**0.02**		**0.03**		**0.03**
Total filtration (mm) Al	**1.5**		**2.5**		**3.5**		**4.5**		**3.5**		**3.5**	
Cu	**0**		**0**		**0**		**0**		**0.1**		**0.2**	

Table 23.28 Normalized organ doses—Upper extremities ap—>17 years

Beam projection:			ap									**Table cells:**
Focus-to-detector distance (cm):			115									(+) = normal font = optimal field setting
Tube voltage (kVp):			80									**(−) = bold font = suboptimal field setting**
Total filtrations:			See last table row.									Empty table cells correspond to values <0.01
Organ or tissue	**Organ dose D_T : Entrance dose K_E**								**(mSv * mGy^{-1})**			
	(+)	(−)	(+)	(−)	(+)	(−)	(+)	(−)	(+)	(−)	(+)	(−)
Skull												
Brain												
Oral mucosa												
Salivary glands												
Cervical spine												
Thyroid												
Thymus										**0.01**		**0.01**
Thoracic spine				**0.01**		**0.01**		**0.01**		**0.02**		**0.02**
Claviculae		**0.09**	0.01	**0.11**	0.01	**0.12**	0.01	**0.13**	0.02	**0.16**	0.02	**0.17**
Scapulae	0.01	**0.05**	0.02	**0.07**	0.02	**0.08**	0.03	**0.09**	0.03	**0.11**	0.04	**0.14**
Ribs	0.04	**0.17**	0.06	**0.21**	0.06	**0.23**	0.07	**0.25**	0.08	**0.29**	0.09	**0.32**
Oesophagus										**0.01**		**0.01**
Trachea												
Lung		**0.02**		**0.03**		**0.03**		**0.04**	0.01	**0.05**	0.01	**0.05**
Mamma		**0.04**		**0.04**		**0.05**		**0.05**		**0.05**		**0.06**
Heart										**0.01**		**0.01**
Stomach												
Spleen												
Liver	0.02	**0.09**	0.02	**0.11**	0.03	**0.12**	0.03	**0.14**	0.03	**0.16**	0.04	**0.19**
Gall bladder		**0.01**		**0.01**		**0.02**		**0.02**		**0.02**		**0.03**
Pancreas												
Small intestine						**0.01**		**0.01**		**0.02**		**0.02**
Large intestine		**0.02**		**0.02**		**0.03**		**0.03**	0.01	**0.04**	0.01	**0.05**
Sigma/rectum												
Lumbar spine										**0.01**		**0.01**
Pelvis		**0.03**	0.01	**0.04**	0.02	**0.05**	0.02	**0.06**	0.02	**0.07**	0.03	**0.08**
Adrenals										**0.01**		**0.01**
Kidneys										**0.01**		**0.02**
Urinary bladder												
Uterus												**0.01**
Ovaries						**0.01**		**0.01**		**0.02**		**0.02**
Prostate												
Testes												
Upper arm	0.39	**0.43**	0.47	**0.52**	0.53	**0.58**	0.57	**0.64**	0.66	**0.73**	0.73	**0.82**
Elbow	0.46	**0.50**	0.55	**0.60**	0.62	**0.68**	0.67	**0.74**	0.77	**0.85**	0.85	**0.94**
Forearm	0.51	**0.55**	0.61	**0.66**	0.69	**0.75**	0.74	**0.81**	0.84	**0.92**	0.92	**1.02**
Upper leg				**0.01**		**0.01**		**0.01**		**0.02**		**0.02**
Knee												
Lower leg												
Lymph nodes				**0.01**		**0.01**		**0.01**		**0.02**		**0.02**
Skin	0.05	**0.07**	0.05	**0.08**	0.05	**0.08**	0.05	**0.08**	0.05	**0.09**	0.06	**0.09**
Muscle	0.01	**0.04**	0.02	**0.04**	0.02	**0.05**	0.02	**0.05**	0.02	**0.06**	0.02	**0.06**
Skeleton	0.07	**0.10**	0.09	**0.12**	0.10	**0.13**	0.11	**0.15**	0.12	**0.17**	0.14	**0.19**
Red bone marrow		**0.02**		**0.02**	0.01	**0.03**	0.01	**0.03**	0.01	**0.04**	0.02	**0.04**
Whole body dose	0.02	**0.04**	0.03	**0.05**	0.03	**0.06**	0.03	**0.06**	0.04	**0.07**	0.04	**0.08**
ED (ICRP 60)		**0.02**		**0.02**		**0.02**		**0.02**		**0.03**	0.01	**0.03**
ED (ICRP 103)		**0.02**		**0.02**		**0.02**		**0.03**		**0.03**	0.01	**0.04**
Total filtration (mm) Al	**1.5**		**2.5**		**3.5**		**4.5**		**3.5**		**3.5**	
Cu	**0**		**0**		**0**		**0**		**0.1**		**0.2**	

Reference

1. Bontrager KL. Textbook of radiographic positioning and related anatomy. St. Louis: Mosby Year Book; 1993.

Hand

24

24.1 Introduction

The type of hand radiography is determined by the clinical indication. Table 24.1 shows indications and Table 24.2 contraindications for hand radiography. Image criteria as required for radiological findings in hand radiographs are listed in Table 24.3. In order to fulfil these image criteria, the field settings listed in Tables 24.4 and 24.5 should be achieved. These field settings go along with the patient preparation as listed in Table 24.6. Exposure parameter combinations for various age groups and hand diameters are listed in Tables 24.7, 24.8 and 24.9. The age groups in Table 24.8 refer to the standard age groups of the mathematical MIRD phantom family which has been used to calculate the normalized organ doses in this book. Table 24.9 shows radiation qualities (i.e. combinations of tube voltage and total filtration) which may be applied in clinical practice. Please note that not all of these radiation qualities are recommended by the authors. The standard projections of hand radiographies of the patient who is correctly positioned are depicted in Fig. 24.1. Please note that the anatomy of the mathematical phantoms has been idealized and somewhat simplified. Therefore, the hand was not inserted into the mathematical phantom, and no normalized organ doses for the hand radiography could be provided.

24.2 Indications and Contraindications

Indications and contraindications for hand radiographs are given in Tables 24.1 and 24.2.

24.3 Image Criteria

The image criteria required for hand radiographs are listed in Table 24.3.

24.4 Field Settings

In Table 24.4, patient positions and beam projections used for hand radiographs are listed, and in Table 24.5, the optimal field settings are listed which are required in order to fulfil the image criteria mentioned in Table 24.3.

Table 24.1 Indications for hand radiographs

Bone age
Trauma, fracture/luxation
Osteonecrosis of carpal bones, e.g. os naviculare, os lunatum
Osteomyelitis
Soft tissue tumours
Bone tumours
Vascular malformation
Congenital malformation

M. Seidenbusch et al., *Imaging Practice and Radiation Protection in Pediatric Radiology*,
https://doi.org/10.1007/978-3-030-18504-6_24

Table 24.2 Contraindications for hand radiographs

None

Table 24.3 Image criteria for hand radiographs

Hand dorsovolar
Visualization of the distal epiphyses and metaphyses of the radius and ulna
Visualization of the carpal bones consistent with age
Visualization of the metacarpals and the fingers including the adjacent soft tissue
Visually sharp reproduction of the physes and epiphyses of the metacarpals and fingers
Visually sharp reproduction of the cortex and the spongiosa of the distal radius and ulna, the carpal bones, the metacarpals, and the fingers
Hand oblique
Visualization of the distal epiphyses and metaphyses of the radius and ulna
Visualization of the carpal bones consistent with age
Visualization of the metacarpals and the fingers including the adjacent soft tissue
Visually sharp reproduction of the physes and epiphyses of the metacarpals and fingers
Visually sharp reproduction of the cortex and the spongiosa of the distal radius and ulna, the carpal bones, the metacarpals, and the fingers

Table 24.4 Patient positions and beam projections for hand radiographs

Hand position	Beam projections
Prone	Dorsovolar
Oblique	Dorsovolar oblique

Table 24.5 Optimal field settings for hand radiographs

Hand dorsovolar	
Upper field edge:	Circa 2–5 cm of the distal forearm
Field centring:	Just distal to the third metacarpo-phalangeal joint
Lower field edge:	Tips of the fingers
Medial lateral field edge:	The thumb must be included
Outer lateral field edge:	The fifth finger must be included

Table 24.6 Patient preparation

General	Remove all clothes, external metallic devices, and jewellery from the hand and forearm of the patient
Positioning	For the dorsovolar view, the thumb and the fingers should be fully extended and put flat on the detector
	For the oblique view, the thumb and the fingers should be kept in slightly flexed positions
Shielding	Lead mask shielding should be used at the entrance side beneath the medial field edges

Table 24.7 Radiographic procedure

X-ray source	Focal spot size between 1.0 and 1.3 dependent on the age of the patient
	Check the use of additional filtration
	Check the focus-to-detector distance
Field setting	Check the field size
	Check the field centring
Image detector system	Check the use of antiscatter grids
Patient	–

24.5 Patient Preparation and Radiographic Procedure

See Tables 24.6, 24.7 and [1].

24.6 Exposure Parameters

In Tables 24.8 and 24.9, exposure parameter combinations used for hand radiographs in various age groups are listed. Dark grey shaded table cells refer to exposure parameter combinations which are not recommended for paediatric hand radiographs. Light blue shaded table cells refer to

Table 24.8 Radiographic techniques of hand radiographs

Age group (years)	Beam direction	Hand position	Beam projection	Typical hand diameter (cm)	Optimal field size (cm x cm)	Focus-to-detector distance (cm)	Voltage (kVp)						Antiscatter grid
							50	60	70	80	90	100	
0	Vertical	Prone	Dorsovolar / oblique	-	-	100 (1)							
1 – 2													
3 – 7						100 (2)							
8 – 12													
13 – 17													
18													
Annotations:													
(1)	The patient is lying on the table top directly on the detector												
(2)	The patient is sitting at the table												
	Parameter combination not recommended, normalized organ doses not provided												
	Parameter combination recommended, normalized organ doses provided												

Table 24.9 Radiation qualities of hand radiographs

Voltage (kVp)	Total filtration					
	Without additional filtration		With additional filtration			
	1.5 mm Al	2.5 Al	3.5 Al	4.5 Al	3.5 Al + 0.1 Cu	3.5 Al + 0.2 Cu
50						
60						
70						
80						
90						
100						
Annotations:						
	Parameter combination not recommended, normalized organ doses not provided					
	Parameter combination not recommended, normalized organ doses provided					
	Parameter combination recommended, normalized organ doses provided					

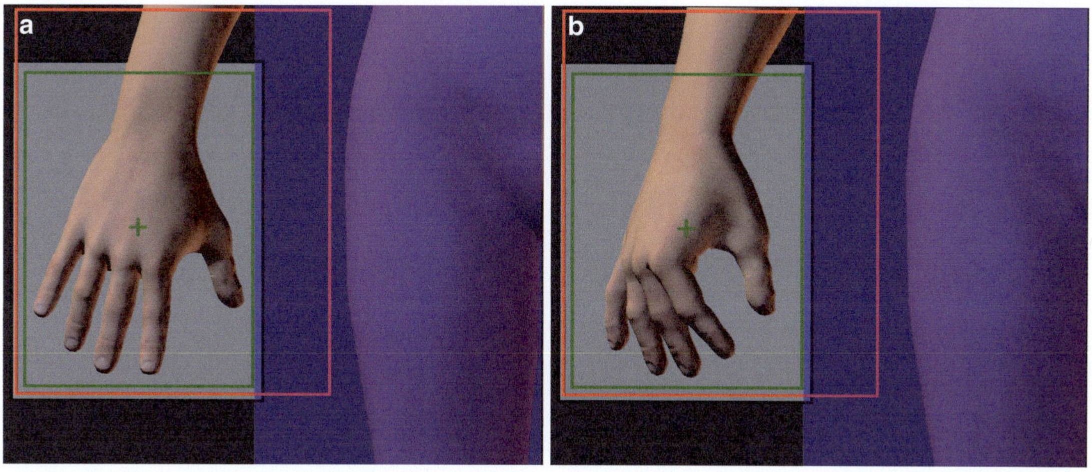

Fig. 24.1 **Field settings for hand radiographs**. (**a**) Dorsovolar projection. (**b**) Oblique projection. (© V. Rösenberger and M. Seidenbusch 2018). Green lines = optimal field settings; green cross = centring point; red lines = suboptimal field settings; grey shaded area = image detector; blue shaded area = lead mask shielding; dark grey shaded area = Bucky table

exposure parameter combinations which are recommended for paediatric hand radiographs.

24.7 Normalized Organ Doses

Please note also that the anatomy of the mathematical phantoms has been idealized and somewhat simplified. Therefore, the hand was not inserted into the mathematical phantom, and no normalized organ doses could be calculated (Table 24.10).

Table 24.10 Inconsistencies between the mathematical MIRD phantoms and real human anatomy

Anatomical region	Inconsistencies
Extremities	No hands have been inserted

For details, see the text

Reference

1. Bontrager KL. Textbook of radiographic positioning and related anatomy. St. Louis, MO: Mosby Year Book; 1993.

25 Lower Extremity

25.1 Introduction

The type of lower extremity radiography is determined by the clinical indication. Table 25.1 shows indications and Table 25.2 contraindications for lower extremity radiography. Image criteria as required for radiological findings in lower extremity radiographs are listed in Table 25.3. In order to fulfil these image criteria, the field settings listed in Tables 25.4 and 25.5 should be achieved. These field settings go along with the patient preparation as listed in Table 25.6. Exposure parameter combinations for various age groups and thigh diameters are listed in Tables 25.7, 25.8, 25.9, 25.10, and 25.11. The age groups in Tables 25.8 and 25.10 refer to the standard age groups of the mathematical MIRD phantom family which has been used to calculate the normalized organ doses in this book. Tables 25.9 and 25.11 shows radiation qualities (i.e. various combinations of tube

Table 25.1 Indications for lower extremity radiographs

Battered child syndrome
Leg length difference (hemihypertrophia)
Bandy legs (coxa vara), knock knees (genua valga)
Congenital malformation of lower extremity, e.g. femur hypoplasia
Skeletal dysplasia
Congenital vascular malformation
Soft tissue tumours
Bone tumours
Vanishing bone disease

Table 25.2 Contraindications for lower extremity radiographs

None

Table 25.3 Image criteria for lower extremity radiographs

Lower extremity ap
Visualization of the entire lower extremity
Reproduction of the hip joint, the knee joint, the ankle joint, and the whole foot
Visualization of all bones of the lower extremity
Visually sharp reproduction of the cortex and spongiosa of the femur, the tibia and fibula, the tarsal bones, and the small bones of the feet

Table 25.4 Patient positions and beam projections for lower extremity radiographs

Patient position	Beam projections
Upright[a]	ap
Supine	ap
Annotation:	
[a]Both lower extremities while standing	

Table 25.5 Optimal field settings for lower extremity radiographs

Lower extremity ap	
Upper field edge:	Iliac bone just below the upper superior spine of the iliac bone
Field centring:	–
Lower field edge:	Tips of the toes
Lateral field edges:	Skin surfaces of the legs

M. Seidenbusch et al., *Imaging Practice and Radiation Protection in Pediatric Radiology*,
https://doi.org/10.1007/978-3-030-18504-6_25

Table 25.6 Patient preparation

General	Remove all clothes, external metallic devices and jewellery from the lower extremities of the patient
	Immobilize infants and uncooperative patients
	Place eventually small X-ray positive markers at palpable tumours or on pain points
Positioning	The lower extremities should be extended in the hip, knee, and ankle joints
Shielding	Lead mask shielding should be used at the entrance side beneath the medial field edges

Table 25.7 Radiographic procedure

X-ray source	Focal spot size between 1.0 and 1.3 dependent on the age of the patient
	Check the use of additional filtration
	Check the focus-to-detector distance
Field setting	Check the field size
	Check the field centring
Image detector system	Check the use of antiscatter grids
Patient	–

voltage and total filtration) which may be applied in clinical practice. Please note that not all of these radiation qualities are recommended by the authors. The standard projections of lower extremity radiographies of the patient who is correctly positioned are depicted in Fig. 25.1. Also the optimal field (marked in green), the suboptimal field (marked in red), and the correct shielding are shown. Tables 25.13, 25.14, and 25.15 outline the virtual field settings on the skin surface of mathematical MIRD phantoms of various age groups and the resulting virtual radiographies with and without bone structures, respectively. Normalized organ doses resulting from these virtual field settings have been calculated for all age groups and have been listed in Tables 25.16, 25.17, 25.18, 25.19, 25.20, 25.21, 25.22, 25.23, 25.24, 25.25, 25.26, 25.27, 25.28, 25.29, 25.30, 25.31, 25.32, and 25.33. Table cells with normal font refer to optimal field settings and table cells with bold font refer to suboptimal field settings. Please note that normalized organ doses have been provided also for exposure parameter combinations, which are not recommended by the authors.

Table 25.8 Radiographic techniques of lower extremity Bucky table radiographs

Age group (years)	Beam direction	Patient position	Beam projection	Typical patient diameter (cm)	Optimal field size (cm x cm)	Focus-to-detector distance (cm)	Voltage (kVp) 50	60	70	80	90	100	Antiscatter grid
0					5 x 25	100[1]							
1–2					7 x 35								
3–7	Vertical	Supine	ap	–	9 x 60	115							
8–12					12 x 84								
13–17					15 x 106								
18					16 x 106								

Annotations:
1 The patient is lying on the table top directly on the detector
Parameter combination not recommended, normalized organ doses not provided
Parameter combination recommended, normalized organ doses provided

Table 25.9 Radiation qualities of lower extremity Bucky table radiographs

Voltage (kVp)	Total filtration: Without additional filtration 1.5 mm Al	2.5 Al	With additional filtration 3.5 Al	4.5 Al	3.5 Al + 0.1 Cu	3.5 Al + 0.2 Cu
50						
60						
70						
80						
90						
100						

Annotations:
Parameter combination not recommended, normalized organ doses not provided
Parameter combination not recommended, normalized organ doses provided
Parameter combination recommended, normalized organ doses provided

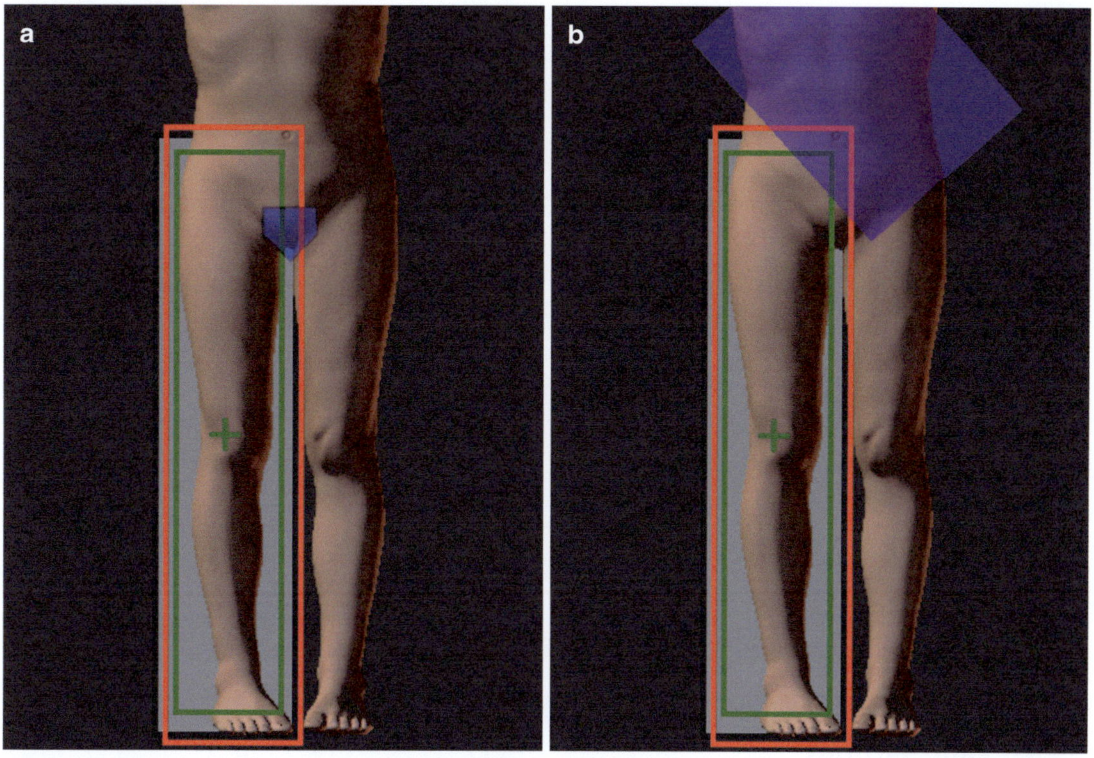

Fig. 25.1 Field settings for lower extremity radiographs in supine position. (**a**) Antero-posterior projection, male. (**b**) Antero-posterior projection, female. (© V. Rösenberger & M. Seidenbusch 2018). Green lines = optimal field settings; green cross = centring point; red lines = suboptimal field settings; grey shaded area = image detector; blue shaded area = lead mask shielding and testicular capsule; dark grey shaded area = Bucky table

Table 25.10 Radiographic techniques of lower extremity vertical stand radiographs

Age group (years)	Beam direction	Patient position	Beam projection	Typical patient diameter (cm)	Optimal field size (cm × cm)	Focus-to-detector distance (cm)	Voltage (kVp)						Antiscatter grid
							50	60	70	80	90	100	
0	This technique is not applied in these age groups												
1–2													
3–7													
8–12	Horizontal	Standing	ap	–	12 × 84	150							
13–17					15 × 106								
18					16 × 106								
Annotations:													
	Parameter combination not recommended, normalized organ doses not provided												
	Parameter combination recommended, normalized organ doses provided												

Table 25.11 Radiation qualities of lower extremity vertical stand radiographs

Voltage (kVp)	Total filtration					
	Without additional filtration		With additional filtration			
	1.5 mm Al	2.5 Al	3.5 Al	4.5 Al	3.5 Al + 0.1 Cu	3.5 Al + 0.2 Cu
50						
60						
70						
80						
90						
100						
Annotations:						
	Parameter combination not recommended, normalized organ doses not provided					
	Parameter combination not recommended, normalized organ doses provided					
	Parameter combination recommended, normalized organ doses provided					

Table 25.12 Inconsistencies between the mathematical MIRD phantoms and real human anatomy

Anatomical region	Inconsistencies
Extremities	No hip joint, no knee, and no ankle joint have been inserted

For details, see the text

25.2 Indications and Contraindications

Indications and contraindications for paediatric lower extremity radiographs are given in Tables 25.1 and 25.2.

25.3 Image Criteria

The image criteria required for lower extremity radiographs are listed in Table 25.3.

25.4 Field Settings

In Table 25.4, patient positions and beam projections used for lower extremity radiographs are listed, and in Table 25.5, the optimal field settings are listed which are required in order to fulfil the image criteria mentioned in Table 25.3.

25.5 Patient Preparation and Radiographic Procedure

See Tables 25.6, 25.7 and [1].

25.6 Exposure Parameters

In Tables 25.8, 25.9, 25.10, and 25.11, exposure parameter combinations used for lower extremity radiographs in various age groups are listed. Dark grey shaded table cells refer to exposure parameter combinations which are not recommended for paediatric lower extremity radiographs and for which no normalized organ doses have been provided. Light grey and light blue shaded table cells refer to exposure parameter combinations for which normalized organ doses have been provided with light grey shaded table cells referring to exposure parameter combinations being not recommended and light blue shaded table cells referring to exposure parameter combinations being recommended for paediatric lower extremity radiographs. Note that due to the fact that the radiation dose is an exponential function of the patient diameter, it may make sense to use the patient diameters instead of the patient's age groups for determining the adequate exposure parameters from Tables 25.8 and 25.10. In this case, for determining the corresponding normalized organ doses from Tables 25.16, 25.17, 25.18, 25.19, 25.20, 25.21, 25.22, 25.23, 25.24, 25.25, 25.26, 25.27, 25.28, 25.29, 25.30, 25.31, 25.32, and 25.33, the age group related to the patient diameter should be used.

25.7 Normalized Organ Doses

For calculating normalized organ doses, the field settings described in Sect. 25.4 were transferred to the mathematical MIRD phantoms. Inconsistencies between the geometrical anatomy of the mathematical phantoms and the real human anatomy have been listed in Tables 25.12 whenever they are of relevance for the validity of the normalized organ doses.

In Tables 25.13, 25.14, and 25.15, optimal and suboptimal field settings are listed for the different age groups and the beam projections given in Tables 25.8 and 25.10. Green lines refer to optimal field settings, red lines refer to suboptimal field settings, respectively. In the tables, the optimal and suboptimal field settings which were used for performing the calculations of the normalized organ doses provided in Tables 25.16, 25.17, 25.18, 25.19, 25.20, 25.21, 25.22, 25.23,

25.24, 25.25, 25.26, 25.27, 25.28, 25.29, 25.30, 25.31, 25.32, and 25.33 were drawn in on the skin surface of the mathematical phantoms. The virtual mathematical radiographs resulting from these field settings were depicted in the tables with and without bone structures, respectively, in order to enable the reader to visually compare the virtual mathematical radiographs with the real radiographs made under clinical conditions. For comparing the virtual mathematical field settings with the real clinical field settings, the virtual mathematical field edges were also listed in the tables.

As the normalized organ doses provided in Tables 25.16, 25.17, 25.18, 25.19, 25.20, 25.21, 25.22, 25.23, 25.24, 25.25, 25.26, 25.27, 25.28, 25.29, 25.30, 25.31, 25.32, and 25.33 refer to entrance kerma values, optimal and suboptimal field sizes were provided in the tables in order to enable the reader to calculate entrance air kerma dose values from clinically measured dose area product values. Please note that these optimal and suboptimal field sizes refer to the mathematical phantoms of an age group only and not to real patients with corresponding ages.

Please note also that the anatomy of the mathematical phantoms has been idealized. Therefore, the virtual field settings for calculating the normalized organ doses have been defined with a view to ensuring that the resulting virtual organ doses would be consistent with the real organ doses which would be achieved during real X-ray examinations being performed with the real field settings recommended in Sect. 25.4. Thus, the field settings shown in the following tables may not be consistent with the clinical field settings as recommended in Sect. 25.4.

Table 25.13 Field settings—lower extremity ap

Age group (years)		Anatomical landmarks			
		Optimal field setting		**Suboptimal field setting**	
0	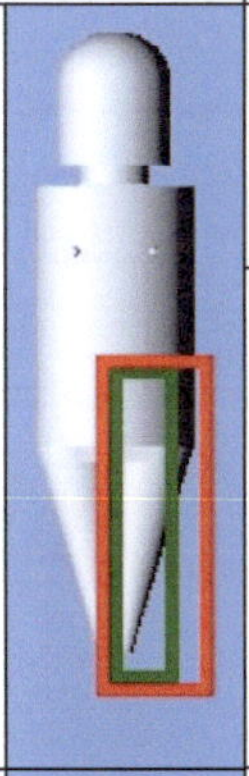	Upper field edge: Just below the spina iliaca ant. sup. Lower field edge: Just below the tips of the toes Medial field edge: About 2 cm lateral to the symphysis pubis Lateral field edge: Just beneath the skin surface Field size = field width × field height: 5 cm × 25 cm = 125 cm²		Upper field edge: Just below the iliac crest Lower field edge: Just below the tips of the toes Medial field edge: Close to the symphysis Lateral field edge: Close to the skin surface of the pelvis including the distal forearm and the hand Field size = field width × field height: 8 cm × 26 cm = 208 cm²	
				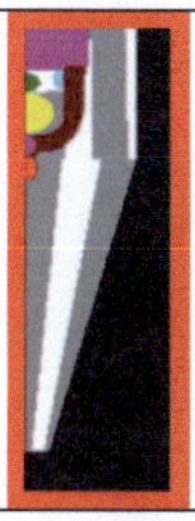	
1–2	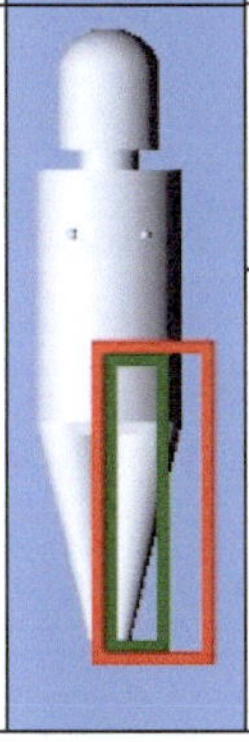	Upper field edge: Just below the spina iliaca ant. sup. Lower field edge: Just below the tips of the toes Medial field edge: About 2 cm lateral to the symphysis pubis Lateral field edge: Just beneath the skin surface Field size = field width × field height: 7 cm × 35 cm = 245 cm²		Upper field edge: Just below the iliac crest Lower field edge: Just below the tips of the toes Medial field edge: Close to the symphysis Lateral field edge: Close to the skin surface of the pelvis including the distal forearm and the hand Field size = field width × field height:12 cm × 36 cm = 432 cm²	

Table 25.14 Field settings—lower extremity ap

Age group (years)		Anatomical landmarks			
		Optimal field setting		Suboptimal field setting	
3–7	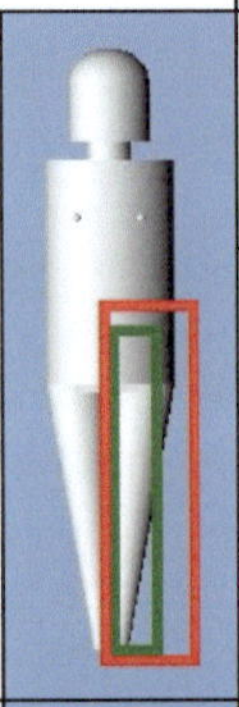	Upper field edge: Just below the spina iliaca ant. sup. Lower field edge: Just below the tips of the toes Medial field edge: About 3 cm lateral to the symphysis pubis Lateral field edge: Just beneath the skin surface Field size = field width × field height: 9 cm × 60 cm = 540 cm^2		Upper field edge: Just below the iliac crest Lower field edge: Just below the tips of the toes Medial field edge: Close to the symphysis Lateral field edge: Close to the skin surface of the pelvis including the distal forearm and the hand Field size = field width × field height:15 cm × 65 cm = 975 cm^2	

8–12	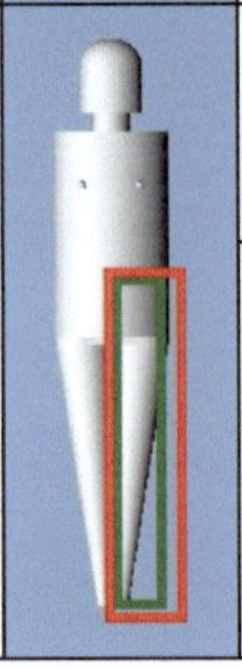	Upper field edge: Just below the spina iliaca ant. sup. Lower field edge: Just below the tips of the toes Medial field edge: About 4 cm lateral to the symphysis pubis Lateral field edge: Just beneath the skin surface Field size = field width × field height: 12 cm × 84 cm = 1008 cm^2		Upper field edge: Just below the iliac crest Lower field edge: Just below the tips of the toes Medial field edge: Close to the symphysis Lateral field edge: Close to the skin surface of the pelvis including the distal forearm and the hand Field size = field width × field height: 18 cm × 90 cm = 1620 cm^2	

Table 25.15 Field settings—lower extremity ap

Age group (years)		Anatomical landmarks			
		Optimal field setting		Suboptimal field setting	
13–17	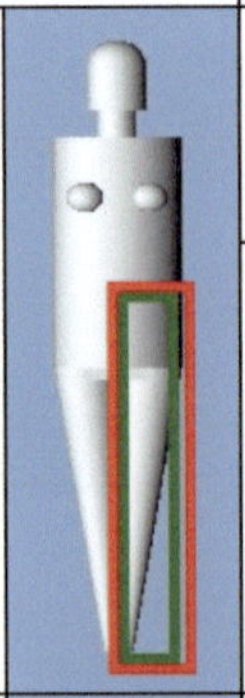	Upper field edge: Just below the spina iliaca ant. sup. Lower field edge: Just below the tips of the toes Medial field edge: About 4 cm lateral to the symphysis pubis Lateral field edge: Just beneath the skin surface Field size = field width × field height: 15 cm × 106 cm = 1590 cm^2		Upper field edge: Just below the iliac crest Lower field edge: Just below the tips of the toes Medial field edge: Close to the symphysis Lateral field edge: Close to the skin surface of the pelvis including the distal forearm and the hand Field size = field width × field height: 20 cm × 110 cm = 2200 cm^2	

18	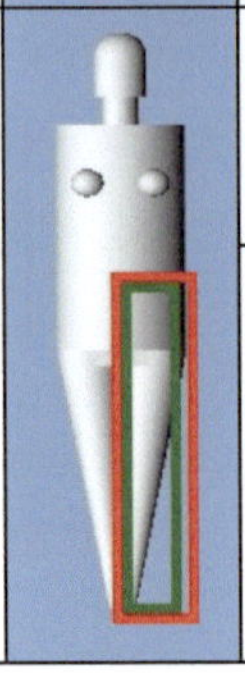	Upper field edge: Just below the spina iliaca ant. sup. Lower field edge: Just below the tips of the toes Medial field edge: About 5 cm lateral to the symphysis pubis Lateral field edge: Just beneath the skin surface Field size = field width × field height: 16 cm × 106 cm = 1696 cm^2		Upper field edge: Just below the iliac crest Lower field edge: Just below the tips of the toes Medial field edge: Close to the symphysis Lateral field edge: Close to the skin surface of the pelvis including the distal forearm and the hand Field size = field width × field height: 22 cm × 112 cm = 2464 cm^2	

Table 25.16 Normalized organ doses—Lower extremities ap—<1 year

Beam projection:			ap									**Table cells:**
Focus-to-detector distance (cm):			100									(+) = normal font = optimal field setting
Tube voltage (kVp):			60									**(−) = bold font = suboptimal field setting**
Total filtrations:			See last table row.									Empty table cells correspond to values < 0.01
Organ or tissue	**Organ dose D_T : Entrance dose K_E**							**(mSv ∗ mGy^{-1})**				
	(+)	(−)	(+)	(−)	(+)	(−)	(+)	(−)	(+)	(−)	(+)	(−)
Skull												
Brain												
Oral mucosa												
Salivary glands												
Cervical spine												
Thyroid												
Thymus												
Thoracic spine										**0.01**		
Claviculae												
Scapulae												
Ribs										**0.01**		
Oesophagus												
Trachea												
Lung												
Mamma												
Heart												
Stomach		**0.02**	0.01	**0.03**	0.01	**0.03**	0.01	**0.03**	0.02	**0.04**	0.02	**0.04**
Spleen		**0.01**		**0.01**		**0.02**		**0.02**	0.01	**0.03**		**0.02**
Liver						**0.01**		**0.01**		**0.01**		**0.01**
Gall bladder		**0.02**	0.01	**0.02**	0.01	**0.03**	0.02	**0.03**	0.02	**0.04**	0.02	**0.04**
Pancreas		**0.01**		**0.01**		**0.02**		**0.02**	0.01	**0.02**		**0.02**
Small intestine	0.08	**0.18**	0.11	**0.22**	0.12	**0.25**	0.13	**0.28**	0.16	**0.32**	0.15	**0.31**
Large intestine	0.02	**0.06**	0.03	**0.08**	0.03	**0.09**	0.04	**0.10**	0.04	**0.11**	0.04	**0.11**
Sigma/rectum	0.27	**0.35**	0.34	**0.44**	0.39	**0.50**	0.43	**0.55**	0.50	**0.64**	0.49	**0.62**
Lumbar spine	0.03	**0.08**	0.03	**0.10**	0.04	**0.12**	0.05	**0.14**	0.06	**0.17**	0.05	**0.16**
Pelvis	0.26	**0.31**	0.34	**0.40**	0.39	**0.46**	0.44	**0.52**	0.52	**0.62**	0.50	**0.59**
Adrenals										**0.01**		
Kidneys		**0.01**		**0.01**		**0.02**		**0.02**	0.01	**0.03**	0.01	**0.02**
Urinary bladder	0.32	**0.33**	0.38	**0.39**	0.42	**0.43**	0.45	**0.47**	0.51	**0.53**	0.50	**0.52**
Uterus	0.19	**0.21**	0.24	**0.27**	0.28	**0.30**	0.31	**0.34**	0.36	**0.39**	0.35	**0.38**
Ovaries	0.15	**0.17**	0.19	**0.21**	0.22	**0.25**	0.24	**0.27**	0.29	**0.32**	0.28	**0.31**
Prostate	0.27	**0.27**	0.33	**0.33**	0.37	**0.38**	0.40	**0.41**	0.46	**0.47**	0.45	**0.46**
Testes	0.58	**0.58**	0.64	**0.64**	0.67	**0.67**	0.69	**0.70**	0.73	**0.74**	0.73	**0.73**
Upper arm												
Elbow								**0.01**		**0.01**		**0.01**
Forearm		**0.48**		**0.57**		**0.63**		**0.68**		**0.76**		**0.76**
Upper leg	0.54	**0.57**	0.68	**0.71**	0.78	**0.81**	0.85	**0.89**	0.99	**1.04**	0.97	**1.02**
Knee	0.63	**0.64**	0.77	**0.79**	0.87	**0.89**	0.95	**0.97**	1.09	**1.11**	1.08	**1.10**
Lower leg	0.82	**0.82**	0.96	**0.97**	1.06	**1.07**	1.13	**1.14**	1.25	**1.26**	1.26	**1.27**
Lymph nodes	0.03	**0.06**	0.04	**0.07**	0.05	**0.08**	0.05	**0.09**	0.06	**0.11**	0.06	**0.10**
Skin	0.13	**0.17**	0.14	**0.18**	0.14	**0.19**	0.15	**0.19**	0.16	**0.20**	0.15	**0.20**
Muscle	0.11	**0.13**	0.12	**0.15**	0.13	**0.16**	0.14	**0.18**	0.16	**0.20**	0.15	**0.19**
Skeleton	0.19	**0.23**	0.23	**0.28**	0.26	**0.32**	0.28	**0.34**	0.32	**0.39**	0.32	**0.39**
Red bone marrow	0.07	**0.09**	0.09	**0.10**	0.10	**0.12**	0.11	**0.13**	0.12	**0.15**	0.12	**0.14**
Whole body dose	0.09	**0.12**	0.11	**0.14**	0.12	**0.15**	0.13	**0.16**	0.14	**0.18**	0.14	**0.18**
ED (ICRP 60)	0.12	**0.13**	0.14	**0.16**	0.15	**0.17**	0.16	**0.18**	0.18	**0.21**	0.18	**0.20**
ED (ICRP 103)	0.08	**0.09**	0.09	**0.11**	0.10	**0.12**	0.11	**0.13**	0.12	**0.15**	0.12	**0.14**
Total filtration (mm) **Al**	**1.5**		**2.5**		**3.5**		**4.5**		**3.5**		**3.5**	
Cu	**0**		**0**		**0**		**0**		**0.1**		**0.2**	

Table 25.17 Normalized organ doses—Lower extremities ap—<1 year

Beam projection:			ap									**Table cells:**
Focus-to-detector distance (cm):			100				(+) = normal font = optimal field setting					
Tube voltage (kVp):			70				**(−) = bold font = suboptimal field setting**					
Total filtrations:			See last table row.				Empty table cells correspond to values < 0.01					
Organ or tissue	**Organ dose D_T : Entrance dose K_E**								**(mSv * mGy^{-1})**			
	(+)	(−)	(+)	(−)	(+)	(−)	(+)	(−)	(+)	(−)	(+)	(−)
Skull												
Brain												
Oral mucosa												
Salivary glands												
Cervical spine												
Thyroid												
Thymus												
Thoracic spine						**0.01**		**0.01**		**0.01**		**0.02**
Claviculae												
Scapulae												
Ribs						**0.01**		**0.01**		**0.01**		**0.02**
Oesophagus												**0.01**
Trachea												
Lung												
Mamma												
Heart												
Stomach	0.01	**0.02**	0.01	**0.03**	0.02	**0.04**	0.02	**0.04**	0.02	**0.05**	0.02	**0.06**
Spleen		**0.01**		**0.02**		**0.02**	0.01	**0.03**	0.01	**0.03**	0.02	**0.04**
Liver				**0.01**		**0.01**		**0.01**	0.01	**0.02**	0.01	**0.02**
Gall bladder	0.01	**0.02**	0.01	**0.03**	0.02	**0.04**	0.02	**0.04**	0.02	**0.05**	0.03	**0.06**
Pancreas		**0.01**		**0.02**		**0.02**	0.01	**0.02**	0.01	**0.03**	0.02	**0.03**
Small intestine	0.10	**0.21**	0.12	**0.25**	0.14	**0.28**	0.15	**0.31**	0.18	**0.35**	0.20	**0.39**
Large intestine	0.03	**0.07**	0.03	**0.09**	0.04	**0.10**	0.04	**0.11**	0.05	**0.13**	0.06	**0.15**
Sigma/rectum	0.32	**0.41**	0.39	**0.49**	0.44	**0.56**	0.48	**0.61**	0.55	**0.70**	0.61	**0.77**
Lumbar spine	0.03	**0.10**	0.04	**0.13**	0.05	**0.15**	0.06	**0.17**	0.07	**0.21**	0.09	**0.24**
Pelvis	0.31	**0.37**	0.39	**0.46**	0.45	**0.53**	0.49	**0.58**	0.57	**0.68**	0.65	**0.77**
Adrenals								**0.01**		**0.01**		**0.02**
Kidneys		**0.01**		**0.02**	0.01	**0.02**	0.01	**0.03**	0.02	**0.03**	0.02	**0.04**
Urinary bladder	0.35	**0.36**	0.41	**0.43**	0.46	**0.47**	0.49	**0.51**	0.54	**0.56**	0.59	**0.61**
Uterus	0.23	**0.25**	0.28	**0.30**	0.31	**0.34**	0.34	**0.38**	0.39	**0.43**	0.44	**0.48**
Ovaries	0.18	**0.20**	0.22	**0.25**	0.25	**0.28**	0.28	**0.31**	0.32	**0.37**	0.36	**0.41**
Prostate	0.30	**0.31**	0.37	**0.37**	0.41	**0.42**	0.44	**0.45**	0.50	**0.51**	0.54	**0.56**
Testes	0.61	**0.61**	0.66	**0.66**	0.69	**0.70**	0.71	**0.72**	0.75	**0.76**	0.78	**0.78**
Upper arm												
Elbow				**0.01**		**0.01**		**0.01**		**0.02**		**0.02**
Forearm		**0.52**		**0.61**		**0.67**		**0.71**		**0.79**		**0.84**
Upper leg	0.62	**0.65**	0.76	**0.79**	0.86	**0.90**	0.93	**0.98**	1.07	**1.12**	1.17	**1.23**
Knee	0.71	**0.72**	0.85	**0.87**	0.95	**0.97**	1.02	**1.04**	1.15	**1.18**	1.25	**1.28**
Lower leg	0.88	**0.89**	1.02	**1.03**	1.11	**1.12**	1.18	**1.19**	1.28	**1.29**	1.35	**1.36**
Lymph nodes	0.04	**0.07**	0.05	**0.08**	0.05	**0.09**	0.06	**0.10**	0.06	**0.12**	0.07	**0.13**
Skin	0.14	**0.18**	0.14	**0.19**	0.15	**0.19**	0.15	**0.20**	0.16	**0.21**	0.17	**0.22**
Muscle	0.12	**0.14**	0.13	**0.16**	0.14	**0.18**	0.15	**0.19**	0.17	**0.21**	0.18	**0.22**
Skeleton	0.21	**0.26**	0.25	**0.31**	0.28	**0.34**	0.31	**0.37**	0.34	**0.42**	0.37	**0.45**
Red bone marrow	0.08	**0.10**	0.10	**0.12**	0.11	**0.13**	0.12	**0.14**	0.13	**0.16**	0.15	**0.18**
Whole body dose	0.10	**0.13**	0.12	**0.15**	0.13	**0.16**	0.14	**0.17**	0.15	**0.19**	0.17	**0.21**
ED (ICRP 60)	0.13	**0.15**	0.15	**0.17**	0.16	**0.19**	0.18	**0.20**	0.19	**0.22**	0.21	**0.24**
ED (ICRP 103)	0.08	**0.10**	0.10	**0.12**	0.11	**0.13**	0.12	**0.14**	0.13	**0.16**	0.14	**0.17**
Total filtration (mm) **Al**	**1.5**		**2.5**		**3.5**		**4.5**		**3.5**		**3.5**	
Cu	**0**		**0**		**0**		**0**		**0.1**		**0.2**	

Table 25.18 Normalized organ doses—Lower extremities ap—1–2 years

Beam projection:		ap										**Table cells:**
Focus-to-detector distance (cm):		100										(+) = normal font = optimal field setting
Tube voltage (kVp):		60										**(−) = bold font = suboptimal field setting**
Total filtrations:		See last table row.										Empty table cells correspond to values < 0.01
Organ or tissue	**Organ dose D_T : Entrance dose K_E (mSv ∗ mGy^{-1})**											
	(+)	(−)	(+)	(−)	(+)	(−)	(+)	(−)	(+)	(−)	(+)	(−)
Skull												
Brain												
Oral mucosa												
Salivary glands												
Cervical spine												
Thyroid												
Thymus												
Thoracic spine												
Claviculae												
Scapulae												
Ribs												
Oesophagus												
Trachea												
Lung												
Mamma												
Heart												
Stomach		**0.01**		**0.02**		**0.02**		**0.03**	0.01	**0.03**		**0.03**
Spleen				**0.01**		**0.01**		**0.01**		**0.02**		**0.02**
Liver												
Gall bladder		**0.02**		**0.02**	0.01	**0.02**	0.01	**0.03**	0.02	**0.04**	0.01	**0.03**
Pancreas						**0.01**		**0.01**		**0.02**		**0.01**
Small intestine	0.05	**0.14**	0.07	**0.18**	0.08	**0.21**	0.09	**0.23**	0.11	**0.27**	0.10	**0.26**
Large intestine	0.01	**0.04**	0.02	**0.06**	0.02	**0.07**	0.03	**0.08**	0.03	**0.09**	0.03	**0.08**
Sigma/rectum	0.20	**0.28**	0.26	**0.35**	0.30	**0.41**	0.34	**0.45**	0.40	**0.54**	0.38	**0.51**
Lumbar spine	0.01	**0.05**	0.02	**0.07**	0.02	**0.08**	0.03	**0.09**	0.04	**0.12**	0.03	**0.10**
Pelvis	0.18	**0.23**	0.24	**0.30**	0.28	**0.35**	0.31	**0.40**	0.39	**0.49**	0.36	**0.45**
Adrenals												
Kidneys				**0.01**		**0.01**		**0.01**		**0.02**		**0.02**
Urinary bladder	0.28	**0.29**	0.34	**0.35**	0.38	**0.40**	0.42	**0.43**	0.48	**0.50**	0.46	**0.48**
Uterus	0.16	**0.18**	0.20	**0.23**	0.24	**0.27**	0.26	**0.30**	0.31	**0.36**	0.30	**0.34**
Ovaries	0.11	**0.13**	0.14	**0.17**	0.17	**0.21**	0.19	**0.23**	0.22	**0.28**	0.21	**0.26**
Prostate	0.21	**0.22**	0.27	**0.28**	0.31	**0.32**	0.34	**0.35**	0.40	**0.42**	0.38	**0.40**
Testes	0.59	**0.57**	0.65	**0.63**	0.69	**0.67**	0.72	**0.70**	0.76	**0.74**	0.76	**0.74**
Upper arm												
Elbow												
Forearm		**0.38**		**0.46**		**0.52**		**0.57**		**0.65**		**0.64**
Upper leg	0.38	**0.40**	0.49	**0.52**	0.58	**0.61**	0.64	**0.68**	0.77	**0.82**	0.73	**0.77**
Knee	0.45	**0.46**	0.58	**0.59**	0.67	**0.68**	0.74	**0.76**	0.88	**0.90**	0.84	**0.86**
Lower leg	0.60	**0.61**	0.74	**0.75**	0.84	**0.85**	0.91	**0.92**	1.05	**1.06**	1.04	**1.05**
Lymph nodes	0.02	**0.05**	0.03	**0.06**	0.03	**0.07**	0.04	**0.08**	0.04	**0.09**	0.04	**0.09**
Skin	0.14	**0.18**	0.15	**0.19**	0.15	**0.20**	0.16	**0.20**	0.17	**0.21**	0.16	**0.21**
Muscle	0.09	**0.11**	0.11	**0.13**	0.12	**0.14**	0.13	**0.16**	0.14	**0.17**	0.14	**0.17**
Skeleton	0.16	**0.19**	0.20	**0.24**	0.23	**0.27**	0.25	**0.30**	0.30	**0.35**	0.29	**0.34**
Red bone marrow	0.04	**0.05**	0.05	**0.07**	0.06	**0.08**	0.07	**0.08**	0.08	**0.10**	0.08	**0.09**
Whole body dose	0.08	**0.10**	0.10	**0.12**	0.11	**0.14**	0.12	**0.15**	0.14	**0.17**	0.13	**0.16**
ED (ICRP 60)	0.11	**0.12**	0.12	**0.14**	0.14	**0.15**	0.15	**0.17**	0.16	**0.19**	0.16	**0.18**
ED (ICRP 103)	0.06	**0.07**	0.08	**0.09**	0.09	**0.10**	0.09	**0.11**	0.11	**0.13**	0.10	**0.12**
Total filtration (mm) **Al**	**1.5**		**2.5**		**3.5**		**4.5**		**3.5**		**3.5**	
Cu	**0**		**0**		**0**		**0**		**0.1**		**0.2**	

Table 25.19 Normalized organ doses—Lower extremities ap—1–2 years

Beam projection:			ap									**Table cells:**
Focus-to-detector distance (cm):			100									(+) = normal font = optimal field setting
Tube voltage (kVp):			70									**(−) = bold font = suboptimal field setting**
Total filtrations:			See last table row.									Empty table cells correspond to values < 0.01
Organ or tissue	**Organ dose D_T : Entrance dose K_E (mSv ∗ mGy^{-1})**											
	(+)	(−)	(+)	(−)	(+)	(−)	(+)	(−)	(+)	(−)	(+)	(−)
Skull												
Brain												
Oral mucosa												
Salivary glands												
Cervical spine												
Thyroid												
Thymus												
Thoracic spine												**0.01**
Claviculae												
Scapulae												
Ribs												**0.01**
Oesophagus												
Trachea												
Lung												
Mamma												
Heart												
Stomach		**0.02**		**0.02**		**0.03**	0.01	**0.03**	0.01	**0.04**	0.02	**0.05**
Spleen		**0.01**		**0.01**		**0.02**		**0.02**		**0.02**	0.01	**0.03**
Liver								**0.01**		**0.01**		**0.02**
Gall bladder		**0.02**	0.01	**0.03**	0.01	**0.03**	0.02	**0.04**	0.02	**0.04**	0.02	**0.05**
Pancreas				**0.01**		**0.02**		**0.02**		**0.02**	0.01	**0.03**
Small intestine	0.07	**0.17**	0.08	**0.21**	0.10	**0.24**	0.11	**0.26**	0.13	**0.31**	0.15	**0.35**
Large intestine	0.02	**0.06**	0.02	**0.07**	0.03	**0.08**	0.03	**0.09**	0.04	**0.11**	0.05	**0.13**
Sigma/rectum	0.25	**0.33**	0.31	**0.41**	0.35	**0.47**	0.39	**0.52**	0.46	**0.61**	0.52	**0.69**
Lumbar spine	0.02	**0.07**	0.03	**0.09**	0.03	**0.11**	0.04	**0.12**	0.05	**0.15**	0.06	**0.18**
Pelvis	0.22	**0.28**	0.29	**0.36**	0.33	**0.42**	0.37	**0.47**	0.45	**0.56**	0.52	**0.65**
Adrenals												**0.01**
Kidneys		**0.01**		**0.01**		**0.02**		**0.02**		**0.02**	0.01	**0.03**
Urinary bladder	0.32	**0.33**	0.38	**0.39**	0.43	**0.44**	0.46	**0.48**	0.52	**0.55**	0.57	**0.60**
Uterus	0.19	**0.22**	0.24	**0.27**	0.27	**0.32**	0.30	**0.35**	0.36	**0.41**	0.41	**0.47**
Ovaries	0.13	**0.17**	0.17	**0.21**	0.20	**0.24**	0.22	**0.27**	0.26	**0.32**	0.30	**0.37**
Prostate	0.25	**0.26**	0.31	**0.32**	0.35	**0.37**	0.39	**0.40**	0.45	**0.47**	0.50	**0.53**
Testes	0.62	**0.60**	0.68	**0.66**	0.72	**0.70**	0.75	**0.73**	0.79	**0.77**	0.82	**0.81**
Upper arm												
Elbow										**0.01**		**0.01**
Forearm		**0.42**		**0.51**		**0.57**		**0.62**		**0.70**		**0.76**
Upper leg	0.46	**0.49**	0.58	**0.61**	0.67	**0.71**	0.74	**0.78**	0.88	**0.93**	1.00	**1.06**
Knee	0.54	**0.55**	0.66	**0.68**	0.76	**0.78**	0.84	**0.86**	0.97	**1.00**	1.09	**1.12**
Lower leg	0.68	**0.69**	0.82	**0.83**	0.92	**0.93**	0.99	**1.00**	1.12	**1.13**	1.22	**1.23**
Lymph nodes	0.03	**0.06**	0.04	**0.07**	0.04	**0.08**	0.04	**0.09**	0.05	**0.10**	0.06	**0.12**
Skin	0.14	**0.18**	0.15	**0.20**	0.16	**0.20**	0.17	**0.21**	0.17	**0.22**	0.18	**0.23**
Muscle	0.10	**0.12**	0.12	**0.14**	0.13	**0.16**	0.14	**0.17**	0.15	**0.19**	0.17	**0.21**
Skeleton	0.18	**0.22**	0.23	**0.27**	0.26	**0.31**	0.28	**0.34**	0.33	**0.39**	0.37	**0.44**
Red bone marrow	0.05	**0.06**	0.06	**0.08**	0.07	**0.09**	0.08	**0.10**	0.10	**0.12**	0.11	**0.13**
Whole body dose	0.10	**0.12**	0.11	**0.14**	0.12	**0.15**	0.13	**0.17**	0.15	**0.19**	0.16	**0.20**
ED (ICRP 60)	0.12	**0.13**	0.14	**0.15**	0.15	**0.17**	0.16	**0.18**	0.18	**0.21**	0.20	**0.23**
ED (ICRP 103)	0.07	**0.09**	0.09	**0.10**	0.10	**0.12**	0.10	**0.13**	0.12	**0.14**	0.13	**0.16**
Total filtration (mm) Al	**1.5**		**2.5**		**3.5**		**4.5**		**3.5**		**3.5**	
Cu	**0**		**0**		**0**		**0**		**0.1**		**0.2**	

Table 25.20 Normalized organ doses—Lower extremities ap—3–7 years

Beam projection:		ap											**Table cells:**
Focus-to-detector distance (cm):		115						(+) = normal font = optimal field setting					
Tube voltage (kVp):		60						**(−) = bold font = suboptimal field setting**					
Total filtrations:		See last table row.						Empty table cells correspond to values < 0.01					
Organ or tissue		**Organ dose D_T : Entrance dose K_E**						**(mSv * mGy^{-1})**					
		(+)	(−)	(+)	(−)	(+)	(−)	(+)	(−)	(+)	(−)	(+)	(−)
Skull													
Brain													
Oral mucosa													
Salivary glands													
Cervical spine													
Thyroid													
Thymus													
Thoracic spine													
Claviculae													
Scapulae													
Ribs													
Oesophagus													
Trachea													
Lung													
Mamma													
Heart													
Stomach			**0.01**		**0.02**		**0.02**		**0.03**	0.01	**0.03**		**0.03**
Spleen							**0.01**		**0.01**		**0.02**		**0.01**
Liver													
Gall bladder			**0.01**		**0.02**		**0.02**	0.01	**0.02**	0.01	**0.03**	0.01	**0.03**
Pancreas									**0.01**		**0.01**		**0.01**
Small intestine		0.07	**0.12**	0.09	**0.15**	0.10	**0.18**	0.12	**0.20**	0.14	**0.24**	0.13	**0.23**
Large intestine		0.01	**0.05**	0.02	**0.07**	0.02	**0.08**	0.03	**0.09**	0.03	**0.11**	0.03	**0.10**
Sigma/rectum		0.19	**0.24**	0.25	**0.31**	0.29	**0.36**	0.33	**0.41**	0.40	**0.49**	0.37	**0.46**
Lumbar spine		0.02	**0.03**	0.02	**0.05**	0.03	**0.06**	0.03	**0.07**	0.04	**0.09**	0.03	**0.07**
Pelvis		0.15	**0.18**	0.20	**0.24**	0.24	**0.29**	0.27	**0.33**	0.34	**0.40**	0.30	**0.36**
Adrenals													
Kidneys							**0.01**		**0.01**		**0.02**		**0.01**
Urinary bladder		0.26	**0.22**	0.32	**0.27**	0.36	**0.31**	0.40	**0.34**	0.46	**0.40**	0.44	**0.38**
Uterus		0.14	**0.10**	0.19	**0.13**	0.22	**0.15**	0.25	**0.17**	0.30	**0.21**	0.28	**0.19**
Ovaries		0.10	**0.11**	0.13	**0.15**	0.15	**0.18**	0.17	**0.20**	0.21	**0.24**	0.19	**0.22**
Prostate		0.19	**0.11**	0.24	**0.14**	0.28	**0.17**	0.32	**0.19**	0.38	**0.23**	0.36	**0.21**
Testes		0.58	**0.16**	0.64	**0.20**	0.69	**0.22**	0.71	**0.24**	0.76	**0.27**	0.76	**0.26**
Upper arm													
Elbow													
Forearm			**0.36**		**0.45**		**0.51**		**0.56**		**0.65**		**0.64**
Upper leg		0.28	**0.30**	0.36	**0.39**	0.43	**0.47**	0.49	**0.53**	0.60	**0.65**	0.55	**0.59**
Knee		0.35	**0.37**	0.46	**0.48**	0.54	**0.56**	0.60	**0.63**	0.73	**0.76**	0.68	**0.71**
Lower leg		0.51	**0.53**	0.64	**0.66**	0.74	**0.76**	0.81	**0.83**	0.95	**0.97**	0.92	**0.95**
Lymph nodes		0.03	**0.04**	0.04	**0.05**	0.04	**0.06**	0.04	**0.07**	0.05	**0.08**	0.05	**0.07**
Skin		0.17	**0.18**	0.19	**0.19**	0.19	**0.20**	0.20	**0.21**	0.21	**0.22**	0.21	**0.22**
Muscle		0.11	**0.12**	0.12	**0.15**	0.14	**0.16**	0.15	**0.17**	0.17	**0.20**	0.16	**0.19**
Skeleton		0.15	**0.17**	0.19	**0.22**	0.22	**0.26**	0.25	**0.29**	0.29	**0.34**	0.28	**0.32**
Red bone marrow		0.04	**0.05**	0.05	**0.06**	0.06	**0.07**	0.07	**0.08**	0.08	**0.10**	0.08	**0.09**
Whole body dose		0.10	**0.11**	0.12	**0.14**	0.13	**0.15**	0.14	**0.17**	0.16	**0.19**	0.16	**0.18**
ED (ICRP 60)		0.10	**0.07**	0.12	**0.08**	0.13	**0.10**	0.14	**0.11**	0.16	**0.13**	0.16	**0.12**
ED (ICRP 103)		0.06	**0.05**	0.07	**0.06**	0.08	**0.07**	0.09	**0.08**	0.11	**0.10**	0.10	**0.09**
Total filtration (mm)	**Al**	**1.5**		**2.5**		**3.5**		**4.5**		**3.5**		**3.5**	
	Cu	**0**		**0**		**0**		**0**		**0.1**		**0.2**	

Table 25.21 Normalized organ doses—Lower extremities ap—3–7 years

Beam projection:			ap									**Table cells:**
Focus-to-detector distance (cm):			115									(+) = normal font = optimal field setting
Tube voltage (kVp):			70									**(−) = bold font = suboptimal field setting**
Total filtrations:			See last table row.									Empty table cells correspond to values < 0.01
Organ or tissue	**Organ dose D_T : Entrance dose K_E**								**(mSv ∗ mGy^{-1})**			
	(+)	(−)	(+)	(−)	(+)	(−)	(+)	(−)	(+)	(−)	(+)	(−)
Skull												
Brain												
Oral mucosa												
Salivary glands												
Cervical spine												
Thyroid												
Thymus												
Thoracic spine												
Claviculae												
Scapulae												
Ribs												
Oesophagus												
Trachea												
Lung												
Mamma												
Heart												
Stomach		**0.02**		**0.03**		**0.03**	0.01	**0.03**	0.01	**0.04**	0.02	**0.05**
Spleen				**0.01**		**0.02**		**0.02**		**0.02**		**0.03**
Liver												**0.01**
Gall bladder		**0.02**	0.01	**0.02**	0.01	**0.03**	0.01	**0.03**	0.02	**0.04**	0.02	**0.05**
Pancreas				**0.01**		**0.01**		**0.01**		**0.02**		**0.02**
Small intestine	0.08	**0.15**	0.11	**0.19**	0.12	**0.21**	0.14	**0.24**	0.16	**0.28**	0.19	**0.33**
Large intestine	0.02	**0.06**	0.03	**0.08**	0.03	**0.09**	0.03	**0.10**	0.04	**0.13**	0.05	**0.15**
Sigma/rectum	0.24	**0.30**	0.30	**0.37**	0.35	**0.43**	0.39	**0.48**	0.46	**0.57**	0.53	**0.65**
Lumbar spine	0.02	**0.05**	0.03	**0.06**	0.04	**0.08**	0.04	**0.09**	0.06	**0.11**	0.07	**0.14**
Pelvis	0.19	**0.23**	0.25	**0.30**	0.29	**0.35**	0.33	**0.40**	0.40	**0.49**	0.47	**0.57**
Adrenals												
Kidneys		**0.01**		**0.01**		**0.02**		**0.02**		**0.02**	0.01	**0.03**
Urinary bladder	0.30	**0.26**	0.36	**0.31**	0.41	**0.36**	0.45	**0.39**	0.51	**0.45**	0.57	**0.51**
Uterus	0.18	**0.12**	0.23	**0.16**	0.26	**0.18**	0.29	**0.20**	0.35	**0.25**	0.40	**0.29**
Ovaries	0.12	**0.14**	0.16	**0.18**	0.18	**0.21**	0.21	**0.24**	0.25	**0.29**	0.29	**0.34**
Prostate	0.23	**0.14**	0.29	**0.17**	0.33	**0.20**	0.36	**0.22**	0.43	**0.27**	0.49	**0.31**
Testes	0.61	**0.19**	0.68	**0.22**	0.72	**0.25**	0.75	**0.27**	0.79	**0.30**	0.83	**0.34**
Upper arm												
Elbow												
Forearm		**0.42**		**0.51**		**0.57**		**0.62**		**0.71**		**0.79**
Upper leg	0.35	**0.38**	0.45	**0.48**	0.52	**0.56**	0.58	**0.63**	0.70	**0.76**	0.82	**0.88**
Knee	0.43	**0.45**	0.54	**0.57**	0.63	**0.66**	0.70	**0.73**	0.83	**0.87**	0.95	**0.99**
Lower leg	0.59	**0.61**	0.73	**0.75**	0.82	**0.84**	0.90	**0.92**	1.03	**1.06**	1.14	**1.17**
Lymph nodes	0.03	**0.05**	0.04	**0.06**	0.05	**0.07**	0.05	**0.08**	0.06	**0.09**	0.07	**0.11**
Skin	0.18	**0.19**	0.19	**0.20**	0.20	**0.21**	0.21	**0.22**	0.22	**0.23**	0.23	**0.24**
Muscle	0.12	**0.14**	0.14	**0.16**	0.15	**0.18**	0.16	**0.19**	0.18	**0.22**	0.20	**0.24**
Skeleton	0.18	**0.21**	0.22	**0.26**	0.26	**0.30**	0.28	**0.33**	0.33	**0.39**	0.38	**0.44**
Red bone marrow	0.05	**0.06**	0.06	**0.08**	0.07	**0.09**	0.08	**0.10**	0.10	**0.12**	0.12	**0.14**
Whole body dose	0.11	**0.13**	0.13	**0.15**	0.15	**0.17**	0.16	**0.19**	0.18	**0.21**	0.20	**0.23**
ED (ICRP 60)	0.11	**0.08**	0.13	**0.10**	0.15	**0.11**	0.16	**0.13**	0.18	**0.15**	0.20	**0.17**
ED (ICRP 103)	0.07	**0.06**	0.08	**0.08**	0.09	**0.09**	0.10	**0.10**	0.12	**0.12**	0.13	**0.13**
Total filtration (mm) Al	**1.5**		**2.5**		**3.5**		**4.5**		**3.5**		**3.5**	
Cu	**0**		**0**		**0**		**0**		**0.1**		**0.2**	

Table 25.22 Normalized organ doses—Lower extremities ap—8–12 years

Beam projection:			ap									**Table cells:**
Focus-to-detector distance (cm):			115									(+) = normal font = optimal field setting
Tube voltage (kVp):			60									**(−) = bold font = suboptimal field setting**
Total filtrations:			See last table row.									Empty table cells correspond to values < 0.01
Organ or tissue	**Organ dose D_T : Entrance dose K_E**								**(mSv * mGy^{-1})**			
	(+)	(−)	(+)	(−)	(+)	(−)	(+)	(−)	(+)	(−)	(+)	(−)
Skull												
Brain												
Oral mucosa												
Salivary glands												
Cervical spine												
Thyroid												
Thymus												
Thoracic spine												
Claviculae												
Scapulae												
Ribs												
Oesophagus												
Trachea												
Lung												
Mamma												
Heart												
Stomach		**0.01**		**0.02**		**0.02**	0.01	**0.02**	0.01	**0.03**	0.01	**0.03**
Spleen								**0.01**		**0.01**		**0.01**
Liver												
Gall bladder		**0.01**		**0.02**		**0.02**	0.01	**0.02**	0.01	**0.03**	0.01	**0.02**
Pancreas										**0.01**		
Small intestine	0.08	**0.12**	0.11	**0.16**	0.13	**0.18**	0.15	**0.21**	0.18	**0.25**	0.16	**0.23**
Large intestine	0.02	**0.06**	0.03	**0.08**	0.03	**0.09**	0.04	**0.10**	0.05	**0.13**	0.04	**0.11**
Sigma/rectum	0.18	**0.21**	0.24	**0.28**	0.28	**0.33**	0.32	**0.37**	0.39	**0.46**	0.36	**0.42**
Lumbar spine	0.02	**0.04**	0.03	**0.05**	0.03	**0.06**	0.04	**0.07**	0.05	**0.10**	0.04	**0.07**
Pelvis	0.13	**0.14**	0.17	**0.19**	0.20	**0.23**	0.23	**0.27**	0.29	**0.34**	0.26	**0.29**
Adrenals												
Kidneys						**0.01**		**0.01**		**0.02**		**0.01**
Urinary bladder	0.23	**0.30**	0.29	**0.37**	0.33	**0.43**	0.37	**0.47**	0.43	**0.55**	0.41	**0.52**
Uterus	0.13	**0.18**	0.17	**0.23**	0.20	**0.28**	0.23	**0.31**	0.28	**0.38**	0.25	**0.35**
Ovaries	0.08	**0.09**	0.11	**0.13**	0.13	**0.15**	0.15	**0.17**	0.18	**0.22**	0.16	**0.19**
Prostate	0.17	**0.25**	0.23	**0.32**	0.27	**0.38**	0.30	**0.43**	0.36	**0.51**	0.34	**0.48**
Testes	0.57	**1.05**	0.64	**1.17**	0.69	**1.24**	0.72	**1.29**	0.77	**1.37**	0.76	**1.35**
Upper arm												
Elbow												
Forearm		**0.31**		**0.39**		**0.44**		**0.49**		**0.58**		**0.56**
Upper leg	0.22	**0.23**	0.28	**0.31**	0.34	**0.37**	0.39	**0.42**	0.48	**0.52**	0.43	**0.46**
Knee	0.28	**0.30**	0.37	**0.40**	0.44	**0.47**	0.50	**0.53**	0.61	**0.66**	0.56	**0.60**
Lower leg	0.43	**0.47**	0.54	**0.60**	0.63	**0.69**	0.70	**0.77**	0.83	**0.91**	0.79	**0.87**
Lymph nodes	0.03	**0.05**	0.04	**0.06**	0.05	**0.07**	0.05	**0.08**	0.06	**0.09**	0.06	**0.09**
Skin	0.21	**0.26**	0.22	**0.28**	0.23	**0.29**	0.24	**0.30**	0.25	**0.32**	0.25	**0.31**
Muscle	0.11	**0.14**	0.14	**0.17**	0.15	**0.19**	0.16	**0.20**	0.19	**0.23**	0.18	**0.22**
Skeleton	0.14	**0.17**	0.19	**0.22**	0.22	**0.26**	0.25	**0.29**	0.30	**0.36**	0.28	**0.33**
Red bone marrow	0.03	**0.03**	0.04	**0.04**	0.04	**0.05**	0.05	**0.06**	0.06	**0.07**	0.06	**0.06**
Whole body dose	0.11	**0.13**	0.13	**0.16**	0.15	**0.18**	0.16	**0.19**	0.18	**0.22**	0.17	**0.21**
ED (ICRP 60)	0.10	**0.16**	0.12	**0.18**	0.13	**0.20**	0.14	**0.22**	0.16	**0.24**	0.15	**0.23**
ED (ICRP 103)	0.06	**0.09**	0.07	**0.11**	0.08	**0.12**	0.09	**0.13**	0.10	**0.15**	0.10	**0.14**
Total filtration (mm) **Al**	**1.5**		**2.5**		**3.5**		**4.5**		**3.5**		**3.5**	
Cu	**0**		**0**		**0**		**0**		**0.1**		**0.2**	

Table 25.23 Normalized organ doses—Lower extremities ap—8–12 years

Beam projection:			ap									**Table cells:**
Focus-to-detector distance (cm):			115						(+) = normal font = optimal field setting			
Tube voltage (kVp):			70						**(−) = bold font = suboptimal field setting**			
Total filtrations:			See last table row.						Empty table cells correspond to values < 0.01			
Organ or tissue	**Organ dose D_T : Entrance dose K_E**								**(mSv ∗ mGy^{-1})**			
	(+)	(−)	(+)	(−)	(+)	(−)	(+)	(−)	(+)	(−)	(+)	(−)
Skull												
Brain												
Oral mucosa												
Salivary glands												
Cervical spine												
Thyroid												
Thymus												
Thoracic spine												
Claviculae												
Scapulae												
Ribs												
Oesophagus												
Trachea												
Lung												
Mamma												
Heart												
Stomach		**0.02**	0.01	**0.02**	0.01	**0.03**	0.01	**0.03**	0.02	**0.04**	0.02	**0.05**
Spleen				**0.01**		**0.01**		**0.02**		**0.02**	0.01	**0.02**
Liver												
Gall bladder		**0.02**	0.01	**0.02**	0.01	**0.03**	0.01	**0.03**	0.02	**0.04**	0.02	**0.05**
Pancreas						**0.01**		**0.01**		**0.02**	0.01	**0.02**
Small intestine	0.11	**0.15**	0.13	**0.19**	0.16	**0.22**	0.17	**0.25**	0.21	**0.30**	0.25	**0.35**
Large intestine	0.03	**0.07**	0.03	**0.09**	0.04	**0.11**	0.05	**0.12**	0.06	**0.15**	0.07	**0.18**
Sigma/rectum	0.23	**0.27**	0.30	**0.34**	0.35	**0.40**	0.39	**0.45**	0.47	**0.54**	0.54	**0.63**
Lumbar spine	0.03	**0.05**	0.04	**0.07**	0.05	**0.09**	0.06	**0.10**	0.07	**0.13**	0.09	**0.16**
Pelvis	0.17	**0.19**	0.22	**0.25**	0.26	**0.30**	0.29	**0.34**	0.36	**0.42**	0.43	**0.50**
Adrenals												
Kidneys		**0.01**		**0.01**		**0.02**		**0.02**	0.01	**0.03**	0.01	**0.03**
Urinary bladder	0.27	**0.35**	0.34	**0.43**	0.38	**0.49**	0.42	**0.54**	0.49	**0.62**	0.56	**0.70**
Uterus	0.16	**0.23**	0.21	**0.29**	0.24	**0.34**	0.27	**0.38**	0.33	**0.46**	0.39	**0.54**
Ovaries	0.11	**0.12**	0.14	**0.16**	0.16	**0.19**	0.18	**0.22**	0.23	**0.27**	0.27	**0.32**
Prostate	0.22	**0.31**	0.27	**0.39**	0.32	**0.45**	0.35	**0.50**	0.42	**0.60**	0.48	**0.68**
Testes	0.61	**1.11**	0.68	**1.22**	0.73	**1.29**	0.76	**1.34**	0.81	**1.43**	0.86	**1.50**
Upper arm												
Elbow												
Forearm		**0.36**		**0.44**		**0.50**		**0.55**		**0.64**		**0.71**
Upper leg	0.28	**0.30**	0.36	**0.39**	0.42	**0.46**	0.48	**0.52**	0.59	**0.63**	0.69	**0.75**
Knee	0.36	**0.38**	0.46	**0.49**	0.53	**0.57**	0.60	**0.64**	0.72	**0.77**	0.84	**0.90**
Lower leg	0.51	**0.56**	0.63	**0.69**	0.72	**0.79**	0.79	**0.87**	0.92	**1.02**	1.04	**1.15**
Lymph nodes	0.04	**0.06**	0.05	**0.07**	0.06	**0.08**	0.06	**0.09**	0.07	**0.11**	0.09	**0.13**
Skin	0.22	**0.27**	0.23	**0.29**	0.24	**0.31**	0.25	**0.32**	0.26	**0.33**	0.28	**0.35**
Muscle	0.13	**0.16**	0.15	**0.19**	0.17	**0.21**	0.18	**0.22**	0.21	**0.25**	0.23	**0.28**
Skeleton	0.18	**0.21**	0.23	**0.27**	0.26	**0.31**	0.29	**0.35**	0.35	**0.41**	0.40	**0.48**
Red bone marrow	0.04	**0.04**	0.05	**0.06**	0.06	**0.07**	0.06	**0.08**	0.08	**0.09**	0.09	**0.11**
Whole body dose	0.12	**0.15**	0.15	**0.18**	0.17	**0.20**	0.18	**0.22**	0.20	**0.25**	0.23	**0.28**
ED (ICRP 60)	0.11	**0.17**	0.13	**0.20**	0.14	**0.22**	0.16	**0.24**	0.18	**0.27**	0.20	**0.29**
ED (ICRP 103)	0.07	**0.10**	0.08	**0.12**	0.09	**0.14**	0.10	**0.15**	0.12	**0.17**	0.13	**0.19**
Total filtration (mm) Al	**1.5**		**2.5**		**3.5**		**4.5**		**3.5**		**3.5**	
Cu	**0**		**0**		**0**		**0**		**0.1**		**0.2**	

Table 25.24 Normalized organ doses—Lower extremities ap—13–17 years

Beam projection:			ap									**Table cells:**
Focus-to-detector distance (cm):			115						(+) = normal font = optimal field setting			
Tube voltage (kVp):			60						**(−) = bold font = suboptimal field setting**			
Total filtrations:			See last table row.						Empty table cells correspond to values < 0.01			
Organ or tissue	**Organ dose D_T : Entrance dose K_E**								**(mSv ∗ mGy⁻¹)**			
	(+)	(−)	(+)	(−)	(+)	(−)	(+)	(−)	(+)	(−)	(+)	(−)
Skull												
Brain												
Oral mucosa												
Salivary glands												
Cervical spine												
Thyroid												
Thymus												
Thoracic spine												
Claviculae												
Scapulae												
Ribs												
Oesophagus												
Trachea												
Lung												
Mamma												
Heart												
Stomach				**0.01**		**0.01**		**0.02**	0.01	**0.02**		**0.02**
Spleen												
Liver												
Gall bladder										**0.01**		
Pancreas												
Small intestine	0.07	**0.08**	0.09	**0.10**	0.10	**0.12**	0.12	**0.14**	0.15	**0.18**	0.13	**0.16**
Large intestine	0.02	**0.04**	0.03	**0.05**	0.04	**0.06**	0.04	**0.06**	0.05	**0.08**	0.04	**0.07**
Sigma/rectum	0.14	**0.16**	0.19	**0.21**	0.23	**0.25**	0.26	**0.28**	0.32	**0.35**	0.29	**0.31**
Lumbar spine	0.01	**0.02**	0.02	**0.02**	0.02	**0.03**	0.02	**0.03**	0.03	**0.04**	0.02	**0.03**
Pelvis	0.09	**0.10**	0.12	**0.13**	0.14	**0.16**	0.16	**0.18**	0.21	**0.23**	0.18	**0.20**
Adrenals												
Kidneys										**0.01**		
Urinary bladder	0.17	**0.19**	0.22	**0.25**	0.26	**0.29**	0.28	**0.32**	0.34	**0.39**	0.32	**0.36**
Uterus	0.23	**0.28**	0.31	**0.37**	0.37	**0.44**	0.42	**0.50**	0.52	**0.62**	0.46	**0.55**
Ovaries	0.06	**0.06**	0.08	**0.09**	0.10	**0.10**	0.11	**0.12**	0.14	**0.15**	0.12	**0.13**
Prostate	0.11	**0.15**	0.14	**0.20**	0.17	**0.24**	0.20	**0.27**	0.24	**0.33**	0.22	**0.30**
Testes	0.39	**0.48**	0.45	**0.56**	0.50	**0.61**	0.53	**0.65**	0.59	**0.72**	0.57	**0.71**
Upper arm												
Elbow												
Forearm		**0.24**		**0.30**		**0.35**		**0.39**		**0.46**		**0.44**
Upper leg	0.15	**0.16**	0.20	**0.21**	0.24	**0.26**	0.28	**0.29**	0.36	**0.37**	0.30	**0.32**
Knee	0.22	**0.23**	0.29	**0.30**	0.35	**0.36**	0.39	**0.41**	0.49	**0.51**	0.44	**0.45**
Lower leg	0.36	**0.38**	0.47	**0.49**	0.54	**0.57**	0.61	**0.64**	0.73	**0.77**	0.69	**0.72**
Lymph nodes	0.03	**0.03**	0.03	**0.04**	0.04	**0.05**	0.04	**0.05**	0.05	**0.06**	0.05	**0.06**
Skin	0.19	**0.23**	0.20	**0.25**	0.21	**0.26**	0.22	**0.27**	0.23	**0.28**	0.23	**0.28**
Muscle	0.10	**0.11**	0.12	**0.14**	0.14	**0.15**	0.15	**0.17**	0.17	**0.19**	0.16	**0.18**
Skeleton	0.12	**0.13**	0.15	**0.18**	0.18	**0.21**	0.20	**0.24**	0.25	**0.29**	0.23	**0.26**
Red bone marrow	0.01	**0.01**	0.02	**0.02**	0.02	**0.02**	0.03	**0.03**	0.03	**0.04**	0.03	**0.03**
Whole body dose	0.10	**0.11**	0.12	**0.13**	0.13	**0.15**	0.14	**0.16**	0.17	**0.19**	0.16	**0.18**
ED (ICRP 60)	0.07	**0.08**	0.09	**0.10**	0.10	**0.12**	0.11	**0.13**	0.12	**0.15**	0.12	**0.14**
ED (ICRP 103)	0.04	**0.05**	0.05	**0.07**	0.06	**0.08**	0.07	**0.08**	0.08	**0.10**	0.08	**0.09**
Total filtration (mm) **Al**	**1.5**		**2.5**		**3.5**		**4.5**		**3.5**		**3.5**	
Cu	**0**		**0**		**0**		**0**		**0.1**		**0.2**	

Table 25.25 Normalized organ doses—Lower extremities ap—13–17 years

Beam projection:			ap									**Table cells:**
Focus-to-detector distance (cm):			115									(+) = normal font = optimal field setting
Tube voltage (kVp):			70									**(−) = bold font = suboptimal field setting**
Total filtrations:			See last table row.									Empty table cells correspond to values < 0.01
Organ or tissue	**Organ dose D_T : Entrance dose K_E**							**(mSv ∗ mGy⁻¹)**				
	(+)	(−)	(+)	(−)	(+)	(−)	(+)	(−)	(+)	(−)	(+)	(−)
Skull												
Brain												
Oral mucosa												
Salivary glands												
Cervical spine												
Thyroid												
Thymus												
Thoracic spine												
Claviculae												
Scapulae												
Ribs												
Oesophagus												
Trachea												
Lung												
Mamma												**0.01**
Heart												
Stomach		**0.01**		**0.02**	0.01	**0.02**	0.01	**0.02**	0.02	**0.03**	0.02	**0.03**
Spleen										**0.01**		**0.02**
Liver												
Gall bladder								**0.01**	0.01	**0.01**	0.01	**0.02**
Pancreas												**0.01**
Small intestine	0.09	**0.10**	0.11	**0.13**	0.13	**0.16**	0.15	**0.18**	0.18	**0.21**	0.21	**0.25**
Large intestine	0.03	**0.05**	0.04	**0.06**	0.05	**0.07**	0.05	**0.08**	0.06	**0.10**	0.08	**0.12**
Sigma/rectum	0.19	**0.21**	0.24	**0.27**	0.28	**0.31**	0.32	**0.35**	0.39	**0.43**	0.46	**0.51**
Lumbar spine	0.02	**0.03**	0.03	**0.03**	0.03	**0.04**	0.04	**0.05**	0.05	**0.06**	0.06	**0.08**
Pelvis	0.12	**0.13**	0.16	**0.17**	0.19	**0.21**	0.21	**0.24**	0.27	**0.30**	0.32	**0.36**
Adrenals												
Kidneys						**0.01**		**0.01**	0.01	**0.02**	0.01	**0.02**
Urinary bladder	0.21	**0.24**	0.26	**0.30**	0.30	**0.34**	0.33	**0.38**	0.40	**0.45**	0.46	**0.51**
Uterus	0.30	**0.36**	0.39	**0.47**	0.47	**0.55**	0.53	**0.62**	0.65	**0.77**	0.77	**0.91**
Ovaries	0.08	**0.09**	0.11	**0.11**	0.13	**0.14**	0.14	**0.16**	0.18	**0.19**	0.21	**0.23**
Prostate	0.14	**0.19**	0.18	**0.25**	0.21	**0.29**	0.24	**0.33**	0.30	**0.40**	0.35	**0.47**
Testes	0.43	**0.53**	0.49	**0.61**	0.54	**0.66**	0.57	**0.70**	0.63	**0.77**	0.68	**0.83**
Upper arm												
Elbow												
Forearm		**0.28**		**0.35**		**0.40**		**0.44**		**0.52**		**0.58**
Upper leg	0.20	**0.21**	0.27	**0.28**	0.32	**0.33**	0.36	**0.38**	0.45	**0.47**	0.54	**0.57**
Knee	0.28	**0.30**	0.37	**0.38**	0.43	**0.45**	0.49	**0.51**	0.60	**0.62**	0.71	**0.74**
Lower leg	0.44	**0.46**	0.55	**0.58**	0.64	**0.67**	0.70	**0.74**	0.83	**0.87**	0.95	**0.99**
Lymph nodes	0.03	**0.04**	0.04	**0.05**	0.05	**0.06**	0.05	**0.06**	0.06	**0.08**	0.07	**0.09**
Skin	0.20	**0.24**	0.21	**0.26**	0.22	**0.27**	0.23	**0.28**	0.25	**0.30**	0.26	**0.31**
Muscle	0.12	**0.13**	0.14	**0.15**	0.15	**0.17**	0.17	**0.19**	0.19	**0.21**	0.21	**0.24**
Skeleton	0.15	**0.17**	0.19	**0.22**	0.22	**0.25**	0.25	**0.28**	0.30	**0.34**	0.35	**0.40**
Red bone marrow	0.02	**0.02**	0.03	**0.03**	0.03	**0.03**	0.03	**0.04**	0.04	**0.05**	0.05	**0.06**
Whole body dose	0.11	**0.13**	0.13	**0.15**	0.15	**0.17**	0.16	**0.19**	0.19	**0.21**	0.21	**0.24**
ED (ICRP 60)	0.08	**0.10**	0.10	**0.12**	0.11	**0.13**	0.12	**0.15**	0.14	**0.17**	0.16	**0.19**
ED (ICRP 103)	0.05	**0.06**	0.07	**0.08**	0.08	**0.09**	0.08	**0.10**	0.10	**0.12**	0.11	**0.13**
Total filtration (mm) **Al**	**1.5**		**2.5**		**3.5**		**4.5**		**3.5**		**3.5**	
Cu	**0**		**0**		**0**		**0**		**0.1**		**0.2**	

Table 25.26 Normalized organ doses—Lower extremities ap—>17 years

Beam projection:	ap											**Table cells:**
Focus-to-detector distance (cm):	115											(+) = normal font = optimal field setting
Tube voltage (kVp):	60											**(−) = bold font = suboptimal field setting**
Total filtrations:	See last table row.											Empty table cells correspond to values < 0.01
Organ or tissue	**Organ dose D_T : Entrance dose K_E**							**(mSv ∗ mGy⁻¹)**				
	(+)	(−)	(+)	(−)	(+)	(−)	(+)	(−)	(+)	(−)	(+)	(−)
Skull												
Brain												
Oral mucosa												
Salivary glands												
Cervical spine												
Thyroid												
Thymus												
Thoracic spine												
Claviculae												
Scapulae												
Ribs												
Oesophagus												
Trachea												
Lung												
Mamma												
Heart												
Stomach								**0.01**		**0.01**		**0.01**
Spleen												
Liver												
Gall bladder												
Pancreas												
Small intestine	0.05	**0.07**	0.06	**0.09**	0.08	**0.11**	0.09	**0.13**	0.11	**0.16**	0.10	**0.14**
Large intestine		**0.03**	0.01	**0.03**	0.01	**0.04**	0.02	**0.05**	0.02	**0.06**	0.02	**0.05**
Sigma/rectum	0.12	**0.14**	0.16	**0.19**	0.19	**0.23**	0.22	**0.26**	0.28	**0.33**	0.24	**0.29**
Lumbar spine		**0.01**		**0.02**	0.01	**0.02**	0.01	**0.03**	0.02	**0.03**	0.01	**0.02**
Pelvis	0.08	**0.09**	0.10	**0.12**	0.12	**0.14**	0.14	**0.16**	0.18	**0.21**	0.15	**0.17**
Adrenals												
Kidneys												
Urinary bladder	0.15	**0.19**	0.19	**0.24**	0.22	**0.28**	0.25	**0.32**	0.30	**0.38**	0.28	**0.35**
Uterus	0.07	**0.10**	0.10	**0.14**	0.12	**0.16**	0.13	**0.19**	0.17	**0.23**	0.14	**0.20**
Ovaries	0.05	**0.06**	0.07	**0.08**	0.09	**0.10**	0.10	**0.11**	0.13	**0.14**	0.11	**0.12**
Prostate	0.08	**0.13**	0.11	**0.18**	0.13	**0.21**	0.15	**0.24**	0.18	**0.30**	0.16	**0.27**
Testes	0.30	**0.45**	0.35	**0.53**	0.39	**0.59**	0.42	**0.63**	0.47	**0.70**	0.45	**0.69**
Upper arm												
Elbow												
Forearm		**0.22**		**0.28**		**0.32**		**0.36**		**0.43**		**0.41**
Upper leg	0.11	**0.12**	0.15	**0.16**	0.18	**0.20**	0.21	**0.23**	0.27	**0.29**	0.23	**0.24**
Knee	0.17	**0.18**	0.23	**0.24**	0.27	**0.29**	0.31	**0.33**	0.40	**0.42**	0.34	**0.36**
Lower leg	0.30	**0.32**	0.39	**0.42**	0.46	**0.49**	0.52	**0.55**	0.63	**0.67**	0.58	**0.62**
Lymph nodes	0.02	**0.03**	0.03	**0.04**	0.03	**0.04**	0.03	**0.05**	0.04	**0.06**	0.04	**0.05**
Skin	0.16	**0.22**	0.17	**0.24**	0.18	**0.25**	0.19	**0.26**	0.20	**0.27**	0.19	**0.27**
Muscle	0.09	**0.11**	0.11	**0.13**	0.12	**0.15**	0.14	**0.16**	0.16	**0.18**	0.15	**0.17**
Skeleton	0.09	**0.11**	0.12	**0.14**	0.15	**0.17**	0.17	**0.20**	0.21	**0.24**	0.18	**0.22**
Red bone marrow		**0.01**	0.01	**0.01**	0.02	**0.02**	0.02	**0.02**	0.02	**0.03**	0.02	**0.02**
Whole body dose	0.09	**0.10**	0.10	**0.12**	0.12	**0.14**	0.13	**0.15**	0.15	**0.18**	0.14	**0.17**
ED (ICRP 60)	0.05	**0.08**	0.07	**0.09**	0.08	**0.11**	0.08	**0.12**	0.10	**0.14**	0.09	**0.13**
ED (ICRP 103)	0.03	**0.05**	0.04	**0.06**	0.05	**0.07**	0.05	**0.07**	0.06	**0.09**	0.06	**0.08**
Total filtration (mm) Al	**1.5**		**2.5**		**3.5**		**4.5**		**3.5**		**3.5**	
Cu	**0**		**0**		**0**		**0**		**0.1**		**0.2**	

Table 25.27 Normalized organ doses—Lower extremities ap—>17 years

Beam projection:			ap									**Table cells:**
Focus-to-detector distance (cm):			115				(+) = normal font = optimal field setting					
Tube voltage (kVp):			70				**(−) = bold font = suboptimal field setting**					
Total filtrations:			See last table row.				Empty table cells correspond to values < 0.01					
Organ or tissue	**Organ dose D_T : Entrance dose K_E**								**(mSv * mGy^{-1})**			
	(+)	(−)	(+)	(−)	(+)	(−)	(+)	(−)	(+)	(−)	(+)	(−)
Skull												
Brain												
Oral mucosa												
Salivary glands												
Cervical spine												
Thyroid												
Thymus												
Thoracic spine												
Claviculae												
Scapulae												
Ribs												
Oesophagus												
Trachea												
Lung												
Mamma												
Heart												
Stomach				**0.01**		**0.01**		**0.02**		**0.02**	0.01	**0.02**
Spleen												**0.01**
Liver												
Gall bladder										**0.01**		**0.01**
Pancreas												
Small intestine	0.06	**0.09**	0.08	**0.12**	0.10	**0.14**	0.11	**0.16**	0.14	**0.20**	0.16	**0.23**
Large intestine	0.01	**0.03**	0.02	**0.04**	0.02	**0.05**	0.02	**0.06**	0.03	**0.08**	0.04	**0.09**
Sigma/rectum	0.16	**0.19**	0.21	**0.24**	0.25	**0.29**	0.28	**0.33**	0.34	**0.40**	0.41	**0.48**
Lumbar spine	0.01	**0.02**	0.01	**0.03**	0.02	**0.03**	0.02	**0.04**	0.03	**0.05**	0.04	**0.06**
Pelvis	0.10	**0.12**	0.14	**0.16**	0.16	**0.19**	0.19	**0.21**	0.23	**0.27**	0.28	**0.33**
Adrenals												
Kidneys										**0.01**		**0.01**
Urinary bladder	0.18	**0.23**	0.23	**0.29**	0.27	**0.34**	0.30	**0.37**	0.35	**0.44**	0.41	**0.51**
Uterus	0.10	**0.13**	0.12	**0.18**	0.15	**0.21**	0.17	**0.23**	0.21	**0.29**	0.25	**0.35**
Ovaries	0.07	**0.08**	0.09	**0.11**	0.11	**0.13**	0.13	**0.15**	0.16	**0.18**	0.19	**0.22**
Prostate	0.11	**0.17**	0.14	**0.22**	0.16	**0.26**	0.18	**0.29**	0.23	**0.36**	0.27	**0.42**
Testes	0.33	**0.50**	0.39	**0.59**	0.43	**0.64**	0.46	**0.69**	0.51	**0.76**	0.56	**0.83**
Upper arm												
Elbow												
Forearm		**0.26**		**0.33**		**0.37**		**0.41**		**0.48**		**0.55**
Upper leg	0.16	**0.17**	0.20	**0.22**	0.24	**0.26**	0.28	**0.30**	0.35	**0.38**	0.43	**0.46**
Knee	0.23	**0.24**	0.30	**0.31**	0.35	**0.37**	0.40	**0.42**	0.49	**0.52**	0.59	**0.62**
Lower leg	0.37	**0.40**	0.47	**0.50**	0.55	**0.58**	0.61	**0.65**	0.73	**0.78**	0.84	**0.89**
Lymph nodes	0.02	**0.04**	0.03	**0.04**	0.04	**0.05**	0.04	**0.06**	0.05	**0.07**	0.06	**0.08**
Skin	0.17	**0.23**	0.18	**0.25**	0.19	**0.26**	0.20	**0.27**	0.21	**0.29**	0.22	**0.30**
Muscle	0.11	**0.12**	0.13	**0.15**	0.14	**0.17**	0.15	**0.18**	0.18	**0.21**	0.20	**0.23**
Skeleton	0.12	**0.14**	0.15	**0.18**	0.18	**0.21**	0.20	**0.24**	0.25	**0.29**	0.30	**0.35**
Red bone marrow	0.01	**0.02**	0.02	**0.02**	0.02	**0.03**	0.03	**0.03**	0.03	**0.04**	0.04	**0.05**
Whole body dose	0.10	**0.12**	0.12	**0.14**	0.14	**0.16**	0.15	**0.18**	0.17	**0.20**	0.19	**0.23**
ED (ICRP 60)	0.06	**0.09**	0.08	**0.11**	0.09	**0.12**	0.10	**0.14**	0.12	**0.16**	0.13	**0.18**
ED (ICRP 103)	0.04	**0.06**	0.05	**0.07**	0.06	**0.08**	0.06	**0.09**	0.08	**0.10**	0.09	**0.12**
Total filtration (mm) Al	**1.5**		**2.5**		**3.5**		**4.5**		**3.5**		**3.5**	
Cu	**0**		**0**		**0**		**0**		**0.1**		**0.2**	

Table 25.28 Normalized organ doses—Lower extremities ap—8–12 years

Beam projection:			ap									**Table cells:**
Focus-to-detector distance (cm):			150									(+) = normal font = optimal field setting
Tube voltage (kVp):			60									**(−) = bold font = suboptimal field setting**
Total filtrations:			See last table row.									Empty table cells correspond to values < 0.01
Organ or tissue	**Organ dose D_T : Entrance dose K_E**								**(mSv ∗ mGy^{-1})**			
	(+)	(−)	(+)	(−)	(+)	(−)	(+)	(−)	(+)	(−)	(+)	(−)
Skull												
Brain												
Oral mucosa												
Salivary glands												
Cervical spine												
Thyroid												
Thymus												
Thoracic spine												
Claviculae												
Scapulae												
Ribs												
Oesophagus												
Trachea												
Lung												
Mamma												
Heart												
Stomach		**0.01**		**0.02**		**0.02**	0.01	**0.03**	0.01	**0.03**	0.01	**0.03**
Spleen								**0.01**		**0.01**		**0.01**
Liver												
Gall bladder		**0.01**		**0.02**		**0.02**	0.01	**0.02**	0.01	**0.03**	0.01	**0.03**
Pancreas										**0.01**		
Small intestine	0.09	**0.13**	0.11	**0.17**	0.13	**0.20**	0.15	**0.22**	0.18	**0.27**	0.17	**0.25**
Large intestine	0.02	**0.06**	0.02	**0.08**	0.03	**0.09**	0.03	**0.11**	0.04	**0.13**	0.04	**0.12**
Sigma/rectum	0.19	**0.22**	0.25	**0.29**	0.29	**0.34**	0.33	**0.39**	0.41	**0.47**	0.37	**0.43**
Lumbar spine	0.02	**0.04**	0.03	**0.05**	0.03	**0.06**	0.04	**0.08**	0.05	**0.10**	0.04	**0.08**
Pelvis	0.13	**0.15**	0.18	**0.20**	0.21	**0.25**	0.25	**0.28**	0.31	**0.35**	0.27	**0.31**
Adrenals												
Kidneys						**0.01**		**0.01**		**0.02**		**0.01**
Urinary bladder	0.24	**0.31**	0.30	**0.38**	0.34	**0.44**	0.38	**0.48**	0.45	**0.57**	0.42	**0.54**
Uterus	0.13	**0.19**	0.17	**0.25**	0.21	**0.30**	0.23	**0.34**	0.29	**0.41**	0.26	**0.37**
Ovaries	0.09	**0.10**	0.12	**0.13**	0.14	**0.16**	0.16	**0.18**	0.20	**0.23**	0.18	**0.20**
Prostate	0.18	**0.26**	0.23	**0.34**	0.27	**0.40**	0.31	**0.44**	0.37	**0.54**	0.34	**0.50**
Testes	0.57	**1.07**	0.64	**1.19**	0.68	**1.26**	0.71	**1.31**	0.76	**1.39**	0.75	**1.38**
Upper arm												
Elbow												
Forearm		**0.32**		**0.40**		**0.46**		**0.51**		**0.60**		**0.58**
Upper leg	0.22	**0.23**	0.29	**0.31**	0.34	**0.37**	0.39	**0.42**	0.49	**0.52**	0.43	**0.46**
Knee	0.28	**0.30**	0.37	**0.39**	0.44	**0.46**	0.49	**0.52**	0.61	**0.64**	0.55	**0.58**
Lower leg	0.43	**0.47**	0.55	**0.59**	0.64	**0.69**	0.71	**0.77**	0.84	**0.91**	0.80	**0.87**
Lymph nodes	0.03	**0.05**	0.04	**0.07**	0.05	**0.08**	0.05	**0.08**	0.07	**0.10**	0.06	**0.09**
Skin	0.21	**0.26**	0.22	**0.28**	0.23	**0.29**	0.24	**0.30**	0.25	**0.32**	0.24	**0.31**
Muscle	0.12	**0.14**	0.14	**0.17**	0.15	**0.19**	0.16	**0.20**	0.19	**0.23**	0.18	**0.22**
Skeleton	0.15	**0.17**	0.19	**0.22**	0.22	**0.26**	0.25	**0.29**	0.30	**0.36**	0.28	**0.33**
Red bone marrow	0.03	**0.03**	0.04	**0.04**	0.05	**0.05**	0.05	**0.06**	0.07	**0.08**	0.06	**0.07**
Whole body dose	0.11	**0.13**	0.13	**0.16**	0.15	**0.18**	0.16	**0.19**	0.18	**0.22**	0.17	**0.21**
ED (ICRP 60)	0.10	**0.16**	0.12	**0.19**	0.13	**0.21**	0.14	**0.22**	0.16	**0.25**	0.15	**0.24**
ED (ICRP 103)	0.06	**0.09**	0.07	**0.11**	0.08	**0.12**	0.09	**0.13**	0.10	**0.15**	0.10	**0.15**
Total filtration (mm) **Al**	**1.5**		**2.5**		**3.5**		**4.5**		**3.5**		**3.5**	
Cu	**0**		**0**		**0**		**0**		**0.1**		**0.2**	

Table 25.29 Normalized organ doses—Lower extremities ap—8–12 years

Beam projection:		ap										**Table cells:**
Focus-to-detector distance (cm):		150						(+) = normal font = optimal field setting				
Tube voltage (kVp):		70						**(−) = bold font = suboptimal field setting**				
Total filtrations:		See last table row.						Empty table cells correspond to values < 0.01				
Organ or tissue	**Organ dose D_T : Entrance dose K_E**							**(mSv * mGy^{-1})**				
	(+)	(−)	(+)	(−)	(+)	(−)	(+)	(−)	(+)	(−)	(+)	(−)
Skull												
Brain												
Oral mucosa												
Salivary glands												
Cervical spine												
Thyroid												
Thymus												
Thoracic spine												
Claviculae												
Scapulae												
Ribs												
Oesophagus												
Trachea												
Lung												
Mamma												
Heart												
Stomach		**0.02**	0.01	**0.02**	0.01	**0.03**	0.01	**0.03**	0.02	**0.04**	0.02	**0.05**
Spleen				**0.01**		**0.01**		**0.02**		**0.02**	0.01	**0.02**
Liver												
Gall bladder		**0.02**	0.01	**0.02**	0.01	**0.03**	0.01	**0.03**	0.02	**0.04**	0.02	**0.05**
Pancreas						**0.01**		**0.01**		**0.02**		**0.02**
Small intestine	0.11	**0.16**	0.14	**0.21**	0.16	**0.24**	0.18	**0.27**	0.22	**0.32**	0.25	**0.37**
Large intestine	0.03	**0.08**	0.03	**0.10**	0.04	**0.12**	0.04	**0.13**	0.06	**0.16**	0.07	**0.19**
Sigma/rectum	0.24	**0.28**	0.31	**0.36**	0.36	**0.42**	0.40	**0.46**	0.48	**0.56**	0.56	**0.65**
Lumbar spine	0.03	**0.06**	0.04	**0.07**	0.05	**0.09**	0.05	**0.10**	0.07	**0.13**	0.08	**0.17**
Pelvis	0.18	**0.20**	0.23	**0.26**	0.27	**0.31**	0.31	**0.35**	0.38	**0.44**	0.45	**0.52**
Adrenals												
Kidneys				**0.01**		**0.02**		**0.02**	0.01	**0.02**	0.01	**0.03**
Urinary bladder	0.28	**0.36**	0.35	**0.44**	0.40	**0.50**	0.43	**0.55**	0.51	**0.64**	0.57	**0.72**
Uterus	0.17	**0.24**	0.21	**0.31**	0.25	**0.36**	0.28	**0.41**	0.34	**0.49**	0.40	**0.57**
Ovaries	0.12	**0.13**	0.15	**0.17**	0.18	**0.20**	0.20	**0.23**	0.25	**0.28**	0.29	**0.33**
Prostate	0.22	**0.32**	0.28	**0.40**	0.32	**0.47**	0.36	**0.52**	0.43	**0.62**	0.49	**0.71**
Testes	0.61	**1.13**	0.67	**1.24**	0.72	**1.31**	0.75	**1.36**	0.80	**1.44**	0.85	**1.50**
Upper arm												
Elbow												
Forearm		**0.38**		**0.46**		**0.52**		**0.57**		**0.66**		**0.74**
Upper leg	0.28	**0.30**	0.36	**0.39**	0.43	**0.46**	0.48	**0.52**	0.59	**0.63**	0.70	**0.75**
Knee	0.35	**0.37**	0.45	**0.48**	0.53	**0.56**	0.59	**0.62**	0.72	**0.76**	0.83	**0.88**
Lower leg	0.51	**0.55**	0.64	**0.69**	0.73	**0.79**	0.80	**0.87**	0.93	**1.01**	1.05	**1.14**
Lymph nodes	0.04	**0.06**	0.05	**0.08**	0.06	**0.09**	0.06	**0.10**	0.08	**0.12**	0.09	**0.13**
Skin	0.21	**0.27**	0.23	**0.29**	0.24	**0.30**	0.25	**0.31**	0.26	**0.33**	0.27	**0.35**
Muscle	0.13	**0.16**	0.15	**0.19**	0.17	**0.21**	0.18	**0.22**	0.21	**0.25**	0.23	**0.28**
Skeleton	0.18	**0.21**	0.23	**0.27**	0.26	**0.31**	0.29	**0.35**	0.35	**0.41**	0.41	**0.48**
Red bone marrow	0.04	**0.04**	0.05	**0.06**	0.06	**0.07**	0.07	**0.08**	0.08	**0.09**	0.10	**0.11**
Whole body dose	0.12	**0.15**	0.15	**0.18**	0.17	**0.20**	0.18	**0.22**	0.21	**0.25**	0.23	**0.28**
ED (ICRP 60)	0.11	**0.18**	0.13	**0.21**	0.15	**0.23**	0.16	**0.24**	0.18	**0.27**	0.20	**0.30**
ED (ICRP 103)	0.07	**0.11**	0.08	**0.13**	0.09	**0.14**	0.10	**0.15**	0.12	**0.17**	0.14	**0.19**
Total filtration (mm) **Al**	**1.5**		**2.5**		**3.5**		**4.5**		**3.5**		**3.5**	
Cu	**0**		**0**		**0**		**0**		**0.1**		**0.2**	

Table 25.30 Normalized organ doses—Lower extremities ap—13–17 years

Beam projection:			ap									**Table cells:**
Focus-to-detector distance (cm):			150									(+) = normal font = optimal field setting
Tube voltage (kVp):			60									**(−) = bold font = suboptimal field setting**
Total filtrations:			See last table row.									Empty table cells correspond to values < 0.01
Organ or tissue	**Organ dose D_T : Entrance dose K_E**								**(mSv * mGy^{-1})**			
	(+)	(−)	(+)	(−)	(+)	(−)	(+)	(−)	(+)	(−)	(+)	(−)
Skull												
Brain												
Oral mucosa												
Salivary glands												
Cervical spine												
Thyroid												
Thymus												
Thoracic spine												
Claviculae												
Scapulae												
Ribs												
Oesophagus												
Trachea												
Lung												
Mamma												
Heart												
Stomach				**0.01**		**0.02**		**0.02**	0.01	**0.02**	0.01	**0.02**
Spleen												
Liver												
Gall bladder										**0.01**		
Pancreas												
Small intestine	0.07	**0.09**	0.10	**0.12**	0.11	**0.14**	0.13	**0.16**	0.16	**0.20**	0.14	**0.17**
Large intestine	0.02	**0.04**	0.03	**0.05**	0.04	**0.06**	0.04	**0.07**	0.05	**0.09**	0.04	**0.08**
Sigma/rectum	0.15	**0.17**	0.20	**0.22**	0.24	**0.27**	0.28	**0.30**	0.34	**0.38**	0.31	**0.33**
Lumbar spine	0.01	**0.02**	0.02	**0.02**	0.02	**0.03**	0.02	**0.04**	0.03	**0.05**	0.02	**0.04**
Pelvis	0.10	**0.10**	0.13	**0.14**	0.16	**0.17**	0.18	**0.20**	0.23	**0.25**	0.19	**0.21**
Adrenals												
Kidneys										**0.01**		
Urinary bladder	0.18	**0.21**	0.23	**0.26**	0.27	**0.30**	0.30	**0.34**	0.36	**0.40**	0.33	**0.38**
Uterus	0.25	**0.30**	0.33	**0.40**	0.39	**0.48**	0.45	**0.54**	0.56	**0.68**	0.49	**0.60**
Ovaries	0.06	**0.07**	0.09	**0.10**	0.10	**0.12**	0.12	**0.14**	0.15	**0.17**	0.13	**0.15**
Prostate	0.11	**0.16**	0.15	**0.21**	0.17	**0.25**	0.20	**0.28**	0.24	**0.34**	0.22	**0.31**
Testes	0.40	**0.48**	0.46	**0.56**	0.50	**0.61**	0.54	**0.65**	0.60	**0.71**	0.58	**0.70**
Upper arm												
Elbow												
Forearm		**0.26**		**0.33**		**0.38**		**0.42**		**0.50**		**0.47**
Upper leg	0.15	**0.16**	0.20	**0.21**	0.25	**0.26**	0.28	**0.29**	0.36	**0.37**	0.31	**0.32**
Knee	0.21	**0.22**	0.28	**0.29**	0.34	**0.35**	0.39	**0.40**	0.48	**0.50**	0.43	**0.44**
Lower leg	0.37	**0.38**	0.47	**0.49**	0.55	**0.57**	0.61	**0.63**	0.73	**0.76**	0.69	**0.72**
Lymph nodes	0.03	**0.03**	0.04	**0.04**	0.04	**0.05**	0.05	**0.06**	0.06	**0.07**	0.05	**0.06**
Skin	0.18	**0.22**	0.20	**0.24**	0.20	**0.25**	0.21	**0.26**	0.22	**0.28**	0.22	**0.27**
Muscle	0.10	**0.11**	0.12	**0.14**	0.14	**0.15**	0.15	**0.17**	0.17	**0.19**	0.16	**0.18**
Skeleton	0.12	**0.13**	0.15	**0.18**	0.18	**0.21**	0.20	**0.24**	0.25	**0.29**	0.23	**0.26**
Red bone marrow	0.01	**0.02**	0.02	**0.02**	0.02	**0.03**	0.03	**0.03**	0.03	**0.04**	0.03	**0.03**
Whole body dose	0.10	**0.11**	0.12	**0.13**	0.13	**0.15**	0.14	**0.16**	0.17	**0.19**	0.16	**0.18**
ED (ICRP 60)	0.07	**0.09**	0.09	**0.11**	0.10	**0.12**	0.11	**0.13**	0.13	**0.15**	0.12	**0.14**
ED (ICRP 103)	0.05	**0.05**	0.06	**0.07**	0.07	**0.08**	0.07	**0.09**	0.09	**0.10**	0.08	**0.10**
Total filtration (mm) **Al**	**1.5**		**2.5**		**3.5**		**4.5**		**3.5**		**3.5**	
Cu	**0**		**0**		**0**		**0**		**0.1**		**0.2**	

Table 25.31 Normalized organ doses—Lower extremities ap—13–17 years

Beam projection:		ap										**Table cells:**
Focus-to-detector distance (cm):		150										(+) = normal font = optimal field setting
Tube voltage (kVp):		70										**(−) = bold font = suboptimal field setting**
Total filtrations:		See last table row.										Empty table cells correspond to values < 0.01
Organ or tissue	**Organ dose D_T : Entrance dose K_E**								**(mSv ∗ mGy^{-1})**			
	(+)	(−)	(+)	(−)	(+)	(−)	(+)	(−)	(+)	(−)	(+)	(−)
Skull												
Brain												
Oral mucosa												
Salivary glands												
Cervical spine												
Thyroid												
Thymus												
Thoracic spine												
Claviculae												
Scapulae												
Ribs												
Oesophagus												
Trachea												
Lung												
Mamma												**0.01**
Heart												
Stomach		**0.01**		**0.02**	0.01	**0.02**	0.01	**0.02**	0.02	**0.03**	0.02	**0.04**
Spleen										**0.01**		**0.02**
Liver												
Gall bladder						**0.01**		**0.01**	0.01	**0.02**	0.01	**0.02**
Pancreas												**0.01**
Small intestine	0.09	**0.12**	0.12	**0.15**	0.14	**0.17**	0.16	**0.19**	0.20	**0.24**	0.23	**0.28**
Large intestine	0.03	**0.05**	0.04	**0.07**	0.05	**0.08**	0.05	**0.09**	0.07	**0.11**	0.08	**0.13**
Sigma/rectum	0.20	**0.22**	0.26	**0.28**	0.30	**0.33**	0.34	**0.37**	0.42	**0.46**	0.49	**0.54**
Lumbar spine	0.02	**0.03**	0.03	**0.04**	0.03	**0.05**	0.04	**0.05**	0.05	**0.07**	0.06	**0.09**
Pelvis	0.13	**0.14**	0.17	**0.19**	0.20	**0.22**	0.23	**0.26**	0.29	**0.32**	0.35	**0.39**
Adrenals												
Kidneys								**0.01**		**0.01**	0.01	**0.02**
Urinary bladder	0.22	**0.25**	0.28	**0.31**	0.32	**0.36**	0.35	**0.40**	0.42	**0.47**	0.48	**0.54**
Uterus	0.33	**0.39**	0.42	**0.51**	0.50	**0.60**	0.56	**0.67**	0.69	**0.83**	0.82	**0.98**
Ovaries	0.09	**0.10**	0.11	**0.13**	0.14	**0.15**	0.15	**0.17**	0.19	**0.22**	0.23	**0.26**
Prostate	0.14	**0.20**	0.18	**0.26**	0.22	**0.30**	0.24	**0.34**	0.30	**0.41**	0.35	**0.48**
Testes	0.44	**0.53**	0.50	**0.61**	0.55	**0.66**	0.58	**0.70**	0.64	**0.77**	0.69	**0.83**
Upper arm												
Elbow												
Forearm		**0.30**		**0.38**		**0.43**		**0.47**		**0.55**		**0.62**
Upper leg	0.21	**0.21**	0.27	**0.28**	0.32	**0.33**	0.36	**0.38**	0.45	**0.47**	0.54	**0.57**
Knee	0.28	**0.29**	0.36	**0.37**	0.42	**0.44**	0.48	**0.49**	0.59	**0.60**	0.69	**0.71**
Lower leg	0.44	**0.46**	0.55	**0.57**	0.64	**0.66**	0.71	**0.73**	0.83	**0.86**	0.95	**0.98**
Lymph nodes	0.03	**0.04**	0.04	**0.05**	0.05	**0.06**	0.06	**0.07**	0.07	**0.08**	0.08	**0.10**
Skin	0.19	**0.23**	0.20	**0.25**	0.21	**0.26**	0.22	**0.27**	0.24	**0.29**	0.25	**0.30**
Muscle	0.12	**0.13**	0.14	**0.15**	0.15	**0.17**	0.17	**0.19**	0.19	**0.21**	0.21	**0.24**
Skeleton	0.15	**0.17**	0.19	**0.22**	0.22	**0.25**	0.25	**0.28**	0.30	**0.34**	0.35	**0.40**
Red bone marrow	0.02	**0.02**	0.03	**0.03**	0.03	**0.04**	0.04	**0.04**	0.05	**0.05**	0.06	**0.06**
Whole body dose	0.11	**0.13**	0.13	**0.15**	0.15	**0.17**	0.16	**0.19**	0.19	**0.21**	0.21	**0.24**
ED (ICRP 60)	0.09	**0.10**	0.10	**0.12**	0.12	**0.14**	0.13	**0.15**	0.15	**0.18**	0.17	**0.20**
ED (ICRP 103)	0.06	**0.07**	0.07	**0.08**	0.08	**0.09**	0.09	**0.10**	0.10	**0.12**	0.12	**0.14**
Total filtration (mm) **Al**	**1.5**		**2.5**		**3.5**		**4.5**		**3.5**		**3.5**	
Cu	**0**		**0**		**0**		**0**		**0.1**		**0.2**	

Table 25.32 Normalized organ doses—Lower extremities ap—>17 years

Beam projection:			ap									**Table cells:**
Focus-to-detector distance (cm):			150									(+) = normal font = optimal field setting
Tube voltage (kVp):			60									**(−) = bold font = suboptimal field setting**
Total filtrations:			See last table row.									Empty table cells correspond to values < 0.01
Organ or tissue	**Organ dose D_T : Entrance dose K_E**								**(mSv * mGy^{-1})**			
	(+)	(−)	(+)	(−)	(+)	(−)	(+)	(−)	(+)	(−)	(+)	(−)
Skull												
Brain												
Oral mucosa												
Salivary glands												
Cervical spine												
Thyroid												
Thymus												
Thoracic spine												
Claviculae												
Scapulae												
Ribs												
Oesophagus												
Trachea												
Lung												
Mamma												
Heart												
Stomach								**0.01**		**0.01**		**0.01**
Spleen												
Liver												
Gall bladder												
Pancreas												
Small intestine	0.05	**0.08**	0.07	**0.11**	0.08	**0.13**	0.09	**0.14**	0.12	**0.18**	0.10	**0.16**
Large intestine		**0.03**	0.01	**0.04**	0.01	**0.04**	0.02	**0.05**	0.02	**0.06**	0.02	**0.05**
Sigma/rectum	0.13	**0.16**	0.18	**0.21**	0.21	**0.25**	0.24	**0.28**	0.30	**0.35**	0.27	**0.31**
Lumbar spine		**0.01**		**0.02**	0.01	**0.02**	0.01	**0.03**	0.02	**0.04**	0.01	**0.03**
Pelvis	0.08	**0.09**	0.11	**0.13**	0.14	**0.16**	0.16	**0.18**	0.20	**0.23**	0.17	**0.19**
Adrenals												
Kidneys												
Urinary bladder	0.16	**0.20**	0.20	**0.26**	0.24	**0.30**	0.26	**0.33**	0.32	**0.40**	0.29	**0.37**
Uterus	0.08	**0.11**	0.10	**0.15**	0.12	**0.18**	0.14	**0.20**	0.18	**0.26**	0.15	**0.22**
Ovaries	0.06	**0.07**	0.08	**0.09**	0.10	**0.11**	0.11	**0.13**	0.14	**0.16**	0.12	**0.14**
Prostate	0.08	**0.13**	0.11	**0.18**	0.13	**0.21**	0.15	**0.24**	0.18	**0.30**	0.16	**0.26**
Testes	0.30	**0.46**	0.35	**0.54**	0.39	**0.59**	0.42	**0.63**	0.47	**0.70**	0.46	**0.69**
Upper arm												
Elbow												
Forearm		**0.24**		**0.31**		**0.35**		**0.39**		**0.47**		**0.44**
Upper leg	0.11	**0.12**	0.15	**0.16**	0.19	**0.20**	0.21	**0.23**	0.27	**0.29**	0.23	**0.24**
Knee	0.17	**0.17**	0.22	**0.23**	0.27	**0.28**	0.31	**0.32**	0.39	**0.40**	0.34	**0.35**
Lower leg	0.31	**0.32**	0.40	**0.42**	0.47	**0.49**	0.52	**0.55**	0.64	**0.67**	0.59	**0.62**
Lymph nodes	0.02	**0.03**	0.03	**0.04**	0.03	**0.05**	0.03	**0.05**	0.04	**0.06**	0.04	**0.06**
Skin	0.16	**0.21**	0.17	**0.23**	0.18	**0.24**	0.18	**0.25**	0.19	**0.27**	0.19	**0.26**
Muscle	0.09	**0.11**	0.11	**0.13**	0.12	**0.15**	0.14	**0.16**	0.16	**0.18**	0.15	**0.17**
Skeleton	0.09	**0.11**	0.12	**0.15**	0.15	**0.17**	0.17	**0.20**	0.21	**0.24**	0.18	**0.22**
Red bone marrow	0.01	**0.01**	0.01	**0.02**	0.02	**0.02**	0.02	**0.02**	0.03	**0.03**	0.02	**0.02**
Whole body dose	0.09	**0.10**	0.10	**0.12**	0.12	**0.14**	0.13	**0.15**	0.15	**0.18**	0.14	**0.17**
ED (ICRP 60)	0.06	**0.08**	0.07	**0.10**	0.08	**0.11**	0.09	**0.12**	0.10	**0.14**	0.10	**0.13**
ED (ICRP 103)	0.03	**0.05**	0.04	**0.06**	0.05	**0.07**	0.06	**0.08**	0.07	**0.09**	0.06	**0.08**
Total filtration (mm) **Al**	**1.5**		**2.5**		**3.5**		**4.5**		**3.5**		**3.5**	
Cu	**0**		**0**		**0**		**0**		**0.1**		**0.2**	

Table 25.33 Normalized organ doses—Lower extremities ap—>17 years

Beam projection:			ap									**Table cells:**
Focus-to-detector distance (cm):			150					(+) = normal font = optimal field setting				
Tube voltage (kVp):			70					**(−) = bold font = suboptimal field setting**				
Total filtrations:			See last table row.					Empty table cells correspond to values < 0.01				
Organ or tissue	**Organ dose D_T : Entrance dose K_E**							**(mSv ∗ mGy⁻¹)**				
	(+)	(−)	(+)	(−)	(+)	(−)	(+)	(−)	(+)	(−)	(+)	(−)
Skull												
Brain												
Oral mucosa												
Salivary glands												
Cervical spine												
Thyroid												
Thymus												
Thoracic spine												
Claviculae												
Scapulae												
Ribs												
Oesophagus												
Trachea												
Lung												
Mamma												
Heart												
Stomach				**0.01**		**0.01**		**0.02**		**0.02**	0.01	**0.02**
Spleen												**0.01**
Liver												
Gall bladder										**0.01**		**0.01**
Pancreas												
Small intestine	0.07	**0.10**	0.09	**0.13**	0.10	**0.16**	0.12	**0.18**	0.14	**0.22**	0.17	**0.26**
Large intestine	0.01	**0.04**	0.02	**0.05**	0.02	**0.06**	0.02	**0.06**	0.03	**0.08**	0.04	**0.09**
Sigma/rectum	0.17	**0.20**	0.23	**0.26**	0.27	**0.31**	0.30	**0.35**	0.37	**0.43**	0.44	**0.51**
Lumbar spine	0.01	**0.02**	0.01	**0.03**	0.02	**0.04**	0.02	**0.04**	0.03	**0.05**	0.03	**0.07**
Pelvis	0.12	**0.13**	0.15	**0.17**	0.18	**0.21**	0.21	**0.23**	0.26	**0.30**	0.31	**0.36**
Adrenals												
Kidneys												**0.01**
Urinary bladder	0.19	**0.24**	0.24	**0.31**	0.28	**0.35**	0.31	**0.39**	0.37	**0.46**	0.43	**0.53**
Uterus	0.10	**0.15**	0.13	**0.19**	0.16	**0.23**	0.18	**0.26**	0.22	**0.32**	0.27	**0.37**
Ovaries	0.08	**0.09**	0.11	**0.12**	0.13	**0.14**	0.14	**0.16**	0.18	**0.21**	0.22	**0.25**
Prostate	0.11	**0.17**	0.14	**0.22**	0.16	**0.26**	0.18	**0.30**	0.23	**0.36**	0.27	**0.43**
Testes	0.33	**0.50**	0.39	**0.59**	0.43	**0.64**	0.46	**0.68**	0.51	**0.75**	0.56	**0.82**
Upper arm												
Elbow												
Forearm		**0.29**		**0.36**		**0.41**		**0.45**		**0.52**		**0.59**
Upper leg	0.16	**0.17**	0.21	**0.22**	0.25	**0.26**	0.28	**0.30**	0.35	**0.38**	0.43	**0.46**
Knee	0.22	**0.23**	0.29	**0.30**	0.34	**0.36**	0.39	**0.40**	0.48	**0.50**	0.57	**0.60**
Lower leg	0.38	**0.40**	0.48	**0.50**	0.55	**0.58**	0.61	**0.65**	0.73	**0.77**	0.84	**0.89**
Lymph nodes	0.03	**0.04**	0.03	**0.05**	0.04	**0.06**	0.04	**0.06**	0.05	**0.07**	0.06	**0.09**
Skin	0.16	**0.23**	0.18	**0.24**	0.18	**0.25**	0.19	**0.26**	0.20	**0.28**	0.21	**0.29**
Muscle	0.11	**0.12**	0.13	**0.15**	0.14	**0.17**	0.15	**0.18**	0.18	**0.21**	0.20	**0.23**
Skeleton	0.12	**0.14**	0.15	**0.18**	0.18	**0.21**	0.21	**0.24**	0.25	**0.29**	0.30	**0.35**
Red bone marrow	0.02	**0.02**	0.02	**0.02**	0.02	**0.03**	0.03	**0.03**	0.03	**0.04**	0.04	**0.05**
Whole body dose	0.10	**0.12**	0.12	**0.14**	0.14	**0.16**	0.15	**0.18**	0.17	**0.20**	0.19	**0.23**
ED (ICRP 60)	0.07	**0.09**	0.08	**0.11**	0.09	**0.13**	0.10	**0.14**	0.12	**0.16**	0.14	**0.19**
ED (ICRP 103)	0.04	**0.06**	0.05	**0.07**	0.06	**0.08**	0.07	**0.09**	0.08	**0.11**	0.09	**0.12**
Total filtration (mm) **Al**	**1.5**		**2.5**		**3.5**		**4.5**		**3.5**		**3.5**	
Cu	**0**		**0**		**0**		**0**		**0.1**		**0.2**	

Reference

1. Bontrager KL. Textbook of radiographic positioning and related anatomy. Mosby Year Book; 1993.

Foot

26

26.1 Introduction

The type of foot radiography is determined by the clinical indication. Table 26.1 shows indications and Table 26.2 contraindications for foot radiography. Image criteria as required for radiological findings in foot radiographs are listed in Table 26.3. In order to fulfil these image criteria, the field settings listed in Tables 26.4 and 26.5 should be achieved. These field settings go along with the patient preparation as listed in Table 26.6. Exposure parameter combinations for various age groups and foot diameters are listed in Tables 26.7, 26.8 and 26.9. The age groups in Table 26.8 refer to the standard age groups of the mathematical MIRD phantom family which has been used to calculate the normalized organ doses in this book. Table 26.9 shows radiation qualities (i.e. combinations of tube voltage and total filtration) which may be applied in clinical practice. Please note that not all of these radiation qualities are recommended by the authors. The standard projections of foot radiographies of the patient who is correctly positioned are depicted in Fig. 26.1. Please note that the anatomy of the mathematical phantoms has been

Table 26.1 Indications for foot radiographs

Bone age (in premature babies)
Trauma, fracture/luxation
Congenital malformation
Osteomyelitis
Soft tissue tumours
Bone tumours
Vascular malformation

Table 26.2 Contraindications for foot radiographs

None

Table 26.3 Image criteria for foot radiographs

Foot dorsovolar
Visualization of the distal epiphyses and metaphyses of the tibia and fibula consistent with age
Visualization of the tarsal bones consistent with age
Visualization of the metatarsals and the toes including the adjacent soft tissue
Visually sharp reproduction of the physes and epiphyses of the metatarsals and toes
Visually sharp reproduction of the cortex and the spongiosa of the distal tibia and fibula, the tarsal bones, the metatarsals, and the toes
Foot oblique
Visualization of the distal epiphyses and metaphyses of the tibia and fibula consistent with age
Visualization of the tarsal bones consistent with age
Visualization of the metatarsals and the toes including the adjacent soft tissue
Visually sharp reproduction of the physes and epiphyses of the metatarsals and toes
Visually sharp reproduction of the cortex and the spongiosa of the distal tibia and fibula, the tarsal bones, the metatarsals, and the toes

Table 26.4 Patient positions and beam projections for foot radiographs

Foot position	Beam projections
Prone	Dorso-plantar

M. Seidenbusch et al., *Imaging Practice and Radiation Protection in Pediatric Radiology*,
https://doi.org/10.1007/978-3-030-18504-6_26

Table 26.5 Optimal field settings for foot radiographs

Foot dorso-plantar	
Upper field edge:	Distal shank
Field centring:	At the base of the second and third metatarsal bone
Lower field edge:	Tip of the toes
Lateral field edges:	Close to the skin of the foot

Table 26.6 Patient preparation

General	Remove all clothes, external metallic devices, and jewellery from the lower extremities of the patient
	The foot is put on the table or the floor. It is fixed in uncooperative patients. The ankle joint is retroflexed by an angle of about 150°
Positioning	Supine or sitting
Shielding	A lead apron should be applied

Table 26.7 Radiographic procedure

X-ray source	Focal spot size between 1.0 and 1.3 dependent on the age of the patient
	Check the use of additional filtration
	Check the focus-to-detector distance
Field setting	Check the field size
	Check the field centring
Image detector system	Check the use of antiscatter grids
Patient	–

idealized and somewhat simplified. Therefore, the foot was not inserted into the mathematical phantom, and no normalized organ doses for the foot radiography could be provided.

26.2 Indications and Contraindications

Indications and contraindications for foot radiographs are given in Tables 26.1 and 26.2.

26.3 Image Criteria

The image criteria required for foot radiographs are listed in Table 26.3.

26.4 Field Settings

In Table 26.4, patient positions and beam projections used for hand radiographs are listed, and in Table 26.5, the optimal field settings are listed which are required in order to fulfil the image criteria mentioned in Table 26.3.

Table 26.8 Radiographic techniques of foot radiographs

Age group (years)	Beam direction	Foot position	Beam projection	Typical foot diameter (cm)	Optimal field size (cm x cm)	Focus-to-detector distance (cm)	Voltage (kVp)						Antiscatter grid
							50	60	70	80	90	100	
0	vertical	prone	dorso-plantar	-	-	100⁽¹⁾							
1–2													
3–7													
8–12						115							
13–17													
18													

Annotations:
(1) The patient is lying on the table top directly on the detector.
Parameter combination not recommended, normalized organ doses not provided
Parameter combination recommended, normalized organ doses provided

Table 26.9 Radiation qualities of foot radiographs

Voltage (kVp)	Total filtration					
	Without additional filtration		With additional filtration			
	1.5 mm Al	2.5 Al	3.5 Al	4.5 Al	3.5 Al + 0.1 Cu	3.5 Al + 0.2 Cu
50						
60						
70						
80						
90						
100						

Annotations:
Parameter combination not recommended, normalized organ doses not provided
Parameter combination not recommended, normalized organ doses provided
Parameter combination recommended, normalized organ doses provided

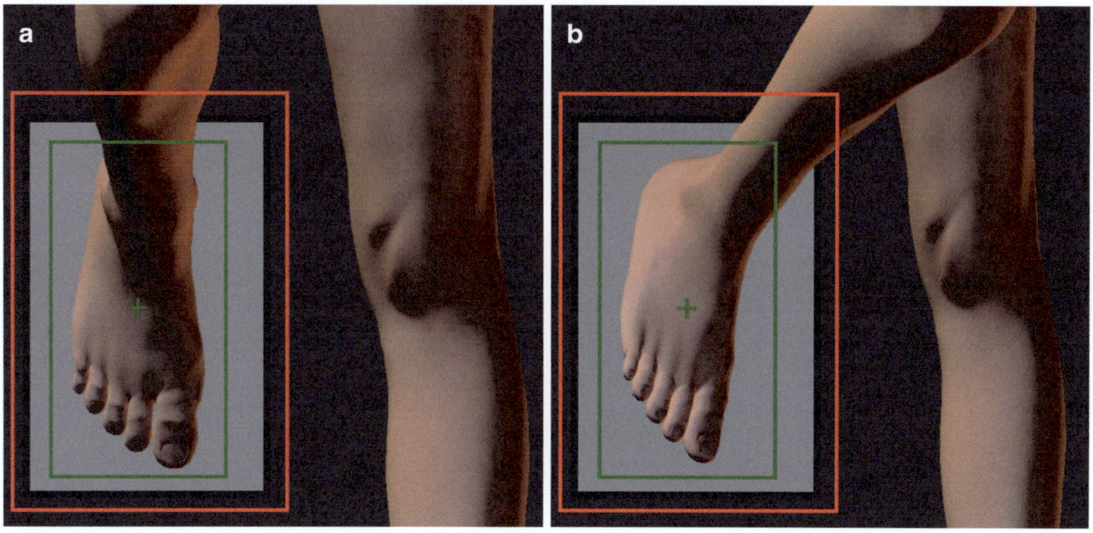

Fig. 26.1 **Field settings for foot radiographs in supine position**. (**a**) Dorso-plantar projection. (**b**) Oblique projection. (© V. Rösenberger and M. Seidenbusch 2018). *Green lines* optimal field settings; *Green cross* centring point; *Red lines* suboptimal field settings; *Grey shaded area* image detector; *Dark grey shaded area* Bucky table

26.5 Patient Preparation and Radiographic Procedure

See Tables 26.6, 26.7 and [1].

26.6 Exposure Parameters

In Tables 26.8 and 26.9, exposure parameter combinations used for hand radiographs in various age groups are listed. Dark grey shaded table cells refer to exposure parameter combinations which are not recommended for paediatric hand radiographs. Light blue shaded table cells refer to exposure parameter combinations which are recommended for paediatric hand radiographs.

26.7 Normalized Organ Doses

Please note also that the anatomy of the mathematical phantoms has been idealized and somewhat simplified. Therefore, the foot was not inserted into the mathematical phantom, and no normalized organ doses could be calculated (Table 26.10).

Table 26.10 Inconsistencies between the mathematical MIRD phantoms and real human anatomy

Anatomical region	Inconsistencies
Extremities	No feet have been inserted

For details, see the text

Reference

1. Bontrager KL. Textbook of radiographic positioning and related anatomy. St. Louis, MO: Mosby Year Book; 1993.

Skeletal Status

27

27.1 Introduction

The skeletal survey is rarely indicated in paediatric patients. Table 27.1 shows indications and Table 27.2 contraindications for the skeletal status. The skeletal survey is a paediatric X-ray examination consisting of several single radiographs: the skull, bony thorax ap and oblique, full spine, pelvis ap, upper and lower extremities ap, hands and feet, see Tables 27.3 and 27.4. The image criteria of the individual radiographs listed in Table 27.4 can be found in the particular chapters. The patient preparation is also listed in the particular tables of each radiographic examination.

Table 27.1 Indications for skeletal survey

Battered child syndrome
Skeletal dysplasia
Congenital lymphangiomatosis of the skeleton[a]
Annotation:
[a]MRI is the preferred diagnostic imaging modality today

Table 27.2 Contraindications for skeletal survey

Skeletal status is no longer indicated in patients with leukaemias, Langerhans cell histiocytosis, Ollier's disease, syndromes, and skeletal dysplasia with known genetic defects

Table 27.3 Image criteria for skeletal survey

X-ray examination	See image criteria in chapter
Skull ap + lat	8
Bony thorax ap + RAO + LAO	10
Full spine lateral	13
Pelvis ap	17
Upper extremities ap	23
Hands dv	24
Lower extremities ap	25
Feet dp	26

Table 27.4 Patient positions and beam projections for skeletal survey radiographs

Patient position	Beam projections
Supine	Skull ap + lat
	Bony thorax RAO/LAO
	Full spine lateral
	Pelvis ap
	Upper extremities ap
	Lower extremities ap
Hand/foot position	**Beam projections**
Prone	Hands dv
	Feet dp

As patients with suspected battered child syndrome are often severely traumatized, they should be appropriately restrained and commonly have to be sedated by a paediatric anaesthesiologist. Lead masking close to the field edge and gonadal protection in males should be applied, if possible.

M. Seidenbusch et al., *Imaging Practice and Radiation Protection in Pediatric Radiology*,
https://doi.org/10.1007/978-3-030-18504-6_27

27.2 Indications and Contraindications

The skeletal status is a paediatric X-ray examination consisting of several single radiographs, usually it consists of X-ray examinations of the skull, chest, full spine, pelvis, upper and lower extremities, hands, and feet (Table 27.3). Indications and contraindications for the skeletal status are given in Tables 27.1 and 27.2.

27.3 Image Criteria

The image criteria required for skeletal status are listed in Table 27.3.

27.4 Field Settings

In Table 27.4, patient positions and beam projections used for skeletal status radiographs are listed, Field settings can be found in the corresponding chapters.

27.5 Patient Preparation and Radiographic Procedure

Patient preparation and radiographic procedures can be found in the corresponding chapters.

Part IV

Supplement

28 Mathematical Phantoms

28.1 Mathematical Phantoms Constructed in Recent Years

The concept of mathematical phantoms representing the human anatomy has been developed by *Fisher* and *Snyder* [11, 13, 40] at the Oak Ridge National Laboratory (ORNL) on behalf of the MIRD[1] commission of the US Society of Nuclear Medicine for determining internal radiation exposure due to incorporation of radionuclides in adults. These phantoms were the predecessors of many mathematical phantoms that have been designed since that time. Overall, there are three generations of mathematical phantoms [47] which have been constructed using different mathematical methods (Table 28.1).

The first generation of mathematical phantoms consisted of mathematical stylized phantoms. In mathematical stylized phantoms, the inner organs and tissues of human body are described by mathematical expressions representing planes or cylindrical, conical, elliptical, or spherical surfaces [34]. Subsequently, several mathematical phantom families of the stylized MIRD type have been constructed. For assessing organ doses in children, a reference phantom family consisting of paediatric phantoms of various age groups has been developed by *Cristy* et al. [7] which were also implemented into the computer algorithm used for calculating the tables in this book [38, 39]. Due to their geometrical construction, mathematical phantoms represent an average anatomy of the corresponding age group as defined by the ICRP data on reference man [21].

In contrast, in the second generation of anthropomorphic voxel phantoms, the phantoms have been constructed from voxel datasets originating from post-mortem computed tomographic (CT) or magnetic resonance tomographic (MRT) slices of real patients. Thus, these phantoms refer to the anatomy of the individual patient. This anthropomorphic type of models was first developed by Gibbs et al. [17] and Williams et al. [45]. Subsequently, a family of anthropomorphic voxel phantoms covering various age groups has been developed by numerous work groups world-wide.

In recent years, mathematical phantoms are based more and more on the representation of anatomical structures through transforming voxel groups into spline functions or polygonal meshes. This approach to anthropomorphic hybrid phantoms as phantoms of the third generation incorporates the best features of both mathematical stylized as well as anthropomorphic voxel phantoms, the mathematical phantoms being economically constructed and the anthropomorphic phantoms being constructed close to reality by

[1] MIRD = Medical Internal Radiation Dose Committee, an expert panel for researching internal radiation exposure through incorporated radionuclides, which was substantially involved in the design of the first mathematical phantoms.

M. Seidenbusch et al., *Imaging Practice and Radiation Protection in Pediatric Radiology*,
https://doi.org/10.1007/978-3-030-18504-6_28

Table 28.1 Generations of mathematical phantoms [47]

Generation	Period	Construction through	Mathematical phantom type	
I	1960–2000s	Geometrical forms	Stylized (S)	Mathematical
II	>Late 1980s	Tomographic voxels	Voxelized (V)	Anthropomorphic
III	>2000	Parametrization of voxels	Hybrid phantoms (H)	

Table 28.2 Paediatric phantom families constructed in recent years—various age groups

Age (years)	Sex	Body height (cm)	Body weight (kg)	Type	Name	Reference
Various	H	Various	Various	S	–	Francois et al. [14]
Various	M	Various	Various	H	–	Segars et al. [37]
Various	F	Various	Various	H	–	Segars et al. [37]
Various	M	Various	Various	H	–	Carver et al. [3]

Table 28.3 Paediatric phantom families constructed in recent years—neonates

Age (years)	Sex	Body height (cm)	Body weight (kg)	Type	Name	Reference
0	H	–	–	S	–	Hwang et al. [20]
0	H	–	3.6	S	–	Cristy et Eckerman [8]
0	F	57	4.2	V	Baby	Zankl et al. [52] Petoussi et al. [34]
0	F	–	3.5	V	–	Nipper et al. [31]
0	M	–	5.4	V	–	Nipper et al. [31]
0	F	–	3.54	V	–	Staton et al. [42]
0	M	–	3.5	H	–	Lee et al. [26]
0	F	–	3.5	H	–	Lee et al. [26]
0	M	Various	Various	H	UFHPDM	Johnson et al. [22]
0	F	Various	Various	H	UFHPDF	Johnson et al. [22]
0	H	51	3.4	H	–	Lee et al. [28]
0	M	–	3.5	H	–	Stabin et al. [41]
0	F	–	3.5	H	–	Stabin et al. [41]
0	H	50.5	3.5	H	P00	Cassola et al. [4]
0	H	–	–	S	–	Gardumi et al. [16]
0	M	52	3.8	H	–	Norris et al. [32]
0	F	52	3.8	H	–	Norris et al. [32]
0	H	–	–	P		Yamauchi-Kawaura et al. [49]
0	M	Various	Various	H	–	Xie et al. [46]
0	F	Various	Various	H	–	Xie et al. [46]

using natural shapes of organs and tissues from CT and MRT datasets [9].

An extensive review of the mathematic phantoms of stylized, voxel as well as hybrid type as developed during the last 50 years has been given by *Xu* [47, 48] and *Varchena* [44].

In this chapter, only phantoms which are intended as or are a part of a phantom family and thus can be used for paediatric dosimetry are listed in Tables 28.2, 28.3, 28.4, 28.5, 28.6, 28.7, 28.8, 28.9, 28.10 and 28.11.

Table 28.4 Paediatric phantom families constructed in recent years—1-year-old children

Age (years)	Sex	Body height (cm)	Body weight (kg)	Type	Name	Reference
1	H	–	9.7	S	–	Cristy et Eckerman [8]
1	M	–	8.9	V	–	Lee et al. [25]
1	H	70	8.9	H	–	Hirata et al. [19]
1	M	Various	Various	H	UFHPDM	Johnson et al. [22]
1	F	Various	Various	H	UFHPDF	Johnson et al. [22]
1	H	76	10.0	H	–	Lee et al. [26]
1	M	–	10	H	–	Stabin et al. [41]
1	F	–	10	H	–	Stabin et al. [41]
1	H	75.8	10	H	P01	Cassola et al. [4]
1	H	–	–	S	–	Gardumi et al. [16]
1	M	Various	Various	H	UFH01M	Geyer et al. [15]
1	F	Various	Various	H	UFH01F	Geyer et al. [15]
1	M	77	10.3	H	–	Norris et al. [32]
1	F	77	10.3	H	–	Norris et al. [32]
1	M	–	–	H	CPP01	Pan et al. [33]
1	M	Various	Various	H	–	Xie et al. [46]
1	F	Various	Various	H	–	Xie et al. [46]

Table 28.5 Paediatric phantom families constructed in recent years—3- to 4-year-old children

Age (years)	Sex	Body height (cm)	Body weight (kg)	Type	Name	Reference
3	H	90	13.0	H	–	Hirata et al. [19]
3	H	90	13	H	–	Nagaoka et al. [30]
3	F	92	13.9	V	Nina	Gosselin et al. [18]
4	F	–	16.7	V	–	Lee et al. [25]

Table 28.6 Paediatric phantom families constructed in recent years—5-year-old children

5	H	–	19.8	S	–	Cristy et Eckerman [8]
5	H	105	17	H	–	Nagaoka et al. [30]
5	M	Various	Various	H	UFHPDM	Johnson et al. [22]
5	F	Various	Various	H	UFHPDF	Johnson et al. [22]
5	H	109	17.8	H	–	Lee et al. [28]
5	F	109	19	H	F05	De Melo Lima et al. [9]
5	M	109	19	H	M05	De Melo Lima et al. [9]
5	M	–	19	H	–	Stabin et al. [41]
5	F	–	19	H	–	Stabin et al. [41]
5	H	–	–	S	–	Gardumi et al. [16]
5	M	Various	Various	H	UFH05M	Geyer et al. [15]
5	F	Various	Various	H	UFH05F	Geyer et al. [15]
5	F	109	17.8	V	Roberta	Gosselin et al. [18]
5	M	110	19.9	H	–	Norris et al. [32]
5	F	110	19.9	H	–	Norris et al. [32]
5	M	Various	Various	H	–	Xie et al. [46]
5	F	Various	Various	H	–	Xie et al. [46]
5	M	110	19	H	USTC-5 M	Pi et al. [35]
5	F	110	19	H	USTC-5F	Pi et al. [35]

Table 28.7 Paediatric phantom families constructed in recent years—6- to 8-year-old children

6	M	–	–	H	–	Alziar et al. [1]
6	M	117	20	V	Thelonious	Christ et al. [6] Gosselin et al. [18]
7	F	115	21.7	V	Child	Williams et al. [45] Zankl et al. [52] Petoussi et al. [34]
7	H	120	23	H	–	Nagaoka et al. [30]
7	M	–	–	H	–	Alziar et al. [1]
8	F	–	28.4	V	–	Lee et al. [25]
8	M	140	26.0	V	Dizzy	Gosselin et al. [18]
8	F	136	30.7	V	Eartha	Gosselin et al. [18]

Table 28.8 Paediatric phantom families constructed in recent years—10-year-old children

10	H	–	33.2	S	–	Cristy et Eckerman [8]
10	H	138	33.5	S	–	Deus et Poston [10]
10	M	Various	Various	H	UFHPDM	Johnson et al. [22]
10	F	Various	Various	H	UFHPDF	Johnson et al. [22]
10	M	–	–	H	–	Taddei et al. [43]
10	H	138	31.9	H	–	Lee et al. [28]
10	F	138	32	H	F10	De Melo Lima et al. [9]
10	M	138	32	H	M10	De Melo Lima et al. [9]
10	M	–	32	H	–	Stabin et al. [41]
10	F	–	32	H	–	Stabin et al. [41]
10	H	–	–	S	–	Gardumi et al. [16]
10	M	Various	Various	H	UFH10M	Geyer et al. [15]
10	F	Various	Various	H	UFH10F	Geyer et al. [15]
10	M	139	33	H	–	Norris et al. [32]
10	F	139	33	H	–	Norris et al. [32]
10	M	Various	Various	H	–	Xie et al. [46]
10	F	Various	Various	H	–	Xie et al. [46]
10	M	139	32	H	USTC-10 M	Pi et al. [35]
10	F	139	32	H	USTC-10F	Pi et al. [35]

Table 28.9 Paediatric phantom families constructed in recent years—11- to 14-year-old children

Age (years)	Sex	Body height (cm)	Body weight (kg)	Type	Name	Reference
11	M	–	33.4	V	–	Lee et al. [25]
11	F	146	36	V	Billie	Christ et al. [6] Gosselin et al. [18]
14	F	–	48	V	Adelaide	Caon et al. [2]
14	M	–	48.7	V	–	Lee et al. [25]
14	M	169	50.4	V	Louis	Gosselin et al. [18]

Table 28.10 Paediatric phantom families constructed in recent years—15-year-old children

15	H	–	–	S	–	Jones et al. [23]
15	F	–	56.8	S	–	Cristy et Eckerman [8]
15	F	161	53.3	H	–	Lee et al. [27] Lee et al. [28]
15	M	167	56.8	H	–	Lee et al. [27] Lee et al. [28]
15	M	Various	Various	H	UFHPDM	Johnson et al. [22]
15	F	Various	Various	H	UFHPDF	Johnson et al. [22]
15	M	–	56	H	–	Stabin et al. [41]
15	F	–	53	H	–	Stabin et al. [41]
15	H	–	–	S	–	Gardumi et al. [16]
15	M	Various	Various	H	UFH15M	Geyer et al. [15]
15	F	Various	Various	H	UFH15F	Geyer et al. [15]
15	M	167	58	H	–	Norris et al. [32]
15	F	162	54	H	–	Norris et al. [32]
15	M	Various	Various	H	–	Xie et al. [46]
15	F	Various	Various	H	–	Xie et al. [46]
15	M	168	55	H	USTC-15 M	Pi et al. [35]
15	F	158	50	H	USTC-15F	Pi et al. [35]

Table 28.11 Paediatric phantom families constructed in recent years—adults

Age (years)	Sex	Body height (cm)	Body weight (kg)	Type	Name	Reference
Adult	H	174	69.9	S	–	Fisher et Snyder [11] Fisher et Snyder [13] Snyder et al. [40]
Adult	M	170	70	S	Adam	Kramer et al. [24] Zankl et al. [51]
Adult	F	160	58	S	Eva	Kramer et al. [24] Zankl et al. [51]
Adult	M	–	73.7	S	–	Cristy et Eckerman [8]
Adult	F	160	58	V	Ella	Christ et al. [6] Gosselin et al. [18]
Adult	M	174	70	V	Duke	Christ et al. [6] Gosselin et al. [18]
Adult	M	176	68.9	V	Golem	Petoussi et al. [34] Zankl et al. [51]
Adult	M	180	103	V	Visible human	Petoussi et al. [34] Zankl et al. [51]
Adult	M	176	69	V	Godwin	Zankl et al. [50]
Adult	M	176	69	V	Rex	Schlattl et al. [36]
Adult	F	170	79	V	Donna	Petoussi et al. [34] Zankl et al. [51] Fill et al. [12]
Adult	F	170	81	V	Helga	Petoussi et al. [34] Zankl et al. [51] Fill et al. [12]
Adult	F	170	79	V	Irene	Zankl et al. [51] Fill et al. [12]
Adult	F	167	59	V	Klara	Zankl et al. [50]
Adult	F	167	59	V	Laura	Zankl et al. [50]
Adult	F	167	59	V	Regina	Schlattl et al. [36]
Adult	M	178	65	H	–	Alziar et al. [1]
Adult	M	Various	Various	H	UFHPDM	Johnson et al. [22]
Adult	F	Various	Various	H	UFHPDF	Johnson et al. [22]

(continued)

Table 28.11 (continued)

Age (years)	Sex	Body height (cm)	Body weight (kg)	Type	Name	Reference
Adult	M	170	60.6	H	CAM	Liu et al. [29]
Adult	F	163	59.7	H	–	Lee et al. [28]
Adult	M	176	73.4	H	–	Lee et al. [28]
Adult	M	–	73	H	–	Stabin et al. [41]
Adult	F	–	60	H	–	Stabin et al. [41]
Adult	H	–	–	S	–	Gardumi et al. [16]
Adult	M	173	65.0	V	Glenn	Gosselin et al. [18]
Adult	M	182	120.0	V	Fats	Gosselin et al. [18]
Adult	M	170	63.1	H	CRAM_S	Chen et al. [5]
Adult	M	170	63	H	USTC-AM	Pi et al. [35]
Adult	F	158	54	H	USTC-AF	Pi et al. [35]

Sex: *F* female phantom; *H* hermaphrodite phantom; *M* male phantom. Type: *H* hybrid phantom; *P* physical phantom; *S* stylized phantom; *V* voxel phantom

References

1. Alziar I, Bonniaud G, Couanet D, Ruaud JB, Vicente C, Giordana G, Ben-Harrath O, Diaz JC, Grandjean P, Kafrouni H, Chavaudra J, Lefkopoulos D, de Vathaire F, Diallo I. Individual radiation therapy patient whole-body phantoms for peripheral dose evaluations: method and specific software. Phys Med Biol. 2009;54:N375–83.
2. Caon M, Bibbo G, Pattison J. An EGS4-ready tomographic computational model of a 14-year-old female torso for calculating organ doses from CT examinations. Phys Med Biol. 1999;44:2213–25.
3. Carver DE, Kost SD, Fraser ND, Segars WP, Pickens DR, Price RR, Stabin MG. Realistic phantoms to characterize dosimetry in pediatric CT. Pediatr Radiol. 2017;47:691–700.
4. Cassola VF, Kramer R, de Melo Lima VJ, de Oliveira Lira CAB, Khoury HJ, Vieira JW, Robson Brown K. Development of newborn and 1-year-old reference phantoms based on polygon mesh surfaces. J Radiol Prot. 2013;33:669–91.
5. Chen Y, Qiu R, Li C, Wu Z, Li J. Construction of Chinese adult male phantom library and its application in the virtual calibration of in vivo measurement. Phys Med Biol. 2016;61:2124–44.
6. Christ A, Kainz W, Hahn EG, Honegger K, Zefferer M, Neufeld E, Rascher W, Janka R, Bautz W, Chen J, Kiefer B, Schmitt P, Hollenbach HP, Shen J, Oberle M, Szczerba D, Kam A, Guag JW, Kuster N. The virtual family—development of surface-based anatomical models of two adults and two children for dosimetric simulations. Phys Med Biol. 2010;55:N23–38.
7. Cristy M. Mathematical phantoms representing children of various ages for use in estimates of internal dose. Oak Ridge, TN: Oak Ridge Laboratory, NUREG/CR-1159, ORNL/NUREG/TM-367; 1980.
8. Cristy M, Eckerman KF. Specific absorbed fractions of energy at various ages from internal photon sources. Oak Ridge, TN: Oak Ridge National Laboratory, Health and Safety Research Division, ORNL/TM-8381/V1; 1987.
9. De Melo Lima VJ, Cassola VF, Kramer R, de Oliveira Lira CAB, Khoury HJ, Vieira JW. Development of 5- and 10-year-old pediatric phantoms based on polygon mesh surfaces. Med Phys. 2011;38:4723–36.
10. Deus SF, Poston JW. The development of a mathematical phantom representing a 10-year-old for use in internal dose calculations. In: Proceedings of the Symposium on Radiopharmaceutical Dosimetry, Food and Drug Administration, HEW Publication (FDA) 76-8044, Oak Ridge, TN; 1976.
11. Fisher HLJ, Snyder WS. Variation of dose delivered by 137Cs as a function of body size from infancy to adulthood. Health Physics Division, Annual Progress Report for Period Ending July 31, Oak Ridge National Laboratory, Report ORNL-4007; 1966 p. 221–228.
12. Fill UA, Zankl M, Petoussi-Henss N, Siebert M, Regulla D. Adult female voxel models of different stature and photon conversion coefficients for radiation protection. Health Phys. 2004;86:253–72.
13. Fisher HLJ, Snyder WS. Distribution of dose in the body from a source of gamma rays distributed uniformly in an organ. Health Physics Division, Annual Progress Report for Period Ending July 31, Oak Ridge National Laboratory, Report ORNL-4168; 1967.
14. Francois P, Beurtheret C, Dutreix A, De Vathaire F. A mathematical child phantom for the calculation of dose to the organs of risk. Med Phys. 1988;15: 328–33.
15. Geyer AM, O'Reilly S, Lee C, Long DJ, Bolch WE. The UF/NCI family of hybrid computational phantoms representing the current US population of male and female children, adolescents, and adults—application to CT dosimetry. Phys Med Biol. 2014;59:5225–42.
16. Gardumi A, Farah J, Desbree A. Creation of ORNL NURBS-based phantoms: evaluation of the voxel effect on absorbed doses from radiopharmaceuticals. Radiat Prot Dosim. 2013;153:273–81.

17. Gibbs SJ, Pujol A, Chen TS, Malcolm AW, James AE. Patient risk from interproximal radiography. Oral Surg Oral Med Oral Pathol. 1984;58:347–54.
18. Gosselin MC, Neufeld E, Moser H, Huber E, Farcito S, Gerber L, Jedensjö M, Hilber I, Di Gennaro F, Lloyd B, Cherubini E, Szczerba D, Kainz W, Kuster N. Development of a new generation of high-resolution anatomical models for medical device evaluation: the virtual population 3.0. Phys Med Biol. 2014;59:5287–303.
19. Hirata A, Ito N, Fujiwara O, Nagaoka T, Watanabe S. Conservative estimation of whole-body-averaged SARs in infants with a homogeneous and simple-shaped phantom in the GHz region. Phys Med Biol. 2008;53:7215–23.
20. Hwang JML, Shoup RL, Poston JW. Mathematical description of a newborn human for use in dosimetry calculations. Oak Ridge, TN: Oak Ridge National Laboratory, Report ORNL/TM-5453; 1976.
21. International Commission on Radiological Protection. Report of the task group on reference man: anatomical, physiological and metabolic characteristics. ICRP Publication 23. Oxford: Pergamon Press; 1975.
22. Johnson PB, Whalen SR, Wayson M, Juneja B, Lee C, Bolch WE. Hybrid patient-dependent phantoms covering statistical distributions of body morphometry in the U.S. adult and pediatric population. Proc IEEE. 2009;97:2060–75.
23. Jones RM, Poston JW, Hwang JL, Jones TD, Warner GG. The development and use of a fifteen-year-old equivalent mathematical phantom for internal dose calculations. Oak Ridge, TN: Oak Ridge National Laboratory, Report ORNL/TM-5278; 1976.
24. Kramer R, Zankl M, Williams G, Drexler G. The calculation of dose from external photon exposures using reference human phantoms and Monte Carlo methods. Part I: The male (Adam) and female (Eva) adult mathematical phantoms GSF-Bericht S-885, Neuherberg: GSF-Forschungszentrum für Umwelt und Gesundheit; 1986.
25. Lee C, Lee C, Williams JL, Bolch WE. Whole-body voxel phantoms of paediatric patients—UF Series B. Phys Med Biol. 2006;51:4649–61.
26. Lee C, Lodwick D, Hasenauer D, Williams JL, Lee C, Bolch WE. Hybrid computational phantoms of the male and female newborn patient: NURBS-based whole-body models. Phys Med Biol. 2007;52:3309–33.
27. Lee C, Lodwick D, Williams JL, Bolch WE. Hybrid computational phantoms of the 15-year male and female adolescent: applications to CT organ dosimetry for patients of variable morphometry. Med Phys. 2008;35:2366–282.
28. Lee C, Lodwick D, Hurtado J, Pafundi D, Williams JL, Bolch WE. The UF family of reference hybrid phantoms for computational radiation dosimetry. Phys Med Biol. 2010;55:339–63.
29. Liu L, Zeng Z, Li J, Qiu R, Zhang B, Ma J, Li R, Li W, Bi L. Organ dose conversion coefficients on an ICRP-based Chinese adult male voxel model from idealized external photons exposures. Phys Med Biol. 2009;54:6645–73.
30. Nagaoka T, Kunieda E, Watanabe S. Proportion-corrected scaled voxel models for Japanese children and their application to the numerical dosimetry of specific absorption rate for frequencies from 30 MHz to 3 GHz. Phys Med Biol. 2008;53:6695–711.
31. Nipper JC, Williams JL, Bolch WE. Creation of two tomographic voxel models of paediatric patients in the first year of life. Phys Med Biol. 2002;47:3143–64.
32. Norris H, Zhang Y, Bond J, Sturgeon GM, Ravin CE, Minhas A, Tward DJ, Ratnanather JT, Miller MI, Frush D, Samei E, Segars WP. A set of 4D pediatric XCAT reference phantoms for multimodality research. Med Phys. 2014;41:033701-1–033701-11.
33. Pan Y, Qiu R, Gao L, Ge C, Zheng J, Xie W, Li J. Development of 1-year-old computational phantom and calculation of organ doses during CT scans using Monte Carlo simulation. Phys Med Biol. 2014;59:5243–60.
34. Petoussi-Henß N, Zankl M, Fill U, Regulla D. The GSF family of voxel phantoms. Phys Med Biol. 2002;47:89–106.
35. Pi Y, Liu T, Xu XG. Development of a set of mesh-based and age-dependent Chinese phantoms and application for CT dose calculations. Radiat Prot Dosim. 2018;179:370–87.
36. Schlattl H, Zankl M, Petoussi-Henss N. Organ dose conversion coefficients for voxel models of the reference male and female from idealized photon exposures. Phys Med Biol. 2007;52:2123–45.
37. Segars WP, Norris H, Sturgeon GM, Zhang Y, Bond J, Minhas A, Tward DJ, Ratnanather JT, Miller MI, Frush D, Samei E. The development of a population of 4D pediatric XCAT phantoms for imaging research and optimization. Med Phys. 2015;42:4719–26.
38. Servomaa A, Rannikko S, Nikitin V, Golikov V, Ermakov I, Madarskyi L, Saltukova L. A topographically and anatomically unified phantom model for organ dose determination in radiation hygiene. STUK-A87. Helsinki: Finnish Centre for Radiation and Nuclear Safety; 1989.
39. Servomaa A, Tapiovaara M. Organ dose calculation in medical x-ray examinations by the program PCXMC. Radiat Prot Dosim. 1998;80:213–9.
40. Snyder WS, Ford MR, Warner CC, Fisher HL. Estimates of absorbed fractions for monoenergetic photon sources uniformly distributed in various organs of a heterogeneous phantom. MIRD pamphlet no. 5. J Nucl Med. 1969;10
41. Stabin MG, Xu XG, Emmons MA, Segars WP, Shi C, Fernald MJ. RADAR reference adult, pediatric, and pregnant female phantom series for internal and external dosimetry. J Nucl Med. 2012;53:1807–13.
42. Staton RJ, Jones AK, Lee C, Hintenlang DE, Arreola MM, Williams JL, Bolch WE. A tomographic physical phantom of the newborn child with real-time dosimetry. II. Scaling factors for calculation of mean organ dose in pediatric radiography. Med Phys. 2006;33:3283–9.

43. Taddei PJ, Mirkovic D, Fontenot JD, Giebeler A, Zheng Y, Kornguth D, Mohan R, Newhauser WD. Stray radiation dose and second cancer risk for a pediatric patient receiving craniospinal irradiation with proton beams. Phys Med Biol. 2009;54: 2259–75.
44. Varchena V. Pediatric phantoms. Pediatr Radiol. 2002;32:280–4.
45. Williams G, Zankl M, Abmayr W, Veit R, Drexler G. The calculation of dose from external photon exposures using reference and realistic human phantoms and Monte Carlo methods. Phys Med Biol. 1986;31:449–52.
46. Xie T, Kuster N, Zaidi H. Computational hybrid anthropometric paediatric phantom library for internal radiation dosimetry. Phys Med Biol. 2017;62:3263–83.
47. Xu XG. An exponential growth of computational phantom research in radiation protection, imaging, and radiotherapy: a review of the fifty-year history. Phys Med Biol. 2014;59:R233–302.
48. Xu XG, Eckerman KF. Handbook of anatomical models for radiation dosimetry. Boca Raton, FL: CRC, Taylor & Francis; 2009.
49. Yamauchi-Kawaura C, Fujii K, Akahane K, Yamauchi M, Obara S, Narai K, Katsu T, Imai K, Ikeda M. Development of age-specific japanese physical phantoms for dose evaluation in infant CT examinations. Radiat Prot Dosim. 2016;171:483–502.
50. Zankl M, Becker J, Fill U, Petoussi-Henss N, Eckerman KF. GSF male and female adult voxel models representing ICRP reference man—the present status. The Monte Carlo method: versatility unbounded in a dynamic computing world. Proceedings of the Monte Carlo 2005 Topical meeting on CD-ROM, Chattanooga, TN; 2005.
51. Zankl M, Fill U, Petoussi-Henss N, Regulla D. Organ dose conversion coefficients for external photon irradiation of male and female voxel models. Phys Med Biol. 2002;47:2367–85.
52. Zankl M, Veit R, Williams G, Schneider K, Fendel H, Petoussi N, Drexler G. The construction of computer tomographic phantoms and their application in radiology and radiation protection. Radiat Environ Biophys. 1988;27:153–64.

29 Normalized Organ Doses in Tables

In this section, tables of normalized organ doses for conventional radiographs are listed that have been published thus far to the knowledge of the authors. The literature listings comprise tables and single publications to normalized organ doses in conventional radiology (Tables 29.1, 29.2, 29.3 and 29.4).

Table 29.1 Systematic tables of normalized organ doses for conventional radiographs in adults

X-ray examination	Reference organ	Age group	Reference
Various	Various	Adults	Drexler et al. [5]
Various	Various	Adults	Hart et al. [14]
Various	Various	Adults	Jones and Wall [18]
Various	Various	Adults	Le Heron [19]
Various	Effective dose	Adults	Hart et al. [11]
Various	Effective dose	Adults	Le Heron [19]

Table 29.2 Articles containing normalized organ doses for conventional radiographs in adults

X-ray examination	Reference organ	Age group	Reference
Various	Various	Adults	Golikov and Nikitin [9]
Various	Various	Adults	Gray et al. [10]
Various	Various	Adults	Jankowski [17]
Various	Various	Adults	Petoussi-Henß et al. [26]
Various	Various	Adults	Servomaa and Kainulainen [37]
Various	Various	Adults	Toivonen et al. [39]
Various	Effective dose	Adults	Huda and Bissessur [15]
Various	Effective dose	Adults	Huda and Gkanatsios [16]
Skull	Various	Adults	Petoussi et al. [25]
Chest	Various	Adults	Petoussi et al. [25]
Chest	Various	Adults	Schultz et al. [27]
Chest	Various	Adults	Wise et al. [41]
Chest	Effective dose	Adults	Biju and Nagajaran [2]
Chest	Effective dose	Adults	Lee et al. [22]
Chest	Red bone marrow	Adults	Archer et al. [1]
Mammography	Breast	Adults	Dance [4]
Mammography	Breast	Adults	Zankl et al. [42]
Abdomen	Various	Adults	Marshall and Faulkner [23]
Abdomen	Effective dose	Adults	Lee et al. [22]
Small intestine	Uterus/gonads	Adults	Weiner et al. [40]
Lumbar spine	Various	Adults	Wise et al. [41]
Whole body	Various	Adults	Fill et al. [7]

M. Seidenbusch et al., *Imaging Practice and Radiation Protection in Pediatric Radiology*,
https://doi.org/10.1007/978-3-030-18504-6_29

Table 29.3 Systematic tables of normalized organ doses for conventional radiographs in children

X-ray examination	Reference organ	Age group	Reference
Various	Effective dose	Various	Hart et al. [12, 13]
Various	Various	Various	Hart et al. [14]
Various	Various	Various	Le Heron [20, 21]
Various	Various	Various	Rosenstein et al. [28]
Various	Various	Various	Rosenstein et al. [29]
Various	Various	Various	Rosenstein et al. [30]

Table 29.4 Articles containing normalized organ doses for conventional radiographs in children

X-ray examination	Reference organ	Age group	Reference
Skull	Various	Various	Seidenbusch et al. [31]
Chest	Various	Various	Seidenbusch et al. [32]
Chest	Various	Neonates	Smans et al. [38]
Chest	Effective dose	Neonates	Elbakri et al. [6]
Babygram	Various	Various	Seidenbusch et al. [34]
Babygram	Effective dose	Neonates	Elbakri et al. [6]
Abdomen	Various	Various	Seidenbusch et al. [33]
Abdomen	Various	Various	Seidenbusch et al. [34]
Spine	Various	Various	Seidenbusch and Schneider [35]
Pelvis/hip joint	Various	Various	Seidenbusch and Schneider [36]
Not specific	Energy imparted	Not specific	Carlsson et al. [3]
Not specific	Energy imparted	Not specific	Gkanatsios and Huda [8]
Not specific	Energy imparted	Not specific	Persliden and Sandborg [24]

References

1. Archer BR, Whitmore RC, North LB, Bushong SC. Bone marrow dose in chest radiography: the posteroanterior vs. anteroposterior projection. Radiology. 1979;133:211–6.
2. Biju K, Nagajaran PS. Computed normalised effective doses to an Indian adult in conventional diagnostic x-ray chest examinations. Radiat Prot Dosim. 2000;88:119–27.
3. Carlsson GA, Carlsson CA, Persliden J. Energy imparted to the patient in diagnostic radiology; calculation of conversion factors for determining the energy imparted from measurements of the air collision kerma integrated over beam area. Phys Med Biol. 1984;29:1329–41.
4. Dance DR. Monte Carlo calculation of conversion factors for the estimation of mean glandular breast dose. Phys Med Biol. 1990;35:1211–9.
5. Drexler G, Panzer W, Stieve FE. Die Bestimmung von Organdosen in der Röntgendiagnostik. Berlin: Hoffmann Verlag; 1993.
6. Elbakri EA. Estimation of dose-area product-to-effective dose conversion factors for neonatal radiography using PCXMC. Radiat Prot Dosim. 2014;158:43–50.
7. Fill UA, Zankl M, Petoussi-Henß N, Siebert M, Regulla D. Adult female voxel models of different stature and photon conversion coefficients for radiation protection. Health Phys. 2004;86:253–72.
8. Gkanatsios NA, Huda W. Computation of energy imparted in diagnostic radiology. Med Phys. 1997;24:571–9.
9. Golikov VY, Nikitin VV. Estimation of the mean organ doses and the effective dose equivalent from rando phantom measurements. Health Phys. 1989;56:111–5.
10. Gray JE, Ragozzino MW, van Lysel MS, Burke TM. Normalized organ doses for various diagnostic radiologic procedures. Am J Radiol. 1981;137:463–70.
11. Hart D, Jones DJ, Wall BF. Estimation of effective dose in diagnostic radiology from entrance surface dose and dose-area-product measurements. Oxon: National Radiological Protection Board, NRPB-R262; 1994.
12. Hart D, Jones DJ, Wall BF. Coefficients for estimating effective doses from pediatric x-ray examinations. ISBN 0-85951-390-4. National Radiological Protection Board, NRPB-R279; 1996.
13. Hart D, Jones DJ, Wall BF. Normalized organ doses for paediatric x-ray examinations calculated using Monte Carlo techniques. Oxon: National Radiological Protection Board (Software Report) NRPB-SR279; 1996.
14. Hart D, Jones DJ, Wall BF. Normalised organ doses for paediatric x-ray examinations calculated using Monte Carlo techniques. Oxon: National Radiological Protection Board (Software Report), NRPB-SR262; 1998.

15. Huda W, Bissessur K. Effective dose equivalents, HE, in diagnostic radiology. Med Phys. 1990;17:998–1003.
16. Huda W, Gkanatsios NA. Effective dose and energy imparted in diagnostic radiology. Med Phys. 1997;24:1311–6.
17. Jankowski J. Organ doses in diagnostic x-ray procedures. Health Phys. 1984;46:234–6.
18. Jones DG, Wall BF. Organ doses from medical x-ray examinations calculated using Monte Carlo techniques. Oxon: National Radiological Protection Board, NRPB-R186; 1985.
19. Le Heron JC. Estimation of effective dose to the patient during medical x-ray examinations from measurements of the dose-area product. Phys Med Biol. 1992;37:2117–26.
20. Le Heron JC. Xdose-software. Christchurch: National Radiation Laboratory, Ministery of Health; 1994.
21. Le Heron JC. Childose-software. Christchurch: National Radiation Laboratory, Ministery of Health; 1996.
22. Lee SC, Wang JN, Liu SC, Jiang SH. Effective dose evaluation for chest and abdomen x-ray tests. Radiat Prot Dosim. 2005;116:613–9.
23. Marshall NW, Faulkner K. Normalized organ dose data measured as a function of field size for abdominal examinations. Phys Med Biol. 1993;38:1131–6.
24. Persliden J, Sandborg M. Conversion factors between energy imparted to the patient and air collision kerma integrated over beam area in pediatric radiology. Acta Radiol. 1993;34:92–8.
25. Petoussi N, Zankl M, Stieve FE, Drexler G. Patient organ doses for proposed technical parameters and their variation. In: Moores BM, Wall BF, Eriskat H, Schibilla H, editors. Optimization of image quality and patient exposure in diagnostic radiology. BIR-Report 20. London: British Institute of Radiology; 1989. p. 246–9.
26. Petoussi-Henß N, Panzer W, Zankl M, Drexler G. Dose-area product and body doses. Radiat Prot Dosim. 1995;57:363–6.
27. Schultz FW, Geleijns J, Zoetelief J. Calculation of dose conversion factors for posterior-anterior chest radiography of adults with a relatively high-energy X-rayspectrum. Br J Radiol. 1994;67:775–85.
28. Rosenstein M. Organ doses in diagnostic radiology. Rockville, MD: U.S. Department of Health, Education, and Welfare, Public Health Service, Food and Drug Administration (FDA), HEW Publication (FDA) 76–8030, Bureau of Radiological Health; 1976.
29. Rosenstein M. Handbook of selected organ doses for projections common in pediatric radiology. Rockville, MD: U.S. Department of Health, Education, and Welfare, Public Health Service, Food and Drug Administration (FDA), HEW Publication (FDA) 79–8079, Bureau of Radiological Health; 1979.
30. Rosenstein M. Handbook of selected tissue doses for projections common in diagnostic radiology. Rockville, MD: U.S. Department of Health, Education, and Welfare, Public Health Service, Food and Drug Administration (FDA), HHS Publication (FDA) 89–8031, Center for Devices and Radiological Health; 1988.
31. Seidenbusch MC, Regulla D, Schneider K. Radiation exposure of children in pediatric radiology. Part 2: the PAEDOS algorithm for computer-assisted dose reconstruction in pediatric radiology and results for x-ray examinations of the skull. Fortschr Röntgenstr. 2008;180:522–39.
32. Seidenbusch MC, Regulla D, Schneider K. Radiation exposure of children in pediatric radiology. Part 3: conversion coefficients for reconstruction of organ doses achieved during chest x-ray examinations. Fortschr Röntgenstr. 2008;180:1061–81.
33. Seidenbusch MC, Regulla D, Schneider K. Radiation exposure of children in pediatric radiology. Part 6: conversion coefficients for reconstruction of organ dose in abdominal radiography. Fortschr Röntgenstr. 2009;181:945–61.
34. Seidenbusch MC, Regulla D, Schneider K. Radiation exposure of children in pediatric radiology. Part 7: conversion factors for reconstruction of organ dose during thoracoabdominal babygrams. Fortschr Röntgenstr. 2010;182:415–21.
35. Seidenbusch MC, Schneider K. Conversion coefficients for determining organ doses in paediatric spine radiography. Pediatr Radiol. 2014;44:434–56.
36. Seidenbusch MC, Schneider K. Conversion coefficients for determining organ doses in paediatric pelvis and hip joint radiography. Pediatr Radiol. 2014;44:1110–23.
37. Servomaa A, Kainulainen E. Calculation of patient dose and somatic dose index due to roentgen examinations. Med Phys. 1985;12:1–4.
38. Smans K, Tapiovaara M, Cannie M, Struelens L, Vanhavere F, Smet M, Bosmans H. Calculation of organ doses in x-ray examinations of premature babies. Med Phys. 2008;35:556–68.
39. Toivonen M, Aschan C, Rannikko S, Karila K, Savolainen S. Organ dose determinations of X-ray examinations using TL detectors for verification of computed doses. Radiat Prot Dosim. 1996;66: 289–94.
40. Weiner GM, Köhler B, Golder W. Organdosen und Konversionsfaktoren für Gonaden und Uterus bei der standardisierten Doppelkontrastuntersuchung des Dünndarms (Enteroklysma nach Sellink). Fortschr Röntgenstr. 2003;175:1551–5.
41. Wise KN, Sandborg M, Persliden J, Carlsson GA. Sensitivity of coefficients for converting entrance surface dose and kerma-area product to effective dose and energy imparted to the patient. Phys Med Biol. 1999;44:1937–54.
42. Zankl M, Fill U, Hoeschen C, Panzer W, Regulla D. Average glandular dose conversion coefficients for segmented breast voxel models. Radiat Prot Dosim. 2005;114:410–4.

30 Abbreviations

Table 30.1 Abbreviations

AEC	Automatic exposure control
ALARA	As low as reasonably achievable
ALARP	As low as reasonably practicable
ARDS	Adult respiratory distress syndrome
CR	Computed radiography
CSF	Cerebrospinal fluid flow (control valves)
CT	Computed tomography
D_T	Organ dose in tissue T
DAP	Dose area product
DI	Dose indicator
DICOM	Digital Imaging and Communications in Medicine
DLR	Digital luminescence radiography
DQE	Digital quantum efficiency
DR	Digital radiography
DRL	Dose reference level
E, ED	Effective dose
ECMO	Extracorporeal membrane oxygenation
ENPR	European Network of Pediatric Radiology
ESPR	European Society for Pediatric Radiology
EU	European Union
FDD	Focus-to-detector distance
FSD	Focus-to-skin distance
GSDF	Grayscale Standard Display Function
gw	Gestational weeks
HVL	Half value length
ICRP	International Commission on Radiological Protection
IRDS	Idiopathic respiratory distress syndrome
K_e	Exit air kerma
K_i	Incidence air kerma
KAP	Kerma area product
Kerma	Kinetic energy released in material
kVp	Peak kilovoltage
mAs	mAs product
LCD	Liquid crystal display
LCH	Langerhans cell histiocytosis
LUT	Look up table
MC	Monte Carlo (simulation)
MMC	Myelomeningocele
MIRD	Medical internal radiation dose (committee)
MR, MRT	Magnetic resonance tomography
MRI	Magnetic resonance imaging
ms	Milliseconds
NCRP	National Counsil on Radiation Protection and Measurements
NIP	Needle image plate
PACS	Picture archiving and communication system
PC	Personal Computer
PCXMC	a dose calculation software provided by STUK
PIP	Powder image plate
ROI	Region of interest
SI	Système international d´unités
SOP	Standard operating procedure
STUK	Säteilyturvakeskus (Finnish Radiation and Nuclear Safety Authority)
TFT	Thin film transistor
US	Ultrasound
VPS	Ventriculo-peritoneal shunt
WHO	World Health Organization

M. Seidenbusch et al., *Imaging Practice and Radiation Protection in Pediatric Radiology*,
https://doi.org/10.1007/978-3-030-18504-6_30

Correction to: Imaging Practice and Radiation Protection in Pediatric Radiology

Correction to:
M. Seidenbusch et al.: *Imaging Practice and Radiation Protection in Pediatric Radiology,* https://doi.org/10.1007/978-3-030-18504-6

The original (print and online) version of the book was inadvertently published with incorrect sentences. These sentences have now been corrected as below:

Chapter 9:
In table 9.5 on p. 129, the phrase "Spinous process of the Sixth cervical vertebra" has been changed to "Spinous process of the sixth cervical vertebra".

In tables 9.15–9.18 on pp. 138–140, the phrase "Above the spinous process of the 6th vertebra" has been changed to "Tip of the mandible, above the clavicles".

In tables 9.19–9.25 on pp. 140–143, the phrase "Above the spinous process of the 6th vertebra" has been changed to "Above the spinous process of the 6th cervical vertebra".

Chapter 10:
In tables 10.11–10.19 on pp. 293–297, the phrase "Lower field edge: Just below the perineum" has been changed to "Lower field edge: Just below the 12th rib".

The phrase "Upper field edge: Orbitae Lower third of the skull" should be changed to "Upper field edge: Lower third of the skull".

The phrase "Lower field edge: Upper third of thighs (in case of extended legs)" has been changed to "Lower field edge: At the upper margins of the iliac crests".

Chapter 11:
In tables 11.13–11.14 on pp. 345–346, the phrase "Upper field edge: Mandibulaspitze" has been changed to "Upper field edge: Tip of the mandible".

The updated online version of these chapters can be found at:
https://doi.org/10.1007/978-3-030-18504-6_9
https://doi.org/10.1007/978-3-030-18504-6_10
https://doi.org/10.1007/978-3-030-18504-6_11

M. Seidenbusch et al., *Imaging Practice and Radiation Protection in Pediatric Radiology*,
https://doi.org/10.1007/978-3-030-18504-6_31

The phrase “Lateral field edges: Beide Humeri proximales Drittel (*)” has been changed to “Lateral field edges: Proximal metaphysis of the humeri”.

The phrase “Upper field edge: Unterrand Orbita” has been changed to “Upper field edge: At the upper margins of the parotid glands”.

The phrase “Lateral field edges: Beide Arme exponiert” has been changed to “Lateral field edges: Both arms included”.